Glossary of Agricultural Biotechnology

Gurbachan Singh Miglani, PhD
Parveen Chhuneja, PhD
Satbir Singh Gosal, PhD

CABI

CABI is a trading name of CAB International

CABI
Nosworthy Way
Wallingford
Oxfordshire OX10 8DE
UK

Tel: +44 (0)1491 832111
E-mail: info@cabi.org
Website: www.cabi.org

CABI
200 Portland Street
Boston
MA 02114
USA

T: +1 (617)682-9015
E-mail: cabi-nao@cabi.org

A catalogue record for this book is available from the British Library, London, UK.

ISBN-13: 9781800627758 (hardback)
9781800627765 (ePDF)
9781800627772 (ePub)

DOI: 10.1079/9781800627772.0000

Commissioning Editor: Ward Cooper
Editorial Assistant: Helen Elliott
Production Editor: Rosie Hayden

Typeset by Exeter Premedia Services Pvt Ltd, Chennai, India
Printed and bound in the UK by Bell & Bain, Glasgow

Contents

MESSAGE

Gurdev S. Khush

This comprehensive book, with over 16,200 terms, is a landmark in agricultural biotechnology literature. Authored by three distinguished scientists devoted to genetics, biotechnology and plant breeding, it is an invaluable resource for students, educators, researchers and industry professionals. As the pioneering glossary in this field, it spans diverse areas including biotechnology, botany, genetics, cytogenetics, bioinformatics, biostatistics, cell biology, epigenetics, molecular biology, tissue culture and artificial intelligence. The authors have meticulously credited those who originally defined and later refined these terms. I applaud their dedication and hard work in creating this extensive and indispensable reference, celebrating the vast and interconnected world of agricultural biotechnology.

Gurdev S. Khush FRS
Adjunct Professor Emeritus
University of California, Davis

Dedication

In heartfelt remembrance of

Dr Satwant Kaur Gosal
(1963–2021)
(wife of Dr Satbir Singh Gosal)
Ex-Professor of Microbiology
Punjab Agricultural University
Ludhiana, Punjab, India,

whose vision birthed the
Glossary of Agricultural Biotechnology

Foreword

P.C. Trivedi

This book, *Glossary of Agricultural Biotechnology*, encompasses two critical concepts: 'glossary' and 'agricultural biotechnology'. A glossary is defined as a 'list of terms in a specialized subject, field or area of usage, accompanied by definitions'. Agricultural biotechnology, or agrobiotechnology, refers to a variety of tools, including traditional breeding and recombinant DNA techniques, which modify agriculturally important plants, animals and microorganisms to meet human agricultural goals. This definition highlights that agricultural biotechnology is a multidisciplinary field, supported by numerous basic and applied sciences. These include artificial intelligence, bioinformatics, biostatistics, biotechnology, botany, cell biology, computer science, cytogenetics, DNA nanotechnology, epigenetics, epigenomics, genetics, intellectual property rights, molecular biology, nanobiotechnology, plant breeding, plant pathology, plant physiology, remote sensing and tissue culture.

A book titled *Glossary of Agricultural Biotechnology* must integrate key terms from all these scientific disciplines. This book adopts a comprehensive approach to honour its promise, presenting a complete and thorough glossary for anyone exploring the vast and intricate realm of agricultural biotechnology.

Three distinguished experts from Punjab Agricultural University, Ludhiana, Punjab, India, collaborated to author the *Glossary of Agricultural Biotechnology*. Dr Gurbachan Singh Miglani, a seasoned educator who retired as Professor of Genetics from the Department of Plant Breeding, Genetics, and Biotechnology, and a former Adjunct/Visiting Professor at the School of Agricultural Biotechnology, is one of them. Dr Miglani has authored 17 widely cited books in areas including general genetics, advanced genetics, molecular genetics, cell biology, recombinant DNA technology, genome editing, epigenetics and epigenomics. He has also written 6 laboratory manuals and 14 book chapters and received the Meritorious Teacher Award from Punjab Agricultural University for 1995–1996.

Dr Parveen Chhuneja, an internationally acclaimed researcher and experienced biotechnology teacher, currently serves as Principal Molecular Geneticist-cum-Director at the School of Agricultural Biotechnology. With 153 publications in refereed journals (100 in the last 8 years), 9143 citations, an H-index of 35 and an i10-index of 82, Dr Chhuneja's contributions to the field are substantial.

Dr Satbir Singh Gosal, currently Vice-Chancellor (Chief Executive Officer) of Punjab Agricultural University, is an accomplished biotechnology educator with numerous academic awards. His publications include 10 books, 38 book chapters, 5 laboratory manuals and 3 innovation papers, with 4200 citations, an H-index of 31, and an i10-index of 84.

The combined expertise of these three authors – a meritorious genetics and biotechnology teacher, an experienced biotechnology researcher and a motivational biotechnology leader – ensures the book's comprehensiveness and authority. Published by CABI, UK, this book is a significant achievement, adding another feather in its cap.

I would like to highlight Dr Gurbachan Singh Miglani's *Dictionary of Plant Genetics and Molecular Biology*, published in 1998 by Food Products Press, an imprint of Haworth Press, USA, later acquired by Taylor & Francis, USA. Due to the overwhelming response from the scientific community, it was reprinted in 2000 by Viva Books Private Ltd., New Delhi, India. Since then, the field of biotechnology has advanced significantly, especially in agriculture and therapeutics.

Today, there is a vast ocean of knowledge in agricultural biotechnology that students, teachers and researchers must grasp. Recognizing this, there has been a pressing need to publish a *Glossary of Agricultural Biotechnology*, covering all scientific areas that contribute to this multidisciplinary field. This book aims to meet the needs of students, educators, researchers, biotechnology-related government agencies and the biotechnology industry involved in agriculture.

Given the success of Miglani's earlier dictionary, I am confident that the *Glossary of Agricultural Biotechnology* will be warmly welcomed by the scientific community. I wish this book great success in becoming an essential resource for all those engaged in the dynamic field of agricultural biotechnology.

P.C. Trivedi PhD, Post-Doc (USA),
FLS (London), FBS, FPSI, FNSI, FBRS, FMA, FES, FNRS, FIAT,
Former Vice-Chancellor of
Jai Narain Vyas University, Jodhpur
M.D.S. University, Ajmer
DDU Gorakhpur University, Gorakhpur
Dr R.M.L. Avadh University, Faizabad
Maharaja Ganga Singh University, Bikaner
Former Head, Botany and Biotechnology, University of Rajasthan, Jaipur

Preface

When specializing in any field of study, one encounters numerous terms that require a thorough understanding of their origin, meaning and application. Precision in defining these terms is essential. In classroom teaching, while teachers often explain concepts, they sometimes fail to provide precise definitions, leaving a gap in our teaching methodology. A glossary addresses this by defining confusing, technical or advanced words, synonymous terms, abbreviations and acronyms, offering both the term and its full name.

Compiling a glossary is a valuable tool for both teachers and students. Given the rapid advancements in the field, regularly updated glossaries are essential. This has been particularly lacking in agricultural biotechnology. Typically, glossaries appear as a section at the end of a book, as a part of a review article, or as an entire book of terms. The *Glossary of Agricultural Biotechnology* is an effort to fill this gap. It aims to provide a comprehensive glossary in reader-friendly language, covering terms from various basic and applied disciplines that have contributed to modern crop improvement approaches. We have included examples wherever possible to enhance understanding. This book is designed to meet the needs of students, teachers, researchers, biotechnology-related government agencies and the biotechnology industry involved in agriculture.

As a glossary, this book does not follow the conventional chapter format found in traditional educational texts. Instead, it features a user's guide, a list of abbreviations and the glossary itself. Terms are organized alphabetically from A to Z, each with a brief description or definition and examples where necessary. Each alphabet section (e.g. 'A') functions like a chapter, making the book effectively 26 chapters long, from Chapter A to Chapter Z.

The book concludes with a bibliography listing dictionaries and glossaries already published on the subject. To keep the book concise, detailed references are omitted, as these can easily be found online using the term, the first author's name and the year of publication.

The *Glossary of Agricultural Biotechnology* will serve as a reference for students, teachers, scientists in agricultural universities, research workers in agricultural biotechnology companies, technical staff in government and non-government organizations and agricultural biotechnology consultants. This book aims to make a lasting impression on the scientific community, providing a valuable resource for agricultural students, educators, researchers and policy makers. We present this glossary in the hope that it will become an indispensable tool in the field of agricultural biotechnology.

In deep gratitude, we humbly acknowledge the pioneering scientists whose groundbreaking discoveries have shaped agricultural biotechnology. We also extend our appreciation to the authors who have compiled and updated essential terminologies in various books and e-books. Drawing from earlier dictionaries, we have meticulously revised and adapted entries to reflect the latest advancements in the field.

Our heartfelt thanks extend to Prof Gurdev S. Khush, distinguished former Principal Plant Breeder at the International Rice Research Institute and esteemed Adjunct Professor at the University of California, Davis. A recipient of numerous prestigious awards including the Borlaug Award, Japan Prize, World Food Prize, Padma Shri, Wolf Prize in Agriculture and VinFuture Prize, Prof Khush has graciously provided his guidance and blessings by contributing a heartfelt 'Message' for our readers.

We are deeply honoured by the participation of renowned Indian botanist Prof (Dr) P.C. Trivedi, former Vice-Chancellor of five state universities and a record holder for his prolific contributions with over 208 books in the subject. His esteemed 'Foreword' enriches this book, reflecting his unparalleled dedication to plant sciences, celebrated by the publication of 20 felicitation books from his students and well-wishers.

The early guidance and support for this book were generously provided by Late Dr Manjit Singh Kang, former Vice-Chancellor of Punjab Agricultural University (PAU), Ludhiana, India, and former Adjunct Professor at Kansas State University, USA. Dr Kang, renowned for his work in quantitative genetics, edited and authored numerous influential books. We also gratefully acknowledge the ongoing inspiration from Late Dr Darshan S. Brar, former Head of the Plant Breeding, Genetics, and Biotechnology Division at the International Rice Research Institute and Honorary Adjunct Professor at SAB, PAU, Ludhiana. Their contributions have profoundly influenced this and many other publications. Our sincere thanks to Dr (Mrs) Sheetal Thapar, Professor of Journalism and Nodal Officer (Content Operations), PAU for critically examining and making useful suggestions in the preliminary pages of the book.

Several colleagues and students from the School of Agricultural Biotechnology, Punjab Agricultural University, Ludhiana, have made significant contributions to this book, directly or indirectly. Dr Naresh Kumar Manda and Ms Ritika Chanan ensured the text's similarity index remained within acceptable limits. Dr (Mrs) Maninder Kaur ensured the inclusion of recent discoveries in agricultural biotechnology. Prabhasis Kaur, a BTech Biotechnology student, and Guranshpal Singh, a BSc Agriculture (Hons.) student, dedicated their weekends from April 2023 to March 2024 to suggest new terms from a student perspective.

Writing this book demanded time we would typically spend with our families, relatives and friends. We are grateful to our families for supporting us to work from home. Their understanding and encouragement made this endeavour possible, and we extend our heartfelt thanks to them all.

Gurbachan Singh Miglani Parveen Chhuneja Satbir Singh Gosal

About the Authors

Gurbachan Singh Miglani

Ex-Adjunct/Visiting Professor, School of Agricultural Biotechnology, Punjab Agricultural University, Ludhiana, Punjab – 141 004, India. E-mail: miglanigs1945@gmail.com

Obtained BSc (Agri & AH) in 1968 and MSc (Genetics) in 1970, from Punjab Agricultural University (PAU), Ludhiana, India; Obtained PhD in 1980, from Howard University, Washington, DC, USA; Retired as Professor of Genetics in 2005 from Department of Plant Breeding, Genetics and Biotechnology, Punjab Agricultural University (PAU), Ludhiana, India, after putting in 35 years of service; Taught undergraduate and postgraduate courses on introductory genetics, fundamentals of genetics, general genetics, animal genetics, elementary evolution, human genetics, cytogenetics, molecular genetics, developmental genetics, biochemical genetics, population genetics, immunogenetics, process of evolution, genetics seminars, advanced genetics, advanced molecular genetics and mutagenesis; Also taught at Howard University, Washington, DC, USA, as a Graduate Teaching Assistant; After superannuation, invited by School of Agricultural Biotechnology (SAB), PAU in March 2010 to teach courses on molecular genetics, epigenetics, genomics and proteomics, molecular biotechnology and genomics, principles of biotechnology, biotechnology seminars, advances in plant molecular biology, advances in microbial genetics and biotechnology to BSc (Biotechnology)/BSc Tech (Biotechnology), MSc (Biotechnology) and PhD (Biotechnology) students, as an Adjunct Professor/Visiting Professor, where worked till August 2020; Guided 12 MSc and one PhD students; Authored 175 publications, including 17 books in the areas of genetics, epigenetics, molecular genetics, biotechnology and so on, and six laboratory manuals; Recipient of the Meritorious Teacher award of the Punjab Agricultural University, Ludhiana, India for the Year 1995–1996.

Parveen Chhuneja

Principal Molecular Geneticist cum Director, School of Agricultural Biotechnology (SAB), Punjab Agricultural University (PAU), Ludhiana, Punjab, India- E-mail: pchhuneja@pau.edu

Obtained BSc (Medical) in 1985 and MSc (Cytogenetics) in 1987 from Punjabi University, Patiala, Punjab, India; PhD (Genetics) in 1992 from Punjab Agricultural University (PAU), Ludhiana, Punjab, India; Serving PAU in different capacities (Assistant Scientist (DST Young Scientist award), Assistant Cytogeneticist, Assistant Cytogeneticist/Geneticist (Senior Scale), Geneticist/Molecular Geneticist, Senior/Principal Molecular Geneticist) since 1999; Presently, Director, School of Agricultural Biotechnology, PAU. International Awards/Recognitions include Jeanie Borlaug Laube Women in Triticum (WIT) Mentor award by Borlaug Global Rust Initiative (BGRI) – 2022; Gene Stewardship Team Award by Borlaug Global Rust Initiative (BGRI) – 2018; Member, International Organising Committee, International Wheat Congress – 2018–2023; and Associate Editor of the International Journal BMC Genomics (2018 till date); Fellow of National Academy of Agricultural Sciences (NAAS) – 2022; National Academy of Sciences, India (NASI) – 2022; Fellowship of Indian Society of Genetics and Plant Breeding – 2020. National awards and recognitions include Dr S. Rajaram Outstanding Woman Scientist Award-2021 by Society for Advancement of Wheat and Barley Research (SAWBAR); Panjabrao Deshmukh Award for Outstanding Woman Scientists – 2018 by Indian Council of Agricultural Research; Dr Gurdev Singh Khush Distinguished Professor Award 2016–2019 by Punjab Agricultural University, Ludhiana, Award of Plaque and Citation for 2013–2014 by Punjab Agricultural University, Ludhiana; Taught UG and PG courses on Cytogenetics, Molecular Biology, Molecular Genetics, Crop Biotechnology, and Genomics; Guided 15 MSc and 9 PhD students as Major advisor.

Satbir Singh Gosal

Vice-Chancellor (Chief Executive Officer), Punjab Agricultural University, Ludhiana (since August 20, 2022) - E-mail: ssgosal@rediffmail.com

Member Board of Management, Punjab Agricultural University, Ludhiana, Punjab, India (July 8, 2015–July 26, 2021); Former Director of Research, Punjab Agricultural University, Ludhiana, Punjab, India; Former Additional Director Research (Agric.), Punjab Agricultural University, Ludhiana, Punjab, India; Founder Director, School of Agricultural Biotechnology (SAB), Punjab Agricultural University, Ludhiana, Punjab, India; Started BSc Biotechnology (Hons), MSc Biotechnology and PhD Biotechnology programmes; Former, Head, Department of Biotechnology, PAU, Ludhiana; BSc (Punjab University, Chandigarh); MSc (Plant Breeding) from Punjab Agricultural University, Ludhiana, Punjab, India; PhD (Plant Breeding) from Punjab Agricultural University, Ludhiana, Punjab, India, with specialization in Agricultural Biotechnology; Postdoctoral Research at University of Nottingham, UK and John Innes Centre, Norwich, England for 2 years and 6 months; Developed Tissue Culture methods for more than 20 crop plants; based on this work, several commercial Tissue Culture Units have been established in Punjab; the Punjab Vidhan Sabha has also passed 'The Punjab Tissue Culture Based Seed Potato Bill, 2020'. Professional recognitions include Award of Royal Society London's Bursary 1983; Award of Rockefeller Foundation's (USA) Career Fellowships 1993–2000 (3 months each year for 6 years); Served as a Consultant FAO/IAEA, Vienna, Austria to undertake tissue expert mission to Iraq during 1997; Honorary member Board of Assessors, Australian Research Council, Canberra; Lifetime Achievement Award by Punjab Academy of Sciences; President Punjab Academy of Sciences (2012–2015); Elected member (Fellow) of Plant Tissue Culture Association (India); Fellow of Indian Society of Genetics and Plant Breeding, New Delhi, India; Distinction award by Society for the Promotion of Plant Science Research, Jaipur, India; Fellow of Punjab Academy of Sciences; Advisory member of several Universities/Institutes in the area of Biotechnology; Served as a member of Review Committee on Genetic Manipulation (RCGM) for three years (March 2009–March 2012) at DBT, Government of India, New Delhi.

User's Guide

- Different terms included in this book have been arranged alphabetically (A–Z) on word-by-word basis.
- Terms that are of limited use, very general in nature or self-explanatory have been avoided.
- Name of the term in bold, lower case (upper case to be used if grammar rules demand) followed by colon (:) and its definition in sentence structure. For example, (i) **feedback:** The influence of the result of a process upon the functioning of the process. (ii) **Klenow fragment:** The larger part of a DNA polymerase.
- When a term is defined by different authors or it may have different meanings in different contexts, such definitions are indicated by (1), (2), (3), etc.
- Sometimes a term is explained by splitting into various parts, indicated as (i), (ii), (iii) and so on; subparts, if any, are indicated as (a), (b), (c) and so on.
- Alternative names of terms, when known, have been given after defining the term; such names are followed by *Also known as* (abbreviated as *Aka*). The alternative names are written in SMALL CAPITALIZED LETTERS. For example, *Aka* CONTRACTION OF CHROMOSOMES.
- Cross-referred terms are provided wherever necessary for economizing space, demonstrating interrelationships and organizing the material in a clear manner, Cross-referred terms are in SMALL CAPITALIZED LETTERS. Thus, information is seldom repeated in this book. For example, *See* SATELLITE DNA; *Compare with* PHENOTYPE; and *Contrasts* HAPLOID.
- Some important words/terms mentioned in the text have been written in SMALL CAPITALIZED LETTERS just to draw attention of the reader.
- Different items listed under one term written in SMALL CAPITALIZED LETTERS which are defined as separate terms are indicted as *See details as separate terms.*
- It has been our sincere effort to name the authors, along with the year, who defined or subsequently modified a term.
- In this area of fast developing science and technology, definitions or descriptions about some terms may be modified with time. The authors, therefore, accept no responsibility for the legal validity, accuracy, adequacy and interpretation of terms given in this book. This is a modest effort to serve the scientific community.
- Errors brought to our notice will be rectified whenever possible.

ABBREVIATIONS USED

Abbreviation	Full form
A	adenine
A site	aminoacyl site
A*	tautomeric adenine
A5′pppp5′A or Ap4A	diadenosine 5′,5″-P1
aaRS	aminoacyl-tRNA synthetase
AAT	analysis and annotation tool
AAV	adeno-associated virus
AAV vectors	adeno-associated virus vectors
ABA	abscisic acid
Abbrev.	abbreviation
ABC transporter	ATP-binding cassette transporter
ABEs	adenine base editors
ABI	Applied Biosystems
ABM	agent-based model
AB-QTL	advanced backcross QTL
ABRC	Arabidopsis Biological Resource Center
ABS	antibody-based screening
ac	acetylation
Aca protein	anti-CRISPR-associated protein
ac4C	N4-acetylcytidine
ACC	1-aminocyclopropane-1-carboxylic acid
ACC oxidase	1-aminocyclopropane-1-carboxylic acid oxidase
ACC synthase	1-aminocyclopropane-1-carboxylic acid synthase
Ac-CoA	acetyl-Coenzyme A
Ac/Ds system	activator-dissociation system
ACLY	ATP citrate lyase
ACMV	African cassava mosaic virus
Aco	aconitase
ACPH enzyme	acid phosphatase enzyme
Acph	*acid phosphatase* gene
Acrs	anti-CRISPRs
acr genes	anti-CRISPR genes
ACSS2	cytosolic acetyl-CoA synthetase
AD	acidic domain; airborne defence

Ad2MLP	Adenovirus-2 major late promoter
ADA	adenosine deaminase
ADAR	adenosine deaminases acting on RNA
ADCC	antibody-dependent cell-mediated cytotoxicity
Adh	*alcohol dehydrogenase* locus
ADI	acceptable daily intake
Adj.	adjective
AdMLP	adenovirus major late promoter
AdoMet (or SAM)	S-adenosylmethionine
ADP	adenosine 5'-diphosphate
ADpr	ADP-ribosylation
ADT	abstract data type
A-EJ	alternative end joining
AER	apical ectodermal ridge; apical epidermis ridge
AFLP	amplified fragment length polymorphism
AFLP markers	amplified fragment length polymorphism markers
AFLP PCR	amplified fragment length polymorphism PCR
AFM	atomic force microscope; atomic force microscopy
AF-M	AlphaFold-Multimer
AFSM	amplified-fragment single-nucleotide polymorphism and methylation
AG	agamous (or AGAMOUS) protein
AG	*AGAMOUS* gene
AgBio	agricultural biotechnology; agrobiotechnology
AGGAGG	Shine-Dalgarno sequence
$AgNO_3$	silver nitrate
Ago	*argonaute* gene
AGO	argonaute protein
agRNA	antigene RNA
AHAS	acetohydroxy acid synthase
AHCs	apical hub cells
AHL	N-acyl homoserine lactone
AHR	assisted human reproduction
AI	artificial intelligence
AID	acquired immune deficiency; activation-induced (cytidine) deaminase
AIDS	acquired immunodeficiency syndrome
AIS	artificial immune system
Aka	also known as
Al	aluminium
AlkB	Alkylation B
ALP	amplicon length polymorphism
α-helix	alpha-helix
ALS	*acetolactase synthase* gene
AMC-MPS	antibody-mediated capture and massively parallel sequencing

AmI	ambient intelligence
AASs	aminoacyl synthetases
amiRNA(s)	artificial microRNA(s)
AMOVA	analysis of molecular variance
AMP	adenosine 5'-monophosphate
Amp	aminopeptidase
AMP-PCR	anchored microsatellite primed polymerase chain reaction
AMR	antimicrobial resistance
Å	Angstrom
Amu or Dalton	atomic mass unit
ANCOVA	analysis of covariance
aNSOM	apertureless near-field optical microscope
ANOVA	analysis of variance
ANT-C	*Antennapedia*
anti-HISG	anti-human immune serum globulin
anti-o	anticoding strand of operator locus
antisense-PTGS	antisense post-transcriptional gene silencing
AO	acridine orange
AOCs	antibody-oligonucleotide conjugates
AOSCA	association of official seed certifying agencies
AP - apyrimidic	apurinic; apyrimidinic
AP endonuclease	apurinic endonuclease; apyrimidinic endonuclease
AP sites	apurinic/apyrimidinic sites
APA	alternative polyadenylation
APC	adenomatous polyposis coli
APC/C	anaphase-promoting complex/cyclosome
APCs	antigen-presenting cells
APHIS	Animal and Plant Health Inspection Service of USDA, USA
API	application programming interface
aPKC	atypical protein kinase C
APLF	aprataxin and polynucleotide kinase-like factor
apoB	apolipoprotein B
APOBEC	apolipoprotein B RNA-editing catalytic component
Ap-PCR	arbitrarily primed PCR
APPFPA	Australian Plant Phenomics Facility Plant Accelerator
AppppA	diadenosine tetraphosphate
AQUA	absolute quantification
AR	augmented reality
ara	arabinose
ara operon	arabinose operon
ARD	asymmetric RNA duplexes
ARF	ADP-ribosylation factor
Arg	arginine

arg operon	arginine operon
ARI	absolute risk increase
ARMGs	antibiotic resistance marker genes
ARP 2/3 complex	actin related protein 2/3 complex
ARR	absolute risk reduction
ARS	autonomously replicating (or replication) sequences
ARTM AFM	attractive regime tapping-mode atomic force microscope/microscopy
ARTS	alignment of RNA tertiary structures
AS	alternative splicing
ASA	allele-specific amplification; acetylsalicylic acid
ASAP	allele-specific associated primers
ASCENDx	acoustic separation and concentration of exosomes for nucleotide detection
ASCII	American Standard Code for Information Interchange
ASCs	antibody-secreting cells
asn	aspargine
ASO	allele-specific oligo; antisense oligonucleotide
AS-PCR	allele-specific PCR
asRNA	antisense RNA
AST	antibiotic susceptibility test
ASTA	American Seed Trade Association
ATAD3A	ATPase family AAA domain-containing protein 3A
ATM	ataxia telangiectasia mutated
ATP	adenosine 5'-triphosphate
ATPase	adenosine 5'-triphosphatase
ATR	ataxia telangiectasia and Rad3-related
ATRIP	ATM and Rad3-related (ATR)-interacting protein
ATRX7	*Arabidopsis* TRITHORAX-RELATED7
AUDPC	area under disease progress curve
AUG	translation initiation codon
AUHTP	AI-based UAS-based High-Throughput Phenotyping
AutoML	automated machine learning
AVP	antiviral protein
AVR	avirulence
Avr	*Avirulence* gene
AWB	Australian Wheat Board
AWU	atomic weight unit
Aza-IP	aza-immunoprecipitation
b	regression coefficient
B cells	B lymphocytes (leukocytes)
b1 locus	*booster 1* locus
BAC	bacterial artificial chromosome
BAC library	bacterial artificial chromosome library
BAC vector	bacterial artificial chromosome vector

BAH domain	Bromo-adjacent homology domain
BAK1	brassinosteroid insensitive 1-associated kinase 1
bar	*bialaphos resistance* gene
BARC	Bioinformatics Applications Research Center
BARNASE	BActerial RiboNucleASE
BC	bisulfite conversion
BCAS	backcross-assisted selection
bcd	bicoid protein
BDI	belief-desire-intention software model
BDMV	bean dwarf mosaic virus
BE	base editor
BEAST	Bayesian Evolutionary Analysis Sampling Trees
BEEM	ballistic electron emission microscopy
BER	base excision repair
BESS	base excision sequence scanning
BET proteins	bromodomain and extraterminal proteins
β-AR	β-adrenergic receptor
β-arr2	β-arrestin 2
BEV	baculovirus expression vector
BeYDV	barley yellow dwarf virus; bean yellow dwarf virus
BGA	blue-green algae
BI	behavior/behaviour informatics
BIB	balanced incomplete block design
BIBAC libraries	binary BAC libraries
BIFS	biologically integrated farming systems
BIK1	botrytis-induced kinase 1
bio	biotin
BioMEMS	biomedical microelectromechanical systems
BioNEMS	biofunctionalized nanoelectromechanical systems
BioPOEMS	biopolymeroptoelectromechanical systems
BIR	break-induced replication
Bk	banded krait (*Bungarus fasciatus*)
Bkm	banded krait minor
Bkm satellite	banded krait minor satellite
bla	*beta-lactamase* gene
BLAST	basic local alignment search tool
BLOSUM	blocks substitution matrix
BLP	best linear prediction; blended learning programs
BLUP	best linear unbiased prediction
BMPs	bone morphogenic (morphogenetic or marrow) proteins
BOD	biochemical oxygen demand; biological oxygen demand
bp	base pair
BPA	bisphenol A

Bq	becquerel
BR	brassinosteroid
BRCA1	*BReast CAncer gene 1*
BRCA2	*BReast CAncer gene 2*
brk	*brinker*
bRNA	bridge RNA
BS	bisulfite/bisulfite sequencing
BSA	bovine serum albumin; bulked segregant analysis
BSA-seq	bulked segregant analysis sequencing
BSC	biological species concept
BSE	bovine spongiform encephalopathy
BsRAD-seq	bisulfite-converted restriction site associated DNA sequencing
BT	behavior/behaviour tree
Bt	*Bacillus thuringiensis*
btd	*button head*
Bti	*Bacillus thuringiensis israelensis*
Btk	*Bacillus thuringiensis kurstaki*
Btt	*Bacillus thuringiensis tenebrionis*; *Bacillus thuringiensis tolworthi*
BTT	balanced tertiary trisomic
bu	butyrylation
bv	biovar
BWS	Beckwith Weidermann Syndrome
BX-C	*Bithorax complex*
BYDV	barley yellow dwarf virus; bean yellow dwarf virus
BYV	beet yellows virus
C	cytosine
C sequences	constant sequences
C*	tautomeric cytosine
C/EBPs	CCAAT/enhancer-binding proteins
CA	combining ability; computer audition
CAAS	Chinese Academy of Agricultural Science
CADD	computer-aided drug design
CAEs	*cis*-acting elements
CAF1 factor	chromatin assembly factor-1
C-AFM	conducting atomic force microscope/microscopy
C-AFM/STM	combined atomic force microscope/scanning tunnelling microscopy
CAGE	cap analysis gene expression
CAK	Cdk-activating kinase
CaM	Calmodium
CAM	cell adhesion molecule; crassulacean acid metabolism
CAM AFM	constant amplitude mode of non-contact atomic force microscope
CAM cambium	crassulacean acid metabolism cambium
CaM kinase	Ca^{2+}/calmodulin-dependent protein kinase

cAMP	3′,5′-cyclic adenosine monophosphate
CaMV	cauliflower mosaic virus
CaMV 35S	cauliflower mosaic virus 35S promoter
candidate SNP	candidate single-nucleotide polymorphism
Canola™	a type of rapeseed variety
CAO	*Chlorophyllide a oxygenase* gene
CAP	cAMP-activated protein; catabolite activator protein
CAPPs	CRISPR-associated primase-polymerases
CAPS	cleaved amplified polymorphic sequence
caRNA	chromatin-associated RNA
CART analysis	classification and regression tree analysis
Cas9	CRISPR-associated 9
Cas 13 an	Ancestral clade of Cas 13
Cas14	CRISPR-associated 14
CAST	CRISPR-associated transposase
CAT	chloramphenicol acetylase or acetyltransferase
CAT	*chloramphenicol acetyltransferase* gene
CATH	Class (C), Architecture (A), Topology (T) Homologous (H)
causative SNP	causative single nucleotide polymorphism
CB	Cajal body
CBC20/CBC80	cap binding complex heterodimer
CBD	Convention on Biological Diversity
CBF1	Centromere Binding Factor 1
CBP	CpG-binding protein; cap-binding protein; CREB binding protein; Conservation Breeding Programme
CC	coefficient of coincidence; central cell; Creative Commons
CCA1	*circadian clock associated 1* gene
CCC	chlorocholine chloride
cccDNA	covalently closed circles of DNA
CCCM	carboxysomal carbon-concentrating mechanism
CCD	charged-coupled device
CCD camera	charged-coupled device camera
CCM	chemical cleavage of mismatch; carbon-concentrating mechanism
CCS	circular consensus sequencing
CCS technology	carbon capture and storage technology
ccSSR	consensus chloroplast simple (single) sequence repeat
CD	catalytic domain
CD site	common docking site
CD/ED	catalytic domain/effector domain
CDC	cell division cycle
CDCA7	cell division cycle associated protein 7
cdc	*cell-division-cycle* gene
CDC protein	cell division cycle protein

CDEs	centromeric DNA elements
CDK11	cyclin-dependent kinase 11
CDKs	cyclin-dependent protein kinases
cDNA	complementary DNA; copy DNA
CDPKs	calcium-dependent protein kinases
CDR	complementarity-determining region
CDS	coding sequence (or region)
CEs	click editors
CE	capillary electrophoresis
Cel operon	cellulose utilization operon of *E. coli* 12
CELI	celery endonuclease
CE-LIF	capillary electrophoresis-laser-induced fluorescence
CEMT project	Centre for Epigenome Mapping Technologies project
CEN	centromere
CENPA	centromere protein A
CER	control event rate
CeU-seq	N_3-CMC-enriched pseudouridine sequencing
CFCGA	computer-facilitated candidate gene approach
CFM	chemical force microscopy
CFTR	cystic fibrosis transmembrane regulator
cfu	colony-forming unit
CG or CpG	cytosine-guanine or cytosine-phosphate-guanine
CGB	Community Gene Bank
CGH	comparative genomic hybridization
CGI	CpG island
CGIAR	Consultative Group on International Agricultural Research
CGIs	CpG islands
cGMP	cyclic guanosine 3′,5′-cyclic monophosphate
CGN	*cis*-Golgi network
cgs	catabolic gene activator site
CGV	cryptic genetic variation
CH	casein hydrolysate
C_H segment	constant heavy segment
CHA	chemical hybridizing agents
CHA chaff	chemical hybridizing agent chaff
CHAPS	3-[(3-cholamidopropyl)-dimethylammonio]-1-propanesulfonate
CHARM	comprehensive high-throughput arrays for relative methylation; coupled histone tail for autoinhibition release of methyltransferase
ChatGPT	Chat Generative Pre-Trained Transformer
cHC	constitutive heterochromatin
CHD	chromodomain helicase-DNA binding; coronary heart disease
CHD proteins	chromosome-ATPase/helicase-DNA binding proteins
CHI	chimera

ChIP	chromatin immunoprecipitation
ChIP-on-chip or ChIP-chip	ChIP technology and genomic tiling arrays
ChIP-SAGE or STAGE	chromatin immunoprecipitation serial analysis of gene expression
ChIP-seq	ChIP and high-throughput sequencing; ChIP-sequencing; Chromatin immunoprecipitation sequencing
χ^2	chi-square
CiA	chromatin *in vivo* assay
CIB	calcium- and integrin-binding protein 1
CID	collision-induced dissociation
CIHR	Canadian Institutes of Health Research
CIM	composite interval mapping
CIMMYT	Centro Internacional de Mejoramiento de Maiz y Trigo
circRNA	circular RNA
CIS element	composite insertion sequence element
cis-TGS	*cis*-transcriptional gene silencing
cit/dea	citrullination/deamination
CITES	Convention on International Trade of Endangered Species
Civet	cluster investigation and virus epidemiology tool
CKI	cdk inhibitor or cyclin-dependent kinase inhibitors
Cl^-	chloride ion
C_L segment	constant light segment
CL3 lab	containment level 3 laboratory
CLERY	combinatorial libraries enhanced by recombination in yeast
CLF	CURLY LEAF
CLIMB	cloud infrastructure for microbial bioinformatics
clk chemistry	click chemistry
clk DNA	click DNA
clone overlap SNP	clone overlap single-nucleotide polymorphism
CLP	chromosome length polymorphism
Clr4	Cryptic loci regulator 4
CLRC	Clr4-containing Clr4 methyltransferase complex
CLS	centrosomal localization of signal
cM	centimorgan
CMAL	chloroplast MraW-like
CMAs	chromatin-modifying activities
CMC	critical micelle concentration; chromatin-modifying complex
CM-cellulose	carboxymethyl-cellulose
CMG assembly	CDC45-MCM2-7-GINS complex
CMGT	chromosome-mediated gene transfer
CML	chronic myeloid leukaemia
CMLM	compressed mixed linear model
CMMs	cellular memory modules
CMS	cytoplasmic male sterility

CMT2	chromomethylase 2
CMT3	chromomethylase 3
CMTs	cortical microtubules
CMTs	chloromethylases
CMV	cucumber mosaic virus; cytomegalovirus
CNL proteins	CC-NBS-LRR proteins
CNS	central nervous system
CNTF	ciliary neurotrophic factor
CNV	copy number variation
C_o	concentration
CoA	coenzyme A
COBRA	combined bisulfite restriction analysis
coding SNP or cSNP	coding single-nucleotide polymorphism
CoIP	co-immunoprecipitation
col factor	colcins factor
common SNP	common single-nucleotide polymorphism
c-onc	cellular oncogene
contig map	contiguous map
COP	coefficient of parentage
COS	conserved orthologous set (of markers)
CoSMoS	colocalization single molecule spectroscopy
CoTC	cotranscriptional cleavage
Co-TGS	cotranscriptional gene silencing
cp	chloroplast
CP	capsid protein (virus coat)
CP	*coat protein* gene (virus)
CP-AFM	conducting probe atomic force microscope/microscopy
CPB	Cartagena Protocol on Biosafety
CPD	contact potential difference; cyclobutane pyrimidine dimer
cpDNA	chloroplast DNA
CPDs	cyclobutane pyrimidine dimers
CpI	cysteine protease inhibitor gene
cPIC	core preinitiation complex
CPK5	calcium-dependent protein kinase 5
CPL2	C-TERMINAL DOMAIN PHOSPHATASE-LIKE 2
CpMV	cowpea mosaic virus
cPP proteins	consensus pentatricopeptide proteins
CPPs	cell-penetrating peptide
CPS	continuation-passing style
CPSF	cleavage and polyadenylation specificity factor
CpTI	cowpea trypsin inhibitor
CpTI	*cowpea trypsin inhibitor* gene
CPU	central processing unit

cr	crotonylation
CRAB	Carbapenem-resistant *Acinetobacter baumannii*
cr/kr	crotonylation/krotonylation
cRdDM	canonical RNA-directed DNA Methylation
CRD	cysteine-rich domain
CRDs	*cis*-regulatory domains
CREB protein	cAMP response element binding protein
CREs	*cis*-regulatory elements
CRISPaint	CRISPR-assisted insertion tagging
CRISPR	clustered regularly interspaced short palindromic repeat
CRISPRa	CRISPR activation
CRISPR-ChAP-MS	clustered regularly interspaced short palindromic repeats-based chromatin affinity purification with mass spectrometry
CRISPRi	CRISPR interference
CRMs	*cis*-regulatory modules
CROO	computerized robotic optimized obtainer
CRP	cAMP receptor protein
crRNA	CRISPR RNA
CRY2	cryptochrome 2 protein
cryo-EM	cryogenic electron microscopy
CsAAT1	*Citrus sinensis* alcohol acyltransferase 1
CSAR	cytoplasmic segregation and recombination
CSb	*citrate synthase* gene
CsCl	caesium chloride
CSF	cleavage stimulation factor; colony-stimulating factor
CSFs	colony-stimulating factors
CSI	cryptic self-incompatibility
CSIRO	Commonwealth Scientific and Industrial Research Organisation, Australia
CSR	class switch recombination
CSVs	comma-separated values
CTAB or C-TAB	cetyltrimethylammonium bromide
CTCF	CCCTC-binding factor
Ctd	C-terminal domain
CTD	canopy temperature depression
C-terminal	carboxyl terminal
Ctf18-RFC	Ctf18 alternative replication factor C
CTGS	co-transcriptional gene silencing
CTP	chloroplast transit peptide
CTV	citrus tristeza virus
CUC	cup-shaped cotyledon
CUTs	cryptic unstable transcripts
CV	coefficient of variation; composition vector; computer vision
cv	cultivar

C-value	genome size
CVD	chemical vapour deposition
CVS	chorionic villus sampling
CWR	crop wild relatives
cybrids	cytoplasmic hybrids
CYCD7	D-type cyclin 7
CyMATE	Cytosine Methylation Analysis Tool for Everyone
Cys	cysteine
cyt	cytochrome
$(CH_2O)_n$	carbohydrate
d	Darwin (unit for the rate of change); map distance
D	dextrorotary; dihydrouridine; genetic distance; linkage disequilibrium; dermatogen stage, diversity sequences
Da	Dalton
dA	deoxyadenosine
dAbs	single-domain antibodies
DAF	DNA amplification fingerprinting
dam	DNA adenine methylation
Dam methylase	DNA adenine methylase
DamID	DNA adenine methyltransferase identification
Dam-MMR	DNA methylase-directed mismatch repair
DAMPs	damage-associated molecular patterns
DARESOME	DNA analysis by restriction enzyme for simultaneous detection of multiple epigenomic states
DArT	diversity arrays technology
DBD	DNA-binding domain
DBM paper	diazo benzyloxymethyl paper
dbSNP	single-nucleotide polymorphism database
dC	deoxycytidine
DC	dosage compensation
dCas9	dead Cas9
DCC	dosage compensation complex
DCL	dicer-like protein
DCL1	dicer-like I protein
DCL2	dicer-like II protein
dcm	DNA cytosine methyltransferase system
DCR	Dicer endoribonuclease
DDB protein	DNA damage binding protein
DDM1	Deficient in DNA methylation 1
ddMMR	dam-directed mismatch repair
ddNTPs	dideoxynucleoside triphosphates
DDR	DNA damage response
DDPs	DNA-dependent polymerases

DDT	DNA damage tolerance
ddTPs	dideoxynucleoside triphosphates; dideoxyribonucleoside triphosphates
DEAD motif	Asp (D)-Glu (E)-Ala (A)-Asp (D) motif
DEAE	diethylaminoethyl
DEAE-cellulose	diethylaminoethyl cellulose
DEDD domain	Asp (D)-Glu (E)-Asp (D)-Asp (D) domain
DEF A	deficiens A
ΔG	actual free energy change
DEMRs	differential epigenetic modification regions
DEMs	DNA demethylases
DES	diethyl sulfate; discrete event simulation
DevRob	developmental robotics
df	degrees of freedom
DFF	DNA fragmentation factor
DFM	dynamic force microscopy
DFP	DNA fingerprint(ing)
DFR	dihydroflavonol 4-reductase
DFS	dynamic force spectroscopy
DFS mutation	dominant female sterile mutation
dG	deoxyguanosine
DGGE	denaturing gradient gel electrophoresis
DGT method	direct gene transfer method
DGTS	direct gene transfer systems
DH	dihaploid
DHA	docosahexaenoic acid
DHCR7	7-dehydrocholesterol reductase
dHPLC	denaturing high-performance liquid chromatography
DHs	doubled haploids
DIB	device-independent bitmap
DIC microscopy	differential interference contrast microscopy
DIF method	differential item functioning method
DigiCGA	digital candidate gene approach
DILPs	*Drosophila* insulin-like peptides
DiMeLo-seq	Directed Methylation with Long-read sequencing
DinR	*Drosophila* insulin receptor
DKO	double knockout
DL	deep learning
dLife	Digital Life Systems
D-loop	displacement loop
DM	dry matter
DMC	dynamic molecular combing
DME	Demeter
DMGs	differentially methylated genes

DMGT	DNA-mediated gene transfer
DMH	differential methylation hybridization
dmin	double minute
DML proteins	Demeter-like proteins
DML2	Demeter-like 2
DML3	Demeter-like 3
DMRs	differentially methylated regions
DMS	dimethyl sulfate
DMSO	dimethylsulfoxide
dN or Ka	nonsynonymous substitution rates
DNA	deoxyribonucleic acid
DNA fibre FISH	DNA fibre-based fluorescence *in situ* hybridization
DNA Pol δ	DNA polymerase δ
DNA Pol ε	DNA polymerase ε
DNA Pol δ-PONA	DNA polymerase δ-PONA
DNA Pol β	DNA polymerase β
DNA Pol γ	DNA polymerase γ
DNA Pol ζ	DNA polymerase ζ
DNA Pol I	DNA polymerase I
DNA Pol II	DNA polymerase II
DNA Pol III	DNA polymerase III
DNA Pol IV	DNA polymerase IV
DNA Pol V	DNA polymerase V
DNAFSMiner	DNA functional site miner
DNase	deoxyribonuclease
DNCs	DNA nanoswitch calipers
dNDPs	deoxyribonucleoside diphosphates
DNFB or FDNB	fluorodinitrobenzene
DNMT1	DNA methyltransferase 1
DNMT2	DNA methyltransferase 2
DNMT3	DNA methyltransferase 3
DNMT3A	DNA methyltransferase 3A
DNMT3B	DNA methyltransferase 3B
DNMT3L	DNA methyltransferase 3L
DNMTs	DNA methyltransferases
DNP	deoxyribonucleoprotein
DNS	Domain Name System
dNTPs	deoxynucleoside triphosphates
dNTPs	deoxynucleoside triphosphates; deoxyribonucleoside triphosphates
DOL	dye-labelled oligonucleotide ligation
DP	docking protein
DPA	diphenylamine
daa (or dpa)	days after anthesis (or days post anthesis)

dpp	decapentaplegic protein
DPP	deep plant phenomics
Dpp	*Decapentaplegic* gene
dPPR proteins	designer pentatricopeptide repeat proteins
DRAG	dinitrogenase reductase-activating glycohydrolase
DRAT	dinitrogenase reductase ADP-ribosyltransferase
DRM2	domain-rearranged methyltransferase 2
DRMs	domain-rearranged methyltransferases
dRNA	de-repressor RNA
DRT	defence-associated reverse transcriptase
dS (or Ks)	nonsynonymous substitutions rate
DSBR	double-strand break repair
DSB	double-strand break
dsDNA	double-stranded DNA
DSDR	double-stranded DNA recombineering
DSGR	daughter-strand gap repair
DSIF	DRB sensitivity-inducing factor
DSP	digital signal processing
dsRNA	double-stranded RNA
dT	deoxythymidine
DUBs	deubiquitin enzymes
DUS testing	testing a variety for distinctness (D), uniformity (U) and stability (S)
DZ	dizygotic twins
E	environmental deviation
E site	exit site
E value	expectation value
E(z)	enhancer of zeste
E. coli	*Escherichia coli*
E2F	transcription factor E2F
EA	evolutionary algorithm
EBI	European Bioinformatics Institute, UK
EBV	Epstein-Barr virus
EcBfr	*Escherichia coli* bacterioferritin
eccDNA	extrachromosomal circular DNA
ECCs	embryonic carcinoma cells
ECE	essential core element
ECF	evolving classification function
ECL	economic trait loci
ECM	extracellular matrix
EcORep	orthogonal DNA replication in *Escherichia coli*
ECs	elongation complexes
ED	effector domain
EDTA	ethylenediamine tetraacetate

EE	embryonic extract
EEA	environment of evolutionary adaptation
EED	embryonic ectoderm development
eEF2	elongation factor 2
eEFSec	eukaryotic sec-specific elongation factor
EER	experimental event rate
EFIIS	elongation factor SII
EFM	electric/electrostatic force microscope/microscopy; early flowering MYB protein
EFR	elongation factor Tu receptor
EFs	elongation factors
EGCs	embryonic germ cells
EGF	epidermal growth factor
EGFR	epidermal growth factor receptor
EGS	external guide sequences
EIA	epigenome isoform analysis; enzyme immunoassay
eIFs	eukaryotic initiation factors
EIL	economic injury level
EIM	expression imbalance map
EIS	epigenetic information system
ELC	expression-linked copy
ELIA	enzyme-linked immuno assays
ELISA	enzyme-linked immunosorbent assay
ELSI	ethical, legal, and social issues
EM	effective microorganisms; electron microscope/microscopy
E_{max}	emission energy
EMBL	European Molecular Biology Laboratory
EMN	engineered meganuclease
EMOs	epigenetically modified organisms
EMP pathway	Embden-Meyerhof-Parnas pathway
ems	*empty spiracles gene*
EMS	ethyl methanesulfonate
EMSA	electrophoretic mobility shift assay
EMT	epithelial-mesenchymal transition
ENA	Europe Nucleotide Archive
enChIP	engineered DNA-binding molecule-mediated ChIP
ENCODE project	Encyclopaedia of DNA Elements Project
ENL protein	eleven-nineteen-leukaemia protein
EOGT	extracellular O-GlcNAC transferase
EP pols	error-prone DNA polymerases
EPA	Environmental Protection Agency, USA
EPIC project	Epigenomics of Plants International Consortium Project
epiQTL	epigenetic quantitative trait loci
epiRILs	epigenetic recombinant inbred lines

epiTOF	epigenetic landscape profiling using cytometry by time-of-flight
EP-PCR	error-prone PCR
EPR	error-prone repair
EPS	extracellular polysaccharide
EPSP synthase	5-enolpyruvylshikimate 3-phosphate synthase
EPSPS	*5-enolpyruvyl shikimate-3-phosphate synthase* gene
EpYVV	eupatorium yellow vein virus
eQTLs	expression-based QTLs
ER	endoplasmic reticulum
ER family	ERECTA family
ERIC sequences	enterobacterial repeated intergenic consensus sequences
ERK	extracellular signal-regulated kinase
ERVs	endogenous retroviruses
ES	electrospray
ESCs	embryonic stem cells
ESI	electrospray ionization
ESI-MS	electrospray ionization mass spectroscopy
eSNP	electronic single-nucleotide polymorphism
ESS	evolutionarily stable strategy
ESTPs	expressed sequence tag polymorphisms
ESTs	expressed sequence tags
ET	ethylene
et al.	and others
EtBr	ethidium bromide
EtBr-CsCl	ethidium bromide-caesium chloride gradient centrifugation
ETC	electron transport chain
ETI	effector-triggered immunity
eTM	endogenous target mimic
ETS	effector-triggered susceptibility
ETS/ITS	external and internal transcribed spacers
EU	5-ethynyl uridine
EvoETR	evolved engineered transcriptional repressor
EVs	extracellular vesicles
EVS	exome variant server database, The National Heart, Lung, and Blood Institute (NHLBI)
Exo-PE	exonuclease-enhanced prime editors
exon SNP	exonic single-nucleotide polymorphism
Eya	eyes absent protein
F factor	fertility or sex factor
F plasmid	fertility plasmid
f. sp.	forma specialis
F_1, F_2, F_3, . . .	first, second, third . . . filial generations
FACS	fluorescence-activated cell sorter
FACT complex	facilitates chromatin transactions complex

FAD	flavin adenine dinucleotide
$FADH_2$	flavin adenine dinucleotide reduced
FAIRE	formaldehyde-assisted isolation of regulatory elements
FAIRE-seq	formaldehyde-assisted isolation of regulatory elements-seq
FAK	focal adhesion kinase
FB DNA	foldback DNA
FCGA	functional candidate gene approach
fCRISPR	fluorogenic CRISPR
FDA	Food and Drug Administration, USA
FDA staining	fluorescein diacetate staining
FDD	floppy disk drive
FDR	first division restitution
FDS	first division segregation
FeMoco	FeMo cofactor
FEN1	flap endonuclease 1
FFDCA	Federal Food, Drug, and Cosmetic Act
fg	Fermtogram (= 10^{-15} g)
FFPE	formalin-fixed paraffin-embedded
FGCMGT	female germ-cell-mediated gene transfer
FGF	fibroblast growth factor
FGF signalling	fibroblast growth factor signalling
FHA	forkhead associated
fHC	facultative heterochromatin
fibre-FISH	fibre fluorescent *in situ* hybridization
FIFRA	Federal Insecticide, Fungicide, and Rodenticide Act
FIS	factor for inversion stimulation
FISH	fluorescent *in situ* hybridization
5-aza-CdR	5-aza-2′-deoxycytidine
5-aza-CR	5-azacytidine
5C	chromosome conformation capture carbon copy
$5^{hm}C$ (or hm^5C)	5-hydroxymethylcytosine
5^mC (or m^5C)	5-methylcytosine
5^mU or (m^5U)	5-methyluridine
5n	pentaploid
5′TOP	5′ terminal oligopyrimidine motif
FLC	FLOWERING LOCUS C
FLC	*FLOWERING LOCUS C* gene
FLP	flippase
FMCA	fluorescence melting curve analysis
f^{Met}	N-formylmethionine
FMN	flavin mononucleotide
FMR1	fragile X mental retardation 1 or fragile X messenger ribonucleoprotein 1
FMUG	find my understudied genes

fo	formylation
FOS	fructoosaccharides
^{14}C dating	radiocarbon dating
4^{ac}C (or ac^{4}C)	N4-acetylcytidine
4C	circular chromosome conformation capture
4E-BPs	eIF4E-binding proteins
4^{m}C (or m^{4}C)	cytosine-N4 methylation
4n	tetraploid
f.w.	formulary weight
FPC	fingerprinted contig
FpPG	*Fusarium phyllophilum* polygalacturonase
FQPA	Food Quality Protection Act
FRET	fluorescence resonance energy transfer
Fri	*FRIGIDA* gene
fRNAs	functional RNAs
FRP	flippase recombination target
FRTs	flippage recognition target sites
FSB	fruit and shoot borer (*Leucinodes orbonalis*)
FWA	*FLOWERING WAGENINGEN* gene
FXS	fragile X syndrome
FYM	farm yard manure
Fz4	Frizzled 4
G	guanine
G protein	GTP-binding protein
G*	tautomeric guanine
G1 phase	gap 1 (or growth 1 phase)
G2 phase	gap 2 or (growth 2 phase)
G3-P	glucose 3-phosphate
G6-P	glucose 6-phosphate
G6-PD	glucose 6-phosphate dehydrogenase
G6-PD	*glucose 6-phosphate dehydrogenase* gene
GA	gibberellic acid; genetic algorithm
GA testing	gibberellic acid (response) testing
GABA	γ-aminobutyric acid
Gadd45a	growth arrest and DNA damage-inducible protein 45 alpha
GAG	glycosaminoglycan
gai	gibberellic acid insensitive genes
GAI	gibberellic acid insensitive
GAL4	galactose 4 regulator
GALT	gut-associated lymphoid tissues
γ-TuRC	γ-tubulin ring complex
GAPIT	genome-associated prediction integrated tool
GAPs	GTPase-activating proteins

GATT	General Agreement on Tariffs and Trade
Gb or GB	gigabase
gbM	gene body methylation
GBS	genotyping by sequencing
GBSS	granule-bound starch synthases
GC	gas chromatography
gca	general combining ability
gcp/cell	genome-containing particles per cell
GDH	glutamate dehydrogenase
GDOs	gene drive organisms
GDPR	General Data Protection Regulation
GDSs	gene drive systems
GDTs	gene drive technologies
GEAC	Genetic Engineering Appraisal Committee, India
GEEN	genome editing with engineered nucleases
GEF	Galinstan expansion femtosyringe; guanine nucleotide exchange factor
GEFs	genetically engineered foods
GEI	genotype × phenotype interaction
GEM	genetically engineered microbe; genetically engineered microorganisms; germplasm enhancement for maize
GEMP	genetically engineered microbial pesticides
GEMS	Gene Expression Mining Server
GEMs	genetically engineered microorganism; Germplasm Enhancement for Maize
gene-based SNP	gene-based single-nucleotide polymorphism
GEOs	genetically engineered organisms
GEPs	gene expression programs
GERU	genomically encoded recording unit
GES-related risks	genetically engineered seeds-related risks
gfp	*green fluorescent protein* gene
GFP	green fluorescent protein
GFs	genetic foods; growth factors
GFSST	gene functional similarity search tool
GG-NER	global genomic nucleotide excision repair
GHGs	greenhouses gases
GI	geographical indications; GenInfo identifier
Gi	inhibitory G protein
GIF	Graphics Interchange Format
GIGO	garbage in, garbage out
GINA	Genetic Information Nondiscrimination Act
GISH	genomic *in situ* hybridization
gl	glutarylation
GlcNAc	N-acetylglucosamine
GLM	generalized linear model

GluProRS	fused glutamyl- and prolyl-tRNA synthetase
GMAC	Genetic Manipulation Advisory Committee
GMAT	genome-wide mapping technique
GMCs	genetically modified crops
GMFs	genetically modified foods
GMMs	genic molecular markers; genetically modified microorganism
GMOP	Generic Model Organism Project
GMOs	genetically modified organisms
GMPP	genetically modified pest protected plants
GMPs	genetically modified plants
GNA	*Glyanthus nivalis aglutenin* gene
GNAT	Gcn5-related N-acetyltransferases
GOGAT	glutamine-2-oxoglutarate aminotransferase
GOI	gene of interest
GOR method	Garnier, Osguthorpe and Robson method
GOS	galacto-oligosaccharides
GPCRs	G-protein-coupled receptors
GPI anchor	glycosylphosphatidylinositol anchor
GPS	Global Positioning System
GPU	graphics processing unit
GR	glucocorticoid receptor
GRA	gel retardation analysis
GRAS	gibberellic acid insensitive repressor of GAI and scarecrow
GRE	glucocorticoid receptor element; glucocorticoid response element
GRK	G protein-coupled receptor kinase
GRK2	G protein receptor kinase 2
gRNA	guide RNA
GRO	genomically recoded organism
GS	gene silencing; glutamine synthetase; genomic selection
Gs	stimulatory G protein
GSCs	germline stem cells
GSK-3β	glycogen synthase kinase-3β
GSS	genome survey sequences
GSS technique	genetic sexing strain technique
GSSM	gene site saturation mutagenesis
GST	glutathione S-transferase
GST silencing	gene-specific translational silencing
GSTs	gene-specific sequence tags
GTFs	general transcription factors; global transcription factors
gTME	global transcription machinery engineering
GTP	guanosine 5'-triphosphate
GTPases	guanosine 5'-triphosphatases
GuHCl	guanidine hydrochloride

GURT	genetic use of restriction technology
GUS	gene of uncertain significance; β-glucuronidase protein
GUS	β-*glucorinase* gene
GWAM	genome-wide association mapping
GWAS	genome-wide association study
H	gene diversity; potential heterozygosity
H^+	hydrogen ions
H_0	null hypothesis
H_1 or H_A	alternative hypothesis
h^2	coefficient of heritability
H_2O_2	hydrogen peroxide
H_3O^+	hydronium ion
HA	haemagglutinin
HACs	human artificial chromosomes
HAC-seq	hydrazine-aniline cleavage sequencing
HAMA	human antimurine antibody response
HAPPY mapping	HAPloid equivalents of DNA and PolYmerase chain reaction mapping
HART	hybrid arrested translation
HAT medium	hypoxanthine, aminopterin and thymidine medium
HATs	histone acetyltransferases
H_b	heritability in broad sense
HBV	Hepatitis B virus
HC	**heterochromatin**
H-chain	heavy immunoglobulin chain
HCI	human-computer interaction
HCN4	*Potassium/sodium hyperpolarization-activated cyclic nucleotide-gated channel 4* gene
HCPs	high CG content promoters
HCRM	host-controlled restriction and modification
hc-siRNAs or het-siRNs	heterochromatic siRNAs
HCT	histone crotonyltransferase
HD proteins	helix destabilizing proteins
HDA	helicase-dependent amplification
HDAC	histone deacetylase
HDACi	histone deacetylase inhibitor
HDCR	histone decrotonylase
HDD	hard disk drive
HDE	histone downstream purine-rich element
HDG	heteroduplex generator
HDGS	homology-dependent gene silencing
HDM	histone demethylase
HDPTGS	homology-dependent post-transcriptional gene silencing
HDR	homology-dependent repair; homology-directed repair
HD-RNAi	host-delivered/derived RNA interference

HDT	horizontal DNA transfer
HDTGS	homology-dependent transcriptional gene silencing
HD-ZIPIII	homeodomain leucine zipper III
HEGs	homing endonuclease genes
HELP	*Hpa*II tiny fragment augmentation by ligation-mediated PCR assay
HEN1	HUA ENHANCER 1
HEPA filter	high-efficiency particulate air filter
hESCs	human embryonic stem cells
HF	hydrogen fluoride
Hfr	high frequency recombination
HGI	Human Genome Initiative
HGP	Human Genome Project
HGPRT	*hypoxanthineguanine phosphoribosyl transferase* gene
HGT	horizontal gene transfer
Hh protein	hedgehog protein
HI	harvest index
hib	2-hydroxylisobutyrylation
HiC	*in situ* high-throughput chromosome conformation capture technology
HICF	high information content BAC fingerprinting
high-density BAC grid	high-density bacterial artificial chromosome grid
HJURP	Holliday Junction Recognition Protein
HI-RNAi	host-induced RNAi
HIV	human immunodeficiency virus
HLA	human leukocyte antigens
HLB	Huanglongbing
HLH	helix-loop-helix
HMG proteins	high mobility group proteins
HMMs	hidden Markov models
HMTs	histone methyltransferases
h_n	heritability in narrow sense
HNO_2	nitrous acid
hnRNA	heterogeneous nuclear RNA
hnRNP	heterogeneous nuclear ribonuclear protein
hnRNPM	heterogeneous nuclear ribonucleoprotein M
H-NS	histone-like nucleoid structuring protein
HOME	histogram of methylation
HOPG	highly oriented pyrolytic graphite
HORs	α-satellite higher-order repeats
hox	homeobox genes; homeotic genes
HP1	heterochromatin protein 1
HPF1	histone PARylation factor 1
hPGCLCs	human PGC-like cells
HPLC	high-performance liquid chromatography; high-pressure liquid chromatography

HPLC-MS	high-performance liquid chromatography coupled with mass spectrometry
hPolι	human DNA polymerase ι
hpRNA	hairpin RNA
hpRNAi	hairpin RNA interference
HR	hypersensitive response; homologous recombination; horizontal resistance
HRM	high resolution melt analysis
HRMDR	homologous recombination-mediated DNA repair
hrp	*hypersensitive reaction and pathogenicity* gene
HRT	hybrid-released translation
HSCs	haematopoietic stem cells
hsd	host specificity locus
HSE	heat shock element
HSF-1	heat shock transcription factor 1
hsp	*heat shock protein* genes
HSP	heat shock protein or high-scoring segment pair
hsr	homogeneously staining region
hSSB1	human SSB protein 1
HSV	herpes simplex virus
HT	herbicide-tolerant
HT stress	high-temperature stress
HTC	homozygous typing cells
HTE	heterogeneity of treatment effect; *Hpa*II tiny fragment
HTF islands	*Hpa*II tiny fragment islands
HTGS	high-throughput genome sequences
HTML	HyperText Markup Language
HTS	high-throughput screening; high-throughput sequencing
htSNP	haplotype single-nucleotide polymorphism
HTST	high temperature short time
http	Hypertext Transfer Protocol
HU	histone-like
HUH	histidine-hydrophobic residue-histidine
HUVECs	human umbilical vein endothelial cells
HV system	host-vector system
HVD	hypervariable domain
HYERs	HYdrolytic Endonucleolytic Ribozymes
HYP effectors	hyper-variable effectors
–OH	hydroxyl group
I^-	iodide ion
I line	inbred line
I&T	integration and testing
IκB	inhibitor kappa B
I/O	input/output
IA	intelligent agent

IAA	indole-3-acetic acid
IADs	integrated access devices
IAP	intracisternal A particle; inhibitors of apoptosis; intracellular protein
IAP retrotransposons	intracisternal A particle retrotransposons
IAPs	intracisternal A type particles
IBA	indole butyric acid
IBSC	Institutional Biosafety Committee, India
IBD	identical by descent; incomplete block design
IBL	ion beam lithography
IBM	International Business Machines
IBM1	Increase in Bonsai methylation 1
ibs	initial binding site
IC	imprinting centre
IC-AFM	intermittent contact AFM
ICE	imprint control element
ICF	centromeric instability, and facial anomalies
ICIM	inclusive composite interval mapping
ICM	inner cell mass
ICNCP	International Code of Nomenclature for Cultivated Plants
ICP	insecticidal crystal protein
ICR	imprint control region; internal control region
ICSI	intracytoplasmic sperm injection
ICTV	International Committee of Taxonomy of Viruses
IDE	integrated development environment
IEC	ion exchange chromatography
IEF	isoelectric focusing
IES	internal eliminated sequences
IETS	inelastic electron tunnelling spectroscopy
IFM	interfacial force microscopy
IFNs	interferons
IFN-α	interferon-α
IFN-β	interferon-β
IFN-γ	interferon-γ
IFs	initiation factors
Ig	immunoglobulin
IgE	immunoglobulin E
IGF-1	insulin-like growth factor 1
IGF-II	insulin-like growth factor II
IGFs	insulin-like growth factors
IgH chain	Ig heavy chain
Ig-like	immunoglobulin-like
IGS	internal guide sequences; intergenic spacers
IGZ	*in gel* zymography

IHF	integration host factor
ihpRNA	intron-containing hairpin RNA
IHS	immunohistochemical staining
IKAP	IκB kinase complex-associated protein
IL	introgression line
IL1	interleukin-1
Ile	isoleucine
IMAGE	integrated molecular analysis of genomes and their expression
IMC systems	isolated microspore culture systems
IMP	inner membrane protein; integral membrane protein
IMSP	impedes mitogenic signal propagation
in silico SNP or *is*SNP	*in silico* single-nucleotide polymorphism
INB	immuno-northern blotting
InDel	insertion/deletion
INIBAP	International Network for the Improvement of Banana and Plantain
init	initialization
INO80	inositol requiring 80
Inr	initiator
Int	integration protein
IOC	ion exchange chromatography
IP	identity preservation; intellectual property
IP3	inositol trisphosphate
IPG gels	immobilized pH gradient gels
IPGRI	International Plant Genetic Resources Institute
IPM	integrated pest management
IPP	isopentenyl-pyrophosphate
IPPC	International Plant Protection Convention
IPR	Intellectual Property Rights
IPRO	iterative protein redesign and optimization
iPSCs (or iPS cells)	induced pluripotent stem cells
IQ	intelligence quotient
IR	intron retention; inverted repeat
I-R system	inducer-reactive system
IRAP	inter-retrotransposon amplified polymorphism
IRES	internal ribosome entry site
IRM	insect resistance management
IRS	insulin receptor substrate
IS	insertion
IS elements	insertion sequence elements
ISAAA	International Service for the Acquisition of Agri-Biotech Applications
ISCGA	*in silico* candidate gene approach
ISH	*in situ* hybridization
iSIT	improved sterile insect technique

ISP	*in situ* PCR
ISR	induced systemic resistance
ISSRs	inter simple sequence repeats
ISWI	Imitation Switch
ISZ	*in situ* zymography
IT security	information technology security
ITCHY	incremental truncation for the creation of hybrid enzymes
ITPGRFA	International Treaty on Plant Resources for Food and Agriculture
ITR	inverted terminal repeat
iTRAQ	isobaric tags for relative and absolute quantitation
ITS	ion torrent sequencing; internal transcribed sequences/spacers
IVF	*in vitro* fertilization
IVS	intervening sequences
IVZ	*in vivo* zymography
J sequences	joining sequences
JA	jasmonic acid
JAK/STAT signalling pathway	Janus kinase/signal transducers and activators of transcription signalling pathway
JFM	joint forest management
JmjC	Jumonji C
JNKs	c-Jun NH_2-terminal kinases; c-Jun N-terminal kinases
K	equilibrium constant; nucleotide sequence difference; Potassium
Ka	affinity constant; association constant
Kac	lysine acetylation
KASP	Kompetitive Allele Specific PCR
kb	kilobase [pairs]
kcal	kilocalorie
Kd	dissociation constant
kDa	kilodalton
KDD	knowledge discovery in databases
kDNA	kinetoplast DNA
KFM	Kelvin probe force microscopy
KI	knock-in
K_m	Michaelis constant
KMTases	lysine methyltransferases
KO	knock-out
KOH	potassium hydroxide
KPCR	kinetic polymerase chain reaction
KPFM	Kelvin probe force microscopy
Kr	*Kruppel*
KRAB	Krüppel-associated boxes
K_s	nonsynonymous substitutions rate
KYP	Kryptonite
L	laevorotary; levorotary

lac operon	lactose operon
LAMP	loop-mediated isothermal amplification
LAMPS	light-activated macromolecular phase separation method
LARP1	La-related protein 1
LASL	linker-amplified shotgun library
LAST-seq	linearly amplified single-stranded RNA-derived transcriptome sequencing
Late-PCR	linear-after-the-exponential PCR
Lb	leghaemoglobin
LB method	Langmuir-Blodgett method
LBD	ligand-binding domain
L-chain	light immunoglobulin chain
LCM	laser capture microdissection
LC-MS	liquid chromatography coupled with mass spectrometry
LCPs	low CG content promoters
LCR	ligase chain reaction; locus control region; low complexity region
LD	lethal dose; linkage disequilibrium
LD mapping	linkage disequilibrium mapping
LD_{50}	lethal dose 50
LDLP	low-density lipoprotein
LDLRs	low-density lipoprotein receptors
LDS	low-dimensional structure
LEA promoter	late embryogenesis abundant promoter
LEC2	*LEAFY COTYLEDON2* gene
LEDs	light-emitting diodes
LFM	lateral force microscopy
LGT	lateral gene transfer
LHC	light-harvesting complex
LHY	*Late Elongated Hypocotyl* gene
LIC	ligase-independent cloning
LIFO	last in, first out
lincRNA	large intergenic noncoding RNA sequences
LINE-1	long interspersed nuclear *elements 1*
LINEs	long interspersed nucleotide elements
liRNAs	long intergenic RNAs
Lit.	literally
LITE system	light-inducible transcriptional effector system
LL-RNAs	long-lived RNAs
LMO	living modified organism
LMPCR	ligation mediated PCR
LMS	learning management systems
lncNATs	long noncoding natural antisense transcripts
lncRNAs	long noncoding RNAs
lnr ncRNA	long nuclear-retained noncoding RNAs

lod	logarithmic odds
LOH	loss of heterozygosity
LOI	loss of imprinting
LPCM	linear pulse code modulation
LPE	local pooled error; lysophosphatidylethanolamine
LRO	low-risk objection
Lrp	leucine-responsive regulatory protein
LRR protein	leucine-rich repeat protein
LRPPRC protein	leucine-rich pentatricopeptide repeat-containing protein
LRR-RLKs	leucine-rich repeat-receptor-like kinases
LPS	lipopolysaccharide
LSC	large single copy
LSD1	lysine-specific demethylase 1
LSU	large subunit of ribosome
LTDRs	long terminal direct repeats
LTIRs	long terminal inverted repeats
LTRs	long terminal repeats
LUCA	last universal common ancestor
lz	*lozenge* locus in *Drosophilla*
m	molal solution
M	molar solution; morgan (crossing over value of 100%)
M Phase	mitotic phase
M strain	maternal strain
m.w.	molecular weight
M13RF	M13 replicative form (double-stranded)
M6P	mannose 6-phosphate
ma	malonylation
MAALs	monosomic alien addition lines
MAbs	monoclonal antibodies
MABC	marker-assisted backcrossing
MAbID	multiplexing Antibodies by barcode Identification
macro ncRNAs	macro noncoding RNAs
MACs	mammalian artificial chromosomes
MADS-box	MCM 1, AG, DEF A and SRF box
MAF	minor allele frequency
MAGE	multiplex automated genome engineering
MAGIC	mating-assisted genetically integrated cloning; multi-parent advanced generation intercross
MAL	*Multiple Aleurone Layer* gene
MALDI	matrix-assisted laser desorption ionization
MALDI-ESI	matrix-assisted laser desorption ionization and electrospray ionization
MALDI-TOF-MS	matrix-assisted laser desorption ionization time of flight mass spectrometry
MAMPs	microbe-associated molecular patterns

MAN	metropolitan area network
MAP	microtubule-associated protein; mitogen activator protein
MAPCs	multipotent adult progenitor cells
MAPK/ERK	mitogen-activated protein kinase/extracellular-regulated kinase
MAPK	mitogen-activated protein kinase
MARS	marker-assisted recurrent selection
MARs	matrix associated regions
MAS	marker-aided selection; marker-assisted selection
MAT	*mating-type* locus
MAT system	multi-auto-transformation system
MAT vector	multi-auto-transformation vector
Mb	megabase (= 1 million nucleotides)
MBD2	methyl-CpG-binding domain 2
MBD4	methyl-DNA-binding protein 4
MBDs	methyl-CpG-binding domains
MBE	molecular beam epitaxy
MBT	mechanical biological treatment
MCB proteins	methylcytosine binding proteins
MCC	mitotic checkpoint complex
Mcc C7	Microcin C7
MCD proteins	multi-consensus design proteins
MCDA	microplate capture and detection assay
M-Cdk	M-phase Cdk
mCIP-Seq	methylcytosine immunoprecipitation and sequencing
MCM	minichromosome maintenance
MCM proteins	minichromosome maintenance proteins
MCM1	*minichromosome maintenance1* gene
Mcm2	minichromosome maintenance 2
MCS	multiple cloning site
MD	molecular dynamics
MDM	mutation data matrix
MDR protein	multidrug resistance protein
ME	mega-environment; molecular electronics
me1	monomethylation
me2	dimethylation
me3	trimethylation
MEA	*MEDEA* gene
MeCPs	methyl-CpG-binding proteins
MeDIP or methyl-DIP	methylated DNA immunoprecipitation
MeDIP-chip	methylated DNA immunoprecipitation chip
MeDIP-PCR	methylated DNA immunoprecipitation PCR
MeDIP-seq	methylated DNA immunoprecipitation sequencing
MED26	Mediator complex subunit 26

mEFs	mouse embryonic fibroblasts
mega-YAC	mega-yeast artificial chromosomes
MEMS	micro-electromechanical systems
meQTL	methylation quantitative trait locus
MeRIP-seq	methylated RNA immunoprecipitation sequencing
MeSA	methyl salicylate
MET1	DNA methyltransferase 1
Methyl-DIP or MeDIP	methylated DNA immunoprecipitation
METTL14	methyltransferase-like 14
METTL3	methyltransferase-like 3
MFM	magnetic force microscope/microscopy
MFSTM	magnetic force scanning tunnelling microscopy
mG	methylated guanine
MGEs	mobile genetic elements
MH	maleic hydrazide
MHC	major histocompatibility complex
MHC genes	major histocompatibility complex genes
MHC I	major histocompatibility complex class I
MHC II	major histocompatibility complex class II
MHC III	major histocompatibility complex class III
MI	mitotic index; machine intelligence; metaphase I
miCLIP	6-methyladenosine individual nucleotide crosslinking and immunoprecipitation
micRNA	mRNA-interfering complementary RNA
μl	microlitre
μm	micron or micrometre or 10^{-6} m
MIDAW tool	MIcroarray Data Analysis Web tool
MII	metaphase II
MIMS	molecular integrated microsystems
miniprep	minipreparation
Mis18 complex	Mis18α-Mis18β-Mis18BP1
mitoDMS-MaPseq	mitochondrial DMS mutational profiling with sequencing
–RNA	negative-strand RNA
–ssRNA viruses	negative-sense single-stranded RNA viruses
–strand	antisense strand
MIP	methylation induced premeiotically
MIPS1	*MYO-INOSITOL-1-PHOSPHATE SYNTHASE 1* gene
MIRA	methylated CGI recovery assay
miRISC	microRNA-induced silencing complex
miRNA	microRNA
miRNP	microRNA-protein; microribonucleoprotein
MIS	maturation-inducing substance
missense SNP	missense single-nucleotide polymorphism
MITE	miniature inverted-repeat transposable element

MJ	methyl jasmonate
MKKs	MPK kinases
ML	machine learning
ml	mainline; millilitre
ML method	maximum likelihood method
MLE	maximum likelihood estimate
MLEE	multilocus enzyme electrophoresis
MLL	mixed lineage leukaemia
MLO	mycoplasma-like organism
MLST	multilocus sequence typing
MLLT1 protein	mixed-lineage leukaemia translocated to 1 protein
MLV	murine leukaemia virus
MMC	megaspore mother cell
MMEJ	microhomology-mediated end-joining
MMP	matrix-degrading metalloproteinase
MMR	mismatch repair
MMS	methyl methanesulfonate
MNase	micrococcal nuclease
MNase-seq	micrococcal nuclease digestion followed by deep sequencing
mncRNAs	medium-sized ncRNAs
MOBE	multiplexed orthogonal base editor
Mop1	*Mediator of paramutation1* gene
MP method	maximum parsimony method
MPF	maturation promoting factor
MP-PCR	microsatellite-primed PCR
MPS	methylation-phosphorylation switch; massively parallel sequencing
MPs	movement proteins
MPSS	massively parallel signature sequencing
mPT	mitochondrial permeability transition
MPX	molecular pixelation
M-QTLs	main-effect QTLs
MR	mutagenic repair
MRFs	male recombination factors
MRI	magnetic resonance imaging
mRNA	messenger RNA
mRRBS	multiplexed reduced representation bisulfite sequencing
MRS	magnetic resonance spectroscopy
MRSA	methicillin-resistant *Staphylococcus aureus*
MS	mass spectrometry
MS/MS	mass spectrometry/mass spectrometry; tandem mass spectrometry
MSA	multiple sequence alignment; mass spectrometry analysis
MSAP	methylation-sensitive amplification polymorphism
MSCI	meiotic sex chromosome inactivation

msDNA	multicopy single-stranded DNA
MSL complex	male-specific lethal complex
MSNP	methylation-sensitive SNP
MSP	massively parallel sequencing; methylation-sensitive PCR
MSRE	methylation-sensitive restriction enzyme
MSSLP	microsatellite simple sequence length polymorphism
msSNP	methylation-sensitive single-nucleotide polymorphism
ms-SNuPE	methylation-sensitive single-nucleotide primer extension
MSSR	microsatellite polymorphism simple sequence repeat
MSSRF	M.S. Swaminathan Research Foundation, Chennai, India
MST	microsystems technology; maternal spindle transfer; microscale thermophoresis
MSUD	meiotic silencing by unpaired DNA
MSV	maize streak virus
MTA	Material Transfer Agreement
MTAs	methylation target arrays
MTases	methyltransferases
MTBF	mean time between failures
mtDNA	mitochondrial DNA
MTHF	methenyl tetrahydrofolate; methylene tetrahydrofolate
MTOC	microtubule-organizing centre
mTORC1	mTOR complex 1 or mammalian target of rapamycin complex 1
MTS	mitochondria-targeting sequences
MTTR	mean time to repair; mean time to recovery
MV	machine vision
MW	molecular weight
MYB	myeloblastosis viral oncogene homologue
MVP	minimum viable population
MYST	MOZ, Ybf2/Sas3, Sas2 and Tip60 complex
MZ	monozygotic twins
n	haploid chromosome number
N protein	nucleocapsid protein
n.	noun
N_2O	nitrous oxide
NAA	naphthalene acetic acid
NAC	N-acetylcysteine
NAC	nucleotide addition cycle
NAD	nicotinamide adenine dinucleotide
NADH	nicotinamide adenine dinucleotide hydrogen
NADPH	nicotine adenine dinucleotide phosphate (reduced form)
NAG	N-acetylglucosamine
NAM	nested-association mapping; N-acetylmuramic acid; no apical meristem
NAPs	nucleoid-associated proteins
NARES-BLP	National Agricultural Research and Education System-Blended Learning Platform

NASC	Nottingham Arabidopsis Stock Centre
NAT10	N-acetyltransferase 10
nat-siRNAs	natural antisense transcript-derived siRNAs
NBD	nested block design
NBPGR	National Bureau of Plant Genetic Resources, New Delhi, India
NBPT	N-(n-butyl) thiophosphoric triamide
NBS	nucleotide-binding site
NBS-LRR protein	nucleotide-binding site–leucine-rich repeat protein
NC	nucleocapsid
NCAM	neural cell adhesion molecule
NCBI	National Centre of Biotechnology Information, USA
NCE	new chemical entity
ncRNAs	noncoding RNAs
ncSMGs	negative conditional selectable marker genes
ncSNP	noncoding single-nucleotide polymorphism
nDNA	nuclear DNA
Ne	effective population size
NE	nuclear envelope
NELF	negative elongation factor
NELF-DSIF	negative elongation factor-DRB sensitivity-inducing factor
NEMS	nanoelectromechanical systems
NER	nucleotide excision repair
NERD pathway	needed for RDR2-independent DNA methylation pathway
NF-κB	nuclear factor kappa B
Nf1	*neurofibromatosis 1* gene
NF-3	neurotrophic factor-3
NFOM	near-field optical microscope
NGF	nerve growth factor
NGOs	non-governmental organizations
NGS	next-generation screening; next-generation sequencing
NH_2OH	hydroxylamine
NH_3	ammonia
NHEJ	nonhomologous end-joining
NHR	nonhomologous recombination
NHS	nuclease hypersensitive site
NIAR	National Institute of Agrobiological Resources
nif/fix genes	nitrogen fixation genes
NIH	National Institute of Health, USA
NIL	nearly isogenic line
NIPBL protein	Nipped-B-like protein
NJU-seq	Nm-Judged-Universally sequencing
NIR spectroscopy	near-infrared spectroscopy
NK cells	natural killer cells

NLM	National Library of Medicine, USA
NLP	natural language processing
NLS	nuclear localization signal
nm	nanometre
Nm	2′-O-methylation
NMD	nonsense-mediated (mRNA) decay
NMR	nuclear magnetic resonance
nmRNAs	non-messenger RNAs
NMS	nuclear male sterility
nncSMGs	negative nonconditional selectable marker genes
NNT	number needed to treat
NO	nitric oxide
NOAEL	no observable adverse effect level
nod genes	nodulation genes
non-LTR retrotransposon	non-long terminal repeat retrotransposons
NORs	nucleolar organizing regions
NOS	nopaline synthase
NO_3	nitrate radicals
NO_x	nitrogen oxide
np	nucleotide pair
NPA	nuclease protection assays
NPC	nuclear pore complex
NPCs	neural progenitor cells
NPD	non-parental ditype
NPP	net primary production
*NPR1**	*Nonexpressor of Pathogenesis-Related 1* gene
NPT	neomycin phosphotransferase
NPTII	*Neomycin phosphotransferase II* gene
nr database	nonredundant database
NRDE pathway	nuclear RNAi defective pathway
NRR	nonhomologous random recombination
nsAA	non-standard amino acid
NSF protein	N-ethylmaleimide-sensitive fusion protein
nSMGs	negative selectable marker genes
NSOM	near-field scanning optical microscopy
nsSNP	nonsynonymous single-nucleotide polymorphism
NT	nuclear transfer
NTD	amino-terminal domain
N-terminal	amino terminal
NTPs	ribonucleoside triphosphates
NTR	nontranslated RNA
NTS	nontranscribed spacer; non-target DNA strand
Nu	nucleosome

Nups	nucleoporins
NuRD	nucleosome remodelling and deacetylase
o; op	operator
O_3	ozone
o^c	*operator-constitutive* mutation
ocDNA	open circular DNA
OD	optical density
ODD	optical disc drives
ODM	oligonucleotide-directed mutagenesis
OECD	Organisation for Economic Co-operation and Development
OG	O-GlcNAcylation
OGT	O-GlcNAc transferase
oh	hydroxylation
OIE	International Office of Epizootics
OLA	oligonucleotide ligation assay
OLED	organic LED
oligos	oligonucleotide
OLS	ordinary least squares
OMGT	ovary-mediated gene transfer
OMIM	Online Mendelian Inheritance in Man
OMM	oligonucleotide-mediated mutagenesis
1^mA	N1-methyladenosine
1^mA-ID-seq	1-methyladenosine-specific RNA immunoprecipitation with demethylase-assisted RNA sequencing
ONM	octanucleotide motif
ONT	Oxford Nanopore Technologies
OOAD	object-oriented analysis and design
OOP	object-oriented programming
ORC	origin recognition complex
ORD	object-relational database
ORF	open reading frame
Ori	origin of replication
OriT	origin of transfer
OrthoRep	orthogonal DNA replication system
OS	operating system
OSI model	one-sided invasion model
OsO_4	osmic acid
OSS	open-source software
otd	*Orthodenticle* gene
O_3	ozone
OTSs	orthogonal translation systems
OTU	operational taxonomic unit
OXPHOS	oxidative phosphorylation

O-DNAP (or O-DNA Pol)	orthogonal DNA polymerase
p	promoter; short arm of a chromosome; frequency of dominant allele
P	perianth; phosphorus; probability
P bodies	processing bodies
P site	peptidyl site
P strain	paternal strain
p65AD	p65 activation domain
PABA	para-aminobenzoic acid
PABP	poly(A)-binding proteins
PacBio	Pacific biosciences
PACs	P1-derived artificial chromosomes; phage artificial chromosomes
PAD	plant airborne defence
PAFc	polymerase-associated factor complex
PAGE	polyacrylamide gel electrophoresis
palm1	palmate-like pentafoliata1
PAM	point-accepted mutation; prediction analysis of microarray; protospacer adjacent motif
PAM sequence	protospacer adjacent motif sequence
PAMAM gene transfer	polyamidoamine-mediated gene transfer
PAMP	pathogen-associated molecular pattern
PAN	peroxyacetyl nitrate
PAP	protein amount polymorphism
PAR	protease-activated receptor; pseudoautosomal region
PARP-1	poly(ADP-ribose) polymerase-1
Pas	polyamides
PASTE	programmable addition via site-specific targeting elements
PAT	phosphinothricin acetyltransferase
pat	*phosphinothricin acetyltransferase* gene
PAZ domain	PIWI-Argonaute-Zwille domain
PB transposon	PiggyBac transposon
PBAT	post-bisulfite adapter tagging
PBG	porphobilinogen
PBIB	partially balanced incomplete block design
PBRs	Plant Breeder's Rights
pBS	pBluescript
PBS	porphobilinogen synthase
PCC	protein-conducting channel
PCD	programmed cell death; premature centromere division
PcG	Polycomb group
PCGA	positional candidate gene approach
PCNA	proliferating cell nuclear antigen
PCR	polymerase chain reaction
PCR-mediated cloning	polymerase chain reaction-mediated cloning
PCR-RFLP	polymerase chain reaction-restriction fragment length polymorphism

Pcsk9	*Proprotein Convertase Subtilisin/Kexin Type 9* gene
PCT	Patent Cooperation Treaty
PD	parental ditype
P-D	proximal to distal
PDB	Protein Data Bank
PDF	pigment-dispersing factor
PDGF	platelet-derived growth factor
PDH45	pea DNA helicase 45
pDMRs	parent-of-origin-dependent DMRs
PDMS	polydimethylsiloxane
PDR	pathogen-derived resistance
PE	prime editing
PE3	prime editor 3
PEDOT:PSS	poly(3,4-ethylenedioxythiophene) polystyrene sulfonate
PEFs	pulsed electrical fields
PEG	parentally expressed imprinted genes; polyethylene glycol
PEG-mediated gene transfer	polyethylene glycol-mediated gene transfer
pegRNA	prime editing guide RNA
PEGs	paternally expressed imprinted genes
PEI	polyethylenimine
PEI-mediated gene transfer	polyethylenimine-mediated gene transfer
PERA	predictive ecological risk assessment
PERK	protein kinase RNA–like endoplasmic reticulum kinase
PERs	pre-edited regions
PERV	porcine endogenous retrovirus
PEs	prime editors
PEST hypothesis	proline (P), glutamic acid (E), serine (S) and threonine (T) hypothesis
PET	paired-end tag; prime editing technology
PETC	photosynthetic electron transport chain
PEV	position-effect variegation
PEX5	peroxisomal biogenesis factor 5
PFC1	paleface 1
PFF	peptide fragment fingerprinting
PFGE	pulsed-field gel electrophoresis
pg	picogram
PGA	polygalacturonase
PGCs	primordial germ cells
PGD	pre-implantation genetic diagnosis
PGIPs	polygalacturonase-inhibiting proteins
PGK	*phosphoglycerate kinase* gene
PGPRs	plant growth promoting rhizobacteria
PGRC	Plant Germplasm Registration Committee
PGRs	plant genetic resources

PGs	polygalacturonases
pgSIT	precision-guided sterile insect technique
ph	phosphorylation
PH domain	pleckstrin homology domain
PHA	polyhydroxyalkanoic acid
phasiRNAs	phased siRNAs
PHB	PHABULOSA
PHD	plant homeodomain
PHD finger	plant homeodomain finger
PHV	PHAVOLUTA
phyB	phytochrome B
pi	proline isomerization
pI	isoelectric point
PI 3-kinase	phosphatidylinositol 3-kinase
PIC	preinitiation complex (for transcription)
PIM-RFLP	PCR-induced mutagenesis-restriction fragment length polymorphism
PIN	protein interaction network
pinna2	*pinnate-like pentafoliata2* gene
piPolB	primer-independent Polymerase B
PIPs	plant-incorporated protectants
piRNAs	Piwi-interacting RNAs
PIs	protease inhibitors
PIV	parainfluenza viruses
PKA	protein kinase A
PKC	protein kinase C
PKS	polyketide synthase
PKU	phenylketonuria
pl.	plural
PLA	proximity ligation assay
PLACE	plant *cis*-acting regulatory DNA elements
PLC-β	phospholipase C-β
PLC-γ	phospholipase C-γ
PlnTFDB	Plant Transcriptional Factor Database
PLK-1	Polo-like kinase 1
PLPs	padlock probes
PLT	primed lymphocyte typing; programming language theory
+RNA	positive-sense RNA
+ssRNA viruses	positive-sense single-stranded RNA viruses
+strand	sense strand
PMC	pollen mother cell
PMF	peptide mass fingerprinting; plant molecular farming
PMP	plant-made pharmaceuticals
PNA	peptide nucleic acid

PNPase	polynucleotide phosphorylase
PNS	positive and negative selection
PNT	pronuclear transfer; plant's novel trait
POI	protein of interest
POLRMT	recombinant human mitochondrial RNA polymerase
pon2	paraoxonase 2
population-specific SNP	population-specific single-nucleotide polymorphism
post-TCs	post-termination complexes
pp	*Proboscipedia*
PPA	Plant Patent Act; Plant Protection Act
PPARγ	peroxisome proliferator-activated receptor γ
PPB	preprophase band
PPBC	Plant Photo Bank of China
PPFM	pink-pigmented facultative methylotroph
ppGpp	guanosine tetraphosphate; guanosine-5′-diphosphate 3′-diphosphate
PPI	protein–protein interactions
PPIN	protein–protein interaction network
PPIX	protoporphyrin IX
PPML	privacy-preserving federated machine learning
PPO	polyphenol oxidase
pppA′p5′A2′p5′A	P4-tetraphosphate (Two-fiveA″)
pppGpp	guanosine-5′-triphosphate 3′-diphosphate
PPR proteins	pentatricopeptide repeat proteins
PPRP	plant pathogenesis-related proteins
PPT	L-phosphinothricim
PPV&FR	Protection of Plant Varieties and Farmers' Rights Act, 2001
pr	propionylation
PR proteins	pathogenesis-related proteins
pRb	retinoblastoma protein
PRC	Polycomb repressive complex
PRC1	Polycomb repressive complexes 1
PRC2	Polycomb repressive complexes 2
PRE	programmable RNA editing
pre-miRNA	precursor microRNA
pre-mRNA	precursor messenger RNA
pre-RC process	pre-replicative process
pre-rRNA	precursor ribosomal RNA
pre-tRNA	precursor transfer RNA
pRi	root-inducing plasmid
PriL	large catalytic subunit of DNA polymerase α
pri-miRNA	primary miRNA
PrimPol	primase-polymerase
PRINS	primed *in situ*

pri-piRNAs	primary piwi-interacting RNAs
PriS	small catalytic subunit of DNA polymerase α
pRNA	RNA primer
Prnp	prion protein
promoter SNP or pSNP	promoter single-nucleotide polymorphism
PROMPTs	promoter upstream transcripts
Protox or PPO	protoporphyrinogen oxidase
PRP	Plant Regulomics Portal; proprietary regulatory property
PrP	prion-related protein
PRR	phytophthora root rot; pattern recognition receptors; postreplication recombination; postreplication repair
PRR disease	phytophthora root rot disease
PRRs	pattern recognition receptors
PRS	polygenic risk score
PRSV	papaya ring spot virus
PSA	prostate-specific antigen
PSC	pseudo-self-compatibility
Pseudo-seq	pseudo-sequencing
Ψ	pseudouridine
pSMP1	positive selectable marker genes, type 1
pSMP2	positive selectable marker genes, type 2
pSMP3	positive selectable marker genes, type 3
pSMP4	positive selectable marker genes, type 4
pSMP5	positive selectable marker genes, type 5
pSMP6	positive selectable marker genes, type 6
pSMP7	positive selectable marker genes, type 7
pSNP	promoter single-nucleotide polymorphism
PSR	phosphatidylserine receptor
PSS	plant stress signalling
PSSM	position-specific scoring matrix
PSTV	potato spindle tuber viroid
Psy	*phytoene synthase* gene
pSym	Sym plasmid
PT	proficiency testing
PTC	phenylthiocarbamide; peptidyltransferase centre; premature termination codon
PTCM (or PTM)	post-transcriptional modification
P-TEFb	positive transcription elongation factor b
P-TEFb complex	positive transcription elongation factor b complex
PTGS	post-transcriptional gene silencing
PTI	PAMP-triggered immunity
pTi	Tumour-inducing plasmid
PTIS	portable translation initiation sequence
PTLM (or PTM)	post-translational modification

PUFA	polyunsaturated fatty acids
PuNNAUGG	Kozak's sequence
pv	pathovar
PVDF	poly(vinylidene fluoride)
PVP	plant variety protection
PVP&FR Act	Plant Variety Protection and Farmer's Rights Act
PVPA	Plant Variety Protection Act
PWWP motif	proline-tryptophan-tryptophan-proline motif
Pyl	pyrrolysine
PYLIS	pyrrolysine insertion sequence
PyV	Polyoma virus
Q	quinone
Q plot	quantile plot
Qb	Q-beta
QBR	Q-beta replicase
QDCMR	quantitative differential chromatin modification region
QDEG	quantitative differential expressed gene
QDMR	quantitative differentially methylated region
QM	quantum microscopy
qMS	quantitative mass spectrometry
Q-PCR/qPCR/qrt-PCR	quantitative real-time polymerase chain reaction
QPM	quantum proximity microscopy
QQ plot	quantile quantile plot
qr	quadriradial
QS	quorum sensing
QSAR	quantitative structure–activity relationship
QSPR	quantitative structure–property relationship
QTL	quantitative trait loci
QTL-seq	quantitative trait loci sequencing
QTN	quantitative trait nucleotide
R	bit rate; selection response
r	correlation coefficient; coefficient off relatedness; recombination frequency
r chromosome	ring chromosome
R genes	*resistance* genes
r mutants	*rapid lysis* mutants
R plasmid	resistance plasmid
R proteins	resistance proteins
R_1, R_2, R_3, . . .	the first, second, third, . . . generations following irradiation
RA	retinoic acid
RAC or RDAC	Recombinant DNA Advisory Committee
RACE	rapid amplification of cDNA ends
RACHITT	random chimeragenesis on transient templates
Radar	RNA Data Analysis and Research

RAFI	Rural Advancement Foundation International
RAHMs	random amplified hybridized microsatellites
RAMs	ribosomal ambiguity mutants
RAMP	random amplified microsatellite polymorphism
RAPD	random amplified polymorphic DNA
rasiRNAs	repeat-associated siRNAs
RATF	ribosome-associated trigger factor
RAV6	*related to abscisic acid insensitive3* (*abi3*)/*viviparous1* (*vp1*) *6* gene
RBCs	red blood cells
RBOH	respiratory burst oxidase homologue
RBH	resequencing by hybridization
RBR1	RETINOBLASTOMA-RELATED1 protein
RCA	rolling circle amplification
RCB design	randomized complete block design
RCC	reciprocal circulating chromatography
R-cell	photoreceptor-cell
RCGM	Review Committee on Genetic Manipulation
RC-PCR	reverse complement PCR
RCT	randomized clinical trial; randomized controlled trial
RDBMS	relational database management systems
RdDM	RNA-directed DNA Methylation
RDMs	random DNA markers
rDNA	ribosomal DNA
rDNA-ITS	ribosomal DNA internal transcribed spacers
RDR, RdR or RdRp	RNA-dependent RNA polymerase
RDR2	RNA-dependent RNA polymerase 2
RDR6	RNA-dependent RNA polymerase 6
ReCODE™	Reconstituting chemically orthogonal directed engineering technology
RECORD method	recombination counting and ordering method
refSNP, rSNP, rsID, SNP ID	reference single-nucleotide polymorphism
regulatory SNP, rSNP	regulatory single-nucleotide polymorphism
REMAP	retroposon microsatellite amplified polymorphism
REMI	restriction enzyme-mediated integration
REML	restricted maximum likelihood
RenSeq	resistance gene enrichment sequencing
REP	repeated extragenic palindrome
RER	rough endoplasmic reticulum; repetitive element repression
REs	restriction endonucleases
RF	recombination-free; replicating form
RF cloning	recombination-free cloning
RF1	release factor 1
RFA	replication factor A
RFC	replication factor C

RFLP	restriction fragment length polymorphism
RFLs	reading frame lengths
RFs	release factors
RG	reporter gene
RGA	resistance gene analogue
RGAI	repressor of gibberellic acid insensitive
RGEN	RNA-guided endonuclease
RGSV1 virus	rice grassy stunt 1 virus
RGSV2 virus	rice grassy stunt 2 virus
ρ factor	rho factor
r^2	coefficient of determination.
Rht	*Reduced height* gene
RI strains	recombinant inbred strains
RIA	radioimmunoassay
RIFF	Resource Interchange File Format
RIGS	repeat-induced gene silencing
RILs	recombinant inbred lines
RING	Really Interesting New Gene
RIP	RNA immunoprecipitation; rearrangements induced premeiotically; ribosome-activating protein
RIP mutation	repeat-induced point mutation
Ri-plasmid	root-inducing plasmid
RIS	recombinant inbred strains
RISC	RNA-induced silencing complex
RITS	RNA-induced transcriptional silencing
RLF	replication licensing factor
R-M system	restriction-modification system
RMG	reporter marker gene
RMGE	recombination-mediated genetic engineering
RNA	ribonucleic acid
RNA Pol I or RNAP I	RNA-dependent RNA polymerase I
RNA Pol II or RNAP II	RNA-dependent RNA polymerase II
RNA Pol III or RNAP III	RNA-dependent RNA polymerase III
RNA Pol IV or RNAP IV	RNA-dependent RNA polymerase IV
RNA Pol V or RNAP V	RNA-dependent RNA polymerase V
RNAi	RNA interference; RNA-dependent interference
RNAi-c	RNA interference due to transcript cleavage
RNAi-gm	RNA interference due to genomic modification
RNAi-r	RNA interference due to repression
RNA-PET	RNA paired-end tag
RNAPs	DNA-dependent RNA polymerases
RNase	ribonuclease
RNase H	ribonuclease H

RNase P	ribonuclease P
RNN	recurrent neural network
RNP	ribonucleoprotein
R_o	RNA concentration
rob translocation	Robertsonian translocation
ROC curve	receiver operating characteristic curve
ROS	reactive oxygen species
ROS1	repressor of silencing 1
RPA	replication protein A
RPC	replisome progression complex; reverse phase chromatography
rpr	*reaper gene*
RRBS	reduced representation bisulfite sequencing
RRF	ribosome recycling factor
RRM	RNA recognition motif
rRNA	ribosomal RNA
RRSV	rice ragged stunt virus
RsmD	ribosomal RNA small subunit methyltransferase D
RSo	orthogonal tRNA synthetases
RSS	recombination signal sequences
RSV	respiratory syncytial virus; Rous Sarcoma Virus
RT	replication timing; reverse transcription
RTBV	rice tungro bacilliform virus
RTF	resistance transfer factor
RTKs	receptor tyrosine kinases
RTL	register transfer language
RTL-PCR	reverse transcription at low deoxyribonucleoside triphosphate concentrations followed by polymerase chain reaction
RT-PCR	real-time reverse transcriptase polymerase chain reaction; real-time PCR; reverse transcribed PCR
RT-qPCR	reverse transcription quantitative polymerase chain reaction
RTS	Rubinstein-Taybi syndrome
RTSV	rice tungro spherical virus
RuBisCO	ribulose-1,5-bisphosphate carboxylase/oxygenase
RuBP	ribulose-1,5-bisphosphate
s	selection coefficient
S	sedimentation coefficient; selection differential; Svedberg unit; raw score
S′	bit score
S locus	*Self-incompatibility* locus
S phase	DNA synthesis phase
SA	salicylic acid
SAGE	serial analysis of gene expression
SAM	scanning acoustic microscope; shoot apical meristem;
SAM (or AdoMet)	S-adenosylmethionine

SAM (or AdoMet) domain	S-adenosylmethionine domain
SAMPLs	selective amplification of microsatellite polymorphic loci
SAND domain	domain made of Sp100 (S), AIRE-1 (A), NucP41/75 (N) and DEAF-1 (D)
SAP	scanning atom probe; specific amplicon polymorphism
SAR	systemic acquired resistance
saRNA	small activating RNA
SASPs	small acid-soluble proteins
SAT chromosome	satellite(d) chromosome
sat DNA	satellite DNA
sat RNA	satellite RNA
SB transposon	sleeping beauty transposon
SBCC	State Biosafety Coordination Committee
SBE	single-base extension
SBP	sequence-based polymorphic; SQUAMOSA promoter binding protein
SBP2	SECIS-binding protein 2
SBT	shifting balance theory
SC	somacolones; scientific computation; scientific computing; synaptonemal complex
sca	specific combining ability
SCAR marker	sequenced characterized amplified region marker
SCARs	sequence characterized amplified regions
SCD proteins	single consensus design proteins
SCD staining	sister chromatid differential staining
S-Cdk	S-cyclin and cyclin-dependent kinase complex
SCE	sister-chromatid exchange
SC5D	sterol C5-desaturase
scFV	single-chain variable fragment
sci-CAR	single-cell combinational indexing-based coassay
SCM	scanning capacitance microscopy
scM&T-seq	single-cell methylation and transcriptome sequencing
SCMR	somatic cell mutation and recombination
SCMV	sugarcane mosaic virus
scnRNA	scan RNA
SCNT	somatic cell nuclear transfer
SCNT technology	somatic cell nuclear transfer technology
SCOP	structural classification of proteins
SCOPE	Suite for Computational identification Of Promoter Elements; structure-based combinatorial protein engineering
scPCR	single-cell PCR
SCPs	single-cell proteins
SCR	scarecrow
SCR protein	*S*-locus cysteine-rich protein
scRNAs	small cytoplasmic RNAs
scRNA-seq	single-cell RNA sequencing

scRNPs or scyrps	snurps-like particles identified in cytoplasm
scRRBS	single-cell reduced representation bisulfite sequencing
Scw	*Screw*
S_d	standard error of difference in means
SD	segregation distorter
S-D sequence	Shine-Dalgarno sequence
SD, s, or σ	standard deviation
SDA	strand displacement amplification
SDLC	software development life cycle
SDM	splicing-dependent microarray; site-directed mutagenesis
SDN	site-directed nuclease; structural DNA nanotechnology
SDR	second division restitution
SDS	second division segregation
SDSA	synthesis-dependent strand annealing
SDS-PAA	sodium dodecyl sulfate-polyacrylamide
SDS-PAGE	sodium dodecyl sulfate-polyacrylamide gel electrophoresis
SE	selection efficiency; standard error
SEC	super elongation complex
SECIS	Sec-insertion sequence
SeCys	selenocysteine
seDSBs	single-ended double-strand breaks
SEM	scanning electron microscope; standard error of difference in means
SE_m	standard error of the mean
SENPs	sentrin-specific proteases
sense-PTGS	sense post-transcriptional gene silencing
SEPM	scanning electrochemical probe microscope
SER	smooth endoplasmic reticulum
ser	serotonylation
Set-1 COMPASS proteins	Complex of Proteins Associated with Set1
7-DHC	7-dehydrocholesterol
7^mG or m^7G	7-methylguanosine or N7-methylguanosine
7TMS receptors	seven-transmembrane-spanning receptors
S-D sequence	Shine-Dalgarno sequence
SF3B1	Splicing factor 3B subunit 1
sfa	sex factor affinity; sex fertility affinity
SFA8	sunflower *albumin 8*
SFM	scanning force microscopy
SFPs	single-feature polymorphisms
SFS	scanning force spectroscopy
SGL1	single leaflet1
sgRNA	single-guide RNA
SH2 domain	Src homology2 domain
Shh	Sonic hedgehog

SHIPREC	sequence homology independent protein recombination
shRNAs	short hairpin RNAs
SI	self-incompatibility; International System of Units; synthetic intelligence
SIB	Swiss Institute of Bioinformatics, Switzerland
SICM	scanning ion-conductance microscopy
SID	mSin3 interaction domains
σ	standard deviation
σ^2	variance
SILAC	stable isotope labelling by amino acids in cell culture
SIM	simple interval mapping
SINEs	short interspersed nuclear elements
sing.	singular
SingleCellGGM	single-cell graphical Gaussian model
single-hit SNP	single-hit single-nucleotide polymorphism
SIRM	Sterile Insect Release Method
siRNA	small/short interfering RNA
siRNP assembly	short interfering ribonucleoprotein assembly
siRNPs	small interfering ribonucleoproteins
SIT	sterile insect technique
SIT AW-IPM	sterile insect technique area-wide integrated pest management
6^mA	6-methyladenine; N6-methyladenosine
6^mA^m	N6,2′-O-dimethyladenosine
6n	hexaploid
SKM	Stress Knowledge Map
SKY	spectral karyotype
SL	shallow learning; spliced leader
SL RNA	spliced leader RNA
SL system	seed-lethal system
SLA	service level agreement
SLG	*S*-locus glycoprotein
SLiCE	seamless ligation cloning extract
SLTS	spliced leader *trans*-splicing
SMA	single-marker analysis
SMART	somatic cell mutation and recombination test
SMC	structural maintenance of chromosome proteins
SmF	submerged fermentation
smFRET	single molecule fluorescence resonance energy transfer
SMG	selection marker gene; selectable marker gene; suppressors with morphogenetic defects in genitalia
SMG7	*Suppressor with Morphogenetic Effects On Genitalia 7i* gene
SMGT	sperm-mediated gene transfer
SMI region	SP11-methylation-inducing genomic region
smi RNA	small molecules interacting with RNA

smRNA	small modulatory RNA
SMRT detection	single-molecule real-time detection
SMRT seq	single-molecule real-time sequencing
SMRT-BS	single-molecule real-time bisulfite sequencing
SMS technologies	single-molecule sequencing technologies
SNAREs	SNAp REceptors; soluble N-ethylmaleimide-sensitive factor activating protein receptors
sncRNAs	small noncoding RNAs
snoRNAs	small nucleolar RNAs
snorps	snurps-like particles identified in nucleolus
SNP	single-nucleotide polymorphism
snRNAs	small nuclear RNAs
snRNPs	small nuclear ribonucleoproteins; small nucleo-ribonucleoproteins
snups	specialized nucleoprotein structures
snurps	specialized nucleoriboprotein structures
SNV	single-nucleotide variant
SOC1	*Suppressor of Overexpression of Constans 1* gene
SOD	superoxide dismutase
SOLiD	sequencing by oligonucleotide ligation and detection
somatic DMRs	somatic differentially methylated regions
SOP	sense organ precursor
SOS	save our souls
SOS3	SALT OVERLY SENSITIVE 3 protein
SP STM	spin-polarized scanning tunnelling microscopy
sp.	species
SpCas9–M-MLVRTΔRNaseH–peg RNA–target DNA complex	Streptococcus pyogenes Cas9 nickase—Moloney murine leukaemia virus reverse transcriptase—pegRNA—target DNA complex
SP11	*S*-locus protein 11
SPAD	soil plant analysis development
SPAR	single primer amplification reaction
SPB	spindle pole body
SPCB	State Pollution Control Board
SPCs	stomodeal precursor cells
SPE	scanning probe electrochemistry
SPF	start promotion factor
S-phase	synthesis phase
spiRNAs	secondary piwi-interacting RNAs
SPLAT	SPlinted Ligation Adapter Tagging
SPM	scanning probe microscope
spRNAPIV	single-polypeptide nuclear RNA polymerase IV
SPS	Sanitary and Phytosanitary Measures
SPT	specific pathogen tested
S-PTGS	PTGS triggered by sense transgenes

SQL	Structured Query Language
Sqv	*squashed vulva* gene
SRA domain	SET- and RING-associated domain
SREs	splicing regulatory elements
SRF	serum response factor
SRK	*S*-locus receptor kinase
sRNA-seq	small RNA-sequencing
SRP	signal recognition particle
SRP RNA	signal recognition particle RNA
SRPR	signal recognition particle receptor
SRS	software requirements specification
SS	simple sequences; splice site; sum of squares
S-S	disulfide bond
SS2PL	strong strict two-phase locking
SSA	single-strand annealing
ssb proteins	single-strand binding proteins
SSBs	single-strand breaks
SSC	small single copy
sscDNA	single-stranded cDNA
SSCP	single-strand conformation polymorphism
SSD	singe-seed descent
ssDB	single-stranded DNA break
ssDNA	single-stranded DNA
SSDR	single-stranded DNA recombineering
SS_E	sum of squares due to environmental differences
SSEA	secondary structure element alignment
SSF	solid state fermentation
SS_G	sum of squares of deviations due to genetic factors
SSH	Secure Shell
SSI	sporophytic self-incompatibility
SS_m	the total sum of squares, $(SS_G + SS_E)$
SSM	slip-strand mispairing
ssp.	subspecies
SSPM	scanning surface potential microscopy
SSR	self-sustained sequence replication; simple sequence repeat; single-sequence repeat
SSR systems	site-specific recombination systems
ssRNA	single-stranded RNA
ssRNA viruses	single-stranded RNA viruses
SSRs	simple sequence repeats
SSs	simple sequences; splice sites
SSS	second-strand synthesis
SSU	small subunit off ribosome
STACs	sirtuin-activating compounds

STATs	signal transducers and activators of transcription
StEP	staggered extension process
SThM	scanning thermal microscope
sTIR	subterminal inverted repeats
STIRs	short terminal inverted repeats
STKE	signal transduction knowledge environment
STM	scanning tunnelling microscopy
STM marker	sequence-tagged-microsatellite marker
STMS	sequence-tagged microsatellite site
STR	short tandem repeat
STRIPAK	striatin-interacting phosphatase and kinase
stRNAs	small temporal RNAs
STRs	simple tandem repeats
STS	sequenced tagged sites; short-tagged sequences; scanning tunnelling spectroscopy
STS marker	sequence-tagged-site marker
STT	speech to text
Su(var)3–9	Suppressor of variegation 3–9
succ	succinylation
SUMO or Sumo	SUMOylation; Small Ubiquitin-like MOdifier
SUTs	stable unannotated transcripts
SVET	scanning vibrating electrode technique
SVMs	support-vector machines
SVT	specific virus tested
SWGA	selective whole genome amplification
SWI/SFN	switching/sucrose non-fermenting; SWItch/Sucrose Non Fermentable
SWR1-C	SWI2/SNF2-Related 1 Chromatin Remodelling Complex
S_x	standard error of sample mean
Sxl	*sex-lethal*
syn	synonymous
Syn-1	F_1 synthetic variety
synPPR proteins	synthetic pentatricopeptide proteins
T	thymine
t	time; translocation
Ta	annealing temperature
Ta Opt	optimal annealing temperature
TβRII	type II TGF β receptor
T*	tautomeric thymine
T_0 generation	transgenic plants directly regenerated from tissue culture
T1SS	secretion system type I
T2SS	secretion system type II
T2T	telomere-to-telomere
T2T assembly	telomere-to-telomere assembly
T3SS	secretion system type III

T4SS	secretion system type IV
T5SS	secretion system type V
T6SS	secretion system type VI
TAB-seq	Tet-assisted bisulfite sequencing
TAD	topologically associating domain
TadA	transfer RNA-specific adenosine deaminase
tag SNP	tag single-nucleotide polymorphism
TAG-KO	transposon-arrayed gene knock out
TAIR	The Arabidopsis Information Resource
TALE	transcription activator-like effector
TALENs	transcription activator-like effector nucleases
TAR	transgenerational acquired resistance
TARDIS	temporal analysis of relative distances
TAS	TALE array sequence
tasiRNAs	*trans*-acting small interfering RNAs
Tat	*trans*-activator of transcription
TBC	transcript-based cloning
TBP	TATA-box binding protein
TBT	Technical Barriers to Trade
T_C cell	cytotoxic T cell
TCA	tricarboxylic acid
TCA cycle	tricarboxylic acid cycle or Krebs cycle
TCEP	tris(2-carboxyethyl)phosphine
TCM	temperature contrast mode
TC-NER	transcription-coupled nucleotide excision repair
TcR	*T-cell receptor* gene
TCS	tentative consensus sequence
TDF	testis-determining factor
TDF	*testis-determining factor* gene
TDG	thymine DNA glycosylase
TDM1	*THREE-DIVISION MUTANT 1*
TDMD model	target-directed miRNA degradation model
T-DNA	transfer DNA
TDRDs	Tudor-domain-containing proteins
TDT	transmission disequilibrium test
TdT	terminal deoxynucleotidyl transferase
TE	transposable element; transformation efficiency
TE buffer	Tris-EDTA buffer
TEI	transgenerational epigenetic inheritance
tel	telomere
TEL patch	TPP1 (T) glutamate (E) and leucine (L)-rich patch
TEM	transmission electron microscope
TERC	telomerase RNA component

TERRA	telomeric repeat-containing RNA
TEs	transposable elements
TESS	Transcription Element Search System
TET1	Ten-Eleven Translocation 1 methylcytosine dioxygenase 1 enzyme
tetra-allelic SNP	tetra-allelic single-nucleotide polymorphism
TETRIS	time-resolved electrochemical technology for plant root environment *in situ* chemical sensing
TFAM	transcription factor A, mitochondrial
TFB2M	transcription factor B2, mitochondrial
TFOs	triplex-forming oligonucleotides
TFs	transcription factors
TGCE	temperature gradient capillary electrophoresis
TGF	transforming growth factor
TGFβ (TGFb)	transforming growth factor β (or transforming growth factor b)
TGF-β superfamily	transforming growth factor-β superfamily
TGGE	temperature gradient gel electrophoresis
TGIRT	thermostable group II reverse transcriptase
TGN	*trans*-Golgi network
TGS	transcriptional gene silencing
T-GURT	trait-level genetic use restriction technology
TGW	thousand-grain weight
TH	true hermaphroditism
T_H cell	helper T cell
THCS	tentative human consensus sequences
Thr	threonine
3′ UTR	3′ untranslated region
3′-ppGpp-5′	guanine tetraphosphate
3C	chromosome conformation capture
3-D	three-dimensional
3GL	third-generation programming language
3^mC	N3-methylcytidine
3n	triploid
3-PGAL	3-Phosphoglyceraldehyde
THUMP domain	THioUridine synthases, RNA Methyltransferases and Pseudo-uridine synthases domain
TI	transcript imaging
Ti plasmid	tumour-inducing plasmid; tumour-induction plasmid
TICO	translation initiation site correction
TILLING	targeting-induced local lesions in genomes
TIM	translocase of the inner membrane
t-IR	transgenerational induced resistance
TIR1	transport inhibitor response 1
TIRs	terminal inverted repeats; Toll-like innate immunity receptors

TIS	translation initiation site
TISM	transgenerational and intergenerational stress memory
TK	thymidine kinase
TKI	tyrosine kinase inhibitors
TLC	thin layer chromatography
TLR	toll-like receptor
TLR3	toll-like receptor 3
TLS	translesion synthesis
T_m	melting temperature of DNA
TM AFM	tapping mode atomic force microscope
TRB proteins	telomere repeat binding proteins
TMGT	testis-mediated gene transfer
tmm	too many mouths protein
TMP-SMZ	trimethoprim and sulfamethoxazole
tmRNAs	transfer-messenger RNAs
TMS	Tandem Mass Spectrometry
TMT	Tandem Mass Tagging
TMV	tobacco mosaic virus
TNA	threose nucleic acid; threofuranosyl oligonucleotides
tncRNAs	tiny non-coding RNAs
TobRV	tobacco ringspot virus
TOC1	Timing of CAB expression 1
TOC1	*timing of CAB expression 1* gene
Tof1	Topoisomerase 1-associated factor 1
TOM	translocase of the outer membrane
TP	transcript profiling
TPE	telomere position effect
TPP	thiamine pyrophosphate
TPS	triisopropyllbenzene sulfonylchloride
tr	triradial
tra	transfer operon
tra-2	*transformer-2*
tracrRNA	*trans*-activating CRISPR RNA
translocase	elongation factor (EF)-G
trans-TGS	*trans*-transcriptional gene silencing
TRAP marker	target region amplification polymorphism marker
TRHs	*trans*-regulatory hubs
tri-allelic SNP	tri-allelic single-nucleotide polymorphism
TRIPs	Trade Related Aspects of Intellectual Property Rights; transcription-replication interaction profiles
tris-EDTA or TE buffer	tris-ethylenediamine tetra-acetic acid buffer
tRNA	transfer RNA
tRNAo	orthogonal amber codon-suppressing tRNA

tRNASec	Sec-specific transfer RNA
trp operon	tryptophan operon
TRS	target recognition site
TrxG	Trithorax group
ts	temperature-sensitive mutant
Ts cells	suppressor T cells
TSG	tumour suppressor gene
TSGFs	tissue-specific growth factors
tsl	*torso*-like gene
tSNPs	tagging single-nucleotide polymorphisms
tsRNAs	sperm transfer RNA-derived small RNA
TSS	transcriptional start site
TSSa-RNAs	transcription start site-associated RNAs
TSWV	tomato spotted wilt virus
TT	tetratype
TTS	transcription termination sequence
TU/cell	transducing units per cell
TULP3	tubby-like protein 3
TuRC	γ-tubulin ring complex
TUs	transcription units
TUTase	terminal uridylyl transferase
2′-*O*-Me	2′-*O*-ribose methylation
2,4-D	2,4-Dichlorophenexy acetic acid
2-D DIGE	two-dimensional differential gel electrophoresis
2-D PAGE	two-dimensional polyacrylamide gel electrophoresis
2-D TLC	two-dimensional thin layer chromatography
2-DE or 2-D GE	two-dimensional gel electrophoresis
2n	diploid
TX	DNA triple crossover molecules
U	uracil
UAFM	ultrasonic atomic force microscopy
UAS	upstream activator sequence
UAS_G	galactose upstream activating sequence
ub	ubiquitination
UCSF	University of California, San Francisco
UDG cloning	uracil-DNA glycosylase cloning
UDGase	uracil-DNA glycosylase
UEC	unequal crossover
UEPs	unique event polymorphisms
UFM	ultrasonic force microscopy
UGMS marker	unigene-derived microsatellite marker
UHV	ultrahigh vacuum
UHV VT SPM	ultra-high vacuum variable temperature scanning probe microscope

UI	user interface
UL	long unique region
UML	Unified Modelling Language
UNG	uracil DNA glycosylase
universal nucleoside	1-(2′deoxy-b-D-ribofuranosyl)-3-nitropyrrole, designated 'M'
Up	probability of ultimate fixation
UPOV	Union for Protection of New Varieties of Plants
URF	unassigned reading frames; unidentified reading frame
URI	Uniform Resource Identifier
URL	Uniform Resource Locator
US	short unique region
USDA	United States Department of Agriculture
UsnRNAs	uridylate-rich small nuclear RNAs
U1snRNA	U1 small nuclear RNA
U2snRNA	U2 small nuclear RNA
U3snRNA	U3 small nuclear RNA
U4snRNA	U4 small nuclear RNA
U5snRNA	U5 small nuclear RNA
U6snRNA	U6 small nuclear RNA
U7snRNA	U7 small nuclear RNA
U8snRNA	U8 small nuclear RNA
U9snRNA	U9 small nuclear RNA
U10snRNA	U10 small nuclear RNA
U11snRNA	U11 small nuclear RNA
U12snRNA	U12 small nuclear RNA
UsnRNPs	uridylate-rich small nuclear ribonucleoproteins
USPTO	US Patent and Trademark Office
UTR	untranslated region
UV	ultraviolet light; ultraviolet radiation
V	variable sequences
V region	variable region
vb.	verb
V_A	additive portion of genetic variance
val	valine
V_D	dominance portion of genetic variance
VDRL	Venereal Disease Research Laboratory
V_E	environmental variance
VECs	valued ecosystem components
V_G	genotypic variance
V-GURT	variety-level genetic use restriction technology
V_I	interaction variance
VIGS	virus-induced gene silencing
VLPs	virus-like particles

V_M	mutational variance
VM	virtual machine
VN	vegetative nucleus
VNTRs	variable number of tandem repeats
VOCs	volatile organic compounds
VPD	Venom phospodiesterase
VPg	potyvirus genome-linked proteins
VPU	vision processing unit
VR	vertical resistance
VRN2	VERNALIZATION2
VSGs	variable surface glycoproteins; variant surface glycoproteins
VSP repair	very short patch repair
VSR	viral suppressor of RNA silencing
VT	virtual terminator
VUS	variant of uncertain significance
w^a	*white-apricot* gene
WAN	wide area network
WB	Western blotting
WBCs	white blood cells (leukocytes)
WBPH	white backed plant hopper
WES	whole exome sequencing
WGA	whole-genome association
WGD	whole-genome duplication
WGS	whole-genome sequencing; whole-genome shuffling
WHB	waxy hull-less barley
WUE	water use efficiency
wHTH domain	winged helix-turn-helix domain
winQTL	win quantitative trait loci
WIPO	World Intellectual Property Organization
without TR	without terminal repeats
WORA	write once, run anywhere
WPGMA	*weighted pair group method with arithmetic mean*
WRKY domain	tryptophan (W)-trginine (R)-tysine (K)-tyrosine (Y) domain
WTO	World Trade Organization
WTSS	whole transcriptome shotgun sequencing
x	basic chromosome number; monoploid number
$X_1, X_2, X_3, \ldots$	the first, second, third, . . . generations following X-irradiation
XCI	X-chromosome inactivation
xDNA	expanded DNA
XHTML	eXtensible HyperText Markup Language
XIC	X chromosome inactivation centre
Xist	X-inactive specific transcript
Xist	*X inactive specific transcript* gene

Xist RNA	X inactive specific transcript RNA
Xite	*X-inactivation intergenic transcription element* gene
Xite RNA	X-inactivation intergenic transcription element RNA
XLB	xylem-limited bacteria (fastidious)
XMA or Xta	chiasma (pl. chiasmata)
Xp	paternal X chromosome
Xpr	X chromosome-pairing region
Y2H	yeast-two hybrid
YAC	yeast artificial chromosome
YAM	yeast artificial mitochondria
yARS	yeast autonomously replicating sequences
Y-CHOPE	Y Chromosome Orthogonal Programmable Endonucleases
YCp	yeast centromere plasmid
YEp	yeast-episomal plasmid
YIp	yeast integrative plasmid
YLBS	Y-ligation based block shuffling
YRp	yeast-replicating plasmids
YTH	YT521-B homology
ZF	zinc finger
ZFN	zinc-finger nuclease
ZFP	zinc-finger protein
ZFY	zinc finger on the Y chromosome
ZMWs	zero-mode waveguides
ZPG	zero population growth
ZTP	5-amino, 4-immidazolecarboxamide riboside 5′-triphosphate
zygDNA	zygotene DNA

A

A chromosome: Any member of the standard diploid or polyploid chromosome complement of a given genome in a species.

A line: The seed-bearing parent line used for producing hybrid seed that is male sterile. A male-sterile parent line is used to produce hybrid seeds in wheat hybrid seed production.

A-tomatine: The main saponin in tomato (*Lycopersicon esculentum*), which displays strong antifungal activity.

A+T/G+C ratio: An expression of the relative number of adenine-thymine pairs to guanine-cytosine pairs in a molecule of DNA.

AAA+ transcription complex structures, DNA opening mechanism in: Gene transcription by RNA polymerase requires the conversion of the initial closed promoter complex, where DNA is double stranded, to a transcription-competent open promoter complex, where DNA is opened up. In bacteria, RNA polymerase relies on σ factors for its promoter specificities. It uses a special form of sigma factor (σ^{54}), which forms a stable closed complex and requires its activator that belongs to the AAA+ ATPases (ATPases associated with diverse cellular activities). F. Ye *et al.* in 2022 reveal a previously unidentified process of DNA melting opening through cryoelectron microscopy structures of transcription initiation complexes. The σ^{54} amino terminus threads through the locally opened-up DNA and then becomes enclosed by the AAA+ hexameric ring in the activator-bound intermediate complex. ATP hydrolysis by the AAA+ activator could remove the σ^{54} inhibition while helping to open up DNA by using σ^{54} amino-terminal peptide as a pry bar (a heavy iron lever with one end forged into a wedge).

***ab initio* gene prediction:** The prediction of exon (coding sequence) structure within a eukaryotic gene using computational algorithms. Several exons present in a pre-mRNA are required to be ligated to enable them to form the structural protein that is encoded by the gene.

ABC exonuclease: A DNA repair enzyme that removes DNA adducts by hydrolysing phosphodiester bonds on both sides of the modified nucleotides.

ABC transporters: Large superfamily of membrane transport proteins that use the energy generated by hydrolysis of adenosine triphosphate (ATP) to transfer peptides and a variety of small molecules (such as sugar molecules, inorganic ions, polypeptides, certain anticancer drugs, certain antibiotics) across membranes. *Aka* ABC TRANSPORTER PROTEINS.

ABCE model of flower development: A, B, C and E are four activities that are present in adjacent whorls of the flower. These four activities are postulated to function combinatorially to specify the identity of the four flower organs – sepals, petals, stamens and carpels.

aberrant ratio: Any departure from Mendelian expectations in F_2 generation.

aberration: Induced or spontaneous variation of chromosome or chromatid structure caused by fragmentation usually followed by union of broken ends into new combinations.

aberration rate: The portion of chromosomal alterations as compared to standard chromosomes.

abiogenesis: Origin of life on earth from non-living matter under a reducing atmosphere when only methane, water vapours, ammonia and hydrogen were present in the environment. This view was given by A. Oparin in 1938 who gave the theory of PRIMARY ABIOGENESIS, which can be summed up as 'abiogenesis at first and biogenesis ever after'. *Aka* CHEMICAL EVOLUTION OF LIFE ON EARTH.

abiological (abiotic) evolution: Evolution of a mechanism that could synthesize macromolecules without the involvement of living systems during origin of life.

abiotic: Physical or chemical (rather than biological) influences or progressions of the non-living environment (climate, geology and atmosphere).

abiotic component: Any non-living chemical or physical component of the environment that distresses living organisms and the functioning of environments, such as the atmosphere and water resources.

abiotic diseases: Diseases that are not triggered by living organisms (pathogens), but by non-living (chemical and physical) factors. *See* NON-INFECTIOUS. *Contrasts* BIOTIC DISEASES and INFECTIOUS DISEASES.

abiotic elicitors: Abiotic elicitors include various chemical and physical stressors such as light and UV-radiation; salts of heavy metals ($Ag_2S_2O_3$, $AgNO_3$, $CdCl_2$, $CuCl_2$, $CuSO_4$, $VOSO_4$, $NiSO_4$, selenium); temperature shift; osmotic stress induced by mannitol, sorbitol, sodium chloride, potassium chloride, cadmium chloride, PVP, etc.; intracellular signalling molecules such as jasmonic acid (JA), methyl jasmonate (MJ), salicylic acid (SA), acetylsalicylic acid (ASA) and systemin.

abiotic stress: The stress (e.g. to crop plants) produced by non-living (physical and chemical) influences such as soil condition, cold, heat, drought, flooding, salinity, metal toxicity, mineral deficiency, adverse pH, air pollution ozone, toxic-to-that-organism metals (e.g. aluminium, for plants) and ultraviolet-B light. Abiotic stress limits crop productivity or destroys biomass. Using biotechnological approaches such as transgenesis and marker-aided selection, attempts are being made to develop crop varieties resistant to various kinds of abiotic stresses.

abiotic stress-related trait: Retort by the plant in terms of resistivity or sensitivity to abiotic stress.

ABO blood group system: Major blood antigen system containing types A, B, AB and O, depending upon the presence or absence of the A and/or B antigens. This method of classification of blood groups was given by Karl Landsteiner in 1901. In this system, individuals with blood group AB can receive blood from individuals with A, B, AB and O; A can receive blood from individuals with A and O; B can receive blood from individuals with B and O; and O can receive blood from individuals with O. So, an individual with O blood group is a universal donor.

abort: To abandon the development of a structure or organ; the collapse or disappearance of seeds or cells.

abortive infection: When pathogenic microorganisms do not become established in the tissue of the host.

abortive transcription initiation: During transcription initiation *in vitro*, prokaryotic and eukaryotic RNA polymerase can engage in abortive initiation, the synthesis and release of short (2–15 nucleotides) RNA transcripts before productive initiation. Abortive initiation involves DNA 'scrunching' in which RNA polymerase remains stationary and unwinds and pulls downstream DNA into itself. Scrunching requires RNA synthesis and depends on RNA length. *See* DNA SCRUNCHING.

abortive transduction: A process by which bacterial DNA carried by the phage is injected into a recipient bacterium but does not replicate because in this case the transducing DNA does not get merged into the recipient chromosome.

abscisic acid (ABA): A growth regulator or phytohormone exemplified by auxins, gibberellins or cytokinins. ABA occurs in various tissues and seeds and is thought to play an important role in controlling various features of seed growth and development, also being involved in fruit growth, rejection of plant organs (such as leaves and fruits) and other physiological phenomena. It also affects the regulation of somatic cell embryogenesis in some plant species. *Aka* ABSCISIN(E).

abscission zone: Zone at the base of a leaf, flower, fruit or other plant part that contains an abscission and/or separation layer and a protecting layer, both being involved in the abscission of the plant part.

abscission: The natural detaching of old leaves or other plant parts such as ripe fruit as the consequence of physical weakness in a specialized layer of cells (called the abscission layer) that grows at the base of the structure. The shedding of flowers, leaves and/or fruit occurs following formation of scar tissue in a plant.

absolute configuration: The arrangement of four different substituent groups around an asymmetric carbon atom, such as relative to dextrorotatory (D) and laevorotatory (L) glyceraldehyde.

absolute quantification (AQUA) of modification states: Absolute quantification of proteins can be determined by using an isotopic synthetic peptide as an internal standard. With a synthetic isotopic modified peptide, individual stoichiometry of a post-translational modification (PTLM) site is also calculated.

absorbance (A): The quantity of light absorbed by a substance suspended in a medium, called matrix. The nature of matrix may be gaseous, liquid or solid. Most biologically active compounds (e.g. proteins) capture light in the ultraviolet (UV) or visible light portion of the spectrum. Absorbance is used to assess the concentration of the substance in question (e.g. a solid substance dissolved in a liquid). *See* OPTICAL DENSITY (OD) and SPECTROPHOTOMETER.

absorption: (i) Acceptance of substances, usually nutrients, water or light, by plant cells or tissue. Here, one substance infuses with another. (ii) In soil science, the physical acceptance of water or ions by a substance (e.g. soil). (iii) In microscopy, the interface of light with matter, ensuing in diminished concentration across whole range or loss of intensity from a portion of the spectrum.

absorption of water: The manner in which water from soil enters roots, particularly to the root xylem.

absorption spectrum: (i) The range of electromagnetic radiation that passes through a medium, which absorbs radiation of certain wavelengths. (ii) A graph that shows the percentage of each wavelength of light absorbed by a pigment (e.g. chlorophyll in plants).

abstract data type (ADT): An accurate model for data types where a data type is defined by its behaviour (semantics) from the point of view of a user of the data, precisely in terms of probable principles, possible operations on data of this type and the behaviour of these procedures. *Contrasts* DATA STRUCTURES, which are concrete representations of data from the point of view of an implementer rather than a user.

abstract method: One with only a name and no implementation body. It is frequently engaged to postulate that a subclass must offer an application of the method. Abstract methods are cast off to stipulate boundaries in some computer languages.

abstraction: (i) In software engineering and computer science, abstraction is the process of eliminating physical, spatial or temporal specifics in the study of substances or systems so as to more meticulously deal with other particulars of interest. It is also very similar in nature to the procedure of generality. (ii) The outcome of this method is an abstract concept object shaped by retaining common features or characteristics to numerous real objects or systems of study.

abundance: The predicted number of entities of a species in an area or entire population.

abundant mRNAs: Individual species of mRNAs present in a large number of copies per cell.

abzyme: An antibody engineered to perform an enzymatic reaction (i.e. a catalytic antibody). Such an antibody is raised up against a transition state analogue for the reaction to be catalysed. It either stabilizes the transition state of a chemical reaction or binds to a specific substrate, thereby enhancing the rate of that chemical reaction.

accelerated breeding: The Consultative Group on International Agricultural Research (CGIAR) initiative on accelerated breeding aims to develop better-performing, farmer-preferred crop varieties and to decrease the average age of varieties in farmers' fields, providing real-time adaptation to climate change, evolving markets and production systems.

accelerated evolution: Orthogonal replication enables rapid continuous biomolecular evolution in *Escherichia coli*. R. Tian *et al.* in 2024 introduced user-defined DNA into an *E. coli* cell in such a way that it is selectively copied and rapidly mutated by distinct replication machinery without affecting the organism's genome.

acceptable daily intake (ADI): A measure of the amount of a chemical (i.e. a pesticide residue) that is believed to be consumed on a daily basis over a lifetime without causing harm. ADI is used by the US government to establish safe levels of potentially hazardous substances in food.

acceptor: An atom that receives a pair of electrons during the formation of a chemical bond.

acceptor junction site: The connection between the right 3′ end of an intron and the left 5′ end of an exon. T is the site where the cleavage of pre-mRNA occurs during RNA splicing.

accession: A discrete sample of germplasm (cultivar, breeding line, population), that is, maintained and deposited in a gene bank for conservation and evaluation. In order to signify the genetic variation of a sample, ideally 4000 seeds are needed for genetically homogeneous lines and about 12,000 seeds for heterogeneous lines.

accession number (in GenBank): (i) An exclusive identifier allotted to the complete sequence record when it is submitted to GenBank. The GenBank accession number is a grouping of letters and numbers that are usually in the format of one letter followed by five digits (e.g. M12345) or two letters followed by six digits (e.g. AC123456). The accession number for a particular record does not change even if the author submits a request to change some of the information in the record. An accession number is a unique identifier for a complete sequence record. On the other hand, a sequence identifier, such as a Version, GI or Protein ID, is a proof of identity, that is, a number assigned only to the sequence data. The NCBI Entrez System is searchable by accession number using the Accession [ACCN] search field. (ii) A unique identification number for a complete RefSeq sequence record. RefSeq accession numbers are also written in the following format: two letters followed by an underscore and six digits (e.g. NT_123456). Here, the first two letters indicate the type of sequence included in the record: NT_123456, constructed genomic contigs; NM_123456, mRNAs (actually, the cDNA sequences constructed from mRNA); NP_123456, proteins and NC_123456, chromosomes.

accessory chromosome: A chromosome that exists in a cell in addition to the normal chromosome complement (e.g. B chromosome).

accessory DNA: Extra DNA present in certain cells or cell stages as a result of gene amplification.

acclimatization: Changes involving the synthesis of proteins, membranes and metabolites that occur in a plant in retort to chilling or freezing temperatures that protect tissues or confer tolerance to the cold. For example, micro-propagated plantlets are usually very delicate and exhibit high mortality rates upon their direct transfer to soil in the field. To thwart this, the plantlets are acclimatized by keeping them on water-moist cotton in open trays in a well-illuminated incubation room for 3–5 days. During this process, plantlets get hardened and resume growth. Thus, acclimatized plants exhibit better survival rates upon their transfer to soil in a greenhouse or field. *Aka* ACCLIMATIZATION.

acclimatized: A condition of physiological adjustment by plants to altered environmental or stress conditions.

accuracy: (i) A measurement or an estimate of measurement; accuracy is the extent to which a measurement represents the true value statistics. (ii) Truthfulness or correctness of a particular value of measurement to the reality. For example, age recorded as 7 years 4 months and 14 days is more accurate than recorded only as 7 years, although this additional accuracy may be redundant.

acellular: Describing tissue or organisms that are not made up of separate cells but often have more than one nucleus.

acentric: (i) Chromosome, chromosome segments or chromatids that have no centromere. (ii) Also applies to a chromosome fragment formed during cell division that lacks a centromere. This fragment is unable to follow the rest of the chromosomes in migration towards one or the other pole, as it does not attach itself to the cell spindle. Consequently, such chromosomes vanish during cell division.

acentric fragment (A): A segment of a chromosome that lacks a centromere.

acetic acid: A carboxylic acid, CH_3COOH, is a simple fatty acid. It is the final product of several fermentation, oxidation or rot processes. It plays a crucial role in energy metabolism. *Aka* ETHANOIC ACID.

aceto-orcein: It consists of the dye orcein dissolved in acetic acid that is used in chromosome staining.

acetocarmine: Used as a dye for staining of chromosomes. For this, cells or tissues are pretreated (fixation) for 12–24 h with a mixture of alcohol and acetic acid (3:1) or alcohol, acetic acid and chloroform after Carnoy. Prior to its squashing, the material is stained with acetocarmine.

acetolactate synthase (ALS): A plant enzyme (also present in some microorganisms), which catalyses one of the early common chemical reaction steps in the synthesis of branched-chain amino acids (isoleucine, leucine and valine), required by plants to sustain life. ALS is the target enzyme for three classes of structurally unrelated herbicides: the sulfonylureas, the imidazolinones and the triazolopyrimidines. Herbicides that deactivate/destroy ALS are effective at killing weeds. *Aka* ACETOHYDROXY ACID SYNTHASE (AHAS).

***Acetolactate synthase* (*ALS*) gene:** A gene that codes the critical-to-plants enzyme ALS in microorganisms or plants' chloroplasts.

acetone: A simple but vital ketone that is frequently used as a lipid solvent.

acetyl group: A chemical group derived from acetic acid that is covalently added and removed off of histones. Acetylation of histones permits transcription and deacetylation hinders transcription.

acetyl-Coenzyme A (Ac-CoA): It consists of a functional acetyl group linked to coenzyme A (CoA) by an easily hydrolysable thioester bond. CoA is the starting product of the citric acid cycle in cells. Acetyl-CoA contributes to many biochemical reactions in protein, carbohydrate and lipid metabolism.

acetylation: A PTLM that results in the addition of a functional acetyl group to organic compounds such as histone and non-histone proteins. *Opposite to* DEACETYLATION.

achiasmate: Meiosis and/or chromosome pairing not involving crossing over and chiasma formation.

achromatic: Parts of the nucleus that are not stainable by common chromosome dyes. *Aka* ACHROMATIN.

acid: A molecule or chemical group that donates a proton (H^+), either to water or to some other base, thus forming a hydronium ion (H_3O^+). Hydrogen ions dissociated from an acid play a role in maintaining acidic pH. The amount of H^+ is expressed by the logarithmic pH scale that ranges from 0 to 14. Chemicals having a pH less than 7 are called acidic. Biochemical reactions take place at a specific level of pH, and the living systems are extremely sensitive to small changes in the pH of the environment. *Contrasts* ALKALINE, BASE.

acid hydrolase: Any of a group of diverse hydrolytic enzymes (such as proteases, nucleases and glycosidases) that have their optimal activity at acidic pH (around 5.0) and are found in the lysosomes.

acid number-acid value. The mass of potassium hydroxide (KOH) in milligrams that neutralizes 1 g of a chemical substance like a fatty acid or in a mixture of compounds. *Aka* NEUTRALIZATION NUMBER, ACID NUMBER or ACIDITY.

acid phosphatase (ACPH): An enzyme that is a member of hydrolases; because of its variability, it is sometimes used as a biochemical marker in genetic studies.

acid precipitation: Low pH precipitation in the form of fog, rain or snow due to the presence of nitric and sulfuric acids formed by the reaction of air pollutants (NO_x and SO_2) with water. *Aka* ACID RAIN.

acid soil: Specifically, a soil with a pH value <7.0, which is caused by the presence of active hydrogen and/or aluminium ions. The pH value of soil decreases as the activity of these ions increases.

acid tolerance: The weathering of soil can result in acidification, limiting the types of plants or microbes that can grow. Repeated fertilization with ammonia can also lead to acidification. To overcome such problems the soil must be limed, or plant species and microbes used that are tolerant of acidity.

acid(ic) dye: An organic anion that stains positively charged macromolecules and acts on protoplasm.

acid-fast stain: A differential stain for the identification of bacteria, which are not decolorized by acid alcohol. It is a stain used to differentiate and distinguish bacteria for classification. This stain strongly binds to those bacteria, which possess a waxy substance in their cell walls. Acid-fast stain is precisely used to recognize *Mycobacterium tuberculosis* that causes tuberculosis and to identify pathogenic strains of genus *Nocardia*. In this procedure, a red dye called carbol fuchsin is applied. The acid-fast microorganisms retain the red colour.

acidity: In soil science, acidity is a measure of the activity of the hydrogen and aluminium ions in wet soil, usually stated as the pH value. Crop plants display precise requirements; some of them grow judiciously well on acid soils (e.g. cowpea pH 5.0–7.0; oats, rye pH 5.5–7.5; maize, sorghum, wheat pH 6.0–7.5; barley pH 6.5–7.5; lucerne pH 7.0–8.0).

acoustic separation and concentration of exosomes for nucleotide detection (ASCENDx): T.D. Naquin *et al.* in 2024 introduced an innovative spaceship-like disc that allows Acoustic Separation and Concentration of Exosomes and Nucleotide Detection: ASCENDx. They created ASCENDx to use acoustically driven disc rotation on a spinning droplet to generate the swift separation and concentration of exosomes from patients' plasma samples. Integrated plasmonic nanostars on the ASCENDx disc enable label-free detection of enriched exosomes with 95.8% sensitivity and 100% specificity.

acquired character: A nonheritable modification of structures or functions enthralled on an individual by environmental stimuli during its development.

acquired immunological tolerance: Insensitivity of the immune system to a given extraneous antigen that can develop in some circumstances.

acquired mutation: A nonheritable genetic change occurring within a somatic cell.

acquired resistance: Plant resistance to a disease conferred after inoculation of the plant with certain microorganisms or treatment with certain chemical compounds.

acquired trait: A phenotypic property, assimilated during growth and development, that is not genetically based and therefore cannot be transmitted to the next generation (e.g. the large muscles of a weightlifter).

acridine dye: A chemical that causes frameshift mutations in the DNA sequence. Several derivatives of acridine, such as ACRIDINE ORANGE, are used as dyes or biological stains.

acridine orange: An acridine dye that works as both a fluorochrome and a gene fluorescence staining mutagen.

acrocentric chromosome: A chromosome whose centromere is located near one end and in this case one arm is very long and the other is very small.

acrosyndesis: Incomplete end-to-end chromosome pairing.

acrylamide: (i) Chemical, C_3H_5NO, that can be chemically persuaded to form a gel matrix. This matrix is used for electrophoretic separation of proteins and nucleic acids, (ii) A compound that forms in some foods during the cooking process (e.g. frying, roasting or baking), due to heat networking with sugars and an amino acid naturally present in some foods.

acrylamide gel: A gel that comprises an acrylamide solution. Commonly used in electrophoresis, the gel provides a mesh-like matrix that makes it possible to accurately separate proteins or DNA based on their length in an easy and inexpensive process.

act: A law made by the parliament or a provincial legislature. The process of making an act of parliament begins with the introduction of a proposed act, or bill, in one of the two houses of parliament (the Senate or the House of Commons). A bill becomes an act if it is passed (approved) by both houses and receives royal assent.

actin: Plentiful protein that forms actin filaments in all eukaryotic cells. The monomeric form is sometimes called GLOBULAR or G-ACTIN, while the polymeric form is called FILAMENTOUS or F-ACTIN.

actin 1 promoter: An efficient promoter for regulating the constitutive expression of a foreign gene in transgenic rice.

actin and myosin in chromatin dynamics and gene expression: The mammalian cell nucleus provides a landscape where genes are regulated through their organization and association with freely diffusing proteins and nuclear domains. In many cases, specific genes are highly dynamic. Actin and myosin have been implicated in chromatin dynamics and gene expression. Dynamics of interphase genome has impact on nuclear organization and gene activity.

actin gene family: Actin genes in eukaryotes represent a multigene family. Here different members are homologous but non-identical, giving rise to slightly different variants. Different members function either at different times or in different tissues at the same time.

actin-binding protein: Protein that connects with either actin monomers or actin filaments in cells and modifies their properties. Examples include myosin, α-actinin and profilin.

actin-related protein 2/3 complex (ARP2/3 complex): A complex of proteins that nucleates actin filament growth from the minus end.

actinobacterium (pl. actinobacteria): A member of a group of prokaryotic microorganisms that are characterized by long filaments. Formerly known as ACTINOMYCETE.

actinomorphic: Regular or radially symmetrical structures or organisms that may be bisected into similar halves in at least two planes. Applies, for example, to steles and flowers in which the perianth segments within each whorl are alike in size and shape. *Compare with* REGULAR. *Contrasts* ASYMMETRICAL, IRREGULAR *and* ZYGOMORPHIC.

actinomycetes: Filamentous bacteria, many of which are treasured in the production of antibiotics.

actinomycin: An antibiotic produced by *Streptomyces chrysomallus* that thwarts the transcription of mRNA.

activated carrier: A small diffusible molecule in cells that stores easily exchangeable energy in the form of one or more energy-rich covalent bonds (e.g. ATP and NADPH). *Aka* COENZYME.

activated charcoal/activated carbon: Charcoal which has been treated to remove hydrocarbons and to increase its adsorptive properties. It acts by condensing and holding a gas or solute on to its surface. Inhibitory substances in the nutrient medium may thus be adsorbed to charcoal included in the medium. Rooting factors such as phenolamines present as pollutants in charcoal may stimulate growth *in vitro*. In addition to rooting medium, they may stimulate root initiation in certain plant species. Activated charcoal may differ in origin and in composition. *Compare with* CHARCOAL *and* PHENOLIC OXIDATION.

activated/deactivated genes: Genes can be activated (where they are turned 'on') or deactivated (turned 'off') depending on how they are chemically marked. An active gene will have its information used in the generation of proteins for cell functions.

activating enzyme: An enzyme that plays a part in translating the genetic code. Each activating enzyme attaches itself to a specific amino acid and also to the appropriate transfer RNA. *See* AMINOACYL-tRNA.

activation domain: The part of a transcription factor that is believed to recruit RNA polymerase to 5′ end of the gene.

activation energy: The minimum collision energy required for a chemical reaction to occur. It is the energy required to overcome the barrier to initiate chemical reaction. It is also the total quantity of energy needed to interrupt the steady electronic arrangement of any molecule so that the electrons can be rearranged. Commonly the rise in temperature increases the reaction rate of a substance. Thus, heat upsurges both the frequency of crashes and the number of molecules that achieve activation energy.

activation-induced deaminase (AID): A member of the family of RNA cytidine deaminases that can deaminate 5-methylcytosine (5^mC) to convert it to thymine.

activation of tRNA: Before translation commences, the activation of amino acids (formylated methionine in prokaryotes and methionine in eukaryotes) takes place in the cytosol at the disbursement of ATP. An amino acid reacts with ATP to become adenylated with associated release of pyrophosphate. As a result of adenylation, the amino acid is attached to adenylic acid via a high-energy ester bond in which the carbonyl group of amino acid is linked to phosphoryl group of adenosine monophosphate (AMP).

activation reaction: A reaction using ATP as energy and conducted by an amino acyl synthetase to activate a specific amino acid. This reaction forms amino acid–AMP complex.

activator: (i) In enzymology, a small protein-like substance that is able to kindle developmental processes. (ii) In molecular biology, a protein upstream from a gene on which the DNA binds; it triggers the transcription of the gene when it binds to an allosteric site.

activator protein: A protein that can stimulate gene expression.

activator RNAs: Component of eukaryotic gene regulation machinery. Activator RNA molecules are synthesized on integrator sites. These RNA molecules or their protein products bind with receptor sites.

activator-dissociator (Ac-Ds) elements: Described genetically in 1948 by Barbra McClintock in maize. Element *Ac* is activator, an autonomous element. Element *Ds* is dissociation locus (chromosomal breaks) – a semi-autonomous element, its mobility depends on *Ac*. *Ds* is derived from *Ac* and contains internal deletions. Both elements contain an 11-bp inverted repeat at the termini (TIR). The *Ac-Ds* system involves an autonomous element (*Ac*) whose insertions are unstable and a non-autonomous element (*Ds*) whose insertions are stable if only *Ds* is present. *Ac* activates *Ds* to transpose or break the chromosome where it is inserted. The *Ac* gene also acts as a gene regulator. It also regulates the appearance and size of colour spots, which are not formed in its absence. The more *Ac* genes present, the smaller the spot size. Apparently, the *Ac* gene delays gene action and therefore not only reduces the ability of the *Ds* gene to induce a break but also regulates movement of breakage. *Aka* ACTIVATOR-DISSOCIATOR (AC-DS) TRANSPOSONS.

activator-dissociator (Ac-Ds)-mediated gene transfer: *Ds* has been successfully utilized in enhancer/gene trap vehicles in *Arabidopsis*. Enhancer/gene traps are advanced versions of the conventional transposon-mediated insertional mutagenesis. 35S CaMV minimal promoter has been commonly used with this system in plants.

activator-dissociator (Ac-Ds)-mediated gene trap system: *Ac* and gene trap *Ds* vectors have been constructed and introduced into the plant genome by *Agrobacterium*-mediated transformation. A doubly enhanced CaMV 35S promoter, untranslated Ω′ sequence and maize *Adh1* intron I were fused to the cDNA of an *Ac* protein-coding sequence. The transcription terminator was from a *PINII* (proteinase Inhibitor II) gene of potato. The *Ac*/*Ds*-mediated gene trap system could prove an excellent tool for the analysis of the functions of genes in rice. A heterologous *Ac*/*Ds* family in rice could be employed in large-scale mutagenesis.

activators: A component of transcriptional processivity. These molecules bind to genes at sites called enhancers. They help to determine which genes will be switched on and speed rates of transcription.

active allele: An allele that is in the 'on' state or is expressed.

active collection: A collection of germplasm cast off for regeneration, multiplication, distribution, characterization and evaluation. Ideally, the germplasm is maintained in sufficient quantities to be made available on request. It is commonly duplicated in a base collection and is often stored under medium- to long-term storage conditions.

active DNA demethylation: An enzymatic process that removes or modifies the methyl group from 5-methylcytosine ($5^{m}C$) to cytosine.

active euchromatin: A fraction of the euchromatin that is actively engaged in transcription.

active histone modifications: Histone modifications that turn on a gene.

active immunity: All methods and reactions that enable a plant to prevent an interaction with a pathogen. Such immunity can be assimilated either by having the disease or through vaccination against it.

active ingredient: In any pesticide product, the component that kills or controls target pests.

active pocket: The region in catalytic proteins (enzymes) that modifies the target (substrate).

active resistance: Resistance resulting from host reactions occurring in response to the presence of the pathogen or its metabolites.

active site: The region on an enzyme that interacts with a specific chemical substance (reactant). In living cells, enzymes serve as biological catalysts. As catalysts, enzymes are very specific. Each enzyme acts on a specific substance, called substrate. For example, sugar is a substrate for the enzyme sucrase that converts sucrose into glucose and fructose. The three-dimensional enzyme molecule possesses an active site that interacts with a specific substrate. Competitive inhibitors plug the active site and compete with the normal substrate for the active site.

active site of spliceosome: Formation of catalytically active RNA structures within the spliceosome requires the assistance of proteins. The number and nature of proteins needed to establish and maintain the spliceosome's active site is not fully known.

active substance: Pesticides or herbicides are generally a blend of different substances; among them the active substance is the most significant one as it attacks the pathogen.

active transport: The passageway of materials (such as a protein or a drug) across a cell membrane against a concentration gradient that requires energy other than that stored in the electrochemical gradient of the transported molecule. Active transport requires an expenditure of energy. In plants, there is an important theory of active transport for the ascent of cell sap. Movement of ions or molecules across the membrane occurs against the concentration gradient. The movement against the concentration gradient or electrochemical gradient is coupled with ATP hydrolysis, and thus, active transport is an energy-requiring process. *Compare with* PASSIVE TRANSPORT.

activity coefficient: The factor by which the concentration of a solute must be multiplied to give its true thermodynamic activity.

activity rhythm: An individual's day-to-day landscape of physiological activity.

Acuron™ Gene: A gene, trademarked by Syngenta AG, that can be inserted into plants via genetic engineering techniques. When inserted into the genome of a plant, this gene confers tolerance to herbicide(s) whose active ingredient is a protoporphyrinogen oxidase (PPO) inhibitor. Thus, such herbicides are known as PPO INHIBITORS.

acute disease: A disease in which symptoms develop rapidly but last only for short time. Every disease distresses the body which changes some structures and functions in a specific way. These are generally associated with some particular symptoms of a particular disease. The scope of the disease is defined in terms of its duration and severity.

acute toxicity: The capability of a single dose of a compound to poison. *Contrasts* CHRONIC TOXICITY.

acute transfection: Short-term infection of cells with DNA.

acute transforming retroviruses: Viruses that have acquired sequences from the host genome, giving them the property of causing tumours or transformation of cells in culture. *Aka* VIRAL ONCOGENES.

acute: (i) Pertaining to symptoms that develop suddenly. (ii) Sharply pointed, but not drawn out, with straight sides approaching the point and converging at an angle of less than 90°. *Contrasts* OBTUSE, ACUMINATE, CHRONIC. *See* LEAF SHAPE.

acyl group: Functional group derived from a carboxylic acid.

acyl-CoA: Acyl derivatives of coenzyme A (acyl-S-CoA). Acyl-CoA is a group of coenzymes that metabolize fatty acids. Acyl-CoAs are susceptible to beta oxidation, ultimately forming acetyl-CoA.

adaptable: Capable of adapting or able to adjust oneself readily to different environmental conditions.

adaptation: Any heritable property of an organism that expands its ability to survive and reproduce in its environment. This term is used to pronounce the mechanism of genetic change within a population, as influenced by natural selection. According to Charles Darwin's theory of evolution by natural selection, organisms that possess heritable traits that enable them to better adapt to their environment compared with other members of their species will be more likely to survive, reproduce and pass more of their genes on to the next generation.

adaptation and domestication of cotton: Transposable elements (TEs) were found to be interwoven with *de novo* epiallele formation, which was considered a source of epigenetic variation that should be further explored in the context of adaptation and domestication.

adaptation of DNA bisulfite methylation mapping to RNA: A critical feature is the portion of intact analyte nucleic acids surviving the sequential treatment under acidic bisulfite (BS) and alkaline conditions.

adaptation to different climates: Variation in various steps of the vernalization mechanism has played a major role in the evolution of different reproductive strategies. Examples: (i) *Arabidopsis thaliana* accessions not requiring vernalization can reproduce multiple times a year; (ii) in *Arabidopsis helleri*, a temperature/time-averaging mechanism integrates temperature over a 6-week period; and (iii) In *Arabis alpina*, a FLOWERING LOCUS C (FLC) homologue (called PEP1) shows interesting regulatory differences with the FLC in *A. thaliana*. Here, unlike

A. thaliana, the silencing is not epigenetically stable. Differences in epigenetic silencing thus seem to have contributed widely to the evolution of reproductive strategies in land plants.

adaptationist programme: A method of understanding evolution that assumes that traits are adaptations.

adapted race: Physiological race.

adapter: A synthetic double-stranded oligonucleotide or a specific type of linker that is used to link sticky ends to a blunt-ended double-stranded DNA molecule.

adaptin: Protein that binds clathrin, a molecular scaffold for vesicular uptake of cargo at the plasma membrane.

adaptive: Changes in a plant that function to preserve its complete development.

adaptive capacity: The genetically established variety or flexibility of reactions in a plant and/or population enabling it to respond to differing conditions in different ways.

adaptive character: A functional or structural characteristic of an organism that enables or enhances the probability of its survival. *Aka* ADAPTIVE TRAIT.

adaptive coloration: Refers to the colour shade or the colour pattern, which seems to camouflage the organism.

adaptive compensation: A process of natural selection in which a relative lack of fitness in one character may be compensated by special suitability in another.

adaptive dynamics: A method for finding the phenotypes that will attack a population. Alleles that differ slightly from the inhabitant alleles are introduced into a population and may invade it; if they do, they will either displace the inhabitant allele, or may coexist with it, in a balanced polymorphism.

adaptive enzyme: An enzyme that is produced in response to a specific environment. *See* INDUCIBLE ENZYME.

adaptive evolution: The process of change in a population driven by variation in reproductive success that is correlated with heritable variation in a trait. In natural selection, individuals with advantageous traits are able to reproduce successfully.

adaptive landscape: A graph of the average fitness of a population related with the frequencies of genotypes in it. Peaks on the landscape match the genotypic frequencies at which the average fitness is high and valleys to genotypic frequencies at which the average fitness is low. *Aka* FITNESS SURFACE or FITNESS LANDSCAPE.

adaptive logic: A behaviour has adaptive logic if it has tendency to increase the number of offsprings that an individual contributes to the next and following generations. If this behaviour is even partly genetically determined, it will be inclined to become widespread in the population. Then, even if circumstances change such that it no longer provides any survival or reproductive advantage, the behaviour will still tend to be displayed – unless it becomes absolutely disadvantageous in the new environment.

adaptive peak: A local maximum in the fitness landscape.

adaptive radiation: The relatively rapid expansion and diversification of an evolving group of organisms, as they adapt to new ecological niches. Adaptive radiation is the process by which one species evolves into two or more species. This occurs as a result of different populations becoming reproductively isolated from each other, usually by adapting to different environments. The branching pattern of evolution resulting from adaptive radiation is known as CLADOGENESIS.

adaptive reaction: Adaptive capacity.

adaptive selection: The evolution of similar types in dispersed but ecologically comparable areas.

adaptive strategies: Any approach dealing with competition or environmental conditions on an evolutionary timescale. Species adapt when subsequent generations underscore beneficial characteristics.

adaptive trials: Multilocational coordinated field experiments across the crop-growing regions in one or more countries undertaken to test the adaptation of varieties or breeding strains under definite ecological conditions.

adaptive value: An assessment of the reproductive efficacy of an organism or genotype compared with other organisms or genotypes.

adaptive zone: A set of ecological niches. A population or taxon makes a transition from one zone to another across an adaptive threshold.

adaptor: A synthetic single-stranded non-self-complementary oligonucleotide used in conjunction with a linker to add cohesive ends to DNA molecule.

adaptor hypothesis: Idea advanced by F.H.C. Crick in 1958 theorizing that there is a molecule that serves as an adaptor to match up amino acids on an mRNA template. The adaptor has one end that matches the mRNA and another end that is attached to the amino acid. This adaptor molecule is tRNA or sRNA.

adaptor protein: General term used for proteins described in intracellular signalling pathways that join different proteins in the pathway directly together.

adding complementary homopolymeric tails to DNA: In this procedure, terminal transferase is used to tail a linearized plasmid vector with G's and the cDNA with C's. When incubated together, the complementary G's and C's anneal to 'insert' the cDNA into the vector, which is then transformed into *E. coli*.

addition haploid: An individual having one alien chromosome more than the genetic chromosome number. It is derived from disomic alien addition line.

addition line: A cell line or line of individuals carrying extra chromosomes or chromosome arms to the normal standard chromosome set.

addition of cap at 5′-end of pre-mRNA: Eukaryotic mRNA invariably carries a 5′ cap composed of a 5′→5′ triphosphate linkage between two modified nucleotides: a 7-methylguanosine and a 2′ O-methyl purine. This cap is a modified guanine (m^7GpppN_1), which is added post-transcriptionally and functions to identify this RNA molecule as an mRNA to the translational machinery.

addition of poly(A) tail at 3′-end: Polyadenylation signal in the train is the sequence AAUAAA in most pre-mRNAs. Polyadenylation involves two steps – cleavage and the addition of poly(A) tail. Multiple copies of polyadenylation signal are a common occurrence in eukaryotic pre-mRNA.

addition rule of probability: The probability of the occurrence of one of two or more mutually exclusive events is the sum of the probabilities of their individual occurrence.

additive allelic effects: Genetic factors that raise or lower the value of a phenotype on a linear scale of measurement.

additive fitness: A selection regime in which the relative fitness of the heterozygote is exactly midway between the fitness of the two homozygotes.

additive gene action: The mode of gene action that causes difference in effect or average character expression associated with the two homozygotes for each gene pair involved. The additive gene effect is also referred to as BREEDING VALUE. This is flexible.

additive genes: Nonallelic genes that interact but do not show dominance (in the case of alleles) or epistasis (if they are not alleles). They do not have individually recognizable phenotypic effects; rather, these genes have cumulative or additive effects.

additive genetic effects: When the combined effects of alleles at different loci are equal to the sum of their individual effects. In this case, effects on a genetic trait are enhanced by each additional gene, either an allele at the same locus or genes at different loci.

additive genetic variance (V_A): Genetic variance due to the difference between homozygotes (for any locus). Heritability in the narrow sense is equal to V_A divided by the total PHENOTYPE VARIANCE (V_P).

additive model: A model in which a quantitative trait is the sum of effects of all the genes involved and a random environment.

additive resistance: Resistance controlled by more than one gene, each of which can be articulated autonomously, but which is strengthened by the appearance of each of the additional genes.

additive tree: A phylogenetic tree in which the subdivision lengths are relative to the evolutionary distance between nodes. *Aka* PHYLOGRAM.

adduct: The covalent complex formed when a chemical is bound with a biomolecule such as DNA or a protein.

adenine (A): A purine base, 6-aminopurine, present in RNA as well as in DNA. It is a component of adenosine diphosphate (ADP) and adenosine triphosphate (ATP). One of the four nucleotide bases in DNA – the other three being cytosine (C), guanine (G) and thymine (T). Adenine pairs with thymine in DNA and uracil in RNA. Its derivatives are involved in a wide variety of biochemical reactions, including cellular respiration. It is added to some tissue culture media as adenine sulfate to promote shoot formation and for its weak cytokinin effect. It is present in plant tissues combined with aminoamide, phosphoric acids and D-ribose.

adenine methylation in mRNA: Post-transcriptional adenine methylation in mRNA impacts the expression of thousands of genes. Thus, a fifth base, N6-methyladenosine (6^mA), infuses the transcriptome. 6^mA sites normally occur in regions of mRNA that are highly conserved across several species of vertebrates. Adenine methylation in RNA is the first demonstration of an epitranscriptomic modification.

adenine sulfate: Functions as a growth factor in some tissue culture media. It enhances the frequency of multiple shoot production and recuperates the chlorotic symptoms of leaves.

adeno-associated virus (AAV) vectors: Viral vectors with small packaging capacity, commonly used in gene therapy, which can infect both dividing and non-dividing cells and do not integrate into the host genome. AAV vectors have been approved for clinical use.

adenosine: A nucleoside holding a heterocyclic nitrogen base (adenine) linked with a pentose sugar.

adenosine 5-diphosphate (ADP): The substance formed from the hydrolysis of ATP. The ribonucleotide containing the purine adenine (A) and two phosphate groups is attached at 5′ end of the sugar moiety to form ADP. The living cells of the body obtain energy for the metabolic processes through the hydrolysis of ATP. Whenever the supply needs refilling, the reaction goes in the opposite direction, that is, conversion of ADP into ATP with the input of energy (ATP + $H_2O \leftrightarrow$ ADP + Energy).

adenosine 5-monophosphate (AMP): The ribonucleotide containing the purine adenine (A) and one phosphate group is attached at 5′ end of the sugar moiety to form AMP. It is one of the four nucleotides in an RNA molecule. AMP is formed by the hydrolysis of ATP or ADP. Two phosphate groups are added to AMP to form ATP.

adenosine 5-triphosphatase (ATPase): An enzyme that hydrolyses ATP to yield ADP, phosphates and energy. The reaction is usually tied to an energy-requiring process. ATP is hydrolysed in the act of shivering and the energy thus produced is transformed into heat to increase body temperature. This type of heat production involves what is known as a futile cycle because the energy is converted to (and wasted as) heat rather than used in motion, and so on. A large number of different proteins have ATPase activity.

adenosine 5-triphosphate (ATP): The ribonucleotide containing the purine adenine (A) and three phosphate groups is committed at the 5′ end of the sugar moiety to form ATP. Nucleoside triphosphate ($C_{10}H_{16}N_5O_{13}P_3$) composed of adenine, ribose and three phosphate groups is the prime carrier of chemical energy in cells. The terminal two phosphate groups are highly reactive in the sense that their hydrolysis, or transfer to another molecule, takes place with the release of a large amount of free energy. ATP and NADPH are the result of light-dependent reactions in plants. ATP is engaged in the reduction and regeneration of ribulose bisphosphate (RuBP). By breaking that one bond and reducing ATP to ADP, the cell can get the energy to carry out its various processes. ATP thus functions as the worldwide medium of biological energy storage and exchange in living cells.

adenosine 5-triphosphate (ATP)-binding cassette (ABC) proteins: Typically contain two nucleotide-binding domains (NBDs) that form two composite nucleotide-binding sites. The transition between the closed ATP-bound to the open ADP-bound state induces a tweezer-like power stroke between NBDs that causes conformational changes in associated domains and/or macromolecules.

adenosine 5-triphosphate (ATP)-dependent chromatin-remodelling complexes: ATP-dependent chromatin-remodelling complexes use the energy of ATP hydrolysis to alter chromatin architecture by repositioning, assembling, mobilizing and restructuring nucleosomes. These complexes are defined by the presence of a conserved SNF2-like, catalytic ATPase subunit that falls into one of four families: SWI/SNF, CHD/Mi-2, ISWI/SNF2L and INO80.

adenosine 5-triphosphate (ATP)-dependent histone exchange: The prototypical remodelling enzyme linked to histone exchange is the Swr1 complex, which directs the replacement of nucleosomal histone H2A/H2B dimers with H2AZ/H2B variant dimers with high specificity.

adenosine 5-triphosphate (ATP) synthase: An enzyme complex that forms ATP from ADP and an inorganic phosphate during oxidative phosphorylation in the inner mitochondrial membrane in animals, thylakoid membrane in chloroplasts in plants and in plasma membranes in bacteria. This is an energy-producing reaction as ATP is employed by cells to sustain their living condition. ATP synthase also occurs on the surface of endothelial cells (the lining of blood vessels), where it assists in building new blood vessels to replace the tissue damaged through injury or disease. Under certain situations, this also produces new blood vessels that provide blood supply to tumours. When separated from the cell's membrane, ATP synthase hydrolyses (breaks down) ATP via a chemical process in which one subunit (designated 'g') of ATP synthase rotates within the other (hollow) part of ATP synthase.

adenosine-to-inosine RNA editing: RNA editing of type conversion of A-to-I is an extensive post-transcriptional mechanism, mediated by adenosine deaminase acting on RNA (ADAR) enzymes to modify the RNA sequence of the primary transcripts.

adenylic acid: The ribonucleotide containing the purine adenine.

adenylyl cyclase: Membrane-bound enzyme that catalyses within the cells and results in the formation of cyclic AMP (cAMP) from ATP. *Aka* ADENYLATE CYCLASE. *See* CYCLIC AMP.

adhesion belt: A belt-like adherens junction that encircles the apical end of an epithelial cell and ascribes it to the adjoining cell. *Aka* ZONULA ADHERENS.

adhesion molecule: Refers to a glycoprotein (oligosaccharide) molecular chain that overhangs from the surface membrane of certain cells. It causes cells possessing matching adhesion molecules to adhere to each other.

adhesion plaque: *See* FOCAL ADHESION.

adhesive end: *See* STICKY END.

adjacent distribution: The positioning and delivery of adjacent chromosomes in the ring or chain configuration of translocation heterozygotes.

adjacent segregation: A reciprocal translocation heterozygote in which during meiosis the segregation of a translocated and a normal chromosome occurs together, producing unbalanced gametes with duplications and deficiencies leading to nonviable zygotes. Adjacent segregation is of two types depending on whether the non-homologous (adjacent-1) or the homologous (adjacent-2) centromeres segregate together. While ADJACENT-1 SEGREGATION is the usual type of adjacent segregation, ADJACENT-2 SEGREGATION is rare.

adjacent-1 segregation: Separation of the non-homologous centromeres during meiosis in a reciprocal translocation heterozygote such that unbalanced gametes with duplications and deficiencies are generated, as opposed to ALTERNATE SEGREGATION and ADJACENT-2 SEGREGATION.

adjacent-2 segregation: Separation of homologous centromeres during meiosis in a reciprocal translocation heterozygote such that unbalanced gametes with duplications and deficiencies are generated, as opposed to ALTERNATE SEGREGATION and ADJACENT-1 SEGREGATION.

adjuvant: A chemical substance added to the culture medium or to a vaccine to enhance its effectiveness. During plant tissue culture for instance certain chemical substances/extracts such as coconut milk, potato extract and bean seed extracts are used to supplement the chemically defined medium, such as used by T. Murashige and F. Skoog in 1962, to improve the cultural response. These enriching compounds are called growth adjuvants.

adnate: Grown from or closely fused to an organ of a different kind, especially along a margin (e.g. a stamen fused to a petal). Adnate anthers have their halves attached to the filament through most of their length. *Contrasts* CONNATE.

adosculation: The fertilization of plants by pollen dropping on the pistils.

ADP-ribosylation factor (ARF) protein: Monomeric GTPase accountable for controlling both COPI coat assembly and clathrin coat assembly at Golgi membranes. COPI is a coatomer, a protein complex that coats vesicles transporting proteins from the *cis* end of the Golgi complex back to the rough endoplasmic reticulum (RER), where they were originally synthesized, and between Golgi compartments.

adsorption: The adhesion of atoms, ions or molecules from a gas, liquid or dissolved solid to a surface. This procedure creates a film of the adsorbate on the surface of the adsorbent. The attachment of phage to host bacterium (e.g. phage lambda adsorbs to a maltose-binding protein of bacterium) is also termed as adsorption.

adult (stage adulthood): Having reached full size and strength, that is, mature.

adult plant resistance: A type of resistance that is not expressed at the seedling stage. Resistance increases with plant maturity (mature-plant resistance or age resistance) and is generally attributed to horizontal resistance.

adult plant resistance genes: The genes that confer resistance to a host in adult stages against a pathogen.

adult stem cell: Self-renewing multipotent stem cells present in the differentiated tissue of an adult organism. *Aka* TISSUE-SPECIFIC STEM CELLS. *Compare with* EMBRYONIC STEM CELL.

advance crop: Forecrop.

advanced: Opposite of primitive. In phylogeny, a plant or trait further removed from an evolutionary divergence than a more primitive one. A feature that displays an actual deviation from the familial trait or type (e.g. an agronomic trait such as brittle rachis).

advanced backcross QTL (AB-QTL) analysis: A method developed by S.D. Tanksley and J.C. Nelson in 1996 that aims at the simultaneous discovery and transfer of valuable QTL alleles from unadapted donor lines (viz. landraces and wild species) into established elite lines. Various steps include crossing an elite line with a donor (wild species, etc.), backcrossing of F_1 to elite line to generate sufficient BC_1 seed (100–200 plants), crossing of phenotypically selected desirable BC_1 plants to the elite parent to generate BC_2 plants. The BC_2 plants are subjected to restriction fragment length polymorphism (RFLP) analysis and are selfed to generate BC_2F_2 families. Also, some desirable BC_2 plants are backcrossed to generate BC_3 plants. The BC_2F_2 families are evaluated and analysed for QTLs. Several QTL-near-isogenic lines (QTL-NILs) are isolated by whole-genome selection. The different QTL-NILs possessing different QTLs for a character can be crossed, and all the desirable QTLs can be pyramided in a single line. *Aka* ADVANCED BACKCROSS QTL (AB-QTL) MAPPING.

advanced generations synthetic variety (beyond Syn1): Descends from an initial intercrossing of a specific set of clones or seed-propagated lines; usually steady for only limited number of generations, including the varieties 'Ranger' and 'Moapa' in lucerne, 'Saratoga' in bromegrass or 'Pennlate' in orchard grass seed.

advancing transcription elongation complex: RNA polymerase combines two contradictory biochemical features: (i) exceptional stability for dissociation and (ii) the ability to easily translocate along DNA. Thus, elongated RNA polymerase simultaneously behaves as a strong DNA-binding protein and as a protein with no affinity for particular DNA sites. The combination of these features ensures the processivity of RNA polymerase.

adventitious: Developing from unusual points of origin, such as shoots or root tissues from callus or embryos from sources other than zygotes (e.g. an adventitious bud produced from a stem rather than from the more typical axil of a leaf). Adventitious roots may develop from nodes of prostrate stems of some plant species, or from the hypocotyl rather than from the radicle of a germinating monocotyledon. This term can also be used to describe agents which contaminate cell cultures.

adventitious bud: A bud forming in an uncommon place (e.g. a bud on leaves).

adventitious embryony: A disorder in a seed in which the embryo ascends from somatic rather than reproductive tissue, such as the progression of a diploid embryo from nucellar or integumentary tissue (sporophyte tissue). It is common in certain grasses and often results in multiple embryos.

adventitious plant: A plant that is produced from somatic rather than reproductive tissue.

adventitious root: Arising from any structure other than a root, for example, from a node of a stem or from a leaf. This phenomenon is quite common in triticale and wheat. The adventitious root development is probably the response to accumulation of auxins in the base of the plant; sometimes this accumulation occurs because the plants are waterlogged and the existing root system is no longer a sink for the auxins due to the lack of oxygen and/or dying off of the roots.

adventive: (i) A plant that has been introduced in a different region but has not yet naturalized. (ii) A plant introduced accidentally, usually referring to a weed.

aeration: Transporting air into a substance, tissue or soil (e.g. by earthworms or digging and turning the soil) to loosen it.

aerenchyma: Plant spongy tissue containing large, intercellular air spaces in the leaves, stems and roots of halophytic plants. It helps in the exchange of gases between the shoot and the root systems.

aerial pathogens: Hostile microorganisms that constrain numerous fungal pathogens of aerial plant parts (e.g. *Tilletiopsis* sp. parasitize the cucumber with powdery mildew fungus *Spaerotheca fuligena*). When present in crop soils, they exert a certain degree of biological control over one or many plant pathogens.

aerial pest control: Pest control *via* application of aircraft and helicopters in order to be more efficient, to prevent damages of the crop or to deal with difficult landscape and soil conditions.

aerial roots: In some epiphytic orchids, the leaves and the shoot axis are reduced or missing; then flatted and green roots take over the fixation of the plant and the function of the leaves.

aerial shoots: Shoots growing high above the ground (e.g. trees and bushes).

aerification: The process of infusing or forcing air, for example, into soil.

aerobe: An organism growing only in the presence of molecular oxygen. These microorganisms use O_2 oxygen as the final electron acceptor for respiration. In aerobic respiration, the ultimate electron acceptor is oxygen. Aerobes produce more energy than an anaerobe. The Kreb's cycle is the best-known example of aerobic respiration. *Aka* AEROBIUM. *Compare with* ANAEROBE.

aerobic respiration: A biochemical process including oxygen in the breakdown of glucose.

aerobic: Refers to a method that requires or occurs in the presence of gaseous oxygen (O_2). *Contrasts* ANAEROBIC.

aerobiology: The study of organic parts, which are inertly transported by the air, including bacteria, fungal spores, very small insects, pollen grains and viruses.

aerotolerant anaerobe: An anaerobe which is not affected by the presence of molecular oxygen. These organisms grow in the absence of oxygen but if oxygen is present, they can tolerate it and can also grow in the presence of oxygen. These organisms ferment carbohydrates to lactic acids which inhibit the growth of other aerobic competitors, which encourages the multiplication of the lactic acid producers. Lactobacilli commonly used in the preparation of pickles and cheese are the best example.

aestivation: Pattern of sepals and petals or their lobes in an unexpanded flower bud. *Contrasts* VERNATION.

aetiology: The study that deals with disease diagnosis and control. Sometimes, aetiology is obvious as trauma from accident. In some other cases the cause may be difficult to assign, for instance, the real cause of cancer.

affected: An individual diagnosed with a disease or disorder who shows signs or symptoms of a condition.

affinity chromatography: A procedure of separating a mixture of proteins or nucleic acids by specific interactions of those molecules with a component known as a LIGAND, which is restrained on a support. If a solution of, say, a mixture of proteins is passed over or through the column, one of the proteins binds to the ligand on the basis of specificity and high affinity; in fact, they fit together like a lock and key. The other proteins in the solution are washed through the column because they are not able to bind to the ligand. Once the column is lacking the other proteins, a suitable wash solution is passed through the column, which causes the protein/ligand complex to dissociate. The protein is finally collected in a highly purified form.

affinity constant (Ka): A measure of the strength of the binding of components in a complex. For components A and B and a binding equilibrium $A + B \rightleftharpoons AB$, the association constant is given by the expression [AB]/[A][B] and is larger the tighter the binding between A and B. *Aka* ASSOCIATION CONSTANT. *See* DISSOCIATION CONSTANT.

affinity matrices: Techniques that isolate mRNA often take advantage of the poly(A) tail typically present at the 3′-end. Affinity matrices use poly(T) oligonucleotides that hybridize to the complementary poly(A) tails.

affinity maturation: Liberal increase in the attraction of antibodies for the immunizing antigen with the passage of time after immunization.

affinity: The tightness of a protein-ligand complex.

afforest: Changing bare or cultivated land into forest.

afforestation: The establishment of forest by natural sequence or by the planting of trees on land where they in the past did not grow.

Affymetrix Genechip Arrays: High-density oligonucleotide microarrays manufactured by using a photolithographic technology with up to 500,000 oligos being synthesized on a single chip. This system allows measurement of the relative concentration of a DNA or RNA sequence in a complex mixture of nucleic acids. Affymetrix (Santa Clara, CA, USA) provides numerous array platforms that are engaged in a number of methylation studies.

Affymetrix Genechip Arrays, uses of: To (i) screen global mRNA abundance in a range of species; (ii) find genes that are uniquely expressed in samples of normal or mutated tissue; (iii) explore changes in gene expression associated with environmental or other changes; (iv) classify putative functions for uncharacterized ESTs; or (v) recognize gene expression 'fingerprints' for efficacy and toxicity.

after-ripening: A term for the collective changes that occur in a dormant seed that make it capable of germination. It is usually considered to denote physiological changes in seed.

agameon: A plant species reproducing solely by apomixis.

agamete: Any nonreproducing germ cell.

agamic: Reproducing asexually.

agamogenesis: Any reproduction without the male gametes.

agamogony: A type of apomixis in which cells undergo abnormal meiosis during megasporogenesis, resulting in a diploid embryo sac rather than the normal haploid embryo sac.

agamospecies: A group of organisms that reproduce asexually but have the characteristics of a species. *Aka* APOMICTIC SPECIES.

agar culture: Cells, organs, tissue or embryos artificially grown on a solid medium composed of agar together with certain nutrients, hormones and so on. *Aka* AGAR MEDIUM.

agar diffusion test: A procedure used to assess the efficacy of antibiotics for specific bacteria. It involves the application of antibiotic solutions of different concentrations to cups, wells or paper discs, placed on the surface of agar plates seeded with a test bacterial strain.

agar-agar: A complex polysaccharide compound, which is produced by red algae. It has long been used as a solidifying agent in culture medium. It is a solid powder that liquefies at about 100°C (212°F) and remains liquid until the temperature drops to about 40°C. In the laboratory, the medium is held in a water bath at below 44°C (123°F) for plating purposes. *Aka* AGAR.

agarose (starch) gel: An inert matrix used in electrophoresis for the separation of proteins or nucleic acids based on their size or conformation. The molecules are visualized in the gel by the ultraviolet fluorescence of ethidium bromide, which is either included in the gel, in the running buffer or used to stain the gel after electrophoresis.

agarose gel electrophoresis: A method for separating nucleic acids or proteins within a gel made of agarose in a suitable buffer under the influence of an electrical field.

agarose: A highly purified form of agar made from seaweed, which is used as a stationary phase (substrate) in some chromatography and electrophoretic procedures for the separation of proteins/nucleic acids by gel electrophoresis.

ageing: Degradative processes which go on in all the cells/organisms and are the cause of decline, in survival or reproduction, with age – and their eventual death. DNA damage and defects in DNA repair mechanisms have been reported to be accountable for ageing. Ageing may be due to the imbalance between rates of DNA repair and damage. Each cell division (mitosis) brings that cell (or organism composed of such cells) closer to its final cell division (i.e. death). *See* SENESCENCE.

ageing in plants: Mitochondrial plasmids that are very similar to mobile genetic elements are involved in the rearrangement of the mitochondrial genome and thus appear to function as mitochondrial mutators. As a result, the inactivation of particular functions may lead to CELL DEATH.

ageing test: A procedure to originally assess seed storability. It exposes un-imbibed seeds to conditions of high temperature (+41°C) and relative humidity (~100%) for short periods (3–4 days). The seeds are then removed from the stress conditions and placed under optimum germination conditions. The two environmental variables lead to rapid seed deterioration. High-vigour seed lots will endure these extreme stress conditions and deteriorate at a slower rate than low-vigour seeds.

agent architecture: A blueprint for software agents and intelligent control systems depicting the arrangement of components. The architectures executed by intelligent agents are referred to as COGNITIVE ARCHITECTURES.

agent-based model (ABM): A class of computational models for simulating the actions and interactions of autonomous agents (both individual and collective entities, such as organizations or groups) with a view to evaluating their impacts on the system as a whole. It combines elements of game theory, complex systems, emergence, computational sociology, multi-agent systems and evolutionary programming. Monte Carlo methods are used to introduce randomness.

agents: Independent, autonomous, software modules that can search the Internet for data or content pertinent to a particular application, such as a gene, protein or biological system.

agglomerate: (i) In biotechnology, a mass of cells grouped together. (ii) In soil science, rock composed of rounded or angular volcanic fragments.

agglutinin: An antibody that can recognize and bind to an immunological determinant on the surface of bacteria or other cells, which causes them to clump (agglutination).

agglutinogen: An antigen interaction with an antibody that causes clumping by the production of agglutinins.

aggregate fruit: Cluster of fruits produced from several pistils in one flower, as in strawberry or blackberry composite fruit. *Compare with* MULTIPLE FRUIT.

aggregate function: In database management, refers to a function in which the values of multiple rows are clustered together to form a single value of more significance or measurement, such as a sum, count or max.

aggressive hybridization: Participation of a subspecies in hybrid formation only as male, not as female or vice versa. The consequence is the reduction in the numbers of species acting as female under specific conditions, which may threaten extinction of the species.

aggressiveness: The ability of a pathogen to contaminate a plant, to break its resistance, to become a parasite on a plant or to use a host plant for reproduction. The degree of fierceness can be estimated only when a pathogen meets a resistant host; when the aggressiveness manages with the resistance of the host, it is termed virulence. *See* VIRULENCE.

agile software development: Under this approach of software development, requirements and solutions evolve through the concerted effort of self-organizing and cross-functional teams and their customer(s)/end user(s). It promotes adaptive planning, evolutionary development, early delivery and continual improvement, and it encourages rapid and flexible response to change.

Agilent array platforms: Agilent (Santa Clara, CA, USA) array platforms are used to examine DNA methylation and methylation-associated modifications in several organisms.

***Agouti* gene**. In mice, the *agouti* (*A*) gene controls the deposition of yellow pigment in developing hairs. There are several alleles (sequence variants), some of which are dominant (e.g. *lethal yellow*, A^y), which result in embryonic lethality if mice inherit two copies. Methylation of the *agouti* reduces its expression, lowering the quantity of yellow pigment produced without altering the sequence of the gene.

agribusiness: A study of the combination of the producing operations of a farm, the making and distribution of farm equipment and supplies and the processing, storage and distribution of farm commodities.

agricultural bioenergy systems (ABSs): ABSs comprise numerous biomasses, various processes and technologies, resulting in different fuels or energy carriers being utilized within agricultural systems and beyond. Key components of ABSs include (i) agricultural biomass; (ii) biomass processing; (iii) biofuels; (iv) biofuel conversion and (v) applications.

agricultural biotechnology (AgBio): A variety of tools, including traditional breeding and recombinant DNA techniques, that alter agriculturally important plants, animals and microorganisms to make or modify products for meeting the human goals in agriculture. *Aka* AGROBIOTECHNOLOGY.

agricultural chemical: A comprehensive term used to cover pesticides, adjuvants, conditioning agents and other chemical tools used in improving agricultural production, protecting crops or controlling pests, diseases and physiological conditions of crop plants.

agricultural variety: A group of similar plants that can be identified from other varieties within the same species through structural features and performance. *See* CULTIVAR.

agricultural weed: *See* WEED.

agriculture: The science and business of producing crops and/or livestock that provides food, fabric and fuel. MODERN AGRICULTURE includes (i) the science of converting sunlight energy into plant and animal products that can be used by humans; (ii) the selective breeding of crops and farm animals has a massive influence on productivity in agriculture; (iii) producing modern varieties of crop plants that have enhanced nutritional value and resistance to disease; and (vi) recent developments in genetic engineering have enabled the potential use of transgenic organisms in agriculture.

agriophyte: Plant species that have invaded native vegetation and can survive there without human intervention. They are established in natural habitats, remaining part of the natural vegetation even after human influence has ceased, and are independent of humans in their continued existence.

***Agrobacterium* plasmid-based vectors:** Vectors based on *Agrobacterium* Ti and Ri plasmids have been designed with the following properties: they do not cause the disease, since they are disarmed by removing gene causing the disease; they carry sites for insertion of foreign gene intended to be transferred and they carry selectable markers, a powerful promoter like 35S derived from cauliflower mosaic virus (CaMV) and a polyadenylation signal required for adding poly(A) to mRNA.

***Agrobacterium rhizogenes*:** A species of Gram-negative, rod-shaped soil bacteria, habitually harbouring large plasmids called Ri plasmids. It can cause a tumourous growth known as hairy root disease in certain plants. *Aka* DEGRADATIVE PLASMIDS.

***Agrobacterium tumefaciens*:** A bacterium that causes crown gall disease in some plants. It infects a wound and injects a short stretch of DNA into some of the cells around the wound; the DNA comes from a large plasmid – the Ti (Tumour induction) plasmid – a short region of which called Ti-DNA is transferred to the plant cell, where it causes the cell to grow into a tumour-like structure. This bacterium is now used to deliberately transfer genetic material into plants through biotechnology. *A. tumefaciens* carries several *Vir* genes (vir region) on its Ti plasmid. Nowadays super-virulent strains of *A. tumefaciens* such as LBA 4404 (p TOK 233) have been developed for the transformation of dicot and monocot plants.

***Agrobacterium*:** A genus of pathogenic bacteria that includes several plant pathogenic species, causing tumour-like symptoms. This bacterium of plants can inject T-DNA of a plasmid DNA into plant cells.

***Agrobacterium*-mediated DNA transfer:** *A. tumefaciens* is a Gram-negative soil bacterium, which results in crown gall formation in the host plant at the site of infection. Gall formation is due to the transfer and stable integration of a well-defined portion (T-DNA) of the Ti plasmid. The transfer of T-DNA from Ti plasmid and its integration into host genome induces tumour formation in plants. An early event in *Agrobacterium*-mediated gene transfer is the nicking of Ti plasmid at two specific sites, between the third and fourth base of the bottom strand of each

25-bp repeat. This initiates DNA synthesis from the nick in the right border in 5′→3′ direction, thus displacing a single T-DNA strand. The single-stranded T-DNA (ssT-DNA) is then released from Ti plasmid. VirD2 protein binds the T-strand at the 5′-end and acts as a pilot protein to guide ssT-DNA to the nucleus of the plant cell. Single-strand binding (ssb) protein, coded by *virE* locus, binds along the whole length of ssT-DNA and perhaps protects it from the action of nucleases. Apart from the *vir* region, there are several chromosomal regions, namely, *chvA*, *chvB*, *cel*, *pscA* and *att* loci, which have been shown to be directly involved in the attachment of *Agrobacterium* to plant cells. Conjugational transfer of DNA between bacteria requires transfer (*tra*) and mobilization (*mob*) gene products. Also, these sequences have been used to transfer T-DNA. Until recently, monocots, particularly the graminaceous crop species were considered outside the *Agrobacterium* host range. Monocots of orders Liliates and Arales were observed to produce tissue swelling in response to inoculation with virulent *Agrobacterium*. Synthesis of *Agrobacterium*-specific opines by *Chlorophytum capense*, Narcissus and *Asparagus officinalis* was considered to be indicative of the integration and expression of the transferred DNA. These studies raised the possibility that monocot species could be transformed by *Agrobacterium*. Ti plasmid of *A. tumefaciens* (causing 'crown gall' disease) and Ri plasmid of *A. rhizogenes* (causing 'hairy root' disease) have been effectively used as vectors for gene transfer in plant cells. The Ti plasmid (and also Ri plasmid) has a region called *vir* region (having six operons, *vir* A, B, C, D, E, G) responsible for virulence towards the host along with the gene intended to be transferred. *Aka* agroinfection.

agrobiodiversity: The variety and variability of animals, plants and microorganisms used directly or indirectly for food and agriculture (crops, livestock, forestry and fisheries). It encompasses the diversity of genetic resources (varieties, breeds, etc.) and species used for food, fuel, fodder, fibre and pharmaceuticals.

agrobiology: The scientific study of plant life related to agriculture, especially with regard to plant genetics and cultivation as a way to upsurge crop yield.

agroeconomic traits: Plant traits such as biotic/abiotic stress resistance, yield, nutritional quality, processing quality and keeping qualities are classified as agroeconomic traits. Better agroeconomic traits in the cultivated crop varieties are required by the farmers, industry and the consumers. Breeding for quantitative traits (controlled by a large number of genes) still is a difficult task. In this regard, biotechnological approaches like molecular marker technology and transgenic technology are now increasingly being used to develop varieties with improved agroeconomic traits. *Bt* cotton is a good example that has revolutionized agriculture in many countries.

agroecosystem: A community of microorganisms, plants and animals, together with their abiotic environment, occurring on farmed land, including the crop species.

agroforestry: A land-use management system in which combinations of trees or shrubs are grown around or among crops or pastureland (together with other land uses such as animal production). Trees in agroforestry systems can also produce wood, fruits, nuts and other useful products with economic and practical values. Agroforestry combines agricultural and forestry technologies to create more diverse, productive, profitable, healthy and sustainable land-use systems; increase farm profitability; help to preserve and protect natural resources such as controlling soil erosions; create habitat for the wildlife and managing animal waste; increase biodiversity, improve soil structure and health and reduce erosion and carbon sequestration.

agroinfection: *Agrobacterium*-mediated plant transformation in which viral DNA is integrated into the T-DNA of *Agrobacterium*. This is the phenomenon in which a virus infects a host as a part of the T-DNA of Ti plasmid of *A. tumefaciens*. After infection, the viral DNA is released to form functional virus that replicates and spreads systemically. In maize, agroinfection with maize streak virus has been demonstrated.

agronomic crops: Staple food crops, such as maize, rice, beans and wheat, which are produced on a large scale and represent the foundation of our human food supply. Agronomic crops are categorized as: (i) green manure crops; (ii) cover crops; (iii) silage crops or (iv) companion crops. About 2% of the 250,000 higher plant species are used in agriculture, horticulture and so on.

agrotain: A product that contains active ingredient N-(n-butyl) thiophosphoric triamide (NBPT), which obstructs the alteration of urea to ammonium carbonate, thereby reducing the potential for ammonia volatilization from urea materials. Similar to 'N-Serve', it diminishes potential nitrogen losses in seasons when cultivation or rain does not integrate the urea into the soil soon after application. It is most useful when urea is applied without incorporation to the surface of fields with high levels of crop residue, such as in no-till situations or fields with high pH levels at the surface.

air layering: A procedure of plant proliferation in which roots are induced to be formed around a stem. For this, a very narrow strip of bark is removed from around the branch or stem and a sliver of wood can be inserted into the cut to keep it open. A bundle of moist sphagnum moss is tied securely around the cut area. The moss must remain moist and the roots of the plant somewhat dry. New roots will sprout from the incision. The new plant is then cut off below the moss, potted and kept in a humid atmosphere until it is established.

air pollution: Air is made up of a number of gases, mostly nitrogen and oxygen and, in smaller amounts, water vapours, carbon dioxide and argon as well as other trace gases. Air pollution occurs when harmful chemicals and particles are emitted to the air (due to human activity or natural forces) at a concentration that affects human health or welfare or that troubles the environment in other ways.

air quality: A measure of the level of pollution in the air.

air pollution in pollination: Chemical pollutants not only reduce animal survival and reproduction, but can also disrupt their senses, changing their behaviour and interactions with other species. Chan *et al.* in 2024 studied the effects of ozone (O_3) and nitrate radicals (NO_3) on nocturnal hawkmoth pollination of a desert plant and found that these air pollutants decreased concentrations of monoterpenes, which potentially affect the insects' ability to locate and pollinate flowers. A global model suggests that many urban areas have sufficient pollution to significantly reduce the distances at which pollinators can sense flowers.

air-screen cleaner: A basic piece of kit for cleaning seed, utilizing airflow and perforated screens seed.

airlock: An airtight chamber allowing passage to or from a space seed.

akaryotic: Cell without a nucleus. The most common type of akaryotes is viruses.

al31 nuclease: An exonuclease that degrades both strands of a DNA molecule at the same time.

alanine (ala): A non-essential amino acid ($C_3H_7NO_2$) of the pyruvic acid family. This amino acid is present in almost all proteins. In its dry, bulk form it appears as a white crystalline solid.

alarmones: Alarmones possibly function as cellular signals of distress as they act on cellular enzymes that are proteins to counteract the problem, whose metabolic consequences they represent, by moving in different sites in the cell, just as hormones. Some examples of alarmones include cAMP (3′–5′ cyclic AMP), ppGpp (guanosine tetraphosphate), pppGpp (guanosine pentaphosphate), AppppA (diadenosine tetraphosphate), ZTP (5-amino, 4-immidazolecarboxamide riboside 5′-triphosphate) and so on.

albinism: A condition in plants, having a deficit of chromoplasts.

albino plant: When a plant is deficient in chromoplasts, it appears white or light coloured, as it has a marked deficiency in green pigmentation.

albuminous seed: A seed that has a well-developed endosperm or perisperm.

albuminous: Seeds containing endosperm.

alcohol fermentation: A process that begins with glycolysis and produces ethyl alcohol. Alcohol fermentation was discovered by L. Pasteur in 1856. It is a process in which yeasts (*Saccharomyces cerevisiae*) convert sugars into alcohol under anaerobic conditions. It begins with the glycolysis of glucose to produce pyruvic acid and ATP. Pyruvic acid further produces acetaldehyde and finally ethanol.

aleurone grain: Small protein grains present in cells of the storage tissue of maturing seeds and tubers.

aleurone layer: A layer of cells below the testa of some seeds (e.g. cereals), which contains hydrolytic enzymes (e.g. amylases and proteases) for the digestion of the food stored in the endosperm. The production of enzymes is activated by gibberellins when the seed is soaked in water prior to germination. Aleurone is a source of nutritionally valuable compounds, such as dietary fibres, proteins, minerals and vitamins, that may exhibit health benefits.

aleuroplast: A leucoplast in which protein granules occur as the main storage product.

alfatoxin: A carcinogenic toxin produced by the fungus *Aspergillus flavus*. Some fungi possess metabolic products that are poisonous to human hosts. For instance, a mould *A. flavus* produces aflatoxin, which when ingested gets altered in the human body to produce mutagens/carcinogens. When ruminants eat feedstuffs containing Aflatoxin B1 (AFB1), it is metabolized and Aflatoxin M1 (AFM1) is excreted in the milk. The International Agency for Research on Cancer (IARC) classifies AFB1 and AFM1 as human carcinogens/mutagens.

alga (pl. algae): Any member of a varied polyphyletic group of photosynthetic, eukaryotic (unicellular, filamentous or multicellular) organism, *aka* kelp. Algae may reproduce sexually or asexually and are often regarded as plants, though they are deficient in most of the complex cell and tissue types that characterize true plants. Alga requires light and air for its growth and development and does not require organic compounds (sugars and vitamins) from the environment unlike photosynthetic plants. Blue-green alga is used in agriculture as a biofertilizer for rice crops. *Spirullina* is used in the industry to produce a variety of organic compounds.

algal biomass: Single-celled plants (such as *Chlorella* spp. and *Spirulina* spp.).

algal blooms: Sudden spouts of algal growth which can affect water quality adversely and can indicate potentially hazardous changes in the local water chemistry.

algorithm design: A method or mathematical process for problem-solving and engineering algorithms. The design of algorithms is part of many solution theories of operation research, such as dynamic programming and divide-and-conquer. Techniques for designing and implementing algorithm designs are *aka* ALGORITHM DESIGN PATTERNS. *See* TEMPLATE METHOD PATTERN and DECORATOR PATTERN.

algorithmic efficiency: A property of an algorithm which relates to the number of computational resources used by the algorithm. An algorithm must be analysed to determine its resource usage, and the efficiency of an algorithm can be measured based on the usage of different resources. Algorithmic efficiency can be thought of as analogous to engineering productivity for a repeating or continuous process.

algorithms, types of: There are several types of algorithms available. The ten important algorithms are: (i) BRUTE FORCE ALGORITHM; (ii) RECURSIVE ALGORITHM; (iii) BACKTRACKING ALGORITHM; (iv) SEARCHING ALGORITHM; (v) SORTING ALGORITHM; (vi) HASHING ALGORITHM; (vii) DIVIDE-AND-CONQUER ALGORITHM; (viii) GREEDY ALGORITHM; (ix) DYNAMIC PROGRAMMING ALGORITHM; and (x) RANDOMIZED ALGORITHM. *See details as separate terms.*

algorithms: Step-by-step procedures designed to solve specific problems such as data processing, data storage, data analysis, visualization and make automated predictions or reasoning tasks from biological data efficiently in computer science and mathematics. These powerful sets of instructions form the backbone of modern technology and govern everything from web searches to artificial intelligence. *See* PROBLEM, INPUT, PROCESSING UNIT and OUTPUT, in relation to algorithms. They are ubiquitous in computing technologies.

alicin: A compound that is produced naturally by the garlic plant when the cells within garlic bulbs are broken open (e.g. during food preparation or consumption). Enzymes present within those garlic cells convert (precursor compound) to alicin. Research indicates that human consumption of alicin confers some specific health benefits (anti-thrombotic, reduces blood cholesterol levels, helps in reducing/avoiding coronary heart disease, enhances the immune system, etc.).

alien: Any plant presented to an area outside its natural range. Often used interchangeably or in combination with foreign, exotic, non-native and/or non-indigenous plants.

alien addition line: A line (strain) of plants with one or more extra chromosomes of an alien species.

alien chromosome: A chromosome from a more or less related species that is transferred to a crop plant.

alien chromosome addition: Incorporation of an additional chromosome from alien species into the normal complement of host species. *See* MONOALLELIC ALIEN ADDITION LINES (MAALs).

alien chromosome substitution: Substitution of homeologous chromosome from an alien species into the genome of other species.

alien chromosome transfer: Cytogenetic procedures that enable the transfer of individual chromosomes from one species to another.

alien gene transfer: The transfer of genes between species or genera by different means.

alien germplasm: Genes presented from a wild relative or non-adapted species.

alien introgression: Refers to the transfer of useful genes from wild species into cultivated crop varieties. Wide hybridization and alien introgression are important areas in plant breeding dealing with the transfer of useful genes from wild relatives into cultivated varieties. Alien genes like *L9, L21, L24* and *L29* conferring resistance to leaf rust have been transferred into wheat from wild species. Molecular marker technology is now being used for monitoring alien gene transfer.

alien species: An organism that has invaded or been presented by humans and is growing in a new region. In cytogenetics, an alien species is a species that serves as the donor of a genome, chromosome or chromosome segment, to be moved into a recipient species or genotype.

alien substitution line: A line of plants in which one or more alien chromosomes from a certain donor species substitute one or more chromosomes of a beneficiary species.

align (vb.): To line up different DNA, RNA or protein sequences in a straight line.

alignment: Depiction of a prediction of the amino acids in tertiary structures of homologues that overlay in three dimensions. Alignments held by setting specific, measurable, achievable, relevant and time-bound (SMART) goals are mostly based on published observations but are updated and edited manually. The process of lining up two or more sequences to achieve maximal levels of identity (and conservation, in the case of amino acid sequences) for the purpose of assessing the degree of similarity, homology, function or other degree of relatedness between two or more genes or gene products. *See* DOMAIN ANNOTATIONS.

alignment of DNA sequences, bioinformatics tools for: DNA sequences are compared by alignments to determine their degree of similarity using various bioinformatic tools. (i) ALIGN Query using sequence data gives very nice output. (ii) LALIGN finds multiple matching subsegments in two sequences. It provides one with a percentage identity for different subsegments of the sequence. (iii) FFAS03: pairwise alignment – nice presentation of results but the server can be very slow. (vi) FASTA compares two sequences. It offers a large number of score matrices. (v) SIM is an alignment tool for proteins that gives fragmented alignments. (vi) Secondary structure element alignment (SSEA) computes the alignments of secondary structures in proteins. The server supports both performing pairwise alignments and searching a secondary structure against a library of domain folds. It can calculate global and local secondary structure element alignments.

alignment of protein sequences, bioinformatics tools for: For protein sequence similarity and identity scores, the tools used are: (i) ToPLign requires a user login. It provides a wide variety of output formats for pairwise alignments; (ii) BLAST2 is useful for DNA sequence comparisons. Provides a small graphic that is only of use with proteins or short DNA sequences; (iii) SUPERMATCHER is part of the EMBOSS group of programs. It uses 10 and 0.5 as the defaults in the gap opening penalty and gap extension penalty, respectively; (iv) MATCHER finds the best local alignments between two sequences (EMBOSS); (v) StretcherN finds the best global alignment between two sequences; (vi) NeedleN for global alignment between two sequences; (vii) zPicture is a DNA or genome alignment and visualization tool based on blastz alignment program. Alignments can be automatically submitted to rVista 2.0 to identify evolutionary conserved transcription factor binding sites; and (viii) FOLDALIGN folds and aligns RNA structures.

alignment of RNA tertiary structures (ARTS): It aligns two nucleic acid structures (RNAs or DNAs) in pdb format and detecting a priori unknown common substructures. The identified common substructures can be either large global folds or small local tertiary motifs with at least two successive base pairs.

alignment, types of: The result of the linear comparison of two or more gene (DNA, RNA or protein) sequences. Sequence alignments are used to determine similarity, homology, function or other degree of relatedness between two or more genes or gene products. Of the two types of alignment, local and global, a local alignment is generally more useful. *See* LOCAL ALIGNMENT and GLOBAL ALIGNMENT.

alkali: A substance having the capability of furnishing hydroxyl (-OH) ions to its solution. The most important alkali metals include potassium and sodium.

alkaline hydrolysis: A chemical method that delivers DNA from a DNA-RNA hybrid.

alkaline method of DNA extraction: Alkaline lysis has been used to isolate plasmid DNA and *E. coli*. The principle of this method is based on selective alkaline denaturation of high-molecular-weight chromosomal (linear) DNA while covalently closed circular DNA remains double stranded. Bacterial proteins, broken cell walls and denatured chromosomal DNA, which are enmeshed into large complexes, are coated with dodecyl sulfate. Plasmid DNA can be isolated from the supernatant after the denatured material has been detached by centrifugation.

alkaline phosphatase: An enzyme that catalyses the hydrolysis of phosphomonoesters of the 5′ nucleotides. Used to dephosphorylate (remove phosphate groups from) the 5′ ends of DNA or RNA molecules, to facilitate 5′ end-labelling with ^{32}P added back by T4 polynucleotide kinase, or to dephosphorylate the 5′ ends of DNA molecules to prevent unwanted ligation reactions during cloning.

alkaline soil: Specifically, a soil with pH value >7.0 caused by the presence of carbonates of calcium, magnesium, potassium and sodium. This method is commonly used for soils showing a pH value of 8.5.

alkaloidity: The alkaloid content of cells, tissue, organs or individuals of plants.

alkaloids: A class of heterocyclic organic toxic compounds that are naturally produced by some organisms (e.g. ants, certain plants such as lupines and certain fungi such as ergot). Alkaloid molecules have one or more alkaline-reacting nitrogen atoms in their carbon structures. Many alkaloids are commercially important as drugs or poisons, such as coniine in hemlock, morphine in poppy fruits, strychnine in seeds of *Strychnos nux* vomica, atropine in nightshades, colchicine in meadow saffron, caffeine in coffee and tea, nicotine in tobacco leaves, theobromine in cacao and quinine in *Cinchona officinalis*. Alkaloids are produced by plants as a defence against herbivores.

alkyl group: A group of covalently linked carbon and hydrogen atoms such as methyl ($-CH_3$) or ethyl ($-CH_2CH_3$) groups. These groups are generally formed by removing a hydrogen atom from an alkane. For example, the removal of a hydrogen atom from methane (CH_4) creates a methyl group ($-CH_3$).

alkylating agents: Alkylating agents carry one (monofunctional), two (bifunctional) or more (polyfunctional) alkyl groups in a reactive form, which are capable of being transferred to other biomolecules where electron density is high. According to the number of reactive groups alkylating agents are capable of transferring, they are called MONOFUNCTIONAL ALKYLATING AGENTS, BIFUNCTIONAL ALKYLATING AGENTS and POLYFUNCTIONAL ALKYLATING AGENTS.

alkylation: All cells experience a biologically significant level of spontaneous DNA methylation. The majority of the bonds in all four bases are susceptible to methylation. Most methylation damage occurs at purine bases.

allele: Short form of the term ALLELOMORPH. One of the alternative forms of a gene. It is a given form of a gene that occupies a specific position or locus on a chromosome. Variant forms of genes occurring at the same locus are said to be alleles of one another. In a diploid cell, each gene possesses two alleles, each occupying the same position (locus) on the homologous chromosomes. Different alleles produce variations in inherited characteristics such as the seed colour of peas. In an individual, one form of the allele (the dominant one) may be expressed more than another form (the recessive one). *Compare with* EPIALLELE.

allele effect: Positive correlation between population growth rate and population density; for example, as can occur when individual reproductive success declines at low population densities owing to difficulty in encountering mates. An allele effect can lead to extinction below a crucial threshold density.

allele exchange: A method of gene replacement that can be used to introduce mutations into the genome of an organism.

allele frequency: A measure of the commonness of an allele in a population; the proportion of the alleles of that gene in a population that are of this specific type. Also called GENE FREQUENCY in population genetics.

allele mining: A method to access new and useful genetic variation in crop plant collections for its use in breeding programmes. Most agronomically important plant traits, such as yield, insect resistance, disease resistance, heading date and grain quality, show continuous phenotype variation governed by QTLs rather than a single gene. For example, an enzyme cytokinin dehydrogenase/oxidase degrades cytokinins (growth regulator), which are needed for increasing the reproductive organs in the inflorescence. Therefore, it has been found in rice that yield can be enhanced by identifying CKX alleles with lower enzyme activity by screening the rice germplasm. The molecular screening of germplasm for desired allele(s) is called allele mining.

allele shift: An alteration of gene frequency caused by selection that can lead to complete loss of an allele in extreme cases, depending on its initial frequency. Similar effects may be a consequence of genetic drift.

allele substitution: The establishment of a previously rare or absent allele as the predominant allele in a population.

allele trend: A directed change in the allele frequency of populations per time unit.

allele-specific amplification (ASA): Use of POLYMERASE CHAIN REACTION (PCR) at a sufficiently high rigour, such that only a primer with exactly the same sequence as the target DNA will be amplified. ASA is a powerful means of genotyping for single-locus disorders characterized at the molecular level.

allele-specific amplification diagnostic assay: Using allele-specific oligonucleotides as primers in PCR reactions. Single-marker alleles are amplified so that the inheritance of these alleles can be followed in the crosses. The marker allele indicative of a resistance allele is correlated with resistance. *Aka* ALLELE-SPECIFIC OLIGONUCLEOTIDE HYBRIDIZATION DIAGNOSTIC ASSAY.

allele-specific associated primers (ASAP): A PCR variant in which the sequence of the decamer-oligo is derived from normal RANDOM AMPLIFIED POLYMORPHIC DNA (RAPD), which produced an absence and/or presence of polymorphism. Such polymorphisms do not need electrophoretic separation of the sample. The presence of an amplification product is noticed by measuring the fluorescence of ethidium-bromide-stained DNA.

allele-specific epigenome editing: This approach of epigenome editing has recovered 'super-specific' variation in which chromatin marks at only one selected allele of the target genomic locus can be modified.

allele-specific methylation: Genetic imprinting, which describes the parent-specific gene expression behaviour of a small set of genes. The methylation pattern of imprinted genes is distinct; the inactive allele is significantly more methylated than the actively expressed allele.

allele-specific methylation on microarrays: This type of arrays has the following methods: (i) methylation-sensitive SNP (MSNP) chip analysis; (ii) profiling on expression microarrays combined with drug-induced demethylation; (iii) profiling methylation in archival formalin-fixed paraffin-embedded samples; and (iv) moderate-throughput quantitative methylation analysis by quantitative MSP.

allele-specific oligo (ASO): A special kind of oligo for an allele-specific PCR. The sequence of the oligo is planned in such a way as to allow and/or inhibit hybridization at the spot where the mutant (resistant) allele differs from its wild-type (susceptible) counterpart.

allele-specific oligonucleotide hybridization: The use of an oligonucleotide probe to determine, which of the two alternative nucleotide sequences is contained in a DNA molecule.

allele-specific PCR (AS-PCR): Refers to the amplification of specific alleles or DNA sequence variants at the same locus. Specificity is achieved by designing the sequence of one or both PCR primers so that they partially overlap the site of sequence difference between the amplified alleles. Variants of this technique have been described under different names such as RAPD (RANDOM AMPLIFIED POLYMORPHIC DNA), AP-PCR (ARBITRARY-PRIMED PCR) or DAF (DNA AMPLIFICATION FINGERPRINTING). They are based on PCR, which comprises randomly synthesized short oligonucleotide sequences as primers. RAPD markers are cheaper and easier to use than RFLP and AMPLIFIED FRAGMENT LENGTH POLYMORPHISM (AFLP) markers. However, RAPDs are dominant and potentially population-specific markers that are very sensitive to laboratory conditions and usually show a low level of polymorphism with PCR amplification using a single random primer.

allele-specific probes: Short-labelled oligonucleotide probes used to distinguish single-nucleotide differences in PCR products through hybridization such that only a completely complementary probe will bind stably to the target sequence.

allelic complementation: The production of a non-mutant phenotype when two independent mutations at the same gene locus, but on different homologous chromosomes, are introduced in a gene.

allelic exclusion: In immunoglobulin-producing cells or plasma cells, only one member of each pair of alleles is involved in the synthesis of the immunoglobulin being expressed. The correct rearrangement of only one chromosome is needed for the generation of each expressed light or heavy chain genes. Since the DNA rearrangement of one chromosome takes place, expression of the alternate allele is excluded.

allelic heterogeneity: Variant alleles found at a single locus that can result in the same phenotype.

allelic imbalance: The epigenomic state of maternal and paternal chromosomes within the nucleus of a diploid organism may differ. An epigenomic difference in a specific locus is referred to as allelic imbalance. An allelic imbalance is experimentally detectable if it spans a heterozygous locus (such as a single nucleotide polymorphism) and if the read-out technology (such as sequencing) provides allelic information. This allelic information allows the separation of an epigenomic signal over a heterozygous locus into signals residing on maternal and paternal chromosomes. An allelic imbalance may be binary (on/off) or quantitative, where the signals from both the maternal and paternal chromosomes are greater than zero but of different magnitudes.

allelic mutations: Two mutations are defined as allelic if they do not show complementation. Two variants are allelic if they are located at the same genetic locus.

allelic scanning: The rapid detection of DNA sequence variations between two strains by analysing total genomic DNA on high-density DNA microarrays.

allelic series: A collection of discrete mutations in a particular gene often associated with multiple phenotypes.

allelic variation: Measurable at a locus as the number of alleles (polymorphism) present, or the proportion of heterozygotes in the population. A NULL ALLELE is a gene variant that lacks the gene's normal function because either it is not expressed or the expressed protein is inactive.

allelism: The common short form of the term ALLELOMORPHISM. One of the two or more forms of a gene arising through mutation and occupying the same relative position (locus) on a homologous chromosome's gene.

allelobrachial: Changes in chromosome structure where the arms of homologous chromosomes are involved.

allelogenous: Females that produce only males or only females in different progenies.

allelomorph: *See* ALLELE.

allelopathic: The ability of one species to inhibit or prevent the growth of another species through the production of toxic substance(s).

allelopathic inhibition: This phenomenon typically results from the combined action of a group of allelochemicals which, collectively, interfere with several physiological processes. Allelopathy is strongly coupled with inherent stresses of the crop environment, including insects and disease, temperature extremes, nutrient and moisture variables, radiation and herbicides; these stress conditions often enhance allelochemical production, thus increasing the potential for allelopathic interference. *See* ALLELOPATHY.

allelopathy: Refers to the release of a chemical substance by an organism that acts as a germination or growth inhibitor to another organism in the environment. Allelopathy offers the potential for selective biological weed management through the production and release of allelochemicals from leaves, flowers, seeds, stems and roots of living decomposing plant materials. A variety of allelochemicals identified include phenolic acids, coumarins, terpenoids, flavonoids, alkaloids, glycosides and glucosinolates. *See* ALLELOMORPHIC INHIBITION.

allelotype: The genetic composition of a breeding population gene in terms of the frequency of different alleles of a gene.

allelozyme: A structurally distinct form of an enzyme due to multiple allelism at a single locus. *See* ISOZYMES.

allergen: A substance, usually a protein, capable of causing an allergy or allergic reaction in the body. It induces a hypersensitive response in the organism that has been previously exposed to an antigen. The antigen in this framework is also called an allergen.

allergenicity: The tendency of a substance to cause allergic response.

allergic reaction: A reaction by the body's immune system after exposure to an allergen, often a protein. Food can contain proteins that induce an immune response. Allergic symptoms may include rash, hives and in extreme cases, shortness or loss of breath or even unconsciousness.

allergy: An abnormal response by the body's immune system after exposure to a particular substance, often a protein, containing a particular food, pollen, fur or dust.

alliaceous: Onion-like in smell or form.

all-in-one biomolecule extraction: When the cellular material is limiting, it is desirable to extract DNA, RNA and protein from the same source. For this, the guanidinium thiocyanate homogenate is extracted with phenol – chloroform at a reduced pH – which allows the preparation of DNA, RNA and protein from tissue or cells. This procedure involves the lysis of cells with guanidine isothiocyanate and phenol in a single-phase solution. A second phase is formed after the addition of chloroform where DNA and proteins are extracted, leaving RNA in the aqueous supernatant. The DNA and proteins can be isolated from the organic phase by precipitation with ethanol or isopropanol, and the RNA is precipitated from the aqueous phase with isopropanol.

alloantigens: Phenotypically definable differences in a single protein that are often created by inherited polymorphisms of amino acid residues and can be detected by antibodies from individuals of the same species. *Compare with* ISOANTIGENS.

allocation ratio: In a parallel group-randomized trial of two treatments, the allocation ratio is the proportion of sample sizes of the two groups.

allocompetition. Cultivation at high plant density infers the presence of strong interplant competition. The individual plants, clones, lines or families are estimated by subjecting them to intergenotypic competition. *Aka* INTERGENOTYPE COMPETITION.

allocycly: Differences in chromosome coiling caused by genotypic or environmental (non-genetic) properties.

allodiploid: Cells or individuals in which one or more chromosome pairs are exchanged by one or more pairs from another species.

allogamous: Cross-fertilizing (in plants); as opposed to autogamous. *See* AUTOGAMOUS.

allogamy: Cross-pollination, that is, the pollination of a flower with the pollen produced on the flower of another plant. Allogamy results in cross-hybridization. The resulting population is highly heterozygous. Plants like maize and papaya are allogamous crops. *Contrast* AUTOGAMY.

allogene: Recessive allele.

allogeneic: Two genetically dissimilar individuals of the same species.

allogenetic: Cells or tissues related but sufficiently dissimilar in genotype to interact antigenically.

allogenic: Applied to successional changes due to any change in abiotic environments.

allograft: A tissue graft from a donor of one genotype to a host of a different genotype but of the same species.

allohaploid: A haploid cell or individual derived from an allopolyploid, composed of two or more different chromosome sets.

alloheteroploid: Heteroploid individuals or cells whose chromosomes derive from various genomes.

alloparapatric speciation: New species arise through populations that are initially allopatric, but later become parapatric before a completely effective reproductive isolating mechanism has evolved.

allopatric speciation: The formation of species in which the evolution of reproductive isolating mechanisms occurs due to the divergence of populations that are geographically (physically) isolated from each other to a level that prevents or interferes with genetic interchange. *Aka* GEOGRAPHICAL SPECIATION. *Compare with* PARAPATRIC and SYMPATRIC.

allopatric: Having separate ranges of distribution. *Contrasts* SYMPATRIC.

allopatry: The complete separation of populations by geographic barriers such as oceans, rivers, lakes, mountains, canyons and islands. *Compare with* SYMPATRY.

allophene: A phenotype which is not due to the mutant genetic constitution of the cell or the tissue in question. Such a cell or tissue develops a normal phenotype if it is transplanted to a wild-type host gene.

allophenic: Characteristics that are produced by intercellular gene action. Sometimes a trait is considered chimeric when it is comprised of cells of two different genotypes. *Aka* HYBRID or CHIMERA HYBRID.

alloplasm: Cytoplasm from an alien species that has been transferred by backcrossing into a cultivated species. Common and readily observable effects of this procedure are male sterility, female sterility and reduced plant vigour. If the alien cytoplasm does not manifest itself, the cytoplasm is not considered alien.

alloplasmic: An individual having the common nucleus, but an alien cytoplasm (e.g. alloplasmic rye containing a wheat cytoplasm). Alloplasmy leads to meiotic disturbances and sterility gene alloploid. *Aka* HETEROPLASMIC, HETEROPLASMONIC. *See* ALLOPOLYPLOID.

allopolyploid: A polyploid formed from sets of chromosomes of different species involving hybridization followed by chromosome doubling. The offspring contains chromosomes from both parental species. Wheat is a good example of allopolyploid. Durum wheat is allotetraploid carrying AABB genomes, whereas bread wheat is allohexaploid carrying AABBDD genomes. In wheat, there is a natural mechanism that prevent pairing among homologous chromosomes. Other examples include *Triticale, Brassica, Raphnobrassica* and cabbage. Triploids are characteristically sterile. Allopolyploidy seems to have played a crucial role in plant speciation.

allopolyploidization: A process of hybridization between two species where an offspring gets the complete diploid chromosome complements of the two parental species.

allopolyploidy: Having an allopolyploid constitution.

all-or-none trait: A trait that is either present or absent in a given individual

allosome: A chromosome deviating in size, form or behaviour from other chromosomes (autosomes). Such a chromosome is called the sex chromosome or B chromosome.

allosteric activator: A ligand that binds to a protein and induces a conformational change such that it increases the protein's activity.

allosteric control: The ability of an interaction at one site of a protein enables it to alter its shape to influence the activity at another site.

allosteric effect: The binding of a ligand to one site on a protein molecule such that the properties of another site on the same protein are impacted by the binding of another small molecule with it.

allosteric enzymes. These are enzymes that possess a special site on their surfaces that is distinct from the enzyme's catalytic site to which specific metabolites (called EFFECTORS or MODULATORS) are reversibly and non-covalently bound. The allosteric binding site is as specific for a particular metabolite as is the catalytic site, but it cannot catalyse a reaction; it can only bind the effector. The binding of the effector causes a conformational change in the enzyme such that its catalytic activity is reduced or stopped. Allosteric enzymes are normally the first enzymes in or near the beginning of a multienzyme system.

allosteric interaction: *See* ALLOSTERY.

allosteric protein: Protein which is modified from one conformation to another when it binds to another molecule or when it is covalently altered. The change in conformation alters the activity of the protein, which can form the basis of directed movement.

allosteric regulation: Regulation of the catalytic activities of all component enzymes in an organism so that it can coordinate its numerous metabolic processes and can grow and differentiate. Allosteric mechanisms can cause large changes in enzymatic activity. An enzyme's catalytic activity can be directly regulated through structural alterations that influence the enzyme's substrate-binding affinity. For example, haemoglobin's oxygen affinity is allosterically regulated by the binding of ligands such as O_2, CO_2, H^+ and BPG; an enzyme's substrate-binding affinity may likewise vary with the binding of small molecules, called allosteric effectors.

allosteric site: The site on an (allosteric) enzyme molecule where the non-covalent binding of a non-competitive inhibitor can upsurge or lessen that enzyme's catalytic activity. Such an effector is called an ALLOSTERIC EFFECTOR because it binds at a site on the enzyme molecule that is other (allo) than the enzyme's catalytic site. This is also called a REGULATORY SITE. The binding of activators or inhibitors (effectors) causes a conformational change that alters an enzyme's activity.

allosteric transition: The change of a protein from one conformation to another.

allosteric: An enzyme whose activity is changed when its structure is modified by an organic compound at a non-substrate site.

allostery: A modification in the shape of a protein or ribozyme due to the binding of a molecule at one site, which then modifies activity at a distant site.

allosubstitution: The substitution of a chromosome or chromosome arm with an alien chromosome or chromosome arm.

allosynapsis: In allopolyploids, the pairing between the genomes of two species; intergenomic pairing.

allosyndesis: Chromosome pairing observed at meiosis of non-homologous chromosomes from the diverse sets of an allopolyploid individual

allotetraploid: A plant that is diploid for two genomes, each one coming from a different species. *Aka* AMPHIDIPLOID.

allotetraploidy: A process through which a plant becomes diploid for two genomes, each contributed by a different species.

allotopic: (i) In botany, a specimen of the sex opposite to that of the holotype. (ii) In immunology, an antibody that functions as an antigen to other antibodies of the same species that possess variant molecular sites.

allotransplantation: Transplantation of cells, tissues or organs taken from another member of the same species.

allotype: Mutant covariant of an immunoglobin gene that follows the rules of simple Mendelian inheritance.

allotypic determinants: Correspond to either specific sequences or configurational sites on different polypeptide chains of an antigen. They can be detected even if the chain is separated or integrated into the molecule.

allotypic monoclonal antibodies: Monoclonal antibodies that are isoantigenic. Typical isoantigens are blood group antigens.

allozygosity: Homozygosity where two alleles are alike but from unrelated genes.

allozygote: A zygote heterozygous for different mutant alleles.

allozymes: Isoenzymes of a protein whose synthesis is usually governed by codominant alleles and inherited by monogenic ratios. They show a specific banding pattern if separated electrophoretically. *Compare with* ISOZYMES.

α-/β-domain: A protein domain composed of beta (β) strands connected by alpha (α) helices.

α-amylase activity: Catalysis of the endohydrolysis of 1,4-α-D-glucosidic linkages in polysaccharides containing three or more 1,4-α-linked D-glucose units.

α-amylase: An enzyme that breaks down starch in germinating grains. It splits starch molecules at random points, forming smaller molecules of widely varying sizes. For example, the so-called dextrinizing enzyme of malt.

α-bromonaphthalene: A chemical agent that is used for the condensation of an artificial chromosome. For this, root tips are treated for several hours with a saturated water solution prior to staining.

α-carbon: The central carbon atom in an amino acid to which side chains (R groups) are bound.

α-complementation: pUC18 and similar molecular vectors contain only a small part of the whole gene for β-galactosidase. This small part gives rise to a shortened protein that forms an enzymatically active heterodimer with a specific mutant β galactosidase.

α-designs: The experimental designs that have a single treatment factor but there is no restriction on the number of levels. The blocking structure has replicates and blocks within replicates.

α-DNA: The clustered repeat sequences of DNA located in the centromeric regions of human chromosomes. *See* DNA HELIX A.

α-helix: Common folding pattern in proteins in which a large repeating linear sequence of amino acids folds into a highly regular right-handed helix pattern stabilized by internal hydrogen bonding between backbone atoms. An α-helix is a tight helix that results from the hydrogen bonding of the carboxyl group of one amino acid to the amino group of another amino acid, four residues away (towards the carboxyl terminus). The α-helix was first discovered in protein molecules by L. Pauling in the late 1940s.

α-ketoglutarate: As anion, α-ketoglutarate (*aka* α-KG, OXOGLUTARATE or 2-OXOGLUTARIC ACID) is one of two ketone derivatives of glutaric acid and is an intermediate in Krebs cycle.

α-lattice design: One of a class of incomplete block designs that cater to a wider range of numbers of entries than 'square' or 'rectangular' lattices.

α-level: Significance level.

α-value: The level of alpha selected in a hypothesis test.

α-satellite higher-order repeats (HORs): Using long-read sequencing, G.A. Logsdon *et al.* (2024) totally sequenced and assembled all centromeres from a second human genome and compared it to the complete reference genome. The two sets of centromeres show at least a 4.1-fold surge in single-nucleotide variation when compared with their inimitable flanks and vary up to 3-fold in size. About 46% of centromeric sequence cannot be dependably aligned using standard methods due to the emergence of new HORs. DNA methylation and CENP-A chromatin immunoprecipitation experiments show one-fourth of the centromeres differing in their kinetochore location by >500 kb. Comparative investigations divulge an almost complete turnover of α-satellite HORs, with distinctive idiosyncratic changes in α-satellite HORs for each species.

alteration in the meaning of code words: In almost all organisms, there seem to be available 'recoding signals' in mRNAs leading to altered meaning of codons. These are exceptions to the universal character of the genetic code. In at least two bacterial genes and three mammalian genes, an internal UGA codon codes for the 21st amino acid selenocysteine (SeCys) which has no unique codon in the genetic code dictionary. A hairpin structure called the 'selenocysteine insertion sequence' (SECIS) is present in the messenger RNA that corresponds to where SeCys is due to be inserted into the growing peptide chain. Pyrrolysine is the 22nd amino acid in proteins. **UAG** is normally the amber-stop codon but encodes pyrrolysine (Pyl) if a pyrrolysine insertion sequence (PYLIS) element is present. *Aka* RECODING.

alteration of generations: The regularly occurring alternation of haploid (gametophyte) and diploid (sporophyte) phases in the life cycle of sexually reproducing plants. Also important are organisms that are haploid for part of their life cycle and diploid for the other part. Such organisms are said to show an alternation of generations, referring to the alternation of the 2n and n stages. Do all these various life cycles show Mendelian genetics? The answer is that Mendelian inheritance patterns characterize any species that has meiosis as part of its life cycle because Mendelian laws are based on the process of meiosis. All the groups of organisms mentioned, except bacteria, undergo meiosis as part of their cycle.

altering epigenetic landscape to transcriptionally reprogramme target immune cells: *Trans*-Activator of Transcription (Tat) is a regulatory protein that severely enhances the effectiveness of viral transcription. Tat protein consists of 86–101 amino acid residues dependent on the subtype. Tat vastly enhances the level of transcription of the HIV dsDNA.

altering photorespiratory metabolism: Photorespiration detoxifies 2-phosphoglycolate (2-PG) and is the main physiological process responsible, in general, for 20–50% yield loss, depending on the environmental conditions and the type of photosynthesis the plant species employs (H. Bauwe, M. Hagemann and A.R. Fernie in 2010). This process involves an energy-expensive metabolic pathway that recycles toxic metabolites of RuBisCO oxygenation reactions. The energy consumption in this futile process reduces assimilate production and increases the cost of crop yield. Research on some alternative pathways is expected to optimize photorespiratory mechanisms.

***Alternaria*:** A genus of fungi. It forms yellowish-brown conidia that are divided by transverse and longitudinal septa. There are many species under this genus, including important plant pathogens (early blight of potato, *Alternaria solani*; black rot of carrot and rape, *A. radicina* and *A. brassica*; Alternaria disease of wheat).

alternate (of leaves of flowers): (i) Borne singly at different levels along a stem, including spiralled parts. (ii) Occurring between something else (e.g. stamens alternating with petals). *Contrasts* OPPOSITE.

alternate forms of DNA: Depending upon base composition and relative humidity, DNA exists in various forms: A-FORM DNA (or A-DNA), B-FORM DNA (or B-DNA), C-FORM DNA (or C-DNA), D-FORM DNA (or D-DNA), E-FORM DNA (or E-DNA), P-FORM DNA (or P-DNA) and Z-FORM DNA (or Z-DNA). These alternative forms of DNA are also, respectively, known as DNA A-HELIX, DNA B-HELIX, DNA C-HELIX, DNA D-HELIX, DNA E-HELIX, DNA P-HELIX and DNA Z-HELIX. *See details as separate terms.*

alternate host: An alternate host in plant pathology is the host belonging to a different family that helps complete the life cycle of a pathogen (e.g. rust fungus) and its survival. These plants are plants other than the primary host on which a pathogen or pest can live. *Aka* ALTERNATIVE HOST, AECIAL HOST or ECONOMICALLY LESS IMPORTANT HOST. *See* TELIAL HOST.

alternate hypothesis (H_1 or H_A): The alternate hypothesis is opposite to our NULL HYPOTHESIS (H_0). If, for example, observed = expected is our null hypothesis, our alternate hypothesis will be observed ≠ expected. If the data you collect demonstrate that your original hypothesis or the null hypothesis was correct, you can reject the alternative hypothesis. *Aka* ALTERNATIVE HYPOTHESIS. *Compare with* NULL HYPOTHESIS.

alternate segregation: At meiosis, in a reciprocal translocation heterozygote, the segregation of both normal chromosomes occurs at one pole and both translocated chromosomes at the other pole, thus giving genetically balanced gametes, or segregation of centromeres during meiosis.

alternating dominance: A change of dominance from one allele to the other (*A1a2 a1A2*) of a pair of alleles during the ontogenetic development of a heterozygous hybrid; the phenotypic expression of the alleles acts as one after another gene.

alternation of generations: The alternation of two or more generations reproducing themselves in different ways (i.e. alternation of gametophyte [sexual reproductive, n] and sporophyte [asexual reproductive, 2n] phases in the life cycle of a plant).

alternation of population density: The population density alternates between high and low values in successive generations if the key factor is density-dependent and strongly overcompensates for a change in population density.

alternative chromatin states: Two states of chromatin with respect to gene expression are the open and closed chromatin states.

alternative disjunction: The distribution of interchange chromosomes at anaphase I of meiosis is resolved by their centromere orientation. In the case of alternative disjunction, chromosomes located alternatively in the pairing configuration are distributed to the same spindle pole. *Opposed to* ADJACENT DISJUNCTION.

alternative dominance: A theory of sex determination that supposes that all individuals are heterozygous for sex, with the only difference being that the male determiners are dominant in male offspring and the female determiners are dominant in female offspring. In fact, the alternative display of parental genomes in two types of hybrid cells echoes the bidirectional reprogramming of the parental genomes.

alternative donor and acceptor splice sites: Introns usually start with dinucleotides GT and the border between exon and intron in a DNA sequence is termed the donor splice site. Introns usually end with dinucleotides AG and the border between intron and exon in a DNA sequence is termed the acceptor splice site.

alternative end joining (A-EJ): The processes of end joining other than HOMOLOGOUS RECOMBINATION (HR) and non-homologous end joining (NHEJ) that repair double-stranded breaks in DNA.

alternative end joining (A-EJ)-mediated V(D)J recombination, regulation of: G0-G1 phase A-EJ is a mutagenic process branded by resected deletion and translocation junctions that are principally straight and are notable from A-EJ in cycling cells that count on much more on MICROHOMOLOGY-MEDIATED END-JOINING (MMEJ). Using chemical and genetic approaches, J. Wang *et al.* (2024) analytically evaluate potential A-EJ factors and DNA DAMAGE RESPONSE (DDR) genes to back this mechanism by mapping the repair destinies of RAG1/2-initiated double-strand breaks in the context of *Igκ* locus V-J recombination and chromosome translocation. A polymerase theta-independent Parp1-XRCC1/LigIII axis is highlighted as central A-EJ machineries, reinforced by 53BP1 in the context of an Ataxia-telangiectasia mutated (ATM)-activated DDR. Mechanically, they reveal varied deviations in short-range resection, MMEJ and translocation, imposed by conceding specific DDR activities, which comprise polymerase alpha, Ataxia-telangiectasia and Rad3-related (ATR), DNA2 and Mre11.

alternative energy: Energy that is not commonly used and is usually environmentally sound, such as solar or wind energy, as opposed to fossil fuels.

alternative fibres: Fibres produced from non-wood sources (such as rice straws, maize stalks, cotton stalks and bagasse) for use in paper making.

alternative fuels: Transportation fuels other than gasoline or diesel. Such fuels are exemplified by natural gas, methanol and electricity.

alternative polyadenylation (APA): A widespread mechanism used to control gene expression. Like alternative splicing, the usage of alternative poly(A) sites allows a single gene to encode multiple mRNA transcripts.

alternative promoters: Alternative promoters in a gene are positively correlated with alternative splicing.

alternative promoters and polyadenylation sites: The presence of alternative promoters and polyadenylation sites in the transcript may also yield different mRNAs molecules of different sizes by differential use of such sites.

alternative RNA splice form: One of the possible alternate combinations of exons into a folded protein that are possible by recombining multiple gene segments during mRNA splicing in higher organisms.

alternative RNA splicing: Refers to the process of the production of diverse mRNAs by differential splicing of the single RNA precursor. It may yield proteins with different compositions of amino acids, or it may involve just the length of 3′ UTR. One reason for alternative/differential splicing is base modification during RNA editing, causing a change in splice sites. This is a fundamental mechanism for encoding genomic complexity in higher eukaryotes without an expansion in gene number. This provides different ways of combining a gene's exons to make variants of the complete protein. *Aka* ALTERNATIVE SPLICING.

alternative RNA splicing, detection of: An 'exon junction array design' uses probes specific to the expected or potential splice sites of predicted exons for a gene. It is of intermediate density, or coverage, to a typical gene expression array (with 1–3 probes per gene) and a genomic tiling array (with hundreds or thousands of probes per gene). It is used to assay the expression of alternative splice forms of a gene. Exon arrays have a different design, employing probes designed to detect each individual exon for known or predicted genes, and can be used for detecting different splicing isoforms.

alternative RNA splicing, functional effects of: Functional effects of alternative RNA splicing include (i) protein localization; (ii) deletion of protein activity; (iii) modification of protein activity; (iv) novel protein activities and (v) RNA stability and translational efficiency.

alternative RNA splicing, mechanisms of: Alternative splicing can occur in several different ways: (i) CONSTITUTIVE SPLICING generates only one type of product; (ii) ALTERNATIVE PROMOTER USAGE generates different products from *Drosophila alcohol dehydrogenase* gene in larvae and adults; (iii) DIFFERENTIAL POLYADENYLATION site usage generates secreted and membrane-bound forms of the immunoglobulin heavy chain; (iv) ALTERNATIVE 3′ SPLICE SITE USAGE produces functional and non-functional forms of *Drosophila* transformer mRNA; (v) EXON SKIPPING generates different forms of *Drosophila double sex* mRNA; (vi) MUTUAL EXCLUSION OF EXONS 16 and 17 generates alternative forms of the rat troponin T mRNA; and (vii) FAILURE TO SPLICE OUT A SINGLE INTRON generates functional and non-functional forms of P-element transposase. This type can be regarded as single intron processing or discard regulation.

alternative RNA splicing, patterns of: The six patterns of alternative RNA splicing are combinatorial exons, mutually exclusive patterns, retained introns, alternative donor and acceptor splice sites, alternative promoter and polyadenylation sites, and splice site compatibility.

alternative RNA splicing, regulation of: Splicing is regulated in a manner analogous to the transcriptional regulation, in that *TRANS*-ACTING PROTEINS (repressors and activators) bind to *CIS*-ACTING REGULATORY SITES (silencers and enhancers) on the RNA. Alternative splicing is controlled by *cis*-regulatory RNA elements that are recognized by RNA BINDING PROTEINS (RBPs). The MALT1 paracaspase, a key factor of signalling pathways, mediates innate and adaptive immune responses. Alternative splicing of *MALT1* is critical for controlling optimal T-cell activation. A.N. Jones *et al.* in 2022 demonstrated that *MALT1* splicing depends on RNA structural elements that sequester the splice sites of the alternatively spliced exon7. The RBPs hnRNP U and hnRNP L bind competitively to stem-loop RNA structures that involve the 5′ and 3′ splice sites flanking exon7. While hnRNP U stabilizes RNA stem-loop conformations that maintain exon7 skipping, hnRNP L disrupts these RNA elements to facilitate the recruitment of the essential splicing factor U2AF2, thereby promoting exon7 inclusion.

alternative RNA splicing, role of: Alternative RNA splicing has a role (i) in developmental control by serving as an on/off switch as well as through alternative RNA splicing of morphogenetic proteins; (ii) in allowing a high degree of protein diversity at a low genetic cost; (iii) in the involvement in the localization of different proteins in different cells, deletion of the function of the gene product and production of isomeric proteins having different functions; (vi) as it is the most efficient mechanism for generating protein diversity due to its reversibility, and it allows genetic variation at a very low genetic cost; (v) as the genes encode proteins with myriad functions and (f) as it is an important regulatory mechanism in metazoans and their viruses.

alternative RNA-splicing activators and inhibitors: Splicing of individual pre-mRNAs is frequently controlled by combinatorial or competitive effects of both activators and inhibitors. *See* ACTIVATORS, INHIBITORS.

alternative RNA-splicing factors: The factors that preferentially regulate the removal of introns harbouring (or closest to) certain clusters. Nova proteins are neuron-specific alternative RNA-splicing factors.

alternative sigma factors: Sigma factor is a vital component for transcription initiation by prokaryotic RNA polymerase. Various forms of sigma factors have been discovered in viruses, prokaryotes and eukaryotes. For a long time, single sigma factor σ^{70} or σ^{D} was thought to exist in *E. coli*. However, in recent years, six new sigma factors in *E. coli* have been discovered which operate under different conditions. σ^{70} accomplishes the vast majority of transcriptional 'housekeeping', which is completed by the RNA polymerase core in a complex with σ^{70}. σ^{S} (σ^{38}) is the general stress factor associated with the stationary phase and a variety of growth-impairing stresses. σ^{H} (σ^{32}) controls heat shock-responsive promoters. σ^{F} (σ^{28}) is used for flagellum-related functions. σ^{E} (σ^{24}) controls responses to extracytoplasmic or extreme heat shock stress. σ^{FecI} (σ^{19}) is required in iron transport for the transcription of the ferric citrate transporter. σ^{N} (σ^{54}) controls promoters for nitrogen assimilation. This repertoire of alternative σ factors reflects the lifestyle of *E. coli* as a gut commensal bacterium.

altitudinal variation: Variation found among individuals of a species or morphological, biochemical or molecular trait along altitude.

altruism: The process through which a gene, trait or behaviour benefits others at a cost to oneself.

altruistic: A gene, trait or behaviour that decreases the fitness of its bearer but surges the fitness of other individuals.

Alu element: A common set of dispersed DNA sequences found throughout the human genome; each is about 300 bases long and is repeated at least 5×10^5–10×10^5 times. Alu sequences are speculated to have originated from viral RNA sequences that integrated into human DNA thousands of years ago. The name has been derived from the *Alu*I restriction enzyme cleavage site that occurs within most Alu elements. *Alu*I is a restriction enzyme that identifies this sequence and cleaves (cuts) it between the G and the C. The repeats arise due to the reverse transcription of RNAs into DNAs. *Aka* ALU FAMILY.

Alu-equivalent family: A set of sequences in a mammalian genome that is related to the human Alu family.

aluminium toxicity: Aluminium is the most abundant metal on earth and has no specific importance in the metabolism of higher plants. Small amounts of its uptake favour the imbibition of the cytoplasm. Its higher concentrations in the soil may result in severe inhibition of plant growth. Aluminium tolerance is a main challenge of plant breeding in several regions of the world. High levels of aluminium can also be toxic to soil microorganisms. Rye, the world's most acid soil-tolerant cereal, contains an enormous diversity in the organization of gene complexes that play a role in controlling aluminium tolerance. The presence of malate transporter (ALMT) genes contributes to the upregulation of aluminium tolerance expression in rye. There is a variation in copy number and chromosome location of ALMT genes in rye. At least two ALMT gene complexes are characterized that vary in gene copy number and expression of rye aluminium tolerance.

alveograph: A dough-testing device for determining the extensibility and resistance to the stretching of dough. It provides a prediction of baking quality. As a standard disc of dough is blown into a bubble, pressure change and bursting pressure are plotted versus time.

***AmA1* gene:** The *AmA1* gene has been cloned from a plant *Amaranthus hypochondriacus*, commonly called 'chalai' and grown for fodder, foliage and seed purposes. It is responsible for the production of seed-specific protein rich in all essential amino acids including lysine, which is an essential amino acid (not produced in the human body). This gene has been introduced into tobacco and potato – the resulting transgenic plants overproduce the lysine-rich protein. Cereals like rice are deficient in lysine. Therefore, attempts are being made to produce transgenic cereals using *AmA1* gene for improving the nutritional quality of rice.

amber codon: The nonsense codon UAG.

amber mutation: Amber suppressor mutation that changes the anticodon of an amino acid–carrying tRNA to UAG, thereby signalling chain termination.

amber suppressors: The mutant genes that code for tRNAs whose anticodons have been altered so that they can respond to UAG codons as well as/instead of to their previous codons.

amber: The chemically altered resin of ancient trees. Soft animal part fossils have been found in amber.

ambidirectional dominance: The dominance at different loci affecting how a character acts in different directions, that is, increasing at some and decreasing at others.

ambiguity: The coding for more than one amino acid by a given codon.

ambiguous codon: Strict one-to-one relationship between codons and amino acids is assumed to be an essential feature of the genetic code. However, it has been observed that one codon can code for two different amino acids with the choice of the inserted amino acid resoluted by a specific 3′ untranslated region structure and the location of the dual-function codon within the mRNA. For example, codon UGA specifies the insertion of selenocysteine and cysteine in the ciliate *Euplotes crassus*. The dual use of this codon can happen within the same gene.

ambisexual: A plant that possesses the reproductive organs of both sexes. *Aka* AMBOSEXUAL.

ambivalent gene: A gene having both advantageous and disadvantageous properties.

amDNA: *See* ANTIMESSENGER DNA.

ameiosis: The failure of meiosis and its replacement by cellular nuclear division that does not reduce the chromosome number to one-half.

ameiotic parthenogenesis: Parthenogenesis in which meiosis has been entirely suppressed in the cell.

amelioration: The process in which DNA that has been acquired by lateral gene transfer changes in composition (e.g. G + C content and codon usage) in order to bear a resemblance to the genome in which it later resides.

American Seed Trade Association (ASTA): One of the oldest trade organizations in the USA. ASTA works on behalf of the seed industry to promote the research, development and movement of quality seeds to meet the world's demand for food, feed, fibre and fuel.

American Standard Code for Information Interchange (ASCII): A character-encoding standard for electronic communications. ASCII codes represent text in computers, telecommunications equipment and other devices. Most modern character-encoding schemes are based on ASCII. However, they are known to support many additional characters.

Ames test: A biological assay, given by B.N. Ames in 1970, widely used to assess the mutagenic potential of environmental chemicals. It banks on the reverse mutations of histidine-requiring (his^-) auxotrophs in *Salmonella typhimurium* to wild-type (his^+) auxotrophs. A positive test shows that the chemical might function as a carcinogen/mutagen although several false positives and false negatives are known. Although this test assesses mutagenesis (causation of mutations) in the DNA of bacteria, its results have also been employed to approve or disapprove various compounds consumed by humans.

amino acid: An organic compound that consists of the basic formula NH_2-CHR-COOH, where 'R' is the side chain, '−COOH' is an acidic carboxyl group and '$-NH_2$' is a basic amino group. Amino acid molecules are linked by amide (also called peptide) linkages to form a polypeptide chain of a protein. Depending on the structure of side group R, they are subdivided into various types: (i) polar or hydrophilic (serine, threonine, tyrosine, asparagine and glutamine); (ii) non-polar or hydrophobic (glycine, alanine, valine, leucine, isoleucine, proline, phenylalanine, tryptophan and cysteine); (iii) acidic (aspartic acid and glutamic acid); and (iv) basics (lysine, arginine and histidine). The sequence of amino acids determines the shape, properties and biological role of a protein. Proteins are the fundamental constituent of living matter. Amino acids are synthesized by autotrophic organisms, such as green plants. There are 20 standard/common/essential amino acids, each coded by a different arrangement of three adjacent DNA nucleotides. Virtually all amino acids (except glycine) possess an asymmetric carbon atom and thus are potentially chiral.

amino acid 'alphabet': The set of amino acids used to build genetically encoded proteins.

amino acid activating enzyme alteration suppression: Suppression of a mutation in a particular triplet, AUU (isoleucine) UUU (phenylalanine) may take place by the occasional attachment of isoleucine to $tRNA^{Phe}$.

amino acid profile: Refers to a quantitative explanation of how much of each amino acid is contained in a given source of (livestock feed or food) protein. For example, of all protein meals, the amino acid profile of soybean meal is matched closest to the profile of amino acids needed for human nutrition. *Aka* PROTEIN QUALITY.

amino acid sequence: Refers to the arrangement of amino acid residues in a polypeptide chain that characterizes the primary structure of a protein. Amino acid arrangement in a protein is usually governed by nucleotide sequence of DNA. Exceptions are those coded for by the RNA of certain viruses, such as HIV. The amino acid sequence is unique to each protein and influences the protein structure (secondary, tertiary and quaternary) and function.

amino acid code, single letter: Alanine (A), Arginine (R), Asparagine (N), Aspartic Acid (D), Cysteine (C), Glutamic Acid (E), Glutamine (Q), Glycine (G), Histidine (H), Isoleucine (I), Leucine (L), Lysine (K), Methionine (M), Phenylalanine (F), Proline (P), Serine (S), Threonine (T), Tryptophan (W), Tyrosine (Y) and Valine (V).

amino acid code, three letter: Alanine (Ala), Arginine (Arg), Asparagine (Asn), Aspartic Acid (Asp), Cysteine (Cys), Glutamic Acid (Glu), Glutamine (Gln), Glycine (Gly), Histidine (His), Isoleucine (Ile), Leucine (Leu), Lysine (Lys), Methionine (Met), Phenylalanine (Phe), Proline (Pro), Serine (Ser), Threonine (Thr), Tryptophan (Trp), Tyrosine (Tyr) and Valine (Val).

amino group: Weakly basic functional group derived from ammonia (NH_3) in which one or more hydrogen atoms are replaced by another atom. In an aqueous solution, it can accept a proton and carry a positive charge.

amino terminus: The end of a polypeptide chain that carries a free α-amino group, by custom drawn at the left of a protein sequence. *Aka* N TERMINUS.

amino transferases: *See* TRANSAMINATION.

aminoacetic acid: The amino acid that has a single hydrogen atom as its side chain. Glycine is the simplest stable amino acid.

aminoacylation: The attachment of an amino acid to the acceptor arm of a tRNA molecule; one of the prerequisites for protein synthesis.

aminoacyl (A) site: The site on the ribosome to which the aminoacyl~tRNA attaches during translation.

aminoacyl~tRNA: A tRNA molecule covalently bound to an amino acid via an acyl bond between the carboxyl group of the amino acid and the 3′-OH of the tRNA.

aminoacyl~tRNA ligase: An enzyme that synthesizes a specific aminoacyl~tRNA molecule, employing a specific amino acid – for example, alanine – its cognate tRNA and ATP to form – for example, alanyl~$tRNA_{ala}$.

aminoacyl~tRNA synthetase: Any one of the diverse family of enzymes responsible for aminoacylation of a specific amino acid with its proper tRNA and ATP to form aminoacyl~AMP (activated amino acid) and pyrophosphate (PPi).

aminocyclopropane-1-carboxylic acid oxidase (ACC oxidase): The enzyme in the ethylene biosynthesis pathway that converts ACC to ethylene.

aminocyclopropane-1-carboxylic acid synthase (ACC synthase): The enzyme in the ethylene biosynthesis pathway that converts S-adenosylmethionine (SAM or AdoMet) to ACC.

aminoglycoside: Antibiotic consisting of amino sugars and an aminocyclitol ring (e.g. streptomycin). Aminoglycosides are the inhibitors of the protein synthesis. Commonly used aminoglycosides include streptomycin, neomycin and gentamycin. These are broad spectrum and also kill mycobacteria.

aminopeptidase (Amp): An enzyme that conducts the hydrolysis of amino acids in a polypeptide chain by acting on the peptide bond adjacent to the essential free amino group.

aminotransferase. *See* TRANSAMINASE.

amitosis: Nuclear division by a process other than mitosis.

amixis: Reproduction in which the essential procedures of sexual reproduction do not occur.

ammonia (NH_3): A compound of nitrogen (N) and hydrogen (H) that plays an important role in the natural nitrogen pathways. It occurs in nature, primarily produced by the anaerobic decay of plant and animal matter. It has also been detected in outer space.

ammonia fertilizer: A material having a high concentration of nitrogen compounds put on soil to stimulate plant growth.

ammonification: The process that releases ammonia from nitrogen-containing organic matter by the action of microorganisms. Microbial decomposition of the dead organisms results in the hydrolytic breakdown of proteins into amino acids. Further microbial ammonification removes the amino groups of the amino acids (deamination) to produce ammonia (NH_3).

ammonium fixation: In soil science, adsorption of ammonium ions by clay minerals. This process makes them insoluble and non-exchangeable.

ammonium phosphate: Chemical formula $(NH_4)_3PO_4$. Used as a mineral fertilizer with a nitrogen content of 28.18%

ammonium sulfate: Chemical formula $(NH_4)_2SO_4$. Used as a mineral fertilizer with a nitrogen content of 21%.

amniocentesis: A prenatal diagnostic technique for the study of metabolic disorders by biochemically testing the cultured cells from the amniotic fluid. A foetus may be checked for any chromosomal abnormality also in early stages of development by karyotyping the cultured cells. Chromosomes in dividing cells can be stained and observed. A sample of fluid is withdrawn with a needle from the amniotic sac. Fetal cells are cultured after a period of 2–3 weeks.

amorph: (i) An inactive gene called an amorphic gene. A mutated allele that has lost the ability of the parent allele (whether wild type or any other type) to encode any functional protein. An amorph mutation, or null, is the loss of genetic information for the synthesis of appropriate mRNA; (ii) In botany, the term refers to something that lacks a discernible shape and can thus be described as AMORPHOUS.

amorphic alleles: The recessive alleles that are not expressed in either homozygotes or heterozygotes compared to recessive alleles that are not expressed in heterozygotes. It is equivalent to the absence of it. *Aka* SILENT ALLELES or ANTIMORPHIC ALLELES.

amorphic mutations: The mutations that lead to almost a total loss of the expression of a trait as they produce non-functional proteins or enzymes (e.g. *white* eye in *Drosophila*).

AMP/ATP ratio: Refers to the ratio of adenosine monophosphate (AMP) to adenosine triphosphate (ATP) in the cell. ATP supplies the energy needed for many biochemical reactions. Therefore, the AMP/ATP ratio reflects the cell's energy level.

amphibivalent: Refers to a ring-like exchange configuration involving four chromosomes.

amphibolic pathway: A pathway that includes both anabolic and catabolic processes. It is a dual-purpose pathway that bridges the reaction dealing with the breakdown (distinctive) and synthesis (constructive) of lipids, proteins, carbohydrates and nucleic acids. Both kinds of reactions occur simultaneously; the breakdown products of one reaction are utilized in another reaction to synthesize different compounds.

amphicarpous: Plants that produce two classes of fruit differing from each other in form or time of ripening.

amphidiploid: An alloploid possessing the complete chromosome complements of two diploid species. Hybridization between two distinct diploid species followed by chromosome doubling is the basic approach for the synthesis of amphidiploids. It is an allotetraploid that appears to be a normal diploid. *Aka* ALLOTETRAPLOID or DIDIPLOID.

amphihaploid: Allohaploid.

amphihaploid: An allopolyploid containing two haploid sets of chromosomes contributed by each of the two species.

amphikaryon: The nucleus of the zygote formed after fertilization.

amphimict: A species or individual reproduced during sexual reproduction by the fusion of two haploid nuclei. *Aka* AMPHIMICTIC.

amphimixis: The copulation of two unrelated haploid cells and nuclei (e.g. egg and sperm) as a part of the sexual reproduction process. *See* AMPHIMICT.

amphimobile: Able to traffic both upward and downward in the phloem, as do some pesticides.

amphipathic molecule: A molecule that bears both polar (hydrophilic) and non-polar (hydrophobic) domains in a phospholipid or a detergent molecule. This term is used to describe large proteins with several domains composed of different amino acid residues. Some examples of such molecules are wetting agents (sodium dodecyl sulfate, SDS) and membrane lipids such as lecithin.

amphiphilic molecules: These molecules have distinct regions of hydrophobic (water-hating) and hydrophilic (water-loving) character within the same molecule. When dissolved in water above a certain concentration, they are capable of forming high-molecular-weight aggregates or micelles. *Aka* AMPHIPHILES or CRITICAL MICELLE CONCENTRATION (CMC).

amphiplasty: Morphological changes of chromosomes after interspecific hybridization. It occurs when the genomes of the parental species are spatially separated in the hybrid nucleus in a concentric fashion. The genome occupying the central position has an active nucleolus organizer region (NOR), while the NOR of the peripherical genome is suppressed (e.g. in hexaploid wheat, NORs on chromosome 1B and

6B function preferentially, although 1A and 5D also carry NORs; however, when the NOR on 1B is deleted, then the NORs on 5D and/or 1A are used to a greater extent). *Aka* NUCLEOLAR DOMINANCE.

amphiploid: *See* ALLOPLOID *and* ALLOPOLYLPOID.

amphiploidy: *See* ALLOPLOIDY *and* ALLOPOLYPLOIDY.

amphitene: *See* ZYGOTENE.

amphitropous ovule: A type of ovule arrangement in which the ovule is slightly curved such that the micropyle comes near to the funicular attachment. *Contrasts* ANATROPOUS, CAMPYLOTROPOUS and ORTHOTROPOUS.

ampholytes: Collection of small amphoteric molecules with an ISOELECTRIC POINT (pI) corresponding to a given pH range. Initially, the buffering molecules are free to diffuse and are distributed evenly, so there is no pH gradient. When an electric field is applied, ampholytes establish a pH gradient and become charge neutral. There is a limitation to the use of ampholytes for establishing a pH gradient.

amphoteric compound: A compound which is capable of both donating and accepting protons and is thus able to act chemically as either an acid or a base.

ampicillin (Amp): A semisynthetic β-lactam antibiotic. In genetic experiments with bacteria, resistance to ampicillin is sometimes utilized as a screening marker. For example, the cloning vector pBR322 carries a gene for ampicillin resistance. Ampicillin interferes with bacterial cell wall synthesis; the *bla* gene is used as a selective marker in many vectors, including pBR322 and pUC18.

amplicon: A specific DNA sequence produced by DNA amplification technology such as PCR. An amplicon is defined by a pair of PCR primers. These primer pairs are sometimes called AMPLIMERS.

amplification at gene level: A mutation at some site produces a new allele. Another mutation at a different site produces another new allele. Intragenic recombination produces another new allele that carries both mutations. Thus, several UNISITE MUTANT ALLELES as a result of intragenic recombination can produce two-site and MULTISITE MUTANT ALLELES. Multisite mutant alleles have been reported to be present in natural populations. This is thus a mechanism for amplifying genetic variation at the gene level.

amplification: Any *in vivo* (spontaneous) or *in vitro* (molecular techniques, e.g. PCR) process that increases the number of copies of a specific DNA fragment. The additional copies of a chromosomal sequence can be seen as intrachromosomal or extrachromosomal DNA.

amplification at genomic level: Variation at the gene level gets further amplified at the genomic level. Independent assortment, crossing over and random union of gametes are the mechanisms for the amplification of genetic variation at the genomic level.

amplification of ribosomal RNA gene repeats: Organisms maintain ribosomal RNA gene repeats (rDNA) at stable copy number by recombination; the loss of repeats results in gene amplification.

amplification-based sequencing: Purified RNA is reverse transcribed into cDNA prior to PCR amplification and DNA sequencing. These sites can be identified by the simple comparison of the genomic DNA and cDNA sequences. A bioinformatics tool distinguishes RNA-editing sites from RNA-sequencing data obtained from multiple samples, without the need for matching the genomic DNA sequence.

amplification-free DNA sequencing: T.D. Harris *et al.* in 2008 reported an amplification-free method for determining nucleotide sequence of more than 28,000 individual DNA molecules simultaneously. A DNA polymerase added labelled nucleotides to surface-immobilized primer-template duplexes in a stepwise fashion, and the asynchronous growth of individual DNA molecules was monitored. They used this method to sequence the M13 virus.

amplified fragment length polymorphism (AFLP): AFLP-PCR or just AFLP is a PCR-based tool used in genetics research, DNA fingerprinting and in the practice of genetic engineering. The technique was originally described by M. Zabeau and P. Vos in 1993 of Keygene (the AFLP technology is covered by patents and patent applications of Keygene N.V.; AFLP is a registered trademark of Keygene N.V.). The procedure of this technique is divided into three steps: (i) Digestion of total cellular DNA with one or more restriction enzymes and ligation of restriction half site–specific adaptors to all restriction fragments. AFLP is based on PCR amplification of a set of restriction fragments, selected from a pool of fragments generated due to digestion with a pair of specific restriction enzymes, one of them being a fragment cutter (e.g. *Mse*I) and the other a rare cutter [e.g. *Eco*RI (6 bp)]. (ii) Selective amplification of some of these fragments with two PCR primers that have corresponding adaptor and restriction site–specific sequences. With the help of these ligated adapters, the number of DNA fragments to be amplified can be restricted. (iii) Electrophoretic separation of amplicons on a gel matrix, followed by visualization of the band pattern. This method generates a large number of bands, facilitating the detection of polymorphism.

amplified fragment length polymorphism (AFLP), advantages of: There are many advantages to AFLP when compared to other marker technologies including RANDOMLY AMPLIFIED POLYMORPHIC DNA (RAPD), RFLP and MICROSATELLITES. AFLP not only has higher reproducibility, resolution and sensitivity at the whole-genome level compared to other techniques, but it can also amplify between 50 and 100 fragments at one time. In addition, no prior sequence information is needed for the amplification. As a result, AFLP has become extremely beneficial in the study of taxa including bacteria, fungi and plants, where much is still unknown about the genomic makeup of various organisms. AFLP can be used to `map' those DNA molecules to speed up plant breeding programmes. AFLP uses selective amplification of a subset of digested DNA fragments to generate and compare unique fingerprints for genomes of interest.

amplified fragment length polymorphism (AFLP) markers, advantages of: (i) Used extensively in genetic mapping for increasing marker density in several species. (ii) Highly polymorphic and reproducible. (iii) Can be used for any organism in primer/probe development. (iv) Provide good genome coverage.

amplified fragment length polymorphism (AFLP) markers, limitations of: (i) The clustering of AFLP markers especially from the methylation insensitive restriction enzyme may aggravate clustering in the highly methylated regions of the genome. (ii) The position of AFLP markers is dependent on the location of the restriction sites of the enzymes used in their production.

amplified fragment length polymorphism (AFLP) variation: A variation on AFLP is cDNA-AFLP, which is used to quantify differences in gene expression levels. Another variation on AFLP is the TE Display, used to detect transposable element mobility. The AFLP technology can detect various polymorphisms in different genomic regions simultaneously. It is also highly sensitive and reproducible. As a result, AFLP has become widely used for the identification of genetic variation in strains or closely related species of plants, fungi, animals and bacteria. The AFLP technology has been used in criminal and paternity tests, to determine slight differences within populations and in linkage studies to generate maps for QUANTITATIVE TRAIT LOCUS (QTL) analysis.

ampligase DNA ligase: This DNA ligase enzyme is active at even much higher temperatures (80–94°C).

amplimer: (i) A piece of DNA formed as the products of natural or artificial amplification events, as in the case of PCR. (ii) PCR-amplified segment of the genome, including SEQUENCE TAGGED SITES (STSs) and EXPRESSED SEQUENCE TAGS (ESTs).

amyl: It means amylose or starch.

amylases: The term refers to a category of enzymes that catalyses the chemical reaction in which amylose (starch) molecules are hydrolytically broken into molecular pieces (e.g. the polysaccharides maltose, maltotriose, a-dextrin, etc.). α-amylase is used to break apart corn starch molecules in the first step of manufacturing fructose (sweetener for soft drinks). Since 1857, amylase has been employed to remove (amylose) starch from woven fabrics in the textile industries. Modern uses of some amylases include the substitution of barley grain for malt in the beer brewing process.

amyliferous: Containing starch.

amylograph: An instrument that determines the change of viscosity with time in a heated mixture of water and a starchy material such as flour.

amylolytic enzymes: The enzymes that hydrolyse starches and similar glucose polymers.

amylopectin: A water-soluble polysaccharide; highly-branched polymer of glucose found in plants.

amyloplasts: Differentiate from LEUCOPLASTS. A colourless plastid type specialized in the storage of starch. Found in the roots and tubers of many plants and are used for starch storage and detecting gravity. Some amyloplasts differentiate into STATOLITHS (various starch grains or other solid bodies in the plant cytoplasm that are held to be responsible for changes in orientation of a part or organ due to changes in their position).

anabiosis: The situation in the life cycle of some plants in which there is no visible metabolic activity; a kind of resting period.

anabolic: An enzymatic reaction leading to the synthesis of a more complex biological molecule from simpler molecule(s).

anabolism: A constructive process in which two or more molecules combine to form a new molecule. For example, joining of (i) sugar molecules to form starch, (ii) amino acids to form protein and (iii) nucleotides to form nucleic acids.

anaerobe: An organism that lives in the absence of molecular oxygen and generally cannot grow in the presence of MOLECULAR OXYGEN (O_2) as the presence of oxygen is toxic for the growth of such organisms. The catabolic metabolism of anaerobic microorganisms reduces a variety of organic and inorganic compounds in order to survive (e.g. carbon dioxide, sulfate, nitrate, fumarate, iron and manganese). Anaerobes produce a large number of end products after metabolism (e.g. acetic acid, propionic acid, lactic acid, ethanol and methane). The genus *Clostridium* that causes tetanus is one of the best examples of an anaerobe. *Aka* ANAEROBIC. *Compare with* AEROBE *or* AEROBIC.

anaerobic respiration: A metabolic progression that does not involve MOLECULAR OXYGEN (O_2) in the breakdown of glucose.

anagenesis: The evolutionary process whereby one species successively evolves into another without any splitting of the phylogenetic tree. For example, species A → species B → Species C. *See* CLADOGENESIS.

analogous: When characters are similar due to convergent evolution.

analogous species: When characters in different species are similar due to convergent evolution.

analogous structures: A set of morphological features in different organisms which have similar form or function but were absent in the organisms' last common ancestor. The cladistic term for the same phenomenon is HOMOPLASY. *Contrasts* HOMOLOGOUS STRUCTURES.

analogue: Also written as 'analog'. A compound (or molecule) that is a (chemical) structural derivative of a parent compound. The word is also used to describe a molecule that may be structurally similar (but not identical) to another, and which exhibits many or some of the same biological functions of the other. For example, the large class of antibiotics known as the sulfa drugs are all analogues of the original synthetic chemical drug, Prontosil, which cures streptococcal infections (discovered by the German biologist Gerhart Domagk). All of the analogue sulfa drugs that were patterned after the original sulfanilamide molecular structure are also called SULFANILAMIDE ANALOGUES. Today, analogues are known for various vitamins, amino acids, purines, sugars, growth factors and many other chemical compounds. Research chemists produce analogues of various molecules in order to ascertain their biological function within a living organism.

analogy: (i) Similarity that is not due to homology (i.e. common ancestry). (ii) Reasoning by which the function of a novel gene or protein sequence may be deduced from comparisons with other gene or protein sequences of known function. Identifying analogous or homologous genes via similarity searching and alignment is one of the chief uses of bioinformatics. *See* ALIGNMENT and SIMILARITY SEARCH.

analysis of covariance (ANCOVA): Analysis of covariance is a tool for evaluating data sets that contain two variables: effect, which is referred to as the variate, and treatment, which is the categorical variable. The categorical variable emerges when one more variable, referred to as the covariate, arises.

analysis of DNA methylation patterns: The following methods are available for this purpose: 5-azacytidine treatment method; shotgun bisulfite sequencing method; degree, position and sequence context of methylation; histogram of methylation (HOME)-based method and RFLP.

analysis of molecular variance (AMOVA): A method for studying molecular variation within a species. A statistical model for the molecular algorithm in a single species, typically biological. The name and model are inspired by ANOVA. The method was developed by L. Excoffier, P. Smouse and J. Quattro at Rutgers University in 1992.

analysis of variance (ANOVA): A statistical method that allows the partitioning of the total variation observed in an experiment among several statistically independent possible causes of the variation, including treatment effects, grouping effects and experimental errors. The statistical test of the hypothesis that the treatment had no effect is the F-test or variance-ratio test. If the ratio of the mean square for treatments to the mean square for error exceeds a certain constant that depends on the respective degrees of freedom of the two mean squares at a chosen significance level, then the treatments are inferred to have been effective. ANOVA usually refers to an analysis of a continuous dependent variable where all the predictor variables are categorical.

analyte: The substance measured by a laboratory test; for instance, a specific mutation or allele.

analytical breeding: In crops such as potato, the genetic improvement through crosses and selection at the dihaploid level and then chromosome doubling of dihaploids.

analytical performance: A term encompassing analytic sensitivity, analytic specificity, assay robustness and assay reliability.

analytical specificity: The probability that a test will be negative when an analyte, or a mutation or an alteration is absent from a specimen. Usually expressed as a percentage. The following hypothetical example illustrates analytical sensitivity and specificity: using DNA samples from 100 known carriers of mutation B for disease A, a PCR/restriction test detects this mutation in 92 of the samples. Thus, the analytical sensitivity is 92%, with eight false negatives. In a parallel test using DNA samples from 100 known unaffected (no mutations, no family history) individuals, the same test detects mutations in five of the samples, in which it is known that there is no mutation. Thus, the analytical specificity is 95%, with five false positives.

analytical validity: The ability of a test to measure the property or characteristic that it was designed to measure (in the above example, specific mutations) by calculating such values as specificity and sensitivity.

anamorph (Adj. anamorphic): The asexual stage in the life cycle of a fungus, when asexual spores (such as conidia) or no spores are produced. *See* IMPERFECT STATE. *Contrasts* HOLOMORPH, TELEOMORPH and PERFECT STATE.

anandrous: Free of anthers, that is, an absence of male sexual organs.

anaphase: A stage that occurs once in mitosis and twice in meiosis; it involves the separation of chromosomal material to a given two sets of chromosomes, which will eventually form part of new cell nuclei. The separation of chromosome material is controlled by the spindle. In the anaphase of mitosis and anaphase II of meiosis, the centromere becomes functionally double and the daughter chromosomes separate from the equator, moving towards the opposite poles of the cell; the spindle then elongates and pushes the two groups of chromosomes further apart. In anaphase II of meiosis, the centromere does not divide; the centromere of each chromosome splits into two chromatids by the replication of its DNA. The two sister chromatids move to opposite poles. Anaphase ends when all the chromatids reach the opposite poles. Composed of ANAPHASE A (chromosomes move towards the two spindle poles) and ANAPHASE B (spindle poles move apart).

anaphase I: From each pair of homologous chromosomes, one chromosome comprising two *chromatids* joined by *centromere* move as a unit to one pole of a spindle and the remaining two chromatids joined by a centromere move to the opposite pole. So, during *anaphase I*, homologous chromosomes of each pair are separated. Division of the centromeres does not take place at this stage. Real reduction in the number of chromosomes occurs here.

anaphase II: The stage of meiosis II is characterized by the division of the centromeres and the separation of sister (single-stranded) chromosomes (individually called CHROMATIDS) from the spindle equator to the spindle pole. It closes with the end of the poleward movement of sister chromosomes. Here reduction of HAPLOIDY in terms of DNA content takes place.

anaphase movement: The movement of chromosomes and/or chromatids towards the opposite poles of the spindle apparatus during mitotic or meiotic anaphase.

anaphase separation: The disjunction of the chromatids of each chromosome during mitotic anaphase and anaphase II of meiosis or the separation of chromosomes in anaphase I of meiosis.

anaphase-promoting complex/cyclosome (APC/C): The anaphase-promoting complex/cyclosome (APC/C) is a multi-component E3 ubiquitin ligase that, by targeting protein substrate for 26S proteasome-mediated degradation through ubiquitination, coordinates the temporal progression of eukaryotic cells through mitosis and subsequent G1 phase of the cell cycle. Ubiquitin ligase endorses the destruction of a set of proteins, some of which initiate the separation of sister chromatids during the metaphase-to-anaphase transition during mitosis. Two APC/C components (APC5 and APC7) interact directly with the coactivators CBP and p300 through protein-protein interaction domains. Anaphase-promoting complex is a ubiquitin ligase, coupled to the Cdc20 adaptor protein, which becomes phosphorylated and thus activated by cyclin-CDK enzymes. Together with ubiquitin proteins, the activated APC transfers ubiquitin to several substrates – including cyclins A and B – whose destruction is necessary to complete DNA segregation.

ancestor: The form or stock from which an organism has descended or the actual or assumed earlier type from which a species or other taxon evolved.

ancestral characteristic: A trait present in an organism being studied; the common ancestor of a group to which the organism under study belongs.

ancestral CRISPR-Cas13 ribonucleases: The RNA-guided ribonuclease CRISPR-Cas13 empowers adaptive immunity in bacteria and programmable RNA operation in heterologous systems. Cas13s reveal in common limited sequence similarity, deterring discovery of related or ancestral systems. P.H. Yoon *et al.* (2024) established an automated structural-search pipeline to recognize an ancestral clade of Cas13 (Cas13an) and further trace Cas13 origins to defence-associated ribonucleases. In spite of being one-third the size of other Cas13s, Cas13an facilitates vigorous programmable RNA depletion and defence counter to assorted bacteriophages. Cas13an practices a single active site for both CRISPR RNA processing and RNA-guided cleavage, showing that the ancestral nuclease domain partakes two modes of activity. Detection of Cas13an increases chances for meticulous RNA editing, highlighting the potential of structure-guided genome mining.

ancestral homology: Homology that evolved before the common ancestor of a set of species, and which also occurs in other species outside that set of species. *Compare with* DERIVED HOMOLOGY.

ancestral: Originating prior in evolution to the derived state.

ancestry: Refers to a series of ancestors, ancestral descent, lineage gene and so on.

anchor gene: A gene that has been located on both the physical map and the linkage map of a chromosome.

anchor sequence: A hydrophobic amino acid sequence that fixes a segment of a newly synthesized, translocating protein within the lipid bilayer membrane of the endoplasmic reticulum.

anchorage-dependent cells or cultures: Eukaryotic cells, or cultures derived from them, which will grow, survive or maintain function only when attached to a surface such as glass or plastic. The use of this term does not imply that the cells are normal or that they are (or are not) neoplastically transformed.

anchored PCR: Anchored PCR allows amplification of full-length mRNA when only a small amount of sequence information lying within the mRNA is available.

anchoring junction: Type of cell junction that attaches cells to the neighbouring cells or the extracellular matrix.

ancient clones: Clones dating from a remote period (i.e. very old or aged). The ancient clones offer proof of the durability of horizontal resistance as such clones may date from centuries or even millennia ago. They are common in figs, olives, date palms, citrus, horseradish, garlic, ginger, turmeric, saffron, rhubarb and so on.

ancient DNA (aDNA): DNA isolated from ancient specimens. Due to degradation processes (including cross-linking, deamination and fragmentation), ancient DNA is more degraded compared to contemporary genetic material. Even under the best preservation conditions, there is an upper boundary of 0.4–1.5 million years for a sample to contain sufficient DNA for sequencing technologies. The oldest sample ever sequenced is estimated to be 1.65 million years old. Genetic material has been recovered from paleo/archaeological and historical skeletal material, mummified tissues, archival collections of non-frozen medical specimens, preserved plant remains, ice and from permafrost cores, marine and lake sediments and excavation dirt. In 2022, The New York Times reported a two-million-year-old genetic material discovered in Greenland, which is currently considered the oldest DNA discovered so far.

ancient DNA methylation reconstruction: A technique to reconstruct high-resolution DNA methylation from prehistoric DNA samples. The method is based on the natural degradation processes that take place in ancient DNA: with time, methylated cytosines are degraded into thymines, whereas unmethylated cytosines are degraded into uracils.

ancillary protein factors: Proteins that are needed for the initiation of transcription. These proteins recognize the sequence on a DNA close to, or overlapping with, the sequence bound by the RNA polymerase itself.

ancillary sites: The term applies to both prokaryotic and eukaryotic genes. The DNA sequences are located between the −30 and the −40 positions. In bacterial genes, it is known as cAMP-ACTIVATED PROTEIN (CAP) site. *Aka* −35 SEQUENCE.

androdioecious: A species bearing male and hermaphroditic flowers on separate individuals. *Datisca glomerata* (Presl.) Baill. (Datiscaceae) is the best-studied example of functional androdioecy in plants. *Contrasts* ANDROMONOECIOUS, POLYGAMODIOECIOUS, POLYGAMOMONOECIOUS *and* POLYGAMOUS.

androdioecy: The condition of a plant population or species in which both male (female sterile) and hermaphrodite individuals are found.

androecium: The male flower consists of stamens. Each stamen is known as a microsporophyll and differentiates into an anther. Each stamen has a bilobed anther, with each anther lobe producing four microsporangia through meiosis.

androecious: Applied to a plant that possesses only male flowers.

androecy: Applied to a plant that possesses only male flowers.

androgamy: Development of embryo neither from egg cells nor from synergid or antipodal cells, but from one of the male gametes itself outside or inside the embryo sac.

androgen: The general term for any substance acting as a male hormone.

androgenesis: Development of haploid embryo and then plant from male gametophyte, that is, microspore/pollen. The cultured anthers/microspores/pollen develop into pollen plants possessing gametophytic chromosome numbers, commonly called haploids.

androgenetic embryo: An embryo that contains two paternally derived sets of chromosomes but lacks maternally derived chromosomes.

androgenetic haploid: A plant possessing chromosomes from the male parent only; chromosomes from the female parent are absent.

androgenous: Producing only males in the progeny.

androgynophore: A stalk bearing both the androecium and gynoecium of a flower above the level of insertion of the perianth.

androgynous: Having male and female flowers in the same inflorescence.

andromonoecious: Having hermaphroditic and male flowers on the same individual plant. *Contrasts* ANDRODIOECIOUS, GYNOMONOECIOUS, POLYGAMODIOECIOUS, POLYGAMOMONOECIOUS *and* POLYGAMOUS.

andromonoecism: Plants bearing only staminate and perfect hermaphroditic flowers.

andromonoecy: The condition of an individual plant or species having individuals that bear both male (i.e. female sterile) and hermaphrodite flowers.

androphore: The stalk or column supporting the stamens of certain flowers usually formed by a union of the filaments (e.g. in plants of the Leguminosae family).

androsome: Any chromosome that is present solely in the nucleus of the male.

anemophilous: Adapted to pollination by wind.

anemophily: Pollination of a flower in which the pollen is carried by the wind, as in grasses.

aneucentric translocation: A translocation involving the centromere so that they result in an acentric chromosome and a dicentric chromosome.

aneucentric: The term is applied to an aberrant chromosome possessing more than one centromere.

aneuhaploid: When the chromosome number diverges from the haploid standard chromosome number of the species or the individual.

aneuploid reduction: Reduction of the genetic variability by reducing the chromosome number.

aneuploid: An individual with a chromosome number other than an exact multiple of a basic number. In aneuploids, only a part of the basic chromosome number, that is, one or few chromosomes of the genome are added or subtracted like $2n-1$, $2n-2$, $2n+1$ and $2n+2$. An aneuploid originates from the non-disjunction of one or more pairs of homologous chromosomes. Aneuploids are used in basic studies dealing with gene location. *Aka* HETEROPLOID.

aneuploidy: The occurrence of one or more extra or missing chromosomes leading to an unbalanced chromosome complement, or any chromosome number that is not an exact multiple of the haploid number. Aneuploidy leads to various genetic disorders in humans. Aneuploidy has been consistently observed in virtually all cancers. A growing body of evidence suggests that aneuploidy is often caused by a particular type of genetic instability, called CHROMOSOMAL INSTABILITY. Aneuploids may involve, in a disomic (2n) individual, deletion of one chromosome ($2n-1$, MONOSOMIC), deletion of two non-homologous chromosomes ($2n-1-1$, DOUBLE MONOSOMIC) or deletion of a pair of homologous chromosomes ($2n-2$, NULLISOMIC). Aneuploids may also arise due to the addition of one chromosome ($2n+1$, TRISOMIC), the addition of two non-homologous chromosomes ($2n+1+1$, DOUBLE TRISOMIC) or the addition of a pair of homologous chromosomes ($2n+2$, TETRASOMIC). Trisomics have been extensively utilized in gene location programmes. Nullisomics and monosomics are particularly useful in pinpointing newly found genes on specific plant chromosomes. *See* MONOSOMICS, NULLISOMICS, TRISOMICS *and* TETRASOMICS.

aneusomatic: Individuals whose cells display variable numbers of chromosomes.

aneusomic: Individuals or cells that possess unequal sets of individual chromosomes.

angiogenesis: Production of new blood vessels from existing ones through tissue engineering. Angiogenesis is being tried to promote the growth of new blood vessels that bypass blockages in arteries. Growth-promoting molecules produced through recombinant DNA technology are being tried in tissue engineering and angiogenesis. Small-scale trials have also been conducted on human beings.

angiosperm: Any vascular plant of the phylum having the seeds enclosed in a fruit, grain, pod or capsule and comprising all flowering plants. They produce seeds bound by fruit (an ovary). There are over 250,000 angiospermous species where flowers are used in reproduction. Angiosperms are divided into two groups: monocots (grasses like wheat and maize) and dicots (beet, beans, etc.). *Aka* ANGIOSPERMOUS.

angiosperms, rise of the: So far, the angiosperm tree of life has been constructed primarily by analysing the plastid genome. However, the inadequate and prejudiced sampling of both taxa and genomes weakens the sureness of the tree and its implications. A.R. Zuntini *et al.* in 2024 have built the tree of life for almost 8000 (~60%) angiosperm genera employing a consistent set of 353 nuclear genes. This 15-fold rise in genus-level sampling relative to comparable nuclear studies offers a judgemental test of previous observations and conveys noteworthy alteration to key groups (especially in rosids) while corroborating many earlier projected relationships. Ascending this tree through time using 200 fossils, they revealed that initial angiosperm evolution was based on high gene tree struggle and explosive divergence, producing more than 80% of the existing angiosperm orders.

Ångstrom (Å): A unit of length equal to 10^{-10} m, 10^{-8} cm or 3.937×10^{-9} inch, earlier used to measure wavelengths and intermolecular distances which has now been replaced by the nanometre (1 Å = 0.1 nm); the unit is named after Swedish pioneer of spectroscopy, A.J. Ångström.

anhydride: A compound formed by eliminating water from a more complex compound.

Animal and Plant Health Inspection Service of USDA, USA (APHIS): Regulations for field trials and releases were first developed and monitored by APHIS in USA. APHIS emphasized the need for a case-by-case and step-by-step evaluation of transgenic products. In 1993, APHIS cleared six transgenic crops (soybean, potato, tomato, maize, cotton and tobacco) for field trials.

anion: An ion with a negative charge. This is the state of an atom in which its outer electron shell is more than half-filled and the atom gains electrons to form a negatively charged ion, commonly called an anion. Examples include chloride ions (Cl^-) and iodide ions (I^-).

anion-exchange method of DNA extraction: Anion-exchange resin uses the anion-exchange principle. It is based on the interaction between positively charged diethylaminoethyl (DEAE) cellulose groups on the resin's surface and negatively charged phosphates of the DNA backbone. The anion-exchange resin comprises defined silica beads with a large pore size, a hydrophilic surface coating and a high charge density. The large surface area of resin permits dense coupling of the DEAE groups. The resin works over a wide range of pH conditions (pH 6–9) and/or salt concentrations (0.1–1.6 M), which can augment the separation of DNA from RNA and other impurities.

anisogametes: Male and female gametes that differ in size, appearance, structure or sex chromosome content from the gamete of the opposite sex (usually known as SPERM and EGG).

anisogamy: Having gametes of different sizes, large eggs and small sperm, fusing during fertilization.

anisogeny: The property of having different inheritance in reciprocal crosses due to cytoplasmic inheritance.

anisomeric: Non-equivalent genes that cooperate to produce particular phenotypes.

anisomery: The condition of having a floral whorl with a different (usually smaller) number of parts compared to the other floral whorls.

anisoploid seeds: A mixture of seeds of different ploidy levels. For example, in different sugarbeet varieties, seeds may be 2×, 3× or 4×.

annealing: The process by which two complementary DNA strands spontaneously re-pair after denaturation to form a double helix. *See* HYBRIDIZATION.

annealing DNA: In molecular genetics, heating causes the separation of the individual strands of any double-stranded nucleic acid helix and cooling leads to the pairing of any molecules that have segments with complementary base pairs; the term annealing is synonymous with hybridization.

annealing temperature (Ta): Ta chosen for PCR relies directly on the length and composition of the primers. Generally, an annealing temperature of about 5°C below the Tm of primers should be used. The optimal annealing temperature (Ta Opt) for a given primer pair on a particular target can be calculated as follows: Ta Opt = 0.3 × (Tm of primer) + 0.7 × (Tm of product) − 14.9, where the Tm of primer is the melting temperature of the less stable primer-template pair, and the Tm of product is the melting temperature of the PCR product.

annotation: The process of identifying the locations of genes in a genome and determining the role of those genes. Annotation is a process that often follows gene sequencing. It is a collection of comments, notations, references and citations, either in free format or precise vocabulary, which describes all the experimental and inferred information about a gene or protein. Annotations are also used to describe other biological systems. Batch, automated annotation of bulk biological sequence, is one of the main uses of bioinformatics tools. Analysis and commentary are added to sequence data in databases. Annotation offers in-context information about coding and non-coding sequences within genes, patterning and motifs, similarities, known or predicted protein structure and function, as well as links to external data, such as morphological comments. It also provides the prediction of genes in a genome, including the location of protein-encoding genes, the sequence of coded proteins, any significant matches to other proteins of known function and the location of RNA-encoding genes. Predictions are based on gene models (e.g. HIDDEN MARKOV MODELS (HMMs) of introns and exons in protein-encoding genes) and models of secondary structure in RNA. *See* BIOINFORMATICS.

annotation software: The priority of any annotation software is to pinpoint the genes by looking for characteristic sequences at the beginning or ends of genes or by comparing new sequences to known genes. Additional computer programs translate those genes into proteins and based on their similarities to other proteins, assign a function to each gene. The National Centre of Biotechnology Information (NCBI)'s BLAST Program compares the new genome data to that from other organisms, such as fruit flies or nematodes.

annotator: An interactive genome annotation tool.

annouline: Refers to the fluorescent protein pigment leaked out by the roots of, for example, ryegrass seedlings. The fluorescent nature of this material is valuable in distinguishing annual and perennial ryegrass seeds.

annual plant: A plant that concludes its life cycle and dies within 1 year. *Contrasts* BIENNIAL *and* PERENNIAL. *Compare with* UNIVOLTINE, PERENNIAL *and* MULTIVOLTINE.

annual ring: In woody plants, the layer of wood formed each year that can be realized when the wood is cut into a cross-section. The number of rings equals the age of the tree.

annulations: The furrowed bands in the cuticle of some nematodes.

annulus: (i) Pappus bristles are sometimes attached to a ring called an annulus or a disk at the top of the achene beak. (ii) In some pollen grains, the exine around the apertures is either thicker or thinner. In pores, this border is termed an annulus. (iii) Certain flowers have ring-like constrictions at the mouth of the flower (e.g. in Huernia and Aristolochia). (iv) A ring of specialized cells on the sporangium. (v) Also applied to any of several ring-shaped nuclear pores.

anomaly: A deviation from the normal.

anomers: A type of stereoisomer and epimer found in carbohydrate chemistry. For example, α-D glucose and β-D glucose are anomers.

anonymous DNA marker: A DNA marker noticeable by variation in its sequence, irrespective of whether or not it occurs in or near a coding sequence. Microsatellites are examples of typical anonymous DNA markers.

anonymous single nucleotide polymorphism (anonymous SNP): Any one of the most frequently existing single nucleotide polymorphisms having no known effect on the function of a gene.

antagonism: A phenomenon of active opposition where one organism does not permit another organism to grow or one drug decreases the effectiveness of the other drug. (i) Microbial antagonism involving competition among different microbes is a common spectacle; (ii) In agriculture, *Trichoderma* is used as a biocontrol agent as it is antagonistic to several other pathogenic fungi; (iii) In the case of drugs, concurrent use of Penicillin and Tetracycline is seen to be less effective compared with when either drug is used alone. *Contrasts* SYNERGISM.

antagonist: An organism or substance that confines or counteracts the action of another.

antagonistic pleiotropy: Pleiotropy occurs when a single gene influences several phenotypic traits that are seemingly unrelated. In the case of antagonistic pleiotropy, expression of the pleiotropic gene has mixed, competing effects; some of these are beneficial but others are detrimental to the organism.

antenatal diagnosis: Diagnosis of a condition before birth.

antenna complex: Part of a photosystem that arrests light energy and channels it into the photochemical reaction centre. It consists of protein complexes that fix large numbers of chlorophyll molecules and other pigments.

antennaria type: Refers to mitotic diplospory where the megaspore mother cell does not enter meiosis.

antephase: *See* PROPHASE.

anterior: Positioned in front of, towards the head end of the body. *Compare with* DISTAL, POSTERIOR.

anterograde regulation: The regulation which occurs or moves in the normal or forward direction of conduction or flow. It coordinates expression levels of nuclear genes encoding photosynthetic proteins and plastid gene expression machineries.

anther culture: Culturing of anthers containing pollen or of single pollen grains. This method is used for obtaining the haploid plants which are subsequently used for the production of doubled haploids (which are homozygous at all the loci) after spontaneous or induced rediploidization, or for breeding on a lower ploidy level.

anther: The name for the terminal portion of a stamen of a flowering plant. The pollen sacs containing pollen are formed on the anther. The number of anthers in a flower varies from 3 to 10 among most species.

anthesis: The period during which flower pollen is presented and/or the stigma is receptive.

anthocyan: *See* ANTHOCYANIN.

anthocyanins: Anthocyanins are naturally occurring glycoside compounds that impart colour to fruit, vegetables and plants. They are thought to play a major role in the high antioxidant activity levels. In acidic conditions, anthocyanin appears as a red pigment while blue pigment anthocyanin exists in alkaline conditions. More than 200 anthocyanins have been discovered.

anthropoid: A member of the group of primates comprising monkeys, apes and humans.

anti-σ factor: Protein that prevents recognition of transcription initiation sites by σ factor of RNA polymerase. An anti-σ factor is synthesized during phage T4 infection.

antiauxin: A chemical that inhibits the auxin response. It may or may not involve the prevention of auxin transport or movement in plants. Some antiauxins, such as 2,3,5-tri-iodobenzoate or 2,4,5-trichlorophenoxy-acetate, may promote morphogenesis *in vitro* and *in vivo*, simulating the growth of some *in vitro* cultures.

antibacterial: Having the ability to kill bacteria.

antibiosis: An association between organisms, or between an organism and a metabolic product of another organism, that is harmful to one of them. For example, certain varieties of corn/maize (*Zea mays* L.) naturally produce chemical substances in their roots that are toxic to the corn rootworm.

antibiotic: Any of a large group of chemical substances, such as penicillin or streptomycin, produced by various microorganisms and fungi, having the capacity (in dilute solutions) to inhibit the growth of or to destroy bacteria and other microorganisms, used chiefly in the treatment of infectious diseases. This term originated from the term antibiosis and was coined by S. Waksman during the 1940s. It has a defensive function and is often toxic to other species. For example, penicillin, originally produced by bread mould, is toxic to numerous human pathogens. Antibiotics generally act by inhibiting protein synthesis, DNA replication, synthesis of cell wall (cytoplasmic membrane) constituents, inhibition of required cell (e.g. bacteria) metabolic processes and nucleic acid (DNA and RNA) biosynthesis, hence killing the (targeted bacteria) cells involved. More than half of the antibiotics in practice are produced by a species of *Streptomyces*. Few antibiotics are produced by the bacterial genus *Bacillus* and others by moulds like *Penicillin* and *Cephalosporium*. Some antibiotics are also toxic to humans.

antibiotic resistance: (i) A heritable trait in microorganisms that empowers them to survive in the presence of an antibiotic; (ii) The ability of bacteria to tolerate an antibiotic and survive being exposed to it. Bacteria may develop this resistance naturally after being exposed to it over many years. (iii) The property of a pathogenic bacteria that allows it to avoid the effect of an antibiotic that had formerly killed or inhibited that cell.

antibiotic resistance gene: A gene that encodes an enzyme that degrades or excretes an antibiotic, thus conferring resistance to that organism against that antibiotic.

antibiotic resistance, occurrence of: Different ways by which antibiotic resistance can occur include: (i) altering the structure of the cell wall (plasma membrane); (ii) synthesis (manufacture) of enzymes to incapacitate the antibiotic (e.g. penicillinases, which inactivate penicillin); (iii) synthesis of enzymes to thwart the entry of an antibiotic into the cell and (iv) active elimination of the antibiotic from the cell. For example, the membrane transporter protein molecules, known as ABC TRANSPORTERS, sometimes aid pathogenic bacteria in resisting certain antibiotics by transporting the antibiotic out before it can kill the bacteria. The ABC transporter is a V-shaped molecule embedded in the bacterial cell's plasma membrane, with the open end of the 'V' pointed towards the interior of the cell. When molecules of certain antibiotics inside the cell contact the ABC transporter molecule, the two 'arms' of the ABC transporter close around the antibiotic molecule. The ABC transporter somersaults over and sends the antibiotic molecule out through the exterior of the cell's plasma membrane, replacing some critical cell metabolic processes with new metabolic processes that bypass the antibiotic's former effect.

antibiotic susceptibility test (AST): A test that governs which antimicrobial drug can successfully inhibit the growth of specific bacteria.

antibiotics as mutagens: Antibiotics, like actinomycin D, mitomycin, aflatoxin B and streptomycin, have chromosome-breaking properties. They affect the biosynthesis of DNA, RNA or proteins. Antibiotics can also enhance or reduce mutagenic effects of other mutagens depending upon the physiological state of the cell.

antibodies: Y-shaped proteins formed by specialized white blood cells (called B lymphocytes or B cells) to recognize and neutralize foreign objects like bacteria, viruses and pollen. When an antibody detects this antigen, it initiates an immune response to rid the body of the antigen. Antibodies produced by the body's immune defence system can bind to foreign molecules and eliminate them. A defence protein is produced in response to the presence of a specific antigen.

antibodies to histone post-translational modification (PTLM): Much of the advancement in our understanding of chromatin modifications and chromatin structure would not have been possible without the development of post-translational modification-specific antibodies. CST H3K27me3 antibody for CUT&RUN and immunocytochemistry and Active motif H3K27ac antibody has been used on human oocytes or embryos. A Histone Antibody Specificity Database has also been prepared.

antibodies, mechanism of action of: The plasma cells are triggered by the foreign molecule's epitope(s) [i.e. group or groups of specific atoms (*aka* a HAPTEN), that are recognized as foreign by the body's immune system] into producing antibody molecules possessing antigen-binding (ab) sites (also called COMBINING SITES or DETERMINANTS). These fit into the foreign molecule's epitope. Thus, via the tips of its arms, the antibody molecule binds specifically to the foreign entity (antigen) that has entered the body. It inactivates that foreign molecule or marks it for eventual destruction by other immune system cells. System marking of the foreign molecule (e.g. pathogen or toxin) for destruction is accomplished by the fact that the stem of the Y (i.e. the Fc) fragment hangs free from the combined antibody-antigen clump, thereby providing a receptor for phagocytes, which roam throughout the body ingesting and subsequently destroying such 'marked' foreign molecules. Antibodies may also kill some pathogens themselves by catalysing the formation of hydrogen peroxide (H_2O_2) from oxygen free radicals (singlet oxygen; formula O=O; also written as $1[O_2]$) and water. H_2O_2 is highly reactive and could potentially kill pathogens when generated by an (attached) antibody.

antibodies, types of: There are five classes of immunoglobulin: IgG, IgM, IgD, IgA and IgE. *Aka* IMMUNOGLOBULIN (IG). *See details as separate terms.*

antibody affinity chromatography: A type of chromatography in which antibodies are immobilized on to the column material. The antibodies bind to their target molecules, while the other components in the solution are not retained. In this way, the purification of antibodies is achieved.

antibody diversity: Organisms have the ability to synthesize millions of different types of immunoglobulins to defend the body against a large number of foreign substances (antigens) that may otherwise harm the body. DNA rearrangements in B lymphocytes hold the key to antibody diversity.

antibody recognition: Certain antibodies can very precisely detect specific histone modifications and can be used to interrogate their positions on the genome. Histone modifications H3S10ph, H3K9me2, H3K27me3, H3K4me3, H3K4me3, H3K9me3, H3K27me3, H3K36me2 and H3K36me3 have been detected through antibody recognition.

antibody structure: This large defence protein consists of two classes of polypeptide chains, LIGHT (L) CHAINS and HEAVY (H) CHAINS. A single antibody molecule consists of two identical copies of the L chain and two of the H chains. They are made by the immune system (B lymphocytes) of the organism. The antibody is composed of four proteins linked together to form a Y-shaped bundle of proteins. The amino acid sequence that makes up the stem (heavy chains) of the Y (i.e. the handles of the taped-together hockey sticks) is similar for all antibodies. The stem is known as the FC REGION of the antibody, which does not bind to antigens but has other regulatory functions. The two arms of the Y are each made up of two side-by-side proteins called light chains and heavy chains (proteins are the chains of amino acids), with identical antigen-binding (ab) sites on the tips of each 'arm'. The antibody is thus bivalent as it has two binding sites for antigen. Taken together, the two arms of the Y are known as the FAB PORTIONS of the antibody molecule, which can be cleaved from the antibody molecule with papain (an enzyme also used as a meat tenderizer) or the Fab portions can be produced by genetically engineered *E. coli*.

antibody titre: It is the amount of antibody in the serum. This is a commonly used method to check the intensity of humoral response. Just after an infection, the serum does not contain detectable antibodies but with some time, there is a sluggish rise in antibody titre; at first, IgM is produced followed by IgG. Further, a steady decrease in antibody titre occurs in this primary infection. Thus, immunoglobulins are produced in higher concentrations after a secondary exposure to an antigen.

antibody-based screening (ABS): ABS is used where functional relationships can be explored by using individual proteins directly as an assay for their building partner in biological samples coated within microwell plates using ELISA test techniques.

antibody-mediated capture and massively parallel sequencing (AMC-MPS): AMC-MPS is perhaps the first protocol to allow a global view of this ubiquitous RNA modification, and it is based on antibody-mediated enrichment of methylated RNA fragments followed by MSP.

antibody-specific methods for RNA modification analysis: Bovine Serum Albumin (BSA), a basic protein possessing many free $-NH_2$ groups, is used as a suitable carrier. Nucleosides are conjugated to carrier proteins. Antibodies enable an unbiased determination of sequences flanking modified nucleotides and their potential consensus motifs. 6-methyl adenosine ($6^{m}A$), 5-methylcytosine ($5^{m}C$), 1-methyladenosine ($1^{m}A$) and 7-methyl guanosine ($7^{m}G$) RNA modifications were thus analysed.

anticoding strand: Refers to the single strand of double-helix DNA that the RNA polymerase uses for transcription. *Aka* ANTISENSE STRAND, TRANSCRIBING or TEMPLATE STRAND. *Compare with* CODING STRAND.

anticodon: The sequence of three nucleotides in a transfer RNA (tRNA) that is complimentary to a codon in the messenger RNA (mRNA). For example, a specific sequence of three nucleotides (CCC) in a tRNA, called anticodon, is complementary to three nucleotides (GGG) for an amino acid (glycine) in a mRNA molecule, called codon. During protein synthesis, base pairing between a codon and an anticodon makes sure that the tRNA carries the corresponding amino acid in a line for addition to the growing peptide chain.

anti-CRISPRs (Acrs): C. Hu *et al.* in 2024 used four cryoelectron microscopy snapshots to define its RNA-guided DNA-binding and cleavage mechanisms in high resolution. The non-target DNA strand (NTS) is accommodated by I-C Cascade in a continuous binding groove along the juxtaposed Cas11 subunits. The binding of Cas3 further traps a flexible bulge in NTS, enabling NTS nicking. They identified two anti-CRISPR proteins (AcrIC8 and AcrIC9) that strongly inhibit *Neisseria lactamica* I-C function. AcrIC8 inhibits PAM recognition through allosteric inhibition, whereas AcrIC9 achieves so through direct competition. Both Acrs potently inhibit I-C-mediated genome editing and transcriptional modulation, providing the first off-switches for type I CRISPR eukaryotic genome engineering.

anti-CRISPR AcrIF25: C.N. Trost *et al.* (2024) describe the anti-CRISPR AcrIF25 and demonstrate that it inhibits the type I-F CRISPR–Cas system by splitting the wholly assembled effector complex. AcrIF25 binds to the principal CRISPR RNA-binding components of this complex, comprising six Cas7 subunits, and removes them from the RNA. AcrIF25 separates one Cas7 subunit at a time, beginning at one end of the complex. Notably, this is accomplished with no apparent enzymatic activity. AcrIF25 is the first protein known to strip a large and stable macromolecular complex in the non-attendance of an external energy source. AcrIF25 founds a paradigm for macromolecular complex inhibitors that may be employed for biotechnological applications.

anti-fuel: A molecular input able to displace the fuel (such as a DNA strand complementary to the fuel strand).

antigen: A foreign substance (a protein, polysaccharide, lipids or nucleic acid) that binds to an antibody and starts an immune response in the body of a vertebrate organism. Antigens bind to antibodies to make antigen-antibody complexes. Also called an IMMUNOGEN. Various sources of antigens include microbes, pollen, white of an egg, serum proteins and surface molecules of transplanted tissues and organs.

antigen-antibody complex: *See* ANTIGEN.

antigen-presenting cells (APCs): The cells that show foreign antigen complex with major histocompatibility complex (MHC) molecules on their surface. PROFESSIONAL APCs include dendritic cells, macrophages and B cells. NONPROFESSIONAL APCs function in antigen presentation for only brief periods include thymic epithelial cells and vascular endothelial cells.

antigene RNA (agRNA): Small RNAs complementary to sequences in the promoter region of mammalian genes that can repress or activate gene expression.

antigenic determinant: Also called an epitope, an antigenic determinant is a specific part of an antigen that is recognized by the immune system, specifically by antibodies and B and T cells. Other immune cells like ANTIGEN-PRESENTING CELLS (APCs) cannot recognize epitopes [PATHOGEN-ASSOCIATED MOLECULAR PATTERNS (PAMPs) and DAMAGE-ASSOCIATED MOLECULAR PATTERNS (DAMPs)]. The interaction between the antigen and antibody depends upon the shape, size and chemical nature of the antigenic determinant as well as the chemical structure of the binding site on the antibody. *Aka* EPITOPE.

antigenic receptors: An antibody-type molecule on B and T cells that enables them to recognize and bind with human antibodies. Antibodies are produced by a special group of lymphocytes – activated B cells divide and differentiate into effector cells called plasma cells, which finally produce antibodies. B cells develop from stem cells and migrate to lymphoid organs such as the spleen. B cells recognize antigens using the antigen receptors present on their cell surfaces.

antigenic variation: Refers to the ability to change the antigens displayed on the cell surface. This is the property of some pathogenic microorganisms that empowers them to escape attack by the immune system.

antihuman immune serum globulin (anti-HISG): An antibody that interacts specifically with human antibodies. For the antigen-antibody complex to be seen in a test, fluorescein-levelled antihuman immune serum globulin (anti-HISG), an antibody that reacts specifically with any human antibody, is added to the slide. Slides carrying known antigens are fixed on it. Anti-HISG will be present only if the specific antibody reacts with its antigen. If the known antigen fixed to the slide emits fluorescence, the antibody specific to the test antigen is present.

antileader: A nucleotide sequence in DNA upstream of the initiation codon of a gene, which is transcribed but not translated. Antileader generally contains the Shine-Dalgarno sequence (L. Shine and J. Dalgarno in 1973), which helps in binding of the mRNA to the ribosome.

antimessenger DNA (amDNA): A DNA strand copied from messenger RNA by RNA-dependent RNA polymerase in a process called REVERSE TRANSCRIPTION. *Aka* ANTISENSE DNA.

antimetabolite: A substance that resembles some naturally occurring compounds in chemical structure and which specifically antagonizes the biological action of such compounds.

antimicrobial drug: A drug that kills microbial pathogens without harming host body cells. A variety of chemical agents are used to regulate the growth of microorganisms. But no single disinfectant is appropriate under all circumstances. Commonly used chemicals include phenol and phenolics, biguanides, halogens, alcohols, heavy metals and their compounds, surfactants, quats, antibiotics and gaseous chaemosterilizer (such as ethylene oxide) and oxidizing agents (such as H_2O_2). *Aka* ANTIMICROBIAL AGENT.

antimicrobial resistance (AMR): An umbrella term used to describe the diverse range of mechanisms that microbes (like bacteria, fungi, viruses) and other parasites use to survive the negative effects of drugs like antibiotics, antivirals, antifungals and antimalarials.

antimitotic: Substances that may cause the termination of mitosis.

antimorph: A dominant mutant allele that acts in a direction opposite to the normal allele. Antimorphs are also called DOMINANT NEGATIVE MUTATIONS.

antimutagens: Non-toxic and non-mutagenic substances that can reduce the rate of mutations. A. Novick and L. Szilard primarily applied this term in 1952 to agents possessing the ability to diminish the rate or frequency of spontaneous mutations but not to chemically induced mutations. Antimutagens are of two types: desmutagens and bioantimutagens. *See* DESMUTAGENS and BIOANTIMUTAGENS.

antimutator gene: A gene, allele or genotype that reduces the spontaneous mutation rate of a gene.

antimutator mutations: Mutations in DNA polymerase that decrease the overall mutation rate of a cell or an organism.

antioncogenes: Those genes whose protein products can avert tumourigenic transformation of cells. These genes act by suppressing malignant growth. *Aka* TUMOUR-SUPPRESSOR GENES (TSGs) *or* RECESSIVE ONCOGENES.

antioxidants: Antioxidants are substances that may protect cells from the damage caused by unstable molecules. Antioxidants interact with and stabilize free radicals and may prevent some of the damage free radicals otherwise might cause, which includes cancer. Examples of antioxidants include beta-carotene, lycopene, vitamins C, E and A and other substances. *Aka* FREE RADICALS.

antiparallel β sheet: A β sheet found in the polypeptide chain of a protein, in which each strand runs in the direction opposite to its immediate neighbours.

antiparallel orientation: The normal arrangement of the two strands of a DNA molecule, and of other nucleic acid duplexes (DNA-RNA, RNA-RNA), in which the two strands are oriented in opposite directions so that the 5′-phosphate end of one strand is aligned with the 3′-hydroxyl end of the complementary strand. The two strands showing antiparallel orientation are called ANTIPARALLEL STRANDS.

antipodal cells: Three of the eight nuclei that result during megasporogenesis by mitotic cell divisions within the developing megagametophyte (embryo sac). These nuclei are usually located at the base of the embryo sac. They may be playing some role in regulating the act of fertilization and double fertilization.

antiporter: Carrier protein that transports two different ions or small molecules across a membrane in opposite directions, either simultaneously or sequentially.

antirepressor for lytic cycle: Gene *cro* codes for a protein (9000 Da) which forms a dimer and acts on promoters P_M (for *cI* gene), P_L (for *N* gene) and P_R (for *cro* gene) itself, so that it stops the expression of repressor gene *cI* as well as that of the early genes, including *N* on the left side and *cro* on the right side. If *cI* is expressed, this will stop the expression of late genes and this will establish lysogeny.

***Antirrhinum* petal asymmetry:** Petal asymmetry in the petal lobe of Snapdragon *Antirrhinum* depends on the direction of growth rather than regional differences in the growth rate. The direction of growth is maintained parallel to the proximodistal axis of the flower, irrespective of changes in shape, implying that long-range signals orient growth along the petal as a whole. Such signals may provide a general mechanism for orienting growth in other growing structures.

antisense: The non-coding DNA strand of a gene. In a cell, antisense DNA serves as the template for producing mRNA, which directs the synthesis of a protein.

antisense and sense strands in double-stranded DNA with respect to a gene: RNA polymerase antisense strand (*aka* TEMPLATE, CODING, TRANSCRIBED or WATSON STRAND, denoted as 3′→5′) sequence of double-stranded DNA is transcribed into RNA (in 5′→3′ direction). Because the first product of a gene is an RNA whose sequence is complementary to the coding strand of the DNA, it is customary to represent the gene by a complementary strand (*aka* SENSE, NON-TEMPLATE, NON-CODING or CRICK STRAND, denoted as 5′→3′) of the duplex DNA. A gene in a double-stranded DNA molecule is located in the antisense strand, but in literature, it is usually represented by the sense strand (in 5′→3′ direction) because its sequence is the same as that of its primary transcript except where there is thymine in DNA there is uracil in RNA.

antisense DNA: A strand of DNA that produces a messenger RNA (mRNA) molecule, which (when reversed end-for-end) has the same sequence as (is complementary to) the unwanted ('bad') messenger RNA. The sense (forward) and antisense (backward) mRNA strands hybridize (are tightly bound to each other), which prevents the bonded pair from leaving the cell's nucleus, so that the bonded pair is rapidly degraded (destroyed) by the nuclei within the cell nucleus. In genetic targeting (to block 'bad' genes), antisense molecules are used to bind to a 'bad' gene's (an oncogene) messenger RNA (mRNA), thus cancelling the (cancer-causing) message of the gene and preventing cells from following its (tumour growth) instructions. Another example would be the use of antisense DNA to block the gene that codes for production of polygalacturonase (an enzyme that causes ripe fruit to soften). Physically, antisense is accomplished by removing a given gene from an organism's genome, reversing it (end-for-end) and reinserting it back into the organism's genome. *Aka* CODING STRAND, TRANSCRIBING STRAND, TEMPLATE STRAND or WATSON STRAND. *Compare with* SENSE DNA STRAND.

antisense gene: A gene construct positioned in inverted orientation relative to a promoter. When it is transcribed, it forms a transcript complementary to the mRNA transcribed from the normal orientation of the gene.

antisense inhibition: A process where the expression of a gene is repressed with the help of an antisense RNA.

antisense oligomers: Single-stranded chains of nucleotides that bind to a specific sequence of RNA, thereby preventing its genetic information from being expressed.

antisense oligonucleotide: A short synthetic nucleic acid sequence that is complementary to an mRNA sequence. Through a variety of mechanisms, antisense oligonucleotides can be used to inactivate gene function by interfering with the ability of an mRNA to produce its corresponding protein product.

antisense post-transcriptional gene silencing (antisense-PTGS): Unintended transcription could produce antisense RNA molecules that participate in the initiation of PTGS triggered by sense transgenes.

antisense RNA (asRNA): According to recent terminology, RNA that is complementary to mRNA. An antisense RNA binds to a naturally transcribed sense mRNA molecule, thus blocking its translation.

antisense RNA (asRNA) approach: Antisense RNA can be produced by reverse transcribing a cDNA copy of an mRNA with respect to the promoter in an expressed vector. This yields a full-length complementary copy of the mRNA sequence. Fragments smaller than full length can also be effective. Antisense RNA molecules are thought to interact with mRNA molecules by base pairing to form double-stranded RNA. Transgenic tobacco plants expressing antisense RNA of Cytomegalovirus (CMV) coat protein gene showed reduced virus accumulation, and prevention of systemic spread of low concentrations of CMV inoculums.

antisense RNA (asRNA) in mRNA export: Precursors of messenger RNAs are transcribed and processed in the nucleus. The processed mRNAs are then transferred to the cytoplasm for translation. The mRNA export is arbitrated by the export receptor heterodimer Mex67-Mtr2 in the yeast *S. cerevisiae* (TAP-p15 in humans). Many long non-coding RNAs (lncRNAs) are known to leave the nucleus. I. Coban *et al.* in 2024 demonstrated that asRNAs hastens mRNA export by binding with its sense counterparts using the helicase Dbp2. These double-stranded RNAs (dsRNAs) govern export compared with single-stranded RNAs (ssRNAs) because they have a higher competence and attraction for the export receptor Mex67. Thus, asRNAs enhance gene expression, which is useful for cells.

antisense RNA (asRNA) inactivation: Antisense RNAs (asRNAs), naturally occurring single-stranded transcripts, can bind their cognate sense RNA (messenger RNA) by sequence complementarity and inhibit mRNA translation to protein.

antisense RNA (asRNA) technology, advantages of: Using this technology, transgenic tomatoes with prolonged shelf life have been obtained through plant genetic engineering. An enzyme, polygalacturonase (PG), is associated with the softening of tomatoes with ripening which is not good for long-time storage. In order to prevent this, transgenic tomatoes have been produced. The antisense gene is produced by inverting the orientation of coding region in relation to promoter, and the antisense gene is then introduced to develop transgenic plants. The presence of endogenous gene and antisense gene produce antisense and sense RNAs which are complementary to each other and thus pair as double-stranded RNA – the mRNA is thus not available for translation leading to the silencing of normal gene. Translation can be regulated by RNA-RNA hybridization. Such antisense RNA that has a regulatory role is *aka* MRNA-INTERFERING COMPLEMENTARY RNA (MICRNA).

antisense RNA (asRNA) technology: Since antisense RNA is complementary to mRNA, they will produce a duplex which will hinder its translation. This technology involves cloning of a particular gene in reverse orientation, with respect to the promoter such that the coding strand acts as a template and the sequence of the mRNA is the same as the coding sense strand. The antisense gene, when transcribed, gives rise to the mRNA, thus inhibiting the target gene expression by forming an RNA duplex. The antisense RNA binds to the complementary mRNA and thwarts its transport to cytoplasm or translation into protein. The antisense strategy has been used by plant breeders to block the synthesis of amylose starch in potatoes.

antisense strand (or primer): Viral RNA that cannot act as mRNA. Polio virus is a single-strand RNA virus. The RNA within the virion is called a sense strand (or + strand) and it can act as an mRNA. After attachment, an enzyme called RNA-dependent RNA polymerase is formed that catalyses the synthesis of another RNA strand complementary to the original strand. The new strand is called an antisense strand (or – strand) that serves as a template to produce additional sense strands.

antisense therapy: *In vivo* treatment of a genetic disease by blocking the translation of a protein with a DNA or an RNA sequence that is complementary to a specific mRNA.

antisense transcript stabilization: There is extensive transcription throughout the eukaryotic genome resulting in both antisense transcripts from coding regions and CRYPTIC UNSTABLE TRANSCRIPTS (CUTs) from intergenic regions. The antisense transcripts, if stabilized by exosome impairment, are able to mediate gene silencing via the recruitment of histone deacetylases.

antisense transcription: The DNA double helix consists of two complementary sequence strands. For any one sequence there is a sense and complementary antisense strand (e.g. 5′ ATTGCGCATT complements 3′ TAACGCGTAA). The sense strand is defined as the normally transcribed strand and translated into mRNA and protein respectively. Antisense transcription refers to the making of RNA from the antisense strand. This RNA will complement the RNA made from the sense sequence, and may bind to it, preventing protein translation.

antisepsis: A chemical method for disinfection. A technique of disinfection using chemical (disinfectant) to treat living tissues is called antisepsis and the chemical used is called ANTISEPTIC.

antiserum (pl. antisera): Blood-derived fluids containing antibodies. Antibodies are present in the serum (plasma – its blood clotting proteins) of the immune animal.

antisilencing mechanisms: DNA methylation-dependent heterochromatin formation is a conserved mechanism of epigenetic silencing of transposons and other repeat elements in many higher eukaryotes. Genes adjacent to repetitive elements are often also subjected to this epigenetic silencing. Consequently, plants have evolved antisilencing mechanisms such as active DNA demethylation mediated by the REPRESSOR OF SILENCING 1 (ROS1) family of 5-methylcytosine DNA glycosylases to protect these genes from silencing. *See* CELLULAR ANTISILENCING FACTORS.

antisocial cells: *See* CANCER CELLS.

antitermination: A process that makes RNA polymerase continue transcription past the terminator sequence, an event called READTHROUGH. Antitermination is used as a control mechanism in both bacterial operons and phage regulatory circuits, and operons controlled by attenuation provide a link between translation and transcription. Termination of transcription ensues when the ribosome is unable to move along a leader segment of the mRNA. During phage infections, different ancillary proteins (antitermination factors) allow the RNA polymerase to bypass specific terminator sequences.

antitermination factor: A protein that when bound at its normal attachment sites lets RNA polymerase read through normal terminator sequences.

antiterminators: At some terminators, an antitermination process causes the enzyme to continue transcription past the terminator sequence. Certain specific ancillary factors interact with the RNA polymerase. NusA protein controls the hairpin formation promoting termination and the stabilization of these contacts by phage lambda N protein leads to antitermination. *Aka* ANTITERMINATION FACTORS.

antitoxin: An antibody with the ability to neutralize a specific toxin. Antitoxins are formed by certain animals, plants and bacteria in response to toxin exposure. Although they are most effective in neutralizing toxins, they can also kill bacteria and other microorganisms.

antitrailer: A nucleotide sequence in DNA downstream of the termination codon of a gene, which is transcribed but not translated. Antileader generally contains signal sequences for the attachment of the poly(A) tail to the RNA polymerase II transcript.

antiviral protein (AVP): The proteins that disrupt virus multiplication. These are proteins produced in response to interferon that stops viral multiplication. For example, oligoadenylate synthetase is an antiviral protein that degrades viral mRNA. Likewise, protein kinase stops protein synthesis.

antiviral response: In mammals, there exists a well-characterized response to dsRNA that acts as an antiviral defence. One obvious role for RNAi/PTGS machinery is a functional homologue of such systems. Indeed, definitive evidence for the use of RNAi as a viral defence comes from studies in plants. *Arabidopsis* mutants that lose the ability to mount a PTGS response are hypersusceptible to viral infections.

antizyme: A protein that renders some other enzyme unstable. For example, in higher animals, an antizyme renders the enzyme ornithine decarboxylase unstable.

antrorse: Directed forward or upward (e.g. of hairs on a stem). *Contrasts* RETRORSE.

anucleolate: Without a nucleus.

apatite: An important inorganic phosphate of soil, but only of restricted use for plant nutrition.

aperture: (i) In an optical instrument, the opening of a lens. (ii) A depressed region in the pollen wall in which the thick intine is covered by the thin exine; the pollen grain appears through the aperture.

apertureless near-field optical microscope (aNFOM): By measuring the modulation of the scattered electric field from the end of a sharp silicon tip which is scanned close to the surface, an image is constructed. Resolutions of the order of a few nm can be achieved.

apetalous: Lacking petals.

apex (pl. apices; Adj. apical): The tip of a root or shoot, containing the apical meristem.

apex culture: The culture of the terminal (0.1–1.0 mm) portion of a shoot comprising the meristem (0.05–0.1 mm) together with primordial and developing leaves as well as adjacent stem tissue. *Aka* SHOOT-TIP CULTURE.

apex separator: A wheat cleaning machine that uses a continuous travelling mesh.

aphibiotic: An organism that can be either parasitic on or symbiotic with a particular host organism.

aphlebia (pl. aphlebiae): Imperfect or irregular leaf endings commonly found on ferns and fern fossils from the Carboniferous Period.

aphyllous: Leafless. The term is applied to flowering plants that are naturally leafless (e.g. several cactus species).

apical: Refers to the tip of a cell, a structure or an organ. The apical surface of an epithelial cell is the bare free surface, opposite to the basal surface. The basal surface relaxes on the basal lamina that separates the epithelium from other tissue.

apical cell: A meristematic initial in the apical meristem of the shoots or roots of plants. As this cell divides, new tissue is formed.

apical dominance: A condition in plants where the stem apex thwarts the development of lateral branches near the apex.

apical meristem: An area of actively dividing cells at the tip of a shoot, branch or root. It is also the precursor of the primary tissue of roots or shoots.

apical placentation: A precentral placentation in fruit where the seeds are attached near the top of the central ovary axis.

apical segment (of spike): The topmost segment of the rachis.

apical spikelet: The spikelet located at the apex (tip) of the ear or panicle.

apiculate: Ending of leaves in a petite, triangular, stretchy point. *See* LEAF SHAPE.

apiculture: Beekeeping for the production of honey and the pollination of crop plants.

apiphily: A system of pollination whereby pollen is dispersed by honeybees.

apoamphimict: A plant that reproduces principally by apomixis but also sexually.

apocarpous: (of a gynoecium) Consisting of one or more carpels which are free from one another (or almost so), as observed in members of the Ranunculaceae and Dilleniaceae.

apocarpy: The condition in which the carpels (female reproductive organs) of a flower are not joined to each other as is the case in buttercup, *Ranunculus* sp.

apochromat: A microscope objective modified for spherical and chromatic aberrations.

apoenzyme: The protein part of an enzyme excluding the co-factors or prosthetic groups. Most of the proteins are made up of two components – a protein component, called an apoenzyme, and a non-protein component, called a co-factor. Apoenzymes are inactive by themselves and the addition of a co-factor (e.g. ions of iron, zinc, magnesium or calcium) activates the enzyme. Apoenzyme plus co-factor is called HOLOENZYME.

apogamety: The self-governing development of a vegetative cell, not the egg cell, into an embryo in an APOMICT.

apogamy: A type of apomixis that includes the suppression of gametophyte formation so that seeds are produced directly from the somatic cells of the parent tissue. *See* APOMIXIS.

apoinducer: A protein that binds to the DNA to switch on transcription by an RNA polymerase.

apolipoprotein B RNA-editing catalytic component (APOBEC): A member of the family of RNA cytidine deaminases. It can deaminate 5^{m}C to convert it to thymine.

apomeiosis: Sporogenesis without the reduction of chromosome number during meiosis, giving rise to apomixis.

apomict: A plant that forms asexual progenies.

apomixis: This term is derived from two words: Apo and mixis. 'Apo' means without and 'mixis' means mixing. So apomixis means without the mixing of male and female genes. A type of asexual reproduction whereby worthwhile seeds or spores genetically identical to the mother are produced asexually (i.e. without fertilization) such that the genetic material they comprise is a clone of the parent's genetic material. The progeny is formed without genetic recombination. A plant produced following this procedure is called an APOMICT. This method of reproduction is used by scientists to propagate (hybrid) plants without having to utilize sexual fertilization. By combining apomixis with tissue culture technology, Cai Detian, Ma Piugfu and Yao Jialin were able to propagate rice varieties in 1994. In 1998, Dimitri Petrov, Phillip Sims and Chester Deald were able to cause apomixis in corn (maize). By 'fixing' hybrid dominance, the need for (sexual) breeding is eliminated and the hybrid vigour is passed down via the seed from generation to generation.

apopetalous: Having separate petals, not fused (sympetalous).

apophase: The period of post-meiotic reconstruction of the cell.

apophysis: (i) The external part of a cone scale. (ii) An outgrowth of an organ or an enlargement of the stem.

apoplast: In a plant, the apoplast is the space outside the plasma membrane within which material can diffuse freely. It is interrupted by the CASPARIAN STRIP in roots, by air spaces between plant cells and by the plant cuticle. The apoplast pathway is where water takes a route going from cell wall to cell wall and not entering the cytoplasm at any point. The apoplast is the 'non-living' extracellular space that surrounds the SYMPLAST. It consists of cell walls and spaces between cells.

apoplastic path: Apoplastic stands for non-living tissue. In apoplastic path, water moves in the cell wall or non-living region between cells without crossing any membranes.

apoptosis: A highly regulated form of PROGRAMMED CELL DEATH (PCD) that transpires in multicellular organisms. It is common in animals. It is conceptualized as a pre-programmed pathway of sequential biochemical events that leads to the physiological cleavage of nuclear chromatin and eventual cell death with characteristic morphological changes. Apoptosis is an active process as compared to passive CELL NECROSIS.

apoptosis and DNA methylation: Pathways linking DNA methyltransferases and methyl-CpG binding proteins to apoptosis, that includes new bioinformatic analyses to characterize the evolutionary origin of two G/T mismatch-specific thymine DNA glycosylases, MBD4 and TDG were discovered by H. X. Meng *et al.* in 2011.

apoptosis due to DNA damage: (i) Mutating the CAACCCCAA sequence in the RNA component of telomerase causes the *in vivo* synthesis of a new telomere sequence corresponding to the mutated RNA sequence, demonstrating that the telomerase contains the template for telomere synthesis. The mutations also lead to nuclear and cell decision defects and senescence, thereby establishing an essential *in vivo* role for telomerase. p63 and p73 are required for p53-dependent apoptosis in response to DNA damage. It has been shown that the combined loss of p63 and p73 results in the failure of cells containing functional *p53* to undergo apoptosis in response to DNA damage. (ii) The Chk2-p53 pathway is a major regulator of DNA damage-induced apoptosis in response to double-strand breaks *in vivo*. Through analysis of 53BP1 complexes a new ubiquitin protease, USP28, was discovered, which regulated this pathway. USP28 is required to stabilize Chk2 and 53BP1 in response to DNA damage.

apoptosis, role of: Apoptotic genes have been reported in mammals. Apoptosis plays a role in development, homeostasis and defence. It is used during early development to eliminate unwanted cells. For example, the regions between the fingers of a developing hand. In adults, apoptosis is used to rid the body of cells that have been damaged beyond repair. It occurs at the early developmental stages (e.g. the loss of the tadpole tail during its metamorphosis into a frog). It is a suicide process that is caused by the fragmentation of nuclear DNA. Generally, it is a natural process but can be induced by DNA damage to the extent that it exceeds the capacity of the repair mechanism of the cell. *Aka* PROGRAMMED CELL DEATH or CELL SUICIDE.

aporepressor: The protein product of a bacterial gene that upon interacting with an amino acid forms a repressor with increased affinity for an operator. For example, the *trpR* gene of *E. coli* forms a repressor with increased affinity for its operator upon interacting with tryptophan.

apospory: The development of a diploid embryo sac in some plants by the somatic division of a nucellus or integument cell without meiosis. A sort of agamospermy in which a seed is produced without fertilization. It occurs in the genera of *Beta*, *Brachiaria*, *Eragrostis*, *Hieracium*, *Hypericum*, *Malus*, *Panicum*, *Purthenium*, *Puspalurn*, *Puspalurn*, *Pennisetum*, *Poa*, *Potentilla*, *Ranunculu* or *Rubus*.

apostatic selection: A selection of very infrequent genotypes or phenotypes that stand out from the norm (e.g. a macromutation).

apparent inducer: Lactose is an apparent inducer of *lac* operon. Actual inducer of lac operon is allolactose.

appendage: A subordinate part attached to a main structure. An external growth that seldom performs any obvious function, hence appendiculate.

applanate: Lying in a plane; flattened; leaves may be presented on the twigs of a plant to give applanated foliage.

application programming interface (API): A set of subroutine definitions, communication protocols and tools for developing a software. In general terms, it is a set of clearly defined methods of communication among various components. A good API makes it simpler to develop a computer program by providing all the building blocks, which are then put together by the programmer.

application software: Computer software designed to perform a group of coordinated functions, tasks or activities for the benefit of the user. Common examples of applications include word processors, spreadsheets, accounting applications, web browsers, media players, aeronautical flight simulators, console games and photo editors. Application software contrasts with SYSTEM SOFTWARE, which is mainly involved in managing the computer's most basic running operations, often without direct input from the user. Collectively, application software refers to all applications. *Aka* APPLICATION or APP.

Applied Biosystems (ABI) SOLiD: The ABI Sequencing Analysis application is designed specifically for the analysis of data produced by the ABI DNA Sequencer. The ABI sequencer is a laser-based instrument that utilizes fluorescent labels to analyse the products of a sequencing reaction as they migrate through a gel. *See* SOLiD.

applied genecology: Plant breeding defined in terms of genetics and ecology; the genetic response of plant varieties to habitat and climate, which has a great bearing on adaptive processes.

applied genetics: A branch of genetics that investigates genetic factors and processes for utilization in agriculture and other branches of science.

aptamers: (i) Oligonucleotide molecules that bind to some other specific molecules (e.g. proteins). Aptamer is derived from the Latin aptus, which means to fit. In 1992, Louis Bock and John Toole isolated aptamers that bind and inhibit the blood-coagulation enzyme thrombin. Since thrombin is vital for the formation of blood clots (coagulation), such aptamers may be useful for anticoagulant therapy (e.g. to prevent blood clots following surgery or heart attacks). (ii) Short, single-stranded DNA or RNA molecules that selectively bind to a specific target, including proteins, peptides, carbohydrates, small molecules, toxins and even live cells. Aptamers assume a variety of shapes due to their tendency to form helices and single-stranded loops. (iii) One of the two components of riboswitches, called DOMAIN, to which a ligand binds, is called APTAMER.

apurinic (AP) DNA: Depurinated DNA.

apurinic/apyrimidinic (AP) endonuclease: AP endonuclease initiates excision repair at apurinic and apyrimidinic sites on DNA. It hydrolyses the phosphodiester bond 5′ or 3′ to an abasic deoxyribose in DNA. AP sites are generated in DNA by spontaneous acid-catalysed hydrolysis, a reaction that becomes facilitated upon base alkylation by radiation, and by the action of DNA glycosylases. These enzymes belong to two classes. Class I AP endonucleases nick the 3′ side of the AP site. Class II AP endonucleases nick the 5′ side of the AP site. Exonucleases remove the nicked-out section, which is then repaired or replaced by DNA polymerase I and ligase.

apurinic/apyrimidinic (AP) sites: Apurinic or apyrimidinic sites that result from the loss of a purine or pyrimidine residue from the DNA.

aquaporin genes: Aquaporin genes are known as MAJOR INTRINSIC PROTEIN (MIP) GENES. In the barley EST database, putative 24 MIPs (contigs) were identified and 11 genes of plasma membrane type aquaporin (PIPs) were detected by PCR. The expression of these 11 PIPs was investigated. One of them, *HvPIP2-1*, was the most abundant and its protein expression was also analysed. *HvPIP2-1* encoded water channel activity in *Xenopus laevis* oocytes injected with *HvPIP2-1* cRNA. Transcripts and proteins of *HvPIP2-1* were reduced in barley roots under salt stress. Overexpression of *HvPIP2-1* increased the shoot/root ratio and raised salt sensitivity in transgenic rice plants, indicating the involvement of *HvPIP2-1* in the cellular mechanism of salt tolerance. The overexpression of the *HvPIP2-1* also increased internal CO_2 conductance and CO_2 assimilation in the leaves of transgenic rice plants, suggesting that *HvPIP2-1* permeates CO_2 in addition to H_2O.

aquatic plant: A plant whose natural habitat is water, living in or on water for all or a substantial part of its lifespan. They are generally restricted to fresh or inland waters.

aqueous: Relating to water (e.g. in an aqueous solution).

aquifer: An area in which groundwater accumulates.

***araBAD* operon in *E. coli*:** Repression of the *araBAD* operon depends upon the operator site to which the regulatory protein AraC binds. AraC protein bound to the upstream operator interacts with another protein bound near the TRANSCRIPTION START SITE (TSS), the intervening DNA looping to allow the interaction. AraC protein binds to sites in *araBAD* operon, which is located in nucleotides lying from −265 to −294 with respect to *araBAD* promoter (P_{BAD}). Previously known sites of protein binding in *ara* operon lie between +20 and −160, whereas the new site is located at −280. The transcription of P_{BAD} is rightwards from +1 and that of P_C is leftwards from −148. To understand the mechanism of repression, the insertions of −16, −8, 0, 5, 11, 15, 24 and 31 bp of DNA between the new site and P_{BAD} were done. Repression was impaired in cases in which half integral turns of DNA helix were introduced, but repression was nearly normal for the insertions of 0, +11 and +31 bp. The data were thus compatible with the DNA looping mechanism.

***Arabidopsis* as a model plant:** The following features make *Arabidopsis* the most commonly used model plant: (i) Short generation time of 8 weeks from seed to seed under normal light conditions and temperature (20°C). It flowers within a month and produces mature seeds within 2 months. The time from seed to seed can be reduced by 2 weeks if plants are grown under continuous light. (ii) It has a small size (adult ~20 cm tall) and is easily grown at a high density in a glasshouse or culture room. The small size permits large numbers of *Arabidopsis* plants to be grown indoors in growth chambers and greenhouses, avoiding the need for researchers to have access to field plots. In addition, the small *Arabidopsis* plants are easy to work with. For example, thousands of seeds can be sown, and plants grown until flowering, on defined agar medium in containers as small as bacteriological Petri plates of 8 cm diameter, which permits the evaluation of specific nutritional requirements or selection of mutants resistant to specific chemicals. (iii) It has a small diploid genome, making the analysis of recessive mutations easy. (iv) It self-fertilizes, so it can isolate seed from a single plant without the need to cross-fertilize. A single plant produces hundreds or thousands of seeds. (v) It possesses a small number of chromosomes (n = 5), simplifying genetic mapping. (vi) The very small *Arabidopsis* seeds are relatively sensitive to chemical mutagens, so 'mutant hunts' are routinely begun by sowing thousands of seeds treated with ethyl methanesulfonate (EMS). (vii) It has a small genome size; around 125 MB is enough for genetic mapping and sequencing. (viii) Its genome was almost completely sequenced in the year 2000 by the Arabidopsis Genome Initiative. (ix) Efficient transformation by *A. tumefaciens.* (x) *Arabidopsis* has a number of natural attributes that collectively make it well suited for molecular genetic studies on flowering plants. (xi) Forward genetics have identified many mutants – over 1500 are freely available from the stock centre. (xii) Its reverse genetic resources are excellent – over 100,000 insertions at precise sequenced locations. (xiii) Most *Arabidopsis* work has been with the (inbred) Columbia (Col) and Landsberg erecta (Ler) ecotypes, but some T-DNA-insertion libraries have been made in the Wassilewskija (Ws) ecotype; there is a large number of ecotypes, native to a variety of latitudes and elevations, that may have interesting differences in genome-maintenance activities.

***Arabidopsis* CEN DNA:** *Arabidopsis* has similar centromere structure to humans but the tandem repeats are 178 bp in length.

***Arabidopsis* chloroplast proteome:** Chloroplasts in most plant species contain 2500 to 5000 proteins. Although only 120 proteins are encoded by the chloroplast genome, several of them are implicated in proteolysis, carbon metabolism and response to oxidation stress. The light-harvesting proteins of photosynthesis I and II were also identified in a number of plant species.

***Arabidopsis* epigenome:** The epigenome of Arabidopsis has been investigated in terms of the following: (i) the distribution of DNA methylation; (ii) the first high-resolution DNA methylome; (iii) highly integrated single-base resolution maps of the *Arabidopsis* epigenome; (iv) integrative epigenomic mapping; (v) stable inheritance of DNA methylation; (vi) exploring the inheritance of DNA methylation; (vii) *Arabidopsis* epigenomics data; (viii) distribution of DNA methylation and (ix) large-scale comparative epigenomics.

***Arabidopsis* epigenomics data:** The following data on *Arabidopsis* are available: (i) BS-seq from silencing mutants; (ii) Whole-genome bisulfite sequencing data; (iii) Whole-genome H3K4m1, H3K4m2 and H3K4m3 profiling; (iv) Whole-genome Dam-ID profiling of LHP1; (v) Whole-genome H3K27m3 and nucleosome density data; (vi) siRNA profiling data from wild-type and nuclear RNA polymerase D mutants; (vii) Whole-genome methylation and expression data and (h) High-throughput RNA-seq.

***Arabidopsis* maps:** (i) Genetic linkage maps reveal the estimated orders and positions of genes along the chromosome. The first linkage groups of *Arabidopsis* were presented by A.D. McKelvie in 1965 at the Gottingen conference for a part of chromosome 1. (ii) M. Koornneef *et al.* in 1983 published a complete genetic map covering all five chromosomes with their 76 markers. (iii) The classical genetic map of *Arabidopsis* contains 462 genes with mutant phenotypes.

(iv) D.W. Meinke *et al.* in 2003 gave a sequence-based map of 620 cloned genes with mutant phenotypes.

***Arabidopsis* mitochondrial proteome:** In *Arabidopsis*, at least 100 different mitochondrial proteins were identified. These proteins were found to be involved in processes such as respiration, citric acid cycle, amino acid metabolism, nucleotide metabolism, protein import, transcription, membrane transport and antioxidant defence. The *Arabidopsis* proteome has also been classified into sub-proteomes like ATP synthase complex and the preprotein translocase of the outer mitochondrial membrane.

***Arabidopsis* nuclear proteome:** Nuclear proteins are implicated in a variety of cellular functions, including signalling, gene regulation, structure, proteolysis and RNA-associated functions. The nuclear proteome was characterized in detail and contained as many as 183 proteins. When alterations in the proteome in response to cold stress were examined, it was found that 54 proteins were up- or downregulated in response to cold treatment.

***Arabidopsis* T-DNA insertion libraries:** *Arabidopsis* has an extremely low frequency of targeted integration of transforming DNA, whether introduced via *Agrobacterium* or biolistic particles. For generation/identification of null mutations, this handicap has largely been overcome by the availability of T-DNA-insertion libraries. Since T-DNA insertion is nearly random (probably with a useful bias towards actively transcribed genes), a few hundred thousand insertions should hit all of the 20,000–30,000 or so *Arabidopsis* genes more than once. Libraries of independent plant lines of that size are available, and the numbers continue to increase. These lines are an outstanding resource for functional genomics which should permit the functions of a large number of genes to be investigated.

***Arabidopsis thaliana* genome:** *A. thaliana* was the first plant whose genome has been completely sequenced (The Arabidopsis Genome Initiative, 2000). It is a small plant belonging to the *Brassicaceae* family whose genome sequencing was completed in 2001. Its genome is one of the smallest among plant genomes and was the first genome to be sequenced. *Arabidopsis* is a popular tool for understanding the molecular biology of many plant traits including flower development and light sensing. *A. thaliana* possesses 70,000 kb in its genome, with very little repetitive DNA. At least two genetic maps have been created for *A. thaliana* (one using yeast artificial chromosomes). The genome of *Arabidopsis* contains 26,000 nuclear genes, 80 chloroplast genes and 60 mitochondrial genes. Nuclear genome contains 120 MB of DNA excluding telomeric and centromeric tandem repeats. The *A. thaliana* genome contains much less repetitive DNA than other angiosperms. Single-copy DNA makes about 80% of *A. thaliana* genome, and an estimated number of 15,000–20,000 genes are present in its five chromosomes. In other words, there is one gene in every 5 kb of genome.

***Arabidopsis thaliana* nucleolus organizer regions:** *A. thaliana* has two ribosomal RNA (rRNA) gene loci – nucleolus organizer regions NOR2 and NOR4 – whose complete sequences are missing in current genome assemblies. Ultralong DNA sequences assembled revealed ~5.5- and 3.9-Mbp sequences for NOR2 and NOR4 in the reference strain, Col-0. Most rRNA gene activity occurs in the central region of *NOR4*, whereas most, but not all, genes of *NOR2* are epigenetically silenced (D. Fultz *et al.* in 2023). Intervals of low CG and CHG methylation overlap regions where gene activity and gene subtype homogenization are high.

***Arabidopsis thaliana*, genetically engineered:** *A. thaliana* was first genetically engineered in 1986. In 1994, researchers succeeded in transferring genes for polyhydroxylbutylate (PHB), a biodegradable plastic, production into *A. thaliana*. The production of PHB requires simultaneous expression of three genes (the PHB production process is 'polygenic'). So far researchers have only been able to insert a maximum of two genes – they have to insert two genes into one plant and one gene into a second plant, then finally get the (total) three genes into (offspring) plants via traditional breeding. In 2001, Eduardo Blumwald and Hong-Xia Zhang inserted a salt-tolerance gene from *A. thaliana* into a tomato (*Lycopersicon esculentum*), and thereby made the tomato plant resistant to salt in concentrations up to 200 mM (far higher than it could previously survive). It has become a model organism in plant biology research because *Arabidopsis* has almost all of the same genes as other flowering plants and has relatively little 'non-coding' DNA. Common names include thale cress, mouse-ear cress or *Arabidopsis*.

***Arabidopsis* transcription factors:** In *A. thaliana*, more than 2000 genes are estimated to encode transcription factors (TFs), which clearly emphasizes the importance of transcriptional control. Nora *et al.* in 2011 developed TF screening platforms, which makes use of publically available GATEWAY®-compatible ORF collections. The *A. thaliana* FOREver-Expression (*At*TORF-Ex) library was thus produced. First, properly recombined vector DNA pools are selected by *E. coli* transformation. The *E. coli* derived vector DNA pools are subsequently used for *Agrobacterium* and *Arabidopsis* flower-dip transformation. After Basta®selection, a seed library of transgenic lines is obtained which overexpresses HA-tagged TF genes. T2 seeds of the transformants are harvested as seed stocks which can be applied for various screening approaches to identify TF genes involved in plant development or stress response. Currently, the *At*TORF-Ex collection harbours transgenic plants overexpressing 650 TFORFs which are available as pools consisting of 30–60 TFs. Developmental phenotypes due to ectopic TF expression have successfully been assayed.

***Arabidopsis* transformation:** *Arabidopsis* transformation has evolved from elaborate tissue culture and regeneration procedures to a technique that is simplicity itself: plants (typically 16 per small pot) that have just begun to set their first seeds (typically 4–5 weeks old) are dipped for a few seconds into growing *Agrobacterium* cultures, then placed upright to mature and produce more seeds. Floral tissues are particularly sensitive to transformation. Transformation efficiencies can reach 0.1%, yielding up to 20 drug or herbicide-resistant progeny from 2000 seeds seen on selective plates. Further, straightforward but laborious screening and segregation analyses yield plants homozygous for single T-DNA insertions when these are required.

arabinose (ara): An aldose sugar that contains five carbon atoms; a member of pentoses. It occurs in sugarbeet. It is metabolized by *E. coli*, which is used as a chemical to induce and activate expression from the promoter pBAD.

***arabinose* (*ara*) operon of *E. coli*:** *Arabinose* (*ara*) operon of *E. coli* has three structural genes – *D*, *A* and *B* – which code for enzymes epimerase, isomerase and kinase, respectively. Gene *C* produces a repressor called C protein, which complexes with arabinose. The corepressor-arabinose complex binds at gene *I*, which induces the production of these enzymes. In *E. coli* cells growing in the absence of arabinose, the three enzymes involved in its breakdown are present in only very small quantities but on the addition of arabinose, all these enzymes are produced in increased amounts. *ara* operon is a glucose-sensitive operon, so *ara* mRNA synthesis is also dependent upon the binding of CAP to promoter.

arbitrary-primed PCR (AP-PCR): PCR-based DNA marker for fingerprinting. This method makes use of a single primer. The length of the primer sequence is ~20 nucleotides. It is similar to RAPDs but gives lesser number of bands as compared to RAPDs because of the larger size of the primers. During AP-PCR, the fragments, which are scored to be similar based on size, may sometimes exhibit sequence variation. It is simple and does not require labour-intensive and expensive procedures, as required by RFLP. Moreover, a very small quantity of genomic DNA is required and there is no need to develop genomic/cDNA libraries.

arboretum (pl. arboreta): (i) A taxonomically arranged collection of trees. (ii) An area, park or garden where trees and shrubs are intentionally grown for educational and/or scientific purposes.

arbuscular mycorrhiza: A symbiotic association between a non-pathogenic or weakly pathogenic fungus and the roots of plants in which fungal hyphae invade cortical cells of the root and produce vesicles and arbuscles. *See* ENDOMYCORRHIZA. *Contrasts* ECTOMYCORRHIZA.

arbuscule (Adj. arbuscular): A branched haustorial structure of some endomycorrhizal fungi that is formed within living cells of the root. The hyphae of the fungus enter the cortical cells of the host where it divides and redivides dichotomously to form a fan-like structure known as arbuscule. In division, branches become thinner and come directly in contact with the host plasmalemma. This is the site of nutrient exchange between the fungus and the host. Arbuscule functions in a bidirectional transfer of nutrients, that is, photosynthates from plant to fungus and nutrients especially P from fungus to plant. A new arbuscule can replace the degenerated arbuscule in the same cortical cell.

Archaea: Also written as archea. One of the three domains of life (the other two being Bacteria and Eukarya) includes species that do not have nuclei and were originally grouped with bacteria into the prokaryotes. They were identified as a separate domain by the analysis of rRNA sequences. Single-celled life forms that can live at extreme ocean depths (high pressure) and in the absence of oxygen. Enzymes sturdy enough for industrial process utilization have been isolated by scientists from some strains of Archaea. Other Archaea strains are sometimes present in the rumen ('first stomach') of cattle and sheep. Those Archaea produce methane gas by breaking down some of the feed consumed by the cattle and sheep.

archaeal histone modifications: Histone modifications do occur in Archaea, but they are rare. Although the proteins that mark histones have a pre-eukaryotic origin, the proteins that read marks and chaperones are eukaryotic innovations. A study by X. Grau-Bové *et al.* in 2022 infers a detailed evolutionary history of eukaryotic chromatin: from its archaeal roots, through the emergence of nucleosome-based regulation in the eukaryotic ancestor, to the diversification of chromatin regulators and their hijacking by genomic parasites.

archaeophyte: A non-native plant that has existed in a particular geographic area for some time.

archegonium: A multicellular structure or organ of the gametophyte phase of certain plants that produce and contain the ovum or female gamete. The corresponding male organ is called the ANTHERIDIUM.

archesporial: The differentiated cell situated in the nucellar tissue of the ovule that is fated to undergo meiosis and produce the haploid generation.

archesporium: Cells formed by mitosis of the microspore and macrospore mother cells.

archiving: Researchers are expected to practice scientific data archiving in compliance with the policies of government funding agencies and scientific journals. Detailed records of their experimental procedures, raw data, statistical analyses and source code are preserved to provide evidence of the effectiveness and integrity of the procedure and assist in reproduction.

Area Under Disease Progress Curve (AUDPC): A measure of the total amount of disease over a period of time, which is determined from graphs of disease vs time, which can be used to compare epidemics quantitatively.

area-effect speciation: When two or more populations characterized by unusual genetic composition expand their ranges, they may meet in a zone across which are sharp differences in gene frequency. If populations whose gene complexes are adaptive produce wee-adapted hybrids, there may be a basis for further divergence and eventual speciation.

arginine (arg): An aliphatic, basic, polar amino acid ($C_6H_{14}N_4O_2$) that contains the guanidino group. In dry, bulk form arginine is colourless, crystalline and water soluble. It is an essential amino acid of the α-ketoglutaric acid family.

arginine methylation: Arginine is one of the building blocks of proteins that can be modified with a chemical tag, in this case, a methyl group (via methylation). *See* LYSINE METHYLATION.

***argonaute* (*Ago*) genes:** A family of evolutionarily conserved genes; their protein products are involved in various RNA interference (RNAi) processes being part of the RNA INTERFERENCE SILENCING COMPLEX (RISC).

argonaute (AGO) proteins: These basic proteins function as catalytic engines. A family of proteins that are characterized by the presence of three functional domains after the N-terminal region: PAZ (which recognizes the 3′-end of small RNAs), MID (which binds the 5′ end of small interfering RNA [siRNA] respectively) and PIWI (an RNase H ribonuclease domain). These proteins are essential for diverse small-RNA pathways. AGOs are effector proteins of small RNA–directed silencing. Small RNAs guide AGOs to their RNA targets. *Aka* PURIFIED PROTEIN DERIVATIVE (PPD). There are ten AGO proteins (AGO1–AGO10) in *Arabidopsis*. Among them, AGO4, AGO6 and AGO9 act in RNA-directed DNA Methylation (RdDM). Argonaute proteins, discovered in plants, are the major actors in RNA-based gene silencing pathways.

arithmetic mean: An average. The number obtained by dividing the sum of a series by the number of items in the series. It is the best measure of the central tendency. Same as MEAN or AVERAGE.

AroA: Refers to the transgene (cassette) initially isolated/extracted from the genome of the *Agrobacterium* bacteria species (strain CP4) and inserted via genetic engineering techniques into a crop plant (e.g. soybean and *Glycine max* L.) to make that (soybean) plant tolerant to glyphosate-based herbicides (and also sulfosate-based herbicides).

aromatic compounds: Compounds that are made up of at least one benzene ring in their structure, commonly having alternating single and double bonds. These compounds have a fragrance or smell, usually restricted to those with pleasant odours of volatile essential oils.

array: (i) Of hairpins, an assembly of alpha helices that cannot be described as a bundle or a folded leaf. (ii) A group of variables that can store multiple values. Each value is retrieved by using an integer index.

array capture: A method for enriching whole genomic DNA for many regions of interest by hybridization to an array containing RNA or DNA sequences complementary to the regions of interest.

array card: A microfluidic card that contains selected primers and probe sets that eases the running of hundreds of real-time PCR reactions simultaneously.

array data structure: A data structure consisting of a collection of elements (values or variables), each identified by at least one array index or key. An array is stored such that the position of each element can be computed from its index tuple by a mathematical formula. The simplest type of data structure is a linear array. *Aka* ONE-DIMENSIONAL ARRAY *or* SIMPLY ARRAY.

array-based genome-wide DNA methylation analysis: Array-based podiums for methylation analysis rely on three basic techniques: (i) restriction digestion; (ii) affinity purification and (iii) bisulfite sequencing. Array-based methodologies have advanced to allow genome-scale DNA methylation analysis through progresses in methylation detection methods, as well as expansion of new high-density arrays.

arrayed library: Individual primary recombinant clones (hosted in phage, cosmid, YAC or other vectors) placed in two-dimensional arrays in microtiter dishes. Each primary clone can be identified by the identity of the plate and the clone location (row and column) on that plate. Arrayed libraries of clones can be used for many applications, including screening for a specific gene or genomic region of interest. *See* LIBRARY, GENOMIC LIBRARY and GENE CHIP TECHNOLOGY.

arrow: The inflorescence of sugarcane.

arsenic (As): A naturally occurring, semi-metallic element that may be found in plants in very low dosages; higher concentrations of As in soil may be toxic for plants.

artefact: Also written as 'artifact'. Something observed in a scientific investigation or experiment that is not naturally present but occurs as a human-made object resulting from the preparative or investigative procedure.

arthropod: A member of the phylum Arthropoda, which consists of animals with articulated bodies and limbs and includes insects, arachnids and crustaceans.

article: (i) In botany, a segment of a jointed stem or a fruit with constrictions between the seeds. (ii) An organ part that separates easily from the rest of the organ at a joint or articulation.

articulate: Jointed; separating freely, leaving a clean scar (e.g. the fronds of certain ferns where they join the rhizome).

artificial chromosome-based vectors: Artificial chromosomes are synthetic chromosomes consisting of fragments of DNA integrated into a host chromosome. These artificial chromosomes are introduced into host cells to propagate and can be used to transfect other cells, introducing new DNA. Artificial chromosomes are useful in cloning larger fragments of DNA anywhere from 300,000 (BACTERIAL ARTIFICIAL CHROMOSOMES, BACs) to 1,000,000 (YEAST ARTIFICIAL CHROMOSOMES, YACs) base pairs, effectively reducing the amounts of runs needed for a large fragment to be analysed. Other artificial chromosome-based vectors are PHAGE ARTIFICIAL CHROMOSOMES (PACs) and MAMMALIAN ARTIFICIAL CHROMOSOMES (MACs). *See details as separate terms.*

artificial chromosomes: Chromosomes experimentally created and constituted, in addition to genetically coding DNA sequences, by ligating the origin of replication, autonomous replicating sequences and telomeric and centromeric sequences. Artificial chromosomes can transmit large numbers of engineered genes and have been projected as a substitute technology for accumulating or recoding genetic information in human cell lines. Early studies revealed that when sequences from human centromeres were incorporated along with a marker gene into cell lines, they moulded steady artificial chromosomes. They are designed to function as vectors, such as BACTERIAL ARTIFICIAL CHROMOSOMES (BACs), YEAST ARTIFICIAL CHROMOSOMES (YACs), MAMMALIAN ARTIFICIAL CHROMOSOMES (MACs) and HUMAN ARTIFICIAL CHROMOSOMES (HACs), that are suitable for large inserts. *See details as separate terms.*

artificial DNA: DNA synthesized in the laboratory using artificial aromatic nitrogen-containing bases.

artificial emotions by robot: Artificial emotions can also be embedded and are composed of a sequence of facial expressions and/or gestures.

artificial gene repressors: The creation of artificial gene repressors is an attractive concept in rational drug design; their activity is based on direct interaction with DNA. Many of these compounds intercalate or bind specifically to DNA. They have contributed in elucidating the function of certain proto-oncogenes in cell differentiation and in understanding the development of cancer.

artificial gene: (i) A double-stranded DNA molecule carrying a specific sequence that will code for a given amino acid sequence and that has been produced *in vitro*. Such a gene may be synthesized by forming a strand of DNA on an mRNA template using reverse transcriptase and then using DNA polymerase to make a double-stranded DNA. (ii) An entirely artificial gene may be created which will code for a novel amino acid sequence by using a template formed from chemically synthesized oligonucleotides followed by enzymatic polymerization.

artificial genetic systems: Artificial genetic systems have artificial DNA molecules that provide deeper insight into how DNA works.

artificial immune system (AIS): A class of computationally intelligent, rule-based machine learning systems enthused by the principles and processes of the vertebrate immune system. The algorithms are typically modelled after the immune system's features of learning and memory for problem-solving.

artificial intelligence (AI): AI is a machine that can think for itself, like a human. John McCarthy defined artificial intelligence in 1955 as the science and engineering of making intelligent machines. Intelligence demonstrated by machines is in contrast to the NATURAL INTELLIGENCE displayed by humans and other animals. In computer science, AI is the study of 'intelligent agents', devices capable of perceiving their environment and taking actions that maximize the chance of successfully achieving their goals. Colloquially, this term is applied when a machine mimics 'cognitive' functions that humans associate with other human minds, such as 'learning' and 'problem-solving'. Frequently confused with MACHINE LEARNING, AI is a procedure for flexibly learning from data, which may be built from elements of ML but is distinguished by the underlying algorithms being created so that the 'machine' can accept new inputs after the developer has completed the initial algorithm. The machine can thus continue to update, refine and teach itself. Currently scientists are working to perfect AI, so in the future, we can utilize robots to do tasks that are being done by humans. *See* MACHINE LEARNING *and* MACHINE INTELLIGENCE.

artificial intelligence (AI) accelerator: A class of microprocessors or computer systems intended as hardware acceleration for artificial intelligence applications, especially artificial neural networks, machine vision and machine learning.

artificial intelligence (AI) control: More than 130 protein science researchers have signed a pledge to use artificial intelligence (AI) safely in their work. Despite AI's potential to help researchers design novel proteins for vaccines, medicines and sustainable materials, terrorists could also use it to build toxins and bioweapons. Among other steps to mitigate those risks, signatories pledge not to use DNA synthesis services from companies that do not adopt standard biosecurity screening protocols. The new guidelines are voluntary and modelled on those adopted for safely using recombinant DNA technology at the Asilomar Conference in 1975. The AI pledge grew out of an October 2023 meeting at the University of Washington (UW). Among the early signatories are 2018 Nobel Prize in Chemistry winner Frances Arnold, UW protein designer David Baker and Harvard University geneticist George Church.

artificial intelligence (AI) in agriculture: (i) AI in agriculture has the potential to provide irrigation, weeding and sprayers through automated sensors to the crops which will boost the manpower efficiency used to increase the crop productivity. Simultaneously, AI covers crop yield, soil content sensing, crop monitoring and crop establishment. (ii) Crop monitoring through AI-based sensors with cameras is achievable through drones to collect information about plants, down to a single leaf. (iii) Information regarding temperature, weather conditions, soil conditions and soil usage can be efficiently controlled using AI-based devices for efficient and precision farming. (vi) AI-based drones installed with robotic lenses to capture high-resolution crop images in wheat sprouts. The aim is to efficiently predict the maturity of crops, which will ease the planning of sowing, harvesting and eventually crop yield. These computerized command-based drones have led to the development of indoor farming, growing crops on a large scale under a packed environment leading to the development of a fully automated indoor growing facility. (v) The manufacturing company John Deere in 2017 developed Blue River Technology for controlling weed population (which has been developed by Forbes to allow the farmers to spray chemicals on weeds while keeping the crop untouched). This robot is based on 'See and Spray' that leverages computer vision for precise spray on weeds to prevent the herbicide resistance problem. (vi) AI software allows sprayers to recognize the difference between the crop (green) and the weeds in cotton plants leading to the development of precise weeding. (vii) An AI machine, Harvest CROO (Computerized Robotic Optimized Obtainer) robotics automated strawberry harvester, was developed in 2013 for strawberry crop harvesting in an efficient and precise manner. Here, the picking wheel is the heart of the tool, which is installed with an AI-based robotic picker to pick up ripe strawberries. Its goal is to lower harvest costs as well as improve fruit quality. (viii) A deep learning (DL) application, Plantix, was developed for the identification of nutritional deficiencies in soil. (ix) Plantix platform 1 harnesses computer vision to enable the early-stage detection of crop diseases with a smartphone camera for precise disease detection. (x) Plantix platform 2 enables the local farmers to visualize symptoms in plants, such as phosphorus deficiency in tomato, giving real-time alerts with the goal of tackling crop failure. (xi) A drone, named SkySquirrel, was also developed based on the algorithms to provide a detailed report on the health of the grapevine leaves. (xii) AI-based tool SkySquirrel-based imaging of vineyards has been developed by SkySquirrel Technologies Inc. (xiii) Researchers combined machine learning (ML) algorithms with satellites for the prediction of weather, analysing crop sustainability and evaluating farms for efficient farming practice. (xiv) AI technology will rescue crops from various factors such as climate changes, population growth and employment and food security issues. (xv) AI technologies prevent the massive use of water, pesticides and herbicides, which ultimately maintains soil fertility. (xvi) Automated irrigation and temperature and moisture sensors ease the decision making to support agriculture. AI-based solution for farmers without the need for experience has the potential to increase their crop productivity. (xvii) Automated systems that depend on ML algorithms, Logistics Regression (LR) and Support Vector Machine (SVM) have achieved an accuracy of 93% and 97%, respectively, across a class of crops, viz., wheat, oats, gram, pea, millets, rice, bajra, maize, potato, groundnut, jute, sugarcane and turmeric with 96% accuracy estimation is major attainment for sustainable development in the farming sector for crop improvement. (xviii) AI has the potential to meet the demand and supply of the market to maintain crop competitiveness and regional crop planning, which will ultimately reduce the labour workforce. (xix) The implementation of AI-based machines and DL will soon become more and more widely adopted for large industrial farms in the agricultural sector to meet the fundamental need for food in the future. Besides AI-based agricultural applications, AI plays a vital role in the collection of immense data in image and video processing, text analysis and speech recognition through the use AI-based powerful computer hardware and algorithms. (xx) AI has the prospect to control the use of harmful pesticides, irrigation, pollution and environmental effects in agricultural practice through automated AI digital world tools. (xxi) AI-based low-cost and effective techniques have the potential to motivate the adoption of green roofs by circumventing water consumption obstacles. (xxii) AI-based weeding is a step towards the reduction in chemical use employed for weeding between target and non-target plants efficiently. An adaptive image segmentation algorithm-based machine vision system is used to detect and locate tomato seedlings and weed plants in a commercial agricultural environment. ANN-based compact airborne spectrographic imager (CASI) identifies weed stress and the nitrogen status of maize. Learning Vector Quantization (LVQ) ANNs distinguish maize from weeds as well as between weed species with an 80–90% accuracy. The major advantage of AI-based weeding tools is that they have the potential to perform site-specific herbicide applications leading to precision farming.

artificial intelligence (AI) in Punjab's agriculture: The AI-based technology linked with Punjab's Agriculture Export Corporation for improving the quality of seed potatoes through real-time monitoring of local weather, crop growth, crop health and pest and crop diseases (indiaai.gov.in). The H_2O model framework has been developed for the determination of evapotranspiration in the Hoshiarpur and Patiala districts of Punjab, based on AI algorithms including deep learning-multilayer perceptrons, generalized linear model, random forest and gradient-boosting machine. In 2019, Punjab Agricultural University, Ludhiana (Punjab) conducted the 43rd Vice-Chancellors' Convention of the Indian Agricultural Universities' Association (IAUA) on the theme 'artificial intelligence for smart agriculture' to bring revolutionary change in the farming community. The emphasis was to implement ML, image recognition and predictive modelling and disease detection in agriculture to boost productivity and efficiency along with reduction in chemical inputs and response to extreme weather conditions. During the year 2020, PhD students from Lovely Professional University (Jalandhar) developed E-parirakshak, based on AI and IoT, for the assessment of fertility, water levels and soil temperature to increase agricultural productivity leading to AI-based farming (www.outlookindia.com). Vinfinet Technologies developed KisanRaja, an AI-based GSM-controlled device for irrigation control using retro-fitted pumps to access farmers on the power supply and water levels through interactive voice response criteria.

artificial intelligence (AI), booming of: AI boomed due to the discovery of AI-based machine learning during the 21st century. It has immense applications in academia and industry that ease the collection of immense data sets through the use of powerful computer hardware systems. In 2020, Quoc Le, a computer scientist at Google, and his colleagues developed an AutoML-Zero program for the development of AI programs with basic mathematical concepts and zero human efforts input (www.sciencemag.org). The implementation of AI-based machine learning, deep learning, big data dealing and data science are now trending in major companies like Google, Facebook, International Business Machines and Amazon for the creation of amazing AI devices that will lead to highly intelligent systems for the future in the coming era. The AI research consists of a 'strong' AI that can autonomously contextualize different specialized problems and 'weak' or 'moderate' AIs that are extremely efficient in their training field.

artificial intelligence (AI), discovery of: The rise of antiquity, myths, stories and rumours of artificial systems equipped with intelligence led to the discovery of the present-day artificial intelligence (AI) era. AI arose during 380 BC–1900 AD when numerous mathematicians, theologians, philosophers, professors and authors contemplated mechanical techniques, machines and numeral systems that led to the concept of mechanized 'human' thought in computer systems. It is a set of sciences, theories and techniques including mathematical logic, statistics, probabilities, computational neurobiology and computer science. The innovation in AI grew significantly from 1900 to 1950 when Alan Turing published 'Computing Machinery and Intelligence' and proposed the idea of The Imitation Game based on the idea that 'machines can think' (www.g2.com). The years from 1952 to 1956 are considered the birth years of AI. During the year 1955, Allen Newell and Herbert A. Simon created the first AI program and named it Logic Theorist. On the other hand, John McCarthy, an American Computer scientist at the Dartmouth Conference, was the first to use the term 'artificial intelligence' in 1956. During the 1980s, AI programs emerged as 'expert systems' throughout the world, providing knowledge based on AI research. It leads to the solution of problems about a specific domain of knowledge through the use of logical rules obtained from expert knowledge.

artificial intelligence (AI), domains of: The natural process in human behaviour is learning, which is also a vital part of machines in the present times, making them more intelligent. The core part of AI includes MACHINE LEARNING (ML) AND DEEP LEARNING (DL).

artificial intelligence (AI), limitations of: AI-based tools provide the flexibility to perform numerous tasks and applications in variable environments, yet these suffer from certain limitations that need to be overcome for better crop management. The major limiting factor is cost involvement in the development of robotics technology, which is a tedious task for poor farmers. The energy and maintenance required also increase the difficulty in implementing robotics-based techniques for crop management. The lack of digital access among farmers and the high cost of research and development hinder the success of AI-based tools. Technological limitations and managerial challenges need to be taken into account for the development of AI-based systems. Moreover, the data recognition in AI-based tools should be noise-free for crop data prediction. AI-based tools depend on climatic factors including rainfall, temperature and humidity, which creates difficulty for farmers; especially in making decisions regarding preparing the soil, sowing seeds and harvesting that need to be monitored during crop data set prediction.

artificial intelligence (AI)-based UAS-based High-Throughput Phenotyping (AUHTP) system: A system that serves as a short-term impact method along with bridging the gap between genomics and phenomics. The combination of remote sensing, simulation models and AI has led to the development of digital agriculture. The integration of AI with genome-wide association studies (GWAS) has the potential to speed up breeding cycles in numerous crops.

artificial intelligence (AI)-complete: In the field of artificial intelligence, the most difficult problems are informally known as AI-complete or AI-hard, implying that the difficulty of these computational problems is equivalent to that of solving the central artificial intelligence problem – making computers as intelligent as people, or strong AI. To call a problem AI-complete reflects the attitude that the problem would not be solved by using a specific simple algorithm.

artificial light: Light other than sunlight; often from fluorescent tubes, used to grow plants in greenhouses or growing chambers, usually out of season.

artificial microRNAs (amiRNAs): The amiRNAs are transcribed from non-coding genes and processed by DCL1 to 21-nucleotide miRNAs. Long dsRNAs derived from inverted repeat constructs (hpRNAi, VIGS) or generated from ssRNA by a member of the RDR family (co-suppression, VIGS) are processed by DCL2, DCL3 or DCL4 to 22-, 24- or 21-nucleotide siRNAs, respectively. 21-nucleotide siRNAs and miRNAs guide the cleavage of target transcripts by AGO1, while 24-nucleotide siRNAs guide AGO4-mediated chromosome modification involving proteins with chromatin-modifying activities (CMAs). Each DCL protein interacts with a specific dsRNA binding protein, DRB (HYL1 for DCL1, others not shown). miRNA and siRNA duplexes are 2′ O-methylated at the 3′-end by HEN1.

artificial plant genome: Researchers in China have synthesized part of the genome of the spreading earth moss, a significant step in synthetic biology for multicellular organisms. This advancement paves the way for the development of artificial genomes in more complex life forms. A research project is focusing on a section of one chromosome, marking progress towards a fully artificial plant genome. The techniques used include trimming and tidying the chromosome, resulting in a 56% reduction in size. Moss plants thus altered grow normally, maintaining their natural stress resistance. This achievement not only advances the field of synthetic biology but also offers insights into the essential elements of multicellular eukaryotic genomes. The researchers aim to continue this work, potentially leading to significant breakthroughs in understanding and utilizing plant genomes.

artificial restriction enzymes: Enzymes generated by fusing a natural or engineered DNA-binding domain to a nuclease domain (often the cleavage domain of the type IIS restriction enzyme *Fok*I). Such enzymes can target large DNA sites (up to 36 bp) and can be engineered to bind to desired DNA sequences. Examples: ZINC-FINGER NUCLEASES (ZFNs), TRANSCRIPTION ACTIVATOR-LIKE EFFECTOR NUCLEASES (TALENs), MEGANUCLEASES (HOMING NUCLEASES) AND CLUSTERED REGULARLY INTERSPACED SHORT PALINDROMIC REPEATS (CRISPR)-CAS SYSTEMS. These artificial restriction enzymes are used to edit genomes. *Aka* GENOME EDITING ENZYMES *or* GENOME ENGINEERING ENZYMES. *See details as separate terms.*

artificial restriction sites diagnostic assay: PCR amplification, either conventional or as site-directed mutagenesis using primers with mismatched 3′-ends, followed by restriction endonuclease digestion, provides rapid, non-isotope assays of known mutations in the human phenylalanine hydroxylase gene.

artificial seeds: Based on embryogenic suspension cultures, embryos may be given a synthetic coat with water-soluble hydrogels (calcium alginate) and other substances (nutrients) to secure proper seed germination even under field conditions. Therefore, artificial seeds serve as an easy delivery system for micropropagation of plants. *Aka* SYNTHETIC SEED.

artificial selection: (i) Plant selection by human or agronomic means; the practice of choosing individuals from a population for reproduction, usually because these individuals possess one or more desirable traits. (ii) The process by which humans breed animals and cultivate crops to ensure that future generations have specific desirable genotypes and characteristics. In artificial selection, breeders select the most desirable variants in a plant or animal population and selectively breed them with other desirable individuals. The forms of most domesticated and agricultural species have been produced by employing artificial selection. It is also an important experimental technique for studying evolution. *Aka* SELECTIVE BREEDING.

artificially acquired active immunity: The production of antibodies by an animal/human body following vaccination. Animals and human beings are protected from several serious viral and bacterial diseases through vaccination, also known as immunization. In this process, antigens (vaccines) are injected into the body. Bacterial toxins, dead microorganisms and living but attenuated microorganisms serve as vaccines. Injected vaccines stimulate an immune response that results in the production of antibodies.

artificially acquired passive immunity: Protection of animals/humans by directly administering antibodies, either orally or through injections. In comparison to active immunization where antigens are injected, in passive immunization antibodies are directly introduced into the body. Using the technique of genetic engineering transgenic, antibody (plantibody) producing plants have been produced for passive immunization.

ascending: (i) A stem initially spreading horizontally, and then directed upwards. An ascending stem is more or less prostrate near its base, then erect. (ii) An ovule attached somewhat above the base.

ascent of sap: The upward transmission of water from the roots to different parts of the plant body.

ascocarp: A sexual fruiting body of an ascomycetous fungus that produces asci and ascospores, including apothecium, ascostroma, chasmothecium/cleistothecium, perithecium and pseudothecium. *See* ASCOMA.

ascogenous: Pertaining to ascus-producing hyphae.

ascogonium (pl. ascogonia): A specialized cell that gives rise to the hyphae that produce asci.

ascoma (pl. ascomata): A sexual fruiting body of an ascomycetous fungus that produces asci and ascospores, including apothecium, ascostroma, chasmothecium/cleistothecium, perithecium and pseudothecium. *See* ASCOCARP.

ascomycete (Adj. ascomycetous): A casual term for a member of Ascomycota.

ascorbic acid: The water-soluble vitamin C ($C_6H_8O_6$) found in large quantities in fruits and vegetables.

ascospore: A haploid (1N) sexual fungal spore enclosed within an ascus produced by ascomycetes. It is a sexual spore, produced by a group of fungi (Ascomycetes) at hyphal tips on the same thalli (homothallic) or on different thalli (heterothallic). Single-celled spores are produced in a sac called an ascus. Eight ascospores per ascus are present and nuclear fusion takes place in the ascus. The diploid zygote nucleus divides by meiosis immediately after fusion to produce four haploid nuclei, which divides once more by mitosis and form eight ascospores per ascus.

ascostroma (pl. ascostromata): A fruiting body containing bitunicate (double-walled) asci in locules; usually dark with multiple locules, but sometimes single. *See* PSEUDOTHECIUM.

ascus (pl. asci): A miniature sac-like structure containing ascospores (typically eight) and usually borne in a fungal fruiting body.

asepsis (Adj. aseptic): This is a condition in which harmful microorganisms are absent. The maintenance of aseptic conditions during *in vitro* culture is a prerequisite and is created by the sterilization of the medium, surface sterilization of the inoculum and performing inoculations in Laminar Air Flow cabinets fitted with HEPA filters. During tissue culture, the maintenance of aseptic/sterile conditions throughout is a prerequisite. In the absence of aseptic conditions, the microbes overgrow the cultured cells/tissues which stop proliferation and get senesced.

aseptate: Having no cross walls and lacking septa. *See* NON-SEPTATE *and* COENOCYTIC. *Contrasts* SEPTATE.

aseptic technique: Procedures used to prevent the introduction of fungi, bacteria, viruses, mycoplasma or other microorganisms into cell, tissue and organ cultures. Although these procedures are used to prevent microbial contamination of cultures, they also prevent cross-contamination of cell cultures. These procedures may or may not exclude the introduction of infectious molecules. *See* ASEPSIS.

asexual reproduction: Any reproduction not involving the union of gametes and meiosis. Often used interchangeably with vegetative reproduction. (i) A propagation without the formation of zygotes by sexual organs and genetic recombination. In plants, there are two types of asexual reproduction – VEGETATIVE PROPAGATION (by stolons, rhizomes, tubers, tillers, bulbs, bulbils or corms) and APOMIXIS (by vegetative proliferation or agamospermy). (ii) A type of reproduction involving a single parent producing offspring genetically identical to the parent. (iii) Asexual reproduction may occur by budding (as in Hydra) by mitotic division of a single cell (as in eukaryotic microorganisms) or by fragmentation of a whole organism into two or more new individuals (as in many plants, as well as some animals like coral, sponges and starfish). It produces an individual genetically identical to the parent. *Contrasts* SEXUAL REPRODUCTION.

asexual spore: A spore produced by mitosis (eukaryotes) or binary fission (actinomycetes). In fungus, spores are formed from aerial hyphae, which can be either asexual or sexual. During asexual spore formation, only mitosis takes place and the organisms developed from spores resemble the parent. Spores are generally produced during unfavourable conditions. Conidiospores produced by *Aspergillus* are asexual spores. Yeast *Candida albicans* also produces thick-walled asexual spores but these are called chlamydospores.

Asilomar Conference: P. Berg suggested in 1975 that international research should continue but under stringent guidelines. These concerns eventually led to the 1975 Asilomar Conference, in which 100 scientists participated to discuss the safety of manipulating DNA from different species.

asparagine (asn): An amino acid ($C_4H_8N_2O_3$) found in the storage proteins of plants (e.g. in peas and beans). The designation was derived from its presence in asparagus.

aspartic acid (asp): A dicarboxylic amino acid (HO_2C-$CH[NH_2]CH_2CO_2H$). In dry, bulk form it appears as a white, crystalline solid. It is found in high amounts in many plants and animals, especially in molasses from young sugarcane and sugarbeets.

***Aspergillus nidulans*:** It is a filamentous fungus. It consists vegetatively of more or less elongated, separate filaments. They are individually called *hyphae* and collectively *mycelium*. The hypha is coenocytic and consists of many nuclei. In this fungus, morphogenesis is controlled by a morphogen. A characteristic feature of fungal morphogenesis is its highly polarized nature. Fungal cells grow by inserting new membranes and cell walls at localized sites in the cell surface. Different types of fungal cells acquire their unique morphogenesis by employing distinctive patterns of polarized morphogenesis. Polarized growths occur in response to either internal or external spatial cues. Internal cues (proteins Bud 3p, Bud 4p) direct bud site selection during the cell cycle whereas external cues (pheromone receptor) dictate the site at which mating projection is formed prior to conjugation. In both cases, spatial information is relayed to morphogenetic complex by a series of GTPase signalling nodules.

aspirator: A seed conditioning (cleaning) machine that uses air to separate seeds according to specific gravity (weight) and resistance to airflow.

assay: (i) A method for determining the presence or quantity of a component. (ii) A method for measuring a biological activity. This may be enzyme activity, binding affinity or protein turnover. Most assays utilize a measurable parameter such as colour, fluorescence or radioactivity to correlate with the biological activity.

assembled epitope: *See* CONFORMATIONAL EPITOPE.

assembler: An all-purpose device for guiding chemical reactions by positioning molecules. A programmable molecular machine, which can build any molecular structure or device from simpler chemical entities. This is similar to an assembly machine in a factory. Originally introduced by Eric Drexler.

assembly PCR: Assembly PCR is a polymerase chain reaction variation that artificially synthesizes long DNA sequences by performing PCR on a pool of long oligonucleotides (primers) with short overlapping segments. The oligonucleotides alternate between sense and antisense directions and the overlapping segments determine the order of the PCR fragments, thereby selectively producing the final long DNA product.

assembly: A compilation of overlapping sequences from one or more related genes that have been clustered together based on their degree of sequence identity or similarity. The sequence assembly may be used to place 'shotgun' sequencing fragments based upon overlapping restriction enzyme digests together, or may be used to identify and index novel genes from single-pass cDNA sequencing efforts.

assertion: In computer programming, a statement that a predicate (Boolean-valued function, i.e. a true-false expression) is always true at that point in code execution. It can help a programmer read the code, help a compiler compile it or help the program detect its defects. For the latter, some programs check assertions by evaluating the predicate as they run and if it is not true – an assertion failure – the program considers itself to be broken and typically deliberately crashes or throws an assertion failure exception.

assertion probability (α), type I: Frequently confusingly labelled as a FALSE POSITIVE PROBABILITY, this is the probability of rejecting H_0 (i.e. declaring 'statistical significance' – not recommended), when the null hypothesis is assumed to be true. The type I assertion probability is often called α and is the probability of asserting an effect when any assertion of effect is by definition false. In common use, the type I probability is the probability that the nominal *P*-value will be <0.05 if there is no effect. It will be 0.05 when (i) only one *P*-value is computed; (ii) all model and experimental design assumptions made by the *P*-value calculation are exactly true and (iii) the *P*-value is computed exactly. *See* ASSERTION PROBABILITIES *and* DECISION ERROR PROBABILITIES.

assertion probability (β), type II: Frequently confusingly labelled as a false negative probability, this is the probability of not asserting an effect (i.e. failing to reject H_0) when there truly is a specific magnitude of effect. The type II probability is referred to as β, which is one minus the power of the test. In other words, the power of the test is $1-\beta$.

assertion probability: The probability of an event that cannot happen or which is impossible is equal to zero.

assessing substrate specificity through microarrays: AbSurance™ Pro microarray from MilliporeSigma can be used to probe the specificity of histone modifications. The AbSurance Pro microarray is a glass microarray containing multiple copies of some 275 modified histone peptides from all four core histone proteins (H2A, H2B, H3 and H4).

assigning function to a gene: Elite genetic stocks from diverse genetic backgrounds provide the raw material for mapping populations such as NEAR-ISOGENIC LINES (NILs), RECOMBINANT INBRED LINES (RILs), DOUBLE HAPLOID (DH) populations, deletion mutants, etc. Newly identified genes are assigned functions with the help of sequence similarities obtained from (i) genomic tools, such as genomic databases, microorganisms and proteomics and (ii) bioinformatic tools, like computer software. Gene functions are linked to the phenotype by knocking out or overexpressing the gene from which a cellular or developmental role can be deduced. Finally, the fitness of plants carrying mutations or natural variants for the gene can be compared in different environments with wild-type plants to study the adaptive function.

assignment test: A test to determine whether a locus is on a specific chromosome by observation of the concordance of the locus and the chromosome in hybrid cell lines.

assimilation: (i) The formation of self-cellular material from small molecules derived from food. (ii) The manufacture of an organic substance from organic elements and compounds via photosynthesis. (iii) The uptake and construction of simple foodstuff.

association: A statistical relationship between two variables. Most commonly an allele found more or less frequently with a specific disease than in the general population.

association analysis: *See* ASSOCIATION MAPPING.

association constant: *See* AFFINITY CONSTANT.

association mapping: In genetics, association mapping, also known as 'LINKAGE DISEQUILIBRIUM MAPPING', is a method of mapping QTLs that takes advantage of historic linkage disequilibrium to link phenotypes (observable characteristics) to genotypes (the genetic constitution of organisms), uncovering genetic associations. 'Association' has a broader meaning that includes the combinations of three or more linked markers, at least some of which are in linkage disequilibrium (LD). These combinations are called haplotypes if specified for a single chromosome. The sex chromosomes in males and certain chromosomal aberrations are monosomic, with each individual carrying only one haplotype. With these exceptions (and some chromosomal aberrations), haplotypes occur in pairs, called diplotypes, consisting of one haplotype from each parent; the two haplotypes of a diplotype cannot be ascertained with certainty if two or more markers are heterozygous, except in special cases that include family studies, physical separation of chromosomes and zero frequency of alternative haplotypes. Association mapping depends on the choice of map taken to represent LD.

association of chlorophyll content with plant traits: Chlorophyll content impacts crop biomass and grain yield. For example, in tartary buckwheat (*Fagopyrum tataricum*), a positive correlation exists between chlorophyll content, plant growth and productivity. The association of leaf chlorophyll content with various physiological traits, such as leaf potassium content, K/Na ratio, plant height, shoot dry weight and survival score, has been observed in *Solanum pimpinellifolium*. Similarly, yield-related traits, such as per cent fruit set, fruit number and yield per plant, were correlated with plant chlorophyll status in *S. pimpinellifolium*. Association of leaf N with chlorophyll content is obvious because N is the chief constituent of the porphyrin ring system in the chlorophyll molecule.

association of DNA methylation status changes with QTL: Methylation pattern variations in epiRILs are constant over the generations (F. Johannes *et al.* in 2009). However, the association of QTL with DNA methylation status exhibits 60–90% heritability in terms of flowering time and primary root length in epiRILs of *Arabidopsis*. The whole-genome bisulfite sequencing and linkage mapping of methylation states uncovered local methyl-QTL.

Association of Official Seed Certifying Agencies (AOSCA): Established in 1919 in the USA, AOSCA sets minimum standards for genetic purity and identity and recommends minimum standards for seed quality. The goal of these agencies is to standardize certification regulations and procedures internationally.

association study: A survey of associations between genetic markers and quantitative traits (including human disease) that aims to locate the QTLs that cause trait variation.

association testing: A method of mapping QTLs that takes advantage of historic LD to link phenotypes to genotypes. Association mapping is based on the idea that traits that have entered a population only recently will still be linked to the surrounding genetic sequence of the original evolutionary ancestor, or in other words, will more often be found within a given haplotype than outside of it. *Aka* LINKAGE DISEQUILIBRIUM (LD) MAPPING.

associative array: An associative array, map, symbol table or dictionary is an ADT composed of a collection of (key, value) pairs, such that each possible key appears at most once in the collection. Operations associated with this data type allow: (i) the addition of a pair to the collection; (ii) the removal of a pair from the collection; the modification of an existing pair and (iii) the lookup of a value associated with a particular key.

associative intron: An intron that comes into being when two unrelated exons join by an unequal recombination between two genes. The introns would render this gene fusion event insensitive to a shift in the reading frame and allow a novel unequal recombination to succeed. It would function evolutionarily by increasing the frequency of successful shifting of gene segments. *Aka* TYPE A INTRON.

associative overdominance hypothesis: Allozyme loci are neutral but are linked to loci that are determinants of the quantitative trait. The linked loci under selection may show overdominance or be subject to an input of deleterious mutations (the partial dominance hypothesis for inbreeding depression).

associative transcriptomics: Associative transcriptomics is a potent new method for rapidly identifying molecular markers associated with trait variation across a diversity panel, exploiting both gene sequence variation and gene expression variation and can be effective even where advanced genomics resources are not available.

assortative mating: The mating of individuals with similar genotype/phenotype. For example, individuals with genotype *AA* (phenotype A) tend to mate with other individuals of the same genotype or phenotypes (e.g. tall individuals tend to mate with other tall individuals). Occurs if the plants that are mating resemble each other, with regard to some traits. Assortative matings also occur in organisms other than humans. An extreme form of assortative mating is self-fertilization, which is the most common form of reproduction in many plants. Thus, assortative mating prevails where matings do not occur at random. *Compare with* DISASSORTATIVE MATING.

aster: Star-shaped system of microtubules originating from a centrosome or from a pole of a mitotic spindle to manipulate the movement of chromosomes during cell division.

astral microtubule: In the mitotic spindle, any of the microtubules radiating from the aster, which are not attached to a kinetochore of a chromosome.

astrobiology: The branch of biology concerned with the effects of outer space on living organisms as well as the search for extraterrestrial life.

asymmetric cell division: A type of cell division that is essential to generate specialized cells of different sizes, different cytoplasmic constituents and with distinct functions. Asymmetric cell division is crucial for forming diverse tissues and organs. *Aka* FORMATIVE CELL DIVISION. *Compare with* SYMMETRIC CELL DIVISION.

asymmetric fusion: A cell formed by the fusion of genetically dissimilar cells, referred to as a heterokaryon.

asymmetric histone segregation: The parental $(H3\text{-}H4)_2$ tetramers are assembled into nucleosomes on both leading and lagging strands, with a slight preference for lagging strands.

asymmetric PCR: Asymmetric PCR is used to preferentially amplify one strand of the target DNA more than the other. Asymmetric PCR is useful in some sequencing and hybridization probing applications where having only one of the two complementary stands is sufficient or required. The asymmetric PCR method is conducted as the standard PCR protocol; however, a great excess of the primers for the chosen strand is used. Due to the slow (arithmetic) amplification later in the reaction after the limiting primer has been used up, extra cycles of PCR are required. In asymmetric PCR, the predominant product is a single-stranded DNA as a result of unequal primer concentrations. As asymmetric PCR proceeds, the lower concentration primer is quantitatively incorporated into double-stranded DNA. The higher concentration primer continues to primer synthesis, but only of its strand.

asymmetric RNA duplexes (ARDs): ARDs consist of 19 to 21-nt antisense and 16-nt sense strands that generate a 16-bp duplex region with 3- to 5-nt long 3′ antisense overhang. The ARDs mediate the cleavage of target mRNAs.

asymmetrical: Irregular or unequal, lacking any plane of symmetry (e.g. flowers of Canna).

asymmetrical sequence: A sequence that is present only on one strand of the DNA.

asynapsis: Chromosomes observed during meiosis I in the which pairing between homologous chromosomes either fails or is incomplete.

asynaptic pachytene: A complete or incomplete pairing of the chromosomes in the pachytene stage of the first meiotic division.

asyndesis: *See* ASYNAPSIS.

atavism: The reappearance of a character after several generations. The character in question being the result of the expression of a recessive gene or complementary genes.

atom: The smallest unit of matter that enters into a chemical reaction. All matter like water, air, soil and organisms are made up of small units called atoms. Different atoms combine in specific amounts to produce molecule. Molecules are the building blocks of the cells. Every atom has a nucleus and particles called electrons that move around the nucleus.

atomic force microscopy (AFM): A form of probe microscopy scanning that delivers atom-level information about molecules. For this, the surface of a molecule is scanned with a microprobe in an x–y grid and the force encountered is measured with piezoelectric sensors. This is a scanning probe microscopic (SPM) device, as it involves the scanning of a probe over a surface to be investigated in atomic steps.

atomic manipulation: Manipulating atoms using SPM techniques.

atomic mass unit (Amu or Dalton): The basic unit of mass on an atomic scale. One dalton is one-twelfth the mass of a carbon-12 atom (i.e. the mass of a hydrogen atom is 1.66×10^{-24} g; therefore, there are 6.023×10^{23} daltons/g).

atomic number: Refers to the number of protons in an atom. The number of protons in the nucleus of the atom ranges from one in the hydrogen atom and more than hundred in the largest atoms known. Atoms are listed by their atomic number. In a sodium (Na) atom, there are 11 electrons and 11 protons. Thus, the atomic number of Na atom = number of electrons = number of protons = 11.

atomic weight: The mass of an atom in atomic weight units (AWU). Protons and neutrons are each 1 AWU. Therefore, we can determine the atomic weight by adding the number of protons and neutrons in the atom. The sum is not exactly the actual atomic weight. For example, sodium has 11 protons and 12 neutrons, but its actual atomic weight is 22.9898, not 23.

atomistic: A theory of inheritance in which the entities controlling heredity are relatively distinct, permanent and capable of independent action. Accordingly, Mendelian inheritance is an atomistic theory because inheritance is controlled by distinct genes.

atomistic simulations: Computations of the properties of nanoscale systems using sophisticated models.

atomizer: Machine for adding water to grain in finely divided spray.

atrazine: A selective herbicide, widely used in maize cropping. It is environmentally significant since it was the second most commonly detected pesticide residue in a US Environmental Protection Agency (EPA) survey of drinking water wells conducted during 1988–1990. Due to concerns about groundwater contamination and worker exposure, the EPA is conducting a special review of atrazine registration.

atrophy: Reduced, contracted or diminished organ size, shape or function, usually a deteriorative change.

atropine: A poisonous crystalline alkaloid ($C_{17}H_{23}NO_3$), it can be extracted from deadly nightshades and other solanaceous plants. It is used in medicine to treat colic, to reduce secretion and to dilate the pupil of the eye.

***att* sites:** Loci on phage and the bacterial chromosome at which site-specific recombination takes place by integration of the phage into, or excision from, the bacterial chromosome.

attached X chromosome: A pair of *Drosophila* X chromosomes joined at one end and inherited as a single unit.

attachment constriction: Centromere; spindle attachment; insertion region; kinetochore; a non-staining localized region in each chromosome that remains single or appears to be attached to the spindle fibre sometime after the rest of the chromosome has divided.

attachment efficiency: The percentage of cells plated (seeded, inoculated) that attach to the surface of the culture vessel within a specified period of time. The conditions under which such a determination is made should always be stated.

attachment technology: A large variety of molecules (including RNA, proteins and many other organic and metallic molecules) are known to be capable of selectively binding directly to short strands of DNA or indirectly binding to DNA via appropriately designed attachment chemistry. It is feasible to bind such molecules to specific sites on DNA tiles within the DNA lattices (these sites might be short stem loops that can be designed to protrude above a DNA tile). This provides for a highly flexible nanostructure construction methodology: by selectively attaching various other types of molecules to the tiles of the lattices, these lattices can be used as superstructures for the placement of nanocomponents composed of a wide variety of materials.

attar: Name used for a perfume from flowers (an essential oil, e.g. the attar of roses). *Aka* ITTAR.

***attB* site:** A 15-nucleotide sequence 5′-GCTTTTTTATACTAAA-3′ in the bacterial genome at which phage chromosome pairs for undergoing site-specific recombination.

attenuated pathogens: Pathogens that have been inactivated and rendered harmless. The disease-causing ability of a pathogen has been sufficiently weakened so that it can still promptly produce an immune response but not the disease. Some ways in which viruses and other pathogens may be attenuated are heat, chemical or radiation treatment. Attenuated pathogens are commonly used for making vaccines. *See* ATTENUATED STRAINS *and* ATTENUATED VIRUSES.

attenuated strains: Strains of pathogenic microorganisms, mainly bacteria and viruses, that have lost virulence.

attenuated vaccine: Vaccine containing weakened, non-pathogenic, live microorganisms. In different vaccines, the nature and source of antigens differ. Some vaccines contain intact but killed microorganisms, whereas others may contain living microorganisms which have been attenuated (weakened, not capable of causing disease).

attenuated viruses: Live viruses are allowed to live on a different host for a large number of generations. Through mutations, these viruses become attenuated which means the viruses cannot cause disease in the original host. These attenuated viruses are used as vaccines. Through back mutations, the viruses may cause disease in the original host.

attenuation: A mechanism by which the expression of an amino acid biosynthetic operon (e.g. the *trp* operon) is regulated by the amino acid levels in the cell. This mechanism relates the supply of an aminoacyl~tRNA to the ability of RNA polymerase to read through a termination site. The terminator is situated at the beginning of the cluster of structural genes coding for the enzymes that synthesize amino acids carried by tRNAs.

attenuation of RNA: Premature termination of an elongating RNA chain.

attenuator: A sequence present downstream a gene that provides the signal for the termination of transcription and the attachment of poly(A) tail and regulates gene activity. A regulator site that acts as a barrier to transcription. The termination event at this site is sensitive to the level of transcription. For example, in tryptophan operon, when tryptophan supply is adequate, transcription stops at about 147 bases after initiation so that no mRNA for *trpE* and other genes of the operon are formed. If tryptophan is in poor supply, transcription continues uninterrupted. During an adequate supply of tryptophan, mRNA for the leader sequence is translated on the ribosome. This synthesis of leader peptide leads to the formation of a hairpin loop. This is an attenuator, which is located in the sequences between 114 and 134. This stops the further synthesis of mRNA. On the other hand, if the synthesis of mRNA is in short supply, two tryptophan codons cannot be translated due to the lack of amino acid tryptophan. The ribosome becomes stalled at this point and makes the formation of a TERMINATOR HAIRPIN LOOP impossible, thus indirectly permitting the transcription of *trpE* and other genes of the operon. This type of termination is possible only in those sequences where transcription and translation are coupled, that is, this is possible only in prokaryotic genes.

attenuator region: A sequence present downstream a gene that provides the signal for the termination of transcription and the attachment of poly(A) tail, thereby regulating gene activity.

attenuator stem: *See* TERMINATOR STEM.

***attP* site:** A site on the phage chromosome, which is represented as the P′OP site. P′ has three binding sites for integration protein, whereas O and P have two sites each. The binding of integration protein molecules at these sites forms a structure called intasome, which is required for site-specific recombination between *E. coli* and λ phage chromosomes.

attractant: A chemical or agent that tricks insects or other pests by stimulating their sense of smell. This is a non-toxic method for attracting insects into traps and is heavily used in orchard crops. Though distinct from toxic baits, attractants are controlled as pesticides.

attractive regime tapping-mode atomic force microscope (ARTM AFM): The ARTM AFM allows the determination of the basic morphologies of the antibodies on the support. More importantly, this regime can resolve the characteristic Y-shaped domain structure of antibodies and the hinge region between domains. It is used for biological imaging.

atypical: Having no discrete distinctive character; not conforming to the type.

AUG: The codon in mRNA that specifies initiation of a polypeptide chain, or within a chain, incorporation of a methionine residue.

auger: (i) A device for boring a hole in the soil used for planting or transplanting seeds, seedlings or bulbs or in fertilizing shrubs, trees and ground covers. (ii) A long pipe with a twist inside that is used to carry the grain up into a granary.

augmented block design: An experimental design that is used to test a large number of germplasm lines in a limited area. In this design, germplasm lines are unreplicated and checks are replicated.

augmented reality (AR): The collaborative involvement of a real-world environment where the substances that exist in the real world are 'augmented' by computer-generated perceptual information, sometimes across multiple sensory modalities, including visual, auditory, haptic, somatosensory and olfactory features.

augmenting gene expression using epigenome-editing technology: Epigenome-editing tool dCas9-HAT can augment abscisic acid-responsive element binding protein1 (*AREB1*) gene expression in *A. thaliana* in response to drought stress.

Australian Wheat Board (AWB): A grower-owned and controlled company operating under Australian corporation laws. This statutory marketing agency controls Australia's domestic marketing of wheat and export marketing of wheat and flour. In the Australian system, farmers take their wheat to elevators designated as official handling agents for the board. Following delivery, the farmers receive an initial payment, then over a period of time they receive additional payments until the full price has been paid.

authorship claim to a variety: *See* SEED BREEDER'S RIGHTS.

autoallopolyploids: Cells or individuals whose genomes show characteristics of both autoploidy and alloploidy.

autoallopolyploidy: A type of ploidy in which one genome is in more than a diploid state (e.g. AAAABB).

autoantibody: Antibody formed by an individual against a protein, or other potential antigen, of its own cells and tissues. Autoantibodies can result in autoimmune disease.

autobivalent: A bivalent in meiosis I that possesses two structurally and genetically identical sister chromosomes.

autocatalysis: The reaction catalysed by one of its products, creating a positive feedback effect on the reaction rate. Thus, this system is self-amplifying.

autocatalytic network: A chemical system that produces a chemical that is a catalyst for the original reaction or that results in other reactions that eventually produce a catalyst for the original reaction.

autocidal insect control: The STERILE INSECT TECHNIQUE (SIT) is not equally effective in all insect species, so other methods need to be investigated for their use in the autocidal control of insect pests. Various approaches to autocidal insect control are inherited sterility, conditional lethal mutations, behavioural changes, hybrid sterility, simply inherited mutations and genetic sexing techniques.

autoclave: An equipment that generates pressurized steam for moist heat sterilization. Autoclave is a bigger version of a home pressure cooker. It is a double thick-walled metal container, that is, fitted with a heating element, safety valve and pressure gauge. Autoclave generates moist heat at 121°C (250°F), which at 15 pounds per square inch (psi) pressure for 20 min is effective in killing all the microorganisms. It is a fact that moist heat is more effective than dry heat.

autocratic plant breeding: The opposite of DEMOCRATIC PLANT BREEDING. In autocratic plant breeding, it is the breeder, rather than the farmer, who decides what kind of cultivar is to be bred and which cultivar is to be grown.

autocrine cell: In animals, a cell that produces hormones, growth factors or other signalling materials for which it also produces the corresponding receptors. *See* ENDOCRINE and PARACRINE.

autocrine signalling: Type of cell signalling in which a cell produces signal molecules that act on itself or on other adjacent cells of the same type.

autofertility: Autogamy; self-fertility.

autogamous: Self-pollinating, self-fertilizing is commonly seen in flowering plants. *See* AUTOGAMY.

autogamy: (i) Obligatory self-fertilization in plants where egg produced in the ovary of a plant fuses with the pollen produced in the same flower, for example, wheat, rice, soybean, chickpea, pigeon pea, pea, groundnut, flax, barley, oats and lupine. Autogamous crops are true breeding plants. This process involves INTERNAL FERTILIZATION, which leads to homozygosity in a single generation. (ii) This occurs in single unpaired animals through a special kind of nuclear organization.

autogenomatic: Genomes that are completely homologous and their chromosomes pair normally in meiosis.

autogenous control: When a gene product (a molecule) either inhibits (NEGATIVE AUTOGENOUS CONTROL) or activates (POSITIVE AUTOGENOUS CONTROL) the expression of the gene that codes for it (Greek *auto*-self). The presence of the product either induces or represses its own production. Several proteins work as repressors for repression of their own production. This is achieved by the binding of repressor protein to the ribosome binding site, that is, SHINE-DALGARNO (S-D) SEQUENCE, or translation initiation codon (AUG) of mRNA. Autogenous regulation occurs among proteins that are incorporated into macromolecular assemblies. The assembled particle may be wobbly as a regulator, because it is large, too numerous or too restricted in its locations. If the assembly pathway is choked for some reason, free subunits accumulate and shut off the unnecessary synthesis of further components. Eukaryotic cells have this type of gene regulation.

autograft: A tissue graft from a donor of one genotype to a host of the same genotype. Autograft is usually done from one part to another part of the same individual.

autoimmune disease: Disorder caused by the immune system of the individual in which the body develops an immunological reaction against its own tissues. It occurs when the immune system loses the ability to distinguish self from non-self. The antibodies produced in the body start damaging its own cells/tissues. For instance, the hepatitis C virus may cause autoimmune hepatitis by this mechanism. Rheumatoid arthritis is another disease caused by the deposition of immune complexes of antibodies such as IgM and IgG in the bone joints causing severe pain.

autoimmune: An organism that makes antibodies against its own tissues.

autoinducer: A chemical used in quorum sensing that is secreted by cells and which is used to quantify cell density.

autoinduction: The induction process of a regulatory cascade in quorum sensing. It is prompted in response to the crossing of a threshold concentration of autoinducer in the environment.

autolysis: The destruction of a cell (or some of its components) through the action of its own hydraulic enzymes.

automata theory: The study of abstract machines and automata, as well as the computational problems that can be unravelled using them. It is a theory in theoretical computer science and discrete mathematics.

automated DNA extraction system: A large, expensive and complex instrumentation designed for high-throughput sample processing that has simplified the isolation of nucleic acids. It is a paramagnetic-particle handling system to process DNA samples and provide consistent yield and purity, as there is no detectable cross-contamination between samples in this system.

automated DNA sequencing through capillary gel systems: In these systems, slab gel electrophoresis is replaced by capillary gel electrophoresis to analyse DNA samples. In these systems, instead of scanning DNA as it migrates through 96 lanes each in a series of poured slab gel, DNA is sent through a set of 96 capillary tubes filled with the polymerized gel. As the automated DNA fragments pass through these capillary tubes, DNA fragments are scanned. Then, the robot moves the DNA samples and reagents through the tubes continuously. The system produces a steady flow of data, with each signal representing one of the four DNA bases.

automated DNA sequencing through slab gel systems: These systems make use of ultra-thin (75 μm) slab gels and involve running at least 96 lanes per gel, the runs being completed twice daily. In these systems, automation in sample loading of sequencing gels has also been achieved by using a plexiglass block – it has wells that are the same distance apart as the comb teeth cut in the drawing samples by the capillary action. Each well in the plexiglass block is filled with a sample (PCR dideoxy-reaction mixture) so that when the porous membrane comb is lowered on to the sample wells in the plexiglass block, the samples are drawn up automatically into the comb teeth by capillary action.

automated DNA sequencing: Automated DNA sequencing reduces the volume of low-level radioactive waste generated on campus while providing a suitable alternative to manual DNA sequencing. Manual DNA sequencing methods utilize radioactive isotopes to label the DNA. Automated DNA sequencing utilizes fluorescent tracers instead of radioisotopes to label the DNA, thereby eliminating or significantly reducing the use of radioactive materials in some research laboratories (M.H. Caruthers in 1985). L.M.

Smith *et al.* in 1986 developed a method for the partial automation of DNA sequence analysis. Fluorescence detection of the DNA fragments was accomplished using a fluorophore covalently attached to the oligonucleotide primer used in enzymatic DNA sequence analysis. A different coloured fluorophore was used for each of the reactions specific for the bases A, C, G and T. Modern automated DNA sequencing instruments (called DNA sequencers) can sequence multiple samples in a batch (run) and perform as many as 24 runs a day. These perform only the size separation and peak reading. The actual sequencing reaction(s), clean up and resuspension in a suitable buffer must be performed separately. The magnitude of the fluorescent signal is related to the number of strands of DNA that are in the reaction. If the initial amount of DNA is small, the signals will be weak. However, the properties of PCR allow one to increase the signal by increasing the number of cycles in the PCR. The presence or absence of a strand of DNA is detected by monitoring the output of the detector. The shorter the strands of DNA moving through the gel matrix, the faster they are detected. There is then a direct correlation between the length of the DNA strand and the time of detection. This relationship is then used to determine the actual DNA sequence. The output of these machines is not perfect, as it may contain reading errors and needs to be processed (*see* SEQUENCE ASSEMBLY). The assembly process for a contig made of ten samples takes about 5–15 min, depending on the quality of the samples. However, modern software can automatically process the output in seconds.

automated machine learning (autoML): AutoML aims to automatically configure a machine learning system to maximize its performance (e.g. classification accuracy).

automated planning and scheduling: A branch of AI that deals with the realization of strategies or action sequences, typically for execution by intelligent agents, autonomous robots and unmanned vehicles. Unlike classical control and classification problems, the solutions are complex and must be discovered and optimized in multidimensional space. Planning is also related to decision theory. *See* DECISION THEORY.

automated reasoning: An area of computer science and mathematical logic, which is dedicated to understanding different aspects of reasoning. The study of automated reasoning is used to produce computer programs that let computers reason completely, or nearly completely, automatically. Although automated reasoning is considered a subfield of AI, it also has connections with theoretical computer science and even philosophy.

automated solid-phase DNA synthesis: DNA can be synthesized by automated solid-phase technique, like polypeptides. A monomer is first attached to an insoluble column, and then activated monomers are supplied in the order of the sequence interested to form an oligonucleotide.

automatic selection: Refers to the increase in frequency of an allele that increases the rate of selfing. It is closely associated with the twofold cost of genome dilution. *Aka* UNCONSCIOUS SELECTION.

automixis: Mandatory self-fertilization.

automorphy: Exclusive derived characteristic. A trait that is present in only one member of a lineage or only one lineage among many genes.

automutagen: A mutagen produced in an organism in the form of a normal or abnormal metabolic product that is mutagenic in the same organism.

automutation: A mutation that arises without exogenic application of mutagens.

autonomic movement: Plant movement as a consequence of internal stimuli.

autonomous apomixis: Agamic seed formation in a plant that does not depend on pollination.

autonomous development: A type of development that is self-controlled, rather than externally based. For example, a *Drosophila* imaginal disc from a mutant for eye colour develops into a mutant eye colour in a wild-type host.

autonomous element: A transposable element that produces all the proteins required for transposition itself. Alleles produced by an autonomous element are mutable alleles, creating mutations that revert when the transposon is excised from the gene. For example, Ac element in maize. *Aka* AUTONOMOUS CONTROLLING ELEMENT. *Compare with* NON-AUTONOMOUS ELEMENTS.

autonomous pathway in flowering: Comprises a combination of factors involved in RNA processing and epigenetic regulation that downregulate the floral repressor, FLOWERING LOCUS C (FLC). However, components of the autonomous pathway are more widely conserved in plant species other than *Arabidopsis* than FLC.

autonomous replication (or replicating) sequence (ARS): A sequence that contains the origin of replication in the yeast genome. It contains four regions

(A, B1, B2 and B3), named in order of their effect on plasmid stability. The A-domain is highly conserved; any mutation abolishes the origin function. Mutations on B1, B2 and B3 diminish, but not prevent, functioning of the origin. *Aka* ARS ELEMENT.

autonomous robot: A robot that performs behaviours or tasks with a high degree of autonomy. Autonomous robotics is usually considered a subfield of AI, robotics and information engineering.

autonomous transposons: Autonomous transposable elements can move on their own. Ac elements of maize, for example, are autonomous because they can move on their own, whereas Ds elements are non-autonomous because they require the presence of Ac to transpose.

autophagy: The degradation of unnecessary or dysfunctional cellular components via the lysosomes, which are compartments inside the cell that digest cellular waste.

autophene: A genetically controlled characteristic that is revealed by the cell's genotype and which shows unusual behaviour in transplants and explants.

autophyte: Any organism such as a photosynthetic plant that manufactures its own food. *Contrasts* HETEROPHYTE.

autopoietic system: A 'system that constructs itself'. The concept of autopoiesis or self-production, pronounces the capability of living systems to make their components and ultimately to manufacture copies of themselves. The bacteria represent true autopoietic systems. *Compare with* CODEPOIESIS.

autoploid: A cell or an individual with more than two complete basic sets of chromosomes, which are derived from the same species. Autoploids are the euploids that contain a multiple of basic chromosome numbers where the same basic number gets multiplied. For instance, triploids (3×), tetraploids (4×), pentaploids (5×), hexaploids (6×) and octaploids (8×) – all are autoploids. Potato, sweet potato, lucerne and banana are natural autoploids.

autopolyploid: A polyploid organism that originates through the multiplication of a single genome of the same species.

autopolyploidization: The phenomenon of doublings of chromosome number by the failure of chromosomes to divide equatorially in mitosis following chromosome replication. Plants have commonly used autopolyploidization as an evolutionary tool.

autopolyploidy: Doubling of one's own entire chromosome set. This may lead to a new tetraploid species. *See* AUTOPOLYPLOID.

autoradiograph: An image displaying the location of radioactive substances in an object. *Aka* AUTORADIOGRAM.

autoradiography: A process to detect radioactively labelled molecules (which usually have been separated in an SDS-PAGE or agarose gel) based on their ability to create an image on photographic or X-ray film. The slab of gel or other material in which the molecules are suspended is placed on top of a photographic film. The two are then tightly fixed together such that no movement is allowed and the film is exposed for a period of time. The exposed film is subsequently developed and the radioactive area is observed as a dark spot. Autoradiography has also been engaged to track the spread of (radioactively labelled) viruses in a living plant. After treatment (the radioactive labelling process), the whole plant (in a slab) is placed on top of a photographic film. The image is called an AUTORADIOGRAPH or AUTORADIOGRAM.

autoreduplication: Biological systems that produce the template for their reproduction to duplicate themselves.

autoregression: The production of a series of observations whereby the value of each observation is partially dependent on the values of those that have immediately preceded it.

autoregulation: A regulatory system of gene expression in which the product of a structural gene moderates its own expression.

autosegregation: The incidence of changes in the chromosome complement during the formation of the egg cell.

autosomal chromosomes: All of the human chromosomes except for the X and Y chromosomes. *Aka* AUTOSOMAL SET.

autosomal dominance: Dominance is a condition in which one member of an allele pair is manifested to the exclusion of the other. An autosomal dominant gene is one that occurs on an autosomal (non-sex-determining) chromosome. As it is dominant, the phenotype it gives will be expressed even if the gene is heterozygous.

autosomal dominant condition: A disease caused by a gene not on a sex chromosome. The phenotype/disease/trait is expressed by individuals who have one copy of the gene variant. Individuals with an autosomal dominant condition have a 50% chance of passing the gene variant to offspring with each pregnancy. *Compare with* AUTOSOMAL RECESSIVE CONDITION.

autosomal dominant disorder: Autosomal dominant is one of many ways that a trait or disorder can be passed down through families. In an autosomal dominant disease, if you get the abnormal gene from only one parent, you can get the disease. Often, one of the parents may also have the disease.

autosomal dominant gene: A gene on one of the autosomes (non-sex chromosomes) that can always be expressed, even if only one copy is present. The chance of passing the gene to offspring is 50% for each pregnancy.

autosomal mutations: Autosomal mutations are positioned in the chromosomes that occur as a homologous pair in both sexes.

autosomal recessive condition: A disease caused by a gene not on a sex chromosome. The phenotype/disease/trait is expressed by individuals who have inherited two genetic variants at the same locus or gene. Most often one changed copy of the gene is inherited from each parent. *Compare with* AUTOSOMAL DOMINANT CONDITION.

autosomal recessive disorder: Autosomal recessive inheritance is a way through which a genetic trait or condition can be passed down from parent to child. A genetic condition can occur when the child inherits one copy of a mutated (changed) gene from each parent. The parents of a child with an autosomal recessive condition usually do not have the condition.

autosomal recessive inheritance: Describes a type of inheritance where an individual must inherit a mutation in both copies of a gene in order to develop the associated trait or disorder.

autosome: A chromosome not involved in sex determination. For example, the diploid human genome consists of 46 chromosomes – 22 pairs of autosomes and 1 pair of sex chromosomes (XX in female and XY in male). Since autosomes are similarly distributed in both sexes, the segregation of autosomes during meiosis does not normally affect the determination of sex. *Compare with* SEX CHROMOSOMES and ALLOSOMES.

autosyndesis: The pairing of complete or partial homologues of chromosomes.

autotetraploid: An autopolyploid is a cell or an individual with four similar genomes from one species. For example, lucerne, coffee, potato and groundnut are autotetraploid plants. If a given gene exists in two allelic forms *A* and *a*, then five genotypic classes can be formed: *AAAA* (quadruplex), *AAAa* (triplex), *AAaa* (duplex), *Aaaa* (simplex) and *aaaa* (nulliplex).

autotetraploidy: The condition in which a cell or an individual has four similar genomes from one single species.

autotransporter system: An alternative name for TYPE V SECRETION SYSTEM (T5SS) of Gram-negative bacteria.

autotriploid: An autopolyploid having three copies of each chromosome.

autotroph: An organism that manufactures its nutritive substances from simple inorganic molecules, carbon and nitrogen sources, such as carbon dioxide and ammonia or plants capable of photosynthesis. Plants are considered autotrophs, as green plants can synthesize organic compounds such as sugars, vitamins and amino acids for themselves through photosynthesis. Therefore, in agricultural practices, these organic compounds are not added to the field. Photosynthetic organisms change light energy into the chemical energy they need to develop, grow and reproduce. *Contrasts* HETEROTROPH.

autotrophic: A living system that obtains complex organic materials from its surroundings and uses them as nutrition.

autotrophy: Autotrophy is the capacity of an organism to use light as the sole energy source for the synthesis of organic material from inorganic elements or compounds. Autotrophic organisms include green photosynthetic plants and some photosynthetic bacteria. *Compare with* HETEROTROPHY.

autotropism: The ability of plants to self-control a stimulated bending or curving of an organ in a way that the previous shape is re-established.

autozygosity: Homozygosity in which the two alleles are identical because they have a common descent (i.e. they are copies of one ancestral gene).

autozygote: A diploid individual in which the two genes of a locus are identical by descent from an ancestral gene. *Aka* HOMOZYGOTE.

auxin-cytokinin ratio: The relative amount of auxin to cytokinin present in plant issue-culture media. It was first documented by C.O. Miller and F. Skoog in 1957. Varying the relative amounts of these two hormone groups in tissue culture formulae was recognized to affect the proportional growth of shoots and roots *in vitro*. As the ratio of these hormones is increased (increased auxin or decreased cytokinin), roots are more likely to be developed, and as the ratio is decreased (decreased auxin or increased cytokinin) root growth drops and shoot initiation and growth are promoted.

auxins: Hormones that encourage longitudinal growth in the cells of higher plants. In combination with cytokinin, auxin is required for the sustained propagation of many cultured plant tissues. Broad-leaved weeds like dandelions are much more susceptible to auxins than narrow-leafed plants like grass and cereal crops. Auxins regulate plant growth (e.g. stimulate cell enlargement, cell division, initiate roots/growth and flowering). Indole acetic acid (IAA), naphthalene acetic acid (NAA), indolebutyric acid (IBA), 2,4-dichlorophenexy acetic acid (2,4-D) and picloram are some of the auxins commonly employed in plant tissue culture. IBA is a rooting auxin, whereas 2,4-D is generally used for initiating callus. They are required in low concentrations ranging from 0.1 to 5.0 mg/l.

auxotroph: A plant unable to grow on a medium containing the minimum nutrients essential for the growth of the wild type. It requires supplementing of the minimal medium by some growth factor(s). Auxotroph mutants are used in basic studies. In plants, auxotroph mutants are used during somatic hybridization to isolate somatic hybrid cells.

availability: Describes the amount of a nutrient and/or water in fertilizers or the soil, respectively, that a plant can immediately absorb. It can be different from the actual amount of the nutrient present.

available water capacity: The amount of water that a soil can store in a form available to plants. It is almost equal to the moisture content at field capacity minus that at the wilting point.

average effect: The effect of an allele on a quantitative trait as predicted by regression. For a random-vvmating population that is in linkage equilibrium, this corresponds to the average excess. *See* AVERAGE EXCESS.

average excess: The difference between the average trait value of individuals who carry a particular allele and the average of the population. For a random-mating population that is in linkage equilibrium, this corresponds to the average effect. *See* AVERAGE EFFECT.

average: A quantity or rating that signifies or comes close to an arithmetic mean.

avidin: A glycoprotein component in egg white that binds to the vitamin biotin with high affinity (kDa = 10^{-15}). Proteins and nucleic acids can be linked to biotin (biotinylated), and the avidin-biotin reaction can then be used in several assay methods, such as antigen-antibody reactions or DNA hybridization (e.g. enzymes conjugated with avidin can be used to bind to biotinylated antibodies).

avidity: The accumulated strength of multiple affinities from multivalent non-covalent binding interactions.

***avirulence* (*avr*) gene:** A gene in a pathogen that makes the pathogen elicit an incompatible (defence) response in a resistant host plant and may increase pathogen virulence in a susceptible host plant. The result of the interaction of an avirulence gene product with its corresponding plant resistance (R) gene product is usually a hypersensitive reaction.

avirulence (Avr): Lack of virulence; the inability of a pathogen to infect.

avirulent bacteria: Benign bacteria that do not cause disease.

avirulent strain: A strain of a parasite that is unable to infect and cause disease in a host plant. *See* NON-PATHOGENIC. *Contrasts* VIRULENT.

Avogadro's number: The number of atoms, 6×10^{23}, in 1 g of hydrogen and thus in the atomic or molecular weight equivalent in grams of any element or molecule.

avoidance of selectable marker genes (SMGs) in plant transformation: Although it is achievable to obtain transgenic plants without using selection, the majority of the screened plants would be untransformed, and thus, not using an SMG at all is very laborious and useful for just a few plant species only.

avoidance: (i) A belief in plant disease control in which plants are grown at times or locations where the pathogen is inactive or not present. (ii) The ability of an organism to avert an injurious stress, pathogen or predator from penetrating its tissues. For example, drought avoidance may be achieved through restriction of water loss or by expansion of the root system to a greater supply of water.

avoiding recircularization during cloning: In order to avoid recircularization after being cut by restriction enzyme, two methods are used – DIRECTIONAL CLONING and PHOSPHATASE TREATMENT. (i) In the case of directional coiling, two cuts are made in two different restriction sites which result in keeping the cuts opened. pBR322 has a unique site for *Bam*-HI and *Hin*dIII very close to each other in the tet^R region. Both generate cohesive ends, but as the sites are different, recircularization is not possible. (ii) Treatment with bacterial or calf intestinal alkaline phosphatase, removes 5′-phosphate group. With the presence of OH groups on both ends, strands remain separate. However, the phosphate group could be added to 5′-OH group by T_4 polynucleotide kinase.

Avr proteins: These are diverse, and many have pathogenicity effector functions that play important roles in enhancing infection. *See* AVIRULENCE PROTEINS.

awn: (i) Any long, bristle-like appendage at the apex of the outer bract of some cereal and grass flowers. (ii) In the Poaceae, awn is an attachment terminating or on the back of glumes or lemmas of some grass spikelets. (iii) In the Geraniaceae, awn is the part of the style that stays attached to the carpel that separates from the carpophore (column). (iv) Awn is a generally straight, stiff pappus element, varying from stiffly and bristle-like to hard and needle-like. In Strophanthus, the awn is the beak of the seed and stipe of the coma hairs.

axenic: A pure culture of one species. It infers that cultures are free of contamination by microorganisms.

axenic culture: A culture without foreign or undesired life forms. An axenic culture may include the purposeful cocultivation of different types of cells, tissues or organisms. Axenic cultures are established by using the technique of surface sterilization of inoculum with appropriate sterilants like $HgCl_2$ or sodium hypochlorite and then culturing the inoculum *in vitro* on a sterile medium.

axil: The upper angle between one part of a plant and another (e.g. the stem and a leaf).

axil placentation: The type of ovule attachment within a fruit in which the seeds become attached along the central axis at the junction of the septa.

axile: (i) Situated in the axis of anything; as an embryo that lies in the axis of a seed. (ii) On an axis; of a placenta, on the central axis of the ovary.

axillary: Borne in or arising from the axil, usually referring to the axil of a leaf.

axillary bud: Develops in the axil of a leaf. The presence of axillary buds distinguishes a leaf from a leaflet. *See* LATERAL BUD.

axillary bud proliferation: (i) Propagation in culture using protocols and media that promote axillary (lateral shoot) growth. (ii) Micropropagation of plantlets in culture, achieved primarily through hormonal inhibition of apical dominance and stimulation of lateral branching.

axillary shoot: Shoot formed axillary to the primary shoot meristem.

axillary tiller: When a tiller is formed at each crown node. Axillary tillers usually begin to emerge when the plant has three leaves; rarely more than five axillary tillers are formed on a cereal plant. *Compare with* COLEOPTILAR TILLER.

axis: The main stem of a whole plant or inflorescence, and the line along which this stem extends. The central column upon which other parts are borne.

axoneme: The bundle of microtubules and accompanying proteins that form the main part of a cilium or flagellum in a eukaryotic cell, that is responsible for their movement.

aza-immunoprecipitation (aza-IP): This method identifies cytosines targeted by $5^{m}C$ methyltransferases and allows the identification of $5^{m}C$ residues even in low abundance RNA species. It is applied to live cells and does not require knocking out any methyltransferase gene.

azacytidine (or 5-azacytidine): A drug ($C_8H_{12}N_4O_5$) that may trigger the expression of rRNA genes through the reduction of their methylation level.

azide: A potent mutagen that induces gene mutations in high frequency. It induces chromosome aberrations in low frequency.

azotobacter: A non-parasitic, nitrogen-fixing diazotrophic bacterium that, due to its high metabolic activity, is a significant component of the nitrogen cycle.

B

B chromosomes: The chromosomes which are found in addition to the normal chromosomes. A heterogeneous group of chromosomes existing in numerous plant species, which differ in their morphology, numerical variation, meiotic pairing and mitotic behaviour from normal A CHROMOSOMES. B chromosomes derive from the A chromosomes by abnormal division processes and successive modifications. Up to 12 B chromosomes have been observed in addition to the diploid A chromosome complement in rye. B chromosomes are not required for normal function in either sex and are present in only some individuals. *Aka* SUPERNUMERARY CHROMOSOMES, ACCESSORY CHROMOSOMES or EXTRA CHROMOSOMES. *Compare with* A CHROMOSOMES.

B line: The fertile counterpart or MAINTAINER LINE of an A line. It does not have fertility restorer genes. B line is used as the pollen parent to maintain the A line and is used in hybrid seed production.

B3 domain: The B3 DBD (IPR003340, SCOP 117343) is found exclusively in transcription factors from higher plants and restriction endonucleases *Eco*RII and *Bfi*I and typically consists of 100–120 residues. It includes seven beta sheets and two alpha helices, which form a DNA-binding pseudobarrel protein fold.

baccate: A fruit looking like a berry that may or may not be a true berry.

***Bacillus*:** A genus that includes a rod-shaped bacterium.

***Bacillus subtilis* (*B. subtilis*):** A (rod-shaped) aerobic bacterium commonly used as a host in recombinant DNA experiments. During the 1990s, research showed that corn (maize) plant tissues infected with the endophyte *B. subtilis* were less likely to become infected with *Fusarium moniliforme* fungus. Other research has indicated the potential for prior infection of maize plant tissues to hinder any subsequent aflatoxin production in that plant by *Aspergillus flavus* fungus.

***Bacillus thuringiensis* (*Bt*):** Discovered by bacteriologist I. Shigetane on a diseased silkworm in 1901, this is a species of Gram-positive bacteria that dwells in soil. Later discovered on a dead Mediterranean flour moth and first named *B. thuringiensis* by E. Berliner in 1915, today *B. thuringiensis* denotes a group of rod-shaped soil bacteria found all over the earth that produce 'cry' proteins which are indigestible by – yet still 'bind' to – specific insects' gut (stomach) lining (epithelium cell) receptors. The 'cry' proteins are toxic to certain classes of insects (corn borers, corn rootworms, mosquitoes, black flies, some types of beetles, etc.) but are harmless to all mammals.

***Bacillus thuringiensis* (*Bt*) gene:** A gene from the bacteria *B. thuringiensis* that confers resistance to lepidopterous insects. *Bt* genes have been successfully transferred to cotton, tobacco, etc. by genetic engineering methodology.

***Bacillus thuringiensis* (*Bt*) genes for production of cry proteins:** At least 20,000 strains of *B. thuringiensis* are known. *B. thuringiensis* is employed as a biological pesticide (biopesticide). Engineered plants possess a gene from *Bt* inserted into their own genetic material. This foreign gene produces a natural protein that kills insects after the *Bt* protein is ingested. The *Bt* toxin degrades rapidly to non-toxic compounds. *Bt* genes that code for the production of these cry proteins that are toxic to insects have been inserted by scientists since 1989 into vectors (i.e. viruses, other bacteria and other microorganisms) in order to confer insect resistance to certain agricultural plants (e.g. via expression of those *Bt* proteins by one or more tissues of the transgenic plant). The targets include beetles, mosquitoes, black flies, caterpillars and moths. Some crops have been genetically engineered to produce *Bt* toxins. *See* *BT* CROPS.

***Bacillus thuringiensis israelensis*:** One of the approximately 30 subspecies groupings within the approximately 20,000 different strains of the soil bacteria known (collectively) as *B. thuringiensis* (*Bt*). When eaten (e.g. due to presence on food), the protoxin proteins produced by *B. thuringiensis israelensis* (*Bti*) are toxic to mosquitoes and black fly (Diptera) larvae.

Bacillus thuringiensis kurstaki: One of the approximately 30 subspecies groupings within the approximately 20,000 different strains of the soil bacteria known (collectively) as *B. thuringiensis* (*Bt*). When eaten (e.g. as part of a genetically engineered plant), the protoxin proteins produced by *B. thuringiensis kurstaki* (*Btk*) are toxic to certain caterpillars (Lepidoptera larvae), such as the European corn borer (pyralis). *Btk* kills borers via perforation of that insect's gut by cry ('crystal-like') proteins that are coded for by the *Btk* gene.

Bacillus thuringiensis tenebrionis: One of the approximately 30 subspecies groupings within the approximately 20,000 different strains of the soil bacteria known (collectively) as *B. thuringiensis* (*Bt*). When eaten (e.g. as part of a genetically engineered plant), the protoxin proteins produced by *B. thuringiensis tenebrionis* (*Btt*) are toxic to certain insects.

Bacillus thuringiensis tolworthi: One of approximately 30 subspecies groupings within the approximately 20,000 different strains of the soil bacteria known (collectively) as *B. thuringiensis* (*Bt*). When eaten (e.g. as part of a genetically engineered crop plant), the protoxin proteins produced by *B. thuringiensis tolworthi* (*Btt*) are toxic to certain caterpillars (Lepidoptera larvae), such as the European corn borer (pyralis).

back mutation: A mutation that reverses the effect of a previous mutation by restoring the original nucleotide sequence in DNA, or original function in the protein, or original phenotype. *Aka* REVERSE MUTATION. *See* REVERTANT.

back mutation-based test: A reverse mutation in which a mutant gene reverts to the original standard form and/or wild type; it is rare to forward mutations, but often strongly selected for. The B.N. Ames' test relies on back mutation for the detection of mutagens gene backward selection. Selection of parent plants is based on results from a progeny test. *See* AMES' TEST.

backbone of an amino acid: Consists of an amide, an alpha carbon and a carboxylic acid or carboxylate group.

backcross: Crossing the F_1 (or individual with dominant phenotype) with either of the homozygous parents. Backcross is used for transferring genes in conventional plant breeding where an otherwise high yielding and popular variety is required to be improved only for a specific character. Also used to describe the breeding protocol of an outcross followed by a backcross. The matings involving a hybrid genotype are used in genetic analyses to determine linkage and crossing-over values. A special combination of backcross with mapping of QTL – gene for quantitative characters in the form of advanced backcross QTL analysis (AB-QTL) – has been used to transfer genes for a quantitative trait (TSS) in tomato. *Compare with* TEST CROSS.

backcross (donor) parent: The parent of a hybrid with which it is again crossed or with which it is repeatedly crossed. Backcrosses may involve individuals of genotype identical to the parent rather than the parent itself.

backcross breeding: A system of breeding whereby recurring backcrosses are made to one of the parents of a hybrid, accompanied by selection for a specific character.

backcross method: See BACKCROSS BREEDING.

backcross ratios: The proportion of heterozygotes to bottom recessives expected in backcross is given by calculating 2^n where n is the number of factors involved (e.g. with 2 factors, $2^n = 4$ and the expected ratio is 3:1, with 3 factors, $2^n = 8$ and the expected ratio is 7:1).

backcross-assisted selection (BCAS): A procedure that permits the selection of plants carrying a favourable recessive allele in each generation, limiting the need for a progeny test, which is common in traditional backcrossing. In cases where the traditional means of selection are limited by environmental conditions (e.g. the presence of an abiotic or a biotic stress such as drought), this selection strategy is superior to conventional ones. Particularly in genetic transformation approaches, where the transgenes can be used as markers, BCAS may display a substantial advantage.

backcrossing: A breeding method used to move one or only a few required genes from an agronomically poor crop line to an elite line. This is done by crossing a donor parent to an elite line and crossing offspring with the 'desired gene(s)' back to the elite parent. *See* BACKCROSS.

background selection: The reduction in genetic diversity caused by selection against deleterious alleles at linked loci.

backtracking algorithm: The backtracking algorithm basically builds the solution by searching among all possible solutions. Using this algorithm, we keep on building the solution following criteria. Whenever a solution fails, we trace back to the failure point and build on the next solution and continue this process till we find the solution or all possible solutions are looked after.

backward elimination: A method to select variables in multiple regression that enters all variables into the regression equations and then eliminates the variable that adds the least to the prediction, followed by the other variables one at a time that decrease the multiple R by the least amount until all statistically significant variables are removed from the equation.

bacteria (sing. bacterium): Members of the Bacteria, one of the two major divisions of prokaryotes, the other being the Archaea. Most bacteria occur as single cells, and some cause disease. An enormous and diverse clade of microscopic, prokaryotic, single-celled organisms which lack a true nucleus. The word comes from the Greek word 'bakterion', which means stick, since the first bacteria viewed by man (via crude microscopes) appeared to be stick shaped. Any of a large group of microscopic organisms having round, rodlike, spiral, filamentous unicellular or non-cellular bodies that are often aggregated into colonies are enclosed by a cell wall or membrane (prokaryotes) and lack fully differentiated nuclei. Bacteria may exist as free-living organisms in soil, water and organic matter, or as parasites in the living bodies of plants and animals.

bacteria signal sequences: In bacteria, proteins fated to be secreted are produced as pre-proteins with N-terminal signal sequences, sometimes termed LEADER PEPTIDES. Bacteria signal sequences are short (25 amino acid residues) and comprise a hydrophobic central core which can acquire an α-helical structure, bordered by a region containing several charged residues.

bacterial artificial chromosome (BAC): A cloning vector derived from a single-copy F-plasmid of *E. coli*. It carries the F replication and partitioning systems that ensure low copy number and faithful segregation of plasmid DNA to daughter cells. Large genomic fragments (100,000–300,000 bases) can be cloned into F-type plasmids, making them useful for constructing genomic libraries.

bacterial artificial chromosome (BAC) clone: A BAC vector that carries a genomic DNA insert. Thousands of BAC clones make up a BAC library, which covers the target genome several-fold (5–10q), such that each genomic sequence is represented at least once.

bacterial artificial chromosome (BAC) contig: A contiguous set of overlapping BAC clones.

bacterial artificial chromosome (BAC) DNA microarray: The ordered alignment of different BAC clones, immobilized on supports of minute dimensions (e.g. nylon membranes, silicon, glass or quartz chips). *Aka* BAC MICROARRAY.

bacterial artificial chromosome (BAC) end pair: The sequence reads from both ends of a BAC clone.

bacterial artificial chromosome (BAC) end sequencing: The estimation of the sequence of one or both ends of a BAC clone. It can actually be used as a first step to study the sequence of whole genomes, starting from a few completely sequenced BAC clones that are extended into overlapping BAC clones selected from a set of end-sequenced BACs.

bacterial artificial chromosome (BAC) ends: Portions of a BAC genomic insert that are near the cloning site of the vector. BAC ends can be easily sequenced and used to anchor the BAC clone to other sequences or to a genetic map.

bacterial artificial chromosome (BAC) fingerprinting: A variant of the restriction fragment length polymorphism (RFLP) technique, which allows discrimination between different clones of a BAC library by digestion of the clones with a set of restriction endonucleases.

bacterial artificial chromosome (BAC) library: A collection of genomic DNA fragments, typically 100–150 kb and maximally 300 kb in length, that are cloned into a bacterial artificial chromosome vector. This molecular tool in plant science is employed to break complex genomes into handy pieces for study; most living organisms have large numbers of genes in their genomes – bacteria have between 5000 and 8000, humans 35,000 and wheat 150,000. The large collection of genes of an entire genome can be broken into small fragments, with one or two genes per fragment, and inserted into pieces of DNA. When the BAC is placed again into bacteria, they synthesize many hundreds of copies of themselves, including the foreign genes introduced. The BAC library consists of 1.2 million BAC clones with a size of 100–200 kb.

bacterial artificial chromosome (BAC) map: The well-ordered alignment of BAC clones such that a physical map of the genome can be is constructed. *Aka* GENOME FINGERPRINT MAP.

bacterial artificial chromosome (BAC) vector: An *Escherichia coli* vector for inserting DNA fragments larger than those that can be inserted by cosmids. BAC vectors are alternative to YAC vectors.

bacterial artificial chromosome (BAC), potential uses of: Some potential uses of BACs include (i) the 'manufacture' of probes (i.e. sequences of DNA utilized to 'find' complementary sequences within large pieces of DNA) via hybridization; (ii) the manufacture of 'DNA sequence markers' for use in marker-assisted selection (e.g. to guide choices made by commercial crop breeders, so they can more quickly select plants bearing gene(s) for a particular trait) to develop future improved crop varieties faster than was previously possible; and (iii) the BAC can accommodate large pieces of DNA up to 1 million base pairs. In principle, BAC is used like a plasmid. BACs can carry DNA from humans or mice or wherever, and the BAC is inserted into a host bacterium. As with the plasmid, when we grow that bacterium, the BACs are replicated as well. Huge pieces of DNA can thus be easily replicated using BACs, usually in the order of 100–400 kb. *See* CLONING VECTOR.

bacterial blight disease: A plant disease spread by bacteria that reasons death of leaves, spotting of bean pods and discolouration of seed. For example, in rice, the causal agent is *Xanthomonas oryzae* var. *oryzae*. Indications are lesions usually starting near the leaf tips or leaf margins or both and spreading down the outer edge. Young abrasions are pale to greyish green, later turning yellow to grey (dead) with time. In very susceptible varieties, abrasions may extend the entire leaf length into the leaf sheath.

bacterial cell: Bacteria are single-celled prokaryotic organisms. Bacterial cells are small, with a typical size range of 0.1–5 μm in diameter. The cell's coiled, circular chromosomal DNA is located in a region called the NUCLEOID, and its non-chromosomal DNA is located in circular plasmids. These structures are not bound by membranes.

bacterial chromosome: A single DNA molecule up to 1 mm long with a molecular weight of over 10^7 daltons, which may contain around 4000–5000 kilobases.

bacterial conjugation: This mechanism requires the presence of a special plasmid called the F PLASMID. F plasmid can give a bacterium new genes that may help it survive in a changing environment. Some plasmids can integrate reversibly into the bacterial chromosome. Bacterial conjugation is the transfer of genetic material (plasmid) between bacterial cells by direct cell-to-cell contact or by a bridge-like connection between two cells. A conjugation event occurs when the male cell extends his sex pili and one pilus attaches itself to the female. This attached pilus is a temporary cytoplasmic bridge through which a replicating F plasmid is transferred from the male to the female. When transfer is complete, the result is two male cells.

bacterial expressed sequence tags (ESTs): Bacterial ESTs are based on sequenced/mapped bacterial genes as an alternative to the genes of ('traditional' EST) *Caenorhabditis elegans* nematode. They are employed to 'label' a given gene in terms of that gene's function/protein.

bacterial filter: A special filter through which bacterial cells cannot pass. Bacterial filters are of immense importance for maintaining sterility of thermolabile chemicals/solutions which cannot be autoclaved (e.g. enzymes, blood plasma, antibiotics, etc.). These are sterilized by passing through nitrocellulose membrane filters with pore size of 2 μm.

bacterial gene bank: It is a set of bacterial clones, each of which carries a small clone of DNA. Altogether, these clones carry the entire genome. It is a kind of gene library, but the term is more common for prokaryotic gene collection. In bacteria, for instance, cloned DNA fragments are about 30 kilo bases (30,000 base pairs) long, so a complete gene bank of an ordinary sized bacterium (4500 kb, *E. coli*) could be carried in 150 bacterial clones.

bacterial growth curve: A curve showing the growth of bacteria in batch culture. To plot the bacterial growth curve, bacterial inoculum is added into liquid culture medium and incubated under optimum conditions. Population is counted at different intervals. A growth curve having four phases as lag, log, stationery and death phase is obtained.

bacteria nodulation sites, regulation of: T. Soyano *et al.* (2024) describe an oscillatory machinery that regulates rhizobium nodulation sites in a legume *Lotus japonicus*. Live-imaging and transcriptome analyses disclose an oscillating gene expression that contests a similar oscillation pattern in response to the plant hormone cytokinin. Disturbing cytokinin signalling leads to changed infection delivery. The size of the root region receptive to infection and the progression of the nodulation process are impacted by this periodic regulation. Cytokinin retort also displays an alike oscillation pattern. Cytokinin signalling is vital to preserving the periodicity, as observed in cytokinin receptor mutants exhibiting changed infection foci distribution.

bacterial lawn: A continuous cover of bacteria on the surface of a growth medium.

bacterial methylase systems: Modification of DNA with a methyl group, catalysed by DNA methyltransferases, is considered to be the most abundant form of post-replicative nucleotide modification found in genomes of both prokaryotes and eukaryotes. Well-studied methyl systems in bacteria include hsd, dam and dcm.

bacterial plasmid conformations: Bacterial plasmid DNA may appear in one of five conformations, namely, nicked open-circular, relaxed circular, linear, supercoiled (or covalently closed-circular), supercoiled denatured plasmids, which (for a given size) run at different speeds in a gel during electrophoresis.

bacterial plasmid subcloning: The insert sequences can derive from essentially any organism, or they may be isolated directly from the genome, from mRNA or from previously cloned DNA segments in which case, the procedure is termed subcloning. Using this method, a gene from a mammalian gene library can be subcloned into a bacterial plasmid (destination platform).

bacterial plasmids: In addition to the very large, circular DNA in the nucleoid, many bacteria contain one or more small, extrachromosomal, covalently closed, circular and independently replicating DNA molecules, called PLASMIDS. Plasmids were discovered by W. Hayes and so named by J. Lederberg in 1952. DNA packaging also occurs in case of plasmids. Plasmids are capable of transmission independent of the nucleoid DNA. A few classes of plasmid DNAs are sometimes temporarily incorporated into the chromosomal DNA. Most plasmids are only a few thousand base pairs long, but some are over 10^5 bp. In many cases, plasmids may not confer any obvious advantage on their host, and their sole function could be self-propagation. However, some plasmids carry genes that make a host bacterium resistant to antibacterial agents. Plasmid DNA is a double-stranded closed circular ring-like structure. Supercoiling is present in plasmid DNA. Plasmid DNA has PLECTONEMIC COILING (twisted thread). Bacterial plasmids are practically useful for cloning DNA inserts less than 20 kb. Some examples of bacterial plasmids are SEX-FACTOR (F AND F' FACTORS), RESISTANCE TRANSFER FACTORS (RTFS OR R PLASMIDS), VIRULENCE PLASMIDS, COLICIN PLASMIDS and METABOLIC PLASMIDS. *See details as separate terms.*

bacterial promoters: Promoters lack any extensive conservation of sequence exceeding 60 bp associated with RNA polymerase, but some short sequences within the promoters seem to be conserved, which could be referred to as signal sequences. A 6-bp sequence upstream from the start point is recognizable in almost all the promoters studied. This signal sequence is TATAAT, known as PRIBNOW BOX. Occurrence of bases in Pribnow box is reflected by consensus sequence $T_{89}T_{89}T_{50}A_{65}A_{65}T_{100}$. The actual location of this hexamer varies from −11 to −5 to −14 to −8. Pribnow box is 10 bp upstream the start points, so it is called −10 sequence. Centre of the similarities of sequence also occur at other locations centred about −35 sequence, sometimes called RECOGNITION REGION. Consensus sequence of −35 region with commonly occurring bases is T_{85} $T_{83}G_{81}A_{61}C_{69}A_{52}$. The distance between −35 and −10 sites varies between 16 and 19 bp in known promoters. The −35 region determines the efficiency of the promoter for polymerase recognition.

bacterial promoters and terminators: (i) Promoter Prediction by Neural Network. (ii) Promoters: (Softberry) – choose from BPROM (bacterial), TSSP (plant) and TSSG and TSSW (human). (iii) PromScan (modified by S. Richards, Queen's University, Canada). Scans small genomes for potential (RpoN) sigma54-RNAP-binding sites. The results will indicate the position of the promoter relative to the nearest gene. (iv) Virtual Footprint offers two types of analyses: (a) Regulon Analysis performs analysis of a whole prokaryotic genome with one regulator pattern; and (b) Promoter analysis performs analysis of a promoter region with several regulator patterns. (v) Prokaryotic promoter analysis using SAK allows analysis of moderately long sequences for prokaryotic Sigma-70 promoters using Sequence Alignment Kernel method. It was tested on a set of 669 known sigma-70 promoters of *E. coli*. (vi) PPP (Prokaryotic Promoter Prediction) determines from input sequences the putative promoter sequences by using multiple HMM models. (vii) SCOPE (Suite for Computational identification Of Promoter Elements), an ensemble of programs aimed at identifying novel *cis*-regulatory elements from groups of upstream sequences.

bacterial resistance: Resistance in a plant to bacterial disease(s). There are several bacterial diseases such as bacterial blight of rice that cause huge crop losses. Using the technique of molecular biology, several genes such as *osmotin* gene, *thionin* gene and *lysozyme* gene have been cloned and are being introduced into crop plants to develop genetic resistance against bacterial diseases.

bacterial RNA polymerase: Bacterial RNA polymerase holoenzyme is composed of six polypeptides ($\alpha_2\beta\beta'\omega\sigma$) of five different kinds. This means five different genes are needed to make *E. coli* RNA polymerase. Core enzyme has five subunits ($\alpha_2\beta\beta'\omega$ of four types). No covalent bond runs between the various chains of RNA polymerase.

bacterial RNA polymerase inhibition: Myxopyronin inhibits bacterial RNA polymerase. The antibiotic binds to a pocket deep inside the RNA polymerase clamp head domain, which interacts with the DNA template in the transcription bubble.

bacterial streaming: Large populations of bacteria that radiate from the cut surface of infected plant tissue when observed with a microscope.

bacterial stressosome: A common strategy used by microorganisms to survive multiple stresses involves a signal transduction cascade that increases the expression of stress-responsive genes. Stress signals can be integrated by multiprotein signalling hub that responds to various signals to affect a single outcome. Different sensory extensions respond to different signals, whereas the conserved domains in the core integrate the varied signals. The architecture of the stressosome provides the potential for cooperativity, suggesting that the response could be tuned dependent on the magnitude of chemophysical insult.

bacterial transcription factors: The protein molecules that bind core RNA polymerase ($\alpha_2\beta\beta'\omega$) and direct transcription initiation at cognate promoter sites.

bacterial transformation: The transformation of a bacterium from one form to another, or the acquisition of a new characteristic by a bacterium, by the uptake of genetic material.

bacterial transposon: Bacterial strains conferring multiple antibiotic resistances are thought to have been produced by activity of transposons. In bacteria, genes move to and out of the chromosome with frequencies as high as 10^{-2}. Transposons in bacteria promote illegitimate recombination. Bacterial transposable elements include insertion sequences, composite insertion sequence (CIS) elements and Tn family of transposons.

bactericidal: Killing or hampering bacteria.

bactericide: An agent that kills the bacteria. Antimicrobial drugs which kill bacteria directly are called bactericide (adjective: bactericidal). It is similar term as fungicide to fungus, insecticide to kill insect and weedicide to kill weed. *Aka* BACTERIOCIDE.

bacterin: Suspension of either killed or attenuated bacteria used for vaccination. Bacterin is the source of antigen that triggers the B cells of the body to manufacture antibodies.

bacteriocins: Bactericidal substances coded by plasmid genes of certain strains of bacteria and are active against some other strains of the same or closely related species.

bacteroid: Bacterial cells that are not normally shaped, usually found in root nodules of legumes.

bacteriology: This branch of science deals with bacteria, which includes classification, isolation/purification, characterization and exploitation.

bacteriolysis: The lysis of bacterial cells, usually induced by antibodies synthesized by the host organism.

bacteriophage (Abbrev. phage): A bacteriophage, or phage, is a virus that infects bacterium. These may be lytic (phage particles multiply in the bacterial cell and get released through lysis) or lysogenic (retention of phage in the host bacterium). Bacteriophage is simple in structure, consisting of either DNA or RNA as genetic material. These are surrounded by a protective protein coat called capsid. There are various types of bacteriophages, including P1, P38, T4, T7, R17, lambda, f1, phi X174 and M13, which have been extensively modified for use as molecular cloning vehicles. Altered forms of phages are employed in DNA cloning work, where they are convenient vectors. The bacteriophages commonly used for this purpose are derived from two 'wild' phages, called M13 and lambda. Lambda phage is used to clone segments of DNA in the range of around 10–20 kb. They are lytic phages – they replicate by rupturing their host cell and releasing more phages. On a bacteriological plate, this results in a small clear zone – a plaque. The M13 system can grow inside a bacterium so that it does not destroy the cell it infects but causes it to make new phages continuously. It is a single-stranded DNA phage and is used in the Sanger dideoxy DNA sequencing method. Both of these phages grow on

E. coli as a host bacterium. Bacteriophages are also utilized to create phage libraries containing human or other genes. *See* DNA SEQUENCING.

bacteriophage λ: A double-stranded DNA virus that infects *E. coli*. Its genome (48.5 kb) contains a central region of 15 kb that is not required for replication or formation of progeny phage in *E. coli*. Modified λ can be used as a cloning vector by replacing the central 15 kb with 10–15 kb of foreign DNA. A maximum of 53 kb can be packed into a phage. To insert larger foreign DNA into phage DNA, phage cloning vectors need to have some non-essential genes deleted. Bacteriophage lambda is frequently employed as a vector in **rDNA** experiments in *E. coli* and attaches to a specific receptor, which in the bacteria also generally functions in sugar transport across the cell wall. There are two kinds of λ phage vectors – INSERTION VECTORS and REPLACEMENT VECTORS.

bacteriophage λ exonuclease: A donut-shaped homotrimeric enzyme that degrades double-stranded DNA in the 5′→3′ direction to create long single-stranded overhangs used in recombination by the bacteriophage λ.

bacteriophage Mu: Temperate bacteriophage Mu (mutator) can cause mutations when it transposes. Its structure includes (i) A 37 kb linear DNA in the phage particle that has central phage DNA and unequal lengths of host DNA at the ends; and (ii) The DNA's G segment can invert and is found in both orientations in viral DNA. Following infection, Mu integrates into the host chromosome by conservative (non-replicative) transposition. (i) Integration produces prophage DNA flanked by 5 bp target site direct repeats; (ii) Flanking DNA from the previous host is lost during integration; (iii) The Mu prophage now replicates only when the *E. coli* chromosome replicates, due to a phage-encoded repressor that prevents most Mu gene expression. Mu prophage stays integrated during the lytic cycle, and replication of Mu's genome is by replicative transposition. Mu causes insertions, deletions, inversions and translocations.

bacteriorhodopsin: Pigmented protein found in the plasma membrane of a salt-loving bacterium, *Halobacterium halobium*. It pumps protons out of the cell in response to light.

bacteriostatic: Antimicrobial drugs which do not kill but prevent the bacteria from growing. A condition in which the growth of bacteria is prevented. *Aka* MICROBISTATIC ANTIMICROBIAL AGENTS.

bacteriostatis: A phenomenon in which antimicrobial agent cause reduction in the bacterial growth. *See* BACTERIOSTATIC.

bacterium (pl. bacteria): A prokaryotic, microscopic, single-celled organism recognized by a peptidoglycan cell wall but without a nucleus. Bacterium reproduces by binary fission. Genetic material (DNA) is present on the chromosomes or plasmid. Bacterial cells exhibit various shapes, such as spherical (coccus), rodlike (bacillus) and spiral.

bacteroid formation and development of mature nodule: When the bacteria are released from the infection thread into the host cell cytoplasm, they get transformed into swollen, irregular-shaped, branched structures called bacteroids, which then become surrounded singly or in small groups by a plant-derived membrane called peribacteroid membrane to form structures called 'symbiosome', the sites of nitrogen fixation. Symbiosomes secrete a hormone which enables the polyploid cells to divide rapidly forming the core of the nodule, and the surrounding diploid cells also to divide and differentiate to cover the nodule in cortical tissue and to form vascular connections with the root. This hormone also enables the production of leghaemoglobin, which protects the nitrogen fixation enzymes from oxygen. The specific nodule components are also produced to complete the nodulation process.

bacteroid: A stage found in symbiotic nitrogen-fixing bacteria after the bacteria are released from the infection thread.

baculiform: Rodlike; longer than wide. *Compare with* CYLINDRICAL.

baculovirus: An insect virus which that specifically infects insect cells and forms the basis of a protein expression system.

baculovirus expression vectors (BEVs): Vectors are used by researchers to insert alien genes into cells. The vectors in which the agent is a baculovirus infects certain types of insects only. These could conceivably be used to make a genetically engineered insecticide that is specific to a targeted insect (would not harm anything but that insect). For example, a BEV might be used to cause a cotton bollworm adult protein to be expressed when the bollworm is a juvenile, thus killing the bollworm before it has a chance to damage a cotton crop.

bait: A material used to trap insects. It is often added to pesticides (e.g. against snails).

bait-and-switch model: This model is based on the supposition that selection pressure favours recognition co-factors that continue to interact with Avr proteins but are not under pressure to avoid targeting by pathogen-derived proteins. A recognition co-factor does not have to be a true virulence target; it just has to look like one.

bait-and-switch model, effector-accessory protein interaction: Interaction of an effector with an accessory protein facilitates direct recognition by the NB-LRR receptor. (i) The NB-LRR protein is in its primed (signalling competent) but autoinhibited (restrained from signalling) state – the 'switch' is in the 'off' position. This state requires a functional nucleotide-binding pocket (presumably bound to ATP or ADP) and multiple intramolecular interactions including a fine-tuned balance between the LRR and ARC2. (ii) The pathogen Avr protein is brought into the NB-LRR system via the bait protein, either through direct binding or through alteration to the bait. (iii) The switch is flipped. Presence of the Avr protein causes perturbation and subsequent conformational changes to the LRR/ARC2 interface. This change is relayed through the ARC2 into an alteration in nucleotide-binding status. This in turn causes further conformational changes, ultimately resulting in signal initiation. Conformational changes within the nucleotide-binding pocket allow signalling motif(s) within the NB domain (green star) to associate with downstream signalling component. (iv) Intramolecular interactions within the NB-LRR protein are probably dissociated and reset, possibly allowing the protein to undergo repeated rounds of recognition, signalling and resetting given the continued presence of the Avr protein.

baking test: A procedure for estimating the overall quality of a flour or meal with respect to its performance in the bakery and the features of bread made from it.

Bal 31 nuclease: An exonuclease that degrades both strands of a DNA molecule at the same time.

balance between lytic cycle and lysogeny: In phage lambda, a very delicate balance is maintained between lytic cycle and lysogeny with a very sophisticated and intricate circuit of events. Repressor's synthesis is first established with the help of a promoter P_E under the control of genes *cII* and *cIII* and is subsequently maintained by the repressor itself through a promoter P_M.

balance in *Drosophila* sex determination: Sex in *Drosophila* is determined by ratio of number of X chromosomes to number of sets of autosomes. X/A ratio of 1.0 is female, 0.5 is male, while 0.67 and 0.75 are intersexes, >1.0 is superfemale and <0.5 is supermale. Zygotes having X/A ratio <0.33 and >1.5 do not develop beyond embryonic stage.

balance model of population structure: According to this model, a typical individual is heterozygous at most and homozygous at a few loci. This model based on studies with natural populations is pluralistic in approach and favours the existence of a variety of genotypes in the population.

balance view: The view that genetic variation is mostly maintained by balancing selection. This view *contrasts* the CLASSICAL VIEW that most variation is due to deleterious mutations.

balanced design: An experimental design in which all treatment combinations have the same number of observations. There are various different kinds of balance designs (variance balance, efficiency balance, etc.), none of which are equivalent to the COMBINATORIALIST'S PAIRWISE BALANCE but become equivalent to it if some extra conditions are satisfied (e.g. in BINARY EQUIREPLICATE UNIFORM BLOCK DESIGNS).

balanced diallelic: The genotype involving a multiple allelic locus in an autotetraploid where two different alleles are represented an equal number of times (e.g. *a1a1a2a2*).

balanced incomplete block design (BIB): A design in which one constant value for the residual variance of the difference between candidates for all pairs of candidates is indicated.

balanced lattice: A special group of balanced incomplete block that allows incomplete blocks to be combined into one or more separate complete replicates.

balanced lethal system: An arrangement of recessive lethal alleles that upholds a heterozygous chromosome combination. Homozygotes for any lethal-bearing chromosome perish.

balanced lethals: Recessive lethals at different loci, such that each homologous chromosome carries at least one lethal and is associated with inversions, so that no recombination occurs between the homologous chromosomes.

balanced lighting: Refers to a light source that provides both red and blue lights. Balanced lighting is important because some plants require both red and blue lights to grow and flower.

balanced load: That which depresses the overall fitness of a population owing to the segregation of inferior genotypes, the component genes of which are maintained in the population because they add to fitness in different combinations (as heterozygotes, for instance).

balanced polymorphism: A genetic polymorphism that is stable and is maintained in a population by natural selection because the heterozygotes for particular alleles have a higher adaptive fitness than either homozygote. It is called OVERDOMINANCE, as opposed to UNDERDOMINANCE, where the heterozygotes have a lower fitness, giving rise to unstable equilibrium. Natural selection maintains a gene even when it is lethal in homozygous state. In this case, more than one forms of a gene are maintained by BALANCING SELECTION due to selective advantage of the heterozygote over either of the homozygote type and where the genotypes are maintained in stable proportions. For instance, in the malaria prone environment, the heterozygotes are at a selective advantage over the two homozygotes. Polymorphism can now be easily detected using molecular markers. *See* POLYMORPHISM.

balanced tertiary trisomic (BTT): Tertiary trisomics constituted in such a way that the dominant allele of a marker gene closely linked with the interchange breakpoint is carried on the tertiary chromosome. The recessive allele is carried on each of the two normal chromosomes. The dominant marker allele for a mature plant character, such as red plant colour (*R*), may be carried on either the centromere portion or on the interchanged segment of the extra chromosome and should be linked with a male fertile gene (*Ms*). The two normal chromosomes should carry the corresponding recessive male-sterile allele (*ms*). All balanced tertiary trisomics would be male fertile and red, and all diploids would be male sterile and green.

balanced incomplete block designs: A statistical experimental design where the experimental units are grouped into blocks such that every pair of treatments occurs in an equal number of blocks. All the comparisons between treatments are made with equal accuracy, such that the design is balanced.

balancer chromosome: A *Drosophila* chromosome that carries multiple inversions, recessive lethals and dominant markers. Such chromosomes cannot recombine and cannot become homozygous and so can be used to keep individual wild-type chromosomes intact.

balancing selection: Selection involving opposing forces in which selective advantages and disadvantages cancel each other out. HETEROZYGOTE ADVANTAGE (or OVERDOMINANT SELECTION) is an example in which an allele selected against in the homozygous state is retained because of the superiority of heterozygotes. Other balanced states may occur including when an allele is favoured: (i) at one developmental stage and is selected against at another (ANTAGONISTIC PLEIOTROPY); (ii) in one sex and selected against in another (SEXUAL ANTAGONISM); and (iii) when it is rare and selected against when it is common (NEGATIVE FREQUENCY-DEPENDENT SELECTION).

Balbiani ring: A large RNA puff on giant chromosomes present during specific times of larval development in the chironomid family of dipteran flies. *Aka* CHROMOSOME PUFF.

ball metaphase: A type of mitosis with typically stomped chromosomes.

ballistic electron emission microscopy (BEEM): Electron tunnelling from the tip to the base electrode injects ballistic electrons. The current owing to these electrons delivers information on the interface electronic structure.

band-pass: A microscopic filter that allows light of a certain restricted range of wavelengths to pass through.

banded krait minor (Bkm) satellite DNA sequences: Repetitious DNA originally isolated from snake, *Bungarus fasciatus* which is localized exclusively in W chromosome of this snake. Bkm DNA-related sequences occur in the Y chromosome of the mouse. Bkm DNA-related sequences are in some way male determining.

banding: A special staining technique for chromosomes, which causes a longitudinal differentiation (e.g. Giemsa staining, which is a complex of stains specific for the phosphate groups of DNA). C-bands signify constitutive heterochromatin. G-banding patterns on mitotic chromosomes match very closely to the chromomere patterns of meiotic chromosome bivalents at pachytene – that is, the chromatin that is densely packed enough to be seen.

banding pattern: The linear pattern of deeply stained bands and weakly staining interbands that results from more or less defined local differences in the degree of DNA compaction along the chromosome.

bands in polytene chromosomes: A part of a chromosome that is distinguishable from its adjacent segments by appearing darker or lighter with one or more banding techniques.

bandwidth: The extreme rate of data transfer across a given path. Bandwidth may be considered as network bandwidth, data bandwidth or digital bandwidth.

bar code: A pattern of light and dark lines on labels that can be read by a light pen for direct entry into a computer. It is used for tagging and labelling of plants, seed accessions and so on.

bar diagram: A graphical representation of appropriate for disjoint categories to show the number of subjects or mean or rates by bars of corresponding height.

***Bar* eye in *Drosophila*:** A phenotype of *Drosophila* characterized by a narrow, oblong, bar-shaped eye in males and homozygous females with fewer facets than wild type. Heterozygous females show kidney bean-shaped eyes. The *Bar* mutation is a tandem duplication of X chromosome region 16A and *double-Bar* a tandem triplication of the region 16A. The *Bar* mutation arose by unequal crossing over between two Roo transposable elements, resulting in a tandem duplication.

***bar* gene in plants:** A gene which confers resistance to the herbicide bialaphos (bar) has been characterized. The *bar* gene was originally cloned from *Streptomyces hygroscopicus*, an organism which produces the tripeptide bialaphos as a secondary metabolite. The *bar* gene is used as a selectable marker in plant transformation, screenable marker in tissue culture and plant breeding and reporter gene in plant molecular biology.

bark: The outer skin of a tree trunk, outside the secondary, vascular cambium. It is composed of phloem tissue, which occurs as living inner and dead outer zones. The outer zone is penetrated by the cork layers formed from the cork cambia.

bark ringing: A procedure used for forcing fruit trees to flower. For this, a complete ring is cut around the trunk below the lowest branch and another ring is cut right below the first. The bark between the two rings is removed. The scar is covered with grafting wax.

barley yellow dwarf virus (BYDV): This virus infects all cereal species. Barley and oats are usually more harshly affected than wheat. Plants are most susceptible to infection early in growth. Infection results in stunting, discolouration and substantial yield loss. The virus is spread by quite a few species of cereal aphids, mainly bird-cherry aphid (*Rhopaosiphum padi*) or grain aphid (*Sitobion avenae*).

BARNASE (BActerial RiboNucleASE): A bacterial protein that consists of 110 amino acids and has ribonuclease activity that kills the cell that the barnase is in. It is synthesized and secreted by the bacterium *Bacillus amyloliquefaciens* but is lethal to the cell when expressed without its inhibitor *Barstar*. When the gene that codes for barnase is introduced via genetic engineering into a given plant and activated only in that plant's pollen (the barnase is synthesized only in its pollen cells), that plant's male parts become sterile. For crop plants possessing both male and female parts (monoecious plants), such male sterility enables the development of hybrids, because self-pollination does not occur. *See* BARSTAR.

barnase/barstar system: Genetic engineering of plants using 'barnase' and 'barstar' genes for developing male-sterile and restorer lines. Dominant nuclear male sterility and restorer genes have been successfully engineered into crop plants. This system utilizes the tapetal-specific transcriptional activity of the tobacco *TA29* gene and an RNase (barnase)/RNase inhibitor (barstar) from bacterium *B. amyloliquefaciens*. Introduction of male sterility gene construct [*TA29*, tapetum-specific promoter-barnase gene for cytotoxic protein] destroys tapetal cells, inhibits pollen formation and causes male sterility, whereas the introduction of restorer gene construct [*TA29*, tapetum-specific promoter-barstar (restorer gene) into another plant leads to the development of restorer line]. Barstar is a specific inhibitor of barnase. Thus, the developed male sterile and restorer lines have been used for production of hybrid seed in maize and oilseed rape.

Barr body: An inactive X chromosome in a cell with more than one X chromosome is made inactive in a process called *lyonization*, in species in which sex is determined by the presence of the Y chromosome (including humans) or W chromosome rather than by the presence of two X chromosomes or two Z chromosomes. The inactive, densely staining, condensed X chromosome is generally located next to the nuclear membrane, in nuclei of somatic cells of XX females. Barr body was so named after its discoverer Murray L. Barr, with graduate student Ewart G. Bertram, in 1948. The number of Barr bodies in such nuclei is one less than the total number of X chromosomes. Barr body is absent in the nuclei of normal males.

barrel structure: The barrel structure of an enzyme is composed of eight catenated strand-loop-helix-turn units. The β-strands are located in the interior of the protein, forming the staves of a barrel, whereas the α-helices pack around the exterior. The active sites of all known (β/α) 8-barrel enzymes are located in the β→α loops.

barren: Sterile or otherwise incapable of reproducing.

barriers to the expression of *E. coli* genes in *B. subtilis*: *E. coli* has the ability to recognize transcription and translation signals from a variety of organisms, e.g. expression of drug resistance genes from Gram-positive bacteria and amino acid biosynthesis genes from yeast. By contrast, *B. subtilis* is restricted in its ability to express genes from other genera. There are barriers to the expression of *E. coli* genes in *B. subtilis*: (i) *B. subtilis* RNA polymerase demands high fidelity to the canonical −35 and −10 promoter hexanucleotides and possibly other sequences as well. Other possible sequences are AT-rich regions upstream from −35 and a PuTTPuTC sequence at positions −18 to −14; (ii) *B. subtilis* ribosomes were not active with any of mRNAs from Gram-negative bacteria. Shine-Dalgarno sequences in mRNAs do not recognize *B. subtilis* ribosomes.

Barstar: A small protein synthesized by the bacterium *B. amyloliquefaciens*. Its function is to inhibit the ribonuclease activity of its binding partner barnase.

basal: Located near the base. The basal surface of a cell is opposite to the apical surface.

basal body: Short cylindrical arrangement of microtubules plus their associated proteins found at the base of a eukaryotic cell cilium or flagellum. It serves as a nucleation site for the growth of the AXONEME, an axial filament is the microtubule-based cytoskeletal structure. Basal body is meticulously comparable in structure to a centriole. *Aka* BASAL GRANULE, KINETOSOME and (in older cytological literature) BLEPHAROPLAST.

basal cell: The lowermost cell produced after the first division of the zygote.

basal factors: A component of transcriptional processivity. These factors are essential in response to activators. These factors position RNA polymerase at the start of the protein-coding region of a gene and send the enzyme on its way. Histones and basal factors (transcriptional factors) complete their binding of TATA BOX, which is a part of the promoter region. Transcription occurs when the basal factor binds at the TATA box site. However, when the site is entrapped in the nucleosome (i.e. bound by the histone), transcription is not initiated.

basal granules: Cytoplasmic organelles of characteristic fine structure to which cilia and flagella are attached.

basal knob: An assembly at the base of a nematode stylet. *See* STYLET KNOB.

basal lamina: Normally thin (20–100 nm) carpet of extracellular matrix secreted by epithelial cells that separates epithelial sheets from various other types of cells such as muscle or fat cells, from connective tissue. It is often incorrectly referred to as the basement membrane.

basal node: The node or joint located at the very bottom of plant stems just above or below the soil. Healing, structural support and biological processes take place at the basal node.

basal placentation: A type of free-central placentation in which the seeds are located at the bottom of the central ovary axis.

Basc *Drosophila* stock: A stock of *Drosophila melanogaster* which carries five mutations – a semidominant marker *Bar*, a recessive marker *white-apricot* (w^a) and three inversions covering entire X chromosome. It is used to detect sex-linked recessive lethal mutations. *See* MULLER-5 TECHNIQUE.

base: A substance that can accept a proton in solution. In a nucleic acid, the aromatic group attached to the pentose sugar of a nucleotide. A cyclic, nitrogen-containing compound that is one of the essential components of nucleic acids. It exists in five main forms: adenine (A), guanine (G), thymine (T), cytosine (C) and uracil (U). A and G have a two-ring structure and are called PURINES; T, C and U have a one-ring structure and are called PYRIMIDINES. A base joined to a ribose sugar joined to a phosphate group is a nucleotide – the building block of nucleic acids. The four bases comprising DNA are adenine (A), cytosine (C), guanine (G) and thymine (T). In RNA, thymine is replaced with uracil (U). In chemistry, an ingredient having a pH of greater than 7. It will react with an acid to form a salt. *See* ALKALINE. *Contrasts* ACID.

base analogue: A base analogue is similar to but differs from its normal form in certain substituents, its electronic structure is modified, and occasional errors may occur in the specificity of hydrogen bonds during incorporation into DNA. Base analogues affect replicating nucleic acids. 5-bromouracil, 5-chlorouracil and 5-iodouracil replace thymine in DNA, and they thus pair with adenine. 5-bromouracil (rare enol state) replaces cytosine. It pairs with guanine and thus causes G-C to A-T transitions and vice versa. 5-fluorouracil, 5-methylcytosine, 5-bromocytosine, hypoxanthine may act as a mutagen when incorporated into DNA. 2-aminopurine can pair in its normal tautomeric form with two bases, with thymine by two hydrogen bonds and with cytosine by one hydrogen bond. More often, 2-aminopurine pairs with thymine by two hydrogen bonds. Thus, 2-aminopurine is incorporated most frequently in place of adenine which itself would not be mutagenic. When 2-aminopurine pairs with cytosine, it is highly mutagenic. It is expected to induce base pair transitions from A-T to G-C and vice versa. Base analogues, 5-bromouracil and 2-aminopurine do not cause transitions. Since these two base analogues cause transitions in both directions (A-T to G-C and G-C to A-T), the mutations induced with them can be reversed by use of these substances.

base collection: A collection of germplasm that is kept for long-term, secure conservation and is not to be used as a routine distribution source. Seeds are usually stored at subzero temperatures and low moisture content.

base editing: A type of genome editing – a way to make targeted changes to the sequence of a piece of DNA. CRISPR, in common parlance, refers to genome editing approaches that use a CRISPR guide RNA molecule to find their target.

base editing in mitochondrial DNA: TALE-fused deaminases enable C-to-T or A-to-G base editing in human mitochondria. S. Cho and others in 2022 presented transcription-activator-like effector (TALE)-linked deaminases (TALEDs), composed of custom-designed TALE DNA-binding arrays, a catalytically impaired, full-length DddA variant or split DddA originated from *Burkholderia cenocepacia*, and an engineered deoxyadenosine deaminase derived from the *E. coli* TadA protein, which induce targeted A-to-G editing in human mitochondria. B.Y. Mok and others in 2022 used phage-assisted non-continuous and continuous evolution to evolve DddA variants with improved activity and expanded targeting scope. Z. Lei and others in 2022 have shown that the mitochondrial base editor induces extensive off-target editing in the nuclear genome.

base editor: Fusion of a Cas protein to a deaminase that enables direct base conversion without a DNA double-strand break.

base editors (BEs): Base editors are effective genome editing apparatuses that produce point mutations without generating double-stranded breaks. Adenine base editors (ABEs) directly alter A•T base pairs to G•C and cytosine base editors (CBEs) convert C•G base pairs to T•A.

base excision repair (BER): A cellular mechanism that repairs damaged DNA and is initiated by the activity of DNA glycosylases. In BER pathway, a damaged DNA base is detached by a glycosylase enzyme and ultimately replaced by a new base.

base excision repair (BER) in eukaryotes: Synthesis of DNA by DNA polymerase β is distributive on single-stranded (ss) DNA templates, but short DNA gaps with a 5′-phosphate in the gap are filled processively to completion. The *RAD3* gene of *Saccharomyces cerevisiae* is required for excision repair of ultraviolet-damaged DNA and is essential for cell viability. The *RAD3*-encoded protein shares a high degree of homology with the human *ERCC2* (*XPD*) gene product.

base excision repair (BER) in prokaryotes: Base excision repair involves the concerted effort of several repair proteins that recognize and excise specific DNA damages, eventually replacing the damaged moiety with a normal nucleotide and restoring the DNA back to its original state. Thus, BER is a repair pathway essential for DNA maintenance.

base excision sequence scanning (BESS): A method that can be used to detect a 'point mutation' in DNA involving rapid DNA sequence scanning.

base J-binding protein: A protein that binds to base J (β-d-glucosyl-hydroxymethyl-U), a modified thymine produced by hydroxylation and glucosylation of the methyl group of thymine (T).

base pair (bp): The hydrogen-bonded structure formed between two complementary nucleotides. Pairing of adenine (A) with thymine (T) or cytosine (C) with guanine (G) in a double-helix DNA. The distance along a piece of DNA is measured in bps. Two strands of DNA are held together in the shape of a double helix. Thymine in DNA replaces with uracil (U) in RNA where A-U base pair is formed. 'Non-canonical' base pairs (e.g. G-U) are common in RNA secondary structure by the bonds between base pairs. 1000 bp = 1 kb. *Aka* NUCLEOTIDE PAIR (NP).

base pair substitution: Nucleotide pair substitution. For example, A=T replaced by G≡C. During translation, it results in an altered protein that may lead to change in the trait even phenotype. These are also called point mutations. Nowadays, synthetic genes are produced in the laboratory by making base substitutions. For example, in *CryAC* gene isolated from *B. thuringiensis* is AT-rich. Whereas, plants are GC-rich. Therefore, for better expression of bacterial gene into plants, some of ATs are replaced with GC's.

base pairing rules: The hydrogen-bonded structure formed between two complementary nucleotides. For example, in double-stranded DNA, adenine [A] = thymine [T], and guanine [G] = cytosine [C].

base population: The initial set of genotypes from which selections are undertaken to establish a breeding population, for example, the wild forest. It is also sometimes referred to as RECRUITMENT POPULATION.

base ratio: The ratio of adenine (A) plus thymine (T) to guanine (G) plus cytosine (C). In double-stranded DNA, the amount of A is equal to the amount of T, and the amount of G is equal to the amount of C, but the amount of A + T does not equal the amount of G + C. The ratio (A + T):(G + C) is constant within a species but varies between species.

base seed: Particularly valuable seeds, usually derived from highly productive single plants (elite plants), which are used for seed production of commercially grown material. It is seed stock produced from breeder's seed by, or under the control of, an appropriate agricultural authority. Also refers to the source of certified seed, either directly or as registered seed.

base selection: Base selection is involved in selecting as to which of the four deoxyribonucleotides is added to the nascent strand. Base selection is done by DNA polymerase more rigorously at monophosphate incorporation step than at triphosphate step.

base sequence: The order of nucleotide bases (A, T, G and C) in DNA or (A, U, G and C) in RNA molecule determines the structure of proteins encoded by that DNA or RNA.

base sequence analysis: A method, sometimes automated, for determining the base sequence.

baseline data: The data that show the status of the subjects at the initial stage, generally at the time of beginning of the study. The baseline is used for evaluating any subsequent change either due to intervention or otherwise.

basic chromosome number: The gametic chromosome number of a true diploid species. It is represented by 'x'.

basic chromosome set: The standard chromosome complement of a given species.

basic dye: Staining salt carrying colour in the positive ion. Basic dyes are used for colouring the microorganisms for initial observations under the microscope. Dyes are salts comprising positive ion (cation) and negative ion (anion). One of these ions is coloured called chromophore. In basic dyes, the colour is carried by positive ion, whereas, in acidic dyes, the colour is carried by negative ion. Commonly used basic dyes include safranin, methylene blue and crystal violet, and common acidic dyes are acid fuchsin and eosin.

Basic Local Alignment Search Tool (BLAST): The most popular local alignment program for similarity search and sequence alignment developed by NCBI. A fast technique for detecting ungapped subsequences that match a given query sequence. A computer program that identifies homologous (similar) genes in different organisms, such as human, fruit fly or nematode. A sequence comparison algorithm optimized for speed used to search sequence databases for optimal local alignments to a query. The initial search is done for a word of length 'W' that scores at least 'T' when compared with the query using a SUBSTITUTION MATRIX. BLAST algorithm generates a list of short word match (default word size is 3 for protein and 11 for nucleotide) in query sequences and then database is searched for the occurrence of these words. Word hits are then extended in either direction in an attempt to generate an alignment with a score exceeding the threshold of 'S'. The 'T' parameter dictates the speed and sensitivity of the search. BLAST finds regions of similarity between biological sequences. The program compares nucleotide or protein sequences to sequence databases and calculates the statistical significance. It includes programs BLAST N (protein), BLAST P (nucleic acid) and BLAST X (protein).

basic proteins as suppressors of gene activity: Basic proteins of the nucleus are inhibitory for the gene action, for example, protamines and histones suppress the activity of special group of genes.

basic research: Research conducted for the advancement of knowledge concerning biological behaviour without concern for the immediate or practical benefits that might result. *Compare with* APPLIED RESEARCH.

basic scientific concepts: There are nine basic scientific concepts: (i) organization, (ii) cause and effect, (iii) systems, (iv) scale, (v) models, (vi) change, (vii) structure and function, (viii) variation and (ix) diversity.

basic: Having the properties of a base.

basidiocarp: A sexual fruiting body of a basidiomycetous fungus.

basidioma (pl. basidiomata): A sexual fruiting body of a basidiomycetous fungus. *See* BASIDIOCARP.

basidiomycetes: A large group of fungi including puffballs, shelf fungi, rusts, smuts and mushrooms that bear sexually produced spores on a BASIDIUM. *Aka* BASIDIOMYCOTA *or* BASIDIOMYCETOUS.

basidiospore: A sexually produced haploid (1N) spore in a basidium in case of basidiomycetes. Basidiospore formation is a common feature of edible fungi (mushrooms). Basidiospores are the meiotic products (haploid) formed on basidium.

basidium (pl. basidia; adj. basidial): A specialized cell or organ, often club shaped, in which karyogamy and meiosis transpire, followed by production of externally borne basidiospores (generally four) that are haploid.

basifixed: Something attached by its base (e.g. an anther attached to the filament). *Compare with* DORSIFIXED.

basipetal: Developing in sequence from the apex towards the base (i.e. with the youngest towards the base). For example, flowers in an inflorescence. Also, moving from leaves to roots (e.g. of molecular signals in plants). *Contrasts* ACROPETAL.

bast A strong, woody fibre, such as flax, hemp, ramie or jute, obtained from phloem tissue.:

bast plant: Crop plants used for fibre production (e.g. flax or hemp).

batch production: A process in which the cells are allowed to grow for a complete cycle (few days) and at the end, the product is collected. In biotechnology, genetically engineered bacteria, plant cells and animal cells are cultured in bioreactors to produce drugs or specialty chemicals. When the product is harvested at the completion of the cycle, this is called batch production.

Batesian mimicry: A kind of palatable mimicry in which one non-poisonous species (the Batesian mimic) mimics another poisonous species.

bathyphyll: A specialized leaf produced at the base of a plant, usually when the plant is immature, and which serves to present the plant to a substrate, especially notable in the fern Teratophyllum. *Contrasts* ACROPHYLL.

Bayes' rule: The conditional probability of an event, based on the occurrence of another event, is equal to the likelihood of the second event given the first event multiplied by the probability of the first event. It is represented by the formula $\Pr(A/B) = \{\Pr(B/A) \bullet \Pr(A)\} \div \{\Pr(B)\}$, where A, B = events; $\Pr(A|B)$ = probability of A given B is true; $\Pr(B|A)$ = probability of B given A is true and Pr(A), P(B) = the independent probabilities of A and B. *Aka* BAYES' THEOREM.

Bayesian Evolutionary Analysis Sampling Trees (BEAST): Software used to carry out Bayesian analysis of molecular sequences related by an evolutionary tree.

Bayesian inference: A branch of statistics based on Bayes' theorem. Bayesian inference does not use *P* values and generally does not test hypotheses. It requires one to correctly specify a probability distribution encapsulating the prior knowledge about, say, a treatment effect. The state of prior knowledge can be specified as 'no knowledge' by using a flat distribution, although this can result in wild and nonsensical estimates. Once the prior distribution is specified, the data are engaged to amend the prior state of knowledge to obtain the post-experiment state of knowledge. Final probabilities computed in the Bayesian framework are probabilities of various treatment effects. The price of being able to compute probabilities concerning the data-generating process is the necessity of specifying a prior distribution to anchor the calculations.

Bayesian programming: A formalism and a procedure for having a technique to postulate probabilistic models and solve problems when less than the essential information is available.

BC_1, BC_2, BC_3 and so on: Symbols signifying the first, second, third and so on backcross generations.

Bce4: The name of a promoter (region of DNA) that regulates an oilseed plant's gene(s) that code for components (such as fatty acids and amino acids) of that plant's seeds. For example, the Bce4 promoter makes such genes to be expressed during one of the earliest stages of canola plant's seed production.

beads theory: The disproved hypothesis that genes are arranged on chromosomes like beads on a necklace, indivisible into smaller units of mutation and recombination. *Aka* BEADS-ON-A-STRING THEORY.

beak: A prominent, pointed terminal projection, especially of a carpel or fruit.

bean dwarf mosaic virus (BDMV): A single-stranded DNA virus (genus *Begomovirus*; family Geminiviridae) that infects common beans (*Phaseolus vulgaris*) and causes stunted plant growth, mosaic and mottle symptoms in leaves. BDMV shows differential pathogenicity in common beans, infecting germplasm of the Andean gene pool (e.g. the snap bean cultivar 'Topcrop') but not that of the Middle American gene pool (e.g. the pinto bean cultivar 'Othello'). Resistance to BDMV in 'Othello' is associated with development of a hypersensitive response in vascular (phloem) tissues.

bean yellow dwarf virus (BeYDV): A disease happening in French beans that results in up to 90% yield losses.

beanstalk: A tower stretching from planetary surface to synchronous orbit. A proposal to generate electricity by tapping solar panels in space and transferring the energy using carbon nanotubes has captivated scientists.

beating up: Replenishing failed areas in a crop or stand by further sowings or plantings. Several other terms in use include BLANKING, FILLING, GAPPING, INFILLING, RECRUITING and REINFORCEMENT PLANTING. In forestry, the term is used to replace dead trees with new ones, especially during the early years of the establishment or reestablishment of a plantation.

Becquerel (Bq): The SI unit of radioactivity. The unit is so named after the discoverer of radioactivity, A.H. BeCquerel, in 1896.

bedding plant: A plant grown for its flowers or foliage that is suited by its habit for growing in beds or masses.

bee-free honey: A food tech start up founded in the Netherlands began investigating honey when making a fruit-based sweetener from peels, the side streams of apple and pear industrial products (C. Turrell in 2024). The Rotterdam-based Fooditive is manufacturing bee-free honey through enzymatic fermentation. The scheme duplicates the metabolic processes in a honeybee's stomach. For doing this, the scientists first recognized an enzyme that changes plant-based nutrients into a honey-like substance. They then engineered *E. coli* to make that enzyme that catalyses sugars into a product with the feeling, palate and viscidness of honey. The team is now working to produce the next version of the honey by reproducing the amino acids found in natural honey.

beet: Any of various biennial plants of the genus *Beta*, of the goosefoot family, especially *Berberis vulgaris*, having a fleshy red or white root and dark-green, red-veined leaves. Sugarbeet derived from *B. vulgaris* by selection has high sugar content.

beet yellows virus (BYV): A virus disease that dramatically reduces yield of sugarbeet crops.

behaviour: How an organism acts in its environment.

behaviour flexibility: All means of plant behaviour allowing temporary adaptation to environmental conditions.

behaviour informatics (BI): The informatics of behaviours so as to obtain behaviour intelligence and behaviour insights.

behaviour tree (BT): A mathematical model of plan execution used in computer science, robotics, control systems and video games. They describe switching between a finite set of tasks in a modular fashion. Their strength comes from their ability to create very complex tasks composed of simple tasks, without worrying how the simple tasks are implemented. BTs present some similarities to hierarchical state machines with the key difference that the main building block of a behaviour is a task rather than a state. Its ease of human understanding makes BTs less error prone and very popular in the game developer community. BTs have shown to generalize several other control architectures.

behavioural changes making specific insects undesirable: Changes in the specific insect behaviours to make them undesirable or destructive would be beneficial for humans. If scientists could genetically change the behaviour of an insect species from multivoltine to univoltine, that would eliminate the economic destructiveness of that pest. Other traits which could be of use are altering host range, fecundity, change in development time, suitable sites for pupation, sex ratio and humidity preferences.

behavioural ecology: The study of the evolutionary basis for animal behaviour due to ecological pressures.

behavioural genetics: (i) A branch of genetics that deals with the inheritance of different types of behaviour. (ii) The study of genes that may influence behaviour.

belief-desire-intention software model (BDI): A software model developed for programming intelligent agents. Superficially characterized by the implementation of an agent's beliefs, desires and intentions, it actually uses these concepts to solve a particular problem in agent programming. In essence, it provides a mechanism for separating the activity of selecting a plan (from a plan library or an external planner application) from the execution of currently active plans. Consequently, BDI agents are able to balance the time spent on deliberating about plans (choosing what to do) and executing those plans (doing it). A third activity, creating the plans in the first place (planning), is not within the scope of the model and is left to the system designer and programmer.

bell curve: The bell curve, *aka* NORMAL DISTRIBUTION, displays the mean, median and mode of the data you collect. It usually follows the shape of a bell with a slope on each side.

benchmark: The act of running a computer program, a set of programs or other operations in order to evaluate the relative performance of an object, normally by running a number of standard tests and trials against it. This term is also commonly used for the purposes of elaborately designed benchmarking programs themselves.

beneficial insect: An insect that serves the best interest of man – for example, insect pest predators and parasitoids and pollinators are friendly insects.

benign missense variants: The missense variants that are not responsible for a disease.

benign polymorphism: A genetic variant for which different phenotypes are clinically normal, such as the ABO blood groups.

benign tumours: Tumour containing cells resembling normal cell. They are non-malignant tumours and stay in the tissue where they originate. In contrast, the MALIGNANT TUMOURS are composed of abnormal multinucleate cells that undergo metastasis and can invade surrounding normal tissue.

benign variant: An alteration of a gene or genome, which does not appear to cause disease.

benzene: Molecule composed of a six-membered ring of carbon atoms, commonly drawn containing three alternating double bonds. The benzene ring occurs as part of many organic molecules of biological relevance.

Bernoulli trial. When a random process or experiment can result in only one of two mutually exclusive outcomes (e.g. dead or alive, sick or well, male or female), the trial is called a Bernoulli trial, and the distribution is called BINOMIAL DISTRIBUTION.

berry: A type of indehiscent fruit with the seeds immersed in the pulp (e.g. tomato).

best linear prediction (BLP): A statistical method that exploits matrix algebra to predict the breeding values for any trait or selection index. In BLP, fixed effects are presumably known. BLP is especially appropriate for analyses of messy or unbalanced data.

best, worst and average case: Expressions of what the resource usage is at least, at most and on average, respectively, for a given algorithm. Usually, the resource being considered is running time (i.e. time complexity), but it could also be memory or some other resource. Best case is the function which performs the minimum number of steps on input data of n elements. Worst case is the function, which performs the maximum number of steps on input data of size n. Average case is the function, which performs an average number of steps on input data of n elements.

β: *See* SELECTION GRADIENT.

β-arrestins: β-Arrestins bind to activated seven-transmembrane-spanning (7TMS) receptors (G protein-coupled receptors) after the receptors are phosphorylated by G protein-coupled receptor kinases (GRKs), thereby regulating their signalling and internalization. β-arrestin 2 (β-arr2) has an unexpected and analogous role for the single transmembrane-spanning type III **transforming growth factor β (TGF-β)** receptor (TβIII, also referred to as beta receptor on its cytoplasmic domain (likely at threonine 841)). However, such phosphorylation was mediated by the type II TGF-β receptor (TβRII), which is itself a kinase, rather than by a GRK. Association with βarr2 led to internalization of both receptors and downregulation of TGF-β signalling. Thus, the regulatory actions of β-arrestins are broader than previously appreciated, extending to the TGF-β receptor family as well.

β-carotene, therapeutic value of: β-carotene aids eyesight in people who consume it and may help prevent lung cancer and heart disease. Because β-carotene is processed into vitamin A by the human body, consumption of this phytochemical can help avoid various human diseases that result from vitamin A deficiency in developing countries where vitamin A is scarce. Examples of such diseases include coronary heart disease, certain cancers (cancer of prostate, lungs, etc.), childhood blindness, macular degeneration (a leading cause of blindness in older people) and various childhood diseases, which often result in death due to a weakened immune system.

β-carotene: A phytochemical (vitamin precursor) which is part of the chemical family called CAROTENOIDS. It is a precursor for vitamin A and has antioxidant properties. It is obviously produced in carrots, other orange vegetables and in the endosperm portion of the maize kernel. If the maize kernel seedcoat is torn (e.g. via insect chewing), the β-carotene inhibits growth of *A. flavus* fungi in the endosperm region of the kernel. In 1970, an orange-fruited cauliflower was discovered growing in a field in Canada. It was the result of a spontaneous mutation that caused β-carotene to be produced in that cauliflower plant at a level that was several hundred times higher than normal for cauliflower.

β-catenin: Multifunctional cytoplasmic protein that is involved in cadherin-mediated cell–cell adhesion, linking cadherins to the actin cytoskeleton. β-catenin can also act independently as a gene regulatory protein. It has an important role in animal development as part of a WNT SIGNALLING PATHWAY.

β-conformation: An extended, zigzag structure of a polypeptide (molecule) chain.

β-DNA: Also called B-FORM DNA. B-DNA is the term given for the canonical right-handed DNA helix that is the most common form of DNA. Canonical B-DNA is a double helix made of two antiparallel strands that are held together via hydrogen bonding in the A•T and G•C base pairs. This is the normal form of DNA which occurs in biological systems as a right-handed helix.

β-galactosidase: The enzyme encoded by the *lacZ* gene of lac operon of *E. coli* that splits lactose into glucose and galactose. It performs two functions: (i) breakdown of disaccharide lactose into two monosaccharides (glucose and galactose), and (ii) formation of inducer by conversion of lactose into allolactose. *Aka* protein 'z'.

β-galactoside acetyltransferase: An enzyme that is involved in lactose metabolism and encoded by gene *lac y* in the *lac* operon. *Aka* THIOGALACTOSIDE TRANSACETYLASE.

β-galactoside permease: An enzyme that is involved in concentrating lactose in the cell and is coded by gene *lac a* in the *lac* operon.

***β-glucuronidase* (*GUS*) gene:** A gene that produces beta-glucuronidase (GUS protein) in *E. coli* bacteria. *GUS* as a reporter gene is commonly used during plant transformation. The gene *uid A* cloned from bacteria *E. coli* codes for an enzyme β-glucuronidase, popularly called GUS, which breaks down glucuronides (X-gluc) giving a blue-coloured reaction. This is a simple but a destructive method.

β-lactam antibiotics: A category of antibiotics (e.g. penicillin G, ampicillin) that kill targeted bacteria by altering their essential cellular function of enzymatic controls that keep cell wall (peptidoglycan) synthesis (creation/repair) in balance with cell wall degradation. This causes cell wall breakdown and death of those bacteria (pathogens).

β-level: The beta level, or simply beta, is the probability of committing a type II error in hypothesis analysis, which entails agreeing to the null hypothesis when it is false.

β-oxidation: Oxidative degradation of fatty acids to form acetyl-CoA by successive oxidations at the β carbon atom.

β-particles: Radioactive particles generated from radioactive decay of heavier elements such as ^{3}H, ^{32}P, ^{35}S. These particles are negatively charged, sparsely ionizing and more penetrating.

β-sheet: A three-dimensional arrangement taken up by polypeptide chains that consist of alternating strands linked by hydrogen bonds between a carboxyl group's oxygen on one strand and the amide group's hydrogen from another strand. The alternating strands together form a sheet that is frequently twisted. Beta sheets may be parallel (both strands oriented in the same direction, amino to carboxyl terminus) or antiparallel (both strands running in opposite directions; for example, if one is oriented amino to carboxyl, the other would be running carboxyl to amino). *Aka* β-PLEATED SHEET.

***Bgl*II:** A type II restriction enzyme from bacterium *Bacillus flobigi*. The enzyme recognizes the DNA sequence 5′ A↓GATCT/3′ TCTAG↑A, where arrows indicate the sites the enzyme cuts. The sticky ends produced by *Bgl*II are complementary to the ends produced by *Bam*H1, *Bcl*I, *Xho*II, *Mbo*I and *Sau*3A. Thus, fragments produced by anyone of these enzymes will have single-strand extensions that can anneal with the sticky ends of any other of the enzymes above.

bias: A systematic error. Bias typically refers to the discrepancy between the average of many estimates over repeated sampling and the true value of a parameter. Therefore, bias is more related to frequentist statistics than to Bayesian statistics. For example, a mis-calibrated machine that reports cholesterol too high by 20 mg% on the average. A satisfaction questionnaire that leads the patients to never report that they are not satisfied with their medical care by using each patient's lowest blood pressure over 24 h to describe a drug's antihypertensive properties.

bicellular: Having composed of two cells.

bi-cropping: A method of growing cereals in a leguminous living mulch. This practice potentially reduces the need for synthetic inputs to cereal production while preventing losses of nutrients and increasing soil biological activity. It is a low-input production system for cereals.

bidirectional DNA replication: DNA replication occurs when two replication forks proceed away from the same origin in different directions. A proof of bidirectional replication came when *E. coli* cells were grown on the medium containing tritiated (^{3}H) thymine at low level, which was abundant to visualize the track of DNA from origin. After some time, the cells were exposed to high level of ^{3}H thymine. Two short tracks of dense silver grains were observed at each end of growing chromosome. This proved a bidirectional mode of DNA replication.

biennial: A plant which finishes its life cycle (including germination, reproduction and death) within 2 years or growing seasons. Biennial plants usually form a basal rosette of leaves in the first year and then flower, bear fruit and die in the second year. *Contrasts* ANNUAL *and* PERENNIAL.

bifloral: Showing two flowers.

bifoliate: A compound leaf consisting of precisely two leaflets, usually in a symmetrical pair (e.g. a leaf of *Colophospermum mopane*). *Compare with* JUGATE LOBED LEAF, found in most species of Bauhinia.

bifunctional alkylating agent: Alkylating agents carry two alkyl groups in reactive form which are capable of being transferred to other molecules where electron density is high, although the actual number of such groups present in the molecule may be more than two (e.g. myleran). *Compare with* MONOFUNCTIONAL ALKYLATING AGENT and POLYFUNCTIONAL ALKYLATING AGENT.

bifunctional DNA glycosylase: An enzyme with both DNA glycosylase and endonuclease/lyase activities.

bifunctional vector (plasmid): A DNA molecule able to replicate in two different organisms (e.g. in *E. coli* and yeast or *E. coli* and Streptomyces). These molecules are thus able to shuttle between the two alternative hosts and are, therefore, *aka* SHUTTLE VECTOR.

bifurcate: Split in two; forked.

bifurcation: A point in a phylogenetic tree in which an ancestral taxon splits into two independent lineages.

big bang theory: The theory that states that the universe began in a state of compression to infinite density, and that in one instant, all matter and energy began expanding and have continued expanding ever since.

big data: Data with many cases (rows) offer greater statistical power, while data with higher complexity (more attributes or columns) may lead to a higher false discovery rate. Big data contain greater variety, arriving in increasing volumes and with more velocity. Put simply, big data comes from new data sources and is larger and more complex. These data sets are so huge that traditional data processing software fail to manage them. These massive volumes of data can, however, be used to deal with the business problems one would have not been able to tackle before. This is referred to as three Vs, namely, VOLUME, VELOCITY and VARIETY. *See details as separate terms.*

big data analytics: Describes the process of uncovering trends, patterns and correlations in large amounts of raw data to help make data-informed decisions. These processes use familiar statistical analysis techniques – like clustering and regression – and applies them to more extensive data sets with the help of newer tools.

big data case study – DTN improves yields and profitability: Digital Transmission Network (DTN), a division of Schneider Electric, offers agricultural information solutions and market intelligence to its customers. Using DTN, farmers and commodity traders can have access to up-to-date weather and pricing data to better manage their business. DTN invests in a modern data integration tool that consolidates data from multiple sources without having to write a ton of custom code. With a clean and consistent set of boundaries, DTN is now able to combine critical weather and agronomic data from fields to provide accurate forecasts. Using DTN, farmers can improve yields and cut costs on the basis of these forecasts.

big data case study – getting started in agriculture: Big data can accurately revolutionize the agricultural sector only by having a cloud-based ecosystem with the right tools and software to integrate various data sources. These tools have potential to consolidate data on climate, agronomy, water, farm equipment, supply chain, weeds, nutrients and so much more to aid the farmer make benefitting decisions.

big data case study – the cloud and the future in agriculture: The introduction of cloud computing and big data has guaranteed that farmers have sufficient data points to make good decisions. Cloud computing has democratized the availability of huge computing power, as data centres and storage are now available on a 'pay-as-you-go' model. This has made it possible to bring together knowledge repositories that contain data such as weather, irrigation practices, plant nutrient requirements and several other farming techniques.

big data empowers precision farming: SMAG is short form for SMART AGRICULTURE. SMAG InVivo have developed many mobile applications to support farmers in their daily operations. SMAG has pooled 30 years of weather data history, satellite and drone images and soil types with a view to make informed decisions faster with an objective of using digitization to solve the food challenges of the 21st century. Using a tool to help process the vast amount of stored and accumulated data, SMAG developed a complex agronomic Data Crop algorithm, allowing for the use of different types of data to improve decision making. For example, Data Crop enables users to track the progress of crops over the year and predict yields – a data point that has led to incredible wheat production results. Currently, 80% of French agricultural land under wheat cultivation is managed through Data Crop. SMAG plans to expand this to other crops and countries as well.

big data in agriculture case studies: Four case studies show how big data solutions can have a real, hard-hitting impact on the ground: (i) DIGITAL TRANSMISSION NETWORK (DTN) uses big data to improve yields and profitability; (ii) SMAG INVIVO uses big data to empower precision farming; (iii) The cloud and the future of big data in agriculture; and (iv) Getting started with big data in agriculture.

big data in agriculture: To counter the pressures of increasing food demand and climate changes, policy makers and industry leaders are seeking assistance from technology forces such as IoT, ANALYTICS and CLOUD COMPUTING. IoT devices help in the first phase of this process – data collection. Sensors plugged into tractors and trucks as well as in fields, soil and plants aid in the collection of real-time data directly from the ground. Analysts integrate the large amounts of data collected with other information available in the cloud, such as weather data and pricing models, to determine patterns. These patterns and insights assist in controlling the problem. They help to pinpoint existing issues, like operational inefficiencies and problems with soil quality, and formulate predictive algorithms that can alert even before a problem occurs. The adoption of analytics in agriculture has been increasing consistently.

big data in feeding a growing population: Big data provides farmers granular data on rainfall patterns, water cycles, fertilizer requirements and more. This enables them to make smart decisions, such as what crops to plant for better profitability and when to harvest. The right decisions ultimately improve farm yields.

big data in managing supply chain issues: McKinsey & Company, an American multinational strategy and management consulting firm, reports that a third of food produced for human consumption is lost or wasted every year, a devastating fact, since the industry struggles to bridge the gap between supply and demand. To address this, food delivery cycles from producer to the market need to be reduced. Big data can help achieve supply chain efficiencies by tracking and optimizing delivery truck routes.

big data in optimizing farm equipment: Companies like John Deere have integrated sensors in their farming equipment and installed big data applications that will help better manage their taskforce. For large farms, this level of monitoring could be a rescuer as it lets users know of tractor availability, service due dates and fuel refill alerts. In essence, this effort enhances usage and ensures the long-term health of farm equipment.

big data in using pesticides ethically: Management of pesticides has been a complicated issue due to its side effects on the ecosystem. Big data allows farmers manage this aspect better by recommending three Ws about pesticides, namely, what, when and by how much to apply. By monitoring it closely, farmers can observe government regulations and avoid overuse of chemicals in food production. Moreover, this leads to an increased profitability, as because of this, crops do not get destroyed by weeds and insects.

big data on the farm, top use cases for: Big data can make a difference in relation to (i) FEEDING A GROWING POPULATION; (ii) USING PESTICIDES ETHICALLY; (iii) OPTIMIZING FARM EQUIPMENT; and (iv) MANAGING SUPPLY CHAIN ISSUES. *See details as separate terms.*

big data, benefits of: (i) Big data makes it possible to gain more complete answers because one has more information available. (ii) More complete answers mean more confidence in the data – which means a completely different approach to tackling problems.

big O notation: A mathematical notation that describes the restraining behaviour of a function when the argument tends towards a particular value or infinity. It is a member of a family of notations invented by Paul Bachmann in 1894, further elaborated by Edmund Landau in 1909 and others, collectively called BACHMANN-LANDAU NOTATION or ASYMPTOTIC NOTATION.

bigerm: Having two seeds.

bilabiate: Having two lips (e.g. the type of petals found in many irregular flowers).

bilateral: (i) Having two distinct sides, such as the two faces of a dorsiventral leaf. (ii) Arranged on opposite sides (e.g. leaves on a stem). *Compare with* DISTICHOUS *and* OPPOSITE. (iii) Bilaterally symmetrical, as in a leaf with a symmetrical outline.

biloculate: Having two loculi (e.g. in anthers or ovaries).

bimitosis: The concurrent occurrence of two mitoses in binucleate cells.

bimodal distribution: A statistical distribution having two modes.

bimodal population: A population containing one or more characteristics, the measurements of which show two peaks of relative frequency.

binary: (i) The term 'binary' refers to the function of interest being divided into two parts encoded by two separate plasmids. (ii) This term has several meanings, associated with the number 2 or the set (0,1). In statistical design theory, a block design is binary if no treatment occurs more than once in a block. In such a situation, a block is akin to a set (rather than a multiset) of treatments, and the block design signifies as an incidence structure.

binary fission: A type of asexual reproduction in which two cells of similar size and shape are formed by the growth and division of one cell. For example, most of the bacteria multiply through binary fission in which the bacterium elongates and then divides to produce two daughter cells of approximately equal size.

binary number: In mathematics and digital electronics, a number expressed in the base-2 numeral system or binary numeral system, which employs only two symbols, typically 0 (zero) and 1 (one).

binary observation: The probability distribution that describes the number of successes X observed in 'n' independent trials, each with the same probability of occurrence.

binary scale: A scale for recording data where there are only two possible responses (e.g. 'yes' or 'no'; 'present' or 'absent').

binary search algorithm: A search algorithm that finds the position of a target value within a sorted array. *Aka* SIMPLY BINARY SEARCH, HALF-INTERVAL SEARCH, LOGARITHMIC SEARCH or BINARY CHOP.

binary switch genes: During development, certain genes act as switches decreasing the number of alternative developmental pathways. Such decision point is usually binary (i.e. there are two alternatives possible), and the action of a switch gene programs the cell to follow only one of these pathways. These genes are defined by their ability to initiate the complete development of an organ or a tissue type. Some well-studied examples of such genes are myoblast-determining genes, eye formation genes in vertebrates and invertebrates and left/right asymmetry in nematode. *Aka* MASTER REGULATORY GENES.

binary switches: The modification of adjacent or nearby histone residues affecting recognition and binding by epigenetic reader proteins.

binary Ti vectors: A T-DNA binary system is a pair of plasmids consisting of a binary plasmid and a helper plasmid. The two plasmids are used together to produce genetically modified plants. They are artificial vectors that have both been created from the naturally occurring Ti plasmid found in *Agrobacterium tumefaciens*. In these vectors, the sequences to be transferred to the plant genome and the *vir* region genes reside on different plasmids.

binary tree: A tree data structure in which each node has at the most two children, referred to as the LEFT CHILD and the RIGHT CHILD. A recursive definition using just set theory notions is that a (non-empty) binary tree is a set of so many elements, called tuple (L, S, R), where L and R represent binary trees (or the empty set), and S is a singleton set. Some authors allow the binary tree to be the empty set as well.

binary variable: A variable or character that is assessed only in two categories such as yes/no, zero/one or male/female. A qualitative variable is mostly binary, but a quantitative variable can also be made binary by dividing into two categories, such systolic blood pressure <140/>140 mmHg.

binary vector system: A two-plasmid system in *A. tumefaciens* for transferring into plant cells a segment of T-DNA that carries cloned genes. One plasmid contains the virulence gene (responsible for transfer of the T-DNA) and the other plasmid contains the T-DNA borders, the selectable marker and the DNA to be transferred. Binary vectors are easy to construct but are difficult to maintain during the bacterial maintenance and multiplication. Different binary vector systems differ mainly in the plasmid region that facilitates replication in *Agrobacterium*.

binding: Interaction of one molecule to another, which causes a change in the shape and function of the target molecule.

binding site: A region on the surface of one molecule (usually a protein or nucleic acid) that is able to interact with another molecule through non-covalent bonding.

binemic: Chromosomes that comprise two DNA helices per metaphase chromatid.

binomial classification: It is the scientific binomial method of classification. *See* SPECIES.

binomial distribution: A common discrete distribution used in statistics, as opposed to a continuous distribution, such as normal distribution. This is because binomial distribution only counts two states, typically represented as 1 (for a success) or 0 (for a failure), given a number of trials in the data. *See* BERNOULLI TRIAL.

binomial distribution of genotype frequencies: Gene frequencies can be computed in decimals and can be treated and manipulated as probabilities. One chromosome carries one allele, and its homologous chromosome carries the other allele. The probabilities of different genotypes in the population can be calculated by combining the chromosomes by chance. In natural populations, the relative frequencies of alleles may vary greatly.

binomial experiment: An experiment with the following properties: (a) Experiment repeated under identical conditions. (b) Dichotomous outcome (success and failure). (c) Probability of success (Prob [Success]) = p is constant from trial to trial, Probability of failure (Prob [Failure]) = 1 − p = q; and (d) Trials are independent.

binomial expression: $(a + b)^n$, where a and b are the probabilities of occurrence and non-occurrence of an event and must total 1, and n is the number of times the event is tried (e.g. Expression $(a + b)^2$ expands to $a^2 + 2ab + b^2$). *Aka* BINOMIAL EXPANSION.

binomial nomenclature of living organisms: The system of naming organisms using a two-part Latinized (or scientific) name that was devised by the Swedish botanist C. Linnaeus (1707–1778). The first part is the generic name (genus), and the second is the specific epithet or name (species). The Latin name is usually printed in italics, starting with a capital. For example, common bread wheat is called *Triticum aestivum* L., where *Triticum* is the name of genus, *aestivum* is the name of species and L. stands for Linnaeus (Authority).

binomial test: When a test has two alternative outcomes, either failure or success, and you know what the possibilities of success are, you may apply a binomial test. People use a binomial test to determine if an observed test outcome is different from its predicted outcome.

binomial theorem: The problem that gives the terms of the expansion of a binomial expression raised to a particular power.

binucleate: Cells with two nuclei.

bioaccumulation: A problem may arise when a stable chemical such as a heavy metal or dichloro-diphenyl-trichloroethane (DDT) is presented to a natural environment. Where there are no agents present to biodegrade it, its concentration can increase as it passes up the food chain and higher organisms may suffer toxic effects. This phenomenon may be employed beneficially for the removal of toxic metals from wastewater and for bioremediation.

bioactive molecule: Showing a direct or assessable effect on living tissues and cells.

bioantimutagens: Bioantimutagens act as suppressors of cellular mutagenicity of mutagens, which is accomplished by interfering with the process of mutagenesis.

bioassay: A procedure for the assessment of a substance by measuring its effect in living cells or on organisms. Animals have been used extensively in drug research in bioassays for the pharmacological activity of drugs. However, bioassays are now usually developed using bacteria or animal or plant cells, as these: (i) are usually much easier to handle than whole animals or plants; (ii) are cheaper to make and keep; and (iii) avoid the ethical problems associated with testing of animals. Sometimes, bioassay is used to detect minute amounts of substances that influence or are essential to growth. If a *Bt* gene is introduced in a plant, then the efficacy of the *Bt* toxin produced by the transgenic plant is tested by feeding the insect on these plants. This is called INSECT BIOASSAY. Similarly, there are DISEASE BIOASSAYS.

bio-assemblies: A precise and adaptable bottom-up approach for the assembly of hierarchical biomaterials. *Aka* BIOMOLECULAR ASSEMBLIES.

bioaugmentation: The process that occurs when microorganisms are imported to a contaminated site to enhance degradation.

biobank: A collection of human biological samples (such as blood and tissue) and medical information about the people who gave their samples for research studies.

bio-based products: Fuels, chemicals, building materials, electric power or heat produced from biological material(s). The term may include any energy, commercial or industrial products other than food or feed that utilizes biological products or renewable domestic agricultural (plant, animal and marine) or forestry materials.

biobleaching: (i) The bleaching of paper using biological agents such as enzymes. (ii) The extraction of valuable metals (such as copper and gold) from ores through the action of microorganisms (such as bacteria).

bioborg: A biological cyborg without non-organic components. Cyborg means cyb(ernetic) org(anism), a person whose physiological functioning is supported by or dependent upon a mechanical or electronic device.

biocatalysis: A near synonym of BIOTRANSFORMATION. The process of catalysis in biological systems. In biocatalytic processes, natural catalysts, such as protein enzymes, perform chemical transformations on organic compounds. *See* BIOTECHNOLOGY.

biocatalyst: A substance of biological origin (e.g. an enzyme) that causes or speeds up a particular chemical or biochemical reaction.

biochemical basis of heterosis: A hybrid between two types, each deficient in the ability to synthesize some chemical, grows normally since each type has a normal allele for the deficiency of the other. Consequently, the F_1 hybrid shows hybrid vigour or positive heterosis in comparison with the parents.

biochemical basis of homeostasis: High developmental homeostasis and high degree of heterozygosity are associated with a low INBREEDING DEPRESSION. At biochemical level, heterozygotes sometimes exhibited better homeostatic ability than the enzyme from either of the two homozygotes. Heterozygote is different from homozygotes by possessing two HOMODIMERIC ALLOZYMES along with a unique HETERODIMERIC ALLOZYME. Thus, homozygotes and heterozygote differ from each other in three respects: (i) heterozygote carries three allozymes where each of the homozygotes carries only one allozyme; (ii) heterozygote possesses the two parental allozymes at the same time; and (iii) it also possesses a unique heterodimeric allozyme.

biochemical genetics: A branch of genetics that deals with the chemical nature of hereditary determinants.

biochemical heterosis: A hybrid between two types, each being deficient in ability to synthesize some substance, grows normally since each type has a normal allele for the deficiency of the other. Consequently, the F_1 hybrid shows HYBRID VIGOUR or POSITIVE HETEROSIS, in comparison with the parents.

biochemical mutants: Term usually used for bacterial or fungal mutants that cannot grow on minimal media (containing sugar, mineral mutants and biotin) unless supplemented by the specific amino acid or vitamin that it is unable to synthesize. They are also called AUXOTROPHS and were first discovered in 1932 by G.W. Beadle, in *Neurospora crassa*. Such biochemical mutants have also been discovered in bacteria *E. coli* and angiosperms such as *Arabidopsis thaliana*.

biochemical oxygen demand (BOD): Measure of biologically degradable organic matter in water. It is an important concept in sewage and waste management. BOD is determined by the amount of oxygen required by bacteria to metabolize organic matter. *Aka* BIOLOGICAL OXYGEN DEMAND.

biochemical pathway: Conversion of one substrate to another involving a series of steps through the use of enzymes.

biochemical profiling: The global investigation of cellular metabolites and other low molecular weight organic molecules formed by cells and organs. The procedures that are used to separate and identify these molecules include liquid chromatography, gas chromatography and mass spectrum analysis. These analytical approaches enable understanding of the impact of various interventions (genes, drugs) on cellular function.

biochemical traits: The features associated with examination of either the impacted metabolite profile or enzymatic activity or biological processes.

biochemistry: (i) The study of chemical processes that comprise living things (systems). (ii) The chemistry of life and living matter. Despite the dramatic differences in the appearances of living beings, the fundamental chemistry of all organisms is strikingly similar. Even tiny one-celled creatures can carry out essentially the same chemical reactions that each cell of a complex organism (such as man) does.

biochip: An electronic device that employs biological molecules as the 'framework' for other molecules acting as semiconductors and functioning as an integrated circuit. Several companies manufactured biochips capable of sequencing the DNA samples. Such biochips have attached to their surfaces all possible DNA PROBES (short sequences of DNA). The DNA sample is passed over the probe-covered surface of the biochip, where each relevant segment (within the large unknown DNA molecule) hybridizes ('pairs') with the short 'DNA probe' attached to a known location on the surface of the biochip. Because the sequence of each DNA probe at each specified location on the biochip is known, the information thus obtained is then used to assemble the complete sequence of the unknown DNA molecule.

biocide: Any chemical that is toxic to living things (systems). Literally, it means 'biokiller' or killer of biological systems. Biocides include weedicide, herbicide, pesticide, insecticide, etc. Most bactericides achieve their task (killing bacteria) via enormous lysis of bacteria cell walls. However, one (triclosan) kills bacteria by inhibiting enoyl-acyl protein reductase, a crucial enzyme used by bacteria in their synthesis of fatty acids. Bavistin is a biocide; more specifically, it is fungicide that kills fungi.

biocoenosis: A community of diverse organisms and its interaction with abiotic factors of habitat.

biocompatibility: The capability of the biological systems to accept foreign objects.

biocontrol: Control of insects and pathogens by using biological organisms. Nowadays, biocontrol is gaining more importance as the traditional chemical control measures are causing environmental pollution and health hazards. Biocontrol methods have been developed for some diseases, as the fungus *Trichoderma* has been recommended to control potato scurf disease. Likewise, NPV viruses are used to kill insect pests. *See* BIOLOGICAL CONTROL.

bioconversion: Process of converting biomass into alternative energy sources by the use of microorganisms. Examples include the production of methane gas from sewage sludge and production of ethanol from agricultural by-products such as molasses.

biocultural evolution: The pattern of human evolution in which the effects of natural selection are altered by cultural inventions. Culture can alter the direction of evolution by creating non-biological adaptations to environmental stresses (e.g. wearing insulating clothes on very cold days). This potentially reduces the need to evolve genetic responses to the stresses. We as human beings have been able to remain essentially tropical animals biologically and live in colder regions of our planet. Biocultural evolution can also involve a mutual, interactive evolution of human biology and culture. An example of this has been the selection favouring sickle-cell trait in Africa. Human agricultural practices altered the environment, which resulted in factors that were advantageous to both the malarial microorganisms and the mosquitoes that transmit them between people. This, in turn, selected for the sickling allele.

biodegradable: Any material that can be broken down by biological processes such as dissimilation, digestion, denitrification. For example, the breakdown of materials (e.g. animal carcasses, dead plants, even man-made chemicals) by microorganisms (bacteria, fungus, etc.). The biodegradation process is often initiated by the actions of animals and insects (e.g. feeding on dead carcasses, which breaks down those carcasses to make their materials more available for microorganisms to 'feed' upon). Nowadays, genes such as *phbA*, *phbB* and *phbC* have been cloned from bacteria and are being introduced into plants with an aim of developing transgenic plants producing biodegradable plastic.

biodegradation: The act of degradation of substances by microorganisms (mainly aerobic bacteria) to simpler chemicals. The materials that are easily biodegraded are colloquially termed BIODEGRADABLE.

biodesulfurization: The removal of organic and inorganic sulfur (a pollution source) from coal by bacterial and soil microorganisms.

biodiesel: An alternative fuel for use in diesel engines that is made from natural renewable sources such as animal fats or vegetable oils. Biodiesel does not contain petroleum. Although it possesses properties similar to petroleum, it releases fewer environmental pollutants in its emissions. Biodiesel can be used in diesel engines with little or no modifications, either as a diesel fuel substitute or added to petroleum-based fuels to reduce their polluting effect. Examples of biodiesel include oils from soybeans, rapeseed, sunflowers or animal tallow.

Biodiversity International: With effect from December 1, 2006, International Plant Genetic Resources Institute (IPGRI) and International Network for the Improvement of Banana and Plantain (INIBAP) function under the name Biodiversity International, 'Bioversity International' for short. This new name resonances the strategy, which pays attention to improving people's lives through biodiversity research.

biodiversity: The Convention on Biological Diversity (CBD) defined diversity as the variability and interrelatedness among living organisms from all sources including terrestrial, marine/aquatic and the complexes of which they are a part. A measure of the variety of life, biodiversity is often described on three levels – ecosystem diversity, species diversity and genetic variability. ECOSYSTEM DIVERSITY describes the variety of habitats present. SPECIES DIVERSITY is a measure of the number of species and the number of individuals of each species present. GENETIC DIVERSITY refers to the total amount of genetic variability present. In agriculture, biodiversity is a production system characterized by the presence of multiple plant and/or animal species. *See* BIOLOGICAL DIVERSITY. *Contrasts* GENETIC SPECIALIZATION OF MONOCULTURE.

bioeconomy: Research and innovation in the biological sciences used to create economic activity and public benefit.

bioelectronics: The branch of biotechnology that deals with the electroactive properties of biological materials, systems and processes, together with their exploitation in electronic devices. Bioelectronics attempts to substitute traditional semiconductor materials (such as silicon or gallium arsenide) with organic materials (such as proteins). *Aka* BIOMOLECULAR ELECTRONICS.

bioenergetics: The discourse of the conversion of energy within and between living organisms.

bioenergy: Energy choices include a wide range of biomass sources (e.g. agriculture, forestry, industry, municipal waste) and conversion technologies (exemplified by fermentation (alcohol production) and co-firing (co-combustion of biomass and coal)). Bioenergy recognizes linkages to wider sustainable development outcomes, critical economic, environmental and security benefits (such as adding value to farm, forestry and other industries) and reducing fossil fuel use (product displacement), waste streams, emission of greenhouse gases and other pollutants. Bioenergy is a form of renewable energy that is derived from recently living organic materials known as biomass, which can be used to produce transportation fuels, heat, electricity and products. *See* AGRICULTURAL BIOENERGY SYSTEMS.

bioengineered food: Food that has been produced through genetic modification using techniques of genetic engineering.

bioengineering: Refers to engineering implicated to biological and medical systems, such as biomechanics, biomaterials and biosensors. Bioengineering also includes biomedical engineering, as in the development of aids or replacements for defective or missing body organs.

bioenrichment: A bioremediation strategy that deals with adding nutrients or oxygen, thereby bolstering the activity of microbes as they break down pollutants. It also investigates the moral implications of biological applications. *Aka* BIOMEDICAL ETHICS.

bioethics: A field of study and counselling dealing with the consequences of certain genetic and medical procedures, such as organ transplants, genetic engineering and care of the terminally ill persons.

biofertilizers. Refer to preparations comprising live or latent cells of efficient microbial strains utilized for application to seed, plant surfaces or soil. The objective here is to accelerate those microbial processes that expand the availability of primary nutrients that can be easily assimilated and adapted by plants. Some living bacteria and fungi are used to serve as fertilizers for nitrogen and phosphorus. Intensive cropping systems require more fertilizers for sustaining the crop yield. As an alternative, some bacteria and fungi serve as good and safe fertilizers (biofertilizers). For example, *Rhizobium*, *Azotobacter*, *Acetobacter* and blue-green algae (BGA) are being used as nitrogen fertilizers. Likewise, fungus vesicular arbuscular mycorrhizae (VAM) helps in mobilization and uptake of phosphorous from the soil. These biofertilizers are gaining importance for organic pharming.

biofilms: A layer of microorganisms growing on a surface, in a bed of polymeric material which they themselves have made. Biofilms tend to form wherever a surface on which bacteria can grow is exposed to some suitable medium and a supply of bacteria.

biofortification: The idea of breeding crops to increase their nutritional value can be realized through conventional selective breeding or genetic engineering to alter the metabolism of a plant to produce higher levels of a specific nutrient. Biofortification refers to enrichment of food crops with micronutrients (Fe, Ca, Mn) and vitamins. Through plant genetic engineering golden rice carrying genes from bacteria *Erwinia* and flower Daffodils, has been developed which produces higher β-carotene (pro-vitamin A) a precursor of vitamin A. Likewise iron dense transgenic rice has also been developed. Some other examples include golden Brassica producing higher vitamin E, golden tomatoes with high flavonols and potatoes with lysine-rich Amaranthus protein. *Ferritin* genes (iron storage protein) cloned from soybean and *Phaseolus* have been transferred to rice to develop 'iron dense' rice, which will allow fortification of iron in grain with enhanced level of its bioavailability.

biofuel: Any fuel that is derived from biomass – that is, plant or algae material or animal waste. Biogas and liquid biofuels must go through a conversion process to become usable fuel. Some biofuel sources are derived from crops that are often used as food and are high in sugar or fat. A second category of biofuel sources are from products that have already been used for their primary purpose or are non-edible, such as crop waste (e.g. hazelnut shells), food waste (e.g. used vegetable oil) or non-edible crops (e.g. poplar trees). Biofuels are being looked at as alternative sources of traditional fuels like diesel and petroleum, as an alternative production of alcohol especially from molasses and Jatropha oil from seeds of Jatropha plant are gaining importance as biofuels.

biofunctionalized nanoelectromechanical systems (bioNEMS): NEMS, which can interact with biology, are a class of devices integrating electrical and mechanical functionality on the nanoscale. NEMS formulate the next logical miniaturization step from the so-called microelectromechanical systems (MEMS).

biogas: A mixture of methane and carbon dioxide resulting from the anaerobic decomposition of waste such as domestic, industrial and agricultural sewage. Gobar gas plants/biogas plants generate huge amounts of biogas that is being used mainly for cooking purposes. Biogas contains methane (50–68%), CO_2 (25–35%), hydrogen (1–5%), nitrogen (2–7%) and oxygen (0.01%).

biogenesis: The maxim that living organisms are formed only by other living organisms. That is basis of the 'theory of generation from pre-existing life'. *Contrasts* ABIOGENESIS or SPONTANEOUS GENERATION.

biogenetic law: Name given to 'recapitulation' in the 1860s by E. Haeckel. Biogenetic law states that ontogeny is the recapitulation of phylogeny. But this definition is not complete (i.e. ontogeny is the shortest and modified recapitulation of phylogeny). Shortening is evident from the fact that what had happened once in millions of years is now performed in only days or weeks. The modifications arise mainly because the embryo at any given time has to be in harmony with its surroundings.

biogeochemical cycle: The flow of chemical elements and compounds between living organisms and the physical environment.

biohardening: Hardening of tissue-cultured plants by applying beneficial microorganisms. Tissue-cultured plants are usually very fragile and show high mortality rate upon their transfer from laboratory to land. Maintenance of aseptic conditions during tissue culture also eliminates the beneficial microbes like *Rhizobium*, *Azotobacter*, *Acetobacter*, *Pseudomonas* and VAM fungi. But these organisms are required by the plants for their better establishment and growth in soil. Therefore, tissue-cultured plants are hardened by applying these organisms to their roots before transfer to soil. Biohardening helps in improving survival rates of tissue-cultured plants.

biohazard: A biological agent, such as an infectious microorganism, or a state that constitutes a threat to humans, especially in biological research or experimentation. It also refers to the potential danger, risk or harm from exposure to such an agent or condition.

bioherbicides: A bioherbicide is a preparation of living inoculum of a plant pathogen formulated and applied in a manner analogous to that of a chemical herbicide in an effort to control or suppress the growth of weed species. The use of bioherbicide is based on the fundamental epidemiological principles of plant pathology. Plant disease is often suppressed by pathogen, environment or plant factors. Pathogen factors such as low inoculum levels, weakly virulent pathogens and poor spore dispersal mechanism; environmental factors such as unfavourable moisture and/or temperature conditions and plant factors such as low susceptibility of the host, host defence mechanisms and widely spread host populations often limit disease. The bioherbicide approach is an attempt to bypass many of these restraints on disease development by periodically dispersing an abundant supply of virulent inoculum uniformly on to a susceptible weed population. The application is timed to take advantage of favourable environmental conditions and/or most susceptible stage of plant growth. Second, it is formulated to avoid unfavourable environmental conditions and to facilitate applications.

bioinformatic tools in genomics and proteomics analysis: There are various programs and tools available for making searches, algorithm analysis modelling and the computer graphics of the databases on genomes and proteomics like BLAST, FASTA, ENTREL, EDITVIEW, DATA-SEQUENCER, FACULTA, AUTOASSEMBLY, GLG Software and GENE EXPLORER 1.4. The first priority of any annotation software is to pinpoint the genes by looking for characteristic sequences at the beginning or ends of genes or by comparing new sequence to known genes. Additional computer programs translate those genes into proteins and based on similarities to other protein to assign a function to each gene. National Centre of Biotechnology Information (NCBI's) BLAST Program compare the new genome data to that from other organisms, such as fruit fly or nematodes.

bioinformatic tools: It includes computer software, etc. These help to collect, compile, analyse, process and represent the information in order to understand and navigate the genome, to look for gene, compare genomes and final information relevant to the stretch of sequence they want to study.

bioinformatics: Bioinformatics is an interdisciplinary field of data management in genomics and proteomics and information on structure and functions of genes and proteins of all forms of life on earth. It syndicates biology, computer science, information engineering, mathematics and statistics to develop methodologies and software tools for *in silico* analyses and interpreting biological data with the goal of revealing new insights and principles in biology. Bioinformatics is especially important in analysing genomic/epigenomic research data by the use of computers. Bioinformatics is a computer-assisted interface discipline dealing with the generation/creation, collection, storage (in databases) and efficient use of data/information from genomics (functional genomics, structural genomics, etc.), combinatorial chemistry, high-throughput screening, proteomics and DNA sequencing research efforts in order to achieve a (research) objective (e.g. to find a new pharmaceutical or a new herbicide). Main public domain bioinformatics facilities include National Centre for Biotechnology Information (NCBI, USA), European Bioinformatics Institute (EBI, UK) and Genome Net (Japan), Swiss Institute of Bioinformatics (SIB, Switzerland). Some important databases thus developed include SWISS-PROT, for proteins, DBBJ for nucleotide sequence, EMBL for DNA sequences. DBBJ gene bank for nucleotide sequence, KEGG for metabolic pathway and PDB for proteins. For database mining, important tools include BLAST, ENTREZ, DNAPLOT, LOCUSLINK, LIGAND and TAXONOMY BROWSER.

bioinformatics and modelling research using *Arabidopsis*: Arabidopsis researchers are supported by the hub provided by TAIR (The Arabidopsis Information Resource). The 'Reactome system' (http://www.reactome.org) provides a framework for describing the molecular biology of an organism (the 'reactome'). Many enzymes of primary metabolism have been carefully annotated and incorporated into specialist databases such as AraCYC. An increasing number of large-scale studies relate biological data to genome features. To manage this work, powerful and useful generic software and database schemas have been developed by the Generic Model Organism Project (GMOD; http://www.gmod.org/).

Bioinformatics Applications Research Center (BARC): A collaboration of North Queensland (NQ, Australia) researchers conducting R&D in the broad area of biotechnology applications for NQ and other tropical regions of the world; BARC is built on the strengths of many component bioscience research programs; thus, many projects under BARC are strongly cross-disciplinary. BARC consolidates complementary biotechnological applications research programs into a cohesive broad-based endeavour, reinforced by the key enabling technologies of supercomputing and advanced networking. This initiative uses advanced software development expertise to create novel software methods and tools for the biotechnology industry.

bioinformatics databases: (i) Diverse data types and sources. (ii) Identification of proteins not present in database. (iii) Large data volume.

bioinformatics, applications of: Some of the applications of bioinformatics related to biological information are: (i) Looking for the sites where restriction enzymes cut a DNA or a protein molecule. (ii) Nucleotide sequences (in DNAs) and amino acid sequences (in proteins) can be compared by aligning corresponding segments and looking for matching and mismatching sites in their sequences. DNAs or proteins sufficiently similar are likely to be related and are therefore said to be 'homologous' to each other. (iii) If a homologue exists then a newly discovered protein may be modelled (i.e. three-dimensional structure of the protein can be predicted without doing laboratory experiments). (iv) Designing of primers uses bioinformatics. Primers are short stretches needed to make many copies of a DNA piece (gene) as done in polymerase chain reaction. (v) Prediction of function of actual gene products. (vi) Information about similarity and – by implication, relatedness – of proteins is used to discover phylogenetic relationships of different molecules during evolutionary time. (vii) Molecular modelling/structural biology is a part of bioinformatics. From primary structure of amino acids of a protein, its secondary structure can be determined. (viii) Bioinformatics handles vast and complex data from X-ray crystallography, nuclear magnetic resonance and electron microscopy investigations to create the three-dimensional models of the molecules. (ix) Medicinal imaging/image analysis might also be considered part of bioinformatics. There is also a whole other discipline of biologically inspired computation: genetic algorithms,

bioinformatics, DNA level problems addressed by: (i) Routine re-sequencing of megabase regions of genomic DNA. (ii) Systematic identification of common variants in genes. (iii) Rapid sequencing of other organisms. (iv) DNA sequence assembly. (v) Repetitive sequences in DNA.

bioinformatics, objectives of: Objectives of science of bioinformatics are: (i) Organizing data in such a way that allows researchers to access existing information and to submit new entries as they are produced; (ii) Developing tools and resources that aid in analysis of data and compare it with previously existing relevant information; and (iii) Interpreting the results in a biologically meaningful manner.

bioinformatics, present involvement of: Bioinformatics now includes the following: (i) DNA microarrays (measure the number of copies of a genetic message at different stages in development or disease, or in different tissues); (ii) Functional genomics (ways of identifying gene functions and associations, e.g. yeast two-hybrid system); (iii) Structural genomics (attempts to crystallize and/or predict the structure of all proteins); (iv) Comparative genomics (comparing genomes of different species to understand evolutionary relationships); and (v) Medicinal genomics (the management of biomedical experimental data associated with particular molecule). *Aka* COMPUTATIONAL BIOLOGY *or* INFORMATICS.

bioinformatics, problems addressed by: Bioinformatics addresses problems at DNA level, RNA level and protein level.

bioinformatics, protein level problems addressed by: (i) Monitoring protein level and its. (ii) Identification of all basic protein shapes.

bioinformatics, RNA level problems addressed by: Simultaneous monitoring of expression of all the genes is now possible. A mRNA signifies expression of a gene. The mRNA levels define the state of a cell with respect to gene expression. The approach is to monitor all mRNAs at quantity sensitive level of one molecule per cell and a qualitative sensitive level sufficient to distinguish alternative splicing. Use of DNA microarrays can help in this. This information can be used for cataloguing of proteins present in different cell types, at different developmental stages and in different environments. Genetic networks and gene expression circuits can be studied in relation to development and response pathways.

bioinoculants: *See* BIOFERTILIZERS.

bioinorganic: Refers to the combination of organic and inorganic materials to create useful materials. For example, Abalone shellfish make their shells via a combination of protein and calcium carbonate. Researchers are working on making semiconductor devices (chips) containing peptides having ability to attach to silicon or gallium arsenide.

bioinsecticide: Any material used in insect control that is derived from living organisms, such as bacteria, plant or animal cells. Examples include *B. thuringiensis* (*Bt*) protein (from bacteria) and Pyrethrum (made from dried flower heads of certain chrysanthemum varieties), both used to control insects.

bioleaching: The bio-mediated recovery of precious metals from their ores. In the recovery of gold, for example, the microorganism *Thiobacillus ferroxidans* may be used to cause the gold to leach out of the ore so it may then be concentrated and smelted. Aluminium may be similarly bioleached from clay ores, using heterotrophic bacteria and fungi.

biolistic method: Refers to a method of introducing DNA into cells by bombarding them with microprojectiles, while carrying the DNA. *See* GENE GUN *and* PARTICLE GUN METHOD.

biolistic transformation: A procedure used for genetic engineering in which plant cells are bombarded with metal particles that are coated with foreign genes using a 'gene gun'. *See* BIOLISTIC GENE GUN.

Biolistic® gene gun: The word 'biolistic' was invented from the words 'biological' and 'ballistic' (in connection with a projectile fired from a gun). This device is used to shoot pellets that are coated with genes (for desired traits) into plant seeds or plant tissues in order to make those plants express the new genes. The gun utilizes an actual explosive (.22 calibre blank) to propel the material. Compressed air or steam may also be used as the propellant. The Biolistic® Gene Gun was invented at Cornell University by J. Sanford, E. Wolf and N. Allen in 1983–1984. The gun and its registered trademark are now owned by E.I. du Pont de Nemours and Company.

biologic isolation: The isolation of some species from its cogneres (byproducts of alcohol being fermented or distilled) by reason of interspecific sterility, incompatibility, preferential mating or differences in time of sexual activity.

biological (scientific) models: Biological models comprising transgenic animals contribute to the study of what happens to genes as function is increased, decreased or completely shut off. Biological models make it easier to look at the functions of genes and what will happen if they are altered.

biological activity: The effect (change in metabolic activity upon living cells) caused by specific compounds or agents. For example, the drug aspirin causes the blood to remain thin in order to clot less easily.

biological assay: *See* BIOASSAY.

biological boundaries: A concept that differentiates one organism from another and suggests that organisms cannot or should not exchange genetic material. An alternative concept is that genes are defined not by the organism from which they came, but by their function. As scientists have identified genes in seemingly non-related organisms such as plants and humans, they have found identical genes in each.

biological containment: One of the precautionary measures taken to prevent replication of recombinant DNA molecule in microorganisms in the natural environment. Biological containment involves the use of vectors and host organisms that have been modified so that they will not survive outside of the laboratory. Biological containment is particularly vital when toxin genes from pathogens are expressed in *E. coli* or other vectors.

biological control: The exercise of using valuable natural organisms to attack and govern harmful plants, animal pests and weeds is called biological control, or biocontrol. This procedure can include introducing predators, parasites and disease organisms or releasing sterilized individuals. The biocontrol methods may be used as an alternative or complementary to chemical and gene-engineered pest control methods. *See* BIODIVERSITY *and* BIOCONTROL.

biological databases: Biological databases are archives of consistent data that are stored in a uniform and efficient manner. These databases contain data from a broad spectrum of molecular biology areas. Primary or archived databases contain information and annotation of DNA and protein sequences, DNA and protein expression profiles. Secondary and derived databases are so called because they contain the result of analysis on the primary resources including information on sequence, patterns or motifs, variants and mutation and evolutionary relationships. These databases help: (i) to make information available globally; (ii) to make biological data available to scientists; (iii) to systematize results from biological experiments and analysis; (iv) to avoid duplication of experiments; and (v) in consistency, accuracy and non-redundancy of data.

biological determinant: A biological factor such as crop species, variety, weed, insect pest or disease that determines the crop conformation and performance of a cropping pattern at a given site or area.

biological diversity: The prevailing genetic variability among living organisms.

biological engineering: Biological engineering is a branch of engineering that focuses on biotechnologies and biological science. It includes different disciplines such as biochemical engineering, biomedical engineering, bioprocess engineering, biosystem engineering and so on. In general, it is an integrated approach of fundamental biological sciences and traditional engineering principles. *Aka* BIOTECHNOLOGICAL ENGINEERING.

biological evolution: Evolution of diverse forms of life from first simple living cell. It refers to change in diversity and adaptation of populations. Population adapts itself to environment. Its rate is slow.

biological information: The information that is carried by the genes of an organism and which directs its development.

biological material: Any chemical compound or structural component unique to a living organism, such as viruses, serums, toxins, antitoxins, vaccines, blood, blood components and derivatives or allergenic products.

biological molecules: Large, complex biomolecules (such as proteins, nucleic acids, lipids and carbohydrates) that are produced only by living organisms. Biological molecules are often referred to as MACROMOLECULES or BIOPOLYMERS.

biological nitrogen cycle: A biogeochemical process through which nitrogen is converted into many forms, consecutively passing from the atmosphere to the soil to organism and back into the atmosphere. It involves several processes such as nitrogen fixation, nitrification, denitrification, decay and putrefaction.

biological nitrogen fixation: Biological nitrogen fixation is reduction of N_2 through symbiotic microorganisms. Biological nitrogen fixation contributes 60% of total nitrogen fixation. Major conversion of N_2 into ammonia, and hence into proteins, is achieved by microorganisms such as bacteria, fungi and algae. In living systems, nitrogen is fixed by enzyme (di)nitrogenase. Nitrogenase is inactivated by O_2, the fixation of N_2 must occur under conditions which are anaerobic at least locally. Fixation of N_2 occurs in special cells known as HETEROCYSTS, which do not photosynthesize but are devoted solely to N_2 fixation.

biological nitrogen fixation, flavonoids in: More than 4000 different flavonoids have been identified in vascular plants. A particular subset of flavonoids is involved in mediating host specificity in the legumes. All flavonoids consist of two benzene rings linked through a heterocyclic pyran or pyrone ring. Specific substitutions on the ring produce flavonols, flavones, flavanones, as well as isoflavonoids, which are derived from a migration of the B ring from the 2 to the 3 position. Isoflavonoids are limited to the legume family. Daidzein and genistein, produced by soybean (*Glycine max*), are effective inducers of *Bradyrhizobium* *japonicum* nod genes but inhibit *Sinorhizobium meliloti* nod gene expression. *S. meliloti nod* genes can be induced by luteolin. This specificity enables rhizobia to distinguish their hosts from other legumes. The specific flavonoid induces not only nod gene expression but also rhizobial chemotaxis. Flavonoids are perceived as aglycones, which induce rhizobial nod genes by interacting with the product of *nodD* gene, a LysR-type regulator. This interaction results in a conformational change in the *nodD* protein such that it binds to nod box elements in the promoters of the nod genes. The concerted expression of these genes leads to the synthesis of Nod factor molecules – lipochitooligosaccharides that usually consist of four or five *N*-acetylglucosamines, β-1–4 linked, with the terminal non-reducing sugar N-acylated with a fatty acid of 16 to 18 carbon residues.

biological nitrogen fixation, importance of: The most important contribution to biological nitrogen fixation comes from the symbiotic association of certain microorganisms with the roots of higher plants. The legumes represent a major direct source of food for man and forage for livestock. A classic example of biological nitrogen fixation is that of the bacteria (*Rhizobium*) which characteristically infect the roots of leguminous plants (e.g. peas, bean, soybean, clover, groundnut) with a high degree of host specificity. Small nodules are formed on the roots, and these become filled with an altered form of the bacteria, called bacteroids, which fix appreciable amounts of nitrogen. The nodulated plant roots supply the bacteria with energy-rich carbon compounds and obtain fixed nitrogen by the bacteroids in return. *Aka* DINITROGEN FIXATION.

biological organization: The pyramid of complex biological structures and systems used to define life using a reductionist approach.

biological oxygen demand (BOD): The oxygen used in fulfilling the metabolic requirements of aerobic organisms in water containing organic compounds. Numerically, BOD is described in terms of the oxygen consumed in water at a temperature of 68°F (20°C) during a 5-day period. The BOD is used as an indication of the degree of water pollution. *Aka* BIOCHEMICAL OXYGEN DEMAND.

biological pesticide: A chemical which is derived from plants, fungi, bacteria or other natural synthetic progressions and which can be utilized for pest control.

biological products: Any virus, therapeutic serum, toxin, antitoxin or analogous product used in the prevention, treatment or cure of diseases or injuries in humans. *Aka* BIOLOGICS.

biological property/character/characteristic/trait: Some attributes of individuals within species for which various heritable differences can be defined.

biological race: Such a race may be said to exist where the individuals of a species can be divided into groups usually isolated to some extent by food preferences occurring in the same locality and runoff showing definite differences in biology but with corresponding structural differences either few, inconsistent or completely absent.

biological software: Bioinformatic tools are software programs that are designed for extracting the meaningful information from the mass of molecular biology/biological databases and to carry out sequence or structural analysis.

biological species: Individuals that can interbreed and produce viable offspring.

biological species concept: In this species concept, emphasis is placed on 'intrinsic' impediments to free exchange of genetic information between populations (i.e. emphasis is on reproductive isolation barriers). According to this concept, a species is a set of organisms that can interbreed among each other. *Compare with* CLADISTIC SPECIES CONCEPT, ECOLOGICAL SPECIES CONCEPT, PHENETIC SPECIES CONCEPT *and* RECOGNITION SPECIES CONCEPT.

biological yield (Ybiol): The total harvest of plant material (i.e. the total biomass including the economic yield – for example, the grain yield). The larger the biological yield, the greater is the photosynthetic efficiency.

Biologically Integrated Farming Systems (BIFS): Integrated (combined) farming systems have been proven to be economical as BIFS reduce the practice of farm chemicals by employing a combination of the best practices from organic, integrated pest management and conventional farming systems.

biology: Branch of science dealing with life and living organisms. Depending upon the nature of organism biology is further divided into different branches like: (i) botany – dealing with plants; (ii) zoology – dealing with animals; (iii) microbiology – dealing with microorganisms; (iv) human biology – dealing with humans; and (v) biochemistry – dealing with chemistry of life.

bioluminescence, uses of: (i) Bioluminescence has been used by man as a genetic marker, such as to cause a genetically engineered plant to glow as evidence that a gene had been successfully transferred into that plant. (ii) Another use of bioluminescence by man is for the rapid detection of foodborne pathogenic bacteria, such as in a food processing factory. One rapid test for bacteria uses two chemical reagents that first break down bacteria cell membranes, and then cause the ATP from those broken cells to luminesce. (iii) Another rapid test uses electrophoresis to first separate the sequences of bacterial DNA, following its extraction from cell and enzymatic fragmentation, and then cause those separated sequences to luminesce. A camera is used to record the sequence-pattern light emission and compare that pattern to those of pathogenic bacteria previously deposited in a database. Some of the examples include bacteria *Vibrio fischeri* that makes symbiotic association with deep-sea fishes and an insect called firefly.

bioluminescence: The enzyme-catalysed production of light by living organisms, typically during mating or hunting. This word literally means 'living light', first identified in 1947 by W. McElroy. Bioluminescence results when the enzyme luciferase comes into contact with adenosine triphosphate (ATP)/luciferin, inside the photophores (organs which emit the light) of the organism. Luciferase picks up electrons from flavoproteins in the electron transport chain and then emits some of the energy of the electrons in the form of light. Such creation of light by living organisms is generally exemplified by fireflies, South America's railroad worm and by many deep ocean marine organisms.

biomacromolecular interactions: Biological macromolecules such as DNA, RNA and proteins function in highly crowded cellular environments. DNA-DNA, DNA-RNA, RNA-RNA, RNA-protein and protein-protein interactions are important biomolecular interactions involved in various biological processes.

biomarker: (i) In food science, a compound that can be identified and assessed, and its presence is related to the potential benefits or risks to health. (ii) In medicine, a biomarker is a quantifiable indicator of the harshness or presence of some disease state. More generally, a biomarker is anything that can be used as a pointer of a particular disease state or some other physiological state of an organism. Diagnostic biomarkers show if a person has a disease. Therapeutic biomarkers are also used to show if a particular therapy is showing the desired effect.

biomass: (i) The total dry weight of all organisms in a specific sample, population, volume or area. (ii) Any organic matter available not only on a renewable or recurring basis such as trees and plants (residues and fibres containing cellulose or lingo-cellulose) but also as poultry litter, animal residues, waste and solid waste from industrial and municipal areas (for example, sawdust, wood chips, paper, grass and leaf compost). Biomass can be above ground or below ground. (iii) All organic matter grown by the photosynthetic conversion of solar energy (e.g. plants) and organic matter from animals. Specifically called LIGNOCELLULOSIC BIOMASS.

biomass concentration: The amount of biological material in a specific volume at a given point in time. It is expressed in g/l.

biomathematics: The theoretical usage of mathematical models and abstractions of living systems to comprehend and envisage biological problems.

biome: Any very large ecological area on the Earth's surface containing fauna and flora (animals and plants) adapting to their environment. Biomes are often defined by abiotic factors such as climate, topographical relief, geology, soils and water resources.

biomechanics: The study of the structure and function of biological systems by means of the methods of 'mechanics', the branch of physics that involves analysis of the actions of forces.

biomedical engineering: The application of engineering principles and design concepts to medicine and biology for healthcare purposes (e.g. diagnostic or therapeutic).

Biomedical Microelectromechanical Systems (BioMEMS): The systems designed to work within biological systems/organisms. Examples of BioMEMS include microfluidic cell sorters or a biochip possessing diverging nanometre-scale etched channels and a fluorescence detector. These systems' function is an electrical field that drives electrophoretic separation of DNA (fragments). The DNA samples could be separated/sorted/identified by fluorescence.

biomedical nanotechnology: Referring to the subject area of nanomedicine.

biomedical research: The area of research that seeks to provide answers to medical questions. These investigations lead to discoveries, which in turn lead to the development of new preventions, therapies and cures for problems in human health.

biomembrane: Semipermeable membrane enveloping the cytoplasm of the cell or the organelles. Biomembranes are made up of phospholipid bilayer, associated membrane proteins and in some cases cholesterol and glycolipids.

biometrical genetics: A branch of genetics that investigates the mechanisms governing the inheritance of quantitative traits using statistical concepts and procedures. *Aka* QUANTITATIVE GENETICS, STATISTICAL GENETICS *or* MATHEMATICAL GENETICS. *See details as separate terms.*

biometrics: The quantitative study involving trait measurement and statistical analysis in medicine and biology. *See* BIOMETRY.

biometry: Branch of mathematical statistics applied to biological investigations.

biomimetic: Imitating, copying or learning from nature. Nanotechnology is practiced by nature, and adaptation of these principles can result in improved technology. *See* WET NANOTECHNOLOGY.

biomimetic chemistry: Chemistry of new molecules, molecular assemblies and materials with biomimetic functions. These materials (such as synthetic cell membranes) are designed for specific applications.

biomimetic materials: Materials that imitate, copy or learn from nature. Employed to describe synthetic analogues of natural materials with advantageous properties. For example, some synthetic molecules function like natural proteins but are not as easily degraded by the digestive system. Other systems such as reverse micelles and/or liposome display certain properties that mimic certain features of living systems.

biomimetics: An interdisciplinary field in which principles from engineering, chemistry and biology are applied to the synthesis of materials, synthetic systems or machines that have functions that mimic biological processes.

biomolecular condensates (BMCs): Biochemical environment is important for fully understanding transcriptional regulation. B.A. Lewis and others in 2023 describe the formation of dense, transcriptionally active bodies *in vitro* with only nuclear extracts and promoter DNA. BMCs are 0.5–1 μm in diameter, have a macromolecular density of approximately 100 mg/ml and are a consequence of a phase transition between promoter DNA and nuclear extract proteins. BMCs are physically associated with transcription as any disruption of one compromised the other. The BMCs contain RNA polymerase II and elongation factors, as well as factors necessary for BMC formation *in vivo*. BMCs are representative of the *in vivo* nuclear environment and a more physiologically relevant manifestation of the preinitiation complex/elongation machinery. BMCs contain elongation factors Positive Transcription Elongation Factor b (P-TEFb), Poly-ADP Ribose Polymerase 1 (PARP-1), RNA Polymerase Associated Factor 1 complex (PAF1c) and Host Cell Factor 1 (HCF-1) as well as Non-POU Domain-Containing Octamer binding (NONO), Splicing Factor Proline- and Glutamine-rich (SFPQ) and Fused in Sarcoma (FUS) proteins required for nuclear membraneless organelle formation *in vivo*.

biomolecule: Molecules and ions that are present in organisms and play an essential role in some typically biological processes such as cell division, morphogenesis or development. Compounds like sugars, various types of proteins and vitamins are important biomolecules in living organisms.

biomotors: Biologically based devices utilized to 'power' nanometre-size machines (e.g. 'nanobots') in one way or another. For example, a molecular machine called 'tweezers' (grasper) consists of three separate strands of DNA; two of them were hybridized separately to small complementary sequences near the two ends of the first DNA strand. The 'tweezers' can then be closed (or opened) by sequentially adding other DNA strands (to the three) which can hybridize to small complementary sequences on the second and third strands, or hybridize to the fourth strand, causing it to unhybridize from the second and the third strands.

bionics: An interscience discipline used for constructing artificial systems that resemble or have the characteristics of living systems. Bionics can involve (in whole or in part) bioelectronics, biosensors, biomimetic materials, biophysics, biomotors and self-assembly (of a large molecular structure).

bionylated cap trapper: This full-length cDNA cloning method is based on chemical introduction of a biotin group into the diol residue of the cap structure of eukaryotic mRNA, followed by RNase I treatment to select full-length cDNA. The selection occurs by trapping the biotin residue at the cap sites using streptavidin-coated magnetic beads, thus eliminating incompletely synthesized cDNAs.

biopesticides. A contraction of 'biological pesticides', includes several types of pest management intervention, through predatory, parasitic or chemical relationships. Biopesticide is a product manufactured from natural sources such as bacteria, animals or plants that is used for pest management. They tend to have less of a harmful effect on the environment and human health because they are less toxic than conventional pesticides and usually affect only one specific pest instead of being broad range. They can also work in low amounts. They breakdown quickly and when used properly, they can reduce the use of conventional pesticides while sustaining crop yields. Biopesticide are in much demand for organic farming. Bacterium *B. thuringiensis*-based formulation have been recommended to kill lepidopteran insects. Similarly, neem cake containing azadrachtin compound is also used as biopesticide. Insect viruses have a market in their natural form as biopesticides, mostly against lepidopterous caterpillar pests of forestry and field crops.

biopharm animals: Transgenic animals modified to produce proteins for extraction, purification and therapeutic use.

biopharmaceuticals: Pharmaceuticals produced by transgenic plants. Genetic transformation has led to the development of designer plants that can produce pharmaceuticals for the drug industry. Transgenic techniques are used to insert therapeutic properties, including vaccines and foods, potentially replacing pills and syringe injections. Transgenic plants are considered as factories for the production of a variety of drugs and specialty chemicals. Agricultural production is cheaper than the laboratory production. Moreover, plants possess natural storage systems such as seeds for storing these compounds. Production of edible vaccines/antibodies is a good example of biopharmaceuticals.

biopharming: Production and use of transgenic plants and animals genetically engineered to produce pharmaceutical substances such as edible vaccines and antibodies in plants or domestic animals. It often involves the insertion of gene constructs derived from humans. *Aka* BIOFARMING.

biophysics: An area of scientific study in which physical principles, physical methods and physical instrumentation are used to study living systems or systems related to life. It overlaps with biophysical chemistry, which is more specialized in scope since it is concerned with the physical study of chemically isolated substances found in living organisms.

biopiracy: The collecting and patenting of genetic stocks and the subsequent privatization of genetic resources collections – formerly held in common – and exploited for profit. The term implies a lack of consent on the part of the originator.

biopoiesis: The origin of organisms from replicating molecules. Biopoiesis is a cornerstone of biogenesis. Deoxyribonucleic acid is the best example of a self-replicating molecule and is found in chromosomes of all prokaryotic and eukaryotic organisms. In some viruses, RNA is a self-replicating molecule. Various chemical and chemical conditions must be met before DNA or RNA is able to replicate. *See* ABIOGENESIS.

biopolymer: A high-molecular-weight organic compound found in nature, whose structure comprises a repeated small unit (i.e. monomer links). Examples of common biopolymers are cellulose (long-chain sugars found in most plants and the main constituent of dried woods, jute, flax, hemp, cotton, etc.) and proteins in general and collagen and gelatine specifically.

biopolymeroptoelectromechanical systems (BioPOEMS): Combining optics and microelectromechanical systems which have found applications in biological sciences.

biopriming: Application of beneficial microbes to seeds. Biopriming is being tried to improve seed germination and subsequent seedling and plant growth. In crops like soybean, lower seed germination is a problem in some varieties. Such methods are being tried to enhance seed germination.

bioprocess engineering: It involves industrial production of pharmaceuticals and other industrial products through engineering bioprocesses to suit the industrial production of a variety of compounds of commercial value. Bioprocess engineering may involve modification of enzymes, enzyme technology, improvement in instrumentation and equipment and control/monitoring of various factors affecting industrial production of pharmaceuticals, organic acids, enzymes, vaccines, etc.

bioprocessing: (i) Any process that utilizes complete living cells or their components (e.g. enzymes, chloroplasts) to bring about anticipated physical or chemical changes. (ii) Processing of foods using biological agents. (iii) Industrial production of pharmaceuticals and speciality chemicals through biotechnology. Modern biotechnology has led to the development of altered genotypes and enzymes. Bioprocessing involving the engineered enzymes, immobilization of cells/enzymes and controlled fermentation processes is becoming important in food processing.

bioprospecting: Searching and sourcing useful genes from other organisms to specifically confer a desired phenotype (for example, medically or industrially relevant pathways and small molecules).

bioprotectants: Microorganisms whose association in the plants protects them from other pathogenic microorganisms. Vesicular Arbuscular Mycorrhizae (VAM) association in the roots of several fields, fruit, vegetable and forest plant species help them in the uptake of phosphorous and water. Besides, in several instances, VAM fungi act as bioprotectant and thus does not allow the other pathogenic bacteria/fungi to invade plant roots.

bioreactor: Refers to fermentation vessel with controlled environmental conditions and aseptic conditions. Vessels usually made up of steel and glass for industrial fermentation are called fermenters. These provide proper aeration, temperature. pH and humidity and also maintain the aseptic conditions. The bioreactors may be of huge size for holding high volumes of as much as 500,000 l.

bioreceptors: Refers to fragments of DNA, antibodies, protein molecules and cellular probes (e.g. adhesion molecule) when attached to a man-made surface (e.g. biochip) for purposes of analysing biological materials.

biorecovery: The use of organisms (e.g. bacteria, plants, fungi, algae) in the retrieval of various metals and/or organic compounds from ores or garbage. *See* BIODESULFURIZATION *and* BIOLEACHING.

bioregulators: Chemicals used in the agriculture to regulate the plant growth. The plant growth in the field is controlled by spraying specific chemicals which are called bioregulators (e.g. in cotton, vegetative growth of the plant can be suppressed by spraying the crop with a growth retarding chemical cycocel).

bioremediation: The use of living organism (plants, bacteria, fungi, etc.), usually microorganisms, to consume, breakdown, eliminate toxic materials from the environment at a polluted site (e.g. the land, ponds, air or groundwater). Bacteria and plants are used to remove toxic heavy metals like mercury, nickel, cadmium, etc. from the environment particularly soil by growing these organisms in polluted environment. Then organisms selectively uptake or degrade the polluting agents. *See* BIOAUGMENTATION *and* BIOENRICHMENT.

bioresponsive hydrogels: Dynamic systems that are capable of responding to or stimulating specific signals through the natural biological processes. The possibility to tailor hydrogel composition, stiffness and degradation rate makes this class of materials promising tools for tissue engineering applications.

biorobotics: An interdisciplinary science that combines the fields of biomedical engineering, cybernetics and robotics to develop new technologies that integrate biology with mechanical systems to develop more efficient communication, alter genetic information and create machines that imitate biological systems.

biosafety: Refers to the avoidance of risk to human health and safety, and to the conservation of the environment, as a result of the use for research and commerce of infectious or genetically modified organisms. Biosafety of transgenics has become as an important concern of the people. There is a fear that transgenic foods may not be safe for humans. The pollen from the transgenic plants may go to normal crop varieties and the marker genes such as kanamycin resistance gene during the development of transgenic may cause resistance to antibiotic kanamycin in humans. Therefore, proper guidelines have been formulated and are being followed by all the institutions dealing with the development of transgenic organisms.

biosafety concerns: Safety of normal organisms from transgenic organisms. With the development and commercialization of transgenic organisms, several biosafety concerns have been raised. These can be grouped into two categories: (i) risk to human health, including risk of toxicity, risk of allergies, antibiotic resistance, eating foreign DNA, change in the nutritional level; (ii) risk to environment, including persistence transgene/transgene product, interaction with non-target organisms, transgene instability, gene flow, resistance in target pests, loss of biodiversity and changes in soil ecology.

biosafety protocol: An internationally approved protocol set up to shield biological diversity from the potential risks posed by the release of genetically modified organisms. It establishes a process for ensuring that countries are provided with the information necessary to make informed decisions before agreeing to the import of such organisms into their territory. *Aka* CARTAGENA PROTOCOL. *See* CONVENTION ON BIOLOGICAL DIVERSITY.

biosecurity: Measures to protect from infection.

bioseeds: Seeds produced via genetic engineering of existing plants.

biosensing: Procedure for the detection of a wide range of chemical and biological agents, including bacteria, viruses and toxins, in the environment and humans.

biosensor: A device that uses a restrained biologically related agent (e.g. an enzyme, antibiotic, organelle or whole cell) to detect or quantify a chemical compound. Reactions between the immobilized agent and the molecule being analysed are transformed into an electric signal by transducer. Biosensor consists of a biological component (for sensing) and a transducer device (for generating electrical signal). Biosensors are of five types: (i) optical; (ii) calorimetric; (iii) amperometric; (iv) potentiometric; and (v) acoustic wave biosensors. Nowadays, transgenic plants have been developed to serve as biosensors to detect environmental pollution and diseases. Besides, biosensors are being used to detect toxins and food contaminants and in medical diagnosis. For example, ExacTech™ blood glucose monitor is now commonly used for measuring blood glucose levels, calorimetrically, and the individual need not go to the clinical laboratory. Genetically modified bacteria or plants are being developed which can be used to detect the pollutants in the environment and diseases of the plants. Biosensor makes use of reporter genes such as *GUS*, *luciferase* gene or *lux* operon from *Vibrio* or photobacterium.

biosensor technology: The use of cells or biological molecules in an electronic system to detect specific substances. Consists of a biological sensing agent coupled with a microelectronic circuit.

biosensors (chemical): Chemically based devices that can detect and/or measure the existence of certain molecules (such as DNA, antigens, pesticides). These devices have been currently developed in the following forms. (i) A two-part diagnostic test that can detect the presence of trace amounts of specific chemicals (e.g. pesticides). The chemical biosensor is made of an immobilized enzyme (which bind the trace chemical) combined with a colour reagent (which indicates the presence of the trace chemical). (ii) A one-part test that can recognize specific DNA segments in 'dirty' complex (multiple component) samples. The biosensor contains 13 nm gold particles on to which numerous nucleotide molecular chains are attached. Each chain contains 28 nucleotides. The 13 nucleotides that are adjoining to each gold particle serve as a spacer, and solutions containing such (spaced) randomly distributed gold particles appear red in colour when illuminated by light. The 15 nucleotides that are farthest from each gold particle are complementary and thus bind to, nucleotide sequences in the target (e.g. DNA) molecule. In the presence of the specific target molecule, a closely linked network of gold particles and double-stranded nucleotide molecular chains forms (overcoming the 13-nucleotide 'spacer' which previously held the gold particles apart). When double-stranded chains form (i.e. target molecule is present), the distance between gold particles becomes less than the size of those particles, making the solution containing (bound) particles appear blue in colour when illuminated by light.

biosensors (electronic): Electronic sensors that are able to detect and measure the presence of biomolecules such as sugars, individual nucleotides or DNA segments. These biosensors are currently created by using various approaches: (i) Fusing organic matter (e.g. enzymes, antibodies, receptors or nucleic acids) to tiny electrodes to make devices that convert natural chemical reactions into electric current to measure blood levels of certain chemicals (e.g. glucose or insulin), control functions in an artificial organ and monitor some industrial processes to act as a robot's nose, etc. (ii) Fusing organic matter (segment of DNA, antibody, enzyme, etc.) on to the surfaces of etched silicon wafers for making devices that alter supramolecular interactions (e.g. nucleotide hybridization, enzyme-substrate binding, lectin-carbohydrate (sugar) interactions, antibody-antigen binding and host-guest complexation) into electric current via a charge-coupled device (CCD) detector. The CCD detector measures the shift in interference pattern caused by change in refractive index that results when the (sensed) molecule firmly binds to the fused (electronic) organic matter. For such an etched-silicon-wafer biosensor, the nucleotide hybridization (binding) enables the detection of femtomolar (10^{-15} mole) concentrations of DNA. If the sensed DNA segment is not complementary to the fused DNA segment, there is no significant change in the interference pattern. Here, a major goal is to build future generations of biosensors directly into computer chips. This would enable low-cost mass production via processes similar to those now used for existing semiconductor chips, with circuits constructed right into the sensor to process data picked up by the biological matter on the chip.

biosilk: A biomimetic, man-made fibre produced by: (i) sequencing the 'dragline silk' protein that is produced by the orb-weaving spider; (ii) synthesizing genes to code for the 'dragline silk' protein (components); (iii) expressing those genes in a suitable host (e.g. yeast, bacteria) to cause production of the protein(s); (iv) dissolving the protein in a solvent and then 'spinning' the protein into fibre form by passing the dissolved protein through a small opening, followed by drying to remove the solvent.

biosome: Any self-governing cell constituent multiplying by autoreduplication.

biosorbents: Microorganisms which, either by themselves or in conjunction with a support/substrate system (e.g. inert granules), impact the extraction (e.g. from ore) and/or concentration of desired (precious) metals or organic compounds by means of selective retention of those entities. Retention of organic compounds (e.g. gasoline) may be used for cleaning the polluted soil. *See* BIOACCUMULATION.

biosphere: Sphere of Earth which includes the lower atmosphere, upper layers of the soil and water. It is a sphere which supports the life of microorganisms, plants, humans and animals.

biostatistics: It is the field of study concerned with collection, organization, summarization, analysis and interpretation of biological data leading to a decision with least error.

biosynthesis: Synthesis of larger molecules from two or more atoms, ions or molecules by living organisms. All the living organisms during anabolism synthesize a variety of macromolecules such as starch from sugars, proteins from amino acids, nucleic acids from nucleotides and so on.

biosynthetic pathway: A pathway involving a series of reactions to synthesize larger molecules from simple molecules. There are several pathways in metabolism, such as Calvin cycle, a photosynthetic pathway that involves a series of reactions for assimilation of CO_2 in plants.

biota: All living organisms collectively of a region, including bacteria, fungi, animals and plants alike.

biotechnologists: Scientists who employ biological processes to develop novel products.

biotechnology: The word biotechnology was coined in 1919 by K. Ereky. It deals with the manipulation of an organism at cellular or molecular level for changing and improving its characteristics efficiently for the welfare of mankind. It is the fusion of 'biology' and 'technology'. There are several definitions of biotechnology given by US National Science Foundation, British Biotechnologists, Japanese Biotechnologists and European Federation of Biotechnology. Office of the Technology Assessment of the United States Congress has also given a comprehensive definition that defines biotechnology as any technique that uses living organisms or substances from these organisms, to make or modify a product, to improve plants or animals or to develop microorganisms for specific uses. Biotechnology further includes plant biotechnology, animal biotechnology, human biotechnology and microbial biotechnology. In particular, biotechnology refers to the use by industry of recombinant DNA technology, protein engineering, nucleotide synthesis, cell fusion, biocatalysis, fermentation, plant tissue culture, cell cultures, cell manipulations, rDNA or gene splicing, enzyme systems, meristem culture, mammalian cell culture, immunology and new bioprocessing techniques for using living organisms or their parts to make or modify products to improve organisms or to make them available for specific uses (e.g. for therapeutics and agriculture).

biotechnology in crop improvement: There are three benefits to agriculture and crop improvement programmes from the use of biotechnology. (i) The reduction of the duration of breeding programmes. This makes use of tissue culture. Use of another culture and molecular markers for selection has speeded up plant breeding. Plant tissue culture also allows creation of new genetic variation, from which useful new varieties can be obtained. (ii) New methods of hybridization can be combined with cell and tissue culture for selection of plants with desired traits. Wide hybridization between wild and weedy relatives are a storehouse for a number of desirable economic traits. Reproductive barriers are a problem in use of this type of variability. Two approaches used to overcome this barrier are EMBRYO RESCUE and SOMATIC HYBRIDIZATION. *In vitro* selection can be done for desirable traits such as disease resistance, tolerance to abiotic stress like drought and excessive salinity or minerals. *In vitro* selection has a distinct advantage in terms of saving space, time and money. Micropropagation of elite genotypes. (iii) Important benefit is from use of recombinant DNA techniques.

biotechnology product: A product developed through genetic engineering (including products where the engineered DNA molecule is itself the 'product', as in an engineered molecule used as a DNA storage medium) or the targeted or *in vitro* manipulation of genetic information of organisms, including plants, animals and microbes. The term also covers some products produced by such plants, animals and microbes or their derived products.

biotechnology, a multidisciplinary approach: Biotechnology draws on the pure biological sciences (genetics, microbiology, animal cell culture, molecular biology, biochemistry, embryology, cell biology). Biotechnology is also dependent on knowledge and methods from outside the sphere of biology (such as chemical engineering, bioprocess engineering, information technology, biorobotics). Thus, biotechnology is a multidisciplinary approach.

biotechnology-driven capabilities: The use of molecular biology and/or recombinant DNA technology, or *in vitro* gene transfer, to develop products or to impart specific capabilities in plants or other living organisms.

biotic elicitors: Biotic elicitors are the substances of biological origin that include polysaccharides originated from plant cell walls (e.g. chitin, pectin and cellulose) and microorganisms. They are complex culture homogenates of fungal or bacterial origin (these involve both pathogenic, e.g. *Botrytis*, and non-pathogenic, e.g. *Aspergillus* microbes). The products which accumulate in plant cell cultures due to elicitation may be antimicrobial in nature, but they should not be confused with phytoalexins unless there is sufficient proof that the source plant responds to pathogens with rapid accumulation of the same product. Therefore, a new term has been coined for the compounds inducible by way of elicitation under the name ELICITATION METABOLITES.

biotic stress: Stress affecting growth and development of plants caused by living organisms. This is the phenomenon that limits crop productivity or destroys biomass caused by living organisms. An adverse condition in which an organism cannot sustain its normal growth due to the interaction with deleterious microorganisms (fungi, bacteria, viruses, viroids, phytoplasma and nematodes). Field-grown plants are prone to several kinds of biotic stresses that affect their growth, development and performance. Major biotic stresses include diseases; caused by mycoplasma, viruses, bacteria, fungi and insects. Weeds also cause stress on the growing plants. The biotic stresses decrease the yield and quality of crops. Using biotechnological approaches like transformation and marker-assisted selection, attempts are being made to develop resistance against biotic stresses in plants. *Bt* cotton (transgenic cotton) is resistant to American bollworm a serious insect pest of cotton. *See* ABIOTIC STRESS.

biotin: A B-complex vitamin, which is vital for life of many grain-eating insects and also for many of the metabolic pathways involved in milk production by cattle. All of the predominant bacteria that breakdown cellulose molecules within the rumen of cattle require biotin to enable them to grow. Biotin is (i) bound to polypeptides or proteins; (ii) important in fat, protein and carbohydrate metabolism; (iii) a common addition to plant tissue culture media; and (iv) acts as a coenzyme in carboxylation reactions, thereby playing a critical role in gluconeogenesis, fatty acid synthesis (manufacture) and protein synthesis reactions occurring within all animals. As biotin enzymes bind tightly to the egg protein avidin, this is used as a marker in a number of nanotechnology applications. *Aka* VITAMIN H. *See* INFECTIOUS. *Contrasts* ABIOTIC and NON-INFECTIOUS.

biotinylated DNA: A DNA molecule labelled with biotin by incorporation of biotinylated dUTP into a DNA molecule. It is used as a non-radioactive probe in hybridization experiments such as Southern transfer. The detection of any hybrids uses a complex of streptavidin-biotin-horseradish, which will produce a fluorescent green colour when hybrids have formed.

biotinylated probe: A DNA sequence in which biotinylated dUTP is incorporated and labelled with biotin. It is used in DNA-DNA hybridization experiments, such as Southern transfer or *in situ* hybridization, with chromosomes. The detection of hybrid molecules is done by a complex of streptavidin, biotin and horseradish-peroxidase. If there is a hybridization, the complex shows a green fluorescence colour.

biotization: Application of beneficial microbes to tissue-cultured plants. *See* BIOHARDENING.

biotope: A portion of a habitat characterized by uniformity in climate and distribution of biotic and abiotic components.

biotoxin: A naturally produced toxic compound which shows marked biological activity and presumably has some adaptive significance to the organism which produces it.

biotransformation: (i) Of a biosynthesized product, refers to POST-TRANSLATIONAL MODIFICATION OF PROTEINS or POST-TRANSCRIPTIONAL MODIFICATION OF TRANSCRIPTS. (ii) Of an introduced compound, refers to biological portion of definition of persistence. (iii) Refers to rapid conversion of one chemical or material into another using a biological catalyst. In biotechnology, biotransformation is becoming an increasingly important technology for converting precursors into end products using culture plant cells, protoplasts or the animal cells. This process helps in MOLECULAR FARMING/PHARMING.

biotreatment: The treatment of a waste or hazardous substance using organisms such as bacteria, fungi and protozoa. *See* BIOREMEDIATION.

biotroph: An organism that lives and multiplies only on another living organism. *See* OBLIGATE PARASITE. *Contrasts* NECROTROPHY.

biotrophic pathogen: A parasitic organism that acquires its nutrient supply only from living host tissue irrespective of whether it can or cannot be cultured artificially.

biotrophy: Refers to obtaining nutrients from living cells. Biotrophs are typically obligate parasites. The term is also used to identify the phase in the infection process where a necrotroph does not destroy the host (hemibiotrophic).

biotype: A subdivision of a species, subspecies or race defined on the basis of some identifiable physiological and biochemical properties such as a specific virulence pattern. This is also called BIOVAR. Existence of various biotypes within a species is a common spectacle in microbes as compared to higher plants and animals. Biotypes generally result from genetic recombination and mutation. *See* BIOVAR.

biotyping: Classification of strains into types (biovars) for epidemiological purposes. Differences in physiological properties are used to classify strains. For instance, *Brucella abortus* can be subdivided into eight types based on growth requirement for CO_2, release of H_2S and their ability to be grown in the presence of some diluted dyes.

biovar (bv): A variant prokaryotic strain that differs physiologically or biochemically from other strains in a particular species.

bioversity: *See* Biodiversity International.

biparental cytoplasmic inheritance: In case of *Plumbago zeylanica*, cytoplasmic genes are transferred to the egg through the male parent (S.D. Russel in 1980). The fertilized egg thus has organelles from the male as well as female parent. Male cytoplasmic organelles were identified in both the egg and central cell immediately after entry of sperm nuclei. Sperm plastids are distinguished by their smaller size, elongated shape, inflated internal lamellae and relatively electron-dense stromata. Egg plastids are larger, circular to elliptical in section, less electron dense and occasionally contain starch grains. Mitochondria of male origin can be distinguished from those of female gametophyte by their size and cellular location. Sperm mitochondria compared to sperm nucleus are circular and female gametophyte mitochondria are elliptical. Presence of extranuclear sperm organelles in the cytoplasm of the egg and central cell supports the view that syngamy in angiosperms is initiated by cellular fusion. Both male organelles and nuclei may contribute to development of the embryo and endosperm.

biparental inheritance: (i) A type of biological inheritance where the progeny inherits a maternal and a paternal allele for one gene. (ii) Plant zygotes that display traits indicating that chloroplast chromosomes from both parents are present and active.

biparental mapping: Biparental mapping populations consist of a set of individuals derived from crosses between two parents often belonging to diverse species of a botanical genus and differing in terms of phenotype and traits to share.

biparental mating: Selective or random intermating or segregants in segregating generations (preferably F_2) of a cross. It provides maximum opportunity for recombination leading to rapid release of latent variability (restricted due to linkage disequilibrium) early.

biparental parentage: An individual having two parents, one male and another female.

biparental populations: The progeny derived subsequent to crossing two genotypes as male and female parents. Such populations include: (i) F_2 genotypes generated from F_1 progeny; (ii) lines generated after doubling the haploids (DHs, obtained from F_1 plants through anther, egg cell or ovule culture or distant hybridization); or (iii) recombinant inbred lines (RILs), which are derived by single-seed descent for at least five or more generations by repeated selfing or sibling mating.

biparental zygote: The zygote which contains mitochondrial or chloroplast genes from both the parents. Two alleles for some cytoplasmic genes may thus occur in the zygote. For example, a *Chlamydomonas* zygote that contains cpDNA from both parents; such cells are generally rare.

bipartite: Consisting of two identifiable parts.

bipolar gene regulation: Transcriptional regulation in yeast seems to be mechanistically bipolar, possibly reflecting a need to balance inducible stress-related responses with constitutive housekeeping functions.

bipolar mitotic spindle: Mitotic spindle having two poles. It is responsible for segregating sister chromatids at anaphase. Microtubule motor proteins generate spindle bipolarity and enable the spindle to perform mechanical work. Central spindle assembly regulates the initiation of cytokinesis and is essential for its completion.

birimose: Opening by two slits (e.g. anthers of plants).

birth defect: Any harmful trait, physical or biochemical, present at birth, whether the result of a genetic or non-genetic (epigenetic) factor. *See* CONGENITAL, GENE, MUTATION, SYNDROME.

bisexual flowers: Flowers bearing both male and female reproductive organs (i.e. the flowers with both stamens and carpels). The term is used synonymously with hermaphrodite, synoecious and monoclinous. Bisexual flowers occur only on monoecious plants. *Aka* HERMAPHRODITE. *See* ANDROGYNOUS, MONOECIOUS *and* PLANT REPRODUCTIVE MORPHOLOGY.

bisphenol A (BPA): A chemical used in some plastics that produces unwanted epigenetic effects.

bistability: A property of a system that results in the system only being stable in two possible states, for example, an active and a silenced state; the system will tend to revert to one of these states if it is initially placed in an intermediate state.

bisulfite conversion: Bisulfite conversion (BC) is a process in which genomic DNA is denatured (made single stranded) and treated with sodium bisulfite ($NaHSO_3$), leading to deamination of unmethylated cytosine into uracil, while methylated cytosines (both 5-methylcytosine and 5-hydroxymethylcytosine) remain unchanged. This method is used for methylation analysis.

bisulfite sequencing (BS-seq), variants of: Certain variants of this method include COMBINED BISULFITE RESTRICTION ANALYSIS (COBRA), TET-ASSISTED BISULFITE SEQUENCING (TAB-SEQ), REDUCED REPRESENTATION BISULFITE SEQUENCING (RRBS), MULTIPLEXED REDUCED REPRESENTATION BISULFITE SEQUENCING (MRRBS), SINGLE-CELL REDUCED REPRESENTATION BISULFITE SEQUENCING (SCRRBS), PADLOCK PROBES and BISULFITE-CONVERTED RESTRICTION SITE ASSOCIATED DNA SEQUENCING (BsRAD-SEQ). *See details as separate terms.*

bisulfite sequencing (BS-seq): In this method, use of bisulfite treatment of DNA before routine sequencing to determine the pattern of methylation. DNA methylation was the first discovered epigenetic mark and remains the most studied. Treatment of DNA with bisulfite converts cytosine residues into uracil but leaves 5-methylcytosine residues unaffected. BS-seq is considered the 'gold standard' of DNA methylation analysis.

bisulfite-converted restriction site associated DNA sequencing (BsRAD-seq): An approach to quantify the level of DNA methylation differentiation across multiple individuals. This reduced representation method is flexible in the extent of DNA sequence interrogated.

bisulfite-treated DNA methylation analysis using microarrays: For high-throughput analysis, bisulfite-treated DNA is analysed using microarrays with two sets of oligonucleotide probes, one of which is complementary to cytosine-containing DNA and the other complementary to thymine-containing DNA.

bit rate (R): In telecommunications and computing, the number of bits conveyed or processed per unit of time. *Aka* BITRATE.

bit score: The value S′ is derived from the raw alignment score S in which the statistical properties of the scoring system used have been taken into account. Because bit scores have been normalized with respect to the scoring system, they can be used to compare alignment scores from different searches. Alignment scores are reported by HMMer and BLAST as bits scores. The likelihood that the query sequence is a bona fide homologue of the database sequence is compared to the likelihood that the sequence was instead generated by a 'random' model. Taking the logarithm (to base 2) of this likelihood ratio gives the bits score.

bit: A basic unit of information used in computing and digital communications. A portmanteau (bag) of binary digit. A binary digit can have one of two possible values and may be physically represented with a two-state device. These state values are most commonly represented as either '0' or '1'.

bitegmic: Of an ovule, covered by two integuments. *Contrasts* UNITEGMIC.

biternate: The structure that with each division divided into three (e.g. six leaflets).

bitunicate: Having a double ascus wall. *Contrasts* UNITUNICATE.

bivalent: (i) In cell division, a duplicated chromosome paired with its homologous duplicated chromosome at the beginning of meiosis. (ii) In electronics, molecules having two binding sites; having two free electrons available for binding.

bivalent formation: The association of two homologous chromosomes as a ring or rod configuration depending on chiasma formation.

bivalent interlocking: Interlocking of bivalent chromosomes during meiosis appears to be of frequent occurrence in both plants and animals. *See* INTERLOCKING.

bivalent, ring: Bivalent with terminal chiasmata or nearly terminal chiasmata at both the ends of the homologous chromosome pair.

bivalent, rod: Bivalent with terminal chiasmata or nearly terminal chiasmata only at one end of the homologous chromosome pair.

bivalent, unequal: Bivalent having two partners of unequal size.

bivariate statistics: Statistical parameters that define the relationship between two variables.

***bla* gene:** *Beta-lactamase* gene responsible for providing resistance to β-lactam ampicillin antibiotics (e.g. ampicillin). *bla* is commonly used as selective marker gene for plasmid vectors.

black-layered (corn): An indicator of a corn (maize) plant's maturity. It refers to a distinctive dark line that is formed in each corn kernel at maturity. *Aka* BLACK-LINED CORN.

blacklist: In computing, a basic access control mechanism that lets through all elements (such as email addresses, users, passwords, URLs, IP addresses, domain names, file hashes), except those explicitly mentioned in a list of prohibited elements. Those items on the list are not allowed access. *Aka* BLOCKLIST. The opposite is a WHITELIST, which means only items on the list are allowed through whatever gate is being used, while all other elements are blocked. A GREY LIST contains items that are temporarily blocked (or temporarily allowed) until an additional step is performed.

blade: The lamina or flattened portion of a leaf, petal or sepal, excluding the stalk or petiole.

blanch: A method to whiten or prevent a plant portion from becoming green by excluding light. Blanching is applied to the stems or leaves of plants (e.g. celery, lettuce and endive). Blanching is done either by banking up the soil around the stems, tying the leaves together to keep the inner ones from light or covering with pots, boxes, etc.

blank codons: Amber nonsense codons and – to a lesser extent, ochre, opal or frameshift codons – have been used to specify unnatural amino acids (UAAs). A sense codon was used to encode a UAA in an auxotrophic strain by exploiting wobble position stability differences and limiting endogenous amino acid concentrations. A general solution that includes the conversion of sense codons into blank codons would be desirable. While the three triplet nonsense codons have been widely explored, quadruplet codons have been examined for the potential of creating additional blank codons for non-canonical amino acid mutagenesis. Two orthogonal quadruplet codons have been demonstrated to simultaneously encode two different non-canonical amino acids within a single protein in bacterial cells.

BLAST: *See* BASIC LOCAL ALIGNMENT SEARCH TOOL.

blasting: A plant symptom identified by shedding of unopened buds leading to failure of producing fruits or seeds.

blastocyst: An early embryonic structure consisting of distinct outer trophectoderm cells, which develop into the placenta, and the inner cell mass (ICM) and the trophoblasts, which will differentiate into embryonic and extra-embryonic tissues, respectively.

blastoderm: A stage of insect embryogenesis in which a layer of nuclei or cells around the embryo surround an internal mass of yolk.

blastomere: One of the cells formed by the cleavage of a fertilized egg forming a BLASTULA.

blastula: Early stage of development of an animal embryo, usually comprising a hollow ball of cells, before beginning of GASTRULATION.

blemishes: Marks on skin of fruit and vegetables, caused by crop parasites. Since the development of synthetic crop protection chemicals, it has become fashionable to see only blemish-free produce on sale. However, blemishes are an indication of freedom from pesticides and are preferred for this reason by lovers of organic food.

blend: A term applied to mechanical seed mixtures of different crop varieties or species that have been mixed together to fulfil a specific agronomic purpose.

blended variety: A variety including mixture of multilines.

blending inheritance: An erroneous theory that organisms contain a blend of their parents' hereditary factors and pass that blend on to their offspring. *Compare with* MENDELIAN INHERITANCE.

blight: A sudden, severe and extensive spotting, discoloration, wilting or destruction of leaves, flowers, stems or entire plants.

blind study: An experimental study in which subjects do not know the treatment they are receiving.

blinding: In a scientific experiment, a blind is where some of the people involved are prevented from knowing certain information that might lead to conscious or subconscious bias on their part, making the results not completely accurate.

block: A number of plots that provide the chance of equal growing conditions. Comparisons among the entries that are tested in the same block offer impartial estimates of genetic differences.

block design: In analysis of variance, a design in which subjects within each block (or stratum) are assigned to a different treatment. This theory of experimental design was developed by R.A. Fisher and F. Yates in the early 1930s. Designs are used to compare the yield of different varieties of grain as it is quite possible that there is an interaction between the environment (type of soil, rainfall, drainage, etc.) and the variety that would alter the yields. So, blocks (sets of experimental plots) are chosen in which the environment is fairly consistent throughout the block. In other types of experiments in which the environment might not be a factor, blocks could be distinguished as plots that receive a particular treatment (e.g. a particular type of fertilizer). In this way, the classification of the experimental plots into blocks and varieties can be used whenever there are two factors that may influence yield. The apparent technique of growing every variety in a plot in every block may, for large experiments, could be too costly or impractical. To deal with this, smaller blocks are used that do not contain all of the varieties. The problem is to minimize the effects of chance due to incomplete blocks and to design the blocks so that the probability of two varieties can be compared; this property is called BALANCE IN THE DESIGN.

block mutation: A term used to denote a change in, or the oversight, of a group of adjacent genes.

blocked DNA replication: A mistake in DNA replication, where an incorrect nucleotide is incorporated. Such a mistake will lead to mutation in next round of DNA replication of the strand with incorrect nucleotide. If DNA replication is, however, not completed, it can lead to aberrant chromosome rearrangements or even cell death.

blocked reading frame: The frame that cannot be translated into protein because it is interrupted by termination codons.

blocker of terminator gene: Between the late promoter and the toxin gene is a piece of DNA called a blocker, which interferes with the ability of the promoter to turn on the toxin gene.

blocking: The process by which experimental units are grouped into homogeneous clusters with a view to improve the comparison of treatments by randomly allocating the treatments within each cluster or block.

Blocks Substitution Matrix (BLOSUM): A substitution matrix in which scores for each position are derived from observations of the frequencies of substitutions in blocks of local alignments in related proteins. Each matrix is tailored to a particular evolutionary distance. In the BLOSUM62 matrix, for example, the alignment from which scores were derived was created using sequences sharing no more than 62% identity. Sequences more similar than 62% are represented by a single sequence in the alignment so as to avoid overweighting closely related family members.

bloom: (i) The white powdery deposit often seen on the surface of the stem, leaves and ears of cereals or sorghum, often of a waxy nature. (ii) In general, the flower of a plant or the state of blossoming.

blooming: In grasses, the period during plant development in which florets are open and anthers are stretched. *Aka* BLOOM PERIOD.

blossom: (i) The flower of a plant, especially of one producing an edible fruit. (ii) The state of flowering.

blot(ting): (i) A procedure in which macromolecules separated on a gel are relocated to a sheet of paper, which immobilizes them. The molecules can be subsequently analysed by several other methods. Blotting is commonly used technique for detecting proteins and nucleic acid sequences in a mixture. Biomolecules of a given sample are first separated using agarose or polyacrylamide gel, separated components are transferred (blotted) to a nitrocellulose sheath which is exposed to a radiolabelled probe which binds to the biomolecule of interest. Finally, it is subjected to autoradiography. (ii) Refers to the autoradiograph produced during the Southern, Northern, Western or Eastern blotting procedure. (iii) As a verb, it refers to transfer DNA, RNA or protein to an immobilizing matrix such as diazobenzyloxymethyl (DBM)-paper, nitrocellulose or biodyne membranes. *Aka* BLOT. *See* NORTHERN BLOTTING, SOUTHERN BLOTTING, WESTERN BLOTTING, EASTERN BLOTTING.

blotch: A necrotic area of tissue irregular in form.

blue biotechnology: It is a term that has been used to describe the marine and aquatic applications of biotechnology, but its use is relatively rare.

Blue Goo: Beneficial nanobots. *Opposite of* GREY GOO. *See* GREY GOO, RED GOO *and* GREEN GOO.

blue-number method: A method of indirect determination of alpha amino nitrogen of proteins using copper reagent (e.g. in sugarbeet). This method was developed by George Papanicolaou in 1942 and is used until now. The method is fast and accurate, and the reagent can be prepared and stored for the whole campaign. A good correlation has been shown between the amount of nitrogen in proteins and the so-called BLUE-NUMBER NITROGEN. The method was used for selection of high-protein fodder beets.

blunt end: End of a DNA fragment produced by a restriction enzyme which cuts both strands of DNA at the same point, leaving no single-stranded sections. DNA with this type of end is more difficult to ligate. Blunt ends are engaged to produce sticky ends of interest. Examples of blunt ends produced by restriction endonuclease *Eco*RV are −GAT↓ATC− / −CTA↑TAG. Hence, both strands are used to produce STICKY ENDS of interest. *Compare with* STICKY END, *aka* COHESIVE END.

blunt-end ligation: In this technique, restriction enzymes are used to cut a duplex DNA at the same place in both the strands. The broken ends are then joined with two ends of another DNA molecule irrespective of sequences present at the end of the two DNA molecules. T_4 ligase is used for this joining reaction. DNA cloned by this method could be inserted at a desired site, and it is very easy to retrieve the inserted DNA when required. Short linkers (short DNA duplexes) that contain *Eco*RI site, which being small, can be synthesized chemically. These can be linked to the blunt ends of vector DNA or an insert. This will allow the insert to be retrieved by cleavage with *Eco*RI without ant restriction on the insert. Ligation of DNA with blunt ends necessitates higher concentration of DNA ligase than sticky-end ligation. It is inhibited by ATP concentrations > 1 mm. *Compare with* STICKY-END LIGATION.

BMP file format: A raster graphics image file format used to store bitmap digital images independently of the display device (such as a graphics adapter), used especially on Microsoft Windows and OS/2 operating systems. *Aka* BITMAP IMAGE FILE, DEVICE INDEPENDENT BITMAP (DIB) FILE FORMAT *or simply* BITMAP.

body methylation: Methylation of DNA bases within coding sequences of actively transcribed genes found within euchromatin.

bole: The trunk of a tree, usually the portion below the lowest branch. *Compare with* CANOPY.

boll: The fruit of cotton.

boll size: Weight in grams of seed cotton from one boll.

Bollgard® insect-protected cotton: A genetically improved cotton that bids defence against the cotton bollworm. The bollworm attack can result in significant damage to a cotton crop and necessitates repeated applications of insecticides. Bollgard® cotton contains a gene that codes for an insecticidal protein from a naturally occurring soil microorganism, *B. thuringiensis*, that provides cotton protection from bollworms.

bond energy: Strength of the chemical linkage between two atoms. Bond energy is measured by the energy in kilocalories or kilojoules required for its interruption.

Bonferroni test: A method for comparing means in analysis of variance. This is a method to counteract the multiple comparisons problem. *Aka* DUNN MULTIPLE-COMPARISON PROCEDURE *or* BONFERRONI CORRECTION.

Boolean algebra: In mathematics and mathematical logic, the branch of algebra in which the values of the variables are the truth values, true and false, denoted usually by 1 and 0, respectively. In elementary algebra, the values of the variables are numbers, and the prime operations are addition and multiplication, but the main operations of Boolean algebra are the conjunction AND (denoted as ∧), the disjunction OR (denoted as ∨) and the negation NOT (denoted as ¬). It is thus a formalism for describing logical relations in the same way that elementary algebra describes numeric relations.

Boolean data type: A data type that has one of two possible values (usually denoted true and false), intended to represent the two truth values of logic and Boolean algebra. It is named after George Boole, who first defined an algebraic system of logic in the mid-19th century. The Boolean data type is principally associated with conditional statements, which allow different actions by changing control flow depending on whether a programmer-specified Boolean condition evaluates to 'true' or 'false'. It is a special case of a more general logical data type (i.e. logic need not always be Boolean).

Boolean expression: An expression used in a programming language that returns a Boolean value when evaluated, that is one of 'true' or 'false'. A Boolean expression may be comprising a combination of the Boolean constants 'true' or 'false', Boolean-typed variables, Boolean-valued operators and Boolean-valued functions.

***booster 1* (*b1*) locus in maize:** This is a well-studied example of paramutation in plants. The booster 1 (*b1*) is a well-studied locus that shows paramutation. The *b1* locus encodes a bHLH transcription factor, which activates genes involved in anthocyanin biosynthesis in maize. The two alleles of the *b1* locus are *B-I* (paramutable allele, active) and *B′* (paramutagenic allele, inactive), and these alleles possess the same DNA sequence but differ in DNA methylation. When *B-I* and *B′* are crossed, the inactive *B′* allele silences the

B-I allele and leads to the lack of expression of the bHLH transcription factor gene in the hybrid plants. The tandem repeats ~100 kb upstream of the *b1* locus are required for paramutation. In maize, MOP1, ZmRPD1 and RMR7 are required for paramutation. RdDM pathway is required for establishment and maintenance of silencing in paramutation.

boot: The lower part of a cereal plant. BOOT STAGE begins following emergence of the flag leaf collar and continues until heading.

booting: The actions implemented in starting up of a computer or computer appliance until it can be used. It can be initiated by hardware such as a button press or by a software command. After the power is switched on, the computer is relatively dumb and can read only part of its storage called READ-ONLY MEMORY. There, a small program is stored called FIRMWARE. It does power-on self-tests and, most importantly, allows access to other types of memory like a hard disk and main memory. The firmware loads bigger programs into the computer's main memory and runs it.

bootstrap: A simulation technique for investigating properties of statistics without having the infinite population available. The most common use of the bootstrap involves taking random samples (with replacement) from the original data set and analysing how some quantity of interest varies. Each random sample should have the same number of observations as the original data set. Some of the original subjects may be lost from the random sample, while some may be sampled more than once. The bootstrap can be applied to compute standard deviations and confidence limits (compatibility limits) without assuming a model. The bootstrap can also be used to internally validate a predictive model without holding back patient data during model development. Bootstrap test allows for a rough quantification of confidence levels. *Aka* BOOTSTRAPPING.

Bootstrap and Jacknife replicates: There is no way of knowing whether tree inferred from the data is truly representative of the evolutionary history of the gene family. Further, not all positions on an alignment are revealing. Ideally one should have a very long alignment with many polymorphic sites (i.e. more the data, less likely it is for an ARTEFACTUAL PHYLOGENY to be produced). The method of bootstrapping is based on the hypothesis that the statistical properties of a sample are same as those of the population from which a sample was drawn. The larger the sample, the more representative it is expected to be of the said population. If the sample is large enough, it should also be possible to have smaller samples from the larger samples and expect that the smaller samples would also be expected to retain most of the statistical properties of the original population. If we repeatedly take random samples of the data set, the resultant small data subsets should give us the same answer as the original large data set. In case of phylogenies, if we create smaller alignments containing only some of the positions from the TOTAL ALIGNMENT and use these MINI-ALIGNMENTS to construct a tree, we should get the same tree each time. If we get the same tree each time, we are confident that all the data are consistent with the tree. Jacknife resampling has now been replaced by BOOTSTRAP RESAMPLING, where sampling is done with replacement. For bootstrap resampling, it is best to create at least 100 bootstrap data sets and redo the phylogeny for each one. A consensus tree can then be built.

bordeaux mixture: This the first and the most enormously successful of all man-made fungicides was discovered in Bordeaux, France by the botanist P.M.A. Millardet in 1882. The mixture, prepared by mixing a solution of copper sulfate with freshly slaked lime, saved the French wine industry from ruin by the newly introduced downy mildew (*Peronospora viticola*), and it also controlled potato blight, caused by *Phytophthora infestans*.

border effect: The environmental effect on plots that are on the boundary of an experimental area.

border strip: A delineation surrounding a plot, usually given the same treatment as the plot. It is arranged in order to curtail border effects.

borer: An insect or insect larva that forms channels or holes in the bark or within the wood of trees.

boring platform: Sterile bottom half of a Petri dish used for preparing explants with a cork borer.

bostrychoid: The inflorescences in which the buds are organized in an almost helical manner on the outside of a long, tapering, conical rachis.

bot: A hardware or software robot or program.

botanical pesticides: Pesticides whose active ingredients are plant-produced chemicals such as nicotine, rotenone or strychnine.

botany: The branch of biology that studies different aspects of plants.

bottleneck effect: A phase in life of an organism when a population becomes reduced due to ecological crisis to only a few individuals. This is a form of random genetic drift that occurs when a population is drastically reduced in size. Consequently, some genes may be lost from the gene pool as a result of chance.

bottom-up: Building larger objects starting from smaller building blocks. Starting from molecules and atoms. It is an iterative approach in chemistry.

bottom-up approach of nanobiotechnology: In bottom-up constructions, many molecules self-assemble in parallel steps, as a function of their molecular recognition properties. DNA nanotechnology is based upon 'bottom-up' approach. As a chemically based assembly system, DNA will be a key player in bottom-up nanotechnology. It is a more biological approach. It finds use in biomimetics, DNA, proteins and atom manipulation. It is helped by SCANNING TUNNELLING MICROSCOPY (STM), ATOMIC FORCE MICROSCOPY (AFM) and TRANSMISSION ELECTRON MICROSCOPY (TEM).

botulism: Food poisoning caused by toxin produced by *Clostridium botulinum. C. botulinum* produces a neurotoxin in contaminated food that causes respiratory failure and a paralytic disease in humans and animals. There are seven biotypes of *Clostridium*, but botulism in human is caused by *C. botulinum* type A, B and E.

boundary between abiological and biological evolution: Abiological evolution means non-living or chemical evolution. Biological evolution means living evolution. Origin of genetic code seems to be a good criterion for distinguishing between living and non-living evolution.

boundary between living and non-living: Scientists now believe in gradual origin of life on earth. Origin of genetic code seems to be a good criterion for distinguishing between living and non-living.

boundary layer: The calm layer of air on surfaces below the layer of more turbulent air.

bouquet stage: A meiotic prophase stage of some organisms during which the chromosome positions by one or both ends towards one point in the nuclear envelope.

brace root: Aerial roots that extend downward from lower above-ground nodes. When coming in contact with the soil, true roots then develop from brace roots and function as support and feeder roots.

brachyomeiosis: An abnormal meiosis recognized by omission of the second meiotic division.

brackling: Refers to the phenomenon of bending over or breaking at the top node. The term is in particular used in the UK, where cereal varieties are rated routinely for normal lodging and brackling. Brackling occurs more common in barley than wheat. The condition is genotype specific, and it is influenced by the environment, frequently associated with wet weather near maturity.

bradytelic evolution: Evolution at much lower rate than horotelic evolution. *Compare with* HOROTELIC EVOLUTION.

brain technology: A technology that employs the latest findings in neuroscience. The term was first introduced by the Artificial Intelligence Laboratory in Zurich, Switzerland, in the context of the ROBOY project. Brain technology can be employed in robots, know-how management systems and any other application with self-learning capabilities. In particular, brain technology applications allow the visualization of the underlying learning architecture often referred to as 'know-how maps'.

branch: Portion of an evolutionary tree diagram connecting two nodes.

branch length: The length of a particular branch in an evolutionary tree. In certain types of evolutionary trees, the branch length is used to represent the amount of evolutionary change or time. A scale bar is used to represent the units of measurement.

branch migration: Once the strand exchange has been initiated, it can move along the duplex. Such mobility is called branch migration. The branch can migrate in either direction as one strand is displaced by the other. The rate of branch migration is uncertain, but probably 30 bp/s. Although the branch point is not fixed, the rate of migration is probably inadequate to support the formation of extensive regions of heteroduplex DNA in natural conditions. Any extensive branch migration *in vivo* must therefore be catalysed by a recombination enzyme. The joint molecule formed by strand exchange must be resolved into two separate duplex molecules. The consequences depend on which pair of strands is nicked. A strand exchange between duplex DNAs always leaves behind a region of heteroduplex DNA, but the exchange may or may not be accompanied by recombination of the flanking regions. Formation of an intermediate involving heteroduplex DNA can be extended by branch migration.

branch point: The region of the pre-mRNA intron containing the branch point. The nucleotide (usually an adenosine) in the branch site sequence that provides the nucleophile for 5′ SS cleavage and becomes the site of lariat formation. In yeast, the branch point is the 3′-most A (underlined) in the UACUAAC branch site consensus sequence. *Aka* BRANCH SITE.

branching process: A process in which individuals produce a random number of offspring independently of each other.

brand: (i) A legal trademark registered by a particular company or distributor for its exclusive use in marketing. A product such as seeds or plants. (ii) In plant pathology, a leaf disease caused by a microscopic fungus (e.g. a rust or smut).

***Brassica*:** A genus within the Brassicaceae (Cruciferae), commonly known as the mustard family. This family of about 375 genera and 3200 species includes crops, ornamentals and many weeds. *Brassica* contains about 100 species, including rapeseed, cabbage, cauliflower, broccoli, Brussels sprouts, turnip, various mustards and weeds.

***Brassica* epigenome:** *Brassica* genus offers a very fascinating plant model for investigation of epigenetic modifications because of their differences in genome size, ploidy level and the organization of heterochromatin blocks. *Brassica rapa* and *Brassica oleracea* are diploid species, and *Brassica napus* is an allotetraploid species that evolved from the hybridization of these two diploids. Landscapes of DNA and histone H3 methylation diverge between *Brassica* species. The most conspicuous variations are reported in the two diploid species. DNA methylation existed entirely in the heterochromatin of only *B. rapa*. In *B. oleracea* and *B. napus*, this modification was spotted in both euchromatin and heterochromatin. A similar pattern was observed for H3K9me2. H3K4me2 is a characteristic marker of euchromatin in *Brassica* species, as it is in other plant species. Complex genome restructuring and epigenetic modifications are suggested to have participated in these allotetrapolyploidization proceedings. Comparison of LONG NON-CODING RNAs (lncRNAs) between *B. rapa* and *Brassica nigra*, *B. oleracea*, *Brassica juncea* and *B. napus* revealed the highest conservation with *B. oleracea*.

brassinosteroids: Indigenous, plant growth-promoting natural products having structural similarities to animal steroid hormones. They impact cell elongation and proliferation, distinct from that of auxins, cytokinins and gibberellic acids, although they interact with them.

breakage and reunion hypothesis: The general mode by which recombination occurs. DNA duplexes are broken and according to the Holliday model.

breakage-first hypothesis: The theory that chromosomes can suffer fragmentation after which the fragments may either remain loose, become reattached to restore the original gene arrangement or unite with other fragments to form new aggregants.

breakage-fusion-bridge cycle: Damage that a dicentric chromosome goes through each cycle.

breakdown point: The breakdown point is a point in which an estimator is no longer useful. A lower breakdown point means the information may not be useful, whereas a higher number means there's less chance of resistance.

breaking: Usually a virus-caused disease symptom involving addition or loss of flower colour to create a variegated pattern. *See* FLOWER BREAK.

breaking dormancy: In cereals, dormancy can be artificially broken by (i) pre-drying seeds at +35–40°C for 5–7 days (with provision for circulation of the air), (ii) prechilling seeds at +10°C for 5 days; after the pre-treatment seeds are germinated on blotters moistened with 500 ppm GA3 for 7 days at +20°C (do not overwater the seeds) or (iii) heat treatment (the grains should be at low moisture content of about 15% or lower) at +40°C for 2–3 days. Wheat dormancy is usually not expressed at low temperatures (<15°C) and does not cause problems for October–November emergence.

breakpoint: The site at which the single or double strand of DNA breaks along a chromosome and/or chromatid when chromosome mutations occur.

breed: (i) Variety in domestic animals. (ii) An artificial mating group derived from a common ancestor or for genetic analysis. (iii) In breeding, a line having the character type .nd qualities of its origin. (iv) In general, a group of plants developed by humans that do not keep their characteristics in the wild breeder. (v) To raise plants primarily for breeding purposes.

breeder: The person who bred or discovered and developed a variety.

breeder's preference: A general impression by the breeder concerning a material that is under experimentation.

breeder's rights: Concerns varietal protection. The legal rights of a breeder, owner or developer in controlling seed production and marketing of crop varieties.

breeder's seed: Seed or vegetative propagating material multiplied by the originating or sponsoring plant breeder or institution. It represents the true pedigree of the variety. It is used for the production of genetically pure, foundation, registered and certified seeds.

breeding: The art and science of propagation and genetic modification of organisms for the purpose of selecting improved offspring; several techniques of hybridization and selection are routinely applied.

breeding (traditional or selective): Making deliberate crosses or matings of plants or animals, so the offspring will have particular desired characteristics derived from one or both of the parents. Practices used in traditional plant breeding may include aspects of biotechnology such as tissue culture, mutational breeding and marker-assisted breeding.

breeding cycle: The shortest period in between successive generations, extending from germination of a seed to reproduction of the progeny. *Aka* SEED-TO-SEED CYCLE.

breeding line: A group of plants with similar traits that have been designated for their special combination of traits from a hybrid or other populations that may be released or used for further breeding approaches.

breeding method: *See* BREEDING SYSTEM.

breeding orchard: A planting of selected trees, usually propagated by cloning or grafting, designed to ease breeding work.

breeding population: A group of individuals phenotypically selected for desirable traits from a wild, experimental or crossing population for use in a breeding programme.

breeding rotation: *See* BREEDING CYCLE.

breeding size: The number of individuals in a population involved in reproduction during particular generations and breeding procedures.

breeding strategy: Prescription for breeding. A sound breeding strategy searches for an optimal compromise between genetic gain, gene diversity, cost, time and other factors.

breeding system: The system by which a species reproduces. More specifically, it refers to the organization of mating that determines the degree of similarity and/or difference between gametes effective in fertilization.

breeding true: Displaying the same characteristic features in the offspring as in the parents, an indication that the trait is genetically determined and that the parents are genetically similar. Homozygous individuals necessarily breed true.

breeding value: The sum of the average effect of each gene. The value of an individual as denoted by the mean value of its progeny, either on the basis of individual traits or a selection index. If an individual is mated at random to others in the population, its offspring will deviate from the population average by one-half of the breeding value.

breeding zone: An area within which a single population of improved trees can be planted without fear of misadaptation.

brevideciduous: A plant that loses all of its leaves only for a briefly period, before growing new ones, so that it is leafless for only a short time (e.g. approximately 2 weeks).

brewing: The process by which beer is made. In the first stage, the barley (or other) grain is soaked in water and allowed to germinate (MALTING), during which the natural enzymes of the grain convert the seed starch to maltose, and then to glucose. The grain is then dried, crushed and added to water at a specific temperature (STEEPING), and any remaining starch is converted to sugar; the resulting liquid (WORT) is the raw material to which yeast is added to convert sugar to alcohol. Hops (dried female flowers of *Humulus lupulus*) are added during this process to give a characteristic flavour.

brick gravel test: This test was given by L. Hiltner in 1917. The principle is that the weak seedlings are not able to generate enough force to overcome the pressure of brick gravels, so this method can be used to differentiate vigour levels in cereal seeds.

brick grit test: A type of seedling emergence test using uniformly crushed brick gravel through which seedlings must emerge to be considered vigourous. The test was originally developed by L. Hiltner and G. Ihssen in 1911 for detecting seed-borne *Fusarium* infection in cereals and it was later modified. The seeds are planted on damp brick grit or in a container of sand shielded with 3 cm of damp brick grit, then germinated in darkness at room temperature.

bridge: Region between the two centromeres of a dicentric chromosome, which at anaphase is being pulled towards both poles of a spindle.

bridge between RNA world and DNA-protein world: Additional RNA functional groups may have been used in amino acids in early DNA-protein world. Thus, 5′-substituted uracils would have been a bridge between RNA and DNA-protein world.

bridge cross(ing): A method of bypassing an incompatibility barrier between two genotypes or species by using a third genotype or species that is partly compatible with each one of them in an intermediate cross. For example, a wild plant is first crossed with another (wild) species, its progeny is selected for desired features which are then crossed with the target variety. This approach is mostly used where the desired trait is easy to select. It is usually a time-consuming process. Once the critical characteristic has been incorporated in the target variety, a number of backcrosses are required to eliminate the undesirable wild characters. *Aka* BRIDGING CROSS.

bridge grafting: When the trunk of fruit trees is damaged by various means and the bark is removed, the bridge graft is a procedure used for repairing a girdled trunk. Although this can be accomplished by inserting the bridges into cuts made in the wood of the trunk, the common method employed is to lift the bark and place the exposed cambiums of scion and stock together.

bridge parent: A parent that is sexually compatible with two reproductively isolated species. This approach is to transfer genes between two species that are reproductively isolated.

bridge RNA (bRNA)-guided recombination: The IS110 synaptic complex contains two recombinase dimers, one of which possesses the target-binding loop of the bRNA and binds to target DNA, whereas the other manages the bRNA donor-binding loop and donor DNA. M. Hiraizumi and co-workers in 2024 discover the formation of a compound RuvC–Tnp active site that lengths the two dimers, positioning the catalytic serine residues adjacent to the recombination sites in both target and donor DNA. A comparison of the three structures revealed that the top strands of target and donor DNA are cleaved at the composite active sites to form covalent 5′-phosphoserine intermediates, the cleaved DNA strands are exchanged and religated to create a Holliday junction intermediate, and this intermediate is finally resolved by cleavage of the bottom strands. A bispecific RNA deliberates target and donor DNA specificity to IS110 recombinases for programmable DNA recombination. *See* BRIDGE RNA.

bridge RNAs (bRNA): M.G. Durrant and co-workers in 2024 report that IS110 insertion sequences produce a structured non-coding RNA – called bridge RNA – that binds precisely to their encoded recombinase. This bridge RNA comprises two internal loops encoding nucleotide stretches that base-pair with the target DNA and the donor DNA, which is the IS110 element itself. The target-binding and donor-binding loops can be autonomously reprogrammed to facilitate sequence-specific recombination between these two DNA molecules. This modularity allows the insertion of DNA into genomic target sites, as well as programmable DNA excision and inversion. The IS110 bridge recombination system enlarges the diversity of nucleic acid–guided systems beyond CRISPR and RNA interference, subscribing an integrated machinery for the three fundamental DNA rearrangements, namely – insertion, excision and inversion – that are essential for genome design.

bridge RNAs in programmable DNA rearrangements: An RNA-guided recombinase system has been reported that is capable of making assorted and user-specific DNA edits. M.G. Durrant *et al.* (2024) display that IS110 elements express a noncoding RNA, called bridge RNA, that comprises two binding loops that separately identify the IS110 DNA donor and its genomic insertion target. These loops can be caused to reprogramme the recombinase IS621, one member of the IS110 family, to invert, excise or insert large DNA sequences at specific section in the genome of *Escherichia coli*. Computational analysis has identified bridge RNAs (bRNAs) in diverse members of the IS110 family, demonstrating that the bridge recombination mechanism is probably a general feature of IS110 elements. The IS110 bridge recombination system inflates the multiplicity of nucleic-acid-guided systems beyond CRISPR and RNA interference, presenting an integrated mechanism for the three central DNA rearrangements—insertion, excision and inversion—that are essential for genome design. M. Hiraizumi *et al.* (2024) describe the cryo-electron microscopy structures of the IS110 recombinase in complex with its bRNA, target DNA and donor DNA in three different stages of the recombination reaction cycle. The top strands of target and donor DNA are cleaved at the composite active sites to produce covalent 5′-phosphoserine intermediates. The cleaved DNA strands are exchanged and religated to generate a Holliday junction intermediate. This intermediate is later resolved by cleavage of the bottom strands. This study divulges the mechanism by which a bispecific RNA deliberates target and donor DNA specificity to IS110 recombinases for programmable DNA recombination.

bridge-breakage-fusion-bridge cycle: A process that can arise from the formation of dicentric chromosomes; daughter cells are formed that differ in their content of genetic material due to duplications and/or deletions in the chromosomes.

bridging cross: A cross made to transfer alleles between two sexually isolated species by first transferring the alleles to an intermediate species by first transferring the alleles to an intermediate species that is sexually compatible with both.

bridging species: A species which is used in a bridging cross in order to bring together the two incompatible species.

bright-field microscope: The light microscope in which the image is normally obtained by simple transmission of light through the object being viewed.

bristle: A straight, stiff hair (smooth or with minute teeth); the upper part of an awn (when the latter is bent and has a lower, stouter and usually twisted part, called the COLUMN).

Britten-Davidson model of gene regulation in eukaryotes, biological significance of: Transcription of the same gene may be needed in different developmental stages. This is achieved by multiplicity of receptor sites and integrator genes. Each producer gene may have several receptor sites, each responding to one activator. Thus, a single activator can recognize several genes, but different activators may activate the same gene at different times. An integrator gene may also fall in cluster with the same sensor gene. A set of structural genes controlled by one sensor site is termed a battery. Sometimes, when major changes are needed, it is necessary to activate several sets of genes. If one sensor site is associated with several integrators, it may cause transcription of all integrators simultaneously, thus causing transcription of several producer genes through receptor sites. The repetition of integrator genes and receptor sites is consistent with the reports that sufficient repeated DNA occurs in the eukaryotic cells. *Aka* GENE-BATTERY MODEL. *See* GENE REGULATORY MECHANISMS IN EUKARYOTES.

Britten-Davidson model of gene regulation in eukaryotes: R.J. Britten and E.H. Davidson developed such a theoretical model in 1969 to explain enormous genome and possible modes of regulation in higher forms. This model was later elaborated by them in 1973. The model assumes the presence of the following four types of sequences: (i) PRODUCER GENE; (ii) RECEPTOR GENE; (iii) INTEGRATOR GENE, (iv) SENSOR GENE. Transcription of a producer gene takes place only if at least one of its receptor sites was activated by forming a sequence-specific complex with activator RNA (or activator protein). This RNA (or protein) is synthesized by an integrator gene in response to the signals received by the sensor site that is sensitive to external or internal development signals. In Britten-Davidson model, it is also proposed that receptor sites and integrator genes may be repeated a number of times to control the activity of a large number of genes simultaneously in the same cell. Repetition of receptor sites ensures that same activator recognizes all of them and in this way several enzymes of one metabolic pathway can be simultaneously synthesized.

broad host range: Broad host range is necessity for cloning foreign genes in organisms other than *E. coli* for the following reasons. (i) It lacks some biochemical pathways that are essential phenotypic expression of certain functions (degradation of aromatic compounds or plant pathogenicity, for example). (ii) It is not used in any of the traditional fermentations. Gram-positive bacteria, *Bacillus*, *Actinomyces* or coryneforms tend to be used. Broad host range plasmids like RP4, RSF1010 and Sa are used for cloning in Gram-negative bacteria. Plasmid RP4 is derived from P group plasmids. It has been used for cloning DNA from *Rhizobium leguminosarum*. Plasmids pRK248 and pRK2501 have been derived from RP4. Plasmid RSF1010 is derived from Q group. It is used in *Pseudomonas aeruginosa* and *P. putida*. A cosmid derivative of RSF1010 is pKT247, which is used for cloning large segments of DNA and for constructing gene backs. Sa is constructed from W group plasmids. It can infect a wide range of Gram-negative bacteria. It has been developed as a vector for use with oncogene bacterium *A. tumefaciens*. Transposons can be used as a broad base host range vector. From Tn7, vector T is developed. Any gene can be inserted into the chromosome by replacing *trpE* gene.

broadcast application: Application of a chemical by spreading or scattering on the soil surface.

broad-sense heritability: *See* HERITABILITY IN BROAD SENSE.

brokens: Pieces of the rice kernel that are less than three-quarters the size of the full kernel.

bromo adjacent homology (BAH) domain: A type of reader module that targets histone lysine methylation marks.

bromodomain and extraterminal (BET) proteins: A family of proteins (BRD2, BRD3, BRD4 and BRDT) characterized by tandem bromodomains that interact with acetylated histones and influence gene expression, cell-cycle regulation and development.

bromodomain: A protein motif that binds acetylated lysine residues. Such motifs are commonly present in proteins that recognize acetylated histones, such as chromatin-remodelling factors. Proteins containing bromodomains are involved in transcription, DNA repair, replication and chromosome condensation.

brooming: The generous branching of woody stems from single stem position.

broth: Liquid medium having all the constituents required for the growth of microorganism devoid of agar.

brown rice: Dehusked paddy, often known as unpolished rice.

brown rot (of wood): A wood degeneration resulting from selective removal of cellulose and hemicellulose, leaving a brown amorphous residue that usually cracks into cubical blocks and comprises largely of slightly modified lignin. *Contrasts* WHITE ROT.

Brownian motion: Inherent motion of particles in a fluid owing to thermal agitation, named after its discoverer, Robert Brown, in 1827. This motion is vital to many of the characteristics of natural systems.

browning: Discolouration that occurs due to phenolic oxidation of freshly cut surfaces of explant tissue. In later culture, this phenomenon may designate a nutritional or pathogenic problem, generally leading to necrosis.

brute force algorithm: Refers to the simplest approach for a problem. A brute force algorithm is the first approach that comes to the thought of a researcher when he/she finds a problem.

bryophyte: In general, any plant that is a moss, hornwort or liverwort. Formally, these plants are placed in three separate divisions: hornworts (Anthocerophyta), liverworts (Marchantiophyta) and mosses (Bryophyta).

BSmooth: An alignment, quality control and analysis pipeline that provides accurate and precise results even with low coverage data, appropriately handling biological replicates.

***Bt* brinjal:** A brinjal variety developed to give resistance against lepidopteran insects, in particular, the Brinjal Fruit and Shoot Borer (*Leucinodes orbonalis*) (FSB) by forming pores in the digestive system. Mahyco, an Indian seed company based in Jalna, Maharashtra, has developed the *Bt* brinjal.

***Bt* corn (maize):** A maize plant that has been developed through biotechnology so that the plant tissues express a protein derived from a bacterium, *B. thuringiensis*, which is toxic to some insects but non-toxic to humans and other mammals.

***Bt* cotton:** A genetically modified pest resistant cotton variety that produces an insecticide to combat bollworm.

***Bt* crops:** Crops that are genetically engineered to carry a gene from the soil bacterium *B. thuringiensis* (*Bt*). The bacterium produces proteins that are toxic to some pests but non-toxic to humans and other mammals. Crops containing the *Bt* gene are able to produce this toxin, thereby providing protection for the plant. *Bt* maize and *Bt* cotton are examples of commercially available *Bt* crops.

***Bt* ICPs:** *B. thuringiensis* insecticidal proteins commonly called delta endotoxins. *See* RESISTANT INSECT and RESISTANCE MANAGEMENT STRATEGIES.

***Bt* soybean:** The first report of the successful introgression and expression of a native cry1Ab *Bt* gene in soybeans was published in 1994 by W.A. Parrott and others.

***Bt* technology:** Use of bacterium *B. thuringiensis*, *Bt* gene(s) or *Bt* toxins to control insect pests. The use of *B. thuringiensis* to control insect pests is not new. What is new in *Bt* crops is that a modified version of the bacterial crystal toxin gene has been incorporated into the plant's own DNA. Transgenic plant's cellular machinery produces the delta-endotoxin, as part of the plant's normal development.

***Bt* toxin genes:** Bacterial toxin genes cloned from a naturally occurring Gram-positive soil bacterium *B. thuringiensis*. *Bt* means 'bacterial toxin'. Several *Bt* genes (*cry* genes) have been cloned from different strains of *B. thuringiensis*. *Bt* cotton and *Bt* maize carry *Bt* genes. Using transgenic approach, *Bt* gene(s) have been introduced into tobacco, cotton, tomato, brinjal and maize, and insect-resistant transgenic varieties have been developed by *Agrobacterium* Ti plasmid-mediated transfer of plants. Development and commercialization of *Bt* cotton has revived cotton in many countries and has revolutionized the agriculture. *Bt* genes are responsible for the production of delta-endotoxin which kill the larvae of insects belonging to order Lapidoptera. Several *Bt* genes like *Cry 1AC*, *Cry 2A*, *Cry 2Ab* and *Cry 9C* have been cloned and now being used in plant transformation to develop insect-resistant crops.

***Bt* toxin in root exudates:** Presence and exudation of *Bt* toxins from roots of transgenic plants. Exudation of *Bt* has been demonstrated for maize and potato but not for cotton, canola and tobacco. It was postulated that it could be linked to proximity of the endoplasmic reticulum to the root cell membranes. Truncated toxins retain activity in soil for at least 234 days (the longest period studied) but at declining rate. Crops grown in soil known to have *Bt* root exudates present from a previous crop do not take up the toxin through their roots. Assessment of effects on earthworms and soil microfauna (nematodes, protozoa and fungi) did not indicate any adverse impact of *Bt* exudates.

bubbles: The nucleic acid configuration during replication in eukaryotic chromosomes, or the shape of heteroduplex DNA at the site of deletion or insertion.

bud: A terminal or axillary structure on a stem consisting of a small mass of meristematic tissue, generally covered wholly or in part by modified scale leaves.

bud eye: A dormant bud located in the axil of a leaf. It is used to propagate a plant through bud grafting.

bud graft: Asexual propagation of stone fruit (peaches, plums, etc.) and many other plants. This method has the advantage of using very little scion material (one bud) and plant survival is usually higher than with other grafting methods. A disadvantage of bud graft is that grafting must be done during a brief period in the spring when the bark slips. To make a graft, a chip of wood containing a leaf with its axillary bud is removed and then a 'T' cut is made in the stock and the leaf blade is clipped off; just the petiole (leaf stalk) is left. It is banged to serve as a convenient handle for manipulating the chip. Then the wood adhering to the bark of the chip is removed, taking care that the cut areas should not be touched with the fingers. A knifepoint is used for that purpose. The bark with its bud and petiole is inserted into the 'T' cut. The whole scion is wrapped with rubber budding tape or a suitable substitute. It is not necessary to use grafting wax, nor is it necessary to remove the wood chip. The type of graft in which a vegetative bud is removed from its parent plant and used as a scion to be grafted on to a stock. The bud is normally removed with a portion of green bark, which is then inserted under the green bark of the stock. This technique is widely used with fruit trees as well as other trees such as rubber, in order to grow a susceptible scion on a resistant rootstock. Using this method, interspecific and intergeneric grafts are often possible.

bud pollination: A procedure applied in maintaining self-incompatible parent lines by self-pollination. Hybrid seed production in plant species having a sporophytic self-incompatibility system is dependent upon the production of inbred lines homozygous for a self-incompatibility allele *S*. In these species, a protein secretion is used to cover the stigmatic surface just prior to anthesis and acts as a barrier to penetration of the stigma by germinating pollen grains. When buds are opened and the pollen applied before the protein barrier is formed, seed set can be obtained (e.g. in rapeseed).

bud scale: A specialized protective leaf of a bud.

bud sport: A somatic mutation occurring in a bud of a plant; it results from local genetic alteration and produces a permanent modification. It is usually retained by grafting. This sort of mutation is often used in fruit tree breeding.

bud wood: Wood consisting of strong, young shoots bearing buds suitable for use in budding.

budding: (i) A method of asexual reproduction in which a new cell/organism is formed as an outgrowth from the parent cell. Budding is a method of asexual reproduction in some microorganisms including bacteria and yeast. In budding yeast, parent cell forms a protuberance (bud) that elongates; parent cell nucleus also divides, and one nucleus migrates into the bud. Subsequently, the newly formed bud gets separated. (ii) A special type of plant grafting procedure using a single bud on the rootstock of another plant. (iii) A method of asexual reproduction in fungi, such as yeasts.

budding yeast: Common name often given to the baker's yeast *S. cerevisiae*, a common experimental organism that divides by budding off a smaller cell.

buffer: A solution mixed by a weak acid and a weak alkali. It prevents changes of the pH value when small quantities of strong acid or base are added. It is a suitable medium for enzyme reactions. Commonly used buffer in the culture medium is the mixture of K_2HPO_4 and KH_2PO_4. Optimum pH in the medium is thus maintained by using pH buffers.

buffering: Resistance to environmental changes leading to improved adaptability, increased stability and reduced genotype by environment interactions.

buffering genes: A complex of polygenes controlling the expression of a major gene and reducing the variability of its phenotype by toning down its reaction to environmental differences.

bulb: A short, flattened, usually globose or disc-shaped, underground, perennial, storage organ composed of concentric layers of overlapping fleshy scale leaves attached to a stem plate at the base.

bulbel: A bulb arising from another bulb. *See* BULBLET.

bulbil: A small, deciduous bulb or tuber formed in the axil of a leaf or pinna. A means of vegetative propagation.

bulblet: A bulb ascending from another bulb. *Aka* BULBEL.

bulbosum technique: *See* *HORDEUM BULBOSUM* PROCEDURE.

bulk breeding: The growing of genetically diverse populations of self-pollinated crops in a bulk plot with or without mass selection, followed by single-plot selection.

bulk generation: In the breeding process, one or more generations in which plants of a genetically diverse population of autogamous crops are grown in a bulk plot with or without mass selection.

bulk population selection: Selection procedure in self-pollinating crops. Segregating populations are propagated as bulks until segregation has virtually ceased, at which time selection is initiated.

bulk screening: A technique for procuring a fair degree of homozygosity for the purposes of late selection. A heterozygous population of an inbreeding species is propagated for several generations in the field with minimal or zero selection in the early stages.Only single gene characters such as marker genes. However, single-seed descent in a greenhouse is usually preferable, because it is faster.

bulk technology: Technology which is based on the manipulation of atoms and molecules in bulk (referring to micrometres).

bulked segregant analysis (BSA): BSA is a technique that allows rapid, cheap development of markers for simply inherited traits in plants. It is a rapid mapping strategy suitable for monogenic qualitative traits. When DNA of a certain number of plants is bulked into one pool, all alleles must be present; two bulked pools of segregants, differing for one trait, will differ only at the locus harbouring that trait. In BSA, a population is screened for a trait of interest and the genotypes at the two extreme ends form two bulks. Then, the two bulks are tested for the presence or absence of molecular markers. Since the bulks are supposed to contrast in the alleles that contribute positive and negative effects, any marker polymorphism between the two bulks indicates the linkage between the marker and trait of interest. BSA is dependent on accurate phenotyping, and the DH population has particular advantage in that they are true breeding and can be tested repeatedly. DH populations are commonly used in bulked segregant analysis, which is a popular method in marker-assisted breeding. This method has been applied mostly to rapeseed and barley.

bulked segregant analysis sequencing (BSA-seq): A high-throughput QTL mapping approach to rapidly identify genomic loci regulating the trait of interest. In contrast to individual segregant analysis (ISA), which classifies segregants according to their marker genotypes, the BSA pools segregants according to their phenotypes.

bullate: Having rounded or globular blisters on the surface.

bulliform cells: Most monocot leaves have bulliform cells, large, round cells grouped on the upper surface, usually near the midrib. These colourless cells absorb water when it is plentiful, which opens the leaf. In dry conditions, these cells become almost empty, causing the leaf to fold or curl in order to reduce the leaf's exposure to light and prevent water loss.

bundle: An array of alpha helices each oriented roughly along the same (bundle) axis. It may have a twist: left-handed if each helix makes a positive angle to the bundle axis or right-handed if each helix makes a negative angle to the bundle axis.

bundle sheath: A bundle sheath surrounds each vascular bundle, or vein, in monocot and dicot leaves. It consists of tightly arranged parenchyma cells in dicots and sclerenchyma cells in monocots. Note: Bundle sheaths also surround the vascular bundles in monocot stems. These cells protect the vascular structures and play a role in photosynthesis.

bunt ball: A smut sorus filled with teliospores that replaces a cereal or grass kernel but is covered by plant tissue at maturity.

buoyant density of DNA: A measure of the density or size of DNA determined by the equilibrium point reached by DNA after density gradient centrifugation.

burl: A deformation or knot in the branches or trunk of a tree, sometimes sought after in woodworking.

burr: (i) A prickly fruit. (ii) A rough or prickly propagule consisting of a seed or fruit and associated floral parts or bracts.

burr knot: A rough outgrowth, often present on the trunks or roots of certain trees.

burst period: Period taken by a virus from its attachment with a cell to release. In case of lytic bacteriophages, the complete cycle comprises different steps such as attachment, penetration, biosynthesis, maturation and release of virus from the cells. The total time taken by phage from attachment to release is called burst period that usually varies around 20–40 min, during which 50–200 phage particles get released from a single cell.

bus: In computer science, a group of lines used to transfer bits between the microprocessor and the components of the computer system.

bush-type: A plant with determinate growth.

butterfly model: DNA replication as envisaged in this model occurs in double-stranded plasmids (e.g. cryptic plasmids). In the 1920s, biologists proposed that butterfly wing pattern diversity evolved as variations of a ground plan of pattern elements that vary in colour, shape and position between different species. George Washington University scientist Anyi Mazo-Vargas and colleagues found that major aspects of this ground plan are determined by an ancient array of 'junk' or non-coding regulatory DNA. The butterfly mode of replication slightly resembles that of theta (θ) replication. Topoisomerases nick one strand and it unwinds as in theta replication. However, in theta replication, the unreplicated portion of the plasmid remains in its usual circular form, while in butterfly, it is supercoiled.

buttress root: A root growing from an above-ground stem or trunk and providing support (e.g. *Ficus macrophylla*).

butyric acid: A carboxylic acid with the formula $CH_3(CH_2)_2COOH$ made by microorganisms. This oily, colourless liquid is responsible for part of the noxious odour of rancid butter or vomit. Epigenetic relevance of butyric acid is that it enhances the transcriptional activity of promoters, which are typically silenced/downregulated due to histone deacetylase activity.

bypass replication: The replication by eukaryotic DNA polymerases past a DNA lesion, such as UV-induced dimers, with only a brief delay. Bypass replication does not require recombination to circumvent the DNA damage that blocks replication.

byte: A unit of digital information that most commonly consists of eight bits, representing a binary number. Historically, the byte was the number of bits used to encode a single character of text in a computer, and for this reason, it is the smallest addressable unit of memory in many computer architectures.

C

C: (i) In molecular biology, for cytosine. (ii) In chemistry, symbol for carbon.

C banding: A staining procedure for chromosomes that labels regions around the centromere with Giemsa stain. A band-like and darkly stained structure consisting of heterochromatin is seen. The technique is intensively employed in chromosome identification and genome characterization, including structural changes and polymorphism.

C genes: The genes that code for the constant regions of immunoglobulin protein in chains.

C mitosis: Mitosis in single-celled organisms results in the production of new individuals (asexual reproduction). Mitosis in multicellular organisms, such as plants, is responsible for the growth of the organism and the repair of damaged tissues. Mitosis is a three-step process including: (i) replication of genetic material in the mother cell; (ii) separation of the replicated genetic material; and (iii) formation of the two daughter cells.

C terminus: Carboxyl terminus, demarcated by the carboxyl (−COOH) group of an amino acid or protein.

C value: Genome is usually expressed as the amount of DNA per haploid cell (usually expressed as picograms) or the number of kilobases per haploid cell and is called the 1C value, and the content of diploids is referred to as the 2C. *See* GENOME SIZE.

C-value paradox: Each human diploid cell has 3.5×10^{12} Daltons (D) of DNA, which measures 174 cm. If a polypeptide has 300 amino acids, this will need 1000 bp for its synthesis, which is 0.67×10^{6} D. If haploid genome content in humans is 2.8×10^{7} bp, which is 1.8×10^{12} D, then this genome size should contain 3×10^{6} genes. However, estimated number of genes in humans is 32,000–35,000. Similarly, in amphibians, the smallest genome is 10^{9} bp, while the largest is 10^{11} bp. If the number of genes is assumed to be roughly similar, most of the DNA in the species with the largest genome cannot be conserved with proteins. There is thus apparent excess of DNA compared with the amount that could be necessary to code for proteins. It is often referred to as PROBLEM OF EXCESS EUKARYOTIC DNA. In reality, most of the eukaryotic genes are much larger than the sequences needed to code for the proteins. This anomalous situation has been described as C-value paradox.

C-value paradox, models of: Two models put forth to explain C-value paradox are: (i) master and slave hypothesis; and (ii) repetitive and unique sequences. *See details as separate terms.*

C-factor: A gene or heterozygous chromosome structural change that reduces the frequency of crossing over.

C3 pathway: Most common photosynthetic pathway of carbon fixation in plants. This photosynthesis produces at first a three-carbon (C3) compound (phosphoglyceric acid). In C3 plants, about 25% of the net carbon uptake is rotated immediately in photorespiration.

C3 photosynthesis: Plants utilizing the 'C3 photosynthesis' process produce a three-carbon molecule when the ribulose bisphosphate carboxylase oxygenase (RuBisCo) enzyme fixes CO_2 into sugar through the Calvin–Benson cycle. Due to photorespiration, it is estimated that C3 plants can lose up to 50% of all CO_2 fixed during photosynthesis. C3 plants produce much lower yields than their C4 cousins. C3 plants include canola, soybeans, rice, potato and wheat, which are all crops of significant value for food and feed production around the world.

C3 plants: A class of plants in which the first product of CO_2 fixation is the three-carbon compound, phosphoglyceric acid. C3 plants are usually temperate plants and are branded by (i) lower dry matter per unit water used; (b) manifestation of photorespiration; and (iii) requirement for greater CO_2 concentration as compared with C4 plants. Photosynthetically, C3 plants are less efficient than C4 plants. The C3 plants are exemplified by wheat, rice and barley. Biofertilizers for C3 plants include *Rhizobium* or *Azotobacter*.

C4 pathway: A carbon fixation pathway in some plants that have high rates of growth and photosynthesis. They are adapted to high temperatures, strong light, low carbon dioxide levels and low water supply. C4 pathway produces at first a four-carbon (C4) compound (phosphoenolpyruvate, PEP). In C4 plants, photorespiration is repressed to a great extent by the existence of a very efficient C2-concentrating machinery.

C4 photosynthesis: C4 plants, which evolved later than C3 plants, reduce photorespiration by sequestering the RuBisCo enzyme in cells where the levels of CO_2 are high and there is very little oxygen. This is achieved with a specialized anatomy of the leaf in C4 plants. A preliminary CO_2 fixation step occurs as carbon enters the leaves forming a four-carbon molecule (thus the name of C4 photosynthesis) that is then shuttled, through a series of steps, to another cell type that contains RuBisCo. CO_2 is released from the four-carbon molecule making it available to RuBisCo for photosynthesis. Unique leaf anatomy allows carbon dioxide to concentrate in 'bundle sheath cells' around RuBisCo. Plants such as sugarcane, maize, sorghum and switchgrass are examples of C4 photosynthetic plants.

C4 plants: Found principally in hot climates whose initial fixation of carbon dioxide in photosynthesis is by the Hatch–Slack–Kortschak (HSK) pathway. The enzyme responsible is PEP carboxylase (carboxylates phosphoenolpyruvate, PEP, to give oxaloacetate), whose products contain four carbon atoms. Subsequently, the carbon dioxide is released and refixed by the Calvin–Benson cycle. The presence of the HSK pathway permits efficient photosynthesis at high light intensities and low carbon dioxide concentrations. Most species of this type have little or no photorespiration; among C4 plants are sugarcane, maize and tropical grasses. *Azospirillum* is the main biofertilizer for C4 plants.

Ca^{2+}/calmodulin-dependent protein kinase (CaM kinase): Protein kinase whose activity is regulated by the binding of Ca^{2+}-activated calmodulin (Ca^{2+}/calmodulin) and which indirectly mediates the effects of Ca^{2+} by phosphorylation of other proteins.

CAAT box: The canonical sequence GGPyCAATCT located within promoter approximately 75 base pairs upstream from eukaryotic transcription start sites. This sequence is one of those that enhance binding of RNA polymerase at the promoter. *Aka* CAT BOX *or* CCAAT BOX.

cadastral gene: A plant gene that controls the expression of floral homeotic genes.

cadherin: A member of a family of proteins that mediates Ca^{2+}-dependent cell-cell adhesion in animal tissues.

caducous: Falling off early, for example, the sepals of poppies, which fall off when the petals begin to open. *Compare with* PERSISTENT *and* FUGACIOUS.

***Caenorhabditis elegans* genome:** *C. elegans* was the first multicellular organism to have its genome completely sequenced. The adult hermaphrodite has 959 somatic nuclei. Its gene density is about 1 gene/5 kb. Introns are 26% of the genome. There are some large intergenic regions containing repetitive DNA sequences. Many genes are arranged in operons: polycistronic series that are transcribed together. *C. elegans* and other nematodes are the only eukaryotes currently known to have operons. The *C. elegans* genome sequence is approximately 100 million base pairs long and contains approximately 20,100 protein-coding genes. The number of known RNA genes in the genome has increased greatly due to the 2006 discovery of a new class of 21U-RNA gene, and the genome is now believed to contain more than 16,000 RNA genes, up from as little as 1300 in 2005. Scientific curators continue to appraise the set of known genes, such that new gene predictions continue to be added and incorrect ones modified or removed.

caesium (Cs): Also written as CESIUM. A rare, highly reactive, soft metallic element of the alkali metal group.

caesium chloride (CsCl): Also written as CESIUM CHLORIDE. A salt that forms very dense solutions in water or aqueous buffers and is used in isopycnic centrifugation to separate DNA molecules of different densities. When spun at a high speed in an ultracentrifuge, Cs^+ atoms distribute themselves (ca. 1.5–1.8 g cm^2) for the separation of DNA molecules.

caesium chloride density gradient centrifugation: A method for purification of DNA by means of centrifugation in a caesium chloride solution, developed in 1958 by M. Meselson and F.W. Stahl.

caffeine: A chemical naturally produced in some plants (e.g. coffee tree) to repel predatory insects. It also acts as a stimulant (when consumed by humans), so it is classified as a phytochemical. S. Diamond during 2000 showed that within the human body, caffeine consumption causes interactions with the synthetic chemical painkiller known as IBUPROFEN. Consuming both together was shown to be more effective in relieving pain, and also brought pain relief faster, than consumption of Ibuprofen alone.

caged molecule: Organic molecule designed to change into an active form when irradiated with light of a specific wavelength. An example is caged ATP.

caging: If safe isolation distances are not available and time isolation is not suited to the plant in order to prevent cross-pollination, caging or bagging techniques are used to insure against crossing. In caging, plants are protected by cages covered with mesh or fabric. The weave of the covering must be small enough to prevent passage of insects or pollen (depending on whether the plant is insect or wind pollinated). Bags over individual flowers or flower heads can be used for self or wind-pollinating plants. In this case, the bags simply act as tiny 'cages'.

Cajal body (CB): A nuclear structure that is associated with the maturation of ribonucleoprotein complexes. CB is found in both plants and animals and contains components of at least three RNA-processing pathways.

calcareous: A soil type or a lichen substrate rock type that is rich in or largely composed of calcium carbonate.

calcicole: A plant that thrives in calcareous soil. *Aka* CALCIPHILE, CALCIPHYTE. *Antonym* CALCIFUGE.

calcium chloride density gradient centrifugation method: Calcium chloride density gradient centrifugation method was used by Meselson and Stahl in 1958 to separate DNA containing ^{15}N from that containing ^{14}N. This method could also be used to separate DNA in which GC:AT ratio differs, since the density of GC is different from that of AT.

calcium oxalate: A crystalline salt normally deposited in the cells of some species of plants. In spinach, the presence of calcium oxalate inhibits absorption of the calcium (present in spinach) by humans. In many animals, calcium oxalate is excreted in urine or retained by the animal's body in the form of urinary calculi.

calcium phosphate–mediated gene transfer: The process of transfection involves the admixture of isolated DNA (10–100 μg) with solution of calcium chloride and potassium phosphate under condition that allows the precipitate of calcium phosphate to be formed. Cells are then incubated with precipitated DNA either in solution or in tissue culture dish. A fraction of cells takes up the calcium phosphate DNA precipitate by endocytosis.

calcium pump (Ca^{2+} ATPase): Transport protein in the membrane of the sarcoplasmic reticulum of muscle cells (and elsewhere) that pumps Ca^{2+} out of the cytoplasm into the sarcoplasmic reticulum using the energy of ATP hydrolysis.

calculus of the genes: Apparent calculation by the genes to determine when a particular altruistic behaviour is beneficial to inclusive fitness and hence worth doing.

calibration: Reliability of predicted values, that is extent to which predicted values agree with observed values. For a predictive model, a calibration curve is constructed by relating predicted to observed values in some smooth manner. The calibration curve is judged against a 45° line. Miscalibration could be called BIAS. Calibration error is frequently assessed for predicted event probabilities.

call back: Also, a call-after function. Any executable code that is passed as an argument to other code that is expected to 'call back' (execute) the argument at a given time. This execution may be immediate, as in a synchronous call back, or it might happen at a later time, as in an asynchronous call back.

callogenesis: A process that forms over a wound or that develops from actively dividing plant tissue in a tissue culture.

callose: An amorphous, hardened, thickened carbohydrate constituent of plant cell walls, callous commonly developing upon injury.

callus: An unorganized growing and dividing mass of plant cells whose formation *in vitro* can be induced by a wound or treatment of tissue explants with phytohormones. In case of plants, the living tissues/organs cultured on tissue culture medium supplemented with 2,4-D (1.0 mg/l–4.0 mg/l) produce a mass of rapidly proliferating cells, commonly called callus. Callus cultures are required to study the process of differentiation, induction of somatic embryogenesis, somaclonal variation and to establish cell suspension cultures.

callus (pl. calli): A protruding mass of undifferentiated tissue formed in response to wounding. Callus formation is a first step in regeneration of plants from tissue culture. It may be grown *in vitro*. In orchids, fleshy outgrowths from the labellum that can be variously shaped from papillae to plates. In grasses, a hardened extension from the base of a floret (formed from the rachilla joint and/or the base of the lemma), which may or may not elongate and is often covered in hairs or bristles.

callus culture: The *in vitro* culture of callus, often as first stage in the regeneration of whole plants in culture.

callus induction: Undifferentiated plant tissue is produced at wound edges. Callus can also be induced and grown *in vitro* by varying the ratio of hormones (e.g. auxin and cytokinin) in the growth medium.

callus, type I: A type of adventive embryogenesis found with gramineous monocots, which has been induced on an explant where the somatic embryos are arrested at the coleoptile or scutellar stage of embryogeny. The embryos are often fused together especially at the coleorhizal end of the embryo axis. The tissue can be subcultured and maintain this morphology.

callus, type II: A type of adventive embryogenesis found with gramineous monocots, which has been induced on an explant where the somatic embryos are arrested at the globular stage of embryogeny. The globular embryos often arise individually from a common base. The tissue can be subcultured and maintain this morphology.

callus, types of: Two types of callus. *See* TYPE I CALLUS *and* TYPE II CALLUS.

calmodulin: Ubiquitous calcium-binding protein whose binding to other proteins is governed by changes in intracellular Ca^{2+} concentration. Its binding modifies the activity of many target enzymes and membrane transport proteins.

caloric restriction: A dietary regimen characterized by a low-calorie intake while maintaining adequate nutrition (i.e. sufficient levels of proteins, vitamins and minerals).

Calvin cycle: The Calvin cycle was first investigated in the late 1940s and early 1950s by the Nobel Prize–winning chemist Melvin Calvin. A series of chemical reactions that occurs as one of two primary phases of photosynthesis in chloroplasts of green leaves of plants, specifically the phase in which carbon dioxide and other compounds are converted into simple carbohydrates such as glucose ($C_6H_{12}O_2$). These reactions occur in the stroma, the fluid-filled area of the chloroplast outside the thylakoid membranes. In the Calvin cycle, the products of previous light-dependent reactions (ATP and NADPH) undergo further reactions that do not require the presence of light and that can be broadly divided into three stages: (i) carbon fixation; (ii) reduction reactions; and (iii) ribulose 1,5-bisphosphate (RuBP) regeneration. The CO_2 fixation is catalysed by an enzyme ribulose-1,5-bisphosphate carboxylase, causing its condensation with five-carbon sugar ribulose 1,5-bisphosphate that produces two molecules of three phosphoglycerate. In this cycle, six molecules of CO_2 are converted into two molecules of glyceraldehyde 3-phosphate. *Aka* LIGHT-INDEPENDENT REACTIONS *or* DARK REACTION.

calyciflorous: Having petals and stamens attached to the calyx.

calycle: A protective structure around a flower formed collectively by the sepals.

calycophyll: Leaf-like structure formed from a sepal or calyx lobe which enlarges, usually manyfold, before or after anthesis, especially when most of the other sepals or calyx lobes retain their original size. More extreme than an accrescent calyx, calycophylls are found in Rubiaceae. *Compare with* SEMAPHYLL *and* PTEROPHYLL.

calyculate: Having an epicalyx. EPICALYX is an additional whorl around the calyx of a flower. It is a group of bracteoles forming the outer whorl of the calyx. Bracteoles are thin leafy or scaly structures on the stalk of the flower. It is present between bracts and calyx.

calyx: (pl. calyces): Collective term for the sepals of one flower; the outer whorl of a flower, usually green. *Compare with* COROLLA.

CaM kinase II: Multifunctional Ca^{2+}/calmodulin-dependent protein kinase found in all animal cells that undergoes autophosphorylation when activated. It is especially abundant in brain and is thought to have a role in learning and memory in vertebrates.

CaM kinase: *See* CA^{2+}/CALMODULIN-DEPENDENT PROTEIN KINASE.

cambium: The steles of dicot roots contain a layer of partially undifferentiated meristem cells, called the *cambium* (or vascular cambium), located between the xylem and phloem. These cells facilitate secondary growth, dividing to create new xylem and phloem cells and thus widening the girth of the root.

cAMP: *See* CYCLIC ADENOSINE 5′-MONOPHOSPHATE.

cAMP-CAP complex: Binding site of cAMP-CAP complex is located proximally to the lac promoter. When cAMP-CAP complex is bound at the promoter, binding of RNA polymerase to the operon is greatly enhanced. Local destabilization of double-helical DNA is induced at RNA polymerase interaction site of the promoter by cAMP-CAP binding. RNA polymerase binds to the destabilized area. RNA polymerase then moves to the operator site and begins transcription. cAMP is synthesized from ATP by adenylate cyclase.

campesterol: A phytosterol produced within the seeds of the soybean plant (*Glycine max* L.), among others. Evidence shows human consumption of campesterol helps reduce total serum (blood) cholesterol and low-density lipoproteins (LDLP) levels, thereby lowering risk of coronary heart disease (CHD). Evidence indicates certain phytosterols (including campesterol) interfere with absorption of cholesterol by the intestines and decrease the body's recovery and reuse of cholesterol-containing bile salts; this causes more (net) cholesterol to be excreted from the body.

campylotropous: When the ovule is oriented transversely (i.e. with its axis at right angles to its stalk) and with a curved embryo sac. *Compare with* AMPHITROPOUS, ANATROPOUS and ORTHOTROPOUS.

CaMV 35S promoter: *See* CAULIFLOWER MOSAIC VIRUS 35S (CAMV 35S) PROMOTER.

Canada balsam: Resin distilled from the bark of *Abies balsamea* (balsam fir) and other similar species; used in cytology for mounting (e.g. chromosome spreads).

canalization: This term, coined by C.H. Waddington, refers to the property of various developmental pathways responsible for producing the standard phenotypes regardless of any genetic and environmental change that could otherwise hamper the normal development of organism. It is the buffering of development by epigenetic networks such that most mutations or environmental conditions will not deflect the genotype from realizing the appropriate phenotype of the cell. The cell begins in a totipotent state and becomes more and more restricted by determinative events. The general properties of this process can be applied to both cytodifferentiation and pattern formation. Canalization allows mutations to build up in the genotype without their being expressed in the phenotype. Thus, it promotes cryptic genetic variation, while preserving the integrity of the differentiating cell.

canalized character: A character whose phenotype is kept within narrow boundaries even in the presence of disturbing environments or mutations.

canalizing selection: The selection of genes that would stabilize the developmental pathways so as to make the phenotype less susceptible to the effect of environmental or genetic disturbances.

canavanine: An uncommon amino acid. It is used in biology as an arginine (another amino acid) analogue. It is a potent growth inhibitor of many organisms.

cancer cells: Certain cells of body can lose contact inhibition of movement and growth. Such cells, called cancerous cells, have a number of properties, including rapid division, invasion of new cellular territories, high metabolic rate, new membrane antigens and altered shapes. These cells obey no community rules and multiply without any regard to tissue organization. *Aka* ANTISOCIAL CELLS *or* MALIGNANT CELLS.

candidate gene: A gene whose function suggests that it may be involved in the genetic variation observed for a particular trait (phenotype, disease or condition), such as the gene for growth hormone is a candidate gene for straw length in cereals or dry matter production. Candidate genes can be divided into two categories: positional and functional. A positional candidate gene is one that might be associated with a trait, based on the location of a gene on a chromosome; a functional candidate gene is one whose function has something in common with the trait under investigation. Positional candidate genes are identified through QTL- and map-based cloning approaches, whereas functional genomics approaches, such as transcriptomics and expression genetics, provide the set of functional candidate genes gene. *See* POSITIONAL CLONING *and* PROTEIN.

candidate gene identification and validation: To identify and validate candidate genes, large insert clones are used for direct complementation analysis via transformation. A special binary BAC (BIBAC) vector system has been introduced for the transfer of large DNA fragments into plants. This strategy may provide evidence that the candidate gene is present on the contig. Candidate genes may also be identified by functional annotation based on sequence comparison to protein sequence databases. The mutagenesis of the target gene by chemical mutagens or mutagenic irradiation is another strategy. Sequence analysis of the candidate gene in independent mutants reveals the presence of functional mutations. This method is based on TARGETING INDUCED LOCAL LESIONS IN GENOMES (TILLING).

candidate gene identification approaches in QTL cloning: Candidate gene (CG) is a gene with known or assumed function that may affect genetic control of a trait and, thus, can be considered a 'candidate gene' for this trait. The CG associates a gene to its phenotypic trait. CG approach is powerful method to detect a quantitative trait locus (QTL). QTL mapping is frequently used to identify genomic regions associated with a phenotypic trait of interest. All genes in the QTL are candidate loci for the status of quantitative trait. The identification of CG can be done using either TRADITIONAL CG APPROACHES (positional and functional) *or* DIGITAL CG APPROACH.

candidate gene mapping of adaptive genes: Candidate gene mapping is based on the assumption that a gene with known or assumed function that may affect genetic control of a trait can be considered a 'candidate gene' for this trait. Furthermore, it is assumed that if this gene is also mapped to the same genomic region as a QTL for this trait, then this gene is very likely to be this QTL that directly controls the trait, although the likelihood depends on marker density, precision of QTL map and genome size.

candidate gene strategy: An experimental approach in which knowledge of the biochemistry and/or physiology of a trait is used to draw up a list of genes whose protein products could be involved in the trait.

candidate population: In forestry, trees that are planted to serve as a base for forward selection or in some way selected from that for further studies. For example, selected phenotypic selections may serve as a candidate population, which is subject to further progeny testing before reselection to the breeding population; the genotypes taken into consideration for the breeding population.

candidate quantitative trait loci, fine mapping of: Plant height (PH) and ear height (EH) are two important traits in maize (*Zea mays* L.), as they are closely related to lodging resistance and planting density. X. Yin *et al.* (2022) created a nested association mapping (NAM) population. The analysis of phenotypic data on PH and EH revealed wide variation among the parents of the NAM population. Genome-wide association study (GWAS) and high-resolution linkage mapping were conducted, and 22 QTL were identified by GWAS and found to be significantly associated with PH and EH. A high-confidence QTL for PH, *Qtl-chr1-EP*, was identified on chromosome 1 *via* GWAS and confirmed by linkage analysis in two RECOMBINANT INBRED LINE (RIL) populations. The triangle heterotic pattern adopted in maize breeding programs by our team is practicable in selecting high-yield crosses based on the low ratio of EH/PH (EP).

candidate single-nucleotide polymorphism (candidate SNP): Any single-nucleotide polymorphism in an exon of a gene that can have an impact on the function of the encoded protein.

candidate tree: A tree that has been tentatively selected for inclusion in a breeding program but has not yet been measured or compared with surrounding trees.

candidate variety: Breeding strains, lines or hybrids of high grade that by a breeder or institution are announced for official national (and international) performance testing in order to be released as a certified variety seed candle; the new shoot growth on needled evergreens before the needles expand.

canker: A plant disease characterized (in woody plants) by the death of cambium tissue and loss and/or malformation of bark, or (in non-woody plants) by the formation of sharply delineated, dry, necrotic, localized lesions on the stem. The term canker also may be used to refer to the lesion itself, particularly in woody plants.

Canola™: A type of rapeseed varieties (*Brassica napus*, *B. rapa* and *B. juncea*) that have been developed and grown in Canada, named after the 'Canadian-Oil-Low-Acid' breeding program. Canola™ is a registered trademark, corresponding to specified low contents in erucic acid in oil and in glucosinolates in meals equivalent to double '0' in the European Union standard. It was initially obtained by conventional breeding; in recent years, herbicide-tolerant varieties have been developed.

canonical code: The genetic code that is used almost universally.

canonical correlation analysis: An advanced method for examining the relationships between two sets of interval or numerical measurements made on the same set of subjects.

canonical nucleosome length: The canonical (peak) distance from the TSS to the midpoint of the −1 nucleosome is −215 bp, and +55 bp for the +1 nucleosome.

canonical RNA-directed DNA methylation (cRdDM): The canonical RdDM pathway involves 24-nt sRNAs, which are specific to that pathway and come predominantly from a single source (the Pol IV-RDR2 complex).

canopy temperature depression (CTD): The cooling effect exhibited by a leaf as transpiration occurs. It gives an indirect estimate of stomatal conductance and is a highly integrative trait being affected by several major physiological processes including photosynthetic metabolism, evapotranspiration and plant nutrition. It has potential for complementing early generation phenotypic selection in plants.

canopy: (i) Branches and expanded foliage of a tree; the crown. (ii) The protective upper layer of a forest. (iii) The vertical projection downward of the aerial portion of plants, usually expressed as percentage of ground so occupied. *Compare with* TRUNK.

cap analysis gene expression (CAGE): (i) A technique that produces a snapshot of the 5′-end of the messenger RNA population in a biological sample (the transcriptome). (ii) Capture of the methylated cap at the 5′-end of RNA, followed by high-throughput sequencing of a small tag adjacent to the 5′ methylated caps. 5′-methylated caps are formed at the initiation of transcription, although other mechanisms also methylate 5′-ends of RNA.

cap site: The probable transcription initiation site of a eukaryotic gene; the primary transcripts of most eukaryotic mRNAs have an adenine in the first position and the cap is added 5′ to it.

cap structure: A structure, which consists of m^7Gp-ppN (where 'm^7G' represents 7-methylguanylate, 'p' represents a phosphate group and N represents any RNA base), is added to the 5′-end of eukaryotic mRNAs during maturation. The cap is introduced by linking the terminal phosphate of 5′ GTP to the terminal base of the mRNA; the added G is methylated. Cap facilitates initiation of translation.

cap-binding protein: A 24-kDa protein with affinity for cap structure at 5′ end of mRNA that probably assists, together with other initiation factors, in binding the mRNA to the 40S ribosomal subunit. Translation of mRNA *in vitro* is faster if it has a cap-binding protein.

cap-independent mRNA translation: Proteins of family 14-3-3 are crucial in a wide variety of cellular responses including cell cycle progression, DNA damage checkpoints and apoptosis. A protein called 14-3-3′ inhibits the cell cycle and may act as a tumour suppressor. It now turns out that it is also involved in regulating cap-independent protein synthesis from messenger RNA during cell division.

capacitation: Poorly understood process that sperm must go through in the female reproductive tract before they are competent for fertilization.

capillary action: It is the movement of water as it is pulled upwards through tubes (xylem) within a plant's roots, stems and leaves. The water (containing minerals and dissolved nutrients) is driven against gravity by adhesion of the water molecules (they stick to the sides of the tubes), cohesion of those molecules (the water molecules sticking together) and surface tension (the forces of the molecules on surface of the upward-moving water).

capillary array: Gel-filled silica capillaries used to separate fragments for DNA sequencing. The small diameter of the capillaries permits the application of higher electric fields, providing high-speed, high-throughput separations that are significantly faster than traditional slab gels.

capillary gel sequencing systems: Capillary gel electrophoresis with fluorescence detection is a technique for rapid DNA sequence analysis. In these systems, slab gel electrophoresis is replaced by capillary gel electrophoresis to analyse DNA samples (L.M. Smith in 1991). In these systems, instead of scanning DNA as it migrates through 96 lanes each in a series of poured slab gel, DNA is sent through a set of 96 capillary tubes filled with the polymerized gel. As the DNA fragments pass through these capillary tubes, DNA fragments are scanned. Then, the robot moves the DNA samples and reagents through the tubes continuously. The systems produce a steady flow of data each signal representing one of the four DNA bases.

capillary matting: A material used to promote the process of capillary action for the purpose of providing plants with the correct amount of water; self-watering devices that use capillary matting.

capped 5′ end: The 5′ end of eukaryotic mRNA that is modified after transcription to form ends with the general structure 7-methylguanosine-(5′) ppm(5′)2-O-methylated nucleoside.

capping: (i) In molecular genetics, process of modification of the 5′ end of messenger RNA, which has consequences for translation efficiency. (ii) In plant propagation, a hardened layer of surface soil sometimes caused by top watering; capping can impede the proper distribution of water through the soil.

capsid: Proteinaceous coat of virus. The outer coat of virus formed by self-assembly of one or more protein subunits and enclosing the viral nucleic acid (DNA or RNA). The capsid containing nucleic acid is called NUCLEOCAPSID. The protein subunits are called CAPSOMERES that can be observed under electron microscope. The capsomeres consist of protein subunits called protomers. *See* COAT PROTEIN.

capsid proteins: Proteins that make up the shell of a virus particle that contains the genetic material. The individual proteins that make up the capsid are called capsomers or protein subunits. It has been discovered that resistance to certain viral diseases may be imparted to some plants by inserting the gene for production of the protein coat into the plants.

capsomere: Protein clusters making up discrete subunits of a viral protein shell as seen under electron microscope. *See* CAPSID.

capsule: (i) Dry fruit formed from two or more united carpels and dehiscing when ripe (usually by splitting into pieces or opening at summit by teeth or pores).

(ii) A gelatinous envelope surrounding many types of microorganisms. The capsule is usually composed of polysaccharides, polypeptides or polysaccharide-protein complexes. Thin layered capsules are called MICROCAPSULES. In case many cells get surrounded by common matrix then the structure is called SLIME. Capsules are not absolutely essential cellular components. The capsule protects the cell from phagocytosis and toxic materials, attaches the cell to surfaces or other cells and helps the cell retain water.

CAPture: This full-length cDNA cloning is an affinity selection procedure using a fusion protein containing the murine cap-binding protein (eukaryotic initiation factor 4E), coupled to a solid support matrix, which allows for the purification of mRNAs via the 5′-cap structure. When combined with a single-strand-specific RNase digestion step, specific retention of complete cDNA-RNA duplexes occurs following first-strand synthesis.

capture probe: Phage or antibody probes that bind proteins in a sample such that their relative expression levels can be detected.

carbohydrate: An organic compound based on the general formula $C_n(H_2O)_n$ or $C_x(H_2O)_n$. The dominating substances of the cell sap are soluble carbohydrates. Carbohydrates are produced by photosynthesis in plants. They occur as disaccharides, such as saccharose and maltose, or as monosaccharides, such as glucose and fructose. The simplest carbohydrates are the sugars (saccharides), including glucose and sucrose. Polysaccharides are carbohydrates of much greater molecular weight and complexity. Examples include starch, which serves as energy store in plant seeds and tubers; cellulose and lignin, which form the cell walls and woody tissue of plants; and glycogen. *Aka* SACCHARIDE.

carbohydrate engineering: The selective, deliberate alteration/creation of carbohydrates (and the oligosaccharide side chains of glycoprotein molecules) by man.

carbon capture and storage (CCS): CCS involves the capture of carbon dioxide (CO_2) emissions from industrial processes, such as steel and cement production, or from the burning of fossil fuels in power generation. This carbon is then transported from where it was produced, via ship or in a pipeline, and stored deep underground in geological formations. It is one of these technologies and can therefore play an important role in tackling global warming. It is a proven technology that has been in safe operation for over 45 years. It adds that all components of CCS are proven technologies that have been used for decades on a commercial scale.

carbon cycle: Processes which convert CO_2 into compounds and back to CO_2. Green plants fix atmospheric CO_2 to synthesize organic compounds carrying carbon (Calvin cycle). These compounds are degraded to release CO_2 (Kreb cycle). After death and decay of the plants the organic compounds are degraded by microorganisms to release CO_2 and returned to the atmosphere through respiration or decomposition. Thus, whole cycle is called carbon cycle.

carbon dioxide (CO_2): A gaseous compound that is formed when carbon combines with oxygen. It is a reactant of photosynthesis, and six CO_2 molecules are needed to produce one glucose molecule. CO_2 molecules move from the surrounding air, entering a plant's leaf through the stomata in the epidermis. Once inside the leaf, these gas molecules move through the spongy mesophyll, entering the chloroplasts of palisade mesophyll cells. Within the stroma of the chloroplast, enzymes, ATP, NADPH and the light energy harvested during the light-dependent reactions drive the light-independent reactions. *Aka* DARK REACTION *or* CALVIN CYCLE that converts CO_2 into sugars.

carbon fixation: The process by which inorganic carbon, particularly in the form of carbon dioxide, is converted to organic compounds by living organisms. ATP and NADPH are used to fix CO_2 into carbohydrates. Carbon fixation takes place in the chloroplast stroma during the second stage of photosynthesis. *See* CALVIN CYCLE.

carbon flow: In plants, the original source of carbon comes from CO_2 in the air. Once the CO_2 enters the chloroplast, it is converted by a series of biochemical reactions into the organic molecule building blocks of the plant to make biomass, fruit and seed.

carbon footprint: Amount of greenhouse gases, specifically carbon dioxide or other carbon compounds, emitted by individuals, companies or countries (i.e. a person's activities or a product's manufacture and transport) during a given period of time. Indicator of air quality often used to measure an entity's environmental impact.

carbon isotope ratio: A measure of the proportion of the carbon-14 (^{14}C) isotope to the carbon-12 (^{12}C) isotope. Living material contains ^{14}C and ^{12}C in the same proportions as exists in the atmosphere. When an organism dies, however, it no longer takes up carbon from the atmosphere, and the ^{14}C it contains decays to nitrogen-14 (^{14}N) at a constant rate. By measuring the ^{14}C-to-^{12}C ratio in a fossil or organic artefact, its age can be determined, a method called RADIOCARBON DATING. Because most ^{14}C will have decayed after 50,000 years, the carbon isotope ratio is mainly useful for dating fossils and artefacts more recent than this. It cannot be used to determine the age of the Earth, for example.

carbon nanocarriers in post-transcriptional gene silencing: Used to deliver siRNA to intact plant cells for efficient gene knockdown. *See* POST-TRANSCRIPTIONAL GENE SILENCING (PTGS).

carbon nanotubes (CNT): A carbon allotrope that resembles a tube of carbon atoms. Carbon nanotubes are extremely robust and difficult to break, but they are still light. Because of their exceptional mechanical, electrical and thermal properties, carbon nanotubes are one of the most investigated nanomaterials. *See* NANOTUBES.

carbon partitioning. As plants grow, the carbon captured by the plant can be used to make biomass, fruit and seed. Efforts to modify carbon partitioning can be used to direct more of the carbon to increase biomass or seed preferentially, thereby increasing crop yield.

carbon-concentrating mechanism (CCM): Synthetic biology provides novel approaches of modifying the C_3 metabolism downstream of the oxygenation step. To enhance photosynthesis, either changes in catalytic properties of RuBisCO or enhancement in CO_2 concentration near the site of RuBisCO can be an appropriate approach. C_3 plants have the disadvantage, under hot-dry conditions, of their photosynthetic efficiency declining because of slow diffusion of CO_2 in apoplastic space and RuBP oxygenation by RuBisCO. The problem of photorespiration is overcome in the plants that possess a CCM around the site of RuBisCO (i.e. C_4 and CAM plants). C_4 photosynthesis has expressively higher conversion efficiency (ε_c) under prevailing atmospheric CO_2 concentration than C_3 photosynthesis.

carbonate: Any member of two classes of chemical compounds derived from carbonic acid or carbon dioxide.

carbonyl group: Pair of atoms consisting of a carbon atom linked to an oxygen atom by a double bond (C=O). The molecules containing a carboxyl group are weak acids – carboxylic acids.

carboxyl (C) terminus: Defined by a free α-carbonyl group (−COOH) of an amino acid or protein.

carboxylase: Any of the class of enzymes that catalyse the release of carbon dioxide from the carboxyl group of certain organic acids.

carboxylation: Introduction of a carboxyl group into an organic compound.

carboxysomal carbon-concentrating mechanism (CCCM) in the cyanelles: Cyanelles are the peculiar plastids of laucocystophyte algae that retained a peptidoglycan wall from the ancestral cyanobacterial endosymbiont. All cyanobacteria and most algae possess an inorganic CARBON-CONCENTRATING MECHANISM (CCM) that involves a microcompartment – 'carboxysomes' in prokaryotes and 'pyrenoids' in eukaryotes – harbouring the bulk of cellular (plastidic) material.

carcinogen: Any physical or chemical agent that may cause cancer by causing changes in a cell's DNA. Some physical radiations and chemicals can cause cancer when organisms get exposed to them either for longer durations or to elevated concentrations. For example, ethidium bromide which is commonly used to stain DNA gels is potentially carcinogenic as it binds to DNA. In this case proper biosafety guidelines should be adhered for our safety in laboratory. *See* MUTAGEN.

cardinal nucleotide: The 3′-nucleotide of anticodon triplet that is considered to be the most important in codon-anticodon recognition.

carmine: Used for preparation of carminic acid, a red dye used for colouring or staining chromosomes and other cytological material. Carmine is prepared from cochineal.

carnoy's fixative: A fixator solution, which consists of ethanol, chloroform and acetic acid in 6:3:1 ratio. It is used in chromosome analysis.

carotene (or carotin): A compound of carbon and hydrogen that occurs in plants. It is a precursor of vitamin A. Its reddish-orange plastid pigment is involved in light reactions in photosynthesis.

carotenoid: A water-insoluble photosynthetic pigment made up of isoprene units. Carotenoids are pigments of yellow, orange or red in colour that are water insoluble but soluble in lipids. It consists of a long aliphatic poly chain composed of isoprene units. Carotenoids are produced by algae, plant and certain bacteria and fungi microorganism. The carotenes and the xanthophylls are the most common. Carotenoids are responsible for the coloration of certain plants (e.g. the carrot) and of some animals (e.g. the lobster). The carotenoid pigments are transferred to animals as an element in their foods. Carotenoids are composed of isoprene units (usually eight), which may be modified by the addition of other chemical groups to the molecule. The carotenes are of importance to higher animals because they are utilized in the formation of vitamin A. Carotenoids act as antioxidants ('quenchers' of free radicals), so consumption of carotenoids apparently thereby reduces the risk of some cancers, coronary heart disease, eyesight loss and cataracts.

carpel: One of the female reproductive organs of the flower in angiosperms, comprising an ovary and usually with a terminal style tipped by the stigma. The GYNOECIUM is the collective term for all of the carpels of a single flower.

carpellary: Referring to carpels or to associated structures or outgrowths of carpels. For example, staminodes attached to carpels in Nymphaeaceae were frequently referred to as carpellary attachments. The current and past usage of the terms CARPELLARY ATTACHMENTS, PARACARPELS and STAMINODES is confusing and meaning varies among authors.

carpopodium: (i) On achenes (Cypselae), an elongation of the base of the gynoecium which looks distinct; the abscission zone, where the achene is separated from the receptacle. (ii) Genus *Carpopodium* in the family Brassicaceae; not to be confused with *Caropodium*, which is a genus of flowering plants belonging to the family Apiacea.

carrier: (i) In disease, a healthy person harbouring pathogenic organisms. Many people harbour pathogens and transmit them to others directly or indirectly, but these people do not exhibit any sign of illness. Such people are called carriers. (ii) In genetics, an individual who has a gene mutation for a recessive disease on one allele while the other allele is normally functioning. This individual most often does not develop symptoms of the disease but is at risk of having a child with the disease if their partner is also a carrier. If both parents are carriers of genetic variants for the same disease-causing gene, each child has a 1 in 4 (25%) chance of inheriting both genetic changes and being affected by the disease. Everyone is a carrier of several genetic variants for recessive disorders.

carrier DNA: DNA of undefined sequence content, which is added to the transforming (plasmid) DNA used in physical DNA-transfer procedures. This additional DNA increases the efficiency of transformation in electroporation and chemically mediated DNA-delivery systems.

carrier protein: Membrane transport protein that binds to a solute and transports it across the membrane by undergoing a series of conformational changes.

carrier screening: A type of genetic test that can tell you whether you carry a gene for certain genetic disorders. When it is done before or during pregnancy, it allows you to find out your chances of having a child with a genetic disorder.

carrier testing: Genetic testing performed to determine if asymptomatic individuals have a genetic variant for an autosomal recessive or X-linked recessive disorder of chromosome rearrangement. Typically performed to determine recurrence risks for a specific disease in a family or population.

carrying capacity: The density of a density-regulated population at equilibrium.

Cartagena Protocol on Biosafety (CPB): The first international treaty dealing with the movement of genetically modified organisms (GMOs) across country borders; the protocol was drawn up under the Convention of Biological Diversity (CBD) and came into force in September 2003. More than 100 countries have ratified the agreement. Although the biosafety protocol was pushed for by the South and drafted as a promise of legal protection against the introduction of GMOs, the weakness of its provisions means that the protocol and the national biosafety laws following its introduction are being steadily turned into tools to facilitate the introduction of GMOs.

cartenoid: A yellow, orange, red or brown pigment that is located in the chloroplast and chromoplast of plants. It acts as photosynthetic accessory pigment.

cartilage: Form of connective tissue composed of cells (chondrocytes) embedded in a matrix rich in type II collagen and chondroitin sulfate.

cartilaginous: Hard and tough; gristly.

caruncle: A small piece of flesh-like tissue, typically lumpy or warty, growing on the testa near the hilum. *Contrasts* ARIL.

caryopsis: A dry, indehiscent, one-seeded fruit in which the seedcoat is closely fused to the fruit wall (e.g. in most grasses).

Cas: General name for CRISPR ASSOCIATED PROTEIN. Includes nucleases like Cas9 and Cpf1. *See* CRISPR/CAS.

Cas9: Cas9 serves as an RNA-guided DNA endonuclease that cleaves DNA upon crRNA-tracrRNA target recognition.

Cas9 nickase: When Cas9 has only one of the two nuclease domains (RuvC or HNH) inactivated, it can cleave only one of the two strands of the target DNA depending on which one of the two domains has been inactivated. Cas9 nickases are of two types – Cas9 D10A (aspartate at position 10 changed to alanine in RuvC domain) and Cas9 H840A (histidine at position 840 changed to alanine in HNH domain). Cas9 D10A will cut the target strand of DNA (which does not contain the PAM sequence), while Cas9 H840A will cut the non-target strand (which contains the PAM sequence).

case-control: An observational study that begins with patient cases who have the outcome or disease being investigated and control subjects who do not have the disease or outcome. It then looks backward to identify possible precursors or risk factors.

case-control study: A study in which subjects are selected on the basis of their outcomes, and then exposures (treatments) are ascertained. For example, to assess the association between race and operative mortality, one might select all patients who died after open heart surgery in a given year and then select an equal number of patients who survived, matching on several variables other than race so as to equalize (control for) their distributions between the cases and non-cases.

case-series study: A simple descriptive account of interesting or intriguing characteristics observed in a group of subjects.

cascade: A sequential series of events (chemical reactions, immune responses, etc.) initiated ('set off') by a specific first event (e.g. a signalling molecule 'docking' at a receptor molecule, an antibody-antigen complex forming in the body, thrombin cleaving fibrinogen).

casein: A protein precipitated from milk; sometimes used for *in vitro* techniques. Casein hydrolysate (CH) is used in the plant tissue culture medium for embryo culture or in the callus induction in otherwise recalcitrant species.

cash crop: A readily saleable crop that is grown and gathered for the market (vegetables or cotton, tobacco, coffee, sisal, etc.).

casparian strips: The material between the cells of the root endodermis that blocks the passive movement of solutes and water.

caspase: Any of a family of intracellular proteases that are involved in initiating the cellular events of apoptosis.

cassette mechanism of mating switch: The mechanism by which homothallic yeast cells alternate mating types. The mechanism involves two silent transposons (cassettes) and a region where these cassettes can be expressed (cassette player).

cassette mutagenesis: A procedure of site-directed mutagenesis that can create specific mutants or families of mutants. Cassette mutagenesis involves removal of a stretch of DNA flanked on either side by a restriction site and insertion of a new cassette of synthetic DNA in its place, which directs various amino acid replacements.

cassette: A 'package' of genetic material (containing more than one gene) inserted into the genome of a cell via gene splicing techniques. May include promoter(s), leader sequence, termination codon, etc.

cassideous: Hood-, helmet- or bonnet-shaped; generally referring to floral anatomy (e.g. in the flowers of Aconitum, Satyrium, etc.).

castaneous: Chestnut coloured, reddish brown.

caste: A group of species in social organisms distinct from other such groups in mating habit and social function. In insects, caste depends on food habit and sex differentiation. In man, one endogamous group is termed caste.

casting: The premature loss of abscised leaves or twigs.

casual alien: An exotic plant that appears with no apparent human assistance but does not develop a sustained population(s), or one that persists only by repeated new introductions. *Compare with* ALIEN.

CAT box: A conserved sequence found within the promoter region of the protein-encoding genes of many eukaryotic organisms. It has the canonical sequence GGPyCAATCT and is believed to determine the efficiency of transcription from the promoter.

catabolic: Pertaining to an enzymatic reaction leading to the breakdown of a complex biological molecule into less complex components, which may either yield energy in the form of ATP or be used in subsequent anabolic reactions.

catabolism: General term for the enzyme-catalysed reactions in a cell by which complex molecules are degraded to simpler ones with release of energy. Intermediates in these reactions are sometimes called CATABOLITES. The reverse of anabolism is catabolism, in which decomposition reactions occur to produce simple molecules. For instance, breakdown of sucrose into glucose and fructose is the function of catabolism. This process releases energy required by the organism.

catabolism energy-yielding pathway: The phase of metabolism involved in the energy-yielding degradation of nutrient (food) molecules.

catabolite: A compound produced by catabolism.

catabolite activator protein (CAP): *Escherichia coli* CAP is a helix-turn-helix motif sequence-specific DNA-binding protein, which plays role in gene regulation. CAP decreases expression of catabolic enzymes as brought about by a catabolite such as glucose. Glucose is the preferred fuel source for certain bacteria, and when present in the culture medium, it represses the formation of enzymes required for the utilization of other fuel sugars, such as β-galactosidase. The protein mediates the action of cyclic AMP (cAMP) on transcription in that cAMP and CAP must first combine. The cAMP-CAP complex then binds to the promoter regions of *E. coli* and stimulates transcription of its operon. Since a cell component increases rather than inhibits transcription, this type of regulation of gene expression is called POSITIVE TRANSCRIPTIONAL CONTROL. Since glucose or other catabolites (other molecules derived from glucose) cause the repression, it is known as CATABOLITE REPRESSION. *Aka* CATABOLITE REGULATOR PROTEIN.

catabolite activator protein (CAP) site: A nucleotide sequence upstream of some bacterial genes and operons. This site serves as the attachment point for the catabolite activator protein (CAP).

catabolite repression: Repression of certain sugar-metabolizing operons in favour of glucose utilization when glucose is present in the environment of the cell.

catalase: Catalase is one in the group of enzymes known as METALLOENZYMES because it requires the presence of a metal in order to be catalytically active. The metal (known as a COFACTOR) is, in the case of catalase, iron. This enzyme is found in nearly all living organisms exposed to oxygen, including bacteria, plants and animals, which catalyses the very rapid decomposition of hydrogen peroxide (H_2O_2) to water and the oxidation by hydrogen peroxide of alcohols to aldehydes during seed germination. Catalase is easily detected by its action on H_2O_2. When a drop of H_2O_2 is added to the bacterial colony producing catalase then oxygen bubbles are released.

catalyse: To induce or accelerate a chemical reaction by a substance that remains unchanged in the process.

catalysis: This term, coined by Jons J. Berzelius in 1838, refers to the act of increasing the rate of a given chemical reaction via use of a catalyst. Almost all chemical reactions in biological systems (e.g. within an organism) are catalysed by molecules known as enzymes. An enzyme typically increases the rate of a given biological/chemical reaction by at least a millionfold without being consumed itself in the reaction.

catalyst: A substance that causes a chemical reaction to happen quicker without being used up in the process. For instance, there is a special catalyst in saliva called amylase that helps break down food so that it can be used in our body faster. Here enzyme amylase is a catalyst. Hydrogen and oxygen do not combine but in the presence of colloidal platinum then get easily combined to produce water. Platinum acts as a catalyst for this reaction. *Aka* CATALYTIC AGENTS.

catalytic antibody: An antibody produced (e.g. via monoclonal antibody techniques) in response to a carefully selected antigen (e.g. target molecule in bloodstream, or molecule involved in chemical reaction of interest) which itself catalyses the 'splitting' of a molecule in the bloodstream (e.g. heroin into two harmless small molecules) or mimics: (i) restriction endonucleases that cleave (cut) proteins or DNA molecules precisely at specific locations on those molecules; (ii) restriction endoglycosidases that are capable of cleaving oligosaccharides or polysaccharide molecules precisely at specific locations on those molecules; (iii) a transition state chemical complex in the chemical reaction that is to be catalysed; the resultant antibody acts both as an antibody (to the selected transition state-complex antigen) and as a catalyst (for the chemical reaction possessing that selected transition state chemical complex). This catalyst (enzyme) thus possesses the remarkable specificity of an antibody (i.e. specific only to the desired transition-state reactant), which holds the potential to yield chemical reaction products of greater purity than those achieved via current (less specific) catalysts. Because the immune system will (in theory) produce an antibody to virtually every molecule of sufficient size to be detected by the immune system (i.e. 6 to 34Å), it should be possible to raise catalytic antibodies for a large number of industrial chemical reactions that are currently catalysed via conventional (less specific) catalysts.

catalytic domain: Epigenetic writers (such as DNA methyltransferases, histone methyltransferases, histone acetylases) or epigenetic erasers (such as DNA demethylases, histone demethylases, histone deacetylases) are used as catalytic domain. *Aka* EFFECTOR DOMAIN (ED).

catalytic RNA: An RNA molecule that acts to cleave any other RNA.

catalytic site: The site (geometric area) on an enzyme molecule (or other catalyst) that is actually involved in the catalytic process. The catalytic site usually consists of a small portion of the total area of the enzyme.

cataphyll: Any plant structure that is morphologically a leaf but has at most an incidental or transient photosynthetic function. They are either shed when their main function has been completed or are incorporated into structures where, when dead, they serve a protective or supportive purpose.

catastrophism: Changes in species that happen because of quick and often violent natural phenomena. This theory *contrasts with* the UNIFORMITARIANISM of Lyell's geology.

catch crop: A method of increasing agricultural or horticultural productivity by filling in the empty spaces; for example, it is created when slower-growing vegetables are harvested with fast-growing crops; in general, a short-duration crop grown in between two main crops in a rotation to maximize cropping intensity (e.g. summer green gram grown in between two main cereal crops (wheat-green gram-maize), usually grown without any extra nutrient application and expected to feed on the residues of nutrients applied to the main crops).

categorical observation: These are those observations that can be sorted into categories or groups. Such observations are further classified into two types, namely, NOMINAL OBSERVATIONS (used to label the variables without providing any numerical value such as letters, words, symbols, gender, etc.) and ORDINAL OBSERVATIONS (that follow natural order differences between which cannot be determined such as those commonly encountered in surveys, questionnaires, finance and economics).

categorical variable: (i) A variable having only certain possible values for which there is no logical ordering of the values. Also called a NOMINAL, POLYTOMOUS, DISCRETE CATEGORICAL VARIABLE or FACTOR. (ii) Represents types of data which may be divided into groups. Examples of categorical variables are race, sex, age group and educational level.

category: Any one of the different levels in the hierarchy of the classification of living things, such as subspecies, species, genus, family and order.

catenane: Interlocked circular DNA molecules that are the exclusive products of bacteriophage λ integrative reaction *in vitro* when substrate is supercoiled molecule containing both attachment sites, *attP* and *attB*.

catenated DNA: Interlocked circular DNA molecules that are the exclusive products of bacteriophage λ integrative reaction *in vitro* when substrate is supercoiled molecule containing both attachment sites, *attP* and *attB*.

catenation: The formation of rings or chains by chromosomes at diakinesis.

catenulate: In the shape of a chain; formed of parts or cells connected as if chained together (e.g. some diatoms, algae and cyanobacteria such as Anabaena). *See* CONCATENATE.

caterpillar: Larva of a moth, butterfly or sawfly.

CATH: It is a novel hierarchical classification of protein domain structures. CATH clusters proteins at four major levels – Class (C), Architecture (A), Topology (T) and Homologous super family (H). Here, CLASS reflects similar secondary structure content and all alpha, all beta, alpha/beta, etc., ARCHITECTURE refers to major structural similarity and secondary structure elements in similar arrangement, TOPOLOGY means probable common ancestry and HOMOLOGOUS SUPER FAMILY denotes clear evolutionary relationship and sequence similarity (>25%).

cathodic drift: As ampholytes themselves migrate to cathode due to electro-osmotic flow, thereby resulting in instability in pH gradient. This is called cathodic drift. To overcome this problem, non-equilibrium pH gradient gels are used.

cation exchange: The exchange between cations in solution and another cation held on the exchange sites of minerals and organic matter.

cation: A positively charged ion. *See* BASE. *Compare with* ANION.

catkin: A conelike cluster of male or female flowers; conelike fruit of angiosperms. A spike, usually pendulous, in which the mostly small flowers are unisexual and without a conspicuous perianth (e.g. in willows, poplars, oaks and casuarinas). The individual flowers often have scaly bracts and are generally wind pollinated. Catkins are usually shed as a unit.

caudate: Having a narrow, tail-like appendage or tip (e.g. a drip tip). *Contrasts* ACUMINATE, CUSPIDATE *and* MUCRONATE.

caudex (pl. caudices): The stem of a plant, especially a woody one; also used to mean a rootstock, or particularly a basal stem structure or storage organ from which new growth arises. *Compare with* LIGNOTUBER.

caudiciform: Stem-like or caudex-like; sometimes used to mean 'pachycaul', meaning 'thick stemmed'.

caulescent: Possessing a well-developed stem above ground, similar to cauline. Antonym: ACAULESCENT (lacking an apparent stem).

cauliflory (Adj. cauliflorous): Having flowers or fruits growing directly from a tree's branches or trunk.

cauliflower mosaic virus (CaMV): A virus that infects cauliflower and other Cruciferae. It is transmitted by insects. The genome size is about 8 kb. It consists of double-stranded DNA with some single-stranded segments. In molecular genetics, it is used as a vector for transformation experiments.

cauliflower mosaic virus 35S (CaMV 35S) promoter: A promoter (sequence of DNA) that is often utilized in genetic engineering to control expression of (inserted) gene (i.e. synthesis of a desired protein in a plant). CaMV 35S promoter from the plant pathogen Cauliflower Mosaic Virus (CaMV) has been instrumental in driving constitutive expression of these transgenes, thus helping elucidate the function of many plant genes and increasing the understanding of plant processes. A 343-bp-long DNA contains a strong transcriptional enhancer, which on duplication results in tenfold increase in expression even at the distance of 2 kbp.

cauline: Borne on an aerial stem or caulis, as with leaves, flowers or fruits (when applied to the latter two organs), usually referring to older stems.

caulirosulate: Borne at the end of the stem or caulis, as with leaves or bracts.

caulogenesis: Process of shoot differentiation from callus. In case of plants, the callus recultured on medium containing low levels of auxins (IAA, NAA) in the presence of cytokinins (kinctin, BAP) exhibit green shoot bud formation, and finally, the shoot differentiation occurs during incubation under florescent light conditions.

causal agent: A general term used to describe an animate or inanimate factor that incites and governs disease and injury.

causal approach: Deals with process and cause of evolution.

causal inference: The study of how/whether outcomes vary across levels of an exposure when that exposure is manipulated. Done properly, the study of causal inference typically concerns itself with defining target parameters, precisely defining the conditions under which causality may be inferred and evaluation of sensitivity to departures from such conditions. In a randomized and properly blinded experiment in which all experimental units adhere to the experimental manipulation called for in the design, most experimentalists are willing to make a causal interpretation of the experimental effect without further ado. In more complex situations involving observational data or imperfect adherence, things are more nuanced.

causal organism: A pathogen of biotic origin. When a pathogenic agent is virulent (the relative aggressiveness of a pathogen), it can cause disease, and if the agent is avirulent, it is a variant of a pathogen that does not cause severe disease. Non-virulent is the preferred synonym to avoid confusion with 'avirulent'.

causation: Causation is a direct relationship between two variables. Two variables have a direct relationship if a change in one's value causes a change in the other variable. In that case, one becomes the cause, and the other is the effect.

causative single-nucleotide polymorphism (causative SNP): Any single-nucleotide polymorphism that is in linkage disequilibrium to a disease phenotype and therefore a responsible candidate for the disease.

cause and effect: Nature behaves in predictable ways. Searching for explanations is the major activity of science; effects cannot occur without causes. Mutation is a cause, and change in phenotype is an effect.

cause-and-effect relationship between an epigenetic mark and its ultimate behaviour: It is a challenge to determine a cause-and-effect relationship between a distinct epigenetic mark and the ultimate behaviour and function of the affected cell. However, CRISPR/Cas9 has provides excellent opportunity to investigators for manipulating the epigenome and observe the effects it may have on cell function, development and differentiation.

CCCTC-binding factor (CTCF): An 11-zinc finger protein that binds insulator elements.

cdc genes: Cell division cycle genes, of which several have now been defined, especially in yeasts.

Cdk activating kinase (CAK): Protein kinase that phosphorylates Cdks in cyclin-Cdk complexes, activating the Cdk.

Cdk inhibitor (CKI): Protein that binds to and inhibits cyclin-Cdk complexes, primarily involved in the control of G1 and S phases.

cell: (i) The basic microscopic structural and functional unit of all living organisms that carries on the biochemical processes of life. A cell may exist as an independent, self-replicating unit (as in the case of unicellular organisms), or in cooperation with other cells, each of which may be specialized for carrying out particular functions within a larger multicellular organism. Cells consist of cytoplasm enclosed within a cell membrane and sometimes a cell wall and serve the fundamental purpose of separating the controlled environment in which biochemical processes take place from the outside world. A eukaryotic cell is compartmentalized into specific functional areas and/or organelles with special tasks. In a cell, the genetic material (DNA) encodes heritable information for the maintenance of life. In eukaryotes, DIPLOID state of a cell is written as '2n' to distinguish it from the HAPLOID state 'n'. Plant cell is surrounded by thick cell wall made up of cellulose, hemicellulose and pectic substances. Inner to the cell wall is the plasma membrane which is lipoprotein bilayer. Membrane surrounds the cytoplasm which carries cell organelles like nucleus, mitochondria and chloroplasts.

cell adhesion: The contact between cells that is involved in cell aggregation and intercellular communication.

cell adhesion molecules (CAMs): (i) Proteins on the surface of an animal cell that mediate cell-cell binding or cell-matrix binding; (ii) The molecules that make cells with similar structure and function to stick together. Cells can change the repertoire of cell-adhesion molecules they express, and thus alter their relationship with the cells around them. The families of CAMs recognized are Ca^{2+}-DEPENDENT CADHERINS and Ca^{2+}-INDEPENDENT CELL ADHESION MOLECULES of the immunoglobulin superfamily (e.g. N-CAM).

cell affinity chromatography: A method of obtaining defined, functionally homozygous cell populations from mixed cultures using affinity chromatography. Affinity absorbents are prepared by immobilizing proteins (antibodies, lectins, etc.) on to macro beads. The sample is applied to a column packed with this material, impurities are washed out, and cells are selectively desorbed by washing them in a buffer with excess competitive agent (e.g. free monosaccharides for elution from immobilized lectins).

cell biology: The branch of biology that studies the structure and function of living cells, including their physiological properties, metabolic processes, chemical composition, life cycle, the organelles they contain and their interactions with their environment. This is done at both microscopic and molecular levels.

cell coat: *See* GLYCOCALYX.

cell competition: Competition among four genetically different spores formed by a female parent during meiosis. *Aka* RENNER EFFECT.

cell cortex: Specialized layer of cytoplasm on the inner face of the plasma membrane. In animal cells, it is an actin-rich layer responsible for movements of the cell surface.

cell counter: An automated device designed to enumerate the number of cells in a sample. The simpler type of cell counters consists of a mechanism to draw a known volume of a suitably diluted cell culture through a detection device. The detector comprises a plate with a small hole in it, across which an electric potential is maintained. As any cell or particle passes through the hole, a pulse is generated, which is applied and logged using an electronic counter. More sophisticated cell counters include flow cytometers.

cell culture: Cultivation of cells *in vitro* in suitable growth medium. In cell cultures, the cells are no longer organized into tissues. Cell cultures are usually grown in the liquid medium in the form of single cells or few cell aggregates in shake cultures using reciprocating shakers. When cultures are established from somatic tissues, these are called SOMATIC CELL CULTURES, and when established from gametic tissues, these are called GAMETIC CELL CULTURES. These are commonly used for basic studies and for the production of secondary metabolites (specialty chemicals for molecular farming/pharming).

cell cycle: It is a series of events that takes place in a cell leading to its division and duplication (replication). In cells with a nucleus (eukaryotes), the cell cycle can be divided into two brief periods: INTERPHASE, during which the cell grows, accumulating nutrients needed for mitosis and duplicating its DNA; and the MITOSIS (M) phase, during which the cell splits itself into two distinct cells. The cell cycle consists of four distinct phases: G1 phase, S phase (DNA synthesis), G2 phase (collectively known as interphase) and M phase (mitosis). Some cells enter G0 phase. G0 phase is often referred to as the resting phase. The G0 phase is the phase in the cell cycle in which the cell is neither dividing nor preparing for division. G1 is the resting stage before DNA synthesis, S is the synthesis phase when DNA replication occurs, G2 phase occurs after the synthesis of DNA and M is the phase when cell division occur. *Aka* CELL DIVISION CYCLE.

cell cycle arrest: The *p53* gene product is a critical human tumour suppressor. In response to DNA damage and oxygen starvation, *p53* induces cell cycle arrest or programmed cell death (PCD, apoptosis).

cell cycle checkpoints: Cell cycle checkpoints are surveillance mechanisms that monitor the order, integrity and fidelity of the major events of the cell cycle. After DNA damage, cell cycle checkpoints are activated. DNA damage checkpoints operate in G1, S and G2 phases of the cell cycle. Checkpoint activation pauses the cell cycle and gives the cell time to repair the damage before continuing to divide. DNA damage checkpoints occur at the G1/S and G2/M boundaries. An intra-S checkpoint also exists. *Aka* DNA DAMAGE CHECKPOINTS.

cell cycle control: Main cell cycle control mechanisms include cyclin-dependent kinases (CDKs), cyclins, cyclin-dependent kinase inhibitors (CKIs), pRb/E2F pathway and protein degradation.

cell cycle control of DSB repair: The choice between homologous recombination (HR) and non-homologous end joining (NHEJ) pathways is governed by cyclin-dependent protein kinases (CDKs), with a major site of control being at the level of DSB resection, an event that is necessary for HR but not NHEJ and that takes place most effectively in S and G2.

cell cycle duration: The duration of these cell cycle phases varies considerably in different kinds of cells. For a typical rapidly proliferating human cell with a total cycle time of 24 h, the G1 phase might last about 11 h, S phase about 8 h, G2 about 4 h, and M about 1 h.

cell cycle progression inhibition: Transforming growth factor-b (TGF-b) potently inhibits cell cycle progression at G1 phase. Smad3 has a key function in mediating the TGF-b growth-inhibitory response.

cell cycle progression: Increased expression of the oncogenic transcription factor c-Myc causes unregulated cell cycle progression.

cell cycle regulation: This phenomenon enables a normal cell to complete specific function before transiting to next phase. Growth factors, oncogene products and cyclins are the agents that determine the regulation of cell cycle by influencing phosphorylation and transcription process. The regulation of cell cycle undergoes two major events as interphase and cell division state. In interphase, the restitutional nucleus enters G1, S and G2 phases. Under favourable conditions, G1 is the phase where the control of cell proliferation is principally determined and the other two landmarks being S and M phases.

cell cycle regulator: Cell cycle regulators are commonly represented by cyclins, CDKs and CDK inhibitors but can also include some of their substrates, interacting partners and upstream regulators.

cell cycle regulatory circuit: It consists of a series of kinases and phosphatases that respond to external signals and checkpoints by phosphorylating or dephosphorylating the next member of the pathway. The ultimate readout of the circuit is to determine the activity of M phase kinase (or the S phase kinase) by controlling its state of phosphorylation.

cell cycle, measuring duration of: The duration of cell cycle and its phases can be measured by the use of radioactive tritiated (^{3}H) thymidine and autoradiography. Roots are grown in the presence of ^{3}H-thymidine for a short period (~30 min). Only those cells that are at S phase will incorporate the ^{3}H-thymidine in their DNA. It is known as PULSE LABELLING. After short pulse labelling (30 min), the roots are grown in absence of ^{3}H-thymidine, and root tip samples are studied after different intervals of time. Mitotic preparations are made from these root tips. Duration of the different phases of cell cycle is worked out by counting the number of labelled dividing mitotic cells out of total number of cells. For an interval of time, no labelled mitotic figures are seen. The time interval after which the labelled prophase cells appear will be equal to the duration of G_2 phase. G_1 can therefore be measured directly. The duration of G_1 is determined by using the formula G_1 = cell cycle − (G_2 + M + S). After some duration, all the dividing cells will appear labelled. The time interval in which all the dividing cells will appear labelled will be equal to the duration of mitosis (M). Gradually, the percentage of labelled cells will drop as the cells that were at G_1 phase at the time of pulse labelling enter into mitosis. Thus, the time interval between the appearance of labelled prophase and their subsequent disappearance is equivalent to the duration of S phase. Gradually, all the dividing cells appear unlabelled. Subsequently, labelled mitotic cells will be seen when earlier labelled cells complete interphase and re-enter mitosis. Thus, the duration of complete cell cycle is estimated by measuring the time interval between successive appearance of labelled cells in mitosis. The duration of complete cell cycle and its component phases vary from one cell type to another. G_1 is the most variable period of cell cycle. Cells have 2C amount of DNA during G_1 and each chromosome has only one chromatid. During G_1, preparation for chromosome duplication takes place. Quantity of nucleotides, histones, tubulins and DNA polymerase increases.

cell determination: It means irreversible commitment of the cell to a specific cell fate. It also means variable gene activity.

cell differentiation: It is the process by which stable differences arise between cells. As all cells arise from a fertilized ovum, which then give rise to various tissues and organs, cell differentiation is a very interesting and complex process. In molecular terms, cell differentiation means variable gene activity in different cells of the same organism, whereas cell specialization involves the preferential synthesis of some specific proteins. Not all genes are tissue-specific. Some genes are called HOUSEKEEPING GENES, which are expected to remain active in all types of cells, such as genes required for building membranes, ribosomes, mitochondria and glycolytic proteins (enzymes), which are components common to all types of cells. Some genes are expressed differentially. Multiple cell types can be produced from one genome through the expression of different groups of genes that regulate cell function and phenotype. Epigenetic modification of genes helps to coordinate the transition of one cell type to another through the developmental process of differentiation. Epigenetic modifications also 'lock down' a terminally differentiated cell to ensure that all of its daughter cells are the same cell type. In particular, developmental gene promoters are regulated by DNA methylation, histone acetylation and H3K4 and H3K27 methylation. Cell differentiation is caused/triggered/assisted by COLONY STIMULATING FACTORS (CSFs), GROWTH FACTORS (GFs) and certain other proteins (e.g. hedgehog proteins). *Aka* CELLULAR DIFFERENTIATION.

cell differentiation, characteristics of: (i) Differentiated state is stable. (ii) Cell determination precedes cell differentiation. (iii) Cell differentiation results from genetically controlled decision.

cell dispersion: A procedure used to liberate the content of a cell. These procedures may be mechanical and result in cell breakage or depend on cell lysis induced by addition of a solvent that affects the cell membrane, antibiotics or antimetabolites that disrupt or disorganize cell wall growth.

cell division: Separation of a cell into two daughter cells. In eukaryotic cells, it entails division of the nucleus (KARYOKINESIS), closely followed by division of the cytoplasm (CYTOKINESIS). In eukaryotes, there are two distinct types of cell division: (i) a vegetative division (mitosis), producing daughter cells genetically identical to the parent cell; and (ii) a cell division that produces haploid gametes for sexual reproduction (meiosis), reducing the number of chromosomes from two of each type in the diploid parent cell to one of each type in the daughter cells. The growth and development of plants and animals and humans is caused by cell division and cell elongation. Here, somatic cells divide and redivide through mitotic divisions. During this division first there is a nuclear division called karyokinesis followed by the division in cytoplasm called cytokinesis. The daughter cells are similar to the parent cells.

***cell division cycle* (*cdc*) gene:** Gene that controls a specific step or set of steps in the eukaryotic cell cycle. Originally identified in yeasts.

cell division axis: In budding yeast, the axis of cell division is defined by the site of bud emergence, which is chosen early in the G_1 phase of the cell cycle. Cells entering the cell cycle have a single microtubule organizing centre called the SPINDLE POLE BODY (SPB), which is embedded in the nuclear envelope from which cytoplasmic microtubules emerge. The SPB duplicates itself and separates, giving rise to a second set of cytoplasmic microtubules and the intranuclear spindle. The initial step in spindle orientation occurs when microtubules from the 'old' SPB are preferentially selected to become attached to the BUD NECK or BUD TIP. This process requires the cortical factor known as Kar9.

cell division controls patterning and growth: C.M. Winter and others in 2024 provide insights into ASYMMETRIC CELL DIVISION and SYMMETRIC CELL DIVISION. These two types of cell division generate cells that are similar to or different from their parents. Analysis of this process in plant roots reveals how this mechanism is regulated. SHORT-ROOT (SHR) and SCARECROW (SCR) are transcription factors that are required for formative divisions in the stem cell niche of Arabidopsis roots. Levels of SHR and SCR early in the cell cycle determine the orientation of the division plane, resulting in either asymmetric (formative) or proliferative cell division. SHR and SCR kinetics do not align with the expected behaviour of a bistable system, and only low transient levels – present early in the cell cycle – are required for formative divisions. These results reveal an uncharacterized mechanism by which developmental regulators directly coordinate patterning and growth.

cell division cycle (CDC) protein: The protein involved directly in the control of the cell cycle For example, CDC2, also known as p34, is a kinase enzyme.

cell division delay: UV light can temporarily delay cell divisions. It also produces a chromatid type of aberration and disrupts H-bonding between strands of natural DNA upon exposure of viral suspensions to UV light.

cell division, priming of: Although the establishment of new walls during plant cell division is inclined to follow maximal tensile stress direction, examination of individual cells over time reveals a much more flexible behaviour. In order to investigate the source of such variability as well as the exact role of interphasic microtubule behaviour before cell division, I. Melogno *et al.* (2024) used the Arabidopsis stem, where the tensile stress pattern is both extremely anisotropic and stable. Even though cortical microtubules (CMTs) usually align with maximal tensile stress, they reveal a specific time window, ca. 3 h before cell division, where cells formulate a radial pattern of CMTs. This microtubule array organization precedes preprophase band (PPB) formation, a transient CMT array predicting the position of the future division plane. It has been detected under different growth conditions and was not linked to cell geometry or polar auxin transport. However, this cortical radial pattern associates with the well-documented upsurge of cytoplasmic microtubule accumulation before cell division. This radial organization is protracted in cells of the *trm678* mutant, where CMTs are unable to form a PPB, whereas division plane orientation in *trm678* is noisier, cell division symmetry is in contrast less variable between daughter cells. This 'radial step' signifies a trade-off in robustness for two essential cell division attributes: symmetry and orientation. This involves a 'reset' stage in G2, where an increased cytoplasmic microtubule accumulation transiently disrupts CMT alignment with tissue stress.

cell division, timer for: A serendipitous observation has revealed that cells make several versions of a key protein needed for cell division. The ratio of these protein isoforms influences how long division can be delayed when errors arise. Cells have a safety mechanism that delays cell division when chromosomes are not properly oriented for their correct distribution into daughter cells. When engaged, this safety mechanism inhibits the protein CDC20, whose activity is crucial for completing division, and this inhibition provides a window of opportunity for errors to be fixed (S. Hauf in 2023). M.-J. Tsang and I.M. Cheeseman (2023) report that cells produce several versions (isoforms) of CDC20 that differ in their susceptibility to inhibition. These isoforms arise when the messenger RNA encoding CDC20 undergoes translation – a process that typically produces only one version of a protein.

cell elongation: The process of cell division is initiated by formation of two complexes of microtubules, the asters. Cells in interphase normally contain one centriole. Duplications of this structure are essential for the formation of two asters. When two asters are present in cell, an amphiester is formed due to the presence of central microtubules, which connect the two centres.

cell envelope: The different surface components of the cell that are present outside the cytoplasmic membrane.

cell fate: In developmental biology, describes what a particular cell at a given stage of development will normally give rise to.

cell fate and pattern formation: Species transmit faithfully their basic characteristic from parents to offspring, which shows that there must be some underlying formation of the general body plan, which encoded the genome. During development, cells adopted specific fates (i.e. capacity to differentiate into particular kinds of cells). A zygotic cell is totipotent, and as the zygote divisions proceed, PERIODIC DECISIONS are made in each cell lineage to specify more exactly the fate of the daughter cells. Some fate decisions are autonomous by the individual cells while others are through communication or collaboration with the neighbouring cells.

cell fate transition: Cell fate transition involves dynamic changes of gene regulatory network and chromatin landscape, requiring multiple levels of regulation, via the crosstalk between epitranscriptomic modification and chromatin signalling. N-acetyltransferase NAT10 controls cell fates via connecting mRNA cytidine acetylation to chromatin signalling. Z. Hu and others in 2024 find that suppression of N-acetyltransferase 10 (NAT10), the writer for mRNA N4-acetylcytidine (ac4C) modification, can notably affect human embryonic stem cell (hESC) lineage differentiation and pluripotent reprogramming. Integrative analysis revealed that NAT10-mediated ac4C modification regulates the protein levels of a subset of its targets, which are strongly enriched for fate-instructive chromatin regulators, and among them, histone chaperone ANP32B is experimentally verified and functionally relevant. Furthermore, NAT10-ac4C-ANP32B axis can modulate the chromatin landscape of their downstream genes (e.g. key regulators of Wnt and TGFβ pathways). NAT10 is thus an essential regulator of cellular plasticity, and its catalysed mRNA cytidine acetylation represents a critical layer of epitranscriptomic modulation and uncovers a previously unrecognized, direct crosstalk between epitranscriptomic modification and chromatin signalling during cell fate transitions.

cell fusion: Fusion of two previously separate cells, occurs naturally in fertilization. It can be induced artificially by the use of FUSOGENS such as polyethylene glycol. Fusion may be restricted to cytoplasm; nuclei may fuse as well. A cell formed by the fusion of dissimilar cells is referred to as a HETEROKARYON.

cell generation time: The interval between consecutive divisions of a cell. This interval can best be determined, at present, with the aid of CINEPHOTOMICROGRAPHY. *Compare with* POPULATION DOUBLING TIME.

cell heredity: Inheritance of a cellular level.

cell hybridization: The fusion of two or more dissimilar cells leading to the formation of a viable cell hybrid. *See* SYNKARYON.

cell immobilization: Cell immobilization is a biotechnology process in which animal and plant cells are fixed in a suitable matrix to immobilize them. It is relatively a new technique of plant tissue culture used for production of secondary plant products. Immobilization of plant cells and protoplasts, among others, is achieved by binding these materials on to or within a solid support. Immobilization plant cells are characterized by large size and sensitivity to shear, and they need cell-to-cell contact for metabolite production. Easier separation of biocatalyst from product occurs by immobilization. Agarose, alginate and carrageenan are the polymers that are used as solid matrix for entrapment. Immobilized cells are capable of maintaining viability for prolonged periods.

cell junction: Specialized region of connection between two cells or between a cell and the extracellular matrix.

cell line/strain: A cell line arises from a primary culture at the time of the first successful subculture. The term cell line implies that cultures from it consist of lineages of cells originally present in the primary culture. The terms finite or continuous are used as prefixes if the status of the culture is known. If not, the term line will suffice. The term CONTINUOUS LINE now replaces the term ESTABLISHED LINE. During the maintenance and multiplication of cell cultures through periodic subculturing, certain cell lines exhibit better growth and performance *in vitro*. These lines are named or numbered that represent a special cell line.

cell lysate: For detection of histone modifications by Western blotting (WB), whole-cell lysates prepared by extraction with SDS (sodium dodecyl sulfate) Laemmli sample buffer are used.

cell manipulation: The collection of means for precise control of the static or dynamic position of individual biological cells with the purpose of observing, probing and/or altering the cellular responses.

cell map proteomics: It deals with construction of physical maps of protein cellular location and interactions. It includes assignment of gene function and identification of protein networks.

cell membrane: The outer membrane of a cell with a discrete structure and function, which separates it from the environment. *Aka* PLASMA MEMBRANE *or* PLASMALEMMA.

cell nucleus: The 'control room' for the cell. The nucleus gives out all the orders.

cell organelles: Various membrane-bound structures that are found within a cell. All the basic functions of the central dogma of molecular biology are found in organelles (chloroplasts and mitochondria). These functions include DNA replication, RNA transcription and protein translation. Thus, certain gene products will result from the expression of the organelle DNA.

cell plate: Flattened membrane-bounded structure that forms by fusing vesicles in the cytoplasm of a dividing plant cell and is the precursor of the new cell wall.

cell polarity: Cell polarity is a fundamental property of all cells. Only now are we beginning to unravel the mechanisms behind a cell's ability to point in any direction and negative effectivity.

cell population: A group of cells that is static (without mitotic activity), expanding (showing scattered mitosis) or renewing (in which mitosis is abundant).

cell proliferation: Broadly speaking, it means uncontrolled cell multiplication. POLYPEPTIDE GROWTH FACTORS control cell proliferation. In multicellular organisms, tissue-specific growth factors are important in coordinating cell proliferation. C-terminal domain (Ctd) is an essential component of the replication complex, which mediates proliferation. The interaction of geminin with Cdt1 inhibits assembly of the MINICHROMOSOME MAINTENANCE (MCM) COMPLEX, so proliferation cannot occur unless geminin is degraded.

cell proliferation and cell cycle relationship: There is intimate relationship between cell proliferation and cell cycle. Antisense-mediated interference with cell cycle affects cell proliferation.

cell proliferation control: If a brink mitotic duration is surpassed, propagation of daughter cells is detained in a way that needs the tumour suppressor p53, the cell cycle inhibitor p21, ubiquitin-specific protease 28 (USP28) and p53-binding protein 1 (53BP1). F. Meitinger *et al.* (2024) elucidate how this procedure functions, with prolonged mitosis resulting in the assembly of a USP28-53BP1-p53 complex – the 'mitotic stopwatch' – which arbitrates daughter cell capture and may avert cancer development.

cell recognition: The mutual recognition of cells due to antigen-antibody or enzyme-substrate reactions.

cell repair machine: Molecular and nanoscale machines with sensors, nanocomputers and tools, programmed to detect and repair damage to cells and tissues, which could even report back to and receive instructions from a human doctor if needed.

cell sap: The inter-organelle fluid of the cell.

cell selection: Selection within a population of genetically different cells *in vitro* by different means and different approaches.

cell sorting: A process utilized (e.g. by researchers) to sort/separate different cells (pathogens, cancerous vs normal cells; sperm that are bearing chromosomes for male vs female; etc.). Some automated means of cell sorting include biochips (utilizing controlled electrical fields to collect specific cell types on to electrodes in the biochip), fluorescence-activated cell sorter (FACS) machines, magnetic particles (attached to antibodies), etc.

cell specialization: It means preferential RNA and hence preferential protein synthesis.

cell strain: A population of cells derived either from a primary culture or from a single cell.

cell suicide: It is a critical process in which the individual's cells undergo SELF-DESTRUCTION when they become superfluous or disordered. Cell suicide is an intrinsic property of all cells. They do this by genetically controlled cell suicide machinery that leads to programmed cell death. Most of the PCD proceeds by apoptosis – a process that includes condensation and segmentation of nuclei and cytoplasm and often extensive fragmentation of CHROMOSOMAL DNA into the nucleosome units. *Aka* PROGRAMMED CELL DEATH or APOPTOSIS.

cell surface: The multi-component structure surrounding a cell.

cell suspension: Cells and small aggregates of cells suspended in a liquid medium.

cell synchrony: When a population of cells proceeds through the stages of cell cycle with synchrony (i.e. all cells divide at one time).

cell theory: The theory that all living things are made up of cells.

cell transformation: A stable heritable alteration in the morphological and biochemical properties of a cell, usually brought about by an oncogenic virus, bacterial infection or experimental means.

cell type: Pattern of gene expression, not genes themselves, defines each cell type. Proteins and RNAs control patterns of gene expression that are passed on to successive generations. Gene extends beyond the sequence to epigenetic modification of that sequence. Addition of methyl groups to DNA plays a role in gene silencing, and the methylation patterns are carried over to next generation.

cell wall: A cell protecting rigid matrix outside the plasma membrane. Cell wall is prominent in plants and most of fungi and prokaryotes but is absent in most of animals. In plants the cell wall is comprised of cellulose, hemicellulose and pectic substances, whereas in fungi, the major components of cell wall include cellulose and chitin. The cell wall balances the outward pressure of the cytoplasm and thus helps in keeping the cell intact. But cell wall is a major barrier for genetic manipulation through somatic hybridization. The cell wall protects the cell, maintains its shape and helps facilitate the passage of water, nutrients, gases and waste into and out of the cell. *See* PROTOPLAST.

cell-cell fusion: Cell-cell fusion is fundamental to the developmental and physiological processes of multicellular organisms. Understanding about cytoplasmic control of nuclear activity comes from cell fusion experiment. Cell fusion is brought about by addition of an inactivated non-infective virus. Virus attaches to the cell membrane and makes the cell sticky. For a period of time the fused cell contains two or more nuclei (i.e. they are multinucleate). If the cells undergoing fusion are derived from different types of tissues or even from different species of animals, they are still capable of fusing and of forming multinucleate cells when the nuclei present in the cytoplasm are genetically dissimilar, a condition termed heterokaryon.

cell-cell interactions: Cell-cell interactions play key role in development. A substance produced by a cell influences developmental fate of another cell, the process is called INDUCTION. Induction is of two types: (i) INSTRUCTIVE INDUCTION and (ii) PERMISSIVE INDUCTION. These two induction mechanisms concern interactions between non-equivalent cells.

cell-cycle control system: Network of regulatory proteins that governs progression of a eukaryotic cell through the cell cycle.

cell-differentiation proteins: The various growth factors and other proteins which cause/assist in cell differentiation.

cell-extrinsic processes of morphogenesis: Cell-extrinsic processes, which derive morphogenesis, include cell-cell adhesion and cell-matrix interactions.

cell-free DNA testing: The cell-free DNA prenatal screening test screens for certain conditions caused by an abnormal number of chromosomes. It does not test for all types of chromosomal disorders.

cell-free extract: A fluid containing most of the soluble molecules of a cell, made by breaking open cells and getting rid of remaining whole cells.

cell-free system: Fractionated cell homogenate that retains a particular biological function of the intact cell and in which biochemical reactions and cell processes can be more easily studied.

cell-intrinsic processes of morphogenesis: The processes which derive morphogenesis. They include the rate of cell growth and proliferation, the nature of cell division, changes in cell shape and cell death.

cell-mediated immune response: That part of an adaptive immune response in which antigen-specific T cells are activated to perform various functions such as killing infected cells and activating macrophages.

cell-penetrating peptide (CPPs): These peptides are able to transverse cell membranes and deliver biological macromolecules to living cells.

cell 'sorting out': When the cells physically interact, they move so as to sort into clusters that maximize contact between cells of the same type. The ability of cells to do this comes from differential cell adhesion. Two well-studied types of cells that sort out are epithelial cells and mesenchymal cells.

cell-specific cDNA libraries: The contents of individual cells are aspirated into micropipette tips filled with RNA extraction buffer. The mRNA from these cells is purified by binding to oligo(dT)-linked magnetic beads and amplified on the beads using reverse transcription and PCR. The cell-specific nature of the isolated mRNA is verified. MONTE CARLO EFFECT is created by small and random differences in amplification efficiency between individual templates in an amplifying cDNA population.

cell-specific gene activation: Equality of DNA content among differentiated cell nuclei means that differentiation cannot in general be explained through selective loss of major fraction of unused genes from the nucleus. Differentiation involves inactivation of DNA coding for properties that are not manifest in a given cell type. Regardless of the process of differentiation, these processes are at least potentially reversible. However, under normal conditions, the processes leading to differentiation are irreversible.

cell-specific genes: *See* LUXURY GENES.

cellomics: A study that combines information from genomics and proteomics with the complex chemical and molecular relationships of cell components. Research about what goes on within a cell can give us valuable information about drug targets and drug development. Microarray technology is an important part of cellomics research.

cellular: Of or relating to a cell.

cellular affinity: Tendency of cells to adhere specifically to cells of the same type. This property is lost in some cancer cells.

cellular antisilencing factors: Some transposons and other repeat elements reside in the introns of genes. ANTI-SILENCING 1 (ASI1), a bromo-adjacent homology domain and RNA recognition motif-containing protein, is required to prevent promoter DNA hypermethylation and transcriptional silencing of some transgenes. Genome-wide DNA methylation analysis reveals that ASI1 has a similar role to that of the histone H3K9 demethylase INCREASE IN BONSAI METHYLATION 1 (IBM1) in preventing CHG methylation in the bodies of thousands of genes. X. Wang *et al.* (2013) found that ASI1 is an RNA-binding protein and ensures the proper expression of *IBM1* full-length transcript by associating with an intronic heterochromatic repeat element of IBM1.

cellular automata: An early model of computing popularized by Conway's *Game of Life* and useful for modelling intercellular interactions and extracellular signalling during tissue growth and remodelling.

cellular basis of Mendel's first principle: In a diploid organism (such as peas), all the cells of the organism contain two chromosome sets. Gametes, however, are haploid, containing one chromosome set. Gametes are produced by meiosis, specialized cell division in the diploid cells in the germinal tissue (ovaries and anthers). The highly programmed chromosomal movements in meiosis cause equal segregation of alleles into separate gametes. The nuclear spindle, a series of microtubules consisting of the protein tubulin, generates the force pulling the chromosome to cell poles. Microtubules attach to the centromeres of chromosomes by interacting with another specific protein located in that area. Orchestration of these molecular interactions is complex yet constitutes the basis of the principles of hereditary transmission in eukaryotes.

cellular defence: A defence response that is mediated by cells. These cells are the non-specific effector cells of the innate immune response. They include scavenger cells, which attack infectious agents directly, and natural killer cells, which attack cells of the body that harbour infectious organisms.

cellular differentiation: A complex multistep biological process by which a cell turns into a different, more specialized, type of cell. However, not every cell has the same ability to differentiate. For instance, some cells called PLURIPOTENT STEM CELLS can turn into almost every cell of the body, while neurons, for example, are terminally specialized cells that can no longer differentiate.

cellular endosperm: A type of endosperm in which the early development is characterized by cell wall formation accompanying each nuclear division.

cellular genomics: A platform for the fusion of genomics and cell biology.

cellular homeostasis: RECEPTOR TYROSINE KINASES (RTKs) control many of the biochemical processes that regulate cellular homeostasis. Given their importance, RTKs are themselves subject to multiple levels of control. A variety of autoinhibition mechanisms have evolved to maintain RTKs in an inactive configuration in quiescent cells. A number of inhibitory mechanisms – such as growth factor – induced receptor endocytosis, recycling and degradation – terminate and fine-tune the signals generated by RTKs after ligand stimulation. Furthermore, growth factor-stimulated protein kinases, guanosine triphosphatases, protein phosphatases and other key components provide another level of control through negative feedback. The correct balance between inhibitory and stimulatory signals is required for control of normal development and cellular homeostasis.

cellular oncogene (c-onc): Altered proto-oncogene (e.g. by mutation), which leads it to acquire an altered cellular function that contributes to carcinogenesis. They are found in the cells of a host organism.

cellular pathways: About 20 μm between the cell membrane and the genetic material in the cell nucleus is the space of the molecules of signal transduction, of the intricate and multifaceted redundancy of the pathways that take signals from the membrane and convert them into the exquisitely selective control of our genes; within these pathways, the regulation of gene transcription is carried out by a multitude of hormones and growth factors.

cellular plasticity: The ability of cells to transition between different phenotypic states.

cellular reprogramming: (i) Conversion of a differentiated cell to an embryonic state. (ii) The process of converting one cell type into another by changing the gene expression programme of the cell.

cellular respiration: A process in which energy is produced from various molecules (like glucose), producing ATP (adenosine triphosphate). During cellular respiration, oxygen is used, and carbon dioxide is produced. Cellular respiration occurs in the mitochondria of eukaryotes and in the cytoplasm of prokaryotes.

cellular responses: p53 is a tumour suppressor protein that regulates the cellular responses to GENOTOXIC STRESSES. p53 is short-lived, and its activity is regulated mostly by stabilization via different post-translational modifications. A novel mechanism of p53 regulation through lysine regulation by Set9 methyltransferase has been proposed. Set9 specifically methylates p53 at one residue within the carboxyl-terminus regulatory region. Methylated p53 is restricted to the nucleus and modification positively affects its stability. Set9 regulates the expression of p53 target genes in a manner dependent on the p53-methylation site.

cellular responses to environmental cues: The signal transduction cascade comprising Raf, mitogen-activator protein (MAP), kinase (MEK) and MAP kinase, is a Ras effector pathway that mediates diverse cellular responses to environmental cues and Ras-dependent oncogenic transformation. Ras effector protein Impedes Mitogenic Signal Propagation

(IMSP) modulates sensitivity of the MAP kinase cascade to stimulus-dependent activation by limiting functional assembly of the core enzymatic components through the activation of Kinase suppressor of Ras (KSR), a scaffold/adaptor protein that couples activated Raf to its substrate MEK.

cellular transformation: One of the steps in cellular transformation may be an epigenetic change in gene activity, such as loss of DNA methylation induced by DNA damage. Cells from large, long-lived species may be effectively buffered against epigenetic changes. One possibility would be an increase in number of methyl groups involved in the control of gene activity. One way to have increased buffer against epigenetic changes would be an increase in the number of methyl groups involved in the control of gene activity. A single $5^{m}C$ in a promoter region can prevent transcription. This $5^{m}C$ can be lost at a frequency of 10^{-3} per cell division. In case of existence of two $5^{m}Cs$ in the same region, either of which can prevent transcription, methyl groups of both $5^{m}Cs$ would be lost at the frequency of 10^{-6} per cell division, and so on.

cellularization: The formation of cells around each nucleus in a multinucleate cytoplasm, transforming it into a multicellular structure.

cellulase: A cell wall degrading enzyme that degrades cellulose to produce cellobiose (simple sugars such as glucose). Cellulase, which is produced by a fungus *Myrothecium verucaria*, is an important plant cell wall degrading enzyme commonly used to obtain protoplasts from different plant tissues. Incubation of peeled leaves, cotyledons, hypocotyls and epicotyls in cellulase solution (2–4%) for 5–10 h yields thousands of protoplasts. Cellulase is sometimes used for maceration of plant tissue in order to improve spreading of chromosomes.

cellulose: A major component of plant cell wall. Cellulose is a polysaccharide, with formula $(C_6H_{10}O_5)_n$, of glucose units linked together by $\beta(1\rightarrow4)$ glycosidic bonds. Cellulose forms long microfibrils, and it is a major component of plant cell wall. Cellulose can be degraded using cell wall degrading enzyme cellulase to obtain plant protoplast. Because of its presence in all plant cells, cellulose is the most abundant biological compound on earth. It provides tensile strength in plant cell walls.

cellulose utilization (cel) operon of *E. coli*: The cel (cellulose utilization) operon of *E. coli* K12 expresses when there are mutations in the operon. Two kinds of mutations are capable of activating the operon. The first type involves the mutations that allow the repressor to recognize the substrate cellulose, arbutin and salicin as inducers. Five different active alleles were studied, and differences were found to be single base pair changes at one of the two lysine codons in the repressor gene. The second type involves the integration of the insertion sequences IS1, IS2 or IS5 into a 108-bp region 72–180 bp upstream of the start of transcription.

CEN: A cloned eukaryotic centromere. Each CEN is given a number corresponding to the chromosome from which it was derived. This notation is confined to yeast molecular biology at the moment but, in principle, could be widely applied.

CEN DNA: The DNA sequence within the centromeric regions is called CEN DNA. CEN DNA can be moved from one chromosome to another and still provide the chromosome with the ability to segregate. These sequences must not provide any other function. Typically, CEN DNA is about 120 bp long. CEN DNA consists of several subdomains: CDE I, CDE II and CDE III. Mutations in the first two subdomains have no effect upon segregation of chromosomes. A point mutation in the CDE III subdomain completely eliminates the ability of the centromere to function during chromosome segregation. CDE III must be actively involved in the binding of the spindle fibres to the centromere.

cenospecies: All the ecospecies that are so related that they are able to exchange gens without the loss of fertility or vigour in the offspring.

censored observation: An observation whose value is unknown, generally because the subject has not been in the study long enough for the outcome of interest, such as death, to occur.

censoring: When the response variable is the time until an event, subjects not followed long enough for the event to have occurred have their event times censored at the time of last follow-up. This kind of censoring is RIGHT CENSORING. For example, in a follow-up study, patients entering the study during its last year will be followed a maximum of one year, so they will have their time until event censored at one year or less. LEFT CENSORING means that the time to the event is known to be less than some value. In INTERVAL CENSORING, the time is known to be in a specified interval. Most statistical analyses assume that what causes a subject to be censored is independent of what would cause an event to occur. If this is not the case, INFORMATIVE CENSORING is said to be present. For example, if a subject is pulled off of a drug because of a treatment failure, the censoring time is indirectly reflecting a bad clinical outcome and the resulting analysis will be biased.

Centegener method: One of the earliest established pure-line systems of plant breeding in which 100 seeds were planted in square plants and plants were selected on the basis of number of seeds per plant.

centimorgan (cM): A unit of measure of recombination frequency. One centimorgan is equal to a 1% chance that a marker at one genetic locus will be separated from a marker at a second locus due to crossing over in a single generation. In human beings, one centimorgan is equivalent, on average, to one million base pairs. Represents GENETIC MAP DISTANCE. *See* MEGABASE.

central body: (i) Centrosome. (ii) The structures at the centre of the aster during mitosis.

central cell: The largest cell in the centre of the embryo sac surrounding the egg apparatus at the micropylar end.

central dogma of molecular biology: Refers to F.H.C. Crick's seminal concept of 1960 that in nature genetic information generally flows from DNA to RNA to protein. This concept holds true for those organisms where DNA is the genetic material. It is, however, not rigorously accurate as illustrated by the facts that DNA (i.e. genes) 'information flow' is influenced (timing, amounts, etc.) by some environmental factors (temperature, humidity, etc.). Prions do not contain any DNA. Central dogma comprises three major processes – replication, transcription and translation. It comprises three major processes in the cellular utilization of genetic information. Replication copies parental DNA to produce complementary daughter DNA molecule. Transcription synthesizes RNA from DNA template. Translation of mRNA produces protein. Any change in the gene makes change in the protein that cause alternation in the trait. Nowadays the organisms are genetically modified by making changes in the base sequence of DNA (gene). *See* MODIFIED CENTRAL DOGMA *and* NEW CENTRAL DOGMA.

central dogma, modified: With the discovery of reverse transcription, this concept was modified.

central dogma, new model of: The new model central dogma of molecular biology is written as Genome (a static information source with a defined gene content) → Transcriptome (dynamic entities, whose content can fluctuate dramatically under different conditions and stages of development due to regulation of transcription, RNA processing) → Proteome (dynamic entities, whose content can fluctuate dramatically under different conditions and stages of development, protein modification) → Metabolome.

central limit theorem: A theorem that states that the sum of a large number of independent variables tends towards a normal distribution. The distribution of means is approximately normal if the sample size 'n' is large enough (n is greater than or equal to 30), regardless of the underlying distribution of the original measurements.

central processing unit (CPU): The electronic circuitry within a computer that carries out the instructions of a computer program by performing the basic arithmetic, logic, controlling and input/output (I/O) operations specified by the instructions. The computer industry has used the term 'central processing unit' at least since the early 1960s. Traditionally, CPU refers to a processor, more specifically to its processing unit and control unit (CU), distinguishing these core elements of a computer from external components such as main memory and I/O circuitry.

central spindle: A miniature spindle that, where centrosomes and asters are present may be located between them, and that, on the disappearance of nuclear membrane (at prometaphase), moves into the middle of the nuclear zone.

central tendency: Among variations, there is still a tendency for a set of values to gather around a central value. Mean, median and mode are measures of central tendency.

central vacuole: The central vacuole is a large membrane-bound organelle located in the cytoplasm of plant cells. In most mature plant cells, there is one central vacuole that takes up 30–80% of the cell's volume. The vacuole stores much of the plant cell's water, storage molecules and enzymes. In addition to storage, its other main functions are maintaining the cell's rigid structure by creating high hydrostatic pressure within the cell, stabilizing pH by removing protons from the cytoplasm and removing harmful substances and waste products from the cytoplasm.

Centre for Epigenome Mapping Technologies (CEMT) project: Establishes the institution of a major Canadian centre of epigenome analysis at the Genome Sciences Centre in Vancouver. These data are presented at epigenomes.ca. CEMT is funded by the Canadian Institutes of Health Research (CIHR) and Genome BC. CEMT aims to generate high-quality reference epigenome maps, incorporating several different epigenomic modifications, as well as matching DNA sequences and gene expression profiles, for several normal and cancerous tissues.

centre of domestication: The area believed to be that in which a particular crop species was first cultivated.

centre of origin: The geographical area from which a given taxonomic group of plants has originated and spread and/or where wild-type species are found in greatest genetic variation; the theory was first published by V.I. Vavilov in 1922.

centre pivot irrigation: A method of agricultural irrigation using a long, wheeled arm with many nozzles that pivots about the centre of a circle; used primarily in arid regions.

centres of diversity: A geographical location or local region where a particular taxon exhibits greater genetic diversity than it does anywhere else. N.I. Vavilov developed this concept, who considered that the centres of diversity are also the centres of origin of a crop species, but the centres of diversity and the centre of origin have subsequently been found to be distinct phenomena. The global centres of origin of crop plants and/or centres of diversity are summarized as follows: (i) China (mountains of Central and Western China and adjoining areas): soybean, *Brassica* spp., radish, poppy, millets, buckwheat, fruit trees, mulberry, naked oat, naked barley; (ii) India and Indo-Malaya (India [without the North-west], Burma, Indochina, Malaysia): rice, sugarcane, banana, cocos palm, pepper, jute; (iii) Central Asia (north-west India, Pakistan, Afghanistan, Tadzhikistan, Tienshan, Uzbekistan): bread wheat, broad bean, pea, lentil, carrot, onion, grape, spinach, apricot; (iv) West Asia (Transcaucasia region, Iran, Turkmenistan, Asia Minor): emmer wheat, einkorn wheat, rye oats, barley, vetches, lucerne, clovers, plums, pea, lentil, fig; (v) Mediterranean coastal and adjacent regions (regions surrounding Mediterranean Sea): vegetables, rape, lupines, beets, clovers, pea, lentil, flax, olive, broad bean, seradella; (vi) South Mexico–Central America (South Mexico, Central America, Antilles): maize, Phaseolus beans, sweet pepper, sweet potato, cotton, sisal, cacao, tomato, cucumber, pumpkins; (vii) South America (Peru, Chile, Bolivia, parts of Brazil): maize, potato, tomato, cotton, groundnuts, bananas, tobacco, rubber tree; (viii) North America: lupines, grape, strawberry, sunflower; (ix) East Africa: coffee, ricinus, sorghum millet, emmer wheat, barley, linseed.

centric constriction: The visible bight along a chromosome that bears the centromere.

centric fission: A chromosomal structural change that results in two acrocentric or telocentric chromosomes from one metacentric chromosome, as opposed to CENTRIC FUSION.

centric fragment: A piece of chromosome containing a centromere.

centric fusion: The whole-arm fusion of chromosomes by the joining together of two telocentric chromosomes to form one chromosome.

centric region: The region where the centromere is placed.

centric: Chromosomes having a centromere as opposed to acentric (having no centromere).

centrifixed: Of a two-branched organ attached by its centre (e.g. a hair or anther).

centrifugal speciation: A concept of speciation in which isolation is not a factor but is believed to involve a sequential series of mutations that radiate from a central point.

centrifugation: Technique of separating macromolecules or cells from suspension using centrifugal force. Centrifugation is common technique routinely used in the laboratories. Centrifugation is done using different kinds of centrifuges such as tabletop centrifuge (microfuge), refrigerated centrifuge, ultracentrifuge, etc. During centrifugation the macromolecules, particles – or cells or protoplasts or cell organelles – get separated based on the differences in mass, shape or density.

centrifuge: A machine that is used to spin liquids in a circular motion at high rates of speed. Particles that are suspended in a liquid medium can be separated according to their density, the heavier particles collecting at the outer rim of the circle and the less dense ones collecting in layers towards the centre.

centriole: A cylindrical cell structure found in most eukaryotic cells, composed mainly of a protein called tubulin. In mitosis, this small spherical body forms the centre of the astral rays. Short cylindrical array of microtubules that are composed of groupings of microtubules arranged in a 9 + 3 pattern and are closely similar in structure to a basal body. A pair of centrioles is usually found at the centre of a centrosome in animal cells. Short cylindrical array of microtubules, closely similar in structure to a basal body. A pair of centrioles is usually found at the centre of a centrosome in animal cells.

centriole DNA: Centrioles seem to contain double-stranded DNA. The temporal relationship between cell cycle events and centriole duplication has been suggested. Centrioles need to duplicate only once per cell cycle. It depended on the timely demolition of a protein that spurred the organelles' replication. Centrioles start reproducing themselves during G1 or S phase.

Centro Internacional de Mejoramiento de Maiz y Trigo (CIMMYT, Mexico DF, Mexico): Responsible for wheat, maize, barley and triticale breeding and research.

centrogene: One of the similar and self-propagating units into which the centromere can be broken by X-rays or can break itself by misdivision.

centromere (CEN): The primary constriction that divides the chromosome into the short arm (p arm) and the long arm (q arm). It is the region of a chromosome that contains the kinetochore, a MICROTUBULE ORGANIZING CENTRE (MTOC) responsible for attachment of the sister chromatids to the spindle apparatus at mitosis or meiosis. During cell division, at late anaphase, the spindle fibres contract and chromosomes segregate towards poles. The genetic loci on chromosomes that link sister chromatids during mitosis and on which kinetochores assemble. A region called CEN on a chromosome that attaches to the spindle at metaphase of mitosis and meiosis and is involved in separating the copies of the chromosome produced during cell division. During this division, paired chromosomes look somewhat like an X, and the centromere is the constriction in the centre. It is a structural feature of a chromosome at which the pair of chromatids of the metaphase chromosome is held together. Each centromere is given a number corresponding to the chromosome from which it was derived. This region does not replicate during S phase as do all the other regions of DNA of the chromosomes. It is absolutely necessary for the stability of the chromosome. Exact position of a centromere is characteristic of a chromosome. The centromeric regions of a chromosome contain special proteins and are characterized by specific DNA sequences. The chromosome regions on either side of a centromere are rich in short, repeated DNA sequences. These sequences often form the satellite DNA. The centromeric region also contains a considerable amount of constitutive heterochromatin. The centromere is generally flanked by repetitive DNA sequences, and it is late to replicate. The centromere is an A–T region of about 130 bp. With it, several proteins bind with high affinity to form the kinetochore. It contains the kinetochore that attaches to the spindle during nuclear division. Thus, the centromere is the DNA region of the eukaryotic chromosome that determines kinetochore formation and sister chromatid cohesion. Centromeres interact with spindle microtubules to ensure the segregation of chromatids during mitosis and of homologous chromosomes in meiosis. The region of centromere in a chromosome is *aka* KINETOCHORE *or* PRIMARY CONSTRICTION.

centromere binding factor 1 (CBF1): A transcription factor that is synthesized (manufactured) within certain plants (*Arabidopsis thaliana*, etc.) when those plants are exposed to cold temperatures. CBF1 then interacts with certain portions, called regulatory sequences, of the plants' DNA to 'switch on' the progression of cold hardening by using proteins coded for by the plant genes.

centromere division: The centromere is the constricted region of the chromosome that separates the shorter 'p arm' from the longer 'q arm'. Prior to cell division, protein complexes called KINETOCHORES form on the centromeres of the replicated chromosome's sister chromatids. During prophase of mitosis or prophase II of meiosis, spindles attach to the kinetochores on each chromatid's centromere. During metaphase, the spindles align the chromosomes on a single plane in the centre of the cell, and during anaphase, they pull the sister chromatids apart to form two separate chromosomes.

centromere identity: The centromere is essential for the inheritance of genetic information on eukaryotic chromosomes. Epigenetic regulation of centromere identity has been implicated in genome stability, karyotype evolution and speciation.

centromere inheritance: Precise chromosome segregation necessitates the addon of microtubules to centromeres, epigenetically defined by the enrichment of CENP-A nucleosomes. During DNA replication, CENP-A nucleosomes experience dilution. To maintain centromere identity, correct amounts of CENP-A must be re-established in a cell cycle-controlled manner orchestrated by the Mis18 complex (Mis18α-Mis18β-Mis18BP1). P. Parashara *et al.* (2024) demonstrate that PLK1 cooperates with the Mis18 complex by recognizing self-primed phosphorylations of Mis18α (Ser54) and Mis18BP1 (Thr78 and Ser93) through its Polo-box domain. Unsettling these phosphorylations perturbed both centromere recruitment of the CENP-A chaperone HJURP and new CENP-A loading. Phosphorylation of Mis18α and PLK1 binding are obligatory to activate Mis18α-Mis18β and encourage Mis18 complex-HJURP interaction.

centromere interference: An inhibitory impact by the centromere on the process of crossing over and the distribution of chiasmata in its vicinity.

centromere markers: Loci located near centromere of a chromosome.

centromere misdivision: A transverse rather than lengthwise division of the centromere, thus producing telocentric chromosomes.

centromere orientation: The process of orientation of centromeres during prometaphase of mitosis as well as meiosis that results in a correct separation of chromatids or chromosomes during anaphase.

centromere position in chromosome: Based on the position of centromere the chromosomes are classified as METACENTRIC (centromere in the middle of chromosome), ACROCENTRIC (centromere towards one end of the chromosome) and TELOCENTRIC (centromere very near to one end).

centromere repulsion: The mutual repulsion of the centromeres of paired chromosomes towards the end of the meiotic prophase.

centromere shift: The displacement of centromeres by structural changes of the chromosomes (e.g. translocations, inversions, etc.).

centromeric chromatin: A type of heterochromatin that is a constituent in the formation of centromeres, which is the region of the chromosome that links sister chromatids to the spindle during cell division. It is flanked by pericentric heterochromatin. CpG islands (CGIs) CG-rich genomic regions, ranging from several hundred bp to 2 kb, are often located at gene promoters.

centromeric constitutive heterochromatin: Chromatin at centromeres is distinct from the chromatin in which the remainder of the genome is assembled. Two features consistently distinguish centromeres: the presence of the histone H3 variant CENP-A and, in most organisms, the presence of constitutive heterochromatin.

centromeric DNA: DNA located in the centromeric regions of chromosomes. It consists of highly repeated short sequences of DNA without a gene-coding function, and due to its anomalous structure, it may be recovered as a centromeric satellite fraction on DNA centrifugation. Centromeric DNA is also highly condensed at all the times and is therefore referred to as constitutive heterochromatin. *Aka* SIMPLE SEQUENCE DNA. *See* CENTROMERE.

centromeric DNA elements (CDEs): Yeast centromere sequence has three elements known as CDEs. (i) *CDEI* is conserved 8-bp palindrome; (ii) *CDEII* is A/T rich region of variable length (78–86 bp); and (iii) *CDEIII* is conserved 25-bp sequence with palindromic characteristics. A 125-bp DNA fragment is necessary and sufficient for formation of complete CDE.

centromeric fission: Creation of two chromosomes from one by splitting the centromere.

centromeric meiotic recombination: The centromeres that contain the kinetochore, which attach chromosomes to the microtubular spindle, are crossover cold spots.

centrosome: The site of spindle fibre association, which provides a polarity to the dividing cell. They are seen only in plant and animal cells and help regulate cell division. In most animal cells, it contains a pair of CENTRIOLES. Centrosome replication precedes the commencement of DNA synthesis. Centrosomes unevenly double in size. A substantial amount of the proteins in centrosomes appears to be tubulins. Then, DNA synthesis transpires. This is called the S PHASE of the cell cycle. Lastly, centrosomes divide and begin movement to the poles. As the centrosomes travel to opposite poles, they play out polar spindle fibres, such that fibres from one pole are interdigitated with fibres from the opposite pole. These polar fibres are engaged during anaphase to push the poles of daughter cells to opposite directions. *Aka* MICROTUBULE ORGANIZING CENTRES.

centrosome cycle: Duplication of the centrosome (during interphase) and separation of the two new centrosomes (at the beginning of mitosis), which produces two centrosomes to formulate the poles of the mitotic spindle.

ceraceous: Having a waxy appearance, colour or texture (e.g. flowers of many species of *Ceropegia*, waxy fruit of some species of *Myrica*).

cereals: Members of the grass family in which the seed is the most important part being used for food and feed.

cernuous: Something nodding, falling headlong or face down, inclined, stooping or bowing forward. The term is applied to many species with a nodding, stooping habit, as found in many *Narcissus* and *Dierama* species. Many plant species bear the specific nickname 'cernua'.

certation: The competition in growth rate between pollen tubes of different genotypes leading to unequal chances of achieving fertilization.

certification: Refers to the seeds, propagative plant material or nursery stock multiplied and sold under inspection of a regulatory body to maintain genetic identity and purity, freedom from harmful pathogens, insect pests and weed seeds. It is approved and certified by an official certifying agency.

certified seed: Seed produced under an officially designated system of maintaining the genetic identity of, and provisions for, seed multiplication and distribution of crop varieties. It is the progeny of breeder, select, foundation or registered seed. The seed is grown in compliance with regulations determining standards of germination, freedom from diseases and weeds and trueness to type.

cessation cassette: A three-gene cassette (genetic sequence construct) that, when inserted into a plant and activated via tetracycline antibiotic, prevents the seeds produced by that plant from germinating. This happens because the 'cessation cassette' stops those resultant seeds from manufacturing a specific protein needed for seed germination.

cetyltrimethyl ammonium bromide (CTAB): A solvent that is widely used to dissolve plant DNA samples (e.g. when a scientist wants to sequence that sample of plant DNA). CTAB solvent helps the scientist to separate out contaminants that commonly exist in samples from plant tissues (polysaccharides, quinones, etc.), because DNA molecules are much more soluble in CTAB than the contaminant molecules. *Aka* HEXADECYLTRIMETHYLAMMONIUM BROMIDE.

CG island: Region of DNA having a greater than average density of CG sequences. These regions generally remain unmethylated.

chain terminator: *See* DIDEOXYNUCLEOTIDE.

chalaza: The base of an ovule bearing an embryo sac surrounded by integuments.

chalky: A colour descriptor describing kernel endosperm of cereal grains (e.g. a rice grain with a high level of chalk is generally undesirable). The chalky look arises from the structure of the endosperm. Voids cause light to be refracted, and hence, the endosperm appears white to reflected light and opaque to transmitted light.

chamber: A cavity of an ovary.

Chambon's rule: The rule that base sequence of an intron begins with GT and ends with AG.

chance agreement: A measure of the proportion of times where two or more raters agree in their measurement or assessment of a phenomenon.

change in the meaning of codon, mechanism of: The encoding of two non-universal amino acids involves dynamic redefinition of 'stop' signals in the genetic code. Mechanisms changing the meaning of codon translation are: (i) stop codons can be reassigned independently of the context wherever they occur in any messenger RNA in a cell. This mode of changing codon meaning occurs in certain specialized genomes, such as those of many mitochondria. In context-dependent reassignment, either (ii) site-specific codon redefinition can occur in response to a colocalized recoding signal (shown as a loop representing an RNA structure); or (iii) mRNA-dependent redefinition of codons might happen in response to a recoding signal embedded outside the coding region.

changing environment during selection process: Shuttle breeding.

changing sowing time during selection process: A change of sowing time (early and late spring; early and late autumn) is useful in selecting for day length insensitivity, reduced vernalization requirement and yield/quality stability in different length of growth period. This method is used in both AUTOGAMOUS and ALLOGAMOUS CROPS.

channel protein: Membrane transport protein that makes an aqueous pore in the membrane through which a specific solute (usually an ion) can pass.

channelled: Sunken below the surface, resulting in a rounded channel.

chaperon: Also spelled as chaprone. Protein that binds to an unfolded target protein. Chaperones may also bind to partially folded target protein which prevents misfolding, degradation or aggregation of the target protein and thus facilitates the proper folding. Chaperons are being investigated for their role in developing resistance to moisture stress (drought) in plants using the process of genetic transformation and development of transgenics. Chaperones also assist in transporting those protein(s) to their ultimate destination(s) in the organism. Later, when cellular protein molecules begin to 'unfold' due to age, heat, viruses or exposure to certain chemicals or ultraviolet light, chaperones frequently cause those unfolded protein molecules to reappear at their correct (initial) conformation. Examples of such chaperones include heat-shock protein 70 (HSP 70) and heat-shock protein 40 (HSP40). *Aka* CHAPERONINS or MOLECULAR CHAPERONE.

chaperoning of the histone octamer: Aprataxin and polynucleotide kinase like factor (APLF) is a DNA repair factor in NON-HOMOLOGOUS END JOINING (NHEJ) repair of DNA double-strand breaks. APLF is recruited to DNA break sites through interactions with DNA end-binding protein Ku and the XRCC4–DNA ligase IV complex to form a scaffold for the NHEJ machinery. APLF has been shown to also have histone chaperone activity via its conserved C-terminal acidic domain ($APLF^{AD}$). $APLF^{AD}$ is intrinsically disordered and can bind H2A-H2B as well as H3-H4 with high affinity. I. Corbeski and others in 2022 proposed that the exposed histone dyad region, where the octamer has highest affinity for DNA and where DNA binding stabilizes the H3-H3′ interface of the $(H3\text{-}H4)_2$ tetramer, allows the interaction with DNA to initiate octamer deposition. The DNA may then displace $APLF^{AD}$ as it wraps around the histone octamer to form the nucleosome.

chaperonins: A class of chaperones. Chaperonins occur in bacteria, chloroplast and mitochondria. Best-known chaperonin is the protein GroE of *E. coli*. This protein in its active form is composed of two components. GroEL and GroES. GroEL (HSP 60) is composed of two disks, each composed of seven copies of a polypeptide. GroES (HSP 10) is a smaller component composed of seven copies of a small subunit. GroEL forms a barrel within which protein folding takes place. As a protein starts to emerge from the ribosome in *E. coli*, two proteins which are products of genes *DnaJ* and *DnaK*, complex with it. A third protein, GrpE, causes release of DnaJ and DnaK proteins leading to interaction with GroEL. Once inside, GroES can cycle on and off as ATPs are hydrolysed, helping GroEL chaperone correct folding of the protein. *See* CHAPERONES.

character: (i) A unit of information that roughly corresponds to a grapheme, grapheme-like unit or symbol, such as in an alphabet or syllabary in the written form of a natural language; (ii) A measurable or observable attribute of a plant ensuing from the interaction of a gene or genes with the environment. *Aka* CHARACTERISTIC or TRAIT.

character displacement: The increased difference between two closely related species where they live in the same geographic region (SYMPATRY) as compared with where they live in different geographic regions (ALLOPATRY). Explained by the relative influences of intra- and inter-specific competition in sympatry and allopatry.

character divergence: The name given by Charles Darwin to the differences developing in two or more related species in their area of sympatry owing to the selective effects of competition.

character methods: These methods are basically of two types: MAXIMUM PARSIMONY (MP) METHOD and MAXIMUM LIKELIHOOD (ML) METHOD. *See details as separate terms.*

character state: The particular form that a character takes (e.g. an A vs T at a particular position in homologous gene sequences).

character state reconstruction: A method used to infer ancestral and derived character states and traits.

characterization/operation/observation: Definitions and measurements of the subject of inquiry are important. Observations can be of structure, colour, pattern or measurements with instruments.

Chargaff's rule: E. Chargaff's generalizations in 1952 that in the base composition of DNA the quantity of adenine equalled the quantity of thymine, and the quantity of guanine equalled the quantity of cytosine. This means equal purine and pyrimidine content.

charge effects: The effect of post-translational histone modifications on altering the electrostatic interaction with DNA.

charged tRNA, formylation of methionine of: In prokaryotes, the first important step in the initiation of translation is charging of $tRNA^{fmet}$ with methionine and methionine is then formylated by enzyme Met-$tRNA^{fmet}$ formyl transferase. Some features of tRNA, called IDENTITY ELEMENTS, and aminoacyl-tRNA synthetases to enable it to discriminate isoaccepting tRNA from rest of 19 amino acids. The accepter stem is an especially important determinant for the specificity of tRNA synthetase recognition. This is a crucial step for formation of translation initiation complex. For initiation of translation in eukaryotes, methionine is not formylated.

charged-coupled device (CCD) camera: A camera that contains a light-sensitive silicon chip. When light falls on that chip, it creates an electrical charge at a specific location. CCD camera is used for digital imaging.

charging transfer RNA: During charging reaction during the initiation of protein synthesis, a specific amino acid is covalently attached to a specific tRNA by a tRNA-specific enzyme called AMINOACYL-TRNA SYNTHETASE. Each tRNA can be charged only with the amino acid for which its anticodon is appropriate. The meaning of tRNA is determined by its anticodon and not by its amino acid. Charging reaction comprises two steps, activation and transfer reactions, which are crucial for functioning of the tRNAs. During charging reaction, each of the 20 different amino acids gets attached to its corresponding tRNA by specific aminoacyl-tRNA synthetases. The amino acid is linked by an ester bond involving its carboxyl group to one of the last bases of the tRNA (which is always adenine).

CHARM: CHARM is an acronym for Coupled Histone Tail for Autoinhibition Release of Methyltransferase. It's a compact, enzyme-free epigenetic editing tool that can silence target genes by recruiting and activating DNA methyltransferases. E.N. Neumann *et al.* (2024) develop a compact epigenetic silencer named CHARM that could proficiently shut off the prion gene throughout the mouse brain when transported systemically by a viral vector without altering the underlying DNA sequence. The epigenetic editor could also be programmed to turn itself off after silencing its target, thus restricting potentially adverse effects from long-term expression. CHARM signifies a therapeutic modality that could be relevant to a range of other diseases caused by the toxic build-up of undesirable proteins.

charomid: A cosmid vector up to 52 kb in length, bearing 1 to 23 copies of a 2 kb spacer fragment linked in head-to-tail tandem arrays. Like cosmids and λ phage, charomids can also be packaged *in vitro* for efficient introduction into bacteria and allow cloning of restriction fragments of any size up to ~45 kb. The number of spacer fragments is variable and determines the lengths of inserts that can be accommodated within the total length of up to 52 kb.

charon phages: A phage that contains two nonsense mutations that prevent phage from growing in *E. coli* strains that do not carry an appropriate nonsense suppressor mutation. This limits phage growth only to special laboratory bacterial strains and thus acting as a safety feature against spreading of undesirable strains.

charon vectors: Charon vectors are modified phage λ plasmids. Their construction incorporates mutations that make them simple to use and also greatly increases their safety for the biological containment of cloned recombinant DNA. Three of the Charon vector phages – 3A, 4A and 16A – have been certified for use as vector-host systems. These INSERTION VECTORS have a deletion that leaves space for an insert of DNA fragments of up to 22 kb (e.g. Charon 16A vector).

chartaceous: Showing a papery texture.

chasmogamous: Of flowers that are pollinated when the perianth is open. *Compare with* CLEISTOGAMOUS.

chasmogamy: Fertilization process that occurs after the opening of a flower, as opposed to CLEISTOGAMY.

chasmophyte: A plant adapted to growth in crevices or hollows, such as in cliff faces. *Compare with* CREMNOPHYTE.

Chat Generative Pre-Trained Transformer (ChatGPT): A chatbot developed by OpenAI and launched on 30 November 2022. Based on a large language model, it enables users to refine and steer a conversation towards a desired length, format, style, level of detail and language. Successive prompts and replies, known as prompt engineering, are considered at each conversation stage as a context.

chatbot: A computer program or an artificial intelligence which conducts a conversation via auditory or textual methods. *Aka* SMARTBOT, TALKBOT, CHATTERBOT, BOT, IM BOT, INTERACTIVE AGENT, CONVERSATIONAL INTERFACE *or* ARTIFICIAL CONVERSATIONAL ENTITY.

check cross: Refers to the crossing of an unidentified genotype with a phenotypically similar individual of defined genotype. F_2 segregation investigation is used to establish whether the phenotype resulted from the action of identical or non-identical alleles of the same gene locus or from the action of non-allelic genes.

check cultivar: A commercial cultivar or experimental strain with well-known characteristics and performance that is normally included for the purpose of comparison with other selections in all testing procedures. *See* STANDARD CHECK.

checkpoint: Point in the eukaryotic cell-division cycle where cycle progression can be halted until conditions are appropriate for the cell to continue to the next stage.

checkpoint proteins: Cyclins are a group of proteins that control the progression of cells through the cell cycle by activating cyclin-dependent kinases (CDKs).

chelate: To combine reversibly, usually with high affinity, with a metal ion such as iron, calcium or magnesium.

chelating agent: A compound which forms more than one bond with metals and helps in keeping them for longer time in solution. Sodium **EDTA** is an important chelating compound that is routinely used in plant tissue culture medium. Sodium EDTA is mixed with ferrous sulfate, and the solution is heated slightly. Interconversion occurs with a result to form Fe-EDTA and sodium sulfate. In this case, the formation of Fe-EDTA complex added to the tissue culture medium, which helps in preventing the oxidation of iron and causes the slow release of iron over a longer period during incubation period of cultures. EDTA tightly and reversibly binds Mg^{2+} and other divalent cations (positively charged ions). If a chelate is allowed to bind to metal ions required for enzyme activity, the enzyme will be inactivated (inhibited).

chelation: The binding of metal cations (metal atoms or molecules possessing a positive electrical charge) by atoms possessing unshared electrons (thus, the electrons can be 'donated' to a bond with a cation). The binding of the metal (cation) to the (electron excess) chelator atom (ligand) results in formation of a chelator/metal cation complex. The intra-atom bonds thus formed are given the name of coordination bonds. The properties of the chelator/metal cation complex frequently differ markedly from the 'parent' cation. Both carboxylate and amino (molecular) groups readily bind metal cations. One of the most widely used chelators is ethylenediamine tetraacetate (EDTA). It has a strong affinity for metal cations possessing two (bi) or more positive (electrical) charges. Each EDTA molecule binds one metal cation. The EDTA molecule can be visualized as a 'hand' (having only four fingers) which grasps the metal cation. Some enzymes (which require metal cations for their activity) are inactivated by EDTA (and other chelators) in that the chelators preferentially remove the metal from the enzyme.

chemical changes in aged seeds: (i) Percentage of leached sugars is somewhat decreased in rye and increased in clover seeds with an increase in moisture content. (ii) Starch content decreases with increase of moisture and temperature of storage. (iii) Quantity of inorganic phosphate is increased. (iv) Quantity of proteins is decreased. (v) Ageing increases the output of carbon dioxide, which helps in reduced respiration. (vi) Decrease in vitamin content is observed. (vii) The enzymes degrade the food in the seed on ageing due to unfavourable conditions. Loss of germination is directly related to enzyme activity. (viii) Auxin content of ageing seeds falls gradually. (ix) Lactic acid was 50 times more in aged dead seeds (under storage conditions) than fresh ones.

chemical cleavage DNA sequencing: In this method of DNA sequencing, given by A.M. Maxam and W. Gilbert in 1977, numerous uniform repeated lengths of DNA are isolated through restriction endonuclease digests which are then broken at various points by four types of chemical treatments that cleave DNA preferentially (i) at guanine and at adenine; (ii) at guanine alone; (iii) at cytosine and at thymine equally; and (iv) at cytosines alone. The fragments differing from one another by one nucleotide are then separated through gel electrophoresis, which are detected by autoradiography.

chemical complexity: The amount of DNA component measured by chemical assays.

chemical composition of chromosomes: Chemical components of chromosomes are DNA, RNA, histone proteins, non-histone proteins, lipids and ions (calcium, magnesium and ferrous). Histones have a fundamental role in chromosome structure gene regulation. DNA double helix polynucleotide chain is very long, but it lies packed in only a few micrometre long chromosomes. Thus, a single human chromosome may have as much as 1 cm of linear DNA. The total length of DNA in the diploid human genome is 1.74 m.

chemical compound: A chemical substance constituted by two or more different chemically bonded elements, with a fixed ratio determining the composition. The ratio of each element is usually expressed by a chemical formula.

chemical desiccation: A method sometimes employed to screen for post anthesis stress tolerance as destruction of the plant's photosynthetic system. The chemical desiccants most commonly used for cereals include magnesium chlorate, potassium iodide and sodium chlorate.

chemical equation for photosynthesis: Sunlight + 6 CO_2 + 6 $H_2O \rightarrow C_6H_{12}O_6$ + 6 O_2.

chemical equilibrium: The state in which both reactants and products are present in concentrations which have no further tendency to change with time in a chemical reaction.

chemical evolution: Chemical reactions that could have produced complex compounds from simple ones prior to the origin of life.

chemical force microscopy (CFM): An AFM technique with molecularly functionalized tip, used to probe the chemical characteristics of surface-bound molecules. Used to study properties such as strength of single molecules and bonds. *Aka* CONTACT FORCE MICROSCOPY.

chemical genetics: Branch of genetics introduced by Rebecca Ward and Timothy Mitchison (1994). It refers to the creation and use of synthetic chemicals that act to either block or enhance the activity of a protein (or gene that codes for protein). Studies under the banner of chemical genetics have enabled scientists to determine the function(s) of specific protein molecules. Sir Archibald Edward Garrod is known as Father of "Chemical Genetics".

chemical group: Set of covalently linked atoms, such as a hydroxyl group (−OH) or an amino group (−NH_2), the chemical behaviour of which is well characterized.

chemical hybridizing agents (CHA): The compounds that are applied to plants prior to anthesis to selectively induce male sterility.

chemical ligation: A chemical reaction that links a fully chemically derived peptide to the end of a recombinant protein.

chemical method of gene transfer: Protoplasts are mixed with vector plasmids containing the gene of interest in the presence of POLYETHYL GLYCOL (PEG).

chemical modification of surface peptides: A new and promising approach which combines small molecules and phage-displayed peptides and enables the isolation of chemically modified bicyclic peptides capable of high-affinity recognition of target proteins.

chemical mutagenesis: The induction of mutations by the use of chemical agents.

chemical mutagens: The chemicals that are capable of causing genetic mutation gene mutagen. A variety of chemicals capable of inducing mutations are inhibitors of nucleic acid precursors, base analogue, dyes, nitrous acid, hydroxylamine, hydrazine, low pH, heat, hydrogen peroxide, maleic hydrazide, antibiotics, azides, alkylating agents, chemicals used in industries, pesticides, cosmetics, food preservatives, etc. The chemical agents that induce mutations are classified into two types depending upon the physiological state of DNA affected by these agents. In first group fall the chemicals that affect replicating nucleic acids. In the second group fall the chemicals that affect resting nucleic acids.

chemical mutagens affecting replicating nucleic acids: Those chemical mutagens which affect replicating nucleic acids. Examples of such chemicals are inhibitors of nucleic acid precursors (e.g. 5-aminouracil, azaserine, benzimidazole, caffeine, 8-ethoxycaffeine, 8-mercaptopurine, 5-nitroquinoxaline, paraxanthine, tetramethyl uric acid, theobromine), base analogues (e.g. 5-bromouracil, 5-chlorouracil, 5-iodouracil, 2-aminopurine, 2,6-diaminopurine) and dyes (e.g. proflavin, acridine orange, 2-aminoacridine, methylene blue, toluidine blue, ethidium bromide, 10-methyl acridinium phenanthridine, propidium diiodide). 5-aminouracil and 8-ethoxycaffeine inhibit thymine, while azaserine, benzimidazole, caffeine, 8-mercaptopurine, 5-nitroquinoxaline, paraxanthine, tetramethyl uric acid and theobromine inhibit purines.

chemical mutagens affecting resting nucleic acids: Examples of such chemicals are nitrous acid, hydroxylamine, hydrazine, low pH, heat, hydrogen peroxide, maleic hydrazide, antibiotics, azide and alkylating agents.

chemical nitrogen fixation: Nitrogen can be artificially fixed for use in fertilizers, explosives or other products. A relatively small amount of ammonia is produced by lightning. The most common method in artificial production is the industrial production by the Haber-Bosch process, wherein very high pressures and fairly high temperature are required.

chemical reaction: A process that results in the transformation of one set of chemical substances to another.

chemical sequencing: *See* DNA SEQUENCING BY MAXAM AND GILBERT METHOD.

chemical vapour deposition (CVD): A technique used to deposit coatings, where chemicals are first vaporized and then applied using an inert carrier gas such as nitrogen.

chemically-defined medium: A nutritive solution for culturing cells in which each component is specifiable and, ideally, is of known chemical structure.

chemiluminescent labelling: A technique used to label DNA probes. Two different probes are constructed that are complementary to adjacent segments of a gene. Each probe carries part of a chemiluminescent system. When brought together, the two probes hybridize in the same region and emit light, which can be detected using a photomultiplier.

chemiosmotic coupling: Mechanism in which a gradient of hydrogen ions (a pH gradient) across a membrane is used to drive an energy-requiring process, such as ATP production or the rotation of bacterial flagella.

chemistry: A branch of science that studies the composition, structure, properties and change of matter. Chemical interactions are at the base of all biological processes.

chemoautotorph: An organism that obtains energy by oxidizing in organic compounds. CO_2 is the sole source of carbon. Chemoautotrophs make use of the electrons from reduced inorganic compounds as a spring of energy, and they use CO_2 as principal source of carbon. For example, inorganic source for Nitrobacter is nitrite ions, and for Nitrosomanas, it is ammonia.

chemoautotrophic: Systems which utilize simple organic molecules as nutrition. The system is thus environment-dependent.

chemofusion of protoplasts: Several chemicals have been used to induce protoplast fusion, such as sodium nitrate, polyethylene glycol and Calcium ions. Chemofusion is a non-specific method. It is inexpensive and can cause massive fusion product. Chemofusion can be cytotoxic and non-selective and having less fusion frequency.

chemokine: Small secreted protein that attracts cells (such as white blood cells) to passage towards its source. Chemokines are important in the functioning of the immune system.

chemolithotrophic bacteria: Bacteria capable of using inorganic, rather than organic, molecules for their fuelling reactions are termed chemolithotrophs, and iron-oxidizing bacteria are a major group in this nutritional category.

chemomechanical conversion: Conversion of chemical energy into mechanical energy.

chemoreceptor: Proteins present on plasma membrane that are specific for various attractants and repellents. Motile bacteria possess receptors on cell membranes. These receptors pick up chemical stimuli as acetosyringone in case of *Agrobacterium tumefaciens*. Stimuli information is passed to the flagella. In case the chemotactic signal is positive, the bacteria move towards stimulus. If the signal is negative, bacteria move away from stimulus.

chemostat: A continuous culture in which growth rate and cell density are held in a steady state by controlling the inputs of growth-limiting nutrient. In this kind of cell culture system, subculturing is avoided, whereas in the close type cell suspension culture system, the cell biomass increases with the incubation period that require periodic subculturing. This device is used for large-scale cultivation of bacteria in the industry. The system regulates the continuous addition of fresh sterile medium and the removal of spent medium containing cells from the culture vessel.

chemosynthesis: Biological conversion of one or more carbon-containing molecules (usually carbon dioxide or methane) and nutrients into organic matter using the oxidation of inorganic compounds (e.g. hydrogen gas, hydrogen sulfide) or ferrous ions as a source of energy.

chemotaxis: Directed movement, growth or sensing of a cell or organism towards or away from changing concentration of a diffusible chemical, often in relation to food or for mating – the process known as CHEMOTROPISM. In nature, wild *A. tumefaciens* infects only the dicots and not monocots, because the dicot plant at the injured site secretes phenolic compounds such as acetosyringone and α-hydroxy acetosyringone, which act as signal molecules to attract *Agrobacterium*.

chemotherapy: Control or treatment of plant disease with chemicals (such as antibiotics and fungicides) absorbed and translocated internally. Chemotherapy is a routine practice for treatment of cancer. Chemicals such as taxols are given to the patient to stop unnecessary cell division.

chemotroph (adj. **chemotrophic**)**:** An organism which obtains energy from chemical compounds. On the basis of energy source organisms are classified as chemotrophs or phototrophs. CHEMOTROPHS derive their energy from oxidation-reduction of organic and inorganic compounds, whereas the PHOTOTROPHS use light as energy source. *See* PHOTOTROPHS.

chemotrophic: *See* CHEMOTROPH.

chiasma (pl. chiasmata) (XMA or Xta): The point where two homologous non-sister chromatids exchange genetic material during the pachytene stage of prophase I meiosis I (sister chromatids also form chiasmata between each other, but because their genetic material is identical, it does not cause any change in the resulting daughter cells). The chiasmata become visible during diplotene stage and represent sites of crossing over.

chiasma frequency: The average number of chiasmata formed per bivalent, or any particular bivalent, in a given organism, under given circumstances.

chiasma interference: The occurrence, less or more frequent than expected by chance, of two or more crossing over and chiasmata events in a given segment of a chromosomal pairing configuration and/or chromosome.

chiasma localization: The physical location of a chiasma in a pairing configuration and/or chromosome.

chiasma terminalization: The progressive shift of chiasmata along the arms of paired chromosomes from their points of origin towards terminal positions due to dissolution of synaptonemal complex.

chiasma theory of pairing: The hypothesis that whenever two chromosomes that have been paired at pachytene remain associated until metaphase, they do so by virtue of the formation of a chiasma or visible exchange of partners among their chromatids.

chi-form (χ-form): An intermediate structure in recombination between DNA molecules.

chimera: (i) An organism or tissue comprising genetically distinct cell sectors. Chimeral plants may originate by grafting, spontaneous mutation, induced mutation, sorting-out from variegated seedlings, mixed callus cultures or protoplast fusion. Chimera may be derived from two separate embryos from the same or different species. (ii) During plant transformation using multicellular target tissues like meristems there are greater chances of obtaining chimeric transgenic plants. (iii) A tissue containing two or more genetically distinct cell types, or an individual composed of such tissue. *Aka* CHIMERIC, CHIMAERA *or* CHIMAERA.

chimeraplasty: A method employed by man to introduce a gene (from the same or another species) into the DNA of a living organism or cell via 'gene repair' mechanism. Scientists add the desired DNA (gene) to a cell, along with RNA, in a paired group known as a CHIMERAPLAST. The chimeraplast attaches itself to the cell's DNA at the site of the specific gene (to be changed), and 'repairs' it by utilizing its (new) chimeraplast-DNA as a 'template'.

chimeric clone: A cloning artefact created by a foreign gene being inserted into a vector in an incorrect orientation resulting in the expression of a protein consisting of a fusion of two different gene products.

chimeric DNA: Recombinant DNA containing spliced genes from two different species.

chimeric gene: A recombinant gene having regulatory sequence from one gene and coding sequence from another gene, which are fused to a promoter and other regulatory sequences derived from a different gene. Most genes used in genetic transformation are chimeric.

chimeric plasmid: Hybrid, or genetically mixed, plasmid used in DNA cloning. A circular plasmid carrying foreign DNA segment and antibiotic resistance genes produced through recombinant DNA technology. Using recombinant DNA technology, a plasmid can be cut open to give a linear DNA molecule. A foreign DNA segment can then be inserted by joining the ends of foreign DNA to obtain chimeric plasmid.

chimeric proteins: Fused proteins from different species, produced from the chimeric DNA template.

chimeric selectable marker gene: A gene constructed from parts of two or more different genes and allows the host cell to survive under conditions where it would otherwise die.

chimeric vector: Bacteriophage or plasmid cloning vector carrying a novel DNA sequence integrated into its DNA. *See* CHIMERIC PLASMID.

Chinese Academy of Agricultural Science (CAAS): Established in 1957, CAAS is China's national agricultural research organization, directly affiliated with the Ministry of Agriculture.

Chinese Spring (wheat): A renowned wheat variety worldwide. It is generally accepted as the standard variety for cytogenetic and molecular investigations in wheat. More than 300 aneuploids were developed from this variety. Agronomically, Chinese Spring has some serious defects, such as shattering, susceptibility to almost all wheat diseases and insects and poor adaptation to the world's major wheat-growing areas. This wheat variety was chosen to produce monosomes, trisomes, tetrasomes and almost all of the other aneuploids. Chinese Spring wheat was used for cytogenetic studies by accident. E.R. Sears made hybrids between wheat and rye in 1936 in attempts to induce chromosome doubling by heat shocks. Chinese Spring variety was known to cross readily with rye. Among the wheat-rye hybrids, a few wheat haploids were obtained. One of these haploids was pollinated by euploid Chinese Spring. Viable seeds (13 in number) were derived from this backcrossing, showing chromosome numbers of 2n = 41, 42, 43 and reciprocal translocations. Nullisomes were eventually obtained from the monosomes, and one nullisome proved to be 3B, which is partially asynaptic and was therefore a good source of additional monosomes and trisomes.

chiral: The property of an asymmetric molecule, the structure of which cannot be superimposed on its mirror image. The molecule exists in two forms called enantiomorphs, which are mirror images of each other. The solutions of these rotate the plane of polarized light in different directions, left (laevo/levo) or right (dextro), making the enantiomers LAEVOROTATORY/LEVOROTATORY or DEXTROROTATORY. There is also a racemic mixture, which has an equal fraction of both and, therefore, will not be optically active.

chiral product: Many chemical substances for crop protection products exist in two forms that are mirror images of each other, just as the right hand is of the left hand. The 'image' and 'mirror image' can have completely different effects (e.g. one form of the amino acid asparagine is used as a sweetness enhancer, while the other form is perceived as bitter).

chiropterophilous: Pollinated by bats.

chi sequence (χ sequence): An octane, non-palindromic sequence that provides a hotspot for recB-CD-mediated genetic recombination in *E. coli.* Wild-type lambda phage lacks a chi sequence. A chi sequence has been added artificially to some lambda cloning vectors, which cannot make concatemers by sigma-type replication. The chi sequence stimulates the formation of lambda dimers through the host recombination functions.

chi site (χ site): Sequence of DNA at which the RecB-CD protein cleaves one of the strands during recombination.

chi-square (chi^2 or χ^2): A statistical procedure that enables researchers to determine how closely an experimentally obtained set of values fits a given theoretical expectation.

chi-square (χ^2) distribution: The distribution used to analyse counts in frequency tables. The distribution of the sum of squares of a number n of normally distributed variables; written as χ2n. The chi-square distribution is a continuous probability distribution with the values ranging from 0 to ∞ (infinity) in the positive direction. The χ2 can never assume negative values. The shape of the chi-square distribution depends on the number of degrees of freedom ('*v*'). When '*v*' is small, the shape of the curve tends to be skewed to the right, and as the 'v' gets larger, the shape becomes more symmetrical and can be approximated by the normal distribution.

chi-square (χ^2) test: (i) A statistical test for testing goodness of fit between expected (on the basis of a gene hypothesis) and observed numbers. (ii) Used to test the null hypothesis (H_0) that proportions are equal or that factors or characteristics are independent or not associated.

chi-square test (corrected): A test for a 2 × 2 table that uses Yates' correction, making it more conservative.

chitin: Chitin $(C_8H_{13}O_5N)_n$ is a complex long-chain polymer of an N-acetylglucosamine (NAG) units. It is a derivative of glucose and is found in many organisms. It is a characteristic component of the cell walls of fungi, animal exoskeletons and nematode eggshells. Besides, chitin is also the main component of exoskeleton of some insects.

chitin magnetic beads method of RNA purification: The poly(A) RNA is extracted by introducing magnetic beads coated with oligo (dT). RNA with a poly(A) tail attach to the oligo (dT). The beads will then be drawn to the bottom of a tube removing mRNA directly from total RNA. The magnetic beads which are specially treated minimize the non-specific binding of other nucleic acids and ensure the purity of mRNA.

chitinase: An enzyme that degrades (breaks down) chitin. It is one of the pathogenesis-related proteins produced by certain plants as a disease-fighting response for entry into plant of pathogenic (disease-causing) fungi. Chitinase does not allow the growth of infecting fungus as it dissolves the fungal wall by degrading chitin that forms a major component of fungal cell wall. Chitinase is also sometimes manufactured by certain fungi and actinomycetes that destroy the eggs (i.e. chitin-containing shells) of harmful roundworms.

chitinase gene: Gene responsible for the production of chitinase enzyme that degrades chitin. Chitinase gene has been cloned from rice and insect. This gene when introduced into a plant produces chitinase enzyme in the transgenic plant. Thus, it is an antifungal gene being tried for developing disease resistant plants. *See* CHITINASE.

chlamydospore: A single-celled thick-walled or double-walled asexual fungal spore produced by the direct differentiation of the cells of mycelium. These are manufactured from the hyphal cells (terminal or intercalary) or by transformation of conidial cells, which are highly resistant to hostile conditions and that can function as an overwintering stage.

***chloramphenicol acetyltransferase* (*cat*) gene:** *cat* gene is utilized as a selective marker for cloning vectors and as a reporter gene.

chloramphenicol acetyltransferase (CAT): Coded by a particular bacterial gene *cat*. It derives from a certain transposon of a plasmid. The resistance to chloramphenicol is commonly used as a reporter gene in genetic experiments for investigating physiological gene regulation and was incorporated in some cloning vectors. β-galactosidase and luciferase genes can also be used for the same purpose.

chloromethylases (CMTs): CMTs are an important class of plant-specific DNA methylation enzymes, which are characterized by the presence of a CHRromatin Organization MOdifier (CHROMO) domain between the cytosine methyltransferase catalytic motifs I and IV. There are three CMT genes encoded in the *A. thaliana* genome: *CMT1*, *CMT2* and *CMT3*. CMT1 is the least studied of the three. It has been suggested that CMT1 is non-essential. The majority of DNA methylation at CHH sites (^{m}CHH) within long transposable elements in pericentromeric regions of the genome is targeted by a *CMT2*-dependent pathway. Allelic variation at *CMT2* has been shown to alter genome-wide levels of CHH DNA methylation (^{m}CHH) and alleles of *CMT2* may play a role in adaptation to temperature. In contrast, DNA methylation at CHG (^{m}CHG) sites is often maintained by *CMT3* through a reinforcing loop with histone H3 lysine 9 di-methylation (H3K9me2) catalysed by the KRYPTONITE (KYP)/SU(VAR)3-9 HOMOLOG 4 (SUVH4), SUVH5 and SUVH6 lysine methyltransferases. In *Z. mays*, ZMET2 is a functional homologue of CMT3 and catalyses the maintenance of ^{m}C.

chlorophyll (Adj. chlorophyllous): One of a group of green pigments found in chloroplasts that are important in autotrophic photosynthesis in plants. It is a magnesium containing porphyrin compound, and by this, plants become capable of manufacturing own food by a process called PHOTOSYNTHESIS. Chlorophyll has a hydrocarbon tail that anchors it to an integral protein in the thylakoid membrane of the chloroplast. Chlorophyll is the source of the green colour of plants and certain other autotrophs. Different types of chlorophyll present in plant kingdom differ in the chemical substituents (Chlorophyll a, $C_{55}H_{72}O_5N_4Mg$; Chlorophyll b, $C_{55}H_{70}O_6N_4Mg$; Chlorophyll c1, $C_{35}H_{30}O_5N_4Mg$; Chlorophyll c2, $C_{35}H_{28}O_5N_4Mg$; Chlorophyll d, $C_{54}H_{70}O_6N_4Mg$; Chlorophyll f, $C_{55}H_{70}O_6N_4Mg$) attached to the main porphyrin ring system.

chlorophyll biosynthesis: The chlorophyll biosynthetic pathway is quite complex. Chlorophyll is synthesized from amino acid glutamate, which is converted into 5-aminolevulinic acid (5-ALA) by enzyme glutamate semialdehyde aminotransferase. Two molecules of 5-ALA are combined to form porphobilinogen (PBG) by enzyme porphobilinogen synthase (PBS), which, by stepwise deamination, produces the resulting large-ring complex known as protoporphyrin IX (PPIX). Mg^{2+} is inserted into PPIX by Mg-chelatase enzyme, followed by light-mediated cyclization of ring E and reduction in ring D of protochlorophyllide molecule. The addition of large phytol tail to the chlorophyllide molecule produces the green-coloured pigment 'chlorophyll'. However, to carry out electron and energy transfer in photochemical quenching, the chlorophyll molecule is stabilized by thylakoid proteins forming the photosystems.

chlorophyll biosynthesis, overexpression of genes regulating: Upregulation of chlorophyll biosynthesis genes, such as *Chlorophyllide a oxygenase* (*CAO*) gene, results in continuous production of chlorophyll, irrespective of normal degradation pathway. Cosmetic stay-green results from mutations that render enzymes in the chlorophyll catabolic pathway non-functional. Transgenics that overproduce chlorophyll and increase their persistence for a longer time are similar to cosmetic stay-green plants. Cosmetic stay-green may also be associated with undesirable effects – mainly in relation to germination and vigour in certain species.

chlorophyll content, impact of environment and abiotic stresses on: Factors, such as climatic conditions, tropical zones, drought stress, osmotic stress, heat stress and low-temperature stress, have been shown to influence leaf chlorophyll content. Because of the direct/indirect association of chlorophyll levels with yield-contributing traits, it would be desirable to increase photosynthetic efficiency via increased biosynthesis of chlorophyll. As photosynthesis generally leads to increase in yield, carbon assimilation pathways can be manipulated in addition to chlorophyll production to enhance yield. Chlorophyll concentration measured at a late developmental stage in maize could serve as an indirect selection criterion for improving grain yield.

chlorophyll-degradation genes, knock out of: The simplest approach of producing a stay-green genotype is to restrict the expression of chlorophyll-breakdown genes. A large number of plastid genes are known, which, when disrupted, result in production of cosmetic stay-greens. Their translated products encode enzymes, such as Chl *b* reductase, Mg-dechelatase, phaeophytinase and pheophorbide *a* oxygenase (PaO), regulate chlorophyll loss during senescence events. In *Arabidopsis*, chlorophyll breakdown was caused by PaO, which is a Rieske-type iron-sulfur protein, encoded by the *accelerated cell death 1* gene.

chlorophyllin: Sodium and copper salts of chlorophyll known to inhibit mutagenic activity of several known mutagens.

chloroplast: A specialized organelle in plant and algae cells that contains chlorophyll pigments and functions in photosynthesis. Chloroplast plays an important role in photosynthesis. It is surrounded by double membrane. Specialized, membrane-bound organelles located in the cytoplasm of green algae and plant cells that carry out photosynthesis. Chloroplast is a specific form of plastid. Plant cells can have anywhere between one and a hundred chloroplasts, which move around the cell and can divide to replicate themselves. Surrounded by two membranes, chloroplasts contain DNA, ribosomes, enzymes, grana and a thick fluid called STROMA. Each granum is a stack of disc-shaped thylakoids that contain the pigment chlorophyll, which converts light energy into chemical energy to produce the food that sustains the plant. Chloroplasts are also involved in the plant's synthesis of fatty acids and amino acids and its immune response. Chlorophyll captures the energy from sunlight and converts and stores it in the molecules ATP and NADPH while freeing oxygen from water. All chloroplasts are evolutionarily derived from cyanobacteria; its function is photosynthesis.

chloroplast aminoacyl-tRNA synthetase: An aminoacyl-tRNA synthetase isolated from chloroplast that is specific for a chloroplast tRNA.

chloroplast chromosome: Circular DNA found in the photosynthesizing organelle (chloroplast) of plants instead of the cell nucleus where most genetic material is located.

chloroplast development regulation by plant hormones: The development and function of the chloroplast are controlled by plant hormones, particularly cytokinin as a positive regulator and brassinosteroid as a negative regulator. Cytokinin is an adenine-derived plant growth regulator with important biological activities, including the promotion of cell division and chloroplast development. Brassinosteroid is a sterol-derived plant growth regulator with important biological activities, including the promotion of cell elongation and negative regulation of chloroplast development.

chloroplast development, plastid DNA copy number increase during: Plastid DNA copy number and ribosome abundance increases dramatically during chloroplast development. A.J. Bendich in 1987 hypothesized that rRNA synthesis was template limited and that the increase in plastid DNA copy number during chloroplast development was required to activate rRNA synthesis.

chloroplast development-dependent nuclear gene expression: In the absence of chloroplast development, expression of nuclear genes encoding chloroplast proteins, such as the *Lhcb* gene for a light harvesting complex protein, are repressed. Molecular genetic studies in *C. reinhardtii* and *Arabidopsis* have suggested that perturbations in the tetrapyrrole pathway around the chlorophyll/haem branchpoint may be involved in generating a distress signal from plastids. The signal may be a tetrapyrrole intermediate or a secondary signal may be generated.

chloroplast division: Plastids contain DNA and are formed by division of existing plastids. In chloroplast division, the PLASTID-DIVIDING RING (PDR) is a key structure of the PD machinery and is a universal structure in the plant kingdom. Proteomic analysis of PD machineries isolated from *Cyanidioschyzon merolae*, Y. Yoshida and others (2010) identified the glycosyltransferase protein plastid-dividing ring 1 (PDR1), which construct the PD ring that is widely conserved from red alga to land plants. PDR1 protein forms a ring with carbohydrates at the chloroplast division site. Fluorometric saccharide ingredient analysis of purified PD ring filaments revealed that only glucose was included, and downregulation of PDR1 impaired chloroplast division. Thus, chloroplasts are divided by the PD ring, which is a bundle of PDR1-mediated polyglucan filaments.

chloroplast DNA packaging: Deoxynucleoprotein (DNP) fibrils released from the chloroplasts after a mild osmotic shock revealed 13–14 nm nucleosome-like globules on their fibrils. The fibrils form rosettes with loops of a contour length of 15–20 μm. Minicircular plasmid-like DNA molecules with nucleosome-like globules were found in the chloroplasts. By means of monoclonal antibodies against the β-subunit of *E. coli* RNA polymerase and protein A-colloidal gold complexes, the sites of DNP fibrils to which chloroplast RNA polymerase are attached were localized. The highest numbers of RNA polymerase molecules were associated with the relaxed smooth DNA fibrils that are almost devoid of nucleosome-like granules. The compactly packed DNP fibrils with a high number of nucleosome-like granules were free of RNA polymerase molecules. The former regions of the chloroplast nucleoid are probably the transcriptionally active sites of the genome, while the latter appear to be transcriptionally inactive. The RNA polymerase molecules were also found to be attached to the minicircular DNA.

chloroplast DNA recombination: DNA recombination has been demonstrated in chloroplasts. Analysis of the cpDNAs from ten randomly selected F_1 hybrids, in each case revealed cpDNA to be recombinant for *Ava*I and *Bst*EII restriction sites, although fragments representative of *C. eugametos*, the mt^+ parent, were typically found in excess of those for *C. moewusii*, the mt^- parent.

chloroplast DNA replication: There is unambiguous evidence for a semi-conservative mode of chloroplast DNA replication and a difference in time of replication of chloroplast DNA and chromosomal DNA during the vegetative growth cycle of *Chlamydomonas reinhardi*.

chloroplast DNA transcription: A template-binding polypeptide has been recognized in pea chloroplast transcriptional complex. It has a molecular weight of 150 kDa and binds to both chloroplast ribosomal (16S rRNA) and messenger (psbA) promoters, and it displays some degree of correlation to total transcriptional element.

chloroplast DNA, maternal inheritance of: To demonstrate that a trait is maternally inherited, specific crosses are necessary to generate the required offspring. A number of crosses were made between cultivated tomato (*Lycopersicon esculentum*) and a number of wild species. Chloroplast DNA was obtained from F_1 plants of a number of crosses in which *L. esculentum* was the female parent in the cross. In each case, the F_1 restriction fragment pattern noted to be identical to the female parent *L. esculentum*. This experiment by J.D. Palmer and D. Zamir in 1982 provided decisive evidence that chloroplast DNA is inherited in a maternal manner.

chloroplast DNA: Plastid DNA occurs as large protein-DNA complexes associated with the inner envelope membrane and called PLASTID NUCLEOIDS. Each plastid generates 10–1000 copies of the circular 75–250 kilobase plastome. The proplastid comprises a single nucleoid located in the centre of the plastid. The number of genome copies per plastid is more than 1000 in rapidly dividing cells, which generally contain 100 or fewer plastids in mature cells, where plastid divisions give rise to a large number of plastids. The plastome comprises about 100 genes encoding ribosomal and transfer RNAs and also proteins involved in photosynthesis and plastid gene transcription and translation. The chloroplast DNA differs from nuclear DNA of same species in nucleotide composition. Plastid DNA replicates through a semi-conservative mode in *Chlamydomonas*. In higher plant species, chloroplast DNA replicates by the rolling circle method. No histone is associated with the chloroplast DNA. The tRNA genes that contain introns.

chloroplast epitranscriptome mapping: Only three 16S rRNA writers – PALEFACE 1 (PFC1) for $6^{m}A$, Chloroplast MraW-like (CMAL) for $4^{m}C^{m}$ and Ribosomal RNA small subunit methyltransferase D (RsmD) for $2^{m}G$ methylations – have been described in chloroplasts, but writers, erasers and readers for these modifications in mRNA are completely unknown. Abundant 2′-*O*-Me have also been identified in chloroplast and mitochondrial rRNAs suggesting that the same or related bacterial-derived, standalone protein enzymes perform the methylation in both organelles.

chloroplast gene introns: Split genes for ribosomal RNA (rRNA), transfer RNAs and some proteins have been reported in the chloroplast genomes of several plants including *Chlamydomonas* and *Nicotiana*. Introns found in chloroplast genes can be classified into three groups on the basis of intron boundary sequences. GROUP I INTRONS (e.g. in *trnL*) are folded in a secondary structure similar to self-splicing rRNA procedure of *Tetrahymena*. These introns can be removed either by self-splicing or by a 'maturase' (as in cytochrome b and cytochrome oxidase mRNA precursors). GROUP II INTRONS (e.g. majority of genes including *trnA* and *trnI*) are generally folded into a complex secondary structure (as in introns of mitochondrial genes for cytochrome oxidase in maize and yeast). GROUP III INTRONS (e.g. *trnG*, *trnK*, *trnV*, *rpl2*, *rps12*, *rps16*) possess conserved sequences at their borders (GTGCGNY at 5′-end and ANCNRYY(N)YYAY at 3′-end where R = purine, Y = pyrimidine and N = any nucleotide) similar to those in the eukaryotic nuclear genes.

chloroplast genes: Chloroplasts of vascular plants contain about 100 genes, most of which encode components of the photosynthetic electron transport machinery and elements of the transcriptional and translational apparatus. Although the progenitor of the chloroplast was a free-living prokaryote, the loss of genetic information to the nucleus to control plastid gene expression has largely placed the chloroplast in a 'receptor' role, where it retorts to nuclear signals. None the less, reverse signalling also occurs, demonstrating the interdependence and need for coordination between the cellular compartments.

chloroplast genetic engineering by electroporation: The chloroplast expression vector pHD203-GUS, which consists of coding regions for β-glucuronidase (*gus*) and chloramphenicol acetyltransferase (*cat*) genes, separated by a double *psbA* promoter fragment from pea (in opposite orientation) is electroporated into chloroplasts.

chloroplast genetic engineering by gene gun: In chloroplast genetic engineering, the recombinant DNA plasmid is bound to small gold nanoparticles that are then injected into the chloroplasts of a leaf using a 'gene gun'. This device requires high pressure to insert the plasmid-coated particles into the cell. The leaf is then grown on a plate containing an antibiotic, which ensures that the only surviving plant cells will be the ones that contain the gene for antibiotic resistance and, therefore, contain the therapeutic gene as well. These cells are then exposed to regenerative factors that induce them to start sprouting shoots and grow into full plants that express the desired protein. Chloroplast genetic engineering technique is environmentally friendly. It permits production of large-scale therapeutic proteins and vaccines. Plastid transformation is becoming more popular and an alternative to nuclear gene transformation because of various advantages such as high protein levels, the feasibility of expressing multiple proteins from polycistronic mRNAs and gene containment through the lack of pollen transmission.

chloroplast genome: The chloroplast genome is highly conserved in size, structure and gene content. Typical chloroplast DNA consists of large single copy (LSC) and small single copy (SSC) regions separated by two duplicated inverted repeat (IR) regions (IR_A and IR_B). Average size of chloroplast genome is 20–30 kb. The IR regions are responsible for the size variation among chloroplast genomes. The boundaries of IR regions with LSC and SSC are also important for expansion and contraction in genome and hence the gene content of this region varies greatly. Genome analysis of chloroplast DNA has revealed 60–200 OPEN READING FRAMES (ORFs).

chloroplast genome analysis: R. Sager and M.R. Ishida in 1963 described a process for isolating relatively intact chloroplasts from *Chlamydomonas*. These chloroplasts contain about 3% of the cellular DNA, of which 25–40% consists of a satellite band with a buoyant density of 1.702 gm/cm^2 and GC content of 39.3%. The enrichment of satellite DNA in the chloroplast extract sevenfold over its concentration in total cell extracts offers strong evidence for its association with the chloroplast *in vivo*.

chloroplast membranes: Chloroplasts are surrounded by outer and inner phospholipid membranes, which are separated by a narrow intermembrane space. The outer membrane is the external layer, which maintains the chloroplast's shape and protects its inner structures. The inner membrane surrounds the stroma, forming a selectively permeable barrier that allows molecules to move into and out of the chloroplast.

chloroplast mRNA: An mRNA transcribed from chloroplast DNA that is only translated on 70S ribosome.

chloroplast proteins: The soluble proteins and the insoluble lamellar proteins. They are encoded by nuclear genes and by chloroplast genes. The enzyme **RuBisCO** is the best studied of the stroma proteins.

chloroplast ribosome: A 70S ribosome found in eukaryotic chloroplasts. Protein synthesis on 70S ribosome is inhibited by chloramphenicol but not by cycloheximide; the reverse is true for 80S cytoplasmic ribosomes. Chloroplast ribosomes disassociate into 50S and 30S subunits.

chloroplast rRNA: The rRNA that consists of various forms of 23S, 16S, 5S and 4.5S with a guanine/cytosine content of 55.8%.

chloroplast structure: In green plants, chloroplasts are surrounded by two lipid bilayer membranes, now believed to correspond to the outer and inner membranes of the ancestral cyanobacterium. The genome is significantly reduced compared to that of free-living cyanobacteria, but the parts that still exist show clear similarities. The fluid within the chloroplast is called the STROMA, corresponding to the cytoplasm of the bacterium, and contains minute circular DNA and ribosomes, though most of their proteins are manufactured by the cell nucleus. Within the stroma are stacks of THYLAKOIDS, the sub-organelle where photosynthesis actually takes place. A stack of thylakoids is called a GRANUM. A thylakoid looks like a flattened disk (stacks of coins), and inside occurs an empty area called the THYLAKOID SPACE or LUMEN. The photosynthesis reaction occurs on the surface of the thylakoid. The photosynthetic proteins in the membrane bind chlorophyll, which is present with various accessory pigments. The photosynthetic proteins give chloroplasts their green colour. Algal chloroplasts may be golden, brown or red and show variation in the number of membranes and the presence of thylakoids.

chloroplast transcription dynamics: Overall transcription rates change dramatically during chloroplast development. In sorghum, RNA polymerase levels increase in parallel with transcription activity. However, in spite of the global changes in transcription rates, the relative ratio of transcription of many plastid genes is constant in different tissues and developmental stages. Three reasons explain this observation. (i) The coding capacity of the plastid genome is restricted, thus preventing the need for regulation in situations where plastid genes are not relevant. (ii) With a few exceptions, plastid genes (e.g. *ndh* genes) have been classified into two major groups: (a) genes encoding proteins and RNAs directly involved in gene expression (i.e. transcription and translation) and (b) genes encoding the photosynthetic apparatus. Expression of these two groups of genes is chronologically linked during chloroplast development such that high levels of the plastid DNA transcription and translation machinery are required only during synthesis and assembly of the photosynthetic apparatus. (iii) Large numbers of plastid genes encode proteins that gather in fixed stoichiometries in a few large protein complexes (RNA polymerase, ribosomes and photosynthetic electron transport complexes). Thus, constant ratios of transcription of genes encoding a basic set of plastid proteins are required in fixed stoichiometry.

chloroplast transformation: Site-specific integration of transgenes into the chloroplast genome differs from random nuclear integration because chloroplast transgenes are integrated via homologous recombination. Chloroplast expression systems may give advantages over routinely nuclear expression. Chloroplast transformation requires (i) A chloroplast-specific expression vector; (ii) a method of delivery of DNA through the double membrane of the chloroplast; and (iii) an efficient selection system. *See* TRANSPLASTOMIC PLANTS.

chloroplast transformation vectors: Vectors for chloroplast transformation are designed with homologous flanking sequences on either side of the transgene. Chloroplast transformation vectors may also carry an origin of replication that facilitates replication of the plasmid inside the chloroplast.

chloroplast transit peptide (CTP): A transit peptide that, when fused to a protein, acts to transport that protein into chloroplast(s) in a plant. Once both the components are inside the chloroplast, the transit peptide is cleaved off the protein and that protein is free to do the intended task. For example, the CP4 EPSPS enzyme in genetically engineered glyphosate-resistant soybean [*G. max* (L) Merrill] plant is transported into the soyabean plant's chloroplasts by N-TERMINAL PETUNIA CHLOROPLAST TRANSIT PEPTIDE (CTP). After both reach the chloroplast, the CTP is cleaved and degraded, so the CP4 EPSPS is set free to perform its task of conferring resistance to glyphosate.

chloroplast tRNA: A tRNA found in the chloroplast. These tRNAs differ in structure from the cytoplasmic tRNAs and are coded by chloroplast DNA. The structure of these molecules is closer to that of prokaryotic tRNA than eukaryotic cytoplasmic tRNA. The chloroplast tRNAs are not amino-acylated by cytoplasmid aminoacyl-tRNA synthetases.

chloroplast-specific expression vector: The generic expression cassette consists of a strong promoter, a selectable marker, gene of interest, 5′- and 3′-untranslated regions to enhance the efficiency of transcription and translation of the introduced gene and two bordering sequences that guarantee that the plasmid is not randomly incorporated into the chloroplast genome. The presence of a chloroplast origin of replication within the chloroplast flanking sequence might facilitate its replication inside the chloroplast, thereby increasing its copy number. This enhances the probability of transgene integration and expedites the process of achieving a homogenous population of transformed plastid genomes, a state called HOMOPLASMY or HOMOPLASTOMY. *See details as separate terms.*

chlorosis (Adj. chlorotic): Refers to the failure of chlorophyll development, instigated by disease or a nutritional disturbance. It is exemplified by fading of green plant colour to light green, yellow or white.

cholesterol: A type of lipid molecule that is made by all animal cells because it is an essential structural component of animal cell membranes, essential for maintaining structural integrity and fluidity of the membrane.

chorionic villus sampling (CVS): An invasive prenatal diagnostic procedure involving removal of villi from the human chorion to obtain chromosomes and cell products for diagnosis of disorders in the human embryo.

chromat: Data file output from most popular DNA sequencers. Chromat files include the fluorescent traces generated by the sequencer for each of the four chemical bases – A, C, G and T – along with the sequence and measures of the error in the traces at each sequence position.

chromatid: One of a pair of homologous chromosomes formed during 'S' phase of cell cycle. In all duplicated chromosomes, each of the two longitudinal subunits becomes noticeable during mitosis or meiosis. During mitosis, the two chromatids separate, each becoming a chromosome of one of the two daughter cells. When the chromosome replicates its DNA prior to cell division, it forms a second DNA molecule that is identical to the original DNA molecule. In the replicated chromosome, the two DNA molecules – now known as SISTER CHROMATIDS – are joined together lengthwise with their strongest connection at the centromere. During cell division, centromere divides into two parts, and consequently, the sister chromatids split apart to form two separate chromosomes.

chromatid aberration: Changes produced in one of the two chromatids of a chromosome as a consequence of spontaneous or induced mutations.

chromatid break: A discontinuity in only one of the two chromatids of a chromosome.

chromatid bridge: A bridge-like configuration produced by a dicentric chromatid having both the centromeres passing to opposite poles during anaphase. The frequency of chromatid bridges in Anaphase II of meiosis is from time to time used as a measure for the level of cytological disturbances (e.g. in induced autopolyploids and allopolyploids).

chromatid conversion: A type of gene conversion that is inferred from the existence of identical sister-spore pairs in a fungal octad that show a non-Mendelian allele ratio.

chromatid exchange: *See* SISTER CHROMATID EXCHANGE.

chromatid gap: A non-staining region of a single chromatid in which there is minimal misalignment of the chromatid.

chromatid interchanges: Interchanges involving breaks in two or more chromatids or different loci of the same chromatid and reunion in new ways.

chromatid interference: A situation in which the occurrence of a crossover between any two non-sister chromatids can be shown to affect the probabilities of those chromatids being involved in other crossover in the same meiosis.

chromatid non-disjunction: The passing of homologous parts of chromatids to the same pole following crossing over between homologous differential segments in a multiple interchange hybrid.

chromatid segregation: Segregation of two sister-chromatid segments of a chromosome.

chromatid tetrad: The four chromatids (two per chromosome) of a bivalent.

chromatin: The term was coined by W. Flemming in 1879. Chromatin is made of DNA, histone proteins, non-histone proteins and some RNA. It is easily stained by basic dyes and can therefore be easily identified and studied under the microscope. In eukaryotes, when the cell is not reproducing, chromatin is seen under compound microscope in the form of threadlike structures called chromatin at interphase stage. Chromatin in the interphase nucleus is structured into discrete domains defined by sites of attachment to the nuclear matrix. Chromatin DNA gradually coils itself around flexible rods of histone proteins during the prophase, forming two parallel compact cylinders (chromatids) connected by a knot-like structure (centromere) at their middles. They appear similar to two rolls of carpeting standing side-by-side that are knotted together with rope at their middles. These cylinders are homologous chromosomes (i.e. the genes of the two chromosomes are linked in the same linear order within the DNA strands of both chromosomes). While they are joined at their middles, these paired chromosomes appear X-shaped. Chromatin is usually not visible during the interphase of a cell but can be made more visible during all phases by reaction with basic dyes specific to DNA. The basic structural unit of chromatin is the nucleosome, consisting of 146 bp of DNA wrapped around an octamer of CORE HISTONES H3, H4, H2A and H2B. Chromatin allows the nearly 2 m of DNA material in each cell of the body to fit in a nucleus measuring just 4–8 μm in diameter. Chromatin provides the context for mitosis, meiosis and gene expression. The DNA carries the cell's genetic information while the proteins, called HISTONE*s*, help package, organize and regulate the genetic material.

chromatin analysis: Chromatin prepared by a method involving restricted nuclease digestion contains the same repeating structure in the nucleus, whereas chromatin prepared by conventional methods involving shear does not. Chromatin displays a series of maxima or bands. Chromatin folding determines the approachability of DNA constituting eukaryotic genomes and subsequently is profoundly important in the nuclear processes, including gene regulation. Nucleosome arrays compact to form a 30 nm chromatin fibre to hitherto disputed structure. Compact nucleosome arrays stabilized by introduction of disulfide cross-links exhibit that the chromatin fibre comprises two stacks of nucleosomes in accordance with the TWO-START MODEL.

chromatin and DSB repair: Repair of DSBs involves both post-translational modification of nucleosomes and concentration of DNA repair proteins at the site of damage. Nucleosome packing and chromatin architecture surrounding the DSB may limit the ability of the DNA damage response to access and repair the break.

chromatin architecture proteins: Core histone variants, post-translational modifications of the histones and linker histones binding to the DNA, all influence the organization and stability of the nucleosome.

chromatin architecture: Higher-order architecture of the genome is similar among cell types with widely variant fates and gene expression patterns. Chromatin architecture includes intra-nucleosome interactions, inter-nucleosome interactions and the influence of non-histone chromatin architectural proteins.

chromatin assembly factor-1 (CAF-1): CAF-1 is a conserved histone chaperone, essential for cell survival in multicellular organisms, that plays a key role in replication-dependent nucleosome assembly and preserves genome stability.

chromatin decondensation: The DNA that is highly coiled gets uncoiled and gets freely dispersed in the nucleus and the chromosomes lose their identity. This leads to the formation of chromatin.

chromatin domain: A region of chromatin, the exact structure and size of which depends on experimental context. It refers to a single nucleosome or an array of more than 100 nucleosomes. The DNA is organized into loop domains by stable attachment to the nuclear matrix at approximately 50,000 bp intervals. Most domains are folded into higher order chromatin structures. The DNA of active domains is extended by multiple sequence-specific dynamic relations with the nuclear matrix. During interphase, the chromatin is attached to the periphery of the nucleus.

chromatin during cell cycle: At interphase, the nucleosome-rich chromatin forms many structural loops, which are anchored together in rosettes in some areas. This initial looping and gathering compresses the genetic material sufficiently to fit into the nucleus, where it appears as a mass of tangled string. As the chromosomes enter prophase of mitosis, looping and gathering increase, and bonding through protein cross-connections begins. By metaphase, the height of looping, gathering and bundling achieves a 250-fold compaction of the roughly 40-fold compacted 300-Å fibre, producing highly condensed, rod-like shapes referred to as mitotic chromosome. Eukaryotic genomes are organized into condensed, heterogeneous chromatin fibres throughout much of the cell cycle. Compaction of eukaryotic genomes into condensed chromatin fibres is necessitated to fit over a metre of DNA within the limited volume of the nucleus. Processes ranging from gene expression to chromosome dynamics during cell division are governed by the folding of DNA into chromatin.

chromatin dynamics: Chromatin displays a spatiotemporal correlation in space and time, and chromatin dynamics are also correlated. Chromatin dynamics are consequence of changes in the physico-chemical attributes of the chromatin constituents themselves or the nuclear environment. DNA is packaged by histones and associated proteins and RNA into chromatin. Several levels of chromatin organization enable accurate cell division.

chromatin folding: Chromatin folding determines the accessibility of DNA constituting the eukaryotic genomes and consequently is greatly important in the mechanism of nuclear processes such as gene regulation. Nucleosome arrays become compacted to form a 30 nm chromatin fibre to previously disputed structure.

chromatin GAGA factor: A multifunctional protein implicated in nucleosome organization and remodelling, activation and repression of gene expression, long distance enhancer-promoter communication, higher order chromosome structure and mitosis.

chromatin immunoprecipitation (ChIP) assay: An experimental procedure designed to determine whether a given protein binds to a particular DNA sequence in the cell. DNA-bound protein is cross-linked to DNA, chromatin is then isolated and the DNA-protein complexes are sheared into small fragments. Antibodies to the proteins are used to pull down fragments of DNA bound to specific proteins. The presence and quantification of specific or global DNA sequences can then be detected by polymerase chain reaction assay or massively parallel sequencing (ChIP-Seq) assay. This procedure that is used to identify the location of DNA-binding proteins and epigenetic marks in the genome. Related techniques – such as methylated DNA immunoprecipitation – use antibodies to recognize DNA modifications directly. Chromatin immunoprecipitation (ChIP)-based assays include: (i) Chromatin immunoprecipitation sequencing (ChIP-Seq); and (ii) Chromatin immunoprecipitation serial analysis of gene expression (ChIP-SAGE (STAGE)) that are used to identify genome-wide profiles of transcription factors, histone modifications, DNA methylation and nucleosome positioning. ChIP is used to investigate a particular protein-DNA interaction, several protein-DNA interactions or interactions across the whole genome or a subset of genes. High-resolution maps of epigenetic features have been generated by using ChIP-on-chip and ChIP-seq methods.

chromatin immunoprecipitation (ChIP): An antibody-based technology used to selectively enrich specific DNA-binding proteins along with their DNA targets. The ChIP method can be used to monitor transcriptional regulation through histone modification (epigenetics) or transcription factor-DNA binding interactions. The ChIP assay method allows analysis of DNA-protein interactions in living cells by treating the cells with formaldehyde or other crosslinking reagents in order to stabilize the interactions for downstream purification and detection. Performing ChIP assays requires knowledge of the target protein and DNA sequence that will be analysed, as researchers must provide an antibody against the protein of interest and PCR primers for the DNA sequence of interest. The antibody is used to selectively precipitate the protein-DNA complex from the other genomic DNA fragments and protein-DNA complexes. The PCR primers allow specific amplification and detection of the target DNA sequence. Quantitative PCR (qPCR technique) allows the amount of target DNA sequence to be quantified. The ChIP assay is amenable to array-based formats (ChIP on chip) or direct sequencing of the DNA captured by the immunoprecipitated protein (ChIP-Seq).

chromatin immunoprecipitation and sequencing (ChIP-seq): A method combining chromatin immunoprecipitation and DNA sequencing to analyse specific DNA-protein interactions. Next-generation DNA sequencing is often performed, resulting in a genome-wide analysis of protein-chromatin interactions. Specific regions of crosslinked chromatin, which is genomic DNA in complex with its bound proteins, are selected by using an antibody to a specific epitope. The enriched sample is then subjected to high-throughput (massively parallel) DNA sequencing to determine the regions in the genome, most often bound by the protein to which the antibody was directed. Most often used are antibodies to any chromatin-associated epitope, including transcription factors, chromatin-binding proteins and specific chemical modifications on histone proteins. ChIP-seq relies on combining of the ChIP experiments with high-throughput sequencing. ChIP-Seq couples ChIP with massively parallel DNA sequencing to recognize the binding sites of DNA-associated proteins. Such methods enable the parallel sequencing of millions of DNA molecules in real time. Building on principles of ChIP-seq and methylated RNA immunoprecipitation (MeRIP), read densities of immunoprecipitated RNA relative to untreated input control are used to identify methylated sites. A consensus motif is deduced, and its distance to the point of maximal enrichment is assessed. ChIP-seq technology is now capable of interrogating the entire human genome at high resolution with only one lane of sequencing. ChIP-seq has recently surpassed ChIP-chip technology for epigenomic analyses.

chromatin immunoprecipitation on chromatin immunoprecipitation chip (ChIP-on-chip): Histone modification was first identified on a genome-wide level through combining the chromatin immunoprecipitation (ChIP) technology with DNA microarrays by locating protein binding sites. For example, proteins to immunoprecipitated are modified histone (e.g. H3K27me3, H3K4me2, H3K9me3), Polycomb-group proteins (PRC2:Suz12, PRC1:YY1) and trithorax group proteins (ASH1) to study the epigenetic landscape or RNA polymerase II to investigate the transcription landscape. It is used to investigate interactions between proteins and DNA *in vivo*. Specifically, it allows the identification of the cistrome, the sum of binding sites, for DNA-binding proteins on a genome-wide basis.

chromatin immunoprecipitation paired end ditag sequencing (ChIP-PET): A high-throughput method where chromatin immunoprecipitation is paired end ditag sequencing (ChIP-PET). High-resolution maps of DNA methylation and H3K4 di- and trimethylation of two entire chromosomes and two fully sequenced centromeres were obtained from rice (*Oryza sativa*) shoots and cultured cells.

chromatin immunoprecipitation serial analysis of gene expression (ChIP-SAGE, STAGE): This method uses ChIP followed by a Sanger sequencing-based SAGE procedure to study the distribution of acetylated (ac) histones H3 and H4 in the yeast genome. The resolution of ChIP-SAGE (STAGE) also depends on the distribution of the restriction enzyme sites in the input ChIP DNA.

chromatin *in vivo* assay (CiA): This system builds on the principle of 'chemically induced proximity', in which a small molecule with two distinct protein-binding domains acts as a scaffold to reversibly bring proteins into proximity.

chromatin insulator: A DNA element that guards a gene from position effects.

chromatin language: DNA methylation and histone post-translational modifications lead to recruitment of protein complexes that regulate transcription. Transcription occurs against a backdrop of mixture of complex modifications, which probably have several roles. Although such modifications were initially thought to be a simple code, a more likely model is that of a sophisticated, nuanced chromatin 'language' in which different combinations of basic building blocks yields dynamic functional outcomes.

chromatin marks: The non-DNA part of a chromosomal locus that affects the nature and stability of gene expression. Covalent modifications of chromatin include DNA methylation ($5^{m}C$) and histone post-translational modifications (methylation, acetylation, ubiquitination, phosphorylation, SUMOylation and poly(ADP)ribosylation). When inherited, stable chromatin marks are referred to as 'EPIGENETIC MARKS'. *Aka* CHROMATIN MODIFICATIONS *or* CHROMATIN TAGS.

chromatin mobility: It contributes to the HOMOLOGY-DIRECTED REPAIR (HDR) of double-strand breaks and replication stress, with some DNA lesions specifically translocated to nuclear hubs of HDR activity.

chromatin modification network: The structural and functional interplay and coexistence of histone and DNA modifications within chromatin.

chromatin organization: Histone proteins help organize DNA into structural units called NUCLEOSOMES, which are then assembled into a compact structure (chromatin) and eventually into very large, high-order structures (chromosomes).

chromatin reconstitution: The rebuilding of chromatin with chromosomal constituents previously removed by chromatin dissociation.

chromatin regulation of flowering: The epigenetic regulation is carried out by the chromatin modification and remodelling activity. These modifications carried out the gene expression regulation and cell identity by transferring the epigenetic information from mother to daughter cells. In *A. thaliana*, chromatin mechanisms perform the flowering time regulation through chromatin modifications and flowering-regulatory gene expression. The chromatin modifiers, plant-specific factors and long non-coding RNAs control the chromatin regulation of *FLC*, a potent floral repressor. The chromatin modifications lead to the *FT* regulation encoding florigen in angiosperm species to control the flowering time in plants.

chromatin regulators: Chromatin, the protein packaging of the genome, harbours information about how the genes it contains are to be regulated. A recipe of histone modification marks, in the absence of a transcription factor, may be sufficient to stably employ a regulatory protein complex to a regulatory site.

chromatin remodellers: The position of the nucleosome influences the access of transcription factors to a gene's regulatory region and therefore is a key determinant of gene expression. Nucleosome positioning is regulated by four subfamilies of chromatin remodelling protein complexes – namely, SWI/SFN (Switching/Sucrose Non-fermenting), ISWI (Imitation Switch), CHD (Chromodomain-Helicase-DNA binding) and INO80 (Inositol Requiring) – which depends on ATP hydrolysis to mediate various functions: (i) change the position of histones within nucleosomes; (ii) removal of histones; and (iii) introduction of histone variants.

chromatin remodellers, functions of: (i) SWI/SFN (Switching/Sucrose Non-fermenting) chromatin modelers are implicated in nucleosome remodelling, DNA repair, differentiation, proliferation. (ii) ISWI (Imitation Switch) chromatin modelers are involved in nucleosome spacing, DNA damage repair, DNA transcription and replication. (iii) CHD (Chromodomain-Helicase-DNA binding) chromatin modelers function in nucleosome structure and density, DNA damage repair, DNA transcription. (iv) INO80 (Inositol Requiring) chromatin modelers have role in transcriptional regulation, DNA repair, recombination and replication.

chromatin remodelling: Refers to a restructuring (at molecular level) of chromatin, that alters specific genes so that DNA subsequently gets expressed. It can be triggered by SHORT-INTERFERING RNA (siRNA) or certain transcription activators. Chromatin remodelling occurs by the action of ATP-dependent remodelling complexes.

chromatin remodelling complexes: Chromatin remodelling is principally carried out by covalent histone modifications by specific enzymes (histone acetyltransferases, deacetylases, methyltransferases and kinases) and ATP-dependent chromatin remodelling complexes, which either move, eject or restructure nucleosomes. *Aka* CHROMATIN REMODELLING FACTORS.

chromatin remodelling in gene regulation: The linker histones (H1, H5 and H10) are involved in the condensation of chromatin into the 30 nm fibre. They have been found to exert both positive and negative effects on gene activity in Protozoa. This SUPRANUCLEOSOME ORGANIZATION correlates with the resting state of chromatin, and it is therefore possible that the linker histones play an active role in the control of chromatin activity. H5 plays role in the control of DNA replication and cell proliferation.

chromatin remodelling mechanisms: The basic mechanism of chromatin remodelling depends on the three dynamic properties of nucleosomes: reconstruction, enzyme-induced covalent modification and repositioning. Regarding reconstruction, nucleosomes can undergo compositional alteration using either canonical histones or special histone variants.

chromatin segregation: Physical model of a chromosome with experimentally measured local interactions segregates into euchromatin- and heterochromatin-like phases.

chromatin states: Different chromatin states are defined by a range of post-translational modifications of core histones, by incorporation of various histone isoforms as well as by DNA methylation.

chromatin stiffening: It is mediated in part by histone H2A phosphorylation. A genome-wide stiffening of the chromatin fibre is a consequence of DNA damage and as a novel mechanism underlying increased chromatin mobility.

chromatin three-dimensional interactions in gene expression: Classically, DNA sequence variants hold the source of heritable phenotypes and epigenetic variation in natural and experimental populations. An association between epigenetics and complex traits has demonstrated dynamic relationship between chromatin and DNA sequence factors among complex traits linked with epigenetics. The natural genetic variation controls numerous molecular phenotypes related to trait of interest among individuals. The EXPRESSION-BASED QTLS (EQTLS) for cell types, tissues and conditions interpret non-coding genetic variations delineating the complex organismal traits and diseases.

chromatin, functional state of: Changes in chromatin structure are affected by modifications in DNA and histones. Eukaryotic genomes are structured into active (euchromatic) and inactive (heterochromatic) domains. Post-translational modifications of histones are the key in defining these functional states, particularly in promoter regions. Asymmetric di-methylation of histone H3 arginine 2 (H3R2me2a) has been observed to counter correlate with di- and tri-methylation of H3 lysine 4 (H3K4me2, H3K4me3) on human promoters.

chromatin-associated proteins: Chromatin is a complex molecule that can be found in either an open decondensed or active state (euchromatin) or a tightly compacted inactive state (heterochromatin). The degree of compaction can be dynamic and is believed to be largely dependent on a variety of chromatin remodelling proteins that influence its structure and gene expression. A variety of histone modification enzymes (histone writers and erasers) impact histones, nucleosomes and the structure of chromatin. These proteins along with the various non-histone proteins modulate chromatin structure and influence the epigenetic regulation of a wide variety of genes. A large number of chromatin-associated proteins are required for normal development and differentiation. Chromatin-associated proteins have role in transcription, development and cancer. Role of chromatin-associated proteins has been determined by developing inhibitors for such proteins. Quinazoline derivatives act as inhibitors of histone methyltransferase (G9a) and G9a-like protein (GLP). There is one quinazoline analogue that inhibits an extra-terminal bromodomain motif (BET). Quinazolines act as dual inhibitors targeting histone deacetylases (HDACs) Zn^{2+}-dependent and kinase receptors for the potential treatment of cancer.

chromatin-associated RNA (caRNA): caRNA acts as a ubiquitous epigenetic layer in eukaryotes and has been reported to be essential in various biological processes, including gene transcription, chromatin remodelling and cellular differentiation. Existing experimental techniques to characterize genome-wide RNA-chromatin interactions to understand their underlying biological functions are generally expensive, time-consuming and limited in identifying all potential sites. Most of the existing computational methods are restricted to detecting only specific types of RNAs interacting with chromatin. Y. Xiong *et al.* (2023) propose a highly interpretable computational framework, named DeepRCI, to identify the interactions between various types of RNAs and chromatin. In this framework, they introduce a novel deep learning component called VARIFORMER and integrate multiomics data to capture intrinsic genomic features at both RNA and DNA levels. DeepRCI can detect RNA-chromatin interactions more accurately, and the sequence features extracted by DeepRCI can be well matched to known critical gene regulatory components, indicating that this model can provide useful biological insights into understanding the underlying mechanisms of RNA-chromatin interactions. They further delineate the relationships between RNA-chromatin interactions and cellular functions, including gene expression and the modulation of cell states.

chromatin-mediated repression: Chromatin is linked to two modes of transcriptional repression in yeast: silencing at the HM loci and telomeres, and repression by the global regulator complex Ssn6/Tup1. In both cases, unique chromatin configurations appear to be orchestrated by multiprotein complexes targeted to particular regions through interactions with specific DNA-binding proteins, such as RAP1, the origin-recognition complex and α2.

chromatin-modifying enzymes: The enzymes which carry out DNA modifications (DNA methyltransferases) and histone modifications (histone acetyltransferases, histone deacetylases, histone methyltransferases, histone demethylases, kinases, E3 ubiquitin ligases, small ubiquitin-related modifier-conjugating enzymes and ADP-ribosyl transferases).

chromatogram: The picture which results from the process of chromatography.

chromatography: Coined by Mikhail S. Tswett in 1906, the term refers to a biochemical technique by which complex mixtures of different molecules may be separated from each other. It is commonly used for purification of proteins and other molecules from a mixture in a solution. This is the most common method of enzyme purification. In chromatography, sample is passed through a column containing solid matrix with which the target molecules interact based on their charge (ion exchange) and their ability to bind specifically to other molecules attached to matrix (affinity). Gel filtration column separates molecules on the basis of their size. During chromatography, the mixture is exposed to many repeated partitioning between a flowing phase and a stationary phase. Commonly used chromatographic methods include adsorption chromatography, ion exchange chromatography, gel filtration chromatography and affinity chromatography. *See* AFFINITY CHROMATOGRAPHY, DNA AFFINITY CHROMATOGRAPHY *or* HIGH-PERFORMANCE LIQUID CHROMATOGRAPHY.

chromatoid bodies: Optical Fourier transforms, made of electron micrographs of the crystalline array of ribonucleoprotein, are called chromatoid bodies, which have been interpreted as the transforms of a helix of 12 nodes in five turns. A model shows that this helix may be built of spheres 180Å in diameter, placed on a radius of 150Å. The optical transforms of the model are very similar to the transform of the original electron micrograph.

chromocenter: Pericentromeric heterochromatic regions of all the polytene chromosomes that are tightly paired such that all chromosomes appear to be fused in this region.

chromodomain: A type of reader module of motif of 40–50 amino acids common to proteins that function in binding DNA, RNA and protein; often binds methylated histones and result in chromatin remodelling by targeting histone lysine methylation marks. Chromodomains are components implicated in the recognition of lysine-methylated histone tails and nucleic acids. CHROMOSOME-ATPASE/HELICASE-DNA BINDING (CHD) PROTEINS control ATP-dependent nucleosome assembly and deployment through their conserved double chromodomains and SW12/SNF2 helicase/ATPase domain. Deletion of chromodomains damages nucleosome binding and remodelling by CHD proteins. Examples of CHD proteins include heterochromatin protein 1 (HP1), which recognizes the trimethylation of lysine 9 on histone H3 (H3K9me3).

chromogene: A hereditary determiner in the nuclear chromosome, in contrast to PLASMAGENES determiners in the cytoplasm. See GENE.

chromomere pattern: The linear order and dispersal of chromomeres along a chromosome. Chromomere pattern has been extensively studied in crops, such as tomato, maize and rye.

chromomeres: The chromomeres are small bead-shaped assemblies along the entire length of a mitotic prophase or meiotic pachytene chromosome at precise regions. They are formed due to local condensation or tight folding of the chromatin. HETEROCHROMATIN CHROMOMERES stain darker and are larger than EUCHROMATIC CHROMOMERES. Number, size and position of the chromomeres on each chromosome are rationally constant and can be used as dependable morphological markers of chromosomes. Chromomeres are also moulded in polytene chromosomes and lamp brush chromosomes.

chromomethylase 3 (CMT3): A plant-specific DNA methyltransferase that is required for the maintenance of DNA methylation at CHG sites. The chromomethylases, which are unique to plants, may preferentially methylate DNA in heterochromatin preferably in CpNpG sequences.

chromonemata (pl. chromonema): The smallest spirally coiled light-microscopically observed central filament of a chromatid along which the chromomeres are aligned. Most contemporary researchers believe that the chromonema is an elementary deoxyribonucleoprotein (DNP) thread 100–200 Å in diameter. During cell division, the chromonema provide support for the spiralization of the chromosome.

chromoplast: The carotinoid-containing plastids responsible for pigment synthesis and storage of (coloured) carotenoids. Chromoplasts give flower petals and fruits their distinctive colour.

chrom-seq: L. Fan *et al.* (2024) established a method termed Chrom-seq to competently capture RNAs associated with various chromatin marks in living cells. Chrom-seq cooperatively applies highly precise chromatin mark reader with APEX2, which catalyses the oxidation of biotin-aniline to label the adjacent RNAs for isolation by streptavidin-coated beads. Using the readers of mCBX7/dPC, mCBX1 and mTAF3, these researchers spotted RNA species significantly linked with H3K27me3, H3K9me3 and H3K4me3, respectively. Chrom-seq outdid other equivalent methods in terms of sensitivity, efficiency and cost of practice. It offers an antibody-free method to analytically map RNAs at chromatin marks with probable regulatory roles in epigenetic procedures.

chromosomal: Relating to the structure, constituents and function of chromosomes.

chromosomal aberrations: All types of changes in chromosome structure (such as deletions, duplications, inversions and translocations) and chromosome number (euploids and aneuploids). *Aka* CHROMOSOME ABERRATIONS or CHROMOSOME MUTATION.

chromosomal chimera: A chimera in which the components do not have the same chromosome number.

chromosomal crossing-over: The exchange of DNA between paired homologous chromosomes in division I of meiosis. It is a sign of genetic recombination, and the crossovers (chiasmata) are cytological evidence of crossing over having been taken place and are visible in the light microscope.

chromosomal cycle: All the changes that take place in the chromosomes of an organism from fertilization to gametogenesis.

chromosomal fibres: Fibre bundles connecting the tiny granules with the poles. These chromosomal fibres have been regarded as intimately involved in chromosomal movement as traction fibres for the chromosomes.

chromosomal fission: When a chromosome fits into two, thereby increasing the number of chromosomes in the karyotype. Fissions are also called DISASSOCIATIONS.

chromosomal fusion: When two non-homologous chromosomes fuse into one, thereby reducing the number of chromosomes in the karyotype.

chromosomal indigestion: Incapability of cells to tolerate genomic multiplicity (polyploidy). Excess chromosomes are eliminated from the nucleus.

chromosomal interference: The increase (negative interference) or decrease (positive interference) in likelihood of a second crossover closely adjacent to another. In most organisms, interference increases with decreased distance between crossovers. *See* COEFFICIENT OF COINCIDENCE *and* COINCIDENCE.

chromosomal mutation: *See* CHROMOSOME MUTATION.

chromosomal polymorphism: A condition where, in the same population, various chromosomal rearrangements exist simultaneously.

chromosomal sex determination: In most of the animals and in some plants, a pair of chromosomes is accountable for the sex determination. These chromosomes exist as a pair in one sex but not in the other. This pair of chromosomes is called SEX CHROMOSOMES or ALLOSOMES. The chromosomes other than the sex chromosomes, which exist in pairs in both the sexes, are called AUTOSOMES. In most diploid individuals, there are 2n−2 autosomes and 2 sex chromosomes, while in some, there are 2n−1 autosomes and 1 sex chromosome. In one of the sexes, two chromosomes are identical or homomorphic (homogametic sex); in the other sex, these are different or heteromorphic (heterogametic sex).

chromosome: The term was proposed by H.W.G. Waldeyer in 1888 for the individual threads within a cell nucleus. This filamentous structure containing many genes in the cell nucleus. It is the structural unit of genetic material consisting of a single, linear double-stranded DNA molecule and associated proteins. During cell division, chromatin composing each chromosome condenses into a compact structure called chromosome, which can be seen under light microscope. Each species has a constant number of chromosomes. The number of chromosomes per cell nucleus is constant for each species, such as maize (*Z. mays* L.) has 20, rice (*O. sativa* L.) 24, bread wheat (*Triticum aestivum* L) 42 and human beings (*Homo sapiens sapiens*) 46 chromosomes/cell. Chromosomes can be seen and counted using compound light microscope in dividing root cells at metaphase and anaphase stages and in pollen mother cells (PMCs) at diakinesis, metaphase and anaphase stages. The chromosomes in prokaryotic cells and eukaryotic cells differ in several key ways. Most eukaryotic cells have multiple chromosomes located in the nucleus, whereas prokaryotic cells have a single chromosome located in a membraneless structure called NUCLEOID and contains a DNA molecule compacted by a series of NUCLEOID-ASSOCIATED PROTEINS (NAPs).

chromosome aberration: *See* CHROMOSOMAL ABERRATION. *Aka* CHROMOSOME MUTATION.

chromosome architecture: Despite the extreme condensation that is necessary to pack the chromosomes into the nucleus, the *Arabidopsis* genome appears to be packed in a predictive manner. Heterochromatin and euchromatin represent two distinct interactomes: (i) interactions between chromosomes correlate with the linear position on the chromosome arm; and (ii) distal chromosome regions have a higher potential to interact with other chromosomes.

chromosome arms: One part of a chromosome apart from the centromere. Each eukaryotic chromosome consists of a linear DNA molecule with two arms, the shorter 'p arm' and the longer 'q arm'. The two chromosomal arms of a chromosome are separated by the centromere.

chromosome banding: The experimental vision of differentially stained regions because of the distribution of different chromatin ingredients along a chromosome.

chromosome break: Induced or spontaneous breaks occurring across the entire cross section of the chromosome.

chromosome breakage and reunion: The general mode by which recombination occurs. DNA duplexes are broken and reunited in a crosswise fashion according to the HOLLIDAY MODEL.

chromosome bridge: A dicentric chromosome that forms a connection between the separating groups of anaphase chromosomes because its two centromeres of the dicentric chromosome are being drawn towards opposite poles.

chromosome cohesion: Cells respond to DNA damage by strengthening chromosome cohesion across the entire genome independently of DNA replication. In S phase, the acyltransferase *Eco*I converts cohesin into cohesion complexes that hold sister chromatids together. A DNA double-strand break in G2 can also trigger cohesin loading and cohesion establishment both at the break site and on undamaged chromosomes.

chromosome coiling: The spiral or helical coiling of the chromonemata of the chromosomes during some stages of cell division (mitosis and meiosis).

chromosome complement: A single complete set of chromosomes derived from a particular gametic or zygotic nucleus of an organism.

chromosome complex: A set of chromosomes having a given segmental arrangement.

chromosome condensation: Dramatic reorganization of the long thin chromatin strands into compact short chromosomes that occurs in mitosis and meiosis.

chromosome configuration: Any association observed by chromosome pairing at meiosis.

chromosome congression: The movement of chromosomes on to the metaphase plate, especially at the first meiotic division.

chromosome conjugation: Association between homologous chromosomes during meiotic prophase.

chromosome constriction: The unspiralized segment of a fixed position found in metaphase chromosome.

chromosome contraction: The coiling and shortening of chromosomes during mitosis and meiosis acquired naturally or experimentally by using specific chemical or cold treatment.

chromosome counting method: The procedure to determine the number of chromosomes per cell.

chromosome counts: *See* CHROMOSOME COUNTING METHOD.

chromosome deletion: The loss of part of a chromosome's DNA.

chromosome diminution hypothesis: (i) Whole genome reduction occurs in all cells undergoing meiosis; two successive rounds of division occur without intervening DNA replication. (ii) In *Ascaris megalocephla*, it was thought that chromosome breakages occur. The segments of acentric chromosomes do not arrange themselves on the equator of the spindle apparatus and are thus lost. Thus, genetic constitution of the descendent cells was not same as the parental cells. It was thought that somatic cells were formed from chromosome diminution. On the other hand, germ cells retained all the genetic material as no chromosome diminution occurred here.

chromosome disorder: A condition due to the gain, loss or rearrangement of an entire or part of a chromosome.

chromosome domain: A region of a chromosome, the exact nature, structure and size of which depend on the experimental context. For example: (i) a domain can be a region of chromosomal packaging, such as a loop extending from two adjacent attachments to a chromosomal axis and can vary in size from an array of less than 100 nucleosomes (~30 kb) to potentially more than 500 nucleosomes and (ii) a chromosomal domain might also represent a functional unit of chromosomal structure defined by boundary elements or insulators at the edges of the domain.

chromosome doubling: Induced or spontaneous doubling of chromosome sets leading to rediploidization or to polyploidization.

chromosome elimination: The loss of chromosomes from nuclei during certain mitotic or meiotic stages. This phenomenon is frequently observed in several artificial autopolyploids and allopolyploids.

chromosome end replication: Extreme 3′-ends of linear DNA are left incomplete during normal DNA replication. So, 3′-ends have single-stranded termini. Because of the redundant terminal sequences, the tails will still have complementary sequence at the 3′-tail that can pair to yield two-unit length concatemer. Gap filling is done by DNA polymerase I and DNA ligase. This process can be replicated to form tetramer and so forth. Specific nucleases then act to produce staggered nicks. These nicks are filled by a DNA polymerase to yield double helices.

chromosome engineering: Manipulation of whole chromosome sets, individual chromosomes or even chromosome segments by different means for scientific analysis or improvement of performance of crop plants.

chromosome fissions: When a chromosome splits up into two, thereby increasing the number of chromosomes in the karyotype. *Aka* DISSOCIATIONS.

chromosome fragmentation: The breakage of chromosomes when acted upon by some mutagen.

chromosome fusion: A mutation in which two chromosomes are connected at their ends. The product is a metacentric chromosome.

chromosome imprinting: An epigenetic system that silences one of the two parental chromosomes such that it results in a heritable (cell-to-cell) change in behaviour of the 'imprinted' chromosome. In some organisms, paternal X chromosome is inactivated in some or all the cells. Chromosome imprinting is obviously heritable but is reversed in germ line. The classical example of epigenetic inheritance is the phenomenon of imprinting, in which the expression standing of a gene depends upon the parent from which it is derived. The X chromosome of paternal and maternal parents act differently probably due to difference in methylation. *Aka* GENETIC IMPRINTING.

chromosome inactivation: Inability of a particular chromosome to transcribe and produce required mRNA transcripts and are unable to produce relevant products in the cell.

chromosome inversion: Chromosome segments that have been turned 180 degrees. The gene sequence for the segment is reversed with respect to the rest of the chromosome.

chromosome jumping: A technique of isolated clones from a genomic library that are not contiguous but hop a region (~200 kb) between known points on the chromosome. This is done generally to evade the regions that are difficult or impossible to 'walk' through or regions known not to be of interest to the investigator. Chromosome jumping is used for large DNA fragments, usually of several hundred kb in size. *Aka* CHROMOSOME HOPPING.

chromosome landing: In plants, richness of repetitive DNA makes chromosome walking impossible. Availability of high-density physical maps based on markers such as Random Amplified Polymorphic DNA (RAPDs) are found close enough to the genes to be included on the same genomic clone. This method is analogous to TRANSPOSON TAGGING.

chromosome length polymorphism (CLP): The phenomenon that chromosomes can considerably vary in length among individuals of a population within a species. This happens usually due to spontaneous structural changes of chromosomes (e.g. in allogamous rye).

chromosome libraries: DNA libraries made from restriction endonuclease cleaved DNA fragments of one chromosome isolated via flow cytometry or pulsed-field gel electrophoresis.

chromosome loss: Failure of a chromosome to become incorporated into a daughter nucleus at cell division.

chromosome map: Refers to the visual appearance of a stained chromosome when examined under a microscope. Particularly important are visually discrete areas – such as light and dark bands, puffs, constrictions, swellings – which give each of the chromosomes a unique appearance. *Aka* CYTOLOGICAL MAP. *See* LINKAGE MAP.

chromosome matrix: A chromophobic membrane enclosing the chromosome proper.

chromosome microarray analysis: A technique using DNA in which very small deletions or duplications of chromosomes can be identified. This is an important test for children with developmental disabilities or autism, with or without additional physical or developmental abnormalities.

chromosome microdissection: The exclusion of relatively large sections (e.g. half a chromosome arm) from an isolated chromosome using atomic force microscopy or laser microdissection. The microdissected pieces still harbour at least 10^6 base pairs and are used for the founding of the subgenomic libraries in bacterial artificial chromosomes or map-based cloning approaches.

chromosome mobilization: The breaking and linearization of the circular bacterial chromosome, establishing the point of origin.

chromosome model: The model which explains organization of chromatin fibres in a chromosome.

chromosome mosaicism: The presence of cell populations of various karyotypes in the same individual.

chromosome movement: The movement of chromosomes during mitosis and meiosis as a prerequisite for the anaphase separation of chromatids and/or chromosomes.

chromosome mutations: Segments of chromosomes, whole chromosomes or even entire sets of chromosomes may be involved in genetic changes. Such mutations include deletions (deficiencies), duplications, inversions and translocations. *See* MACROMUTATIONS.

chromosome number: The specific chromosome number (2n) in somatic cell of a given species or its derivative.

chromosome number, changes in: Chromosomal mutations that involve (i) an incomplete set of chromosomes (ANEUPLOIDS) or (ii) a complete set of chromosomes (EUPLOIDS).

chromosome orientation: *See* CENTROMERE ORIENTATION.

chromosome packaging in eukaryotic nucleus: In eukaryotic chromosomes, a DNA-histone protein thread, usually associated with non-histone (acidic) proteins and RNA, occurring in the nucleus of a cell. A single DNA molecule complexed with chromatin proteins is organized to facilitate a hierarchical packing scheme: (i) DNA helix is wound twice around a core particle of histone proteins; (ii) 30 nm fibre, six histone core particles per turn; and (iii) loops of 30 nm fibres are formed by attachment of chromatin to the nuclear matrix roughly every 30–100 kb. It is important to make the distinction between decondensed interphase chromosomes and condensed mitotic chromosomes. During interphase, most of the chromosomal material needs to be in an open configuration to allow for gene expression to occur. During mitosis, the chromatin needs to be condensed. Chromosomes are involved in the transmission of hereditary information. It is self-replicating genetic structure of cells. They work in tandem with other nucleic acids to build proteins and are also involved in cell division (mitosis or meiosis).

chromosome painting: Fluorescent *in situ* hybridization of a precisely labelled DNA probe or probes that hybridizes to the entire chromosome of the probe's origin. Using different fluorescent dyes and probes, a pattern of multicoloured chromosomes or chromosome segments appear.

chromosome pairing: The highly specific side-by-side association of homologous chromosomes during meiotic prophase. *See* SYNAPSIS.

chromosome polymorphism: The presence of one or more chromosomes in two or more alternative structural forms within the same population.

chromosome puffing: Despiralization of the deoxyribonucleoprotein of distinct regions of a chromosome during particular cell stages.

chromosome puffs: During certain stages of development, diameter of specific sites of chromosome is greatly increased. These swellings indicate sites of gene activity. *Aka* BALBIANI RINGS *or simply* PUFFS.

chromosome pulverization: The destruction of chromosome structure, varying from an apparently total fragmentation of the chromatin to various degrees of defective condensation and erosion.

chromosome rearrangement: The structural changes (such as deletions, addition, duplication or substitution) of the chromosome regions.

chromosome reduplication: The synthesis of all compounds that result in formation of an identical copy of the original chromosome gene.

chromosome region p: A designation for the short arm of a chromosome.

chromosome region q: A designation for the long arm of a chromosome.

chromosome scaffold: Basic architectural form after histones have been removed from chromosomes. In this form, revealed by electron microscopy of spread films of such chromatin, long loops of DNA appear to be subtended by a persisting framework of non-histone proteins. These are loosely referred to as the SCAFFOLD PROTEINS.

chromosome segregation: Separation of the members of a pair of homologous chromosomes in eukaryotes in a manner that only one member is present in any post-meiotic nucleus. Chromosome segregation phase follows a single DNA replication phase.

chromosome separation in yeast: The budding yeast Polo-like kinase Cdc5 was found to be instrumental in establishing the meiosis I chromosome segregation program. Cdc5 was required to phosphorylate and remove meiotic cohesion from chromosomes. An absence of CDC5, kinetochores were biorient-ed during meiosis I. Mam1, a protein essential for coorientation, failed to associate with kinetochores. Thus, sister-kinetochore coordination and chromosome segregation during meiosis I are coupled. During meiosis I, the centromeres of a pair of sister chromatids work in unison so that they face, and attach to, the same spindle pole A. A cohesion complex containing the Rec8 protein holds sister chromatids together at centromeres and along chromosome arms. Anaphase I phase is triggered by cleavage of arm Rec8 by separase, allowing the homologous pairs to disengage and move to opposite poles. Sgo1 protects Rec8 at centromeres during anaphase I, so that sister chromatids remain tethered. Sgo1 is degraded after anaphase I (telophase I), so that, at metaphase II, centromeric rec8 is no longer protected from separase. Cleavage of this centromeric Rec8 at anaphase II allows sister chromatids to separate.

chromosome set: The minimum viable complement of indispensable chromosomes (each represented only once) of an individual.

chromosome size: The physical size of a complete chromosome.

chromosome sorting: See SORTING.

chromosome staining: The pre-treatment and treatment of chromosomes with different dyes in order to make them most suitable for chromosome counting or specific analyses.

chromosome stickiness: Chromosome agglutination that gives a pycnotic or sticky appearance to chromosomes. Sometimes it may be caused by gene mutations or by treatment with certain chemical or physical agents.

chromosome structure, changes in: Translocations, inversions, deletions and duplications are included in this category. These structural changes in chromosomes produce semi-sterility by generating unbalanced meiotic products that may themselves be lethal or result in lethal zygotes. *Aka* CHROMOSOMAL ABERRATIONS.

chromosome substitution: The replacement of one or more chromosomes by others from another source by spontaneous events or a crossing scheme.

chromosome theory of inheritance: The theory states that the genes are located in the chromosomes. Chromosomes thus constitute physical inheritance. W. Sutton and Th. Boveri in 1902 independently pointed out similarities between Mendelian factors (genes) and chromosomes in their behaviour during meiosis and fertilization and suggested relationship between them. T.H. Morgan in 1910 provided evidence to this theory by mapping the gene responsible for *white* eye colour in *Drosophila melanogaster* on the X chromosome.

chromosome thread: The protein framework of a chromosome.

chromosome translocation: Translocations may involve transfer of a part of chromosome to the same chromosome or to a non-homologous chromosome, or two non-homologous chromosomes may mutually exchange their parts; in the latter case, two simple translocations are simultaneously achieved.

chromosome walking: A technique developed by W. Bender, P. Spierer and D.S. Hogness in the early 1980s for studying segments of DNA, larger than can be individually cloned, by using overlapping probes. It is used for the determination of a gene on a particular DNA clone of a DNA library. The total DNA of a chromosome has to be available as a series of overlapping DNA fragments. Such fragments are constructed either by DNA shearing or by cleavage using restriction enzymes. These fragments are used for series of hybridizations. This technique starts with a cloned gene, which is already identified on the same chromosome. This known gene serves as a probe for detection of clones (fragments) that are contained in neighbouring DNA sequences. During the following hybridization, this DNA sequence is used as a probe for detecting the next neighbouring sequences, and so on. Through each hybridization, there is progress away from the known gene towards the unknown chromosomal site. This method is practically more appropriate for plants with small genomes than for crop plants such as wheat, because the ratio of kilobases of DNA to the genetic map units is roughly proportional to the size of the genome. The cloning of genes by this method turns out to be much easier if the entire genome is already represented by a contiguous array of ordered DNA clones. *Aka* JUMPING LIBRARIES.

chromosome-mediated gene transfer (CMGT): A chromosome from a mammalian cell can be isolated by differential centrifugation of carefully broken cells whose mitotic cycles have been arrested at metaphase by the drug colchicine, since metaphase chromosomes are relatively robust and are easy to be isolated. These isolated chromosomes are allowed to be taken up by cells in culture, and success has been achieved in some cases where individual human chromosome 17, or a segment, could be transferred to a mouse cell line. This chromosome 17 thus transferred carried the gene for thymidine kinase, and therefore, its transfer could be checked on HAT medium, discussed for chromosome mapping. The chromosome segment with markers was later found to be translocated to a mouse chromosome. This process is parallel to the bacteriophage-mediated transduction in bacteria. CMGT allows a number of closely related genes to be transferred from one cell to another.

chromosomes classified based on centromere position: Depending on the position of centromere, chromosomes are classified as METACENTRIC, SUBMETACENTRIC, ACROCENTRIC, TELOCENTRIC and ACENTRIC. *See details as separate terms.*

chromosomes classified based on number of centromeres: Depending upon number of centromeres in a chromosome, chromosomes are classified as MONOCENTRIC, DICENTRIC and POLYCENTRIC. *See details as separate terms.*

chromotype: The chromosome set.

chronic toxicity: Poisoning caused by low levels of exposure to a compound over a long period of time. *Contrasts with* ACUTE TOXICITY.

chronic: Pertaining to slow-developing, persistent or recurring symptoms. *Contrasts with* ACUTE.

chronoculture: Improving crop yield and sustainability by exploiting the plant circadian cycle.

Chytridiomycota: A group of fungi that yield motile zoospores and resting sporangia. The plant pathogenic species are all biotrophs that are restricted to the cells of their host. *Aka* CHYTRIDIOMYCETES.

cI repressor: One of the regulatory proteins responsible for controlling the expression of γ genes.

cilium: Hairlike extension of a eukaryotic cell containing an essential bundle of microtubules and capable of performing repeated beating movements. If these projections are numerous and short then they are called CILIA. Cilia are found in large numbers on the surface of many cells and are responsible for the swimming of many single-celled organisms.

cipher: In cryptography, an algorithm for performing encryption or decryption. A series of well-defined stages that can be followed as a procedure. *Aka* CYPHER.

circadian clock: A circadian clock is an internal biochemical mechanism that oscillates with a period of 24 h and is coordinated with the day-night cycle. Circadian clocks are the central mechanisms which drive circadian rhythms. For example, the sleep-wakefulness cycle in humans. *Aka* CIRCADIAN OSCILLATOR.

***Circadian Clock Associated 1* (*CCA1*) gene:** A gene that is central to the circadian oscillator of angiosperm.

circadian clock genes: A set of core clock genes that forms a feedback loop of gene transcription and translation. The core genetic clock generates circadian rhythms in cells throughout the body.

circadian clock in *Arabidopsis*: Circadian clocks are believed to confer an advantage to plants, but the nature of that advantage has been unknown. A substantial photosynthetic advantage is conferred by correct matching of the circadian clock period with that of the external light-dark cycle. In wild-type and in long- and short-circadian period mutants of *Arabidopsis thaliana*, plants with a clock period matched to the environment contain more chlorophyll, fix more carbon, grow faster and survive better than plants with circadian periods differing from environment. This explains why plants gain advantage from circadian control.

circadian clock in cyanobacteria: An autoregulatory transcription-translation feedback loop is thought to be essential in generating circadian rhythms in any model organism. In the cyanobacterium *Synechococcus elongates*, the essential clock protein KaiC is proposed to form this type of transcriptional negative feedback. Nevertheless, temperature-compensated, robust circadian cycling of KaiC phosphorylation has been demonstrated even without KaiBC messenger RNA accumulation under continuous dark conditions. This rhythm persisted in the presence of a transcription or translation inhibitor. Thus, cyanobacterial clock can keep time independent of *de novo* transcription and translation processes.

circadian clock in *Drosophila*: Output from the circadian clock controls rhythmic behaviour through poorly understood mechanisms. In *Drosophila*, null mutations of the *neurofibromatosis 1* (*Nf1*) gene produced abnormalities of circadian rhythms in locomotor activity. Mutant flies showed normal oscillations of the clock gene *period* (*per*) and *timeless (tim)* and of their corresponding proteins but altered oscillations and levels of a clock-controlled reporter. Mitogen-activated protein kinase (MAPK) activity is increased in *Nf1* mutants, and the circadian phenotype is rescued by loss-of-function mutations in the Ras/MAPK pathway. Thus, *Nf1* signals through Ras/MAPK in *Drosophila*. Immuno-histochemical staining revealed a circadian oscillation of phospho-MAPK in the vicinity of nerve terminals containing pigment-dispersing factor (PDF), a secreted output from clock cells, suggesting a coupling of PDF to Ras/MAPK signalling. Full-length mammalian protein *mTim* exhibited a 24-h oscillation.

circadian clock in mitosis regulation: Cell division in many mammalian tissues is associated with specific times of day. In the regenerating liver of mice, the circadian clock controls the expression of cell cycle-related genes that in turn modulate the expression of active cyclin B1-Cdc2 kinase, a key regulator of mitosis. Among these genes, expression of *wee1* was directly regulated by the molecular components of the circadian clockwork. In contrast, the circadian clockwork oscillated independently of the cell cycle in single cells. Thus, the intracellular circadian clockwork can control cell-division cycle directly and indirectly in proliferating cells.

circadian clock in *Neurospora*: Recent identification of a large number of putative antisense transcripts in *Neurospora crassa* strengthens the view that antisense RNAs might affect a wider variety of processes than previously thought. In this model, organism entrainment of the circadian clock, which is critical for the correct temporal expression of genes and their products, is controlled partly by an antisense RNA arising from a clock component locus.

circadian oscillator: Networks of biochemical feedback loops that generate 24-h rhythms in organisms from bacteria to animals.

circadian plasticity of chromatin states: Chromatin regulation is a necessary component of clock-mediated transcriptional oscillations and negative feedback inhibition.

circadian rhythm: A circadian rhythm, or circadian cycle, is a natural, internal process that regulates the sleep–wake cycle and repeats roughly every 24 h. It can refer to any process that originates within an organism (i.e. endogenous) and responds to the environment (entrained by the environment). These 24-h rhythms are driven by a circadian clock that seems to be independent of outside stimuli. They have been widely observed in animals, plants, fungi and cyanobacteria.

circadian rhythmicity: The molecular machinery that governs circadian rhythmicity is based on clock proteins organized in regulatory feedback loops. Although post-translational modification of clock proteins is likely to finely control their circadian functions, only limited information is available to date. BMAL1, an essential transcription factor component of the clock mechanism, is SUMOylated on a highly conserved lysine residue (Lys^{259}) *in vivo*. BMAl1 shows a circadian pattern of SUMOylation that parallels its activation in the mouse liver. SUMOylation of BMAL1 requires and is induced by CLOCK, the heterodimerization partner of BMAL1. Ectopic expression of a SUMO-deficient BMAL1 demonstrates that SUMOylation plays an important role in BMAL1 circadian expression and clock rhythmicity. This reveals an additional level of regulation within the core mechanism of the circadian clock.

circular consensus sequencing (CCS): A sequencing technology with long read length and high accuracy that enables comprehensive variant discovery in a single experiment. CCS derives a consensus sequence from multiple passes of a single template molecule, producing accurate reads from noisy individual subreads.

circular fork model of DNA replication: Circular fork model has only been studied in prokaryotes. This model is bidirectional – DNA replication proceeds in both the directions with respect to the point of replication origin which gives rise to θ structure. The θ structure alternates around its axis to give rise to 8-structure, which further differentiates to form two daughter strands.

circular permutation: In a population of phage chromosomes, the chromosomes are redundant at the ends, and the redundancies are different for each chromosome. This causes physically linear chromosomes to yield a circular chromosome.

circular restriction map: A pictorial representation of the position of the sites recognized by various restriction enzymes on circular genomes from bacteriophages, viruses, chloroplasts mitochondria, bacteria or cyanobacteria.

circularization of DNA: A DNA fragment produced by digestion with a single restriction endonuclease will have complementary 5′ and 3′ extensions (sticky ends). If these ends are strengthened and ligated, the DNA fragment is changed into a covalently closed circle (or circularized).

circulative transmission: A type of virus transmission categorised by a long period of acquisition of the virus by a vector (typically an insect) for a latent period of several hours before the vector is able to transmit the virus, and the retention of the virus by the vector for a long period, usually several days. The virus circulates in the body of the vector. *See* PERSISTENT TRANSMISSION *and* PROPAGATIVE TRANSMISSION. *Contrasts with* NON-PERSISTENT TRANSMISSION *and* STYLET-BORNE TRANSMISSION.

cirrhose: (of a leaf) Ending in a tendril at the apex.

cirrhus (pl. **cirrhi**)**:** A curled, tendril-like form of exuded spores that are held together by a slimy matrix.

***cis*:** The term when used in molecular biology denotes an interaction between two sites which are located within the same molecule. However, a *cis*-acting protein can either be one that acts only on the molecule of DNA from which it was expressed or a protein that acts on itself (e.g. self-proteolysis).

***cis* face:** Face of a Golgi mass at which material enters the organelle. It is adjacent to the *CIS*-GOLGI NETWORK.

***cis*/*trans* isomerism:** A type of geometrical isomerism (the phenomenon in which more than one compounds have the same chemical formula but different chemical structures) found (e.g. in alkenic systems), in which it is possible for each of the doubly bonded carbons to carry two different atoms or groups. Two similar atoms or groups may be on the same side (*cis*) or on opposite sides (*trans*) of a plane bisecting the alkenic carbons and perpendicular to the plane of the alkenic systems.

***cis*-acting function:** The ability of a DNA sequence or transcript to regulate the expression of one or more genes on the same chromosome. *Aka* GENE REGULATION IN *CIS*. *Contrasts* *TRANS*-ACTING FUNCTION.

***cis*-acting gene regulation:** The gene regulation which involves proteins that influence regulation locally on the genome.

***cis*-acting locus:** A regulatory sequence that effects activity of DNA sequences located on its own molecule of DNA.

***cis*-acting ncRNAs:** Non-coding RNAs known to be involved in the *cis*-regulation of target genes located at or near the same genomic locus. Their expression is often coordinated with that of neighbouring protein-coding genes, and in many cases, related transcripts can influence each other at one step or another during their biogenesis.

***cis*-acting protein:** A *cis*-acting protein has the extraordinary property of acting only on the molecule of DNA from which it was expressed.

***cis*-configuration:** Two sites on the same molecule of DNA.

***cis*-dominant mutation:** A mutation in the operator that affects regulation of genes of an operator only under *cis* arrangement.

***cis*-elements:** The regions of non-coding DNA which regulate the transcription of neighbouring genes. *cis*-regulatory elements (CREs) are vital components of genetic regulatory networks, which in turn control morphogenesis, the development of anatomy, and other aspects of embryonic development, studied in evolutionary developmental biology. *Aka* *CIS*-ACTING ELEMENTS (CAEs), *CIS*-REGULATORY MODULES (CRMs) *or* *CIS*-REGULATORY ELEMENTS (CREs).

***cis*-genics:** Transfer of gene(s) from one plant line to another using recombinant DNA technology – but only using sequences from sexually compatible species.

***cis*-Golgi network (CGN):** A network of interconnected cisternae and tubules that receives vesicles from the endoplasmic reticulum and transfers material to the *cis* face of the Golgi apparatus.

***cis*-heterozygote:** This term is synonymously used for COUPLING LINKAGE (+ +/ *a b*). However, this is more appropriate term for a heterozygote in which two dominant genes are contributed by one parent and their corresponding recessive alleles by the other parent. *Compare with* *TRANS*-HETEROZYGOTE (e.g. + *b*/ a +).

***cis*-inactivation:** It involves inactivation (or silencing), frequently by methylation, of multiple linked copies of transgenes at a single locus, irrespective of the copy number, which can be arranged as inverted/ directed repeats. These are the results of transformation techniques, particularly direct gene transfer methods.

***cis*-position pseudoallele:** The location of two closely linked alleles in one chromosome with the two normal alleles in the homologous chromosome.

***cis*-position:** When two mutant alleles or elements are located in the same DNA strand.

***cis*-regulation:** Refers to the regulation with respect to when and where a gene is transcribed by DNA sequences that lie to either side (5′ or 3′) of the gene or within the introns of the gene.

***cis*-regulatory modules (CRMs):** *Aka* *CIS*-REGULATORY DOMAINS (CRDs) or CIS-REGULATORY ELEMENTS (CREs) that can be divided into three classes: enhancers, insulators and silencers.

***cis*-RNAs:** The *cis*-acting small RNAs that direct DNA methylation of genomic DNA homologous to a transcript that is cleaved by a miRNA.

***cis-trans* complementation test:** A genetic analysis that tests whether two mutations lie in the same or different genes, which can also provide information on dominance and recessiveness. The test involves introducing the two mutated genes into a single cell, for example, by introducing an F′ plasmid carrying one mutated gene into a recipient bacterium with a chromosomal copy of the second gene. The phenotypic distinction is referred to as the POSITION EFFECT. *Aka* *CIS-TRANS* TEST.

***cis*-transcriptional gene silencing (*cis*-TGS):** Transcriptional gene silencing occurrence affecting single or multiple copies inserted at one locus (i.e. it does not require the presence of homologous sequences in the genome). *Compare with* *TRANS*-TRANSCRIPTIONAL GENE SILENCING (*TRANS*-TGS).

cisgenesis: the genetic modification of a recipient plant with a natural gene from a crossable – sexually compatible – plant. Such a gene includes its introns and is flanked by its native promoter and terminator in the normal sense orientation. Cisgenic plants can harbour one or more cisgenes, but they do not contain any transgenes.

cisterna (pl. cisternae): Compressed membrane-bounded compartment, as commonly found in the endoplasmic reticulum or Golgi apparatus.

cistrome: The term refers to the set of *CIS*-ACTING TARGETS of a *TRANS*-ACTING FACTOR on a genome-wide scale, *aka* the *in vivo* genome-wide location of transcription factors or histone modifications.

cistron: Coined by S. Benzer in 1955, a genetic unit that encodes single polypeptide. Cistron is a nucleotide sequence in DNA specifying a single genetic function as defined by the *CIS-TRANS* COMPLEMENTATION TEST. It is also defined as a nucleotide sequence coding for a single polypeptide or ‘a gene’. The term cistron became common instead of ambiguous term ‘gene’. Later, the term gene again become popular, and cistron is being used less. Thus, terms gene and cistron are now considered equivalent. Two mutations are said to belong to the same cistron if they do not reveal complementation in *cis-trans* test or in a simple *trans* complementation test. A cistron codes for a specific gene product, either a protein or an RNA. Cistron is a sequence of bases in DNA that specifies one polypeptide, including the relevant translational start (initiation) and stop (termination) codons.

citrate synthase: The enzyme used (by plants) to synthesize (create) citric acid.

***citrate synthase* gene:** A bacterial gene utilized by certain bacteria (*Pseudomonas*) to code for (i.e. cause to be produced by bacterium possessing that gene) the enzyme known as citrate synthase. That enzyme is used to synthesize (create) citric acid. In 1996, L. Herrera-Estrella discovered that inserting the *CSb* gene from *Pseudomonas aeruginosa* into certain plants caused those plants to produce up to ten times more citrate in their roots and to release up to four times more citric acid from those roots into the surrounding soil (thus decreasing aluminium toxicity by chemically ‘binding’ aluminium ions that are present in some soils). Such soil aluminium, which slows plant growth and decreases crop yields, is present to a certain degree in approximately one-third of the Earth’s arable land (e.g. in the country of Colombia, it affects 70%).

citric acid: A tricarboxylic acid present naturally in plants, especially citrus fruits. It is used as a flavouring agent, an antioxidant in foods, an animal feed ingredient and a sequestering agent. The commercially produced form of citric acid melts at 153°C (307°F). Some plants naturally release citric acid from their roots into the surrounding soil so that it can chemically ‘bind’ aluminium ions that are present in some soils. For example, 70% of the agricultural land in the country of Colombia possesses harmful amounts of aluminium to damage crops. Maize yields are reduced up to 80% by such aluminium in soils. Soybeans, cotton and field bean yields are also reduced.

citric acid cycle: An enzyme system that converts pyruvic acid to carbon dioxide in the presence of oxygen with the release of energy in the form of ATP molecules. Its central involvement in many biochemical pathways indicates that it was one of the earliest established components of cellular metabolism and that it may have originated abiogenically. In eukaryotic cells, it occurs in the mitochondria. *Aka* KREBS CYCLE (named after discoverer, H.A. Krebs in 1937) *or* TRICARBOXYLIC ACID CYCLE (TCA CYCLE).

citrus tristeza virus (CTV): One of the most important citrus pathogens. It causes the death of millions of trees grafted on sour orange (*Citrus aurantium*); however, this rootstock is very well adapted to the Mediterranean and semiarid conditions. Some species confer resistance, for example, *Poncirus trifoliata*.

cl repressor: One of the regulatory proteins responsible for controlling expression of genes.

CL3 laboratory: Laboratory classified to have containment level 3. This is required for laboratories that work with high-risk biological agents, including human pathogens that have the potential to be transmitted via inhalation.

clade: A branch of the evolutionary tree that contains all the species descended from a single common ancestor (and therefore having similar features). A clade is always monophyletic on a phylogenetic tree. *Aka* MONOPHYLETIC GROUP *or* NATURAL GROUP. *See* MONOPHYLETIC.

cladistic species concept: The concept of species, according to which a species is a lineage of populations between two phylogenetic branch points (or speciation events). *Compare with* BIOLOGICAL SPECIES CONCEPT, ECOLOGICAL SPECIES CONCEPT, PHENETIC SPECIES CONCEPT *and* RECOGNITION SPECIES CONCEPT.

cladistics: Initially popularized by W. Hennig's 1950 book *Phylogenetic Systematics*, cladistics is a procedure of classifying organisms on the basis of common ancestry and the branching of the evolutionary family tree. Organisms that share common ancestors (and therefore have similar features) are grouped into taxonomic groups called CLADES. Cladistics can also be used to predict properties of yet-to-be discovered organisms.

cladode: A photosynthetic branch or stem, often leaf-like and usually having foliage leaves either absent or much reduced. *Aka* CLADOPHYLL. *Compare with* PHYLLODE.

cladogenesis: An evolutionary process whereby one species splits into two or more species. *Aka* TRUE SPECIATION. *Compare with* ANAGENESIS.

cladogenetic: Evolutionary change occurring during speciation events. Usually applied to fossils.

cladogenetic asymmetry: The 'nowhere to go but up' process whereby a group (clade) originates at a character state (e.g. size) that is physically restricted to expansion mainly in one direction (e.g. a mammalian only gets so small before encountering metabolic problems); the descendent radiation is therefore asymmetric in that character state (e.g. the larger size). The resulting trend is an increase in variance.

cladogram: A branching diagram that illustrates hypotheses about the evolutionary relationships among groups of organisms. Cladograms can be considered as a special type of phylogenetic tree that concentrates on the order in which different groups branched off from their common ancestors. A cladogram branches like a family tree, with the most closely related species on adjacent branches.

cladophyll: *See* CLADODE.

claim: In patenting biological material, a complete and exact description that defines the scope of an invention.

clamp connection: A bridge or clasp protrusion found at the septa of hyphae in basidiomycetous fungi and linked with cell division.

clasping: To clutch parts together.

class: In biological classification, class (Latin: classis) is a taxonomic rank, as well as a taxonomic unit, a *taxon*, in that rank. It is a group of related taxonomic orders. Other well-known ranks in descending order of size are life, domain, kingdom, phylum, order, family, genus and species, with class fitting between phylum and order.

class discovery: Most early expression profiling experiments, and many current ones, have the form that is known as class discovery. A popular approach to class discovery encompasses grouping similar genes or samples together. The simplest form of class discovery would be to separate out all the genes that altered by more than a certain amount between two experimental conditions. This approach helps in GENE ANNOTATION.

class I elements: Intrinsically duplicative elements that transpose using an RNA intermediate, in which a

DNA copy of the element transcript is integrated into a new position.

class II elements: The elements that transpose via a 'cut and paste' mechanism, in which the element is excised and then reintegrated at a new position.

class of seed and seed crop: Refers to the generations of pedigreed seed and seed crops, such as breeder seed, select seed, foundation seed, registered seed and certified seed, which have met the standards prescribed by recognized seed and seed crop certification agencies.

class switch recombination: A mechanism that changes the class or isotype of antibody produced by an activated B cell. This does not change the affinity towards an antigen but instead allows for interaction with different effector molecules.

class switching: The change in the expression of the C region of an immunoglobulin heavy chain during lymphocyte differentiation from making one class of immunoglobulin molecule (e.g. IgM) to making another class of immunoglobulin molecule (e.g. IgG) that many B cells undergo during the course of an immune response.

class-based programming: A style of OBJECT-ORIENTED PROGRAMMING (OOP) in which inheritance occurs via defining 'classes' of objects instead of via the objects alone. *Aka* CLASS-ORIENTATION PROGRAMMING. *Compare with* PROTOTYPE-BASED PROGRAMMING.

classes or class limits: The subdivisions of a numerical feature (or the widths of the classes) when it is exhibited in a frequency table or graph (an example is ages by decades).

classical breeding: Classical plant breeding uses deliberate interbreeding (crossing) of closely or distantly related individuals to produce new crop varieties or lines with desirable properties. Plants are crossbred to introduce traits/genes from one variety or line into a new genetic background.

classical genetics: Study of genetics on an organismal rather than cellular level. *Aka* MENDELIAN GENETICS *or* TRANSMISSION GENETICS.

classical model of population structure: This model assumes that a typical diploid individual is homozygous at most and heterozygous at a few loci. This model is based on laboratory mutants whose fitness is much lower than that of normal alleles and considers a genotype homozygous for normal alleles as the best.

classical pathway: A pathway for activating the complement system that is initiated by immunoglobulin IgG or IgM antibodies bound to the surface of a microbe.

classification and regression tree (CART) analysis: A multivariate method used to detect significant relationships among variables, which are then used to develop predictive models for classifying future subjects.

classification of a disease: The classification of a disease can be done on the basis of the pathogen, the host, the age of the host, the name of the disease, a plant part, symptoms, location, causal agent, geography or by order of importance within a given location. Taxonomic classification is the systematic ordering of plants and animals.

classification: Important in the study of organisms. Cladistics is a method of classification based on common ancestry. The Linnaean system is based on a simple hierarchical structure.

clathrate: Shaped like a net or lattice (framework). Pierced with apertures, as with a cage.

clathrin: Protein that accumulates into a polyhedral cage on the cytosolic side of a membrane so as to make a clathrin-coated pit, which buds off by endocytosis to form an intracellular clathrin-coated vesicle.

clathrin-coated pit: Area of plasma membrane of animal cells that is coated with the protein clathrin on its cytosolic face. Such regions are continually formed and budded off by endocytosis to form intracellular clathrin-coated vesicles containing extracellular fluid and the materials dissolved in it.

claw: (i) A narrow, stalk-like, basal portion of a petal, sepal or bract. (ii) In Melaleuca, the united portion of a stamen bundle.

***ClB* method:** A technique developed by H.J. Muller in 1920 to rapidly screen adult fruit flies for detection of sex-linked recessive lethal mutations. The *CIB* chromosome carries a recessive lethal (*l*), a semi-dominant marker *Bar* (*B*) and inversions, which act as crossover suppressor (*C*). *See* BASC METHOD.

clean gene technology: Refers to the methods utilized in the production of marker-free transgenic plants. Several strategies have been developed to remove SELECTABLE MARKER GENES (SMGs) from GM products while retaining the transgenes of interest involves integration of an SMG and gene of interest (GOI) on unlinked genomic loci during transformation so that the SMG can be segregated away from the GOI, integration of an SMG and GOI on the same locus. Molecular tools are used for SMG removal, using plant-derived marker genes, and using non-antibiotic resistance genes as markers are considered safe SMGs. Marker-free transformation is the most useful and desirable approach used for obtaining marker-free transgenics.

clean transgenic plants: Transgenic plants that contain the desired GOI but lack the SMG used in its production are termed 'clean'.

cleanroom software engineering: A software improvement procedure intended to produce software with certifiable level of reliability. The cleanroom process was originally developed by H. Mills and several of his colleagues, including A. Hevner in 1986 at International Business Machines (IBM). The focus of the cleanroom process is on defect prevention, rather than defect removal.

cleavage and polyadenylation specificity factor (CPSF): Protein involved in the cleavage of the 3′ signalling region from a newly synthesized PRE-MESSENGER RNA (PRE-mRNA) molecule in the process of gene transcription.

cleavage division: One of the early divisions of the zygote that leads to the formation of an embryo in animals.

cleavage map: *See* RESTRICTION MAP.

cleavage or translational repression: miRNAs play important regulatory roles in animals and plants by targeting mRNAs for cleavage or translational repression. For example, microRNA398 (miR398) is a conserved miRNA of plants that targets two of the three copper/zinc superoxide dismutases (SOD) of Arabidopsis (CSD1 and CSD2) by triggering cleavage or inhibiting translation of their mRNAs.

cleavage sites: The sites at which a restriction enzyme can cleave DNA. For a particular restriction endonuclease, there may be a number of cleavage sites in DNA from different organisms. For example, for *Bam*HI isolated from *Bacillus amyloliquefaciens* H, in DNA of λ, Ad2 and SV40, there are 5, 3 and 1 sites, respectively.

cleavage stimulation factor (CSF): A heterotrimeric protein, made up of the proteins CSTF1 (55kDa), CSTF2 (64kDa) and CSTF3 (77kDa), totalling about 200 kDa. It is involved in the cleavage of the 3′ signalling region from a newly synthesized pre-mRNA molecule.

cleavage, mechanism of: Physical splitting of a cell into two is called cleavage. First, there is nuclear division (karyokinesis) and then there is cytoplasmic division (cytokinesis). If nucleus is removed after mitosis, cytoplasm still divides. If nucleus is removed before mitosis, cytokinesis does not occur. This suggests that the division of the nucleus in some way causes division of the cytoplasm. If the nucleus is removed during mitosis, there is a critical time at which even if the nucleus is removed, cytokinesis occurs. Mitotic apparatus acts like a seed. Cleavage begins with a cell that has a greatly bloated volume and a distorted nucleus-to-cytoplasm ratio in comparison with cells of the adult body. Genes are replicated. Genes and cytoplasm are partitioned during this process.

cleave: Refers to producing in DNA a double-stranded cut with the help of a restriction endonuclease.

cleaved amplified polymorphic sequences (CAPS): PCR-amplified DNA that is digested with restriction endonucleases to discover polymorphisms in restriction sites. *See* SEQUENCE TAGGED SITE (STS), EXPRESSED SEQUENCE TAGS (ESTs) *or* SEQUENCE CHARACTERIZED AMPLIFIED REGION (SCAR).

cleft grafting: Refers to split grafting cleistogamous (a self-pollinated) plant that produces inconspicuous flowers that never open (e.g. wheat or tomato).

cleistogamous: Having flowers that self-pollinate and never open fully, or that self-pollinate before opening. *Compare with* CHASMOGAMOUS.

click (clk) chemistry: Click chemistry is a two-step process that practises quantitative chemical reactions of alkyne and azide moieties to produce covalent carbon-heteroatom bonds between biochemical species. The reaction uses copper(I) as a catalyst and constitutes a 1,2,3-triazole between an azide and terminal alkyne. The methodology is dependable and stable, which brands it a perfect oligonucleotide labelling method.

click (clk) DNA: Editing efficiency can be enhanced by speedily screening clkDNA oligonucleotides with numerous modifications, including repair-evading substitutions.

click (clk) editing: J. Ferreira da Silva *et al.* (2024) establish click editing, a genome writing platform that combines the beneficial properties of DNA-dependent polymerases (DDPs) with RNA-programmable nickases to enable the setting up of a series of edits, including substitutions, insertions and deletions. Through iterative optimization of the modular components of click editors (CEs) and their clkDNAs, they establish the capability to install precise genome edits with minimal indels in diverse immortalized human cell types and primary fibroblasts with accurate editing proficiencies of up to around 30%. Click editing is thus an accurate and adaptable genome editing approach for diverse biological applications.

click (clk) editors (CEs): CEs influence the 'click'-like bioconjugation capability of histidine-hydrophobic residue-histidine (HUH) endonucleases with single-stranded DNA substrates to covalently rein 'click DNA' (clkDNA) templates encoding user-specifiable edits at targeted genomic loci.

client: A piece of computer hardware or software that accesses a service made available by a server. The server is often (but not always) on another computer system, in which case the client accesses the service by way of a network. The term applies to the role that programs or devices play in the client-server model.

climate chamber: A successional community of plants capable of optimal development under the prevailing environment and in dynamic equilibrium with its environment. It uses phytotron system.

climate: Statistics of temperature, humidity, atmospheric pressure, wind speed, precipitation, atmospheric particle count and other meteorological elemental measurements in a given place or region over long periods. Takes into consideration the variations of cold and heat, dryness and moisture, calm and wind in a given region or country. In general, the combined result of all the meteorological phenomena of any region affects its crop or vegetable production. *Compare with* WEATHER *and* ENVIRONMENT.

climate change: While the term generally refers to a significant change from one climatic condition to another, 'climate change' has been used interchangeably with the term 'global warming' to refer to broad, long-term changes in the earth's climate and weather patterns due to long-term changes in air, soil or water temperature; precipitation regimes; wind speed; or other climate-related factors. The main cause of climate change is burning fossil fuels such as oil, gas and coal. When burnt, fossil fuels release carbon dioxide into the air, causing the planet to heat up. Agriculture is very sensitive to weather and climate. It also relies heavily on land, water and other natural resources that climate affects. Climate changes (such as in temperature, precipitation and frost timing) could lengthen the growing season or allow different crops to be grown in some regions and can make conditions better or worse for growing crops in different regions.

climber: A plant growing more or less erect by leaning on or twining around another structure for support, or by clinging with tendrils.

climbing plants: A creeper or other plant that willingly grows up a support (climbing rose), twines up a slender support (hop, honeysuckle) or grips the support by special organs such as adventitious aerial roots (e.g. English ivy, poison ivy, trumpet creeper), tendrils, hook-tipped leaves (e.g. gloriosa lily, rattan) or stipular thorns (catabrier); some climbing plants when not supported become trailing plants (e.g. English ivy).

clinal propagation: Asexual reproduction of plants that are considered to be genetically uniform and originated from a single individual or explant.

clinal selection: Selection that changes steadily along a geographical slope.

cline: (Adj. clinal): (i) A continuous morphological dissimilarity in form within a species or sometimes between two species. (ii) A spatial incline in any measurable characteristic – for example, allele frequency or the mean of a quantitative trait. (iii) An environmental incline and a corresponding phenotypic incline in a population of plants; when clines are assessed by provenance tests, they are often found to have a genetic basis.

clinical heterogeneity: Different clinical presentations within the same disorder.

clinical performance: A term encompassing clinical sensitivity, clinical specificity, penetrance and gene/environmental modifiers.

clinical sensitivity: The proportion of individuals with a specified clinical disorder whose test values indicate that the disorder is present (e.g. the mutation associated with the disorder is identified). Alternatively, the proportion of individuals who do not have a specified clinical disorder and whose test results indicate that the disorder is not present. The following hypothetical example illustrates CLINICAL SENSITIVITY and SPECIFICITY: A clinician determines that 100 patients have recessively inherited disease C, based on symptoms, family history and ethnicity. Upon genetic testing, 60 of the patients are shown to have two mutations in gene *D*. Thus, the clinical sensitivity is 60%. The remaining 40 patients may have only one or no mutations in *D*. The same clinician sees 100 patients who do not have disease C. When tested, five of the individuals are shown to have two mutations in *D*. Thus, the clinical specificity is 95%.

clinical specificity: The proportion of individuals who do not have a specified clinical disorder.

clinical trial: Medical research undertaken with informed and consenting human subjects in a controlled environment. The intent of a clinical trial is for the sponsoring company or research institution to gather information on the safety and effectiveness of new drugs or therapies before seeking approval of a procedure or product for use by the public. Though almost always clinical trial is utilized to represent a randomized experiment, a clinical trial refers to any type of future study of human subjects in which therapies or clinical strategies are compared. Treatments may be allocated to individual patients or to groups, the latter including cluster randomized trials. For a RANDOMIZED CLINICAL TRIAL or RANDOMIZED CONTROLLED TRIAL (RCT), the choice and timing of treatments is beyond the control of the physician and patient but is (usually) set in advance by a randomization device. RCTs do not need representative patients but do require representative treatment effects. The RCT is the gold standard for establishing connection.

clinical trials, phases in: Biotechnology companies typically engage in clinical trials to assess the efficacy and safety of new therapies and to answer scientific questions. Typically, there are three phases during a clinical trial. PHASE I is designed to evaluate the safety of the product in humans. PHASE II analyses the influences of dose escalation. PHASE III conclusively estimates the clinical efficacy of the product.

clinical utility: The value of a test in diagnosing/ruling out a disease in suggesting treatment or prevention strategies and in evaluating risks and benefits associated with the test.

clinical validity: The ability of a test to distinguish affected and unaffected populations, including a determination of the probability of being affected. The clinical sensitivity, specificity and predictive value of a test.

clock-controlled genes: Genes are linked to the circadian oscillator. Circadian output genes are linked to the oscillator by E boxes, RORE and/or D-elements. Per1 inhibits the activity of the Clock (C) or Npas2 (N) and Bmal1 (B) heterodimers.

clonal expansion: Refers to the population of cells produced from a single cell. The term is identical with clone but is utilized in particular context. For example, a cell with a particular chromosomal abnormality can, by clonal expansion, produce a population of the same type of cell within the organism.

clonal multiplication: The clonal multiplication reproduces clones, which contain all the genetic information of the parent tree. Clonal multiplication is done in four stages (I–IV). *See* CLONAL MULTIPLICATION STAGE I, CLONAL MULTIPLICATION STAGE II, CLONAL MULTIPLICATION STAGE III *and* CLONAL MULTIPLICATION STAGE IV.

clonal multiplication stage I: A step during *in vitro* propagation characterized by the establishment of an aseptic tissue culture of a plant.

clonal multiplication stage II: A step during *in vitro* propagation characterized by the rapid numerical increase of organs or other structures.

clonal multiplication stage III: A step during *in vitro* propagation characterized by preparation of propagule for successful transfer to soil – a process involving rooting of shoot cuttings, hardening of plants and initiating the change from the heterotrophic to autotrophic state.

clonal multiplication stage IV: A step during *in vitro* propagation characterized by the establishment in soil of a tissue culture derived plant, either after undergoing a stage III pretransplant treatment or, in certain species, after the direct transfer of plants from stage I into soil.

clonal propagation: Asexual reproduction of plants that are considered to be genetically uniform and originated from a single individual or explant. In vegetative propagated plants, the propagation is generally done through cutting, budding, grafting, layering, tubers, sets, bulbs, corms and micropropagation. This is known as CLONAL PROPAGATION as resulting population is a clone (i.e. a group of genetically similar plants).

clonal selection: A scientific theory in immunology that explains the functions of cells (lymphocytes) of the immune system in response to specific antigens invading the body. The theory has become the widely accepted model for how the immune system responds to infection and how certain types of B and T lymphocytes are selected for destruction of specific antigens.

clonal selection theory: A theory that elucidates how the adaptive immune system can reply to millions of different antigens in a highly specific way. From a population of lymphocytes with a vast collection of randomly generated antigen specificities, a given foreign antigen selects and activates only those cells that possess the corresponding antigen specificity.

clonal test: Evaluation of genotypes by comparing clones in a plantation.

clone: (i) A cell population derived from a single progenitor cell or organism. (ii) A clone comprises two mitotic cells with the same gain or structural aberration or three cells with the same chromosome loss. (iii) A genetically engineered exact copy of biological material such as a DNA segment (e.g. a gene or other region), a whole cell or a complete organism. (iv) A group of genetically identical individuals. Clones are the result of asexually reproduction. For example, bacteria, viruses and also some metazoan, such as aphids, rotifera and cladocera (water fleas), can reproduce clonally. In a clone, all the molecules/cells or organisms are true copies of the parent. This kind of population, being genetically alike, does not exhibit much phenotypic variation and is exploited in the basic and applied research.

clone bank: *See* GENOMIC LIBRARY.

clone by clone approach for genome sequencing: This method involves splitting the genome into relatively large chunks, called clones. Each clone is then cut into smaller overlapping pieces and sequenced. The pieces and the overlaps are then assembled to recreate the sequence of the whole clone. Researchers in human genome sequence consortium combined approach this with bacterial artificial chromosomes (BACs)-based clone-by-clone approach.

clone coverage: The degree to which a genome is represented in clones of, for example, a BAC library or plasmid library. Clone coverage is the amount of physical DNA coverage of the target genome.

clone library: A collection of clones that contain a number of different genes. *See* GENOMIC LIBRARY.

clone overlap single-nucleotide polymorphism (clone overlap SNP): Any single-nucleotide polymorphism perceived by sequencing the overlaps of two (or more) BACs and comparison of the resultant sequences.

clone variety: Refers to a crop variety that consists of individuals originating from a single clonal genotype (MONOGENOTYPIC), such as in potato, cassava, sweet potato, rubber, mango, avocado, apple, pear, banana, pineapple, strawberry, brambles, grape, peach, cherry, almond, citrus, date, Jerusalem, artichoke, yams, black pepper, olive, fig, pistachio or edible aroids.

clone-based map: Any physical map of a genome that is created on the alignment and sequencing of overlapping BAC clones.

cloned DNA in *B. subtilis*, fate of: When DNA from an organism, which has little or no homology with *B. subtilis*, is cloned in *B. subtilis*, the chimeric plasmid appears to be stable. The extent of homology of pCD1 with *B. subtilis* can be greatly reduced by excising a 1.5 MDAl *Bgl*II fragment carrying the *thy* gene. Thus, pCD2 was formed.

cloned gene: A gene that has been inserted into a cloning vector.

cloned gene analysis: Cloned genes are analysed through ELECTROPHORESIS, RESTRICTION ENZYME MAPPING, SOUTHERN BLOTTING, DNA SEQUENCING, RNA SEQUENCING *or* POLYMERASE CHAIN REACTION (PCR). *See details as separate terms.*

cloned gene analysis through RNA sequencing: It is sometimes necessary and more useful to sequence RNA molecules. Various methods used for RNA sequencing include GENERATING DNA THROUGH REVERSE TRANSCRIPTION, ION EXCHANGE CHROMATOGRAPHY, RNA SEQUENCING WITH RADIOACTIVE CHAIN-TERMINATING RIBONUCLEOTIDES, NEXT-GENERATION SEQUENCING TECHNOLOGIES FOR RNA SEQUENCING, GENOME ALIGNMENT AND READS ASSEMBLY *and* DIRECT RNA SEQUENCING. *See details as separate terms.*

cloning: (i) The process to vegetatively propagate a certain crop and/or plant, creating an organism that carries the same genes as the original. (ii) In molecular genetics, the cloning of DNA molecules from prokaryotic or eukaryotic sources as part of a bacterial plasmid or phage replicon biomes of spent medium. (iii) The process of producing many copies of a type of cell. Because reproduction is asexual, the progeny is genetically identical to the original ancestor. For example, the vegetative production of new plants or the propagation of DNA molecules by insertion into plasmids. (iv) In biotechnology, cloning refers to the artificial creation of copies of cells, DNA fragments or other biomolecules by various laboratory techniques. *See* MOLECULAR CLONING.

cloning DNA: Specialized DNA technology is employed to produce multiple, exact copies of a single gene or other segment of DNA to obtain enough material for further study. The resulting cloned (copied) collections of DNA molecules are called CLONE LIBRARIES.

cloning efficiency: The percentage of cells plated (seeded, inoculated) that form a clone. One must be certain that the colonies formed arose from single cells in order to properly use this term. *See* COLONY-FORMING EFFICIENCY.

cloning in a supersize genome: The large size and large amounts of repetitive DNA and the presence of three homologous genomes result in major difficulties during chromosome walking and are thus challenge for map-based cloning. Due to the complexity and the size of the supersize genome (such as allopolyploid bread wheat), an alternative is to use diploid or tetraploid wheat genomes that are closely related to the genomes of modern hexaploid wheat. For physical mapping, lower-ploidy wheat species are used as models for the three different hexaploid wheat subgenomes A, B and D.

cloning in *Bacillus subtilis*: *Bacillus* species are Gram-positive and generally obligate aerobes compared with *E. coli*, which is a Gram-negative facultative anaerobe. *Bacillus* can sporulate and therefore cannot be used as a model for prokaryotic differentiation. *Bacillus* species are widely used in fermentation industry, particularly for the production of exozymes. They can be tailored to secrete products of cloned eukaryotic genes. It is an extremely safe organism. Cloning in *Bacillus* is done with bifunctional vectors that replicate in both *E. coli* and *B. subtilis*. Such bifunctional vectors can be transferred by replica plating *E. coli* containing them on to a lawn of competent *B. subtilis* cells.

cloning in streptomyces: Streptomyces is the only Gram-positive bacteria other than *B. subtilis* in which there has been extensive gene manipulation. This species produces most of the known antibiotics. Three vehicles are available: (i) a low copy number, narrow host range plasmid; (ii) a high copy number, broad range host range plasmid; and (iii) a phage vector. Streptomyces promoters do not appear to function in *E. coli*, whereas *E. coli* promoters and *B. subtilis* promoters permit transcription in *S. lividans*.

cloning in yeast: By incorporating eukaryotic DNA sequences into various prokaryotes and by altering *in vivo* by bacterial genetic techniques and *in vitro* by specific enzyme modifications, it has been possible to study effects of induced changes on function and expression of eukaryotic genes. Eukaryotes have certain systems like: (i) ATP-generating systems that are localized to mitochondria; (ii) DNA associated with histones; (iii) mitosis and meiosis; and (iv) obligate differentiation of cells that are absent in prokaryotes, which have been studied. Yeast spheroplasts (wall-less cells) were fused with polyethylene glycol in the presence of DNA and $CaCl_2$. Then spheroplasts were allowed to regenerate walls. The transforming DNA was plasmid pYeLeu10. This hybrid plasmid was composed of ColE1 and yeast DNA containing the *Leu2*$^+$ gene. Spheroplasts from a stable *Leu2*$^-$ were transformed.

cloning of human telomere: Although eukaryotic chromosomes vary in size over five orders of magnitude and are constituents of diverse genetic systems, the fundamental features of their telomeres appear to be almost completely conserved. This has enabled molecular cloning of human telomere in yeast, which contributes reagents for mapping mammalian chromosomes and that they provide one set of elements for construction of artificial mammalian chromosomes.

cloning of quantitative trait loci: Refers to the identification of the DNA sequences (coding or non-coding) responsible for QTLs. Classical methods include INSERTIONAL MUTAGENESIS and POSITIONAL/MAP-BASED CLONING. Reverse genetics strategies of QTL cloning include ECO TILLING and GENOMIC METHODS. GENOMIC METHODS FOR CLONING QTLS ARE RECOMBINATION FINE MAPPING, GENE EXPRESSION VARIATION, HYBRIDIZATION POLYMORPHISM, ASSOCIATION TESTING – LD MAPPING, QUANTITATIVE COMPLEMENTATION and TRANSGENIC ALLELIC COMPLEMENTATION. *See details as separate terms.*

cloning sequences adjacent to the site of insertion: Once a recombinant plasmid has been integrated into the chromosome it is relatively easy to clone adjacent sequences.

cloning site: Restriction site, usually exclusive in a vector, where DNA can be inserted (cloned).

cloning technique: A method to produce genetically identical copies of a biological entity.

cloning vector: A cloning vector is a small piece of DNA that is skilled to undergo replication in a host organism, into which a foreign DNA fragment (or gene) can be introduced to construct a recombinant DNA molecule. This is carried out by treating the vehicle and the foreign DNA with a restriction enzyme that creates the same overhang, then ligating the fragments together. There are many types of cloning vectors, such as plasmids, cosmids, BACs and YEAST ARTIFICIAL CHROMOSOMES (YACs). Genetically engineered plasmids (such as pBR322) and bacteriophages (such as phage λ) are most commonly used for this purpose. Cloning vectors, which are very similar to expression vectors, involve the same process of introducing a new gene into a plasmid, but the plasmid is then added to bacteria for replication purposes. *Aka* VECTOR. *See* EXPRESSION VECTORS.

cloning vectors, common features of: (i) Main purpose of expression vectors in these vehicles is the controlled expression of a particular gene inside a convenient host organism (e.g. *E. coli*). (ii) For control of gene expression, it is usually desirable to insert the target DNA into a site that is under the control of a particular promoter. Some commonly used promoters include T7 promoters, lac promoters (*bla* promoter) and cauliflower mosaic virus's 35S promoter (for plant vectors). (iii) To allow for convenient and favourable insertions, most cloning vectors have had nearly all their restriction sites engineered out of them and a synthetic MULTIPLE CLONING SITE (MCS) inserted that contains many restriction sites. MCSs allow for insertions of DNA into the vector to be targeted and possibly directed in a chosen orientation. (iv) A selectable marker, such as an antibiotic resistance (e.g. beta-lactamase), is often supported by the vector to permit the selection of positively transformed cells. (v) All plasmids must carry a functional origin of replication. (vi) Some other possible features present in cloning vectors are *vir* genes for plant transformation, integrase sites for chromosomal insertion, lacZα fragment for α complementation and blue-white selection, and/or reporter genes in setting with and flanking the MCS to facilitate the production of recombinant proteins (e.g. fused to the GREEN FLUORESCENT PROTEIN [GFP] or to the glutathione S-transferase). (vii) Many general-purpose vectors such as pUC19 generally include a system for perceiving the presence of a cloned DNA fragment, based on the loss of an easily scored phenotype. The most commonly used is the gene coding for *E. coli* β-galactosidase, whose integrity can easily be recognized by the ability of the enzyme it codes for to hydrolyse the soluble, colourless substrate X-gal (5-br omo-4-chloro-3-indolyl-beta-d-galactoside) into an insoluble, blue product (5,5′-dibromo-4,4′-dichloro indigo). Cloning a fragment of DNA within the vector-based gene encoding the β-galactosidase thwarts the production of an active enzyme. If X-gal is included in the selective agar plates, transformant colonies are generally blue in the case of a vector with no inserted DNA and white in the case of a vector containing a fragment of cloned DNA.

cloning yeast telomeres, linear vectors for: Only a linear plasmid can model the eukaryote chromosome. Linear vectors have been used to clone yeast telomeres. Linear vector was constructed from two components: (i) pSZ213, a yeast-replicating plasmid; and (ii) a linear rRNA-encoding plasmid from *Tetrahymena*. Telomeres from yeast cloned in linear vector were analysed. Three unusual features of this vector were noted: (i) ends have a large number of short 5′-CCCCAA-3′ repeat units; (ii) specific single-strand interruptions within the repeated sequences; and (iii) a cross-linked terminus.

cloning, types of: There are two recognized forms of cloning related to humans – REPRODUCTIVE CLONING and THERAPEUTIC CLONING. *See details as separate terms.*

close breeding: *See* INBREEDING.

closed chromatin: Highly condensed chromatin whose regulatory sequences of the genes are not accessible to transcriptional machinery. Only 50% of DNA in nucleosome is protected by digestion with *Staphylococcus* nuclease regardless of concentration of nuclease used. This is called 'closed' chromatin. Major effect of chromatin proteins is to restrict transcription by decreasing the number of available sites. *Aka* TIGHT CHROMATIN.

closed continuous culture: A type of cell suspension culture in which cells are added back to the culture. A continuous-type cell suspension culture in which inflow of the fresh medium is balanced by the outflow of equal volume of the used medium containing cells. Thereafter, the cells are mechanically separated from the medium and added back to the culture.

closed mating system: A system in which no outside individuals are permitted to introduce gametes into the population.

closed population: A group of interbreeding plants (occurring in a certain area or in an experimental design) or plants originating from one or more common ancestors, where there is no immigration of plants or pollen (i.e. no genetic input by forces other than mutation).

closed promoter complex: The initial complex formed between RNA polymerase and a promoter, in which the double helix is still completely base paired.

closed reading frame: The reading frame contains termination codons that prevent its translation into protein.

closed, partly opened and opened: For all alpha structures, describes the extent in which the hydrophobic core is screened by the comprising alpha-helices. Opened means there is space for at least one more helix to be easily attached to the core.

closed-ended questions: Questions that provide respondents with a fixed set of alternatives from which they have to choose the answer.

cloud computing: Shared pools of configurable computer system resources and higher-level services that can be rapidly provisioned with minimal management effort, often over the Internet. Cloud computing relies on sharing of resources to achieve coherence and economies of scale, similar to a public utility.

cloud infrastructure for microbial bioinformatics (CLIMB): UK's largest cloud-based system for the analysis of microbial genomes.

cloud robotics: A field of robotics that attempts to invoke cloud technologies, such as cloud computing, cloud storage and other Internet technologies cantered on the benefits of converged infrastructure and shared services for robotics. When connected to the cloud, robots can benefit from the powerful computation, storage and communication resources of modern data centre in the cloud, which can process and share information from various robots or agent (other machines, smart objects, humans, etc.). Humans can also delegate tasks to robots remotely through networks. Cloud computing technologies enable robot systems to be endowed with powerful capability while reducing costs through cloud technologies. Thus, it is possible to build lightweight, low-cost, smarter robots that have intelligent 'brain' in the cloud. The 'brain' consists of data centre, knowledge base, task planners, deep learning, information processing, environment models, communication support, etc.

cloverleaf tRNA structure: A convenient two-dimensional representation of the structure of a tRNA molecule.

CLUSTAL W: A fully automatic programme for global multiple alignments of DNA and protein sequences. It improves the sensitivity of progressive multiple sequence alignment through sequence weighting, position-specific gap penalties and weight matrix choices. This programme can create multiple alignments, manipulate existing alignments, perform profile analysis and create phylogenetic trees. The input file for CLUSTAL W is a file containing all sequences in one of the formats: NBRF/PIR, EMBL/SWISS Pnt, Clustal.

cluster: The grouping of similar objects in a multidimensional space. Clustering is used for creating new features that are abstractions of the prevailing features of those objects. The quality of the clustering relies crucially on the distance metric in the space. In bioinformatics, clustering is achieved on sequences, high-throughput expression and other experimental data. Clusters of partial or complete gene sequences can be engaged to identify the complete (contiguous) sequence and to better identify its function. Clustering expression data allows the researcher to determine patterns of coregulation in groups of genes.

cluster analysis: A statistical analysis method in which similar variances are grouped or clustered. The results of statistical calculations are frequently represented as dendrograms. In cross-breeding, the cluster analysis is employed for selecting most varied parents for crossing. It is an innovative method that determines a classification or taxonomy from multiple measures of a set of objects or subjects. Cluster analysis is a main task of exploratory data mining and a common technique for statistical data analysis, used in many fields, including machine learning, pattern recognition, image analysis, information retrieval, bioinformatics, data compression and computer graphics.

cluster analysis in phylogenetic tree construction: Cluster analysis is the simplest method for phylogenetic tree construction. The tree is built by a sequential clustering algorithm. In the matrix, the two OPERATIONAL TAXONOMIC UNITS (OTUs) with the smallest distance are clustered into a COMPOSITE OTU, which will be further treated as a new single OTU. A new matrix is created, where the distances of other OTUs to the composite OTU are calculated. This process is repeated until only two OTUs remain. These are clustered together, and the root is placed at half the calculated distance between the two clusters. The UNWEIGHTED PAIR GROUP METHOD WITH ARITHMETIC MEAN (UPGMA) calculates the distances to a composite OTU as the arithmetic mean of the distances between the constituent OTUs in each composite OTU. If a simple average of the distances between the composite OTUs is used, the method is called WEIGHTED PAIR GROUP METHOD WITH ARITHMETIC MEAN (WPGMA). *Aka* CLUSTERING.

cluster clone theory: L. Bogorad in 1975 put forth the cluster clone theory. A prokaryote having a single duplicated strand of DNA has several genes (say *a b c d e*). DNA strand breaks into two parts such that they have different genes (*a b c* and *d e*). Some of these genes (say *a b c*) were retained in the nucleus, and the others (say *d e*) were converted into mitochondria by acquisition of new membranes. Neither the method nor the source of these membranes is explained.

cluster investigation and virus epidemiology tool (CIVET): Software suite that works on CLIMB to generate a report for a set of genome sequences of interest, such as for an outbreak investigation.

cluster random sample: A two-stage sampling procedure in which the population is split into clusters, a random sample of clusters is selected and then random samples of subjects within the clusters are designated.

cluster sampling: Sampling by distributing the population into groups or clusters and drawing samples from only some groups.

clustered regularly interspaced short palindromic repeat (CRISPR): An important part of a prokaryotic adaptive immune system that uses short RNAs to guide the CRISPR-associated 9 (Cas9) nuclease to specific targets, which cleaves foreign DNA elements such as plasmids and phage genomes. A genomic locus in bacteria or archaea where protospacers and direct repeat sequences are arrayed in tandem. It is associated with adaptive immunity against invading phages and plasmids. This natural mechanism has been manipulated by researchers to develop genome-editing techniques. Programmable nucleases comprised of bacterially derived endonuclease (Cas9) and a single-guide RNA (sgRNA) are used to edit the base pairs of a gene.

clustered regularly interspaced short palindromic repeat (CRISPR)-associated transposase (CAST) systems: Samuel Sternberg of Columbia University in 2023 used CAST systems to engineering bacterial genomes. This protocol includes guidelines on available vectors, customization of guide RNAs and DNA payloads, selection of common delivery methods and genotypic analysis of integration events. The method allows the isolation of clonal strains containing a novel genomic integration event of interest within 1–2 weeks using available plasmid constructs and standard molecular biology techniques.

clustered regularly interspaced short palindromic repeat (CRISPR)-based gene drive: A powerful technology that promotes the inheritance of the gene drive tool itself via sexual reproduction and can, therefore, spread quickly through a population. Reversal drives can undo the effects of a prior drive.

clustered regularly interspaced short palindromic repeat (CRISPR)-Cas: A bacterial adaptive immune system comprising clustered regularly interspaced short palindromic repeats (CRISPRs) and CRISPR-associated (Cas) endonucleases that has been repurposed for genome editing.

clustered regularly interspaced short palindromic repeat (CRISPR)-Cas systems, classification of: The CRISPR-Cas systems are partitioned into two classes: (i) class 1 with multi-subunit effectors; and (ii) class 2 in which the effector consists of a single, large protein. The class 2 effectors originate from nucleases encoded by different mobile genetic elements (MGEs), whereas the origin of the class 1 effector complexes remains murky. Class I CRISPR associated systems are divided into three types: type I, III and IV. Class II CRISPR associated systems are also divided into three types: II, V and VI. *See details as separate terms.*

clustered regularly interspaced short palindromic repeat (CRISPR)-Cas system, type I: CRISPR-Cas systems encode multiple Cas proteins to multi-Cas effector machineries, which is termed the CRISPR-associated complex for antiviral defence (Cascade). Cascade was initially used for characterizing the type I-E CRISPR-Cas effector complex of *E. coli.*

clustered regularly interspaced short palindromic repeat (CRISPR)-Cas system, type II: CRISPR-Cas system that encodes cas9, cas1 and cas2 within the CRISPR-Cas loci, in addition to a tracrRNA, which is partially complementary to repeats in the CRISPR array.

clustered regularly interspaced short palindromic repeat (CRISPR)-Cas system, type III: Despite primary targeting of RNA, Type III CRISPR-Cas systems protect cells from viruses with DNA genomes and interfere with plasmid transformation, as long as viral or plasmid DNA is transcribed with the production of RNA molecules complementary to the crRNA spacers.

clustered regularly interspaced short palindromic repeat (CRISPR)-Cas system, type III-B: CRISPR-Cas is the strategy of an infected host to sacrifice itself to save the population from being infected. Type III CRISPR-Cas systems use highly diverse signal transduction mechanisms to achieve this objective. J.A. Steens and others in 2024 studied this sophisticated type III system from the myxobacterium *Haliangium ochraceum*, which caused a cascade of proteolytic cleavages. They found that target recognition triggered the allosteric activation of a caspase-like protease (SAVED-CHAT) that specifically cleaved and activated yet another caspase-like protease (PCaspase). The latter enzyme had a broad substrate specificity that resulted in dormancy or death of the infected host cell.

clustered regularly interspaced short palindromic repeat (CRISPR)-Cas systems, type IV: All known type IV CRISPR-Cas systems are encoded on plasmids, integrative and conjugative elements (ICEs) or prophages and are thought to contribute to competition between these elements, although the mechanistic details of their function remain unknown.

clustered regularly interspaced short palindromic repeat (CRISPR)-Cas systems, type V: Type V CRISPR-Cas systems are distinguished by a single RNA-guided RuvC domain-containing effector, Cas12. Cas12c, Cas12g, Cas12h and Cas12i. Cas12c, Cas12h and Cas12i subtypes have been identified and characterized. The distinct domain architectures and diverged RuvC sequences of uncharacterized Cas12 proteins suggest unexplored functional diversity. RNA-guided double-stranded DNA (dsDNA) interference activity has been demonstrated.

clustered regularly interspaced short palindromic repeat (CRISPR)-Cas systems, type VI: Type VI CRISPR-Cas systems contain a single RNA-guided effector protein, Cas13, that specifically targets RNA to protect bacteria from mobile genetic elements. Cas13 exhibits two distinct ribonuclease activities: one for pre-crRNA processing and another for *cis*- and *trans*-RNA-target cleavage.

clustered regularly interspaced short palindromic repeat (CRISPR) locus: A bacterial or archaeal DNA locus constituted of a CRISPR array, its corresponding Cas.

clustered regularly interspaced short palindromic repeat (CRISPR) RNA (crRNA): CRISPR RNA base pairs with tracrRNA to form a two-RNA structure that guides the Cas9 endonuclease to complementary DNA sites for cleavage.

clustered regularly interspaced short palindromic repeat (CRISPR)/CRISPR-associated systems (CRISPR/Cas): Clustered regulatory interspaced short palindromic repeats are loci that contain multiple short direct repeats and provide acquired immunity to bacteria and archaea. CRISPR systems rely on crRNA and tracrRNA for sequence-specific silencing of invading foreign DNA.

clustered regularly interspaced short palindromic repeat (CRISPR)-associated primase-polymerases (CAPPs): Enzymes that are physically associate with the Cas proteins to form bespoke CRISPR-Cas complexes.

clustered regularly interspaced short palindromic repeat (CRISPR)-associated protein (Cas): Includes nucleases like Cas9 and Cpf1.

clustered regularly interspaced short palindromic repeat activation (CRISPRa): Dead Cas9 (dCas9) is attached with transcriptional activators, including the transactivation domain of the VP16, VP64 or VP160 with an appropriately designed sgRNA. Using a dCas9 or dCas9-activator with a gRNA increases transcription of a target gene.

clustered regularly interspaced short palindromic repeat interference (CRISPRi): KRAB-dCas9 fusion-mediated transcriptional repression was verified to be a commanding way of downregulating the endogenous gene with an appropriately designed sgRNA. Using a dCas9 or dCas9-repressor with a gRNA represses/decreases transcription of a target gene by blocking RNA Pol II binding.

clustered regularly interspaced short palindromic repeats (CRISPR)-associated Cas 9 (CRISPR-Cas9): A procaryotic immune system that protects bacterium against foreign DNA such as plasmids and phages. Mechanistically, in its simplest form, a nuclease (Cas9) binds to an appropriate small guide RNA molecule of the CRISPR class which targets the entire complex to its complementary target DNA sequence. It is an RNA-guided DNA endonuclease from the type II CRISPR system of *Streptococcus pyogenes* that induces double-strand breaks in DNA and has been adapted for use in genome-editing applications.

clustered regularly interspaced short palindromic repeats (CRISPR)-associated Cas14 (CRISPR-Cas14): David F. Savage identified in 2019 CRISPR-Cas enzyme C14. C14 has small size (400–700 amino acids). Activity assays revealed that Cas14 cleaved only **single-stranded DNA (ssDNA)** and not double-stranded DNA or single-stranded RNA. Researchers could detect no PAM binding requirement for Cas14. H. Altae-Tran and others in 2023 showed that Cas14 cuts RNA precisely.

clustered regularly interspaced short palindromic repeats-based chromatin affinity purification with mass spectrometry (CRISPR-ChAP-MS): This is a CRISPR-based approach for identification of histone modifications. PrA-tagged Cas9 bound to gRNA is targeted to a specific region of chromatin. Following chemical cross-linking, the chromatin is trimmed to ~1 kb in size and subjected to affinity separation with IgG-coated beads. Isolated chromatin containing PrA-tagged Cas9/gRNA is then analysed with high-resolution MS to identify specifically related proteins and histone PTLMs.

CMG assembly: A key event during replication initiation is the CMG assembly by CMG- helicase, which unwinds DNA at the fork and is composed of the hexameric MCM2-7 ATPase and two accessory factors, CDC45 and GINS. After the MCM2-7 double hexamers are recruited to origins in G1 and cells enter S phase, CDC45 and GINS associate sequentially with MCM2-7 double hexamers, leading to CMG assembly. DONSON scaffolds are a pre-Loading complex that delivers GINS to CDC45-MCM2-7 for CMG assembly.

C_nH_{2n}: an alkene with only one carbon-carbon double bond.

coacervate: Term coined by A. Oparin in 1980. An aggregate of varying degrees of complexity, formed by interaction of two or more colloidal suspensions. Originally called MICROSPHERE by S.W. Fox in 1977.

coactivator proteins: Gene activation in higher eukaryotes necessitates the determined action of transcription factors and coactivator proteins. Coactivators occur in multiprotein complexes that anchorage on transcription factors and modify chromatin, permitting effective transcription to happen. These adaptor molecules integrate signals from activators and repressors and relay the result to basal factors.

coadaptation: (i) Beneficial interaction between (a) a number of genes at different loci within an organism; (b) different parts of an organism; or (c) organisms belonging to different species. (ii) The process of selection by which pleasantly interacting genes become amassed in the gene pool of a population.

coagulate: When antigens and antibodies clump together.

coalescence time: The time back to when two genes share a common ancestor.

coalescence: (i) A process occurring during the calculation of how long ago two alleles shared a common ancestor: the branches of a genealogy coalesce as one works back in time. (ii) The merging of two genetic lineages into a single common ancestor.

coalescent: Having plant parts glued or grown together to form a single unit.

coalescent process: A model in which as one moves back in time, each pair of lineages merges at a rate ½Ne.

coancestry: A quantification of relatedness between two individuals; the probability that genes taken from two individuals are identical by descent. The coancestry of mates becomes the inbreeding of the progeny. Coancestry of full sibs (from unrelated non-inbred parents) is 0.25, of half-sibs 0.125 and of first cousins 0.0625. It is possible to talk about self-coancestry. Thus, an individual's relatedness with itself for non-inbred individuals is 0.5. Coancestry and inbreeding are relative concepts. Their values depend on how far back ancestry is traced. Thus, it is convenient to use them against some reference.

coancestry matrix: Square matrix with coancestry among individuals in the cells. On the diagonal appears self-coancestry, in the other cells pairwise coancestry; the matrix is symmetric.

coat protein: The protein molecules that make up the shielding layer of the nucleic acid core of a virus. *See* CAPSID.

coat(ing): In the context of virus capsid, it refers informally to the defined geometric structure of a capsid or the membrane of an endosome containing an integral virion.

coated vesicle: Small membrane-bounded organelle with a cage of proteins (the coat) on its cytosolic surface. It is formed by the pinching off of a coated region of membrane (coated pit). Some coats are made of clathrin, whereas others are made from other proteins.

cobots: Collaborative robots designed to work alongside human operators. Prototype cobots are being used on automobile assembly lines to help guide heavy components like seats and dashboards into cars so that they don't damage auto body parts as workers install them.

coccus (pl. cocci): A globular (or near-globular) bacterial cell. *See* BACTERIUM.

cochaperonin: A protein molecule residing inside the living cells that 'works together' with applicable chaperonin(s) to assist in proper folding of the (new) protein molecules that are produced in the cell's ribosomes.

cochineal carmine staining: *See* ACETO-CARMINE STAINING.

cochleariform: Bowl- or spoon-shaped.

cochleate: Coiled like a snail's shell.

cocloning: The additional or accidental cloning (i.e. copying) of extra molecular fragments, other than the desired one, that occasionally happens when a researcher is trying to clone a molecule.

coconservation: The simultaneous correction of two sites during gene conservation.

coconut water: Liquid endosperm from the centre of the coconut seed; a complex, undefined addendum of variable quality and effects in some nutrient solutions (2–15% v/v) for plant tissue culture. It shows growth-promoting effects and cell division factors. It is replaceable in some cases by cytokinins and/or sugar.

coconversion: The concurrent correction of two sites during gene conversion.

cocultivation: Growth of cultured cells together. During *Agrobacterium*-mediated gene transfer, cocultivation is the step, in which *Agrobacterium* and target plant tissues are grown together in the petri dish containing plant tissue culture medium in total dark at 28°C for 3–5 days. During this cocultivation, the gene transfer from *Agrobacterium* to plant cell occurs. This technique is used for transforming protoplasts and other explants, or for *in vitro* selection by incubating them with a low density of transformed bacteria or a certain concentration of selective substances. *Aka* COCULTURE.

code: *See* CODON, GENETIC CODE.

code dictionary: A listing of the 64 possible codons and their translational meanings (the corresponding amino acids). *Aka* GENETIC CODE DICTIONARY.

code letter: In standard genetic code, code letters are represented as sequence of nucleotides A, T, G and C in DNA and A, U, G and C in mRNA. Codon sequence of nucleotides specifies an amino acid in triplets.

code library: A collection of non-volatile properties used by computer programs, often for software development. These may include configuration data, documentation, help data, message templates, pre-written code and subroutines, classes, values or type specifications. In IBM's OS/360 and its successors, these are referred to as partitioned data sets.

code word: A word used for secrecy or convenience instead of the usual name for something. For example, in genetics, three adjacent nucleotides in DNA or RNA code for one amino acid. *Aka* GENETIC CODE.

codepoiesis: The source of the first cells was founded on the skill of the inherited systems to produce the rules of the genetic code, and the successive evolution of the cells was founded on two matching processes: (i) the production of new organic codes; and (ii) the other was the preservation of the prevailing ones. Taken together, these two progressions are the two faces of a biological marvel that can be referred to as 'codepoiesis'. Codepoiesis is not autopoiesis. The descendants of the first cells and the eukarya are codepoietic systems. *Compare with* AUTOPOIETIC SYSTEM.

coding: Computer programming is the process of designing and building an executable computer program for accomplishing a specific computing task. Programming involves tasks such as analysis, generating algorithms, profiling algorithms' accuracy and resource consumption and implementing algorithms in a chosen programming language (commonly referred to as coding). The source code of a program is written in one or more programming languages. The purpose of programming is to find a sequence of instructions that will automate the performance of a task for solving a given problem. The process of programming thus often requires expertise in several different subjects, including knowledge of the application domain, specialized algorithms and formal logic.

coding capacity: The amount of protein that can be specified by a given DNA or RNA sequence. Estimates of coding capacity usually require assumptions about the absence of introns or overlapping genes. As a rough guide, 1 MDa (megadalton) of double-stranded DNA can encode 60,000–70,000 Da of protein.

coding exonic regions and non-coding exonic regions: Exons and introns were named by W. Gilbert in 1978. Exons are regions of DNA that are transcribed to RNA and retained after introns are spliced out. However, the term 'exon' is often misused as synonymous to 'protein coding'. Not all exons are protein coding. In contrast, only a fraction of exonic sequences are protein coding (<30% in humans). Both exons and introns are also present in UNTRANSLATED REGIONS (UTRs) and non-coding RNAs. Misuse of the term exon is problematic. For example, 'whole-exome sequencing' technology targets <25% of the human exome, primarily regions that are protein coding. We need to recognize the importance of the original definition of an exon for making functional distinctions in genetics and genomics. J.L. Aspden and others in 2023 recommend the use of clearer language referring to coding exonic regions and non-coding exonic regions. They propose the use of CODING EXOME SEQUENCING (CES) to more appropriately describe sequencing approaches that target primarily protein-coding regions rather than all transcribed regions.

coding ratio: The number of bases in nucleic acids divided by the number of amino acids whose sequences they determine in a particular polypeptide.

coding region: All the exons of a gene that result in the protein product of the gene. *Aka* OPEN READING FRAME.

coding RNA: RNA that serves as a template during translation. Of all the types, mRNA is the only coding RNA.

coding sequence (CDS): The part of a gene (DNA) or mRNA confined by start and stop codons that determines the sequence of amino acids of a protein, as opposed to non-coding sequences, such as promoter, operator, intron or terminator regions. *Aka* CODING REGION.

coding single-nucleotide polymorphism (coding SNP or **cSNP):** A term for a single-nucleotide polymorphism that is positioned within an exon of a gene.

coding strand: The DNA strand that is used as a template for transcribing mRNA. *Aka* ANTISENSE STRAND, CODOGENIC STRAND, TRANSCRIBING STRAND *and* TEMPLATE STRAND. *Compare with* ANTICODING STRAND.

coding theory: The theory describing the properties of codes and their respective fitness for specific applications. Codes are used for data compression, cryptography, error detection and correction, data transmission and data storage. Codes are investigated in various scientific disciplines – such as information theory, electrical engineering, mathematics, linguistics and computer science – for the purpose of designing efficient and reliable data transmission methods. This theory typically involves the elimination of redundancy and the correction or detection of errors in the transmitted data.

coding versus non-coding DNA: In prokaryotes, DNA as a template for the synthesis of mRNAs, tRNAs or rRNAs. Average gene size is 1000 bp. Non-coding DNA functions mostly as promoters and other regulatory sequences. Generally, no intron exists in prokaryotes. In eukaryotes, most DNA is non-coding (introns). Average gene size is 27,000 bp.

codominance: In this case, phenotype of F_1 hybrid reveals characteristics of both the parents. Thus, in the heterozygote, both the alleles are expressing themselves. Example is provided by AB blood group of ABO system in man. *Compare with* DOMINANCE *and* RECESSIVENESS.

codominant alleles: Both the alleles present at a locus in a diploid individual contribute to the phenotype.

codominant marker: Any molecular marker that notices both alleles of a particular genomic locus (e.g. RESTRICTION FRAGMENT LENGTH POLYMORPHISM (RFLP) or sequence-tagged microsatellite site markers).

codominant: (i) In genetics, a heterozygote that displays fully the phenotypic manifestations of both alleles at a gene locus. (ii) In forestry, a tree receiving full light from above but comparatively little from the sides. Such trees usually have medium-sized crowns.

codon: A sequence of three adjacent nucleotide bases in the nucleic acid which codes for specific amino acid. Sequence of three nucleotide bases in DNA or mRNA which specifies a particular amino acid during protein synthesis. It is also called triplet. In total of 64 possible codons, three are stop codons which do not specify amino acids. Set of all codons constitute the genetic code. *Aka* TRIPLET CODON, CODE, GENETIC CODE and CODE WORD.

codon ambiguity: An extreme form of mistranslation in which a codon can be translated as two different amino acids.

codon bias: The idea that for amino acids with several codons, one or a few are preferred and are used disproportionately. Synonymous codons do not occur with equal frequency. They would match with tRNAs that are abundant. An important determinant of codon bias is the **guanosine-cytosine (GC) content** of the genome. An organism that has a relatively low G + C content of 30% will be less likely to have G or C at the third position of a codon (wobble position) than A or T to specify an amino acid that can be characterized by more than one codon. Selection of synonymous codons in a gene is strictly non-random. This is dependent upon tRNA availability in the cell and non-random choices between pyrimidine-ending codons. *Aka* CODON PREFERENCE, CODON USAGE BIAS and CODON CHOICE.

codon length: When only as many amino acids are coded as there are code words in end-to-end sequence. For example, UUUCCC has two code words, and they code for only two amino acids. Code length is three code letters.

codon optimization: A multifactorial process that utilizes synonymous codons to reflect the preferred codon frequency in order to optimize protein expression in a target species.

codon-anticodon misreading suppression: A change in ribosomal configuration that enables codon-anticodon misreading to occur and thereby suppress some mutations through misreading.

codon-anticodon pairing affinity: The genetic code is degenerate. Many amino acids are encoded by more than one codon. For example, CCU, CCA, CCC and CCG all encode proline. In some of these cases, different tRNAs recognize the different codons, but certain tRNAs recognize several different codons. The base at the 5′-end of the anticodon does not have as strict base-pairing requirements as the other two bases, permitting it to form hydrogen bonds with several bases at the 3′-end of the codon. This is especially true when the base at the 5′-end of the anticodon is inosine (abbreviated as I); it is particularly 'wobbly'.

codon-anticodon relationship: This is alignment on the ribosome of an anticodon of aminoacyl-tRNA molecule against its complementary codon on mRNA.

codon-restriction theory of ageing: The idea that with the passage of time, different sets of codons come into use, and others are permanently repressed. This could lead to a more limited set of proteins being synthesized and thus to a more limited capacity to adapt to changing environmental conditions.

coefficient: A coefficient measures a variable by using a multiplier. When conducting research and computing equations, the coefficient is often a numerical value that multiplies by a variable, giving you a coefficient of the variable. If a variable doesn't have a number, the coefficient is always one.

coefficient of coancestry: *See* COANCESTRY.

coefficient of coincidence (CC): A measure of INTERFERENCE in the formation of chromosomal crossovers during meiosis. CC is calculated by using the formula: Observed frequency of double crossovers (O_f)/Expected frequency of double crossovers (E_f), where O_f = DCO/N and E_f = [(SCO I + DCO)/N] × [(SCO II + DCO)/N]. See whether the value of CC is less or more than one. It expresses strength of interference. *Aka* COINCIDENCE COEFFICIENT.

coefficient of consanguinity: The probability that a related mate is a carrier for a particular trait if the other partner is known to be a carrier for the trait.

coefficient of correlation: A numerical value that measures the linear dependence between two random variables. Limiting values of −1 and +1 indicate perfect negative and perfect positive correlation, respectively. A correlation of zero suggests a complete lack of association between the two variables.

coefficient of determination (r^2): The percentage of variation in a variable explained by one or more of the others. In a simple linear regression setup, this is the square of the correlation coefficient. In a multiple linear regression setup, this is the square of the multiple correlation coefficients. In some other setups, this parameter is explained as a proportion of total sum of squares.

coefficient of heritability: The degree of correspondence between genotypic and phenotypic values is measured by the coefficient of heritability (h^2), usually expressed as the ratio of genetic over phenotypic variance ($h^2 = \sigma_g^2/\sigma_p^2$). The higher the coefficient of heritability approaches unit, the closer the correspondence between genotypic and phenotypic values, and therefore, the greater the expected response through phenotypic selection.

coefficient of inbreeding: *See* INBREEDING COEFFICIENT.

coefficient of integration: The relationship between the frequency of transfer of a donor marker to recipients and frequency to its transmission from recipients to recombinants following F-X FR conjugal mating.

coefficient of kinship: Defined as the probability that two homologous alleles drawn from each of two individuals are identical by descent (IBD), is a classic measurement of relatedness. *See* COANCESTRY.

coefficient of migration: *See* MIGRATION COEFFICIENT.

coefficient of parentage (COP): COP measures the genetic relationship between strains of germplasm according to the proportion of identical alleles that they share by descent through their pedigrees. COP values are therefore important measures of genetic diversity. *Aka* COEFFICIENT OF KINSHIP.

coefficient of relatedness (r): The probability that two individuals have received a certain gene from a common ancestor, represented by formula r = n × (0.5) L, where n denotes the alternative routes between the related individuals along which a particular allele can be inherited and L indicates the number of meiosis or generation links. *Aka* COEFFICIENT OF RELATIONSHIP.

coefficient of replacement: *See* MIGRATION COEFFICIENT.

coefficient of selection (Abbrev. s or t): A measure of the relative change in gene frequency between generations as a result of differential selection.

coefficient of variation (CV): Standard deviation (σ) divided by mean ($\bar{x}$), generally multiplied by 100. It is used to obtain a measure of relative variation. This unit-free parameter is used to compare dispersion of one variable with another, such as dispersion of cholesterol level with dispersion of body temperature in a group of cases.

coenocyte: A single cell with multiple nuclei. This happens formed when nuclear division is not followed by cytokinesis, such as in some protoplast or cell fusion products.

coenocytic: Condition in which an organism consists of filamentous cells with large central vacuoles and whose nuclei are not divided into separate compartments. This condition results in a long tube containing many nuclei, with all the cytoplasm at the periphery. *See* ASEPTATE *and* NON-SEPTATE. *Contrasts* SEPTATE.

coenocytic endosperm: Multinucleate condition in which nuclei divide within a common cytoplasm without being detached from each other by a plasmalemma or cell wall.

coenospecies: A group of individuals having common evolutionary origin comprising more than one taxonomic species.

coenzyme A (CoA): A water-soluble vitamin that is required as a cofactor in all living cells by certain condensing enzymes and functions in acyl-group transfer and in fatty acid metabolism. A cofactor that is a small organic or organometallic non-protein molecule or ion that participates in an enzyme-catalysed reaction and functions to transfer atoms or electrons between itself and various molecules. While some cofactors are bound covalently, others are not. Examples of cofactors include biotin, NAD^+, $NADP^+$ and coenzyme A. Most of the enzymes consist of both a protein portion called an APOENZYME and a non-protein portion called cofactor. Ions of Fe, Mg and Zn are examples of COFACTORS. If the cofactor is organic molecule, it is called coenzyme. Many coenzymes are derivative of vitamins. Sometimes, the same coenzyme is needed by different enzymes involved in the catalysis of different reactions. Coenzymes have been classified into two large groups: fat soluble and water soluble. Examples of a few water-soluble vitamins are thiamin, biotin, folic acid, vitamin C and vitamin B12 and those of fat-soluble vitamins are vitamins A, D, E and K. *Aka* PANTOTHENIC ACID. *See* ACETYL CoA.

coevolution: Cooperative evolution of two unrelated species that have a close ecological relationship resulting in reciprocal adaptations as happens between host and parasite, and plant and insect. *Aka* CONCERTED EVOLUTION.

coevolution of epigenetic and genetic variation: Natural epigenetic and genetic variations coevolve. The inheritance of majority of methylation variants is linked to Mendelian modes of inheritance but some epigenetic variations do not obey the Mendelian principles of inheritance. The major potentials of epialleles include: (i) determination of epigenetic variation; (ii) effect of epi-marks on phenotype; and (iii) the inheritance pattern of epigenetic mark-linked superior phenotypes.

coevolutionary conflict in resistance (*R*) and *avirulence* (*Avr*) genes: Antagonistic relationship between *R* and *Avr* genes results in coevolutionary conflict as selection favours evolution of resistance in plants and virulence in their pathogens.

coexistence: Refers to a state where different primary production systems, including non-GM systems such as organic production, conventional agriculture can each working smoothly alongside one another.

cofactor: A non-protein component required by some enzymes for activity. The cofactor may be a metal ion or an organic molecule called a COENZYME. Cofactors are generally heat stable. Cofactors may be either tightly or loosely bound to the enzyme.

cofactor recycle: The regeneration of a consumed cofactor by a supplementary reaction such that it may be reprocessed many times over by a cofactor-requiring enzyme during a reaction.

cog region: Any group of similar sequences of nucleotides in DNA molecules that may specifically be recognized by endonucleases or other enzymes.

cognate: In molecular biochemistry, the term refers to two biomolecules that typically interact with each other. For example, interaction between anticodon on transfer RNA and codon on messenger RNA, on the small subunit of the ribosome. On the basis of such a relationship, there are three types of codon-anticodon relationship: COGNATE CODON-ANTICODON PAIRING, NEAR COGNATE CODON-ANTICODON PAIRING *and* NONCOGNATE CODON-ANTICODON PAIRING. *See details as separate terms.*

cognate codon-anticodon pairing: Occurs when there occurs complementary base pairing between anticodon and codon at (i) all the three positions (1, 2 and 3) as, for example, between AAA/UUU; or (ii) complementary base pairing at first two positions (1 and 2) and Wobble at position 3 (e.g. between AAG and UUA).

cognate genes: Genes that are related or analogous in nature, character or function. For example, *AvrXa21* (*Xanthomonas*) and *Xa21 R*? (Rice) are cognate genes. The first LRR-kinase found to be involved in plant defence was an *R*-gene product, Xa21 of rice. It is debatable whether *Xa21* is an *R* gene. PRR (XA21)-Elicitor (Ax21) are related in function.

cognate tRNAs: All the tRNAs that are recognized by a particular aminoacyl-tRNA synthetase.

cognitive science: The interdisciplinary, scientific study of the mind and its processes. It examines the nature, the tasks and the functions of cognition (in a broad sense). Cognitive scientists study intelligence and behaviour, with a focus on how nervous systems represent, process and transform information. Mental faculties of concern to cognitive scientists include language, perception, memory, attention, reasoning and emotion. To understand these faculties, cognitive scientists borrow from fields such as linguistics, psychology, artificial intelligence, philosophy, neuroscience and anthropology.

cognotechnology: Convergence of nanotech, biotech and INFORMATION TECHNOLOGY (IT) for remote brain sensing and mind control.

coherent spore dispersion via drop-leaf interaction: The dispersion of plant pathogens, such as rust spores, is responsible for more than 20% of global crop yield loss annually. Z. Wu *et al.* (2024) investigated the interplay between leaf elasticity and rainfall, revealing how a flexible leaf structure can generate a lateral flow stream, with embedded coherent structures that enhance transport. They modelled the linear coupling between drop momentum, leaf vibration and the stream flux from leaf surfaces. With Lagrangian diagnostics, they then mapped out the nested coherent structures around the fluttering profile, providing a dynamical description for local spore delivery. This model can facilitate the construction of physically informed analytical models for local crop disease management.

cohesin: A protein responsible for the formation of cohesion complex that holds the sister chromatids together to form a chromosome during mitosis, facilitate spindle attachment to chromosomes and are involved in DNA repair through recombination. *See* COHESION COMPLEX.

cohesin assembly factor 1 (CAF-1): CAF-1 is a conserved histone chaperone, essential for cell survival in multicellular organisms, that plays a key role in replication-dependent nucleosome assembly and preserves genome stability.

cohesin complex: Complex of proteins that clasps sister chromatids together along their length before their separation during cell division.

cohesin loading: Cohesin rings are topologically loaded on to chromosomes at sites that are occupied by the cohesin loader and from where they slide away once the loading reaction is complete. NIPPED-B-LIKE PROTEIN (NIPBL) is a CAF that recruits histone deacetylases to mediate local chromatin modifications.

cohesion complex: The cohesion complex, formed by cohesin molecules, is best known for its role in cell division. During cell division, this complex mediates the pairing of sister chromatids. Cohesion complex containing the Rec8 protein holds sister chromatids together at centromeres and along chromosome arms. The centromeres of a pair of sister chromatids work in unison so that they face, and attach to, the same spindle pole. Cohesion complex also has roles in interphase cells. It does not rest between divisions and, instead, participates in regulating gene expression.

cohesion generation in response to the DSBs: Faithful chromosome segregation and repair of DOUBLE-STRANDED BREAKS (DSBs) require cohesin, the protein complex that mediates sister-chromatid cohesion. Cohesion between sister chromatids is thought to be generated only during ongoing DNA replication by an obligate coupling between cohesion establishment factor such as Eco1 (Ctf7) and the replisome.

cohesive ends: In molecular biology, a fragment of DNA (often produced by a staggered cut on the DNA using restriction enzymes) in which the strands are of unequal length, and an end that is cohesive due to an overhang or a stretch of unpaired nucleotides that easily matches up with another overhang in the complementary region. Cohesive ends are employed in building of recombinant DNA molecules. *Aka* STICKY ENDS. *Contrasts* BLUNT ENDS.

cohort: Refers to an inclusive group of all genotype selections inflowing the trials at one time. Some kind of peer grouping is essential for valid comparisons in breeding programmes, because only the selected material enters at the initial stage (e.g. yield nurseries in seasonal or yearly groups). With the passage of time each cohort is unique in its number of possible data years (e.g. cohorts may enter the system on an annual basis and may be melded into a conglomerate cohort, which each year may contain one plot of every selection undergoing evaluation). The cohort (master nursery) is rationally stable in size. The addition of the annual new cohort almost balances the abrasion by discard from all older cohorts. This basic nursery serves as a resource pool from which outstanding selections periodically are chosen to become entries in replicated advanced trials.

cohort study: (i) A study in which all subjects meeting the entry criteria are included. Entry criteria are defined at baseline (e.g. at time of diagnosis or treatment); (ii) An observational study that begins with a set of subjects who have a risk factor and a second set of subjects who do not have the risk factor. Both sets are followed prospectively through time to learn how many in each set develop the outcome or consequences of interest.

coiled coil: Especially stable rodlike structure in proteins that is shaped by two α helices coiled around each other.

coiling: When CHROMOSOME CORES first become visible in late prophase of mitosis, SISTER CORES are adjacent to one another and run along the inner sides of sister chromatids. As prophase proceeds, the PROTEINACEOUS CORES separate, and sister chromatids coil. The chromatids coil because cores actively shorten. Probably the shrinkage of a core located to one side of a chromatid causes the chromatid to coil with the core to the inside and chromatin to the outside.

coimmunoprecipitation (CoIP)/pulldown experiments: CoIP is a robust method by which a target protein is extracted from a mixture (e.g. cell lysate or serum) via its affinity to a specially prepared solid support.

coincidence: The chance of a simultaneous occurrence of two events.

coincidence coefficient: *See* COEFFICIENT OF COINCIDENCE (CC).

coincidental evolution: *See* CONCERTED EVOLUTION *and* COEVOLUTION.

cointegrate plasmid: A plasmid carrying both genes for antibiotic resistance and genes for transfer.

cointegrate structure: A fusion of two elements; an intermediate structure in transposition.

cointegrating vectors: Vectors constructed by homologous recombination of a bacterial plasmid with the T-DNA region of an endogenous Ti plasmid in *Agrobacterium*. Integration of the two plasmids requires a region of homology in both. Three vectors necessary in this system are disarmed *Agrobacterium* Ti plasmids, intermediate vectors and helper vectors. A cointegrated plasmid assembled by *in vitro* manipulation normally contains the *vir* genes, the left and right T-DNA borders, an exogenous DNA sequence between the two T-DNA borders and plant and bacterial selectable markers. *Aka* COINTEGRATED VECTORS or HYBRID TI PLASMIDS.

coisogenic: Nearly identical strains of an organism; they vary at only a single locus. *Aka* CONGENIC.

cojugative plasmids: Bacterial plasmids that can be used to transfer genes to bacteria outside their own species.

Col plasmids: These plasmids carry genes that confer the ability to the host bacterium to kill other bacteria by secreting bacteriocins, a type of proteins. Bacteriocins often kill cells by creating channels in the plasma membrane, thus increasing its permeability. They also may degrade DNA or RNA or attack peptidoglycan and weaken the cell wall. Bacteriocins act only against closely related strains. Col E1 plasmid of *E. coli* code for the synthesis of bacteriocin called colicins, which kill other susceptible strains of *E. coli*. Col plasmids of some *E. coli* code for the synthesis of bacteriocin, namely cloacins that kill *Enterobacter* species. Lactic acid bacteria produce bacteriocin NisinA, which strongly inhibits the growth of a wide variety of Gram-positive bacteria and is used as a preservative in the food industry.

colcemid: A synthetic equivalent of colchicine. *See* COLCHICINE.

colchicine: A poisonous alkaloid drug ($C_{22}H_{25}NO_6$) that is attained from meadow saffron (*Colchicum autumnale*). It has a disruptive effect on microtubular activity. Thus, it affects tissue metabolism generally, and mitosis and meiosis in particular; it is used for the induction of polyploidy.

colchiploidy: Polyploidy induced by application of an alkaloid, colchicine.

cold dilute soup: Belief that first life on earth originated in a cold solution of simple organic compounds.

cold hardening: A process of adaptation in which certain organisms manufacture specific proteins that guard them from freezing to death during the winter. Among other organisms, the common housefly, the *Arabidopsis thaliana* plant, the fruit fly *Drosophila* and 'no-see-ums' (*Culicoides variipennis*) can produce these proteins (during the gradually decreasing temperatures of a typical autumn season in North America). The amount of such proteins produced within their bodies is proportional to the severity and duration of the cold experienced. For example, prior to cold hardening, *Culicoides variipennis* insects usually die after exposure for 2 h to a temperature of 14°F (−10°C). If those insects are first exposed for 1 h to a temperature of 41°F (5°C), approximately 98% of these insects can then survive exposure for 3 days to a temperature of 14°F (−10°C).

cold temperature as trigger of epigenetic switch: Vernalization provides a classic example of environmentally mediated epigenetic gene silencing. Vernalization is the acceleration of the transition to flowering by a prolonged period of cold temperatures (usually winter).

cold-sensitive: When a mutant is defective at low temperature but functional at normal temperature.

coleoptilar tiller: When a tiller forms a bud located at the coleoptilar node. The coleoptilar tiller can emerge at any time, independent of the number of leaves on the main stem. *Compare with* AXILLARY TILLER.

coleoptile: One type of sheath in the structure of monocotyledonous seeds. The coleoptile is a protective sheath or cap (pileus), generally more or less pointed, that covers the monocotyledonous plumule as it emerges from the soil. It generally turns green and contributes to photosynthesis until its function is superseded by the main growth of the seedling. *Contrasts* COLEORHIZA.

coleorhiza: One type of sheath in the structure of monocotyledonous seeds. The coleorhiza connects the coleoptile to the radicle and protects the monocotyledonous radicle during germination. Unlike the coleoptile, the coleorhiza is associated with the root and does not emerge from the soil during germination. *Contrasts* COLEOPTILE.

colicin E1: A protein that kills *E. coli*.

colicin plasmids: These bacterial plasmids carry genes that confer ability to the host bacterium to kill other bacteria by secreting bacteriocins, which often kill cells by creating channels in the plasma membrane thus increasing its permeability. They also may degrade DNA or RNA or attack peptidoglycan and weaken the cell wall. ColE1 plasmid genes of *E. coli* code for the synthesis of bacteriocin called COLICINS that kill other susceptible strains of *E. coli*.

colicins: Proteins produced by *E. coli* that are toxic (primarily) to other closely related strains of bacteria. The particular *E. coli* that produce a given colicin are generally unaffected by the colicin that they produce.

coliform bacteria: Gram-negative rods resembling *E. coli* and similar species that normally inhabit the colon.

collinearity: (i) Gene is a linear sequence of nucleotides. Polypeptide is a linear sequence of amino acids. The linear relationship between the gene and its protein product is termed collinearity which means that there is a point-for-point correspondence between gene and its protein product with the 3′-end of the template strand of the gene corresponding to the amino-terminus of the polypeptide. (ii) Conserved order of genes or markers in different species belonging to same group. Such collinearity is common in rice and other grass species, *Arabidopsis* and tomato.

collinearity in eukaryotes: In eukaryotes, the nuclear codon and amino acid sequences coded by them are similar to those in prokaryotes. However, the distance between any two amino acids is not always proportionate to the distance between the codons coding for them. In many eukaryotic proteins, two specific amino acids may be located close together, but the codons coding for them may in fact be located far apart in the gene. This is due to non-split nature of most of the eukaryotic genes.

collinearity in prokaryotes: In prokaryotes, there is perfect collinearity between a gene, its RNA transcript (mRNA) and the polypeptide chain (product). The sequence of the amino acids in a polypeptide chain matches with the sequence of the codons coding for those amino acids. Moreover, the distance between any two amino acids is proportionate to the distance between the codons coding for them. This is because prokaryotic genes are generally non-split. A prokaryotic gene and its protein product have a point-for-point correspondence. In other words, a prokaryotic gene and its protein product are perfectly colinear. Thus, the sequence of nucleotides in a prokaryotic gene determines the sequence of amino acids in a protein/polypeptide. This conclusion was result of a study carried out by S.D. Yanofsky and others in 1987 on many mutants of bacterial tryptophan synthetase A polypeptide.

collagen: Fibrous protein rich in glycine and proline that is a major constituent of the extracellular matrix and connective tissues. Collagen exists in many forms: type I, the most common, is found in skin, tendon and bone; type II is found in cartilage; and type IV occurs in basal laminae.

collagen fibril: Extracellular structure shaped by self-assembly of secreted fibrillar collagen subunits.

An abundant component of the extracellular matrix in many animal tissues.

collar: The portion of the seedling or plant near the surface of the soil. In grafted woody plants, the scion portion of the plant near the soil surface.

collection: A collection is a grouping of some variable number of data items (possibly zero) that have some shared significance to the problem being solved and need to be operated upon together in some controlled fashion. Generally, the data items will be of the same type or, in languages supporting inheritance, derived from some common ancestor type. A collection is a concept applicable to abstract data types and does not prescribe a specific implementation as a concrete data structure, though often, there is a conventional choice. *Aka* CONTAINER.

collenchyma: In dicot stems and leaves, collenchyma cells can be found in tissues where growth is occurring. Alive at maturity, these elongated cells have thick cellulose and pectin walls. They provide flexible structural support.

colleter: A multicellular, glandular hair that usually manufactures a mucilaginous substance and is positioned on sepals, stipules or petioles or on nearby parts of stems. It is commonly found on plants in the order Gentianales.

colloid: A state of matter in while particle dimension changes from that of true molecules and bulk suspensions. The dimensions vary from 1 nm to 100 nm or larger. Such solutions possess a range of properties. The particles may have charge and can be separated by electrophoresis, except at their isoelectric point (when no net charge exists). Originally used to refer to amorphous state, in *contrast* to CRYSTALLOIDS.

colocalization single-molecule spectroscopy (CoSMoS): A single-molecule fluorescence microscopy technique in which biochemical reactions can be studied by watching multiple biomolecules, each labelled with a unique fluorophore, colocalize with a surface-tethered substrate.

colonial: Condition in which many unicellular organisms live together in a somewhat coordinated group. Unlike true multicellular organisms, the individual cells retain their separate identities and usually their own membranes and cell walls.

colonization: The establishment and implication of a pathogen within a host plant.

colonization of a host: The establishment, growth and reproduction of the pathogen on or in infected plant.

colonize: To infect, multiply and spread through plant tissue with the growth of a pathogen.

colony: (i) A growth of a group of microorganisms derived from one original organism, specifically as a pure culture. After a sufficient growth period, the growth is visible to the eye without magnification. Sometimes, organisms can be identified from the colour and forms of the colony. (ii) In biology, a colony is composed of two or more conspecific individuals living in close association with, or connected to, one another. This association is usually for mutual benefit such as stronger defence or the ability to attack bigger prey. (iii) In tissue culture, a visible mass of cells. (iv) A group of plants where all plants arise from one root system (e.g. in sunflower). Cells in the centre are usually in the stationery or death phase, whereas the cells on periphery are in exponential phase.

colony hybridization: A method for screening of genomic and cDNA libraries. Using this method bacteria containing chimeric vectors are grown into colonies followed by preparation of replica plate using nitrocellulose membrane. Their DNA is denatured on replica plate, hybridized with labelled probe and then detected using autoradiography. Colony hybridization is one of the techniques used to select and identify particular clones containing particular DNA fragments. Autoradiography then identifies those colonies or plaques that bear desired DNA sequences complementary to the probe and thus allows further cloning and isolation of appropriate fragments. *Aka* PLAQUE HYBRIDIZATION.

colony PCR: This protocol is designed to quickly screen for plasmid inserts directly from *E. coli* colonies. The plasmid should be high copy number such as pUC18, pUC19 or pBluescript. Even though blue/white screening can be used to determine if inserts are present, this technique can be used to determine insert size and/or orientation in the vector. Alternately, the presence of an insert and its size can be determined by growing each colony in liquid, the plasmid purified by a boiling or alkaline preparation protocol, digestion of the plasmid with restriction enzyme(s) that excises the insert, followed by separation by agarose gel electrophoresis.

colony-forming unit (CFU, cfu or Cfu): Plate count of microorganisms is reported as a colony-forming unit. Plate count assumes that one colony develops from one microorganism on solid agar medium. The cell or aggregates of cells that give rise to a single colony in the plate count technique is called colony-forming unit. It is a commonly used term during culturing of single cells or protoplasts or pollen. A known number of these are plated on the culture medium. After optimum incubation the number of colonies thus formed are counted and the colony-forming efficiency is calculated by dividing the number of colonies obtained with total number of cells/protoplasts/pollens cultured.

colorimeter: An instrument that estimates the concentration of a chemical element with the aid of a colour reagent.

colorimetry: A technique that measures the concentration of a chemical element with the aid of a colour reagent.

colour blindness: Refers to the inability to see certain colours as they normally appear to others. The most common form of this vision deficiency is X-linked genetically inherited red-green colour blindness or deficiency inherited as a sex-linked recessive gene in human beings.

colour inhibitor: One of the numbers of dominant genes that preclude colour formation.

colour separator: A machine that is used to separate seed on the basis of their surface colour. It is useful for seed cleaning or type separation.

columella: In flowering plants, the central axis of the cone or fruit (e.g. in Callitris).

column: (i) A structure extending above the ovary and incorporating the style and stamens (e.g. in orchids and milkweeds). (ii) In grasses, the lower, firmer and usually twisted part of an awn, which is distinct from the lean upper part or bristle. *Aka* GYNOSTEGIUM.

column chromatography: The separation of organic compounds by infiltrating a liquid containing the compounds through a porous material (e.g. ion exchange resin) in a cylinder.

column diagram: *See* HISTOGRAM.

column-complete Latin square: A Latin square is column-complete if each ordered pair of discrete symbols occurs precisely once in successive positions in a column of the square.

column-quasi-complete Latin square: A Latin square is column-quasi-complete if each unordered pair of discrete symbols occurs precisely twice in successive positions in a column of the square.

coma: (i) A tuft of hairs from testa or funiculus at one or both ends of some seeds (e.g. in *Strophanthus*, *Asclepias* or *Alstonia)*. (ii) Sterile bracts (e.g. in *Curcuma*, *Ananas* or *Eucomis*). (iii) Sterile flowers (e.g. in Muscari and Leopoldia, at the apex of some inflorescences). (iv) A tuft of hairs at the base of some flowers (e.g. in *Pfaffia gnaphalioides*). (v) A tuft of hairs at the apex or base of some spikelets. (vi) An axil tuft of hairs in inflorescences in some Poaceae (e.g. in *Eragrostis comata*).

combination: A formula in probability that gives the number of ways a specific number of items can be selected from the total number of items in the entire population or sample.

combination breeding: A breeding method that employs the genetic diversity of individuals or varieties in order to generate and select new phenotypes on the basis of genetic recombination of useful characteristics of parental material.

combinatorial biology: A term used to describe the set of DNA technologies used to generate a large number of samples of new chemicals (metabolites) via the creation of non-natural metabolic pathways. The collection of samples thus generated is called a LIBRARY, and the samples are then tested for potential use (e.g. for therapeutic effect, in the case of a pharmaceutical). These technologies enable greater efficiency in a pharmaceutical researcher's screening process for drug discovery.

combinatorial chemistry: The use of chemical methods to create all possible combinations of chemicals starting with a subset of compounds. The building blocks may be peptides, nucleic acids or small molecules. The libraries of compounds produced by this methodology are used to probe for new pharmaceutical reagents. *See* HIGH-THROUGHPUT SCREENING.

combinatorial control: Refers to the rheostat of a step, stage of phase in a cellular process, such as the initiation of DNA transcription by a specific combination of proteins rather than by any individual constituent.

combinatorial determinants: The antigenic determinants which result from the interaction of different polypeptide chains of a molecule. The determinants are absent in either of the parents *Aka* HYBRID DETERMINANTS.

combinatorial exons: A true stimulation of 3′ splice site activity by the downstream binding of U1 snRNP and suggest a possible mechanism by which combinatorial patterns of exon selection are achieved for alternatively spliced pre-mRNAs. In this type of alternative RNA splicing, inclusion or exclusion of a particular exon does not dependent on any of the other exons existing in the transcript.

combinatorial explosion: A fundamental problem in evolutionary optimization and computing. As the size of the investigated system and the number of corresponding parameters increase, the number of combinations that one has to examine grows exponentially, which requires an intolerable amount of time to examine them.

combinatorial libraries enhanced by recombination in yeast (CLERY): A PCR- and *in vivo* recombination-based strategy for recombining multiple parental genes in yeast.

combinatorial library: A collection of small molecules synthesized by combinatorial chemistry.

combinatorial mutagenesis: Generally, simple mutations introduced by site-directed mutagenesis are not expected to have as drastic effects on altering an enzyme's substrate recognition pattern, as many amino acid residues in the enzyme affect the binding pocket of the substrate in the enzyme. Thus, several amino acid residues may need to be altered simultaneously to achieve the goal of altering substrate specificities. However, as any amino acid residue may be altered into 19 other ones, the number of amino acid combinations that can be made if mutations are introduced at various residues simultaneously can become very large. It is difficult to predict on the basis of a primary sequence whether the 3D structure of a protein (and of the active site) is good for the defined purpose. The best approach in this situation would be to have essentially all different combinations made at the DNA level in different plasmids (can be done using degenerate oligonucleotides), use all these different plasmids as a mixture to transform *E. coli*, have *E. coli* express the different proteins (each *E. coli* cell and its clones will express one particular protein, assuming it has taken up one plasmid molecule) and select the *E. coli* cell(s) that may be able to convert a new substrate. One chooses selection methods depending on the properties of the protein that has been altered. Peptides can also be synthesized from a degenerate mix of amino acid analogues, and the resulting mix of peptides can be screened for desired properties, in particular pharmaceutical applications. Similarly, RNA can also be synthesized combinatorically, and degenerate RNA mixtures can be used to study features that are needed to provide RNA with catalytic properties. Thus, combinatorial mutagenesis provides virtually limitless possibilities for genetic engineering and has become an important tool in biotechnology.

combined atomic force microscope/scanning tunnelling microscopy (C-AFM/STM): C-AFM/STM uses qPlus sensors and discusses the stability and thermal drift. By using a material with a low thermal expansion coefficient, one can perform constant height measurements and achieve atomic resolution in both AFM and STM on various samples. Moreover, the design allows a wide-angle optical access to the sensor and the sample that is of interest for combining with optical microscopes or focusing optics with a high numerical aperture.

combined bisulfite restriction analysis (COBRA): A quantitative technique for the identification of methylated DNA in which DNA is subjected to bisulfite conversion and digestion with restriction endonucleases that are specific for sequences containing CpG sites (and thus are subject to methylation); the digestion products are a direct reflection of DNA methylation at the restriction sites. COBRA assay consists of three major steps: (i) treatment of genomic DNA with sodium bisulfite; (ii) PCR amplification; and (iii) restriction digestion. It allows for the sensitive quantification of DNA methylation levels at a specific locus.

combined water: Water chemically bound to soil surface.

combining ability (CA): Combining ability in crosses is defined as the cultivars or parents' ability to combine among each other during hybridization process such that desirable genes or characters are transmitted to their progeny. The average performance of a strain in a series of crosses is called general combining ability (GCA), while deviation in a particular cross from performance predicted on the basis of GENERAL COMBINING ABILITY is known as SPECIFIC COMBINING ABILITY (SCA). *Aka* COMBINING PRODUCTIVITY. *See* GENERAL COMBINING ABILITY (GCA) *and* SPECIFIC COMBINING ABILITY (SCA).

combining site: The site on an antibody molecule that adheres on to an EPITOPE (hapten).

commaless code: Said of a genetic code in which successive codons are contiguous and not separated by non-coding bases or groups of bases. Genetic code is actually commaless. For example, UUUCCCGGGAAA has four code words, and upon translation, we will have a tetrapeptide chain of Phe-Pro-Gly-Lys. So, all the letters are used to code for one or the other amino acid.

comma-separated values (CSV): A delimited text file that uses commas to separate values. A CSV file stores tabular data (numbers and text) in plain text. Each line of the file is a data record. Each record comprises one or more fields, separated by commas. The use of the comma as a field separator is the source of the name for this file format.

commensal bacteria: Bacteria that obtain food or other benefits from its host without causing harm.

commensalism: A symbiotic relationship in which one partner is benefited and the other is neither benefited nor harmed. A kind of relationship among the members of different species living together in the same environment in which one organism gets benefit from the association but the other is neither benefited nor harmed. This type of association exists between humans and some members of microbiota. For example, the silverfish live with the army ants, participate in the raids and share their prey. Silverfish are of no harm or benefit to the ants.

commercial name: A name generally of no botanical standing and not governed by the International Code of Nomenclature for Cultivated Plants (ICNCP). The term generally refers to names such as Trademark Names, names covered by Plant Breeders Rights, Patents and Promotional Names, which are often coined to boost the sale of a plant.

commissure: The seam or face at which two carpels stick to. *See* FISSURE *and* SUTURE.

commodity: A product of agriculture. Examples of agricultural commodities include wheat, rice, beet, maize, beef, soybean and coffee.

common allele: At a given locus, the nucleotide seen most frequently in a given population.

common ancestor: The most recent ancestral form or species from which two different species evolved.

common single-nucleotide polymorphism (common SNP): Any single-nucleotide polymorphism whose minor allele exists in more than 10% of the genomes of a population.

commonsense knowledge: In artificial intelligence research, commonsense knowledge consists of facts about the everyday world, such as 'Lemons are sour', that all humans are expected to know. The first AI program to address commonsense knowledge was Advice Taker in 1959 by John McCarthy.

commonsense reasoning: A branch of artificial intelligence concerned with simulating the human ability to make presumptions about the type and essence of ordinary situations they encounter every day.

Commonwealth Scientific and Industrial Research Organization (CSIRO) Australia: This organization conducts research in science, industry and agriculture and is partly funded by industry.

communicating junction: Type of cell junction that facilitates the route of chemical or electrical signals from one cell to another.

communication, community and culture in scientific research: Frequently, a scientific method is employed not only by a single person but also by several people cooperating directly or indirectly. Such cooperation can be regarded as one of the defining elements of a scientific community. Various techniques have been developed to ensure the integrity of that scientific method within such an environment.

community: (i) Any ecological assemblage (naturally existing group) of organisms that occupy a common environment. The term is a general one as it covers groups of various sizes. (ii) A grouping of interacting populations in a particular habitat.

community gene bank: *Ex-situ* gene bank conservation stores seeds at very low temperature and moisture levels to preserve seed viability for many years. Gene bank conservation is static. It gives safety against genetic erosion and natural calamities. M.S. Swaminathan Research Foundation (MSSRF) maintains a gene bank in Chennai. It was established in 1993. Its mission is to empower farming and tribal communities for releasing the farmers' rights provided in the PPV&FR Act 2021. Hence, this gene bank is called Community Gene Bank (CGB). The CGB is thus uniquely different from the national gene bank, the National Bureau of Plant Genetic resources (NBPGR), New Delhi and International Agricultural Research Centres. CGB also maintains a herbarium of most of the specimen varieties.

community genomics: Community genomics emphasizes the analysis of natural coexisting species populations through cultivation-independent environmental genome sequencing. The approach enables postgenomic functional assays to be carried out to understand the ecology and evolution of microbial consortia.

compact epigenetic editor: Neumann and others (2024) report a new class of epigenetic editor, called

COUPLED HISTONE TAIL FOR AUTOINHIBITION RELEASE OF METHYLTRANSFERASE (CHARM), that can competently silence the expression of gene coding for PRION PROTEIN (PrP) in the brains of mice when transported systemically by a viral vector without altering the underlying DNA sequence and provides a new method to deal with neurodegenerative diseases. It has upgraded delivery, durability, tunability and safety. CHARM epitomizes a therapeutic procedure that could be useful to a range of other diseases produced by the toxic build-up of unwanted proteins. *See* COUPLED HISTONE TAIL FOR AUTOINHIBITION RELEASE OF METHYLTRANSFERASE (CHARM).

compact operons: All the components of operon are located in a single (contiguous) region of the chromosome, such as lactose (*lac*) operon.

companion cell: (i) Type of cell presents in phloem tissue, associated with sieve cells and pumps nutrients (sugars) for the process of translocation. (ii) Cells in the germ line of plants that will not contribute genetically to progeny but are produced by meiosis. These are the VEGETATIVE NUCLEUS (VN) in the male germ line and the CENTRAL CELL (CC) in the female germ line. The CC is fertilized to produce the endosperm that acts as a supportive tissue to the developing embryo.

companion crop: Two or more different crops grown side-by-side for some gain to one or more of them. It is particularly used of the small grains with which forage crops are sown. Acceptable companion crops are oat (*Avena sativa* L.), flax (*Linum usitatissimum* L.) and pea (*Pisum sativum* L.).

comparative analysis: The relations of two or more traits among higher taxa (e.g. species, genera, families) are analysed with proper control for phylogeny and covariates such as body weight.

comparative biology: The use of comparative methods to investigate the similarities and differences between two or more biological organisms (e.g. two organisms from the same time period but different taxa, or two organisms from the same taxon but different times in evolutionary history). Biologists infer the organisms' genetic relatedness and their natural histories from side-by-side comparison of morphological or molecular characteristics of different organisms. It is a fundamental method in many biological disciplines, including anatomy, physiology, palaeontology and phylogenetics.

comparative candidate positional cloning: The molecular cloning of candidate genes responsible for a specific phenotype and the detection of SINGLE-NUCLEOTIDE POLYMORPHISM(S) (SNPs) within these genes in different individuals.

comparative epigenomics of potato and tomato: Potato (*S. tuberosum*) and tomato (*S. lycopersicum*) are the first two sequenced genomes of asteroids and share a common polyploidization event. The overall levels of DNA methylation are different, although patterns of DNA methylation are similar in potato and tomato. Different types of duplicated genes can display different methylation designs in potato and tomato.

comparative gene mapping: Mapping of a common set of DNA probes on to linkage maps of different species is done and compared to indicate substantial conservation of blocks of genes and even large segments of chromosomes between species. This method shows synteny of markers among related species or genera. Such maps have been established for cereals (such as wheat, maize, oats, rye, rice, sorghum, millet) and Solanaceae (such as potato, tomato and paprika). Since appreciable similarity prevails in the arrangement of genes, groups of genes (MICROSYNTENY) or whole chromosomal regions (MACROSYNTENY), RFLP probes are used to map the corresponding sequences on the wheat genome, especially since they are sufficiently homologous to cross-hybridize.

comparative genomic hybridization (CGH): A DNA microarray technique to detect gains and losses at chromosomal and sub-chromosomal levels. The test and a well-characterized reference sample are hybridized to the same microarray, and the relative intensity between the two samples is used to infer copy number variation all over the genome in a single hybridization. *Compare with* CIS-ACTING FUNCTION.

comparative genomics: A comparison of gene numbers, gene locations and biological functions of genes that play a unique biological role in different species such as mice, the fruit fly and the bacterium *E. coli*. Comparative genomics analyses different groups of plants using molecular maps. Comparative genomics has been applied to grasses and members of Solanaceae family. All genomes of grass family can be described in terms of their relationships to a single reference (i.e. rice genome). Less than 30 linkage blocks are needed to represent genomes of all grass species, including rice, maize, wheat, rye, barley, sorghum, oat, sugarcane and millets.

comparative mapping: Refers to mapping a common set of probes on to linkage maps of different species. This approach aids in establishing synteny of markers and unscrambling the relationship among related genera. Comparative maps have been prepared for cereals such as wheat, maize, rice and sorghum and some dicots such as tomatoes, potatoes and peppers. *See* ORTHOLOGOUS GENE MAPPING.

comparative mapping of adaptive genes: Comparative mapping is a very gifted genomic method for determining adaptive genes. This method takes advantage of high similarity in gene location in chromosomes of closely related species and applies it across different species to search for functional genes and their genomic organization. Comparative mapping in various species has revealed that gene content and gene order are preserved over long chromosomal regions among related species of animals or angiosperm plants. The genetic maps of closely related species can be directly compared due to synteny (i.e. cooccurrence of two or more genes on the same chromosome) among the genomes of these species.

comparative positional candidate gene: A gene that is likely to be positioned in the same region as a DNA marker that has been shown to be linked to a single-locus trait or a QUANTITATIVE TRAIT LOCUS (QTL), where the gene's likely position in the genome of the species in question is based on its identified location in the map of another species (i.e. is based on the comparative map between the two species).

comparative trial: Trials with two or more treatment groups, planned with sufficient power or precision to detect pertinent clinical differences in treatment effectiveness among the groups.

comparator: A known non-biotechnology organism that is similar to the engineered organism, except for the engineered trait.

comparing multiple sequences: (i) ClustalW does multiple sequence alignment. This provides one with a number of options for data presentation, homology matrices and presentation of phylogenetic trees. (ii) DbClustal aligns sequences from a BlastP database search with one query sequence. The alignment algorithm is based on ClustalW modified to incorporate local alignment data in the form of anchor points between pairs of sequences. (iii) PROBCONS is a novel tool for generating multiple alignments of protein sequences. (iv) PRALINE is a multiple sequence alignment program with many options to optimize the information for each of the input sequences (e.g. global or local pre-processing, predicted secondary structure information and iteration capabilities). (v) Gene Context Tool is an incredible tool for visualizing the genome context of a gene or group of genes (synteny). (vi) Consensus (EMBL, Universitat Heidelberg) performs multiple alignments and calculates the consensus. (vii) MultAlin performs Multiple sequence alignment. (viii) DiAlign is a novel program for multiple alignment. It was developed by Burkhard Morgenstern and others. While standard alignment methods rely on comparing single residues and imposing gap penalties, it constructs pairwise and multiple alignments by comparing whole segments of the sequences. (ix) T-COFFEE is reported to be more accurate than ClustalW for sequences with less than 30% identity. (x) LocARNA performs multiple alignment of RNAs and is a tool for multiple alignment of RNA molecules. It requires only RNA sequences as input and will simultaneously fold and align the input sequences. It outputs a multiple alignment together with a consensus structure.

compartment: Regions in the embryo that are formed exclusively from the descendants of a few founder cells. There is no cell movement between the compartments once delimited (e.g. subdivision of molecules, cells or genetic functions into discrete spatial or temporal units). This refers to the grouping together of a protein into distinct parts of the cell or of networks of genetic interactions into diverse functional units.

compartmentalization: The demarcation of a specific tissue area by host barrier tissues.

compatibility: In molecular biology, compatible plasmids have different replication functions and can coexist in the same cell. Incompatible plasmids are very similar or identical in their replication functions, and, in the absence of selection, only one plasmid survives in the cells, while the other is lost instinctively from a culture. The smaller, faster-replicating plasmid generally survives in this process.

compatibility group: When plasmids contain membrane to exist in the same bacterial cell.

compensating trisomic: An aneuploid individual with an extra chromosome in which a missing standard chromosome is compensated for by two novel interchange chromosomes. The two novel chromosomes carry two different arms of the missing chromosome.

competence factor: A surface protein that binds extracellular DNA and enables the cell to be transformed.

competence: The physiological growth stage of microorganism at which it can take up foreign DNA leading to transformation. Competence usually results from changes in the cell wall that make it permeable to foreign DNA molecules. Usually, the naturally occurring bacteria say *E. coli* are not amenable to transformation. But the simple chemical treatments make these competent to take up foreign DNA leading to transformation. Therefore, in the routine cloning work the competent bacterial strains are used instead of normal (wild) types.

competency: An ephemeral state, induced by treatment with cold cations, during which bacterial cells are capable of taking up foreign DNA.

competent: (i) In tissue culture, able to function or to develop. (ii) In molecular biology, competent cells have the ability to take up foreign extrachromosomal exogenous DNA. *E. coli* is artificially made competent by washing in the cold with $CaCl_2$, and some bacteria become naturally competent during certain growth phases.

competent cells: The bacterial cells that can take up the foreign DNA from the surroundings by a process called transformation are known as competent cells. F. Griffith in 1928 first reported it in *Streptococcus pneumoniae*. *E. coli* cells are more likely to uptake the DNA if their cell walls are altered.

competition: When two species seek simultaneously an essential resource of environment that is in limited supply.

competitive asymmetry: Organisms that, by virtue of a particular trait or a set of traits, have a disproportionate advantage over others. For example, organisms with larger body size often have a competitive advantage over smaller organisms because they can eat food that smaller ones cannot, consume more and have foraging advantages.

competitive exclusion: The principle that no two species can coexist in the same place if their ecological requirements are identical. *Aka* GAUSE PRINCIPLE.

competitive inhibitor: A chemical that competes with the normal substrate for the active site of an enzyme. Competitive inhibitors mimic the substrate, as their chemical structures resemble with that of substrate. Thus, these inhibitors block the active site and prevent the binding of substrate and block the reaction. For example, sulfanilamide is a competitive inhibitor of paraminobenzoic acid.

compiler: A computer program that alters computer code written in one symbolic programming language (the SOURCE LANGUAGE) into another programming language (the TARGET LANGUAGE) so that the advice can be executed by a computer. Compilers are a type of translators that support digital devices, primarily computers.

compilospecies: A genetically hostile species that integrates genomes from related species. In some cases, it may completely integrate genome of a species, causing it to become extinct. Examples of compilospecies include *Helianthus* (sunflowers), *Draba* (whitlow-grasses) and *Armeria villosa* Girard.

complement: (1) A soluble protein found in blood serum that is responsible for killing bacteria. The protein was originally called 'alexin' when it was discovered in the late 19th century, but the term was subsequently changed to reflect the fact its action 'complements' that of an antibody in the destruction of bacteria. (2) In genetics, a genetic process when two strains of an organism with different homozygous recessive mutations that produce the same mutant phenotype (for example, a change in wing structure in flies) have offspring that express the wild-type phenotype when mated or crossed.

complement fixation test: A classic test based on the phenomenon of quantitative binding of complement by a complex of antigen and its corresponding antibody.

complement system: System of serum proteins activated by antibody-antigen complexes or by microorganisms. Helps eliminate pathogenic microorganisms by directly causing their lysis or by promoting their phagocytosis.

complementarity: The specific binding of adenine to thymidine (or uracil in RNA) and cytosine to guanine on opposite strands of DNA or RNA.

complementary bases: Bases which are hydrogen bonded specifically in a DNA duplex or in DNA/RNA heteroduplex. In DNA, since guanine (G) pairs with cytosine (C), they are complementary bases; similarly, since adenine (A) pairs with thymine (T), these bases are also complementary.

complementary chiasmata: Chiasmata formed between four chromatids in such a way that the first chiasma occurs between two of the chromatids and the second between their homologues.

complementary determining region (CDR): The hypervariable regions of an antibody molecule, consisting of three loops from the heavy chain and three from the light chain, that together form the ANTIGEN-BINDING SITE.

complementary DNA (cDNA): A DNA molecule produced from mRNA molecule by using enzyme reverse transcriptase. cDNA lacks introns present in the genomic DNA. The amino acid sequence of the target protein can be easily deduced from the cDNA sequence. cDNA helps in gene cloning.

complementary DNA (cDNA) chips: cDNAs of 0.5–2.0 kb are amplified by PCR, and about 1 ng of cDNA is deposited by robots at intervals of 100–300 μm either by mechanical microspotting or by inkjetting. In this case, a single, long double-stranded DNA probe per gene is used. In both cases, probes are generally planned from sequences located nearer to the 3′ end of the genes near the poly(A) tail in eukaryotic mRNA and different probes are used for different exons. This system consists of a chip, hybridization chamber to control hybridization conditions, a reader and software to analyse the chip data. The array is placed in a temperature-controlled hybridization chamber. A target sequence is fluorescently tagged and then injected into the chamber and left overnight, where the target hybridizes to its complementary sequences on the array. Laser excitation enters through the back of the array fixated at the interface of array surface and the target solution. Fluorescence emission is collected by a confocal lens and passes through a series of optical filters to a sensitive detector. A quantitative fluorescence image proportional to hybridization intensity is attained by CHARGED COUPLED DEVICE (CCD). Quantitative estimates of number of transcripts per cell can be attained directly by averaging the signal from multiple probes. Rare mRNA transcripts of 1:100,000 can be detected.

complementary DNA (cDNA) clone: A DNA molecule constructed from an mRNA sequence via sequential use of reverse transcriptase (acting on mRNA) and DNA polymerase.

complementary DNA (cDNA) cloning: A method of cloning the coding sequence of a gene starting with its mRNA transcript. It is normally used to clone a DNA copy of a eukaryotic mRNA. The cDNA copy, being a copy of a mature messenger molecule, will not contain any intron sequences and may be readily expressed in any host organism if attached to a suitable promoter sequence within the cloning vector.

complementary DNA (cDNA) libraries, applications of: (i) Discovery of novel genes. (ii) Cloning of full-length cDNA molecules for *in vitro* study of gene function. (iii) Study of the repertoire of mRNAs expressed in different cells or tissues. (iv) Study of alternative splicing in different cells or tissues. (v) cDNA libraries are useful in reverse genetics, but they only represent a very small (less than 1%) portion of the overall genome in a given organism. (vi) Since cDNAs do not carry introns, they therefore can be expressed in prokaryotic cells.

complementary DNA (cDNA) library: Collection of DNA copies of all the mRNAs produced by a cell type, tissue or organism that is formed using reverse transcriptase, representing all the genetic information expressed by such cells inserted into a cloning vector. cDNA libraries are produced for gene isolation and cloning purposes. The size of the cDNA library depends upon a number of expressing genes for a particular trait. For instance, if the protein gene is to be cloned, then mRNAs are isolated and purified from developing seeds of grain legumes, which are the rich source of proteins. mRNA is converted into cDNA using an enzyme reverse transcriptase. cDNA segments are inserted into suitable vector. Collection of cDNA segments is called cDNA library. Different steps of cDNA library construction include (i) MRNA EXTRACTION, (ii) CDNA CONSTRUCTION, (iii) LIGATION TO VECTOR and (iv) SCREENING THE CLONES. *See details as separate terms.*

complementary DNA (cDNA) library vs genomic DNA library: A cDNA library absences the non-coding and regulatory elements found in genomic DNA and is much less resource-intensive to generate and maintain. Genomic DNA libraries provide much more detailed information about the organism but are much more resource-intensive to generate and maintain.

complementary DNA (cDNA) of satellite RNA: Some RNA viruses have small RNA molecules called satellites. These satellites depend on viral genome for their replication but are not necessary for viral function. In many cases, satellites either increase or decrease the severity of disease produced by the virus carrying it. The complementary copies of satellites that reduce disease severity have been integrated into host genomes. The expression of satellites has been shown to reduce disease symptoms. For example, tobacco plants expressing satellites of CUCUMBER MOSAIC VIRUS (CMV) showed reduced disease symptoms.

complementary DNA (cDNA), construction of: Once mRNA is purified, oligo-dT (a short sequence of deoxy-thymine nucleotides) is marked as a complementary primer, which binds to the poly-A tail providing a free 3′-OH end that can be extended by reverse transcriptase to create the complementary DNA strand. The mRNA is then detached by using a RNase enzyme saving a SINGLE-STRANDED cDNA (sscDNA). This sscDNA is converted into a double-stranded DNA with the help of DNA polymerase. However, for DNA polymerase to synthesize a complementary strand, a free 3′-OH end is needed. This is offered by the sscDNA itself by generating a hairpin loop at the 3′ end by coiling on itself. The polymerase extends the 3′-OH end, and later, the loop at 3′ end is opened by the scissoring action of S_1 nuclease. Restriction endonucleases and DNA ligase are then used to clone the sequences into bacterial plasmids. The cloned bacteria are then selected, often through the use of antibiotic selection.

complementary DNA (cDNA)-amplified fragment length polymorphism (AFLP) transcript profiling technique: This technique has been utilized to analyse gene expression during flower development in *Petunia hybrid* (F. Cnudde and others in 2003). Reproductive and vegetative floral organs were sampled at five developmental stages and gene expression profiles were compared. This allowed assembly of an inventory of genes expressed mainly in anthers during microspore development and in ovaries during macrosporogenesis. About 6000 transcript tags were generated, 354 of which revealed a modulated and/or organ-specific expression pattern. Stamen-specific transcripts exhibiting an upregulation in gene expression were well represented in our screening. Ovary-specific transcripts were less frequently observed and often displayed a constant level of gene expression. Of 194 fragments characterized further by sequencing, 35% showed homology with known genes in a database search. They belong to a wide range of gene classes, such as proteases, transcription factors and genes involved in metabolism, cell cycle and disease resistance. The usefulness of cDNA-AFLP transcript profiling as a tool to unravel complex developmental processes at the molecular level is discussed.

complementary DNA (cDNA)-amplified fragment length polymorphism (AFLP): This is one of the simplest methods to get an idea about the differences in transcriptome of: (i) plants subjected to different stresses; (ii) plants that differ for specific genes; and (iii) different organs or developmental stages of the same plant. The objective here is to identify DIFFERENTIALLY TRANSCRIBED FRAGMENTS (DTFs). The mRNA is isolated from two or more target plants or tissues and reverse transcribed into cDNA, which is then subjected to AFLP to identify polymorphic fragments. DNA from each polymorphic band is eluted, cloned and sequenced to identify the genes involved in differential expression.

complementary effect: Genes that by interaction their products produce a phenotype qualitatively distinct from the phenotype of any of them separately.

complementary event: An event opposite to the event being investigated. Complementary events occur when there are only two outcomes, such as clearing an examination or not clearing an examination.

complementary gene interaction: Either gene, when homozygous recessive, is epistatic to the other. Such genes are called COMPLEMENTARY GENES because two dominant genes complement each other to produce the wild-type phenotype. Complementary genes are similar in phenotypic effect when present separately. When existing together they interact to produce a wild-type phenotype. The genes showing this type of interaction are known as complementary genes because two dominant genes interact to produce the wild-type phenotype. In this case, wild-type and mutant phenotypes appear in 9:7 ratio in F_2.

complementary genes: Genes that interact to produce an effect distinct from the effects of an individual gene. Complementary genes yield same mutant phenotype when present separately, but when present together, they interact to produce a wild-type phenotype.

complementary homopolymeric tailing: The process of adding complementary nucleotide extensions to different DNA molecules, for example, dG (deoxyguanosine) to the 3′-hydroxyl ends of one DNA molecule and dC (deoxycytidine) to the 3′-hydroxyl ends of another DNA molecule to facilitate, after mixing, the linking of the two DNA molecules by base pairing between the complementary extensions.

complementary nucleotides: A property of nucleotides, whereby adenine (A) always pairs with thymine (T) or uracil (U) in DNA and RNA, respectively, while cytosine (C) always pairs with guanine (G). The DNA double helix is made up of two strands of DNA held together by complementary base pairs.

complementary RNA: Synthetic RNA produced by transcription from a specific DNA single-stranded template.

complementary sequence: Nucleic acid base sequence that can form a perfectly paired double-stranded structure with another DNA fragment by following Watson-Crick base-pairing rules (A pairs with T and

C pairs with G). The complementary sequence to GTAC, for example, is CATG.

complementation: The ability of two different genetic defects to compensate for one another. The production of a wild-type phenotype when two haploid genomes bearing different recessive mutations are united in the same cell. Demonstration of the recessive nature of individual mutants is a crucial result that allows us to proceed with a complementation test. Usually, a protein product is involved, as this is a diffusible molecule that can exert its effect away from the DNA itself. For example, a *lacZ*$^+$ gene on a plasmid can complement a mutant (*lacZ*$^-$) gene on the chromosome by enabling the synthesis of β-galactosidase. Dominant mutations cannot be used in a complementation test. This phenomenon is exploited to isolate somatic hybrid cells in plants. For instance, in case of chlorophyll-deficient mutants, the plants are albino, whereas the somatic hybrid among them becomes green because of complementation. Therefore, green plants are selected as somatic hybrids.

complementation assay *in vitro*: The assay that identifies a component of a wild-type cell that can confer activity on an extract prepared from a mutant cell. The assay identifies the component reduced inactive by the mutation.

complementation group: Cistron determined by the *cis-trans* complementation test.

complementation map: A map developed from the complementation relationships between alleles, normally in a small segment of the chromosome.

complementation mapping: According to M.M. Green and K.C. Green in 1949, *Lozenge* (*lz*) in *D. melanogaster* is a complex locus, which consists of four groups, each containing one or more different lozenge alleles. Alleles of different groups recombine. All these mutants are recessive. When two mutations at two subloci are present in *cis*-position, they have their wild-type alleles on the other homologue. However, when two mutations are placed in *trans*-position, many heterozygous combinations of different lozenge alleles do show mutant effects. From this, E.B. Lewis (1951) hypothesized that this effect arose from sequential nature of gene product synthesis. S. Benzer used this method between 1954 and 1962 to study fine structure of the gene. *See* FINE STRUCTURE OF THE GENE.

complementation matrix: A tabular representation of complementation tests involving a number of phenotypically similar mutants.

complementation test: A genetic test to investigate if two mutations with similar phenotypes are allelic (i.e. occur in the same gene) or are non-allelic (i.e. occur in separate genes). Individuals homozygous for each mutation are crossed. If the mutations are allelic, the offspring will have the mutant phenotype. If they are nonallelic, the offspring will be heterozygous for both the alleles and will have the wild-type phenotype. The molecular explanation of such results is often in relation to biochemical pathways in the cell.

complete block design: A simple experimental design in which all testers are included in each replication of the experiment and are placed in a random order within replication. For accuracy, this design should be used with small numbers of entries only. Replications may be placed end to end or opposite each other, so that the total area covered by the experiment is as nearly square in shape as possible. Entries with apparent weakness may be rejected before harvest, and data may still be analysed by an ANALYSIS OF VARIANCE (ANOVA).

complete copy choice hypothesis of recombination: J. Lederberg in 1955 gave a modified version of J. Belling's hypothesis. The entirety of recombinant chromosomes arise from newly synthesized sections that have copied a part of their sequence from a section of one parental and the other part from a section of the other parental chromosome. This hypothesis relates recombination to replication of the hereditary material. This hypothesis requires that homologous chromosomes shall be paired at least at certain points along their length at the time of replication and that new strands may be replicas of one parental strand in one region and of the other in another region. This theory signifies that replication occurs at the chromatid level. It also provides an explanation of how within a tetrad on site may show a 2:2 ratio and a neighbouring one a 3:1 ratio (gene conversion).

complete diallel: A mating design and subsequent progeny test ensuing from the crossing of a certain number of parents in all possible combinations including selves and reciprocals. Because of severe inbreeding depression in the selves, these are often hopped; nevertheless, the test is still called a FULL DIALLEL. *See* DIALLEL CROSS.

complete digest: The treatment of a DNA preparation with an endonuclease for sufficient time so that all the potential target sites within that DNA are cleaved.

complete dominance: Comprehensive resemblance of F_1 with one of its parents. *See* DOMINANCE.

complete flower: A flower that has all the four essential parts – pistils, stamens, petals and sepals.

complete genome sequencing for gene discovery, an alternative to: An alternative to complete genome sequencing for discovering genes is being used in trees and many other organisms. The procedure is based on identifying only the DNA that includes the genes that are expressed in the genome. They are partial or complete sequences of COMPLEMENTARY DNA (cDNA) obtained from mRNA isolated from different tissues and therefore represent genes expressed in these tissues with often known or suggested function. EXPRESSED SEQUENCE TAGS (ESTs) are compared to all other sequences in gene databases to identify matches likely representing highly homologous genes. If there is noticed a high similarity (homology) to some other gene sequence whose identity has been determined, then the identity of the EST can be immediately inferred. ESTs can be used as a resource for identifying candidate genes for QTLs involved in genetic control of adaptive traits. Large libraries of partial or complete sequences of thousands of expressed genes have already been obtained, and large databases of ESTs are accessible for many animal and plant species, including several forest tree species, such as Monterey or radiata pine (*Pinus radiata*), loblolly pine (*P. taeda*), Norway spruce (*Picea abies*), *Eucalyptus* and *Populus*.

complete imprinting: Some of the imprinted genes from plants show differential rather than complete (binary) imprinting. Quantification of Allele-Specific Expression by Pyrosequencing® is a tool suitable for observing whether or not imprinting is complete. *Aka* BINARY IMPRINTING.

complete Latin square: A Latin square is considered to be complete if it is both row-complete and column-complete.

complete lethal: In F_2 progeny obtained in Basc procedure, four phenotypic classes, namely Muller-5 females, heterobar females, Bar eyed males and normal males are observed in 1:1:1:1 ratio statistically, when no recessive lethal was present in X-chromosome of treated *Drosophila* males. But this ratio is observed to be 1:1:1:0 when complete recessive lethal mutation was induced in the treated *Drosophila* male.

complete linkage: A condition where no crossovers are observed in a test cross or F_2 progeny. Closely linked genes may show complete linkage. In some species, crossing over is absent in one or the other sex. For example, crossing over is absent in male *Drosophila* and female silkworm.

complete medium: A culture medium that is enriched to contain all the growth requirements of a strain of organisms.

complete penetrance: A situation in which a dominant gene always produces a phenotypic effect.

complete resistance: A resistant reaction in which some aspect of disease development, usually symptom expression or pathogen reproduction, is completely inhibited. *Contrasts* PARTIAL RESISTANCE.

completely randomized design: The structure of the experiment in a completely randomized design is assumed to be such that the treatments are allocated to the experimental units completely at random.

completeness of genome test: Direct test of completeness of genome in a differentiation cell is to take out the nucleus and put it into an enucleated egg. If this nucleus directs the development of a normal embryo, it clearly carries all necessary genetic information in a readily available form. Experimental test of totipotency of nuclei in differentiated cells provided evidence of genome constancy in differentiated cells.

completion of the first survey of the entire human genome: 26 June 2000 was a historic event at the White House, when US President Bill Clinton announced that international Human Genome Project and Celera Genomics Corporation had both completed an initial sequencing of the human genome. He underscored that this genetic information must never be used to stigmatize or discriminate against any individual or group. Such scientific advances must always incorporate our most cherished values, and the privacy of this new information must be protected. Coding of the human genome is expected to lead to new ways to prevent, diagnose, treat and cure disease.

complex: Assembly of molecules that are seized together by non-covalent bonds. Protein complexes perform most cell functions.

complex adaptive traits via QTL mapping: The procedure adopted for detecting and locating QTLs is called QTL mapping. The conceptual basis of this method is comparatively simple, but it requires relatively dense genetic maps with constantly distributed markers covering the entire genome, appropriate statistical tools and an appropriate experimental population of sufficient size segregating for both genetic markers and phenotypic traits. It involves numerous steps. First, multi-locus genotypes (molecular markers) and phenotypes (quantitative traits) are measured on all individuals of segregating population. Then, phenotypic values are statistically related with genotypic values, usually using multiple regression or maximum likelihood methods to recognize markers that have a strong connotation (joint segregation) with the quantitative trait. Such association can be the consequence of either tight linkage of the genetic marker and QTL (because they are positioned in the same region of the chromosome) or direct involvement of this particular marker(s) in genetic control of the trait. QTLs have already been detected and mapped in forest trees for such adaptive traits as growth rhythm, phenology, form, wood quality, disease resistance, cold hardiness and drought tolerance. Once a QTL controlling an adaptive trait has been precisely mapped, it then may become possible to clone the gene underlying the QTL based merely on the knowledge of its genetic map position and without knowing its function or DNA sequence. *Aka* POSITIONAL CLONING *or* MAP-BASED CLONING.

complex character: One whose difference from an alternative is not transmitted as single unit of heredity.

complex disease: A disease or disorder which does not follow a mendelian inheritance pattern. Typically, complex diseases are the result of multiple susceptibility alleles along with environmental factors. Examples include heart disease, obesity, diabetes and cancer. This stands in contrast to a SIMPLE GENETIC DISEASE that is more directly caused by mutations in a single gene. *Aka* MULTIFACTORIAL DISEASES.

complex gene: Complex genes are so called because either there is rearrangement of DNA at different levels before 3′←5′ strand of DNA is available for transcription (as is the case of IMMUNOGLOBULIN GENES), or pre-mRNA is cleaved through a series of steps before mRNA is available in translatable form (as is the case of DIMORPHIC GENES), as exemplified by the kallikrein gene, or these genes have a mysterious structure in that the ultimate active product or products are carried within the precursor protein (as is the case of CRYPTOMORPHIC GENES), exemplified by yeast SEX PHEROMONE GENE (*MF1*). Mammalian pancreatic glucagon genes, which are released after enzymatic breakdown of the precursor, become functional following further processing. *See* IMMUNOGLOBULIN GENES, DIMORPHIC GENES *and* CRYPTOMORPHIC GENES.

complex genetics of mtDNA disorders: Human diseases related with defective mitochondrial function exhibit a broad array of clinical problems. Typically, a single mitochondrial mutation leads to a complex medical syndrome.

complex heterozygous: Special type of genetic system based on the heterozygosity for multiple reciprocal translocations.

complex locus: A bunch of two or more closely linked and functionally related genes comprising a pseudoallelic series.

Complex of Proteins Associated with Set1: Also called Set1-COMPASS proteins. H. Woo *et al.* (2024) show that N-terminal acetylation of Set1-COMPASS proteins by N-terminal acetyltransferases (NATs) can moderate H3K4 methylation patterns. Precisely, deleting NatA substantially reduces global H3K4me3 levels and causes the H3K4me2 peak in the 5′ transcribed regions to move to the promoters. N-terminal acetylation of Set1-COMPASS proteins can thus directly fine-tune the purposes of this complex, thereby significantly influencing H3K4 methylation patterns.

complex oligosaccharide: Chain of sugars attached to a glycoprotein that is produced by trimming of the original oligosaccharide attached in the endoplasmic reticulum and subsequent addition of further sugars.

complex promoters: These promoters have interdigitated array of constitutive and regulatory elements. Both constitutive and regulatory elements are necessary for precise enhancer function.

complex regulatory network formation: Chemical modifications to DNA and histone proteins form a complex regulatory network that modulates chromatin structure and genome function.

complex trait: (i) Trait that has a genetic component that does not follow strict Mendelian inheritance. May involve the interaction of two or more genes or gene-environment interactions. (ii) The traits determined by quantitative trait loci (QTLs) are also called complex traits. *See* MENDELIAN INHERITANCE *and* ADDITIVE GENETIC EFFECTS.

complex transcription units: The transcription units in which a primary transcript may yield two or more mRNAs that encode different proteins. Complex transcription units may have two or more poly(A) sites or two or more splice sites.

complexity: Total length of different sequences of DNA present in a given preparation.

complexity of an algorithm: Describes the number of steps required by the algorithm to solve a problem and the amount of data (e.g. the length of sequences to be aligned).

complexity of DNA: It is the total length of DNA of different sequences DNA present in a given preparation.

complexity of gene sequence: The term 'low complexity sequence' may be thought of as synonymous with regions of locally biased amino acid composition. In these regions, the sequence composition deviates from the random model that underlies the calculation of the statistical significance (P-value) of an alignment. Such alignments among low complexity sequences are statistically but not biologically significant (i.e. one cannot infer homology (common ancestry) or functional similarity).

complexity of phenotypic variation: QTL analysis based on phenotypic variation rather than QTL mapping based on DNA sequence or chromatin information holds strength. The initial QTL scan at time point (T = 0) founded on DNA or chromatin markers did not detect any QTL at any of the two loci (L1 or L2). This was due to the occurrence of 'suppresses' or 'buffers' for differential gene expression at non-polymorphic epigenomic loci (epi-loci) that halt the DNA sequence variation. Later on, the environmental trepidation at T = 0 affects L1 (T = 1) and a sequence-dependent (*cis*) chromatin change occurs in the next generation due to buffered sequence variation release. At T = 1, QTL analysis deciphered $phQTL^{dna}$ and $phQTL^{epi}$ linked with chromatin markers. The hidden sequence variants enhanced the gain in phenotypic variation among population. At T = 2, the generation undergoes attenuation and locus L1 trigger induction of chromatin changes at locus L2 (*trans*). The combined approach of chromatin level data in quantitative genetic investigations has the potential to provide scrutiny to the complex phenotypes in populations over time.

complexity of transcriptome, proteome and genome: The transcriptome and proteome are much more complex than genome. A single gene can produce many different mRNAs and hence proteins. Different transcripts from the same gene can be generated by alternative splicing, alternative promoter usage, polyadenylation site usage or special RNA processing strategies (like RNA editing). Different proteins can be generated from the same transcript by alternate use of translation start and stop codons. Proteins synthesized from one mRNA can be variously modified during or after translation.

complon: A subunit of complementation map of a cistron.

component of fitness: *See* FITNESS COMPONENT.

component of variance: *See* VARIANCE COMPONENT.

component weight: For each stage of sampling, the component weight is equal to the reciprocal of the probability of selecting the unit at that stage.

components of translation: (i) mRNA serves as a template for production of protein which in turn performs structural or functional role. (ii) On the ribosome (ribonucleoprotein), mRNA in conjunction with covalently linked amino acids with tRNAs aggregate accomplishes protein synthesis and thus function as the sites of polypeptide synthesis. (iii) Twenty essential amino acids found in proteins are required. The three types of RNAs (mRNA, tRNAs. rRNAs) are important components of the protein synthesizing machinery. (iv) All three of them are found together at the ribosome during protein synthesis. (v) For binding of each amino acid and a specific tRNA, one specific aminoacyl synthetase (AAS) enzyme is required. tRNA is an adaptor molecule with dual functions of being used to recognize both the codon and the amino acid.

composite: A mixture of genotypes derived from diverse sources and maintained by normal pollination. *See* SYNTHETICS.

composite cross: A population resulting from the hybridization of several parents, either by hand pollination or by the use of male sterility.

composite insertion sequence (CIS) elements: Composite IS elements are 3000–10,000 bp long. These elements are composed of parts of two or more different IS elements.

composite interval mapping (CIM): This method of QTL mapping combines interval mapping with linear regression and includes additional genetic markers in the statistical model in addition to an adjacent pair of linked markers for interval mapping. *Compare with* INCLUSIVE COMPOSITE INTERVAL MAPPING.

composite karyotype: A karyotype containing all clonally occurring abnormalities in a tumour.

composite mixture: Breeder seed attained by mechanically combining seed from two or more strains. The mixture is increased through successive steps in a certified seed programme and distributed as a synthetic variety.

composite transposons: Bacterial transposons carry genes (e.g. antibiotic resistance) flanked on both sides by insertion sequence (IS) elements (IS modules). The IS elements are of the same type and are called IS_L (left) and IS_R (right). IS_L and IS_R may be in direct or inverted orientation to each other. Insertion elements only carry enzymes needed for transposition (transposase, resolvase) composite transposon probably evolved from IS elements by the chance location of a pair in close proximity to one another. Inactivation of one element by mutation would not harm ability to transpose and would assure continued transposition of the entire transposon. Example of composite transposon is Tn10.

composite variety: A plant population in which at least 70% of progeny comes from cross of the parent lines.

composite-cross population: A population produced by hybridization more than two varieties and/or lines of normally self-fertilizing plants and proliferating successive generations of the segregating population in bulk in specific environments so that natural selection is the principal power functioning to produce genetic change. Artificial selection may also be executed. The resulting population is expected to have a continually changing genetic makeup.

composition of nuclear compartment: Four prominent components make up typical nucleus. These components are (i) nucleolus; (ii) nuclear sap, *aka* FLUID MATRIX *or* NUCLEOPLASM; (iii) nuclear envelope or surrounding membrane; and (iv) chromosomes.

compost: A rich soil-like material produced from decayed plants and other organic matter, such as food and animal waste, that breaks down naturally. Most food waste can be put into compost, but you should not include meat, bones, cheese, cooking oils and fish. These may take a long time to break down and attract unwanted pests.

composting: Preparing plant material waste through microbial decomposition for manuring. It is a common practice by which the plant parts such as leaves, stems and roots are cut into small pieces and buried in the pits for degradation using microbial consortia. Composting is usually done for preparing organic manures.

compound: Composed of several parts (e.g. a leaf composed of multiple leaflets); a gynoecium composed of multiple carpels, or an inflorescence made up of multiple smaller inflorescences.

compound antigen: Formation of unique serological determinant by combination of two common antigens.

compound chromosome: A chromosome formed by the amalgamation of two separate chromosomes.

compound cross: An amalgamation of desirable genes from more than two inbred lines, breeding strains or varieties.

compound genes: Genes in which coding sequences (exons) are separated by non-coding sequences (introns). Discovery of split genes was reported independently by two groups led by P.A. Sharp and R.J. Roberts in 1977. *Aka* SPLIT GENES.

compound heterozygote: When an individual has two different variant alleles in a given gene which results in disease.

compound leaf: A leaf that is divided into many separate parts (called leaflets) along a midrib (the rachis). All the leaflets of a compound leaf are slanted towards the same plane. When a compound leaf falls from the tree, it does so as a unit. A DOUBLE COMPOUND LEAF is one in which each leaflet of a compound leaf is also made up of SECONDARY LEAFLETS.

compound leaf morphogenesis: Compound leaves are initiated as simple primordia at the flanks of the shoot apical meristem (SAM), which subsequently undergo a series of coordinated morphogenetic events to develop into a complex structure with multiple leaflets. The two most important events of this process are leaflet initiation and boundary formation. Some progress in understanding the regulatory pathways for each event has been made. L. He and others in 2024 identified the *PINNATE-LIKE PENTAFOLIATA2* (*PINNA2*) gene encoding a newly identified GRAS transcription factor in *Medicago truncatula*. PINNA2 transcripts are preferentially detected at organ boundaries. Its loss-of-function mutations convert trifoliate leaves into a pinnate pentafoliate pattern. PINNA2 directly binds to the promoter region of the LEAFY orthologue SINGLE LEAFLET1 (SGL1), which encodes a key positive regulator of leaflet initiation, and downregulates its expression. PINNA2 synergizes with two other repressors of SGL1 expression, the BEL1-like homeodomain protein PINNA1 and the C_2H_2 zinc finger protein PALMATE-LIKE PENTAFOLIATA1 (PALM1), to precisely define the spatiotemporal expression of SGL1 in compound leaf primordia, thereby maintaining a proper pattern of leaflet initiation. The enriched expression of PINNA2 at the leaflet-to-leaflet boundaries is, however, positively regulated by the boundary-specific gene MtNAM, which is essential for leaflet boundary formation.

Thus, boundary-expressed transcription factor PINNA2 regulates leaflet initiation. The expression of PINNA2 at leaflet boundaries is positively regulated by a boundary-specific protein NO APICAL MERISTEM (NAM)/CUP-SHAPED COTYLEDON (CUC), which is crucial for leaflet boundary formation. See GRAS GENE FAMILY.

compound light microscope: A microscope with visible light as source of illumination with magnification up to 1000×. Compound light microscopes are commonly used in the laboratories to examine the cells, protoplasts, fungi or even the bacteria. The eye pieces are usually of 10×, whereas objective lens may be of 10×, 40× or 100×. The illumination source is usually the electric bulb. Advanced models are fitted with phase contrast systems to study the unstained cells/organisms.

compound palmate: Having leaflets that radiate from a central point (usually at the top of a petiole), like spread out fingers radiating from the palm of a hand. *Compare with* PALMATE.

comprehensive high-throughput arrays for relative methylation (CHARM): CHARM has been developed from the *Mcr*BC assay by improving statistical procedures and the array design algorithm.

computability theory: A branch of mathematical logic, computer science and the theory of computation that came into light in the 1930s with work of K. Gödel, A. Church, R. Péter, A. Turing, S. Kleene and E. Post on computable functions and Turing degrees. The field has since expanded to include the study of generalized computability and definability. In these areas, recursion theory overlaps with proof theory and effective descriptive set theory. *Aka* RECURSION THEORY. *See* TURING MACHINES.

computation: Any type of calculation that includes both arithmetical and non-arithmetical steps and follows a well-defined model (e.g. an algorithm). The study of computation is paramount to the computer science.

computational biology: The field of biology that involves the development and application of data-analytical and theoretical methods, mathematical modelling and computational simulation techniques to the study of biological, ecological, behavioural and social systems. The field is broadly defined to encompass foundations in biology, applied mathematics, statistics, biochemistry, chemistry, biophysics, molecular biology, genetics, genomics, computer science and evolution. *See* BIOINFORMATICS.

computational chemistry: A branch of chemistry that uses computer simulation to facilitate solving of chemical problems. It uses methods of theoretical chemistry, incorporated into efficient computer programs, to calculate the structures and properties of molecules and solids.

computational complexity theory: A subfield of computational science that focuses on categorizing computational problems on the basis of their inherent difficulty and linking these classes to each other. A computational problem is a task solved by a computer. A computation problem can be solved by mechanical application of mathematical steps, such as an algorithm.

computational forecasts of histone methylation: For the establishment of an improved histone occupancy and modification landscape for multiple histone markers, M.N. Asim and others in 2023 presented an end-to-end computational multi-paradigm framework 'Histone-Net'. To facilitate researchers and practitioners, Histone-Net multi-paradigm framework is deployed as an interactive and user-friendly web server at https://histone.opendfki.de/.

computational intelligence (CI): Usually refers to the ability of a computer to learn a specific task from data or experimental observation.

computational model: A mathematical model in computational science that needs extensive computational resources to investigate the behaviour of a complex system by computer simulation.

computational nanotechnology: Computational studies of modelling and simulation of nanomaterials and structures. Almost any property of a material can be simulated. The predictive and analytical power of computation is critical to success in nanotechnology: nature required several hundred million years to evolve a functional 'wet' nanotechnology. The insight provided by computation should allow us to reduce the development time of a working 'dry' nanotechnology to a few decades. It will have a major impact on the 'wet' side as well.

computational prediction of DNA methylation: H. Zheng *et al.* (2023) developed CpGIMethPred, the support vector machine-based models to predict the methylation status of the CpG islands in the human genome under normal conditions. CpGIMethPred is available at http://users.ece.gatech.edu/~hzheng7/CGIMetPred.zip.

computational resources and bioinformatics tools in deep sequencing: Various programs/softwares with Web address/availability are CLC Genomics Workbench (commercial), Galaxy (academic, freely available), Genomatix genome analyser (commercial), NEXTGENe (commercial), SeqMan NGen (commercial), RMAP (academic, freely available), ELAND (commercial), Exonerate (academic, freely available), SSAHA2 (academic, freely available), Edena (academic, freely available), SHARCGS (academic, freely available) and EagleView (academic, freely available).

computational science: An interdisciplinary field that employs advanced computing capabilities to understand and solve complex problems. It is an area of science that spans many disciplines, but at its core, it involves the development of computer models and simulations to understand complex natural systems. *Aka* SCIENTIFIC COMPUTING *or* SCIENTIFIC COMPUTATION (SC).

computational steering: The practice of manually overruling with an otherwise autonomous computational process, to change its outcome.

computational thinking: The breaking down of a complex problem into a step-by-step solution in a way that a computer could execute. This type of thinking leads computer programming.

computer: A device that can be tutored to carry out sequences of arithmetic or logical operations automatically via computer programming. Modern computers have the ability to follow generalized sets of operations, called PROGRAMS. These programs enable computers to perform an extremely wide range of tasks.

computer architecture: A set of rules and methods that are followed for the functionality, organization and implementation of computer systems. It defines the capabilities and programming model of a computer but not a particular implementation. In some cases, computer architecture involves instruction set architecture design, microarchitecture design, logic design and implementation.

computer data storage: A technology comprising computer components and recording media that are used to retain digital data. Data storage is a core function and fundamental component of all modern computer systems. *Aka* STORAGE *or* MEMORY.

computer ethics: A part of practical philosophy concerned with how computing professionals should make ethical decisions regarding professional and social conduct.

computer graphics: Pictures and films created using computers. Normally, the term refers to computer-generated image data created with the help of specialized graphical hardware and software. It is a vast and recently developed area of computer science.

computer network: A digital telecommunications network which permits nodes to share resources. In computer networks, computing devices share or exchange data with each other using connections (data links) between nodes. These data links function over cable media, such as wires or optic cables, or wireless media, such as Wi-Fi. *Aka* DATA NETWORK.

computer programming: The process of designing and building an executable computer program for achieving a specific computing task. Programming involves responsibilities such as analysis, generating algorithms, profiling algorithms' accuracy and resource consumption and implementing algorithms in a chosen programming language, commonly referred to as coding. The source code of a program is written in one or more programming languages. The purpose of programming is to find a sequence of instructions that will automate the performance of a task for solving a given problem. The process of programming thus often requires expertise in several different subjects, including knowledge of the application domain, specialized algorithms and formal logic.

computer science: The concept, investigation and engineering that form the basis for the design and use of computers. It is the study of algorithms that process, store and communicate digital information. A computer scientist specializes in the theory of computation and the design of computational systems. *See* COMPUTATIONAL SCIENCE.

computer security: The protection of computer systems from theft or damage to their hardware, software or electronic data, as well as from disruption misuse or misdirection of the services they provide. *Aka* CYBERSECURITY *or* INFORMATION TECHNOLOGY SECURITY (IT SECURITY).

computer vision (CV): CV is a field of ARTIFICIAL INTELLIGENCE (AI) that enables computers and systems to derive meaningful information from digital images, videos and other visual inputs and take actions or make recommendations based on that information. If AI enables computers to think, computer vision enables them to see, observe and understand. Computer vision trains machines to perform these functions the human visual system can do, but it has to do it in much less time using cameras, data and algorithms rather than retinas, optic nerves and a visual cortex.

computer vision (CV) in agriculture: CV in agriculture is a great way for farmers to identify product defects, sort the produce by weight, colour, size, ripeness and many other factors. CV solutions offer significant benefits when used in conjunction with appropriate machinery.

computer-aided drug design (CADD): A broad range of theoretical and computational approaches that are part of modern drug discovery. CADD can contribute to selecting candidate drugs with favourable bioactivity and side-effect profiles. This is applicable in both designing novel drugs or replacing existing ones.

computing: It is any goal-oriented activity demanding, benefiting from or creating computing machinery. It comprises study of algorithmic processes and development of both hardware and software. It has scientific, engineering, mathematical, technological and social aspects. Major computing fields comprise computer engineering, computer science, cybersecurity, data science, information systems, information technology and software engineering.

concatamer: A DNA molecule that comprises a series of smaller DNA molecules linked head-to-tail via cohesive ends. Concatamers are formed during the replication of some viral and phage genomes. For example, Lambda phage DNA must be concatemers in order to be packaged.

concatenate: Interlocked circles (e.g. plasmids), joined together in a chain-like form. *See* CATENANE.

concatenation: In formal language theory and computer programming, string concatenation is the operation of joining character strings end-to-end. For example, the concatenation of 'snow' and 'ball' is 'snowball'. In certain formalizations of concatenation theory, also called STRING THEORY or STRING CONCATENATION, string concatenation is a primitive notion.

concept: Describing a concept requires a bit more elaboration than simply naming a principle or topic. An example of a scientific concept would be the idea that all matter is composed of atoms, which are themselves composed of smaller particles (protons, neutrons and electrons), which are further made of fundamental particles, such as quarks. Some other such examples include concept of gene structure and concept of dominance. These concepts have changed with new knowledge in the subject.

concept drift: In predictive analytics and machine learning, the concept drift means that the statistical properties of the target variable, which the model is trying to predict, change over time in unforeseen ways. This causes problems because the predictions become less accurate as time passes.

concept of dominance: Earlier, a gene was considered dominant if it was expressed in the heterozygous condition and recessive if it was masked. Sometimes, heterozygous individuals for a particular locus reveal an intermediate phenotype (incomplete dominance or mixture of two phenotypes; codominance). These genes can't be classified simply dominant or recessive without referring to the condition(s) at which the relationships exist. Classification of alleles as dominant or recessive ultimately depends on the function of the gene products produced and level of analysis applied to the phenotypes in question. Dominance is not a universal property of a gene. It is relative property of a gene to produce a particular phenotype in a particular genetic and environmental background.

conception: The penetrance of the egg by a sperm cell and the formation of a zygote.

concerted evolution: (i) The evolution of repeated sequences, which tend to persist homogeneous because of processes such as unequal crossing over and gene conversion. (ii) The tendency of the different genes in a gene family to evolve in concert; that is, each gene locus in the family comes to have the same genetic variant. (iii) The ability of two related genes to evolve together as though constituting a single locus.

concordance: When both members of a twin pair have a given trait. *Compare with* DISCORDANT.

concordant regulation: Perturbation of antisense RNA can alter the expression of sense messenger RNAs, suggesting that antisense transcription contributes to the control of transcriptional outputs. Gene expression profiling in mammals reveals frequent concordant regulation of sense/antisense pairs of transcripts.

concurrency: The ability of different parts or units of a program, algorithm or problem to be executed out-of-order or in partial order without affecting the final outcome. This permits parallel execution of the concurrent units, which can significantly enhance overall speed of the execution in multiprocessor and multicore systems. In more technical terms, concurrency refers to the decomposability property of a program, algorithm or problem into order-independent or partially ordered components or units.

concurrent controls: Control subjects assigned to a control condition during the same period that an experimental treatment or procedure is being evaluated.

condensation of chromosomes: The spiralization and charging of a chromosome with nucleic acid. *Aka* CONTRACTION OF CHROMOSOMES.

condensation reaction: A reaction in which a covalent bond is formed between two molecules through –OH groups with loss of a water molecule, as in the formation of a peptide bond between two adjacent amino acids during synthesis of a polypeptide chain.

condensin complex: This complex consists of five proteins, including two SMC proteins (Cut3 and Cutl4) and three non-SMC proteins (Cnd1, Cnd2 and Cnd3). Condensin complex is involved in condensation of chromosomes prior to mitosis and is a target for the M-Cdk.

condensins in bacteria: Condensins are conserved proteins containing STRUCTURAL MAINTENANCE OF CHROMOSOME (SMC) MOLECULES that are involved in establishment of compact chromosomes. MukBEF, the condensin in *E. coli*, cooperatively compacts a single DNA molecule into a filament with an ordered, repetitive structure in an ADENOSINE TRIPHOSPHATE (ATP) binding-dependent manner. MukBEF complexes are dumbbell- shaped with a central pivot. In an ATP-dependent manner, free MukBEF molecules with various pivot angles cooperatively polymerizes along the length of DNA, trapping and supercoiling variable lengths of DNA. The binding propagates along the length of DNA, creating a compressed, microheterogeneous and repetitive structure. Intermolecular head-head binding provides force-resistant interactions, whereas intramolecular head-head binding traps DNA in a force-sensitive manner. When the filament is extended, the force-sensitive intramolecular interactions are sequentially broken from one end, releasing variable lengths of sequestered DNA. When the force is lowered, recondensation begins, with a random nucleation event that cooperatively propagates along the extended filament by reforming the trapped DNA supercoils. Nucleation can occur at different positions along the extended filament, which results in multiple recondensation pathways.

conditional: A property of a programming language that performs different computations or actions depending on whether a programmer-specified Boolean condition assesses to true or false. Apart from the case of branch predication, this is always achieved by selectively altering the control flow based on some condition. *Aka* CONDITIONAL STATEMENT, CONDITIONAL EXPRESSION *or* CONDITIONAL CONSTRUCT.

conditional expression: Gene expression that occurs only in response to certain stimuli or specific conditions.

conditional lethal genes: Genes that are expressed in response to a specific environmental stimulus such as a specific chemical. When the lethal genes are expressed, they are lethal to the organism.

conditional mutation: A mutation that alters a protein or RNA molecule such that its role is altered only under some conditions. A mutation that has the wild-type phenotype under certain environmental conditions (e.g. temperature, age, nutrition) and a mutant phenotype under other conditions. For example, some of the male-sterile mutants in maize are male-sterile in one season (summer) and fertile in other (winter).

conditional probability: The probability (P) of the veracity of a statement or of an event A given that a specific condition B holds or that an event B has already occurred, denoted by $P(A \mid B)$. This is a probability in the presence of knowledge captured by B. For example, if the condition B is that a person is male, the conditional probability is the probability of A for males. It could be argued that there is no such thing as a completely unconditional probability. In this example, one is indirectly conditioning on humans even if not considering the person's sex.

conditional RNA interference: Specialized laboratory techniques have also been developed to improve the utility of RNAi in mammalian systems by avoiding the direct introduction of siRNA, for example, by stable transfection with a plasmid encoding the appropriate sequence from which siRNAs can be transcribed, or by more elaborate lentiviral vector systems, allowing the inducible activation or deactivation of transcription, known as conditional RNAi.

conditional-lethal mutant: A mutant that is lethal under one condition but not so under another condition.

conditioned dominance: Dominance impacted by the presence of other genes or by environmental variation.

conducting atomic force microscope (C-AFM): Contact mode AFM with conductive tip. The tip-surface current is monitored.

conducting probe AFM (CP-AFM): This is used to measure I–V characteristics of a nanoscale sample while doing AFM.

conduction: Plasmid deployment involving cointegrate formation.

cone: A type of fruit, usually woody, ovoid to globular, including scales, bracts or bracteoles arranged around a central axis (e.g. in gymnosperms, especially conifers and Casuarina).

confidence bands: Lines on each side of a regression line or curve having a given probability of containing the line or curve in the population.

confidence belt: A term for a pair of numbers that predict the range of values (confidence interval) within a particular parameter. *Aka* CONFIDENCE LIMIT.

confidence coefficient: Refers to the formula for a confidence interval that determines probability level associated with the intervals, such as 90%, 95% and 99%.

confidence interval (CI): The interval within which results of other similar studies are expected to lie with a certain confidence level.

confidence level: The degree of assurance that other studies of similar type will have the same results as obtained by the participants in the current study.

confidence limits: (i) To state that the 0.95 confidence limits for an unknown quantity are [a, b] means that 0.95 of similarly constructed confidence limits in repeated samples from the same population would contain the unknown quantity. Many confidence intervals are asymmetric (e.g. intervals for probabilities, odds ratios and other ratios). (ii) Another way to define a confidence interval is the set of all values that if null hypothesized would not be rejected at one minus the confidence level by a specific statistical test. For that reason, confidence intervals are better known as COMPATIBILITY INTERVALS.

confidentiality: In genetics, the expectation that genetic material and the information gained from testing that material will not be available without the donor's consent.

configuration: (i) The complete ordering and detailed description of all parts of a software or database system. (ii) The three-dimensional arrangement in space of substituent groups in stereoisomers.

configuration management: The use of software to identify, catalogue and maintain the component modules that collectively constitute one or more systems or products.

confirmatory testing: Repeat testing generally performed to corroborate a positive test result.

conflict: Evolutionary conflict arises when two genes that interact with each other have different transmission patterns and therefore different evolutionary interests. Conflict arises, because the genes have only partially overlapping interests.

conflorescence: A term describing substantial differences between the overall structure of an inflorescence and that of its individual branches (e.g. the bottlebrush multiple-flower head of members of the genus *Callistemon*).

confocal microscope: A type of light microscope that produces a clear image of a given plane within a solid object. It uses a laser beam as a source of illumination and scans across the plane to produce a two-dimensional 'optical section'. Thus, it can be used to determine the exact length, breadth and height of the specimen.

confocal optics: A microscope optical system in which the condenser and objective lenses both focus on to one single point in the specimen.

conformation: The precise three-dimensional structure of atoms and bonds in a macromolecule (such as a protein or nucleic acid) describing its geometry and hence its molecular function.

conformational epitope: An epitope which is dependent upon folding of a protein. Amino acid residues existing in the antibody binding site are often located at sites in the primary sequence of the protein, which are at some distance from each other. The massive majority of B-cell (antibody binding) epitopes are conformational.

confounded: A term used to describe a study or observation that has one or more nuisance variables present that may lead to incorrect interpretations.

confounder: A variable measured before the exposure (treatment) that is a common reason of (or is just linked with) the retort and the exposure variable. A confounder, when accurately controlled for, can explain away an apparent association between the exposure and the response. A formal definition of confounder is 'a pre-exposure covariate C for which there exists a set of other covariates X such that effect of the exposure on the outcome is unconfounded conditional on (X, C) but such that for no proper subset of (X, C) is the effect of the exposure on the outcome unconfounded given the subset'.

confounding variable: A variable more likely to occur in one group of subjects than another that is related to the outcome of interest and thus potentially confuses or confounds the results.

congenic strain: A variant plant strain that is produced by backcrossing a donor plant strain to an inbred parental strain for at least eight generations while maintaining by appropriate selection the presence of a small genetic region derived from the donor strain.

congenital: (i) Any trait that is present at birth, whether the result of a genetic or non-genetic factor especially the uterine environment. (ii) Common term for diseases that occur by birth. These diseases usually affect the B cells and T cells of the body, and finally, the immune system, which leads to infection and death in early life. *See* BIRTH DEFECT.

congression: The movement of chromosomes on to the metaphase plate, especially at the first meiotic division.

congruency: A measure of degree of similarity among different phylogenetic trees constituted of the same OPERATIONAL TAXONOMIC UNITS (OTUs).

congruent crossing: A type of cross-breeding when parents match precisely in morphology, physiological and/or cytogenetic behaviour.

conidia: External asexual fungal spores that are produced singly or in chains.

conidiogenesis: The production of asexual spores (conidia).

conidiogenous: Producing and bearing asexual spores (conidia).

conidioma (pl. conidiomata): A dedicated conidia-bearing structure (e.g. acervulus, pycnidium, sporodochium, synnema).

conidiophore: A simple or branched hypha on which conidia are formed.

conidium (pl. **conidia**): An asexual, non-motile fungal spore that develops externally or is liberated from the cell that had formed it.

conjugate: (i) (of a Latin square) A Latin square obtained by permuting the roles of 'rows', 'columns' and 'symbols'. (ii) Molecule formed by fusing together (via recombination or chemically) two different molecules. The purpose of conjugation is to produce a molecule in which one of the original molecules has one function (e.g. a toxic, cell-killing function) while the other original molecule has another function (e.g. targeting the toxin to a specific site that might include cancerous cells).

conjugated protein: A protein containing a metal or an organic prosthetic group or both (e.g. a glycoprotein is a conjugated protein bearing at least one oligosaccharide group).

conjugated vaccine: A vaccine consisting of more than one component of the antigen. In case of cellular vaccines, purified cellular components act as vaccine. Sometimes, these components are too weak to serve as an effective vaccine. This problem has been solved for instance by combining (conjugating) the polysaccharide antigen with a protein such as diphtheria toxoid to make it strong antigen.

conjugation: (i) A process analogous to sexual reproduction in bacteria where plasmid-mediated transfer of genetic material takes place from a 'male' donor bacterium to a 'female' recipient through mating in bacteria. (ii) A process that comprises cell-to-cell contact and the one-way transfer of DNA from the donor to the recipient. This process involves the formation of conjugation tube to transfer DNA from one bacterial cell to other. This transfer of DNA always occurs from F^+ to F^- cell or from Hfr to F^- cell. Bacteria can be stimulated to conjugate by transposons and plasmids, which thereby achieve horizontal transmission. Though conjugation, both plasmids and chromosomal DNA can be exchanged. (iii) In contrast to some other DNA-transfer processes of bacteria, conjugation may involve the transfer of large portions of the genome. (iv) A progression whereby organisms of identical species, but opposite mating types, pair and exchange genetic material (DNA) gene.

conjugation mapping: The technique that lets the relative positions of bacterial genes to be mapped by determining the time it takes for each gene to be transferred during conjugation.

conjugative plasmid: Double-stranded circular DNA molecule possessing ability to transfer a copy of its own to another bacterial cell through conjugation. Plasmids are extra chromosomal, double-stranded circular DNA, chiefly found in bacteria and also in several eukaryotes such as yeast. The F factor is a conjugative plasmid that carries genes for sex pili and for transfer of plasmid from donor to recipient bacterial cell during conjugation. These plasmids contain so-called *tra* genes, which perform the complex process of conjugation, the transfer of plasmids to another bacterium. Conjugative plasmids in bacteria can be transferred even to bacteria outside their own species. For example, F-factor of *E. coli* can be transferred during conjugation to bacteria of other genera within the family Enterobacteriaceae. Similarly, plasmid from *P. aeruginosa* (R1822) has an extraordinary broad host range, as it can be transferred to many different families of Gram-negative bacteria. Bacteriophages and retroviruses can be used as vectors for gene transfer.

conjugative transposons: Those transposons which are transferred between different bacterial cells through conjugation.

connective: The part of an anther that attaches the anther cells.

connexon: Water-filled pore in the plasma membrane moulded by a ring of six protein subunits. Part of a gap junction (i.e. connexons from two adjoining cells) join so as to form a continuous channel between the two cells.

consanguineous mating: The mating of closely related individuals, who share a common ancestor in the prior two or three generations, such as brothers, sisters or cousins. Another name for this mating pattern is INBREEDING. This is an extreme form of POSITIVE ASSORTATIVE MATING.

consanguinity: Coancestry (i.e. related by a common ancestor). A consanguineous marriage is a marriage between two close relatives.

consensus chloroplast simple (single) sequence repeat (ccSSR): Average or most typical form of a sequence that is generated with minor variations in a group of chloroplast DNA sequences. The consensus sequence of chloroplast DNA reveals the nucleotide or amino acid most often found at each position. The conservancy of a consensus implies that the sequence is functionally vital. The simple or single sequence repeat (i.e. microsatellite sequences repeat) is seen repeatedly in sequence within the DNA molecule in a way that enables them to be utilized as markers.

consensus map: *See* MOLECULAR GENETIC MAPS.

consensus sequence: A single sequence demarcated from an alignment of multiple constituent sequences in a group of related DNA, RNA or protein sequences that characterizes a 'best fit' for all those sequences. A 'voting' or other selection procedure is used to determine which residue (nucleotide or amino acid) is placed at a given position in the event that not all of the constituent sequences have the identical residue at that position. The term is also utilized for genes that encode the same protein in different organisms. The preservation of a consensus indicates that the sequence is functionally important. An example is the CAAT BOX in the promoter sequence.

conservation: Substitution of one amino for another to preserve the physicochemical properties of the original residue (e.g. when a hydrophobic amino acid residue is replaced by another hydrophobic residue).

conservation biology: The scientific study of nature and of the Earth's biodiversity with the aim of shielding species, their habitats and ecosystems from excessive rates of extinction and the erosion of biotic interactions.

conservation breeding: Refers to the practice of keeping endangered plants and animals in a controlled environment. *Aka* CAPTIVE BREEDING.

conservation breeding programmes (CBPs): The programmes that are governed by policy statements of national and international organizations for conserving the endangered species. Conservation over long term requires management to reduce risks, including *ex situ* conservations that could support and interact demographically and genetically with wild population. The World Association of Zoos and Aquariums (WAZA) follows IUCN red list and prioritized the funding opportunity for the conservation breeding of such species worldwide. Conservation Planning Specialist Group (CPSG) is a Specialist Group of the International Union for Conservation of Nature (IUCN) Species Survival Commission (SSC).

conservation of plant genetic resources: Plant genetic resources are conserved by two important methods. They are on-farm and off-farm conservations. On-farm conservation is the traditional and most effective method of conserving varieties and wild species. This method involves continuous re-propagation and selection of varieties under cultivation. On-farm conservation alone is dynamic in enhancing genetic variability. On-farm conservation is of two kinds. When a variety or wild species is conserved in its habitat region, it is called *IN-SITU* ON-FARM CONSERVATION. When the conservation is done away from this region, it is *EX-SITU* ON-FARM CONSERVATION. A combination of *in-situ* and *ex-situ* on-farm conservation is the best safeguard against genetic erosion. OFF-FARM CONSERVATION is *aka EX-SITE* GENE BANK.

conservation tillage: Refers to farming techniques or practices such as low-tillage crop production and no-tillage crop production that circumvent or diminish the disturbance of topsoil. Cumulatively, time and money are saved, and environmental impact is reduced (e.g. conserving and enhancing the quality of topsoil, reducing pesticide runoff into groundwater and reducing fossil fuel use).

conservation-based strategies: These strategies originate from the homology shown between miRNAs in different species and are useful to predict miRNAs positioned to intergenic regions that were conserved across related species.

conservative DNA transposons: A DNA-based transposable element that itself travels from one (donor) site to another (target) site in the genome but does not leave a copy in the original location (e.g. Tn5, Tn10, P elements). *Aka* CONSERVATIVE TRANSPOSITION.

conservative recombination: The recombination that involves breakage-and-reunion of pre-existing strands of DNA without the synthesis of new stretches of DNA.

conservative replication: A hypothesized mode of DNA replication in which an intact double helix acts as a template for synthesis of a new double helix, known to be incorrect.

conservative substitution: A nucleotide mutation which changes the amino acid sequence of the protein, but which causes the substitution of one amino acid with another that has a side chain with similar charge or polarity characteristics. The size of the side chain of amino acid may also be an important consideration. Conservative mutations are generally thought improbable to greatly change the structure or function of a protein, but there are many exceptions. It depends upon the site of replacement. Certain biochemical or kinetic properties of a protein may be affected by conservative mutations. Such substitutions cannot, however, be detected electrophoretically. *See* AMINO ACID *and* NON-CONSERVATIVE SUBSTITUTION.

conservative test: A term used to describe a test if it reduces the chances of a type I error.

conserved: (i) The number of genes present within the DNA of more than one species. For example, ~25% of the genes found within the human nuclear DNA are also found within the nuclear DNA of plants. (ii) A particular domain of a molecule on the surface of a rapidly mutating microorganism (e.g. the influenza virus, the AIDS virus) that remains the same in all, or most, variations of that microorganism. If that conserved region is appropriate to function as an antigen (HAPTEN, EPITOPE), it may be possible to produce a successful vaccine against that microorganism, that would otherwise be unsuccessful due to the fact that the rapid mutation would cause it (e.g. the AIDS virus) to appear to be different than the one (antigen) against which the vaccine was designed.

conserved function: A property of paralogous genes with high DNA sequence homology that code for proteins with similar function in distantly related organisms.

conserved orthologous set (of markers) (COS): COS is used for comparative mapping between closely related plant species. For a given group of species, a COS is formed by identifying genes from each species that are orthologous to genes of other species in the set.

conserved sequence: A base sequence in a DNA molecule (or an amino acid sequence in a protein) that has remained essentially unchanged throughout evolution.

consistency: A measure of whether, given sufficient data, a method will generate a correct answer.

consistency index: A measure of how well the character states for a specific character trait map on a phylogenetic tree. It is calculated as the ratio of the minimum number of evolutionary state changes (i.e. steps) that could be estimated from all conceivable trees divided by the actual number of changes on the tree being examined.

conspecific: Belonging to the same species.

constant (C) region of immunoglobulin (Ig): The region in light as well as heavy chains of immunoglobulins that are coded by C genes and are parts of a chain that vary the least in amino acid sequence. The constant regions of heavy chains identify the type of immunoglobulin.

constant-amplitude mode of non-contact atomic force microscope (CAM AFM): Constant-amplitude mode of non-contact AFM. Here the tip is oscillating with a constant amplitude of 1–10 nm at a given frequency. The interactions with the surface change the frequency. The amplitude of oscillation is kept constant. When the cantilever oscillation is damped as a result of interaction, the excitation voltage to keep the amplitude constant (V_{exc}) increases. The shift in the frequency and the shift in the excitation voltage can be used for imaging the surface and/or studying the strength of interaction.

constant-current mode for STM: In constant-current mode, the feedback keeps the tip-surface current constant, and the distance between the sample and the tip is monitored.

constant-force mode for AFM: The feedback keeps the force between the cantilever and the sample constant.

constant-height mode for scanning probe microscope: Similar to the constant-current or constant-force modes, but the feedback mechanism is turned off, and so the technique is faster.

constitutional deletion: Deletion inherited in the germ line from one or other parent and present in every cell of the body (usually examined in lymphocytes). *Contrasts* SOMATIC.

constitutive: A plant is said to be constitutive for the production of an enzyme or some other protein if that protein is always manufactured by the cells at constant rate under all physiological conditions. The constitutive expression of a transgene is required where the stress (biotic or abiotic) can affect any part of the plant. For example, *Bt* gene is made to express in the cotton in constitutive manner in all the cells of transgenic cotton plant irrespective of tissue does not allow the bollworm (insect) to attack the crop. This kind of expression is accomplished by using constitutive promoter like CaMV 35S along with *Bt* gene in the gene construct. If desired, using tissue-specific promoters, the transgene can also be expressed in tissue-specific manner (e.g. in roots, anthers or fruits). *Contrasts* INDUCED.

constitutive elements in eukaryotic gene regulation: Constitutive elements are common in a large number of different genes and are required for a basal level of expression. One such constitutive element is Goldberg- Hogness box or TATA box which is positioned 20–30 bases upstream of the transcription initiation site and is involved in determining the precise position of initiation of transcription. Constitutive sequences upstream of TATA box, such as CAAT and GC boxes, play role in controlling the frequency of transcription initiation.

constitutive enzymes: Enzymes that are part of the basic metabolic machinery of the cell. They are formed at a constant rate and in constant amounts regardless of the metabolic state of the organism. Presence or absence of an inducer does not affect constitutive enzyme production. Constitutive enzymes are products of constitutive (housekeeping) genes. Hence, they are called the HOUSEKEEPING ENZYMES. For example, constitutive enzymes include enzymes that are involved in the (i) production of cell-usable energy (such as ATP); (ii) the enzymes of the glycolytic sequence (which is the most ancient energy-yielding catabolic pathway); and (iii) enzymes of the monophosphate pathway. Constitutive enzymes are produced by all the body cells of an organism.

constitutive expression: Continual production of RNA or protein by an organism. *Aka* CONSTITUTIVE SYNTHESIS.

constitutive gene: A gene that is continually expressed in all the cells of an organism and its expression is not regulated. The product of a constitutive gene is produced regularly and is invariably present in the cell at roughly the same concentration regardless of whether substrate to that enzyme is ever presented to the cell or not. Such genes thus have invariant control. *Aka* HOUSEHOLD GENES or HOUSEKEEPING GENES.

constitutive heterochromatin (cHC): Heterochromatin that is heritable and is found at a particular locus at all times. Highly repetitive DNA sequences, such as centromeric (*aka* PERICENTRIC HETEROCHROMATIN, which is irreversibly silenced) and telomeric domains are examples of cHC. cHC is found in hypoacetylated nucleosomes. It is also associated with H3K9me. DNA within constitutive heterochromatin is typically AT-rich. cHC is always condensed during interphase. Permanently non-expressed sequences, usually SATELLITE DNA. *Compare with* FACULTATIVE HETEROCHROMATIN.

constitutive mutant: A mutant whose transcription is no longer under regulatory control.

constitutive mutations: Mutations that cause non-constitutive genes to become constitutive genes.

constitutive promoter: An unregulated promoter that causes continual transcription of its associated gene.

constitutive resistance: Genetically controlled (inherited) resistance to a disease.

constitutive RNA splicing: The process of intron removal and exon ligation of all the exons in the order in which they appear in a gene.

constitutive secretion: The proteins which are continuously secreted from the bacterial cells.

constitutive secretory pathway: Pathway existing in all cells by which molecules such as plasma membrane proteins are constantly transported to the plasma membrane from the Golgi apparatus in vesicles that fuse with the plasma membrane. *See* DEFAULT PATHWAY.

constitutive strain: A strain which constitutively produces a gene product.

constitutive synthesis: *See* CONSTITUTIVE EXPRESSION, HOUSEKEEPING GENES.

constitutively active gene: *See* CONSTITUTIVE GENE.

constriction: An unspiralized segment of fixed position in the metaphase chromosomes (e.g. nucleolar constriction, primary constriction or centric constriction, secondary constriction).

construct validity: A demonstration that the measurement of a characteristic is related to similar measures of the same characteristic and is not related to measures of other characteristics.

constructed language: A language whose phonology, grammar, and vocabulary are consciously devised, instead of having developed naturally. Constructed languages may also be referred to as artificial, planned or invented languages.

constructing an alignment and phylogeny: The steps in constructing an alignment and phylogeny are illustrated using programmes from the BIRCH SYSTEM. Assume a set of *GENBANK* entries has been retrieved, all of which represent homologous genes from several species. In GenBank entries, protein-coding sequences are annotated as 'CDS' feature. The FEATURES PROGRAMME can extract CDS sequences from a group of GenBank entries automatically. Next, coding sequences must be translated. Two multiple alignment programmes are available. CLUSTALX aligns by successive addition of sequences, using a phylogenetic tree as a guide. PIMA uses a clustering method based on hierarchical groupings of amino acids according to chemical structure. To display the alignment for evaluation or for final publication, it is best to try several of the programmes listed to tailor the output to your needs. REFORM generates straight ASCII TEXT, which can be easily imported into a word processor or drawing programme. The other programmes listed generate PostScript output of direct viewing or printing. If an aligned DNA sequence is desired, MRTRANS from the FASTA package can be read in an aligned protein sequence and generate a file containing DNA alignment.

constructing DNA lattices with complex patterning: The assembly process can be used to perform computations. This entails the building up of superstructures from the starting units such that the assembly process itself performs the actual computation.

constructive research: The area of that which develops solutions to a problem.

consultand: Anyone who consults a healthcare provider/genetic counsellor for genetic information.

Consultative Group on International Agricultural Research (CGIAR): An informal association of 58 public and private sector members supporting 16 international agricultural research centres. A group of donors established the CGIAR in the early 1970s to fund agricultural research around the world. The institutes under its auspices are: Africa Rice Center (ARC, formerly WARDA); Center for International Forestry Research (CIFOR); Centro Internacional de Agricultura Tropical (CIAT); International Center for Agricultural Research in the Dry Areas (ICARDA); International Crops Research Institute for the Semi-Arid Tropics (ICRISAT); International Food Policy Research Institute (IFPRI); International Institute of Tropical Agriculture (IITA), Ibadan, Nigeria; International Livestock Research Institute (ILRI), Nairobi, Kenya and Addis Ababa, Ethiopia; International Maize and Wheat Improvement Center (Centro Internacional de Mejoramiento de Maiz y Trigo, CIMMYT); International Plant Genetic Resources Institute (IPGRI); International Potato Center (CIP), Lima; International Rice Research Institute (IRRI), Los Banos, Philippines; International Water Management Institute (IWMI), Colombo, Sri Lanka; World Agroforestry Centre (ICRAF), Nairobi, Kenya; World Fish Center (WFC), Penang, Malaysia. These 16 International Agricultural Research Centres, which now call themselves FUTURE HARVEST CENTRES, comprising more than 8500 scientists and support staff working in more than 100 countries.

contact atomic force microscope (C-AFM): The scanning probe tip is in contact with the sample surface being investigated.

contact fungicide: A fungicide that sticks on the surface where it is applied and prevents infection often through the inhibition of spore germination; no after-infection activity of the fungicide is observed. *See* PROTECTANT FUNGICIDE. *Contrasts* SYSTEMIC FUNGICIDE.

contact hypothesis: The theory that chromosomes occasionally undergo illegitimate crossing over between non-homologous sections, resulting in an interchange of blocks of genes.

contact inhibition of locomotion: A phenomenon characterizing certain cells in which two cells meet, locomotory activity diminishes and the forward motion of one cell over the surface of the other is stopped.

contact mode: A technique used in ATOMIC FORCE MICROSCOPY (AFM).

contact points: The points at which the chromosomes first come in contact with each in pairing at zygotene.

contact-dependent signalling: Cell-cell communication in which the signal molecule remains bound to the signalling cell and only influences cells that physically contact it.

contagious disease: A disease which easily spread across the individuals. The contagious diseases are communicable diseases that spread from one person to another (e.g. chicken pox and measles).

container: A class, a data structure or an ABSTRACT DATA TYPE (ADT) whose occurrences are collections of other objects. They store objects in a systematized way that follows specific access rules. The size of the container depends on the number of objects (elements) it contains. Underlying (inherited) implementations of various container types may vary in size and complexity and provide flexibility in choosing the right implementation for any given scenario.

containment: Refers to the measures taken to prevent release of recombinant DNA molecules into the natural environment. For achieving this objective, biological and physical methods are applied. *See* BIOLOGICAL CONTAINMENT *and* PHYSICAL CONTAINMENT.

contaminant: Growth of microorganisms alongside of the plant tissue *in vitro*.

contamination: Growth of microorganisms in tissue cultures. During plant tissue cultures, contamination is a serious problem. Contamination is generally caused by bacteria and fungi. Contaminated cultures exhibit declined growth, and severe contamination causes death of culture tissues. Contamination can be minimized by maintaining aseptic conditions in the laboratory.

contelligence: The combination of awareness and computational power required in an ARTIFICIALLY INTELLIGENT (AI) network before we could, without loss of anything essential, upload ourselves into them.

content validity: A measure of the degree to which the items on a test or measurement scale are illustrative of the characteristic being measured.

contig: A length of contiguous sequence amassed from partial, overlapping sequences, generated from a 'shotgun' sequencing project. Contigs are typically created computationally by comparing the overlapping ends of several sequencing reads generated by restriction enzyme digestion of a segment of genomic DNA. The creation of contigs in the presence of sequencing errors, ambiguities and repeats is one of the most computationally challenging aspects of the role of bioinformatics in genome analysis. *Aka* CONTIGUOUS.

contig map: A map depicting the relative order of a linked library of overlapping clones representing a complete chromosomal segment. *Aka* CONTIGUOUS MAP.

contiguous gene syndromes: Refer to disorders caused by microdeletions or microduplications in neighbouring functional genes. Inheritance is usually infrequent, but reappearances are possible.

contiguous genes: Genes physically close on a chromosome that, when acting together, express a phenotype.

contiguous: Adjoining, touching, but not united. *See* CONTIG.

contiguous map: *See* CONTIG MAP.

con-till: An abbreviation that refers to CONSERVATION TILLAGE farming practices.

continental drift: The process by which the continents move as part of large plates floating on the Earth's mantle. *See* PLATE TECTONICS.

contingency chi-square test: Sometimes, it is desirable to compare one set of observations under particular conditions to those of a similar nature taken under different conditions. In this case, there are no expected values. The question to be answered is whether the results are dependent (contingent upon) or independent of the conditions under which they were observed. *Aka* TEST FOR INDEPENDENCE.

contingency loci: Loci made up of microsatellite repeats (e.g. ATATATAT) in which, when the number of copies of the repeat changes, the phenotype of the cell is drastically altered. These loci are of common occurrence in the genomes of some bacterial pathogens.

contingency table: A table of frequencies showing two or more nominal or quantitative variables simultaneously and used in testing their independence. *See* CROSS-TABULATIONS.

continuation-passing style (CPS): A style of functional programming in which control is passed explicitly in the form of a continuation. CPS *contrasts with* DIRECT STYLE, which is the usual style of programming. Gerald Jay Sussman and Guy L. Steele, Jr. coined the phrase in AI Memo 349 in 1975, which sets out the first version of the Scheme programming language.

continuity correction: An adaptation to a test statistic when a continuous probability distribution is employed to evaluate a discrete probability distribution.

continuity of life: Germplasm is continuous between all descent generations. It occurs through reproduction.

continuous cell culture: A culture which is apparently capable of an unlimited number of population doublings; often referred to as an immortal cell culture. Such cells may or may not express the characteristics of *in vitro* neoplastic or malignant transformation. *See* IMMORTALIZATION.

continuous cropping: The practice of growing the same crop in the same location repeatedly.

continuous culture: Cell suspension cultures continuously maintained by adjusting the inflow of fresh medium and out flow of the spent medium. Unlike batch cultures in flasks in the experimental laboratory, the continuous culture system is preferred in the industry where the cell mass generally keeps increasing in the culture vessel.

continuous replication: In DNA, uninterrupted replication in the 5′→3′ direction using a 3′←5′ template. *Compare with* DISCONTINUOUS REPLICATION.

continuous scale: A scale for keeping a tally of quantitative data for which the number of potential values is not predefined and is theoretically limitless (e.g. age, seed weight in grams). Quantitative traits show MULTIGENIC INHERITANCE. *Aka* POLYGENIC INHERITANCE.

continuous trait: Quantitative trait; variation that cannot be represented by discrete classes and requires measurement data. Multigenic or polygenic inheritance represents this type of trait.

continuous variable: A variable that can theoretically have infinite number of possible values within a short range. Age is continuous since within 8 and 12, it can be 8.17, 10.874, 9.756 years, etc. Age can be measured in terms of days, hours and minutes, although practically, there is no need to do this. Blood pressure is a continuous variable but is measured in integers for convenience. Parity is not a continuous variable, because there is no possibility of it being 2.75 or 1.6.

continuous variation: Variation in the expression of inherited traits in which a series of non-discrete, intermediate types, which cannot be divided into separate categories, connect the extremes with their distribution on a continuum. Variation that might be produced by a multitude of individual genes, each with a small effect on the measured character. Continuous variation is also governed by Mendelian genes. Continuous variation noted for quantitative traits is result of both genotype and environment. *Aka* QUANTITATIVE VARIATION.

contort: (of sepals or petals) A type of imbricate aestivation in which one side of each segment overlaps one of the adjacent segments and the other side is overlapped by the other adjacent segment. *See* CONVOLUTE.

contractile ring: Ring containing actin and myosin that is formed under the surface of animal cells undergoing cell division and contracts to tweak the two daughter cells apart.

contractile vacuole: In many protists, a specialized vacuole with associated channels is designed to collect excess water in the cell. Microtubules periodically contract to force this excess water out of the cell, thereby regulating the cell's osmotic balance.

contrasting genetic character: A genetic character with noticeable phenotypic differences.

control elements: DNA sequences in genes that interact with regulatory proteins (such as transcription factors) to determine the rate and timing of expression of the genes as well as the beginning and end of the transcript.

control event rate (CER): The number of subjects in the control group who develop the outcome being studied.

control flow: The order in which individual statements, instructions or function calls of an imperative program are executed or evaluated. The stress on explicit control flow distinguishes an imperative programming language from a declarative programming language. *Aka* FLOW OF CONTROL.

control pollination: In horticulture and forestry, to deliberately pollinate the female flowers of a tree with pollen from a known source. Usually, the flowers are secluded from undesirable pollen by covering them with a pollen-tight cloth or paper bag before they have become receptive. It is a way to produce FULL-SIB FAMILIES.

control sequences: Those sequences of DNA that are positioned adjacent to a gene (in genome) and which 'turn on' and/or 'turn off' that gene.

control subjects: In a clinical trial, subjects assigned to the placebo or control condition; in a case-control study, subjects without the disease or outcome.

control variables: Variables whose value is apprehended constant under all conditions of an experiment.

control: Regulation of expression of genes by other genes and/or elements of chromosomes.

controlled breeding: The skilful reproduction of desired characteristics.

controlled for: Term used to describe a confounding variable that is taken into consideration in the design or the analysis of the study.

controlled trial: Subjects are assigned to a control condition as well as to an experimental condition.

controlling cell proliferation: Cultured cells can be grown on well-defined media containing hormones, nutrients and growth factors. In the past, 5–10% of serum was essential protein factor providing cell proliferation. In the last 10 years, synthetic culture media have come which provide cell proliferation even in the absence of serum. In multicellular organisms, TISSUE-SPECIFIC GROWTH FACTORS (TsG-Fs) are required.

controlling elements: Mobile (autonomous or non-autonomous) genetic component that is capable of producing an unstable, mutant target gene. The sequences that are in fact involved in regulation of gene expression. They have been shown to regulate phenotypic expression, being capable of moving to different locations and exerting their influence on a variety of genes. For example, controlling elements of maize are transposable units originally recognized solely by their genetic properties. They may transpose only in the presence of an autonomous element.

controlling factor: With reference to competition, any factor the effect of which becomes more severe as the density of the population increases.

controlling gene: A gene that has a role in turning on or off the transcription of structural genes. Two types of genetic elements exist in this process – a REGULATOR ELEMENT and a RECEPTOR ELEMENT. The receptor element is one that can be inserted into a gene, making it a mutant, and can also exit from the gene; both of these functions are under control of the regulator element.

controlling region and genes of *lactose* operon: Nucleotide length of different components of *lac* operon of *E. coli*: *lac i* promoter; 40 nucleotides; *i* gene, 111 nucleotides; *lac i* operator, 26 nucleotides; *lac z* gene, 3063 nucleotides; *lac y* gene, 800 nucleotides; *lac a* gene, 800 nucleotides. *E. coli lac* promoter has three components: cgs, catabolic gene activator (CAP) site; ibs, initial binding site; and op, operator.

convenience sampling: Used in exploratory research, where the researcher is focused on getting an inexpensive approximation of the truth. For this, the samples are selected because they are convenient. This is a non-probability method, which is often used during preliminary research efforts to get a gross approximation of the results, without incurring the cost or time required to select a random sample.

Convention on Biological Diversity (CBD): The international treaty governing the preservation and use of biological resources around the world. CBD was signed by more than 150 countries at the 1992 United Nations Conference on Environment and Development. Article 19.4 of the CBD called for establishment of a PROTOCOL ON BIOSAFETY to govern the transnational-boundary movement of non-indigenous living organisms.

Convention on International Trade of Endangered Species (CITES): This convention provides guidelines/ regulations for the international trade in endangered species of plants and animals.

conventional agriculture: Crop production from elite plant lines, maximized with inputs such as insecticides, herbicides or fertilizers.

conventional breeding: Propagating plants or animals sexually, selecting for certain traits. Using selective cross-breeding, people can develop different varieties of plants and breeds of animals. The selection process is used to improve the characteristics of a plant. Genetically superior parents are selected for breeding in a cyclic process until the desired trait is received. The goal of conventional breeding is to produce genetically improved populations. *Aka* SELECTIVE BREEDING.

conventional genetics and development: The DNA methylation model for the control and heritability of gene expression can provide the molecular link between conventional genetics and the known features of development. The necessary changes in DNA are specific, stable and heritable. External influences could directly or indirectly influence proteins that impose or remove methylation and thereby alter the pattern of gene expression. It is proposed that signal receptor transmits information to one or more specific sites in DNA, which are then methylated or demethylated. This modification is heritable and can alter the gene expression. These changes could occur in single cells or in groups of cells acting in concert. The pathway of information flow would be applicable to any situation where an induced modification to chromatin structure and gene expression is transmitted through cell division. Whatever the epigenetic mechanisms be, the major conclusion is that the processes known to occur during development look extremely different from those that might be expected from classical molecular genetics. The rule of clonal expansion is broken; the rule of cell lineage inheritance is broken; there is good reason to believe that external influences can induce heritable changes in DNA or chromatin, and the central dogma of molecular biology is more complex than thought.

conventional pseudogene: A DNA sequence that resembles a gene, but which has become inactive owing to the accumulation of mutations.

convergence: (i) The end point of any algorithm that uses iteration or recursion to guide a series of data-processing steps. An algorithm is usually said to have reached convergence when the difference between the steps computed and those observed fall below a predefined threshold. (ii) The process by which similar characters evolve independently even in two different taxonomic groups. *Aka* ANALOGY *or* CONVERGENT EVOLUTION.

convergence-divergence selection: A breeding scheme in which selection of promising genotypes is performed in a bulk population at different locations, followed by massing of selection and permitting mating among them in a pollination field. The harvested bulk seeds thus constitute the basis for the next propagation cycle.

convergent cross mapping (CCM): A method to analyse causality of non-linear time series variables. Different from the traditional linear system analysis method, CCM gets historical information based on their state space reconstruction.

convergent evolution: An evolutionary process by which two or more species of different lineages independently develop similar characteristics, often to the point that these species appear to be more closely related than they actually are. Evolution of similar phenotypes in different populations or species results from adaptation to similar environments or ecological niches.

convergent extension: Cellular rearrangement within a tissue that causes it to extend in one dimension (e.g. length) and shrink in another (e.g. width).

convergent improvement: Double backcross method for simultaneous improvement of two parents involved in a single cross for improving per se performance of lines without interfering with their combining ability.

convergent transcription: Sister chromatids, the products of eukaryotic DNA replication, are held together by the chromosomal cohesion complex after their synthesis. This lets the spindle in mitosis recognize pairs of replication products for segregation into opposite directions. Cohesin forms large protein rings that may bind DNA strands by encircling them. Cohesin positions almost exclusively between genes that are transcribed in converging directions. Active transcription positions cohesins at these sites, not the underlying DNA sequence. Cohesin localization to regions of convergent transcription is conserved in fission yeast, suggesting that it is a common feature of eukaryotic chromosomes. *Compare with* DIVERGENT TRANSCRIPTION.

convolute: (i) (of flowers) Referring to the organization of floral or foliar organs in a bud when each organ or segment has one edge overlapping the adjacent organ or segment. This is a type of imbricate arrangement. *See* CONTORT. (ii) (of leaves) A type of vernation in which one leaf is rolled up inside another. (iii) A type of vernation of two leaves at a node, in which one half of each leaf is exposed and the other half is wrapped inside the other leaf.

cool morning pulse: Trying to keep a greenhouse in negative **DIFFERENTIAL ITEM FUNCTIONING (DIF)** is a difficult task. Another way to reduce a plant's height is 'cool morning pulse'. This method involves dipping the greenhouse temperature −15 to −12°C lower than the night temperature for 2–3 h, starting 30 min before dawn, reduced plant height as effectively as negative DIF and was easier to do. For example, night temperature of 20°C, 2-h drop to 15°C (30 min before dawn), and then 18°C maintained during the day, or for warmer plants 22°C at night; 17°C predawn for 2–3 h; 18°C d. There are thermostatically controlled devices that can be set up to automatically change these temperatures at the correct times. Salvia, rose, snapdragon and fuchsia have shown very encouraging responses to this method, compared to aster, French marigold, tulip and squash, which show little or no response.

cool-season crop: A crop that grows best during the cool temperatures of spring and fall.

cooperation between nuclear and chloroplast DNAs: One clear-cut case of cooperation between nuclear DNA and chloroplast DNA is known. Small subunit of fraction I protein is coded by nuclear DNA and is synthesized in the cytoplasm, which is then transported to the chloroplast. The large subunit of fraction I protein is synthesized in the chloroplast. The small and large subunits of fraction I protein join each other in the chloroplast.

cooperation between nuclear and mitochondrial DNA: There are many proteins in mitochondria that have revealed dual origin, some polypeptides having nuclear gene origin and others having mitochondrial gene origin. Two examples are documented: (i) Mitochondrial cytochrome C oxidase comprised 7 different polypeptides. Four polypeptides are synthesized on cytoplasmic ribosomes while three are produced on mitochondrial ribosomes. (ii) Mitochondrial ATPase is made up of several polypeptides. Two polypeptides are synthesized on mitochondrial ribosomes while the remaining polypeptides are made at cytoplasmic ribosomes.

cooperation response genes: Transformation of normal cells into cancer cells entails concerted changes in the expression of many genes. Oncogenic mutations in the transcription factor p53 and in the small GTPase protein RAS, which individually have limited effects on promoting cancer, cooperate to transform normal cells into cancer cells. A large proportion of such genes are controlled synergistically by loss-of-function.

cooperativity: Phenomenon in which the binding of one ligand molecule to a target molecule promotes the binding of successive ligand molecules. Seen in the assembly of large complexes, as well as in enzymes and receptors composed of multiple allosteric subunits, where it sharpens the response to a ligand.

coordinate gene regulation: The common control of a group of genes.

coordinate repression: Interrelated regulation of the structural genes in an operon utilizing a regulatory molecule that interacts with the operator sequence.

coordinated framework for regulation of biotechnology: The regulatory 'framework' through which the US evaluates and approves new products derived via biotechnology. The Coordinated Framework assigns specific regulatory tasks to each of the US government's applicable agencies. For example, the US Environmental Protection Agency (EPA) is assigned to evaluate and regulate all GENETICALLY MODIFIED PEST PROTECTED (GMPP) new plants, in terms of their impact on pests. The US Food and Drug Administration (FDA) is assigned to evaluate and regulate all new food crops derived via biotechnology, in terms of their potential impact on food safety (allergenicity, toxicity, etc.). The US Department of Agriculture (USDA) is assigned to evaluate and regulate all new plants derived via biotechnology, in terms of field (outdoor) testing and of potential impact on the environment such as weediness.

coordinated gene expression: A three-dimensional investigation of gene regulation suggests that portions from different chromosomes 'communicate' with each other and bring related genes together in the nucleus to coordinate their expression, which is required for homeostasis, growth and development in all organisms. Such coordination may be partially achieved at the level of messenger RNA stability, in which the targeted destruction of subset of transcripts produces the potential for cross-regulating metabolic pathways.

coordination of development and proliferative pool: The interaction between geminin and Hox proteins prevents Hox proteins from binding to DNA, inhibits Hox-dependent transcriptional activation of reporter and endogenous downstream target genes and displaces Cdt1 from its complex with geminin. By establishing competitive regulation, geminin functions as a coordinator of developmental and proliferative control.

coorientation: *See* CENTROMERE ORIENTATION.

***Copia* elements:** *Copia* elements are just like proviruses. Retroviruses might have evolved from transposable elements or transposons can be degenerate proviruses.

***Copia* gene family:** This gene family is 4–5 kb long and seems to consist of genes of very similar structure positioned at multiple sites that are categorized by different flanking sequences.

***Copia*-like elements:** *Copia*-like elements may range from 5 to 8.7 kb in length, and their number may vary from 10 to 80. Each element carries a long direct terminal repeat ranging from 268 to 512 bp and a short, imperfect inverted repeat ranging from 0 to 18 bp in length.

copper fist: Configuration of a DNA-binding protein that resembles a fist closed around a penny. In this case, the penny is copper ions. The knuckles of the fist of the yeast ACE1 protein interact with the promoter of the metallothionein gene, enhancing its transcription.

coppice: The regrowth of woody plants by sprouts derived from stumps or roots.

copulation: The union of sexual elements.

copy correction: If the gene is introduced in the inverted repeat (IR) region of the chloroplast DNA, integration in one inverted repeat is followed by the phenomenon of copy correction that duplicates the introduced transgene into the other inverted repeat as well.

copy DNA (cDNA): *See* COMPLEMENTARY DNA.

copy error: A mistake in the DNA replication process producing to a gene mutation.

copy gene: Genetic material integrating the genetic code for a desirable trait, which has been copied from DNA of the donor to the host organism.

copy number: The number of copies of a plasmid, gene, transposon or repetitive sequence in the genome of a cell. Each plasmid has a typical copy number value ranging from 1 to >50. Higher copy number has been associated with a higher yield of the protein encoded for by the plasmid gene in each cell.

copy number variation (CNV): A condition when the number of copies of a gene differs from one individual to the next. CNVs occur when sections of the genome are repeated, and the number of repeats varies between individuals. CNVs can occur due to insertions, duplications or deletions in the genome and are collectively referred to as copy number variants.

copy-choice hypothesis: (Of genetic recombination) During replication of the paired chromosomes, the new DNA (RNA, in case of RNA viruses) strand being formed along the parental chromosome switches to maternal one and it thereafter copies. If complementary replica of the maternal strand also switches templates when it reaches the same point, two reciprocal recombinant strands would be formed. This is now a discarded hypothesis which states that recombination results from the swapping of the DNA-replicating enzyme from one homologue to the other. This hypothesis provides an interpretation of intrachromosomal genetic recombination that is not regarded as a physical exchange of preformed genetic strands.

copyright: Copyright is a right given by the law to the creators of literary, dramatic, musical and artistic works and producers of cinematograph films and sound recordings. In fact, it is a bundle of rights including *inter alia* rights of reproduction, communication to the public, adaptation and translation of the work. Copyright ensures certain minimum safeguards of the rights of the author over their creations, thereby protecting and rewarding creativity.

Copyright Act: Copyright Act 1957 protects original literary, dramatic, musical and artistic works and cinematograph films and sound recordings from unauthorized use. Unlike the case with patent, copyright protects the expression and not the ideas. There is no copyright in an idea. Ordinarily, the author is the first owner of copyright in a work. The owner of the copyright in an existing work or the prospective owner of the copyright in a future work may assign to any person the copyright either wholly or partially and either generally or subject to limitations and either for the whole term of the copyright or any part thereof. Copyright of the works of the countries mentioned in the International Copyright Order is protected in India, as if such works are Indian works. It is not necessary to register a work to claim a copyright. Acquisition of copyright is automatic, and it does not require any formality. However, the certificate of registration of copyright and the entries made therein serve as *prima facie* evidence in a court of law with reference to disputes relating to the ownership of copyright.

copyright infringements: The following commonly known acts involve infringements of copyright: (i) making infringing copies for the sale or hire or selling or letting them for hire; (ii) permitting any place for the performance of works in public where such performance constitutes infringement of copyright; (iii) distributing infringing copies for the purpose of trade or to such an extent so as to affect prejudicially the interests of the owner of the copyright; (iv) public exhibition of infringing copies by way of trade; and (v) import of infringing copies into India. A copyright owner can take legal action against any person who infringes the copyright in the work. The copyright owner is entitled to remedies by way of injunctions, damages and accounts. Copyright infringement is a criminal offence.

copyright office: The National Copyright Office established under the provisions of the Copyright Act 1957 provides registration facilities for all types of works indicated in the Act and is headed by a Registrar of Copyrights. The office is located at B2/W3, Curzon Road Barracks, Kasturba Gandhi Marg, New Delhi.

cordycepin: Chemically, 3′-deoxyadenosine, a derivative of the nucleoside adenosine, differing from the latter by the replacement of the hydroxy group in the 3′ position with a hydrogen. It acts as an inhibitor of polyadenylation of RNA.

core collection: The basic sample of a germplasm collection. It is premeditated to epitomize the wide range of diversity in terms of morphology, geographic range or genes. It contains, with a minimum of repetitiveness, the genetic diversity of a crop species and its wild relatives. It is not envisioned to replace existing gene bank collections but to include the total range of genetic variation of a crop in a relatively small and practicable set of germplasm accessions.

core DNA: The 146 bp of DNA contained on a core particle of a nucleosome.

core enzyme: RNA polymerase having subunit constitution $\alpha_2\beta\beta'\omega$, which is used in elongation of a transcript. *See* RNA POLYMERASE CORE ENZYME. *Compare with* HOLOENZYME.

core histones: The histone proteins around which DNA is wound tight, ordering DNA into nucleosomes. Core histones include H2A, H2B, H3 and H4.

core octamer: The assembly containing two subunits each of the four histones H2A, H2B, H3 and H4, which forms the essential component around which DNA is wound to form a nucleosome. *Aka* CORE PARTICLE.

core particle: It is composed of two molecules of each of the core histones H2A, H2B, H3 and H4 with approximately 146 base pairs (bp) of DNA wrapped in 1.67 left-handed superhelical turns around the histone octamer.

core promoter: Eukaryotic promoter much similar in function to bacterial promoter, located close to coding region of a gene.

corepressor: A metabolite that in combination with a regulatory molecule, called repressor, binds to the operator present in an operon and prevents the synthesis of a repressible enzyme. In repressible gene control, end product is termed corepressor. Effectors of repressible operons are known as corepressors.

cork: The external protecting tissue of a stem or root, impermeable to water and gasses. It is the primary component of bark.

cork cambium: A cylinder of meristematic cells (lateral meristem) in the stems of woody plants that makes cork (bark).

corm (Adj. cormose, cormous): A fleshy, swollen stem base, usually underground and operational in the storage of food reserves, with buds naked or enclosed in very thin scales. A type of rootstock.

cormel: A small corm (or cormlet), establishing at the base of a growing larger corm.

corolla: A collective term used to represent the petals of a flower. *Compare with* CALYX.

corona: (i) In flowering plants, a ring of assemblies that may be united in a tube, arising from the corolla or perianth of a flower and located vertical between the perianth lobes and the stamens. The trumpet of a daffodil is an example of a corona. (ii) In grasses, a hardened ring of tissue riding over the lemma in some species.

correction: The production (possibly by excision and repair) of a properly paired nucleotide pair from a sequence of hybrid DNA that contains an illegitimate pair.

correlated response: Change of phenotype occurring as an incidental consequence of selection for a seemingly independent character, such as sterility, resulting from selection for high bristle number.

correlation: The degree or strength of relationship between two quantitative variables. Loosely used for qualitative variables as well. For linear relationships, it is measured by PEARSONIAN CORRELATION COEFFICIENT that varies from −1 to +1. A negative correlation informs that increase in the value of one variable is accompanied by linear decrease in the other, and vice versa. A positive correlation signifies that the two values move together in the same direction. A correlation close to zero means that increase or decrease in one does not linearly impact the other. Correlation coefficient can be close to zero when a strong relationship is present but is non-linear.

correlation between a metric trait and heterozygosity: Three principal hypotheses have been advanced for a correlation between a metric trait and heterozygosity: TRUE OVERDOMINANCE HYPOTHESIS, ASSOCIATIVE OVERDOMINANCE HYPOTHESIS and INBREEDING DEPRESSION HYPOTHESIS. *See details as separate terms.*

correlation coefficient (r): The most commonly used amount of correlation between two variables (x, y) estimated on the same set of subjects. It was formulated by Karl Pearson in 1890 and is defined as the ratio between the covariance and the square roots of the variances. It ranges from −1 to +1, with 0 demonstrating no relationship. *Aka* PEARSON PRODUCT.

correlation correspondence analysis: A complementary analysis to genetic distances and dendrograms. It displays a global view of the associations among populations. This type of analysis is more enlightening and precise than dendrograms, especially when the number of loci is less than 30. Using the same allele frequencies that are used in phylogenetic tree construction, correspondence analysis can be performed on the VISTA or MVSP computer programs.

correndonucleases: These DNA repair endonucleases are of two types: (i) CORRENDONUCLEASE I is specific for damaged DNA possessing monoadduct derivatives. (ii) CORRENDONUCLEASE II is specific for diadduct modified regions of DNA.

corresponding gene pair: A pair of genes in a parasite that parallels with a pair of genes in a host, which function together to bring about a specific interaction.

cortex (Adj. cortical, pl. cortexes or cortices): The cortex is a ground tissue region found in monocot and dicot roots, located between the outer epidermis and the inner vascular structures. It is primarily composed of parenchyma cells, but it may contain an outer layer of sclerenchyma cells in monocot roots. The primary functions of the cortex are diffusing water, nutrients and other substances into the inner vascular structures and storing starch. The endodermis regulates the flow of water, ions and hormones into and out of the vascular structures within the stele.

cortical granule: A specialized secretory vesicle occurring under the plasma membrane of unfertilized eggs, including those of mammals. After fertilization, this structure is involved in preventing the entry of further sperm.

corticolous: Manufactured on bark or on wood with the bark uncovered. *Compare with* LIGNICOLOUS.

corymb (Adj. corymbose): An inflorescence with branches ascending at different points but reaching approximately the same height, giving the flower cluster a flat-topped look.

***cos* site:** The site of the circular form of λ phage DNA or others that is sliced by the terminase to generate the cohesive 12 bp 5′ overhang ends of the linear phage as it is wrapped into the capsid.

cosduction: The transduction of foreign DNA by use of cosplasmid (a plasmid on to which the cohesive ends form). The transducing phage particles are called CONDUCING PARTICLES and recipients COSDUCTANTS.

cosegregation: In *Chlamydomonas*, parallel behaviour of different chloroplast markers in a cross due to their close linkage with chloroplast DNA.

cosexual: The individual that produces both male and female gametes. It is equivalent to hermaphrodite.

cosmid: A cloning vector carrying λ *cos* sites inserted into a bacterial plasmid. Cosmid cloning is a common method for cloning larger fragments 28–45 kb long. *Cos* sites are recognized during head filling of lambda phages. Cosmids have high transformation efficiency. Cosmids are useful for cloning large segments of foreign DNA. Cosmids can also be used to build genomic libraries. Examples of cosmids are pBR322 and pJB38.

cosmid cloning: (i) A hybrid vector made up of plasmid sequences and the cohesive ends of the bacteriophage lambda. (ii) A genetically engineered plasmid encompassing bacteriophage λ packaging signals and potentially very large pieces of inserted foreign DNA (30–50 kb) which can be replicated in bacterial cells. The vector contains a selectable marker, plasmid origin of replication, polylinker or restriction site into which DNA can be inserted. Cosmid cloning permits for isolation of DNA fragments that are larger than those that can be cloned in conventional plasmids.

cosmid library: Cosmid libraries are used for cloning genes with large introns and for sequencing larger chunks of the genome. The insert sizes are of the order of 35–45 kb. Cosmid cloning vector can be used for genomic library but not for cDNA libraries.

cosmopolitan cultivars: Cultivars that have an extensive geographical and environmental array.

cost of genome dilution: The disadvantage to a female that reproduces sexually and therefore dedicates resources to proliferating her mate's genes.

cost of meiosis: *See* COST OF GENOME DILUTION.

cost of selection: The proportion of adaptively inferior genotypes that are produced in a population in the course of producing adaptively superior genotypes. *See* SUBSTITUTION LOAD.

cosuppression in metabolic engineering: The phenomenon of cosuppression is useful for reduction in catabolic enzymes. Genetic engineering of highly nutritional food crops requires both GENE SILENCING and COUNTER-SILENCING TECHNOLOGIES.

cosuppression: It involves introduction of extra copies of an endogenous gene to boost expression of endogenous or introduced gene, which may cause the coordinate silencing not only of the introduced transgenes but also of the endogenous genes. The mechanisms of suppression of sense genes may involve interference of RNA strands with the transcription process itself. For example, high-oleic oil soyabeans result when the *GmFad2-1* gene (which codes for native Δ12 desaturase enzyme) is inserted and expressed in traditional varieties of soyabeans. That happens because the inserted gene silences itself and the endogenous *GmFad2-1* gene (i.e. the one naturally/originally present in the soyabean plant) and prevents formation of the Δ12 desaturase enzyme (which normally causes most oleic acid within soybeans to be converted into polyunsaturated acid/linoleic acid). Cosuppression can mirror either transcriptional repression (i.e. TRANSCRIPTIONAL GENE SILENCING [TGS] or POST-TRANSCRIPTIONAL GENE SILENCING [PTGS]). The latter process is the plant version of RNA interference. *Aka* SENSE-SUPPRESSION.

cosuppression in plants: R.A. Jorgensen coined this term in 1990 to define the loss of mRNAs of both the endogene and the transgene. Now, this term is used in plants *C. elegans* and *Drosophila* to designate the reciprocal silencing of transgenes and (partially) homologous endogenous genes. Cosuppression occurs either at the transcriptional level, when the homology is within the promoter, or at the post-transcriptional level, when the homology is within the coding sequence.

C_0t: The product of DNA concentration (C_0) and time (t) of incubation in a reassociation reaction.

C_0t curve: Graphic representation of the progress of a (liquid) hybridization experiment. It is used to determine the complexity of DNA mixtures (e.g. the size of the genome) gene. In the C_0t curve, the y-axis represents the DNA that remains single-stranded. This is expressed as a ratio of the concentration of single-stranded DNA (C) to the total concentration of the starting DNA (C_0). The x-axis is a log-scale of the product of the initial concentration of DNA (in moles/litre) multiplied by length of time the reaction proceeded (in seconds) (C_0t). The curve itself is called a C_0t curve. The curve appears rather smooth, which reflects that reannealing occurs, slowing gradually over a period of time.

C_0t curve of a eukaryotic genome: Eukaryotes actually have a wide array of sequences that are represented at different levels of repetition. Genomes that contain these different classes of sequences reanneal in a different manner than genomes with only single copy sequences. Instead of having a single smooth C_0t curve, three distinct curves can be observed, each representing a different repetition class: (i) The first sequences to reanneal represent highly repetitive sequences because so many copies of them exist in the genome, and because they have a low sequence complexity. (ii) The second portion of the genome to reanneal denotes the middle repetitive DNA. (iii) The final portion to reanneal signifies the single-copy DNA. Repeated DNA sequences have distinctive effects on C_0t curves. If a specific sequence is represented twice in the genome, it will have two complementary sequences to pair with and as such will have a C_0t value half as large as a sequence represented only once in the genome.

C_0t curve shape: The shape of a C_0t curve for a given species is a function of two factors: (i) the size or complexity of the genome; and (ii) the amount of repetitive DNA within the genome. If we plot the C_0t curves of the genome of three species such as bacteriophage lambda, *E. coli* and yeast, we will see that they have the same shape, but the Cot½ of the yeast will be the largest, *E. coli* next and lambda smallest. Physically, the larger the genome size, the longer it will take for any one sequence to encounter its complementary sequence in the solution. This is because two complementary sequences must encounter each other before they can pair.

C_0t plot: A graph showing the extent of reannealing of previously denatured molecules as function of C_0t value. C_0t plots indicate the relative amounts of unique, moderately repetitive and highly repetitive DNA sequences in a genome, giving some idea of the total complexity of the sequences.

C_0t value: An expression for the rate of DNA renaturation (annealing-reannealing). DNA renaturing at low C_0t is composed of highly repetitive sequences, and DNA renaturing at high C_0t values is minimally repeated or non-repetitive.

$C_0t½$: Midpoint values on C_0t curves (C_0t values plotted against concentration of remaining single-stranded DNA), which estimate the length of unique DNA in a sample. Renaturation of DNA depends on random collision of the complementary strands and follows second-order kinetics. A greater $C_0t½$ value implies slower reaction.

cotranscriptional cleavage (CoTC): This primary cleavage event within β-globin pre-messenger RNA, downstream of the poly(A) site is crucial for efficient transcriptional termination by RNA polymerase II. CoTC process involves a self-cleaving activity. Autocatalytic core of the CoTC ribozyme has a functional role in efficient termination *in vivo*. CoTC may be a general phenomenon, and functionally, it may provide an entry point for exonuclease involved in mRNA maturation, turnover and, in particular, transcriptional termination.

cotranscriptional gene silencing (Co-TGS): The coupling of repressive epigenetic modification with transcription by an RNA polymerase that produces a nascent RNA molecule targeted by small RNA.

cotransduction: The transfer of two or more genes on a single DNA molecule during transformation of a bacterium.

cotransfection: The concurrent transfection of two markers.

cotransformation: The concurrent transformation of two or more genes. Only the genes located in the same chromosomal vicinity can be transformed. The closer together the genes lie, more frequently they will be cotransformed.

cotranslational: Describes importation of a protein into the endoplasmic reticulum before the polypeptide chain is completely synthesized.

cotranslational protein translocation: Secreted and membrane proteins travel across or into cell membranes through a protein-conducting channel (PCC) when they are being manufactured. This translocating

PCC forms connections with ribosomal RNA hairpins on two sides and ribosomal proteins at the back, leaving a frontal opening. For entry of these proteins into the endoplasmic reticulum, leader sequence is necessitated at the N-terminal end of the protein. *Aka* COTRANSLATIONAL TRANSFER.

cotranslational transfer: The process through which membrane-bound polysomes synthesized proteins enter the endoplasmic reticulum when they are being synthesized. A protein that resides within a membrane is called an INTEGRAL MEMBRANE PROTEIN. Secretion of proteins is explained by SIGNAL HYPOTHESIS.

cotransport: Membrane transport process in which the transfer of one molecule depends on the concurrent or sequential transfer of a second molecule. *Aka* COUPLED TRANSPORT.

cotton epigenome: Cotton genus, *Gossypium*, diploids (AA and DD) allotetraploid (AADD), is a very suitable material for investigation of impacts of polyploidization and subsequent domestication on the epigenome, specifically the methylome. Many alterations in the CG methylation of genes were associated with changes in gene expression. A small percentage of methylated CG (^{m}CG) DMRs were tossed between the A and D genomes when occurring in the tetraploid species *G. hirsutum*; that is, if a locus was ^{m}CG in the A genome of one ancestral diploid and not methylated in the D genome of the other, in *G. hirsutum*, the locus was ^{m}CG in the D genome and unmethylated in the A genome. Genome-wide 5^{m}C levels are often related with TE content. Despite the increased level of overall methylation in the A genome, the D genome had increased levels of gene-body ^{m}CG. The cotton anther genome revealed almost varied methylation levels at all sequenced C sites and in all the three sequence contexts (CG, CHG and CHH).

cotyledon: The leaf-forming part of the embryo in a seed. It may act as a storage organ from which the seedling draws food, or it may absorb and pass on to the seedling nutrients stored in the endosperm. Once it is exposed to light, it manufactures chlorophyll and functions photosynthetically as the first leaf.

cotyledonary node: The point of attachment of the cotyledons to the embryonic axis.

coumarin couple method: A breeding procedure exclusively utilized in breeding of allogamous plants. From an original population, single plants are selected and, subsequently, pairwise crossed, thwarting unwished pollination. The crossing partners should be as similar as possible in spite of colour, growth habit, etc. The offspring is grown in separate plots in the following year. The selection of individuals from the plots and a repeated pairwise crossing can be performed during the fourth year. During the fifth year, offspring is grown in plots and again selected for progeny testing.

coumarin: A white crystalline compound ($C_9H_6O_2$) with a vanilla-like odour, which gives sweet clover its characteristic odour. It also acts as a chemical growth inhibitor that has germination-inhibiting ability.

counteracting replication-associated shortening of telomeres: Functional telomeres are required to provide chromosome stability and to ensure faithful segregation of genetic material into daughter cells upon cell division. Telomeres conceal linear chromosome ends from detection and inappropriate repair and thus provide a buffer to counteract replication-associated shortening.

countercurrent exchange: The crossover of some property, usually heat or some component, between two fluids flowing in opposite directions to each other. In biological systems, countercurrent exchange mechanism is used by fish in their gills to transfer oxygen from the surrounding water into their blood.

counter-silencing: A phenomenon opposite to gene silencing. A logical counter-defence mechanism by many viruses has been to evolve RNAi inhibitors, which allows them to infect plants productively even in the presence of an active RNAi pathway. This event seems to have happened often during evolution, and some 20 different RNA silencing suppressors have been identified from different plant viruses. These suppressors of RNA silencing might prove to be useful tools not only for dissecting the biochemical steps of RNAi but also for promoting the expression of plant transgenes.

counter-transcript: Any RNA species that is initiated from a single start within the leader region and transcribed from the opposite strand, one terminating very close to one mRNA start and the other further upstream. One or both these counter transcripts serve as negative control elements.

counting X chromosomes: In mammals, X-inactivation establishes X-chromosome dosage partly between males and females. X chromosome counting and choice are regulated not only by 'blocking factor' but also by both a 'blocking' and a 'competence' factor. Homologous associations driven by X-pairing region (*Xpr*) of the *Xic* enable a cell to sense that more than one X chromosome is present and coordinate reciprocal *Xist/Tsix* expression.

couplant: A material (usually liquid) that facilitates the transmission of ultrasonic energy from the transducer into the test specimen. Couplant is generally necessary because the acoustic impedance mismatch between air and solids (e.g. the test specimen) is large.

coupled histone tail for autoinhibition release of methyltransferase (CHARM): CHARM is an epigenetic editor produced for performing targeted DNA methylation and gene silencing. E.N. Neumann and co-workers in 2024 engineered CHARM, which contains a DNA-binding domain (DBD) glued to the Dnmt3l domain (D3L) and histone H3 tail. CHARM is targeted to the prion protein (*Prnp*) promoter, D3L engages the endogenous methyltransferase DNMT3A and the H3 tail stimulates DNMT3A to silence *Prnp* gene through methylation of the cytosine-guanine dinucleotide (CpG) island (CGI). Adeno-associated virus (AAV)-delivered CHARM silences *Prnp* all over the mouse brain. Self-silencing of CHARM transiently limits its expression. *See* COMPACT EPIGENETIC EDITOR.

coupled reaction: Linked pair of chemical reactions in which the free energy released by one reaction is utilized to drive another reaction.

coupling: The phase state in which either two dominant or two recessive alleles of two different genes occur on the same chromosome (i.e. linked recessive alleles occur in one homologous chromosome and their dominant alternatives occur in the other chromosome). It is also called *CIS*-CONFIGURATION; opposed to repulsion in which one dominant and one recessive occur in each member of the pair of homologous chromosomes.

coupling linkage: Tendency of two dominant characters, inherited by F_1 from one parent stay together in F_2.

coupling of factors (*cis*-arrangement): Linkage between two genes in which both dominant alleles are contributed one parent, or the condition in which a double heterozygote has received two linked mutations from one parent and their wild-type alleles from the other parent, such as *a b* / + + (called COUPLING LINKAGE) as opposed to *a* + / + *b* (called REPULSION LINKAGE).

covalent bond: Stable chemical link between two atoms formed by sharing one or more pairs of electrons.

covalent catalysis: A catalytic mechanism in which the formation of a transient covalent bond between catalyst and reactant lowers the free energy of reactions transition state. Generally, the covalent bond is formed by the reaction of a nucleophilic group on the catalyst with an electrophilic group on the substrate. Hence, this type of catalysis is also known as nucleophilic catalysis. The nucleophilicity of a substance is closely related to its basicity. In covalent catalysis, the more stable the covalent bond is, less easily can it decompose in the final steps of a reaction.

covalent cross-linking of DNA bases: Some antibiotics (e.g. mitomycin C) and some reagents can form covalent linkages between a base in one strand and an opposite base in the complementary DNA strand. This prevents strand separation during DNA replication. For example, THYMINE DIMERS, CYTOSINE DIMERS or CYTOSINE-THYMINE DIMERS. *See details as separate terms.*

covalent modification: The interconversion of a and b forms of glycogen synthase and glycogen phosphorylase is accomplished through enzyme-catalysed phosphorylation and dephosphorylation, a process that is under hormonal control. If enzymes that covalently modify a target enzyme are themselves under allosteric control, it is possible for a small change in concentration of allosteric effector of modifying enzyme to cause a large change in activity of its modifiable target enzyme.

covalent trapping of proteins for 5^{m}C mapping: This approach involves the covalent trapping of proteins that interact with modified nucleotides, providing adducts that are detected by reverse transcription. This is utilized by Aza-IP, which was recently used to map 5^{m}C.

covalently closed circular DNA (cccDNA). A double-stranded DNA that originates in a linear form that is ligated by means of DNA ligase to a covalently closed ring.

covalently modulated enzymes, types of: A covalently modulated enzyme occurs in two forms: (i) 'phosphorylase a' is the more active form; and (ii) 'phosphorylase b' is the less active form. In order for the enzyme to acquire the maximal catalytic activity (i.e. to be 'phosphorylase a') certain serine residue on all four subunits must have a phosphate covalently attached. If, due to other regulatory signals it has received, the enzyme phosphorylase phosphatase hydrolytically cleaves and removes the phosphate group from the four subunits, the tetramer dissociates into the inactive (or much less active) dimer, phosphorylase b. Another enzyme, phosphorylase kinase, is able to re-phosphorylate the four specific serine residues of the four subunits at the expense of ATP and regenerate the active phosphorylase a tetramer.

covalently modulated enzymes: These regulatory enzymes can be interconverted between active and inactive (or less active) forms by the covalent attachment (or removal) of a modulating metabolite by other enzymes. Hence, the activity of one enzyme can, under certain conditions, be regulated by other enzymes. Glycogen phosphorylase, an oligomeric protein with four major subunits (tetramer), is a classic example of a covalently modulated enzyme.

covariance: Measurement of degree of relationship between two quantitative variables (x, y). It is defined as the expected product of their deviations from the mean. It is used during the calculation of correlation coefficient.

covariance analysis: An analysis of the mean of the product of the deviation of two variates from their individual means. Covariance analysis measures the interrelationship between two variables.

covariance matrix: An n × n matrix giving the covariances between a set of n variables. The diagonal gives the covariances of each variable with itself (i.e. the variances).

covariate: A potentially confounding variable that is controlled for in the analysis of covariance. See PREDICTOR.

covirus: One of a pair of viruses that have complementary functions, and a cell must be coinfected with both the members for successful viral transmission.

cowpea mosaic virus (CpMV): A virus that infects cowpea (*Vigna unguiculata*) plants, known as black-eyed peas in the US, but it does not infect animals. Researchers have discovered how to cause CpMV to express certain animal virus proteins (i.e. antigens) on its surface through genetic engineering. These viral antigens hold the ability to replace the antigens currently utilized in vaccines, which are faced with problems due to their production in animal cells, bacterial cells or yeast cells. In addition, CpMV acts as an intrinsic, natural adjuvant to the (animal virus) antigens, since it provokes an immune response itself.

cowpea trypsin inhibitor (CpTI): A chemical that is naturally coded for by a certain cowpea (*Vigna unguiculata*) plant gene. CpTI kills certain insect larvae by inhibiting digestion of ingested trypsin, thereby starving the larvae to death.

Cox model: The Cox proportional hazards regression model is used to relate a set of patient descriptor variables to time until death or the other event. Cox analyses are based on the entire survival curve. The time-to-event may be reduced due to loss to follow-up or by another event, as long as the censoring is independent of the risk of the event under study. Descriptor variables may be used in two ways: (i) as part of the regression model and (ii) as stratification factors. For variables that enter as regressors, the model specifies the relative effect of a variable through its impact on the hazard or instantaneous risk of death at any given time since enrolment. For stratification factors, no assumption is made about how these factors affect survival (i.e. the proportional hazards assumption is not made). Separately shaped survival curves are permissible for these factors. The LOG RANK TEST for comparing two survival distributions is a special case of the Cox model. Cox models are used to estimate adjusted hazard ratios. *Aka* COX PROPORTIONAL HAZARD MODEL. *See* SURVIVAL ANALYSIS.

CP4 EPSPS: The enzyme 5-enolpyruvylshikimate-3-phosphate synthase, which is naturally manufactured by an *Agrobacterium* species (strain CP4) of soil bacteria. CP4 EPSPS is crucial for the functioning of that bacterium's metabolism biochemical pathway. CP4 EPSPS remains unaffected by glyphosate-containing or sulfosate-containing herbicides, so introduction of the *CP4 EPSPS* gene into crop plants (e.g. soyabeans) makes those plants essentially impervious to glyphosate-containing or sulfosate-containing herbicides.

CpG dyad in the context of a non-B DNA: Cell division cycle associated protein 7 (CDCA7), a protein with a carboxyl-terminal cysteine-rich domain (CRD). CDCA7 guides DNA methylation to juxtacentromeric regions. S. Hardikar *et al.* (2024) show that the CDCA7 CRD accepts a unique zinc-binding structure that identifies a CpG dyad in a non-B DNA shaped by two sequence motifs. CDCA7, but not immunodeficiency, centromeric instability and facial anomalies (ICF) mutants, favourably binds the non-B DNA with strand-specific CpG hemi-methylation. The unmethylated sequence motif is extremely enriched at centromeres of human chromosomes, whereas the methylated motif is dispersed throughout the genome. At S phase, CDCA7, but not ICF mutants, is focused in constitutive heterochromatin foci, and the formation of such foci can be repressed by exogenous hemi-methylated non-B DNA bound by the CRD. Binding of the non-B DNA designed in juxtacentromeric regions during DNA replication offers a mechanism by which CDCA7 defines the specificity of DNA methylation.

CpG (or CG) island (CGI): The 'p' in CpG refers to the phosphodiester bond between the cytosine and the guanine. A CG-rich region of 200–500 bp DNA, usually at the 5′-end (promoter region) of approximately 40% of the mammalian genes, of a gene that is relatively rich in unmethylated CpG dinucleotides. They often overlap TRANSCRIPTION START SITES (TSS). No standard definition of CpG islands has been agreed on. The definition usually takes the following form: (i) 'an X-bp stretch of DNA with a C+G content of Y% and an observed CpG/expected CpG in excess of Z'. An early definition suggests X = 200, Y = 50, Z = 0.6. A revision to X = 500, Y = 55, Z = 0.65 has been proposed more recently. (ii) CG-rich DNA has a G+C content of higher than 50% and an observed-to-expected CpG ratio equal to or greater than 0.6 or 0.65 and a minimum 50% or 55%. CpG islands are typically 300–3000 bp long, located upstream of gene coding regions and usually protected from DNA methylation. CGI are often associated with genes and possibly involved in transcriptional regulation. In primordial germ cells, these imprints are erased and then reset later during gamete formation. With all this accomplished, the development programme – a cascade of changes in gene regulation – unfolds normally.

CpG (or CG) island shores: Regions of DNA that lie at the outskirts of CpG islands rather than in the islands itself. They have variable methylation and have been proposed to be enriched in methylation differences between tissue types or between normal and cancer samples.

CpG (or CG) methylation maintenance: DNA methyltransferase 1 (Dnmt1) is the key enzyme that is responsible for maintenance of CpG methylation.

CpG (or CG) methylation: A heritable chemical modification of DNA (replacement of cytosine by 5-methyl cytosine) that, when present in a control region, usually suppresses expression of the corresponding gene. CpG methylation has role in the regulation of gene expression, silencing of parasitic DNA elements, genomic imprinting and embryogenesis.

CpG (or CG) motif: A genetic element contained in DNA vaccines that stimulates the immune system.

CpG (or CG) sequences/site (CG sequence/site): Cytosine followed by a guanine in the 5′→3′ directions on the same strand of DNA linked by phosphate (p). It is the major substrate for DNA methylation in mammals.

cranium: The part of the skull that protects the brain in vertebrates.

Crassulacean acid metabolism (CAM): A metabolic pathway for adaptation of certain plants that lets them take up CO_2 at night instead of during the day. This activity greatly reduces transpirational water loss during photosynthesis. This type of metabolism is commonly found in xerophytes.

crassulacean acid metabolism (CAM) cambium: In the stem and roots of vascular plants, CAM cambium is a single layer of cells lying between the xylem and phloem.

Cre: A site-specific recombinase that recognizes and binds to specific sites called loxP. Two loxP sites recombine at nearly 100% efficiency in the presence of Cre, allowing DNA cloned between two such sites to be removed by Cre-mediated recombination.

creation science: An assortment of many different, non-scientific attempts to disprove evolutionary theory and efforts to prove that the complexity of living things can be explained only by the action of an 'intelligent designer'.

creationism: The religious doctrine that all living things on Earth were created separately, in more or less their present form, by a supernatural creator, as stated in the Bible; the precise beliefs of different creationist groups vary widely. *See* SEPARATE CREATION.

Creative Commons (CC): An American non-profit organization dedicated to escalating the range of creative works available for others to build upon legally and to share. The organization has released several copyright licenses, *aka* CREATIVE COMMONS LICENSES, free of charge to the public.

Cre-*loxP* system of site-specific recombination: Cre-*loxP* is a site-specific recombination system of bacterio-phage P1 controlling resolution. This system is entirely independent of Cin-cixL/R system. Its primary function is to resolve plasmid dimmers.

crisis: A stage of the *in vitro* transformation of cells. It is characterized by reduced proliferation of the culture, abnormal mitotic figures, detachment of cells from the culture substrate and the formation of multinucleated or giant cells. During this massive cultural degeneration, a small number of colonies usually, but not always, survive and give rise to a culture with an apparent unlimited *in vitro* lifespan. This process was first described in human cells following infection with an oncogenic virus (SV40).

See CELL LINE, *IN VITRO* TRANSFORMATION *and* *IN VITRO* SENESCENCE.

CRISPaint homology-independent knockin method: Targeted genomic knockin is a valuable tool to probe gene function. CRISPaint homology-independent knockin method is based on the work first described as CRISPR-assisted insertion tagging (CRISPaint) to insert universal donor plasmids into the genome. An adapted version of this method for *D. melanogaster* uses CRISPR/Cas9 and NHEJ. This method is a simple and fast alternative to HDR for certain strategies, such as C-terminal tagging and gene disruption. Existing collection of CRISPaint universal donor plasmids is compatible with insect cells. In addition, they inserted the transgenesis marker *3xP3-RFP* into seven genes in the fly germ line, producing heritable loss-of-function alleles that were isolated by simple fluorescence screening.

CRISPR: Acronym of CLUSTERED REGULARLY INTERSPACED SHORT PALINDROMIC REPEAT. A bacterial genomic region used in pathogen defence.

CrisprStitch: Y. Han *et al.* (2024) describe CrisprStitch, a server-less web application designed to provide non-bioinformaticians with a user-friendly platform for quantification and visualization of genome-editing efficiency statistics within a target region. CrisprStitch accepts raw FASTQ sequencing files as input and generates histograms that illustrate genome-editing efficiency. Depending on the intended use, it can be accessed by visiting either https://zhangtaolab.org/software/crisprstitch or https://bioinfor.yzu.edu.cn/software/crisprstitch. They also provide an install package for a local computer system. *See* FASTQ.

criss-cross inheritance: Transmission of a trait or an X chromosome from heterogametic sex (father XY male) to homogametic sex (daughter XX female) and then back to heterogametic sex (her son XY male).

criss-cross nanoseed detection: An enzyme-free pathogen detection platform that can target specific nucleic acid sequences and proteins. The technology leverages biomolecular nanotechnology to assemble nucleic acid-based 'nanoseeds' that only form in the presence of specific pathogen-derived molecules. The nanoseeds then trigger downstream signal amplification via the rapid (within minutes) self-assembly of much larger, ribbon-like DNA structures that can easily be detected using low-cost readout methods. The method is highly specific and ultrasensitive due to the suppression of non-specific interactions that cause unwanted false-positive results.

criss-crossing: A continuous, rotational crossbreeding system interchangeably using males or pollinators of two different breeds. This system is simple to manage and breeds its own replacements. It utilizes the benefit of hybrid vigour; compared to the common F_1, some hybrid vigour can be lost, but that loss is more than compensated for by reduced management effort and cost.

criss-crossing cristae: The multiply-folded inner membrane of a cell's mitochondrion that has finger-like projections. The walls of the cristae are the site of the cell's energy production where ATP is produced.

crista (pl. cristae): One of the folds of the inner mitochondrial and plastid membrane.

criterion validity: An indication of how well a test or scale forecasts another related characteristic, ideally a 'gold standard'.

criterion variable: The outcome (or dependent variable) that is forecast in a regression problem.

critical breed: A breed where: (i) the total number of breeding females is less than 100 or the total number of breeding males is less than or equal to five; or (ii) the overall population size is close to, but slightly above, 100 and decreasing, and the percentage of purebred females is below 80% gene.

critical chi-square: A chi-square value for a given degree of freedom and probability level to which an experimental chi-square is to be compared.

critical concentration: Concentration of a protein monomer, such as actin or tubulin, that is balanced with the assembled form of the protein (i.e. assembled into actin filaments or microtubules, respectively).

critical difference: A value indicating least significant difference at values greater than which all the differences are significant.

critical ratio: The term refers to the Z SCORE used in tests.

critical region: The region or set of values in which a test statistic must occur for the null hypothesis to be rejected.

critical set (in a Latin square): A set of entries in a square grid which can be fixed in precisely one Latin square, with the property that if any entry of the critical set is deleted, the remaining set can be fixed in more than one Latin square.

critical value: The value of a test statistic (e.g. t, F, χ^2, z) that if exceeded by the observed test statistic would result in statistical significance at a chosen α level or better. For a z-test (normal deviate test), the critical level of z is 1.96 when $\alpha = 0.05$ for a two-sided test. For t- and F-tests, critical values decrease as the sample size increases, as one requires less penalty for having to estimate the population variance as 'n' gets large.

Cro repressor: One of the regulatory proteins responsible for controlling expression of genes.

cron: The time unit for evolutionary processes (1 cron = 10^6 years; 1 kilocron = 10^9 years; 1 millicron = 10^{12} years).

Cronbach's alpha coefficient: Cronbach's alpha coefficient is a measurement of internal consistency. It shows the nature of the relationship between multiple variables in a group of data. Additionally, Cronbach's alpha coefficient increases when you increase the number of items and vice versa.

crop: Refers to the plants and fungi that are grown and then gathered and harvested for some specific use. Its yield depends on the environment and other factors such as how much water it gets and the type of soil where it grows. Some examples of the types of crops include fruits, vegetables, cotton, coffee and hay.

crop diversification: Cropping system where a number of different crops are grown in the same general area and may be rotated from field to field, year after year.

crop ferality: Feral plants are often semidomesticated, escaped from the field a long time ago, having been domesticated and now growing wild (e.g. *Cyperus rotundus*, *Cynodon dactylon*, *Echinochloa crus-galli*, *Echinochloa colonum*, *Eleusine indica*, *Sorghum halepense*, *Imperata cylindrica*, *Portulaca oleracea*, *Chenopodium album*, *Digitaria sanguinalis*, *Convolvulus arvensis*, *Avena fatua*, *Amaranthus hybridus*, *Amaranthus spinosus*, *Cyperus esculentus*, *Paspalum conjugatum* and *Rottboellia exaltata*). Growing wild has to be confined properly. Only in exceptional cases a feral crop really grows in the wild.

crop improvement: The science that deals with the development of new crop varieties with superior quality and quantity. Crop improvement makes use of modern molecular techniques with the following objectives in mind: (i) to increase crop yield per given area; (ii) to improve the quality of produce; (iii) to improve crop resistance to pest and pathogens; (iv) to adapt to adverse climatic conditions; and (v) to improve harvesting quality.

crop plant: A plant explicitly cultivated for some specific use. The majority of crops can be classified as: (i) root and tuber crops (e.g. potato, yams); (ii) cereal crops (e.g. wheat, oats, barley, rye, rice, maize); (iii) oil and protein crops (e.g. rapeseed, pulses); (iv) sugar crops (e.g. sugarbeet, sugarcane); (v) fibre crops (e.g. cotton, jute); or (vi) forage crops (e.g. grasses, legumes).

crop rotation: The practice of successive growing a series of different types of crops in the same area across a sequence of growing seasons. This practice reduces the reliance of crops on one set of nutrients, pest and weed pressure, along with the probability of developing resistant pests and weeds. A well-designed crop rotation can reduce the need for synthetic fertilizers and herbicides by better using ecosystem services from a diverse set of crops. Additionally, crop rotations can improve soil structure and organic matter, which reduces erosion and increases farm system resilience. *Compare with* MONOCROPPING.

cropping intensity (R): Refers to the number of years of cropping multiplied by 100 and divided by the number of years of the rotation (e.g. 4 years crop, 6 years fallow = 10-year rotation; thus $R = (4 \times 100)/10 = 40$).

cropping pattern: The yearly sequence and spatial arrangement of the crops or of crops and fallow on a given area. It includes sequential or multiple cropping, intercropping, mixed cropping, relay cropping, etc. (e.g. rice followed by wheat, maize followed by wheat followed by green gram).

cross: Bringing together genetic material from different individuals through mating in order to achieve genetic recombination.

cross back: *See* BACKCROSSING.

cross-classification: Classification according to more than one characteristic at the same time.

cross-fertilization: A process in which sexual reproduction ensues as a result of the fusion of male and female sex cells (gametes) from different individuals from different populations (lines, breeds, races or species). *Contrasts* SELF-FERTILIZATION.

cross-hybridization: (1) The annealing of a single-stranded DNA sequence to a single-stranded target DNA to which it is only partially complementary. Often, this refers to the use of a DNA probe to detect homologous sequences in species other than the origin of the probe. (2) In biotechnology, the hydrogen bonding of a single-stranded DNA sequence that is partially but not entirely complementary to a single-stranded substrate. Often, this requires hybridizing a DNA probe for a specific DNA sequence to the homologous sequences of different species.

cross-inoculation groups: Groups of legumes, any one of which is capable of forming nodules when inoculated with rhizobia isolated from another legume in the group.

cross-licensing: In patenting biological material, it refers to agreement in which two or more firms with competing and similar technologies strike a deal to reduce need for legal actions to clarify who is to profit from applications of a particular technology.

cross-linking ChIP: Cross-linking ChIP engages chromatin fixed with formaldehyde and fragmented by sonication.

cross-linking of DNA: When two strands of DNA are connected through covalent bonds, situation is described as cross-linking. Alkylating agents may cause cross-linking of DNA and thus cause lethal effect by inhibiting DNA duplication because due to cross-linking DNA strands do not separate.

cross-pollination: Fertilization of a plant with pollen from another plant, usually from the same species. Pollen may be transferred by wind, insects, other organisms or humans. *Contrasts* SELF-POLLINATION.

cross-pollination efficiency: The ease with which cross-pollination can be achieved. Generally measured by the number of hybrid progeny generated per flower pollinated.

cross-protection: The process by which a normally susceptible host is infected with a less virulent pathogen (usually a virus) and thereby becomes resistant to infection by a second, usually related, more virulent pathogen.

cross-resistance: Resistance associated with a change in one genetic factor that causes resistance to different chemical pesticides that were never applied.

cross-sectional study: An observational study that estimates a characteristic or set of characteristics in a set of subjects at one point in time; a 'snapshot' of a characteristic or condition of interest. *Aka* SURVEY *or* POLL.

cross-sterility: The failure of fertilization due to genetic or cytological conditions (incompatibility) in crosses between individuals crossing barrier.

cross-tabulations: For cross-classified data, the frequency for the values that fall into each possible combination of levels from two different factors is recorded. One of the factors is associated with the columns of the contingency table. The other factor is associated with the rows of the contingency table. Individual qualitative data values are simply grouped by the values of the classification variable. For individual quantitative data values, a classification may be created by defining intervals or cut points to place the data into separate groups. *See* CONTINGENCY TABLES.

cross-validation: This technique involves leaving out 'm' patients at a time, fitting a model on the remaining 'n − m' patients and obtaining an unbiased evaluation of predictive accuracy on the 'm' patients. The estimates are averaged over ≥n/m repetitions. Cross-validation provides estimates that have more variation than those from bootstrapping. It may require >200 model fits to yield precise estimates of predictive accuracy.

crossability: The ability of two individuals, species or populations to cross or hybridize.

crossbreed: Fertilization between separate individuals.

crossbreeding: Outbreeding or the breeding of individuals from genetically unrelated populations (lines, breeds, races or species). This may necessitate the transfer of pollen from one individual to the stigma of another of a different genotype.

crossbreeding barrier: A pre- and/or postfertilization condition (i.e. PROGAMOUS and/or POSTGAMOUS INCOMPATIBILITY) that prevents or reduces crossbreeding or any form of gene transfer. It is caused by genetic, environmental, physical or chemical influences.

crossing: The deliberate mating of two parental types of organisms in genetic analysis.

crossing barrier: Any of the genetically governed processes that either entirely prevent or at least significantly reduce the ability of individuals of a population to hybridize with individuals of other populations.

crossing block: A crop plant nursery containing the parental stocks for a breeder's crossing programme.

crossing groups: Any group of individuals that includes a unique set of parents. Accordingly, there are three types of crossing groups: (i) DIALLEL CROSSING GROUP – controlled crosses are done between each pair of parents in the group but crosses with parents outside the group are omitted. (ii) FACTORIAL CROSSING GROUP – a limited number of parents are employed as male testers in controlled crosses with an unlimited number of female parents. (iii) OPEN-POLLINATED CROSSING GROUP – all parents in a breeding population are involved in a progeny test or series of tests.

crossing over: Exchange of chromosome segments between homologous non-sister chromatids by breakage and reunion during meiosis. Crossing over leads to recombination of genes. Crossing over takes place among the maternal and paternal chromatids resulting in new combination of genes. Combined with random segregation, crossing over is the source of new combination of genes in the interbreeding populations. In plant breeding in self-pollinated crops, for instance, when two true breeding lines are crossed (hybridized), there is a huge variation in the F_2 generation that is a result of crossing over and recombination. During meiosis at pachytene stage crossing over takes place which is facilitated by tight synapse of the four chromatids and involves the formation of a zipper-like structure along the length of chromatids called the synaptonemal complex resulting in a recombinant chromosome. In other words, crossing over is a mechanism for genetic recombination. This process can result in an exchange of alleles between chromosomes. *See* RECOMBINATION, CHIASMA *and* CHROMOSOMAL CROSSING-OVER.

crossing over map: A genetic map constructed by using crossing-over frequencies as a measure of the relative distances between genes in one linkage group (chromosome).

crossing over, significance of: Crossing over is a means of introducing new combinations of genes and hence traits. It thus increases genotypic and phenotypic variability, which is useful for natural selection and for adjustment under the changed environment. Frequency of crossing over depends on the distance between two genes. This serves as a basis for preparing linkage chromosome maps. Linkage maps that use the phenomenon of crossing over as a basis prove that genes lie in linear fashion in the chromosome. Useful recombinants produced by crossing over are picked by breeders to produce new varieties of crop plants and animals.

crossing over, stage of occurrence of: Tetrad analysis in *Neurospora* and attached X chromosome experiment in *Drosophila* have shown that crossing over takes place at the four-strand stage.

crossing-over suppressor: In a heterozygote for an inversion, crossing over within the inversion loop produces dicentric and acentric chromosomes. Individuals producing crossover chromatids do not survive and thus do not produce recombinant gametes. Thus, crossovers in such cases are not detected. In genetics experiments, inversions are used as suppressors of crossing over.

cross(ing)-over unit: A 1% crossing-over value between a pair of linked genes. *See* MORGAN UNIT.

crossover: (i) In genetics, a recombination event between two sites within a chromosome at meiosis. (ii) In genetic algorithms and evolutionary computation, a genetic operator used to combine the genetic information of two parents to generate new offspring. It is one way to stochastically generate new solutions from an existing population and analogous to the crossover that happens during sexual reproduction in biological organisms. Solutions can also be generated by cloning an existing solution, which is analogous to asexual reproduction. Newly generated solutions are typically mutated before being added to the population.

crossover fixation: A possible consequence of unequal crossing over that lets a mutation in one member of a tandem cluster to spread through the whole cluster or to be eliminated.

crossover study: A clinical trial in which each group of subjects receives two or more treatments, but in different sequences.

crossover suppression: The apparent lack of crossing over within an inversion loop in heterozygotes. It occurs due to mortality of zygotes carrying defective crossover chromosomes rather than actual suppression.

crossover unit: The percentage of recombination of linked genes. One per cent of recombination equals one crossover unit in a linkage map. *Aka* MAP UNIT *or* CENTIMORGAN.

crossover value: The percentage of crossing over in hybrid population. The term is used mostly in determining linkage percentage, particularly in chromosome mapping.

crosstalk at composite response elements: This model allows for cell-specific effects by ubiquitous factors and divergent effects by closely related factors from a given family. For example, the GLUCOCORTICOID RECEPTOR (GR) acts through a zinc finger DNA-binding domain in the presence of the hormones, assuming a dimer conformation. In the absence of DNA, the receptor is predominantly monomeric. Thus, heterodimerization and post-translational modifications of transcription factors and the organization of composite promoter elements are strategies by which diverse sets of genes can be regulated in a specific manner using a finite number of protein factors.

crosstalk between chloroplasts and mitochondria: Several pieces of evidence demonstrate the reality of crosstalk between chloroplasts and mitochondria. Mutants either showing defect in photorespiration activity or having abrasions in a gene encoding a component of mitochondrial protein complexes (such as

NADH dehydrogenase) revealed altered chloroplast functional states, including photosynthetic activity, while, in chloroplast development mutants such as ALBOSTRAINS of *Arabidopsis*, development of mitochondria was also affected. Furthermore, chloroplast-mitochondria crosstalk appears to contribute to chloroplast-to-nucleus retrograde signalling.

crosstalk between histone modifications: Crosstalk between histone modifications occurs when one or more histone modifications modulate the addition, removal or recognition of another modification, or synergistically function to promote or repress the transcription of a gene.

crown gall tumour: A tumour formed, usually, on the stems of broad-leaved plants when infected with *A. tumefaciens* containing a Ti plasmid. The bacterium is only necessary for the initiation of the tumour. The genome of the affected plant cells contains several copies of a segment of the Ti plasmid (the T-DNA). Crown galls can be of two types, OCTOPINE or NOPALINE, depending on the type of Ti plasmid which initiated the tumour. Whole plants can be regenerated from crown-gall tissue and some of these will contain the T-DNA. Galls are moulded at the crown (stem-root junction) or, less frequently, on roots, stems or branches.

crown: (i) The upper dome of tree, bearing leaves, flowers and fruits. (ii) Junction of root and stem of a plant, usually at the soil line. (iii) In grafted woody plants, the rootstock portion of the plant near the soil surface. The term is also used for the TREETOP. *See* CANOPY.

cruciferae: A taxonomic group ('family') of plants that includes canola, mustard, oilseed, rape, etc.

cruciform DNA: One structural distortion of DNA is cruciform, which is cross-shaped structure. This structure generates due to intramolecular pairing within a double-stranded DNA molecule that contains an inverted repeat. When such sequences have a length of 10 or more base pairs with a centre of symmetry, double-stranded DNA can assume cruciform shape like a hairpin, or a single-stranded DNA can assume stem-loop structure. Such regions may be present in regulatory segments for recognition by specific protein factors. Cruciform DNA is produced in the presence of (i) inverted repeats such as CGATCTGG – CCAGATCG, and (ii) minor repeats such as GGTTGGCC – CCGGTTGG and direct repeats such as GGTTGGCC – GGTTGGCC.

cruciform structure: The cross-shaped structure that can arise by intramolecular pairing within a double-stranded DNA molecule that contains an inverted repeat.

crude rate: A rate for the entire population that is not specific or adjusted for any given subset of the population.

***Cry* genes:** The insecticidal *crystal protein* (*Cry*) genes of *Bacillus thuringiensis* that are a key gene resource for generating transgenic crops with pest resistance. Different strains of *B. thuringiensis* are distinguishable on the basis of crystal (delta endoxin) shapes and sizes. For instance, *Cry1*, *Cry2* have been cloned from *B. thuringiensis* strain 'Kurstaki', *Cry9* from strain 'Japonensis' and *Cry25* from strain 'Jegathesan' and so on. *Cry* genes produce delta endotoxins which are highly toxic to lepidopteran larvae at very low concentrations.

Cry proteins: A class of proteins produced by *B. thuringiensis* (*Bt*) bacteria (or plants into which a *Bt* gene has been inserted). Cry ('crystal-like') proteins are toxic to certain categories of insects (corn borers, corn rootworms, mosquitoes, black flies, armyworm, tobacco hornworm, some types of beetles, etc.) but are harmless to mammals and most beneficial insects.

Cry1A protein: A protein derived from the bacterium *B. thuringiensis* that is toxic to some insects when ingested. This bacterium transpires widely in nature and has been used for decades as an insecticide, although it constitutes less than 2% of the overall insecticides used.

Cry1A(b) protein: One of the cry ('crystal-like') proteins, a protoxin that when eaten by certain insects (e.g. Lepidoptera larvae such as the armyworm or tobacco hornworm or European corn borer) is toxic to those crop pest insects. However, if eaten by a mammal, the Cry1A(b) protein is digested harmlessly within 1 min.

Cry1A(c) protein: One of the cry ('crystal-like') proteins produced by the Gram-positive bacterium, *B. thuringiensis* (*Bt*) during sporulation, which act as insecticides.

Cry1F protein: One of the cry ('crystal-like') proteins, a protoxin that, when eaten by the European corn borer, south-western corn borer, black cutworm and fall armyworm, is toxic to those insects.

Cry3B(b) protein: One of the cry ('crystal-like') proteins, a protoxin that, when eaten by certain insects (e.g. larvae of corn rootworm *Diabrotica virgifera virgifera*), is toxic to those insects.

Cry9C protein: One of the cry ('crystal-like') proteins, a protoxin that, when eaten by the European corn borer, south-western corn borer, black cutworm and some species of armyworm, is toxic to those insects.

CRY-CIB1 heterodimerizing domains: Here, CRY stands for Cryptochrome and CIB1 stands for cryptochrome-interacting basic helix–loop–helix 1. A light-inducible protein interaction between the blue light-sensitive cryptochrome 2 protein (CRY2) and its interacting partner CIB1 from *Arabidopsis thaliana*.

cryobank: Place for the preservation or storage of plant material under very low temperatures, usually in liquid nitrogen (−196°C).

cryobiological preservation: Preservation of cells/tissues/organs at very ultralow temperature, usually at −196°C. This has become a common method for long-term storage (ranging from months to several years) of germplasm especially of the vegetatively propagated plant species. Wherever required, these are regenerated into entire plants in tissue culture. The technique of cryopreservation is also used for storage of human and animal semen. *Aka* CRYOPRESERVATION *and* FREEZE PRESERVATION.

cryobiology: The branch of biology that studies the effects of low temperatures on living things within the Earth's cryosphere or in laboratory experiments.

cryodamage: Damage caused by exposure to cold conditions.

cryoelectron microscopy: A type of electron microscopy technique in which the objects to be viewed (e.g. macromolecules and viruses) are rapidly frozen.

cryoprotectant: A substance (chemical) which is used to protect seeds, cultured material, tissue, organs or cells from the low temperature in cryopreservation (e.g. glycerol).

cryoscopy: A technique for determining the molecular weight of a substance by dissolving it and measuring the freezing point of the solution.

cryostat: A device designed to provide low-temperature environments in which experiments may be carried out under controlled conditions.

cryptic colouration: Colouration that permits an organism to match its background and hence become less vulnerable to prediction or recognition by prey.

cryptic element: A transposable element that is epigenetically inactivated for a period of time and has lost the ability to mobilize.

cryptic error: The biological error observed when plots in an experiment differ from the fields they are meant to represent.

cryptic genetic variation (CGV): CGV is defined as genetic variation that does not affect the normal phenotype observed in the population but can modify the phenotype which appears after environmental changes or the introduction of new alleles. Electrophoresis is also unable to detect this type of variation. Cryptic variation is detected at protein level in context of some biochemical (kinetic) properties. Cryptic variation is suggested to compensate in response to varying environmental conditions (temperature) during different stages of development (larvae, pupae and adults). Such variation was suggested to play role in adaptation.

cryptic genetic variation in *Drosophila*: 16 natural populations of *D. melanogaster* were collected from different sites of Mexico by eminent Drosophilist Sarah B. Pipkin of Howard University, Washington, DC, who studied with J.T. Patterson and H.J. Muller at University of Texas at Austin and was a student of J.B.S. Haldane at King's College London. One of us, G.S. Miglani (as Pipkin's doctorate student only for less than a year, September 1976–June 1977 and then for next 3 years under the supervision of Franklin R. Ampy), used these populations for the detection and analysis of cryptic variation for alcohol dehydrogenase (ADH). For this purpose, isochromosomal lines were constructed using genetic markers so that these lines were homozygous for chromosome 2, which carries the *alcohol dehydrogenase* (*Adh*) locus. These isochromosomal lines were homozygous either for Adh^{I} or Adh^{II} allele. Crude extract was immediately used for measuring ADH activity spectrophotometrically. Statistical analysis revealed cryptic variation in the activity of alcohol dehydrogenase (ADH I and ADH II) in 12 Adh^{I}/Adh^{I} and 40 Adh^{II}/Adh^{II} isochromosomal lines. The existence of additional genetic variation in ADH activity was demonstrated in these isochromosomal lines subsequent to treatment with three denaturants: guanidine hydrochloride (GuHCl), urea and heat. A significant increase in the coefficient of variation (CV) values observed after treatment with these denaturants provided evidence for the existence of cryptic genetic variation in isochromosomal lines having approximately the same native ADH activity. Mean alcohol dehydrogenase (ADH) activity levels within ADH I and ADH II isochromosomal lines varied significantly among the Mexican populations of *D. melanogaster* used. Also, the relative amount of genetic variation differed among and within the ADH I and ADH II strains examined. The quantitative nature of the differences in ADH activity supported the concept of the additive gene action of the *Adh* locus. *See* REGULATORY PROPERTIES OF CRYPTIC VARIANTS.

cryptic genetic variation storing through polyploidization: As a result of the current polyploidization, most genes in tetraploid and hexaploid wheat species occur in multiple functional copies, referred to as homeologs. These duplicated genes buffer the rapid natural modifications going on in the giant and dynamic wheat genomes. These duplicate genes alleviate the rapid natural changes generated by the great genomes of wheat. This variation remains hidden to nature and human choice, since changes in loss of functions in a single wheat homeolog are often masked by redundancy of other homologs. This becomes an advantage in the way that redundancy confers tolerance to mutations, thereby creating mutant populations.

cryptic genetic variation, evolutionary role of: The accumulation of CGV occurs under weakened selection, which can serve as an ultimate source of evolutionary innovations. However, this kind of selection could probably lead to accumulation of strongly deleterious or lethal alleles, which can swamp the effect of any potentially adaptive alleles when they are revealed. Revealing and assimilating CGV might be a powerful force in evolution, as adaptive combinations of mutations can appear more often if the weakened selection acts on such variation. Therefore, these combinations of mutations may be relevant to the evolution of novel adaptations.

cryptic genetic variation, significance of: CGV can fuel evolutionary alternate. C.H. Waddington's classic experiment, which dates back to 1942, confirmed this. He used ether vapor to set off a severe reaction in the developing embryos of *Drosophila* (C.H. Waddington in 1942). In response, a share of embryos developed a change in their phenotype, a second thorax. He made selections in *Drosophila* with bithorax character for over 20 years and eventually resulted in fixation of the trait. Similarly, Waddington in 1953 exposed *Drosophila* pupae to heat shock, ensuing in the development of a character known as *crossveinless* (lacking a characteristic crossvein in adults). Both the phenomena laid the initial foundation of the notion referred to as genetic assimilation that environmentally triggered modifications ought to grow to be fixed in a population.

cryptic genetic variation, storing of: The hidden mutations don't require canalization as a prerequisite, as any genetic system with epistatic relationships between loci can store mutations. More is the number of genes contributing to a trait, more will be the alternative phenotypes of that trait. The idea of hiding and storing of mutations in this way has been suggested in various situations.

cryptic plasmids: Plasmids to which phenotypic traits have yet not been ascribed.

cryptic polymorphism: A type of polymorphism which is controlled by recessive genes.

cryptic satellite: A satellite DNA sequence not identified as such by a separate peak on a density gradient (i.e. it remains detectable in the main-band DNA).

cryptic self-incompatibility (CSI): Botanical expression that is used to describe a weakened self-incompatibility (SI) system. CSI is one expression of a mixed mating system in flowering plants.

cryptic species: The species not distinguishable by external morphology.

cryptic splice sites: The splice sites that are not noticeably used in wild-type pre-mRNA, but are only selected as a consequence of a mutation elsewhere in the gene, most frequently at the reliable splice site.

cryptic splicing: Deep intronic splicing procedures such as recursive splicing or poison exon splicing that employ intronic motifs that benevolently participate with the spliceosome. The regions implicated may be more vulnerable to exonization than other intronic areas would be. *See* EXONIZATION.

cryptic transposable element: A transposable element that is epigenetically inactivated for a period of time and has lost the ability to mobilize.

cryptic variants, regulatory properties of: 35 cryptic variant lines that revealed statistically significant differences in mean ADH activity levels subsequent to treatment with GuHCl, urea or heat were used to study regulatory properties of cryptic variants and examine the mechanisms involved in genetic modulation of alcohol metabolism in *Drosophila*. For this purpose, studies were performed at three developmental stages (i.e. late third-instar larval, preemergence pupal and adult) of *Drosophila* cultured in different environments (i.e. at 18 and 28°C). Spectrophotometric analyses for native ADH activity and residual ADH activity after treatment with a denaturant were performed. The three denaturants affected various developmental stages different at the two temperatures. Differential response of cryptic variants to treatment with the denaturants during development suggested that this variation may have an adaptive significance by compensating in metabolism for changes in the environmental conditions. *See* CRYPTIC GENETIC VARIATION IN *DROSOPHILA*.

cryptic variation: Differences among individuals in terms of some biochemical properties (e.g. denaturation) that are not revealed at morphological level or even electrophoretically.

cryptobiosis (hidden life): The ability to enter a state of suspended metabolic activity during unfavourable environmental conditions (drying, cold). It is a survival strategy of some plant pathogenic nematodes.

cryptochrome: Flavoprotein responsive to blue light, found in both plants and animals. In animals, it is involved in circadian rhythms. *See* PHOTOMORPHOGENESIS.

cryptogam: Any of the 'lower plants' which produce spores and do not have stamens, ovaries or seeds. These are literally the plants whose sexual reproductive organs are not conspicuous. This group typically includes ferns, bryophytes, algae and sometimes fungi (including lichenized fungi). *Compare with* PHANEROGAM.

cryptogamous: Reproduction by spores or gametes rather than seeds.

cryptogenes: In trypanosome mitochondrial genes, AUG (translation initiation codon) is absent, frameshifts and most conserved genes like ATP synthetase subunits 6, 8 and 9 and tRNA genes are absent. The genes whose transcripts get edited are called cryptogenes and portions of cryptogenes that are edited are called PRE-EDITED REGIONS (PERs).

cryptogenes, types of: Depending on the position of editing in protein-encoding region, three types of cryptogenes are found. TYPE I CRYPTOGENES are edited in the internal portion (e.g. *COII* in *Trypanosoma brucei*, *Crithidia fasiculata* and *Leishmania tarentolae*). TYPE II CRYPTOGENES are edited at 5′-end (e.g. *Cyb* of *T. brucei*, *C. fasiculata* and *L. tarentolae*). TYPE III CRYTOGENES are extensively edited throughout the protein-encoding region (e.g. *COIII* in *T. brucei*). Editing proceeds in a 3′→5′ direction. In the 3′ poly(A) tails of transcripts of maxicircle genes in the trypanosomatid mitochondria, uridines (U's) are added.

cryptomere: (i) A gene which by itself has no visible effect but whose existence can be demonstrated by means of suitable crosses (i.e. a complementary factor). (ii) A recessive factor.

cryptomeric gene: *See* CRYPTOMERISM.

cryptomerism: The phenomenon in which a gene or an allele does not display a phenotypic effect unless it is activated by another genetic factor, which leads to a sudden change of qualities in the progeny not recognized among the ancestors.

cryptomorphic gene: The gene has a cryptic structure such that the ultimate active product or products are carried within the precursorial molecule. The active protein is released after enzymatic breakdown of the precursor followed by processing. *See* COMPLEX GENES.

crystal (cry) proteins in biotic stress: The *cry* gene of *B. thringiensis* produces a protein, which forms crystalline inclusions in the bacterial spores. The crystalline protein is responsible for insecticidal activities of the bacterial strains. *Cry* genes (and Cry proteins), on the basis of size, have been grouped into 18 distinct groups, which code for 130 kDa or a 70 kDa protein. On the basis of insecticidal activities, Cry proteins have been divided into four groups – CryI, CryII, CryIII and CryIV. Later on, some cry proteins were placed in other groups – CryV and CryVI. The insecticidal crystalline protein δ-endotoxin is characterized by relatively narrow range of activity. *Bt* crystalline proteins are ingested by insects and subsequently dissolved in the insect midgut under alkaline conditions, resulting in protoxin of 130 kDa. The protoxins are subsequently trimmed by gut proteases to an N-terminal 65–70 kDa truncated form which represents activated toxin. Cry proteins require an alkaline pH of 9.5 or above for processing into an active molecule. The toxin binds to specific receptors on the membranes of midgut epithelial cells, inserts itself into the membrane, forms membrane pores that kills the epithelial cells and eventually the insect by osmotic lysis. Cry protein cannot act in human or animal intestine because their intestine is acidic, pH 1.5. Also, there are no receptors. The cry proteins produced in *Bt* crops have been shown to degrade when crop residues are incorporated into soil, like any other protein added to soil. Thus, they are safe for human and environment.

crystal (cry) proteins, domains of: The toxic fragment of cry proteins has three domains – I, II and III. Domain I (33–253 amino acids) functions in pore and ion channel formation. Domain II (265–461 amino acids) is involved in receptor recognition. Domain III (463–609 amino acids) receptor binding, ion channel formation and target insect specificity. The specificity of cry proteins may possibly be due to the proteolytic processing of Cry proteins, which may be cleaved differentially in different insect species. The presence of specific receptors in the midgut epithelium is most likely to be the chief reason for Cry toxin specificity.

crystal (cry) proteins, increasing toxicity of: The toxicity of Cry proteins can be increased by following two approaches: (i) Site-directed mutagenesis within the toxin encoding region. For example, a mutation in *cry1c* gene may lead to a 100-fold increase in toxicity. (ii) Production of hybrid *cry* genes by fusing selected toxic domains of two or more different *cry* genes.

crystal lattices: In crystallography, crystal structure is a description of the ordered arrangement of atoms, ions or molecules in a crystalline material. Ordered structures occur from the intrinsic nature of the constituent particles to form symmetric patterns that repeat along the principal directions of three-dimensional space in matter.

crystal structure: A high-resolution molecular structure derived by x-ray crystallographic analysis of protein or other biomolecular crystals.

crystallography: The study of a protein or some other biomolecular structure by crystallizing the protein and examining the crystals using X-rays. The diffraction angles of the X-rays are used to compute the relative positions of components of the protein and thus its structure.

CryX protein: One of the cry ('crystal-like') proteins that, when eaten by corn rootworm larvae (*Diabrotica virgifera virgifera*), is toxic to those insects.

cuckoo chromosome: An alien chromosome that displays a preferential transmission during generative reproduction. Cuckoo chromosome is found in certain wheat-*Aegilops* crossing progenies.

cuckoo gene: Refers to a gene conferring preferential transmission from the maternal parent (e.g. in *Lablab purpureus*).

cucullate: Hood-like or hooded appearance, commonly referring to the shape of leaves or petals (e.g. *Pelargonium cucullatum*). Similarly derived terms include CUCULLIFORM and CUCCULARIS.

Cucumber Mosaic Virus (CMV): One of the most widely occurring plant viruses that infects more than 800 plant species and has a considerable harmful effect on agriculture worldwide. In addition to an RNA genome, certain strains of CMV naturally harbour an RNA of 330–400 nucleotides, called sat -RNA that depends on the CMV for its entire cycle, has no noteworthy similarity with the viral genome and seemingly does not encode protein. The sat-RNAs are capable of weakening CMV strains, resulting in a sharp decrease in virus titre in infected plants and a nearly comprehensive lack of symptoms (e.g. lethal tomato necrosis).

culling: The postharvest elimination of pathogen-infected or damaged fruit, seeds or plants by screening procedures. The culled or off-graded material can later be individually analysed or discarded.

culm: In grasses, sedges, rushes and some other monocotyledons, an aerial stem bearing the inflorescence, extending strictly from the base of the plant to the lowest involucral bract (or base of the inflorescence).

cultigen: A cultivated plant or group of plants for which there is no known wild ancestor. Their origin or selection is primarily due to intentional human activity.

cultivar (cv): A cultivar is a taxon that has been designated for a particular feature or combination of features and that is clearly discrete, uniform and stable in its morphological, physiological, cytological, chemical or other features, such as colour, shape of flowers, fruits, seeds, height or form and that, when proliferated by appropriate means, retains those characteristics. Refers to a crop variety produced by scientific breeding or farmer's selection methods. According to International Code of Nomenclature for Cultivated Plants (ICNCP-1995), term cultivar is used synonymously with *Sorte* (German), *variety* (English) or *variété* (French). Cultivar names are written with single quotation marks around them (e.g. 'Blue Carpet' or 'Alba'). All new names established after 1 January 1959 must be written in common language (i.e. not in Latin), but names established in Latin prior to this date are retained in their original form. *See* VARIETY.

cultivar development: Any activity of crossing, transformation and/or selection (including marker-assisted selection) among plants, which has the direct purpose of releasing a crop variety.

cultivar epithet: The defining part of a name that denominates a cultivar. Cultivars are designated by fancy (q.v.) epithets appended either to the scientific name or to the common name of the taxon to which they belong. They are not italicized but placed in single quotation marks (e.g. *Rubus nitidoides* 'Merton Early'). 'Merton Early' is the cultivar epithet.

cultivar identification system: A classification system based on sequence-tagged microsatellite loci analysis with fluorescent primers and appropriate computer software. This system allows unequivocal identification of varieties, paternity testing and duplicate identification.

cultivar mixture: A mixture of different varieties prepared for improving the environmental adaptability or the resistance to pathogens.

cultivation: (i) The art or process of agriculture. (ii) The practice of periodic soil tillage to eliminate all vegetation other than the crop being grown. All plant residues are taken off the field and are not recycled *in situ*.

cultural evolution: Change in culture (i.e. information passed on by learning and imitation rather than by biological inheritance).

cultural practices: The manner in which plants are grown (e.g. application of nutrients, irrigation practices, type of cultivation); may be used for disease management.

culture: Growing of living cells/tissues/organs on/in culture medium *in vitro*. The culture technique is commonly practiced in bacteria, algae, protozoa, virus, yeasts, plant cells, mammalian cells, etc. for the purpose of production, trade and utilization or experimentation. Typically used to refer to a population of the cells of a single species or a single strain. The living cells are aseptically cultured on/in chemically defined medium for basic studies to increase their number. This requires aseptic conditions, optimum growth medium and suitable incubation conditions. All clone specimens are exactly alike and identical to the original. Further, depending upon the nature of inoculum, it can be of various types, such as EMBRYO CULTURE, CELL CULTURE, ANTHER CULTURE, POLLEN CULTURE, LEAF CULTURE, BACTERIAL CULTURE and FUNGUS CULTURE.

culture medium: Any nutrient system for the artificial cultivation of bacteria or other cells. The medium usually consists of a complex mixture of organic and inorganic materials. For example, the classic culture (growth) medium used for bacteria consists of nutrients (required by that bacteria) plus agar to solidify or semi-solidify the nutrient containing mass. Also refers to medium on or in which tissues, organs or cells are cultured. It supplies inorganic salts, organic compounds, vitamins, mineral, carbohydrates and hormonal requirements for the growth of a particular type of organism.

culture room: Room or chamber designated for incubating cultures with controlled light, temperature and humidity.

culture tube: A tube in which tissue, organs, cells or organisms are cultured.

culture, pure: A medium which contains only one specific type of organism (e.g. *E. coli* bacteria). A culture may be preserved (stored alive) by freezing, drying (in which the cells go dormant), subculturing on agar medium or other methods. *See* TISSUE CULTURE.

culturing: The growth of a particular microorganism in the laboratory in isolation from other organisms.

cumulative distribution: The probability that a random variable will be less than a given value is known as its cumulative distribution.

cumulative effect: Action of two alleles of a gene giving a more pronounced effect than one in the heterozygous conditions. Probably one allele supplies insufficient enzyme for the reaction normally conditioned by two. The hybrids are distinguishable from parents. Cumulative effect is sometimes erroneously referred to as INCOMPLETE DOMINANCE.

cumulative genes: Polymeric non-allelic genes. *See* POLYGENES.

cumulative population doublings: *See* POPULATION DOUBLING LEVEL.

curative pesticide: A pesticide that has an ability to inhibit or eradicate a disease-causing organism after it has become established in the plant.

curie: Named after the discoverer of radium, Marie Curie, the term denotes the quantity of any radioactive material giving 3.7×10^{10} disintegrations/s.

curing: The elimination of a plasmid from its host bacterial or eukaryotic cell due to action of agents (e.g. ethidium bromide), which inhibits DNA replication.

curing agent: A substance that enhances the rate of loss of plasmids during bacterial growth.

curl: Distortion, puffing and/or crinkling of a leaf resulting from unequal growth.

***Curly-Lobe-Plum Drosophila*.** A balanced lethal system in chromosome 2 of *Drosophila*. Used to detect autosomal recessive lethal mutations of chromosome 2.

curved and bent DNA: During transcriptional initiation or activation, the upstream regions of DNA are observed to be bent and curved so that the regulatory proteins bound at far regions, by curving or bending of the DNA, bring distant regions near to each other for interaction. Such curved DNA is also observed during initiation of replication.

cut: A DNA double-strand break (DSB); as a result of action of physical or chemical mutagens or the wild-type function of most Cas nucleases or restriction enzymes.

cut-and-paste transposition: Type of movement of a TRANSPOSABLE ELEMENT (TE) in which TE is cut out of the DNA and inserted at a new site by a special TRANSPOSASE enzyme.

cuticle (Adj. cuticular): (i) The non-cellular outer layer of an insect or a nematode, composed of the polymers cutin and/or cutan and waxes. (ii) Water-repellent, waxy layer of epidermal cells of plant parts, such as leaves, stems and fruit. The leaf's dermal cells produce and secrete a waxy substance, composed of lipid polymers, that forms a protective covering. It protects the leaf from light damage and invading pathogens, such as viruses, bacteria and fungi.

cutting: A section of a plant that is removed and used for propagation. Several types of cuttings can be taken from the parent stock, which depends on the point on the parent stock where the cutting is taken; four major categories of cuttings are: (i) stem cuttings; (ii) leaf cuttings; (iii) leaf-bud cuttings; and (iv) root cuttings. *See* ROOT CUTTING.

Cyanelle DNA: Cyanelles are photosynthetically active organelles, which resemble the cyanobacterium for many morphological and physiological features. Cyanelles are found in a few eukaryotes like *Cynophora paradoxa*. Cyanelle DNA is similar to both chloroplast DNA and cyanobacterium DNA. Cyanelles have been suggested to represent a bridge between cyanobacteria and chloroplasts.

Cyanobacteria: One of the major phyla of bacteria. Referred to as blue-green algae. Most species are photosynthetic, and chloroplasts are hypothesized to have been derived from cynobacteria.

cyathium (pl. cyathia): An inflorescence of unisexual flowers surrounded by involucral bracts, especially the flowers of Euphorbia.

cyathophyll: In Euphorbia, the bract-like structure on which the involucre sits, usually but not always occurring in pairs. They may sometimes be brightly coloured and confused with petals.

cyberbullying: A form of intimidation or pestering using electronic means. *Aka* CYBERHARASSMENT *or* ONLINE BULLYING.

cybernetic: The one involving a control or governing mechanism operated by feedback mechanism.

cyberspace: Widespread, interconnected digital technology.

cybrid: A eukaryotic cell produced by the fusion of a whole cell with an enucleated cell (cytoplast). Thus, cybrids combine the cytoplasmic DNA (traits) from two parents and nuclear DNA (trait) from one parent only. Cybrid is a unique combination as during conventional hybridization we combine nuclear genes from two parents and cytoplasm is contributed by the female parent only. Cybrids are useful in single-step transfer of cytoplasmic traits. *Aka* CYTOPLASMIC HYBRID.

cyclic adenosine 5′-monophosphate (cAMP): A specialized nucleotide generated from ATP by adenylyl cyclase in retort to stimulation by many types of cell-surface receptors. cAMP functions as an intracellular signalling molecule by activating cyclic-AMP-dependent kinase (protein kinase A, PKA). It is hydrolysed to AMP by a phosphodiesterase. cAMP is very important for CELL RESPONSE to protein hormones. cAMP can also induce formation of nervous system. cAMP also acts as a MORPHOGEN. It is essential for activation of glycoprotein hormones, alpha polypeptide (cga) protein. cAMP-cga protein system is influenced by level of glucose. This exercises a positive control on expression of lactose operon. cAMP is the key metabolite that impacts glucose effect. *Aka* SECOND MESSENGER.

cyclic adenosine 5′-monophosphate (cAMP)-dependent protein kinase A (protein kinase A or PKA): An enzyme that phosphorylates target proteins in reply to a rise in intracellular cyclic AMP level.

cyclic guanosine 5′-monophosphate (cGMP): A specialized small, soluble intracellular signalling molecule nucleotide formed from GTP by guanylyl cyclase in response to photoreceptor stimulation in the retina. cGMP is a common regulator of ion channel conductance, glycogenolysis and cellular apoptosis.

cyclic phosphorylation: The reaction that involves the movement of electrons along the electron transport chain in the mitochondria, which generates a proton gradient that drives the production of ATP in plants through the action of the ATP synthase enzyme. *Aka* PHOTOSYNTHETIC PHOSPHORYLATION (PHOTOPHOSPHORYLATION).

cyclical changes in DNA methylation status: Replication of methylation patterns by DNA methyltransferases maintains genome stability through cell division. There is evidence of an unanticipated dynamic role for DNA methylation in gene regulation in human cells. Periodic, strand-specific methylation/demethylation occurs during transcriptional cycling. Cyclical changes in the methylation status of promoter CpGs may thus represent a critical event in transcriptional achievement.

cyclin: Protein that periodically increases and decreases in concentration in step with the eukaryotic cell cycle. Cyclins activate crucial protein kinases (*Aka* CYCLIN-DEPENDENT PROTEIN KINASE, CDK) and thereby control progression from one stage of the cell cycle to the next.

cyclin-Cdk complex: Protein complexes that are moulded periodically during the eukaryotic cell cycle as the level of cyclin increases and in which the cyclin-dependent kinase (CDK) becomes incompletely activated.

cyclin-dependent kinase inhibitors (CKIs): CKIs inactivate Cdk-cyclin complexes and promote cell cycle arrest. Three different CKIs – p21Cip1, p27Kip1 and p57Kip2 – form the Cip/Kip family.

cyclin-dependent kinases (CDKs): CDKs are small serine/threonine protein kinases that require association with a cyclin subunit for their activation. The CDKs limit the activation of DNA replication origins to once per cell cycle by preventing the assembly of pre-replicative complexes during S, G_2 and M phases of the cell cycle in the budding yeast. CDKs inhibit each pre-RC component (ORC, Cdc6, Cdt1/Mcm2-7) by different mechanisms. Mitotic CDK, Clb2/Cdc28, binds tightly to an amino-terminal domain (NTD) of Cdc6, and Cdc6 in this complex is unable to assemble pre-RCs. The progression through the cell-division cycle is driven by activation and inactivation of CDKs by phosphorylating specific target proteins that trigger the transition to subsequent phases of the cycle.

cyclins: A diverse group of proteins ranging in size from 35 to 130 kDa, all with a similar core structure based on two symmetrical domains of five α-helices. One of these domains, the cyclin box, is highly conserved and is the defining structural feature of these proteins. At least 16 different cyclins have been identified in humans. Four different classes of cyclins defined on the basis of the stage of the cell cycle in which they bind and activate CDKs are G1-cyclins, G1/S-cyclins, S-cyclins and M-cyclins.

cyclobutane pyrimidine dimers (CPDs): UV-induced structures that are repaired much more efficiently in an actively transcribed gene than in non-coding regions, and this repair is more efficient in the transcribed strand of such genes than in the non-transcribed strand.

cyclobutyl dimer: A dimeric structure formed between two adjacent pyrimidine bases in a polynucleotide, resulting from UV irradiation.

cycloheximide: A chemical that impedes protein synthesis by the 80S eukaryotic ribosomes. It does not, however, inhibit the 70S ribosomes of prokaryotes. This chemical inhibits peptide bond formation by binding to the large eukaryotic ribosomal subunits. *Aka* ACTIDIONE.

cylindrical: Rod-like in shape and is two to three times as long as wide. *Compare with* BACULIFORM.

cyme (Adj. cymose): A type of inflorescence in which the main axis and also all lateral branches end in a flower (each lateral branch may be repeatedly branched).

cynobacteria: A simple prokaryote which in terms of its biochemistry and metabolism seems to occupy a middle position between anaerobes and eukaryotes because they can tolerate oxygen, make use of oxygen both metabolically in aerobic respiration and in synthetic pathways and can make unsaturated fatty acids by oxidative desaturation. Some cynobacteria under anaerobic conditions can use H_2S instead of water releasing S instead of oxygen.

cypsela: A type of dry, one-seeded, indehiscent fruit formed from the inferior ovary.

cyst: A protective capsule around some protozoa during unfavourable conditions. During unfavourable conditions some protozoa develop a protective capsule around the body which is called a cyst. The term cyst is also used to describe a sac with a distinct wall containing fluid or other material. Cysts are also detected in human body through radio-diagnostic techniques. In fungi, cyst is a resting structure in a protective membrane or shell-like enclosure; in nematode females, the egg-laden carcass of a female nematode; in bacteria, a specialized type of bacterial cell enclosed in a thick wall, often dormant and resistant to environmental conditions.

cysteine (Cys): An aliphatic, polar alpha-amino acid that contains a sulfhydryl group.

***Cystic Fibrosis Transmembrane Regulator* (*CFTR*) gene:** The gene responsible for cystic fibrosis.

cystic fibrosis: An autosomal recessive genetic condition that affects not only the lungs but also the pancreas, liver, kidneys and intestine, and usually appears shortly after birth. They include faulty digestion, breathing difficulties and respiratory infections due to mucus accumulation and excessive loss of salt in sweat. In the past, cystic fibrosis was almost always fatal in childhood, but treatment is now improved such that patients commonly live into their 20s and beyond.

cystine: Two cysteine amino acids covalently joined by a disulfide bond. Cystines are important in biochemistry, as the disulfide bridges represent one important way in which the conformation of a protein is maintained in the active form. Cystine bridges lock the structure of the proteins in which they occur in place by disallowing certain types of (molecule) chain movement. When the disulfide bond is with a free cysteine (i.e. one that is not a part of the same protein molecule's amino acid backbone), the free cysteine is known as a THIOL GROUP. Cystine can be metabolized from methionine by certain animals (e.g. swine) but not vice versa.

CystX: Refers to a naturally occurring gene present in the genome (DNA) in some varieties of soyabean plant that confers on those particular soyabean varieties (some) resistance to the soyabean cyst nematode. Discovered and developed during the 1990s by Jamal Faghihi, John Ferris, Virginia Ferris and Rick Vierling.

cytidine: The ribonucleoside containing the pyrimidine cytosine (C).

cytidine 5′-diphosphate (CDP): The ribonucleotide containing the pyrimidine cytosine (C) and two phosphate groups are attached at 5′ end of the sugar moiety.

cytidine 5′-monophosphate (CMP): The ribonucleotide containing the pyrimidine cytosine (C) and one phosphate group is attached at 5′ end of the sugar moiety. *Aka* CYTIDYLIC ACID *or* CYTIDYLATE.

cytidine 5′-triphosphate (CTP): The ribonucleotide containing the pyrimidine cytosine (C) and three phosphate groups are attached at 5′ end of the sugar moiety.

cytidylic acid: The ribonucleotide containing the pyrimidine cytosine and one phosphate group attached to sugar moiety.

cytochemistry: The analysis, often by determination of staining properties, of the chemical composition of cell components.

cytochimera: Different tissues or parts there of that differ in chromosome number.

cytochrome b-c1 complex: Second of the three electron-driven proton pumps in the respiratory chain. It accepts electrons from ubiquinone.

cytochrome (Cyt): Any of the complex protein respiratory pigments (enzymes) occurring within plant and animal cells. They usually occur in mitochondria and function as electron carriers in biological oxidation during respiration and photosynthesis. Cytochromes are haem-containing proteins involved in the 'handing off' of electrons to each other in a stepwise fashion. In the process of 'handing off', other events take place, which result in the production of energy that the cell needs and is able to use.

cytochrome oxidase complex: Third of the three electron-driven proton pumps in the respiratory chain. It accepts electrons from cytochrome c and generates water using molecular oxygen as an electron acceptor.

cytodifferentiation: Cytodifferentiation is the sum total of all processes by which, during the development of the individual, the zygote, specialized cells, tissue and organs are formed. This is the process whereby cell acquires those biochemical and morphological properties necessary to perform a function. The process of transformation from undifferentiated to fully differentiation state is a gradual one. It begins with a shift in developmental potential from totipotent to more restricted. As development proceeds the repertoire of types into which a given cell can differentiate becomes gradually limited. The process of cytodifferentiation in embryo and continued process of tissue differentiation in adult is accomplished by a population of stem cells, each capable of providing organism with a limited variety of specialized cells. Each differentiated cell type reaches its final stage of differentiation via a predictable and unique series of events. An outstanding feature in a multicellular organism is that each differentiated cell falls into clearly recognizable discrete category sharply set out for other cell type. All such specializations are recognizable in the general features of undifferentiated cell although each may be unique and having certain proteins not present in other. In a sense, general property of cell represents keyboard on which various types and patterns of selection are made. One event that seems to occur in a wide variety of differentiating tissue is the stabilization of mRNAs for the tissue-specific proteins at some stage during differentiation. The stability of dedifferentiation state of a cell is questionable. A cell can be differentiated and redifferentiation. After a process of dedifferentiation into another cell type seems to occur.

cytodifferentiation during tissue culture: Differentiation of callus cells into a variety of cells during regeneration of plants. True callus is a homogeneous mass of undifferentiated cells. Prolonged culturing, physical and chemical factors induce cytodifferentiation in which the callus cells exhibit morphological, cytological, cytogenetical and biochemical differentiation. Cells start organizing into different kinds of tissues and then organs. Among various factors, the kind and concentration of growth regulators play dominant role in cytodifferentiation.

cytogamy: Refers to the fusion or conjugation of cells.

cytogenetic breeding: Breeding techniques that comprise the manipulation or alteration of genetic material in cells (e.g. exposure to radiation or application of chemicals such as an alkaloid colchicine).

cytogenetic disorders: Disorders involving the cellular constituents concerned in heredity (i.e. chromosome abnormalities).

cytogenetic map. A genetic map giving the position of gene loci relative to chromosome bands. The characteristic appearance of a chromosome when stained and examined under a microscope. In this case, cytological and genetic map are aligned. Molecular cytogenetic mapping uses fluorescent *in situ* hybridization (FISH) and primed *in situ* (PRINS) labelling techniques. *See* GENETIC MAP.

cytogenetics: Scientific discipline that fuses cytology with genetics. It includes: (i) the organization of chromosomes in the nucleus with the necessities of gene expression in different developmental and environmental contexts; (ii) the relationship between the behaviour of chromosomes with transmission of phenotypes to progeny; (iii) changes in chromosomal structure and number with speciation; and (iv) the evolution of the genome relative to the evolution of the species.

cytogenic male sterility: Male sterility controlled by interaction between cytoplasmic and nuclear genes. *See* CYTOPLASMIC MALE STERILITY.

cytohet: A cell containing two different cytoplasmic genomes (e.g. mitochondria) that differ in one or more genes contributed by two parents. Thus, the individual is CYTOPLASMATICALLY HETEROZYGOUS.

cytokine receptor: Type of cell-surface receptor whose ligands are cytokines such as interferons, growth hormone and prolactin, and which acts through the Jak-STAT pathway. *See* JAK-STAT PATHWAY.

cytokines: Substances produced by cells of the immune system of the animal that function to stimulate or modify the immune response (e.g. interferon and interleukins). A large class of glycoproteins similar to lymphokines but produced by non-lymphocytic cells such as normal macrophages, fibroblasts, keratinocytes and a variety of transformed cell lines. They participate in regulating immunological and inflammatory processes and can contribute to repair processes and to the regulation of normal cell growth and differentiation. Although cytokines are not produced by glands, they are hormone-like in their intercellular regulatory functions. They are active at very low concentrations and for the most part appear to function non-specifically. For example, the cytokines stimulate the endothelial cells to express (synthesize and present) P-selectins and E-selectins on the internal surfaces (of blood vessels). These selectins protrude into the bloodstream, which causes passing white blood cells (leukocytes) to adhere to the selectins, then leave the bloodstream by 'squeezing' between adjacent endothelial cells. Cytokines are exemplified by the interferons.

cytokinesis: During the division of a cell, the division of the constituents of the cytoplasm. It usually begins in early telophase with the formation of a cell plate, which is assembled within the phragmoplast across the equatorial plane. The PHRAGMOPLAST is a complex array of Golgi-derived vesicles, microtubule, microfilaments and endoplasmic reticulum that assembles during the late anaphase and is dismantled upon completion of the new cell wall. During division of cytoplasm equal distribution of cytoplasmic organelles is ensured. Animal cells divide their cytoplasm by furrowing. A circular constriction appears round the middle of the cell. Plant cells divide their cytoplasm by cell plate formation. It starts from the centre of the cell and grows towards the periphery. Cytokinesis is delayed until both nuclear divisions (meiosis I and meiosis II) are completed so that four cells are simultaneously formed owing to crossing over during pachytene. The four haploid cells produced by division of a single diploid cell are genetically non-identical. *Aka* CYTOPLASMIC DIVISION.

cytokinesis I: The division of the cytoplasm after meiosis I, giving rise to two haploid cells. Cytoplasm divides at its middle, but the two daughter cells formed from one parental cell remain intact.

cytokinesis II: The division of the cytoplasm after meiosis II giving rise to four haploid cells.

cytokinins: One of a group of phytohormones (e.g. kinetin, benzylaminopurine, zeatin) that acts synergistically with auxins to promote cell division but, unlike auxins, that endorse lateral growth. Cytokinins are usually found in the roots, young fruits and seeds. Kinetin is commonly used for callus production as these chemicals at concentrations ranging from 0.2–3.0 mg/ml cause active cell division. Further, during micropropagation, the excessive shoot multiplication is also a game of cytokinins.

cytokinins in *Arabidopsis*: Cytokinins are a class of phytohormones involved in various physiological events of plants. The *Arabidopsis* sensor histidine kinase CRE1 was reported to be cytokinin receptor. Transcription factor-type response regulator ARR1 directs transcriptional activation of the *ARR6* gene, which responds to cytokinins without *de novo* protein synthesis. This fact, together with characteristics of *ARR1*-overexpression plants and *arr1* mutant plants, indicated that phosphorelay to ARR1, probably from CRE1, constitutes an intracellular signal transduction occurring immediately after cytokinin reception.

cytological band: An area of the chromosome that stains differently from areas around it. *See* CYTOLOGICAL MAP.

cytological map: A type of chromosome map whereby genes are located on the basis of cytological findings by observation of chromosomal morphology obtained with the aid of chromosome mutations.

cytology: The branch of biology dealing with the structure, function and life history of the cell.

cytolysis: The dissolution of cells, particularly by destruction of their surface membranes.

cytomixis: The extrusion or passage of chromatin from one cell into the cytoplasm of an adjoining cell.

cytopathic: Damaging to cells.

cytopathology: The study of changes induced by disease at the cellular level.

cytophotometry: Photometric analysis of different macromolecules (such as DNA, RNA, proteins, polysaccharides) in cells or tissue sections.

cytoplasm: The matrix (ground substance) of a cell composed of mainly of water and proteins and surrounded by plasma membrane. Cytoplasm of a cell lies inside the cytoplasmic membrane and external to the nucleus. It primarily consists of water (90%). Most of the metabolic reactions take place in the cytoplasm. It possesses some internal structures such as chloroplasts, mitochondria, ribosomes, free-floating granules of nutrients and other biomolecules. Cytoplasm is a continuous viscous material that exhibits cytoplasmic streaming movements in the living cells that can be seen under light microscope using phase contrast system. Cell organelles like chloroplasts and mitochondria possess their own DNA called chloroplast DNA (cp DNA) or mitochondrial DNA (mt DNA); collectively, these are called CYTOPLASMIC GENES. During hybridization, the cytoplasm is always contributed by the female parent, with some exceptions, as in Pelargonium. Cytoplasmic hybrids (cybrids) are only possible through protoplast fusion. *Aka* CYTOSKELETON.

cytoplasmic: Pertaining to or centred in the cytoplasm.

cytoplasmic DNA: The DNA within an organism (e.g. plant) that is not inside the cell's nucleus. Cytoplasmic DNA refers to the DNA found in the cell's mitochondria and the chloroplasts. Unlike nuclear DNA, cytoplasmic DNA is not transferred from plant to plant via pollen.

cytoplasmic factor: A hereditary trait transmitted by way of cytoplasm rather than the nuclear material of a cell.

cytoplasmic gene control: Wide range of half-lives for different specific mRNAs in the same cells exists and different half-lives for the same mRNA in the same cell exist under different circumstances. Some pieces of evidence show that cytoplasm can control nuclear activity.

cytoplasmic genetic system: Refers to the genetic systems of cell organelles.

cytoplasmic hybrid: *See* CYBRID.

cytoplasmic inheritance: A non-Mendelian (extranuclear) inheritance via genes in cytoplasmic organelles (mitochondria, plastids). Some traits (e.g. male sterility, herbicide resistance) are governed by the genes present in the mitochondria and chloroplast DNA. *Aka* UNIPARENTAL INHERITANCE. *See* CYTOPLASMIC TRAIT, EXTRACHROMOSOMAL INHERITANCE *and* MATERNAL INHERITANCE.

cytoplasmic male recombination: When recombination is induced in males of *D. melanogaster* under the influence of cytoplasmic factors.

cytoplasmic male sterility (CMS): A condition under which a flowering plant is unable to produce functional pollen. It is widespread among higher plants. Pollen abortion due to cytoplasmic factors, which are maternally transmitted but which act only in the absence of pollen-restoring genes. This type of sterility can also be transmitted by grafting. CMS systems represent a valuable tool in the production of hybrid seed in crop species, including maize, rice, cotton and a number of vegetable crops.

cytoplasmic mutations: Mutations in the cytoplasmic genes. All the F_1s show the characters of the female parent. Chloroplast and mitochondrial mutations fall in this category.

cytoplasmic protein synthesis: It is the translation of mRNAs representing nuclear genes. It occurs via ribosomes attached to the cytoskeleton.

cytoplasmic segregation and recombination (CSAR): A process suggested to explain assortment and recombination of organelle-based genes. From the rate and pattern of segregation of these genes, it has been inferred: (i) hybrid cells are cytoplasmic diploids containing one segregating copy of each parental allele; (ii) the distribution of cytoplasmic genes to daughter cells is spatially oriented and controlled by some cytoplasmic structure so that as a rule both daughters of cytoplasmic diploids are themselves diploid; and (iii) replication of one cytoplasmic chromosome instead of two is the mechanism of restoring the haploid state.

cytoplasmic streaming: The movement of cytoplasm in the protoplast or cell of higher organisms. Cytoplasmic streaming movements can be seen under high-power compound microscope and cytoplasmic streaming is an indication that the protoplast or the cell under the microscope is living as dead protoplasts/cells do not exhibit streaming movements.

cytoplasmic trait: A genetic characteristic in which the genes are found outside the nucleus, in chloroplasts or mitochondria. Results in offspring inheriting genetic material from only one parent.

cytoplasmon: All cytoplasmic hereditary constituents of a cell excepting those localized in the plastids and mitochondria.

cytoplasm-restorer: *See* CYTOPLASMIC MALE STERILITY.

cytoplast: Intact cytoplasm with degenerated nucleus but possessing the outer plasma membrane. The cytoplasts are produced by degenerating the protoplast nucleus with γ-irradiation (10–50 Kr). The cytoplasts are then used in making fusions with intact protoplasts to obtain cybrids.

cytoplastic inheritance: *See* CYTOPLASMIC INHERITANCE.

cytosine (C): A pyrimidine base with chemical name is 2-oxy, 6-amino pyrimidine. It is one of the bases found in DNA and RNA. It occurs in both DNA (along with adenine guanine, thymine) and RNA along with adenine guanine, thymine and uracil. It is paired with guanine (G) in double-stranded DNA. *See* BASE PAIR *and* NUCLEOTIDE.

cytosine hydroxymethylation: DNA modification generated by oxidation of methylcytosine. This modification is mediated by the 10-11 translocation (TET) family proteins and is considered both stable and an intermediate in DNA demethylation.

Cytosine Methylation Analysis Tool for Everyone (CyMATE): CyMATE quantifies and illustrates general and pattern-specific methylation at CG, CHG and CHH (H = A, C or T) sites, both per sequence and per position. CyMATE amplifies and sequences DNA after Sodium bisulfite treatment of genomic DNA deciphers the methylation state of cytosine residues at the highest resolution possible.

cytosine methylation in DNA: In mammalian DNA, 2–7% of the total cytosine is changed to 5-methylcytosine ($5^{m}C$). Methylation occurs enzymatically after DNA synthesis by methyl transfer from S-adenosylmethionine (SAM or AdoMet) to position 5 of cytosine. Cytosine methylation occurs in CpG doublets. *See* DNA METHYLATION.

cytosine methylation in DNA in different contexts: Plants methylate DNA in three different sequence contexts: dinucleotide CpG (or CG) sites and trinucleotide CpHpG (or CHG) and CpHpH (or CHH) (H = A, T or C) sites. This is in contrast to mammals where methylation is primarily found in the CpG context and in specific cell types, the CpH context.

cytosine methylation in RNA: Post-transcriptional RNA modification is a characteristic feature of non-coding RNAs and has been described for rRNAs, tRNAs and miRNAs. (Cytosine-5) RNA methylation has been detected in stable and long-lived RNA molecules. Cytosine methylation in RNA can be reproducibly and quantitatively detected by bisulfite sequencing.

cytosine-to-uracil RNA editing: Three RNA-editing mechanisms are potentially responsible for C-to-U conversion: (i) deamination/amination; (ii) base exchange; and (iii) nucleotide exchange. *See details as separate terms.*

cytoskeleton: Microscopic integrated system of molecules within eukaryotic cells that delivers them with shape, internal spatial organization, motility and may assist in communication with other cells and the environment. Its most abundant components are actin filaments, microtubules and intermediate filaments. Cytoskeleton provides capacity for directed movement and that participates cell.

cytosol: The matrix of the cytoplasm. It is the fluid portion of the cytoplasm. The cytosol present inner to the plasma membrane and outer to the nuclear envelop of eukaryotic cell carry important organelles such as mitochondria and chloroplast. It excludes membrane-bounded organelles such as endoplasmic reticulum, chloroplasts and mitochondria. Operationally defined as the cell fraction remaining after membranes, cytoskeletal components, and other organelles have been detached by low-speed centrifugation.

cytostatic: Any physical or chemical agent accomplished with inhibiting cell growth and cell division.

cytosterility: A genetically controlled condition of male sterility in breeding lines of a self-pollinating crop variety. A breeding line with cytosterility is employed as the seed-producing parental line in the production of hybrid seed.

cytotaxonomy: The study of natural relationships of organisms by fusion of cytology and taxonomy.

cytotoxic: Poisonous to cells.

cytotoxic T cell: Type of T cell accountable for killing infected cells.

cytotoxic T lymphocytes (Tc): T cells that are responsible for attacking host cells that have been infested with an invading bacterium or virus. They directly react with the antigens.

cytotype: Any variety of a species whose chromosome complement differs quantitatively or qualitatively from the standard complement of that species.

D

D: This symbol is used to denote several terms such as DERMATOGEN STAGE, DEXTROROTATORY, DIVERSITY SEQUENCES, GENETIC DISTANCE and LINKAGE DISEQUILIBRIUM.

D-loop: Configuration found during DNA replication of chloroplast and mitochondrial chromosomes wherein the origin of replication is different on the two strands. The first structure formed is a DISPLACEMENT LOOP.

D-loop model: Initiation of DNA replication requires separating the DNA strands and commencing bidirectional DNA synthesis. Replication starts at a specific origin in the circular duplex DNA. However, initially, only one of the two parental strands (the H-strand in mammalian mitochondrial DNA) is used as a template for synthesis of a new strand. Synthesis proceeds for only a short distance, displacing the original partner (L) strand, which remains single stranded. The condition of the region gives rise to the DISPLACEMENT LOOP (D-LOOP).

daemon: In multitasking computer operating systems, a daemon [/ˈdiːmən/ or /ˈdeɪmən/] is a computer program that runs as a background process, rather than being under the direct control of an interactive user. Traditionally, the process names of a daemon end with the letter 'd' for clarification that the process is in fact a daemon and for differentiation between a daemon and a normal computer program. For example, System Logging Protocol daemon (syslogd) implements system logging facility and Secure Shell Daemon (sshd) serves incoming Secure Shell (SSH) connections.

dalton (Da): A unit of mass very approximately equal to that of a hydrogen atom and precisely equal to 1.0000 on the atomic mass scale. Named after John Dalton, who developed the atomic theory of matter. One amu (an arbitrary unit used for the atomic scale) is the mass of 1/12th of one atom of Carbon 12 and equals to 1.66057×10^{-24} g.

***dam* gene:** DNA adenine methylation gene of *Escherichia coli*. *dam* gene product methylates the sequence GATC (*Sau*3A cleaves methylated and unmethylated DNA, *Mbo*I cleaves only unmethylated DNA, *Dpn*I cleaves only the methylated sequence).

dam system: DNA adenine methyltransferase (dam) system discriminates the strands of newly replicated DNA by methylating adenine. It may be involved in regulation of replication and marking DNA strands for repair.

dam-directed mismatch repair (ddMMR) system: This repair system of *Escherichia coli* removes potential mutations arising as replication errors (mutation avoidance). Correction of biosynthetic errors by MMR occurs directly after replication, just behind the replication fork, where the parental DNA strand is fully methylated at GATC (dam) sequences, and the newly synthesized strand is not yet methylated.

damping-off: The death of a seedling before or shortly after emergence, caused by decomposition of the root and/or lower stem. It is possible to distinguish between preemergence and postemergence damping-off.

***Danio rerio* genome:** Researchers at the Institute of Genomics and Integrative Biology, Delhi, India (2009), announced the sequencing of a wild-caught strain of zebrafish (*Danio rerio*), containing 1.7 billion genetic letters. Although the zebrafish has approximately 30,000–50,000 genes, only a small proportion of them are necessary for normal development.

dark field microscopy (DFM): A microscope designed such that the entering centre light rays are blacked out, and the peripheral rays are directed against the object from the side. As the result, the object being observed appears bright upon a dark background.

dark reaction: The phase of photosynthesis that does not require light. In this phase, carbohydrates are synthesized from carbon dioxide.

dark repair: The repair of DNA damage by DNA-repairing enzymes that do not require light for their activation.

dark respiration: *See* DARK REACTION.

Darwin (d): A unit used to denote the rate of change of morphology. It was introduced by J.B.S. Haldane and is equal to a change by a factor e = 2.71828 per million years or 1 part per million per year.

Darwin's paradox resolved: Raghavendra Gadagkar working on social wasps found in 2010 and 2011 that when creatures are solitary and start becoming social, they come together for mutual benefit. In the course of social evolution, some individuals become slaves with the hope that one day they would eventually be queens.

Darwin's paradox: In his theory of natural selection, Charles Darwin stated that worker bees were sterile and did not reproduce; therefore, they should have disappeared, but they have not. This is a paradox to his theory.

Darwinian demon: A hypothetical organism having indeterminately high survival and fertility. This is an idea used to emphasize the belief that all real organisms must face constraints on their fitness.

Darwinian evolutionary theory: This theory of evolution proposed by Charles Darwin states that biological species have evolved with 'modification by descent'. It is based on the observation that all species are built up of populations exhibiting natural variation, some of which is selected for because it enhances the fitness of an individual under a particular set of environmental conditions. Natural variation is generated by mutation, whose degree and kind can be influenced by epigenetic traits.

Darwinian fitness: The genetic input of an individual to the gene pool of the next generation relative to the average for the population. It is usually measured by the number of offspring or close kin that survive to reproductive age.

Darwinian medicine: Refers to the application of evolutionary principles to medicine.

Darwinism: Charles Darwin's evolutionary theory stating that species originated by evolution from other species and that evolution is mainly driven by natural selection. The term is commonly used as a synonym for evolution but differs from neo-Darwinism mainly because Darwin did not know about Mendelian inheritance.

data: A set of observations, generally in numerical format but can be in text format also.

data centre: A dedicated space used to house computer systems and associated components, such as telecommunications and data storage systems. It generally includes redundant or backup components and infrastructure for power supply, data communications connections, environmental controls (e.g. air conditioning and fire suppression) and various security devices.

data cleaning: A process whereby automated or semi-automated algorithms are used to process experimental data, including noise, experimental errors and other artefacts, in order to generate and store high-quality data for use in subsequent analysis. Data cleaning is typically required in high-throughput sequencing, where compression or other experimental artefacts limit the amount of sequence data generated from each sequencing run or 'read'.

data management: The organized preparation of the recording of data for its analysis and its interpretation.

data mining: The process of looking at a large amount of information to see if there are any patterns or connections. For instance, once an experiment has been done, scientists have to make sense of the information they have collected so they can determine the reason the experiment worked the way it did. The ability to query very large databases in order to satisfy a hypothesis ('top-down' data mining); or to interrogate a database in order to generate new hypotheses based on rigorous statistical correlations ('bottom-up' data mining). Data mining is the step analysis for the 'KNOWLEDGE DISCOVERY IN DATABASES' (KDD) process. Aside from the raw analysis step, data mining also involves database and data management aspects such as data preprocessing, model and inference considerations, interestingness metrics, complexity considerations, postprocessing of discovered structures, visualization and online updating.

data processing: The systematic performance on data of such operations as handling, merging, sorting and computing. The semantic content of the original data should not be changed, but the semantic content of the processed data may be changed.

data science: An interdisciplinary field that makes use of scientific methods, processes, algorithms and systems to abstract knowledge and insights from data in various forms, both structured and unstructured, similar to data mining. Data science is a concept that unifies statistics, data analysis, machine learning and their related methods in order to understand and analyse actual phenomena. It employs techniques and theories drawn from many related fields within the context of mathematics, statistics, information science and computer science.

data sharing: When additional information is necessitated before a study can be reproduced, the author of the study is expected to provide it promptly on request – although a small charge may apply. If the author, however, refuses to share data, appeals can be made to the editors of the journal who published the study or to the institution which funded the research.

data sheet: A specially prepared pro forma on paper or computer for recording the primary data.

data structure: A format specifying data organization, management and storage that empowers effective access and modification. More precisely, a data structure is a collection of data values, the relationships among them and the functions or operations that can be applied to the said data.

data type: A feature of data which tells the compiler or interpreter how the programmer intends to use the data. Most programming languages support common data types of real, integer and Boolean. A data type coerces the values that an expression, such as a variable or a function, might take. This data type specifies the manoeuvres that can be done on the data, the meaning of the data and the way values of that type can be stored. *Aka* simply TYPE.

data warehouses: Vast arrays of heterogeneous (biological) data, stored within a single logical data repository, which are accessible to different querying and manipulation methods.

databanks (GenBank): A comprehensive database that contains publicly available nucleotide sequences for more than 2,40,000 named organisms.

database: Refers to store of a large amount of information in a form that can be accessed and handled electronically by a computer. Recently, in breeding, numerous databanks are in use, such as field registration, nursery plans, selection data, data of international observation trials, statistical values and data recording. Where databases are more intricate, they are often developed by means of formal design and modelling techniques.

database search: Once an open reading frame or a partial amino acid sequence has been resolved, the investigator compares the sequence with other similar databases using a computer and a search algorithm. This is usually done in a protein database such as protein information resource (PIR) or Swiss-Prot. Nucleic acid sequences are in GenBank and EMBL databases. The search algorithms most commonly used are BLAST and FASTA.

databases, types of: (i) Protein databases (primary databases, composite databases, secondary databases). (ii) Databases of macromolecular structures, nucleotide and genome sequences. (iii) DNA microarray data. (iv) Other databases.

dauermodification: *See* PERSISTENT MODIFICATION.

daughter cells: Young cells produced from the replication and division of a single parent cell.

daughter chromosome: Refers to either of the two chromatids of which the replicated chromosome consist after mitotic metaphase or anaphase II of meiosis.

daughter nucleus: The nuclei that are produced from the division of a single nucleus.

daughter-strand gap repair (DSGR): DSGR is a DNA repair mechanism that promotes recombination to repair the daughter-strand gap, not the dimer, and is a means to manage the problem of a noncoding lesion continuing in DNA.

days post anthesis (dpa): Refers to the number of days after flowering and fertilization.

days to flower: Refers to the number of days required to 50% blooming or flowering in a field.

days to harvest: The least number of days between the last pesticide application date and the harvest date, as established by law, to prevent contact with a hazardous residual amount of pesticide.

days to heading: Number of days required to 50% panicles or tassels emergence from the flag leaf of grasses or cereals in a field or study.

days to maturity: Number of days required from seeding to seed/grain ripening.

dbSNP: NCBI database of genetic variation. This database has served as a central, public repository for genetic variation. Once such variations are identified and catalogued in the database, additional laboratories can use the sequence information around the polymorphism and the specific experimental conditions for further research applications. As with all NCBI resources, the data within dbSNP are available freely and in a variety of forms.

dCas9: *See* DEAD CAS9.

dCas-based programmable RNA editing, limitations of: The dCas-based PROGRAMMABLE RNA EDITING (PRE) is a tedious approach compared with the PENTATRICHOPEPTIDE REPEAT (PPR) protein-based PRE as it suffers from numerous limitations: The major disadvantages of dCas13-based PRE include the small RNA target system (<22 nucleotides), high level of off-target effects and toxicity of the cleavage of RNA target to the eukaryotic cells. The overexpression of deaminase enzymes during recruitment of endogenous transcripts hinders the specificity of PRE tools. The overexpressed proteins or nucleic acid sequences generally hamper the editing efficiency. Any defect in the construct delivery system could also reduce the editing efficiency. Limitations still exist in the existing RNA-editing tools in relation to identification and manipulation of specific RNA molecules at the single-base level. The property of Cas13 to specifically cleave single-stranded RNA transcript might limit the *in vivo* cleavage specificity of the unstructured targeted RNA regions. The targeting of a transcript for single-base conversion with RNA's secondary or tertiary structure and RNA-binding proteins could limit the PRE approach. The effects of editing on RNA are limited on account of the natural half-life of RNA targets because of the rapid turnover of mRNAs in mammalian cells.

dCasRx-based tool for epitranscriptome editing of $6^{m}A$ modification: A bidirectional dCasRx epitranscriptome-editing podium composed of a nuclear-localized dCasRx conjugated with either a methyltransferase, METTL3, or a demethylase, AlkB homologue H5 (ALKBH5), to operate methylation events at targeted $6^{m}A$ sites. Leveraging this platform, they precisely and proficiently edited 6-methyladenine ($6^{m}A$) modifications at targeted sites, reflected in gene expression and cell proliferation.

Dcm system: DNA cytosine methylase (Dcm) is suggested to play a role in cellular differentiation and development. In *Escherichia coli*, internal cytosine residues in the sequence CCWGG are converted to 5-methylcytosine ($5^{m}C$) by Dcm. This methylation system might shield the genome from the action of restriction enzymes.

de facto: In fact; in reality. Something that exists or occurs de facto is not the result of a law but because of circumstances.

***de novo*:** A Latin term that means arising 'anew', 'from the beginning', 'once more', or 'afresh'.

***de novo* DNA methylation:** A method by which methyl groups are added to unmethylated DNA at specific CpG sites, catalysed by DNMT3A and DNMT3B. This type of DNA methylation occurs rarely. It is not clear whether or not *de novo* DNA methylase is sequence specific. CG doublets do not by themselves provide specificity. Surrounding sequence may be important for protein recognition during gene expression. Most $5^{m}C$ may play a regulatory role.

***de novo* DNA methylation in animals:** In mammals, DNA methylation patterns are established early in development via a highly orchestrated progression that involves genome-wide demethylation and *de novo* methylation. During preimplantation development, DNA methylation marks inherited from gametes are mostly erased. On implantation, the embryo undergoes a wave of *de novo* methylation that establishes a new methylation pattern. DNMT3A and DNMT3B are mainly responsible for *de novo* methylation. This *de novo* methylation pattern is copied during division of somatic cells. DNMT1 is the major enzyme for maintaining DNA methylation patterns. Another developmental stage that exhibits substantial *de novo* DNA methylation in mammals is gametogenesis. DNA methylation in both male and female germ cells plays a critical role in the establishment of genomic imprinting. Genomic imprinting is an epigenetic process that marks alleles according to their parental origin. Genome imprinting results in monoallelic expression of a small subset of genes. DNMT3A and DNMT3L are vital for setting up DNA methylation imprints in germ cells. DNMT3L has no enzymatic activity, but it has been demonstrated to interact with DNMT3A and inspire its activity. DNMT3L binds the N-terminal tail of histone H3, signifying a role for DNMT3L in engaging DNMT3A to specific genomic regions. The interaction between DNMT3L and histone H3 is repressed by methylation at H3K4, indicating that H3K4 methylation may control germline imprinting. KDM1B (*aka* AOF1 or LSD2), a histone H3K4 demethylase highly expressed in growing oocytes, is needed for *de novo* DNA methylation of maternally imprinted genes. DNA methylation in mammalian germ cells is also indispensable for suppression of transposons. *DNMT3L*-deficient male mice fail to launch *de novo* methylation of transposons. This results in unrestrained transposon expression and spermatogenesis failure. This phenotype is unusually parallel to that of mice deficient for the Piwi family members MILI or MIWI2. This advocates that the Piwi-piRNA complexes and DNA methylation machinery act together to overpower transposons in germ cells.

***de novo* gene synthesis by an antiviral reverse transcriptase:** Reverse transcriptase of bacterial origin produces extrachromosomal genes through rolling-circle amplification method to deliberate effective antiviral immunity.

***de novo* genome sequencing:** Determination of the DNA sequence of the full genetic complement (genome) of an organism.

de novo **methyltransferases**. They recognize something in the DNA that allows them to newly methylate cytosines. These enzymes are expressed mainly in early embryo development, and they set up the pattern of methylation.

de novo **mutation:** A genetic alteration which arises in an individual that was not inherited from a parent or was present for the first time in the egg or sperm from the parent.

de novo **strategy:** *De novo* peptide sequencing is reconstitution of the peptide sequence from the fragmentation spectra based on peptide fragmentation rules. It is a straightforward approach for protein identification when protein identification becomes elusive. This task was traditionally done by Edman sequencing, but with the advent of the latest technologies and bioinformatics tools, the data produced is searched for protein homology using the BLAST tool.

de novo **sequencing:** The base-by-base sequencing of a complete genome for which neither sequence data nor any fragment libraries (e.g. bacterial artificial chromosome libraries) are yet available.

de novo **transcriptome assembly:** The *de novo* sequence assembly method of generating a transcriptome without the aid of a reference genome. *De novo* transcriptome assembly is most favoured method for studying non-model organisms, since it is cheaper and much easier than building a genome. The transcriptomes of these organisms can thus disclose novel proteins and their isoforms that are occupied in such unique biological phenomena.

deacetylation: (i) In biochemistry, the removal of acetyl groups from organic compounds. (ii) In epigenetics, the removal of the acetyl group from both histone and non-histone proteins, a form of PTLM. *Opposite to* ACETYLATION.

dead Cas9 (dCas9): A mutated wild-type Cas9 nuclease which is an enzymatically inactive form of Cas9. It can bind but cannot cleave DNA. The nuclease domains of Cas9 have been altered by mutating H840A in the HNH domain and D10A in the RuVC1 domain. It does not break DNA but still binds with DNA. It is used as a component of gene activation or gene repression tool along with an effector molecule. *Aka* DCAS9 NULL MUTANT, DEFICIENT CAS9 *or* INACTIVE CAS9.

DEAD-box RNA helicase: An RNA helicase that contains the DEAD (Asp-Glu-Ala-Asp) or DEXD/H (Asp-Glu-X-Asp/His domain, where X represents any amino acid) motif. These proteins use the energy produced by ATP hydrolysis to unwind RNA.

DEAD-box RNA helicase domain: An evolutionarily conserved domain in a family of enzymes that use ATP hydrolysis to unwind RNA duplexes. The domain is named after the DEAD (Asp-Glu-Ala-Asp) motif.

DEAE-cellulose: The complete name is DIETHYLAMINOETHYL CELLULOSE. It is the ion exchange material that binds anions, used in chromatography especially for proteins.

DEAE-Dextran-mediated gene transfer: DEAE-Dextran is used in the transfection medium in which DNA is present. DEAE-Dextran is polycationic, high molecular weight substance and convenient for transient assays in *cos* cells. It is not efficient method for the production of stable transfectants. If DEAE-Dextran treatment is coupled with DIMETHYLSULFOXIDE (DMSO) shock, then up to 80% of transformed cell express the transferred gene.

deamination: (i) The process in which the amino group of an amino acid is removed and converted to ammonia. Upon the death of an organism microbial decomposition takes place which causes hydrolytic breakdown of proteins into amino acids. Further deamination causes the conversion of amino acids into ammonia by removing the amino groups. (ii) The removal of amino groups from molecules via the energy-consuming metabolism of excess amino acid in take by that animal. For example, when livestock are fed more lysine than their body needs per day, the excess lysine is metabolized to urea and then excreted.

debugging: The process of identifying and resolving defects or problems within a computer program that prevents accurate working of computer software or the system as a whole. Debugging strategies can involve collaborative debugging, control flow analysis, unit testing, integration testing, log file analysis, monitoring at the application or system level, memory dumps and profiling.

decapentaplegic (dpp) protein: Dpp protein produced by boundary cells induces different levels of response in cells near to or distant from the boundary.

decarboxylation: Removal of carbon dioxide from an amino acid or acid. Pyruvic acid cannot enter the Krebs cycle until it loses one molecule of CO_2 to become two carbon compounds. This process of removal of CO_2 is called decarboxylation.

decay: The gradual decomposition of organic matter.

decay of genetic variability: The reduction of heterozygosity because of genetic drift leading to loss and fixation of alleles at various loci.

decentralized plant breeding: A defined set of breeding experiments performed in a variety of local sites (e.g. communities, farmer fields) that characterize real farming conditions, as opposed to a single, central, research station site that does not embody a real farming context.

deciduous plant: A plant whose leaves are shed as a habit, at a season or growth stage, by abscission. The function of this practice is usually to escape an adverse season, such as a winter or a tropical dry season. The deciduous habit also has advantages in the control of leaf parasites by providing a discontinuous pathosystem in which a gene-for-gene relationship can function as a system of biochemical locking. The term also refers to the shedding of other plant structures, such as petals after flowering or fruits when ripe. *Contrasts* PERSISTENT.

decision error probabilities: In statistics, the term 'error' arises in two ways. First, it arises in the context of decision making, where the probability of error may be considered as being the probability of making a wrong decision and which would have a different value for each type of error. Second, it arises in the context of statistical modelling (for example, regression) where the model's predicted value may be in error regarding the observed outcome and where the term probability of error may refer to the probabilities of various amounts of error occurring.

decision support system (DSS): An interactive software-based system used to help decision makers compile useful information from a combination of raw data, documents and personal knowledge and identify and solve problems to make an optimized decision. The DSS architecture consists of the database (or knowledge base), the model (i.e. the decision context and user criteria) and the user interface. The main advantages of using a DSS include examination of multiple alternatives, better understanding of the processes, identification of unpredicted situations, enhanced communication, cost-effectiveness and better use of data and resources. The application of DSS in agriculture and environment has rapidly increased in the past decade, which allows rapid assessment of agricultural production systems around the world and decision making.

decision theory: A branch of applied probability theory and analytic philosophy concerned with the theory of making decisions based on assigning probabilities to various factors and assigning numerical consequences of the outcome.

declining vitality: Refers to seeds that are aged or have been subjected to unfavourable storage conditions. Such seeds usually show a slow germination. Some of the essential plant parts are frequently undersized or missing. Saprophytic fungi may also negatively affect the growth of the seedlings.

decoding: *See* RECODING.

decoding region: Located on the small subunit (SSU) of rRNA and contains highly conserved nucleotides that are involved in interactions with transfer RNAs and messenger RNA. The A-site binds to aminoacyl-tRNA. The P site binds to peptidyl-tRNA.

decomposer: Any organism that eases the breakdown of dead or decaying organisms by decomposing complex biomolecules into simpler substances. Decomposers are heterotrophs which obtain energy and nutrition for their own growth and reproduction by recycling the chemical compounds captured from organic substrates. Microorganisms such as bacteria and fungi are the biosphere's chief decomposers, but invertebrates such as earthworms are also sometimes considered decomposers.

decomposition: Decomposition is the reverse of synthesis. It splits large complex molecules into simpler organic or inorganic matter such as carbon dioxide, water, simple sugars and mineral salts. These reactions are catabolic in nature and occur naturally by both biotic means (biodegradation, such as that practiced by many bacteria and fungi) and abiotic means (basic physical and chemical processes, such as hydrolysis). Decomposition recycles matter present in the biosphere, making it an essential part of the nutrient cycle. Organisms that facilitate decomposition are known as DECOMPOSERS. The scientific study of decomposition is known as TAPHONOMY.

decompound: Divided to more than one level (e.g. in bipinnate leaves, in which the leaflets of what would otherwise be a pinnate leaf are themselves pinnately divided).

decondensation stage: A stage between interphase and prophase of mitosis in which heterochromatin is normally decondensed only for a short period.

decontamination: Clearing of the surface by removing contaminating microorganisms. It is removal of unwanted/contaminating microorganisms to encourage the growth of desired microbe.

deconvolution: A mathematical procedure to separate out the overlapping effects of molecules, such as mixtures of compounds in a high-throughput screen or mixtures of cDNAs in a high-density array.

decorticate: (i) To shed the outer bark of a tree, usually seasonally as part of the natural growth cycle. (ii) To strip the peel, crust, bark or other surface tissues from a plant or from harvested material, such as in extracting fibre from harvested Agave leaves.

DEcoTILLING: A modification of TILLING and EcoTILLING that uses an inexpensive method to identify fragments. Since the advent of NEXT-GENERATION SEQUENCING (NGS) technologies, TILLING by sequencing has been developed based on Illumina sequencing of target genes amplified from multidimensionally pooled templates to identify possible single-nucleotide changes.

decoy: Decoy specializes in perception of the effector by the R protein but itself has no function either in the development of disease or resistance. Many pathogen effectors have multiple targets in the host. Additional targets of AvrPto and AvrBs3 prompted the concept that some host targets of effectors act as decoys to detect pathogen effectors via R proteins. Host targets of effectors act as decoys/baits to detect pathogen effectors via R proteins. Effector target with no function in host defence or susceptibility in the absence of its cognate R protein. Effector alteration of the bait does not result in enhanced pathogen fitness in plants that lack the R protein and triggers innate immunity in plants that carry the R protein.

decoy crop: A crop that stimulates germination of seeds of a parasitic plant, such as witchweed (*Striga* spp.), but is not susceptible to infection by the parasitic plant. It aids in reducing seed populations of the parasite in soil. This way, a susceptible crop can be planted.

dedifferentiated calli: The process in which tissues that have undergone cell differentiation can be made to reverse the process so as to become a primordial cell again.

dedifferentiation: Refers to the loss of specialization of a cell. It can be observed when differentiated cells are placed *in vitro* culture and these cells grow in reverse (i.e. from a partially or terminally differentiated stage to a less differentiated stage within their own lineage).

deep contrastive learning: Based on the unproven assumption that no two fingerprints, even from different fingers of the same person, are alike, fingerprint biometrics are integral to digital authentication and forensic science. This renders them useless in scenarios where the presented fingerprints are from different fingers than those on record. However, G. Guo and others (2024) find that these similarities hold across all pairs of fingers within the same person, even when controlling for spurious factors like sensor modality. They also find evidence that ridge orientation, especially near the fingerprint centre, explains a substantial part of this similarity, whereas minutiae used in traditional methods are almost non-predictive. These experiments suggest that, in some situations, this relationship can increase forensic investigation efficiency by almost two orders of magnitude.

deep intronic splicing: Transcriptomic analyses have detected numerous deep intronic mis-splicing variants in the causal genes governing Mendelian disorders. These deep intronic mutations frequently cause pseudoexon inclusion by creating *de novo* splice sites (SSs), thereby producing loss of gene function by generating premature termination codons (PTCs). *See* PSEUDOEXON INCLUSION.

deep learning (DL): DL is the deep neural networks learning. The DL is one of the domains of AI composed of network model with neurons, but it serves as a subgroup of MACHINE LEARNING (ML). The term deep implies the numerous layers between the input and output of a neural network. AI based on neural networks has application in computer science, mathematics, physics, engineering image/signal processing, economic/finance, philosophy, linguistics and neurology.

deep plant phenomics (DPP): A tool based on the AI deep convolutional neural networks is used for the purpose of plant phenotyping to ease the leaf counting, mutant classification and age regression tasks for *Arabidopsis thaliana*. AI has eased the quantification of field scale phenotypic information accurately and initiate the integration of big data into predictive and prescriptive management tools accessible through the implementation of UNMANNED AIRCRAFT SYSTEMS (UAS).

deep-freezing: Process of storing microbial cultures at super low temperature ranging from −70°C to −90°C. Microbial cultures are stored in the deep freezers maintained at super low temperatures such as −70°C in glycerol stocks for years together. Whenever required, the glycerol stocks are taken out of the deep freezers and re-cultured on the fresh medium.

There is no risk of mutations during storage at such a low temperature.

default pathway: Constitutive secretory pathway that automatically transports material from the Golgi apparatus to the plasma membrane if no other sorting signals exist.

defective interfering virus (DIV): A virus that has lost some function and that depends on coinfection with intact virus for transmission.

defective viral genome: Defective or deleted genome of some RNA and DNA viruses disrupt the replication of complete genome of those viruses with which they are associated. Plants expressing such subgenomic DNA show decreased disease severity due to the concerned virus. For example, African cassava mosaic virus (ACMV) genome consists of two single-stranded DNA molecules designated as A- and B-DNAs. A 50% deleted B-DNA integrated into their genome showed reduced systemic spread when they were infected with ACMV.

defence against spread of transposable elements: The hypermethylation of repetitive DNA in plants and some fungi may signify a defence mechanism to thwart the spread of transposable elements (as it is in vertebrates). The involvement of repetitive sequences in both co-suppression and paramutation appears to back this theory. Loss of methylation reflects a deregulation of epigenetic programming rather than increased mutagenesis.

defence against viruses: High levels of dsRNA produced by viral infections or highly transcribed transgenes can provoke DNA methylation. However, RNA-DIRECTED DNA METHYLATION (RdDM) can also affect endogenous genes expressed at lower levels. RdDM can be triggered by levels of trigger RNAs lower than those needed for RNAi, as long as the source of RNA is continuously expressed over several plant generations.

defence priming: Priming is an adaptive strategy that improves the defensive capacity of plants. This phenomenon is marked by an enhanced activation of induced defence mechanisms.

defence response: Refers to active response of a plant to pathogen attack. It comprises the elements that inhibit pathogen development. A defence response is triggered in compatible and incompatible interactions.

defence responses leading to resistance: In gene-for-gene resistance, the products of host resistance (*R*) genes determine recognition of pathogen 'avirulence' (*Avr*) gene products to initiate defence responses leading to resistance. There is high degree of specificity between corresponding *R* and *Avr* genes.

defence-associated reverse transcriptase (DRT): DRT schemes offer an interesting counterpoint to antiviral immune approach by empowering DNA synthesis. S. Tang and others (2024) reveal that DRT2 systems accomplish an unparalleled immunity apparatus that comprises *de novo* gene synthesis through rolling-circle reverse transcription of a non-coding RNA (ncRNA). Unbiased profiling of RT-associated RNA and DNA ligands in DRT2-expressing cells exposed that reverse transcription produces concatenated cDNA repeats via programmed template jumping on the ncRNA. The occurrence of phage at that time generates second-strand cDNA synthesis, resulting in the manufacture of long double-stranded DNA. This double-stranded DNA product is proficiently transcribed, producing messenger RNAs that encode a stop codon-less, never-ending OPEN READING FRAME (ORF (neo)) whose translation reasons strong growth arrest.

deficiency: The absence or deletion of a slice of genetic material. *See* DELETION.

deficiency diseases: Disease caused by the absence or inadequacy of some nutrient, element or compound (e.g. copper, zinc or iron deficiency in cereals).

deficiency loop: The loop formed by the non-synapsing portion of an unaltered chromosome, caused by a deficiency in the homologous chromosome. *Aka* DELETION LOOP.

defined medium: A culture medium in which each component is of known chemical structure and the specified quantity. In plant tissue culture, for instance, the medium defined by T. Murashige and F. Skoog in 1962 is called the defined medium. But when we add growth adjuvants like coconut water or potato extract, then it becomes undefined.

definite: Of a constant number, such as (i) twice as many stamens as petals or sepals (or less) or (ii) an inflorescence ending in a flower or an aborted floral bud, typically a cymose inflorescence. *Contrasts* INDEFINITE.

definitive nucleus: A nucleus formed in the embryo sac by the fusion of two polar nuclei. The definitive nucleus later fuses with a male nucleus to form triploid endosperm nucleus.

defoliation: The loss of leaves from a plant, whether normal or premature.

deforestation: A permanent conversion of forest into some other kind of ecosystem, such as agriculture or urbanized land use.

degeneracy and evolution: Within biological systems, degeneracy refers to circumstances where structurally dissimilar components/modules/pathways can perform similar functions (i.e. are effectively interchangeable) under certain conditions, but perform distinct functions in other conditions. Degeneracy is thus a relational property that requires comparing the behaviour of two or more components. In particular, if degeneracy is present in a pair of components, then there will exist conditions where the pair will appear functionally redundant but other conditions where they will appear functionally distinct. Examples of degeneracy are found in the genetic code, when many different nucleotide sequences encode the same polypeptide; in protein folding, when different polypeptides fold to be structurally and functionally equivalent; in protein functions, when overlapping binding functions and similar catalytic specificities are observed; in metabolism, when multiple, parallel biosynthetic and catabolic pathways may coexist. More generally, degeneracy is observed in proteins of every functional class (e.g. enzymatic, structural or regulatory), protein complex assemblies, ontogenesis, the nervous system, cell signalling and so on.

degeneracy: In genetics, coding of a particular amino acid by more than one codon. Since numerous distinct codons define the same amino acid, the genetic code degenerates. This allows for greater permutations and combinations to code for a particular amino acid.

degenerate code: When there is more than one codon for a particular amino acid. For example, there are six codons for leucine specified by UUA, UUG, CUU, CUC, CUA, CUG codons, six codons for serine specified by UCA, UCG, UCC, UCU, AGU, AGC, four codons for alanine specified by GCU, GCC, GCA and GCG and three codons for isoleucine specified by AUU, AUC and AUA. This is explained like this: the first two positions of the triplet codon on mRNA pair precisely with the first two nucleotides of anticodon of tRNA but pairing at third position may not be precise. *Aka* DEGENERACY OF GENETIC CODE *or* DEGENERACY.

degenerative genes: Genes that in homozygous condition have an effect ranging from degenerative to lethal.

degradation: (i) Breaking up of DNA into usually non-coding molecules. During DNA isolation, sometimes due to some chemical and physical factors, the DNA gets degraded. The hydrogen bonding between the bases disrupts and the linkage between deoxyribose of one molecule and phosphate group of the next also gets disrupted to yield smaller non-functional DNA molecules. Degraded DNA loses its native confirmation and function by exposure to destabilizing agents such as high temperature. During DNA handling, degradation should be avoided, otherwise degraded DNA cannot be used for POLYMERASE CHAIN REACTION (PCR) and also for genetic transformation. (ii) In a technique developed by Fred Sanger, a large molecule is broken down into smaller fragments and then pieced together to work out the sequence of the molecule's building blocks. Due to physical shearing or exposure to endogenous or exogenously added nucleases, DNA and RNA can become hydrolysed or degraded to the oligonucleotide or single-nucleotide level.

degree day: The departure of the average daily temperature from a defined base (e.g. the minimum documented temperature for the growth of a plant species). The number of degree days may be added up to assess the amassed warmth of a particular year's growing season.

degree of dominance: Square root of ratio of H/D, where $H = (d)^2$ and $D = (h)^2$. 'd' represents difference in performance between homozygous parents and their mid-parent, and h represents difference in performance between F_1 and mid-parent. If this ratio is =1, the trait shows complete dominance, if the ratio is >1, the trait shows overdominance, and if the ratio is <1, the trait shows incomplete dominance.

degree of genetic determination: The portion of total variance that is genetically determined.

degrees of freedom (df): The number of objects of data that are free to vary independently. In a set of quantitative data, for a specified value of the mean, only $(n - 1)$ items are free to vary, since the value of the nth item is then determined by the values assumed by the others and by the mean. For a continuous variable Y, there are two types of df values: NUMERATOR DF and DENOMINATOR DF. *See details as separate terms.*

dehiscent: Breaking open into parts at maturity to release contents. For example, the opening of fruits to release seeds, of anthers to release pollen and of sporangia to release spores. *Contrasts* INDEHISCENT.

dehydration: Dehydration is a common term used for depletion of water from body. Excessive dehydration may cause the death of an organism as more than 15% loss of body weight through dehydration is usually fatal. It can be treated by fluid therapy.

dehydration reaction: A chemical reaction that includes the loss of a water molecule from the reacting molecule.

dehydrins: Proteins formed during the late stages of plant embryo development and following any environmental provocation involving dehydration.

dehydrogenase: An enzyme that catalyses the exclusion of hydrogen from a substrate.

dehydrogenation: The removal of hydrogen atoms from molecules. When those molecules are the components of vegetable oils/fats, a lower content percentage of saturated fats is produced.

deism: The view that God works through fixed laws of nature (i.e. natural events occur without supernatural intervention).

delayed inheritance: The inheritance whereby each generation manifests with regard to particular characters, the genotype of the female parent.

delayed maturity: A breeding goal for fruits and vegetables. Many fruits like tomatoes, melons and bananas are usually harvested before ripening. The firmness of unripe fruit facilitates mechanical harvesting and spreads the period of time allowed for transportation and storage. For several fruits and vegetables, transgenic methods have been employed to develop cultivars with delayed maturity. It is achieved by suppressing the synthesis of polygalacturonase, an enzyme responsible for fruit softening, either by obstructing the production of ethylene (a plant hormone that triggers the ripening process) by introducing genes for enzymes that degrade ethylene precursors or by obstructing the polygalacturonase gene using antisense technology. Transgenic tomatoes with extended shelf life were the first to be introduced to the US market.

delayed segregation: A condition where segregation occurs in F_3, rather than in F_2 since the phenotype of the individual is determined by the genotype of the mother (e.g. inheritance of sinistral and dextral coiling in snails).

delayed senescence carnation: Flower senescence is promoted by ethylene. Ethylene biosynthesis involves two enzymes, viz. 1-aminocyclopropane-1-carboxylate (ACC) synthase (encoded by *acsI*) and ACC oxidase (encoded by *aco*). Gene *aco* was suppressed by using the antisense construct of its cDNA. Such transgenic carnation flowers produce little detectable ethylene. The vase life of transgenic flowers is increased by 200%.

deleterious effects: Substitutions that have negative influence on phenotype or function of an individual. Such mutations are eliminated by natural selection. Many mutations taking place in coding DNA have deleterious effects. The observed frequency of surviving mutations in coding DNA is thus much less than that in non-coding DNA, with the result that coding DNA sequences are conserved over evolutionary time.

deletion: A deletion may involve part of a gene, whole gene, a group of genes, part of a chromosome, complete chromosome, more than one chromosome or even a complete set of chromosomes. Deletions are usually lethal in homozygous state. Cytologically, deletions can be recognized by failure of a segment of a chromosome to pair properly. Deletions are usually lethal in homozygous state. This is referred to as DELETION LOOP. Chromosome with a deletion can never revert to a wild-type condition. Deletions may show pseudodominance because in this case a recessive gene present in a single dose expresses itself. Some deletions may have phenotypic effects. *Aka* DEFICIENCY.

deletion analysis: The identification of regulatory sequences for a gene determining by determining effect on gene expression of specific deletions in the upstream rejoin.

deletion heterozygote: An individual that is heterozygous for the absence of an internal segment of one of its chromosomes.

deletion loop: *See* DEFICIENCY LOOP.

deletion map: A description of a specific chromosome that uses defined mutations – specific deleted areas in the genome – as 'biochemical signposts' or markers for specific areas.

deletion mapping: Mapping mutation by use of overlapping deletion mutants to determine whether or not a mutation includes the site of a mutant gene. Process in which different deletions in a region of DNA are created and used to map the functionally critical areas of that DNA. For example, the minimal region of DNA required for a test promoter can be ascertained by systematic deletions in the region of interest.

deletion mutants in protein engineering: In addition to the substitution of amino acids at specific positions, amino acids can be deleted from the sequence, either individually or in groups. These proteins are referred to as deletion mutants. Deletion mutants may or may not be missing one or more functions or properties of the full, naturally occurring protein. Proteins from eukaryotes (e.g. mammals, plants and fungi) can have very long amino acid sequences and, as a consequence, tend to be organized as several modular protein domains linked together. Deletion mutants have been a useful way to create smaller proteins that contain only one or a few of these domains so that the individual properties can be studied.

deletion mutation: A mutation in which one or more bases are removed from the DNA sequence of a gene.

deletion series: A set of clones derived from the same initial recombinants but in which the insert lacks sequences at one of its ends because of treatment with an exonuclease.

deletion/duplication in imprinting: Deletions involving imprinted regions can affect imprinting by either: deletion of an expressed gene on that allele (thus abolishing gene expression) or deletion of the imprinting control centre, which results in loss of regulatory control of imprinting. Duplications involving the imprinted regions can double the expression of imprinted genes expressed from that allele.

DELLA proteins: DELLA proteins are nuclear-localized growth inhibitors that are key components of the mechanism that allows the phytohormone gibberellin (GA) to promote plant growth. They are produced by a family of genes (of which '*reduced plant height* (*Rht*)' in cereals is one) that suppress the activity of important genes involved in plant growth. Gibberellin destabilizes DELLA proteins, reducing the amounts of the proteins in cells and their repressive effects on cell growth. For example, in dwarf wheat, a mutation makes the DELLA proteins produced by the *Rht* gene insensitive to gibberellin, and so they are not destabilized by gibberellin, and cell growth is repressed.

Δ12 desaturase: An enzyme present within the soyabean plant and in other oilseed crops (canola, maize/corn, etc.) that is involved in the pathway used by oilseed crops to manufacture polyunsaturated fatty acids (e.g. linoleic acid) from monounsaturated fatty acids (e.g. oleic acid) in seeds while they are developing.

ΔGp: Actual free-energy change of ATP hydrolysis under a specified set of conditions.

deme: A discrete local, randomly mating population connected with other such populations by gene flow. A set of demes makes a metapopulation. Often used in the same context as population.

Demeter (DME): A $5^{m}C$ DNA glycosylase/DNA demethylase that is important for gene imprinting in *Arabidopsis*.

Demeter-like proteins (DML): Paralogues of DML1 and ROS1 – DML2 and DML3 – are $5^{m}C$ DNA glycosylases that are expressed in a wide range of plant organs.

demethylase-facilitated RNA-seq: Alkylation B (AlkB)-facilitated RNA methylation sequencing (ARM-seq) involves pre-treatment of RNA with *Escherichia coli* AlkB dealkylating enzyme to demethylate 1-methyladenosine ($1^{m}A$), 3-methylcytidine ($3^{m}C$) and 1-methylguanosine ($1^{m}G$). The demethylated RNA together with an untreated sample is subjected to steps similar to the ones used in RNA-seq.

demethylation model of gene regulation: A model that elucidates the role of undermethylation of cytosine in gene regulation and differentiation in eukaryotes. In undifferentiated cells, DNA is fully methylated at all sites that ever will be methylated. During development, sequence-specific proteins would obstruct methylation during DNA replication, leading to methylation patterns specific for each tissue. Once the specific demethylation events occur, the differentiated methylation pattern is inherited clonally as a result of the maintenance methylase system, provided that *de novo* methylation is slow. Starting from a uniform, fully-methylated ground state and proceeding unidirectionally by demethylation is logically attractive model and is consistent with the studies showing sperm and embryonic DNA to be highly methylated. Demethylation alone is not capable of turning on any gene. If appropriate *trans*-acting factors are present to repress transcription, or if the tissue is missing specific *trans*-acting factors that positively regulate transcription, demethylation may have little or no effect.

demic diffusion: The process of colonization of an area by a population. This term is used in contrast to cultural diffusion, which involves movement of ideas rather than petition.

demic selection: Special type of intergroup selection that does not essentially involve direct competition. It affects the general genetic composition of a population if subsets of a population have different gene frequencies.

demicyclic: A rust fungus that does not have the urediniospore (repeating) stage (e.g. many species of Gymnosporangium). *Contrasts* MACROCYCLIC *and* MICROCYCLIC.

democratic plant breeding: The opposite of AUTOCRATIC PLANT BREEDING. In democratic plant breeding, as many breeders as possible are growing as many cultivars as possible so that farmers have a wide choice of cultivars. This approach is possible with the use of horizontal resistance because breeding for this kind of resistance is so easy. In many cases, farmers can do their own plant breeding.

demographic transition: A change in human population parameters from a condition of high birth and death rates to one of low birth and death rates.

demography: The study of populations and how they grow, survive, reproduce and die.

demonstration strip design: The simplest form of design for a field trial. Varieties or breeding lines, herbicide or fertilizer applications are compared using demonstration strips on a field. In this case, the stress is on visual impact, not on measured results that are critically compared. Usually, the comparison is two to four treatments, up to a maximum of ten. Although this design is designed for visual comparison, some measurements may, however, be assembled for general comparison. The simplest measurement method is to take one randomly located sample for a variable, such as yield, from near the centre of each strip. The measured data from individual strips are compared to one another.

denaturant: A chemical capable of unfolding a protein in solution at ordinary temperatures.

denaturation: (i) A reversible or irreversible process in which proteins unfold their higher-order (quaternary, tertiary and secondary) structure present in their native state, when exposed to some external stress, such as heat or chemical denaturants (such as guanidine hydrochloride, urea, a strong acid or base, a concentrated inorganic salt or an organic solvent). (ii) With respect to nucleic acids, the term refers to the conversion from double-stranded to the single-stranded state, often achieved by heating or alkaline conditions. During PCR denaturation of target DNA is the first step at which denaturation (melting of double-stranded DNA) is done at 95°C to yield single-stranded DNA that acts as template for the synthesis of another complementary strand. *Aka* MELTING DNA.

denaturation map: A map of a stretch of DNA showing the locations of local denaturation loops, which correspond to regions of high AT content.

denatured DNA: DNA converted from double-stranded to single-stranded form by some denaturation process such as heating the DNA solution. In the case of heat denaturation, the solution turns very gelatinous and viscous.

denatured protein: A protein whose properties have been changed by treatment with physical or chemical agents.

denatured state: The partially or completely unfolded form of a biological macromolecule, in which it is incapable of carrying out its biochemical and biological functions.

denaturing gel: An agarose or acrylamide gel run under conditions which destroy secondary or tertiary protein or RNA structure. For protein, this usually means the inclusion of 2-ME (which reduces disulfide bonds between cysteine residues), SODIUM DODECYL SULFATE (SDS) and/or urea in an acrylamide gel. For RNA, this usually means the inclusion of formaldehyde or glyoxal to destroy higher-order RNA structures. In DNA-sequencing gels, urea is included to denature ds-DNA to ssDNA strands. In denaturing gels, macromolecules tend to be separated on the basis of size and (to some extent) charge, while shape and oligomerization of molecules are not important. *Contrasts* NATIVE GEL.

denaturing polyacrylamide gel electrophoresis (PAGE): The use of PAGE in order to separate and analyse DNA fragments (sequences) after DNA has been denatured. This methodology can be utilized to scan DNA in order to detect point mutations.

dendrimers: Large molecules. From the Greek word 'dendra' (meaning tree) a dendrimer is a branching polymer. The diversity of the branches varies depending on the generation of its growth; several of them are of nanometre dimensions. Dendrimers are used in nanoparticle synthesis.

dendrite: An extension of a nerve cell that receives stimuli from other nerve cells.

dendrogram: A graphical procedure for representing the output of a hierarchical clustering method. A dendrogram is strictly defined as a binary tree with a distinguished root that has all the data items at its leaves. Conventionally, all the leaves are shown at the same level of the drawing. The ordering of the leaves is arbitrary, as is their horizontal position. The heights of the internal nodes may be arbitrary or may be related to the metric information used to form the clustering. *Aka* ULTRAMERIC TREE. *See* PHENOGRAM.

dendroid: Tree-like; branching like a tree.

Denhardt's solution: A solution used in DNA hybridization procedures such as the Southern technique. It is used to coat a nitrocellulose filter, thus blotting to inhibit non-specific binding of the probe.

denitrification: The alteration of nitrate (NO_3^-) or nitrite (NO_2^-) to gaseous products, chiefly nitrogen (N_2) and/or nitrous oxide (N_2O), owing to the loss of nitrogen into the atmosphere and is therefore undesirable in agriculture. Denitrification is performed by a large group of heterotrophic facultative anaerobic bacteria (e.g. *Pseudomonas*) and is a fundamental component of the nitrogen cycle.

denominator degree of freedom (df): It is also called ERROR DF and is the sample size minus the number of parameters that need to be estimated. It is the denominator of a variance estimator.

denominator elements: Genes (or elements) on the autosomes of *Drosophila* or worms that regulate the sex switch (*sxl*) to the off condition (maleness). Such elements must exist as autosomal genes encoding the general machinery of the cell. Refers to the denominator of the X/A genic balance equation.

densitometer: A device that directs a collimated light beam through a film and into a photoelectric cell, which converts the remaining unattenuated light into an electric current.

density gradient centrifugation: The isolation or purification of macromolecules or subcellular particles by sedimentation through a gradient of increasing density under the effect of a centrifugal force. The density gradient may either be designed before the centrifugation run by mixing two solutions of different densities (e.g. in sucrose density gradients) or it can be formed by the process of centrifugation itself (e.g. in CsCl and Cs_2SO_4 density gradients).

density-dependent inhibition of growth: Mitotic inhibition correlated with increased cell density.

density-dependent selection: Selection that favours different trait values at different population densities. The limiting of the size of a population (e.g. a vertical pathotype) by mechanisms that are also controlled by the size of the population. This is a probable genetic mechanism for controlling the system of locking of the n/2 model, ensuring that all the n/2 biochemical locks and keys occur with an equal frequency. The rarity of a vertical pathotype or pathodeme is a reproductive advantage that leads to commonness. Commonness is a reproductive disadvantage that results in rarity. In contrast, under DENSITY-INDEPENDENT SELECTION the same trait value is favoured under all densities.

deoxyadenosine: The deoxyribonucleoside containing the purine adenine (A).

deoxyadenosine 5′-diphosphate (dADP): A deoxyribonucleotide having adenine (A) and two phosphate groups are attached at 5′ end of deoxyribose.

deoxyadenosine 5′-monophosphate (dAMP): A deoxyribonucleotide having adenine (A) and one phosphate groups is attached at 5′ end of deoxyribose. *Aka* ADENYLIC ACID *or* ADENYLATE.

deoxyadenosine 5′-triphosphate (dATP): A deoxyribonucleotide having adenine (A) and three phosphate groups are attached at 5′ end of deoxyribose.

deoxyadenylic acid: The deoxyribonucleotide monophosphate containing the purine adenine (A). *Aka* DEOXYADENOSINE 5′-PHOSPHATE.

deoxycytidine: The deoxyribonucleoside containing the pyrimidine cytosine (C).

deoxycytidine 5′-diphosphate (dCDP): A deoxyribonucleotide having cytosine (C) and two phosphate groups are attached at 5′ end of deoxyribose.

deoxycytidine 5′-monophosphate (dCMP): A deoxyribonucleotide having cytosine (C) and one phosphate group is attached at 5′ end of deoxyribose. *Aka* CYTIDYLIC ACID *or* CYTIDYLATE.

deoxycytidine 5′-triphosphate (dCTP): A deoxyribonucleotide having cytosine (C) and three phosphate groups are attached at 5′ end of deoxyribose.

deoxycytidylic acid: The deoxyribonucleotide monophosphate containing the pyrimidine cytosine (C). *Aka* DEOXYCYTIDINE 5′-PHOSPHATE.

deoxyguanosine: The deoxyribonucleoside containing the purine guanine (G).

deoxyguanylic acid: The deoxyribonucleotide monophosphate containing the purine guanine (G). *Aka* DEOXYGUANOSINE 5′-PHOSPHATE.

deoxygunosine 5′-diphosphate (dGDP): A deoxyribonucleotide having guanine (G) and two phosphate groups are attached at 5′ end of deoxyribose.

deoxygunosine 5′-monophosphate (dGMP): A deoxyribonucleotide having guanine (G) and one phosphate group is attached at 5′ end of deoxyribose. *Aka* GUANYLIC ACID *or* GUANYLATE.

deoxygunosine 5′-triphosphate (dGTP): A deoxyribonucleotide having guanine (G) and three phosphate groups are attached at 5′ end of deoxyribose.

deoxynucleoside 5′-triphosphates (dNTPs): Tri-phosphorylated ('high energy') nucleotide precursors required for synthesis of DNA, where N refers to one of the four bases (A, G, T or C).

deoxyribonuclease (DNase): Enzyme that hydrolyses a DNA molecule. Nucleases cut or degrade the DNA molecule by breaking the phosphodiester bonds that link one nucleotide to the next in a DNA strand. Some DNases may function as endonucleases, while some others may work as exonucleases. Some DNases cleave double-stranded DNA, while some others cleave single-stranded DNA molecules. For example, an enzyme Bal 31 removes nucleotides from both strands. Another enzyme *Escherichia coli* exonuclease 111 degrades just one strand leaving single-stranded DNA as the product. Likewise, S1 endonuclease cleaves only single strand, whereas DNase 1 prepared from cow pancreas cuts both single- and double-stranded molecules. The two main types of DNase found in animals are known as DEOXYRIBONUCLEASE I (DNASE I) family and DEOXYRIBONUCLEASE II (DNASE II) family. In addition, DEOXYRIBONUCLEASE III FAMILY and DEOXYRIBONUCLEASE IV FAMILY have also been found. *See details as separate terms.*

deoxyribonuclease (DNase) I family: The DNase I Family includes DNase I, DNase1L1, DNase 1L2 and DNase1L3. The first set of DNases is DNase I. This family consisted of DNase I, DNase1L1, DNase 1L2 and DNase1L3. DNase I cleaves DNA to form two oligonucleotide-end products with 5′-phospho and 3′-hydroxy ends and is produced mainly by organs of the digestive system. The DNase I family requires Ca^{2+} and Mg^{2+} cations as activators and selectively expressed. In terms of pH, the DNases I family is active in normal pH of around 6.5–8.

deoxyribonuclease (DNase) II family: The DNase II Family includes DNase IIα and DNase IIβ. The second set of DNases is DNase II. This family consisted of DNase IIα and DNase IIβ. Like DNase I, DNase II cleaves DNA to form two oligonucleotide-end products with 5′-hydroxy and 3′-phospho ends. This type of DNase is more widely expressed in tissues due to high expression in macrophages but limited cell-type expression. Unlike DNase I, they do not need Ca^{2+} and Mg^{2+} cations as activators. In terms of pH, the DNase II family is expressed in acidic pH. The cleavage pattern of DNase II is altered in the presence of Dimethyl sulfoxide (DMSO), which significantly affects the structure of DNA.

deoxyribonuclease I (DNase I): A nuclease that cleaves DNA preferentially at phosphodiester linkages adjacent to a pyrimidine nucleotide, yielding 5′-phosphate-terminated polynucleotides with a free hydroxyl group on position 3′, on average producing tetranucleotides. It acts on single-stranded DNA, double-stranded DNA and chromatin. It has a role as a waste-management endonuclease and in DNA fragmentation during apoptosis.

deoxyribonuclease II (DNase II): An endonuclease that hydrolyses phosphodiester linkages of deoxyribonucleotide in native and denatured DNA, yielding products with 3′-phosphates and 5′-hydroxyl ends, which occurs as a result of single-strand cleaving mechanism. It functions optimally at acid pH because it is commonly found in low pH environment of lysosomes. The action of DNase II occurs in three phases: The initial phase introduces multiple nicks in the phosphodiester backbone. The second phase produces acid-soluble nucleotides. The third phase, which is the terminal phase, consists of hyperchromic shift resulting from reduction of oligonucleotides.

deoxyribonuclease III (DNase III): DNase III is localized in the cell nucleus, and it appears to be the predominant DNA exonuclease in many different mammalian tissues. It hydrolyses denatured DNA four times more rapidly than native DNA, with no apparent base specificity,

deoxyribonuclease IV (DNase IV): Phage-induced deoxyribonuclease IV has both an exonucleolytic and an endonucleolytic activity. It functions at abasic or apurinic-apyrimidinic sites when the cell is undergoing nucleotide excision repair. In addition, the endonuclease IV consists of several activities such as apurinic-apyrimidic (AP) endonuclease, 3′-diesterase, 3′→5′ exonuclease and 3′ phosphatase.

deoxyribonucleic acid (DNA): DNA is the genetic material in all forms of life and some viruses. DNA was discovered by Frederick Miescher in 1869. DNA carries genetic information in living systems. In 1944, O.T. Avery, C.M. Macleod and M. McCarty established that DNA is the genetic material. Later, J.D. Watson and F.H.C. Crick (1953) proposed DNA model for the physical and chemical structure of DNA. Accordingly, DNA is the long linear polymer composed of four kinds of deoxyribose nucleotides linked by phosphodiester bonds. In its native state, DNA is a double helix of two anti-parallel strands. The chemical substance from which genes are made. Each DNA molecule is made of long sequences of nucleotides, each of which includes one of four nitrogenous bases – adenine (A), cytosine (C), guanine

(G) and thymine (T) – attached to a sugar-phosphate complex which acts as a 'backbone' for the long-chain polymer. DNA most commonly occurs in 'double-stranded' form (i.e. as a pair of nucleotide polymers bound together by COMPLEMENTARY BASE PAIRING). In nature, base pairs in DNA are formed only between A and T and between G and C bases. Thus, the base sequence of each single strand can be inferred from that of its partner. The sequence order of the four bases in the DNA strands determines all the genetic information contained within it. DESOXYRIBONUCLEIC ACID and DESOXYRIBONUCLEATE are archaic terms for DNA.

deoxyribonucleic nucleic acid (DNA) replication: When J.D. Watson and F.H.C. Crick (1953) put forth the double-helical model of DNA, the salient feature that excited the world of genetics was the self-complementary nature of DNA (i.e. the relationship between the sequences of the bases on its intertwined polynucleotide chains). They also suggested semi-conservative mode of DNA replication.

deoxyribonucleoside: A portion of a DNA molecule composed of one deoxyribose molecule plus either a purine or a pyrimidine. There are four kinds of deoxyribonucleosides in DNA – DEOXYADENOSINE, DEOXYGUANOSINE, DEOXYCYTIDINE and DEOXYTHMIDINE.

deoxyribonucleotide: A portion of a DNA molecule composed of one deoxyribonuceoside bonded with 1, 2 or 3 phosphate groups attached with the sugar moiety at the 5′ end and a deoxyribonucleotide may have one of the four standard bases – adenine, cytosine, guanine or thymine. Thus, there are following 12 types of deoxyribonucleotides: DEOXYADENOSINE 5′-MONOPHOSPHATE, DEOXYGUANOSINE 5′-MONOPHOSPHATE, DEOXYCYTIDINE 5′-MONOPHOSPHATE, DEOXYTHMIDINE 5′-MONOPHOSPHATE, DEOXYADENOSINE 5′-DIPHOSPHATE, DEOXYGUANOSINE 5′-DIPHOSPHATE, DEOXYCYTIDINE 5′-DIPHOSPHATE, DEOXYTHMIDINE 5′-DIPHOSPHATE, DEOXYADENOSINE 5′-TRIPHOSPHATE, DEOXYGUANOSINE 5′-TRIPHOSPHATE, DEOXYCYTIDINE 5′-TRIPHOSPHATE and DEOXYTHMIDINE 5′-TRIPHOSPHATE. *See details as separate terms.*

deoxyribose: A 5-carbon sugar derived from ribose by replacement of a hydroxyl group by hydrogen at position 2′ contained in DNA molecule. Deoxyribonucleotides are the structural units of nucleic acids. Each deoxyribonucleotide consists of three parts: a deoxyribose sugar, a nitrogen-containing aromatic base and a phosphate group.

deoxythymidine 5′-diphosphate (dTDP): A deoxyribonucleotide having thymine (T) and two phosphate groups are attached at 5′ end of deoxyribose.

deoxythymidine 5′-monophosphate (dTMP): A deoxyribonucleotide having thymine (T) and one phosphate group is attached at 5′ end of deoxyribose. *Aka* THYMIDYLIC ACID *or* THYMIDYLATE.

deoxythymidine 5′-triphosphate (dTTP): A deoxyribonucleotide having thymine (T) and three phosphate groups are attached at 5′ end of deoxyribose.

deoxythymidine: *See* DEOXYTHYMIDYLIC ACID *and* DEOXYTHYMIDINE 5′-MONOPHOSPHATE.

deoxythymidylic acid: The deoxyribonucleotide monophosphate containing the pyrimidine thymine (T). *Aka* DEOXYTHYMIDINE 5′-MONOPHOSPHATE *or* DEOXYTHYMIDINE.

dependent variable: A dependent variable is a value that depends on another variable to exhibit change. When computing in statistical analysis, you can use dependent variables to make conclusions about causes of events, changes and other translations in statistical research.

dephosphorylation: The removal of a phosphate group from an organic compound (e.g. changing of ATP to ADP).

deployment: The physical movement of clones (ramets) or other genetic units from one site (usually a nursery) to another (usually plantations), often including their spatial configuration on the recipient site.

depolarization: The process of retreating the charge across a cell membrane (such as that of a neuron), thereby causing an action potential. In depolarization, the inside of the membrane, which is normally negatively charged, becomes positively charged and the outside becomes negatively charged. This is transpired by positively charged sodium ions rapidly passing into the axon.

depth of coverage (depth): The number of sequencing reads generated during massively parallel sequencing which align to and cover a specific genomic region or base pair. Higher depth of coverage generally increases confidence in variant calls.

depurination/depyrimidination: It means loss of a purine (complete removal of A or G) or a pyrimidine (complete removal of T or C) base leading to apurinic site. The glycosidic bond of a purine/pyrimidine nucleotide is spontaneously broken at physiological temperatures, though at a very low rate and in this process purine/pyrimidine is lost from the DNA.

deregulation: The process or act of removing government restrictions or regulations on planting, import and/or export. Plant commodities are deregulated once the government receives and evaluates scientific research demonstrating food, feed and human safety and minimal impact on the environment.

derepressed: The condition of an operon that is transcribing because repressor controlled has been lifted.

derepression: (i) Initiation of or increase in transcriptional activity of a gene sequence resulting from the removal of a repressor molecule from the sequence or a neighbouring control sequence. (ii) The mechanism by which repression is alleviated, as when a repressing metabolite is removed, resulting in the increased level of a protein or an enzyme.

derivative chromosome: A structurally rearranged chromosome generated either by a rearrangement involving two or more chromosomes or by multiple chromosomes. The term always refers to a chromosome that has an intact centromere.

derivative hybrid: A hybrid originating from a certain cross between two hybrids.

derived characteristic: A feature noticed in an organism that was not present in the common ancestor of a group of organisms being investigated.

derived homology: Homology that first appeared in the common ancestor of a set of species and is unique to those descendent species. *Compare with* ANCESTRAL HOMOLOGY.

dermatogen: A specialized meristem in flowering plants in which floral instruction begins. It also gives rise to the epidermis.

dermatogen (D) stage: The stage at which the embryo has experienced three mitotic divisions and contains eight cells. The outer protodermal cells give rise to the epidermis through a series of anti-clinal divisions. Development of the suspensor is complete at this stage.

desaturase: An enzyme family that is present within the soybean plant and other oilseed crops (e.g. canola, corn/maize). One or more desaturases are involved in the synthetic 'pathway' through which oilseed crops produce unsaturated fatty acids (e.g. linoleic acid). A desaturase is also involved in production of β-carotene in some plants.

descendant: An individual originating from the sexual reproduction of one parental pair of individuals.

descent with modification: (i) Darwin's term for evolution. (ii) Passing down traits that might change over time.

descriptive statistics: Methods that assist in organizing, summarizing and interpreting the sample data at hand. This may include a representation of the total population or a sample population.

descriptor: An identifiable and measurable feature used to facilitate data classification, storage, retrieval and use.

desensitization: *See* ADAPTATION.

desert: A temperate or tropical biome characterized by prolonged drought, usually receiving less than 25 cm of precipitation per year.

deserticolous: Inhabiting a desert.

desertification: (i) The formation of deserts in areas that were previously productive. (ii) The increasing aridity of drylands. (iii) An environmental change that can render agriculture difficult or impossible.

desiccate: To dry out.

desiccation: Drying up or the removal of water. Desiccation of the agar solidified culture medium is a common phenomenon, particularly in those culture vessels, which are plugged with cotton. The water from the medium evaporates causing desiccation and drying up of the medium. In plant tissue culture, desiccation sometimes helps in plant regeneration from the callus cultures. Desiccation can be prevented to some extent by using airtight plastic lids on the culture vessels.

design of experiment: The format of collection, compilation, analysis and interpretation of observations. A procedure, which can be employed interactively to form experimental designs of various types. There exist several types of designs: ORTHOGONAL HIERARCHICAL DESIGNS, FACTORIAL DESIGNS (WITH BLOCKING), FRACTIONAL FACTORIAL DESIGNS (WITH BLOCKING), LATTICE DESIGNS, LATTICE SQUARES, LATIN SQUARES, LATIN SQUARES BALANCED FOR CARRYOVER EFFECTS, SEMI-LATIN SQUARES, ALPHA DESIGNS, CYCLIC DESIGNS, BALANCED INCOMPLETE BLOCK DESIGNS, NEIGHBOUR-BALANCED DESIGNS, CENTRAL COMPOSITE DESIGNS and CENTRAL COMPOSITE DESIGNS.

designated host: A genetically stable host (i.e. a clone or pure line) that has been selected for use in the one-pathotype technique in a horizontal resistance breeding programme. It possesses a resistance that is matched by the designated pathotype, which is cultured on that host for the entire duration of the breeding programme. All the original parents of the breeding population are chosen on the basis of their susceptibility to the designated pathotype, which is used to inoculate every screening population. It ensures that all the vertical resistances are corresponding during the screening for horizontal resistance, regardless of how the vertical resistance genes may have recombined during the crossing process.

designer chromosome: Gene cloning and transformation can be utilized to circumvent LINKAGE DRAG EFFECTS that can plague conventional interspecific gene transfers. These techniques can also be employed to create desirable genetic linkages (e.g. the use of *Nicotiana glutinosa N*-gene mediated TOBACCO MOSAIC VIRUS (TMV) resistance in flue-cured tobacco, *Nicotiana tabacum*, has been controlled due to linkage drag effects). Transformation was used to introduce the cloned *N*-gene into 'NC152', a chromosome addition line possessing a chromosome pair from *N. Africana*. This chromosome has been projected to be used as a 'designer chromosome' into which numerous transgenes could be inserted to form a desirable linkage package. This system was proposed to shuttle a large number of transgenes from genotype to genotype.

designer crops: Designer crops are the crops designed by the breeder through genetic manipulation of crop over native genotype of plant using advanced techniques of molecular biology.

designer pentatricopeptide repeat (dPPR) proteins: Engineering or synthesizing artificial PPR domains through a consensus-based approach can be programmed for sequence-specific RNA binding in an effective manner. The presence of each protein with multiple PPR domains, ranging from 2 to 30 per protein, provides a useful reserve of sequence data to derive consensus sequence for dPPR proteins. The engineered PPR proteins include CONSENSUS PPR (cPPR), DESIGNER PPR (dPPR), SYNTHETIC PPR (synPPR), SINGLE CONSENSUS DESIGN (SCD) as well as MULTI-CONSENSUS DESIGN (MCD) PROTEINS.

desmosome: A plaque-like site on cell surfaces characterized by dense plaques of protein formed between two epithelial cells, into which intermediate filaments in the two adjoining cells get inserted. Desmosomes function in maintaining cohesion with an adjacent cell. Cells in this case adhere firmly, and they anchor by a line. No cell communication occurs in this case. *Aka* MACULA ADHAERENS.

desmutagen: An antimutagen that suppresses mutagenic activity of a mutagen by directly inactivating the mutagen by destroying oxygen radicals produced.

desynapsis: The premature separation of paired chromosomes during diplotene or diakinesis, which paired during zygotene and remained paired at pachytene of meiotic prophase. It is often genetically controlled but can also be induced by special environmental conditions (e.g. heat).

desyndesis: *See* DESYNAPSIS.

detasseling: Artificially removing (cutting or pulling) the tassel of the female parent to thwart selfing during hybrid seed maize production.

detectable difference: The value of a true population treatment effect (difference between two treatments) that, if held, would result in a statistical test having exactly the desired power.

detecting presence of a cloned gene: Many general-purpose vectors such as pUC19 usually include a system for detecting the presence of a cloned DNA fragment, based on the loss of an easily scored phenotype. The most widely used system is the gene coding for *Escherichia coli* β-galactosidase, whose activity can easily be detected by the ability of the enzyme it encodes to hydrolyse the soluble, colourless substrate X-gal into an insoluble, blue product.

detection and quantitation of messenger RNA: Methods commonly used for detection and quantitation of messenger RNA are the following: NORTHERN BLOTTING, NUCLEASE PROTECTION ASSAYS, *IN SITU* HYBRIDIZATION, REAL-TIME PCR and AFFINITY MATRICES. *See details as separate terms.*

detection and quantitation of small RNAs: An improved northern blot method enhances detection of small RNA molecules such as microRNA (miRNA), short-interfering RNA (siRNA) and Piwi-interacting RNA. Conventional UV cross-linking of RNA to nylon membranes is replaced with a novel, 1-ethyl-3-(3-dimethylaminopropyl) carbodiimide (EDC)-mediated, chemical cross-linking step that enhances detection of small RNA by up to 50-fold.

detection of linkage: Statistical analysis in which total χ^2 is partitioned into various components to determine presence of linkage in F_2 or test cross progeny. If linkage is complete, in a dihybrid cross, only two parental classes of offspring are produced. If linkage is incomplete and some recombination occurs, then recombinant classes are also produced but these will not be as frequent as the parental classes. The consequence of incomplete linkage is departure from the frequencies of phenotypes expected according to independent assortment. Linkage between two gene pairs is not the only cause influencing frequencies of certain classes. Abnormal segregation for alleles at a locus can be another reason for departure from the expected ratio. So, for detection of linkage, CHI-SQUARE must first test whether each gene pair segregates and then whether two gene pairs segregate independent of each other again as expected in Mendelian fashion. If the gene pairs singly show Mendelian segregation but jointly fail to show independent assortment, it can be concluded with certainty that the genes in question are linked. On the same principle, for detection of linkage in a three-point cross, four χ^2 values must be calculated: one each for segregation of the three individual gene pairs and the fourth for their joint segregation.

detection of non-coding RNAs: The following approaches have been used for detection of non-coding RNAs: (i) direct labelling and separation on polyacrylamide gels, (ii) sub-fractionation of nuclear extracts, (iii) association with specific proteins and (iv) mutations or phenotypes resulting from overexpression.

detection of promoters: A wide variety of algorithms has been developed to facilitate detection of promoters in genomic sequence, and promoter prediction is a common element of many gene prediction methods. A promoter region is located before the −35 and −10 consensus sequences. The closer the promoter region is to the consensus sequences, the more often transcription of that gene will take place.

detection of ribosomal RNA: 28S rRNA is detected by cytometry and fluorescence *in situ* hybridization (FISH) method. 16S rRNA is detected by SURFACE PLASMON RESONANCE (SPR) IMAGING.

detection of RNA-protein interactions using tRSA method: This method uses an RNA aptamer, scaffolded to a tRNA, to create an affinity matrix that efficiently pulls down transcript-specific RNA-binding proteins from cell lysates. The addition of tRNA scaffold to a Streptavidin aptamer (tRSA) increased binding efficiency by ~10×.

detection of transfer RNA: Method for tRNA detection is based on search for sequences that have the ability to form acceptor stem, T arm, D arm, anticodon stem, check invariant and semi-invariant bases of a tRNA in the found sequence and detect tRNA-like structures through RNase P cleavage of the IRES domain.

detergent: (i) A type of small amphipathic molecule that has a tendency to coalesce in water, with its hydrophobic tails hidden and its hydrophilic heads exposed. It is extensively used to solubilize membrane proteins. (ii) Any synthetic organic cleaning agent that is liquid or water soluble and has wetting and emulsifying properties.

deterioration of genetic purity (of a variety): Genetic purity of a variety can decline due to several factors during production cycles, such as development variations, mechanical mixtures, mutations, natural crossing, minor genetic variations, selective influence of diseases, selective influence of diseases and techniques of plant breeding.

deterioration of genetic purity due to development variations: When the seed crops are cultivated in a difficult environment, under different soil and fertility conditions, different climatic conditions, under different photoperiods or at different elevations for several successive generations, the development variations may ascend sometimes as differential growth response. To diminish the opportunity for such shifts to happen in varieties, it is advisable to grow them in their ranges of adaptation and growing seasons.

deterioration of genetic purity due to mechanical mixtures: This is the utmost imperative source of variety deterioration during seed production. They may often happen at the time of sowing, if more than one variety is sown with the same seed drill, through helper plants of the same crop in the seed field or through different varieties cultivated in adjacent fields. Two varieties growing along with each other in the field are often mixed to some extent during harvesting and threshing operations. Often the seeds produced by all the varieties are kept on the same threshing floor, leading to considerable varietal mixtures. To avoid this type of mechanical contamination, it would be essential to reprobate the seed fields and exercise the extreme care during seed production, harvesting, threshing and further handling.

deterioration of genetic purity due to minor genetic variations: This may occur even in the varieties looking phenotypically uniform and homogeneous at the time of their release. During later production cycles, some of these variations may be vanished because of selective elimination by the environment. To overwhelm these variations, yield trials of lines are propagated from breeders' seeds for the maintenance of self-fertilized crop varieties.

deterioration of genetic purity due to mutations: This is not a grave factor of varietal deterioration. In the majority of the cases, it is difficult to distinguish minor mutations. In the case of vegetatively propagated crops, periodic increase of true-to-type stock would eradicate the deteriorating influences of mutations.

deterioration of genetic purity due to natural crossing: In sexually propagated crops, natural crossing is a significant cause of varietal decline due to introgression of genes from dissimilar stocks. This can only be solved by deterrence.

deterioration of genetic purity due to selective influence of diseases: New crop varieties often become susceptible to new races of diseases, often caused by obligate parasites, and are out of seed programmes. Similarly, the vegetatively propagated stocks decline fast if infected by viral, fungal or bacterial diseases. During seed production, it is, therefore, imperative to produce disease-free seeds/ stocks.

deterioration of genetic purity due to techniques of plant breeder: In certain instances, serious issues may occur in varieties due to cytogenetic irregularities not properly evaluated in the new varieties prior to their release. Premature release of varieties still segregating for resistance and susceptibility to diseases or other factors may also be central to the deterioration of varieties. Other factors that may significantly lower the genetic purity include break down in male sterility, certain environmental situations and other heritable dissimilarities.

determinant: A factor that is responsible, fully or partially, for an outcome.

determinate: Ceasing vegetative growth when the first flower or reproductive structure forms. *Contrasts* INDETERMINATE.

determinate growth: The sort of development in which a plant matures all of its seeds at around the same time.

determination: It is cell division without growth. Potential of the cell becomes restricted. The process of commitment to a particular fate.

determination of blastomeres: Initial determination of blastomeres follows a particular path of differentiation, which involves the interaction of nuclear genome with cytoplasmically localized information. T.H. Morgan first gave this hypothesis in 1934.

determination of homoplasmy: Southern analysis is performed to determine the copy number and degree of homoplasmy of the introduced transgene. The transgenic plants possess a higher molecular weight plastid genome that is distinguishable on the autoradiogram from the smaller molecular weight fragment representing the untransformed plastid genome. If the transgenic plants are heteroplasmic, a wild-type fragment is visible along with the larger transgenic fragment. Absence of the wild-type fragment confirms homoplasmy.

determination of the genomic loci for specific histone modifications: Various methods used for this purpose include (i) CHROMATIN IMMUNOPRECIPITATION (two types of ChIP – native ChIP and cross-linking ChIP), (ii) CHIP-ON-CHIP (ChIP-chip), (iii) CHIP-SEQUENCING (ChIP-seq), (iv) CHIP-PET, (v) HALO-CHIP™ SYSTEM, (vi) PROXIMITY LIGATION ASSAY and (vii) READING DOMAINS. *See details as separate terms.*

determined: In developmental biology, an embryonic cell is thought to be determined if it has turned out to be committed to a particular specialized pathway of development. This determination echoes a change in the internal character of the cell, and it leads to the much more readily detected process of cell differentiation.

determining the size of a DNA fragment: A particular DNA molecule or a DNA sample is digested with a specific restriction endonuclease. A series of bands appear. The position of different bands would depend upon the size of the fragments. Smaller fragments move faster than the longer fragments from the loading site. Using DNA fragment of known length, the gels are calibrated, and the length of the experimental fragment could be known. A rough estimation of the size of an unknown DNA sample can be made from examination of the gel but for more accurate sizing the distances of the bands from the wells must be measured and a graph drawn. From a graph of migration distance vs log size of standard fragments, the exact size of unknown fragments is extrapolated.

deterministic: The process where the present state completely determines the future state. *Contrasts* STOCHASTIC PROCESS.

deterministic algorithm: An algorithm that, given a particular input, will always produce the same output, with the underlying machine always passing through the same sequence of states. Deterministic algorithms are by far the most studied and familiar kind of algorithm, as well as one of the most practical, since they can be run on real machines efficiently.

deterministic events: Events that have no random or probabilistic aspects and proceed in a fixed, predictable fashion.

detoxication: The metabolic progression by which toxins are altered into less toxic or more willingly excreted materials.

detrimental alleles: Any gene form that imparts a disadvantage to the individual carrying it and decreases the carrier's chance of reproduction or survival compared with that of an individual not possessing that allele.

detritivore: A heterotroph that feeds on dead organic matter.

development: Sequence of changes that occur in an organism as a fertilized egg progresses to form an adult plant or animal. It usually begins with a single cell, a fertilized egg, but in certain species, it may well be initiated asexually, for example, by parthenogenesis or by budding to form an adult. A simple and crude pattern is formed in the early embryo and filled with detail as the embryogenesis progresses. Three overlapping and interdependent processes that describe development include DIFFERENTIATION, REGIONAL SPECIFICATION and MORPHOGENESIS.

development of reproductive isolation: The evolutionary changes of regulatory genes are primarily responsible for development of reproductive isolation and, thus, for the origin of new species. The non-coding sequences, transposable elements, repetitive DNAs and other such structural features of various genomes have also been proposed to carry out important functions in the regulation of gene expression and thus contribute effectively to the process of speciation.

development, genes necessary for: To have an idea about number of genes necessary for development, let us consider case of *Drosophila* development. Total number of genes in *Drosophila* genome is 5000. Estimated number of total maternal effect genes in *Drosophila* is 117. There are 617 genes necessary for early embryogenesis. Total number of genes needed for imaginal disc development is 1000. Thus, there are 1617 genes necessary for normal embryogenesis and metamorphosis, which is roughly 30% of the entire genome. This estimate, however, omits genes that exist in multiple copies, such as those for histones, ribosomal RNAs and transfer RNAs. This estimate probably omits genes whose functions are strictly devoted to metabolic housekeeping throughout the life of the fly.

development, main key to: Development is not just more than growth or maturation, requiring constant negotiation with the environment. The question is, how do 30,000 human genes determine 10^{11} cells with 10^{15} connections? Obviously, they cannot do it in the same way that the roundworm's 18,000 genes govern its 959 cells. The cells are not quite genetically identical – the same set of genes is always there, but only some are switched on. Which genes switch on and which off in any given cell – and when, how and why – determined the cell's character and function. A main key to development is this on-off pattern, a pulsing, embryo-wide light show that turns genetic instructions into animals. Elucidating the control of these switches – by signals inside the cell, beyond it or even outside the body – is the main task of biology in the twenty-first century. And the switches are not flipped just in early life. Understanding the process of development helps in understanding processes that are responsible for evolution. The crucial role of growth factors, morphogens or other effectors is completely accepted in development. This means that external agents can change the properties of the genes, and in many contexts, such changes are inherited. If such effects are processed by signal transduction and alter the DNA in the chromosomes, then in this sense, it could inheritance of acquired characteristics in somatic cells.

development, regulation of: The developmental programme is regulated by intrinsic (internal) and extrinsic (external) influences. In most organisms, the first facet of differentiation and patterning ensues through pre-existing (intrinsic) asymmetry in the egg. Internal controls comprise metabolism and behavioural pattern. External controls include temperature and light. Further diversification and pattern formation are accomplished by extrinsic processes, such as cell-to-cell communication and the interaction between cells-environment interaction.

development, steps in: Well-recognized steps in development are gametogenesis, fertilization, cleavage, determination, differentiation, gastrulation, organogenesis and morphogenesis. During all these above-mentioned stages of development, no growth takes place. Growth comprises increase in size of the embryo. Growth can be with or without ell division. This completes developmental cycle.

development, stresses in: Molecular signals are not the only forces that pattern and shape the developing *Drosophila* embryo. Mechanical stresses sensed by cells also seem to be involved in creating the body plan. It has been shown that mechanical compression of STOMODEAL PRECURSOR CELLS (SPCs) can affect the expression of *twist* gene. Pressure causes the β-catenin protein to move from cell membrane to cytoplasm, increasing its concentration there. This movement in turn allows β-catenin to accumulate in the nucleus, where it activates *twist* and probably other developmental genes. The product of *twist* gene is required for the invagination and development of the stomodaeum.

development, types of genes in: On the basis of their role in development, genes can be categorized into three groups: (i) MATERNAL GENES, (ii) SEGMENTATION GENES and (iii) HOMEOTIC GENES. These three types of genes act sequentially to demarcate the properties of the progressively restricted parts of the embryo.

development-related processes: In addition to various stages of development, an individual goes through other processes, which are not part of developmental cycle. Senescence (ageing) is one such property of the cell. Cancer is another property of the cell. Regeneration may also be regarded as a type of developmental processes, although these properties of the cell are not normal.

developmental biology: The branch of biology that investigates the processes by which living organisms grow and develop over time. This field of biology also encompasses the other fields such as reproduction, regeneration, metamorphosis, growth and differentiation of stem cells in established tissues.

developmental constraints: The ontogenetic rules that determine (i) the course of evolution, (ii) the degree of 'intrinsic' control over evolutionary direction (in contrast to 'extrinsic' natural selection) and (iii) confines on what kinds of organism can develop.

developmental correlation: The pleiotropy resulting from the influence of the same gene upon the character differences that are necessarily affected by the same developmental or metabolic pathway during ontogeny.

developmental cycle: The gradual progression of phenotypic alterations of an organism during development.

developmental genetics, major tool in. Major tool in developmental genetics is mutation. Development of mutant is compared with that of normal. Whatever is not being accomplished in the mutant is assigned to the normal allele of the mutant gene. This is how specific genes are allotted specific functions. Analysis of normal and mutant development informs the time of epigenetic events. The fact that none of the many thousands of genes that are left can take over the tasks of missing factor is evidence of the high degree of individuality in the structure and function of gene. Most of the studies in developmental genetics are, however, based on lethal genes, which kill the individual in homozygous state. Mutations disturb the developmental process.

developmental genetics: Transmission genetics must be understood at a molecular level to understand development because development has a genetic basis. Genes play a very important role during development. Thus, it is an integrated study of genetics and development. Development interprets a genotype into a phenotype.

developmental homeostasis: The capacity of the developmental pathway to produce a normal phenotype in spite of developmental or environmental disturbances. The developmental pathways of organisms are genetically adjusted to produce the characteristic morphology of the species regardless of variations in internal and external conditions of development. This is termed genetic homeostasis or developmental homeostasis.

developmental machinery: Because developmental pathways are often highly integrated, some developmental mechanisms or intermediate stages may be highly conserved. Integration in this sense means that particular mechanisms (e.g. signal transduction modules) may be reused for different purposes at different developmental times in different regions, implying high conservation is required to maintain the integrity of the system. There may be many mechanisms that converge at a particular stage that is important for continued development. Integration and redundancy may allow a form to be retained despite changes in some of the developmental pathways.

developmental noise: The influence of random molecular processes on development.

developmental programme, factors regulating: The developmental programme is controlled by intrinsic (internal) and extrinsic (external) influences. In most organisms, the first facet of differentiation and patterning occurs through pre-existing (intrinsic) asymmetry in the egg. Internal controls comprise metabolism and behavioural pattern. External controls include temperature and light. Further divergence and pattern formation are achieved by extrinsic processes, such as cell-to-cell communication and the interaction between cell and their environment. Processes in embryo and adult are regulated in the same way.

developmental robotics (DevRob): A scientific field which aims at studying the developmental mechanisms, architectures and constraints that allow lifelong and open-ended learning of new skills and new knowledge in embodied machines. *Aka* EPIGENETIC ROBOTICS.

developmental stitches: Phenotypic sources of variation among organisms can be described by developmental stitches and reaction norms. A reaction norm is a mirror that reflects environmental effects into phenotypes.

developmental trial: Mini-kit trials that are conducted by a large number of farmers under their own socio-economic conditions to determine whether they can grow a new variety. It is a sort of farmers' participation in varietal evaluation before the new varieties are finally released by the breeders for cultivation.

developmentally regulated gene rearrangements: (i) A type of gene rearrangements that take place during differentiation of heterocysts in cyanobacteria or in mother cell of *Bacillus subtilis* during sporulation. (ii) Such rearrangements in DNA take place during the differentiation for regulation of gene expression

devernalization: The decline of vernalization by non-vernalizing temperatures or other means.

deviation: The parting of a quantity from its expected value.

deVriesism: The hypothesis that evolution in general and speciation in particular are the results of drastic mutations (saltations). *See* MUTATION THEORY OF EVOLUTION.

DexD/H-box ATPases: RNA-binding enzymes that hydrolyse ATP to facilitate a number of biochemical processes. These enzymes can be divided into subfamilies based on conserved sequence motifs. The spliceosome contains three DEAD-box ATPases (Sub2, Prp5 and Prp28), four DEAH-box ATPases (Prp2, Prp16, Prp22 and Prp43) and one DEIH-box ATPase (Brr2). These ATPases are believed to facilitate structural transitions that occur during splicing, and some of these (Prp5, Prp16, Prp22 and Prp43) take part in discard pathways that enhance splicing fidelity.

dextrorotatory (D) isomer: A compound that interchanges the plane of polarized light clockwise as it reaches the observer. *Compare with* LAEVOROTATORY (L) ISOMER.

dextrose: An aldohexose monosaccharide ($C_6H_{12}O_6$) that is a main intermediate compound in cellular metabolism. It is the dextrorotatory form of glucose, found in fruits and commercially attainable from starch by acid hydrolysis.

dG:dC tailing: A method used for molecular cloning of random fragments of DNA by adding complementary homopolymer tails to the insert and the vector DNA. Oligo dG and oligo dC tails are added by a terminal transferase.

DH10B: A strain of *Escherichia coli* that has been modified and selected to accept large **BAC** clones by transformation. DH10B is defective for RecA recombination.

diacylglycerol: Lipid produced by cleavage of inositol phospholipids in reply to extracellular signals. Composed of two fatty acid chains linked to glycerol, it works as a signalling molecule to activate protein kinase C.

diadelphous: Referring to a class of adelphous structure in which anthers remain free and filaments are united.

diagnosis: Concerned with the development of algorithms and techniques that are able to determine whether the behaviour of a system is correct. If the system is not functioning correctly, the algorithm should be able to determine – as accurately as possible – which part of the system is failing and which kind of fault it is facing. The computation is based on observations, which provide information on the current behaviour.

diagnostic (n. diagnosis): Relating to a distinctive characteristic important in the diagnosis of a disease or other condition.

diagnostic assays: Once mutations have been defined, they can be searched for repeatedly by methods referred to as diagnostic methods. Such methods include (i) allele-specific oligonucleotide hybridization, (ii) allele-specific amplification, (iii) ligation, (iv) primer extension and (v) artificial introduction of restriction sites.

diagnostics: The determination of the nature and/or cause of a disease or disordered condition.

diakinesis: A substage of prophase I of meiosis I during which chromosomes condense further and shorten in size and become thicker. Chromosomes have separated except at the ends. The homologous chromosomes are present as bivalents. The nucleoli begin to disappear. The nuclear envelope starts breaking down.

diallel: In either the complete or incomplete diallel, individualities of both seed and pollen parents are upheld for each family.

diallel crossing: The crossing in all possible amalgamations of a series of genotypes. *Aka* DIALLEL MATING.

diallel selective mating: Enlarging the initial hybrid base (otherwise non-existent in self-fertilizing plants) by developing three or four hybrid pools (populations) from multiple crossing in diallel fashion.

diallelic map: A genetic map that is prepared on the basis of molecular markers (e.g. single-nucleotide polymorphisms) of which both parental alleles are recognized. *Aka* BIALLELIC MAP.

dialogue system: A computer system intended to converse with a human with a coherent structure. Dialogue systems have employed text, speech, graphics, haptics, gestures and other modes for communication on both the input and output channels.

diamond code: An early view of the genetic code in which diamond-shaped pockets in the DNA molecules were hypothesized to serve as sites for specific alignment of amino acids.

Diamond vs Chakrabarty case: Diamond vs Chakrabarty US Department of Commerce, 1980, is a landmark case in which the US Supreme Court held that 'the inventor of a new microorganism whose invention otherwise met the legal requirements for obtaining a patent could not be denied a patent solely because the invention was alive'. It essentially allowed the patenting of life forms.

diamondoid: Structures resembling diamond with three-dimensional network of covalent bonds. Structures of this kind are formed mostly with second row atoms. Most useful diamondoid structures are carbon rich. Structures have some of the properties of diamond such as high strength.

diapause: A period of unprompted dormancy, independent of environmental conditions, interrupting developmental activity in an embryo, larva or pupa. Diapause is in a way a resting stage that lets organisms to survive hostile conditions.

diaspore: Any reproductive part of a plant adapted for dispersal and for establishing new plants; may be a disseminule, such as a seed, or other parts, such as specialized buds, branches, inflorescences or fruits.

diatomaceous earth method of DNA extraction: Diatomaceous Earth, *aka* KIESELGUHR or DIATOMITE, has silica content as high as 94%, is used in filtration and chromatography and is useful for the purification of plasmid and other DNA by immobilizing DNA onto its particles in the presence of a chaotropic agent.

dicentric chromosome: An aberrant chromosome with two centromeres that forms when two chromosome segments (from different chromosomes or from the two chromatids of a single chromosome), each with a centromere and a fuse and without the resulting acentric fragments.

dicer: The term was initially coined by Greg Hannon to describe a *Drosophila* multidomain enzyme of the RNase III family. *Drosophila* Dicer cuts long dsRNA into small dsRNAs of 21–23 nucleotides with a 30 overhang, known as siRNAs. In plants, at least two types of detectable Dicer activity are responsible for the production of siRNAs with distinct sizes of 21 and 24 nucleotides. Because plants have several Dicer homologues, they are sometimes referred to individually as Dicer-like I and Dicer-like II.

dicer-like (DCL) protein: Key components in small RNA biogenesis. A small protein family in plants whose diversification time dates to the emergence of mosses (*Physcomitrella patens*). DCLs are ubiquitously but not evenly expressed in tissues at different developmental stages and in response to environmental stresses.

dicer-like I protein (DCL1): Catalyses cleavage of intergenic/intronic fold-back precursors to release miRNAs. These molecules incorporate into an RNA-INDUCED SILENCING COMPLEX (RISC) to promote cleavage or translational repression of cellular transcripts carrying miRNA complementary sites.

dicer-like II protein (DCL2): Synthesizes a 24-nt natural antisense transcript siRNA that regulates stress responses.

dichogamy: The condition in which male and female portions of a flower mature at different times.

dichophase: The phase of the mitotic cycle in which a cell is resolved for further mitotic differentiation or for distinct cell functions.

dichotomous: Forking into two more or less equal branches. This may result from an equal division of the growing tip, or may be sympodial, in which the growing tip is aborted and replaced. Typically refers to mode of branch growth, as in *Aloidendron dichotomum*, but also to other organs, such as the venation patterns on leaves, the thorns of various species of Carissa (which morphologically are branches) and the thalli or hyphae of various algae and fungi.

dichotomous key: A tool used in the documentation of unknown plants by stating the noticeable features by which the various taxa can be recognized. Two contrasting choices are presented at each step – one choice in the couplet is accepted and the other rejected – thereby knitting a path to the identity of the plant.

dichotomous outcomes: An experiment with two outcomes, generally referred to as 'success' or 'failure'.

dichotomous regulators: Dual function regulators that activate transcription in one circumstance and repress in another.

dichotomous transcription elongation: Elongation of transcription progresses in alternating laps of monotonous and inch-worm-like movement with the elastic RNA polymerase configuration being subject to direct sequence control.

dicot: An abbreviated name for DICOTYLEDON, which means a plant with two cotyledons or seed leaves. Dicots are also called BROADLEAVES and usually show leaf veins in a netlike pattern and a tap root. *Contrasts* MONOCOT.

dicot epidermis: The stem's outer dermal tissue layer is the epidermis, a single layer of cells that prevents damage caused by sunlight, pathogens and herbivores. In dicot stems, multicellular trichomes (hairs) extend from the epidermis. In woody dicot stems, the epidermis is replaced by periderm, composed of cork and other tissues. In some stems, the dermal cells secrete a waxy substance that forms a cuticle, a protective covering that helps the stem retain water.

dicot hypodermis: The hypodermis, composed of collenchyma cells in dicots, forms the stem's outer layer of ground tissue, located interior to the epidermis. It provides structural support to the stem.

dicot leaf: Dicot leaves have an upper and lower epidermis, and the lower epidermis contains several small pores, called STOMATA, which facilitate gas exchange and allow water vapour to exit the leaf. Having stomata only on its lower surface helps the dicot leaf conserve water while having most stomata open. The upper and lower epidermis are separated by ground tissue, called MESOPHYLL, which contains parenchyma cells that carry out photosynthesis as well as sclerenchyma and collenchyma cells that provide structural support. In dicot leaves, the ground tissue is often arranged into two distinct types, the palisade and spongy mesophyll. The dicot leaf's vascular structures form netlike veins that are connected to the stem's vascular structures.

dicot root: Dicot roots have a taproot structure, meaning they form a single thick root, with lateral branches, which grows deep into the soil. The ground tissue of dicot roots, primarily composed of parenchyma cells, surrounds the roots' central vascular structures.

dicot stem: The majority of the dicot stem is composed of ground tissue, which primarily consists of parenchyma cells. Sclerenchyma and collenchyma cells are also found in regions that require extra strength. Dicot stems have a ring of vascular bundles, composed of xylem and phloem, which divide the ground tissue into the outer cortex and central pith. Dicot vascular bundles have an additional component, not present in monocot stems, called CAMBIUM. The stem's outer dermal tissue layer is the epidermis.

dicot stomata: Stomata (sing. stoma) are small epidermal pores that allow gases and water vapour to move between the interior leaf structures and the surrounding air. In dicot leaves, the lower epidermis contains several stomata. Having stomata only on its lower surface helps the dicot leaf conserve water while having most stomata open.

dicotyledon: A flowering plant whose embryo has two or more cotyledons (seed leaves). Abbreviated as DICOT.

dideoxynucleotide: A modified deoxynucleotide that lacks the 3′-hydroxyl group and thus prevents further chain elongation when incorporated into a growing polynucleotide in enzymatic method of DNA sequencing.

dideoxynucleotide DNA-sequencing method: *See* DNA-SEQUENCING METHODS.

dideoxyribonucleotide 5′-triphosphate (ddNTP): A triphosphate deoxyribonucleotide which sans both 3′ and 2′ hydroxyl groups. Such dideoxynucleotides can be added to a growing nucleic acid chain but do not then present a 3′-OH group to support further propagation of the nucleic acid chain. ddNTPs are useful in DNA and RNA-sequencing reactions. *Aka* CHAIN TERMINATORS. *Compare with* DEOXYRIBONUCLEOTIDE.

didiploid (also **allotetraploid**): Two different diploid chromosome sets occurring in one cell or organism.

dieback: The progressive death beginning at the tips of shoots, leaves or roots.

differential affinity: The failure of two chromosomes to pair at meiosis in the presence of a third chromosome, although they pair in its absence.

differential amphiplasty: A reversible effect resulting from the interaction of alien chromosomes.

differential centrifugation: A method of separating subcellular particles by centrifugation of cell extracts at successively higher speeds. It is based on differences in sedimentation coefficients that are roughly proportional to particle size. In other words, large particles (nuclei, chloroplasts or mitochondria) sediment at lower speeds than small particles (ribosomes).

differential display: In this approach of screening cDNA libraries, mRNA is reverse transcribed using a series of primers. mRNAs that are homologous to the randomly chosen sequence should be reverse transcribed, producing a single-stranded cDNA. The addition of another primer, again, randomly chosen, will allow amplification of a subset of the reverse transcribed cDNAs. Depending on the distance between the two primers, fragments of varying molecular weights will be obtained. By following this procedure with mRNA that has been isolated from two different cell populations, the pattern of expression between the two cell types or cell states can be determined.

differential display proteomics: 2D-PAGE correlative analysis, antibody-based screening and protein chip technology are the tools used in functional analysis of proteins. Antibody-based screening is used where functional relationship can be explored by using individual proteins directly as assay for their building partner in biological samples coated within microwell plates using ENZYME-LINKED IMMUNOSORBENT ASSAY (ELISA) test techniques. In protein chip technology, a variety of bait proteins, such as antibodies, are immobilized in an array format onto a specially treated surface. The surface is then probed with the sample of interest, and only the proteins that bind to the relevant antibodies remain bound to the chip. These are then placed in a chip reader, which includes a mass spectrophotometer for analysis of the proteins. The protein chip is probed with fluorescently labelled proteins from two different cell states. Cell lysates are labelled by different fluorophores and mixed such that the colour acts as a reader for the change in abundance of the protein bound to the antibody. Phage presentation utilizes fusion proteins encoded by chimeric sequences of bacteriophage viral coat protein and gene of interest. cDNA libraries encoding fusion protein are created. Individual clones in the library can be selected by panning with any ligand of interest (e.g. a herbicide whose mode of action is unknown). If this ligand is immobilized to an inert support, the bacteriophage clones that bind can be purified by elution. The bacteriophage particles can then be used as template for PCR with a set of fluorescent-labelled primers that flank the cDNA inserts. The cDNA fragments in this enriched population could then be characterized by microarray hybridization. Three tools used in functional analysis of proteins are the following: (i) 2D-PAGE correlative analysis, (ii) antibody-based screening and (iii) protein chip technology.

differential ELISA test: Detection of antibodies specific to either the vaccine or the disease agent by the enzyme-linked immunosorbent assay.

differential gene action: According to this hypothesis, in different cells, different genes are active at different times. Along with differential gene activation, there was also differential gene inactivation. Some genes become inactivated irreversibly. In insects, there are several pieces of evidence for differential gene action. Puffs on a particular chromosome constitute a puffing pattern, which is different in different cells at a certain specific time, and the puffing pattern is also different at different times in a developing system. Different sets of genes are active in the same developing system at different times. Gene activity at molecular level means synthesis of MESSENGER RNA (MRNA) by a particular gene. Thus, synthesis of mRNA was studied to demonstrate differential gene activity.

differential gene alteration*:* Some genes are altered, while others are not. There are two types of gene alteration – differential gene inactivation and differential gene replication.

differential gene replication: Some genes are replicated, while others are not. This phenomenon can be illustrated by studying replication of rRNA in the oocytes of amphibians.

differential host: A plant host that helps in distinguishing various strains or races of a given plant pathogen on the basis of disease reactions. *Aka* DIFFERENTIAL CULTIVAR.

differential imprinting: Differential sex-specific methylation of the alleles of a gene. Such methylation distinguishes male and female genomes and is inherited in a parent-of-origin-specific manner.

differential item functioning (DIF) method: A method in the industry that controls plant height in greenhouses by temperature. DIF was developed by researchers from Michigan State University back in the 1980s, which showed that the average day plus night temperature affects plant's growth rate, with higher averages leading to more rapid growth and development. The difference between day and night temperatures affects stem elongation and height. DIF is calculated as the day temperature minus the night temperature, and this value can be either positive (day temperature is higher than night), zero (day temperature is equal to night) or negative (day temperature is less than night).

differential medium: A differential medium contains specific ingredients or chemicals that let the observer visually distinguish which species possess or species lack a specific biochemical process.

differential methylation hybridization (DMH): In this method, one set of the genomic DNA is digested with methylation-sensitive Res, and a parallel set of the DNA is left undigested. Both sets of DNAs are subsequently amplified, and each one is labelled with fluorescent dyes and used in two-colour array hybridization. The level of DNA methylation at a given loci is determined by the relative intensity ratios of the two dyes.

differential processing: Production of more than one functional mRNA from one type of complex transcription unit.

differential reproduction: Reproduction in which different types do not contribute to the next generation in the proportion to their numbers.

differential RNA processing: It refers to the production of more than one functional mRNA from one type of complex transcription unit through differential processing. For example, during B-lymphocyte development and immunoglobulin synthesis, the μ heavy chain is first introduced as an integral protein in plasma membrane. Later, a similar μ chain is noticed as a part of secreted immunoglobulins. Analysis of cDNA of the two mRNAs showed that carboxyl termini of the μ membrane were different and amino termini were the same for the two μ chains. Further examination revealed that genomic DNA possessed two poly(A) sites in the same transcript unit and same primary transcript appeared to cover both of them. Depending upon choice of poly(A) site, two heavy chain mRNAs with different 5′-most sequences can be designed, one for the membrane chain and other for the μ secreted chain.

differential screening: A method of screening cDNA libraries whereby two probes, differing in one or just a few sequences, are used as probes in two parallel *in situ* hybridizations using duplicate lifts of the library.

differential segments: Portions of chromosomes that do not pair in meiosis.

differential sensitivity: A difference between two organisms in sensitivity to environmental conditions (e.g. one may be killed by a temperature that is optimum for the other).

differential splicing: A form of splicing wherein the combination of introns that are removed varies between cell types, allowing a single gene to produce transcripts with different primary structures and therefore in coding potential.

differential staining: Chromosomes, when denatured and subsequently renatured and stained with some dye (such as quinacrine mustard or Giemsa), show some darkly stained bands along the chromosome. This technique is very useful for identifying individual chromosomes.

differential variety: A host variety, consisting of a set differing in disease reaction, used to identify physiologically specialized forms of pathogens.

differentially methylated genes (DMGs): Some of the TISSUE-SPECIFIC DIFFERENTIALLY METHYLATED REGIONS (TDMRs) are within 5′ promoter CpG islands of genes that exhibit a tissue-specific expression pattern that is consistent with methylation status and a role in tissue differentiation.

differentially methylated regions (DMRs): Genomic regions that have different levels of methylation between the two chromosomes of a diploid cell or sample groups; often associated with genomic imprinting. DMRs can be context specific (CpG, CpHpG or CpHpH) or can refer to overall methylcytosine content. CG-dinucleotide-rich genomic regions that, in diploid cells, are methylated on one parental chromosome and unmethylated on the other. Thus, in an organism's genome, DMRs are methylated differentially between disease phenotypes. Gametic DMRs acquire their parental-specific DNA methylation during gametogenesis, either in the developing haploid oocyte or sperm, whereas somatic DMRs acquire their parental-specific DNA methylation in somatic diploid cells.

differentiated: Cells that maintain, in culture, all or much of the specialized structure and function typical of the cell type *in vivo*.

differentiated substances: Those substances that are not needed by the cell for its maintenance, but they are characterizing its differentiated state. Such substances are synthesized by non-dividing cells. The synthesis of extra ribosomes is a differentiation function. Synthesis of extra ribosomes is accompanied by specific gene replication.

differentiation: Differentiation of organs from a mass of undifferentiated cells. During plant tissue culture, callus is considered as a homogeneous mass of undifferentiated cells. Upon subculturing on the regeneration medium, it starts differentiating into structures resembling embryos (somatic embryos), roots (rhizogenesis) and shoots (caulogenesis). Differentiation is process of specialization of cells. Special properties of the cell appear during this stage. Specific protein synthesis takes place. This leads to formation of different cell types with specialized structures and functions.

differentiation and genetic cascades: DNA serves as a blueprint from which RNA is transcribed. Proteins come from the amino acid chains that are translated from the RNA. The levels of transcription and translation of a gene determine how much of that gene's protein will be present in the cell. Gene expression, which encompasses transcription and translation, is the general term to describe the processes in which DNA produces RNA and proteins. It can also include other factors, such as the rate at which RNA is degraded before it can be translated. Differential gene expression will result in varying concentrations and kinds of proteins in cells, causing them to look and function differently. This differential transcription and translation of genes ultimately allows for cellular differentiation. Development is a programme that regulates gene expression at the appropriate locations and times.

diffraction pattern: A pattern set up by wave interference between radiation communicated or dispersed by different parts of an object.

diffusible signals: Small molecules often function as diffusible signals. In unicellular organisms, diffusible signals may be environmental in origin or may be released from other cells. Examples of diffusible signals include yeast mating-type pheromones and cAMP in *Dictyostelium*.

diffusion: (i) The process by which molecules (e.g. of oxygen) move inertly from a region of high concentration to a region of low concentration. (ii) Dispersal due to the cumulative effect of small random movements. As nutrients are captivated, a lower concentration is observed at the root interface. To even out the concentration throughout the soil, nutrients traffic from areas of higher concentration to areas of lower concentration around the roots.

diffusion approximation: A mathematical estimate that describes diffusion (dispersal) using a differential equation. It describes the blowout of chemical concentration or allele frequency through physical space and also the spread of a probability distribution through the space of allele or genotype frequencies.

diffusion equation: An equation that defines how a probability distribution spreads out with time. Diffusion equation is applied when accidental fluctuations are small.

diffusion pressure deficit: Net capacity of plant cell for absorbing water.

digenic inheritance: Refers to inheritance of a trait that is determined by two genes. Inheritance pattern observed in a cross between two individuals identically heterozygous at two gene pairs gives 9:3:3:1 ratio in F_2 in case of COMPLETE DOMINANCE. *See* DIHYBRID CROSS.

digestive enzymes: The peptidases that are synthesized in the pancreas and secreted into the duodenum. These enzymes catalyse the hydrolysis of peptide bonds but with different specificities for side chains – the scissile (to be cleaved) peptide bond. CHYMOTRYPSIN is specific for bulky hydrophobic residue preceding the scissile bond, whereas TRYPSIN is specific for a positively charged residue.

digital agriculture: Refers to the use of computers and other electronic devices in farming and other agricultural endeavours. Digital agriculture enables farmers and other stakeholders in the agriculture value chain to increase food production using advanced technologies incorporated into a single system. *Aka* EAGRICULTURE.

digital candidate gene approach (DigiCGA): DigiCGA is product of bioinformatics. This approach enables the systematic identification of genes underlying biological traits. Gene functional similarity search tool (GFSST) is a digital resource that makes it possible to identify candidates by functional similarity. *Aka* IN SILICO CANDIDATE GENE APPROACH (ISCGA) *or* COMPUTER-FACILITATED CANDIDATE GENE APPROACH (CFCGA).

digital data: In information theory and information systems, the discrete, discontinuous representation of information or works. Numbers and letters are commonly used representations.

digital imaging: Digital imaging is the creation of a digital representation of the visual characteristics of an object, such as a physical scene or the interior structure of an object. The term is often assumed to imply or include the processing, compression, storage, printing and display of such images. A key advantage of a digital image, vs an analogue image such as a film photograph, is the ability to digitally propagate copies of the original subject indefinitely without any loss of image quality. Digital image processing technology is also more applied to agriculture, such as pesticide detection of vegetables, pest control of crops, growth trend detection of crops, colour and classification of crops. *Aka* DIGITAL IMAGE ACQUISITION.

digital life systems (dLife): This concept proposed by X. Zhang and others in 2023 is a new paradigm to leash the potential of genomics and artificial intelligence in revolutionizing medicine. It includes the multiscale and multi-granule measure and representation of life in the digital space, the mathematical and/or computational modelling of the biology behind physiological and pathological processes and, ultimately cyber twins of healthy or diseased human body in the virtual space that can be used to simulate complex biological processes and deduce effects of medical treatments.

digital signal processing (DSP): Employing digital processing, such as by computers or more specialized digital signal processors, to perform a wide variety of signal processing operations. The signals processed in this way are a sequence of numbers that characterize samples of a continuous variable in a domain such as time, space or frequency.

digitate: Refers to segments dispersing from a common centre, like the fingers of a hand. *See* PALMATE, PALMATISECT *and* LEAF SHAPE.

digitiform: Designed like a finger.

digynic triploid: An embryo that contains two maternal genomes and one paternal genome.

dihaploid: A haploid cell or individual containing two haploid chromosome sets. For instance, haploids of tetraploid species are called dihaploid. A dihaploid occurs as a result of the spontaneous or induced chromosome doubling in haploid cells during embryogenesis; not to be confused with DOUBLED HAPLOID.

dihaploid (DH) populations: DH populations have become standard resources in genetic mapping for species in which DHs are readily available. Double-haploid populations are ideal for genetic mapping. It is possible to produce a genetic map within 2 years of the initial cross regardless of the species. Map construction is relatively easy using a DH population derived from a hybrid of two homozygous parents, as the expected segregation ratio is simple (i.e. 1:1). DH populations have now been used to produce genetic maps of barley, rapeseed, rice, wheat and pepper. DH populations played major role in the generation of molecular marker maps in eight crop species.

dihybrid: A cross between individuals that differ with respect to two specified gene pairs.

dihybrid cross: A sexual cross in which the inheritance of two pairs of alleles is followed (tall, yellow × dwarf, green). For this, a cross is made between individuals differing for two pairs of genes affecting two different characteristics. The principle of independent assortment (independent segregation) was deduced from Mendel's observations on dihybrid crosses.

diiso-compensating trisomic: A compensating trisomic in which a missing chromosome is compensated by two isochromosomes, one for each arm of missing chromosome.

diisosomic: A cell or an individual which possesses a pair of homologous isochromosomes for one arm of a particular chromosome.

diisotrisomic: A cell or an individual in which one chromosome is absent but carries two homologous isochromosomes of one arm of a particular chromosome.

dikaryon: *See* DIKARYOTIC.

dikaryosis: The existence of two nuclei in each cell.

dikaryotic (n. dikaryon): Possessing two different and distinct sexually compatible haploid nuclei per cell that divide simultaneously. This pair of haploid nuclei differing in some character which can be genetically complementary to each other. This situation is observed in the fungi. A dikaryotic individual is called a DIKARYON. This phase is called DIKARYOPHASE. *Contrasts* HAPLOID, DIPLOID *and* POLYPLOID.

dilution plating: A method to attain pure colonies of bacteria and fungi in which infected plant material or infested soil is diluted in sterilized water. Small samples of the water are spread on the medium surface of several petri plates to find which dilution will be appropriate for producing pure colonies.

dilution streaking: The procedure that involves repeated streaking of bacteria on the surface of a nutrient medium with a sterile metal loop to allow pure colonies to grow.

dimensionality: The spatial and temporal scales of a risk assessment.

dimer: A composite molecule formed by the binding of two identical (homo-) or different (hetero-) molecules. Similarly, trimers and tetramers are bunches of three and four molecules, respectively. *See* HOMODIMERS, HETERODIMERS, TRIMERS *and* TETRAMERS.

dimerization: Chemical union of two identical molecules.

dimethyl sulfoxide (DMSO): A liquid solvent (C_2H_6OS) permitted for improved penetration of specific substances through the cell wall.

dimethylsulfoxide (DMSO)-mediated gene transfer: DMSO increases membrane permeability and enhances uptake of DNA. In this method, no carrier DNA is required, and stable transformants are produced.

dimonoecious: Having perfect (bisexual) flowers as well as pistillate and staminate flowers on one plant.

dimorphic: Occurring in two different states (with respect to shape and/or size; e.g. stamens, fronds, or leaves). *See* MONOMORPHIC *and* POLYMORPHIC.

dimorphic gene: A gene whose protein product exists in two forms – inactive and active; active form is produced by cleaving a part of polypeptide chain of the inactive protein. *See* COMPLEX GENES.

dimorphism: The existence of a morphological distinction between organisms of the same population or species, such that individuals of that species ensue in one of two distinct forms which diverge in one or more characteristics, such as colour, size, shape or any other phenotypic trait. Dimorphism based on sex (e.g. male vs female) is frequent in sexually reproducing organisms such as plants and animals.

dinucleotides and trinucleotides in creation of chromatin marks: 5′-CG-3′ dinucleotides and 5′-CNG-3′ trinucleotides are found in published sequences of small interfering RNA and microRNA more often than they should be in random DNA sequences. Small RNAs containing these di- and trinucleotides participated in the creation of chromatin marks of epigenetic information through a highly specific search for repressible DNA sequences and through the initiation of the methylation *de novo* of 5′-CG-3′ and 5′-CNG-3′ sites in DNA fragments appearing to be bound complementary to small RNAs.

dioecious: Commonly used term for unisexual plants. The plant is either male or female. In such type of plants cross pollination is a rule rather than exception. In natural progeny, about 50% of plants carry male flowers and rest 50% carry female flowers. Only the female plants bear fruits and seeds. For example, hemp, hops, asparagus, spinach, papaya and date palm are unisexual plants. *Contrasts* MONOECIOUS.

dioecism: The phenomenon of plants showing either male or female sex organs.

dioecy: Having separate sexes; individuals are either males or females; used for plants. *See* GONOCHORISM.

dioxygenases: Enzymes that catalyse the addition of the two oxygen atoms from its molecular free form (O_2) to one or two organic substrates.

dip pen nanolithography: An AFM-based soft-lithography technique.

diphasic: Chromosomes that display both euchromatic and heterochromatic sectors.

diphenylamine (DPA) method for DNA detection: DPA indicator confirms the presence of DNA. This procedure involves chemical hydrolysis of DNA. When heated (e.g. ≥95°C) in acid, the reaction requires a deoxyribose sugar and therefore is specific for DNA.

diplo-chromosome: A chromosome that has divided twice instead of once since the preceding effective mitosis, its centromere being undivided.

diploid (2n): An organism or cell that contains two sets of homologous chromosomes in the same nucleus of each cell and hence two copies of each gene or genetic locus, one from each parental set. One set of chromosomes is contributed by each parent. When a human somatic cell undergoes mitosis, or cell division, the sister chromatids separate, resulting in two daughter cells with diploid chromosomes each. During mitosis, a parent cell splits into two genetically identical, diploid daughter cells. Most animal cells except the gametes have a diploid set of chromosomes. For example, the diploid human genome has 46 chromosomes. Commonly grown high-yielding rice is diploid with chromosome number 2n = 24. *See* HAPLOID *and* POLYPLOID. *Compare with* HAPLOID.

diploid number: Refers to the full component of chromosomes normally found in somatic cells. In humans, the diploid number (2n) is 46. *Compare with* HAPLOID NUMBER.

diploidization: In polyploids, a natural or induced mechanism in which the chromosomes pair completely or partially as bivalents, even though polyploid sets of chromosomes are present. Diploidization may be caused by a structural differentiation of homologous chromosome sets or by genetic control. For example, in bread wheat, three homoeologous genomes are available (*AABBDD*); they do not pair as HEXAVALENTS but exclusively as BIVALENTS. This diploidization behaviour in wheat is due to *Ph* locus on chromosome 5B of hexaploid wheat.

diploidy: The presence of two homologous sets of chromosomes in somatic cells.

diplonema: *See* DIPLOTENE.

diplontic: *See* DIPLOID.

diplontic life cycle: A life cycle in which diploid somatic adults produce haploid gametes by meiosis that fuse to form diploid zygotes that develop into somatic adults.

diplontic selection: Selection of somatic mutants under the competition between mutated and surrounding non-mutated cells forming chimera. This is true for colchiploids as well.

diplophase: A phase in the life cycle of an organism in which the cells of the organism have two copies of each gene. When this state exists, the organism is said to be DIPLOID.

diplosis: Establishment of the zygotic chromosome number, usually by syngamy and karyogamy.

diplospory: A mode of apomixis in plants in which a diploid gametophyte is formed mitotic or partly meiotic divisions of the spore-forming cells.

diplostemonous: Having stamens arranged in two whorls, with the outer whorl alternating with the petals, while the inner whorl is opposite the petals. *Compare with* OBDIPLOSTEMONOUS *and* HAPLOSTEMONOUS.

diplotene: Fourth stage of division I of meiosis I. At this stage, homologous chromosomes start separating. Separation of chromosomes does not take place at all the points as the sites where crossing over has already occurred show cross-type structures known as chiasmata (sing. chiasma), which are observable under the microscope.

direct chemical method for sequencing RNA: Four different base-specific chemical reactions offer a method of directly sequencing RNA terminally labelled with ^{32}P. After a partial, specific modification of each kind of RNA base, an amine-catalysed strand break produces labelled fragments whose lengths determine the position of each nucleotide in the sequence. Dimethyl sulfate modifies guanosine. Diethyl pyrocarbonate attacks primarily adenosine. Hydrazine attacks uridine and cytidine, but salt suppresses the reaction with uridine. In all these cases, aniline induces a subsequent strand break. The electrophoretic fractionation of the labelled fragments on a polyacrylamide gel is followed by autoradiography, which determines the RNA sequence.

direct cloning vector for *B. subtilis*: Plasmid pBD214, which encodes Cm^R and carries a *thy*$^+$ gene and plasmid BD393, which is a highly competent thymine autotroph for the basis of direct selection.

direct cross: A cross between two individuals.

direct DNA repair: DNA repair that involves neither removal nor replacement of bases or nucleotides; rather, the covalent modification in DNA is simply reversed. *Aka* DNA DAMAGE REVERSAL.

direct DNA sequencing using PCR: Direct sequencing of PCR-generated templates is a commonly used technique in molecular biology laboratories. E. Iannelli and others in 1998 described an improved method for direct sequencing of PCR fragments longer than 20 kb obtained with a commercial mixture of *Taq* and Pwo DNA polymerases. The sequencing protocol was optimized for an automated infrared DNA sequencer, consistently yielding long reads (500–600 bases). Direct DNA sequencing using PCR is *aka* LIGATION-MEDIATED PCR (LMPCR). PCR has also been used for DNA product. This method of DNA sequencing is faster and more reliable and can utilize either the whole genomic DNA or cloned fragments for sequencing a particular DNA segment. The DNA sequencing using PCR involves two steps. (i) Generation of sequencing template. (ii) Sequencing PCR products either with the thermolabile DNA polymerase or with the thermostable *Taq* DNA polymerase. Thus, the DNA-sequencing method using PCR eliminates the need for cloning the DNA in single-stranded DNA phage vector (i.e. M13).

direct DNA transfer: Refers to methods of inserting a gene directly into a cell's DNA without the engaging a vector (e.g. electroporation).

direct embryogenesis: Embryoid formation directly on the surface of zygotic or somatic embryos or on seedling plant tissues in culture without an intervening callus phase.

direct gene transfer (DGT) methods: Gene transfer through direct DNA (gene) uptake by cell or through certain physical and chemical processes. DGT methods are broadly classified as PHYSIOCHEMICAL, PHYSICAL AND ELECTRICAL. *Aka* VECTORLESS METHODS OF GENE TRANSFER, DNA-MEDIATED GENE TRANSFER (DMGT) *or* DIRECT GENE TRANSFER (DGT) SYSTEMS. *See* VECTOR METHODS.

direct interaction between Avr and R proteins: Co-localization of Avr and R proteins in the flax cytoplasm and genetics of the gene-for-gene interactions are consistent with direct interaction between these proteins.

direct intracellular interactions: *R* gene products are receptors for ligands encoded by *Avr* genes. Ligand (Avr) binding to receptor (R) triggers a signalling cascade leading to activation of defence mechanisms, including HYPERSENSITIVE RESPONSE (HR). The seat for recognition is intracellular.

direct lineage reprogramming: The conversion of fully differentiated cells of a certain type into another cell type, while bypassing the intermediate pluripotent state.

direct organogenesis: Organ formation directly on the surface of relatively large intact explants without an intervening callus phase.

direct penetration: The penetration of plant tissues by a pathogen through a barrier (e.g. leaf cuticle) by chemical and physical means (e.g. INFECTION PEG and PENETRATION PEG). *Contrasts* INDIRECT PENETRATION.

direct read-off RNA sequencing: A two-dimensional direct read-off sequencing procedure uses *Escherichia coli* 5S ribosomal RNA as a model compound. Radioactive bands are transferred from an acrylamide gel fractionation in the first dimension onto a DEAE-cellulose thin-layer plate. After *in situ* enzymatic digestion with RNase T2, mononucleoside 3′,5′-diphosphates were separated in the second dimension by electrophoresis at pH 2.3.

direct regeneration: In direct regeneration, *in vitro* organs are directly induced from explant tissues. *Compare with* INDIRECT REGENERATION.

direct repeats: Identical or related sequences occurring in two or more copies in the same orientation in the same molecule of DNA that may or may not necessarily be adjacent.

direct RNA sequencing: Two methods of direct RNA sequencing are as follows. (i) RNA that is both polyadenylated and 3′ deoxy-blocked with poly(A) polymerase is seized on poly(dT)-coated surfaces. A 'fill-and-lock' step is performed, in which the 'fill' step is done with natural thymidine and polymerase, and the 'lock' step with fluorescently labelled A, C and G Virtual Terminator (VT) nucleotide and polymerase. Then, chemical cleavage of the dye-nucleotide linker is performed to release the dye and prepare the templates for nucleotide incorporation. Incubation of this surface with one labelled nucleotide and a polymerase mixture is done. After this step, imaging is achieved to locate the templates that have incorporated the nucleotide. Chemical cleavage of the dye allows the surface and DNA templates to be ready for the next nucleotide-addition cycle. Nucleotides are added in the C, T, A, G order for 120 total cycles (30 additions of each nucleotide). (ii) Radioactive bands are transferred from an acrylamide gel fractionation in the first dimension onto a Diethylaminoethyl cellulose (DEAE-C) thin-layer plate. After *in situ* enzymatic digestion with RNase T2, mononucleoside 3′,5′-diphosphates are separated in the second dimension by electrophoresis at pH 2.3. Using this two-dimensional method, the entire sequence of 163 residues of the previously unknown *Vicia faba* (broad bean) 5.8S ribosomal RNA has been worked out.

direct selection: A change in genotype frequency that occurs by the effects on fitness of the alleles themselves.

direct sequencing of candidate gene for a quantitative trait locus (QTL): In candidate gene screening approach, candidate genes are selected based on the following criteria: those QTLs known to cause or be associated with condition with phenotypes similar to the genes with known function that emerged from expression microarray data and those QTLs expressed that are also related to condition that have increased prevalence.

directed alternate disjunction: Regular movement of alternate centromeres to the same pole of the spindle in nuclei heterozygous for one or more reciprocal translocations:

directed dominance: *See* DIRECTIONAL DOMINANCE.

directed evolution: A laboratory process used on isolated molecules or microbes to cause mutations and identify subsequent adaptations to novel environments. Random mutagenesis is applied to a protein, and a selection regime is used to pick out variants that have the desired function or behaviour. Further rounds of mutation and selection are then applied. This method mimics natural evolution and generally produces superior results to rational design. The great advantage of directed evolution is that it requires no prior structural knowledge of a protein, nor is it necessary to be able to predict what effect a given mutation will have. Indeed, the results of directed evolution experiments are often surprising; desired changes are often caused by mutations that were not expected to have that effect. The drawback is that they require high throughput, which is not feasible for all proteins. Directed evolution has two approaches: (i) CLASSICAL or *IN VITRO* APPROACH and (ii) SYNTHETIC GENETIC SYSTEMS or *IN VIVO* APPROACH (R.L. Williams and C.C. Liu in 2024). *See details as separate terms.*

directed evolution, classical approach: This approach to directed evolution relies on a repeated sequence of labour-intensive steps where genes encoding biomolecules of interest are diversified *in* *vitro*, then transformed into cells, expressed and subjected to selection or screening to achieve desired activities. As a result, the evolutionary searches that these manually staged processes constitute are limited in scale and depth.

directed evolution, synthetic genetic systems approach: There has been a growing effort to develop synthetic genetic systems that autonomously diversify user-defined genes *in vivo* so that biomolecules of interest quickly and autonomously evolve, as cells are grown under selection for the gene's function. R. Tian and others (2024) report the establishment of an orthogonal DNA replication system in *Escherichia coli* (EcoRep) that can allow *in vivo* rapid continuous directed evolution of RNAs, enzymes, proteins and genetic circuits towards new activities without passing on catastrophic defects to offspring. This approach massively accelerates the evolution of new function from user-defined DNA sequences.

directed gene amplification: A type of gene amplification, which is genetically programmed. It occurs at some specific stage during development. For example, programmed amplification occurs in rRNA genes of many amphibian oocytes.

directed mutagenesis: Alteration of DNA at a specific site and its reinsertion into an organism to study any effects of the change.

directed mutagenesis and protein engineering: Directed mutagenesis has been used in protein engineering in (i) protein engineering of laundry detergent, (ii) combinatorial mutagenesis, (iii) reconstituting chemically orthogonal directed engineering (ReCODE™) technology, (iv) ProteinGPS™ engineering technology and (v) transforming Protein Engineering with Infologs™.

directed mutation: Selective mutation of a particular gene sequence or a class of sequence. Such mutations may be achieved by the use of drugs that have a particular affinity for the sequence in question (e.g. active as compared to inactive gene sequencing) or that tend to alter permanently the stage of differentiation of cells in culture.

directed protein evolution: A protein engineering method to evolve proteins with desirable properties. It mimics and accelerates natural evolutionary processes by applying *in vitro* diversification-selection-amplification cycles.

directed selection: The course of controlling genetic change in domesticated plants through manipulation of the plants' environment and their breeding process.

directed sequencing: Successively sequencing DNA from adjacent stretches of chromosome.

direction of replication: J. Cairns in 1963 proved that only one origin of replication existed in prokaryotes. DNA replication is of two types with respect to direction of replication – UNIDIRECTIONAL and BIDIRECTIONAL. The question of direction of DNA replication was resolved by D.M. Prescott and P.L. Kuempel in 1972 and 1973, when *Escherichia coli* cells were grown on the medium containing ^{3}H Thymine at low level which was enough to visualize the track of DNA from origin. After some time, the cells were exposed to high level of ^{3}H Thymine. As a result, two short tracks of dense silver grains were observed at each end of growing chromosome. This proved BIDIRECTIONAL MODE OF DNA REPLICATION. DNA replication proceeds outwardly from the point of start in both the directions. Evidence for bidirectional replication in *Drosophila* was provided by H.J. Kreigstein and D.S. Hogness in 1974. Thus, bidirectional replication is characteristic of prokaryotes and eukaryotes. UNIDIRECTIONAL REPLICATION is seen in φX174 and some other single-stranded DNA viruses.

directional cloning: In biotechnology, DNA inserts and vector molecules are digested with two different restriction enzymes to create non-complementary sticky ends at either end of each restriction fragment. It permits the insert to be ligated to the vector in a specific orientation and stops the vector from recircularizing.

directed assembler: A specific type of assembler that makes use of directed assembly such that the assembly process requires external energy or information input.

directional dominance: A type of dominance in which the majority of dominant alleles show positive effects in one direction.

directional gene frequency: Greater frequency of dominant alleles than that recessive alleles, or vice versa, in a population. This leads to asymmetry in the distribution of allele frequency.

directional mutation theory: This theory predicts that when the mutational bias between A/T and G/C nucleotide pairs is equilibrated with the base composition of a neutral set of DNA nucleotides, the mutation frequency per gene will be much lesser than the frequency observed immediately after the mutator mutation takes place.

directional selection: Refers to the following. (i) Selection resulting in a consistent directional change in the form of a population through time (e.g. selection for larger body size). (ii) Selection favouring one allele over another or favouring increased values of a quantitative trait. It is comparable to positive selection. Selection for or against one extreme of a trait. (iii) Selection resulting in a shift in the population mean in the direction desired by the breeder.

directionality problem of replisome: In all organisms, the protein machinery is responsible for the replication of DNA. The replisome is confronted with directionality problem. The anti-parallel nature of duplex DNA permits the leading-strand polymerase to advance in a continuous fashion but forces the lagging-strand polymerase to synthesize in opposite direction. By extending RNA primers, the lagging-strand polymerase restarts at short intervals and produces Okazaki fragments. At least in prokaryotic systems, this directionality problem is solved by the formation of a loop in the lagging strand of the replication fork to reorient the lagging-strand DNA polymerase so that it advances in parallel with the leading-strand polymerase.

dirty crop approach: Partially resistant multiline where each component of the multiline carries a single gene for resistance but none of the resistances are effective against all known races of the pathogen (D.R. Marshall in 1977). It has been supported as a dynamic, natural biological system for the control of crowd diseases in crop plants. Multilines are a way to stabilize the race structure of the pathogen population.

disarmed *Agrobacterium* Ti plasmids: Oncogenes located in the T-DNA region are replaced by exogenous DNA. The left and right border sequences as well as the nopaline synthase gene of the Ti plasmid are conserved.

disassembler: An instrument able to take apart structures a few atoms at a time, recording structural information at each step. This could be used for uploading, copying objects (with an assembler), a dissolving agent or a weapon. *See* ASSEMBLER.

disassortative mating: Occurs if the plants that are mating resemble each other less than plants belonging to pairs of random plants, with regard to some trait. *Compare with* ASSORTATIVE MATING.

disc gel electrophoresis: A technique in which a gel, usually of agarose or polyacrylamide, is cast in a vertical tube instead of a slab. The nucleic acid or protein bands are resolved as a series of discs in the cylindrical gel. This method is seldom used now mainly because of the difficulty of ensuring that each individual gel is run in an identical manner. However, disc gels and their variants still find application in preparative procedures.

discontinuous character: Variation in which distinct classes can easily be visually recognized (e.g. flower colour, straw length). *Aka* QUALITATIVE CHARACTER.

discontinuous gene: A gene in which the biological information is divided between two or more exons, separated by introns. *Aka* SPLIT GENE.

discontinuous replication: DNA replication in short 5′ to 3′ segments using the 5′ to 3′ strand as a template while going backwards, away from the REPLICATION FORK. *See* OKAZAKI FRAGMENTS.

discontinuous trait: Variation in which discrete classes are easily recognized (e.g. tall vs dwarf plants). *Aka* QUALITATIVE TRAIT.

discontinuous variation: Variation that falls into discrete categories (e.g. flower colour of garden peas).

discordance: In twin studies, if both the individuals do not exhibit the same form of the trait in question, they are said to be discordant. *Compare with* CONCORDANCE.

discovery: In biology, a finding where a biological substance or its properties already existed in nature but were noticed for the first time individually or collectively by humans.

discovery rights: In patenting biological material, selling only the research findings while keeping rights to all the knowledge that is uncovered along the way with the discoverer.

discrete control points in cell progression: There are two points at which a decision may be taken on whether to proceed through another cell cycle. (i) Commitment to chromosome replication occurs in G_1 phase. If conditions to pass a 'commitment point' are satisfied, there is a lag period, and then a cell enters S phase. The commitment point has been defined most clearly in yeast cells, where it is called START. The comparable feature in animal cells is called the RESTRICTION POINT. (ii) Commitment to mitotic division occurs at the end of G_2. If the cell does not divide at this point, it remains in the condition of having twice the normal complement of chromosomes.

discrete distribution: A probability distribution that only reveals a finite or countably infinite number of possible values.

discrete event simulation (DES): A model of the operation of a system as a discrete sequence of events in time. Each event transpires at a particular instant in time and marks a change of state in the system. Between consecutive events, no change in the system is assumed to occur. Thus, the simulation can directly jump in time from one event to the next.

discrete generations: Generations that have no overlapping reproduction. All reproduction takes place between individuals of the same generation.

discrete resistance: The ability to struggle disease or the agent of disease (the vector) and so stay healthy. Resistance or tolerance to disease is a topic of intensive interest in plant tissue-culture work. Screening or selection at the cellular level is followed by plant regeneration and screening of these and their progeny.

discrete variable: A variable that can take only finite, practically small, number of possible discrete values. Parity for a woman is a discrete variable. Deaths in an area with population of 200,000 are theoretically discrete but can be considered continuous for statistical purposes, because it can take a large number of possible values.

disease: A state in which the use or structure of any part of a living organism is not normal. It is a harmful departure from normal operationing of physiological processes. Diseases are often identifiable by specific signs and symptoms. Disease may occur due to external factors such as infectious pathogens or by internal dysfunctions such as immune deficiency or senescence. Six types of causal agents considered as cause of disease include (i) fungi, (ii) bacteria, (iii) viruses, (iv) nematodes, (v) insects and (vi) plant parasites.

disease Δ: Spoken as DISEASE TRIANGLE. Refers to the three conditions – a susceptible host, a suitable pathogen and an appropriate environment – required for a disease to occur.

disease avoidance: Avoiding the disease by growing the crop sufficiently early so that the vulnerable part of the plant's growing cycle is over before the disease-causing organism arrives in the area. This stops the disease from starting. This phenomenon is sometimes referred to as PASSIVE RESISTANCE.

disease control: Several types of disease control include (i) disease resistance, (ii) disease protection, (iii) disease avoidance, (iv) disease exclusion, (v) disease eradication and (vi) disease therapy. *See details as separate terms.*

disease cycle: The sequence of all of events and interactions that occur among the host, parasite and environment that transpire in occurrence of a disease, from initial infection of the plant by a causal agent, through pathogenesis to over-seasoning until another infection occurs.

disease eradication: This pragmatic control measure is applied to a situation in which the disease is present in the area. It involves removing the disease (e.g. by burning all stubble of the diseased crop) in order to prevent transfer of the disease from the previous crop to the next season.

disease escape: For a variety of reasons, some individuals in a screening population may escape pests or disease. This phenomenon is, however, very easily confused with resistance. *Aka* CHANCE ESCAPE.

disease-free: An organism free from pathogen. In agriculture, disease-free planting material is preferred to obtain higher yields and better quality. In seed-propagated crops, the gametes are the sieves against pathogens, and most of pathogens do not pass through seed. However, in case of vegetatively propagated crops, there is no such sieve, and pathogens keep accumulating generation after generation, causing decline in the yield and quality. Therefore, availability of disease-free planting material is the key factor in enhancing the crop production. In this regard, meristem culture is being commercially exploited for generation of disease-free clean planting material in the vegetatively propagated plant species.

disease incidence: Refers to the number of plants affected by a disease within a population.

disease progress curve: A graph representing a measure of disease (i.e. severity) over time.

disease protection: Involves protecting the plant with a chemical. The chemical is applied before the disease starts and prevents the beginning of problem. Systemic fungicides can penetrate and move inside the plant. They therefore have a greater exposure to the pathogenic organism. Some are taken up by roots, others by leaves.

disease pyramid: It describes how disease can ultimately destroy a plant. It is comprised of the offerings of the pathogen that cause the disease, the plant or host and the environmental conditions that set up the pathogen to go after the plant and time.

disease quadrangle: *See* FUNDAMENTAL ELEMENTS FOR DISEASE IN PLANTS.

disease resistance and pathogen-associated molecular pattern (PAMP), link between: There is a direct link between disease resistance and PAMP. Immediately after ligand binding, the PATTERN RECOGNITION RECEPTOR (PRR) forms an active complex with BRASSINOSTEROID INSENSITIVE 1-ASSOCIATED KINASE 1 (BAK1). This results in transphosphorylation of the respective kinase domains of the PRR and BAK1. Signalling through this active complex can be mediated directly by BOTRYTIS-INDUCED KINASE 1 (BIK1), or by MITOGEN-ACTIVATED PROTEIN KINASES (MAPKs) or CALCIUM-DEPENDENT PROTEIN KINASES (CDPKs). Transphosphorylation is involved in initiation of these signals.

disease-resistance plant: Plants are susceptible to viral, bacterial and fungal diseases. Much progress has been made in evolving transgenic plants resistant to viruses. For example, expression of a gene that encodes the coat protein of TMV in transgenic tobacco plants has been shown to cause the plants to resist TMV infection. A number of other viral-resistant plants species have been developed, including squash and potatoes. Genetic engineering of crop plants for resistance to fungal and bacterial infections has been more difficult. However, by studying the protective genes that are expressed in naturally disease-resistant plants, encouraging progress has been made. The proteins encoded by these so-called PATHOGENESIS-RELATED PROTEINS (PR proteins) can, in some cases, provide limited disease protection in transgenic plants. There are several strategies for engineering plants for viral resistance, and these utilize the genes from virus itself (e.g. the viral coat protein gene). The virus-derived resistance has given promising results in number of crop plants (e.g. tobacco, tomato, potato, lucerne and papaya). Some viral-resistant transgenic plants (e.g. papaya resistant to papaya ring spot virus) have been commercialized in some countries. Plants respond to pathogens by inducing a variety of defence responses, like PR proteins, enzymes that degrade/destroy fungal cell wall (chitinase), antifungal proteins and compounds, phytoalexins and so on. Several transgenic crop plants showing increased resistance to fungal pathogens are being raised with genes coding for the different compounds mentioned above.

disease responsive genes: Histone acetylation/deacetylation has an important role in the pathogenic response in plants. HDAC19, involved in several developmental processes, is induced by JASMONIC ACID (JA) and by 1-AMINOCYCLOPROPANE-1-CARBOXYLIC ACID (ACC), an ETHYLENE (ET) precursor. Overexpression of the *HDAC19* gene increases the expression of chitinase and glucanase genes, as well as resistance against the pathogenic fungus *Alternaria brassicicola*.

disease severity: Refers to amount of damage caused by a disease.

disease-associated genes: Alleles carrying particular DNA sequences associated with the presence of disease.

disinfect: To eradicate a pathogen from infected plant tissues.

disinfectant: A chemical treatment used to sterilize seed for planting; it is commonly useful for surface-borne pathogens.

disinfest: To destroy pathogens that have not yet initiated disease or other contaminating microorganisms that occur in or on inanimate objects (e.g. soil or tools) or that occur on the surface of plant parts (e.g. seed).

disjunct: Some event occurring in broadly disconnected geographic areas; distinctly separate. The term applies to a discontinuous range in which one or more populations are separated from other potentially interbreeding populations with sufficient distance so as to prevent gene flow between them.

disjunct populations: Populations whose ranges do not overlap.

disjunction of chromosomes: The parting of homologous chromosomes after they paired and undergone crossing over. The separation of homologous chromosomes occurs during meiotic anaphase I (FIRST-DIVISION SEGREGATION) of meiosis and of chromatids of a chromosome during anaphase II (SECOND-DIVISION SEGREGATION).

disk floret: A floret existing most characteristically in the disk of the capitulum of flowers in the family Asteraceae and to some degree in other plants that bear a flowering head with a disk (e.g. *Scabiosa*, also known as Pincushion flower).

disk storage: A general type of storage mechanisms where data are recorded by various electronic, magnetic, optical or mechanical changes to a surface layer of one or more rotating disks. A DISK DRIVE is a device implementing such a storage mechanism. Notable types of disc drives are the following: (i) HARD DISK DRIVE (HDD) containing a NON-REMOVABLE DISK, (ii) FLOPPY DISK DRIVE (FDD) and its REMOVABLE FLOPPY DISK and (iii) various OPTICAL DISC DRIVES (ODD) and ASSOCIATED OPTICAL DISC MEDIA. *Aka* DRIVE STORAGE.

dislocation: The movement of a chromosome segment away from its original location in the chromosome.

disome: A cell or an individual possessing two homologous sets of chromosomes. *Aka* DIPLOID *or* DISOMIC.

disomic: An individual with euploid chromosome complement.

disomic alien addition line: A line with an extra chromosome pair from a related species.

disomic haploid: An aneuploid that is disomic for one chromosome out of the total haploid complement; (x + 1, 2x + 1, 3x + 1). It is derived from a tetrasomic.

disomic inheritance: Normal inheritance dependent upon two or more alleles for a gene, customarily found in diploid species with no extra chromosomes or chromosome sets (genomes).

disorder: Any deviation from the normal structure or function of any part, organ or system of the body that is manifested by a characteristic set of symptoms and signs whose pathology and prognosis may be known or unknown. Also sometimes known as DISEASE.

dispermy: The inflowing of two sperm cells into one egg cell.

dispersal: The spread of infectious material (inoculum) from diseased plants to healthy plants. *Aka* DISSEMINATION.

dispersion: The degree of variability or scatter of the values of a variable from the mean when measured for different subjects or at different times.

dispersive replication: A postulated mode of DNA replication combining aspects of conservative and semi-conservative replication; known to be incorrect.

displacement loop: *See* D-LOOP.

disposable soma theory: The somatic cells of an organism are expendable. Ageing results from destruction caused by molecules (including oxygen-free radicals) produced in the normal course of living. When these wear and tear processes outpower the repair systems, ageing occurs.

disposal: In this guide, getting rid of waste by discarding it into a bin and, when it is collected, by incinerating it or sending it to landfill.

disputed inheritance: Between 1856 and 1863, an Austrian monk named Gregor John Mendel performed a succession of ground-breaking experiments by breeding plants. Using garden peas (*Pisum sativum*), he established that traits such as seed colour and plant height are transmitted to the next (F_2) generation in a 3:1 ratio (three plants with yellow peas for everyone with green peas, for instance). Mendel anticipated that unknown particles, which he called 'factors' or 'elements', were responsible for this pattern. The type of factor inherited by each offspring determined whether it would have the 'dominant' majority or 'recessive' minority trait. The monk died before the term GENE was coined, but his theories came to be known as Mendelian genetics (B.K. Hall in 2023). From mid to late 19th century, divergent thinking about evolution and inheritance was prevailing at the time. Mendel's conclusions drew attention after they were reproduced by a trio of botanists (Hugo de Vries, Carl Correns and Erich von Tschermak) independently in 1900. The focus of many biologists turned to a newly proposed term 'gene' in 1909 to replace terms such as 'factor' or 'element'. The Mendelian way of thinking left no place for the influence of the environment, natural selection and other considerations in heritability. William Bateson at the University of Cambridge was an influential advocate for Mendelian genetics. But Raphael Weldon at the University of Oxford viewed variation in traits as a spectrum – influenced by genes, yes, but also by development, the environment and the ancestral history of each species. Since Weldon's death, many biologists have pursued to find alternatives to Mendelian genetics. In *Disputed Inheritance: The Battle over Mendel and the Future of Biology*, published by University of Chicago Press in 2023, historian of science Gregory Radick provided a scholarly perceptive analysis of why Mendelian genetics has long conquered our approach to understanding inheritance, despite valiant attempts to propose alternatives. Conrad Waddington in 1942 proposed that traits can change before genes do. This occurs, for instance, when fruit flies (*Drosophila melanogaster*) exposed to ether develop a second thorax, complete with wings, where a wingless abdominal segment should be. Selective breeding of these flies produced offspring that developed a body with two winged segments without further ether exposure. Waddington named this phenomenon as 'epigenetic inheritance'. The word epigenetic means 'above the gene'. That this idea was rejected for decades is a testament to the tenacious hold of Mendelian inheritance. It is now evident that epigenetic inheritance occurs because of changes that affect gene expression without altering the DNA sequence, such as the addition of heritable molecular 'tags' to histone proteins around which the DNA is wrapped. These tags can be induced by environmental changes. This development brings us back to W. Weldon's emphasis on the environment. The discovery of epigenetic inheritance reveals a more complete theory of inheritance than William Bateson would have approved.

disruptive selection: Selection for both extremes of a trait and against the middle. In the case of polygenic traits that are expressed as a continuum of phenotypes (e.g. human stature), it would be selection for both very tall and very short people and against those who are average in height. The result would be a progressive increase in both of the extreme forms of this trait and a reduction in the middle range. In the case of a trait controlled by only two alleles, it would be selection or fixation for both recessive and dominant homozygotes and against heterozygotes. The result would be a progressive reduction in the number of people who are heterozygous for the trait in the population. *See* DIRECTIONAL SELECTION *and* STABILIZING SELECTION.

disruptive technology: Technology that is significantly cheaper than current, is much higher performing, has greater functionality, is frequently more convenient to use and will revolutionize markets by superseding existing technology. 'Paradigm shifting' is a well-worn connotation. Although the term may sound negative to some, it is in fact neutral. It is only negative when businesses who are unprepared for change fail to adapt, only to fall behind and fail. The results are not evolutionary; they are revolutionary.

dissected: Deeply divided; cut into many segments.

disseminate: Usually refers to long-distance distribution and spread usually refers to local distribution. The dissemination of pathogenic organisms can also be accomplished by wind, rain, irrigation, contaminated seeds and transplants. A few pathogens have the ability to move short distances on their own. Nematodes, zoosporic fungi, oomycetes and some bacteria can move from host to host if they are close enough to one another and the conditions are favourable.

dissemination: The action or process of spreading or sending out freely or widely, as though sowing or strewing seed; the state of being dispersed throughout in small particles.

dissepiment: A barrier or septum in a plant part, usually denoting to septa between the loculi of capsules or of other fruits with multiple partitions.

dissipative DNA nanotechnology: DNA systems that incorporate a chemical fuel-to-waste conversion. Chemical energy stored in the fuel is used to drive the system out of equilibrium.

dissociation constant (Kd): Measure of the tendency of a complex to dissociate. For components A and B and the binding equilibrium $A + B \rightleftharpoons AB$, the dissociation constant is given by [A][B]/[AB], and it is smaller the tighter the binding between A and B. *Compare with* ASSOCIATION CONSTANT.

distal protein: With reference to the assembly of complex structures (e.g. ribosomes), there are proteins that bind late in the assembly process and whose binding is dependent on the prior binding of proximal proteins. *See* PROXIMAL PROTEIN.

distance: In taxonomy, refers to the quantitatively measured difference between the phenetic appearance of two groups of individuals, such as populations or species (PHENETIC DISTANCE) or the difference in their gene frequencies (GENETIC DISTANCE).

distance matrix: A table presenting the evolutionary distance between sets of operational taxonomic units(e.g. species or genes).

distance methods: In these methods of phylogenetic analysis, all possible pairs of sequences are aligned to determine which pairs are the most similar or closely related. These alignments provide a measure of genetic distance between the sequences. These distance measurements are then used to predict the evolutionary relationship.

distichous: Arranged in two opposite rows and hence in the same plane.

distinctness, uniformity and stability (DUS) testing: Refers to the procedures and standards of the identification and description of varieties, where D = distinctness, U = uniformity and S = stability. This is elaborated by different teams of an international organization, the Union for Protection of New Varieties of Plants (UPOV) established for the protection of plant varieties. The so-called TG/01/2/1/ contains the most important prescriptions.

distinctness, uniformity and stability (DUS) testing, characteristics involved in: The characteristics implicated in DUS testing can be separated into two main types: (i) MEASURED or QUANTITATIVE CHARACTERISTICS and (ii) VISUALLY OBSERVED, BONITATED or QUALITATIVE CHARACTERISTICS. The measured characteristics have values on the continuous scale. Visually observed data have values in the interval 1, 2 to 9. In other words, visually observed traits are of the so-called 'ordinal type', while measured data are on the so-called 'interval or ratio scale'. Phenological data must also be considered a special type of measured characteristic (e.g. number of days until flowering).

distinctness, uniformity and stability (DUS) testing, legal protection: For the legal protection of varieties, the candidates (i) have to have names that have not yet been used for the registration of recognized varieties, (ii) must have been discovered distinct of the other 'known' varieties (D), (iii) have to be uniform and homogeneous (U) and (iv) have to be stable (S).

distinguishable hybrid: A type of hybrid in which intermediate inheritance is phenotypically expressed(i.e. the heterozygous gene constitution is evident from the phenotype).

distributed computing: A field of computer science that deals with distributed systems. A distributed computing is a system whose components are located on different networked computers that communicate and coordinate their actions by passing messages to one another. The components interact with one another in order to achieve a common goal. Three significant characteristics of distributed systems are: concurrency of components, lack of a global clock and independent failure of components. Examples of distributed systems vary from SOA-based systems to massively multiplayer online games to peer-to-peer applications.

distributed intelligence: An intelligent entity that is distributed over a large volume (or inside another system, like a computer network) with no distinct centre. This is the opposite to the strategy of CONCENTRATED INTELLIGENCES. Distributed intelligences have much longer communications lags but are more flexible in their structure and can survive damage to their parts.

distribution: *See* STATISTICAL DISTRIBUTION.

distribution function: A graph of some precise quantitative measure of a character against its frequency of occurrence.

distributive DNA replication: *See* DISPERSIVE REPLICATION.

distyly: A polymorphism possessing two different arrangements of anther and stigma that encourages outcrossing. Flowers of some species occur in two forms that differ only by the length of the style and stamens, and flowers of only one of these forms appear on any one plant. For example, the two crosses, pin × pin and thrum × thrum, are incompatible due to alleles at a single locus. The thrum morphology is controlled by a dominant allele *S* and the pin morphology by the recessive alleles. *Compare with* HETEROSTYLY.

disulfhydryl bridge: A particular type of covalent bond, incorporating two sulfur atoms that can form between two cysteines in a single or on different polypeptides.

disulfide (S-S) bond: Covalent linkage moulded between two sulfhydryl groups on cysteines. For extracellular proteins, a common way of joining two proteins together or linking different parts of the same protein. S-S bonds are formed in the endoplasmic reticulum of eukaryotic cells. In this way, S-S bonds contribute to holding proteins together and also help provide the internal structure (conformation) of the protein. *Aka* DISULFIDE BRIDGE *or* DISULFIDE LINKAGE.

ditelocentric: *See* DITELOSOMIC.

ditelomonotelosomic: A cell or an individual that possesses a pair of telocentric chromosomes for one arm and a single telocentric chromosome for the other arm.

ditelosomic: A cell or an individual that has possesses telocentrics of one chromosome arm.

ditelotrisomic: A cell or an individual that possesses two telocentric chromosomes of one arm plus one complete chromosome of the homologue.

ditertiary compensating trisomic: A cell or an individual with a compensating trisomic chromosome in which a lost chromosome is compensated by two tertiary chromosomes.

ditype: Two types of ascospores in the same ascus sac.

diurnal rhythm: *See* CIRCADIAN RHYTHM.

diurnal plants: Having leaves or flowers that open in daylight and close at night. The morning glory and crocus are examples of diurnal plants.

divalent: *See* BIVALENT.

divaricate: In biology, refers to mean branching or having separation or a degree of separation. The angle between branches is wide.

divergence: (i) Related species no longer resemble each other because evolution has adapted them to different ecological conditions. (ii) The attainment of differences after evolutionary separation (e.g. of species).

divergent: Spreading in different directions, generally upward.

divergent evolution: A kind of evolutionary change that leads to increasing morphological difference between initially more similar lineages.

divergent multigene family: Different members of the family have originated from concerted evolution to the construction of divergent sets of duplicates (e.g. immunoglobulin genes, globin genes, actin genes, albumin genes, α-fetoprotein genes, serine protease genes, the interferon genes (for defence against viral infection) genes and chorion protein-making genes).

divergent transcription: Transcription initiation by RNA polymerase II is generally assumed to occur unidirectionally from most of the genes. The initiation of transcription occurs at two promoters facing in the opposite direction, so that transcription proceeds away in both directions from a central region. TRANSCRIPTION START SITE-ASSOCIATED RNAs (TSSA-RNAs) non-randomly border active promoters, with peaks of antisense and sense short RNAs at 250 nucleotides upstream and 50 nucleotides downstream of TSSs, respectively. Divergent transcription over short distances is commonly observed for active promoters. *Compare with* CONVERGENT TRANSCRIPTION.

diversifying selection: Selection in which two or more genotypes reveal optimal adaptiveness in different sub-environments.

diversity: This is the most obvious characteristic of the natural world. For example, there are many types of objects and organisms.

diversity (within a species): Refers to the genetic variation that occurs within a population (of organisms) in a species. This diversity is the result of one or more SINGLE-NUCLEOTIDE POLYMORPHISMS (SNPs) in each individual's genome (DNA) within the population of organisms.

diversity arrays technology (DarT) marker: A DarT marker is a segment of genomic DNA, the occurrence of which is polymorphic in a defined genomic representation. DarT markers are diallelic and perform in a dominant (present vs absent) or codominant (two doses vs one dose vs absent) manner. To identify the polymorphic markers, a complexity reduction method is used on the metagenome, a pool of genomes representing the germplasm of interest. The genomic representation accomplished from this pool is then cloned, and individual inserts are arrayed on a microarray resulting in a 'DISCOVERY ARRAY'. Labelled genomic representations prepared from the individual genomes included in the pool are hybridized to the discovery array. Polymorphic clones (DarT markers) show variable hybridization signal intensities for different individuals. These clones are subsequently assembled into a 'GENOTYPING ARRAY' for routine genotyping. The method was developed to provide a practical and cost-effective whole-genome fingerprinting tool. DarT has three key attributes of interest to plant breeders and scientists studying and managing genetic diversity: (i) it is independent from DNA sequence, (ii) the genetic scope of analysis is defined by the user and easily expandable and (iii) the method provides for high-throughput and low-cost data production. The discovery of polymorphic DarT markers and their scoring in subsequent analysis does not necessitate any DNA sequence data.

diversity arrays technology (DarT): A procedure that can detect and type DNA variation at several hundred genomic loci in parallel without relying on sequence information. It can be effectively applied to genetic mapping and diversity analyses of barley, a species with a 5000-Mbp genome. DarT has been used in genetic mapping and fingerprinting studies in wheat, Arabidopsis and barley. It is a high-throughput, quick and highly polymorphic and reproducible method. Their dominant inheritance is still a limitation for mapping.

diversity panel: A collection of genetic material or individual samples taken from a diverse population of a certain species. The idea is to illustrate the genetic and phenotypic diversity of the species. Diversity panels exist for human populations, mouse and other organisms.

divide-and-conquer algorithm: An algorithm design paradigm which is founded on multi-branched recursion. A divide-and-conquer algorithm works by recursively breaking down a problem into two or more subproblems of the same or related type until these become simple enough to be solved directly. The solutions to the subproblems are then combined to give a solution to the original problem. 'Divide-and-conquer' algorithm thus consists of the following three steps: divide, solve and combine.

divided gene: A gene whose exons are located far away in the chromosome, such as a gene for ribosomal protein of smaller subunit (12 reps) in chloroplast. This gene consists of three exons, with five exons located far away from the other two in all land plants. Mature RNA from this gene is supposed to be produced by *trans*-splicing.

divided operons: When structural genes of an operator are located at different positions in the chromosome, the operon is said to be divided. Each of these structural genes lies next to its operator. Examples include tryptophan (*trp*) operon, arabinose (*ara*) operon and arginine (*arg*) operon.

divine creation: *See* SPECIAL CREATION.

division: A taxonomic rank below kingdom in the standard taxonomic hierarchy, generally used only for plants as the approximate botanical equivalent of the term 'phylum', which is used for animals and other kingdoms.

divisive introns: An intron which arises as a divisive element separating a previously contiguous exon into two halves. Here, random insertions of DNA (bearing rudimentary splice signals) get incorporated into previously intact genes. *Aka* TYPE B INTRONS. Such introns could arise by aberrant recombination, retrovirus or transposon integration, or retrovirus-mediated insertion of cellular mRNA sequences.

dizygotic twins (DZ): Twins that developed from two separate fertilized eggs and therefore related in the same way as siblings. *Aka* FRATERNAL TWINS *or* NON-IDENTICAL TWINS.

dN: An alternative notation for non-synonymous substitution rate (Ka).

dN/dS ratio: In molecular phylogenetic studies, the ratio of the number of non-synonymous nucleotide substitutions (dN) to the number of synonymous nucleotide substitutions (dS) is determined. In the case of functionally important (or otherwise constrained) genes, dN is expected to exceed dS (Dn/Ds <1), because most amino acid changes will disrupt protein structure and those non-synonymous substitutions (dN) causing them will not be maintained. In a non-functional pseudogene, there will be no discrimination between them, and equal numbers of dN and dS are expected (Dn/Ds = 1); when natural selection is acting to favour changes at the amino acid level, it is predicted that dN will exceed dS, hence a high dN/dS ratio will be observed. In classical loci, in the peptide binding regions (allele-specific sequences), because of heterozygote advantage/frequency-dependent selection, there is always a high dN/dS ratio (>1), whereas in the remainder of the gene, dN/dS <1 (due to functional constraints). This approach thus suggests that balancing selection is acting on peptide binding regions.

DNA A helicase: An *Escherichia coli* enzyme which binds to replication origin site *OriC*. 20–30 molecules are required for the formation of DNA replication initiation complex.

DNA adenine methylase (dam) system: An enzyme that adds a methyl group to the adenine of the sequence 5′-GATC-3′ in newly synthesized DNA. Immediately after DNA synthesis, the daughter strand remains unmethylated for a short time. It is an orphan methyltransferase system that is not part of a restriction-modification system and regulates gene expression.

DNA adenine methyltransferase identification (DamID): A protocol used to map the binding sites of DNA- and chromatin-binding proteins in eukaryotes. DamID detects binding sites by expressing the proposed DNA-binding protein as a fusion protein with DNA methyltransferase. Analogously to chromatin immunoprecipitation (ChIP), genomic regions bound by a protein of interest can be isolated and used to probe a microarray to determine binding site occupancy. Unlike ChIP, DamID does not require antibodies but makes use of adenine methylation near the protein's binding sites to selectively amplify those regions, introduced by expressing minute amounts of protein of interest fused to bacterial DNA adenine methyltransferase.**DNA affinity chromatography:** A technique used for purifying sequence-specific DNA-binding proteins by their binding to a matrix to which the appropriate DNA fragments are attached.

DNA amplification fingerprinting (DAF): PCR-based DNA marker for fingerprinting. This method makes use of single primer. The length of the primer sequence is around 6–8 nucleotides. It is similar to RANDOM AMPLIFIED POLYMORPHIC DNA (RAPD), but in DAF, using smaller primers, the number of amplified products increases many folds, which make DAF more useful for DNA fingerprinting. DAF is dominant marker, but the possibility of being codominant cannot be ruled out. It is simple and does not require labour-intensive and expensive procedure as required by RFLP. Moreover, a very small quantity of genomic DNA is required, and there is no need of developing genomic/cDNA libraries.

DNA analysis by restriction enzyme for simultaneous detection of multiple epigenomic states (DARESOME): 5-methylcytosine ($5^{m}C$) and 5-hydroxymethylcytosine ($5^{hm}C$) are the most abundant DNA modifications that play roles in gene regulation. Current methods are unable to simultaneously measure both these epigenetic modifications, particularly in samples with limited quantities. This limitation hampered detailed studies of these different epigenetic marks aimed at understanding their combined effects and dynamic interconversion. R. Viswanathan and others in 2023 present DARESOME, an assay based on modification-sensitive restriction digest and sequential tag ligation that can concurrently perform quantitative profiling of unmodified cytosine, $5^{m}C$ and $5^{hm}C$ in CCGG sites genome-wide. Observations reveal the opposing roles of $5^{m}C$ and $5^{hm}C$ in gene expression regulation as well as their interconversion during ageing in a mouse brain. Pronounced $5^{hm}C$ strand bias reflects the semi-conservative replication of DNA.

DNA and histone methylation in recombination: A meiotic recombination model predicts half-methylated sites in sperm and oocyte DNA. By restriction analysis, it has been observed that the DNA of sperm is indeed methylated differently from somatic DNA. *Arabidopsis* chromosomes are highly DNA methylated in the repetitive centromeres, which are also crossover-suppressed. RNA-directed DNA methylation is sufficient to locally silence *Arabidopsis* euchromatic crossover hotspots and is associated with increased nucleosome density and H3K9me2. Loss of CG DNA methylation maintenance in *met1* triggers epigenetic crossover remodelling at the chromosome scale, with pericentromeric decreases and euchromatic increases in recombination. Crossover remodelling is driven by loss of CG methylation within the centromeric regions. DNA methylation is sufficient to silence crossover hotspots and plays a key role in establishing domains of meiotic recombination along chromosomes.

DNA array: A group of technologies in which specific DNA sequences are either deposited or synthesized in a 2D (or sometimes 3D) array on a surface in such a way that the DNA is covalently or non-covalently attached to the surface.

DNA as genetic material, evidence of: Evidence for DNA as genetic material have come from (i) TRANSFORMATION, (ii) TRANSDUCTION and (iii) VIRAL INFECTIVITY experiments. Transformation and transduction displayed that DNA is the genetic material in bacteria. Viral infectivity experiment established that genetic material of phage T_2 is DNA, not protein. *See details as separate terms.*

DNA as the genetic material, transformation as evidence for: Transformation experiments conducted in 1928 by F. Griffith displayed that bacteria *Diplococcus pneumonia* can transfer genetic information through a process called transformation. However, it failed to explain the biochemical nature of genetic material. A group of scientists, Oswald Avery, Colin MacLeod and Maclyn McCarty (1944), continued the Griffith experiment in search of biochemical nature of the hereditary material. Their discovery revised the concept of protein as genetic material to DNA as genetic material.

DNA B helicase: An enzyme that causes unwinding of DNA double helix prior to replication in phage and *Escherichia coli*; this enzyme is localized in these systems by proteins P and DnaC, respectively.

DNA banding: A topological state in which a DNA molecule is vent so that two distant sites on the double helix are brought close together.

DNA bank: A service that stores DNA extracted from blood samples or other human tissue.

DNA banking: In addition to the *in situ* and *ex situ* conservation of plant genetic resources, DNA banking is a complementary conservation strategy and an effective tool for more efficient management of gene banks and conservation research. However, it is still not a common practice, since technical capacity and financial resources are currently limited, particularly in developing countries. DNA banks can be the libraries for reference sequences that can be utilized to measure genetic diversity and relative change within populations or to identify samples when genotypes mix up morphological-based identification. DNA is an unusually stable biomolecule that often outlasts the organism it encodes. DNA banks provide an accessible means for genetic characterization and electronic integration of many existing biomaterial collections.

DNA barcoding: A taxonomic method that uses a short genetic marker in an organism's DNA to identify it as belonging to a particular species. It differs from molecular phylogeny in that the main goal is not to determine classification but to identify an unknown sample in terms of a known classification. Applications include, for example, identifying plant leaves even when flowers or fruit are not available, identifying insect larvae (which typically have fewer diagnostic characters than adults), identifying the diet of an animal based on stomach contents or faeces and identifying products in commerce (e.g. herbal supplements or wood).

DNA base sequence: A chain of repeating units of deoxyribonucleotides (adenine, guanine, cytosine and thymine) arranged in a particular pattern.

DNA box: Refers to a short nucleic acid consensus sequence or motif that is universal within kingdoms of organisms. Examples of DNA boxes are the PRIBNOW BOX (TATAAT) for RNA polymerase, the HOGNESS BOX (TATA) that has a similar function in eukaryotic organisms, and the HOMEOBOX.

DNA bricks: A Lego-like DNA block that is used in DNA technology to build 2D and 3D nanostructures.

DNA bridges: Large segments of DNA whose sequence is known and mapped in toto. These sequences are then employed by scientists to bridge the DNA segments and thereby assemble a more complete map of an organism's chromosome or genome.

DNA bubbles: The nucleic acid configuration during replication in eukaryotic chromosomes or the shape of HETERODUPLEX DNA at the site of a deletion or insertion.

DNA C: An *Escherichia coli* protein required to localize DNA replication proteins.

DNA characterizations: (i) G.J. Mendel's work in 1865 led to particulate nature of unit of heredity. (ii) In 1950, it was known that genetic inheritance had a mathematical description. (iii) Researchers in Sir Lawrence Bragg's laboratory at Cambridge University made X-ray diffraction pictures of various molecules. (iv) Using clues that were painstakingly assembled over the course of decades, beginning with its chemical composition, it was determined that it should be possible to characterize the physical structure of DNA and X-ray images would be the vehicle.

DNA chimera: One DNA molecule composed of DNA extracted from two different species.

DNA chip: A purpose-built microchip used to identify mutations or alterations in a gene's DNA. *Aka* GENE CHIP *or* DNA MICROCHIP. *See* DNA CHIP TECHNOLOGY.

DNA chip technology: *See* MICROARRAYS.

DNA chips and functional genomics: The collection of genes that are expressed or transcribed for genomic DNA are referred to as expression profile or the transcriptome. Microarrays are used to: (i) understand the function of genes knowing when, where and to what extent is a gene expressed; (ii) find expression markers by determining the causes and consequences of diseases (e.g. by analysing multiple samples obtained from individuals with or without acute leukaemia gene expression was discovered that can be used in classification of these cancers); (iii) understand mechanism of action of drug candidates in cell; and (iv) finding potential therapeutic drug targets. Gene expression behaviour is observed over multiple experiments to first cluster genes together into groups. Genes with similar expression behaviour (e.g. increases or decreases together under similar circumstances) are likely to be related functionally.

DNA circularization: A DNA fragment generated by digestion with a single restriction endonuclease will have complementary 5′ and 3′ extensions (STICKY ENDS). If these ends are annealed and ligated, the DNA fragment will have been converted into a covalently closed circle, thus becoming circularized.

DNA classification based on centrifugation density gradient: When eukaryotic DNA is centrifuged on a density gradient machine, two types of DNA bands may be distinguished on the basis of optical absorbance – MAIN BAND DNA and SATELLITE DNA. Sometimes, an additional smaller peak(s) is/are seen at a different value. This material is called satellite band and the DNA in this band is termed SATELLITE DNA.

DNA classification based on sequences: Based on their sequences, DNA can assume specific structural configurations, (e.g. CRUCIFORM DNA, TRIPLE HELICAL DNA, QUADRUPLEX DNA, SLIPPED DNA, SINGLE-STRANDED DNA and SUPERCOILED DNA). *See details as separate terms.*

DNA classification based on the number of strands: On the basis of number of strands, DNA can be classified as follows: (i) DOUBLE-STRANDED DNA (as found in many viruses, bacteria and all organisms) or (ii) SINGLE-STRANDED DNA (as found in ϕX174, s13, F1 and M13 viruses). TRIPLEX DNA and QUADRUPLEX DNA structures are also known. *See details as separate terms.*

DNA classification based on the shape of the DNA molecule: DNA can be classified as (i) CIRCULAR DNA (as in most of the viruses (e.g. SV40) and bacteria (e.g. *Escherichia coli*, *Salmonella typhimurium*)) or (ii) LINEAR DNA (as in phage T7 and eukaryotic chromosomes).

DNA cleavage of unmodified complexes: With unmodified complexes, ATP triggers a transformation of *Eco*K* to a new form *Eco*K$^+$. *Eco*K$^+$ loosens enzyme-bound S-adenosine methionine (AdoMet or SAM), and it forms a complex that has smaller diameter. ATP serves as an energy source for DNA cleavage.

DNA clone: A section of DNA that has been inserted into a vector molecule (e.g. a plasmid or a phage chromosome) and then replicated to form many copies.

DNA cloning: In order to be useful, the recombinant DNA requires to be replicated many times (i.e. cloned). Cloning can be done *in vitro*, via the PCR, or *in vivo* (inside the cell) using unicellular prokaryotes (e.g. *Escherichia coli*), unicellular eukaryotes (e.g. yeast) or mammalian tissue-culture cells.

When a fragment of DNA cloned happens to be a gene, the term used is GENE CLONING or MOLECULAR CLONING.

DNA computing: A developing branch of unconventional computing that uses DNA, biochemistry and molecular biology hardware instead of the traditional electronic computing. Research and development in this area deals with theory, experiments and applications of DNA computing. This field originally began with the demonstration of a computing application by Len Adleman in 1994.

DNA concatemer: A DNA molecule that comprises a series of smaller DNA molecules linked head to tail. Concatemers are formed during the replication of some viral and phage chromosomes.

DNA concatenates: Multiple circular DNA molecules linked in a chain-like arrangement. Concatenates can arise as a result of the normal processes of DNA replication with circular genomes and cannot be separated without strand breakage. Plasmids in bacteria and kinetoplast DNA in trypanosomes frequently form concatenates.

DNA concatenation: A single, large DNA molecule composed of two or more phage chromosomes.

DNA condensation: The nucleocapsid (NC) protein NCp7 of the immunodeficiency virus type 1 is a small basic protein with two zinc-finger motifs. NCp7 has key roles in virus replication and structure, which rely on its interactions with nucleic acids. Although most interactions involve RNAs, binding to the viral DNA is thought to be important to achieve protection of the DNA against cellular nucleases and its integration into the host genome. The histone-like property of NCp7 leads to DNA condensation and thus seems to contribute to viral DNA stability *in vivo*.

DNA conductivity: Electric current can pass through DNA molecules.

DNA content: Usually, the total DNA amount per nucleus, given as picograms.

DNA cross-linking: Interstrand thymines of DNA form dimers thus blocking replication.

DNA cytosine methyltransferase (dcm) system: DNA cytosine methyltransferase methylates the second cytosine in 5′CCWGG3′ sequences. The *dcm* gene is in an operon with the *vsr* gene which codes for a protein that repairs T:G mismatches caused by deamination of $5^{m}C$.

DNA damage: Any modification that alters the chemical structure of DNA double helix. DNA damage can occur at the bases, at the sugar or affect the integrity of sugar-phosphate backbone. Mutation is defined as a sudden change in the DNA at a particular locus in an organism. The main types of chemical damage arise spontaneously from endogenous processes, such as mitochondrial respiration leaking out oxygen radicals (superoxide and peroxides). These can cause oxidation, alkylation or hydrolysis of bases. At sites of base loss, the deoxyribose residues exist in equilibrium between the closed furanose form and the open aldehyde form. The 3' phosphodiester bonds are readily hydrolysed by a β-elimination reaction producing single-strand breaks. Further, double-strand breaks (i.e. two nearby single-strand breaks) can be caused by replication fork collapse, endonuclease activity and radiation.

DNA damage and ageing: DNA damage may reduce the expression of selectively vulnerable genes involved in learning, memory and neuronal survival, initiating a programme of brain ageing that starts early in adult life.

DNA damage checkpoint pathway: The checkpoint kinases transduce genomic stress signals to halt cell-cycle progression and promote DNA repair. ATAXIA TELANGIECTASIA MUTATED (ATM) and ATM and RAD3-RELATED (ATR)-INTERACTING PROTEIN (ATRIP) has been identified that is phosphorylated by ATR, regulates ATR expression and is an essential component of the DNA damage checkpoint pathway. ATRIP and ATR are mutually dependent partners in cell-cycle checkpoint signalling pathways.

DNA damage checkpoints: The DNA damage checkpoints, like other signal transduction pathways, have four components: damage sensors, mediators, signal transducers and effectors. A brief description of components of DNA damage checkpoints follows.

DNA damage removal: To repair damage to one of the two helical domains of DNA, the other strand must remain intact so that a replacement of damaged information can be made by the information from the undamaged copy. This type of DNA damage is referred to as single-strand damage. The cells that divide have an additional means of DNA repair via DNA polymerases. Cells that do not divide (such as brain and heart cells) cannot use this important DNA repair mechanism. Excision and mismatch repair are the mechanisms that remove DNA damage.

DNA damage response (DDR): The processes by which cells repair damage to DNA and coordinate repair with cell-cycle progression. Structural changes to DNA severely affect its functions, such as replication and transcription. DDR results in activation of complicated and entangled network of DNA repair mechanisms, including multiple DNA repair pathways, damage tolerance processes and cell-cycle checkpoints safeguard genomic integrity. DDR is a chromatin-associated process that is generally tightly controlled in time and space.

DNA damage response, early: RNF8, an forkhead associated (FHA/ Really Interesting New Gene (RING) domain-containing protein, played a critical role in the early DNA damage response. RNF8 facilitates the accumulation of checkpoint mediator proteins BRCA1 and 53BP1 to the damaged chromatin, on one hand through the phosphor-dependent FHA domain-mediated binding of RNF8 to MDC1, on the other hand via its role in ubiquitylating H2AX and possibly other substrates at damaged sites.

DNA damage sensors: DNA damage checkpoints require the recognition of DNA damage to initiate subsequent events. For example, REPLICATION PROTEIN A (RPA) is an important damage sensor for both nucleotide excision repair and the DNA damage checkpoints.

DNA damage tolerance: DNA damage tolerance (DDT) has been employed loosely to include a collection of mechanisms by which cells survive replication-blocking lesions with or without associated genomic instability.

DNA damage-related transcription factors: Failure of cells to respond to DNA damage is a primary event associated with mutagenesis and environmental toxicity. Genome-wide binding locations for 30 damage-related transcription factors (TFs) after exposure of yeast to METHYL METHANESULFONATE (MMS) were measured to map the transcriptional network controlling the damage response. The resulting 5272 TF-target interactions revealed extensive changes in the pattern of promoter binding and identified damage-specific binding motifs.

DNA delivery for chloroplast transformation: Different methods used so far for chloroplast transformation are called GENE GUN. *Aka* BIOLISTIC METHOD, BOMBARDMENT METHOD, POLYETHYLENE GLYCOL-MEDIATED TRANSFORMATION, MICROINJECTION *or* ELECTROPORATION.

DNA demethylases (DEMs): Glycosylases that nick the DNA, leading to repair processes that replace methylated cytosines with unmethylated cytosines.

DNA demethylation: Removal of methyl group from modified cytosine. DNA demethylation occurs through two different pathways: active and passive. *See* ACTIVE DNA DEMETHYLATION *and* PASSIVE DNA DEMETHYLATION.

DNA denaturation/renaturation: When double-stranded DNA is exposed to heat or treated with some denaturants, DENATURATION (separation of DNA strands) takes place. DNA becomes single stranded. When after heat treatment, DNA is allowed to cool down or in case of treatment with a denaturant, the denaturant is removed, RENATURATION, also called REANNEALING (repairing), of DNA strands commences.

DNA denaturing polyacrylamide gels: To look at smaller DNA molecules with much higher resolution, DNA is denatured via heat and run it through a thin polyacrylamide gel that is also kept near the denaturing temperature. These gels usually contain additional denaturing compounds (e.g. urea). *Aka* DNA DENATURING GELS *or* DNA-SEQUENCING GELS. *Compare with* NATIVE GELS.

DNA detection using silver stain method: Silver stain demonstrates a reproducible curvilinear relationship between silver density and the amount of protein or DNA over an averaged concentration range from 1 to 300 ng for proteins and 10 to 710 ng for DNA.

DNA double helix: DNA is a macromolecule that consists of two polynucleotide strands – composed of units called NUCLEOTIDES – twisted around a common axis in a double helix shape. The DNA double helix looks like a twisted ladder due to the way the components of the nucleotides bind together. Pyrimidines pair with purines, cytosine and guanine are joined by three hydrogen bonds, and adenine and thymine are joined by two hydrogen bonds.

DNA effectors: Working in coordination with various structural proteins, DNA effectors change the three-dimensional structure of DNA and allow for compression/condensation and relaxation of the DNA double helix.

DNA elasticity: DNA can bend itself to allow DNA-protein and protein-protein interactions involved in various biological phenomenon (e.g. replication, transcription, transposition, repair, gene regulation).

DNA end extension: The single-stranded DNA tail found at the end of a restriction fragment. Different restriction endonucleases generate extensions at either the 5′ or 3′ end of the DNA strand.

DNA end ligation: The ends of DNA fragments are joined by the formation of phosphodiester bonds between the 3′-hydroxyl of one DNA terminus with the 5′-phosphoryl of another.

DNA erasers: Enzymes that remove chemical tags from modified DNA. For example, all the DNA demethylase enzymes function as DNA erasers.

DNA excision repair: The removal or excision of the dimer and the subsequent repair of the damaged DNA strand is referred to as the excision repair system and requires the action of several enzymes. EXCISION REPAIR is *aka* DARK REPAIR.

DNA excision repair, pre-excision step in: An intermediate step between incision and excision has been suggested. The *uvrC* gene product prevents resealing of the correndonuclease II incision before the excision of the pyrimidine dimer.

DNA experiments: J.D. Watson and F.H.C. Crick presented an initial (and incorrect) proposal for the structure of DNA to a team from Kings College – Rosalind Franklin, Maurice Wilkins and Raymond Gosling. R. Franklin – immediately pointed out the flaws that concerned the water content. Later, J.D. Watson saw R. Franklin's detailed X-ray diffraction images, which revealed an X shape, and became sure that the structure of double-stranded DNA was helical. This rekindled J.D. Watson and F.H.C. Crick's model building in 1953 and led to the correct structure. E. Chargaff's rules helped J.D. Watson and F.H.C. Crick in building of double-helical model of DNA.

DNA extension: The 5′ and 3′ extensions are ssDNA regions at the ends of ds-DNA. Sometimes, they are called OVERHANGS or STICKY ENDS.

DNA extraction: There are three basic and two optional steps in a DNA extraction. (i) Breaking the cells open, commonly referred to as cell disruption or cell lysis, to expose the DNA within. This is commonly achieved by chemical and physical methods – blending, grinding or sonicating the sample. (ii) Removing membrane lipids by adding a detergent or surfactants such as SDS and by centrifugation. (iii) Removing proteins is optional but almost always done using a protease such as pronase and proteinase K and denatured protein is separated from the cell extract. (iv) Removing RNA by adding an RNase (often done), which rapidly degrades RNA into ribonucleotide subunits. (v) Precipitating the DNA with an alcohol – usually ice-cold ethanol or isopropanol. Since DNA is insoluble in these alcohols, it will aggregate together, giving a pellet upon centrifugation. This step also removes alcohol-soluble salt. Refinements of the technique include adding a chelating agent to sequester divalent cations such as Mg^{2+} and Ca^{2+}, which prevents enzymes like DNase from degrading the DNA. Cellular and histone proteins bound to the DNA can be removed either by adding a protease or by having precipitated the proteins with sodium or ammonium acetate or extracted them with a phenol-chloroform mixture prior to the DNA precipitation. If desired, the DNA can be resolubilized in a slightly alkaline buffer or in ultra-pure water.

DNA extraction using cetyltrimethylammonium bromide (C-TAB) method: For plant DNA extraction, sample is grinded after freezing it with liquid nitrogen. Sample is then resuspended in a suitable buffer such as C-TAB, a non-ionic detergent that precipitates nucleic acids and acidic polysaccharides from low ionic strength solutions, while proteins and neutral polysaccharides remain in solution under these conditions. In solutions of high ionic strength, C-TAB does not precipitate nucleic acids and forms complexes with proteins.

DNA extraction using ethidium bromide-caesium chloride (EtBr-CsCl) gradient centrifugation method: Genomic DNA can be purified by centrifugation through a CsCl density gradient method. Cells are lysed using a detergent, and the lysate is alcohol precipitated. Resuspended DNA is mixed with caesium chloride (CsCl) and ethidium bromide (EtBr) and centrifuged for several hours. EtBr can insert or interpose between bases of DNA if the DNA is flexible. This occurs in case of linear DNA but not with DNA that is circular. Ethidium intercalation makes DNA dense, which can be separated from DNA not containing ethidium. This technique is used to separate circular DNA of episomes, including mitochondria, chloroplasts and plasmids. Bacterial DNA, although circular, is sufficiently long, as it breaks during isolation and hence behaves if it were linear DNA.

DNA extraction using solid-phase method: Solid-phase system absorbs nucleic acid in the extraction process depending on the pH and salt content of the buffer. The absorption process is based on hydrogen-binding interaction with a hydrophilic matrix under chaotropic conditions, ionic exchange under aqueous conditions by means of an anion exchanger, and affinity and size-exclusion mechanisms. Solid-phase purification is normally performed by using a spin column, operated under centrifugal force.

DNA fibre-based fluorescence *in situ* hybridization (DNA fibre FISH): This single-molecule strategy allows direct identification of whole-genome $5^{m}C$ by the combination of DNA fibre FISH and ATOMIC FORCE MICROSCOPY (AFM). Using extended DNA fibres and anti-$5^{m}C$ antibody for the detection of $5^{m}C$, it is possible to map the physical location of $5^{m}C$ within the genomic DNA.

DNA filter assay: An analytical procedure used to recognize recombinant DNA in cloned cells. The transformed cells only represent a small percentage of the total population of cells present in the culture. The culture is diluted and placed on petri dishes at such a concentration that individual colonies form from each cell. A replicate of the pattern of cells on the plate is taken using a cellulose nitrate filter. The cells are lysed, and the filter is heated to bind DNA onto the filter in a denatured form in a position equivalent to that of the cell colonies on the petri dish. The filter is then treated with the probe, which combines with DNA hybridization only at the positions of transformed DNA. These are then located using autoradiography or florescent antibodies, as appropriate.

DNA fingerprint(ing) (DFP): A technique developed by A. Jeffreys and others in 1985 that is valuable in establishing near-perfect identity of an unidentified body based on the nucleotide sequences of certain regions of human DNA that are unique to individuals. The technique is based on the observation that every human being is conferred with DNA variations in form of VARIABLE NUMBER OF TANDEM REPEATS (VNTRs). Accuracy of this method is 1-in-75 billion error probability ratio. The unique pattern of DNA fragments is identified by Southern hybridization (using a probe that binds to a polymorphic region of DNA) or by PCR using primers flanking the polymorphic region. DNA fingerprinting is used in a variety of situations, such as criminal investigations, other forensic purposes and paternity testing. In these situations, one aims to 'match' two DNA fingerprints with one another (i.e. a DNA sample from a known person and one from an unknown person). The technique has applications in forensics, paternity testing, anthropology, conservation biology and ecological research. DNA fingerprinting helps to distinguish even closely related individuals. In case of plants, using DNA markers like SSRs, AFLPs and VNTRs, DNA fingerprints have been established in various crops including wheat, rice, maize, barley, oats, brassica, potato and soybean. These fingerprints help in the identification of breeding lines and varieties for research and IPR issues. DNA fingerprinting is also used in forensic medicine to determine paternity or to prove that blood on murder suspect's clothes come from the murder victim. *Aka* DNA TYPING, DNA PROFILING *or* GENETIC SIGNATURES.

DNA flexibility: DNA flexibility is an important property of DNA, as DNA bending is important in packaging, recombination and transcription.

DNA footprinting: A method for determining the sequence specificity of DNA-binding proteins. It utilizes a damaging agent that cleaves DNA at all regions except where protein is bound to DNA.

DNA forensics: With the entry of functional genomics, this area of research is now completely modernized at the scene. It identifies potential suspect whose DNA may match the evidence left at the crime scenes, establish paternity and other family relationships and identifies protected species.

DNA forms: Different forms of DNA, such as B-DNA, Z-DNA, A-DNA, C-DNA and D-DNA, have been reported to exist. All these forms of DNA play significant roles in the regulation of gene expression. The Watson and Crick right-handed double helix of DNA is commonly called B-DNA. Another type of DNA that is known as Z-DNA, possesses left-handed helical structure and gives zigzag appearance of sugar-phosphate backbones. B-DNA and Z-DNA are the two major forms of DNA. Besides, other forms of DNA such as A-DNA, C-DNA and D-DNA have been found in natural and synthetic DNA. These are right-handed helices with 11.0, 7.9–9.6 and 8.0 base pairs per turn, respectively (B-DNA possesses 10 base pairs per turn of the helix).

DNA four-arm junction: A series of four-arm junctions are used as summits within a network of double-helical DNA fragments. Careful distortion of the junctions from their most relaxed conformations guarantees that a scaffold strand can traverse through individual summits in multiple directions. These four strands associate into a DNA four-arm junction because this structure maximizes the number of correct base pairs, with A matched to T and C matched to G.

DNA fragmentation: The pathway that leads to DNA fragmentation has been identified. Caspase-3 cleaves one subunit of a dimer called DNA FRAGMENTATION FACTOR (DFF). The other subunit then activates a nuclease that degrades DNA.

DNA G primase: An enzyme which synthesizes primer required for λ phage and *Escherichia coli* DNA replication; this enzyme is localized in these systems by proteins P and Dna C, respectively.

DNA glycosylases: Endonucleases that initiate excision repair at the sites of various damaged or improper bases in DNA. Bases that have been modified by alkylation or deamination may be removed from DNA by special DNA glycosylases. Each type of modified base has a corresponding DNA glycosylase that removes the base, leaving an apurinic or apyrimidinic site in the DNA.

DNA grooves, major and minor: In DNA, the glycosidic bonds between sugars and bases of given base pairs are not precisely opposite to each other, and two grooves of unequal width are formed around double helix. These two types of grooves are called MAJOR GROOVE (measuring more than 180°) and MINOR GROOVE (measuring less than 180°). The functional utility of grooves is that proteins interact with particular bases, thus disrupting the DNA double helix.

DNA gyrase: A type II topoisomerase first reported in *Escherichia coli*, which is responsible for relaxing the tension on the double helix owing to supercoiled twists brought about by the unwinding of the double helix without rotation. M. Vayssières and others (2024) resolve the structure of *E. coli* DNA gyrase bound to a negatively supercoiled minicircle DNA. They reveal how DNA gyrase captures a DNA crossover, displaying both conserved molecular grooves that lodge the DNA helices. Together with molecular tweezer experiments, the assembly reveals that the DNA crossover is of positive chirality, unifying the binding step of gyrase-mediated DNA relaxation and supercoiling in a single structure. This study offers evidence for a model projected nearly four decades ago. *Aka* HELIX-UNWINDING PROTEIN.

***dna*:** Mutants of bacteria that are temperature sensitive, as they cannot synthesize DNA at 42°C but can do so at 37°C.

DNA: *See* DEOXYRIBONUCLEIC ACID.

DNA helicases: Enzymes responsible, during DNA replication, for breaking the hydrogen bonds that hold the double helix together.

DNA hybridization: Base pairing of DNA from two different sources. In biotechnology, a technique for selectively binding specific segments of single-stranded (ss) DNA or RNA by base pairing to complementary sequences of ssDNA molecules that are trapped on a nitrocellulose membrane.

DNA hydrogels: DNA hydrogels comprising three-dimensional networks of DNA polymeric chains, have received considerable attention as a new class of polymeric materials particularly as biomaterials, showing great potential in a wide range of promising applications. DNA chain composed of four deoxyribonucleotide monomers (abbreviated as A, T, C and G) can be designed rationally and synthesized precisely, thus endowing a hydrogel with desirable functions and properties, such as responsiveness upon pH, enzyme, ions and biomolecules. DNA hydrogels can obtain multiple stimulus responsiveness and tuneable mechanical property via rational design of DNA sequences. DNA hydrogels provide an attractive vehicle for delivering CRISPR/Cas system to achieve effective gene editing.

DNA hydroxylation: Oxidation of the 5-methylcytosine ($5^{m}C$) to 5-hydroxymethylcytosine ($5^{hm}C$) and higher oxidation states. This process is the first step in DNA demethylation, and the modified bases function as a chromatin modification. *Aka* PASSIVE DNA DEMETHYLATION.

DNA hypotheses: Linus Pauling proposed in 1952 that DNA might be a triple helix. When J.D. Watson and F.H.C. Crick learned of L. Pauling's hypothesis, they understood from existing data that Pauling was wrong and that Pauling would soon admit his difficulties with that structure. Thus, the race started to figure out the correct structure. In 1953, Watson and Crick proposed a double helix model of DNA structure for which they, along with M. Wilkins, got a Nobel Prize in Physiology or Medicine in 1962.

DNA insertases: A family of enzymes that directly insert missing bases into DNA. At apurinic and apyrimidinic sites, in a single step.

DNA isolation: DNA isolation is a routine procedure to collect DNA for subsequent molecular or forensic analysis. However, these terms may also be used to represent different steps in DNA analysis. *See* DNA EXTRACTION *and* DNA SEPARATION. *Aka* DNA EXTRACTION, DNA PURIFICATION *and* DNA SEPARATION.

DNA iterations: After considerable fruitless experimentation, being dejected by their superiors from continuing and numerous false starts, in 1953, J.D. Watson and F.H.C. Crick were able to deduce the essential structure of DNA by real modelling of the physical shapes of the nucleotides that comprise it. They were directed by the bond lengths that had been deduced by L. Pauling and by R. Franklin's X-ray diffraction images.

Dna J: One of the two phage proteins (the other being Dna K) required to free Dna B helicase (unwinding enzyme) from the specialized nucleoprotein (snup) complex.

Dna K: One of the two phage proteins (the other being Dna J) required to free Dna B helicase (unwinding enzyme) from the specialized nucleoprotein (snup) complex.

DNA lattices: This entails the building up of superstructures from starting units consisting of single strands of DNA. The binding is through Watson-Crick base-pairing DNA sequences and is highly selective and specific. This ensures that with the appropriate choice of these multiple component strands of DNA, in many cases, predictable nanostructures self-assemble.

DNA libraries: A set of collection of DNA clones in a certain vector. The goal is to have each gene (DNA fragment) represented in the library at least once. DNA library is very important for gene cloning. In the library, DNA fragments are kept in a vector, usually in *Escherichia coli*. Depending upon the nature of DNA, it may be GENOMIC DNA LIBRARIES, CHROMOSOME LIBRARIES and COMPLEMENTARY DNA (cDNA) LIBRARIES. *See details as separate terms.*

DNA ligase: An enzyme that produces a phosphodiester bond between the 5′-PO_4 end of one polynucleotide and the 3′-OH end of another, thereby making a single, larger polynucleotide. This enzyme (usually from the T4 bacteriophage) facilitates the joining of DNA strands together by catalysing the formation of a phosphodiester bond. It plays a role in repairing single-strand breaks in duplex DNA in living organisms, but some forms of ligases (e.g. DNA ligase IV) may specifically repair double-strand breaks. In the cell, ligases are involved in DNA replication to join Okazaki fragments to produce a continuous DNA strand. Purified ligases are used in building of recombinant DNA molecules and *in vitro* DNA replication. *Aka* POLYNUCLEOTIDE LIGASE.

DNA ligase (bacteriophage T4): The most commonly used ligase in laboratory research can ligate cohesive ends of DNA, oligonucleotides, as well as RNA and RNA-DNA hybrids, but not single-stranded nucleic acids. It can also ligate blunt-ended DNA with much greater efficiency than *Escherichia coli* DNA ligase. T4 DNA ligase has an absolute requirement for ATP as a cofactor.

DNA ligase (*Escherichia coli*): The *Escherichia coli* DNA ligase is encoded by the *lig* gene. It uses energy gained by cleaving nicotinamide adenine dinucleotide (NAD) to create the phosphodiester bond. It normally does not ligate blunt-ended DNA.

DNA ligase activity measurement units: Two units used for expressing ligase activity: (i) *Weiss unit.* The amount of ligase that catalyses the exchange of 1 nmole of ^{32}P from inorganic pyrophosphate to ATP in 20 min at 37°C. This is the most commonly used unit for expressing ligase activity. (ii) *Modrich-Lehman unit.* Defined as the amount of enzyme required to convert 100 nmoles of $d(A\text{-}T)_n$ to an exonuclease III resistant form in 30 min under standard conditions. Many commercial suppliers of ligases use an arbitrary unit based on their ability to ligate cohesive ends.

DNA ligase I (mammalian): Ligates the nascent DNA of the lagging strand after the Ribonuclease H has removed the RNA primer from the Okazaki fragments.

DNA ligase II (mammalian): Alternatively spliced form of DNA ligase III, found in non-dividing cells.

DNA ligase III (mammalian): DNA ligase III complexes with DNA repair protein XRCC1 to aid in sealing DNA during the process of nucleotide excision repair and recombinant fragments. The only DNA ligase found to be present in mitochondria.

DNA ligase IV (mammalian): It complexes with XRCC4 to catalyse the final step in the non-homologous end-joining (NHEJ) DNA double-strand break repair pathway. It is also required for V(D)J recombination, the process that generates diversity in immunoglobulin and T-cell receptor loci during immune system development.

DNA ligases (mammalian): Four types of DNA ligases exist in mammals: DNA LIGASE I, DNA LIGASE II, DNA LIGASE III and DNA LIGASE IV. *See details as separate terms.*

DNA ligases (thermostable): These DNA ligases are active at higher temperatures. Examples of thermostable DNA ligases are *Taq* DNA ligase NEB and the Ampligase DNA ligase.

DNA ligation: Chemical linking of two DNA strands. It is made use of in DNA repair, replication and other molecular progressions.

DNA loop extrusion: Many protein complexes that drive key processes in cells are 'molecular motors' – assemblies that consume (electro)chemical energy to produce mechanical work. Examples include the FoF1 synthase rotary motor that catalyses adenosine triphosphate (ATP) production, kinesin and myosin motors that 'walk' along cytoskeleton filaments or polymerases and helicases that move along DNA. STRUCTURAL MAINTENANCE OF CHROMOSOMES PROTEIN COMPLEXES (SMCs) have only recently been identified as an entirely distinct class of DNA-translocating motors, although their key role in folding the linear DNA double helix into intricate three-dimensional structures, such as X-shaped mitotic chromosomes, was known for decades. C. Dekker and others (2023) discuss how insights from biophysical, biochemical and structural studies are starting to yield an understanding of the mechanism by which these motors extrude loops of DNA to structure genomes.

DNA looping: Looping of DNA seems to be the most likely mechanism, enabling protein-protein interactions involved in gene regulation. This mechanism envisages that proteins bound at widely separated sites act in the same way, with the intervening DNA looping or bending to allow occurrence of protein-protein interactions. According to this idea, in reality, it is the interaction between DNA-bound proteins, not the looping per se, that controls gene expression.

DNA loss: A special incidence of DNA loss is the extrusion the pushing out the nucleus from mammalian reticulocytes, which differentiate into nucleate erythrocytes (nulliploidy). Selective DNA loss is seen in many lower eukaryotes and is associated with the differentiation of the somatic cells and germ cells. Generally, the germ cells retain the entire complement of DNA, while there is selective DNA loss or chromatin diminution in the somatic cell lineage. This process involves site-specific DNA cleavage and segregation. Significance of DNA loss is unclear, but it appears to be an essential aspect of somatic cell development and is controlled by the regional distribution of cytoplasmic components.

DNA machine: A molecular machine constructed from DNA. Research into DNA machines was pioneered in the late 1980s by Nadrian Seeman and co-workers from New York University. DNA is used because of the numerous biological tools already found in nature that can affect DNA and the immense knowledge of how DNA works.

DNA machines assemble complex and tiny electrical and mechanical devices: Its precise structure and ability to bind with other molecules make DNA an attractive scaffolding material for nanotech researchers. Scientists have already used DNA to construct two-dimensional patterns, three-dimensional objects and simple shape-changing devices. Now, researchers have separately made complex programmable machines using DNA molecules. Ned Seeman has combined multiple DNA devices to make an assembly line. They walk. They work. New DNA robots strut their tiny stuff. The nano-contraption picks up gold nanoparticles as it tumbles along a DNA-patterned surface. A computer-generated conception of nanorobots, microscopic machines made from DNA molecules, mimics the work of living cells.

DNA marker, properties: Properties sought in perfect DNA markers include highly polymorphic nature, preferably codominant inheritance, frequent occurrence in genome, neutral behaviour to the environmental conditions and management practices, easy and fast assay, high reproducibility and easy exchange of data between laboratories.

DNA marker, type I: Any molecular marker derived from gene sequences. Such genic markers are usually highly conserved and therefore exhibit little sequence polymorphism and are less suitable for the analysis of pedigrees and population diversity.

DNA marker, type II: Any molecular marker derived from hypervariable microsatellite sequences. Such markers are highly informative in forensic analysis, pedigree and population studies. Type II markers are almost evenly dispersed throughout eukaryotic genomes, and each locus exists as many alleles as possible in a population.

DNA marker, type III: Any one of a series of biallelic single-nucleotide polymorphism markers from sequences within or between coding regions that are highly informative for pedigree, family or population screens within species but are usually not informative for comparative ortholog identification between orders.

DNA marker: A piece of DNA molecule that is associated with a certain trait in an organism. These are linear landmarks in the DNA or chromosome where genotypic differences arising from point mutation, insertions, deletion or transpositions can be detected and visualized by various molecular tools. A large number of acronyms for DNA markers are now commonly used for variety of purposes. Certain markers – RESTRICTION FRAGMENT LENGTH POLYMORPHISMS (RFLPs) – consist of segments of DNA that can be identified on autoradiographs (produced after digestion of the DNA by restriction enzymes and segregation of the resulting fragments through gel electrophoresis). Molecular marker is any site (locus) in the genome of an organism at which the DNA base sequence varies among the different individuals of a population. Such markers generally have no apparent effect on the phenotype of the individual, but they can be determined by biochemical analysis of the DNA. The advent of such genetic tools as restriction enzymes and the polymerase chain reaction, plus the growing abundance of DNA sequence data, coupled with automated high-throughput assays, have revealed several classes of molecular markers. DNA markers have been employed to help or fasten the process of plant breeding since the mid-1970s, via MARKER-ASSISTED SELECTION (MAS). Many molecular marker techniques are available today. DNA markers are generally classified into HYBRIDIZATION-BASED (NON-PCR) MARKERS and PCR-BASED MARKERS. *Aka* MOLECULAR MARKER. *See separate terms for details.*

DNA markers, mechanisms basing the: The molecular processes that form foundations of DNA markers include (i) single-base substitution in the restriction sites or PCR priming sites, (ii) arrangement within the DNA intervening the two restriction sites or PCR priming sites, (iii) error in replication of arrays of tandemly repeated DNAs and (iv) mutation in DNA sequence.

DNA methylation: In biological systems, it refers to the addition of a methyl group ($-CH_3$) to a DNA base (adenine or cytosine) by endogenic methylases. It can regulate the expression of genes. Methylation of specific nucleotides within a target site of a restriction enzyme can defend the DNA against attack by that enzyme. Typically, such addition occurs on to the cytosine bases of DNA. Some methylation changes are heritable. *See* MATERNAL IMPRINTING.

DNA methylation analysis: The assessment of DNA methylation within particular regulatory regions/genes of interest. Vast majority (about 90%) of the m^5C residues in eukaryotic DNA are found in the dinucleotide sequence CpG context. Fortunately, several restriction enzymes include CpG in their recognition sequence. Some of these 'CpG enzymes' (and cutting sites) serve as the recognition sites of several restriction endonucleases, such as *Hpa*II (CCGC), *Msp*I (CCGG), *Hha*I (GCGC), *Xho*I (CTCGAG), *Ava*I (CPyC GPuG), *Sal*I (GTCGAC) and *Sma*I (CCCGGG). Most of these enzymes do not cut the DNA if the CpG sequence is methylated. Thus, these enzymes can be used as a tool to probe for methylation.

DNA methylation analysis, methods of: The routine methods for DNA methylation analysis are primarily based on CHROMATOGRAPHIC TECHNIQUES, THIN-LAYER CHROMATOGRAPHY, MASS SPECTROMETRY and DENSITOMETRIC SCANNING OF ELECTROPHEROGRAPHS. Relatively recent methods include METHYLATION-SENSITIVE RESTRICTION ENZYMES, BISULFITE CONVERSION, HIGH-RESOLUTION MELT ANALYSIS, ENTROPY-BASED QUANTITATIVE TOOLS and PCR-BASED METHODS.

DNA methylation analysis using deep sequencing: Refers to the concept of sequencing a genomic region multiple times, sometimes hundreds or even thousands of times. This NGS approach allows researchers to detect rare clonal types, cells or microbes comprising as little as 1% of the original sample. This technology allows sequencing of billions of nucleotides in a single run. Main deep sequencing technologies in use are Roche/454, Illumina/SOLEXA, ABI SOLiD, Polonator, Heliscope and Ion Torrent.

DNA methylation analysis using different array platforms: Four different array platforms available for DNA methylation analysis are NimbleGen, Affymetrix, Agilent and Illumina Arrays.

DNA methylation analysis using methylation-sensitive restriction enzymes: METHYLATION-SENSITIVE RESTRICTION ENZYMES (MSREs) play a role in the analysis of methylated DNA, as they are used to analyse the methylation status of cytosine residues in CpG sequences. These restriction enzymes, as their name implies, are not able to cleave methylated cytosine residues, leaving methylated DNA intact. MSREs cleave DNA at specific unmethylated cytosine residues and DNA is then amplified by PCR following digestion. Because unmethylated DNA is digested by MSREs, the only amplification products detected are methylated DNA.

DNA methylation analysis using PCR-based methods: PCR-based methods are routinely used to study DNA methylation on a gene-specific basis, after bisulfite treatment. Variations of this method include bisulfite sequencing, methylation-specific PCR, real-time PCR-based MethyLight and methylation-sensitive high-resolution melting PCR.

DNA methylation analysis, applications of: Investigating DNA methylation patterns quantitatively across the genome using high-throughput sequencing and array-based techniques yields various benefits. The results obtained can help researchers gain valuable insight into gene regulation and identify potential biomarkers. DNA methylation analysis empowers (i) separating plant nuclear DNA from organellar DNA, (ii) cytosine methylation analysis tool for everyone, (iii) understanding role of heritable changes in gene expression and (iv) ancient DNA methylation reconstruction.

DNA methylation and epigenetics: DNA methylation is used for long-term epigenetic silencing of imprinted genes and can either increase or decrease level of transcription. Numerous types of epigenetic modifications of DNA include methylation of cytosine at position 5, a process that is known to inactivate eukaryotic genes. Methylated cytosine ($5^{m}C$) is the only modified base yet found in DNA of higher organisms. Methylation occurs through the enzymatic reaction. This is a postreplication chemical modification of a base in DNA. These DNA modification mechanisms offer a molecular basis for inheritance of a particular pattern of gene activities. A particular pattern of unmodified or modified cytosines could be inherited. Maintenance methylase distinguishes hemi-methylated DNA (*aka* HALF-METHYLATED DNA) formed after replication and that methylates the nascent strands. Such an enzyme does not act on sequences that contained non-methylated cytosine. Cytosine methylation occurs in CPG DOUBLETS.

DNA methylation and gene expression: DNA methylation is one mechanism out of the four bases chemically that can alter at least two by attaching methyl groups to them. Adenine gets methylated at 6th carbon position and cytosine at 5th carbon position. Both adenine and cytosine are amino bases. Methylation of either coding or non-coding DNA strand was sufficient to block expression of the hemi-methylated chromatin. DNA methylation is known to have important role in gene expression. Methylation of DNA leads to B→Z transformation which affects interaction of regulatory proteins with DNA. There are several examples which show that methylation at specific sites in a gene has regulatory role, particularly the methylation at 5′-end of a gene. The pattern of DNA methylation in the genome changes as a result of a dynamic process involving both *de novo* DNA methylation and demethylation. As a consequence, differentiated cells develop a stable and unique DNA methylation pattern that regulates tissue-specific gene transcription.

DNA methylation and gene silencing: The integration of extra copies of an endogenous gene or multiple copies of a transgene into a plant genome often results in epigenetic silencing of both endogenous and exogenous genes. The silencing mechanism can occur at two levels, either transcriptionally (HOMOLOGY-DEPENDENT TRANSCRIPTIONAL GENE SILENCING (HDTGS), which may involve in *cis* or *trans* DNA pairing), or post-transcriptionally HOMOLOGY-DEPENDENT POST-TRANSCRIPTIONAL GENE SILENCING (HDPTGS) or co-suppression, which may involve antisense RNA (DEGRADOSOMES), but there is no change in DNA sequence. In many cases, epigenetic silencing correlates to hypermethylation of the repetitive sequences.

DNA methylation detection, transformer-based: A number of analytical tools, such as Nanopolish, Guppy/Tombo and DeepMod, have been developed to detect DNA methylation on Nanopore data. Current limitations for DNA methylation detection that are inherent in short-read bisulfite sequencing or methylation microarrays. X. Wang and others (2023) apply TRANSFORMER ARCHITECTURE to detect DNA methylation on ionic signals from Oxford Nanopore sequencing data. Transformer is an algorithm that adopts self-attention architecture in the neural networks and has been widely used in natural language processing. Transformers have the ability to detect methylation on ionic signal data.

DNA methylation in animals: In mammals, DNA methylation and demethylation are involved in diverse processes, including (i) early embryogenesis, (ii) stem cell differentiation, (iii) genomic imprinting, (iv) X chromosome inactivation, (v) silencing of repetitive elements, (vi) regulation of neuronal development and (vii) development of cancers.

DNA methylation in chromosome folding, packing and sorting: The effects of DNA methylation on chromatin structure require that nuclear factors be able to distinguish methylated and non-methylated DNA. METHYL-CPG-BINDING PROTEIN (MeCP) complexes with a variety of unrelated DNA sequences when they are methylated at CpG. Unmethylated CpG sequences do not bind MeCP. The complex secondary and tertiary folding of eukaryotic chromosomes probably comprises specific protein-DNA interactions that could be impacted by the distribution of $5^{m}C$ in the DNA. The high methylation of centromeric regions suggests a possible role in the mitosis and chromosome sorting. DNMT1 interacts with hSNF2H chromatin remodelling enzyme, and DNMT1 binds mononucleosomes with higher affinity in the presence of hSNF2H. A chromatin remodelling enzyme can alter the biological properties of a DNA methyltransferase.

DNA methylation in early embryogenesis: In mammals, complex alterations in DNA methylation levels occur during embryonic development. Immediately after fertilization, the paternal and maternal genomes in zygotes differ in levels of DNA methylation. For example, the DNA methylation level is initially higher in the paternal genome than in the maternal genome. Within 3–6 h after fertilization, the maternal genome is rapidly methylated through *de novo* DNA methylation. The paternal genome is actively demethylated in an elongator-dependent manner. Demethylation of the paternal genome continues in preimplantation embryos. After rapid *de novo* DNA methylation in the maternal genome of zygotes, passive DNA demethylation commences gradually during DNA replication cycles. The replication-dependent demethylation is due to exclusion of maintenance methyltransferase DNMT1 from the nucleus of embryos. At the morula stage, the parental and maternal genomes have equally low DNA methylation levels. Embryonic DNA methylation patterns are established after implantation through lineage-specific *de novo* methylation. DNA methylation levels increase rapidly in the primitive ectoderm, which gives rise to the entire embryo, whereas methylation is either inhibited or not maintained in the trophoblast and the primitive endoderm lineage, which give rise to the placenta and yolk sac membrane, respectively. Demethylation and *de novo* methylation also occur during gametogenesis, which are critical events for the institution of genomic imprinting and for the suppression of transposons in germ cells.

DNA methylation in eukaryotes: A heritable, reversible epigenetic modification in which a methyl group ($-CH_3$) is covalently added to a DNA sequence, usually the 5th carbon of the cytosine pyrimidine ring, in a CpG dinucleotide, although CpHpG and CpHpH sequences can be methylated in plants, where H is any DNA base other than G. DNA methylation is used by healthy cells to switch off certain genes when their expression is no longer needed (e.g. switch off genes involved in juvenile development after organism reaches adulthood). DNA methylation (of cell genes that would normally prevent inappropriate cell division/proliferation) also occurs in some cancers. Extensive methylation of the cytosine base in CG sequences is used in vertebrates to keep genes in an inactive state. However, $6^{m}A$ has also been reported in eukaryotes. DNA methylation is often associated with epigenetic gene silencing. *Aka* DNA MODIFICATION.

DNA methylation in gene regulation: Adding methyl groups to DNA is a means of regulating some genes and genomic sequences. Structural analysis shows that the enzyme complex that mediates this process shows unexpected sequence specificity. In these regions, CG dinucleotides occur with a periodicity of 8–10 bp. Different patterns of CG periodicity might reveal different functional specificities within the genome. An antisense transcript not naturally occurring but induced by genetic mutation leads to gene silencing and DNA methylation. DNA methylation is an important epigenetic mark for TRANSCRIPTIONAL GENE SILENCING (TGS) in a variety of organisms. Correlation has been observed between undermethylation and gene activity. Ethionine (an analogue of methionine) and 5-azacytidine (an analogue of cytosine) are potent inhibitors of DNA methylation. Consequently, they cause undermethylation of cytosine in DNA and gene activity. Gene expression is regulated at various levels. Cytosine methylation at different stages of DNA packaging: transcription factor-binding dynamics at 1-nt resolution, nucleosome positioning at 20-nt resolution, RNA splicing at 1000-nt resolution, gene expression at 1–5-kb resolution and genome imprinting at 10-kb resolution. The most important role of DNA methylation in gene regulation is in transcription initiation. DNA methylation in general recognized as repressor of gene expression. However, if some CpGs in the promoter are unmethylated, transcription is initiated, but if these sites in the promoter are methylated, there is no transcription.

DNA methylation in gene silencing and genome stability: DNA methylation is essential for gene silencing and genome stability in many organisms. Gadd45a is a nuclear protein involved in maintenance of genomic stability, DNA repair and suppression of cell growth. This protein has a key role in active DNA demethylation. Gadd45a overexpression activates methylation-silenced reporter plasmids and promotes global DNA demethylation. Gadd45a knockdown silences gene expression and leads to DNA hypermethylation. During active demethylation of oct4 in *Xenopus laevis* oocytes, Gadd45a is specifically recruited to the site of demethylation. Active demethylation occurs by DNA repair and

Gadd45a interacts with and requires the DNA repair endonuclease XPG. Gadd45a relieves epigenetic gene silencing by promoting DNA repair, which erases methylation marks.

DNA methylation in human health: Hypermethylation happens at sites already repressed in precancerous tissue and so is unlikely to contribute to gene silencing. In addition to hypermethylation, hypomethylation at some promoters has also been found to be associated with cancer.

DNA methylation in mutation: Methylase-deficient *Escherichia coli* mutants have been isolated that have reduced methylation of GATC sequences. These mutants show increased spontaneous mutagenesis. In *Escherichia coli*, methylated sites can be mutation 'hotspots'. A major hotspot in the *lac* repressor gene is the sequence CC(A/T)GG which, in wild-type *Escherichia coli*, is always methylated. Deamination of cytosine produces uracil, which is recognized and preferentially removed. Deamination of m^5C generated thymine, which is not preferentially corrected. DNA hypomethylation leads to elevated mutation rates at *Hprt* and *tk* genes. Mammalian DNA methylation has an important role in maintaining genome stability.

DNA methylation in plants: Cytosine DNA methylation is an important mechanism in regulating gene expression and in silencing transposons and other repetitive sequences. Genomic studies in *Arabidopsis thaliana* have revealed that many endogenous genes are methylated either within their promoters or within their transcribed regions. Gene methylation is also highly correlated with transcription levels. However, plants have different types of methylation controlled by different genetic pathways.

DNA methylation in prokaryotes: In prokaryotes, DNA adenine-N6 is the major methylated base, with methyl group on the C5 position of cytosine (5^mC) and cytosine-N4 methylation (4^mC) occurring less frequently. The DNA of *Escherichia coli* contains minor amounts of 6-methyladenine (6^mA) and 5-methylcytosine (5^mC). These bases are produced by postreplicative modification of residues in specific DNA sequences by the action of three methylases: (i) hsd (host-specificity) – bifunctional enzymes – restriction and modification of DNA; (ii) dam (DNA adenine methylation)-directed mismatch repair, which corrects replication errors in the newly synthesized daughter-strand DNA and (iii) dcm (DNA cytosine methylation) – required for VSP (very short patch) mismatch repair, which is probably required to correct base-pair mismatches formed by cytosine deamination. These methylated bases are not essential for viability.

DNA methylation in recurrent endogenization of giant viruses: L.A. Sarre *et al.* (2024) display that the protist strictly interrelated to animals *Amoebidium appalachense*]structures both transposon and gene body methylation (5^mC), a pattern suggestive of invertebrates and plants. Unexpectedly, hypermethylated genomic regions in *Amoebidium* arise from viral insertions, including hundreds of endogenized giant viruses. Using a combination of inhibitors and genomic assays, they demonstrate that 5^mC silences these colossal virus insertions. 5^mC is life-threatening for the meticulous coexistence of newly acquired viral DNA into eukaryotic genomes, making *Amoebidium* an exclusive model to understand the hybrid origins of eukaryotic DNA.

DNA methylation in repair: Dam system, which methylates adenine, facilitates postreplicative mismatch repair. Dcm system, which methylates cytosine, is involved in very short-patch mismatch repair.

DNA methylation in replication and repair: Robin Holliday (1979) proposed a model, whereby DNA replication – if uncoupled from DNA methylation – could result in changes in the methylation pattern of the gene and subsequent gene expression after two rounds of DNA replication. DNA-damaging agents also could apparently turn on genes, if the methylation pattern is not restored after DNA repair. This model provides a testable framework to investigate the relationship between DNA methylation and DNA replication and repair. In the phage ϕX174, DNA methylation at one specific site is known to be involved in replication, blocking methylation blocks DNA synthesis. The replication of the *Escherichia coli* chromosome also stops after one round of synthesis in the absence of methylation. Dam methylation system also plays role in DNA replication. Methylation of GATC sites by the dam methylase is required for efficient initiation of DNA replication at the replication origin, *oriC*, of *Escherichia coli*. This is demonstrated by the inability of minichromosomes to be maintained in *dam* mutant strains. The requirement for methylated GATC sites is less stringent *in vitro* than *in vivo*. The time required for complete methylation of the origin region apparently determines the minimal spacing of replication forks on the chromosome.

DNA methylation in transposable element silencing: DNA methylation is an epigenetic mark that has been found to be associated with transposable element silencing and gene imprinting in flowering plants and mammals. Imprinting in plants has been suggested to evolve from targeted methylation of transposable element insertions near genic regulatory elements followed by positive selection when the resulting expression change was advantageous.

DNA methylation in various cellular processes: (i) Initiation of replication, (ii) cell division, (iii) regulation of transcription initiation, (iv) restriction and modification, (v) recombination, (vi) mutation, (vii) DNA repair and (viii) regulation of gene expression.

DNA methylation maintenance: Specialized DNA methyltransferase enzymes can recognize hemi-methylated DNA templates (e.g. a ds-DNA molecule) immediately after synthesis of the new strand. Maintenance methyltransferases then duplicate the DNA methylation onto the new strand, ensuring that $5^{m}C$ patterns can be inherited reliably over several cell-division cycles.

DNA methylation occurrence: Model nematode species *Caenorhabditis elegans* (roundworm) is a genuine methylation-free zone. In nematode species *Trichinella spiralis*, the existence of DNA methylation was experimentally demonstrated. *Trichinella spiralis* seems to have retained a DNA methylation machinery that might be missing in the ancestral nematode.

DNA methylation of imprinted genes: Genome imprinting, found in flowering plants and placental mammals, employs DNA methylation to produce gene expression that is dependent on the parent of origin.

DNA methylation patterning, conservation and divergence of: Shotgun bisulfite sequencing DNA methylation profile in eight eucaryotic organisms, including *Arabidopsis thaliana* (thale cress), *Oryza sativa* (rice), *Populus trichocarpa* (poplar), *Chlamydomonas reinhardtii* (green algae), *Ciona intestinalis* (sea squirt), *Apis mellifera* (honeybee), *Danio rerio* (zebrafish) and *Mus musculus* (mouse), showed demonstrate wide variations in methylation levels and targets. Methylation in genes, exons, repetitive DNA and transposons shows distinct maintenance and deviation among the organisms tested, serving to define the evolution of DNA methylation patterning and function.

DNA methylation patterns, heritability of: Pattern of methylation is inherited through DNA synthesis and cell division.

DNA methylation sequencing strategies: Different DNA methylation sequencing strategies include *Mcr*BC-seq, methylcytosine immunoprecipitation and sequencing (mCIP-Seq) and sodium bisulfite conversion followed by sequencing (BS-seq).

DNA methylation theory of development: A theory put forward to explain developmental reprogramming by reprogramming of gametes prior to fertilization that depends upon specific *de novo* methylation of chromosomal DNA.

DNA methylation variants in RILs, inheritance of: Epigenomic analysis of soybean RECOMBINANT INBRED LINES (RILs) deciphered that DMRs mostly co-segregate with the genotype and solved the theory of uncoupling in genotype and epigenotype. The epi-allelic transgressive segregation also forms the basis of variation between parental lines and EPIGENETIC RECOMBINANT INBRED LINES (EPI-RILs).

DNA methylation, characteristics of: (i) Methylation sites exist in three states. (ii) Methylation patterns are symmetrical. (iii) Methylation patterns are clonally inherited. (iii) DNA methylases have limited specificity. (iv) Distribution of methylated sequences is non-random. (v) Methylation patterns are tissue specific. (vi) Overall levels of methylation changes are very small.

DNA methylation, detection and quantification of: Global quantification of DNA methylation is crucial for understanding the roles that gene expression and silencing play in the development of cancer and other diseases. Various EPIGENTEK kits include MethylFlash Methylated DNA Quantification Kit (Colorimetric), MethylFlash Global DNA Methylation ($5^{m}C$) ELISA Easy Kit (Colorimetric), MethylFlash Global DNA Hydroxymethylation ($5^{hm}C$) ELISA Easy Kit (Colorimetric) and MethylFlash 5-Formylcytosine ($5^{f}C$) DNA Quantification Kit.

DNA methylation, functions of: (i) Host restriction and modification; (ii) DNA replication; (iii) mutation; (iv) gene silencing and genome stability; (v) recombination; (vi) repair; (vii) chromosome folding, packing and sorting and (viii) epigenetic modifications in DNA.

DNA methylation, luciferase-based sensors of: In this method, the firefly luciferase reporter protein (Fluc) serves to assess DNA methylation at the genome level or at specific DNA sequences. Split-luciferase biosensors comprise of two fusion proteins: (i) a DNA-binding domain fused to the

N-terminal portion of Fluc and (ii) a DNA-binding domain fused to the Fluc C-terminus. To assess levels of global DNA methylation, both fusion proteins are assembled using the DNA-binding domain of a methyl-CG-binding domain protein, such as MBD2, which has a 100-fold preference for methylated CG sites over unmethylated CG sites.

DNA methylome: The set of nucleic acid methylation modifications in an organism's genome or in a particular cell.

DNA methyltransferase 1 (DNMT1): The chief maintenance methyltransferase in mammals. It predominantly methylates hemi-methylated CpG dinucleotides. DNMT1 is 7- to 100-fold more active on hemi-methylated DNA as compared with unmethylated substrate *in vitro*. It is still more active at *de novo* methylation than other DNMTs. The recognition motif for the human enzyme involves only three of the bases in the CpG dinucleotide pair: a C on one strand and CpG on the other. The enzyme is ~1620 amino acids long. The first 1100 amino acids constitute the REGULATORY DOMAIN of the enzyme. The remaining residues constitute the CATALYTIC DOMAIN. These domains are joined by Gly-Lys repeats. Both domains are needed for the catalytic function of DNMT1. Somatic DNMT1 is always found in the nucleus of somatic tissues. DNMT1 has several isoforms. Splice variant (DNMT1B) null mutant embryonic stem cells are viable. Mouse embryos homozygous for deletion in Dnmtl die at 10–11 days of gestation. Oocyte-specific isoform (DNMT1o) is synthesized and stored in the cytoplasm of the oocyte and translocated to the cell nucleus during early embryonic development.

DNA methyltransferase 2 (DNMT2): While the biological function of DNMT2 is not yet known, the strong binding to DNA suggests that DNMT2 may mark specific sequences in the genome by binding to DNA through the specific target-recognizing motif.

DNA methyltransferase 3 (DNMT3): A family of DNA methyltransferases that could methylate hemi-methylated and unmethylated CpG at the same rate. The structure of DNMT3 enzymes is similar to that of DNMT1, with a regulatory region attached to a catalytic domain. There are three known members of the DNMT3 family: DNMT3A, DNMT3B and DNMT3L.

DNA methyltransferase 3A (DNMT3A): A member of DNMT3 family of DNA methyltransferases that can mediate methylation-independent gene repression. DNMT3A can co-localize with heterochromatin protein (HP1) and methyl-CpG-binding protein (MeCBP). Can also interact with DNMT1, which might be a cooperative event during DNA methylation. DNMT3A prefers CpG methylation to CpA, CpT and CpC methylation. DNMT3A methylates CpG sites at a rate much slower than DNMT1 but greater than DNMT3B.

DNA methyltransferase 3B (DNMT3B): A member of DNMT3 family of DNA methyltransferases. DNMT3B can mediate methylation-independent gene repression. This enzyme can also interact with DNMT1, which might be a cooperative event during DNA methylation. There appears to be some sequence preference of methylation for DNMT3A and DNMT3B.

DNA methyltransferase 3L (DNMT3L): A member of DNMT3 family of DNA methyltransferases. DNMT3L contains DNA methyltransferase motifs and is required for creating maternal genomic imprints, despite being catalytically inactive. Expressed during gametogenesis when genomic imprinting takes place. Loss of DNMT3L results in biallelic expression of genes normally not expressed by the maternal allele. Interacts with DNMT3A and DNMT3B and co-localized in the nucleus. DNMT3L appears to be incompetent for methylation reaction, but it is, however, participating in transcriptional repression.

DNA methyltransferases (DNMTs): DNA methyltransferases (DNMTs) are responsible for the transfer of a methyl group from the universal methyl donor, SAM, to DNA. MTases can be divided into three different groups on the basis of the chemical reactions they catalyse: (i) 6^{m}A methyltransferases (EC 2.1.1.72) generate 6^{m}A, (ii) 4^{m}C methyltransferases (EC 2.1.1.113) generate N4-methylcytosine and (iii) 5^{m}C methyltransferases (EC 2.1.1.37) generate C5-methylcytosine. These enzymes work throughout the life of the organism to preserve the methylation pattern that had been initially established by the *de novo* methyltransferases. DNMTs do not methylate an unmethylated site. DNMTs are essential for mammalian development. *Aka* MAINTENANCE METHYLASES.

DNA methyltransferases (DNMTs), functions of: DNA methyltransferases (DNMTs) have a key role in epigenetic regulation. DNMT activity is highly regulated. Key regulatory mechanisms of DNMTs include molecular interactions, post-translational modifications, alternative splicing, gene duplication and gene loss. Molecular functions of DNMTs include gene silencing, transcriptional activation and post-transcriptional regulation (mediated by DNMT2-dependent RNA methylation). DNMT mutations are associated with several human diseases, including cancer.

DNA methyltransferases (DNMTs) in mammals: Methyl group for methylation comes from the cofactor S-ADENOSYL-L-METHIONINE (SAM) to the -O, -C or -N position of a target nucleotide. Mammalian DNA methyltransferases (DNMTs: DNMT1, DNMT3, DNMT3a, DNMT3b, DNMT3L) have been classified into two functional families – DNMT1 and DNMT3 – as they are encoded by their own single gene and consist of catalytic and regulatory regions. Molecular functions of DNMTs include gene silencing, transcriptional activation and post-transcriptional regulation. DNMT1 is required for maintenance of DNA methylation. DNMT2 does not participate in genome methylation in mammals. DNMT2 binds DNA to form a denaturant-resistant complex *in vitro*. While the biological function of DNMT2 is not yet known, the strong binding to DNA suggests that DNMT2 may mark specific sequences in the genome by binding to DNA through the specific target-recognizing motif. DNMT3A/B are involved in *de novo* methylation for mammalian development.

DNA methyltransferases (DNMTs) in plants: In plants, three DNA methyltransferases have been characterized: MET1, CMT3 and DRM2. MET1, a homologue of mammalian DNMT1, is responsible for maintenance of symmetric CpG methylation. CMT3 is a plant-specific DNA methyltransferase that is required for the maintenance of DNA methylation at CpHpG sites. DRM2 is responsible for *de novo* DNA methylation at all sequence contexts, but its role in CpHpH methylation is most prominent, since CpHpH methylation cannot be maintained and thus must rely entirely on *de novo* methylation. The DRM2 activity is highly regulated by the RdDM pathway. *See* RNA-DIRECTED DNA METHYLATION.

DNA microarray: A multiplex technology, initially established by Patrick Brown during the 1980s, which consists of an arrayed series of thousands of microscopic spots of DNA oligonucleotides, called features, each containing picomoles (10^{-12} moles) of a specific DNA sequence, known as *probes* (or *reporters*). These can be a short section of a gene or other DNA element that are used to hybridize a cDNA or cRNA sample (called *target*) under high-stringency conditions. Probe-target hybridization is usually detected and quantified by detection of fluorophore-, silver- or chemiluminescence-labelled targets to determine relative abundance of nucleic acid sequences in the target. DNA microarrays can be used to measure changes in expression levels (e.g. between diseased and non-diseased tissues) to detect SNPs or to genotype or resequence mutant genomes. DNA microarrays is a CLOSED SYSTEM OF GENE EXPRESSION ANALYSIS. Its analysis is limited to only those species previously identified and assigned to an array. With this method, transcript levels for many genes at once (expression profiling) can be measured. Recent advances in microarray technology allow for the quantification, on a single array, of transcript levels for every known gene in several organism's genomes, including humans.

DNA microarray platforms: In standard microarrays, the probes are attached to a solid surface by a covalent bond to a chemical matrix. The solid surface can be glass or a silicon chip, in which case they are colloquially known as an AFFY CHIP when an Affymetrix chip is used. Other microarray platforms, such as Illumina, use microscopic beads, instead of the large solid support. DNA arrays are different from other types of microarrays only in that they either measure DNA or use DNA as part of its detection system.

DNA microarray, applications of: (i) Gene expression profiling, (ii) comparative genomic hybridization, (iii) GeneID, (iv) chromatin immunoprecipitation on ChIP, (v) DamID, (vi) SNP detection, (vii) alternative splicing detection and (viii) fusion genes microarray.

DNA microarray, major types of: Two major types of DNA microarrays exist: (i) The traditional solid-phase array is a collection of orderly microscopic 'spots', called features, each with a specific probe attached to a solid surface, such as glass, plastic or silicon biochip (commonly known as a GENOME CHIP, DNA CHIP or GENE ARRAY). Thousands of them can be placed in known locations on a single DNA microarray and (ii) The alternative bead array is a collection of microscopic polystyrene beads, each with a specific probe and a ratio of two or more dyes, which do not interfere with the fluorescent dyes used on the target sequence. The broadest distinction is whether they are spatially arranged on a surface or on coded beads.

DNA microarray, making of: For the production of DNA microarray, cellular mRNA is employed to make segments of complementary DNA (cDNA) of approximately 500–5000 bp long, using the reverse transcriptase polymerase chain reaction (RT-PCR). These cDNA segments are then adhered to a nylon or glass surface at known spots, so when hybridization of sample DNA occurs, the location of the spot tells what DNA was in the sample. Another way to build a type of DNA microarray is to similarly attach oligonucleotides or peptide nucleic acids of known sequence (composition) at known spots on the nylon or glass surface and pass the biological sample containing DNA (e.g. in liquid) over that surface to identify the DNA in the sample by the spot to which it hybridizes. *Aka* DNA CHIP.

DNA microarray, methodology: Microarray technology evolved from Southern blotting, where fragmented DNA is attached to a substrate and then probed with a known gene or fragment. The step required in a microarray experiment include (i) purification of sample, (ii) reverse transcription, (iii) coupling, (iv) hybridization and washes, (v) scanning and (vi) normalization and analysis.

DNA microarray, principle of: The core principle behind microarrays is hybridization between two DNA strands. Fluorescently labelled target sequences that bind to a probe sequence generate a signal that depends on the strength of the hybridization determined by the number of paired bases. Total strength of the signal, from a spot (feature), depends upon the amount of target sample binding to the probes present on that spot. Microarrays use relative quantization in which the intensity of a feature is compared to the intensity of the same feature under a different condition, and the identity of the feature is known by its position. An alternative to microarrays is serial analysis of gene expression, where the transcriptome is sequenced allowing an absolute measurement.

DNA microarray, uses of: DNA microarrays can be used to detect DNA (as in comparative genomic hybridization) or RNA (most commonly as cDNA after reverse transcription) that may or may not be translated into proteins. DNA microarray technique is used for study of polymorphisms in genes, *de novo* sequencing or molecular diagnosis of disease. Since an array can contain tens of thousands of probes, a microarray experiment can accomplish many genetic tests in parallel. Therefore, arrays have dramatically accelerated many types of investigation. The process of measuring gene expression via cDNA is called EXPRESSION ANALYSIS or EXPRESSION PROFILING.

DNA microchip: DNA chip is a microchip that searches and distinguishes DNA from the sample of half of DNA double helix. *See* DNA CHIP.

DNA modification: (i) Includes all changes made to the deoxyribonucleotides after their initial incorporation into the deoxypolynucleotide chain and (ii) Modifications that impact gene activity without changing the DNA sequence (e.g. modification of cytosine to 5-methylcytosine). Alteration of DNA by DNA methylases ensues after DNA replication. Site-specific methylation protects the DNA (e.g. of bacteria).

DNA modification, role of: Methylation of adenine of prokaryotic DNA plays role in host restriction and modification and cytosine in eukaryotic DNA is involved in maternal inheritance of chloroplast DNA, DNA replication, recombination, mutation, chromosome folding and packing, mammalian chromosome inactivation, conformational changes in DNA, ageing, epigenetic modification and so on. DNA methylation protects the host DNA from its restriction enzymes. *See* DNA METHYLATION.

DNA motifs: While one can use established lists of motifs to search one's DNA sequence, one can also discover them directly. In order to do this, one has to derive a consensus sequence or probability matrix. In the case of bacterial proteins for which the binding sites have been determined, a good place to start is the *Escherichia coli* DNA-Binding Site Matrices. (i) DNA Motifs Gibbs Sampler algorithm finds DNA motifs of unknown length and complicated structure in a set of unaligned DNA sequences. It uses an improved motif length estimator and careful Bayesian analysis of the possibility of a site absence in a sequence. (ii) PromScan scans small genomes for potential factor-binding sites including IHF-binding sites. (iii) FindTerm is a tool for mapping rho-independent terminators. (iv) TransTerm searches for rho-independent terminators in the vicinity of annotated genes. (v) ZLAB finds interaction between regulatory proteins and their DNA/RNA target sites, including the tools Cluster-Buster, Comet, Cister. (vi) Motif Sampler tries to find over-represented motifs (*cis*-acting regulatory elements) in the upstream region of a set of co-regulated genes. (vii) Motif Elucidator in Nucleotide Sequence Assembly

(Melina) helps one extract a set of common motifs shared by functionally related DNA sequences. (viii) BioProspector discovers conserved DNA motifs in upstream regulatory regions of co-expressed genes and examines the upstream region of genes in the same gene expression pattern group and looks for regulatory sequence motifs. (xi) Suite for Computational identification Of Promoter Elements (SCOPE) is aimed at identifying novel cis-regulatory elements from groups of upstream sequences.

DNA moving pictures: Spectacular images of intact chromosomal DNA molecules as long as 100 Mb have been taken by M.P. Cosma and M.V. Neguembor in 2023.

DNA mutation: A long-lasting alteration to the nucleotide sequence of DNA that can occur during replication and recombination.

DNA nanomachine: A motif that changes its structure in response to an input, demonstrated in 1999 by N. Seeman. An improved system, which was the first nucleic acid device to make use of toehold-mediated strand displacement, was demonstrated by B. Yurke in 2000. The next advancement in DNA nanotechnology was to translate this into mechanical motion.

DNA nanostructures: Individual nucleic acid strands rationally assemble into the desired structures. This process begins with specification of a desired target structure or function. Then, the overall secondary structure of the target complex is determined, specifying the arrangement of nucleic acid strands within the structure and which portions of those strands should be bound to each other. The last step involves the primary structure design, which is the specification of the actual base sequences of each nucleic acid strand.

DNA nanoswitch calipers (DNCs): The technology enables researchers to perform distance measurements on single peptides with high precision by applying small amounts of force. By rapidly making many of these distance measurements on the same molecule, DNC creates a unique 'fingerprint' that can be used to identify the same molecule in subsequent experiments.

DNA nanoswitch catenanes: A structure where both ends of a longer synthetic DNA strand are linked to two antibody fragments that each specifically bind different parts of the same biomarker molecule of interest, thus allowing for high target specificity and sensitivity.

DNA nanotechnology: A branch of nanotechnology that uses artificial nucleic acids for the design and construction of nanomachines for technological purposes. In this field, nucleic acids are used as non-biological engineering materials for nanotechnology rather than as the carriers of genetic information in living cells. The devices thus made are called DNA NANOTECHNOLOGICAL DEVICES.

DNA oligonucleotide microarrays: In this microarray technology, oligos are synthesized *in situ* and enable the direct synthesis of oligonucleotide probes on the chip surface using methods such as photolithography. These oligonucleotide microarrays can also be designed in the form of TILING MICROARRAYS to permit sequencing of cDNA clones, thus providing signature sequences of the cDNA that is being examined.

DNA oozing: In this case, binding of a regulatory protein to its operator helps binding of another protein to adjacent sequences, which in turn helps another protein to bind next to it, and so on, until a procession of proteins has oozed out from the control sequences to the gene where transcription is to be initiated. It seems unbelievable that the oozing effect works over thousands of base pairs. The idea of oozing should not be confused with non-specific wrapping of DNA around histone-like proteins. Oozing is not important for site-specific recombination. For the case involving ends of Tn3, the amount of protein sufficient to catalyse the reaction is insufficient to cover DNA between sites. Regulatory proteins recognize specific DNA sequences in DNA using structures that are complementary to ordinary helix. DNA-bound regulatory proteins influence transcription by excluding binding of other proteins. These descriptions hold whether the interacting proteins are lying adjacent to each other or at a distance.

DNA organization, analysis of: Restriction enzymes offer a powerful analytical tool, particularly in higher organisms in which a large amount of DNA in each cell has been an impediment to investigate one particular region of DNA. If a particular piece of DNA can be identified, it can be isolated and studied. One method of identification is to develop appropriate DNA probe. A common technique for this is to purify mRNA product of a particular gene and to form a radioactive labelled copy of cDNA with the help of reverse transcriptase. The labelled cDNA is the probe for the gene from which the original mRNA was produced. By denaturing and renaturing together the cDNA and the restriction fragment of DNA containing the gene under study, the complementary strands from these two sources can be made to hybridize. The identity of the hybrid can be ascertained by radioactive label contributed by the probe. *Escherichia coli* RY13 from which *Eco*RI type II RE was isolated which produces sticky ends. *Streptomyces marcescens*, from which *Sma*I restriction endonuclease was isolated, produces blunt ends. With the use of restriction endonucleases, the recombinant DNA molecules can be generated combining DNA segments from any two species.

DNA origami: DNA origami is an example of folding-based nanostructure. It makes use of a long natural virus strand as a 'scaffold' strand and computationally designed shorter 'staple' strands that bind to portions of the scaffold strand and force it to fold into the desired shape. This method has the advantage of being simple to design, as the base sequence is predetermined by the scaffold strand sequence, and it also does not require high strand purity and accurate stoichiometry, as most other DNA nanotechnology methods do. DNA origami was first demonstrated for two-dimensional shapes.

DNA packaging: The length of the DNA as an extended molecule surpasses the dimensions of the compartment that contains it. The DNA must be compressed exceedingly tightly to fit into the space available. So, in contrast with the customary picture of DNA as an extended double helix, structural deformation of DNA to bend or fold it into a more compact form is the rule rather than exception. DNA packaging has been investigated in viruses/bacteriophages, prokaryotes (nucleoids and plasmids) and eukaryotes (nuclear, mitochondria and chloroplast chromosomes). *Aka* DNA PACKING.

DNA packaging in bacteria: *Escherichia coli* has around 3000 genes on a single double-stranded circular molecule. Once believed to be 'naked' DNA, simple basic amino acids are found to be associated with the bacterial chromosome. The bacterial chromosome is a circular DNA molecule called a NUCLEOID, which lacks a membrane enclosure. Supercoiling and associated basic proteins influence the folding of the bacterial DNA molecule. It forms looping domains. Bacterial chromosome adopts its shape due to supercoiling of DNA molecule. DNA gyrase is an enzyme that introduces super-helical turns into bacterial DNA. The chromosome of *E. coli* consists of 4,639,221 bp and has a contour length of about 1.7 nm, some 850 times the length of the *E. coli* cell. When *E. coli* cells are lysed, fibres are released in the form of loops attached to the broken envelope of the cell. The DNA of these loops does not exist in the extended form of a free duplex but is compacted by association with proteins. Bacterial chromosomes are in a highly folded and coiled state. Such a chromosome is known as folded genome. The bacterial nucleoid consists of hundreds of DNA loops, shaped in part by non-specific DNA bridging proteins such as histone-like nucleoid-structuring (H-NS) protein, leucine-responsive regulatory protein (Lrp) and structural maintenance of chromosomes (SMC) protein. H-NS protein is dynamically organized between two DNA molecules in register with their helical pitch. Dimeric structure of H-NS protein is hypothesized. H-NS interacts with DNA duplexes as a dimer in either a parallel or anti-parallel configuration. A folded genome of *E. coli* has about 100 loops, folds or domains. These folds are stabilized by RNA molecules. The loops are 50–100 kbp in length. Each fold in the chromosome is independently supercoiled. The formation of loops and of supercoils within the loops packs the >1000 μ long *E. coli* chromosome into a structure of 1–2 μ diameter. The folded genome is stabilized by RNA and proteins associated with it.

DNA packaging in eukaryotic organelles: The mitochondria and chloroplasts are the eukaryotic organelles that are classified as SEMI-AUTONOMOUS ORGANELLES. These organelles also contain their own circular DNA molecules. Both organelles have evolved from prokaryotes by endocytosis into proto-eukaryotic cells. That is why DNA packing of organelles DNA is somewhat similar to that of bacterial DNA, but the amount of DNA is much smaller.

DNA packaging in nucleus of diploid eukaryotes: Diploids possess two complete sets of chromosomes, except for the sex chromosomes in one of the sexes. Number of base pairs in a eukaryotic chromosome is very large. Humans have 2.2 m of DNA in each diploid cell, typically only 6 μ wide and 2 μ deep. Basic proteins are protamines (in spermatozoa) and histones. Acidic (non-histone) proteins perform structural and regulatory functions. In eukaryotes, chromosomes are diffused throughout the nucleus during interphase. These appear in form of darkly stained bodies during cell division. To achieve the overall packing ratio, DNA is not packaged directly into final structure of chromatin.

Instead, it contains several hierarchies of organization. The nucleosome is only the first level of packaging nuclear DNA. The BEADS-ON-A-STRING fibres are 10 nm in diameter, which can be compacted further into a '30-nm' chromatin fibre. Three levels of DNA packaging in eukaryotic chromosome are nucleosome, solenoid structure and scaffolding protein complex. The DNA is wound around four core histones (H2A, H2B, H3 and H4) to form a NUCLEOSOME. From 2 nm of the double helix to 11 nm in thickness, H1 acts as a linker histone as it links together nucleosome beads and then influences them to form a SOLENOID-LIKE STRUCTURE. This makes a 30-nm-thick structure. There is additional structure and packaging that is not yet understood. There is some evidence for a central axis SCAFFOLDING PROTEIN COMPLEX, which is thought to be required for the amount of packaging found in the highly compact chromosomes like mitotic chromosomes.

DNA packaging in viruses/bacteriophages: Viruses/bacteriophages have a single chromosome, which has a single molecule of nucleic acid. This viral chromosome may be linear or circular. The nucleic acid may be DNA or RNA and single stranded or double stranded. DNA-containing viruses are known as DEOXYVIRUSES, and RNA-containing viruses are known as RIBOVIRUSES. Viral genomes are packaged into capsids which are either symmetrical or quasi-symmetrical structures assembled from only one or a few types of protein monomers.

DNA packaging of bacterial plasmids: Plasmids are capable of independent transmission. Plasmid DNA is a double-stranded closed circular ring like assembly. Supercoiling is present in plasmid DNA. Plasmid DNA has PLECTONEMIC COILING (twisted thread).

DNA photolyase: It is a monomeric protein of 55–65 kDa with two chromophore cofactors: a pterin in the form of methenyl tetrahydrofolate (MTHF) and a flavin in the form of flavin adenine dinucleotide (FADH). There are two types of photolyases: DNA photolyase and 6-4 photoproduct photolyase. DNA photolyase repairs cyclobutane pyrimidine dimmers induced by UV light by splitting the cyclobutane ring. 6-4 photoproduct photolyase repairs the second major UV photoproduct by light-initiated reaction.

DNA pistons: Researchers in the UK and Germany have managed to assemble tetrahedrons out of DNA 'struts' with some chemical trickery and then fed the shape DNA 'fuel' to get the tetrahedron to contract. Some 'anti-fuel' expands the shape again, creating a sort of piston with all sorts of potential. The researchers are currently working to assemble larger structures using the tetrahedrons as building blocks. Possible applications of the technology range from drug delivery to the motors of nanoscale robots.

DNA polymerase δ: DNA polymerase δ (pol δ) is exceedingly processive and has proofreading 3′→5′ exonuclease activity. It is thought to be the main polymerase involved in lagging-strand synthesis.

DNA polymerase ε: DNA polymerase ε (pol ε) is extremely processive and has proofreading 3′→5′ exonuclease activity. It is related to pol δ and is thought to be the main polymerase involved in leading-strand synthesis. It also has a role in gap filling of lagging strands.

DNA polymerase β: DNA polymerase β (pol β) is implicated in base-excision repair and gap-filling synthesis. MW of DNA polymerase β is <50,000. It is a basic protein resistant to reagents that block sulfhydryl groups. It copies activated DNA well and, to somewhat lesser extent, the synthetic template $A_n dT_{15}$. It contains no detectable nuclease activity. DNA pol β is present in eukaryotic cells in resting and differentiated states.

DNA polymerase α: DNA polymerase α (pol α) is supposed to carry out lagging-strand replication. Its synonyms are DNA primase, DNA polymerase. It forms a complex with a small catalytic (PriS) and a large non-catalytic (PriL) subunit, with the Pri subunits acting as a primase (synthesizing an RNA primer) and then with DNA pol α elongating that primer with DNA nucleotides. After around 20 nucleotides, elongation is taken over by pol ε (on the leading strand) and pol δ (on the lagging strand). Molecular weight of pol α is >100,000. It copies activated double-stranded DNA. It has almost no ability to copy oligo-homopolymer $A_n dT_{15}$. Polα is involved in the replication of cellular DNA and some mammalian viruses (e.g. SV40 and polyoma). Polα is believed to carry out lagging-strand replication.

DNA polymerase γ: DNA polymerase γ (pol γ) is an acidic protein that has molecular weight >100 kDa. It needs sulfhydryl-containing compounds for maximal activity. Pol γ is also seen in eukaryotic cells in resting and differentiated states but is usually localized in mitochondria, and it is believed to be involved exclusively in mitochondrial DNA replication. Pol γ is the sole DNA polymerase in animal mitochondria. It also has proofreading 3′→5′ exonuclease activity.

DNA polymerase ζ: DNA polymerase ζ (pol ζ) is involved in the bypass of DNA damage in eukaryotes.

DNA polymerase Family D: The family D of DNA polymerase was discovered by Y. Ishino in 1998 from *Pyrococcus furiosus* and *Methanococcus jannaschii*. The PolD complex is a heterodimer of two chains, each encoded by DP1 (small proofreading) and DP2 (large catalytic). Unlike other DNA polymerases, the structure and mechanism of the DP2 catalytic core resemble that of multisubunit RNA polymerases. The DP1-DP2 interface resembles that of Eukaryotic Class B polymerase zinc finger and its small subunit. DP1, a Mre11-like exonuclease, is likely the precursor of small subunit of pol α and pol ε, providing proofreading capabilities now lost in eukaryotes. Its N-terminal HSH domain is similar to AAA proteins, especially pol III subunit δ and RuvB, in structure. DP2 has a Class II KH domain. *Pyrococcus abyssi* polD is more heat stable and more accurate than *Taq* polymerase but has not yet been commercialized. It has been proposed that family D-DNA polymerase was the first to evolve in cellular organisms and that the replicative polymerase of the LAST UNIVERSAL CELLULAR ANCESTOR (LUCA) belonged to family D.

DNA polymerase I: Like all known DNA polymerases, DNA polymerase I requires a primer from which to initiate replication and polymerizes deoxyribonucleotides into DNA in the 5′→3′ direction using the complementary strand as a template. Newly synthesized DNA is covalently attached to the primer but only hydrogen bonded to the template. The template provides the specificity according to Watson-Crick base pairing. Each incoming dNTP is positioned by base pairing with the appropriate template nucleotide, and a phosphodiester bond is created by nucleophilic attack of the primer-strand 3′ hydroxyl group onto the alpha phosphate of the incoming dNTP. Only the alpha phosphate of the dNTP is incorporated into newly synthesized DNA. The rate of DNA synthesis by DNA polymerase I is only 20 nucleotides/second, much slower than the rate of 1000 nucleotides/second measured for the replication of *Escherichia coli* DNA. P. De Lucia and J. Cairns in 1969 isolated a viable mutant (*polA*) lacking DNA polymerase I activity, an indication that pol I is not the main enzyme used to replicate DNA. Interestingly, *polA* mutants are defective in repairing DNA damage. DNA polymerase I is involved in termination of DNA replication. DNA polymerase I from *E. coli* was the first DNA polymerase characterized. There are around 400 molecules of the enzyme per cell. *E. coli* DNA polymerase I (abbreviated as pol I) is a single large protein with a molecular weight of approximately 103 kDa. The enzyme requires a divalent cation (Mg^{++}) for activity and has three enzymatic activities connected with it: (i) 5′→3′ DNA polymerase activity, (ii) 3′→5′ exonuclease (proofreading) activity and (iii) 5′→3′ exonuclease (nick translation) activity. Used for nick translation and for the production of the KLENOW FRAGMENT. *Aka* KORNBERG ENZYME. *See* NICK TRANSLATION.

DNA polymerase II: It can add new nucleotides at the 3′-end. DNA polymerase II (pol II) has 100 molecules/cell. Enzyme is around 90,000 kDa in size. It is coded by *pol b* gene. Synthesis is induced during S phase. It can re-initiate DNA synthesis downstream of gaps. It has low error rate but is too slow and hence cannot be used during normal DNA synthesis. The crystal structure of a target fragment of yeast DNA pol II reveals a heart-shaped dimeric protein with a large central hole. It provides a molecular model of the enzyme as ATP-modulated clamp with two sets of jaws at opposite ends connected by multiple joints. DNA pol II is implicated in repair of damaged DNA and has 3′→5′ exonuclease activity.

DNA polymerase III: DNA polymerase III (pol III) is the main polymerase for *in vivo* replication and has 3′→5′ exonuclease proofreading ability. It is also referred to as replicase, pol III, polC, DnaE or the alpha subunit. This is a major enzyme for DNA replication. The DNA polymerase III has 10 molecules/cell. DNA pol III from *Escherichia coli* is a single protein of molecular weight 130 kDa. Though the molecule has DNA polymerase activity by itself, it works to replicate DNA in the bacterial cell in conjunction with other proteins. This multiprotein complex is referred to as the pol III holoenzyme. The proteins (called subunits) that associate with DNA pol III in the holoenzyme perform several functions. The most interesting subunit is called beta.

DNA polymerase IV: The SOS-inducible, error-prone DNA polymerase IV (pol IV), encoded by *dinB*, is required for adaptive point mutation in the *Escherichia coli lac* operon. A non-polar *dinB* mutation reduces adaptive mutation frequencies by 85% but does not affect adaptive amplification, growth-dependent mutation or survival after oxidative or UV damage. Pol IV, together with the major replicase pol III, can account for all adaptive point mutations at *lac*. These results have thus identified a role for pol IV in inducible genetic change.

DNA polymerase V: DNA polymerase V (pol V) participates in bypassing DNA damage. When *Escherichia coli* is exposed to high levels of radiation or of a mutagen, major damage to the bacterial DNA can occur. The cell responds by inducing a special 'last resort' repair pathway called the SOS REPAIR PATHWAY. Among the SOS, genes that are induced are UmuC and UmuD. The products of these two genes form pol V. Actually, one copy of UmuC and two copies of a truncated form of UmuD are required to form this polymerase.

DNA polymerase variant: A DNA polymerase with a single mutation and a divalent calcium cofactor catalyses the synthesis of unnatural N3′→P5′ phosphoramidate (NP) bonds to form NP-DNA. However, this template-directed phosphoryl transfer activity remains orders of magnitude slower than native phosphodiester synthesis. V.S. Lelyveld and others in 2023 used time-resolved X-ray crystallography to show that NP-DNA synthesis proceeds with a single detectable calcium ion in the active site. Using insights from isotopic and elemental effects, they propose that one-metal-ion electrophilic substrate activation is inferior to the native two-metal-ion mechanism. Scandium (III) in particular confers highly specific NP activity with kinetics enhanced by more than 100-fold over calcium (II), yielding NP-DNA strands up to 100 nucleotides in length.

DNA polymerase, eukaryotic: One of a group of enzymes primarily involved in copying a single-stranded DNA molecule to make its complementary strand. Eukaryotic DNA polymerases contribute to chromosomal replication, repair, crossing-over and mitochondrial replication. To initiate replication, DNA polymerases require a priming RNA molecule. They extend the DNA using deoxyribonucleotide triphosphates dATP, dGTP, dCTP and dTTP as substrates and releasing pyrophosphates. The dNMPs are added to the 3′-OH end of the growing strand (i.e. DNA replication proceeds from 5′ to 3′ end). There are numerous types of DNA polymerase. Some help in replication of DNA when a cell divides, and others help in the day-to-day repair and maintenance of DNA.

DNA polymerases: Enzymes that synthesize a new strand of DNA complementary to an existing DNA or RNA template. DNA polymerases are commonly used during DNA amplification using PCR.

DNA polymerases in prokaryotes: A group of enzymes required for DNA synthesis. Scientists have found five DNA polymerases in *Escherichia coli*. DNA POLYMERASE I removes the RNA primer by 5′→3′ exonuclease activity and replaces the primer by its polymerase activity in the lagging strand. DNA POLYMERASE II has main role is in repair and also a backup of DNA polymerase III. It has 3′→5′ exonuclease activity. DNA POLYMERASE III is the main enzyme responsible for replication in prokaryotes. DNA polymerase III is the main enzyme for replication. The polymerization and processivity rate are maximum in DNA polymerase III. It also assumes proofreading 3′→5′ exonuclease activity. DNA POLYMERASE IV has main role is in DNA repair during SOS response, when DNA replication is stalled at the replication fork. DNA POLYMERASE V is also involved in translesion synthesis during SOS response and DNA repair. It is made up of UmuC monomer and UmuD dimer. *See details as separate terms.*

DNA polymerases, eukaryotic, type A: Type A DNA polymerases include DNA polymerase γ, whose main function is to replicate and repair mitochondrial DNA. It also functions by proofreading 3′ to 5′ exonuclease activity. Mutations on pol γ significantly affect the mitochondrial DNA causing autosomal mitochondrial disorders. *See* DNA POLYMERASE γ.

DNA polymerases, eukaryotic, type B: Type B DNA polymerases include three members: DNA polymerase α, DNA polymerase δ and DNA polymerase ε. These three enzymes are the main polymerases employed in DNA replication. Pol α works by binding to the primase enzyme, forming a complex, where they both play a role in initiating replication. Primase enzyme creates and places a short RNA primer which allows pol α to start the replication process. Pol δ starts the synthesis of the lagging strand from pol α, while pol ε is believed to synthesize the leading strand during replication. Studies indicate that pol δ replicates both the lagging and leading strand. Pol δ and ε also have a 3′ to 5′ exonuclease activity. *See* DNA POLYMERASE α, DNA POLYMERASE δ *and* DNA POLYMERASE ε.

DNA polymerases, eukaryotic, type C: Type C (or Family X) includes three members: DNA polymerase β, DNA polymerase μ and DNA polymerase λ. Pol β has a short-patch base-excision repair mechanism where it repairs alkylated or oxidized bases. Pol λ and pol μ are important for rejoining DNA double-strand breaks due to hydrogen peroxide and ionizing radiation, respectively. *See* DNA POLYMERASE β, *and* DNA POLYMERASE λ.

DNA polymerases, eukaryotic, type D: Type D (or family Y) includes three members: DNA polymerases η, DNA polymerase ι and DNA polymerase κ. These DNA polymerases are majorly used in repairing of DNA by a mechanism known as translesion synthesis. They are prone to errors during DNA synthesis. Pol η functions by accurately ensuring the translesion synthesis of DNA damages that is caused by ultraviolet radiation. Pol κ is still understudies but one of its known functions is to extend or insert specific bases at certain DNA lesions. Translesion synthesis polymerases are activated by stalled replicative DNA polymerase. In addition to above four types, there are some present other important DNA polymerases.

DNA polymerases, types of eukaryotic: Eukaryotic DNA polymerases are classified into various types: TYPE A, TYPE B, TYPE C, TYPE D, TERMINAL DEOXYNUCLEOTIDYL TRANSFERASE (TDT) and TELOMERASE. *See details as separate terms.*

DNA polymorphism: One of two or more alternate forms (alleles) of a chromosomal locus that differ in deoxyribonucleotide sequence or have variable numbers of repeated deoxyribonucleotide units. Two or more kinds of DNA sequence variations that are maintained in a breeding population.

DNA predictions: J.D. Watson, F.H.C. Crick and others theorized in early 1950s that DNA had a helical structure. This implied that DNA's X-ray diffraction pattern would be 'X shaped'. This prediction was followed from the work of W. Cochran, F.H.C. Crick and V. Vand (and independently) by A. Stokes in 1952. The Cochran-Crick-and-Stokes theorem provided a mathematical explanation for the empirical observation that diffraction from helical structures produces 'x-shaped' patterns. Also, in their first paper, J.D. Watson and F.H.C. Crick (1953) predicted that the double-helix structure provided a plausible mechanism for DNA replication.

DNA primase: Enzyme that synthesizes a short strand of RNA on a DNA template, producing a primer for DNA synthesis.

DNA primer: A short DNA piece that is required for a polymerization reaction (e.g. DNA synthesis) and is structurally similar to the product of the reaction itself. A primer pairs with one strand of the template DNA in the POLYMERASE CHAIN REACTION (PCR) technique. In PCR testing (e.g. in a paternity test), the primer is selected to be complementary to the analytically relevant sequence of DNA. It is the growing end of the DNA chain, and it merely provides a free 3′-OH end at which the enzyme DNA polymerase adds on deoxyribonucleotide units (monomers). Which deoxyribonucleotide is added is dictated by base pairing to the template DNA chain. Without a DNA primer sequence, a new DNA chain cannot form, as DNA polymerase is not able to initiate DNA chains.

DNA primer synthesis: A primer must be synthesized by an enzyme called PRIMASE, which is a type of RNA polymerase, before DNA replication can occur. The synthesis of a primer is necessary because the enzymes that synthesize DNA, which are called DNA POLYMERASES, can only attach new DNA nucleotides to an existing strand of nucleotides.

DNA probe: A fragment of single-strand DNA that contains a specific nucleotide sequence that is used to detect a complementary sequence in a sample genome. The probe is labelled with a radioactive or chemical tag, which provides a signal when it binds to the complementary DNA. Probe DNA can come from cDNA from highly expressed mRNA from a tissue, homologous gene from a related organism or DNA obtained from 'reverse genetics' (protein → DNA). When a DNA probe of known composition and length is mixed with pieces of DNA (genes) from a chromosome, the probe will pair with its exact counterpart in the chromosomal DNA pieces (genes), forming a stable double-stranded hybrid. The presence of this 'labelled' probe is detected visually or with the help of another detection instrument. Because the composition of the DNA probes is known, scientists can scan through a chromosome, noticing segments of DNA (i.e. genes) that seem to be linked to genetic diseases. DNA probes are used in DNA-DNA and DNA-RNA hybridization experiments. *Aka* GENE PROBE *or* GENETIC PROBE.

DNA profiling: Invented in 1985 by A. Jeffreys, this technique is used by forensic (i.e. crime-solving) chemists to match biological evidence (e.g. a bloodstain) from a crime scene to the person (e.g. the assailant) involved in that particular crime. DNA profiling involves the use of restriction fragment length polymorphism (RFLP) analysis or allele-specific oligonucleotide (ASO)/PCR analysis to identify the specific sequence of bases (i.e. nucleotides) in a piece of DNA taken from the biological evidence. Since the specific sequence of bases in DNA molecules is different for each individual (due to DNA polymorphism), a criminal's DNA can be matched to that of the evidence to prove guilt or innocence. Biological evidence may include, among other things, blood, hair, nail fragments, skin and sperm. *Aka* GENETIC FINGERPRINTING *or* DNA FINGERPRINTING.

DNA protective mechanism: The cells that are destined to become gametes in multicellular organisms are almost always shielded both physically and chemically from excessive attack from outside world. This process thus serves as a defence of the DNA that is destined to be transmitted to next generation.

DNA puff: Cross band in a polytene chromosome swollen due to the relaxation of chromatin of banded region at the time of high activity.

DNA pull-down assay: Pull-down assays are used to selectively extract a protein-DNA complex from a sample. Typically, the pull-down assay uses a DNA probe labelled with a high-affinity tag (e.g. biotin), which allows the probe to be recovered or immobilized. A biotinylated DNA probe can be complexed with a protein from a cell lysate in a reaction similar to that used in the ELECTROPHORETIC MOBILITY SHIFT ASSAY (EMSA) and then used to purify the complex using agarose or magnetic beads. The proteins are then eluted from the DNA and detected by Western blot or identified by mass spectrometry. Alternatively, the protein may be labelled with an affinity tag, or the DNA-protein complex may be isolated using an antibody against the protein of interest (similar to a super shift assay). In this case, the unknown DNA sequence bound by the protein is detected by Southern blotting or through PCR analysis.

DNA pull-down assay, limitations of: (i) Long DNA probes can show significant non-specific binding, (ii) requires very specific antibodies for native proteins, (iii) nuclease-free conditions required and (iv) assay must be performed *in vitro*.

DNA pull-down assay, strengths of: (i) Enrichment of low abundant targets, (ii) end-labelled DNA can be generated by several methods, (iii) isolation of intact complex and (iv) compatible with immunoblotting and mass spectrometry analysis.

DNA readers: The proteins that recognize and bind with DNA modifications and impact their functional outcome. Methyl-CG-binding proteins are examples of DNA readers. MeCP2 has been identified as a microsatellite DNA-binding protein that targets the 5^{hm}C-modified CA-rich strand of DNA.

DNA recombination: (i) The exchange of chromosomal segments that occurs when homologous DNA sequences become closely associated. This is how the genetic information from mother and father changes in every generation to give genetically unique offspring. (ii) Molecular process representing double-strand break (DSB) DNA repair. Homologous recombination (HR) involves two homologous DNA double strands that may or may not result in crossing over of the initial DNA strands. *Aka* GENETIC RECOMBINATION *or* GENERALIZED RECOMBINATION.

DNA recombination in eukaryotes: DNA recombination in eukaryotes comprises both mitotic and meiotic recombination. Mitotic recombination takes place in somatic cells, while meiotic recombination occurs in germ cells. Both mitotic and meiotic recombination involve exchange of segments. Mitotic recombination repairs damaged chromosomes. Meiotic recombination, more important in sexually reproducing organisms, leads to creation of new combination of genes (i.e. new linkages).

DNA recombination, types of: DNA recombination is categorized into various types: HOMOLOGOUS (OR GENERAL) RECOMBINATION, DNA TRANSPOSITION, ILLEGITIMATE (or NON-HOMOLOGOUS) RECOMBINATION, ARTIFICIAL RECOMBINATION and SITE-SPECIFIC RECOMBINATION.

DNA repair: Collective name given to various biochemical processes that repair unintended changes in the DNA. Most damage to DNA is repaired by removal of the damaged bases, followed by resynthesis of the excised region. Some lesions in DNA, however, can be repaired by direct reversal of the damage, which may be a more efficient way of dealing with specific types of DNA damage that occur frequently. Repair of damaged DNA is nature's method to maintain the integrity of genetic information. DNA repair systems make use of enzymes that reverse the chemical modification of nucleotide bases as well as more complicated multienzyme systems that depend upon the inherent redundancy of the information in duplex DNA to restore the damaged molecule.

DNA repair genes: DNA repair genes encode enzymes that fix DNA damage. Different repair genes function in homologous recombination or DSB repair, including *RecA*, Rad50/SMC complex, encompassing NHEJ, break-induced replication (BIR), single-strand annealing (SSA) and synthesis-dependent strand annealing (SDSA). Further, gene-encoded enzymes are used in nucleotide excision repair

(NER) (e.g. eroderma pigmentosum group D; cyclobutane pyrimidine dimers; XPD), base-excision repair (BER) (e.g. thymine DNA glycosylase; TDG-glycosylase), mismatch repair (e.g. msh2), damage reversal (e.g. cyclobutane pyrimidine dimer (CPD) photolyase), damage tolerance (e.g. Rad6), checkpoint-regulatory proteins (e.g. RecQ helicases, Rad17, ATM, PARP), DNA polymerases and others (e.g. p53).

DNA repair pathways: At least five major DNA repair pathways – (i) BASE-EXCISION REPAIR (BER), (ii) NUCLEOTIDE EXCISION REPAIR (NER), (iii) MISMATCH REPAIR (MMR), (iv) HOMOLOGOUS RECOMBINATION (HR) and (v) NON-HOMOLOGOUS END JOINING (NHEJ) – are active throughout different stages of the cell cycle, allowing the cells to repair the DNA damage. *See details as separate terms.*

DNA replication: Reproduction of a DNA molecule (inside a cell). This process can be viewed as occurring in stages, in which the first stage consists of an enzyme 'unwinding' the double helix of the DNA molecule at a replication origin, forming a replication fork. At the replication fork, the two separated (DNA) strands serve as templates for new DNA synthesis. The template strand is not destroyed. That new DNA synthesis is accomplished on each strand via enzymes, known as DNA polymerases, which travel along each (single) strand making a second complementary strand by catalysing the addition of DNA bases to the newly growing strands. The end result is two new double helices (DNA molecules), each of which has one chain from the original DNA molecule and one chain that was newly synthesized by the DNA polymerase enzymes. This is termed SEMI-CONSERVATIVE DNA REPLICATION. In humans and other eukaryotes, replication occurs in the cell nucleus. Epigenetic reprogramming occurs during DNA replication. *See* EPIGENETIC REPROGRAMMING.

DNA replication errors in spontaneous mutations: Replication errors made by DNA polymerases, if not corrected, result in spontaneous mutations, of which transition mutations are most common. Mispairing during DNA replication may lead to mutation.

DNA replication factories: DNA is recruited to replication factories, and sister replication forks initiated from the same origin are held together within a single replication factory. Formation of replication factories is a consequence of DNA replication itself. Sister replication forks generated from the same origin stay associated with each other within a replication factory, while the entire replicon is replicated.

DNA replication fidelity: The ability of a DNA polymerase to correct misincorporated nucleotides as a result of its 3′→5′ exonuclease activity. The most recently added nucleotide is excised if it is non-complementary. In prokaryotes as well as eukaryotes, this mechanism is contributing significantly to the total fidelity of DNA replication. The processes recognized in this process are base selection, proofreading and mispair correction.

DNA replication fork coupling: Dynamic chromatin interactions accompanying fork movement. DNA replication forks are spatially coupled in human and murine cells. This coupling arises from sister forks at the same origin or tightly associated chromatin loop anchors, ensuring precise replication, elongation and termination. Y. Liu and others in 2024 developed a replication-associated *in situ* high-throughput chromosome conformation capture (HiC) method to capture chromatin interactions involving nascent DNA. Hierarchical structure of the DNA replication machinery and suggests that replication termination is established during the replication initiation stage rather than being triggered by DNA polymerase collision.

DNA replication in bacteria: In prokaryotes, the process of replication has mostly been studied in *Escherichia coli*, where length of the chromosome is 1 mm, and one round of DNA replication is completed in 30 min. This time is almost equal to the cell doubling time of the bacterial cells. Rate of replication in prokaryotes is 25,000–30,000 nucleotides per minute. Division of the cell rapidly follows replication of DNA, and replication instantly restarts in the daughter cells. DNA replication in bacteria is semi-conservative, as evidenced by M. Meselson and F.H. Stahl in 1958.

DNA replication in eukaryotic cells: At chromosomal level, DNA replication was shown to be semi-conservative by J.H. Taylor, P.S. Woods and W.L. Hughes in 1957. They labelled dividing root-tip cells of the broad bean *Vicia faba* with tritiated (^{3}H)-thymidine which is specifically incorporated into newly synthesized DNA. It was noticed that a new chromosome was replicating semi-conservatively because it consisted of half radioactive newly formed chromosome and half parental chromosome. When the new chromosomes (hybrid) were allowed to grow and divide on unlabelled media, half of the new chromosomes were labelled, and the other half were unlabelled. Thus, random labelling of chromosomes did not occur, and the mode of replication was consequently not dispersive. Semi-conservative replication of the eukaryotic chromosome is based essentially on an underlying structure that is a continuous length of DNA double helix. Although the basic mechanism of DNA replication is same in prokaryotes and eukaryotes, due to complexity, there lie differences. Using the same technique, P. Filner in 1965 proved semi-conservative mode of DNA replication in higher plant *Nicotiana tobaccum*.

DNA replication in phage T2: The first study at the molecular level that replication might be semi-conservative was conducted by C. Levinthal in 1956 in an autoradiographic study of the distribution of ^{32}P-labelled DNA from phage T2. They suggested that a large piece of DNA replicated in complementary manner, as suggested by J.D. Watson and F.H.C. Crick in 1953.

DNA replication in prokaryotes and eukaryotes, differences in: (i) DNA replication in prokaryotes commences in cytoplasm, but in eukaryotes, it takes place in the nucleus. (ii) In prokaryotes, DNA replication occurs continuously during their life, but in eukaryotes, it occurs almost only in S phase of the cycle. (iii) DNA replication in prokaryotes proceeds from single origin mostly in both the directions, while in eukaryotes, it proceeds from many origins in both the directions. (iv) In prokaryotes, both lagging and leading strands are synthesized by DNA polymerase III, while in eukaryotes, the leading and lagging strands are synthesized by DNA polymerases α and δ, respectively. (v) RNA primers are removed in prokaryotes by DNA polymerase I and in eukaryotes by DNA polymerase β. (vi) DNA replication in prokaryotes does not require PROLIFERATING CELL NUCLEAR ANTIGEN (PCNA), while DNA replication in eukaryotes PCNA protein is associated with DNA polymerase. (vii) In prokaryotes, there is no telomerase activity, while in eukaryotes, telomerase completes the lagging strand in the telomere regions of chromosome. (viii) 2000 nucleotides are replicated per second in prokaryotes, but only 100 nucleotides are replicated per minute in eukaryotes. (ix) Okazaki fragments are 1000–2000 nucleotide long in prokaryotes but less than 200 nucleotide long in eukaryotes. (x) Prokaryotes have only one replicon in their genome, while eukaryotes have as many as 500–60,000 replicons per chromosome. (xi) In prokaryotic cells, nucleosomes do not bind with DNA, while in eukaryotes, DNA binds immediately after replications associates with histones to form nucleosomes.

DNA replication in study of cancer: The study of the mechanism of DNA replication is crucial for the understanding of cancer development, because every cancerous cell must replicate its DNA before division. The discovery of initiation protein Noc3p will not only advance our study on DNA replication but also help the fight against cancer.

DNA replication inhibitor: Compounds that bind DNA covalently and inhibit DNA replication (e.g. acridine, acriflavine, 8-aminoquinoline, echinomycin and proflavine).

DNA replication initiation: In a cell, DNA replication begins at specific locations, or origins of replication, in the genome. Unwinding of DNA at the origin and synthesis of new strands, accommodated by an enzyme known as helicase, results in replication forks growing bidirectionally from the origin. A number of proteins are associated with the replication fork to help in the initiation and continuation of DNA synthesis.

DNA replication licensing system: MATURATION PROTEIN FACTOR (MPF) has two components: kinase (34 kDa) and cyclin (45 kDa). Two types of cyclins have role in control of mitosis. Members of cyclin A play an important role in DNA synthesis and are thus S-phase cyclins. Cyclin B are mitotic cyclins. REPLICATION LICENSING FACTOR (RLF) restricts DNA replication to once during cell division.

DNA replication once per cell cycle: Eukaryotic DNA replication is regulated to ensure all chromosomes replicate once and only once per cell cycle.

DNA replication origin: A unique location on the chromosome where DNA replication begins. Supercoiling is necessary at the replication origin to allow first separation of the two DNA strands and the formation of a replicating growing point.

DNA replication proteins and enzymes: Major ones are: (i) HELICASE (opens up the DNA at the replication fork), (ii) SINGLE-STRAND BINDING (SSB) PROTEINS (coat the DNA around the replication fork to prevent rewinding of the DNA), (iii) TOPOISOMERASE (works at the region ahead of the replication fork to prevent supercoiling), (iv) PRIMASE (synthesizes RNA primers complementary to the DNA strand), (v) DNA POLYMERASE III (extends the primers, adding on to the 3′-end, to make the bulk of the new DNA), (vi) DNA POLYMERASE I polymerizes deoxyribonucleotides into DNA in the 5′→3′ direction using the complementary strand as a template, (vii) newly synthesized DNA is covalently attached to the primer, but only hydrogen bonded to the template, (viii) DNA POLYMERASE II has main role is in repair and also a backup of DNA polymerase III. It has 3′→5′ exonuclease activity. RNA primers are removed and replaced with DNA by the gaps between DNA fragments and (ix) DNA LIGASE seals the gaps.

DNA replication timing programme: Highly organized temporal sequence of locus replication events that occur during the synthesis phase (S phase) of the cell cycle from early-to-late S phase.

DNA replication, activities in: DNA replication comprises the following activities: (i) unwinding of double-helical DNA, (ii) stabilization of single-stranded DNA, (iii) synthesis of pre-primer, (iv) formation of primer, (v) extension of primer by DNA polymerase III, (vi) excision of primer, (vii) gap filling by DNA polymerase I and (viii) zipping.

DNA replication, circular fork model: A common type of replication that takes place in circular DNA (e.g. in *Escherichia coli* and other bacteria) is called THETA REPLICATION, because it generates a structure that resembles the Greek letter theta (θ). In theta replication, double-stranded DNA begins to unwind at the replication origin, producing single-stranded nucleotide strands that then serve as templates on which new DNA can be synthesized. The unwinding of the double helix generates a loop, termed a REPLICATION BUBBLE.

DNA replication, different modes of: M. Delbruck and G.S. Stent in 1957 anticipated three modes of DNA replication: SEMI-CONSERVATIVE, CONSERVATIVE and DISPERSIVE. In order to discover the actual mode of replication of DNA, chemical labelling of DNA was done. In later duplications, the label was detected. Expectations in the three modes of replication were predicted. In case of conservative replication, both strands of newly synthesized DNA double helix should be labelled, while the old double helices remain unlabelled. In case of semi-conservative replication, the replicated double helixes carry one old and one new stand, and all are, therefore, labelled. In dispersive mode of replication as well, label in both the strands of both the helixes will be expected. Semi-conservative and dispersive modes can be differentiated in next cycle of DNA replication.

DNA replication, fork model of: The most widely used mechanism of DNA replication comprises formation of replication forks, moving in one direction in unidirectional replication and in both directions in bidirectional replication. This model applies to both linear and circular DNA molecules. This model of DNA replication is of two types: LINEAR FORK MODEL and CIRCULAR FORK MODEL.

DNA replication, linear fork model: Eukaryotic DNA involves a linear mode of replication. Eukaryotes have longer linear chromosomes and initiate replication at multiple origins within these.

DNA replication, proofreading enzymes in: Mistakes during DNA replication can be identified and repaired by proofreading enzymes. A mismatched nucleotide may occur at the rate of one per 100,000 base pairs, causing a pause in replication. DNA repair enzymes accomplish a proofreading function and reduce the error rate to one per billion. Many complex mechanisms have evolved to repair damages in DNA, which can transpire as a result of damage from ultraviolet radiation, X-rays and so on, as well as from errors during replication. Over 50 genes have been implicated in yeast to be involved in repair mechanisms.

DNA replication, proofreading in: Any mechanism for correcting errors in nucleic acid synthesis that involves inspection of individual bases after they have been added to the DNA chain. This function is carried out by a 3′ to 5′ exonuclease, among others. Proofreading intensely increases the fidelity of the base-pairing mechanism.

DNA replication, rolling circle: In some cases (e.g. F factor of *Escherichia coli*), DNA replicates in a rolling-circle fashion.

DNA replication, steps in: For the sake of discussion, DNA replication comprises three steps: initiation, elongation and termination. Initiation includes activation of deoxyribonucleotides, exposure of unwinding of double helix and synthesis of RNA primer. Elongation includes formation of leading and lagging strands, chain growth, removal of RNA primers, gap filling and nick sealing. Termination is the last activity in DNA replication.

DNA robots as molecular machines: Such molecular machines could bind to a cell surface, maybe carry a cargo and release something. They combined a programmable DNA track and a squad of mobile robotic walkers with a set of independently controlled molecular forklifts that can deliver parts on command. The result was a functioning nanofactory.

DNA robots spin gold: US scientists have developed microscopic robots composed of DNA that can follow instructions and work together like an assembly line to make products such as particles of gold. Nadrian Seeman and colleagues describe a tiny DNA factory consisting of a DNA track for assembly, three molecular forklifts that can deliver parts and a DNA WALKER that moves around like a car on an assembly line. The team had produced the first such DNA walker in 2004, knitting together strands of DNA to form a mobile molecule. The research takes advantage of DNA's unique ability to store information, which the team manipulated to adjust the structure of the molecular robots and how they connect to other molecules. Adding DNA strands to the walker and the forklifts allows them to move. Another study appearing in *Nature* featured a programmable DNA walker that could start, stop and turn.

DNA robots, future of: (i) Imagine if your phone behaves like a remote that controls your whole life. Imagine if the remote has the ability to save your life history to be the device that you can use to bring your memory back to life when you need it. These devices are currently being continuously developed by scientists and researchers around the world. Cellular telephone technology is expected to be far more advanced for 10 years or so. (ii) It is not entirely unreasonable to imagine seeing a cellular phone that can memorize your travel schedule and make sure everything is ready on the road, booking airline tickets and even train and give you the absolute best route to reach your destination on time. (iii) Imagine a phone that will control the central heating in your home to keep you warm even before you get out of bed in the morning or provide emergency authorities with information on your home if your home security system is triggered. (iv) Technology of the future remains to be seen, but we can assume from the way that things have changed in the past, things would be far more advanced in the future. Future technology may not yield any of these things, but they would likely be far better than earlier technologies. (v) Given how the world is now compared to what it was 100 years ago, it stands to reason that the future technology will bring us the things that we never felt. (vi) A robot could be repairing DNA inside the cell. (vii) Artificial DNA could power future computers. DNA uses just four basic building blocks, known as bases, to code proteins used in cell functioning and development. (viii) Findings in the area of DNA nanotechnology could lead to improvements in gene therapy, futuristic nanosized computers and other high-tech advances, as Japanese researchers say. Scientists have tried for years to develop artificial versions of DNA in order to take advantage of its amazing information storage capabilities. Already, DNA has been harnessed to create simple electronic circuits. Like natural DNA, the new rip-offs were right-handed, and some easily formed triple-stranded structures. The unique chemistry of these structures and their high stability offer unprecedented possibilities for developing new biotech materials and applications.

DNA safeguards: To safeguard the genetic material within a cell, organisms have evolved various DNA protection mechanisms. Antimutagens play role in protecting DNA against the mutagenic effects of radiation and environmental chemicals. Non-DNA repair (an intrinsic feature of an organism) also serves as means for protection of the genetic material.

DNA self-replication: The self-replication of palindromic duplex DNA-like nucleotides, 24-monomer long, in the absence of enzymes by means of a cycle that transfers information from template to copy is potentially capable of extension to include non-symmetrical sequences, selection and mutation. Replication proceeds by a chemical process involving the formation of an intermediate triplex structure and is sequence selective in the sense that mismatches impair its efficiency.

DNA separation: Following methods are used for DNA separation: (i) CALCIUM CHLORIDE DENSITY GRADIENT CENTRIFUGATION METHOD, (ii) ETHIDIUM BROMIDE METHOD, (iii) HYDROXYAPATITE COLUMNS AND NITROCELLULOSE FILTERS and (iv) DNA SEPARATION BY SILICA ADSORPTION. *See details as separate terms.*

DNA separation by silica adsorption, steps in: There are four fundamental steps of this DNA purification/isolation/purification method: (i) The sample is run through a microchannel. (ii) DNA binds to the channel and all other molecules remain in the buffer solution. (iii) The channel is washed of impurities. (iv) An elution buffer removes the DNA from channel walls, and the DNA is accumulated at the end of the channel. A sample (which may be anything from purified cells to a tissue specimen) is placed onto the chip and lysed. The resultant mixture of proteins, DNA, phospholipids and so on is then run through the channel, where the DNA is adsorbed by silica surface in the presence of solutions with high-ionic strength. The highest DNA adsorption efficiencies are seen to occur in the presence of buffer solution with a pH at or below the pKa of the surface silanol groups. The mechanism for this interaction possibly involves reduction of the surface of silica's negative charge owing to the high ionic strength of the buffer. This decrease in surface charge results in a decline in the electrostatic repulsion between the negatively charged DNA and the negatively charged silica. In the meantime, the buffer also decreases the activity of water by formatting hydrated ions. This step leads to the silica surface and DNA becoming dehydrated. These conditions result in an energetically favourable situation for DNA to adsorb to the silica surface.

DNA separation by silica adsorption: DNA separation by silica adsorption is an important method of DNA separation that is used in novel technologies that use microchannels. The principle behind this type of separation relies on DNA molecules binding to silica surfaces in the presence of certain salts and under certain pH conditions. Conventional methods for DNA extraction, such as centrifugation with ethanol or preparations using commercial purification kits, cannot be integrated onto microchips because they require multiple hands-on processing steps. In addition, they also require large equipment and high volumes of reagents and samples. DNA extraction on microchips provides a fast, cost-effective and effective for high-throughput screening, which also has a very small footprint. This new method has useful applications for biosensors, 'lab-on-a-chip' devices and other new technologies that require high-quality DNAt.

DNA separation from proteins, lipids, RNA and carbohydrates: For separating phage DNA, aqueous suspension of phage is shaken in phenol. Phenol breaks down the protein coat, which precipitates at the aqueous phenol coating. DNA gets precipitated in ethanol and can be separated. For separating bacterial DNA, bacterial cell wall and inner cell membrane can be dissolved by phenol. They are degraded by enzyme lysozyme and some detergents like SDS. Cell contents thus released are excluded from RNA by RNase and proteins by phenol. DNA is precipitated in ethanol.

DNA separation using salting out procedure: This procedure involves salting out of the cellular proteins by dehydration and precipitation with a saturated NaCl solution.

DNA sequence: The relative order of base pairs, whether in a DNA fragment, gene, chromosome or an entire genome. Sequence of bases in 5′→3′ direction (printed left to right) along a strand of DNA. There are four organic bases – adenine (A), thymine (T), guanine (G) and cytosine (C) – that define the sequence in a single-strand DNA. The DNA sequence of a gene determines the amino acid composition of a protein in a linear manner, the phenomenon known as collinearity. *See* BASE SEQUENCE ANALYSIS.

DNA sequence assembly: In order to sequence a genome of an organism, nucleotide sequences of overlapping fragments involving the entire genome are determined. The objective is to assemble them into the original DNA sequence from which the fragments were derived. One approach is the PAIRWISE SEQUENCE ALIGNMENT. For a given pair of sequences and a method for scoring similarity of sequences, one has to determine how similar two sequences are (best similarity score) and show where two sequences match (best alignment).

DNA sequence divergence: Nucleotide sequencing techniques made it possible to know the primary structure of many genes, which made it possible to compare pairs of homologous nucleotide sequences of a wide variety of genes in different organisms. This enabled us to evaluate the extent of sequence divergence in coding regions (exons) and non-coding regions (introns) of a gene. From the comparison of DNA sequences in coding regions, it became possible to evaluate the extent of sequence divergence for two types of substitutions: (i) a nucleotide replacement leading to an amino acid change (mis-sense mutations) and (ii) a nucleotide replacement leading to a synonymous codon change (same-sense mutations). In molecular terms, at DNA level, evolutionary rate is expressed as difference in nucleotide per site per year.

DNA sequencing: Refers to sequencing methods for determining the order of the nucleotide bases – adenine (A), guanine (G), cytosine (C) and thymine (T) – in a molecule of DNA. It is an important part of genetic analysis, not only for basic research into fundamental biological processes but also for applied research such as diagnostics, forensics and gene therapy. There are two main methods for DNA sequencing: Sanger's (dideoxy) method and Maxam-Gilbert method. These are different techniques but allow the sequencing of DNA of several kb in length. The full genomes of *Arabidopsis*, rice and human have been sequenced. *See* GENE SEQUENCING. Large-scale DNA sequencing requires quick, high-quality and cost-effective DNA sequencing. It requires facilities to process hundreds of thousands of samples per day. The sequence of bases tells scientists the type of genetic information that is carried in a particular segment of DNA. DNA sequencing is now done as a matter of routine using automated process.

DNA-sequencing automation and sample preparation: Automated DNA-sequencing instruments (DNA sequencers) can sequence up to 384 DNA samples in a single batch (run) in up to 24 runs a day. DNA sequencers carry out capillary electrophoresis for size separation, detection and recording of dye fluorescence and data output as fluorescent peak trace chromatograms. Sequencing reactions by thermocycling, clean up and resuspension in a buffer solution before loading onto the sequencer are performed separately. A number of commercial and non-commercial software packages can trim low-quality DNA traces automatically. These programs score the quality of each peak and remove low-quality base peaks (generally located at the ends of the sequence). The accuracy of such algorithms is below visual examination by a human operator but sufficient for automated processing of large sequence data sets. Large-scale sequencing aims at sequencing very long DNA pieces (e.g. whole chromosomes). Common approaches consist of cutting (with restriction enzymes) or shearing (with mechanical forces) large DNA fragments into shorter DNA fragments. The fragmented DNA is cloned into a DNA vector and amplified in *Escherichia coli*. Short DNA fragments purified from individual bacterial colonies are individually sequenced and assembled electronically into one long, contiguous sequence.

DNA sequencing by analysis of cloned genes: DNA sequencing is an approach used for analysis of cloned genes. Various methods of DNA (nucleotide) sequencing include MANUAL AND AUTOMATED METHODS. Manual methods use radioactive probes that are harmful while automated methods use fluorescence dyes that are safe.

DNA sequencing by chemical-cleavage method: DNA can be sequenced by a chemical procedure that breaks a terminally labelled DNA molecule partially at each repetition of a base. The lengths of the labelled fragments then identify the positions of that base. A.M. Maxam and W. Gilbert in 1977 described four reactions that cleave DNA preferentially at guanine, at adenine, at cytosine and thymine equally and at cytosine alone. The reagents used are highly toxic and very dangerous. When the products of these reactions are resolved by size, using electrophoresis on a polyacrylamide gel, the DNA sequence can be inferred from the pattern of radioactive bands. The technique allows sequencing of at least 100 bases from the point of labelling. In this method, many uniform repeated lengths of DNA are separated through restriction endonuclease digests. Copies of the DNA fragments are labelled at their 3′ or 5′ end with radioactive ^{32}P. These fragments are then cleaved at various points by four separate chemical treatments, each treatment removing on an average either one purine or one pyrimidine from any particular chain. The result of each such break in the DNA is to generate a ^{32}P-labelled fragment of a specific length that bands at a specific position on a gel subjected to an electric gradient. This process of gel electrophoresis can separate molecules that differ by a single-nucleotide length, and the ^{32}P-labelled DNA fragments on the gel are then detected by autoradiography. The radioactive DNA molecules are split into four groups. These four groups are treated separately. These treatments are adjusted in such a way that, on an average, only one nucleotide is removed from each DNA strand. A particular treatment will generate fragments of varying lengths, depending upon which nucleotide was removed. Products of the four different treatments are placed in parallel lanes on a polyacrylamide gel at the negative pole of an electrophoretic apparatus. Pore size of the gel governs the mobility of DNA molecules. The smaller the nucleotide fragment, the faster is the migration rate towards the positive pole. After electrophoresis, the gel is covered with a sensitized film, and an autoradiograph is taken. From the banding pattern, nucleotide sequence of the DNA is then read in ascending order from the 5′ terminal end at the positive pole. *Aka* MAXAM-GILBERT METHOD OF DNA SEQUENCING.

DNA sequencing by clone-by-clone approach: It consists of breaking the genome into relatively large chunks called clones, then cut each clone into smaller overlapping pieces, sequence the pieces and use the overlaps to reconstruct the sequence of the whole clone. Researchers in Human Genome Sequence Consortium combined this with bacterial artificial chromosomes (BAC)-based clone-by-clone approach, starting by breaking genome into overlapping fragments 100–200 kbp in size, plugging these into BACs, cloning in *Escherichia coli*, mapping each of these clones to its correct position in the genome using a restriction enzyme-generated 'fingerprint' of clone to match a spot on a restriction map of entire genome. Then, one takes each BAC, shears it into pieces and determines the sequence using the shotgun approach. Eventually, the BACs are pieced together to make the whole genome. Analogy: First separate the copies of book into individual pages and then shred each page into its own separate pile. Then work on one page at a time. The clone-by-clone approach produced the first draft of the human genome, which was published in June 2000, February 2001 and April 2003 in *Nature* and *Science*.

DNA sequencing by dideoxy method: F. Sanger and others (1977) developed a sequencing method that is similar to 'plus or minus' method of F. Sanger and A.R. Coulson (1975) but makes use of 2′,3′-dideoxy and arabinonucleoside analogue of the normal deoxynucleoside triphosphates, which act as specific inhibitors of DNA polymerase. They determined complete sequence of ϕX174 DNA of approximately 5375 nucleotides that codes for nine proteins in this virus. The scheme employed comprises the following steps: DNA polymerase-generated single-stranded oligonucleotides are electrophoresed on polyacrylamide gels. Conditions are set such that the nucleotide sequences can be read directly from an autoradiograph of the electrophoresed polyacrylamide gel. Primer is elongated at 3′, the end towards the region of bases ATGCTG using one nucleotide radioactively labelled with ^{32}P. This comprises the plus series of four experiments. Synthesis of the new strand is halted at various points by the insertion of dideoxy analogues. Dideoxy analogues lack 3′-OH group of deoxyribose sugar. In the minus series of experiments, the oligonucleotide primer is elongated using only three of the deoxynucleoside triphosphates (e.g. dGTP, dCTP, dTTP are added, but dATP is missing). Each primer is elongated until the missing nucleotide is specified by the template. The termination point thus lies before dA. A minus experiment that leaves out each of the four nucleotides is performed. DNA products of each of the four minus experiments are denatured and electrophoresed on a polyacrylamide gel. Then, autoradiography is employed to locate the radioactive DNA bands. The shortest fragments move the fastest in this analysis. The sequence 3′-TACGAC-5′ can be read directly from the left group of the four minus lanes in the autoradiograph. The sequence is simultaneously confirmed using the plus series of experiments. The T_4 DNA polymerase will degrade double-stranded DNA from its 3′ end. If one type of deoxynucleoside triphosphate (e.g. dATP) is added during this process, the exonuclease action of the T_4 DNA polymerase stops at the nucleotides containing the same base as the free nucleotide (at nucleotides with the base A). The minus series stops just before an A, whereas the plus series degrades until an A is reached. The ϕX174 DNA sequence identifies many of the features responsible for the production of proteins of the nine known genes of the organism, including initiation and termination sites for the proteins and RNAs. Two pairs of genes are coded by the same region of DNA utilizing different reading frames. Of the several techniques available for sequencing cloned DNA fragments, Sanger's dideoxy sequencing offers significant advantages in terms of rapidity and simplicity. F. F. Sanger won Nobel prize twice – once in 1958 for determining amino acid sequence of a protein (insulin) and then in 1980. F. Sanger and W. Gilbert were awarded Nobel prize in 1980 for discovering the method for DNA base sequencing. *Aka* CHAIN TERMINATION METHOD *or* DIDEOXY CHAIN TERMINATION METHOD.

DNA sequencing by hybridization using microarrays on DNA chips: Sequencing by hybridization was developed for detection of SNPs using DNA microchips to meet this objective. A sequencing microchip is developed either by immobilizing or by parallel synthesis of a complete set of 8-mers (i.e. $4^8 = 65536$) on a solid support. Using printing technology and robotics can further reduce the dimension of the microchip. The technique involves hybridization of a short DNA fragment of unknown sequence with a set of short oligonucleotides, particularly 8-mers of known sequences. Identification and analysis of a set of overlapping oligomers that form perfect duplexes with the DNA of interest permits reconstruction of the target DNA sequence. For example, to determine the sequence of a 17-base target DNA fragment and hybridize it with complementary overlapping 8-mers to form perfect duplexes, the overlap of seven or fewer adjacent 8-mers showing perfect match enables reconstruction of the complementary DNA sequences of the target.

DNA sequencing by PCR: A method for direct sequencing of PCR fragments sequences longer than 20 kb obtained with a commercial mixture of Taq and Pwo DNA polymerases. The sequencing protocol was optimized for an automated infrared DNA sequencer, consistently yielding long reads (500–600 bases). Direct DNA sequencing using PCR is *aka* LIGATION-MEDIATED PCR (LMPCR).

DNA sequencing by transcription: Recently, DNA sequencing was successfully achieved using RNA polymerase enzyme for transcription, instead of using sequenase or thermosequenase used in Sanger's dideoxynucleotide method of chain termination. This method of sequencing described as transcriptional sequencing makes use of 3′ dNTPs for terminating transcription in the same manner as ddNTPs are used for terminating DNA synthesis in Sanger's method (whenever 3′ dNTP is available will be for the formation of phosphodiester bond). The 3′ dNTP can be labelled with fluorescent (rhodamine) dyes so that four dNTPs with four colours can be used along with sufficient quantities of all the four NTPs. This will allow RNA synthesis to proceed normally with the help of RNA polymerase until, occasionally, a 3′ dNTP is incorporated. This will yield RNA chains of different lengths, each having at its terminus a specific dNTP that will be identified by its colour on an automatic sequencer. The method is believed to be superior because (i) incorporation of NTP during transcription of NTP during transcription does not require a primer, thus making it faster (240 bases/s) than DNA replication by *Taq* polymerase (60/s), (ii) a large amount of product can be obtained from small template (600 molecules of RNA from one DNA molecule) and (iii) direct PCR sequencing can be done, without the elimination of primers and 2′-dNTPs.

DNA sequencing by transcriptional motion of RNA polymerase: W.J. Greenleaf and S.M. Block (2006) reported a method for DNA sequencing that relied upon resolving the motion of a processive nucleic acid enzyme. They used a newly developed assay for tracking transcription by single molecules of *Escherichia coli* RNA polymerase in which a pair of opposite traps levitates polystyrene beads: one attached to the RNA polymerase enzyme and the other to the distal end of a DNA template. Transcriptional motion of RNA polymerase along the template changes the length of the DNA tether joining the two beads, leading to the displacements that can be registered with angstrom-level precision, affording single-base-pair resolution.

DNA sequencing by wandering spot analysis: F. Sanger and co-workers gave this method in 1973, which comprises four steps. First of all, DNA is labelled with ^{32}P at the 5′ end of the molecule. Then fragment is partially digested with snake venom phosphodiesterase, which sequentially removes nucleotides from the 3′ end. All the fragments obtained upon digestion contain radioactive label. These fragments are then subjected to electrophoresis, then the strip that results from the electrophoresis is applied to a sheet of chromatography medium, and HOMOCHROMATOGRAPHY is performed. Two-dimensionally separated nucleotides are autoradiographed. Partial digests are applied to cellulose acetate strip, and on the basis of shift in mobility, the results are inferred. Dark spots indicate the location of radioactive nucleotide fragments when X-ray film is developed. Finally, a sequence is obtained.

DNA-sequencing manual methods: (i) CHEMICAL-CLEAVAGE METHOD of A.M. Maxam and W. Gilbert in 1977, (ii) PLUS AND MINUS METHOD of F. Sanger and others in 1978 and (iii) WANDERING SPOT ANALYSIS METHOD of F. Sanger and others in 1973. *See details as separate terms.*

DNA-sequencing methods: (i) WANDERING SPOT ANALYSIS, (ii) CHEMICAL-CLEAVAGE METHOD, (iii) ENZYMATIC METHOD, (iv) CHAIN TERMINATION DIDEOXY PROCEDURE FOR LARGE-SCALE DNA SEQUENCING, (v) AUTOMATED DNA SEQUENCING, (vi) SLAB GEL SEQUENCING SYSTEMS, (vii) CAPILLARY GEL SEQUENCING SYSTEMS; (viii) DIRECT DNA SEQUENCING USING PCR, (ix) DNA SEQUENCING THROUGH TRANSCRIPTION; (x) DNA SEQUENCING BY HYBRIDIZATION USING MICROARRAYS ON DNA CHIPS, (xi) PYROSEQUENCING, (xii) RNA-SEQ METHOD, (xiii) DNA SEQUENCING THROUGH TRANSCRIPTIONAL MOTION OF RNA POLYMERASE, (xiv) AMPLIFICATION-FREE DNA SEQUENCING, (xv) HIGH-THROUGHPUT DNA SEQUENCING, (xvi) THIRD-GENERATION GENOME SEQUENCING, (xvii) LYNX THERAPEUTICS' MASSIVELY PARALLEL SIGNATURE SEQUENCING (MPSS), (xviii) POLONY SEQUENCING, (xix) 454 PYROSEQUENCING, (xx) ILLUMINA (SOLEXA) SEQUENCING and (xxi) SOLiD SEQUENCING. *See details as separate terms.*

DNA sequencing of entire genomes: Complete sequencing of genomes of several important and model species is a significant achievement of genomics, which provides the basis for comparative and functional analysis. Answers to questions such as (i) the number, location and distribution of genes in genome, (ii) gene organization and their function, (iii) what genes are the same or highly conserved across different species and (iv) what genes are responsible for species adaptation and evolution can now be obtained. Complete genome sequences have been determined for the bacterium *Escherichia coli* K12, yeast *Saccharomyces cerevisiae* (1997), the nematode *Caenorhabditis elegans* (1998), the fruit fly *Drosophila melanogaster* (2000), the annual plant *Arabidopsis* (2000), humans (2001), mouse (2002), rice (whole-genome shotgun sequenced genomes made were available from Monsanto in 2000 and Syngenta in 2002 [Goff and others in 2002] and was published in 2006 by International Rice Genome Sequencing Project 2005), forest tree *Populus* (2006), maize (2009) and *Eucalyptus* (2014).

DNA-sequencing technology: The instrumentation and procedures used to determine the order of nucleotides in DNA.

DNA-sequencing tools: (i) The DNA Sequence Quality provides base calling, chromatogram display and high-quality sequence region evaluation and presentation for up to five sequences simultaneously. (ii) Sequencing errors is used if your DNA sequence does not match the expected protein sequence you can check for errors at ERR_WISE or Wise2: Intelligent algorithms for DNA searches. (iii) SEQERR is used in detection of Frameshift Errors in Coding Regions. (iv) http://www.insilico.com/ allows in silico experiments including theoretical PCR amplification, AFLP-PCR, restriction analysis and pulsed-field gel electrophoresis (PFGE) with bacterial and archael genomes.

DNA sequencing, applications of: (i) For organo-chemical synthesis of a gene, knowing its nucleotide sequences is a prerequisite. (ii) Understanding a particular DNA sequence can shed light on a genetic condition and offer hope for the eventual development of treatment. (iii) Most frequently used to determine genome sequence. (iv) Many emerging science and technological fields rely heavily on DNA sequencing. (v) Important for planning the procedure and method of gene manipulation. (vi) Used for construction of restriction endonuclease map. (vii) Used to find tandem repeats or inverted repeat for the formation of hairpin. (viii) Used to find whether any ORF coding for a polypeptide exists. (ix) Used to find a polypeptide sequence from the data bank or to compare with DNA sequences from other related organism for phylogenetic analysis. (x) Used to construct the molecular evolution map. (xi) Useful in identifying exons and introns.

DNA sequencing, disadvantages of: (i) Genetic testing will expand quickly and soon adding highly targeted data to peoples' medical records. (ii) As these records go electronic, outsiders will find it increasingly easy to pursue an individual's health information. (iii) Private details could be uncovered. (iv) Insurers could deny coverage to someone with a complex medical condition. (v) Employers could also use this private information to disadvantage of an employee. (vi) Existing laws offer weak protection. (vii) Legislation is needed to give individuals more control over their own data to limit unauthorized disclosures by others.

DNA shuffling: It mixes and matches pieces of successful variants in order to produce better results. This process mimics the recombination that occurs naturally during sexual reproduction. Large amounts of recombinant DNA must be mutated, and the products must be screened for desired qualities by PCR-mediated reassembly process. The sheer number of variants often requires expensive robotic equipment to automate the process. Furthermore, not all desired activities can be easily screened for. DNA shuffling methods have been applied to single genes, metabolic pathways, groups of homologous genes in family shuffling formats and entire genomes of organisms.

DNA sleds: Something small that can slide along DNA by itself or carry different cargos (especially proteins) along DNA.

DNA sliding: This mechanism of DNA-protein interaction in gene regulation envisages that a protein recognizes a specific site on DNA and then moves (slides, tracks) along the DNA to another specific sequence where, perhaps, by interacting with another protein, it initiates transcription. This idea also envisions that the regulatory protein remains bound at the original site, and the DNA is thread past or through the bound protein until the second critical site is encountered. Regulatory proteins are not known to act by sliding, but this process is important for many other precise DNA transactions. For example, tracking has been suggested as an explanation for various site-specific recombination systems, but R. Craige and K. Mizuuchi in 1986 established with the help of two intertwined circles that recombination is possible without tracking. In replication, initiation of *Escherichia coli* and λ phage helicases cause unwinding of double-stranded DNA by tracking. Polymerases replicate nucleic acids through tracking. Type I restriction enzymes bind their unique recognition sequences and cleave the DNA at sites that can be thousands of base pairs away. This is possible only through tracking. *Aka* DNA TRACKING.

DNA strand: One of the two strands in a double-helical DNA molecule.

DNA structure: R. Altman in 1861 coined the term nucleic acid. The earliest work to determine the structure of nucleic acids was done by P. Levene in 1929, who determined 2-deoxyribose sugar and proposed tetranucleotide theory. E. Chargaff in 1950 concluded that deoxypentose nucleic acids from animal and microbial cells contained varying proportions of the same four nitrogenous constituents. Their composition appeared to be characteristic of the species but not of the tissue from which they are derived. There are four types of nucleotides in DNA: DEOXYADENOSINE-5′-PHOSPHATE, DEOXYGUANOSINE-5′-PHOSPHATE, DEOXYCYTIDINE-5′-PHOSPHATE and DEOXYTHMIDINE-5′-PHOSPHATE. Each deoxyribonucleotide contains phosphate, deoxyribose sugar and one of the four nitrogen-containing bases. The four bases are: ADENINE (A), GUANINE (G), CYTOSINE (C) and THYMINE (T). Adenine and guanine have a two-ring structure and are known as PURINES, whereas thymine and cytosine have one-ring structure and are known as PYRIMIDINES. Chargaff gave generalizations that the total amount of pyrimidine (T + C) nucleotides is always equal to total amount of purine (A + G) nucleotides in DNA. Also, amount of T always equals amount of A and C always equals G. The width of the DNA molecule was found to be 20 Å units, and each nucleotide pair was spaced 3.4 Å apart. Keeping these observations in mind, J.D. Watson and F.H.C. Crick in 1953 proposed a double-helix structure for DNA.

DNA supercoiling: Additional twisting of the DNA helix that occurs in response to the superhelical tension created when, for example, a circular DNA is partly unwound.

DNA supercoiling and transcription: Many prokaryotic genes are responsive to DNA supercoiling. Transcription itself can be major contribution to the level of supercoiling.

DNA synthesis: The building (i.e. polymerization manufacture) of a known sequence of nucleotides into a chain called an oligonucleotide (of which genes are made) or DNA (deoxyribonucleic acid). Invented by H.G. Khorana and his colleagues at the University of Wisconsin, Madison, in 1968, this process enables scientists to create genes or gene fragments for use in research. In 1973, R.B. Merrifield developed a means to partially automate the oligonucleotide assembly process. This led to automated machines that can now rapidly manufacture a gene fragment, gene or DNA probe. *See* GENE SYNTHESIS.

DNA synthesis *in vitro*: Synthesis of DNA outside a cell (in a test tube) involving four types of deoxynucleoside triphosphates (dNTPs) can be done organo-chemically, without the use of any template or polymerase. DNA synthesis can also be done by using DNA- or RNA-based templates.

DNA synthesis, organo-chemical method of: H.G. Khorana and co-workers in 1979 developed a method for synthesis of a given gene after the discovery of polynucleotide ligase and polynucleotide kinase. This method does not make use of any template or polymerase. However, this method requires that nucleotide sequence of the DNA fragment (gene) be known. The technique involves the principle of blocking these reactive groups of the triphosphate nucleotides that are not required to participate in the reaction and removing blocks of the groups that are required to participate in the reaction.

DNA tags: DNA tags provide a quick, inexpensive way to identify species. A small segment of DNA is selected from the mitochondria, the same short strand for each species is used for the identification of animal species. The segment chosen comes from a gene called *CO1*. It contains only 648 bp of nucleic acids, making for quick reading of its DNA sequence. However, the small piece varies enough from creature to creature for the differences to be distinguished from one species from another.

DNA tiles: These are quite rigid and stable DNA nanostructures that are formed from multiple DNA anti-parallel crossovers. DNA tiles typically have a geometry that is roughly rectangular. These tiles come in multiple varieties that differ from one another in the geometry of strand exchange and the topology of the strand paths through the tile.

DNA topoisomerase: An enzyme that generates positive supercoiling ahead and leaves negative supercoiling behind. This enzyme introduces or removes turns from the double helix by momentary breakage of one or both polynucleotides. These enzymes are vital during transcription. They prevent DNA knotting during replication. Two types of DNA topoisomerases – type I and type II – are known to occur. *See* DNA TOPOISOMERASE TYPE I *and* DNA TOPOISOMERASE TYPE II.

DNA topoisomerase type I: Type I DNA topoisomerase rectifies supercoiling of DNA by producing a temporary break in one strand of DNA. *Compare with* DNA TOPOISOMERASE TYPE II.

DNA topoisomerase type II: Type II DNA topoisomerase reduces negative supercoiling during transcription by producing a temporary break in both the strands of DNA. *Compare with* DNA TOPOISOMERASE TYPE I.

DNA transcription: A process in which a portion of the double-stranded DNA template gives rise to a single-stranded RNA molecule. In some cases, the RNA molecule itself is a 'finished product' that serves some important function within the cell. Messenger RNAs act as template for translation. tRNAs take a specific amino acid to the ribosome, rRNAs are an important constituent of ribosome, which is the site of protein synthesis, and some small non-coding RNAs have a regulatory role in gene expression. *See* TRANSCRIPTION.

DNA transfection: The transfer of DNA into cells in culture.

DNA transposons: DNA transposable elements that do not use a reverse-transcription step to integrate copies into the genome. These transposons transpose via DNA intermediates that have short, inverted repeats that flank this element. The enzyme transposase recognizes these sequences, creates a stem/loop structure and excises the loop from the region of the genome. The excised loop can then be inserted into another region of the genome (e.g. Ac/Ds and Spm in plants, P elements in animals, Tn in bacteria).

DNA tumour virus: A general term for a variety of different DNA viruses that can produce tumours.

DNA twisting: In case of DNA twisting, the regulatory proteins are considered to bind to some altered form of DNA (e.g. left-handed DNA or single-stranded DNA), or the regulatory protein possesses an enzymatic activity that changes DNA conformation (e.g. by unwinding it). Gene activation in this case would be a result of conformational change propagated through the DNA, which permits other proteins to bind and begin transcription. All known or suspected regulatory proteins isolated from prokaryotes and eukaryotes are supposed to recognize ordinary helical DNA. λ repressor protein bears protruding α-helices that fit into the DNA major groove and have sequence-specific contacts with functional groups exposed along the edges of the base pairs. These proteins do not greatly change the structure of DNA on binding. The idea of twisting has been protracted to a eukaryotic case. The DNA-binding and gene activation roles of a yeast transcription regulator have been discovered to be separated. Yeast GAL4 regulator has been found to consist of two domains: (i) the amino domain, which identifies sequences in the galactose upstream activating sequence (UAS_G) and (ii) the carboxyl domain, which, when bound to DNA through the amino domain, contacts another protein to mediate positive control. When the carboxyl domain is removed, this fragment identifies UAS_G but fails to activate transcription. When DNA-binding domain is replaced with that of the bacterial protein LexA, this protein activates transcription in yeast if LexA operator is located near the start site of a gene.

DNA twisting, contradictions about There is no indication that GAL4 can itself twist DNA. None of the experiments rules out the possibility that GAL4 helps some other proteins to bind adjacent to UAS_G and that GAL4 activates transcription by untwisting or twisting. TAILED CIRCLE experiment performed by S. Plon and J. Wang in 1986 on a mammalian virus argues against possibility of twisting.

DNA typing: A catch-all term for a wide range of methods for studying genetic variations. Each method has its own advantages and limitations, and each is at a different state of technical development. Each DNA typing method involves three steps. (i) Laboratory analysis of samples to determine their genetic marker types at multiple sites of potential variation. (ii) Comparison of the genetic marker types of the samples to determine whether the types match and thus whether the samples could have come from the same source. (iii) If the types match, statistical analysis of the population frequency of the types to determine the probability that such a match might have been observed by chance in a comparison of samples from different persons.

DNA unwinding enzyme: An ATP-dependent enzyme involved in the early stages of DNA replication. It unwinds the parental strand to create two template strands for DNA replication.

DNA unwinding, problem of: Contradiction was raised with respect to the amount of time DNA would require for such replication to occur in a very long molecule. T2 phage DNA is 200,000 nucleotide pairs. Since these are 10 base pairs to each rotation, such a molecule would have 20,000 rotations. The molecules will be $20{,}000 \times 34 = 68\ \mu$ in length. Such a molecule would have to be folded a number of times to fit into cells or nuclei. This problem of unwinding of DNA and replication appeared profound.

The answer to this question comes from the fact that replication proceeds before the parental DNA is completely unwound. J. Cairns in 1963 presented evidence to this thought. The *Escherichia coli* chromosome is circular and has a double-stranded structure. Two component strands separate during replication, with each strand duplicating individually, as shown by an autoradiograph. The fact that this duplication produces a Y-type of joint indicates that unwinding of two complementary strands of DNA is not completed before replication begins. In the second round of replication, θ structure appears, which shows that one of the arcs was twice as heavily labelled as the other arc. It clearly supported the semi-conservative mode of DNA replication.

DNA vaccine: A vaccine made up of 'naked' DNA. Using gene transfer approaches, the foreign DNA (gene) is introduced into an organism that stimulates an immune response (e.g. either a cellular immune response, humoral immune response or otherwise raise antibodies against the pathogen from which the naked genes have ascended) and produces specific protein which acts as a vaccine. The concept of 'edible vaccines' is based on DNA vaccine.

DNA vector: A vehicle (such as a virus) used for transferring genetic information (DNA) from one cell to another.

DNA viruses: Those viruses that contain DNA as genetic material. These viruses can be further divided into following two classes. One class comprises those viruses that have their genes in a double-stranded DNA molecule (ds-DNA). Examples of ds-DNA viruses are smallpox, vaccinia, varicella-zoster, adenoviruses, SV40, T2, T4, lambda, herpes viruses, KSHV, cytomegalovirus (CMV) and Epstein-Barr virus (EBV). Other class includes those viruses that have their genes in a molecule of single-stranded DNA (ssDNA). Well-known examples of ssDNA viruses are ϕX174 and **adeno-associated virus (AAV)**.

DNA walker: The scientists got the walker to move down a path 100 nm long, an unprecedented feat. The robot could follow a complex curve or a straight line. Though random walkers had been developed before, the team made their robot move around by manipulating how it interacts with the landscape. In effect, chemical commands to 'stop', 'start' or 'turn' were programmed into the DNA environment. Scientists are one step closer to creating molecular robots that may eventually perform complex tasks, such as building nanomolecules or delivering drugs to target tissues. Scientists have constructed DNA-based robots that can walk along a specific path unaided or collect various nanoparticles along an assembly line. This work will absolutely pave the way for how you build molecular robots. The spider is fuelled by the chemical interactions its single-stranded DNA 'legs' have with the origami surface. In order to take a 'step', the legs first cleave a DNA strand on the surface, weakening its interaction with that part of the origami surface. This encourages the spider to move forward, pulled towards the intact surface, where its interactions are stronger. When the spider binds to a part of the surface that it is unable to cleave, it stops. In essence, the researchers created a DNA spider that can sense the environment. The next step, he added, is to increase the complexity of movements performed by such autonomous robots by compiling a collection of rules of interactions between molecules and between molecules and environment.

DNA world: Life systems became so complex during origin of life that RNA, which was initially the genetic material, remained no longer suitable to function as genetic material. RNA is far more vulnerable to spontaneous hydrolysis than DNA. Thus, RNA had to be replaced with DNA as the genetic material.

DNA writers: The enzymes that write epigenetic modifications on DNA. All the DNA methyltransferases are DNA writers.

DNA, A-form: The more dehydrated form of right-handed helical DNA than the typical B-form obtained under non-physiological conditions. It is more compact, with 11 nitrogen bases per turn of the helix and a pitch of 28 Å, which gives A-form DNA an axial hole. In A-form DNA, the planes of its base pairs are fitted at 20° with respect to the helix axis. A-form DNA has very deep major groove and shallow minor groove. The base pairs are pushed towards the minor groove. In A-form DNA, there is no water spine. A-form DNA is found in fibres at 75% relative humidity and necessitates the presence of sodium, potassium or caesium as the counterion. Instead of lying flat, the bases are tilted with regard to the helical axis. The A-form DNA is biologically interesting because it is probably very close to the conformation adopted by DNA-RNA hybrids or by RNA-RNA double-stranded regions. The reason for this is that the presence of the 2′2 hydroxyl group prevents RNA from lying in the B-form DNA. *Aka* A-DNA. *See* B-FORM DNA.

DNA, B-form: The typical right-handed structure of DNA, which has ten nitrogen bases per turn of the helix. Stacked bases are regularly spaced 0.34 nm apart, and the helix makes a complete turn every 3.4 nm. The form of DNA described by J.D. Watson and F.H.C. Crick in 1953 is B-form. These conformations diverge from one another in relative rotation per base pair, repeating helix unit, mean base pair per turn, inclination of bases, rise per base pair along helix axis, glycosyl torsion angle and sugar-pucker conformations. Out of all the DNA forms, B-form is the most stable structure in the physiological conditions of the cell. The minor groove spine of hydration stabilizes the B-form DNA at high humidity levels. *Aka* B-DNA.

DNA, C-form: C-DNA is found under even greater dehydration conditions than those observed during the isolation of A- and B-forms of DNA. This form of DNA arises when DNA fibres are maintained in 66% relative humidity in the presence of lithium ions. It has only 9.3 bp per turn and is, thus, less compact. It is also right-handed. Its helical diameter is 19 Å. Like A-form DNA, C-form DNA does not have its base pairs lying flat; rather, they are slanted relative to the axis of the helix. *Aka* C-DNA.

DNA, D-form: It is right-handed helix of DNA. It occurs in helix deficient in guanine in their base composition. It possesses 8 base pairs per turn. *Aka* D-DNA.

DNA, E-form: E-form DNA is right-handed helix. It occurs in helix lacking guanine in their base composition. It contains 7 base pairs per turn. *Aka* E-DNA.

DNA, P-form: Jean-Francois Allemand and co-workers have shown in 1998 that if DNA is artificially stretched, still another form of the molecule is assumed, which they have called P-form DNA. It is longer and nanomer than B-form DNA. The phosphate groups are present inside the molecule. The nitrogenous bases are found closer to the external surface of the helix. There are 2.62 bases per turn. *Aka* P-DNA.

DNA, replicating form: The double-stranded form of single-stranded DNA viruses (for example, φX174 phage).

DNA, Z-form: Z-DNA is a left-handed higher-order structure known to form in alternating purine-pyrimidine or pyrimidine-purine (APP) dinucleotide repeat sequences, with the propensity of Z-DNA to form in the order GC>CA>TA. Evidently, the Z-form DNA conformation is not readily assumed by DNA segments with alternating purines and pyrimidine base sequences. This helix has 12 bp/turn, having a pitch of 45 Å. This DNA has a deep minor groove and no discernible major groove. The base pairs are tossed 180° relative to those in B-DNA. The methylation of cytosine residues at C-5 also promotes Z-DNA structure. The reversible conversion of specific segments of B-form DNA to Z-form DNA under appropriate circumstances acts as a kind of switch in regulating genetic expression. *Aka* Z-DNA.

DNA/RNA chip technology: Nucleic acid arrangements are based on hybridization of sample (labelled RNA or DNA) in solution to the immobilized DNA fragments on a solid surface. The arrayed DNA fragments often come from cDNA, genomic DNA or plasmid libraries. Usually, an array is designed on the basis of specific sequence information, a process sometimes referred to as DOWNLOADING THE GENOME ONTO A CHIP. There are several dissimilarities in this basic technical theme, which as hybridization reaction may be driven by an electric field. Other detection methods besides fluorescence can be used, and the surface may be made of materials other than glass, such as plastic, silicon, gold, gel or membrane or even comprised of beads at the ends of fibre optic bundles. *Aka* DNA/RNA microarray.

DNA/RNA FISH: A fluorescence *in situ* hybridization technique that uses a complementary DNA or RNA strand to determine the localization of DNA sequences or transcripts in cell nuclei.

DNA/RNA hybridization: If RNA is added to denaturized DNA, RNA/DNA hybrids are formed. Renaturation depends on copy number of the segment present in a unit DNA solution. The greater the copy number, the higher the chances of collision.

DNA/RNA microarray: *See* DNA/RNA CHIP TECHNOLOGY.

DNA/RNA packaging in viruses/bacteriophages: Viruses/bacteriophages have a single hereditary unit (i.e. viral chromosome), which has a single molecule of nucleic acid. This viral chromosome may be linear or circular and rodlike or spherical. The nucleic acid may be DNA or RNA and single stranded or double stranded. DNA-containing viruses are known as DEOXYVIRUSES, whereas RNA-containing viruses are known as RIBOVIRUSES. Viral genomes are packaged into capsids that are either symmetrical or quasi-symmetrical structures assembled from only one or a few types of protein monomers. There are two general patterns of viral DNA packaging: (i) The protein shell is assembled around the nucleic acid, condensing the DNA or RNA by protein-nucleic acid interactions during this process. (ii) The capsid is first formed into which the nucleic acid is introduced being condensed as it enters the capsid.

DNA-amplified fingerprinting: A technology based on amplification of random genomic DNA sequences attained by a single short (5–8 bases) oligonucleotide primer of arbitrary sequence. It generates a characteristic spectrum of short DNA pieces of varying complexity that are resolved on polyacrylamide gel (PAGE) following silver staining. It is used to detect genetic differences between genotypes as well as to detect polymorphism even between organisms that are closely related, such as near-isogenic lines.

DNA-based processes, orchestration of: To facilitate DNA-based functions, modifications orchestrate the unravelling of chromatin to help the execution of a given function. This may be a very local function, such as transcription of a gene or the repair of DNA, or it may be a more genome-wide function, such as DNA replication or chromosome condensation. All these biological tasks require the ordered recruitment of the machinery to unravel DNA, manipulate it and then put it back to the correct chromatin state.

DNA-binding domain (DBD): A DBD is an independently folded protein domain that contains at least one motif that recognizes double- or single-stranded DNA. A DBD can recognize a specific DNA sequence (a recognition sequence) or have a general affinity to DNA. Some DNA-binding domains may also include nucleic acids in their folded structure.

DNA-binding domain (DBD) for epigenome editing: For epigenome editing, a DBD can be taken from a transcription factor (e.g. zinc-finger protein (ZFP), TALE or dCas-guide RNA ribonucleoprotein).

DNA-binding domain (DBD) interaction with DNA nucleotides: The DBD interacts with the nucleotides of DNA in a DNA sequence-specific or non-sequence-specific manner, but even non-sequence-specific recognition encompasses some sort of molecular complementarity between protein and DNA. DNA recognition by the DBD can transpire at the major or minor groove of DNA, or at the sugar-phosphate DNA backbone. Each specific type of DNA recognition is tailor made to the protein's function. For example, the DNA-cutting enzyme DNase I cuts DNA almost randomly, and so it must bind to DNA in a non-sequence-specific manner. Even so, DNase I enzyme recognizes a certain 3D DNA structure, yielding a somewhat specific DNA cleavage pattern that can be useful for studying DNA recognition by a technique called DNA FOOTPRINTING. Many DNA-binding domains must recognize specific DNA sequences, such as DBDs of transcription factors that activate specific genes, or those of enzymes that modify DNA at specific sites, like restriction enzymes and telomerase. The hydrogen bonding pattern in the DNA major groove is less corrupt than that of the DNA minor groove, providing a more attractive site for sequence-specific DNA recognition. The specificity of DNA-binding proteins can be studied using many biochemical and biophysical techniques, such as gel electrophoresis, analytical ultracentrifugation, calorimetry, DNA mutation, protein structure mutation or modification, nuclear magnetic resonance, X-ray crystallography, surface plasmon resonance, electron paramagnetic resonance and cross-linking and microscale thermophoresis (MST).

DNA-binding domain (DBD) without nuclease activity: DNA-binding proteins without nuclease activity, such as ZFPs, TALEs or ribonucleoproteins – including dead Cas9 (dCas9) complexed with SINGLE-GUIDE RNA (sgRNA) – act as DBDs. DBDs can also be synthesized to favourably bind to DNA sequences at will chemically as well as biotechnologically. The chemical approaches include TRIPLEX-FORMING OLIGONUCLEOTIDES (TFOs), PEPTIDE NUCLEIC ACIDS (PNAs) and POLYAMIDES (Pas). TFOs are single-stranded DNA molecules, which can be synthesized to bind particular DNA sequences in the DNA major groove. PNAs are derivatives of TFOs. Polyamides are small chemicals, which combine predetermined sequences in the minor groove.

DNA-binding domain (DBD), functions of: One or more DNA-binding domains are frequently part of a larger protein consisting of additional domains with differing function. The additional domains often regulate the activity of the DNA-binding domain. The function of DNA binding is either structural or regulatory (involving transcriptional control), with the two roles sometimes overlapping. DNA-binding domains with functions involving DNA structure have biological roles in the replication, repair, storage and modification of DNA, such as methylation.

Many proteins involved in the regulation of gene expression contain DBDs. For example, proteins that regulate transcription by binding DNA are called TRANSCRIPTION FACTORS. The final output of most cellular signalling cascades is gene regulation.

DNA-binding domain (DBD), types of: (i) Helix-turn-helix, (ii) zinc finger, (iii) leucine zipper, (iv) winged helix, (v) winged helix-turn-helix, (vi) helix-loop-helix and (vii) HMG-box. II. *See details as separate terms.*

DNA-binding domains (unusual): Three types of unusual DNA-binding domains are the following: (i) IMMUNOGLOBULIN FOLD DOMAINS, (ii) B3 DOMAIN and (iii) TRANSCRIPTION ACTIVATOR-LIKE EFFECTOR (TALE) DOMAIN. *See details as separate terms.*

DNA-binding motif: Common sites on different proteins that facilitate their binding to DNA (e.g. leucine zipper and zinc-finger proteins).

DNA-binding protein: Any protein that attaches to DNA as a part of its normal function (e.g. histone, RNA polymerase, *lac* repressor).

DNA-dependent RNA polymerases in plants: Eukaryotes have three multisubunit DNA-dependent RNA polymerases – RNA polymerase I, RNA polymerase II and RNA polymerase III – with discrete functions. Higher plants have two extra RNA polymerases – RNA polymerase IV (*aka* NRPD1) and RNA polymerase V (*aka* NRPE1) – that have a role mainly in RdDM pathway. RNA Polymerase IV functions upstream in RdDM to generate and amplify the small RNA trigger for silencing. RNA polymerase V acts downstream to transcribe non-coding RNAs that deliver scaffolds for attracting silencing complexes and could also be involved in the reinforcement of silencing through a positive feedback loop.

DNA-DNA hybridization: The annealing of two complementary DNA strands to produce hybrid nucleic acid molecules; it is used to identify the base sequences in two polynucleotide chains from different sources.

DNA-driven hybridization: It involves the reaction of an excess of DNA with RNA.

DNA-driven reaction: A reaction in which nucleic acid hybridization is utilized in an experimental situation and complementary DNA is greatly in excess of a radioactive tracer RNA. Therefore, RNA is forced to with a much greater proportion of hybridizing DNA and thus will bind to highly repetitious DNA sequences.

DNA-histone methylation switch: Eukaryotes regulate the effective dosage of their ribosomal RNA (rRNA) genes, expressing fewer than half of the genes at any one time. rRNA gene dosage control occurs through an epigenetic switch in which concerted changes in promoter cytosine methylation density and specific histone modifications dictate the 'on' and 'off' states of the rRNA genes. A key component of the off switch is HDT1, a plant-specific histone deacetylase that localizes to the nucleolus, and is required for H3K9 deacetylation and subsequent H3K9 methylation.

DNA-mediated gene transfer (DMGT): A vectorless method of gene transfer in which a purified DNA fragment carrying gene of interest is mixed with a carrier DNA and is precipitated out of solution with calcium phosphate. When target cells are incubated with the precipitate, there is increased uptake of DNA, and some of the cells are transformed by the desired gene. DMGT methods are classified as PHYSIOCHEMICAL, PHYSICAL and ELECTRICAL. These vectorless methods have been successfully tried in a wide range of plants and animals. *Aka* VECTORLESS METHODS OF GENE TRANSFER *or* DIRECT GENE TRANSFER SYSTEMS (DGTS).

DNA-modified histone interactions: Histone acetylation can loosen the tightly wound DNA structure and allow for increased DNA access to transcription factors, thus allowing gene transcription to occur.

DNA-only transposon: Type of transposable element that exists as DNA throughout its life cycle. Many types of transposons move by cut-and-paste mechanism.

DNA-protein interactions: Proteins interact with DNA through electrostatic interactions (salt bridges), dipolar interactions (hydrogen bonding, H-bonds), entropic effects (hydrophobic interactions) and dispersion forces (base stacking). DNA-protein interactions are required for precise DNA transactions, including initiation of DNA replication, initiation of transcription and site-specific recombination. In fact, DNA-protein interactions are involved in almost all biological functions of DNA. These interactions occur when a protein binds a molecule of DNA, often to regulate the biological function of DNA, usually the expression of a gene. Among the proteins that bind to DNA are transcription factors that activate or repress gene expression by binding to DNA motifs and histones that form part of the structure of DNA and bind to it less specifically. In general, proteins bind to DNA in the major groove; however, there are exceptions. DNA-binding proteins include transcription factors, which modulate the process of transcription, various polymerases, nucleases that cleave DNA molecules and histones that are involved in chromosome packaging in the cell nucleus. DNA-binding proteins can incorporate such domains as the zinc finger, the helix-turn-helix and the leucine zipper (among many others) that facilitate binding to nucleic acid.

DNA-protein interactions in gene regulation: Regulatory proteins recognize and bind to specific DNA sequences using structural features that are complementary to ordinary helix. DNA-bound regulatory proteins impact transcription by excluding binding of other proteins. These descriptions hold whether the interacting proteins are lying adjacent on DNA or whether they are separated; in the latter case, the DNA in between the separated sites loops out to permit protein-protein interactions.

DNA-protein interactions, models of: Four types of models have been suggested to allow DNA-protein interactions for gene regulatory mechanisms to operate: (i) DNA LOOPING, (ii) DNA TWISTING, (iii) DNA SLIDING and (iv) DNA OOZING. *See details as separate items.*

DNA-protein interactions, types of: Protein-DNA interactions are of mainly two types: (i) sequence-specific DNA-protein interactions and (ii) non-sequence-specific DNA-protein interactions.

DNA-RNA hybrid: A double helix that involves one chain of DNA hydrogen bonded to a chain of RNA by following complementary base pairs.

DNA-RNA hybridization: When a mixture of DNA and RNA is heated and then cooled, RNA can hybridize with DNA form a double helix that has a complementary nucleotide sequence.

DNase: *See* DEOXYRIBONUCLEASE.

DNase C: An *Escherichia coli* protein required to localize replication proteins.

DNase I: An endonuclease that introduces random single-stranded nicks in DNA. This enzyme is engaged for nick translation. *See* NICK TRANSLATION.

DNase-seq: The DNase I enzyme preferentially cuts live chromatin preparations at sites where there are specific (non-histone) proteins nearby. The resulting cut points are then sequenced using high-throughput sequencing to determine those sites 'hypersensitive' to DNase I, corresponding to open chromatin. The technique involves adaption of established regulatory sequence assay.

DNA-silica binding: A better description of how DNA binds to silica is based on the action of guanidium hydrochloride (GuHCl), which functions as a CHAOTROPE. A chaotrope denatures biomolecules by disrupting the shell of hydration around them. This permits positively charged ions to construct a salt bridge between the negatively charged silica and the negatively charged DNA backbone in high salt concentration. The DNA is then washed with high salt and ethanol and finally eluted with low salt. After the DNA is adsorbed to the silica surface, all other molecules pass through the column. Most likely, these molecules are sent to a waste section on the chip, which can then be closed off using a gated channel or a pressure- or voltage-controlled chamber. The DNA is then washed to eliminate any excess waste particles from the sample and then eluted from the channel using an elution buffer for further downstream processing. Methods using silica beads and silica resins have already been developed that successfully isolate DNA molecules, which are then PCR amplified. However, a few problems accompany these methods. First, beads and resins are highly variable depending on how well they are packed and are thus hard to reproduce. Each loading of a microchannel can lead to a different amount of packing and thus change the amount of DNA that adsorbed to the channel. Furthermore, these methods form a two-step manufacturing process. Silica structures are a much more accurate method of packing material because they are engraved into the channel during its fabrication and is thus the outcome of one-step manufacturing processes via soft lithography. Silica structures are therefore simpler to use in highly parallelized designs (since beads or resins do not need to be introduced post manufacturing the chip).

DNA-to-DNA transposition: Transposons of this type are class II introns. Transposition in this case may be conserved or replicative.

DNA-to-protein translation: A number of excellent sites exist, all of which allow translation in all six reading frames. 'ORF Finder' is recommended tool because of its visuals, as are Pipeline or GeneMark, if one is seriously interested in finding genes within the sequence. The latter two programs permit the analysis of long sequences. (i) For frameshift errors in DNA sequence, three sites are available: (a) GENIO/frame, (b) FrameD and (c) AMIGene. (ii)

ORF Finder presents the potential ORFs as coloured bars with the coordinates. If a bar is clicked, the amino acid sequence is presented in single letter code underlying the nucleotide sequence – Batch ORF Finder. (iii) Glimmer provides access to ORF analysis as well as the capability to design ORF-specific primer pairs – Translation Initiation site (TIS) Correction. It offers a boundary for direct postprocessing of the predictions obtained from Glimmer to improve the accuracy of annotated Translation Initiation Sites (TIS). (iv) Genome Analysis Pipeline (Genome Analysis and System Modelling Group, Life Sciences Division, Oak Ridge National Laboratory, USA) offers both prokaryotes as well as higher eukaryotes and *Saccharomyces* genome analysis. In the case of prokaryote analysis, there is a wide choice of model organisms. (v) GeneMark Homepage offers a family of programs for ORF analysis. This site links one to a growing number of programs for modelling phage, bacterial and eukaryotic data. Extensive control is possible with the data output (i.e. one can request the nucleotide and protein sequence of the ORFs). (vi) Generation – Microbial Gene Prediction System – can train itself on microbial and model organisms to produce a set of data, which can be used to recognize genes in these organisms. (vii) EasyGene is a prokaryotic gene finder tool that ranks ORFs by statistical significance. It produces a list of predicted genes given a sequence of prokaryotic DNA. Each prediction is attributed with a significance score (R-value), indicating how likely it is to be just a non-coding open reading frame rather than a real gene. The user needs only to specify the organism hosting the query sequence. If you are interested in the analysis of existing bacterial genomes, consult EasyGene 1:2. (viii) GeneHacker is a system for gene structure prediction in microbial genomes. It can model eight bacterial species, detect overlapping/nested genes and frameshift errors within protein-encoding regions. Interesting output which can be modified. (ix) FrameD is a noise-resistant gene finder for prokaryotic and matured eukaryotic sequences offering considerable flexibility in search strategies and output format. (x) FgenesB (SoftBerry) is fast Pattern/Markov chain-based bacterial operon and gene prediction. Somewhat limited range of model bacteria and archaea. (xi) Gene Identification – one can modify the minimum ORF size and obtain both the DNA and protein sequence of the ORF of interest by clicking on the diagram. (xii) ExPASy – Translate tool (ExPASy, University of Geneva, Switzerland). This site is useful if there is a gene which begins with an alternative start codon. An alternative site is Translate Nucleic Acid Sequence Tool which permits choice of reading frame(s) and genetic code.

Dobzhansky-Muller model: A simple model for the evolution of reproductive isolation in which two populations accumulate different alleles. These alleles cause no loss of fitness within the genetic background in which they ascended, but alleles derived from different populations are incompatible with each other.

docking modules: Docking is a molecular modelling technique that is used to envisage how a protein (enzyme) interacts with small molecules. Some epigenetic modifications may function as docking modules for related factors.

docking protein: Responsible for attaching (docking) a ribosome to a membrane by interacting with a signal particle attached to a ribosome destined to be membrane bound.

documentation: Written text or illustration that accompanies computer software or is implanted in the source code. It either describes how it operates or how it is used and may mean different things to people in different roles.

domain: (i) In a chromosome, this term may refer either to a discrete structural entity defined as a region within that supercoiling is independent of other domains, or to an extensive region, including an expressed gene that has heightened sensitivity to degradation by the enzyme DNase I. (ii) In a protein, a discrete folded continuous part of the amino acid sequence that can be equated with a particular function (e.g. ligand-binding domain). The tertiary structure of larger proteins is usually subdivided into distinct globular or rodlike structures called domains. It is the folded region of polypeptide. Protein domains can be seen using electron microscopy and X-ray crystallography (e.g. haemagglutinin contains a globular domain and a fibrous domain). (iii) In software engineering, the targeted subject area of a computer program. It characterizes the target subject of a specific programming project, whether narrowly or broadly defined. Conserved structural entities with distinctive secondary structure content and a hydrophobic core. In small disulfide-rich and Zn^{2+}-binding or Ca^{2+}-binding domains, the hydrophobic core may be provided by cystines and metal ions, respectively. Homologous domains with common functions usually show sequence similarities. *See* PROTEIN DOMAIN *and* CHROMOSOME DOMAIN.

domain composition: Proteins with the same domain composition have at least one copy of each of domains of the query.

Domain Name System (DNS): A tiered and dispersed naming system for computers, services or other resources connected to the Internet or to a private network. It links various information with domain names assigned to each of the participating entities. Most prominently, it translates more readily memorized domain names to the numerical IP addresses needed for locating and identifying computer services and devices with the underlying network protocols. By providing a worldwide, distributed directory service, the DNS has been an indispensable component of the functionality of the Internet since 1985.

domain of attraction: The domain of attraction of a stable equilibrium is the set of states from which a system evolves towards that equilibrium.

domain organization: Proteins having all the domains as the query in the same order.

domestication: Refers to the selective breeding by humans of species in order to accommodate human needs. There have been significant changes in plant characters, such as synchrony, free threshing, gigantism, increased harvest index and altered genetic population.

dominance (genetic): An allele (*A*) is DOMINANT if the phenotype of the heterozygote (*Aa*) is the same as that of the dominant homozygote (*AA*). The allele (*a*) does not influence the heterozygote's phenotype and is called RECESSIVE. An allele may be partly, rather than fully, dominant. In that case, the heterozygous phenotype is nearer to, rather than identical with, the homozygote of the dominant allele.

dominance and epistasis: In the case of dominance, one member of the allelic pair masks the expression of the other member (interallelic interaction), whereas epistasis is the suppression of action of a gene or genes by a gene or genes not allelomorphic to those expressed (INTERGENIC INTERACTION).

dominance deviation: The difference between the trait value of a genotype and the value expected with no dominance.

dominance gene effects: The effects representing the dominance of one allele over another so that heterozygote is more like one parent than the other.

dominance genetic variance: Refers to the phenotype deviation caused by the interactions between alternative alleles that control one trait at one specific locus. Epistatic variance involves an interaction between different alleles in different loci.

dominance hypothesis: The theory that heterosis is caused by masking of harmful recessive alleles by dominant alleles. *See* OVERDOMINANCE HYPOTHESIS.

dominance model of heterosis: This model states that heterosis is caused by the complementation of deleterious recessive alleles. It predicts that an inbred line of equal performance to the F_1 hybrid could be achieved by eliminating all deleterious alleles and accumulating favourable alleles.

dominance modifiers: The genes that, through relative strength of action of the two alleles at a locus or action of other genes (EPISTASIS), affect dominance-recessiveness relationship.

dominance not a law: Dominance is not a law since incomplete (partial) dominance exists. Codominance also shows that dominance is not a law.

dominance portion of genetic variance (V_D): That portion of the genetic variance for a given trait that results from the fact that heterozygotes do not always score exactly midway between the homozygotes.

dominance theory: A theory proposed by D.D. Commins in 1996 that infers the specific social cognitive functions as adaptations to the pressures of living in a dominance (or status) hierarchy. This provides an explanation for Haldane's rule, which assumes that F_1 sterility (or inviability) is caused by recessive alleles.

dominance variance (V_D): The variance in a quantitative trait that is caused by dominant gene effects. It is defined as the variance of DOMINANCE DEVIATIONS.

dominance, analysing type of: The terms dominance, incomplete dominance and codominance are somewhat arbitrary unless analysed critically. The type of dominance inferred depends on the phenotypic level at which the observations are being made – organismal, cellular or molecular. Indeed, the same caution can be applied to many of the categories that scientists use to classify structure and processes; humans devise these categories for convenience of analysis. The type of dominance is determined by the molecular functions of the alleles of a gene and level of analysis.

dominance, scale of: Let us consider a locus *A* with two alleles, A_1 and A_2. Let us assume that the phenotype of homozygote A_1A_1 is P_1 and that of A_2A_2 is P_2. If the phenotype of heterozygote A_1A_2 is (i) same as P_1, A_1 is dominant over A_2, (ii) same as P_2, A_2 is dominant over A_1, (iii) ranges between P_1 and P_2, A_1 and A_2 show incomplete dominance, (iv) exhibits properties of both P_1 and P_2 phenotypes, codominance is the result and (v) lies beyond the range P_1–P_2, overdominance is said to exist between alleles A_l and A_2. When the phenotype of F_1 falls at the mid-parent value, some workers call it NO DOMINANCE.

dominance: If the heterozygote (e.g. *A/a*) is exactly intermediate between the two homozygotes (*A/A* and *a/a*), there is no dominance. Any deviation from this additive model is called as dominance. The condition in which one member of a pair of allelic genes expresses itself in whole (complete dominance) or in part (incomplete dominance) over the other member

dominance/recessiveness relationship: Dominant and recessive relationships are the best-known forms of gene interaction between alleles within a single locus. *See* DOMINANCE *and* RECESSIVENESS..

dominant: In genetics, it refers to an allele of a gene expressed in the phenotype of heterozygote. In Mendel's experiments, when he crossed a true breeding red flowered plant with a true breeding white flowered one, the progeny produced red flowers. The whiteness seemed suppressed, and the redness seemed to dominate. Mendel, therefore, called such traits (as redness of flowers) dominant and their alternative (such as whiteness) recessive.

dominant allele: In diploids, an allele is dominant if phenotypic effect is fully expressed in a heterozygote. An allele that disguises the presence of a recessive allele. Discovered by Gregor J. Mendel in the 1860s, this gene produces the same phenotype when it is heterozygous as it does when it is homozygous (i.e. trait, or protein, is expressed even if only one copy of the gene is present in the genome). For example, if *A* is a dominant over *a*, then *A/A* and *A/a* have the same phenotype. In the heterozygote, the protein produced by the FUNCTIONAL ALLELE suffices for the normal needs of the cell. Thus, the functional allele acts as a DOMINANT ALLELE. *Compare with* RECESSIVE ALLELE.

dominant epistasis: When a dominant allele conceals the expression of both recessive and dominant alleles at other loci, it is called dominant epistasis (e.g. one dominant factor A is epistatic to another factor B, or B is hypostatic to A, provided A and B factors are positioned at different loci). Such genes are called DUPLICATE GENES.

dominant gene effects: Gene action with deviations from the additive such that the heterozygote is more like one parent than the other.

dominant inheritance: *See* AUTOSOMAL DOMINANT.

dominant lethal: Action of a gene lethal in both the homozygous and heterozygous conditions; a gene that frequently kills in a single dose. *See* LETHAL.

dominant negative mutation: Mutation that dominantly influences the phenotype by means of a defective protein or RNA molecule that interferes with the function of the normal gene product in the same cell.

dominant phenotype: When an individual showing one phenotype is crossed with another of the opposite sex, having another contrasting phenotype for the same trait, the F_1 generation consisted of progeny having the same phenotype as one of the parents. The phenotype that appears in the F_1 generation is termed DOMINANT. *Aka* DOMINANT TRAIT.

dominant S-haplotype: Known as class I haplotype. In *Brassica rapa*, class I haplotypes include S8, S9, S12 and S52. These haplotypes are codominant when present in the same plant but are always dominant to class II haplotypes in pollen.

dominant species: The species with the greatest influence on both the biotic and abiotic components of its community.

dominant vs recessive: Please note that these terms should be applied only to phenotypes, not to alleles. Phenotype 1 is dominant to phenotype 2 if the heterozygote has phenotype 1. Here, phenotype 1 is dominant, and phenotype 2 is recessive.

dominigenes: Modifying genes that are able to alter the dominance of another gene.

domino theory: According to this theory, cell cycle is a linear assembly line.

donor: An individual, line, population or variety from whom pollen or genetic material is used for transfer to another.

donor cell: The cell that donates genes during a genetic cross between bacteria.

donor junction: The connection between the left 5′ end of an exon and the right 3′ end of an intron.

donor parent: The parent from which one or a few genes are shifted to the recurrent parent in backcross breeding.

donor plant: The source plant used for propagation, crossing and so on, whether donor is a simple individual, an explant, graft or cutting.

donor splicing site: *See* LEFT SPLICING JUNCTION.

DONSON: An enigmatic protein which was previously implicated in DNA replication. DONSON is the missing link in the assembly of the complex that unzips the DNA double helix during replication. Working in the roundworm *Caenorhabditis elegans*, Y. Xia and others in 2023 showed that the worm homologue of DONSON, DNSN-1, is essential for the assembly of the replication machinery. Homologues of DNSN-1 are absent from yeast but present in plants and animals. Mutations of the equivalent protein in humans cause a form of microcephalic dwarfism often seen when assembly of the chromosome duplication machinery is defective. Y. Lim and others in 2023 applied AlphaFold-Multimer (AF-M), an artificial intelligence–based system technology to predict which proteins expressed in human cells bind DONSON.

dopeyballs: Superconducting Buckyballs have the highest critical temperature of any known organic compound.

dormancy (Adj. dormant): A situation of deferred growth and reduced metabolism of an organism seed's inability to sprout for a given time after it is produced.

dormant: To remain inactive when conditions are not suitable. Seeds are often dormant until conditions are right for germination.

dorsiventral: Possessing structurally and visibly different upper and lower surfaces (e.g. some leaves). *Compare with* BILATERAL *and* ISOBILATERAL.

dorsoventral: Refers to the axis running from the back to the belly of an animal or from the upper side to the underside of a structure.

dosage compensation: A phenomenon in which the activity of a gene is increased or decreased according to the number of copies of that gene in the cell.

dosage compensation complex (DCC): The DCC in *Caenorhabditis elegans* is engaged to specific regions of the X chromosome and blows out along the chromosome from these initial binding sites. Multiple distinct recognition factors on the X chromosomes recruit DCCs to initiate their chromosomal spreading. Association of DCCs with both X chromosomes results in transcriptional repression of genes on those chromosomes.

dosage compensation in *Caenorhabditis elegans*: In this worm, dosage compensation is achieved by the activity of the DCC, which localizes to both X chromosomes in hermaphrodites to downregulate gene expression by half.

dosage compensation in *Drosophila*: Dosage compensation in *Drosophila* increases the transcription of genes on the single X chromosome in males to equal that of both X chromosomes in females. Site-specific histone acetylation by the MALE-SPECIFIC LETHAL (MSL) COMPLEX is alleged to perform a fundamental role in the amplified transcriptional output of the male X.

dosage compensation in mammals: To ensure gene dosage equivalence between male and female mammals, one X chromosome is inactivated in all somatic cells in females. The silencing is initiated in the early post-implantation embryo when one chromosome is randomly chosen to be inactivated. The inactive chromosome is epigenetically marked with DNA methylation at gene promoters, transcriptionally repressive histone modifications, and the silencing requires the ncRNAs Xist and Tsix. High levels of H3K27 trimethylation are present on the inactive X chromosome.

dosage compensation in plants: In a minority of flowering plants, separate sexes are genetically determined by sex chromosomes. The Y chromosome has a non-recombining region that degenerates, causing a reduced expression of Y genes. In some species, the lower Y expression is accompanied by dosage compensation (DC), a mechanism that re-equalizes male and female expression and/or brings XY male expression back to its ancestral level.

dosage compensation mechanisms: To compensate for X chromosome dosage difference between sexes, different mechanisms have evolved to equalize X-linked transcript levels in males and females. Recent genomic research into the mechanisms of X chromosome dosage compensation in three model organisms – *Mus musculus* (mouse), *Drosophila melanogaster* (fly) and *Caenorhabditis elegans* (worm) – suggest that these animals use different strategies. Dosage compensation has also been investigated in some plant systems (e.g. white campion *Silene latifolia*, formerly *Melandrium album*). Various mechanisms of dosage compensation include transcriptional repression, hypertranscription and hypotranscription, which exist in different species. In general, a genetic procedure that compensates for genes that occur in two doses in the homozygous dominants, so that the heterozygotes yield the same amount of gene product as the homozygotes. In XX-XY or XX-XO system of sex determination, it refers to the equalization of most X-linked gene products between males, which have one X chromosome and a single dose of X-linked genes, and females, which have two Xs and two doses of such genes. It is mediated by the MSL complex consisting of a core of five protein subunits encoded by male-specific lethal 1, 2 and 3 (msl1, msl2, msl3), males absent on the first (mof), and maleless (mle), as well as one of two non-coding RNAs [RNA on the X1 and X2 (roX1 and roX2)]. Mechanisms involved in equalizing the expression of genes encoded on the X chromosome between the two sexes; some examples are X chromosome inactivation in female mammals, X chromosome upregulation in male *Drosophila* flies and partial repression of both X chromosomes in hermaphrodite *C. elegans* worms.

dosage compensation, epigenetic basis of: An epigenetic regulatory mechanism observed in mammals, flies and worms equalizes the expression of genes on the X chromosome between XY/X0 males and XX females. To ensure gene dosage equivalence between male and female mammals, one X chromosome is inactivated in all somatic cells in females. The inactive chromosome is epigenetically marked with DNA methylation at gene promoters, transcriptionally repressive histone modifications, and the silencing requires the ncRNAs Xist and Tsix. High levels of H3K27 trimethylation are present on the inactive X chromosome.

dosage effect: A type of inheritance in which two like dominant alleles have a stronger effect than a single one of the same type. *Aka* DOSE EFFECT.

dosage effects in polyploids: Most genes are compensating for gene dosage (83% with no differences between diploid and tetraploid plants).

dosage repetitions of DNA sequences: In case of dosage repetition, there are many copies of a DNA sequence per nucleus. The rRNA, 5SRNA, tRNAs and histone genes show dosage repetition.

dot (slot) blot: The immobilization of DNA or RNA to a sample spot (slot) on a filter that will be subsequently probed by hybridization. Usually, many samples are applied, each to a different location on the same filter, for mass screening. Unlike Southern or Northern blots, there is no parting of the target DNA or RNA by electrophoresis (size), and thus, potentially much less specificity occurs.

dot blot(ting): (i) Technique in which small spots, or dots, of nucleic acid (DNA or RNA) are immobilized on a nitrocellulose or nylon membrane for hybridization. (ii) Hybridization of cloned DNA without electrophoretic separation. Dot and slot blots differ only in the geometry of the blot, a series of spots giving a hybridization pattern that is amenable to analysis by densitometric scanning. Protein, DNA and RNA samples are dotted through circular templates directly onto the membrane or paper substrate. The dot representing the sequences related to probe will light up by autoradiography. This is a simple method, as there is no need of DNA digestion, gel electrophoresis and transfer of DNA from gel to membrane. Dot blotting is more advanced, less time-consuming, accurate and applicable to a wide variety of gene/source simultaneously.

dot plot: A graphical method of comparing two sequences corresponding to regions of sequence similarity.

dot-blot analysis: A variant of the Southern transfer; different concentrations of RNA or DNA may be identified. Non-radioactive DNA will be denatured and with different concentrations transferred to nitrocellulose filters. Only small dots are transferred. That DNA may hybridize with radioactively labelled probe. After autoradiography, the strength of blackness is used as a simple measure of DNA concentration or DNA homology. *Aka* DOT-BLOT ASSAY.

double back cross: Mating of an individual who is heterozygous at two loci with an individual who is homozygous recessive at the same two loci.

double *Bar Drosophila*: A more extreme bar phenotype, manifests as a narrower bar with reduced facet number, brought about by the duplication of a segment of the X chromosome at the *Bar* locus. Double (or homozygous) Bar *Drosophila* has 68–70 facets compared with 810 in wild-type.

double cross: A hybrid between two single crosses of different pure inbred lines (e.g. a cross between two F_1 hybrids). Four different lines (A, B, C, D) are used for making two single crosses: A × B = AB hybrid and C × D = CD hybrid; the single-cross hybrids (AB and CD) are then crossed, and the double-cross hybrid (ABCD) seed is used for the commercial crop. This method is used for producing hybrid seed. Most commercial hybrid corn hybrids are double crosses. *Aka* DOUBLE HYBRID.

double crossing over: The situation in which two crossing-over events occur in a defined chromosomal region within a tetrad. *Aka* DOUBLE CROSSOVER.

double-digest DNA: The product formed when two different restriction endonucleases act on the same segment of DNA.

double ditelocentric: The situation in which both arms of a certain chromosome are present in a cell as telocentrics, and each telocentric with its homologue.

double fertilization: One sperm nucleus of the male gamete fuses with the egg nucleus of the egg cell, and the second male gamete fuses with the secondary nucleus of the central cell. The former fusion is known as SYNGAMY, which gives the zygote, and the latter fusion is known as TRIPLE FUSION, which results in the formation of primary endosperm nucleus.

double flower: Flowers that have more than one row of petals. Stamens and sometimes pistils are transformed into petals, or sometimes, the petals split to form several more. Completely double flowers have usually lost their reproductive organs and are therefore unable to produce seeds. These varieties of plants are bred to stay fresh longer than single flowers when cut. Commonly, plants do not show double flowers. They can only be formed when three specific genes are simultaneously mutating. These genes are the main regulators for flower formation. Their DNA sequences are almost identical. By induced mutations and subsequent combination of those genes, double flowers can be induced (e.g. in *Arabidopsis*).

double haploid (DH) population: DH populations have become standard resources in genetic mapping for species in which DHs are readily available. Doubled-haploid populations are ideal for genetic mapping. It is possible to produce a genetic map within 2 years of the initial cross regardless of the species. *See* DOUBLED HAPLOID.

double helix: *See* DOUBLE-HELICAL MODEL OF DNA.

double heterozygote: An individual who is heterozygous at two loci under investigation (e.g. *AaBb*).

double infection: Infection by two genetically different phages.

double isocompensating trisomic: An individual deficient in one chromosome, which is compensated by two isochromosomes, one for each arm of the missing chromosome.

double membrane: In mitochondria and plastids, there exists a two-layered membrane which environs the organelle. This is believed to be the result of endosymbiosis, with the outer membrane coming from the eukaryotic cell and the inner membrane belonging to the original prokaryote that was 'swallowed'.

double minute (dmin): Extra pairs of chromosomal fragments visualized on metaphase spreads, associated with amplification (e.g. multidrug resistance, *myc* oncogene).

double monoisosomic: The presence of two isochromosomes, each derived from one chromosomal arm. It can be derived from a double monotelosomic individual.

double monosomic: An individual/cell that lacks two different (non-homologous) chromosomes in a disomic chromosome complement ($2n - 1 - 1$).

double monotelosomic: The presence of two telocentrics, one derived from each arm.

double recessive: An individual who is homozygous for a recessive allele (e.g. *aa*), as opposed to homozygous for a dominant allele (e.g. *AA*).

double recombination: Two recombination events between the same DNA molecules, which may result in insertion of a short linear molecule into a longer linear or circular molecule.

double reduction: The genetic consequence of chromatid segregation, as opposed to chromosome segregation, whereby two sister-chromatid segments are part of the same meiotic product.

double sex: An allele that converts fruit fly males and females into developmental intersexes.

double telotrisomic: The incidence of two telocentrics, one for each arm of a missing whole chromosome, but together with the complete homologue.

double tertiary compensating trisomic: A compensating trisomic in which a missing chromosome is rewarded by two *trans*-located chromosomes.

double trisomic: An individual/cell that has two different (non-homologous) chromosomes extra in a disomic chromosome complement ($2n + 1 + 1$).

double-crossover (DX) array: The earliest example of lengthy lattices that used double-crossover (DX) complexes as the basic tiles, each containing four sticky ends designed with sequences that caused the DX units to combine into periodic two-dimensional flat sheets that are principally rigid two-dimensional crystals of DNA.

doubled haploid (DH): A diploid plant derived from spontaneous or induced chromosome doubling of a haploid cell or plant, usually after anther or microspore culture by using different methods.

doubled-haploid (DH) lines: Homozygous lines derived from haploidization and doubling again the chromosome number.

doubled-haploid (DH) method: A procedure used to speed up the production of homozygotes and to decrease the population size for selection. In other words, generating haploid plants by parthenogenesis or by another culture followed by doubling of the number of their chromosomes (spontaneously or induced).

double-helical model of DNA: Proposed by J.D. Watson and F.H.C. Crick in 1953. Accordingly, two polynucleotide chains give double-helical shape. Its width is 20 Å. One complete turn has 10 base pairs. The length of one turn is 34 Å, and the distance between two base pairs is 3.4 Å. Two strands of DNA are complementary (adenine pairs with thymine and guanine pairs with cytosine). Two strands are anti-parallel. It stores information for biological messages. The model proposed semi-conservative mode of DNA replication and explained that mutation arises due to a change in nucleotide sequence. *Aka* DOUBLE HELIX.

double-hit single-nucleotide polymorphism (double-hit SNP): Any SNP for which each allele occurs in two (or more) samples from a discrete population.

double-precision floating-point format: A computer number format that signifies a wide lively range of numerical values by using a floating radix point.

double-strand break (DSB): A deleterious form of DNA damage that occurs when the covalent bonds of both strands of a double helix are broken at a locus. Double-strand breaks in DNA are produced by reactive oxygen species ionizing radiation and chemicals that produce reactive oxygen species. Double-strand breaks are also a normal consequence of V(D)J RECOMBINATION, but they occur abnormally during replication as a result of replication fork arrest and failure. Failure to rejoin breaks properly can lead to lose of portions of chromosome or to rearrangements. DSBs can be repaired by homologous recombination (HR) or error-prone NHEJ.

double-strand break repair (DSBR) model: This model elucidates meiotic recombination in plants and intrachromosomal recombination in somatic cells (i.e. with inverted repeats and gene conversion). It describes the conservative mode of DNA recombination.

double-stranded DNA: DNA that consists of two polynucleotide chains whose nitrogenous bases are connected by hydrogen bonds. Within this arrangement, each strand mirrors the other as a result of the anti-parallel orientation of the sugar-phosphate backbones, as well as the matching nature of the A-T and C-G base pairing.

double-stranded DNA (ds-DNA): A DNA molecule that consists of two bound strands, each that complements the other. DNA is usually double stranded. *Compare with* SINGLE-STRANDED DNA.

double-stranded DNA as genetic material: Double-helical DNA is now known to be the genetic material in many viruses, bacteria and all organisms. Depending upon shape of the DNA molecule DNA can be categorized as (i) CIRCULAR (as in most of the viruses, like SV40, and bacteria like *Escherichia coli*, *Salmonella typhimurium*) or (ii) LINEAR (as in phage T_7 and eukaryotic chromosomes).

double-stranded DNA recombineering (DSDR): Recombineering based on double-stranded linear DNA fragments. DSDR requires the expression of a phage-derived exonuclease (Exo//or RecE) and an ssDNA-binding protein (Beta or RecT). In the case of the l-Red system, Gam (vii) can also be added, which inhibits the DNA exonucleases of the hosts.

double-stranded DNA viruses: These viruses have double-stranded DNA as the genetic material (e.g. smallpox, vaccinia, varicella-zoster, adenoviruses, SV40, T2, T4, lambda herpes viruses, human herpes virus 8, cytomegalovirus, Epstein-Barr virus).

double-stranded RNA: Refers to high molecular weight RNA in the 'A' form, having the following properties: (i) a base composition expected for an RNA duplex composed of two complementary, anti-parallel strands stabilized by hydrogen bonds and hydrophobic interactions, (ii) a molar absorbance (per phosphodiester group) lower than that of single-stranded RNA (ssRNA), (iii) an absolute hypochromicity remarkably larger than that of ss-RNA, (iv) temperature transition profiles of a co-operative type, with ionic strength-dependent Tm values and (v) sedimentation coefficients (s20,w), above 8–9S.

double-stranded RNA spray: A spray consisting of crude preparations of dsRNA which are efficient for silencing genes of plant viruses.

double-stranded RNA viruses: Double-stranded RNA (dsRNA) is the genetic material in reovirus and several plant viruses. In wound tumour (plant) virus and retrovirus (animal virus) also, dsRNA is the genetic material. These virus particles comprise enzymatic machinery that transcribes each of the double-stranded RNA molecules into a mRNA (complete with cap) and exports these into the cytosol of the infected cell.

doubling dose: The dose of a mutagenic substance to induce as many mutations as can occur spontaneously in one generation.

doubling time: The average time required to double the number of individuals of a population.

down mutations: Mutations with lost or reduced transcription of the adjacent gene which could result from deletion of an extragenic part of the promoter.

Down mutations can be produced involving only a single-base pair.

down promoter mutations: Mutations in promoters that decline the frequency of initiation of transcription. They result in the production of less mRNA than is done in the case of non-mutated state.

down-regulating: Refers to regulatory sequences, chemical compounds (e.g. transcription factors), mutations (e.g. down promoter mutations) and so on, that result in a given gene to express less of the protein that it normally codes for.

downgrading regulation: A mechanism in dosage compensation where two X chromosomes in one sex decrease expression of its genes to equalize the expression of a single X chromosome in the other sex (e.g. in nematode, HYPOTRANSCRIPTION exists).

downloading: In computer networks, to receive data from a remote system, typically a server, such as a web server, an FTP server, an e-mail server or other similar systems. This *contrasts with* UPLOADING, where data are sent to a remote server. A download is a file offered for downloading or that has been downloaded or the process of receiving such a file.

downregulation: Process by which a cell loses its sensitivity to growth factor or hormone stimulation (often by endocytosis of growth factor receptors).

downstream: (i) Recognizes sequences proceeding farther in the direction of expression (e.g. the coding region is downstream from the initiation codon, towards the 3′ end of an mRNA molecule). In double-stranded DNA, downstream is towards the 3′ end of a polynucleotide of an antisense (coding or template) DNA strand. downstream (or 3′ to) is in the direction of transcription (or translation). (ii) Sometimes, the term refers to a position within a protein sequence, in which case downstream is towards the carboxyl end, which is synthesized after the amino end during translation. *Compare with* UPSTREAM.

downstream gene: A gene under the control of a regulatory gene; the genes located downstream of a regulatory gene constitute a regulatory pathway.

downstream processing: Separation and purification of useful products from bioprocesses. The bioprocesses release a variety of undesirable metabolites along with the useful products. Separation of useful products from undesirable compounds is described as 'downstream' processing. The cost of downstream processing is sometimes more than 50% of the manufacturing cost that requires more research.

downy mildew: A plant disease in which the pathogen gives the impression as a downy growth on the host surface. This condition is caused by a member of the Oomycota (the oomycetes).

DQAsomes: DQAsomes are mitochondriotropic bolasomes that are derived from the compound dequalinium, a dicationic compound resembling bolaform electrolytes and is a symmetrical molecule with two charge centres separated at a relatively large distance. This symmetric bipolar compound can self-associate into mechanically very stable spherical monolayer aggregates, called DQAsomes, upon sonication in aqueous medium. DQAsomes migrate and localize with mitochondria within living cells. These liposomal-like vesicles are capable of binding plasmid DNA and form DQAsome-DNA complexes in which the plasmid DNA is protected from nucleases. DQAsomes have been employed to deliver both oligonucleotides and plasmid DNA conjugated with a MITOCHONDRIA-TARGETING SEQUENCE (MTS) to mitochondria of living cells.

draft sequence: The sequence generated by the HUMAN GENOME PROJECT (HGP) as of June 2000 that, while incomplete, offers a virtual road map to an estimated 95% of all human genes. Draft sequence data are mostly in the form of 10,000 base-pair-sized fragments whose approximate chromosomal locations are known. See SEQUENCING, FINISHED DNA SEQUENCE *and* WORKING DRAFT DNA SEQUENCE.

drastic amino acid substitutions: When amino acids that have dissimilar charges, chemically reactive and non-reactive or bulky and light side chain amino acids replace each other. These substitutions may have drastic effects on functioning of the protein.

drastic mutations: These are the mutations in which amino acids that have different charges substitute each other. Chemically reactive amino acids may be substituted by non-reactive amino acids or amino acids that have high molecular weight may be substituted by amino acid having low molecular weight, or vice versa. These mutations cause differences in mass/charge ratio of protein and hence can be identified electrophoretically. The consequence of such mutations depends upon the type of replacement or site of substitution, which may vary from 'no' effect to 'full' effect. *Aka* RADICAL MUTATIONS.

drift: (i) Of pesticides, the movement of airborne particles of a spray, dust or vapour away from the target area during or shortly after an application. (ii) In genetics, *see* RANDOM GENETIC DRIFT.

drip tip: A long, narrow, acuminate, caudate or cuspidate extension at the tip of a leaf or leaflet. Commonly an adaptation to rainy conditions, as it promotes shedding of water by its dripping from the narrow tip. The term drip tip is not anatomically descriptive in the way that acuminate or cuspidate are. Rather, it is an explanation of the functional shape that aids dripping, regardless of the specific geometry of the shape itself.

driver DNA: Excess of DNA used in a DNA-driven reaction.

driver mutation: A genetic mutation that gives cancer cells a growth advantage, leading to rapid reproduction and growth.

domains rearranged methyltransferase 2 (DRM2): A methyl transferase responsible for *de novo* DNA methylation at all sequence contexts; its role in CHH methylation is most prominent, since CHH methylation cannot be maintained and, thus, must rely entirely on *de novo* methylation. The DRM2 activity is highly regulated by the RdDM pathway.

Drosha: The first human RNase III enzyme identified and cloned. A novel pathway of small guide RNA maturation and the first example of a host factor (RNase III) required for bacterial RNA-mediated immunity against invaders. Drosha is responsible for performing pri-RNA cropping.

***Drosophila*:** The dipteran fly that reproduces rapidly and that is commonly utilized in classical genetics experiments and development studies due to its short life cycle (14 days) and simple genome (four chromosome pairs). Because of these factors, a large base of knowledge about Drosophila genetics has been accumulated by the world's scientific community. For example, of the nearly 300 'disease-causing' genes in the human genome, more than half have an analogous gene in the *Drosophila* genome. *Drosophila* was one of the first organisms to have its entire genome sequenced by man. Experiments on *Drosophila* led to the chromosome theory of inheritance by T.H. Morgan. C.B. Bridges, A.H. Sturtevant and H.J. Muller used this fly extensively. *Drosophila* continues to be an important experimental organism even in molecular genetics experiments. Most commonly used species is *Drosophila melanogaster*.

***Drosophila* epigenome:** The *E*(*z*) histone methyltransferase heterozygous mutation in *Drosophila* increases lifespan and stress resistance. Age-dependent modifications in the expression of genes involved in these pathways related to immune response, cell cycle and ribosome biogenesis.

***Drosophila* epigenomic landscape analysis:** Several genes in the pericentromeric regions of *Drosophila melanogaster* depend on heterochromatic environment for their expression. A comprehensive analysis of the epigenetic landscape of heterochromatic genes across all the developmental stages of *Drosophila* was performed using the available histone modification and expression data from modENCODE. Heterochromatic genes exhibited combinations of active and inactive histone marks that corresponded to their level of expression during development. Nuclear Matrix Associated Regions (MARs) were reported to occur prominently in the intergenic regions of heterochromatic genes during embryonic stages suggesting their reasonable role in pericentromeric genome organization. Meta-analysis of the various genomic data sets suggested that the epigenomic and genomic landscape of the heterochromatic genes were discrete, which could be contributing to their unusual regulatory features as opposed to the surrounding heterochromatin, which is repressive in nature.

Drosophila melanogaster: Species of small fly, commonly called a fruit fly, commonly used in genetic, cytogenetic and molecular investigations because of its favourable features (e.g. small size, short life cycle, large progeny, clearly defined wing shape, wing size, wing, veins, body colour, eye shape, eye colour, easy growth on simple medium, small genome size, small chromosome number, salivary chromosomes).

***Drosophila* retrotransposon:** The *copia* retrotransposons include several families, each highly conserved and present in 5–100 widely scattered copies per genome. An RNA intermediate and reverse transcriptase are used for transposition. Virus-like particles (VLPs) occur with copia. Integration results in target site duplication (3–6 bp).

***Drosophila* transposons:** It is estimated that 15% of the Drosophila genome is mobile. These transposons fall into different classes: (i) Retrotransposons (e.g. copia) and (ii) P-element.

drosopterin: A red pigment in *Drosophila* eye.

drought: Insufficient soil water for normal plant growth.

drought resistance: Drought resistance plants synthesize and accumulate Osmoprotectants – low-molecular-weight compounds like sugars alcohols, amino acid proline, quaternary ammonium salts (like glycine betaine) that lower osmotic potential and help maintain turgor.

drought tolerance: Tolerance of the crop to an extended dry period that otherwise results in crop stress and harvest reduction. Drought resistance is one of the most important problems in agriculture. Severe droughts have caused reduction in global grain production by 5% or more in key countries. Plants tolerate drought through escape, avoidance or resistance mechanisms. Conventional breeding for drought tolerance has been slow because of the complexity of the trait. Using biotechnological techniques, hundreds of genes that are induced under drought have been identified, and some of these have been cloned. Using transgenic approach, different genes encoding for protein factors, enzymes for osmolytes, proteins involved in scavenging the reactive oxygen species and membrane proteins are being used to develop drought tolerance in plants.

drought-responsive genes: Plants may become drought tolerant by activating specific drought-responsive genes. Many such genes are modified by chemical marks, called EPIGENETIC MARKS, such as H3K9ac. Two key steps are needed to coordinate the activation of the drought-responsive gene: (i) an increase in the H3K9ac mark and (ii) a TF that binds to the drought-responsive gene for activation.

drought stress in Brassica: Drought stress is a multi-dimensional encounter that intensely influences crops at diverse levels, including morphological, physiological and molecular responses. Plants undergo structural modifications (e.g. growth of deeper root systems, higher root hair density and reduced leaf surface area) as a tactic to improve water uptake while minimizing water loss. Plant cells accrue osmoprotectants and antioxidants to combat oxidative stress. A.R. Sheoran *et al.* (2024) suggest approaches for Brassica crop resilience including Brassica-specific genes, regulatory networks associated with the drought signalling pathway and drought-responsive proteins. Management approaches like planting techniques, breeding, transgenics, nanotechnology, microorganisms and chemical priming are proposed to battle the negative possessions of drought stress in Brassica crops.

drug addicts: DNA modified by a covalently bound drug or the reactive side group of a drug.

drug discovery: METTL3 a good drug target for cancer therapy. Moreover, tissue samples taken from acute myeloid leukaemia patients revealed high levels of the enzyme fat mass and obesity-associated protein (FTO), which is a $6^{m}A$ eraser.

drug discovery cycle: The cycle of events required to develop a new drug. Typically, this involves research, preclinical testing and clinical development and can take 5–12 years.

drug resistance factor: Duplex ring DNA plasmid or episome, which confers upon its bacterial host resistance to one or more antibiotics.

drug resistance gene: A gene that codes for an enzyme that enables the host cell to resist the effects of a drug. In general, the enzyme hydrolyses the drug or modifies its structure in some way. Drug resistance genes include the genes conferring antibiotic resistance that are carried on the plasmids.

drupe: A type of succulent fruit formed from one carpel. The single seed is enclosed by a stony layer of the fruit wall (e.g. in peaches and olives). *Aka* KERNEL.

dry nanotechnology: Derives from surface science and physical chemistry, focuses on fabrication of structures in carbon (e.g. fullerenes and nanotubes), silicon and other inorganic materials. Unlike the 'wet' technology, 'dry' techniques admit use of metals and semiconductors. The active conduction electrons of these materials make them too reactive to operate in a 'wet' environment, but these same electrons provide the physical properties that make 'dry' nanostructures promising as electronic, magnetic and optical devices. Another objective is to develop 'dry' structures that possess some of the same attributes of the self-assembly that the wet one's exhibit. The two terms, DRY NANOTECHNOLOGY and WET NANOTECHNOLOGY, were originally introduced by Prof. R.E. Smalley in 2004. *Compare with* WET NANOTECHNOLOGY.

dry waste: Waste that can be collected by the containers for valuable materials (e.g. glass, paper, cardboard, plastics, metals and similar recyclable wastes).

dry weight: Weight of any plant (or other organism) part after all its water content has been removed by drying.

dS: An alternative notation for synonymous substitution rate (Ks).

dual nickase: A method to decrease off-target effects by using a single Cas9 nickase and two different gRNAs that bind in close proximity on opposite strands of the DNA to create a DSB. *Aka* DOUBLE NICK *or* DOUBLE NICKING.

dual-specificity splice sites: These splice sites can be alternatively recognized as either 5′ or 3′ splice sites, and the dual splicing is conceptually similar to a pair of mutually exclusive exons separated by a zero-length intron. The dual-splice-site sequences are essentially a composite of canonical 5′ and 3′ splice-site consensus sequences, with a CAG|GURAG core.

dummy variable: A device used in a multivariable regression model to describe a categorical predictor without assuming a numeric scoring. Indicator variable might be a better term. For example, treatments A, B, C might be described by the two dummy predictor variables X_1 and X_2, where X_1 is a binary variable taking on the value of 1 if the treatment for the subject is B and 0 otherwise, and X_2 takes on the value 1 if the subject is under treatment C and 0 otherwise. The two dummy variables completely define three categories, because when $X_1 = X_2 = 0$ the treatment is A.

dumping: Disposing of waste illegally by not using bins or official recycling centres, civic amenity sites or landfills.

duplex: A nucleic acid molecule in which two strands are base paired with each other.

duplex type: A polyploid plant that possesses two dominant alleles for a given locus.

duplicate dominant epistasis: Dominant alleles at either of two loci can mask the expression of recessive alleles at the two loci resulting in 15:1 ratio. *Aka* DUPLICATE EPISTASIS.

duplicate genes: Two or more pairs of genes in one diploid individual that alone or together produce identical effects. In this case, wild-type and mutant phenotypes occur in 15:1 ratio in F_2.

duplicate loci: A pair of non-allelic loci having the same genetic meaning; two isoloci.

duplicate recessive epistasis: Recessive alleles at either of the two loci can mask the expression of dominant alleles at the two loci resulting in 9:7 ratio. *Aka* COMPLEMENTARY EPISTASIS.

duplicated operators: Chromosomal changes can, therefore, connect the structural genes of operon to new operator genes affected by different repressors, or conversely, an operator gene affected by the same repressor may be duplicated in various operons throughout the genome.

duplication: A type of genetic mutation that involves the duplication of one or more copies of a single nucleotide, gene or region of a chromosome, a complete chromosome or a complete set of chromosomes without loss of the original and increasing the size of the genome by inserting the copied sequence elsewhere. When the duplicated sequence is a gene, the incident is called GENE DUPLICATION. Duplication is important in the evolution of genomes. A duplication, like deletion, may include a part of a gene, whole gene, a group of genes, complete chromosome, more than one chromosome or even a complete set of chromosomes. Duplications are usually not lethal in homozygous state. The reciprocal of a deletion is called duplication, which at pachytene gives a loop-like structure. This is called a DUPLICATION LOOP. There may be a tandem duplication of part or the whole gene. Duplications have played a crucial role in the evolution of new genes during evolution. Duplications may have visible effects on phenotypes. Duplication of 16A segment of X chromosome of *Drosophila melanogaster* results in reduction in eye size. Flies homozygous for wild-type allele (B^+/B^+) have 779 facets. Those heterozygous for *Bar* allele (B^+/B) have 358 facets. Those homozygous for *Bar* allele (B/B) have 68 facets. *Ultrabar* females have three 16A segments on one X chromosome and only one 16A segment on the other (B^U/B^+); the eye has 45 facets.

duplication of germplasm: A duplicated seed sample that is prepared for safety reasons. The two samples are usually stored at different locations.

duplicative transposition: A mechanism of transposition that results in a copy of the element at both the excision and acceptor sites. This can occur even after excision of the element by the process of gap repair.

durable resistance: A resistance that stays effective during prolonged and widespread use in an environment favourable to disease. It thus remains effective and stable during the agronomic life of a crop. This type of resistance is usually determined by several genes and is a main target of resistance breeding. *See* HORIZONTAL RESISTANCE *and* RACE-NON-SPECIFIC RESISTANCE. *Contrasts* SPECIFIC RESISTANCE *and* VERTICAL RESISTANCE.

DUS testing: Testing for distinctiveness (D), uniformity (U) and stability (S) is called as DUS. These are well-defined UPOV guidelines for DUS in different crops that can be adopted as such or may be modified as per experience and need of a member country. Government of India has selected 35 crops to start with. Out of these, 25 are field crops (rice, wheat, maize, sorghum, pearl millet, chickpea, pigeon pea, mungbean, urdbean, lentil, rajmah, field pea, rape seed, mustard, groundnut, soybean, sunflower, safflower, castor, sesame, linseed, cotton, jute and sugarcane), 8 vegetable crops (potato, brinjal, okra, tomato, cauliflower, cabbage, onion and garlic) and 2 flower crops (rose and chrysanthemum). Guidelines for all the 35 crops have been developed. There is a need for developing detection methods for genetically modified organisms.

DUST: A computer program for filtering low-complexity regions from nucleic acid sequences.

***DWARF1(D1)* gene in rice:** This is a well-studied example of paramutation in plants. In rice, the spontaneous mutant *Epi-d1* produces a metastable dwarf phenotype. The phenotype is mitotically and meiotically heritable and corresponds to the metastable epigenetic silencing of the *DWARF1(D1)* gene. The silenced state is correlated with DNA hypermethylation in the *D1* promoter region. The epigenetic state of *D1* is bidirectionally mutable, from active to repressed and from repressed to active. The bidirectional epigenetic state might signify that the epigenetic regulation of *D1* is due to *de novo* DNA methylation. Epigenetic alleles such as *Epi-d1* could offer a mechanism of rapid adaptation to changing environmental conditions.

dwarfing: The underdevelopment of a plant or plant organs, which may be caused by disease, inadequate nutrition or unfavourable environmental conditions.

dwarfing genes in maize: One class of genes that control the height of the plant. According to D.S. Multani and co-workers (2003), compact stalks of maize and dwarf sorghum mutants are the result of a loss of an auxin transporter class of P-glycoproteins (multidrug resistant). While not currently used as a source of height reduction in maize, *br2* may offer an agronomic advantage to dwarf maize. Interestingly, *dw3* of sorghum was observed to have a direct repeat in one of its exons (exon V). Sporadic unequal crossing over in this exon produces mutant reversions (tall plants). The *dw3* locus has often been utilized in elite sorghum lines and the infrequent tall plant has, until now, been something of a mystery.

dwarfing genes in wheat: The dwarfing alleles *Rht-B1b* and *Rht-D1b* – previously known as *Rht1* and *Rht2*, respectively – have been utilized in wheat improvement programmes worldwide. Dwarf and semi-dwarf cultivars usually have higher grain-yield potential and improved lodging resistance under irrigated and high input conditions. These characteristics empowered the Green Revolution in wheat. The deployment of the *Rht* genes by International Maize and Wheat Improvement Center (CIMMYT), Mexico and by many other public breeding programmes has led to the replacement of taller wheat with shorter cultivars. In the late 1990s, over 80% of registered wheat cultivars globally possessed at least one dwarfing allele. The *Rht-B1b* and *Rht-D1b* alleles – which are located at homoeologous loci on chromosomes 4B and 4D of wheat, respectively – reduce sensitivity to gibberellic acid, which is necessary for stem elongation. Gibberellic acid response testing (GA testing) is done to recognize various dwarfing alleles.

dyad: (i) A pair of cells (i.e. one of the products of the disjunction of the tetrads at the first meiotic division, contained in the nuclei of secondary gametocytes). (ii) Two sister chromatids attached to the same centromere.

dye-labelled oligonucleotide ligation (DOL): A homogeneous DNA detection method that requires no further manipulations after the initial reaction is set up. This assay combines the PCR and the oligonucleotide ligation reaction in a two-stage thermal cycling sequence with FLUORESCENCE RESONANCE ENERGY TRANSFER (FRET) detection monitored in real time. Because FRET occurs only when the donor and acceptor dyes are in close proximity, one can infer the genotype or mutational status of a DNA sample by monitoring the specific ligation of dye-labelled oligonucleotide probes.

dyes as mutagen: A coloured substance that chemically bonds to the substrate to which it is being applied. Dyes affect replicating nucleic acids. Caffeine owes its genetic effect to its interference with one of the steps concerned with the repair of genetic damage. Caffeine-induced chromosome breaks do not join. Acridine dyes attach directly to nucleic acids and thus cause mutations. The presence of acriflavin in the growth medium of *Escherichia coli* removes the F′ factor. The presence of proflavin prevents assembly of phage T_4. In darkness, acridine dyes induce deletions or additions of bases in DNA. Methylene blue, toluidine blue and proflavin, when exposed to light, oxidize some DNA bases and have a strong lethal effect on phages, provided the dyes are inside the phage. Proflavin induces frameshift mutations as it gets intercalated between two adjacent purines. Acridine orange can stack along nucleic acids between two bases of a DNA strand and apparently attaches itself to a phosphate backbone. The two bases are separated by 6.8 Å. Thereafter a base will be missing. X-ray data show that acridine can be sandwiched between purine bases.

Acridines induced only deletions or insertions of one base pair in DNA. These dyes thus cause FRAMESHIFT MUTATIONS. Ethidium bromide and propidium diiodide cause likewise frameshift mutations.

dynamic analysis of histone variants using proteomics. Some PTLMs on histone variants play crucial roles. Quantification analysis of the histone variant of interest can be accomplished by using one or a few variant-specific peptides produced from an enzymatic digestion.

dynamic DNA nanotechnology: DNA systems that dynamically respond to the addition of fuel strands by adapting in a thermodynamically controlled fashion to the new energy landscape.

dynamic flux analysis of histone modifications: Three methods are suggested to measure the kinetic flux of histone proteins (e.g. protein turnover experiments): (i) Incorporation of metabolic analogues, for example, the methionine surrogate azidohomoalanine (Aha), in nucleosome turnover experiments. Aha can be conjugated to alkynyl group containing reagents via click chemistry, which permits facile isolation of new synthesized histones. (ii) metabolic labelling of histone PTLMs with small-molecule metabolite precursors, which uses radio isotope-labelled metabolites to determine protein and PTLM turnover. (iii) Methodologies coupling metabolic labelling and synchronised cell populations can be used to pulse-chase label and therefore distinguish the newly synthesized histones.

dynamic force microscopy (DFM): While the static, or contact mode, AFM is a widespread technique to obtain nanometre resolution images on a wide variety of surfaces, true atomic resolution imaging is routinely observed only in the dynamic mode. Dynamic AFM comprises a whole family of operational modes. Two modes of operation dominate the application of dynamic AFM: (i) AMPLITUDE MODULATION MODE (also called TAPPING MODE) is shown to exhibit an instability, which separates the purely attractive force interaction regime from the attractive-repulsive regime. (ii) SELF-EXCITATION MODE is derived, and its experimental realization is outlined. While the first is primarily used for imaging in air and liquid, the second dominates imaging in UHV (ultrahigh vacuum) for atomic resolution imaging. *Aka* DISSIPATION FORCE MICROSCOPY.

dynamic force spectroscopy (DFS): Here, the resonance frequency of the cantilever as a function of the resonance amplitude is measured.

dynamic instability: The property of sudden conversion from growth to shrinkage, and vice versa, in a protein filament, such as a microtubule or actin filament.

dynamic molecular combing (DMC): DCM technique is more reproducible and to make measurements directly on human genomic DNA, as done in the previous techniques (FISH on chromosomes, DNA halo, decondensed chromatin, etc.). In order to avoid breakage of the DNA fibres, which occurs during the manipulation of drop of DNA solution, the DNA was protected by agarose blocs, a polymer that behaves like a solid gel at room temperature and melts above 60°C. Cells containing the DNA to comb are mixed at high temperature with melted agarose and, after solidification of the agarose blocks, disrupted and deproteinized using enzymes. The DNA fibres remain entangled into the polymer mesh. DNA-containing blocks are put in solution and melted, in order to release the DNA. In order to prevent resolidification of the agarose, it is digested by an enzyme called β-agarose. DNA is then free to decondense and fluctuate freely and eventually to bind to a treated glass surface.

dynamic programming: A program that allows a computer to explore efficiently all possible solutions to certain types of complex problems; it divides a problem into reasonably sized subproblems and uses parts to compute the final answer.

dynamic programming algorithm: This algorithm uses the concept of using the already found solution to avoid repetitive calculation of the same part of the problem. It divides the problem into smaller overlapping subproblems and solves them.

dynamic view of epigenome dynamics: The DNA sequence variation occurs either in *cis* or in *trans*. This forced link undergoes relaxation in the facilitated category that induces changes in epiallelic states that can be passed on to succeeding generations.

dysgenic: Exercising a harmful consequence on later generations through the inheritance of undesirable characteristics.

dysplasia: A change in cell growth and behaviour in a tissue in which the structure becomes disordered.

dysploid: A plant or species in which the chromosome number is more or less than the expected normal euploid number.

dysploidy: Abnormal ploidy (e.g. the appearance of diploid or triploid individuals in a normally tetraploid population or of triploid and tetraploid ones in a normally diploid population).

E

early development in phage: The period of a phage infection preceding the start of DNA replication.

early generation test: Selection schemes in which poor recombinants are rejected already in F_2 and F_3 generations.

early genes: Bacteriophage genes that are expressed during the early stages of infection of a bacterial cell; usually, the products of early genes are involved in replication of the phage genome.

early ideas about heredity: PREFORMATION, PANGENESIS, INHERITANCE OF ACQUIRED TRAITS, EPIGENESIS, BLENDING THEORY and GERMPLASM THEORY. *See details as separate terms.*

early intron theory: This theory explains that exons are descents of ancient minigenes and that the introns are descendants of the spacers between them. Genes large enough to encode proteins were first assembled from sets of exons. The machinery of splicing originated in an ancient RNA world. Introns were completely lost from both kingdoms of bacteria as well as several protist groups.

early selection: It is performed on highly heterozygous individuals that are used as the parents of the next screening generation.

early stages of flower development: The early stages of flower development requires AGAMOUS protein (AG) from *Arabidopsis* homeotic gene *AGAMOUS* (*AG*) that encodes transcription factor of the MADS-box family for reproductive specification of stamens and carpels. During later flower developmental stages, *AG* is expressed in the distinct reproductive organs. Therefore, the ultimate function of *AG* is the maturation of stamens and carpels and organogenesis.

early steps in processing 5S rRNA: After transcription, the primary transcript is slashed into two parts: one part carries the genes for the minor rRNA and one or two tRNAs present on the intergenic spacer, and other part contains the major and the supplemental rRNAs, depending upon whether or not a tRNA occurs on the 3′-end. Neither RNase III nor RNase E is implicated in this initial enzymatic reaction.

early vs late genes: Those genes transcribed early in a bacteriophage-mediated infection process as compared to those genes transcribed sometime later. May require different sigma factors for recognition of promotors. Such genes fall in three categories: immediate-early (IE), early (E) and late (L) genes.

early vs late proteins: During viral infection, viral-specific proteins are produced at distinctive times after infection, frequently under positive control of different bacterial and viral sigma factors. Such proteins fall in three categories: immediate-early (IE), early (E) and late (L) proteins.

early warning system (EWS): Represents the set of capacities required to produce and transmit timely and meaningful warning information to allow individuals, communities and organizations threatened by a hazard to prepare and to perform appropriately and in sufficient time to reduce the possibility of harm or loss. On the one hand, an EWS includes the gathering, processing and presentation of information in a consistent and meaningful manner to allow the generation of alert messages and, on the other hand, the generation and transmission of alert messages to the citizens at risk by means of warning communication. *Aka* FOREWARNING SYSTEM.

early λ phage genes: Soon after phage infection, transcription starts with the help of two promoters – P_R and P_L, on the right and left side, respectively. The transcription is initiated on both DNA strands. One strand is transcribed on the left side and the other on the right side. Early gene *cro* is transcribed on the right side under the control of promoter P_R, and *N*-gene is transcribed on the left side under the control of the promoter P_L. *Aka* IMMEDIATE-EARLY GENES.

Eastern blot(ting): This term has been fought over extensively. Many different techniques have been variously described as Eastern blotting, most of which involve using different types of probes to detect POSTTRANSLATIONAL MODIFICATIONS (PTLMs or PTMs) of proteins including the addition of lipids, phosphates and glycoconjugates, for instance the use of lectins to probe a protein-blotted membrane in search of specific carbohydrates. Eastern blot is considered an extension of the biochemical technique of Western blot. Multiple techniques have been described by the term 'Eastern blot(ting)', and most use phosphoprotein blotted from SODIUM DODECYL SULFATE-POLYACRYLAMIDE GEL ELECTROPHORESIS (SDS-PAGE) gel on to a polyvinylidene fluoride or nitrocellulose membrane. Transferred proteins are analysed for PTLMs using probes that may detect lipids, carbohydrate, phosphorylation or any other protein modification. This technique probes proteins blotted to polyvinylidene difluoride (PVDF) membrane with lectins, cholera toxin and chemical stains to detect glycosylated, lipoylated or phosphorylated proteins. Multiple variants of Eastern blotting technique have been described in literature. Eastern blotting can be considered an extension of the biochemical technique of WESTERN BLOTTING to detect PTLMs, including the addition of lipids, phosphates and glyconjugates. It is generally engaged to detect carbohydrate epitopes. *Compare with* WESTERN BLOTTING.

Ebola virus: The Ebola virus can cause severe and often fatal haemorrhagic fever in humans and other mammals. *Aka* ZAIRE EBOLAVIRUS.

eclipse period: The period following viral infection when no intact infection virions are present. After infection, the virion ceases to exist as an entity. New virions are assembled just before and during the burst period.

ecogenomics: N.M. Van Straalen and D. Roelofs (2012) define ecogenomics as a scientific discipline that studies the structure and functioning of a genome with the aim of understanding the relationship between the organism and its biotic and abiotic environments.

ecological amplitude: An array of environmental conditions in which an organism can survive.

ecological conversion: A long-term alteration in the character of the ecosystem at some place (e.g. when a natural forest is converted into an agricultural land use).

ecological efficiency: The efficiency with which energy is shifted from one trophic level to the next. It is determined by a combination of efficiencies concerning the organismic resource acquisition and assimilation in an ecosystem.

ecological footprint: The area of ecoscape (i.e. landscape and seascape) necessitated to supply a human population with the necessary food, materials, energy, waste disposal and other crucial goods and services.

ecological genetics: The study of evolution in action in nature, by a combination of fieldwork and laboratory genetics.

ecological integrity: A concept related to environmental quality but focusing on changes in natural populations and ecosystems, rather than effects on humans and their economy. *Aka* ECOSYSTEM HEALTH. *See* ENVIRONMENTAL QUALITY.

ecological justice: A worldview in which all species (i.e. not just humans) have a right to equitable access to the necessities of life and happiness.

ecological pyramid: A model of the trophic structure of an ecosystem systematized with plant productivity on the bottom, that of herbivores above and carnivores above the herbivores. *Aka* TROPHIC PYRAMID, ELTONIAN PYRAMID, ENERGY PYRAMID *or* FOOD PYRAMID. *See* SOCIAL JUSTICE.

ecological race: A local race that owes its most conspicuous attributes to the selective effects of a specific environment.

ecological risk assessment of genetically modified plants (GMPs): There are a very large number of ecological potential risks associated with the release of GMPs into the farming system. Perhaps the two most important risks are the following: (i) GMP establishes in semi-natural or natural habitats and (ii) inserted transgenes incorporate into other species, thereby affecting non-target organisms in farms or natural habitats.

ecological species concept: A concept of species, according to which a species is a set of organisms adapted to a particular, discrete set of resources (or 'niche') in the environment. *Compare with* BIOLOGICAL SPECIES CONCEPT, CLADISTIC SPECIES CONCEPT, PHENETIC SPECIES CONCEPT *and* RECOGNITION SPECIES CONCEPT.

ecological subspecies: A distinctive population whose distribution in space is in response to vegetational or other narrow ecological influences, rather than to the broader influences of climate, as in geographical subspecies.

ecological succession: The more or less predictable and methodical set of changes that transpires in the composition or structure of an ecological community over time.

ecological values: Broader useful values that are based on the needs of humans but also on those of other species and natural ecosystems.

ecologically sustainable development: It considers the human requirement for resources within an ecological context and includes the need to withstand all species and all components of Earth's life support system. *Compare with* SUSTAINABLE DEVELOPMENT.

ecology: The scientific analysis and investigation of interactions between organisms and their environment. It is an interdisciplinary field that combines concepts from biology, geography and Earth science.

economic heterosis: Superiority of the F_1 hybrid over the standard commercial cultivar. *Aka* HETEROBELTIOSIS *or* STANDARD HETEROSIS.

economic threshold: The pathogen abundance or disease level at or above which the value of crop losses (in the absence of management efforts) would exceed the cost of management practices.

economic trait loci (ECL): Sites in the genome that govern the traits of economic significance.

ecophene: The range of different phenotypes produced by one genotype within a certain environment.

ecophenotype: A non-genetic modification of the phenotype in response to an environmental condition.

ecophysiology: A discipline of biology that studies the adaptation of an organism's physiology to environmental conditions.

***Eco*RI:** A type II restriction endonuclease from *Escherichia coli* RY13. *Eco*RI recognizes the below-mentioned DNA sequence and arrows indicate the cuts – 5′ G↓A A T T C / 3′ C T T A A↑G.

ecospecies: A locally adapted species that displays minor changes in morphology and physiology to another species that are related to a habitat and are genetically determined.

ecosystem: A community of living organisms in concurrence with the non-living components of their physical environment, interacting as a system.

ecosystem disruption: Any perturbation to either the structure or function of an ecosystem.

EcoTILLING: A high-throughput, low-cost technique for quick discovery of polymorphisms in natural populations of plants. It is comparable to TILLING, but it differs from TARGETING INDUCED LOCAL LESIONS IN GENOMES (TILLING) in that natural polymorphisms are spotted rather than being induced through chemical mutagenesis. On the other hand, SINGLE-NUCLEOTIDE POLYMORPHISMS (SNPs) – small insertions and deletions and variations in microsatellite repeat number – can be efficiently identified using the EcoTILLING technique. The sequence of a cloned QTL gene offers the perfect marker for MARKER-ASSISTED SELECTION (MAS) and provides the template for the identification of potentially superior allelic variants in the crop species or wild progenitors via EcoTILLING. Combining mutagenesis with DNA sequence information, EcoTILLING is utilized to investigate the functions of specific genes and to create novel variation. This technique is also employed for studying natural variation for the target genes (e.g. loci governing grain number per panicle in wheat and rice). *See* TILLING.

ecotype: A locally adapted population of a widespread species. It displays minor changes in morphology and physiology that are associated with a habitat and are genetically determined. The individuals of an ecotype are only uniform with regard to the traits that provide them with special adaptation to specific environments. All the other characters may vary. Ecotypes may be noticed in perennial clover, lucerne, grasses or other forage crops. In evolutionary ecology, a genetically distinct geographic variety, population or race within a species which is adapted to specific environmental conditions. *Aka* ECOSPECIES.

ectopic gene expression: Expression of a (trans)gene in a tissue where, or developmental stage when, such expression is not expected.

ectopic gene expression, confirmation of: To further confirm that the ectopic expression (an abnormal gene expression in a cell type, tissue type or developmental stage in which the gene is not usually expressed) of *Os02g45850* is responsible for the developmental defects in *Epi-rav6*, M.F. Zhang and others in 2015 overexpressed *Os02g45850* in the wild type under the control of the maize (*Zea mays* L.) *Ubiquitin* promoter. The positive transformants with high expression of *Os02g45850* displayed large leaf inclination and small grains. It was concluded that the abnormal phenotype in this mutant resulted from the high expression of *Os02g45850*. *Os02g45850* encodes a B3 DNA-binding domain-containing protein, which belongs to the RAV family and was named RAV6, and the allele was named *Epi-rav6*. OV-1 and OV-2 indicate transgenic plants overexpressing RAV6.

ectopic recombination: Recombination between repetitive DNA elements observed in different regions of the genome (e.g. between transposable elements at different sites). This can cause chromosome rearrangements and deletions.

ectosymbiont: The organism that resides outside a host organism.

ectotherm: An organism in which internal physiological sources of heat are of comparatively small or quite negligible importance in controlling body temperature compared to ambient sources of heat (e.g. invertebrates, reptiles, amphibians and plants). Ectotherms generally experience alterations in body temperature that closely match changes in the temperature of their environment. Colloquially, these organisms are often referred to as 'cold-blooded'. *Contrasts* ENDOTHERM.

edible vaccine: Production of vaccines in the edible plant parts through transgenic approach. Conventionally, vaccines are produced through mammalian cell culture that is very expensive and requires cool chain from the site of production to the site of use. Charles J. Amtzen, early during 1990s, gave the concept of edible vaccines from transgenic plants. Antigens have been produced in plants, for example, against rabies virus in potato and tomato, against respiratory syncytial virus in tomato, and antibodies have been produced against *Streptococcus* bacteria in tobacco plants. Crop plants offer cost-effective bioreactors to express antigens, which can be used as edible vaccines. The genes encoding antigenic proteins can be isolated from the pathogens and expressed in plants, and such transgenic plants or their tissues producing antigens can be eaten for vaccination/immunization. The expression of such antigenic proteins in crops like banana and tomato is useful for immunization of humans, since banana and tomato fruits can be eaten raw. The edible vaccines that are produced in transgenic plants have great advantages, like the alleviation of storage problems, easy delivery system by feeding and low cost as compared to recombinant vaccines produced by bacterial fermentation. Edible vaccines are also derived from eggs via the activation of the hen's immune system to cause that hen to secrete desired molecule(s) into the eggs it lays.

editing reactions during translation: Synthesis of proteins containing errors (mistranslation) is prohibited by aminoacyl-tRNA synthetases through their accurate aminoacylation of cognate tRNAs and their ability to correct occasional errors of aminoacylation by editing reactions. A principal source of mistranslation comes from misidentifying glycine or serine for alanine.

edition enzyme: Enzyme responsible for proofreading the DNA and correcting errors in replication or following from mutation.

effect size: Effect size is a statistical term that quantifies the degree of a relationship between two given variables. For example, we can learn about the effect of therapy on anxiety patients. The effect size aims to determine whether the therapy is highly successful or mildly successful.

effective breeding size: The breeding size of a population, attuned mathematically to permit comparisons with others.

effective microorganisms (EM): Natural microorganisms occurring freely in nature and, when applied as single strains or mixtures, may serve to improve the quality and fertility of soil as well as the growth and quality of crops.

effective population size (Ne): The size of the ideal Wright-Fisher population that might yield the same rate of random drift as the actual population (of size N) in question. This defines effective population size for drift. There are other definitions for unequal family sizes, subdivided populations, differences in male vs female reproductive success and variable population sizes.

effective population size in crop improvement: Effective population size is a measure of genetic diversity, representing the number of individuals donating to the next generation. J. Duminil and others in 2019 estimated the effective population size of African mahogany in the Amazon, finding it smaller than the actual population size, thereby indicating reduced genetic diversity and increased inbreeding risk. M.S. Islam and others in 2020 estimated the effective population size of Indian mackerel in the Bay of Bengal, also observing a smaller effective population size and reduced genetic diversity. Such studies aid in developing effective crop improvement programmes and strategies. A comprehensive understanding of population genetics principles is essential for breeders and crop improvement experts.

effectively neutral: An allele that alters fitness by less than approximately 1/Ne. Its probability of fixation is approximately the same as if it were absolutely neutral.

effector: A class of usually small molecules that facilitates the activity of a specific protein (e.g. enzyme) molecule by binding to a specific site on the protein. Control of (existing) enzyme activity may be achieved by combination of the effector with the enzyme. The effector molecule may either physically block the active site on the enzyme molecule or alter the three-dimensional conformation of the enzyme molecule. Such conformational change results in a change in the enzyme's catalytic activity. Effector is a general term. Effector molecules may be activators (cause an increase in the enzyme's catalytic activity) or inhibitors (cause a decrease in the enzyme's catalytic activity). A special class of effectors, *aka* ALLOSTERIC EFFECTORS, binds to the enzyme molecule at a site other than the enzyme's active site (thereby activating or inhibiting). Effectors also pathogen molecules, usually proteins, which are translocated into host cells where they may function to directly manipulate host innate immunity.

effector cell: A cell that brings out the final response or function of a particular process. The main effector cells of the immune system, for example, are activated lymphocytes and phagocytes – the cells involved in destroying the potential threat.

effector domains, types of: The EDs used for DNA methylation and histone modifications are categorized as repressors and activators of gene expression. The original VP16 domain is derived from Herpes Simplex Viral Protein 16 and comprises amino acid residues 437–447 (DALDDFDLDML). VP16 was then engineered into VP64 domain (which is a fusion containing four tandem copies of VP16 each connected by dipeptide glycine-serine (GS) linker (DALDDFDLDML)-GS-(DALDDFDLDML)-GS-(DALDDFDLDML)-GS-(DALDDFDLD ML)) or into VP160 (which is a fusion of ten tandem copies of VP16). VP16 is the most commonly used transactivation domain. One consequence of using the VP64 transactivation domain is that it recruits p300, which results in activation of H3K27ac to intensify at the targeted locus over time and represents an example of transcription-driving domain. Transcriptional repression, on the other hand, utilizes repression domains and is characteristically accomplished by variants of a 45-aa segment from KRÜPPEL-ASSOCIATED BOXES (KRAB) or repressive epigenomic modifications.

effector gene: The gene which drives transcription of another gene with the help of a promoter from another gene.

effector molecule: A molecule (a sugar, an amino acid or a nucleotide) that can bind to a regulator protein and thereby alter the capability of the regulator molecule to interact with the operator. Effectors in inducible operons are called INDUCERS.

effector plasmid: Effector plasmids carry a gene that expresses a regulatory protein which in turn controls the expression of another gene carried in a reporter plasmid.

effector proteins: (i) Microbial proteins (virulence factors) that are secreted into the host cells and promote infection, such as protein products of *Avr/hrp* genes (e.g. enzymes and toxins). (ii) Biological responses to histone methylation critically depend on the faithful readout and transduction of the methyllysine (^{me}K) signal by 'effector' proteins.

effector-triggered immunity (ETI): Race/cultivar-specific resistance based on the perception of pathogen effector proteins by host resistance proteins, leading to a quick and strong induction of specific defence responses.

effector-triggered immunity (ETI), invocation of: Invocation means to invoke someone or something. PRR protein FLS2 produces basal defence (PTI). R protein Xa21 is likely involved in basal (PTI) as well as HR-based defence (ETI). Host proteins (e.g. Pto, RIN4) are involved in HR-based defence (ETI).

effector-triggered susceptibility (ETS): The successful infection of a plant by a pathogen through the release of effector molecules, which interfere with PTI, resulting in ETS.

efficiency: A measure of the rapidity at which a method performs its function.

efficiency elements: Eukaryotic regulatory elements that act to increase the overall rate of genetic transcription but do not directly confer tissue specificity of expression.

efficient multigene editing: Z.P. Li *et al.* (2024) have advanced an optimized and modular approach enabling the targeting of several genes simultaneously in conjunction with the co-ordinated expression of the Cas9 endonuclease in the egg cell. This scheme permits substantial editing efficiency while evading mosaicism. The adaptable system they offer enables adaptation to inducible and/or tissue-specific edition in accordance with the promoter chosen to drive the expression of the *Cas9* gene.

egg: The mature female gamete in sexually reproducing organisms. It is usually a large and immobile haploid cell.

eIF1: This eukaryotic translation initiation factor ensures the fidelity of initiation codon selection, promotes ribosomal scanning, stimulates binding of eIF2/GTP/Met-tRNAi to 40S subunits and prevents premature eIF5-induced hydrolysis of eIF2-bound GTP and Pi release.

eIF1A: This eukaryotic translation initiation factor stimulates binding of eIF2/GTP/Met-tRNAi to 40S subunits and cooperates with eIF1 in promoting ribosomal scanning and initiation codon selection.

eIF2: This eukaryotic translation initiation factor forms an eIF2/GTP/Met-tRNAi ternary complex that binds to the 40S subunit, thus mediating ribosomal recruitment of Met-tRNAi.

eIF2B: This eukaryotic translation initiation factor is a guanosine nucleoside exchange factor that promotes GDP/GTP exchange on eIF2.

eIF3: This eukaryotic translation initiation factor binds 40S subunits, eIFs 1, 4G, 5; stimulates binding of eIF2/GTP/Met-tRNAi to 40S subunits; promotes attachment of 43S complexes to mRNA and subsequent scanning; possesses ribosome dissociation and anti-association activities preventing joining of 40S and 60S subunits.

eIF4A: This eukaryotic translation initiation factor acts as DEAD-box ATPase and ATP-dependent RNA helicase. Two paralogues (eIF4AI and eIF4AII), encoded by different genes, are functionally indistinguishable, but eIF4AIII has no activity as an eIF.

eIF4B: This eukaryotic translation initiation factor is an RNA-binding protein that enhances eIF4A's helicase activity.

eIF4E: This eukaryotic translation initiation factor binds to the m7GpppG 5′-terminal 'cap' structure of mRNA.

eIF4F: This eukaryotic translation initiation factor is a Cap-binding complex, comprising eIF4E, eIF4A and eIF4G; unwinds 5′-proximal region of mRNA and mediates attachment to it of 43S complexes; assists ribosomal complexes during scanning.

eIF4G: This eukaryotic translation initiation factor binds eIF4E, eIF4A, eIF3, PABP, SLIP1 and mRNA and enhances eIF4A's helicase activity. Two paralogues (eIF4GI and eIF4GII), encoded by different genes, are functionally similar but show some selectivity towards different mRNAs. eIF4GI is generally the more abundant.

eIF4H: This eukaryotic translation initiation factor is an RNA-binding protein that enhances eIF4A's helicase activity; homologous to a fragment of eIF4B.

eIF5: This eukaryotic translation initiation factor is a GTPase-activating protein specific for eIF2•GTP that induces hydrolysis of eIF2-bound GTP upon recognition of the initiation codon.

eIF5B: This eukaryotic translation initiation factor is a ribosome-dependent GTPase; mediates ribosomal subunit joining.

eIFs: Eukaryotic initiations factors.

elaiosome: An external structure attached to the seed of many species of plants. Elaiosomes appear fleshy and, in some species, are rich in oils or other nutritious materials. Their functions diverge and are not always obvious. Commonly, they attract ants or other animals that aid in dispersal, but they may also deter other animals from eating the seed.

elastin: Hydrophobic protein that forms extracellular extensible fibres, called ELASTIC FIBRES, which give tissues their elasticity and resilience.

electrical methods of gene transfer: These methods include ELECTROPORATION and ELECTROFUSION. *See details as separate terms.*

electro-blotting transfer: The electrophoretic transfer of macromolecules (DNA, RNA or protein) from gel in which they have been separated to support matrix such as nitrocellulose sheet. This is an alternative to the capillary transfer usually used in techniques such as Southern and Northern blotting.

electrochemical gradient: A gradient of electrochemical potential, usually for an ion that can move across a membrane. The gradient consists of two parts: (i) the electrical potential and (ii) the difference in chemical concentration across the membrane.

electrochemical proton gradient: The consequence of a combined pH gradient (proton gradient) and the membrane potential.

electrofusion: A phenomenon by which neighbouring cells are induced to fuse by applying pulse of an electric field. Electrofusion is also used to make tetraploid embryos.

electrofusion of protoplasts: Mild electric stimulation can be used to fuse protoplast. An electric field of low strength (10 kv/m) gives rise to dielectrophoretic dipole generation within the protoplast suspension. High strength of electric fields (100 kv/m) for some microseconds is applied.

electromorphs: Allozymes that can be recognized by separation through electrophoresis.

electron: Negatively charged subatomic particle that mostly occupies orbitals surrounding the nucleus in an atom.

electron acceptor: Any chemical entity (atom or molecule) that receives electrons transferred to it from another chemical entity. It is an oxidizing agent that, by virtue of its accepting electrons, is itself reduced in the process. *Contrasts* ELECTRON DONOR.

electron carrier: Any of various molecules that are capable of receiving one or two electrons from one molecule and then donating them to another in the process of electron transport. As the electrons are transferred from one electron carrier to another, their energy level decreases, and energy is released (e.g. cytochrome c transfers an electron from a donor molecule to an acceptor molecule).

electron donor: A chemical entity that easily gives electrons to another chemical entity. It is a reducing agent that, by virtue of giving up its electrons, is itself oxidized in the process. *Contrasts* ELECTRON ACCEPTOR.

electron microscope (EM): A type of microscope that uses a beam of electrons to create an image of structures of viruses, bacteria, pollen grains, molecules, etc. The technique uses beams of electrons instead of light rays. EMs are capable of much higher magnifications and have greater resolving power than conventional light microscopes, allowing them to see much smaller objects in finer detail. EMs are becoming more important for nanobiotechnology, as they ensure 1–3 hundred thousand times magnification. They are further equipped with scanning and transmission-type facilities for obtaining better images.

electron transport chain (ETC): The process of oxidative phosphorylation, by which the NADH and succinate generated by the citric acid cycle get oxidized and electrons are transferred sequentially down a long series of proteins, ultimately to the enzyme ATP synthase, which uses the electrical energy to catalyse the synthesis of ATP by the addition of a phosphate group to ADP. The process takes place in the cell's mitochondria and is the primary means of energy generation in most eukaryotic organisms.

electronic Northerns: The use of an electronic database of cDNA sequences (or probes derived from them) in order to measure the relative levels of mRNAs expressed in different cells or tissues. An example of the use of an electronic Northern might be to identify the differences in the genes expressed in prostate cancer and those in benign prostate hyperplasia by subtracting the database of one from the other and seeing which cDNAs remain.

electronic single-nucleotide polymorphism (eSNP): Any SINGLE-NUCLEOTIDE POLYMORPHISM (SNP) resulting from EXPRESSED SEQUENCE TAG (EST) databases by data mining. The EST sequences from many different individuals of a species are aligned, and the sequence is screened for SNPs. These SNPs are discovered solely *in silico*.

electrophoresis: A procedure for separating molecules (such as DNA or proteins) based on the differential movement of charged particles through a matrix when exposed to an electric field depending on the charge or binding affinity and size of the molecule. The term is usually useful for large ions of colloidal particles dispersed in water. The most important application of electrophoresis is in the analysis of proteins, for which a technique known as GEL ELECTROPHORESIS has been developed. The mixture of molecules is added to one end of a gel-like medium. Agarose and acrylamide gels are the media commonly used for electrophoresis of proteins and nucleic acids. Since the proportion of proteins varies widely in different diseases, electrophoresis can be used for diagnostic purposes. Electrophoresis is a common way to separate, identify and purify plasmid DNA, DNA fragments ensuing from digestion (of DNA) with restriction endonucleases and RNA. Electrophoresis is also used to study bacteria and viruses, nucleic acids and some types of biomolecules, including amino acids. Once the molecules are separated, the gel can be used in a blot (Southern, Northern or Western). Molecules displaying variation can be shown to be heritable and their transmission traced between various pedigree generations. Not all variant forms of a protein are detectable by this method.

electrophoretic karyotypes: Refers to details of chromosome number and size produced by pulse field gel electrophoresis in species with chromosomes that are too small to be seen under the light microscopy.

electrophoretic mobility shift assay (EMSA): The EMSA is used to study proteins binding to known DNA oligonucleotide probes and can be used to assess the degree of affinity or specificity of the interaction. The technique relies based on the observation that protein-DNA complexes migrate more slowly than free DNA molecules when subjected to non-denaturing polyacrylamide or agarose gel electrophoresis. Because the rate of DNA migration is shifted or retarded upon protein binding, the assay is also referred to as a gel shift or gel retardation assay. Adding a protein-specific antibody to the binding components creates an even larger complex (antibody-protein-DNA), which migrates even slower during electrophoresis. This is known as a super shift and can be used to confirm protein identities. Until conception of the EMSA, protein-DNA interactions were studied primarily by nitrocellulose filter-binding assays using radioactively labelled probes.

electrophoretic mobility shift assay (EMSA), limitations of: (i) Analyses protein-DNA interactions *in vitro*, (ii) difficult to quantitate and (iii) need to perform supershift assay with antibody to be certain of protein identity in a complex.

electrophoretic mobility shift assay (EMSA), strengths of: (i) Detect low abundance DNA-binding proteins from lysates, (ii) test binding site mutations using many probe configurations with the same lysate, (iii) test binding affinity through DNA probe mutational analysis and (iv) non-radioactive EMSA possible using biotinylated or fluorescently labelled DNA probes.

electroporation: Electroporation is electron-mediated gene transfer technique for transfecting cells. A high-voltage DC electric pulse is applied for this purpose. At high-voltage, cell membrane gets electrochemically compressed which leads to pore formation. Through these pores intracellular and extracellular exchange of components occurs. This step renders the cells permeable to the nucleic acid. Also useful for introducing synthetic peptides into eukaryotic cells. In 1995, the US company Dekalb Genetics Corp secured a patent for producing genetically engineered corn via introduction of a foreign gene into corn cells via electroporation. In this vectorless method, the foreign DNA is placed in the solution, which enters through these pores and reaches the chromosomes in the nucleus, resulting in transformation through which foreign DNA enters and integrates with the native DNA. Also called ELECTRO-PERMEABILIZATION. *See* TRANSFECTION.

electrospray ionization mass spectroscopy (ESI-MS): Principal used for this method is the same as the one used in MALDI. Contrary to MALDI, ESI produces multiply charged ions that can increase the complexity of the obtained spectra especially for samples containing multiple molecular species. If the components of a complex mixture are to be identified, using ESI-MS alone is not practicable, as spectra interpretation may become challenging. A chromatographic separation technique is frequently used in conjunction with ESI-MS for such experiments.

electrostatic force microscopy (EFM): Electrostatic force between the DC-biased conductive tip and the surface is measured by measuring the change in the resonance frequency of the cantilever. Also used to refer to another technique, ELECTRIC FORCE MICROSCOPY.

electrostatic interactions: Non-covalent interaction between atoms or groups of atoms due to attraction of opposite charges.

element: The particulate hereditary determiner of Mendel. *Aka* FACTOR. Later named GENE.

elements of a scientific method: Essential elements of scientific method are iterations, recursions, interleavings and orderings of characterizations, hypotheses, predictions and experiments.

elephophily: A procedure of pollination whereby pollen or spores are distributed by the feet of elephants, as in *Rafflesia arnoldii*.

elicitation metabolites: Compounds inducible by way of elicitation. *See* ELICITORS.

elicitor-receptor model: A model proposed to explain gene-for-gene hypothesis. Accordingly, either primary virulence gene products or metabolites resulting from their metabolic activity are recognized by specific plant receptors encoded by disease genes.

elicitor-suppressor model: A model proposed to explain gene-for-gene hypothesis. Accordingly, either primary virulence gene products or metabolites resulting from their metabolic activity are recognized by specific plant receptors encoded by disease genes. Pathogens are thought to produce general elicitors that initiate host defence unless a specific suppressor substance is also liberated by a particular pathogen. This scheme is understood to work where virulence rather than avirulence is dominant.

elicitors: These are specific signal molecules produced by the pathogen. Elicitors are identified by the host through receptors encoded by genes conferring resistance in plants. Elicitors are thus involved in plant-microbe interaction. In context of product accumulation by plant cells in suspension cultures, they are mediator compounds. There are two types of elicitors: ABIOTIC ELICITORS and BIOTIC ELICITORS. *See details as separate terms.*

elimination reaction: A reaction that comprises the ejection of a specific group from a molecule, often leading to the formation of a carbon-carbon double bond.

ELISA: *See* ENZYME-LINKED IMMUNOSORBENT ASSAY.

elite germplasm: Germplasm that has been selectively bred and optimized to a specific environment is referred to as 'elite' germplasm. New traits developed through biotechnology are typically bred into elite germplasm plant lines prior to regulatory testing and commercialization. For example, corn/maize (*Zea mays* L.), which is native to Mexico, has been adapted and optimized to grow in field conditions in many of the world's countries.

elite material: The material that is apparently superior over the other material.

ellipsoid: A three-dimensional shape that is elliptical in all sections through the long axis.

elliptical: Planar, oval, shaped like a flattened circle, symmetrical along both the long and the short axes, tapering equally to both the tip and the base. *Aka* ELLIPTIC.

elongation factor Tu receptor (EFR): The epitopes (the part of an antigen molecule to which an antibody attaches itself) recognized by the plant as PAMPs are highly conserved domains of the proteins, defined by the peptides flg22 and elf18, respectively. EFR is the most abundant and most conserved PAMP. EFR recognizes elf18, consisting of the first 18 aa of EF-Tu. EFR is exclusive to *Brassicae* family.

elongation factors (EFs): The proteins (EF-Ts, EF-Tu, EF-G) that play an ancillary role in the elongation step during translation process for the adding of amino acids to growing polypeptide chains on ribosomes.

elongation of DNA replication: Elongation is a nucleotidal group-transfer reaction because the 3′-OH of the pre-existing nucleotide makes a nucleophilic attack on the 5′-phosphate of the incoming nucleotide. The deoxyribonucleotide triphosphates are produced in phosphorylation step of initiation. Hence, phosphodiester bond is formed, liberating two inorganic phosphates. Enzymes that replicate DNA using a DNA template are called DNA polymerases. However, there are also enzymes that synthesize DNA using an RNA template (reverse transcriptases) and even enzymes that make DNA without using a template (terminal transferases). Most organisms have more than one type of DNA polymerase (e.g. *Escherichia coli* has five DNA polymerases), but all work by the same basic rules: Polymerization occurs in only 5′→3′ direction. Polymerization requires a template to copy: the complementary strand. Polymerization requires four dNTPs: dATP, dGTP, dCTP and dTTP (TTP is generally not designated with a 'd' since there is no ribose containing equivalent). Different types of DNA polymerases differ with respect to mass (kDa), molecules/cell, turnover number, affinity for nucleoside triphosphate at 37°C (molecules/minute), polymerization 5′→3′, exonuclease activity 3′→5′, exonuclease 5′→3′, use of nicked duplex, use of primed single strands, intact duplex and gapped duplex or single-stranded 5′ ends.

elongation of transcription in prokaryotes: Ribonucleotides are linked together on DNA template through phosphodiester bonds. Nus A factor plays role in elongation of transcription.

elongation of translation in prokaryotes: This process starts when 'A' site is occupied by an incoming amino acid~tRNA and peptide bond is formed between F-met and the incoming amino acid. The amino acid~tRNA that binds to the 'A' site is specifically determined by the triplet codon of the mRNA that occupies the 'A' site on the ribosome. tRNA deacylase is responsible for cleaving the bond between N-terminal of first amino acid (formylated-methionine). So, elongation cannott take place at this end. The other end (−COOH) is available for chain elongation. The factors required for elongation are Tu, Ts, G and GTP. Tu and Ts are involved in exchange reaction. Peptidyl transferase, the peptide-forming enzyme, is needed one copy for one 50S subunit. On completion of peptide bond formation, the 'A' site is engaged by F-met amino acid-$tRNA_f^{met}$ A1, and the 'P' site is employed by t-RNA.F-met. Then, the translocation step follows. The $tRNA_f^{met}$ is discharged from the ribosome. F-met amino acid 1-$tRNA_{A1}$ dipeptide is shifted from 'A' to 'P' site as ribosome moves by length of one codon along the mRNA in a 5′→3′ thus bringing a new codon to 'A' site. The $tRNA_{f\text{-}}^{met}$ finally occupies 'E' site. Thus, polypeptide assembly starts at the amino acid and ends at carboxyl end. mRNA is read in a 5′→3′ direction. After the initiation triplet, the bases are read in group of three in sequence, until a chain-terminating codon is reached (UAA, UGA, UAG) at which time a polypeptide is released from the mRNA and ribosome.

elongation step in PCR: The last temperature step, which occurs at 72°C in PCR procedure, during which a DNA polymerase adds nucleotides to the growing DNA strand.

elongator complex: A protein complex originally identified in budding yeast to be associated with the elongating and hyperphosphorylated RNA polymerase II. It has also been implicated in tRNA modification, exocytosis and neuronal maturation.

elution: Washing out or collection of liquid from a chromatography column. The liquid, which may contain the fractionated samples, is called the ELUTE.

emarginate: Typically refers to leaf margins that are notched or sunken at some part of the edge, such as the apex. The recess is usually broad and shallow. The location of a leaf's margination(s) might be one or more of apical, lateral or basal.

emasculation: Removing the anthers from a bud or flower before the pollen is shed. Emasculation is a normal preliminary step in crossing to prevent self-pollination.

Embden-Meyerhof-Parnas (EMP) pathway: Allows the metabolic use of glucose to generate ATP, NADH and several biosynthetic precursors such as 3-phosphoglycerate or pyruvate.

embiont: A heterotrophic protocell that became autotrophic through evolution of metabolic pathways.

EMBL3 vector: A replacement vector for the cloning of large (~20 kb) DNA fragments; it derives from phage lambda.

embroid: Non-zygotic embryo formed in culture.

embryo: An initial stage of development of a multicellular organism. In organisms that reproduce sexually, embryonic development is the part of the life cycle that begins just after fertilization of the female egg cell by the male sperm cell.

embryo culture: (i) A procedure of inducing the artificial growth of embryos by excising the young embryos under septic conditions and placing them on suitable nutrient media. Developing embryos are excised and cultured *in vitro* to produce plant callus or transgenic plants. In plant breeding, embryo culture is a common practice to develop hybrids from wide crosses involving genetically diverse parents. For instance, embryo culturing is required to produce hexaploid triticale from a cross involving tetraploid wheat (*Triticum durum*) and diploid rye (*Secale cereal*). This method is often applied when POSTGAMOUS INCOMPATIBILITY exists (e.g. in wide crosses). (ii) *In vitro* development or maintenance of isolated mature or immature embryos.

embryo development: A process that marks the beginning of the formation of the embryo. It is the beginning of development, the most important part of development, because this is the period during which very important changes take place. After fertilization, plant zygote undergoes cleavage, which distributes cytoplasm of egg among numerous smaller cells. In plant cells, cellular walls are present, because of which cleavage mechanism is different than that in animals. In plants, cells separate from one another by furrowing after mitosis. Plant cells maintain in position with rigidity. Hence, cleavage of plant cells is deliberate. Each division plan occurs at a precise site so as to fix the location of every cell. These cells may in turn divide and differentiate to produce different regions of the plant body. Thus, in plants, cleavage is directly linked to the formation of body regions. The plant body has polarity. The rigid positioning of the plant cells means polarity must be accomplished by division pattern rather than by rearrangement of cells after mitosis. *Aka* EMBRYONIC PERIOD OF DEVELOPMENT, EMBRYOGENESIS *or* EMBRYONIC DEVELOPMENT.

embryo gene expression programmes: Embryo gene expression programmes occur during a development period that is critical for establishing the polarity and morphological patterns of the plant in which major plant tissues and cell types are formed.

embryo gene therapy: A type of gene therapy in which the genetic constitution of embryo at the postzygotic level is altered so that the inheritance will be altered.

embryo microinjection: Transfer of genetic material by microinjection into early embryos has been extensively used for production of transgenic animals. Foreign genes have been reported to express in tissues of mice by using a DNA-coated microprojectile. Microinjection into pronucleus has the efficiency of generating transgenic lines. Transfer of injected ova into recipient females has produced transgenic animals capable of transcribing foreign genes to produce authentic gene products.

embryo rescue: When cross-pollination is done or occurs between genetically widely different plants, the ensuing embryo may be aborted because of parental mutual incompatibility. Such embryos may be excised during an early stage and grown on a congenial medium, such as nutrient agar. During wide hybridization, for instance, making cross between *Hordeum vulgare* with *Secale cereale* (intergeneric cross), the endosperm of the developing grain degenerates, and subsequently, the developing embryo starts degenerating. In such a situation, the developing embryos (before degeneration) are excised from the grain and cultured *in vitro* on suitable medium to make them to grow into entire plants.

embryo sac: The megaspore of a seed-bearing plant, situated within the ovule, giving rise to the endosperm and forming the egg cell or nucleus from which the embryo plant develops after fertilization.

embryogenesis: The formation of an embryo. After double fertilization, seed development begins with endosperm, embryo growth and differentiation. During subsequent seed maturation, the seed storage proteins, starch and lipids accumulate before the seed undergoes desiccation and dormancy. During late embryogenesis, endosperm and mature embryo express seed storage reserve genes. Many other hydrophilic proteins with certain functions that share similar sequence repeats also are accumulated. They are called LATE EMBRYOGENESIS-ABUNDANT PROTEINS, protecting cells during seed desiccation (e.g. abscisic acid).

embryogeny in dicots: On the basis of the plane of cell division in the apical cell at 2-celled proembryo and the contribution of the basal cell and apical cell in the formation of embryo proper, embryogeny in dicots is classified into various types: crucifer, asterad, soland, caryophyllad, chenopodial and piperad type.

embryogeny in monocots: In monocots, the development of embryo varies from species to species. Among the monocots, graminae family is highly evolved. Regular occurrence of oblique divisions in early embryogeny is characteristic feature of monocot embryo.

embryoid: Non-zygotic (somatic) embryo formed in culture. During plant tissue culture, many plant species exhibit frequent direct or indirect somatic embryogenesis. During culture under the influence of growth regulators, the cultured tissues like callus, roots, leaves and cotyledons exhibit the formation of nodular structures called embryoids, which behave like zygotic embryos and germinate into plants.

embryology: Scientific study of the early stages of growth and development in organisms in which a single highly specialized cell, the egg, is transformed into a complex multicelled organism resembling its parents.

embryonic: Related to an embryo or being in the state of an embryo.

embryonic callus: Callus cultures that under suitable conditions are capable of producing embryos. Conversely, callus cultures that have lost this capability are termed 'non-embryonic'.

embryonic development: In multicellular organisms, embryo development is the most important part of development because this is the period during which very important changes take place. This period is called EMBRYONIC PERIOD.

embryonic stem cell (ESC)-mediated gene transfer: This method involves prior insertion of the desired DNA sequence by homologous recombination into an *in vitro* culture of ES cells. These cells are then incorporated into an embryo at the blastocyst stage of development. It is the method of choice for gene inactivation, the so-called KNOCKOUT METHOD.

embryonic stem cells (ESCs): Self-renewing cells secluded from the vertebrate inner cell mass of a developing embryo. As ESCs are totipotent, they are capable of differentiating into any somatic cell or tissue, but they also preserve their ability to remain within the developmental pathway of germ cells. Thus, they are part of the germline.

emergence: The growth of the seedling shoot through the surface of the soil.

emigration: The movement of organisms out of an area.

empiric risk: Risk estimate based on observation.

empirical: Determined by experimentation.

empirical research: The area of research which tests the feasibility of a solution using empirical evidence.

enabling technologies: Technical processes that allow genetic modifications in organisms. Enabling technologies confer the ability to activate genes specifically or to influence them in their activity or the use of marker genes.

enantiomers: Two optical stereoisomers that are related to each other by a reflection. They rotate polarized light. For example, dextro (D)-glucose and levo (L)-glucose. *Aka* OPTICAL ISOMERS.

enantiostyly: The condition in which the gynoecium bulges laterally, to the right (dextrostyly) or to the left (sinistrostyly) of the androecium (e.g. Senna).

enation: Leaf-like extension from a surface.

encapsidate: To cover virus nucleic acid with a protein coat.

encapsidation (of viruses): The process in which a virus synthesizes the protein shell that surrounds the virus nucleic acid.

encapsulation: (i) Enclosing of fragile organic material in a protective casing, sometimes of a semisolid nature. It is used for planting or moving somatic embryos. (ii) The theory believing that a female includes all hereditary information for all her immediate and remote progeny.

enChIP: Engineered DNA-binding molecule-mediated ChIP, using a tagged-dCas9+gRNA to purify specific genomic regions to identify molecules associated with the genomic regions.

encrypted genes: The genes occurring as separate segments around the genome, so that, for example, all building blocks of a given mRNA molecule can be located, as modules, on separate chromosomes. In the organelles of microbial eukaryotes and in the prokaryotes, many examples of so-called encrypted genes are known.

encryption: In cryptography, encryption is the process of encoding information. This process converts the original representation of the information, called PLAINTEXT, into an alternative form, known as CIPHERTEXT. Ideally, only authorized parties can decipher a ciphertext back to plaintext and access the original information. Encryption does not itself avert interference but denies the intelligible content to a would-be interceptor. For technical reasons, an encryption scheme normally employs a pseudo-random encryption key created by an algorithm. It is possible to decrypt the message without having the key, but for a well-designed encryption scheme, considerable computational resources and skills are required. An authorized recipient can easily decrypt the message with the key provided by the originator to recipients but not to unauthorized users. Historically, various types of encryption have been utilized to aid in cryptography. Early encryption techniques were frequently employed in military messaging. New techniques have since then emerged and become commonplace in all areas of modern computing. Modern encryption schemes utilize the concepts of public-key and symmetric-key. Modern encryption techniques ensure security, because modern computers are inefficient at deciphering the encryption.

encyst: To make a cyst or defensive layer.

end free-space alignment: This model of sequence analysis detects maximum similarity between substrings. It is assumed that at least one of these substrings must be a prefix of the original string and one (not necessarily the other) must be a suffix.

end-filling: Conversion of a sticky end (5′ overhang only) into a blunt end by enzymatic synthesis of the complement to the single-stranded overhang.

end-joining reaction catalysed by DNA ligases: The end-joining reaction catalysed by DNA ligases is necessitated by all organisms and functions as the decisive step of DNA replication, repair and recombination processes. One of three well-characterized mammalian DNA ligases, DNA ligase I, joins Okazaki fragments during DNA replication.

end-labelling: Addition of a radioactively labelled group to one end (5′ or 3′) of a DNA strand by the use of specific enzymes as one means of making labelled probes.

end-product inhibition: Describes the ability of a product of a metabolic pathway to inhibit the activity of an enzyme that catalyses an early step in the pathway. *Aka* FEEDBACK INHIBITION.

end-to-end warning system: A complete set of components that connects those who need to hear messages to others who compile and track the hazard information of which messages are composed. An effective end-to-end early warning system includes four core elements: (i) risk knowledge, (ii) monitoring, (iii) response capability and (iv) warning communication, where each component must function efficiently for the system to be successful. RISK KNOWLEDGE builds the baseline understanding about risks (hazards and vulnerabilities) and priorities at a given level. MONITORING oversees how those risks and vulnerabilities change over time. RESPONSE CAPABILITY reduces risk once trends are spotted and announced – this may be through pre-season mitigation activities, evacuation or duck-and-cover reflexes, depending on the lead time of a warning. WARNING COMMUNICATION packages the monitoring information into actionable messages understood by those that need, and are prepared, to hear them.

endangered species: Any species which is highly probable to become extinct in the near future, either worldwide or in a particular area. Such species may be susceptible to factors such as habitat loss, hunting, disease and climate change, and most show a declining population or a very limited range.

endemic disease: A disease that is always present in a population at about the same level. The best example of endemic disease is the common cold. An epidemic disease that quickly spreads in a population worldwide is called PANDEMIC DISEASE.

endemic species: Species confined to a given geographic region. *Compare with* NATIVE SPECIES

endemism: The ecological state of an organism or species being unique to a defined geographic location, such as an island, nation, country, habitat type or other defined zone.

endergonic reaction: A chemical reaction that consumes energy. During chemical reactions, whenever bonds between atoms are formed, energy is required, and the reaction absorbs more energy than it releases. For instance, when KNO_3 is dissolved in water, the solution becomes cold, and it becomes difficult to dissolve more KNO_3 in this solution. Slight warming helps in making concentrated solution. *Aka* NON-SPONTANEOUS REACTION *or* UNFAVOURABLE REACTION.

endless summer tomato: The antisense RNA technology has been utilized to overpower ethylene biosynthesis in tomato. The construct of ACC synthase gene was transferred in tomato, which reduces ethylene production. The fruit remains green and hard on the plants so that they amass more sugars and organic acids that render the fruits sweeter and richer in taste. The fruit are harvested and ripened in warehouse by exposing them to ethylene. The new transgenic variety is called 'Endless Summer'. It has longer shelf life.

endocarp: The innermost layer of the wall of a fruit; in a drupe, the stony layer surrounding the seed.

endocytic-exocytic cycle: The processes of endocytosis and exocytosis that, respectively, add and remove plasma membrane from the cell, resulting in no overall change in the cell's surface area and volume.

endocytosis: The importation of substances (e.g. hormones, viruses and toxins) into a cell via specific receptor/ligand binding. The entity under consideration binds to a receptor(s) positioned in the plasma (cell) membrane, which then invaginates (infolds), hence taking up the entity via 'endosomes' (formed by pinching off an infold to form a 'bag') into vesicles located within the cell. It is one route to deliver essential metabolites to cells (e.g. low-density lipoprotein), and it is a means to modulate the cell's responses to many protein hormones and growth factors (e.g. insulin, epidermal growth factor and nerve growth factor). It is a route by which certain proteins targeted for destruction can be taken up and transported to the cell's lysosomes. For example, phagocytic cells have receptors, enabling them to take up antigen-antibody complexes for subsequent destruction by the phagocytic cell. This route is also a means used by certain viruses and toxins to gain entry into cells through the otherwise impervious cell membranes (e.g. used by the AIDS virus and the Semliki Forest Virus). Disorders of endocytosis can lead to disease states (e.g. high cholesterol levels in the blood of people whose low-density lipoprotein receptors are impaired). Drugs (e.g. certain painkillers) can be targeted to specific receptors via RECEPTOR MAPPING (RM) and RECEPTOR FITTING (RF) for greater efficacy. *Aka* RECEPTOR-MEDIATED ENDOCYTOSIS. *See* PINOCYTOSIS *and* PHAGOCYTOSIS.

endocytotic vesicles: Membranous particles that transport proteins through endocytosis, being transported into the cell within membranous vesicles.

endodeoxyribonuclease: Endodeoxyribonuclease are both endonucleases and deoxyribonucleases. They catalyse cleavage of the phosphodiester bonds in DNA. Examples of endodeoxyribonuclease include DEOXYRIBONUCLEASE I, DEOXYRIBONUCLEASE II, DEOXYRIBONUCLEASE IV, RESTRICTION ENZYMES, UVRABC ENDONUCLEASE and MICROCOCCAL NUCLEASE. *See details as separate terms.*

endoderm: One of three cell layers found innermost in bilaterian embryos, the other two being the ECTODERM (outside layer) and MESODERM (middle layer). The endoderm forms structures, such as the lining of the digestive system and portions of organs such as the liver, lung and pancreas.

endodermis: The endodermis is a single layer of cylindrical parenchyma cells that separates the root's cortex and stele. Endodermal cells have thick walls, and their radial and transverse walls contain Casparian strips, waxy bands that prevent water from passing between the cells. The endodermis regulates the flow of water, ions and hormones into and out of the vascular structures within the stele.

endoduplication: The doubling of the haploid chromosome complements due to the failure of cell wall formation.

endogenous: (of a substance or process) Originating from within a system (such as an organism, tissue or cell), as with endogenous cannabinoids and circadian rhythms. *Contrasts* EXOGENOUS.

endogenous cytokinin and senescence genes: The senescence-delaying effect of cytokinin, famously known as the RICHMOND-LANG EFFECT, occurs in plants. Investigations performed on *Arabidopsis*, sorghum and soybean have confirmed a decline in expression of cytokinin biosynthetic genes during senescence with upregulated expression of cytokinin-degradation genes. With the discovery of cytokinin receptor AHK3 and type B response regulator ARR2 in *Arabidopsis*, the enhanced activity of cell wall invertase, mediated by cytokinin, was observed to be accountable for delayed senescence. Maintenance of cytokinin levels in source organ (i.e. leaf) through metabolic engineering aids in retaining the normal carbon fixation to improve photosynthetic yields.

endogenous DNA damage: Lesions can be caused by endogenous DNA damage like depurination, deamination oxidative damage and alkylation. Endogenous damage affects the primary rather than secondary structure of the double helix. It can be subdivided into four classes: depurination/depyrimidination, deamination, oxidative damage and alkylation.

endogenous gene expression reduction by using antisense gene: An antisense gene is generated by inverting (i.e. reversing the orientation of the protein-encoding region of a gene in relation to its promoter). As a result, the natural sense strand of the gene becomes oriented in the 3′→5′ direction with reference to its promoter and is transcribed. The normal antisense strand is not transcribed since now its orientation is 5′→3′. The RNA produced by this gene has the same sequence as the antisense strand of the normal gene and is, therefore, called ANTISENSE RNA (ASRNA). When the antisense gene is present in the same nucleus as the normal endogenous gene, the transcription of the two genes yields antisense and sense RNA transcripts, respectively. Since the sense and antisense RNAs are complementary to each other, they would pair to produce double-stranded RNA molecules. The mRNA is unavailable for translation. The dsRNA molecules are attacked and degraded by ds RNA-specific RNases. These events lead to methylation of promoter and coding regions of the normal gene resulting in silencing of the endogenous gene. Some of the examples of application of antisense RNA technology include FlavrSavr tomato, 'Endless Summer' tomato and Delayed senescence carnation. *See* ENDLESS SUMMER TOMATO.

endogenous metabolites: The metabolites produced by the host organism.

endogenous protein-encoding gene modification for seed protein quality in plants: This approach is based on the isolation and modification of the concerned protein-encoding gene sequence either by (i) replacing one or few codons with selected codons or (ii) inserting one or few selected additional codons at appropriate sites. For example, prolamine storage proteins (e.g. zein) of cereals are deficient in the essential amino acids (e.g. lysine and tryptophan). Single lysine replacements in the N-terminal coding sequence as well as within and between the peptide repeats and double lysine replacement constructs of prolamine genes have been prepared. In addition, short oligonucleotides encoding lysine- and tryptophan-rich peptides were inserted separately at several different points in the coding sequence. These constructs were shown to express well in *Xenopus* oocytes, and their polypeptides were able to form normal aggregates of protein bodies.

endogenous provirus: *See* ENDOGENOUS RETROVIRUS.

endogenous retrovirus: Integrated retrovirus DNA (provirus) derived from infection of the germline of an ancestral animal. All animals are thought to carry numerous endogenous (but non-functional) retroviruses, some of which were inserted many millions of years ago.

endogenous rhythm: A type of rhythmic plant response or growth capability that is not impacted by exterior stimuli.

endomembrane system: Successions of intracellular membrane compartments observed in eukaryotic cells.

endomitosis: A doubling of the chromosomes within a nucleus that does not divide, thus producing a polyploid. The doubling may be repeated a number of times, which results in endopolyploidy.

endomycorrhiza (pl. endomycorrhizae): A symbiotic link between a non-pathogenic or weakly pathogenic fungus and the roots of plants in which fungal hyphae invade cortical cells of the root. *See* ARBUSCULAR MYCORRHIZA. *Contrasts* ECTOMYCORRHIZA.

endonuclease: The enzyme that cleaves the phosphodiester bond within a polynucleotide chain (namely DNA or RNA) at specific locations corresponding to specific base sequences within a polynucleotide chain. Each endonuclease cleaves at its own specific site. It is used to identify genomes, genotypes and chromosomes, and it functions as a genetic marker. A restriction enzyme is a type of endonuclease. It does not need a free end to act. *See* RESTRICTION ENZYME *and* RESTRICTION ENDONUCLEASE.

endonucleases, type of: There are three categories of restriction endonucleases that relatively contribute to the cleavage of specific sequences. The types I and III are large multisubunit complexes that include both the endonucleases and methylase activities. TYPE I endonucleases can cleave at random sites of about 1000 base pairs or more from the recognition sequence, and it requires ATP as source of energy. TYPE II endonucleases behave slightly differently and were first isolated by Hamilton Smith in 1970. They are simpler versions of the endonucleases and require no ATP in their degradation processes. Some examples of type II restriction endonucleases include *Bam*HI, *Eco*RI, *Eco*RV, *Hind*III and *Hae*III. TYPE III endonucleases, however, cleave the DNA at about 25 base pairs from the recognition sequence and also require ATP in the process.

endoparasite: A parasitic organism that lives in the internal organs or tissues of its host and feeds from inside its host. *Contrasts* ECTOPARASITE.

endoparasitic sequences: DNA sequences that are repeated multiple times in the genome; a type of transposable element.

endophyte: A plant emerging inside another organism. The term is also used for endoparasitic fungi living in grass species.

endoplasm: The granular central material of the cytoplasm.

endoplasmic reticulum (ER): A double-membrane system in the cytoplasm, continuous with the nuclear membrane and bearing numerous ribosomes. This system is involved in translation, folding and transport of proteins. *See* ROUGH ENDOPLASMIC RETICULUM (RER) *and* SMOOTH ENDOPLASMIC RETICULUM (SER).

endoplasmic reticulum (ER) lumen: The space bounded by the membrane of the endoplasmic reticulum.

endoplasmic reticulum (ER) resident protein: Protein that occurs in the endoplasmic reticulum or its membranes and performs its function there, as opposed to proteins that are present in the ER only in transit.

endoplasmic reticulum (ER) retention signal: Short amino acid sequence on a protein that thwarts its movement out of the endoplasmic reticulum. Found on proteins that reside in the ER and function there.

endoplasmic reticulum (ER) signal sequence: N-terminal signal sequence that guides proteins to enter the endoplasmic reticulum and is cleaved off by signal peptidase after entry.

endopolyploid: Diploid individuals whose cells contain 4C, 8C, 16C, 32C, etc. amounts of DNA in their nuclei *See* ENDOMITOSIS *and* ENDODUPLICATION.

endopolyploidy: An increase in the number of chromosome sets caused by replication without cell division.

endoreduplication: Duplication of the genome without mitosis, giving rise to four-stranded chromosomes at prophase and metaphase.

endoreplication: Replication of the nuclear genome in the absence of mitosis, which leads to elevated nuclear gene content and polyploidy.

endoribonucleases: These ribonucleases cleave either single-stranded or double-stranded RNA, depending on the enzyme. Example includes both single proteins such as RNase III, RNase A, RNase T1, RNase T2 and RNase H and also complexes of proteins with RNA such as RNase P and the RNA-induced silencing complex.

endosmosis: In plants, diffusion of solvent inside the cell.

endosome: Membrane-bounded nucleus-like organelle in animal cells that receives materials newly ingested by endocytosis and transports many of them on to lysosomes for degradation.

endosperm: In grasses, the reserve food material in the caryopsis lies outside the embryo. It is a starchy nutritive tissue that is formed during grain development. It delivers nourishment for the developing embryo and for the seedling after germination until it can establish itself in the supporting media. Usually, the endosperm is triploid (3n) and originates during the double fertilization when one of the two sperm nuclei fuses with two polar nuclei.

endosperm development: Primary endosperm nucleus divides and gives rise to undifferentiated food-laden parenchymatous tissue around the developing embryo. It stores food materials in form of carbohydrates, proteins and fats. Functionally, it is comparable to the female gametophyte of gymnosperms. In gymnosperms, endosperm is formed before fertilization and is haploid, whereas in angiosperms, it is formed after fertilization and is generally triploid. The developing embryo either consumes entire endosperm or only a part of it. The endosperm that is composed of two or three different coloured patches that are arranged in irregular pattern (mosaic pattern) is known as MOSAIC ENDOSPERM. These patches also differ in genetic make-up. In *Zea mays* L., the endosperm is made up of red and white patches, so its endosperm is morphologically and genetically mosaic. Most commonly, the endosperm is triploid. However, in some species, it may be diploid or polyploid. The highest level of ploidy reported for endosperm is 24576n in case of *Arun maculatum*. In general, endosperm is composed of morphologically, physiologically and genetically similar type of cells. However, in cereals, the outermost 2–5 layers of cells become highly specialized morphologically and physiologically. Such layers are known as aleurone layers. Their cells contain aleurone granules (protein storage organelles). Endosperm regulates precise mode of embryo development. If the endosperm is not formed properly, it adversely affects the development of embryo. Some endosperms (coconut) contain regulatory substances such as auxins, cytokinins, gibberellins. Immature endosperm is a potent mitogen. That is why coconut milk is used in formation of callus and differentiation of embryoids.

endosperm genetics: Double fertilization results in the formation of endosperm tissue, which partakes, of both paternal and maternal inheritance. In a plant in which the ovary wall and seed coat are thin and transparent (e.g. a kernel of corn), a direct effect of the male gamete on the character of the endosperm is evident. Geneticists studying endosperm traits can keep one generation ahead in the game of genetic analysis compared to some sporophytic character.

endosperm mutation: A change that is detected by a difference in the endosperm of the corn kernel (e.g. starchy to sweet).

endosperm proliferation: Flowering plants undergo two fertilization events to form the embryo and the nutritive endosperm. After gametes fuse, they must then resume cell-cycle progression to allow them to divide and develop. S. Simonini and others (2024) found that in *Arabidopsis* a protein called RETINOBLASTOMA RELATED1 (RBR1) holds the female endosperm precursor cell in stasis, preventing full DNA replication. During fertilization, the sperm cell delivers a core cell-cycle component D-type cyclin CYCD7 to the female cell to reactivate the cell cycle. Upon fertilization, delivery of the CYCD7 causes RBR1 degradation and thus S-phase progression, ensuring the formation of functional endosperm and, consequently, viable seeds.

endospore: A resistant, thick-walled asexual spore produced within a bacterial or fungal cell. A resting, exceptionally hardy structure formed inside some bacteria. During unfavourable conditions (stress condition) some bacteria produce resting structures within their bodies which can survive up to hundreds of years. During favourable conditions, these endospores germinate to produce actively growing bacterial cells. Bacterial endospores are structurally different from fungal spores.

endospory: The production of spores that germinate into a reduced multicellular gametophyte contained within the spore wall. *Contrasts* EXOSPORY.

endosymbiont: The organism that resides inside a host organism. The term also applies to those organisms that reside within the cells of another organism.

endosymbiosis: An interdependence in which one organism lives within cells of another.

endosymbiotic theory: An evolutionary theory regarding the origin of eukaryotic cells from a hypothetical internal symbiosis between prokaryotic organisms, first articulated by the Russian botanist K. Mereschkowski in 1905 and 1910, advanced and authenticated with microbiological evidence by L. Margulis in 1967. This is the currently accepted theory as to how cells evolved. A prokaryotic cell having a flexible cell membrane was capable of ingesting bacterial cell, which evolved into mitochondria. Spirochetes became ingested and were converted into nuclear fibres, centrioles and flagella of a primitive eukaryotic cell. Blue-green algae cells became incorporated into the colourless unicellular eukaryotes to form simple green flagellate. *Aka* ENDOSYMBIOGENESIS.

endothelial cell: Flattened cell type that forms a sheet (the endothelium) lining all blood vessels.

endotoxin: Complex molecules (lipopolysaccharides) that compose an integral part of the cell wall and are released only when the integrity of the cell is disturbed. Lipopolysaccharides of outer membrane (lipid A) of most Gram-negative bacteria. Endotoxins are toxic to humans, animals and insects. All Gram-negative bacteria produce endotoxins. It is the lipopolysaccharide component of their outer membrane. Lipid portion (lipid A) of lipopolysaccharide acts as endotoxin. *Bacillus thuringiensis* produces delta endotoxin, which is highly toxic to the larvae of insects belonging to order Lepidoptera. *Bt* gene has been commercially exploited for controlling insects mainly through transgenic approach.

ends-in gene targeting: In this method, a break is made within the region of homology. Recombination with the target results in a tandem duplication of all the homologous sequence carried on the donor, separated by any additional sequences that are between the FLIPPAGE RECOGNITION TARGET SITES (FRTs).

ends-out gene targeting: This method involves a simple replacement of the genomic sequence with the homologous sequence. This is a strategy for knocking out a gene with a heterologous sequence.

energy consumption: Exploitation of the free energy released from the conversion of fuel into waste.

enforced outbreeding: Deliberate avoidance of mating between relatives.

engineered gene drive, pest management using: RNA-guided engineered nuclease CRISPR-Cas9 has potential to serve as a general method for spreading altered traits through wild populations over many generations. The ability to edit populations of sexual species would offer substantial benefits to humanity and the environment. RNA-guided gene drives could potentially prevent the spread of insect-borne disease, curtailing the spread of vector-borne infectious diseases, support agriculture by reversing pesticide and herbicide resistance in insects and weeds, and control damaging populations of environmentally and economically invasive species.

engineered gene drives: Engineered gene drives use artificial endonuclease genes or tools for gene drive process in which the homing process was carried out using genome-editing tool CLUSTERED REGULARLY INTERSPACED SHORT PALINDROMIC REPEATS (CRISPR) and the associated RNA-guided endonucleases (such as Cas9 or Cpf1) or TRANSCRIPTION ACTIVATOR-LIKE EFFECTOR NUCLEASES (TALENs).

engineered meganuclease (EMN): Microbially derived meganucleases that are modified, fused or rationally designed to cause site-directed DSB. *Aka* LAGLIDADG ENDONUCLEASES *or* HOMING NUCLEASES.

engineered proteins: Protein Top7 with a novel fold as well as sensors for unnatural molecules have been designed. The engineering of fusion proteins has yielded rilonacept, a pharmaceutical which has secured **FDA** approval for the treatment of cryopyrin-associated periodic syndrome.

engineered strains of insect pest species: Two approaches have been used to engineer strains of insect pest species: (i) Developing lethality that only operates on females and (ii) Using transposable elements as potential vectors for transformation.

engineered vaccines: The vaccines prepared by recombinant DNA technology. These vaccines avoid dangers of incomplete inactivation. There are four steps involved in the preparation of vaccines by recombinant DNA technology: (i) Identify the surface antigens of the parasite, (ii) Clone the gene for the antigen, (iii) Gene is got expressed in *Escherichia coli* or some other system and (iv) Purified antigen is used as a vaccine.

engineering complex properties of a highly ordered nano-protein assembly: Computation-aided design has been used to engineer complex properties of a highly ordered nano-protein assembly. A protein cage, *Escherichia coli* bacterioferritin (EcBfr), which naturally shows structural instability and an incomplete self-assembly behaviour by populating two oligomerization states, has been used as the model protein. This protein has a smaller than average dimeric interface on its twofold symmetry axis mainly due to the existence of an interfacial water pocket centred around two water-bridged asparagine residues. To engineer EcBfr for modified structural stability, a semi-empirical computational method is used to virtually explore the energy differences of the 480 possible mutants at the dimeric interface relative to the wild-type EcBfr. This computational study also converges on the water-bridged asparagines. Replacing two asparagines with hydrophobic amino acids results in proteins that fold into alpha-helical monomers and assemble into cages. All redesigned proteins possess increased stability. One mutation shifts the population in favour of the higher-order oligomerization state in solution.

engineering herbicide resistance in plants: Three approaches are used for engineering herbicide resistance in plants: (i) overproduction of herbicides sensitive biochemical target, (ii) structural alteration of biochemical target resulting in reduced herbicide affinity and (iii) detoxification by degradation of herbicide before it reaches biochemical target inside the plant cell. Glufosinate is an inhibitor of glutamine synthase, which is required for N_2 metabolism for NH_3 assimilation. The inhibitor of this enzyme will result in accumulation of ammonia and death of plant cells. This enzyme basically carries out assimilation of N_2 into amino acid glutamine.

engineering mitochondria for disease-free babies: A child directly inherits only its mother's mitochondrial DNA and in some cases will be certain to inherit a debilitating mitochondrial mutation. Because of the lack of a viable treatment for these patients, and their families, preventing the transmission of MITOCHONDRIAL DNA (mtDNA) disease is a priority. THREE-PARENT EMBRYOS may provide us its solution.

engineering novel metabolic pathways: Multigene engineering has been used to introduce completely novel pathways and thus to produce completely alien products. Multipoint metabolic engineering is now beginning to supersede single-point engineering as the best way to manipulate metabolic flux in transgenic plants. Several points in a given metabolic pathway can be controlled simultaneously, either by overexpressing and/or suppressing several enzymes or through the use of transcriptional regulators to control several endogenous genes.

engineering of light-harvesting antenna size: A continuous stock of energy equivalents (ATP and NADPH) from LIGHT-HARVESTING COMPLEX (LHC) is always required for enhancing operational photosynthetic efficiency in the plant system. However, the quantum flux gotten through sunlight should be optimum to mediate light-induced photosynthesis. To permit maximum light interception by whole plant canopy, there should be minimum antenna size in upper and large antenna size in lower regions of the canopy.

engineering T-DNA: Sequences within T-DNA carrying opines and oncogenes, not required for gene transfer, can be removed without affecting gene transfer. Left border 25-bp repeat is not required for gene transfer. A small synthetic plasmid with only right border sequence is in regular use for gene transfer. The virulence genes of the Ti plasmid are not required in *cis*. It means virulence genes can work even when present in some other plasmid. This leads to the development of binary vectors. The major advantage of binary vectors is their lack of dependence on an engineered Ti plasmid. This can broaden the host species of range of *Agrobacterium*.

enhanced photosynthesis to plants, benefits of: Plants possessing enhanced photosynthesis cope with fluctuating environmental conditions, reveal improved abiotic stress tolerance, possess delayed senescence and are in general better than those having less efficient photosynthesis in manufacturing heterotic biomass.

enhanced thermotolerance in rice: Zhang and others in 2022 identified a quantitative trait locus, THERMOTOLERANCE *3* (*TT3*), consisting of two genes – *TT3.1* and *TT3.2* – that interact together to enhance rice thermotolerance and reduce grain-yield losses caused by heat stress. Upon heat stress, plasma membrane-localized E3 ligase TT3.1 translocates to the endosomes, on which TT3.1 ubiquitinates chloroplast precursor protein TT3.2 for vacuolar degradation, implying that TT3.1 might serve as a potential thermosensor. Lesser accumulated, mature TT3.2 proteins in chloroplasts are essential for protecting thylakoids from heat stress.

enhanceosome: An assembly that educates recruitment programme of chromatin modifiers/remodellers and general transcription factors to the promoter. This programme terminates with sliding of a nucleosome, blocking the core promoter to a downstream position, a prerequisite for transcriptional activation. This assembly is, for example, needed for transcriptional activation of the IFN-β gene in response to virus infection. The identity of a gene expression programme is achieved and maintained by the dynamic interplay between specific enhanceosomes and specific local chromatin structure.

enhancer: A special type of eukaryotic regulatory sequence that can increase the rate of transcription of a gene located some distance away in either direction. They are mostly *cis*-acting, but *trans*-acting ones are also recognized. They are operative whether lying upstream or downstream. They are lively whether they lie in the same or opposite polarity as the gene. They are equally effective irrespective of the organism or the gene from which they are derived when attached to foreign DNA. A common feature of enhancers is GGTGTGGAAAG. First enhancer investigated has 72-bp tandemly repeated sequence positioned 100 nucleotides upstream of the ancillary site in DNA of SV40. Enhancer is a distant element activated by protein factor binding, and a signal is transmitted to promoter, where a number of different constitutive and regulatory transcription factors are bound. Unlike promoter sequences, the position and orientation of the enhancer sequence is generally not important to its activity, as enhancers can act on promoters over many tens of kilobases of DNA and be 5′ or 3′ to the promoter that they regulate. Generally, enhancers and promoters are on the same chromosome. Together, they control transcription and hence the expression of genes. Binding of specific proteins to an enhancer either stimulates or decreases the rate of transcription. Enhancer sequences work well regardless of organism or gene from which they are taken when attached to foreign gene. ENHANCER TRAPS are now being used to enhance the expression of transgene(s) in the transgenic organisms.

enhancer dynamics: L. Belver and others in 2021 report that enhancers control dynamic changes in gene expression and orchestrate the tightly controlled transcriptional circuitries that direct and coordinate cell growth, proliferation, survival, lineage commitment and differentiation during lymphoid development.

enhancer element: *Cis*-acting sequence that increases the utilization of eukaryotic promoters. *See* *CIS*-ACTING LOCUS.

enhancer function and activity analysis using epigenome-editing technology: Editing of gene enhancer regions in the genome via targeted epigenetic modification has been skilled by utilising a TALE-LSD1 effector fusion protein in order to target enhancers of genes to persuade enhancer silencing in order to have enhancer activity and gene control.

enhancer of zeste (E(z)): Originally identified in genetic screens for homeotic transformations in *Drosophila melanogaster* and later shown to encode a histone H3 lysine 27 (H3K27) methylating enzyme.

enhancer RNAs: A class of relatively long non-coding RNA molecules (50–2000 nucleotides) transcribed from the DNA sequence of enhancer regions. They were first detected in 2010 independently by T.K. Kim and others and F. De Santa and others through the use of genome-wide techniques such as RNA-seq and ChIP-seq. eRNAs can be subdivided into two main classes – 1D eRNAs and 2D eRNAs – which differ primarily in terms of their size, polyadenylation state and transcriptional directionality. The expression of a given eRNA correlates with the activity of its corresponding enhancer in target genes. Increasing evidence suggests that eRNAs actively play a role in transcriptional regulation in *cis* and in *trans*, and while their mechanisms of action remain unclear, a few models have been proposed.

enhancer-driven genomic recording of transcriptional activity in multiplex (ENGRAM): Dimensions of gene expression or signal transduction activity are conservatively performed using approaches that necessitate either the destruction or live imaging of a biological sample within the timeframe of interest. W. Chen, *et al.* (2024) demonstrate a paradigm in which such biological activities are stably recorded to the genome. ENGRAM relies on the signal-dependent production of prime editing guide RNAs that arbitrate the insertion of signal-specific barcodes (symbols) into a GENOMICALLY ENCODED RECORDING UNIT (GERU). This approach can be employed for multiplex recording of the cell-type-specific activities of dozens to hundreds of *cis*-regulatory elements with extreme fidelity, sensitivity and reproducibility. Using ENGRAM, leveraging signal transduction pathway-responsive *cis*-regulatory elements, time- and concentration-dependent genomic recording of WNT, NF-κB and Tet-On activities have been accomplished.

enhancer-promoter encounters: A crucial step in gene regulation is the physical encounter of dispersed enhancer-promoter pairs across the genome. D.B. Brückner and others in 2023 visualized the three-dimensional motion of pairs of DNA loci of varying separations along the chromosome and their transcriptional output in developing fly embryos. They found an unexpected combination of dense packing and rapid diffusion, leading to encounter times with a weak dependence on genomic separation. Transcriptional contacts are possible across large genomic distances, with crucial implications for gene regulation.

enhancing photosynthesis by metabolic acclimation: Plants control their photosynthetic capacity in response to their external environment to maximize their fitness and yield. This acclimation in photosynthetic response to environment comprises fluctuations in leaf metabolome, such as RuBisCO production, ribulose-1,5-bisphosphate (RuBP) regeneration, kinetics and association between other enzymes of the Calvin cycle and antioxidant metabolites to maximize carbon gain. However, regulatory networks involved in plant photosynthetic acclimation to particular environment need to be investigated.

enhancing plant yield by leveraging photosynthetic efficiency: Photosynthetic systems have so far been unimproved in crops, and utilization of photosynthetic efficiency still falls far short of its biological limit. The yield outcome resulting from manipulation of photosynthetic machinery may be limited by feedback effects arising from interactions with crop growth, development dynamics and the changing climate. In this connection, a reliable model has been given in wheat and sorghum that enlightens which photosynthetic modifications in plants can enhance the yield in these crops. This model can predict the yield gains linked with photosynthesis in true-to-life crop environments. Simply enhancing photosynthesis will not always improve crop yield. Thus, root and canopy architecture must be considered in yield-improvement programmes aimed at enhancing photosynthetic efficiency. Three major photosynthesis manipulation targets have been examined by an Australian team in wheat and sorghum to increase yield: (i) improving the activity of the key photosynthetic enzyme RuBisCO, (ii) improving the capacity of plant leaves to transport electrons and (iii) improving the movement of CO_2 via the internal layers of the leaf involved in yield formation.

enhancing seed germination: Smoke derived from burning plant material increases germination of a wide range of species from Australia, North America and South Africa. The compound present in plant- and cellulose-derived smoke that promotes germination of a variety of smoke-responsive taxa at a level similar to plant-derived smoke water has been identified. The structure of the compound was confirmed as butenolide 3-methyl-2H-furo[2,3-c]pyran-2-one (1) (Scheme 1). The results could provide benefits for horticulture, agriculture, mining and disturbed land restoration.

enhancing stress tolerance to improve production: Multiple approaches are being employed to attain improved photosynthesis with improved water use efficiency in plants, metabolic acclimation, increased N for faster plant growth in low-light environments and improved interspecific root interactions.

enhancing the expression of transgenes: Two approaches can be used: (i) Heterologous introns can enhance expression of transgenes. (ii) Post-transcriptional regulatory element enhances expression of transgenes.

enhansons: The enhancer elements collaborate with one another or duplicates of themselves to improve transcription. These elements are bipartite, being composed of subunits, called enhansons, which can be duplicated or interchanged to generate new enhancer elements. Enhansons vary from the enhancer elements as they are very sensitive to changes in spacing.

enneaploid: A polyploid plant with nine sets of chromosomes.

enriched histone fraction: The isolation of nuclei from cultured metazoan cells or yeast is very straightforward and requires three essential steps: (i) hypotonic swelling, (ii) cell membrane lysis by mechanical shearing and (iii) isolation of nuclei by centrifugation.

enriched medium: *See* COMPLETE MEDIUM.

enrichment: The process of increasing the proportion of mutant cells in a culture. Enrichment is used when the desired mutants cannot be selected for directly. The technique is essentially the converse of selection. Conditions are established such that the desired mutants will not grow. Wild-type (growing) cells are then killed by some physical or chemical treatment which does not affect non-growing (mutant) cells. Such methods include treatment with drugs such as penicillin (for bacteria) or nystatin (for fungi), inositol starvation and heat shock.

enrichment of small RNAs: Two methods used for enrichment of small RNAs are HIGH-AFFINITY BINDING PROTEINS METHOD and CHITIN MAGNETIC BEADS METHOD. *See details as separate terms.*

Enterobacteriaceae: Large family of Gram-negative bacteria that includes *Escherichia coli*, *Klebsiella*, *Salmonella*, *Shigella* and *Yersinia pestis*.

enterotoxin: An exotoxin (secreted proteins) that causes gastroenteritis. Enterotoxin is a type of exotoxin produced and released by some bacteria, which affect cells lining the gastrointestinal tract Exotoxins are among the most lethal substances. One milligram of botulinum exotoxin can kill 1 million (10 lakh) guinea pigs.

entomophily: A type of pollination whereby pollen or spores are dispersed by insects.

Entrez: An online resource provided by the National Center for Biotechnology Information (NCBI). It organizes GenBank sequences and links them to the literature sources in which they originally appeared.

entropy: A measure of the disorder or randomness in a molecule or system. The second law of thermodynamics explains that the entropy of a closed system can never decline; the higher the entropy, the greater the disorder.

entropy-based quantitative DNA methylation tools: Integrated software platforms that support quantification of epigenetic difference and identification of DIFFERENTIAL EPIGENETIC MODIFICATION REGIONS (DEMRs) by Shannon entropy. Two main modules – QUANTITATIVE DIFFERENTIAL CHROMATIN MODIFICATION REGION (QDCMR) and QUANTITATIVE DIFFERENTIALLY METHYLATED REGION (QDMR) – are provided for bioinformatic analysis of chromatin modifications and DNA methylation data, respectively. The third module – QUANTITATIVE DIFFERENTIAL EXPRESSED GENE (QDEG) – is used to identify differentially expressed genes. The platform-free and species-free nature of EpiDiff makes it potentially applicable to a wide variety of epigenomes at an unprecedented scale and resolution.

entry time: The time when a patient starts contributing to the study. In randomized studies or observational studies where all patients have come under observation before the study starts (for example, studies of survival after surgery), the entry time and time origin of the study will be identical. However, for some observational studies, the patient may not start follow-up until after the time origin of the study, and these patients contribute to the study group only after their 'late entry'.

enucleate: (of cells) Eradicating the nucleus from a cell.

enucleated oocyte: An egg cell from which the nucleus has been removed mechanically. *Aka* CYTOPLAST.

enveloped virus: Virus with a capsid surrounded by a lipid membrane (the envelope), which is derived from the host cell plasma membrane when the virus buds from the cell (e.g. retroviruses, herpesviruses, influenza viruses and many more).

envelopes: Lipid bilayer surrounding some organelles (e.g. nucleus and mitochondria).

environment: It is an outer media in which human or another living being withstands own interactions during their life. The physical elements comprising this media are air, water and soil. The biological elements involved are producers (plants), utilizers (animals) and decomposers (bacteria and fungus). The economical environmental elements include the activities of humans, which are related with utilization and operation of natural, human, economical and materialistic sources. The social elements include demographic structure, historical and cultural infrastructure of human and the type of life with respect to accommodation, health, education and culture demands. The physical, chemical and biotic conditions surround a living organism.

environment of evolutionary adaptation (EEA): Refers to a group of selection pressures occurring during an adaptation's period of evolution responsible for producing the adaptation. A selection pressure can be any factor in a population that impacts reproductive success.

environment, external: The aggregate of all external conditions and influences affecting the life and development of an organism.

environment, internal: The cellular environment affecting the expression of manifestation of certain genes (e.g. hormonal constitution affects expression of the gene for baldness in men and women).

environmental biology: The branch of biology concerned with the relations between organisms and their environments.

environmental determination of sex: A phenomenon in which the sex of sexually undifferentiated organism is determined by the environment.

environmental deviation (E): The difference between the expected and actual trait value of a given genotype.

environmental diseases: The diseases that are caused by persistent unfavourable environmental conditions, including factors such as temperature, moisture, wind, light, soil pH, soil structure, host nutrition, herbicides, chemicals and air pollutants.

environmental effects: Measured value of contribution of environment towards the total phenotypic variation.

environmental effects on QTL mapping: Environmental effects influence the expression of QTL. Experiments that are replicated across sites and over time may enable the researcher to investigate such effects. Recombinant inbred (RI) or DOUBLED-HAPLOID (DH) populations are ideal for such studies.

environmental epigenetics: (i) A study of the molecular mechanisms of environmental factors interacting with the epigenome and impacting the biology in specific organ and tissue and cells, resulted in adaptive evolutionary changes or disease development. (ii) Branch of epigenetics which investigates how the environment impacts phenotype is performed under the umbrella of environmental epigenetics.

environmental factors: The conditions that encourage or discourage the susceptibility of the host and the pathogenicity of the pathogen. The environmental conditions can also affect the interaction between the host and the pathogen. As related to genetics, the term refers to exposures to substances (e.g. pesticides or industrial waste) where we live or work, behaviours (e.g. smoking or poor diet) that can increase an individual's risk of disease or stressful situations (e.g. racism). Genetic studies often take environmental factors into consideration, as these exposures can increase an individual's risk of genetic damage or disease.

environmental genomics: Predominantly consists of a range of molecular components, including DNA and RNA in (meta) genomes and (meta) transcriptomes collected directly from the environment. *Aka* MICROBIAL ECOGENOMICS. *See* METAGENOMICS.

environmental mutagenesis: Branch of genetics that deals with studies on the mutagenic effects of environmental agents using a battery of test systems. *See* GENETIC TOXICOLOGY.

environmental mutagenic hazards: A variety of industrial and agricultural chemicals are in use, and millions of workers are regularly exposed to them. Applying fertilizers, insecticides, fungicides, bactericides, etc. is industrializing even our agriculture. Persons not working in industry and agriculture expose themselves to chemicals in other ways (e.g. medicines, food dyes, food preservatives, dyes). Therefore, chemicals have become an important part of our life. A dangerous consequence is their effect on the genetic material. Our technologically oriented society has greatly increased the amounts of various gases (e.g. CO, CO_2, SO_2, NO, NO_2, H_2S) that can be quite hazardous. For example, sulfurous gases enter the atmosphere through industrial and automobile combustion. All these compounds are mutagenic.

environmental mutagens: Chemicals, sun and radiation that can cause mutations in DNA.

environmental protection: The practice of protecting the natural environment by individuals, groups and governments. Its objectives are to conserve natural resources and the existing natural environment and, where it is possible, to repair damage and reverse trends. It is that part of resource management that deals with the discharge into the environment of substances that might be harmful or that might have harmful physical effects and with safeguarding beneficial uses.

Environmental Protection Act (EPA): Established in 1970, The EPA is an independent executive agency of the United States federal government tasked with environmental protection matters. All US states have their own state-level departments of environmental protection, which may issue regulations more stringent than the federal ones. Commercial releases of transgenic crops in USA relating to environmental aspects are approved by EPA. For example, the transgenic crops carrying herbicide, insecticide resistance genes need EPA approval. The Constitution of India has a number of provisions demarcating the responsibility of the Central and State governments towards environmental protection. The state's responsibility with regard to environmental protection has been laid down under article 48-A of the constitution, which stated, 'The states shall endeavour to protect and improve the environment and to safeguard the forest and wildlife of the country'.

environmental quality standards: The maximum confines or concentrations of pollutants that are allowed in specific media.

environmental resistance: All genetically based features that guard the reproduction of plants against negative impacts of the environment.

environmental risk: A hazard or probability of suffering damage or misfortune because of exposure to some environmental circumstance.

environmental risk assessment: A quantitative evaluation of the risks associated with an environmental hazard.

environmental science: An interdisciplinary branch of science that investigates questions related to the human population, resources and damages caused by pollution and disturbance.

environmental security: The protection of people and the public interest from environmental risks, particularly those associated with anthropogenic activities and accidents but may also include natural dangers.

environmental stress adaptation: Chromatin modifications, nucleosome positioning and DNA methylation have been recognized as important components in environment stress adaptations.

environmental stress memory: Environmental stress memory is a mechanism for organisms to store the message from an initial stress and becomes more robustness in response to future stresses, leading to the tolerating, priming or acclimation affects after prestressing.

environmental studies: An extremely interdisciplinary strategy that examines the scientific, social and cultural aspects of environmental problems.

environmental teratogen: A teratogenic impact that comes upon in the environment. *See* TERATOGEN.

environmental toxicology: The investigation of environmental factors that impact exposures of organisms to potentially toxic levels of chemicals. *Compare with* TOXICOLOGY *and* ECOTOXICOLOGY.

environmental values: Perceptions of the worth of environmental components, divided into two broad classes: utilitarian (useful rather than decorative) and intrinsic (extremely important and basic characteristic).

environmental variance: The variance due to environmental variation.

environmental variation (V_E): When individuals with the same genotype develop in different environments, their phenotypes may be quite different (e.g. identical twins raised in different environments show different phenotypes due to environmental variation).

enzymatic: Activity of an enzyme that is a substance produced by a living organism and acting as a catalyst to promote a specific biochemical reaction.

enzymatic DNA sequencing: *See* DNA SEQUENCING BY DIDEOXY METHOD.

enzyme: A functional protein that catalyses a chemical reaction but is itself neither consumed nor altered. Enzymes control the rate of metabolic processes in an organism; they are, for instance, the active agents in the fermentation process. Each enzyme is highly specific with regard to the type of chemical reaction it catalyses and to the substances (called SUBSTRATES) upon which it acts. This specific catalytic activity and its control by other biochemical constituents are of primary importance in the physiological functions of all organisms. Although all enzymes are proteins, they may, and usually do, contain additional non-protein components called COENZYMES that are essential for catalytic activity. Most enzymes are proteins, but some RNAs, known as ribozymes, also act as catalyst. *See* COENZYMES.

enzyme catalysis: Enzyme-based catalysis of biochemical reactions. Enzymes achieve their enormous rate accelerations via the same catalytic mechanism used by chemical catalysts. Enzymes, like other catalysts, reduce the free energy of the transition state; they stabilize the transition state of the catalysed reaction.

enzyme engineering: *See* PROTEIN ENGINEERING.

enzyme immunoassay (EIA): EIA is used to measure the amount of a particular substance by virtue of its binding to a specific antibody. Examples of EIA include ENZYME-LINKED IMMUNOSORBENT ASSAY (ELISA) and Western blotting.

enzyme kinetics: The rate of a reaction controlled by an enzyme. In low substrate concentrations, the reaction rate (V) is always proportional to the substrate concentration ([S]), but when the substrate is present at high concentrations, enzyme saturation will occur (V_{max}). The Michaelis-Menten equation expresses the rate of reaction as $V = V_{max}[S]/(K_m + [S])$, where K_m is the Michaelis constant and is equal to the concentration of the substrate when the speed of the reaction is half its maximum.

enzyme regulation: Enzymes may be regulated at two levels: (i) at the level of gene expression and protein synthesis through induction and repression and (ii) at the enzyme level through enzyme inhibition or activation as a result of the binding of effector molecules at allosteric sites on the enzyme.

enzyme repression: A mechanism that prevents the synthesis of an enzyme by the formation of repressors that bind to DNA, thus preventing transcription.

enzyme-linked immunosorbent assay (ELISA): A highly sensitive immunoassay based on enzymatic reactions. Usually two antibodies are used. The first antibody binds to an antigen and directs itself towards the second antibody, the antiglobulin. The antiglobulin is linked to an enzyme (e.g. horseradish peroxidase). The enzyme activity is detected (e.g. by a staining reaction). This test can readily measure less than a nanogram (10^{-9} g) of a protein. The degree of enzyme reaction measured quantitatively by fluorometric or colorimetric method estimates the number of antibodies of the first type (i.e. of the original antigen) present. ELISA is used to test for the presence of a particular genetically engineered organism. ELISA test is commonly used to detect virus particles in the plants. Plant tissue extract is mixed with specific antibodies and then the formation of antigen-antibody complexes is studied using ELISA plate reader, thus offering a simple test to check the presence or absence of disease-causing virus in the plant.

enzyme-linked receptor: Major type of cell-surface receptor in which the cytoplasmic domain either has enzymatic activity itself or is associated with an intracellular enzyme. In both cases, the enzymatic activity is stimulated by ligand binding to the receptor.

enzyme-substrate complex: A temporary molecule formed when the substrate binds to the enzyme. When an enzyme binds to its substrate, it undergoes a conformational change or temporary change in shape. However, once the product(s) is released, the enzyme regains its original shape.

epecophyte: Species of recent appearance, usually numerous and constant in a geographical area, but confined to artificial habitats (e.g. meadows and ruderal vegetation) and are dependent on humans for existence.

ephemeral resistance/performance: Impressive resistance or performance when the variety is new or in breeder's trials, but it fades when the variety is released or becomes popular (under commercial cultivation).

epialleles: (i) Alleles with differential chromatin states that are mitotically and/or meiotically inherited and are a good source of variation that can lead to phenotypic diversity. Spontaneous epialleles arise independent of changes in DNA sequence. (ii) Alternative chromatin states at a given locus, defined with respect to individuals in the population for a given time point and tissue type. Epialleles vary greatly in their stability, and they affect gene expression levels or patterns rather than gene products. (iii) Chromatin differences at a locus between different individuals or cells. (iv) Versions of a gene generated by an epimutation. Alleles that are genetically identical (no sequence difference) but present distinct epigenetic profiles due to differences in chromatin marks (e.g. DNA methylation), which may be associated with changes in gene expression depending on the location of the modification. Epialleles vary greatly in their stability, and they affect gene expression levels or patterns rather than gene products. Thus, some epialleles may reflect epigenetic variation, but others may reflect genetic variation. Note that an epiallele may occur due to genetic differences (at *cis*-genomic or *trans*-genomic locations). Epialleles at a given locus are defined with respect to individuals in the population for a given time point and tissue type.

epialleles, classes of: There are three major classes of epialleles on the basis of underlying genetic variants – OBLIGATE, PURE and FACILITATED. *See details as separate terms.*

epialleles, transmission of: Epigenetic variation is based on the stability in the next generation. However, the epialleles are not always as stable as the primary DNA sequence. The sites of epiallelic conditions occur locally and distantly to other chromosomes. Epialleles confer flexibility in transcriptional behaviour for the generation of novel phenotypic variation. RNA-dependent DNA Methylation (RdDM) leads to the formation of random epialleles that have higher significance than genetic variation. About 97% of epialleles were transmitted through five meiosis.

epibreeding: External and internal stimuli may cause epigenetic changes such as DNA methylation, histone modifications and the action of non-coding RNAs that change chromatin structure, which leads to alterations in gene expression. If heritable, these changes can lead to stable epialleles impacting phenotypic variation and can be exploited for the selection of desired genotypes in plant breeding programmes. The comparison of different epigenomes arising from different genotypes/tissues/cell types/environmental conditions can provide valuable information for the development of biomarkers paving the way to what is nowadays termed PLANT EPIBREEDING.

EPIC Informatics Hub: This Hub intends to serve as a chief community resource. This idea includes collaborating with iPlant (funded by the US National Science Foundation), the nascent *Arabidopsis* Information Portal and other computational and informatics resource centres.

epicarp: Refers to the outer layer of the wall of a fruit; the 'skin'.

epicotyl: The portion of the stem of a plant embryo or seedling above the node to which the cotyledons are attached. *Contrasts* HYPOCOTYL.

epidemic: An unarrested, widespread upsurge of disease in a population. An overall and serious outbreak of disease. An epidemic may extend over a single season or many years and over a wide or relatively small area. *See* EPIPHYTOTIC.

epidemic disease: *See* ENDEMIC DISEASE.

epidemic rate: The increase or decrease per unit of time in a given plant population.

epidemiology (Adj. epidemiologic): Science dealing with the studies regarding the factors affecting the disease initiation, development and spread in plants, animals and humans. It also deals with the study of when and where diseases occur and how these are transmitted from one organism to another. Epidemiological studies help in the forecasting and control of diseases.

epidermal growth factor (EGF): A growth factor involved in cell growth, proliferation and differentiation.

epidermis (Adj. epidermal): (i) An organ's outermost one-cell thick layer of cells of leaves and other plant parts. (ii) Epithelial layer covering the outer surface of the body. It has dissimilar structures in diverse animals. The root's outer dermal tissue layer is the epidermis, a single layer of cells that protects the root and controls water and mineral absorption. Their large surface area allows root hairs to efficiently absorb water via osmosis and minerals via diffusion.

epigenator: In an epigenetic process, epigenator makes available a signal that triggers changes in the environment of the cell to create an epigenetic phenotype. It can be anything – nutrition, toxin, radiation, hormones, etc. The best example of an epigenator comes from plants, where environmental signals such as temperature affect the epigenetic process of paramutation.

epigene: A novel chromatin segmentation method for the identification of active transcription units (Tus) using transcription-associated histone modifications. Geneticists study the gene; however, for epigeneticists, there is no obvious 'epigene'. So epigene is a hypothetical term corresponding to gene.

epigenesis: A concept according to which process of development consists of more than growth. Development of an individual is epigenetic. Genes act in epigenetic systems, and adult structures develop from uniform embryonic tissues. The development of an organism occurs from a zygote through cell differentiation and formation of morphology. This view is a *counterargument to* PREFORMATIONISM, which holds that the whole organism is already present in the sex cells and merely gets bigger over time.

epigenetic: (i) Relating to or arising from reversible non-genetic influences on gene expression.

(ii) Heritable changes in gene expression without changes in primary DNA sequences. (iii) Refers to the mitotic or meiotic transmissibility of chromatin variation, independent of DNA sequence variation. (iv) Any change in the phenotype that is not heritable. During plant cell culture, sometimes we get variants at cellular/plant level that exhibit changed phenotypes, but the traits are not heritable to the next generation through sexual reproduction. This kind of variation is called EPIGENETIC VARIATION.

epigenetic analysis: The study of heritable changes that are distinct from changes in DNA sequence. Epigenetic changes include covalent modifications to the DNA bases or histone proteins that make up chromatin; they can be influenced by the environment and transmitted through DNA replication.

epigenetic analysis using next-generation methods: Next-generation methods include WHOLE-GENOME BISULFITE SEQUENCING, ChIP-SEQUENCING (ChIP-SEQ), MICROCOCCAL NUCLEASE (MNASE) DIGESTION WITH DEEP SEQUENCING (MNASE-SEQ) and SMALL RNA SEQUENCING (sRNA-SEQ).

epigenetic cell memory: During development and differentiation, cellular phenotypes are stably propagated through numerous cell divisions. This epigenetic cell memory helps to maintain stable pattern of gene expression. DNA methylation and the propagation of specific chromatin structure may both contribute to cell memory.

epigenetic changes: Include covalent modifications to the DNA bases or histone proteins that make up the alterations that influence the phenotype without altering the genotype in terms of DNA sequence. Methylation of cytosine at position 5 ($5^{m}C$) in DNA at certain positions is known to inactivate eukaryotic genes.

epigenetic changes cause physical variation: In eukaryotes, the genome is compacted into chromatin, and the chromatin structure plays an important role in gene expression. Gene expression can be controlled by changes in the structure of chromatin without changing the DNA sequence. Epigenetic change can cause phenotypic variation, and thus, epigenetic change can be considered an important factor in understanding phenotypic change. DNA methylation and histone modification are well-known epigenetic modifications.

epigenetic changes causing changes in gene regulation: Epigenetic regulation is achieved by specific mechanisms involving DNA methylation, histone PTLMs and the action of NON-CODING (NC) RNAs, which result in open or closed chromatin states related with gene activation or gene silencing, respectively. Extensive investigations have offered a wealth of information on epigenetic regulation at specific loci both in model and crop plants and the effect it may have on various aspects of plant development such as proper vegetative growth, successful reproduction and viability, effects on yield and efficiency in coping with stress. Transgenerational natural or induced epigenetic variation can be a new source of phenotypic diversity, especially for species with low genetic variation. The comparison of different epigenomes generated from different genotypes/tissues/cell types/environmental conditions offer valuable information for the development of biomarkers flagging the way to PLANT EPIBREEDING. Many of the environmentally induced epigenetic changes in plants are reset during gametogenesis. However, some epigenetic changes persist through gametogenesis and can be stable through many generations. The plant germline cells descend from somatic cells, and they carry epigenetic marks as persistent remnants of earlier environmental stimuli. Epigenetic erasure during embryogenesis is more or less incomplete in plants.

epigenetic code: The central dogma rose from studies in molecular genetics, and the central statement is that genetic information flows from nucleic acid to protein, but not in the reverse direction. In modified central dogma of molecular biology, reverse transcription was also added. Currently, the molecular biology of gene expression in eukaryotes has been found to depend in the large part on the study of protein-DNA interactions. These interactions provide the key to an understanding of developmental processes, including the heritability of determined or differentiated state of cells. In case of DNA methylation, the enzymes involved actually alter DNA structure, thereby imposing an epigenetic code over and above the genetic code and thus specifically altering the information content of the DNA. Both DNA methylation and histone modification are involved in establishing patterns of gene repression during development. Certain forms of histone methylation cause local formation of heterochromatin, which is readily reversible, whereas DNA methylation leads to stable long-term repression. DNA methylation and histone modification pathways can be dependent on one another and that this crosstalk can be mediated by biochemical interactions between SET domain HISTONE METHYLTRANSFERASES (HMTs) and DNA METHYLTRANSFERASES (DNMTs). Relationships between DNA methylation and histone modification have implications for understanding normal development as well as somatic cell reprogramming and tumourigenesis.

epigenetic code of cellular memory: As each of the histone proteins can be modified with different PTLMs at different amino acids, the logical question that arises is whether these modifications in different combinations constitute a form of information, language or histone code. *See* CELLULAR MEMORY.

epigenetic control of biological processes: Non-genetic regulation of molecular and cellular functions, eventually causing functional phenotype changes (e.g. diseases).

epigenetic control of development: Modified DNA base 5-METHYLCYTOSINE ($5^{m}C$) is strongly involved in control of gene expression in higher organisms. Gene transcription is associated with the absence of methylation at one or more locations in promoter or sometimes other regions of the gene. In cells not expressing a gene, some sites are found to be methylated. It is not yet clear whether methylation is a cause or an effect of gene activation. Now, sufficient evidence is available to show that inactivation is a result of methylation.

epigenetic control of developmental programme: There are two major landscapes of epigenetic control of the developmental programme: (i) the segregation of gene activities and (ii) the stability of cells that are differentiated. Pattern of methylation is inherited through DNA synthesis and cell division.

epigenetic control of effectors in plant pathogens: Pathogen effector proteins facilitate disease but can become avirulence (Avr) factors when the host acquires discrete recognition capabilities that trigger immunity.

epigenetic control of gene expression: Chromatin-based control of gene expression that tells a gene, in response to internal or endogenous stimuli, when to perform its function, using genetically determined epigenetic mechanisms such as DNA modifications, RNA modifications, histone modifications and small RNAs. This is in contrast with GENE REGULATION, which explains the mechanism that initiates production of transcript on DNA sequence of a gene.

epigenetic crosstalk: Communication between/among different epigenetic marks located on histones, DNA and RNA have been reported to crosstalk. For example, crosstalk has been observed between/among: (i) DNA methylation marks in different contexts; (ii) histone modifications; (iii) DNA and histone modifications and (iv) RNA and histone modifications.

epigenetic databases: The EpiGenie team has searched and compiled a list of the most commonly used free tools and databases that epigenetics researchers use.

epigenetic defects theory of ageing: There is a constant likelihood that a methyl group will be lost in each cell division. *De novo* methylation either does not occur or occurs at a very low rate. This follows that level of methylation should progressively decrease. Thus, the possibility of genetic errors through loss of methyl groups should be added as a cause of ageing. Rate of loss of methyl groups could be correlated to lifespan of the organisms. In fact, observed rate of methyl group loss is dependent upon cell division rather than chronological time. A single treatment of cultured diploid cells with 5-AZACYTIDINE (5-aza-CR) or 5-AZA-2′-DEOXYCYTIDINE (5-aza-CdR) inhibits growth to some extent, but cells rapidly recover and have normal phenotype and growth rate. However, they retain a memory of the treatment, because their growth potential is substantially reduced. Multiple treatments have a progressively strong effect in shortening lifespan. This means that the physiological age of cells is advanced by the loss of methylation induced by 5-aza-CR. Permanent lines may escape senescence, because they have the ability to restore methyl groups by *de novo* methylation; decline in the level of $5^{m}C$ in DNA may reactivate silent genes. Changes associated with ageing accrue more and more rapidly as the end of normal span is approached. This appears to be true of a whole organism, as well as of cultured human diploid cells.

epigenetic defects, repair of: Purpose of meiosis is reprogramming gametes prior to the formation of fertilized eggs. This programming, according to DNA methylation theory of development, would depend at least in part on specific *de novo* methylation or demethylation of chromosomal DNA. Transmethylase maintenance activity is also essential in germline cells to prevent transcription of genes with specialized somatic functions. Unprogrammed loss of methyl groups creates a problem for any *de novo* methylase. The methylase will have to be sequence specific, not to be recognized by all the demethylated sites, or it will have to be non-specific to methylate a large number of sites in meiosis. Recombination in meiosis may play an essential role in the repair of epigenetic defects that occur due to loss of methylation.

epigenetic differentiation: It is believed that certain development areas are present in the egg cytoplasm. Fertilization then provides an inducer which leads to differentiation within the existing genes and initiation of a new differentiated area. The inducer may act within these areas by establishing a gradient or by influencing gradients that are already present. Further inducers then arise causing further differentiated areas or new areas. The uniformity of the pattern in different individuals lies in the fact that its history is identical in each individual. When the history of the pattern is changed, such as change in the relative positions or a change in the effect of one of the inducers, a changed phenotype is the result. It has been suggested that each new set of positional coordinator stimulates special selector genes, which in turn causes cells to commit themselves to a particular developmental direction. Subsequent exposure to further positional influence stimulates a different battery of selector genes to a further commitment, etc.

epigenetic disruptors: Molecules that could modify gene activity and in the case of stem cells, alter cell fate or number, causing an increased risk of disease (e.g. cancer).

epigenetic diversity in pairs of sympatric, closely related plants: Epigenetic diversity can back adaptation. Epigenetic diversity compensates for the loss of genetic diversity. Genetic and epigenetic diversities and their possible co-variation were evaluated in three populations of each focal species using AMPLIFIED FRAGMENT LENGTH POLYMORPHISM (AFLP) and METHYLATION-SENSITIVE AMPLIFIED POLYMORPHISM (MSAP). All species and populations displayed modest to high levels of genetic polymorphism regardless of their ecological features. Epigenetic diversity was greater than genetic diversity in all the cases studied.

epigenetic diversity in productivity and stability of plant populations: Mutants of 'Fuji' apple are used as the assessment material. Considering the genetic variation within 'Fuji' as the control, the topographies of epigenetic variation at different levels in both varieties and mutant groups have been examined. Vast majority of loci maintained the original methylation pattern. CHG methylation variation was the main form of epigenetic variation observed. The variation in methylation patterns between the mutant groups was higher than that of methylation levels. The variation in CHG methylation patterns (including CHG hypermethylation and CHG demethylation) was expected to be dominant. The variation in methylation patterns was more apparent in both the *Early maturation* and *Spur* mutant groups.

epigenetic drugs: The compounds that target the proteins involved in writing, reading or erasing epigenetic marks are attractive targets for drug development. Epigenetic drugs elicit a 'specific' effect on cancerous cells by reverting unique expression changes.

epigenetic effects on genome structure: Epigenetic mechanisms are linked in various ways to genome structure and there are several indications that environmental stimuli induce epigenetic mechanisms with genetic consequences. There is evidence that DNA repair is affected by RNA silencing in plants. Some genetic effects could be related to the epigenetic control of transposons. A study of nutrient stress-induced heritable changes in flax may also illustrate the potential of the environment to induce DNA rearrangements that influence gene expression. Flax plants with a 'plastic' genotype grew as large plants under high nutrient conditions and were small under low nutrient conditions. Many of the induced differences persisted for several generations in progeny grown under standard nutrient conditions. The inherited differences include changed height and weight, nuclear DNA content and methylation, ribosomal gene number and seed capsule characters. These phenotypes were not graft-transmissible, but they are stable in subsequent generations.

epigenetic erasers: Enzymes that remove epigenetic chemical/modifications, such as methyl group from DNA, RNA or histones. *Aka* CHROMATIN ERASERS.

epigenetic event: Any change in a phenotype that does not result from an alteration in DNA sequence. This change may be stable and heritable and includes alteration in DNA methylation, transcriptional activation, translational control and PTLMs.

epigenetic gatekeepers: Epigenetic silencing genes (e.g. *p16*, *SFRPs*, *GATA-4*, *GATA-5* and *APC*) in stem/precursor cells of adult cell-renewal systems that may assist to abnormally lock these cells into stem-like states that adopt abnormal clonal expansion.

epigenetic gene silencing maintenance: The institution and preservation of epigenetic gene silencing is central to cell determination and function. The essential epigenetic systems implicated in heritable repression of gene activity are Polycomb group (PcG) proteins and the DNA methylation systems.

epigenetic hybrid plants: Hybrid systems show altered DNA methylation and siRNA patterns that potentially contribute to the hybrid vigour phenotype.

epigenetic imprinting: An epigenetic form of gene regulation that determines the parent-dependent gene expression of marked or imprinted genes during gametogenesis and embryonic development.

epigenetic influences on DNA replication origins: DNA replication in multicellular animals begins at specific origins to which a PRE-REPLICATIVE PROCESS (pre-RC) binds. Origin identity can change during development, suggesting that there are epigenetic influences. A notable example of developmental specificity occurs in *Drosophila*, where somatic follicle cells of ovary transition from genomic replication to exclusive replication at origins that control amplification of chorion protein genes. Chromatin acetylation is critical for this developmental transition in origin specificity. They have found that the histones at active origins are hyperacetylated, coincident with binding of the ORIGIN RECOGNITION COMPLEX (ORC). Mutation of the HISTONE DEACETYLASE (HDAC) Rpd3 induced genome-wide hyperacetylation, genomic replication and a redistribution of the origin-binding protein ORC2 in amplification-stage cells, independent of effects on transcription. Tethering Rpd3 or Polycomb proteins to the origin decreased its activity, whereas tethering the Chameau acetyltransferase increased origin activity. These results have suggested that nucleosome acetylation and other epigenetic changes are important modulators of origin activity in metazoa.

epigenetic inheritance: (i) Germline-mediated inheritance of epigenetic information between generations in the absence of direct environmental influences that leads to phenotypic variation without that information being encoded in the nucleotide sequence of the gene. (ii) Somatic inheritance of the differentiated state of the cell through cycles of cell division. *Aka* EPIGENETIC TRANSGENERATIONAL INHERITANCE.

epigenetic inheritance system (EIS): (i) A mitotically and/or meiotically heritable trait that is not accompanied by a change in the DNA sequence. (ii) A system that enables the phenotypic expression of the information in a cell or individual to be transmitted to the next generation. EISs described so far are DNA methylation, histone modification and SMALL NON-CODING RNAs (SNCRNAs) involved in RNA INTERFERENCE (RNAI).

epigenetic initiator. In an epigenetic process, epigenetic initiator translates the epigenator signal to facilitate the epigenetic effect on chromatin. Priming of epigenetic initiator by epigenator makes initiator identify location on a chromosome where epigenetic state is to be established. Initiator could be a DNA-binding protein, a non-coding RNA or any other entity that can define the coordinates of the chromatin structure to be accumulated. Examples of initiator are non-coding X-inactive specific transcript (Xist) RNA, which is sufficient for silencing the mammalian X chromosome and DNA-binding factors that lead to reprogramming of differentiated cells into stem cells in metazoans.

epigenetic interactions: Various approaches used for studying DNA-protein interactions include: (i) DNA-protein interaction analysis; (ii) digital DNase; (iii) DNase-seq; (iv) chromosome conformation technologies (such as chromosome conformation capture (3C), circular chromosome conformation capture (4C) and chromosome conformation capture carbon copy (5C)); (v) target classes and technologies in histone methylation analysis; (vi) target classes and technologies in acetylation analysis; (vii) links between metabolism and epigenetic modification of DNA and (viii) epigenetics screening library.

epigenetic landscape in *Arabidopsis*: In *Arabidopsis*, 26,852 regions were detected with a significant level of DNA methylation, covering 22,554,840 bp and representing 18.9% of the entire sequenced nuclear genome.

epigenetic landscape in rice: The first high-resolution genome-wide map of DNA methylation in rice was obtained using the methylcytosine immunoprecipitation (mCIP)-chip method. Histone modifications H3K4me1 and H3K9me2 were distributed across the entire transcribed region while the histone marks H3K4me2, H3K4me3, H3K9ac, H3K9me3, H3K27ac, H3K27me3 and H3K36me3 were enriched at the 5′-end of the gene body. H3K4me1 and H3K4me2 did not show correlation with transcript richness.

epigenetic landscape of *Arabidopsis* centromeres: Patterns of higher-order repetition within centromeres and many satellite variants are personal to each chromosome, which has implications for the recombination pathways acting in the centromeres. CENH3 preferentially occupies the satellites with the least amount of divergence and that show higher-order repetition.

epigenetic landscape profiling using cytometry by time-of-flight (epiTOF): A method that measures the overall cellular levels of chromatin marks at a single-cell level and in high-throughput format. This method measured the levels of 8 classes of histone PTLMs and 4 histone variants in 22 subsets of immune cells.

epigenetic landscape: Refers to a mechanism proposed by C.H. Waddington in 1942 in which a cell proceeds through development by transversing (i.e. moving to and fro over or along). Waddington's visual represents a ball rolling down a valley as an analogy for the way cells make decisions for following a path during development. The cell can thus opt for one of several allowable pathways leading to different cell fates (or cell types). The cell starts from a totipotent state and becomes more and more restricted by determinative events. The general properties of this process can be applied to both cytodifferentiation and pattern formation. Extrinsic forces such as hormonal stimuli or induction can influence the decision points, and the decisions will be dependent upon the genetic response of the cell to that stimulus. The cell moves from high to low end of the plane through the valley system. Depths and directions of this landscape are controlled by genes. Waddington called this process CANALIZATION.

epigenetic landscapes and transcriptomes, relationship between: Genomic landscapes of representative epigenetic modifications were associated with mRNA and sRNA transcriptomes in maize shoots and roots. The epigenetic patterns differed intensely between genes and TRANSPOSABLE ELEMENTS (TEs), and two repressive marks (H3K27me3 and DNA methylation) were usually observed to be mutually exclusive.

epigenetic learning and memory: Learning is the process of acquiring new understanding, knowledge, behaviours, skills, values, attitudes and preferences. The changes induced by learning often last a lifetime. Memories are recollections of actual events stored within our brains. Epigenetic mechanisms influence genomic activities in the brain to produce long-term changes in synaptic signalling, organization and morphology, which in turn support learning and memory.

epigenetic maintainer: In an epigenetic process, epigenetic maintainer sustains the chromatin state created by initiators in the first and subsequent generations but is not sufficient to initiate it. This signal encompasses many different pathways, including DNA methylation, histone modifications, histone variants and nucleosome positioning.

epigenetic maintenance of centromeres: Fruitful cell division involves precise separation of chromosomes with near perfect fidelity to the daughter cells. Fundamental to this essential process is attachment of the chromosomal centromeres to microtubules of the mitotic spindle. After DNA replication, properly timed regeneration of the centromere and its associated protein complex is crucial. D. Conti *et al.* (2024) and Parashara *et al.* (2024) reveal that Polo-like kinase 1 (PLK-1) coordinates this progression at the end of mitosis. PLK-1 has complex interactions with proteins at the centromere, first phosphorylating associated proteins and then itself binding to those sites, instigating conformational changes that endorse the recruitment of the histone H3-like centromeric protein A (CENP-A) and its chaperone protein Holliday Junction Recognition Protein (HJURP). These consequences assist to elucidate how the protein complex at the centromere is placed and regulated to construct in a timely manner once each cell cycle.

epigenetic mapping vs linkage disequilibrium: The epigenetic mapping is better than linkage disequilibrium mapping (gap) conferring the causative genes contributing towards phenotypic variation. Epigenetics mapping has the potential to disentangle the complex genetic and epigenetic regulatory network of candidate genes governing complex traits. The epi-QTL studies reveal that candidate genes have an association with multiple SINGLE-NUCLEOTIDE POLYMORPHISMS (SNPs), which shows that the expression of these candidate genes is controlled by complex genetic mechanisms. Stable and newly formed epigenetic variants over generations have significant utility for crop improvement.

epigenetic maps and epi-QTL: With the prevailing biotechnology-based advanced tools, researchers nowadays construct the epigenetic maps (E-maps) and detection of epi-QTL to target the trait of interest that confer the scrutiny in the study of complex traits. The E-maps have been prepared through SSR and MSAP that have the potential to efficiently detect the DNA methylation polymorphisms to construct the methylation-based linkage maps in *Sorghum bicolor* and *Brassica napus*, respectively. AMPLIFIED FRAGMENT SINGLE-NUCLEOTIDE POLYMORPHISM AND METHYLATION (AFSM) marker have been used to generate the highly dense cassava genetic linkage map with 18 linkage groups. E-maps have been constructed with a total of 19 major linkage groups (>50 cM) and 2463

MSAP markers that constitute the overall framework for the linkage E-map in *Populus* species.

epigenetic markings: Changes in how the expression of a gene is regulated that is not caused by a change in the gene sequence.

epigenetic marks: Covalent modifications of DNA (e.g. cytosine methylation) or chromatin proteins (e.g. histone methylation, acetylation, phosphorylation, ubiquitination and SUMOylation) that affect gene expression, which is mitotically and/or meiotically heritable. Epigenetic marks are the landscapes not directly controlled by the genetic code. Epigenetic marks modify the role of the tagged proteins and influence the way genes are expressed. *Aka* CHEMICALS TAGS *or simply* TAGS.

epigenetic marks, stability of the: For transgenerational stability, the epigenetic marks pass through germline for their transmission to the subsequent generations without being erased by any inherent mechanism, which is thus efficiently maintained. More than 99% of methylated regions in DNA undergo inheritance across each generation in *Arabidopsis thaliana*. The transgenerational inheritance is mainly acquired through environmental changes that employ all non-sequence-based effects which get transferred to the subsequent generations.

epigenetic mechanisms: Molecular processes affecting cell behaviour via changes in gene expression that do not involve genetic alterations. Epigenetic mechanisms result in changes in chromatin structure which renders genes differentially competent for transcription. These mechanisms include epigenetic modifications such as DNA methylation, histone modifications, chromatin remodelling and regulatory ncRNAs. This forms the basis of development of a multicellular organism with multiple cell types.

epigenetic mechanisms at genome-wide levels: Advances in NGS technologies have provided an opportunity to investigate the epigenetic mechanisms at genome-wide levels. This will further help in development of methods for manipulating these pathways using genetic and biochemical approaches. This technology could prove to be a useful research tool for displaying the systematic relationships of genetic and epigenetic variations, especially in terms of cytosine methylation onto the genomic region in a specific cell or tissue.

epigenetic meiotic inheritance: Epigenetic meiotic inheritance is caused by imperfect reprogramming in the early embryo, which results in the dissemination of epigenetic information from parent to offspring. This phenomenon gives rise to patterns of phenotypic inheritance that obey Mendelian rules. The mechanism that guarantees that methylation marks at promoter DNA persist through meiosis, gametogenesis and early embryogenesis are not fully understood. It is also not understood why some epigenetic marks are heritable whereas others are not.

epigenetic memory: An essential process of life that governs the inheritance of predestined functional characteristics of normal cells and the newly acquired properties of cells affected by cancer and other diseases from parental to progeny cells. Epigenetic modifications of histones are thought to act as a layer of epigenetic information, because of their strong association with gene expression and their potential role in transcriptional memory.

epigenetic mitotic inheritance: This type of epigenetic inheritance is judgementally involved in cellular differentiation and cell fate decisions.

epigenetic modifications: Changes in DNA, histones or non-coding RNAs that influence the phenotype without altering the genotype. Epigenetic modifications are heritable chemical or physical changes in chromatin. The term is commonly used to describe a change in nucleosome structure caused by histone modifications, histone variants or modification (methylation) of the DNA. These changes are not necessarily epigenetic in the sense that they are stable through cell divisions, but some (e.g. symmetrical DNA methylation) might be. There are two broad types of epigenetic modifications – DNA methylation and histone modifications. *See* EPIGENETIC PHENOMENON *and* PTLMs.

epigenetic modification in expression divergence: Epigenetic changes were thus shown to contribute to the expression divergence of three *TaEXPA1* homoeologues during wheat development.

epigenetic modifications in phenotypic variation: Identification of genome-wide epi-markers has proved to be useful in exploring the impact of epigenetic modifications on phenotype. The preliminary criterion is the selection of plants with the desired trait of interest, controlling phenotype based on the adaptability under extreme environmental conditions. It comprises three major steps. The foremost step in such studies is the phenotyping of a population with respect to the target gene of interest in diverse germplasm. Next, precise phenotyping leads to the identification of genome-wide methylation that delineates the genetic basis underlying the phenotypic variation among the genotypes under-studies. The last step includes the identification of methylated and unmethylated cytosine loci among different genotypes, which interprets the epigenetic polymorphism. The combined approach of identifying epi-polymorphisms with phenotypic variation through statistical analysis confers epi-genotyping, which leads to the identification of genome-wide methylation among number of targeted genotypes.

epigenetic modifications, functional associations of coexisting: Functional associations between coexisting histone and DNA modifications are illustrated by the following examples: (i) H3K4me2/3 + H4K16ac have an association with transcriptionally active homeotic genes. (ii) H3K4me2/3 + H3K9/14/18/23ac has an association with transcriptionally active chromatin. (iii) H3S10ph + H3K14ac has an association with mitogen-stimulated transcription. (iv) H3S10ph + H3K14ac has an association with mitogen-stimulated transcription. (v) H3K4me3 + H3K27me3 has an association with bivalent domains. (vi) H3K9me3 + H3K27me3 + $5^{m}C$ has an association with silent loci. (vii) H3K27me3 + H2AK119ub1 has an association with silent homeotic genes. (viii) H3K9me3 + H4K20me3 + $5^{m}C$ has an association with heterochromatin. (ix) H3K9me2/3 + H4K20me1+ H4K27me3 + $5^{m}C$ has an association with inactive X chromosome.

epigenetic moonlighting: Recent molecular and genetic studies revealed that the function of many histone-modifying enzymes extends independently and beyond their catalytic activities. M.A.J. Morgan and A. Shilatifard in 2023 highlight original and recent advances in the understanding of the non-catalytic functions of histone modifiers. Histone-modifying enzymes have 'moonlighting' activities in transcriptional regulation independent of their catalytic activities. The development of CRISPR-Cas9 technology now allows for the rapid mutagenesis of multiple genes, a crucial advantage when examining the many potentially redundant, closely related modifiers in mammalian cells. Thus, emerging evidence that histone modifications have only minor or context-specific roles in transcriptional regulation should not diminish their potential biological importance but, instead, should guide chromatin research into new areas such as cell-cycle control, DNA damage responses, and genome replication or other chromatin templated processes.

epigenetic mutations: Aberrant methylation levels that can lead to unusual gene expression.

epigenetic patterns: The patterns involving spatial arrangement of tissues and organs are formed through a complex network of reactions and interactions influenced by genes.

epigenetic phenomenon: A stable and heritable (through cell divisions) change in gene expression that is independent of DNA sequence changes and is, in principle, reversible. *See* EPIGENETIC MODIFICATIONS.

epigenetic profiling: Revealing aetiologically distinct patterns of epigenetic modifications related with a phenotype.

epigenetic quantitative trait loci (epi-QTL): Epigenetic variants that are quantitative in nature, are associated with a trait and do not have any change in the DNA sequence.

epigenetic readers: Protein complexes and domains capable of detecting the precise temporal and spatial patterns of chromatin modifications and subsequently initiating regulatory processes based on these patterns. Readers are capable of initiating or silencing transcription, DNA repair and other vital processes. Some readers are the products of tumour-suppressor genes. Epigenetic readers produce a reversible change in gene expression (increasing or decreasing gene transcription). *Aka* CHROMATIN READERS.

epigenetic recombinant inbred line (epi-RIL): A quasi-homozygous line that is almost identical at the genetic level but that segregates at the DNA methylation level. EpiRILs are classically obtained by crossing parents with hypomethylated genomes (mutants in genes necessary to maintain DNA methylation) with isogenic wild-type parents. Plants of the F_1 progeny are genetically identical but have hybrid epigenomes. After outsegregation of the mutation in the F_2 generation, plants are selfed across seven to eight generations, which results in a population with wild-type genotypes but mosaic epigenomes.

epigenetic recombinant inbred lines in crop improvement: Construction and investigation of EPIGENETIC RECOMBINANT INBRED LINES (EPI-RILS) using *Arabidopsis thaliana* drastically altered the rate at which epialleles could be produced and related to phenotypic traits.

epigenetic regulation: It involves revocable changes in DNA methylation and/or histone modification patterns. Short-interfering RNAs (siRNAs) can guide DNA methylation and heterochromatic histone modifications, resulting in sequence-specific transcriptional gene silencing. For example, H2B deubiquitination by SUP32/UBP26 is needed for heterochromatic histone H3 methylation and DNA methylation.

epigenetic regulation of gene expression: Epigenetic processes, including DNA methylation, histone modifications and small non-coding RNAs, seem to permit an organism to reply to the environment through changes in gene expression. Epigenetic regulatory processes do not involve changes in DNA sequence.

epigenetic regulation of photoperiodic flowering: The DNA demethylating reagent promotes flowering in *Perilla frutescens*, which shows that epigenetic mechanisms control photoperiodic flowering. The coevolution of *VIN3* and *FLC* families results in the proper epigenetic regulation of vernalization. The imprinted memory in the plant epigenome, accompanied by the condition-dependent reversible alteration of chromatin architecture, regulates flowering time mechanism. The epigenetic mechanisms, including diverse transcription factors and chromatin modifiers, regulate the vernalization response, photoperiodic control and flowering genes including *FLC* and *FT* in response to seasonal signals in the *Arabidopsis thaliana*.

epigenetic regulation of stomatal development: P.J. Tricker and others in 2012 have suggested epigenetic regulation of stomatal development that allows for anatomical and phenotypic plasticity and may help to explain at least some of the plant's resilience to fluctuating relative humidity. *Arabidopsis thaliana* 'Landsberg erecta' plants grown under low relative humidity have a reduced stomatal index and that two genes in the stomatal development pathway, SPEECHLESS and FAMA, become *de novo* cytosine methylated and transcriptionally repressed. These environmentally induced epigenetic responses were abolished in mutants lacking the capacity for de novo DNA methylation, for the maintenance of CG methylation and in mutants for the production of SHORT-INTERFERING NON-CODING RNAs (SIRNAs) in the RNA-directed DNA methylation pathway.

epigenetic reprogramming: There must be an epigenetic reprogramming mechanism by which cells in some of C.H. Waddington's valleys (those that result in sperm and eggs) can return to the beginning. There is some evidence of epigenetic reprogramming in action in early embryos and in gametes. Sperm DNA experiences a drastic loss of DNA methylation only hours after fertilizing an egg, and before its DNA is replicated, but the egg-derived nucleus remains highly methylated. Further demethylation and reorganization of histone modifications take place as cells in the early embryo divide, before the first differentiation steps begin. After that, and concomitant with differentiation, high levels and cell-type-specific patterns of DNA methylation and histone modifications are acquired. Epigenetic reprogramming results in the resetting of epigenetic tags so they look like those of other cells from earlier developmental stages. This point is of particular relevance for germline cells after the fusion of gametes, when the genome is brought back into a kind of 'zero-state' of gene expression. Two major epigenetic reprogramming events occur in mammalian development during gametogenesis and early embryogenesis.

epigenetic reprogramming, *in vitro* reconstitution of: Epigenetic reprogramming reconstitutes parental epigenetic memories and differentiates PRIMORDIAL GERM CELLS (PGCs) into mitotic pro-spermatogonia or oogonia, guaranteeing sexually dimorphic germ-cell development for totipotency. Y. Murase and others (2024) launch a vigorous approach for persuading epigenetic reprogramming and differentiation of pluripotent stem cell (PSC)-derived human PGC-like cells (hPGCLCs) into mitotic pro-spermatogonia or oogonia, united with their extensive amplification (~>1010-fold). Bone morphogenetic protein (BMP) signalling is a main driver of these processes. BMP-driven hPGCLC differentiation comprises an attenuation of the mitogen-activated protein kinase/extracellular-regulated kinase (MAPK/ERK) pathway and both *de novo* and maintenance DNA methyltransferase (DNMT) activities, possibly promoting replication-coupled passive DNA demethylation. On the other hand, hPGCLCs deficient in 10–11 translocation methylcytosine dioxygenase 1 (TET1), an active DNA demethylase abundant in human germ cells, differentiate into extra-embryonic cells, including amnion, with de-repression of crucial genes bearing bivalent promoters; these cells do not fully activate genes crucial for spermatogenesis and oogenesis, with their promoters remaining methylated.

epigenetic signals: As long as a transcriptional response is self-sufficing in the absence of the initiating stimulus, it can be characterized as epigenetic. Epigenetic signals are accountable for the institution, preservation and revocability of metastable transcriptional states that are primary for the cell's ability to 'remember' past events, such as changes in the external environment or developmental cues.

epigenetic signals *in cis*: Molecular signs that are physically linked with the DNA sequence they regulate.

epigenetic signals *in trans*: Self-proliferating *trans*-acting mechanisms.

epigenetic silencing: It involves chemical modifications to DNA and the histone proteins that remodel the chromatin surrounding DNA, rendering genes inaccessible to transcription-related proteins. Epigenetic silencing also targets 'transposons', genetic elements that can move around the genome. Both paramutagenic alleles and transposons contain tandem or inverted repetitive DNA sequences. Such repeats can induce heritable silencing when they trigger the production of double-stranded RNAs, which are then processed into small interfering RNAs that can inactivate genes through DNA methylation and other mechanisms. For example, embryonic cells silence transcription by retroviruses, but how? They recognize viral DNA. A TRIM28 corepressor complex binds to the retrovirus promoter-binding site. Epigenetic silencing of retrovirus transcription is achieved by 'writing' a dimethyl mark on lysine 9 of histone H3 (H3K9me2) that is read by the heterochromatin protein HP1γ.

epigenetic stability: A broader phenomenon that incorporates maintenance of phenotypic traits in both non-replicating and replicating cells independent from DNA sequence.

epigenetic state: Genes, transposable elements, heterochromatin and so on are marked by a particular combination of epigenetic marks that define their epigenetic state.

epigenetic stress memory: Epigenetic mechanisms are indispensable for stress memories and adaptation in plants. Plants adapt to environmental stress by developing the ability to 'remember' a stress episode and to react more efficiently (faster and more strongly) upon subsequent exposures to stress, that is, with increased resistance. Two types of 'memories' seem to coexist: (i) at the level of a population, multiple lines of evidence point towards a transgenerational transmission of the stress memory to the progeny and (ii) at the level of the plant lifetime, short-term memory (*aka* PRIMING), persisting for days, can also increase the potential of plants to respond to environmental challenges. Epigenetic stress memory provides the necessary information for the cellular machineries to assume the right epigenome at the right time under stress conditions. Targets of RNA-dependent DNA methylation (RdDM), such as retrotransposon LTRs, can be derepressed by different types of stress and provide promoter/enhancer activities for adjacent plant genes.

epigenetic switches: Small-molecule probes may not have full drug-like properties, but their communications with target proteins can help to understand the relationship between epigenetic marks and the underlining genes that are being turned on or off.

epigenetic templating: By preferentially modifying hemi-methylated DNA, DNA methyl transferase DNMT1 transmits patterns of methylation to a newly synthesized strand after DNA replication and, therefore, is often referred to as MAINTENANCE METHYLATION. To emphasize the difference between this molecular mechanism of inheritance and the canonical Watson-Crick base-pairing mechanism of transmission of genetic information, the term epigenetic templating was coined.

epigenetic therapy: Refers to the use of drugs or other epigenome-influencing techniques to treat medical conditions. Many diseases (e.g. cancer, heart disease, diabetes and mental illnesses) are influenced by epigenetic mechanisms, and epigenetic therapy offers a potential way to influence those pathways directly. For example, chemical compounds, such as DNA methyltransferase inhibitors (e.g. 5-azacytidine, 5-aza-2′-deoxycytidine), to target epigenetically regulated mechanisms in patients.

epigenetic tools: Data summarization and visualization is extremely important where quantity of data becomes overwhelming, particularly for sequencing approaches.

epigenetic trait: A distinct feature of a developing or developed individual that has appeared as a consequence of mechanisms regulating the expression of the genes rather than differences in the gene sequence. Epigenetic variation is likely to contribute to the natural variation. Epigenome analysis helps explain how natural epigenetic variation causes phenotypic differences in plants. Stable inheritance of complex traits such as flowering time has been noticed in the EPI-RECOMBINANT INBRED LINE (EPI-RIL) populations, thereby providing important evidence that epigenetic variation can contribute to simple as well as complex traits.

epigenetic transgenerational inheritance: Germline-mediated inheritance of epigenetic information between generations in the absence of direct environmental influences that leads to phenotypic variation.

epigenetic variance, performing cooperative structure with genetics: AFLP and MSAP techniques, and association analysis and sequencing were used to survey epigenetic variation and its association with genetic variance, environment factors and traits in about 100 accessions of ornamental tree *Prunus mume*.

epigenetic variation: Heritable variation that is not a consequence of differences in primary DNA sequence. Epigenetic variation has a non-genetic basis. Differences in the DNA methylation, histone modifications (and variants) and small RNAs are often related with epigenetic variation but can be impacted by genetic variation or development/environment.

epigenetic variation and phenotypic diversity in plants: The epigenetic variation is responsible for phenotypic diversity in plants followed by transmission to offspring for numerous generations leading to epigenetic inheritance in the form of epialleles. Environment-dependent chromatin modifications lead to the development of epigenetic changes in gene function.

epigenetic variation contributes to complex traits: Epigenetic variation contributes to complex traits through the establishment of stable inheritance of complex traits such as flowering time, as investigated in epi-RIL populations. The presence of two divergent genomes and epigenomes in the same nucleus has the perspective towards alteration in gene expression and the phenotype of an organism. Epigenome is clearly a possible contributor to heterosis, especially in intraspecies hybrids with increased vigour and limited genetic diversity. The interaction between the epigenomes contributed by the two parental sets of chromosomes results in altered gene expression in F_1 hybrids. This approach has the potential to discover the epigenetic features of the chromatin inheritance system to study the epigenome associated with sequence-dependent and sequence-independent regions of variation and constancy.

epigenetic variation exploitation: Because the epigenome is dynamic, it remains a difficult task to portray active epigenomes during different growth stages and in response to various environmental stimuli in plants. In spite of some preliminary investigations on the function of the epigenome or different regions of the epigenome in plant growth, the machineries for the interaction between the epigenome and signal transduction networks, such as light and plant hormones, are distant from being adequately explained. How epigenomic information becomes available between different cell types and transferred to the next generation during plant development merits thorough examination.

epigenetic variation exploration: Occurrence and distribution of epigenetic marks is associated with eukaryotic developmental progressions and specifically recognise where this may be pertinent to crop improvements both in terms of specific growth stages and in relation to functional responses to ecological alteration.

epigenetic writers: Enzymes that add chemical groups to DNA, RNA and histones. *Aka* CHROMATIN WRITERS.

epigenetically modified organisms (EMOs): (i) The organisms produced by making changes in the epigenome by changes in control of gene expression whose bases lie not in the changes in the DNA sequence. (ii) The organisms produced by introducing epigenetic modifications through the use of epigenome-editing technology.

epigenetically inherited chromatin domains: Eukaryotic DNA is organized into structurally distinct domains that regulate gene expression and chromosome behaviour. Epigenetically heritable domains of heterochromatin control the structure and expression of large chromosome domains and are required for proper chromosome segregation. Recent studies have identified many of the enzymes and structural proteins that work together to assemble heterochromatin. The assembly process appears to occur in a stepwise manner, involving sequential rounds of histone modification by silencing complexes that spread along the chromatin fibre by self-oligomerization, as well as by association with specifically modified histone amino-terminal tails. Finally, an unexpected role for non-coding RNAs and RNA interference in the formation of epigenetic chromatin domains has been uncovered. DDM1 encodes a protein similar to the chromatin remodelling factor SW12/SN2, and there is other evidence suggesting an intimate link between DNA methylation and chromatin structure; thus, if DNA methylation really is part of a host control system, it seems likely that chromatin must be implicated too. Intriguingly, it has been suggested that other epigenetic phenomena, including genomic imprinting in placental mammals and X chromosome dosage compensation, may themselves have evolved from such host defence mechanisms directed against parasitic DNA.

epigenetics: Epigenetics deals with cellular, physiological and phenotypic trait variations caused by external or environmental factors that, by adding or removing epigenetic tags on DNA, histones and/or RNA affect how cells express genes, as opposed to those caused by changes in the DNA sequence. In some cases, epigenetic modifications are stable and passed on to future generations, but in other instances, they are dynamic and change in response to environmental stimuli. Epigenetics deals with the following. (i) The study of factors outside of the genetic code that affect how the code (called DNA) is expressed. All cells in the body contain an almost identical set of genes, which are specific stretches of DNA that act as a 'recipe' for building an organism. The epigenome tells these genes when to switch on or off and to what extent, altering how genes are expressed and, ultimately, the physical traits that result from this expression. (ii) Changes in gene expression or cellular phenotype that are transmitted during mitosis and/or meiosis or that are stably maintained in cells in the absence of any triggering signal and are not due to a change in the underlying DNA sequence. (iii) Modifications of DNA or chromatin proteins that alter the ability of DNA to respond to external signals. Epigenetic processes are involved in the unfolding development of an organism. (iv) The study of biochemical modifications to the genome, either to the DNA itself through DNA methylation, or to the DNA packaging proteins called HISTONES, to regulate gene expression. The term most commonly refers to HERITABLE CHANGES in gene expression that do not result from changes in actual gene sequences. The word 'heritable' here refers not only to inheritance between parents and offspring via meiosis, but also between parent and daughter cells via mitosis. Nearly every aspect of biology is influenced by epigenetics, making it one of the most important fields in science.

epigenetics and photosynthetic efficiency: Epigenetic regulation of the activity of genes involved in chlorophyll biosynthesis and photosynthetic efficiency has yet not been explored. Epigenetic analysis would prepare molecular plant breeders with highly useful information on epigenetic regulation of the photosynthetic process.

epigenetics compared with artificial selection: Epigenetics is less time consuming as compared to artificial selection. The methylated regions have the potential to map genomic regions controlling the trait of interest. The epigenetics associated with the trait provides avenues for the identification of the defective clones using DNA methylation profiling. The perturbation of DNA methylomes of crop plants results in the creation of novel epiallelic variation in plant species. The first site-specific epigenome editing in plants was performed in *Arabidopsis thaliana* using a **Zinc finger nuclease (ZFN)** fused to SUVH9 protein, which is integral to RdDM. Due to the stability of epigenetic variants in epigenetics, it would be readily acceptable for the stable improvement of agronomic traits.

epigenetics databases content: MethDB (DNA Methylation Database), PubMeth (Pubmeth.Org), REBASE, MeInfoText, MethPrimerDB, The Histone Database, CREMOFAC, The Krembil Family Epigenetics Laboratory and MethyLogiX DNA methylation database.

epigenetics of ageing: Ageing is related with reflective epigenetic changes, ensuing alterations of gene expression and conflicts in broad genome architecture and the epigenomic landscape.

epigenetics on precision treatment and medicine: Epigenetic dysregulation is now a hallmark of several complex pathologies, including cancer, metabolic disorders, cardiovascular and neurological diseases. Disease-specific epigenetic signatures (e.g. DNA methylation, hydroxymethylation and non-coding RNAs) are now being utilized clinically for prognostics and diagnostics, while an expanding collection of genetically aberrant, abnormally expressed or chromatin-interacting epigenetic enzymes is positioned as a promising target for therapeutic intervention.

epigenetics with long-read sequencing: The unique chemistry of long-read HiFi sequencing technology reveals the epigenetic landscape of samples. On-instrument 5-base HiFi sequencing detects 5-methylcytosine ($5^{m}C$) methylation in standard sequencing runs without any changes to library prep or sequencing workflows required.

epigenome: A form of the genome that has been modified, without alteration in DNA sequence, by changes like DNA methylation, histone modifications, small interfering RNAs (siRNAs) that are guiding *de novo* DNA methylation, nucleosome occupancy and/or histone variants. The epigenome consists of chemical compounds that alter, or mark, the genome in a way that tells it what to do, where to do and when to do. Epigenomes of the same individual can differ among different cell types. Thus, an organism can have several epigenomes, but only one genome. The epigenome controls the differential expression of genes in specific cells. Changes to the epigenome can result in changes to the structure of chromatin and changes to the function of the genome. The epigenome provides the complete portrayal of these potentially heritable changes across the genome. The composition of the epigenome within a given cell is a function of genetic determinants, lineage and environment.

epigenome changes: All cells in the body encompass in essence the same genome. However, the chemical tags on DNA and histones get reorganized when cells become specialized. The epigenome can also alter during a person's lifetime. Lifestyle and environmental factors can expose a person to stresses that stimulate chemical responses in DNA and histones.

epigenome comparison: (i) A pairwise comparison of two epigenomes to determine loci with specific differences and similarities in any of the epigenomic components. (ii) a comparison of an individual epigenome against a database such as the Human Epigenome Atlas to determine the most similar epigenomes. (iii) any multiway comparison, such as clustering of epigenomes to determine their mutual relationship.

epigenome death match: A defensive barrier is suggested to be formed by H3K4me3 and CCCTC-binding factor (CTCF) enrichment, with H3K4me3 (typically representative of euchromatin) even pushing into the endogenous retrovirus element itself.

epigenome diversity: Epigenetic variation can substitute genetic diversity in the lab. Using epimutations, the range of phenotypes generated can indeed be enormous. A phenotype is the physical manifestation of a gene. There is good potential for creating epigenome diversity.

epigenome diversity in crop breeding: Plant breeders have been considering genetic diversity or genetic variation as the most significant donor of phenotypic variation and have been innocently neglecting yet an unexplored rich wealth of epigenetic variation present in the crop populations for a number of traits.

epigenome dynamics: New experimental strategies and analysis models care used to apprehend the theoretically dynamic interaction between chromatin and DNA sequence factors in complex traits. Two views of epigenome dynamics include static view and dynamic view.

epigenome editing: A type of engineering in which the epigenome is modified at specific sites using engineered molecules targeted to those sites. TRANSCRIPTION ACTIVATOR-LIKE EFFECTOR (TALE)- or CLUSTERED REGULARLY INTERSPACED SHORT PALINDROMIC REPEATS (CRISPR)-based genome-editing technologies have been explored beyond genome editing to develop genome-editing systems. Epigenome-editing systems rely on the targeting capabilities of a genome-editing system but are modified to have the nuclease inactivated, so it is incapable of cleaving DNA but can still reach its target. The novel functionality is given by a fusion of the altered genome-editing system with an effector domain (e.g. TET1, LSDA1, HAT, HADC), which is a modular domain that gives the system the power of a transcription factor or epigenome modifier. Epigenome editing with the CRISPR-Cas9 platform is a promising technology for modulating gene expression to direct cell phenotype and to dissect the causal epigenetic mechanisms of gene regulation. *Aka* EPIGENOME ENGINEERING.

epigenome editing linked to pathology and therapeutics: Scientific community endures to document and expand the knowledge related to the extensive catalogue of pathologies associated with mis-regulation of the epigenome, which arise as a consequence of abnormal patterns of DNA and histone epigenetic modifications, as well as defects in regulation by ncRNAs.

epigenome editing, *in vitro* tissue culture based: Epigenetics is a highly sensitive regulatory mechanism largely impacted by the environmental factors. During plant development, the genome is constant as it shows little change. Epigenetic modifications are highly sensitive to *in vitro* conditions. Under varied growth conditions, epigenetics plays an important role in the final outcome of a phenotype.

epigenome editing, prerequisites for: If the objective of using epigenome-editing tools is to produce universal or arbitrary modifications in the epigenomic landscape, then random approaches to epigenome editing are engaged, but if the objective is to produce modification(s) at a specific locus or region (e.g. promoter, enhancer, silencer and insulator), then precise methods of epigenome editing are required.

epigenome functions: Three types of investigations: (i) The epigenome helps control how genes are switched on and off, (ii) epigenome effects on genome structure and (iii) epigenome effects plant genome evolution.

epigenome isoform analysis (EIA): EIS systematically discovers and analyses coordination patterns using a hierarchical latent mixture model. The *cis*-regulatory element isoforms revealed by EIA provide a brief but biologically meaningful description of the highly complex epigenome landscape.

epigenome mapping initiatives: Rapid progress of experimental technologies provoked several epigenome mapping initiatives, such as The Encyclopaedia of DNA Elements (ENCODE) project, Epigenomics of Plants International Consortium (EPIC), Centre for Epigenome Mapping Technologies (CEMT) project and Plant Regulomics Portal (PRP).

epigenome mapping: Process of diagramming the positions of different epigenetic modifications (such as DNA methylation, histone modifications, non-coding RNAs) across the genome of an organism, DNA sequencing-based mapping technologies, such as ChIP-seq, Bisulfite-seq, MINT-Chip, RNA-seq and chromatin conformation, capture assays are used.

epigenome mapping using next-generation sequencing (NGS): DNA methylation and histone modification using NGS technologies have led to new discoveries and thus enhanced our familiarity with human biology and diseases.

epigenome modifiers, dietary agents as: Numerous bioactive dietary components are of special interest in the field of human epigenetics. A number of nutritional compounds have epigenetic targets in cancer cells. Emerging evidence strongly advocates that consumption of dietary agents can modify normal epigenetic states as well as reverse abnormal gene activation or silencing. Epigenetic modifications influenced by bioactive dietary compounds are suggested to be advantageous. Considerable evidence is accumulating to show that normally consumed bioactive dietary factors play a role in altering the epigenome and may be incorporated into an EPIGENETIC DIET. Dietary polyphenols (i.e. tea polyphenols, resveratrol, curcumin, isoflavones (genistein) and isothiocyanates) can meaningfully modify the epigenome of cancer cells and are regarded as attractive possibilities for anticancer therapeutics.

epigenome prediction: Bioinformatic research has been dedicated to the estimation of epigenetic information. An extensive volume of bioinformatic research has been dedicated to the prediction of epigenetic information from the characteristics of the genomic sequence. Promoter prediction is viewed as the first attempt to forecast epigenetic states from the DNA sequence. CGI prediction has an approximate overlap with promoter prediction, because the majority of promoters in mammalian genomes co-localize with CGIs. DNA methylation prediction is theoretically simpler than the prediction of more volatile epigenetic mechanisms because DNA methylation patterns reveal relatively low tissue specificity compared to other epigenetic information. Forecast of nucleosome positioning is based on the reasoning that the sequence composition of DNA molecules strongly marks their nucleosome affinity.

epigenome-editing systems, chemical and light inducible: Chemically inducible epigenome editing includes the use of small molecules that function as chemical inducers. For example, rapamycin-sensitive dimerization domains have been demonstrated to chemically induce CRISPR-Cas9-based genome editing and transcription regulation when Cas9 is torn apart into two fragments that, upon chemical induction, reconstruct a full-length catalytic nuclease.

epigenome-editing technologies, overcoming challenges faced in: The Duke researchers engaged themselves in histone acetylation by fusing dCas9 with a HAT domain. The scientists were in a position to induce acetylation of the histone at a particular locus. This provides researchers with unparalleled control over the workings of specific enhancers or gene promoters. One talented and alluring technique to amend the problem of antibody specificity and reproducibility that is often faced in epigenetic experiments is the practicability of using recombinant epigenetic readers as reliable substitutes for histone PTLM antibodies.

epigenome-editing technology in determining the function of specific methylation sites: Functional role that specific methylation sites perform in controlling gene expression can be deciphered by using a TALE-TET1 fusion protein to demethylate a single CG methylation site. This strategy necessitates many controls to guarantee specific binding to the target loci.

epigenome-editing technology in determining the role of epigenetic modifications directly: An optogenetic two-hybrid system integrates the sequence-specific TALE DNA-binding domain with a light-sensitive cryptochrome 2 protein (CIB1). Once expressed in the cells, the system could inducibly edit histone modifications and control their function in a specific context.

epigenome-editing technology induced phenotypic changes: DNA methylation changes using epigenome-editing technology to target DNA sequences have resulted in expected phenotypic consequences.

epigenome-editing technology, applications of: Epigenome-editing technology have been used to produce changes in DNA or histones with regard to (i) phenotypic change, (ii) targeting chromatin-modifying enzymes, (iii) production of variety in endogenous gene expression, (iv) studying enhancer function and activity, (v) determining the function of specific methylation sites, (vi) augmenting gene expression in response to drought stress and (vii) defining the role of epigenetic modifications directly.

epigenome-editing technology, challenges in the use of: The major challenges of editing technologies include their editing efficiency, delivery and cytotoxicity. The restrictions flanking substrate specificity of HAT and other epigenetic effectors pose an immense hindrance to the development of effective and programmable CRISPR-mediated epigenome-editing tool.

epigenome-editing technology, limitations of: Numerous thresholds of epigenome-editing technology, such as poor specificity of histone modifiers towards histone substrates, off-target effects, non-specific hyperacetylation of nucleosomal substrates and functions of other than transcriptional repression. With respect to the claim that highly specific epigenome editing can be accomplished across promoters and enhancers genome-wide, the data presented by the CRISPR-dCas9-p300Core experiment verify only targeting of desired DNA loci, not the deposition of unique epigenetic marks on histone tails.

epigenome-editing tool, principal components of an: A DNA-BINDING DOMAIN (DBD) binds to a defined DNA sequence and aids in the recruitment of FUNCTIONAL CATALYTIC DOMAIN (CD) or EFFECTOR DOMAIN (ED) to the demarcated target loci in the genome, where it can position itself and modify the chromatin state, and thereby modify the epigenetic state of DNA or histones, depending upon the specificity of the DNA-BINDING DOMAIN (DBD) and nature of the ED. In a genome/epigenome-editing tool, DBD is either directly attached to ED or joined indirectly to ED through a 22 amino acid linker peptide. A DBD binds to a defined DNA sequence and aids in the recruitment of functional CD/ED to the demarcated target loci in the genome, where it can position itself and alter the chromatin state, and thereby modify the epigenetic state of DNA or histones, depending upon the specificity of the DBD and the nature of the ED.

epigenomic diversification within the genus *Lupinus*: Several instances of epigenetic changes exemplify the diversity of lupin genomes, which could be valuable to explain further epigenetic changes in the evolution of the lupin genome.

epigenomic diversity in *Arabidopsis*: More than 1000 high-quality single-base resolution methylomes (and in higher number transcriptomes) were constructed from the collection of 1001 genomes of *Arabidopsis thaliana*. Although the genetic basis of methylation variation is highly intricate, geographic origin is a major predictor of genome-wide DNA methylation levels and of modified gene expression triggered by epialleles.

epigenomic landscape survey: Technological advances have permitted base-pair resolution of various epigenomic features, resulting in novel insights into epigenetic control of gene expression. Many approaches, such as MNase-seq based methods, DNase-seq based methods, ChIP protocols, Formaldehyde-assisted isolation of regulatory elements (FAIRE) and Sono-seq for single-base-pair resolution mapping, have been put forth, which have empowered probing numerous features of the epigenome.

epigenomic map: Diagrammatic illustration of the gene expression, DNA methylation and histone modification, and small RNA status of a particular genomic region or genome.

epigenomic modification compartmentalization in plant genomes: Plant genomes are decidedly compartmentalized, with distinct regions of the genome related to discrete patterns of epigenomic modifications. For example, TEs and other repetitive sequences are richly DNA methylated in all three sequence contexts (i.e. CpG, CpHpG and CpHpH, where H is A, T or C) and are also targeted by the siRNA and H3K9me2 pathways.

epigenomic studies, future prospects of: Epigenetic regulation is thought to be another coating of genetic control of the complex traits that are predisposed by an environmental stimulus. Unlike other regulatory mechanisms, many of the epigenetic modifications may be remembered/inherited over time/generations as epigenetic memory.

epigenomic studies, rice as a model plant for: Recent results on rice epigenome profiling and three-dimensional chromatin structure studies reveal specific features and implications for gene regulation during rice plant development and adaptation to environmental changes. Results on rice chromatin regulator functions shed light on mechanisms of establishment, recognition and resetting of epigenomic information in plants. The cloning of several rice epialleles associated with important agronomic traits highlights the importance of epigenomic variation in rice plant growth, fitness and yield.

epigenomics: The study of epigenetic marks, such as DNA methylation, PTLMs and small RNAs of chromatin, on a genome-wide scale. It also deals with the study of complex expression networks or linkages both spatially (within the body) and temporally (at different times in development).

epigenomics essential data sets: Essential genome-wide data sets include the following categories. (i) Deep-sequence data identifying all RNA transcripts, including protein-coding mRNAs, lncRNAs and sncRNAs (e.g. microRNAs and siRNAs). (ii) Positions and frequencies of methylated cytosines, investigated at single-nucleotide resolution, genome-wide. (iii) Positions of nucleosomal histones showing informative modifications, including, but not limited to, histone marks typical of silent chromatin. For example, histone H3 that is monomethylated, dimethylated or trimethylated on Lys9 (abbreviated as H3K). (iv) Positions of histone variants. (v) Positions of DNase-hypersensitive sites, corresponding to likely gene regulatory regions. (vi) Positions of nucleosomes, which can make known nucleosome-free or nucleosome-depleted regions that are hotspots of gene regulation.

epigenomics in *Brassica* improvement: Treatment with the demethylating agent 5-azaC produced a range of developmental changes at a high frequency in *Brassica oleracea*.

epigenomics in climate-smart crop breeding: Modifications of chromatin states comprise a wide range of DNA methylation, histone PTLMs, histone variants and activity of non-coding RNAs, which can epigenetically administer specific transcriptional outputs.

epigenomics in cotton improvement: Complete epigenomic and functional analyses were done on nearly 12 million differentially methylated cytosines in domesticated allotetraploid cottons and their tetraploid and diploid kin. To improve cotton, the barricade of restricted epigenetic variation in plants is certainly required to be overcome through some unconventional technology. After conventional plant breeding, mutational and targeted genome-editing approaches, epigenome editing holds great promise for crop improvement.

epigenomics in maize improvement: Work on maize in progress to determine how stress-induced epigenetic variation can be useful in enhancing stress tolerance and improving yields of crops.

epigenomics in tomato improvement: The Cnr phenotype in tomato appears due to a spontaneous epigenetic change in the SQUAMOSA promoter-binding protein (SBP)-box promoter.

epigenomics of plants' responses to environmental stress: Genome-wide epigenetic changes in plants during development and environmental stresses, which are often associated with gene expression at the transcriptional level, are under investigation. The sum total of the biochemical changes in nuclear DNA, PTLMs in histone proteins and variations in the biogenesis of non-coding RNAs in a cell constitutes an epigenome. These changes frequently produce variations in the expression of the gene. These alterations might also cause variations in chromatin structure, thereby resulting in changes in function/activity of the genome.

Epigenomics of Plants International Consortium (EPIC) project: EPIC project intends to decipher and control the 'second code' (i.e. 'epigenetic code' of biology). A key step towards this ambitious goal is the CoGe Epigenomics Browser, a web-based comparative genomics system that offers access to 20,000 genomes from 15,000 organisms, and users can take advantage of over 30 tools for the analysis, comparison and visualization of genomic data from the scale of whole genomes to individual nucleotides.

epigenomics research, challenges in: (i) Investigating the critical influence of the microbiome on health and disease. (ii) Studying genome-epigenome interaction involved in the biology, character and extent of an established disease and not just in disease development. (iii) Studies on the impacts and the mechanistic influence of epi-drugs and their effect on the genome, development and validation of new epigenetic drug candidates and rational design of combination therapies of genetic and epigenetic drugs. (iv) Detecting functional associations between epigenetics and genetics necessitates the analysis of *ex vivo* samples of primary cells, and therefore, the sampling, sorting and analytical procedures require to be optimized and adapted. (v) Performing data analysis by integrating data from different repositories is problematic because of the different approaches employed to obtain the data sets; (vi) Integrating the findings from omics research into clinical practice. (vii) Improving partnerships need to be fostered by the establishment and harmonization of typical operating procedures for sample processing, data acquisition and formatting. (viii) Development of software that is user-friendly.

epigenotype: (i) The epiallelic constitution of a locus. (ii) Includes the specific set of epigenetic marks peculiar to different cells, which determines their fate (i.e. whether they remain germline or will be brain or bone cells). (iii) The pattern of epimutations in an individual or in a cell.

epimerization: The reaction that alters the steric arrangement around one atom, as in a sugar molecule.

EpiMix: An integrative tool for epigenomic subtyping using DNA methylation. It detects abnormal DNA methylation (DNAme) present in small patient subsets. It enables DNAme analysis of distal enhancers and non-coding RNAs (Y. Zheng and others in 2023). Non-coding RNAs regulated by abnormal DNAme are biomarkers for lung cancer prognosis. The process is called EPIGENOMIC SUBTYPING.

epimutant: Any organism formed by epimutation.

epimutation: (i) A heritable change in phenotype that is not the result of an altered DNA base sequence but due to a change in the chemical structure of DNA. (ii) A specific loss of methylation (hypomethylation) or gain of methylation (hypermethylation) at an imprinting control centre without any change in DNA sequence. This alters the expression of imprinted genes on the same allele. (iii) Environmentally induced differential presence of epigenetic alterations that can lead to altered genome activity when compared with organisms not having the exposure. (iv) Epigenetic modification that results in heritable or stable changes in gene expression (activation of a silent gene or silencing of an active gene) but is not associated with modifications to the underlying DNA sequence. (v) Alterations to normal epigenetic marks that can be passed down from parents to offspring via their germline. (vi) Changes in the structure of amino acids in histone proteins (*aka* EPIMUTATIONS). (vii) A heritable change in gene activity that is not due to a change in base sequence. (viii) The term is generally applied to an abnormal change in gene activity. Note that, in development, normal changes in gene activity occur, and these are often somatically inherited. Such changes are not epimutations.

epiphenomenon (pl. epiphenomena): Epiphenomena are those processes that ostensibly have no precedent at lower levels of scientific organization. An epiphenomenon is a secondary phenomenon that is a by-product of and accompanies another phenomenon. In this review, it is argued that many genetic processes, including ploidy, dominance, heritability, pleiotropy, epistasis, mutational load and recombination, all are at least analogous to biochemical events that were requisite features of the RNA world. Most, if not all, of these features of contemporary whole organisms, and populations may have their ultimate evolutionary roots in the chemical repertoire of catalytic RNAs. Some of these phenomena will eventually prove to be not only analogous but also homologous to ribozyme activities.

epiphyte: A plant, alga or fungus that grows on another plant without deriving nourishment from it but just using it for support.

epiphytic: Living on the surface of plants, but not as a parasite. *Compare with* EPILITHIC *and* LITHOPHYTIC.

epiphytotic: An epidemic in a plant population. *See* EPIDEMIC.

epiproteome, applications of: Any given genome can potentially result in an infinite number of proteomes. A single gene can give rise to multiple gene products. One of the most important applications of proteomics will be the characterization of PTLMs.

epiproteome: The term epiproteome was first coined by D.B. Dai and T.P. Rasmussen in 2007 to refer to proteomic changes directly associated with epigenetic modifications, namely histone acetylation. The term currently describes the set of all PTLMs made to the proteins comprising a cell or organism.

epi-QTL mapping in crops: The epi-QTL mapping has the potential to detect the effect of epigenetic loci linked with phenotypic variations to target trait of interest. It has a vital role in unravelling the complex genetic and epigenetic regulatory network of candidate genes that control the intricate characteristics. DMRs that act as epi-QTL are also responsible for 60–90% of the heritability for flowering time and primary root length traits in the epi-RIL populations of isogenic *Arabidopsis* lines. Moreover, these epi-QTLs that segregate according to differences in DNA methylation are reproducible in nature and are independent of changes in the DNA sequences. Epigeneticists have employed MSAP and retrotransposon epi-markers and identified epi-QTLs linked with various agronomic traits (i.e. yield, seed quality and plant development) in *Brassica napus*.

epiQTLdna: Refers to a QTL influencing chromatin states (epi) in either *cis* or *trans*, which can be demonstrated to be due to DNA sequence (DNA).

episome: Those genetic elements (DNA molecules called plasmids) that have the ability to live independent of the bacterial chromosome (i.e. AUTONOMOUS GENETIC ELEMENT) or as an integral part (covalently inserted) of the main chromosome. The F (fertility) factor is an episome.

epistasis: The non-reciprocal interaction of non-allelic genes. The situation in which one gene makes the expression of another; more specifically, the gene interaction in which one gene interferes with the phenotypic expression of another non-allelic gene so that the phenotype is determined effectively by the former. The latter is described as a hypostatic gene.

epistasy: The property of non-reciprocal conditioning of the manifestation of one gene difference, said to be *hypostatic*, by the action of another, said to be EPISTATIC.

epistatic: A gene whose effect overrides that of another gene with which it is not allelic.

epistatic genes: The genes that suppress expression of other genes.

epistatic interactions: Interactions between two mutations whereby the phenotypic effect of one mutation depends on the presence of another mutation. In this case, one gene modifies the expression of another gene when the two are not alternate alleles.

epistatic QTLs (E-QTLs): E-QTLs are defined as loci at which trait values are determined by interactions between alleles at two or more loci and are detected by associations between trait values and multilocus marker genotypes using epistatic models. Three types of epistatic QTLs are recognized. Type I epistasis is well described in classic quantitative genetics theory, in which two M-QTLs are involved in epistasis and affect the same phenotype. Type II epistasis involves interactions between alleles at an M-QTL and a background (or modifying) locus. Type III epistasis represents epistasis between two complementary loci that do not have detectable main effects.

epistatic variance: What remains unaccounted for by additive and dominance variance components in genetic variance (i.e. failure of the summation of the intra-locus genetic variance to account for the total variation among genotypes).

epithelial-like: Resembling or characteristic of, having the form or appearance of epithelial cells. In order to define a cell as an epithelial cell, it must possess characteristics typical of epithelial cells. Often, one can be certain of the histologic origin and/or function of the cells placed into culture, and under these conditions, one can be reasonably confident in designating the cells as epithelial. It is incumbent upon the individual reporting on such cells to use as many parameters as possible in assigning this term to a culture. Until such time as a rigorous definition is possible, it would be most correct to use the term epithelial-like.

epithelium (pl. epithelia): Articulate cell pane formed from one or more layers of cells covering an external surface or coating a cavity.

epithet: The adjectival element in a binomial scientific name, usually more specifically called a SPECIFIC EPITHET. The final word or combination of words in a name of more than one word (other than a term denoting rank) that denominates an individual taxon. The simplest and commonest example is the second word in a two-word name of a species, such as '*mirabilis*' in *Welwitschia mirabilis*.

epitope: A molecular region on the surface of an antigen capable of eliciting an immune response and of combining with the specific antibody produced by such a response. There may be several epitopes per molecule. *Aka* ANTIGENIC DETERMINANT. *See* ANTIGENIC DETERMINANT.

epitope mapping: Approaches used for investigating the interactions of antibodies with specific regions of protein antigens. Applications of epitope mapping are found within the area of immunochemistry.

epitranscriptome: Includes all the biochemical modifications of the RNA (the transcriptome) within a cell.

epitranscriptome deciphering: The epitranscriptomic changes play vital roles in developmental processes and stress responses. Besides being non-genetically encoded, they mostly have reversible nature and are involved in fine tuning the expression of gene. Advances in chemogenetic RNA-labelling and high-throughput NGS techniques are enabling functional analysis of epitranscriptomic modifications to reveal their roles in different aspects of mRNA biology.

epitranscriptome engineering in boosting agricultural production: The possibility of engineering the epitranscriptome in crop improvement was established where epitranscriptome engineering was engaged to upsurge crop yield.

epitranscriptome in relation to plant growth: The epitranscriptome establishes a crucial and extensive regulatory mechanism fundamental to almost all aspects of plant growth and development, including root development, SHOOT APICAL MERISTEM (SAM) proliferation, leaf growth, trichome development, flowering, sporogenesis, seed and fruit development.

epitranscriptome manipulation for crop improvement: Three major strategies concerning epitranscriptome engineering for crop breeding are the following: (i) manipulating the activities of RNA MODIFICATION-RELATED PROTEINS (RMPs); (ii) eliminating RNA modification on a specific target through precise base editors, such as the adenine base editor composed of the catalytically impaired CRISPR-Cas9 protein and an engineered adenosine deaminase (Kim 2018), which might result in reduced or enhanced function of the target gene(s) useful for agricultural application; and (iii) creating or removing RNA modifications on selected target sites with the catalytically inactivated Cas13 (dCas13)-RMPs.

epitranscriptome mapping: Detection of the primary epitranscriptome can be accomplished with dot blot and high-performance liquid chromatography (HPLC) coupled with mass spectrometry (MS). Dot blot is a simple method to detect the presence of a PROTEIN OF INTEREST (POI) in a sample. With high-throughput sequencing combined with antibodies against a specific type of methylation, the position and richness of RNA modification can be interpreted precisely.

epitranscriptome mapping, challenges in: Antibodies can cross-react with other modifications, so researchers should use at least two different antibodies and cross-correlate the hits. Chemical methods can cut, bind or enrich some areas more frequently than others, thereby generating biased fragmentation patterns. Sequencing depth and choice of bioinformatics algorithms can impact the detection of modification sites. Even the length of time for which cells are kept alive in culture could also influence RNA modification levels.

epitranscriptome mapping, overcoming challenges in: Scientists at Oxford Nanopore Technologies in the United Kingdom established one such method, which extends to RNA the capability to thread through a nanopore embedded in a membrane. Researchers at Pacific Biosciences have also developed a direct RNA sequencing method using SINGLE-MOLECULE REAL-TIME (SMRT) sequencing technology. SMRT sequencing uses DNA polymerase to replicate strands of DNA, capturing the addition of fluorescent nucleotides in real time.

epitranscriptome, applications of: Research in this area deals with drug discovery, dCasRx-based tool for epitranscriptome editing of 6^{m}A modification, epitranscriptome engineering boosts agricultural production and role of epitranscriptome in plant growth.

epitranscriptomics: Epitranscriptomics is composed of many kinds of modification of RNA; some examples include N1-methyladenosine (1^{m}A), pseudouridine (Ψ), 5-methylcytidine (5^{m}C) and N6-methyladenosine (6^{m}A). *Aka* RNA EPIGENETICS.

equational division: The partition of each chromosome during the metaphase of mitosis or meiosis II into two equal longitudinal halves resulting in the formation of sister chromatids, which are dispersed to the two cell poles and then merged into separate daughter nuclei. *Aka* HOMOTYPIC DIVISION.

equatorial plate: A preparation of the chromosomes in which they lie nearly in one plane, at the equator of the spindle. It is seen during metaphase of mitosis and meiosis I and meiosis II.

equilibrium: State where there is no net change in a system. For example, equilibrium is attained in a chemical reaction when the forward and reverse rates are equal, the concentration of all the reactants and products becomes constant, and no further net change occurs.

equilibrium constant (K_{equ}): For a chemical reaction, the equilibrium constant can be defined as the ratio between the amount of reactant and the amount of product, which is used to determine chemical behaviour. At a particular temperature, the rate constants are constant.

equilibrium gene frequency: In a population, gene frequency that varies non-directionally about a mean by an amount described by the standard deviation under conditions of unchanging selection pressure and mutation rate, with no intermixing from other populations.

equivalent reversion: The type of reversion in which the original codon is not restored. Equivalent reversions are of two types. In one type, a different codon coding for the same amino acid is reconstituted. For example, AGG (Serine) mutates to ACG (Cysteine) which reverse mutates to TCC

(Serine). Thus, a same-sense codon is generated. In the second type, a different codon coding for a similar amino acid is reconstituted. For example, reverse mutation produces a mis-sense codon, but the amino acid substituted has a similar charge (e.g. Arginine (AGG) → Lysine (AAG); both amino acids are positively charged) and thus does not affect wild-type activity of the protein.

equivocal result: A test result that cannot be interpreted as either negative or positive.

eradication: The management of plant disease by abolishing the pathogen after it is established or by removing the plants that are infected by the pathogen.

erect: Standing straight, more or less perpendicular to the ground or point of attachment. *Compare with* PATENT (spreading) *and* ERECTO-PATENT (between erect and patent).

ergot: A disease of certain grasses and cereals, especially rye, caused by *Claviceps* spp. and other closely related fungi in infected flowers of parasitized grain plants.

erosion: The corporal removal of rocks and soil through the combined actions of flowing water, wind, ice and gravity.

error variance: Variance originating from unrecognized or uncontrolled influences in an experiment with which the variance of recognized factors is equated in tests of significance.

error, type I: If our NULL HYPOTHESIS (H_0) is correct then the decision to accept it will be correct. If we reject a true null hypothesis, we commit an error called type I error. This type I error is the rejection of a true hypothesis.

error, type II: The error of accepting a false NULL HYPOTHESIS (H_0) (i.e. the error of wrongly concluding that there is no difference when actually some difference is present). This error leads to a false negative result.

error-catastrophe model of ageing: Mutations in the protein-synthesizing apparatus have a cascading effect on a widening series of cell products, causing cell malfunction. If an error occurs in the proteins that are needed for the synthesis of other proteins, a wrong enzyme will synthesize many more wrong proteins before it has degraded itself. Consequently, the activity of the cell gets so slow that it dies due to error catastrophe.

error-prone DNA repair: A type of DNA repair in which DNA chain growth occurs across damaged DNA segments, but some of these new strands are themselves defective. Such error-prone repair appears to be the chief source of genetic mutation in the bacteria in which it occurs.

escape: The failure of characteristically susceptible plants to become diseased, even though the disease is prevalent.

***Escherichia coli* (*E. coli*):** A type of rodlike aerobic bacteria that commonly live in the intestines of most vertebrates. There are many types of *Escherichia coli*. Most pose no harm to human health, but one group produces a potent toxin which can cause food poisoning. Because of its rapid multiplication and best genetic characterization, it is extensively employed in molecular genetics and genetic engineering. One of the more exotic uses of genetically engineered *E. coli* was to make indigo dye (originally discovered in 1983, using indole or tryptophan as starting materials). *E. coli* is also used in many microbiological experiments. It has historically been considered the workhorse of genetic engineering research, and genetically engineered versions have been engaged to produce human proteins (e.g. insulin).

***Escherichia coli* (*E. coli*) Ada:** One of the most mechanistically unusual DNA repair enzymes, which acts by direct reversal of the damage to the target lesion. The N-terminal half (20kDa) of the Ada protein repairs the relatively benign methylphosphotriester lesion and is primarily involved in regulating the expression of *ada* and other alkylation repair genes. The C-terminal half (19kDa) repairs the highly promutagenic O^6-methylguanine, which base pairs promiscuously with cytosine and thymine and which, if not repaired, generates a G-C→A-T transition.

***Escherichia coli* (*E. coli*) genome:** F.R. Blattner and others in 1997 worked out the 4,639,221-base-pair sequence of *Escherichia* coli K-12. Of 4288 protein-coding genes annotated, 38% have no attributed function. The genome as a whole is strikingly organized with respect to the local direction of replication; guanines, oligonucleotides possibly related to replication and recombination, and most genes are so oriented. The genome also contains INSERTION SEQUENCE (IS) elements, phage remnants and many other patches of unusual composition indicating genome plasticity through horizontal transfer.

***Escherichia coli* (*E. coli*) growth on different sugars:** When *Escherichia coli* is cultured on a medium containing both glucose and lactose, only glucose is utilized and during this time the lac operon is inactive. The same situation holds good in case of glucose and galactose where galactose operon remains inactive. The glucose blocks the activity of other operons through cAMP. When glucose is available as the energy source, cAMP level decreases, which prevents the functioning of other operons involved in the metabolism of other sugars (e.g. lactose, arabinose, galactose, maltose). When glucose levels decrease, cAMP levels increase, and other sources of sugar can be used by *Escherichia coli*.

***Escherichia coli-Baciillus subtilis* hybrid replicons, construction of:** Plasmid pBR322 was ligated with pC194 to get hybrid replicons. The beauty of a hybrid plasmid is that the expression of the cloned gene can be tested in either host. Two such hybrid clones are pHV14 and pHV15. These hybrid plasmids do not express tetracycline resistance because the continuity of the promoter for the tetracycline-resistance gene is destroyed. PHV32 is a very useful hybrid plasmid.

***Escherichia coliform* (*E. coliform*):** Gram-negative, facultative anaerobic rod-shaped bacteria commonly found in faeces of warm-blooded animals, in soil and aquatic environments. These bacteria are mostly harmless, but they are used as a hygiene indicator of faecal contamination. High levels indicate unsanitary food processing conditions or poor hygiene practices. For factories that have effective kill steps, they indicate postprocessing contamination.

essential amino acids: Those amino acids that cannot be synthesized by humans and most other vertebrates and therefore must be obtained from the diet. They are phenylalanine, valine, threonine, tryptophan, isoleucine, methionine, histidine, leucine and lysine. The essential amino acids for poultry are arginine, glycine, histidine, leucine, isoleucine, lysine, methionine, cystine, phenylalanine, threonine, tryptophan and valine. Out of these, the ones critical in practical diets are arginine, lysine, methionine, cystine and tryptophan.

essential element: Any chemical element which is essential to the normal growth and reproduction of a plant. There are 16 essential elements, namely, boron (B), calcium (Ca), carbon (C), chlorine (Cl), copper (Cu), hydrogen (H), iron (Fe), magnesium (Mg), manganese (Mn), molybdenum (Mo), nitrogen (N), oxygen (O), phosphorus (P), potassium (K), sulfur (S) and zinc (Zn).

essential gene: A gene whose deletion is lethal to the organism.

essential nutrients: A nutrient necessitated for normal physiological function which cannot be synthesized by a particular organism, either at all or in sufficient quantities, and which therefore should be dutifully supplied from external sources such as food. In humans, a set of nine amino acids (*see* ESSENTIAL AMINO ACIDS), two fatty acids (omega-3 fatty acids and omega-6 fatty acids), thirteen vitamins A, C, D, E, K, B thiamine, riboflavin, niacin, pantothenic acid, biotin, B6, B12 and folate) and sixteen minerals (*see* ESSENTIAL ELEMENTS) are considered essential nutrients.

established cell lines: The eukaryotic cell lines that have been adapted to indefinite growth in culture. *Aka* IMMORTALIZED CELL LINES.

establishing linkage groups: Markers are assigned to linkage groups using the odds ratios. *Odds ratio* is defined as the ratio of the probability that two loci are linked with a given recombination value over the probability that the two are not linked. This ratio is called a LOGARITHM OF ODDS (LOD) value or LOD SCORE. The critical LOD scores used to establish linkage groups and calculate map distances are called 'LINKLOD' and 'MAPLOD'. Researchers generally use a LINKLOD value of 3 as the minimum threshold value in order to decide whether or not loci were linked. A LOD value of 3 between two markers indicates that linkage is 10^3 (i.e. 1000) times more likely than no linkage.

ester: Molecule shaped by the condensation reaction of an alcohol group with an acidic group. Phosphate groups typically form esters when linked to a second molecule. For example, two nucleotides in nucleic acids are connected by ester bonds.

estimate: A statistical estimate of a parameter based on the data. Examples include the sample mean, sample median and estimated regression coefficients. *See* PARAMETER.

estimation of linkage: Calculating distance between genes from F_2 or test cross (TC_1) progeny on the basis of crossing-over frequency.

estuary: A coastal, semi-enclosed ecosystem that is open to the sea and has habitats transitional between marine and freshwater conditions.

eternal molecular biology: In 1938, Warren Weaver (1894–1978) coined the term 'molecular biology,' and this is why the date is frequently referred to as its foundation (W. Weaver in 1970). Weaver in fact did much more than devising this name; as a division head at the Rockefeller Foundation, he provided support for budding research directions.

ethelochoric: Thoughtful introduction by seedlings, seeds or plants in a new habitation by humans.

ethics: The study of fundamental principles that define values and determine moral duty and obligation.

ethidium bromide (EtBr) method: Ethidium can inset or intercalate between bases of DNA if the DNA is elastic. This so happens in the case of linear DNA but not with DNA that is circular. Ethidium intercalation makes DNA dense which could be separated from DNA not containing ethidium. This procedure is used to separate circular DNA of episomes, including mitochondria, chloroplasts and plasmids. Bacterial DNA, although circular, is sufficiently long, as it breaks during isolation and hence works as if it were linear DNA.

ethidium bromide (EtBr): Luminous molecule that inserts itself between base pairs of DNA and RNA in such a way that they fluoresce under UV light. Ethidium bromide staining is commonly used to visualize RNA or DNA in agarose gels placed on UV light boxes. Proper precautions are required, because the ethidium bromide is highly mutagenic and the UV light damaging to the eyes. Ethidium bromide is also included in caesium chloride gradients during ultracentrifugation, to separate supercoiled circular DNA from linear and relaxed circular DNA. *Aka* ETHIDIUM.

ethionine: A methionine analogue with chemical formula $C_6H_{13}NO_2S$ that inhibits most methyltransferases, is an effective inducer of globin gene expression in Friend erythroleukemia cells. It has been suggested that the induction of globin production is caused by undermethylation of the DNA. More definitive results have been obtained from studies with 5-azacytidine. 5-azacytidine is a cytosine analogue with a nitrogen atom replacing the carbon atom at position 5 of the pyrimidine ring and thus cannot accept a methyl group. With a short time with ethionine or 5-azacytidine, it was observed that several generations later, foci of differentiated cells (muscle, chondrocytes and adipocytes) were apparent.

ethnic group: A population group united by manners and customs and an association (present or past) with a particular geographical region.

ethological barrier: Isolating mechanisms caused by behavioural incompatibilities of potential mates.

ethyl ($-CH_2CH_3$): Hydrophobic chemical group derived from ethane (CH_3CH_3).

ethyl methanesulfonate (EMS): An alkylating compound that functions as a mutagen.

ethylene: A gaseous plant hormone that influences various aspects of vegetative growth and stimulates fruit ripening, dropping of leaves and senescence of flowers. *Aka* RIPENING HORMONE.

ethylenediamine tetraacetate (EDTA): An organic molecule which, due to the chemical groups it contains and their juxtaposition within that molecule, is able to chelate with certain other molecules such as divalent metal cations. EDTA thus impedes the activity of some enzymes requiring such ions for activity.

etiolation: The weak growth resulting in elongation of stems and yellowish colour caused by reduced light intensities.

etioplasts: Photosynthetically inactive plastids that accumulate when light levels are too low for chloroplast maturation. The etioplast inner membrane consists of a paracrystalline tubular lattice and peripheral, disk-shaped membranes, respectively, known as the prolamellar body and prothylakoids.

etoposide: A chemical agent that blocks topoisomerase and therefore introduces random DNA breaks, which results in S-phase arrest.

eubacteria: Comprise the major lines of prokaryotes. True bacteria such as *Escherichia coli* and *Salmonella typhimurium*. Eubacteria are physiologically distinct from archaebacteria.

eucell: A eukaryotic cell showing a nucleus, nuclear envelopes, chromosomes and nuclear divisions.

eucentric: A chromosomal exchange by which the translocated segment does not change its position in relation to the centromere.

euchromatin (EC): Light-staining, decondensed and transcriptionally accessible regions of the chromosomes (genome), allowing access for nuclear factors and nuclear repressors, acetylated nucleosomes, H3K4me, H3K36me and gene expression. Euchromatin is rich in gene concentration. It is uncoiled during interphase and condenses during mitosis, reaching a maximum density at metaphase. Euchromatin is extended, transcriptionally active in nucleosomes which show HIGH MOBILITY GROUP PROTEINS (HMGs). HMGs are typically found in place of histone H1 in transcriptionally active chromatin. Euchromatin has a fibrillar internal nuclear matrix, replicating during early S phase. Single copy and middle repetitive DNA sequences are less methylated. Euchromatin is conformationally favourable for transcription. It has less DNA methylation than heterochromatin, and its associated histones have modifications that favour gene transcription. *See* CHROMATIN. *Compare with* HETEROCHROMATIN.

euchromatization: The induced or spontaneous alteration of heterochromatin into euchromatin.

euchromosome: A chromosome displaying the distinctive characteristics of the standard complement of a given species.

eugenic: Favourable to the inherent quality of a human population, as opposed to DYSGENIC.

eugenics: The strategy of social control (such as selective breeding) to improve the inherited abilities of future generations characteristically in reference to humans.

euhaploid: A haploid genome displaying no differing number of chromosomes compared with the standard genome of the species.

eukaryon: The highly organized nucleus of a eukaryotic cell.

eukaryote: An organism containing a membrane-bound nucleus and other organelles, such as mitochondria and chloroplasts. Eukaryotes include all higher plants, animals, humans, yeast, fungi, algae and protists. Eukaryotic cells are much larger and more complex than prokaryotic cells. The genetic material (DNA) is enclosed in a membrane-bound nucleus. Besides, the cytoplasm also contains organelles such as chloroplasts and mitochondria, which have their own DNA. Yeast is an example of a single-celled eukaryote. *Contrasts* PROKARYOTE.

eukaryotes lack operon system: The operon system of gene regulation has proved not to be applicable in eukaryotes. Crossing-over and chromosome structural changes (such as translocations, inversions) disrupt clusters of genes that otherwise might become operons.

eukaryotic: Having cells with 'good' or membrane-bound nuclei. Plants are known for their cell walls of cellulose, as are green algae and certain protists, while fungi have cell walls of chitin.

eukaryotic alkyl DNA transferases: Methylguanine DNA methyltransferase is a small eukaryotic protein of 24 kDa in humans. After forming a low-stability complex with the DNA backbone at the damage site, it is thought to flip out the O^6-methylguanine base into the active site cavity, wherein the methyl group is transferred to an active site cysteine. The protein then dissociates from the repaired DNA, but the C-S bond of methylcysteine is stable, and therefore, after one catalytic event, the enzyme becomes inactivated.

eukaryotic AP endonuclease: The abasic site (i.e. the site of base loss) that is formed by DNA glycosylase activity is subsequently recognized by Ape1 (the major AP endonuclease), which incises the phosphodiester backbone immediately 5′ to the lesion leaving behind a strand break with a normal 3′-hydroxyl group and an abnormal 5′-abasic terminus.

eukaryotic cell: A cell with a distinct nucleus.

eukaryotic chromosome replication: J.H. Taylor and co-workers in 1957 labelled dividing root-tip cells of the broad bean *Vicia faba* with tritiated (^{3}H)-thymidine, which is specifically combined into newly synthesized DNA. The new chromosome was observed to replicate semi-conservatively because it consisted of half radioactive newly formed chromosome and half parental chromosome.

eukaryotic chromosome, functional domains in: Eukaryotic chromatin is separated into functional domains differentiated by histone PTLMs, histone variants and DNA methylation. DNA methylation is related to the expression of transcriptional initiation in plants and animals and is frequently observed in transposable elements.

eukaryotic chromosomes, structural components of: Various structural components of eukaryotic chromosomes are centromere, telomeres, kinetochore, primary constriction and secondary constriction. No eukaryotic chromosome can survive without a centromere and telomeres.

eukaryotic DNA: With respect to the copy number of a sequence, there are three classes of DNA in genomes of the eukaryotes – single copy/unique, moderately repetitive and highly repetitive. Thus, the whole genome of eukaryotic DNA is made up of three separate cot curves. On the basis of the C_0t values and the part of the genome that each segment comprises, the degree of repetitiveness of each segment can be determined. *See* C_0T CURVE.

eukaryotic DNA photolyases: Photolyase enzyme is found in many bacteria, lower eukaryotes, insects and plants. It seems to be absent in mammals (including humans). The gene is present in mammals but may code for a protein with an accessory function in another type of repair.

eukaryotic DNA polymerases: Eukaryotic DNA polymerase are named in a different way as compared to those in prokaryotes. Major eukaryotic DNA polymerases are DNA polymerases α, β, γ, δ, ε and ζ. In addition, other eukaryotic DNA polymerases, η, ι, κ, Rev1. θ, λ, φ, σ and μ are not well characterized yet.

eukaryotic DNA replicates only once per cell cycle: In several eukaryotes, CYCLIN-DEPENDENT

KINASES (CDKs) have been implicated in promoting the block to reinitiation of DNA replication within a cell cycle. This is very crucial to maintain constancy in chromosome number from one-cell cycle to the other. B-type CDKs in *Saccharomyces cerevisiae* prevent reinitiation through multiple overlapping mechanisms, including phosphorylation of the ORIGIN RECOGNITION COMPLEX (ORC), downregulation of Cdc6 activity and nuclear exclusion of the Mcm2-7 complex. Only when all three inhibitory pathways are disrupted do origins re-initiate DNA replication in G2/M cells.

eukaryotic DNA replication: Because of the intricacies of eukaryotic chromosomes, the rate of replication of eukaryotic chromosomes is slow (2600 bp/min) compared to *Escherichia coli* (50,000 bp/min). A single chromosome of eukaryotes has a multitude of replicons. A replicon has three parts of origin where replication begins and two forks that diverge from replicons in opposite direction. Fork movement continues until it meets another fork from the neighbouring replicon. DNA polymerase α is solely responsible for DNA synthesis at replication forks while DNA polymerase γ alone or together with DNA polymerase carries out the synthesis of mitochondrial DNA.

eukaryotic DNA replication, end-joining reaction in: The end-joining reaction performed by DNA ligases is required by all organisms and serves as a decisive step of DNA replication, repair and recombination processes. DNA ligase I joins Okazaki fragments during DNA replication.

eukaryotic DNA replication, non-enzymic proteins in: (i) PROLIFERATING CELL NUCLEAR ANTIGEN (PCNA). (ii) REPLICATION FACTOR A (RFA). (iii) REPLICATION FACTOR C (RFC).

eukaryotic gene activation, mechanism of: The interactions involving four classes of molecules, namely, SENSOR SITE, INTEGRATOR SITE, RECEPTOR SITE and PRODUCER GENE, regulate gene expression in eukaryotes. Transcription of a producer gene occurs only if at least one of its receptor sites is activated by forming a sequence-specific complex with activator RNA. This RNA (or protein) is synthesized by the integrator gene in response to the signals by the sensor site that are sensitive to external or internal development signals.

eukaryotic gene organization: A eukaryotic gene consists of ten regions: (i) recognition region (~50 kb), (ii) transcription initiation site, (iii) 5′-untranslated region, (iv) translation initiation, (v) alternating exons and introns, (vi) splice donor and acceptor sites, (vii) translation stop site, (viii) 3′-untranslated region, (ix) polyadenylation signal and (x) transcription stop site. *Aka* EUKARYOTIC GENE REGIONS.

eukaryotic gene regulation at the level of translation initiation: Assembly of the eIF4E/eIF4G complex has a central role in the regulation of gene expression at the level of eukaryotic translational initiation. This complex is regulated by the eukaryotic translation initiation factor 4E-binding proteins (4E-BPs), which compete with eIF4G for binding to eIF4E and which have tumour-suppressor activity.

eukaryotic gene regulation model, feature of: A model designed to portray gene regulation in eukaryotes must account for numerous observations. A simple external signal mediates a change in the activity of a cell. Integrated activity of numerous non-contiguous genes is necessary for a particular state of cellular differentiation. Heterogeneous nuclear RNA (hnRNA) is produced only in nuclei of higher organisms. Genomes of higher organisms are very large as compared to bacterial genomes. A high proportion of DNA in higher organisms does not code for mRNA. Some DNA may have regulatory and some protective functions. Repetitive DNA sequences are found in higher (but not in bacterial) cells. The repetitive DNA sequences are transcribed according to cell-specific patterns in differentiated cells. All these conditions require great flexibility in gene regulation. It is an accepted fact that portions of DNA involved in regulatory functions would prevail over those carrying structural information.

eukaryotic gene regulation through DNA rearrangements: Gene regulation in eukaryotes through DNA rearrangements is exemplified by expression of immunoglobulin genes, yeast mating-type switching, trypanosome surface antigen switching and synthesis of mRNA in pieces in VARIANT SURFACE GLYCOPROTEIN (VSG) GENES in trypanosome.

eukaryotic gene regulation, constitutive elements in: Constitutive elements are of common existence in a large number of different genes and are required for a basal level of expression. One such constitutive element is the Goldberg-Hogness (or TATA) box, identified by R.P. Lifton and others in 1978, which is positioned 20–30 bases upstream of the transcription initiation site and is implicated in determining the precise location of initiation. Constitutive sequences upstream of the TATA box, such as CAAT and GC boxes, are known to play the dominant role in controlling the frequency of transcription initiation in eukaryotic genes.

eukaryotic genes, recognizing intron/exon boundaries in: Because many genes in eukaryotes are interrupted by introns it is difficult to identify the protein sequence of the gene. Various computer programs have been designed for recognizing intron/exon boundaries for a particular organism or group of organisms: (i) DNA functional site miner (DNAFSMiner) contains two software tools: TIS Miner which can be used to predict TRANSLATION INITIATION SITE (TIS) in vertebrate mRNA, cDNA or DNA sequences. The other is called Poly(A) Signal Miner which can be used to predict polyadenylation (poly(A)) signal in human DNA sequences. (ii) FirstEF is a first exon and promoter prediction program for human DNA. (iii) AUGUSTUS uses gene prediction in eukaryotic (Human, *Drosophila*, *Arabidopsis*, Brugia, *Aedes*, *Coprinus* and *Tribolium*) sequences that are based on a generalized hidden Markov model, a probabilistic model of a sequence and its gene structure. (iv) TWINSCAN can be used for the analysis of mammalian (human and non-human), plant (Arabidopsis and non-Arabidopsis dicots), worm (*Caenorhabditis elegans*) and fungus (*Cryptococcus neoformans*) gene analysis. (v) GeneMark offers plenty of model systems (*Arabidopsis thaliana*, *Caenorhabditis elegans*, *Drosophila melanogaster*, *Gallus gallus*, *Homo sapiens*, *Takifugu rubripes*, *Mus musculus*, *Rattus norvegicus* and *Saccharomyces cerevisiae*) with many options. (vi) GlimmerM provides the following model organisms: *Arabidopsis*, *Aspergillus*, *Brugia*, *Cryptococcus*, *Entamoeba*, *Oryza*, *Plasmodium*, *Schistosoma* and *Theileria*. (vii) Genome Analysis Pipeline offers both prokaryote (Generation and Glimmer) as well as higher eukaryotes and *Saccharomyces* genome analysis. (viii) FGENESH (SoftBerry) is a very fast HMM-based gene structure prediction program. (ix) geneid is used for the prediction of human and *Drosophila* genes. (x) GenLang. is used for the prediction of vertebrate, Drosophila and dicot genes. (xi) HMMgene is used for the prediction of vertebrate and *Caenorhabditis elegans* genes. (xii) NetPlantGene is for neural network predictions of splice sites in *Arabidopsis thaliana* DNA. (xiii) Genie is finder based upon generalized Hidden Markov Models. Human and *Drosophila* genes. (xiv) GeneSeqer is used for humans, mice, rats, chicken, *Drosophila*, nematode, yeast, *Aspergillus*, *Arabidopsis* or maize. (xv) GrailEXP is used for humans, mice, *Arabidopsis*, maize and many others. (xvi) FrameD is a noise-resistant gene finder for prokaryotic and matured eukaryotic sequences offering considerable flexibility in search strategies and output. (xvii) SplicePort is an interactive splice-site analysis. (xviii) JCat (Codon Adapter Tool) offers a complete range of eukaryotic and prokaryotic cells; and the ability to select against rho-independent terminators and restriction sites. These programs may not recognize all the intron/exon boundaries.

eukaryotic genome complexity analysis: For calculating eukaryotic genome complexities, each genome fraction must be treated as an independent kinetic component whose reassociation is compared with standard DNA. For example, the slow component represents 45% of DNA, so its concentration in the reassociation reaction is 0.45 of the measured concentration (C_0). Thus, the $C_0t½$ of the slow fraction alone is 0.45 × 630; 283. Thus, if the slow component is isolated from other components, it renatures with $C_0t_{½}$ value of 283. If under such conditions the *Escherichia coli* DNA reassociates with $C_0t½$ value of 4.0, the complexity of the slow component comes out to be 3.0 × 10^8 bp. Similarly, the complexity of intermediate and fast components comes out to be 6.0 × 10^5 and 340, respectively. Thus, the faster the component reassociates, the lower its complexity.

eukaryotic genome kinetics: When the DNA of the eukaryotic genome is characterized by reassociation kinetics, usually the reaction occurs over a range of C_0t values having up to eight orders of magnitudes. This is much broader than the 100-fold range expected. The reason for this is that the expectations are based on a single kinetically pure reassociating component. A genome actually includes several such components, each reassociating with its own characteristic kinetics.

eukaryotic genome organization: When DNA of eukaryotic genome is characterized by reassociation kinetics, usually the reaction occurs over a range of C_0t values. A eukaryotic genome actually includes several such components, each reassociating with its own characteristic kinetics. The reaction falls into three distinct phases, called fractions I, II and III, which are named as three components – fast, intermediate and slow. The FAST COMPONENT represents 25% of the total DNA renaturing with C_0t value of 10^{-4} and ~2 × 10^{-2} with a $C_0t½$ value of 0.0013. The INTERMEDIATE COMPONENT represents the renaturation of

30% of the DNA. It renatures between C_ot values of ~2×10^{-2} and 10^2 with a Cot½ value 1.9. The SLOW COMPONENT represents 45% of total DNA. It extends over C_ot range from ~10^2 to 10^4 with C_ot½ value of 630. A plateau separates phases I and II but phases II and III overlap slightly.

eukaryotic gyrase: In eukaryotes, gyrase relaxes negatively supercoiled DNA in the presence of ATP and Mg^{2+}.

eukaryotic primases: Eukaryotic primases form a complex with DNA polymerase α and its accessory β subunit. The small eukaryote primase subunit contains the active site for RNA synthesis, and its activity correlates with DNA replication during the cell cycle.

eukaryotic promoters and terminators, prediction of: (i) Promoter 2.0 Prediction Server predicts transcription start sites of vertebrate Pol II promoters in DNA sequences. (ii) PROMOSER Human, Mouse and Rat promoter extraction service – maps promoter sequences and transcription start sites in mammalian genomes. (iii) Promoter and gene expression regulatory motifs search (Softberry, USA) offers a variety of promoter-scanning programs.

eukaryotic ribosomes: Ribosomes of eukaryotes and prokaryotes diverge in size and other details. The cytoplasmic ribosomes of eukaryotes are of 80S, size, contain 60% rRNA and 40% protein and dissociate into a smaller (40S) subunit and a larger (60S) subunit. 60S subunit has 5S, 5.8S and 28S rRNA and 40S subunit has 18S rRNA.

eukaryotic RNA polymerases: Eukaryotic RNA polymerases are characterized by the type of RNA they synthesize. Five types of eukaryotic RNA polymerases are well known: *See* RNA POLYMERASE I, RNA POLYMERASE II, RNA POLYMERASE III, RNA POLYMERASE IV *and* RNA POLYMERASE V. *See details as separate terms.*

eukaryotic single-strand DNA-binding proteins (SSBs): In eukaryotes, the major SSB, replication protein A (RPA), is a heterodimer. Another human SSB protein is hSSB1. Upon induction of DNA damage, hSSB1 accumulates in the nucleus and forms distinct foci independent of cell-cycle phase.

eukaryotic transcription precision: Prokaryotic transcription regulators must be highly accurate but are likely to operate in a 'moderate precision' mode because an error per cell generation is likely to be a tolerable number. However, to achieve the same biological accuracy, the localization mechanism controlling eukaryotic transcription is likely to be much more stringent because some 10^{10} base pairs must be scanned by the regulators instead of the roughly 10^7 of their prokaryotic counterparts. Two troublesome *pseudosites* in the prokaryotic genome expand to 2000 in the eukaryotes. Thus, the biological accuracy problem is the ratio of sites scanned to sites selected. From this point of view, regulation of eukaryotic transcription can be considered a high-precision interaction.

eukaryotic translation elongation: Elongation factor eEF3 is an ATPase that, in addition to the two canonical factors eEF1A and eEF2, serves an essential function in the translation cycle of fungi. eEF3 is required for the binding of the aminoacyl-tRNA-eEF1a-GTP ternary complex to the ribosomal A-site and has been suggested to facilitate the clearance of deacyl-tRNA from the E-site.

eukaryotic translation initiation: In eukaryotes, at least 11 eukaryotic translation initiation factors (eIF1, eIF2, eIF3, eIF4A, eIF4B, eIF4C, eIF4 D, eIF4E, eIF4F, eIF5, eIF6) are known to exist. There are two cap-binding proteins (CBPI and CBP2). CBP1 binds to the 5′ cap (m^7GpppX) of mRNA and facilitates formation of a complex between the mRNA and the 40S ribosomal subunit. CBP2 performs some yet unknown function. The 5′ cap is required for efficient translation. The eIF1 assists mRNA binding, eIF2 binds met-tRNA. It has three subunits – α (binds ATP), β (may be a recycling factor) and γ (binds met-$tRNA_f^{Met}$). The eIF3 binds mRNA. The initiation factor eIF4A assists mRNA binding and also binds ATP, eIF4B assists mRNA binding and unwinding, eIF4C binds 60S subunit. The function of eIF4D and eIF4E is unknown. Initiation factor eIF4F mediates the function of cap. Initiation factor eIF4E binds the cap. This step is thought to be controlled by phosphorylation. Initiation factor eIF5 releases eIF2 and eIF3 while eIF6 prevents 40S–60S joining. The eukaryotic initiation factor 4G (eIF4G) is the core of a multicomponent switch controlling gene expression at the level of translation initiation. It interacts with the small ribosomal subunit interacting protein, eIF3 and the eIF4E/cap-mRNA complex in order to load the ribosome onto mRNA during cap-dependent translation. Initiation factor eIF5A seems to have a hand in every step. Mammalian eIF6 is required for efficient initiation of translation, *in vivo*.

eukaryotic translation termination: In the eukaryotic system, only one release factor (eRF) is known to exist. GTP seems to be essential for the activity of eRF. The released peptidyl-tRNA comes in the cytoplasm.

eukaryotic translational release factors (RFs): In eukaryotes, translation termination requires two eukaryotic release factors, eRF1 and eRF3. eRF1 is required for recognition of the stop codon and eRF3 supports the polypeptide chain release in a GTP-dependent manner.

euphenics: A branch of genetics which deals with the control of hereditary diseases, specially, inborn errors of metabolism.

euploid: A cell carrying an exact multiple of the haploid chromosome number. For example, Individuals having one, two, three, four, five, six (and so on) complete sets of chromosomes are called HAPLOIDS, DIPLOIDS, TRIPLOIDS, TETRAPLOIDS, PENTAPLOIDS, HEXAPLOIDS (and so on). Polyploids are commonly found in plants but in animals, polyploids are rare and restricted to groups that reproduce asexually or are hermaphroditic (an individual with both male and female reproductive organs) or parthenogenetic (an individual developed from an unfertilized egg). Examples are earthworms, some types of shrimps and parthenogenetic species of insects and lizards. A plant alkaloid, colchicine, is used to produce polyploids, as it acts by dissolving the spindle apparatus thereby doubling the number of chromosomes in the cell. A practical application of sterility associated with triploidy is the production of seedless varieties of watermelons, bananas, etc. These varieties are more palatable to consumers. For instance, in wheat basic chromosome number is $x = 7$; the diploid wheat is $2x = 14$; tetraploid wheat is $4x = 28$ and hexaploid wheat is $6x = 42$. But all these wheats are euploids.

euploidy: Variation in chromosome number by whole sets or exact multiples of monoploid (haploid) number. Euploidy above the level of diploidy is collectively referred to as polyploidy.

Europe Nucleotide Archive (ENA): Repository for uploading and sharing annotated DNA and RNA sequences.

European Bioinformatics Institute (EBI) databases: A variety of general and specialized databases that are mentioned at EBI. These include databases for primary structure collections, such as the European Molecular Biology Laboratory (EMBL), Nucleotide Sequence Database, the protein databases SWISS-PORT and TrEMBL, amalgamated into the Universal Protein Resource UniPort, in addition to higher-order structure databases available in the Macromolecular Structure Database (MSD). Information on these and other data collections can be inspected at http://www.edi.ac.uk/Databases. The detailed description is given in the *EMBL Nucleotide Sequence Database User Manual* which can be inspected at http://www.ebi.ac.uk/embl/Documentation/user-manual/usrman.html.

European Bioinformatics Institute (EBI): It is an outstation of the European Molecular Biology Laboratory (EMBL) and maintains nucleotide sequence database. It can be accessed and searched through the SRS system at EBI, or one can download the entire database as flat files.

evaporation: The change of state of water from a liquid or solid to a gas.

evapotranspiration: The process by which water is transported from the land to the atmosphere by evaporation from the soil and other surfaces and by transpiration from plants. *See* TRANSPIRATION.

even-pinnate: Having an even number of leaflets in a compound leaf. *Aka* PARIPINNATE.

event: An 'event' in genetic engineering is a specific genetic modification in a specific species. For example, the same gene incorporated by man into a given plant genome at two different locations (loci) along that plant's DNA would be considered two different events. Alternatively, two different genes incorporated into the same locus of two same-species plants would also be considered two different events. Generally speaking, the world's regulatory agencies confer new biotech-derived product approvals in terms of events.

event-driven programming: A programming model in which the flow of the program is determined by events such as user actions (mouse clicks, key presses), sensor outputs or messages from other programs or threads. Event-driven programming is the dominant model used in graphical user interfaces and other applications (e.g. JavaScript web applications) that are run on performing certain actions in response to user input. This is also true of programming for device drivers (e.g. P in USB device driver stacks).

evergreen: Not deciduous; having leaves all the year.

evo-devo: The informal name for EVOLUTIONARY DEVELOPMENTAL BIOLOGY, a branch of evolutionary investigations that seeks to offer insights from genetics, embryology and microbiology to help scientists better know how new species originate from existing life forms. The year 2000 might be considered the birth of evo-devo when two journals arose to publish the results of evo-devo research. If 2000 saw the birth of evo-devo as a discipline, then 1977 must have been the year of its conception. In that year, three publications (S.J. Gould in 1977; F. Jacob in 1977; A.M. Maxam and W. Gilbert in 1977) paved the way for evolutionary developmental biology. J.T. Bonner in 1982 and R.A. Raff and T.C. Kaufman in 1983 set forth many of the paradigms that we are presently using. Evo-devo is shifting the balance of developmental biology from the DIFFERENTIATION question towards the MORPHOGENESIS question.

evolution: (i) The change in the heritable traits of interbreeding populations of biological organisms over successive generations, which may occur by mutation, gene flow, natural selection or random chance. (ii) The transformation of a group of interbreeding organisms in a way that descendants differ from their predecessors. (iii) Darwin defined evolution as 'descent with modification'. It is the change in a lineage of populations between generations. (iv) In general terms, biological evolution is the process of change by which new species develop from pre-existing species over time. (v) In genetic terms, evolution can be defined as any change in the frequency of alleles in populations of organisms from generation to generation. It is a population of organisms that evolves, not individual organisms.

evolution at molecular level: Most studies of evolution at the molecular level have been conducted on structural genes as these are the ones that are transcribed to generate RNAs that either function as mRNA and are translated to produce proteins, or have functional roles, like rRNAs or tRNAs.

evolution at protein level: The number of amino acid substitutions between two species was approximately proportional to the time since the divergence of the species. Amino acid substitutions occurred less frequently in the functionally important proteins on protein regions than in less important proteins or protein. It was thus assumed that amino acid substitutions in the non-conserved regions of proteins are nearly neutral or slightly positively selected and that the amino acids in functionally important sites do not change easily to maintain the same function.

evolution by natural selection – a scientific law and not just a theory: The inheritance of biological properties of sexual macroorganisms is explained by Mendelian genetics in such a manner that fundamental heritable information is transmitted unaltered to later generations, but with occasional instability in the form of mutations. This procedure sets two vital situations which, together with a capability for reproduction and variation in longevity or reproductive success, assure the accomplishment of natural selection resulting in adaptation. This, according to D.J.M. Crouch and W.F. Bodmer in 2024, launches adaptive evolution by natural selection as a scientific law by any judicious definition. Given that natural selection is then certain to happen for biological beings and so is a law in that context. It can only endure to be viewed as a theory in the logic that there might be other mechanisms for adaptive evolution to take place without it, but such processes are not so far known to occur.

evolution by random walk: A hypothesis that believes in the evolution of species through forces acting at random rather than in a specific direction.

evolution of cellular pathways: The regulation of many different cellular processes requires the use of protein interaction domains to direct the association of polypeptides with one another and with PHOPHOLIPIDS, small molecule or nucleic. The modular nature of these domains, and the flexibility of their binding properties, have facilitated the evolution of cellular pathways. Conversely, aberrant interactions can induce abnormal cellular behaviour and disease.

evolution of complexity in DNA replication: DNA replication has been keeping pace with evolution. In core complex, the numbers of subunits (polypeptides) present in T7 phages are 4, T4 phages are 7, *Escherichia coli* are 13, and yeast and mammals are 27. Thus, polypeptide subunits involved in the machinery of DNA replication increased giving rise to complexity during evolution.

evolution of DNA as genetic material: Proteins are more versatile as enzymes than RNA. Heritable mechanism for mass production of proteins was needed by the protocell. RNA is very unstable as a rule. This severely limited the complexity of RNA-based living systems. DNA bonds in a double-helical molecule are much more stable. Selection for greater stability in the stored code led to the evolution of DNA. DNA fulfils the need for a stable genetic material.

evolution of double-stranded genomes: Double-stranded RNA genomes evolved before the birth of the DNA world. This expedited the transition from RNA to DNA genomes as this step was chemically more feasible. The key event in ushering the DNA world was the ability of the enzyme ribonucleoside diphosphate reductase to convert ribonucleoside diphosphates (rNDPs) to deoxyribonucleoside diphosphates (dNDPs).

evolution of new genes: Origin of new genes involves a number of steps: (i) duplication of an existing gene, (ii) elimination of its existing function through a nonsense mutation (pseudogene), (iii) accumulation of more mutations over the passage of time thus the duplicate gene acquires a new function and (iv) a gene with a new function is added to the genome. Duplications in this way provide additional genetic material potentially capable of giving rise to new genes during the process of evolution.

evolution of organisms, steps in: Heterotrophic → Chemoautotrophic → Photoautotrophic → Photosynthesis → Respiration. *See details as separate terms.*

evolution of photosensitive cells: Hypothetical evolution of photosensitive cells containing rhodopsin as a light receptor and monophyletic evolution of the various eye types, starting from a Darwinian prototype eye consisting of a single photoreceptor cell and a pigment cell assembled under the control of *Pax 6*, has been suggested.

evolution of speech: The ability to speak evolved from our non-speaking under the force of natural selection under the force of natural selection. We all are using the same genetic capacity to symbolize the same concept. Different languages and scripts evolved in different regions.

evolution pressure: The joint action of mutation, immigration, hybridization and selection pressure.

evolution, forces of: Mutation, recombination, migration, random genetic drift, Founder effect, bottlenecks, assortative mating, selection, etc. are regarded as forces of evolution.

evolutionarily conserved: *See* CONSERVED SEQUENCE.

evolutionarily important genes: Evolution at the organismic level depends predominantly on regulatory changes and structural gene changes have a secondary role in organismal evolution. Simple gene rearrangements such as deficiencies, duplications, inversions and translocations can markedly change regulatory control over gene function. Evolutionary changes of regulatory genes are primarily responsible for the development of reproductive isolation and, thus, for the origin of new species.

evolutionarily stable strategy (ESS): (i) A strategy, which, if adopted by individuals in a population, prevents the invasion into the population of a mutant adopting an alternate strategy. (ii) A strategy that cannot be replaced by any alternative. Individuals that adopt the ESS must be at least as fit as any feasible alternative when the rest of the population is adopting the ESS.

evolutionary algorithm (EA): A subset of evolutionary computation, a generic population-based metaheuristic optimization algorithm. An EA uses mechanisms inspired by biological evolution, such as reproduction, mutation, recombination and selection. Candidate solutions to the optimization problem play the role of individuals in a population, and the fitness function determines the quality of the solutions. Evolution of the population then takes place after the repeated application of the above operators.

evolutionary biology: The subfield of biology that investigates evolution and the evolutionary processes that produced the diversity of life on Earth from a hypothesized single common ancestor. These processes include (i) descent of species and (ii) origin of new species.

evolutionary breeding: Breeding procedure in which a variety is developed from an unselected progeny of a cross or multiple crosses that have experienced evolutionary changes.

evolutionary character state reconstruction: *See* CHARACTER STATE RECONSTRUCTION.

evolutionary classification: Method of classification that uses both cladistic and phenetic classificatory principles. To be precise, it allows PARAPHYLETIC GROUPS (which are permitted in phenetic but not in cladistic classification) and MONOPHYLETIC GROUPS (which are permitted in both cladistic and phenetic classification) but excludes POLYPHYLETIC GROUPS (which are barred from cladistic classification but allowed in phenetic classification).

evolutionary clock: This is defined by the rate at which mutations accumulate in a given gene. By comparing the amino acid sequence of the resultant protein, it is possible to discover how much the proteins and the DNA of various species differ from each other. This information is used to estimate the rate of evolutionary change.

evolutionary computation: A family of algorithms for global optimization inspired by biological evolution and the subfield of artificial intelligence and soft computing studying these algorithms. In technical terms, they are a family of population-based trial and error problem solvers with a metaheuristic (strategies that guide the search process) or stochastic (involving a random variable) optimization character. *Aka* EVOLUTIONARY COMPUTING. *See* GENETIC PROGRAMMING *and* GENETIC ALGORITHM.

evolutionary developmental biology: Evolutionary biology, medicine and ecology areas are themselves interacting with one another to highlight questions of developmental biology that had become peripheral – genetic assimilation and life history strategies; teratology, endocrine disruption; developmental constraints and origin of *Homo sapiens*. Together, these constitute a larger evolutionary developmental biology that is shifting the balance of developmental biology from the differentiation question towards the morphogenesis question. The linkage of these areas will provide the second stage in the expansion of developmental biology and its reconciliation with every other area of biological sciences. If genetics is 'Darwin's missing evidence', then it must include developmental genetics as well as population genetics and molecular genetics.

evolutionary developmental mechanisms: Evolutionary developmental mechanisms, HETEROCHRONY and HETEROTROPHY, at various levels of the biological hierarchy have been described. Heterochrony and heterotrophy are evolutionary developmental mechanisms that span all levels.

evolutionary divergence: Divergent evolution represents the evolutionary pattern in which species sharing a common ancestry become more distinct due to differential selection pressure which gradually leads to speciation over an evolutionary time period.

evolutionary divergence, stages in: F. Nie in 2018 suggested five stages of increasing evolutionary divergence during speciation: (i) geographic species, (ii) subspecies, (iii) semi- or incipient species, (iv) sibling species and (v) full species.

evolutionary ecology: Study of the evolutionary penalties of interactions between species and between a species and its environment.

evolutionary epigenetic clock: Heritable changes in epigenetic markers, known as epimutations, occur at higher rates than genetic mutations. N. Yao and others (2023) develop a tool that allows for the estimation of phylogenies based on these epimutations. Stochastic DNA methylation changes at a subset of cytosines in plant genomes display a clock-like behaviour. They located regions of the genome that experience neutral, clock-like epimutations in the classic plant model *Arabidopsis thaliana* and were able to recapitulate known phylogenies of very recent timescales. This will help to provide the tools and theoretical basis for high-resolution temporal studies of plant biodiversity. The new tool may have potential applications in plant breeding and the study of ecology, evolution and archaeology.

evolutionary game: An interaction between individuals in which the payoff depends on the strategy played by each of them.

evolutionary genetics: Offers useful information for guiding gene mapping (e.g. understanding the phylogeny helps to manipulate horizontal gene transfer between microorganisms).

evolutionary novelty: A newly acquired structure or other property that permits the performance of a new function.

evolutionary plasticity: The degree of genetic adaptability of a species or other taxa.

evolutionary rates: The rate of divergence between taxa measurable as amino acid substitutions per million years.

evolutionary responses: Interacting populations form an environment for each other and mutually impose selective forces that stimulate evolutionary changes to alter their relationship by making all the interacting populations respond.

evolutionary robotics: A method of robotics in which a number of differing robots are submitted to tests. Those which perform best are used as a model to create a subsequent 'generation' of robots.

evolutionary species concept: In this species concept, emphasis is laid on different evolutionary lineages. *Aka* PHYLOGENETIC SPECIES CONCEPT.

evolutionary synthesis: The integration during the 1930s and 1940s of principles of population genetics with those in the other fields of biology (e.g. palaeontology, systematics and botany).

evolutionary trees, properties of: (i) The root is the common ancestor of all taxa. It is defined by including a taxon that we are reasonably sure branched off earlier than other taxa under study but should be related to the remaining taxa. A unique path leads from one node to any other node and the direction indicates evolutionary time. (ii) If we do not have taxa to define the root, we can predict relationships by an uprooted tree. (iii) Leaves represent things (genes, proteins, species) being compared. (iv) Internal nodes are hypothetical ancestral units. (v) In a rooted tree, the path from the root to a node represents an evolutionary path. (vi) An uprooted tree specifies relationships among things, but evolutionary paths. (vii) Paralogues are genes that diverged within the same species. These are multiple genes (tree leaves) per species. (viii) Orthologous are the genes that diverged with species.

evolvability: Ability to produce a heritable variation that can be exploited by selection.

evolve: To develop a population using the process of evolution and changing in some way as an adaptation to the environment.

evolved engineered transcriptional repressor (EvoETR): EvoETR is characterized by a high specificity profile, further reducing the circulating levels of PCSK9 in mice with an efficiency comparable with that obtained through conventional gene editing, but without causing DNA breaks. Permanent epigenetic silencing using programmable editors equipped with transcriptional repressors holds great promise for the treatment of human diseases.

evolving classification function (ECF): Evolving classifier functions or evolving classifiers are used for classifying and clustering in the field of machine learning and artificial intelligence, typically employed for data stream mining tasks in dynamic and changing environments.

EVS: Short form of NHLBI exome variant server database. The EVS annotation source contains exome sequencing variants retrieved from the Exome Variant Server (EVS) for the NHLBI Exome Sequencing Project (ESP). The EVS annotation data was generated from approximately 2500 exomes and evs_5400 from approximately 5400 exomes.

***ex situ*:** Out of place or not in the original habitat (e.g. seeds stored in a gene bank).

***ex situ* conservation:** A conservation method that requires the removal of seed, pollen, sperm or individual organisms from their original habitat, keeping these resources of biodiversity alive outside of their natural environment. The storage of plant genetic resources as seeds in gene banks at sub-zero temperatures is the most widely applied method. Nearly 90% of more than 6 million accessions stored *ex situ* worldwide are maintained as seeds. Other methods of *ex situ* conservation include: (i) maintenance of whole plants in field gene banks; (ii) as tissue culture in *in vitro* gene banks; and (iii) immersion of tissue, embryo or seeds in liquid nitrogen; the latter two methods are particularly used for conservation of vegetatively propagated species and for those with recalcitrant seeds.

***ex vitro*:** *In vitro* grown plants transferred to *in vivo* conditions. The term *ex vitro* is used for *in vitro* grown plants when these plants are grown outside the *in vitro* conditions usually in the greenhouse or the field.

***ex vivo*:** (Latin meaning 'out of the living') literally means that which takes place outside an organism. In science, *ex vivo* refers to experimentation or measurements done in excised living cells or on tissue from an organism in an external environment with minimal alteration of natural conditions. *Ex vivo* experiments are done to determine the effect of the substance (e.g. pharmaceutical) on the biochemistry of the cell.

***ex vivo* gene therapy:** The delivery of a gene or genes to the isolated cells of an individual. After culturing, the transformed cells are introduced back into the individual by transfusion, infusion or injection to assuage a genetic disorder.

exact reversion: The type of reversion (or back mutation) where the same codon is reconstituted. For example, codon UUU (phe) through a forward mutation → CUU (leu) which through exact reverse mutation → UUU (phe).

exaggeration: Expression of a hypomorphic gene placed opposite to a deficiency.

exalbuminous: In seeds of a given species, having no endosperm (i.e. no albumen) (e.g. in Fabaceae and Combretaceae).

excavate: Member of a kingdom of eukaryotes. All are single-celled species, and none are known to have mitochondria.

exception handling: The process of responding to the occurrence, during computation, of exceptions – anomalous or exceptional conditions requiring special processing – often disrupting the normal flow of program execution. It is provided by specialized programming language constructs, computer hardware mechanisms like interrupts or operating system IPC facilities like signals.

exceptions to genome constancy: Random and programmed changes, gene amplification, chromatin diminution and other cases of DNA loss and programmed recombination are a few instances which question the concept of genome constancy in development.

exceptions to Mendelian principles: Except for the basic principles of inheritance, the whole of modern genetics is an exception to Mendelism. Modern genetics seems to defy if not question the universality of Mendelian inheritance. MULTIGENE INHERITANCE, LINKAGE, GENE–GENE INTERACTIONS, MULTIPLE ALLELISM, QUANTITATIVE CHARACTERS, INCOMPLETE DOMINANCE, EPISTASIS, CYTOPLASMIC INHERITANCES, GENE MUTATIONS, CHROMOSOME MUTATIONS, EXPANDED TRIPLET REPEATS, GENOME IMPRINTING, COMPLEX TRAITS, etc. are by some authors considered exceptions to Mendelism. *See details as separate terms.*

exchange: Refers to either CHROMATID EXCHANGE or CHROMOSOME EXCHANGE. Exchange is the result of two or more chromatid or chromosome lesions and subsequent rearrangement of chromatid or chromosome material.

exchange pairing: The type of pairing of homologous chromosomes that allows genetic crossing-over to take place.

exchange reaction: The reaction involved in polypeptide elongation where the aminoacyl located at A-site is translocated to P site on 50S subunit of the ribosome. This step involves two protein factors Tu and Ts in prokaryotes.

excision: (i) The cutting out of a piece of damaged or defective DNA by enzymes. DNA damage might be constituted by the presence of a thymine dimer which inactivates that part of the DNA. The region of the dimer is cut out along with a few flanking bases and then repaired. (ii) Excision of phage or episome or other sequence describes its release from the host chromosome as an autonomous DNA molecule.

excision (Xi) protein: A protein required for excision of phage chromosome from bacterial chromosome through site-specific recombination.

excision and reinsertion mechanisms: Two types of exonucleases, DNA polymerase I-associated and DNA polymerase I-unassociated, act on the incised DNA for short-patch repair and long-patch repair, respectively.

excision of thymidine dimers: Thymine dimers are excised, and the gap is patched by the sequential action of four enzymes. The pyrimidine dimers are excised from a DNA molecule through the sequential action of a variety of enzymes, namely, endonucleases, exonucleases, polymerases, ligases, insertases and glycosylases.

excision repair: A DNA repair system that removes a single-stranded DNA sequence containing damaged or misplaced bases and replaces them in a duplex with a complementary strand.

exclusion: The management of disease by eliminating the pathogen or infected plant material from crop production areas (e.g. by quarantines and embargoes).

exclusion principle: The principle that states that two species cannot coexist at the same locality if they have identical ecological requirements.

executable: Also refers to EXECUTABLE CODE, EXECUTABLE FILE and EXECUTABLE PROGRAM which causes a computer 'to perform indicated tasks according to encoded instructions, as opposed to a data file that must be analysed by a program to be meaningful. The exact interpretation depends upon the use, while 'instructions' is traditionally taken to mean machine code instructions for a physical CPU, in some contexts a file containing bytecode or scripting language instructions may also be considered executable.

executable module: A process in computer and software engineering in which a computer or virtual machine implements the instructions of a computer program. Each instruction of a program is a description of a particular action which is to be carried out as per instructions of a program for solving a specific problem. Therefore, the actions they describe are carried out by an executing machine; specific effects are produced in accordance with the semantics of the instructions being executed. *Aka* EXECUTION.

exergonic reaction: A chemical reaction that releases energy. During chemical reactions, bonds between atoms are formed or broken, and some change of energy occurs. When a bond is broken energy is increased. A chemical reaction that releases more energy than it uses is called an exergonic reaction. For example, when NaOH is dissolved in water the solution becomes hot.

exhaustion test: A type of vigour test that evaluates the ability of seeds to grow rapidly under rigidly controlled conditions of high temperatures, relative humidity and moisture content in continuous darkness.

exine: The outer coat of a pollen grain.

existence detection: An existence check before reading a file can catch and/or prevent an incurable error.

exit (or E) site: Third site on large subunit of ribosome where tRNA lies before it exits the ribosome.

exocarp: The outer layer of the pericarp; often refers to the skin of fleshy fruits.

exoconjugant: Each of the two cells that separate after conjugation has taken place.

exocytosis: Process by which most molecules are secreted from a eukaryotic cell. These molecules are packaged in membrane-bounded vesicles, called ENDOSOMES, that fuse with the plasma membrane, releasing their contents to the outside.

exocytotic vesicles: Membranous particles that transport and store proteins during exocytosis. *Aka* SECRETORY VESICLES.

exodeoxyribonuclease: An enzyme that is capable of detaching the terminal nucleotide from a deoxyribonucleic acid (DNA) chain. An exonuclease releases one nucleotide at a time (serially) beginning at one end of a nucleic acid.

exogenote: DNA that a bacterial cell has taken through one of its sexual processes.

exogenous: A substance or process originating outside of or external to a system (such as an organism, tissue or cell), as with drugs and many pathogens. *Contrasts* ENDOGENOUS.

exogenous DNA: DNA originating outside an organism that has been introduced into the organism.

exogenous metabolites: Metabolites of foreign substances, such as drugs. *Aka* XENOMETABOLITES.

exome: Sum total of all of the protein-coding regions (exons) in the genome constituting (e.g. approximately 1% of the human genome). Currently, most known disease-causing variants are located in the exome.

exome sequencing: A DNA sequencing technique focused on the portion of the genome that codes for proteins, known as the exome. The exome comprises only 1% of the genome. Uses sequence-specific probes in solution or fixed to microarrays to purify coding regions of the DNA. These regions are then amplified, sequenced and compared with a reference to identify variations in the coding regions. *Aka* WHOLE EXOME SEQUENCING (WES).

exon: One of the coding regions of a discontinuous gene. The sequence present in a gene (DNA) is complementary to that present in its processed mRNA, tRNA or rRNA. This is the part of a eukaryotic gene whose DNA sequence is preserved in post-transcriptional splicing and is represented in the spliced (processed/mature) mRNA. Exons abundantly occur in eukaryotic genes but rarely in prokaryotic genes. These regions of genes code for amino acids, with other parts that do not code for amino acids interspersed among them called INTRONS. An exon usually lies adjacent to an intron. The exons may or may not be coding (i.e. translatable). The exons which are translatable are called TRANSLATABLE EXONS while others that are not translatable are called UNTRANSLATABLE EXONS. Untranslatable exons occur at both the ends of mRNAs. Typically, a mature eukaryotic protein is composed of several domains coded by different exons within a single gene.

exon length: A computer analysis of large-sized random DNA sequences revealed that an exon has an upper limit of 600 nucleotides in most of the eukaryotic genes.

exon sharing: Likely mechanism for the sharing of exons coding for different proteins is through the duplication and migration of exons during evolution.

exon shuffling: Mixing up of exons during evolution. This process can lead to substantial genome rearrangements due to meiotic division. It is believed to be responsible for the evolution of split genes in nuclear genes. During the splicing stage, different coding sequences (segments or exons) could splice in a large number of different combinations, thus generating a large number of different genes. *In vitro* exon shuffling is performed using a mixture of oligonucleotides to allow combinations of exons to be spliced together to generate mosaic proteins.

exon theory of genes: Split genes arise not by insertion of introns into unsplit genes but from combinations of primordial MINIGENES separated by spacers.

exon trapping: This method is most often used to isolate exons from new DNA, rather than simply to identify exon-intron boundaries. In exon trapping, a restriction fragment from a new DNA sequence is cloned into a cognate restriction site in an intron of a cloned gene. This chimeric DNA is introduced into an appropriate eukaryotic host, usually yeast, and the cloned gene is expressed. During the processing of the initial transcript, introns are spliced out, leaving only the exon from the cloned restriction fragment behind. DNA from this mRNA is obtained via reverse-transcription-PCR, and its sequence is determined. Comparison of this sequence, containing only the exon from the cloned restriction fragment, with the sequence of the restriction fragment itself shows where the intron-exon boundaries are located.

exonic genes: See SPLIT VS NON-SPLIT GENES.

exonic single-nucleotide polymorphism (exon SNP): Any single-nucleotide polymorphism that occurs in an exon of a gene. *Aka* EXPRESSED SINGLE-NUCLEOTIDE POLYMORPHISM.

exonic variant: A DNA alteration located in an exon which may or may not alter an amino acid.

exonization: A procedure, whereby genes attain new exons from non-protein-coding, principally intronic, DNA sequences. Genomic insertions or point mutations in DNA sequences frequently produce substitute splice sites, triggering the splicing system to embrace new sequences as exons or to extend existing exons.

exonuclease: One of the enzymes that removes nucleotides from the free ends of the DNA and not within the strand. These are present in both eukaryotic and prokaryotic cells. They are also found in venoms of lizards and snakes which work as toxins and cleave DNA encoding essential proteins. The majority of exonucleases act in the 3′←5′ direction of the DNA/RNA strand and biochemically catalyse the removal of mononucleotides. As an exonuclease removes nucleotides from terminal positions of nucleic acids one at a time, this creates overhangs called sticky ends. Exonucleases mainly function in the control of genetic quality, DNA proofreading during replication, homologous combination and DNA repair. Therefore, they are known to be the protectors of genome stability and thus exonucleases are central to many DNA metabolism processes. *See* DNASE.

exonuclease I: This is DNA-specific exonuclease catalyses the removal of nucleotides from linear single-stranded DNA in the 3′ to 5′ direction and is ideal for removal of single-stranded primers in PCR reactions prior to Sanger DNA sequencing or SNP analysis, removal of single-stranded primers for nested PCR reactions and removal of linear single-stranded DNA, leaving behind double-stranded DNA in the sample.

exonuclease II: This enzyme is associated with DNA polymerase I, which contains a 5′ exonuclease that clips off the RNA primer contained immediately upstream from the site of DNA synthesis in a 5′→3′ manner.

exonuclease III: This is a double-stranded DNA-specific exonuclease that initiates at the 3′ termini of linear double-stranded DNA with 5′ overhangs or blunt ends and 3′ overhangs containing less than four bases and also initiates at nicked sites in double-stranded DNA. It catalyses the removal of nucleotides from linear or nicked double-stranded DNA in the 3′ to 5′ direction. Exonuclease III is ideal for (i) site-directed mutagenesis, (ii) preparation of single-stranded DNA for dideoxy sequencing and (iii) preparation of nested deletions in double-stranded DNA.

exonuclease IV: This enzyme adds a water molecule, so it can break the bond of an oligonucleotide to nucleoside 5′ monophosphate. This exonuclease requires Mg^{2+} in order to function and works at higher temperatures than exonuclease I.

exonuclease V: This enzyme is a 3′ to 5′ hydrolysing enzyme that catalyses linear double-stranded DNA and single-stranded DNA, which requires Ca^{2+}. This enzyme is extremely important in the process of homologous recombination.

exonuclease VI: This enzyme has an extra nuclease activity related with DNA polymerase. Like exonuclease II it is an exonuclease and releases deoxynucleoside-5′-monophosphates. Among the few substrates so far studied, it appears to degrade d(A-T) preferentially. Like exonuclease II, this new activity might perform a normal polymerizing function of the enzyme. In contrast to exonuclease II, this enzyme catalyses sequential degradation beginning at the 5′-terminus of polydeoxynucleotides.

exonuclease VII: This DNA-specific exonuclease cleaves linear single-stranded DNA in both the 3′ to 5′ and 5′ to 3′ directions. Its preferred substrate is linear single-stranded DNA and is **ideal for** removal of primers with or without 3′ or 5′ terminal phosphorothioate bonds, mapping positions of introns in genomic DNA and removal of single-stranded DNA, leaving behind the double-stranded DNA in a sample.

exonuclease VIII: This enzyme is a 5′ to 3′ dimeric protein that does not require ATP or any gaps or nicks in the strand but requires a free 5′ OH group to carry out its function.

exonucleases, classification of: In prokaryotes and eukaryotes, exonucleases are of three types: (i) Decapping 5′ to 3′ exonuclease, (ii) Independent 5′ to 3′ exonuclease and (iii) Poly A specific 3′ to 5′ exonuclease.

exonucleases, examples of: EXONUCLEASE I, EXONUCLEASE II, EXONUCLEASE III and λ EXONUCLEASE. *See details as separate terms.*

exopeptidase protein sequencing, engineered nanopores for: Proteins, unlike DNA, do not have a uniform charge, and the electrophoretic force cannot generally be used to translocate or linearize them. A. Sauciuc and others (2023) show that the introduction of sets of charges in the lumen of the CytK nanopore spaced by around 1 nm creates an electroosmotic flow that induces the unidirectional transport of unstructured natural polypeptides against a strong electrophoretic force. Molecular dynamics simulations indicate that this electroosmotic-dominated force has a strength of around 20 pN at −100 mV, which is similar to the electric force on single-stranded DNA. Unfolded polypeptides produce current signatures as they traverse the nanopore, which may be used to identify proteins. Nanopores can identify individual amino acids cleaved by a peptidase. This approach may eventually lead to a method for nanopore protein sequencing (A. Bonini and others in 2024). In two independent publications in *Nature Methods*, K. Wang and others in 2023 and Y. Zhang and others in 2023 show that nanopores can identify the 20 proteinogenic amino acids and some of their PTLMs. Coupled with an exopeptidase, a bottom-up approach to protein sequencing using nanopores is on the horizon.

exoribonuclease: The enzymes that degrade RNA by removing terminal nucleotides from either the 5′-end or the 3′-end of the RNA molecule. Accordingly, there are two types of exoribonucleases: 5′–3′ exoribonucleases and 3′–5′ exoribonucleases. Exoribonucleases exist in all kingdoms of life – bacteria, archaea and eukaryotes. Exoribonucleases are involved in the degradation of many different RNA species, including messenger RNA, transfer RNA, ribosomal RNA and miRNA. Exoribonucleases can be single proteins (such as RNase D or RNase PH) or complexes of multiple proteins, such as the exosome complex in which four of the major exoribonuclease families are represented.

exoribonuclease I: Degrades single-stranded RNA from 5′→3′, exists only in eukaryotes. Exoribonuclease I from yeast is a 175 kDa protein that is responsible for the majority of cytoplasmic mRNA degradation.

exoribonuclease II: A close homologue of Exoribonuclease I. It causes exonucleolytic cleavage in the 3′→5′-direction to yield nucleoside 5′-phosphates. This enzyme has a preference for single-stranded RNA.

exoribonuclease III: This enzyme, isolated from nuclei of mouse liver, kidney, embryo, mammary tumour and Ehrlich ascites tumour, degrades single-stranded RNA to nucleoside 5′-monophosphates. This enzyme does not hydrolyse DNA.

exoribonuclease-mediated mRNA transcription termination: Eukaryotic organisms utilize a conserved exoribonuclease-mediated machinery to terminate the mRNA transcription by RNA polymerase II (Pol II). Y. Zeng and others in 2024 describe two cryogenic electron microscopy structures of *Saccharomyces cerevisiae* Pol II pre-termination transcription complexes attached to the 5′-to-3′ exoribonuclease Rat1 and its partner Rai1. These structures reveal that Rat1 displaces the elongation factor Spt5 to dock at the Pol II stalk domain. Rat1 safeguards the RNA exit channel of Pol II, directs the nascent RNA towards its active centre and loads three nucleotides at the 5′ terminus of the nascent RNA. The structures further reveal that Rat1 rotates towards Pol II as it shortens RNA. These results describe the structural mechanism for the Rat1-mediated termination of mRNA transcription by Pol II in yeast and the exoribonuclease-mediated termination of mRNA transcription in other eukaryotes.

exoribonucleases, examples of: Exoribonuclease are classified as polynucleotide phosphorylase (PNPase), RNase II, RNase T, PH, RNase II, RNase R, RNase D, Oligoribonuclease, Exoribonuclease I and Exoribonuclease II. *See details as separate terms.*

exosmosis: In plants, diffusion of solvent outside the cell. *Contrasts* ENDOMOSIS.

exosome: Exosomes are 50–90 nm vesicles secreted by a wide range of mammalian cell types. Exosomes possess a mechanism for selective removal of many plasma membrane proteins. A major eukaryotic ribonuclease is located in both the nucleus and the cytoplasm, which contributes to the processing, quality control and/or turnover of a large number of cellular RNAs.

exospory: The production of spores that germinate into free-living multicellular gametophytes. *Contrasts* ENDOSPORY.

exotesta: The outer layer of the testa (seed coat) derived from the outer integument of the ovule. The exotesta functions as the mechanical layer.

exotic: Not native. Introduced from another geographical region or country.

exotic germplasm: Germplasm that has not been adapted by selective breeding to the environment intended for its offspring via selective breeding by man.

exotoxic: Refers to damage to the environment.

exotoxins: Proteins produced by bacteria that are able to diffuse out of the cells; generally, more potent and specific in their action than endotoxins.

expanded DNA: Slipping in an extra benzene ring creates a broader DNA double helix that is similar to, but different from natural DNA. A normal base and an expanded base (with an extra benzene ring) pair up according to hydrogen bonding and size complementarity to form a duplex with four possible base pairs. There are four viable partnerships in the expanded duplex: xG-C, G-xC, xA-T and A-xT. Expanded DNA can code more genetic information as it would provide an unexplored avenue towards expanding the genetic alphabet. In expanded duplex DNA, the so-called major and minor grooves of xDNA duplex are markedly wider and swallower. *Aka* xDNA.

expanded DNA repeats: These DNA repeats seem to be predisposed to such expansion because they have unusual structural features, which disrupt the cellular replication, repair and recombination machineries. The presence of expanded DNA repeats alters gene expression in human cells, resulting in disease.

expanded DNA repeats and human diseases: Nearly 30 hereditary disorders in humans result from an increase in the number of simple repeats in genomic DNA. These DNA repeats seem to be predisposed to such expansion because they have unusual structural features, which disrupt the cellular replication, repair and recombination machineries. The expanded DNA repeats alter gene expression in human cells, leading to disease. Many of these diseases are caused by repeat expansions in the non-coding regions of their resident genes. Peculiar structures of repeat-containing transcripts are at the heart of the pathogenesis of these diseases.

expansion of genetic alphabet: A new Watson-Crick base pair with a hydrogen pattern different from that in the A·T and G·C base pairs is inserted into the duplex DNA and RNA by DNA and RNA polymerases and expands the genetic alphabet from four to six letters. The genetic code can thus be extended artificially.

expectation value (E-value): The number of different alignments with scores equivalent to or better than that expected to occur in a database search by chance. The lower the E-value, the more significant the score. This represents the number of sequences with a score greater than, or equal to, X, expected absolutely by chance. The E-value connects the score ('X') of an alignment between a user-supplied sequence and a database sequence, generated by any algorithm, with how many alignments with similar or greater scores that would be expected from a search of a random sequence database of equivalent size. Version 2.0 E-values are calculated using Hidden Markov Models, leading to more accurate estimates than before.

expected heterozygosity: *See* GENE DIVERSITY.

expected lifetime fecundity: *See* R_0.

expected ratio: In statistical analysis, evaluation of an observed ratio (the ratio by which deviations are calculated) to determine whether they are significant.

expected results: Known outcomes for a specific test for a particular disease or condition, obtained from published data, reports, etc.

experiment: A controlled test or investigation designed to deliver evidence for, or preferably against, a hypothesis about the natural or physical world.

experimental alteration of germplasm: The changing of the germplasm so that the inheritance is altered. It can be produced by ionizing radiation, ultraviolet light, heat, chemicals and novel methods such as transduction and transformation.

experimental design: The planning of a process of data collection. Also refers to the information necessary to describe the interrelationships within a set of data. It includes considerations, such as the number of cases, sampling methods, identification of variables and their scale types, identification of repeated measures and replications.

experimental error in QTL detection: Marker genotyping and missing data can affect the order and distance between markers within linkage maps. A reliable QTL map can only be produced from reliable phenotypic data.

explanation: Preparation of an explant by isolating tissue or a group of cells from an organism or embryo.

explanatory variable: It explains variations in the response variable. In an experimental study, it is manipulated by the researcher. *Aka* INDEPENDENT VARIABLE *or* PREDICTOR VARIABLE.

explant: Tissue/organ excised from its original site and transferred to a culture medium for growth and maintenance. During plant tissue culture, the organs like leaves, meristems, shoot apices, embryos, anthers, cotyledons and roots are excised (cut) from the mother plant and then aseptically cultured on the growth medium. Excised organs are called explants.

explant culture: The maintenance or growth of an explant in culture.

exploitation competition: A form of competition that revolves around the superior ability to gather resources, rather than an active interaction among organisms for these resources.

exploiting gene regulation: In biology, exploitation is the act of employing some biological information, process, phenomenon, concept or principle to the greatest possible advantage. Various approaches in use for exploiting gene regulatory mechanisms are gene regulatory networks; exploiting network-based approaches for understanding gene regulation and function; exploiting regulatory variation to identify genes; and gene expression regulation by miRNAs.

exploitive adaptation: The genetic capacity of an organism to bring out phenotypic changes in response to an altered environment so as to achieve acclimatization.

exploratory research: The area of research which structures and identifies new problems.

exploratory system: A system that shapes originally random variation so as to produce a well-informed functional consequence.

exploring epigenome: Changes in the epigenome also can lead to, or be a consequence of, disease. Epigenomics has now become an important part of attempts to better comprehend the human body and to improve human health.

explosion method: A technique for gene-into-cell introduction in which the gene or foreign DNA is driven into plant cells by the force of an explosion (vaporization) of a drop of water to which the gene and gold particles have been added. The explosion is caused by the application of high-voltage electricity to the drop of gene-laden water. The water is then vaporized explosively, driving the shot (gold particles) and genetic material through the cell membrane. The plant cell then heals itself, incorporating the new gene into its genetic complement.

explosive evolution: Refers to the evolution of a large number of species from a single closed species. *Aka* EXPLOSIVE SPECIATION.

exponential growth: It is revealed when the rate of change of the value of a mathematical function is proportional to the function's current value, resulting in its value at any time being an exponential function of time. *Aka* GEOMETRIC GROWTH *or* LOGARITHMIC GROWTH.

exponential growth phase: *See* LOG PHASE.

export: rRNA is used in the nucleus for assembly of ribosomal subunits. Hence, rRNA and unprocessed mRNA are not exported. For this, export of processed mRNA and snRNA depends upon 5′-CAPPING with a methyl group (5′-monomethyl cap). Several cap-binding proteins have been identified. Exportins play a role in the export process.

exposure: This term in ecotoxicology refers to the interaction of organisms with an environmental stressor at a particular place and time.

exposure assessment: A study of the resources by which organisms may come across a potentially toxic level of a chemical or other environmental stressor.

express: To translate the cell's genetic information stored in the DNA (gene) into a specific protein (synthesized by the cell's ribosome system). Certain proteins (i.e. when present in relevant cells) regulate the expression (e.g. increase/decrease/timing) of some genes.

expressed fraction of genome: Several supposedly silent intergenic regions in the genome of budding yeast are actually transcribed by RNA polymerase II, suggesting that the expressed fraction of the genome is higher than anticipated.

expressed gene: *See* GENE EXPRESSION.

expressed proteins, purification of: After expression of the gene product, the purification of the protein is essential. Since the vector is introduced to a host cell, the protein of interest has to be purified from the proteins of the host cell. Therefore, to make the purification process easy, the cloned gene should have a tag (histidine tag or any other marker peptide). Expression vectors must have expression signals such as a strong promoter, a strong termination codon, adjustment of the distance between the promoter and the cloned gene, and the insertion of a TRANSCRIPTION TERMINATION SEQUENCE (TTS) and a PORTABLE TRANSLATION INITIATION SEQUENCE (PTIS).

expressed sequence tag (EST) clustering: Clustering or assembling of EST data. EST is useful information in a sense that it is a profile of expressed gene sequences. It usually does not contain full-length gene sequences, because about 600 bp sequences are generated from 5′ and 3′ end of cDNA clones. At the same time, EST permits low-quality bases due to single-pass sequencing, and often, some sequences are highly redundant in certain genes. In order to overcome these disadvantages and collect more unique sequences (called UniGene), clustering ESTs is necessary. *Phrap*, *TIGR* assembler and *CAP3* are used to cluster or assemble EST data.

expressed sequence tag (EST) method: In this method, given by P. Liang and A.B. Pardee in 1992, mRNAs from desired cells are isolated, cDNAs are made by reverse transcription and a single-pass sequence of 200–500 bp are determined from 5′ or 3′ or both ends of randomly chosen cDNA clones from various libraries. These sequences are known as EXPRESSED SEQUENCED TAGS. A major problem with genomic sequences is how to distinguish coding regions from non-coding introns. From comparisons between EST and cDNA sequences and sequences similar to known coding sequences, one can assign intron position for many genes. Sequencing only the beginning portion of the cDNA produces what is called a 5′ EST. A 5′ EST is obtained from the portion of a transcript that usually codes for a protein. A GENE FAMILY is a group of closely related genes that produces similar protein products. These regions tend to be conserved across species and do not change much within a gene family. Sequencing the ending portion of the cDNA molecule produces what is called a 3′ EST. Because these ESTs are generated from the 3′ end of a transcript, they are likely to fall within non-coding, or untranslated regions (UTRs), and therefore tend to exhibit less cross-species conservation than do coding sequences. A UTR is that part of a gene that is not translated into protein. For the human genome, computer programmes, such as GRAIL, can predict the beginning and ends of gene and intron positions with high reliability. The NetPlantGene programme is used for the same purpose in plants. By 1998, the database of private companies, like Incyte and Human Genome Sciences, claimed to have 2,000,000 human EST, and the public database contained 127,000 EST Sequences from 19 plant species.

expressed sequence tag polymorphisms (ESTPs): ESTPs are derived from ESTs. Using EST sequences polymerase chain reaction (PCR) primers are designed to amplify ESTs from individual genomic DNA. Allelic polymorphism in the amplification product (ESTPs) can be revealed using different modern methods for the detection and visualization of DNA alterations. ESTPs mostly reveal genetic variation within genes, although variation can be found in both coding and non-coding regions of genes. Thus, ESTPs are the most informative markers in terms of gene function among the most recently developed ones and are the first genetic markers that offer real potential for detecting adaptive genetic diversity broadly.

expressed sequence tags (ESTs): ESTs are PCR-based DNA markers using a pair of primers. These are like STS markers. An EST is a DNA sequence from a cDNA clone that corresponds to an mRNA or a part thereof. ESTs in most of the genomes are 150–400 bp long and are useful in search of similarity and mapping of genomes. EST databases are a wonderful resource for discovering genes. They act as physical markers for cloning and full-length sequencing. Now EST databases exist for a number of crop plants including tomato, soybean, maize, rice, wheat, cotton, potato and oilseed rape. STS markers can be developed from EST databases. EST-derived microsatellites and SNPs have been identified to facilitate the mapping of ESTs. ESTs are being used for synteny mapping and gene cloning.

expressed sequence tags (ESTs), disadvantages of: EST technology however has certain disadvantages: (i) incompleteness of sequence information and (ii) high cost.

expressed sequence tags (ESTs), limitations of: (i) Low copy number genes, (ii) rare tissues, (iii) enrichment of 3′ ends and (iv) incomplete coverage of genes.

expressed sequence tags (ESTs), uses of: ESTs have been extensively utilized for (i) development of DNA-based molecular markers representing the functional region of the genome (e.g. EST-SSR and EST-SNP); (ii) annotation of genomic sequences and (iii) analysis of differential expression of genomic sequences.

expression: (i) In a programming language, expression is a combination of one or more constants, variables, operators and functions that the programming language interprets (according to its particular rules of precedence and of association) and computes to produce ('to return', in a stateful environment) another value. This process, as for mathematical expressions, is called EVALUATION. (ii) In gene expression, it is the production of an observable phenotype by a gene – usually by directing the synthesis of a protein.

expression analysis of cloned genes: Methods used in expression analysis of cloned genes are based on RNA product (NORTHERN BLOTTING and RNA SEQUENCING) and protein product (WESTERN BLOTTING and EASTERN BLOTTING) of the genes.

expression component: One of the two components of riboswitches. This component regulates gene expression, and contrary to the other component (aptamer), it can vary greatly in order to affect the different processes of transcription, translation and RNA processing. *See* APTAMER.

expression genomics: It deals with the preparation of quantitative maps of the proteins expressed from cells or tissues. It includes co-expression as a means of identifying biological pathways and their perturbation by disease, drug action and effects of biological stimuli, and identification of disease markers.

expression imbalance map (EIM): A visualization procedure for detecting mRNA expression imbalance regions, reflecting genomic losses and gains at a much higher resolution than conventional technologies such as comparative genomic hybridization. Simple spatial mapping of the microarray expression profiles on chromosomal location offers little information about genomic structure, because mRNA expression levels do not completely reflect genomic copy number and some microarray probes would be of low quality.

expression mapping: Creation of quantitative maps of protein expression from cell or tissue extracts (e.g. expressed sequence tag maps). The approach relies on 2-D gel maps and image analysis and opens up the possibility of studying cellular pathways and their perturbation by disease or other biological stimuli at the whole-proteome level.

expression marker: Any expressed sequence (e.g. a cDNA), a tag derived from serial analysis of gene expression (SAGE), or an expressed sequence tag (EST), that has been recognized by high-throughput expression profiling and serves as a diagnostic (or even prognostic) marker for a disease.

expression microarray profiling combined with drug-induced demethylation: This method banks on the functional property of transcriptional activation to discover genes that have been epigenetically silenced in cancer cells. Demethylating drugs, such as 5-aza-2′-deoxycytidine (5′), can reactivate the expression of genes silenced via promoter hypermethylation by combining 5′ treatment with standard messenger RNA (mRNA) expression profiling on oligonucleotide or cDNA microarrays.

expression of cloned genes in yeast: Most yeast and even bacterial genes are expressed in yeast. Yeast promoters and translation initiation signals are required for the expression of genes. Factors affecting expression are (i) level of transcription is affected by distance between promoter and initiation ATG codon and (ii) differences in codon usage between donor and host organism. 5′ flanking region TATA box was found in most of the yeast genes. All yeast genes had adenine residue at position −3 and usually thymine at position +6. Ideally, cloned genes should be placed downstream from a controllable promoter. Secretion of proteins by yeast can be achieved by addition of a signal sequence. Yeast cells can recognize and correctly process a single sequence from a higher eukaryote.

expression of cloned genes, maximizing the: Following factors affect the expression of cloned genes: promoter strength, translational initiation sequences, codon choice, plasmid copy number, transcription termination, plasmid stability, segregative instability, structural stability and host cell physiology:

expression of *Cry* genes in plants: The *cry1* genes have been successfully transferred into tobacco, potato and tomato. Typically, truncated *cry* genes are used for the production of transgenic plants since the level of expression of the complete genes in transgenic plants is extremely low.

expression of foreign genes: Following transformation, we need to establish whether foreign genes remain intact in the host genome and that they are expressed at desired levels in target tissues and in response only to expected chemical or environmental stimuli. The foreign gene's local environment within the genome can affect its expression. Transgene inactivation, sometimes due to methylation of DNA, can also complicate the assessment of foreign gene expression.

expression of immunoglobulin genes: Cells that manufacture antibody molecules, called IMMUNOGLOBULINS, experience DNA rearrangements during cell differentiation, suggesting that although a genome usually remains constant in different cells, it does undergo a change with regard to immunoglobulin genes in mammals.

expression on the cell surface with antibody screening: Cell-surface receptors can sometimes be isolated using an expression strategy. Cell-surface molecules on lymphocytes, for example, have been identified by the isolation of specific monoclonal antibodies. Likewise, the ligand for many receptors has been isolated before the nature of the receptor is established. In both cases, a cDNA for the cell-surface molecule when expressed will lead to the presence of a binding site on the cell surface. This binding site can be used to screen a library either by a method analogous to the antibody screening or by using the ligand or antibody as an affinity reagent to 'pan' for cells that express the binding site.

expression profiling: The first step in converting genetic information to an observable trait involves the transfer of information from gene to protein through an intermediate molecule called RNA. Measuring all of the RNA produced in one definable moment, to create a global picture of cellular function produces a profile of total expression activity. The level and duration of expression of one or more genes, selected from a particular cell or tissue type, generally obtained by a variety of high-throughput methods, such as sample sequencing, serial analysis or microarray-based detection. DNA microarrays are used to determine the expression level of thousands of genes simultaneously. Expression profiling is a logical next step after sequencing a genome: the sequence tells us what the cell could possibly do, while the expression profile tells us what it is actually doing now.

expression profiling experiments: Expression profiling experiments often involve measuring the relative amount of mRNA expressed in two or more experimental conditions. This is because altered levels of a specific sequence of mRNA suggest a changed need for the protein coded for by the mRNA, perhaps indicating a homeostatic response or a pathological condition.

expression profiling, limitations of: Expression profiling studies generally report those genes that show statistically significant differences under changed experimental conditions. This is typically a small fraction of the genome for several reasons. (i) Different cells and tissues express a subset of genes as a direct consequence of cellular differentiation so many genes are turned off. (ii) Many of the genes code for proteins that are required for survival in very specific amounts so many genes do not change. (iii) Cells use many other mechanisms to regulate proteins in addition to altering the amount of mRNA, so these genes may stay consistently expressed even when protein concentrations are rising and falling. (iv) Financial constraints limit expression profiling experiments to a small number of observations of the same gene under identical conditions, reducing the statistical power of the experiment, making it impossible for the experiment to identify important but subtle changes. (v) It takes a great amount of effort to discuss the biological significance of each regulated gene, so scientists often limit their discussion to a subset. (vi) Newer microarray analysis techniques automate certain aspects of attaching biological significance to expression profiling results, but this remains a very difficult problem.

expression profiling, uses of: Sometimes, a scientist has an idea (a hypothesis) about what is going on and he or she performs an expression profiling experiment to potentially disprove this hypothesis. Here the scientist is making a specific prediction about levels of expression that could turn out to be false. More commonly, expression profiling takes place before enough is known about how genes interact with experimental conditions for a testable hypothesis to exist. With no hypothesis, there is nothing to disprove but expression profiling can help to identify a candidate hypothesis for future experiments.

expression proteomics: This branch of proteomics is dedicated to the analysis of protein abundance and involves the separation of complex protein mixture, the identification of individual components, their systematic quantitative analysis and characterization of protein modification.

expression screening: A method of screening for a specific cDNA clone where the cDNA is inserted next to a promoter active in the host cell and an immunoassay or bioassay is used to detect the required clone.

expression stability: The tendency for gene expression levels in transgenic plants to remain stable over multiple generations (i.e. not affected by epigenetic silencing phenomena).

expression system: A system explicitly designed for the production of a gene product of choice. This is normally a protein although may also be RNA or a ribozyme. An expression system consists of a gene, normally encoded by DNA, and the molecular machinery required to transcribe the DNA into mRNA and translate the mRNA into protein using the reagents provided. In the broadest sense, the expression system includes every living cell, but the term is more normally used to refer to expression as a laboratory tool.

expression variegation: A type of disparity in gene expression during development that causes streaks or patches of cells to have different phenotypes.

expression vector: A cloning vector that is engineered to allow the expression of protein from a cDNA. The expression vector provides appropriate promoter and restriction sites that allow the insertion of cDNA. The plasmid also contains regulatory sequences that function as enhancer and promoter regions and result in efficient transcription of the gene carried on the expression vector. The goal of a well-designed expression vector is the making of large amounts of the protein product. Nowadays, overexpression vectors are available, which results in the overexpression of the desired protein or toxin for bioassays. Expression vectors or overexpression vectors are also used to check the efficacy of various *Bt* genes by overexpressing them in *Escherichia coli* and using the protein thus formed in insect bioassays. This technique is highly rewarding as the effective gene(s) can be shortlisted for further transformation. *Bt* genes are attached to promoters, such as CaMV 35S, to achieve their expression in plants. The cloning vehicles which provide these signals and that can be used in the production of recombinant protein are called expression vectors.

expression vector for *Bacillus subtilis*: Specific *Bacillus subtilis* expression vectors have been constructed. Plasmid pPL603 was generated which can be used to clones active in *Bacillus subtilis* and specific Km^R and carries structural gene CAT, which lacks a promoter. Plasmid pPL608 was derived from pPL603. *Escherichia coli trpC* gene and mouse *DHFR* gene was cloned in pPL608. Eight N-terminal amino acids of foot and mouth disease virus VP1 antigen was fused into the Em^R gene of *Bacillus subtilis* vector. *Escherichia coli* lac promoter was placed on the 3′ side of a hybrid promoter controllable by *Escherichia coli* lac repressor. *Escherichia coli* lac repressor gene was placed under the control of a promoter and binding site functional in *Bacillus subtilis*.

expression vectors, limitations of: There has been controversy over the use of expression vectors to modify crops due to the fact that there might be unknown health risks, possibilities of companies patenting certain genetically modified food crops and ethical concerns. None the less, this technique is still being used and heavily researched. For example, baculovirus happens to be commonly used as expression vector in insect cells.

expression vectors, uses of: (i) Important basic tools for biotechnology and the manufacture of proteins that are central for medical treatments of specific diseases. (ii) Expression vectors are also employed in molecular biology techniques such as site-directed mutagenesis. (iii) Expression vectors have been engaged in the introduction of specific genes in organisms, especially plants used in agriculture. Expression vectors have also been employed to introduce genes such as for vitamin A precursor, beta-carotene, into rice plants. This product is called 'golden rice'. (iv) This process has also been used to introduce a gene into plants that manufactures an insecticide, called *Bacillus thuringiensis* toxin or *Bt* toxin which reduces the need for farmers to apply insecticides since it is produced by the modified organism. (v) Expression vectors have been successfully used to extend the ripeness of tomatoes by altering the plant so that it produces less ethylene, the chemical that causes the tomatoes to rot.

expression-based studies on designer PPR proteins: Gene expression modification studies have been conducted using DESIGNER PENTATRICOPEPTIDE REPEAT (DPPR) proteins in the mitochondria of *Arabidopsis*, maize, soybean and sugarbeet. The development of customized PPR assembly would be used for plant organelle RNA manipulations through targeting of organellar transcripts.

expression-linked copy (ELC): A copy of the gene which is active.

expressivity: In genetics, expressivity is the degree to which a phenotype is expressed by individuals having a particular genotype. Although a gene is stable, its phenotypic consequence is variable. Some genetic disorders exhibit variable expressivity. *Compare with* PENETRANCE.

expressivity modifiers: Those genes which are responsible for variable expressivity of a phenotype.

exserted: Projected beyond something (e.g. stamens beyond the corolla tube).

exstipulate: Devoid of stipules.

extein: The remaining portions of the protein after inteins (protein introns) have been spliced out. *Compare with* INTEINS.

extended anticodon hypothesis: The structure of the anticodon loop and the proximal anticodon stem are associated with the sequence of the anticodon. In other words, the anticodon is extended into the nearby structure and consists of (i) two nucleosides at the 5′ end of the anticodon loop, (ii) three nucleosides of anticodon, (iii) two nucleosides at the 3′ end of the anticodon loop and (iv) five nucleoside pairs in the anticodon stem. Extended anticodons are involved in translation. The 3′ nucleotide of anticodon triplet is considered to be the most important and is therefore called CARDINAL NUCLEOTIDE.

extended heredity: August Weismann proposed that environmentally caused changes in an organism are not inherited. Russell Bonduriansky and Troy Day in their book *Extended Heredity* published in 2018 believe that 'Weismann's barrier' is remarkably porous, with epigenetics representing 'only the tip of the iceberg'.

extended lattices: Small nucleic acid complexes can be equipped with sticky ends and combined into larger two-dimensional periodic lattices containing a specific tessellated pattern of the individual molecular tiles.

extended phenotype: The phenotype of all the individuals affected by a gene.

eXtensible HyperText Markup Language (XHTML): Part of the family of XML markup languages. It mirrors or extends versions of the widely used HyperText Markup Language (HTML), the language in which web pages are formulated.

extension (in nucleic acids): (i) A single-stranded tail found at the end of a restriction fragment. Different restriction enzymes generate extensions at either 5′ or 3′ end of the DNA strand. (ii) The nucleic acid strand elongation (lengthening) that transpires in a polymerization reaction.

extensive crop: A crop that has low production costs and profit margins. For example, soybeans, maize and wheat are typical extensive crops in North America.

external fertilization: A type of fertilization in which a sperm unites with an egg external to the body or bodies of the parent organisms. For example, liverworts develop an umbrella-shaped structure, both male and female. The sperm released from the male umbrella swim along with the plant's moist surface and fertilize the egg. *Contrasts* INTERNAL FERTILIZATION.

external stimuli causing epigenetic changes: External and internal stimuli may result in epigenetic changes such as DNA methylation, histone modifications and the action of non-coding RNAs that change chromatin structure causing changes in gene expression. If heritable, these changes can be a source of stable epialleles, impacting phenotypic variation, and can be exploited for the selection of desired genotypes in EPIBREEDING PROGRAMMES. Ultimately the combinatorial application of epibreeding and breeding could pave the way for the development of efficient strategies for crop improvement.

extinction: The termination of the existence of a particular kind of organism or a particular taxon, often a species, as a result of the death of the last individual of the taxon, though the capacity to breed and recover may have been lost much before this point, rendering the taxon functionally extinct.

extracellular: Of or occurring in the space outside the plasma membrane of a cell. *Contrasts* INTRACELLULAR.

extracellular matrix (ECM): Region outside of metazoan cells which includes compounds attached to the plasma membrane, as well as dissolved substances attracted to the surface charge of the cells. ECM is a complex network of polysaccharides (such as glycosaminoglycans or cellulose) and proteins (such as collagen) secreted by cells. ECM serves as a structural element in tissues and also impacts their development and physiology. The ECM plays a dual role in keeping animal cells adhered together and well as buffering them from their environment.

extracellular microRNA biomarkers: Extracellular circulating miRNAs offer several potential advantages as informative biomarkers when compared to protein-based blood biomarkers. For instance, while low abundance can significantly hinder the detection of some protein-based biomarkers, most circulating miRNAs can be readily detected by PCR. Circulating extracellular miRNAs clearly possess many desirable properties when compared to protein biomarkers.

extracellular polysaccharide (EPS): A sugar polymer that aids in movement, helps prevent desiccation and contributes to the slimy appearance of bacteria, fungi and nematodes. EPS probably also promotes colonization of plant tissues and disease development in bacteria.

extracellular vesicles: Lipid bilayer-delimited particles that are naturally released from almost all types of cells and, unlike a cell, cannot replicate.

extrachromosomal: Structures or processes outside the nuclear chromosomes.

extrachromosomal circular DNA (eccDNA): A type of double-stranded circular DNA that is derived and free from chromosomes. It is found in various organisms, including humans. It ranges in size from tens to millions of base pairs.

extrachromosomal element: All genetic elements that are not at all times part of a chromosome, for example, plasmids, phages, transposons, insertion sequences, plastids (mitochondrial and chloroplast) genomes.

extrachromosomal gene: Any gene that is not carried by the prokaryotic cell's chromosome(s). For example, genes carried by plasmids. *Compare with* EXTRANUCLEAR GENES.

extrachromosomal inheritance: Inheritance that is not governed by chromosomal determinants (located in the nucleus) but by cytoplasmic determinants (i.e. those in mitochondria and chloroplasts). *See* CYTOPLASMIC INHERITANCE *and* MATERNAL INHERITANCE.

extragenic suppressor: In this case, mutations in tRNA gene cause suppression of a protein-encoding gene. These suppressors may be further classified as NONSENSE SUPPRESSORS, MIS-SENSE SUPPRESSORS or FRAMESHIFT SUPPRESSORS.

extranuclear genes: Genes that are residing within the cell, but outside the nucleus. Generally, extranuclear genes are located in the organelles such as mitochondria and chloroplasts. *Compare with* EXTRACHROMOSOMAL GENES.

extranuclear genetic factors: All genetic factors that lie outside the nucleus are called extranuclear genetic factors or extranuclear genomes. E. Baur in 1908 and C. Correns in 1909 offered the first genetic evidence for the existence of cytoplasmic genes regulating chloroplast development. Extranuclear genomes are present in mitochondria, chloroplasts, kinetoplasts and centrioles. These genes were further described by R. Sager in 1977 after the name of the organelle (e.g. CHLOROPLAST GENES and MITOCHONDRIAL GENES).

extranuclear inheritance: A transmission of genes that takes place outside the nucleus. It is characterized by differences in reciprocal crosses, progeny shows characters of the female parent and displays no linkage with the nuclear genes. *See* CYTOPLASMIC INHERITANCE.

extrastaminal: Outside the stamens or androecium, usually referring to the location of a nectary disk.

extremophile: An organism that flourishes in environments that are at the extremes of conditions where life is normally found.

extremophilic bacteria: Bacteria that live and reproduce outside (either colder or hotter) the typical temperature range of 40°F (4°C) to 140°F (60°C) and that tend to be found inside the Earth. Other extremes are high pressure (e.g. at the ocean bottom), salt saturation (e.g. the Dead Sea), pH lower than 2 (e.g. coal deposits), pH higher than 11 (e.g. sewage sludge), high levels of radiation, etc.

extremozymes: Enzymes that function under extreme conditions. Some organisms (microbes) live under extreme temperature conditions ranging from freezing point to as high as 140°C. These organisms are called extremophiles and the enzymes that function under these conditions are described as extremozymes. Because extremozymes can catalyse reactions under high pressure, high temperatures, etc., they are increasingly being employed as catalysts for industrial processes. Examples of extremozymes include cellulases, amylases, xylanases, proteases, pectinases, keratinases, lipases, esterases, catalases, peroxidases and phytases, which have great potential for application in various biotechnological processes.

extrinsic ethical concerns in genetic engineering: These concerns are based on doubts about the technology, its potentiality, newness and applicability to all life forms.

extrorse: Of anther locules, opening towards the outside of the flower. *Contrasts* INTRORSE *and* LATRORSE.

exudate: A liquid excreted or discharged from diseased tissues, from roots and leaves or by fungi.

eyes: In DNA replication, it refers to the configuration of replicating DNA in eukaryotic chromosomes.

eyespot: Light-sensitive organelle found in many groups of protists and in some metazoans.

F

F^- bacteria: Bacteria that do not have an F plasmid. *Aka* FEMALE BACTERIA or F^- CELL.

F′ cell: A bacterium that carries a modified F plasmid, a small piece of DNA derived from the host bacterial DNA molecule.

F factor: A type of transmissible (i.e. can enter other cells) plasmid that consists of double-stranded DNA having about 10^5 bp carrying 25 genes that direct conjugal transfer of DNA between bacteria (e.g. *Escherichia coli*). Its presence confers donor ability (maleness). The F factor is 40 times smaller than the bacterial chromosome. In F^+ stain, the F factor is circular, and one copy is present in each bacterial cell. It is known as an EPISOME of bacteria, which has three regions. One region of the plasmid contains genes involved in the regulation of the DNA replication (*rep* genes), the other region contains transposable elements (IS3 and Tn1000, IS3 and IS2) – genes involved in its ability to function as an episome – and the third large region, the *tra* region, consists of *tra* genes and possess the ability to promote the transfer of plasmids during conjugation. *Aka* FERTILITY FACTOR, F PLASMID, SEX FACTOR or SEX PARTICLE. *See* CONJUGATION and EPISOME.

F pili: Hair-like projections of an F′ or Hfr bacterium involved in attaching during conjugation and presumably through which DNA passes. *Aka* SEX PILI.

F plasmids: These are the first described plasmids that play a major role in conjugation in bacteria. It is a circular ds-DNA molecule of 99,159 base pairs. One region of the plasmid contains genes involved in regulation of the DNA replication (*rep* genes), the other region contains transposable elements: two copies of the insertion sequence IS3, one IS2 and one transposon Tn1000 (also called γδ) involved in its ability to function as an episome. The third large region, the *tra* region, consists of *tra* genes and possesses the ability to promote the transfer of plasmids during conjugation (e.g. F plasmid of *Escherichia coli*). *Aka* F-FACTORS.

F-Box proteins: Proteins manufactured within some eukaryotic cells that have a crucial role in the degradation of cellular regulatory proteins, after those regulatory proteins have 'completed their job' in the cell.

F-duction: The transfer of bacterial genes from donor to recipient bacterium via an F′ plasmid as the basis for a gene mapping technique. *Aka* F-DUCTION, SEXDUCTION or F-MEDIATED TRANSDUCTION.

F-test: An F-test is any test that uses F-distribution with a null hypothesis. Researchers may employ the F-test to evaluate the parity of two population variances. They use it to assess if the two independent samples selected from a normal population have similar variability or not. Also, a statistical procedure for the test of hypothesis on the equality of three or more means.

F^+ bacteria: Bacteria that have an F plasmid. *Aka* MALE BACTERIA or F^+ CELL.

F_0, F_1, F_2, F_3: Various generations (parent, offspring, grand-offspring, great-grand-offspring, etc.)

F_1: Refers to the first filial generation. F_1 is obtained by crossing two different parents. The parents are quite distinct: one is homozygous dominant for both traits, and the other is homozygous recessive for both traits. The parental generation is referred to as P_1.

F_1 hybrids: The first-generation offspring of crossbreeding two pure-bred lines. These are sometimes observed to be more healthy, productive and uniform than their parents. *Aka* FIRST FILIAL HYBRIDS.

F_1 synthetic variety (Syn-1): A variety derived by intercrossing a specific set of clones or seed-propagated lines. These may include normally cross-fertilizing or self-fertilizing crop varieties into which mechanisms have been introduced to maximize cross-fertilization.

F_2: Refers to the second filial generation (i.e. the progeny produced by the intercrossing or self-fertilization of F_1 individuals).

F_2 hybrid: A plant breeding term for the result of a plant arising from a cross between two F_1 hybrids. This also refers to self-pollination in a population of F_1 hybrids.

F_2 variety: Refers to the next-generation seed derived from the F_1 hybrid generations. The F_2 variety cannot be perpetuated by growing additional generations. An example of an F_2 variety is the tomato variety 'Foremost F_2'.

F_3: Refers to the third filial generation or the progeny obtained by self-fertilizing F_2 individuals.

facilitated diffusion: *See* PASSIVE TRANSPORT.

facilitated epialleles: Facilitated epialleles have a dependence upon genetic variant breakdowns.

facilitated recurrent selection: A type of recurrent selection in which genetic male sterility is sustained in the population to sustain heterozygosity and genetic diversity, and to allow the recombination and shifting of gene frequencies.

fact: In science, a 'fact' typically refers to an observation, measurement or other form of evidence that can be expected to occur the same way under similar circumstances. However, scientists also employ the term 'fact' to refer to a scientific explanation that has been tested and confirmed repeatedly such that there is no longer a compelling reason to keep testing it or looking for additional examples.

factor: The Mendelian unit of inheritance; the determining agents responsible for each trait. An independent variable under investigation in an experiment as a probable reason for variation. In a diploid organism, there are two factors for a trait. Each factor is present on the two homologous chromosomes at the same corresponding position. *Aka* ELEMENT. Now this term has been replaced by 'GENE'.

factorial: The product of all integers from the specified number down to one (unity).

factorial analysis: A multivariate statistical analysis in which the independent variables are clustered into a small number of factors describing the variance of the dependent variables. It is used in cluster analysis. This method pulls the largest common variance across all variables and converts it into a single number. The additional analysis uses this number as an index of all factors. *Aka* FACTOR ANALYSIS.

factorial design: In a factorial design, the effect of two or more factors can be concurrently observed (e.g. planting time, fertilizing, watering regime) – the design offers evidence of the average effect of the individual factors as well as the interaction between these factors. This design type also permits a wider application of the deductions reached for the effect of each factor as each factor is verified over a wide range of conditions of other pertinent factors. The statistical analysis used is the ANALYSIS OF VARIANCE (ANOVA) followed by Duncan's multiple range test. Because the total number of treatment combinations (the product of the numbers of levels of the factors) may be large, thus it is often essential to test only some of these combinations, giving a 'fractional factorial design'. A suitable fraction may be an orthogonal array or a subgroup of the direct product of abelian groups of orders equal to the number of levels of each factor.

factorial designs (with blocking): These experimental designs have numerous treatment factors and a single blocking factor (giving strata for blocks and plots within blocks). The blocks are so small that they cannot contain a complete replication of the treatment combinations, and so various interactions are confounded with blocks.

factorial trial: An experimental design in which the influences of several different factors are investigated at the same time. The factorial set of treatments comprises all combinations that can be fashioned from different factors. The trial can include randomized complete blocks and Latin squares (e.g. treatments A and B at levels x and y). The sources of variance are replicates A, B, A × B and error.

facultative: Able to accomplish a particular life function, or to live generally, in more than one way. These are capable of changing lifestyles (e.g. from saprophytic to parasitic or vice versa). *Compare with* OBLIGATE.

facultative anaerobe: An organism that is capable of producing energy through aerobic respiration as well as switching to anaerobic respiration depending on the amounts of oxygen and fermentable material in the environment.

facultative apomictic: An apomictic plant, though some seeds develop due to the fusion of male and female gametes (e.g. sorghum).

facultative cells: Cells that can survive in either the presence or absence of oxygen.

facultative heterochromatin (fHC): (i) Heterochromatin that switches to euchromatin at points in a few developmental or physiological states, and heterochromatin in others. fHC includes imprinted (silenced) regions or transcriptionally repressed genes, such as inactive X chromosome or imprinted regions, or transcriptionally repressed genes, hypoacetylated nucleosomes and H3K27me. (ii) fHC is present in only one of a pair of homologues or not permanently present. (iii) The chromatin that becomes inactive in certain parts of the life cycle and resumes activity before entering the germline. For example, one X chromosome gets characteristically condensed and inactivated.

facultative parasite: (i) A chiefly saprophytic organism with feebly pathogenic properties. (ii) A parasite having the ability, or 'faculty', to adapt to an alternative mode of living (i.e. parasitic or saprophytic; rotten plant). Facultative parasites gain their nourishment by digesting dead organic material.

facultative parthenogenesis: A process that enables sexually reproducing organisms to switch between sexual and asexual parthenogenetic reproduction. A.L. Sperling *et al.* (2023) sequenced the genomes of sexually reproducing and parthenogenetic strains of *Drosophila mercatorum* and identified differences in the gene expression in their eggs. They then identified a polygenic system where the increased expression of the mitotic protein kinase *polo*, and decreased expression of a desaturase, *Desat2*, caused facultative parthenogenesis in the non-parthenogenetic species that was enhanced by increased expression of *Myc*. The genetically induced parthenogenetic *Drosophila melanogaster* eggs exhibit *de novo* centrosome formation, fusion of the meiotic products and the onset of development to generate predominantly triploid offspring. They thus demonstrated a genetic basis for sporadic facultative parthenogenesis in an animal.

facultative pathogen: A microorganism capable of growth and reproduction in either a non-living medium or a living host; such that in later instances, a disease condition of the host may arise. *Compare with* OBLIGATE parasite.

facultative saprophyte: An organism that is customarily parasitic but is capable of obtaining food by absorbing dissolved organic material (saprophytic).

facultative weed: A weed observed growing wild as well as in association with humans.

failure of centromeres to perform function: Non-disjunction occurs when the centromere fails to function properly. This leads to the occurrence of aneuploids.

fallow: Cultivatable land kept free from a crop or weeds during the normal growing season.

false negative result: The result of a diagnostic test that wrongly comes out as normal when the disease is present. *Aka* TYPE II ERROR.

false positive result: The result of a diagnostic test that came back positive or abnormal when the disease is not present. False positives can aggravate assessing the performance of a principal substance. *Aka* TYPE I ERROR.

falsifiability: A measure of whether the consequences of a method allow one to determine if the assumptions used have been violated.

familial trait: A trait that occurs more frequently in relatives than in the general population. The trait can be due to genetics, the environment or both.

family: (i) A group of individuals directly related by descent from a common ancestor, with at least one parent in common. Progenies with one parent in common are called half-sibs. Progenies with both parents in common are called full-sibs. A selfing family is a family obtained by self-pollinating a genotype. (ii) A taxonomic group of one or more genera. It is the term for the rank between order and genus.

family (Latin: **familia, pl. familiae**): One of the eight major hierarchical taxonomic ranks in Linnaean taxonomy. It is classified between the order and genus. A family may be divided into subfamilies, which are intermediate ranks between the ranks of family and genus.

family analysis of genetic maps: Different approaches include: (i) bulked segregant analysis, (ii) identity by descent and (iii) whole-genome association.

family selection: The selection of progeny families on the basis of their mean performance. In addition, the best individuals are usually selected in the best families.

family shuffling: A format of DNA shuffling in which nucleotide sequences from multiple homologous genes are recombined to form highly diverse genetic variants.

far-infrared (FIR) imaging system: Utilizes light in the FIR region (spectrum: 15 μm to 1 mm) to study the temperature differences and their effects on salinity tolerance, photosynthesis efficiency, water usage, etc.

farmer selection: In several plant breeding programmes the farmers make the final selection of cultivar. Each farmer is given a different group of new clones or pure lines of a crop emerging from a breeding programme. Farmers then may grow them and select those they like best. Their favourites are usually their property, with the sole facility that the breeder may have some of them for further breeding. The farmers can then grow that material for their own use, and give or sell propagating material to their friends and neighbours. *Aka* PARTICIPATORY PLANT BREEDING.

Farmer's fixative: Comprises anhydrous ethanol and glacial acetic acid in a 3:1 ratio. It is used as a fixing and dehydrating agent in histology. It is also used in conjunction with Feulgen stain and carbolfuchsin stain for chromosome analysis. *Aka* CARNOY'S SOLUTION I.

farmer's privilege: A clause in the plant breeder's rights legislation of most countries that allows a farmer to use some of their own crop of a registered cultivar for seed only on their own farm. A farmer may not sell any of that crop for seed unless qualified to do so. Some seed companies deny this right, particularly concerning genetically modified organisms, by a separate clause in the sale contract.

farmer's rights: The appreciation of the rights of farmers (past, present and future) as *in situ* agricultural innovators who jointly conserve and develop agricultural genetic resources around the world. Farmers are recognized as innovators entitled to intellectual integrity and compensation whenever their innovations are commercialized. In the Protection of Plant Variety and Farmers Right (PPV&FR) Act, there are nine rights of farmers – (i) right to seed, (ii) right to register traditional varieties, (iii) right to award and recognition, (iv) right for benefit sharing, (v) right to get compensation for the loss suffered from the registered variety, (vi) right to receive compensation for undisclosed use of traditional varieties, (vii) right to the seed of registered varieties, (viii) right to receiving free services and (ix) right to protection against innocent infringement.

farming (or pharming): The manufacture of medical products from genetically modified plants or animals.

farnesyl transferase: An enzyme used by the *ras* gene to help by signalling certain cells to divide/grow.

Fas protein: Membrane-bound receptor that initiates apoptosis in the receptor-bearing cell after binding to its ligand (Fas ligand).

fast component of eukaryotic genome: First fraction of DNA (I) to reassociate during renaturation represents 25% of the total DNA renaturing with C_0t value of 10^{-4} and ~2×10^{-2} with a $C_0t^{1/2}$ value of 0.0013.

fast neutrons: The high velocity, density ionizing particles generated from the radioactive decay of heavier elements in atomic reactors or cyclotrons

FASTA: A powerful international tool for storing biosequences and scanning databases to find sequences that are similar to a query sequence. Usually it is a program for making protein-protein comparisons, but it can also compare DNA sequence to DNA databank. If a significant number of exact matches are found, FASTA uses the dynamic programming algorithm to compute optional alignments. Current FASTA programs can be found at URL http://fasta.bioch.virginia.edu/. FASTA searches molecular databases, the first line begins with '>' and a short description followed by the sequence. The program looks for optimal local alignments by scanning the sequence for small matches called WORDS. Initially, the scores of segments in which there are multiple word hits are calculated (*init1*). Later the scores of several segments may be summed to generate an *initn* score. An optimized alignment that includes gaps is shown in the output as 'opt'. The sensitivity and speed of the search are inversely related and controlled by the *k-tup* variable which specifies the size of a 'word'.

FASTA format: A sequence in FASTA format begins with a single-line description, followed by lines of sequence data. The description line is distinguished from the sequence data by a greater than symbol (>) in the first column. It is recommended that all lines of text be shorter than 80 characters in length. Sequences are expected to be represented in the standard IUB/IUPAC amino acid and nucleic acid codes, with these exceptions: lowercase letters are accepted and are mapped into uppercase letters; a single hyphen or dash can be used to represent a gap of indeterminate length and in amino acid sequences, U and * are acceptable letters. Invalid characters (digits, blanks, etc.) are removed automatically.

fastidious: (In reference to prokaryotic organisms) having special growth and nutritional requirements.

fastidious xylem-limited bacteria (XLB): The bacterial pathogens of plants found only in the xylem, causing wilting, scorching and stunting symptoms; vectored by insects that feed on xylem fluid; not sap transmissible; require complex culture media.

FASTQ: A text-based sequencing data file format that stores both raw sequence data and quality scores. FASTQ files have become the standard format for storing NEXT-GENERATION SEQUENCING (NGS) data from Illumina sequencing systems and can be used as input for a wide variety of secondary data analysis solutions.

fate map: Diagrammatic representation of the type of tissue the cells in developing embryos will become. Scientists have been able to create fate maps for several organisms (such as the sea urchin) since the early decades of the twentieth century. Researchers use various methods including removing cells from embryos. If the adult developed from these embryos is missing specific tissues, researchers infer that the removed cells would have become those missing tissues. Researchers can also use a variety of stains to trace cells in the developing embryo. The analyses of numerous gynandromorphs have enabled various workers to establish fate maps in which many adult structures are localized to their relative positions on the blastoderm surface. Note that, in contrast to the one-dimensional linear linkage map derived from the linear chromosome, the fate map is two-dimensional since it derives from the surface of the blastoderm embryo. Thus, cross distances between different structures are also part of the map.

fate mapping: A technique used to show the movement of a cell or tissue and what it will become during normal development. Fate mapping was developed as a means of tracing the development of specific regions in the early embryo. Mitotic recombination has been used for fate mapping in *Drosophila*. A precise fate mapping of blastoderm cells was performed by a technique originally invented by A.H. Sturtevant. This technique depends upon the occurrence of mosaics in which one organ or structure is genetically different from the other because of a chromosomal loss during one of the early mitotic divisions. Fate mapping is then performed by noting a simple correlation: the greater the physical distance between two organs whose cells derive from the blastoderm, the greater the changes that a mosaic boundary will pass between these organs.

fate of completed proteins: In both prokaryotes and eukaryotes, mRNAs that code for particular proteins may be translated at special locations and perhaps by special ribosomes. In the bacterium Bacillus, synthesis of neutral protease (which is secreted into the medium) seems to occur on the ribosomes located at the cell periphery. In eukaryotes, ribosomes may be free (SMOOTH ENDOPLASMIC RETICULUM, SER) or attached to the endoplasmic reticulum (ROUGH ENDOPLASMIC RETICULUM, RER). Membrane-bound ribosomes manufacture proteins that are exported from the cell into the surrounding serum. 60S subunit of the ribosome is attached to the endoplasmic reticulum (ER). The ER membrane is used to form a vesicle containing the protein. This vesicle becomes a secretary message whose membrane fuses with the cell membrane, emptying its contents outside the cell. On the other hand, free ribosomes synthesize specific non-serum liver proteins. Leader sequences on mRNAs may determine which ribosomes shall translate them. Some short proteins are not synthesized on ribosomes. Such proteins are antibiotic polypeptides, such as gramicidin and tyrocidine.

father plant: The individual or species from which pollen was obtained for the cross made to create a hybrid.

fatuoid: A mutant commonly occurring in cultivated oats that resembles wild oats (*Avena fatua*).

fault-tolerant computer system: A system designed around the concept of fault tolerance. In essence, it must be able to continue working to a level of satisfaction even in the presence of errors or breakdowns.

fauna: The animals of an area or a period. The term is often used to distinguish between plant life, FLORA.

favoured mutations: The mis-sense mutations that disturb less critical sites of a polypeptide chain such that the basic functional feature of the polypeptide does not change. Such mutations may, however, alter some kinetic properties (like K_m, K_i or V_{max}) of the protein/enzyme and may be more efficient under one environment or the other. Such mutations may accompany the process of evolution.

Fc receptor: One of a family of receptors specific for the invariant constant region (Fc region) of immunoglobulins (other than IgM and IgD); different Fc receptors are specific for IgG, IgA, IgE and their subclasses.

feasibility study: An investigation aimed to objectively and rationally discover the strengths and weaknesses of an existing business or proposed venture, opportunities and threats existing in the natural environment, the resources needed to carry through and the projections for success. In its simplest terms, the two chief criteria to judge feasibility include the cost required and the value to be attained.

feature-based approaches: Gene recognition programs work by searching for various ad hoc features of genes and identifying regions that score high enough. Typical features include codon bias, donor/acceptor sites and coding frame length. Since stop codons should occur every 20 codons or so, long OPEN READING FRAMES (ORFs) without stop codons are strongly suggestive of genes. ORFs are easy to find with computers.

features of cloning vectors: Small size, semi-conservative replication, transferability, recombination either with another vector DNA or with bacterial DNA, INSERTION SEQUENCES (IS) which are homologous in vectors and bacteria, MULTIPLE CLONING SITE (MCS), easily scored phenotype, high copy number, selectable marker(s), reporter gene(s) and elements for expression are important features of cloning vectors.

fecundity: The probable number of offspring produced in a unit of time.

fecundity selection: The forces working to make one genotype more fertile than another genotype.

Federal Food, Drug and Cosmetic Act (FFDCA): The US federal law that decides the legal tolerances of pesticides on food products at harvest.

Federal Insecticide, Fungicide and Rodenticide Act (FIFRA): The US federal law that enacts the conditions for the use of pesticides.

feedback: The effect of the consequence of a process upon the functioning of the process.

feedback inhibition: Feedback inhibition in eukaryotic cells comprises time delays of gene expression from transcription, transcript splicing and processing and protein synthesis. In principle, such a phenomenon could lead to the oscillation of gene expression. The production of a particular substance in the cell is repressed when that substance has accumulated to a certain level, thereby balancing the amount available with the amount needed. This type of control is generally observed in metabolic pathways. *Aka* END-PRODUCT INHIBITION. Well-known examples of feedback inhibition include (i) isoleucine synthesis in *Escherichia coli*, (ii) production of protease kallikrein in rats and (iii) maturation of insulin hormone.

feedback mechanism: A control manoeuvre in a system. Homoeostatic systems comprise numerous negative feedback mechanisms that tend to counterbalance positive changes, thus maintaining stability.

feeder layer: A layer of living cells positioned between the solid culture medium and the target cell/tissues/organs to be grown *in vitro*. A feeder layer, also called 'nurse cells', is used to grow the recalcitrant cells/tissue that otherwise do not exhibit good growth on the synthetic medium. Feeder cells secrete certain natural growth-promoting substances that, in combination with medium components, help in improving the growth of target cells. Feeder cells are either embedded in the growth medium or separated from target cells/tissues by a membrane filter. Feeder layer cells may be from the same species (HOMOLOGOUS SYSTEM) or different species (HETEROLOGOUS SYSTEM). In case of animal cell culture, a layer of lethally irradiated cells is used. *See* NURSE CULTURE.

feeder root: A fine root that absorbs water and dissolved nutrients.

female flower: *See* PISTILLATE FLOWER.

female gametophyte: Functional haploid megaspore represents gametophytic generation. The megaspore produces a female gametophyte or female megagametophyte or embryo sac. The development of female gametophyte in angiosperms is endosporic (i.e. formed within the megaspore wall). The embryo sac is a three-celled structure. The three-celled egg apparatus occupies the micropylar region of the embryo sac whereas the chalazal region is occupied by three antipodals. The remaining portion of the embryo sac is known as the central cell, which has a central larger vacuole and two polar nuclei. The embryo sac has its own wall. It is composed of cellulose and pectin material. The micropylar portion of the embryo sac is occupied by the big vacuole. The egg nucleus and other cell organelles such as mitochondria, dictyosomes, plastids, ribosomes and abundant starch are present at the chalazal portion of the embryo sac. If synergids are absent in the egg apparatus, the egg cell has hook-like structures and wall ingrowths, as in synergids at the micropylar region. The antipodals generally degenerate before or after fertilization or remain persistent.

female germ-cell-mediated gene transfer (FGCMGT): FGCMGT is a novel and useful tool for efficient transgenesis. Oocytes collected from ovaries are cultured and transfected with a virus or DNA-liposome complex *in vitro*, embryos are generated by *IN VITRO* FERTILIZATION (IVF) and then the viable ones are reimplanted to the pseudopregnant females. *Aka* OVARY-MEDIATED GENE TRANSFER (OMGT).

female sterility: Some crops (e.g. banana) do not produce true seed because of female sterility. However, male sterility is much more common and is more useful in plant breeding as a technique for achieving cross-pollination.

FeMo cofactor (FeMoco): The primary cofactor, containing iron (Fe) and molybdenum (Mo), of nitrogenase that acts during nitrogen fixation.

femto: A prefix meaning Femtometer, 10^{-15} m. Unit suitable to express the size of atomic nuclei against Ångstrom (10^{-10} m), used for measuring atomic dimensions.

femtosecond: 10^{-15} s.

femtotechnology: Manipulating materials on the scale of elementary particles (leptons, hadrons and quarks). A technology after picotechnology, that is below the length scale of nanotechnology. *Aka* FEMOTECH.

feral: Refers to an individual or population that has returned to the wild after a history of domestication.

feralization promoting cryptic variation: The term feralization is defined as the phenomenon of conversion of cultivated species back to wild form. The food we have comes from domesticated species, that are the result of rounds of artificial selection performed on their wild relatives during evolution under the process known as domestication. De-domestication or feralization has been confirmed in various crops like wheat, rice and barley, as well as livestock. In a simplified way, it is not a complete conversion but rather the acquisition of phenotype similar to the wild crop under loosened artificial selection or by omitting intensive farming practices. Domestication is responsible for several genetic bottlenecks that dramatically reduce genetic variations and evolution potential under changing conditions due to genetic homogeneity. Feralization has turned this fact upside down by confirming that cultivated crops can undergo adaptive evolution. Despite low genetic variation, cryptic beneficial alleles were released upon selection pressure in different environments that pre-existed in cultivated progenitor populations. Feral plants live in the wild but are descended from domesticated individuals. Feral plants thus offer a proofreading concept for studying convergent evolution as well as the above-discussed concept of conditional neutrality.

fermentation: (i) A process in which chemical changes are brought about in an organic substrate through the actions of enzymes produced by microorganisms. (ii) The enzyme-catalysed, energy-yielding pathway in cells by which 'fuel' molecules such as glucose are broken down anaerobically (in the absence of oxygen). A guaranteed product of this pathway is the energy-rich compound ADENOSINE TRIPHOSPHATE (ATP). Other products can be alcohol, glycerol and carbon dioxide from the yeast fermentation of various sugars; butyl alcohol, acetone, lactic acid and acetic acid from various bacteria and citric acid, gluconic acid, antibiotics, vitamins B12 and B2 from mould fermentation. (iii) The enzymatic transformation of organic substrates (feedstocks), especially carbohydrates, generally accompanied by the evolution of gas. (iv) A physiological counterpart of oxidation, permitting certain organisms to live and grow in the absence of air. It is used in various industrial processes for manufacturing products such as alcohols, acids and cheese by the action of yeasts, moulds and bacteria. Alcoholic fermentation is the best-known example. (v) Anaerobic energy-yielding metabolic pathway in which pyruvate produced by glycolysis is converted, for example, into lactate or ethanol, with the conversion of NADH to NAD^+. *Aka* ZYMOSIS.

fermentor: A glass or steel vessel used for large-scale culture of microorganisms or plant cells/organ cultures. The fermentor is generally used for the large-scale commercial production of microorganisms such as those acting as biofertilizers, biopesticides or plant cell cultures/hairy root cultures for the production of secondary metabolites or speciality chemicals. Fermentor ensures aseptic conditions and optimum incubation conditions such as temperature and humidity.

ferrochelatase: A mitochondrial enzyme that catalyses the incorporation of iron into the protoporphyria molecule.

ferrodoxin: A non-heme iron- and sulfur-containing protein vital in the electron-transfer processes of photosynthesis and nitrogen fixation in plants. It also plays a role in the metabolism of some bacteria and was first found in an anaerobic bacterium.

ferroptosis: A non-apoptotic form of cell death that can be triggered by small molecules or conditions that inhibit glutathione biosynthesis or the glutathione-dependent antioxidant enzyme glutathione peroxidase 4 (GPX4). Ferroptosis has received considerable attention not only as a means to eradicate defined tumour entities but also because it provides unforeseen insights into the metabolic adaptation that tumours exploit to counteract phospholipid oxidation.

ferroptosis, endogenous suppressor of: 7-dehydrocholesterol (7-DHC) is an endogenous suppressor of ferroptosis.

fertile: A plant that produces seed capable of germination or that produces viable gametes (like anthers containing pollen).

fertility: Reproductive potential measured by the total quantity or relative proportion of developing eggs or of mating resulting in the progeny.

fertility restoration: Restorer gene restores fertility in cytoplasmic male sterile lines.

fertility restorer: Cytoplasmic male sterile lines that produce viable pollen in certain genotypes. Restorers have genes that restore fertility to a cytoplasmic male sterile line.

fertility-restoring genes: Cytoplasmic male sterile lines produce viable pollen in certain genotypes. Restorers have nuclear genes that restore fertility to a cytoplasmic male sterile line.

fertilization: The fusion of male and female gametes to form a zygote. Plants, animals and humans produce haploid male and female gametes that unite to form a diploid zygote – giving rise to a new individual organism or offspring and initiating its development. There are two stages in the process of fertilization – gamete approach and gamete fusion (fertilization). Fertilization scripts the beginning of the development of a new organism via cellular differentiation. *Aka* GENERATIVE FERTILIZATION, SYNGAMY and IMPREGNATION.

fertilization in angiosperms: It provides a very useful system for the study of double fertilization, embryo development and endosperm development in angiosperms. About 95% of the ovules on a shoot are cross-pollinated and 5% are self-pollinated. It is a monoecious plant as male and female flowers are in the same plant. A single tassel of this plant produces as many as 25,000,000 pollen grains. Each kernel has about 25,000 pollen grains and an ear contains about 1000 kernels. Hot and dry weather hastens pollen shedding, which goes on for many days. On an ear shoot, spikelets are borne as pairs, which is why there is an even number of kernel rows on the ear. In the process of fertilization, the pollen grain makes contact with the stigma (silk) of the pistil and then germinates into a long pollen type carrying the two male sperm nuclei to the embryo sac. One male gamete fertilizes the female gametic nucleus to form the diploid zygote, and the other male gamete combines with the diploid endosperm nucleus to form the triploid (diploid + haploid) tissue that will nourish the embryo. Thus, double fertilization forms two types of tissues: embryonic and endosperm.

fertilization wave: It begins when the fertilizing sperm is attached to the egg surface as a wave. Ca^{2+} ion release is an important event in the process of fertilization. Calcium stimulates all the events to follow. A chemical can cause the release of Ca^{2+} in the cell. This is an ionophore, which comprises group of molecules that transport ions across the membrane. The egg is then activated. Events that follow after Ca^{2+} release are the emission of the acid (H^+) from the egg and the prevention or block of polyspermy. Due to the effect of Ca^{2+}, there is the release of enzymes (which increase respiration) in the next 20 s. The unfertilized egg is inactive while the fertilized egg is active in respiration. The early response usually happens in under a minute. When eggs are treated with ammonia, it causes late responses bypassing all early responses. Ammonia can initiate late responses without the release of Ca^{2+} ions. This results in the increase in transport of nucleotides, amino acids and inorganic phosphate. DNA synthesis, mRNA activation and protein synthesis follow, which lead to cleavage. The sperm has to bind at the surface of the egg for all the events to occur. Injection of sperm into the egg will not lead to the occurrence of any of the above events.

fertilized egg microinjection: The transfection of fertilized egg may comprise the transfer of the whole nuclei, whole chromosomes (or fragments) or DNA segments. For the transfer of whole nuclei, the egg cells are treated with cytochalasin-B and subjected to centrifugation causing enucleation. Incubation of desirable karyoplasts with these enucleated eggs results in the transfer of whole nuclei in the presence of POLYETHYLENE GLYCOL (PEG). For the transfer of whole chromosomes, the chromosomes are first isolated from the metaphase cells by hypotonic lysis and may be fractionated using density centrifugation or flow cytophotometry. Individual chromosomes or fragments thus isolated are then incubated with whole cells (eggs) for the incorporation of chromosomes into nuclei. Microinjection of DNA segments into fertilized eggs is the most commonly used method. Several hundred copies of the DNA segments are used for this purpose and after stable integration, the eggs are used for transgenic animals. This method is commonly used for the production of transgenic mice.

fertilizer: Any substance that is applied to the land as a source of nutrients for plant growth. It may be recycled waste (e.g. farmyard manure, crop residues or compost) or produced industrially.

fetus (or **foetus**)**:** An animal embryo after 8 weeks of development, excluding any time during which its development has been suspended, and ending at birth.

Feulgen reaction: A cytochemical test that uses Schiff's reagent as a stain and DNA hydrolysis. This method, discovered in 1912 by Robert Feulgen, is highly specific for DNA detection. It permits a wide range of chromosome studies and quantitative determination of DNA contents of nuclei by applying the so-called CYTOPHOTOMETRY. *See* FEULGEN STAIN.

Feulgen stain: A histochemical and/or cytochemical reagent for quantifying nuclear DNA content or for staining chromosomes in cells. For preparing this stain, 900 ml distilled water is boiled in a 2-l flask and then 5.0 g basic fuchsine is slowly added. It is swirled for 1 min and then vacuum filtered through two layers of filter paper in a Buchner funnel into a 1-l flask. The contents are cooled to 50°C. While swirling 100 ml hydrochloric acid (1.0 N), 10.0 g potassium metabisulphite ($K_2S_2O_5$) is added. The swirling is done for 2 min. The cloudy red solution will become clear blood red. It is stored in the refrigerator for 24 h, removed, warmed to room temperature and supplemented by 3.75 g activated charcoal. It is shaken vigorously for 1 min and vacuum filtered. The reagent should be clear and colourless. If not, the charcoal step must be repeated. The reagent has to be stored in the refrigerator and used within 48 h. The stain remains stable for months.

Feulgen test: It is a chemical test for DNA. The test is made by first hydrolysing the chromosomes with Schiff's reagent ($C_{20}H_{19}N_3{\cdot}HCl$) and then treating it with Feulgen stain. It forms a characteristic colouration (magenta or deep purple).

FGENEH: A dynamic programming algorithm that uses linear discriminant functions. It is used to identify the presence of genes.

fibre cell: A type of cell found in sclerenchyma. It is flexible, strong and stringy with an elongated structure. It dies at maturity soon after an extensive modification of its cell wall. The cell wall is normally thickly lignified but is sometimes gelatinous.

fibre fluorescent *in situ* hybridization (fibre FISH): A variant of the conventional FISH technique for the detection and quantification of target sequences in which different probes labelled with different fluorochromes are hybridized to chromosome fibres.

fibrillar collagen: A type of collagen molecule that assembles into rope-like structures. Collagens type I (common in skin), II, III, V and XI belong to this type.

fibroblast: A common cell type found in connective tissue. It secretes an extracellular matrix rich in collagen and other extracellular matrix macromolecules. Migrates and proliferates voluntarily in wounded tissue and tissue culture.

fibroblast-like: Resembling or characteristic of, having the form or appearance of fibroblast cells. To define a cell as a fibroblast cell, it must possess characteristics typical of fibroblast cells. Often one can be certain of the histologic origin and/or function of the cells placed into culture and under these conditions, one can be reasonably confident in designating the cells as fibroblast. It is incumbent upon the individual reporting on such cells to use as many parameters as possible in assigning this term to a culture. Until a rigorous definition is possible, it would be correct to use the term fibroblast-like.

fibronectin: Extracellular matrix protein that has a role in the adhesion of cells to the matrix and the guidance of migrating cells during embryogenesis. INTEGRINS on the cell surface are receptors for fibronectin.

fidelity of DNA replication: Refers to various processes responsible for the faithful replication of DNA through generations of dividing cells. The processes recognized in this process are base selection, proofreading and mispair correction.

field: Data have several parts, known as a RECORD, that can be divided into fields. Relational databases arrange data as sets of database records, so-called ROWS. Each record consists of several fields. The fields of all records form the columns. Examples of fields are name, gender, hair colour, etc.

field capacity: Field capacity refers to the amount of soil moisture or water content held in the soil after excess water has drained away and the rate of downward movement has decreased.

field inversion gel electrophoresis (FIGE): A chromatographic method for the separation of a mixture of molecules by using a two-dimensional electrical field, applied across a gel matrix containing those molecules. For example, FIGE is commonly used to separate mixtures of large DNA molecules up to 2000 Kbp in length by their mass and charge.

field resistance: The term is synonymously used with GENERAL RESISTANCE. It is under polygenic control and is characterized by severe injury in the laboratory but shows resistance under normal field conditions. In general, field resistance is longer lasting than RACE-SPECIFIC RESISTANCE. Field resistance slows down the rate at which disease rises in the field.

field test: A test of a new crop variety, including biotechnology-derived, conducted outside the laboratory with specific requirements on location, plot size, methodology, etc. *Aka* TRIAL or FIELD TRIAL.

field trial: In science, the term refers to a test of the performance of some new product under the conditions in which it will be used. In agriculture, field trials are woven into its history and future. They have been heavily used in formal plant breeding for advancing agricultural practices, products and ideas. There are five key steps to use this as an effective on-farm, boots-on-the-ground practice: identify the why, design the trial, gather data, interpret results and draw conclusions.

figure eight and chi structure: A practical consequence of recombination between circular DNAs is that the recombination intermediate should take the form of a figure eight (8). The nature of the figure eight can be distinguished by cleaving the isolated molecule with a restriction enzyme that cuts each monomeric circle only once. The two parts of the figure eight then will fall apart unless they are covalently connected. If the figure eight is a recombination intermediate, cleavage will generate a structure in which the parental duplex molecules are held together by a region of heteroduplex DNA at the point of fusion. The cleaved molecule has four arms, each pair of the arms corresponding to each of the original monomeric circles. In the cleaved form, this molecule is called a CHI STRUCTURE or HOLLIDAY INTERMEDIATE. The four arms of duplex DNA are connected by a region in which the heteroduplex segments have been pulled apart into their constituent single strands (probably by the conditions of preparation). Holliday junctions are used in the formation of complex DNA molecules in the development of nanomachines.

figure eight: Two circles of DNA linked together by a recombination event that has not yet been completed.

filament: (i) The stalk of a stamen. (ii) Any very narrow, thread-like structure that is one or a few cells thick. (iii) Long chain of proteins, such as found in hair, muscle or flagella.

filamentous: Consisting of filaments or fibres; filiform; hair-like.

filename extension: An identifier specified as a suffix to the name of a computer file. The extension indicates a characteristic of the file contents or its intended use.

filial generation: Each generation of offspring in a breeding programme, designated F_1, F_2, etc.

filiform: Thread-like (e.g. stamen filaments or leaf shapes).

filing date: In the patenting of biological material, the date on which a complete patent application is received by the US Patent and the Trademark Office.

filling the regulatory gaps: The criteria that are presently in the use for regulating GENE DRIVE ORGANISMS (GDOs) are the same as are applied to GENETICALLY MODIFIED ORGANISMS (GMOs). This seems to be logically unacceptable as the natural selection pressure is not prevalent in the case of GDOs to the extent existing for GMOs. Thus, it would be desirable to formulate separate regulations for controlling research on GDOs.

filopodium (pl. filopodia): Thin, spike-like protrusion with an actin filament core, generated on the leading edge of a crawling animal cell.

filter (software): A computer program or subroutine to process a stream, producing another stream. While a single filter can be used individually, they are frequently laced together to form a pipeline.

filter bioreactor: Cells are grown on an open mesh of an inert material, which allows the culture medium to flow past it but retains the cells. This is similar in idea to membrane and hollow fibre reactors but can be much easier to set up, being similar to conventional tower bioreactors, but with the meshwork replacing the central reactor space. *Aka* MESH BIOREACTOR.

filter enrichment: A technique used for recovering auxotrophic mutants in filamentous fungi.

filter hybridization: Hybridization of nucleic acid fragments on a filter (e.g. nitrocellulose) as a carrier.

filtering strategy: A series of decisions an investigator makes after performing MASSIVELY PARALLEL SEQUENCING (MPS) to narrow down a list of candidate genetic variants from thousands to a few variants that may cause disease in an individual or family.

filtering: The process of hiding regions of (nucleic acid or amino acid) sequence having characteristics that frequently lead to spurious high scores. *Aka* MASKING. *See* SEG and DUST.

fimbral antigen: Proteins that comprise the hair-like organelles on the surface of bacteria responsible for motility.

fimbria (pl. fimbriae): All Gram-negative bacteria and many Gram-positive bacteria have slender, short, hair-like projections that support and move them. Fimbriae are short protein tubes that extend from the cell wall and cover the cell's surface. Bacterial cells can have up to a thousand fimbriae, which attach them to substrate or other cells.

fimbriate: Fringed (e.g. where the ends of a petal are split into two or more divisions).

find my understudied genes (FMUG): R. Richardson *et al.* (2024) proposed that understudied genes are methodically unrestrained in favour of better-studied genes between the completion of -omics experiments and the reportage of consequences. To encourage the study of understudied genes, they condensed their insights into a tool, FMUG, that empowers scientists to contribute with probable bias during the selection of hits. They reveal the usefulness of FMUG via the documentation of genes that remain understudied in the ageing process of vertebrates. FMUG is developed in Flutter and occurs at fmug.amaral.northwestern.edu as a MacOS/Windows app. *See* FLUTTER.

fine mapping: The process by which a trait-associated region from a GENOME-WIDE ASSOCIATION STUDY (GWAS) is analysed to identify the particular genetic variants that are likely to causally influence the examined trait.

fine structure genetic mapping: *See* FINE STRUCTURE OF GENE.

fine structure of an oncogene: In an attempt to locate the sequence responsible for oncogenicity (with restriction enzymes), the proto-oncogene and oncogene were cleaved at the same site, the fragments recombined, and the recombinants tested for transforming activity. The exact position was narrowed to the 350-bp sequenced region. The cause of oncogenic activity was found to be a single-base change (G→A) that results in a valine substitution for glycine.

fine structure of gene: S. Benzer in 1955 and 1956 used extensively cis-trans complementation tests in his study of *rII* locus in T4 phage. He observed a group of non-complementary mutants falling within a short map segment, corresponding to a unitary function. Since the *cis-trans* complementation test was used by Benzer, he proposed the term cistron. The term cistron and gene are now used synonymously. Benzer located 2400 independent mutants at over 300 sites within *rII* locus. The existence of many mutational sites in the *rII* locus implied that this locus possessed an intricate fine structure. Different units of mutational sites within a gene showed recombination. Gene was, therefore, found to be not a unit of recombination. The unit of recombination is a much smaller unit, termed by Benzer as a recon. Operationally, a cistron is demarcated by COMPLEMENTATION TEST. Two mutations belong to the same cistron if they do not display complementation in a *cis-trans* test or a simple *trans*-complementation test. Both muton and recon are defined by RECOMBINATION TEST. Two mutations belong to the same recon or muton if they do not reveal recombination. In the context of the molecular structure of DNA, given by J.D. Watson and F.H.C. Crick in 1953, a gene can be defined as a part of the DNA molecule possessing a specific base-pair sequence and a mutation as a change in the base-pair sequence. Comparing Benzer's observations with general estimates of the size of the phage chromosome, the T4 chromosome has a linear length of 200,000 nucleotides and 1500 map units (MU). Thus, 1 MU = 200,000/1500 = 133 nucleotides. Cistron *rIIA* = 6 MU and cistron *rIIB* = 4 MU. Number of nucleotides in rII region = 133 (6 + 4) = 1330. Within this length, recombination values range from 10% to 0.02%. In other words, there are 10/0.02 = 500 recombinational units within the *rII* locus. Benzer estimated 400–500 mutational sites in the rII region. The number of nucleotides per mutational site = 1330/500 ≅ 3. The same was suggested to be the size of the recombinational unit. These estimates were not considered precise. In the light of insights into the structure of the gene, from a geneticist's point of view, a gene is a discrete chromosomal region that is responsible for the synthesis of a specific cellular product (mRNA, tRNA or rRNA), which consists of a linear collection of potentially mutable sites each of which can exist in several alternative forms and between which crossing over can occur. See *CIS-TRANS* COMPLEMENTATION TEST.

fine tuning gene expression via histone modification: The binding of a methyl group to histone H_3 creates a platform for the binding of another protein HP1 that prevents transcription. Chemical modifications after the proteins sometimes clear the way for transcription and other times block it. C.D. Allis in 2000 suggested that methylation, acetylation, phosphorylation and other changes most likely occur in combinations that define histone code that fine tunes gene expression.

fingerprint: (i) The characteristic spot pattern formed by electrophoresis of the polypeptide or DNA fragments through the denaturation of a particular protein or DNA with a proteolytic enzyme or other means. (ii) A fingerprint is a set of motifs utilized to predict the occurrence of similar motifs, in either an individual sequence or a database. Fingerprints are refined by iterative scanning of a composite protein sequence database. A composite or multiple-motif fingerprint contains numerous aligned motifs from different parts of a multiple alignment. True family members are easy to recognize by virtue of owning all elements of the fingerprint, while subfamily members may be recognized by possessing only part of it.

fingerprinted contig (FPC): A software package that allows the assembly of clones into contigs by using either the end-labelled double-digest method or the complete-digest method.

fingerprinting: In genetics, the procedure used for the identification of multiple specific alleles on a person's DNA to produce a unique identifier for that person.

The method combines electrophoresis and chromatography to separate the components of a protein or DNA. The protein is denatured using a proteolytic enzyme and the resulting polypeptide fragments produce a characteristic spot pattern, referred to as a fingerprint, after electrophoresis. It is also used with hydrolysed fragments of nucleic acids. It is a common and very sensitive method for the identification of different plant genotypes. *See* FORENSICS *and* DNA FINGERPRINTING.

finished DNA sequence: High-quality, low-error, gap-free DNA sequence of the human genome. Achieving this ultimate 2003 HGP goal requires additional sequencing to close gaps, reduce ambiguities and allow for only a single error every 10,000 bases, the agreed-upon standard for the HGP finished sequence. *See* SEQUENCING *and* DRAFT SEQUENCE.

finite cell culture: A culture capable of only a limited number of population doublings after which it ceases proliferation. *See* *IN VITRO* SENESCENCE.

first eukaryotic cell: The time of appearance of the first eukaryotic cell is somewhat 1.5×10^6 million years ago. In terms of their biochemistry and metabolism, cyanobacteria seem to occupy a middle group between anaerobes and eukaryotes. Cyanobacteria can tolerate oxygen, make use of oxygen metabolically in both aerobic respiration and synthetic pathways, and can make unsaturated fatty acids by oxidative desaturation.

first living cell: The first living cell was probably a heterotroph lacking chlorophyll. The atmosphere was strongly reducing. The first living cell was also anaerobic and slow growing. It was enucleated, had a genetic code and RNA replicase which functioned as both genetic material and enzyme. The first cell was delimited by a permeable membrane capable of withstanding abuse.

first self-replicating molecule: There are four general requirements for primordial genetic material, regardless of its chemical nature: (i) informational properties (some type of heteropolymer, composed of at least two monomeric subunits), (ii) capable of directing the ordered assembly of monomeric starting materials to form additional copies of itself, (iii) the monomeric starting materials must have been readily available on the primitive Earth and (iv) sufficiently stable for its rate of reproduction to exceed its rate of decomposition.

first-degree relative: A family member who shares about 50% of their genes with a particular individual in a family.

first-division restitution (FDR): FDR occurs when meiosis fails to separate homologous chromosomes. It results from an abnormal orientation of the spindles right before meiotic anaphase II. Non-sister chromatids end up in the same nucleolus. FDR has been reported in several crop species, such as potato, rye, etc. *Compare with* SECOND-DIVISION RESTITUTION.

first-division segregation (FDS): The segregation of two members of an allelic pair at a locus at meiosis I. FDS gives a 1:1 ratio during tetrad analysis. *Compare with* SECOND-DIVISION SEGREGATION (SDS).

first-generation molecular markers: DNA markers such as RAPDs and RFLP. See RAPDs and RFLPs.

fission: (i) Centric fission refers to breakage through the centromere resulting in two derivative chromosomes. (ii) Division of single-celled organisms, especially prokaryotes, in which mitosis does not occur. (iii) Also refers to mitosis in certain unicellular fungi. (iv) In genetics, the division of one cell by cleavage into two daughter cells, or a chromosome into two arms.

fission yeast: A common name generally given to the yeast *Schizosaccharomyces pombe*, a common experimental organism. It divides to give two equal-sized cells.

fitness (W): (i) Relative lifetime reproductive success of a genotype, which includes the probability of surviving to reproduce. The fitness of a genotype is manifested through its phenotype, which is also affected by the developmental environment. (ii) The number of offspring left by an individual after one generation. (iii) The fitness of an allele is the average fitness of individuals that carry that allele.

fitness component: Traits such as survival, mating success and reproduction, that determine fitness.

fitness landscape: Refers to either a graph of fitness as a function of individual genotype or phenotype, or of population mean fitness as a function of allele frequencies or trait means. In evolutionary biology, fitness landscapes or adaptive landscapes (types of evolutionary landscapes) are used to visualize the relationship between genotypes and reproductive success. It is assumed that every genotype has a well-defined replication rate (often referred to as fitness). This fitness is the 'height' of the landscape.

5′ and 3′ non-coding regions: Untranslated regions at the 5′ and 3′ ends of mRNA sequences.

5′ cap: Modified 5′ end of eukaryotic mRNA, consisting of an extra methylated nucleotide and methylated 2′ position of the sugar in one or more subsequent nucleotides. The cap plays a role in the binding of the ribosome to mRNA and impacts mRNA stability and the removal of introns.

5′ DNA end: Describes the different, complementary ends of a DNA single strand; ends with a phosphate group.

5′ P-DNA strand terminus: The end of a polynucleotide that terminates with a mono-, di- or triphosphate attached to the 5′ carbon of the sugar.

5′ splice site: The 5′ end of an intron where cleavage takes place in RNA splicing.

5′ untranslated region (UTR): Sequence of nucleotides at the 5′ end of mRNA; does not code for the amino acids of a protein. *Compare with* 3′ UNTRANSLATED REGION.

5′→3′ DNA growth: The synthesis of a nucleic acid chain by joining the 5′ (PO_3) end of a nucleotide to 3′ (OH) end of the last nucleotide already in the uncompleted chain.

5′→3′ exoribonucleases: Enzymes that remove nucleotides from the 5′ end of RNA.

5′TOP mRNAs, regulation of: A central mechanism of the mammalian target of rapamycin complex 1 (TOR complex 1 or mTORC1) signalling is the coordinated translation of ribosomal protein and translation factor mRNAs mediated by the 5′-terminal oligopyrimidine motif (5′TOP). La-related protein 1 (LARP1) has been proposed to be the specific regulator of 5′TOP mRNA translation downstream of mTORC1 while eIF4E-binding proteins (4EBP1/2) have been suggested to have a general role in the translational repression of all transcripts. T. Hochstoeger *et al.* (2024) used single-molecule translation site imaging of 5′TOP and canonical mRNAs to study the translation of single mRNAs in living cells. 4EBP1/2 has a dominant role in the repression of translation of both 5′TOP and canonical mRNAs during pharmacological inhibition of mTOR. LARP1 selectively protects 5′TOP mRNAs from degradation in a transcriptome-wide analysis of mRNA half-lives.

5-azacytidine: A cytosine analogue with a nitrogen atom replacing the carbon atom at position 5 of the pyrimidine ring, making it unable to accept a methyl group. A short treatment with it produces foci of differentiated cells in mammalian cell lines by undermethylation of cytosine.

5-bomouracil: A base analogue used as a mutagen. This compound can be incorporated into DNA sequences in place of thymine and can cause incorrect base pairing after a period.

5-bromodeoxyuridine: Growing cells readily incorporate this base analogue into nucleic acids and can be killed by exposure to light. It suppresses the utilization of deoxythymidine. This procedure is a selection mechanism utilized in mutagenic studies to eliminate wild-type cells and recover the auxotrophs.

5-enolpyruvylshikimate 3-phosphte (EPSP) synthase: An enzyme in plants' metabolic pathways that leads to the production of aromatic amino acids necessary for developing proteins required for plants' growth.

5-hydroxymethylcytosine ($5^{hm}C$): An epigenetic modification of DNA that can result from the oxidation of 5-methylcytosine by the Tet family of enzymes.

5-methylcytosine ($5^{m}C$): An epigenetic modification of DNA that usually occurs at CpG dinucleotides; the $5^{m}C$ modification usually correlates with repressed gene expression. *Aka* the 5TH BASE OF DNA.

5-methylcytosine ($5^{m}C$) deaminase: The AID/APOBEC family enzyme or other enzyme that can deaminate 5-methylcytosine to thymine.

5-methylcytosine ($5^{m}C$) DNA glycosylase: DNA glycosylase that removes methyl group from 5-methylcytosine bases from DNA.

5-methylcytosine ($5^{m}C$) mapping and quantification in RNA: Preliminary tRNA BS-seq experiments showed a substantial variation in the cytosine methylation patterns of tRNAs. The robust quantification of cytosine methylation levels at specific positions therefore needs a substantial sequencing depth of 100 (or more) sequence reads. This can be accomplished by NGS technologies, which allow the MPS of hundreds of thousands of single DNA molecules.

5-methylcytosine ($5^{m}C$) methyltransferases: Those enzymes that generate C5-methylcytosine (EC 2.1.1.37). In mammalian cells, cytosine-specific methyltransferases methylate certain CpG sequences. In bacteria, these enzymes are a component of restriction-modification systems. $5^{m}C$ MTases comprise of two domains – a larger catalytic domain-containing catalytic and cofactor binding sites, and a smaller DNA recognition domain.

5S ribosomal RNA: 5S ribosomal RNA (5S rRNA) is an integral component of the large ribosomal subunit in all known organisms. It has approximately 120-nucleotide–long ribosomal RNA molecules. The term 5S refers to the molecule's sedimentation velocity in an ultracentrifuge. The exact function of 5S rRNA is unclear, but it is believed to enhance protein synthesis by stabilizing the ribosome scaffold.

5S RNA gene family: This family of genes occurs independently of rDNA, which is localized in the NUCLEOLAR ORGANIZER REGION (NOR) in eukaryotes. In prokaryotes and yeast, the *5S rRNA* gene family exists in close vicinity of rDNA. The 5S rRNA genes are transcribed by RNA polymerase II.

***fix* genes:** *fixABCX* genes are involved in electron transport to nitrogenase. *fix NOQP* genes are required to support bacteroid respiration under conditions of low oxygen present in root nodules. *fix GHFS* genes are involved in transmembrane proteins. We do not know the functions of *fixLKJ* genes. We also do not understand why *fixN* gene is present in both clusters.

fixable genetic variance: The genetic variance component arising from additive gene action including additive × additive and its higher interactions. The variance helps selection to be most effective. *Aka* FIXABLE GENETIC VARIATION.

fixation: (i) If an allele at a polymorphic locus goes to fixation, it reaches a frequency of 100%. This means that it has replaced all other allelic variants. (ii) In creationism, species are described as 'fixed' in the sense that they are believed not to change their form, or appearance, through time. (iii) The first step that is essential in making permanent preparations of tissue, etc. for microscopic study. The procedure aims at killing cells and preventing subsequent decay with the least distortion of structure.

fixation agent: Any chemical that has the ability to preserve the structure or chemical composition of animal or plant tissues is considered to be a fixing agent. A solution is used for the preparation of tissue for cytological or histological studies. It precipitates the proteinaceous enzymes and prevents autolysis, destroys bacteria, etc. (e.g. acetic acid, formalin, Farmer's fixative, etc). *Aka* FIXATIVE.

fixation of alleles: The establishment of a single allele within a population through random chance loss of the other member or members of the same set.

fixation probability: The probability that a new mutation will reach fixation.

fixation time: The time (number of generations) required for new mutations to become fixed or reach fixation.

fixative: Chemical reagents such as formaldehyde or osmium tetroxide used to preserve cells for microscopy. Samples treated with these reagents are said to be 'fixed', and the process is called FIXATION.

fixed allele: An allele for which all members of the population under study are homozygous, such that no other allele for this locus exists in the population.

fixed breakage point: In the heteroduplex recombination model, the point from which the unwinding of the DNA double helices starts as a runup to the formation of heteroduplex DNA.

fixed rearrangement: When a chromosomal rearrangement is present in different species.

fixing: The process of attaching specimen on the surface of a slide for long-term microscopic studies. In case of plants, the cells are fixed on the surface of the slide by treating them with acetic alcohol, passing them through the alcohol xylene series and finally mounted on CANADA BALSAM. Whereas the microorganisms are fixed by passing through the flame or by covering the slide with methanol for 1 min. Air drying and flaming fix microorganisms on the slide.

fixity-of-species doctrine: An old doctrine that argues that species were specially created and do not undergo any change over time. *Opposite to the* THEORY OF ORGANIC EVOLUTION.

flag: An isolated, wilted or necrotic branch with dead leaves attached.

flag tag: A short peptide sequence that is added to a protein to allow the protein to be recognized by antibodies raised against the flag tag.

flagellum (pl. flagella): A lash-like, hair-like, whip-like or tinsel-like appendage of a motile cell, bacterium or zoospore that protrudes from the cell body and provides locomotion. Eukaryotic flagella are a longer form of cilia. Bacterial flagella are shorter and completely different in structure and mechanism of action. The eukaryotic flagellum is sometimes referred to as an UNDULIPODIUM.

flaming: Sterilization of instruments (forceps, scalpel) by using an open flame after dipping in alcohol.

flanking regions: Non-coding DNA sequences on both ends of a mature gene.

flanking sequence: A segment of DNA molecule that either precedes or follows the region of interest on the molecule.

flap endonuclease 1 (FEN1): Structure-specific nuclease with 5′-flap endonuclease and 5′–3′ exonuclease activities involved in DNA replication and repair.

flavanoids: Large group of water-soluble plant pigments comprising antioxidant, anti-inflammatory and antiviral properties that are beneficial to health.

flavin: One of a group of pale yellow, greenly fluorescing biological pigments widely spread in small quantities in plant and animal tissues. Flavins are manufactured only by bacteria, yeast and green plants. For this reason, animals rely on plant sources for riboflavin (vitamin B2), the most predominant member of the group. *Aka* LYOCHROME.

flavin adenine dinucleotide (FAD): (i) A redox-active cofactor, more specifically a prosthetic group of a protein, associated with various other proteins, involved in different important enzymatic reactions in metabolism. (ii) The coenzyme of some adenine dinucleotide oxidation-reduction enzymes; it contains riboflavin.

flavin adenine dinucleotide reduced (FADH$_2$): Activated carrier molecule that is produced by the citric acid cycle.

flavin mononucleotide (FMN): Riboflavin phosphate, a coenzyme of certain oxidoreduction enzymes. A biomolecule derived from vitamin B2 (riboflavin) through the catalytic activity of the enzyme riboflavin kinase. FMN is a mononucleotide that acts as a cofactor. In particular, it assists certain oxidoreductases (e.g. NADH dehydrogenase) in various oxidation-reduction reactions. It also functions as a cofactor in blue-light photoreceptors.

flavin nucleotides: Biomolecules derived from riboflavin (vitamin B). These are commonly used as cofactors in biological blue-light photoreceptors. For example, FMN and FAD.

flavin-linked dehydrogenases: Dehydrogenases are enzymes (involved in removing hydrogen atoms from their substrate) which require one of the riboflavin coenzymes, FMN or FAD, in order to perform their function.

flavonoids: A category of phytochemicals produced by plants that are typically beneficial to the health of humans that consume them. For example, the three isoflavones (genistein, daidzein and glycitein) produced in seeds of the soybean plant (*Glycine max* (L.) Merrill) are flavonoids, and confer several health benefits to humans. Coffee, tea and chocolate products contain several antioxidant flavonoids (i.e. polyphenols). Because oxidation of lipids (low-density lipoproteins) in the bloodstream is the initial step in atherosclerosis disease, consumption of large amounts of coffee may help to prevent atherosclerosis. High coffee ingestion by humans reduces the oxidation of lipids in the bloodstream by 30%. Cranberries (*Vaccinium macrocarpon*) encompass many antioxidant flavonoids and their intake in large amounts regularly may impede the development of breast cancer. Blueberries (genus *Vaccinium*) comprise a number of flavonoids, and their ingestion in large amounts on a regular basis helps to strengthen eyesight, improve memory and inhibit some physical aspects of the ageing process. Other subcategories of flavonoids are flavones, flavonols, flavanols, aurones, chalcones, etc.

flavonols: A group of phytochemicals, consisting of a subcategory of the flavonoid 'family' of phytochemicals. Flavonols are present in citrus fruits such as grapefruit, oranges, etc. and are characteristically advantageous to the health of humans that consume them. However, at least one flavonol (quercetin glycoside) occurs in tomato peels.

flavoprotein: An enzyme containing a flavin nucleotide as a prosthetic group.

FlavrSavr tomato: The enzyme polygalacturonase (PG) degrades pectin, which is a major component of the fruit cell wall. Transgenic tomatoes have been produced which contain an antisense construct of the gene coding for PG. These transgenics have reduced the expression of PG and have a 2-week longer shelf life. The FlavrSavr tomatoes have improved flavour and soluble solids in addition to increased shelf life.

***FLC* expression, activation of:** The flowering repression occurs due to the hyper-trimethylation of H3K4 *FLC* chromatin in winter-annual *Arabidopsis*. The H3K4 trimethylation and ATP-dependent nucleosome remodelling are interrelated. A model summarizing the putative roles of several of the players required for *FLC* expression by the PAF1-complex-mediated trimethylation of H3K4 has been proposed. When the *Arabidopsis* PAF1 complex (whose known members include ELF7, ELF8 and VIP4) associates with RNA polymerase II, an H3K4 methyltransferase (EFS) is recruited to *FLC* chromatin by Pol-II–PAF1 complex, leading to H3K4 trimethylation in the 5′ portion of transcribed *FLC* chromatin. The binding of PIE1 to the trimethylated H3K4 causes the *FLC* chromatin remodelling to an active state for *FLC* transcription. Vernalization converts active *FLC* chromatin into a heterochromatin-like state so that the PAF1 complex is not able to access the *FLC* locus. FRIGIDA acts as an *FLC* activator that enables *flowering locus T (FT)* induction under unfavourable conditions.

***FLC* expression, repression of:** The vernalization condition is inherited through mitotic divisions, but this vernalized state is not inherited from one generation to the next. *FLC* repression occurs through histone modifications involving the Vernalization 2 (VRN2) protein complex in *Arabidopsis*. Other protein complexes participate in the activation of the flowering time genes. The vernalization regulates the flowering time genes in numerous plant species by preventing the *FT* induction of the floral promoter during autumn but permitting *FT* induction in spring for flowering at the proper time in the plants.

fleck: A minute, discoloured spot in green tissue.

flexible DNA: Contrary to the curved DNAs, certain sequences (like CTG_nCAG_n and CGGnCCGn), called flexible DNA, exhibit 20% faster movement than the normal B-DNA in a gel. This unusual mobility is impacted by the length of repeats, temperature and percentage of the acrylamide gel.

flexible populations: These populations react by genetic changes to seasonal as well as altitudinal environmental alterations.

flexistyle: Depending on the degree of maturation of the stamens, the style moves up (cataflexistyle) or down (hyperflexisyle).

flip-flop model of mating switch: Assumes that two mating-type alleles are present, one on each side of a common regulatory site. Mating switch is brought about by recombination within the common regulatory site of the mating-type locus.

floating chromosomal rearrangements: When a chromosomal rearrangement is present within a species, it is called floating or polymorphic.

floccose: Having a soft and woolly covering of hairs.

floppy disc: A thin plastic disc coated with magnetic oxide and used for information storage.

flora: An enumeration of all the plants found growing in a certain region or country or period, listed by species and considered a whole, generally with a guide to their identification (e.g. the Flora of North America, Flora of China, Flora of Victoria, Flora of New South Wales and so on). In this case, flora is written with a capital 'F'.

floral bud: A plant bud that only produces a flower.

floral diagram: A graphical means to describe flower structure, usually a schematic cross-section through a young flower.

floral envelope: The calyx and corolla of a flower. *See* PERIANTH.

floral formula: A description of flower structure using numbers, letters and various symbols. The parts of the flower are represented by K (for calyx), C (for corolla), P (for perianth), A (for androecium) and G (for gynoecium). These are followed by numbers indicating the number of parts in each whorl (e.g. K5 indicates a calyx of five sepals); ∞ indicates an indefinite number of parts (more than 12).

floral integrators: Floral pathway integrator gene expression mediates the gradual transmission of environmental and endogenous cues to flowering time. The floral integrators and light wavelength play a crucial role in activating or repressing the flowering. The function of floral integrators is the upregulation of floral meristem identity genes.

floral leaves: The upper leaves located at the base of the flowering branches.

floral meristem identity gene *leafy* (*lfy*): A gene that functions within the floral primordium to promote the expression of genes that specify floral organ identity.

floral organ identity genes: The function of floral meristem identity genes is to activate floral organ identity genes. These genes have been divided into three classes – A, B and C. The class A genes, *AP1* and *AP2*, initiate the sepal and petal formation in whorls 1 and 2, respectively. Another function of these genes is the repression of class C activity in whorls 1 and 2. The class B genes, *AP3* and *PI*, are essential for petal development in whorl 2 and stamens in whorl 3 while the class C gene, *AG*, marks the identity of stamens and carpels in whorls 3 and 4, respectively. The additional function of class C genes is the repression of class A genes in whorls 3 and 4 during flower development.

floral organs, initiation of: Several genes belonging to the MADS-box genes serve as the transcription factors to regulate the gene expression for the initiation of each floral organ. The MADS-box genes are present in flowering plants, viz. *Arabidopsis*, *Petunia* and non-flowering plants, viz. mosses, ferns and green algae. MADS-box genes control sperm motility, cell division and elongation of the stem of gametophores.

floral tube: A vague term sometimes used as a synonym of hypanthium, corolla tube or calyx tube.

floret: A small flower, usually referring to the individual true flowers clustered within an inflorescence, particularly those of the Poaceae grasses and the pseudanthia of the family Asteraceae.

florigen: Also called FLOWERING HORMONE. The universal hormone that supposedly causes plants to undergo a transition from the vegetative to the reproductive state. In 2005, after 70 years of this hypothesis, O. Nielson and co-workers identified mRNA of the *FT* gene of thale cress (*Arabidopsis thaliana*) that is produced in leaves and induces flowering when transported to apex tissue.

Floury-2: A gene in corn/maize (*Zea mays* L.) that (when present in the DNA of a given plant) causes that plant to produce seeds that contain higher-than-traditional levels of the amino acids – methionine and tryptophan.

flow cytometry: A method of counting/separating cells/protoplasts using flow cytometry that makes use of fluorescent dyes. During protoplast fusion, the hybrid cells are detected and separated from the parental protoplasts by a machine called fluorescence-activated cell sorter (FACS) which is based on flow cytometry. One type of protoplasts is stained with red dye, and the second type with green dye. The hybrids give two colours, viz. red and green. The FACS separates the hybrid cells based on the double colour. *See* SORTING.

flow densitometry: Multiphase flow measuring unit for performing X-ray density measurements on a fluid flow, the unit comprising an X-ray source generating an X-ray beam within a chosen spectral range and a first detector aimed at the beam for detecting radiation scattered by or transmitted through the flow, the spectral range being a narrow, essentially monochromatic range. *See* SORTING.

flow karyotyping: Use of flow cytometry to analyse and separate chromosomes according to their DNA content.

flow-sorted chromosomes: Chromosomes sorted by flow cytometry based on size.

flower: The sexual reproductive structure of the angiosperms, typically with a gynoecium, androecium, perianth and an axis. A typical flower of angiosperms or plants whose seeds are enclosed in an ovary is composed of petals, sepals, stamens and a pistil. The flower morphology contributes to the relative importance of self- and cross-pollination. The two structures directly involved in sexual reproduction are the male stamina and the female pistil. A stamen consists of an anther, which contains the pollen grains, and a filament on which the anther is borne. Some flowers, called PERFECT FLOWERS, have both male and female reproductive organs while others, called IMPERFECT FLOWERS, are characterized by only male reproductive organs (STAMENS) or only female reproductive organs (OVARY, STYLE, STIGMA). Some plants have both male and female flowers, while others have males on one plant and females on another.

flower break: A break or stripe in flower colour. *See* BREAKING.

flower development: Flowers are derived from SHOOT APICAL MERISTEM (SAM). Those plant species that are determinate in nature, SAM finally recognizes itself directly into a floral meristem as in sunflower (*Helianthus annus*) or through the inflorescence meristem that finally gives rise to the floral meristem as in rice (*Oryza sativa*), whereas in case of indeterminate plant species, flowers are formed on the lateral branches or through lateral inflorescence called COINFLORESCENCE. Hence, in determinate plant species, SAM persists and continues to produce floral meristem laterally. Thus, both vegetative and reproductive phases occur simultaneously. The flower development has four temporal phases. The first phase includes the environmental and endogenous signals for switching into the reproductive development from the vegetative stage through the flowering time genes. The second phase includes the flowering time pathways signals to activate the meristem identity genes. The third and fourth phases comprise the activation of the floral organ identity genes and organ-building genes, respectively. There are several HOMEOTIC MUTATIONS in flowers where different parts replace others. For example, in one class of mutations, sepals develop where petals should, and carpals develop where stamens should. These mutations have been identified as defects of a family of genes that all encode a particular class of transcription factor, called the MADS-box family. MADS-box transcription factors occur in both plants and, to a lesser extent, in animals and contain a conserved 58 amino acid sequence.

flower development in *Arabidopsis*: *Arabidopsis* has been extensively studied as a model for flower development. The developing flower has four basic organs: sepals, petals, stamens and carpels (which go on to form pistils). These organs are arranged in a series of whorls: four sepals on the outer whorl, followed by six petals inside them, six stamens and a central carpel region. Homeotic mutations in *Arabidopsis* result in the change of one organ to another – in the case of the Agamous mutation, for example, stamens become petals and carpels are replaced with a new flower, resulting in a recursively repeated sepal-petal-petal pattern. Observations of homeotic mutations led to the formulation of the

ABC model of flower development by E.S. Coen and E.M. Meyerowitz in 1991.

flower development, ABC model of: Class A genes are active in whorls 1 and 2, class B genes are active in whorls 2 and 3, while class C genes are active in whorls 3 and 4. This is referred to as the 'ABC model' to describe the gene activity in meristem towards sexual reproduction during flower development.

flower development, ABCE model of: The *SEP* gene discovery, termed class E gene lead to the emergence of revised ABC model. These are referred to as E genes and not as D genes because earlier researchers characterized D genes in *Petunia*. The newly emerged 'ABCE model' specifies the sepals by A activity, petals by A1B1E, stamens by B1C1E and carpels by C1E. In *Petunia*, *FBP7* and *FBP11* specifies the placenta and ovule identity. Due to co-suppression in plants, *FBP7* and *FBP11* genes undergo downregulation that results in ovule-to-carpel transformations.

flower development, epigenetic control of: The epigenetic mechanisms of gene regulation include DNA methylation, histone modifications including repressive marks (H3K9me, H3K27me, H2Aub), active marks (H3K4me, H3K36me, H3Ac) and chromatin remodelling (nucleosome composition, occupancy and location). Plant environmental epigenetics mechanisms include chromatin-mediated covalent histone, DNA modifications, nucleosome positioning, remodelling, chromatin configuration, histone variants regulation and environmentally induced chromatin changes transmitted through cell divisions (mitosis and/or meiosis) for regulating flowering.

flower induction: The process where the leaf generates a flowering stimulus in response to appropriate photoperiods. *Aka* FLORAL INDUCTION *and* FLOWER INITIATION.

flower structure: A flower consists of two essential parts: the vegetative part (calyx and corolla) and reproductive part (androecium and gynoecium). In *Arabidopsis* and other plants, the developing flower is composed of four concentric whorls. The outermost whorl (whorl 1) becomes the sepals, the outer floral leaves. It surrounds whorl 2, which becomes the petals, the white inner floral leaves. Whorl 3 becomes stamens, which contain the male organs. The innermost whorl (whorl 4) becomes the carpels, which form the ovary.

flowering genes, categories of: There are three classes of flowering genes, namely – Class A, Class B and Class C genes. These genes are active in the three regions (sepals, petals and stamens) of the flower meristem.

flowering hormone: *See* FLORIGEN.

***flowering locus C* (*flc*):** A flowering gene that encodes a MADS-box protein that may act as a transcriptional repressor of flowering and plays a dominant role in the vernalization response.

***FLOWERING LOCUS T* (*FT*) repression molecular switch:** FT is also a crucial component of the florigen that is transported from leaves to shoot apices to induce flowering. *FT* is specifically transcribed in the leaf vascular tissues, where its production is suppressed by many flowering repressors, including the myeloblastosis viral oncogene homologue (MYB) transcription factor EARLY FLOWERING MYB PROTEIN (EFM). Y. Zhang *et al.* (2024) revealed that a plant CTD phosphatase [C-TERMINAL DOMAIN PHOSPHATASE-LIKE 2 (CPL2)] suppresses *FT* expression in leaf vascular tissues by modifying the binding activity of EFM. CPL2 interacts with, and dephosphorylates, EFM to enable the binding of dephosphorylated EFM to *FT* chromatin, thereby constraining flowering. Thus, CPL2-mediated dephosphorylation of the floral repressor EFM serves as a molecular switch, adding one more layer of regulation to fine-tune *FT* transcription and ensure the timely occurrence of flowering.

flowering plant sperm compaction: T. Buttress *et al.* (2022) demonstrated that a histone variant, H2B.8, mediates sperm chromatin and nuclear condensation in *Arabidopsis thaliana*. The loss of H2B.8 causes enlarged sperm nuclei with dispersed chromatin, whereas ectopic expression in somatic cells produces smaller nuclei with aggregated chromatin. This result demonstrates that H2B.8 is sufficient for chromatin condensation. H2B.8 aggregates transcriptionally inactive AT-rich chromatin into phase-separated condensates, which facilitates nuclear compaction without reducing transcription.

flowering plants: All flowering plants are collectively called ANGIOSPERMS. They form seeds enclosed in fruit (an ovary). Their flowers are used in reproduction. These are the dominant types of plants today. There are over 250,000–300,000 angiosperm species. Angiosperms evolved about 140 mya, during the late Jurassic period. They became the dominant land plants about 100 mya (edging out conifers, a type of gymnosperm).

flowering signal: Florigen comprising numerous genes, including *Constans* (*CO*), *Flowering Locus C* (*FLC*) and *FLOWERING LOCUS T* (*FLT*), is required for the flower development. Florigen in leaves is responsible for the induction of several different physiological and morphological changes during flower development.

flowering switch: The switch to flowering is a response to the length of light and dark periods in many plants. The timing of the flowering switch is vital for reproductive success. The transition of the vegetative phase to floral parts depends upon the seasonal variation of the seed production under favourable environmental conditions. The flowering-regulatory genes regulate the floral gene expression in varying seasons through the photoperiod pathway and the vernalization pathway for proper floral development.

flowering time control: Multiple environmental and endogenous inputs determine when plants flower. The molecular genetic dissection of flowering time control in *Arabidopsis* has recognized an integrated network of pathways that quantitatively regulate the timing of this DEVELOPMENTAL SWITCH. This framework offers the foundation to understand the evolution of different reproductive strategies and how floral pathways interact through seasonal progression. The inputs encourage or suppress the activation of genes termed FLORAL PATHWAY INTEGRATORS. Depending upon the wavelength of flight, the light quality pathway can block or endorse flowering. Floral pathway integrators upregulate the function of floral meristem identity genes. Deciding when to flower is very critical to plants. A gene has been identified that functions in measuring the duration of cold exposure and establishing the vernalization state. The plant homeodomain (PHD) finger protein Vernalization Insensitive 3 (VIN3) mediates vernalization in *Arabidopsis thaliana*.

flowering time genes: The function of *FT* gene is to perform the long-day photoperiod, gibberellin (GA), autonomous and vernalization pathways. The MADS (MCM1, AGAMOUS, DEFICIENS and SERUM RESPONSE FACTOR) domain transcription factor FLC promotes or represses flowering during vernalization in *Arabidopsis*. The downregulation of *FLC* occurs through chromatin modifications under cold conditions to respond to flowering in spring. The accessions requiring vernalization are termed as 'winter annuals', germinating in the fall, overwintering as rosettes and flowering in the spring. The presence of *Arabidopsis* floral meristem identity gene *AP1* and the florigen gene *FT* is required for vernalization in cereals. The gibberellin and autonomous pathways promote flowering under non-inductive photoperiodic conditions.

flowering time under stress, epigenetic processes control: The number of genes controls the expression pattern of flowering under variable environmental conditions to ensure the transgenerational seed production. The epigenetic modifications control the floral initiation and stress tolerance mechanism. Understanding the relationship between epigenetic pathways, flowering and stress-responsive genes is expected to lead to the discovery of novel epigenomic players. The dynamic and stable structure changes in the chromatin results in the regulation of stress response and memory storage. Histone modifications serve as an epigenetic blueprint during stress responses in plants.

flowering time, chromatin-mediated regulation of: Chromatin modification is the base of the transcriptional activation and repression of flowering. These modifications include histone modification and chromatin remodelling for the transcriptional activation. Both the transcriptional activation and repression of the major floral repressor gene *FLOWERING LOCUS C* (*FLC*) are under the control of numerous chromatin modifiers. Some of these modifiers are involved in histone modification or chromatin remodelling, which is generally linked to transcriptional activation. The *FRI* gene initiates high-level *FLC* through multiple chromatin modifications at the *FLC* locus. The modifications including histone deacetylation and methylation processes undergo *FLC* repression. The histone modification in *FLC* chromatin serves as an epigenetic memory of vernalization. The proteins homologous to chromatin modification regulate gene expressions in the photoperiodic flowering pathway.

FLOWERING WAGENINGEN (FWA): FWA is silenced in wild-type plants and the reverse of the imprinted silencing causes a late flowering phenotype.

FLOWERING WAGENINGEN C (FLC): A MADS-box transcriptional repressor involved in silencing the genes that are required for the switch to flowering.

flowering, initiation of: Initiation of flowering requires a switch from vegetative to reproductive phase. The switch in apical meristem from a vegetative to an inflorescence meristem and then floral meristem could be a two-phase transition or directly from vegetative meristem to floral meristem could be a single-phasetransitiondependingoncriticalcues,both environmental and genetic.

flowering, repression of: The repression of flowering is an important mechanism during plant developmental under specific environmental conditions (I. del Olmo *et al.*, 2010). The gene regulatory networks behind the flower repression of floral promoter (*FT*) in non-inductive photoperiodic conditions are important in flowering mechanism.

FlowerMate 2.0: The 20K model has been combined into the app 'FlowerMate 2.0' and its corresponding platform (www.iPlant.cn), both of which were established by G. Xie and co-workers in 2024. Serving as the professional version of 'FlowerMate', the 'FlowerMate 2.0' app and related platform combine wide-ranging data on plant specimen images and species information amassed by generations of researchers in China. This amalgamation enables users to engage in complete and specialized studies of plants.

Flpe: Flpe is a genetically enhanced, site-specific Flp recombinase that recognizes and binds to FRT sites.

fluctuation test of mutations: An investigational process that measures mutation rates and defines whether mutations arise before or in reply to selection.

fluidity: A property of membranes that reveals the ability of lipids to move laterally within their particular monolayer.

fluid-phase endocytosis: A type of endocytosis in which small vesicles bud off from within the plasma membrane, transporting extracellular fluid and dissolved material into the cell. *See* PINOCYTOSIS.

fluorescein: Fluorescent dye emitting light when irradiated at the right frequency. Several biologically friendly fluorescent molecules, especially fluorescein isothiocyanate (FITC), are used for probing biological processes *in situ*.

fluorescence: The property of certain chemicals to absorb light of one wavelength and produce light of a higher wavelength. The light is emitted by the excited molecules when they revert to their ground state. Fluorescein isothiocyanate (fluorescent compound) when irradiated with invisible UV light fluorescence's visible green light. Likewise, there are fluorescent compounds that fluoresce red, yellow and orange light.

fluorescence-activated cell sorter (FACS): A machine utilized to sort cells from a mixed group of cells (e.g. separating only the cells into which a new gene has been inserted via genetic engineering techniques). The desired cells are first labelled with a specific fluorescent dye and then passed through a flow chamber that is illuminated by a laser beam, which causes the labelled cells to glow. The molecules of the fluorescent dye, which 'stick' to only one type of cell in the mixture, contain chromophores that can be raised to an excited, unstable state via irradiation with specific wavelength(s) of light. Those chromophores remain in that excited state for a maximum of 10^{-9} s before releasing their energy by emitting light and returning to their unexcited 'ground' state. This fluorescence is a measurable property, and the FACS machine uses it to separate the desired cells from the rest of the mixture.

fluorescence imaging system: Uses a computer program to convert the fluorescence generated when an object absorbs light into false-colour signals for plant health analysis.

fluorescence *in situ* hybridization (FISH): (i) A physical mapping approach that uses fluorescein tags to detect the hybridization of probes with metaphase chromosomes and the less-condensed somatic interphase chromatin. (ii) A method for perceiving the presence of a particular gene (e.g. in a biological sample), which uses a fluorescein-tagged DNA probe. When the DNA probe hybridizes to that particular gene, the 'tag' glows, thereby demonstrating its presence in the sample of that particular gene positively. This physical mapping method uses fluorescein tags to detect hybridization of probes with metaphase chromosomes and with the less-condensed somatic interphase chromatin. This approach identifies specific sequences of intact chromosomes by hybridization with complementary nucleic acid probes that are covalently bound to fluorochromes. (iii) A procedure for visual detection in the microscope of specific DNA sequences on cytological fixed chromosomes, after hybridization with DNA probes labelled with a fluorochrome. This can be done on both interphase and metaphase chromosomes. *Aka* FLUORESCENT *IN SITU* HYBRIDIZATION.

fluorescence microscope: Microscope fitted with UV light to illuminate specimens stained with fluorescent dyes. Fluorescence preparations are self-luminous. It is intended to view material stained with fluorescent dyes. These are similar to a light microscope, but the illuminating light in this case is passed through one set of filters before the specimen, to select those wavelengths that excite the dye, and through another set of filters before it reaches the eye to select only those wavelengths emitted when the dye fluoresces. The fluorescing parts of the stained object then appear bright against a dark background. The staining technique is extremely sensitive. *See* FLUORESCENCE.

fluorescence resonance energy transfer (FRET): A distance-dependent physical process by which energy is transferred non-radioactively from an excited molecular fluorophore (the donor) to another fluorophore (the acceptor) by means of intermolecular long-range dipole-dipole coupling.

fluorescence staining: Very few biological samples are characteristically fluorescent such that they can be imaged directly. Instead, fluorescence microscopy principally uses a fluorochrome that is introduced through some form of staining procedure called fluorescence staining for flow cytometric, fluorescence or light microscopic analysis. The fluorescence can be directly applied, such as 4′-6-diamidino-2-phenylindole (DAPI). It forms fluorescent complexes with natural double-stranded DNA, showing a fluorescence specificity for AT, AU and IC clusters. Because of this property, DAPI has been a useful tool in various cytochemical investigations. When DAPI binds to DNA, its fluorescence is strongly enhanced, which has been interpreted in terms of a highly energetic and intercalative type of interaction, but there is also indication that DAPI binds to the minor groove, stabilized by hydrogen bonds between DAPI and acceptor groups of AT, AU and IC base pairs. Other applications of this technique use single-stranded DNA sequences with a fluorescent label to hybridize with its complementary target sequences in the chromosomes, permitting it to be visualized under ultraviolet light. Fluorescence staining technique offers several advantages, such as high resolution, live cell staining and the possibility of dual or multiple labelling. Since there is no radioactive decay, long shelf life for probes and nucleotides eliminates time-consuming handling, monitoring and clean up, does not require designated special 'hot' areas in lab and is often quicker than doing autoradiography.

fluorescent dye: Refers to molecule that absorbs light at one wavelength and responds by emitting light at another wavelength. The emitted light is of a longer wavelength (and hence of lower energy) than the light absorbed.

fluorescent *in situ* hybridization (FISH), techniques not using: Present techniques do not allow for the precise positioning of probes down to a single-nucleotide base. Therefore, it became necessary to use overlapping clones, which are clones that share a common segment. Covering of a region of the genome can then be accomplished by a set of partially overlapping clones, also called a CONTIG (for contiguous clones). Building a contig of clones for a given region is thus the first and crucial step of physical mapping. For this, one principally picks up clones out of a clone library or bank obtained by systematic cloning of pieces ensuing from the enzymatic digestion of the whole genome. These clones are selected based on their positivity for markers specific to the studied region and must be organized through physical mapping. Following this approach, one thus acquires a minimal continuous string of overlapping clones that can ultimately be sequenced. One method used for organizing these clones utilize a matrix organization of these clones on a membrane, which are then hybridized to one another and detected by autoradiography. A combinatorial analysis then explicitly gives the order of the overlapping clones.

fluorescent *in situ* hybridization (FISH), techniques using: The contig organization does not provide a precise information on the size of the overlap between two clones. Even less information is provided on their distance (or gap between clones). In the last few years, different techniques have been developed in order to precisely measure the respective position of clones on to a partially linearized DNA fibre. All these techniques use FISH, which was initially developed for sensing probes on chromosomes in metaphase. This involves first preparing cells on a slide, then partially or fully decondensing and/or releasing the DNA from the cell nucleus. The DNA is then hybridized with labelled probes, and detection with fluorescent systems allows for direct observation of the relative position of the probes using epifluorescence microscopy.

fluorescent resonance energy transfer (FRET): Procedure for monitoring the closeness of two fluorescently labelled molecules (and thus their interaction) in cells.

fluorescent-labelled antibodies: Antibodies attached with fluorescent chemicals. It is a diagnostic tool that makes use of fluorescent antibody, that are seen using a fluorescent microscope. Fluorescent antibodies are made by covalently binding the fluorescent compound with the antibodies. It is also called immunofluorescence.

fluorescein diacetate (FDA) staining: Living cells are stained with FDA fluorescence in the presence of UV light to assess the viability of cell cultures, etc.

fluorodinitrobenzene (DNFB or FDNB): A chemical that reacts with the N-terminal amino acid of polypeptides. It is used to sequence proteins. *Aka* SANGER'S REAGENT *or* DINITROFLUOROBENZENE.

fluorogenic CRISPR for genomic DNA imaging: Conventional CRISPR-based imaging approaches image genomic loci in living cells expressing constitutively fluorescent proteins, resulting in high background and non-specific nucleolar signal. Z. Zhang in 2024 constructed fluorogenic CRISPR (fCRISPR) with dCas9, an engineered single-guide RNA (sgRNA) and a fluorogenic protein. Fluorogenic proteins are degraded unless they are bound to specific RNA hairpins. These hairpins are inserted into sgRNA, resulting in dCas9: sgRNA: fluorogenic protein ternary complexes that enable fluorogenic DNA imaging. With fCRISPR, they imaged various genomic DNA in different human cells with high signal-to-noise ratio and sensitivity. fCRISPR offers a high-contrast and sensitive platform for imaging genomic loci.

Flutter: An open-source framework shaped by Google that enables a single code base to generate complex user interfaces on multiple devices irrespective of whether it is an Android or iOS operating system. Flutter's 'hot-reload' function permits developers to present changes in almost real time making it effective, and Google's Material Design enables developers to adapt each and every pixel on screen using 'widget' to generate reliable UI strategies for all platforms during cross-platform development. *See* WIDGET.

fmet: Short form of N-formyl-methionine. The modified amino acid carried by the $tRNA_f^{met}$ that initiates translation in bacteria.

focal adhesion: A type of anchoring cell junction, forming a small region on the surface of a fibroblast or other cell that is anchored to the extracellular matrix. Attachment is mediated by transmembrane proteins such as integrins, which are linked, through other proteins, to actin filaments in the cytoplasm. *Aka* FOCAL CONTACT *or* ADHESION PLAQUE.

focal adhesion kinase (FAK): Cytoplasmic tyrosine kinase present at cell-matrix junctions (focal adhesions) in association with the cytoplasmic tails of integrins.

focus (pl. foci): A small zone of diseased plants within a population.

focus formation: The ability of transformed eukaryotic cells to nurture in dense clusters, piled up on one another.

focus-forming unit (ffu): The determined concentration of virus is expressed as focus-forming units FFU/ml.

focus map: A fat map of areas of the *Drosophila* blastoderm destined to become specific adult structure, based on the frequencies of specific kinds of mosaics.

***Fok*I:** A type IIS restriction enzyme found in *Flavobacterium okeanokoites* that is composed of a separable DNA-binding domain and a nuclease domain that is used to construct ZFNs and TALENs.

Fokker-Planck equation: An equation introduced by R. Metzler and co-workers in 1999 that describes diffusion processes. It is used by population geneticists to describe random genetic drift.

fold-back DNA: It comprises inverted repeats that have renatured by intrastrand reassociation of denatured DNA. When DNA from any eukaryote is denatured and allowed to reanneal at low concentrations, the first sequences to become double stranded are those that occur as inverted repeats. There occur 2000–4000 pairs of inverted repeats or potential fold-back structures in *D. melanogaster*, which cover about 3% of the total genome. The repeats may be closely adjacent (FB1 and FB5) or separated by up to several thousand base pairs (FB4, FB5, FB7, FB8 and FB9), and as the reaction is intramolecular, the structures thus moulded are called SNAP-BACK or STEM-AND-LOOP. *Aka* FOLD-BACK (FB) ELEMENT *or* SNAP-BACK (SB) ELEMENTS.

folded leaf: A layer of alpha-helices wrapped around a single hydrophobic core but not with the simple geometry of a bundle.

foliar: Pertaining to leaves.

foliate: Preceded by a number to signify having a certain number of leaflets (e.g. 3-foliate signifies three leaflets).

foliole: A small, leaf-like addition on the front or back.

foliose: Flattened like a leaf.

follicle: A dry fruit arisen from one carpel splitting along a single suture to which the seeds are attached (e.g. from the pod of a legume).

follicle cell: One of the cell types that surround an emerging oocyte or egg.

follicle-stimulating hormone (FSH): A protein hormone used in traditional medical therapy in an attempt to enhance the production of sperm inside the follicles of the testes.

food: Any substance, whether processed, semi-processed or raw, intended for human consumption. It includes drinks, chewing gum and any substance used to manufacture, prepare or treat 'food'. It excludes cosmetics, tobacco and substances used only as drugs.

Food and Drug Administration (FDA): The federal US agency empowered with approving all pharmaceutical and food ingredient products sold within the country. In 1992, prior to the approval of any of the biotechnology-derived food crop plants, the FDA decided that food crops produced via 'biotechnological (i.e. recombinant) technologies' must meet the same rigorous safety standards as those created via 'traditional breeding methods', both of which are regulated by the FDA. Historically, new food crops created via TRADITIONAL BREEDING TECHNOLOGIES (e.g. crossing with wild type in order to confer disease resistance, increased yield, etc. on the resultant domesticated plant varieties/strains) have sometimes contained unexpectedly high levels of known (and naturally occurring) toxins (e.g. solanine, a naturally occurring toxin in potatoes and some other plants; psoralene, a naturally occurring toxin in celery, etc.). Any transgenic crop variety of a food/feed crop to be released in the USA requires an additional clearance from the FDA. The FDA is also responsible for ensuring the safety, efficacy and security of human and veterinary drugs, foods, vaccines, cosmetics, biological products and medical devices.

food biotechnology: The application of biotechnology approaches to the production of food.

food chain: A hierarchical model of feeding relationships among species in an ecosystem. The chain of eating and getting nutrition starts from a small herbivorous animal and ends at a big carnivorous organism. This flow of energy happens through plants to animals as they eat their food.

food engineering: A scientific, academic and professional field that interprets and applies principles of engineering, science and mathematics to food manufacturing and operations, including the processing, production, handling, storage, conservation, control, packaging and distribution of food products. Some of the applications and practices used in food engineering to produce safe, healthy, tasty and sustainable food are refrigeration and freezing, evaporation, packaging, energy for food processing, heat transfer in food processing and food-safety management systems (FSMS).

food processing enzyme: Enzyme used to provide food texture, flavour, appearance and to a certain extent, nutritional value. Amylases break down complex polysaccharides in to simple sugars. Proteases tenderize meat proteins. Biotechnology can play a role in the development of new food enzymes by searching for or engineering enzymes that better fit with the other processes that the food must undergo, like cooking or canning.

Food Quality Protection Act (FQPA): The US federal law that defines a health-based standard regulating pesticide tolerance levels in food.

food safety disputes between member nations: (i) Codex Alimentarius Commission – for foods and food ingredients. (ii) International Plant Protection Convention (IPPC) – for plants. (iii) International Office of Epizootics (OIE) – for animal diseases.

food security: Availability of and access to sufficient, nutritious food consistently, as well as the knowledge and ability to select and prepare foods to ensure safety and adequate nutrition. *Antonym*: FOOD INSECURITY.

food web: A complex model of feeding relationships that describes the connections among all food chains within an ecosystem. It is a complex way of showing how energy is transferred between organisms through eating. A food web shows the prey and predator relationships of animals, and it could show more than one prey per organism. The arrows on the web are drawn from the prey to the predator to show the flow of energy.

food-web magnification: The tendency for top predators in a food web to have the highest residues of certain chemicals, especially organochlorines. *Aka* FOOD-WEB ACCUMULATION *or* FOOD-WEB CONCENTRATION. *Compare with* BIOACCUMULATION.

footprinting: A technique for identifying DNA regions that bind protein. Radiolabelled DNA/RNA digested with DNase/RNase in the presence or absence of a DNA/RNA binding protein and then subjected to gel electrophoresis. Regions of DNA/RNA bound to a protein do not get digested. Therefore, the patterns of fragment bands from bound and unbound samples will differ, which allows the identification of the protein binding regions of DNA. Footprinting determines the length of nucleic acid in contact with a protein.

for-loop: Also written as '*for loop*'. A control flow statement for specifying iteration, which allows code to be executed repeatedly. Various keywords are used to specify this statement. For example, descendants of ALGOL use 'for', descendants of Fortran use 'do' and COBOL use 'PERFORM VARYING'. In perform varying, a paragraph will be executed till the condition in Until phrase becomes true.

forb: Any non-woody flowering plant that is not a grass, sedge or rush.

forbidden mutations: The mutations that affect the function of a gene drastically such that the individual cannot tolerate them (e.g. FRAMESHIFT MUTATIONS). Such mutations do not accompany the process of evolution.

force modulation microscopy (FMM): FMM is based on contact mode **atomic force microscope (AFM)** with an additional mechanical modulation applied to the cantilever during the contact scan. The FMM amplitude and phase are sensitive to changes in the local elasticity, friction and adhesion force between the tip and sample surface. Therefore, FMM acquires images of the sample topography simultaneously to local nanomechanical properties that allow a qualitative differentiation of material components in composite samples, such as polymer blends.

forces of evolution: Mutation, recombination, migration, random genetic drift, selection, etc. are regarded as forces of evolution.

foreign DNA: DNA that is not found in the normal chromosome complement of the organism concerned. It may be introduced into a vector by complexing with native DNA using techniques of gene manipulation and, hence, entering the new host. Foreign DNA may also be introduced into the host cells following a viral infection.

forensic biology: Individualization of biospecimens encountered during a criminal investigation, paternity dispute or immigration requirement, personal identification, etc.

forensics: The use of DNA for identification. Some examples of DNA use are to establish paternity in child support cases, establish the presence of a suspect at a crime scene and identify accident victims.

forest: Vegetation subjugated by trees with single trunks, including closely arranged trees with or without an understorey of shrubs and herbs. Forests are usually conducive to wide biodiversity.

forest decline: The combination of biotic factors (such as insect pests and pathogens) and abiotic factors (such as drought and air pollution) that contribute to poor growth and the death of tree populations.

forest floor: Litter and other organic debris deposited on top of the mineral soil of a forest.

forestry: The harvesting of trees and management of postharvest succession to bring about the regeneration of another forest.

form genus: A genus based on morphology and not on evolutionary relationships (such as for imperfect fungi).

forma (in common usage, **form**): A taxonomic category lower in rank to species and within the taxonomic hierarchy, below variety (varietas) and usually differentiated by a minor character.

forma specialis (f. sp.): A taxonomic group within a pathogenic fungal species defined in terms of host range (i.e. members of different formae speciales infect different groups of plants). Forma specialis used for fungi is equivalent to pathovar (pv) for bacteria.

formal methods: A set of mathematically based techniques for the specification, development and verification of software and hardware systems. The use of formal methods for software and hardware design is inspired by the anticipation that, as in other engineering disciplines, performing suitable mathematical analysis can add to the reliability and robustness of a design.

formal verification: The act of proving or disproving the correctness of intended algorithms underlying a system with respect to a certain formal description or property, using formal methods of mathematics.

formaldehyde-assisted isolation of regulatory elements (FAIRE)-seq: FAIRE isolates nucleosome-depleted genomic regions by exploiting the difference in cross-linking efficiency between nucleosomes (high) and sequence-specific regulatory factors (low). FAIRE consists of cross-linking, phenol extraction and sequencing the DNA fragments in the aqueous phase.

formation and extension of D-loop: The formation of heteroduplex DNA produces a D-loop in which one strand of the donor duplex is displaced. The D-loop is protracted by repair synthesis, using the free 3′ end as a primer. Ultimately, the D-loop becomes large enough to correspond to the entire length of the gap on the recipient chromatid. When the extruded single strand spreads to the far side of the gap, the complementary single-stranded sequences anneal. Heteroduplex DNA exists on either side of the gap, with the gap itself represented by the single-stranded D-loop. The duplex integrity of the gapped region can be re-established by repair synthesis using the 3′ end on the left side of the gap as a primer. Overall, two individual rounds of single-stranded DNA synthesis repair the gap. Branch migration converts this structure into a molecule with two recombinant joints. The joints must be resolved by cutting. (i) If both joints are resolved in the same way, for example, the inner strands are cut at each joint, the original non-crossover molecules will be released, each with a region of altered genetic information that is a footprint of the exchange event. (ii) If the two joints are resolved in opposite ways, one is cut on the inner strand and the other on the outer strand resulting in a genetic crossover. Following the double-strand break, heteroduplex DNA is formed at each end of the region involved in the exchange. Between the two heteroduplex segments is the region corresponding to the gap, which has the sequence of the donor DNA in both molecules. Thus, the arrangement of the heteroduplex sequence is asymmetric, with part of one molecule having been converted to the sequence of another, which is why the initiating chromatid is called the RECIPIENT.

forward epimutations: *De novo* methylation leads to forward epimutations. For example, (i) in DNA, cytosine → 5-methylcytosine or (ii) in histone protein, lysine → methylated lysine. So far, there is no evidence for shutting off of genes by *de novo* methylation in diploid somatic cells apart from X chromosome inactivation. *See* EPIMUTATIONS.

forward genetics: Genetic analysis that proceeds from trait (phenotype) to DNA (gene) through positional cloning or a candidate gene approach. *See* REVERSE GENETICS.

forward mutation: Any mutation away from the standard or wild type is defined as forward mutation. For example, $y^+ \rightarrow y$ or $R \rightarrow r$. *Compare with* BACK MUTATION *or* REVERSE MUTATION.

forward phenomics: It uses phenotyping tools to distinguish most promising genotypes having the most desirable traits within a large collection and this allows selection of the 'best of the best' genotype/elite line. Thus, it enables rapid identification of traits at pre-stage and makes it less necessary to grow plants up to the maturity stage in field. The application of forward phenomics facilitates the screening of a large number of pots (containing plants) that run on a conveyor belt and pass through a chamber comprising automated imaging systems.

forward selection: Choosing individuals with desirable attributes out of a progeny test for possible use in seed orchards and/or subsequent generations of breeding.

fosmid library: Fosmids are commonly used for preparing genomic libraries when a smaller insert size is desired. The inserts have an average size of 40 kb and are produced by random shearing, yielding a more uniform coverage of the genome than other library types.

fossil: Most commonly, an organism, a physical part of an organism or an imprint of an organism that has been preserved from ancient times in rock, amber or by some other means. New techniques have also revealed the existence of cellular and molecular fossils.

fossil fuel: A fuel, such as organic-rich geological materials (coal, petroleum oil and natural gas), formed by the decomposition of ancient (fossilized) plants and animals. *Compare with* ALTERNATIVE ENERGY.

fossil water: The water in aquifers (a body of rock and/or sediment that holds groundwater) that has accumulated over long periods of time and is not quickly replaced through infiltration from recent rainfall.

foundation seed: Seed stocks increased from breeder seed (i.e. the progeny of breeder, select or foundation seed). Foundation seed is handled to closely maintain the genetic identity and purity of the variety. It is a sort of certified seed, either directly or through registered seed.

founder effect: Extreme cases of random genetic drift occur when a new population is established by only very few individuals. This has been called the founder effect by E. Mayr in 1942.

founder principle: A small population size effect in which the genes of a few people (the population founders) are inherited over time by a large number of descendants. *Aka* FOUNDER EFFECT *or* SEWALL WRIGHT EFFECT. *See* GENETIC DRIFT.

^{14}C dating: Radiocarbon dating.

***412* gene family:** This gene family at each of the 30 or so sites contains a complete *412* gene. Each is 7.0 kb in length. It is terminally redundant.

454 pyrosequencing: A parallelized version of pyrosequencing was developed by 454 Life Sciences. The method amplifies DNA inside water droplets in an oil solution (emulsion PCR), with each droplet containing a single DNA many picolitre-volume wells each containing a single bead and sequencing enzymes. Pyrosequencing uses luciferase to produce light for detection of the individual nucleotides added to the nascent DNA, and the combined data are used to create sequence read-outs. This technology offers intermediate read length and price per base compared to Sanger sequencing on one end and

Solexa and SOLiD on the other. 454 Life Sciences has since been acquired by Roche Diagnostics.

4^{m}C methyltransferases (MTases): Those enzymes that generate N4-methylcytosine (EC 2.1.1.113). The N- and C-terminal domains form a cleft that accommodates the DNA substrate. The type I restriction-and-modification system is composed of three polypeptides: The M (hsdM) and S (hsdS) subunits together form a methyltransferase that methylates two adenine residues in complementary strands of a bipartite DNA recognition sequence. In the presence of the R (hsdR) subunit, the complex can also act as an endonuclease. Whether the DNA is cut or modified depends on the methylation state of the target sequence. When the target site is unmodified, the DNA is cut. When the target site is hemi-methylated, the complex acts as a maintenance methyltransferase, modifying the DNA so that both strands become methylated. 4 ^{m}C Mtases specifically methylate the amino group at the C-4 position of cytosines in DNA. Found as components of type II restriction-modification systems in prokaryotes. Recognize a specific sequence in DNA and methylate a cytosine in that sequence. In this way, they protect DNA from cleavage by type II restriction enzymes that recognize the same sequence.

four-strand stage crossing over: Crossing over between chromatids during the occurring pachytene stage, evidenced by chiasma seen at the diplotene stage in meiotic metaphase when four chromatids are present.

foxo transcription factors: These transcription factors have been implicated in diverse biological processes, including metabolism, cellular stress response and ageing.

fractal: A mathematical object having fractional dimension.

fractional factorial designs (with blocking): Experimental design containing several treatment factors but the design does not comprise every treatment combination and therefore some interactions are misidentified. A blocking factor may also be present, which can confound some interactions with blocks.

fractional replication: As the name indicates, fractional replication of factorial designs involves eliminating some of the treatment combinations from a factorial design instead of simply confounding them with groups. A fractional replication of a factorial design might possess only a small fraction of the number of possible treatment combinations. For example, in a fractional replication of a 2 × 2 × 2 × 2 design, only a half or only a fourth of the treatment combinations might be included in the experiment. In a fractional replication of a 3 × 3 design, only a third of the treatment combinations might be included. Fractional replication designs are like confounded factorial designs in that modular arithmetic is used to divide a factorial design into independent sets of treatment combinations. Fractional replication designs differ from confounded factorial designs in that in fractional replication some of the sets of treatment combinations are removed from the design instead of being confounded with groups.

fractionation: A separation process that divides a certain quantity of a mixture into several smaller quantities.

Fragmid: A toolkit for the rapid assembly and assessment of CRISPR technologies. McGee and co-workers in 2024 leveraged Golden Gate cloning to create Fragmid, a toolkit composed of modular fragments that allows users to create customized CRISPR plasmids in a matter of days. Approximately, 200 Fragmid modules work together like building blocks to allow users to mix and match individual modules to create a desired targeting construct. *See* GOLDEN GATE CLONING.

fragile chromosome site: Heritable chromosomal sites that exhibit gaps or constrictions on chromosomes when exposed to partial replication stress. Fragile sites are classified as either COMMON or RARE, depending on their frequency.

fragile X syndrome (FXS): A genetic condition that causes a range of developmental problems including learning disabilities and cognitive impairment. Usually, males are more severely affected by this disorder than females. Affected individuals have delayed development of speech and language by age 2. The genetic disorder FXS is caused by changes in a gene named *Fragile X Messenger Ribonucleoprotein 1* (*FMR1*). *FMR1* usually codes for a protein called FMRP needed for brain development. People who have FXS do not make this protein. FXS has symptoms that include chromosome instability and intellectual disabilities. Its aetiology involves the expansion and methylation of a CGG repeat in the fragile X mental retardation 1 (FMR1) five-prime (5′) UTR and promoter methylation.

fragmented ions, analysis of: This allows the masses of polypeptide fragments to be determined which are used in correlative database searching or the derivation of *de novo* sequences.

frameshift: A deletion, substitution or duplication of one or more bases that makes the reading frame of a structural gene to shift from the normal series of triplet codons.

frameshift intergenic suppressors: A mutation in an RNA gene suppresses the mutagenic consequence of a frameshift mutation.

frameshift mutation: A mutation resulting from addition or deletion of one or more base pairs, which is not a multiple of three, in a gene. Frameshift mutation shifts the mRNA's reading frame so that incorrect amino acids are added to the polypeptide chain after the mutation locus. Unless a frameshift mutation transpires close to the C-terminus, it reduces the function of the gene. *Aka* PHASESHIFT MUTATION.

frameshift suppression: The reversion of the effects of a frameshift mutation by a second mutation in the same gene, thus restoring the reading frame.

frameshift suppressor: The suppressor that involves a change in the size of the code word. In any case, a single tRNA contains a four-nucleotide anticodon that can read a four-letter codon resulting from a single-nucleotide pair insertion.

frameshift variant: The insertion or deletion of a nucleotide, or nucleotides, which is not a multiple of three that results in an alteration of the codon reading frame and may result in a premature stop codon or truncated protein product. May be denoted as FRAMESHIFT (INDEL). *Aka* FRAMESHIFT MUTATION, OUT-OF-FRAME DELETION *or* OUT-OF-FRAME INSERTION.

fraternal twin: Siblings born at the same time as the result of fertilization of two ova by two sperm. They share the same genetic relationship as any other siblings and are not necessarily of the same sex or more similar in appearance than other siblings. They share half of their genomes, just like any other siblings. *Aka* DIZYGOTIC TWINS *and* NON-IDENTICAL TWINS. *Compare with* MATERNAL TWINS.

free: Not united with other organs of the same type (i.e. not attached at one end).

free amino acids: Singular molecules not bound by peptide bonds to other amino acids.

free central (of placentation): Ovules attached to a free-standing column in the centre of a unilocular ovary.

free energy (G): A function, designed to yield a criterion for spontaneous change, that combines the entropy (the degree of disorder or uncertainty in a system) and enthalpy (the measurement of energy in a thermodynamic system) of a molecule or system. Free energy decreases for a spontaneous process and remains unchanged at equilibrium.

free nuclear division: Refers to the mitotic division of nuclei without accompanying cytokinesis (i.e. nuclei divide in a common cytoplasm with the cell walls only forming around each later).

free radical: An atom or group of atoms possessing an unpaired electron. The homolytic cleavage of covalent bonds to generate free radicals has now been found in a range of biochemical processes such as reactions of methyl-CoA mutase, ribonucleotide reductase and DNA photolyase. The term also refers to an oxygen (atom) bearing an 'extra' electron. Because of that, it holds a large amount of energy, and in a biological system, it can damage body tissues when it releases that energy. In animals, free radicals are known to cause cancer by damaging DNA but they also play an important role in the immune system where they are used to kill viruses and bacteria invading the body. They are known to have a similar role in plant growth. All cell growth is controlled by the production of these highly reactive and therefore very toxic free radicals. A gene *Rhd2* has been identified that codes for a protein that produces free radicals. Controlled production of free radicals by *Rhd2* stimulates calcium channels in the membranes of cells resulting in calcium being taken up by the cells. The accumulation of calcium in turn activates cell expansion. In plants where the *Rhd2* gene was inactivated by a mutation, the roots and root hairs were stunted. *Aka* REACTIVE OXYGEN SPECIES (ROS), SINGLET OXYGEN *or* OXYGEN-FREE RADICAL.

free radical theory of ageing: Free radical theory has been proposed to explain the basic cause of ageing. Free radicals have unpaired electrons, which increase in concentration with age. If mice are fed with antioxidants like 2-mercaptoethyl amine and 2,6-di-tertary butyl hydroxy toluene, their lifespan increases by 30–40%. When mice are fed with amino acids, tyrosine and histidine, their lifespan decreases. It was generalized that animals with high metabolic rate have short lifespan because they produce free radicals more rapidly. Humans and other long-lived species produce more superoxide dismutase overall and are more resistant to oxidation. It has also been observed that oxidized inactive proteins accumulate in cells because the degradation ability of the body declines.

free ribosome: Ribosome that is free in the cytosol, unattached to any membrane. It is the site of synthesis of all proteins encoded by the nuclear genome other than those destined to enter the endoplasmic reticulum.

free water: Unbound water. The term is frequently used to describe a film of water on a plant surface.

free-energy change (ΔG): Change in the free energy (ΔG) during a reaction calculated by using the formula: the free energy of the product molecules (change in enthalpy, ΔH) minus the free energy of the starting molecules (change in entropy, TΔS; here T is the temperature in Kelvin). A large negative value of ΔG indicates that the reaction has a strong tendency to occur. *See* ΔG.

freeze drying: A method for preserving microbial cultures for long periods. It is also called lyophilization. In this process, microbial suspension is quickly frozen at temperatures ranging from −54°C to −72°C, and water is removed by a high vacuum. Under vacuum, the container is sealed with a high-temperature torch. Powder-like residue that contains living microbes can be stored for years. Whenever required these can be revived by adding a suitable nutrient medium.

freeze preservation: Conditioning and preservation of plant cells, tissues or organs at extremely low temperatures, usually in liquid nitrogen.

freeze squeeze: A method of recovering DNA from an agarose gel. The portion of the gel containing the desired class of molecules is cut out and frozen. It is then placed between two sheets of Parafim™ and squeezed until a drop of DNA solution is ejected. Approximately, 60% of the DNA contained in the gel fragment may be recovered this way.

freeze-fracture electron microscopy: A technique for studying membrane structure, in which the membrane of a frozen cell is cracked along the interior of the bilayer, splitting it into two monolayers with the interior faces exposed.

frequency: (i) The number of times something happens in a specific group. (ii) The frequency of occurrence per unit of time (per month, per year, etc.).

frequency distribution: A synthetic presentation of a series of observations, obtained by specifying the number of observations falling in each 'class'; the frequencies of individuals with values of x between x_1 and x_2, between x_2 and x_3 and so on, define a class. May be graphically presented as frequency diagrams, histograms or line graphs.

frequency histogram: A 'step curve' in which the frequencies of various arbitrarily bounded classes are graphed.

frequency polygon: Frequency polygons are a graphical device for understanding the shapes of distributions. They serve the same purpose as histograms but are especially helpful for comparing sets of data. Frequency polygons are also a good choice for displaying cumulative frequency distributions.

frequency spectrum: The distribution of allele frequencies.

frequency-dependent fitness: Fitness differences whose intensity changes in the relative frequency of genotypes in the population.

frequency-dependent selection: Selection where the fitness of a type diverges with its frequency (i.e. whereby a genotype is at a benefit when rare and at a loss when common). In this type of natural selection, either rare types (negative frequency-dependent selection) or common types (positive frequency-dependent selection) are favoured. *See* NEGATIVE FREQUENCY-DEPENDENT SELECTION *and* POSITIVE FREQUENCY-DEPENDENT SELECTION.

frequency-independent fitness: Fitness that is not determined by interactions with other individuals of the same species.

frequency-independent selection: Selection in which the fitness of genotypes is independent of their relative frequency in the population.

frequentist statistical inference: This is currently the most commonly used statistical philosophy. It uses hypothesis testing, type I and II assertion probabilities, power, P-values, confidence limits (compatibility intervals) and adjustments of P-values for testing multiple hypotheses from the same study. Probabilities calculated using FREQUENTIST METHODS, P-values, are probabilities of finding values of statistics. The FREQUENTIST APPROACH is also called the SAMPLING APPROACH as it considers the distribution of statistics over hypothetical repeated samples from the same population. The frequentist approach is linked with long-run operating characteristics of statistics and estimates. Because of this and the backwards time/information ordering of P-values, frequentist testing requires complex multiplicity adjustments but provides no supervisory principles for exactly how those adjustments should be derived.

frequentist statistics, confusion in: Frequentist statistics involves the confusion of two ideas: (i) a priori probability that an experiment will generate misleading information (e.g. the chance of an assertion of an effect when there is no effect) – type I assertion probability (α) and (ii) *a posteriori* probability, the evidence for an assertion after the experiment is run. The latter should not involve a multiplicity adjustment, but because the former does, frequentists do not know how to interpret the latter when multiple hypotheses are tested or when a single hypothesis is tested sequentially. Frequentist statistics as typically practiced places emphasis on hypothesis testing rather than estimation.

friability: A term indicating the tendency of plant callus cells to separate from one another. Depending upon the plant species, composition of culture medium, primary explant and growth conditions, callus cultures exhibit either smooth, compact and hard type growth or nodular and friable type of growth. The latter growth is preferred for establishing cell suspension cultures.

frictional force microscopy (FFM): FFM is a derivative of the AFM, which is optimized to measure the lateral forces acting between a sharp, nanoscopic tip and a surface during the scanning of the tip. C.M. Mate and co-workers demonstrated that the frictional force can be measured at the atomic scale in 1987.

Friedman's two-way analysis of variance: Researchers use Friedman's two-way analysis of variance test to examine if there are statistically meaningful variations between groups when comparing them, using various parameters for each group.

friendly bacteria: A common reference to those bacteria that offer some benefit to human hosts, such as *Lactobacillus* species, which convert milk protein to lactic acid in the gut. The presence of such bacterial colonies also inhibits the growth of potentially pathogenic bacteria.

***FRIGIDA* (*Fri*):** A plant gene that controls whether or not a plant needs a winter period before it will flower. It guarantees that flowering is delayed until after winter so that the plant flowers in the favourable conditions of spring. This process of vernalization (the acceleration of flowering by a cold period of 3–8 weeks at +4 to +8°C) is an important characteristic in many plants. *FRIGIDA* is a single-copy gene in *Arabidopsis thaliana* and encodes a protein product of 609 amino acids. This protein has no significant match with any other protein or fragment sequence of known function. The action of an active *Fri* is dependent on an active *Flc* gene and *Fri* is known to increase the level of RNA from *Flc*. *FRIGIDA* is dominant. Summer annual types are recessive.

frond: A leaf of a fern, cycad or palm.

frozen-accident theory of genetic code: A theory given by F.H.C. Crick in 1968 to explain the relationship between an amino acid and its codon. There was a chance selection of code words that were then frozen. Once this association took place, it perpetuated as such. This theory can explain the particulate nature of the codon but not the pattern of degeneracy. This theory was later modified. Accordingly, the choice of the first codon was by accident but then the choice of synonymous codons is by selection. This explains the degeneracy of code.

fructification: A general term used for a spore-bearing organ of a fungus.

fruit: A seed-bearing structure existing in all angiosperms. It is formed from the mature ovary and sometimes accompanying floral parts upon fertilization.

fruiting body: Any of numerous, complex, spore-bearing fungal structures.

frutescent: Shrub-like (fruticose) or becoming shrub-like. *Aka* FRUTICOSE.

fuel: (i) In the context of dissipative DNA nanotechnology, a molecular input (such as nucleic acid strand, ATP) with high chemical potential that activates a DNA-based system. (ii) In the context of dynamic DNA nanotechnology, a molecular input (such as nucleic acid strand, pH or small molecule) that serves as 'input' for the activation of the system by initiating a strand displacement reaction or other fuel-induced downstream processes.

fuel gate: A DNA system able to process input molecules to release secondary fuels.

fugacious: In botany, disappearing, falling off or withering plant parts. *Compare with* PERSISTENT *and* CADUCOUS.

full gene sequence: The complete order of bases in a gene. This order determines which protein a particular gene will produce.

full-length cDNA cloning: The three methods commonly used for full-length cDNA cloning are OLIGO-CAPPING, CAPTURE and BIONYLATED CAP TRAPPER. *See details as separate terms.*

full-sib (FS): The members of a progeny with both parents in common.

fully-methylated DNA site: When both complementary strands of DNA carry a methyl group at a site. *Aka* HOMODUPLEX MODIFIED DNA. *Compare with* HETERODUPLEX MODIFIED DNA *and* UNMODIFIED DNA.

fumigant (vb. fumigate): A gas or volatile substance used to kill or inhibit the growth of microorganisms or other pests.

function: A subroutine that returns a value.

functional alleles: Alleles determined/defined based on complementation test.

functional candidate gene approach (FCGA): The FCGA works on the identification of traits associated with the expressed gene. Candidate genes are statistically detected from the genes controlling large components of inheritable gene expression variation.

functional complementation: Specific clones in a bacterial or phage library can be detected through their ability to confer a missing function on a mutant line of the donor organism, which acts as the transformation recipient. This method depends on the ability to transform the donor organism, often a eukaryote. The transforming vector contributes something that the recipient genome lacks, and the recipient genome contributes something that the vector lacks.

functional domains in eukaryotic chromosome: Eukaryotic chromatin is separated into functional domains differentiated by DNA methylation, post-translational histone modifications and histone variants. Methylation is associated with the expression of transcriptional initiation in plants and animals and is frequently found in transposable elements.

functional food: Ordinary food that has components or ingredients added to give it a specific medical or physiological benefit, other than a purely nutritional effect. For improving nutritional quality of foods, transgenic plants are being developed to obtain functional foods. Functional foods contain higher amounts of health-promoting nutrients such as quantity and quality of proteins, starches, flavonols and reduced concentrations of life-threatening substances such as allergens, 'toxins' caffeine, etc. *Aka* NUTRACEUTICAL.

functional genomics: Study dealing with assigning functions to every bit of the genome of an organism. The timing and location of the expression of those genes are also influenced by external/environmental factors such as temperature, sunlight, humidity, the presence of SIGNAL TRANSDUCERS, ACTIVATORS OF TRANSCRIPTION (STATs), etc. Also influencing the functions/traits are gene interactions, signalling cascades and response/reaction mechanisms within the body of that organism. Typically, functional genomic study follows the discovery of complete genome sequences found via structural genomics approaches. Some methods utilized to determine which traits/functions result from which gene(s) include: (i) SITE-DIRECTED MUTAGENESIS (SDM), to compare two same-species organisms possessing two different genes at the same site on the genome; (ii) ANTISENSE DNA SEQUENCE, to compare two same-species organisms (one of which has the gene at the same site 'turned off' via antisense DNA); (iii) REPORTER GENE, to compare two same-species organisms (with two different genes at the same site on the genome) via a 'reporter' gene located next to the gene/site, to detect the presence of the desired trait/function; (iv) CHEMICAL GENETICS, to compare two same-species organisms (one of which has a gene at the same site on the DNA molecule at least partially inactivated by a specific chemical); (v) 'SILENCING' or 'KNOCKING OUT' a particular gene via other methods than antisense or chemical genetics, to compare.

functional genomics research, impact of: The genomics age will change biology forever providing a sequence blueprint for numerous bacteria, plants, animals and human beings. Gene hunts have been dramatically accelerated by functional genomics. Traditional Research approaches have become more productive, fast and efficient, which help to give a deep insight into understanding biological systems in totality. Of course, the threat of patents battles over key aspects of the technology and over the novel genes may limit its expansions.

functional genomics techniques, classification of: Experimental techniques used to measure gene expression are divided into two groups based on their degree of multiplexity. (i) LOW-TO-MID-PLEX TECHNIQUES: (a) REPORTER GENE; (b) NORTHERN BLOT; (c) WESTERN BLOT; (d) FLUORESCENT IN SITU HYBRIDIZATION and (e) REVERSE TRANSCRIPTION PCR. (ii) HIGHER-PLEX TECHNIQUES: (a) SERIAL ANALYSIS OF GENE EXPRESSION (SAGE); (b) DNA MICROARRAY; (c) TILING ARRAY and (d) RNA-SEQ. *See details as separate terms.*

functional genomics tools: Elite genetic stocks from diverse genetic backgrounds provide the raw material for mapping populations such as NEAR-ISOGENIC LINES (NILs), RECOMBINANT INBRED LINES (RILs), DOUBLE HAPLOID (DH) POPULATIONS, DELETION MUTANTS, etc. Newly identified genes are assigned functions with the help of sequence similarities obtained from genomic tools, such as genomic databases, microorganisms, proteomics, etc. and bioinformatic tools, like computer software. Gene functions are linked to the phenotype by knocking out or overexpressing the gene from which a cellular or developmental role can be deduced. Finally, the fitness of plants carrying mutations or natural variants for the gene can be compared in different environments with wild-type plants to study the adaptive function.

functional genomics, methods of: For large-scale expression studies, recently devised methods used in functional genomics approach are as follows: (i) EXPRESSED SEQUENCE TAGS (ESTs), (ii) SERIAL ANALYSIS OF GENE EXPRESSION (SAGE), (iii) MICROARRAYS OR DNA/RNA CHIP TECHNOLOGY and (iv) PROTEOMICS. *See details as separate terms.*

functional genomics, potential applications of: (i) Looking into the fundamental causes of diseases, which improves disease diagnosis, helps in earlier detection of genetic predisposition to disease, leads to rational drug design and improves gene therapy and control system for drugs. (ii) Being applied in risk assessment where it quantifies DNA damage caused by radiation, mutagens and carcinogens. (iii) Bioarchaeology, anthropology, evolution and human migration are some other fields where functional genomics is useful. (iv) Studies evolution through germline mutations, studies the migration of different population groups based on female genetic inheritance and Y-chromosome lineages. (v) DNA forensics is now completely modernized with the entry of functional genomics of the scene. It identifies potential suspects whose DNA may match the evidence left at the crime scenes, establishes paternity and other family relationships and identifies protected species. (vi) Agriculture, livestock breeding and bio-processing areas in which functional genetics actively involve themselves include disease, insect and drought-resistant crops, healthier, more productive, disease-resistant farm animals, more nutritious produce, biopesticides, edible vaccines incorporated into food products, etc. (vii) Designer babies is a futuristic concept. Genes once found responsible for better resistance against diseases, for better IQ, longevity, etc. could be cloned or synthesized from known sequences from published human genome maps.

functional group: A portion of a molecule that participates in chemical interactions with other substances. The common functional groups are acyl, carbonyl, amino, carboxyl, ester, ether, hydroxyl, phosphoryl and sulfhydryl.

functional hemizygosity: An autosomal gene is functionally hemizygous if only one of the two copies in a diploid cell is expressed.

functional linkages, detection of: The advent of fully sequenced genomes has facilitated the development of computational methods for establishing functional linkage between proteins. Phylogenetic profiles describe the pattern of the presence or absence of a particular protein across a set of organisms whose genomes have been sequenced. If two proteins have the same phylogenetic profile (i.e. the same pattern of presence or absence in all surveyed genomes), it is inferred that the two proteins have a functional link. If there are 'n' fully sequenced genomes there are up to '2^n' phylogenetic profiles.

functional map: A physical map of a genome in which the positions of genes are known and their functions described.

functional markers (FMs): The DNA markers that have been directly derived from functionally characterized genomic regions of a trait-controlling gene. FMs are more useful in plant breeding compared with RANDOM DNA MARKERS (RDMs) and GENETIC MOLECULAR MARKERS (GMMs). FMs have been called 'perfect markers' in contrast with RDMs as 'non-perfect markers' such as RFLPs, AFLPs or SSR markers.

functional polymorphism: Any polymorphism in a gene or a promoter (or non-coding regulatory sequence) that either changes the underlying codon, and hence, the amino acid composition of the encoded protein, or alters the sequence of a transcription factor–binding site in a promoter such that binding and subsequent activation of the adjacent gene is prevented.

functional programming: A programming model of building the structure and elements of computer programs that consider computation as the evaluation of mathematical functions and evades changing-state and mutable data. It is a declarative programming pattern in that programming is accomplished with expressions or declarations instead of statements.

functional proteins: These are the proteins (enzymes) that conduct various biochemical processes.

functional proteomics: A straightforward approach to finding out the function of a gene.

functional regions of a flower: There are three functional regions of flower meristem, namely – A, B and C.

functional RNAs: Different types of RNA species (e.g. tRNA, rRNA and mRNA) that perform specific functions during protein synthesis. *Aka* NON-GENETIC RNAs.

functional state of chromatin: Eukaryotic genomes are organized into two types of domains: (i) active (euchromatic) and (ii) inactive (heterochromatic) domains. Posttranslational modifications of histones are the key epigenetic modifications to define these functional states, particularly in promoter regions. Asymmetric di-methylation of histone H3 arginine 2 (H3R2me2a) counter correlates with di- and tri-methylation of H3 lysine 4 (H3K4me2, H3K4me3) on human promoters.

functional unknomics: Refers to the organized screening of conserved genes of unknown function.

functionalized/functionalization: The attachment of a chemically active moiety to an inert molecule/entity.

functions of non-coding RNAs: The roles of ncRNAs are as diverse as their structures and range from the purely structural to the purely regulatory. Non-coding RNAs play a role in diverse cellular processes such as (i) CHROMOSOME REPLICATION, (ii) TRANSCRIPTIONAL REGULATION, (iii) RNA PROCESSING AND MODIFICATION, (iv) MRNA STABILITY AND TRANSLOCATION, (v) PROTEIN DEGRADATION AND TRANSLOCATION, (vi) GENE SILENCING and (vii) DEVELOPMENT. *See details as separate terms.*

functions of repeated DNA: Assumed functions of repeated DNA are suggested to be many, including involvement in chromosome pairing, control of gene expression, processing of RNA (excision of introns) and participation in DNA replication. So far none has been fully established.

fundamental elements for disease in plants: The four fundamental elements required for disease in plants are: a susceptible host, a pathogen capable of causing disease, a favourable environment and adequate time. This is referred to as the DISEASE QUADRANGLE.

fundamental niche: All available roles an individual can play in an ecosystem.

fundamental number: The number of chromosome arms in a somatic cell of a particular species.

fundamental theorem of natural selection: The upsurge in mean fitness due to the selection in allele frequencies is equal to the additive genetic variance in fitness divided by the mean fitness.

fungal resistance: Resistance in a plant to fungal disease(s). Fungal diseases cause huge crop losses. Rust and Karnal bunt of wheat, sheath blight of rice, powdery mildew of grapes and Alternaria of brassica are important fungal diseases. Fungal diseases are controlled either by using fungicides or by growing resistant crop varieties. Many antifungal genes such as the *chitinase* gene, *glucose oxidase* (H_2O_2 producing) gene, *glucanase* gene, *RB* gene from wild potato and *peroxidase* gene have been cloned and are used in plant transformation to develop fungal resistance.

fungi imperfecti: An archaic term for a group of fungi without a sexual stage. Precisely, it refers to the asexual stage of members of Ascomycota (the ascomycetes) and Basidiomycota (the basidiomycetes). *See* DEUTEROMYCETES *and* IMPERFECT FUNGI.

fungicide (Adj. fungicidal): A chemical or physical agent that kills or constrains the growth of fungi.

fungistasis: Refers to the inhibition of fungal growth, sporulation or spore germination but not its death. The term is used to describe the non-specific phenomenon in natural soils where spore germination is inhibited and often overcome by rhizosphere nutrients.

fungus (pl. fungi): A eukaryotic organism that is usually filamentous (forming a mycelium) and heterotrophic, has cell walls composed of chitin and reproduces by sexual and/or asexual spores. Fungi are responsible for many plant diseases and a relatively small number of animal diseases. It lacks chlorophyll. It includes microorganisms such as yeasts and moulds, as well as the more familiar mushrooms.

furrow formation: Posterior positioning of the cell-division plane during mitosis is essential for determining the size and position of the two daughter cells. This is a critical step in development and cell differentiation. A bipolar microtubule array has been proposed to be the minimum requirement for furrow positioning in mammalian cells, with furrow forming at the site of the microtubule plus-end overlap between the spindle poles. Cytokinesis occurs at high frequency in monopolar cells.

furrowing: A mechanism through which animal cells undergo cytokinesis.

fusion gene: A gene derived from parts of two other genes, either by unequal crossing over, by deletion of the part of the chromosome segment between two adjacent genes or by recombinant DNA technology.

fusion genes microarray: A fusion gene microarray can detect fusion transcripts (e.g. from cancer specimens). The principle behind this method is the ALTERNATIVE SPLICING MICROARRAYS. The oligo design strategy empowers combined measurements of chimeric transcript junctions with exon-wise measurements of individual fusion partners.

fusion of species: Complete breakdown of reproductive isolating barriers between species. Consequently, some well-defined species merge into one species.

fusion pattern analysis of protein domain: By analysing the fusion patterns of protein domains, functionally linked polypeptides can be discovered. When two free-standing polypeptides A and B are functionally active in one organism but are expressed as a fused protein in some other species, a successful search through other genome sequences for the corresponding fused protein offers powerful evidence that polypeptides A and B are linked functionally. A is the fused protein that is similar to both A and B and is termed a Rosetta Stone Sequence. Because A and B have unrelated sequences, this type of functional linkage cannot be detected by a homology search. If the genes that encode two proteins are neighbours in several genomes, the proteins tend to be functionally linked. This is called the GENE NEIGHBOURHOOD method, which is useful in uncovering functional linkages in prokaryotes, where operons are common but also display promise for analysing interacting proteins in eukaryotes.

fusion protein: (i) The protein resulting from the genetic joining and expression of two different genes. (ii) A product of recombinant DNA in which the foreign gene product is juxtaposed ('fused') to either the carboxyl-terminal or amino-terminal portion of a polypeptide encoded by the vector itself. Use of fusion proteins often facilitates the expression of otherwise lethal products and the purification of recombinant proteins. For instance, by fusing a difficult-to-study gene with the β-galactosidase gene, the (protein) product that results can easily be identified and measured using chromatography. *See* CHIMERIC PROTEIN.

fusion protein purification by affinity chromatography: Fusion proteins facilitate purification by affinity chromatography. A short sequence of amino acids or a protein domain is fused to one end of the polypeptide. Six histidines can be used to purify a protein through nickel affinity. The Fc region of an antibody (region of an antibody molecule that binds to cell-surface antibody receptors) can be fused to a protein to purify it by affinity to protein A, a protein from bacteria that specifically binds antibodies.

fuzzy logic: A simple form for the many-valued logic, in which the truth values of variables may have any degree of TRUTHFULNESS that can be represented by any real number in the range between 0 (as in COMPLETELY FALSE) and 1 (as in COMPLETELY TRUE) inclusive. Consequently, this logic is employed to handle the concept of partial truth, where the truth value may range between completely true and completely false. In contrast to Boolean logic, where the truth values of variables may have the integer values 0 or 1 only.

fuzzy logic artificial intelligence (AI): Fuzzy logic AI explains human intelligence and decision-making behaviour. In logic, fuzzy logic is a form of many-valued logic in which the truth value of variables may be any real number between 0 and 1 both inclusive.

G banding: A special staining technique for chromosomes that results in a longitudinal differentiation by Giemsa stain, which is a complex of stains specific for the phosphate group of DNAs. The characteristic bands thus produced are called G bands. These bands are generally produced in AT-rich heterochromatic regions.

G bands: Eukaryotic chromosome bands produced by treatment with Giemsa stain. For G banding, the chromosome preparations on a slide are air dried, subjected to denaturation-renaturation and then stained with Giemsa. The representative bands produced are called G bands. These bands are generally formed in A/T-rich, heterochromatic regions.

G-overhang: A telomeric single-stranded DNA 3′ tail consisting of TTAGGG repeats, which is the substrate of telomerase.

G-proteins: A large group of guanine-nucleotide-binding trimeric proteins with GTPase activity that act as molecular switches in many different cellular pathways, controlling processes such as sensory perception, intracellular transport, protein synthesis and cell growth and differentiation. They undergo a large conformational change when a bound guanosine 5′-triphosphate (GTP) is hydrolysed to guanosine 5′-diphosphate (GDP).

G-protein–coupled receptors (GPCRs): An organism's body is a fine-tuned system of interactions between billions of cells. Each cell has tiny receptors that enable it to sense its environment, so it can adapt to new situations. About a thousand genes code for such receptors, for example, for light, flavour, odour, adrenalin (also called epinephrine), histamine, dopamine and serotonin. About half of all medications, including beta-blockers, antihistamines and several psychiatric medications, achieve their effect through GPCRs.

G-protein–linked receptor: Cell-surface receptor that associates with an intracellular trimeric GTP-BINDING PROTEIN (G-protein) after receptor activation by an extracellular ligand. These receptors are seven-pass transmembrane proteins.

G-quadruplex RNA: In addition to serving as a messenger, RNA regulates the process of transcription itself. Polycomb repressive complex 2 (PRC2) is an epigenetic gene-silencing complex that silences genes through trimethylation of histone H3K27. PRC2 is essential for cell differentiation and is often deregulated in cancers, is a prominent system for studying this RNA-mediated regulation. J. Song *et al.* (2023) used cryo-electron microscopy to solve a structure of PRC2 bound to precursor messenger RNA and long non-coding RNA, with a binding preference for G-quadruplex RNA, revealing that the RNA promotes dimerization of the protein complex.

G-quadruplexes (G4): Tertiary structures in which groups of four guanines in single-stranded DNA or RNA form tetrads through hydrogen bonds.

G/T mismatch repair: GT mismatches are corrected by the VSP repair pathway (very short-patch DNA repair synthesis), which depends on the genes *vsr* and *polA* and is strongly stimulated by MutL and MutS.

G0 phase: State of withdrawal from the eukaryotic cell-division cycle by entry into a quiescent Gap 1 phase. A common state for differentiated cells. Gap 0 cells have exited the cell cycle and entered into a resting phase. *Aka* G0 STAGE.

G1 checkpoint: The G1/S checkpoint is the main decision point for a cell – that is, the primary point at which it must choose whether or not to divide. Once the cell passes the G1 checkpoint and enters S phase, it becomes irreversibly committed to division at a stage called the restriction point in mammalian cells and starts in budding yeast.

G1 phase (Gap 1): The first of four phases of the cell cycle that takes place in eukaryotic cell division. The Gap 1 cells accomplish most of their growth; they get bigger in size and make proteins and organelles needed for normal functions of DNA synthesis. Here, proteins and RNAs are synthesized, and, more especially, the centromere and the other components of the centrosomes are made. G1 phase of the eukaryotic cell-division cycle occurs between the end of cytokinesis and the start of DNA synthesis. G1 phase is the most variable in length, depending on physiological conditions it may last for days, months or years. *Aka* GAP 1 PHASE *or* GROWTH 1 PHASE. *See* GAP PERIOD.

G1/S-Cdk: Complex formed in vertebrate cells by a Gap 1/S-cyclin and the corresponding cyclin-dependent kinase (Cdk).

G2 checkpoint: At this stage, the cell checks DNA integrity and DNA replication. If errors or damage are detected, the cell will pause at the G2 checkpoint to allow for repairs.

G2 phase (Gap 2): Gap 2 is the third subphase of interphase in the cell cycle directly preceding mitosis. It follows the successful completion of S phase, during which the cell's DNA is replicated. Cytoplasmic organelles (centrioles, mitochondria and Golgi apparatus) are doubled. Proteins for the spindle apparatus are synthesized. The cell is metabolically active and stores energy required for the division of the cell. In G2, the amount of DNA is double (i.e. 4C). During this phase, p34cdc2 dephosphorylates and combines with M-specific cyclins to form maturation promotion factor (MPF). It phosphorylates many substrates, which push the cell into M phase. G2 phase marks the beginning of mitosis. *Aka* GAP 2 PHASE *or* GROWTH 2 PHASE. *See* GAP PERIOD.

***GA21*:** A naturally occurring gene (i.e. expressed at low levels in some plants) which confers resistance to glyphosate-containing herbicides. When the *GA21* gene is inserted by man into crop plants (e.g. maize/corn) in a way that causes high expression, those crop plants are subsequently unaffected when glyphosate-containing herbicides are applied to fields to control weeds in those crops.

gain-of-function mutation: Gain-of-function mutation alters the gene product such that it gains a new function. These mutations usually show dominant phenotypes and are often called NEOMORPHIC MUTATIONS.

GAL4: A sequence-specific DNA-binding protein that activates transcription in yeast.

galactose (gal): A monosaccharide existing in both LAEVOROTATORY or LEVOROTATORY (L) and DEXTROROTATORY (D) arrangements as a constituent of plant and animal oligosaccharides (lactose and raffinose) and polysaccharides (agar and pectin). *Aka* CEREBROSE.

galactose repression in *Escherichia coli*: For galactose repression in *Escherichia coli*, two operators, O_E and O_I, exist. O_E is located immediately upstream of two overlapping *gal* promoters, while O_I is located inside first structural gene. PG1 is cAMP-dependent promoter, whereas PG2 is cAMP-independent promoter. Cat is cAMP.CRP complex binding site. E_N is the start site of the translation of the *gal* gene. Six natural or engineered endonuclease sites. Purified repressor shows concentration-dependent binding to both O_E^+ and O_I^+ but not O_E^C and O_I^C sequences. O_E and O_I are separated by more than 90 bp. Negative control of the *gal* operon requires repressor binding at both O_E and O_I. How does the repressor binding to operator sites inhibit transcription, and why two operators are needed? There are two possibilities: (i) Repressor dimer binds independently to O_E and O_I. These two repressor-DNA complexes then associate. This could either stabilize the repressor-O_E complex or alter the conformation of DNA in the resulting loop. (ii) The *gal* repressor dimer binds simultaneously to O_E and O_I, generating a DNA loop.

galactose upstream activating sequence (UAS_G): A piece of DNA recognized by the amino domain of yeast GAL4 regulator.

galactose-1,3-galactose: A carbohydrate found as a modification to cell-surface glycoprotein on all mammals (and many other organisms) except for old-world primates (including humans). The presence of naturally occurring antibodies to this modification in humans is a major (but not the only) cause of rejection of xenotransplanted organs.

galactose-responsive transcription factor (GAL4): A sequence-specific DNA-binding protein that activates transcription in yeast.

galactoside permease: Product of *y* gene of *lac* operon. It facilitates the entry of lactose into the cell. *Aka* GALACTOSE PERMEASE.

galactosyl transferase: The enzyme, lacking in old-world primates, responsible for adding galactose-1,3 galactose to proteins.

galaxy: A web-based platform that makes command-line tools available to biologists, is flexible, sharable and can be run from (almost) any computer.

gall: Irregular localized outgrowth, often roughly spherical, on external plant tissues, in response to attack by viruses, fungi, bacteria, nematodes, plants, insects and mites. *See* KNOT *and* TUMOUR.

game theory: The study of mathematical models of strategic interaction between rational decision makers. It has applications in all fields of social science, as well as in logic and computer science. Originally, it addressed zero-sum games, in which each participant's gains or losses are exactly balanced by those of the other participants. Today, game theory applies to a wide range of behavioural relations and is now an umbrella term for the science of logical decision making in humans, animals and computers.

gametangium (pl. gametangia): An organ or cell in which gametes are produced that is found in many multicellular protists, algae, fungi and the gametophytes of plants. In contrast to gametogenesis in animals, a gametangium is a haploid structure, and the formation of gametes does not involve meiosis.

gamete: A male or female reproductive cell. In plants, animals and humans, the organism body is diploid. Meiosis takes place in the reproductive organs. As a result, haploid gametes are produced. In animals (and humans), the functional, mature, male gamete is called a SPERMATOZOON; in plants, it is called a SPERMATOZOID. In animals and plants, the female gamete is called the OVUM, or EGG. Later male and female gametes combine to produce a diploid zygote, which further grows into a diploid organism. *Aka* SEX CELLS *or* GAMETIC CELL.

gametic: Tissue or generation that has haploid (n) number of chromosomes in contrast to zygotic tissue with diploid (2n) number of chromosomes. Some microorganisms for the most part are haploids.

gametic (phase) equilibrium: In relation to any two loci, the existence of gametes with a frequency equals the product of the frequency of the two relevant alleles (e.g. loci A and B are in linkage equilibrium if the frequency of the gamete A_nB_n is equal to the product of the frequencies of alleles A_n and B_n).

gametic differentially methylated regions (DMRs): DMRs acquire their parental-specific DNA methylation during gametogenesis, either in the developing haploid oocyte or sperm.

gametic imprinting: The expression of a gene depends on its parental origin in the offspring.

gametic mutation: A mutation occurring during gamete formation.

gametic selection: The impacts exerted to result in differential reproductive success of one allele over another in a heterozygote.

gametocidal gene: A gene encoding a product that abolishes cells that divide to yield the gametes.

gametocide: A chemical agent used to specifically kill either male or female gametes. Such a chemical is used in hybrid seed production of autogamous crops (e.g. barley or wheat).

gametoclonal variation: Differences among regenerants obtained from pollen and/or anther culture. This variation in phenotype could be due to either genetic or epigenetic causes and is expressed by gametoclones.

gametoclone: Plants regenerated from cell cultures resulting from meiospores, gametes or gametophytes. *Aka* GAMETOCLONAL.

gametocyte: A cell that will undergo meiosis to form gametes. In protozoa, the male or female cell is called a gametocyte. Some protozoa produce gametes, haploid sex cells that are called gametocytes. During reproduction, two gametes combine to yield a diploid zygote, which further grows into an organism.

gametogamy: The combining of the sexual gametes and the production of the zygotic nucleus.

gametogenesis: The process of formation of gametes from gametocytes that participate in fertilization. In this process, genes are processed by meiosis and stored in the form of gametes. During oogenesis, yolk, tRNAs and mitochondria are stored.

gametophore: Specialized structures on the gametophytes of some bryophyte species. For example, in many species in the order Marchantiales, the gametes are produced on the gametophores.

gametophyte: The haploid multicellular phase in the alternation of generations of plants and algae that bear gametes. In bryophytes, the gametophyte is the dominant vegetative phase; in ferns and their allies, it is a small free-living plant known as the PROTHALLUS; in gymnosperms and angiosperms, the gametophytes are reduced to microscopic structures dependent on the sporophyte, male gametophytes contained in pollen grains and females contained within the ovules. Gametophyte alternates with a diploid, sporophyte generation. *Contrasts* SPOROPHYTE.

gametophytic apomixis: Agamic seed formation in which the embryo sac originates from an unreduced initial cell. It includes both DIPLOSPORY and APOSPORY. *See details as separate terms.*

gametophytic differences: Differences observable in the gametophyte (e.g. *waxy* versus *starchy* pollen in maize and some other cereals). Most, if not all, of the differences in lower forms (haploids) are gametophytic.

gametophytic generation: The haploid microspores and megaspores represent the gametophytic generation. The microspore germinates in microsporangium and produces male gametophyte. The megaspore on germination produces a female gametophyte or embryo sac in the megasporangium. The embryo sac has an egg apparatus, antipodals and a secondary nucleus. During pollination, pollen alight on the stigma of the pistil. Pollen grain germinates in the presence of stigmatic fluid and produces the pollen tube, which penetrates through the stylar tissue and reaches the embryo sac of the ovule. The pollen tube carries two male gametes, which are released into the embryo sac.

gametophytic incompatibility: A botanical phenomenon controlled by the complex *S* locus in which a pollen grain cannot fertilize an ovule produced by a plant that carries the same *S* allele as the pollen grain (e.g. S_1 pollen cannot fertilize an ovule made by an S_1/S_2 plant).

gametophytic self-incompatibility: Self-incompatibility relies on the genotypic and phenotypic relationship between the female and male reproductive system. Alleles in cells of the pistil govern its receptivity to pollen. The phenotype of the pollen, expressed as its inability to effect fertilization, may be resolved by its own alleles, the phenomenon referred to as GAMETOPHYTIC INCOMPATIBILITY (e.g. S_1 pollen cannot fertilize an ovule made by an S_1/S_2 plant).

γ-retroviruses: A species of retrovirus that includes porcine endogenous retrovirus (PERV) and Murine leukaemia virus (MLV).

γ-tubulin ring complex (γ-TuRC): Protein complex containing γ-tubulin and other proteins that is an efficient nucleator of microtubules.

gamophyllous (syn. symphyllous): With joined or fused petals. A single perianth whorl of united segments. *Compare with* APOPHYLLOUS *and* POLYPHYLLOUS.

gap: A space introduced into an alignment to compensate for insertions and deletions in one sequence relative to another. To prevent the accumulation of too many gaps in an alignment, the introduction of a gap causes the deduction of a fixed amount (the gap score) from the alignment score. Extension of the gap to encompass additional nucleotides or amino acids is also penalized in the scoring of an alignment.

gap genes: Mutations in these genes lead to the deletion of a group of several adjacent segments from the final pattern.

gap junction: There is a very small gap between two adjacent cell membranes. Certain proteins fill in the gap, which allows the passage of some ions and small molecules from one cell to another. This permits coordination of cell behaviour.

gap penalty: Numeric score used in sequence alignment to penalize the presence of gaps within an alignment. The price applied to a similarity score for the introduction of an insertion or deletion gap, the extension of a gap, or both. Gap penalties are generally deducted from a cumulative score being worked out for a comparison of two or more sequences via an optimization algorithm that attempts to maximize that score.

gap period: One of the two intermediate periods within the cell cycle. Gap 1 (G_1) occurs between mitosis (M) and the DNA synthesis (S) phase and gap 2 (G_2) occurs between DNA synthesis (S) and next mitosis (M).

gap repair: A linear plasmid vector (gapped vector) can be circularized by homologous recombination between its ends and a target DNA.

gap-filling repair in eukaryotes: Replication by template switch is thought to mediate DNA damage–bypass and filling of gaps. Gap-filling repair requires homologous recombination as well as Rad18- and Rad5-mediated proliferating cell nuclear antigen (PCNA) polyubiquitylation.

gapped DNA: A duplex DNA molecule with one or more interior single-stranded sections.

garbage in, garbage out (GIGO): Describes the concept that flawed or nonsense input data produces nonsense output or 'garbage'. It can also refer to the unforgiving nature of programming, in which a poorly written program might produce nonsensical behaviour.

Garnier, Osguthorpe and Robson (GOR) method: This method assumes that amino acids up to eight residues on each side influence the secondary structure of the central residue. This method uses information, theory and values in tables to calculate the probability that the central residue is one type of secondary structure and not another.

gas-liquid chromatography: A chromatographic method in which the materials to be analysed are vaporized and carried in an inert gas (e.g. nitrogen) through a stationary liquid phase. The method is suitable for analysing complex fatty acid mixtures that are volatile at reasonably low temperatures.

gastrulation: The stage in animal embryogenesis during which the embryo is transformed from a ball of cells to a structure with a gut (a gastrula). During gastrulation, cell rearrangement and cell-cell interactions take place. These interactions form the basis of further differentiation.

gated transport (of a protein): One of three ways for a protein molecule to pass between compartments within eukaryotic cells. The compartment 'wall' (membrane) owns a 'sensor' (receptor) that senses the existence of a correct protein (e.g. after that protein has been synthesized in the cell's ribosomes) and then opens a 'gate' (pore) in the membrane to permit that protein to pass from the first compartment to the second compartment.

gateway recombinational cloning: A recombinational cloning procedure based on the site-specific recombination system of the λ bacteriophage. Gateway is comprised of the BP reaction and LR reaction, both *in vitro* recombinations.

Gause law: An ecological principle stating that different species occupying identical ecological niches (i.e. ecological homologues) cannot coexist in the same habitat.

Gaussian curve: *See* NORMAL DISTRIBUTION.

Gaussian derivative wavelets: A computational approach to investigate epigenetic changes using the shape of the signals for histone modification. Employing a Gaussian Derivative Wavelet (CGD-Wavelet) method, the proposed procedure models a nucleosome with a Gaussian and detects the peak and the edges of the Gaussian.

Gaussian distribution: *See* NORMAL DISTRIBUTION.

GC box: A constituent of many eukaryotic promoters, particularly those from constitutively expressed genes. The consensus sequence for the GC box is 5′-GGGCGG-3′.

GC content: The degree of the abundance of G+C nucleotides relative to A+T nucleotides within DNA sequences.

GC-rich area: Many DNA sequences carry long stretches of repeated G and C which often indicate a gene-rich region.

geitonogamy: The fertilization of a flower by pollen from another flower on the same (or a genetically identical) plant. *Compare with* XENOGAMY.

gel: The inert matrix used for the electrophoretic separation of nucleic acids or proteins.

gel electrophoresis: A technique by which molecules are separated by size or charge by passing them through a gel under the influence of an external electric field. *See* ELECTROPHORESIS.

gel filtration chromatography: In this process, separation is based on the differing ability (due to differing molecular size) of molecules in the sample to enter the pores of the gel filtration medium. The stationary phase in this technique consists of beads of a hydrated, sponge-like material that has pores of molecular dimensions and with a narrow range of sizes. *Aka* SIZE-EXCLUSION CHROMATOGRAPHY *or* MOLECULAR-SIEVE CHROMATOGRAPHY.

gel retardation analysis (GRA): A technique that identifies a DNA fragment that has a bound protein by virtue of its decreased mobility during gel electrophoresis.

gel shift: A method by which the interaction of a nucleic acid (DNA or RNA) with a protein is detected. The mobility of the nucleic acid is monitored in an agarose gel in the presence and absence of the protein: if the protein binds to the nucleic acid, the complex migrates more slowly in the gel (hence 'gel shift'). A 'supershift' allows determination of the specific protein, by virtue of a second shift in mobility that accompanies binding of a specific antibody to the nucleic acid-protein complex.

gel-based assays: The presence of SINGLE-NUCLEOTIDE POLYMORPHISM (SNP) can be detected by RESTRICTION FRAGMENT LENGTH POLYMORPHISM (RFLP), or AMPLIFIED FRAGMENT LENGTH POLYMORPHISM (AFLP) conducted on PCR products, whenever such an SNP produces or abolishes a specific restriction site for an enzyme. SINGLE-STRAND CONFORMATION POLYMORPHISM (SSCP) can also be used for the detection of SNPs.

gelling agent: A chemical compound that causes gelling and solidification of the culture medium. The gelling agent does not play any nutritional role, but when added into medium before autoclaving, it causes solidification of autoclaved medium kept at room temperature. Commonly used gelling agents include, Agar, Agarose, Gelrite and Phytagel.

geminivirus: A single-stranded DNA virus that results in serious diseases in cereals, vegetables and fibre crops worldwide. The important diseases include maize streak virus (MSV), wheat dwarf virus, cassava latent virus, tomato yellow leaf curl virus and bean golden mosaic virus. Like nanoviruses, it is spread by either whiteflies or leafhoppers. All Whitefly-transmitted geminiviruses are placed in the genus *Begomovirus,* which typically have two-part genomes (DNA A-form and DNA B-form) embracing circular DNAs of around 2700 nucleotides, although some have been found to be monopartite, possessing only a DNA A-form component. In contrast, the nanoviruses are a recently established group of plant viruses that are spread by either aphids or planthoppers and have multi-part genomes comprising circular DNAs of around 1000 nucleotides. These viruses replicate via double-stranded DNA forms. These viruses are also used to develop expression vectors for plants.

gemmule: An old and no longer appropriate term used to describe hypothetical units in the body that carried the essence of specific body traits to the gonads, from which they were transmitted to the offspring.

gen: A branch of a cladogram.

GenBank: A data bank of genetic sequences governed by a division of the National Institutes of Health, USA.

gender: Differences observed between any two complementary organisms, defined as males and females, of the same species that make them capable of mating.

gene: (i) Classically, a unit of inheritance which is too small to be seen under a microscope, a segment of DNA on a chromosome that encodes a protein and includes all the regulatory sequences (e.g. promoter, operator, terminator) required to control the expression of that protein. (ii) A gene is an ordered sequence of nucleotides located in a particular position on a particular chromosome that encodes a specific functional product (i.e. a protein or RNA molecule). (iii) From a geneticist's point of view, a gene is a discrete chromosomal region which is responsible for the synthesis of a specific cellular product (mRNA, tRNA, rRNA or regulatory RNA (small RNAs)), has one or more specific effects on the phenotype and can mutate to various allelic forms, each of which can exist in several alternative forms and between which crossing over can occur. (iv) Modern functional definition of gene is a DNA sequence coding for a specific polypeptide but could be a split gene where DNA sequence also includes introns (non-coding segments) and exons (coding segments that make proteins) and other regulatory DNA elements (e.g. promoters, enhancers, regulator genes, operators). (v) Some genes code for rRNA, tRNA and snRNPs. (vi) From an evolution point of view, gene is any portion of chromosomal material that potentially lasts for enough generations to serve as a unit of natural selection. (vii) A DNA segment known to produce a product. *See* ALLELE.

gene action: The expression of the gene based on the transcription into complementary RNA sequences and the subsequent translation of mRNA into polypeptides, which may ultimately form a specific protein.

gene activation: Diverse mechanisms involved in the repression and activation of genes.

gene activator protein: A gene regulatory protein that when bound to its regulatory sequence in DNA activates transcription.

gene activity theory of cell differentiation, evidence of: Some of the evidence is as follows. (i) Only a minor fraction of DNA sequences is transcribed in any given differentiated cell type. (ii) Diverse sets of sequences are transcribed in each differentiated cell type. (iii) Differentiated cell nuclei contain inactive (nontranscribed) DNA transcription in isolated chromatin, which is restricted to a small fraction of DNA. Therefore, much of the chromatin DNA is inactive. (iv) There is also cytological evidence that supports variable genomic activity. In many cell types, easily stained and densely packed heterochromatin region of the interphase chromatin is designated. These regions tend to be inactive in RNA synthesis.

gene activity theory of cell differentiation: In this theory, cell differentiation is a two-step mechanism at each stage of development. (i) Gene expression is unstable. It occurs stochastically and produces different cell types. (ii) Gene expression is stabilized by means of cellular interactions.

gene amplification: A process by which a cell increases the copies of a particular gene within the genome, in some specific tissue, disproportionately greater than their representation in the parent molecules, during development. During prolonged cell culturing, gene amplification sometimes takes place, which leads to somaclonal variation. It may occur in the presence of elevated concentrations of growth regulators in the culture medium which cause DNA duplication and excessive cell division.

gene amplification, modes of: Two modes of programmed amplification can be distinguished *in vivo* – WHOLE-GENOME AMPLIFICATION and SELECTIVE WHOLE-GENOME AMPLIFICATION. When a mutant copy is present in a multigene family, spontaneous amplification of functional or non-functional mutant copy may take place. When gene amplification involves the functional copy in some cases of the mutant gene, then the mutant gene will be expressed, but when the non-functional copy is involved, the individual may get eliminated. In either case, it will lead to homogenization. *See* GENE *and* ONCOGENE.

gene amplification, techniques of: The five primary techniques used by man to perform gene amplification are (i) POLYMERASE CHAIN REACTION (PCR), (ii) LIGASE CHAIN REACTION (LCR), (iii) SELF-SUSTAINED

SEQUENCE REPLICATION (SSR), (iv) Q-BETA REPLICASE TECHNIQUE and (v) STRAND DISPLACEMENT AMPLIFICATION (SDA).

gene and enzyme, relationship between: A. Garrod in 1909 observed that several hereditary human defects are produced by recessive mutations. A relationship between the unit of heredity (gene) and its protein product (enzyme), which is accountable for directing a biochemical reaction, was suggested.

gene annotation: Statistics reliably identifies which gene products undergo alteration under experimental conditions, deducing that the biological logic of expression profiling relies on knowing which protein each gene product forms and what function this protein performs. Gene annotation offers functional and other information (e.g. the location of each gene within a particular chromosome). Some functional annotations are more consistent than others while some are absent. Gene annotation databases update regularly, and various databases refer to the same protein by different names, reflecting a fluctuating understanding of protein function. The use of standardized gene nomenclature helps address the naming aspect of the problem, but precise matching of transcripts to genes remains a significant deliberation.

gene arrangement: A particular distribution of genes along a chromosome, typically identified by the distribution of bands along a polytene chromosome.

gene as functional entity: From a functional point of view, a gene is a specific linear sequence of nucleotides performing a primary cellular function. Now gene function is better appreciated in terms of one gene-one primary cellular function hypothesis. Some recent concepts include one gene many functions and many genes one function hypotheses.

gene augmentation: A number of copies of a normal gene for the defective gene are introduced into the cell and are made to express at a high level.

gene bank: (i) A group of recombinant DNA molecules containing inserts which when together contain the entire genome of an organism. (ii) An establishment in which both somatic and hereditary genetic material are conserved (seeds, pollen, whole plants, extracted DNA). It stores, in a viable form, material from plants that are at risk of extinction in the wild and cultivars that are not currently in popular use. The stored genetic material can be asked for when required. The normal method of storage involves reducing the water content of seed material to around 4% and storing it at 0°C or less (−20°C). All stockpiled stocks are periodically checked by germination tests. *See* GENOMIC LIBRARY.

gene battery model: In the Britten-Davidson model of gene regulation in eukaryotes, a set of structural genes regulated by one sensor is termed as a BATTERY.

Gene Center: The Gene Center is a research facility of the Ludwig Maximilian University (LMU) of Munich that includes the Department of Biochemistry, independent junior groups and research groups.

gene centre: Any of a number of areas on the Earth from which arose important crop plants and domestic animals. *Aka* GENE POOL CENTRES.

gene chip: A microchip that holds DNA probes that form half of the DNA double helix and can recognize DNA from samples being tested. The covalent attachment of oligonucleotides or cDNA directly on to a small glass or silicon chip in organized arrays. Over 50,000 different DNA fragments can be presented on a single chip providing a high-throughput parallel method of probing gene expression, genotype or gene function. *Aka* GENE ARRAYS. *See* DNA CHIP.

gene chip technology: Development of cDNA microarrays from a large number of genes. Used to monitor and measure changes in gene expression for each gene represented on the chip. This technology permits the analysis of thousands of genes at once. For this, cheap and simple customized chips capable of deconstructing long segments of DNA have been manufactured. This technique enables biologists to screen huge chunks of animal and plant genomes in search of the genes that promote disease, the genetic switches that govern such biological phenomena as ageing, and the DNA codes that permit microorganisms to make antibiotics.

gene cloning: Insertion of DNA piece (gene) into a cloning vector (plasmid) and subsequent multiplication of recombinant DNA molecule in a host organism. The discovery of two types of enzymes (i) restriction enzymes (cut DNA into pieces) and (ii) ligases (insert DNA pieces into plasmid) has made gene cloning a reality. Recombinant DNA molecules thus produced are introduced into bacterial cells (*Escherichia coli*), which are multiplied, and all descendants carry the desired gene. Now, gene cloning is also possible using PCR. Basic steps in gene cloning are (i) generation of a DNA fragment (isolation/synthesis/digestion); (ii) joining to a vector molecule or a carrier molecule (select DNA capable of self-replication); (iii) moving recombinant DNA into the host cell, which provides the machinery for DNA replication (by techniques such as transformation, electroporation or transfection) and (iv) select host cells that contain recombinant DNA. *Aka* MOLECULAR CLONING, DNA CLONING *or* MOLECULAR CLONING.

gene cloning and polymerase chain reaction: With DNA polymerase, DNA can be amplified to have 10^6 copies. With recent modifications that use DNA polymerase from thermophilic bacteria, a given DNA fragment can be amplified up to 4×10^6 times in two cycles. The length of the target sequence is determined by the distance between the two primer sites and can be up to 2 kb. The sensitivity of the technique (error rate/replication) is about 2×10^{-4}, which means an early error will be multiplied enormously. Therefore, multiple samples must be taken. The technique offers a powerful approach to distinguish different alleles of a gene and diagnosis of disease can be done at the sequence level. PCR can be used to examine the sequence in the genome of a particular individual to determine whether the allele at a given locus is wild type or mutant. Amplification by PCR is so sensitive that the target sequence in an individual cell can be characterized. This allows DNA to be amplified from a very small sample, which is useful for diagnostic and forensic purposes.

gene cloning host-vector system: A host (could be bacteria, fungus, higher organisms) should be *rec*$^-$. A vector (could be plasmid, 4–5 kb; phage, 20–25 kb; cosmid, 40–45 kb; YACs, 100–400 kb) must be *tra*$^-$. For replication of foreign DNA segment in recipient DNA, there are two requirements: (i) The presence of a region on the DNA which serves as point of origin of DNA replication. (ii) Controlling or regulating element required for the expression of the inserted gene.

gene cloning, basic steps in: (i) Generation of a DNA fragment (isolation/synthesis/digestion). (ii) Joining to a vector molecule or a carrier molecule (select DNA capable of self-replication).

gene cloning, obtaining DNA fragments for: In gene cloning, DNA is required in micrograms. For gene cloning, a DNA fragment suitable for cloning is obtained first through one of the following methods available: (i) If a DNA fragment is identified and characterized, gene cloning is straightforward. (ii) Sometimes, DNA carries a function, which allows DNA fragment through a biological test. (iii) Where mRNA is available, cDNA could be obtained first, which could be cloned later on. (iv) In cases where only protein is available, knowing its amino acid sequence is important. From amino acid sequence, the DNA sequence is inferred using the coding dictionary. The nucleotide sequence of the gene thus deciphered can be synthesized chemically, which can then be used for cloning.

gene cluster: A group of adjacent genes that are identical or related.

gene complement: The set of genes characteristic of any one species of organism.

gene complementation: The process by which one defective gene complements the other defective gene to produce the desired phenotype. Gene complementation takes place among non-allelic mutants, and this phenomenon is exploited for the selection of somatic hybrid cells. For instance, during protoplast fusion, the protoplasts of one albino (chlorophyll less) mutant are fused with protoplasts of a second non-allelic albino mutant. After fusion due to gene complementation, the somatic hybrid cell can synthesize chlorophyll and become green.

gene concept: Understanding genes in terms of their structure and function.

gene concept, modern phase of: A phase where a gene is defined on the basis of *cis-trans* complementation test and is termed as a cistron which is made up of a large number of mutable sites (MUTON) which undergo recombination (RECON).

gene concept, Pontecorvo's contribution to: G. Pontecorvo in 1952 claimed that if progeny obtained for detecting recombination is studied in thousands (10^5), recombination can be observed between alleles of multiple series. He suggested that the term pseudoallelism should be discarded and the concept of gene should be suitably modified. A gene may be regarded as a unit of function but not as a unit of recombination. Now it is established that recombination can take place within a gene.

gene conservation population: A population engaged to maintain original genetic variation in species.

gene construct: An experimentally engineered gene that have functional and non-functional properties. For successful transformation, a gene construct is introduced in a cell rather than the single gene itself. The construct carries a selectable marker/reporter gene, specific promoter, gene of interest and terminator (usually *nos*) sequences. The construct is maintained in *Escherichia coli* (for particle bombardment) or in *Agrobacterium* for *Agrobacterium*-mediated transformation. The whole construct gets integrated into the recipient genome.

gene control region: DNA sequences necessitated to begin transcription of a given gene and control the rate of start of transcription.

gene conversion. The alteration of all or part of a gene by a homologous donor DNA that is itself not altered in the process. This is thus a specific type of meiotic homologous recombination that comprises the unidirectional transference of genetic material from a 'donor' sequence to a highly homologous 'acceptor'. In this non-reciprocal transfer of genetic information, one DNA sequence replaces a homologous DNA sequence such that the two sequences become identical. At a molecular level, the non-reciprocal exchange of genetic information occurs as a result of heteroduplex (hybrid DNA) formation between non-sister chromatids. Thus, heterozygous sites are converted to sites homozygous for one or the other allele. Gene conversion can be a consequence of the repair of damaged DNA as described by the double-strand break repair model. In Ascomycete fungi, a 2:2 ratio of alleles is expected after meiosis, yet a 3:1 ratio is sometimes observed. The mechanism of gene conversion is explained by the repair of heteroduplex DNA produced by the Holliday model of recombination. *Aka* NON-RECIPROCAL RECOMBINATION.

gene conversion in lower eukaryotes: When recombination between alleles was discovered, the natural assumption was that it takes place by the same mechanism of reciprocal recombination that applies to more distant loci. That is to say, an individual breakage and reunion event occurs within the locus to generate a reciprocal pair of recombinant chromosomes. However, in the close quarters of a single gene, formation of the heteroduplex DNA itself is usually responsible for the recombination event. Individual recombination events can be studied in the *Ascomycetes* fungi, because the haploid products of a single meiosis are held together within a large cell, the ascus, in a linear order. Mitosis occurs after the production of these four nuclei, giving a linear series of eight haploid nuclei. Each of these nuclei effectively represents the genetic character of one of the eight strands of the four chromosomes produced by meiosis. Meiosis in a heterozygote should generate four copies of each allele. This is seen in the majority of the spores. However, there are some spores with abnormal ratios. They are explained by the formation and correction of heteroduplex DNA in the region in which the alleles differ. Some asci display ratios of 3:5 or 2:6 in which one or two spores, respectively, that should have been of one allelic type are actually of another type. These ratios are a consequence of the process of recombination. An uneven ratio (3:5 or 5:3) can result only from the segregation of two mismatched strands in one DNA duplex; only the other heteroduplex must have been corrected. Even ratios (2:6 or 6:2) could in principle result from independent correction of two heteroduplexes but are more likely to result from a repair mechanism such as the copying event resulting from a double-strand break. Gene conversion does not depend on crossing over but correlates with it. Large proportions of aberrant asci show genetic recombination between two markers on either side of a site of interallelic gene conversion.

gene copy number variation: Variation in gene copy number is recognized as a common heritable source of inter-individual differences in genomic sequences. Non-homologous end-joining mechanisms are well-known, recent models focus on perturbation of DNA replication, and replication of non-contiguous DNA segments, including the proposal that repair of broken replication forks switches under stress from high-fidelity homologous recombinational to non-homologous repair that promotes gene copy number variation. *Aka* COPY NUMBER VARIATION (CNV).

gene delivery: In gene therapy, this term refers to putting the new gene into correct target cells. *See* GENE THERAPY.

gene discoveries: Refers to phase in the understanding of gene function when many new different types of genes, including repeated genes, moveable genes (transposable genetic elements), pseudogenes, retrogenes and related phenomena, such as multiple polyadenylation sites, protein *trans*-splicing, were discovered. With these discoveries, the neoclassical concepts of gene structure and gene function became obsolete.

gene disruption: A mechanism of transcriptional gene silencing in which the physical nature of a gene is disrupted by insertion of a piece of DNA of known sequence, such as T-DNA or transposon elements.

gene diversity (H): The probability that two randomly chosen genes will carry different alleles. It is equal to the heterozygosity for a randomly mating diploid population with no selection. Genetic diversity is critical for a population to adapt to changing environments.

gene dosage: The number of times an allele is present in a particular genotype.

gene drive: Genes in sexually reproducing organisms normally show a 50% chance of being inherited, but some genes have an even higher chance of being inherited. These genes can increase in relative frequency in a population even if they reduce the odds that each organism will reproduce. Scientists are now experimenting on how populations can be changed by adding, disrupting or editing genes or suppressed by proliferating traits that reduce reproductive capacity. The release of just a few individuals within a population may result in a complete invasion of the gene drive cassette within 15–20 generations. Potential beneficial uses of such 'gene drives' include reversing the development of pesticide resistance and locally eradicating invasive species.

gene drive organisms (GDOs): GDOs are organisms that are supposed to reliably force one or more genetic traits on to future generations of their own species. The term for gene drives used by French scientists, 'Forcage Genetique' (which means genetic forcer), makes the intention clear: to force a human-crafted genetic change through an entire population or even an entire species. GDOs are an uncontrollable and potentially irreversible technology, designed to override the natural rules of inheritance. GDOs have been proposed or are being developed to address public and environmental health issues not addressed easily by conventional means.

gene drive research, code of ethics for: Gene drives hold promise for use in controlling insect vectors of diseases and agricultural pests and for conservation of ecosystems against invasive species. At the same time, this technology comes with potential risks that include unknown downstream effects on entire ecosystems as well as the accidental or nefarious spread of organisms that carry the gene drive machinery. A code of ethics can be a useful tool for all parties involved in the development and regulation of gene drives and can be used to help ensure that a balanced analysis of risks, benefits and values is taken into consideration in the interest of society and humanity. G.J. Annas *et al.* (2021) developed a code of ethics for gene drive research with the hope that this code will encourage the development of an international framework that includes ethical guidance for gene drive research and is incorporated into scientific practice by gaining broad agreement and adherence.

gene drive systems (GDSs): GDSs are capable of altering the traits of wild populations and associated ecosystems. A gene drive biases the transmission of a particular allele of a gene such that it is inherited at a greater frequency than by random assortment. The endonuclease-containing cassette that is copied is referred to as a 'gene drive' or simply a 'drive'. Gene drives are systems of biased inheritance that enhance the likelihood a sequence of DNA passes between generations through sexual reproduction and potentially throughout a local population and ultimately all connected populations of a species. A consequence of gene drives is an increased frequency of specific genetic elements or alleles and their accelerated spread throughout populations over successive generations. On the basis of occurrence, there exist two types of gene drives – natural and engineered.

gene drive technologies (GDTs): The technologies that promote the rapid spread of a particular genetic element within a population of non-human organisms. Approaches of such gene drives include (i) CRISPR-based gene drive, (ii) *Medea* gene drive system, (iii) Safe gene drive, (iv) Y-chromosome shredding and (v) Resistance evolution in a Cas9-based sex conversion-suppression gene drive.

gene drive technology, limitations of: (i) Gene drives necessitate many generations to spread through populations. The total time needed to spread the process to all members depends on the number of drive-carrying individuals that are released, the generation time of the organism, the efficiency of homing, the impact of the drive on individual fitness and the dynamics of mating and gene flow in the population, but in general it will need several dozen generations. (ii) Gene drives cannot influence species that exclusively practice asexual reproduction through clonal division or self-fertilization. This category includes all viruses and bacteria as well as most unicellular organisms. (iii) Drive-mediated genome alterations are not everlasting on an evolutionary timescale. While gene drives can spread traits through populations even if they are costly to each individual organism, harmful traits will eventually be outcompeted by more fit alleles after the drive has gone to fixation. Maintaining deleterious traits within a population is likely to necessitate scheduled releases of new RNA-guided gene drives to periodically overwrite the broken versions in the environment. (iv) Our knowledge of the risk management and containment issues associated with gene drives comes from the efforts of researchers focused on mosquito-borne illnesses. While these examples provide an invaluable starting point for investigations of RNA-guided gene drives targeting other organisms, investigations examining the particular drive, population and associated ecosystem in question will be required.

gene drive technology, safety issues of: Gene drives may present environmental and security challenges as well as benefits. Transforming the genetics of an entire population may have unintended consequences on the environment. Extensive testing should precede any release of gene-edited individuals into the environment to transform the entire (e.g. mosquito) population. It is very difficult to predict how gene drive would affect population dynamics and ecosystems. In some cases, the purpose of gene drives would be to reduce population sizes of an organism, which could influence processes like pollination and transmission of parasites. In other cases, gene drives can be used to weed out disease by driving the population that carries that disease to extinction. Eliminating an organism (e.g. one plant species) or reducing its numbers can greatly cause the proliferation of others, and this leads to a series of changes in the ecosystem.

gene drive, mechanism of: At the molecular level, endonuclease gene drives function by cutting the chromosomes that do not encode the drive at a specific site, inducting the cell to repair the damage by copying the drive sequence on to the damaged chromosome. This is derived from genome-editing techniques and similarly relies on the fact that DSBs are most frequently repaired by HOMOLOGOUS RECOMBINATION (HR) if a template is present, and less frequently by NON-HOMOLOGOUS END JOINING (NHEJ). The cell then has two copies of the drive sequence. To achieve this behaviour, endonuclease gene drives consist of two nested elements: (i) a HOMING ENDONUCLEASE (HE) that cleaves the target sequence in recipient cells and (ii) a template sequence used by the DNA repair machinery after the target sequence is cleaved. To attain the self-propagating nature of gene drives, this repair template holds at least the endonuclease sequence. Because the template must be used to repair a DSB at the cutting site, its sides are homologous to the sequences that are located next to the cutting site in the host genome. By targeting the gene drive to a gene-coding sequence, this gene will be inactivated. Additional sequences can be presented in the gene drive to encode new functions. As a result, the gene drive insertion in the genome will reoccur in each organism that inherits one copy of the modification and one copy of the wild-type gene. If the gene drive is already present in the egg cell (e.g. when received from one parent), all the gametes of the individual will carry the gene drive (instead of 50% in the case of a normal gene).

gene drive, synthetic: Genetic engineering makes the manipulation of genetic drive achievable in a process termed 'GENETIC WELDING'. This process is similar to but distinct from artificial selection, is extremely powerful and profoundly capable of promoting the evolution of complex and novel phenotypic changes to both captive and wild populations by biasing allele transmission. A.D. Cutter in 2023 suggests that the usage of synthetic homing gene drives (the human-mediated analogue of endogenous genetic drives) permits the description of 'genetic welding' as an anthropogenic evolutionary force. The emergent reputation of genetic welding also forces their obvious appreciation of genetic drive as an addition to the other four fundamental forces: gravity, electromagnetism and the strong and weak nuclear forces. According to E. Stokstad (2024), the first synthetic 'gene drive' for plants could assist tame weeds – or transform them.

gene drives in nature: There are four types of gene drives occurring in nature: (i) homing endonuclease genes (HEGs), (ii) segregation distorters, (iii) transposons and (iv) heritable microbes.

gene drives, characteristics of: Homing, evolutionary stability, reversibility and wide applicability are some important characteristics of gene derive.

gene drives, evolutionary stability of: Gene drives are evolutionarily stable in the sense that the mutant forms are most likely to arise as they spread through a population that will be selected against and lost.

gene drives, regulatory aspect of: A Food and Drug Administration (FDA) guidance issued in 2009 states that genetically engineered DNA constructs proposed to influence the structure and function of an animal, regardless of their use, meet the criteria for veterinary medicines and are regulated as such. Developers are required to prove that such constructs are safe for the animal. Existing international conventions support international movements of gene drives, but do not outline standards for assessing effects, estimating damages or vindicating harms. International movements of living modified organisms are treated under the 2003 Cartagena Protocol on Biosafety (CPB), ratified by 167 nations not including the USA and Canada. Article 17 of the Protocol necessitates parties to notify an International Biosafety Clearinghouse and affected nations of releases that may lead to the movement of living modified organisms with adverse effects on biological diversity or human health.

gene drives, wide applicability of: In sexual reproduction, each of the two alleles of any gene is transmitted to 50% of offspring. Gene drives are genetic elements that circumvent this rule, significantly increasing the probability that the offspring will inherit the allele containing a gene drive element rather than a wild-type allele. Because of this, a gene drive can spread through a population even if it carries a fitness cost to the organism, as individuals with a gene drive element will produce more offspring with the gene drive allele than without it.

gene duplication: A process in evolution in which a gene is copied twice and the two copies lie side by side along the same chromosome. *See* DUPLICATION.

gene editing: Gene editing is a type of genetic engineering in which DNA is inserted, deleted or replaced in the genome of an organism using ENGINEERED NUCLEASES, *aka* MOLECULAR SCISSORS. The CRISPR/cas9 technology is one system to achieve genome editing, and it has received considerable attention as a means to modify DNA in plants and other organisms. This relatively new technology has been of considerable interest to plant scientists as it may represent a way to modify and improve plants without introducing any additional DNA by precisely deleting small pieces of the host plant's DNA. Yield10 has identified a number of gene targets that may be useful as gene-editing targets.

gene editing in Tephritid flies: Targeted gene editing by CRISPR/Cas9 system was carried out in three Tephritid fruit fly species: *Anastrepha*, *Bactrocera* and *Ceratitis* using homology-driven repair (HDR) or NHEJ repair. The *white* gene was targeted in *Anastrepha ludens*, *Bactrocera dorsalis* and *Ceratitis capitata* using SINGLE-GUIDE RNAS (SGRNAS). Successful mutant lines were established with individuals having *white-eye* mutation arise in each of these species. The *white* gene was deleted in the mutant *white-eye* phenotypes due to point mutations, frameshift mutation and premature stop codons. This indicated that technique optimization is critical even among closely related species.

gene emigration: The movement or flow of genes out of a population, caused by emigrating individuals, which do not interbreed with the residents of the original population. *Compare with* GENE IMMIGRATION.

gene eviction: A method, originally developed in yeast molecular biology, which permits the retrieval of a chromosomal copy of any gene which has previously been cloned. Thus, if the wild-type copy of a gene has been cloned gene eviction may be used to obtain a mutant copy and vice versa. A feature of the gene eviction procedure is that DNA sequences adjacent to the gene of interest are retrieved at the same time and so the method may be used for CHROMOSOME WALKING.

gene evolution: New genes in the genome evolve through gene duplication and diversification. At first, the function of the second copy of the gene is knocked out through a FORBIDDEN MUTATION and becomes a PSEUDOGENE. Then, with the help of new mutations, the duplicate gene assumes a new function.

gene expression: A measure of the presence, amount and time course of one or more gene products in a particular cell or tissue. The process of converting genetic information into RNA and protein for use in the cell. Gene expression includes transcription, translation, post-translational modifications and possibly transport reactions. In molecular genetics, gene expression means transcription taking place. In biochemistry, gene expression usually refers to the making of an observable phenotype by a gene, usually via directing the synthesis of a protein. Gene expression can be turned on and off, or the level of gene expression can be regulated, through epigenetic processes. Many genes are expressed only at specific times during the life period of an organism. Some genes are expressed in a 'cascade' of related expressions. The gene product is then translocated to the site of function. Used RNA and protein molecules are then degraded. On the basis of gene expression genes may be divided into several classes: tRNA genes synthesize pre-tRNA, rRNA genes synthesize pre-rRNA and structural genes synthesize mRNA (in prokaryotes) or pre-mRNA (in eukaryotes); these transcripts after processing become functional molecules, tRNA, rRNA and mRNA, respectively. Whereas tRNA and rRNA do not encode polypeptides, mRNA contains codons and acts as a template for translation. Every gene is not expressed at the same level and at the same time. Expression patterns, easily analysed using microarray technology, can give a lot of information about the role genes play in different situations, such as disease and health. Some genes that may be upregulated or downregulated during a cellular process, in response to an external stimulus or in sickness or disease. Gene chips and proteomics now allow the study of expression profiles of sets of genes or even entire genomes.

gene expression analysis through protein quantification: The most commonly used method of protein quantification is to perform a Western blot against the protein of interest. This gives information on the size of the protein in addition to its identity.

gene expression analysis: Generally accomplished via the use of 'biochips' (which have numerous detection/analytical devices invented on to their silicon surface) or 'microarrays', gene expression analysis comprises evaluation of the expression (and expression levels) of numerous genes in a biological sample, to analyse/compare any differences between gene expression products in (i) normal cells vs diseased cells, (ii) normal cells vs those responding to a stimulus, (iii) cells from the same organism, at different stages of development (e.g. embryo versus adult), (iv) normal (historic wild-type) cells vs genetically engineered cells (those that have been engineered to cure a disease or resist an herbicide, etc.) and (v) normal cells vs the same cells treated with a given pharmaceutical (candidate). Analysis generally involves the measurement of gene expression markers (i.e. molecules synthesized or cellular consequences such as apoptosis) to determine which genes are expressed (and when/how much, etc.).

gene expression and evolution: During evolution, novel phenotypes emerge through changes in gene expression. Expression divergence between species was generally due to changes in *cis*. Divergence in *trans* reflected a differential response to the environment and explained the tendency of certain genes to diverge rapidly.

gene expression cascade: A sequential series of individual gene expressions (i.e. each gene leading to the synthesis of a separate/different protein) that is set off by the first gene expression. For example, a gene expression cascade is often initiated by the first gene causing the expression of a transcription factor (i.e. protein that itself interacts with the cell's DNA to either cause or speed up yet another gene expression). The protein resulting from the second gene expression could yet be another transcription factor that triggers another (i.e. third) gene expression, and so on.

gene expression control at transcription initiation level: The process of transcription initiation is the most important step of gene expression as most of the genes are regulated at the level of transcription initiation.

gene expression control at translation level: For lac operon of *Escherichia coli*, there is 10:5:2 ratio of the products of genes *lacZ*, *lacY* and *lacA*. The relative amounts of polycistronic mRNA depend upon the primary structure of mRNA. At the end of each cistron on mRNA, there is at least one chain termination codon followed by an intercistron segment prior to the next translation initiation codon. The intercistron nucleotide sequences govern whether or not a ribosome 'out of gear' will fall off the message before initiating the next polypeptide chain. Presumably, in the lac operon, half (50%) of the ribosomes fall off between *lacZ* and *lacY* and another more than half (60%) fall off between genes *lacY* and *lacA*.

gene expression control, quantitative understanding of: R. Balakrishnan *et al.* (2022) measured promotor on-rates, messenger RNA abundance, and protein abundance for more than 1500 genes in the bacterium *Escherichia coli* under many different growth conditions. A balancing of transcription with translation has been proposed through Rsd, a factor that controls the availability of RNA polymerase. A simple mathematical formula relates promoter activities and protein concentrations across growth conditions, enabling quantitative inference of gene regulation from omics data.

gene expression divergence: A phenotypic trait reflecting the evolution of gene regulation and characterizing dissimilarity between species and between cells and tissues within the same species.

gene expression markers: Refers to molecules (e.g. synthesized as a result of a specific gene's expression) or consequences (e.g. cell apoptosis due to a specific gene's expression) that can be evaluated as an evidence of the gene's expression in gene expression analysis.

gene expression networks: Genes have sometimes been viewed as NODES in a network, with INPUTS being proteins such as transcription factors, and OUTPUTS being the level of gene expression. The node itself accomplishes a function, and the procedure of these functions has been construed as executing a kind of information processing act within cells and controlling cellular behaviour.

gene expression profiling: Refers to the analysis of gene expression levels of many genes concurrently. Technologies that can be employed to extract this information include Northern blot, DNA chips, high-density arrays spotted on glass or membranes and quantitative techniques. Expression profiling is a logical next step after sequencing a genome: the DNA sequence tells us what the cell could possibly do, while the expression profile tells us what it is actually doing now. Genes contain the instructions for making messenger RNA (mRNA), but at any moment, each cell makes mRNA from only a fraction of the genes it carries. If a gene is used to produce mRNA, it is considered 'on', otherwise 'off'. Many factors determine whether a gene is on or off, such as the time of day, whether or not the cell is actively dividing, its local environment and chemical signals from other cells. Skin cells, liver cells and nerve cells turn on (express) somewhat different genes and that is in large part what makes them different. Therefore, an expression profile allows one to deduce a cell's type, state, environment and so forth. In an mRNA or gene expression profiling experiment, the expression levels of thousands of genes are simultaneously monitored to study the effects of certain treatments, diseases and developmental stages on gene expression. For example, microarray-based gene expression profiling can be used to identify genes whose expression is changed in response to pathogens or other organisms by comparing gene expression in infected to that in uninfected cells or tissues.

gene expression profiling, class discovery in: Most early expression profiling experiments, and even many current ones, have this form which is known as class discovery. A popular approach to class discovery comprises grouping similar genes or samples together. The simplest form of class discovery would be to include all the genes that change beyond a certain amount between two experimental conditions.

gene expression profiling, limitations of: (i) In general, expression profiling experiments report only those genes that show statistically significant differences under changed experimental conditions. This is characteristically a small fraction of the genome for several reasons. (ii) Different cells and tissues express a subset of genes as a direct consequence of cellular differentiation so many genes are turned off. (iii) Many of the genes code for proteins that are needed for survival in very specific amounts, so many genes do not change. (iv) Cells use many other mechanisms to regulate proteins in addition to altering the amount of mRNA; such genes may be constantly expressed even when concentrations of their protein products are rising and falling. (v) Financial restraints may limit expression profiling experiments to a small number of observations of the same gene under identical conditions, thereby reducing the statistical power of the experiment, making it impossible for the experiment to identify important but not obvious changes. (vi) It takes a great amount of effort to infer the biological significance of each regulated gene, so scientists often limit their discussion to a subset. Newer microarray analysis techniques automate certain aspects of attaching biological significance to expression profiling results, but this remains a very challenging problem.

gene expression profiling, principle of: A genomic technique is based on the fact that only a fraction of the genes encoded in the genome of a cell is expressed by being transcribed into messenger RNA (mRNA). The complement of mRNAs in a cell largely dictates its complement of proteins. Changes in gene expression underlie many biological phenomena. The use of DNA microarrays provides insight into tissue- and developmental-specific expression of genes and the response of gene expression to environmental stimuli. The correlation of transcript profiles, probe clustering and expression patterns in a particular tissue or developmental stage may help to assign biological relevance to many of the unknown genes.

gene expression profiling, quantitative analysis of: In gene expression analysis, the quantitative amounts of each protein being expressed are resolved via the use of such technologies as 2D gel electrophoresis, Southern blot analysis, fluorescence tagging, radiolabelling, RT-PCR, qPCR, plane polarimetry, etc.

gene expression profiling, uses of: Typical uses of gene expression profiling include the following: (i) equating diseased cell with normal cell, (ii) describing quantitatively the 'normal' state, (iii) associating a given drug's impact (i.e. treated cell with normal cell) and (iv) associating old cell with young cell.

gene expression regulation: Epigenetics regulates gene expression in a reversible manner by chemically modifying DNA and histone proteins, which prevent permanent mutations or alterations within the gene themselves.

gene expression variation: The molecular basis underlying allelic variation at QTL is similar to the identified variation for simple Mendelian loci. Alterations in gene expression or protein function form the basis underlying allelic variation at QTL. Our ability to identify the molecular basis of quantitative traits is being enhanced by genomic methodologies such as transcriptomics, metabolomics and proteomics.

gene expression, levels of: Gene regulatory procedures operate at various different levels such as DNA, transcriptional, post-transcriptional, RNA transport, translational and post-translational levels.

gene expression, measurement of: Measuring gene expression is an important part of many life sciences – the ability to quantify the level at which a particular gene is expressed within a cell, tissue or organism can give a huge amount of information. Analysis of the location of expression protein is a powerful tool and this can be done on an organism or cellular scale. Investigation of localization is particularly important for the study of development in multicellular organisms and as an indicator of protein function in single cells. Ideally, the measurement of expression is done by detecting the final gene product (for many genes this is the protein) however it is often

gene family: A group of genes that has arisen by duplication of an ancestral gene. Subsets of genes comprising homologous sequences usually relate to a common function, as they make similar products. Members of a gene family may be functionally very similar or differ widely. The globin gene family is an example. Some members of the family may act together; others may be silenced and become PSEUDOGENES.

gene flow: (i) The movement of genes from one individual or population to another genetically compatible individual or population by way of interbreeding. (ii) The exchange of genetic traits between populations by movement of individuals, gametes or spores. It involves the spread of new variants among different populations through dispersal. *Aka* GENE MIGRATION *or* ALLELE FLOW.

gene flow in crop improvement: The transfer of genes through migration and interbreeding influences genetic diversity and population structure. It reduces genetic differences between populations. M. Lehnert *et al.* (2020) studied wild soybean populations in China and observed gene flow between regions, leading to increased genetic diversity and reduced differentiation. Similarly, X. Yang *et al.* (2019) investigated maize populations in China and found evidence of gene flow between regions and varieties, resulting in enhanced genetic diversity and reduced differentiation. These examples highlight the impact of gene flow on genetic diversity and the homogenization of populations through gene exchange.

gene frequency: The frequency in the population of a particular gene relative to other genes at its locus. Expressed as a proportion (between 0 and 1) or percentage (between 0% and 100%). If there are 100 individuals in a population of diploid individuals, and we consider one locus (one gene) that is present in two forms (two alleles), *A* and *a*, then if 20 of the individuals carry two copies of *A* (they are *AA* homozygotes), 60 of the individuals are *Aa* heterozygotes and the remaining 20 individuals are aa homozygotes, then the gene frequencies are calculated as the number of each allele divided by the total number, in this case, (40 + 60) / 200 = 0.50 for both alleles.

gene function: Gene function can be considered from several points of view: (i) biochemical function (e.g. protein kinase), (ii) cellular function (e.g. a role in a signal transduction pathway), (iii) developmental function (e.g. a role in pattern formation) and (iv) adaptive function (the contribution of the gene product to the fitness of the organism). Having identified a new sequence, the comparison with sequence databases is the simplest way to obtain functional information.

gene function analysis: Refers to the determination of which protein is caused to be 'manufactured' by each gene in an organism's genome/DNA. Typically, gene function analysis follows after discovery of gene sequences found via a structural genomics study.

gene function analysis, methods of: Some methods utilized to determine which proteins result from which gene(s) are as follows: (i) SITE-DIRECTED MUTAGENESIS (SDM) to compare two same-species organisms possessing two different genes at the same site (SNP) on the genome (i.e. on organism's DNA); (ii) antisense DNA sequences to compare two same-species organisms, one of which has a gene at the same site 'turned off' (silenced) via antisense DNA; (iii) reporter gene, to compare two same-species organisms (possessing two different genes at the same site on genome/DNA) via a reporter gene adjacent to the gene/site, to detect the presence or absence of the desired trait/function; (iv) comparison of same organism (e.g. crop plant) when one of the two is 'challenged' by a specific plant disease; (v) chemical genetics, to compare two same-species organisms (one of which has gene at the specific site at least partially inactivated by a specific chemical); (vi) 'silencing' or 'knocking out' a particular gene via other methods than antisense or chemical genetics, to compare and (vii) use of already-known 'model organisms' such as *Drosophila* for comparing insect genes, *Arabidopsis thaliana* for plant genes and *Caenorhabditis elegans* for animal genes.

gene fusion: Refers to the technology/methods applied to fuse together two or more genes. Mutations that cause portions of two genes to fuse together and form a hybrid gene are of frequent occurrence in blood-related and lung cancers. Certain exons can be members of at least two, and possibly even more, transcripts. These chimeric transcripts are formed by the transcription of two consecutive genes into one RNA. When such a 'fused gene' is inserted into a genome (e.g. the DNA of a plant), it causes production (in plant's ribosomes) of protein(s) consisting of all or part of the amino acid sequences (known as the DOMAIN) of the two proteins typically coded for by those two genes. This fusion is often done in order to put expression of the second (fused) gene under the control of the (strong) promoter of the first gene. In 2001, Rajbir Sangwan *et al.* inserted a fused gene into a potato plant (*Solanum tuberosum*), a major source of plant starch. That fused gene coded for the production of the two proteins α-amylase and glucose isomerase; both are enzymes. α-amylase catalyses the conversion of potato starch into glucose (a sugar), and glucose isomerase catalyses the conversion of glucose to fructose (a more valuable sugar).

gene gun: A device used to bombard plant cells with metal particles coated with foreign genes to achieve genetic engineering. It is a direct gene transfer method; no vector is used for gene delivery. A device invented at Cornell University that allows genetic material to be introduced into a new organism. The genetic material from the donor is bound to small gold or tungsten nanoparticles (0.2–0.7 μm) which are then 'shot' into cells or organelles (e.g. chloroplasts) of the recipient cells and the material is incorporated into its DNA. Gene gun is a very popular method for nuclear as well as chloroplast transformation. Following bombardment, transgenic cells are selected and regenerated into entire plants through tissue culture. Gene gun is also called 'particle gun' and this method is called biolistic method. *Aka* BIOLISTIC METHOD, BIOLISTIC TRANSFORMATION *or* BOMBARDMENT METHOD.

gene homology: The degree of similarity between DNA sequences of genes.

gene immigration: The movement or flow of genes into a population, by immigrating individuals, which interbreed with the residents. *Compare with* GENE EMIGRATION.

gene imprinting: An inheritance process that is independent of the classical Mendelian inheritance. An epigenetic mechanism that determines the expression or repression of a subset of genes according to their parent of origin. When an allele coming from one parent is silenced, the allele coming from the other parent is expressed. For example, imprinted genes are either expressed only from the allele inherited from the mother (e.g. *H19* or *CDKN1C*) or in other cases from the allele inherited only from the father (e.g. *IGF-2*). Various forms of genomic imprinting have been demonstrated in insects, mammals and flowering plants. Thus, whether a gene in a cell lineage is expressed or not depends on which parent contributed the gene. A process of silencing genes occurs through DNA (cytosine) methylation and/or histone modifications (H3K4, H3K27 and H3K9).

gene imprinting, beginning of: The distinctive mechanisms of gene imprinting in the endosperm, which comprise DNA methylation and histone methylation, begin in the central cell and sperm preceding fertilization. Gene imprinting is widespread in flowering plants. In *Arabidopsis*, parent-of-origin-specific gene expression is primarily found in the endosperm. Imprinted genes (e.g. *FWA*, *MEA*, *FIS2* and *PHERES1*) are specifically expressed from the maternal genome of the endosperm while the alleles from the paternal genome are silenced. *FWA*, a suppressor of flowering, was initially identified from late-lowering epigenetic mutants that show ectopic *FWA* expression. The hypomethylation of the DNA repeat region in the *FWA* promoter is required for the normal expression of *FWA*. *FWA* is expressed in the endosperm but not in any other tissues. The *FWA* imprint depends on *DME*. Unlike the constitutive expression of *ROS1*, the expression of *DME* is concentrated in the central cells and endosperm during seed development. DME specifically demethylates the maternal genome in central cells (progenitors of endosperm) and leads to the expression of imprinted genes. The molecular regulation of imprinted gene expression varies between imprinted regions. Appropriate expression of imprinted genes is important for normal development, with numerous genetic diseases associated with imprinting defects including Beckwith-Wiedemann syndrome, Silver-Russell Syndrome, Angelman Syndrome and Prader-Willi Syndrome. *Aka* GENETIC IMPRINTING, GENOME IMPRINTING, GENOMIC IMPRINTING *or* PARENTAL IMPRINTING.

gene index: Refers to listing of the number, type, label and sequence of all the genes identified within the genome of a given organism. Gene indices are usually created by assembling overlapping EXPRESSED SEQUENCE TAGS (ESTs) into clusters and finding out if each cluster matches a unique gene. Methods using which a cluster can be identified as representing a unique gene include identification of long open reading frames, comparison to genomic sequence and detection of single-nucleotide polymorphisms or other features in the cluster that are known to exist in the gene.

gene inference: Genes are inferred by observing precise mathematical ratios in the filial generations descending from two parental individuals.

gene inhibition therapy: For the treatment of dominant negative loss-of-function mutations, and gain-of-function mutations, GENE AUGMENTATION therapy is less powerful. Additional functional copies of dominantly malfunctioning genes are unlikely to affect the phenotype. In principle, a valid approach to the treatment of such disease would be targeted correction (i.e. allele replacement) or gene knockout to remove the mutant allele. Targeted inhibition of gene expression can also take effect at the protein level, by the expression within a cell of genetically engineered antibodies (INTRABODIES) which bind to inactive mutant proteins.

gene insertion: The process of addition of one or more copies of a normal gene into a defective chromosome.

gene interaction: Modification of gene action by the interaction between products of non-allelic genes. In genetics experiments, type of gene interaction is reflected by modified F_2 ratios (e.g. 9:7, 9:3:4, 15:1, 13:3) compared with the normal 9:3:3:1 ratio.

gene introduction: Introduction of transgene(s) into the cell using gene transfer methods. *See* TRANSGENESIS.

gene isolation: Refers to isolating a specific region of DNA and producing millions of identical copies of the DNA (gene) within a microbial cell culture. Gene isolation is the first step in the genetic engineering process. Molecular cloning involves several tools and techniques to introduce DNA as an insert into a vector molecule.

gene knockdown: Refers to an experimental technique by which the expression of one or more of an organism's genes are reduced. The reduction can occur either through genetic modification or by treatment with a reagent such as a short DNA or RNA oligonucleotide that has a sequence complementary to either gene or an mRNA transcript. Since RNA interference may not totally abolish the expression of the gene, this technique is sometimes referred to as a 'knockdown', to distinguish it from 'knockout' procedures in which expression of a gene is entirely eliminated. Extensive efforts in computational biology have been directed toward the design of successful dsRNA reagents that maximize gene knockdown but minimize 'off-target' effects. Off-target effects arise when an introduced RNA has a base sequence that can pair with and thus reduce the expression of multiple genes at a time. Such problems occur more frequently when the dsRNA contains repetitive sequences. A multitude of software tools have been developed implementing algorithms for the design of general, mammal-specific and virus-specific siRNAs that are automatically checked for possible cross-reactivity.

gene knockin (KI): Involves the insertion of a new functional gene. Foreign genes knocked into the mouse genome have produced site-specific transgenic mice.

gene knockout (KO): Leads to disruption of function of the existing gene. The yeast Flp/FRT recombinase system and bacteriophage recombinases such as Cre and its recognition sequence, *lox*P, allow spatial and temporal control of knockouts. The two most popular methods employed for knocking out the genes are INSERTIONAL INACTIVATION (of which homologous recombination is a specific technique) and ANTISENSE INACTIVATION.

gene library: A collection of cloning vectors containing a complete set of DNA fragments of the entire genome of an organism. The development of gene libraries (i.e. the collection of all genes in the vector) has become very important for gene cloning. From the gene library, the gene of interest is isolated using appropriate DNA probes. Depending upon the nature of the DNA segment (gene), the library may contain the genomic DNA (genomic library), or it may contain cDNA (cDNA library). Gene libraries are usually maintained in *Escherichia coli*. *See* DNA LIBRARY *or* GENOMIC LIBRARY.

gene location: Estimation of physical or relative distances among different genes on a particular chromosome.

gene location in double-stranded DNA: The leader is actually located before the 3′-end of the functional region of the coding strand and the train follows the 5′-end. One given strand of DNA does not necessarily bear the coding sequences of all the genes located in any single sector of the genome; sometimes some genes are located on one strand while others are located on the opposite one. Since cistrons are always read (transcribed) in 3′to 5′ direction on the coding strand, those on opposite strands also differ in polarity. Thus, the train of one gene may be located under a mature gene of opposite polarity or even below its leader.

gene locus: Position of a gene on a chromosome. Also refers to the DNA at that position. The use of locus is sometimes restricted to mean expressed DNA regions.

gene machine: (i) An instrument which, when fed information about the amino acid sequence of a protein (usually via a protein sequencer), will automatically produce polydeoxynucleotide genes (or segments) to code for that protein. (ii) In common literature, it refers to an idiomatic description of an automated oligonucleotide synthesizer.

gene manipulation: The formation of a new combination of heritable material by the insertion of nucleic acid molecules, produced by whatever means outside the cell, into any virus, bacterial plasmid or any other vector system so as to allow their incorporation into a host organism in which they do not naturally occur, but in which they are capable of continued propagation. *See* GENETIC TRANSFORMATION.

gene map: (i) In classical genetics, a graphic demonstration of the linear arrangement of genes along a chromosome or segment. It shows the relative distance between loci in linkage experiments. (ii) In molecular genetics, descriptive representation of the structure of a gene. It includes the DNA sequence of a gene with introns and exons, 3′ or 5′ transcribed-untranslated regions, termination/polyadenylation signal and regulatory elements such as promoters and enhancers, and it may include known mutations defining alternative alleles of the same gene.

gene mapping: Refers to the process of locating genes within a genome. The regions of a genome that directly encode proteins, or the protein-coding genes, are usually of great interest to scientists. Therefore, it is often a priority to locate each gene in a genome. Gene maps may be based on classical genetic recombination analysis or on direct DNA data obtained by DNA sequencing. *Aka* GENOME MAP *or* LINKAGE MAPPING.

gene mapping in yeast and *Chlamydomonas*: Yeast and *Chlamydomonas* have unordered tetrads. It is not possible in such cases to distinguish between tetrads resulting from first- and second-division segregation. For detection and estimation of linkage, the tetrads are classified as parental ditype (PD), NONPARENTAL DITYPE (NPD) and TETRATYPE (TT). Detection of linkage in yeast and *Chlamydomonas* is done the same way as in *Neurospora*. In the case of unordered tetrad data, distances are measured between genes but not between centromere and gene. Per cent recombination is (NPD + ½TT)/(PD + NPD + TT).

gene migration: Gene migration is of two types: GENE IMMIGRATION and GENE EMIGRATION. *See* GENE FLOW. *See details as separate terms.*

gene mining: *See* ALLELE MINING.

gene mutation: A heritable alteration of a gene detected and estimated by resulting phenotypic modifications. As a result of mutation, a gene can mutate from one allelic form to another. *Aka* POINT MUTATION.

gene name: The official name assigned to a gene. According to the Guidelines for Human Gene Nomenclature developed by the HUGO Gene Nomenclature Committee (HGNC), it should be brief and describe the function of the gene.

gene neighbour method: If in several genomes the genes that encode two proteins are neighbours, the proteins tend to be functionally linked. This is useful in uncovering functional linkages in prokaryotes, where operons are common but also show promise for analysing interacting proteins in eukaryotes. Once a function has been assigned to an individual protein, one can search for other proteins with related functions by seeking proteins whose amino acid sequences are similar to the original protein. This HOMOLOGY METHOD is used widely to extend knowledge of protein function from one protein to its cousins, which are presently described from the same common ancestral protein.

gene networks: The networks of genes, proteins and their regulatory interactions.

gene of uncertain significance (GUS): Refers to a gene that has been identified through genome sequencing but whose connection to a human disease has not been established.

gene ontology: A controlled vocabulary of terms relating to molecular function, biological process or cellular components developed by the Gene Ontology Consortium. A controlled vocabulary allows scientists to use consistent terminology when describing the roles of genes and proteins in cells.

gene organization: There are differences in gene organization between prokaryotes and eukaryotes. (i) In prokaryotes, the nature of genes is predominantly non-split and no processing is necessitated for a transcript to become mRNA. In eukaryotes, most of the genes are split in nature. Coding sequences (introns) are separated by non-coding sequences (introns). Here transcript is required to be processed in the nucleus before mature mRNA is available for translation. (ii) Metabolically related genes are linked together in prokaryotes but in eukaryotes, metabolically related genes are not contiguous; in fact, they may even be spread across different chromosomes. (iii) Messenger RNAs are polycistronic in prokaryotes; an mRNA may code for two or more different polypeptide chains. On the other hand, eukaryotic mRNAs are unicistronic; one mRNA molecule codes for only one polypeptide chain. (iv) In prokaryotes, all the genes are always active, and, in this case, regulatory mechanisms are designed to turn off those genes whose product is not required by the cell. In eukaryotes, all genes are normally inactive unless turned on by gene regulatory mechanisms. Other features investigated in gene organization include (a) genes may be overlapping or non-overlapping; (b) genes may be functional or non-functional; (c) genes may be repeated or unique; and (d) genes may occupy a fixed position or shift from one place to another in the genome. Present gene organization in different organisms is the net result of concerted efforts of genome evolution.

gene pairs: In eukaryotes, genes occur in pairs. The two copies of a gene occur in a diploid, one on each homologous chromosome.

gene patenting: Protection provided by governmental or non-governmental institutions to the discoverer of new genes, genotypes, strains or testing procedures so that the detailed information can be declared publicly. Currently, a synthetic gene (but not a natural gene itself) can be patented. However, a sequenced functional unit or specific utilization of a natural gene can be a matter of patent. For instance, the *CryI Ac* gene, which imparts resistance to insects in plants, has been patented in the USA by Monsanto. Patenting protects the intellectual property rights. According to patenting laws, patented discovery cannot be exploited by anybody else without signing the agreement.

gene pool: A set of all genes, or genetic information, in any population, usually of a particular species or lower and higher taxa available for crossing and selection. It may be differentiated between (i) PRIMARY GENE POOLS, (ii) SECONDARY GENE POOLS and (iii) TERTIARY GENE POOLS. *See details as separate terms. See* ALLELE, GENE *or* POLYMORPHISM.

gene prediction: Prediction of genes in genome projects. Gene prediction is one of the important subjects in genome projects. In eukaryotes, gene prediction and annotation are not a simple process mainly because of the various sizes of introns (non-coding sequences) located between exons (coding sequences). In addition, many genes have alternative splice variants. In other words, every eukaryotic gene shows different structures and lengths to be predicted. Many gene prediction programs mainly three groups have been developed for genome-wide annotation: (i) *Ab initio* program: FGENESH, Genscan, Grail, MZEF and HMMgene; (ii) Similarity-based program: CRASA, AAT and AgenDa and (iii) Combined methods: GenomeScan, Procrustes and FGENESH+.

gene prediction, methods of: Major methods of gene prediction are laboratory-based, feature-based, homology-based and statistical and HMM-based approaches.

gene probe: A specific single-stranded DNA that is usually labelled with radioactive phosphorus to help the detection of a gene with a complementary sequence to which the probe can bind.

gene product: The biochemical material, either RNA or protein, resulting from the expression of a gene. The amount of gene product is used to measure how active a gene is; abnormal amounts can be correlated with disease-causing alleles.

gene proper: The portion of the template strand containing coded information for a functional product.

gene protection technology for hybrid seed production: A different strategy, utilizing only two genes (terminator and recombinase), one in each of the two parents of the hybrid is used and no repressor gene is needed.

gene protection technology for pure lines seed production: In actual commercial production and sale of seeds of pure lines, the crop for commercial seed production will be grown by the breeder or Seed Company without chemical treatment. The harvested seeds from this crop will be treated with the chemical before it is sold to the farmer so that in crops grown by the farmer, the toxic substance is produced at the time of embryo development, and the embryos die or fail to develop. A seed thus produced will carry the endosperm, but not the embryo, so that it can be used or sold as grain but cannot be used for sowing. Tetracycline is one such chemical trigger that can be used to treat the seeds before selling them in the market.

gene protection technology, reduction of transgene introgression by: A repressible seed-lethal (SL) system aimed at reducing the probability of transgene introgression into a population of sexually compatible plants was developed. To evaluate the potential of this method, tobacco plants were transformed with an SL construct comprising gene 1 and gene 2 from *Agrobacterium tumefaciens* whereby gene 1 was controlled by the seed-specific phaseolin promoter modified to contain a binding site for the *Escherichia coli tet* repressor (R). The expression of this construct allows normal plant and seed development but inhibits seed germination. Plants containing the SL construct were crossed with plants containing the tet^R gene to derive plant lines where the expression of the SL construct is repressed. Plant lines that contained both constructs allowed normal seed formation and germination, whereas seeds in which the SL construct was separated from the *R* gene through segregation did not germinate.

gene pyramiding: Incorporation of more than one gene of similar effect for durable performance. Gene pyramiding is an attractive approach in plant breeding for developing varieties possessing better resistance to diseases and insects for extended periods. For instance, *Xa* genes impart resistance to bacterial blight of rice. Nowadays, more than one *Xa* gene (i.e. *Xa5*, *Xa13*, *Xa21*) has been incorporated using marker-assisted selection to develop superior varieties possessing durable resistance. Gene pyramiding can also be achieved through genetic transformation using a 'particle gun' and multiple genes can be introduced together.

gene redundancy: The presence of a gene in multiple copies due to POLYPLOIDY, POLYTENY, GENE AMPLIFICATION *or* CHROMOSOMAL DUPLICATIONS.

gene reduplication: The duplication of genes in the chromosomes at cell division so that each daughter gene and chromosome is an exact replica of its original.

gene regulation: The process of turning genes on and off, which ensures that the appropriate genes are expressed at the proper times. The modulation of gene expression. Genes are regulated at the level of DNA, RNA and protein, the interactions between which are a direct consequence of their immediate biochemical environment.

gene regulation at DNA level in eukaryotes: DNA arrangements are necessary for the regulation of some eukaryotic genes. Some examples are expression of immunoglobulin genes, yeast mating-type switching, trypanosome surface antigen switching and synthesis of mRNA in pieces in VARIANT SURFACE GLYCOPROTEIN (VSG) genes in trypanosome.

gene regulation at translation level in eukaryotes: Differential translation of mature mRNA takes place in the cell cytoplasm. For this process to occur, the dormant mRNA must have characteristics, such as structural determinants and subcellular location that permit their selection and timely translational activation.

gene regulation at translational level in bacteria: For *lac* operon of *Escherichia coli*, there is 10:5:2 ratio of protein products of *lacZ*, *lacY* and *lacA* genes. The relative amounts of polycistronic mRNA depend upon the primary structure of mRNA. At the end of each cistron on mRNA, there is at least one chain termination codon followed by an intercistron segment prior to the next translation initiation codon. The intercistron nucleotide sequences govern whether or not a ribosome 'out of gear' will fall off the message before initiating the next polypeptide chain. Presumably, in *lac* operon, half of the ribosomes fall off between *lacZ* and *lacY* and another more than half (60%) fall off between *lacY* and *lacA*.

gene regulation by hormones: Various hormones regulate gene expression by acting as signals. Peptide hormones, such as insulin and epinephrine, are too large to enter into the cell. Their effect appears to be mediated by receptor proteins located in target cell membranes and by the intracellular levels of cyclic adenosine monophosphate (cAMP). The cAMP activates a protein kinase, which phosphorylates and thus activates many specific enzymes. Steroid hormones (e.g. oestrogen, progesterone, testosterone) are small molecules which readily enter through the plasma membrane. Once inside the appropriate target cells, the steroid hormones get tightly bound to specific receptor proteins, which are present only in the cytoplasm of the target cells. The hormone-receptor complexes activate the transcription of specific genes in two possible ways: (i) The hormone-receptor protein complexes interact with specific non-histone chromosomal protein and this interaction inspires the transcription of correct genes. (ii) The hormone-receptor complexes activate the transcription of target genes by binding to specific DNA sequences present in *cis*-acting regulatory regions (the enhancers and promoters) of genes. In both cases, hormone-receptor complexes activate transcription as positive regulators (activators) of transcription much like the cAMP-CRP complexes in prokaryotes. Here CRP denotes cAMP receptor protein. Genes control the production of hormones and physiology of an organism. Hormones play a role in development by controlling transcription.

gene regulation by non-histone proteins: The non-histone (acidic) proteins are thought to be activators of gene expression as they can take out histones and thereby open some regions of DNA for transcription. Certain characteristics of non-histone proteins that resemble DNA are (i) high rate of synthesis and metabolism, ((ii) high concentrations of the chromatin of metabolically active tissues, (iii) localization on those parts of chromatin which reveal especially high levels of RNA synthesis and (iv) phosphorylation during the gene activation period. The acidic proteins promote the transcription of free DNA, and they also govern transcriptional specificity.

gene regulation in cellular differentiation and development: Gene regulation plays a very important role in cellular differentiation and development of multicellular organism. Initiation of transcription is the first important level of gene expression of a prokaryotic and eukaryotic cell. When the regulatory protein binding sites are close to RNA polymerase binding sites, initiation of transcription is controlled by physical contact. When the protein binding site is at a distance from the RNA polymerase binding site, then there are several mechanisms to explain gene regulation. Various such mechanisms will now be discussed. Gene regulation in both prokaryotes and eukaryotes occurs through DNA-protein and protein-protein interaction.

gene regulation in *cis*. *See* CIS-ACTING FUNCTION.

gene regulation in eukaryotes, mated model of: According to the mated model of gene regulation in eukaryotes, de-repressor RNA (dRNA) binds to the anticoding strand of operator locus (*anti-o*), and structural gene (*sg*) loci. The dRNA of a given operon is complementary to the base sequence to the operator portion of the direct transcription product. The direct transcriptional product is cleaved to form messenger RNA (mRNA) and operator RNA (oRNA). Cleavage can occur directly or after the formation of heterodimeric duplex RNA by base pairing of de-repressor RNA with the operator portion of the transcription product. By such duplex formation, de-repressor RNA is selectively removed from the operator locus, thus providing feedback inhibition of transcription of operon. With the consumption of mRNA and with the degradation of operator RNA, de-repressor RNA is released from the duplex, thus providing positive feedback de-repression of transcription of the operon. Different structural genes may share operators with common base sequences, and thus be equally sensitive to given species of de-repressor RNA, but during gene transcription and during its selective inhibition.

gene regulation in eukaryotes: In general, genes of eukaryotes are apparently 'turned off' and must be activated. Differentiation and growth/development take place only in eukaryotes. No cell division occurs in prokaryotes. At any one time, in a eukaryotic cell, only 2–15% of the total genome is expressed. It means some sort of regulation is occurring. In chromatin, two types of proteins present are histones and nonhistones. Histones are basic in nature and are involved in a structural role of chromosomes. Recently histones have been found to play a regulatory role also. Nonhistones are acidic in nature and are thought to act as controlling elements for gene transcription.

gene regulation in eukaryotic cells, levels of: Various possible control points recognized in eukaryotes include regulation at the level of gene structure, initiation of transcription, processing of transcript (including capping, addition of poly(A) tail, RNA splicing, RNA editing), mRNA transport from the nucleus to cytoplasm, mRNA stability, time and frequency of translation in the cytoplasm, post-translational control and protein targeting.

gene regulation in three dimensions: Genes from different chromosomes are in close proximity until a developmental signal stimulates the cells when the genes split apart. One moves to a region that represses gene expression (heterochromatin) and the other relocates to an area with many active genes (euchromatin). Genes from different chromosomes might come into contact when the chromatin containing them loops out from their chromosome 'territory'.

gene regulation *in trans*: *See* TRANS-ACTING FUNCTION.

gene regulation theory of ageing: Sequential changes in the expression of genes occur from the beginning of the lifespan of an organism until the attainment of reproductive maturity and then ageing results. M.S. Kanungo proposed the gene regulation theory of ageing in 1971. Ageing may be due to regulatory changes in the activation of genes. This theory is supported by changes in enzyme patterns known to occur during development and growth. Various types of chemical modifications (i.e. phosphorylation, acetylation, methylation and ADP-ribosylation) are involved in altering the expression of genes. Some of these chemical modifications have been correlated with transcription. Conformational changes in chromatin, particularly in postmitotic cells, occur as a function of age. The gene regulation theory is further supported by age-related decline in RNA synthesis. Methylation of cytosine has been shown to be correlated with gene activity. Methylation of cytosine/adenine has been correlated with age in *Enterobacter cloacae*. Regulation of gene expression is important during ageing because the key difference between normal-lived and long-lived strains is the different manners in which the otherwise identical genes express themselves. Ageing resulted in differential gene expression patterns indicative of marked stress response and lower expression of metabolic and biosynthetic genes. Most alterations were completely or partially prevented by caloric restriction. Transcription of animals under caloric restriction suggested that it retards ageing by causing a metabolic shift towards increased turnover and decreased macromolecular damage. Ageing occurs due to changes in the expression of genes after reproductive ability and adulthood activities are attained.

gene regulation, linear view of: The promoter (p) near the start of a gene provides the minimal information needed for gene expression. The function of the promoter is supplemented by enhancers or silencers, farther away, where regulatory proteins bind to activate or repress transcription of the gene.

gene regulation, meaning of: In *Escherichia coli*, the estimated number of genes is 3000–4000 and the number of proteins found at a given time is from 600 to 700. This means that all the genes are not always active. Only those genes are active at a particular time whose products are needed by a cell. Further, in a bacterial cell, there are about 1,000,000 protein molecules, and the number of copies of various protein molecules varies from 10 to 500,000. This suggests that all the active genes do not produce their products at the same rate. The genes synthesize their products in the quantities they are needed and at a time when they are needed by the cell/organism. This shows that the activity of genes is regulated by some mechanism(s). This is true not only for *E. coli* but for all the organisms – prokaryotes as well as eukaryotes. Why is regulation of gene expression necessary? It would be most efficient for the cell and in the interest of the organism to synthesize only those proteins it needs and in the needed proportions at a time and place they are needed. Some regulatory mechanisms have evolved to achieve the above objective.

gene regulatory circuitries: Study of evolution of genetic regulatory systems that are responsible for differences in gene expression at different developmental stages, in different tissues of the organism and in different species.

gene regulatory mechanism in eukaryotes: Transcription of a PRODUCER GENE occurs only if at least one of its RECEPTORS is activated by forming a sequence-specific complex with ACTIVATOR RNA. The activator RNA is synthesized from INTEGRATOR GENES in response to signals by SENSOR GENES which are sensitive to external/development signals.

gene regulatory mechanism in prokaryotes: They are based on the presence or absence of substrate (INDUCER), end-product (CO-REPRESSOR) or product of a regulatory gene (REPRESSOR). Based on whether control is inducible or repressible, or whether control is positive or negative, several models of gene regulation in prokaryotes are conceived. Basically, gene regulation involves interaction between RNA polymerase and promoter to initiate transcription, repressor and operator thus blocking passage of RNA polymerase, and repressor and inducer to inactivate the repressor.

gene regulatory protein: Any protein that binds to a specific DNA sequence to alter the expression of a gene.

gene repair (by man): The 'repair' of a damaged gene (e.g. mutation) or replacement of a given gene, via a process developed by Eric Kmiec in 1993. The desired DNA (gene) is added to a cell, along with RNA, in a paired group known as a CHIMERAPLAST. The chimeraplast attaches itself to the cell DNA at the site of the specific gene (i.e. the one that is to be changed), and 'repairs' it by using its (new) chimeraplast DNA as a 'template'.

gene repair (natural): Refers to the natural processes through which all cells in an organism continually repair their DNA damaged by ultraviolet light, various chemicals, etc. In the natural gene repair processes, first, an enzyme complex recognizes the damaged site on DNA (e.g. on one of the two strands of the DNA molecule). Then, an enzyme cuts out the damaged portion of the DNA (on that one strand, leaving the other (good) strand intact). Next, a DNA polymerase enzyme enters the gap and synthesizes the new DNA (to replace the portion that was cut out), using the intact (good) DNA strand as a template. Finally, the new DNA is joined to the old DNA with the help of the DNA ligase enzyme.

gene replacement: A method of substituting a cloned gene or a part of a gene, which has been mutated *in vitro*, for the wild-type copy of the gene within the host's chromosome.

gene repression: The inhibition of gene repression.

gene repressor protein: A regulatory protein that prevents the initiation of transcription of a gene.

gene rescue: The method allowing the reisolation of a transferred gene from stably transformed cell lines by way of its activity in *Escherichia coli*. The recovery of plasmids is called PLASMID RESCUE.

gene safety: In gene therapy, this term is used to emphasize that a new gene introduced into the target cells of the patient should not harm the cells or, by extension, the organism.

gene scanning: A method in which the insertion of mutations at specific sites allows the determination of those DNA sequences needed for the activity of a gene. A series of mutants are constructed in which the position of the block is systematically moved. In this way, a length of DNA can be scanned rapidly for regions of importance for the protein product of a gene or for regions involved in the regulation of gene expression.

gene sequencing: Technique of determining the base sequence in the DNA. In order to determine the base sequences small segment of DNA is used as a primer for polymerization of DNA up to specific nucleotides ('minus') or for degradation of DNA down to a specific nucleotide ('plus'). This generates very small segments of DNA of various sizes, all ending with known nucleotides. These segments are separated using gel electrophoresis and analysed 'plus and minus' method of Sanger and Coulson is very important for sequencing. *See* DNA SEQUENCING.

gene shears: A genetically engineered molecule which can seek and destroy unwanted genetic messages within a cell. It may play a role in the prevention of plant and animal diseases, and the destruction of viruses, which invade human, plant and animal cells.

gene silencing (GS): A general term describing epigenetic processes of gene regulation that prevent the expression of a certain gene by a mechanism other than genetic modification. That is, a gene which would be expressed (turned on) under normal circumstances is switched off by machinery in the cell. An endogenous gene can also be suppressed by the introduction of a related transgene. Gene silencing has mostly been applied to plants and insects. Endogenous genes are controlled by TRANSCRIPTIONAL GENE SILENCING (TSG) or post-transcriptional GENE SILENCING (PTGS). During transformation, sometimes the introduced gene expresses very well in the T_0 or T_1 generation, but it stops expressing in the following generations (i.e. transgene is present but does not express). Methylation (attachment of CH_3 groups to DNA) and co-suppression (in case of multicopy transformants) are the two major causes of gene silencing in transgenic plants. *See details as separate terms.*

gene silencing (GS), mechanisms of: At a molecular level, gene silencing may occur via (i) DNA-DNA pairing, (ii) DNA-RNA pairing, (iii) autoregulation, (iv) antisense-mediated RNA degradation, (v) cellular mechanisms and (vi) chromatin-mediated repression.

gene silencing (GS), applications of: Gene silencing is used as a tool to prevent the occurrence of other genetic events (e.g. transposition of mobile elements), which have the potential to disturb the normal structure and function of the host genome. Gene silencing has been used in crop improvement. The suppression of gene expression, for example, of the gene for polygalacturonase which causes fruit to ripen, or the gene for P34 protein in soybeans, via a variety of methods, such as RNA interference (RNAi), chemical genetics, effect of certain viruses, zinc-finger proteins or sense or antisense genes. *See* RNA INTERFERENCE.

gene site saturation mutagenesis (GSSM): GSSM uses sets of degenerate oligonucleotides to introduce all 19 amino acid substitutions at every position of the gene.

gene space: Long gene-rich regions on a chromosome that contain the vast majority of genes, separated by long gene-poor regions in the genome of a given species. It is a general feature of plant species which have a large genome size owing to the abundance of repetitive DNA sequences (transposons and retrotransposons) in their genome.

gene splicing: (i) The isolation of a gene from one organism and then the introduction of that gene into another organism using techniques of biotechnology.

(ii) It also refers to the removal of introns and the joining of exons during pre-mRNA maturation to functional mRNA. (iii) The enzymatic joining of one gene or part of a gene to another.

gene squelching: In eukaryotes, most transcription factors directly interacting with the pre-initiation complex are transcriptional activators. The analysis of transcription factor domain structure has shown that the activation domains (which contact the basal apparatus to facilitate transcriptional activation) can be assigned to a small class of families and also to DNA-binding domains. The acidic activation domain is highly interchangeable between transcriptional factors from different species, suggesting that all transcriptional factors of a given activation family interact with the same, highly conserved component of basal apparatus. The phenomenon of squelching supports this conclusion. A gene is said to have been squelched when a transcription factor is overexpressed and not only activates all its target genes, but other sequences also, by protein-protein interactions, components of the basic apparatus of genes, it does not normally regulate, causing global downregulation.

gene stacking: In general, the process of transferring two or more (possibly synthetic) genes into the genome. The usefulness of gene stacking lies in the combination of more than one gene for plant disease resistance, crop productivity or other horticultural traits. Gene stacking, also called GENE PYRAMIDING, is becoming an attractive alternative for developing durable resistance and for simultaneous and effective control of more than one insect together. For instance, in cotton, a new transgenic event 'Bollgard2' carries two *Bt* genes: *Cry1Ac* (against American bollworm) and *Cry2Ab2* (against tobacco budworm). *See* GENE PYRAMIDING.

gene string: The linear functional gene group of a chromosome. A chromonema with the genes it carries.

gene structure: The basic genetic material is fundamentally the same in all living organisms. It consists of chain-like molecules of nucleic acids – deoxyribonucleic acid (DNA) in most organisms and ribonucleic acid (RNA) in certain viruses – and is usually associated in a linear arrangement that (in part) constitutes a chromosome. The segment of DNA forms a gene that is involved in producing a polypeptide chain. It includes regions preceding and following the coding region (leader and trailer) as well as intervening sequences (introns) between individual coding segments (exons). More than one protein can be expressed (made) from a given gene (i.e. the particular protein expressed is determined by factors such as the cell's temperature or any other environmental variable, or the presence of SIGNAL TRANSDUCERS AND ACTIVATORS OF TRANSCRIPTION (STATs), some of which themselves are proteins).

gene structure concept, classical phase of: Gene was regarded as a hypothetical particle, performing a specific function and recombining with other genes.

gene structure concept, modern phase of: Gene was a unit of function but not a unit of recombination or that of mutation. The fine structure of the gene was defined in terms of cistron, recon and muton. The gene has a well-defined internal structure.

gene structure concept, transitional phase of: Some alleles may occupy complex locus which can be divided into subloci.

gene structure, classical phase of: In the beginning, the gene was regarded as a hypothetical particle, performing a specific function and recombining with other genes. A gene was considered a unit of function, and the unit of recombination was regarded as no larger than a gene. After the independent discovery of W. Sutton and Th. Boveri in 1903, a gene was considered a concrete particle – a small part of the chromosome. There are several genes carried by a chromosome. Genes located in the same chromosome do not show independent assortment. Genes showing their location on the same chromosome can also show recombination by crossing over. Studies by Morgan showed that multiple allelism arises from alternative modifications of the same gene. Multiple alleles do not show complementation in F_1 (i.e. F_1 is not wild type) and do not show recombination in F_2. The unit of recombination was regarded as no smaller than a gene. A gene thus was considered a unit of recombination.

gene substitution: The replacement of one allele by another allele in a population by natural or directed selection.

gene surfing: The identification of genic DNA sequences in an unidentified DNA by comparing it against sequences in genome and/or protein databases that already have assigned (or proven) functions (e.g. coding sequences).

gene switch: A change in a cell or an organism from the expression of one type of gene, or set of genes, to another related one, or set. Classical examples of gene switching occur during the development processes of higher organisms.

gene symbol: Designating a gene, usually by an abbreviation of the name or description of a given gene. In the past, genes have been described by Latin names and subsequently the one- to three-letter abbreviations. Currently, several systems of gene naming and symbolization are in use, although a comparable, uniform symbolization is sought. Symbols for human genes are usually designated by scientists who discover the genes. The symbols are created using the Guidelines for Human Gene Nomenclature developed by the HGNC. Gene symbols usually consist of no more than six uppercase letters or a combination of uppercase letters and Arabic numbers. Gene symbols start with the first letters of the gene name. For example, the gene symbol for insulin is 'INS'. A gene symbol must be submitted to HGNC for approval before it can be considered an official gene symbol.

gene synteny: Conserved gene order in the chromosomes of related species. Using the techniques of gene mapping it has been observed that there are certain DNA sequences common in the chromosomes of some cereal crops like rice, maize and wheat. This conserved gene order is called gene synteny. Gene synteny helps in evolutionary studies. *See* SYNTENY.

gene synthesis: Organo-chemical synthesis of a gene involving blocking of certain reactive groups of nucleotides and joining the same one by one. No template or polymerase is used. For using this method, the nucleotide sequence of the gene to be synthesized needs to be known. Now the genes can be synthesized in the laboratory, and these are called synthetic genes. For instance, if the base sequence of *CryI Ac* gene is known, then it can be synthesized by joining together the nucleotides using the DNA synthesizer. *See* DNA SYNTHESIS.

gene synthesis machines: These are the automated machines which can rapidly synthesize oligonucleotides chemically. This method is based on the development of silica-based supports and stable deoxyribonucleoside phosphoramidites called SYNTHONS. These machines work under the control of the microprocessor.

gene tagging close to marker gene: Mapping of genes of agronomic/economic importance close to known markers. A molecular marker closely linked to a gene acts as a 'tag', which is used for the indirect selection of genes in breeding programmes. Gene tagging is a prerequisite for marker-aided selection and map-based gene cloning. For example, in rice bacterial blight resistance genes x_a1, x_a2, X_a3, X_a4, x_a5, x_a8, x_a13 and X_a21 have been tagged.

gene targeting: (i) A genetic technique that uses homologous recombination to change an endogenous gene. The method can be used to delete a gene, remove exons, add a gene and introduce point mutations. Based on the resultant function of the targeted gene, gene targeting methods have opened two lines of investigation: the GENE KNOCKOUT (KO) and the GENE KNOCKIN (KI). (ii) The insertion of antisense DNA molecules *in vivo* into selected cells of the body to block the activity of undesirable genes. These genes might comprise oncogenes or genes crucial to the life cycle of parasites.

gene targeting by homologous recombination: The gene targeting technique, which can specifically modify the host genome, is based on EMBRYONIC STEM CELLS (ESC) and homologous recombination. Through the homologous recombination of the desired gene construct and the specific chromosomal DNA, gene targeting is capable of precisely modifying or changing the specific chromosome sequence. The resulting new gene is stably inherited and expressed. There are two general methods of gene targeting, called ENDS-IN and ENDS-OUT. They differ in whether the double-strand break is within the region of homology (ends-in) or at the ends (ends-out). These two methods produce very different products.

gene targeting by recombination of a replacement vector: This method improves the efficiency of gene correction in mouse embryonic stem cells using homologous recombination of a replacement vector. The protocol uses microelectroporation chambers and a modified electroporation procedure that does not cause significant cell death. Plating and growth of the electroporated cells at an optimum density to maintain viability significantly increases the recovery of targeted cells.

gene targeting by site-specific methods using germ cells: Female germline mosaics are generated by inducing site-specific homologous mitotic recombination with a yeast recombinase (FLP) driven by a heat shock promoter. These germline mosaics are produced in flies heterozygous for the agametic, germline-dependent, dominant female sterile (DFS) mutation *ovoD1*, where only flies possessing germline clones are able to lay eggs. The 'FLP-DFS' technique is very efficient. This heat-inducible, site-specific mitotic recombination system does not affect viability, and the germline clones recovered are physiologically the same as those created by X-ray-induced mitotic recombination.

gene targeting by site-specific methods using somatic cells: The wild-type ADENO-ASSOCIATED VIRUS (AAV) can insert its genome at a specific site of the human chromosome 19 (AAVS1) through the activity of a specific replicase/integrase protein (Rep) binding both the AAVS1 and the viral INVERTED TERMINAL REPEATS (ITRs). Novel hybrid vectors developed carry an integration cassette flanked by AAV ITRs and a tightly regulated, drug-inducible *Rep* expression cassette in the framework of a helper-dependent, fully deleted adenoviral (Ad) vector.

gene targeting by site-specific methods: Gene targeting by site-specific methods can be done by using somatic cells or germ cells.

gene technology: In a broad sense, it is the artificial transfer of genes between cells or individuals by means of molecular and *in vitro* techniques. Prerequisites for gene technology are (i) the availability of a gene as a DNA fragment, (ii) a cloneable DNA, (iii) the fragment has to be transferable by different systems, (iv) the incorporation of the DNA fragment into a recipient cellular genome has to be feasible, (v) the transformed cells have to be regenerable into a normal plant and (vi) the gene thus transferred must be expressed in the alien genetic background.

gene testing: A type of medical test that identifies changes in genes, chromosomes or proteins. The results of a genetic test can confirm or rule out a suspected genetic condition or help determine a person's chance of developing or passing on a genetic disorder. More than 77,000 genetic tests are currently in use, and others are being developed. *See* GENETIC TESTING *and* GENETIC SCREENING.

gene theory: One of the five basic principles of biology, stating that traits are inherited through gene transmission.

gene therapy: A technique that modifies a person's genes to treat or cure disease. An experimental procedure aimed at replacing, manipulating or supplementing non-functional or misfunctioning genes with healthy genes. The therapeutic gene is typically delivered using recombinant virus or liposome-based delivery systems. First, gene therapy experiment began in 1990 to cure human ADENOSINE DEAMINASE DEFICIENCY (ADA), which is an immunodeficiency that can cause death during childhood. A retroviral vector carrying the normal ADA gene was introduced into T lymphocytes isolated from the patient. Altered cells were returned to the patient and produced immunoglobulins for immune response. Human cells/embryos are being transferred with the normal gene to replace the defective gene. The first gene therapy to treat haemophilia in humans was done in 1999. Nowadays, antisense RNA technology or RNA interference techniques are increasingly being tried to block the expression of certain specific genes, whose products cause disease. *See* GENE, INHERIT, SOMATIC CELL GENE THERAPY *and* GERMLINE GENE THERAPY.

gene tracking: The procedure of following the inheritance of a particular gene from generation to generation.

gene transcript: An RNA molecule of a defined size synthesized over the length of a gene. A cell uses a DNA strand as a template to make a transcript (or mRNA molecule). This transcript is further processed and then translated to make a protein. Different types of transcripts fulfil regulatory, structural or catalytic functions.

gene transcription: The process by which cells read out the genetic instructions in their genes to RNAs, such as messenger RNA, which is the first phase in the synthesis of proteins.

gene transfer methods: There are many methods of gene transfer. *See* CHEMICAL METHOD, MICROPROJECTILES METHOD, AGROBACTERIUM-MEDIATED DNA TRANSFER, MICROINJECTION, COTRANSFORMATION, TRANSFECTION, ELECTROPORATION, CHROMOSOME-MEDIATED GENE TRANSFER, LIPOSOME-MEDIATED GENE TRANSFER *and* DNA-MEDIATED GENE TRANSFER.

gene transfer: Incorporation of new DNA into an organism's cells, usually by a vector such as a modified virus. Used in gene therapy. *See* MUTATION, GENE THERAPY *and* VECTOR. *See* GENETIC TRANSFORMATION.

gene transfer methods in animals: (i) P-element-mediated gene transfer, (ii) Transfer of genes to cultured cells, (iii) Chromosome-mediated gene transfer, (iv) Transformation of cells with free DNA, (v) Gonadal microinjection, (vi) Gene targeting and (vii) Restriction enzyme-mediated integration. These are most commonly used methods for gene transfer in animals. *See details as separate terms.*

gene transfer methods in plants: These methods can be broadly classified into two groups: (i) *Agrobacterium* Ti or Ri plasmid-mediated gene transfer and (ii) Physical delivery or DNA-mediated gene transfer (DMGT). Gene transfer methods in plants make use of a variety of target cell types, which include protoplasts or cultured cells, meristem cells from immature embryos, shoots or flowers and pollen or zygotes.

gene translocation: The transfer or movement of a gene or gene fragment from one chromosomal location to another. A translocation often alters or abolishes gene expression. *See* POSITION (or POSITIONAL) EFFECT.

gene-based single-nucleotide polymorphism (gene-based SNP): Any SNP that is positioned in an exon, an intron or a promoter of a gene.

gene-body methylation (gbM): Refers to CG methylation in the body of the gene. It is evolutionarily conserved, occurring in animal species as diverse as humans and honeybees. Its enhancement within exons suggests a potential role in pre-mRNA splicing.

gene-deleted organism: An organism with one or more genes removed from the genome to render it non-pathogenic so it can be used as a vaccine.

gene-environment interaction: A gene produces a particular effect only in the presence of a particular environment. The term refers to the interaction of genes (and, more broadly, genome function) and the physical and social environment. Examples include (i) organophosphate pesticides and Parkinson's disease, (ii) abuse and antisocial behaviour, (iii) asthma and air pollution and (iv) oesophageal cancer and alcohol.

gene-enzyme relationship: The first hint to the nature of the primary gene function came from the studies on humans. A. Garrod in 1909 reported that several hereditary human defects are caused by recessive mutations. Such disorders were ascribed to an absence or a defect in an enzyme. Thus, a relationship between genes and enzymes, which are responsible for performing various biochemical reactions, was suggested.

gene-for-gene hypothesis: This hypothesis states that for every gene for resistance (R) in the host there is a corresponding and specific gene for avirulence (Avr) in the pathogen. A disease fully develops only if the resistance gene in the host is overcome by a corresponding and specific virulence gene in the parasite. The inheritance of resistance in the host and the ability of a parasite to cause disease is controlled by these pairs of matching genes. This hypothesis was given by H.H. Flor in 1956 which states that for every gene for resistance in the host, there is a corresponding and specific gene for avirulence in the pathogen. Genes for avirulence in the pathogen and resistance in the host are genetically dominant A disease fully develops only if the resistance gene in the host is overcome by a corresponding and specific avirulence gene in the parasite. Flax has 30 resistance (R) genes, of which about 30 corresponding flax rust avirulence (Avr) genes are reported.

gene-for-gene theory: In certain plant-pathogen interactions, a gene for resistance in the host matches to and is directed against a gene for virulence in the pathogen. In flowering plants, four families of disease-resistance genes confer gene-for-gene resistance to a wide array of pathogens by recognizing the products of the corresponding pathogen avirulence genes. *See* FLOR'S HYPOTHESIS.

gene-gene interactions: Genes are not merely separate elements producing distinct individual effects but may interact with each other to give completely novel phenotypes. Gene interactions may give modified F_2 ratios. This so happens because two or more gene pairs, which interact, influence many characters. In the case of independent assortment, we deal with two gene pairs or more, all governing different characters, whereas in gene-gene interactions we deal with two or more gene pairs affecting one and the same character. Variations of the latter type show multigene inheritance. In order to avoid complexity, in textbook examples, we deal only with the cases where both gene pairs show complete dominance. Even in the absence of any type of gene interaction, the F_2 ratios are modified to give a higher number of phenotypic classes when there is incomplete dominance for two gene pairs (1:2:2:4:1:2:1:2:1) or for one gene pair (3:6:3:1:2:1).

gene-gene interactions, conclusions from: The discovery of gene-gene interaction dispelled the notion that each gene produced a single, non-overlapping individual effect and that these effects fitted together like a mosaic to produce the organism. This viewpoint is called PREFORMATIONISM. The present view is that the development of an organism results from the interaction of gene products with each other and with the environment. This viewpoint is called EPIGENESIS. Multiple genes govern the inheritance of a trait. In the final expression of a trait, no single gene acts by itself and the phenotype is an outcome of several integrated reactions. Gene interactions will change the expression of a phenotype beyond the limits of a gene

GeneCalling: It is an mRNA profiling technology that has found increased use in the field of genomics and invented by Dr Jonathan M. Rothberg, R.A. Shimkets *et al.* (1999) and developed by *CuraGen* Corporation. GeneCalling is an open architecture system capable of assaying more than 95% of the genes expressed in a tissue. GeneCalling method is not dependent upon an existing sequence or clone database. It uses as low as 50 pg of cDNA from samples and identifies cDNA fragments that are differentially modulated within a set of samples. With the use of 96 pairs of restriction enzymes, more than 30,000 cDNA fragments are routinely assayed to identify those that are differentially modulated. GeneCalling has been successfully applied to expression profiling of several plant and fungal species, which resulted in the identification and characterization of genes that are useful in commercial applications towards improving agriculturally important traits in plants. This genomics technique swiftly identifies candidate genes for use in drug discovery and development. Known as well as novel genes in any organism – most commonly humans, animals, plants and pathogens – can be detected by this open expression system. This mRNA profiling technique for determining differential gene expression uses but does not require, prior knowledge of gene sequences. This method permits high-throughput reproducible detection of most expressed sequences with a sensitivity of greater than 1 part in 100,000. Gene identification by database query of a restriction endonuclease fingerprint, confirmed by competitive PCR using gene-specific oligonucleotides, facilitates gene discovery by minimizing isolation procedures..

GeneCalling methodology: There are various steps involved in GeneCalling methodology. First, mRNAs are purified from a given sample and processed into fragments for analysis. This usually involves the synthesis and subdivision of double-stranded cDNAs from poly(A) RNA. Distinct sets of restriction enzymes are then used to digest sets of the divided cDNAs and resulting fragments are ligated to labelled adapters to be amplified by PCR. PCR products are then purified and subjected to gel electrophoresis on a mounted platform employing stationary laser excitation and a multi-colour charge-coupled device imaging system. A fluorescent label at the 5′ end of one of the PCR primers permits visualization of the PCR fragments, and the cDNAs are exposed to several isolated and identical restriction digests to generate a merged profile based on peak height and variance. The merged digestion profiles from the cDNA preparations are then compared to locate differentially expressed fragments (such as between normal tissue and diseased or drug-responsive tissue). These profiles are then compared by various internet-ready databases such as GeneScape.

GeneCalling, advantages of: The GeneCalling process is vigorous and efficient. It is well validated and needs only minimal-sized samples. Furthermore, it is widespread in identifying nearly all genes in a given area of research. Beyond this, it identifies subtle changes in gene expression while maintaining reproducibility. Because it is an open expression system, no prior knowledge of sequence information is necessary, and the ability to share information obtained through databases permits for a complete analysis of gene and protein function.

GeneCalling, patent of: On 17 February 1999, *CuraGen* was granted a patent covering its open architecture system of gene expression technology. The patent was entitled 'Method and Apparatus for Identifying, Classifying or Qualifying DNA Sequences in a Sample Without Sequencing' and covered the newly developed process of GeneCalling. While most methods of analysing gene expression to this point were closed systems such as ChIP, the open nature of this system was ground-breaking in its provision of greater flexibility through analysis of gene expression without prior sequence information as well as the ability to compare these results to a sequence database.

genecology: *See* ECOLOGICAL GENETICS.

GeneDesign: An excellent means for designing synthetic genes. Characteristically, it includes tools for codon optimization and removal of restriction sites.

Genefinder: A popular tool to detect coding regions and splice sites.

GeneID: Small microarrays to check IDs of organisms in food and feed (like GMO, mycoplasma in cell culture) or pathogens for disease detection, mostly combining PCR and microarray technology.

genelet: (i) A very small gene, especially one whose function can be switched. (ii) In immunogenetics, the term refers to a coding region of immunoglobulin genes. These genelets recombine in precursor immunoglobulin-producing cells to generate a functional immunoglobulin gene ready to undergo transcription. (iii) In computer science, genelet is a next-generation web development framework. It aims to build secure, scalable and high-performance websites with minimal programming effort. It is currently available in GO Lang, PHP, Java and Perl.

general amphiplasty: The overall changes in chromosomal morphology (shortening, thickening or lengthening of chromosomes) from one species in the interspecific hybrids or amphidiploids.

general combining ability (GCA or gca): The average performance of a parent in a hybrid combination and is primarily recognized as a measure of additive gene action, whereas general combining ability describes those cases in which certain hybrid combinations do relatively better or worse than what would be expected. *Compare with* SPECIFIC COMBINING ABILITY.

General Data Protection Regulation (GDPR): European Union regulation designed to protect personal data and privacy.

general recombination: Recombination that occurs between two homologous chromosomes (as in meiosis). *Aka* GENERAL GENETIC RECOMBINATION.

general resistance: Refers to resistance that is effective against all biotypes of the pathogen. *See* DURABLE RESISTANCE, HORIZONTAL RESISTANCE *and* RACE-NON-SPECIFIC RESISTANCE. *Contrasts* SPECIFIC RESISTANCE *and* VERTICAL RESISTANCE.

general shutdown of protein synthesis: In response to binding viral double-stranded RNA by-products within a cell, the RNA-dependent protein kinase, PKR, phosphorylates α subunit of the translation initiation factor eIF2 on a regulatory site, Ser51. This triggers the general shutdown of protein synthesis and inhibition of viral propagation.

general transcription factors (GTFs): Any of the proteins whose assembly around the TATA box is required for the initiation of transcription of most eukaryotic genes. GTFs are also called BASAL FACTORS, which are required for the synthesis of all mRNAs, select the transcription initiation site and deliver RNA polymerase II to it thereby forming a complex that initiates transcription at a basal rate. The transcription initiation of most structural genes requires the presence of seven GTFs, named TFIIA, TFIIB, TFIID, TFIIE, TFIIF, TFIIH and TFIIJ, many of which are multiprotein complexes. These GTFs combine in an ordered pathway with the promoter and RNA polymerase near the transcription start site to form a so-called PRE-INITIATION COMPLEX (PIC) that supports a basal level of transcription.

general waste: No special treatment is necessary for this waste as it can be disposed of with municipal waste. Food waste from tuberculosis or similar category treatment areas is not general waste and it should be autoclaved before disposal.

generalizability: *See* REPLICATION, REPRODUCTION, ROBUST *and* GENERALIZABLE.

generalization: To operationalize the experiment and analysis differently, use new data and get largely the same result (e.g. using different genetics, proteomics or an imaging platform or translating a questionnaire to a different language and doing a survey in a different country). Generalization is also taken to mean validating that a treatment works similarly on patients who are different from those in a clinical study.

generalized lattice design: Each replicate is subdivided into incomplete blocks typically of 5–6 plots. Sometimes referred to as 'alpha design', given by H.D. Patterson and E.R. Williams in 1976.

generalized linear model (GLM): A model that has the same right-hand side form as a linear regression model but whose dependent variable can be categorical or can have a continuous distribution that is not normal. Examples of GLMs include BINARY LOGISTIC REGRESSION, PROBIT REGRESSION, POISSON REGRESSION and MODELS FOR CONTINUOUS Y HAVING A γ DISTRIBUTION plus the GAUSSIAN DISTRIBUTION SPECIAL CASE OF THE LINEAR REGRESSION MODEL. GLMs can be fitted by MAXIMUM LIKELIHOOD, QUASI-LIKELIHOOD or BAYESIAN METHODS.

generalized recombination: Recombination between two double-stranded DNA molecules which share extensive nucleotide sequence similarity.

generalized transduction: A phage attaches to the cell wall of a bacterium and injects DNA. The bacterial chromosome is broken down and biosynthesis of phage DNA and protein occurs. Sometimes bacterial DNA can be packaged into the virus instead of phage DNA. This phage is defective (cannot destroy another host cell) because it does not carry its own genetic material. The cell lyses, releasing viruses. The phage carrying bacterial DNA infects another cell. Crossing over between donor and recipient DNA can occur producing a recombinant cell. In generalized transduction, any bacterial genes can be transferred because the host's chromosome is broken down into fragments. Whatever piece of bacterial DNA happens to get packaged within the phage is the genetic material that will be transferred between cells. *Compare with* SPECIALIZED TRANSDUCTION.

generating new phenotypic variations using epigenomics technologies: (i) A large number of epigenome-editing tools based on small RNAs have been employed for controlling plant gene expression in thale cress (*Arabidopsis thaliana*), tobacco (*Nicotiana benthamiana*), rice (*Oryza sativa*), potato (*S. tuberosum*) and *Zea mays*. (ii) RNAi-based tools have been used in *Arabidopsis thaliana*; RNA-directed DNA Methylation (RdDM) in *Arabidopsis thaliana*. (iii) RdDM involving transgenically produced exogenous small interfering RNAs in *Oryza sativa*. (iv) Site-specific DNA-binding proteins that have been successfully repurposed to function as DNA-binding domains of epigenome-editing tools include the following. (a) Zinc-finger proteins, transcription activator-like effectors and dead Cas9 complexed with single-guide RNA. Zinc-finger-based epigenome-editing tools have been applied in *Arabidopsis thaliana* and *Oryza sativa*. (b) Transcription activator-like effector and dead Cas9-based tools in *Oryza sativa* and *Arabidopsis thaliana*, respectively.

generating new phenotypic variations using genome-editing technologies: The genome-editing tools used in plants to generate new phenotypic variations for utilization in plant breeding programs are (i) zinc-finger nucleases in maize (*Zea mays* L.); (ii) transcription activator-like effector nucleases in rice (*Oryza sativa* L.), sugarcane (*Saccharum* spp.) and wheat (*Triticum aestivum* L.); and (iii) clustered regularly interspaced short palindromic repeats (CRISPR)/CRISPR-associated protein 9 (Cas9) in *Oryza sativa* and tomato (*Solanum lycopersicum* L.).

generation: (i) A body of living beings constituting a single step in the line of descent from an ancestor. (ii) A group of individuals born and living contemporaneously with the younger generation. (iii) A group of individuals having contemporaneously a status (such as that of students in a school) which each one holds only for a limited period. (iv) A type or class of objects usually developed from an earlier type.

generation time: Time required for a cell (bacteria) or population to double in number. The living cells or microorganisms divide, grow and increase in number. However, this process is highly affected by the genotype of the cell/organism and the prevailing environmental conditions such as nutrients, temperature, light, moisture, etc. Most bacteria have a generation time of 1–3 h. The *Escherichia coli* generation time under favourable conditions is 20 min (i.e. doubling occurs after every 20 min). Therefore, under optimum cultural conditions in about 10 h (i.e. 30 generations), a single *Escherichia coli* cell will produce one billion *Escherichia coli* cells.

generative artificial intelligence: Generative artificial intelligence is artificial intelligence capable of generating text, images or other media in response to prompts. Generative AI models learn the patterns and structure of their input training data and then generate new data that has similar characteristics, typically using Transformer-based deep neural networks.

generative nucleus: A haploid nucleus of a pollen grain that yields two sperm nuclei by pollen grain mitosis.

generic diversity: The differences observed between individuals belonging to different genera.

generic event: An event that harbours the same inserted genetic sequences as the corresponding commercialized Event, but that is made *de novo* with nucleotide sequences and technologies available in the public domain.

generic name: (i) In taxonomy, the name of a taxonomic genus, such as *Acacia* and *Eucalyptus*. (ii) Consumer products identified by product characteristics rather than brand name. It refers to a product that is not protected by intellectual property rights and that is freely available for use by third parties for commercial and development purposes.

genes as the blueprint for life: In genetics, genes are considered blueprints of life. This view, according to D. Noble in 2024, is oversimplified and out of date. Philip Ball in his book *How Life Works*, published in 2024, states that genes are not the blueprint for life. He does not consider the human genome sequenced in 2001 as an 'instruction manual' for life. Organisms are extremely robust, and a particular function can often be performed even when key genes are removed. For instance, although the *potassium/sodium hyperpolarization-activated cyclic nucleotide-gated channel 4* (*HCN4*) gene encodes a protein that acts as the heart's primary pacemaker, the heart retains its rhythm even if the gene is mutated.

genes of terminator technology: The three genes of terminator technology are named Gene *I*, Gene *II* and Gene *III*. Gene *I* is a repressor gene that produces a repressor protein that interacts with a binding site near Gene *II*. Gene *II* is a recombinase gene that is controlled by a promoter. Between the gene and the promoter is a binding site for the repressor from Gene *I*. The recombinase gene produces a recombinase protein, which is an enzyme that snips out pieces of DNA. Gene *III* is a toxin (lethal) gene which produces a toxin that is lethal to embryos. This gene is controlled by a late promoter, which is active only during the late stage of seed development when the embryo is developing.

genes on the basis of gene expression, classification of: On the basis of gene expression, genes may be divided into several classes: (i) NON-CODING RNA, including tRNA genes – synthesize pre-tRNA and rRNA genes – synthesize pre-rRNA, small RNAs – play role in gene regulation; and (ii) PROTEIN-ENCODING or STRUCTURAL GENES – synthesize mRNA (in prokaryotes) or pre-mRNA (in eukaryotes); these transcripts after processing become functional molecules tRNA, rRNA and mRNA, respectively. Whereas tRNA and rRNA do not encode polypeptides, mRNA is the only RNA that contains codons and acts as a template for transcription. *See details as separate terms.*

genes, class I: In molecular genetics, genes transcribed by RNA polymerase I. For example, the genes for 18S, 5.8S and 28S rRNA form a single transcription unit that is transcribed by RNA polymerase I to give a single large transcript (45S pre-rRNA). The extra eukaryotic 5.8S rRNA, which is absent in prokaryotes, lies between the 18S and the 28S rRNA.

genes, class II: In molecular genetics, genes transcribed by RNA polymerase II. Precursors of mRNAs and most snRNA and micro RNAs belong to this class. Many class II genes have a characteristic region situated between 19 and 27 bp upstream from the TRANSCRIPTION START SITE (TSS) whose consensus sequence TATANA is similar (except in location) to the prokaryotic TATA BOX, *aka* HOGNESS BOX, where N represents any nucleotide. Class II promoters include yet another upstream element, called GC box, with consensus sequence GGGCGG which functions in either direction and often exists in multiple copies and binds the transcription factor SP1. Class II genes carry still other binding sites for transcription factors at a variety of positions. Certain sequences are present only in some classes of genes (e.g. those for heat shock proteins); these sequences bind to regulatory factors.

genes, types of: From a structural point of view, genes are classified into three types: (i) SIMPLE GENES (most of the prokaryotic genes), (ii) COMPOUND GENES (most of the eukaryotic genes) and (iii) COMPLEX GENES.

gene-specific marker: DNA markers that are associated with a specific gene. For example, a *Ph-3* on-gene CLEAVED AMPLIFIED POLYMORPHIC SEQUENCE (CAPS) marker, Ph3.gsm/*Hinc*II, was developed by Y.-Y. Wang in 2016 based on the published gene sequence of *Ph-3*.

gene-specific translational silencing: Gene-specific translational (GST) silencing is known to be a novel function of the fused glutamyl- and prolyl-tRNA synthetase (GluProRS). GluProRS is released from a multisynthetase translation complex in response to γ-interferon and forms a four-protein Gamma Activated Inhibitor of Translation (GAIT complex) that silences translation of ceruloplasmin (Cp), a protein linked to the inflammatory response. The four proteins of the GAIT complex are: glutamyl-prolyl-tRNA synthetase (EPRS), NS1-associated protein 1 (NSAP1), glyceraldehyde-3-phosphate dehydrogenase (GAPDH) and ribosomal protein (RP) L13

genetic: Related to genes, the origin or common ancestor or ancestral type, heredity; referring to heritable characteristics..

genetic ablation: A method of modifying DNA in order to disrupt the production of a specific gene.

genetic advance: The expected gain in the mean of a population for a particular quantitative character by one generation of selection of a specified per cent of the highest-ranking plants.

genetic algorithm (GA): A metaheuristic inspired by the process of natural selection that belongs to the larger class of EVOLUTIONARY ALGORITHMS (EA). A population of initial solutions is encoded and the algorithm searches through these by applying a predefined fitness measurement to each solution, selecting those with the highest fitness for reproduction. Genetic algorithms are commonly used to generate high-quality solutions to optimization and search problems by relying on bio-inspired operators such as mutation, crossover and selection.

genetic analysis, conditions for model organism: The organism used in genetic experiments should fulfil certain conditions: (i) have a number of well-defined contrasting differences, (ii) be sexually reproducing and have biparental parentage, (iii) undergo inherent recombination, (iv) controlled matings should be possible, (v) have a short life cycle, (vi) yield a large number of offspring, (vii) be convenient in handling and (viii) its culturing should be relatively inexpensive.

genetic analysis. It includes (i) finding out the number of genes that control the difference between mutant and wild-type phenotype; (ii) finding out the location of gene(s) (i.e. nuclear or cytoplasmic); (iii) in animals, finding out whether genes are sex linked or autosomal; (iv) mapping position of the gene(s) in particular chromosome(s); and (v) knowing nucleotides sequence of the genes, marking particularly regulatory sequences. It deals with the analysis of a gene in terms of its (a) nature (i.e. nuclear or cytoplasmic; mitochondrial or chloroplast), dominant or recessive, autosomal or sex linked (if applicable); (b) location on a particular chromosome; (c) position on the chromosome; and (d) nucleotide sequence. Several prokaryotes and haploid and diploid eukaryotes are very useful organisms in genetic analysis.

genetic ancestry: Refers to information about the people that an individual has biologically descended from. It includes their genetic relationships. This genetic information can be combined with historical information to infer where an individual's distant ancestors lived.

genetic and physical maps, relationship between: In a high-density genetic map, the order of markers remains the same as that of the physical map.

genetic architecture: The dispersal of genetic variation in a species, usually described hierarchically as variation at the regional, local, family and individual levels. It is also related to the proportions of additive and nonadditive inheritance.

genetic architecture of complex traits: Genetic architecture refers to the numbers and genome locations of genes that affect a trait, the magnitude of their effects and the relative contributions of additive, dominant and epistatic gene effects. Quantitative trait locus (QTL) mapping techniques are commonly used to investigate genetic architectures of complex traits.

genetic assimilation: Genetic assimilation is the process by which a phenotypic response to the environment through the process of selection. It is taken over by the genotype so that it becomes independent of the original environmental inducer. For genetic assimilation to work, four things have to be shown. (i) The genome must be responsive to environmental inducers. (ii) The competence to be induced must be transferred from an external inducer to an internal, embryonic inducer. (iii) There has to be cryptic variation within a population so that the physiological induction can be taken over by embryonic inducers. (iv) There must be selection for the phenotype. Genetic assimilation has experimental support. (v) Ultimate extinction of a natural species as massive pollen flow takes place from another related species and the natural species turns out to be more like the related species.

genetic association: Refers to an agreement greater than predicted by chance between a specific allele and another trait that may or may not have a genetic basis. Evaluation of connotation requires the study of unrelated individuals. Association analysis may prove useful in identifying a genetic factor in a trait; except when linkage disequilibrium exists, the association is not due to genetic linkage and should not be confused with it.

genetic background: The set of genes with which a gene of interest is associated. Recombination moves genes from one genetic background to another.

genetic balance: The ideal interaction of coadapted genes, alleles or genetic systems within a given individual gene.

genetic barrier: A physical or psychological condition that hinders or prevents the movement of individuals and thus of genetic information.

genetic base: Amount of genetic variability available for selection at a plant breeder's disposal.

genetic basis of determination: The specific nature of changes that accompany determination is unknown. Changes during cleavage reflect differential action of the genome among the cells of the embryo. Evidence shows that differentiating cells engage in differential cell programme. RNA population extracted from different tissues contains widely different RNA species. Chromatin isolated from one tissue synthesizes RNA *in vitro* not made by chromatin from other tissue. Fetal-level chromatin experiment produces global mRNA *in vitro*, while chromatin from brain cells does not.

genetic block: In biochemical genetics, a block is a step in the synthesis or segregation of a biochemical substance brought about by a mutant gene. The decrease or complete suppression of an enzyme activity results from a specific gene mutation.

genetic bottleneck: A temporary reduction in population size that causes alleles to be lost, while others are fixed. *See* FOUNDER EFFECT.

genetic cascade system: A system in which the completion of one event triggers the initiation of a second event. In genetics, usually a system for controlling the order in which genes are expressed.

genetic code: The biochemical basis of heredity consists of codons (base triplets along the DNA sequence) that determine the specific amino acid sequence in proteins. Under normal conditions, the code is not ambiguous – each codon always designates the same amino acid. There are more than 20 different amino acids and only 4 bases in DNA (adenine, thymine, cytosine and guanine). The mRNA code is a triplet code, that is, each successive 'frame' of three nucleotides (called a CODON) of the mRNA corresponds to one amino acid of the protein. This rule of correspondence is the genetic code. The genetic code consists of 64 entries as this number of triplets is possible when there are four possible nucleotides, each of which can be at any of the three places (4 × 4 × 4 = 64). Since only a little over 20 amino acids exist, 44 codons seem to be superfluous which in fact is not the case as there is some redundancy in the system. Hence, certain amino acids are coded for by two, three, four or six different triplets. A gene's DNA sequence can be used to predict the mRNA sequence, and the genetic code can in turn be used to predict the amino acid sequence. The processes of genetic code are the language of life, or it may be considered a fundamental secret of nature. A gene's DNA sequence can be employed to predict the mRNA sequence, and the genetic code can in turn be used to predict the amino acid sequence. *See* GENETIC CODE DICTIONARY *and* CODE DICTIONARY.

genetic code at work: If there is a change in the nucleotide sequence of DNA, it would lead to a corresponding change in the mRNA. The change in nucleotide sequence of DNA may be brought about by a point mutation, insertion of nucleotide(s), deletion of nucleotide(s), etc. A change in nucleotide sequence of DNA may be reflected by a change of amino acid in protein via a change in the nucleotide sequence of mRNA.

genetic code dictionary: In tabular form, it shows which triplet DNA/RNA codon codes for which amino acid. For example, AUG is an initiation codon and codes for formyl-methionine (in prokaryotes) or methionine (in eukaryotes) and UAA, UGA and UAG are termination/nonsense codons.

genetic code expansion: Modification of the genetic code to enable site-directed incorporation of amino acids beyond the canonical 20 into proteins.

genetic code in nuclear and yeast mitochondrial genes: Genetic code assignments in nuclear and yeast mitochondrial genes have been found to be different. This questions the universality of the genetic code. Phenomena of RNA EDITING and RECODING to some extent resolve this dilemma.

genetic code specificity: Modification of a particular base in a tRNA which is central for reading its specific codon for an amino acid for the specificity of its aminoacylation.

genetic code, evolution of precise: Primitive code was structurally triplet but functionally doublet. With functional differentiation, doublet code became triplet. The triplet code is a very small jump from a doublet code. F.H.C. Crick in 1968 gave the 'frozen-accident' theory. Chance selection and then freezing of the selected code words took place. Once this association took place it perpetuated as such. This explains the particularity and degeneracy of the codon.

genetic code, properties of: (i) Genetic code is triplet – three code letters form one code word which codes for one amino acid. (ii) Codon is commaless. UUUCCCGGGAAA has four code words, and upon translation, it will produce a tetrapeptide chain phe-pro-gly-lys. So, all the letters are used to code for one or the other amino acid. (iii) Code is degenerate; there exists more than one codon for a particular amino acid (e.g. there are four codons for alanine: GCU, GCC, GCA and GCG). This is explained like this: the first two positions of the triplet codon on mRNA pair precisely with the first two nucleotides of the anticodon of tRNA but pairing at the third position may not be precise. (iv) Code is largely unambiguous; one code word codes for only one amino acid. In only one case the code seems ambiguous in the sense that codon AUG specifies two kinds of tRNA ($tRNA_f^{met}$ at initiation site and $tRNA^{met}$ in the reading frame). (v) Nonsense codons exist. For example, UAA, UAG and UGA do not code for any amino acid. They serve as termination codons. (vi) A synthetic ribonucleotide chain directs incorporation of the same amino acid into the polypeptides in cell-free extracts derived from almost any organism. This shows that code is universal. Even nonsense codons appear to be universal in prokaryotes and eukaryotes. (vii) Code is non-overlapping. There will be only two codons from six code letters if the code is non-overlapping. (viii) Initiation codon is AUG or GUG.

genetic collapse: When the concentration of lethal genes in a population becomes great enough to cause the population to die out.

genetic complement: The genetic complement of an organism is called genotype. *See* GENOME.

genetic complementation: Refers to a relationship between two different strains of an organism which both have homozygous recessive mutations that produce the same phenotype (e.g. a change in wing structure in flies) but which do not reside on the same (homologous) gene.

genetic condition: A disorder (disease) caused or influenced by genes that may or may not be passed on from parent to child.

genetic consultation: An opportunity to meet with a geneticist to deal with genetic issues. The visit usually entails a complete medical history, family history, sometimes a physical examination and a review of previous laboratory testing. Any indicated testing is ordered, and when all testing has been completed, the geneticist provides appropriate genetic counselling and a written report to the referring physician.

genetic control of biochemical reactions: Lawrence and co-workers investigated pigment inheritance in cape primrose (*Streptocarpus*) and showed the effect of genes in controlling specific biochemical steps.

genetic control of cell cycle: Genes control the cell cycle in two very general ways. (i) There are genes that control the production of proteins whose functions are vital for the cell cycle to take place, for example, replication of DNA is under the control of many enzymes. The lack of any of these enzymes may prevent DNA replication, thus interrupting the cell cycle. (ii) More interesting are genes that control the initiation of each phase. Initiation of mitosis involves two major gene products one of which oscillates in quantity during the cell cycle and one of which is constant. The constant gene product is an enzyme controlled by *cdc2* gene. *Cdc* stands for cell-division cycle. The enzyme product by this gene is called $p34^{cdc2}$, where p stands for protein, and 34 indicates that the size of the protein is 34 kDa. This protein is a kinase. *See* CELL-DIVISION CYCLE.

genetic copy-and-paste machinery: Long interspersed nuclear *elements 1 (*LINE-1) DNA elements self-duplicate, inserting the copy into new regions of the genome. This is a key process in chromosome evolution. Structures of the machinery that performs this process in humans have been reported (G. Cristofari 2024). A. Thawani *et al.* (2023) and E.T. Baldwin *et al.* (2023) deciphered the X-ray and cryo-electron microscopy structures of the remarkable complex that enables L1 to propagate itself, thus providing insight into how and where this DNA element is duplicated.

genetic correlation: The correlation between the genotypic values of two characters with respect to the genetic/breeding character. This may be due to linkage disequilibrium between genes affecting the different traits or to pleiotropy, in which alleles affect both traits.

genetic counselling: The educational process that helps individuals, patients, couples or families to understand genetic information and issues that may have an impact on them. Education includes information regarding the disorder and its consequences, the probability of developing or transmitting a disorder (risk assessment) and ways to prevent or ameliorate a disorder.

genetic counsellor: A person who provides patients and their families education regarding genetic-related conditions and advocacy and support. Genetic counsellors are healthcare professionals with graduate training in medical genetics and counselling that typically work as part of a healthcare team to provide patient education and support.

genetic death: Complete elimination of an allele from a population without reproducing. It is caused by mutationally derived alleles that reduce the fitness of a genotype and/or taxon. It is difficult to eliminate an allele. According to an estimate, it needs 30,000 generations for the genetic death of an allele.

genetic decoding: *See* RECODING.

genetic deterioration: Refers to the fact that the genetic health of the human population is deteriorating day by day because of our increased influence on natural processes through environmental pollution and advances in medical science.

genetic discrimination: Genetic discrimination occurs when people are treated differently by their employer or insurance company because they have a gene mutation that causes or increases the risk of an inherited disorder. Fear of discrimination is a common concern among people considering genetic testing.

genetic disease: A disease or condition caused by a change or mutation in a gene or change in the chromosomes.

genetic dissection: The use of recombination and mutation to piece together the various components of a given biological function.

genetic distance (D): A measure of gene differences between individuals or populations measured by the differences of several characters. Such distance may be based on phenotypic traits, allele frequencies or DNA sequences (e.g. genetic distance between two populations that have the same allele frequencies at a particular locus and based solely on that locus is zero). The distance for one locus is maximum when the two populations are fixed for different alleles. When allele frequencies are estimated for many loci, the genetic distance is calculated by averaging over these loci. The most widely used measure of genetic distance is M. Nei's 'D'. There exists a clear-cut relationship between reproductive isolation and genetic distance.

genetic distancing: The collection of the data on phenotypic traits, marker allele frequencies or DNA sequences for two or more populations and estimation of the genetic distances between each pair of populations. From these distances, the best representation of the relationships among all the populations may be obtained.

genetic diversity: Differences or similarities at the genetic level. Genetic diversity is commonly measured by genetic distance or genetic similarity, both of which imply that there are either differences or similarities at the genetic level. MOLECULAR MARKER–BASED GENETIC DIVERSITY ANALYSIS (MMGDA) has the potential to assess changes in genetic diversity over time and space. Population genetic diversity and phylogenetic relationships may not only illustrate the evolutionary process and biological conservation but also provide useful information for the selection of parents for conventional breeding and hybrid breeding programmes. The grand totality of all such genetic diversity among all cultivated crop plants is called PLANT GENETIC RESOURCES (PGRs). Farmers have created much or all of the PGR. Scientist breeders always develop new varieties by using available genetic diversity. Conservation and enrichment of existing genetic diversity in India are more vital to the future of agriculture than to the present. The Biological Diversity Bill (2000 is in the process of enactment and the revision of Seeds Act) 1966 is also receiving attention. The need to provide for protection in the areas specific to the farm animal sector is also being released. All these steps in India for implementation of intellectual property rights (IPR) will have a significant impact on the course of agricultural R&D.

genetic drift: Changes in the frequencies of alleles in small population that occur by random chances due to sampling error, rather than because of natural selection. Genetic drift occurs most rapidly in small populations. In large populations, random deviations in allele frequencies in one direction are more likely to be cancelled out by random changes in the opposite direction. Variation in genes is not correlated with variation in reproductive success. *See* RANDOM GENETIC DRIFT.

genetic drift in crop improvement: Genetic drift refers to chance-driven fluctuations in gene/allele frequencies within a population, particularly impactful in small populations. It can reduce genetic variation and lead to allele fixation. L. Gao *et al.* (2019) demonstrated in rice the impact of genetic drift on genetic diversity and structure. Genetic drift in small populations resulted in the loss of genetic variation and fixation of alleles.

genetic effects of inversions: In a heterozygote for a paracentric inversion, crossing over within the inversion produces dicentric and acentric chromosomes. Individuals producing crossover chromatids do not survive and hence do not produce recombinant gametes. Thus, crossovers cannot be detected. Further, in a heterozygote for a pericentric inversion crossover products are not recovered. Thus, inversions in many genetic experiments are used as suppressors of crossing over.

genetic engineering: A technology that involves the introduction, elimination or rearrangement of specific genes to alter the genetic constitution of an organism to enable them to make new substances or perform new functions. It uses the methods of modern molecular biology, particularly those techniques referred to as recombinant DNA techniques. The technique allows the bypassing of all the biological restraints to genetic exchange and mixing and even permits the combination of genes from widely differing species. Genetic engineering lets scientists isolate a specific gene for a particular trait, such as resistance to insect attack, in a plant or animal, and transfer it into another plant. The technique permits the bypassing of all the biological hurdles to genetic exchange and mixing and even allows the combination of genes from widely differing species. *Aka* GENETIC MODIFICATION, GENETIC MANIPULATION *or* GENE SPLICING. *See* RECOMBINANT DNA TECHNOLOGY.

Genetic Engineering Approval Committee (GEAC): India's GEAC must approve a recombinant DNA (rDNA) product (e.g. a genetically engineered crop plant that earlier received its 'biosafety clearance' from the Indian Department of Biotechnology) before that rDNA product is allowed to be commercially planted. The GEAC functions under the supervision of the Ministry of Environment, Forest and Climate Change (MoEF&CC), Government of India. As per Rules, 1989, it is responsible for the appraisal of activities involving large-scale use of hazardous microorganisms and recombinants in research and industrial production from the environmental angle. The committee is also responsible for the appraisal of proposals relating to the release of genetically engineered (GE) organisms and products into the environment including experimental field trials.

genetic engineering plasmid: Development of recombinant DNA molecules (new genes) using *in vitro* techniques. For instance, the Ti-plasmid of *Agrobacterium tumefaciens* does not carry *Bt* gene. Thus, a *Bt* (bacterial toxin) gene is isolated and cloned from another bacteria, *Bacillus thuringiensis* and then it is inserted into the Ti-plasmid using restriction (to cut open) and ligation (to join) enzymes *in vitro*. Ti-plasmid carrying *Bt* gene thus created is called a genetically engineered plasmid, and the process is called genetic engineering. This term is more common for microorganisms.

genetic engineering, explosion method of: A technique for gene-into-cell introduction in which the gene is introduced into plant cells by the force of an explosion (vaporization) of a drop of water (to which the gene and gold particles have been added). The explosion is the consequence of the application of high-voltage electricity to the drop of gene-laden water. The water is then vaporized explosively, driving the 'shot' (gold particles) and genetic material through the cell membrane. The plant cell then heals itself (reseals the hole form where the gene entered), incorporates the new gene into its genetic complement and produces whatever product (e.g. a protein) for which the newly introduced gene codes.

genetic engineering, intrinsic ethical concerns in: These concerns are based on how people view life, nature, religion, their personal emotions and values. There is a feeling that mixing up genes in the organisms for our use is 'playing God' and human beings should not intervene in God's realm.

genetic engineering, moral issues in: Transgenic animals raise several particular moral issues: Are animals that combine species an unethical alteration of the natural order of the universe? Is it unethical to modify an animal's genetic make-up for a specific purpose, without knowing in advance if it will have any side effects that will cause suffering to the animal? Does 'creating' animals by genetic engineering amount to treating the animals entirely as commodities? Is it unethical to create 'diseased' animals that are very likely to suffer?

genetic engineering, religious issues in: Religious groups may have specific reasons for objecting to genetic engineering. A Muslim would object to pig genes being inserted into vegetables and fruits. Vegetarians would object to animal genes being inserted in fruits and vegetables, as they could no longer eat those products if they felt strongly about not eating meat. Religious attitudes to GM are mainly concerned with what is done with the technology rather than whether it should be done.

genetic engineering, safety issues in: The main objections to the safety of GM techniques have generally fallen into two categories: (i) risk and safety; and (ii) issues that arise not from the technology itself but from the ways in which it is commercialized.

genetic enhancement: Genetic engineering can also be used to enhance or improve normal human traits. The concept of 'designer babies', for example, leads to debates over whether it is significantly worse for parents to try and give their children the best opportunities by selecting certain genes using *IN VITRO* FERTILIZATION (IVF) and PREIMPLANTATION GENETIC DIAGNOSIS (PGD), versus hiring sperm and egg donors with coveted genetic traits. Germline genetic enhancement is the most controversial form of genetic intervention.

genetic equilibrium: The condition of a population in which successive generations consist of the same genotypes with the same frequencies, in respect of particular genes or arrangements of genes. A propensity for the population to equilibrate its genetic composition and resist sudden change is called GENETIC HOMEOSTASIS.

genetic erosion: The loss of genetic information that transpires when highly adaptable cultivars are developed and threaten the survival of their more locally adapted ancestors, which form the genetic base of the crop.

genetic exceptionalism: The point of view that DNA is unique, but uniqueness does not itself imply any moral duty. Every human being is 'unique' by virtue of DNA, environment and upbringing. The uniqueness of DNA has never forbidden us implicitly or explicitly to modify what we encounter in nature.

genetic fine structure: The structure of the gene analysed at the level of the smallest units of recombination and mutation (nucleotides). S. Benzer used *cis-trans* complementation test in 1962 to study the fine structure of the gene.

genetic flexibility: The degree to which networks can be reprogrammed by mutations.

genetic gain: The change attained by artificial selection in a specific trait. The gain is usually estimated in terms of change per generation or the change per year. It is affected by selection intensity, parental variation and heritability of the genetic factors.

genetic gain due to environmental effects: The strong genetic gain in rice grain production under varying environmental conditions is achievable through the utilization of *Os*SPL14WFP epiallele. Chromatin, histone and DNA changes based on epigenetic processes enable the plant to survive under stressful environments to cope up with the stress in future through the establishment of plant stress memory at the intergenerational and transgenerational levels. Epigenetics has the potential to delineate the G × E interactions that lead to the generation of different phenotypes. The impact of chromatin modification and DNA methylation depends upon the environmental changes that cause epigenetic modifications in gene function. The QTL analysis on epigenetic RILs performed under different environments revealed their power to give scrutiny to the underlying DNA methylated loci linked with any observed variation and significant heritable variation.

genetic heterogeneity: A similar phenotype caused by alterations of different genes or by a different genetic mechanism. *Aka* HETEROGENEITY. *Compare with* ALLELIC HETEROGENEITY, CLINICAL HETEROGENEITY *and* LOCUS HETEROGENEITY.

genetic homeostasis: A mechanism for promoting the stability of phenotypic expression of a genotype when grown over a wide range of environments. The populations under selection have the tendency to regress toward the original mean. *Aka* DEVELOPMENTAL HOMEOSTASIS.

genetic homology: Two genes are said to be homologous if they derive from the same ancestral gene. Homology lets Darwinian evolutionary theory to applied precisely across the biosciences.

genetic identity (I): The proportion of genes the products of which are not distinguishable by their electrophoretic behaviour.

genetic illness: Sickness, physical disability or other disorder resulting from the inheritance of one or more deleterious alleles.

genetic imprinting: *See* GENE IMPRINTING.

genetic induction: Stimulation of a specific gene by an inducer molecule. In bacterial operons, more than one gene may be induced simultaneously by the same inducer. Inducers may act directly or indirectly, usually through an effect on the RNA polymerase or on the promoter region upstream from the gene-coding sequence.

genetic informatics: *See* BIOINFORMATICS.

genetic information: Genetic information is contained within the DNA or RNA sequence.

genetic information, continuity of: The specificity in the order of nucleotide sequence in DNA maintains its continuity because genetic material has the ability of undergoing a process called REPLICATION to synthesize more copies like themselves. DNA is a double-helical structure. Each of the two original strands is complementary to the other. When replication occurs, the hydrogen bonds between the complementary bases break and the strands replicate as they unwind. Each strand functions as a template for the formation of a new complementary chain. Two pairs of chains thus appear after the completion of one cycle of replication whereas before replication only one pair existed. The mechanism of synthesis of DNA was proposed by J.D. Watson and F.H.C. Crick in 1953 suggesting that every single strand is a template or mould for its complement, and a new helix has one old strand and one strand that is newly synthesized. Thus, they proposed semi-conservative mode of DNA replication.

genetic instability: Diverse mechanisms that produce phenotypic variation. For this, genes mutate and back mutate with a very high frequency. Certain controlling elements are considered to be responsible for this activity.

genetic interaction: The interaction between genes resulting in different phenotypic expressions.

genetic isolation: Isolation of a variety or species by genetic means, such as cross sterility in maize and cross incompatibility of amphidiploid species with parents or other species.

genetic load: (i) A measure of all of the harmful recessive alleles in a population or family line. A high genetic load would be one in which there is a high frequency of deleterious recessive alleles. (ii) A reduction in the average fitness of the members of a population because of the deleterious genes, or gene combinations, in the population. It has many particular forms, such as 'mutational load', 'segregational load' and 'recombinational load'. (iii) The relative decrease in the mean fitness of a population due to the presence of genotypes (usually comprising lethal genes) that have less than the highest fitness.

genetic locus: A particular defined site of a chromosome. At a given locus, there can be several or many alleles (i.e. alternative versions of the gene). *See* LOCUS.

genetic male sterility: Male sterility controlled by nuclear genes.

Genetic Manipulation Advisory Committee (GMAC): A government expert advisory committee that offers leadership to the government and industry on the safe and responsible development and use of gene technology in Australia, since June 2001, before the commencement of the Gene Technology Act 2000.

genetic map: A diagram depicting the linear arrangement (i.e. the relative positions) of molecular markers and/or morphological markers along so-called linkage groups, based on their frequency of crossovers or recombinations. The genetic distance is given in centimorgans. One centimorgan equals a recombination frequency of 1%. Molecular markers utilized as 'signposts'/'guideposts' in such maps include SNPs, restriction sites (i.e. the specific locations where each restriction endonuclease 'cuts' a DNA strand) and microsatellites. Such markers located in or close to the gene of interest (e.g. a disease-causing gene within a chromosome) to a researcher are more likely to be inherited along with that gene. Three strategies used to construct genetic maps are: LINKAGE MAP (*aka* RECOMBINATION MAP *or* MEIOTIC MAP), CYTOGENETIC MAP *and* PHYSICAL MAP. *See details as separate terms.*

genetic mapping: A technique for the determination of the linear order of molecular markers or genes along a stretch of DNA. The result of genetic mapping is the construction of a genetic map of a genome. Formerly, genetic mapping was generally done on a gene-by-gene basis. Before the discovery of molecular markers, there were no explicit projects to map entire genomes; therefore, because of greater precision (<0.5 cM) has been achieved by fewer crosses, use of F_2 data, fewer breeding generations, more reliance on maps, mapping kits, etc.

genetic mapping, functions of: Recombination frequencies are translated into map distances.

genetic mapping, principle of: When two genes are close together on the same chromosome, they do not assort independently and are said to be linked. Genes that are closer together or tightly linked will be transmitted together from parent to progeny more frequently than those genes located far apart.

genetic mapping, prospects of: (i) The development of genetic maps based on markers that are simple to generate, highly reproducible, codominant and specific for known linkage groups are highly desirable for their application in breeding. (ii) The development of high-density maps that incorporate EST-derived RFLP, SSR and SNP markers will provide researchers with a greater arsenal of tools for identifying genes or QTLs associated with economically important traits. (iii) Genetic maps with good genome coverage and confidence in locus order require not only large numbers of DNA markers but also the analyses of large numbers of individuals. (iv) The ultimate physical map is the whole-genome DNA sequence.

genetic mapping, steps in: (i) Develop appropriate mapping population. (ii) Decide the sample size. (iii) Decide the type of molecular marker(s) for genotyping the mapping population. (iv) Screen parents for marker polymorphism and then genotype the mapping population (parents plus all progenies). (v) Perform linkage analyses using statistical programs: (a) Calculate pairwise recombination frequencies between markers. (b) Establish linkage groups. (c) Estimate map distances. (d) Determine map order.

genetic maps, obtaining data for: A genetic map is as good as the data that were used to construct it. Researchers construct a linkage map assuming no errors present in the data. Data must be critically checked for all possible errors: typographical error, missing data, genotype coding error and order of genotypes along all loci, etc. It is absolutely essential that the order of the individuals is identical over all loci in the data file.

genetic marker: A genetic marker is defined as a gene or DNA sequence with a known position on a chromosome that can be used to recognize individuals or species. It can be described as a variation (which may appear due to mutation or alteration in the genomic loci) that can be investigated and whose inheritance can be followed successively. A genetic marker may be a short DNA sequence, such as that surrounding a single-base-pair change (SNP), or a long one, like minisatellites. A marker may or may not be part of a gene. Markers are crucial for use in linkage studies and preparation of the genetic maps to help scientists precisely know the possible location of new genes and to determine the relationship between genetic mutations and disease. Genetic markers include such traits as (i) expression of luciferase-catalysed bioluminescence in leaf cells (causing leaves to glow when illuminated by certain light sources), (ii) resistance to specific antibiotics, (iii) the nature of the cell wall and capsule characteristics, (iv) requirements for a particular growth factor and (v) carbohydrate utilization. The first genetically engineered plants bearing a marker gene were field-tested in 1986.

genetic material: The hereditary substance in the cell of an organism. It carries all information specific to an organism. Genetic material has been found to be linear double-stranded DNA (deoxyribonucleic acid) in all eukaryotes, circular double-stranded DNA in all prokaryotes, as well as some viruses, linear or circular single-stranded DNA or RNA in some viruses.

genetic material, nature of: 'Genes lie in the chromosomes' is the proposition of the CHROMOSOME THEORY OF INHERITANCE. Apparently, some component of the chromosome must be functioning as genetic material. Chromosome is made up of proteins and nucleic acids. Proteins are composed of polypeptides, which are made up of amino acids linked together. Twenty kinds of essential amino acids are known to exist in proteins. Thus, proteins are quite complex, and they appear to qualify as possible genetic substances. Nucleic acids are of two kinds – deoxyribonucleic acid (DNA) and ribonucleic acid (RNA). Nucleic acids are long molecules made up of four types of nucleotides. In nucleic acids, nucleotides are arranged together in the form of sugar-phosphate-sugar-phosphate . . . linkage to make a polynucleotide chain. Nitrogen-containing aromatic bases are attached to sugar molecules as side chains. DNA molecules are very complex, and therefore, DNA is also a candidate for genetic material. This is strong evidence to show that DNA is the genetic material in all eukaryotes and prokaryotes. Some viruses, however, have RNA as the genetic material.

genetic material, organization of: Genetic material occurs as a compact mass, occupying a very limited volume in the cell. Various activities of genetic material, such as replication and transcription, must be accomplished within this confined area. The organization of this genetic material must accommodate conversions between inactive and active states. The structure of the nucleoprotein complex is determined by the interactions of the proteins with the DNA (or RNA). Packaging of DNA transpires in bacteriophages/viruses, bacterial cells, as well as eukaryotic nuclei, chloroplasts and mitochondria. Organization of genetic material chiefly refers to the arrangement of genetic material in a cell. The organization of prokaryotic and eukaryotic genomes diverges with respect to split versus non-split nature of gene, coding versus non-coding DNA, application of operon concept, polycistronic versus unicistronic mRNAs, nature of regulatory control, unicellularity vs multicellularity, singularity vs plurality of chromosomes, DNA packaging, haploidy vs diploidy, unique vs repeated sequences and collinearity between gene and its product. Prokaryotic and eukaryotic chromosomes are similar with respect to two basic features – unineme structure and semi-conservative DNA replication.

genetic material, properties of the: Storage of genetic information, mutation, recombination and continuity (through replication) are fundamental properties of genetic material. DNA also possesses the capability of damage repair, transposition, elasticity, conductivity and elasticity.

genetic material, single-stranded DNA as: Single-stranded DNA (ssDNA) molecules reveal $5' \rightarrow 3'$ polarity. When ssDNA is suspended in an aqueous solution, they show randomly coiled features because each of the nucleotides has bases linked to their sugars by glycosidic bonds at C1′ position in such a way their sugars show C2′-endo and C2′-exo puckering; hence, they display anti-configuration. The ssDNA strands also show duplex, hairpin or stem-loop structural features in the regions where the sequences are complementary to each other. Replication of ssDNA is via ds replicative form of DNA. Some viruses possess single-stranded DNA as genetic material. Examples are ϕX174, M13 and AAV.

genetic material, single-stranded RNA as: Some viruses possess single-stranded RNA as genetic material. Examples include Tobacco Mosaic Virus (TMV), Influenza Virus, Rous Sarcoma Virus (RSV), Poliomyelitis and bacteriophages (MS2, F2, r17).

genetic material, threose nucleic acid (TNA) as: TNA is based on sugar called threose, instead of the deoxyribose found in DNA and the ribose in RNA. Threose is similar but simpler sugar than ribose. It was postulated that TNA could be a long-lost precursor to RNA. There are certain properties that favour TNA as genetic material. It forms stable double helices. Its backbone is composed of sugar molecules with only 4-carbon atom. It pairs up with complementary strands of RNA and DNA.

genetic medicines: It is the use of snippets of genetic material which can block the expression of defective genes. These blocks may be at transcription level (triplex approach) or at translation level (antisense approach). These techniques are being perfected.

genetic modification: Refers to any genetic perturbation, including GENETIC KNOCKOUT, GENE ACTIVATION or GENE REPRESSION. Some countries (other than the USA) employ this term to specifically mention genetic engineering. It includes processes such as duplicating, deleting or inserting one or more new genes or changing the activities of an existing gene. Genetic modification can be performed on microbes, plants or animals (humans included). When this is done in humans, it is called GENE THERAPY. *Aka* GENETIC MANIPULATION.

genetic modification of host cells: Several approaches for the genetic modifications of the host cells include the removal of the enzyme or its inhibition to eliminate a competitive pathway or a toxic by-product, the amplification of a gene or group of genes to improve the synthesis of existing products, the expression of heterologous enzyme(s) to extend the substrate range, to produce a novel product, to provide pathways for the degradation of toxic compounds or to design a more environmentally resistant plant. Alternatively, the deregulation of existing enzymes might be necessary to overcome the existing control mechanisms of a rigid node. A combination of modifications may be needed to achieve the goal. Additional manipulation of the central metabolic pathway may also be required to generate the precursor, cofactors and energy needed to sustain the modified pathway.

genetic modifiers: Other genes that can influence the phenotype associated with the gene of interest.

genetic mosaic: An organism in which different cells contain different genetic sequences. This can be the result of a mutation during development or fusion of embryos at an early developmental stage.

genetic mutagen: A gene or a DNA sequence that causes other genes to mutate (e.g. the *Ac-Ds* system in maize or the *Dt* gene in maize).

genetic mutation: A change that occurs in the DNA sequence. Such a change can occur either when mistakes are made in the process of DNA replication or result from environmental factors (e.g. ultraviolet radiation from the sun or cigarette smoke). Some mutations are only carried in an individual organism. A change can be carried over into the organism's offspring when the mutation happens in a germline cell.

genetic networks: Genetic networks are important in gene regulation. In genetic networks, the purpose is regulation of other processes. There is a component of regulation also in metabolic networks. The enzyme activities are regulated through interactions with substrates, products or other regulatory molecules so that appropriate conditions are upheld. Bioinformatics helps in understanding metabolic and signalling networks in terms of their function in the organism and in relation to the data we already have.

genetic nomenclature: The designation of genes by abbreviated gene descriptions or symbols, usually a beginning capital letter of the abbreviation represents a dominant allele, while a small beginning letter refers to a recessive allele.

genetic operator: An operator used in genetic algorithms to guide the algorithm towards a solution to a given problem. There are three main types of operators (mutation, crossover and selection), which must work in conjunction with one another in order for the algorithm to be successful.

genetic perspective: The approaches used to study subject matter in genetics.

genetic polymorphism: An incidence in a population of two or more genotypes in frequencies that cannot be explained by recurrent mutation (i.e. when the most common allele has a frequency of not >99%).

genetic predisposition: A genotype that increases the risk but is insufficient to result in disease. Impaired expression of alleles at other gene loci and/or environmental factors is needed before the disease appears.

genetic privacy: It deals with the idea of regulating the collection, maintenance, use and dissemination of genetic information collected from individuals. The freedom from unlawful interruption. Often mentioned as the right to be let alone, it protects regional, bodily, psychological and informational integrity and decision making. Many of these safeties are directly concerned by genetic testing. Informational privacy protects the access, control and spread of personal information. Privacy is essential to maintain the relations of trust. The Supreme Court of Canada has stated that confidential therapeutic relations appreciate some Charter protection, but they are not absolute. Recognized exceptions include those authorized in law, an individual's consent or waiver of privacy and overriding duties to third parties.

genetic prognosis: It illustrates some basic principles underlying genetic predictions in terms of the probability of a particular genetic disorder appearing in an individual.

genetic program: The transcriptional profiles of mouse embryonic, neural and hematopoietic stem cells were compared to define a genetic program for stem cells. A total of 216 genes are enriched in all three types of stem cells, and several of these genes are clustered in the genome. Compared to differentiated cell types, stem cells express a significantly higher number of genes (represented by expressed sequence tags) whose functions are unknown. Embryonic and neural stem cells have many similarities at the transcriptional level. These results provide a foundation for a more detailed understanding of stem cell biology.

genetic programming: Generating computer programs by selection on a population of variant programs. It is similar to the use of GENETIC ALGORITHMS.

genetic properties of QTLs: Refers to the magnitude of the effect of individual QTLs. Only QTLs with sufficiently large phenotypic effects will be detected; QTLs with small effects may fall below the significance threshold of detection. Another genetic property is the distance between linked QTLs. QTLs that are closely linked (approximately 20 cM or less) will usually be detected as a single QTL in typical population sizes (<500).

genetic recombination: Physical exchange of chromosome regions (genes) among homologous chromosomes by crossing over during meiosis. During meiosis, genetic recombination occurs through chiasma formation and recombinant gametes/individuals are formed. Whereas recombination of the unlinked genes is based on the mechanism of free combination of chromosomes in meiosis, recombination of linked genes takes place through an exchange of segments of homologous chromosomes, a process known as CROSSING OVER. In crossing over, a reciprocal exchange of information takes place between homologous chromosomes, while in GENE CONVERSION, the exchange is predominantly non-reciprocal in character. *See* RECOMBINATION.

genetic resistance: Resistance against pathogens or pests due to specific or general gene action.

genetic resources: The gene pool in natural and cultivated stocks of organisms that is available for human exploitation. Genetic resources are the properties of the sovereign states to which they are indigenous. Future accessions of such resources would require consent from the states. The Convention on Biological Diversity (CBD) promulgates ensuring the conservation and sustainable use of biological diversity, besides fair and equitable sharing of benefits from their utilization. Having abundant natural resources, developing countries have an edge over developed nations.

genetic restoration: Organisms can sometimes rewrite their DNA on the basis of RNA messages inherited from generations past. Plants and animals can inherit allele-specific DNA sequence information that was not present in the genome of their parents but was present in previous generations. First found in *Arabidopsis thaliana*, later, the same phenomenon was also found in mammals, suggesting a common occurrence.

genetic risk: Children, parents and grandparents often share similar health problems. If a particular disease runs in your family, you may have inherited factors that put you at risk. Inherited risk factors are passed down from parent to child by way of genes.

genetic risk assessment: Quantitative estimate of probable impact on the gene pool of subsequent generations from a specific mutagenic exposure.

genetic RNA: RNA, which functions as genetic material. RNA viruses may carry single-stranded RNA (ssRNA) or double-stranded RNA (dsRNA) as genetic material.

genetic screen: A search through a large collection of mutants for a mutant with a particular phenotype.

genetic screening: Testing a group of people to identify individuals at high risk of having or passing on a specific genetic disorder. An example is newborn screening for PHENYLKETONURIA (PKU).

genetic segregation: When an organism makes gametes, each gamete receives just one gene copy, which is selected randomly. *See* SEGREGATION.

genetic sexing techniques: Autocidal insect control techniques, which involve separation of males and females before release, are considered effective for the supervision of insect pests. Separation is done to (i) evade assortative mating, (ii) to avert any possibility of an increase in the wild population size during a genetic control programme and (iii) to abolish females as disease vectors, extremely obnoxious or detrimental to the crop. In autocidal control programmes, the main goal is the exclusion of females during the rearing process of the Mediterranean fruit fly *Ceratitis capitata*. With genetic sexing, wild-type females that have brown pupae are separated from black male pupae by electronic colour recognition circuitry in a mechanical or air-driven sorter. The autocidal control techniques are most effective and money saving, when the population of insect pests is low. Since, these techniques include release of genetically transformed laboratory-reared insects, the size of the area in which the release is to be made will depend upon the number of insects reared in the laboratory.

genetic slippage: The loss of a potential genotype that may sometimes fail to reach an optimum level of phenotype due to genotype × environment interaction (hence selected against).

genetic sterility: A type of male sterility habituated by nuclear genes. *Opposed to* CYTOPLASMIC STERILITY.

genetic stock: A variety or strain identified to carry specific genes, alleles or linkage groups.

genetic structures responsible for triggering gene expression: The grouping and activity of regulatory elements across individuals have revealed two genomic structures – *cis*-regulatory domains (CRDs) and *trans*-regulatory hubs (TRHs) – responsible for triggering gene expression. This has the potential to link the non-coding genetic variation and gene expression within and across chromosomes. Relationship has been observed between genetic variation, regulatory elements (REs) and gene expression to discover *cis*- and *trans*-regulatory harmonization. The population-scale data set combines the activity of REs for monomethylated histone 3 at lysine 4 (H3K4me1), trimethylated histone 3 at lysine 27 (H3K4me3) and acetylated H3 at lysine 27 (H3K27ac).

genetic switches: According to this theory, the cell cycle is a linear assembly line whereas according to the CLOCK THEORY, the cell cycle is an oscillator. The key components of an embryonic and somatic cell cycle are the products of homogenous genes. Different control mechanisms act as switches which regulate different developmental phases of an organism. MATURE PROMOTING FACTOR (MPF) is capable of inducing the same process in immature frog oocytes as in fully-grown oocytes without receiving the hormonal signals. The activity of MPF was found to be fluctuating and was detectable only during mitosis and was never found in interphase. Different events of the cell cycle are thus sequentially executed and the completion of one gives the signal for the initiation of the next phase.

genetic system: The organization of genetic material in a given species and its method of transmission from the parental generation to its filial generations.

genetic targeting: Refers to the addition of antisense DNA molecules *in vivo* into specific cells of the body in order to inhibit the activity of undesirable genes. These genes might include oncogenes, or genes critical to the life cycle of parasites such as trypanosomes (which reason sleeping sickness). *See* GENE TARGETING.

genetic testing: A laboratory test, done most often on a blood sample, but also on cheek cells, skin cells, bone marrow, amniotic fluid or a placenta sample. It looks at a particular gene for changes, or mutations, which might confirm the diagnosis of a genetic disease or that show a tendency to a genetic disease.

genetic toxicology: A field of research in which genetic samples from a living organism (including humans) are placed on a DNA microarray (gene chip) and tested in a computerized device to detect the presence of toxic substances from the environment. It is done to determine if the organism providing the sample has been exposed to specific chemicals which have caused problems such as mutations, cancer and birth defects. The study of the pattern of occurrence of such biomarkers in a sample of individuals or a community is termed GENETIC EPIDEMIOLOGY.

genetic trait: A distinct phenotypic feature, characteristic or quality of character, of a developing or developed individual that is associated with a physical or genetic marker.

genetic transformation: The process of introduction, stable integration and expression of foreign gene(s) in an organism. Genetic transformation of microbes, animals and plants is now being commonly and widely attempted to develop transgenic organisms. For instance, *Bt* cotton (transgenic cotton) has been developed through the genetic transformation of cotton using *Bt* gene(s) cloned from bacterium *Bacillus thuringiensis*. Likewise, in agriculture, this process has now been commercially exploited by developing *Bt* cotton, *Bt* maize, *Bt* tomato and herbicide-resistant crops like soybean, maize and disease (virus)-resistant papaya. *See* TRANSGENESIS.

genetic trespass: The movement of undesirable genes into a crop by cross-pollination from a crop in a neighbouring field.

genetic uniformity: Homogeneity accruing from a narrow or single genetic base (e.g. an F_1 hybrid developed from a single male sterile line). This may lead to genetic vulnerability of the resultant crop, which may become highly prone to epidemics.

genetic use restriction technologies (GURTs): An all-purpose term referring to numerous different technologies envisioned to control the expression (or non-expression) of the gene(s) for specific (e.g. valuable) traits. TERMINATOR GENE TECHNOLOGY, TERMINATOR TECHNOLOGY and SUICIDE SEEDS are other names given to GURT methods for restricting the use of genetically modified plants by causing second-generation seeds to be sterile. The strategy behind this technology is to kill only the embryos while leaving other important seed components such as oils and proteins intact.

genetic variability: Individuals diverging in their genotypes as a result of mutational, recombinational or selective mechanisms.

genetic variance: A component of total phenotypic variance due to the presence of different genotypes in the population. It has three major components: ADDITIVE GENETIC VARIANCE, DOMINANCE VARIANCE and EPISTATIC VARIANCE. *See details as separate terms.*

genetic variant: Any alteration of genetic material ranging from an entire chromosome to a single nucleotide. If the frequency or function of the alteration is unknown, it is called a VARIANT instead of a POLYMORPHISM or MUTATION.

genetic variation: (i) A phenotypic variability of a trait in a population attributed to genetic heterogeneity. (ii) Variations of genomes between members of species, or between groups of species thriving in different parts of the world as a result of genetic mutation. Genetic diversity in a population or species is a result of new gene combinations (e.g. crossing over of chromosomes), genetic mutations, genetic drift, etc.

genetic variation in *Arabidopsis thaliana*: This model plant also exhibits significant natural genetic variation, enabling research on traits such as flowering time, architecture and stress tolerance. This variation aids in understanding plant development, stress response and evolution.

genetic variation in maize: Maize (*Z. mays* L.) exemplifies genetic variation in plants, leading to diverse varieties with distinct traits. This variation has improved crop yields, nutritional quality and resistance to pests and stressors.

genetic variation, predictive power of: Variation within populations can be utilized to predict how traits evolve under natural selection. A. Holstad *et al.* (2024) publish analyses of a massive database of traits from both natural populations and the fossil record, revealing that within-population variation forecasts how traits evolve across population and species boundaries. They test several rival hypotheses that may elucidate this relationship and reach the amazing inference that the propensity of natural selection to change in its direction and intensity may clarify why genetic variation leaves such a long-lasting stamp.

genetic variation, usefulness of: Variation in genetic sequences and the detection of DNA sequence variants genome-wide allow studies relating the distribution of sequence variation to a population history. This in turn allows one to determine the density of SNPs or other markers needed for gene mapping studies. Quantitation of these variations together with analytical tools for studying sequence variation also relate genetic variations to phenotype.

genetic vulnerability: The defencelessness of genetically uniform crops to damage or destruction resulting from outbreaks of a disease or pest or unusually poor weather conditions or climatic change.

genetic welding: The idea given by A.D. Cutter in 2023 that genetically engineered gene drives inserted into wild populations would form a new evolutionary force. This view is also known as HUMAN-CONTROLLED EVOLUTION.

genetically engineered food: *See* GENETICALLY MODIFIED FOOD.

genetically engineered male sterility: Induction of male sterility through recombinant DNA technology.

genetically engineered microbe (GEM): A microorganism that is altered by genetic engineering. The term specifically refers to microorganisms (i.e. bacteria or fungi, including yeasts) that humans have modified using *in vitro* molecular biology techniques (*aka* MODERN BIOTECHNOLOGY) to perform a specific function. *Aka* GENETICALLY MODIFIED MICROORGANISM (GMM).

genetically engineered microbial pesticide (GEMP): One or more microbes that have been genetically engineered to be operative in fighting pest(s) that attack crops or livestock. For example, a microbe that attacks in nature a crop pest could be genetically engineered to make the microbe more potent or more durable in field environments when applied via a selected method of microbe application.

genetically engineered microorganism (GEM): *See* GENETICALLY ENGINEERED MICROBE (GEM).

genetically engineered organism (GEO): The organism whose genome has been engineered in the laboratory in order to favour the expression of desired physiological traits or the production of desired biological products. In genetic modification, however, recombinant genetic technologies are employed to produce organisms whose genomes have been precisely altered at the molecular level, usually by the inclusion of genes from unrelated species of organisms that code for traits that would not be obtained easily through conventional selective breeding. LIVING MODIFIED ORGANISMS (LMOs) and TRANSGENIC ORGANISMS are other terms often used instead of GMOs. GMOs include GM MICROORGANISMS, GM PLANTS and GM ANIMALS. GM animals are produced using methods that include REPRODUCTIVE CLONING TECHNOLOGY and RECOMBINANT DNA TECHNOLOGY. *Aka* GENETICALLY MODIFIED ORGANISMS, a term currently in use most often in official discussions that designates crops, which carry new traits that have been inserted through advanced genetic engineering methods (e.g. flavour-saver ('FlavrSavr') tomato, Roundup™-ready soybeans, *Bt* cotton or *Bt* maize).

genetically enhanced: An organism that has experienced the progression of inserting new genetic information into existing cells in order to adapt the organism for the purpose of improving one or more of its undesirable characteristics.

genetically heterogeneous: A genetic condition where more than one genetic factor can result in similar characteristics. Within the clinic, genetic heterogeneity means the presence of a variety of genetic defects which cause the same disease. These can be due to mutations at different sites on the same gene. Many cancers and genetic disorders (e.g. type 1 diabetes) are genetically heterogeneous.

genetically identical: Two organisms with the same DNA are genetically identical. Two different people are very similar but not exactly the same. Identical twins (*aka* MATERNAL TWINS) are genetically identical whereas fraternal twins are born at the same time but do not share the same DNA.

genetically linked resistance gene analogues (RGAs): The RGA pic19 in maize, a candidate for sugarcane mosaic virus (SCMV) resistance gene (R gene) Scmv1, was used to screen a maize BAC library to identify homologous sequences in the maize genome and to investigate their genomic organization. Fifteen positive BAC clones were identified and could be classified into five physically independent contigs consisting of overlapping clones. Genetic mapping clustered three contigs into the same genomic region as Scmv1 on chromosome 6S. The two remaining contigs mapped to the same region as a QTL for SCMV resistance on chromosome 1. Thus, RGAs mapping to a target region can be successfully used to identify further-linked candidate sequences. The pic19 homologous sequences of these clones revealed a sequence similarity of 94–98% on the nucleotide level. The high sequence similarity reveals potential problems for the use of RGAs as molecular markers. Their application in MARKER-ASSISTED SELECTION (MAS) and the construction of high-density genetic maps is complicated by the existence of closely linked homologues resulting in 'ghost' marker loci analogous to 'ghost' QTLs. Therefore, the implementation of genomic library screening, including genetic mapping of potential homologues, seems necessary for the safe application of RGA markers in MAS and gene isolation.

genetically modified (GM) material in non-GM products: The adventitious presence of GM material is the unintended occurrence of plant material from crops improved through modern biotechnology in other seed, food and feed. It occurs through natural pollen flow or from co-mingling of grain that occurs in the production/distribution system. It is the logical and unavoidable consequence of the development of GM crops in many parts of the world. In contrast to off-types mentioned in the previous section, which are checked on phenotypic characteristics, the adventitious presence of GM material is in general checked based on DNA characteristics.

genetically modified banana: The genetically modified banana has been developed by James Dale, leader of the banana biotechnology program at the Queensland University of Technology after 20 years of work as an industry backup in case the Panama Tropical Race (TR4) breaks out but will not replace the Cavendish variety. Panama TR4 was first recorded in the Tully Valley in 2015. For the first time, food-safety regulators have given the green light for a genetically engineered banana. QCAV-4 has been bioengineered with a single gene, *RGA2*, from a south-east Asian banana. *Musa acuminata ssp malaccensis*, a wild banana. Variety, QCAV-4, is resistant to the devastating fungal disease Panama TR4, which has spread worldwide and for which there is currently no treatment or cure. It has been approved in Australia and New Zealand as a 'safety net' in case biosecurity efforts fail to restrict the spread of the disease.

genetically modified crop (GMCs): Genetically modified crop, sometimes referred to as 'GM CROP', are plants used in agriculture containing DNA that has been modified using genetic engineering methods. Genes for beneficial traits such as herbicide resistance or pest tolerance have been successfully introduced into plants such as maize and soybean. *Aka* GENETICALLY ENGINEERED CROPS.

genetically modified food (GMF): A food item obtained from transgenic organisms. The main objective of these foods is the value addition to our food. Genetically modified foods include high lycopene tomatoes, Flavr-Savr® tomato, antioxidant tomatoes, high flavanol tomatoes. Carotene, vitamin A and iron-pumping golden rice; golden brassica; iron-dense rice; cavity-fighting apples; proteinaceous potatoes; decaffeinated coffee; and cavity-fighting apples and tea are some leading examples of genetically modified foods for the future.

genetically modified microorganism (GMM): *See* GENETICALLY ENGINEERED MICROORGANISM (GEM).

genetically modified mustard: DMH-11 is an indigenously developed transgenic mustard. It is a genetically modified variant of Herbicide-Tolerant (HT) mustard. DMH-11 is a result of a cross between Indian mustard variety 'Varuna' and East European 'Early Heera-2' mustard.

genetically modified organism (GMO): *See* GENETICALLY ENGINEERED ORGANISM *and* TRANSGENICS.

genetically modified organism (GMO) testing: PCR detects the presence of a certain DNA sequence and is used to test maize, soybean, rice, cotton and Canola™ tissue resulting in qualitative results. Tests for the 35S promoter gene are associated with GMO. DNA is mined from ground material. A certain section of the DNA is replicated numerous times to accomplish sufficient DNA to recognize the promoter through electrophoresis. Electrophoresis separates different-sized particles, allowing the 35S promoter to be detected when compared to a known 35S sample.

genetically modified pest-protected (GMPP) plant: A plant that has been genetically engineered so it resists (or is more tolerant to) attacks by pests (e.g. insects).

geneticist: A physician or PhD with special training in some specific area of genetics.

genetics: W. Bateson coined the term genetics in 1905. A branch of biology that deals with the mechanisms responsible for the transmission of physical and behavioural attributes, including medical conditions, from one generation to the next. Genetics deals with the study of genetic variation, genes and heredity of organisms. In 1865, Mendel defined what FACTORS (ELEMENTS or GENES) are and that they can be dominant or recessive (within the offspring's genome/DNA, which has two 'copies' of each gene). Genetics is also a scientific study of the way the development of living things is controlled by assets that have been passed on from parents to children. Genetics also includes the study of the mechanism of gene action – the manner in which the genetic material (DNA) affects physiological reactions within the cell. Genetics deals with understanding the nature, molecular structure, organization, biological function, regulation and manipulation of hereditary particles called GENES, which play a central role in life processes. Genetics also covers finding out ways and means to use knowledge of genetics for the welfare of mankind.

genetics in biological control of pests: With the help of genetic knowledge, biological control of insect pests can be enhanced in a very systematic manner in view of the influence on natural enemies as well. Applications of molecular genetics in the management of insect pests include (i) developing resistant varieties of crops and trees expressing insect-specific toxins, (ii) enhancing potential insecticidal activity of microbial biocontrol agents and (iii) developing engineered insects that have the ability to transfer lethal genes to wild populations after their field release and with comparative genomic analyses. It is also possible to identify and target insect-specific genes using cell-based, high-throughput screening assays.

genic and plasmagenic interaction: Interaction between a plasma gene and a gene in the chromosome to produce an effect. For example, in *Paramecium*, plasma gene *kappa* and nuclear dominant gene *K* interact to produce a killer phenotype.

genic balance theory: The theory of Bridges that the sex of a fruit fly is determined by the relative number of X chromosomes and autosomal sets (X/A).

genic control of sex: One or two gene differences may control sex in some plant species. *Asparagus* is dioecious with distinct male and female plants. Here the maleness is determined by a dominant gene (*P*) while the femaleness is attributed to recessive homozygous condition (*pp*). Maize (*Z. mays* L.) is monoecious with male inflorescence (tassel) and female inflorescence (silk) located on the same plant. Two gene pairs (*Ts, ts* and *Sk, sk*) have been identified to play a role in sex determination in maize. When a female *tsts sksk* (having seeds on tassels but silkless) is crossed with a male *Tsts sksk* (with normal tassels but silkless), it produces male and female plants in 1:1 ratio. This will be a stable dioecious condition. *Ts* locus affects tassel formation. *Ba* locus affects both male and female gamete production.

genic disharmony: When genes contributed by two parents fail to work together in a hybrid or its progeny. It can be expressed at various stages of development and is associated with the process of nuclear metabolism, including DNA replication and transcription. Such individuals are at a selective disadvantage. It serves as a POSTZYGOTIC REPRODUCTIVE ISOLATING MECHANISM.

genic equilibrium: The tendency of a particular gene frequency in a panmictic population to remain constant generation after generation.

genic explanation of hybrid vigour: Hybrid vigour is widely explained as the result of a dominance of linked genes. Many dominant genes are involved. Different inbreds contribute different dominant genes so that the F_1 has more than either parent.

genic male sterility: Male sterility generally characterized by the impairment of the male reproductive development as a result of underlying genetic causes and leads to the malformation of male gametes.

genic regions: The genic region is defined as from the transcription start site to the end of 3′ untranslatable region (UTR).

genie: An artificial intelligence (AI) combined with an assembler or other universal constructor, programmed to build anything the owner wishes. Sometimes called a SANTA MACHINE. This assumes a very high level of AI and nanotechnology.

genocide: The destruction of individuals having an undesirable genotype.

genocopy: The term 'genocopy' was coined by H. Nachstheim in 1957. A gene or genotype produces the same phenotype as another gene or genotype. Genocopies are the basis of genetic heterogeneity and are central to genetic diagnosis and counselling. Genocopies may also be referred to as 'GENETIC MIMICS' *See* PHENOCOPY.

genome: In eukaryotes, a complete set (n) of chromosomes (hence, of genes) inherited as a unit from one parent plus one sex chromosome from the other parent in heterogametic individuals. Each cell of an organism contains the entire genetic material in the form of DNA in its nucleus. For example, the human genome is comprised of 23 pairs of chromosomes and approximately three billion base pairs. Animals have two genomes, nuclear and mitochondrial. Plants have three genomes – nuclear, mitochondrial and chloroplast. The organization of all of the genetic information of an organism is in the form of double-stranded DNA.

genome activity patterns: When puffing patterns from different stages of development and from different tissues are compared, one can determine whether genomic activity patterns change during development and whether they are tissue specific. Puffs appear and disappear ranging from a very low number of 20 bands. They begin to form at a single band but progressively incorporate neighbouring bands, which suggests that during puff development, more and more cistrons are stimulated to deliver their information in a sequential way.

genome analysis: The study of the genome by a combination of cytogenetics, karyotyping and crossing methodologies.

genome analysis using arrays: Detection of polymorphism and genotyping is possible through arrays. An overwhelming majority (99.9%) of genes perfectly match in human population but 0.1% of genes vary. It is this single-base-pair difference in the DNA sequences of individuals in a population, which is termed SNP. These are biallelic markers present throughout the human genome with an average frequency of approximately one per thousand base pair. The SNP consortium was established to produce an ordered high-density SNP map with 200,000–300,000 SNPs evenly distributed in the genome by 2001. In future, a simple genetic test may determine whether you are likely to be treated effectively by a given drug or whether you face the risk of being killed by that drug. At this moment, personalized medicines are still on the laboratory bench. Locations of hundreds of chromosomal origins of replication in yeast are also possible through arrays. Fragmented chromatin can be cross-linked to a protein and then immunoprecipitated with an antibody to that protein. The DNA fraction of the immunoprecipitate can be labelled and hybridized to identify the approximate location of the binding sites. Full genome analysis allows the chromosomal landscape of silencing to be mapped as in the case of telomeric genes and centromeric proximal genes.

genome annotation: The process of identifying functional elements along the sequence of a genome, thus giving meaning to it.

genome annotation tools: (i) BASys Bacterial Annotation Tool performs automated, in-depth annotation of bacterial genomic sequences. (ii) ORF provides one of the choices of Glimmer, ZCurve or GeneMark predictions coupled with GenBank or Fasta-formatted output. It works fine and quickly with phage-sized genomes. (iii) BAGEL reports the presence of bacteriocins based on a database containing information on known bacteriocins and adjacent genes involved in bacteriocin activity. (iv) MICheck (MIcrobial genome Checker) allows rapid verification of sets of annotated genes and frameshifts. (v) RibEx (Riboswitch Explorer) images <40 kb DNA for potential genes (which are linked to BLASTP) and a few hundred regulatory elements, including riboswitches. (vi) tRNAscan-SE site is extremely sensitive and also delivers secondary structure diagrams of the tRNA molecules.

genome assembly: Genome assembly refers to the process of putting nucleotide sequences into the correct order. Assembly is required, because sequence read lengths – at least for now – are much shorter than most genomes or even most genes.

genome comparison tools: (i) GeneOrder 2.0 is perfect for comparing small GenBank genomes (up to 0.25 Mb), while GeneOrder 3.0 outspreads the limits to around 2.0 Mb. Each gene from the query sequence is compared to all of the genes from the reference database using BLASTP. (ii) CoreGenes is intended to analyse two to five genomes simultaneously, producing a table of related genes – orthologues and putative orthologues. (iii) CoreGenes 3 counts the total number of genes in common between the two genomes being compared, displays the per cent value of genes that are in common with a specific genome and determines the unique genes contained in a pair of proteomes. (iv) WebACT is a DNA sequence comparison viewer based on Artemis. (v) WebGMAP is used for annotating and mapping individual cDNA sequences to the genomes of many eukaryote species, currently including *Arabidopsis thaliana*, *Chlamydomonas reinhardtii*, *Glycine max*, *Oryza sativa*, *Physcomitrella patens* and *Populus trichocarpa*. (vi) Panseq is a group of tools used for the analysis of the 'pan-genome' of a group of genomic sequences. The pan-genome of a bacterial species consists of a core genome and an accessory gene pool, the latter of which enables subpopulations of the organism to adapt to specific environments.

genome constancy: Gene content of all the cells of an organism is more or less the same. The phenomena that question the concept of genome constancy in development include random and programmed changes, gene amplification, chromatin diminution and other instances of DNA loss and programmed recombination.

genome coverage: The percentage of all base pairs or loci of the genome covered by sequencing. The average coverage for a whole genome can be calculated from the length of the original genome (G), the number of reads (N) and the average read length (L) as N × L/G. For example, a hypothetical genome with 2000 base pairs reconstructed from 8 reads with an average length of 500 nucleotides will have 2× redundancy.

genome cutter: A peptide nucleic acid (PNA) 'clamp', consisting of two predesigned 8-bp sequences of PNA linked together, can bind strongly and sequence specifically to DNA, hence known as BIS-PNA, protecting the binding region from methylation. After the removal of bis-PNA, methylated DNA can be cut quantitatively at restriction sites selected by the PNA clamp. *See* PEPTIDE NUCLEIC ACID.

genome doubling: *See* AUTOPOLYPLOIDIZATION.

genome duplications in the origin of vertebrates: According to the 2R hypothesis, a large proportion of gene duplicates in vertebrate genomes are the result of at least one complete genome duplication event that occurred at the origin of vertebrates. Most studies published so far testing the validity of the 2R hypothesis include a limited number of genes from organisms phylogenetically close to the point of the hypothesized genome duplications.

genome editing: Any procedure with which a target-specific DNA sequence can be changed. The resultant DNA sequence may include an insertion, deletion or substitution of at least a single nucleotide. The edited DNA sequence may produce a changed gene product (e.g. a protein with an amended amino acid (aa) sequence, a non-coding RNA with an altered nucleotide sequence), may perform a modified function (e.g. changed promoter function or changed enhancer function) or may not give rise to a gene product. Genome editing is used to produce changes in the genome at will. Genome-editing technology uses engineered nucleases and DNA repair and recombination enzymes.

genome editing with engineered nucleases (GEEN): Genetic engineering where DNA is inserted, replaced or removed from a genome using SDN.

genome editing with retroelements: Genome-editing methods generally involve two critical steps: targeting a DNA sequence of interest and introducing the desired genetic modification. Capabilities for programmable DNA targeting have advanced rapidly over the past decade, owing primarily to the retooling of bacterial RNA-guided CRISPR-Cas systems. Efficient and precise modification of the target site particularly for large-scale perturbations such as the insertion of gene-sized DNA payloads remains challenging. Research is now drawing inspiration from MOBILE GENETIC ELEMENTS (MGEs), DNA segments that have exploited diverse targeting and integration strategies to propagate in the genomes of their hosts for billions of years. Of particular interest are retroelements, which mobilize using an RNA intermediate. Recent investigations by S. Tang and S.H. Sternberg in 2023 into the structure and function of retroelement-encoded protein-RNA complexes have resolved key molecular details of their targeting and integration mechanisms, unveiling exciting opportunities to re-engineer them for programmable DNA insertion.

genome engineering: The use of tools that allow rapid and precise changes directly across chromosomes of living cells instead of limiting modifications to single genes. *Aka* GENOME EDITING.

genome engineering in chloroplasts: Techniques for the transformation of plastid genome have been developed relatively recently to explore the plastid gene function and expression. Plastid transformation can be used as an important tool to investigate and alter the functions of plastid gene products. Two methods developed for chloroplast genome engineering are GLASS BEADS METHOD and PARTICLE BOMBARDMENT METHOD. Transformation of the chloroplast genome has several potential problems that nuclear transformation does not have. It is more difficult for DNA to cross the chloroplast double membrane than the nuclear membrane. Chloroplast genomes are present in much higher copy number than the nuclear genomes. *Chlamydomonas* is a favourable organism for chloroplast genome transformation because it has a single large cup-shaped chloroplast, which lies adjacent to the plasma membrane along most of the periphery of the cell. Photosynthetic mutants are viable if reduced carbon is provided in the form of acetate. *Chlamydomonas* cells without cell walls are stable due to active contractile vacuoles.

genome evolution: The process by which a genome changes in structure over time, through mutation, horizontal gene transfer and sexual reproduction in viruses and bacteria to higher eukaryotes. In general, there is no correspondence of genomic evolution with morphological evolution.

genome formula: A mathematical representation of the number of genomes (sets of genetic instructions) in a cell or organism. Examples: n for haploid gamete, x for monoploid somatic cell, 2n for diploid, 3n for triploid, 4n for tetraploid, 2n − 1 for monosomic, 2n + 1 for trisomic, 2n − 2 for nullisomic, etc. *See* GENOME SYMBOL.

genome imprinting: The process that temporarily and reversibly marks the genes passed on by females and males in different ways. Offspring that receive marked genes from their mothers are consequently different from those that receive the genes from their fathers. *Aka* GENOME PROFILING.

genome imprinting in plant development: Transposable elements in the *Arabidopsis* genome are normally methylated and silenced. Pollen grains contain one vegetative cell and two accompanying sperm cells. Transposable elements can be reactivated and transpose in the pollen vegetative cell but not in the sperm cells, which provide DNA to the fertilized zygote. DNA in the pollen vegetative nucleus has reduced methylation, which correlates with the reactivation and transposition of transposable elements. The expression of transposable elements in the vegetative cell may be important to ensure the silencing of these elements in sperm cells. The silencing status of transposable elements in the male gametes is essential for genome stability and integrity. Many genes involved in the RdDM pathway have reduced expression in mature pollen. This reduced expression coincides with the downregulation of 24-nt siRNAs in the vegetative cell nucleus. Reactivation of transposable elements in the pollen vegetative nucleus leads to the accumulation of 21-nt siRNAs from transposable elements. These 21-nt siRNAs could move to the sperm cell to post-transcriptionally silence transposons that may have escaped transcriptional gene silencing. Passive DNA demethylation contributes to the reactivation of transposable elements in the pollen vegetative nucleus.

genome instability: Defects in cellular DNA repair processes have been linked to genome instability, heritable cancers and premature ageing syndromes. Defects in some repair processes manifest themselves primarily in neuronal tissues.

genome interpretation: Refers to the mechanisms involved in negotiating the genome for transcription initiation. Spontaneous preferences of DNA-binding proteins are a primary mechanism by which cells interpret the genome.

genome mapping: *See* GENE MAPPING.

genome organization: The arrangement of the genetic material in the cells of an organism. Most of the well-characterized prokaryotic genomes consist of double-stranded DNA organized as a single circular chromosome 0.6–10 Mb in length and one or more circular plasmid species of 2 kb–1.7 Mb. The structural organization of the prokaryotic genes within the genome was thought of as simple, but the discovery of overlapping genes in bacteriophages, animal viruses, bacterial insertion sequences and split genes in prokaryotic genomes suggests that the organization of the prokaryotic genome is quite complex. Although the length of the DNA molecule of *Escherichia coli* is 1 mm, it occupies a diameter of at most only 1 μ. The bacterial chromosome is packaged in order to remain within this confined space in the absence of a nuclear membrane. *Escherichia coli* chromosome has about 100 loops of DNA held together mostly by RNA and some proteins. Prokaryotes package their DNA in bacterial chromosomes and plasmids. The bacterial chromosome is localized in the nucleoid region of the cell (no nucleus) and is looped into negative coils. The loops are 50,000–100,000 bp in length (similar to eukaryotic chromosomes), which are held in place by RNA and small basic (histone-like) proteins. RNA and protein binding to DNA molecule help in stabilizing the folded structure of DNA molecule. In eukaryotic chromosome, DNA is complexed with basic (histone) and acidic (non-histone) proteins. *See* NUCLEOSOME.

genome profiling: *See* GENOME IMPRINTING.

genome project: Research and technology-development effort aimed at mapping and sequencing the genome of human beings and certain model organisms. *See* HUMAN GENOME PROJECT.

genome resetting model: This model emphasizes that at least some of the DNA involved in species evolution should be functional. In particular, these sequences must be regulatory elements whose positions and sequences influence the development pathways in a coordinated manner. At some instances, such changes may lead to species divergence in beneficial ways. Genome resetting may have a role in GENETIC PROGRAMMING (e.g. during development). The evolutionary consequence of successful genome reorganization is speciation. These organizations that are the result of genome turnover serve to establish new genetic programmes in development and differentiation, which define the morphological and physiological characteristics of new species.

genome selection: A form of marker-assisted selection in which genetic markers covering the whole genome are used so that all QTL are in linkage disequilibrium with at least one marker.

genome sequencing: Determining the exact DNA sequence of an individual's genome including all coding and non-coding regions.

genome sequencing strategies: There are two major approaches to genome sequencing: the SHOTGUN APPROACH and HIERARCHICAL SHOTGUN or CLONE-BY-CLONE APPROACH. *See details as separate terms.*

genome sequencing, progresses in: Significant progress has been made in the field of DNA sequencing. As many as 259 genomes of both prokaryotes and eukaryotes are completely sequenced. The number of ongoing DNA sequencing of prokaryotic genomes is 655 and that of eukaryotic genomes is 483. There are some important organisms like *Haemophilus influenzae*, *Helicobacter pylori*, *Methanococcus janaschii*, *Escherichia coli*, *Sachharyomyces cereviseae*, *Bacillus subtilis*, *Drosophila*, *Caenorhabditis elegans*, *Arabidopsis thaliana*, *Homo sapiens* and *Oryza sativa*, whose genomes have been completely sequenced. About 200 organisms have been completely sequences so far.

genome size (C-value): The size of a single haploid genome. It is sometimes expressed as the mass of a single haploid genome generally expressed as its total number of base pairs or measured in picograms (1 pg = 10^{-12} g). One picogram corresponds to approximately 1000 Mb of DNA.

genome size of some model organisms: The number of nucleotides in a complete haploid complement of genetic information is called genome size. The genome size varies with the species. For example, the genome size in Mb of rice *Oryza sativa* L. is 430.0, *Arabidopsis thaliana* is 130.0, *Escherichia coli* is 4.8, *Drosophila melanogaster* is 180.0, humans is 320.0, Mycoplasma is 0.58, yeast is 12.1 and bread wheat is 17,000.

genome stability: A feature of every organism to preserve and faithfully transmit the genetic material from generation to generation or from one somatic cell to another. One of the important functions of RNA-dependent DNA polymerase is to hit transposons to protect the genome from their harmful effects. Transposable elements may possess some structural features that render them more distinguishable from other sequences and hence more susceptible to the RdDM. Several lines of evidence have pointed out that transposable elements could possess long untranslated regions or specific secondary structures that make them more recognizable to the methylation machinery. Transposons are methylated at the ends as well as in the middle of their sequence.

genome structure maintenance: Telomeres and centromeres possess unique epigenetic structures that facilitate chromosome integrity and organization of replication and recombination. Epigenetic modifications also facilitate DNA damage repair.

genome survey sequences (GSS): DDBJ/EMBL/Gen Bank division dealing with sequences which are genomic in origin rather than cDNA. GSS division deals with the following types of data: random 'single-pass read' genome survey sequences; single-pass read from cosmid/BAC/YAC ends, which could be chromosome specific – but need not be – exon-trapped genomic sequences.

genome symbol: The description of a specific genome by a symbol, usually a capital letter with or without a specification gene. For example, hexaploid bread wheat *Triticum aestivum* (AABBDD) evolved from three wheat species: diploid wheat *Triticum monococcum* (AA genome), diploid wheat *Aegilops speltoides* (BB genome) and diploid wheat *Aegilops squarrosa* (DD genome). Here, A, B and D are three genome symbols.

genome type: A set of genomes that confer a specific genetic risk for a given disease.

genome visualization tools: (i) CGView Server is a comparative genomics tool for circular genomes that lets sequence feature information to be envisioned in the context of sequence analysis results. (ii) GenomeVx makes editable, publication-quality maps of mitochondrial and chloroplast genomes and of large plasmids. These maps reveal the location of genes and chromosomal features as well as a position scale. (iii) DNAPlotter is used for making circular and linear representations of genomes. (iv) GeneWiz makes linear or circular genome altases. They have ready name ones for most bacteria, but by uploading custom data in GenBank format (.gbk), one can make one's own diagram showing the genetic and physical properties of your genome. (v) SLEP is used for predicting the localization of bacterial proteins starting from genome sequences. It combines the results of several tools.

genome-assisted breeding (GAB): GAB can help the breeders select suitable parents for different crossing programs so as to have novel combinations leading to elite breeding lines. GAB is applied through MAS as well as genomic selection (GS).

genome-associated prediction integrated tool (GAPIT): An R package that performs a GENOME-WIDE ASSOCIATION STUDY (GWAS) and genome prediction (or selection). GAPIT implements advanced statistical methods including the COMPRESSED MIXED LINEAR MODEL (CMLM) and CMLM-based genomic prediction and selection. The GAPIT package can handle large data sets in excess of 10,000 individuals and 1 million single-nucleotide polymorphisms with minimal computational time, while providing user-friendly access and concise tables and graphs to interpret results.

genome-editing enzymes: *See* ARTIFICIAL RESTRICTION ENZYMES.

genome-editing mutagenesis: New allelic variation can be produced by targeted or random mutagenesis. Seemingly, random mutagenesis is outdated because clustered regularly interspaced short palindromic repeats (CRISPR) – Cas technology is much more precise and potentially faster. Various genome-editing approaches used in insect pest control include (i) Using gene-editing Tephritid flies, (ii) Recreating mutant alleles in *Drosophila*, (iii) CRISPaint homology-independent knockin method, (iv) genome-editing mutagenesis in Queensland fruit fly and (v) Sex reversal by a gene-edited mutation.

genome-editing mutagenesis in Queensland fruit fly: CRISPR/Cas mediated mutagenesis in *Bactrocera tryoni* generated a series of frameshift mutations in the ATP-dependent binding cassette transporter, *white*, causing a classic white-eye phenotype. This work has established methods for CRISPR/Cas genome editing in Tephritid fruit flies and demonstrated its potential for developing genetic sexing strains which could be used for SIT-based pest-control methods.

genome-editing principles, technologies using: GENE DRIVE and GENE-EDITING MUTAGENESIS are two methodologies developed to manage pests of diverse crops.

genome-editing-based methods: Genome-editing approaches rely on the induction of a site-specific double-strand break (DSB) in DNA. DSB switches on the natural DNA maintenance mechanism of the cell to repair a DNA strand. It is during the repair of the DSB that a site-specific change in DNA can be made. DSB repair pathways used in genome editing are HOMOLOGY-DIRECTED/DRIVEN REPAIR (HDR) and NON-HOMOLOGOUS END JOINING (NHEJ). They can generate a range of genetic modifications (such as insertions, deletions and replacements). CRISPR/Cas technologies have been applied to the fruit fly, *Drosophila melanogaster*, investigating into the details of established and cutting-edge methods and their potential applications to future studies in this and other insect models. Genome editing has also been applied to several plant species.

genome-wide association mapping (GWAM): GWAM technique recognizes QTLs by studying the marker-trait associations which could be credited to the power of linkage disequilibrium between different molecular markers and functional polymorphisms through a pool of diversified germplasm. GWAM can be used to identify the genomic regions related to yield- and stress-related traits under multi-location stress conditions.

genome-wide association study (GWAS): A method that maps DNA polymorphisms in affected individuals and controls matched for age, sex and ethnic background with the aim of identifying underlying genetic or epigenetic variants. GWAS is utilized to recognize genomic variants that are statistically linked with a risk for a disease or a particular trait. The method includes surveying the genomes of many people, looking for genomic variants that occur more frequently in those with a specific disease or trait compared to those without the disease or trait. Once such genomic variants are identified, they are then typically employed to search for nearby variants that contribute directly to the disease or trait.

genome-wide DNA methylation profiling: Assessing the methylation status for a large number of genes or the entire genome to facilitate the understanding of the nature of gene regulation in cells and the epigenetic mechanism of interactions between cells and the environment. Genome-wide DNA methylation profiling methods include (i) METHYLATION-SENSITIVE RESTRICTION-BASED DNA MICROARRAY, (ii) ALLELE-SPECIFIC METHYLATION ON MICROARRAYS, (iii) METHYLIGHT ASSAY DESIGN and (iv) METHYLATION TARGET ARRAYS. *See details as separate terms.*

genome-wide high-resolution DNA methylation mapping: Two methods are available for this purpose: *In situ* method for DNA methylation analysis and Cytosine extension assay.

genome-wide histone modification profiling: Various methods available for genome-wide profiling of histone modifications include the following assays: (i) ChIP assays, (ii) mass spectrometry such as top-down proteomics, (iii) cytometry by time-of-flight, (iv) histone modifications and protein-DNA-binding analyses, (v) histone acetylation profiling, (vi) histone methylation profiling, (vii) histone phosphorylation profiling, (viii) histone SUMOylation profiling and (ix) Spatial-CUT&Tag method.

genomic: An adjective, usually applied to purified DNA from a particular source. Pertaining to a genome.

genomic allopolyploid: Polyploid species with two or more homologous genomes.

genomic analysis: Once we have sequenced entire genomes, the next step in genomics is genomic analysis. Various tools (computer software) are used for different types of genomic analyses, including alignments, comparing multiple sequences, DNA motifs, homology searches, comparison of homology between two small genomes, designing PCR primers, short-interfering RNA (siRNA) design, real-time PCR primer design, phylogeny, promoters and terminators, repeats and secondary structure, restriction endonuclease digestion, translation: DNA to protein, eukaryotic genes, transcription factor–binding sites and RNA analysis (including structural RNAs, RNA folding, pseudoknots, ribosomal RNAs and siRNAs).

genomic clone: The term genome is taken to mean the entire complement of genetic information as it is found in the chromosomes. A genomic clone is the one containing a fragment of chromosomal DNA rather than a copy of mRNA (i.e. cDNA).

genomic conflict: *See* CONFLICT.

genomic disease model: This model postulates that at times an isolated population loses or fails to acquire a transposable element that otherwise has spread throughout the population, and as a consequence, these isolated individuals on mating with the population lead to abnormalities that end up with sterility in F_1 and F_2 progenies, thereby establishing postzygotic reproductive isolation. To avoid wastage of gametes, natural selection might then act to establish some prezygotic barriers. There is enough evidence in support of this model, especially in *Drosophila*.

genomic DNA: DNA sequence typically obtained from mammalian or other higher-order species, which includes exon sequence (coding sequence) and intron and other non-coding regulatory sequences such as promoter and enhancer sequences. In case of plants, the genomic DNA can be easily isolated following standard methods like the cetyltrimethylammonium bromide (C-TAB) method. The quantity and purity can be checked by using UV-visible spectrophotometer. Pure DNA is then utilized for genomic studies using PCR and Real-Time PCR. Often used in contrast to cDNA (DNA prepared by reverse transcription from messenger RNA). Genomic DNA clones represent DNA cloned directly from chromosomal DNA, and a collection of such clones from a given genome is a genomic DNA library. *Contrasts* MITOCHONDRIAL DNA and CHLOROPLAST DNA.

genomic evolution, mechanism of: A major mechanism that has shaped the genome of many organisms is gene duplications. As the number of completely sequenced genomes increases, the impact of gene duplications on their structure becomes obvious. The extent of gene duplications in a variety of organisms, the mechanisms that create gene duplicates as well as the dynamics that lead to their preservation within genomes need to be understood. A major duplication event affecting the entire genome of the early vertebrates is one of the most discussed hypotheses.

genomic footprinting: A technique used to investigate areas of DNA-protein interaction. Cells in suspension are treated with dimethylsulphate (DMS) or UV light to modify specific DNA bases that undergo a quantitative cleavage reaction. The modified sites are identified by genomic sequences. Proteins bound to DNA protect these areas from modification by DMS as subsequent chemical cleavage, leaving their 'footprint' – the DNA. In contrast, UV light cleaves at sites of DNA-protein interaction.

genomic imprinting: A phenomenon in which the phenotype of a heterozygote depends on which allele came from the father and which from the mother.

genomic *in situ* hybridization (GISH): An *in situ* hybridization technique that uses the total genomic DNA of a given species as a probe and the total genomic DNA of another species as a blocking DNA. It is based on fluorescence *in situ* hybridization. It is a useful method to detect interspecific or intergeneric genome differentiation, chromosome rearrangements (translocations) and substitutions or additions.

genomic instability: Caused by DNA damage, aberrant DNA replication or uncoordinated cell division, which can lead to chromosomal aberrations and gene mutations.

genomic islands tools: (i) MobilomeFINDER is used for *in silico* and experimental discovery of bacterial genomic islands. (ii) Prophage Finder predicts potential prophage loci in prokaryotic genome sequences. However, it does not make any predictions about whether the identified prophage is functional. Moreover, the identified prophage region will most likely not represent the entire prophage. (iii) Phage_Finder identifies prophage regions in completed bacterial genomes. (iv) Prophinder is a similar program.

genomic library: A population of host bacteria, each of which contains a DNA molecule that was inserted into a cloning vector such that the collection of cloned DNA molecules represented the entire genome of the source organism. Once a genomic library is produced, a researcher can work with it in a number of different ways. The genomic library can be stored (frozen) for future reference. As long as it is kept in stable conditions, it can last for an extremely long time. *Aka* GENE LIBRARIES *or* GENE BANKS. *See* LIBRARY *and* ARRAYED LIBRARY.

genomic library construction: Desired DNA molecules of an organism of interest are isolated. DNA molecules are then partially digested by an endonuclease restriction enzyme. The digested DNA molecules are separated by size using agarose electrophoresis. A suitable range of lengths of DNA pieces are isolated and ligated into vectors using DNA ligase. This creates a large pool of recombinant molecules.

genomic library, applications of: (i) Determining the complete genome sequence of a given organism (*see* GENOME PROJECT). (ii) Serving as a source of genomic sequence for the generation of transgenic animals or plants through genetic engineering. (iii) Study of the function of regulatory sequences *in vitro.* (iv) Study of genetic mutations in cancer tissues. (v) Exploring the genome of an organism to learn more about genomic structure and function. (vi) Mapping the genome, identifying the locations of specific genes. (vii) Developing the tests which can be used to locate genetic variations including mutations which may be related to congenital conditions.

genomic library, construction of: The DNA molecules of an organism of interest are isolated. The DNA molecules are then partially digested by an endonuclease restriction enzyme. Sometimes, the DNA molecules are digested for different lengths of time in order to ensure that all the DNA has been digested to manageable sizes. The digested DNA molecules are separated by size using agarose electrophoresis, and a suitable range of lengths of DNA pieces are isolated and ligated into vectors. The vectors can then be taken up by suitable hosts. The hosts are kept in liquid media and can be frozen at −80°C for a long period of time. Usually, the hosts are bacteria that do not contain any plasmids, so as to be sensitive to antibiotics. A computer-based collection of genetic information from these biomolecules can be a 'virtual genomic library'.

genomic library, creating a: The process of subdividing genomic DNA into clonable elements and inserting them into hosts is called creating a GENOMIC LIBRARY, A CLONE BANK or a GENE BANK.

genomic medicine: An emerging medical discipline that encompasses using genomic information about an individual as part of their clinical care (e.g. for diagnostic or therapeutic decision making) and the health outcomes and policy implications of that clinical use. Already, genomic medicine is making an impact in the fields of oncology, pharmacology, rare and undiagnosed diseases and infectious diseases.

genomic non-equivalence: The discovery of the phenomenon of genomic imprinting in mammals showed that the parental genomes are functionally non-equivalent.

genomic selection (GS): A method in which the number of polymorphic bands resembling the recurrent parent is used as the selection criterion. It is a form of MAS in which genetic markers covering the whole genome are used so that all QTL are in linkage disequilibrium with at least one marker. This approach has become feasible due to the large number of SNPs discovered by genome sequencing and new methods to efficiently genotype a large number of SNPs.

genomic shock: Release of genome-wide chromatin constraints of gene expression including activation of transposons in response to genomic, environmental and physiological changes.

genomic technology: Comprises genomics (DNA synthesis, sequencing, genotyping and expression profiling) and proteomics (peptide synthesis, protein sequencing and mass spectrometry) approaches. More recently, it also embraces applications of nanobiotechnology, isolation, imaging and characterization of single molecules.

genomic tools: Genomic tools include genomic database, microorganisms, proteomics, etc.

genomic variation: Refers to DNA sequence differences among individuals or populations. Some variants impact biological functions (such as a mutation that causes a genetic disease), while others have no biological influences.

genomically extinct: The 'genetic swamping' of the Scottish wildcat is so extreme that the most endangered mammalian carnivore in Britain is now 'GENOMICALLY EXTINCT'. The species' only hope may lie in a captive population of Scottish wildcats, which researchers have begun releasing into the wild, far from domestic cats. The team hopes that as the animals adapt to their environment, they will begin to shed their domestic DNA.

genomically recoded organism (GRO): An organism in which codons have been reassigned to create an alternate genetic code. For example, an *Escherichia coli* strain in which all instances of the TAG stop codon have been mutated to TAA and release factor 1 (RF1) has been deleted, enabling the use of TAG as an open coding channel. Three steps used in production of GROs are (i) identification of suitable loci within essential genes, (ii) site-specific incorporation of TAG codons into essential genes and (iii) isolation of synthetic auxotrophs. *See details as separate terms.*

genomics and marker technology: This sequence-based polymorphic (SBP) marker technology allows the development of molecular markers for targeted genomic regions of *Arabidopsis*. It should facilitate the isolation of codominant molecular markers for targeted genomic regions of any animal or plant species, whose genomic sequences have been assembled.

genomics: The term genomics has been derived from the word 'genome' and was coined in 1986 by Thomas H. Roderick of the Jackson Laboratory, Bar Harbor, Maine. It includes structural genomics and functional genomics. The mapping, sequencing, tagging and molecular characterization of genomes is called structural genomics, whereas various studies relating to the gene function (expression analysis, revalidation of transgene(s), etc.) are included in functional genomics. Genomics also examines the molecular mechanisms and the interplay of genetic and environmental factors in disease. The whole-genome sequences are often analysed through the analysis of ORFs. On the basis of such a study, genomic sequences are first classified into coding and non-coding sequences, so that coding sequences may be used for identification of genes, albeit without known functions. Analysis of genomic sequences does not allow identification of those non-coding sequences that are transcribed, but not translated, thus being involved only in the regulation of gene expression at the post-transcriptional level. Genomics studies have extensive reliance on powerful computer technologies. Evolution and mapping of genomes are also included in genomics.

genomics and proteomics, simultaneous studies in: Genomics and proteomics studies simultaneously investigate large numbers of genes or proteins in one single experiment. Autoimmune diseases are characterized by a coordinated attack of the immune system against molecules of the human body, causing chronic inflammation and leading to the destruction of tissues. While the causes of autoimmune diseases are unknown, many genes and proteins are involved in this 'attack against self'. Thus, genomic and proteomic studies are of particular interest to researchers in the field of autoimmunity.

genomics approach toward pest management: Parasitic wasps attack and kill pest insects, keeping pest numbers low. Parasitoid wasps are useful natural enemies that find and kill only specific kinds of insects. If their full potential is harnessed, they would be much preferred over chemical pesticides. The Nasonia Genome Working Group (2010) sequenced genomes of three parasitoid wasp species (*Nasonia vitripennis*, *Nasonia giraulti* and *Nasonia longicornis*) to identify the genes that determine which insects a parasitoid will attack, the dietary needs of parasitoids to assist in economical, large-scale rearing of parasitoids and parasitoid venoms that could be used in pest-control programmes. In addition to being advantageous for controlling pests and offering promising venoms, the wasps could function as a new genetic system for this purpose. Male *Nasonia* have only one set of chromosomes (haploid), instead of two sets (diploids). A single set of chromosomes is a simple and handy genetic tool, particularly for studying how genes interact with each other.

genomics approaches for improvement of abiotic stress: A number of genes have been identified, isolated, cloned and expressed in plants, which are potential sources of resistance to abiotic stresses. These genes include *Rab* (responsive to abscisic acid) and *SalT* (induced in response to salt stress) genes of rice, genes for enzymes involved in proline biosynthesis in bacteria (*proBA*, *proC* in *Escherichia coli*).

genomics approaches for improvement of biotic stresses: Such approaches include (i) insect resistance transgenic plants, (ii) viral resistance and (iii) resistance to bacterial and fungal genes.

genomics approaches for improvement of quality characteristics in crop plants: With the help of genetic engineering such varieties of crop plants can be developed that are suited to specific consumer, industrial, etc. needs. Such varieties are popularly termed as 'designer crops'. Designer crops may have a specific protein, fat or starch quality; nutritional features like vitamin content, elimination of an antinutritional factor, colour, flavour, taste, etc. of fruit/grain; keeping quality, may produce a specific novel biochemical of pharmaceutical/industrial value or may have some other features that make them more useful to mankind.

genomics in agriculture: Applications of genomics in agriculture include food security, food processing and animal husbandry. Control and agricultural management of infection and plant-pathogen interaction through the application of genome sequencing and analysis. Maximizing soil fertility is an important area for application of genomics. Organisms present at the root-soil interface mediate the availability and uptake of soil nutrients; metagenomic investigations can identify those organisms. Crops producing high yields from low inputs must be utilized to make bioenergy economically viable. Interpretation of the genomes of energy crops will enable breeders to produce new varieties to exploit this valuable market.

genomics in DNA fingerprinting: (i) Repeated DNA sequences in the human genome are sufficiently variable among individuals that they can be used in human identity testing. The FBI uses a set of 13 short tandem repeat (STR) DNA sequences for the Combined DNA Index System (CODIS) database, which contains the DNA fingerprint or profile of convicted criminals. Investigators of a crime scene can use this information in an attempt to match the DNA profile of an unknown sample to a convicted criminal. (ii) DNA fingerprinting can also identify victims of crime or catastrophes, as well as many family relationships, such as paternity. (iii) While we think of forensics in terms of identifying people, it can also be used to match donors and recipients for organ transplants, identify species, establish pedigree or even detect organisms in water or food. (iv) In Buenos Aires in the 1970s and 1980s, children of activists 'disappeared' during the military dictatorship. The children were placed in orphanages or illegally adopted when their parents were killed. Now Mary-Claire King is using mitochondrial DNA, which is inherited only maternally, to reunite the children with their grandmothers.

genomics tools: (i) STRUCTURAL GENOMICS discovers what particular gene sequences are present and where they are located within an organism's DNA. (ii) GENE FUNCTION ANALYSIS is used to determine which protein is expressed (i.e. caused to be 'manufactured') by each gene in an organism's genome. Typically, gene function analysis follows after a structural genomics study. (iii) FUNCTIONAL GENOMICS is the study or discovery of what traits/functions are conferred to an organism by a given gene sequence(s). (iv) CHEMICAL GENETICS is used to compare two same-species organisms (one of which has a given gene, or genes, inactivated by a specific chemical or site mutation). (e) GENE EXPRESSION ANALYSIS determines the product(s) resulting (such as an enzyme or other critical protein) when a given gene is 'switched on', by measuring the fluorescence of individual messenger RNA (mRNA) molecules (specific to which particular gene is 'switched on' at the time), when that mRNA hybridizes (with DNA pieces corresponding to proteins produced/analysed, that were attached to hybridization surface on biochip).

genomics, applications of: (i) DNA FINGERPRINTING, (ii) MEDICINE and (iii) AGRICULTURE. *See details as separate terms.*

genomics, cascade of: Research in genetics works in a cascade. The genome project reads the sequence of a gene and annotates it. Structural genomics checks out the shape of the protein the gene dictates and proteomics checks out the functional relevance of the protein and how different proteins work together. This is precisely what is termed proteomics. There is no strict linear relationship between gene activity and protein abundance in a cell. The existence of an OPEN READING FRAME (ORF) in genomic data does not necessarily imply the existence of a functional gene. Modifications of the proteins that are not apparent from the DNA sequences, such as isoforms and post-translational modifications, can be determined only by proteomic methodologies. The localization of gene products which is often difficult to predict from the sequences can only be determined through proteomics.

genomics, fields that interface with: The fields that are included within the field of genomics include bioinformatics, epigenetics, epigenomics, metabolomics, proteomics and transcriptomics.

genomics, various branches of: (i) FUNCTIONAL GENOMICS, (ii) STRUCTURAL GENOMICS, (iii) COMPARATIVE GENOMICS, (iv) EPIGENOMICS (EPIGENETICS) and (v) MICROBIAL GENOMICS. *See details as separate terms.*

genomics-assisted breeding (GAB): Mapping of some important traits, including resistance to *Fusarium* wilt and sterility mosaic disease, fertility restoration and determinacy with other agronomically important traits, have paved the way for applying GAB through MAS as well as genomic selection (GS). GAB can help the breeders select suitable parents for different crossing programmes so as to have novel combinations leading to elite breeding lines.

genophore: The chromosome (genetic material) of prokaryotes and viruses.

genosensors, automated: Automated genosensors allow rapid and non-destructive sequencing of DNA molecules. *Aka* GENOSENSORS.

genosome: The physical part of a chromosome at the point where a gene is located.

genospecies: A group, all members of which are genotypically identical.

genotoxic carcinogens: Compounds that act directly on the genetic material (i.e. DNA) of an organism, thus causing cancer in that organism. Of the numerous chemicals that have been documented to be human carcinogens, the majority of them are genotoxic.

genotoxic: Refers to substances and circumstances inducing mutants and damage of the heritable material. For example, the antitumour antibiotic family of duocarmycin drugs.

genotype by environment interaction: Refers to the interplay of the effects of genetic and non-genetic factors on development. Besides allelic and non-allelic interactions, this is another type of interaction between genes and their environment. Genes determine the potential, and the environment determines whether that potential is to be reached or not. The extent and nature of this interaction varies with each genotype and environment. The variance due to this interaction (V_I) is part of the total phenotypic variance of a trait. Several statistical methods (e.g. analysis of variance, the Azzalini/Cox test, the Hildebrand procedure, the Kubinger approach and the De Kroon/Van der Laan technique) can be applied for the analysis of genotype by environment interactions in cross-classified data sets from cultivar performance yield trials with rows = cultivars and columns = environments (locations and/or years). The procedures Hildebrand/Kubinger and De Kroon/Van Der Laan are nonparametric methods based on ranks, while analysis of variance and the Azzalini/Cox test proceed from the original absolute yield data. The Azzalini/Cox and De Kroon/Van Der Laan methods are based on the crossover concept of interaction (different rank orders), while the other methods are based on the usual concept of interaction (deviations from additivity of main effects). For an analysis of usual interactions, the procedures Hildebrand/Kubinger and the analysis of variance are approximately equivalent. For the crossover concept of interaction, the Azzalini/Cox approach is recommended, especially if one is particularly interested in rank changes between environments within genotypes. *Aka* GENOTYPE × ENVIRONMENT INTERACTION.

genotype by yield*trait (GYT) approach: A novel approach for genotype selection based on multiple traits where 'trait' can be any breeding objective other than yield; it may be an agronomic trait, a grain quality, processing quality, nutritional quality trait or disease resistance. This approach tackles the problem of genotype evaluation on multiple traits. It is based on the following two concepts. (i) Yield is the most important trait and all other target traits are important only when combined with high yield. (ii) The superiority of a genotype should be judged by its levels in combining yield with other target traits, rather than by its levels in individual traits. In this approach, the GENOTYPE BY TRAIT (GT) two-way table from a variety trial(s) is first transformed into a GYT two-way table, in which each column is the combination of yield and a trait. The GYT table is then displayed in a GYT biplot.

genotype frequency: The proportion of individuals in the population that carry a given genotype at the locus or loci of interest.

genotype-environment correlation: The non-random assortment of particular genotypes among particular environments within a given population.

genotype: Concept given by W. Johannsen in 1911. (i) The total DNA content of a cell's nucleus, irrespective of whether a trait is externally observable or not. (ii) The complete genetic make-up of an individual organism that is determined by the specific variants (i.e. alleles) of each gene carried by the individual. Differences in alleles among individuals account for the differences in phenotype observed among those individuals. Genotype is relatively stable throughout the life of an organism except when there occurs a somatic mutation. (iii) The sum total of heredity (i.e. the genetic constitution an organism receives from its parents). The genotype represents potential properties but not the properties themselves. In molecular terms, the genotype is the collection of genes – the entire DNA of the organism. *Compare with* PHENOTYPE.

genotype-environmental interaction: The interplay of a specific genotype and a specific environment, affecting the phenotype. The extent and nature of this interaction varies with each genotype and environment. The variance due to this interaction (V_I) is part of the total phenotypic variance of a trait.

genotype-phenotype correlation: The association of a DNA variant (genotype) with a particular disease or trait (phenotype).

genotype-phenotype relationship: G. Beadle and B. Ephrussi (1937) performed eye disc transplantation experiments in *Drosophila melanogaster*, which provided a strong link between phenotype (eye colour) and genotype (*v* and *cn*). The phenotype of a trait is governed by a genotype. The process of development interprets a genotype for its phenotypic manifestation. Gene controls the synthesis of a protein, which eventually determines a phenotype. There is a big gap between a genotype and the phenotype that it determines. Epigenetics seems to fill in the gap between genotype and phenotype.

genotypic: Phenomenon and processes that are linked with the genotype.

genotypic control: The control of chromosome behaviour by the genotype.

genotypic environment: The aggregate of all the genes considered acting on one or more of them.

genotypic ratio: Ratio of different genotypes in a segregating generation (e.g. genotypic ratio of F_2 of a monohybrid cross is 1*AA* : 2*Aa* : 1*aa*).

genotypic value: The average trait value G of individuals with a particular genotype.

genotypic variance (V_G): The variance of the genotypic value. *See* COMPONENTS OF VARIATION.

genotypic variation: When two or more individuals develop in the same environment and come to possess different phenotypes, these individuals have different genotypes. Genotypic variation arises due to mutations in the germplasm.

genotyping: Molecular characterization of the individuals in a population by using protein and DNA markers to assess their similarities. Variations in the DNA sequence such as copy number variations, insertions, deletions and SNPs are referred to as the genotype of the sample and can be detected using comparative genome hybridization, whole-genome sequencing and exome sequencing.

genotyping by sequencing (GBS): This genotyping technique includes (i) amplicon-based targeted sequencing, (ii) hybridization-based enrichment sequencing and (iii) restriction enzyme-reduced representation sequence-based genotyping. It is a genetic screening method for discovering novel plant and animal SNPs and performing genotyping studies. GBS identifies variants other than SNPs, including small insertions, deletions and microsatellites. It also enables comparative analyses across samples in the absence of a reference genome. *Aka* NEXT-GENERATION GENOTYPING.

Genquest: Program for sequence assembly, analysis and comparison.

GENSCAN: Prediction program for coding regions and exon/intron splice sites.

genuine pleiotropy: When the same gene does different things in different tissues, which it produces directly and by the use of different mechanisms. *Compare with* SPURIOUS PLEIOTROPISM.

genus (pl. genera): A group of one or more species with landscapes or ancestry (or both) in common. The genus is the principal category of taxa intermediate in rank between family and species in the standard nomenclatural hierarchy. The genus or generic name is the first name in a Latin binomial. (e.g. in *Pisum sativum*, *Pisum* is the name of the genus). *See* SPECIES.

geocarp: A fruit that ripens beneath the ground (e.g. groundnut pod).

geographic barrier: Any type of geographical feature, whether mountain, river or other type of terrain that prevents gene flow between populations.

geographic isolate: A population that is isolated from other populations of a given species by geographical distance or barrier.

geographic isolation: The prevention of gene flow between by geographical barriers. *See* REPRODUCTIVE ISOLATION.

geographic speciation: *See* ALLOPATRIC SPECIATION.

geographical indications (GI): Geographical Indication is an indication or appellation of origin. It is used to identify agricultural, natural or manufactured goods originating in the said area. It originates from a definite territory. It should have a special quality, characteristic or reputation based on the climatic or production characteristics unique to the geographical location.

Geographical Indications Act: In December 1999, the Indian Parliament passed the Geographical Indications of Goods (Registration and Protection) Act, 1999. This act seeks to provide for the registration and protection of Geographical Indications relating to goods in India. This act is administered by the Controller General of Patent and Designs and Trade Marks, who is the Registrar of Geographical Indications.

geographical indications registration infringement: When unauthorized use indicates or suggests that such goods originate in a geographical area other than the true place of origin of such goods in a manner, which misleads the public regarding their geographical origin.

geographical speciation: The formation of new species during a period of geographical isolation. Any type of geographical feature whether mountain, river or other type of terrain that prevents gene flow between populations is known as GEOGRAPHIC BARRIER. *See* GEOGRAPHICAL SUBSPECIES.

geographical subspecies: A distinctive population whose distribution is correlated to a geographical region; one that results from broad climate selective forces.

geographical variation: Population differences correlated to geographical distribution.

geologic time: The timescale used to describe events in the history of Earth.

geometric mean: The average value or mean which signifies the central tendency of the set of numbers by finding the product of their values. An average is defined by the nth root of the product of n values. Basically, we multiply the numbers altogether and take the nth root of the multiplied numbers, where n is the total number of data values. For example: for a given set of two numbers such as 3 and 1, the geometric mean is equal to $\sqrt{(3 \times 1)} = \sqrt{3} = 1.732$.

germ cell: Any precursor cell that gives rise to gametes. For instance, in plants the germ cells are present in the flowers which undergo meiosis to produce male gamete (microspore/pollen) and female gamete (egg cell). In animals, germ cells include sperm and egg cells and their precursors. In sexually reproducing higher organisms, diploids, germ cells are haploid and have only one set of chromosomes, while all somatic cells have two sets of chromosomes. *See* GAMETE.

germ-cell migration: Lipid phosphates can act as signalling molecules to influence cell division, apoptosis and migration. *Wunen* and *wunen2* encode *Drosophila* lipid phosphate phosphohydrolases, integral membrane enzymes that dephosphorylate extracellular lipid phosphates. Genes *wun* and *wun2* act redundantly in somatic tissues to repel migrating germ cells. Gene *wun2 also* functions in germ cells, enabling them to perceive the *wun*/*wun2*-related signals from the soma. Upon *wun2* expression, cultured insect cells dephosphorylate and internalize exogenously supplied lipid phosphate. They have proposed that *Drosophila* germ-cell migration and survival are controlled by competition for hydrolysis of a lipid phosphate between germ cells and soma.

germline: The cell lineage in a multi-tissued eukaryote that – in normal development – gives rise to gametes for passing the genetic material from one generation to the next, which then bridges the gaps between generations, unlike somatic cells in the body of an organism.

germline gene therapy: An experimental process of inserting genes into germ cells or fertilized eggs to cause a genetic change that can be passed on to offspring. May be used to alleviate effects associated with a genetic disease. *See* GENOMICS *and* SOMATIC CELL GENE THERAPY.

germline mutation: A DNA sequence alteration in the egg or sperm which can be passed to subsequent generations. *See* MUTATION.

germ theory: The theory that infectious or contagious diseases are triggered by microorganisms (germs).

germ track: A layer of germline cells which produce gametes.

germ tube: A hypha ensuing from an extension of the spore wall and cytoplasm after germination.

germinal mutations: These mutations transpire in those tissues that eventually form sex cells. Mutations induced in the cells of these tissues may be transmitted to the next generation.

germinal selection: The selection occurring during gametogenesis counter to induced mutations that reduce the spread of mutant cells.

germinal vesicles: The nuclei of amphibian eggs during early oogenesis.

germinate: To begin growth of a seed, spore, sclerotium or other reproductive body.

germination: (i) Of seeds; describing the complex sequence of physiological and structural alterations that occur from resting to growth stage. Germination of seeds follows the uptake of water. This is referred to as IMBIBITION. In some cases, a cold period or high heat is also required for germination. For germination, starch is metabolized by amylase. Transcription of the *amylase* gene is stimulated by GIBBERELLIC ACID (GA). The initial stages in the growth of a seed to form a seedling. The embryonic shoot (plumule) and embryonic root (radicle) emerge and grow upward and downward, respectively. Food reserves for germination come from tissue within the seed and/or from the seed leaves (cotyledons). (ii) Of a pollen grain; formation of a pollen tube upon contact with a stigma receptive to it.

germline: A subset of cells in an organism that gives rise to reproductive (sperm or egg) cells. Mutations in the DNA of germline cells are therefore passed on to future generations. Non-germline cells are called SOMATIC CELLS or SOMA.

germline cell: A reproductive cell (egg or sperm cell) that comprises half the set number of chromosomes. These cells contain inherited material that comes from the eggs and sperm and that is passed on to offspring.

germline determination: Refers to irreversible segregation of the germ-cell lineage from somatic lineages. It can take place early or late in development.

germline editing of *Drosophila* by CRISPR-Cas9-based base editors: Germline-editing tools comprising of Target-AID, BE3 and ABE7.10 base editors fused to the catalytically modified Cas9 and xCas9(3.7) were tested for germline editing of the fruit fly *Drosophila melanogaster*. For this purpose, N. Thakkar *et al.* (2023) developed a guide RNA-expressing construct, *white*-4gRNA, targeting splice sites in the *white* gene, an X chromosome–located gene. Using *white*-4gRNA flies and transgenic lines expressing Target-AID, BE3 and ABE7.10 base editors, they tested the efficiency of stable germline gene editing at three different temperatures. Classical Cas9 generating insertions/deletions by NHEJ served as a reference. Gene editing was most efficient at 28°C, the highest temperature suitable for fruit flies.

germline gene therapy: An altered gene is inserted into the sperm or egg cells (germ cells), and this ultimately leads to a change in not only the individual receiving the treatment but also future offspring. It is also possible to insert the altered cells into an early-stage embryo that would affect both the germline and somatic cells.

germline nuclear RNAi: In plants, germline cells arise late in development from somatic stem cells whereas in animals, the germline is specified early in development. Transposons must be silenced extensively throughout development. Generally, chromatin marks that exist during somatic development must be reset in the germline. In somatic cells, both the RdDM pathway and maintenance DNA methyltransferases keep transposons silent; however, such a change in the companion cells of the germline does not contribute genetically to the next generation.

germline sexual development: A process that determines germline sex of an individual.

germline theory of antibody diversity: The genome comprises thousands of related but unique DNA sequences, each of which specifies one of the *V* regions synthesized by an organism. A large percentage of the mammalian genome is devoted to carrying variable sequence families and each person inherits and transmits all *V* regions through his or her germline. Each lymphocyte will selectively express only one of these regions (genes) for the *V* portion of a light chain and only one of the *V* portions of a heavy chain. Because most lymphocytes will be selecting different combinations of sequences to express, the organism will happen to express its vast diversity of antibody-producing cells.

germline-specific genes: Genes of unique gametogenic function that are tightly regulated outside gametes or the early embryo by DNA methylation. They are specifically demethylated during **primordial germ cell (PGC)** specification.

germplasm: (i) The genetic make-up of an organism – the sum total of its genes, both dominant and recessive. (ii) The potential heredity materials within a species, taken collectively. (iii) The total genetic variability to an organism, represented by the total available pool of germ cells or seed. (iv) The term germplasm is frequently slackly used for cultivars and breeding lines.

germplasm collection: A collection of genotypes of a particular species, from different sources and geographic locations, used as source materials in plant breeding. *See* WORLD COLLECTION.

germplasm conservation: There are raw germplasm identified by scientists with one or more outstanding traits, such as resistance to diseases, high protein content, bold grain, etc. but these do not qualify to get notified as a variety because of poor agronomic performance. Such material could be used in plant improvement. Realizing the importance of these materials, the Indian Council of Agricultural Research constituted the Plant Germplasm Registration Committee (PGRC) in 1996 to document information for the germplasm developed/identified with potentially valuable trait(s).

germplasm enhancement: Any activity that includes gene transfer via sexual and asexual means from germplasm accessions or increasing the frequencies of desirable genes in crop gene pools that will be used for developing parents or cultivars.

Germplasm Enhancement for Maize (GEM): A project conducted under the patronage of the US Department of Agriculture, in cooperation with 16 American universities and 20 maize (corn) seed companies. GEM's commitment is to cross exotic (not in current use) germplasm with commercial maize lines in order to improve maize yield.

germplasm resource: The sum total of all allelic resources of a species.

germplasm theory: A. Weismann in 1893 established through his famous experiment on mice that multicellular organisms give rise to two types of tissues – SOMATOPLASM and GERMPLASM. Somatoplasm consists of tissues essential for the functioning of the organism but lacking the property of entering into sexual reproduction. Changes that take place in somatic cells are not transmitted from parents to offspring. Thus, somatic cells ultimately die as changes in somatic tissues are not heritable. The germplasm is set aside for reproductive purposes as the cells of these tissues divide meiotically to form gametes. Any change in the germplasm is transmitted from parents to offspring and could lead to altered inheritance. This work thus rejected Lamarck's theory of inheritance of acquired traits. However, epigenetic inheritance reveals attributes that seem to favour Lamarck's theory. The Lamarckism is being revisited in light of epigenetic inheritance.

gerontoplasts: These plastids control the dismantling of the photosynthetic apparatus during senescence.

gestation: The period in animals bearing live young (especially mammals) from the fertilization of the egg and its implantation into the wall of the uterus until the birth of the young (parturition), during which the young develop in the uterus. In humans, gestation is known as PREGNANCY and takes about 9 months (40 weeks).

GI (in GenBank): A GI (GenInfo identifier) is a sequence identifier that can be allocated to a nucleotide sequence or protein translation. Each GI is a numerical value of one or more digits. The protein translation and the nucleotide sequence confined in the same record are assigned different GI numbers. Every time the sequence data for a particular record is revised, its version number increases, and it receives a new GI. However, although each new version number is based on the previous version number, a new GI for an altered sequence may be completely different from the previous GI. For example, in the GenBank record M12345, the original GI might be 7654321, but after a revision in the sequence is submitted, the new GI for the changed sequence could be 10529376. Individuals can search for nucleotide sequences and protein translations by GI using the UID search field in the NCBI sequence databases. NCBI's Sequence Revision History page can be used to view all the various GI numbers, version numbers or update dates linked with a particular GenBank record.

giant cell: An enlarged, multinucleate cell formed in plant roots by repeated nuclear division without cytoplasm division induced by secretions of certain inactive plant-parasitic nematodes.

giant chromosomes: *See* POLYTENE CHROMOSOMES.

gibberellic acid (GA): A gibberellin (group of growth-promoting chemical substances) that regulates many growth retorts and seems to be a widespread component of seeds and plants. *See* GIBBERELLINS.

gibberellin pathway: Gibberellin (GA), a phytohormone, plays a major role in affecting flowering in the model plant *Arabidopsis thaliana*. The GA pathway interacts with other flowering genetic pathways and phytohormone signalling pathways through either DELLA proteins or mediating GA homeostasis. Gibberellin independently promotes flowering through the activation of *SOC1* and other floral activator genes. Plants integrate multiple internal and external flowering cues through gene expression changes.

gibberellins: Plant growth regulators that, among other functions, regulate the stem growth of grass species, including rice (after the relevant gibberellin is activated by an enzyme). In 1996, L. Mander and R. Pharis discovered an analogue to grass gibberellin that does not cause grass growth. When this analogue is sprayed on to grass, it mixes into the naturally occurring grass gibberellin and significantly slows down grass growth (thus potentially reducing the amount of mowing required for lawns, golf courses, etc.), thus working as a weedicide.

Giemsa stain: *See* BANDING.

gigabase (Gb): A unit of measurement used to designate the length of DNA. One gigabase is equal to 1 billion bases.

gigabyte (GB): A multiple of the unit byte for digital information which is equal to one billion bites.

gigapreparation of plasmid DNA: This method combines time-tested consistency of alkaline-SDS lysis of bacterial cells with innovative high-efficiency DNA-binding technology to recover large-scale, high-quality plasmid DNA. The product is suitable for automated fluorescent DNA sequencing, restriction endonuclease digestion, transfection of mammalian cells and other manipulations.

Gin system of site-specific recombination: Gin is an enzyme that can introduce topological changes in circular DNA molecules. Gin and its bound DNA are modelled as a 2-string tangle which undergoes changes during recombination, thereby changing the topology of the DNA substrate. The Gin system is one of the few cases where tangle analysis can be carried out systematically and rigorously, yielding a single, biologically reasonable solution.

Gini's mean difference: A measure of variability (dispersion) that is much more interpretable than the standard deviation and more robust to outliers and also applies to non-symmetric distributions. It is the mean absolute difference between all possible pairs of observations. There is a fast-computing formula for the index, and the index is highly statistically efficient.

girdle: Act to circle and cut through a stem or the bark and outer few rings of wood, thus disrupting the phloem and xylem.

GitHub: GitHub is an internet platform that hosts open-source software development projects.

gland: A secretory structure located within or on the surface of a plant.

glandular hair: A hair with a gland at a tip. They produce essential oils which apparently protect against herbivores and pathogens.

glass beads method: In this method, chloroplast transformants can be recovered by agitation of cell wall deficient cells in the presence of glass beads and DNA. The shear force generated during agitation damages the cells transiently, allowing the uptake of DNA before the damage is repaired. To get chloroplast transformants by this method, we need a mutant that has deletion in one of its genes.

glass particle method of DNA extraction: The adsorption of nucleic acid on to the glass substrate is based on a mechanism and principle similar to adsorption chromatography. Nucleic acid purification can also be done on silica gel and glass mixture. A mixture of silica gel and glass particles can be used to separate nucleic acid from other substances in the presence of a chaotropic salt solution.

glasshouse: *See* GREENHOUSE.

glaucous: Describes the external surface of a plant part that has a whitish covering, in some cases with a blueish cast. Often applied to plants with a woolly or arachnoid surface, but properly referring to pruinose surfaces, meaning those with a waxy bloom. The surface of the young leaves of many eucalypts provides a good example, as does that of some xerophytes.

gliadin: A major component of wheat gluten made up of single-chain polypeptides, with an average molecular weight of 25–100 kDa, linked by intramolecular disulphide bonds. It is essential for giving bread the ability to rise properly during baking.

global alignment: Attempts to match as many characters as possible, from end to end, in a set of two or more nucleic acid or amino acid sequences.

global chromatin environment: Modifications help partition the genome into distinct domains such as euchromatin, where DNA is kept 'accessible' for transcription and heterochromatin, where chromatin is 'inaccessible' for transcription. Grossly speaking, there are two different types of chromatin environments in the genome, silent heterochromatin and active euchromatin. Each of these is associated with a distinct set of modifications. In mammals, demarcation between the different environments is set up by boundary elements, which recruit enzymes to modify the chromatin. The CCCTC-BINDING FACTOR (CTCF) is an example of a boundary element binding protein that delivers the modifying enzymes. Heterochromatin boundaries are maintained by the presence of methylation at H3K4 and H3K9 in adjacent euchromatic regions. Thus, one critical function of chromatin modifications is that they dictate the different chromatin environments and preserve these two types of domains.

global genomic nucleotide excision repair (GG-NER): GG-NER repairs damage in both transcribed and untranscribed DNA strands in active and inactive genes throughout the genome. This pathway employs several 'damage sensing' proteins including the DNA DAMAGE BINDING (DDB) and XPC-HR23B COMPLEXES that constantly scan the genome and recognize helix distortions. Upon identification of a damaged site, subsequent repair proteins are then recruited to the damaged DNA to verify the presence of DNA damage, excise the damaged DNA surrounding the lesion then fill in the repair patch. The GGR process is not dependent on transcription.

global RNA modification mapping strategies: Global RNA modification mapping approaches trust the fact that some RNA modifications can serve as handles for attaching chemical tags. Modified RNA building blocks called PSEUDOURIDINES or PSEUDOUS are well known in other classes of RNA but not in mRNA. Pseudouridylation is the most abundant internal POST-TRANSCRIPTIONAL MODIFICATION (PTCM) of stable RNAs, with fundamental roles in the biogenesis and function of spliceosomal snRNAs and ribosomal RNAs. A chemical labelling and pull-down method for enriching the modification on transcripts. PseudoUs are much more abundant in human and mouse mRNA than was previously thought.

global transcription factors (GTFs): In plants, global transcription factors act as 'master switches' for regulating the expression of certain genes. Modification of the expression of GTFs represents promising new targets for achieving yield improvements in that they control the expression of multiple genes in different biochemical pathways.

global transcription machinery engineering (gTME): An approach for reprogramming gene transcription to elicit cellular phenotypes relevant to technological applications.

global variable: In computer programming, a variable with global scope, meaning that it is visible (hence accessible) throughout the program, unless shadowed. The set of all global variables is known as the GLOBAL ENVIRONMENT or GLOBAL STATE. In compiled languages, global variables are generally static variables, whose extent (lifetime) is the entire runtime of the program, though in interpreted languages (including command-line interpreters), global variables are generally dynamically allocated when declared, since they are not known ahead of time.

global warming: Retention of solar heat by substances/gases in the atmosphere. Global warming is a phenomenon due to which there is an ever increase in the atmospheric temperature by 0.2°C to 0.5°C/year. This is mainly because of the depletion of the ozone layer and the production of gases from the fuels. Global warming is affecting crop yields, as several winter crops are susceptible to heat stress.

globin gene family: In adult humans only single genes, each specifying alpha (α) and beta (β) polypeptide chains of haemoglobin are expressed. This is not sufficient during development, as a whole family of related α globin genes and another family of β globin genes are also required to function. These genes are turned off and on in a coordinated and sequential manner during embryo development till the fetus is born, depending upon the intracellular milieu.

globular protein: Any protein with an approximately rounded shape. Such proteins are in contrast with highly elongated, fibrous proteins such as collagen.

globulins: (i) Proteins whose molecules have a globular shape. The globulins found in blood serum are classified as α, β and γ globulins, in order of decreasing electrophoretic mobility. They play an important role in liver function, blood clotting and fighting infection. (ii) Globulins are a chief class of seed storage proteins in cereals and legumes, which were thought to be enzymatically inactive. These proteins belong to the most ancient cupin superfamily. Based on their sedimentation coefficients, globulins can be graded into 11S legumin type, which may play a role in auxin homeostasis and 7S vicilin type that have a role in sucrose binding, desiccation, defence against microbes and oxidative stress.

globulose: Approximately spherical.

glucocorticoid receptor (GR): A sequence-specific DNA-binding protein required for activation of mammalian genes. It binds at GLUCOCORTICOID RESPONSE ELEMENTS (GREs) to perform its function.

glucocorticoid receptor element (GRE): A mammalian gene sequence at which glucocorticoid receptors bind for activating transcription.

glucocorticoid regulation: A 25-bp element that resides upstream of the mouse *proliferin* gene conferred on linked promoters either positive or negative glucocorticoid regulation, depending upon physiological context. The element called GLUCOCORTICOID RECEPTOR ELEMENT (GRE) was bound selectively *in vitro* both by glucocorticoid receptor and by c-Jun and c-Fos, components of the phorbol ester-activated AP-1 transcription factor. A general model for composite GRE action is proposed that involves both DNA-binding and protein-protein interactions by receptor and nonreceptor factors.

glucocorticoids: Steroid hormones produced from the cortex of adrenal glands. Glucocorticoids have a pivotal role in the glucose, protein and fat metabolism of the body.

glucose ($C_6H_{12}O_6$): Sugar that is the product of photosynthesis. Glucose is six-carbon sugar formed from two 3-Phosphoglyceraldehydes (3-PGALs). 3-PGAL is an isomer of 3-Phosphoglyceric acid (PGA) formed during the Calvin cycle. Glucose is stored in polymeric form as glycogen in animal cells and starch in plant cells.

glucose 3-phosphate (G 3-P): It occurs as a by-product in the biosynthesis pathway of tryptophan, an essential amino acid that cannot be produced by the human body.

glucose as a product of photosynthesis: A glucose ($C_6H_{12}O_6$) molecule is the product of each light-independent reaction that occurs in the stroma of the chloroplast during photosynthesis. Glucose is made by joining two of the three-carbon sugar products of the Calvin cycle into a six-carbon sugar. Glucose can be made into larger sugars (sucrose) or carbohydrates such as starch and cellulose. Sugars exit the leaf through the phloem, travelling to the roots for storage or to other parts of the plant, where they are used as energy to fuel the plant's activities.

glucose oxidase gene: Gene responsible for the production of an enzyme glucose oxidase that converts glucose into hydrogen peroxide (H_2O_2). The cloned glucose oxidase gene is being used in plant genetic engineering to develop plants resistant to fungal diseases. Plants produce glucose through photosynthesis. A glucose oxidase gene when introduced, converts glucose into hydrogen peroxide in the transgenic plant. H_2O_2 is antimicrobial. This approach may help in developing transgenic crop varieties resistant to fungal diseases.

glucose-sensitive operons: Those operons whose functioning is blocked by the presence of glucose. Glucose indirectly lowers the cyclic AMP level thereby blocking a necessary positive control signal.

glucoside: A substance that, on decomposition, produces glucose and certain other compounds. Some glucosides function as defence compounds produced by plants (e.g. cyanogenic or phenolic glucosides).

glutamic acid (glu): A dicarboxylic amino acid of the α-ketoglutaric acid family having chemical formula $HOOC(CH_2)_2CH(NH_2)COOH$. It is involved in purine biosynthesis and is occasionally added to plant tissue-culture media. It may replace ammonium ions as the nitrogen source. It is of vital importance in pollen growth *in vitro*.

glutamine: An amino acid with chemical formula $C_5H_{10}N_2O_3$. The monamide of glutamic acid. Glutamine is of central importance for amino acid biosynthesis in all forms of life.

glutamine synthetase (GS): An enzyme that plays an essential role in the metabolism of nitrogen by catalysing the condensation of glutamate and ammonia to form glutamate.

glutamine-2-oxoglutarate amino-transferase (GOGAT): An enzyme that manufactures glutamate from glutamine and α-ketoglutarate and also plays a central role in the regulation of nitrogen assimilation in photosynthetic eukaryotes and prokaryotes.

glutaraldehyde: Small reactive molecule with two aldehyde groups that is often used as a cross-linking fixative.

glutathione: A tripeptide, containing glutamic acid, cysteine and glycine, capable of being alternately oxidized and reduced. It plays a crucial role in cellular oxidation.

gluten: A naturally occurring mixture of two different proteins – glutenin and gliadin – in the seeds of bread wheat (*Triticum aestivum*). In flour made from traditional varieties of wheat, glutenin proteins constitute approximately 50% of the total gluten. The relative content of these two proteins determines one of the most commercially important characteristics of the wheat (strength and elasticity of the flour made from that particular wheat). For example, more of the high molecular weight glutenin is 'stretchy' and provides physical strength to a dough made from such flour, so that dough holds together while rising. It provides a flour that is better suited to manufacture higher-quality yeast-raised bread products. *See* GLUTENIN *and* GLIADIN.

glutenin: Protein naturally present in the gluten within seeds of wheat (*Triticum aestivum*). Glutenin is responsible for providing the strength and elasticity of dough.

glutinous: Sticky in texture.

***Glyanthus nivalis aglutenin* (*GNA*) gene:** A lectin-producing gene cloned from snowdrop (*Glyanthus* nivalis) plant. GNA means *G. nivalis* aglutenin gene. GNA is a plant gene that produces lectin protein which is an insect control protein. It has been cloned from an ornamental flowering plant commonly called snowdrop. Following the transgenic approach, GNA has been introduced into rice plant, and transgenic rice possessing a good level of resistance against sucking type of insect – White Backed Plant Hopper (WBPH) – has been developed. *Bt* gene works against chewing-type insects (borers), whereas GNA works against juice-sucking insects; moreover, it is a plant-derived gene.

glycerol: Small organic molecule from which are derived many small molecules in the cell, including phospholipids.

glycerol gradient centrifugation: A technique for estimating the sedimentation coefficient (Svedberg unit or S value) of proteins or protein complexes. The sample is centrifuged through a gradient of glycerol, which provides a near-constant rate of movement through the gradient.

glycine (gly): The simplest (and smallest) of the amino acids, with chemical formula $C_2H_5NO_2$, found in proteins. It is the only amino acid that does not possess an asymmetric carbon atom within its molecule. Thus, glycine is not optically active.

glycocalyx: Carbohydrate-rich coating that forms the outer layer of a eukaryotic cell. Composed of the oligosaccharides linked to (i) intrinsic plasma membrane glycoproteins and glycolipids and (ii) glycoproteins and proteoglycans that have been secreted and reabsorbed on to the cell surface. *Aka* CELL COAT.

glycogen: A multi-branched tree-like polysaccharide of glucose that serves as a form of energy storage in humans, animals and fungi. Large granules of glycogen are especially plentiful in liver and muscle cells. Its molecular weight may be as high as several million.

glycolipid: Membrane lipid molecule possessing a sugar residue or oligosaccharide linked to the polar headgroup.

glycolysis: A linear metabolic pathway that breaks glucose into two molecules of pyruvate in the presence of oxygen. Glycolysis produces the energy sources – adenosine 5-triphosphate (ATP), nicotinamide adenine dinucleotide (NADH) – and water. In the absence of oxygen, glucose is anaerobically broken into two molecules of pyruvate which are converted into lactate and synthesizes two molecules of ATP.

glycopeptide: Chains of amino acids with attached carbohydrate (sugar).

glycoprotein: Any protein with one or more oligosaccharide chains covalently linked to amino acid side chains. Most secreted proteins and most proteins exposed on the outer surface of the plasma membrane are glycoproteins.

glycosaminoglycan (GAG): Long, linear, highly charged polysaccharide molecule composed of a repeating pair of sugars, one of which is always an amino sugar. GAG is mainly found covalently linked to a protein core in extracellular matrix proteoglycans. Examples include chondroitin sulphate, hyaluronic acid and heparin.

glycosidic bond: A bond joining a sugar molecule (S) with an aromatic base (B). Such bonds are involved in the formation of nucleosides, a constituent of nucleic acids.

glycosylation: The covalent addition of sugar moieties to N or O atoms present in the side chains of certain amino acids of certain proteins, generally occurring within the Golgi apparatus during secretion of a protein. *See* O-LINKED GLYCOSYLATION *and* N-LINKED GLYCOSYLATION.

glycosylphosphatidylinositol (GPI) anchor: Type of lipid linkage by which some membrane proteins are bound to the membrane. It is formed as the proteins travel through the endoplasmic reticulum.

glyphosate: An herbicide used to kill weeds, such as those that compete with commercial crops. Also known under the trade name *Roundup*®. Farmers favour glyphosate for its ability to control many types of weeds and its low toxicity compared with other herbicides.

GMO unique identifier: Refers to an international GMO identification code. Every genetically modified organism that obtains approval is assigned a unique identifier comprising nine letters and/or numbers; the first two or three characters designate the company submitting the application, while the final five or six characters signify the corresponding transformation event; the last digit indicates a verifier. The unique identifier was established by the Organization for Economic Co-operation and Development (OECD). The OECD is now supervised by the European Union and formulates the basis for the legally authorized system of traceability.

Goldberg-Hogness box: Sequences discovered by the team led by M.L. Goldberg and D.S. Hogness (R.P. Lifton, M.L. Goldberg, R.W. Karp, D.S. Hogness in 1978) in promoters of eukaryotic genes. Sequences similar to Pribnow box. The sequence of the Hogness box is $T_{82}A_{97}T_{93}A_{35}A_{43}/T_{27}A_{83}A_{50}/T_{37}$. Hogness box is surrounded by GC-rich sequence. *Aka* TATA box.

Golden Gate cloning: A molecular cloning method that allows a researcher to assemble multiple DNA fragments simultaneously and directionally into a single piece using Type IIS restriction enzymes and T4 DNA ligase. This assembly is performed *in vitro*. Most commonly used Type IIS enzymes include *Bsa*I, *Bsm*BI and *Bbs*I. *Aka* Golden Gate assembly.

golden rice: Vitamin A deficiency is the single most important cause of blindness among children in developing countries. This deficiency occurs where rice is the staple food, since rice grain does not contain vitamin A (i.e. β-carotene). Rice endosperm does synthesize geranyl-geranyl pyrophosphate, which can be converted into β-carotene by the following four enzyme activities: (i) phytoene synthase (ii) phytoene desaturase, (iii) ζ-karotene and (iv) lycopene cyclase. A biotechnology-derived rice (*Oryza sativa*) created in the 1990s by I. Potrykus and P. Beyer, which contains large amounts of beta-carotene (precursor of vitamin A) in its seeds. These researchers utilized *Agrobacterium tumefaciens* bacteria to genetically engineer the rice plant by inserting *phytoene synthase* (*PSY*) gene from daffodil (narcissus), which codes for phytoene synthase, which converts geranylgeranyl-diphosphate into phytoene, *CRTL* gene from *Erwinia uredovora*, which codes for phytoene desaturase, causes the rice plant to convert phytoene (a 'light-harvesting' carotenoid involved in photosynthesis) into lycopene (a carotenoid which is then utilized by the rice plant in the production of beta-carotene) and *Lycopene beta-cyclase* (*Lcy-b*) gene from daffodil, which synthesizes *lycopene beta-cyclase*, which converts lycopene into beta-carotene. All these transgenes were introduced together in a single cotransformation experiment. The resulting transgenic rice, popularly called 'golden rice' contains good quantities of β-carotene content, which gives rice a golden colour. Iron-deficiency anaemia is the most common nutritional disorder in the world. Rice grain has the lowest iron content among crops; it also has phylate, which reduces iron absorption in the human intestine. Phylate is used for phosphate storage in seeds.

golden tomato: Transgenic tomato with enhanced vitamin A and lycopene. Golden tomato has been developed through transgenic approach, and thus, developed tomato produces more vitamin and lycopene, which improve the nutritional quality of tomato. Golden tomato is being evaluated under containedconditionsforitsstudyingdifferentbiosafety aspects.

Goldschmidt's models: R. Goldschmidt during 1940–1946 gave his own version of evolutionary developmental biology that he called physiological genetics. He presented two models relating gene activity, development and evolutionary dynamics. In the first model, he argued that new species might originate as 'hopeful monsters' that result from mutations in developmentally important loci (developmental macromutations). In the second model, he argued that chromosomal rearrangements (systemic mutations) would have an effect on many developmental macromutations and cause even larger phenotypic changes. His view of systemic mutations did not win much favour, he did provide the influential idea that 'a single mutational step affecting the right process at the right moment can accomplish everything, providing that it is able to set in motion the ever-present potentialities of embryonic regulation'.

Golgi bodies: First described by Camillo Golgi in 1898, these are the primary 'sorting centres' of cells, and the mechanism for glycosylation of (i.e. adding oligosaccharide and polysaccharide branches on to) proteins, before the proteins are transported by transfer vesicles to lysosomes, secretory vesicles or the plasma membrane. In plant cells, Golgi complexes are the sites where complex polysaccharides are 'sorted' and assembled in preparation for making the cell wall (located just outside the cell's plasma membrane). Visually, a Golgi complex is a stack of flattened membranous sacs (usually 6 in mammal cells and 20 in plant cells). Animal cells usually have one or several Golgi bodies, whereas plant cells can have hundreds. Each Golgi body contains stacks of small, flattened, membrane-bound sacs called *cisternae*. Cisternae are arranged in parallel with a distance of 20–30 nm and surrounded by numerous vesicles. This is a feature of almost all eukaryotic cells. This structure is involved in the packaging of many products of cell metabolism. The Golgi body receives proteins, synthesized by ribosomes on the rough endoplasmic reticulum, via transport vesicles. Enzymes in the cisternae modify these proteins and pack them into vesicles that bud off the Golgi body and transport the proteins to different areas in the cell. The Golgi body also processes and distributes lipids and complex carbohydrates. It is responsible for manufacturing, warehousing and shipping certain cellular products. *Aka* GOLGI COMPLEX *and* GOLGI APPARATUS.

gonadal dysgenesis: A condition where the gonads in an individual are not clearly differentiated into a male or female sex. This condition is due to an X-linked recessive allele which does not allow differentiation of gonads to testes in 46 XY embryos.

gones: The four cells, or their nuclei, which are normally the immediate product of meiosis.

good genes: Genes that increase fitness for reasons that are not due to sexual selection on a signal trait. Such components of fitness are most conveniently thought of as viability, but fertility and mating success could also be included.

goodness of fit: Assessment of the agreement of the data with either a hypothesized pattern (e.g. independence of row and column factors in a contingency table or the form of a regression relationship) or a hypothesized distribution (e.g. comparing a histogram with expected frequencies from the normal distribution). Goodness of fit methods are also employed to test the conformity of an observed empirical distribution function of data with a postulated theoretical distribution function (e.g. Chi-square test) by comparing observed and expected frequency counts. The Kolmogorov-Smirnov test calculates the maximum vertical distance between the empirical and postulated distribution functions.

Google for DNA: A device that works like a Google for DNA has established its promise for making all of the world's biological sequence data cheaply and readily searchable, according to the Swiss team that established it. In a proof of principle study, the researchers claim they have magnificently indexed 10% of the world's known DNA, RNA and protein sequences.

governance: A set of values, norms, processes and institutions through which society manages technology development and deployment and resolves conflict formally or informally. Governance includes oversight, which is defined more narrowly as watchful and responsible care or regulatory supervision.

***GPA1* gene:** A gene responsible for controlling water retention and cell division in plants. The *GPA1* gene codes for a G-protein, which sends/controls signals (light, temperature, phytohormones, nutrients, etc.) controlling the plant's development. In 2001, Alan Jones and colleagues reported that silencing the *GPA1* gene caused the (then-resultant) G-protein to be insensitive to ABSCISIC ACID (AA). Because AA is a phytohormone utilized by plants to regulate the size of stomatal pores [i.e. the openings in leaves through which plants exchange oxygen and carbon dioxide (and also water inadvertently) with the atmosphere], the 'knocked-out *GPA1*' plants wilted due to uncontrolled water loss to the atmosphere.

Gq: Class of receptor-coupled G-protein that activates phospholipase C-β and originates the inositol phospholipid signalling pathway.

gradient: A gradual change in some quantitative feature over a specific distance.

gradients in egg: Gradients are established in the egg along two axes, posterior-anterior and dorsal-ventral. The gradients consist of molecules that are differentially distributed in the common cytoplasm. A nucleus can, in principle, detect its position in the cytoplasm by the level of each gradient.

gradual speciation: True speciation which occurs slowly over time in small steps in populations.

gradualism: Changes in species that happen over long periods of time.

graft chimaera: A taxon whose members consist of tissue from two or more different plants in intimate association originated by grafting. The addition sign '+' is used to indicate a graft chimaera either as a part of a formula (e.g. *Crataegus monogyna* + *Mespilus germanica*) or in front of an abbreviated name (e.g. + *Crataegomespilus* 'Dardarii'). The nomenclature of graft hybrids is ruled by the International Code of Nomenclature for Cultivated Plants. *Aka* GRAFT HYBRID.

graft transmission: The transmission of a pathogen from one host plant to another through blend of living tissue from the diseased host with living tissue of a healthy host.

graft(ing): The procedure of plant propagation used to seam together parts of plants by holding cut surfaces in position until a union of living cells is formed. The fused parts will endure their growth as one plant. This technique is frequently used to asexually propagate fruit trees and other woody plants that do not readily root from cuttings. While nearly all of the plants that are commonly grafted can be produced from seed, the inconsistency in the plants produced is so high that it is more practical to clone these plants by grafting. Grafting has been a horticultural art for several thousand years. In the past three centuries, the potential of grafting was realized for using selected rootstocks to affect growth and performance of scions of plants. Grafting is presently in use to create novelty items such as potato plants that produce tomatoes, moon cactus and fruit trees that bear more than one variety. Grafting also is employed to incorporate both maleness and femaleness into the same dioecious plant, such as holly, so that it bears fruit. It is used to accelerate a breeding program, to test for compatibility and to determine if symptoms are virus caused. Four common graft methods are bridge, bud, cleft and whip. The stem or shoot that is inserted into a rooted plant is called the SCION. The plant or part of a plant into which the scion is inserted is called the STOCK or UNDERSTOCK. There are many different methods of grafting, including WOOD GRAFTING, CLEFT GRAFTING, STUB GRAFTING, BARK GRAFTING, AWL GRAFTING, VENEER GRAFTING, BUD GRAFTING, FLAT GRAFTING, SPLIT GRAFTING *and* SIDE GRAFTING. Sometimes, this term refers to the point where a scion is inserted in the stock.

grafting experiment in *Acetabularia*: The important point that this experiment brings forth is that the nucleus controls the synthesis of products at points far removed from its own location. At molecular level these results are interpreted in terms of mRNA being transported from nucleus to cytoplasm. It was observed in *Acetabularia* that simple structure changed into complex structure. There are some morphological and biochemical changes responsible for differentiation. As development proceeds, it assumes certain specific form. This is called MORPHOGENESIS. Message in the stalk is stable because cap can grow even for several days. The message has ability to form a characteristic cap. It was concluded that cytoplasmic factors govern the developmental state of nucleus. When the cap is fully developed, it signals the nucleus at the other pole of the cell to start dividing. So, development results from nucleocytoplasmic interactions. Nucleus synthesized certain messages for controlling the development and the cytoplasm determines as to what message is to be synthesized.

graft-versus-host reaction: A reaction in which foreign lymphocytes (such as those which are passengers in a tissue transplant) mount an immune response against the cells of the host organism.

Grail × GRAIL/Grailcnt: Predicts exons, putative genes, detects promoter regions, poly(A)s, CpG islands, similarities in ESTs and repetitive segments.

grains: The seeds of cereal grasses, such as wheat, maize, oats, barley, rye and rice. Grain foods include bread, cereals, rice and pasta.

Gram stain: A differential stain for bacterial classification. Gram stain is commonly used stain that was developed by Danish Bacteriologist Hans Christian Gram. This staining procedure classifies bacteria into two groups: Gram-negative and Gram-positive. Following staining, bacteria which retain violet or purple colour after decolourization with alcohol are called Gram-positive, whereas those that lose the colour are classified as Gram-negative.

Gram staining: A staining method devised by H.C. Gram in 1884 to illuminate the composition of the physical structure of the cell wall of bacteria being tested. It is employed to estimate the effectiveness of a given chemical compound (e.g. an antibiotic) against bacterium types. It is used to differentiate two large groups of bacteria based on the chemical and physical properties of their cell walls. This method involves staining the cells with crystal violet dye and then adding iodine and potassium iodine. GRAM-POSITIVE BACTERIA show up as purple while GRAM-NEGATIVE BACTERIA show up as pink. An antibiotic is classified in terms of the group of (pathogenic) bacteria that it is effective against, which is known as an antibiotic's 'spectrum of activity'. An antibiotic is said to have a spectrum of activity against Gram-positive bacteria, Gram-negative bacteria or the bacteria of both the groups. An antibiotic that is effective against both groups of bacteria is termed 'BROAD SPECTRUM' or 'WIDE SPECTRUM'.

Gramicidin S: It was first isolated from the soil bacterium *Bacillus brevis* by the French-born American microbiologist René Dubos in 1939 with the help of Rollin Hotchkiss. The antibiotic inhibits the growth of GRAM-POSITIVE BACTERIA. An antibiotic that was discovered in 1942 by the Russian microbiologist G.F. Gause and his wife M. Brazhnikova. The antibiotic is effective against some GRAM-POSITIVE and GRAM-NEGATIVE bacteria and some fungi.

Gram-negative (G^-) bacteria: It is also variously abbreviated as GN, GrN, GM^- or gr^-. Pertaining to one of the most important ways of classifying bacteria by means of the differences in the way they stain. The set of bacteria that are not able to be stained (blue) when treated with the Gram staining procedure. Gram negativity (and Gram positivity) is conferred not by the chemical constituents of the bacteria, but rather by the physical structure of the bacteria cell wall. The staining procedure involves the staining of all cells in a sample with a blue dye. Gram-negative bacteria have a very thin peptidoglycan cell wall (capsule). Hence, the washing procedure, which is an integral part of the overall staining procedure, washes out the blue dye (known as crystal violet). This leaves the Gram-negative bacteria colourless. The cells are then stained with a red acidic counterstain (dye) such as acid fuchsin or safranine. After treatment with counterstain, the Gram-negative cells are red, and the Gram-positive cells are blue. *See* GRAM STAIN.

Gram-positive (G+) bacteria: It is also variously abbreviated as GP, GrP, GM+ or gr+. Pertaining to bacteria, holding the colour of the primary stain (blue) when treated with Gram's stain or Gentian violet solution. In contrast to the Gram-negative bacteria, the Gram-positive bacteria possess a much thicker peptidoglycan cell wall (capsule). Because of this, the blue crystal violet dye (with which the bacteria were stained) does not wash out of the cell and the bacteria appear blue under the microscope. See GRAM STAIN.

grana (sing. granum): Stacked membrane discs (thylakoids) in chloroplasts that contain chlorophyll and are the site of the light-trapping reactions of photosynthesis. *See* STROMA.

graph of phyletic gradualisma: Progressive straight line of change over time.

graph theory: In mathematics, the study of graphs, which are mathematical structures used to model pairwise relations between objects. A graph in this context is made up of vertices (*aka* NODES or POINTS) which are connected by edges (*aka* LINKS or LINES). A distinction is made between undirected graphs, where edges link two vertices symmetrically, and directed graphs, where edges link two vertices asymmetrically.

graphical genotype: The portrait of restriction fragment polymorphism data in a graphical form for description of an entire genome in a single graphic image. It describes the parental origin and allelic composition throughout the entire genome.

Graphics Interchange Format (GIF): The Graphics Interchange Format (GIF; /gɪf/ GHIF or /dʒɪf/ JIF) is a bitmap image format that was developed by a team at the online services provider CompuServe led by American computer scientist Steve Wilhite and released on June 15, 1987. It is in widespread usage on the World Wide Web (www) due to its wide support and portability between applications and operating systems.

graphics processing unit (GPU): Specialized electronic processor designed to accelerate graphics rendering.

GRAS gene family: A plant-specific transcription factor family, the name GRAS is derived from the three functional members, gibberellic acid insensitive (GAI), repressor of GAI (RGA) and scarecrow (SCR), which are widely distributed in plants.

gratuitous inducers: A type of inducers that resemble authentic inducers of transcription but are not substrates for the induced enzymes.

greedy algorithm: In this type of algorithm, the solution is built part by part. The solution for the next part is built based on the immediate benefit of the next part. The one solution giving the most benefit will be chosen as the solution for the next part.

Greek-key: A topology for a small number of beta-sheet strands in which some interstrand connections going across the end of barrel, or in a sandwich, fold between beta sheets.

green algae: Algae that are clear green in colour.

green biotechnology: It is biotechnology applied to agricultural processes with the aim of improving the nutritional quality, quantity and production economics. An example would be the selection and domestication of plants via micropropagation. Another example is the designing of transgenic plants to grow under specific environments in the presence (or absence) of chemicals. One hope is that green biotechnology might produce more environmentally friendly solutions than traditional industrial agriculture. An example of this is the engineering of a plant to express a pesticide, thereby ending the need of external application of pesticides. An example of this would be Bt maize. Whether or not green biotechnology products such as this are ultimately more environmentally friendly is a topic of considerable debate.

***green fluorescent protein* (*gfp*) gene:** A gene cloned from jellyfish used as reporter gene in plant transformation. Unlike *GUS* gene, GFP allows non-destructive *in vivo* assay. The *gfp* gene was isolated from jellyfish (*Aequorea victoria*). GFP is a 27 kDa monomer that emits green light when exposed to UV light. This is species independent as has shown expression in bacteria, nematodes, *Drosophila*, mice and several plant species.

green fluorescent protein (GFP): Protein from the jellyfish, *Aequorea victoria* (Scyphozoa). The *gfp* gene is used as a reporter gene. GFP fluoresces in UV light. Several variants have been developed, each exhibiting a characteristic spectrum. Significant advantages are that the protein can be seen in living tissue and is not toxic.

green liver concept: Plants not only function as 'green lungs' due to photosynthetic oxygen production but also as 'green liver' because of their metabolism of xenobiotics.

green manure: Living plant biomass that is grown and then incorporated into the soil by tillage.

green revolution: A dramatic enhancement in agricultural yields that occurred in the 1950s through 1960s. The Green Revolution was based on many advancements in plant science, including the genetic improvement of many plants (including new, high-yield hybrid varieties), improved irrigation, more efficient machinery, new fertilizers and pest controls that increased plants' disease resistance, improved their hardiness and increased their productivity (especially rice, wheat and maize).

green seed: A Canola™ production problem in which an early frost inhibits the degradation of the green pigment chlorophyll. The presence of chlorophyll in the oil lowers the yield and grade of oil obtained from the seed.

greenhouse: A building, room or area, usually of glass, in which the temperature is maintained within a desired range. It is used for cultivating tender plants or growing plants out of season.

greenhouse effect: The heating of the atmosphere as radiation from the sun is transformed to infrared radiation but is trapped by so-called atmospheric GREENHOUSE GASES (i.e. water vapours, carbon dioxide, methane, ozone), preventing its release. This condition is similar to a greenhouse, in which glass traps infrared radiation heated air. Greenhouse effect raises planetary temperatures above those expected from a simple model of heat radiation received from the sun and re-radiated into space. Greenhouse effect is a big challenge in sustainable agriculture.

greenhouse gas: The molecules in greenhouse gases remain in the atmosphere and can trap heat from the sun against the earth's surface. This makes the overall temperature of the Earth go up. We want some natural greenhouse gases to keep our Earth warm, but any extras will make it too warm which can harm environments like the polar ice caps. Carbon dioxide is a common example of a greenhouse gas.

greenhouse gases (GHGs): Gases that absorb and re-radiate infrared radiation. When present in the atmosphere, these gases contribute to the greenhouse effect, trapping heat near the surface of the planet. On Earth, the main greenhouse gases are carbon dioxide, water vapour, methane, nitrous oxide, ozone and some halocarbon compounds. *Aka* RADIATIVELY ACTIVE GASES.

Grey Goo: Destructive nanobots, opposite of Blue Goo. Several other Goo's have been proposed. *Aka* GRAY GOO.

GridION: Benchtop nanopore sequencing device designed to run multiple experiments simultaneously with five independent flow cells. Produced by Oxford Nanopore Technologies.

Griffith effect: F. Griffith in 1928 published his work on bacterium *Diplococcus pneumonia* which is associated with certain types of pneumonia in mice. This bacterium occurs in two forms. The first type has smooth (S) cells which secrete a covering capsule of polysaccharide materials causing the colonies on agar to be smooth and shiny. This type is virulent. It produces pneumonia. The second type has rough (R) cells which lack a polysaccharide capsule. Colonies appear rough and dull. This type is nonvirulent and does not cause pneumonia in mice. Smooth (S) and rough (R) traits are genetically determined. When laboratory mice were injected with S strain of bacteria, mice died. When he injected mice with R strain of bacteria, mice lived. When mice were injected with heat-killed S strain of bacteria, mice lived. However, when mice were injected with R strain plus heat killed S strain of bacteria, mice died.

This experiment showed that the heat killed S strain somehow transformed R strain to virulence. This observation was named as GRIFFITH EFFECT. This process was called TRANSFORMATION. The agent that was thought to be responsible for transformation was called TRANSFORMING PRINCIPLE or TRANSFORMING AGENT. *See* TRANSFORMING PRINCIPLE.

groundcover: A plant that covers the soil surface so densely that it smothers all beneath it.

group coancestry: The probability that two genes taken from a gene pool of a population are identical by descent, or the average of the cells in a coancestry matrix for the population concerned.

group I introns: *See* INTRONS, GROUP I.

group II introns: *See* INTRONS, GROUP II.

group selection: The selection operating between groups of individuals rather than between individuals. It would produce attributes beneficial to a group in competition with other groups rather than attributes beneficial to individuals.

grouped data: Quantitative values in categories such as age into 0–4, 5–9, 10–14, etc.

grouping plants according to their low-temperature tolerance: Plants can be grouped into three different classes based on their low-temperature tolerance. (i) FROST TENDER PLANTS are sensitive to chilling injury and can be killed by short periods of exposure to temperatures just below freezing point. They fail to tolerate ice in their tissues and readily exhibit frost injury symptoms that include a water-soaked flaccid appearance with loss of turgor followed by rapid drying upon exposure to warm temperatures (e.g. beans, maize, rice and tomato). (ii) PLANTS TOLERANT FOR THE PRESENCE OF EXTRACELLULAR ICE IN THEIR TISSUES have frost resistance ranges from the broad-leafed summer annuals, which are killed at temperatures slightly below freezing point, to perennial grasses that can survive exposure to −40°C. As temperature decreases, the outward migration of intracellular water to the growing extracellular ice crystal causes dehydration stress that eventually results in irreversible damage to the plasma membranes, which is the primary site of low-temperature injury. (iii) VERY COLD HARDY PLANTS are predominantly temperate woody species. Their lower limits of cold tolerance depend on the stage of acclimatization, the rate and degree of temperature decline and the genetic capability of tissues to accommodate extracellular freezing and the accompanying dehydration stress.

group-transfer reaction: Reaction (excluding oxidations or reductions) in which molecules exchange functional groups.

growth: This comprises of increase in size of the embryo. Growth can be with or without cell division. This completes developmental cycle.

growth arrest and DNA damage-inducible protein 45 alpha (Gadd45a): A small acidic nuclear protein induced by stress; a putative regulator of active DNA demethylation in mammals.

growth cone: Migrating motile tip of a growing nerve cell axon or dendrite.

growth factor: Extracellular, usually secreted polypeptide signal molecule that can stimulate a cell to grow or proliferate. Signals can be autocrine (self-derived), paracrine (from another cell but local in origin), endocrine (from a distant cell/organ) or neurocrine (from nerve cells). Examples are EPIDERMAL GROWTH FACTOR (EGF) and PLATELET-DERIVED GROWTH FACTOR (PDGF). Most growth factors also have other functions.

growth hormone: A substance that stimulates growth, especially a secretion of the anterior pituitary, that directly influences protein, carbohydrate and lipid metabolism and controls the rate of skeletal and visceral growth.

growth regulator: A chemical substance produced in one part of an organism and transported in minute quantities to induce a growth response in another part. These chemicals can regulate the plant growth under *in vivo* or *in vitro* conditions. Since the discovery of 'chemical regulation of plant growth' by C.O. Miller and F. Skoog (1957), growth regulators are commonly used to regulate plant growth under *in vitro* and *in vivo* conditions. Plant growth regulators are broadly classified into four groups: auxins, cytokinins, gibberellins and growth retardants. Auxins induce cell elongation, rooting and callusing. Cytokinins help in cell division, shoot bud development and shoot growth. Gibberellins help in breaking dormancy and quick elongation of shoots. Retardants help in minimizing callusing during micropropagation.

GT-AG rule: Refers to the fact that with introns present in nuclear protein-coding genes, the first two nucleotides of the intron are GT and the last two AG.

GTPase: Also, the common name used for monomeric GTP-binding proteins. Refers to enzyme activity that converts GTP to GDP. *See* GTP-BINDING PROTEIN.

GTPase-activating protein (GAP): Protein that binds to a GTP-binding protein and inactivates it by motivating its GTPase activity so that it hydrolyses its bound GTP to GDP.

guanidinium thiocyanate-phenol-chloroform method of DNA extraction: Cells are lysed, and cell debris is removed by centrifugation. Single or multiple separation and/or purification steps are needed to desalt the sample comprising the nucleic acid. RNase activity is inhibited by using a mixture of phenol:chloroform:isoamyl alcohol (25:24:1). Proteins, lipids, carbohydrates and cell debris are removed through extraction of the aqueous phase with the organic mixture of phenol and chloroform.

guanine (G): A purine; one of the bases found in DNA and RNA. Its chemical name is 2-amino, 6-oxypurine. In double-stranded DNA or RNA, it pairs with cytosine (C).

guanine nucleotide exchange factor (GEF): Protein that binds to a GTP-binding protein and activates it by stimulating it to release its tightly bound GDP, thereby allowing it to bind GTP in its place.

guanosine: A nucleoside containing purine guanine (G)

guanosine 5′-diphosphate (GDP): The ribonucleotide containing the purine guanine (G) and two phosphate groups are attached at 5′ end of the sugar moiety.

guanosine 5′-monophosphate (GMP): The ribonucleotide containing the purine guanine (G) and one phosphate group is attached at 5′ end of the sugar moiety. *Aka* GUANYLIC ACID *or* GUANYLATE.

guanosine 5-triphosphatases (GTPases): These are G-proteins (enzymes) which are critical for growth, movement and maintenance of the cell's shape. When active, GTPases are bound to cell membrane surfaces by an isoprene molecule, a receptor.

guanosine 5-triphosphate (GTP): The ribonucleotide containing the purine guanine (G) and three phosphate groups are attached at 5′ end of the sugar moiety. Nucleoside triphosphate is produced by phosphorylating GDP (guanosine diphosphate). Like ATP, it releases a large amount of free energy on hydrolysis of its terminal phosphate group. It plays a special role in microtubule assembly, protein synthesis and cell signalling.

guanylic acid: A ribonucleotide constituent of ribonucleic acid that is chemically the phosphoric acid ester of nucleoside guanosine. *Aka* GUANOSINE MONOPHOSPHATE.

guanyl-nucleotide-binding proteins (G-proteins): Discovered by Martin Rodbell *et al.* at America's National Institutes of Health (NIH), and Alfred G. Gilman and co-workers at the American University of Virginia, Charlottesville, during the 1970s–1980s. For this work, they were awarded Nobel Prize in 1994. These proteins are entrenched in the surface membrane of cells. G-proteins receive chemical signals' from outside the cell (e.g. hormones) and pass the signal into the cell, so that cell can retort to the signal. For example, a hormone, drug, neurotransmitter (or other signal) binds to a receptor molecule on the surface of the cell's exterior membrane. That receptor then activates the G-protein, which allows an effector inside cell to produce a second signal chemical inside the cell, which causes the cell to react to the original external chemical signal. The G-proteins are so called because they become GTP and GDP forms alternately, as part of their reaction cycle (i.e. in 'passing the signal'). Dysfunction of G-proteins in plants causes rapid water loss (wilting). G-proteins have intrinsic GTPase activity eventually converts the GTP to GDP which inactivates the protein. These GTPases thus act as molecular switches in intracellular signalling pathways. One family of G-proteins is composed of three different subunits (heterotrimeric GTP-binding proteins). The members of the other, very large family are monomeric GTP-binding proteins which are commonly referred to as monomeric GTPases.

guard and decoy models: An effector that modifies an accessory protein, which may be its virulence target (guard model), or a structural mimic of such a target (decoy model).

guard cell: A specialized type of plant epidermal cells, two of which surround each stoma and change their turgidity and causes stomatal opening and closing. These cells regulate photosynthesis and respiration, facilitating gas exchange while open and preventing excess water loss while closed.

guard hypothesis: This represents a model of indirect recognition of *Avr* proteins by *R* proteins. R proteins recognize the action of virulence factors as they modify host cellular targets in the cytoplasm and suppress basal defence, thus resulting in disease. When avirulence proteins and R proteins interact with the target, virulence is defeated. Release of R protein signals resistance. In *Arabidopsis thaliana*, effectors have specific targets in the host, and R proteins act by guarding these effector targets. Modification of this target by the effector results in the activation of the R protein, which activates disease resistance in the host. Guard model explains how multiple effectors could be perceived by a single R protein, thus allowing a relatively small R gene repertoire to target the broad diversity of pathogens that attack plants. Guarded effector target (also called the GUARDEE) is vital for the virulence function of the effector protein in the absence of the cognate R protein. Effector target has a role in host defence or susceptibility in the absence of its cognate R protein. Effector alteration of the guardee results in enhanced pathogen fitness in plants that lack the R protein and triggers innate immunity in plants that carry the R protein.

guardee: Intracellular nucleotide-binding, leucine-rich repeat receptors (NLR) proteins mediate recognition of AVR effectors either by direct physical interaction or in an indirect manner. In cases of indirect recognition, the NLR responds to modifications of another plant protein induced by the AVR protein. This plant protein may be the operational target of the effector, in which case it is called a 'guardee', since it is guarded by the NLR.

guardee protein, evolution of: Classical guardee proteins are often expendable for the virulence activities of effectors in plants lacking the R protein. Guardee attempts to escape detection by modification by the effector in the absence of a functional *R* gene. In the presence of a functional *R* gene, guardee evolves with improved interaction with an effector. This places guardee in an evolutionarily unstable situation that could be relaxed upon the evolution of a host protein to become a BAIT. In the absence of a functional *R* gene, natural selection is expected to make the guardee lower its binding affinity with the effector and thereby escape detection and modification by the effector. In the presence of a functional *R* gene, natural selection is expected to favour guardees with improved interaction with an effector to enhance pathogen perception. These two conflicting selection pressures on the same effector interaction surface of the guardee lead to an evolutionarily unstable situation that could be relaxed upon the evolution of a host protein, termed here 'decoy'.

guest: With respect to symbiosis, guest is PARASITIC, MUTUALISTIC or COMMENSALIST ORGANISM in its interactions with the host. The guest typically receives shelter and nourishment from the host.

guide RNA (gRNA): (i) An RNA molecule (or a part of one) which hybridizes to eukaryotic mRNA immediately downstream of the pre-edited regions (PER). gRNAs are of around 60 nucleotide length and assist in the splicing of intron sequences. Guide sequences may be either external (EGS) or internal (IGS) to the mRNA being processed and may hybridize to either intron or exon sequences close to the splice junction. Guide RNA is implicated in RNA editing. RNA interference and related RNA silencing phenomena utilize short antisense guide RNA molecules to repress the expression of target genes. The guide RNAs are specific for each edited mRNA and encode additional U residues as complementary A or G residues. The mRNAs are edited most commonly, whereas edits of tRNAs and rRNAs are extremely rare.

guide RNA (gRNA) scaffold sequence: The sequence within the guide RNA that is responsible for Cas9 binding, it does not include the 20-bp spacer/targeting sequence that is used to guide Cas9 to target DNA. *Aka* GRNA TARGETING SEQUENCE.

guide RNA (gRNA) targeting sequence: The 20 nucleotides that precede the PAM sequence in the genomic DNA.

guide RNA targeting sequence: Refers to the 20 nucleotides that precede the protospacer adjacent motif (PAM) sequence in the genomic DNA. This sequence is cloned into a gRNA expression plasmid but does not include the PAM sequence or the gRNA scaffold sequence. *See* PAM

guide RNAs in gene silencing: In effector step, the antisense siRNAs produced by a DICER serve as guides for a different ribonuclease complex, called RNA-INDUCED SILENCING COMPLEX (RISC), which cleaves the homologous single-stranded mRNA. RISC cuts the mRNA approximately in the middle of the region paired with antisense siRNA, after which the mRNA is further degraded. Although the most protein component of RISC has not yet been identified, they might include endonuclease, a helicase, exonuclease and a homology-searching activity.

guide sequence: An RNA molecule (or a part of one) which hybridizes to and aids in the splicing of intron sequences. Guide sequences may either be external (EGS) or internal (IGS) to the mRNA being processes and may hybridize to either intron or exon sequences close to the splice junction.

guidelines by NIH: Guidelines are formed to bring about uniformity in a process in terms of dos and do nots in research. NIH guidelines for recombinant DNA research, guidelines for writing of theses/research papers/manuscripts/books; guidelines for student seminars.

guild: A group of closely related species presumed to be in competition for the same resources and using the same foraging strategy.

gum: A gelatinous (sugary) aggregate synthesized and displayed by plant tissues.

gummosis (pl. gummoses): A pathologic condition and the products of cell degeneration recognized by unwarranted formation of gums.

***GUS* gene:** A gene that codes for production of β-glucuronidase (i.e. GUS protein) in *Escherichia coli.* The *GUS* is generally utilized as a 'marker gene' for genetically engineered plants. β-glucuronidase causes a colour change, in the presence of 5-bromo-4-chloro-3-indoyl-beta-Dglucuronic acid by 'cutting' a glucuronic acid molecule off the 5-bromo-4-chloro-3-indoyl-beta-D-glucuronic acid. The residual molecule is an insoluble blue dye.

guttation: Exudation of water usually through structures called HYDATHODES present at the tip of veins of leaves. *See* HYDATHODE.

gymnosperm: A seed-bearing plant with naked ovules arisen on the surface of a sporophyll. Gymnosperms are among the oldest clades of vascular plants. They are today represented by approximately 1000 extant species worldwide, including, among others, conifers, Ginkgo, Gnetum and cycads. *Compare with* ANGIOSPERM.

gymnospermous: A type of plant that produces seeds but not fruits. The seeds are not borne within an ovary and are called unenclosed. Also called GYMNOSPERMS.

gynandromorph: An individual displaying both male and female sexual differentiation.

gynodioecious: Of a species in which some plants bear only bisexual flowers and others bear only female flowers. *Aka* GYNODIOECY.

gynoecium: The term collectively refers to the female reproductive parts of a flower or for the carpels of a flower, whether united or free. For instance, G designates a superior ovary and G(5) specifies having five fused carpels. *Aka* GYNOECIOUS. *Contrasts* ANDROECIUM.

gynogenesis: Parthenogenesis in which the embryo contains only maternal chromosomes owing to the failure of the sperm to fuse with the egg nucleus.

gynomonoecious: Of a species bearing bisexual flowers and female flowers on the same plant.

gynophore: A stalk supporting the gynoecium and located above the level of insertion of the other floral parts.

gyrase: A type of topoisomerase II that relaxes supercoiling caused by unwinding of DNA double helix.

gyrase, type I: This enzyme can catalyse catenation of double-stranded circular DNA or catenanes can be separated into simple circles provided that at least one circle contains a single-stranded break.

gyrase, type II: This enzyme can carry out catenation reaction. In this case, no single-stranded break is required. It relaxes supercoiling caused by unwinding of DNA double helix.

H

H: H is the relative entropy of the target and background residue frequencies. H can be thought of as a measure of the average information (in bits) available per position that distinguishes an alignment from chance. At high values of H, short alignments can be distinguished by chance, whereas at lower H values, a longer alignment may be necessary.

H DNA: *See* DNA, H-FORM.

H-Y antigen: A histocompatibility factor determined by the Y chromosome. It is thought to be the major male-determining factor in mammals. In humans, the gene for the H-Y antigen is on the short arm of the Y chromosome.

H^+: *See* PROTON.

H2AK119 ubiquitination-dependent inheritance: Polycomb repressive complexes 1 and 2 (PRC1 and 2) are obligatory for heritable repression of developmental genes. The *cis*- and *trans*-acting factors donate to epigenetic inheritance of mammalian Polycomb repression. A. Tiasha *et al.* (2024) demonstrate that, in human cells, ectopically induced Polycomb silencing at originally active developmental genes, but not near universally expressed housekeeping genes, is inherited for many cell divisions. Surprisingly, silencing is heritable in cells with mutations in the H3K27me3 binding pocket of the Embryonic Ectoderm Development (EED) subunit of PRC2, which are known to interrupt H3K27me3 recognition and results in loss of H3K27me3. This method of inheritance is less stable and necessitates intact PRC2 and recognition of H2AK119ub1 by PRC1. The upkeep of Polycomb silencing is sensitive to local genomic context and can be facilitated by PRC1-dependent H2AK119ub1 and PRC2 autonomously of H3K27me3 recognition.

H4Kacme associated with marks of active chromatin: Lysine residues in histones and other proteins can be modified by post-translational modifications that encode regulatory information. Lysine acetylation and methylation play role in regulating chromatin and gene expression. Pathways involving these post-translational modifications are targets for clinically approved therapeutics to treat human diseases. Lysine methylation and acetylation are generally assumed to be mutually exclusive at the same residue. W.J. Lu-Culligan *et al.* (2023) observe cellular lysine residues that are both methylated and acetylated on the same side chain to form Nε-acetyl-Nε-methyllysine (Kacme) on histone H4 (H4Kacme) across a range of species and across mammalian tissues. Kacme is associated with marks of active chromatin, increased transcriptional initiation and is regulated in response to biological signals. H4Kacme can be installed by enzymatic acetylation of monomethyllysine peptides and is resistant to deacetylation by some HDACs *in vitro*. Kacme can be bound by chromatin proteins that recognize modified lysine residues, as demonstrated with the crystal structure of acetyllysine-binding protein BRD2 bound to a histone H4Kacme peptide. These results establish Kacme as a cellular post-translational modification with the potential to encode information distinct from methylation and acetylation alone and demonstrate that Kacme has all the hallmarks of a post-translational modification with fundamental importance to chromatin biology.

habit: The general look of a plant, with respect to its size, shape, texture and orientation.

habitat: (i) An area with a particular kind of environment within an ecosystem which is suitable for the growth of an organism. (ii) The living place of an organism or community characterized by its physical or biotic properties. (iii) The place, in an ecosystem, where animals, people and plants and non-living things are together found.

habituation: The acquired ability of cells, tissues and organs to divide and grow independently of exogenously supplied growth regulators. During tissue culture, the transfer of *in vivo* grown plants to *in vitro* conditions is transitory step and all the explants do not resume growth. But certain explants, tissues or cell sectors slowly and slowly get acclimatized to *in vitro* conditions and start proliferation. The acquired ability is termed habituation.

***Hae*III:** A type II restriction enzyme from the bacterium *Haemophilus aegyptius* that recognizes the DNA sequence 5′-G G↓C C / 3′-C C↑G G, where arrows indicate the sites of cleavage in the two strands, thus producing blunt ends.

haemagglutinin (HA): A special protein that some viruses make use of to gain entry into the cells they have besieged. HA protein assists the virus stick to the cell it targets. HA also refers to specific plant cell proteins (lectins) that are naturally manufactured by certain plants such as the soybean plant (*Glycine max* (L) Merrill). The presence of those lectin molecules (e.g. on surfaces of root cells of the soybean plant) helps nitrogen-fixing *Rhizobium japonicum* bacteria to stick to soybean plant roots, where they begin to 'fix nitrogen' (i.e. create natural nitrate fertilizer), which improves the soil and helps plants to grow.

haemoglobin: The major protein in red blood cells that associates with O_2 in the lungs by means of a bound haem group. Haem is an iron-containing molecule.

haemophilia: A metabolic disorder characterized by free bleeding from even slight wounds because the blood does not clog properly. This occurs due to the absence of blood clotting factor. Haemophilia is of two types – haemophilia A and B. Both haemophilia A and B are inherited in the same way, because both the genes for factor VIII and factor IX are located on the X chromosome (chromosomes are structures within the body's cells that contain the genes).

hair: A single, elongated cell or row of cells borne on the surface of an organ. Hair is a protein filament.

hairpin: (i) A helical (duplex) region formed by base pairing between adjacent (inverted) complementary sequences within a single strand of RNA or DNA. (ii) A protein naturally formed by the *Erwinia amylovora* bacteria, which usually causes the plant disease called fire blight in apple trees, pear trees and some ornamental plants of the rose family.

hairpin loop: A section of highly curving, single-stranded DNA or RNA formed when a long piece (string) of the DNA or RNA bends back on itself and hydrogen bonds (is able to base pair) in some regions to form double-stranded regions. The structure can be visualized by taking a human hair, bending it back on itself and holding it in such a way as to half its original length. The section where the two ends of hair lie next to each other represents the section of double-stranded DNA or RNA. At one end the hair will have to make a sharp turn and will form a loop. This loop represents the single-stranded hairpin loop.

hairpin bisulfite PCR: Ligation of a hairpin oligonucleotide to genomic DNA prior to bisulfite conversion and POLYMERASE CHAIN REACTION (PCR) amplification physically links the two complementary DNA strands. This additional step in the conversion procedure overcomes the limitations of conventional bisulfite sequencing where information of the cytosine methylation status is only obtained from one of the two strands of an individual DNA molecule.

hairpin RNA (hpRNA): Ribonucleotide strand with complementary sense and antisense sequences that folds to produce double-stranded RNA with a loop at one end. *Aka* SHORT HAIRPIN RNA *or* SMALL HAIRPIN RNA (SHRNA).

hairpin, uses of: Discovered in 1992 by Z.-M. Wei *et al.*, harpin reasons numerous species of plants to recruit a protective/defensive response (cascade) against bacteria, viruses, fungi, some insects and nematodes. Harpin also causes plants (i.e. that it is sprayed on to) to enhance their photosynthesis and to have increased root growth/proliferation which can result in greater crop yields.

hairy root cultures: *Agrobacterium rhizogenes* induced transformed roots grown in culture. Root organ culture is exploited in the production of secondary metabolites through tissue culture. In comparison, to normal roots, the hairy roots (transformed roots) exhibit faster growth even in the absence of rooting hormones. Therefore, such roots are now preferred for secondary metabolite production.

half chromatid: One of the two DNA strands in a chromatid.

half genes: Genome sequence analysis of the small hyperthermophilic archaeal parasite *Nanoarchaeum equitans* has not shown genes encoding the glutamate, histidine, tryptophan and initiator methionine transfer RNA species. A computational approach was employed to look for widely separated genes encoding tRNA halves that, on the basis of structural prediction, could form intact tRNA molecules. The search discovered nine genes that encode tRNA halves; collectively they explain for missing tRNA genes.

half-chromatid conversion: A type of gene conversion that is inferred from the existence of non-identical sister spores in a fungal octad showing a non-Mendelian allele ratio.

half-inferior ovary: An ovary located partly below and partly above the level of attachment of the other floral parts. *Compare with* INFERIOR OVARY *and* SUPERIOR OVARY.

half-life: The amount of time taken by one-half of the atoms of a radioactive material to decay to a stable form. For example, the half-life of carbon-14 is 5568 years.

half-methylated site: When only one strand of DNA carries a methyl group whereas the other complementary strand does not do so at a site. *Aka* HETERODUPLEX MODIFIED DNA.

HaloTag method: A modified haloalkane dehalogenase designed to covalently bind to synthetic ligands (HaloTag ligands). The synthetic ligands comprise a chloroalkane linker attached to a variety of useful molecules, such as fluorescent dyes, affinity handles or solid surfaces. This method has the utility for cellular imaging and protein immobilization by analysing multiple molecular processes associated with NF-κB-mediated cellular physiology, including imaging of subcellular protein translocation and capture of protein-protein and protein-DNA complexes.

handle: In computer programming, the term handle is used for an abstract reference to a resource that is used when application software reference blocks of memory or objects that are managed by another system like a database or an operating system.

hand-pollination: The controlled activity of pollination that eliminates the possibility of open pollination.

haplodiploid: A genetic system of sex determination in which females develop from fertilized (diploid) eggs and males from unfertilized (haploid) eggs.

haplodiploidy: Sex differentiation mechanism in which males are haploids and females are diploids. For example, ants, bees and wasps.

haploid: A cell or an individual with one set of chromosomes (n). Egg and sperm are haploid cells. Once they join during fertilization, the resulting cell will be diploid with two sets (2n) of chromosomes. Haploids occur as a single set of chromosomes during almost the entire part of their life and thus monoploids are distinguishable from true haploids. *Compare with* DIPLOID, DIKARYOTIC *and* POLYPLOID. *See* MONOPLOID.

haploid breeding: To obtain homozygous breeding lines, originally self-pollination was needed over 6–8 generations, which is a very time-consuming and costly process. The production of homozygous lines for true-breeding lines is best done through haploid breeding. The desired trait will be transmitted unchanged to subsequent generations only if the trait is present in homozygous form. Homozygous lines of some plant species (e.g. tobacco, barley, potatoes, rape, wheat, etc.) can be produced from gametes, which contain only one set of chromosomes. In most cases, unripe pollen is placed on a suitable culture medium, where it develops into plants with a single set of chromosomes (HAPLOID ANDROGENESIS). Similarly, ovaries may also be utilized as the source material, although this is done less frequently (HAPLOID PARTHENOGENESIS). Following a cultivation period of 3–4 weeks, the haploid plantlets are treated with colchicine in order to induce duplication of the chromosomes. The resulting cells produce 'double haploids', which are fully homozygous plants that produce identical offspring.

HAPloid equivalents of DNA and PolYmerase chain reaction (HAPPY) mapping: An *in vitro* technique for the determining the arrangement (order and spacing) of DNA markers directly on genomic DNA and the establishment of physical maps of any density, without need for parental lines, polymorphic markers and segregating populations.

haploid number (n): Referring to the number of chromosomes in a gamete. A somatic cell is diploid and has twice that many chromosomes (2n). Haploid number is the result of the halving of chromosome pairs during meiosis. *See* DIPLOID NUMBER.

haploid number and monoploid number, relationship between: In diploid organisms, haploid number (n) is equal to monoploid number (x). For example, in human, $x = n = 23$, which is also denoted as $2n = 2x = 46$. In polyploids, x and n differ. For example, bread wheat (*Triticum aestivum*) has six sets of chromosomes as it is a hexaploid. In this case, $2n = 6x = 42$; $x = 7$; and $n = 21$.

haploid parthenogenesis: The situation in which an unfertilized egg hatches and develops normally to produce a viable male adult whose cells contain only the haploid number of chromosomes.

haploidization: Production of haploid from a diploid by progressive chromosome loss. For example, in barley (*Hordeum vulgare*) by crossing with wild (bulb) barley (*Hordeum bulbosum*) or in bread wheat (*Triticum aestivum*) by crossing with pearl millet (*Pennisetum glaucum*).

haploids, production of: This has been made possible through: (i) anther and microspore culture; (ii) ovule culture; and (iii) chromosome elimination. Haploids have been used in plant breeding programmes for: (a) production of homozygous lines, (b) hybrid sorting in haploid breeding and (c) production of spontaneously doubled haploids for breeding. Haploids have been found to be useful in: (a) mutation research, (b) cytogenetics research, (c) evolutionary studies and (d) genetic studies.

haploidy versus diploidy: Prokaryotic chromosomes occur in single copy and are, therefore, called haploids. In eukaryotes (except some haploid eukaryotes like fungi), chromosomes occur in pairs and are, therefore, called diploids. Doubling the chromosome number in eukaryotes is probably the reason for their greater complexity.

haplo-insufficiency: Reduced gene dosage. A situation where the presence of only one normal copy of a gene is not sufficient to maintain its normal function. It is observed as a dominant mutation on one allele (or deletion of it) resulting in total loss of function in a diploid cell because of the insufficient amount of the wild-type protein encoded by the normal allele on the other haplotype gene hemizygote. Haplo-insufficiency may cause some syndromes. For example, haplo-insufficiency may be one of the reasons of RUBINSTEIN-TYABI SYNDROME (RTS). Greg syndrome is also a consequence of haplo-insufficiency.

haplontic: Refers to organisms in which meiosis occurs in the zygote resulting in the formation of four haploid cells. In life cycle of a haplontic, the multicellular phase is haploid.

haplontic life cycle: A life cycle in which haploid adults produce haploid gametes by mitosis; the diploid zygotes immediately undergo meiosis to produce haploid individuals.

haplontic selection: Selection of haploid cells at gametic level. The DIPLONTIC SIEVE is more rigorous than HAPLONTIC SIEVE.

haplophase: A phase in the life cycle of an organism in which it possesses only one copy of each gene. The organism is then said to be haploid. Yeasts can occur as true haploids. Humans are haploid for only a few genes and cannot occur as true haploids.

haplosis: The meiotic division ensuing in haploid cells and/or gametes.

haplosomic: The situation in which the homologue of a pair of chromosomes is absent in somatic cells.

haplostemonous: Having a single series of stamens equal in number to the proper number of petals and alternating with them. *Compare with* DIPLOSTEMONOUS *and* OBDIPLOSTEMONOUS.

haplotype: (i) A set of genes at more than one locus inherited by an individual from one of its parents. It is the multilocus analogue of an allele. (ii) The particular combination of alleles in a defined region of some chromosome, in effect the genotype in miniature. Originally used to describe combinations of MHC alleles, it is now also used to describe particular combinations of RFLPs.

haplotype block: A specific arrangement of adjacent alleles in a given region of genomic DNA, usually in the range of 10,000–100,000 bp, that is inherited as a 'block', probably because its recombination frequency is lower than in other parts of the genome. *Aka* HAPBLOCK.

haplotype characterization: The characterization of SINGLE-NUCLEOTIDE POLYMORPHISMS (SNPs) by coherent packages (i.e. SNPs that are usually transmitted together).

haplotype mapping of human genome: Haplotypes are stretches of DNA that travel as a unit, each carrying a group of SNPs, when chromosomes recombine during inheritance. Human genome contains about 10 million common SNPs. Finding those SNPs that differ between healthy people and those, with, say, diabetes or cancer, began to look impossibly costly. Its solution came from haplotypes. If disease researchers see one haplotype that predominates in their patients, they can narrow their SNP search so that haplotype typically around 10,000–20,000 bases of DNA. Two SNPs are sufficient to identify each of these three haplotypes. In 2004, 9 million SNPs were available for HapMappers.

haplotype mapping of human genome: If disease researchers see one haplotype that predominates in their patients, they can narrow their SNP search so that haplotype typically around 10,000–20,000 bases of DNA.

haplotype mapping: The process of creating a HAPLOTYPE MAP. Haplotype map is modification of a genome map in which the HAPLOTYPE BLOCKS of the genome of an organism are depicted.

haplotype signature: Any characteristic configuration of specific alleles in an organism. Haplotype signatures can be prepared for a collection of organisms and a disease allele can be recognized.

haplotype single-nucleotide polymorphism (htSNP): Any single-nucleotide polymorphism that is identified or 'tagged' from larger SNP databases, located in a specific genomic region and used to define haplotypes and haplotype structure in that region. *Aka* HAPLOTYPE-TAGGED SINGLE-NUCLEOTIDE POLYMORPHISM *or* HAPLOTYPE TAG SNP.

haplotyping: The process of determination of the haplotype of an individual.

haplo-viable mutants: These prevent normal functioning of pollen but are transmitted through the egg carrying them. *Aka* MALE GAMETE ELIMINATORS.

hapten: A substance that can combine with a specific antibody but lacks antigenicity of its own. Many small molecules of <1000 Da such as toxins, drugs and hormones are not capable of invoking immune response when injected directly into animals. They are thus not immunogenic by themselves.

haptoglobins: Proteins in blood serum that bind haemoglobin and transport it from broken red cells to sites where it is further metabolized.

hard disk: A disk fixed in a microprocessor. Such disks have more storage capacity and faster in reading and writing than floppy disks.

hard problem: Computational complexity theory which focuses on (i) classifying computational problems on the basis of their inherent difficulty and (ii) relating these classes to each other. A COMPUTATIONAL PROBLEM is a task solved by a computer. A computation problem is solvable by mechanical application of mathematical steps, such as an algorithm.

hard selection: Selection where numbers produced by a patch are directly associated with the relative survival of the individuals of that patch. *See* SOFT SELECTION.

hardening: Preparation and acclimatization of tissue-cultured plants for their transfer to soil. The plants grown in the field are autotroph and get exposed to various kinds of biotic and abiotic stresses, whereas the *in vitro* grown plants are heterotroph and very delicate. The transfer of tissue-cultured plants from *in vitro* to *in vivo* conditions is a transitory step that requires greater care and skill. The direct transfer leads to high mortality. Therefore, the plantlets are first acclimatized (hardened) in the laboratory to make them hard and survive under *in vivo* conditions.

hardiness: The ability to withstand stress.

Hardy-Weinberg equilibrium: Occurrence in a population of three genotypes at a locus, two homozygous and one heterozygous (e.g. *AA*, *Aa* and *aa*), where the frequencies of the three genotypes are p^2 for homozygous dominants, 2pq for heterozygotes, q^2 for homozygous recessives, p = frequency of *A*, q = frequency of *a*. At these frequencies, there is no change in genotypic frequencies as long as isolation mechanisms, selection, mutation and drift are absent. The equation ($p^2 + 2pq + q^2 = 1$) is used by population geneticists to determine probable genotype frequencies of a population for specific traits. By comparing these frequencies for subsequent generations, it is possible to track the direction and rate of evolution.

Hardy-Weinberg equilibrium in crop improvement: Hardy-Weinberg equilibrium is a mathematical model that predicts the distribution of genetic variation within a population in the absence of evolutionary forces such as mutation, selection and gene flow. The Hardy-Weinberg equilibrium assumes that the population is large, randomly mating and not subject to any evolutionary forces. A study by H. Chen *et al.* (2018) tested for genetic equilibrium in a population of Chinese jujube trees. The population was not in Hardy-Weinberg equilibrium, suggesting that evolutionary forces such as selection and gene flow were impacting the genetic structure of the population. A study by X. Li *et al.* (2017) revealed that the Chinese cabbage population was in Hardy-Weinberg equilibrium, suggesting that the population was not subject to significant evolutionary forces such as selection or gene flow.

Hardy-Weinberg law: In 1908, G.H. Hardy and W. Weinberg independently reported the basic principles of population genetics. Their discovery is known as the Hardy-Weinberg law which states that in a large population mating at random, after one generation of such mating the frequencies of alleles and of genotypes remain constant generation after generation, in an infinitely large, interbreeding population in which mating is at random and there is no selection, migration or mutation. The population must be large enough so that sampling errors can be disregarded. This implies that there should be no random genetic drift. There can be no mutations, either $A \rightarrow a$ or $a \rightarrow A$. If there are the $A \rightarrow a$ mutation, they must be the same in frequency as $a \rightarrow A$. If this is not the case, frequencies of alleles *A* and *a* will change over time. Mating must be random; that is all mating combinations must be equally probable. For example, let us consider three genotypes – *AA*, *Aa* and *aa*. Various mating combinations *AA* × *AA*; *AA* × *Aa*; *AA* × *aa*; *Aa* × *Aa*; *Aa* × *aa* and *aa* × *aa* should be in the same proportion. All genotypes should be equally viable and equally fertile. No selection should be operative. All gametes must have equal probabilities of forming zygotes. The population must be isolated to avoid individuals, and hence genes, from migrating into and out of it. If these conditions are met, genetic equilibrium will be established; if not, gene frequencies will change. The changing of gene frequencies over time is the entire basis of evolution. *Aka* HARDY-WEINBERG EQUILIBRIUM *or* HARDY-WEINBERG THEOREM.

Harlequin chromosome: A chromosome whose sister chromatids stain differentially such that that one appears dark (usually red-violet) and the other light. *See* SISTER CHROMATID EXCHANGE (SCE).

harmomegathy: A process by which the pollen grains in arid environments shut their apertures to avoid losing water.

harnessing epigenetic variation: Proper yoking of epigenetic variation may offer novel prospects for crop improvement. Technological developments have provided novel insights into the sources and inheritance of epigenetic variation.

harnessing landrace genetic diversity in wheat breeding: S. Cheng *et al.* (2024) studied the genetic and phenotypic diversity of the A.E. Watkins landrace collection of bread wheat (*Triticum aestivum*), a major global cereal, through whole genome resequencing (827 Watkins landraces and 208 modern cultivars) and comprehensive field assessment spanning a decade. They observed that modern cultivars have been derived from just two of the seven ancestral groups of wheat and preserve very long-range haplotype integrity. The remaining five groups exemplify untapped genetic sources, offering access to landrace-specific alleles and haplotypes for breeding. LINKAGE DISEQUILIBRIUM (LD) based haplotypes and association genetics analyses link Watkins genomes to the thousands of high-resolution QUANTITATIVE TRAIT LOCI (QTL), and significant marker-trait associations are recognized. Using the structured germplasm, genotyping and informatics resources, these researchers discovered many Watkins-unique advantageous haplotypes that can arbitrate superior traits in modern wheat. They also assessed the phenotypic effects of 44,338 Watkins-unique haplotypes, introgressed from 143 ranked QTL in the context of modern cultivars, bridging the gap between landrace diversity and current breeding. This approach thus launches an outline for systematically exploiting genetic diversity in crop improvement to attain sustainable food security.

harvest index (HI): Determination of crop yield (i.e. the weight of a harvested product, such as a grain) as a percentage of the total plant weight of a crop or, more general, the proportion of the biological yield to economic yield.

hash function: Any function that can be utilized to map data of arbitrary size to data of a fixed size. The values reimbursed by a hash function are called HASH VALUES, HASH CODES, DIGESTS or simply HASHES. Hash functions are generally used in combination with a hash table, a common data structure used in computer software for rapid data lookup. Hash functions accelerate table or database lookup by detecting duplicated chronicles in a large file.

hash table: In computation, a hash table (or map) is a data structure that implements an associative array abstract data type, a structure that can map keys to values. A hash table uses a hash function to compute an index into an array of buckets or slots from which the desired value can be found.

hashing algorithm: Hashing algorithms work similarly to the searching algorithm. But they contain an index with a key ID. In hashing, a key is assigned to specific data.

HAT selective medium: A medium containing hypoxanthine, aminopterin and thymidine which is used to select somatic hybrid cells.

haulm: Refers to the stems or stalks collectively.

haustorium (pl. haustoria): (i) In parasitic plants, haustorium is a structure developed for penetrating the host's tissues. (ii) A specialized outlet of a parasite moulded inside host cells to absorb nutrients.

Haworth projection: A Haworth projection is a common way of writing a structural formula to represent the cyclic structure of monosaccharides with a simple three-dimensional perspective.

Hayflick limit: The observation that cells have a specific number of cell divisions that they can go through.

hazard: Any potential cause of harm irrespective of how likely or unlikely that potential harm.

hazard rate: The instantaneous risk of a patient experiencing a particular event at each specified time. The instantaneous rate with which an event occurs at a single point in time. It is the probability that an event occurs between time t and time t + δ given that it has not yet occurred by time t, divided by δ, as δ becomes vanishingly small. The rates, unlike probabilities, can exceed 1.0 because they are quotients.

hazard ratio: The ratio of hazard rates at a single time t, for two types of subjects. Hazard ratios are in the interval $[0, \infty]$, and they are frequently good ways to summarize the relative effects of two treatments at a specific time t. Like odds ratios, hazard ratios can apply to any level of outcome probability for the reference group. Note that a hazard ratio is distinct from a risk ratio, the latter being the ratio of two simple probabilities and not the ratio of two rates.

head: A type of inflorescence in plants in which plentiful small flowers are densely packed on a disk or around a central stalk.

heap: A specialized tree-based data structure which is in essence an almost complete tree that placates the heap property: if P is a parent node of C, then the key (the value) of P is either greater than or equal to (in a max heap) or less than or equal to (in a min heap) the key of C. The node at the 'top' of the heap (with no parents) is termed the ROOT NODE.

heapsort: A comparison-based sorting algorithm. Heapsort can be assumed as an improved selection sort: like that algorithm, it splits its input into a sorted and an unsorted region, and it iteratively shrinks the unsorted region by extracting the largest element and moving that to the sorted region. The improvement consists of the use of a heap data structure rather than a linear time search to find the maximum.

HEAT domain: HEAT (comprising proteins Huntington, Elongation Factor 3, PR65/A, TOR) has a domain of 37–47 amino acids consisting of tandemly repeated pairs of antiparallel alpha helices. HEAT domains have a super-helical structure and often function as protein-protein interaction surfaces.

heat labile: Easily destroyed by heat. Antibiotics commonly used in biotechnology research are usually heat labile. Therefore, these are always stored in the refrigerator. During their use these are not co-autoclaved but added separately by filtering their solutions using membrane filters (0.2 µm).

heat mutagenesis: Induction of mutations at above melting temperature by strand separation, depurination and chain breakage.

heat killed viruses: These are used as vaccines because heat destroys pathogenic property of virus, but heat may even destroy antigenic property of the virus.

heat-shock elements: Two copies of a consensus sequence CNNGAANNTTCNNG, occurring in *Drosophila* heat-shock protein genes (e.g. hsp70) required for their maximal transcription. Heat-shock transcription factors bind to these elements.

***heat-shock protein* (*hsp*) genes:** Genes responsible for the production of specific proteins in response to heat stress. Heat-shock is an important abiotic stress affecting the crop yield. For example, wheat is susceptible to high temperature. Now heat-shock protein genes have been cloned from bacteria, yeast and plants. and are now being introduced into crop plants for developing heat tolerance. *Aka* HEAT-SHOCK GENES.

heat-shock proteins (HSPs): A family of proteins synthesized in increased amounts in response to an elevated temperature or other stressful treatment, which usually help the cell to survive the stress. Prominent examples are hsp60 and hsp70. HSPs also function as accessory molecules in antigen presentation. High levels of HSPs prevent stress-induced apoptosis and may have a transforming potential. *Aka* STRESS-RESPONSE PROTEIN.

heat-shock proteins in cell differentiation: HSP synthesis is induced by heat-shock. HSPs also play a role in the normal lifespan of the cell in cellular differentiation. HSP70 is transiently expressed during normal cell cycle. Microorganisms utilize HSP genes during normal life cycle as well as during a stress response. Bacteriophages have co-opted the cell HSPs for their own development, employing the proteins in DNA replication and phage assembly. HSPs in higher eukaryotes may function during DNA replication and can be affected by multiple stimuli, ranging from viral infection to hormones to cell cycle activity. HSPs are essential component in cell homeostasis. True function of these proteins is not well understood.

heat-shock response: A ubiquitous and highly conserved defence mechanism for protection of cells from harmful conditions such as heat-shock, UV irradiation, toxic chemicals, infection, transformation and appearance of mutant and misfolded proteins. Enhanced expression of specific genes heat-shock protein genes in response to high temperature. This response is widespread in the prokaryotic and eukaryotic organisms. This is accompanied by decreased transcription of other genes and decreased translation of other mRNAs. Under heat stress, heat-shock genes express and produce special HEAT-SHOCK PROTEINS (HSPs) which protect the organism from heat stress. Cloned heat-shock protein genes are now being used for transformation of crop plants to develop high-temperature tolerant transgenic crops.

heat-shock transcription factor 1 (HSF-1): A specific transcription factor which regulates the heat-shock response and also influences ageing. Reducing *hsf-1* gene activity accelerates tissue ageing and shortens lifespan. Overexpression of *Hsf-1* gene activity extends lifespan. HSF-1 like DAF-16 is needed for *daf-2*-insulin/IGF-1 receptor mutations to extend lifespan. This is because

HSF-1 and DAF-16 together activate expression of specific genes, including genes encoding small heat-shock proteins, which in turn promote longevity. The small heat-shock proteins also suspension the onset of polyglutamine expansion protein aggregation, suggesting that these proteins couple the normal ageing process to this type of age-related disease.

heat-shock transcription factors: A family of regulatory proteins that bind to heat-shock elements of the heat-shock genes.

heat stress memory: Heat stress-related transcription factors and histone-modifying enzymes function in the sensitized expression of heat memory genes via the deposition and removal of histone modifications.

heat stress memory gene: The genes that produce heat stress-related transcription factors and histone-modifying enzymes. Chromatin remodelling complexes and miRNA accumulation also trigger the sustained expression of heat memory genes.

heavy chain: Larger of the two polypeptide chains existing in an immunoglobulin molecule. *Aka* H CHAIN.

heavy chain genes: Genes that code for heavy chain of immunoglobulin molecules. *Aka* H CHAIN GENES.

heavy chain variable (V_H) domains: The domains of the antibody molecule's heavy chain that vary in their amino acid sequence. The 'chains' comprising the antibody (immunoglobulin) molecule include a region of variable (V) amino acid sequence and a region in which the amino acid sequence remains constant (C). An antibody molecule possesses two antigen-binding sites, and it is the variable domains of the light (V_L) and heavy (V_H) chains which contribute to this antigen-binding ability.

Heisenberg uncertainty principle: A quantum-mechanical principle, according to which the position and momentum of an object cannot be precisely determined. W. Heisenberg stated in 1927 that more precisely the position is determined, the less precisely the MOMENTUM is known. *Aka* UNCERTAINTY PRINCIPLE.

HeLa cell line: This cell line is derived from cervical cancer cells taken on February 8, 1951, from Henrietta Lacks, a 31-year-old African American mother of five, who later died of cancer and after whom they are named. The cell line is durable and prolific, allowing it to be used extensively in scientific study.

helicase: An enzyme that unwinds the double DNA helix near the replication fork before DNA polymerase acts on it. The replication fork moves from 3′ to 5′ of the leading strand. Unwinding is also necessary for DNA repair.

helicase-dependent amplification (HDA): HAD is a method which is different but is based on the PCR. HDA uses a constant temperature instead of cycling temperatures through denaturation and annealing/extension cycles. DNA helicase, an enzyme that unwinds DNA, is used in place of thermal denaturation.

helitrons: The eucaryotic rolling-circle transposable elements that are most prevalent among plant and animal genomes. Studies have identified three additional coding and structural variants of helitrons called HELENTRONS, PROTO-HELENTRON and HELITRON2.

helix: A helix is a three-dimensional spiral, like the shape of a spring or the railing on a spiral staircase. A DNA molecule consists of two helixes intertwined. *See* DNA.

helix-destabilizing (HD) proteins: HD proteins prevent the reannealing of complementary DNA strands by binding specifically to single-stranded DNA and unwinding enzymes that are single-strand-dependent ATPases that catalyse unwinding of DNA duplex. *Aka* SINGLE-STRAND BINDING (SSB) PROTEINS.

helix-loop-helix (HLH): The helix-loop-helix domain is found in some transcription factors and is characterized by two α helices connected by a loop. One helix is typically smaller and, due to the flexibility of the loop, allows dimerization by folding and packing against another helix. The larger helix typically contains the DNA-binding regions. Should not be confused with the HELIX-TURN-HELIX.

helix-turn-helix (HTH)-domain-containing anti-CRISPR-associated (Aca) protein: All organisms recurrently involve HTH domain proteins for regulation of gene expression which must be attuned to meet cellular requirements. For example, in the arms race between bacteria and bacteriophages, rapid expression of phage anti-CRISPR (acr) genes upon infection empowers evasion from CRISPR–Cas defence; transcription is then repressed by an HTH-domain-containing anti-CRISPR-associated (Aca) protein, possibly to reduce fitness costs from excessive expression. A single HTH regulator adjusts anti-CRISPR production to cope with increasing phage genome copies and accumulating acr mRNA. N. Birkholz *et al.* (2024) reveal that the HTH domain of the regulator Aca2, in addition to repressing Acr synthesis transcriptionally through DNA binding, impedes translation of mRNAs by binding conserved RNA stem-loops and blocking ribosome access. The cryo-electron microscopy structure of the close to 40 kDa Aca2–RNA complex establishes how the versatile HTH domain precisely discriminates RNA from DNA binding sites. These combined regulatory modes are extensive in the Aca2 family and facilitate CRISPR-Cas inhibition in the face of rapid phage DNA replication without toxic acr overexpression. The Aca proteins are proposed to elicit regulatory control by dual DNA and RNA binding.

helix-turn-helix motif: Configuration observed in DNA-binding proteins consisting of a recognition helix and a stabilizing helix separated by a short turn. Originally discovered in bacteria, the helix-turn-helix motif is generally found in repressor proteins and is about 20 amino acid long. In eukaryotes, the homeodomain comprises two helices, one of which recognizes the DNA (*aka* RECOGNITION HELIX). They are common in proteins that regulate developmental processes.

helper cells: Cells engineered to express retrovirus packaging proteins to produce retroviruses capable of infecting cells when the packaging genes have been deleted from the retrovirus genome. Such cells are widely used for the production of retrovirus vectors. The retroviruses produced are incapable of making progeny that can infect cells. *Aka* PACKAGING CELLS.

helper phage: A virus that, when it infects a cell with a defective virus to complete its infective cycle, by providing gene products for which the defective genome does not code.

helper T (T_H) cell: A specialized mammalian T cells that interacts with an antigen before B cells interact with antigen. Helper T cells play a central role in immune response. T_H cells produce cytokines with the help of which they induce the formation of cytotoxic T cells. Once activated by an antigen T_H cells use these chemical messengers to influence the activity of other immune system cells.

helper vectors: Small plasmids maintained in *Escherichia coli* that contain transfer (*tra*) and mobilization (*mob*) genes, which allow the transfer of the conjugation-deficient intermediate vectors into *Agrobacterium*.

haeme: Cyclic organic molecule containing an iron atom that carries oxygen in haemoglobin and carries an electron in cytochromes. The iron-porphyrin prosthetic group of a class of proteins is called HAEME PROTEINS.

hemerochory: A plant that has been transported voluntarily or involuntarily by humans in a territory which it could not have colonized by its own natural mechanisms of dissemination or at least much more slowly.

hemialloploid: Not a full alloploid but a SEGMENTAL ALLOPLOID, in which some (half) parts of the unified genomes show some degree of structural ploidy.

hemiautoploid: Autopolyploids with a certain degree of differentiation between the diploid sets of chromosomes, either by a subsequent differentiation of previously homologous sets of chromosomes or by spontaneous or induced intervarietal or subspecies hybridization.

hemichromosome: A chromosome divided into chromatids without previous reduplication at interphase.

hemidesmosome: Specialized anchoring cell junction between an epithelial cell and the underlying basal lamina.

hemigamy: A condition where gametes fuse but the nuclei do not, thereby forming a dikaryon.

hemihaploid: Individuals with half of the normal haploid chromosome number.

hemi-legume: A legume fruit in which the seed or seeds and one valve of the pod are dispersed as a unit. The valve catches the wind and blows away with the seeds, as in *Acacia tenuifolia* and *Peltogyne paniculata*.

hemi-methylated DNA: Duplex DNA in which only one of the two strands is methylated. Not be confused with monoallelic methylation, in which one allele of a locus is methylated in a diploid organism. *Compare with* MONOALLELIC METHYLATION.

hemi-methylated sites: DNA sequences (mostly CpGs) that are methylated only on one of the two DNA complementary strands.

hemi-methylated state: The status of a symmetrical DNA sequence (such as CG or CHG) that is methylated on only one strand.

hemi-methylation: The process through which a DNA duplex is methylated in one strand but remains unmethylated in the other.

hemiparasite: A parasitic flowering plant that contains chlorophyll when mature (photosynthetic) and obtains water with dissolved nutrients by connecting to the host xylem via the haustorium, such as true mistletoe (*Phoradendron* spp. or *Viscum* spp.) or witchweed (*Striga* spp.). *Contrasts* HOLOPARASITE.

hemiploid: Individuals with half of the somatic chromosome number.

hemizygote: A diploid individual that has lost its copy of a particular gene (e.g. because a chromosome has been lost) and which therefore has only a single copy.

hemizygous: A term that describes a person who has only one copy of a gene rather than the usual two copies. Hemizygosity can occur under normal conditions. For example, (i) most males have one X chromosome and one Y chromosome, they are hemizygous for genes on those chromosomes, (ii) monosomics and (iii) haploids.

heptane: A straight-chain alkane with seven carbon atoms. It has been found in Jeffrey pine (*Pinus jeffreyi*). It has a role as a non-polar solvent and a plant metabolite. It is a volatile organic compound and an alkane.

hepten: Small molecules that cannot induce production of antibodies but interact with antibodies.

herb: Any vascular plant that does not develop a woody stem at any point during its life cycle (e.g. wheat, paddy, cabbage, tomato, mustard, radish, sunflower, carrot, ginger, daffodil, turnip).

herbaceous: Refers to primary, soft, nonwoody tissue, as a plant or plant part or having the characteristics of an herb.

herbarium (pl. herbaria): (i) A collection of preserved, usually pressed and dried, plant material used for identification and comparison. (ii) A building in which such collections are stored.

herbicide: Class of crop protection and specialty chemicals used to control weeds in agriculture, horticulture, gardening and forests, as well as in non-agricultural applications such as golf courses, public properties and home lawns. Herbicide can be selective (kill selected species) or non-selective (broad spectrum, i.e. kill all plants).

herbicide resistance in transgenic plants: Herbicides normally affect process like photosynthesis or biosynthesis of essential amino acids. Two approaches have been used for development of resistant plants. In the first approach, it is seen that either the target molecules should become insensitive to herbicide, or the target protein should be overproduced. In the second approach, a pathway is introduced that will detoxify the herbicide. Target has been modified for developing resistance against three herbicides (glyphosate, sulfonylureas and imidazolinones). Resistance against glyphosate has been achieved in petunia by over expression of *5-enolpyruvyl shikimate-3-phosphate* (*EPSPS*) gene. Transgenic plants expressing a mutant *acetolactase synthase* (*ALS*) gene showed resistance to sulfonylureas and imidazolinones. A pathway is introduced that will detoxify the herbicide. A number of enzymes have been identified that detoxify herbicides. Glutathione-S-transferase (GST) (in maize and other plants) detoxifies the herbicide atrazine. This herbicide inhibits photosynthesis. Another enzyme nitrilase (coded by gene *bxn* in *Klebsiella pneumoniae* detoxifies) the herbicide bromoxynil. This herbicide also inhibits photosynthesis. Phosphinothricin acetyl transferase (PAT), coded by *bar* gene in *Streptomyces*, detoxifies the herbicide L-phosphinothricin (PPT). Transgenic plants using *bxn* gene from *Klebsiella* and *bar* gene from *Streptomyces* have been produced in tomato, potato, oilseed rape and sugarbeet.

herbicide resistance: Resistance in the crop to herbicides. Chemical weed control is now common practice in the agriculture where weeds are controlled by using herbicides such as Butachlor 50 EC (Machete) in rice and Atrataf 50 WP (Atrazine) in rice and 2,4-D in wheat. These herbicides are selective killers for weeds. Due to their continuous use, the weeds are becoming tolerant to herbicides. When higher doses of herbicides are used these affect the crop also. Therefore, we need crop varieties resistant to herbicides so that weeds can be killed even by using higher doses of herbicides without affecting the main crop. Using transgenic approach, genes such as *bar*, *bxn* and *tfda* have been introduced in several plant species to develop herbicide-resistant crop varieties. 'Round up' ready transgenic soybean is resistant to a common herbicide 'Round up' in the USA.

herbicide tolerance: Genetic adjustment of plant structures or metabolism that interferes with action of compounds toxic to plants. One can therefore apply the specific herbicide directly to the field without damaging crop.

herbicide-resistant plant: Many effective broad-spectrum herbicides do not discriminate between weeds and crops. Crop plants can, however, be modified genetically to make them resistant to herbicides, so as to eliminate weeds more selectively. For example, the herbicide Roundup™ contains the active ingredient glyphosate, which kills plants by binding to the active site of enzymes enolpyruvylshikimate phosphate (EPSP) synthase. This enzyme is critical for the synthesis of aromatic amino acids. Roundup is an extremely effective herbicide, but it kills almost all species of plants. On the other hand, it is very safe for humans and animals because they do not have EPSP synthase. By using rDNA technology, modified

EPSP synthase gene (that produced enzymes that are functional but are not inhibited by glyphosate) has been introduced into crop plants such as cotton and soybean. These genetically modified plants have been found to be highly resistant to treatment with Roundup. Genes that provide resistance to other herbicides, such as sulfonylureas and glufosinates, have also been developed and transferred to produce various transgenic plants.

herbicide-tolerant (HT) crops: Crop plants that have been modified to be able to survive application(s) of one or more herbicides by the integration of certain gene(s) via either genetic engineering, natural mutation or mutation breeding. Because it has been utilized for decades, most relevant national laws consider mutation breeding to be one of the so-called TRADITIONAL PLANT BREEDING techniques. For example, European laws that require special labelling of food products containing genetically engineered (via rDNA) crops do not require such special labelling for food products that contain crops created via mutation breeding. Several crops (soybean, canola, cotton, etc.) are made tolerant to glyphosate- or sulphosate-containing herbicides by the insertion (via genetic engineering techniques) of the *aroA* transgene (cassette) for CP4 EPSPS. Corn is made tolerant to glyphosate-containing herbicides by integration (via genetic engineering techniques) of the *mEPSPS* or *GA21* transgene (cassette). Some soybean varieties are made tolerant to sulfonylurea-based herbicides by adding (via traditional breeding methods) the *ALS* gene (which confers the sulfonylurea-tolerance trait). Corn and rice are made tolerant to imidazolinone-containing herbicides by adding (via traditional breeding techniques) the imidazolinone-tolerance trait. That trait is imparted by the T-Gene, imidazolinone resistant (IT)-Gene or the imidazolinone resistance (IR)-Gene.

herbicide-tolerant (HT) maize: HT maize is essentially a new weed-control approach. The potential agronomic changes from the use of HT maize are multidimensional and subtle. It must be understood within the context of maize farming systems and the inherent crop-weed competition. HT maize can change the weed-control systems of maize production, in general, as well as their economics.

herbicide-tolerant (HT) soybean: Genetically modified (GM) glyphosate-resistant (GT) soybeans are tolerant to glyphosate. This means that farmers can kill weeds by spraying glyphosate, without killing their soybean crops too. Glyphosate targets the shikimate pathway enzyme (EPSPS) which is essential for plant life.

hereditary cancer: Cancer that occurs due to the inheritance of an altered gene within a family. *See* SPORADIC CANCER.

hereditary determinant: Any genetically acting unit of an organism that is replicated and conserved, transferred from generation to generation.

hereditary factor: *See* HEREDITARY DETERMINANT.

hereditary univalents: Chromosomes that have evolved a system of approximately even distribution in segregation without going through the process of pachytene pairing. *Compare with* SPONTANEOUS UNIVALENTS.

heredity: (i) The passage of characteristics from parents to offspring. It is the cause of similarities between parents and offspring. It operates primarily by the germ cells in sexually reproducing species. (ii) The passing on of phenotypic traits from parents to their offspring, either through sexual or asexual reproduction. Offspring cells and organisms are said to inherit the genetic information of their parents. For this reason, brothers and sisters with the same parents resemble each other for certain characteristics.

heritability (h^2): A portion of phenotypic variance that is consequence of the varying genotypes of the individuals (V_G) in a population. Together with the environmental variance (V_E), it adds up to the total phenotypic variance (V_P) observed among individuals in a population. V_G comprises additive (resulting from differences between homozygotes, V_{add}) and dominance variance (resulting from specific effects of various alleles in heterozygotes, V_{dom}). Thus, $V_G = V_{add} + V_{dom}$. The quotient V_{add}/V_E is termed HERITABILITY IN NARROW SENSE (h_n^2), while V_G/V_E refers to HERITABILITY IN BROAD SENSE (h_b^2).

heritability in broad sense (h_b^2): The proportion of total phenotypic variation that can be ascribed to all the genetic differences among individuals in a population ($h_b^2 = V_G/V_P$, where V_G is total genotypic variance and V_P is phenotypic variance. h_b^2 is used to estimate the degree of genetic control of a trait in a population, and it is useful for predicting response to clonal selection. *See* HERITABILITY.

heritability in narrow sense (h_n^2): An estimate of heritability measuring contribution of only additive genetic effects contributing towards the phenotype

($h_n^2 = V_A/V_P$, where V_A is ADDITIVE GENETIC VARIANCE and V_P is PHENOTYPIC VARIANCE.

heritable: A genetic trait that can be inherited or passed on to future generations.

heritable basis of complex traits: Epigenetic differences in natural and experimental populations display the dynamic relationship between chromatin and DNA sequence factors in complex traits. The chromatin variation in experimental and natural populations serves as an added source of phenotypic variation. Indeed, chromatin differences independent of DNA sequence polymorphisms have the potential to transmit information across mitosis and meiosis in both mammals and plants at the cell, tissue and whole organism level.

heritable changes in gene expression: Methylation-sensitive enzymes have been used to explore epigenetics – heritable changes in gene expression – that affect flowering in *Arabidopsis*.

heritable epigenetic diversity and morphological diversity: Genetically identical East African Highland banana clones are epigenetically diverse with heritable epialleles that can result in morphological diversity. Heritable epigenetic variation can support agronomic traits in crops and should be considered in germplasm conservation.

heritable epigenetics: Stably heritable phenotype resulting from changes in a chromosome without alterations in the DNA sequence.

heritable legacy of induced epigenetic changes: Epigenetic marks show a tendency to undergo spontaneous gain or loss, as indicated by the analysis of a 30-generation pedigree in *Arabidopsis*. Why some loci should be prone to spontaneous epigenetic change? Epigenetic marks are gained and lost more easily than in other regions of the genome. In the allotetraploids of *Arabidopsis*, for example, there is a loss of repressive histone marks from the CCA1 and LHY circadian clock regulators and a consequent up-regulation of 130 downstream genes that are regulated by these factors. There is also evidence in hybrids of cultivated and wild tomato of altered production of siRNAs and, at one locus at least. A change was observed in the methylation status at the associated DNA.

herkogamy: Pollination by the neighbour individual, population individual or species individual.

hermaphrodite (Adj. hermaphroditic): An organism possessing both male and female reproductive systems. Most of the higher plants are hermaphrodite plants (e.g. wheat, rice, soybean, cotton, citrus, etc.) possess both male reproductive organs (andrœcium) and female reproductive (gynoecium) organs on the same plant even in the same flower. This kind of situation encourages self-population. In contrast to this, some plant species like papaya and date palm are unisexual, as their male and female parts are on different plants.

hermaphroditism: A condition in which an individual has both male and female sexes.

herpes viral vectors: Herpes viruses include infectious human viruses including herpes simplex virus (HSV). An infection by an HSV is marked by watery blisters in the skin or mucous membranes of the mouth, lips or genitals. HSV-1 and HSV-2 are two species of the herpes virus family whose genomes are complex and contain two unique regions called the LONG UNIQUE REGION (ULR) and the SHORT UNIQUE REGION (USR). HSV-derived vectors have very high transgenic capacity of the virus. Three different classes of vectors can be derived from HSV: REPLICATION-COMPETENT ATTENUATED VECTORS, REPLICATION-INCOMPETENT RECOMBINANT VECTORS and DEFECTIVE HELPER-DEPENDENT VECTORS (*aka* AMPLICONS).

hetero: A chemical nomenclature prefix meaning 'different'. For example, a HETEROCYCLIC COMPOUND is one with a (ring) structure made up of more than one kind of atom. A HETEROKARYON refers to a cell containing nuclei of different species.

heteroallele: An allele that differs from other alleles of the same gene by nucleotide differences at different sites within the gene. This is in contrast with TRUE ALLELES, of which only four are possible at each site within the gene. *Compare with* HOMOALLELE.

heterobrachial aberration: Chromosomal aberration involving both arms of a chromosome.

heterobrachial duplication: When a duplicate segment is present in another arm of the chromosome.

heterochromatic: Entire chromosomes or portions of the chromosomes that do not manifest the usual prophase-telophase transformations and appear to lack genes with major phenotypic effects such as a dense, compact structure in telophase, interphase and early prophase.

heterochromatic barriers: Heterochromatin (HC) that acts as a barrier to DNA double-strand break repair. Ataxia telangiectasia mutated (ATM) is essential for repair of HC-DSBs.

heterochromatic siRNAs (hc-siRNAs): These are the 24-nt-long small RNAs that are drawn in

TRANSCRIPTIONAL GENE SILENCING (TGS) which leads to silencing of endogenous genes or transgenes through the inactivation of promoter sequences. The siRNAs transcriptionally control gene expression by RNA-DIRECTED DNA METHYLATION (RdDM) pathway. The biogenesis of most hc-siRNAs depends on the plant-specific RNA polymerase IV.

heterochromatin (HC): Heterochromatic regions are generally gene poor, replicate late, have a low level of transcription as it is compacted, dark-stained interphase portion of chromatin that is inaccessible to transcription factors and conformationally unfavourable for transcription. HC is associated with repressive histone modifications, DNA methylation and gene silencing and consists primarily of epigenetically silenced pseudogenes and transposable elements. Certain chromosome sections, particularly those containing the centromeric and nucleolus organizers, may stick to form a chromocenter (the point at which the *Drosophila* polytene chromosomes join) and knobs (as in maize at pachytene). Some chromosomes are composed primarily of heterochromatin. These chromosomes are termed heterochromosomes, such as the Y chromosome in some species. It has a peripheral nuclear matrix, late-replicating, highly repetitive DNA. Heterochromatin replicates during latter half of the S phase. Heterochromatin is characterized by histone hypoacetylation (in all eukaryotes) and by methylation of histone H3 on lysine 9 (H3K9) in higher eukaryotes, but not in single-celled eukaryotes such as yeast. Histone methylated at H3K9 is bound by heterochromatin protein 1 (HP1), a highly conserved protein that is directly associated with pericentric heterochromatin. Heterochromatin may exist in two states – CONSTITUTIVE HETEROCHROMATIN and FACULTATIVE HETEROCHROMATIN. *See details as separate terms. Compare with* EUCHROMATIN.

heterochromatin assembly: The organization of eukaryotic genomes requires a harmony between efficient compaction and accessibility. It is characterized by the establishment of H3K9me and HP1/Swi6 proteins bound to H3K9me.

heterochromatin protein 1 (HP1): Conserved component of silent heterochromatic regions, which contains a chromodomain that binds nucleosomes containing methylated histone H3K9.

heterochromatinization: The conversion of euchromatin into heterochromatin during the inactivation of the X chromosomes in the embryogenesis of mammalian females. Inactive state of X-linked genes, namely, HYPOXANTHINE GUANINE PHOSPHORIBOSYL TRANSFERASE (HGPRT), PHOSPHOGLYCERATE KINASE (PGK) and GLUCOSE-6-PHOSPHATE DEHYDROGENASE (G6-PD), in female mammalian cells has been correlated with methylation in *Hpa*II tiny fragment (HTF) islands that are associated with these genes.

heterochromosome: (i) In animal systems, any chromosome that differs from the autosomes in size, shape and behaviour. (ii) In plants, M. Tahara (1910) discovered in *Morus indica*, the wild mulberry, a differentiation among the chromosomes comparable to that earlier discovered among animals. During the early prophase stages in nuclei of the sporophyte and still more noticeably as the chromosomes enter the equatorial plate, two pairs of the chromosomes are larger than the others. In the spore mother cells (X-generation), there are 14 chromosomes (bivalent), one pair of which shows larger than the rest.

heterochrony: An evolutionary developmental mechanism responsible for change in timing of development. It is a mechanism operating during embryonic development but that can be modified over the course of evolution and is thus instrumental in effecting evolutionary change in the phenotype.

heterocysts: Specialized cells of some cyanobacteria in which nitrogen fixation occurs. Some nitrogen-fixing and oxygen-producing cyanobacteria protect their nitrogenase by separating it into specialized cell known as heterocyst which do not produce oxygen and are impermeable to it. Thus, in this way they protect their nitrogenase while growing in the presence of oxygen.

heterodimer: Protein complex composed of two different polypeptide chains.

heteroduplex: A base-paired structure moulded between two polynucleotides that are not entirely complementary. They appear when primary pairing region, together with adjacent region of branch migration, encompasses the site(s) of genetic differences between the two parental chromosomes. Heteroduplex DNA model describes both crossing over and gene conversion by assuming the formation of a short stretch of heteroduplex DNA (shaped from both the parental DNAs) in the vicinity of a chiasma. Hybridization between an RNA strand and a DNA strand may produce heteroduplexes. Heteroduplexes may also be consequence of mutation or annealing DNA single strands *in vitro*.

heteroduplex analysis: A technique of detecting gene mutation by mixing PCR-amplified mutant and wild-type DNA followed by denaturation and re-annealing. The resultant products are identified by gel electrophoresis, with single-base substitutions detectable under optimal electrophoretic conditions and gel formulations. Large base pair mismatches may also be identified by using electron microscopy to envision heteroduplex regions.

heteroduplex DNA model: A model that explains both crossing-over and gene conversion by assuming the production of a short stretch of heteroduplex DNA (formed from both parental DNAs) in the vicinity of a chiasma.

heteroduplex DNA: Generated by base pairing between complementary single strands derived from different parental duplex molecules; heteroduplex DNA molecules occur during genetic recombination *in vivo* and during hybridization of different but related DNA strands *in vitro*. Since the sequences of the two strands in a heteroduplex differ, the molecule is not perfectly base-paired; the melting temperature of a heteroduplex DNA is dependent upon the number of mismatched base pairs.

heteroduplex mapping: Heteroduplex is a double-stranded nucleic acid molecule when two strands are not entirely complementary. Heteroduplex mapping is a type of mapping where heteroduplex analysis is used to determine the location of various insertions, deletions or heterogeneities between two DNA molecules. Heteroduplex mapping is a laboratory technique used to locate a particular sequence of nucleotides. It involves making a heteroduplex by pairing the unknown nucleic acid strand with a reference strand which has a known sequence. A HETERODUPLEX GENERATOR (HDG) is constructed by site-directed mutagenesis to contain a short deletion of 1–3 bases from the variable site. This construct is coamplified with target DNA, and after the last cycle of PCR, the amplicons are melted and allowed to reanneal. Apart from correctly annealed homoduplexes (target amplicons and HDG amplicons), heteroduplexes are formed, with a short bubble due to the deletion in the HDG. The size of the bubble depends on the surrounding sequences including the variant base. Thus, the heteroduplexes containing the variation are seen by their differential electrophoretic migration. Heteroduplexes are prepared by mixing the denatured DNA preparations (e.g., one type of a wild type of phage and other from a deletion phage). In case of such a heteroduplex, a loop can be seen under the electron microscope at the site of deletion, which could be mapped. Heteroduplex mapping makes use of the CHEMICAL CLEAVAGE OF MISMATCH (CCM) method to assess whether a region of DNA contains mutations and to localize them. CCM has a higher diagnostic sensitivity suited to the detection of mutations in tumour genes and can analyse amplicons ≤2 kb in length. To detect mutations, PCR heteroduplexes are incubated with two mismatch-specific reagents. Hydroxylamine modifies unpaired cytosine and potassium permanganate modifies unpaired thymine. The samples are then incubated with piperidine, which cleaves the DNA backbone at the site of the modified mismatched base. Cleavage products are separated by electrophoresis, revealing the identity and location of the mutation. The CCM method can efficiently detect point mutations as well as insertions and deletions. This protocol can be completed in 10 h.

heteroduplex mapping, uses of: S. Benzer (1962) located hundreds of deletion mutants. It permits determining size of the deletion and mapping of finding location of a deletion. This technique when extended to DNA-RNA heteroduplexes has permitted the discovery of demarcations of introns and exons in split genes. Heteroduplex mapping is also used to locate a particular sequence of nucleotides. Heteroduplex mapping has applicability of DNA heteroduplex patterns in establishing identity/non-identity of alleles in the major histocompatibility complex.

heteroduplex modified DNA: DNA having no methyl group on a particular adenine or cytosine on one DNA strand included in a host-specificity site.

heteroecious: (i) A species that forms male and female gametes on different individuals. (ii) The requirement of a pathogen for two host species to complete its life cycle. *Contrasts* AUTOECIOUS.

heterofertilization: Fertilization of the nuclei of endosperm and embryo-forming cells by genetically different gametes.

heterogametic: A species that sexually reproduces by two types of gametes. *Aka* HETEROGAMIC.

heterogametic sex: The sex that carries distinct sex chromosomes. An individual that forms two types of gametes with respect to sex determination. For example, in mammals, males are the heterogametic sex. Males are XY in most mammals, but in birds the females are the heterogametic sex. *Compare with* HOMOGAMETIC SEX.

heterogamic mating: Mating between individuals of different genotypes.

heterogamy: Reproduction process involving union of two types of gametes.

heterogeneity: Lack of agreement of different bodies of data in regard to the value of one or more parameters of which they are capable of yielding estimates.

heterogeneity index: A measure for degree of genetic differences within populations.

heterogeneous: (i) Means composed of different components. For example, tumours are described as being a heterogenous mixture of cells – non-cancerous cells and cancerous cells with a variety of different genetic mutations. This term is also commonly used during the cell culturing when the cell population is not uniform, which may differ genetically, physiologically or morphologically. Even at organism level if individuals in a population are not uniform, it is called heterogeneous population. (ii) Catalysis occurring at a phase boundary, usually a solid-fluid interface. (iii) A chemical reaction in which the reactants are of different phases: for example, gas with liquid, liquid with solid or a solid catalyst with liquid or gaseous reactants. (iv) One that consists of two or more phases, such as liquid-vapour or liquid-vapour-solid.

heterogeneous biodiversity restoration: To succeed, large-scale restoration requires the development of national and regional policies that promote the supply of considerable species sets, accounting for local and regional diversity. Ideally, the heterogeneity found in reference ecosystems should guide goal setting and species selection.

heterogeneous nuclear ribonuclear protein (hnRNP): Any of a group of proteins that assemble on newly synthesized RNA, organizing it into a more compact form. hnRNPs are involved in several aspects of the cell cycle by recruiting, splicing and co-regulating certain cell cycle control proteins.

heterogeneous nuclear RNA (hnRNA): The nuclear RNA fraction that comprises the unprocessed transcripts synthesized by RNA polymerase II, converted to mRNAs after capping, splicing and polyadenylation.

heterogenetic: Chromosome pairing between more or less different genomes in amphiploids.

heterogenic: Gametes or populations differing in alleles or genes.

heterogenote: A partially diploid bacterium, with different alleles at the diploid locus, resulting from transduction.

heterograft: The scion and rootstock derived from different species. *Aka* HETEROLOGOUS GRAFT.

heterokaryon (Adj. heterokaryotic): A cell possessing two or more than two genetically different nuclei in the common cytoplasm. During protoplast fusion the parental nuclei in the beginning coexist in the common cytoplasm of newly formed somatic hybrid cell. This kind of coexistence of the nuclei is called heterokaryon and the cell containing heterokaryon and cytoplasm from both parents is known as HETEROKARYOCITE. *Aka* HETEROCARYON.

heterokaryon test: A test for cytoplasmic mutations, based on new associations of phenotypes in cells derived from specially marked heterokaryon.

heterokaryosis: The presence of two or more genetically different nuclei within the single cell of a mycelium.

heterokaryotic vigour: The hybrid vigour brought about by two different nuclei in a heterokaryon in a single cell of a mycelium.

heterokaryotype: A chromosome complement that is heterozygous for any type of chromosome mutations.

heterolabelling: A chromosomal labelling pattern resulting from induced or spontaneous exchange of labelled and non-labelled half-chromatids.

heterologous DNA: Refers to a double-stranded DNA molecule in which each of the strands is derived from different sources (e.g. different species).

heterologous duplication: When a duplicate segment is present in the nonhomologous chromosome.

heterologous encapsidation: The formation of 'new' viruses by surrounding one virus with the envelope protein of another virus. It is a natural process that can occur when plants are co-infected by different strains of viruses. It has also been observed under laboratory conditions for transgenic, virus-resistant plants that produce viral envelope (capsid) proteins. Since envelope proteins are responsible for virus transmission (e.g. by insects like aphids), heterologous encapsidation can change mode of transmission of a virus. This can allow animals to transmit plant viruses in ways that were not previously possible. It is only effective for one virus generation because in case of heterologous encapsidation the virus genome is not modified, the original envelope proteins will be produced by these viruses in the next generation. *Aka* TRANSCAPSIDATION.

heterologous gene expression: Expression of a gene in another host.

heterologous genes for metabolic engineering: This approach has been used for production of new compounds. Completion of partial pathways to obtain new product, transfer of an entire biosynthetic pathway, heterologous host for producing novel products, altering nutrient uptake and metabolite flow are some of the applications.

heterologous introns in enhancing the transgene expression: Certain genomic constructs were expressed more efficiently in transgenic mice than constructs that were identical except for the lack of introns. Using the mouse metallothionein promoter-rat growth hormone gene construct as a model, it has been shown that the *rrst* intron of the rat growth hormone gene is essential for high-level expression, whereas the other three introns are less effective. Several heterologous introns placed 3′ of the coding regions of an intronless rat growth hormone gene are also ineffective. However, insertion of some heterologous introns between the metallothionein promoter and the growth hormone gene improves expression. To determine whether addition of heterologous introns would provide a general strategy for improving expression, they tested them in conjunction with other intronless genes and with different promoters.

heterologous probing: Searching at low stringency with a DNA fragment that originates from another organism and thus does not have an identical counterpart in the target DNA. Frequently it gives significant signals because of sequence conservation.

heterology: A sequence of amino acids in two or more proteins that are not identical to each other.

heteromeric: Genes that govern a trait by joint gene action, but each of them shows a definitely different influence on the final product.

heteromorphic: Chromosomes that differ in size and/or shape. The term also refers to meiotic pairing configuration, which is composed of different chromosomes and/or chromosome segments. *Compare with* ISOMORPHIC.

heteromorphic bivalent: A bivalent comprising of nonhomologous chromosomes or segments.

heteromorphic chromosomes: Homologous pair of chromosomes members of which do not show morphological identical, for example, the sex chromosomes X and Y in a heterogametic sex.

heteromorphic incompatibility: *See* INCOMPATIBILITY.

heteromorphic self-incompatibility: In this type of incompatibility, same species produces more than one morphological type of flowers. For example, relative length of styles and stamens in flowers of same species.

heteromorphs: The occurrence of plants having flowers of different forms with respect to the length of the stamens and pistils.

heteromultimeric protein: A protein which consists of non-identical subunits (coded by different genes).

heteronuclear ribonucleoprotein M (hnRNPM): An indispensable RNA-binding protein that suppresses cryptic splicing via binding to deep introns, maintaining human transcriptome integrity. *See* CRYPTIC SPLICING *and* CRYPTIC SPLICE SITES.

heteroplasmy: A mixture of mitochondria in a single cell. Some of the mitochondria or chloroplast may contain wild-type DNA and some may contain mutant DNA. *Aka* HETEROPLASTOMY.

heteroplastidic: Cells that carry plastids different in shape.

heteroploid: The term given to a cell culture when the cells comprising the culture possess nuclei containing chromosome numbers other than the diploid number. This is a term used only to describe a culture and is not used to describe individual cells. Thus, a heteroploid culture would be one which contains aneuploid cells.

heteropolymer: An artificial nucleic acid molecule made up of a mixture of different nucleotides.

heteropyknosis: A condition in which chromosomes chromosome regions are out of phase with respect to staining properties and the coiling cycle.

heteropyknotic: Chromosomes or chromosomal segments that show a different coiling or staining pattern.

heterosis: The foundation of a comprehensive understanding of *heterosis* was laid by G.H. Shull (1908) and E.M. East (1936). The concept was further amplified by G.H. Shull (1952). The term heterosis has different connotations for different workers. According to Th. Dobzhansky (1952), true heterosis is the mutational EUHETEROSIS which results from sheltering deleterious recessive mutants by the adaptive superior dominant alleles in the population of sexually reproducing and cross-fertilizing organisms. He also described 'THE BALANCED EUHETEROSIS' which arises out of a balanced type of adaptive mutations and these mutations confer on the heterozygote a better adaptive value than is found in corresponding homozygotes. The condition is sometimes called OVERDOMINANCE. The term heterosis is usually applied to size, rate of growth, survival, fertility or general fitness. Partial dominance, complete dominance and overdominance are considered three different aspects of the phenomenon of heterosis.

heterosis, biochemical basis of: K.S. Trehan and K.S. Gill (1987) at the Punjab Agricultural University, Ludhiana, Punjab, India investigated the biochemical basis of heterosis using acid phosphatase gene-enzyme system of *Drosophila* as a model. They discovered through partial purification of the individual electrophoretic bands that superior performance of the heterozygote compared to the two parents and the F_1 was due to the presence of two homodimeric (formed by two identical proteins) and one heterodimeric (made up of two similar but not identical subunits) enzyme compared to the two parents which had only one type of homodimeric enzyme.

heterosis, types of: In breeding experiments, three types of heterosis are distinguished: (i) F_1 yielding more than the mean of the two parents (J.L. Brewbaker, 1964); (ii) F_1 yielding more than the best yielding parents; and (iii) F_1 yielding more than the best yielding variety. For describing the genetic basis of heterosis, two hypotheses have received the most attention: DOMINANCE HYPOTHESIS and OVERDOMINANCE HYPOTHESIS. *See details as separate terms. See* HYBRID VIGOUR.

heterosomal: A chromosome mutation that involves nonhomologous chromosomes.

heterosome: A chromosome that departs from the standard chromosomes in size, shape or behaviour.

heterospory: The production of spores of two different sizes (small and large) by the sporophytes of land plants. *Compare with* HOMOSPORY.

heterostyly: A species possessing flowers with different style and stamen lengths. However, all the flowers of any one plant are identical. *See* DISTYLY.

heterothallism (Adj. heterothallic): The condition in which sexual reproduction occurs only in the presence of genetically different mycelia. *Contrasts* HOMOTHALLISM.

heterotroph (Adj. heterotrophic): Organism that cannot manufacture its own vitamins and other organics. The field-grown plants are considered autotrophs as they can synthesize organics for their growth through the process of photosynthesis, whereas the same plants cultured *in vitro* and kept for their growth in the growth room under fluorescent light are considered heterotrophs as they cannot synthesize sufficient organics in the absence of photosynthesis under tissue culture conditions. *Contrasts* AUTOTROPH.

heterotrophic: See HETEROTROPH.

heterotrophy: An evolutionary developmental mechanism responsible for alteration in spatial location of one or more developmental processes. One of E. Haeckel's primary examples of heterotrophy given in 1866 was the origin of germ cells from endoderm or mesoderm in different animal groups.

heterozygosity: The presence of different alleles in a population at one or more loci on homologous chromosomes.

heterozygosity index: A measure of genetic variance in a population, with respect to one locus, stated as the frequency of heterozygotes for that locus.

heterozygote: Diploid organism/cell possessing two different alleles. When two homozygotes are crossed, the F_1 is heterozygote. For example, homozygote (*AA*) homozygote (*aa*) resulted in F_1 (*Aa*) heterozygote. Heterozygotes do not breed true as the resulting F_2 population exhibit variation with respect to various characters. In crops, F_1 hybrids are given to farmers that give higher yield. Due to hybrid vigour the hybrids are higher yielding.

heterozygote advantage: A condition in which the fitness of a heterozygote is higher than the fitness of either of the homozygotes. *See* OVERDOMINANCE *and* HETEROSIS.

heterozygous: Carrying two different variants (i.e. alleles) of a given gene. An individual having inherited different forms of a particular gene from each parent. *Compare with* HOMOZYGOUS.

hexaploid: (Usually) a diploid with duplicate sets of three different genomes (6x). Common bread wheats (6x = 2n = 42 chromosomes) are hexaploids in contrast to primitive diploid wheats with 2n = 14.

hexasomes: Noncanonical nucleosomes with six instead of eight histones. Hexasomes are linked to actively transcribed genes. M. Zhang *et al.* (2023) report structural and mechanistic findings on the remodelled subfamily of ATP-dependent chromatin remodelling complex INO80, which prefers hexasomes over canonical nucleosomes. Hexasomes not only modify the packaging of DNA but may also alter how enzymes and other factors interpret the regulatory information of chromatin. H. Wu *et al.* (2023) determined a structure showing how INO80's remodelling subunit binds at a location called super-helical location-2 (SHL-2) on a hexasome, rotated 180° compared with its position on a nucleosome. Both hexasome and nucleosome sliding by INO80 necessitate action from SHL-2, suggesting additional steps that slow nucleosome sliding. INO80's highly regulated mechanism explains its functional versatility. *Aka* NONCANONICAL NUCLEOSOME REMODELLING.

hexasomic: A cell or individual displaying one chromosome six times.

***Hha*I:** Type II restriction enzyme isolated from *Haemophilus haemolyticus* that recognizes the tetranucleotide target sequence. Cleavage of the DNA duplex occurs indicated by arrows, 5′-G C G↓C 3′ / 3′-C↑G C G 5′, to yield a fragment of DNA having sticky ends.

hidden genes: These are short open reading frames (i.e. sequences initiated by the alternative translation initiation codons ACG, AUA and GUG in the 5′ terminal extra-cistronic region).

hidden Markov model (HMM): A joint statistical model for an ordered sequence of variables. The result of stochastically perturbing the variables in a Markov chain (the original variables are thus 'hidden'), where the Markov chain has discrete variables that select the 'state' of the HMM at each step. The perturbed values can be continuous and are the 'outputs' of the HMM. A hidden Markov model is equivalently a coupled mixture model where the joint distribution over states is a Markov chain. Hidden Markov models are valuable in bioinformatics because they allow a search or alignment algorithm to be trained using unaligned or unweighted input sequences and because they allow position-dependent scoring parameters (e.g. gap penalties), thus more accurately modelling the consequences of evolutionary events on sequence families.

hidden transcription: CRYPTIC UNSTABLE TRANSCRIPTS (CUTs) have been described as a main class of RNA Polymerase II transcripts in budding yeast. These transcripts are targeted for degradation immediately after synthesis by the action of the Nrd1-exosome-TRAMP complexes.

hierarchical classification: A system of classifying organisms in a nested series of levels including life, domain, kingdom, phylum, class, order, family, genus and species.

hierarchical shotgun DNA sequencing: Hierarchical shotgun approach includes breaking the genome into relatively large chunks, called CLONES. Each clone is then cut into smaller overlapping pieces and sequence the pieces and use the overlaps to reconstruct the sequence of the whole clone. *Aka* CLONE-BY-CLONE DNA SEQUENCING, TOP-DOWN SEQUENCING *or* MAP-BASED SEQUENCING.

hierarchy: Any system of organization based on level or ranks.

high information content BAC fingerprinting (HICF): A technique for the detection of minimum overlaps between neighbouring BAC clones that is based on the 1–5 base terminal sequences of the clones. HICF increases the power to detect clone overlaps and helps to accurately assemble contigs, to close gaps, to verify existing contigs and to establish minimal tiling paths with fewer clones.

high-temperature-induced promoter methylation modifications: Anther development in flowering plants is extremely sensitive to HIGH-TEMPERATURE (HT) STRESS. Understanding the probable epigenetic mechanism of anther infertility caused by HT stress in cotton (*Gossypium hirsutum* L.) is vital for the real use of genetic resources to guide plant breeding.

high-amylose corn: Refers to those corn hybrids that produce kernels in which the starch is contained within those kernels is at least 50% amylose, versus the average of 24–28% amylose in traditional corn starch.

high-density bacterial artificial chromosome grid (high-density BAC grid): Any nitrocellulose or nylon membrane on to which different BAC clones are speckled at high density.

high-density molecular maps: The SATURATED MAPS prepared with the help of molecular markers such that markers are spaced out at recombination frequencies of 1–5cM or about 100–500 kb physical distances. Some laboratories consider a map a high-density map if marker density is 0.5cM or less (genetic map) and 5–10 kb or less (physical map). The molecular markers provide capacity for complete coverage of nuclear, mitochondrial and chloroplast genomes.

high-density molecular maps, molecular markers: Molecular markers used for this purpose are of several kinds. *Aka* HIGH-RESOLUTION MAP, HIGH-DENSITY GENETIC MAP, HIGH-DENSITY LINKAGE MAPS, HIGH-RESOLUTION GENETIC MAP *or* HIGH-RESOLUTION PHYSICAL MAP. *See* AMPLIFIED FRAGMENT LENGTH POLYMORPHISM (AFLP), AMPLICON LENGTH POLYMORPHISM (ALP), ARBITRARILY PRIMED PCR (AP-PCR), ALLELE-SPECIFIC PCR (AS-PCR), DNA AMPLIFICATION FINGERPRINTING (DAF), RANDOM AMPLIFIED POLYMORPHIC DNA (RAPD), RESTRICTION FRAGMENT LENGTH POLYMORPHISM (RFLP), SATURATED LINKAGE MAP, SPECIFIC AMPLICON POLYMORPHISM (SAP), SEQUENCE-CHARACTERIZED AMPLIFIED REGION (SCAR), SINGLE-STRAND CONFORMATION POLYMORPHISM

(SSCP), MICROSATELLITE SIMPLE SEQUENCE LENGTH POLYMORPHISM (MSSLP), SIMPLE SEQUENCE REPEATS (SSR) *and* SEQUENCED TAGGED SITES (STS). Of all these markers, RFLP and RAPD are the principal classes of markers. High-density molecular maps are helpful in demonstrating SYNTENY. *See details as separate terms.*

high-efficiency particulate air (HEPA) filter: A filter used in the Laminar Air Flow cabinets that clean the air by removing particles (microbes, dust particles) larger than 0.2 μm from the air. HEPA is abbreviation for High-Efficiency Particulate Air filter which is fitted in the Laminar Air Flow cabinets, growth rooms and operation theatres fitted with positive pressure modules. HEPA filters are of pore size ranging from 0.2 to 0.4 μm. Out of these the one with pore size of 0.2 or 0.3 is common in use. Most of the microbes are bigger and get retained, whereas viruses and mycoplasmas are smaller and can pass through, but these do not create problem being obligates and thus do not grow on the culture medium.

high-efficiency transformation of *Penicillium nalgiovense*: *Penicillium nalgiovense* is a filamentous fungus that is gaining increasing biotechnological importance in the food industry owing to its widespread use as starter culture for cured and fermented meat products. Strains of *P. nalgiovense* can be improved by genetic modification to remove the production of penicillin and other potentially hazardous secondary metabolites, to improve its capacity to control the growth of undesirable fungi and bacteria on the meat product and other factors that contribute to the ripening of the product in order to get safer and better-quality foods. Genetic manipulation of *P. nalgiovense* has been restricted by the lack of molecular genetics tools that have been available for this fungus, particularly for 'self-cloning' avoiding the use of exogenous DNAs. A series of vectors, selectable markers and transformation methods are now available that can be engaged for efficient transformation of *P. nalgiovense*, gene cloning and expression. A uridine auxotrophic *P. nalgiovense* mutant with an inactive *pyrG* gene has been isolated. The *P. nalgiovense* wild-type *pyrG* gene has been cloned and sequenced, and vectors carrying the gene shown to complement the *pyrG* mutant. Autonomously replicating plasmids carrying the AMA1 region from *Aspergillus nidulans* transformed *P. nalgiovense* very efficiently (F. Fierro *et al.*, 2004). These plasmids can be maintained as stable extrachromosomal elements in *P. nalgiovense* and rescued in *Escherichia coli*. The mitotic stability of self-replicative AMA1 plasmids in *P. nalgiovense* is higher than that for *Penicillium chrysogenum*.

high-energy bond: Covalent bond whose hydrolysis produces an unusually large amount of free energy under the conditions existing in a cell. A group linked to a molecule by such a bond is willingly transferred from one molecule to another. Examples of high-energy bond include the phosphodiester bonds in ATP and the thioester linkage in acetyl-CoA.

high-frequency recombination (hfr) bacteria: The F plasmid can behave as an episome when the F^+ plasmid is integrated within the bacterial chromosome. Hfr means high frequency of recombination cell. The Hfr cell still behaves as a F^+ cell, transferring F genes to a F^- cell but can take some of the bacterial chromosome with it.

high-isoflavone soybeans: Developed in the USA in the 1990s, these are soybean varieties which contain greater proportion of isoflavones than do the traditional soybean varieties (i.e. isoflavones constitute 0.15–0.3% of a traditional variety soybean's dry weight). Consumption of isoflavones helps to reduce the blood level of low-density lipoproteins ('bad cholesterol') in humans.

high-lactoferrin rice: Refers to rice plants (*Oryza sativa*) that have been genetically engineered to yield substantial amounts of lactoferrin in the grain they yield. Lactoferrin is a compound that is naturally manufactured in human breast milk. Consumption of lactoferrin by infants helps to strengthen their immune system. Consumption of lactoferrin (e.g. from genetically engineered rice) by older humans helps their immune systems resist some infectious diseases. Lactoferrin 'binds' free iron (e.g. in body fluids), thereby denying that iron to pathogenic bacteria (which need free iron to grow/infect).

high-laurate canola: Refers to canola (*Brassica napus/campestris*) varieties genetically engineered (e.g. via insertion of gene for lauroyl-ACP thioesterase) to have at least 40% laurate (lauric acid) in their oil (in seed).

highly conserved sequence: DNA sequence that is very similar across several different types of organisms. *See* GENE *and* MUTATION.

high-lysine corn: Developed in the USA in the mid-1960s, these were initially corn (maize) varieties possessing the *opague-2* gene. This gene is responsible for corn to contain 0.30–0.55% lysine (i.e. 50–80%

more than traditional No. 2 yellow corn). Other genes were later been discovered that when inserted into the corn genome (e.g. via genetic engineering techniques) cause production of larger amounts of lysine than in traditional corn varieties. High-lysine corn is particularly advantageous for feeding of swine.

high-mannose oligosaccharide: Refers to chain of sugars linked to a glycoprotein which contains many mannose residues. It is produced by a trimming of the original mannose-rich oligosaccharide that leaves most of the mannose residues with no subsequent addition of sugars.

high-methionine corn: Developed in the USA in the mid-1960s. These were initially corn varieties possessing the *floury2* gene which causes corn to contain slightly higher levels of methionine than traditional No. 2 yellow corn. Other genes were also subsequently discovered that, when inserted into corn/maize genome (e.g. via genetic engineering techniques), could reason the production of larger amounts of methionine than in traditional corn/maize varieties. High-methionine corn is particularly beneficial for feeding of poultry, since traditional No. 2 yellow corn does not contain enough methionine for optimal poultry (especially feather) growth.

high-mobility group (HMG) proteins: Non-histone proteins (e.g. HMG-14 and HMG-17) that are involved in the switch from an inactive to an active chromatin structure by associating themselves with nucleosomes.

high-oil corn: Conceived in 1896 at the University of Illinois in the USA, high-oil corn (HOC) is defined to be corn (maize) possessing a kernel oil content of 5.8% or greater. Traditional No. 2 yellow corn varieties tend to contain 4.5% or less oil content.

high-oleic oil soybeans: Soybeans from plants which have been genetically engineered to yield soybeans having oil that contains more than 70% oleic acid, instead of the typical 24% oleic acid content of soybean oil produced from traditional varieties of soybeans. Cosuppression, via inserted gene for Δ 12 desaturase (an enzyme that normally converts oleic acid to linoleic acid as part of the oil creation process in traditional varieties of soybean plants), results in the higher than traditional amount of oleic acid in the soybean oil. High-oleic soybean oil would be disposed to possess greater oxidative stability (especially at elevated temperatures) than soybean oil from traditional varieties of soybeans. Because of that, nuts that were fried in high-oleic oil have been shown to possess a longer shelf life than nuts fried in traditional oils. A human diet containing a large amount of oleic acid causes lower blood cholesterol level and consequently lower risk of coronary heart disease (CHD).

high-performance liquid chromatography (HPLC): Chromatography carried out in a vacuum, in which the solvent phase is forced through the soli phase under high pressure, improving speed and under certain circumstances, resolution.

high-performance ultrafiltration (UPUF): A type of high-performance liquid chromatography that uses a gel filtration column.

high-phytase corn and soybeans: Crop plants that have been genetically engineered to contain in their grain/seed high(er) levels of the enzyme phytase (which aids digestion and absorption of phosphate in that grain/seed). High-phytase grains or oilseeds are particularly suitable for the feeding of swine and poultry, since traditional No. 2 yellow corn (maize) or traditional soybean varieties do not contain phytase in amounts required for complete digestion/absorption of phosphate naturally contained in those traditional soybeans and corn varieties in the form of phytate.

high-precision DNA transactions: These are those genetic processors that involve interaction between DNA and proteins and between proteins and proteins. Such interactions are involved in site-specific recombination, initiation of DNA replication and regulation of gene action. DNA topoisomerases are enzymes that carry out vital cellular processes like DNA strand scission, rejoining of strands and modulation of DNA supercoiling inside cells. These enzymes solve various topological problems required for high-precision DNA transactions such as transcription, replication, recombination, chromosomal condensation and segregation. *See* TOPOISOMERASES.

high-resolution genome-wide mapping of histone modifications: GENOME-WIDE MAPPING TECHNIQUE (GMAT) is used for this purpose. A combination of SERIAL ANALYSIS OF GENE EXPRESSION (SAGE) and CHROMATIN IMMUNOPRECIPITATION (ChIP) protocols has been used to determine the distribution of hyperacetylated histones H3 and H4 in the *Saccharomyces cerevisiae* genome.

high-resolution melt analysis (HRM) of DNA methylation: DNA from forensic stains can be isolated, processed using bisulfite-modified PCR and detected by real time PCR with HRM capability.

high-resolution QTL mapping: Aim of QTL mapping is to produce a comprehensive framework that covers all chromosomes evenly in order to identify markers flanking those QTLs that control traits of interest. Even the closest markers flanking a QTL may not be tightly linked to a gene of interest. This means that recombination can occur between a marker and QTL, thus reducing the reliability and usefulness of the marker. By using larger population sizes and a greater number of markers, more tightly linked markers can be identified.

high-scoring segment pair (HSP): Local alignments with no gaps that achieve one of the top alignment scores in a given search.

high-stearate canola: The seeds of canola varieties which have been genetically engineered contain a higher percentage of stearate (*aka* STEARIC ACID) in the canola oil than the typical stearate content in canola oil produced from traditional canola varieties. Cosuppression, via inserted gene for D-stearoyl-ACP desaturase (i.e. enzyme that normally converts stearic acid to oleic acid in the oil creation process in traditional varieties of canola), causes the higher than traditional amount of stearic acid in the canola oil.

high-stearate soybeans: Soybean plant varieties which have been bred or genetically engineered so their beans contain at least 12% stearate (*aka* STEARIC ACID) within their soybean oil (i.e. more than four times the typical 3% stearic acid content in the soybean oil produced from traditional soybean varieties). Some high-stearate soybeans contain more than 20% stearate. Cosuppression, via inserted gene for D-stearoyl-ACP desaturase (i.e. enzyme that normally converts stearic acid to oleic acid in the oil creation process in traditional varieties of soybeans), is the primary way to cause the higher than traditional amount of stearic acid in the resultant soybean oil. A human diet containing stearate instead of alternative saturated fatty acids does not cause an increase in blood cholesterol levels.

high-sucrose soybeans: Another name for low-stachyose soybeans because the soybeans replace the (reduced) stachyose with (additional) sucrose.

high-temperature adaptation: An allele of the nitrate transporter gene, *OsNRT2.3*, is required to maintain high yield and high nitrogen use efficiency under high temperatures. Y. Zhang *et al.* (2022) demonstrate that this tolerance to high temperatures in rice accessions harbouring the *HTNE-2* (*high-temperature resistant and nitrogen efficient-2*) alleles from enhanced translation of the *OsNRT2.3b* mRNA isoform and the decreased abundance of a unique small RNA (sNRT2.3-1) derived from the 5′ untranslated region of *OsNRT2.3*. sNRT2.3-1 binds to the *OsNRT2.3a* mRNA in a temperature-dependent manner. The allelic variation in the 5′ untranslated region of *OsNRT2.3* leads to an increase in OsNRT2.3b protein levels and higher yield during high-temperature stress. These results provide a breeding strategy to produce rice varieties with higher grain yield and lower N fertilizer input suitable for a sustainable agriculture that is resilient against climate change.

high-throughput antibody-independent method for RNA modification analysis: Method depends on the anti-6^{m}A antibody, but its use is limited by its poor reproducibility and poor resolution.

high-throughput approach: An approach in the molecular marker technology for genotyping of larger populations in minimum time. High-throughput approach is usually required while genotyping large number of entries (breeding populations) using efficient DNA marker systems such as AFLPs and microsatellites. This involves the mass DNA extraction using Matrix Mill or FastPrep systems, followed by DNA amplification through DNA Engine Tetrad (Tetrad PCR machine) and laser detection through fluorescent labelling and automated sequencing machines.

high-throughput DNA sequencing: The process of determining the exact order of the four-letter code (bases, G, A, T or C) that comprises all genetic material. Because a representative higher organism contains more than 20,000 genes, each with several thousand bases of DNA, data storage and its analysis have been among the most important challenges of this process. J.C. Venter *et al.* (2004) applied whole-genome shotgun sequencing to microbial populations from seawater samples collected from the Sargasso Sea near Bermuda and identified over 1.2 million previously unknown genes represented in these 1800 genomic species. They harnessed the power of high-throughput DNA sequencing and computational genomics to produce a massive data set of large DNA fragments from total microbial genomes. They discovered an enormous number of new genes from a few samples extracted from the subtropical North Atlantic Ocean of the Bermuda coast. These efforts are critical to understanding how life evolved.

high-throughput genome sequences (HTGS): High-throughput genome sequences DDBJ/EMBL/Gen Bank division dealing with high-throughput genome sequences. Now several genome sequencing centres worldwide have started the large-scale sequencing of human and other higher eukaryotic genomes. The databases have deemed at beneficial to put the unfinished sequences that are the result of such sequencing efforts in a separate division. The unfinished records, in most cases, are notable for important numbers of gaps in the nucleotides, low accuracy and no annotations on the record.

high-throughput identification: Determination of the identification of a given chemical compound (e.g. within a mixture), the desired impact (CELL APOPTOSIS, etc.), a specific segment (sequence) of DNA (i.e. a specific gene), a specific ligand or receptor (e.g. 'attaching' itself to a given molecule), etc. within the overall process known as HIGH-THROUGHPUT SCREENING.

high-throughput measurements: Real time polymerase chain reaction is also called quantitative real time polymerase chain reaction (Q-PCR/qPCR/qrt-PCR) or kinetic polymerase chain reaction (KPCR). Both DNA microarrays and Q-PCR exploit the preferential binding or 'base pairing' of complementary nucleic acid sequences, and both are used in gene expression profiling, often in a serial fashion. While high-throughput DNA microarrays lack the quantitative accuracy of Q-PCR, it takes about the same time to measure the gene expression of a few dozen genes via Q-PCR as it would to measure an entire genome using DNA microarrays. So, it often makes sense to perform semi-quantitative DNA microarray analysis experiments to identify candidate genes, then perform Q-PCR on some of the most interesting candidate genes to validate the microarray results. Other experiments, such as a Western blot of some of the protein products of differentially expressed genes, make conclusions based on the expression profile more persuasive, since the mRNA levels do not necessarily correlate to the amount of expressed protein.

high-throughput screening (HTS): A methodology applied to rapidly screen large numbers of compounds for use as pharmaceuticals or agrochemicals (e.g. herbicides). For example, when finding chemical compounds for potential use as a pharmaceutical, the goal frequently is to evaluate differences between diseased and treated cells. HTS enables identification of a pharmaceutical candidate that favourably influences change in protein level (i.e. gene expression), which typifies a diseased state or some other gene expression marker (e.g. apoptosis). Although whole living cells or whole microscopic animals such as nematodes could be utilized in HTS, it is better to use a proxy (e.g. receptors, enzymes or STATs from applicable cells) whose interaction with candidate compounds can be inferred to cell (and/or organism) effects. Characteristically, these screenings are carried out in 96-well plates using automated, robotic station-based technologies or in higher-density array (chip) formats. HTS is a fast way of determining the order of bases in DNA. *See* SEQUENCING.

high-throughput sequencing technologies (H-TST): H-TST generate millions of short reads from a library of nucleotide sequences, whether they come from DNA, RNA or a mixture, the sequencing mechanism of each platform does not vary. There are many tools available for aligning genomic reads to a reference genome (sequence alignment tools); however, special attention is needed when alignment of a transcriptome to a genome, mainly when dealing with genes having intronic regions. Several software packages exist for short read alignment, and recently, specialized algorithms for transcriptome alignment have been developed (e.g. *TopHat* and *Cufflinks*).

high-velocity microprojectile transformation: A technique for introduction of DNA into plant cells. For example, gold particles are layered with DNA and driven at high speed through the target cell walls by means of an electrical or gunpowder expulsion.

high-voltage electron microscopy: Type of transmission electron microscopy that accelerates electrons through 10^6 volts and can be used for examining sections as thick as 1 μm.

high-yielding variety: The phrase refers to 'modern' (post-1960) crop varieties having a much higher grain yield potential (5–10 t/ha) than traditional tall varieties. For example, in maize, hybrids or composites were developed. In wheat and rice, usually stiff-strawed, short-statured varieties or hybrids were developed. High-yielding varieties are more responsive to fertilizer application than local tall varieties and their large-scale cultivation triggered the Green Revolution.

higher category: A taxonomic unit larger than a species of genus, such as order, class and phylum.

highly oriented pyrolytic graphite (HOPG): HOPG is widely used in scanning tunnelling and force microscopies as the substrate material because of its atomic flatness and cleanness.

highly repetitive DNA sequences: Highly repetitive DNA sequences (10^6–10^7) have simple sequence organization comprising a basic unit of about ten nucleotide pairs. As these basic units are repeated in tandem, these sequences are also termed CLUSTERED REPEATS. These are generally not transcribed and in chromosome apparently arise from sudden replication of pre-existing sequences. Divergence of family members proceeds by nucleotide substitution in individual members and the extent of divergence, more or less, reflects age of a family. Highly repetitive DNAs have been demonstrated to have an important part in the process of speciation. These sequences are observed in regions of the chromosome such as heterochromatin, centromeres and telomeres and tend to be organized as a tandem repeat. The tandemly repeated sequence is exemplified by ATTATA ATTATA ATTATA // ATTATA. Highly repetitive DNA sequences are five to a few hundred-bp long. Highly repetitive sequences are believed to be structural component of chromosomes as they are located mainly at centromere and telomeric positions, but their function is, however, unknown. Tandemly repetitive sequences exist as satellite DNA. Unequal crossing-over is mainly accountable for spreading of this DNA. For dispersed repetitive sequences, integration mechanism (such as in bacterial transposons) is thought to be responsible for evolution of this DNA.

highly repetitive DNA, functions of: Postulated functions for highly repetitive DNA include: (i) structural or organizational role in chromosome; (ii) participation in chromosome pairing during meiosis; (iii) contribution in crossing-over or recombination; (iv) 'protection' of important structural genes, like histone, rRNA or ribosomal protein genes; (v) a source of unessential DNA sequences for use in the future evolution of the species and (vi) no function at all – just 'junk' DNA that is carried along by the processes of replication and segregation of chromosomes. The validity of any of these postulated roles still remains to be established.

highly unstable RNAs: The bulk of eukaryotic genomes is transcribed. A class of short, polyadenylated and highly unstable RNAs has been reported. These promoter upstream transcripts (PROMPTs) are produced around 0.5–2.5 kb upstream of active transcription start sites. PROMPT transcription occurs in both sense and antisense directions with respect to the downstream gene. It necessitates the presence of the gene promoter and is positively associated with gene activity. PROMPT transcription is a common feature of RNA polymerase II transcribed genes with a possible regulatory potential.

HighSeq 2500: High-throughput sequencing system developed by Illumina.

Hill-Robertson effect: Refers to the interference between selection at linked loci. In a fundamental computer simulation study, B. Hill and A. Robertson in 1966 investigated multilocus interactions under realistic conditions, including selection and recombination in finite populations, where the possibility of fixation of a beneficial mutation in the presence of another segregating beneficial mutation. The results confirmed R.A. Fisher's predictions and revealed that selection at one locus hinders with selection at a second beneficial mutation reducing its probability of fixation. That is, in natural populations (with finite size), selection at more than one site causes an overall reduction in the efficacy of selection. The magnitude of this effect depends on selection intensities, initial frequencies of the alleles and more importantly, the degree of interference increases with genetic linkage between the loci under selection.

hilum: The scar on a seedcoat from where it separates from its stalk (funicle).

Hin system of site-specific recombination: Hin system of *Salmonella typhimurium* has recombinase genes and enhancers are located within the invertible segment, while the genes involved in variation lie outside. DNA invertase protein Hin switches the expression of the surface antigens of *Salmonella*.

hinge DNA: *See* DNA, H-FORM.

hip: The fruit of a rose plant.

hirudin: It is a polypeptide. Recombinant hirudin is encoded by a synthetic gene and is expressed in fusion with oil-body protein oleosin, which greatly facilitates the purification of heterologous (different from those endogenous to the plant species) polypeptides. The hirudin provides a successful example of transgene's expression in plants for the isolation of a polypeptide of interest at a commercial scale.

His3: A yeast selectable marker that encodes an enzyme required for histidine (His) biosynthesis. HIS3 yeast mutants cannot grow in media without His. On HIS-deficient medium, recombinants that restore the wild type gene are able to grow again.

histidine (his): A basic amino acid that is essential in the nutrition of the rat. It is formed by the decomposition of most proteins (as globin).

histidine-kinase-associated receptor: Type of transmembrane receptor existing in the plasma membrane of bacteria, yeast and plant cells and involved, for example, in sensing stimuli that result in bacterial chemotaxis. These receptors are associated with a histidine protein kinase on its cytoplasmic side.

histocompatibility: Tissues can be transplanted between genetically identical animals without anxiety for immunologic rejection, whereas transplants between genetically non-identical animals are generally rejected with time. *Aka* TISSUE COMPATIBILITY.

histocompatibility antigens: Those antigens that regulate the acceptance or rejection of a tissue graft. These antigens are produced by HISTOCOMPATIBILITY GENES. These genes are involved in the manufacture of cell surface antigens that are identified by the rejection or tolerance of tissue transplants.

histocompatibility genes: The genes that code for the histocompatibility antigens.

histogram: A diagram showing the number or percentage of subjects with values in different intervals by means of contiguous bars. It is used only for quantitative variables.

histology: The study of the microscopic anatomy of cells and tissues of plants and animals.

***HI*stome.** A freely available histone infobase, specialist, electronic database aimed to display information about human histone variants, sites of their PTLMs and about various histone-modifying enzymes.

histone acetylation (ac): The chemical process whereby a histone acetyltransferase enzyme adds an acetyl group ($-COCH_3$) to a specific lysine or arginine residue in the 'tail' of a histone protein. When present on histones it neutralizes their positive charge allowing chromatin decompaction and enhanced gene expression. Addition of an acetyl group to an amino acid of a histone protein. Histone acetyltransferases (HATs) cause acetylation reaction. Acetyl coenzyme A (acetyl-CoA) synthetase enzyme uses acetate to produce acetyl-CoA. Lysine is the only amino acid which is acetylated in histones. HATs control expression of genes and control cell cycle. Histone deacetylases play role in transcriptional silencing. Acetylation of histones reduces their ability to inhibit RNA synthesis *in vitro*. The rate of histone acetylation is enhanced in the types of cells where RNA synthesis is encouraged. Acetylation within the globular core domain of histone H3 on lysine 56 has been observed to have a critical role in packaging DNA into chromatin following DNA replication and repair in budding yeast. *See* HISTONE ACETYLTRANSFERASES.

histone acetylation analysis: Eleven-nineteen-leukaemia (ENL), also named mixed-lineage leukaemia translocated to 1 (MLLT1) protein in humans, is encoded by the *MLLT1* gene. This protein functions as a chromatin reader, interacts with acetylated histones and recruits/stabilizes the transcriptional machinery to enhance gene expression.

histone acetylation profiling: Histone acetyltransferases (HATs) and histone deacetylases (HDACs) play an important role in controlling histone acetylation levels. EpiQuik Histone Acetylation Quantification Kits developed by Epigentek are ELISA-based histone acetylation quantification tools which measure total levels of histone H3 or H4 acetylation.

histone acetyltransferases (HATs): Enzymes that add an acetyl group to specific amino acids on histone tails, promoting an 'open' chromatin state and transcriptional activation. Enzyme Rtt109 is known to mediate acetylation of H3K56 (H3K56ac) is fungal specific. In response to DNA damage, histones bearing acetylated K56 are assembled into chromatin in *Drosophila* and human cells, forming foci that colocalize with the site of DNA repair. Sirtuins, a group of conserved NAD^+-dependent deacetylases or ADP-ribosyltransferases, promote longevity in diverse organisms.

histone ADP-ribosylation (ADPr): Among histones, arginine, lysine and serine are ADP-ribosylated. For lysine, ADP-ribosylation involves four discrete reaction intermediates involving enzymes poly(ADP-ribose) polymerase and poly(ADP-ribose) glycohydrolase. Serine ADPr is stringently dependent on histone PARylation factor 1 (HPF1), a regulator of PARP-1. Arginine ADP-Ribosylation is catalysed by a subfamily of ADP-ribosyltransferases.

histone ADP-ribosylation removal: ADP-ribose removing depends almost exclusively on poly(ADP-ribose) glycohydrolase.

histone biotinylation (b): The process that covalently attaches biotin ($C_{10}H_{16}N_2O_3S$), *aka* VITAMIN H or B7, to a protein, nucleic acid or other molecule. Biotin is one of the B complex vitamins that help the body convert food into energy. Biotinylation of lysine in histones involves acetylation of unmodified amino acid. Acetylated lysine is then deacetylated by sirtuin. Deacetylated lysine is then biotinylated using sulfo-NHS-biotin as a donor of biotin. It is rapid, specific process and it is unlikely to disturb the natural function of the molecule due to the small size of biotin (MW = 244.31 g/mol). Biotinylation of histones plays a role in cell proliferation, gene silencing and cellular response to DNA damage.

histone butyrylation (bu) and 2-hydroxylisobutyrylation (ib): Butyrylation refers to a biochemical interaction process where the butyryl functional group covalently modifies the lysine amino acid of a protein. Four lysine butyrylation (Kbu) sites (H3K14, H4K12, H2BK42 and H2BK134) on rice histones have been identified. Butyrylation has been associated with transcriptional misregulation. Lysine 2-hydroxyisobutyrylation (Khib) is a newly identified histone mark conserved from yeast to humans; specifically, H4K8 2-hydroxyisobutyrylation (H4K8hib) has been detected in actively transcribed genes in mouse meiotic and post-meiotic cells. Khib has been suggested to be involved in transcription of meiotic and post-meiotic cells.

histone cassettes: Short sequences in histone proteins with clustered histone modifications that direct the biological readout in a combinatorial fashion.

histone citrullination/deimination (cit/dea) analysis: Citrullination of arginine has been revealed to impede additional methylation of some arginines thus changing the transcriptional activity at target genes. While arginine is positively charged, citrulline has no net charge thus increasing the hydrophobicity – the relation of non-polar groups or molecules in an aqueous environment which ascends from the predisposition of water to exclude non-polar molecules. This change from basic to neutral isoelectric point may modify the tertiary structure of or impede binding ability of histone 'readers' at the protein domain.

histone citrullination/deimination (cit/dea): Citrullination or deimination is the post-translational conversion of the amino acid arginine in a protein into the amino acid citrulline by the enzyme peptidylarginine deaminases (PADs). Peptidyl arginine deiminase 4 (PADI4) unambiguously deiminates arginine residues at positions R2, R8, R17 and R26 in the H3 tail. Deimination by PADI4 thwarts arginine methylation by CARM1. Arginine is positively charged at a neutral pH, whereas citrulline has no net charge. This increases the hydrophobicity of the protein, which can lead to changes in protein folding, affecting the structure and function. This increases the hydrophobicity of the protein, which can lead to changes in protein folding, affecting the structure and function. The immune system can attack citrullinated proteins, leading to autoimmune diseases such as rheumatoid arthritis (RA) and multiple sclerosis (MS). Fibrin and fibrinogen may be favoured sites for arginine deimination within rheumatoid joints.

histone citrullination/deimination (cit/dea) prevention: Dimethylation of arginine stops deimination by PADI4 although monomethylation still permits deimination to occur.

histone code: This term has been insecurely used to describe the role of histone modifications to enable DNA functions. This term, although valuable in signifying the necessity for a specific set of modifications for a given task, is unlikely to truly echo the presence of a predictable 'code' in the strictest sense of the word. The hypothesis that the locations and types of histone modifications, through chromatin remodelling and/or recruitment of transcription factors, predicts the effects of those modifications on gene expression. Specific histone-tail modifications effected by enzymes, control the activity of genes in the nucleus, by providing binding sites for regulatory proteins.

histone crotonylation (cr)/krotonilation (kr): Crotonylation occurs on the ε-amino group of lysine residues and modifies histone charge. This histone PTLM is specifically enriched at active gene promoters and potential enhancers in mammalian cell genomes. Crotonylation can be catalysed reversibly by protein crotonyltransferases and decrotonylases. Similar to acetylation, the substrate for crotonylation is a donor molecule linked by a thioester to the sulfhydryl group of coenzyme A, namely crotonyl-CoA. The crotonylation of lysine has been identified on histones H1, H2A, H2B, H3 and H4. The histone acetyltransferases (HATs) also have histone crotonyltransferase (HCT) activities. Examples of such HATs are p300/CBP (CREB-binding protein), GNAT (Gcn5-related N-acetyltransferases) and MYST (MOZ, Ybf2/Sas3, Sas2 and Tip60) families. The p300/CBP and MYST families are identified only in eukaryotic cells. GNAT family is present and conserved in all domains of life. Histone deacetylases (HDACs) were also reported to have histone decrotonylase (HDCR) activity. The role of lysine crotonylation in the regulation of gene transcription, acute kidney injury, spermatogenesis, depression, telomere maintenance, human immunodeficiency virus (HIV) latency and cancer.

histone deacetylase inhibitors (HDACi): Certain chemical agents (viz. benzoylhydrazide scaffold, lead compound, UF010) inhibit acting as inhibitors of histone deacetylases promote growth arrest, differentiation and apoptosis of tumour cells, with minimal effects on normal tissue.

histone deacetylases (HDACs): Enzymes that remove acetyl groups from specific N(6)-acetyl-amino acid residue of histone proteins. This has the general effect of increasing the positive charge of histones and increasing their attraction for the negatively charged phosphate groups in DNA. Histone deacetylation results in a 'closed' chromatin state and transcriptional repression.

histone deacetylation: Deacetylases remove acetyl groups from acetylated histones and help in transcriptional silencing.

histone decrotonylation (dcr)/dekrotonylation (dkr): Reverse of histone crotonylation (cr)/krotonilation (kr).

histone demethylases (HDMs): These enzymes remove methyl groups from histones.

histone demethylation: Removal of methyl groups from methylated lysines or arginines. Deimination and demethylation are two processes that reverse methylation, but they are unlikely to be redundant. Specific histone lysine or arginine demethylases function to remove the methyl groups.

histone dephosphorylation: Histones H1 and H3 are both dephosphorylated after ionizing radiation (IR) in a mutated ataxia telangiectasia (ATM)-dependent manner.

histone deubiquitination. Ubiquitin modification is removed via the action of ISOPEPTIDASES, called DEUBIQUITIN ENZYMES (DUBs).

histone dimethylation: Some amino acids (e.g. lysine and arginine) have ability to attach two methyl groups. H3K9me2 is generally associated with heterochromatin and silenced transposons.

histone effectors: Usually, small molecule that selectively binds to a specific histone and regulates its biological activity. Effector proteins can also be recruited and stabilized by some histone modifications such as methylation. There are dozens of histone effectors that have been usefully employed in epigenetic editing approaches.

histone erasers: Enzymes that remove histone epigenetic marks from chromatin. Some categories of this group of enzymes are arginine deiminases, arginine demethylases, lysine deacetylases, lysine debiotinases, lysine demethylases, lysine deribosylases, lysine deubiquitinases, serine phosphatases, threonine phosphatases, tyrosine phosphatases. Examples of histone erasers are histone deacetylases (HDAC) and histone demethylases (HDMs). HDACs belong to three classes: Class I (HDAC1, HDAC2, HDAC3, HDAC8); II (HDAC4, HDAC5, HDAC6, HDAC7, HDAC9, HDAC10); III (Sirtuin) to IV (HDAC11). HDMs are exemplified by Jumonji, LSD1.

histone exchange: The SWR1 complex catalyses an ATP-dependent histone exchange with specificity for the histone variant H2A.Z. The histone variants are incorporated into the nucleosome outside of S phase relationship between histone exchange and activator binding, preinitiation complex (PIC) assembly.

histone formylation (fo): This histone modification has been revealed to modulate chromatin conformations and gene activation. Oxidative stress produces a significantly different environment in which acetyllysine can be speedily outcompeted by the formation of formyllysine due to the high reactivity of formylphosphate species.

histone gene family: Five types of histones – H1, H2A, H2B, H3 and H4 – are regularly involved in DNA packaging for chromatin formation. Genes synthesizing these histones are present in multiple copies.

histone glutarylation (gl): The cellular dynamics of H4K91glu is controlled by Sirt7 as a deglutarylase and KAT2A as a glutaryltransferase. H4K91glu as a new regulatory mark for chromatin dynamics. H4K-91glu is regulated by KAT2A and Sirt7, its 'writer' and 'eraser', respectively.

histone H1: H1 histone is lysine rich. It is called LINKER HISTONE as it is found on the linker DNA, the DNA that connects two adjacent nucleosomes.

histone H3 expression in cellular plasticity: R.J. Gleason *et al.* (2023) used CRISPR-Cas9–mediated gene editing in *Caenorhabditis elegans* to investigate the expression pattern and role of individual replication-coupled histone H3 genes and compared them with the histone variant, H3.3. They report a tightly regulated epigenome landscape change from the germline to embryos that are regulated through differential expression of distinct histone gene clusters. Together, this study reveals that a change from a H3.3- to H3-enriched epigenome during embryogenesis restricts developmental plasticity and uncovers distinct roles for individual H3 genes in regulating germline chromatin.

histone H3 tail loss in gene expression: Histone H3 has a chunk of its tail, which when trimmed off influences the expression of genes with which the histone is associated. The amino acid sequence of the mammalian histone H3 tail is shown along with the major sites of methylation (circles) and acetylation (triangles) that have been mapped to arginine (R) and lysine (K) residues. Enzyme cathepsin L in mouse and a serine protease in yeast cleave H3 after alanine 21. H3 clipping could set the transcriptional state of a particular cell lineage during differentiation. At cellular level, H3-tail shortening could simply clear all repressive marks from chromatin, thereby enabling the binding of transcription-activator complexes to the affected DNA.

histone hydroxylation (oh): Jumonji domain containing 6 (JMJD6) hydroxylates histone lysyl-5 residues of histones 3 and 4 tails. 5-hydroxylation of lysyl residues inhibited N-acetylation and N-methylation by an acetyltransferase and a methyltransferase, respectively, *in vitro*. Histone 5-hydroxylation may have central roles in epigenetic regulation of gene transcription or chromosomal rearrangement.

histone isomerization: Isomerization is the chemical process by which a compound is transformed into any of its isomeric forms (i.e. forms with the same chemical composition but with different structure or configuration) and, hence, generally with different physical and chemical properties such as A-B-C → B-A-C. Proline isomerization histone H3 is an innovative noncovalent histone modification which controls lysine methylation and gene expression. Proline isomerase Fpr4 is a member of the FK506 binding protein family in *S cerevisiae*. Fpr4 binds the amino-terminal tail of histones H3 and H4 and catalyses the isomerization of H3 proline P30 and P38 *in vitro*. P38 is required for methylation of K36. Isomerization by Fpr4 negatively impacts the ability of Set2 to methylate H3K36 *in vitro*. The conformational state of P38, controlled by Fpr4, is thus crucial for methylation of H3K36 by Set2. Agreeing with such an aggressive role, retraction of Fpr4 catalytic activity *in vivo* leads to increased levels of H3K36 methylation and delayed transcriptional induction kinetics of specific genes in yeast. Proline isomerization thus controls transcription and also offers evidence for crosstalk between histone lysine methylation and proline isomerization.

histone malonylation (ma): Malonylation is an evolutionarily conserved modification wherein malonyl-CoA is used as a substrate to add a malonyl group to the amino acid lysine, changing its charge from +1 to −1. In histone H2A at lysine 119, malonylation has been identified which may function in chromosome segregation during mitosis and meiosis.

histone methylation: Histones may be methylated on either lysine (K) or arginine (R) residues. Histone methylation process adds one, two (in case of arginine) or even three (in case of lysine) methyl groups. It may induce alterations in chromatin architecture, either condensing or relaxing its structure. Lysine or arginine methyltransferases add methyl groups to these amino acids. Specific kinds of methylation (e.g. H3K9me) are associated with inactive chromatin where genes are relatively inactive. H3K9me acts as a binding site for heterochromatin Protein 1 (HP1, Swi6 in *Schizosaccharomyces pombe*) the presence of which is the defining feature of heterochromatic loci. Methylation at certain histone amino acid residues may also be important for activation, whereas at other sites it is a signal for inactivation. Methylation of histones is another reaction involved in a relatively late affair in the cell cycle. Methylation of histones is related with changes that take place in the nucleus before mitosis, when chromatin condensation occurs, and the nucleus becomes synthetically less active. Histone methylation is crucial for regulating chromatin structure, gene transcription and the epigenetic state of the cell. Histones package DNA and post-translational modifications of histones can regulate access to DNA.

histone methylation analysis: Histone arginines are the most frequent hydrogen bond donors in the linking of histones with the DNA. While the methylation does not alter the charge of arginine or lysines significantly, methyl groups are hydrophobic and bulkier; therefore, the main change they impart to histones may be to inhibit or promote the binding of proteins.

histone methylation and facultative heterochromatin: Distinct histone lysine methylation marks are involved in transcriptional repression linked to the formation and maintenance of facultative heterochromatin. Two models have been proposed: (i) In BRIDGING MODEL, one L3MBTL1 monomer or dimer binds to two nucleosomes moving them into close vicinity. (ii) In ASSOCIATION MODEL, each homodimeric or monomeric L3MBTL1 binds one nucleosome and facilitates linker DNA bending or internucleosomal interactions.

histone methylation profiling: The global histone methylation can be changed by inhibition or activation of histone methyltransferases (HMTs). BioCat offers Epigentek's tools for the quantification of H3 and H4 methylation *in situ* and *in vitro*. The EpiQuik™ Global Histone H3 Methylation Assay Kits are appropriate for determining site-specific histone H3 methylation. ChIP protocols for detecting DNA interaction partners of methylated histones are also available. A kit for quantifying total H3 is also available.

histone methylation state association with different functions: In plants, DNA methylation, histone deacetylation and histone methylation in H3K9 (9th lysine of H3) and H3K27 are associated with gene repression. DNA demethylation, histone acetylation and histone methylation in H3K4 and H3K36 are associated with gene activation. Histone lysine residues are able to be mono-, di- or trimethylated, and each methylation state is associated with different functions. Epigenetic modifications play important roles in various aspects of the plant life cycle such as genome integrity, transgene silencing, nucleosome arrangement, nucleolar dominance, paramutation, flowering and parent-of-origin-specific gene expression (IMPRINTING).

histone methyltransferase Meisetz: Epigenetic modifications of histones regulate gene expression and chromatin structure. Meisetz (meiosis-induced factor containing a PR/SET domain and zinc-finger motif) is a histone methyltransferase that is important for the progression of early meiotic prophase.

histone methyltransferases (HMTs): A class of enzymes that add methyl groups to specific histone residues; members include histone-lysine N-methyltransferase and histone-arginine N-methyltransferase. These enzymes, which contain SET domains, can add up to two methyl groups to arginines or up to three methyl groups to lysine.

histone modification analysis: Histone modifications, produced post-translationally, can be detected, quantified and profiled using a variety of techniques. General methods used for histone modification analysis include cell lysates, effector proteins, gaussian derivative wavelets, enriched histone fractionation and histone purification.

histone modification binding partners: For identification of histone modification binding partners, the methods used include co-immunoprecipitation (CoIP)/pulldown experiments, peptide pulldown assays and peptide microarrays.

histone modification detection using western blot (WB): WB is highly searching and specific for analysis of histone marks. Both pan-(sequence-independent) and sequence-specific anti-PTLM antibodies can be utilized for analysis of histone marks. Despite its wide applications, WB has several limitations for analysis of histones and histone marks.

histone modification detection with high sequence coverage and sensitivity: Major approach to advance sequence coverage is chemical derivatization by acylation reaction at amine groups. The derivatization usually reacts with the ε-amine group of lysine residues, blocking trypsin digestion at the lysines. Deuterated acetic acid is one of the earliest substances employed to derivatize histones. The derivatization adds a deuterated acetyl group to originally unmodified lysine residues.

histone modification imaging: Several techniques for *in vitro* and *in vivo* imaging of modified histones have been developed. Most of these techniques use a split-reporter complementation system.

histone modification in cell type-specific gene expression: Multicellular eukaryotes consist of diverse cell types with discrete functions. Lineage specification depends on cell-specific gene expression, which in turn is determined by promoters, enhancers, insulators and other *cis*-regulatory DNA sequences for each gene. Chromatin state at promoters and CTCF-binding at insulators is largely constant across diverse cell types. In contrast, enhancers are marked with highly cell-type-specific histone modification patterns, strongly associated with cell-type-specific gene expression programmes on a global scale and are functionally active in a cell-type-specific manner.

histone modification nomenclature: The common system of nomenclature of histone modifications includes five components: (i) The name of the histone (e.g. H2A, H2B, H3, H4); (ii) The single-letter amino acid abbreviation (e.g. K for lysine, R for arginine); (iii) The amino acid position in the protein from N-terminal (e.g. 27, 36); (iv) The type of chemical tag (methyl, phosphate, acetyl, ubiquitin) attached; and (v) The number of copies of a chemical group attached. For example, only methyl (me) group occurs in one, two or three copies per residue. Thus, me1, me2 and me3, respectively, represent mono-, di- and trimethylation. So, histone modification H3K4me1 denotes the monomethylation of the 4th residue, a lysine, from the start (i.e. the N-terminal tail) of the H3 protein.

histone modification pattern: This pattern changes very little through the cell cycle, including metaphase. At microscopic resolution (10–50 Mb), histone modification patterns are suggested to be dependent primarily on genomic features embedded in DNA sequence, such as gene density or frequency of repetitive elements.

histone modification reversal: Histone package DNA and post-translational modifications of histones can regulate access to DNA. Until recently, histone methylation – unlike all other histone modifications – was considered a permanent mark. The discovery of enzymes that reverse the methylation of lysine and arginine challenges our current thinking on the unique nature of histone methylation and subsequently increases the complexity of histone modification pathways.

histone modification switching: H3K9me2 is documented by a chromodomain of HP1 protein. The binding of HP1, which is a product of the *SU(VAR)25* gene, employs other proteins which change local chromosomal structures and gene expression.

histone modification, simultaneous detection and quantification of: Various methods developed for this purpose include: (i) histone modification detection with high sequence coverage and sensitivity, (ii) detection of multiple histone modifications, (iii) quantitative proteomics for defining changes in the levels of histone modifications, (iv) stoichiometry of histone modifications, (v) dynamic flux analysis of histone modifications, (vi) mass spectrometry for identifying histone modifications, (vii) Western blot for identifying histone modifications, (viii) assessing substrate specificity through microarrays, (ix) dynamic analysis of histone variants using proteomics approaches; (x) quantification of global alterations of histone modifications and (xi) identification of novel types of histone modifications.

histone modifications: In addition to DNA modification, histone modification also plays an important role in regulation of gene expression. Post-translational covalent modifications of histone tails, which include reversible methylation, acetylation, phosphorylation, ubiquitylation, SUMOylation, citrullination and ADP-ribosylation, are involved in transcriptional regulation and epigenetic silencing. Lysine (K), arginine (R), serine (S), threonine (T) and proline (P) are chiefly the amino acids that have been found to be modified in core histones (H2A, H2B, H3 and H4).

histone modifications and protein-DNA binding analyses: A robust ChIP protocol, allowing *in vivo* analyses of multiple chromatin modifications and binding of histone modifiers in different plant organs and tissues. This method is quantitative and provides a technique to study the dynamic state of chromatin during plant development and also in response to different environmental stimuli. The EPI-MODIFIER™ screening system offers a platform for expediently investigating modulators that target site-specific histone modifications.

histone modifications and variants in diversification of a chromatin landscape: Histone modifications and variants collectively contribute to divergence of a chromatin landscape shaped by self-motivated processes that are driven principally by transcription and nucleosome remodelling systems.

histone modifications in chromatin structure and transcription: The mechanism of 'repression' and 'depression' of various genes is associated with an interaction of chromosomal DNA with proteins. Basic proteins of the nucleus are inhibitory for the gene action (e.g. protamines and histones suppress the activity of special group of genes).

histone modifications in DNA packaging control: Histone modifications and the proteins that bind to nucleosomes have revealed a mechanism that might be controlling DNA packaging. Three contrasting mechanisms have been found to regulate binding of proteins that target methylated histones. (i) Phosphorylation of a serine or threonine amino acid that lies next to a methylated lysine creates a 'methyl-phospho' module. This module can no longer associate with methyl-binding proteins. Phosphorylation is readily reversed by a phosphatase. This mechanism leaves the original pattern of methylation intact. (ii) Passage of RNA polymerase II may result in nucleosome replacement, erasing the original pattern of methylation. (iii) Site-specific patterns of methylated nucleosomes can be erased enzymatically by histone demethylase.

histone modifications in epigenetic control: Chromosomal regions can assume stable and heritable alternative states resulting in bistable gene expression without changes to the DNA sequence. Such epigenetic control is often related with alternative covalent modifications of histones. The stability and heritability of the states are believed to involve positive feedback where modified nucleosomes engage enzymes that similarly modify nearby nucleosomes. The mechanism can give strong bistability that is resistant both to high noise due to random gain or loss of nucleosome modifications and to random partitioning upon DNA replication. Robust bistability requires cooperativity, the activity of more than one modified nucleosome, in the modification reactions and that nucleosomes occasionally stimulate modification beyond their neighbour nucleosomes. Basic ingredients of this model include the relevant nucleosome types – methylated (M), unmodified (U) or acetylated (A) – can be interconverted by recruitment of histone-modifying enzymes by nearby M or A nucleosomes or by random 'noisy' transitions.

histone modifications in gene expression control: The post-translation histone modifications (PTLMs) impact gene expression by altering chromatin structure or recruiting histone modifiers. Histone modifications and the enzymatic machineries that set them are crucial regulators that can control cellular proliferation, differentiation, plasticity and malignancy processes.

histone modifications, class I: The histone modifications that indirectly control chromatin structure through enrolment of chromatin-associated proteins.

histone modifications, class II: The histone modifications that function through direct chemical intrusion with histone-histone or histone-DNA interactions.

histone modifications, consequences of: There are two major consequences of histone modifications: (i) establishment of global chromatin environment and (ii) orchestration of DNA-based processes.

histone modifications, detection of: Various approaches used for this purpose include mass spectrometry, antibody recognition, anti-PTLM histone antibody quality issue and validation of histone modification antibodies.

histone modifications, effects of: These modifications can dramatically affect the electrical and other properties of the chromatin, and they play a major role in gene regulation. These histone modifications are an important component of chromatin and can be involved in transcriptional regulation and epigenetic silencing.

histone modifications, global detection of: Active-Motif has for several years offered a series of ELISA kits specifically targeted to histone H3 which can indicate whether a population of cells has altered its histone-modification profile on a global level in response to some treatment, which can then be followed up with ChIP to gain a sequence-level perspective. Abcam also offers ELISA assays for histone modifications, including kits to quantify up to 21 histone H3 modifications and 10 histone H4 modifications simultaneously.

histone modifications, identification of novel types of: Various approaches used for this purpose include: (i) uncovering of a novel mass shift that is caused by a new PTLM; (ii) inference of chemical structure for a novel PTLM; and (iii) authentication of novel types of histone modifications.

histone modifications, problems faced in the detection of: (i) PTLMs on histones are lively as they are rapidly changing. (ii) Problems of antibody specificity. (iii) Problems in detection of PTLMs that are specific to MS. (iv) Peptide coverage is variable for different parts of the histone, and this reduces. (v) Modification of other substrates by the same enzyme will play a parallel role in the biological response.

histone modifications, types of: Specific lysines in core histones (H2A, H2B, H3 and H4) may undergo monomethylation (me1), dimethylation (me2), trimethylation (me3), acetylation, phosphorylation (ph), ubiquitination, Small Ubiquitin-like Modifier (SUMO)lyation or ADP-ribosylation (ADPr), O-GlcNAcylation (og), citrullination/deimination (cit/dea), crotonylation/krotonylation (cr/kr), proline isomerization (pi), biotinylation (b) propionylation (pr), butyrylation (bu) and 2-hydroxylisobutyrylation (hib), malonylation (ma), succinylation (succ), glutarylation (gl), formylation (fo), hydroxylation (oh) and serotonylation (ser). Serines and threonines have been observed to get phosphorylated. Lysines can be mono- (me1), di- (me2) or trimethylated (me3). Arginines can be monomethylated (H3R2me1) or dimethylated asymmetrically (H3R2me2a, H3R17me2a,) or symmetrically (H4R3me2s). Phosphorylation occurs at serine/threonine/tyrosine residues.

histone modifications, typical patterns of: Typical patterns of histone methylations displayed at promoters, insulators, enhancers and transcribed regions have been identified. The monomethylations of H3K27, H3K9, H4K20, H3K79 and H2BK5 are all related with gene activation, whereas trimethylations of H3K27, H3K9, and H3K79 are associated with repression. H2A.Z is linked with functional regulatory elements, and CCCTC-binding factor (CTCF) scripts boundaries of histone methylation domains. Histone modifications and the proteins that bind to nucleosomes have shown a mechanism that might control DNA packaging. Phosphorylation of histones is another mechanism known for neutralization of basic proteins. This reaction occurs after the completion of the histone synthesis. It is a reversible reaction. Phosphorylation of lysine and arginine-rich histones precedes an increase in RNA synthesis. Histone modifications are implicated in impacting gene expression. High-resolution maps for the genome-wide distribution of 20 histone lysine and arginine methylations as well as histone variant H2A.Z, RNA polymerase II and the insulator binding protein CTCF across the human genome have been constructed.

histone modifications, working of: The surface of the nucleosomes is dotted with a multiplicity of modifications. At least eight different classes have been categorized to date and many different sites have been identified for each class. Operationally, histone modifications function either by disrupting chromatin contacts or by affecting the recruitment of non-histone proteins to chromatin. The presence of non-histone proteins in chromatin can command their high-order chromatin structure in which DNA is packaged and can orchestrate the ordered recruitment of enzyme complexes to manipulate DNA. Thus, histone modifications have the capability to impact many fundamental biological processes, some of which may be epigenetically inherited.

histone modifications: Post-translational modification (PTLM) of proteins plays an important part in mediating protein interactions and/or the recruitment of specific protein targets. PTLM can be mediated by the addition of functional groups (for example, acetylation or phosphorylation), peptides (for example, ubiquitylation or SUMOylation) or nucleotides (for example, poly(ADP-ribosyl)ation). Histone modifications are integral to choreographing the expression of the genome. Modifications on histones control important processes through their effects on chromatin structure. Post-translational modification of chromatin has profound effects on many biological processes including transcriptional regulation, heterochromatin organization and X chromosome inactivation. Histone modifications are drawn in influencing gene expression.

histone modifiers: Post-translational modifications (PTLMs) of histone proteins are central to the control of chromatin structure, playing essential functions in modulating the activation and repression of gene transcription. Different types of histone modifications and their different types of modifiers influence chromatin structure during transcription. Histone modifiers identified in litchi include 6 histone acetyltransferases (HATs), 11 histone deacetylases (HDACs), 48 HMTs and 22 histone demethylases (HDMs).

histone modifiers as oxygen sensors: Hypoxia induces histone lysine hypermethylation, a form of epigenetic chromatin modification. However, whether this represents a direct sensing of oxygen tension or an indirect effect, perhaps through the HIF pathway, has not been established. Activity of the lysine-specific demethylases (KDMs) KDM5A and KDM6A is oxygen sensitive, thereby identifying them as oxygen sensors.

histone monomethylation (hme1). Addition of one methyl group to lysine or arginine amino acid of a histone.

histone octamer: A histone octamer comprises two copies of each of the four core histone proteins – H2A, H2B, H3 and H4. This octamer gathers when a tetramer consists of two copies of H3 and H4 composites with two dimers of H2A/H2B. They assist in the nucleosome-formation. Histone octamer interacts with DNA to give rise to a structure called NUCLEOSOME. Linker histone H_1 resides on DNA in between two nucleosomes.

histone O-GlcNAcylation removal: O-GlcNAc is removed by a single O-GlcNAcase (OGA) by hydrolysis of the O-linked β-N-acetyl glucosamine (O-GlcNAc) modification, an essential protein glycosylation event that modulates the function of numerous cellular proteins in response to nutrients and stress.

histone O-GlcNAcylation: A reversible enzymatic PTLM characterized by a β-glycosidic bond between the hydroxyl group of serine or threonine side chains and N-acetylglucosamine (GlcNAc). O-GlcNAC transferase (OGT) adds O-GlcNAC at serine/threonine of histone. Reverse reaction is performed by GlcNAcase. The addition of GlcNAC has been compared with phosphorylation, and there may be significant interplay between the two histone modifications. O-GlcNAcylation has putative involvement in a range of cellular activities, including the stress response, transcription, translation, cell signalling and cell cycle regulation. Changes in O-GlcNAc modification have also been implicated in a host of diseases such as diabetes, Alzheimer's disease and cancer. O-GlcNAcylation is regulated by a discrete set of enzymes. It was originally thought to exist solely on nuclear and cytosolic proteins catalysed by a single OGT. However, O-GlcNAcylation has lately been revealed on extracellular proteins, where O-GlcNAc addition is catalysed by extracellular OGT (EOGT). O-GlcNAcylation on histone H2A at Thr101 can alter chromatin structure through destabilizing H2A-H2B dimer.

histone phosphorylation (ph): Addition of a phosphate group to an amino acid of a histone protein. Protein Kinases perform the function of phosphorylation. H1 and H3 are maximally phosphorylated in metaphase. Phosphorylation of histones is a mechanism for neutralization of basic proteins. This reaction transpires after the completion of the histone synthesis. It is a reversible reaction. Phosphorylation of lysine and arginine-rich histones precedes an increase in RNA synthesis. Histones and protamine are phosphorylated at the final stages of spermatogenesis. Specific kinases participate in transportation of phosphorous group from ATP to histones. The reaction is stimulated by cyclic AMP. Phosphorylation of H1 affects its interaction with DNA and it is seen that chromatin with phosphorylated histones is a better template for RNA synthesis. Histone phosphorylation can transpire at two levels: molecular and microscopic. MOLECULAR LEVEL PHOSPHORYLATION includes the nucleofilaments and supercoil formation. MICROSCOPIC-LEVEL PHOSPHORYLATION includes chromosomal condensation. Histone phosphorylation has role in transcription regulation, mitosis, apoptosis and DNA damage repair.

histone phosphorylation analysis: Detection of phosphorylation incidences on histones is accomplished by mass spectroscopy (MS) or phosphor-specific antibodies. In tandem MS, during the Collision Induced Dissociation (CID), the fragmentation of the ionized peptide that holds the phosphor group results in the loss of a H_3PO_4 molecule and the neutral loss is sensed by mass spectrometers.

histone phosphorylation profiling: Inactivation or activation of these protein kinases can result in modification in intracellular histone phosphorylation. Histone H3S10 and H3S28 are well-kept-up in eukaryotes, and an increase in phosphorylation has been shown to correlate with gene activation and cell growth.

histone pre-mRNA processing: Processed histone mRNAs contain a stem-loop structure at their 3′ ends. This end is formed by specific machinery that recognizes features of the histone pre-mRNA, the stem-loop as well as HISTONE DOWNSTREAM PURINE-RICH ELEMENT (HDE) 15–20 nucleotides (nt) downstream from the stem-loop.

histone presenters: Histone presenters are thought to convey substrate specificity for histone-modifying enzymes by binding with them. The H3K4 presenter WDR5 forms protein complexes with H3K4 methyltransferases MLL1-MLL4 and binding partner proteins (e.g. RBBP5, ASH2L and DPY30) and plays a central role in histone H3K4me3, chromatin remodelling, transcriptional activation of target genes, normal biology and diseases such as MLL-rearranged leukaemia.

histone proline isomerization (pi): Isomerization is the chemical process by which a compound is transformed into any of its isomeric forms. Symbolic representation of this process can be done by the reaction A-B-C → B-A-C. The dihedral angle of a peptidyl proline's peptide bond interconverts between the *cis* and *trans* conformations, the states differing by 180°. Proline isomerases enable this interconversion.

histone propionylation (pr): *In vivo* lysine propionylation has been identified in histone H4 at Lys-5, Lys-8 and Lys-12. Bacterial propionyl-CoA synthetase and human histone H4 are propionylated at specific lysine residues that have been previously acknowledged to be acetylated. Propionylation of H3 lysine at position 23 (H3K23pro) was detected in the leukaemia cell line U937. Histone propionylation might be produced by the same set of enzymes as for histone acetylation and that selection of donor molecules (propionyl-CoA versus acetyl-CoA) may determine the key difference of modifications.

histone proteins: Histone are relatively small proteins with a preponderance of the basic, positively charged amino acids lysine and arginine. The histones' strong positive charge enables them to bind to and neutralize the negatively charged DNA throughout the chromatin. Histones make up half of all chromatin proteins by weight. Two copies of each of the four core histones, H2A, H2B, H3 and H4, form an octamer, which serves as a core particle in the formation of the nucleosome. Histone H1 links two nucleosomes. H1 resides on LINKER DNA.

histone purification: Histones are extracted from nuclei using a dilute sulfuric acid solution and then purified by column chromatography. Nucleic acids and many of the non-histone proteins are easily detached by centrifugation because of their insolubility at acidic pH. The soluble histone fraction is then precipitated with trichloroacetic acid (TCA) and, if necessary, purified over a reversed-phase HPLC column.

histone readers: Readers distinguish and bind to the epigenetic marks laid down by writers, thereby determining their functional outcome. They include proteins containing bromodomains, chromodomains, Tudor and PHD finger. Histone readers are chromatin-associated proteins that recognize specific epigenetic modifications of histones and influence transactivation via different mechanisms, including induction of nucleosome cross-linking, modulation of transcription factor and RNA polymerase complex access, promotion of additional chromatin post-translational modifications, readers interact specifically with modified-histone sequences as part of multiprotein complexes and readers help modulate transactivation, either activating or repressing expression. Different families of histone readers are identified by their unique modified-histone-binding domains or 'binding modules': BROMODOMAIN has representative associated proteins SNF2, TAF-(II)250, HRX/MLL and PCAF. CHROMODOMAIN has representative associated proteins Polycomb, and HP1. *Tudor* has representative associated protein 53BP1. SET has representative associated proteins Suv39H1, HRX/MLL and G9a. (v) PHD finger has representative associated protein RAG2.

histone recycling: To restore chromatin on new DNA during replication, recycling of histones evicted ahead of the fork is combined with new histone deposition. The Asf1 histone chaperone, which buffers excess histones under stress, is a key player in this process.

histone repressive mark in flower development: Trimethylation at lysine of histone H3 at position 27 (H3K27me3) is a repressive mark in flower development, which not only functions during floral induction through regulating *FLC* expression and floral meristem identity through *LFY*, it also participates in the organogenesis during flower development.

histone serotonylation: Serotonylation (ser) of glutamine at position 5 on histone H3 (H3Q5ser) has been reported in organisms that produce serotonin (*aka* 5-HYDROXYTRYPTAMINE (5-HT)). Transglutaminase 2 can serotonylate H3K4me3-marked nucleosomes, resulting in the presence of combinatorial H3K4me3Q5ser *in vivo*. H3Q5ser functions as an accommodating marker for gene activation in collaboration with H3K4me3 during neuronal cell differentiation.

histone shuttling: A mechanism proposed to guide specific proteins to sites of repair. In addition, histone shuttling driven by the poly ADP-ribosylation system seems to be involved in nucleosomal unfolding of chromatin in DNA excision repair. Two nuclear enzymes, poly(ADP-ribose) polymerase and poly(ADP-ribose) glycohydrolase, may cooperate to function as a histone shuttle mechanism on DNA.

histone shuttling, mechanism of: Two nuclear enzymes – poly(ADP-ribose) polymerase and poly(ADP-ribose) glycohydrolase – may cooperate to function as a histone shuttle mechanism on DNA. The mechanism involves four distinct reaction intermediates that were analysed in a reconstituted *in vitro* system: (i) The enzyme poly(ADP-ribose) polymerase is activated in the presence of histone-DNA complexes and converts itself into a protein carrying multiple ADP-ribose polymers. These polymers attract histones that dissociate from the DNA as a histone-polymer-polymerase complex. (ii) The DNA assumes the electrophoretic mobility of free DNA and becomes susceptible to nuclease digestion. (iii) Poly(ADP-ribose) glycohydrolase degrades ADP-ribose polymers and thereby eliminates the binding sites for histones. (iv) Histones reassociate with DNA, and the histone-DNA complexes exhibit the electrophoretic mobilities and nuclease susceptibilities of the original complexes prior to dissociation.

histone succinylation (succ): Succinylation is an acylation that alters the cationic lysine side chain to an anion (from +1 to −1) but its exact role remains unknown. Histone succinylation has large potential impacts on protein structure and function. Role of H3K122 succinylation has been suggested in chromatin function and transcription.

histone SUMO-deconjugation: For histone SUMO-deconjugation, sentrin-specific proteases (SENPs) are accountable for processing SUMO peptides and for deconjugating SUMOylated species.

histone SUMOylation: Covalent attachment of SUMO molecules to specific lysines of all the four histones occurs via the accomplishment of E1, E2 and E3 enzymes. This modification is involved in trafficking proteins to the nucleus among many other processes.

histone SUMOylation profiling: SUMO paralogues control different mitotic processes in mammalian cells.

histone tail clipping: Histone tail clipping plays very important biological role. Perhaps the most fundamental way to eliminate histone modifications is to remove the histone N-terminal tail in which they reside.

histone tail loss: Histone H3, with its tail trimmed off, affects the expression of genes with which the histone is associated. At cellular level, H3-tail trimming could simply clear all repressive marks from chromatin, thereby permitting the binding of transcription-activator complexes to the affected DNA.

histone tails: N-termini of histones which are in particular highly modified.

histone trimethylation (hme3): Addition of three methyl group to an amino acid of a histone protein.

histone trimethylation as transcriptional signal: Diverse histone modifications are catalysed and recognized by various specific proteins, establishing unique modification patterns that act as transcriptional signals. In particular, histone H3 trimethylation at lysine 36 (H3K36me3) is associated with actively transcribed regions and has been proposed to provide landmarks for continuing transcription.

histone turnover: Chromatin plays role in processes governed by different timescales. Histone H3 turnover was measured in G1-arrested *S. cerevisiae* at single-nucleosome resolution over 4% of the genome, and at lower (around 265 bp) resolution over the entire genome. Nucleosomes at promoters are replaced more rapidly than at the coding regions and that replacement rate over coding regions correlates with polymerase activity. In addition, rapid histone turnover is observed at known chromatin boundary elements. Rapid histone turnover assists to functionally separate chromatin domains and prevent spread of histone states.

histone ubiquitination (ub): The addition of one or a chain of ubiquitin proteins to a histone. Ubiquitin (ub) is a small regulatory 76-amino acid polypeptide existing in almost all tissues of eukaryotic organisms. Ubiquitination results in a much larger covalent modification. Ubiquitination is a PTLM in which carboxylic acid of the terminal glycine from the diglycine motif in the activated ubiquitin makes an amide bond to the epsilon amine of the lysine in the modified protein. Histones H2A and H2B are two of the most highly ubiquitinated proteins found in the nucleus. Monoubiquitination (ub1) of H2A is catalysed by Polycomb group proteins. The main enzyme accountable for monoubiquitinated H2B is Brel in yeast and its homologues RNF20/RNF40 in mammals. **Histone ubiquitination** often targets the histone for degradation but can also change its localization, conformation or binding.

histone variants: Histones that substitute for the linker (H1) and core (H2A, H2B, H3 and H4) canonical histones are PARALOGOUS HISTONES which may have intrinsic gene regulatory functions. Histones characterized by the distinct protein sequences that can be inserted into nucleosomes and may have intrinsic gene regulatory functions. They carry out specialized functions in DNA repair, chromosome segregation and regulation of transcription initiation or perform tissue-specific roles. The sequence and domains of histone variants are not overtly different from that of canonical nucleosome histones. The histone variants alter nucleosome structure or chromatin organization. For example, (i) Centromere Protein A (CENPA) is a H3 variant with centromere function. (ii) H2A.X is a H2A variant which marks damaged DNA and induces repair mechanisms. (iii) H2A.Z is another H2A variant which is frequently found in association with euchromatin; associated with heterochromatin during embryonic development. (iv) H3.3 is a H3 variant found in nucleosomes of non-dividing cells and is involved in senescence epigenetic events; important in the brain for neuronal plasticity. *Aka* REPLACEMENT HISTONES. *See* PARALOGOUS HISTONES.

histone variants in male germ cell development: Male germ cells contain many histone variants, which are expressed at different stages of male germ cell development.

histone writers: All the histone-modifying enzymes are known as writers. Writers add epigenetic marks to histones. They have characteristic distinction wherein some are ubiquitously expressed in all cell types and some show tissue-specific expression. These enzymes have associated domains to recognize histone substrates. Hundred unique enzymes that belong to different categories of writers are known: arginine deiminases, arginine methyltransferases, lysine acetyltransferases, lysine biotinases, lysine methyltransferases, lysine ribosylases, lysine ubiquitinases, serine kinases, threonine kinases and tyrosine kinases. The important categories of this large group are: Many of these writers have other non-histone targets as well. Writers may contribute towards regulation of any cellular process by more than one mechanism. Specific examples of writers are Histone Acetylases (HATs): CBP/p300, MYST and GNAT (GCN5) and HMTs: Set1, SMYD3, MLL, Suv39H1 and G9a.

histone writers and erasers, functions of: Writers and erasers help determine chromatin's transcriptional state through a dynamic process of post-translational covalent tag addition and removal. Specificity for the addition or removal of marks varies among family members, however HMT and HDM members generally target specific histones and residues.

histone-dependent activation: This is the first stage in the activation of genes in which activator proteins, at the UAS, directly or indirectly cause the core histone particles to dissociate from TATA BOX and form and form a preinitiation complex and thereby generate basal level of transcription.

histone-dependent inactivation: This is the second stage in activation of genes in which activators stimulate the transcription preinitiation complex to maximize production of mRNA.

histone-independent activation: This is the second stage in the activation of genes in which activators stimulate the preinitiation complex to maximize the production of mRNA. A Histone H_1 has been found out to be a potent repressor of transcription by RNA polymerase II.

histone-dependent activation stage. At this stage, activator proteins at the UPSTREAM ACTIVATOR SEQUENCE (UAS) directly or indirectly cause the histone core particle to disassociate from the TATA box. This displacement enables basal factors to assemble on the TATA box and form a preinitiation complex thereby generating basal level of transcription.

histone-like (HU) proteins in bacteria: Protein HU is the most histone-like; it has an amino acid composition similar to eukaryotic histone H2A. HU proteins are required to stimulate reaction leading to DNA replication in *Escherichia coli*. Several DNA-binding proteins isolated from *Escherichia coli* include HU, H, IHF, H1, HLP1, FirA and P. Many of these proteins resemble the eukaryotic histones. HU is an abundant DNA-binding protein capable of wrapping DNA, and its primary structure is highly conserved among bacterial species. DNA is wrapped around tetramers of HU. Each turn consists of 38–39 bp, and each linker is about 19-bp long. Thus, adjacent tetramers are inverted relative to each other. IHF is similar to protein HU, while protein P is similar to protamines that bind to DNA in the sperms of some animal species. HLP1 is small basic protein and binds to DNA. Protein H1 binds very strongly to DNA and may be able to compact it.

histone-mediated repair restriction: In sex-separated *Caenorhabditis elegans* strains, paternal, but not maternal, exposure to ionizing radiation leads to transgenerational embryonic lethality. The offspring of irradiated males displayed various genome instability phenotypes, including DNA fragmentation, chromosomal rearrangement and aneuploidy. Paternal DNA double-strand breaks were repaired by maternally provided error-prone polymerase theta-mediated end joining. S. Wang *et al.* (2022) show that depletion of an orthologue of human histone H1.0, HIS-24 or the heterochromatin protein HPL-1, could significantly reverse the transgenerational embryonic lethality. Removal of HIS-24 or HPL-1 reduced histone 3 lysine 9 dimethylation (H39me2) and enabled error-free homologous recombination repair in the germline of the F_1 generation from ionizing radiation-treated P0 males, consequently improving the viability of the F_2 generation.

histones: A class of proteins that associate with DNA in the eukaryotes. These are five small basic proteins, viz. H_1, H_2A, H_2B, H_3 and H_4 which are rich in arginine and lysine and function to package, condense and organize DNA are highly conserved in the eukaryotic cells. These proteins associate with DNA in the nucleosome. The combination of about 146 base pairs of DNA and an octamer of four histones (H2A, H2B, H3 and H4) is called NUCLEOSOME. H1 binds to the linker DNA sequence of 54 base pairs. The N-terminal regions of histone proteins are largely unstructured and are subject to various chemical modifications. The linker histone H1 locks the DNA into place and allows the formation of a higher-order structure. The histone-DNA interactions are thus involved in packaging of DNA in the nucleus. Each core histone has a short tail that contains sites for methylation and acetylation, chemical processes that adjust how tightly the DNA winds around the histone core proteins. Thus, these histone tails play an important role in differential gene expression, because only loosely compacted genes are available for transcription.

histones as gene regulators: Histones not only are important basic proteins which impart structure to eukaryotic chromosome but also play an important role in cell cycle by undergoing reversible modifications, such as acetylation of lysine, phosphorylation of serine and threonine and ubiquitination of lysine. They act as repressors of gene activity.

histones as repressors of gene activity: Since histones act as repressors of gene activity, it is important to understand mechanism of inhibition of DNA transcription by histones. The mechanism of 'repression' and 'depression' of various eukaryotic genes is linked to an interaction of chromosomal DNA with histones. Acetylation, phosphorylation and methylation of histones are considered to be important in repression of gene activity in eukaryotes. There are two hypotheses put forth to explain histones as repressors of gene activity. According to the first hypothesis, histones, which are assembled with DNA, prevent RNA polymerase motion during transcription. The second hypothesis suggests that the structure of DNA is altered as a result of histone-DNA association and supercoiling of DNA occurs. RNA polymerase is not able to transcribe those regions, which are supercoiled. Different fractions of histones have different mechanisms of action, for example, lysine-rich histones inhibit transcription when associated with DNA. The arginine-rich histones inhibit RNA polymerase activity when two are associated. Histones and basal factors (TFIIIA; TFIID) compete for the TATA region of DNA. Activator molecules probably have a disruptive influence on nucleosomes. They may alter some basal factors in the preinitiation complex, which in turn disrupts the nucleosome and gains access to the TATA box.

histones H2A and H2B: Both these histones are slightly lysine rich. The dense, inactive chromatin of thymus lymphocyte nuclei is constituted by lysine-rich histone cross-linking the chromatin fibrils.

histones H3 and H4: Both these histones are arginine-rich and are bound to the chromatin fibrils but do not cross-link them. Methylation of histone H3 lysine 9 generates a binding site for HETEROCHROMATIN PROTEIN 1 (HP1). It contributes to higher-order chromatin organization by propagating heterochromatin-specific affinity in the histone H3 amino terminus. This histone in combination with other proteins compacts the 11-nm fibre into higher-order structures such as the 30-nm fibre. H3 is accountable for the stable inheritance of the heterochromatic state.

hit-and-run epigenome editing: Durable epigenetic silencing after the delivery of transient delivery of epigenetic editors *in vivo* has been confirmed. M.A. Cappelluti *et al.* (2024) targeted *Proprotein Convertase Subtilisin/Kexin Type 9* (*Pcsk9*), a gene expressed in hepatocytes that is involved in cholesterol homeostasis. *In vitro* screening of different editor designs indicated that zinc-finger proteins were the best-performing DNA-binding platform for efficient silencing of mouse *Pcsk9*. A single administration of lipid nanoparticles loaded with the editors' mRNAs almost halved the circulating levels of PCSK9 for nearly 1 year in mice. *Pcsk9* silencing and accompanying epigenetic repressive marks also persisted after forced liver regeneration, further corroborating the heritability of the newly installed epigenetic state.

hitch-hiking: The increase in the frequency of a neutral allele that is associated with a selectively favourable allele at another locus. It is used in broad sense to refer to any type of indirect selection.

hitch-hiking effect: Genes are favoured in a population by close linkage with other genes, which are positively selected.

HIV reverse transcriptase: The reverse transcriptase of HIV performs a series of reactions to change the single-stranded RNA genome of HIV into double-stranded DNA for host-cell integration. This task necessitates the reverse transcriptase to discriminate a variety of nucleic acid substrates such that the active sites of the enzyme are correctly positioned to support one of three catalytic functions, namely, RNA-directed DNA synthesis, DNA-directed DNA synthesis and DNA-directed RNA hydrolysis.

HLA complex: Another name for the MHC in humans. Refers to the HUMAN LEUKOCYTE ANTIGEN (HLA) COMPLEX. HLA complex comprises at least four closely linked loci, in the sequence *D–B–C–A*, located on chromosome 6, that control the synthesis of HLAs that are located on the surfaces of the leukocytes and which affect tissue compatibility in organ transplants and skin grafting.

HLA genes, endogenous functions of: These genes may have played some role in our evolutionary history in response to some environmental selection pressure and/or may be intimately involved in cancer. Presumably, one of the normal functions of the HLA genes is to recognize cancer cells as a kind of 'foreign' agent within the body. Some HLA alleles seem to confer resistance or susceptibility to certain specific diseases. The HLA genes also play an important role in normal immune response.

HMG-box: HMG-box domains are found in high-mobility group proteins which are involved in a variety of DNA-dependent processes like replication and transcription. The domain consists of three alpha helices separated by loops.

hnRNA: *See* HETEROGENEOUS NUCLEAR RNA.

Hogness-Goldberg box: A nucleotide sequence that makes up part of the eukaryotic promoter. *Same as* TATA BOX.

holandric genes: Genes located in the differential region of the Y chromosome are known as holandric genes. Genes for maleness in man may also exist in differential region of the Y chromosome. Y-linked genes are transmitted from father to son and then to his son and so on.

holandric inheritance: The inheritance of traits governed by genes linked to the differential region of the Y chromosome.

holandric traits: The traits controlled by the genes located in the differential region of Y chromosomes. In humans, hairy ear rim is a holandric trait.

holism: The idea that only the entire organism should be studied because to separate the parts from the whole does not reveal how the separated parts function in the intact organism.

Holliday junction: A branched DNA structure comprising four double-stranded arms observed during DNA recombination, held together at the site of crossing-over, also called a CROSS-STRAND EXCHANGE, which is eventually resolved to produce recombinant products. Holliday junctions are used in DNA nanotechnology.

Holliday model: According to R. Holliday put forth during 1962–1968, the first event in recombination is recognition and alignment of homologous nucleotide sequence in two duplex chromosomes. Breaks are next made at comparable position in two of the polynucleotide chains having the same polarity. Each broken chain then invades the opposite helix, establishing base pairs with complementary nucleotides. Ligase enzyme then seals the discontinuities to produce a molecule with the shape of the Greek letter Chi (χ), which is referred to as a HOLLIDAY INTERMEDIATE. Once created, connection is free to move to the right or to the left. Such shifts are called BRIDGE MIGRATIONS. The HYBRID DNA thus formed may carry a mismatched base pair. It is, therefore, said to contain heterologous base sequence and is called a heteroduplex. Holliday model is the most satisfactory explanation of recombination at molecular level. In a reciprocal exchange, all four polynucleotide chains participating in the exchange events are conserved. Reciprocal exchange is thought to accompany most general recombinational events, but there are important cases like bacterial transformation and transduction where exchange appears to be nonreciprocal event, a piece of donor DNA is incorporated into a recipient chromosome and the 'extra' or dangling DNA is degraded by endonuclease. Recombination involving reaction between homologous sequences of DNA is called GENERALIZED RECOMBINATION. Recombination between chromosomes comprises a physical exchange of parts. The structure created by this exchange is visible at meiosis in the form of a chiasma. The chiasma represents the results of breakage and reunion, in which two non-sister chromatids (each containing a duplex of DNA) are broken and then linked with each other in a new combination. Recombination occurs between precisely corresponding sequences, so that not a single base pair is added to or lost from the recombinant chromosomes.

Holliday structure: An intermediate structure believed to be formed during recombination between two DNA molecules.

holocentric: Applied to chromosomes with diffuse centromeres in such a way that the properties of the centromere are distributed over the chromosome. Also called HOLOKINETIC.

holocentric chromosome: A chromosome that possesses multiple kinetochores along their length rather than the single centromere typical of other chromosomes. In holocentric chromosomes, centromeric activity is spread over the entire chromosome.

holoenzyme: An enzyme consisting of an apoenzyme and co-factor that is metabolically active. Some of the enzymes consist of only proteins, whereas others consist of both protein part (apoenzyme) and dialyzable non-protein part called co-factor (if it is a metal ion) or coenzyme (if it is organic molecule). Apoenzymes are inactive and these get activated by co-factors. Together the apoenzyme and co-factor form holoenzyme. For example, the version of the *Escherichia coli* RNA polymerase that has the subunit composition $\alpha_2\beta\beta\omega'\sigma$ is involved in efficient recognition of promoter sequences. Only holoenzyme is capable of efficiently initiating transcription. *Compare with* CORE ENZYME.

holomorph: The whole fungus in all its states. *Contrasts* ANAMORPH, IMPERFECT STATE, PERFECT STATE *and* TELEOMORPH.

holo-omics: It refers to the combined investigation of non-targeted molecular data layers from host-microbiota systems or holobionts, which is progressively used to unravel the complex interactions between the elements that compose them. Navigating through the generation, analysis and integration of omics data, focusing on the commonalities and main differences to produce and analyse the various types of omics, with a special focus on optimizing data generation and integration, I. Odriozola *et al.* (2024) support careful generation and distillation of data, followed by independent exploration and analyses of the single omic layers to obtain a better understanding of the study scheme, before the integration of multiple omic layers in a final model is attempted.

holoparasite: A parasitic flowering plant that lacks chlorophyll (nonphotosynthetic) and must depend totally on the contents of the xylem and the phloem of the host plant, such as broomrape (*Orobanche* spp.). *Contrasts with* HEMIPARASITE.

holophyletic: Consisting of species and all of its descendants without species.

holorepressor: A functional repressor protein consisting of an aporepressor plus corepressor complex. *See* APOREPRESSOR *and* COREPRESSOR.

homeobox genes: Pattern formation in animal development requires that genes be expressed differentially according to position in sheets of cells that make up the early embryo. The homeobox genes of *Drosophila* are control genes active both in the establishment of segmentation patterns and in the specification of segment identity. The homeobox genes of mammals seem to be important in pattern formation. Genes that control appendages are called homeotic genes. These genes produce proteins that function as transcription factors, which bind to DNA, switching them on or off. The genes that contain homeoboxes can control the expression of other genes. There are at least 24 homeobox genes, out of which only some but not all are also homeotic in their outcome. In general, 'homeotic' genes are genes that control the identity of body parts. They are active in the early stages of embryonic development of organisms. 'Hox' genes are a subgroup of homeotic homeobox genes that determine positional cell differentiation and development. *Aka* HOMEOBOX-CONTAINING GENES, HOMEOTIC SELECTOR GENES, HOMEOTIC GENES *or* HOX GENES.

homeobox: A 180 base pair sequence, discovered by W. Gehring during the 1970s, in some regulatory genes in *Drosophila*, that codes for a 60 amino acid protein segment that binds to DNA as a key part of a transcription factor, called the HOMEODOMAIN, which plays role as a master genetic regulatory element in pattern formation and cell differentiation during development in species as diverse as worms, fruit flies and humans. Mutations in the homeobox can have HOMEOTIC EFFECTS. In the 1980s, J.C. Nusslein-Volhard revealed that one homeobox was associated (in adjacent exon) with each of the genes that were responsible for embryonic development (i.e. 'switched on' only in an embryo that is developing into an adult). Thus, it is now possible to locate many embryonic-development genes in many species by using a DNA probe (made via a *Drosophila* homeobox DNA sequence) to find homeobox sequences attached to those embryonic-development genes. Because of such a role, the respective homeobox sequences attached to each gene are known as DNA MARKERS.

homeodomain: A 60 amino acid protein domain coded for by the 180-bp DNA sequence (HOMEOBOX) region of a homeotic gene. This protein motif forms a DNA-binding domain that occurs in many transcription factors that play a role in pattern formation and cell differentiation.

homeodomain protein: A homeodomain protein consists of three main regions: (i) VARIABLE REGION, which regulates the protein's specific activity; (ii) HINGE REGION, a small connective region and (iii) HOMEODOMAIN, which binds homeoprotein to the DNA or RNA.

homeogenetic induction: Process by which a differential cell can induce differentiation in an adjacent cell not previously so committed. The term is used chiefly in relation to plant cells.

homeologous: Having partial homology.

homeologous chromosome: Chromosome with similar set of genes but in another genome contained by an allopolyploid plant. In allopolyploid as in case of bread wheat (*Triticum aestivum*) there are three genomes namely AABBDD. Some of the chromosomes of A genome have similarity with other B or D genomes that can result in homologous pairing, multivalent formation and sterility. However, allopolyploid systems have evolved strategies (*ph* locus in wheat) that prevent homologous pairing and thus the polyploids behave like diploids.

homeologous chromosomes: Partially homologous chromosomes or genomes that are believed to have originated from ancestral homologous chromosomes. Such chromosomes found in related species that are derived during evolution from the same ancestor and are found to coexist in an allopolyploid.

homeologous genes: Genes in related species that are derived from the same ancestor and coexist in an allopolyploid.

homeologous group: Series of two or more chromosomes with similar but not homologous chromosomes. This situation is commonly found in allopolyploids (e.g. in hexaploid wheat or tetraploid cotton).

homeologous pairing: Partial pairing between chromosomes due to their partial homology.

homeologous plastid DNA transformation: The use of the incompletely homologous (homologous) plastid DNA as donor results in plastid transformation frequency comparable with other experiments where completely homologous plastid DNA was introduced. The integration of the cloned plastid DNA into other plastid genome involved multiple recombination events.

homeologue-expression bias: Refers to the preferential expression of one homologue relative to the other in resynthesized *B. napus* (A and C homologues contribute unequally to the total gene expression).

homeosis: The alteration of one organ or a segmental or homologous series from its own characteristic to that of some other member of the series.

homeostasis: The developmental pathways of organisms are genetically adjusted through internal regulatory mechanisms to produce the characteristic morphology of the species regardless of variations in internal and external conditions of development. Homeostasis may operate in cross-pollinated crops. Homeostasis in self-pollinated populations cannot be explained on the basis of heterozygous superiority. Perhaps homeostasis in such populations is brought about by selection in favour of genes that can perform optimally in a range of environments, rather than in a single environment. This is termed HOMEIOSIS, GENETIC HOMEOSTASIS or DEVELOPMENTAL HOMEOSTASIS.

homeostatic genotypes: Those which interact with variable environments in such a way as to maintain relative uniformity of an observable character. Stability of development in variable environments implies physiological adjustments and the capacity of such adjustment is a property of homeostatic genotypes.

homeostatic mechanism: A physiological procedure that plays role in the maintenance of a relatively stable internal environment in a multicellular organism.

homeotic (homoeotic) mutations in flower development: There are several homeotic mutations in flowers where different parts replace others. For example, in one class of mutations, sepals develop where petals should and carpals develop where stamens should. These mutations have been identified as defects in a family of genes that all encode a particular class of transcription factors, called the MADS-BOX family. MADS-box transcription factors occur in both plants and, to a lesser extent, in animals and contain a conserved 58 amino acid sequence. The *Arabidopsis* homeotic gene *agamous* (*ag*) is necessary for the specification of reproductive organs (stamens and carpels).

homeotic effects: A homeotic protein can activate one gene but repress another, producing effects that are complementary and necessary for the ordered development of an organism. Homeotic genes contain a sequence of DNA known as a HOMEOBOX, which encodes a segment of 60 amino acids (homeodomain) within the homeotic transcription factor protein.

homing: Refers to ability of certain animals to return to a given place when displaced from it, often over great distances. *See* HOMING GENES.

homing genes: Genes that have the ability to copy themselves on to the opposite chromosomes that do not contain gene drive alleles. The copying process is termed HOMING.

***Homo sapiens* genome:** The entire human genome is over three billion base pairs. Generally, a locus is located on a particular chromosome by somatic cell hybridization. Loci could be pinpointed further using aberrant chromosomes. Since the number of human genes known in this way is limited, genetic mapping is continued with RESTRICTION FRAGMENT LENGTH POLYMORPHISMS (RFLPs). This saturates our map with RFPLs markers, referred to as a modern linkage map as compared to a linkage map of known genes, referred to as classical linkage map. With the help of the human genome project, a physical map of our chromosome is being created which marks off chromosomes not in recombinational distances but in base pairs. This is done by creating libraries of overlapping, contiguous clones, referred to as contig. This is being done chromosome by chromosome. Molecular geneticists begin with overlapping clones obtained, for example, by chromosome walking. The cloned pieces are then isolated and totally digested.

Clones are then compared to see which segments they have in common. From shared segments it can be inferred which parts of the clones are in common. Contigs of parts of the chromosome can thus be built up. Later contigs comprising part of a chromosome can be placed in order by taking an end clone of a completed contig and using it as a probe to begin chromosome walking to find an end clone of a nearby contig. Physical maps are integrated with linkage maps by locating various RFPLs in their appropriate clones within contigs. Clones from contigs can be sequenced to add to the information of the base sequence of that chromosome. With haploid genome of 3 billion bases, the average chromosome has about 140 million base pairs. If a cosmid can hold about 50,000 base pairs, the average chromosome needs about 3000 cosmids to contain it. By 1993, the following had been obtained: a complete contiguous set of overlapping clones of human genome in yeast artificial chromosome; 1 cM genetic map; integration of two maps and correlation with cytogenetic map. By 1994, one megabase of contiguous human DNA sequenced on chromosome 4 and X chromosome was prepared. In 1996, ten megabases were sequenced on chromosome Xq and 21q (q is the short arm). In 1997, complete chromosome 21 was sequenced. Recently, the completion of the first survey of the entire human genome has been announced. By the end of 2005, the complete human genome was expected to be sequenced and complete genetic and physical map would be generated. Human DNA has 3.1 billion letters, which contains 30,000–40,000 genes. However, nearly 40% of genes are still a total mystery, with no known function.

homoalleles: The alleles that arise due to base pair replacement in the same muton; recombination is theoretically not possible between homoalleles. *Compare with* HETEROALLELES.

homobrachal duplication: When a duplicate segment is present in the same arm of the chromosome.

homochromatography: Type of chromatography that rests on the displacement of oligonucleotides (small groups of nucleotides) fixed on some ion-exchange paper by some unlabelled oligonucleotides. In this process, a solvent moves up the chromatography medium separating the labelled oligonucleotide fragments according to length. The shorter the fragment, the faster the migration. The resulting two-dimensionally separated nucleotides are then autoradiographed. Dark spots indicate the location of radioactive nucleotide fragments when X-ray film is developed.

homodimer: A dimeric protein having two identical polypeptides.

homoduplex DNA: A DNA duplex that has precisely complementary bases in the two strands.

homoduplex modified DNA: DNA that has methyl group on a particular adenine or cytosine of both strands included in a host-specificity site. *Aka* FULLY METHYLATED DNA.

homoduplex unmodified DNA: DNA having no methyl group on a particular adenine or cytosine on both strands included in a host-specificity site. *Aka* UNMETHYLATED DNA.

homoeosis: *See* HOMEOSTASIS.

homoeostasis: The tendency of a biological system to struggle against change and to maintain itself in a state of stable equilibrium even under the changed environment.

homogametic sex: The sex having two similar sex chromosomes; for organisms with chromosomal sex determination; females are XX in humans. *Aka* HOMOGAMETIC. *Compare with* HETEROGAMETIC SEX.

homogamic: Refers to mating between individuals belonging to the same population or species.

homogamous flowers: Hermaphroditic flowers showing coordinated formation and function of male and female sex organs.

homogamy: The preference of individuals to mate with others of a similar genotype or phenotype. In botany, the condition in which male and female parts of the flower mature simultaneously.

homogeneous population: Uniform; all of a single category.

homogeneous: All of a single type (uniform or clone). During cell culturing, if all the cells in a population look alike then the population is called homogeneous population. Likewise, at organism level, for instance, in case of self-pollinated plants the progeny is homogeneous.

homogeneously staining region (hsr): Intrachromosomal segments of various length and uniform staining intensity after G-banding that house amplified genes because the chromosomes are not attached at the centromere; for example, premature centromere division (PCD) with chromatid puffing in areas of constitutive heterochromatin is found in Robert syndrome.

homogenic: Refers to chromosome pairing between morphologically identical genomes in amphiploids. *Compare with* HETEROGENIC.

homogenization of DNA sequences: The uniformity of sequence within families of satellite DNA, and within families of dispersed repeats too, in any one species is striking. There must be mechanisms that homogenize the sequences against the constant propensity to deviate. As a consequence, the clustered and dispersed classes of repeated sequence display a remarkable constancy of sequence against a background of steady modification in much of the single-copy sequence DNA. When homologous chromosomes pair, there is only a small chance of perfect matching of tandem arrays, one against the other. Imperfect pairing leads to unequal crossing-over and the ultimate consequence is reciprocal amplification and deletion of sequences in tandem arrays. Another process of homogenization is known as gene conversion, and this is not a consequence of tandem repeats.

homogenote: A partially diploid bacterium with the same alleles at the diploid locus, resulting from transduction.

homogenotization: A genetic procedure used to replace one copy of a gene, or one DNA sequence within a genome, with an altered copy of that sequence. For this, the DNA is first cloned and then changed in some way, for example, when a transposon is inserted into a gene; the mutated copy of the gene can be used to replace the original copy by recombination *in vivo*. The integration of the mutated gene is usually selected, for example, by virtue of it containing a transposon-encoded antibiotic resistance.

homogenous libraries: Those libraries that cover a maximal sequence space. For construction of such libraries, focus is set on the methods that enable the exclusion of disturbing sequences. The parameters that define the variability, the minimal numbers of copies per library are considered and the use of alternating panning cycles is made to avoid the loss of selected hits.

homogentisic acid: A reducing substance excreted in the urine of persons with a genetic metabolic deficiency known as ALKAPTONURIA. Upon oxidation it turns black, also called ALKAPTON.

homograft: A graft of tissue taken from a donor of the same species as the recipient.

homokaryon: A hybrid cell possessing two or more genetically identical nucleic in the common cytoplasm, derived as a result of cell-to-cell fusion. During somatic cell hybridization through protoplast fusion if the two parental protoplasts are of the same species or variety/cultivar particularly of a self-pollinated crop then resulting hybrid cell immediately after its formation is called HOMOKARYOCYTE, and the nuclear situation is known as HOMOKARYON.

homologous: (i) Chromosomes or chromosome segments (genes) that are identical with respect to their integral sequence, genetic loci and/or their visible structure (in the case of chromosomes), regardless of whether the similarity is evolutionarily shared from a common ancestor or convergent. In molecular biology, when comparing two sequences, the corresponding sites are called 'homologous' if they have the same nucleotide. They likewise talk about percent homology between the two sequences. Homology in this context simply means similarity. For example, a gene of 'unknown' function in humans could be compared (in a database) with genes of a simpler organization (e.g. *Caenorhabditis elegans*). If the human gene is homologous, and the function of the *C. elegans* gene is known, the function of the human gene could be inferred by comparison. (ii) Refers to biomolecules (DNA, RNA, proteins, etc.), chromosomes, structures, characters or organs that look similar because of their common evolutionary origin.

homologous chromosome: One of the two copies of a particular chromosome in the diploid cell. A diploid chromosome complement consists of a number of pairs of chromosomes that varies species to species. The members of each pair look alike under microscope. They are derived from maternal and paternal sex cells, and these are called homologous (similar) chromosome, which start pairing lengthwise during zygotene of meiosis I.

homologous genes: Genes with a common ancestor. Term is generally used to describe genes from different species but are similar and have the same function.

homologous protein: A protein having identical functions and similar properties in different species. For example, the haemoglobins that perform identical functions in the blood of different species.

homologous recombination (HR): A reciprocal genetic recombination process in which nucleotide sequences are exchanged between two similar or identical molecules of DNA. The exchange of sequence between two related but different DNA (or RNA) molecules, with the result that a new 'chimeric' molecule is created. HR is used to repair DSBs when a complementary strand is available to act as a template. In DNA recombination, breakage of single strands of DNA in the two recombination partners is followed by joining of strands present in opposing molecules and may involve specific enzymes. Recombination of RNA molecules may occur by other mechanisms. Homologous recombination occurs naturally in eukaryotic organisms, bacteria and certain viruses. HR is a powerful tool in genetic engineering. *See* HOMOLOGOUS RECOMBINATION MEDIATED DNA REPAIR (HRMDR) *and* HOMOLOGY-DIRECTED REPAIR (HDR).

homologous recombination (HR)-based methods: HR-based methods of directed evolution include: (i) DNA-SHUFFLING, STAGGERED EXTENSION PROCESS (STEP); (ii) GENE SITE SATURATION MUTAGENESIS (GSSM); (iii) RANDOM CHIMERAGENESIS ON TRANSIENT TEMPLATES (RACHITT); (iv) COMBINATORIAL LIBRARIES ENHANCED BY RECOMBINATION IN YEAST (CLERY); (v) ASSEMBLY OF DESIGNED OLIGONUCLEOTIDES (ADO); (vi) MUTAGENIC AND UNIDIRECTIONAL REASSEMBLY (MURA); (vii) Y-LIGATION BASED BLOCK SHUFFLING (YLBS) and (viii) WHOLE-GENOME SHUFFLING (WGS). *See details as separate terms.*

homologous recombination mediated DNA repair (HRDMR): The fundamental reaction in homologous recombination is the exchange of strands between a single-stranded DNA and a homologous double-stranded DNA. Homologous recombination plays essential roles in the repair of DNA DSBs and collapsed replication forks.

homologous recombination repair: When damage occurs to both DNA strands, the only way that it can be repaired is by HR using the intact chromosome copy. This allows a damaged chromosome to be replaced, using the sister of the chromosome pair as the template. The proteins facilitating this process are not sequence-specific but are homology-dependent.

homologous RNA recombination: Homologous RNA recombination involves two identical or similar molecules and is called PRECISE if recombinant junction sites are located accurately at the corresponding nucleotides or IMPRECISE when junction sites occupy different positions within the recombining molecules.

homologous species: When characters in different species are similar due to common ancestry.

homologous structures: The structures shared by a set of related species because they have been inherited, with or without modification, from their common ancestor. For example, the bones that support a bat's wing are similar to those of a human arm.

homologue: Also spelled as HOMOLOG. A member of a chromosome pair in diploid organisms or a gene that has the same origin and functions in two or more species. The term covers both ORTHOLOGUES and PARALOGUES. Real homologies are based on a common evolutionary descent in contrast with CONVERGENT EVOLUTION or ANALOGIES. Genes discovered initially in fruit flies often have homologues in mammals. *See* HOMOLOGOUS CHROMOSOME.

homologue/orthologue identification: Z. Xu *et al.* (2023) presented a comprehensive pipeline for homologous sequences identification, phylogenetic relationship inference and potential functional profiling of genes in plants. Identification of orthologues uses large-scale genomic and transcriptomic data. This protocol is generalized for analysing the evolution of plant genes.

homology: (i) In phenotypic studies, it refers to a character shared by a set of species and present in their common ancestor. (ii) In genetic studies, it refers to genes that are present in the same locus in the genome. (iii) In evolutionary studies, homology refers to similarity in DNA or protein sequences between individuals of the same species or among different species.

homology assessment between natural and designer PPR proteins: The consensus PPR coding sequence has eight tandem repeats, with the most enriched amino acid at each position for efficient RNA specificity interaction with the designed cPPR protein.

homology between two small genomes, tools for comparison of: (i) SCAN2 provides one with a colour-coded graphical alignment of genome length DNAs in Java. (ii) Advanced PipMaker aligns two DNA sequences and returns a percent identity plot of that alignment, together with a traditional textual form of the alignment. (iii) JDotter is a dot matrix plotter for Java that produces similar diagrams to the above-mentioned programmes, but with better control on output. (iv) multi-zPicture is a multiple sequence alignment tool that provides nice dotplot graphs and dynamic visualizations. If simple gene locations are provided in the form (e.g. >2000–5000 RNA polymerase; indicates that the RNA polymerase gene is found on the plus strand between bases 2000 and 5000), these data will be added to the dynamic visualization. zPicture alignments can be automatically submitted to rVista to identify conserved transcription factor binding sites. (v) GeneOrder

3.0 is ideal for comparing small GenBank genomes (up to 2 Mb). Each gene from the query sequence is compared to all of the genes from the reference sequence. (vi) CoreGenes has been designed to analyse two to five genomes simultaneously, generating a table of related genes – orthologues and putative orthologues. These entries are linked to their GenBank data. It has a limit of 0.35 Mb, while the newer version CoreGenes 2.0 spreads the limit to around 2.0 Mb. (vii) CoreGenes 3.0 determines unique genes contained in a pair of proteomes a single pairs of genomes. This has proved extremely valuable in determining unique genes in comparisons between large Myoviridae.

homology map: Any genetic or physical map that compares gene sequences in homologous segments of two (or more) genomes (or organisms).

homology of transcription mechanisms: TATA-BINDING PROTEIN (TBP) is a classless factor essential for all transcription in eukaryotic cell nuclei and Archaebacteria showing that transcription mechanisms of Archaebacteria and eukaryotes are fundamentally homologous.

homology screening: Once a gene of interest has been isolated from one species, it is relatively straightforward to use a low stringency hybridization strategy to isolate cDNAs from another species. Likewise, additional family members from the same species can frequently be identified.

homology searches, tools for: (i) BLASTN compares the given DNA sequence with all the DNA sequences in the nonredundant (nr) database. (ii) BLASTP compares your protein sequence with all the protein sequences in nr database. (iii) BLASTX translates your nucleotide sequence in all six reading frames and the products compared with the nr protein database. (iv) EMB BLAST is very convenient since it permits one to specifically search databases such as prokaryote, bacteriophage, fungal and 16S rRNA using BLASTN and specific bacterial genomes or SwissProt using BLASTX or BLASTN. (v) ParAlign employs a heuristic method for sequence alignment. In essence, ParAlign is about as sensitive as Smith-Waterman but runs at the speed of BLAST. (vi) GTOP Sequence Homology Search offers BLASTP search capability against individual archaea, bacteria, eukaryota and viruses. (vii) T4-like Phage NCBI MegaBLAST comprises a growing list of T4-like completed phage sequences as well as those in the draft and contig stages of completion. (viii) WU-BLAST categorically finds regions of sequence similarity quickly, with minimum loss of sensitivity. This tool yields functional and evolutionary clues about the structure and function of a novel sequence. (ix) Batch BLAST has been developed specifically for DNA or protein BLAST sequence analysis against the NCBI databases. It enables one to submit a file that contains multiple sequences and then will organize the results by each individual sequence contained in the file. (x) FASTA is a pairwise sequence alignment tool which takes input as nucleotide or protein sequences and compares it with existing databases. It is a text-based format and can be read and written with the help of text editor or word processor.

homology through evolutionary descent: Evolutionary descent from a common ancestor may be due to gene duplication. Homology is typically measured by the amount of identity (in the case of DNA) or conservative replacements (in the case of protein) that they register along their lengths. Sequence homology searches are typically performed with a query DNA or protein sequence to identify known genes or gene products that share significant similarity and hence might inform on the ancestry, heritage and possible function of the query gene. It is probably better to use more quantitative and descriptive terms such as nucleotide 'identity' or, in the case of proteins, amino acid 'identity' or 'relatedness'. *Compare with* ANALOGY.

homology versus similarity: Homology is not synonymous with similarity. Homologous is judgement, similarity is a measurement.

homology-based approaches: Searching for a known homologue is the most widely understood means of identifying new protein-encoding genes. Such searches depend only on evolutionary relatedness and so are widely applicable. Homology-based gene prediction systems scan databases to find similarities to previously identified coding regions. Such homology-based approaches can only identify previously known genes.

homology-dependent gene silencing (HDGS): Genes can be silenced due to sequence homology either between the transgene or the endogenous gene or among the transgenes themselves. It can occur either due to *cis*-inactivation, *trans*-inactivation or cosuppression. Presently known three general modes of HDGS include: (i) *CIS*-INACTIVATION, (ii) *TRANS*-INACTIVATION and (iii) COSUPPRESSION. *See details as separate terms.*

homology-dependent post-transcriptional gene silencing (HDPTGS): The integration of extra copies of an endogenous gene or multiple copies of a transgene into a plant genome often leads to epigenetic silencing of both endogenous and exogenous genes. The silencing mechanism can operate at two levels, either transcriptionally (homology-dependent transcriptional gene silencing, HDTGS), which may involve in *cis* or *trans* DNA pairing, or post-transcriptionally (homology-dependent post-transcriptional gene silencing, HDPTGS) or cosuppression, which may involve antisense RNA (degradosomes), but there is no change in DNA sequence. In many cases, epigenetic silencing correlates to hypermethylation of the repetitive sequences.

homology-directed repair (HDR): A mechanism of DNA repair that uses a homologous DNA as template to rebuild sites of genomic damage (DNA nicks and double-stranded breaks). HDR can be leveraged in genome editing experiments to make precise genomic alterations by supplying the desired sequence for insertion flanked by segments of DNA that are homologous to the sequence surrounding the Cas9-induced DSB. This approach enables the insertion of single or multiple transgenes, as well as single-nucleotide substitutions.

homomeric: Genes that govern the determination of a trait by joint gene action, and each of them shows a similar contribution to the final phenotype.

homomorph: Plants with flowers which are identical for size of stamens and pistil.

homomorphic bivalent: A bivalent that is made by two homologous chromosomes (i.e. in size and shape).

homomorphic incompatibility: *See* INCOMPATIBILITY.

homomorphic self-incompatibility: The type of self-incompatibility which is not connected with morphological differences of flowers. The incompatibility may be due to the genotype of the plant on which it is produced (SPOROPHYTIC CONTROL) or by its own genotype (GAMETOPHYTIC CONTROL).

homomultimeric proteins: Proteins that consist of identical subunits.

homoplasmic: Cells or individuals that contain two or more different types of cytoplasmic components. *Aka* HOMOPLASMONIC.

homoplasmy: Refers to a eukaryotic cell whose copies of mtDNA or chloroplast DNA are all identical. The similarity that has evolved independently and is not indicative of common phylogenetic origin (i.e. a similarity of traits not due to homology but instead to convergence or parallel evolution). *Aka* HOMOPLASTOMY.

homopolymer: An artificial nucleic acid molecule where one strand comprises just one nucleotide.

homopolymer tailing: In this case, random DNA fragments, which are obtained by mechanical shearing or sonicating genomic DNA and then treated with λ exonuclease, can be cloned in a vector. The DNA fragments have 3′-single-stranded ends or tails, which are extended by adding adenines using enzyme terminal deoxynucleotide transferase. The vector molecule is cut at a unique site to generate 3′-tails again. For example, staggered ends which are extended by terminal transferase adding thymines. Now foreign and vector DNA have 3′ complementary tails which could be annealed by ligase. *Aka* A'S AND T'S METHOD *or* dA:dT TAILING METHOD.

homosequential linkage map: A linkage map of one genome (say A) that shares a similar or even identical marker order with a linkage map (or parts of it) from another genome (say B).

homosequential species: The group of related species having identical gene arrangements in all their chromosomes as indicated by their banding patterns.

homospory: A condition in which an organism produces spore of only one type and size (e.g. microspores of only one size by the sporophytes of land plants). *Compare with* HETEROSPORY.

homothallic strains: In yeast, strains that are capable of switching mating type and thus of mating with themselves, forming zygotes, asci and spores under appropriate conditions.

homothallism (Adj. homothallic): A condition in which sexual reproduction transpires with a single thallus (i.e. self-fertile). *Contrasts* HETEROTHALLISM.

homozygosity: The presence of identical alleles at one or more loci in homologous chromosomal segments.

homozygosity mapping: Recessive characters necessitate two copies of an allele for expression. Because of linkage disequilibrium, loci surrounding the trait locus will tend to be homozygous in affected individuals. Searching for homozygous segments in mutated individuals helps geneticists to locate the trait gene.

homozygote: A diploid cell or organism having two identical alleles at a given location/locus in the genome, one on each chromosome of a homologous pair. For example, an individual may be homozygous dominant (*AA*) or homozygous recessive (*aa*).

Homozygotes always breed true. Traits controlled by recessive gene appear in the individuals homozygous for recessive allele. *Compare with* HETEROZYGOUS.

homozygous: A condition in which a diploid cell or organism contains two identical copies (or versions) of a given gene or gene marker, chromosome or chromosome region/arm from each biological parent. Individuals having identical rather than different alleles in the corresponding loci of a pair of chromosomes therefore breeding true. *Contrasts* HETEROZYGOUS.

homozygous typing cells (HTC): A technique used to type cells of an individual on the basis of class II HLA molecules. Lymphocytes of the individual to be typed are stimulated with a panel of cells each of which is homozygous for one of the known alleles. The individual carrying one or both alleles identical with a particular allele will not react with an individual. If an individual's cells respond to all known cells, it must carry a new allele.

honeycomb method: To minimize adverse effects of interplant competition, plants may be grown in a hill surrounded by six other hills which form a hexogen, thus sharing a common microenvironment (one hexagon encompasses seven hills).

Hoogsteen base pairing: Type of base pairing that occurs in TRIPLEX-STRANDED DNA (H DNA) and RNA structures. Hoogsteen base pairing in DNA is seen as T=A=T and C=G=C triple base pairing.

***Hordeum bulbosum* procedure:** A method for producing zygotic haploids, also called double haploids (DHs) in barley by crossing *H. bulbosum* with *H. vulgare* genotypes. After formation of zygotes, the wild *H. bulbosum* chromosomes are afterward eliminated during embryogenesis, which results in haploid *Hordeum vulgare* plants. The procedure involves: (i) day 0: emasculation of florets inside the spike; (ii) day 2: pollination with *H. bulbosum* pollen; (iii) day 3 (to 5): 40% of the embryonic cells are haploid, endosperm abortion occurs, GA3 treatment enhances retention of florets; (iv) day 11: 94% of the embryonic cells are haploid; (v) day 14 (to 16): embryos are dissected and cultured in the dark at 18–22°C, embryos develop *in vitro*; and (vi) day 22 (to 28): embryos are transferred to light for seedling development. Similarly, this technique has been used to produce doubled haploid (DH) wheat and barley plants by crossing it with *Triticum aestivum* or *Hordeum vulgare*, followed by the elimination of the *H. bulbosum* chromosomes from the offspring.

horizontal DNA transfer (HDT): Transmission of DNA between species, involving close contact between the donor's DNA and the recipient, uptake of DNA by the recipient and stable incorporation of the DNA into the recipient's genome. HGT includes the spread of antibiotic resistance genes among bacteria (except for those from parent to offspring), fuelling pathogen evolution.

horizontal gene transfer: Gene transfer (exchange) between two organisms living together in the same environment. It is a rare phenomenon as compared to the vertical gene transfer in which genes get transferred from an organism to its offspring. While growing transgenic crops in the soil, there is an apprehension that there could be horizontal gene transfer from transgenic plants to soil microorganisms, but it is a remote possibility. *See* VERTICAL GENE TRANSFER.

horizontal resistance (HR): This type of resistance is equally effective against all the races (biotypes) of the pathogen. Variation in host resistance is primarily due to differences between varieties and between isolates, rather than to specific variety × isolate interactions. The type of resistance is non-specific in nature, which is governed by MINOR GENES (POLYGENES or QUANTITATIVE GENES) which act additively together to control resistance which is non-specific in nature and does not obey gene-for-gene hypothesis. This type of resistance is difficult to identify. *Aka* MANY-GENE RESISTANCE *or* NON-SPECIFIC RESISTANCE. *See* DURABLE RESISTANCE, GENERAL RESISTANCE *and* RACE-NON-SPECIFIC RESISTANCE. *Contrasts* SPECIFIC RESISTANCE *and* VERTICAL RESISTANCE.

horizontal transmission: Transmission of genetic information between different individuals other than from parent to offspring. *See* VERTICAL TRANSMISSION.

hormonal control of sex: A fertile hen (female) changed to a fertile cock (male). This resulted due to damaged ovary in female. The ovary in the female secretes a male suppressing hormone. Therefore, in the absence of ovary, the testis could develop. In cattle, when twins are produced, one male and another female, the female is sterile, and male is normal. These are known as Free-Martin and would be produced only when there is a vascular connection between two embryos. The male hormone, perhaps, suppresses the development of the ovary in the female, thus the sex is changed.

hormone: A class of signalling molecules that are produced in one part of an organism and transported in minute quantities to induce through their mitogenic activity a growth response in another part. It has one or more of four basic functions: (i) enables and promotes normal development, (ii) enables and promotes the adjustment of performance level, (iii) plays a role in keeping certain physiological functions in balance and (iv) the 'messenger' molecules for the body that help coordinate the actions of various tissues, they produce a specific effect on the activity of cells remote from their point of origin. For example, (a) in plants, auxin, cytokinin and gibberellin and (b) insulin produced by pancreas. *See* GROWTH REGULATOR. Hormone is the general term used in animals and humans where growth related substances are produced by ductless glands. Similar substances that regulate the plant growth are called PLANT GROWTH REGULATORS.

hormone net: The integration and control of various interacting hormones in an individual.

hormone-induced increases in transcription: Hormones induce increases in transcription. For example, oestrogen treatment stimulates production of egg white proteins in chicken oviduct.

hormone-receptor protein complexes: Specific receptor proteins which activate the transcription of specific genes.

hormones and transcription: Sex differentiation mechanism and function of hormones shows that comprehension of genotypic capability rests on and can be modified by hormones, while the production of hormones is itself directed by the genetic composition of the individual.

hormones in gene regulatory sequence: The hormone (H) attaches itself to a specific cytoplasmic receptor (R) which is then further processed into a form (R′) enabling the H-R′ complex to cross the nuclear membrane to enter the nucleus. In the nucleus, the H-R′ segment is bound to a non-histone protein acceptor to which phosphate groups are then added by specific phosphorylating kinase enzymes. The negatively charged phosphorylated non-histone protein can combine with the positively charged basic histones to cause removal of histones from DNA and thereby allow transcription to proceed. Some phosphorylated non-histones may also act as sigma factor components of RNA polymerase, allowing the enzyme to recognize promoters and transcribe specific DNA sections.

hormonic mean: The reciprocal of the average of the reciprocals of the data values. It is based on all the observations, and it is rigidly defined. Harmonic mean gives less weightage to the large values and large weightage to the small values to balance the values correctly. In general, the harmonic mean is used when there is a necessity to give greater weight to the smaller items. It is applied in the case of times and average rates.

horotelic evolution: A normal or average rate of evolution per million years of genera within a given taxonomic group. Thus, slowly or rapidly evolving lines may be horotelic during certain episodes in their history. *Compare with* TACHYTELIC EVOLUTION.

horticulture: The cultivation of a garden, orchard, nursery, flowers, fruits, vegetables or ornamental plants.

host: (i) In molecular biology, a host is the organism into which a gene from another organism is introduced. (ii) In plant pathology, an organism that becomes infested or infected with other organisms. This is a common term used during the association of two organisms in which host is usually larger partner which usually provides shelter/nutrient to the other small organism. Host may get benefit or harm from the other organism. For instance, in plants the pathogen causes disease in the host plant, whereas, in symbiosis host gets benefit from symbiont. Plant hosts can be broadly classified by IMPORTANCE (a) primary or principal hosts; (b) secondary hosts; (c) intermediate hosts; (d) accessory hosts; (e) accidental hosts; and (f) definite or final hosts; SEASON (a) winter hosts and (b) spring hosts; and OTHER FUNCTIONS (a) alternative, alternate or differential hosts and (b) transport or transfer hosts.

host cell: (i) A cell whose metabolism is employed for growth and reproduction by a virus. (ii) The cell into which a plasmid is introduced during recombinant DNA experiments. (iii) A cell whose metabolism is utilized for the growth of a pathogen.

host-cell physiology: Host-cell physiology can affect the level of expression. There has been no systematic study of the effect of different growth conditions on synthesis of foreign proteins in *Escherichia coli*. Certain factors (viz. choice of nutrients, way in which they are supplied to culture, temperature and dissolved oxygen tension) constitute physiological conditions of the host cell.

host genes: These are the genes that are responsible for resistance to diseases in the host. They are usually dominant. Resistance could be monogenic, oligogenic or polygenic depending upon whether it is controlled by one, a few or many genes. On the basis of effectiveness, resistance may be classified into VERTICAL and HORIZONTAL.

host genomics: The genetic makeup of a host organism.

host plant: A living plant attacked by or harboured a parasite or pathogen and from which the invader obtains part or all of its required nourishment.

host plant resistance: A procedure of pest control in which resistant, tolerant or unattractive host organisms are utilized. The inherited qualities of resistance impact the extent of pest damage.

host races: Genetically distinct populations that specialize on different hosts.

host range: (i) The spectrum of genotypes that can infect a specific pathogen or pest. (ii) All different types of hosts in which a phage or plasmid can replicate. Restriction is one issue that can limit the host range of plasmids or phages. Most of the viruses are specific as they infect specific host species, whereas there are certain viruses which possess a wider host range as these can infect, fungi, insects, plants and humans. In case of bacteria, for example, naturally occurring *Agrobacterium tumefaciens* can infect a wide range of dicot plants, such as tobacco, soybean, tomato, aubergine and poplar, whereas these wild types of strains do not infect monocot plants under natural conditions.

host resistance: The consequence of genetic manipulation of the host which makes it less susceptible to pathogens that would or do attack the host.

host restriction-modification in prokaryotes: Host-controlled restriction is the process by which certain strains of bacteria recognize and degrade foreign DNA. The same strains are able to carry out modification (i.e. methylation) of their own DNA to protect it from restriction. 'Modification' used in strict sense here means DNA methylase activity where adenine or cytosine gets methylated by transfer of methyl groups from S-adenosylmethionine (AdoMet or SAM). The phenomenon of restriction and modification was originally noticed during the lytic phage infection of bacteria. When a strain of bacteriophage λ infects *Escherichia coli* K λ, phage infection and replication were normal, leading to lysis of the cell and release of progeny phage. By contrast, when bacteriophage λ.B (carrying the B modification) or λ.O (lacking modification) infects *Escherichia coli* K, the host's restriction and modification system distinguishes the DNA as foreign and slashes it before it has a chance to replicate. This modification protects the bacterial DNA from its restriction enzymes. Thus, 'host restriction and modification' system offers a defence against foreign DNA, such as that borne by bacteriophages.

host restriction-modification system, biological significance of: The host restriction-modification system acts as a bacterial protection system against the foreign DNA. In fully modified DNA both strands are methylated. This DNA represents the homologous bacterial DNA which is not substrate for methylation (modification) or restriction (cleavage). Heteroduplex DNA is the product of semi-conservative replication and as such must be protected from restriction and newly synthesized daughter strand must be modified. This enzyme system does not cleave heteroduplex DNA but methylates it. Unmodified DNA represents foreign DNA, which must be restricted. The enzyme remains bound to the recognition site.

host restriction-modification system, type I: Type I restriction and modification (R-M) systems are complex. A single multifunctional enzyme can respond to the modification state of its target sequence with the alternative activities of modification or restriction. In the absence of DNA modification, a type I R-M enzyme behaves like a molecular motor, translocating vast stretches of DNA towards itself before eventually breaking the DNA molecule.

host restriction-modification system, type II: Type II RE is responsible for restriction only and a separate enzyme is accountable for modification (methylation) of the same target sequence. Structures of these proteins are also complex. Best understood are the components of *Eco*RI system where restriction enzyme is a dimer of identical subunits and methylase is a monomer. The target sites for type II REs are the palindromic sequences of 4–6 bp. Because of symmetry, the bases to be methylated occur in both the strands. So, a target site could be fully methylated, hemi-methylated or non-methylated. A fully methylated site is not used as a target for restriction or methylation. A hemi-methylated site is not recognized by RE but may be converted by methylase into fully methylated condition. It happens during the perpetuation of fully methylated DNA which on replication forms hemi-methylated DNA. This is primarily the usual mode of action of methylase *in vivo*. A non-methylated target site may be substrate for either restriction or modification *in vivo*. It is rare for unmodified DNA to survive by gaining the modification pattern of new host. Most of the type II REs cut DNA at an unmethylated target site by slashing one bound in each strand. *Aka* HOST-CONTROLLED RESTRICTION.

host restriction-modification system, type III: Type III systems are typically composed of two functionally distinct proteins, Mod and Res. The Mod subunit is responsible for specific DNA sequence recognition and host-protective DNA methylation. The Res subunit contains the endonuclease domain for DNA cleavage and an ATP hydrolysis domain related to SF2 helicases. Restriction requires both Mod and Res subunits, whereas methylation generally requires only the mod enzyme.

host restriction-modification system, type IV: They are not true RM systems because they only contain a restriction enzyme and not a methylase. Unlike the other types, type IV restriction enzymes recognize and cut only modified DNA. McrA, McrBC and Mrr are prototype *Escherichia coli* enzymes in this class that show different restriction spectra.

host-cell reactivation: The repair of damaged DNA of viruses by host-cell repair mechanisms.

host-controlled modification: After DNA replication, the host methylates its own DNA with the help of its methylase enzyme. This enzyme is composed of polypeptide products of *hsdM* and *hsdS* genes in ratio 2:1.

host-controlled restriction: DNA synthesized in one cytoplasm that undergoes cleavage when it enters the host cytoplasm. The restriction endonuclease is made up of three different polypeptides, R, M and S, in ratio 2:2:1, synthesized by three genes, *hsdR*, *hsdM* and *hsdS*, respectively.

host-controlled restriction and modification (HCRM): The mechanism by which certain bacteria attempt to protect themselves from phage infection. Incoming phage (incoming) DNA is destroyed by site-specific restriction endonucleases. The bacterial genome is protected from destruction by its own restriction enzymes by the modification of some of the nucleotides in the recognition site of the enzyme. Base methylation is the usual method of DNA modification.

host-delivered/derived RNA interference (HD-RNAi): Host-derived RNAi is achieved by genetically transforming host plants with double-stranded RNA constructs targeted at essential parasite genes leading to generation of siRNAs. Si RNAs formed in the host are then delivered to the parasite and transported to target cells.

host-induced RNAi-based pest control: The RNAi technique can be successfully utilized as a control tactic against insect and mite pests. In most of the functional genomic studies, microinjection or feeding of bacteria expressing dsRNA was used as a method of delivery to silence different target genes. Gene silencing caused distressing effects on the growth, development and survival of insects. These investigations suggested the prospect of employment of approach based on feeding bioassay for insect pest control through RNAi. In host-induced RNAi (HI-RNAi) or plant-mediated approach, a crop plant is engineered with HAIRPIN RNAI (hpRNAi) vector for the production of dsRNA against the target gene of insect pest. The dsRNA enters into the insect gut during feeding on plant parts lead to the induction of RNAi machinery and then, result in silencing of the target gene in concerned insect pest.

host-mediated assay: In this mutagenicity test, microorganisms (bacteria or yeast cells) are introduced into a host organism such as a rat (or a mouse). The host is then treated with the substance being tested. After a period of time, the microorganisms are recovered, checked for mutations and compared to the control. It is presumed that experimental microorganism comes in contact with the substance after it has been metabolized by the host.

host-mediated restriction: A mechanism by which bacteria prevent infection by phages originating from other bacteria. Restriction also acts against unmodified plasmid DNA in transformation experiments. Restriction endonucleases cleave the foreign DNA while the host DNA is protected from cleavage by specific methylation.

host-parasite specificity: Refers to ability of a pathogen to affect distinct group of plants. It is known to operate at genus, species and genotypic level.

host-specificity (*hsd*) locus: It synthesizes enzyme having methylase and/or restriction activity. It is composed of three genes in the order *hsdR* (responsible for restriction), *hsdM* (responsible for modification) and *hsdS* (required for recognition of host-specificity site).

host-specificity (*hsd*) promoters: The *hsd* locus has two promoters, P_1 and P_2. P_1 is located upstream of the first gene *hsdR* and is responsible for transcription of this gene. P_2 is located upstream of the second gene *hsdM* and controls transcription of *hsdM* and *hsdS* genes.

host-specificity sites: These are those sites in host DNA where recognition and methylation occur. The cleavage may take place elsewhere. Thus, restriction enzyme recognizes two sites on DNA: the RECOGNITION SITE (HOST-SPECIFICITY SITE) and the

CLEAVAGE SITE. The sK and sB are the recognition sites of *Eco*K and *Eco*B strains, respectively. These recognition sites are very similar architecturally: sB 5′ T-G-A*-N-N-N-N-N-N-N-N-T-G-C-T 3′/3′ A-C-T-N-N-N-N-N-N-N-N-A*-C-G-A 5′; sK 5′ A-A*-C-N-N-N-N-N-N-G-T-G-C 3′/3′ T-T-G-N-N-N-N-N-N-C-A*-C-G 5′. Recognition sequence is hyphenated, with two constant domains of three and four bases, respectively, separated by a spacer of eight bases for sB and six bases for sK. Alteration in sequence of constant domains or in the length of spacer results in non-functional site. Neither sB nor sK sequences show twofold symmetry. Two methyl groups per sB site were incorporated, one for each strand. Adenines that may become methylated are marked with asterisks (*).

host-vector (HV) system. In recombinant DNA technology, host is the organism into which a gene from another organism is introduced. The GUEST GENE is carried by a vector (i.e. a larger DNA molecule, such as a plasmid or a virus into which that gene is inserted) which then propagates in the host. Vectors (plasmids and phages) thus transmit man-made and natural genetic information. A host cell develops, expresses and multiplies the transmitted information.

host-virus interactions: Define as to how viruses sustain themselves within the host organisms on a molecular, cellular or population level. Virus-host interaction is often virus- and host cell-specific.

hot dilute soup: Belief that early earth was hot, and first life originated in a hot solution containing simple organic compounds.

hot spots: Sites in genes at which events, such as mutations, happen with unusually high frequency.

hot-start PCR: Hot-start PCR is a technique that reduces non-specific amplification during the initial set up stages of the PCR. It may be performed manually by heating the reaction components to the melting temperature (e.g. 95°C) before adding the polymerase. Specialized enzyme systems have been developed that inhibit the polymerase's activity at ambient temperature, either by the binding of an antibody or by the presence of covalently bound inhibitors that only dissociate after a high-temperature activation step. Hot-start/cold-finish PCR is achieved with new hybrid polymerases that are inactive at ambient temperature and are instantly activated at elongation temperature.

household waste: Waste that contains paper, cardboard, textiles (for example, fabric or carpet), timber, food, garden clippings, glass, plastic and other manufactured materials.

housekeeping enzymes: Enzymes occurring in all cells capable of routine metabolism. They are indispensable for the synthesis or breakdown of proteins, nucleic acids and lipids for glycolysis and respiration and for many normal metabolic pathways.

housekeeping genes: The genes that code for proteins involved in standard metabolic processes common to all cell types, regardless of their specific role, and as such seem to comprise the major class of mammalian genes. Many such genes lack a TATA box and, presumably as a consequence of this, have multiple initiation sites. They all possess CG-rich regions with multiple binding sites. Further, promoters of housekeeping genes possess the ability to support bidirectional transcription. A house keeping gene promoter has a simple array of control elements. Such genes are destined to express continuously. These genes are needed throughout development for the smooth running of the cell's metabolism. *See* CONSTITUTIVE GENES.

Hox complex: Two tightly linked clusters of genes in *Drosophila* (the bithorax and Antennapedia complexes) that govern the differences between the different segments of the body. Homologous Hox complexes exist in other animals, where they also determine pattern along the anteroposterior axis.

Hox genes: A particular subgroup of homeobox genes that act to pattern the axis of an organism's body and govern where limbs and other body segments will grow as the embryo develops.

Hpa tiny fragment (HTF) islands: HTF islands are methylated CpG-rich regions in the genome. About 80% of these regions occur at or near genes, particularly the housekeeping genes. The HTF islands are not methylated on active X chromosomes.

***Hpa*I:** A type II restriction enzyme from bacterium *Haemophilus parainfluenza* that recognizes the DNA sequence 5′-G T T↓A A C / 3′-C A A↑T T G where cuts at the sites are indicated by arrows.

***Hpa*II tiny fragment augmentation by ligation-mediated PCR assay (HELP)**. This method can be used to determine genome-wide patterns based on the combined activity of *Hpa*II and *Msp*I.

***Hsa*I:** A type II restriction enzyme from bacterium *H. sapiens* that recognizes the DNA sequence 5′-G C G↓C 3′ / 3′-C↑G C G 5′ where cuts at the sites are indicated by arrows.

***hsd* locus of *Escherichia coli* K:** The *hsd* locus of *Escherichia coli* K includes restriction and modification genes, *hsdR*, *hsdM* and *hsdS* which code for hsdR, hsdM and hsdS subunits. Restriction is carried out by hsdR subunit and modification by means of methylation by the hsdM subunit. Subunit hsdS is responsible for recognition of the host-specificity site. This locus has two promoters, P_1 and P_2, which control transcription of the genes: P_1 for *hsdR* gene and P_2 for both *hsdM* and *hsdS* genes. Translation of the transcripts of these three genes, respectively, yields polypeptides with molecular weights of 135,000 (hsdR), 62,000 (hsdM) and 55,000 (hsdS). These three polypeptides, in the ratio of 2:2:1, form the enzyme *Eco*K. The transcription pattern controlled by two promoters enables the cell to synthesize a methylase composed of *hsdM* and *hsdS* gene products independently from synthesis of the restriction endonuclease which contains all the three gene products.

hsd system: A system that deliberates specificity by methylating adenine. There are counter parts to this system in many bacteria. The specificity is characteristic of each system.

***hsdM* gene:** The *Escherichia coli* gene which encodes the DNA methylase involved in the modification system which prevents its DNA being attacked by its own restriction endonucleases. Host strains in gene cloning experiments usually have a functional *hsdM* gene facilitate gene transfers.

***hsdR* gene:** The *Escherichia coli* gene that encodes restriction endonuclease activity. The initials stand for host-specified defence restriction. Host strains for gene cloning experiments are often *hsdR* mutants so that the foreign DNA carried by the vector is not attacked by the host's restriction system.

***hsdS* gene:** The *Escherichia coli* gene that encodes host-specificity site recognition activity. The initials stand for host-specified defence restriction. Host strains mutant for this gene produce enzyme that does not recognize host-specificity site on DNA.

human artificial chromosomes (HACs): A minichromosome that can function as a new chromosome in a population of human cells that have been synthesized from chemicals that are identical to chromosomes within human cells. HACs have been used to create transgenic animals for use as animal models of human disease and for production of therapeutic products. HACs formed when sequences from human centromeres were incorporated along with a marker gene into cell lines were often large concatenated (linked together) mixtures of the incorporated DNA, dipping their utility for precision genome engineering. C.W. Gambogi *et al.* (2024) pronounce a significantly enhanced method that forms HACs that are small and structurally well defined. HAC is around 750 kb, much larger than prior HACs and adequate to accommodate the multidomain chromatin at the centromere needed for inheritance through cell divisions. Along with streamlined cellular delivery methodologies, these advances offer the methods to improve chromosome engineering in mammals and many other eukaryotes. These HACs are also known as SINGLE-COPY ARTIFICIAL CHROMOSOMES.

human CEN DNA: In humans, one of the most recognized satellite DNA sequences is the alphoid family which is found mainly in the centromeric regions. Alphoid sequences, each 171-bp in length, are present in tandem arrays of up to one million bp. Each alophoid repeat in humans acts as a binding site for a single molecule of centromere-specific protein CEN-B.

human diseases due to triplet repeat DNA amplification: A number of human diseases that are caused by virtue of having triplet repeat amplification such as in Huntington's disease where 11–34 repeats of CAG in the Huntington's disease gene is normal, but 50–100 results in the disease.

human DNA polymerase ι (hPolι): hPolι incorporates the correct nucleotide opposite to a template adenine with a several hundred to several thousand-fold greater efficiency than it incorporates a correct nucleotide opposite a template thymine. Its efficiency for correct nucleotide incorporation opposite a template guanine or cytosine is intermediate between these two extremes.

human epigenome atlas: A compendium of reference epigenomes produced by the NIH Epigenomics Roadmap Initiative (http://www.roadmapepigenomics.org) which is suitable for integrative analysis, visualization and epigenome comparison. The atlas is accessible at http://www.epigenomeatlas.org. Human epigenome maps have been prepared in recent years owing to rapid development of high-throughput epigenome-mapping technologies. Many international consortia have been constituted to tackle the fundamental problems in epigenomics by sharing resources and protocols. Exonic intermediate methylation associates with exon inclusion at a level between that of fully methylated and unmethylated exons, highlighting gene context-dependent functions. The human epigenome mapping includes detailed descriptions of the epigenetic markers in 111 types of cells and tissues. Partial epigenome mapping has also been achieved for many other cell types. The analysis of DNA methylation levels within the major histocompatibility complex, including regulatory exonic and intronic regions associated with 90 genes in multiple tissues and individuals, revealed a bimodal distribution of methylation profiles.

human exposure to *Bt* proteins: Exposure of humans to *Bt* toxins in the food and environment. As the acreage of *Bt* crops increases, more are the chances of our exposure to *Bt* protein. *Bt* crops or not, we already are exposed to *Bt* proteins. Our exposure may be ancient as well as unavoidable. Studies of the natural ecology of *Bt* show it is abundant in the foliage of numerous plant species, and its presence in the soil may result from wash-off with rainfall. Even more curious is the occurrence of *Bt* in stored grain that has not been specifically sprayed. Stored grain is commonly infested by moths and beetles that may be susceptible to naturally occurring *Bt*. Exposure to *Bt* by organisms feeding on the grain is confirmed by finding Bt spores in the faeces of birds and rodents collected from the feed mill. It might be argued that *Bt* sprays do not leave as much insecticidal protein as is present in a transgenic plant, in which every tissue makes large quantities of the protein. Documents compiled by seed producers of transgenic plants filed with various agencies responsible for the regulation of transgenic plants in the USA – United States Department of Agriculture (USDA), Animal and Plant Health Inspection Service (APHIS); the Environmental Protection Agency (EPA) and the Food and Drug Administration (FDA) – have included tests of oral toxicity to rodents, birds, honeybees, insect predators, earthworms and Daphnia. FDA has ruled that products derived from *Bt* cotton, corn and potatoes are safe for human and animal consumption.

Human Genome Project (HGP): An international effort officially launched in October 1990 and completed in April 2003 to determine all three billion bases of DNA within the entire human genome and all the alleles of all the genes. Through HGP, a physical map (more precisely DNA sequence) of our chromosomes has been created which marks off chromosomes not in recombinational distances but in base pairs. This has been done by creation of libraries of overlapping, contiguous clones, mentioned in literature as contig. This is done chromosome by chromosome. The process begins with overlapping clones obtained, for example, by chromosome walking. The cloned pieces are then isolated and totally digested. Clones are then compared to mark the segments that they have in common. From shared segments it can be inferred which parts of the clones are in common. Contigs of parts of the chromosome can thus be constructed. Later, contigs comprising part of a chromosome can be assembled by taking an end clone of a completed contig and using it as a probe to begin chromosome walking to find out an end clone of a nearby contig.

human leukocyte antigen (HLA) complex: A complex encompassing the genes that are responsible for synthesis of antigens that determine acceptance or rejection of a tissue graft. HLA system is based on three classes of HLA genes on chromosome number 6 which control immunological acceptability. Each gene may possess 8–40 alleles, each allele specifying a particular antigen. A particular chromosome 6 may have one of 75,000 theoretically possible combinations of HLA alleles; each particular combination being called a HAPLOTYPE. HLA complex encodes cell surface molecules specialized to present antigenic peptides to the T-cell receptor (TCR) on T cells. HLA antigens have a significant role in transplant-mediated rejection. HLAs are of three main types. CLASS I HLA ANTIGENS include HLA-A, B and C molecules; CLASS II HLA ANTIGENS, which includes HLA-DR, -DQ and -DP loci, are on antigen-presenting cells; and CLASS III HLA ANTIGENS contains proteins that have immune functionality.

human leukocyte antigens (HLA): A very complex array of six proteins that cover the surface of leukocytes (and the bone marrow cells that produce leukocytes). These HLA are usually different (i.e. a nonmatch) for individuals that are not genetically related to each other (e.g. a father-son or a father-daughter), so have been used in the past to prove paternity. HLA must also be matched (as nearly as possible) for successful bone marrow transplants, to prevent the donated bone marrow (and the marrow recipient) from 'rejecting' each other. *See* MAJOR HISTOCOMPATIBILITY COMPLEX.

human methylome atlas: N. Loyfer *et al.* (2023) describe a human methylome atlas, based on deep whole-genome bisulfite sequencing, allowing fragment-level analysis across thousands of unique markers for 39 cell types sorted from 205 healthy tissue samples. Replicates of the same cell type are more than 99.5% identical, demonstrating the robustness of cell identity programmes to environmental perturbation. Unsupervised clustering of the atlas recapitulates key elements of tissue ontogeny and identifies methylation patterns retained since embryonic development. Loci uniquely unmethylated in an individual cell type often reside in transcriptional enhancers and contain DNA-binding sites for tissue-specific transcriptional regulators. Uniquely hypermethylated loci are rare and are enriched for CpG islands, Polycomb targets and CTCF-binding sites, suggesting a new role in shaping cell-type-specific chromatin looping. The atlas provides an essential resource for study of gene regulation and disease-associated genetic variants and a wealth of potential tissue-specific biomarkers for use in liquid biopsies.

human mitochondrial genomes: The human mitochondrial genome is a circular double-stranded DNA molecule with a size of 16,569 bp. The mtDNA has no intron but retains compactly arranged 37 genes (13 proteins, 22 tRNAs and 2 rRNAs) critical for producing energy through OXIDATIVE PHOSPHORYLATION (OXPHOS).

human telomeric chromatin structure: A. Soman *et al.* (2022) used negative stain electron microscopy and single-molecule magnetic tweezers to characterize 3-kbp-long telomeric chromatin fibres. They also obtained the cryogenic electron microscopy structure of the condensed telomeric tetranucleosome and its dinucleosome unit. The structure displayed close stacking of nucleosomes with a columnar arrangement, and an unusually short nucleosome repeat length that comprised about 132 bp DNA wound in a continuous superhelix around histone octamers. This columnar structure is primarily stabilized by the H2A carboxy-terminal and histone amino-terminal tails in a synergistic manner. The columnar conformation results in exposure of the DNA helix, which may make it susceptible to both DNA damage and the DNA damage response.

human telomeric DNA: Human telomeric DNA comprises around 2–20 kb of tandem repeat $(TTAGGG)_n$ oriented 5′®3′ towards end. An α thalassaemia mutation associated with terminal truncation of the short arm of chromosome 16 showed that $(TTAGGG)_n$ has been added directly to the site of the break.

humoral immune response: A type of immune response in which antigen stimulates the production of antibodies which react with specific antigens so as to destroy them.

humoral immunity: Immunity conferred by circulating antibodies produced by B lymphocytes and plasma cells. It is also called antibody-mediated immunity. B cells produce antibodies which are released into extracellular fluids such as blood plasma, lymph and mucus secretions. Humoral immunity protects from bacteria, viruses and bacterial toxins which circulate freely in the body's fluid.

hundred-dehulled grain weight: A parameter to express weight of the 100-dehulled grain having pericarp (seedcoat).

hundred-seed weight: Measurement in grams of 100 well-developed whole grains (seeds with the hull), dried to 13% moisture content, weighed on a precision balance.

husk: Protective outer covering of certain seeds, including the leafy outer covering of an ear of maize, leathery covering of the walnut or spiky covering of the chestnut.

***hut* operon of *Salmonella typhimurium*:** In *S. typhimurium*, histidine is converted into glutamate and formate by four enzymes (histidase, urocanase, imidazolone propionate hydrolylase and farminino glutamate hydrolylase). *HUT* operon controls the synthesis of all these enzymes. This operon carries two operators and two promoters. HUT repressor mediates the negative control. Only when HUT repressor is prevented from binding to respective operators can the signal of either C or N starvation be transmitted.

hyaline: Any of several translucent nitrogenous substances related to chitin, found especially around cells, and readily stained by eosin. It gives a glassy appearance. It could be transparent or nearly so; translucent.

hybrid: (i) Seed or plants produced as the result of controlled cross-pollination between two genetically unlike parents (from lines, races, breeds, varieties, species or genera) as opposed to seed produced as the result of natural pollination. Hybrid seeds are selected to have higher quality traits (e.g. yield or pest tolerance). (ii) Across species, fusing two different cell types *in vitro*. *Aka* CHIMAERA.

hybrid breakdown: A type of reproductive failure that appears after the F_2 generation of crosses between different species or subspecies produces weak or sterile F_2 progeny by vigorous fertile F_1 hybrids. It is caused by incompatibility between interacting genes.

hybrid breeding: The discovery of heterosis has been acknowledged as one of the major landmarks of plant breeding. In comparison to inbred lines and homozygous material, the phenotypic superiority of heterozygotes is the foundation of hybrid breeding. It is employed for production of hybrid, synthetic and composite varieties. Hybrid breeding can be achieved using traditional breeding techniques, or the process can be accelerated by using gene marker technology to rapidly recognize parents with desired genes for certain attributes. For hybrid breeding, two genetically different parent lines are produced and crossed. The offspring (hybrids) are bigger, more fruitful and more resilient than their parental lines (hybrid vigour) but the heterosis effect is maintained for only one generation. Seeds from a hybrid variety, if planted, will not deliver the same benefits as the original seeds and after several offspring will have lost the desired qualities from the original hybridization. *Aka* HETEROSIS BREEDING.

hybrid cell: Uninucleate cell formed after the fusion of two different cells. During somatic cell hybridization, when two protoplasts from different species are fused the immediate product is heterokaryon, it is followed by spontaneous fusion of the two nuclei form SYNKARYON (one fused nuclei). The newly formed cell containing fused nucleus is called a hybrid cell.

hybrid chlorosis: Plant and/or leaf chlorosis due to interacting genes and/or cytoplasm of parental lines in a hybrid gene. Hybrid chlorosis was linked to two recessive interacting loci: one on chromosome 1 and one on chromosome 5, of which one, originating from Lov-5 on chromosome 5, appears to encode an RNA helicase (*AtRH18*). The other gene located on chromosome 1 (*At1g08210*) encodes a protein that belongs to the A1 family of pepsin-like aspartic proteases).

hybrid complex: Concealing morphological differences of parental lines in a hybrid plant.

hybrid DNA: DNA whose two strands have different origins. Hybrid DNA is initially formed on one chromatid by asymmetrical strand transfer and subsequent isomerization and branch migration can generate hybrid DNA on both chromatids.

hybrid dysgenesis: The inability of certain strains of *D. melanogaster* to interbreed, because the hybrids are sterile (although otherwise they may be phenotypically normal).

hybrid formula: The names of the parents of a hybrid joined by a multiplication sign (e.g. *Cytisus ardonoi* × *C. purgans*).

hybrid gene: A gene composed of a fragment of the gene of interest ligated to some other heterologous DNA sequence that permits regions to respond to regulatory signals.

hybrid inferiority: When a hybrid is inferior to either of the parents.

hybrid inviability: Reduced vigour of hybrid plants compared to their crossing parents.

hybrid lethality: The failure of hybrids to produce viable offspring. *Aka* HYBRID STERILITY. *See* HYBRID NECROSIS.

hybrid necrosis: Hybrid necrosis is a common type of incompatibility found in F_1 progeny of many crosses within species and between species, which suggests it could be a model for understanding factors that are important at various stages in the processes of genetic differentiation and perhaps speciation.

hybrid plasmid: A plasmid that carries an inserted piece of foreign DNA.

hybrid proteins: Protein sequences can be joined or fused to that of another protein. The resulting protein is called a hybrid, fusion or chimeric protein, which generally has characteristics that combine those of each of the joined partners. Protein fusions have been extensively used to study interactions between two or more proteins. An application of protein fusion methods called the yeast two-hybrid screen identifies proteins that interact with each other. The system was developed by separating a yeast transcription-activating protein into two functional domains. The first domain of the transcription-activating protein is fused to a protein that is being studied for interactions. The second domain of the transcription-activating protein is fused to a collection of proteins encoded in a protein library. The first fusion protein and one from the protein library are expressed together in yeast. When the two fusion proteins interact, they can bring together the two halves of the transcription activator and turn on the expression of a reporter gene. Another example is the fusion of a protein to an inherently fluorescent protein, such as the GREEN FLUORESCENT PROTEIN (GFP) from the jellyfish *Aequorea victoria*. Fluorescent fusion proteins have been used to study the location, movement, appearance and degradation of proteins in living cells. *Aka* FUSION PROTEINS *or* CHIMERIC PROTEINS.

hybrid rescue allele: An allele that assuages hybrid sterility or hybrid inviability.

hybrid resistance: The rejection of parental tissue grafts by the F_1 hybrids of two inbred strains.

hybrid screening: Radioisotope technique used to determine whether a hybrid plasmid contains a particular gene or DNA region.

hybrid seed: The seeds produced by combination of more or less defined pure parental lines. The hybrid seeds are generally more productive or more appropriate than pure lines. Hybrid seed is produced with the intention of moving all positive traits into a common background. They are used for subsequent growing and commercial production of a crop seed.

hybrid selection: The process of choosing plants possessing desired traits among a hybrid population.

hybrid sorting: The selection of superior plants among haploids derived from F_1 hybrids through another culture.

hybrid sterility: Refers to the failure of hybrids to produce viable offspring.

hybrid sterility in insect pest management: Species are reproductively isolated populations of insects and scientists trust that reproductive barriers may exist between different races of the same species inhabiting different geographical areas or between sibling species. Therefore, hybrids can be produced between two closely related species in the laboratory which would be sterile and then released into wild populations of insects.

hybrid variety: A variety produced from the cross-fertilization of inbred lines with favourable combining ability; the progeny is homogeneous and highly heterozygous. It can be produced by (i) two inbred lines, (ii) single crosses, (iii) a single cross and an open-pollinated or a synthetic variety, or (iv) two selected clones, seed lines, varieties or species. *Aka* HYBRID LETHALITY. *See* HYBRID NECROSIS. *See* VARIETY.

hybrid vehicle: An episome or plasmid containing an inserted piece of foreign DNA. *Aka* HYBRID VECTOR.

hybrid vigour: W.G. Whaley (1952) used terms heterosis and hybrid vigour synonymously. However, hybrid vigour precisely refers to the developed superiority of hybrid and heterosis refers to the mechanism by which superiority is developed. By this definition, hybrid vigour is a manifestation of heterosis. Thus, heterosis is the cause, and hybrid vigour is an effect. Hybrid vigour is *aka* POSITIVE HETEROSIS. *See* HETEROSIS.

hybrid weakness: The decrease in vigour of hybrids below their parental inbred types. For example, in rice, a hybrid weakness phenomenon is governed by a set of complementary *hybrid weakness c* genes, *Hwc1* and *Hwc2*. The *Hwc2* gene is prevalent among temperate japonica rice but not among tropical japonica or indica rices. The chromosomal location of the *Hwc2* locus was determined from the segregation in the F_1 hybrids made between recombinant inbred lines and the cultivar 'Jamaica'; *Hwc2* was located between two restriction fragment length polymorphism loci, XNpb264 and XNpb197 on chromosome 4.

hybrid zone: A narrow region in which genetically distinct populations meet, mate and hybridize.

hybrid-arrested translation: A means of identifying recombinant DNA clones by their ability to hybridize to, and thus prevent the translation of, a specific messenger RNA in a cell-free system. Hybrid-arrested translation is founded on the fact that an mRNA will not direct the synthesis of a protein in a cell-free system when it is in a hybrid from with its complementary DNA. The approach is thus similar to HYBRID SELECTION, *aka* HYBRID RELEASE, in which messenger RNAs homologous to cDNA clones are specifically trapped as RNA/DNA hybrids on nitrocellulose filters. By washing the filters at specific temperatures and salt concentrations, RNA molecules bound non-specifically are removed leaving only the RNA homologous to the DNA bound. This RNA can then be released by boiling the filter, precipitated with ethanol and translated *in vitro*. *Compare with* HYBRID-RELEASED TRANSLATION.

hybrid-released translation (HRT): HRT enables a cloned DNA to be correlated with the protein(s) which it encodes. HRT is a direct method in which cloned DNA is bound to a cellulose nitrate filter and hybridized with an unfractionated preparation of mRNA or even total cellular RNA. The filter is washed and hybridized mRNA is eluted by heating in low salt buffer. Recovered mRNA is then translated in a cell-free translation system. The proteins encoded by the message can be analysed by electrophoresis through a polyacrylamide gel. *Compare with* HYBRID-ARRESTED TRANSLATION.

hybridization: (i) The pairing of a single-stranded, labelled probe (usually DNA) to its complementary sequence. (ii) The creation of RNA-DNA hybrids by a heating process, so that the RNA becomes associated with the complementary DNA. (iii) The process of complementary base pairing between sense and antisense RNA. (iv) In plant genetics, the mating of two plants from different species or genetically very different members of the same species to yield hybrids (first filial hybrids) possessing some of the characteristics of each parent. Hybrids can also be produced from more than two ('parent') species. (v) Some hybrids have occurred spontaneously in nature. For example, wheat (*Triticum aestivum*) arose centuries ago from a naturally occurring interbreeding of three

Middle East grasses. Hybrids can contain two DNA strands, two RNA strands or one DNA or one RNA strand. In molecular biology, hybridization forms the basis of powerful technique, such as SOUTHERN BLOTS, NORTHERN BLOTS, MICROARRAY, for detecting specific nucleotide sequences.

hybridization fingerprinting: A technique used to identify clone overlaps. It is designed to combine that high data rates achievable by hybridization techniques and the favourable scaling behaviour and the repeat intensity of the gel fingerprinting approaches. In this approach, filters carrying a large number of DNA spots are generated by colony lysis. These DNA spots are hybridized to oligonucleotide probes or more complex probe mixtures complementary to many positions in the genome.

hybridization *in situ*: Finding the location of a gene by adding specific radioactive probes for the gene and detecting the location of the radioactivity on the chromosome after hybridization.

hybridization kinetics: It is the speed with which a probe binds with complementary strands. It gives an estimate of number of copies of a sequence present.

hybridization probe: Labelled nucleic acid molecule used to detect complementary DNA sequences after hybridization.

hybridization probing: A method that uses a labelled nucleic acid molecule to identify complementary or homologous molecules through the formation of stable base-paired hybrids.

hybridization stringency: The percentage of nucleotides, which must match on two unrelated single-stranded nucleic acid molecules before they base pair with each other to form a duplex, under a given set of physical and chemical conditions. The hybridization stringency is used to determine when a hybridization probe and a target nucleic acid will come together and can be set by varying the conditions. In general, if the percentage of matching nucleotides is lower than 70%, the two single-stranded nucleic acid molecules are considered non-homologous, and any hybridization is considered non-stringent.

hybridization surfaces: Various physical substrates (surfaces) on to which genetic materials (DNA, RNA, oligonucleotides, etc.) have been 'attached'. Relevant complementary genetic materials are hybridized on to those attached-to-surface genetic materials for various specific purposes (e.g. detection of the presence of those unattached genetic materials, in the case of biosensor's hybridization surface). One of the technologies that can be engaged to assay DNA from hybridization surfaces is matrix-assisted laser desorption ionization time of flight mass spectrometry (MALDI-TOF-MS).

hybridize: (i) To form a hybrid performing a cross. (ii) To anneal nucleic acid strands from different sources.

hybridoma: The term refers to fused cells resulting from fusion of cells from two different sources. A cell hybrid between an antibody-producing B cell and a tumour cell, which divides indefinitely and produces a single antibody (monoclonal antibody) in culture. The role of the myeloma cell in this process is to deliver the normal B lymphocytes with the cancerous property of uninhibited growth, thereby permitting hybridomas to grow easily both in tissue culture and after injection into mice. Normal non-cancerous B cell would rapidly die in either situation.

hybridoma technology: A method for producing large numbers of identical antibodies (also called monoclonal antibodies). This process starts by injecting a mouse (or other mammal) with an antigen that provokes an immune response. A type of white blood cell, the B cell, produces antibodies that bind to the injected antigen. These antibody-producing B cells are then harvested from the mouse and, in turn, fused with immortal myeloma cancer cells (myeloma cells are cancerous plasma cells; plasma cells are derived from activated B cells), to produce a hybrid cell line called a hybridoma, which has both the antibody-producing ability of the B cell and the longevity and reproductivity of the myeloma. The hybridomas can be grown in culture, each culture starting with one viable hybridoma cell, producing cultures each of which consists of genetically identical hybridomas which produce one antibody per culture (monoclonal) rather than mixtures of different antibodies (polyclonal).

hydathode: (i) An epidermal leaf structure (pore) specialized for secretion or exudation of water. (ii) Leaf opening at terminus of vein.

hydrazine ($_2$HN-NH$_2$): Hydrazine breaks rings of uracil, cytosine and thymine. Breakage of pyrimidine is much more frequently lethal than mutagenic. Mutagenic effect of hydrazine is A-T to G-C transition.

hydrazine-aniline cleavage sequencing (HAC-seq): This method analyses ribosomal RNA (rRNA)-depleted total RNAs in human cells and found that tRNAs are the major 3^{m}C-modified RNA species, with

17 3^{m}C modification positions on 11 cytoplasmic and 2 mitochondrial tRNA isoacceptors in MCF7 cells.

hydrocarbon: An organic compound consisting entirely of hydrogen and carbon atoms. Hydrocarbons from which one hydrogen atom has been removed are functional groups called hydrocarbyls.

hydrogels: Cross-linked 3-D networks containing hydrophilic polymer chains able to adsorb a significant amount of water with a high versatility degree of chemical and physical properties. Hydrogel has been introduced in urban agriculture in the past few decades. Hydrogels act as a reservoir of water and/or nutrients with gradual release according to plant requirements and in a controlled manner. Thus, a favourable environment for the development of roots and plants is created and maintained throughout their growth.

hydrogen bond: A noncovalent interaction between the donor atom (e.g. nitrogen or oxygen), which is bound to a positively polarized hydrogen atom, and the acceptor atom, which is negatively polarized. Though not covalent, the hydrogen bond holds the donor and acceptor atom close together, maintains the secondary structure (α-helices and β-sheets) of proteins and is also the primary interaction between water molecules.

hydrogen fluoride (HF): A primary pollutant from volcanoes and several industrial processes such as brick factories, aluminium smelters and phosphate fertilizer factories. HF is used to make refrigerants, herbicides, pharmaceuticals, high-octane gasoline, aluminium, plastics, electrical components and fluorescent light bulbs.

hydrogen peroxide (H_2O_2): Hydrogen peroxide and organic peroxides are decomposed into OH radicals (written as $OH^{\bullet}$). These may react with pyrimidines at 5–6 double bonds thus attaching −OH group at position 6, initially. Hydroperoxide is thus produced which leads to transitions G-C to A-T transition.

hydrogenosome: An organelle which is devoid of genomes and found in some anaerobic ciliates, flagellates and fungi that produces hydrogen gas and ATP. It is conceivably derived from mitochondria.

hydrological cycle: The cycling of water molecules among its different states – gaseous form (water vapour), liquid and ice.

hydrolyse: To 'cut' or 'break' a chemical bond (i.e. with a molecule) via hydrolysis.

hydrolysis: Cleavage of a band such as anhydride by addition of elements of water molecules (H^+ and OH^-), yielding two or more products. Hydrolysis (addition of water) is the reverse of dehydration. Using this process an organism breaks down macromolecules into monomers to obtain nutrients. As the covalent band is broken one part of the molecule combines with a hydrogen ion while the other combine with a hydroxide ion. The net outcome of the hydrolysis is the splitting of macromolecules in the body cells. Cleavage of a covalent bond with accompanying addition of water, −H being added to one product of the cleavage and −OH to the other. *Aka* HYDROLYTIC REACTION.

hydronium ion (H_3O^+): Water molecule linked with an additional proton.

hydropathy plot: A measure of the hydrophobicity of a protein region and therefore occurrence of the likelihood that it will reside in a membrane.

hydrophilic: Describes a polar molecule or part of a molecule that forms enough energetically favourable interactions with water molecules to dissolve readily in water. Literally, 'water loving'. The molecules that get dissolved in the water are called hydrophilic, whereas those that do not dissolve are called HYDROPHOBIC.

hydrophilicity: The degree to which a molecule is soluble in water. Hydrophilicity depends to a large degree on the charge and polarizability of the molecule and its ability to form transient hydrogen bonds with (polar) water molecules. Hydrophilic groups interact with water, so that hydrophilic regions of protein or biomolecule tend to reside in water.

hydrophily: Form of pollination whereby pollen is dispersed by the flow of waters.

hydrophobic effect: The tendency of non-polar groups in water to self-associate and thereby minimize their contact surface area with the polar solvent.

hydrophobic: A non-polar molecule or part of a molecule that cannot make energetically favourable interactions with water molecules and therefore does not dissolve in water. Literally, 'water hating'. *Contrasts* HYDROPHILIC.

hydrophobic force: Force exercised by the hydrogen-bonded network of water molecules that brings two non-polar surfaces together by excluding water between them.

hydrophobicity: Literally, water hating. the degree to which a molecule is insoluble in water and hence is soluble in lipids. If a molecule lacking polar groups is placed in water, it will be driven to find a hydrophobic environment (such as the interior of a protein or a membrane).

hydroponics: The cultivation of plants by placing the roots in liquid nutrient solutions rather than in soil.

hydrothermal vents: Very hot areas in the ocean where primitive life might have begun.

hydroxyapatite: A form of calcium phosphate that binds double-stranded DNA.

hydroxyapatite columns: This method is used for separating double-stranded DNA from single-stranded DNA. Double-stranded DNA binds to hydroxyapatite whereas single-stranded DNA does not. Double-stranded DNA is separated from single-stranded DNA by placing them into a column containing hydroxyapatite. Denatured DNA when being annealed, the repetitive sequences could become double-stranded and hence would be separated from single-stranded DNA. Mixture of double-stranded and single-stranded DNAs is passed through a column of hydroxyapatite crystals, which retains the double-stranded DNA and permits single-stranded DNA to pass through. Double-stranded DNA is then eluted by raising temperature and its quantity is measured. NITROCELLULOSE has the same properties as hydroxyapatite. With nitrocellulose also, double-stranded DNA can be separated from single-stranded DNA.

hydroxyl (-OH): Chemical group consisting of a hydrogen atom linked to an oxygen, as in an alcohol.

hydroxylamine (NH_2OH): Hydroxylamine reacts with pyrimidine bases. Its effect is strong on cytosine while on uracil, the effect is less. It breaks and removes pyrimidine ring of uracil. It causes base pair transitions from G-C to A-T only. Mutations induced with hydroxylamine cannot be reversed with hydroxylamine.

hydroxylation reaction: A chemical reaction in which one or more hydroxyl groups (i.e. the −OH group) is chemically attached to a molecule.

hydroxyproline-based DNA mimics: A type of DNA mimic that has improved hybridization ability and functions as efficient gene silencers.

hydroxyurea: Causes depletion of the dNTP pool by inhibition of ribonucleotide reductase and thereby stalls DNA replication. Removal of hydroxyurea might stimulate viral second-strand synthesis.

hygroscopic water: Water held lightly by soil particles.

hylogram: A phylogenetic tree in which the branch lengths are proportional to the evolutionary distance between nodes. *Aka* ADDITIVE TREE.

hyperchromacity: The increase in optical density that ensues when DNA is denatured.

hypercycle: (i) Cooperation between a set of replicating molecules (A, B, . . ., Z), in which A helps replication of B, B assists C, . . . and Z supports A. It was proposed as a way for early life to replicate genetic information despite having high mutation rates. (ii) The metabolome forms a large network of metabolic reactions, where outputs from one enzymatic chemical reaction are inputs to other chemical reactions.

hypermethylation: Increase in the level of DNA methylation in a population of cells relative to a reference or normal sample; may be used to describe a specific nucleotide or a group of nucleotides.

hypermorph: A type of mutation in which the altered gene product possesses an increased level of activity or in which the wild type gene product is expressed at an increased level. The enlargement of tissues occurs via an increase in the number of cells by cell division.

hypermorphic allele: An allele that increases gene expression or function beyond wild-type levels.

hypermorphic mutations: These mutations lead to an increased expression compared with that of the wild type (e.g. *rII* mutants of bacteriophage T4 yield relatively large plaques and are called RAPID LYSIS MUTANTS).

hyperparasite: An organism that is parasitic upon another parasite.

hyperplasia (Adj. hyperplastic): The uncharacteristic upsurge in the number of cells which often results in the formation of galls or tumours.

hyperploid: Aneuploid containing a small number of extrachromosomes.

hyperploidy: A condition having additional chromosome complements compared to the standard chromosome.

hyper-resupinate: In botany, describing leaves or flowers that are in the usual position but are borne on a petiole or pedicel that is twisted 360°. The term is used to describe organs (e.g. orchid flowers) that are usually resupinate. *Compare with* RESUPINATE.

***hypersensitive reaction and pathogenicity* (*hrp*) gene:** A type of avirulence gene necessitated for elicitation of the hypersensitive response (HR) in resistant plants and causation of disease in susceptible plants.

hypersensitive response (HR): A protective/defensive retort by certain plants to 'infection' by plant pathogens (bacteria, fungi, etc.), in which the plant cells that are immediately adjacent to the infected area of plant are 'instructed' to undergo self-destruct via apoptosis, in order to barricade the infected area. This is done to avoid further spread of the infection. The initiation of the hypersensitive response is often triggered by signalling molecules that are produced by the pathogens themselves. This is a kind of non-host resistance which plant possesses to defend it.

For example, one particular protein produced by the soil fungus triggers a hypersensitive response that frequently is so severe that the entire plant dies. *Aka* SYSTEMATIC ACQUIRED RESISTANCE (SAR).

hypersensitive: Extremely or excessively sensitive; often refers to an extreme reaction to a pathogen.

hypersensitive site: A region of DNA located in a chromatin structure that renders it more sensitive to attack by endonucleases than DNA sites located elsewhere in the chromatin. The presence of hypersensitive sites is associated with transcription of adjacent DNA sequences in eukaryotic cells. Hypersensitive site probably includes an area from which nucleosomes are excluded.

hypersensitivity: The response to attack by a pathogen of certain host plants in which the assaulted cells die quickly and prevent further spread of infection.

hyperspectral remote sensing data analysis for agricultural crops: Hyperspectral remote sensing provides many ways to analyse the crop health to develop prediction models for crop health parameters as well as modelling for monitoring of timely water deficit stress in crop etc.

hyperthermophile: An organism that flourishes at temperatures above 80°C.

hypertonic: A solution which possesses a higher concentration of the solute than isotonic solution. Hypertonic solution is used for plasmolysis of the plants to obtain protoplasts. For example, 10% solution of sodium chloride w/v acts as hypertonic solution. Plant cells dipped in this solution exhibit removal of the water from the cytoplasm and vacuole leading to the shrinkage of protoplast that can be taken out mechanically by rupturing the cell wall.

hypertranscription: The coordinate increase in the nascent output of the majority of the transcriptome, including 'housekeeping' genes such as those coding for ribosomal proteins.

hypertrophy (Adj. hypertrophic): The abnormal increase in the size of cells in a tissue or organ, often causing in the formation of galls or tumours.

hypervariable (HYP) effectors: HYP effectors of plant-parasitic nematodes are single-gene loci that possibly harbour thousands of alleles. HYP effectors are essential for parasitism in plant-parasitic potato cyst nematodes. U. Sonawala *et al.* (2024) exposed an unparalleled variation. The alleles differ in the arrangement as well as the number of motifs within a central hypervariable domain (HVD). They enlarge the HYP collection of two plant-parasitic nematodes and define discrete species-specific 'rules' fundamental to the apparently faultless genetic rearrangements. They suggest a mechanism of programmed, locus-specific, somatic genome editing, termed HVD editing, where changes are locus specific, stringently oversaw by rules and apparently produce thousands of error-free variants.

hypervariable loci: Loci with many alleles; especially those whose variation is due to variable numbers of tandem repeats.

hypervariable region: Any of three small regions (V, D and R) within the variable region of an immunoglobulin light or heavy chain that show the highest level of variability from molecule to molecule. These regions determine the specificity of the antigen-binding site.

hypha (pl. hyphae; adj. hyphal): A single, tubular filament of a fungal thallus or mycelium which is considered the basic structural unit of a fungus.

hypocarpium: Enlarged plump structure that is formed below the fruit from the receptacle or hypanthium.

hypomethylation: Decrease in the level of DNA methylation in a population of cells relative to a reference or normal sample; may be used to describe a specific nucleotide or a group of nucleotides.

hypomethylation of the promoter region of *RAV6* in *Epi-rav6* plants: X. Zhang *et al.* (2015) found no nucleotide sequence difference between the wild-type and variant *Epi-rav6* rice. They did observe changed level of *RAV6* expression in the *Epi-rav6* plants. This suggested that the difference in expression may result from an epigenetic modification. They therefore investigated the DNA methylation status of the *RAV6* locus. The promoter region of *RAV6* contained many TEs, including an *adh5*-like miniature inverted-repeat transposable element (MITE), a short interspersed nuclear element (SINE), two unclassified retrotransposons and a Snabo-like MITE. Some TEs were especially close to genes. They acted as epigenetic mediators to influence nearby gene expression. They analysed methylation in a 2358-bp genomic region consisting of 573 bp of gene body, 1321 bp of upstream region and 464 bp of 5′ distal retrotransposon sequence. In the wild type and homozygous *Epi-rav6* mutants, they found no DNA methylation in all three sequence contexts in the 573-bp gene body of *RAV6*. They, however, observed higher CG and CHG but not CHH DNA methylation in proximal and distal upstream regions of *RAV6* in the wild type compared with the homozygous *Epi-rav6* mutant. All the changed methylation sites occurred in a contiguous 96-bp region at −600 to −504 bp relative to the transcriptional start site, including the closest MITE (*adh5*-like MITE). This region is hypermethylated in the wild type but hypomethylated in *Epi-rav6*, containing four CG sites and two CHG sites. To further confirm that the ectopic expression of *RAV6* in *Epi-rav6* results in hypomethylation of the promoter region, they treated the wild-type seeds with 5-aza-2′ deoxycytidine (5-aza-dC), an inhibitor of DNA methylation. The expression levels of *RAV6* were measured in 7-day-old seedlings with or without 5-aza-dC treatment. Treatment with 5-aza-dC upregulated *RAV6* expression to varying degrees in three cultivated rice strains, including two *japonica* (cv Zhonghua11 and Nipponbare) and one *indica* (cv Kasalath) accessions. These three cultivated accessions also contain the MITE; therefore, these results indicate that the DNA methylation mediated by the TE in the 5′ upstream sequence of *RAV6* plays an essential role in the regulation of *RAV6* expression.

hypomorphic alleles: An allele whose gene product possesses a reduced level of activity or in which the wild type gene product is expressed at a reduced level.

hypomorphic mutants: Mutant cells or individuals in which the altered gene product possesses a reduced level of activity or in which the wild-type gene product is expressed at a reduced level.

hypomorphic mutations: The mutations which result in a partial loss in the expression of a trait because the enzyme produced by the mutant allele is partially functional (e.g. *chlorina* mutants in crop plants).

hypophysis: The uppermost cell or suspensor from which part of the root and root cap in the embryo of angiosperms is derived.

hypoploid: An aneuploid with a small number of chromosomes missing.

hypoploidy: A condition representing missing chromosome complements compared with the standard chromosome set.

hypoproduction: Dosage compensation in mammals is achieved by hypoproduction, called DOWNREGULATION, due to inactivation of one of the two X chromosome of homogametic sex. This situation is observed in case of mammals.

hypostasis: Interaction between nonallelic genes in which one gene does not express in the presence of a second.

hypostatic epistasis: Genes whose expression is affected (blocked or masked) are hypostatic to the interfering genes.

hypostatic genes: Hypostatic genes are those whose action is suppressed. Gene interactions are classified on the basis of the manner in which the concerned genes influence or modify the expression of each other. Some of the common interactions are described below.

hypotheses: Includes theoretical, hypothetical explanations of observations and measurements of the subject. A hypothesis is a suggested explanation of a phenomenon, or alternately, a reasoned proposal suggesting a possible correlation between or among a set of phenomena. It is a limited statement regarding cause and effect in specific situations. It also refers to our state of knowledge before experimental work has been performed and perhaps even before new phenomena have been predicted. A hypothesis is an educated guess, based on observation. Usually, a hypothesis can be supported or refuted through experimentation or more observation. New hypotheses are formed with increasing knowledge in the subject. Generally, a hypothesis is used to make predictions that can be tested by observing the outcome of an experiment. If the outcome is inconsistent with the hypothesis, then the hypothesis is rejected. However, if the outcome is consistent with the hypothesis, the experiment is said to support the hypothesis. Examples of hypotheses are different stages in understanding gene function – ONE GENE–ONE ENZYME HYPOTHESIS, ONE GENE–ONE CHROMOMERE HYPOTHESIS, ONE GENE–ONE POLYPEPTIDE HYPOTHESIS, ONE CISTRON–ONE POLYPEPTIDE HYPOTHESIS, ONE GENE–ONE PRIMARY CELLULAR FUNCTION HYPOTHESIS. A hypothesis can change with more knowledge in the area of research. *See details as separate terms.*

hypothesis testing: Hypothesis testing is an act in statistics whereby an analyst tests an assumption regarding a population parameter. The methodology employed by the analyst depends on the nature of the data used and the reason for the analysis. Hypothesis testing is used to infer the result of a hypothesis performed on sample data from a larger population. The determination of whether to accept or reject a proposed null hypothesis based on the likelihood of the experimental results.

hypotranscription: Reduced levels of transcription. For example, in *Caenorhabditis elegans*, dosage compensation takes the form of hypotranscription.

hypovirulence: The reduced ability to cause disease.

I

I_1, I_2, I_3 and so on: Symbols used to denote first, second, third and so on, inbred generations. *See* S_1, S_2, S_3 AND SO ON.

I-value: The total concentration of DNA terminal in a ligation reaction, which increases with increasing DNA concentration. I-values and J-values influence the formation of products in a ligation reaction. By choosing different I- and J-values (i.e. various concentrations of vector and insert DNAs) one can influence the structures of the recombinant molecules formed. These may be linear oligomers, circularized vectors or vector plus insert.

I-R system: This is a system of hybrid dysgenesis in which when a female from a REACTIVE (R) STRAIN mates with a male from an INDUCER (I) STRAIN, specific female (SF) sterility is induced. The reactive and inducer strains can be arranged from 'strong' to 'weak'. The stronger the paternal strain, the higher the reduction of female sterility. Reactivity corresponds to a cytoplasmic state and the inducer is a factor chromosomal in nature.

identical: When a corresponding character is shared between two species or populations, the character is said to be identical.

identical by descent: When two individuals share an allele inherited from a common ancestor. Two or more alleles are identical by descent (IBD) if they are identical copies of the same ancestral allele. This characteristic is frequently used in genetic linkage to recognize alleles that are potential candidates for harbouring mutations accountable for a genetic disease. A common way to identify alleles as identical by descent is using SINGLE-NUCLEOTIDE POLYMORPHISM (SNP) data. If enough SNPs in two alleles are observed as the same, the two alleles are inferred as IBD. However, this does not guarantee that the two alleles are identical. Even if they are indeed derived from the same ancestral allele, they could harbour different mutations that arose during the inheritance process, like single-nucleotide mutations, insertion, deletions or others. Given the new importance that IBD alleles are assuming in genetic association studies, a new definition of IBD that would better suit the underlying inheritance property would be that two or more alleles are IBD if they have been inherited from the same ancestral allele without recombination events.

identical DNA repeats: All copies of DNA repeat exact copies of one another. The advantage of such repeats is that they can make millions of identical products, such as ribosomes, as are needed for the mass production of a protein. For example, identical repeats of rDNA genes are tandem repeats of DNA coding for rRNA. Here, a primary transcript is processed. *Compare with* NON-IDENTICAL DNA REPEATS.

identical multigene families: Different members of the gene family are multiple copies of the same gene.

identical species: When a corresponding character is shared between two species or populations, the character is said to be identical.

identical twins: Identical twins result from the fertilization of a single egg by a single sperm, with the fertilized egg then splitting into two. Identical twins share the same genomes and are always of the same sex. The epigenetic profiling of identical twins discordant for a disease is a unique experimental design as it eliminates the DNA sequence – age – and sex differences from consideration. *Aka* MONOZYGOTIC TWINS. *See* DIZYGOTIC TWINS *and* MATERNAL TWINS. *Compare with* FRATERNAL TWINS.

identification of gene of interest: This can be done through probing for the gene, complementation, positional cloning or tagging.

identification of the 5′-end and intron-exon junctions of the gene: For this purpose, S1 nuclease mapping and primer extension methods are used. In the S1 NUCLEASE MAPPING METHOD, a DNA probe labelled at its 5′-end overlapping the gene 5′-end or 5′-end of an exon is hybridized to the gene DNA. The S1 nuclease, which is specific for either single-stranded DNA or RNA, is utilized to digest the single-stranded DNA. The resulting 5′-labelled probe is then 'sized' via a Southern blot. Its size pinpoints the 5′-end of the gene or exon. In the PRIMER EXTENSION METHOD, the DNA probe labelled at the 5′-end is contained within the gene and hybridized to the mRNA from the gene. The probe DNA is used as a primer for reverse transcriptase, which extends this primer using mRNA as a template, synthesizing **cDNA** to the end of the mRNA. The resulting 5′-labelled DNA probe, identical to the one produced in the S1 nuclease mapping approach, is then 'sized' via a Southern blot to pinpoint the 5′-end of the gene.

identification via hybridization to mRNA or cDNA: The method employed includes Northern blots and zoo blots. Northern blots are the same as Southern blots except that the mRNA is run out on the gel. Zoo blots are simply Southern blots of a labelled probe from the new DNA sequences against genomic DNA. Restriction fragments are recognized from different organisms (the Zoo). These are used to find out if DNA sequences are highly similar and hence possibly homologous to the new DNA present in one or more of these other organisms.

identifier: In computer languages, identifiers are tokens that name language entities. The kinds of entities an identifier might denote include variables, types, labels, subroutines and packages. *Aka* SYMBOLS

identity preservation (IP): The segregation of one crop type from another at every stage from production and processing to distribution. This process is usually performed through audits and site visits and provides independent third-party verification of the segregation.

identity: The extent to which two (nucleotide or amino acid) sequences are invariant.

ideochromatin: Nuclear chromatin controlling cell division but otherwise dormant.

ideotype breeding: A method of breeding to augment genetic yield potential based on modifying individual traits, where the breeding goal for each trait is stated.

ideotype: The term 'ideotype' was coined by C.M. Donald in 1968. The literal meaning is a form denoting an idea, and in the broadest sense, it is a biological model that is expected to perform or behave in a predictable manner within a defined environment. More specifically, a crop ideotype is a plant model that influences photosynthesis and yields a greater quantity of grain, oil or other useful products when developed as a cultivar.

idioblast: A cell, especially of a leaf, differing markedly from surrounding cells. These often synthesize specialized products such as crystals.

idiochromosome: A chromosome that contributes to the determination of sex.

idiogram: A photograph or diagram of the chromosomes of a cell arranged in an orderly fashion. *See* KARYOTYPE.

idioplasm: All hereditary elements of a plant including genotype and plasmotype.

idiotype: Idiotype is an established term, first used by H.W. Siemens in 1921. The 'idio' in idiotype is derived from the Greek word 'idios', denoting one's own, personal, separate, distinct gene. The region of the antibody molecule that enables each antibody to recognize a specific foreign structure (i.e. epitope or hapten) is said to have an idiotype for that epitope or hapten. The determinants are composed of the antigen-combining site or COMPLEMENTARY-DETERMINING REGIONS (CDRs). Every unique antigenic determinant has a specific antibody with its own unique idiotype. *See* EPITOPE and HAPTEN.

idiotypic variation: Differences in the variable parts of immunoglobulin genes.

idling reaction: The production of guanosine tetraphosphate (3′-ppGpp-5′) by the rigorous factor when a ribosome encounters an uncharged tRNA in the A-site.

ignorant DNA: *See* SELFISH DNA.

illegitimate crossing-over: *See* UNEQUAL CROSSING-OVER.

illegitimate end joining: This reaction occurs frequently in eukaryotes. It repairs double-strand chromosome breaks by direct re-ligation of the ends. It can result in translocations and other rearrangements when ensuing on illegitimate ends. It also enables transfected linear DNA to be integrated into the genome.

illegitimate recombination: Recombination between DNA fragments that do not share extensive DNA sequence homology between the transposon and the target DNA. Illegitimate recombination includes TRANSPOSON IMMUNITY, TRANSPOSON TAGGING and TRANSPOSON-MEDIATED GENE TRANSFER.

illegitimate replication: This process transpires in repetitive DNA and DNA that forms stabilizing secondary structures like hairpins and cruciforms. The primer strand jumps out of register with the template and produces insertions and deletions.

illegitimate strand exchange: Illegitimate strand exchange takes place when the free strands become joined to the wrong partner (e.g. a single strand that happens to be in the vicinity of the reaction).

illegitimate V-(D)-J joining: Illegitimate recombination between recombination signal sequences (RSS)-elements in the immunoglobulin and T-cell receptor (TCR) gene loci and cryptic elements positioned elsewhere in the genome is responsible for a spectrum of chromosome aberrations often associated with lymphoid tumours.

Illumina (Solexa) sequencing: Solexa, now part of Illumina, developed a sequencing technology based on reversible dye terminators. DNA molecules are first attached to primers on a slide and amplified so that local clonal colonies are formed (bridge amplification). Four types of ddNTPs are added and non-incorporated nucleotides are washed away. Unlike pyrosequencing, the DNA can only be extended one nucleotide at a time. A camera takes images of the fluorescently labelled nucleotides, and then the dye along with the terminal 3′ blocker is chemically removed from the DNA, allowing the next cycle.

Illumina: Illumina (San Diego, CA, USA) BeadArrays has provided one of the most advanced array-based methods for both custom and large-scale DNA methylation analysis. It has been used by various investigators for addressing DNA methylation-related questions.

immediate early gene: A viral gene that is expressed soon after infection.

immobilization: Imprisonment of a biocatalyst in a distinct phase that allows exchange with, but is separated from, the bulk phase in which the substrate, effector or inhibitor molecules are dispersed and monitored. In case of enzymes, these are physically entrapped or covalently bonded by chemical means to an inert and usually insoluble matrix, where it can act upon its natural substrate. The matrix is usually polyacrylamide, cellulose, starch and glass beads. Likewise, the cells, protoplasts and microbial agents are also immobilized.

immobilized pH gradient (IPG) gels: IPG gels solve the issue of both cathodic drift and poor reproducibility. IPG is established using immobile (a collection of non-amphoteric) molecules that contain weak acid or base buffering at one end and acrylic double bonds to facilitate the immobilization reaction at the other. No movement of the buffering molecules transpires when an electric field is applied, but the proteins are separated.

immortal cell culture: *See* CONTINUOUS CELL CULTURE.

immortalization: The production of a cell line that has the capability of undergoing an unlimited number of cell divisions, whether by perturbation or intrinsically, of the attributes of a continuous cell line. It can be a consequence of a chemical, a viral transformation or a fusion of the original cells with cells of a tumour line.

immortalized cells: Cells that continue to divide indefinitely, unlike normal cells that cease to divide with normal ageing. An immortalized cell is not necessarily one that is neoplastically or malignantly transformed.

immune: Host not affected by pathogens; host exempt from infection. The condition of a host possessing qualities that do not allow the development of a disease.

immune deficiency: The absence of an adequate immune response. There are several causes such as cancers, drugs or pathogens of immune deficiency. Many viruses are capable of infecting and killing lymphocytes and thus lowering the immune response. HIV infection causes ACQUIRED IMMUNODEFICIENCY SYNDROME (AIDS), which affects T cells (the virus destroys T_H cells), and the body becomes susceptible to other viral, bacterial, fungal and protozoan diseases.

immune electrophoresis: A procedure for the differentiation of proteins in solution, based on both their electrophoretic and immunological properties. Initially, the proteins are separated by gel electrophoresis. They are then reacted with specific antibodies by double diffusion through the gel. The pattern of precipitin arcs thus formed can be successfully used to identify the proteins.

immune memory: The increased response of the immune system to an antigen that had been encountered before. This forms the basis of vaccination.

immune reaction: The reaction occurring between a specific antigen and antibody. When plants are inoculated (e.g. with the BUK strain of tomato blacking nepovirus), they are subsequently protected against secondary infection with a similar viral strain but not against a dissimilar strain of virus.

immune response: The immune response refers to how your body recognizes and responds to defend itself against bacteria, viruses and other substances that appear foreign and harmful. The response of the immune system when a foreign substance or microorganism enters its body. *See* INNATE IMMUNE RESPONSE, ADAPTIVE IMMUNE RESPONSE, PRIMARY IMMUNE RESPONSE, SECONDARY IMMUNE RESPONSE *and* IMMUNE REACTION.

immune system: (i) The active defence of plants against infections and other invasive hostilities. It detects antigens in the invading entities and creates new antibodies to destroy them. (ii) The population of lymphocytes and other white blood cells in the vertebrate body that defend it against infection.

immunity: A natural or acquired resistance of a plant to a pathogenic microorganism or its products. This means that a host cannot be parasitized by a particular species of parasite. For example, coffee is immune to wheat rust, and wheat is immune to coffee rust. Immunity is non-variable. The maximum level of horizontal resistance may be a deceptive immunity, but it is not true immunity because it is variable and can be eroded. Vertical resistance has often been called immunity, but it too is a superficial immunity because it operates only against the non-matching strains of the parasite. The human body develops an immune response against foreign substances (ANTIGENS). Immune response involves the production of antibodies and specialized lymphocytes that can destroy antigens. Immunity may be naturally attained during a lifetime, artificially developed (ACTIVE IMMUNITY) through vaccination or artificially acquired (PASSIVE IMMUNITY) through the introduction of antibodies.

immunity breeding: *See* RESISTANCE BREEDING.

immunity in phages: The ability of a prophage to avert another phage of the same type from infecting a cell. It usually results from the synthesis of a phage repressor by the prophage genome.

immunity in plasmids: The ability of a plasmid to prevent another of the same type from becoming established in a cell. It results from interference with the ability to replicate.

immunity in transposons: The ability of certain transposons to stop others of the same type from transposing to the same DNA molecule. It results from a variety of mechanisms.

immunization: Artificially stimulating the body's defence system. Immunization, also called vaccination, introduces vaccines (antigens) into the body and the human body then produces antibodies. This is active immunization. Antibodies (rather than antigens) can be introduced directly, which is passive immunization.

immuno: Therapies and/or treatments that kindle the immune system.

immuno blot: *See* WESTERN BLOT.

immunoassay: The use of antibodies to recognize and quantify substances by a variety of methods. The binding of antibodies to antigen (the substance being measured) is frequently followed by TRACERS, such as fluorescence or radioisotopes, to enable the measurement of the substance.

immunocompromised: A state in which the immune system is defective and unable to mount an immune response against disease pathogens or other foreign cells.

immunoconjugate: A molecule formed by attachment to each of two originally different molecules. One of these is generally an antibody; hence, the word 'immunoconjugate'. Classic organic drug molecules such as methotrexate, adriamycin chlorambucil, radionuclides, enzymes, toxins and ribosome-inhibiting proteins may be conjugated to antibodies. The antibody portion of the conjugate functions to direct the biologically active molecule to its target.

immunodeficiency: An inborn, acquired or induced inability to produce a normal immune response.

immunodiffusion test: A test involving the precipitation reaction of antigens and antibodies mixed by diffusion through a gel. This test is usually done in a Petri plate of agar. The preparations of antigens and antibodies (without dilution) are put in separate wells that diffuse. These result in a sharp line of cloudiness at the zone of equivalence due to precipitation between the two wells.

immunodominant epitopes: Regions of a protein that stimulate the production of antibodies.

immunoelectrophoresis: A test in which antigens and antibodies are diluted and mixed by electrophoresis through a gel. It is used to identify various protein antigens in human serum. Antigens are first separated by electrophoresis, after which, the antibody mixture is added to a trough that runs parallel. Antigens and antibodies diffuse towards one another forming a series of precipitation lines. The location of lines identifies various antigens.

immunofluorescence: Any of various methods used for detecting an antigen or antibody in a sample by coupling its specifically interactive antibody or antigen to a fluorescent compound, mixing with the sample and perceiving the reaction under an ultraviolet light microscope. An animal is infected with a known bacterium (antigen). That results in the production of antibodies that are labelled with fluorescent dyes and added to a microscopic slide containing an unknown bacterium. If the unknown bacterium is the same as that was injected, the antibodies bind on the surface of the bacterium and cause fluorescence.

immunogen: *See* ANTIGEN.

immunogenetics: The study of the inheritance of those molecules that can trigger an immune response. The study of inheritance of antigens and antibodies.

immunoglobulin (Ig): An antibody molecule secreted by derivatives of B lymphocytes (plasma cells) that acts as a vital part of the immune response by specifically identifying and binding to particular antigens, such as bacteria or viruses, and aiding in their destruction. These are a major component of the group of immune defence molecules, collectively called antibodies. A member of the globulin protein family contains two light and two heavy chains linked by disulfide bonds. All antibodies are immunoglobulins. A group of different types of soluble antibodies are collectively known as immunoglobulins. Higher vertebrates have five classes of immunoglobulin – IgA, IgD, IgE, IgG and IgM – each with a different role in the immune response. *See details as separate terms.*

immunoglobulin A (IgA): The class of antibodies found in secretions.

immunoglobin allotype: Refers to the allele of the antibody chains occurring in an individual.

immunoglobulin classswitch recombination (CSR): Activation-induced (cytidine) deaminase (AID) is needed for the DNA cleavage step in immunoglobulin class switch recombination. AID is anticipated to deaminate cytosine to produce uracil (U) in either mRNA or DNA. DNA cleavage depends on the enzyme uracil-DNA glycosylase (UNG) for the removal of U. UNG is involved in the repair step of CSR through a yet unknown mechanism. Uracil exclusion introduces immunoglobulin classswitch recombination.

immunoglobulin D (IgD): The class of antibodies found on B cells.

immunoglobulin domain: Characteristic protein domain of about 100 amino acids that exists in immunoglobulin light and heavy chains. Similar domains, known as IMMUNOGLOBULIN-LIKE (IG-LIKE) DOMAINS, occur in many other proteins involved in cell-cell interactions and antigen recognition and define the Ig superfamily.

immunoglobulin E (IgE): The class of antibodies involved in hypersensitivity and the expression of allergies.

immunoglobulin fold domains: The immunoglobulin domain (IPR013783) consists of a beta-sheet structure with large connecting loops that serve to recognize either DNA major grooves or antigens. Usually found in immunoglobulin proteins, these are also present in the Stat proteins of the cytokine pathway. This is likely because the cytokine pathway evolved relatively recently and made use of systems that were already functional, rather than creating its own.

immunoglobulin G (IgG): The most abundant class of antibodies in serum.

immunoglobulin genes: The complex of GENELETS responsible for the synthesis of immunoglobulins. These genes undergo DNA rearrangement at different levels before the DNA acts as a template for transcription. If the convention that one polypeptide is coded for by one gene is accepted, each of the final DNA segments coding a complete heavy or light chain should be considered a single gene. Various alternative sequences within each set (V, J, D or C) should be termed simply as DNA sequences according to this definition since they are all parts of a single gene coding for one polypeptide chain. With respect to immunoglobulin genes, the genome of somatic cells is not constant. These constitute a multigene system of three unlinked multigene families: λ, k and H. *See* GENELETS.

immunoglobulin heavy chain genes: These genes code for heavy chains of immunoglobulins. There are five types of heavy chains of constant region (C_H) known: μ, δ, γ, ε and α, which define five types of immunoglobulins, IgM, IgD, IgG, IgE and IgA, respectively. Each of the five heavy chain classes may be associated with either k or λ light chains. L (leader), V (variable), J (joining), D (diversity) and C (constant) sequences are also present in heavy chain genes.

immunoglobulin light chain genes: These genes code for light chains of immunoglobulins. Only two types of light chain genes are known: k and λ, each differing in amino acid sequence of constant light (C_L) segment and includes light chains (L) with many different variable light chain (V_L) sequences. L (leader), V (variable), J (joining) and C (constant) sequences are also present in light chain genes.

immunoglobulin M (IgM): The first class of antibodies to appear after exposure to an antigen.

immunoglobulin superfamily: A large family of proteins that comprise immunoglobulin domains or immunoglobulin-like domains. Most immunoglobulins are implicated in cell-cell interactions or antigen recognition.

immunogold electron microscopy: An electron microscopy technique in which cellular structures or molecules of interest are labelled with antibodies tagged with electron-dense gold particles. These labels show up as black spots on the electron microscopy (EM) image.

immunological distance: A measure of phylogenetic relationship based on differences in antibody-antigen binding affinity between species. The magnitude of the difference is used as an approximate evolutionary clock.

immunological memory: A long-lived state that occurs after a primary immune response to many antigens, in which subsequent encounters with that antigen will aggravate a rapid secondary immune response.

immunological tolerance: If the body of an individual does not respond to an antigen, the antigen may be inducing tolerance.

immuno-northern blotting (INB): This method combines two commonly used molecular biology techniques: Western and Northern blotting. With this method, antibodies against the modified nucleotides are used for detection instead of the radiolabelled DNA probes used in the classical Northern blotting protocols.

immunopotentiating: Compounds strengthening the immune system.

immunoprecipitation: A process whereby a particular protein of interest is isolated by the addition of a specific antibody, followed by centrifugation to pellet the resulting immune complexes. Often, staphylococcal protein A or G, bound to Sepharose or some other type of macroscopic particle, is added to the reaction mix to increase the size and ease the collection of the complexes. Usually, the precipitated protein is examined by SDS-PAGE.

immunoprecipitation-based methods: Immunoprecipitation-based methods for DNA methylation analysis are relatively forthright, as they neither require digestion of genomic DNA nor bisulfite treatment. Also, since there is no chemical conversion of epigenetic information to genetic information, downstream data processing does not necessitate reference bisulfite-converted genomes.

immunoproteins: All the proteins that are part of the immune system (including antibodies, interferon and cytokines).

immunopurification: The process includes the separation of a specific antigen from a mixture of very similar antigens. This purified antigen is then used for developing a vaccine against a pathogen.

immunoscreening: The use of an antibody to detect a polypeptide synthesized by a cloned gene. The immunological screening approach uses procedures like ENZYME-LINKED IMMUNOSORBENT ASSAY (ELISA) and specific antibodies to detect antigenic expression products (usually polypeptides) produced by clones, and to identify and select the ones of interest. The label used in this method is often an enzyme such as alkaline phosphatase or horseradish peroxidase, but it can also be ^{125}I.

immunomodulators: Compounds capable of stimulating or downregulating the immune response.

immunostimulating: Compounds capable of stimulating an immune response.

immunosuppressants: A substance that results in or affects immunosuppression.

immunosuppression: Suppression of the immune response (e.g. by irradiation or administration of certain substances). During organ transplants in humans, the recipient's body starts rejecting an allograft due to the active immune response. Therefore, the immune response is suppressed by a drug called cyclosporine. Tacrolimus (FK-506) is another commonly used immunosuppressant drug. These drugs disrupt cell-mediated immunity by cytotoxic T cells. Some new drugs like Sirolimus (rapamycin) affect cell-mediated response and antibody reproduction as well.

immunotherapy: A therapy based on the immune system to suppress tumour cells (e.g. in vaccine development). May also refer to the therapy of immune system caused diseases. *See* IMMUNOTOXIN *and* CANCER.

immunotoxicity: The toxicity of a therapeutic agent as it could cause immune reactions or allergies.

immunotoxin: A therapeutic agent combining a monoclonal antibody antitoxic agent and a toxic agent. Immunotoxins consisting of a poison such as the drug 'ricin' or radioactive compound combined with a monoclonal antibody are directed against a particular type of tumour-associated antigen. The antibody selectively locates the cancer cells and the attached toxic agent destroys the cell. It causes little or no damage to healthy tissue.

immunotyping: The process of screening patients' specimens to identify the specific viral antigen on antigen-presenting cells or identifying specific viral antibodies.

imperfect flowers: Unisexual flowers; flowers lacking either male or female part.

imperfect state (of fungi): The asexual state of a fungus (i.e. the state in which no sexual reproduction occurs).

implant: A material artificially positioned in an organism.

implication of reverse transcription: The most important implication of reverse transcription is to understand cancer induction in human beings. The DNA transcripts of viral RNA are attached to the genome of the cell and can, when genetically active, cause spontaneous cancer. Hepatitis B virus (HBV), Cauliflower mosaic virus (CaMV) and retroviruses have been thought to have the same origin. CaMV is being successfully used as one of the vectors for gene transfer in plants.

import of a nuclearly encoded tRNA into mitochondria: The import of tRNAs encoded in the nucleus is essential for proper protein translation within mitochondria of a variety of organisms. An *in vitro* assay has been developed for the import of nuclearly encoded tRNAs into plant mitochondria that utilizes isolated mitochondria and synthetic tRNAs transcribed from cloned nuclear tRNA genes.

importin: Receptor on the pore to which a signal sequence binds.

imprint control element (ICE): Genomic region that acts, in a methylation-sensitive manner, to determine whether imprinted genes are expressed or not according to the parent of origin. A short DNA sequence that controls the imprinted expression of multiple genes in *cis*. All known ICEs are also gametic DMRs; however, their identification requires *in vivo* analysis of a deletion of the DMR. *Aka* IMPRINT CONTROL REGION (ICR) or IMPRINTING CENTRE (IC).

imprinted control regions (ICR): The grouping of imprinted genes within clusters enables them to share common regulatory elements, such as non-coding RNAs and DIFFERENTIALLY METHYLATED REGIONS (DMRs). When these regulatory elements control the imprinting of one or more genes, these are known as imprinting control regions. The expression of non-coding RNAs, such as *Air* on mouse chromosome 17 and *KCNQ1OT1* on human chromosome 11p15.5, has been revealed to be crucial for the imprinting of genes in their corresponding regions. DMRs are the segments of DNA rich in cytosine and guanine nucleotides, with the cytosine nucleotides methylated on one copy but not on the other. Contrary to expectation, methylation does not necessarily mean silencing; instead, the effect of methylation depends on the default state of the region.

imprinted gene (or locus): (i) Genomic region with a methylation mark that is present only on the maternally or paternally derived copy of an allele. (ii) A gene (or an allele) that is expressed in a parent-of-origin-specific way. For example, the *INSULIN-LIKE GROWTH FACTOR 2 (IGF2)* copy of the gene that is contributed by the mother is silent, whereas the copy that is donated by the father is active and is expressed into protein. Abnormal imprinting can cause embryonic lethality.

imprinted gene expression: Humans inherit two alleles from the mother and the father. Both are functional for the majority of the genes, but sometimes one is turned off or 'stamped' and does not show in offspring. That gene is imprinted. Imprinting means that a particular gene is silenced, and the gene from the other parent is expressed.

imprinted genes in plants, regulation of: The regulation and function of the imprinted genes *MEDEA (MEA)* and *FLOWERING WAGENINGE (FWA)* during the life cycle of *Arabidopsis* are well known. Constitutive repression of the late-flowering gene *FWA* is maintained by the methylation of its promoter region through the activity of a methyltransferase 1 (MET1) and a SWItch/Sucrose Non-Fermentable (SWI/SNF) ATP-dependent chromatin-remodelling factor DDM1. The activity of MET1 and Deficient in DNA methylation 1 (DDM1) is possibly dependent on siRNAs that correspond to retrotransposon-derived tandem repeats present in the 5′ regulatory region of the *FWA* locus. MET1 plays a crucial role in maintaining methylation patterns during gametogenesis and directs the methylation of the *MEA* promoter. In contrast, DEMETER (DME) acts defiantly to MET1 and activates the maternally inherited alleles of *MEA* and *FWA* in the central cell, probably by removing 5′ methylcytosines. The function of *FWA* in the endosperm is still unknown, whereas *MEA* function is necessary to repress cell proliferation by inhibiting the activity of the MADS-box gene *PHE1*. During endosperm formation, *FIE* and *FIS2* are not regulated by DME. *MEA* is expressed from both parental alleles during late embryo and vegetative development; a subsequent mechanism to re-establish imprinted states in *MEA* is likely to act at some point during flower development. Because the endosperm does not transmit any genetic material to the next generation, imprinting genes that act in this tissue do not need subsequent reprogramming during later stages of the plant life cycle.

imprinted genes in plants: Like eutherian mammals, the gene imprinting phenomenon has also been described in flowering plants (angiosperms). During fertilization of the embryo in flowers, a second separate fertilization occurrence gives rise to the endosperm, an extraembryonic structure that nourishes the seed similar to the mammalian placenta with the embryo. Unlike the embryo, the endosperm often contains two copies of the maternal genome and fusion with a male gamete results in a triploid genome. This uneven ratio of maternal to paternal genomes seems to be critical for seed development. Some genes are found to be expressed from both maternal genomes, while others are expressed exclusively from the lone paternal copy.

imprinting: (i) A cellular process in which certain genes within an organism's cells are 'disabled' during the earliest stage(s) of the organism's development. For example, the embryo of a female mammal which receives two copies of the X chromosome – one from each parent – disables one of those copies at random in each of its cells, so the female becomes a genetic mixture of its two parents. (ii) Epigenetic regulation in which maternally and paternally inherited alleles are differentially expressed owing to *cis*-acting modifications of DNA or histones inherited from parental chromosomes. For instance, both Prader-Willi and Angelman syndromes are inherited when the same part of chromosome 15 is missing. When the father's complement of 15 is missing, the child has Prader-Willi, but when the mother's complement of 15 is missing, the child has Angelman syndrome. *See* GENOMIC IMPRINTING.

imprinting epigenetics: The inheritance of epigenetic tags from cell generation to cell generation and from the parents to their offspring. The epigenetic tags on imprinted genes usually stay put for the life of the organism. But they are reset during egg and sperm formation. Regardless of whether they came from the mother or the father, certain genes are always silenced in the egg, and others are always silenced in the sperm.

imprinting mechanisms: Imprinting is a dynamic process. It must be possible to erase and re-establish the imprint through each generation. The nature of the imprint must therefore be epigenetic (modifications to the structure of the DNA rather than the sequence). In germline cells, the imprint is erased, and then re-established according to the sex of the individual, that is in the developing sperm, a paternal imprint is established, whereas in developing oocytes, a maternal imprint is established. This process of erasure and reprogramming is necessary such that the current imprinting status is relevant to the sex of the individual. In both plants and mammals, there are two major mechanisms that are involved in establishing the imprint; these are DNA methylation and histone modifications.

imprints: Allele-specific DNA methylation marks at ICRs. *See* GENOME IMPRINTING.

improved site-specific recombination strategies for SMG removal: Improvements in site-specific recombination strategies for SMG removal include constitutive expression of a recombinase gene using plant hybridization or re-transformation, transient expression of the recombinase gene, induced expression of a recombinase gene, developmentally programmed recombinase gene expression and the use of developmentally controlled promoters.

improved sterile insect technique (iSIT): After decades of genetic research, a range of classical and molecular genetics tools have been developed that can be used to develop improved strains for use in SIT programmes. The problem of inbreeding occurring in mass-reared strains of *Bactrocera tryoni* can be overcome by four-way crossing strategies. Hybrids produced between *Bactrocera tryoni* and the other two species in the laboratory have proved useful for the development of genetically marked strains. Similarly, male and female embryos can be easily distinguished in any strain that carries a *Bactrocera jarvisi* Y chromosome. This enabled scientists to study homologues of the sex-determination genes during the development of *Bactrocera jarvisi* and *Bactrocera tryoni*, necessary for the development of genetic sexing strains.

improved vectors for cloning *Bacillus subtilis*: An ideal vector pHV11 for cloning *Bacillus subtilis* was constructed. It confers more than one selectable phenotype on host cells and in addition has unique restriction endonuclease cleavage sites in the corresponding genes.

improving crop performance through 6^{m}A modulation: The FAT MASS AND OBESITY-ASSOCIATED (FTO) transgene leads to 6mA demethylation of mRNAs and repeat RNAs in rice nuclei, inducing an open chromatin state and transcriptional activation. The presence of FTO kindles root meristem cell proliferation and tiller bud formation, and promotes photosynthetic efficiency and drought tolerance, but has no consequences on mature cell size, shoot meristem cell proliferation, root diameter, plant height or ploidy.

improving plant yield: In addition to increasing crop productivity by using built-in protection against diseases, pests, environmental stresses and weeds to minimize losses, efforts are being made to use biotechnology to improve crop yields directly. Researchers at Japan's National Institute of Agrobiological Resources (NIAR) added maize photosynthesis genes to rice to increase its efficiency of converting sunlight to plant starch and increase yields by 30%. Other scientists are altering plant metabolism by blocking gene action to shunt nutrients to certain plant parts. Yields increase as starch accumulates in potato tubers and not leaves or oilseed crops (such as canola) allocate most fatty acids to the seeds. The needs of the crops that have better accessibility to micronutrients are also being developed. Mexican scientists have genetically modified plants to secrete citric acid, a naturally occurring compound, from their roots. In response to the slight increase in acidity, minerals bound to soil particles, such as calcium, phosphorous and potassium, are released and made available to the plant. Nitrogen is the grave limiting element for plant growth, and researchers from many scientific disciplines are working on the details of the symbiotic relationship that allows nitrogen-fixing bacteria to capture atmospheric nitrogen and provide it to the plants that harbour them in root nodules.

***in-frame* deletion**: A deletion of a multiple of three nucleotides that does not alter the reading frame of the protein.

in-fusion cloning: In-fusion protocol utilizes either some form of recombination or the production and annealing of single-stranded overhangs to avoid the need to restrict the digestion of the input DNA insert. In both cases, the starting DNA is a PCR product corresponding to a complete OPEN READING FRAME (ORF), or domains thereof, produced from either a genomic or cDNA template. The PCR primers incorporate either restriction enzyme recognition sites or the sequences required for LIGATION-INDEPENDENT CLONING (LIC) REACTIVITY.

***in-gel* zymography (IGZ)**: For in-gel zymography, samples are prepared in a standard, non-reducing loading buffer for SDS-PAGE. No reducing agent or boiling is necessary, since these would interfere with the refolding of the enzyme. A suitable substrate (gelatine or casein) is embedded in the resolving gel during the preparation of the acrylamide gel. Following electrophoresis, the SDS is removed from the zymogram by incubation in unbuffered Triton X-100, followed by incubation in an appropriate digestion buffer, for an optimized length of time at 37°C. The zymogram is subsequently stained commonly with Amido Black or Coomassie Brilliant Blue. Areas of digestion appear as clear bands against a darkly stained background.

in planta: Refers to the direct transformation of the plant without involving any tissue culture step.

***in planta* transformation**: In planta transformation involves the transformation of intact plants or plant tissues without callus culture or regeneration. The most well-known and widely used in planta transformation method is the floral dip method developed by S.J. Clough and A.F. Bent in 1998 for *Arabidopsis thaliana*.

in silico: A modern term to describe biological experiments carried out entirely on a computer.

***in silico* (biology)**: Refers to computer-mediated biology. A set of computer-modelling technologies, through which researchers can construct computer models of (i) specific cells to see how a given disease influences that cell and how a given pharmaceutical impacts that cell, (ii) specific organs to see how a given disease influences that organ and how a given pharmaceutical impacts that organ, (iii) specific organisms to see how a given disease influences that organism and how a given pharmaceutical then influences that disease within that organism, (iv) specific organisms that possess a given genome to see how a given disease impacts that specific organism/phenotype and how a given pharmaceutical then impacts that disease within that organism/phenotype and (v) protein 'digestion' (i.e. breaking apart into constituent peptides), for comparison with the actual peptides (fragments) that have resulted from chemical digestion of those protein molecules (e.g. via immersion in trypsin).

***in silico* map**: A physical map prepared by *IN SILICO* MAPPING.

***in silico* mapping**: A variant of the comparative mapping method that makes the most use of genome sequence databases or the establishment of physical maps of an unsequenced genome (or part of it). *In silico* mapping adventures the macrosynteny and/or microsynteny between related (and also less related) genomes.

***in silico* screening**: A set of computer-modelling technologies through which researchers can (vicariously) screen chemical compounds for their potential as pharmaceutical candidate compounds, pesticide candidate compounds and so on. The chemical compounds are 'generated' (e.g. from data available about compounds actually created in a laboratory in the past) and computer modelling is then employed to (i) evaluate their influence on 'generated' specific cells, tissues and so on (from data available about that chemical type of molecule's effect on that type of cell/tissue when actually tested on it in a laboratory in the past); (ii) produce an analogous chemical compound that is likely to be more efficacious or have fewer undesirable side effects; (iii) repeat the process, for example, when screening compounds for potential usefulness as a pharmaceutical, the goal is to assess (modelled/predicted) differences between diseased (untreated) and treated cells; thus, enabling prediction of (better) pharmaceutical candidate compounds for ultimate actual testing on real cells/tissues and (v) some of the more classy *in silico* screening software can even model Absorption, Distribution, Metabolism and Excretion (ADME) properties for selected pharmaceutical candidate compounds.

***in silico* single-nucleotide polymorphism (*in silico* SNP, isSNP)**: Any single-nucleotide polymorphism that is recognized *in silico* by pulling out overlapping sequences in EXPRESSED SEQUENCE TAGS (ESTs) or genomic databases.

in situ: In the natural, existing or original, position (e.g. inside the body).

***in situ* conservation**: Refers to the maintenance of plant populations in the territories where they naturally occur and have evolved. In crop species, conservation can be achieved in home gardens or farms, and in wild species through the designation of protected areas such as natural parks and nature reserves.

***in situ* hybridization (ISH)**: A methodology used to detect the segments complementary to specific nucleic acid molecules. Chromosomes, which are treated to denature the DNA and remove RNA and proteins, are then incubated with radioactively labelled nucleic acids (DNA or RNA) or non-radioactive labelled probes of special properties. The hybridized segments are then visualized by autoradiography or directly under the microscope and in photography of fluorescent signals. ISH is mainly used to detect a gene or a messenger RNA molecule in a cell or tissue by hybridization. *See* FLUORESCENCE *IN SITU* HYBRIDIZATION.

***in situ* hybridization (ISH) procedure**: The pairing of a labelled probe to its complementary sequence within intact, banded chromosomes. The procedure begins by fixing samples in neutral-buffered formalin and embedding the tissue in paraffin. The samples are then sliced into thin sections and mounted on to microscope slides (Alternatively, tissue can be sectioned frozen and postfixed in paraformaldehyde). After a series of washes to dewax and rehydrate the sections, a Proteinase K digestion is performed to increase probe accessibility, and a labelled probe is then hybridized to the sample sections. Radiolabelled probes are visualized with liquid film dried on to the slides, while non-isotopically labelled probes are conveniently detected with colorimetric or fluorescent reagents. *Aka* *IN SITU* NUCLEIC ACID HYBRIDIZATION.

***in situ* localization of histone methylation**: A procedure performed to visualize the distribution of H3K4me3 and H3K9me2 POST-TRANSLATIONAL MODIFICATIONS (PTLMs) without the loss of histological integrity of different tissues from numerous plant species.

***in situ* PCR sensitivity**: The two factors that affect PCR sensitivity are (i) the strandedness of the target molecule and (ii) the lack of a complementary sequence proximal to the target sequences. Reverse transcriptase-catalysed *in situ* transcription has even been employed to detect RNAs which exist at relatively high copy number. Initially, the stoichiometry of the nucleic acid components makes the reaction primer-driven. As the reaction proceeds, the PCR product itself begins to function as a target for further amplification, as well as a primer for further elongation. It is the accumulation of sequences elongated in multiple reactions that is retained by the fixed cell and acts as the target for the radiolabelled probe in the detection phase. The complex physical nature within a fixed cell impedes both the efficiency of the polymerase as well as diffusion and annealing and thus elongation times are much longer than those conducted under aqueous (standard PCR) conditions. The optimum signal is usually produced using a 15-min elongation.

***in situ* PCR (ISP)**: This is a polymerase chain reaction that actually occurs inside the cell on a slide. *In situ* PCR amplification can be performed on fixed tissue or cells. *IN SITU* HYBRIDIZATION (ISH) applies the procedure of nucleic acid hybridization at the cellular level. Combining cytochemistry and immunocytochemistry, *in situ* PCR permits the identification of cellular markers and further allows their localization to cell-specific sequences within cell populations, such as tissues and blood samples. ISP is limited to the detection of non-genomic material such as RNA, genes or genomes, as the detection limit in most conditions is several copies of the target nucleic acid per cell. Therefore, due to copy number limitations, RNA hybridization is more sensitive than DNA detection.

***in situ* zymography (ISZ)**: A unique laboratory technique that enables the localization of matrix-degrading metalloproteinase (MMP) activity in histological sections. Frozen sections are placed on glass slides coated with fluorescently labelled matrix proteins. After incubation, MMP activity can be observed as black holes in the fluorescent background due to the proteolysis of the matrix protein. Alternatively, frozen sections can be incubated with matrix proteins conjugated to quenched fluorescein. Proteolysis of the substrate by MMPs leads to the release of fluorescence. This technique can be combined with immunohistochemistry to enable the co-location of proteins such as cell type markers or other proteins of interest.

in utero: The term means 'in the uterus' or 'before birth'.

in vitro: Plant tissue culturing is done under *in vitro* conditions (in an unnatural position). For example, the testing of a substance, or the experimentation in (using) a 'dead' cell-free system, in an artificial, controlled environment outside the body, such as in a Petri dish or test-tube. The term is also used by biochemists to describe a process taking place in an isolated cell-free extract. After autoclaving, the inoculum is aseptically cultured on/in this medium and the cultures are then incubated in the growth room to obtain the desired growth. This type of culturing is called IN VITRO CULTURING. *Compare with* IN VIVO.

***in vitro* cell fate switch**: Plant mature cells can reverse their state of differentiation and produce new organs under specific culture conditions. Two phases, dedifferentiation and redifferentiation, are commonly characterized during *in vitro* organogenesis. In these processes, cells undergo fate switch several times regulated by extrinsic and intrinsic factors, associated with re-entry to the cell cycle, the balance between euchromatin and heterochromatin, reprogramming of gene expression and so forth.

***in vitro* collection**: An assembly of germplasm maintained as plant tissue grown in active culture on solid or liquid medium. It can be sustained as plant tissue ranging from protoplast and cell suspensions to callus cultures, meristems, shoot tips and embryos.

***in vitro* complementation**: Complementation of allelic sequences demonstrated in a test-tube.

***in vitro* culture**: The cell, organ or tissue culture performed under artificial conditions in tubes, glasses, dishes and so on.

***in vitro* fertilization**: (i) Pollination done aseptically *in vitro* by direct application of the pollen to the ovule. It is used to overcome prezygotic incompatibility. *Aka* IN VITRO POLLINATION. (ii) A procedure to help infertile couples conceive. Eggs are removed from the woman and fertilized with the man's sperm in the laboratory. Fertilized eggs can then be transferred to the woman's uterus to try to establish a pregnancy or they can be frozen for future use.

***in vitro* marker**: A mutation that enables identification *in vitro* of a cell line possessing the marker.

***in vitro* mutagenesis**: Any one of several procedures employed to produce a specified mutation at a predetermined position in a DNA molecule outside the host cells. Mutagenesis can be random or site specific, and the base change depends on the method used.

***in vitro* neoplastic transformation.** The acquisition, by cultured cells, of the property to form neoplasms, benign or malignant, when inoculated into animals. Many transformed cell populations that arise *in vitro* intrinsically or through deliberate manipulation by the investigator, produce only benign tumours that show no local invasion or metastasis following animal inoculation. If there is supporting evidence, the term '*in vitro* malignant neoplastic transformation' or '*in vitro* malignant transformation' can be used to indicate that an injected cell line does, indeed, invade or metastasize.

***in vitro* packaging**: Synthesis of infective λ phage particles from a preparation of λ capsid proteins and a concatemer of DNA molecules separated by *cos* sites.

***in vitro* pollination**: Pollination performed aseptically *in vitro*. Pollen is applied directly to ovules in an attempt to overcome various types of prezygotic incompatibility that otherwise impede fertilization.

***in vitro* propagation**: The propagation of plants on synthetic medium under meticulously controlled and sterilized conditions, using plastic or glass culture vessels, aseptic techniques and a defined growing media. In vegetatively propagated crops such as sugarcane and potato, the conventional rate of multiplication is rather slow, usually 1 to 10/year. Using the technique of *in vitro* propagation through tissue culture, thousands of plants can be generated from one plant within a year.

***in vitro* propagation, stages in:** Four stages occur in *in vitro* propagation. STAGE I is characterized by the establishment of an aseptic tissue culture of a plant. STAGE II is characterized by the rapid numerical increase of organs or other structures. STAGE III is characterized by the preparation of propagules for successful transfer to soil, a process involving rooting of shoot cuttings, hardening of plants and initiating the change from the heterotrophic to the autotrophic state. STAGE IV is characterized by the establishment in the soil of a tissue culture–derived plant, either after undergoing a Stage III pretransplant treatment or, in certain species, after the direct transfer of plants from Stage II in to the soil.

***in vitro* protein synthesis:** The incorporation of amino acids into polypeptide chains in a cell-free system.

***in vitro* recombination:** *See* RECOMBINANT DNA TECHNOLOGY.

***in vitro* replication techniques:** Common amplification techniques include POLYMERASE CHAIN REACTION (PCR), LIGASE CHAIN REACTION (LCR), SELF-SUSTAINED SEQUENCE REPLICATION (SSR), Q-BETA REPLICASE (QBR) TECHNIQUE and STRAND DISPLACEMENT AMPLIFICATION (SDA). *See details as separate terms.*

***in vitro* screening:** The search and selection for particular characters in cells, organs or tissues, performed under artificial conditions in tubes, glasses, dishes and so on, usually in combination with special nutritional media, which enable a differentiated growth of the cells and so on.

***in vitro* selection:** (i) Selection at the cellular or callus stage of individuals possessing certain traits, such as herbicide resistance or salt tolerance. (ii) A search process (e.g. for a new pharmaceutical) that first involves the construction of a large 'pool' of polynucleotide sequences (at least some of which are likely to possess the desired pharmaceutical properties) that are synthesized by a totally random process. This is followed by repeated cycles of screening (for those sequences possessing desired properties) and/or enriching and amplification (of the screened/enriched sequences).

***in vitro* senescence:** In vertebrate cell cultures, the property attributable to finite cell cultures; namely, their inability to grow beyond a finite number of population doublings. Neither invertebrate nor plant cell cultures exhibit this property. *See* STAGES IN *IN VITRO* PROPAGATION.

***in vitro* synthesis of DNA:** M. Goulian *et al.* (1967) took tritium-labelled (^{3}H) ϕX174 plus (+) strand (parental) and used it as a template for the synthesis of a minus (−) strand. In the minus (−) strand, the usual T was replaced by its heavier analogue 5-bromouracil (%-BU) and ^{32}P. This step was performed to differentiate minus (−) strands from the parental plus (+) strands. The minus (−) strand was permitted to form a circular DNA molecule with DNA ligase. The double helixes were briefly exposed to DNase and then denatured by heat to allow isolation through centrifugation of the heavier circular minus (−) strand. Again, a reaction mixture was prepared with DNA polymerase and ligase. The product thus obtained had a completely synthetic DNA molecule just like its natural replicative form. The treatment with DNase permitted the separation of newly synthesized (+) strands, which were later shown to be fully infective when exposed to bacterial protoplasts.

***in vitro* synthesis of genes from mRNA:** RNA-directed DNA polymerase (also called reverse transcriptase) has the ability of synthesizing DNA on an RNA template. This procedure enabled molecular biologists to synthesize complementary DNA using mRNA template. If mRNA transcribed from a specific gene is made available in purified form, the COMPLEMENTARY DNA (cDNA) synthesized with its help will represent the synthesized gene.

***in vitro* transcription:** The specific and accurate synthesis of RNA in the test -tube using purified DNA preparations as a template. A so-called 'COUPLED SYSTEM' may be obtained from *Escherichia coli* that carries out both mRNA synthesis and its translation into protein. For eukaryotes, separate cell-free systems have to be set up to demonstrate the activity of the three functionally different RNA polymerase complexes. *Aka* CELL-FREE TRANSCRIPTION.

***in vitro* transformation:** A heritable change, occurring in cells in culture, either intrinsically or from treatment with chemical carcinogens, oncogenic viruses, irradiation, transfection with oncogenes and so on and leading to the acquisition of altered morphological, antigenic, neoplastic, proliferative or other properties. This expression is distinguished from '*in vitro* neoplastic transformation' in that the alterations occurring in the cell population may not always include the ability of the cells to produce tumours in appropriate hosts. The type of transformation should always be specified in any description.

***in vitro* translation:** The synthesis of proteins in the test-tube from purified mRNA molecules using cell extracts containing ribosomal subunits, the necessary protein factors, tRNA molecules and aminoacyl-tRNA synthetases. ATP, GTP, amino acids and an enzyme system for regenerating the nucleoside triphosphates are added to the mix. Prokaryotic translation systems are usually prepared from *Escherichia coli* or the thermophilic bacterium *Bacillus stearothermophilus*. Eukaryotic systems usually employ rabbit reticulocyte lysates or wheat germ.

in vivo: Studies and propagation of whole, living organisms on intact organ systems therein or on populations of microorganisms.

***in vivo* engineering of mammalian mitochondrial genomes:** An efficient method has been developed for replicating and stably maintaining entire mitochondrial genomes in *Escherichia coli*. These mitochondrial DNA genome clones could be engineered using standard molecular biological techniques. This was accomplished by inserting an *Escherichia coli* replication origin and selectable marker into isolated, circular mtDNA at random locations using an *in vitro* transposition reaction and then transforming the modified genomes into *Escherichia coli*.

***in vivo* nitrogenase Fe protein regulation model:** The activities of two enzymes – dinitrogenase reductase ADP-ribosyltransferase (DRAT) and dinitrogenase reductase-activating glycohydrolase (DRAG) – are required. DRAT requires Mg^{2+} and ADP to attach Adenosine diphosphate ribose (ADPR) to the active oxidized form of Fe protein and in this reaction, NAD^{+} is converted into Nicotinamide. This reaction requires increased darkness or NH_4^{+}. DRAG requires MgATP and Mn^{2+} to remove ADPR from the inactive reduced form of Fe protein-ADPR and in this reaction, ADP-ribose is released. This reaction requires decreased darkness or NH_4^{+}. The oxidized form of active Fe protein is converted into a reduced active form by the addition of an electron. The reduced form of inactive Fe protein-ADPR is converted into an oxidized inactive form by the addition of an electron.

***in vivo* zymography (IVZ):** A type of zymography in which living tissue and whole organisms are used. Fluorescent substrate, such as modified type IV collagen, is injected into the tissue. The substrate either contains a fluorophore and a quencher or is highly quenched, resulting in increased fluorescence upon substrate hydrolysis. IVZ gives real-time, 3D information about MMP-9 activity. Artefacts associated with sample preparation are minimized.

inactivated agent: A virus, bacterium or other organism that has been treated to prevent it from causing a disease.

inactivating DNA alteration: Any DNA damage that prevents DNA replication. If not healed by DNA repair processes, such damage is usually lethal to cells. Occasionally, it may produce gene mutations, and commonly chromosome structural changes by mispairing.

inactive X chromosome: In females, one of the two X chromosomes is inactivated to prevent the overexpression of X-linked gene products to equalize gene dose between females (with two X chromosomes) and males (with one X chromosome). X chromosome inactivation is achieved through the formation of heterochromatin. The highly condensed inactive X chromosome observed in mammals is called a BARR BODY.

inborn errors of metabolism: Genetically caused metabolic dysfunctions.

inbred: Plants resulting from successive self-fertilization of parents through several generations.

inbred (I) lines: Involves the inbreeding of annual seed-propagated material resulting in homogeneous and homozygous lines that are isolated by the selection of desired recombinants or segregated in F_2 to F_7 generations of crosses between parental pure lines. Generally, mono-genotypic lines can be blended to form multilines, for example, in tomato, lettuce, soybean, pea, cowpea, snap bean, field bean, Arabian coffee, Capsicum pepper, aubergine, okra, lentil and papaya. *Aka* INBRED PURE LINE.

inbred-variety cross: The cross-producing F_1 of an inbred line with a variety.

inbreeding: A system of mating where genetically related individuals are mated with each other. It leads to an increase in homozygosity. Self-pollination is an extreme form of inbreeding. In cross-pollinated crops forced selfing is done to achieve inbreeding for the development of inbred lines for the production of hybrid seeds or synthetic composite varieties. Thus, the developed inbreds are less vigorous and difficult to maintain because of inbreeding depression. Inbreeding decreases the genetic and phenotypic variability and increases homozygosity and inbreeding depression among individuals in a population. *See* CONSANGUINEOUS MATING.

inbreeding coefficient: The probability that the two genes at any locus in a diploid individual are identical by descent; they were created from the replication of one gene in a previous generation.

inbreeding depression: Reduced offspring fitness as a result of breeding between related parents. Inbreeding depression is a major force in maintaining self-incompatibility (SI) systems in plants. The reduction in the vigour, survival or reproduction potential of inbred individuals is caused by homozygote deleterious recessive genes that were present in the parents as heterozygotes. The effect opposite to heterosis is usually severe in open-pollinated outcrossing species.

inbreeding depression hypothesis: Heterozygosity at the allozyme loci is generally an indicator of the overall level of heterozygosity and individuals with high levels of homozygosity suffer inbreeding depression.

inbreeding load: The extent to which a population is weakened by inbreeding.

inbreeding minimum: The stage reached after a long period of inbreeding. Because of the homozygosity attained, vigour is not further decreased but remains constant if inbreeding is continued.

incertae sedis: Of unknown taxonomic affinity; having obscure relationships.

incestuous mating: A mating between blood relatives that are more closely related than the law of the land permits. *See* CONSANGUINEOUS MATING.

inch-worming model of abortive transcription initiation: It invokes a flexible element in RNA POLYMERASE (RNA POL). In each cycle of abortive initiation, a module of RNA pol containing the active centre detaches from the remainder of RNA polymerase and translocates downstream; upon release of the abortive RNA, this module of RNA polymerase reverse translocates.

incidence: The number or proportion of new cases of a specified condition in a population over a specified period.

incidence rate: Incidence per unit of time and per unit of population. If the incidence of benign prostatic hyperplasia in a population of 100,000 adults in 6 months is 16 cases, the incidence rate is 0.32 per 1000 adults per year. This can also be calculated per person-year or per 100 person-years and so on.

incineration: Burning of garbage in special facilities by utilizing inherent thermal values of solid wastes.

incinerator: A device in which solid, semi-solid, liquid or gaseous garbage is burnt as a means of disposal. If the material is not itself supporting combustion, auxiliary fuel is added. Many types of industrial and domestic wastes are incinerated and there are many types of incinerators to deal with different types of wastes.

incipient species: Populations that are too diverse to be considered subspecies of the same species, but not sufficiently differentiated to be regarded as different species. Sometimes called SEMISPECIES.

incised: Cut deeply and usually unevenly. A condition in between toothed and lobed.

incision step in DNA excision repair: An incision is caused close to the damage on the DNA strand by a specific correndonuclease (i.e. correndonuclease I or correndonuclease II depending upon the mono or diadduct formation and acts in the absence of divalent cations Mg^{2+}). If there is no incision, further steps in excision repair are not followed.

inclusion body: A structure developed within a plant cell as a consequence of infection by a virus that is often useful in classifying the virus.

inclusive composite interval mapping (ICIM): ICIM includes methods for QTL mapping with populations derived from biparental crosses. Two steps were included in the ICIM. In the first step, stepwise regression was applied to identify the most significant regression variables. In the second step, a one-dimensional scanning or interval mapping was conducted for detecting additive (and dominance) QTL and a two-dimensional scanning was conducted for detecting digenic epistasis. The ICIM provides intuitive statistics for testing additive, dominance and epistasis, and can be used for most experimental populations derived from two inbred parental lines. The EM algorithm used in ICIM has a fast convergence speed and is therefore less computing intensive. The ICIM retains all the advantages of the CIM over interval mapping, and avoids the possible increase of sampling variance and the complicated background marker selection process in the CIM. *Compare with* COMPOSITE INTERVAL MAPPING.

inclusive fitness: A measure of individual fitness that comprises the effects of that individual on its neighbours' fitness, weighed by their relatedness, and reduces the effects of neighbours' on the individual's fitness. *See* KIN SELECTION.

incompatibility: A genetically governed inability to obtain fertilization and seed formation after self-pollination or cross-pollination. There are numerous types of PROGAMOUS INCOMPATIBILITY (*aka* POSTGAMOUS INCOMPATIBILITY); in contrast to HETEROMORPHIC INCOMPATIBILITY (e.g. heterostyly in *Primula* sp.), HOMOMORPHIC INCOMPATIBILITY is not associated with morphological differences.

incompatibility group: Plasmids that show incompatibility with each other are said to belong to the same incompatibility group.

incompatibility of plasmids: The inability of certain types of plasmids to survive together in the same cell.

incomplete antibodies: Those antibodies that react with individual erythrocytes but not agglutinate with them. *Aka* BLOCKING ANTIBODIES because they cover the antigens and make them inaccessible to antibodies that otherwise would produce agglutination.

incomplete block design (IBD): A block design in which not every treatment occurs in a block. It is preferred where large numbers of cultivars are compared in a single yield trial. The entries in each replication are subdivided into smaller blocks, in a manner designed to reduce the error caused by soil variation. Usually, IBD refers to a lattice design, considering the restriction that (i) the number of testers must be harvested, (ii) inferior strains cannot be discarded prior to harvest to reduce harvest expenses and (iii) researchers still must analyse the experiment as a lattice design.

incomplete cell division promotion: In many vertebrate and invertebrate organisms, gametes develop within groups of interconnected cells called GERMLINE CYSTS formed by several rounds of incomplete divisions. The loss of the deubiquitinase *USP8* gene in *Drosophila* transforms incomplete divisions of germline cells into complete divisions.

incomplete diallel: A partial sampling in which any individual family or type of family may be omitted.

incomplete digest: *See* PARTIAL DIGEST.

incomplete dominance: In this case neither of a pair of alleles displays dominance and the F_1 progeny is phenotypically intermediate between the two homozygous parents with contrasting pair of traits. Flower colour in four o' clock plant, *Mirabilis jalapa*, provides a good example of incomplete dominance. Neither of the alleles R1 and R2 are dominant in the hybrid, as the homozygote R1R1 produces red, homozygote R2R2 produces white and their F_1 heterozygote R1R2 produces pink flowers. Incomplete dominance increases the number of phenotypes in the segregating progeny. *Aka* SEMIDOMINANCE. *See* PARTIAL DOMINANCE.

incomplete flower: An incomplete flower lacks one of the parts (stamen, a pistil, petals and sepals) of a flower.

incomplete karyotype: The karyotype present is incomplete, usually because of poor chromosome quality.

incomplete Latin squares: Row-and-column designs for n − 1 replicates of n treatments arranged in n − 1 rows and n columns. These designs can be constructed by ignoring any row from an n × n Latin square. The n − 1 rows comprise a set of replicate blocks of size n, whereas the n columns comprise a set of balanced incomplete blocks of size n − 1. The designs have a balanced partitioning of treatment effects and have an orthogonal factorial structure for factorial treatments design. If n is too large for a simple incomplete Latin square, a design of size m × n where m < n − 1 can sometimes be beneficial. If the rows are complete replicate blocks and the columns are balanced incomplete blocks, the designs are called a YOUDEN RECTANGLE and have similar properties to incomplete Latin squares. YOUDEN DESIGNS can be very useful for row-and-column-type designs but the handiness of balanced Youden designs is extremely limited. Also, if m is substantially smaller than n, a substantial amount of treatment information will be confounded between columns and some method for recovery of information from the columns stratum of the analysis will be necessary for an efficient analysis of variance.

incomplete linkage: Linked genes that reassociate through genetic crossing-over.

incomplete penetrance: Means the genetic trait is expressed in only parts of the population. The per cent penetrance also may change with the age range of the population.

incomplete resistance: A kind of resistance that is not complete and shows slow susceptibility to the pathogen.

incompletely sex-linked gene: Genes located on homologous portions of the X and Y (or Z and W) chromosomes. *See* PSEUDOAUTOSOMAL GENE.

increasing nitrogen for faster plant growth in low-light environments: N supply is essential for the growth of all plants. Increased N supply stimulates tomato plants to grow faster in low-light environments, resulting in enhanced light-harvesting efficiency. The evaluation of plant dry-weight build-up with N stimulation shows that high N doses are related to high number of grana per chloroplast, improved thylakoid piling and high electron density. Such a strong photosynthetic machinery results in improved green-light absorption of light-harvesting machinery in tomato but not blue- or red-light absorption. The increased quantum yield and larger palisade parenchyma cell area for green-light absorption results in increased plant growth under natural-light-limited photosynthesis. Such experimental observations have confirmed the role of additional N supply in improving light-harvesting efficiency under low-light conditions.

increasing the breadth of resistance: Many R genes have a narrow range of resistance. Broad-spectrum resistance induction occurs by coordinate expression of an *R* gene and a corresponding *Avr* transgene controlled by pathogen-inducible promoter. There is no need of pyramiding numerous R transgenes. The avirulence (AVR) protein in presence of R protein leads to activation of defences.

incremental truncation for the creation of hybrid enzymes (ITCHY): ITCHY is a method for generating chimeric gene libraries by non-homologous recombination through a process of nested gene deletions and fusions.

incubation period: The time during which microorganisms inoculated on to a medium are allowed to grow. The period between the infection and the appearance of the symptoms of a disease. Incubation period in some diseases is always the same, whereas, in others, it is variable. Incubation period is determined by the host resistance, number of infecting microorganisms and the virulence of pathogens. For example, incubation period for influenza is 18–36 h, chicken pox (14–16 days), cholera (1–3 days), typhoid fever (14 days), AIDS (10 days), tetanus (3–21 days) and malaria (2 weeks).

incurved: In general, bent or curved inward. Of leaf margins, when bent or curved towards the adaxial side.

indefinite: Variable in number and, as a rule, numerous. For example, more than twice as many stamens as petals or sepals, but no particular standard number of stamens. Also used as a synonym for the term INDETERMINATE, meaning the condition in which an inflorescence is not terminated by a flower, but continues growing until limited by physiological factors. *See* INDETERMINATE. *Compare with* NUMEROUS. *Contrasts* DEFINITE.

indehiscent: Not opening, or without any special method of opening, at maturity as in anthers, asci, fruiting bodies, fruits, sporangia and so on. *Contrasts* DEHISCENT.

indel: A general term that may refer to insertion, deletion or insertion and deletion of nucleotides in genomic DNA. Indels ≥50 bases in length are classified as STRUCTURAL VARIANTS. Indels can be employed as genetic markers in natural populations, especially in the phylogenetic studies.

independent characters: Characters governed by genes showing independent assortment.

independent contrasts, method of: A comparative method that controls for the fact that character states in related organisms are not statistically independent because of shared ancestors. The basic idea is that differences between one pair of species are independent of differences between another pair of species even if both pairs are related.

independent genes: Genes that show independent assortment.

independent segregation (assortment), principle of: Different segregating genes behave independently during meiosis. *See* MENDEL'S PRINCIPLES, PRINCIPLE OF INDEPENDENT SEGREGATION (ASSORTMENT).

independent t-test: The independent t-test analyses the averages of two independent samples to see if there is statistical proof that a related population average or mean differs substantially.

independent variable: In a statistical experiment, an independent variable is one that you modify, control or manipulate in order to investigate its effects. It is called independent since no other factor in the research affects it.

indeterminate: Descriptive of an inflorescence in which the terminal flower is the last one to open. The flowers arise from axillary buds and the floral axis may be indefinitely prolonged by a terminal bud. Some crops, such as haricot beans, can have either the determinate or the indeterminate habit. With the former, they are self-supporting, bushy plants and with the latter, they grow as vines. On the other hand, potatoes are determinate plants but when grafted on to tomatoes, they become indeterminate. This is a very valuable technique when many flowers are needed for the production of true seed for breeding purposes.

index of prevalence: Proportion of plant stands affected by disease or infestation in a defined area.

index selection: A type of intentional simultaneous selection. With the index selection, some index value is allocated to each candidate. The index value specifies the aggregate value of each candidate across several traits. The index selection consists of truncation selection with regard to the index values. In population improvement, it is important to understand that the breeding material undergoes selection as well as recombination in recurrent cycles. Separate simulation is used to resolve the numerical problem of model calculations to take recombination into account. Two negatively correlated traits with high, as well as high and low heritabilities, are measured in a population with initially equal gene frequencies for the favourable and unfavourable allele. When the linear Smith-Hazel, Pesek-Baker and heritability indices, as well as the non-linear Elston index, are compared with another type of non-linear selection index over 40 cycles of recurrent selection, the new type of selection index shows characteristics for population improvement. A genetic variation is maintained and less favourable alleles are lost in a recurrent selection programme compared to other indices. It can be shown that situations exist where the classical SMITH-HAZEL INDEX is not optimal in medium and long-term population improvement. Moreover, if two negatively correlated traits have considerably different heritabilities, the selection response is negative for the low heritable trait for many recurrent selection cycles with all selection indices except for the PESEK-BAKER INDEX and the new type of selection index.

indexing: (i) The process used to scrutinize vegetatively reproduced plants for freedom from viral diseases before multiplying those plants. (ii) The testing of a plant for infection, often by mechanical transmission or by grafting tissue from it to an indicator plant.

indicator organism: An organism whose presence indicates infection/contamination. Coliform bacteria are used as indicator organism. The predominant faecal coliform is *Escherichia coli*. Its detection is a good indication that human wastes are entering the water. The common method for checking the presence of coliform in water is largely based on the lactose fermenting ability of coliform bacteria. In case of plants, *Chenopodium* is used as an indicator plant to check for viral infection in crop plants.

indicator plant: A plant that responds to a pathogen or an environmental factor with specific symptoms and is employed to detect or identify the pathogen or determine the effects of the environmental factor.

indigenous: Native to the area, not introduced, and not necessarily confined to the region discussed or present throughout it. It is hardly distinct from NATIVE but is usually applied to a smaller area. For example, the Cootamundra Wattle (*Acacia baileyana*), a shrub or tree in the flowering plant family Fabaceae, is native to Australia but indigenous to the Cootamundra region of southern New South Wales. *Compare with* ENDEMIC.

indirect end labelling: It is a technique for examining the organization of DNA by making a cut at specific site and isolating all fragments containing the sequence adjacent to one side of the cut; it reveals the distance from the cut to the next break(s) in DNA.

indirect penetration: The penetration of plant tissues by a pathogen via natural openings (e.g. stomata) or wounds. *Contrasts* DIRECT PENETRATION.

indirect regeneration: In indirect regeneration, a *de novo* organ is typically formed from an intermediate tissue, the callus. *Compare with* DIRECT REGENERATION.

indirect selection: Selection of a trait that ascends by virtue of its association with other traits rather than because of its direct causal effect on fitness. Also refers to the selection of a trait that arises from the effects of a gene in one individual on the fitness of the other individuals. *Compare with* GROUP SELECTION.

individual buffering: A mechanism promoting genotypic adaptation of homogeneous populations (e.g. pure line or F_1 hybrids) that tend to stabilize their production under different environments. It operates at allelic level within an individual. *Aka* DEVELOPMENTAL HOMEOSTASIS.

indole-3-acetic acid (IAA): A substance ($C_{10}H_9NO_2$) that functions as a growth hormone or auxin in plants, where it regulates cell enlargement and, through interaction with other plant hormones, also influences cytokinesis.

indoor culture: Growing plants indoors using natural and/or artificial light and additional heating. It is generally used for subtropical or tropical plants or for plant propagation.

induced: Produced in response to a stimulus; regulated. *Contrasts* CONSTITUTIVE.

induced chromosome break: Chromosome break caused by some agent usually external to the chromosome, such as radiations or chemicals. It can also be produced genetically, e.g. by the *Ac-Ds* system in maize.

induced epigenetic variation: Induced epigenetic variation includes mutations in the epi-RILs, chemical treatment with inhibitors of DNA methylation and chromatin modifications, directed epigenome editing and tissue culture-based treatments.

induced fit: A substrate-induced change in the shape of an enzyme molecule that causes the catalytically functional groups of the enzyme to assume positions that are optimal for catalytic activity to occur.

induced fit hypothesis: An interaction between a protein and its ligand that produces a conformational change in the protein that enhances the protein's affinity for the ligand. Usually, the active site consists of a cleft on the surface of an enzyme molecule that is complementary in shape to the substrate. Molecules that differ from substrate in shape or with respect to functional group cannot bind to the enzyme. The substrate binding site of most enzymes is preformed but may undergo some conformational change on substrate binding, a phenomenon called induced fit.

induced mutation: Mutation that is produced under changed growth conditions; as a result of exposure of an organism to some agent that does not form normal growth condition for the organism. An agent that has the ability to produce mutation is called a MUTAGEN. A mutagen can be a physical agent (UV light, X-rays, γ-rays, α-particles, β-particles, etc.) or a chemical agent (nitrogen mustard, hydrazine, hydrogen peroxide, nitrous acid, base analogues, alkylating agents, dyes, etc.). An individual possessing a mutation is termed as a mutant. These are those mutations whose cause is known. *See* PHYSICAL MUTAGENS *and* CHEMICAL MUTAGENS. *Compare with* SPONTANEOUS MUTATION.

induced pluripotent stem cells (iPSCs): Differentiated cells reprogrammed to pluripotency by the ectopic expression of reprogramming factors such as Oct3/4, Sox2, Klf4 and c-Myc.

induced protoplast fusion: This type of protoplast fusion needs a fusion-inducing agent (e.g. MECHANICAL FUSION, CHEMOFUSION and ELECTROFUSION).

induced resistance: A form of resistance activated in plants by an external stimulus (e.g. elicitor treatment or pre-inoculation with a microorganism). It can be a local or systemic acquired resistance.

induced responses: A change in phenotype that occurs in response to a specific environmental signal (or cue) and that improves survival and/or reproduction. Without the signal, the change does not take place. *See* PHENOTYPIC PLASTICITY.

induced somatic cell haploidy: A. Mikhalchenko *et al.* (2024) developed and explored a model of premature cell division, where nonreplicated, G0/G1-stage somatic cell nuclei are transplanted to the metaphase cytoplasm of mouse oocytes. Subsequent cell division generates daughter cells with reduced ploidy. High occurrence of somatic genome haploidization was observed in the nuclei from inbred genetic backgrounds but not in hybrids, emphasizing the importance of sequence homology between homologues.

induced sterility technique: In SIT, and other similar techniques, males and females can be easily distinguished by GENETIC SEXING STRAIN (GSS) TECHNIQUE facilitating the release of male-only strains. It is now a known fact that the efficacy of many STERILE INSECT TECHNIQUES (SITs) has been increased by this method.

induced systemic resistance (ISR): The reduced disease symptoms on a portion of a plant distant from the area where the inducing agent is active and resistance is caused by the triggering of active plant defences against a variety of pathogens. ISR is also used to describe increased resistance in plants induced by certain rhizobacteria. *See* SYSTEMIC ACQUIRED RESISTANCE (SAR).

inducer: A molecule that induces the expression of a gene or operon by binding to a REPRESSOR protein and thereby preventing the repressor from attaching to its OPERATOR site. EFFECTORS in inducible operons are known as INDUCERS. A small molecule that binds to a repressor that triggers the transcription of a gene, such as *LacZ*. Inducers may be enzyme substrates or modified substrates such as isopropyl β-d-1-thiogalactopyranoside (IPTG).

inducer of terminator gene: In terminator gene technology, before selling to farmers, a seed company treats the seeds with a chemical inducer – probably tetracycline – to initiate terminator gene interactions. If the seed company does not want to initiate the terminator genes, it will not apply the inducer. This allows the repressor protein to bind to the binding site on Gene II, preventing the production of recombinase.

inducible and repressible operons, similarities between: In both these systems, inducers and corepressors are usually small molecules described as effectors whose presence leads to changes in the function of regulatory repressor proteins that control transcription. In the case of *lac* systems, the actual effector (allolactose) is different from the apparent effector (lactose), and this can also be true for some repressible systems where the corepressor is an amino acid–charged tRNA rather than the amino acid itself. Similarities between inducible and repressible systems also extend to the occurrence of both polar and promoter mutations.

inducible control: In this case, substrate functions to induce the production of the enzymes; repression occurs in the absence of the substrate.

inducible enzymes: Enzymes whose rate of production in an organism can be increased by the presence of certain chemical molecules.

inducible system: A system comprising a coordinated group of enzymes involved in a catabolic pathway is inducible if the metabolite upon which it works causes transcription of the genes controlling these enzymes. These systems primarily occur in prokaryotic operons.

inducible transcription factors: These factors function similarly to upstream transcription factors but must be activated by phosphorylation or by specific ligands in order to bind to their target DNA sites and influence transcription initiation. These are synthesized and/or activated in specific tissues at particular times and therefore mediate gene expression in a positionally and temporally specific manner.

inducible versus repressible gene regulation: Substrates and end-products play a role in the regulation of gene action. Accordingly, two types of controls are recognized – INDUCIBLE CONTROL and REPRESSIBLE CONTROL.

inducing tissue: This is the tissue that produces a protein and releases it into extracellular spaces and induces differentiation.

induction: (i) A chemical or physical treatment that results in excision from the host genome of the integrated form of a lysogenic phage, followed by the switch to the lytic mode of infection. (ii) An increase in gene expression due to a regulatory signal. The transcription of genes is induced by inactivation of a repressor or by the action of an activator. (iii) In developmental biology, a change in the developmental fate of one tissue due to an interaction with another tissue. Such an interaction is known as INDUCTIVE INTERACTION. (iv) It is the initiation of structure or process or organ during *in vitro* culturing. During plant tissue culture, when elevated levels of a growth regulator 2,4-D are added into the culture medium, it results in callus formation in the majority of plant species. This phenomenon is called callus induction. Likewise, when Indole Butyric Acid (IBA) is used in the medium, the culture produces roots and this process is called root induction.

induction of flowering: The initiation of the process of the production of flowers, possibly stimulated by florigen.

induction of mutation: The process of producing a variation or mutation via treatment with a physical or chemical agent.

induction of prophage: A chemical or physical treatment that results in excision from the host genome of the integrated form of a lysogenic phage, followed by the switch to the lytic mode of infection.

industrial crops: Crops that are processed on a large scale by industrial means, for example, potato, pea, lupine, dwarf bean or cereals for starch production; sugarbeet, beets, sweet sorghum, chicory or Jerusalem artichoke for sugar processing; linseed, false flax, common marigold, crambe, caper spurge, *Cuphea* sp., meadowfoam or jojoba for oil processing; flax, hemp or nettles for fibre processing; and foxglove, poppy, yellow bark or cocoa for pharmaceutical utilization.

industrial melanism: Industrial melanism is the darkening of moths during the industrial revolution and is an interesting case study of natural selection. It illustrates the type of selection that can be caused by human intervention in natural systems. It is a spectacular example of a very rapid evolution in nature.

inelastic electron tunnelling spectroscopy (IETS): The process of inelastic electron tunnelling occurs when electrons tunnelling through a thin barrier lose part of their energy in exciting phonon modes in the barrier or vibrational modes of proximate molecules. The onset of such inelastic tunnelling process is indicated by a kink in the current versus voltage characteristics.

inert: A chromosomal segment that is assumed to be genetically inactive or without coded genetic information.

inert gene: A gene that, through apparent lack of mutation or effect in disturbing the balance, has been assumed to be inactive. Often associated with heterochromatin.

inert matter: A component of seed purity investigation. It includes non-seed and seed materials that are classified as inert according to the rules framed for testing seeds. For example, (i) seed units in which it is clearly known that no true seed is apparent; (ii) pieces of broken or damaged seed units half or less than half the original size; (iii) those appendages not classed as being part of the pure seed, in accordance with the definitions of the pure seed, must be removed and included in the inert matter; (iv) the seeds of Leguminosae, Cruciferae, Cupressaceae, Pinaceae and Taxodiaceae with the seedcoat completely removed; (v) unattached sterile florets, empty glumes, lemmas, paleas, chaff, stems, leaves, cone scales, wings, bark, flowers, nematode galls, fugues bodies – such as ergot – sclerotia and smut balls, soil, sandstones and all other non-seed matter and (vi) all material, except other seeds, left in the light fraction when the separation is done by the uniform blowing method. *See* TESTING SEEDS FOR PURITY.

infanticide: The killing of an offspring by its parent.

infect (vb.): To enter, invade or penetrate and develop a parasitic association with a host plant.

infection (n): The invasion of an organism's cells or tissues by a disease-causing pathogen, its growth and/or multiplication, and the reaction of the host organism to the infectious agent and the toxins it produces. The variety of biological pathogens includes certain bacteria, viruses, fungi, protists, parasitic worms and arthropods.

infection court: A certain part of a given plant that is susceptible to a particular pathogen or pathogens.

infection cushion: An organized mass of hyphae produced on the surface of a plant from which numerous infective hyphae develop.

infection focus: An initial site of infection, which is followed by secondary spread, generally with reference to a population of plants.

infection peg: The specialized, narrow hyphal strand on the bottom of an appressorium that penetrates host cells. *See* PENETRATION PEG.

infection period: The time required for the infection to ensue under conducive environmental conditions (e.g. usually hours of leaf wetness and temperature for foliar pathogens).

infectious: Relating to a disease that is caused by a biotic agent skilled of spreading from plant to plant. *See* BIOTIC. *Contrasts* ABIOTIC *and* NON-INFECTIOUS.

infectious transfer: The rapid transmission of free episomes (plus any chromosomal genes that it may carry) from donor to recipient cells in a bacterial population.

infective: Referring to an organism capable of attacking a host and causing infection. It also refers to a vector carrying or containing a pathogen that can be transferred to a host plant.

inference: A conclusion drawn from evidence.

inferential statistics. Inferential statistics is a test you use to compare a certain set of data within a population in a variety of ways. Inferential statistics include parametric and nonparametric tests. When conducting an inferential statistical test, you take data from a small population and make inferences about whether it will provide similar results in a larger population.

infertile: Organism not able to reproduce or not able to make viable gametes.

infertility: Infertility is the inability of a person, animal or plant to reproduce by natural means.

infest (n. infestation): (i) To attack as a pest (used especially in case of insects and nematodes), (ii) To contaminate, as with microorganisms. (iii) To exist in large numbers.

infestation: Refers to the establishment (or 'running over') on the surface of a host by a large number of insects or other animal pests. With infestation there is no implication that infection has occurred.

infinite-alleles model: A model that assumes that every mutation produces a new allele.

infinitesimal model: A model that accepts that quantitative trait variation is caused by a large number of loci, with infinitesimally small additive effects. Selection changes the mean, but the genetic variance remains unchanged. In a nutshell, it is equivalent to believing that offspring are normally distributed around the mid-parental value with constant variance.

inflammation: The body's response to injury or infection observable by pain, redness, heat, swelling or sometimes the loss of function. Inflammation can transpire anywhere in the body. But it is easily identified when occurring in the exposed tissue. It develops stepwise as the infection stimulates the release of inflammatory intermediators. Some of these intermediators act on capillaries causing their enlargement and leakiness that results in redness, swelling, heat and pain.

inflammatory response: The local response of a tissue to injury or infection – characterized by tissue redness, swelling, heat and pain. It is caused by the invasion of white blood cells, which release various local mediators such as histamine.

inflorescence: Several flowers closely grouped and arranged together to form an efficient structured unit.

influence of natural variation on phenotypic differences: Epigenetic variation also contributes to phenotypic variation. Epigenetic change, which does not involve alteration to the nucleotide sequence, can also cause changes in gene activity by changing the structure of chromatin through DNA methylation, histone modifications or small non-coding RNAs.

informatics pipeline for profiling and annotating RNA modifications: A protocol to identify and annotate RNA modifications based on the informatic analysis of METHYLATED RNA IMMUNOPRECIPITATION AND SEQUENCING (MeRIP-seq) data, using RNAmod, a convenient one-stop online interactive platform for the annotation, analysis and visualization of mRNA modifications.

informatics: *See* BIOINFORMATICS.

informational gene family: It comprises individual members that may differ markedly in sequence from one another although all are homologous and obviously reveal an ancient ancestry.

informational genes: Genes implicated in core 'informational' processes, including DNA replication and repair, transcription and translation. These are believed to be less disposed to lateral gene transfer than operational genes.

informational molecules: Molecules that contain information in the form of specific sequences of different building blocks. Proteins and nucleic acids are informational molecules.

informed consent: A communication process during which one learns the risks, benefits, limitations, potential implications, possible results and any necessary information about a test or research study so the individual can make an informed decision regarding their care or desire to participate in research. By giving permission to proceed with the test or to refuse the test, the individual acknowledges they understand the test and all of their queries have been answered. *See* PRIVACY.

informosome: A complex of mRNA and protein that protects the mRNA from degradation.

infrageneric: Representing taxonomic ranks below the genus level (e.g. subgenera, sections and series).

infraspecific: Denotes taxonomic ranks below the species level (e.g. subspecies).

infructescence: A fruiting structure that comprises more than a single fruit.

inherit: In genetics, to receive genetic material from the parents through biological processes.

inheritance: The process by which genetic information is passed on from parent to offspring.

inheritance (in computer science): In object-oriented programming, the mechanism of basing an object or class upon another object (prototype-based inheritance) or class (class-based inheritance), retaining similar implementation. Also defined as deriving new classes (subclasses) from existing ones (superclass or base class) and forming them into a hierarchy of classes.

inheritance of acquired characters: Jean-Baptiste Lamarck in 1889 tried to explain the extraordinary ability of hereditary factors to respond directly to the environment. He assumed that each hereditary particle had a spiritual conscious–like property that could absorb and interpret messages from outside. He believed that variations in organisms are induced in response to an urgent need and he regarded these variations as heritable. This theory could not be proven experimentally. However, in the light of epigenetic inheritance, this theory is being revisited.

inheritance of epigenome: Much of the epigenome is reorganized when parents pass their genomes to their offspring. Under some circumstances, however, some of the chemical tags on the DNA and histones of eggs and sperm may also be passed on to the next generation. When somatic cells divide mitotically, often much of the epigenome passes on to the daughter cells and epigenetic modification patterns are conserved as such, thus helping the cells remain specialized during their multiplication.

inheritance of *Epi-rav6*: To determine the inheritance of *Epi-rav6*, Zhang *et al.* in 2015 examined the phenotypes of 260 progeny of self-pollinated *Epi-rav6* plants. Three phenotypes were segregated in these progenies: 62 were like the wild type; 143 were like *Epi-rav6*, and 55 had severe defects, including fewer tillers, very large leaf angles and sterility. The ratio of the wild type to *Epi-rav6*-like to severe phenotypes was 1:2.3:0.89. In addition, the 27 F_1 plants derived from the cross between *Epi-rav6* and cv Zhonghua11 showed nearly a 1:1 ratio of wild-type progeny and *Epi-rav6*-like progeny. This genetic evidence demonstrated that the mutant is controlled by a single, semi-dominant locus and the derived allele is heterozygous. They regarded *Epi-rav6*-like progeny as heterozygotes and referred to them as *Epi-rav6* (+/−). They also regarded the progeny with severe defects as homozygotes and refer to them as *Epi-rav6* (−/−).

inheritance of modified histones: Parental histones H3-H4 segregate to both daughter DNA strands with a weak leading-strand bias, skewing partition at **topologically associating domain (TAD)** borders and enhancers proximal to replication initiation zones. The segregation of parental histones to the leading strand increased markedly in cells with histone-binding mutations in mini-chromosome maintenance protein, MCM2, part of the replicative helicase, exacerbating histone post-translational modification sister chromatid asymmetry.

inheritance of new phenotypes of a differentiating cell: At gene level, the requirements are (i) response to a signal, (ii) transduction of the signal to one or more genes in the nucleus, (iii) alteration of gene activities and (iv) inheritance through mitosis of the new pattern of gene activity. If this is correct, the external influences change the epigenotype. This raises another paradox, namely that information flows from DNA to proteins, but other information flows back to DNA, which can be changed to a new heritable state. In case of dividing differentiated cells, this inheritance could be extremely stable.

inheritance of transgenes: Genetic studies relating to the transfer of transgene(s) to following generations. Transgene(s), depending upon their number have been found to be inherited in the Mendelian fashion and can be transferred from transgenic crop varieties to other varieties through conventional cross-hybridization. In T_0 generation, the transgene is in a hemizygous condition that gets fixed following selfing. Although the expression levels are similar under hemizygous or homozygous conditions, the homozygous true breeding lines are developed for field evaluation.

inherited: A trait or gene that is passed down from parents to their offspring.

inherited sterility: Inherited sterility involves the transmission of aberrant chromosomes from the released genetically transformed insect population to the wild population. There are two methods to achieve this. (i) By treatment of laboratory-reared insect populations with a dose of radiation that is sufficient to break chromosomes, but with the frequency low enough to produce direct sterility. (ii) By rearing those insect populations that contain homozygous translocations. When the aberrant chromosomes are passed on to the wild populations, the individuals with aberrations in the heterozygous state will be sterile. The number and size of the translocations will further control the level of sterility. *Aka* DELAYED STERILITY, PARTIAL STERILITY *or* F_1 STERILITY.

inherited sterility, advantages of: (i) The partially sterile released individuals have improved reproductive competitiveness in comparison with the fully sterile individuals used in Sterile Insect Release Method (SIRM). (ii) The F_1 individuals are produced in the wild population. So, insect rearing facilities can be engaged for the higher production of the individuals for primary release.

inhibition: The destruction of the biological function of an enzyme or system by chemical or physical methods.

inhibitor: As it pertains to DNA manipulation *in vitro*, contamination in the preparation or sample can inhibit the biochemical processes involved in the manipulation, generally polymerases. Can be endogenous (from the sample matrix) or exogenous.

inhibitors of apoptosis (IAP): A group of negative regulators of both caspases and cell death. Although best known for their ability to regulate caspases and cell death, it is now clear that these function as arbiters of diverse biological processes.

inhibitors of nucleic acid precursors: Inhibitors of nucleic acid precursors that induce mutations are 5-aminouracil, azaserine, urethane, benzimidazole, caffeine, 8-ethoxycaffeine, 8-mercaptopurine, 5-nitroquinoxaline, paraxanthine, tetramethyl uric acid and theobromine. 5-aminouracil and 8-ethoxycaffeine inhibit thymine while azaserine, benzimidazole, caffeine, 8-mercaptopurine, 5-nitroquinoxaline, paraxanthine, tetramethyl uric acid and theobromine inhibit purines.

inhibitory epistasis: One gene when dominant is epistatic to the other, but the other when recessive is epistatic to the first. These two genes give 13:3 phenotypic ratio in F_2 progeny.

inhibitory G protein (Gi): G protein that can regulate ion channels and inhibit the enzyme adenylyl cyclase.

inhibitory neurotransmitter: Neurotransmitter that opens transmitter-gated Cl^- or K^+ channels in the postsynaptic membrane of a nerve or muscle cell and thus tends to inhibit the generation of an action potential.

initial inoculum: The inoculum usually obtained from an overwintering source that initiates disease in the field, as opposed to an inoculum that spreads disease during the season. *Aka* PRIMARY INOCULUM. *Contrasts* SECONDARY INOCULUM.

initiation codon: The first codon in translation. Initiation codon is mostly but not exclusively AUG as in certain cases, UUG or GUG also act as initiation codons. At initiation positions, UUG, AUG and GUG are recognized by $tRNA_f^{met}$. At internal positions, UUG is recognized by $tRNA^{leu}$. AUG is recognized by $tRNA^{met}$, and GUG is recognized by $tRNA^{val}$. *Aka* START CODON.

initiation complex in prokaryotes: The formation of translation initiation complex is the first step in translation initiation. It necessitates mRNA, 30S subunit of ribosome, a special initiating species of amino-acyl-tRNA that apparently starts all polypeptide chains and three initiating protein factors (IF_1, IF_2 and IF_3). The first amino acid incorporated is formylated at N-methionine. At this stage, the formation of 30S initiation complex is complete. After the entry of the 50S subunit of the ribosome at the scene, the 70S initiation complex is moulded.

initiation factors: Protein molecules that play an ancillary role in the initiation stage of translation.

initiation factors in eukaryotes: Initiation factors required for eukaryotic translations are eIF1, eIF1A, eIF2, eIF2B, eIF3, eIF4A, eIF4B, eI4E, eIF4F, eIF4G, eIF4H, eIF5, eIF5A and eIF5B. *See* TRANSLATION INITIATION IN EUKARYOTES.

initiation factors in prokaryotes: Protein molecules that play an ancillary role in the initiation stage of prokaryotic translation (IF1, IF2 and IF3). Initiation factors stimulate the proper association of ribosomes with messenger RNA, which is required for the initiation of protein synthesis.

initiation of bacterial DNA replication: DnaA is an origin-binding protein. It binds cooperatively to the four 9-bp repeats in *Ori*C region. The initial closed complex contains the origin DNA wrapped around an assembly of 10–20 monomers of DnaA complexed with ATP. An open complex is formed when the three AT-rich 13-bp repeats in OriC unwind as a consequence of the DNA wrapping around the assembly of DnaA. The OriC plasmids must be negatively supercoiled in order to form open complexes; the relaxed OriC plasmids can form only closed complexes. DnaA then guides the DnaB (helicase) hexametric protein from a DnaB-DnaC complex in a solution to its place around each strand. The DnaB unwinds the DNA strands using ATP energy and moves processively (i.e. it does not leave the DNA until replication is complete). DnaB encircles the DNA strand in the 5′→3′ direction along the DNA. DnaA together with the use of ATP energy is required to load the DnaB (helicase) on to the DNA in the form of a DnaB-DnaC complex. After loading the DnaB on to the replication fork, DnaC is released from the DnaB-DnaC complex and leaves the DNA. Thus, DnaC is needed for loading DnaB on to the DNA. The DnaB helicase activity is enhanced more than tenfolds by making contact with the DNA polymerase III holoenzyme. Helicases and gyrases (topoisomerases) are involved in the exposure of the two DNA strands. Strand separation, in case of double helical DNA, is the first step in DNA replication. The three proteins required in strand separation are helix destabilizing protein (HDP), unwinding protein and gyrase. Helicases use the energy of nucleoside 5′-triphosphate (NTP) hydrolysis to translocate along a nucleic acid strand and catalyse reactions such as DNA unwinding. Helicases unwind the double helix by providing a single strand so the DNA polymerase can work upon it. It uses the energy of NTP hydrolysis to break the H-bonds. These are processive in nature. It translocates along 5′→3′ or 3′→5′. Their activity is affected by the presence of SINGLE-STRAND BINDING PROTEINS (SSBs) and the concentrations of TOPOISOMERASES. Topoisomerases relax the topological stress in the positively supercoiled regions of the DNA. These change the linking number of DNA by lysing. These enzymes are of two types – topoisomerase I produces transient nicks in only one strand of DNA while topoisomerase II produces transient nicks in both the strands of DNA. It revolves one cut end by 360° before re-joining the two strands.

initiation of eukaryotic DNA replication: For eukaryotic DNA replication to start, histones are displaced to one side. The initiation of replication needs attachment of the tumour antigen (T-ag) to the helix for unwinding it and single-strand binding (SSBs) proteins to stabilize single-stranded DNA, DNA polymerase α primase to synthesize RNA primer and the polymerization of deoxyribonucleotides during initiation and the synthesis of Okazaki fragments and DNA polymerase δ-PONA for the polymerization of nucleotides on the lagging strand of DNA.

initiation of genetic exchange: Genetic exchange is initiated by a double-strand break. Then, one of the single strands migrates to the other duplex so that the duplex molecules become connected by a stretch of heteroduplex DNA. Recombination is initiated when an endonuclease makes a double-strand break in one chromatid, the 'recipient'. The cut is enlarged to a gap, probably by exonuclease action. The exonucleases nibble away one strand on either side of the break, generating 3′ single-stranded termini. One of the free 3′ ends then invades a homologous region in the other, 'donor', duplex.

initiation of patterns: The initiation of patterns begins with the egg. Differences, such as those imposed by the uneven deposition of maternal substances in the egg cytoplasm or by external factors such as GRAVITY, help to initiate localized differences in development. Superimposed upon these differences are new localized influences that occur once development is initiated, such as the effect of substances that diffuse from particular points, or differentiated cells that migrate from special areas. Their effects cause further differentiation, which, in turn, causes the evocation of further inducer substances and still further differentiation. It is thus self-evident that a particular pattern seen in an individual does not arise *de novo* but is based upon previous patterns. It is because the previously existing patterns are not identical to those that arise later that development is considered epigenetic.

initiation of reverse transcription: During reverse transcription, the synthesis of DNA requires the transfer of nascent DNA strand twice between templates. The synthesis of negative (minus) strand is initiated at 3′-end of U5 sequence of viral RNA with the help of the host template RNA primer. The first jump occurs and forms a bridge-like structure with the 3′-end of the second RNA subunit extended over the major expanse of the genome. The synthesis of positive (plus) strand starts at the 5′ boundary of U3 at the purine-rich tract. AATG and its extension are also limited by the amount of available template. For the second jump to occur, the duplex can be formed with the 3′ end of the negative strand. This results in the extension of both strands, which produces a blunt-ended linear duplex of the DNA molecule.

initiation of translation in eukaryotes: In eukaryotes, only AUG is recognized as the initiation codon. Also, methionine is not formylated in eukaryotes. $tRNA^{met}$ is present in eukaryotes compared to $tRNA_f^{met}$ in prokaryotes. In eukaryotes, there are at least eight initiation factors (eIF1, eIF2, eIF3, eIF4A, eIF4B, eIF4C, eIF4D and eIF5). Their roles are not yet clearly understood.

initiation of translation in prokaryotes: Initiation involves reactions prior to formation of the peptide bond between the first two amino acids of the protein, which is a relatively slow step in protein synthesis. The 30S initiation complex is first formed. It requires mRNA, 30S subunit of ribosome, a special initiating species of amino-acyl-tRNA (that apparently starts all polypeptide chains) and three initiation factors – IF_1, IF_2 and IF_3. The first amino acid incorporated is formylated at N-methionine. At this stage, the formation of the 30S initiation complex is complete. Then the 70S initiation complex is formed. Three functional sites exist on the 50S subunit: (i) acyl ('A') site for the attachment of aminoacyl (amino acid tRNA); (ii) peptidyl ('P') site for the growing peptide chain attached to tRNA, F-met-$tRNA_f^{met}$ sits in the P site and (iii) exit ('E') site, involved in the exiting of the tRNA from the ribosome. The ribosome moves in a 5′ to 3′ direction, synthesizing polypeptide in an H_2N- to -COOH direction. In bacteria, RNA transcription and translation occur simultaneously.

initiation site: The site on the DNA molecule where the synthesis of RNA (transcription) begins.

initiator proteins: Proteins that identify and bind to the origin of replication on a replicon and take part in PROMOSOME formation.

initiator tRNA: Special $tRNA^{f\text{-}met}$ that initiates translation in prokaryotes, which carries the amino acid formylated methionine (in prokaryotes). $tRNA^{met}$ initiates translation in eukaryotes, which carries methionine. The formylated methionine or methionine, which is essential for such initiation, is removed from most proteins before or shortly after protein synthesis is terminated. *Aka* INITIATING TRNA.

innate immune response: The immune response (of both vertebrates and invertebrates) to a pathogen that involves the pre-existing defences of the body – the innate immune system – such as the barriers formed by skin and mucosa, antimicrobial molecules and phagocytes. Such a response is not specific for the pathogen.

inner cell mass (ICM): Pluripotent cells in very early developing embryo, when the embryo is merely a ball of cells. *Aka* EMBRYOBLAST.

inner membrane: The innermost of the two membranes surrounding an organelle. In the mitochondrion, it encloses the matrix and contains the respiratory electron transport chains.

inner nuclear membrane: The innermost of the two nuclear membranes. It contains binding sites for the chromatin and the nuclear lamina on its internal surface.

inoculate: (i) To place inoculum in an infection court. (ii) To insert a pathogen into healthy tissue.

inoculation: The process of applying inoculum to a host. The inoculum must be on some part of the host that can be invaded. This part of the host body is called the infection court. A repeating cycle is a series of secondary infections that endure for a specific period of time during the growing season. For example, the addition of effective rhizobia (bacteria) to legume seeds prior to planting for the purpose of promoting nitrogen fixation.

inoculum: Explant, callus, aliquot of cell suspension culture, microbial culture, inoculated for *in vitro* culture and multiplication. Inoculum is a small tissue or aliquot of a suspension or a microbial colony taken for inoculation and raised in an *in vitro* culture for further multiplication. For instance, in plants, if the leaf is excised from the donor plant and cultured, then the leaf is the inoculum. In the case of bacterial culture, the bacterial colony transferred to a liquid medium serves as the inoculum.

inoculum density: A measure of the number of propagules of a pathogenic organism per unit area or volume. *See* PROPAGULE.

inosine (I): A deaminated product of adenosine that occurs in tRNA (anticodon) and can pair with A, U or C in the codon in mRNA, resulting in a wobble during a codon-anticodon interaction.

inositol phospholipids: One of a family of lipids containing phosphorylated inositol derivatives. Although minor components of the plasma membrane, these are important in signal transduction in eukaryotic cells. *Aka* PHOSPHOINOSITIDES.

input: After designing an algorithm, the algorithm is given the necessary and desired inputs.

input device: A device that transfers information from the outside world to the computer (e.g. a keyboard).

input load: The load of inferior alleles in a gene pool. This is caused by mutation and immigration.

input/output (I/O, io or IO): The communication between an information processing system, such as a computer, and the outside world, possibly by a human or another information processing system. Inputs are the signals or data received by the system and outputs are the signals or data sent from it. The term can also be utilized as part of an action. To 'perform I/O' means to accomplish an input or output operation.

insect: A member of the class Hexapoda (phylum Arthropoda) possessing three sets of limbs attached to a central body segment.

insect pest control through virus-induced gene silencing (VIGS): VIGS is one of the enormous reverse genetics tools for the analysis of gene function that uses viral vectors carrying a target gene fragment to produce dsRNA, which triggers RNA-mediated gene silencing. Insect-specific viruses can be exploited as vectors for VIGS to control insect pests.

insect pest control, simply inherited mutations in: Genetic mutations in a single gene may cause decreased fitness of the individuals in the wild populations if the gene could be transferred to the wild populations in enormous numbers. These mutations could be recessive lethal mutations, like changes in eye colour to increase or decrease light sensitivity, deformities in antennae, legs or wings and changes in pupal or adult body colour. G. Farnz in 2005 achieved autocidal control of *Ceratitis capitata* by using genetic sexing methods by a Y-autosome translocation and using a temperature-sensitive lethal (*tsl*) mutation.

insect pest management, population genetics methods of: By including the analysis of adaptive history and the responses of pest species, scientists hope to discover what exactly makes a species a good invader and how to best predict the future susceptibility of an environment to invasion by an insect. The better a population of an invasive species is understood from a genetic perspective, the more informed management efforts can be made and it is more likely that insect problems can be addressed.

insect resistance development through Cry proteins: Can be done by (i) combining two or more kinds of Cry proteins, (ii) altering two or more kinds of Cry proteins, (iii) reducing the selection pressure on insects by limiting *Cry* gene expression to only the economically important plant parts, (iv) introducing *Cry* gene expression only when insect starts feeding on the plant, (v) lowering the levels of expression of *Cry* genes, (vi) the premature termination of transcription, (vii) the splicing of *Cry* gene mRNA, (viii) mRNA instability elements and (ix) biased codon use. Cry proteins do not bioaccumulate causing any delayed effect. Although there are reports of soil binding under certain conditions, the bound Cry proteins are also reported to be rapidly degraded by microbes upon elution from soil.

insect resistance in plants: To minimize crop damage by insects, mites and nematodes, farmers use synthetic pesticides extensively, which cause severe effects on human health and the environment. The transgenic technology provides an alternative and innovative method to improve pest control management that is eco-friendly, effective, sustainable and beneficial in terms of yield. This involves genetic incorporation of toxic gene (the production of which is lethal to insects) in to the plant. This kills the insects without using dangerous insecticides. Genes available for genetic engineering of crop plants for pest resistance were *Cry* genes (popularly known as *Bt* genes) derived from the bacterium *Bacillus thuringiensis*. These *Cry* genes are specific to a particular group of insect pests, and are not harmful to other useful insects like butterflies and silk worms. Transgenic crops (e.g. cotton, rice, maize, potato, aubergine, cauliflower, cabbage, etc.) with *Bt* genes have been developed and such transgenic varieties have proved effective in controlling the insect pests, and it has been claimed worldwide that it has led to significant increase in yield along with dramatic reduction in pesticides use. The most notable example is *Bt* cotton (which contains *Cry1Ac* gene) that is resistant to a notorious insect pest bollworm (*Hellicoperpa armigera*). *Bt* cotton was adopted in India in 2002.

insect resistance in transgenic plants: The genes for *Bt* toxins and genes for Protease Inhibitors have been transferred to develop insect resistance in transgenic plants.

insect resistance management (IRM): The most effective strategy to combat insecticide resistance is to do everything possible to prevent it from occurring in the first place. To this end, crop specialists recommend IRM programmes as one part of a larger IPM approach covering three basic components: monitoring pest complexes in the field for changes in population density, focusing on economic injury levels and integrating multiple control strategies. An approach for delaying the development of pesticide resistance by maintaining a part of the pest population in a shelter that is free from contact with the insecticide. For *Bt* crops, this enables the insects feeding on the *Bt* toxin to mate with insects not exposed to the toxin produced in the plants.

insect resistance to Cry proteins: Several insects have developed a resistance to Cry proteins. For example, *B. thuringiensis* formulations developed resistance against Diamondback moth, and several species of Lepidoptera, Coleoptera and Diptera. Recent evidence has indicated that transgenic plants expressing *Cry* genes markedly delay the development of resistance as compared to spraying *B. thuringiensis* formulations. The resistance to Cry proteins may be due to the reduced binding of the toxin to the midgut epithelium or the complete degradation of Cry proteins.

insect-protected crops: Plants with the ability to withstand, deter or repel insects, thereby preventing them from feeding on the plant. The traits (genes) determining resistance may be selected by plant breeders through cross-pollination with other varieties of this crop or through the introduction of genes such as *B. thuringiensis* (*Bt*) through genetic engineering.

insect-resistant crops: Plants with the ability to endure, dissuade or prevent insects and thereby avert them from feeding on the plant. The traits (genes) determining resistance may be carefully chosen by plant breeders through cross-pollination with other varieties of this crop or through the introduction of novel genes such as *Bt* genes through genetic engineering.

insect-resistant plants: Plants that exhibit resistance to insect pests. Various kinds of insect attack crops in field, orchards and forests. These insects cause heavy losses in the yield and quality of the crop. One way to control insects is through chemical control in which insecticides are applied to plants/field to kill the insects. But these methods add to the cost of cultivation, cause environmental pollution and are hazardous to humans and animals. Another strategy is host plant resistance that involves the development of insect-resistant crop varieties that are not attacked by insects and thus require little to no insecticide. Using transgenic approach, the development and commercialization of *Bt* cotton is the best example of insect-resistant crop that can resist the attack of American bollworm, a serious insect pest of cotton.

insect-resistant transgenic plants: Insect-resistant plants contain either a gene (e.g. *Cry* genes) from the bacterium *B. thuringiensis* or some other genes.

insecticide: A class of crop protection and specialty chemicals used to control or kill insects on farms and forests, as well as non-agricultural applications such as residential lawn care, golf courses and public properties.

insecticide resistance: The development or selection of heritable traits (genes) in an insect population that allow individuals expressing the trait to survive in the presence of levels of an insecticide (biological or chemical control agent) that would otherwise debilitate or kill this species of insect. The presence of such resistant insects makes the insecticide less useful for managing pest populations.

insectivorous: Feeding largely or exclusively on insects.

inseparable disciplines in biology: Evolution, genetics, development (embryology) and cell biology are inseparable disciplines. Genetics explains how developmental differences are inherited, the processes of evolution and development. Evolution explains evolution from genes to complex organisms. Evolution acts on phenotypes and has a genetic basis. Development explains how a genotype is interpreted into a phenotype. Cells play a key role in development. Development is genetically as well as epigenetically controlled. Thus, these four disciplines of life sciences are integrated and only an integrated approach will help in the growth of development and evolution as sciences.

insert: A section of foreign DNA introduced into a phage, plasmid or another vector DNA.

insert orientation: Descriptive of the situation in which the vector and the inserted fragment of DNA have the same or opposite orientation.

insertion: A chromosomal segment is displaced (two breaks) and relocated into a different chromosomal region (necessitating a third break). The orientation of the inserted segment may be retained in its original or inverted orientation. An insertion may involve different chromosomes or may be intrachromosomal. It usually disrupts the gene's normal function. *See* CHROMOSOME, DNA, GENE *and* MUTATION.

insertion (IS) elements: Sequences of DNA that specify the location of insertion of different episomes into the genome. The IS elements are small (about 750–1400 bp long) in simple transposons found in bacteria and archaea. The target DNA repeats and terminal repeats are variable in length in different transposons. These have the ability to transpose. These promote recombination between non-homologous chromosomes. IS elements encode only genes for mobilization and insertion of its DNA. These contain two inverted repeats and the transposase at the centre. These have TERMINAL INVERTED REPEATS (TIRs) and generate flanking direct repeat at insertion site. These are 768–5 kb in length and are an autonomous unit – these sponsor their own transposition. The recognition of the ends of an inverted repeat (IR) is critical in transposition; point mutations abort it. *See* ILLEGITIMATE RECOMBINATION, TRANSPOSABLE ELEMENT, TRANSPOSABLE GENETIC ELEMENT.

insertion mutagenesis: Alteration in gene action owing to an insertion event that either changes a gene directly or disturbs regulatory mechanisms.

insertion mutation: A mutation caused by the insertion of an extra base or mutagen between two successive bases in DNA.

insertion point: The point at which one organ or structure (such as a leaf) is joined to the structure (such as a stem) that bears it.

insertion region: The centromere, spindle attachment or kinetochore; a non-staining, localized region in each chromosome to which the spindle 'fibre' appears to be attached at metaphase. The centromere remains single for some time after the rest of the chromosome has divided and, at anaphase, starts to move towards the pole before the rest of the chromatid.

insertion sequences: *See* TRANSPOSABLE ELEMENTS, IS ELEMENTS.

insertion site: (i) A unique restriction site in a vector DNA molecule into which foreign DNA can be inserted. (ii) The position of integration of a transposon or IS element. *Aka* CLONING SITE.

insertion sort: A simple cataloguing algorithm that shapes the final sorted array (or list) one item at a time.

insertion vector: Bacteriophage λ vector that contains a unique cleavage site in a non-essential region whereby foreign DNA with a size of 5–11 kb may be inserted. *Contrasts* REPLACEMENT VECTOR where a piece of DNA is replaced in the process of cloning.

insertion/deletion (InDel) polymorphisms: DNA sequence variations that involve the insertion or deletion of one or more nucleotides.

insertion/deletion (InDel): Following the creation of a DOUBLE-STRANDED BREAK (DSB) by Cas9, the cell initiates DNA damage repair. Repair by the error-prone NONHOMOLOGOUS END-JOINING (NHEJ) pathway can result in small insertions and/or deletions at the site of cleavage. These indels can cause frameshift mutations or premature stop codons resulting in a genetic knock-out.

insertional duplication: Inserting additional homologous base pairs into a recipient genome, which produces a mutation (by duplicated segments).

insertional inactivation: In this technique, the foreign DNA is cloned in a restriction site that lies within the coding sequence of a gene of the vector (e.g. amp^R gene). The insertion interrupts the gene's sequence and the function of the gene is no longer expressed. It permits the detection of the foreign DNA following transformation. For example, cloning in *Pst*I site of plasmid pBR322 inactivates amp^R gene but leaves the tet^R gene intact. Following transformation, recombined molecules will have tet^R amp^S phenotype.

insertional mutagenesis: Transformation approach for the identification and isolation of new genes. Insertional mutagenesis deals with the insertion of foreign DNA segment in the genome that causes change in the phenotype after integration. The insertion is done either through transposon mutagenesis or T-DNA mutagenesis. The insertion generally leads to a loss of function, mutations or 'gene knock-outs' and only rarely leads to the gain of function mutations. A change in phenotype helps in the identification of the gene. It has a potential of cloning QUANTITATIVE TRAIT LOCI (QTL). Insertional mutations occur naturally, mediated by virus or transposon, or can be artificially created.

insertional mutation: *See* INSERTION.

insertional RNA editing: It involves the editing of kinetoplastid mitochondrial mRNA via base pairing to specific guide RNAs with poly(U) tails. The mechanism includes the identification of editing sites in the mRNA by mis-matches with the guide RNA, cleavage of the mRNA and re-ligation of the mRNA halves with added or deleted U residues by TUTase. This process results in the insertion/deletion of multiple uridine residues in trypanosome kinetoplast mRNA by a transcriptional strand slipping mechanism.

insertional translocation: The insertion of a segment from one chromosome into another non-homologous one.

insertions/deletions/rearrangements: Insertions and deletions within a protein-coding sequence can disturb protein function. Insertions or deletions whose length is not a multiple of 3 cause a frameshift mutation, meaning that incorrect amino acids will be incorporated downstream from the mutation. Such rearrangements in a protein-coding sequence most probably cause loss of function mutations.

inside marker: The middle locus of three linked loci.

insignificant mutation: Referring to a gene variant that has no impact on the health or function of an organism.

instantaneous speciation: When a single individual is reproductively isolated from the species to which the parents belong and is reproductively and ecologically capable of establishing a population that forms a new species.

Institutional Biosafety Committee (IBC): The IBC in India is responsible for reviewing projects that involve, but are not limited to, recombinant DNA, RNAi, pathogens, human materials and other potentially infectious material, as well as transgenic animals. The IBC provides recommendations to the intramural community in matters pertaining to the control of biohazards associated with the use of microbiological agents and their vectors.

institutional plant breeding: Refers to plant breeding conducted or directed by large institutes. This kind of breeding is usually cost-intensive as it aims to develop cultivars with a wide climatic adaptability.

instruction cycle: The cycle that a CENTRAL PROCESSING UNIT (CPU) follows to process instructions from the boot-up until the computer has shut down. It is composed of three main stages: the FETCH STAGE, the DECODE STAGE and the EXECUTE STAGE. *Aka* FETCH-DECODE-EXECUTE CYCLE or simply FETCH-EXECUTE CYCLE.

instructive induction: In this case, responding cells adopt different fates in the presence or absence of the inducer.

instructive tissue interaction: Instructive tissue interactions are those in which cells actually change their behaviour. This is called INDUCTION.

insulators: *Cis*-regulatory elements (regions of DNA) that work indirectly by interacting with other nearby *cis*-regulatory modules to separate (or insulate) the expression of one region from that of the next. *Aka* INSULATOR ELEMENT.

insulin: Polypeptide hormone that is secreted by β cells in the pancreas and helps regulate glucose metabolism in animals.

insulin/IGF-1 signalling pathway: This pathway influences longevity, reproduction and diapause in many organisms. Because of the fundamental importance of this system in animal physiology, the question is when during an animal's life is it necessary to regulate these different processes. This pathway acts during adulthood to relatively advanced ages to influence ageing. In contrast, it regulates diapause during development. In addition, the pathway controls longevity and reproduction independently of one another. Lifespan regulation can be dissociated temporally from phenotypes that might seem to decrease the quality of life.

insulin-expressing beta cells: Insulin-expressing beta cells, found in pancreatic islets, are capable of generating more beta cells even in the adult. Fibroblast-like cells derived from adult human islets donated post-mortem proliferate readily *in vitro*. These mesenchymal-type cells, which exhibit no hormone expression, can then be induced to differentiate into hormone-expressing islet-type cell aggregates, which re-establishes the epithelial character typical of islet cells.

insulin-like growth factor II: IGF-II is a mitogenic polypeptide for most of the cells. The IGF-II gene uses at least three promoters and expresses several transcripts in many tissues during the embryonic and neonatal periods, whereas its expression in adult animals is confined to the choroid plexus and the leptomeninges through disrupting one of the alleles by gene targeting. The progeny is heterozygous and smaller compared to its wild-type counterparts. Heterozygous animals are apparently normal and fertile. This provides the first direct evidence for a physiological role of IGF-II in embryonic growth.

intact peptide ions, analysis of: This type of analysis lets masses of intact peptides to be calculated for identifying proteins in a sample by database searches.

intasome: Nucleoprotein structure formed by phage-encoded integration (Int) protein and *att*P site required for pairing with *Escherichia coli att*B site and site-specific recombination.

integer: A datum of integral data type; a data type that signifies some range of mathematical integers. Integral data types may include different sizes and may or may not be permitted to contain negative values. Integers are commonly embodied in a computer as a group of binary digits (bits). The size of the grouping varies so the set of integer sizes available varies between different types of computers. Computer hardware, including virtual machines, nearly always provide a method to represent a processor register or memory address as an integer.

integral host factor (IHF): A protein required for the precise formation of intasome. It is required both for integration and excision through site-specific recombination.

integral membrane protein (IMP): A protein that is seized tightly in a membrane that can only be detached by treatments that interrupt the lipid bilayer.

integrase: A viral enzyme that catalyses site-specific recombination involving the insertion of the genome of a temperate bacteriophage into a bacterial host chromosome.

integrated development environment (IDE): A software application that offers comprehensive services to computer programmers for software development. An IDE normally comprises at least a source CODE EDITOR, BUILD AUTOMATION TOOLS and a DEBUGGER.

integrated forest management: Forest management programmes that house the need to harvest timber from landscapes, while also supporting other values, such as hunting wildlife, outdoor recreation and biodiversity.

integrated genetic maps: Maps prepared by combining the maps of different molecular markers. Integrated genetic maps have been prepared by integrating microsatellite maps into RESTRICTION FRAGMENT LENGTH POLYMORPHISM (RFLP) maps in rice, barley, tomato, potato, maize and soybean. These maps are more dense than individual maps.

integrated nutrient management (INM): Nutrient management in the soil using different approaches. Intensive crop cultivation has deteriorated the soil health; the use of chemical fertilizers alone is not good for the soil health. Therefore, an integrated approach involving chemical fertilizers, biofertilizers, farm yard manure (FYM), crop residues, vermicompost and so on is being used for effective nutrient management.

integrated pest management (IPM): The broad-based, coordinated, safe and economical approach that assimilates a range of practices for the economic control of pests. IPM aims to overpower pest populations below the ECONOMIC INJURY LEVEL (EIL). Different pest control methods are classified as cultural, biological, genetic, chemical, biotechnological, biopesticides, *Bt* transgenic varieties, insect-resistant varieties, clean cultivation, sanitary, baits and traps.

integrating cryptic genetic variation (CGV) into plant breeding programmes: In crop plants, different threshold is required to express the new state of character as not all alleles are universally penetrant; some show dominance while others reveal epistasis or G × E relationship. Living systems are crafted in a robust way that makes the trait persistent under perturbations through canalization or accumulation of genetic variation without having a phenotypic effect. Genetic vulnerability affects the standing genetic variation due to long-term directional selection. Therefore, to adapt to novel environments, CGV can play a significant role and contribute in adaptive evolution. This in turn will open up an avenue for standing genetic variation either with beneficial or deleterious effects to respond to selection. That will lead to two forms of phenotypes in the population: one will be pre-adapted to new conditions and another part will carry the burden of deleterious alleles known as maladaptive phenotypes.

integration: The physical covalent insertion of DNA (like a provirus) into the host cell genome. The process is used by retroviruses, where a specific enzyme catalyses the process or can occur at random sites with other DNA (e.g. transposons).

integration efficiency: The frequency with which a particular DNA sequence is inserted into the genome of a recipient bacterium following transformation, or into a recipient animal or plant cell or ovum following its artificial introduction.

integration host factor (IHF): A protein required for the precise formation of intasome. It is required both for integration and excision through site-specific recombination.

integration of an SMG and GOI on unlinked genomic loci: The various systems for integration of a SELECTABLE MARKER GENE (SMG) and GENE OF INTEREST (GOI) on unlinked genomic loci during transformation so that the SMG can be segregated away from the GOI are as follows: COTRANSFORMATION-BASED SYSTEM, CO-TRANSFORMATION AND SEGREGATION, TRANSPOSON-BASED SMG REMOVAL and MULTI-AUTOTRANSFORMATION (MAT) SYSTEM. *See details as separate terms.*

integration of flowering signals: The integration of flowering signals occurs through the genes *FLC*, *SUPPRESSOR OF OVEREXPRESSION OF CONSTANS 1* (*SOC1*), *FT* and *LFY*. The key floral integrator is the *FLC* expression upregulated by *FRI*, *FRL1* and H3K4 trimethylation mediated by the PAF1 complex (ELF7, ELF8 and VIP4). Autonomous and vernalization pathways repress the FLC expression. The function of *FLC* is to repress the 'flowering-time integrators' *SOC1* and *FT*, while the photoperiod pathway enhances the expression of these integrators. The integrators *SOC1* and *FT* activate the meristem identity genes *LEAFY* and *AP1* to initiate flowering. The linking signal *SOC1* initiates flower development through long day, autonomous, vernalization pathway, plant age and gibberellin signals. The positive regulation of *SOC1* and *AGL24* in the shoot apex results in the activation of the meristem identity gene; *LFY* for the floral organ development through the activation of class A genes. These genes are also involved in floral repression through the activation of class B and C genes to repress the *SEP3* gene. The major function of *SOC1* and *AGL24* is floral induction and flower development at a specific time. CO-FT induction and floral evocation occur.

integration protein: λ phage integrase from phage λ (Intλ) that binds with the *attP* site on DNA to form a nucleoprotein structure called INTASOME necessary for integration and excision through site-specific recombination with bacterial chromosome. *Aka* INT PROTEIN.

integration testing: A phase in software testing in which individual software modules are united and evaluated as a group. Integration testing is done to evaluate the compliance of a system or component with specified functional requirements. Integration testing is done after unit testing and before validation testing. Integration testing involves taking modules that have already undergone unit testing, grouping them into larger aggregates and applying tests outlined in an integration test plan to these aggregates. The result is an integrated system that is prepared for system testing. *Aka* INTEGRATION AND TESTING (I&T).

integrative analysis of haplotype-resolved epigenomes: The resultant huge assembly of haplotype-resolved epigenomic maps exposes extensive allelic biases in both chromatin state and transcription, which reveal considerable variation across tissues and between individuals and enable us to examine *cis*-regulatory relationships between genes and their control sequences.

integrative biology: Refers to various forms of cross-disciplinary and multitaxon investigation.

integrative transformation, proof of: Refers to providing suitable proof that transgenic plants have been produced. Proof of integrative transformation requires controls for treatment and analysis, a tight correlation between treatment and predicted results, a tight correlation between physical (e.g. Southern blot) and phenotypic (e.g. enzyme assay) data, complete Southern analysis (containing the predicted signals in high-molecular-weight DNA, including hybrid fragments between host DNA and foreign gene, and the presence of the complete gene and evidence for the absence of contaminating DNA fragments or identification of such fragments), data that allow discrimination between false positives and correct transformants in the evaluation of the phenotypic evidence, correlation of the physical and phenotypic evidence with transmission to sexual offspring, and molecular and genetic analysis of offspring populations.

integrator gene: The integrator gene is comparable to the regulator gene. A component of eukaryotic gene regulation that responds to the signals provided by SENSOR GENE. These genes are transcribed to produce ACTIVATOR RNAS.

integrator site: In eukaryotic gene regulation, integrator sites transcribe activator RNA molecules in reply to the activation of a sensor site. The integrator site is akin to the prokaryotic regulator gene and is accountable for the synthesis of an activator RNA molecule that may or may not produce a regulatory protein before it stimulates the receptor site.

integrator-mediated transcription regulation: The integrator binds RNA Pol II and the pausing factors DRB sensitivity-inducing factor (DSIF) and negative elongation factor (NELF) to exclude binding of the elongation factors SPT6 and PAF1 complex. The integrator also binds the C-terminal domain of RNA Pol II and positions PP2A to counteract RNA Pol II phosphorylation and elongation.

integrin: A member of a large family of transmembrane proteins implicated in the adhesion of cells to the extracellular matrix and each other.

integrin activation: Control of the integrin affinity for ligands (integrin activation) is essential for normal cell adhesion, migration and the assembly of extracellular matrix. Integrin activation is usually mediated through the integrin β subunit cytoplasmic tail and can be regulated by many different biochemical signalling pathways. Specific binding of the cytoskeleton protein TALIN to the integrin β subunit cytoplasmic tail leads to the conformational rearrangements of integrin extracellular domains that increase their affinity. Thus, regulated binding of talin to the integrin β tails is a final common element of cellular signalling cascades that control integrin activation.

integron integrases: Integron integrases are proteins that mediate site-specific DNA recombination between a proximal primary site (*att*I) and a secondary target site (*att*C) found within mobile gene cassettes encoding resistance or virulence factors.

integron recombination: Integrons are bacterial genetic elements capable of incorporating and expressing promoterless genes structured as cassettes.

integument: Any covering but particularly the covering of an ovule.

intein: An intein is a segment of a protein that is capable of excising itself and joining the remaining portions (called the exteins) with a peptide bond during protein splicing. Inteins have also been called PROTEIN INTRONS, by equivalence with RNA INTRONS.

intellectual property (IP): A category of legal property that includes intangible creations of the human intellect. There are many types of intellectual property and some countries recognize more than others. The most well-known types are copyrights, patents, trademarks and trade secrets. The patent system provides the only protection available for the intellectual products of research. There is increasing pressure on the patent system to patent life forms.

intellectual property rights (IPR): A system of patents that permits ownership over applications of research like patents, copyrights and trademarks. Industrial design protection, a special protection for new plant varieties, has been available since 2007. The International Union for the Protection of New Plant Varieties (UPOV) and the Trade Related Aspects of Intellectual Property Right (TRIPS) Agreement offer this protection. The Act of UPOV Convention in 1991 provided this important protection to plant breeders. The UPOV office is located in Geneva and coordinates the Protection of New Plant species. There is a debate in many countries that the plants developed do not satisfy the non-obvious requirement in a patent application system, as existing techniques are used and the new breed is obvious. Therefore, a unique *sui generis* system is employed for the Protection of New Plant Varieties. The TRIPS Agreement protects for 25 years in the case of trees and vines, while the protection for other plants is 20 years. There are seven forms of IPRs recognized in the TRIPS Agreement – COPYRIGHT AND RELATED RIGHTS, TRADEMARKS, GEOGRAPHICAL INDICATIONS, INDUSTRIAL DESIGNS, PATENTS, LAYOUT-DESIGNS (TOPOGRAPHICAL) OF INTEGRATED CIRCUITS and PROTECTION OF UNDISCLOSED INFORMATION. Perhaps the best-known form of IPR is patent. Examples of IPR include patents, domain names, industrial design, confidential information, inventions, moral rights, database rights, works of authorship, service marks, logos, trademarks, design rights, business or trade names, commercial secrets and computer software.

intellectual property rights (IPR) – key to new wealth: One-of-its-kind multimedia interaction guide on a CD-ROM. It unravels the mysteries of converting intangible intellectual property into tangible assets. This product has been developed with support from the Department of Scientific and Industrial Research.

intelligence quotient (IQ): The mental age of an individual (as measured by standardized tests) divided by his or her real age and multiplied by 100. This formulation establishes the average IQ as 100. The usefulness and reliability of IQ as a measure of intelligence has been questioned, in part because of the difficulty of devising standardized tests that are free of cultural biases.

intelligent agent (IA): In artificial intelligence, it refers to an autonomous entity that acts, directing its activity towards achieving goals (i.e. it is an agent), upon an environment using observation through sensors and consequent actuators (i.e. it is intelligent). Intelligent agents may also learn or use knowledge to achieve their goals. These may be very simple or very complex. A reflex machine, such as a thermostat, is considered an example of an intelligent agent.

intelligent decision support system (IDSS): A decision support system that makes extensive use of ARTIFICIAL INTELLIGENCE (AI) techniques. Examples of specialized intelligent decision support systems include FLEXIBLE MANUFACTURING SYSTEMS (FMS), intelligent marketing decision support systems and medical diagnosis system.

intelligent design: The argument that organisms are irreducibly complex and so must have been created or designed by an intelligent designer, unidentified supernatural or extra-terrestrial intelligence. It is a version of the argument from design.

intense selection: Selection for the traits of interest with little or no retention of lines not having the traits of interest. In a population comprising overlapping generations, intense selection can disturb the age distribution and thus impact the rate of increase of an advantageous allele.

intensifying screen: A plastic sheet saturated with a rare-earth compound, such as calcium tungstate, which absorbs beta radiation and emits light. When placed on one side of a piece of X-ray film with a radioactive sample on the other side, the intensifying screen captures some of the beta emissions that pass through the film, blackening the X-ray film and thus greatly improving the sensitivity of the detection. An intensifying screen is valuable in Southern and Northern blotting procedures.

intensive crop: A crop that has high-profit margins and consequently justifies considerable expense in its production. Horticultural crops are classified as intensive crops, while cereals are usually counted as EXTENSIVE CROPS.

intention-to-treat: Subjects in a randomized clinical trial are analysed according to the treatment group to which they were assigned, even if they did not receive the intended treatment or received only a portion of it. If, in a randomized study, the analysis fails to assign all patients to the groups they were originally randomized into, the study can no longer be strictly considered a randomized trial, as the randomization process is effectively compromised. Intention-to-treat analyses are pragmatic in that they reflect real-world non-adherence to treatment.

inter-organelle cross-talk in acetyl-coenzyme A homeostasis and lipogenesis: Proliferating cells rely on acetyl-CoA to support membrane biogenesis and acetylation. Several organelle-specific pathways provide acetyl-CoA as nutrient availability fluctuates, so understanding how cells maintain acetyl-CoA homeostasis under such stresses is critically important. R.S. Kuna *et al.* in 2023 applied ^{13}C isotope tracing cell lines deficient in these mitochondrial [ATP-citrate lyase (ACLY)]–, cytosolic [acetyl-CoA synthetase (ACSS2)]– and peroxisomal [peroxisomal biogenesis factor 5 (PEX5)]–dependent pathways. ACLY knock-out in multiple cell lines reduced fatty acid synthesis and increased reliance on extracellular lipids or acetate. Double knock-out (DKO) of both ACLY and ACSS2 severely stunted but did not entirely block proliferation, suggesting that alternate pathways can support acetyl-CoA homeostasis. Metabolic tracing and PEX5 knock-out studies link the peroxisomal oxidation of exogenous lipids as a major source of acetyl-CoA for lipogenesis and histone acetylation in cells lacking ACLY, highlighting a role for inter-organelle cross-talk in supporting cell survival in response to nutrient fluctuations.

inter-quartile range: The range between the outer quartiles (25th and 75th percentiles). It is a measure of the dispersion of the data distribution – a central interval containing half the sample.

inter-retrotransposon amplified polymorphism (IRAP): Any difference in DNA sequence between two genomes detected by PCR-mediated amplification between two neighbouring retrotransposons using a left and right outward-facing primer complementary to the conserved regions within the LONG TERMINAL REPEATS (LTRs). Any observed polymorphism is the consequence of mutations primarily in the region between the two retrotransposons as well as in one or both long terminal repeats.

inter-simple sequence repeat (ISSR) PCR: This PCR-based method is used for DNA fingerprinting by amplifying regions between some SIMPLE SEQUENCE REPEATS (SSRs) to produce a unique fingerprint of amplified fragment lengths. It can replicate itself and catalyse chemical reactions as RNA does. It can serve as a template both for its own replication and for the formation of RNA from its subcomponents. A polycyticidine decamer of PEPTIDE NUCLEIC ACID (PNA) can function as a template for the oligomerization of activated guanosine mononucleotides.

inter-simple sequence repeats (ISSRs), uses of: (i) The technique was found to be stable across a wide range of PCR parameters. (ii) Non-anchored ISSR markers are a good choice for DNA fingerprinting. (iii) ISSRs are advantageous because no prior genomic information is required for their use. (iv) ISSRs have wide applicability in a variety of plants. (v) ISSR is a recently developed modification of SSR-based marker systems, which circumvents the requirement for flanking sequence information. ISSR analysis involves PCR amplification of genomic DNA using a single primer that targets the repeat per se, with 1–3 bases that anchor the primer at the 3′ or 5′ end. (vi) In addition to freedom from the necessity of obtaining flanking genomic sequence information, ISSR analysis is technically simpler than many other marker systems. The method provides highly reproducible results and generates abundant polymorphisms in many systems.

inter-simple sequence repeats (ISSRs): A general term for a genome region between microsatellite loci. The complementary sequences to two neighbouring microsatellites are used as PCR primers; the variable region between them gets amplified. The limited length of amplification cycles during PCR prevents excessive replication of overly long contiguous DNA sequences, so the result will be a mix of a variety of amplified DNA strands that are generally short but vary in length. Sequences amplified by ISSR-PCR can be used for DNA fingerprinting. Since an ISSR may be a conserved or a non-conserved region, this technique is not useful for distinguishing individuals, but rather for phylogeography analyses or maybe delimiting species; sequence diversity is lower than in SSR-PCR, but still higher than in actual gene sequences. In addition, microsatellite sequencing and ISSR sequencing are mutually assisting, as one produces primers for the other. Non-anchored ISSRs are arbitrary multi-loci markers produced by PCR amplification with a microsatellite primer. Polymorphisms were abundant among seven dicot species tested with two tri-nucleotide and two tetra-nucleotide primers.

inter-SSR-PCR: A technique closely related to MP-PCR (MICROSATELLITE-PRIMED PCR) that uses 5′ or 3′ anchored satellites of the dinucleotide type as PCR primers (e.g. GC[AT]$_7$). This technique is also called AMP-PCR (ANCHORED MICROSATELLITE-PRIMED PCR), during which complex fingerprint-like patterns are obtained. The banding pattern complexity and informativeness of AMP-PCR are considerably higher as compared to MP-PCR.

inter-subunit bridges: RNA-RNA and RNA-protein interactions at the interface formed upon association of the ribosomal LARGE SUBUNIT (LSU) and SMALL SUBUNIT (SSU) during translation initiation.

interaction of genes: The process by which one gene difference affects the expression of another gene difference.

interaction of transcription factors with DNA as dimers: Although the ability to dimerize is not a universal phenomenon among transcription factors, homodimerization and heterodimerization are frequently observed. Several classes of dimerization interfaces have been identified. The leucine zipper motif is the most extensively studied transcription factor. First detected in C/EBP (CCAAT/enhancer-binding protein), many proteins (e.g. Jun, Fos and Myc) have been found to contain a heptad repeat of leucines within a stretch of 30 amino acids, capable of forming an α-helix. The 'zipping' of two such helices, facilitated by the attractive interactions between the leucine residues partitioned along the hydrophobic face to each helix, facilitates dimerization. The basic regions adjacent to the zipper region make contact with the DNA.

interaction proteomics: This branch of proteomics considers the genetic and physical interactions among proteins as well as interactions between proteins and nucleic acids or small molecules. This approach provides information about cellular localization, biochemical functioning of individual proteins and also about how various proteins function in pathways, networks and complexes.

interaction theory: A theory of quantitative factor inheritance that assumes that the effect of each factor on the genotype is dependent upon all other factors present. The smaller the visible effect of a certain factor, the greater the number of factors acting in the same direction.

interaction variance (V_I): Refers to the variance in a quantitative trait caused by epistatic interactions between loci. It is the difference between the genotypic (V_G) and the additive (V_A) plus dominance variance (V_D) and is $V_I = V_G - V_A - V_D$.

interactome: Interactome is defined as the whole set of molecular interactions in cells. Molecular interactions can occur between molecules belonging to different biochemical families (proteins, nucleic acids, lipids, carbohydrates, etc.) and within a given family. When spoken in terms of proteomics, interactome refers to PROTEIN-PROTEIN INTERACTION NETWORK (PPIN), or PROTEIN INTERACTION NETWORK (PIN). Another extensively studied type of interactome is the protein–DNA interactome (network formed by transcription factors and DNA or chromatin regulatory proteins) and their target genes. The word INTERACTOME was originally coined in 1999 by a group of French scientists headed by Bernard Jacq.

interallelic complementation: The change in the properties of a heteromultimeric protein brought about by the interaction of subunits coded by two different mutant alleles; the mixed protein may be more or less active than the protein consisting of subunits only of one or the other type.

interbands: The region of chromosomes that are lightly stained. These regions mostly have unique DNA sequences.

intercalary: (i) Of chromosome, chromosomal segments located beside terminal regions. (ii) Of growth, occurring between the apex and the base of an organ. *Aka* INTERSTITIAL.

intercalary evolution: For the evolution of the eye morphogenetic pathway, a mechanism of intercalary evolution has been proposed. Since both the master control genes at the top of the gene cascade, *Pax 6* and *sine oculis*, as well as the essential photoreceptor genes of the rhodopsin family at the bottom of the hierarchy, are present already in the primitive eyes of flatworms, it has been proposed that the more elaborate eyes have evolved from these primitive eyes by intercalation or the recruitment of additional genes into the eye developmental pathway. The two best-understood genetic mechanisms for intercalation are gene duplication and subsequent divergence, and enhancer fusions.

intercalary heterochromatin: Refers to a heterochromatin other than the centromeric heterochromatin, dispersed throughout the eukaryotic chromosomes.

intercalary meristem: An internodal meristem positioned between differentiated tissues. It produces cells perpendicular to the growth axis thereby causing internode elongation.

intercalary regeneration: A type of regeneration that fills in the missing tissues upon grafting together two mismatched parts of a structure.

intercalary segments: The segments of chromosomes between the terminal segment and centromere in each arm.

intercalary translocation: A chromosome arrangement involving the transfer of a part of the chromosome to another site in the same chromosome.

intercalating agents: Hydrophobic heterocyclic ring molecules that resemble the ring structure of base pairs (e.g. ethidium bromide, acridine orange and actinomycin D). Insertion of these agents distorts the DNA double helix, thereby interfering with DNA replication, transcription and repair or causing a frameshift mutation. *Aka* INTERCALATE.

intercalation: The process in which certain drugs or dyes, such as ethidium bromide, can be inserted into DNA or double-stranded RNA between adjacent base pairs. Molecules with this property are called INTERCALATING AGENTS. The binding of such dyes reduces the buoyant density of the DNA. The DNA duplex increases in length. The insertion of these agents distorts the DNA double helix, thereby interfering with DNA replication, transcription and repair or causing a frameshift mutation.

interchange: An exchange of segments between non-homologous chromosomes resulting in translocations.

interchange complex: Association of four or more chromosomes in a translocation at diakinesis or metaphase I.

intercellular junctions: Plasma membrane, in addition to acting as a selective barrier for different biomolecules, also has another functional capacity. These serve as intercellular junctions, which provide the cells with a way to communicate with each other.

intercellular: Between or among cells.

interchange trisomic: An additional chromosome to the diploid set, made of two different chromosomes via translocation.

interchromosomal: Effects and processes between chromosomes.

interchromosomal interactions in coordinated gene expression: Some eukaryotic genes (e.g. cytokine genes), located in separate chromosomes, associating physically in the nucleus via interactions that may function in coordinating gene expression.

interchromosomal mutations: Changes in chromosome structure between two chromosomes (e.g. translocations).

interchromosomal pairing: The interchromosomal pairing between HAS1pr-TDA1pr alleles in saturated cultures of *Saccharomyces* yeast is mediated by three transcription factors, Leu3, Sdd4 (Ypr022c) and Rgt1.

interchromosomal recombination: Recombination resulting from independent assortment.

interchromosome balance: The condition in which whole chromosome are adjusted in proportions that give satisfactory development of the organism.

intercistronic region: The distance between the termination codon of one gene and the initiation codon of the next gene. *Aka* INTERGENIC SPACERS.

intercrop: To grow two or more crops at the same time on the same area of land.

intercropping: Intercropping is the exercise of growing two or more crops next to each other at the same time, with the benefit of growing more crops in a given area. This practice has found favour with farmers because it improves plant productivity and resource-use efficiency. Below-ground interactions in these species-diverse agroecosystems play an essential part in enhancing crop yields.

intercropping, interspecific root interaction improvement during: There has been an incomplete understanding of how plant roots interact to influence crop biomass, photosynthetic rates and the control of diverse proteins involved in CO_2-fixation and photosynthesis. A pot experiment involved three root-barrier treatments – full, partial and no root interactions between foxtail millet [*Setaria italica* (L.) P. Beauv.] and groundnut (*Arachis hypogea* L.) across two growing seasons. The biomass of foxtail millet and groundnut plants in the treatment favouring full root interaction in both the growing seasons was at least threefold higher in comparison to the treatment when there was no root interaction. Net photosynthetic rates were also meaningfully enhanced in millet and groundnut when full root interaction was compared with no root interaction. Not all combinations of crops in an intercropping system can produce mutual benefits. The advantages of the best intercropping combinations can probably be derived from high LIGHT-USE EFFICIENCY (LUE) aboveground and nutrients belowground (e.g. N). Intercropping farming systems significantly enhanced soil fertility, resource-use efficiency, carbon sequestration, phosphorus uptake and availability, water conservation in soils, and crop and fodder productivity. *Compare with* MONOCULTURE FARMING SYSTEMS.

intercrossing: Mating between two heterozygotes for the same allele. *Aka* INTERCROSS.

interdeme selection: Selection at the level of a local population.

interested and affected parties: People, groups or organizations that decide to become informed about and involved in a risk characterization or decision-making process. Interested parties may or may not be affected parties, who are people, groups or organizations that may experience benefit or harm as the results of a hazard or the processing leading to a decision about risk.

interface: A collective boundary across which two or more different components of a computer system exchange information. The exchange can be between software, computer hardware, peripheral devices, humans and any combination of these. Some computer hardware devices, such as a touchscreen, can send and receive data through the interface, while others such as a mouse or microphone may only provide an interface to send data to a given system.

interfacial-force microscopy (IFM): The interfacial-force microscope can measure the mechanical and adhesive properties of a single monolayer of organic molecules. Scanning microscopy can be implemented at all levels of interfacial force, both attractive and repulsive.

interference (in genetics): An effect in which the occurrence of a crossover in a certain region affects the probability of a crossover in the adjacent region. Mapping functions are of two types: (i) HALDANE'S MAPPING FUNCTION – assumes the absence of interference between crossovers in meiosis. (ii) KOSAMBI'S MAPPING FUNCTION – assumes a certain degree of interference. Such mapping functions convert recombination fractions into map units called centimorgans (cM). For recombination frequencies above 10%, Haldane mapping functions give a higher map distance than Kosambi. Interference reflects the frequency of double crossovers, which fail to be recovered in relation to their expected frequency. One crossover interferes with the coincident occurrence of another crossing-over in the same pair of chromosomes. Interference is calculated using the formula interference = 1 − Coefficient of coincidence (CC). Depending upon whether the value of the CC is zero, less than one or more than one, interference can be no interference, positive interference or negative interference. *See* NO INTERFERENCE, POSITIVE INTERFERENCE *and* NEGATIVE INTERFERENCE.

interference (in patenting biological material): When two or more patent applications or issued patents claim the same invention. For example, V. Šikšnys (20 March 2012), the J.A. Doudna/E. Charpentier team of UC (25 May 2012) and Feng Zhang of The Broad Institute (12 December 2012) all filed patent applications at the US Patent and Trademark Office (USPTO) claiming some variation of the CRISPR system using the Cas9 enzyme. Though the UC team had priority on its side, the patent application by Feng Zhang claimed that, unlike the UC application, it was the only one not restricted to simple life forms. Therein lies the essence of a hostile patent battle that lasted until October 2019.

interference competition: A form of competition that involves a fight or active interaction among organisms.

interference distance: The distance within which further crossing-overs may occur after the previous one happened away from the centromere.

interference microscope: A light microscope that measures the refractive index of microscopic objects, including parts of living cells. Like the phase microscope, the interference microscope is employed for witnessing transparent structures and provides guidance to dry mass or thickness.

interference range: The distance large enough for forming two crossing-overs without mutual interference.

interferon regulatory factors: These are the molecules that control the expression of the interferon genes by activation of kinases that phosphorylate substrate proteins called SIGNAL TRANSDUCERS AND ACTIVATORS OF TRANSCRIPTION (STATs).

interferons (IFNs): Small antiviral proteins (glycoproteins) first recognized in animals for their role in inhibiting viral replication and inducing resistance in host cells. Virus-infected cells produce and release interferon that are taken up by neighbouring healthy cells. These start producing **antiviral proteins (AVPs)**, which are enzymes that interfere with viral protein synthesis and stop the spread of infection. Interferons are virus non-specific as the same interferon is active against many different viruses. The IFNs are a highly conserved family of multi-functional, species-specific, secreted proteins originally categorized on the basis of cellular origin, including leucocyte IFN (alpha), beta fibroblastic IFN and immune gamma IFN. IFNs have been reclassified based on their recognition of cell-surface receptors.

interferons (IFNs), major classes of: In humans, there are three major classes of IFNs: (i) Type I consists mainly of the original types – alpha (including various isoforms) and beta, (ii) Type II consists of IFN gamma and (iii) Type III consists of IFN lambda. The IFNs have multiple biological roles demonstrated to a greater or lesser degree, including (i) the induction of intracellular mechanisms having antiviral effects (affecting viral protein synthesis) and anti-proliferative effects (affecting cell replication), (ii) the stimulation of cellular immune responses against viruses, bacteria and tumours and (iii) the regulation of immune and inflammatory responses.

interferon-α (IFN-α): A type I interferon belonging to a family of structurally related, species-specific proteins exclusive to vertebrates. The interaction of IFN-alpha with its receptor component results in the activation of a number of signalling pathways that are mainly involved in innate immune response against viral infection.

interferon-β (IFN-β): A type I interferon IFN-β is an approved treatment for multiple sclerosis, and antithyroid autoantibodies and thyroid dysfunction have been reported in patients who are treated with IFN-β. IFN-β has two subtypes: Interferon β-1a and Interferon β-1b. Interferon β-1a is a cytokine in the interferon family used to treat multiple sclerosis (MS). It is produced by mammalian cells. Interferon β-1b is produced by modified *Escherichia coli*.

interferon-γ (IFN-γ): IFN-γ is a cytokine secreted by certain types of T cells after activation, which enhances the antiviral response and macrophage activation.

interfission period: The period between these two divisions in unicellular organisms where development is not as well defined as in multicellular organisms. In this case, changes occur between two division cycles.

intergenerational stress memory: (i) Stress memory that extends in duration to the next generation. (ii) When only the first stress-free generation has a detectable memory effect. *Compare with* TRANSGENERATIONAL STRESS MEMORY.

intergeneric cross: Spontaneous or experimental crosses of individuals of different genera (e.g. wheat [*Triticum aestivum*] and rye [*Secale cereale*]); this cross resulted in a human-made new crop plant 'triticale'.

intergeneric hybrid: Intergeneric cross.

intergenic (extragenic) suppression: A mutation at a second locus that apparently restores the wild-type phenotype to a mutant at a first locus. Intergenic suppressors are of three types: NONSENSE SUPPRESSORS, MISSENSE SUPPRESSORS and FRAMESHIFT SUPPRESSORS. *See details as separate terms.*

intergenic changes: Mutational changes involving more than one locus (e.g. inversions, deletions, translocations and duplications).

intergenic effects: The effects and phenomena between genes.

intergenic mutation: Mutations that involve more than one gene.

intergenic region: A stretch of DNA sequences located between genes. Intergenic regions may contain functional elements and junk DNA.

intergenic regions: The stretches of DNA located between genes. In humans, intergenic regions are non-protein coding and comprise a large majority of the genome. Some intergenic DNA is known to regulate the expression of nearby genes.

intergenic sequences: The DNA sequence between the genes in the human genome.

intergenic spacers (IGS): (i) The genomic regions that are not directly involved in the final product and are observed between actual coding sectors of two adjacent genes. These spacers range in length from a single base pair to many thousand base pairs. These are frequently referred to as FLANKING REGIONS. Certain parts of these sequences reveal specialization of function. The sector of the non-coding sequence preceding 5′-end of gene proper (mature gene, antisense strand) is called LEADER, while the one that follows the 3′-end is the TRAIN. (ii) Each member of repeat unit in a ribosomal gene family has a coding region with genes that specify 18S, 5.8S and 28S rRNA molecules, a spacer region called INTERGENIC SPACERS (IGS). The variation in the number and size of the subrepeats in IGS is accountable for variation in length of the repeat unit (IGS + coding region).

intergenic suppressor: A mutation that overpowers the phenotype of another mutation in a gene other than the one in which the suppressor mutation occurs.

intergenic transcription: Transcription by RNA polymerase II in budding yeast and in humans is widespread, even in genomic regions that do not encode proteins. The function of such intergenic transcription is largely unknown, although it can be regulatory. Transcription of *SER3* is tightly repressed during growth in rich medium. The regulatory region of this gene is highly transcribed under these conditions and produces a non-protein-coding RNA (*SRG1*). The expression of SR transcription across the G1 RNA is required for the repression of *SER3*. Repression occurs by a transcription-interference mechanism in which *SRG1* transcription across the *SER3* promoter interferes with the binding of activators.

intergenic/extragenic suppressors: Mutation in one gene suppresses the effect of a mutation in another gene. Intergenic suppressors fall into six categories, namely, nonsense suppressors, missense suppressors, frameshift suppressors, physiological suppressors, codon-anticodon misreading suppressors and suppression through change in amino acid activating enzyme.

intergenotypic competition: *See* ALLOCOMPETITION.

intergroup selection: An evolutionary process resulting from the differences in growth rate or survival of different competing or non-competing groups.

interjections: Non-human vocables that reflect the individual's physical or emotional state. They do not, as true language does, name, discuss, abstract or symbolize. These are emotional signals, not a discourse.

interkinesis: A resting stage that occurs between the first and second meiotic division.

interleukin: A secreted protein or peptide produced naturally by our bodies to stimulate our immune systems. There are at least 18 known kinds of interleukins. It chiefly mediates local interactions between white blood cells (leucocytes) during inflammation and immune responses.

interlocking: The intertwisting of non-homologous chromosomes and/or chromosome configurations observed during meiotic pairing.

intermated recombinant inbreds: A different structure of mapping population – like F_2, backcross or near-isogenic lines. The intermating of F_2 individuals results in new recombination events. The intermated-recombinant-inbred populations seem to possess improved genetic resolution.

intermediary metabolism: The chemical reactions that take place in the cell that transform the complex molecules derived from food into the small molecules needed for the growth and maintenance of the cell.

intermediary trait: A plant trait controlled by a heterozygous pair of alleles, which result in an intermediate phenotype as compared to the corresponding homozygous genotypes.

intermediate complexes in eukaryotic transcription initiation: Five sets of complexes containing distinct components have been resolved through native gel electrophoresis DNA assay on the adenovirus major late promoter (Ad2MLP) by general transcription factors and RNA polymerase II. The sequential binding of transcription factors TFIID, TFIIA, TFIIB, RNA polymerase II and TFIIE generates INTERMEDIATE COMPLEXES 1, 2, 3, 4 and 5, 6 and 7, and 8 and 9. *See details as separate items.*

intermediate complexes in eukaryotic transcription initiation complex 1: The first step in transcription initiation is possibly the recognition of the TATA element by (transcription factor) TFIID. TATA element sequences −37 to −17 seem to be involved in this binding reaction.

intermediate complexes in eukaryotic transcription initiation complex 2: It comprises both transcription factors TFIID and TFIIA bound specifically to the −42 to −17 region of adenovirus major late promoter (AdMLP). That is why TFIIA is positioned on the upstream side of TFIID. Addition of TFIIA slightly improves the binding ability of TFIID to a cryptic TATA sequence at −70 than to MLP TATA element.

intermediate complexes in eukaryotic transcription initiation complex 3: After binding to transcription factors TFIID and TFIIA to TATA element, TFIIB possibly binds to the initiation complex. The sequences −10 to +10 seem to be implicated in the formation of this complex. TFIIB acts as a 'bridging' molecule between the TFIIA-TFIID-TATA element complex and RNA polymerase II.

intermediate complexes in eukaryotic transcription initiation complexes 4 and 5: The addition of purified RNA polymerase produces complexes 4 to 7. Complexes 4 and 5 represent the binding of RNA polymerase to complex 3. These complexes probably cover upstream sequences up to −47.

intermediate complexes in eukaryotic transcription initiation complexes 6 and 7: The addition of transcription factor TFIIE to complexes 4 and 5 produces complexes 6 and 7. TFIIE seems to bind downstream of the polymerase at sequences +20 to +30. The TFIIA binding then undergoes ATPase activity, which generates 'activated' transcription complexes. This causes the dissociation of complexes 6 and 7 and generates complexes 6* and 7*. In case of these newly generated complexes, downstream boundary of +30 is retarded to +25. This change may be ascribed to a helicase activity, which would unwind DNA making the template strand available to the polymerase.

intermediate complexes in eukaryotic transcription initiation complexes 8 and 9: The addition of NTPs that allow transcript elongation to +11 results in the formation of complexes 8 and 9. These complexes contain precisely initiated transcripts. Complexes 8 and 9 migrate near complexes 4 and 5, respectively. This event is consistent with the view that transcription factor TFIIE dissociates from the activated transcription complexes. TFIIE in fact may act as an inhibitor of elongation but not initiation of transcription.

intermediate components: The components of reassociation reaction that stand between fast (satellite) and slow (non-repetitive DNA) components. These components comprise moderately repetitive DNA.

intermediate filament: Fibrous protein filament (~10 nm in diameter) that forms rope-like networks in animal cells. One of the three most prominent types of cytoskeletal filaments.

intermediate host: A host required for the completion of the life cycle of a parasite, but in which it does not achieve sexual maturity. Thus, it harbours the asexual stage of the disease-causing organism. Malaria is caused by a protozoan *Plasmodium vivax*. Mosquitoes that harbour its sexually reproducing stage are the definitive hosts of *Plasmodium*, whereas the *Plasmodium* undergoes asexual reproduction in the human body. In this case, the human is the intermediate host.

intermediate inheritance: In the inheritance of quantitative character, the F_1 generation is often intermediate between the parents and the F_2. This intermediate character may apparently prevail. This phenomenon is due to polymeric genes. *See* POLYMERISM.

intermediate lesions: When there is either an addition or a deletion of one or a few nucletotide pairs from a gene. This leads to a change in the number of nucleotide pairs in a gene. These changes may cause a frameshift mutation.

intermediate vectors: The small pBR322-based plasmids (*Escherichia coli* vectors) containing a T-DNA region. These are used to overcome problems derived from the large size of disarmed Ti plasmids and their lack of unique restriction sites. The intermediate vectors are replicated in *Escherichia coli* and are transferred into *Agrobacterium* by conjugation. These cannot replicate in *A. tumefaciens* and, therefore, carry DNA segments homologous to the disarmed T-DNA to permit recombination to form a co-integrated T-DNA structure.

intermembrane space: The sub-compartment formed between the inner and outer membranes of mitochondria. and chloroplasts.

intermitosis: The resting stage that often occurs between two mitotic divisions; interphase.

intermittent contact AFM (IC-AFM): Also called the HIGH-AMPLITUDE RESONANCE MODE or TAPPING MODE.

intermolecular triple helical DNA: Structures that are easily formed with polypurine-polypyrimidine tracts. In this case, the third strand can pair in two different variations that depend on its strand nature. However, the base pairing requires protonation, or low pH of cytosine. The base pairing is by HOOGSTEEN BODING (K. Hoogstein in 1963). This type of triple-strand formation is possible during homologous mode of recombination, where one of the nicked strands from the homologous segment intercalates into a helical duplex that is paired and thus produces triple-strand structures. These helices are characterized by short oligonucleotides binding in the grooves of a longer DNA duplex. However, the base pairing requires protonation, a fundamental chemical reaction and is a step in many stoichiometric and catalytic processes.

internal balance: The harmonious epistatic interactions of genes at different loci.

internal control region (ICR): In all eukaryotic genes that are transcribed by RNA polymerase III, a small DNA sequence guides upstream inhibition by the enzyme. ICRs ensue within the DNA encoding the mature RNA. Eukaryotic tRNA gene transcription is regulated by two ICRs, the D-control region (which in the DNA codes for the D-stem and D-loop) and T-control region (which in tRNA codes for the T-loop).

internal control signals: In this case, promoter sequences are located within a mature gene itself (e.g. the genes coding for small pieces of RNA like tRNAs and 5S rRNA have internal control signals).

internal differentiation in gene: H.J. Muller (1927) reported that mutations could be induced with X-ray treatment in *Drosophila* that provided evidence that the gene is particulate in nature. New alleles of a gene are produced by mutations. The discovery of multiple allelism revealed that a gene has multiple mutable sites indicating that the gene is differentiated internally.

internal documentation: Computer software is thought to have internal documentation of the notes on how and why various parts of code operate are included within the source code as comments. It is often united with meaningful variable names with the intention of providing potential future programmers a means of understanding the workings of the code. This term *contrasts* EXTERNAL DOCUMENTATION, where programmers retain their notes and explanations in a detached document.

internal fertilization: A kind of fertilization that transpires inside the egg-producing individual. *See* EXTERNAL FERTILIZATION.

internal genetic balance: The harmonious epistatic interactions of genes at different loci.

internal guide sequence (IGS): In group I introns, it refers to a sequence that pairs with exon sequences adjacent to both the 5′ and 3′ splice sites. The IGS usually begins with 20 nucleotides of the 5′ splice site. IGS 5′ exon pairing is important in both steps of RNA splicing.

internal membrane: A eukaryotic cell membrane other than the plasma membrane. The membranes of the endoplasmic reticulum and the Golgi apparatus are examples of internal membrane.

internal promoter: Promoter sequence that exists within the coding sequence of a gene. The 5S genes of *Xenopus* are known to have an internal promoter in addition to the conventionally placed upstream promoter at 5′ end of the gene.

internal ribosome entry site (IRES): A structure that is located in the 5′ UTR or open reading frame of some mRNAs of cellular or viral origin. It mediates translation initiation independently of the mRNA's 5′-end by recruiting the ribosome directly to an internal position on the mRNA.

internal transcribed sequences/spacers (ITS): Each member of a repeat unit in a ribosomal gene family possesses a coding region comprising the genes that specify 18S, 5.8S and 28S rRNA molecules; this spacer region is called INTERGENIC SPACERS (IGS) or INTERNAL TRANSCRIBED SEQUENCES (ITS). These genes appear to be nearly identical among a wide variety of species, but the DNA spacers between them are quite variable and may be species specific. This makes them potential targets for the binding of 'primers' for POLYMERASE CHAIN REACTION (PCR) amplification, which is used to identify a wide variety of organisms.

International Committee of Taxonomy of Viruses (ICTV): Set up in 1966, ICTV authorizes and organizes the classification and naming of viruses.

International Service for the Acquisition of Agri-Biotech Applications (ISAAA): ISAAA is maintaining and updating the most comprehensive database regarding the commercialization of transgenic crops in the world. The data regarding the total countries growing transgenic crops, total area under transgenics in the world, crop-wise area, trait-wise area and so on, can be obtained from the ISAAA website, www.isaaa.org.

International System of Units (SI): Refers to the modern form of the metric system of units and measurements. This is the system of measurement formally implemented for use in the physical and natural sciences.

internet: The global system of interconnected computer networks that use the Internet protocol suite (TCP/IP) to connect devices worldwide. It is a network of networks that comprises private, public, academic, business and government networks of local to global scope, associated by a broad array of electronic, wireless and optical networking technologies.

internet bot: A software application that runs automated tasks (scripts) over the internet, usually with the intent to imitate human activity on the internet, such as messaging, on a large scale. Typically, bots perform jobs that are both simple and structurally repetitive, at a much higher rate than would be possible for a human alone. The largest use of bots is in web spidering (web crawler), in which an automated script fetches, analyses and files information from web servers at many times the speed of a human. *Aka* WEB ROBOT, ROBOT or simply BOT.

internode (Adj. internodal): Refers to the portion of a stem between two consecutive nodes.

internucleosomal DNA cleavage: Internucleosomal DNA cleavage, visualized as ladders, is thought to occur during apoptosis. Also occurs during necrosis and is an early event but appears to be triggered by proteolytic mechanisms significantly different from those documented in apoptosis.

interpetiolar stipules: Stipules located between the petioles of opposite leaves (e.g. in Rubiaceae).

interphase: The period between two cell divisions. Misnamed as RESTING STAGE. Metabolically very active stage as DNA synthesis occurs during this phase. It comprises first gap (G_1), synthetic (S) and second gap (G_2) subphases. The duration of these subphases varies in different organisms. Cells spend most of their time in interphase, also called resting phase but as a matter of fact this is a very active phase biochemically as it is a period of intense biosynthetic activity. Some cells do not divide at all after birth and remain in the interphase for the entire lifespan of the person (e.g. nerve cells or neurons).

interphase chromatin: V.I.P. Keizer *et al.* in 2022 developed a technique to measure how a genomic locus, in its native nuclear context, responds to a point force of physiological magnitude. Interphase chromatin has been found to be liquid like, with moderate hindrance and topological effects when probed in a living cell, contrasting with the common view of a crowded and entangled nuclear environment. These results have implications for the overall compaction of the genome and the coupling between mechanical and biochemical signalling needed for genome integrity and function (W.Y. So and K. Tanner in 2022).

interphase heterochromatin organization: Heterochromatin positioning in the interphase nucleus is, to some extent, influenced by the chromosome arrangement during cell division.

interphase nucleus: A nucleus during the interphase stage in which there is no visible activity related to the division of the nucleus, but in which metabolic and synthetic activities are in progression.

interphase nucleus mapping: A variant of the conventional fluorescent *IN SITU* HYBRIDIZATION (ISH) technique for the imagining of specific genes in interphase nuclei (in which the chromosomes are invisible). The gene is detected by a fluorochrome-labelled gene probe that hybridizes to the target gene. The location of the hybridized probe is visualized by laser-induced excitation of the fluorochrome.

interplanting: The practice of planting a fast-growing crop between a slower-growing one.

interplay between morphogenetic proteins: The interplay between BMPs and their antagonists governs developmental and cellular processes as diverse as the establishment of the embryonic dorsal-ventral axis, the induction of neural tissue, the formation of joints in the skeletal system and neurogenesis in the adult brain. So far, the three-dimensional structures of BMP antagonists and the structural basis for inactivation have remained unknown. The crystal structure of the antagonist Noggin bound to BMP-7 shows that Noggin inhibits BMP signalling by blocking the molecular interfaces of the binding epitopes for both type I and type II receptors.

interpreter: A computer programme that directly executes instructions written in a programming or scripting language, without necessitating them to have been previously assembled into a machine language programme.

interrupted mating: A mapping technique in which bacterial conjugation is interrupted after specified time intervals for measuring the distance between two genes in terms of the time taken to transfer from one bacterial cell to another.

intersex: A class of individuals of a bisexual species that have sexual characteristics intermediate between the male and the female.

interspecific: Effects and phenomena between species.

interspecific cross: A crossing between two species.

interspecific hybrid: A hybrid between two or more species.

interspecific hybridization: Crossing between species.

interspecific selection: A selection that operates to improve the competitive power of one species in relation to the other species, as distinct from INTRASPECIFIC SELECTION.

interspersed DNA: Even though the genomes of higher organisms contain single-copy, middle-repetitive and highly repetitive DNA sequences, these sequences are not organized similarly in all species. In some species, repeated DNA is interspersed. Interspersed DNA is of two types: SHORT-PERIOD INTERSPERSION and LONG-PERIOD INTERSPERSION.

interspersion pattern of DNA: The pattern of dispersal of coding and non-coding sequences within a genome. In many genomes, especially of higher eukaryotes, there is a tendency for coding sequences of structural genes to be interspersed with long lengths of DNA that are not transcribed. Such DNA is partly regulatory but much of this fraction may be redundant.

interstitial chiasma: A chiasma that has a length of chromatid on either side of it.

interstitial region: The chromosomal region between the centromere and the side of a rearrangement.

interstitial segment: A chromosome region between the centromere and a site of rearrangement. *Aka* INTERSTITIAL REGION.

interstock: An intermediate plant part that is compatible with both the scion and the rootstock.

intervarietal: Effects and phenomena between varieties (cultivars).

interveinal: Between (leaf) veins.

intervening sequence: A non-coding nucleotide sequence in eukaryotic DNA, separating two portions of nucleotide sequence found to be contiguous in cytoplasmic mRNA. *Aka* INTRON.

intoxication: Poisoning resulting from the ingestion of microbial toxin. Some human pathogens cause disease through their toxins. Intoxication is caused by the ingestion of such toxins that affect the gastrointestinal tract. Infection and intoxication together cause diarrhoea and dysentery – a major cause of infant mortality. These are caused by rotaviruses of the *Escherichia coli* and *Shigella* species.

intra-allelic competition: The restoration of activity of an enzyme consisting of subunits in a heterozygote of two mutants that when homozygous do not have any activity. In this case, activity is caused by interactions of non-identical subunits in the protein.

intra-allelic complementation: Synergistic effect visible in some proteins that are mixed polymers with subunits coded by more than one allele or more than one gene. Thus, mixed tetramers of haemoglobin and lactate dehydrogenase are more efficient in some situations than the homopolymers.

intrabreeding: A mating type in which only the individuals from the same populations are combined.

intracellular (adv. **intracellularly):** Of or happening inside or within the enclosed interior of a cell. *Contrasts* EXTRACELLULAR.

intracellular domains: Domain families that are most prevalent in proteins within the cytoplasm.

intracellular membranes: Intracellular membranes surround organelles. One such membranes is ENDOPLASMIC RETICULUM (ER). It is a network of a group of membranes. Its important part is the Golgi complex, which involves secretion. Another type of intracellular membrane is lysosomes, which contain hydrolytic enzymes that break down lipids, proteins and so on. Lysosomes are derived from ER. There are certain regions of the body where the cells stop dividing after a particular stage of development. For example, the cells of the tips of the fingers stop dividing due to lysosomes.

intracellular promiscuity: Intrachromosomal gene transfer of DNA. *See* PROMISCUOUS DNA.

intracellular protein inhibitors of apoptosis (IAP) family: A group of negative regulators of both caspases and cell death. Although best known for their ability to regulate caspases and cell death, it is now clear that they function as arbiters of diverse biological processes.

intracellular signal for DNA replication initiation: Initiation of DNA replication in eukaryotes presumably occurs in response to some intracellular signal. During S phase of the cell cycle, each section of DNA replicates only once. Cell cycle is controlled in two phases, one at G1 and another at G2. G1 cells initiate replication, S phase cells have activators of

DNA replication and G2 nuclei do not re-replicate DNA.

intracellular signalling: The process of communication of a molecular message from the surface of a cell to the nucleus via the participation of a series of molecules, including receptors, enzymes, proteins and small molecules. The end result of the signalling process is the up- or downregulation of a particular series of genes that may be involved in cell growth, division or differentiation.

intrachromosomal mutations: Structural changes within a chromosome, such as deletions, duplications and inversions.

intrachromosomal recombination: Recombination resulting from crossing over two gene pairs.

intrachromosome balance: A condition produced by the adjustment of genes within one chromosome.

intracisternal A type particles (IAPs): Mouse-specific class II LTR-containing retroelements that retain high DNA methylation levels throughout development. These transposons are virus-like particles. These give rise to poly(A) RNAs that give rise to IAB proteins of 13,000 Daltons. There are a 1000 copies of this particle per genome of *Mus muscuclus* (mouse).

intracistronic complementation: A type of complementation between mutant variants of certain genes coding catalytic enzymes, receptor-type proteins or membrane proteins. Complementation is the consequence of the combination of mutant polypeptides in which each mutant chain delivers or modifies the conformation of the site(s) of activity of protein.

intraclass correlation: A measure of phenotypic resemblance between relatives that is determined by the ratio of covariance of full-sib (FS) or half-sib (HS) to the total phenotypic variance (V_p) − t(FS)/V_p and t(HS) = Cov(HS)/V_p. The higher the size of t, the greater the resemblance between relatives.

intracodon recombination: Genetic recombination between adjacent nucleotides within a codon.

intracytoplasmic sperm injection (ICSI): Fertilization by direct injection of sperm into the cytoplasm of an egg. ICSI can also be used as a means of transfection.

intragenic: Effects and phenomena within a gene.

intragenic changes: Mutational alterations in the nature of the individual gene, as opposed to INTERGENIC CHANGES.

intragenic chromatin modifications: Both neuronal cell depolarization and the use of double-stranded small RNAs targeting intragenic regions can modulate alternative splicing through chromatin remodelling, taking advantage of the kinetic coupling between splicing and transcription.

intragenic inversion. An inversion within a gene that changes the amino acid sequence of a polypeptide encoded by that gene. The phenotype may be altered or the function may be lost. Inversions do have some genetic effects.

intragenic mutations: Mutations that occur within the same gene.

intragenic recombination: Intragenic recombination is defined as recombination within a gene. New alleles are added to natural populations through intragenic recombination.

intragenic suppression: A frameshift of opposite sign at a second site can act as an intragenic suppressor. In the mutant protein, amino acids incorporated are different. If these amino acid changes are in less important sites, the function of the protein will not be affected. A second-site missense mutation in the same gene may also serve as an intragenic suppressor.

intragenic suppressor: A compensating mutation that restores the original reading frame after a frameshift mutation.

intragenotype competition: When only a single genotype is used in a series of monocultures and the character is defined in such a way that its expression shows a linear regression on the density of the culture, the regression coefficient, b_m, provides a measure of the strength of intragenotypic competition. *Aka* ISOCOMPETITION.

intragonadal microinjection: The syncytial nature of the *Caenorhabditis elegans* gonad and the overall transparency of this worm, allowing easy visualization of this region by differential interference contrast (DIC) microscopy, has made gonadal microinjection as the most widely used approach to transgenesis in this model organism.

intramarginal: Something occurring inside but close to the margin (e.g. a vein in a leaf).

intramolecular triplex helical DNA: The triplex DNA structures form easily if the DNA segment contains minor repeat symmetry in addition to PuPy sequences. Intramolecular triplex DNA may contain repeats of G, GA, GGAA, GGGA or AAA-GXN. Third-strand pairing requires protonation of cytosine for pairing with guanine (requires lower pH). It can also occur in four different isoforms.

Supercoiling of DNA favours intramolecular triplex form. The third strand can be produced during replication by a process called SLIPPAGE PROCESS, where the polymerase in certain regions transverses back and replicates the same strand a second time after replicating a segment, and it can be repeated several times.

intra-S checkpoint: It monitors the process of DNA replication and stops it if the DNA breaks or stalled replication forks are detected. It delays the onset of mitosis until the replication of the genome is complete. The damage sensors for the intra-S checkpoint encompass a large set of checkpoint and repair proteins.

intraspecific: Effects and phenomena within a species.

intrastaminal: Usually refers to the location of a nectary disk – inside the stamens or androecium.

intrinsic networks: Mutants and metabolically changed plants are also analysed using transcriptomics, proteomics and metabolomics to understand how different intrinsic networks are influenced.

intrinsic rate of increase (r): A measure of population growth. It is the instantaneous rate of increase of a population or genotype. It is used as a measure of fitness. *See* MALTHUSIAN PARAMETER.

introduced species: Any species living outside its native geographic range and has arrived there either by accidental or deliberate human activity. Such human-caused introduction of species to foreign environments is distinguished from biological colonization, by which species spread to new areas through 'natural' means (i.e. without the involvement of humans). *Aka* EXOTIC SPECIES, FOREIGN SPECIES, ALIEN SPECIES, NON-NATIVE SPECIES *or* NON-INDIGENOUS SPECIES.

introducing exogenous DNA into mitochondria: Exogenous DNA can be imported into isolated mammalian mitochondria *in vitro* by electroporation, bacterial conjugation or the natural competence mechanism.

introgression: Transferring of a gene or few genes controlling a desired character through backcrossing from wild species to a cultivated variety. In plant breeding, the useful traits from related wild species are transferred into cultivated crop varieties through backcross breeding where the useful genes get introgressed into the genome of recipient variety through recombination. For instance, several gene controlling biotic and abiotic stresses have been transferred from *Oryza nivara* (wild species of rice) into *Oryza sativa* (cultivated rice) following this approach.

introgression library: Increasing the genetic diversity of elite breeding materials with exotic germplasm necessitates techniques that minimize negative side effects attributable to genetic interactions between recipient and donor. This seems attainable by an introgression library approach involving the systematic transfer of a limited and/or restricted number of short donor chromosome segments from an agriculturally unadapted source into an elite line (recipient or recurrent parent). Established introgression libraries signify a dynamic resource that can substantially foster breeding programmes and provide an opportunity to advance towards functional genomics.

introgression line population for QTL mapping: A novel INTROGRESSION LINE (IL) population is used for the mapping of interspecific variation for quantitative traits associated with yield. Among the lines there is a complete representation of the wild species genome. The ILs are nearly isogenic to the recipient genotype, and therefore all the genetic variation that differentiates them can be associated with the introgressed segment. Since each line carries only a small fraction of the wild species genome, most of the fertility problems can be eliminated, and yield-associated traits can be measured.

introgression, uses of: Introgression uses crosses of closely related plant species, not to combine complete genomes but to transfer chromosome fragments (blocks of genes), which are usually translocated to distal chromosome regions of the receptor genome. Tracking and tracing of translocations is done with molecular markers nowadays.

introgressive hybridization: Crossbreeding of plants from different species that lead to introgression.

intron: An intervening sequence/segment of deoxyribonucleic acid (DNA) that is transcribed but is removed from within the transcript before it is used for translation, by splicing together the sequences (exons) on either side of it (in the molecule). It is generally considered a non-coding portion of the molecule. Putative functional roles of introns have been recognized in various cellular processes such as splicing, mRNA transport, NONSENSE-MEDIATED DECAY (NMD) and expression regulation. Besides, introns may provide some advantages as a mutational buffer in eukaryotic genomes protecting coding sequences from being affected by randomly occurring deleterious mutations. Some introns also code for proteins that act as helpers in RNA splicing. Yeast pre-mRNAs usually contain a single intron of less than 1000 nts in length. Human pre-mRNAs can contain many introns ranging from a few dozen to >400,000 nucleotides in size. Introns are mostly present in eukaryotic genes but rarely in prokaryotic genes. *Compare with* EXON.

intron retention (IR): An alternative splicing mode whereby introns, rather than being spliced out as usual, are retained in mature mRNAs.

intron types based on length: Depending on the length, introns are of three types: (i) SHORT or VERY SHORT INTRONS are found in many tRNA genes. (ii) Much LONGER INTRONS are found in nuclear RNA and mitochondrial RNA genes and many mitochondrial mRNA genes. (iii) The LONGEST INTRONS (a few hundred to tens of kb long) are found in genes specifying proteins.

intron types based on mode of splicing: Depending on the mode of splicing, there are four types of introns: (i) TYPE 1 INTRONS are self-splicing and occur in nuclear, mitochondrial and chloroplast rRNA, tRNA and mRNA. In type 1 introns, the 3′ OH group of a free guanosine acts a nucleophile, attacking the 5′ phosphate at the splice site, displacing it from the exon. The 3′ OH at the exon, usually a U, acts as a nucleophile, attacking the phosphodiester bond, removing the other end of the intron and reforming the phosphodiester bond with itself. (ii) TYPE 2 INTRONS are also self-splicing and occur in genes in mitochondria and chloroplasts of fungi algae and plants. In type 2 introns, a similar mechanism is used, except that an internal 2′ OH of an adenosine is used as the nucleophile, attacking the 5′ splice site of the exon to produce a lariat structure. In the final step, the 3′ OH of the 5′ exon attacks the phosphodiester bond of 3′ uracil, to remove the lariat. (iii) TYPE 3 INTRONS require a spliceosome and SMALL NUCLEAR RIBONUCLEOPROTEINS (SNRNPS) and are responsible for exclusively eukaryotic splicing. In type 3 introns, a lariat structure is formed, but the snRNPs are used to form the secondary structures and to mediate the reaction, it still occurs without free energy input. There are six snRNP subunits that make up the spliceosome. (iv) TYPE 4 INTRONS require ATP and an endonuclease.

intron-exon junctions: Consensus sequences at the boundaries of intron-exon junctions have shown that GT is always found at the 5′-side of the intron and AG at 3′-side. *Aka* CHAMBON'S RULE.

intronic gene: A gene positioned inside an intron and flanked by non-coding regions.

intronic variant: A nucleotide change in an intronic region that may alter splicing and result in a mature messenger RNA sequence with retained intron sequence.

introns (chloroplast gene), group I: Self-splicing introns that do not require an external nucleotide for splicing. The intron is released in a lariat form.

introns (chloroplast gene), group II: Introns that are not self-splicing. They need small guide RNAs or U RNAs for their splicing.

introns (chloroplast gene), group III: Introns that have conserved sequences at their borders [GTGCGNY at 5′ end, and ATCHRYY(N)YYAY at 3′ end], similar to those in the eukaryotic nuclear genes. For example, *trnG*, *trnK*, *trnV*, *rpl2*, *rps12* and *rps16* have group III introns.

introns of chloroplast genes: Split genes for ribosomal RNA (rRNA), transfer RNAs and some proteins have also been reported in the chloroplast genomes of several plants including *Chlamydomonas* and *Nicotiana*. Such introns can be classified into three groups on the basis of intron boundary sequences. GROUP I INTRONS (e.g. in *trnL*) can be folded in a secondary structure similar to self-splicing rRNA procedure of *Tetrahymena*. These can be removed either by self-splicing or by a 'maturase' (as in cytochrome b and cytochrome oxidase mRNA precursors). GROUP II INTRONS (e.g. majority of genes including *trnA* and *trnI*) can be folded into a complex secondary structure (as in introns of mitochondrial genes for cytochrome oxidase in maize and yeast). GROUP III INTRONS (e.g. *trnG*, *trnK*, *trnV*, *rpl2*, *rps12*, *rps16*, etc.) have conserved sequence at their borders [GTGCGNY at 5′-end and ATCHR YY(N)YYAY at 3′-end], similar to those in the eukaryotic nuclear genes.

introns of fungal mitochondrial genes: Introns of split genes in fungal mitochondria are classified into two groups on the basis of their internal organization. GROUP I INTRONS are found in a majority of the fungal mitochondrial split genes, do not carry any conserved sequence at intron-exon junctions but carry a short conserved sequence called INTERNAL GUIDE SEQUENCE (IGS) internally. GROUP II INTRONS resemble nuclear gene introns and have consensus sequences (GT and APy) and a branch sequence that resembles the TACTAAC BOX.

introns, type A: The introns that result from the original assembly of the gene from smaller units. Type A introns are further classified according to whether they derive from the creation of a domain from smaller units (A1), from the duplication of a complete domain (A2) or from the joining of two dissimilar domains (A3).

introns, type B: The introns that result from DNA sequences that never had any functions.

introrse: Of anther locules, with opening towards the centre of flower (at least in bud). *Compare with* EXTRORSE *and* LATRORSE.

intumescence: The blisters formed on leaves and other plant parts under conditions of high moisture and restricted transpiration. *See* OEDEMA.

inulin: A polysaccharide in which about 32 β-fructose units are joined in a chain by glyosidic linkages between the first and second carbon atoms on neighbouring sugar units. Insulin occurs as a storage compound in roots, rhizomes and tubers of many species of Compositae.

invagination theory: Given by R.A. Raff and H.R. Mahler in 1972. A prokaryotic cell having a flexible membrane evolves extensive invaginations that are derived from the internal membranes of cell. Enzymes responsible for glycolysis and respiration became associated with these invaginations. The invaginated membranes lose contact with cell membranes and become converted into mitochondria.

invariance: The reciprocal of the variance.

invariant: One can come across invariants that can be trusted upon to be true during the execution of a programme, or during some portion of it. It is a logical statement that is always held to be true during a certain phase of execution. For example, a loop invariant is a state that is true at the beginning and the end of every execution of a loop.

invariant DNA code: Variant code means that a mutation in the genetic code would place new amino acids in certain loci and entirely eliminate amino acids from the other loci of practically all proteins in an organism. This actually does not occur as DNA code in fact is invariant.

invasion: The spreading of a pathogen through the tissues of a diseased plant.

invasive alien: Non-native species that survive in wild habitats and perhaps aggressively out-compete native species or cause other kinds of ecological damage.

invasiveness: the ability of a species to spread beyond its introduction location and become settled in new locations where it may provide a deleterious effect on organisms already existing there.

invention: Invention is defined as an act of inventing. In Patent Laws, it means the conception of an idea and the means or apparatus by which the result is obtained.

inverse metabolic engineering: Inverse metabolic engineering begins by identifying the desired phenotype, then determining the environmental or genetic conditions that produce this phenotype and finally modifying the phenotype of the chosen host through genetic manipulation.

inverse PCR (IPCR): IPCR was first described by H. Ochman *et al.* (1988). A restraint of standard PCR is that 5′ and 3′ flanking regions of the DNA fragment of interest must be known. IPCR is a variant of PCR and is used when only one internal sequence of the target DNA is known. It is therefore very useful in identifying flanking DNA sequences of genomic inserts. Similar to other PCR methods, IPCR amplifies target DNA using DNA polymerase. IPCR uses standard PCR; however, it has the primers oriented in the reverse direction of the usual orientation. The template for the reverse primers is a restriction fragment that has been ligated upon itself to form a circle. The IPCR method comprises a series of digestions and self-ligations with the DNA being cut by a restriction endonuclease. This cut results in a known sequence at either end of unknown sequences.

inverse PCR, steps in: (i) The target DNA is lightly cut into smaller fragments of several kilobases by restriction endonuclease digestion. (ii) Self-ligation is induced under low concentrations causing the phosphate backbone to reform. This gives a circular DNA ligation product. (iii) The target DNA is then restriction digested with a known endonuclease. This generates a cut within the known internal sequence generating a linear product with known terminal sequences. This can now be used for PCR. (iv) Standard PCR is conducted with primers complementary to the now-known internal sequences. Inverse PCR functions to clone sequences flanking a known sequence. Flanking DNA sequences are digested and then ligated to generate circular DNA. PCR primers pointing away from the known sequences are then employed to amplify the flanking sequences. Inverse PCR has numerous applications in molecular biology including the amplification and identification of sequences flanking transposable elements, and the identification of genomic inserts.

inversion: A type of chromosomal abbreviation that may involve a portion of a gene or a group of genes that have been cut out, turned through an angle of 180° and reattached in a reverse order. In case of an inversion, there is no net loss or gain of genetic material observed and thus heterozygotes are perfectly viable. An inversion that may involve only one chromosomal arm and not include a centromere are called PARACENTRIC INVERSION. The inversion that involves both the chromosomal arms and thus includes a centromere is called PERICENTRIC INVERSION. An intragenic inversion may reverse amino acid sequence of polypeptide encoded by that gene. Cytologically, a heterozygous inversion is easily detectable as an inversion loop. Consequently, phenotype may be changed or function may be lost. Inversions have some genetic effects. In a heterozygote for a paracentric inversion, crossing-over within the inversion loop yields dicentric and acentric chromosomes. Individuals producing crossover chromatids do not survive and thus do not produce recombinant gametes. Thus, crossovers are not detected. In a heterozygote for a pericentric inversion, crossover products are not recovered. Inversions in many experiments in genetics are used as suppressor of crossing-over. For genes present in the inverted chromosome, the linkage relationship is reversed. *See* PARACENTRIC INVERSION *and* PERICENTRIC INVERSION.

inversion mutation: Alteration of the sequence of a DNA molecule through the removal of a segment, followed by its reinsertion in reverse orientation.

inversion polymorphism: The occurrence of two or more chromosome sequences, diverging by inversions, in the homologous chromosomes of a population gene.

inverted repeat sequences: The repeats that comprise two copies of the same sequence of DNA repeated in opposite orientation on the same molecule. Commonly found in transposons and viruses. Adjacent inverted repeats constitute a PALINDROME. *Contrasts* DIRECT REPEATS.

inverted terminal repeats: The short, related or identical sequences present at the ends of some transposons.

investigation: Investigation is a process of carefully making observations and studying something in more detail. You are gathering information! Investigation requires planning, observing and outcomes. With investigation, you can test your theory, strengthen your theory or sometimes find something that does not fit your theory.

involute: Rolled inward, for example, when the margins of a leaf are rolled towards the adaxial (usually upper) surface. *Compare with* REVOLUTE.

iodine (I): A non-metallic halogen element existing as a greyish-black crystalline solid substance that sublimes to a dense violet vapour when heated. Iodine is used in radiolabelling.

iodine number: The mass of iodine in grams that is consumed by 100 g of a chemical substance. Iodine numbers are often used to determine the amount of unsaturation in fatty acids. *Aka* IODINE VALUE, IODINE ADSORPTION VALUE *or* IODINE INDEX.

***iojap*:** An idiomatic portrayal of a mutant locus in maize that results in a variegation gene.

ion: An atom or molecule with a net electric charge resulting from the loss or gain of one or more electrons (e.g. Na^+ and Cl^-).

ion beam lithography (IBL): Lithography using ions rather than light beams.

ion channel: Transmembrane protein complex that forms a water-filled channel across the lipid bilayer through which specific inorganic ions can diffuse down their electrochemical gradients.

ion exchange chromatography (IEC): Separation of ionic compounds (such as nucleic acids and proteins) in a chromatographic column containing a polymeric resin (i.e. the stationary phase) having fixed charge groups. In this process, the charges of the column (stationary phase) interact with the opposite charges of the material dissolved in the solution that is flowing through the column (mobile phase). The charge interaction between the column material and the protein slows down the rate of movement of the protein through the column. The other molecules, meanwhile, which do not interact with the column, flow right on through. This leads to the formation of a separation process. IEC includes digesting RNAs with sequence-specific RNase fractionating the resultant oligoribonucleotides by ion-exchange chromatography or two-dimensional homochromatography. This process establishes the order of bases within each fragment by exonuclease digestion by employing two different methods of fragmentation.

ion exchange column: A column filled with materials suitable for ion exchange chromatography. Such materials are resins, cellulose or other supporting compounds to which charged groups have been attached. For example, diethylaminoethyl-cellulose (DEAE-cellulose) is positively charged at pH 7.0 and therefore acts as an anion exchange material. Carboxymethyl-cellulose (CM-cellulose) is negatively charged and used for ion exchange chromatography. The mixtures of DEAE-cellulose and CM-cellulose, having different i soelectric points, and thus charges at a given pH value, is passed down the column material and eluted sequentially by a gradient of buffer of increasing ionic strength or by a gradient of changing pH value.

ion torrent semiconductor sequencing: Ion torrent sequencing uses a 'sequencing by synthesis' approach, in which a new DNA strand, complementary to the target strand, is synthesized one base at a time. A semiconductor chip detects the hydrogen ions produced during DNA polymerization. Following polony formation using emulsion PCR, the DNA library fragment is flooded sequentially with each **deoxynucleoside triphosphate (dNTP)**, as in pyrosequencing. The dNTP is then incorporated into the new strand if complementary to the nucleotide on the target strand. Each time a nucleotide is successfully added, a hydrogen ion is released, and it is detected by the sequencer's pH sensor. As in the pyrosequencing method, if more than one of the same nucleotides is added, the change in pH/signal intensity is correspondingly larger.

ionic bond: A type of non-covalent chemical bond that involves the complete transfer of valence electron(s) between two atoms. Such bonds typically ensue between elements characterized as metals and non-metals, and thus generate two oppositely charged ions. The metal loses electrons to become a positively charged cation and the non-metal accepts those electrons to become a negatively charged anion.

ionizing radiations in induction of mutations: Ionizing radiations, such as X-rays, alter nucleic acid in both direct and indirect manner. In direct manner, there is a rupture of the chemical bonds of the bases and the deoxyribose as well as of the backbone of the DNA molecule. In indirect effect, ionizing radiation produces free radicals either from water or from organic molecules. Ionizing radiation causes three types of damage to DNA: (i) Single-strand breaks that are mostly sealed by DNA ligase so do not contribute to lethality. (ii) Double-strand breaks that are often lethal because these breaks cannot be resealed by ligase, so degraded by nucleases. (iii) Alteration of bases that is an oxidative type of damage and is usually lethal because it forms a replication barrier at that site. X-rays may have a direct effect on the chromosome (e.g. directly breaking it or altering one of the DNA bases). It may have an indirect effect, since along the track of each particle, a large number of ions or radicals remain that can initiate a chain of chemical reactions. Chromosomes are extensively sensitive to breakage in meiotic prophase. During mutation breeding of plants, X-irradiation and γ-irradiation are preferred as these radiations cause ionization resulting in higher frequencies of mutations.

ionome: The mineral nutrient and trace element composition of an organism that represents the inorganic component of cellular and organismal systems.

ionomics: The study of elemental build-up in living systems using high-throughput elemental profiling. This approach is applied extensively in plant studies focusing on forward and reverse genetics, screening diversity panels and the modelling of physiological states.

ionophore: Small hydrophobic molecule that dissolves in lipid bilayers and upsurges their penetrability to specific inorganic ions.

iontophoresis: A separation process that relies on charge-charge interactions between proteins in a sample and charges immobilized on resin. *Aka* ION EXCHANGE CHROMATOGRAPHY (IOC).

ion torrent sequencing (ITS) technology: Ion Torrent™ technology directly translates chemically encoded information (A, C, G, T) into digital information (0, 1) on a semiconductor chip. This approach marries simple chemistry to proprietary semiconductor technology.

IPR recommendations: Several recommendations have been made concerning (i) harmonization of IPR system, (ii) awareness generation and literacy in IPR, (iii) IPR education, training and human resource development, (iv) strengthening the institutional mechanism – legal, regulatory and administrative, (v) Strengthening the policy area, (vi) harnessing IP-linked technical opportunities in agriculture and (vii) linkages and cooperation.

IPRs in agriculture: Agriculture is a rule-bound enterprise of investment and profit and profit making. This became obvious with the inclusion of agriculture in the intergovernmental negotiations for the General Agreement on Tariffs and Trade (GAAT) for the first time in the Uruguay Round (1986–1994). This round led to the establishment of the World Trade Organisation (WTO) in January 1995. WTO has several intergovernmental agreements that directly affect agriculture. These are the Agreement on Agriculture (AoA), Applications of Sanitary and Phytosanitary Measures (SPS), Technical Barriers to Trade (TBT), Anti-Dumping, Subsidies and Countervailing

Measures, Safeguards and Trade Related aspects of Intellectual Property Rights (TRIPs). Following the establishment of international institutional mechanisms, such as the Convention of Biological Diversity (CBD), and the WTO, signing of International Treaty on Plant Resources for Food and Agriculture (ITPGRFA), the growing importance and scope of IPR in agriculture were well realized. IPR-related issues in agriculture include (i) protectable subject matter in agriculture, (ii) technical opportunities in agriculture and (iii) enabling the environment for accelerated R&D and global competitiveness in Indian agriculture.

iron (Fe): An element required by plants. It is used in reactions in which rapid reductions occur by the transfer of electrons as in photophosphorylation and oxidative phosphorylation reactions. Iron-deficient plants produce chlorotic young leaves. At first, the veins are green but later they too become chlorotic. Fertilizers containing iron chelates are added to the soil or sprayed on the leaves to make iron accessible to the roots and foliage.

iron deficiency chlorosis: Iron is one of the more abundant minerals on earth, but it may often occur in forms that are not readily available to plants. In alkaline soils, particularly those rich in limestone, low-lying plants cannot access the iron that is present. The plant may consequently become deficient in this element and chlorotic.

iron-sulfur centre: Electron-transporting group consisting of either two or four iron atoms bound to an equal number of sulfur atoms, found in a class of electron transport proteins.

irradiate: The act of exposing an organism (or callus) to radiation that may surge the mutation rate of some genes and enhance genetic variation. Radioactive irradiation can directly and indirectly cause damage to the DNA molecule. Not only the energy of the electrons but also the charge of the electrons can cause damage to the molecules. Slow and fast electrons may cause mutations.

irradiated callus: Callus that has been exposed to radiation.

irradiation breeding: *See* MUTATION BREEDING.

irreducibly complex: A system that cannot function if any one of its components is absent.

irregular: A structure not divided into two equal halves through any vertical plane. *See* ASYMMETRICAL. *Compare with* ZYGOMORPHIC, ACTINOMORPHIC *and* REGULAR.

irreversible increase: Increase in cell numbers by cell division and an increase in cell size.

IS element: *See* INSERTION ELEMENT.

island model: This model deals with a species subdivided into a number of discrete finite populations, races or subspecies, between which some migration (m) occurs. A fraction '*m*' of the genes in a deme immigrates from outside. Immigration may be from the mainland population or other demes.

islets of Langerhans: The insulin-generating portion of the pancreas.

iso-Seq method RNA sequencing: Iso-Seq method sequences the entire cDNA molecules – up to 10 kb or more – without the need for bioinformatics transcript assembly, so one can characterize novel genes and isoforms in bulk and single-cell transcriptomes and further: (i) characterize ALTERNATIVE SPLICING (AS) events, including alternative start sites, end sites, intron retention and exon-skipping events; (ii) find gene fusions in tumour samples; (iii) identify allele-specific isoforms; (iv) detect differentially expressed isoforms and isoform switching events and (v) predict functional impact of novel isoforms through ORF prediction.

isoacceptor tRNAs: Two or more tRNA molecules specific for the same amino acid.

isoalleles: Several alleles may exhibit themselves within the same phenotypic range, whether wild type or mutant. Such alleles are called isoalleles. Isoalleles may be wild type or mutant type. Detailed allelic studies of any trait would classify multiple alleles into these two categories.

isoallelic mutations: These mutations do not influence the power of expression of the concerned trait. The enzymes encoded by such alleles are comparable in activity to those produced by their wild-type alleles. However, the proteins produced by isoalleles can be differentiated electrophoretically, as these have different mass/charge ratios (and hence mobility) under specific conditions.

isoantigens: An antigen derived from one member of a species that can cause the production of antibodies in some other members of the same species. *Compare with* ALLOANTIGENS.

isobifacial: Of flat structures, especially leaves. Having both surfaces similar, usually with respect to cell types or to the number and distribution of stomata.

isobrachial: A metacentric chromosome, with a centromere position resulting in two chromosome arms with equal length.

isocaudomers: Pairs of restriction enzymes that have slightly different recognition sequences, but upon cleavage generate identical termini.

isochromatid break: A break induced by X-rays or other ionizing radiation at the same locus in two already divided chromatids.

isochromocentric: Nuclei revealing as many chromocentres as chromosomes.

isochromosomal line: Strains of organisms that are uniform for a particular chromosome.

isochromosome: An abnormal chromosome with two genetically and morphologically identical arms; usually derived from telocentric chromosomes.

isocompetition: Cultivation at high plant density implies the presence of strong interplant competition. However, when there is no genetic variation, the competition is called ISOCOMPETITION.

isoderivative chromosome: Designates an isochromosome formation from one of the arms of a derivative chromosome.

isodicentric chromosome: A structurally abnormal chromosome that encompasses a duplication of a part of the chromosome including the centromere. The resulting chromosome contains two centromeres and a point of symmetry that rests on the location of the breakpoint.

isoelectric focusing (IEF): In standard gel electrophoresis, no pH gradient exists because the electrophoresis buffer has a uniform pH. Each protein reaches either the anode or the cathode. In the case of IEF, the charge density of each protein declines as it moves along the pH gradient towards its isoelectric point. When the isoelectric point is reached, the protein charge is zero and it comes to a halt. Diffusion still functions against this tendency to focus at a single position in a gel, but a protein diffusing away from its isoelectric point (pI) becomes charged and therefore moves back to its focus. Two-dimensional (2D)-gel electrophoresis includes the purification of proteins from the desired organelle, tissue or cells. Protein mixtures are separated first in one dimension (1-D) by pI and then in two dimensions (2-D) by molecular weight (MW). The gels are then stained and data analysed. Protein identification is done through mass spectrometry. Proteins with different pI values, therefore, focus at different positions in the pH gradient. Although there may be an initial sieving effect that separates the proteins based on size, running the gel for a suitably long period ensures that all proteins reach their isoelectric point and size-independent separation is successfully achieved. There are two ways used to establish pH gradients in the gel – (i) with ampholytes and (ii) IPG gels.

isoelectric point: The pH at which a charged molecule in a solution possesses no net electric charge and therefore does not travel in an electric field.

isoenzyme (or isozyme): A species of enzyme that exists in two or more structural forms, which are functionally active at different speeds and easily identified by electrophoretic methods.

isoflavone: An oestrogen-like substance made by some plants, including the soybean. Soy isoflavones are being studied for their role in the prevention of cancer, hot flashes that occur with menopause and osteoporosis (loss of bone density).

isogametes: Male and female gametes cells look similar although biochemical differences do exist; can be classified as '+' or '–'.

isogamous: Refers to a single kind of gamete produced by the two parents.

isogamy: A reproduction process in which the two mating partners produce gametes of the same size and morphology. *Compare with* ANISOGAMY.

isogeneic: Applied to a graft that involves a scion and stock that are genetically identical.

isogenic: A group of individuals showing the same genotype (except for sex).

isogenic lines: Two or more lines differing from each other genetically at one locus only. *Compare with* RANDOM LINES.

isogenic stocks: Strains of organisms that are genetically uniform; completely homozygous.

isogeny: The situation when a group of individuals shows the same genotype.

isolate (n.): (i) In microbiology, a culture or subpopulation of a microorganism separated from its parent population and maintained in some controlled condition. (ii) In reproductive biology, a segment of a population within which assortative mating takes place. (iii) In plant pathology, as a verb, it means to remove an organism (e.g. a fungus) from the plant in a pure form.

isolated microspore culture (IMC) systems: IMC systems offer an unlimited supply of embryogenic cells and developing embryos, which facilitate investigations of key developmental processes related to seed formation and metabolism.

isolating chromosomal mechanisms: A difference in chromosome number or morphology that maintains genetic isolation.

isolating mechanisms: Any structural, physiological, behavioural or other features of an individual, or any geographical or geological barrier, that prevents the individuals of one population from successfully interbreeding with those of other populations.

isolation by distance: Disagreement between allele frequencies in different spaces within a spatially continuous population. The term typically refers to divergence caused by random genetic drift.

isolation distance: Distance that splits one field of plants from another. It is used to preserve seed purity in seed production fields and minimize cross-pollination between different kinds of plants (e.g. transgenic crops and organically grown crops). Distance is needed to isolate pedigreed seed crops from other crops that may be a source of pollen or seed contamination. It is used by most seed certification agencies as one of the requirements to maintain the varietal purity of pedigreed seed crops.

isolation of DNA from microbes: For DNA extraction, bacterial cells are grown in a liquid medium until they reach a maximum density of 2–3 × 10^9 cells/ml, and then harvested. The growth of the cells can be monitored by analysing the turbidity of the culture and quantified by measuring the optical density (OD) at 600 nm and correlating it with the growth curve. The collected cells can then be processed to extract DNA. Cell lysis for DNA extraction from bacterial cells is often done using chemical methods. This step is followed by the removal of cellular components from DNA using an organic extraction method or silica-based technology. The last step involves DNA precipitation to obtain pure DNA at high concentrations.

isolation of genomic DNA: In boiling preparations, the proteins and linear DNA fragments from the genome irreversibly denature and precipitate. For the alkaline lysis technique, the alkaline conditions also denature the DNA and some protein, which then precipitates upon the addition of a large amount of salt and a drop in pH (both accomplished through the addition of potassium acetate). In both cases, the plasmid does not denature and precipitate because the two strands are unable to separate in the closed, circular, supercoiled DNA. The precipitate is then removed by centrifugation.

isolation of plasmid DNA: This method uses neutral pH for lysing the cells. Non-supercoiled plasmid species are recovered more efficiently in this method compared with other methods.

isolation of polyomavirus DNA: Cells infected by polyomavirus are lysed by the addition of the detergent SDS. The detergent is precipitated with sodium chloride at 4°C. The DNA of the cells, if not sheared, precipitates too. The viral DNA however, which is smaller in size by several orders of magnitude, stays in the supernatant.

isoleucine (Ile): A crystalline monocarboxylic amino acid ($C_6H_{13}O_2$) present in most dietary proteins.

isoleucine synthesis in *Escherichia coli*: Studies on isoleucine synthesis in *Escherichia coli* have demonstrated that the addition of isoleucine (the end-product of a five-step conversion of threonine) into a culture of bacterium results in immediate blocking of the isoleucine threonine pathway. The addition of isoleucine inhibits the action of the enzyme responsible for the deamination of threonine to α-ketobutyrate. The end-product binds to the enzyme so that it competes with the substrate for a site on the enzyme molecule. The competition is not for the same site. The enzyme has two specific recognition sites, one for its substrate and the other for the inhibition of the end-product. Attachment at the inhibitor site affects the substrate site. It alters the interaction between the substrate site and the substrate. The term allosteric interaction is applied to such changes in enzyme activity that occur due to a change in the shape of the enzyme molecule produced by the binding of a second substance at a different and non-overlapping site.

isolines: Isogenic lines.

isomer: One of the two or more chemical substances having the same elementary percentage composition (i.e. same atoms) and molecular weight, but differing in structure and therefore in properties. There are many ways in which such structural differences (between the two or more isomeric molecules) can occur, such as n-butane [$CH_3(CH_2)_2CH_3$] and isobutane [$CH_3CH(CH_3)_2$].

isomerase: An enzyme catalysing conversion of a compound into its positional isomer.

isomeric genes: Different genes, each capable of producing the same or similar phenotype.

isomerization: In chemistry, isomerization is the process in which a molecule, polyatomic ion or molecular fragment, is transformed into an isomer with a different chemical structure. Enolization is an example of isomerization, where ketones are generally converted into enols. Isomerization does not change the atomic make-up of a molecule, but rather changes its geometric conformation, yielding a slightly different molecule.

isomerous: Of flowers, having an equal number of parts or markings in the whorls.

isomers: Any two molecules having the same molecular formula but different arrangement of the functional groups. Isomers differ with respect to structure and properties but possess the same chemical formula. For example, glucose ($C_6H_{12}O_6$) and fructose ($C_6H_{12}O_6$) have the same chemical formula, but the position of oxygens and carbons differ in the two different molecules that also causes changes in their physical and chemical properties. Thus, glucose and fructose are isomers.

isometric: The term is usually used to designate virus particles that are icosahedral in structure and look almost round. *See* ICOSAHEDRAL.

isomorphic: A structure with all features morphologically similar – of similar size and shape. *Compare with* HETEROMORPHIC.

isomorphs: *See* HYPOMORPH.

isophene: Lines on geographic maps that connect points of equality in the phenotypic expression of a clinically varying character.

isoprene (2-methyl butadiene): A 5-carbon compound (C_5H_8) that formulates the structural basis of many biologically important compounds, such as terpenes.

isoprenoid: Member of a large family of lipid molecules with a carbon skeleton based on multiple five-carbon isoprene units (e.g. retinoic acid and dolichol). *Aka* POLYISOPRENOID.

isoschizomeric enzymes: Pairs of structurally distinct restriction enzymes with the same recognition sequence and the same cleavage positions but their cutting is methylation specific. *Aka* ISOSCHIZOMERS.

isoschizomers: Two different restriction enzymes that recognize and cut DNA at the same recognition site. For example, *Sma*I and *Xma*I both recognize and cut the sequence CCCGGG.

isosome: A chromosome having morphologically and genetically identical arms.

isosomic: Cells or individuals characterized by isosomes, usually derived from telocentric chromosomes.

isotelo compensating trisomic: A compensating trisomic in which a missing chromosome is compensated for by the presence of one telocentric and one tertiary chromosome.

isotertiary compensating trisomic: A compensating trisomic. A missing chromosome is compensated for by the presence of one isochromosome and one tertiary chromosome.

isotomic: Having branches of the same diameter. *Compare with* ANISOTOMIC.

isotonic solution: A solution containing a cell, where the osmotic pressures of the cell and solution are the same. Isotonic solution is also called isoosmotic solution (iso=equal). Bacterial or other cells can be subjected to three kinds of osmotic solutions, viz. hypertonic, hypotonic or isotonic. An isotonic solution is that in which the concentration of solutes equals that of the cell. Therefore, the cell's content remains in equilibrium with the outside solution (Also see hypotonic and hypertonic).

isotope labelling *in vitro*: PTLM stoichiometry is determined in this method by introducing an isotope into the modified residues. To determine the phosphorylation stoichiometry of protein phosphorylation on a large scale, the proteolytic protein lysate is divided into two identical aliquots. One aliquot is treated with a phosphatase to remove the phosphate group. Treated and untreated samples are then chemically labelled by reductive dimethylation with isotopic 'heavy' and 'light' formaldehyde, respectively, and then mixed. The resulting mixture is analysed by HPLC/MS/MS analysis.

isotope labelling: Isotopic tracer.

isotopes: Refers to one of the several 'varieties' of atoms that exist for the same element, which differ from each other in the number of neutrons in the atom's nucleus but have the same atomic mass. Also defined as the stable or radioactive forms of an element. For example, the element chlorine exists primarily in two forms (isotopes) in nature, with 18 neutrons (76% of the time) and 20 neutrons (24% of the time). The chemical properties of isotopes of a given element are virtually identical. Oxygen has three isotopes, viz., 16_8O, 17_8O and 18_8O. Isotopes are being used as tracer elements.

isotopic dating: An approach to determining the age of certain materials by reference to the relative abundances of the parent isotope and the daughter isotope. If the decay constant and the concentration of the daughter isotope are known, it is possible to calculate the age.

isotopic tracer: Isotopically labelled precursors of nucleic acids. The labelled compounds are injected or fed to plants; subsequently, the excretions (solid, liquid or gas) or tissues are analysed to determine by detection of radioactive tracer how the original compound has been changed. Many tracers have been used, but the most common are ^{3}H, ^{14}C, ^{32}P, ^{35}S, ^{125}I and ^{131}I.

isotrisomic: When the extra chromosome shows identical arms.

isozyme: Alternate forms of an enzyme with different molecular weights but with the same substrate specificity and activity. In humans for instance an enzyme called lactate dehydrogenase (LDH), involved in metabolic activities of muscles and other tissues exhibit five types, viz. A_4, A_3B, A_2B_2, AB_3 and B_4, are very similar in structure and function but possess different electrical charges so they can be easily separated by gel electrophoresis.

isozyme markers: Markers that detect allelic variants of specific enzymes.

isozymes: Different forms of the same enzyme that catalyse the same reaction but differ from each other concerning their structure and regulatory properties. Aka ISOENZYMES.

iteration: Repetition of a series of steps in an algorithm whereby the processing of data is performed repetitively until the result exceeds a particular threshold. Iteration is often used in multiple sequence alignments whereby each set of pairwise alignments is compared with every other set, starting with the most similar pairs and progressing to the least similar, until there are no longer any sequence pairs remaining to be aligned. Each repetition of the process is a single iteration, and the result of each iteration then provides the starting point of the next iteration. At any stage, some consideration may lead the scientist to repeat an earlier part of the process. For example, (i) if the scientist nosedives to develop an interesting hypothesis, they re-define the subject; (b) if the scientist fails to produce interesting and testable predictions, he reconsiders the hypothesis and (c) if the scientist fails to produce interesting results, he might reconsider the experimental method.

iterative cycles of design-build-test-learn: Plant synthetic biology is still in its infancy. C. Tian and others in 2023 discuss genetic parts libraries and quantitative characterization approaches have been developed. Synthetic switches and circuits have been created for plant systems and of synthetic products produced in plant chassis. Challenges exist in the iterative cycles of design-build-test-learn for introducing novel traits into plants. The quantitative characterization of genetic parts is limited; the orthogonality and the transfer functions of circuits are unpredictable and also, the mathematical modelling-assisted circuit design still needs to improve predictability and reliability. Plant synthetic biology promises to provide important solutions to many issues in agricultural production, human health care and environmental sustainability.

iterative protein redesign and optimization (IPRO): This computational method redesigns proteins to increase or give specificity to native or novel substrates and cofactors. This is done by repeatedly randomly perturbing the backbones of the proteins around specified design positions, identifying the lowest energy combination of rotamers and determining if the new design has a lower binding energy than previous ones. The iterative nature of this process allows IPRO to make additive mutations to the protein sequence that collectively improve the specificity towards the desired substrates and/or cofactors. Details on how to download the software implemented in Python and experimental testing of predictions are outlined. IPRO has successfully engineered the switching of cofactor specificity of *Candida boidinii* xylose reductase.

iteroparity: Repeated breeding. *Compare with* SEMELPARITY.

iteroparous: Having several discrete reproductive events per lifetime.

ixeris type: Diplospory where a syndetic prophase leads to a restitution nucleus, which divides. It is not followed by a cytoplasmic division. This resultant megaspore contains two unreduced nuclei. Two further mitotic divisions lead to an eight-nucleate embryo sac.

J

jackknifing: A statistical procedure in which new data sets are produced by resampling an original data set shorn of replacement.

jackknife resampling: This method has the disadvantage that the sub-replicates of size smaller than the original data set may alter the statistical attributes of the sample. Jackknife resampling has now been substituted by bootstrap resampling, where sampling is done with replacement. For bootstrap resampling, it is best to generate at least 100 bootstrap data sets and rebuild the phylogeny for each one. A consensus tree can then be built.

Jacob-Monod model: The operon model of gene regulation in *lac* region of *Escherichia coli* was given by F. Jacob and J. Monod in 1961. This model explains the mechanism of gene expression in prokaryotes. *See* GENE REGULATORY MECHANISM IN PROKARYOTES.

Jacob-Monod model, confirmation of: W. Gilbert (1966) isolated purified *i* gene product, repressor of lactose operon. Repressor was present in i^+ but not in i^- cells. Repressor protein binds to o^+ but not o^c DNA. This confirmed the model of F. Jacob and J. Monod, put forth in 1961.

Janus kinase/signal transducers and activators of transcription (JAK-STAT) signalling pathway: Some signal transduction paths have huge numbers of apparatuses (thereby permitting a high degree of amplification) and many responding back circuits (thus allowing delicate regulation of the period and intensity of the signal). The JAK-STAT pathway is much simpler and consists of three components. JAK-STAT pathways are triggered by numerous cytokine receptors. These receptors do not enjoy intrinsic kinase activities. However, binding of a cytokine reasons its receptor to dimerize, which offers the signal to assistant and activate a JAK kinase. Originally, Janus kinases got their name from the typical presence of two kinase domains in each molecule. Several members of the family are recognized (JAK 1,2,3, etc.); each one contacts a specific set of cytokine receptors. The interaction between the activated (dimeric) cytokine receptor and JAK kinase(s) consequently produces the same result as the ligand-induced dimerization of a tyrosine kinase receptor. The difference, however, is that the receptor and kinase activities are possessed in different proteins rather than by the same protein. The JAK kinases are tyrosine kinases whose major substrates are TRANSCRIPTION FACTORS (TFs) are termed SIGNAL TRANSDUCERS AND ACTIVATORS OF TRANSCRIPTION (STATs). There occur more than seven STATs, and each STAT is phosphorylated by a particular set of JAK kinases. The phosphorylation occurs while the JAK is accompanying the receptor at the plasma membrane. A pair of JAK kinases is connected with an activated receptor, and both may be required for the pathway to function. An example of JAK-STAT pathway is that stimulation by the interferon - IFNγ which requires both JAK1 and JAK2. STAT phosphorylation results in the creation of both homodimers and heterodimers. The basis for dimerization is a reciprocal interaction between a Src homology 2 (SH2) domain in one subunit and a phosphorylated Tyr in the other subunit. The STAT dimers get transported to the nucleus and in some cases associate with other proteins. They bind to definite recognition elements in target genes, whose transcription is then activated.

jarovization: Means vernalization. *See* VERNALIZATION.

jasmonic acid (JA): JA has chemical formula $C_{12}H_{18}O_3$. It is distributed throughout higher plants, synthesized from linolenic acid through the octadecanoid pathway. An essential role of JA likely is its operation as a 'master switch', in SYSTEMIC ACQUIRED RESISTANCE (SAR) when SAR is activated in plants through (i) spray application of harpin protein to various plants, (ii) chewing of insects on the leaves of certain plants and/or (iii) via the entry into plant of certain pathogenic bacteria/

fungi. Proteins determined by jasmonate-induced genes include enzymes of alkaloid and phytoalexin synthesis, storage proteins, cell wall ingredients and stress protectants (e.g. the wound-induced formation of proteinase inhibitors, in which JA associates with abscisic acid (ABA) and ethylene to defend the plant from predation).

Java: A general-purpose, class-based, object-oriented, although not a pure OBJECT-ORIENTED (OO) PROGRAMMING language that is designed to possess as few implementation dependencies as possible. It is envisioned to let application developers 'WRITE ONCE, RUN ANYWHERE' (WORA), meaning that assembled Java code can run on all platforms that support Java without the requirement for recompilation.

jiffy pots: Pots made of wood pulp and peat and commonly utilized for transplanting tissue-culture-derived plants into soil medium.

joining segment: A small DNA segment that links coding parts of genes (called genelets) to generate a purposeful immunoglobulin gene encoding an immunoglobulin molecule. For example, through early B-cell differentiation in the bone marrow (BM), the variable (V), diversity (D) and joining (J) gene segments of the immunoglobulin (Ig) genes are rearranged in an ordered fashion to generate the primary Ig repertoire. Formation of Ig heavy (IgH) chain gene rearrangement precedes Ig light (IgL) chain gene rearrangement and D_H to J_H assembly precedes V_H to DJ_H assembly. *See* GENELET.

JoinMap: An MS Windows® programme used for construction of genetic linkage maps in experimental populations of diploid species. The software can deal with the common types of experimental population, including the full-sib family (F_1) of a cross between two individuals of an outbreeding species. It provides high quality tools that permit detailed study of the experimental data.

joint DNA molecule: Any DNA molecule comprising material contributed by two different DNA duplexes, as a consequence of breakage and reunion recombination (*in vivo*) or owing to recombinant DNA technology (*in vitro*).

joint single-nucleus (hydroxy)methylcytosine sequencing (Joint-sn^{hm}C-seq): E.B. Fabyanic *et al.* (2024) present an accessible and quantitative method that concurrently profiles $5^{hm}C$ and true $5^{m}C$ in single cells by yoking differential deaminase activity of APOBEC3A towards $5^{m}C$ and chemically protected $5^{hm}C$.

juggernaut polymerase: A transcriptase (named DNA-dependent RNA polymerase) that disregards all signals normally used for termination of transcription due to the rho (ρ) factor. This polymerase is assumed to form by attachment of other proteins conferring antitermination ability.

jumbling sequence addition order: Most methods used for phylogeny construction are sensitive to the order in which sequences are added to the tree. The simplest way developed to test a phylogeny is to repeat the analysis several times with different addition orders. Most of the phylogeny software have an option called JUMBLE that uses a random number generator to choose which sequence is to be added at each step.

Jumonji C (JmjC): An evolutionarily conserved motif which when present in a protein is predicted to be protein hydroxylases or histone demethylases.

jumping DNA/genes: *See* TRANSPOSABLE ELEMENTS.

jumping gene: The gene that keeps on changing its position in a chromosome and also between the chromosomes in a genome. These genes are associated with transposable elements. *Aka* TRANSPOSON *or* TRANSPOSABLE ELEMENT. *See details as separate terms.*

Juncea: Refers to a group of related plants, often commonly called WILD MUSTARD.

junctional diversity: Variability in immunoglobulins caused by variation in the exact crossover point during V-J, V-D and D-J joining.

junctional proteins: Proteins specifically associated with cell functions.

junctional sequence: Refers to any of the two terminal regions of the intron in an RNA precursor molecule.

junk DNA: Presently, only the function of a small fraction of the genomic DNA is known in eukaryotes, and the rest (over 95%) is classified as 'junk' which supposedly has largely unknown function. The most exhaustive knowledge is about the genes responsible for the morphological structures, the structural genes, which are the simplest part of the system. The knowledge about the most important part of this system, the regulator genes and their incomplete genetic code language is only partially known. The sequences of this unknown DNA are inherited and some of them are repetitive. The DNA has been found to mutate. A key feature of biological organisms is expending optimal energy for performing a function. To synthesize and retain massive amounts of superfluous molecules seems conflicting to this fundamental energy saving feature of biological organisms. It has now been discovered that at least some of what was formerly thought of as junk DNA (e.g. INTRONS) allows more than one specific protein molecule to be made from certain genes. Growing evidence now indicates many important functions of this DNA, including various regulatory roles. Some of the so-called junk DNA seems to serve as a reservoir for the future evolution. *Aka* SELFISH DNA, INNOCENT DNA *or* NON-CODING DNA.

just so stories: Untestable explanations for adaptations.

juvenile: (i) A stage in the sexual cycle of a plant recognized by differences in appearance from the adult and which lacks the ability to retort to flower-inducing provocations. Some horticultural and forest plants after their planting do not flower for several years for instance pear takes 7–8 years and bamboo takes many decades to flower. This non-flowering phase is called juvenile phase. (ii) An immature form that appears similar to but usually smaller than the adult and is not sexually mature (e.g. insects with gradual metamorphosis, nematodes).

juvenile leaves: Leaves growing on a young plant that characteristically differs from the adult plant leaves.

juvenile resistance: The typically vertical resistance of plants in the seedling stage. *Aka* SEEDLING RESISTANCE.

juvenile *in vitro* embryo technology: A technology involving collection of undeveloped ova from young animals, *in vitro* maturation of those ova, *in vitro* fertilization and then transfer of the resultant embryos into recipient females. This method is adopted for achieving rapid generation turnover.

juxtacrine signalling: When signals are often associated with the extracellular matrix or presented on the exterior of neighbouring cells.

K

K: A statistical parameter used in computing Basic Local Alignment Search Tool (BLAST) scores that is believed to be a natural scale for search space size. The value K is useful in converting a raw score (S) of a bit score (S').

K^+ leak channel: A K^+-transporting ion channel that functions in the plasma membrane of animal cells. It stays open even in a 'resting' cell.

K-selection: Selection acting on the density-dependent fitness parameter, K. This concept is not used anymore in evolutionary biology. *Compare with* R-SELECTION.

Ka: *See* DN.

kanamycin-resistant tissue: A tissue that is resistant to the lethal influences of kanamycin. Some cloning vectors carry a kanamycin-resistant (*Kan^r*) gene as a selectable marker.

***Kan^r*:** *Kanamycin resistance* gene.

kappa particle: A particle present in the cytoplasm of some Paramecia (killers) that secretes a substance that kills.

Karnal bunt: A fungus disease of wheat that diminishes yields and results in a foul tasting, but inoffensive flavour in flour pulverized from infected grains.

karyoevolution: Evolutionary modifications in the chromosome set, measured as alterations in number and gross structure of the chromosomes. In a broad sense, evolutionary associations between taxa are shown by karyotypic modifications.

karyogamy: Union of the haploid nuclei of two cells to form a diploid (2n) cell. During somatic cell hybridization, the fusogen induces fusion of two protoplasts. Cytoplasmic fusion first results in the heterokaryon situation, where two different kinds of nuclei coexist in the same cytoplasm. It is followed by karyogamy resulting into the fusion of two nuclei. It is also a frequent phenomenon during the sexual cycle of fungal reproduction. *See* PLASMOGAMY.

karyogenesis: Refers to biogenesis during mitosis of the nucleus, the dominant structure of a cell. Cell is the smallest, most basic unit of life that is capable of existing by itself. The division of the cell nucleus (karyokinesis) is distinct from cytoplasmic division or cytokinesis. It represents a system by which the genetic information delimited in the chromosomes of eukaryotes is disseminated to the daughter nuclei, which are generally identical to each other and the mother cell nucleus.

karyogram: A systematic array (picture or figure) of chromosomes. *See* IDIOGRAM.

karyokinesis: Division of nucleus. It comprises prophase, metaphase, anaphase and telophase stages. *See* CELL DIVISION.

karyological races: Races of species with all the members possessing the same chromosome number, which is characteristic of that species.

karyology: The study of the nucleus and its components.

karyolymph: A clear fluid or nuclear sap within the nuclear membrane.

karyolysis: During karyogenesis, nuclear membrane visible in interphase and prophase nucleus vanishes during metaphase.

karyomorph: A clear fluid or nuclear sap within the nuclear membrane.

karyoplast: A cell nucleus obtained from the cell by enucleation. It is surrounded by little cytoplasm bounded by plasma membrane. Karyoplasts are formed from the cells or protoplasts. These are smaller than the parental cell/ protoplast and are used in fusion and transformation experiments.

karyosome: Any of several masses of chromatin in the reticulum of a cell nucleus.

karyosphere: The part of the nucleus into which the chromosomes have contracted.

karyostasis: The stage of the cell cycle in which there is no visible chromosome dividing activity in the nucleus but there occurs abundant metabolic and synthetic activity.

karyotype: A systematic diagrammatic representation of the number, sizes and shapes of the entire set of metaphase chromosomes. The karyotype is generally species specific. In the majority of the organisms of a species, all cells have the same karyotype. But similar looking species may have different karyotypes, demonstrating that similar genetic potential can be planned on chromosomes in several different ways.

karyotyper: A scientist or more commonly an automated analytical machine that (i) captures a video picture of a given cell under a microscope, (ii) digitizes that picture within a computer, (iii) 'cuts out' the individual chromosomes contained within that cell's genome and (iv) arranges the cell's chromosomes in pairs by size into a chart, popularly called a KARYOTYPE.

karyotyping: A laboratory procedure that permits scientists to see all of the chromosomes present in the cell at one time in black and white. It is beneficial for observing the number, size and shape of all the chromosomes. Inferring these karyotypes needs a skilled person, who might need a few hours to examine a single chromosome.

karyotyping, spectral: In this case, each pair of chromosomes is painted in a different fluorescent colour. Even non-experts can see instances where a chromosome, painted in one colour, has a small piece of a different chromosome, painted in another colour, attached to it. Spectral karyotyping offers depiction of certain numerical and structural changes and highly informative analysis of sample materials with only single or few cells available for investigation.

Kb/cM rate: The average chromosome distance (measured as DNA in kilobases) per unit of genetic distance (measured in cM). It denotes a general mean of a species (computed as genome dimension in kb/total genetic map) or to a precise chromosome segment.

keel (Adj. keeled): A protruding longitudinal crest like the keel of a boat (e.g. in flowers, the erection of the corolla formed by the fusion of the lower edge of the two abaxial anterior petals of flowers in the Fabaceae).

Kelvin probe force microscopy (KPFM): In the Kelvin method (or vibrant capacitor method), two conductors are organized as a parallel plate capacitor, and the spacing is kept small. The CONTACT POTENTIAL DIFFERENCE (CPD) between the two materials is sensitive to the changes in the surface electronic structure of the plates. Various ways exist by which measurements can be obtained to understand the structure of interfaces.

kernel: (i) In botany, a whole grain or seed of a cereal plant or the part of the seed inside the pericarp. *See* DRUPE. (ii) In computer science, the first section of an operating system to load into memory. As the centre of the operating system, the kernel needs to be small, efficient and loaded into a protected area in the memory so that it cannot be overwritten. It may be responsible for several essential tasks, such as disk drive management, file management, memory management and process management.

key gene: A gene with a major effect on a phenotype. *See* OLIGOGENE.

key innovation: A novel phenotypic attribute that permits successive evolutionary radiation and accomplishment of a taxonomic group.

keystone species: The species that are very important to their ecosystem. They are the MINIMUM VIABLE POPULATIONS (MVPs). Without the keystone species, many other species are disturbed, changing the ecosystem. If most of the animals in an area depend on blackberries to live, blackberries will be a keystone species, because without them, a lot of animals will go hungry.

killer yeast: Yeast cells comprise a double-stranded RNA plasmid that is transcribed to encode genes required for production of a killer substance that is lethal to other cells. The toxic substance is closely similar in its action and synthesis to a bacterial colicin.

kilobases (kb): A unit of measure of DNA or RNA sequence 1000 nucleotides in length.

kilobase pairs (kbp): An abbreviation for 1000 base pairs in double-stranded DNA **or** RNA.

kilocalorie (kcal): Unit of heat energy equal to 1000 calories, which is frequently used to denote the energy content of food or molecules (e.g. bond strengths are measured in kcal/mole). An alternative unit in wide use is the KILOJOULE, equal to 0.24 kcal.

kilodalton (abbr. **kDa**): A unit of mass equal to 1000 Da. *See* DALTON.

kilojoule: Standard unit of energy equal to 1000 joules, or 0.24 kilocalories.

kin discrimination: The skill to distinguish between related and unrelated individuals.

kin selection: (i) A change in the frequency of an allele, triggered by the influence of that allele on the fitness of other individuals who carry the allele. (ii) Adaptive evolution of genes owing to relatedness. An allele instigating an individual to act to benefit its relatives will increase in frequency if that allele is also found in the relatives and if the benefit to the relatives is higher than reimbursing the cost to the individual.

kinase: A type of enzyme that speeds up chemical reactions in the body by adding a phosphate group (PO_4^{3-}) to other molecules. This may cause other molecules in the cell to turn out to be either active or inactive. Kinases are a part of many cell processes. It catalyses reactions related with the transfer of phosphates from a nucleoside triphosphate (e.g. adenosine triphosphate (ATP)) to another substrate. Kinase $p34^{cdc2}$ thus controls the events that initiate changes of phase in cell cycle, whereas cyclin B seems to be clock within changes quantity marking time when changes take place. Since $p34^{cdc2}$ requires cyclin before it can function, it is referred to as *cyclin-dependent kinase* (*cdc13* gene produces cyclin). The synchronized action of *cdc2* and *cdc13* genes is controlled by two genes, *cdc25* and *wee1*. cdc25 is a phosphatase which dephosphorylates *cdc2* to make it functional, Wee1 produces a kinase which phosphorylates cdc2. Another gene, *nim1* (never in mitosis), exercises a negative control on cell by phosphorylation. *Cdc25* gene produces a phosphatase. Its activity removes phosphate from tyrosine molecule at position 34 of *cdc2*. An increased amount of Cdc25 causes early entry into mitosis, thus causing lethality. Increase in the ratio of Cdc25 and Wee1 produced increases cell size needed for the entry into mitosis, whereas reduction in the ratio results in reduction in critical cell size. *See* PHOSPHORYLATION.

kinases in eukaryotic DNA replication: Cyclin-dependent kinases (CDKs) control major cell cycle steps, such as the start of chromosomal DNA replication. P. Zagerman and J. Diffley (2007) recognized two S phase CDK (S-CDK) phosphorylation positions in the budding yeast Sld3 protein that, together, are crucial for DNA replication. When phosphorylated, these sites bind to the amino-terminal BRCA1 C-terminal (BRCT) repeats of Dpb11. A Sld3-Dpb11 fusion construct bypasses the recruitment for both Sld3 phosphorylation and the N-terminal BRCT repeats of Dpb11. Co-expression of this fusion with a phospho-mimicking mutant is a second essential CDK substrate, Sld2, promotes DNA replication in the absence of S-CDK. Therefore, Sld2 and Sld3 are the minimal set of S-CDK targets necessary for DNA replication. DNA replication in cells lacking G1 phase CDK (G1-CDK) requires expression of the Cdc7 kinase regulatory subunit, Dbl4, as well as Sld2 and Sld3 bypass. This describes how G1- and S-CDKs promote DNA replication in yeast.

kinesin: One type of motor protein that uses the energy of ATP hydrolysis to move along a microtubule.

kinetic complexity: The complexity of a DNA component measured by the kinetics of DNA reassociation.

kinetic constriction: A non-staining, restricted region in each chromosome that stays single for some time after the remaining chromosome has divided and that, at metaphase, appears to be attached to the spindle fibre.

kinetics: A dynamic process involving motion.

kinetin: A synthetic cytokinin. A group of plant growth substances, natural or synthetic, that inspires cell division.

kinetochore microtubule: In a mitotic or meiotic spindle, a microtubule having one end attached to the kinetochore on a chromosome.

kinetochore: A dense, plaque-like part of the centromere region of a chromatid, to which the microtubules of the spindle become attached during cell division. Kinetochore is shaped from proteins on a mitotic chromosome to which microtubules attach and which plays a dynamic part during the movement of chromosomes to the poles. Two chromatids when joined together by a centromere make a complete chromosome. Centromeric DNA at the kinetochore has a disc- or plate-like structure. Kinetochore is composed of both DNA and protein. Small fibres are attached to the kinetochore during cell division and help in movement of chromosomes.

kinetoplast: These are adapted mitochondria positioned near the base of flagella. A kinetoplast may be bigger than a nucleus.

kinetoplast DNA: The kinetoplast DNA (kDNA) consists of a large quantity of interlocked 'minicircles'. They are equivalent to small bacterial plasmids. The kinetoplast DNA of trypanosomes and other parasitic members of the order Kinetoplastida is systematized into a complex system containing thousands of catenated circular DNA molecules.

kinetoplast DNA maxicircle: Maxicircle in Trypanosome kinetoplast DNA comprises 23 to 36 kbp. Maxicircle molecules are homologues of informational mitochondrial DNA molecules in animals and fungal cells and carry various genes. Several maxicircle genes represent CRYPTOGENES (i.e. incomplete genes) whose transcripts are edited to produce translatable sequences. *Compare with* KINETOPLAST DNA MINICIRCLE.

kinetoplast DNA minicircle: Minicircles in trypanosome kinetoplast DNA have 465 to 2500 bp. No noticeable protein-encoding genes are known to occur in minicircles.

kinetoplastid: Member of a phylum of eukaryotes in the *Excavata* kingdom identified by the existence of a kinetoplast organelle.

kingdom: In biology, a kingdom is the second-highest taxonomic rank, just below domain. Kingdoms are divided into smaller groups called phyla. The plant kingdom comprises vascular plants, bryophytes and green algae (*Aka* the CLADE Viridiplantae). Current molecular data have generally strengthened the evolutionary significance of the kingdoms Animalia, Plantae and Fungi. The single-celled eukaryotes once placed into the kingdom Protista are now known to be very diverse and not closely related to one another. The prokaryotic organisms once placed into the kingdom Monera are now considered to belong to separate domains, Eubacteria and Archaea. *See* TAXON.

kink: A sudden and significant bend between two helical segments of DNA achieved by unstacking one base pair and twisting (kinking) the polynucleotide backbone.

Kirby-Bauer test: *See* DISC DIFFUSION METHOD.

kissing in RNA pairing: The first step in RNA pairing, which involves initial loop-loop interaction.

Klebsiella: An inoffensive type of Gram-negative bacteria exist in the gut but can be cause of several different forms of infections, together with pneumonia, bloodstream infections, wound or surgical site infections and meningitis.

Klenow fragment: The large portion of *E. coli* DNA polymerase I which is devoid of 5'→3' exonuclease activity and can therefore not be used for nick translation but is very valuable for filling-in reactions and DNA sequencing reactions by the SANGER method, which proceeds in a 5'→3' fashion.

Klinefelter's syndrome: A genetic disease in man due to the XXY karotype. It produces sterile males with some intellectual disability.

knee: In plants, a sudden bend in a root or stem, commonly observed at a node. A cypress knee, or pneumatophore, is a type of bend or knob in the root of some plants, especially conifers, such as some of the Taxodioideae that reveal a projection of the root above ground level or mud level.

knob: A heavily stainable and quite big chromomere observed along a chromosome of some plants, for example, in maize, it is used as a marker in pachytene analysis.

knock-in: Replacement of a gene by a mutant version of the same gene using homologous recombination.

knock-out: A laboratory organism (a plant or an animal) in which a gene has been made non-functional. By eliminating the gene's function, scientists can deduce its involvement in the organism's physiology from the phenotype. Knock-out procedure is usually performed via homologous recombination, following transfection with an appropriate DNA construct.

knock-out genes: Transgenes that have been designed and incorporated to silence an existing gene of a host.

knock-out mice: Mice which have been engineered to lack activity of a chosen gene. The specific gene is inactivated in the embryonic stem cells using the procedure of homologous recombination. These cells are then transferred into an early-stage embryo (blastocyst), and this is then transplanted into a recipient mouse. The successive progeny lacks the targeted gene in some cells. This technique is used to know the function of the chosen gene. *See* GENE TARGETING.

knock-out technique: O. Smithies (1985) studied gene function by knocking off genes in mice. Knock-out mice are genetically modified as they carry one or more genes that have been silenced. Knock-out mice are utilized to reconstruct human diseases in mice to study the role of individual genes in an organism's development.

knot: An abnormal swelling or confined outgrowth, often roughly spherical, formed by a plant as a result of attack by a fungus, bacterium, nematode, insect or other organism. *See* GALL *and* TUMOUR.

Koch's postulates: The procedure used to demonstrate the pathogenicity of an organism (i.e. its role as the contributory agent of a disease).

Kompetitive Allele-Specific PCR (KASP): KASP is a homogenous, fluorescence-based genotyping variant of polymerase chain reaction which is based on allele-specific oligo extension and fluorescence resonance energy transfer for signal generation. KASP is a high-throughput genotyping procedure established on the basis of SINGLE-NUCLEOTIDE POLYMORPHISM (SNP). Because of its high flux, low cost and strong operability, this method has revealed a great application potential in the field of crop improvement.

Kornberg enzyme: *See* DNA POLYMERASE I.

Kosambi formula: Recombination fractions and map distances match well only over relatively short recombinational distances. As genetic distance increases, the probability of a second (and correcting) recombination also increases. Hence, the measured recombination for two loci is less than would have been apparent if a third intervening locus were present. Various mapping functions seem to permit single recombination fractions to be converted to map distances. D.D. Kosambi (1944) presented the simplest and probably most general functions, which is given as $x = 25 \log_n [(1 + 2y)/(1 - 2y)]$ where x is the map distance (in cM) corresponding to the recombination fraction, y (e.g. if the recombination value is 0.05, then the distance is 5 cM; if the value is 0.1 and then the distance is 10.1 cM). *Aka* KOSAMBI FUNCTION.

Kozak's scanning hypothesis. In eukaryotic mRNA, first translation initiation codon AUG in 90% of the cases occurs in the form of consensus sequence CCRCCAUGG (where R = purine), known as M. Kozak's sequence (1983). In 5% cases, one or more AUG codons have been reported to be present upstream to first AUG codon. These extra AUG codons are also read according to Kozak's hypothesis. Such initiation codons make false starts and consequently extended protein is obtained. If there is a termination signal between the false and real starts irrespective of reading frame, ribosome does not leave mRNA but reading of the real message starts if AUG is encountered within a reasonable distance. An AUG codon can be missed if it poorly matches Kozak's sequence.

Krebs cycle: A biochemical pathway that converts two carbon compounds to CO_2. It is also called TRICARBOXYLIC ACID (TCA) CYCLE or CITRIC ACID CYCLE. It operates in most of the body cells. In this cycle, every two molecules of acetyl-CoA that enter the cycle, four molecules of CO_2 are released. Besides, six molecules of NADH and two molecules of $FADH_2$ and two molecules of ATP are also formed. It is a catabolic pathway. CO_2 thus formed is released into the atmosphere. *See* CITRIC ACID CYCLE *and* TCA CYCLE.

Krüppel-associated box (KRAB): A conserved domain of a transcription repressor that can be fused to DNA-binding proteins for targeted transcription repression.

Ks: *See* Ds.

L

lab on a chip: A microdevice that allows rapid, microanalytical analysis of DNA or protein in a single, fully integrated system. Typically, these devices are miniature surfaces, made of silicon, glass or plastic, which carry the necessary microdevices (pumps, valves, microfluidic controllers and detectors) that allow sample separation and analysis. These devices are used in drug discovery, genetic testing and separation science.

label: (i) As a noun, a chemical group, radioactive atom or fluorescent dye added to a biomolecule (DNA, RNA or protein) for following its progress through a biochemical reaction or to detect it spatially. (ii) As a verb, to add such a group or atom to a cell or molecule. Incorporation of an easily detectable signal into a DNA molecule. Currently, radioactive labelling is being increasingly replaced by non-radioactive methods. *See* LABELLING.

label-free quantification: In this method, all forms of a peptide are required to be identified, and the ionization efficiency needs to be normalized because it varies among differently modified peptides. This method is used to have a rough estimation of histone post-translational modification stoichiometry.

labelled oligonucleotide probe from known peptide sequence: Micropeptide sequencers are available that can rapidly generate sequence information. From this sequence, you can use reverse translation of the amino acid sequence to obtain the nucleic acid sequences of an amino-terminal fragment and a carboxy-terminal fragment. In this case, the fragment complementary to the anticoding strand will be a direct conversion of the genetic code for the amino acids in the amino-terminal fragment of the protein. The strand complementary to the coding strand will be complementary to the derived nucleic acid sequence of the carboxy-terminal fragment. Screening of a cDNA library can be done with a labelled oligonucleotide probe based on a known peptide sequence.

labelling: The process of marking a DNA molecule using radioactive or non-radioactive labels.

labelling of foods: The process of developing a list of ingredients contained in foods. Labels imply that the list of ingredients can be verified. The US Food and Drug Administration has jurisdiction over what is stated on food labels.

labelling seeds: Adding labels to seed lots with information on variety identity, purity and seed quality.

labelling the 3′ ends of DNA: The substrate for this reaction is a fragment of DNA generated by digestion with a restriction enzyme that leaves a 3′ overhang, but oligodeoxynucleotides can also be used. When such DNA is incubated with tagged nucleotides and terminal transferase, a string of the tagged nucleotides will be added to the 3′ overhang or to the 3′ end of the oligonucleotide.

labellum: Lip, one of three or five petals which are normally different from the others, such as in Orchidaceae, Zingiberaceae, Cannaceae and Stylidiaceae.

labile gene: Genes that are constantly mutating.

laboratory populations: The relative frequency of alleles at a locus in a laboratory population is usually fixed at some convenient ratio by the design of the experiment, such as when two homozygous parents are crossed ($TT \times tt$) to begin an experiment, the alleles are thereby introduced in equal frequency. In laboratory populations, the system of mating is usually sharply defined, such as crossing $F_1 \times F_1$ or F_1 to one of the parents. Thus, the breeder entertains a wide variety of crosses.

laboratory-based approaches for gene prediction: Experimental procedures for gene prediction in new DNA are basically of three types: (i) identification via hybridization to mRNA or cDNA; (ii) identification of the 5′ end and intron-exon junctions of the gene and (iii) exon trapping.

***Lactis aerogenes*:** Harmless organism sometimes involved in the souring of milk. Genus *Lactis* is also called *Aerobacter*.

lactose (lac) operon in *Escherichia coli*, working of: The controlling region of lac operon is composed of three sites, each of which serves to bind a specific protein. In the absence of a repressor protein, RNA polymerase begins transcription at the operator, which is also the site at which the repressor attaches. When the repressor, a product of the regulator gene, binds to the operator, RNA polymerase is physically blocked. Transcription is not allowed to take place. This happens when substrate lactose is absent. Lac operon is thus an inducible operon. Since the final product of the regulatory gene acts to suppress transcription, it is undergoing a negative control. Lac operon is, therefore, under inducible negative control. In the presence of substrate lactose, the repressor is inactivated. It does not bind to the operator. If the repressor is already bound with the operator, it gets released. Now, there is no physical block to stop the activity of RNA polymerase. All three genes *lacZ*, *lacY* and *lacA* are cotranscribed and the mRNA contains messages from all three genes. It is, therefore, a polycistronic mRNA, which is then used as a template to produce rapidly and simultaneously all three enzymes, β-galactosidase, galactoside permease and thiogalactoside transacetylase, which are used in lactose metabolism. β-galactosidase accomplishes two functions: (i) breakdown of disaccharide lactose into two monosaccharides, glucose and galactose; and (ii) formation of inducer by conversion of lactose into allolactose. Galactose permease facilitates the entry of lactose into the cell. Thiogalactoside transacetylase is involved at some point in lactose metabolism.

lactose (lac) operon in *Escherichia coli*: F. Jacob and J. Monod in 1961 studied the regulation of lactose (lac) operon in *Escherichia coli*. The lac operon in *E. coli* contains a promoter (p), an operator (o) and three structural genes *lacZ*, *lacY* and *lacA*. The three cistrons (genes) are closely linked, cotranscribed and controlled coordinately. Lengths (in nucleotides) of different components of lac operon of *E. coli* are as follows: lac i promoter, 40; lac i operator, 26; *lacZ*, 3063; *lacY*, 800 and *lacA*, 800. Polypeptide size (in number of amino acids) of repressor and three enzymes produced by lac operon is: repressor, 360; β-galactosidase, 1021; galactose permease, 275; thiogalactoside transacetylase, 275. *Escherichia coli* lac promoter has three components: catabolic gene activator site (cgs); initial binding site (ibs); and operator (op). CYCLIC ADENOSINE MONOPHOSPHATE (cAMP) is essential for the activation of the cga protein. cAMP-cga protein system is impacted by the level of glucose. This exerts a positive control on the lac operon.

***lactose* (lac) promoter mutations:** Promoter in the lactose operon was found to be present to the left of the operator.

lactose (lac) repressor of *Escherichia coli*: A tetramer protein that can bind simultaneously to two lac operator DNA fragments. This binding seems to be of physiological importance. Cooperative DNA binding of regulatory proteins (e.g. lactose repressor) is a beneficial approach for (i) enhancing the specificity of DNA-protein interactions and (ii) building delicate genetic switches to regulate transcription of the *lac* operon in retort to the levels of lactose in the cellular environment.

lacuna: An empty space, hole, cavity, pit, depression or discontinuity.

ladder system approach: A systematic approach in plant breeding commencing from developing a genetic base, exploitation of variability, removal of bottleneck genes, boosting productivity and quality improvement adaptive design, finally leading to the release of an improved variety.

laevorotatory or **levorotatory (L) isomer:** An isomer of an optically active compound which rotates (when illuminated) the plane of plane-polarized light to the left. *Compare with* DEXTROROTATORY (D) ISOMER.

lag phase: When microbes or plant/animal cells are inoculated into the fresh liquid medium, they do not exhibit any growth for a period ranging from one hour to several days. During this phase, the cultured cells undergo a high rate of metabolism, DNA and enzyme synthesis required for cell division.

lagging chain: The strand of the double helix that, during DNA replication, is synthesized in a discontinuous fashion in the form of Okazaki fragments. *See* OKAZAKI FRAGMENT.

lagging chromosome: Deferred movement of a chromosome from the equator to the poles at anaphase so that it becomes left out from the daughter nuclei after anaphase. *Aka* LAGGARD.

lagging strand: During DNA replication, the term refers to the strand that is semi-continuously synthesized in the 3′ to 5′ direction by ligation of short DNA strands, called OKAZAKI FRAGMENTS, manufactured discontinuously in the 5′ to 3′ direction. Synthesis of each Okazaki fragment is primed using an RNA primer. Subsequent to Okazaki fragment synthesis, RNA primer is removed, and the Okazaki fragments are joined later by DNA ligase to make the lagging strand continuous.

lagging strand DNA synthesis coupling to parental H3-H4 transfer: Inheritance of epigenetic information is crucial for preserving cell identity. The transfer of parental histone H3-H4 tetramers, the principal transporter of epigenetic modifications on histone proteins, embodies a vital step in the inheritance of epigenetic information. A. Serra-Cardona *et al.* (2024) reveal that the lagging strand DNA polymerase, Pol δ, interrelates directly with H3-H4 and that the interaction between Pol δ and the sliding clamp PROLIFERATING CELL NUCLEAR ANTIGEN (PCNA) controls parental histone transfer to lagging strands, most likely self-defining their functions in DNA synthesis. When united, mutations at Pol δ and Mcm2 that compromise parental histone transfer lead to a greater reduction in nucleosome occupancy at nascent chromatin than mutations in either of them alone. PCNA donates to nucleosome positioning on emerging chromatin. PCNA–Pol δ complex connects lagging strand DNA synthesis to parental H3-H4 transfer, easing epigenetic inheritance.

Lamarckism: The idea that an organism can pass on characteristics that are acquired during its lifetime to its offspring. To date, there is no clear-cut evidence showing the inheritance of acquired traits. The concept of epigenetic inheritance, however, seems to support this idea. *See* THEORY OF INHERITANCE OF ACQUIRED characteristics.

λ: A statistical parameter used in computing BLAST scores that can be considered a natural scale for the scoring system. λ is used for converting a RAW SCORE (S) to a BIT SCORE (S').

λ bacteriophage (phage): A temperate bacteriophage of size 48.5 kb that infects *Escherichia coli*. It is commonly utilized as a vector in recombinant DNA construction.

λ bacteriophage lytic cascade: One of the most sophisticated examples of gene regulation known so far includes numerous finely coordinated interactions between λ genes. When sensitive *Escherichia coli* strains are infected by the temperate λ virus, a choice occurs between two alternative pathways: LYSIS or LYSOGENY. If the lytic pathway is followed, ~100 viral particles are formed within one hour by an infected bacterial cell and the cell is destroyed for release of virus particles. On the other hand, the lysogenic pathway results in the integration of the λ chromosome into the bacterial chromosome, λ and bacterial chromosome develop harmony, and the cell now continues to live and replicate, the bacterial cell becomes immune from lysis by λ until induction occurs. The choice of which pathway to follow is determined by factors in both host and virus, such as the nutritive state of the host and the number of λ viruses infecting a cell. Lysogeny is augmented both in starved cells and when there is a high multiplicity of infection.

λ exonuclease: This DNA-specific exonuclease catalyses the removal of nucleotides from linear or nicked double-stranded DNA in the 5′ to 3′ direction. It shows highly processive degradation of double-stranded DNA from the 5′ end, and its preferred substrate is 5′-phosphorylated double-stranded DNA, although non-phosphorylated substrates are degraded at a greatly reduced rate. Lambda Exonuclease is ideal for conversion of the linear double-stranded DNA to single-stranded DNA via preferred activity on 5′-phosphorylated ends.

λ phage repression: Most direct display of interaction between DNA-bound proteins and between separated proteins with concomitant DNA looping has come from the studies of the λ phage repressor. The λ repressor is known as both a negative and positive controller of gene expression. When bound to the right operator (O_R) on the phage chromosome, the suppressor switches off (rightward) transcription of the *cro* gene essential for lytic growth and concurrently switches on (leftward) transcription of its own gene, *cl*. Repressor bound to the strong site O, O_R^1, assists another repressor to bind to the other weaker adjacent site O_R^2 and this arrangement triggers expression of the repressor gene through contact of O_R^2-bound repressor with RNA polymerase and thereby assists it to bind and begin transcription in the adjacent leftward promoter. At the same time, rightward transcription is suppressed because the bound repressors stop polymerase from binding to the rightward promoter. Interaction between carboxyl domains is also involved in cooperative repressor binding to adjacent operator sites. An amino domain of the repressor dimer O_R^2 contacts polymerase at the adjacent promoter. If the concentration of repressor enhances transiently, site O_R^3 becomes occupied by a third repressor dimer and consequently transcription of the repressor gene is turned off. The λ repressors bind cooperatively to separated operator sites just as they do to the adjacent sites, the DNA in between the sites bending smoothly to accommodate the interaction. Electron microscopy provides an excellent piece of evidence for DNA looping induced by λ repressor binding.

λ phage system of site-specific recombination: Phage λ integrates into and excises from the host *Escherichia coli* genome by a regulated site-specific recombination event. The recombination reaction requires three proteins: (i) λ Int protein is used for integration and excision. (ii) λ Xis protein is required for excision only. (iii) INTEGRATION HOST FACTOR (IHF protein) is required for both directions.

λ protein: A protein of λ phage required to localize replication proteins.

lamina: The blade of a leaf or the expanded upper part of a petal, sepal or bract. *Contrasts* PETIOLE.

laminar air flow cabinet: A machine that generates clean air free from bacteria, fungus and dust particles. Laminar air flow cabinets are commonly used in tissue culture laboratories. This is fitted with a HEPA filter (0.2–0.3 μm pore size) that does not allow the bacteria and fungi to pass through. Therefore, the air thus generated by this machine is free from these microorganisms and good for performing inoculations under aseptic conditions. *Aka* LAMINAR FLOW CABINET. *See* HEPA FILTER.

laminin: Extracellular matrix protein found in basal laminae, where it forms a sheetlike network.

lampbrush chromosome: It is a type of puffed chromosome. A chromosome that has paired loops extending laterally to reveal a lampbrush-like shape occurs in primary oocyte nuclei of amphibians; they represent sites of active RNA synthesis. Puffed chromosome regions may also exist in plants (e.g. *Phaseolus* beans).

lanate: Enclosed in or consisting of woolly hairs.

land use: The deployment of land for any purpose. Competition for limited land necessitates the establishment of priorities among claims, which is the primary thing of land-use planning.

landfill: The activity involving dumping of refuse by tilting it on land. Often the refuse is used to fill in old mine workings or low-lying land, to reclaim land from water or to create a feature on flat land. If the refuse is dropped in prepared trenches or holes, over which earth can be heaped at the end of each day, this is called CONTROLLED TIPPING in the UK and sanitary landfill in the USA.

landmark: A cytological feature of a chromosome that aids in the identification of that specific chromosome (e.g. the centromere, p arm, q arm, telomere or certain defined bands).

landraces: The local varieties or populations developed as products of natural selection, which were cultivated by farmers over a long period of time on account of their adaptability and resistance to biotic and abiotic stresses. These are also called primitive cultivars. The original material might have been selected by each farmer according to his requirements and the environmental factors splitting it into different types. These are considered the best sources of resistance genes. In fact, these are the farmer's varieties. The availability of high-yielding varieties 'super varieties' has led to large-scale replacement of landraces. But still, these are available in some inaccessible areas and specific niches.

Langmuir-Blodgett (LB) method: A technique used to make ultrathin films. To do so, first, a Langmuir film is made (at the air-liquid border), which is then compressed and transferred to a substrate.

large insert libraries: Gene libraries constructed in Bacterial Artificial Chromosome (BAC) or Yeast Artificial Chromosome (YAC) possessing large inserts. Libraries are exploited for the development of detailed genetic and physical maps with higher levels of resolution appropriate for positional cloning of genes of interest. BAC libraries are preferred over YAC libraries due to various reasons: (i) BAC DNA is easy to purify in contrast to YAC DNA purification; (ii) BAC clones are stable in the host strains; (iii) the ease of using the physical maps constructed with BAC in genome research and (iv) low level of chimerism. The BAC library can permanently serve as a genomic research source. In the case of wheat, three BAC libraries from diploid wheat, *Triticum monococcum* (A genome) and *Aegilops tauschii* (D genome), one BAC library from tetraploid, wheat, *Triticum dicoccoides* (AB genome) and four BAC libraries from hexaploid wheat, *Triticum aestivum* (ABD genome) have been constructed.

large language models (LLMs): A type of language model notable for its ability to achieve general-purpose language understanding and generation. LLMs acquire these abilities by using massive amounts of data to learn billions of parameters during training and consuming large computational resources during their training and operation. LLMs are artificial neural networks (mainly transformers and are (pre-) trained using self-supervised learning and semi-supervised learning). Protein science has been one of the pioneering areas adopting the use of LLMs

(Tang, 2023). Using diverse protein sequence data sets compiled throughout the tree of life, protein language models with hundreds of millions to billions of parameters. A. Madani *et al.* (2023) describe ProGen, a language model that can generate protein sequences with a predictable function across large protein families, akin to generating grammatically and semantically correct natural language sentences on diverse topics. Z. Lin *et al.* (2023) trained transformer protein language models with up to 15 billion parameters on experimental and high-quality predicted structures and found that information about atomic-level structure emerged in the model as it was scaled up. They created ESMFold, a sequence-to-structure predictor that is nearly as accurate as alignment-based methods and considerably faster.

large non-coding RNA (lncRNA): Non-coding RNA larger than 200 nucleotides can have roles in epigenetic regulation of gene expression.

large X effect: The reflection that the X chromosome is implicated in hybrid incompatibilities more often than would be anticipated from its size.

large-scale double-stranded DNA isolation: This method employed for the isolation of large-scale cosmid and plasmid DNA is actually a modification of an alkaline lysis procedure tailed by equilibrium ultracentrifugation in caesium chloride-ethidium bromide gradients. Cells containing the anticipated plasmid (or cosmid) are harvested by centrifugation, raised in a lysozyme buffer and treated with an alkaline detergent. Detergent solubilized proteins and membranes are precipitated with sodium acetate, and the lysate is cleared first by filtration of precipitate using cheesecloth and then by centrifugation. The DNA-containing supernatant is transferred to a new tube, and the plasmid (or cosmid) DNA is precipitated by the addition of polyethylene glycol and collected by centrifugation.

large-scale double-stranded M13RF DNA isolation: Double-stranded M13 REPLICATIVE-FORM (M13RF) DNA is isolated for usage in M13 *Sma*I cut, dephosphorylated vector preparation. The growth conditions of M13-infected bacterial cells seem complicated but result in a maximal amount of M13 RF molecules per cell. After the M13RF-containing bacterial cells are collected by centrifugation, the double-stranded molecules are isolated using the caesium chloride method for large-scale plasmid isolation.

large-scale gene function analysis: Microarray transcript profiling and RNA interference are two new technologies critical for large-scale gene function studies in multicellular eukaryotes. Both these techniques count on sequence-specific hybridization between complementary nucleic acid strands. For this, GENE-SPECIFIC SEQUENCE TAGS (GSTs) are generated representing almost all the genes. GST resources offer innovative and commanding tools for functional genomics.

largest eukaryotic genome: P. Fernández *et al.* (2024) report the unearthing of the largest eukaryotic genome in *Tmesipteris oblanceolata*, a New Caledonian fork fern. At 160.45 Gbp/1C, this record-breaking genome contests existing records and unlocks new paths to discover the evolutionary forces at work of genomic gigantism.

lariat RNA: Refers to an RNA intermediate produced during the splicing of mRNA precursors. The lariats have a circular component with an extending tail and are shaped by a branch where an adenosine residue is linked to the 5′ end of an intron and to the remainder of the intron.

laser beam method of gene transfer: Laser beam method has been used for the direct transfer of alien genetic material into human cultured cells. Described A protocol for gene transfer in plants using a laser beam was developed by G. Weber *et al.* (1989).

last universal common ancestor (LUCA): Hypothesized common ancestral cell from which all the three domains of life – the Bacteria, the Archaea, and the Eukarya – originated. LUCA probably was a cellular organism that had a lipid bilayer and used DNA, RNA and protein.

***Late Elongated Hypocotyl (LHY)* gene:** An oscillating gene found in plants that function as part of their circadian clock. *LHY* encodes components of mutually regulatory negative feedback loops with Circadian Clock Associated 1 (CCA1) in which overexpression of either results in dampening of both of their expression. This negative feedback loop affects the rhythmicity of multiple outputs creating a daytime protein complex.

late embryogenesis abundant (LEA) promoter of toxin gene: The toxin gene is placed under the control of the LEA protein promoter, permitting *RIP* to express only during late embryogenesis, thus affecting only the embryo development. The promoter used becomes active only in the later stages of seed maturation.

late genes: A group of bacteriophage genes that are expressed during the later stage of the infection cycle. Late genes usually code for proteins needed for the synthesis of new phage particles.

late intron view: Split genes arise from uninterrupted genes by insertion of introns. Genes large enough to encode proteins first arose (presumably from smaller genes) without the participation of introns. The machinery of spliceosomal splicing arose from fragmented self-splicing introns. Spliceosomal introns were never observed in the ancestors of those organisms that now lack them.

late lambda (λ) phage genes: Late genes for lysis, *S* and *R*, are positioned on the right end while those for tail and head (*A* to *J*) are situated on the left end. Genes for recombination are also located on the left side of *cIII* and are transcribed as late genes. Soon after the infection, the two ends of the DNA molecule join to produce a ring, so that all late genes are arranged in a single group containing *S*-*R* genes from the right end and the head and the tail genes *A* to *J* from the left end. There is a promoter gene P_R' between genes *Q* and *S*. In the absence of the product pQ of gene *Q*, the transcription from P_R' is constitutive but terminates at tR3 lying close to P_R' forming a product 6S RNA (194 bases long). However, when pQ is present, it suppresses tR3 and 6S RNA is prolonged with the consequence that late genes are heavily expressed. Thus, gene *Q* induces transcription of late genes, which continue to be transcribed. Transcription from both sides stops on the ring molecule before the RNA polymerase molecules from the two sides clash.

late replicating: Often, the heterochromatic regions of chromosomes reveal a later replication than the euchromatic ones.

late replication: The heterochromatic regions of chromosomes, which show a later replication of DNA than the euchromatic ones.

late selection: Applied in self-pollinating crops, that is, in the variable progeny of a cross for three to four generations, by means of either the bulk breeding method or single-seed descent and producing a mixed population of relatively homozygous individuals. The selection is done among these homozygous individuals. It is effective because it produces plants with a reduced hybrid vigour, which can be misleading during the screening process, and it also yields a greater expression of recessive alleles that are exhibited only in the homozygous state. Late-selected plants have a higher heritability than those of early-selected plants. The advantage of late selection has been equated with the longer breeding cycle required.

late-acting self-incompatibility: The occurrence of SELF-INCOMPATIBILITY (SI) in flowering plants where pollen tubes from self-pollen successfully reach the ovary, but ovules fail to develop.

latent: (i) A pathogen present but not manifested or visible (e.g. a symptomless infection by a pathogen). (ii) Currently, the pathogen is inactive but is capable of becoming active.

latent infection: A long-lasting infection in which host-pathogen equilibrium is recognized without any visible symptoms of disease. In this case, a pathogen lives in the host for a long period without triggering disease. Latent infection is very common in viral diseases of humans. The virus can live within the body for years together without causing the disease. HIV can live up to 10 years or more. In the case of herpes simplex virus, many times, the individual carries the virus but signs never appear. Likewise, chicken pox virus can also occur in a latent state.

latent period: The time between infection and the formation of new inoculum.

latent virus: A virus that does not reveal symptom development in its host.

later steps in processing 5S rRNA: The 10S RNA contains six STEM-AND-LOOP STRUCTURES produced by the action of RNase III. Then, by the action of RNase P, fragment 7S RNA is produced, which contains 5S rRNA and a termination stem. Removal of this stem is achieved by RNase E.

lateral bud: A bud that progresses in the axil of a leaf. *See* AXILLARY BUD.

lateral force microscopy (LFM): LFM measures the forces parallel to the surface plane. The feedback loop must be slowed down, as always, when a force channel is measured. Changes of F_y and F_z deflect the light beam in the same direction. The cantilever is then buckled, which must distort the topography signal.

lateral gene transfer (LGT): The spread of DNA from one evolutionary lineage to another. *Aka* HORIZONTAL GENE TRANSFER.

lateral inhibition: The mechanism responsible for creating differences within the population of equivalent cells.

latex: A milky fluid that radiates from plants such as spurges, figs and dandelions.

laticiferous: Latex-bearing; manufacturing a milky juice.

Latin square design: An experimental which simultaneously controls fertility variation in two directions.

Latin squares: Designs available for any number of treatments, where feasible, more than one orthogonal treatment factor can be created to form Graeco-Latin squares.

Latin squares balanced for carryover effects: Pertinent when the same plots or subjects are treated during several successive time periods, and there exists interest in investigating both (i) the direct effect of treatment during the period in which it is applied and (ii) its carryover (or 'residual') effect during later periods.

latrorse: A type of anther dehiscence in which the anthers open laterally towards adjacent anthers. *Compare with* INTRORSE *and* EXTRORSE.

lattice design: An incomplete block design in which the number of treatments must be an exact square. There are numerous types of lattice designs (e.g. ALPHA LATTICE and PARTIALLY BALANCED SQUARE LATTICE). The lattice design is 'simple' if r = 2. SQUARE LATTICE DESIGNS are also called 'NETS'; they are incompletely balanced with respect to an association scheme of Latin square type; first associates harmonize once in blocks, second associates, never. For a 'rectangular' lattice design, there are n(n–1) treatments in r × n blocks of size n–1. The construction is similar to that for a square lattice design, except that n cells are removed from the square array. For a 'cubic' lattice design, there are n^3 treatments in $3n^2$ blocks of size n; the treatments form a cube of order n; the blocks are the 1-dimensional slices.

law: A law generalizes a body of observations. At the time it is made, no exceptions have been found to a law. Scientific laws explain things, but they do not describe them. One way to tell a law and a theory apart is to ask if the description gives you a means to explain 'why'. Consider Newton's Law of Gravity. Newton could use this law to predict the behaviour of a dropped object, but he could not explain why it happened. Some other such examples include Newton's Laws of Motion, Boyle's Law, Charles' Law, law of land. A law remains so only as long as it is not refuted by further experimentation.

law of ancestral inheritance: The idea that an individual contains, in equal amounts, genetic material from each of the four grandparents, each of the blend of the previous generation and each ancestor contributes a specific proportion of an individual's genome.

lawn: A continuous layer of bacteria on the surface of agar medium.

lazy learning: In machine learning, lazy learning is a learning method in which generalization of the training data is, in theory, delayed until a query is made to the system, as opposed to in eager learning, where the system tries to generalize the training data before receiving queries.

LD_{50}: *See* LETHAL DOSE 50.

leader DNA: The untranslated segment that lies upstream of the initiation codon on a mRNA molecule.

leader peptidase: An enzyme used to slash leader peptide from the preprotein to be secreted in bacterial cells.

leader peptide gene: A small gene within the attenuator control region of repressible amino acid operons. Translation of the gene is used to test the concentration of amino acids in the cell.

leader peptides: In bacteria, proteins fated to be secreted are synthesized as pre-proteins with N-terminal signal sequences which are short (25 amino acid residues) and include a hydrophobic central core that can assume an α-helical structure, flanked by a region containing several charged residues.

leader segment: Sector of the non-coding DNA sequence preceding 5′-end of gene proper (mature gene, antisense strand). These are the sequences between the initiation codon and the promoter-operator region. For example, in tryptophan operon of *Escherichia coli*, mRNA of gene *trpE*, which codes for enzyme anthranilate synthetase has a sequence of 162 bases between the initiation codon and the promoter-operator region of this gene. A nucleotide sequence of the mRNA to which the ribosomes bind; in other words, the non-translated sequence at 5′ end of mRNA, or N-terminal sequence of a protein constituting a signal for transport through a membrane, which is later removed. The leader segment determines post-transcriptional control. This post-transcriptional control is in addition to normal transcriptional control provided in the operon model. *Aka* LEADER SEQUENCE.

leader transcript: The untranslated segment of mRNA that lies upstream of the initiation codon on an mRNA molecule. Leader transcript includes SHINE-DALGARNO SEQUENCE, which assists in the binding of mRNA with the ribosome.

leading strand: During DNA replication, one strand is produced continuously and the other discontinuously.

The continuous strand is called the leading strand in which 5′ → 3′ synthesis proceeds in the same direction as replication for K movement. The discontinuous strand is called the lagging strand. This DNA strand is synthesized in short pieces called Okazaki fragments.

leading strand, continuous synthesis of: SINGLE-STRAND BINDING PROTEINS (SSBs) do not unwind DNA, but they bind to and stabilize unwound single-stranded DNA. Thus, SSB protein holds denatured *Ori*C open. DnaB helicase is loaded onto replication forks. DnaG primase is recruited to the replication fork by protein-protein interactions with DnaB helicase. Since β polymerase is needed to rein core polymerase to DNA, the end result is a primase-to-polymerase switch. Tethered DNA polymerase III now synthesizes the leading strand in a continuous fashion at the growing 3′ tip. DNA synthesis proceeds at a rate of 1000 nucleotides/second. Protein-protein contact between DNA polymerase III and DnaB helicase kindles helicase activity.

leaf: An outgrowth of a stem, usually flat and green. The main function of the leaf is manufacturing food via photosynthesis.

leaf apex: Leaf tip.

leaf dip: The method involving inspection of a sample of sap from a suspected virus-infected leaf using an electron microscope.

leaf gap: A parenchymatous zone in the stele above (distal to) a leaf trace.

leaf hopper: A mobile insect with sucking mouthparts (order Hemiptera).

leaf morphogenesis in *Arabidopsis thaliana*: Long-distance movement of RNA through the phloem is known to occur. Grafting experiments with a naturally occurring dominant gain-of-function leaf mutation in tomato have been used to demonstrate long-distance crusade of mutant mRNA into wild-type scions. The stock-specific pattern of mRNA expression has been graft-transmissible, indicating that the mRNA accumulation pattern is inherent to the transcript and not accountable to the promoter. The translocated mRNA caused modifications in the leaf morphology of the wild-type scions, indicating that the translocated RNA is functional. *KNOTTED*-like homeobox (KNOX1) genes regulate develop of the leaf from SHOOT APICAL MERISTEM (SAM) and may regulate leaf form. KNOX1 expression was correlated with complex leaf primordial. However, complex primordial may mature into simple leaves. Therefore, not all simple leaves develop similarly, and final leaf morphology may not be an adequate predictor of homology.

leaf scar: A healing layer established on a stem where a leaf has fallen off.

leaf senescence: A type of programmed cell death (PCD), during which leaf cells undergo coordinated changes in cell structure, metabolism and gene expression, resulting in a sharp deterioration in photosynthetic capacity. A cytokinin class of plant hormones plays a part in regulating leaf senescence because a decline in the cytokinin level occurs in senescing leaves. The external application of cytokinin often delays senescence.

leaf sheath: The lower, tubular portion of a grass leaf that hooks the culm.

leaf spot: A plant disease lesion that typically restricts development in the leaf after reaching a distinctive size.

leaf stomata: The epidermis is a single layer of epidermal cells that shelters all the parts of plants, including roots, stems, leaves and flowers. Unlike the other plant parts, leaves have two epidermal layers, the upper and lower epidermis, which border the inner mesophyll. In the epidermis, small pores, called STOMATA, permit leaves to exchange gases with the surrounding air. During photosynthesis, carbon dioxide (a reactant) is admitted into the leaf and oxygen (a by-product) is released through the stomata.

leaf trace: A vascular bundle connecting the stele to a leaf.

leaflet: One of the separate blades or divisions; the last segments of a compound leaf.

leafroll: A disease symptom in which the edges of the leaf roll or turn, either up or down. It often serves as an indication of virus infection.

leaky mutants: A mutant in which the mutated gene product, such as an enzyme, still retains a fraction of its normal biological activity.

leaky mutations: The mutations that lead to only partial loss of a characteristic of the organism possessing it.

least squares estimate: The value of a regression coefficient that gives the minimum sum of squared errors, where an error is defined as the difference between an observed and a predicted dependent variable value.

least squares method: A method of approximation based on the minimization of sums of squares.

leavening agent: Substance causing expansion of doughs and batters by the release of gases within such mixtures, producing baked products with porous structure. Such agents include air, steam, yeast, baking powder and baking soda.

lectins: A class of proteins that bind carbohydrates with high affinity and specificity. Lectins are found in all organisms. Lectins play an important role in the adhesion process, cell-to-cell recognition and signalling process. The presence of lectins in plants makes them resistant to some sucking type of insects. A lectin gene (*GNA*) cloned from a snowdrop (*Galanthus nivalis*) is being exploited in plant genetic engineering for developing insect-resistant crops.

left splicing junction: Refers to the boundary between the right end of an exon and the left end of an intron.

left-handed DNA: *See* DNA, H-FORM.

leghaemoglobin: A haem-containing, oxygen-carrying phytoglobin found in the nitrogen-fixing root nodules of leguminous plants.

legume: A fruit typical of the family Fabaceae, moulded from one carpel and either dehiscent along both sides or indehiscent. Also, a crop species in the family Fabaceae. Legumes are an important type of vegetable. They are commonly used for food and agriculture. Their seeds are used as food too. Common examples are beans and lentils. They play a very important role in the nitrogen cycle because they are home to important nitrogen-fixing bacteria. They are also a good source of protein!

lentic ecosystem: A freshwater ecosystem categorized by non-flowing water, such as a pond or lake. *Compare with* LOTIC ECOSYSTEM.

lenticel: Typically lens-shaped, porous tissue in bark with large intercellular spaces that allow direct exchange of gases between the internal tissues and atmosphere through the bark.

lentivirus: A genus of slow-growing viruses capable of integrating exogenous genetic information into the genome of dividing and non-dividing host cells and thus acts as one of the most efficient gene delivery vectors. Examples include HIV, SIV and FIV.

leptonema: *See* LEPTOTENE.

leptotene: The first substage of prophase I of meiosis I during which the chromosomes seem thread-like. Centrioles move apart. The DNA of each of the chromosomes has already replicated, and each chromosome consists of two identical members (each called CHROMATID). Chromosomes are thin and long and are distinguishable individually. *Aka* LEPTONEMA.

lesion: A localized diseased area or wound.

lethal dominant mutation: Even if one occurs in a single dose in a diploid organism, it causes the death of the organism before reaching the adult stage.

lethal dose (LD): The concentration of a treatment that eliminates a certain number of cells, individuals, etc. from the test population.

lethal dose 50 (LD_{50}): The dose of a compound that causes death in 50% of the test population treated.

lethal equivalent alleles: Alleles whose total effect is that of lethality. For example, four alleles, each of which would be lethal 25% of the time (or 25% of their bearers), are considered equivalent to one lethal allele.

lethal gene: A gene that renders an organism or a cell possessing it inviable in proper arrangement for expression. Dominant lethal alleles kill heterozygotes, whereas recessive alleles kill homozygotes only.

lethal load: It happens in many organisms that normal individuals are carriers of genes, which have no effect in heterozygotes, but are recessive and detrimental or harmful in homozygous form. For example, conifers are often carriers of a genetic load that is released by inbreeding (in particular selfing) and actually reduce the amount of selfing (inbreeding) in vital seeds compared to the frequency of selfing.

lethal mutation: Mutation of a gene to yield no, or a totally defective, gene product (protein), thereby rendering it unable to function and hence unable to sustain the life of the organism. It causes the death of the organism before it reaches the adult stage. A lethal mutation can be dominant or recessive in action.

lethal recessive mutation: The case in which inheriting two recessive alleles of a gene results in the death of the organism.

lethality that only operates on females: Engineered strains of some agricultural pest species have been developed with lethality that only operates on females of those species. This type of lethality was achieved in the Mediterranean fruit fly *Ceratitis capitata* by female-specific excision of cassette exon from *tra* gene.

leucine (leu): An aliphatic, non-polar, neutral, monocarboxylic, essential amino acid with the chemical formula ($HO_2CCH(NH_2)CH_2CH(CH_3)_2$) that, unlike most amino acids, is sparingly soluble in water.

leucine zipper: A structural motif occurring in several DNA-binding proteins. For example, eukaryotic transcriptional factors are rich in leucine (e.g. C/EBP, Fos, Jun and GCN4). The basic leucine zipper (bZIP) domain (motif), which binds DNA contains four to five leucines, are found at seven amino acid intervals. This motif occurs typically in transcription factors and other proteins that bind DNA. A structural motif is seen in many DNA-binding proteins in which two α helices from distinct proteins are joined together in a coiled-coil (rather like a 'teeth in a zipper'), forming a protein dimer. When binding to the DNA, basic amino acid residues bind to the sugar-phosphate backbone while the helices are seated in the major grooves. It plays a role in the regulation of gene expression.

leucine-rich repeat protein (LRR protein): Common type of receptor serine/threonine kinase in plants. Characterized by a tandem array of leucine-rich repeat sequences in the extracellular portion.

leucoplasts. The sites for monoterpene synthesis, leucoplasts are food-storage bodies, which store large amounts of starch. Leucoplasts sometimes differentiate into more specialized plastids, namely, AMYLOPLASTS, ELAIOPLASTS and PROTEINOPLASTS. In algae, the term leucoplast is used for all unpigmented plastids and their function differs from the leucoplasts of plants.

level of *Bt* expression: Concentration of *Bt* protein in the transgenic plant. *Bt* protein concentration in the transgenic plant varies from 0.1% to 0.5% of the total soluble plant protein, and this concentration is sufficient to kill the larvae of insects. The exception is chloroplast expression of *Cry2Aa2* in tobacco that reaches 2–3% of total soluble leaf protein. Expression levels in leaves, stems, roots and plant reproductive tissues vary both with the promoter employed and with the age of the plant.

level of significance: The probability value in statistics which is used to reject the null hypothesis (H_0). The maximum tolerable probability of type I error or rejecting a true null hypothesis. It is fixed in advance, such as 5% or 1% and denoted by α. In statistical terms, the agreed threshold of probability of rejecting a true null hypothesis.

levels of gene expression: Gene regulatory mechanisms operate at various different levels, namely, DNA, transcriptional, post-transcriptional, RNA transport, translational and post-translational levels.

levels of packaging: To accomplish the overall desired packing ratio, DNA is not packaged directly into the final structure of chromatin. Instead, it contains several hierarchies of organization. For example, three levels of DNA packaging in eukaryotic chromosomes are nucleosome, solenoid structure and scaffolding protein complex.

Levene model: H. Levene in 1953 proposed a model of a structured population in which individuals from a single gene pool devote part of their lives to competing within small reinforcements. This model describes the influence of spatially varying selection on a population's genetic structure. It covers migration between a finite number of demes and the evolution of gene frequencies from one generation to the next. Therefore, it is formulated as a discrete-space, discrete-time migration–selection model.

lexicon: In bioinformatics, it refers to a pre-defined list of terms that together completely define the contents of a particular database. The component in the grammar is in the bare form of a list of words or lexical entries.

***lfy*:** Floral meristem identity gene.

liana (or liane): A woody ascending plant, rooted in the ground.

library: (i) In computer science, a collection of non-volatile resources used by computer programmes, often for software development. These may include configuration data, documentation, help data, message templates, pre-written code and subroutines, classes, values or type specifications. (ii) In biotechnology and molecular genetics, an assembly of cells, usually bacteria or yeast, which have been transformed with recombinant vectors carrying DNA inserts from a single species (e.g. cDNA), expression or genomic library. *See* COMPLEMENTARY DNA LIBRARY, GENOMIC LIBRARY. (iii) In molecular biology, a large collection of compounds, peptides, cDNAs or genes, which may be screened in order to isolate cognate molecules.

lichen: A beneficial relationship between a fungus and green algae. Lichen is a combination of green algae and a fungus, which is placed in the kingdom fungi and classified according to the fungal partner. Both partners live in a mutualistic relationship in which each partner benefits. These are found on rocks, trees and rooftops and are the slowest-growing organisms on the Earth. Lichens are important actors in nutrient cycling and act as producers which many higher trophic feeders feed off of (e.g. reindeer, gastropods, nematodes, mites and springtails).

life: Any entity that is capable of making a reasonable accurate reproduction of itself and duplicates being able to produce the same task and subjected to a low rate of alteration, and these changes being heritable.

What establishes 'life' is infamously difficult to define, and there is at present no agreement definition of life, though some popular standards are that living things are self-possessed of cells, undergo a life cycle, experience metabolism, maintain homeostasis, acclimatize to environments, answer to stimuli, reproduce and evolve.

life cycle: The entire series of development stages undergone by an individual from zygote to maturity and death.

life cycle of an infectious disease: The sequence of distinct events (e.g. sexual reproduction) that occur between the appearance and reappearance of the causal organism. The stages of the disease cycle are the appearance, development and perpetuation of a pathogen and the effect of the disease on the host. Because the advancement of the disease involves the host, the pathogen and in some cases biological vectors, the life cycle of the pathogen as well as environmental factors are involved in the disease cycle.

life, certain facts about: (i) Life comes only from pre-existing life. (ii) Life exists in diverse forms. (iii) Like begets like. (iv) Traits (i.e. biological characteristics) show recombination. (v) Variation exists in living forms. (vi) Living things have individuality. (vii) Biological individuality has patterns, limitations and characteristics that depend at least in part on the parents of an individual. (viii) Similarities and differences are heritable.

life, continuity of: Continuity of life is the perpetuation of a species to repeat its genetic possession in every subsequent generation. A species preserves its biological information in DNA. Each cell of a given species has a fixed amount of DNA that is doubled by replication. Cell division guarantees equal delivery of parent cell DNA to daughter cells. Gametes carry half the amount of DNA, while the zygote holds the amount of DNA characteristic of the species. Replication of DNA is the molecular basis of reproduction and hence perpetuation of life. Replication devotedly upholds genetic information intact over generations. The continuity of germplasm between all the descendant generations of a species explains most of the biological similarities that are inherited.

life, origin of: Includes the processes by which the first living cell evolved from the chemicals present on the primitive Earth. The first living systems could not have evolved unless the most essential building blocks of life (purines, pyrimidines, sugars, amino acids, etc.) had already been synthesized by prebiotic pathways.

life-history trait: Traits (e.g. impermanence rate, fertility or age of reproduction) that are meticulously related to fitness.

ligand: A small molecule or ion that can bind to or interact with a protein or receptor molecule. For example, a pharmaceutical that binds to a receptor protein molecule on the surface of a cell may be called a LIGAND, which may be attained by passing unbound coenzymes through the column. The ligand may be hormones (i.e. used to isolate receptor molecules) or any other type of molecule that is capable of binding specifically and reversibly to the desired protein or protein complex.

ligand-binding site: A site on the surface of a protein at which another molecule binds.

ligase chain reaction (LCR): An *in vitro* nucleic acid amplification method that uses DNA ligase, an enzyme that joins two pieces of DNA together.

ligase-independent cloning (LIC): A method that uses T4 DNA polymerase or exonuclease III and does not involve the traditional ligation procedure. Recombinants are generated between PCR products and a PCR-amplified plasmid vector. The procedure does not require the use of restriction enzymes, T4 DNA ligase or alkaline phosphatase. *Aka* RESTRICTION DIGESTION *or* LIGATION-INDEPENDENT CLONING (RLIC).

ligases: A group of enzymes that catalyse a reaction that seams two ligates or substrates using energy derived from the synchronized hydrolysis of a nucleotide triphosphate. In general, a ligase seals single-strand breaks in DNA through phosphodiester bonds. DNA ligases are formed by *Escherichia coli* bacteria infected with T_4 phage. DNA ligases in the Eppendorf tube join together individual DNA molecules or two ends of the same DNA molecule. A foreign gene is integrated into a vector through a ligase enzyme to produce a recombinant DNA molecule.

ligation: The reaction involving the formation of a phosphodiester bond to link two adjacent bases separated by a nick in one strand of a double helix of DNA (deoxyribonucleic acid). The term applies to blunt-end ligation and to the joining of RNA (ribonucleic acid) strands.

light (L) chain: One of the smaller polypeptide chains present in every immunoglobulin or myosin molecule. There are two types of light chains, kappa (κ) and lambda (λ). *Compare with* HEAVY (H) CHAINS.

light (L) chain genes: Genes that determine the light chain of immunoglobulin molecules. Two types of light-chain genes known are λ and *k*.

light-activated macromolecular phase separation (LAMPS) method: Y.J. Kim *et al.* (2023) devised a biotinylated CRISPR-dCas9–based optogenetic method, LIGHT-ACTIVATED MACROMOLECULAR PHASE SEPARATION (LAMPS), to enable inducible formation, affinity purification and multi-omic dissection of nuclear condensates at the targeted genomic loci. LAMPS-induced condensation at enhancers and promoters activates endogenous gene transcription by chromatin reconfiguration, causing increased chromatin accessibility and de novo formation of long-range chromosomal loops. Proteomic profiling of light-induced condensates by dCas9-mediated affinity purification uncovers multivalent interaction-dependent remodelling of macromolecular composition, resulting in the selective enrichment of transcriptional coactivators and chromatin structure proteins.

light-chain variable (V_L) domains: The regions of the antibody (molecule's) 'light chain' that vary in their amino acid sequence. The chains comprising the antibody (immunoglobulin) molecule consist of a region of variable (V) amino acid sequence and a region in which the amino acid sequence remains constant (C). An antibody molecule has two antigen binding sites, and it is the variable domains of the light (V_L) and heavy (V_H) chains which contribute to the antigen binding ability.

light-dependent reactions: During the light-dependent reactions that transpire in the thylakoids of the chloroplast, the water molecules are split to convert $NADP^+$ to NADPH and to produce oxygen, a by-product that is released through the leaf's stomata into the surrounding air. Additionally, ATP is made from ADP and a free phosphate, using ATP synthase.

light-harvesting complexes: Multi-protein unit in the thylakoid membrane that absorbs light to function as energy for reactions. *Aka* PHOTOSYSTEMS COMPLEXES.

light-independent reactions: In the light-independent reactions (*aka* DARK REACTIONS or CALVIN CYCLE) that happen within the stroma of the chloroplast, enzymes, ATP and NADPH drive the conversion of CO_2 into sugars. Three carbon dioxide molecules enter each round of the Calvin cycle, and it takes two rounds to produce a glucose molecule. Two rounds of the Calvin cycle will produce two G3Ps that the plant cell converts into one glucose molecule, or it can use them to make sucrose and other organic molecules. ADP and $NADP^+$ exit the Calvin cycle and return to the thylakoids to start the next set of light-dependent reactions. *See* CALVIN CYCLE.

light-inducible transcriptional effector (LITE) system: This system employs a modular design consisting of two independent components: The first constituent is the genomic anchor and includes a customizable DNA-binding domain, based on TRANSCRIPTION ACTIVATOR-LIKE EFFECTORS (TALEs) from *Xanthomonas sp.*, fused to the light-sensitive CRYPTOCHROME 2 (CRY2) protein from *Arabidopsis thaliana*. The second component includes CRY2's interacting partner CALCIUM AND INTEGRIN-BINDING PROTEIN 1 (CIB1) fused to a desired effector domain (CIB1-effector). In the absence of light (inactive state), TALE-CRY2 binds the promoter region of the target gene while CIB1-effector remains free within the nuclear compartment. Illumination with blue light (peak~450 nm) triggers a conformational change in CRY2 and subsequently enrolls the CIB1-effector.

light-sensing studies in *Arabidopsis*: The photoreceptors comprising phytochrome A, B, C, D and E mediate red light-based phototropic response. Understanding the function of these receptors has assisted plant biologists in understanding the signalling cascades that regulate photoperiodism, germination, detiolation and shade avoidance in plants. Light response has even been observed in roots, which were believed not to be particularly sensitive to light. While the gravitropic response of *Arabidopsis* root organs is their predominant tropic response, specimens treated with mutagens and selected for the absence of gravitropic action revealed a negative phototropic response to blue or white light, and positive response to red light, suggesting that the roots also reveal positive phototropism.

lignans: A type of phytochemicals that have defensive roles within land plants. Lignans are also referred to as 'phyto-oestrogens' and are typically beneficial to the health of humans that consume them. Lignans are found within the seedcoats, stems, leaves or flowers in virtually all fruits, vegetables and cereals (grains). One of the beneficial lignans, commonly consumed by humans, is sesamin, found in seeds of the sesame plant (*Sesamum indicum*), which acts as an antioxidant.

ligneous: Hard lignified tissues or woody parts of a plant.

lignicolous: Growing on wood tissue after the bark has fallen or been stripped off. *Compare with* CORTICOLOUS.

lignification: The hardening of tissue due to the deposition of lignin in the plant cell wall.

lignins: A group of phenolic ('ring-shaped' molecules) polymeric compounds formed by land plants within the cell walls (i.e. exterior of cell's plasma membrane) of xylem vessels and certain other plant cells. A major constituent of wood to reinforce/strengthen those cell walls.

lignotuber: A woody swelling of the stem below or just above the ground. It contains adventitious buds from which new shoots can develop (e.g. after fire).

likelihood: Given a hypothesis, the probability of observing certain data.

likelihood function: The probability of the observed data as a function of the unidentified parameters for the data distribution. Here 'probability' is used in a loose sense and called 'likelihood' so that it can apply to both discrete and continuous outcome variables. When the outcome variable Y can take on only discrete values (e.g. Y is binary or categorical), given a statistical model, one can compute the exact probability that any given possible value of Y can be observed. In this case, the joint probability of a set of such occurrences can easily be computed. When the observations are independent, this joint probability is the product of all the individual probabilities. The likelihood function is then the joint probability that all the observed values of Y would have occurred, as a function of the unidentified parameters that create the whole distribution of an individual observation's Y. When Y is continuous, the probability elements constituting the likelihood function are the probability density function values assessed at the observed data. Because joint probabilities of many observations are very small, and for another reason about to be assigned, it is customary to state natural logs of likelihoods rather than using the original scale. The log-likelihood accomplished by a model – the log-likelihood at the maximum likelihood estimates of the unidentified parameters is a gold standard information measure and is employed to compute various statistics including R^2, Akaike Information Criterion (AIC) and likelihood ratio χ^2 tests of association. *See* MAXIMUM LIKELIHOOD ESTIMATE.

likelihood ratio: A ratio of the likelihood of creating a particular data set given one hypothesis relative to the probability of generating the same data given another hypothesis (Prob(D | H1)/Prob(D | H2)). This is the value calculated in likelihood-based phylogenetic rebuilding methods.

limited backcrossing: Contrary to complete backcrossing, which requires at least six cycles, only two or three cycles are coupled with rigorous selection to gain the advantage of transgressive segregation.

limited chromosomes: Heterochromatic chromosomes restricted to the germline in some species.

line: A group of individuals of shared ancestry and more narrowly defined than a strain or variety. In breeding, it represents any group of genetically uniform individuals formed from the selfing of a common homozygous parent. *See* CELL LINE.

line diagram: A diagram showing the trend by a line.

line varieties: *See* VARIETY.

line-breeding: A system of breeding in which a number of genotypes, which have been progeny tested in respect to some character or a group of characters, are composited to form a variety.

lineage: (i) A chart that traces the flow of genetic information from generation to generation. (ii) Used to describe a group of closely related viruses with a common ancestor.

lineage analysis: Drawing the ancestry of individual cells in a developing embryo.

lineage sorting: The process by which, following the parting of two species, the ancestry of every gene converges to the overall phylogeny of the species. This process takes about 2Ne generations, and INCOMPLETE LINEAGE SORTING implies discordance between genealogies.

lineage-specific developmental mechanisms: Developmental mechanisms found within all organisms of one lineage but not in other lineages, responsible for the morphology that characterizes the lineage. They limit the genetic variation that can be expressed in the lineage.

lineal consanguinity: Marriage of persons, one of whom is descended from the other.

linear DNA: The DNA has free ends, either because both strands have been cut or because the DNA was linear *in vivo*. You can model this with an electrical extension cord that is not plugged into itself.

linear: A term used to describe the physical state of nucleic acid molecules. Two strands at the end of a linear double-stranded DNA can be free, as in a restriction fragment, bound to a specific protein, as in some eukaryotic extra chromosomes. Eukaryotic chromosomes are linear and their ends, called telomeres, have special sequences which promote their replication.

linear epitope: An epitope formed by a series of amino acids which are adjacent to each other within the primary structure of the protein. Such epitopes can be successfully modelled by synthetic peptides but comprise only a small proportion of all epitopes. The minimal epitope size is about five amino acid residues. *Aka* SEQUENTIAL EPITOPE.

linear fork model of DNA replication: Unidirectional DNA replication; DNA replication proceeds in one direction with respect to origin or replication. This results in a Y-fork shape structure (e.g. DNA replication in col E1, puc plasmids and pBR322).

linear regression model: Refers to regression for a continuous dependent variable and generally to the case where the residuals are presumed to be Gaussian. The linear model is now and then called GENERAL LINEAR MODEL, not to be confused with GENERALIZED LINEAR MODEL, where the distribution can take on many non-Gaussian forms. *Aka* ORDINARY LEAST SQUARES (OLS).

linear search: A process for discovering an element within a list. It sequentially checks each element of the list until a match is found or the whole list has been searched. *Aka* SEQUENTIAL SEARCH.

linear tetrad: A tetrad that results from the occurrence of the meiotic and post-meiotic nuclear divisions in such a way that sister products remain adjacent to one another (with no passing of nuclei).

linear-after-the-exponential-PCR (late-PCR): A modification of asymmetric PCR is late-PCR. The asymmetric PCR is valuable in some sequencing and hybridization probing applications where having only one of the two complementary stands is sufficient or mandatory. Late-PCR uses a LIMITING PRIMER with a higher melting temperature (T_m) than the EXCESS PRIMER to maintain reaction effectiveness as the limiting primer concentration decreases mid-reaction.

linearly amplified single-stranded RNA-derived transcriptome sequencing (LAST-seq): J. Lyu and C. Chen (2024) develop the LAST-seq method, which unswervingly intensifies the original single-stranded RNA without preceding REVERSE TRANSCRIPTION (RT) and SECOND-STRAND SYNTHESIS (SSS) and delivers high-sensitivity RNA detection and a low level of technical noise in single-cell transcriptome analysis. LAST-seq has been pragmatic in quantifying transcriptional bursting kinetics in human cells, enriching our understanding of the role of chromatin organization in controlling gene expression. *See* SINGLE-CELL RNA SEQUENCING (SCRNA-SEQ).

linkage: A phenomenon discovered by English Scientists W. Bateson and R.C. Punnet in 1906 in Sweet Pea (*Lathyrus odoratus*). However, it was put forward as a regular concept by T.H. Morgan in 1910 from his work on (*Drosophila melanogaster*). Genes or markers that are carried on the same chromosome are stated to be linked. The physical association between two genes results from them being present in the same chromosome. The tendency of the parental combination of genes to stay together in F_2. Linkage is shown by virtue of the fact that a chromosome contains many genes. Genes located in the same chromosome are linked. Thus, linkage is an exception to independent segregation and instead of 9:3:3:1 phenotypic ratio in a dihybrid cross a different ratio will be observed which will depend upon a degree of linkage. A gene located in the same chromosome 50 or more map units apart does not show linkage; rather it shows independent assortment. Linkage can, however, be demonstrated in such cases with the help of another gene located in between the two genes. Marker genes that are linked to a gene (e.g. for a given trait or disease) of interest can be utilized to predict the presence of that (trait or disease-causing) gene. *See* COUPLING LINKAGE *and* REPULSION LINKAGE.

linkage analyses using computer software: Linkage analyses and mapping are computerized. Computer packages available are as follows: (i) JoinMap (P. Stam in 1993), (ii) MAPMAKER/EXP (E.S. Lander *et al.*, 1987), (iii) GMENDEL (C. Echt *et al.*, 1992), (iv) LINKAGE (K.A. Suiter *et al.*, 1983) and (v) Map Manager QTX (K.F. Manly *et al.*, 2001). Manuals with each computer package are available.

linkage analysis: The estimation of the frequency of crossovers or recombinations between DNA sequences, called RECOMBINATION FREQUENCY, is used to establish the position of a particular sequence (e.g. a gene) on a genetic map.

linkage analysis under unusual situations: Distorted segregation, differential viability and partial manifestation leading to misclassification are considered unusual situations in linkage. Detection and estimation methods under these unusual situations are based on the partitioning of chi-square.

linkage detection: Deals with partitioning of a total χ^2 into different components to find out the presence of linkage.

linkage disequilibrium (LD): (i) The level of non-random assortment of alleles at two or more different loci in a population. (ii) A state in which the haplotype frequencies in a population deviate from the values they would have attained if the genes at each locus were combined at random. It can offer a route to the identification and isolation of genes controlling quantitative sources of phenotypic variation. (iii) Where alleles occur together more often than can happen by chance. This occurrence indicates that the two alleles are physically close on the DNA strand. *Compare with* LINKAGE EQUILIBRIUM *and* MENDELIAN INHERITANCE.

linkage disequilibrium (LD) decay: Linkage disequilibrium (LD) is the non-random co-segregation of alleles at two or more loci. In other words, it is the difference between observed and expected allelic frequencies, assuming random distribution due to independent assortment. This non-random co-segregation could occur between loci on the same chromosome or between loci on different chromosomes. During prophase I of meiotic cell division, recombination tends to shuffle genetic material between chromosomes resulting in the decay of LD. As a result, after many meiotic events, only loci tightly linked to a functional mutation will show significant association with the trait of interest.

linkage disequilibrium (LD) mapping: LD mapping is founded on seeking relations between phenotype and allele frequencies. The principal methods used to identify marker-trait associations in crop species have been based on the construction of linkage maps or the use of BULKED SEGREGANT ANALYSIS. Both procedures are based around the production and phenotyping of special populations, which are generally developed from two varieties that depict a major difference in the traits targeted for mapping. This leads to several complications in existing mapping strategies, which can be dealt with association or LD mapping. *See* ASSOCIATION MAPPING.

linkage disequilibrium (LD) mapping, advantages of: There are four potential rewards of this method in mapping genes in crop species: (i) offers a new perspective for trait mapping; (ii) yields detailed fingerprinting information on a large number of lines and varieties; (iii) uses real breeding populations, the material is diverse and relevant and the most important genes (for adaptation, etc.) should be segregating in such populations and (iv) pattern analysis of marker data has the possibility of detecting intricate combinations (even epistatic interactions) between alleles at several loci, which inspire the superior individuals in a breeding population.

linkage drag: The transfer of something, which is not desirable/required along with the gene of interest. During wide hybridization, the wild species are hybridized with cultivated varieties to transfer useful genes (disease resistance, insect resistance) from the wild into cultivated ones. Often, these useful traits are tightly linked to some undesirable (e.g. seed shattering) traits. In this situation, the undesirable traits accompany the desired ones and get introgressed into cultivated ones. This is an undesirable phenomenon and is commonly called linkage drag.

linkage equilibrium: Absence of linkage disequilibrium, so that haplotype frequencies are equal to the product of allele frequencies.

linkage group homology: The presence of identical markers on linkage groups from two different crosses, so that a homology can be inferred.

linkage group: Any group of genes or DNA markers that are contiguous on a linear chromosome map (i.e. genes are located in the same chromosome) and show a high degree of linkage.

linkage map: A genetic map of a chromosome or genome described by mapping the locations of genes, genetic markers or DNA sequences to their chromosomes by their linkage to willingly identifiable genetic loci. Distance between loci is measured in CENTIMORGANS (cM). The genetic maps prepared through recombination frequency do not correspond to physical portions of the chromosomes, though recombination (crossing-over) frequency is a function of the distance between genes and is not completely independent of the mutants used, its location of chromosome, genetic background, environmental conditions and so on. So, a crossing-over value in one system may not give the same picture of the gene structure in another system. Moreover, distances between genes on a genetic map may not correspond to the distances between them in the DNA molecule. A gap between them may arise due to non-availability of the mutant. Genes are situated on chromosomes in a linear order such that their position with respect to each other along the chromosome can be determined. All the genes situated on a specific chromosome constitute a linkage group out of which any one gene serves to identify the concerned chromosome. The true position of a gene along the DNA in 'kbp' constitutes the physical map, whereas the genetic map refers to the relative position of genes determined from the frequency of recombinants of alleles of two genes. Now with the use of various molecular markers (e.g. RAPD and Microsatellites), it has become possible to develop high-density saturated linkage maps using a computer software like MAPMAKER. *See* GENETIC MAP.

linkage map, construction of: The mechanism responsible for the origin of new phenotypic combinations is crossing-over. RECOMBINATION FREQUENCY is a function of the distance between genes in the chromosomes. The greater the distance between two genes, the greater the recombination frequency observed. The distance between linked genes is measured in map units. One map unit equals 1% crossing over, which is equal to one centimorgan (cM). Show the symbols of the mutant genes indicating the distance between them in map units along a straight line. Arrangements of genes in a linkage group in such a way that the distance between any two of them reflects the per cent crossing-over between them is known as a linkage map, which uses the symbols of the mutant genes.

linkage mapping: Development of comprehensive molecular genetic linkage maps. The most common use of DNA markers is to develop genetic linkage maps in which the linear orders, and the whole genome in an organism are determined genetically and represented graphically. This approach is particularly suited to crops in which dense molecular marker / genetic linkage maps with a variety of DNA markers are already available, the whole genome has been claimed to be sequenced and where rapid progress has been made in the field of genomics. For example, in rice, several high-density molecular linkage maps have been constructed that have greatly facilitated efforts in QTL mapping.

linkage maps based on cytosine methylation: The first step to explore the epigenetic variation is the determination of the chromosomal distribution of 5-methylcytosine ($5^{m}C$) which is possible through the construction of methylation-based linkage maps. The cytosine methylation depicting linkage maps shows polymorphism on the basis of DNA methylation patterns rather than variation among DNA sequences, conferring epigenetics phenomenon.

linkage number: The number of times the two strands of a closed DNA duplex cross over each other.

linkage number paradox: The inconsistency between the occurrence of –2 supercoils in the path of DNA on the nucleosome compared with the measurement of –1 supercoil released when histones are removed.

linkage of genes: The relationship of genes that results from their being located on the same chromosome. Linkage is detected by the greater association in the inheritance of two or more non-allelic genes than would be expected from the independent assortment. The nearer such genes are to each other on a chromosome, the more closely linked they are considered to be and the less often they are likely to be separated in future generations by crossing over.

linkage value: Recombination fraction stating the proportion of non-parental or recombinant versus parental types in a progeny. In diploids, the recombination fraction can vary between zero and one-half.

linkage, estimation of: Estimation of linkage means measuring the distance between the linked genes. For estimation of linkage, per cent recombination is calculated. In the F_2 progeny of a dihybrid cross, there are two parental classes and the same number of recombinant classes. The recombinant types in the F_2 progeny most likely originated from new combinations of the linked genes. In order to measure linkage in a two-point cross, mark the parental pair of complementary classes as 'P_1' and 'P_2'. The pair showing the maximum number of flies belongs to the parental types. Mark the pair of complementary classes that arise due to crossing-over as 'R_1' and 'R_2'. The pair showing the least number of flies belongs to this category. To calculate per cent recombination, use the following formula: Number of recombinants/Total progeny × 100. Estimation of linkage is done from F_2 data by using this method mostly in the case of plants in which crossing-over occurs in both sexes in almost equal frequency. The following steps are followed for estimation of linkage in a three-point cross. The first step is to ascertain the order of the three genes. For this, arrange the eight phenotypic classes in four complementary pairs. Work out the total number of flies in each pair of classes. Mark the parental pair of complementary classes as 'P'. The pair showing the maximum number of flies belongs to the parental types. Mark the pair of complementary classes that arise due to double-crossing over as 'DCO'. The pair showing the least number of flies belongs to this category. Compare parental and double crossover pairs of complementary classes and determine the gene that changes position. The gene that changes position lies in the middle. Rearrange the order of genes in the eight phenotypic classes according to the gene order worked out above in all the eight phenotypic classes if the given arrangement is different. Write the genotype of the heterozygous parent of the TC_1 progeny in proper gene order. By assuming a single crossing-over between the first two genes, work out the resultant phenotypic classes. Mark these classes as 'SCOI'. Similarly, assuming a single crossing-over between the second and third genes, work out the resultant classes and mark them 'SCOII'. Attempt double-crossing over, one event between the first two genes and the other between the second and third genes. What are the resultant phenotypes? Do they match with the classes marked 'DCO' in the beginning as those resulting from double-crossing over? If this matching does not occur, you have committed some mistake. Go over the above steps again to correct yourself. Calculate per cent recombination between the first two genes by using the formula: % Recombination = (SCOI + DCO)/N × 100. Calculate per cent recombination between second and third genes by using the formula: % Recombination = (SCO II + DCO)/N ×100.

linked genes. Any set of one or more genes that are adequately close together on the same chromosome that they are very unlikely to assort independently and therefore are normally inherited together.

linked list: A linear assembly of data elements, whose order is not shown by their physical placement in memory. Instead, each element points to the next. Thus, it is a data structure comprising a collection of nodes which together represent a sequence.

linked marker: Any molecular marker located near a target gene on the same chromosome such that the recombination frequency between them approaches zero. Linked markers are exploited for the isolation of the linked gene(s) via positional cloning.

linker: (i) In computer science, a computer utility programme that takes input from one or more object files created by a compiler or an assembler and combines them into a single EXECUTABLE FILE, LIBRARY FILE or another OBJECT FILE. A simpler version that writes its output directly to memory is called the LOADER, though loading is characteristically thought a separate process. *Aka* LINK EDITOR. (ii) In molecular biology, linker refers to a short synthetic duplex oligodeoxynucleotide comprising the target site for some restriction enzyme. It may be added to the ends of a DNA fragment prepared by cleavage with some other enzyme reconstructions of recombinant DNA. Following the ligation reaction, the product is digested with the endonuclease, generating a DNA fragment with the desired sticky or blunt ends.

linker DNA: In molecular biology, linker DNA is double-stranded DNA (38–53 base pairs long) in between two nucleosome cores that, in association with histone H1, hold the cores together. They look like beads on a string.

linker histone and DNA methylation in gene regulation: The linker histones can participate in epigenetic regulation of gene expression by contributing to the maintenance or establishment of specific DNA methylation patterns.

linker histones: A linker histone connects two adjacent nucleosomes. Linker histones are involved in the condensation of chromatin into the 30-nm fibre. H1, H5 and $H1^0$ are well-known linker histones. They have been found to exert both positive and negative effects on gene activity in Protozoa. This supra-nucleosome organization correlates with the resting state of chromatin and it is therefore possible that the linker histones play an active role in the control of chromatin activity. H5 plays a part in the regulation of DNA replication and cell proliferation. *Compare with* CORE HISTONES.

linker insertion mutagenesis: An *in vitro* mutagenesis technique employed to produce a large library of mutants with specific interchanges that are then dispersed throughout a cloned DNA to identify these mutants for a desired phenotype.

linker oligonucleotide: A synthetic oligodeoxyribonucleotide that contains a restriction site. Linkers may be blunt-end ligated onto the ends of DNA fragments to create restriction sites that can be used in the successive cloning of the fragment into a vector molecule. It may be necessary to protect the fragment to be cloned from the action of the restriction endonuclease by treating them with DNA methylase.

linker peptide: A 22-amino acid protein is used as a linker protein to connect a DNA-BINDING DOMAIN (DBD) with an EFFECTOR DOMAIN (ED) in an epigenome-editing tool.

linker scanner mutations: The mutations that are identified by recombining two DNA molecules *in vitro* at a reaction fragment are added to the end of each. This procedure results in the insertion of the linker sequence at the site of recombination.

linking: The process of reversibly 'attaching' a drug or a toxin to a monoclonal antibody, or another homing molecule of the immune system. As this attachment must be reversible, so that the homing molecule can release the drug or toxin after delivering the drug or toxin to the desired site in the body (e.g. delivery of a toxin to a tumour in order to kill it), linking is rather a difficult process to reliably achieve its objective.

linking number: The number of times the two strands of DNA molecules *in vitro* at a reaction fragment are added to the end of each; the result is to insert the linker sequence at the site of recombination.

Linnaean classification: A hierarchical method of naming classificatory groups, invented by the 18th-century Swedish naturalist Carl von Linné, or Linnaeus. Each individual is assigned to a species, genus, family, order, class, phylum, kingdom and some intermediate classificatory levels. Species are referred to by a Linnaean binomial of its genus and species, such as *Magnolia grandiflora*.

lipase: An enzyme that catalyse the hydrolysis of triglycerides into glycerol and fatty acids. Some microbes produce extracellular enzymes that break fats into fatty acids and glycerol. Such enzymes are called lipases. Many bacteria that hydrolyse fatty acids can use the same enzymes to degrade petroleum products.

lipid: A biomolecule that is insoluble in water and soluble in non-polar organic solvents such as alcohol, ether and chloroform. They are an important component of living cells. Together with carbohydrates and proteins, they are the key constituents of plant and animal cells. Cholesterol and triglycerides are lipids. A special class of lipids, the phospholipids, provides the structural basis of biological membranes.

lipid bilayer: Thin bimolecular leaf of mainly phospholipid molecules that constitutes the core structure of all cell membranes. The two-layer sheet of phospholipid molecules is packed with its hydrophobic tails pointing inwards and their hydrophilic heads outward, exposed to water.

lipid raft: Small area of the plasma membrane augmented in sphingolipids and cholesterol.

lipofection: A method of transfection in which DNA is incorporated into lipid vesicles (liposomes), which then are fused to the membrane of the target cells.

lipophilic: *See* HYDROPHOBIC.

lipopolysaccharide: *Aka* ENDOTOXIN. A lipopolysaccharide is a large molecule found on the outer membrane of the bacterial cell wall, which can activate the immune system and cause harm.

lipoprotein: A biochemical assemblage that contains both proteins and lipids, bound to the proteins, which enable fats to move through the water inside and outside cells. The proteins function to emulsify the lipid molecules.

liposome: A lipid vesicle containing foreign DNA for transformation. During plant transformation, the liposomes are being used to deliver cloned DNA into protoplasts. The cloned DNA is encapsulated in the lipid vesicle for DNA protection from nucleases. Liposomes are taken up by protoplasts through pinocytosis, lipase enzyme within the protoplast dissolves the lipid vesicle and DNA gets released for integration.

liposome-mediated gene transfer: A method of gene transfer which involves encasement of DNA in lipid bags, which can be made to fuse with protoplasts with the help of a peg. This technique has been used successfully for gene transfer in a number of plant species. For example, stable transformation of mouse L cells deficient in THYMIDINE KINASE (TK) was achieved by liposome-mediated transfer of a recombinant plasmid carrying the *Tk* gene. Ten per cent of the recipient cells expressed TK activity. The transformed phenotype (e.g. 200 out of 10^6 cells) was stable under selective and non-selective conditions.

liquid chromatography coupled with mass spectrometry (LC-MS): This method relies on nuclease P1 and alkaline phosphatase digestion of RNA to nucleosides, isolation of the digestion products by reverse phase chromatography and analysis by MS and UV light (254 nm) absorption.

liquid crystal display (LCD): LCD is the most important technology used in flat panel displays. Alignment of liquid crystals can be changed with electric current. Light is transmitted through a crystal in one alignment and in another alignment caused by the electric field, the transmission is affected. This can be employed to generate a show if alignment at individual positions can be changed. By packing red, blue and green emitting crystals appropriately, a full-colour display can be obtained.

liquid nitrogen: Nitrogen gas condensed to a liquid form with a boiling point of about –196°C. Very commonly, it is used as the medium in which containers of genetic material are stored. *See* CRYOPRESERVATION.

list: An abstract data type that represents a countable number of ordered values, where the same value may occur more than once. An instance of a list is a computer representation of the mathematical concept of a finite sequence; the (potentially) infinite analogue of a list is a stream. Lists are a basic example of containers, as they contain other values. If the same value occurs multiple times, each occurrence is considered a distinct item.

lithification: A geological process in which live materials are captured, aggregated, densified and cemented into new sedimentary rocks.

lithophytic: A plant growing on rocks. An epilithic plant.

litmus: Litmus is a special dye that we get from algae. Litmus is extremely useful because it turns red when it is in an acid and blue when it is in a base. Thus, we use this to see if something is acidic or basic.

live vaccine: A living, non-virulent form of a microorganism or virus that is used to elicit an antibody response that will protect the inoculated organism against infection by a virulent form of the microorganism or virus. Also, a living, non-virulent microorganism or virus that expresses a foreign antigenic protein and is used to inoculate humans or animals. The latter organisms are *aka* LIVE RECOMBINANT VACCINES. For example, large DNA viruses, such as Vaccinia virus, are used as a biological delivery agent. Vaccinia virus has a linear double-stranded DNA genome of approximately 185,000 bp and replicates in the cytoplasm of infected cells.

living modified organism (LMO): A living organism is a biological entity that can transfer or replicate genetic material. LMO is defined in the Cartagena Protocol on Biosafety as any living organism that possesses a novel combination of genetic material obtained through the use of modern biotechnology. In everyday usage, LMOs are usually considered to be the same as GMOs (GENETICALLY MODIFIED ORGANISMS), but definitions and interpretations of the term GMO vary widely. Common LMOs include agricultural crops that have been genetically modified for greater productivity or for resistance to pests or diseases. This term is being used for living transgenic organisms to distinguish them from dead transgenics. *See* TRANSGENIC.

loader: The part of an operating system that accounts for loading programmes and libraries. It is one of the crucial stages in the process of starting a programme, as it places programmes into memory and prepares them for execution. Loading a programme involves reading the insides of the executable file containing the programme instructions into memory and then carrying out other necessary preparatory tasks to prepare the executable instructions for running. Once loading is over, the operating system starts the programme by transferring control to the loaded programme code.

local alignment: Attempts to align regions of sequences with the highest density of matches. In doing so, one or more islands of subalignments are created in the aligned sequences. This model of sequence analysis means two strings may not be highly similar to a whole but may be similar only in certain regions. The task is to find and extract a pair of regions, one from each of the two given strings, which exhibit high similarity. *Aka* LOCAL SIMILARITY.

local lesion: A small, restricted lesion, often the distinctive reaction of differential cultivars to specific pathogens, especially in retort to mechanical inoculation with a virus.

local mediator: Secreted signal molecule that functions at short range on adjacent cells.

localization: Numbers of domains that are thought from SwissProt annotations to be present in different cellular compartments (cytoplasm, extracellular space, nucleus and membrane-associated) are shown in annotation pages.

localize: Determination of the original position (locus) of a gene or other marker on a chromosome.

localized cell death: In this case, mitosis stops in certain regions. For example, at the end of a flower bud, localized cell death leads to the formation of petals.

localized cell division: In this case, the area underlying rapid mitosis becomes bigger and bigger.

localized deposition: In certain cases, localized deposition of extracellular material or matrix takes place. For example, in case of development of bone, proteins and calcium ions are deposited. When matrix components are appropriate, cells will react with the matrix carbohydrate component. They induce changes in the matrix, and successive changes become possible.

localized protein secretion: A signal sequence may contain information that directs the secretion of a protein to one subcellular region, in addition to its classical role in promoting secretion. The signal sequences of two surface proteins, PrtM (M) and PrtF (F), direct secretion to different subcellular regions. The signal sequence of M protein promotes secretion of the division septum, whereas that of F protein preferentially promotes secretion at the old pole.

lockdown: Policy used in emergency situations to restrict the movement of people to prevent the spread of disease.

locule: A compartment or cavity encompassing seeds within an ovary, pollen within an anther or spores in a sporangium.

locus (pl. loci): A specific position on a chromosome where a gene is located; in diploids, loci pair during meiosis and, unless there have been translocations, inversions and so on, the homologous chromosomes contain identical sets of loci in the same linear order; at each locus is one gene; if that gene can take several forms (alleles), only one of these will be present at a given locus. It may refer to a single-nucleotide site, a substantial length of DNA sequence, a chromosome marker or a gene. The use of this term is sometimes made with reference to expressed DNA regions. At a given locus, any one of the variant forms of a gene may exist. The variants are said to be alleles of that gene. The term locus (plural loci) is used both to indicate the location of a gene on a chromosome map and to designate the unit whose variants act as alleles. In the latter sense, 'locus' is almost equivalent to 'gene'. *See* GENE EXPRESSION.

locus control region (LCR): A genomic region that enhances the expression of genes from a distance. It is defined by their capability to augment the expression of linked genes to physiological levels in a tissue-specific and copy number-dependent way at ectopic chromatin sites. The perception that developmental and cell lineage-specific control of gene expression relies not only on gene-proximal elements such as promoters, enhancers and silencers but also on long-range interactions of numerous *cis*-regulatory elements and dynamic chromatin alterations. Segments of DNA are important for the correct and coordinated expression of large regions (e.g. those encoding haemoglobins).

locus heterogeneity: Genes at different chromosomal positions (loci) resulting in the same phenotype.

locus-specific: Plant characters or genetic activity that is solely correlated with a particular chromosomal locus.

locus-specific, co-dominant markers from RAPDs: For developing locus-specific, co-dominant markers, the RANDOM AMPLIFIED POLYMORPHIC DNA (RAPD) marker band is isolated from the gel. It is amplified in the PCR reaction. The PCR product is cloned and sequenced. New longer and specific primers are designed for the DNA sequence, which is called the SEQUENCED CHARACTERIZED AMPLIFIED REGION (SCAR) marker.

LOD score: Log_{10} ratio of two probabilities for recombination ratio (i.e. observed recombination and free recombination). The higher the LOD value, the stronger the linkage. A LOD score of three means that the probability of linkage value is 10^3. *Compare with* THE PROBABILITY OF FREE RECOMBINATION.

LOD score method: A technique (LOGARITHMIC ODDS) for determining the most likely recombination frequency between two loci from pedigree data.

LOD value in establishing linkage groups: The higher the critical LOD value, the greater is the number of fragmented linkage groups. The smaller the LOD values, the fewer are the linkage groups. The number of linkage groups arrived at should be the same as the haploid chromosome numbers of the species under study. The choice of 'maplod' value is arbitrary, and it can be as low as 0.01 to as high as 3.0.

lodging resistance: Plants that can fight back lodging by an optimal root system, stiffer straw or other characteristics. For example, in cereal breeding, the introduction of 'semi-dwarf genes' contributed to shorter plants and thus higher lodging resistance even when nitrogen fertilization is increased.

log phase: A period of exponential growth of bacteria or cells in culture. The log phase is the period when cells are most active, and generation time is minimal. Therefore, a logarithmic plot of growth during the log phase is a straight line. During this phase, the cells are highly sensitive to adverse conditions.

logarithm of odds (LOD) score: A significance measure used in maximum likelihood statistics. In interval gene/marker mapping, the LOD score is log10 of the ratio of the probability that a QTL is present to the probability that a QTL is absent, for example, if the presence of a QTL is 1000 (10^3) times more probable than its absence, then the LOD score = 3, because log_{10} of $10^3 = 3$.

logic error: In computer programming, a bug in a programme that makes it operate incorrectly, but not terminate abnormally. A logic error generates unintentional or undesirable output or other misbehaviour, although it may not immediately be predictable as such.

logic programming: A type of programming model that is mainly based on formal logic. Any programme written in a logic programming language is a set of sentences in logical form, expressing realities and directions about some problem domain. Major logic programming language families comprise Prolog, answer set programming (ASP) and Datalog.

logistic growth: A pattern of population growth in which the growth rate is reduced with a growing number of individuals until it turns out to be zero when the population achieves a saturation size. When the number of individuals is small, the population increases almost exponentially, but as the growth rate decreases the growth curve (population size versus time) slowly levels out towards a constant value at the saturation size, thus forming a characteristic 'S-shaped' curve.

logistic regression model: A multivariable regression model relating one or more predictor variables to the probabilities of various outcomes. The most frequently utilized logistic model is the binary logistic model, which forecasts the probability of an event as a function of several variables. There are several types of ordinal logistic models for estimating an ordinal outcome variable, and there is a polytomous logistic model for categorical retorts. The binary and polytomous models employ the χ^2 test for testing for relationship between categorical variables. One commonly used ordinal model, the PROPORTIONAL ODDS MODEL, uses the Wilcoxon 2-sample rank test. Binary logistic models are suitable for forecasting events in which time is not very important. They can be utilized to forecast events by a specified time, but this can cause a loss of information. Logistic models are utilized to estimate adjusted odds ratios as well as probabilities of events. The actual probability of a specific value for a continuous variable is zero. The probability that they actually occur is now moot, since the Y values have already been observed.

long interspersed nuclear elements (LINEs): Each long, interspersed sequence is 5000–7000 bp long. Each sequence is bordered on either side by short direct repeats of 255–400 bp. Thus, they resemble pro-retroviruses. It is probable that retroviruses evolved from them. They are also moveable. The mammalian genome has 20,000–50,000 copies of LINES, which are 6–7 kb long and spread throughout the genome. They make up around 21.1% of the human genome. LINEs are one of the most ancient and successful inventions in eukaryotic genomes. The LINE machinery is believed to be responsible for most reverse transcription in the genome. Each species has a specific LINE-1 (L-1) family on the basis of which it can be categorized. Each LINE comprises one or two OPEN READING FRAMES (ORFs). ORF has an initiation as well as termination codons but its product is not known. L-1 elements are capable of transposition and make target site duplication.

long non-coding RNA (lncRNA): Non-protein-coding RNA, more than 200 nucleotides. Certain regulatory activities like controlling genome dynamics, cell biology and developmental programming have been related to them. Many of their functions remain to be characterized. Examples include X INACTIVE SPECIFIC TRANSCRIPT (Xist), which is the master regulator of X chromosome inactivation, and HOX transcript antisense intergenic RNA, which is highly upregulated in liver cancer and high expression in hepatocellular carcinoma. The primordial lncRNA 'signals' may often contain functional seeds to become decoys, guides, and scaffolds. *See* MACRO NON-CODING RNA, DECOYS, GUIDES *and* SCAFFOLDS.

long nuclear-retained non-coding RNAs (lnrncRNA): One of the ncRNAs, which controls synapse formation by moderating the expression of genes important in synapse formation and/or maintenance.

long PCR: Efficient long PCR results from the use of two polymerases: a non-proofreading polymerase is the main polymerase in the reaction, and a proofreading polymerase (3′ to 5′ exo) is present at a lower concentration.

long-period interspersion of DNA sequences: The second type of arrangement is long-period interspersion. This is characterized by 5000 bp stretches of repeated sequences interspersed within regions of 35,000 bp of single-copy DNA. *Drosophila* is an example of a species with this uncommon sequence arrangement. In both cases, the repeated sequences are usually from the middle repetitive class.

long terminal repeat (LTR) retrotransposons: A type of retrotransposon that repeats in a direct orientation at either end. *Contrasts* NON-LTR RETROTRANSPOSONS.

long terminal repeat (LTR): A DNA sequence directly repeated at the ends of the provirus (integrated DNA) of all retroviruses, derived during reverse transcription by duplication of sequences at the ends of the genome RNA. It contains most of the control elements necessary for the expression of the provirus.

long-branch attraction: A phenomenon in phylogenetic analyses when taxa or operational taxonomic units (OTUs) that lie at the end of long branches are inferred to be closely related, regardless of their true evolutionary relationships.

long-day plant: A plant in which flowering is favoured by long days (>14 h daylight) and corresponding short dark periods. There occur two types of long-day plant species: (i) those in which there is a total requirement for these conditions and (ii) others in which flowering is simply hastened by them.

long-lived nuclear RNAs: S. Zocher *et al.* (2024) reveal that newly synthesized RNA labelled with 5-ETHYNYL URIDINE (EU) in early postnatal mice still occurred in many brain cells 2 years later. The complex pattern of when and which cells are labelled suggests that EU that is incorporated into RNA in NEURAL PROGENITOR CELLS (NPCs) frequently remains in adult neurons. This suggests that a diversity of long and repeat-rich RNAs, together called LONG-LIVED RNAs (LL-RNAs), can be stable fixtures in postmitotic and quiescent neural cells. These RNAs could be critical for lifelong functions of the brain but might also contribute to its age-dependent decline.

long-patch base excision repair (BER) in eukaryotes: DNA polymerase I-unassociated exonuclease may act on incised DNA where patches are long (around 3000 nucleotides) in length. Long-patch BER is dependent on DNA polymerase II and DNA polymerase III. A gap expansion step is necessary, which is catalysed by ATP-stimulated exonuclease activity of exonuclease V (recB, recC). Since Pol II and Pol III lack $5'\rightarrow3'$ exonuclease activity, they (specifically Pol III) must have a role in reinsertion reactions.

long-range restriction mapping: Restriction enzymes are proteins that cut DNA at precise locations. Restriction maps depict the chromosomal positions of restriction-enzyme cutting sites. These are used as biochemical 'signposts', or markers of specific areas along the chromosomes. The map will detail the positions where the DNA molecule is cut by particular restriction enzymes.

long-short-day plants: The plants which flower when first grown under long-day conditions and then short-day conditions.

long-term gene pool: A population with extensive genetic variability developed for long-term breeding objectives. Lower selection pressure is exerted when it is improved through recurrent selection. It can also offer genetic variability to other gene pools.

long-term gene regulation: It includes phenomena connected with determination, differentiation or more generally development. Short-term regulation is a characteristic of both developing and fully-differentiated eukaryotic cells; even when a cell is undergoing differentiation, it responds to its environment. During differentiation, certain characteristics of a short-term regulatory process may be modified melodramatically. For example, DNA synthesis in early embryos occurs rapidly, and the cell cycle exhibits very short, if any, G1 and G2 periods. As differentiation proceeds, these periods lengthen and ultimately assume cell-specific proportions. *Aka* REVERSIBLE GENE REGULATION.

longevity: The perseverance of an individual for longer than most members of its species, or of a genus and/or species over a lengthy period of geological time.

longitudinal study: A research study involving repeated observations of the same variables in the same individuals over long periods of time.

loop: A single-stranded area at the end of a hairpin in RNA (or single-stranded DNA). It corresponds to the sequence between inverted repeats in duplex DNA.

loop rolling model of DNA replication: Rolling circles are used in the replication of some phages. The A protein that nicks the ϕX174 origin has the rare asset of *cis*-action. It performs only on the DNA from which it was synthesized. It remains connected to the displaced strand until an entire strand has been replicated and then nicks the origin again, freeing the displaced strand and starting another cycle of replication. Rolling circles also are implicated in bacterial conjugation when an F plasmid is moved from a donor to a recipient cell, following the initiation of contact between the cells by means of the F-pilli. *Aka* ROLLING CIRCLE MODEL OF DNA REPLICATION.

loop-mediated isothermal amplification (LAMP): A single-tube technique that enables the amplification of DNA, which unlike conventional PCR does not require a thermocycler.

looping in DNA-protein interactions: Looping of DNA appears to be the most likely apparatus enabling protein-protein interactions involved in gene regulation. This mechanism envisages that proteins bound at widely parted sites act in the same way, with the intervening DNA looping or bending to permit protein-protein interactions. According to this view, it is the interaction between DNA-bound proteins, not the looping *per se*, that controls gene expression. The examples which offer support to the case for looping are as follows: (i) galactose repression in *Escherichia coli*, (ii) *araBAD* operon in *Escherichia coli*,

(iii) early gene promoter of mammalian virus SV40, (iv) λ phage repression and (v) *lac* repressor.

looping-scanning model of gene regulation: The locus control region regulates several genes. Proteins binding here scan through large portions of DNA, looping the intervening region out, until they find the relevant gene.

loss of heterozygosity (LOH): A common genetic event in cancer development and is known to be involved in the somatic loss of wild-type alleles in many inherited cancer syndromes. The wider involvement of LOH in cancer is assumed to relate to unmasking a somatically mutated tumour suppressor gene through the loss of the wild-type allele.

loss of imprinting (LOI): Activation of an allele normally silenced by genomic imprinting; LOI causes excess gene product to be produced and is often associated with tumorigenesis.

loss-of-function mutation: A DNA alteration that results in reduced or abolished protein activity. When the allele has a comprehensive loss of function (NULL ALLELE), it is often known as AMORPHIC MUTATION.

lost gene: A situation in which all the genes at a certain locus are the same, and the other allele has been permanently eliminated from the population.

low complexity region (LCR): Areas of biased composition including homopolymeric runs, short-period repeats and more indefinable over-representation of one or a few residues. The SEG programme is utilized to disguise or filter LCRs in amino acid queries. The DUST programme is used to disguise or filter LCRs in nucleic acid queries.

low-pressure liquid chromatography: An automated system for carrying out liquid chromatography.

low-density lipoprotein (LDLP): A large complex structured by a single protein molecule and many esterified cholesterol molecules, together with other lipids. LDLP is the state in which cholesterol is transported in the blood and taken up by the cells.

low-dimensional structures (LDSs): LDSs include quantum wells, quantum wires and quantum dots.

low-input variety: A crop variety with low claims at macro- and micronutrient fertilizers, pest control and agronomic measures.

low-linolenic oil soybeans: Soybeans from *Glycine max* plant varieties that have been produced specifically to harvest soybeans bearing oil that contains less than 4% linolenic acid, instead of the typical 8% in soybean oil produced from traditional varieties of soybeans. Low-linolenic soybean oil tends to have greater flavour stability (especially at elevated temperatures utilized in frying foods) than soybean oil from traditional varieties of soybeans.

low-phytate corn: Developed in the USA during the 1990s, these are corn hybrids possessing the *Lpa1*, *Lpa2* or the *HAP* (*highly available phosphorous*) gene (which was discovered by Victor Raboy). That gene causes corn (maize) hybrids possessing it to yield much less phytate than the 0.15% typically present in traditional varieties of corn (maize). Because phytate is not digestible in humans and other monogastric animals (swine, poultry, etc.), substituting low-phytate corn in place of traditional corn varieties of those animals' diets assists reduce the hostile environmental influence of animal feeding (e.g. phosphorous emissions in excess of annual cropland requirements). Swine fed a diet in which traditional corn varieties have been replaced by low-phytate corn produce up to 30% less phosphorous in their manure, thereby lessening the phosphorous impact of those swine on the environment. Humans consuming a diet based profoundly on corn/maize (e.g. tortillas) absorb 50% more iron when traditional corn varieties are replaced by low-phytate corn varieties. That is because the phytate (inositol hexaphosphate) molecule binds iron (and also some other metals) within the digestive system and prevents its absorption into the body.

low-phytate soybeans: Developed in the USA during the 1990s, these are soybean varieties having less than 0.30% (of total soybean weight) phytate, versus the typical 0.45% phytate content of soybeans from traditional varieties. Because phytate is not digestible in humans and other monogastric animals (swine poultry, etc.), substituting low-phytate soybeans in place of traditional soybean varieties of those animals' diets assists to reduce the adverse environmental impact of animal feeding. Swine fed a diet in which traditional soybean varieties have been replaced by low-phytate soybeans produce up to 20% less phosphorous in their manure, thereby lessening the phosphorous influence of those swine on the environment. Due to the fact that the amino acids lysine, methionine, cysteine, arginine and threonine all become more 'bioavailable' (i.e. available for the animal to build its body tissue or otherwise utilize) in a low-phytate diet, low-phytate diets also aid in reducing excess nitrogen emissions.

low-risk objection (LRO): LRO deals with the preservation of life. Because GENETICALLY ENGINEERED

SEEDS (GES)-related risks are low, they are no threat to Lockean preservation of life. Logically, LRO confuses absence of evidence (for GES-related risks) with evidence of absence. Even if GES risks are low, stakeholders (those affected) have the right to decide whether the risks are worth the benefits or whether associated benefits can be realized in alternative ways.

low-stachyose soybeans: Those soybean varieties that comprise lower than 1% levels of the relatively indigestible stachyose carbohydrate (and thus higher levels of easily digestible other nutrients) than traditional varieties of soybeans (which typically contain 1.4–4.1% stachyose in traditional soybean varieties). Compared to traditional varieties of soybeans, low-stachyose soybeans have ~10% more energy content metabolizable by animals and a 3% increase in amino acid digestibility. Low-stachyose soybeans are principally beneficial for feeding monogastric animals (swine, poultry, etc.), since their single stomach cannot digest stachyose. Thus, stachyose attempts to promote excess bacterial growth in their intestines, causing them to feel impulsively full.

low-temperature acclimatization: Ability of plants to cold acclimatize when exposed to gradually declining temperatures below a specific threshold. This is the general process that plants have evolved for acclimatizing to low-temperature stress. It is a gradual process during which there are changes in just about every measurable morphological, physiological and biochemical parameters. These changes are determined by genotype-environment interactions.

low-tillage crop production: A procedure of crop production in which the farmer employs a minimum of mechanical cultivation (i.e. only two to four passes over the field with tillage equipment instead of the conventional five passes per year for traditional crop production). This abridged mechanical tillage leaves more carbon in the less disturbed soil, leaves more earthworms (*Eisenia foetida*) per unit area living in the topsoil and decreases soil compaction (i.e. the reduction in interstitial spaces between individual soil particles), thereby enhancing the fertility of 'low till' farm fields. The plant residue lasting on the field's surface assists in controlling weeds and reducing soil erosion; it also offers sites for insects to shelter and reproduce, leading to a need for increased pest insect control via techniques such as inserting a *Bacillus thuringiensis* (*Bt*) gene into certain crop plants. But if a farmer needs to apply synthetic chemical pesticides, the plant residue remaining on the field's surface assists in breaking down (into substances such as carbon dioxide and water) of those pesticides. That is why the plant residue aids to maintain moisture in the field surface environment, thereby enhancing the growth of the microorganisms that help in breaking down the pesticides.

luciferase: An enzyme that acts on substrate (luciferin) and emits a photon of light. Some bacteria and insects exhibit bioluminescence (light emission). These organisms have an enzyme called luciferase which picks up electrons from the flavoproteins in the electron transport chain and then emits some of the electron's energy as a light. Firefly is a good example of an insect that exhibits bioluminescence. The luciferase gene from this insect has been cloned and introduced into tobacco that produced light when supplied with luciferin and ATP. The luciferase gene is also used as a reporter gene during plant transformation.

Ludwig theorem: The theory that new genotypes can be supplemented to a population if they can make use of new components of the environment by occupying new subniches even if they were inferior in the ancestral niche.

lumen (pl. lumina): (i) Region within the thylakoid membrane where water is torn apart to attain oxygen. The oxygen diffuses out of the cell, while the protons remain inside to enhance the positive electrical charge inside the thylakoid. (ii) Epithelial tissues are widespread throughout the body. They form the covering of all body surfaces, line body cavities and hollow organs, and are the major tissue in glands. They perform a variety of functions that include protection, secretion, absorption, excretion, filtration, diffusion and sensory reception.

Luria and Delbrück experiment: An experiment in genetics designed by S.E. Luria and M. Delbrück (1943) to reveal whether a specific mutation could be environmentally induced or a random occurrence. It involves the use of a fluctuation test, in which numerous bacterial cultures are simultaneously screened for resistance to lysis by bacteriophage. If mutation to phage resistance were random, then considerable fluctuation in the numbers of mutant bacteria would be expected to occur in the parallel cultures, depending upon precisely when the mutation had occurred. If it occurred early in the life of the culture, many progenies would accumulate; if late, no progeny would accumulate. Since marked fluctuations in the numbers of surviving resistant colonies were obtained, it was obvious that mutation was occurring at a constant, random rate and was not being induced by or exposure to the phage or other environmental factors.

luxuriance: Hybrids that are larger, faster growing or otherwise outdo the parental forms in some traits. Luxuriance usually ensues due to complementary gene action occurring in the parents and combined in the hybrid.

luxury genes: The genes coding for specialized functions; their products are synthesized (usually) in large amounts in particular cell types.

lyase: An enzyme that catalyses non-hydrolytic reactions in which groups are either detached or supplemented to a substrate, thereby generating or abolishing a double bond, especially between carbon atoms or between carbon and oxygen atoms.

lycopene: An antioxidant carotenoid ('light-harvesting' pigment used by plants in the photosynthesis process) that is an obviously occurring phytochemical in tomatoes, watermelon, guava and pink grapefruit (and some other fruits). Ingesting of significant amounts of lycopene by humans results in an increase in the concentration of lycopene in the blood plasma.

Lyell's notion of gradual change: Also called uniformitarianism or gradualism, C. Lyell's concept was that Earth has been shaped by the same forces and processes (that operate even today), acting uninterruptedly over very long periods of time. For example, the ongoing erosion caused by flowing water in a river could, if given enough time, carve out the Grand Canyon (a National Park, located in north-western Arizona, USA).

Lynx Therapeutics' massively parallel signature sequencing (MPSS): The first of the next-generation sequencing technologies, MPSS was developed in the 1990s at Lynx Therapeutics, a company founded in 1992 by S. Brenner and S. Eletr. MPSS is a bead-based method that utilizes a complex approach of adapter ligation followed by adapter decoding, reading the sequence in increments of four nucleotides. This method made it susceptible to sequence-specific bias or loss of specific sequences. Because the technology was so complex, MPSS was only performed 'in-house' by Lynx Therapeutics and no machines were sold; the union with Solexa later led to the advancement of sequencing-by-synthesis, a more simple approach with numerous benefits, MPSS became outdated. However, the essential properties of the MPSS output were typically for later 'next-gen' data types, together with hundreds of thousands of short DNA sequences. In the case of MPSS, these were typically employed for sequencing cDNA for measurements of gene expression levels.

Lyon hypothesis: M.F. Lyon in 1961 suggested that Barr body represented an inactive X chromosome, which in females becomes tightly coiled into heterochromatin (condensed chromatin) and therefore characteristically visible form of chromatin. For example, in any given cell of a female, one X chromosome is active and the other is inactive. Lyon hypothesis states that in cells with multiple X chromosomes, all but one is inactivated during mammalian embryogenesis.

lyonization: Lyonization (also called X-INACTIVATION) refers to the normal phenomenon in which one of the two X chromosomes in every cell of a female individual is inactivated during embryonic development. This inactivation prevents females from having twice as many X chromosome gene products as males, who possess only a single copy of the X chromosome. Lyonization is named after Mary F. Lyon, the British geneticist who discovered the phenomenon in 1961. *See* FREEZE DRYING.

lysate: A liquid containing the components of disrupted cells that is used as the source of DNA for testing patient samples.

lyse: To abolish, rupture or jumble cells by enzymes, viruses or other means.

lysine (lys): An essential nutritionally important amino acid ($HO_2CCH(NH_2)(CH_2)_4NH_2$) that can be attained from many proteins by hydrolysis (i.e. cutting apart the protein molecule). Lysine is found in proteins of legumes, whole grains and nuts. Specific lysines within histone proteins are often post-translationally tagged by acetyl, methyl and other chemical moieties.

lysine acetylation: A post-translational modification in which an acetyl group reacts with the primary amine on the side chain of a lysine residue.

lysis: The term is applied when a bacterial cell is killed for the release of phage progeny. When a virulent bacteriophage, such as lambda (λ), T2, T4 or SPO, infects a host bacterium, it completely depends on the host for its replication. It takes over the host's function and utilizes host machinery for producing a large number of phage particles. Enzyme lysozyme coded by phage gene is synthesized that results the bacterial cell wall to break open. The bacterial cell undergoes lysis and dies to liberate the phage particles, which are each then ready to restart another cycle by infecting new bacterial cells. This cycle is known as LYTIC CYCLE.

lysogen: A bacterium that possesses a repressed prophage as part of its genome.

lysogenic: The pattern of bacteriophage infection that involves the integration of the phage genome into the host DNA molecule.

lysogenic cycle: Life cycle of a phage in which phage DNA becomes part of the host cell's genome, and it is called prophage. This is in contrast to lytic cycle. In the lysogenic cycle, DNA of the phage is retained in the host through bacterium thousands of cell divisions. The phage DNA is inserted into the bacterial genome. The integrated form of phage DNA is called prophage. Eventually, prophage is released from the host genome and lyses the host cell. Lambda (λ) phage is lysogenic.

lysogenic immunity: The capability of a prophage to prevent another phage from re-entering the lytic cycle.

lysogenic repressor: The protein responsible for averting a phage from re-entering the lytic cycle.

lysogeny: State of a bacterium in which it transports the DNA of an inactive virus integrated into its genome. The virus can later be activated to replicate and lyse the cell.

lysophosphatidylethanolamine (LPE): LPE is one of the phospholipids naturally found in soybean oil. In plants, LPE functions as a signalling molecule (e.g. speeding the ripening process). *Aka* PHOSPHATIDYL ETHANOLAMINE.

lysosome: A membrane-surrounded organelle in the cytoplasm of eukaryotic cells which comprises many hydrolytic enzymes. The lysosome internalizes and digests foreign proteins as well as cellular debris. The protein fragments (epitopes) are 'presented' to T cells by the MAJOR HISTOCOMPATIBILITY COMPLEX (MHC) proteins on the surface of the eukaryotic cell. Lysosomes contain lysozyme, lactic acid, oxidants and other digestive enzymes. The hydrolytic enzymes are typically most active at the acid pH found in the lumen of lysosomes.

lysozyme: It is an enzyme that destroys Gram-positive bacterial cell walls. It does so by hydrolysing the β(1→4)-glycosidic linkages from N-ACETYLMURAMIC ACID (NAM) to N-ACETYLGLUCOSAMINE (NAG) in cell wall peptidoglycans. Lysozyme gene has been cloned from human beings and it is being transferred into crop plants for developing resistance against bacterial diseases.

lytic cycle: Steps of lytic cycle are as follows: attachment of bacteriophage to receptors on bacterial cell wall, penetration of the cell wall by tail core, DNA is injected into the host, bacterial DNA is hydrolysed, phage DNA directs biosynthesis of viral parts using the host cell's machinery, the phages mature as the parts are assembled, and lyses of host bacteria and release of the new phages.

lytic cycle control: A mechanism whereby the pathway of transcription of phage DNA is controlled in such a way that either the lytic pathway or the lysogenic pathway is followed. If protein (cl repressor) is formed, lysogeny results; if protein cro is formed, phage progeny develops.

lytic infection: A viral infection in which the ultimate act of the infection is to lyse (i.e. burst or destroy) the cell. This releases the new (progeny) viruses, so they can go on to infect other cells.

M

M phase: It is composed of two tightly coupled processes: (i) mitosis, in which the cell's chromosomes are divided between the two daughter cells; and (ii) cytokinesis, in which the cell's cytoplasm divides in half, forming distinct cells.

M phase kinase: This enzyme comprises two types of subunits: (i) Subunit Cdc2 is the catalytic subunit that phosphorylates serine and threonine residues in target proteins. It is named after the homologous protein in Schizosaccharomyces pombe. (ii) Subunit Cyclin is a regulatory subunit, which is necessary for the kinase to function with appropriate substrates. There are (at least) two general forms of the M phase kinase: Cdc2-cyclin A and Cdc2-cyclin B. The common properties of the cyclins suggest that they perform a similar function (i.e. impact the activity of the catalytic subunit of M phase kinase). The two classes of cyclins have only weak similarity and follow a different temporal and spatial pattern of behaviour. Two general models could be proposed for its role: (a) It could be a master regulator, phosphorylating target proteins that in turn act to control other necessary functions (a classic cascade); or (b) it may be a workhorse, directly phosphorylating the crucial substrates required to execute the regulatory events or cell reorganization involved in the cycle.

M-checkpoint: A checkpoint in mitosis, where the cell examines whether all the sister chromatids are correctly attached to the spindle microtubules.

M-cyclin: Type of cyclin existing in all eukaryotic cells that promotes the events of mitosis.

M1, M2, M3, etc.: The symbols used to label first, second, third, etc. generations after treatment with a mutagen.

M13 phage vector: Since 7.2-kb-long M13 phages occur in both single- and double-stranded forms, replication form (RF) can be digested with restriction endonucleases, and inserts can be cloned in it like plasmids. '+' strands from phage particles are a convenient source of single-stranded DNA and are used for Sanger's method of dideoxy chain termination sequencing and site-directed mutagenesis. Different-sized DNA molecules are packaged as phage particles. Phages with inserts >2 kb replicate slower.

M_2 population: The progeny derived by selfing M_1 plants, which themselves comprise progeny that arise by selfing plants grown from a mutagenized seed. Recessive mutations, resulting from the seed mutagenesis, are detected in M_2 plants, which are homozygous for the mutation.

m^6A-dependent translational regulation: A.M. Vicente *et al.* (2023) identified epitranscriptomic players as modulators in acclimation processes. During cold treatment, the expression of components of the primary nuclear m^6A methyltransferase complex was upregulated, accompanied by a significant increase in cellular 6-methyldenossine (m^6A) mRNA marks. In the cold, the presence of FIP37, a core component of the writer complex, played an important role in positive regulation of thylakoid structure, photosynthetic functions, and accumulation of photosystem I, the Cytb_{6f} complex, cyclic electron transport proteins, and Curvature Thylakoid1, but not that of photosystem II components and the chloroplast ATP synthase. Downregulation of FIP37 affected abundance, polysomal loading and translation of cytosolic transcripts related to photosynthesis in the cold, suggesting m^6A-dependent translational regulation of chloroplast functions.

m^6Anet: A neural network-based method that leverages the multiple instance learning framework to specifically handle missing read-level modification labels in site-level training data. m^6Anet outperforms existing computational methods, shows similar accuracy as experimental approaches, and generalizes with high accuracy to different cell lines and species without retraining model parameters. C. Hendra *et al.* (2022) demonstrate that m^6Anet captures the underlying read-level stoichiometry, which can be used to approximate differences in modification rates. Thus, m^6Anet offers a tool to capture the transcriptome-wide identification and quantification of m^{6}A from a single run of direct RNA sequencing.

macerate: (i) To cause the disintegration of tissues by separating the cells. (ii) To make it softer by soaking.

machine intelligence (MI): Machine intelligence is advanced computing that enables a technology (a machine, device or algorithm) to interact with its environment intelligently, meaning it can take actions to maximize its chance of successfully achieving its goals.

machine learning (ML): An algorithmic procedure for prediction or classification that tends to be empirical, nonparametric and flexible, and it does not capitalize on the additivity of predictors. A. Samuel defined machine learning in 1959 as a 'field of study that gives computers the ability to learn without being explicitly programmed'. Machine learning does not use a data model (i.e. a probability distribution for the outcome variable given the inputs) and does not place emphasis on interpretable parameters. Examples of machine-learning algorithms include neural networks, support vector machines, bagging, boosting, recursive partitioning and random forests. Ridge regression, the lasso, the elastic net and other penalized regression techniques, which have identified parameters and make heavy use of additivity assumptions, fall under statistical models rather than machine learning. By allowing high-order interactions to be potentially as important as main effects, machine learning is 'data hungry', as sample sizes needed to estimate interaction effects are much larger than sample sizes needed to estimate additive main effects. Machine learning is not to be confused with artificial intelligence. ML algorithms build a mathematical model based on sample data (*aka* TRAINING DATA) to make forecasts or conclusions without being explicitly programmed to perform the task. Currently, ML is providing benefits to farmers worldwide at the pre-harvesting, harvesting and postharvesting stages of crops to minimize losses in farming. The major application of ML will allow the farmer to attain more efficient and precise farming in agriculture with less human manpower and high crop quality production.

machine listening: A general field of study of algorithms and systems for audio understanding by machines. *Aka* COMPUTER AUDITION (CA).

machine perception: The capability of a computer system to interpret data in a manner that is similar to the way humans use their senses to relate to the world around them.

machine vision (MV): The know-how and procedures utilized to deliver imaging-based automatic inspection and analysis for such applications as automatic inspection, process control and robot guidance, usually in industry. Machine vision, in fact, refers to many technologies, software and hardware products, integrated systems, actions, procedures and expertise. Machine vision as a systems-engineering discipline can be thought of as distinct from computer vision, a form of computer science. It attempts to assimilate existing technologies in new ways and broadly apply them to solve real-world problems. MV is the prevalent term for these functions in industrial automation environments, but it is also used for these functions in other environments such as security and vehicle guidance.

machine-learning strategies: The machine-learning approaches utilize previously confirmed miRNA hairpins as a positive training data set and a negative training data set with other hairpin structures, containing, for example, mRNAs, tRNAs and rRNAs.

macro noncoding RNAs (macro ncRNAs): Noncoding RNAs that can be as short as a few hundred nucleotides or as long as several hundred thousand nucleotides, the function of which does not depend on processing into short interfering RNAs or microRNAs. *See* LONG NONCODING RNA.

macroclimate: Climatic conditions affecting an extensive area. *Compare with* MICROCLIMATE.

macroconidium (pl. macroconidia): The larger of two types of conidia produced by certain fungi. *Contrasts* MICROCONIDIUM.

macrocyclic: Pertaining to a rust fungus that typically exhibits all five stages of the rust life cycle: spermatia (0), aeciospores (I), urediniospores (II), teliospores (III) and basidiospores (IV). *Contrasts* DEMICYCLIC *and* MICROCYCLIC.

macroevolution: Evolution occurring on a scale of separated gene pools. Macroevolutionary studies focus on change that transpires at or above the level of species (e.g. genera, families or classes). The origin or adaptive radiation of a higher taxon (e.g. vertebrates) could be called a MACROEVOLUTIONARY EVENT. *Contrasts* MICROEVOLUTION.

macrogenesis: The sudden origin of new types by saltation.

macroinjection: Needles used for injecting DNA have a diameter greater than the cell's diameter. DNA is injected with a conventional syringe into a region of the plant, which will develop into floral tillers. DNA solution is injected at a point above the tiller node until several drops of solution come out from the top of the young inflorescence.

macrolesions: (i) Mutation exerts a large phenotypic effect; one that produces a phenotype well outside the range of variation previously existing in the population. (ii) Include changes in chromosome structure and number. These changes can be observed at high resolution with a light microscope. (iii) A mutation that causes a philosophical change in an organism, for example, a change in a regulatory gene that controls the expression of many structural genes. *Aka* MACROMUTATIONS, CHROMOSOME MUTATIONS *or* CHROMOSOME REARRANGEMENTS. *Contrasts* MICROLESIONS *and* MICROMUTATIONS.

macromolecule: A large organic molecule having a molecular weight ranging from a few thousand to many millions. Proteins, nucleic acids and oligosaccharides are examples of macromolecules. During the metabolism, growth and development of an organism, these compounds play very important roles. Macromolecules are commonly generated by the polymerization of smaller subunits (monomers). After the death of the organism, macromolecules are degraded into smaller molecules (monomers) by microorganisms.

macromutation: A mutation with distinct morphological changes in the phenotype, usually observed in qualitative characters.

macronucleus: The larger of the two nuclei instituted in ciliate cells. It functions in ways similar to those of animal somatic cells. *Compare with* MICRONUCLEUS.

macronutrient: An element necessitated in relatively large quantities for plant growth, metabolism and other body functions, including macronutrients nitrogen (N), phosphorus (P), and potassium (K) provide calories or energy. There are three basic types of macronutrients: (i) fats, (ii) proteins and (iii) carbohydrates. *Contrasts* MICRONUTRIENT.

macrophagocytosis: A process involved in swallowing the largest immune cells, which means it functions by literally swallowing up other particles or smaller cells. Macrophages engulf and digest debris (e.g. dead cells) and foreign particles through the process of phagocytosis. Thus, macrophages act like scavengers. It also has antigen-presenting functions in immune responses. Phagocytic cells, derived from blood monocytes, reside in most tissues.

macrorestriction map: Map depicting the order of and distance between sites at which restriction enzymes cleave chromosomes.

macrosynteny: The conserved order of large genomic blocks in the genomes of related or unrelated species, as detected by chromosomal *in situ* suppression hybridization or fluorescent *in situ* hybridization.

Megaclone technology: This technology allows both the simultaneous cloning of millions of DNA molecules or fragments in a sample, and the parallel probing or assaying of the millions of clones in the resulting sample. Furthermore, Megaclone™ technology enables all this without requiring prior separation, purification, individual amplification or identification of any of the original molecules or fragments in the sample. This technology is based on a highly proprietary, patented, parallel cloning process that simultaneously replaces each individual molecule or fragment in a sample, whether known or unknown, by a microbead carrying about 100,000 copies of that molecule or fragment. In other words, a single sample containing 1,000,000 or more DNA molecules or fragments is converted into a single sample containing 1,000,000 or more clones (each about 100,000 copies strong) concentrated and segregated on as many microbeads as possible.

magnetic bead-based DNA purification method: This method is based on reversibly binding DNA to magnetic solid surfaces, beads or particles, which have been coated with a DNA-binding antibody or a functional group that interacts specifically with DNA. After DNA binds, beads are separated from other contaminating cellular components, washed and finally the purified DNA is eluted using ethanol extraction.

magnetic force microscopy (MFM): An atomic force microscopy (AFM)-based technique in which an AFM tip with a magnetic coating is employed to investigate local magnetic fields with the typical AFM spatial resolution, thus enabling one to obtain images reflecting the local magnetic properties of the samples at the nanoscale.

magnetic force scanning tunnelling microscopy (MFSTM): The MFSTM technique is shown to be a powerful technique for generating images of magnetization patterns. Images of the overwritten data on commercial magnetic discs and images of deliberately written patterns that serve as benchmarks are presented. The usefulness of this method for probing the subtle effects of insufficient erasure during data overwriting.

magnetic resonance imaging (MRI): A medical imaging technique using the property of nuclear magnetic resonance to image the nuclei of atoms inside the body. MRI can create more detailed images of the human body than are possible with X-rays.

magnetic resonance spectroscopy (MRS): A research technique using nuclear magnetic resonance to provide information about the chemical properties of atoms and molecules.

magnificent megabacterium: Sampling a mangrove swamp, J.M. Volland *et al.* (2022) found an unusually large, sulfur-oxidizing bacterium with a complex membrane organization and predicted life cycle. Using a range of microscopy techniques, the authors observed highly polyploid cells with DNA and ribosomes compartmentalized within membranes. Single cells of the bacterium, dubbed *Candidatus Thiomargarita magnifica*, although thin and tubular, stretched more than a centimetre in length. This bacterium has an average cell length greater than 9000 μm, is visible to the naked eye and is thus not a microbe by definition.

main band DNA: Upon centrifugation on the density gradient, most of the eukaryotic genomes form a variety of fragments that appear as a rather broad peak cantered on the buoyant density corresponding to the average guanine-cytosine (G-C) content of the genome.

main-effect QTLs (M-QTLs): M-QTLs are defined as single Mendelian factors at which additive and/or dominance effects on a given phenotype arise from allelic substitution and are detected by marker-trait associations using single-factor ANOVA QTL mapping or interval mapping models.

mainline (ML): A quantitative term referring to the most frequent chromosome constitution of a tumour cell population.

maintainer: A line used for maintenance and multiplication of a cytoplasmic male-sterile line. The line usually comprises genotypes containing the normal cytoplasm and recessive at the restorer locus.

maintaining genetic purity: Various steps involved in maintaining the genetic purity of a line include the following: (i) Providing satisfactory isolation to prevent contamination by natural crossings or mechanical mixtures. (ii) Rouging (the act of identifying and removing plants with undesirable characteristics from agricultural fields) of seed fields prior to the stage at which they could contaminate the seed crop. (iii) Periodic testing of varieties for genetic purity. (iv) Avoiding genetic shifts by growing crops in areas of their adaptation only. (v) Certification of seed crops to maintain genetic purity and quality of seed. (vi) Adopting the generation system. (vii) Grow-out tests seed puroindoline proteins: small, basic, cysteine-rich proteins found in bread wheat (*Triticum aestivum*). By engineered introduction of a puroindoline gene, the hard-textured grain of durum wheat (*Triticum durum*) and other cereals, including maize and barley, can be converted into soft texture. The puroindoline genes, *puroindoline a* and *puroindoline b*, represent the grain hardness gene on chromosome 5DS. In hexaploid wheat, soft kernel texture is a wild-type trait, whereas hard texture (hardness) results from any one of several mutations in the puroindolines.

maintenance methylase: An enzyme that methylates a half-methylated site on DNA obtained after replication, thus converting it into a fully methylated site. This enzyme does not methylate an unmethylated site.

maintenance methylation: A process where a half-methylated site in DNA is fully methylated through the use of maintenance methylase enzyme, which transfers a methyl group from S-adenosylmethionine (SAM or AdoMet) to cytosine in eukaryotes and adenine in prokaryotes.

maintenance of CpG methylation: DNA methyltransferase 1 (Dnmt1) is the principal enzyme responsible for maintenance of cytosine methylation in CpG context and is essential for regulation of gene expression, silencing of parasitic DNA elements, genomic imprinting and embryogenesis.

maintenance of genome integrity: The maintenance of genome integrity is key to preserving the functions of ESCs and reducing the risks of cancer development, cell cycle arrest and abnormal replication. During the self-renewal or differentiation of ESCs, abundant HOMOLOGOUS RECOMBINATION (HR) factors promote the repair of DNA damage and recovery from replication stress, thereby maintaining genome integrity.

maintenance substances: Those substances that are needed by the cell for its maintenance. The synthesis of ribosomes in somatic cells is a maintenance function.

maintenance-type DNA methylation: Most DNA methylation is maintenance-type following DNA replication. Maintenance-type DNA methylation is not sequence-specific. Half-methylated sites of double-stranded DNA are the substrates for this type of DNA methylation. Maintenance-type DNA methylation functions to restore hemi-methylated DNA generated during DNA replication to the fully methylated state.

maize epigenome: The maize genome contains a class of 22-nucleotide siRNA that is absent in a number of other monocots and introduced the possibility of two distinct biogenesis pathways for siRNA in maize. Landscapes of typical epigenetic changes and their associations with mRNA and sRNA transcriptomes have been investigated in maize shoots and roots. The genic DNA methylation pattern in maize resembles that of rice in that it reveals a peak around ATG sites but differ substantially from that of *Arabidopsis*, which shows higher DNA methylation in transcribed regions. Genome-wide high-resolution allele-specific maps of DNA methylation and H3K27me3 have been constructed in maize endosperm. For DNA methylation, thousands of PARENT-OF-ORIGIN-DEPENDENT DMRs (pDMRs) were identified. All pDMRs were paternally hypermethylated and maternally hypomethylated in a uniform manner. The maize embryo genome contained more DNA methylation than the endosperm. Among the numerous imprinting genes found, two-fifths were similarly methylated, and three-fifths were differentially methylated between embryo and endosperm.

maize diversification and domestication by RNA interference: Introgression of *mexicana* into early cultivated maize is believed to have been critical to its geographical dispersal throughout the Americas, and a tightly linked inversion in *mexicana* spans a major domestication sweep in modern maize.

maize streak virus (MSV): A geminivirus, which carries a single-stranded DNA genome that replicates in the nucleus of the host cells to offer a double-stranded replication intermediate and transcription template.

major gene resistance: Races of the pathogen are controlled by major genes in the host.

major gene: A gene with marked effects on phenotype. *Contrasts* MODIFIER GENES.

major groove of DNA: In DNA, the glycosidic bonds between sugars and bases of given base pairs are not located exactly opposite each other, and two grooves of unequal width are formed around the double helix. The major groove measures more than 180°. Thus, this is the larger of the two grooves that spiral around the surface of the DNA double helix. The major groove is much wider than the other groove, called the MINOR GROOVE. *Compare with* MINOR GROOVE OF DNA.

major histocompatibility antigens: These antigens act with high intensity and reject a graft fast.

major histocompatibility complex (MHC): Proteins of the MHC complex are highly polymorphic in vertebrates. They appear on the surface of cells, imparting to them the property of 'self' (belonging to that organism). Some other functions of the MHC complex are also suggested. In humans, a group of similar proteins is known as the HUMAN LEUKOCYTE ANTIGEN (HLA) COMPLEX, *aka* MHC PROTEINS. MHC proteins are of three types: MAJOR HISTOCOMPATIBILITY COMPLEX CLASS I, MAJOR HISTOCOMPATIBILITY COMPLEX CLASS II and MAJOR HISTOCOMPATIBILITY COMPLEX CLASS III. *See details as separate terms.*

major histocompatibility complex class I (MHC I): Proteins present on almost all cell types that present viral peptides on the surface of virus-infected cells, where they are recognized by cytotoxic T cells.

major histocompatibility complex class II (MHC II): Proteins present on 'professional' antigen-presenting cells that present foreign peptides to helper T cells.

major histocompatibility complex class III (MHC III): Proteins comprising a mixture of complement, chemokine and cytokine proteins that are not involved in antigen presentation.

major histocompatibility genes (MHC genes): A chromosomal region, ~3000 kb in size, that encodes for three classes of transmembrane (cell) proteins. MHC genes must be matched (between an organ donor and organ recipient) to avert rejection of organ transplants.

major satellite: Tandem repeating DNA sequences of 234 bp that are primarily located in the pericentromeric regions of the mouse genome.

major spliceosome: Major spliceosomes are assembled from U1, U2, U4, U6 and U5 snRNPs (which are named according to the U snRNA(s) they contain). *Compare with* MINOR SPLICEOSOME.

major splicing pathway: The major splicing pathway takes away U2-type introns permitting the adjacent exons to be spliced together in the nucleus before export of the fully spliced transcripts to the cytoplasm. The transcripts that also contain U12-type introns are exported as partially spliced transcripts and their exons are spliced together by the U12 pathway in the cytoplasm. Although U2-type introns are found globally, U12-type introns may occur preferentially in particular sets of genes, such as those involved in cellular proliferation.

male flower: *See* staminal FLOWER.

male gametophyte: A microspore may be oval or fusiform in shape and contain denser cytoplasm. It has a haploid nucleus in the centre. All kinds of cell organelles, such as dictyosomes, ribosomes, mitochondria, endoplasmic reticulum, plastids and microtubules, are present. The food is stored in the form of carbohydrates, proteins and lipid granules. The mature pollen grain is surrounded by a thick and highly stratified wall, the sporoderm. The sporoderm comprises two principle layers: the outer thick EXINE and the inner thin INTINE. Exine plays a protective role. Exine again consists of two sublayers: the outer thick SEXINE and the inner thin NEXINE. The sexine is further differentiated into the inner foot layer, the central bacula and the inner cellulose layer. Intine has uniform thickness, but at the pores region, it has thick callose material. Intine is differentiated into an outer pectic layer and an inner cellulose layer. Sporoderm has some weak (thin) areas known as apertures, which are directly or indirectly associated with the germination of pollen grains. The gametophytic development starts in microsporangium. Some striking changes occur in the microspore. In angiosperms, the microsporangium is known as the NUCELLUS. Generally, one or two outer coverings surround these integuments. The microsporangium with its protective coat is known as the ovule.

male recombination factors (MRFs): Genetic factors responsible for spontaneous recombination in males of *Drosophila melanogaster* where this phenomenon is normally known to be absent.

male sterility: The condition in which pollen is not produced or produced pollen is non-functional for pollination. Male sterility is an important tool in plant breeding for the development of hybrids. Male sterility may exist as pollen sterility (i.e. non-functional pollen is produced); staminal sterility, in which stamens are malformed or even absent; and functional sterility, where viable pollen is trapped in indehiscent anthers. Among these, pollen sterility is the most common. Male sterility may be controlled by nuclear genes (nuclear male sterility), cytoplasmic genes (cytoplasmic male sterility) or both (cytoplasmic – genetic male sterility). Nowadays, transgenic approaches are also used to develop male-sterile lines by introducing the *barnase* gene, which causes the degeneration of tapetal cells in the anther, leading to the formation of non-functional pollen.

male sterility and fertility restoration: Transgenic plants for male sterility and fertility restoration have been produced in *Brassica napus*. Another specific promoter from TA29 of *Bacillus amyloliquefaciens* and a bacterial coding sequence from *the barnase* gene (which codes for ribonuclease) were used to produce a transgenic plant, which prevented normal pollen development, leading to male sterility.

male sterility through recombinant DNA technology: The development of recombinant DNA technologies has opened new possibilities for creating and manipulating male sterility in crops for pollination control. There are four approaches based on biotechnological methods: (i) DOMINANT MALE-STERILITY GENES – expression of a gene encoding a cytotoxin is placed under the control of an anther-specific promoter, resulting in male-sterile plants of otherwise normal phenotype. (ii) RECESSIVE MALE-STERILITY GENES – male-sterile phenotypes are normally recessive; therefore, most male-sterility genes are simply mutations in genes essential for male fertility. (iii) TARGETED GAMETOCIDE – a transgene with a suitable anther-specific promoter converts an applied chemical into a phytotoxic substance in a confined area, such as the tapetum cells, thus conferring sterility. (iv) DUAL METHOD – two different versions of the female parent are created, one with an anther-specific promoter and one with the sequence for a protein, which will disrupt the formation of viable pollen. Crosses of the two females assemble the male-sterility gene. The benefits and limits of genetically engineered pollination versus cytoplasmic male sterility are discussed.

male sterility, cytoplasmic: Male sterility resulting from specific cytoplasmic-genic interactions. Sterile cytoplasm is denoted by S or cms, in contrast to normal cytoplasm N.

male sterility, genetic: Male sterility resulting from the action of nuclear genes.

maleic anhydride: An important organic chemical used in the manufacture of synthetic resins, in fungicides, in the dyeing of cotton textiles and to prevent the oxidation of fats and oils during storage and rancidity.

maleic hydrazide (MH): A herbicide, fungicide, growth inhibitor and growth regulator. It is a structural isomer of uracil. It induces chromosomal breaks frequently in heterochromatic segments near the centromere.

malignant: Refers to tumours and tumour cells that are aggressive and/or able to undergo metastasis. A malignant tumour is a cancer. *See* BENIGN.

malignant tumour: The tumours from which cells detach and migrate to other parts of the body, giving rise to secondary tumours. They invade the tissue. The phenomenon is known as METASTASIS. *See* METASTASIS.

Malpighian layer: A protective layer or layers of cells present in the coats of many seeds. It is typically made up of close-packed, radially placed, heavy-walled columnar cells without intercellular spaces. The cells are often heavily cutinized or lignified and are relatively resistant to moisture and gases.

Malthusian parameter: The parameter that determines the exponential rate of growth is called the Malthusian parameter. Simple explicit formulae are available both for the age-dependent branching process and for Green's model of asymmetrical cell division. *See* R *and* INTRINSIC RATE OF INCREASE.

malting: The conversion of grain starch into glucose and maltose. Almost all alcoholic beverages, like beer, are produced from grain starches (barley grains) using microorganisms. First, the starchy grains are sprouted, dried and ground to prepare malt. The malt contains starch-degrading enzymes like amylase that convert starch into carbohydrates (glucose + maltose), which are further fermented by yeast to produce alcohol.

mammalian artificial chromosome (MAC): MACs are functional, add-on chromosomal elements that behave as normal chromosomes, replicating and partitioning into daughter cells at each cell division. They are deployed as useful gene expression vectors as they remain independent from the host genome. To construct MACs, telomere repeats and selectable markers were introduced into a 100-kb YEAST ARTIFICIAL CHROMOSOME (YAC) containing human centromeric DNA. This YAC, which has a regular repeat structure of alpha-satellite DNA and CENTROMERE PROTEIN B (CENP-B) boxes, efficiently formed MACs that segregated accurately. The MACs appear to be about 1–5 Mb in size and contain YAC multimers. MACs have the potential to be used as vectors for introducing genes into mammals.

mammalian DNA methyltransferases: Active DNA methyltransferases identified in mammals are DNMT1, DNMT3a, DNMT3b and DNMT3L. *See details as separate terms.*

mammalian mitochondrial rRNA modifications: Mitochondrial ribosomes are unique in having a set of methylated residues that includes universally conserved ones (i.e. those that could be present either in bacterial or archaeal/eukaryotic cytosolic ribosomes and those that are present exclusively in mitochondria).

mammalian mitochondrial tRNA modifications. Faults in RNA modification, arising from either MITOCHONDRIAL (MT)-DNA mutations or mutations in nuclear-encoded mitochondrial modification enzymes, underlie multiple mitochondrial diseases. Associated advances in the identification of the mitochondrial RNA modification machinery and recent structural views of the mitochondrial translation apparatus now allow the molecular basis of such mitochondrial diseases to be understood on a mechanistic level.

mammalian RNA polymerase II gene promoters: These promoters are characteristically arranged in the following order: upstream sequence motif(s)/ TATA box/initiation site. The polarity of transcription is principally determined by the linear order of an upstream sequence relative to a TATA box rather than by the individual locations of either of these two elements.

mammalian spot test: In this mutagenicity test, embryos that are heterozygous for different recessive coat-colour genes are treated *in utero* during the 7th–10th day of fetal development by injection of a mutagen into the peritoneal cavity of the mother animal. If this treatment, in a pigment precursor cell, leads to an alteration of the wild-type allele of one of the genes under study, a colour spot can be seen in the adult coat.

mammalian virus-based vectors: Some of the mammalian virus-based vectors used in gene therapy include retroviruses, adenoviruses, parvoviruses, herpesviruses and poxviruses.

mammalogy: The branch of biology that studies mammals, a class of vertebrates with features such as homoeothermic metabolism, fur, four-chambered hearts and complex nervous systems.

mammary glands: Specialized glands that can produce milk for feeding young. Only found in mammalian females.

Manhattan plot: A scatter plot that offers a useful grand-scale visualization of the GENOME-WIDE ASSOCIATION STUDIES (GWAS) results. The assayed

SNPs are sorted by chromosome locations on the x-axis. The y-axis shows the negative logarithm of P-values, and as a result, the higher dots represent significant hits.

manipulating regulatory genes for metabolic engineering: To increase flux through biochemical pathways, there is interest in manipulating regulatory genes such as kinases or transcription factors to upregulate entire pathways. The unresolved issue, in this respect, concerns the degree to which all genes in biosynthetic pathways are coordinately controlled. Several examples suggest that at least some pathways are under coordinated control.

manipulation force microscope: The manipulation force microscope is an AFM adapted to measuring the force necessary to displace micron-size samples sticking to surfaces. *Aka* MANIPULATION FORCE MICROSCOPY.

mantle: A thick hyphal mass of an ectomycorrhizal fungus enclosing the short feeder roots of plants. *See* HYPHAL SHEATH.

many genes-one polypeptide hypothesis: The immunoglobulin genes, which are also called ASSEMBLED GENES, do not fit into any classical or neoclassical definition of the gene since the genetic unit in the germline and in the mature immune cell is completely different. Functional genes of immunoglobulins mature by the somatic recombination process from a few units in the germline during the maturation of immune cells. The somatic recombination in immunoglobin genes led to many genes-one polypeptide hypothesis.

map: (i) As a verb, to determine the relative or physical position of a gene, DNA molecule or chromosome segment. (ii) As a noun, a map is a pictorial depiction of the position of genes on the chromosomes of an organism. A linkage map is constructed from percent crossing-over between the linked genes.

map distance: The distance between any two gene markers on a genetic map, based on the percentage of crossing-over. The distance between linked genes is 1% to <50%. If two genes are located on the same chromosome and are far enough apart that they undergo recombination at least 50% of the time, such genes show independent assortment and no linkage is detected in such analysis. Linkage between such genes can be indirectly demonstrated by showing linkage between each one of these genes and another gene located in between the two genes in question. One map unit corresponds to a recombination frequency of 1%. *See* SYNTENY.

map distance (d) and recombination frequency (r), relationship between: When map distances are small (<10 cm), d = r. This relationship does not apply for map distances that are >10 cm.

map marker: Three different categories of molecular markers are first-generation markers, second-generation markers and new-generation markers. These types of markers may also identify dominance and co-dominance within the genome. Identifying dominance and co-dominance with a marker may help identify heterozygotes from homozygotes within the organism. Co-dominant markers are more beneficial because they identify more than one allele, thus enabling someone to follow a particular trait through mapping techniques. These markers allow for the amplification of particular sequences within the genome for comparison and analysis.

map unit: A unit used to describe the distance between two genes on a chromosome, now superseded by centiMorgan. One map unit is equal to one per cent recombination, which is equal to one centiMorgan (cM). *See* GENETIC MAP DISTANCE.

map-based cloning: A method for the isolation of an important gene by cloning the gene in question on the basis of molecular maps, where the physical location in the genome is known but the biochemical function is not known. It involves the identification of a mutant phenotype for the trait of interest (obtained by mutagenesis or from natural variation) and genetic fine-tuning using many progeny plants. The method of isolation of genes solely based on their map location is applicable to any gene whose phenotypic effect can be followed in populations segregating for natural variants or mutant forms of the gene. This map is then used for CHROMOSOME WALKING or CHROMOSOME LANDING, with the help of large-insert DNA libraries or physical maps to isolate the gene. *Aka* POSITIONAL CLONING.

map-based cloning, steps in: Map-based cloning involves four steps: (i) Development of high-resolution molecular linkage maps; (ii) Physical mapping of the region of interest; (iii) Identification of appropriate YAC or BAC clones for isolating putative clones harbouring the gene of interest; (iv) Transgene validation. G.B. Martin *et al.* (1993) were the first to clone the disease resistance gene *Pto* in tomatoes using map-based cloning. Since then, several genes conferring resistance to viruses, bacteria, fungi and nematodes have been cloned.

MapChart: A computer package for the MS-Windows ® platform that produces charts of genetic linkage maps and QTL data. These charts are composed of a sequence of vertical bars representing the linkage groups or chromosomes.

MAPMAKER: A computer programme developed for the detection and estimation of linkages. This package is based on the maximum likelihood method and is mostly used for the construction of RFLP maps. MAPMAKER/EXP performs full multipoint linkage analysis for dominant, recessive and co-dominant (e.g. RFLP-like) markers in BC_1 backcrosses, F_2 and F_3 (self) intercrosses and recombination inbred lines. MAPMAKER/QTL is a companion programme to MAPMAKER/EXP that permits one to map genes controlling polygenic quantitative traits in F_2 intercrosses and BC_1 backcrosses relative to a genetic linkage map.

mapping: A general term in genetics for the plotting of specific positions of genes along a chromosome or strand of DNA. *Aka* GENE MAPPING, GENOME MAPPING *or* GENETIC MAPPING. *See* GENE MAPPING, LINKAGE MAP *and* PHYSICAL MAP.

mapping function: The mathematical relationship between measured map distance and actual recombination frequency.

mapping histone modifications: Histone modifications, which are located mainly in the elastic N-terminal 'tail' regions of the proteins, impact the likelihood that bound DNA will be transcribed. Certain modifications, such as lysine acetylation (Kac), can directly relax chromatin structure to increase DNA accessibility to transcriptional machinery, whereas other modifications employ specific transcriptional activators or repressors to turn gene expression on or off.

mapping kit: A set of molecular markers that are evenly spread out on the chromosomes. If a minimal set of markers can be defined, one can detect linkage by testing a minimal number of probes. For example, if the genome is 1468 cM in length and markers are evenly spaced at 15 cM distance, then at least 98 markers are needed to detect any gene ($N = \ln(1 - P)/\ln(1 - f)$), where N = the number of markers essential to give a probability P of finding at least one marker linked within d map units of the gene and f = the fraction of the genome represented by a region of d map units on either side of the gene (i.e. 2d/genome size; e.g. a marker is required in the genome of 1468 cM and linked within 10 cM of any gene), then $N = \ln(1 - 0.99)/\ln[1 - (2 \times 10\ \text{cM})/1468\ \text{cM}]$ = 337 markers. When a marker should be linked to a gene of interest, then screening of randomly chosen markers necessitates that a large number of markers be investigated to be sure that at least one is linked to the critical gene. However, once a genome has been saturated with markers, it is essential to only search for a small set of evenly spaced markers that together cover the entire genome. That way, no matter where the gene is located, at least one of the markers screened must be closely linked with the gene in question.

mapping methods: (i) FAMILY ANALYSIS and (ii) POSITIONAL GENE IDENTIFICATION. *See details as separate terms.*

mapping mRNA modifications for functional studies: C. Liu *et al.* (2022) introduced a quantitative, transcriptome-wide $6^{m}A$ mapping method unbiased by the sequence and structure of surrounding bases, whereas Q. Dai *et al.* (2022) presented a robust, quantitative Ψ mapping platform that builds on previous RNA bisulfite-sequencing (RBS-seq) methods. Together, these studies achieve both the quantification and resolution necessary to move beyond measuring global distributions of modifications to mechanistic studies of individual sites at high throughput.

mapping population: A population used for gene mapping is called a mapping population. Controlled crosses are made to develop a mapping population. Sufficient variation for the traits must exist between the parents at the DNA sequence and at the phenotypic level. The variation at the DNA level is essential to tracing the recombination events. Different kinds of mapping populations that are often used in linkage mapping include: (i) The F_2 population, (ii) The F_2derived F_3 (F_2:F_3) population, (iii) backcrosses, (iv) doubled haplotypes (DHs), (v) recombinant inbred lines (RILs) and (vi) near-isogenic lines (NILs).

mapping population, sample size of: Significance is a function of sample size. It is important to have a sufficient sample size and to do the appropriate statistical tests. A larger sample size also avoided a skewed representation.

mapping populations for self-pollinating species: Backcross progenies (BC1, BC1F_2), F_2 and double haploids.

mapping protein-DNA interactions: Directed methylation with long-read sequencing (DiMeLo-seq), which methylates DNA near each target protein's DNA-binding site *in situ*, then leverages the ability to distinguish methylated and unmethylated bases on long, native DNA molecules using long-read and single-molecule sequencing technologies.

mapping the chloroplast epigenomes: A number of studies provided support for the existence of both cytosine and adenine methylation in plastid genomes and anticipated a role for cytosine methylation in the transcriptional regulation of plastid genes. Some other studies suggest that at least cytosine methylation may be missing in higher plant plastid genomes. Using DNA cleavage with methylation-sensitive and methylation-dependent restriction endonucleases, it was revealed that the plastid genomes in the transplastomic plants were proficiently methylated.

mapping the methylated DNA: Examination of the expression of genes related to the analysed CG ISLANDS (CGIs) revealed that active genes had unmethylated promoters but methylated coding regions. This discovery is contrary to our conventional perception that DNA methylation always silences genes and underlines the significance of analysing CG sites away from the promoter regions.

mapping the mitochondrial epigenomes: Mitochondrial epigenetics appears to be similar with respect to DNA methylation and, to some extent, in the role of ncRNAs, but differs significantly in changes in components coiling DNA. Nuclear DNA is coiled around histones, but mitochondrial DNA, together with associated proteins, resides in mitochondrial pseudocompartments called NUCLEOIDS. Mitochondrial epigenetic mechanisms impact cell fate, transcription regulation, cell division, cell cycle, physiological homeostasis, bioenergetics and even pathologies. Methylation or hydroxymethylation in the mtDNA is proposed to play a pivotal epigenetic role in regulating the replication and transcription of the mtDNA.

mapping-by-sequencing in *Arabidopsis*: Mapping-by-sequencing combines genetic mapping with whole-genome sequencing in order to accelerate mutant identification. The combination of bulk segregant analysis and whole-genome resequencing has proven to radically speed up this process. This speed-up was gained by whole-genome sequencing of bulked recombinants and a subsequent analysis for local skews in the parental allele frequencies, which are introduced through phenotypic selection for mutant phenotypes. This directs the analysis to an approximate mapping interval that can be screened for underlying mutations using the exact same sequencing data. A guideline for mapping-by-sequencing for *Arabidopsis* was made available by T.Y. James *et al.* (2013) following an experimentally established recombination landscape. A simulation pipeline, called pop-seq or next-generation sequencing, of >400,000 mapping-by-sequencing experiments is simulated to analyse the differences in the design of mapping populations in relation to the number of candidate mutations identified in the course of such an experiment. This can be applied to other species as well as other mapping or sequencing strategies.

marginal likelihood: The marginal likelihood is a parameter variable's likeliness to marginalize. Marginalize, in this sense, refers to tracking the probabilities of new propositions with already existing probabilities.

mariner: A transposon originally isolated from insects, but related elements have been found in many other animals, including humans.

marker: A locus or allele whose phenotype provides information about a chromosome or chromosomal segment during genetic analysis. A marker may be morphological, genetic, biochemical or molecular. *See* GENETIC MARKERS.

marker chromosome: A structurally abnormal chromosome in which no part can be explicitly identified by conventional banding techniques.

marker DNA: A DNA fragment of known size used to calibrate an electrophoretic gel.

marker gene: A gene that recognizes the plants that have been successfully transformed. A gene of known function and location, or a mutation within a gene that permits investigating the inheritance of that gene. Phenotype-based, metabolic-based (polyphenol profiles, flavonoids, carbohydrates, oils, secondary products), protein-based (isozymes, seed storage proteins, total soluble proteins) and DNA-based (RESTRICTION FRAGMENT LENGTH POLYMORPHISM (RFLP), by HYBRIDIZATION, and RANDOM AMPLIFIED POLYMORPHIC DNA (RAPD), SIMPLE SEQUENCE REPEATS (SSR), SHORT TANDEM REPEAT (STR), SEQUENCE-TAGGED MICROSATELLITE SITE (STMS), AMPLIFIED FRAGMENT LENGTH POLYMORPHISM (AFLP), CLEAVED AMPLIFIED POLYMORPHIC SEQUENCES (CAPS), EXPRESSED SEQUENCE TAGS (EST), INTER-SIMPLE SEQUENCE REPEATS (Inter-SSR), SINGLE-NUCLEOTIDE POLYMORPHISMS (SNPs), SEQUENCE-CHARACTERIZED AMPLIFIED REGION (SCAR) and PCR-SEQUENCING (PCR)) markers can be distinguished.

marker retention: This a technique used in yeast (a unicellular fungal plant) to test the degree of linkage between two mitochondrial mutations.

marker stock: A stock having some genetically characterized markers.

marker trait: A genetically well-characterized trait in a stock that is useful for genetic analysis.

marker vaccine: A recombinant organism containing a foreign gene in which, when used as a vaccine, the foreign gene or antibodies to the expressed protein from the foreign gene can be detected. Marker vaccines or animals receiving marker vaccines can be detected with specific diagnostic tests.

marker-assisted backcrossing (MABC): It is one of the most promising approaches to making use of molecular markers to identify and select genes controlling resistance to those factors. MABC can contribute to developing resistant, high-yielding or quality rice varieties by incorporating a gene of interest into an elite variety that is already well adapted by farmers. MABC is a newly developed, efficient tool by which, using large population sizes (400 or more plants) for the backcross F1 generations, it is possible to recover the recurrent parent genotype using only two or three backcrosses. So far, many high-yielding, biotic and abiotic stress tolerance, quality and fragrance rice varieties have been developed in rice-growing countries through MABC within the shortest timeframe. Nowadays, MABC is widely used in plant breeding programmes to develop new variety/lines especially in rice.

marker-assisted breeding: An indirect selection process where a trait of interest is selected based on a marker (morphological, biochemical or DNA/RNA variation) linked to a trait of interest (e.g. productivity, disease resistance, abiotic stress tolerance and quality) rather than on the trait itself. This process is used in both plant and animal breeding.

marker-assisted introgression: The use of DNA markers to increase the speed and efficiency of the introgression of a new gene or genes into a population. The markers will be closely linked to the genes in question.

marker-assisted recurrent selection (MARS): It is an important breeding strategy that involves the use of molecular markers to aid in the selection of desirable traits in plants. MARS is also a type of recurrent selection that includes choosing and breeding the best individuals from a population across multiple generations in order to improve the population's overall performance.

marker-assisted selection (MAS): Indirect selection manipulating the association between the qualitative variation in a trait (isoenzymes, DNA markers) and the quantitative variation in another trait. It is a strategy permitting plant selection at the juvenile stage in early generations. The vital requirements for MAS in plant breeding are: (i) markers should cosegregate or be closely linked (1 cM or less) with the desired trait; (ii) an effective method of screening large populations for molecular markers should be available; and (iii) the screening methodology should have high reproducibility across laboratories, be economical to use and should be user-friendly. MAS has been utilized in many plant (e.g. crop) breeding programmes since the mid-1970s. It involves the use of genetic markers to follow regions of the genome that encode specific characteristics of a plant. For example, a marker genetically linked to a disease resistance locus can be used to predict the presence of the resistant or susceptible allele. The reliability of the prediction will depend on the closeness of the genetic linkage. Markers that cosegregate with the target trait are absolutely reliable and can be regarded as diagnostic. MAS is now being used in plant breeding programmes particularly for GENE PYRAMIDING (GENE STACKING). *Aka* MARKER-AIDED SELECTION.

marker-gene systems: Tools for studying the transfer of genes into an experimental organism. A marker allows the researcher to determine whether the transgene has been transferred, where it is located and when it is expressed. The marker gene is placed next to the transgene to make a single piece of DNA, which is then transferred.

markerless transformation: Transformation without SELECTABLE MARKER GENES (SMGs) is an ideal way to attain marker-free GM plants. Two methods for markerless transformation include the POLLEN-TUBE PATHWAY and the OVARYDRIP. In most of the cases, putatively transformed plants were screened by PCR to confirm the presence of the transgene. Selection of transformants may also be facilitated by expressing a screenable marker by challenging the putative transformants to a phenotypic property associated with the expression of the GENE OF INTEREST (GOI) or based on the direct screening of the GOI expression product.

markerless transformation through the ovary dip method: The ovary drip method involves the complete removal of the styles and the subsequent application of a DNA solution directly to the ovaries of transgenic plants. The movement of exogenous DNA is monitored using fluorescein-isothiocyanate labelled DNA.

markerless transformation through the pollen-tube pathway: Via the pollen-tube pathway, exogenous DNA is taken up by the egg cell or zygote after fertilization and then integrated into the host genome at mitosis.

market-basket survey: The systematic assemblage and testing of foods offered for purchase by consumers for pesticides and other likely contaminants.

Markov chain: Any multivariate probability density whose independence diagram is a chain. The variables are ordered, and each variable depends on its neighbours only in the sense of being in principle independent of the others. Markov chains are an integral component of HIDDEN MARKOV MODELS.

marsh: A productive wetland, typically dominated by species of monocotyledonous angiosperm plants that grow as tall as several meters above the water surface.

mask: A method for identifying the positions in a multiple-sequence alignment that can be used for phylogenetic reconstruction.

masked (blind) testing: Use of a specimen whose contents are unknown to the laboratory or to the laboratory technician to assess the ability of the laboratory to perform a test correctly.

masked mRNA: An mRNA complexed with protein so that it is not translated or enzymatically degraded.

masked symptoms: Plant symptoms (e.g. caused by a virus) that are not observable under some environmental conditions but do appear when the host is exposed to certain conditions of light and temperature.

masked testing: Use of a specimen whose contents are unknown to the laboratory or to the laboratory technician to assess the ability of the laboratory to perform a test correctly. *Aka* BLIND TESTING.

masking: (i) The removal of repeated or low-complexity regions from a sequence in order to improve the sensitivity of sequence similarity searches performed with that sequence. (ii) Preventing the subject, treating physician, patient interviewer, study director, or statistician from knowing which treatment a patient is given in a comparative study. A SINGLE-MASKED STUDY is one in which the patient does not know which treatment he/she is receiving. A DOUBLE-MASKED STUDY is one in which neither the patient nor the treating physician or other personnel involved in data collection know the treatment given. A TRIPLE-MASKED STUDY is one in which the statistician is unaware of which treatment has been given. *Aka* BLINDING *or* FILTERING.

mass emasculation: In hybrid (e.g. maize) breeding, the emasculation of the male flowers by mechanical means.

mass extinction: An event in which large numbers of species died out completely.

mass pedigree method of selection: A system of breeding in which a population is proliferated in bulk until circumstances favourable for selection arise. Usually, after mass pedigree selection, pedigree selection trails.

mass selection: A form of breeding in which individual plants are carefully chosen based on their individual advantages and the next generation is proliferated from the combined collection of their seeds. The easiest method for doing this is to select and multiply together those individuals from a mixture of phenotypes that correspond to the breeding aim (positive mass selection). The likelihood of success through mass selection rest on the amount of genetic variation and the strength of the relationship between genotype and phenotype. Mass selection in plant breeding is implemented for two purposes – the development of new varieties and the purification of established varieties.

mass spectrometer (MS): An instrument used to identify nucleic acids, proteins, oligonucleotides, peptides or peptide fragments in a substance by their mass-to-charge (m/z) ratio. The basic components of a mass spectrometer are the inlet, ion source, mass analyser, detector and instrument control system. The instrument control system is attached to a vacuum system and a data system. MS is an instrument that can measure the mass/charge (m/z) ratio of ions in a vacuum. From these data, the molecular weight (mass) of proteins and nucleic acids, the sequence of (composition and order of amino acids comprising) protein molecules, the chemical composition of virtually any material, and the rapid identification of intact Gram-negative and Gram-positive microorganisms (the latter using matrix-assisted laser desorption ionization time of flight mass spectrometry) can be determined with a high degree of accuracy, allowing molecular comparison of a given sample or analyte to be determined. Two types of analysis that can be carried out are (i) ANALYSIS OF INTACT PEPTIDE

IONS and (ii) ANALYSIS OF FRAGMENTED IONS. *See detail as separate terms.*

mass spectrometry (MS): MS allows the measurement with high precision and sensitivity of atomic and molecular masses. Material is vaporized in a vacuum (ionized) and then passed first through a strongly accelerating electric potential, and then through a powerful magnetic field. This process obliges us to separate the ions in order of their charge (i.e. mass ratio). Detection is made using an electrometer, which measures the force between charges and, hence, the electrical potential. Steps in mass spectrometry include reproducibility and separation of proteins with extreme photoionization (pI) values. Further, with the availability of ionization techniques like matrix-assisted laser desorption/ionization (MALDI) and electrospray (ES), mass spectrometry has been strengthened. *See* MATRIX-ASSISTED LASER DESORPTION/IONIZATION (MALDI).

mass spectrometry applications of MS: Approach proved effective in the identification and quantification of histone PTLMs and their binding proteins, outside the limitations of specific antibodies. MS has proven beneficial in the investigation of host-pathogen interactions. A sample to be analysed may be solid, liquid, or gaseous. It is ionized by bombarding it with a beam of electrons. Some of the sample's molecules break up into positively charged fragments or simply become positively charged without fragmenting. These ions (fragments) are then separated according to their mass-to-charge (m/z) ratio.

massively parallel sequencing (MPS): A next-generation DNA sequencing technology that allows millions or billions of base pairs of DNA or RNA to be sequenced simultaneously. *Aka* NEXT-GENERATION SEQUENCING (NGS), SECOND-GENERATION SEQUENCING or HIGH-THROUGHPUT SEQUENCING (HTS).

massively parallel signature sequencing (MPSS): This method is similar to serial analysis of gene expression (SAGE) in so far as it also measures short-tag sequences (STSs, 17–20 bp tags per transcript), but it allows the sequencing of millions of tags in parallel. It involves the following two activities: (i) an *in vitro* cloning process for loading cDNA on microbeads using a technique named MEGACLONE TECHNOLOGY and (ii) identification of each microbead-loaded cDNA through a 17-mer signature sequence to be obtained through hybridization-ligation-based parallel sequencing. In MPSS, in addition to transcript assignment, the frequency of each transcript is a measure of transcript abundance. The biggest advantage of MPSS is its parallel capacity, where millions of transcripts can be assessed simultaneously. This is one order of magnitude greater than SAGE.

mast: Edible fruit and nuts produced by woody species of plants (e.g. acorns and beech mast), which are expended on the ground by wildlife species and some domestic animals.

mast cell: A cell filled with basophil granules, found in good numbers in connective tissue and releasing histamine and other substances during inflammatory and allergic reactions.

master chromosome: A plant mitochondrial genome that exists in the form of circular DNA molecules of different sizes. It is known as a MULTIPARTITE GENOME ORGANIZATION.

master gene: A gene that controls other genes, particularly when the genes are characteristically expressed in a particular type of differential cell. *See* MASTER-AND-SLAVE HYPOTHESIS.

master regulators: The genes that respond most directly to the X/A ratio. These regulators are located on the X chromosome in both the fruit fly and the nematode.

master regulatory genes: *See* BINARY SWITCH GENES.

master-and-slave hypothesis: According to this hypothesis, a cell has several copies of a gene, one of which is the master gene and the other is the slave gene. This hypothesis was put forth to explain the 'C-value paradox'. However, this hypothesis did not get experimental support. *See* C-VALUE PARADOX.

material transfer agreement (MTA): The transfer of material from one laboratory to another laboratory is now done through a kind of agreement known as 'Material Transfer Agreement' (MTA). In India, MTAs have been in use essentially for the supply of biological material required by plant breeders in its research system. This has been facilitated by the National Bureau of Plant Genetic Resources (NBPGR), New Delhi, which has a rich collection of germplasm. In view of the increasing use of biological material, NBPGR was called upon to develop an MTA for use by members of the SAARC countries. Recently, when a new law relating to biodiversity, the 'Biological Diversity Act' of 2002, came into force in 2004, the need for evolving appropriate MTAs was felt. If this law is enacted, almost every university in India, be it general or agricultural, will have an office of intellectual property protection. Each university will have its own standard

MTA. Thus, MTAs are going to be extremely important for scientific organizations engaged in research in India in the future. MTAs will have to be designed in such a way that material taken from NBPGR is not misused by the recipient.

maternal chromosome: In bisexual organisms, the chromosome of a homologous pair that is inherited from the mother.

maternal effect genes: Those genes that affect egg polarity and the special coordinates of the egg and future embryo.

maternal effect: The maternal effect occurs when the phenotype of the individual is dependent on its mother's genotype, not its own. In cases of maternal effect, the transmission pattern of the alleles is the same as in standard Mendelian genetics, but the action of the gene occurs a generation later. A maternal effect gene where the mutant allele (m) is recessive to the wild-type allele. In the cross of homozygous, wild-type females to homozygous, mutant males, all the F_1 offspring are heterozygotes and appear normal. In the reciprocal cross, all of the F_1 offspring are heterozygotes but have the mutant phenotype. Although the F_1 offspring are genotypically identical in the reciprocal crosses, they are phenotypically different. This is because the phenotype is due to the action of the mother's genotype. Maternal effects in snails of the genus *Limnaea* are well known.

maternal genes: The genes that are expressed during oogenesis by the mother. They function in the nurse cells that surround the oocyte. Some of the maternal gene products behave as morphogens.

maternal imprinting of genes: The process by which only one copy of a gene from their mother is expressed, while the other copy coming from the father is suppressed.

maternal influence: The influence of the genotype of a heterozygous mother on the eggs and larvae. The dominant colour of the mother persists for a short time in eggs and larvae, even with recessive genotype.

maternal inheritance: The inheritance of nonnuclear genes (i.e. those in mitochondria and chloroplasts). *Compare with* CYTOPLASMIC INHERITANCE *and* EXTRACHROMOSOMAL INHERITANCE.

maternal inheritance of human mitochondrial DNA: In many species, including humans, the sperm mitochondria are introduced to the oocyte during fertilization. The mechanisms hypothesized to prevent paternal mtDNA transmission include ubiquitination of the sperm mitochondria and mitophagy. Experiments show a lack of mitophagy in mammalian oocytes. W. Lee *et al.* (2023) found that mitochondria in human spermatozoa are devoid of intact mtDNA and lack mitochondrial transcription factor A (TFAM), the major nucleoid protein required to protect, maintain and transcribe mtDNA. During spermatogenesis, sperm cells express an isoform of TFAM, which retains the mitochondrial presequence ordinarily removed upon mitochondrial import.

maternal origin: Derived from the mother.

maternal somatic genes: These genes are expressed in the somatic cells of the mother that affect egg development, for example, follicle cells.

maternal spindle transfer (MST): This method for gene/DNA transfer into mitochondria, developed by D.M. Turnbull, has three steps: (i) Doctors use standard IVF procedures to collect eggs from the mother. (ii) They then pluck the nucleus from one of the eggs and drop it into a healthy donor egg that has its own nucleus removed. The new egg has all the mother's chromosomes but the donor's healthy mitochondria, apart from a tiny portion of faulty ones that inevitably carry over with the mother's nucleus. (iii) The egg is then fertilized with the father's sperm, and the embryo is implanted as a standard IVF procedure.

maternal twins: IDENTICAL TWINS, also called MONOZYGOTIC TWINS, result from the fertilization of a single egg by a single sperm, with the fertilized egg then splitting into two. Identical twins share the same genomes and are always of the same sex.

maternal X chromosome: The X chromosome was contributed by the mother.

maternally expressed imprinted genes: These genes are generally expressed only when inherited from the father.

maternal-specific DNA demethylation: Maternal endosperm chromosomes inherited from the central cell display a pattern of extensive demethylation, which is essential for seed development in *Arabidopsis thaliana*.

mating: In biology, mating is the pairing of either opposite-sex or hermaphroditic organisms for the purposes of sexual reproduction.

mating types: In microorganisms, the two strains of opposite types are usually designated as '+' and '−'; 'α' and 'a'; or 'A' and 'B', necessary for sexual reproduction in heterothallic fungi. *See* SELF-INCOMPATIBILITY.

mating-assisted genetically integrated cloning (MAGIC): This method facilitates the rapid construction of recombinant DNA molecules. MAGIC uses bacterial mating, *in vivo* site-specific endonuclease cleavage and homologous recombination to catalyse the transfer of a DNA fragment between a donor vector in one bacterial strain and a recipient plasmid in a separate bacterial strain. Recombination events are genetically selected and result in the placement of the gene of interest under the control of new regulatory elements with high efficiency. MAGIC eliminates the need for restriction enzymes, DNA ligases, preparation of the DNA and all *in vitro* manipulations required for subcloning and allows the rapid construction of multiple constructs with minimal effort.

mating-type loci: Genetic elements in yeast containing mating-type information (a or α) that is activated by recombination from heterochromatic copies of one of the two mating-type alleles.

mating-type switch: Haploid yeast cells occasionally switch their mating type. At a genetic level, such a switch has been shown to involve the excision of a particular mating-type gene sequence from the genome, with its replacement by a gene of the alternative mating type.

mating-type switching in fission yeast: Fission yeast exhibits a homothallic life cycle in which the mating type of the cell mitotically alternates in a highly regulated fashion. Only one cell (of the two sister cells) switches mating types.

matriclinous inheritance: Phenotypic differences due to factors such as chloroplasts and mitochondria transmitted by the female gamete.

matrix genetics: A branch of bioinformatics and mathematical biology that studies the matrix arrangements of presentations of the genetic code.

matrix space: The central subcompartment of a mitochondrion, bounded by the inner mitochondrial membrane. The corresponding compartment in a chloroplast, which is more commonly known as the *stroma*.

matrix-assisted laser desorption ionization (MALDI): The most commonly used method for uncovering of modified nucleotides in RNA. Gas phase ions produced by MALDI are singly charged ions that are either protonated or deprotonated, depending on the ion detection mode used (positive or negative).

matrix-assisted laser desorption ionization and electrospray ionization (MALDI-ESI): These two techniques are most commonly used to volatize and ionize peptides and proteins in MS analysis. In the case of ESI, the analyte is dissolved and forced through a narrow needle held at high voltage. A fine spray of charged droplets emerges from the needle and is directed to the vacuum chambers of the MS through a small orifice. As they enter MS, the droplets are dried using a stream of inert gas, resulting in gas phase ions that are accelerated through the analyser towards the detector. Because ESI produces gas phase ions from solution, it is readily integrated with upstream protein separation by liquid phase methods, especially high-performance liquid chromatography, sometimes referred to as HIGH-PRESSURE LIQUID CHROMATOGRAPHY (HPLC). This method is more suited to the study of complex samples.

matrix-associated laser desorption ionization time of flight mass spectrometry (MALDI-TOF-MS): A mass spectrometry procedure that can find, in seconds, the identity, purity, etc. of a sample of proteins, oligonucleotides or (poly)peptides. MALDI-TOF-MS can also identify Gram-positive microorganisms and classify genetic materials (DNA, RNA, etc.) on hybridization surfaces. MALDI-TOF-MS utilizes measurement of the time for particles (e.g. proteins) to transit a specific distance after being 'dislodged' from the 'adhered' surface by a specific amount of energy to precisely determine the molecular weight (of proteins, etc.).

matroclinal: Having hereditary characteristics that are more maternal than paternal (e.g. in certain banana hybrids).

matromorphy: Resembling the female parent in morphology, the plants are called MATROMORPHS, MATROMORPHICS or MATROMORPHIC plants. Matromorphy should not be confused with matrocliny, as the matroclinous plants are hybrids (formed after fertilization of the egg cell), which resemble the mother much more closely than the father. This condition may be caused by dominant gene action or dose effects caused by different numbers of genomes from the mother and the father in the hybrid individual. Besides MATROMORPHY and MATROCLINY, PATROMORPHY (diploid androgenesis) and patrocliny may occur. PATROCLINOUS plants are hybrids that resemble the father much more closely than the mother. *Aka* DIPLOID PARTHENOGENESIS.

maturase: Any of the proteins encoded by self-splicing introns that are, together with nuclear-encoded proteins, needed for splicing. All the maturases function only by splicing the intron in which they are encoded (or closely related introns).

maturation: (i) The completion of development and the process of ripening. (ii) The processing of messenger RNAs, transfer RNAs, ribosomal RNAs or small nuclear RNAs.

maturation division: *See* MEIOSIS.

maturation of insulin hormone: Rat preproinsulin peptide has 86 (proinsulin) plus 24 (signal peptide) – 110 amino acids. Insulin is initially synthesized as an 86 amino acid single chain. Disulfide bonds are formed, one between cysteines at positions 7 and 72 and the other between cysteines at positions 19 and 85. The internal cuts excise 36 amino acids. Mature insulin hormone carries chain A (30 amino acids) and chain B (21 amino acids).

maturation phase in oogenesis: Maturation means meiosis. By the time the growth phase is completed, the chromosomes of the oocytes reach the diakinesis stage. On entering the maturation phase, the nuclear membrane disintegrates, liberating the chromosomes and the nuclear sap. Nuclear sap intermingles with the cytoplasm. Certain substances dissolved in the nuclear sap are important for the normal development of the embryo in the later stages. When the chromosomes are liberated by the rupture of the nuclear envelope, they undergo metaphase and so on. Finally, the first polar body is separated out as a polocyte. This completes the first maturation phase and the secondary oocyte is formed, which is haploid (n) in chromosome number. Meiosis II is achieved by sperm entry. When a mature egg is fertilized with an incoming sperm to give rise to a zygote (2n), it undergoes mitotic division to give rise to an embryo.

maturation phase in spermatogenesis: After the last spermatogonial division, the cell grows in volume and becomes primary spermatocytes. It enters meiosis. At the end of meiosis I, secondary spermatocytes are formed. One primary spermatocyte produces two secondary spermatocytes. Then meiosis II occurs. After the completion of meiosis I, spermatids are formed. Each secondary spermatocyte produces two spermatids. Thus, at the end of meiosis II, four spermatids are produced. As much as 99.7% of total DNA is synthesized during the S phase preceding the first meiotic prophase. DNA synthesis, which occurs during the zygotene and pachytene stages, involves all the chromosomes. It mainly replicates those parts of chromosomes that had not been previously replicated during interphase.

maturation-promoting factor (MPF): MPF initiates metaphase by breaking up the nuclear envelope. It phosphorylates the lamina proteins of the nuclear envelope. Phosphorylated lamina subunits are dissociated, leading to the breakdown of the nuclear envelope. MPF also induces chromosome condensation and the assembly of the spindle. During anaphase, the cyclin component of MPF is phosphorylated and becomes a substrate for protease.

mature: Fully differentiated and functionally competent RNA molecules, cells, tissues or organisms.

mature unfertilized egg stage: In between the activated egg and the growing oocyte stage, the mature unfertilized egg is a sandwich stage. This stage is characterized by a lack of both DNA and RNA synthesis. If this is true, a nucleus injected into a mature, unfertilized egg should stop both synthetic activities. This is what has actually been observed.

Maxam-Gilbert DNA sequencing: *See* CHEMICAL SEQUENCING METHOD OF DNA SEQUENCING.

maxi-cell: An *Escherichia coli* or *Bacillus subtilis* cell in which plasmid genes are preferentially expressed because the chromosomal genes have been activated by ultraviolet (UV) irradiation. A bacterial strain carrying mutations that render it defective in both DNA repair and recombination is exposed to UV light. Because of these deficiencies, the chromosomal DNA will be severely damaged and degraded. If such a cell contains a multicopy plasmid-cloning vector, some of the plasmids will not be hit by UV and these will continue to be produced. A radioactively labelled amino acid can be added to the irradiated culture and will be specifically incorporated into plasmid-encoded proteins.

maximum equational segregation: If the crossing-over always takes place between the gene and the centromere in a multivalent (tri- or tetrasomic), the separation of sister chromatid segments carrying the locus will always take place during the first meiotic division.

maximum likelihood estimate (MLE): An estimate of a statistical parameter (e.g. a regression coefficient, mean, variance, or standard deviation) that is the value of that parameter, making the data most likely to have been observed. MLEs have excellent statistical properties in general, such as converging to population values as the sample size increases and having the best precision among all such competing estimators when the statistical model is correctly specified. When the data are normally distributed, maximum likelihood estimates of regression coefficients and means are equivalent to least squares estimates. When the data are not normally distributed (e.g. binary outcomes or survival times), maximum likelihood is the standard method to estimate the regression coefficients (e.g. logistic regression, Cox regression). Unlike Bayesian estimators, MLEs cannot take extra-study information into account. MLEs can be overfitted when the data's information content does not allow a reliable estimation of the number of parameters involved (*see* OVERFITTING). Penalized MLEs can solve this problem by maximizing a penalized log likelihood function. When extra-study information is not allowed to be utilized, MLE is considered a gold standard estimation technique. *See* LIKELIHOOD FUNCTION.

maximum likelihood method: A statistical, phylogenetic procedure of linkage estimation depending on maximizing the log likelihood in which probabilities are considered for individual nucleotide substitution in a set of sequence alignments. A purely statistically based method of phylogenetic reconstruction. This method relies on first obtaining a reliable multiple-sequence alignment and then examining the changes in each column in the alignment. In this case, however, the likelihood of a particular tree may be miscalculated using an unexpected model of change in the sequences. For example, all nucleotides are presumed to be equally frequent and the probability of a change of any nucleotide to any other nucleotide is supposed to be the same in the Jukes-Cantor model. For each possible tree, the likelihood of finding the actual sequence alterations at each column in the aligned sequences is calculated. The probabilities for each aligned position are then multiplied to make available likelihood for each tree. The tree that provides the maximum likelihood value is the most probable tree.

maximum parsimony (MP) method: In this method, a multiple-sequence alignment is produced in order to predict which sequence positions are likely to correspond. These positions will appear in vertical columns in the multiple-sequence alignment. For each aligned position, phylogenetic trees that require the smallest number of evolutionary changes to produce the observed sequence changes are identified. This analysis is continued for every position in the sequence alignment. Finally, those trees that produce the smallest number of changes overall for all sequence positions are identified.

maxipreparation of plasmid DNA: The level of endotoxin contamination in purified plasmid DNA is critical for many applications. For this reason, a specific procedure reduces endotoxins to a very low level. A modified alkaline/SDS lysis procedure prepares the bacterial cell pellet for plasmid purification. The bacterial lysate is cleared by filtration and loaded on to the equilibrated column. After efficient washing of the column, the purified plasmid DNA is eluted from the column.

mCIP-seq: In this protocol, an antibody against 5^mC is used to immunoprecipitate the methylated DNA. The method can be scaled up to perform genome-wide DNA methylation analysis.

***Mcr*BC methodology:** A reverse approach by using genomic representations made by digesting with *Mcr*BC, an unusual restriction enzyme that digests methylated rather than unmethylated DNA. *Mcr*BC cleaves between two closely spaced (55–100 bp) methylated cytosines in the $(G/A)^mC$ context, and therefore preferentially digests densely methylated regions of DNA, such as abnormally methylated CGIs. After the two-step restrictions and subsequent PCR, amplicons were labelled with different fluorescent cyanine dyes, Cy5 or Cy3, for either sample or reference DNA, mixed and hybridized to oligonucleotide microarrays.

***Mcr*BC-seq:** McrBC is an endonuclease that specifically cleaves DNA containing methylcytosine (5-methylcytosine or 5-hydroxymethylcytosine) on one or both strands. Sites on the DNA recognized by *Mcr*BC consist of two half-sites of the form $(G/A)^mC$, which can be separated by up to 3 kb, but the optimal separation is 55–103 bp. *Mcr*BC requires GTP for DNA cleavage.

mean: The sum of all the values divided by the number of observations (for n values, $z_1, \ldots, z_n, = (\Sigma iz_i)/n$). The mean of a binary variable is equal to the proportion of ones because the sum of all the zero and one values equals the number of ones. The mean can be heavily influenced by outliers. When the tails of the distribution are not heavy, this influence of more extreme values is what gives the mean its efficiency compared to other estimators, such as the median. When the data distribution is symmetric, the population mean and median are the same. The sample mean is a better estimator of the population median than the sample median when the data distribution is symmetric and Gaussian-like. *Aka* ARITHMETIC AVERAGE *or* ARITHMETIC MEAN. *Compare with* HARMONIC MEAN *and* GEOMETRIC MEAN.

mean fitness: The average fitness of a population.

mean square: The square of the mean variation of a set of observations around the sample mean.

measures of variability: Measures of variability, also referred to as measures of dispersion, denote how scattered or dispersed a database is. The four main measures of variability are the interquartile range, range, standard deviation and variance.

measuring DNA purity: Measuring the intensity of absorbance of the DNA solution at wavelengths of 260 and 280 nm is used as a measure of DNA purity. DNA absorbs UV light at 260 nm and aromatic proteins absorb UV light at 280 nm; a pure sample of DNA has a ratio of 1.8 at 260/280 and is relatively free from protein contamination. A DNA preparation that is contaminated with protein will have a 260/280 ratio lower than 1.8.

measuring RNA purity: When using a spectrophotometer, a ratio of absorbances at 260 and 280 nm (A260: A280) greater than 1.8 is usually considered an acceptable indicator of RNA purity.

meat-rice: S. Park *et al.* (2024) use manufacturing methods similar to those for other cultured meat products, in which animal cells are grown on a scaffold in a laboratory, bathed in a growth medium. Using rice as the scaffold has the benefit of adding nutrition to the rice, with the beef–rice having a slightly higher fat and protein content than standard rice. The hybrid rice could be a cheaper, more efficient source of nutrition than large pieces of lab-grown meat, such as patties or steaks. The food properties and production value of this rice-based meat show its potential as a sustainable food that guarantees safety from food crises and global warming.

mechanical biological treatment (MBT): A way of sorting and treating waste. The waste is first sorted mechanically into materials that can and cannot be recycled. Any waste that can be recycled is then broken down biologically, often through composting, while the rest is usually sent to landfill. *See* COMPOSTING.

mechanical genome incompatibility model: This model postulates that species formation is an outcome of processes like unequal crossing-over and gene conversion that alter the positions, numbers or sequences of clustered and dispersed repetitive DNAs. It explains that turnover processes, like transposition, unequal crossing-over and gene conversion, lead to disruption in the chromosomal interaction processes between individuals affected by such processes. Affected and unaffected individuals, on mating, produce effectively sterile hybrids.

mechanical injury: Includes any activity that damages the roots, root collar, stem, branches or leaves due to cuts and abrasions. Plants can sense the injured tissue as an altered self and induce responses similar to those activated by pathogen infection.

mechanical inoculation: A method of spreading the pathogen from plant to plant; for example, sap from diseased plants or a defined inoculum is rubbed on test-plant leaves that usually have been sprinkled with carborundum or other abrasive materials. This method is applied for experimental testing of plant resistance.

mechanical isolation: Interbreeding between members of different populations is not taking place as a consequence of the structural incompatibility of male and female secondary reproductive organs.

mechanical protoplast fusion: The isolated protoplasts are brought into intimate physical contact mechanically under a microscope and using a micromanipulator or perfusion micropipette.

mechanical purity: Refers to the amount of freedom a seed lot has from the seeds of other crops, weed seeds and inert matter seeds.

mechanical transmission: The spread or introduction of inoculum to an infection court (wound) by human intervention, accompanied by physical disruption of host tissues.

mechanics of inheritance: Study of the manner in which characters are transmitted from one generation to the next. A satisfactory answer to the mechanics of inheritance came from the work of an Austrian monk, Gregor J. Mendel, the Father of Genetics, who in 1865 provided the first proof for a theory that explained heredity by transmission of units in the reproductive cells. This branch of genetics is termed TRANSMISSION GENETICS, CLASSICAL GENETICS or MENDELIAN GENETICS.

mechatronics: Electronics and mechanical engineering combined to make intelligent machines.

mecrotroph: A parasite that typically kills host cells and obtains its energy from them. *Contrasts* BIOTROPH.

***Medea* gene:** A maternally expressed imprinted gene that makes a Polycomb group protein involved in cell proliferation during seed development in *Arabidopsis*.

***Medea* gene drive system:** A synthetic *Medea* gene drive system in *Drosophila suzukii* is based on an engineered maternal 'toxin' coupled with a linked embryonic 'antidote' is capable of biasing Mendelian inheritance rates with up to 100% efficiency. However, drive resistance, resulting from naturally occurring genetic variation and associated fitness costs, could be selected to hinder the spread of such a drive. This gene drive could maintain itself at high frequencies in a wild population and spread to fixation if either its fitness costs or toxin resistance were reduced, providing a clear path forward for developing such systems in this pest.

median: Median is the 'middle item' in an array (i.e. the middle point of data). Typically, if you have a data set with an odd number of items, the median appears directly in the middle of the numbers. When computing the median of a set of data with an even number of items, you can calculate the simple mean between the two middle most values to achieve the median.

median centromere: A centromere that is located midway on chromosomes, resulting in two equal-length arms. *See* METACENTRIC CHROMOSOME.

median test: A median test is a nonparametric test that tests two independent groups that have the same median. It follows the null hypothesis that each of the two groups maintains the same median.

mediator complex: A multiprotein complex (comprising up to 26 subunits in humans) required for transcription of most RNA polymerase II transcripts. The complex binds to the C-terminal domain of RNA polymerase II and bridges the enzyme to transcription factors.

mediators: These proteins simultaneously associate the damage sensors and signal transducers at certain phases of the cell cycle and as a consequence, provide signal transduction specificity. Several mediator proteins, such as γ-H2AX, BRCA1, MDC1, Claspin and 53BP1, work to coordinate the localization of various factors in DNA damage repair, promote their activation, and regulate substrate accessibility.

medium (pl. media): (i) A mixture of organic and/or inorganic chemical compounds and water that delivers the nutrients needed for the growth of a microorganism *in vitro*. (ii) For higher plants, a mixture of fertilizers and other components in which a plant can sustain its growth.

medium formulation: In tissue culture, it refers to the particular formula (content, quantity and procedure) for the preparation of a culture medium. It commonly contains macroelements and microelements, some vitamins, plant growth regulators (auxin, cytokinin and sometimes gibberellin), a carbohydrate source (usually sucrose or glucose), and often other substances, such as amino acids or complex growth factors. Media may be liquid or solidified with agar. The pH of the medium is adjusted between 5 and 6, and the solution is sterilized (usually by filtration or autoclaving). Some formulations are very specific in the kind of explant or plant species that can be maintained, while others are very general.

medullary rays: Cellular structures found in some species of wood. They appear as radial planar structures, perpendicular to the growth rings, which are visible to the naked eye. In a transverse section, they appear as radiating lines from the centre of the log. In an axial section, they may appear as a variety of transverse markings, depending on how close the section is to the plane of the ray. In a tangential section, they may be hard to see at all. *Aka* VASCULAR RAYS *and* PITH RAYS.

mega-: Prefix denoting 10^6. From Greek 'megas', meaning huge, powerful.

megabase (Mb): A unit of measurement used to help designate the length of DNA. One megabase is equal to 1 million bases. 1 mb is approximately equal to 1 cM. *See* CENTIMORGAN.

megabase cloning: A technique for the cloning of extremely large fragments of DNA (ranging from one to several megabases) into suitable vectors.

megabase mapping: The construction of a linear genetic or physical map using markers that are within one million bases from each other.

mega-environment (ME): A broad, not necessarily contiguous area, occurring in more than one country and frequently transcontinental, defined by similar biotic and abiotic stresses, cropping system requirements, consumer preferences and a volume of production. The concept of ME was introduced by CIMMYT in 1988 to address the needs of diverse wheat-growing areas of the world.

mega-evolutionary shift: Refers to new general adaptations during trans-specific evolution.

megagametogenesis: The process of developing the female gametophyte from a functional megaspore.

megagametophyte: Megagametophyte is the female gamete, *aka* EMBRYO SAC. In heterosporous plants, the megagametophyte develops from the megaspores. The megagametophyte acts as an embryo sac in angiosperms. *See* EMBRYO SAC.

meganucleases: Artificial restriction enzymes characterized by a large recognition site (double-stranded DNA sequences of 12–40 bp). Meganucleases are considered to be the most specific naturally occurring restriction enzymes. *Aka* HOMING NUCLEASES.

megapreparation of plasmid DNA: The procedure used is the same as that used for MAXIPREPARATION OF PLASMID DNA. Each prep yields up to 2000 μg of endotoxin-free plasmid DNA.

megasporangium: The larger of two kinds of sporangium produced by heterosporous plants. Large spores contain the female gametophytes. *Compare with* MICROSPORANGIUM.

megaspore: (i) One of the four haploid spores produced from the megaspore mother cell through meiosis in the ovary. The megaspore is the female gamete, which fuses with male gamete (i.e. pollen nuclei in plants and sperm nuclei in humans, to produce diploid zygote). (ii) The larger of two kinds of spores produced by a heterosporous plant, giving rise to the female gametophyte. *Compare with* MICROSPORE.

megaspore mother cell: A diploid cell in the ovary that produces, through meiosis, four haploid megaspores. *Aka* MEGASPOROCYTE.

megasporocyte: A diploid cell that undergoes meiosis to produce megaspores as part of MEGASPOROGENESIS. *Aka* MEGASPORE MOTHER CELL. *See* EMBRYO SAC.

megasporogenesis: The development of the megaspore from the archesporial cell (the cell or group of cells from which spore mother cells develop). In the young ovule in the micropylar region, a single hypodermal cell becomes more prominent. This cell is larger than the surrounding cells and contains a large active nucleus and denser cytoplasm. It is known as an archesporial cell. In the ovules of some species, the archesporial cell divides transversally and produces the outer primary parietal cell and the inner PRIMARY SPOROGENOUS CELL. The parietal cell divides and produces a mass of parietal tissue. It pushes the sporogenous cell into a deeper portion of the nucellus. The sporogenous cell functions as a megaspore mother cell (MMC). In some species, archesporial cells directly act as MMC. Plasmodesmatal connections are present between MMC and surrounding cells. At first, MMC has a thin cellulose wall, but before the onset of meiosis, cellulose material is deposited. The MMC undergoes meiosis I division and gives rise to a dyad. The dyad undergoes meiosis II division and gives birth to four haploid megaspores. The megaspores are arranged in a linear tetrad. Out of four megaspores, generally the micropylar three are smaller in size and non-functional. The chalazal megaspore is larger and more functional. The functional megaspore germinates and produces female gametophytes, whereas the non-functional megaspores degenerate eventually.

megasporophyll: Refers in heterosporous plants to a modified leaf bearing one or more megasporangia. *Compare with* MICROSPOROPHYLL.

mega-yeast artificial chromosomes (mega-YAC): A large (>500 bp) fragment of DNA cloned inside a living yeast cell. While most bacterial vectors cannot carry DNA inserts that are >50 bp and standard YACs characteristically cannot carry DNA pieces that are >500 bp, mega-YACs can carry DNA pieces (or chromosomes) as large as 1 million bp.

meiocyte: The cells that undergo meiosis to produce gametes. They are diploid and one meiocyte yields four haploid cells after meiosis.

meiome: The term refers in functional genomics to the meiotic transcriptome. Meiosis is a key feature of all sexually reproducing eukaryotes in which homologous chromosome pairing, synapse and recombination are unique. Since meiosis in most organisms occurs in a short time period, the study of meiotic transcript profiling is extremely hard due to the challenge of the isolation of meiotic cells. There is a list of groups studying meiotic transcriptomes using high-throughput techniques such as microarrays and sequencing technologies. Meiome is the procedure to investigate transcript profiling using mRNA isolated from enriched meiocytes.

meiosis: Discovered by E. Van Beneden in the 1870s. A process of nuclear division in which the number of chromosomes per nucleus is halved (i.e. converting the diploid state to the haploid state). Meiosis is confined to certain types of cells (oocytes, spermatocytes and sporocytes) and occurs at some stage in the life cycle of sexually reproducing organisms. There are two successive divisions – MEIOSIS I and MEIOSIS II – with only one round of DNA replication. During meiosis I, genetic material can be exchanged between homologous chromosomes and the diploid chromosome number is halved (REDUCTIONAL DIVISION). The subsequent division (meiosis II) virtually corresponds to mitosis (EQUATIONAL DIVISION). The duration of meiosis is positively associated with DNA content per nucleus in diploid species and with mitotic cycle time. Meiotic duration is also impacted by chromosomal organization, DNA structure and the developmental pattern of the species. Meiosis reduces the starting number of 4n chromosomes to

1n in each of the four product cells. The cells thus produced may mature to form germ cells (sperm or eggs). Four cells produced at the completion of meiosis are genetically non-identical. *Contrasts* MITOSIS. *See details as separate terms.*

meiosis I: During meiosis I, the two homologous chromosomes of each pair separate from each other and go to separate daughter cells, thus reducing the number of chromosomes from diploid to haploid. Meiosis I is, therefore, known as reductional division. Meiosis I comprises four phases: prophase I (with five subphases, namely, LEPTOTENE, ZYGOTENE, PACHYTENE, DIPLOTENE and DIAKINESIS), METAPHASE I, ANAPHASE I and TELOPHASE I. *See details as separate terms.*

meiosis II: Meiosis II is an EQUATIONAL DIVISION. The four stages known in this division are PROPHASE II, METAPHASE II, ANAPHASE II and TELOPHASE II. Stages in meiosis II are similar to those in mitosis. The difference between meiosis II and mitosis lies with respect to chromosome number in the cell: haploid (n) in the former and diploid (2n) in the latter. *See details as separate terms.*

meiosis, significance of: Meiosis produces gametes for sexual reproduction. Meiosis halves the chromosome number in gametes and fertilization restores the original chromosome number. Crossing-over during meiosis results in the formation of chromosomes with new combinations of hereditary units and genes. Random distribution of paternal and maternal chromosomes during anaphase I gives new combinations of chromosomes.

meiosis-to-mitosis shift: In angiosperm plants, meiosis leads to a haploid but multicellular reproductive structure. A. Cairo *et al.* (2022) found that a germline-specific protein THREE-DIVISION MUTANT 1 (TDM1) is incorporated into P-bodies through interaction with SUPPRESSOR WITH MORPHOGENETIC EFFECTS ON GENITALIA 7 (SMG7). TDM1 attracts eIF4F, the main translation initiation complex, temporarily sequestering it in P-bodies and inhibiting translation. This recognition process suggestively remodels the TRANSLATOME and facilitates the transition to mitosis. *See* TRANSLATOMICS.

meiospore: Cell that is one of the products of meiosis in plants.

meiotic crossing-over: *See* CHIASMA.

meiotic cycle: *See* MEIOSIS.

meiotic drive: Distortion of the fairness of meiosis by nuclear genes to increase their representation in the gametes at the expense of other alleles. A form of GENOMIC CONFLICT.

meiotic duration: The time required for the completion of two meiotic divisions (meiosis I and meiosis II) under certain conditions. The duration of meiosis in *Arabidopsis* is about 26 h.

meiotic male recombination: Recombination may rarely occur under natural conditions or may be induced with the help of a physical or chemical agent in the meiotic germ line cells in males of *Drosophila melanogaster*.

meiotic mutation: These mutations are used to elucidate several areas of genetic research. It enables gene-centromere mapping by half-tetrad analysis with 4x–2x crosses (potato), where the 2x parent forms 2n pollen by either first-division restitution (FDR, ps/ps) or second-division restitution (SDR, pc/pc). Since two of the four strands of a bivalent are recovered together in 4x progeny by crossing a 4x nulliplex (*aaaa*) with a 2x heterozygote (*Aa*), the frequency of 4x nulliplex progeny offers an estimate of gene-centromere map distance through half-tetrad analysis.

meiotic restitution: Failure of meiosis to produce gametes with a reduced chromosome number. It can lead to the restoration of fertility in allohaploids. Two main types of restitution are distinguishable by different genetic consequences: FIRST-DIVISION RESTITUTION (FDR) and SECOND-DIVISION RESTITUTION (SDR). *See details as separate terms.*

meiotic segregation: During meiosis, cohesins are lost from the chromosomes in a step-wise manner. Loss of cohesins from chromosome arms is required for homologous chromosomes to segregate during meiosis I.

meiotic sex chromosome inactivation: Epigenetic regulation of gene expression is critical during spermatogenesis. We are beginning to understand epigenetic regulation and the consequences of its aberrant regulation during mitosis, meiosis and spermiogenesis (N.M. Zamudio *et al.*, 2008). Epigenetic effects result in impaired fertility and potentially long-ranging effects. Male germ cells are epigenetically distinct from somatic cells and they undergo several important epigenetic transitions (O. Meikar *et al.*, 2013). In PRIMORDIAL GERM CELLS (PGCs), the epigenome is reprogrammed by genome-wide resetting of epigenetic marks, including the sex-specific imprinting of certain genes. Postnatal spermatogenesis is characterized by drastic chromatin rearrangements during meiotic recombination, sex chromosome silencing and compaction of sperm nuclei, which is accomplished by replacing nearly all histones by sperm-specific PROTAMINES. Small RNAs, including MICRO RNAs (miRNAs), ENDOGENOUS SMALL INTERFERING RNAs (endo-siRNAs) and PIWI-INTERACTING RNAs (piRNAs), are also involved in the control of male gamete production.

meiotic silencing by unpaired DNA (MSUD): The mechanism that repressed the expression of genes in the regions contained in the diploid phase unpaired sister chromosomes.

melanin: A dark brown pigment found in the cells of the skin, hair and iris of the eye. This pigment is also found in many plant species.

melanism: An increased frequency of dark-coloured black or deep-brown individuals in a population.

melting curve: A graphical plot of the change in absorbance at 260 nm as a function of temperature that occurs when a solution of DNA is heated. Heating results in denaturation due to the breaking of the hydrogen bonds between the bases and the dissociation of the two strands of the double helix. The absorbance of double-stranded DNA is always lower than that of single-stranded DNA, which is lower than the equivalent number of free nucleotides in solution.

melting of DNA: Melting of DNA means to heat-denature it. When this transpires, the hydrogen bonds holding the two strands of the DNA double-helix molecule originally together in the normal way are disrupted. This disruption causes a more random polymer structure of DNA to exist.

melting profile: A profile, commonly represented on a graph, showing the percentage of a given sample of DNA that melts at a specific time on a gradually increasing temperature scale. Since some DNA, such as imprecisely base-paired sequences, low-molecular-weight material and molecules rich in A/T bases, will melt out at lower temperatures than other DNA, some useful information can be derived from melting profiles.

melting: The dissociation of a duplex nucleic acid molecule into single strands, usually by increasing temperature. *See* DENATURATION.

melting temperature (T_m) of DNA: The temperature at which the two strands of a double-stranded DNA molecule come apart; a short (<18 nucleotides) oligonucleotide Tm value (in °C) is estimated by the formula: T_m = (number of A + T) × 2 + (number of G + C) × 4. Tm is represented as the midpoint of the temperature range over which DNA is denatured.

membrane: A sheetlike, semi-fluid structure, 7–10 nm wide, that forms the boundary between a cell and its environment and also between various compartments and organelles within the cell; it is composed of lipids, proteins and some carbohydrates. It functions as a selective barrier and also as a structural base for enzymes.

membrane bioreactors: Bioreactors where cells grow on or behind a permeable membrane, which lets the nutrients for the cell through but retains the cells themselves.

membrane channel: Transmembrane protein complex that permits inorganic ions or other small molecules to diffuse passively across the lipid bilayer. Passive transport of solutes occurs through an aqueous diffusion pore.

membrane filter: A filter with small pores (0.2–0.45 μ) used to filter thermolabile chemicals/solutions. The membrane filters are usually made up of nitrocellulose material with varying pore sizes. These are commonly used in laboratories for filtering thermolabile chemicals (e.g. antibiotics and cell wall degrading enzymes). Membrane filters are also used in nurse culturing of cells, protoplasts and pollen.

membrane potential: Voltage difference across a membrane due to a slight surplus of positive ions on one side and of negative ions on the other. A typical membrane potential for an animal plasma membrane is 60 mV.

membrane protein: A protein that is normally associated closely with a cell membrane.

membrane transport: Passage of molecules across a membrane mediated by a membrane transport protein.

membrane transport protein: A membrane protein that smoothens the passage of ions or molecules across a membrane. Examples include ion channels and carrier proteins.

membrane-bound ribosome: Ribosome attached to the cytosolic face of the endoplasmic reticulum. It presents the site for the synthesis of proteins that are destined to enter the endoplasmic reticulum.

membranes of a cell: Refers to the thin 'skin-like' structures that envelope the exterior of a cell (plasma membrane) and also enclose various specialized bodies (nucleus, mitochondria, etc.) within the cell itself. The membrane that surrounds the cell's nucleus is called the NUCLEAR ENVELOPE. Membranes are lipoidal (i.e. made of fat-like material) in which proteins and protein complexes are embedded. For example, protein molecules known as RECEPTORS are embedded in the plasma membrane (i.e. the outermost membrane of the cell) and in the nuclear envelope.

membranous: Thin, translucent and flexible; seldom green.

memory: In computer science, the term refers to computer data storage, often called STORAGE. It is a technology comprising computer components and recording media that are utilized to recall digital data. It is a core function and a fundamental component of computers. A process of maintenance of the primed state over time, which may be apparent, such as in altered gene activation patterns (→ transcriptional memory).

memory cells: Some B cells may not secrete antibodies and are called memory cells. They go back to become plasma cells.

memory genes: Genes that show transcriptional memory.

Mendel's principle of independent assortment: The principle of independent assortment (segregation) status that unlinked or distantly linked segregating gene pairs behave independently. Now known as the PRINCIPLE OF INDEPENDENT SEGREGATION.

Mendel's principle of segregation: The two members of a gene pair segregate from each other during meiosis such that each gamete has an equal probability of obtaining either member of the gene pair. Now known as the PRINCIPLE OF SEGREGATION.

Mendel's principles of inheritance: The inheritance of chromosomal genes is based on the chromosome theory of heredity and two principles are considered: (i) PRINCIPLE OF SEGREGATION and (ii) PRINCIPLE OF INDEPENDENT ASSORTMENT (more appropriately INDEPENDENT SEGREGATION). *Aka* MENDELIAN GENETICS, TRANSMISSION GENETICS *or* CLASSICAL GENETICS. *See details as separate terms.*

Mendelian character: A character that follows the principles of inheritance formulated from Gregor J. Mendel's work. *Aka* MENDELIAN TRAIT.

Mendelian genetics: A set of principles that attempts to explain inheritance and biological diversity according to the tenets of Gregor John Mendel given in 1866 regarding the transmission of genetic characters from parent organisms to their offspring based on his statistical analysis and scientific breeding experiments on pea plants. His significant contributions served as the cornerstone of genetics. Mendel's work on *Pisum sativum* formed the basis for two principles of inheritance: the PRINCIPLE OF SEGREGATION and the PRINCIPLE OF INDEPENDENT ASSORTMENT (SEGREGATION). These principles explain the mechanism of inheritance of biological properties of qualitative nature from one generation to the next. The principle of segregation states that genes occur in pairs and segregate from each other during meiosis. Mendel's Principle of independent assortment (segregation) states that members of one pair of genes do not influence the way in which other pairs of genes are distributed. The inheritance of quantitative traits, explained through segregation and an independent assortment of genes, is also included in Mendelian genetics.

Mendelian inheritance: One mechanism by which genetic traits are passed in all diploid multicellular organisms from parents to offspring. Named after Gregor John Mendel, who first studied and recognized the existence of 'factors' or 'elements', now called 'genes'. At each locus, an individual carries two genes: one inherited from its father and the other from its mother. These two genes are represented in equal proportions in its gametes. Female gametes are fertilized by male gametes to produce different genotypes and phenotypes in the Mendelian ratio. *See* MENDELIAN RATIO, AUTOSOMAL DOMINANT, RECESSIVE GENE, SEX-LINKED *and* MENDEL'S PRINCIPLES OF INHERITANCE.

Mendelian lottery: A particular allele will or will not be represented in the offspring because of the segregation of alleles at meiosis and the random chance that any particular gamete will form a zygote.

Mendelian population: Specifically, a Mendelian population is a group of individuals who interbreed among themselves according to a certain system of mating and form a breeding community with equal survival of all genotypes. These individuals share a common gene pool, which is the total genic content of the group.

Mendelian ratio: The segregation ratios of genotypes and phenotypes expected in accordance with Mendel's principles of inheritance.

Mendelism: Generally, refers to Mendel's two principles of inheritance: (i) The PRINCIPLE OF SEGREGATION and (ii) THE PRINCIPLE OF INDEPENDENT ASSORTMENT (SEGREGATION). *See* MENDELIAN GENETICS.

Mendelism and its rediscovery, the impact of: An important consequence of Mendel's work and its subsequent rediscovery was to replace the blending theory of inheritance with a PARTICULATE THEORY OF INHERITANCE. Mendel's observation that one of the parental characteristics was absent in the F_1 hybrid and reappeared in unchanged form in the F_2 generation was inconsistent with the BLENDING THEORY. It was concluded from Mendel's work that traits from the parental lines were transmitted as two different elements of particulate nature that retained their purity in the F_1 hybrids.

Mendelism or the rediscovery of: An Austrian monk, Gregor J. Mendel, whose plant hybridization experiments in garden peas (*Pisum sativum*), begun in 1856, led to visions into the mechanisms of heredity in 1866 that are considered the foundation of genetics today. His work was ignored in his lifetime and only independently rediscovered in 1900 by Hugo DeVries, C. Correns and E. von Tschermak. W. Bateson (1905) confirmed Mendel's work through a series of hybridization experiments. The rediscovery of Mendel's work went a long way towards understanding Mendelism in its real sense. This rediscovery marks the beginning of the science of formal genetics. *See* MENDELIAN INHERITANCE.

mendelize: To segregate according to Mendel's principles of inheritance.

mericlinal: Refers to a chimaera in which the inner tissue possesses a different genetic constitution than the surrounding outer tissue. For example, in the case of a red mutant on a delicious yellow fruit, the surface of the fruit may have longitudinal streaks or sectors of red on an otherwise yellow surface. *Aka* INCOMPLETELY PERICLINAL.

mericloning: A plant proliferation method that uses shoot tips in culture to multiply buds, which can then be separated, rooted and planted out.

meristem (Adj. meristematic): Any actively dividing plant tissue. Tip of a plant's growth. A plant tissue is characterized by frequent cell division, producing cells that become differentiated into specialized tissues. Meristematic tissues produce certain antimicrobial substances and lack vascular connections. The meristem of a plant is generally free, even from systemic pathogens. Meristematic cells are found at the growing tip of a root or stem (apical meristem), in the cambium (lateral meristem), and also within the stem and leaf sheaths (intercalary meristem) of grasses. The meristem culture in plants is being exploited for the production of disease-free planting material, particularly in vegetatively propagated plants.

meristem cloning: Artificial propagation of a plant using cells taken from the meristem of a parent plant and yielding genetically identical offspring.

meristem culture: The aseptic *in vitro* culture of a generally shiny, dome-like plant structure measuring less than 0.1 mm in length when excised, most often excised from the shoot apex. This method is used to produce pathogen-free plants.

meristematic: Pertaining to the meristem.

meristematic tip culture: Refers to the meristematic vault and one pair of leaf primordia used as explants, particularly to produce virus-free plant materials.

meristic mutations: Mutations that alter the number or structure within a floral whorl.

meristomoid: A confined group of cells in callus tissue, categorized by a build-up of starch, RNA and proteins, producing adventitious shoots or roots.

merodiploid: A haploid organism that is diploid for a small region of the chromosome (i.e. a partial diploid). *Aka* MEROZYGOTE.

merodiploidy: Temporary partial diploidy in bacteria due to the presence of transferred genes. *Aka* MEROZYGOTES.

merogony: (i) An individual with the egg cytoplasm coming from one parent and the egg nucleus from the other. (ii) An enucleated egg fertilized by a sperm. (iii) A form of asexual reproduction whereby a parasitic protozoan replicates its own nucleus inside its host's cell and then induces cell segmentation.

merozygote: *See* MERODIPLOID.

Meselson-Stahl experiment: M. Meselson and F. Stahl (1958) labelled the DNA of *Escherichia coli* bacteria with 'heavy' nitrogen (^{15}N) by culturing them on a ^{15}N-containing medium that was the only source of nitrogen for many generations. The DNA extracts of such cells gave a distinctive UV-light pattern, showing a dark band near one end of a tube that had been spun at high speed in an ultracentrifuge. When these labelled ^{15}N cells were grown on unlabelled media (containing ^{14}N), the DNA extracted comprised a hybrid DNA carrying both ^{14}N and ^{15}N at the same time (i.e. DNA had not replicated in the separately labelled and unlabelled conservative form). Conservative replication was thus excluded as the mechanism of DNA replication. To distinguish between semi-conservative and dispersive replication, Meselson and Stahl produced the next generation of growth again on unlabelled media. Unlabelled DNA was produced in amounts equal to partially labelled hybrid DNA. Additional generations of growth on unlabelled media gave a relative increase in the amount of unlabelled DNA. Since unlabelled DNA was always formed despite the presence of labelled hybrid DNA, duplication evidently did not involve random dispersive labelling of all newly formed DNA. The dual $^{15}N/^{14}N$ composition of the hybrid DNA was also confirmed by heating this band of DNA to 100°C and then reanalysing it by ultracentrifugation. Denaturation with heat broke the A=T and G≡C hydrogen bonds between two complementary DNA strands and two single strands of DNA were produced. Two separate bands were formed in the ultracentrifuge density gradient, one heavy (^{15}N) and one light (^{14}N). This experiment thus supported the semi-conservative mode of DNA replication. *See* DNA REPLICATION IN EUKARYOTIC CELLS.

mesocarp: The middle layer of the fruit wall.

mesophyll: (i) The parenchyma tissues between the upper and lower epidermis. They vary in function in the storage, starch and photosynthesis tissues of a leaf. (ii) In ecology, the blade of a leaf or leaflet that has a surface area of 4500–18,225 mm^2.

mesophyll cell: A type of plant cell located between the upper and lower epidermis that is the site for photosynthesis.

mesophyll explant: An explant arising from internal parenchyma tissue.

mesophyte: A plant flourishing under intermediate environmental conditions of moderate moisture and temperature without major seasonal fluctuations.

messenger DNA (mDNA): A single-stranded DNA that acts as a template for the synthesis of messenger RNA, which in turn functions as a messenger of protein biosynthesis. Coding DNA sequences of genes may be termed mDNA. This term is, however, not in common use.

messenger RNA (mRNA): A complementary RNA copy transcribed from a single-stranded DNA template during transcription. It functions as an intermediate molecule between DNA and ribosomes (in a cell), which acts as a template for the synthesis (manufacture) of those proteins coded for by the cell's DNA. mRNA is transported from the nucleus to the cytoplasm, where it is processed into a sequence carrying the information to code for a polypeptide domain. mRNA contains an initiation codon (AUG or GUG), at least one of the termination codons (UAA, UGA or UAG) and a base sequence in the form of triplet codons that dictate the order of amino acids in a polypeptide chain. mRNA also includes certain trailer and leader sequences that are not translated. The mRNA fraction is heterogeneous in size, ranging from 500 to 6000 nucleotides in *Escherichia coli*. Protein-encoding regions of mRNA are composed of contiguous, nonoverlapping strings of codons called OPEN READING FRAMES **(ORFs)**, each of which specifies a single protein. Translation starts at the 5′-end of the open reading frame and proceeds one codon at a time to the 3′-end.

messenger RNA (mRNA) caps: Addition of 7^mG group(s) at the 5′ end of majority of the eukaryotic primary transcript. These caps are of three types: (i) CAP 0: Cap with a single methyl group, found in 100% of cases; (ii) CAP 1: methyl group may be present on the penultimate base at the 2′ O position of the sugar moiety, which is present in most cases; and (iii) CAP 2: methyl group may be present in the third base also at the 2′ O position of sugar, present in 10–15% of cases.

messenger RNA (mRNA) cleavage: Ribosome stalling during translation caused by various reasons leads to mRNA cleavage, resulting in non-stop mRNAs that are eliminated in a TRANSFER-MESSENGER RNA (TMRNA)-dependent manner. mRNA cleavage is induced by amino acid starvation, resulting in the accumulation of truncated mRNAs in cells lacking tmRNA. The truncated mRNAs are eliminated in wild-type cells, indicating that the tmRNA system rapidly degrades the truncated mRNAs.

messenger RNA (mRNA) decay: mRNA decay is a major step in the regulation of gene expression. After messenger RNA has been used as a template for translation, it is protected till needed by the cell or degraded by the cellular machinery when not required. mRNA has noncoding nucleotides at either end of the molecule. These segments contain information about the number of times mRNA can be transcribed before being destroyed by ribonucleases. Hormones stabilize certain mRNA transcripts. This process is responsible for the differential survival of mRNAs in the cytoplasm. Site-specific endonucleases seem to control this process.

messenger RNA (mRNA) in molecular investigations: For the identification and characterization of structural genes, mRNA is considered the ideal intermediate molecule because one can move in both directions from this molecule, either to investigate a gene or a polypeptide. A population of mRNAs, established as the spectrum of the sequences found in the polygenes, defines the entire set of genes expressed in a cell or tissue. Thus, the constitution of mRNA displays both nature and number of structural genes.

messenger RNA (mRNA) modifications: Modifications of mRNA include N6-methyladenosine (6^mA), N1-methyladenosine (1^mA), N6,2′-O dimethyladenosine (6^mA^m), N3-methylcytidine (3^mC), N7-methylguanosine (7^mG), psudouridylation (Ψ), 5-hydroxymethylcytosine ($5^{hm}C$), G-qaudruplexes and RNA-loops (RGQs).

messenger RNA (mRNA) quantification: Measurement of the abundance of different transcripts in a transcript profile.

messenger RNA (mRNA) splicing: The editing of the primary transcript of a gene from a higher organism to excise noncoding intervening sequences. The leaves contain a continuous coding sequence specifying the protein in question.

messenger RNA (mRNA) targeting mitochondria: The crosstalk between the mitochondrial and nuclear genomes ensures mitochondrial biogenesis, dynamics and maintenance. Mitochondrial proteins are partly encoded by the nucleus, synthesized in the cytosol and partly in the mitochondria, coded by the mitochondrial genome. The transport systems allow the transport of nuclear-encoded gene products (e.g. proteins and mRNAs) to the mitochondrial vicinity to allow for their translation and/or import.

messenger RNA (mRNA) transport: After mRNAs are synthesized, processed and become associated with a number of different proteins at the transcription site, they are released into the nucleoplasm. The mechanism by which these large mRNA-protein (mRNP) complexes then move through dense nucleoplasm to reach the nuclear pores. The coupling between mRNA maturation and nuclear export is an important mechanism for providing only fully functional and competent mRNA to the cytoplasmic translational machinery, thereby ensuring accuracy and swiftness of gene expression.

messenger RNA (mRNA), translation of: Upon receiving the message encoded in the DNA, the mRNA passes through the ribosomes like a reel of punched paper passes through an old piano (pianola), giving the ribosomes the specifications for making the coded-for proteins. This process is aided by **transfer RNA (tRNA)** molecules, which quest for amino acids that float around in the cell (outside of the cell's nucleus and ribosomes). The tRNA molecules attach to and escort individual amino acids to the ribosome as and when the mRNA directs. Each of the 20 different amino acids has at least one of its own purpose-built tRNA molecules, which possess a three-letter anticodon at the stem of the cloverleaf-shaped rRNA molecule. The ribosome has room for only two tRNA molecules at a time. The mRNA molecule (which itself is passing through the ribosome) calls over the first tRNA molecule, which fetches the specified amino acid. Short sections of the mRNA and tRNA molecules pair together inside the ribosome at a region where these two molecules meet as their three nucleotides of tRNA (anticodon) and mRNA (codon) are complementary. The whole (locked together) apparatus shifts along by three notches (i.e. nucleotides), and a second tRNA molecule (bearing another amino acid) slips in next to the first tRNA molecule. Next, the first amino acid (brought in by the first tRNA molecule) jumps over to the second tRNA molecule, joining the amino acid that was brought in by the second tRNA molecule, thus initiating the synthesis of a POLYPEPTIDE or PROTEIN MOLECULE. The empty (first) tRNA molecule falls out of the ribosome, and the whole (locked together) apparatus (i.e. mRNA plus the second tRNA molecule) moves three more notches (i.e. nucleotides) along the mRNA molecule to make room for a third tRNA molecule bearing another amino acid, and so on. This process of creating ever-longer chains of amino acids continues to repeat itself inside the ribosome until the protein (coded for by the DNA, which code was transferred to mRNA, which transferred it to the ribosome) is completed. The translation of mRNA stops at the termination codon. *See* TRANSLATION.

messenger RNA processing: The collective term for the modifications to eukaryotic RNA that are necessary before the RNA can be transported to the cytoplasm for translation. It includes various steps such as CAPPING, INTRON REMOVAL and POLYADENYLATION. *See details as separate terms.*

meta research: A collection of apparatuses for handling crowdsourcing responsibilities from the commencement through to data packaging for research usage.

metabolic control of transcription: Mechanisms that link environmental stimuli to organismal transcriptional responses facilitate adaptation and survival. Because it signals energy availability through photosynthesis, light is among the most important environmental signals for plants. F. Huang *et al.* (2023) demonstrate that, in the plant *Arabidopsis thaliana*, light exposure stimulates chromatin-local depletion of the **tricarboxylic acid (TCA) cycle** metabolite α-ketoglutarate to orchestrate global gene expression by impairing demethylation of nucleosomal histone tails. Jumonji C domain–containing histone demethylases (JMJs) dynamically control the level of histone methylations.

metabolic engineering: Improving cellular activities through the management of enzymatic transport and regulatory functions of cells using recombinant DNA technology. The performance of metabolic pathways or bioprocesses can be altered by genetic manipulation so that a particular metabolite is overproduced. Unlike conventional approaches, cell function can be altered using targeted changes in normal cellular activities. The introduction of foreign gene(s) causes the production of heterologous proteins/enzymes, leading to the production of new compounds. Genes can be transferred for partial or entire biosynthetic pathway, creating new products and new reactions. For example, a number of carotenoid genes have been cloned from the bacteria *Rhodobacter* and *Erwinia* and expressed in *Escherichia coli* and rice (golden rice). *See* PROTEIN ENGINEERING.

metabolic engineering and secondary products: Plant biotechnology has produced improved plant sources for the production of valuable secondary metabolites. Overexpression of the gene that encodes for the first enzyme in a pathway generally results in higher levels of the desired end product. This has been successfully achieved in the improvement of taxol production from the transformed tissue cultures of *Taxus* sp. Another strategy involves the use of *Agrobacterium rhizogenes* to induce the excessive formation of secondary roots in plants that normally produce useful secondary metabolites. Transgenic plants can be used as factories to produce polyhydroxy butyrate (PHB, biodegradable plastic). Genetically engineered *Arabidopsis* plants produce PHB globules exclusively in their chloroplasts without effecting plant growth and development. The large-scale production of PHB may be easily accomplished in tree plants like populus, where PHB can be extracted from leaves. The industry has already started to explore the production of biodegradable plastics from transgenic plants.

metabolic engineering for novel carotenoid biosynthesis: Microorganisms and plants produce more than 600 carotenoids, which are of interest for pharmaceuticals, food colourants, animal feed and nutrient supplements for human food. There is enough scope for developing novel carotenoids on an industrial scale (e.g. C_{40} carotenoid phytoene). Phytoene is used as a substrate for the desaturase enzyme crtl to bring about desaturation. This leads to the production of the following six compounds: (i) phytofluene, (ii) ξ carotene, (iii) neurosporene, (iv) lycopene, (v) 3,4, didehydrolycopene and (vi) 3,4, 3′,4′ tetrahydrolycopene.

metabolic engineering for novel polyhydroxyalkanoates biosynthesis: Polyhydroxyalkanoates (PHAs) provide biodegradable alternatives to petrochemical plastics. Polyhydroxybutyrate (PHB) and polyhydroxyvalerate (PHBV) are two such polymers. PHBV is more useful, but its production is difficult. Metabolic engineering of *Arabidopsis thaliana* and *Brassica napus* for production of PHBV has been useful through the use of four genes (*ilvA*, *bktB*, *phbB* and *phbQ*) and by diverting metabolic pools of acetyl-CoA and threonine.

metabolic engineering, reasons for: Some reasons for performing metabolic engineering of an organism include changing cell behaviour and organismal metabolic designs to (i) induce production of proteins/polypeptides and/or metabolites that are desired by mankind (e.g. 'golden rice'); (ii) enable a given organism to consume or gather toxic wastes or valuable materials (e.g. gold) that occur at a site in low concentration or highly dispersed; and (iii) cure disease.

metabolic engineering, uses of: Metabolic engineering has been successfully used for improving the nutritional value of crops like rice and maize. Different approaches used in metabolic engineering include: (i) heterologous genes for metabolic engineering; (ii) redirecting metabolic flow; and (iii) the development of novel biosynthetic pathways.

metabolic mutant: A mutant having a lesion in a gene that codes for an enzyme involved in metabolism. Nonmetabolic mutants must, for example, have alterations in parts of their DNA that are not expressed as enzyme proteins.

metabolic pathway: Refers to a consecutive series of enzymatic reactions involving the synthesis, degradation or transformation of a metabolite. The pathway can be linear, branched or cyclic and it can be directly or indirectly reversible.

metabolic plasmids: These bacterial plasmids possess genes to code enzymes that degrade unusual substances such as toluene (aromatic compounds), pesticides (2, 4-dichlorophenoxy acetic acid) and sugars (lactose). The TOL (= pWWO) plasmid of *Pseudomonas putida* is an example. However, some metabolic plasmids occurring in certain strains of *Rhizobium* induce nodule formation in legumes and carry out the fixation of atmospheric nitrogen.

metabolism: The entire set of life-sustaining enzyme-catalysed pathways of organic energy-producing (nutrient) molecules in living cells. The conversion of food and water into nutrients that can be used by the body's cells and the use of those nutrients by those cells to sustain life, grow, etc. The three main functions of metabolism are: (i) the conversion of the energy in food to energy available to run cellular processes; (ii) the conversion of food to building blocks for proteins, lipids, nucleic acids and some carbohydrates; and (iii) the elimination of metabolic wastes. These enzyme-catalysed reactions allow organisms to grow and reproduce, maintain their structures and respond to their environments. There are two kinds of metabolism – CONSTRUCTIVE METABOLISM (or ANABOLISM), which concerns the synthesis of proteins, carbohydrates and fats that form tissue and store energy and DESTRUCTIVE METABOLISM (or CATABOLISM) and the breakdown of complex substances, producing energy and waste matter.

metabolite, primary: The term was adopted by F. Czapek during 1922–1925. A kind of metabolite that is directly implicated in normal growth, development and reproduction. It usually performs a physiological function in the organism (i.e. an intrinsic function). A primary metabolite is typically present in many organisms or cells. It is also referred to as a CENTRAL METABOLITE, which has an even more restricted meaning (present in any autonomously growing cell or organism). Some common examples of primary metabolites include lactic acid and certain amino acids.

metabolite, secondary: The term secondary metabolite was first coined by A. Kossel, the 1910 Nobel Prize laureate for medicine and physiology. Secondary metabolites, also called specialized metabolites, toxins, secondary products or natural products, are organic compounds produced by any life form that are not directly involved in the normal growth, development or reproduction of the organism. Instead, they generally mediate ecological interactions, which may produce a selective advantage for the organism by increasing its survivability or fecundity. Secondary metabolites often play an important role in plant defence against herbivory and other interspecies defences. Humans use secondary metabolites as medicines, flavourings, pigments and recreational drugs.

metabolites: Metabolites are intermediate products of metabolism. Examples include antibiotics and pigments. These are small molecules that exhibit enormous variation. A metabolite is usually defined as any molecule less than 1 kDa in size. In plant-based metabolomics, it is common to refer to PRIMARY METABOLITES and SECONDARY METABOLITES. *See details as separate terms.*

metabolome: The quantitative complement of all the low-molecular-weight molecules that are present in cells in a particular physiological or developmental state. All metabolites in a complete set make a metabolome (e.g. hormones and other signalling molecules and secondary metabolites of medicinal or aromatic value). Although the metabolome can be defined readily enough, it is not currently possible to analyse the entire range of metabolites by a single analytical method. Over 50,000 metabolites have been characterized from the plant kingdom, and many thousands of metabolites have been identified and/or characterized from single plants.

metabolome analysis: The analysis of the metabolome involves (i) The alignment of chromatographic and spectroscopic data using specific bioinformatics tools and (ii) the identification of different metabolites. While mRNA gene expression data and proteomic analyses do not tell the complete story of what might be happening in a cell, metabolic profiling can give an instantaneous snapshot of the physiology of that cell. One of the challenges of systems biology and functional genomics is to integrate proteomic, transcriptomic and metabolomic information to give a more complete picture of living organisms. The metabolome forms a large network of metabolic reactions, where outputs from one enzymatic chemical reaction are inputs to other chemical reactions. Such systems have been described as hypercycles.

metabolomics: (i) Systematic global analysis of non-peptide small molecules (e.g. vitamins, sugars, hormones, fatty acids and other metabolites) in a cell or an organism at different developmental stages and under different conditions. (ii) The quantitative measurement of the dynamic multiparametric metabolic response of living systems to pathophysiological stimuli or genetic modification. (iii) It involves descriptions of biological pathways and current metabolomic databases, such as the Kyoto Encyclopaedia of Genes and Genomes (KEGG). Metabolomics is a similar term to 'genomics'. As genomics deals with the study of genes, metabolomics deals with the study of metabolites (gene products). It provides a better understanding for the elucidation of gene function. It possesses special significance in integrated genomics involving the genome, transcriptome, proteome and metabolome. Using this approach, it has been possible to propose a new metabolic pathway for citramalate synthesis using *Arabidopsis*. Citramalate was not reported earlier in plants. More information can be found at www.metabolomics.nl. *Aka* METABONOMICS.

metabolomics and metabonomics, differences between: Two major differences exist: (i) Metabolomics is more associated with MASS SPECTROMETRY (MS)-BASED TECHNIQUES, while metabonomics is associated with NUCLEAR MAGNETIC RESONANCE (NMR) SPECTROSCOPY. (ii) Metabolomics lays a greater emphasis on metabolic profiling at the cellular or organ level and primarily deals with normal endogenous metabolism. 'Metabonomics' extends metabolic profiling to include information about perturbations of metabolism caused by environmental factors (including diet and toxins), disease processes and the involvement of extragenomic influences (e.g. gut microflora).

metabolomics, analytical technologies in: For separation of metabolites, GAS CHROMATOGRAPHY (GC), HIGH-PERFORMANCE LIQUID CHROMATOGRAPHY (HPLC) and CAPILLARY ELECTROPHORESIS (CE) are used. For the detection of metabolites, MASS SPECTROMETRY (MS) and NUCLEAR MAGNETIC RESONANCE (NMR) SPECTROSCOPY are used. *See details as separate terms.*

metabolomics, applications of: The metabolomics research is applied in (i) characterization of genotypes; (ii) integration with transcript and protein profiling; and (iii) identification of regulated key sites in the network.

metabonomics and metabolomics: These very similar terms have arisen at about the same time in different areas of bioscience research, mainly animal biochemistry and microbial/plant biochemistry, respectively. Although both involve the multiparametric measurement of metabolites, they are not identical. METABONOMICS deals with integrated, multicellular and biological systems, including communicating extracellular environments. METABOLOMICS deals with simple cell systems and, at least in terms of published data, mainly intracellular metabolite concentrations.

Metabuli: Metagenomic taxonomic classifiers analyse either DNA or amino acid (AA) sequences. Metabuli (https://metabuli.steineggerlab.com), developed by J. Kim and M. Steinegger (2024), however, together analyses both DNA and AA to impact AA conservation for sensitive homology detection and DNA mutations for specific differentiation of closely related taxa. In the Critical Assessment of Metagenome Interpretation 2 plant-associated data set, Metabuli covers 99% and 98% of classifications of state-of-the-art DNA- and AA-based classifiers, respectively.

metacentric: A chromosome having a centromere in the middle.

metacentric chromosome: A chromosome with a centrally located centromere.

metadata: In genomics, metadata provides information on the time and origin of the sample that the genetic material was taken from, as well as certain characteristics (e.g. a person's age), which is essential to correctly interpret the genomic data and understand its context. Data about data or specifically, in our current context, data describing web resources. One of the major issues with the World Wide Web (www) is that it is really hard to automate any task. So far, the web is mainly built as a forum for human interaction. The only available form of searching on the web is to simply match words or sentences contained in documents. Anyone who has used a service like Acta Vista knows that typing in a few keywords and receiving thousands of 'hits' is not very useful. A possible solution to the search problem is to provide a mechanism that allows more precise descriptions of things on the web. It could elevate the status of the web from machine-readable to machine-understandable.

metafemale: Abnormal females in *Drosophila*, usually sterile and weak, with an overbalance of X chromosomes with respect to autosomes; X/A ratio greater than 1.0 but less than 1.5, as an X/A ratio greater than 1.5 is lethal. *Aka* SUPERFEMALE.

metagenesis: Alternation of generation; alternation of a sexual with an asexual form; or cytologically, the alternation of a haploid with a diploid stage.

metagenomic approach: Characterization of soil-extracted nucleic acids for investigating the vast portion of soil bacterial communities that are not culturable.

metagenomics: Large-scale sequencing of the function of entire genomic DNA bulks sampled directly from a whole community of organisms inhabiting an environment. Metagenomics is often used to study a specific community of microorganisms, such as those residing on human skin, in the soil or in a water sample. *Aka* ENVIRONMENTAL GENOMICS, ECOGENOMICS *or* COMMUNITY GENOMICS.

metagon: An RNA molecule required for the maintenance of mu particles in *Paramecium*.

MetaGraph: The quantity of biological sequencing data existing in public repositories is rising exponentially, yet making it full-text searchable and easily available to researchers in life and data science has been an unsolved problem. M. Karasikov *et al.* (2024) harvest the benefits of recently established, very efficient data structures and algorithms for representative sequence sets. They style Petabases (10^{15}) of DNA sequences across all clades of life, including viruses, bacteria, fungi, plants, animals and humans, fully searchable and render the indexes accessible to the research community. They present a methodological framework called MetaGraph that enables us to index huge sets of DNA or protein sequences using annotated De Bruijn graphs.

metal-regulated transcription in eukaryotes: Metal-regulated gene transcription has been well documented in diverse forms of organisms, including bacteria, fungi, plants, insects, fish, rodents and humans. Eukaryotic organisms can sense changes in environmental metal levels and respond to this consciousness by regulating the transcription of specific cellular target genes. In higher eukaryotes, the metallothionein (MT) genes have been most intensively investigated and are the best-known examples of metal-regulated transcription units.

metalloenzyme: Metalloproteins structurally. An enzyme has a metal ion as its prosthetic group. Some notable examples of metalloenzymes (and their metal coenzymes) include ascorbic acid oxidase (copper), alcohol dehydrogenase (zinc), glutathione peroxidase (selenium) and xanthine oxidase (molybdenum). Metalloenzymes promote a diverse range of reactions, including hydrolytic processes and oxidation/reductions.

metamale (supermale): Abnormal males in *Drosophila* with an overabundance of autosomes on X chromosomes, having an X/A ratio <0.5 but >0.33. The X/A ratio of <0.33 is lethal.

metamodel methods of bioinformatics: Metamodel methods refer to the methods utilized to integrate data that have been independently generated/created and generally stored in separate database models via independent genomics research projects, combinatorial chemistry projects and high-throughput screening projects through biochip use. Metamodel methods sometimes display imperative interrelationships that were not obvious in the individual models: (i) created solely for the genomics project data, (ii) created solely for the combinatorial chemistry project data or (iii) created solely for the high-throughput screening project data.

metamorphosis: One or more alterations in form during the life cycle of an organism (e.g. an amphibian or insect) in which the juvenile stages differ from the adult. An example is the transition from a tadpole to an adult frog. The term COMPLETE METAMORPHOSIS is applied to insects such as butterflies, in which the caterpillar stage is distinct from the adult stage. The term INCOMPLETE METAMORPHOSIS describes the life histories of insects such as locusts in which the young go through a series of larval stages, each of which bears similarities to the adult. Metamorphosis in both insects and amphibians is controlled by hormones and often involves considerable destruction of larval tissues by enzymes.

metaphase: A stage of mitosis at which centrioles push each other apart to lie at opposite poles. Chromosomes get arranged on the equatorial or metaphase plate. The nucleolus and nuclear membrane are not visible. The spindle apparatus is very prominent.

metaphase arrest: The stopping of cell division at mitotic or meiotic metaphases, usually by application of specific agents.

metaphase chromosome: A chromosome at the metaphase stage of cell division, when the structure is in its most organized state and features (e.g. the banding pattern) can be visualized.

metaphase I (mI): A stage of meiosis I at which the nuclear membrane disappears, the spindle apparatus is completely formed and the homologous chromosomes, in the form of bivalents, are arranged on the equatorial (metaphase) plate of the spindle apparatus. A single spindle fibre instead of two, as in mitosis, joins each centromere.

metaphase II (mII): A stage of meiosis II at which the nuclear membranes in the two haploid cells seen at prophase II disappear, the spindle apparatus is completely formed in each haploid cell and the chromosomes are arranged on the equatorial plates of the spindle apparatus. Two spindle fibres join each centromere, one from each pole of the spindle pole. The chromosomes take positions at the centre of the spindle, forming a metaphase plate.

metaphase II (mII) transgenesis: Metaphase II oocytes can efficiently promote transgenesis when co-injected with sperm and small (<5 kb) ubiquitously expressed transgenes. The mII transgenesis can readily be applied to a range of larger transgenes (11.9–170 kb), including BACs and MACs.

metaphase mapping: The *IN SITU* HYBRIDIZATION (ISH) of labelled DNA probes to the mitotic metaphase chromosomes. It is important for rapid localization to a small chromosome region. The limit of resolution in the metaphase mapping is 1 mb.

metaphase plate: The arrangement of the chromosomes in a horizontal plane at the equator of the spindle apparatus during the metaphase stage of mitosis and meiosis.

metatranscriptomics: Retrieves and sequences environmental mRNAs from a microbial ecosystem to assess which genes may be expressed in that community.

metaxenia: The influence of pollen on the maternal tissue of a fruit outside the embryonic sac (i.e. edible fruit tissues). The phenomenon is seen on dates and is sometimes called ZENIA. In maize (*Zea mays* L.) with normal endosperm, the relative advantage in weight of cross-fertilized to self-fertilized kernels can reach 13% when inbred lines are used as female parents.

methicillin-resistant *Staphylococcus aureus* (MRSA): A staphylococci species that is resistant to many widely used antibiotics.

methionine (met): An essential amino acid with the chemical formula $C_5H_{11}NO_2S$. Donates to the organism both labile methyl groups and sulfur necessary for normal metabolism. Methionine is an intermediate in the biosynthesis of cysteine, carnitine, taurine, lecithin, phosphatidylcholine and other phospholipids. The only one of two amino acids encoded by a single codon (AUG) in the standard genetic code (tryptophan, encoded by UGG), is the other. The codon AUG is also significant as it carries the 'start' message for a ribosome that signals the initiation of protein translation from mRNA.

method: In OBJECT-ORIENTATED PROGRAMMING (OOP), a procedure is associated with a message and an object. An object consists of data and behaviour. The data and behaviour comprise an interface that specifies how the object may be applied by any of the various consumers of the object.

methodology: In software engineering, a software development process separates software development work into distinct steps to improve design, product management and project management. The methodology may include the pre-definition of specific deliverables and artefacts that are created and completed by a project team to develop or maintain an application. *Aka* SOFTWARE DEVELOPMENT LIFE CYCLE (SDLC).

methyl ($-CH_3$) group: Hydrophobic chemical group derived from methane (CH_4).

methyl blue: A dye, normally dissolved in ethanol, that is used in histology for staining cell nuclei.

methyl green: The basic dye used in histochemistry to stain DNA. It is often used in conjunction with pyronine.

methyl jasmonate: The volatile chemical compound that results when methyl groups ($-CH_3$) are chemically added to a molecule of jasmonic acid. Methyl jasmonate alleviates the stress of salinity on plant growth by positively impacting plant morphology and physiology, including plant growth, plant hormones, plant photosynthesis rate, chlorophyll and proline content and rate of transpiration.

methyl methanesulfonate (MMS): A frequently used, very efficient chemical mutagen ($C_2H_6O_3S$), which acts by adding a methyl ($-CH_3$) group to guanine. Thus, it causes base-pairing errors as methylated guanine pairs with adenine.

methyl/phospho switch: Histone methylation and phosphorylation is referred to as 'methyl/phospho' switch. The generation of H3K9me3S10ph rests on Suv39h and Aurora B and transpires at pericentric heterochromatin during mitosis in different eukaryotes. Most heterochromatin protein 1 (HP1) is typically dissociated from chromosomes during mitosis, but if phosphorylation of H3S10 was inhibited, HP1 remained chromosome-bound throughout mitosis. H3 phosphorylation by Aurora B is therefore part of a 'methyl/phospho' switch mechanism that displaces HP1 and perhaps other proteins from mitotic heterochromatin.

methyl-CpG-binding domain 4 (MBD4): A protein that in humans is encoded by the *MBD4* gene. The human MBD4 protein has 580 amino acids with a methyl-CpG-binding domain at amino acids 82–147 and a C-terminal DNA glycosylase domain at amino acids 426–580. These domains are separated by an intervening region that interacts with UHRF1, an E3 ubiquitin ligase and USP7, a deubiquinating enzyme.

methylase: A DNA modification enzyme that covalently attaches a methyl ($-CH_3$) group to a specific nucleotide base of a DNA molecule.

methylase reaction: Though methylase and cleavage reactions are fundamentally different from each other, the first step of both reactions appears to be the same. The *Eco*K enzyme and S-adenosylmethionine (AdoMet) or SAM produce the EcoK-AdoMet complex through noncovalent binding. The *Eco*K enzyme has a minimum of 2–3 AdoMet binding sites. The *Eco*K-AdoMet complex undergoes a transition into an activated form (*Eco*K*). This is the rate-limiting step of this enzyme. *Eco*K* interacts with any DNA, regardless of the presence of sK sites. Initial complexes are unstable and can be dissociated by exposure to heparin (a polyanion that mimics DNA). *Eco*K* then specifically binds to sK sites, regardless of their methylated state. It is the *Eco*K* and sK site interactions that determine whether *Eco*K* undergoes a transition to restriction or modification mode.

methylated cap: A methylated guanine nucleotide that is added to the 5′ end of eukaryotic hnRNA and mRNA after transcription but prior to later processing and translation.

methylated CpG island recovery assay (MIRA): MIRA is based on the high affinity of the MBD2b/MDB3L1 complex for double-stranded methylated DNA, which allows for the recovery of methylated DNA without the use of bisulfite conversion or antibody recognition.

methylated DNA: Refers to a DNA molecule that is saturated with methyl groups. When cytosine in DNA is methylated, the genes comprising that DNA molecule are generally 'turned off' (inactivated).

methylated DNA immunoprecipitation (Methyl-DIP or MeDIP): This method makes use of non-specific fragmentation of the genomic DNA followed by anti-5^{m}C antibodies to enrich for methylated DNA fragments. The resulting immunoprecipitated DNA, enriched in hypermethylated sequences, and total genomic DNA (input) are labelled with fluorescent cyanine dyes Cy5 and Cy3, respectively and co-hybridized on to microarray chips. The ratio of fluorescent intensity (Cy5 to Cy3) specifies the methylation status at each particular gene.

methylated DNA immunoprecipitation chip (MeDIP-chip): This method uses the MeDIP approach in combination with a tiling array for the investigation of genome-wide DNA methylation patterns (or DNA methylomics).

methylated DNA immunoprecipitation PCR (MeDIP-PCR): This method combines the MeDIP technology with rtqPCR technology to interrogate the exact value of the methylation level of specific regions over a wide range.

methylated DNA immunoprecipitation sequencing (MeDIP-seq): This method is commonly used to study 5^{m}C or 5^{hm}C modifications. Specific antibodies can be used to study cytosine modifications.

methylated genomic DNA fragments, enrichment of: Affinity enrichment is a technique that is often used to isolate methylated DNA from the rest of the DNA population. This is usually accomplished by antibody immunoprecipitation methods or with METHYL-CPG-BINDING DOMAIN (MBD) proteins. Various methods under affinity enrichment include METHYLATED DNA IMMUNOPRECIPITATION (MEDIP), METHYLATED DNA IMMUNOPRECIPITATION PCR (MEDIP-PCR), METHYLATED DNA IMMUNOPRECIPITATION CHIP (MEDIP-chip), METHYLATED DNA IMMUNOPRECIPITATION SEQUENCING (MEDIP-seq) and METHYLATED CGI RECOVERY ASSAY (MIRA).

methylated RNA immunoprecipitation sequencing (MeRIP-seq): N(6)-methyladenosine-sequencing (6^{m}A-seq), as this method is used to localize 6^{m}A in the transcriptome at high resolution. This technique is based on RNA fragmentation and immunoprecipitation with 6^{m}A specific antibodies, followed by rigorous washing and release of the remaining molecules. The modified nucleotides are marked throughout the transcriptome by accumulations of aligned sequence reads, *aka* PEAKS.

methylated states and restriction/modification: There are three possible states of methylated sites: (i) HOMODUPLEX-MODIFIED DNA has a methyl group on each strand. *Aka* FULLY METHYLATED DNA. (ii) HETERODUPLEX DNA carries a methyl group on only one strand. *Aka* HALF-METHYLATED or HEMI-METHYLATED DNA. There are two variants for heteroduplex DNA. (iii) HOMODUPLEX, UNMODIFIED DNA has no methyl groups at all. *Aka* UNMETHYLATED DNA. The methylated state of DNA controls whether DNA will be a substrate for restriction or modification. Unmodified DNA is susceptible to endonucleolytic action but is methylated very slowly in the presence of AdoMet and *Eco*K. Heteroduplex DNA is rapidly methylated but undergoes neither single- nor double-stranded cleavage. Modified DNA is neither methylated nor cleaved.

methylation induced premeiotically (MIP): A form of reciprocal *trans*-transcriptional gene silencing (*trans*-TGS) observed in the fungus *Ascobolus immerses*, resulting in a block of transcription elongation within the duplicated sequences.

methylation of telomere cytosines: A decrease in repressive chromatin marks (e.g. levels of cytosine methylation in subtelomeric regions), which results in telomere elongation in mouse cells.

methylation patterns: The pattern of how particular amino acids are methylated within an array of nucleosomes (e.g. lysine can be mono-, bi- or trimethylated or arginine can be mono- or bimethylated), deciding about the degree of heterochromatinization. DNA also becomes directly methylated by DNA methyltransferases. DNA methylation occurs at CpG, CpHpG and CpHpH sites. The pattern of histone or DNA methylation may alter gene expression and can be a reason for developing cancer.

methylation profiling in archival formalin-fixed paraffin-embedded (FFPE) samples: Defining the epigenetic changes that transpire at the earliest stages of tumour initiation (e.g. *in situ* carcinomas and dysplasias) can only be analysed in micro-dissected material, which is often attained from FFPE specimens.

methylation quantitative trait locus (meQTL): One mechanism through which expression QTLs act is epigenetic (i.e. by modulating DNA methylation). In such cases, there are quantitative differences in DNA methylation between the two alleles of the causal polymorphism and the association signal. *Aka* EPIGENETIC QUANTITATIVE TRAIT LOCUS (EPIQTL).

methylation specificity: Most DNA methylation is maintenance-type methylation following DNA replication. The enzyme is not sequence-specific. *De novo* DNA methylation occurs rarely. Here, non-methylated DNA is the substrate. It is not known whether or not *de novo* DNA methylase is sequence-specific. CG doublets do not, by themselves, provide specificity.

methylation target arrays (MTAs): An approach for methylation profiling that reverses the normal process of hybridizing an array of genes with a probe prepared from a single experimental sample.

methylation: The attachment of a methyl group to some molecule (e.g. histone, DNA or RNA).

methylation-induced premeiotically (MIP): A form of reciprocal *trans*-TGS observed in the fungus *Ascobolus immersus*, resulting in a block of transcription elongation within the duplicated sequences.

methylation-phosphorylation switch (MPS): Polo-like kinase 1 (Plk1) activity is controlled by a balanced methylation and phosphorylation switch. The methyltransferase G9a monomethylates Plk1 at Lys209, which antagonizes phosphorylation of T210 to inhibit Plk1 activity.

methylation-sensitive amplification polymorphism (MSAP)-based genetic diversity analysis: The analysis exhibited differences in the DNA methylation level among the samples. However, the samples collected from saline areas and some species hybrids displayed precise DNA methylation patterns. AMPLIFIED FRAGMENT LENGTH POLYMORPHISM (AFLP) data were utilized in combination with MSAP data to prepare a phylogenetic tree, showing higher efficiency in distinguishing the samples. This combined analysis distinguished samples formerly grouped in the AFLP analysis.

methylation-sensitive PCR (MSP): a technique used to determine the methylation status of specific DNA sequences by PCR amplification of a bisulfite-converted template with different primer sets that distinguish methylated DNA from non-methylated (C→T converted) DNA.

methylation-sensitive restriction-based DNA microarray: Three methods in methylation-sensitive restriction-based DNA microarrays are: (i) DIFFERENTIAL METHYLATION HYBRIDIZATION; (ii) MCRBC METHODOLOGY; and (iii) COMPREHENSIVE HIGH-THROUGHPUT ARRAYS FOR RELATIVE METHYLATION.

methylation-sensitive single-nucleotide polymorphism (msSNP) chip analysis: This method used restriction endonuclease-based methylation profiling on Affymetrix SNP arrays to measure net methylation and allele-specific methylation genome-wide.

methylation-sensitive single-nucleotide primer extension (ms-SNuPE): A technique used to query the methylation status of a targeted base bisulfite conversion followed by primer extension with labelled dCTP or dTTP to distinguish methylated and non-methylated DNA.

methylation-specific PCR: This method is based on a chemical reaction of sodium bisulfite with DNA that converts unmethylated Cs to Us thus altering CG dinucleotides to UG, followed by traditional PCR. However, methylated Cs are not converted in this process, and primers are designed to overlap the CG site of interest, which enables the determination of methylation status as methylated or unmethylated.

methylene blue: An important staining agent used to dye chromosomes.

MethyLight assay design: MethyLight assay, a TaqMan® probe-based methylation profiling approach, can be efficiently used for DNA methylation profiling, measuring DNA methylation or detecting methylated alleles in the presence of unmethylated ones.

methylmercuric hydroxide: A powerful denaturing agent that may be incorporated into agarose gels in order to determine the single-strand molecular weight of an RNA or DNA molecule.

methylome: Refers specifically to the distribution of 5-methylcytosone (5^mC) over the entire genome.

methylomes and evolution: A 'methylome' comprises a list of genes and genetic elements that have been switched on or off through a natural process called DNA methylation. The density of possible methylation sites and the distribution of 5-methylcytosine (5^mC) were not uniform in the genome contexts, indicating that the role of DNA methylation in promoters, gene bodies, regulatory features and transposable and repetitive elements can be unusually different. The evolutionary history of DNA methylation in gene bodies and transposons is, in fact, independent.

methylomes, phylogeny of: Human DNA shows dense and genome-wide methylation at CG sites, including methylation of transposable elements, which suppress their transcription and recombination. Unmethylated CG sites are mainly positioned in so-called CpG islands (CGIs), the regions with large CG content. These islands are found in the promoter regions of genes, and if methylated, they repress gene expression. Plants share features with mammalian methylation (e.g. methylation of transposons and lack of methylation at promoters of expressed genes), but they possess three methylation systems specific for CG, CHG and CHH.

methyltransferases (MTases): The enzymes that methylate substrates (e.g. DNA, RNA, histone and non-histone proteins). Enzymes that catalyse highly specific transfer of a methyl group from the co-factor S-adenosyl-l-methionine to the -O, -C or -N position of a target nucleotide or -C of an amino acid.

METI family enzymes: Most likely function as maintenance methyltransferases but may also play a role in *de novo* methylation. *Arabidopsis* METI methyltransferase shows a preference for cytosines in CG sequences.

metric character: A trait that varies more or less continuously among individuals and is therefore placed into classes according to measured values of the trait. *Aka* QUANTITATIVE CHARACTER.

metrical variation: *See* CONTINUOUS VARIATION.

Michaelis constant (K_m): For an enzyme that follows the Michaelis-Menten model, $K_m = (k_1 + k_2)/k_1$; K_m is equal to the substrate concentration at which the reaction velocity is half-maximal (i.e. $V_0 - V_{max}/2$). Therefore, if an enzyme has a small value of k_m, it achieves maximal catalytic efficiency at low substrate concentrations. Different enzymes that act on a single substrate have different km values. The magnitude of km varies widely with the identity of enzyme and the nature of the substrate. It is also a function of temperature and pH. *See* AFFINITY CONSTANT.

Michaelis-Menten hypothesis: A mathematical expression that describes the activity of an enzyme in terms of the substrate concentration [S], the enzyme's maximal velocity (Vmax), and its Michaelis constant (K_m). V_0 = Vmax [S]/Km + [S]. The Michaelis – Menten equation describes the rate of enzymatic reaction as a function of substrate concentration. In this kinetic scheme, the formation of products from ES is a first-order process. Thus, the rate of function of the product can be expressed as the product of the rate constant of the reaction yielding the product and the concentration of its immediately preceding intermediate.

micro-: Prefix denoting 10^{-6}.

microaerophile: An organism that sustains best in the presence of a small amount of oxygen.

microarrays: High-density miniaturized arrays of molecular samples for screening genomic or cDNA samples. It is also called DNA chip technology, having submicrolitre spots on silica chips on glass slides. It can detect the presence of 1 in 100,000 or more DNA sequences. Generally, two types of DNA chips are available: (i) cDNA-based chips and (ii) oligonucleotide-based chips. The former is used for studying gene expression pattern, whereas the latter is generally used for DNA sequencing by hybridization. Microarrays are generally used for the detection of single-nucleotide polymorphisms (SNPs), the detecting of human diseases, functional genomics and proteomics. Generally, an array is planned on the basis of specific sequence information, a process occasionally referred to as DOWNLOADING THE GENOME ON

TO A CHIP. There are several variations of this basic technical theme, which, as a hybridization reaction, may be driven by an electric field. Other detection methods besides fluorescence can be used and the surface may be made of materials other than glass, such as plastic, silicon, gold, gel, membrane or even beads at the ends of fibre optic bundles.

microarray data, statistical analysis of: Microarray data sets are generally very huge, and analytical accuracy is impacted by a number of variables. Statistical problems include taking into account the effects of background noise and a suitable normalization of the data. Normalization methods may be appropriate for definite platforms and, in the case of commercial platforms, the analysis may be exclusive. Algorithms that affect statistical analysis include: (i) IMAGE ANALYSIS: gridding, spot recognition of the scanned image (segmentation algorithm), removal or marking of poor-quality and low-intensity landscapes (called FLAGGING). (ii) DATA PROCESSING: background subtraction (based on global or local background), determination of spot intensities and intensity ratios, visualization of data (*See* MA PLOT), log-transformation of ratios, global or local normalization of intensity ratios. (iii) IDENTIFICATION OF STATISTICALLY SIGNIFICANT CHANGES: t-test, ANOVA, Bayesian method Mann–Whitney test methods tailored to microarray data sets, which take into account multiple comparisons or cluster analysis. These methods evaluate statistical power based on the difference present in the data and the number of experimental replicates and can help minimize Type I and Type II errors in the analyses. (iv) NETWORK-BASED METHODS: Statistical methods that take the fundamental assembly of gene networks into account, representing either associative or causative interactions or dependencies among gene products. (v) MICROARRAY DATA may necessitate further processing aimed at reducing the dimensionality of the data to assist comprehension and more focused analysis. (vi) OTHER METHODS permit analysis of data consisting of a low number of biological or technical replicates; for example, the Local Pooled Error (LPE) test pools standard deviations of genes with similar expression levels in an effort to compensate for insufficient replication.

microarray experimental design: Three main elements are to be considered while designing a microarray experiment: (i) Replication of the biological samples is indispensable for drawing conclusions from the experiment. (ii) Technical replicates (two RNA samples obtained from each experimental unit) aid to guarantee precision and enable testing differences within treatment groups. The technical replicates may be two independent RNA extractions. (iii) Spots of each cDNA clone or oligonucleotide are present as replicates on the microarray slide to deliver a measure of technical accuracy in each hybridization.

microarray technology: A general laboratory approach that involves binding an array of thousands to millions of known nucleic acid fragments to a solid surface, referred to as a 'chip'. The chip is then bathed with DNA or RNA isolated from a study sample (e.g. cells or tissue). Complementary base pairing between the sample and the chip-immobilized fragments produces light through fluorescence that can be detected using a specialized machine. Microarray technology can be used for a variety of purposes in research and clinical studies (e.g. measuring gene expression) studying intertwined interactions among genes and their products and detecting specific DNA sequences (e.g. single-nucleotide polymorphisms, or SNPs). *Aka* BIOCHIPS, DNA CHIPS *and* DNA/RNA CHIP TECHNOLOGY.

microarrays and bioinformatics: Gene expression data from microarray experiments can be represented as heat maps to envision the outcomes of data analysis. The advent of inexpensive microarray experiments created numerous specific bioinformatics challenges: (i) the multiple levels of replication in experimental design (EXPERIMENTAL DESIGN); (ii) the number of platforms and independent groups and data format (STANDARDIZATION); (iii) the treatment of the data (STATISTICAL ANALYSIS); (iv) accuracy and precision (RELATIONSHIP BETWEEN PROBE AND GENE); and (v) the sheer volume of data and the ability to share it (DATA WAREHOUSING).

microarrays and next-generation sequencing: With the advent of microarray and NGS technology, truly high-resolution and genome-wide DNA methylation analysis became possible. BSmooth and Kismeth are the two tools that have thus been developed.

microarrays, fabrication of: Microarrays can be manufactured in different ways, depending on the number of probes under examination, costs, customization requirements, and the type of scientific question being asked. Arrays may have as few as ten probes or up to 2.1 million micrometre-scale probes from commercial vendors. Microarrays can be fabricated using a variety of technologies, including printing with fine-pointed pins on to glass slides, photolithography using pre-made masks, photolithography using dynamic micromirror devices, ink-jet printing, or electrochemistry on microelectrode arrays.

microarrays, salient features of: (i) PARALLELISM – this method permits parallel acquisition and analysis of enormous data in a single reaction. (ii) MINIATURIZATION – this step results in less consumption of DNA probes and reagents. (iii) MULTIPLEXING – multiple samples can be analysed in a single assay. Labelling with multicolour fluorochrome comparison of multiple samples can be accomplished on a single DNA chip. This removes chip-to-chip variation and discrepancies in reaction conditions. (iv) AUTOMATION – advanced manufacturing techniques permit the mass production of DNA chips. The major weaknesses of microarrays include: (a) its cost and necessity for a specialized arraying robot and scanner and (b) the arrays cannot be reused, which further increases the cost.

microarrays, types of: (i) DNA microarrays (e.g. cDNA microarrays, oligonucleotide microarrays and SNP microarrays); (ii) MMChips, for surveillance of microRNA populations; (iii) protein microarrays; (iv) tissue microarrays; (v) cellular microarrays (also called transfection microarrays); (vi) chemical compound microarrays; (vii) antibody microarrays; (viii) carbohydrate arrays (glycoarrays); and (ix) spotting-microarray technology.

microbe: The term was coined in 1878 by M. Sedillot, a colleague of L. Pasteur. A non-scientific and very general term with no taxonomic significance, sometimes used for microscopic organisms that may or may not be disease-causing. *Aka* MICROORGANISM. The term microorganism is applied to the simple, unicellular and structurally similar representatives of the plant and animal kingdoms. With few exceptions, unicellular organisms are invisible to the naked eye and generally have dimensions between a fraction of a micron and 200 microns. Microbiology also studies how microorganisms interact with the environment and other organisms, including their host responses.

microbe becoming pathogen: The four-phased ‘Zigzag’ model explains how a microbe becomes a pathogen. The four phases of this model are: (i) PAMPs/MAMPs are directly recognized by PATTERN RECOGNITION RECEPTORS (PRRs) which leads to PTI and the infection is halted. (ii) Only successful pathogens survive PTI, which means that they need to have EFFECTORS which can interfere with PTI; this results in EFFECTOR-TRIGGERED SUSCEPTIBILITY (ETS). (iii) The used effector is recognized by a SPECIFIC DISEASE RESISTANCE GENE. Most of these genes encode for an NBS-LRR protein. The effect resulting from this is called *ETI*. ETI is basically an accelerated and amplified PTI response, which results in disease resistance. (iv) Pathogens with modified or new effectors have a relative advantage against the host. The hosts that are able to develop resistance against the new or changed effector have an advantage against their conspecifics (belonging to the same species).

microbe-associated molecular patterns (MAMPs): The molecular signals or elicitors from microbes, including pathogens, which are recognized by plant or animal receptors and can influence the innate immunity of the host. *Aka* PATHOGEN-ASSOCIATED MOLECULAR PATTERNS (PAMPs).

microbe-induced growth promotion: These salinity-tolerant plant growth-promoting bacteria (PGPB) assist the plants in withstanding saline conditions. The naturally associated microbiome of plants has the potential to protect the host through stress avoidance, tolerance and resistance strategies.

microbial: Referring to microbes or microorganisms.

microbial diversity: Variety of microorganisms in an environmental sample. Microbial diversity is a general term used to include genetic diversity, that is, the amount and distribution of genetic information, within microbial species; diversity of bacterial and fungal species in microbial communities; and ecological diversity, that is, variation in community structure, complexity of interactions, number of trophic levels, and number of guilds. Microorganisms dominate the biosphere. Traditional methods for culturing microorganisms limit analysis to those that grow under laboratory conditions. Recent research in molecular microbial ecology provides compelling evidence for the existence of many novel types of microorganisms in the environment and comparatively few microorganisms are amenable to laboratory cultivation. The genomes of the total microbiota found in nature, which is termed the metagenome, contain vastly more genetic information than is contained in the culturable subset. Direct analysis of 16S rRNA gene sequences in the environment is widely used to describe the diversity of microorganisms in an environmental sample without culturing. Several PCR-based techniques, including DENATURING GRADIENT GEL ELECTROPHORESIS or TEMPERATURE GRADIENT GEL ELECTROPHORESIS (TGGE), have been developed to rapidly analyse microbial communities basedonthesequence-specificseparationof16SrDNA amplicons. Analysis of stable markers such as the beta subunit of RNase (*rpo*B), gyrase (*gyr*B) and rDNA internal transcribed spacer is also frequently employed for microbial diversity analysis.

microbial genetics: The study of genes and gene function in bacteria, archaea and other microorganisms. Often used in research in the fields of bioremediation, alternative energy and disease prevention. *See* MODEL ORGANISMS, BIOTECHNOLOGY *and* BIOREMEDIATION.

microbial genomics: This area of genomics endeavours to sequence the genomics of bacteria useful in energy production, environmental remediation, toxic waste reduction and industrial processing. This helps in the development of fuel energy sources (biofuels), the detection of pollutants, monitoring of the environment and protection from biological and chemical warfare.

microbial physiology: Deals with the cell structure, growth factors, metabolism and genetics of microorganisms.

microbiology: A life science dealing with the biology, structure, function, classification, physiology and distribution of microorganisms (e.g. viruses, bacteria, algae, fungi, slime moulds and protozoa) and with their technical and medical significance.

microbiome: The collection of all microbes (e.g. bacteria, fungi, viruses and their genes) that naturally live on our bodies and inside us. Although microbes are so small that they require a microscope to see them, they contribute in big ways to human health and wellness.

microbiostatic antimicrobial agents: Those drugs that stay active on a surface in their dry state long after they have been sprayed or wiped on. They adhere to the surface in a monomolecular layer that both attracts and destroys microbes by puncturing through the cell wall.

microcapsules: Heterogenous particles where a membrane shell is surrounding the core, forming a reservoir.

microcell: A cell fragment, containing one to a few chromosomes, that is formed by the enucleation or disruption of a micronucleated cell.

microchromosomes: Tiny dot-like chromosomes that are too small for their centromeres and individual arms to be clearly observed under the light microscope.

microclimate: (i) The atmospheric physiognomies prevailing within a small space eco-macroclimate. (ii) Weather conditions on a small scale (e.g. at the surface of the plant or within a crop).

micrococcal nuclease: An endoexonuclease that preferentially digests single-stranded nucleic acids. The rate of cleavage is 30 times greater at the 5′ side of A or T than at G or C and results in the production of mononucleotides and oligonucleotides with terminal 3′-phosphates. The enzyme is also active against double-stranded DNA and RNA and all sequences will be ultimately cleaved.

microcomputer: The computer that is built with a microprocessor and meant for use by one person at a time. *Aka* PERSONAL COMPUTER.

microconidium (pl. microconidia): The smaller of the two types of conidia formed by some fungi. *Contrasts* MACROCONIDIUM.

microcyclic: Relating to a rust fungus that produces only teliospores and basidiospores. *Contrasts* DEMICYCLIC *and* MACROCYCLIC.

microelectrode, micropipette: Piece of fine glass tubing pulled to a smooth, finer tip. It is used to pierce a cell to study its physiology or to inject electric current or molecules.

micro-electromechanical systems (MEMS): Microscopic devices incorporating both electronic and moving parts. Americans use the term MEMS, while Europeans utilize the term MICROSYSTEMS TECHNOLOGY (MST). *Aka* NANOTECHNOLOGY.

microevolution: Refers to small evolutionary events, typically described as changes in allele frequencies within a species or population. Small evolutionary changes within a species result from the differential survival of the constituent individuals in response to natural selection. The accumulation of microevolutionary changes over a very long period of time can result in macroevolution. *Compare with* MACROEVOLUTION.

microfibres: Tiny fibres composed primarily of actin involved in the structural integrity and movement of eukaryotic cells. Together with microtubules, these fibres make up the cytoskeleton. *Aka* MICROFILAMENTS. *See* ACTIN FILAMENT.

microfilaments: Thread-like organelles involved in cell motion. *See* MICROFIBRES.

microflora: The sum total of all microorganisms in a particular environment.

microgametogenesis: The development of the microgametophyte (i.e. pollen grain) from a microspore. *Compare with* MEGAGAMETOGENESIS.

micrografting: Grafting of plants under *in vitro* conditions. Grafting a scion on an appropriate rootstock is a common method for propagation of several fruit plants. For instance, kinnow is grafted on rough lemon rootstock that is relatively more tolerant to abiotic and biotic stresses. But this technique does not eliminate pathogens from scion cultivars (kinnow). Micrografting is an *in vitro* technique in which the meristem of a scion cultivars (kinnow) is grafted on the rootstock (rough lemon) of a seedling grown *in vitro*.

micrograph: Photograph of an image taken by a light or electron microscope, called a LIGHT MICROGRAPH or an ELECTRON MICROGRAPH depending on the type of microscope employed.

microhomology-mediated end joining (MMEJ): DNA double-strand breaks pose a threat to genome integrity, and during interphase, they are repaired by non-homologous end joining (NHEJ) and homologous recombination (HR). A. Brambati *et al.* (2023) showed that during mitosis, when these pathways are suppressed, mutagenic repair by MMEJ prevails. In MMEJ, DNA ends are joined via microhomologous sequences flanking the breakpoint. Ku proteins are essential in MMEJ *in vivo*. A partner protein named RHINO plays a critical role in restricting MMEJ to mitosis. RHINO accumulates exclusively in the M phase, and its phosphorylation by PLK1 facilitates the recruitment of DNA polymerase μ to condensed chromosomes to promote end joining. Ultimately, mitotic MMEJ ensures that cellular division does not proceed with unrepaired DNA lesions, thus preserving the stability of our genome.

microinjection: A vectorless method for gene transfer in plants. Glass micropipettes (0.5–10 μ diameter tip) are used to inject foreign DNA into protoplasts or cells. The recipient cells are immobilized on a solid support or held by a suction pipette. In general, specially designed equipment called a micromanipulator is used for transformation. Introduction of foreign DNA (gene) into a living cell or protoplast using a micropipette (0.2 μ). Microinjection is a vectorless method for direct gene transfer into cells or protoplasts. It is a device attached to a compound microscope. But this method is very labour-intensive and thus not in common practice. Embryos, oocytes or fertilized eggs can be microinjected with recombinant DNA. The recombinant DNA plasmid can also be injected into the chloroplasts of a leaf using a Galinstan Expansion Femtosyringe (GEF). This method can be used on plants as well as animals.

microlesions: These are those mutations in which only one nucleotide pair is replaced. These changes are too small in size to be observed even under a microscope. Such mutations can be detected by comparing the nucleotide sequence of wild-type and mutant DNA/RNA. When there is a base-pair substitution, the number of bases does not change. Microlesions are best detected through DNA sequencing. *Aka* POINT MUTATIONS *or* GENE MUTATIONS.

microlitre (μl): A unit of measurement of volume. 1 μl is equal to 10^{-6} l; 1 ml = 1000 μl.

micromachining: The technology, tools or methods employed to generate the very small parts, grooves (in chips/arrays), etc. in NANOELECTROMECHANICAL SYSTEMS (NEMS), BIOCHIPS, MICROARRAYS and other devices of nanotechnology.

micromanipulation: Manipulation or surgery performed while viewing the object under a microscope and often carried out with the help of injection or dissection of substances or particles. *Aka* MICROSURGERY *or* MICRURGY.

micromutation: Mutations without visible phenotypic changes, generally observed for quantitative characters.

micronucleated cell: A cell that has been mitotically arrested and in which small groups of chromosomes function as foci for the reassembly of the nuclear membrane, thus forming micronuclei, the maximum of which could be equal to the total number of chromosomes.

micronucleolus: A small nucleolus that arises by nucleolar budding in the process of nucleolar degeneration.

micronucleus: (i) The smaller nucleus, distinct from the larger nucleus, formed during the telophase of mitosis or meiosis by lagging chromosomes or chromosome fragments derived from spontaneous or induced chromosome aberrations. (ii) The smaller of the two nuclei found in ciliate cells. It functions in ways generally similar to the germ cells of animals.

micronucleus test: The micronucleus test is a procedure that employs small animals and has been developed primarily for recognizing chemicals that have chromosome-breaking effect. During anaphase, acentric chromatids and chromosome fragments lag behind and are not included in any of the spindle poles. Nuclear membranes are formed at two poles around the damaged but not undamaged chromosomes.

Chromosome elements, which are not included in any of the regular nuclei, are transformed into one or several secondary nuclei that are much smaller than the principal nucleus. These are called MICRONUCLEI.

micronutrient: An essential element required by plants for growth in very small amounts (i.e. in concentrations < 0.5 mM/l). *Contrasts* MACRONUTRIENT.

microorganism: A living organism too small to be seen with the naked eye. Microorganism are seen under a microscope and include bacteria, fungi, protozoa, viruses and microscopic algae. The branch of science that deals with microorganisms is called MICROBIOLOGY. *See* MICROBE.

microparticles: Spherical particles formed by a polymer matrix with typical sizes of 1–1000 μm. Also refers to the metal particles (R gene gun).

microphotography: Photography requiring optical enlargement.

micropipette: *See* MICROELECTRODE.

microplate capture and detection assay: A hybrid of the DNA-pulldown assay and ELISA, microplate capture assays use immobilized DNA probes to capture specific protein-DNA interactions and confirm protein identities and relative amounts with target-specific antibodies. Typically, a biotinylated DNA probe is immobilized on the surface of 96- or 384-well microplates coated with streptavidin. A cellular extract is prepared in binding buffer and added for a sufficient amount of time to allow the putative binding protein to bind to the oligonucleotide. The extract is then removed and each well is washed several times to remove non-specifically bound proteins. Finally, the protein is detected using a specific antibody labelled for detection. This method can be extremely sensitive when performed with enzyme-labelled antibodies and a chemiluminescent substrate, detecting less than 0.2 pg of the target protein per well. The microplate format is efficient and high-throughput compatible, allowing statistical mutational and activation assays to be performed. This method may also be utilized for oligonucleotides labelled with other tags, such as primary amines that can be immobilized on microplates coated with an amine-reactive surface chemistry.

microplate capture and detection assay, limitations of: (i) Requires antibodies with affinity for DNA-bound native proteins (i.e. supershift antibodies). (ii) Data only provide relative changes in transcription factor-DNA affinity or abundance. (iii) Assay kits are available for only a few targets.

microplate capture and detection assay, strengths of: (i) The use of ELISA-based technology increases speed and throughput compatible with drug screening. (ii) It is possible to optimize sensitive non-radioactive assays.

microprocessor: A semiconductor device manufactured by using either large or very large integration technique. It includes an arithmetic logic unit (ALU), register arrays and control circuits on a single chip.

microprojectile bombardment: High-velocity microprojectile transformation.

microprojectile method: A method of gene transfer where gold or tungsten particles coated with DNA are carried by a macroprojectile and are accelerated into target cells, where they penetrate the cell wall, leaving DNA to be incorporated.

microprojectiles: Microprojectiles have been used to introduce exogenous DNA into intact cells without killing the cells. Microprojectiles (gold or tungsten particles coated with DNA, 1–3 μ in diameter) are carried by a macroprojectile (or bullet) and are accelerated into target cells, where they penetrate the cell wall, leaving DNA to be incorporated. This technique has been successfully used for transformation in soybean, tobacco, maize, wheat, oats and tall fescue. This method is now believed to be universal in its application, as it is being used for gene transfer in plants, irrespective of species and genotype. *Aka* PARTICLE BOMBARDMENT, GENE GUN *or* BIOLISTICS.

microprojectiles method of gene transfer: Small tungsten or gold particles are coated with DNA to be transferred. These are bombarded into cells with the help of a microgun or particle gun.

micropropagation: Propagation of plants through tissue culture, starting from very small tissues or explants measuring less than 1.0 mm. Micropropagation of plants is the best example of commercial exploitation of tissue culture. Micropropagation ensures, true to type, rapid and mass multiplication under disease-free conditions. Starting from one corm, tuber or plant, thousands of plants can be produced in a year. In vegetative propagation, four stages (I-IV) are known. STAGE I deals with the institution of small fragments of stock plants in tissue culture. STAGE II comprises multiplication of propagules through stimulation of branching and subsequent division of shoot clumps on smaller explants, which are then placed on a fresh medium. By repeating this process, the initial number of propagules can be increased a million times in 1 year. STAGE III deals with the preparation of propagules for transfer to normal growing conditions through rooting or elongation of shoots. STAGE IV establishes stage II or III propagules in normal growing conditions – usually in soil or potting mix in a greenhouse.

micropyle: A channel in the coverings of the nucellus through which the pollen tube usually passes during fertilization. Later, when the seed matures and starts to germinate, the micropyle functions as a minute pore through which water enters.

MicroReact: An open-source web tool for visualizing and sharing genomic epidemiology, phylogenetic trees and metadata.

microRNA (miRNA): A tiny piece of RNA, around 21–24 bases in length, which binds to matching pieces of messenger RNA to make it double-stranded and decrease the production of the corresponding protein. MicroRNAs were first discovered in the roundworm *Caenorhabditis elegans* in the early 1990s and are now known to exist in many species, including plants. Functional miRNA is derived from primary microRNAs (pri-miRNAs), which form a stem-loop structure that is processed to form another precursor (pre-miRNA) before being exported to the cytoplasm. Further processing by the Dicer protein generates the mature miRNA, one strand of which is incorporated into the RNA-INDUCED SILENCING COMPLEX (RISC). In general, miRNAs differ from small interfering RNAs (siRNAs) in that they are usually produced from double-stranded RNA (dsRNA). MiRNAs are also defined by having sequences that are only partially complementary to the target mRNAs. Whereas plant miRNAs have extensive complementarity to their target mRNAs, animal miRNAs are much less homologous and, when paired to their targets, form bulges that are proposed to block the translation of the target mRNA. *Compare with* SHORT INTERFERING RNAS (SIRNAS).

microRNA (miRNA) decay pathways: There are multiple ways mature miRNAs decay, as well as degradation mechanisms for their different precursor forms, acting together with transcriptional regulation to control the levels of these small molecules.

microRNA (miRNA) genes: A large class of small non-coding genes that produce microRNAs involved in the regulation of protein production by binding to partially complementary sites on mRNA transcripts.

microRNA (miRNA) precursor: An imperfect stem-loop RNA structure generated from a long transcript, known as primary miRNA, which is encoded by a non-protein-coding region in the genome. Dicer or Dicer-like enzyme cleaves a miRNA precursor and produces mature miRNA.

microRNA (miRNA) prediction: Strategies used for this purpose include: (i) Conservation-based strategies and (ii) Machine-learning strategies.

microRNA (miRNA) processing pathway: MicroRNAs (miRNAs) are short (21–24) noncoding RNAs that regulate gene expression by base pairing with 3′-untraslated regions (3′-UTRs) of their regulated transcripts. miRNA biogenesis involves two main RNase III endonucleases, Drosha and Dicer. Precursor miRNAs (pre-miRNAs) are around 70 nucleotide-long RNA molecules with a characteristic hairpin structure. They originate in longer primary transcripts (pri-miRNAs) that are cleaved in animals by the Drosha endonuclease in the nucleus. Following the export of pre-miRNAs to the cytoplasm by Exportin-5, the loop region of the hairpin is removed by the Dicer endonuclease to produce a short, double-stranded RNA (dsRNA). One of the strands is preferentially incorporated in the RISC, producing a biologically active mature miRNA (generally the −5p miR), while the inactive strand (the −3p miR) is degraded. The coupling of the active miRNA to the 3′ UTR of its target gene facilitates mRNA degradation or translation inhibition. MiRNAs regulate many biological processes and have critical roles in cell proliferation, differentiation and death. The only difference between plant and animal miRNAs is the extent of their complementarity to target mRNAs. Whereas plant miRNAs have extensive complementarity to their target mRNAs, animal miRNAs are much less homologous and, when paired to their targets, form bulges that are proposed to block the translation of the target mRNA. *Compare with* SHORT INTERFERING RNA (SIRNA) PROCESSING PATHWAY.

microRNA (miRNA), biogenesis of: Pri-miRNAs are transcribed from miRNA genes and cleaved into pre-miRNAs. The pre-miRNAs are further processed by DICER-LIKE1 (DCL1) to form an RNA duplex consisting of a 5p and a 3p miRNA. Any of the miRNAs could then be incorporated into AGOs involved in the RISC complex and target single-strand transcripts through base pairing.

microRNA (miRNA)-induced silencing complex (miRISC): RISC with incorporated miRNA.

microRNA-protein (miRNP) complex: An AGO-containing ribonucleoprotein (RNP) complex incorporating an endogenously (miRNA)-derived single-stranded miRNA.

microRNAs (miRNAs) and organismal complexity: The basis of evolution of organismal complexity at a cellular level and the mechanism through which a genome encodes may lie in differences, not in the number of genes an organism has but rather in the regulation of gene expression. Small RNAs have been identified from animal phyla that diverged before the emergence of the bilateria. Animal evolution seems to have been relatively dynamic, with precursor sizes and mature miRNA sequences differing greatly between poriferans, cnidarians and bilaterians. None the less, miRNAs and piRNAs have been available as classes of riboregulators to shape gene expression throughout the evolution and radiation of animal phyla.

microRNAs (miRNAs) as activators during stress: MicroRNAs (miRNAs) are instrumental in a variety of stress responses in addition to their more recognized role in development. miRNAs, which normally suppress the expression of target transcripts, may become activators of expression during stress.

microsatellites: A series of short (2–10 bp) middle-repetitive, tandemly arranged and highly variable (hypervariable) sequences. $(A)_n$, $(AT)_n$, $(ATT)_n$ and $(AATTA)_n$ are such microsatellites. Hypervariability of certain microsatellites leads to human diseases but can also be exploited for the generation of so-called sequence-tagged microsatellite markers. Microsatellites are typically co-dominant. They are used as molecular markers in genetics, for kinship, population and other studies. They can also be used to study gene duplication or deletion. Another common example of microsatellite is a $(CA)_n$ repeat, where n varies between alleles. These markers often present high levels of inter- and intra-specific polymorphism, particularly when the number of repetitions is ten or greater. Microsatellites can be amplified for identification by the PCR process, using the unique sequences of flanking regions as primers. DNA is repeatedly denatured at a high temperature to denature the double strand, then cooled to enable annealing of primers and the extension of nucleotide sequences through the microsatellite. This procedure causes enough DNA production to be visible on agarose or polyacrylamide gels. Only small amounts of DNA are required for amplification because, in this way, thermocycling creates an exponential increase in the replicated segment. With the abundance of PCR technology, primers that flank microsatellite loci are simple and quick to use, but the development of correctly functioning primers is often a complicated and expensive process. Microsatellites are pieces of the same small DNA segment that appear repeatedly in sequence within the DNA molecule adjacent to a specific gene. Thus, these 'microsatellites' are linked to that specific gene. *Aka* SHORT TANDEM REPEAT (STR), SIMPLE SEQUENCE REPEAT (SSR), REPETITIVE SIMPLE SEQUENCE (RSS) *or* SIMPLE REPETITIVE SEQUENCE (SRS).

microsatellite marker: These are tandem arrays of short nucleotide repeats from di- to hexa-nucleotide bases per unit flanked by unique sequences. These are dispersed throughout the genome and are highly variable. These were first independently discovered by two groups – (i) J.L. Weber and P.E. May and (ii) M. Litt and J.A. Lutty – in 1989. These are flanked by unique sequences that are highly conserved. These are PCR-based, and flanking unique sequences are analysed and their complementary primers synthesized. These markers provide co-dominant landmarks that are ubiquitous in the eukaryotic genomes. SSRs have been widely used for fingerprint accessions, genotype identification, varietal protection, diversity analysis, identification of introgressions, pedigree analysis, location of genes and QTLs on chromosomes, marker-aided selection and evaluation of seed purity. SSRs have been developed for a number of crops including maize, rice and wheat. *Aka* SIMPLE SEQUENCE REPEATS (SSRs).

microsatellite polymorphism: Any difference in the number of microsatellite repeats at corresponding genomic loci in two (or more) different genomes. *Aka* SHORT TANDEM REPEAT POLYMORPHISM (STRP).

microsatellite primers, creation of: If searching for microsatellite markers in specific regions of a genome, for example, within a particular exon of a gene, primers can be designed manually. This involves searching the genomic DNA sequence for microsatellite repeats, which can be accomplished by eye or by employing automated tools such as repeat masker. Once the potentially useful microsatellites are resolved by removing non-useful ones such as those with random inserts within the repeat region, the flanking sequences can be utilized to design oligonucleotide primers that will augment the specific microsatellite repeat in a PCR reaction.

Random microsatellite primers can be established by cloning random segments of DNA from the focal species. These random segments are introduced into a plasmid or bacteriophage vector, which is in turn implanted into *Escherichia coli* bacteria. Colonies are then developed and screened with fluorescently labelled oligonucleotide sequences that will hybridize to a microsatellite repeat, if present on the DNA segment. If positive clones have been attained by employing this procedure, the DNA is sequenced and PCR primers are chosen from sequences flanking such regions to determine a specific locus. This process involves significant trial and error on the part of researchers, as microsatellite repeat sequences must be predicted and primers that are randomly isolated may not display significant polymorphism. Microsatellite loci are widely distributed throughout the genome and can be isolated from the semi-degraded DNA of older specimens, as all that is needed is a suitable substrate for amplification through PCR. More recent techniques involve using oligonucleotide sequences consisting of repeats complementary to repeats in the microsatellite to 'enrich' the DNA extracted (MICROSATELLITE ENRICHMENT). The oligonucleotide probe hybridizes with the repeat in the microsatellite and the probe/microsatellite complex is then pulled out of solution. The enriched DNA is then cloned as normal, but the proportion of successes will consequently be much higher, thus drastically reducing the time needed to develop the regions for use. However, deciding which probes to use can be a trial-and-error process in itself.

microsatellite repeat sequences: Sequences of 2–5 bp repeated up to 50 times such as a TA dinucleotide repeat polymorphism in the DNA molecule adjacent to a specific gene. Thus, the microsatellites are linked to a specific gene. The variable number of repeats creates polymorphism. They may occur at 50,000–100,000 locations in a certain genome. The microsatellites mutate faster than nonrepeat polymorphisms and can be used to estimate evolutionary relationships over shorter timescales gene.

microsatellite-primed PCR (MP-PCR): A technique producing RAPD-like patterns after agarose gel electrophoresis and ethidium bromide staining. The MP-PCR technique is more reproducible than RAPD analysis because of its higher strictness.

microsatellites in forensics: The microsatellites in use today for forensic analysis are all tetra- or penta-nucleotide repeats, as these repeats yield a high degree of error-free data while being robust enough to survive degradation in non-ideal conditions. Shorter repeat sequences usually suffer from artefacts (e.g. PCR STUTTER and PREFERENTIAL AMPLIFICATION), as well as the fact that several genetic diseases are linked with tri-nucleotide repeats (e.g. Huntington's disease). Longer repeat sequences suffer more extremely from environmental degradation and do not get amplified by PCR as well as shorter sequences.

microsatellites in gene regulation: Length alterations of microsatellites within promoters and other *cis*-regulatory regions can also change gene expression quickly between generations. The human genome contains many (>16,000) short-sequence repeats in regulatory regions, which provide 'tuning knobs' on the expression of many genes. Length changes in bacterial SSRs can impact fimbriae formation in *Haemophilus influenza* by modifying promoter spacing. Microsatellites in control regions of the *Vasopressin 1a* receptor gene in voles affect their social behaviour and level of monogamy.

microsatellites within introns: Microsatellites within introns also influence phenotype through means that are not currently understood. For example, a GAA triplet expansion in the first intron of the *X25* gene appears to interfere with transcription and cause Friedreich. Tandem repeats in the first intron of the *Asparagine synthetase* gene are linked to acute lymphoblastic leukaemia. A repeat polymorphism in the fourth intron of the NOS3 gene is linked to hypertension in a Tunisian population. Reduced repeat lengths in the EGFR gene are linked with osteosarcomas.

microsatellites within transposons: Microsatellites are distributed throughout the genome. Almost 50% of the human genome is contained in various types of transposable elements (also called transposons or jumping genes), and many of them contain repetitive DNA. It is probable that short-sequence repeats in those locations are also involved in the regulation of gene expression.

microsatellites, features of: Microsatellites are (i) highly variable due to a variable number of tandem repeats (VNTR)-type of polymorphism; (ii) more or less evenly distributed throughout genome; and (iii) probably non-functional and, therefore, selectively neutral. Microsatellites are widely used as genetic markers.

microsatellites, limitations of: Microsatellites are multi-purpose molecular markers, particularly for population analysis, but they have some limitations. Microsatellites established for a particular species can often be applied to closely related species, but the percentage of loci that successfully amplify may, however, decrease with increasing genetic distance. Point mutations in the primer annealing sites in such species may result in the occurrence of 'null alleles', where microsatellites fail to amplify in PCR assays. Null alleles can be attributed to several phenomena. Sequence divergence in flanking regions can cause poor primer annealing, especially at the 3′ section, where extension commences. Preferential amplification of particular size alleles owing to the competitive nature of PCR can result in heterozygous individuals being scored for homozygosity (partial null). PCR failure may be the consequence when particular loci fail to amplify, whereas others amplify more efficiently and may appear homozygous on a gel assay, when they are in reality heterozygous in the genome. Null alleles confuse the interpretation of microsatellite allele frequencies and thus make estimates of relatedness faulty. Mutation in microsatellite alleles is biased in the sense that larger alleles contain more bases and therefore may be mistranslated in DNA replication. Smaller alleles have a tendency to increase in size, whereas larger alleles have a tendency to decrease in size, as they may be subject to an upper size limit. This limitation has been determined, but possible values have not yet been specified. If there is a large size difference between individual alleles, then it may cause increased instability during recombination at meiosis. In tumour cells, where controls on replication may be damaged, microsatellites may be gained or lost at an especially high frequency during each round of mitosis. Hence, a tumour cell line might show a different genetic fingerprint from that of the host tissue.

microscope: An optical instrument having a magnifying lens or a combination of lenses for visualizing objects too small to be seen distinctly by the unaided eye.

microscopic robots: Robots made from DNA molecules can walk, follow instructions and work together to assemble simple products on an atomic-scale assembly line, mimicking the machinery of living cells. These experimental devices are advances in DNA nanotechnology, in which bioengineers are using the molecules of the genetic code as nuts, bolts, girders and other building materials on a scale measured in billionths of a meter. The effort, which combines synthetic chemistry, enzymology, structural nanotechnology and computer science, takes advantage of the unique physical properties of DNA molecules to assemble shapes according to predictable chemical rules.

microscopic-level phosphorylation: It includes chromosomal condensation. H1 phosphorylation has been reported to control the initiation step of mitosis through chromosome condensation.

Microsnoop: A generalist tool for microscopy image representation developed by D. Xun *et al.* (2024). It is a deep learning tool for profiling heterogeneous microscopy images, provides generalist pipelines for processing various types of images, achieves cutting-edge microscopy image representation ability with great potential for expansion, and is highly scalable for studies from small scale to high throughput.

microsome: A small vesicle that results from fragmented endoplasmic reticulum produced when cells are homogenized.

microspheres: Characterized as matrix systems in which the drug is homogeneously dispersed, either dissolved or homogenously suspended.

microsporangium: The smaller of two types of sporangium formed by a heterosporous plant, producing microspores that encompass the male gametophyte. *Compare with* MEGASPORANGIUM.

microspore: An uninucleate, haploid cell that develops into a pollen (male gamete). Microspores are the product of meiosis. In case of plants, pollen mother cell (PMC) undergoes meiosis that leads to the development of four microspores/PMC. The newly formed microspores are thin-walled and subsequently develop into pollen grains surrounded by thick walls (intine and exine) in an uninucleate or binucleate condition. The smaller of two types of spores generated by a heterosporous plant. *Compare with* MEGASPORE.

microspore culture: Refers to the *in vitro* culture of pollen grains to harvest haploid callus or haploid plantlets directly from the pollen grains. Microspore cultures differ from pollen cultures by the stage of development in gametogenesis.

microspore mother cell: One of the many cells in the microsporangium (anther) that experience microsporogenesis to produce four microspores; *Contrasts* MEGASPORE MOTHER CELL.

microsporocyte: *See* POLLEN MOTHER CELL (PMC).

microsporogenesis: The formation of haploid microspores from diploid microspores mother cells (PMC) by meiotic divisions. At the onset of meiosis, the plasmodesmatal connections disappear between tapetal cells and PMC. The pollen mother cell wall becomes very thick due to the deposition of additional callose wall material. During this process, cytoplasmic channels replace plasmodesmatal connections. These channels help in the movement of cytoplasmic contents from one PMC to the other. Due to this, a close synchrony is maintained among the PMC during meiosis and finally, the channels disappear. Each pollen mother cell, after reductional division, produces four pollen grains. The four pollen grains remain inside the common wall and are known as *pollen tetrads*. After the disintegration of the tetrad wall, the spores become free from each other.

microsporophyll: In heterosporous plants, a modified leaf bears one or more microsporangia. *Compare with* MEGASPOROPHYLL.

microstrobilus: The smaller of two types of cones or strobilus formed by gymnosperms, being male and producing the pollen.

microsynteny: Genomic relationships at the gene level.

microtome: A machine for cutting thin slices of embedded tissue. These sections may be stained and examined with a light or electron microscope.

microtubule: A long hollow cylindrical structure composed of the protein tubulin (α and β proteins) with a string-like structure. Some microtubules are the primary structural components of the eukaryotic flagellum. Microtubules function primarily to support and shape the cell. Other microtubules are the 'tow ropes' utilized to move proteins within cells via vesicular transport (vesicles are small hollow structures that contain those protein molecules). They periodically contract to force excess water out of the cell, regulating the cell's osmotic balance. It is one of the three major classes of filaments in the cytoskeleton, the other two being actin filaments and intermediate filaments.

microtubule-associated protein (MAP): Any protein that binds to microtubules and alters their properties. Many different kinds of MAPs have been found, including structural proteins (e.g. MAP-2) and motor proteins (e.g. dynein).

microtubule-organizing centre (MTOC): Bundles of protein tubes (e.g. a centrosome or a basal body) that may be found at the base of a eukaryotic flagellum, from which microtubules grow. In animals, they also function by creating the arrays of microtubules that pull the chromosomes apart during mitosis.

microwave sensors: A microwave radar sensor is a device that uses microwaves to detect the presence of objects. Remote sensing of earth resources using spaceborne sensors is becoming a useful tool for application scientists to provide input to natural resource monitoring and management systems. Sensors operating in the microwave and millimetre wave frequency bands have also been used on a limited scale for a variety of applications. These frequencies have certain unique capabilities (e.g. propagating through clouds, penetrating soils, vegetation cover, etc.). Also, they do not depend on the sun's illumination and therefore can operate during the day and night. Microwave sensors can be broadly classified as PASSIVE and ACTIVE MICROWAVE SENSORS. *See details as separate terms.*

microwave sensors, active: Active sensors illuminate the targets with its own energy. The energy thus scattered in the direction of the sensor is processed to study their physical properties and activities as well as to generate imageries of the targets as a function of their physical and electrical properties like size, shape, conductivity and dielectric constant as well as frequency, polarization and the illumination angle of the sensor. Active microwave sensors are generally divided into two distinct categories: IMAGING and NON-IMAGING. *See details as separate terms.*

microwave sensors, active imaging: The most common form of imaging active microwave sensors is RADIO DETECTION AND RANGING (RADAR), which essentially characterizes the function and operation of a radar sensor. The sensor transmits a microwave (radio) signal towards the target and detects the backscattered portion of the signal. The strength of the backscattered signal is measured to discriminate between different targets and the time delay between the transmitted and reflected signals determines the distance (or range) to the target. The two primary advantages of radar are all-weather and day or night imaging.

microwave sensors, active non-imaging: Non-imaging microwave sensors include ALTIMETERS and SCATTEROMETERS. In most cases, these are profiling devices that take measurements in one linear dimension as opposed to the two-dimensional representation of imaging sensors. Radar altimeters transmit short microwave pulses and measure the round-trip time delay for targets to determine their distance from the sensor.

microwave sensors, passive: All the natural materials emit electromagnetic radiation, which is a complex function of the physical properties of the emitting surface. Passive sensors are radiometers that detect this radiated energy in the microwave spectrum. Useful information regarding the physical properties as well as the activities of the targets can be derived from the energy received. Passive microwave measurements from earth orbits provide global data on a wide range of geophysical and meteorological phenomena.

mid-parent value: In quantitative genetics, the middling of the phenotypes of two mates.

mid-prophase: The meiotic I stage following zygotene when paired chromosomes are somewhat shortened and thickened. This stage, called PACHYTENE, has been studied extensively.

middle lambda (λ) phage genes: *cII*, *cIII* and *Q* are the middle genes that exercise influence on lysogeny and the lytic cycle. *cII* (on the left side) and *cIII* (on the right side) are the genes that control synthesis of repressor by *cl* to enable phages to enter lysogeny. Gene *Q* (on the right side) controls by acting as an anti-terminator and enabling the transcription of late genes meant for lysis. The two pathways – lytic and lysogeny – are so closely related that it is difficult to predict which pathway will be followed and how the decision for alternative pathways is actually exercised. *Aka* DELAYED EARLY GENES.

middle lamella: The layer, consisting chiefly of pectic substances, between the walls of neighbouring plant cells.

middle repetitive DNA: Middle repetitive DNA is found in more than 1000–100,000 copies per haploid genome. Middle repetitive DNA varies from 100–300 bp to 5000 bp in size and can be spread throughout the genome. It is suggested that these sequences are regulatory in nature. Also, they have parasitic DNA. Examples of middle repetitive DNA comprise rRNA and tRNA genes and storage proteins in plants such as maize. Middle repetitive DNAs are of two types with respect to size and frequency of repetitive DNA. Some families of moderately repetitive sequences are spread throughout the genome, while others are bunched in tandem at the centres (centromeres) and ends (telomeres). In addition, there are other members of functional genes that can also be properly positioned in this group (e.g. those for histones, actin, β-globin and immunoglobulins). Moderately repetitive DNA sequences can be transcriptionally active or inactive, in the form of clustered or dispersed repeats. Transcriptionally active clustered repeats are of four types, including rRNA genes, tRNA genes, histone genes and small nuclear RNA (snRNA) genes. Transcriptionally active dispersed-type repeats include retroposons and retrotransposons. Some moderately repetitive DNA sequences match with highly conserved multigene families. Middle repetitive sequences are also found as short and long interspersed elements that are intermingled with unique sequences; these elements are, respectively, known as SINEs and LINEs. In addition to these, some middle repetitive DNA sequences, known as NOMADIC SEQUENCES, are able to migrate from one position to another in the genome to regulate the expression of other genes. *Aka* MODERATELY REPETITIVE SEQUENCES *or* DISPERSED REPEATS.

midiprep double-stranded DNA isolation: A midiprep double-stranded DNA isolation generates a sufficient amount of template DNA for several Sequenase[TM] catalysed fluorescent terminator reactions. Here, one bacterial colony which harboured the plasmid of interest is picked into 3 ml of liquid media containing ampicillin and incubated in a 37°C shaker for 8–10 h. At this time, the culture is transferred into 50 ml of ampicillin-containing media and incubated further for 10–12 h. After harvesting the cells by centrifugation, a diatomaceous earth-based alkalinelysis purification method is performed. The purified DNA is crudely assayed for concentration and purity by agarose gel electrophoresis against known standards.

midipreparation of plasmid DNA: System designed to isolate transfection-quality plasmid DNA. The starting *Escherichia coli* culture volume is 15–25 ml of LYSOGENY BROTH (LB) and the expected DNA yield is 100–350 μg. Plasmid DNA can be purified in as little as 30 min with the vacuum protocol, greatly reducing the time spent on purification.

migration: The movement of an organism from one habitat to another. It may or may not involve a return. The term is used synonymously with GENE FLOW.

migration coefficient: The constant indicating the proportion of individuals migrating from one population to another.

migratory: Moving from one place to another on a plant or from one plant to another when feeding. *Contrasts* SEDENTARY.

mildew: A thin coating of mycelial growth and spores on the exteriors of infested plant parts.

milli: Prefix denoting 10^{-3}.

Millipore™ filter: A disc-shaped synthetic filter with holes of diameter (0.005–8 μ) through its surface.

milpa: Small farmers in Mexico and parts of Central America have practiced a form of agriculture, known as milpa, that traces back to the ancient Maya. Tradition holds that mixing crops—typically the three sister crops, maize, beans and squash—in a single field improves yields. Scientists now know that beans fix nitrogen into the soil, enriching it for all plants; the maize stalks provide a structure for the beans to climb; and the squash covers the ground, suppressing weeds and keeping the soil moist. Researchers also suspected that this 'intercropping' has an additional benefit: helping decrease damage by plant pests. A study showed that planting genetically different tomatoes together only protected them against pests when the set of varieties produced the right blend of volatile chemicals. Thus, it is important which crop variety you are using for intercropping.

MIM number: A unique six-digit number allotted to each entry listed in the catalogue of human genes and genetic disorders, 'Online Mendelian Inheritance in Man' (OMIM). The first digit of a MIM number refers to a gene's mode of inheritance. *Aka MIM#*, OMIM NUMBER *or* MCKUSICK CODE.

mimetics: Compounds that mimic the function of other molecules via their high degree of structural (conformational) similarity and hence physiochemical properties.

mimic: The organism that mimics another organism.

mimic gene: Two or more non-allelomorphic genes that produce similar or identical effects.

mimicry: An adaptive resemblance in which one species appears more or less similar to another species. Some kinds of mimicry result in a selective advantage for concealment and protection from predators. Another type of mimicry enables protection for the mimic through its resemblance to another species that is toxic or in some other way, dangerous. *See* BATESIAN MIMICRY *and* MÜLLERIAN MIMICRY.

mimivirus: Mimivirus is the largest known virus with a 400-nm particle size. It is a double-stranded DNA virus with a 1,181,404-bp genome sequence consisting of 1262 putative open reading frames.

mineralization: The conversion of organic tissue to an inorganic state by decomposition of the organic content.

mini-Ti plasmid: A derivative of the wild-type, tumour-inducing plasmid of *Agrobacterium tumefaciens* in which part of the DNA sequence that is not needed for infection and replication is excised. This leaves a smaller plasmid that may be used as a cloning vehicle for recombinant gene technology in higher plants.

miniature inverted-repeat transposable element (MITE)-mediated DNA methylation and *RAV6* expression: MITEs are short (less than 600 bp), nonautonomous DNA transposons. As the TEs with the highest copy numbers in the rice genome, MITEs mainly occur in the chromosomal arms, especially in the vicinity of genes. The *japonica* rice 'Nipponbare' genome contains 2984 copies of the adh5-like MITE. The adh5-like MITE in the *RAV6* promoter region is 100 bp long and located 198 bp upstream of the *RAV6* gene body. To further explore the evolutionary significance of the epigenetic regulation of *RAV6*, X. Zhang *et al.* (2015) investigated the natural variation of the MITE in the promoter region of the *RAV6* locus in 40 accessions of cultivated rice. The results showed that the TE insertion in the promoter region of *RAV6* is conserved in all cultivated rice accessions. They aligned the sequences of the MITEs, as well as the 5′ region of *RAV6*, in four known assembled rice genomes, including three cultivated rice strains, cv Nipponbare, 93-11 and Kasalath, and the wild rice *Oryza brachyantha*. The MITE insertion was conserved in the cultivated rice genomes but was absent in *O. brachyantha*. Consistent with this, the sequence of the 5′ promoter region of *RAV6* is also more conserved in cultivated rice, especially the six cytosine methylation sites, which are essential for regulation of *RAV6* expression.

miniature inverted-repeat transposable elements (MITEs): Small interfering RNAs (siRNAs) 21–26-nt in size and are often associated with transposon silencing. MITEs are so small that they cannot encode any proteins. The larger transposons that do encode the necessary enzyme and recognize the same inverted repeats are perhaps accountable for copying and moving MITEs to new sites. There are over 100,000 MITEs in the rice genome (representing some 6% of the total genome).

minichromosome maintenance (MCM) proteins: MCM proteins include only the six proteins Mcm2 through Mcm7 (with the recent addition of an Mcm8 in some species). These proteins in the eukaryotic cell that bind to origin recognition complexes in DNA in early G1 and are involved in forming the pre-replicative complex. All these proteins contribute to viability.

minichromosome: A very small chromosome, usually formed as a consequence of chromosome aberrations. Engineered minichromosomes provide an opportunity to enhance crop performance. Unlike conventional gene transformation technologies, minichromosomes can be utilized simultaneously to transfer and stably express multiple sets of genes. As they segregate independently of host chromosomes, they offer a platform for accelerating plant breeding. Approaches for constructing artificial chromosomes may include deliberate engineering of endogenous chromosomes or *de novo* assembly from chromosomal constituents.

minimal genomes: Genomes that carry only the minimal genetic information necessary for life in a given environmental condition. Reduction towards a minimal essential gene set can occur either naturally (for example, in symbionts) or by genetic engineering. The smallest known genome of a free-living Cosmopolitan Oceanic bacterium *Pelagibacter ubique*, reported by S.J. Giovannoni *et al.* (2005), is 1.3 Mb with ~1100 genes. However, significantly more reduced genomes are commonly observed in naturally occurring symbiotic and parasitic organisms.

minimal life: Sequencing by A Goffeau (1995) of the *Mycoplasma genitalium* genome represents the first complete molecular definition of minimal life with 482 genes. The genome is only 580 kb long.

minimal medium: A chemically defined medium that contains only the minimum number of nutrients needed for the growth of the organism. In the case of plant tissue culture (T. Mrashige and F. Skoog, 1962), basal medium is minimal medium. Further, it is supplemented with auxins and cytokinins to obtain the desired growth.

minimal promoter: The minimum size of the promoter is 12 bp. Consensus sequences are made known by aligning all identified examples so as to maximize their homology. Some conclusions in this regard come from work on *Escherichia coli*. The startpoint in >90% of cases is a purine. Just upstream of the start point, a 6-bp section is recognized in almost all the promoters. The centre of the hexamer is usually located close to 10 bp upstream of the start point. This hexamer is, therefore, often called the −10 sequence. This –10 sequence is TATAAT; conservation of these bases varies from 45% to 96%. Capital letters in the sequence T80 A95 t45 A60 and T96 are used to indicate bases conserved, which are more than 75.4%. At the −35 sequence of the promoter and the consensus sequence is T82 T84 G78 A65 C54 A45. The distance between the −10 sequence and the −35 sequence in 90% of cases is 16–18 bp. The sequence of these 16–18 bp *per se* may not be important. Actually, the distance between two sites may be crucial for appropriate separation for the geometry of RNA polymerase. Thus, an optimal promoter is defined as a sequence consisting of the −35 hexamer, separated by 17 bp from the −10 hexamer, lying 7 bp upstream of the start point. A typical promoter can use the −35 and −10 sequences to be recognized by RNA polymerase.

minimal tiling path: Any map or table showing the placement and order of a set of clones (e.g. bacterial artificial chromosome clones) that completely and contiguously cover a specific segment of DNA.

minimal tillage practices: Practices that allow farmers to reduce the tilling of the land in order to conserve topsoil and its nutrients.

minimalist proteins: Proteins are usually large molecules composed of several hundred or even more than a thousand amino acids. Yet the portion of the protein responsible for a specific biological activity is usually concentrated in a small part of the structure. In these circumstances, there are advantages to creating a minimalist protein that has enough structure to retain the desired activity. A smaller protein is often superior for analysing and manipulating structure-function relationships, and the reduced cost of materials for manufacturing a small protein can also be an important motivation for reducing the size. To obtain minimalist proteins with novel properties, libraries with large numbers of different amino acid sequences have been created. These sequences are often so short (less than 50 amino acids) that they are referred to as peptides, oligopeptides, or polypeptides and not as proteins. (A protein is regarded as a long polypeptide, generally at least over 80 amino acids in length and usually much longer.) Those with novel properties, such as the ability to bind to a specific ligand (a unique molecule that is usually small) are then selected. In these instances, the sequences of the peptides are created de novo and are not based on those of a naturally occurring protein.

minimized proteins: The domain/active site of a (former) native protein after all or most of its extraneous (unneeded) portions (peptides) have been removed. In 1995, B. Cunningham and J.A. Wells reduced the 28-residue (peptide) protein (hormone) Atrial

Natriuretic Factor to 15-residue (peptide) size without reducing its potency (biological activity). Minimized proteins that retain their potency hold the potential for medicines possessing a greater serum lifetime (when injected into a patient's body) and as 'models' for the creation of organic-chemical-synthesized mimetic drugs possessing the same therapeutic effect as the native protein did.

minimum descriptor: The minimum number of distinct morpho-physiological features that can effectively discriminate among genotypes for evaluation of germplasm collections.

minimum number of nucleotide substitutions: Changes in amino acid sequences are used to estimate the rate of evolutionary change. Many amino acid substitutions have occurred between the two known groups of organisms. The genetic code dictionary allows us to estimate the minimum number of nucleotide substitutions required for this change. For example, if one protein contains a phenylalanine in position 7 (codons UUU, UUC) and the same protein in a different species has an isoleucine in the same position (AUU, AUC, AUA), we can see that the minimum number of substitutions to convert a phenylalanine codon to an isoleucine codon (UUU→AUU) is one.

mining the epigenetic landscape: Various types of epigenomic data are available in public repositories, epigenomic tools for comparative analysis, and address challenges faced while mining epigenomic data sets.

MinION: A portable nanopore sequencing device produced by Oxford Nanopore Technologies.

miniprep double-stranded DNA isolation: The standard method for the miniprep isolation of plasmid DNA includes the same general strategy as large-scale isolation. However, smaller aliquots of antibiotic-containing liquid media inoculated with plasmid-containing cell colonies are incubated in a 37°C shaker for 12–16 h. After collecting the plasmid-containing cells by centrifugation, the cell pellet is resuspended in a hypotonic sucrose buffer. The cells are successively incubated with an RNase-lysis buffer, alkaline detergent and sodium acetate. The lysate is cleared of precipitated proteins and membranes by centrifugation, and the plasmid DNA is recovered from the supernatant by isopropanol precipitation. The DNA is crudely checked for concentration and purity using agarose gel electrophoresis against known standards.

minipreparation (Abbrev. miniprep): A small-scale preparation of plasmid or phage DNA commonly utilized after cloning to analyse the DNA sequence inserted into a cloning vector.

minipreparation of plasmid DNA: It is a rapid, small-scale isolation procedure of plasmid DNA from bacteria. It is based on the alkaline lysis method. Minipreps are used in the process of molecular cloning to analyse bacterial clones. A typical plasmid DNA yield of a miniprep is 50 to 100 μg depending on the cell strain.

minisatellites: Highly polymorphic DNA markers, including a variable number of tandem repeats that tend to cluster near the telomeric ends of chromosomes. First reported by A.J. Jeffreys *et al.* (1985), their utility through PCR was suggested later. The microsatellites are more evenly dispersed in the genome than minisatellites, which are generally confined to telomeres. For example, a dinucleotide SSR $(CA)_n$ occurs in the human genome, about 50,000 times, with n equal to 10 to 60. The tri- and tetranucleotide repeats are commonly found in the human genome. Multiple copies of short sequences, from 9 base pairs up to several hundred base pairs are dispersed over many sites, with 10–100 repeats, and are highly variable and hence valuable as genetic markers. These are families of about 15 bp repeats forming 0.5 to 30 kb sequences distributed throughout the eukaryotic genome. These show allelic differences in the number of repeats called VNTRs. The occurrence of many highly polymorphic DNA loci simultaneously is responsible for the specific DNA fingerprints of different individuals. Minisatellites in gene regulation are also linked to abundant variations in *cis*-regulatory control regions in the human genome.

minor allele frequency (MAF): The frequency of the rare allele at a given locus. Variants with MAF >5% are typically considered common variants. The frequency can be derived from a number of public variant databases such as dbSNP, 1000 Genomes Project or the NHLBI.

minor gene: A gene that individually exerts an effect on the phenotype that is so small that it is not easily detectable. *See* MODIFIER GENE *or* MODIFYING GENE.

minor gene resistance: Host resistance is governed by a number of genes with small effects.

minor groove of DNA: In DNA, the glycosidic bonds between sugars and bases of given base pairs are not exactly opposite to each other and two grooves (minor and major) of unequal width are moulded around a double helix. The minor groove measures less than 180°. Thus, this is the smaller of the two grooves that spiral around the surface of the DNA double helix.

minor histocompatibility antigens: These antigens act with low intensity and reject the graft slowly.

minor satellite: Tandem repeating AT-rich DNA sequences of 120-bp that are located in the centromeric regions of the mouse genome.

minor spliceosomes: In eukaryotes, 99% of introns (U2 type) are excised by a major spliceosome. The remaining introns (U12 type) are excised by the minor spliceosome. Minor spliceosomes are assembled from U11, U12, U4atac, U5 and U6atac snRNPs. R. Bai *et al.* (2024) reconstituted a fully assembled human minor spliceosome before splicing activation and uncovered how the U11 small nuclear ribonucleoprotein engages the U12-type intron through specific recognition of its 5′-splice site and how the minor spliceosome is put together. The 3.3-Å cryo-electron microscopy structure of the fully assembled human minor spliceosome pre-B complex. The atomic model includes U11 SMALL NUCLEAR RIBONUCLEOPROTEIN (snRNP), U12 snRNP, and U4atac/U6atac.U5 tri-snRNP. U11 snRNA is recognized by five U11-specific proteins (20K, 25K, 35K, 48K and 59K) and the heptameric Sm ring. The 3′ half of the 5′-splice site forms a duplex with U11 snRNA; the 5′ half is recognized by U11-35K, U11-48K and U11 snRNA. Two proteins, CENATAC and DIM2/TXNL4B, specifically associate with the minor tri-snRNP. A structural analysis uncovered how two conformationally similar tri-snRNPs are differentiated by the minor and major pre-spliceosomes for assembly. *Compare with* MAJOR SPLICEOSOMES.

minor splicing pathway: The minor spliceosome is a ribonucleoprotein complex that catalyses the removal of an uncharacteristic class of spliceosomal introns (U12-type) from eukaryotic messenger RNAs. This pathway was first identified and characterized in animals, where it was observed to contain several unique RNA constituents that share structural similarity with and seem to be functionally analogous to the small nuclear RNAs (snRNAs) contained in the major spliceosome. Subsequently, minor spliceosomal components and U12-type introns were reported in plants but not in fungi.

minus end: The end of a microtubule or actin filament at which the addition of monomers occurs least willingly is the slow-growing end of the microtubule or actin filament. The minus end of an actin filament *aka* the POINTED END.

−1 and −3 rule: In mature proteins, amino acids with small residual groups are often located adjacent to the cleavage site (the −1 position) and the next but one upstream residue (the −3 position).

−10 consensus sequence: Consensus sequence (TATAAT) found in most bacterial promoters is approximately 10 bp upstream of the transcription start site. *See* PRIBNOW BOX.

−25 box: A component of the nucleotide sequence that makes up the prokaryotic promoter.

−35 consensus sequence: Consensus sequence (TTGACA) found in many bacterial promoters is approximately 35 bp upstream of the transcription start site.

misalignment generated by DNA palindromes: The crucial aspect of misalignment schemes inducing frameshift mutations is their reliance on correct base pair in an incorrect context. This type of misalignment cannot explain many frameshift and deletion mutations. Many frameshift mutations and deletions do not arise within or between repeated sequences. Certain such mutations can be explained by misalignment generated by DNA palindromes.

misalignment mutagenesis: Spontaneous mutation induction process in which the bases are correctly paired but the pairing occurs out of register.

misdivision: An aberrant chromosome division in which transversal but no longitudinal separation of the centromere occurs. It may result in telocentric chromosomes.

misdivision haploid: Individual product due to the misdivision of the telocentric or isochromosomes in the haploid.

MiSeq: Benchtop genome sequencer produced by Illumina.

misfolded proteins: After insertion into the endoplasmic reticulum, proteins that fail to properly fold there are destroyed. Through a process termed DISLOCATION, such misfolded proteins move to the cytosol, where modifications such as ubiquitination, deglycosylation and finally proteasomal proteolysis dispense with the unwanted polypeptides. Some abnormally folded proteins result from mutations in genes encoding disease-associated proteins (for example, repeat-expansion diseases). Low levels of mischarged tRNAs can lead to an intracellular accumulation of misfolded proteins.

mismatch assay A method for detection of indel mutations following Cas9 cleavage. Targeted genomic DNA is amplified by PCR. The PCR products are melted and reannealed to allow heteroduplexes to form between wild-type and mutant DNA. The hybridized products are then incubated with an enzyme that cleaves heteroduplexes but not perfectly matched DNA. The resulting DNA fragments are analysed by electrophoresis to determine the percentage of cleavage events that result from indel formation.

mismatch correction *in vitro*: *In vitro* assays for mismatch repair have utilized heteroduplex constructs containing a mismatched base pair within the recognition sequence for type II restriction enzyme. Such sites which are resistant to cleavage by endonucleases are rendered sensitive to the enzyme by mismatch correction on the appropriate DNA strand.

mismatch distribution: The distribution of numbers of differences between random pairs of sequences that have been sampled from a population.

mismatch DNA repair: A form of excision repair initiated at the sites of mismatched bases in DNA.

mismatch repair (MMR) in eukaryotes: The general mechanism of MMR is the same in prokaryotes and eukaryotes. MMR in eukaryotic cells serves to maintain genetic fidelity following replication errors in recombination and in genome stability. Homologues of *mutS* and *mutL* are found in yeast and mammalian cells.

mismatch repair (MMR) in humans: Five human genes encoding proteins specifically involved in MMR have been identified. These genes are *MutS* homologues (*hMSH2*), GT-binding protein GTBP/p160 (*hMSH6*) and *MutL* homologues (*hMLH1, hPMS1* and *hPMS2). hMSH2*, in consort with GTBP recognizes various single-nucleotide mismatches and short displaced loops that are created in the intermediate stage following replication slippage. The recognition of the incorrect strand may be methyl-directed or directed by transient nicks in the daughter strand, with the nascent strand being transiently hypomethylated. hMLH1 and hPMS2 dimerize and are coupled to the hMSH2/GTBP heterodimer bound to the DNA strand. Subsequent recruitment of DNA exonuclease and helicase results in unwinding, nicking and excision of the incorrect strand. This is followed by repair with a DNA polymerase using sequences from the parent strand as a template.

mismatch repair (MMR): DNA repair process that corrects mismatched nucleotides inserted during DNA replication. A short stretch of newly synthesized DNA, including the mismatched nucleotide, is removed and replaced with the correct sequence with reference to the template strand. *Aka* STRAND-DIRECTED MISMATCH REPAIR.

mismatch repair (MMR) in prokaryotes: Prokaryotic cells possess mismatch repair systems that are characterized by short excision repair tracts (typically 20 nucleotides or less). They have a relatively restricted specificity with respect to the mismatches that they repair and are operational only within a limited sequence context.

mismatch repair, strand selection in: The adenine residues in GATC sites in DNA are methylated by the DNA-adenine methylase (Dam methylase) in a postreplication DNA modification reaction. Thus, immediately after replication, GATC sites are hemi-methylated (i.e. the adenine in the GATC site in the original), parental DNA strand is methylated, but the adenine in the daughter DNA strand is not methylated. This difference in the presence of a methyl group between parental and daughter DNA strands provides the basis for strand selection for mismatch repair in these organisms.

mismatch score: The penalty allocated by an algorithm when non-identical restudies are aligned.

mismatch-induced stalling of polymerase: A mechanism by which high-fidelity DNA polymerases maintain replication accuracy involves stalling of the polymerase in response to covalent incorporation of mismatched base pairs, thereby favouring subsequent mismatch excision.

mismatching: A region of DNA in a heteroduplex where bases do not pair.

mispairing: The presence in one chain of a DNA double helix of a nucleotide that is not complementary to the nucleotide occupying the corresponding position in the other chain gene.

missense intergenic suppressor: A gene coding for a mutant tRNA that retorts to one or more of the missense mutations.

missense mutation: A nucleotide substitution (point mutation) within a protein-coding region of a gene that results in the replacement of one amino acid by another amino acid. For example, AGU (Ser) → AGA (Arg) or

AGG (Arg). Many, and perhaps most missense mutations, have little or no effect on the function of the protein. Evidence for this comes from the fact that comparison of amino acid sequences from homologous proteins from many species often shows that some regions of a protein are highly conserved, while other regions appear to tolerate a great deal of amino acid substitution. Some missense mutations may cause a radical effect on protein function. *Contrasts* SILENT MUTATION.

missense single-nucleotide polymorphism (missense SNP): Any SNP that occurs in the coding region of a gene and changes the amino acid sequence of the encoded protein.

missense suppressors: A tRNA gene undergoes a mutation in its anticodon region such that it becomes complementary to some other amino acid. For example, UGC (Cys) tRNA has anticodon for Cys but carries Ser (UCC).

missense variants: These variants occur when a missense mutation potentially changes the structure of the protein they encode. Not all missense variants lead to a disruption of downstream function and organism fitness, and therefore missense variants can be further classified as pathogenic or benign, depending on their effect.

missing link: An absent member needed to complete an evolutionary lineage.

mistranslation: There is some error rate for normal translation. This value is estimated to be one codon out of 300 reads. Under conditions of amino acid limitation and tRNA imbalance, the error rate increases.

mitochondria (sing. mitochondrion): Granular or rod-shaped bodies (organelles), whose length averages 2μ, are found in the cytoplasm of cells of most eukaryotes, such as animals, plants and fungi that have a double membrane structure and use aerobic respiration to generate adenosine triphosphate (ATP), which is used throughout the cell as a source of chemical energy for many key metabolic reactions. They are semi-autonomous organelles containing its own DNA and ribosomes and reproducing by binary fission. They contain the enzyme systems needed in the citric acid cycle, electron transport, beta oxidation of fatty acids and the synthesis of ATP via oxidative phosphorylation. The mitochondrion is a site where cellular respiration occurs. Mitochondria have their own genomes, derived from bacterial ancestors (e.g. α-proteobacteria). *See* DOUBLE MEMBRANE.

mitochondria as inducer of apoptosis: The mitochondrion is emerging as a crucial control point in the induction of apoptosis. The release of cytochrome *c* is preceded by changes in the permeability of the mitochondrial membrane. Bcl2 family members act at the mitochondrial membrane, and although their mode of action is not known, one possibility is that they form channels in the membrane. Apoptosis involves localization (or perhaps increased concentration) of Bcl2 family members at the mitochondrial membrane, including Bid (required to release cytochrome *c*) and Bax (perhaps involved in membrane permeability changes).

mitochondria-associated membraneless compartment: Mammalian oocytes stop transcribing DNA into messenger RNA (mRNA) during the final stages of their development. The oocyte's meiotic divisions and early embryonic development occur in the absence of transcription and rely instead on maternal mRNAs that are stored in the oocyte. S. Cheng *et al.* (2022) discovered that mammalian oocytes, including those in humans, store maternal mRNAs around the mitochondria in a membraneless compartment with hydrogel-like properties. The RNA-binding protein ZAR1 drives the assembly of this compartment, which clusters the mitochondria and protects the mRNAs against degradation.

mitochondria-targeting sequences (MTS): The intracellular sorting of newly synthesized precursor proteins (preproteins) to mitochondria depends on the MTS, which are located at the amino termini of the preproteins. The use of MTS is considered to be the most promising candidate for the delivery of DNA to mitochondria when complexed with cationic liposomes, including DQASOMES or POLYETHYLENIMINE (PEI) polymers.

mitochondrial complementation: The heterotic effect expressed in terms of greater mitochondrial activity (ATPase, oxidative activity) of the mixture of mitochondria from two different inbreds than that of the inbreds themselves. This may be a laboratory exercise to identify potential inbred parents for hybridization.

mitochondrial dimethyl sulfate (DMS) mutational profiling with sequencing (mitoDMS-MaPseq): The secondary structure of mRNA stages crucial roles in gene expression. J.C. Moran *et al.* (2024) developed a chemical labelling approach named mitoDMS-MaPseq and used a clustering algorithm to divulge the folding patterns of mitochondrial mRNA (mt-mRNA) in wild-type and leucine-rich pentatricopeptide repeat-containing (LRPPRC) protein-deficient cells. LRPPRC is a protein vital for mt-mRNA maintenance and translation and functions as an mRNA holdase, impacting mRNA folding. LRPPRC is a crucial regulator of mt-mRNA stability, polyadenylation and translation. A high-throughput, genome-wide RNA structure probing strategy that makes use of a high-fidelity and processive thermostable group II reverse transcriptase (TGIRT) enzyme that converts DMS modifications in the RNA to mismatches in the complementary DNA. DMS reactivity information is then utilized as a restriction in RNA folding algorithms based on thermodynamics to obtain highly accurate secondary structure models. Analyses of genome-wide and mt-mRNA-specific folding landscapes revealed mRNA-programmed translational pausing and programmed ribosomal frameshifting. These findings have established a crucial layer in mitochondrial gene expression, providing mt-mRNA folding maps for varied physiological and pathological studies.

mitochondrial DNA (mtDNA): In most eukaryotic cells, mtDNA is double-stranded and circular (as in bacteria) in structure. It replicates independently of nuclear DNA in a semi-conservative manner. Molecular weight = 9×10^6 to 10×10^6; length = 5 μ; size = 14,000 bp. Although mitochondria contain their own DNA, they depend substantially on the nucleus for their structure and function. Because mitochondria are generally carried in egg cells but not in sperm, mitochondrial DNA passes to offspring from mothers, but not fathers. Animal mtDNA is widely used as a genetic marker because it does not recombine. The maternally inherited nucleic acid is found in the cytoplasm whose homologue in plants is chloroplastic DNA. This small circular DNA codes for tRNAs, rRNAs and some mitochondrial DNAs. It is more closely related to bacterial DNA than to eukaryotic nuclear DNA. Mitochondrial DNA mutates 10–20 times faster than nuclear DNA. It is much more abundant than nuclear DNA and this is why most ancient DNA studies use mtDNA.

mitochondrial DNA damage theory of ageing: Irreparable damage to genes does occur in mitochondria also. This organelle serves as the cell's powerhouse. Mitochondria attracted attention because it was observed that the ability of mitochondria to generate energy declined with age. It was suspected that free radicals attacked mitochondrial DNA, as mitochondria are a major source of free radicals in the body. The radicals form when energy-producing machinery in the mitochondria uses oxygen and nutrients to synthesize ATP. These radicals attack and may injure the machinery itself and the mitochondrial DNA that is needed to construct parts of the mitochondria. Thus, radicals may harm mitochondria and cells. Indeed, the rate of DNA oxidation is much higher in mitochondria than in the nucleus. The present theory of ageing is based on the possibility that a relentless assault on mitochondrial DNA may lead to the accumulation of damage and slowly interfere with the functioning of the mitochondria in normal ageing adults, causing ATP production decline. As the cells become starved of energy and damaged, they function less efficiently. This may lead to the impairment of cells. There is evidence that at least one respiratory chain complex falls with age in the brain, skeletal muscles, skin, heart and liver. Further, various rearrangements in the mitochondrial DNA of many tissues, especially the brain, have been observed. *In vitro* studies have shown that cells growing old in culture exhibit similarities with the cells undergoing senescence in the body.

mitochondrial DNA diseases: Refers to human diseases associated with mutations in mitochondrial DNA. Diseases due to single deletions or duplications, multiple deletions or point mutations in mitochondrial DNA are known.

mitochondrial DNA packaging: Mitochondrial DNA (mtDNA) is packed into highly organized structures called MITOCHONDRIAL NUCLEOIDS. A novel mtDNA packaging protein, Glom (a protein inducing agglomeration of mitochondrial chromosomes), was extracted from the highly condensed mitochondrial nucleoids of the true slime mould, *Physarum polycephalum*. This protein binds to the entire mtDNA and packages it into a highly condensed form *in vitro*. Glom is specifically localized throughout the mt-nucleoid. Glom has a lysine-rich region with proline-rich domain in the N-terminal half and two HMG-boxes in the C-terminal half. Deletion analysis of Glom revealed that the lysine-rich region was adequate for the intense mtDNA condensation *in vitro*.

mitochondrial DNA phylogeny: Mitochondrial DNA evolves ten times faster than nuclear DNA and thus is very useful for studying the phylogenetic or genealogical relationships of closely related species. A restriction endonuclease analysis of mtDNA is performed. The genetic similarity index between mtDNA from different organisms is based on the proportion of fragments shared in the digestion profiles. The positions of the restriction sites are mapped in relation to one another or to some known landmarks and this process is called RESTRICTION MAPPING. The digestion profile data of mtDNA are transformed into quantitative estimates of mtDNA.

mitochondrial DNA recombination: mtDNA recombination has long been recognized but only in laboratory experiments and under conditions in which HETEROPLASMY is ensured. Recombination in mtDNA transpires in yeast. There is also some evidence for the occurrence of recombination in human mtDNA.

mitochondrial DNA replication: DNA replication in mitochondria occurs through the formation of D-loops. About 100 min are necessary for the completion of mtDNA replication.

mitochondrial DNA transcription: The basal machinery for the initiation of mtDNA transcription has been molecularly defined.

mitochondrial DNA transfer, issues concerning: The strategy involving mtDNA transfer for handling mitochondrial disease is fascinating, important and ethical, but it certainly crosses the line of engineering genes. The prospect of eradicating mitochondrial disease by replacing the mother's faulty mitochondrial DNA with healthy mtDNA from a donor egg is another issue. The resulting egg would contain the birth mother's nuclear DNA and the donor mother's mtDNA and would be fertilized *in vitro* by the father's sperm – three-parent babies!

mitochondrial DNA transformation: In yeast *Saccharomyces cerevisiae*, genetic transformation of mitochondrial DNA has been reported as a mean to generate predetermined alterations in mitochondrial DNA (mtDNA). Mitochondrial transformation has been reported for single-celled *Chlamydomonas* as well.

mitochondrial epitranscriptome mapping: Mammalian mitochondrial tRNA and rRNA modifications have been investigated.

mitochondrial gene introns: Introns of split genes in fungal mitochondria are classified into two types on the basis of their internal organization. GROUP I INTRONS have been identified in the majority of the fungal mitochondrial split genes. They do not carry any conserved sequence at intron-exon junctions but carry internally a short, conserved sequence called the INTERNAL GUIDE SEQUENCE (IGS). GROUP II INTRONS resemble nuclear genes and have consensus sequences (GT and APy where Py is pyrimidine T or C) and a branch sequence that resembles the TACTAAC box.

mitochondrial genetic code: It is similar to the universal genetic code, with some deviations due to RNA editing. UGA codes for tryptophan (stop codon in standard code), AUA for methionine (isoleucine in standard code), AGA and AGG for termination codons (arginine in standard code), and CGC for tryptophan (arginine in standard code).

mitochondrial inheritance: Mitochondria are inherited from the ova, as sperm do not contain mitochondria. Mitochondria contain their own genome, which is maternally inherited. Alterations in the mitochondrial genome can result in disease.

mitochondrial mRNA structurome: The human mtDNA is a double-stranded 16.569-Kb molecule with heavy (H) and complementary light (L) strands. Transcription distances almost their entire length, producing two long polycistronic transcripts that are administered to yield 11 mature mRNAs, as two transcription units remain unprocessed as bicistronic elements containing overlapping and –1-shifted open reading frames (ORFs) encoding ATP8 and ATP6 or ND4L and ND4.

mitochondrial mutations: Mitochondrial DNA (mtDNA), which is located inside mitochondrial organelles, exists in single or multiple copies of a circular loop without any histone association. Consequently, mtDNA is far more prone to damage than nuclear DNA because it lacks the structural protection afforded by histone proteins. In addition, the highly oxidative environment inside mitochondria that exists due to the constant production of adenosine triphosphate (ATP) makes mtDNA even more prone to damage.

mitochondrial permeability transition: Mitochondria play an important role in energy production, Ca^{2+} homeostasis and cell death. In apoptosis and necrosis, the mitochondrial permeability transition (mPT), which leads to disruption of the mitochondrial membrane and mitochondrial dysfunction, is considered to be one of the key events in cell death. Cyclophilin D-dependent mitochondrial permeability transitions regulate some necrotic but not apoptotic cell death. Loss of cyclophilin D revealed a critical role for mitochondrial permeability transition in oxidative damage-mediated cell death, but not Bcl-2 family member-regulated death.

mitochondrial plasmid: In mitochondria of several higher plants and in some fungi, any of the small plasmid-like DNA molecules (minicircles and mini-linears).

mitochondrial precursor protein: Any mitochondrial protein that is encoded by a nuclear gene, synthesized in the cytosol and eventually transported into mitochondria.

mitochondrial ribosome: A ribosome associated with the mitochondrion having a sedimentation constant of 70S.

mitochondrial RNA: RNA plasmids have been identified in corn mitochondria. Four species of RNA plasmids have been discovered. The first two species are relatively short (900 and 850 nucleotides) and about one-tenth of their concentrations are long (2850 and 2150 nucleotides).

mitochondrial signal sequence: A thread of amino acids that causes a eukaryotic protein to be delivered to a cell's mitochondria.

mitochondrial transcription factors: The initiation of transcription by recombinant human mitochondrial RNA polymerase (POLRMT) is aided by two initiation factors: transcription factor A, mitochondrial (TFAM) and transcription factor B2, mitochondrial (TFB2M) for finer control of experimental components by A.H. Hsieh *et al.* (2023), during transcription by PORMT in a minimalistic system.

mitochondrial transcription initiation: Mitochondria recruit a single-subunit RNA polymerase (RNAP) and one or two auxiliary factors to initiate transcription. Q. Goovaerts *et al.* (2023) solve high-resolution cryogenic electron microscopy structures of yeast mitochondrial RNAP and the transcription factor Mtf1 catalysing two- to eight-nucleotide RNA synthesis at single-nucleotide addition steps. The growing RNA-DNA is accommodated in the polymerase cleft by template scrunching and non-template reorganization, creating stressed intermediates. During early initiation, non-template strand scrunching and unscrunching destabilize the short two- and three-nucleotide RNAs, triggering abortive synthesis. Subsequently, the non-template reorganizes into a base-stacked staircase-like structure supporting processive five- to eight-nucleotide RNA synthesis. The expanded non-template staircase and highly scrunched template in initiation complex IC8 destabilize the promoter interactions with Mtf1 to facilitate initiation bubble collapse and promoter escape for the transition from initiation to the elongation complex (EC). The series of transcription initiation steps, each guided by the interplay of multiple structural components, reveal a finely tuned mechanism for potential regulatory control.

mitochondrial transformation method: DNA can be delivered into yeast mitochondria by microprojectile bombardment, in which DNA coated on microbeads composed of tungsten or gold is introduced into living cells at high speeds. The DNA delivered into mitochondria is subsequently incorporated into mtDNA by the highly active homologous recombination machinery operating in the yeast organelle. This strategy allows introducing the desired mutations into endogenous mitochondrial genes and to insert new genes, inserting new genes into mtDNA and providing a powerful *in vivo* tool for the study of mitochondrial biogenesis in yeast.

mitochondrion phosphorylation: The addition of a phosphate group to a compound, involving the formation of an ester bond between the reactants.

mitogen: A substance that initiates cell division within an organism's body. Mitogens include temperature, most angiogenic growth factors (e.g. fibroblast growth factor), phytohormones (like cytokinin, auxin and gibberellin) and animal hormones (like insulin and steroids).

mitogen-activated protein kinases (MAPKs): These kinases are crucial components for many signalling pathways that regulate cellular activities as diverse as cell motility, gene expression, cell division and programmed cell death. MAPK cascades are well-studied signal transduction systems present in a wide variety of eukaryotes. MAPK signalling pathways transduce signals for processes as diverse as mating, cell proliferation and organ development. Depending on the cellular context, the MAPK cascade may provide a switchlike all-or-none decision between two different responses or a graded response over a wide range of stimulus strengths. MPK kinases (MKKs) are selective in recognizing specific MAPKs and phosphorylate both the tyrosine and threonine residues in the tyrosine/threonine (TXY) motif of the MAPK activation loop. In mammals, MAPKs guide cellular maturation and can induce inflammation and apoptosis. The MAPK family includes extracellular signal-regulated kinases (ERKs), which are activated by mitogens; c-Jun NH_2-terminal kinases (JNKs) and p38 MAPKs, which are primarily activated by cytokines and in response to cellular stress.

mitogen-activated protein kinases (MAPKs), prevention of cross-talk among: With all of these related kinases in the same place, how does the cell keep the pathways distinct from one another? It turns out that many mechanisms exist to prevent cross-talk between pathways, thereby ensuring that the cell responds correctly to each environmental challenge. One way that MAPKs maintain specificity is to physically bind to other proteins in their information cascade through highly specific docking sites. Many MAPKs contain the common docking (CD) site and the Glu-Asp (ED) site; MAPK-interacting proteins have D domains. The structural basis for these precise interactions remained elusive until Chang and colleagues resolved the crystal structure of p38 MAPK bound to the docking domains of a p38 substrate and a p38-activating enzyme. The MAPK signalling cascades transmit signals from the cell surface to its interior. There are at least four MAPK cascades in yeast, and in response to a specific extracellular signal, one of these cascades 'fires'. Each cascade contains three kinases, and sometimes a given kinase is a component of more than one cascade. Several cascades are triggered by a common kinase Ste20.

mitogenomes: Refers to mitochondrial genomes. The mitochondrial genome may include vital data for understanding phylogenetic and molecular evolution.

mitogenomics: The study of the complete genome of the mitochondrion of an organism.

mitomycin C: A family of antibiotics ($C_{15}H_{18}N_4O_5$) synthesized by *Streptomyces caespitosus*. It inhibits DNA replication by cross-linking the complementary strands of the DNA double helix.

mitonuclear communication: The communication of mitochondria with the nucleus by metabolites that are produced by the mitochondria provides the cells with a dynamic regulatory system that is able to respond to changing metabolic conditions.

mitosis: The process of nuclear division in the diploid somatic cells by which two diploid daughter nuclei are formed, each identical to the other and the parent nucleus. Mitosis is a continuous process, but for sake of discussion, it is divided into various phases, namely, prophase, metaphase, anaphase and telophase. Mitosis is generally preceded by the synthesis (S) stage of interphase, when the cell's DNA is replicated. These chromosomes then separate, by division of the centromere, during mitosis, so that one duplicate goes into each daughter nucleus. In contrast to the PROPHASE of meiosis I, the prophase of mitosis does not involve pairing of chromosomes or crossing-over between the homologous chromosomes. During METAPHASE, the individual chromosomes line up at the equatorial plate of the cell and a spindle fibre develops that links each of their chromatids to one of the two poles in the cell. The chromatids of each chromosome separate and move to opposite poles at ANAPHASE. At TELOPHASE, a nuclear membrane develops around the chromosomes to define the nucleus of the cell. Mitosis is followed by cytokinesis, when the cytoplasm and cell membrane are divided into two new daughter cells. It is an equational division in which the chromosome number of the daughter cell remains the same as that of the mother cell. This phenomenon led to the development of the concept of totipotency in plant cells by G. Haberlandt in 1902. *Compare with* MEIOSIS.

mitosis, exiting and entering: Entry and exit from the cell cycle are determined by several processes, like synthesis of nucleoproteins (e.g. cyclins, phosphorylation/dephosphorylation of pre-existing proteins and degradation of proteins like cyclins).

mitosis, significance of: It helps maintain the size of the cell. A cell, when fully grown, divides by mitosis instead of growing further. A fertilized egg produces an embryo, which develops into an adult through repeated mitotic divisions. It keeps the number of chromosomes equal in all the cells of an individual. Mitosis provides new cells to replace the old, worn-out and dying ones. Mitosis produces new cells for the healing of wounds and regeneration. It plays a role in asexual as well as sexual reproduction. Mitosis being essentially similar in different kinds of organisms provides evidence of the basic relationship between organisms.

mitotic (M) phase: Mitosis consists of five morphologically distinct phases: prophase, prometaphase, metaphase, anaphase and telophase. Each phase involves characteristic steps in the process of chromosome alignment and separation.

mitotic analysis: Examination of the mitotic division cycle. The purpose might be to count the number of chromosomes at metaphase or to calculate the rate of cell division, etc.

mitotic apparatus: An organelle consists of three components: (i) the ASTERS, which form the centrosome; (ii) the GELATINOUS SPINDLE; and (iii) the TRACTION FIBRES, which connect the centromeres of the various chromosomes to either centrosome. *See* SPINDLE.

mitotic checkpoint complex: When chromosomes are not attached correctly to the mitotic spindle, a MITOTIC CHECKPOINT COMPLEX (MCC) is formed that prevents chromosome separation by inhibiting the action of the ANAPHASE-PROMOTING COMPLEX/CYCLOSOME (APC/C) UBIQUITIN LIGASE.

mitotic chromosome: Highly condensed duplicated chromosome with the two new chromosomes, which are held together at the centromere as sister chromatids.

mitotic crossing-over: *See* SOMATIC CROSSING-OVER.

mitotic cycle: The sequence of steps by which the genetic material is equally split before the division of a cell into two daughter cells happens.

mitotic death: Cells fatally damaged by ionizing radiation may not die until the next mitosis, at which point the radiation damage to the DNSA becomes evident, particularly when there is fragmentation of chromosomes.

mitotic fidelity: Cyclin-dependent kinase 1 (CDK1) is the only cyclin-dependent kinase (CDK) partner of cyclin B1 for driving mitotic progression. X.-F. Zheng *et al.* (2024) demonstrate that CDK5 is active during mitosis and is essential for maintaining mitotic fidelity. CDK5 is an atypical CDK due to its high expression in post-mitotic neurons and activation by non-cyclin proteins p35 and p392. They specifically repealed CDK5 activity during mitosis, and observed mitotic defects, nuclear atypia and substantial alterations in the mitotic phosphoproteome. Cyclin B1 is a mitotic co-factor of CDK5. CDK5-cyclin B1 form a functional complex. Disruption of this complex phenocopies CDK5 retraction in mitosis. Cyclin B1 thus partners with both CDK5 and CDK1, and CDK5-cyclin B1 works as a canonical CDK-cyclin complex to safeguard mitotic fidelity.

mitotic hypomethylation clock: Global hypomethylation represents a 'mitotic clock' that counts divisions in somatic cells and functions to restrain ageing cells and limit malignant progression. Therapies that modulate the pace of methylation loss or eliminate hypomethylated cells could alleviate ageing-associated diseases or cancers (S.E. Johnstone *et al.*, 2022). DNA hypomethylation arises from inefficient maintenance of CG methylation in late-replicating genomic regions with low CpG density.

mitotic index (MI): MI is generally expressed as the fraction of a total of 1000 somatic cells that are undergoing mitosis in a given sample at one time.

mitotic inhibition: In many systems, exposure to γ-radiation gives rise to, among other effects, the temporary disappearance of mitotic figures from the cell population. The cause of mitotic inhibition is a block in the cell division cycle immediately prior to prophase I.

mitotic nondisjunction: This occurs when sister chromatids fail to migrate to opposite poles of the cell during mitotic anaphase. The result is daughter cells with hyperploid and hypoploid chromosome counts.

mitotic poison: Any substance that interferes with proper mitosis.

mitotic recombination: Mitotic recombination between homologous chromosomes can lead to mitotic segregation, where heterozygous loci segregate into somatic daughter cells, generating homozygous clones identifiable by their distinct phenotypes. Mitotic recombination may be a by-product of the DNA repair process. Mitotic recombination between homologous chromosomes can be used to derive genetic maps. Some developmental processes (e.g. immunoglobulin gene rearrangements) require recombination in somatic cells to make an immunoglobin gene ready for transcription. Mitotic recombination serves as a mechanism for the generation of variation in heterokaryons. *Aka* SOMATIC RECOMBINATION.

mitotic spindle: A bipolar structure composed of microtubules (MTs) and associated motor proteins segregates sister chromatids from daughter cells. Initially, some MTs emanating from one centrosome attach to the kinetochore at the centromere of one of the duplicated chromosomes. This attachment allows rapid poleward movement of the bound chromosome. Subsequent attachment of the sister kinetochore to MTs growing from the other centrosome results in the bi-orientation of the chromosome, in which interactions between kinetochores and the plus ends of MTs are formed and stabilized. These processes ensure the alignment of chromosomes during metaphase and their correct segregation during anaphase. Small GTPases of the Rho family regulate cell morphogenesis by organizing the actin cytoskeleton and regulating MT alignment and stabilization. Cdc14 promotes localization of microtubule-stabilizing proteins to the anaphase spindle and dephosphorylation of the kinetochore component. Ask1 contributes to both the silencing of microtubule turnover and successful anaphase A. During unequal cell divisions, a mitotic spindle is eccentrically positioned before cell cleavage. It has been found that the force imbalance is due to a larger number of force generators pulling on the astral microtubules of the posterior aster. Moreover, activation of a heterotrimeric guanine nucleotide-binding protein (G protein) α subunit is required to generate these astral forces.

mitotic stopwatch: Mitosis is a challenging time for cells owing to the potential for structural and numerical chromosomal abnormalities, termed chromosomal instability, which is a trademark of cancer. In M phase, chromosomal instability is prohibited by the spindle assembly checkpoint, which postpones cell cycle progression until chromosomes are properly committed to spindle microtubules, ensuring proper separation into daughter cells.

mitotic-meiotic switch: Refers to critical biochemical events in the cell that lead to a shift from their mitotic to meiotic behaviour.

mixed families: Groups of four codons sharing their first two bases and coding for more than one amino acid.

mixed farming: A type of farming that involves both the growing of crops and the raising of livestock. For example, a mixed farm may grow cereal crops (e.g. wheat or rye) and also keep cattle, sheep, pigs or poultry. Such agriculture occurs across Asia and in countries such as India, Malaysia, Indonesia, Afghanistan, South Africa, China, Central Europe, Canada and Russia. Though in the beginning, it mainly served domestic consumption, countries such as USA and Japan now use it for commercial purposes.

mixed strategy: A probability distribution someone uses to randomly choose among available actions in order to circumvent being predictable. In a mixed strategy equilibrium, each player in a game uses a mixed strategy, one that is best for him against the strategies the other players are using.

mixed-function oxygenases: Enzymes catalysing concurrent oxidation of two substances by oxygen, one of which is usually NADPH or NADH.

mixoploid: Cell populations in which different cells reveal different chromosome numbers.

mobile CRISPRi: A high-throughput method designed by K. Rachwalski *et al.* (2024) to screen genetic interactions in bacteria using a conjugative CRISPR INTERFERENCE (CRISPRi) plasmid. The method enables systematic studies of gene essentiality in diverse genomic and environmental contexts and is applicable to *Escherichia coli* as well as other bacteria. The mobile CRISPRi collection combines the benefits of an ordered CRISPRi collection with the advantage of being compatible with existing genetic collections, including plasmid-borne collections (e.g. transcriptional reporter or protein overexpression collections). Similar to other ordered CRISPRi collections, it enables systematic studies of the phenotypic consequences of silencing essential genes while avoiding interactions via secreted metabolites and proteins that may occur in pooled CRISPRi screens. The co-expression of dcas9 and sgRNA from the same conjugative plasmid makes the approach versatile and compatible with high-throughput genomic experiments.

mobile dispersed genetic element: *See* TRANSPOSON, MOBILE GENETIC ELEMENT *and* TRANSPOSABLE GENETIC ELEMENT.

mobile restriction-modification system: This system provides phage defence and resolves an epigenetic conflict with an antagonistic endonuclease. Such systems are frequently mobile, influencing horizontal gene transfer by altering the compatibility of the host for foreign DNA uptake.

mobile RNAs: Small RNAs are passed as consignment in EXTRACELLULAR VESICLES (EVs) but can also be found in body fluids outside EVs. Maintenance of their stability outside EVs necessitates binding proteins, specific secondary structures or RNA modifications, which might also permit their transport into and out of cells.

mobile silencing: Maintenance of epigenetic effects in plants is effective if the initiator stimulus is applied to the meristems. The silenced state can then be passed to new cells, tissues, and organs as they arise. In vernalization, for example, the cold stimulus is perceived in meristems and the silenced state of FLC is propagated with the replication of chromatin as the cells divide. There are also instances in which initiation and maintenance are in separate cells and there are mobile signals that move between cells. Mobile signals were illustrated by experiments in which shoots from one genotype of plant were grafted on to another. One illustrative experiment was based on plants in which a GFP target transgene was expressed under the control of a meristem-specific promoter and enhancer, and in which a silencer transgene was designed to produce siRNA to target the necessary enhancer component. The target transgene was methylated at C residues and silenced in root meristems if they were grafted to shoots carrying the silencer. However, this promoter

DNA methylation and silencing did not occur if the silencer shoots were defective for the 24-nt siRNA-specific Dicer. Correspondingly, the 24-nt siRNAs were present in the grafted root if the shoot was from the wild type but not the Dicer mutant shoot. 24-nt siRNAs are the mobile signals in plants and they can initiate epigenetic effects in recipient meristems.

mobilization: The transfer of genes from one place to another (in the same or a different cell or organism) mediated by a retrovirus or transposable element.

modal number: The chromosome count, which is characteristic of cells from a particular individual, cell line or tumour cell population. The modal number may be expressed as a range.

mode: Mode refers to the value in a database that repeats the greatest number of times. If none of the values repeats, there is no mode in that database.

mode of action: The mechanism by which a pesticide, drug, or other biologically active chemical performs its functions.

mode of action of specification: The basic axes of the embryo are determined by maternally supplied positional amplification, specifically anterior versus posterior and dorsal versus ventral. The first event involves the partitioning of the presumptive germ band into a group of large domains, some of which are determined by gap loci. The pair-rule genes further subdivide these large regions into domains. This is followed by the action of the homeotic genes.

mode of reproduction: Two general modes of reproduction are distinguished: (i) sexual and (ii) asexual. SEXUAL REPRODUCTION involves the union of male and female gametes derived from the same or different parents. ASEXUAL REPRODUCTION ensues by multiplication of plant parts or by seed production that does not involve the union of sexual gametes. The breeding procedures are fundamentally dependent on the mode of reproduction.

model: A model of an object is a physical representation that shows what it looks like or how it works. The word model is reserved for situations where it is known that the hypothesis has at least limited validity. A model could be mathematical, graphical, structural, etc. We can create or design objects that represent other things. Well-known examples of models are: (i) double-helical model of DNA proposed by J.D. Watson and F.H.C. Crick (1953); (ii) the F. Jacob-J. Monod model of gene regulation in prokaryotes; and (iii) the R.J. Britten-E.H. Davidson model of eukaryotic gene regulation.

model building: An experimental approach in which possible structures of biological molecules are assessed by building scale models of them.

model organisms/systems: Selection of the lowest level of organism/system that will allow researchers to answer a research question. They reproduce quickly in large numbers, are easy to handle and represent particular experimental properties not able to handle experientially with human research systems. Most model organisms have a known genome sequence. Among bacteria, *Escherichia coli* has been used intensively over a long period and thus serves as a 'model' for the study of the biology of a particular type of organism. Examples of eukaryotic organisms used in classical genetics, molecular genetics, developmental genetics and epigenetic studies are mouse *Mus musculus*, the worm *Caenorhabditis elegans*, the budding yeast *Saccharomyces cerevisiae* and fission yeast *Saccharomyces pombe*, mouse-ear cress plants *Arabidopsis thaliana* and the fruit fly *Drosophila melanogaster*.

modelling: (i) Generally, modelling is represented in the form of a series of mathematical equations or procedures that simulate a real-life process given a set of assumptions, boundary parameters and initial conditions. (ii) In bioinformatics, the term refers to molecular modelling, a process whereby the three-dimensional architecture of biological molecules is interpreted, predicted, visually represented and manipulated in order to govern their molecular features. (iii) The use of statistical analysis, computer analysis or model organisms to predict the outcomes of research.

models of protein degradation: Two well-known models of protein degradation include the N-END RULE and the PEST HYPOTHESIS. *See details as separate terms.*

moderate-throughput quantitative methylation analysis by quantitative-MSP: This approach allows measurement of the degree of methylation by real-time PCR of bisulfite-converted genomic DNA. This strategy has an experimentally proven sensitivity capable of detecting one methylated DNA molecule in a background of ten thousand unmethylated ones.

moderately repetitive sequences: When the number of copies of a sequence in eukaryotes is 10^3 to 10^5. These sequences are interspersed with unique sequences and are also called dispersed repeats.

modern biotechnology: Used to distinguish newer applications of biotechnology (e.g. genetic engineering and cell fusion) from more conventional methods (e.g. breeding or fermentation). Most often, the term BIOTECHNOLOGY is used interchangeably with MODERN BIOTECHNOLOGY.

modern farming practices: Farming practices that maximize the amount of production per unit (either per acre or per animal) while conserving soil and water resources. May include the use of modern government-approved aids (e.g. fertilizers, insecticides, herbicides and antibiotics), which undergo extensive safety testing before approval. *See* ANTIBIOTICS, HERBICIDES, INSECTICIDE *and* PESTICIDES.

modern genetics-based tools for insect pest management: Various such tools are sterile insect technique (SIT), using insects homozygous for a repressible dominant lethal, engineering insect pests, transgenic crops expressing insecticidal toxins, primary resistance management method, gene silencing (transcriptional and post-transcriptional gene silencing/RNA interference), genome-editing-based methods (gene drive technology and mutagenesis through gene editing), autocidal insect control techniques, genomics approaches and using microbes for precision insect pest control that have been developed with a goal of managing agricultural crop pests.

modern synthesis of the process of development: The modern synthesis would involve supplanting embryology with genetics. Embryology was left out of the modern synthesis. Cells differentiate according to which subset of their total complement of genes is expressed. Master control genes are genes that initiate differentiation pathways. These genes encode transcription factors that activate a cascade of gene expression, resulting in differentiation. Master control genes can be activated by induction (a signal from another cell) or by autonomous regulation (inheritance of a determinant – usually a mRNA that encodes a master control gene product). Evolutionary developmental biology (evo-devo, devo-evo or EDB) seeks as a discipline to identify those developmental mechanisms that bring about evolutionary changes in the phenotypes of organisms. The origins of evo-devo are multiple, and evolutionary developmental mechanisms are many and varied, reflecting the hierarchical organization of organisms and the many levels at which evolutionary change can occur. Evo-devo reflects a long search to find and understand relationships between the transformation of an organism within a single generation (development, ontogeny, ontogenetic change) and those transformations that occur between generations (evolution, phylogeny and phylogenetic change).

modern synthesis of theory of evolution: Combining natural selection and Mendelian inheritance into a synthesis of a novel theory of evolution. This neo-Darwinian theory of evolution was introduced by a number of evolutionary biologists, such as T. Dobzhansky, J.B.S. Haldane, R.A. Fisher, Sewall Wright, G.L. Stebbins, Ernst Mayr. *Aka* NEO-DARWINISM.

modes of selection: Various modes of selection in natural populations are directional, disruptive, diversifying, balance, frequency-dependent (density-dependent), normalizing and stabilizing (centripetal).

modifiability: The capability of phenotypic variation of a particular genotype to respond to varying environmental conditions.

modification enzyme: An enzyme that recognizes the same nucleotide sequence as a corresponding restriction endonuclease and methylates certain bases within the sequence, thus conferring protection from the endonuclease activity.

modification of bases in DNA/RNA: This includes all changes made to the bases in nucleotides after their initial incorporation into the polynucleotide chain. All the bases except the usual four in DNA (T, C, A, G) or RNA (U, C, A, G) are synthesized; they result from postsynthetic changes in nucleic acids.

modification of DNA/RNA: This includes all changes made to the nucleotides after their initial incorporation into the polynucleotide chain.

modified bases: All those except the usual four bases in DNA (T, C, A and G) and RNA (U, C, A and G) are synthesized; they result from postsynthetic changes in the nucleic acids.

modified central dogma of molecular biology: Crick's suggestion of central dogma (DNA → RNA → protein) was modified with the emerging knowledge that in some plant viruses, RNA is the genetic material where genetic information flows in reverse direction (RNA → DNA) with the help of an enzyme called RNA-DEPENDENT DNA POLYMERASE, *aka* REVERSE TRANSCRIPTASE. Another modification in Crick's central dogma was that RNA can self-replicate. Protein cannot self-replicate, nor can it use amino acid sequence information to reconstruct RNA or DNA. However, under laboratory conditions, DNA can directly be used as a template for translation. All these points were added to the classical central dogma to have the 'modified central dogma of molecular biology'.

modified nucleosides in origin of life: Compartmentalization of RNA in biopolymer-rich membraneless organelles (coacervates) is pervasive and critical for the function of extant biology, which has been suggested as a prebiotically plausible way to accumulate RNA. However, compartment-RNA interactions that drive encapsulation have the potential to influence RNA structure and function in compartment- and RNA sequence–dependent ways. M.O. Meyer *et al.* (2023) performed NGS experiments in coacervates to characterize the fold of many different transfer RNAs (tRNAs) simultaneously under the potentially denaturing conditions of these compartments. Natural modifications favour the native fold of tRNAs in these compartments, suggesting that covalent RNA modifications could have played a critical role in metabolic processes at the origin of life.

modifier gene: A gene that affects or moderates the expression of another gene. Modifier genes can affect the penetrance, pleiotropy, dominance and expressivity of a trait. *Aka* MODIFYING GENE.

modifying gene expression: In biology, modification means: (i) A nonhereditary change in an organism (e.g. one that is acquired from its own activity or environment). (ii) A chemical or structural alteration in a molecule, typically a macromolecule (e.g. a DNA, RNA) or protein molecule, by chemical means; often, the covalent addition by some reagent or chemical group. Such modification may result in a change in behavioural, chemical or physiological properties of that macromolecule.

modifying single-cross hybrids: As a substitute for using an inbred line as the female parent, sisterlines (SLs), the F_1 between two highly related inbred lines (A3A*), have been utilized in corn seed production. Hybrids thus produced using SLs are referred to as modified single-cross (MSC) hybrids.

MODOMICS: A database of PTCMs that offers comprehensive information about the chemical structures of modified ribonucleosides, their biosynthetic pathways, RNA-modifying enzymes and the location of modified residues in RNA sequences.

modular enzymes: An assembly having several component enzyme domains. A molecular enzyme is a multi-component system. The order of enzymatic domains specifies which monomer substrates are activated, condensed and elongated. Individual components can be changed to obtain products with the altered desired structure. Polyketide synthase (PKS) and non-ribosomal peptide synthetase (NRPS) are examples of modular enzymes.

modularity: In biology, modularity refers to the assembly of living organisms into distinct parts that function together to ensure individual and species' well-being. For example, plant hormones can be classified into different modules, each with a specific purpose in the reproduction process.

modulating codon: DNA-coding triplets (codons) that code for rare transfer RNA (tRNA) molecules. Since the tRNA that carries the appropriate amino acid to the mRNA is in short supply in the cell, the presence of such codons tends to slow down the translation of the mRNA containing them.

modulation of plant cell proliferation: G PROTEIN-COUPLED RECEPTORS (GPCRs) at the cell surface activate heterotrimeric G proteins by inducing the G protein alpha (Gα) subunit to exchange guanosine diphosphate for guanosine triphosphate. Regulators of G protein signalling (RGS) proteins accelerate the deactivation of Gα subunits to reduce GPCR signalling. An RGS protein (*At*RGS1) in *Arabidopsis* has a predicted structure similar to a GPCR as well as an RGS box with GTPase accelerating activity. Expression of *At*RGS1 complemented the pheromone supersensitivity phenotype of a yeast RGS mutant, *sst2Δ*. Loss of *At*RGS1 increased the activity of the *Arabidopsis* Gα subunit, resulting in increased cell elongation in hypocotyls in darkness and increased cell production in roots grown in light. Thus, *At*RGS1 is a critical modulator of plant cell proliferation.

module: (i) In proteins or nucleic acids, a unit of structure or function that occurs in a variety of different contexts in different molecules. (ii) In computer science, a software package that can readily be used by other programmers. Modules can comprise methods, classes, constants and other modules. The term MEMBER or METHOD is used to signify something that a module contains. Some modules are written in Pike, but C and C++ can also be used for this purpose.

moiety: Referring to a part or portion of a molecule, generally complex, having a characteristic chemical or pharmacological property.

moisture stress tolerance: *See* DROUGHT TOLERANCE.

molal: A solution that contains 1 molecular weight of solute in a kg of solvent (abbreviated as 1m).

molar: One mole of solute dissolved in solvent (water) to give a total volume of 1000 ml (1 l). For example, one molar solution of potassium hydroxide (KOH) is prepared by dissolving 56 g of KOH (39 + 16 + 1) in about 250 ml of water and then adding more water to get the final volume of 1000 ml. Whereas to prepare one molar solution of sodium hydroxide (NaOH), 40 g of NaOH (23 + 16 + 1) is dissolved in about 250 ml of water and a final volume of 1000 ml is made by adding more water.

mole: (i) Amount of chemical equal to the atomic weights of all the atoms in a molecule of the chemical. The term mole is used to relate the molecular level to the laboratory level. One mole of the substance refers to its molecular weight expressed in grams. For instance, 1 mole of water weighs 18 g because the molecular weight of H_2O is 18 [(2 × 1) + 16]. (ii) The amount of a substance that contains exactly $6.02214076 \times 10^{23}$ 'elementary entities' of the given substance. The number $6.02214076 \times 10^{23}$ is popularly known as the AVOGADRO CONSTANT and is often denoted by the symbol 'N_A'.

molecular assembler: K. Eric Drexler (1991) defines molecular assembler as a 'proposed device able to guide chemical reactions by positioning reactive molecules with atomic precision'. A molecular assembler is a kind of molecular machine. Some biological molecules such as ribosomes fit this definition. This is because they receive instructions from messenger RNA and then assemble specific sequences of amino acids to construct protein molecules. *Aka* ASSEMBLER.

molecular barriers to the development of interspecific hybrids: Although the exact nature of molecular barriers to the development of interspecific hybrids is not known, the phenomenon of allelic repression occurring in hybrids shows that regulatory barriers may be very important. Evolutionary loss of the ability of two species to hybridize appears to result from the accumulation of incompatibilities between genes during embryonic development. Mammals appear to have certainly undergone both rapid regulatory evolution and rapid reorganization of genes. Thus, gene rearrangement offers an important method of accomplishing new patterns of regulation.

molecular beacon: This term refers to specific oligonucleotides possessing a 'hairpin loop' and bearing a fluorescent dye. A QUENCHER DYE located on a nearby portion of the hairpin loop thwarts fluorescence until the hairpin loop is opened up. Molecular beacons (*aka* FLUOROGENIC PROBES) are utilized (e.g. in high-throughput screening or identification) to perceive the presence of a desired TARGET molecule that possesses the desired functional group or desired property. When the target is present within a given sample being evaluated, the 'hairpin loop' opens up because a portion of it makes a stronger bond to the 'target', than to the rest of the loop thereby enabling the fluorescent dye to emit light.

molecular beacon assay: In this assay, an oligonucleotide probe (called molecular beacon) is used, which is made of the target SNP sequence, with its two ends being complementary to each other. The two ends of this molecular beacon are labelled: 5′-end with REPORTER and 3′-end with QUENCHER. When the probe fails to make a duplex with the template DNA due to the presence of SNPs, it results in the formation of a hairpin structure due to the self-annealing of its two ends, thus quenching the reporter. But when the probe anneals with the template, it gets linearized, thus separating the reporter from quencher and allowing fluorescence signal (as digestion will occur in PCR). These fluorescent signals can thus be detected by appropriate sensing devices.

molecular beam epitaxy (MBE): Process used to construct compound semiconductors involving the deposition of multiple layers, one by one.

molecular biology: Molecular biology emerged as a distinct scientific discipline largely as a result of a decision to provide (i) support for the application of new physical and chemical methodologies to biology during the 1930s by W. Weaver, Director of the Biology (Funding) Program at America's Rockefeller Foundation and (ii) platform for the study of the structure, function, and interaction between proteins and nucleic acids in biological systems. The term molecular biology, actually coined by V. Bush during the 1940s, referred to the study and manipulation of molecules that institute or interact with cells. Thus, molecular biology deals with the structures and functions of different nucleic acids, amino acids and proteins and their modifications and manipulations to improve the characteristic of living organisms for the welfare of mankind.

molecular breeding: The application of molecular biology tools, often in plant breeding. The areas of molecular breeding include QTL mapping or gene discovery, marker-assisted selection and genomic selection and genetic engineering. This term also sometimes refers to the use of molecular genetics methods to select the organisms (e.g. crop varieties) that possess gene(s) for a particular trait (higher yield, disease resistance, etc.).

Molecular Breeding™: A trademarked term that refers to certain 'molecular evolution' skills developed by Maxygen Company.

molecular chaperone: A molecular chaperone is a protein that is required for the assembly or proper folding of some other protein but which is not itself a component of the target complex. The molecular chaperones provide a structure on which proper folding of proteins occurs. The chaperones do not seem to provide the three-dimensional structure of the proteins they help, but rather the chaperones seem to bind to a protein during its early stages of folding and prevent unproductive folding. Thus, molecular chaperones enable proteins to find a functional, stable state by providing the proteins with opportunities to fold into a thermodynamically stable and functional state. Each cycle of refolding requires ATP energy. A well-studied class of chaperones is known as CHAPERONINS. *See* CHAPERONE.

molecular clock hypothesis: The approximately constant rate of nucleotide or amino acid substitutions for particular genes and classes of genes within particular lineages. The difference between the structure of a molecule in two species is then proportional to the time since the species diverged from a common ancestor and molecules became of great importance in the inference of phylogeny.

molecular cloning: DNA segments of different sizes of prokaryotic or eukaryotic origin are identically multiplied as a part of bacterial plasmids or phages. The alien DNA is incorporated into the host DNA ring molecule. Through the rapid division of the host cells, the alien DNA segments are simultaneously cloned. Subsequently, those cloned DNA segments can be excised, separated and purified for further utilization. The process of *in vitro* recombination makes it possible to cut different strands of DNA, *in vitro* (outside the cell), with a restriction enzyme and join the DNA molecules together via complementary base pairing. *Aka* DNA CLONING *or* GENE CLONING. *See details as separate terms.*

molecular combing and fluorescent *in situ* hybridization: The method developed in collaboration with J.W. Gray's group (University of California) improves the mapping measurements both qualitatively and quantitatively. It is based on molecular combing, which allows a regular stretching of the target DNA. Protocols used in standard fluorescent hybridization assays work perfectly in these new conditions. This technique yields physical mapping results whose precision is in principle only limited by the optical resolution (i.e. roughly 0.5 microns) corresponding to approximately 1000 bases.

molecular contact memory: A potential future memory technology that stores data at the molecular level. Using lasers and the bacteriorhodopsin protein molecule, scientists have built a molecular memory device holding several hundred megabytes. The advantage of such a memory is its small size and stability. It holds its content without power.

molecular diversity: Alternatively, sometimes referred to as IRRATIONAL DRUG DESIGN, this term refers to the drug design methodology of creating large numbers of diverse candidate molecules (e.g. pieces of DNA, RNA, proteins, or other organic moieties) at random (via a variety of molecular biology methods). These diverse candidate molecules are then tested to find the best for working against a disease/condition (e.g. fitting a cell receptor or category of receptors relevant to the disease in question). Molecular candidates that display potential (e.g. via a 'pretty good fit' to receptor) are then manufactured in larger quantities (e.g. via polymerase chain reaction techniques) along with additional molecules that are similar (though slightly different) in structure (e.g. via site-directed mutagenesis), with the goal of producing a molecule that is a 'perfect fit' (e.g. to a receptor).

molecular docking: A method that predicts the preferred orientation of one molecule to another when a ligand and a target are bound to each other to form a stable complex. Knowledge of the preferred orientation in turn may be used to predict the strength of association or binding affinity between two molecules using, for example, scoring functions.

molecular drive: A molecular mechanism that explains evolution in multiple gene families. It includes unequal recombination, transposition and gene conversion. A large amount of DNA sequence may have evolved by multiplication of already existing sequences inside the genome by a process of molecular drive. Detailed analysis of eukaryotic genomes indicates that the molecular drive, which is an outcome of turnover processes of repetitive DNA, is a unique phenomenon that has led to the formulation of models describing speciation. These models were described to be of three kinds: the genomic disease model, the mechanical genome incompatibility and the genome resetting model.

molecular dynamics (MD): A computer simulation method for analysing the physical movements of atoms and molecules. The atoms and molecules are allowed to interact for a fixed period of time, giving a view of the dynamic 'evolution' of the system.

molecular evolution: Refers not to one but to several research lines in the continuum comprised between the study of the evolution of macromolecules (e.g. nucleic acids and proteins) and the use of these molecules in the study of organismic evolution. Molecular evolution encompasses the study of the forces shaping the structure and function of macromolecules as well as their use in studying the evolution of the organisms possessing them.

molecular evolution, epochs of: The molecular evolution can be classified into four epochs: prebiotic world, RNA world, ribonucleoprotein world and DNA/protein world.

molecular evolution, forces of: The production and redistribution of variation is produced by mutation, genetic drift and gene flow. Natural selection, in turn, acts on the variation produced by these agents. While mutations and genetic drift are random forces, natural selection is not, as it preferentially selects for different mutations based on differential survival.

molecular evolutionary clock: A measurement of evolutionary time in nucleotide substitutions per year. With the minimum number of substitutions known, we can calculate molecular evolutionary rates or nucleotide substitutions per million years. In a sense, these rates provide us with a molecular evolutionary clock, measuring evolutionary time in nucleotide substitutions.

molecular farming: Certain recombinant proteins and pharmaceuticals are being produced in the milk, urine, blood, etc. of transgenic animals, which can be used to extract drugs. Transgenic plants can yield a variety of proteins for use in diagnostics for detecting human diseases and therapeutics for curing human and animal diseases on a large scale at a low cost. Monoclonal antibodies, blood plasma proteins, peptide hormones and cytokinins are being produced by transgenic plants. Plants are amazing and cheap chemical factories. Production of value-added products like nutrition supplements, pharmaceuticals, fuels, etc. using transgenic crops. Molecular farming also refers to the application of biotechnology for the selected production of pharmaceutical compounds or other health or industrial compounds within a microbe. Examples of molecular farming include: (i) hirudin; (ii) phytase; (iii) polyhydroxybutyrate; (iv) research reagents; and (v) post-translational protein modifications. *Aka* MOLECULAR PHARMING.

molecular fossil: RNA editing has been nicknamed 'molecular fossil' by T. Cech (1992). It has been suggested to be a living relic that provides a look backwards to life's origin.

molecular genetic maps: Genetic maps prepared using molecular markers. The genetic maps (linkage maps and cytogenetic maps) were inadequate for the identification of candidate genes and for positional cloning. Therefore, using different kinds of molecular markers, molecular genetic (high-resolution) maps have now been prepared for several crops. Such maps form the basis of future applications of molecular markers in crop breeding. Several efforts have gone in to develop genetic maps of hexaploid wheat (A, B and D genomes), tetraploid wheat (A, B genomes) and their progenitor diploid genomes namely *Aegilops tauschii* (D genome) and *Triticum monococcum* (A genome), at the level of whole genomes using various populations comprising F2 populations resulting from narrow or wide crosses, single-seed descent-derived recombinant inbred lines (RILS) or doubled haploids (DHs). A consensus molecular genetic map has been prepared in order to carry out studies on macrosynteny between cereals and establish reference maps. The availability of a consensus map provides a basis for selecting alternative markers in a chromosome region of interest as well as for compiling gene-rich regions.

molecular genetics: A branch of genetics concerned with the structure and activity of genetic material at the molecular level. It deals with issues such as the nature, structure, properties and organization of genes as well as non-gene DNA. It attempts to answer at the molecular level some fundamental questions of life, such as how a gene is copied, how a mutation arises and how genes are turned on and off. Thus, it is the study of the expression, regulation and inheritance of genes at the level of DNA and its transcription products. It is the area of knowledge concerned with the genetic aspects of molecular biology, especially with DNA, RNA and protein molecules.

molecular genetics, birth of: The seminal discovery of double-helical structure of DNA by J.D. Watson and F.H.C. Crick (1953). With this, genetics entered the DNA age. This was a very exciting phase in the development of genetics.

molecular glue: Gram-negative bacteria such as *Acinetobacter baumannii* contain lipopolysaccharide (LPS) on their outer membrane that blocks the entry of most small-molecule drugs, which makes the bacteria hard to kill. These antibiotics are often derivatives of existing classes, and no new classes of antibiotics have been developed against these Gram-negative bacteria for over 50 years (I. Marchal, 2024). In 2024, two teams of scientists (M.S.A. Gilman *et al.* and C. Zampaloni *et al.*) identify and characterize a new class of antibiotics that targets the LPS transport machinery, which provides insights into lipid transport inhibition as a druggable target. Assembly of the outer membrane requires transport of LPS across a protein bridge that spans from the cytoplasmic membrane to the cell surface. Maintaining outer membrane integrity is essential for bacterial cell viability, and its disruption can increase susceptibility to other antibiotics. Thus, inhibitors of the seven lipopolysaccharide-transport (Lpt) proteins that form this trans-envelope transporter have long been sought. K.S. Pahil *et al.* (2024) show that these new antibiotics trap a substrate-bound conformation of the LPS transporter that stalls this machine. The inhibitors accomplish this by recognizing a composite binding site made up of both the Lpt transporter and its LPS substrate. They identify an unusual mechanism of lipid transport inhibition, reveal a druggable conformation of the Lpt transporter and provide the foundation for extending this class of antibiotics to other Gram-negative pathogens. Carbapenem-resistant *Acinetobacter baumannii* (CRAB) is a major global pathogen with limited treatment options. C. Zampaloni *et al.* (2024) report the identification and optimization of tethered macrocyclic peptide (MCP) antibiotics with potent antibacterial activity against CRAB. The mechanism of action of this molecule class involves blocking the transport of bacterial lipopolysaccharide from the inner membrane to its destination on the outer membrane, through inhibition of the $LptB_2FGC$ complex. A clinical candidate derived from the MCP class, zosurabalpin (RG6006), effectively treats highly drug-resistant contemporary isolates of CRAB both *in vitro* and in mouse models of infection, overcoming existing antibiotic resistance mechanisms. This chemical class represents a promising treatment paradigm for patients with invasive infections due to CRAB.

molecular hybridization: (i) The formation of duplex molecules from complementary single strands (DNA-DNA, DNA-RNA, RNA-RNA). (ii) The annealing of previously purified and denatured DNA strands by different means. Hybridization is used to detect sequence homologies between two different nucleic acid molecules.

molecular imprinting: The phenomenon in which there is differential expression of a gene depending on whether it was maternally or maternally inherited.

molecular integrated microsystems (MIMS): Microsystems in which functions investigated in biological and nanoscale systems are integrated with materials.

molecular map: In recent years, with the advent of recombinant DNA and PCR technologies, successful efforts have been made to prepare genetic, cytogenetic and physical maps of animal genomes. These maps utilize molecular markers in the form of RESTRICTION FRAGMENT LENGTH POLYMORPHISMS (RFLPs), RANDOM AMPLIFIED POLYMORPHIC DNAs (RAPDs or 'rapids'), MINISATELLITES or VARIABLE NUMBER OF TANDEM REPEATS (VNTRs) and MICROSATELLITES or SIMPLE SEQUENCE REPEATS (SSRs). Preparation of physical maps is also facilitated by the preparation of contiguous sets of overlapping clones (called 'contigs') in cosmid and YEAST ARTIFICIAL CHROMOSOME (YAC) vectors. *See details as separate terms.*

molecular maps, uses of: (i) Molecular maps are useful in linking a genetic marker to disease. For example, a genetic marker linked to Huntington disease was found on chromosome 4 in 1983, making Huntington disease (HD) the first genetic disease mapped using DNA polymorphisms. HD is inherited as an autosomal dominant disease. (ii) LARGE-INSERT BACTERIAL ARTIFICIAL CHROMOSOME (BAC) LIBRARIES, PLANT-TRANSFORMATION-COMPETENT BINARY BAC (BIBAC) LIBRARIES and SIMPLE SEQUENCE REPEAT (SSR) MARKERS are essential for many aspects of genomics research. The BAC and BIBAC libraries and new SSR markers provide valuable resources for chickpea genomics research and breeding. *See details as separate terms.*

molecular marker: Any specific DNA segment whose base sequence is different (polymorphic) in different organisms and whose role is therefore diagnostic. Molecular markers can be visualized by either hybridization or PCR techniques. A molecular marker (identified as a genetic marker) is associated with a certain location within the genome.

Molecular markers are used in molecular biology and biotechnology to identify a particular sequence of DNA in a pool of unknown DNA. Isozymes (enzymes with the same catalytic activity but a different molecular weight) have been tagged with genes of agronomic importance. For instance, nematode resistance in tomatoes is associated with a rare allele of the acid phosphatase locus *Aps-1* that has been commercially exploited in tomato breeding. But the number of useful protein markers is limiting factor. Now, the use of DNA markers has revolutionized molecular breeding. Specific spots on DNA molecules, in both coding and noncoding regions, can be identified as markers. DNA markers are independent of stage and environment. Particular DNA sequences and/or segments that are closely linked to a gene locus and/or morphological or other traits of a plant can be identified and visualized by molecular techniques.

molecular marker, types of: Different types of molecular markers are AMPLIFIED FRAGMENT LENGTH POLYMORPHISM (AFLP), AMPLICON LENGTH POLYMORPHISM (ALP), ARBITRARILY PRIMED PCR (AP-PCR), ALLELE-SPECIFIC PCR (ASPCR), DNA AMPLIFICATION FINGERPRINTING (DAF), DIVERSITY ARRAYS TECHNOLOGY (DArT), MICROSATELLITE POLYMORPHISM SIMPLE SEQUENCE REPEAT (MSSR), MICROSATELLITE SIMPLE SEQUENCE LENGTH POLYMORPHISM (MSSLP), RESTRICTION-SITE ASSOCIATED DNA MARKERS (RAD MARKERS), RANDOM AMPLIFICATION OF POLYMORPHIC DNA (RAPD), RESTRICTION FRAGMENT LENGTH POLYMORPHISM (RFLP), SPECIFIC AMPLICON POLYMORPHISM (SAP), SEQUENCE-CHARACTERIZED AMPLIFIED REGION (SCAR), SINGLE FEATURE POLYMORPHISM (SFP), SINGLE-NUCLEOTIDE POLYMORPHISM (SNP), SINGLE-STRAND CONFORMATION POLYMORPHISM (SSCP), SHORT TANDEM REPEAT (STR), SEQUENCED TAGGED SITES (STS) and VARIABLE NUMBER TANDEM REPEATS (VNTRs). Of all these markers, RFLP and RAPD are the principle classes of markers. *See details as separate terms.*

molecular markers for genetic mapping: (i) RESTRICTION FRAGMENT LENGTH POLYMORPHISMS (RFLPs). (ii) MICROSATELLITES OR SIMPLE SEQUENCE REPEATS (SSRs). (iii) EXPRESSED SEQUENCE TAGS (ESTs). (iv) CLEAVED AMPLIFIED POLYMORPHIC SEQUENCE (CAPS). (v) RANDOM AMPLIFIED POLYMORPHIC DNA (RAPD). (vi) AMPLIFIED FRAGMENT LENGTH POLYMORPHISMS (AFLPs). (vii) INTER-SIMPLE SEQUENCE REPEAT (ISSR). (viii) DIVERSITY ARRAYS TECHNOLOGY (DArT). (ix) SINGLE-NUCLEOTIDE POLYMORPHISMS (SNPs). *See details as separate terms.*

molecular markers, advantages of: (i) The precision (<1cM) by which molecular markers can be used to map a site on a genome is often better than phenotypic markers because one can use more of them, and there are fewer errors in scoring. (ii) Linkage can be easily established; when a set of markers is carefully chosen from different parts of the genome, it is possible to map any locus to a marker. (iii) The markers do not even have to be genes; they can be any piece of DNA that detects a small number of bands, and for which there is polymorphism in the population; it makes them selectively neutral. (iv) For RFLPs, the polymorphism detected does not have to fall inside the region of the probe itself; it only has to be between the probe and the next site for whichever restriction enzyme is used.

molecular markers, groups of: Roughly, three groups of markers can be classified: (i) HYBRIDIZATION-BASED DNA MARKERS such as RESTRICTION FRAGMENT LENGTH POLYMORPHISMS (RFLPs) and OLIGONUCLEOTIDE FINGERPRINTING; (ii) PCR-BASED DNA MARKERS such as RANDOM AMPLIFIED POLYMORPHIC DNAs (RAPDs), which can also be converted into SEQUENCE-CHARACTERIZED AMPLIFIED REGIONS (SCARs), SIMPLE SEQUENCE REPEATS (SSRs) or MICROSATELLITES, SEQUENCE-TAGGED SITES (STS), AMPLIFIED FRAGMENT LENGTH POLYMORPHISMS (AFLPs), INTER-SIMPLE SEQUENCE REPEAT AMPLIFICATION (ISA), CLEAVED AMPLIFIED POLYMORPHIC SEQUENCES (CAPs) and AMPLICON LENGTH POLYMORPHISMS (ALPs) and (iii) DNA CHIP AND SEQUENCING-BASED DNA MARKERS such as SINGLE-NUCLEOTIDE POLYMORPHISMS (SNPs). *Aka* DNA MARKER. *See details as separate terms.*

molecular markers, limitations of some: AFLPs, RAPDs and ISSRs have the following limitations: (i) The chromosomal location of markers scored from one cross cannot be extrapolated to others unless indirectly determined using other known markers, such as RFLPs and microsatellites. (ii) They are all dominant markers and are less valuable for mapping F_2 or BC populations.

molecular markers, uses of: (i) Each marker system has advantages and disadvantages. (ii) Various factors are considered in selecting one or more of these markers. (iii) The first large-scale efforts to produce genetic maps were performed mainly using RFLP markers.

molecular mechanics: A method by which the molecular potential energy function is calculated to predict and understand various properties of the system.

molecular medicine: This area of functional genomics looks into the fundamental causes of diseases, which improves disease diagnosis, helps in earlier detection of genetic predisposition to disease, leads to rational drug design and improves gene therapy and control systems for drugs.

molecular models of recombination: (i) The WHITEHOUSE-HASTINGS MODEL and (ii) HOLLIDAY'S MODEL. *See details as separate items.*

molecular morphogenesis, events in: The development of body structure depends upon two types of events. First, gradients are established in the egg along two axes, posterior-anterior and dorsal-ventral. These gradients comprise molecules that are differentially dispersed in the common cytoplasm. A nucleus can, in principle, detect its position in the cytoplasm by the level of each gradient. Second, distinct regions are resolved in the embryo that correspond to parts of the adult body. In terms of the posterior-anterior axis, these regions form a striped pattern along the embryo. Ultimately, they result in the production of segments of the adult insect. Single mutations that influence the structure of the entire body part have aided in the identification of potential regulator genes that may switch or select between developmental pathways.

molecular morphogenesis: Process of morphogenesis explained in molecular terms. The analysis of the process of morphogenesis in molecular terms. At fertilization, the egg possesses two parental nuclei and is distinguished at the posterior end by the presence of region called POLAR PLASM. A nucleus determines its position in the embryo with reference to the anterior-posterior and dorsal-ventral gradients and behaves accordingly. Four groups of genes concerned with the development of particular regions of the embryo can be identified by mutations in maternal genes. The genes in each group can be ordered into a pathway that reflects their sequence of action, either by conventional genetic tests, such as comparing the properties of double mutants with the individual mutants, or by biochemical assays, showing which mutants contain components that can bypass the stages that are blocked in other mutants. Each pathway is initiated by localized events, which may exist either within or outside the egg. This results in the localization of a signal within the egg. This signal takes the form of a protein (MORPHOGEN) with an asymmetric distribution.

molecular pathway: A chain of chemical reactions occurring within a cell.

molecular pharming: *See* MOLECULAR FARMING.

Molecular Pharming™: A trademark of the Groupe Limagrain company, it refers to the production of pharmaceuticals and certain other chemicals (e.g. intermediates utilized to manufacture pharmaceuticals) in agronomic plants that have been genetically engineered.

molecular phylogenetics: A relatively new scientific discipline that involves the comparative analysis of the nucleotide sequences of genes and the amino acid sequences and structural features of proteins from which evolutionary histories and relationships, and in some cases also functions, can be inferred.

molecular pixelation (MPX): F. Karlsson *et al.* (2024) provide MPX, an optics-free, DNA sequence-based procedure for spatial proteomics of single cells using antibody-oligonucleotide conjugates (AOCs) and DNA-based, nanometre-sized molecular pixels. The comparative positions of AOCs are estimated by sequentially correlating them into local neighbourhoods using the sequence-unique DNA pixels, forming >1000 spatially linked zones per cell in 3D. For each single cell, DNA sequencing reads are computationally organized into spatial proteomics networks for 76 proteins. By studying immune cell dynamics using spatial statistics on graph representations of the data, they identify known and new patterns of spatial organization of proteins on chemokine-stimulated T cells, revealing the potential of MPX in defining cell states by the spatial arrangement of proteins.

molecular recognition: Molecules stick in a highly specific way though noncovalent bonds, recognizing each other, in an organized fashion. The shape of the molecules plays a vital role in the strength of binding.

molecular recombination: The physical cutting and joining of two double-stranded DNA molecules.

molecular scissors (restriction enzymes): An enzyme produced chiefly by certain bacteria, having the property of cleaving DNA molecules at or near a specific sequence of bases.

molecular self-assembly: The essentially automatic ordering and assembly of certain molecules into large structures, often called SUPRAMOLECULAR

STRUCTURES or SUPRAMOLECULAR ASSEMBLIES, include micelles, reverse micelles, ribosomes, peptide nanotubes and TOBACCO MOSAIC VIRUS (TMV). The first discovery of a self-assembling active biological structure occurred in 1955 when H. Fraenkel-Conrat and R.C. Williams showed that TMV can reassemble into functioning, infectious virus particles after TMV has been dissociated into its components via immersion in concentrated acetic acid.

molecular self-assembly, goals of: In the near future, human-directed self-assembly of molecular structures is expected to serve as: (i) cages to carefully protect and deliver sensitive/unstable pharmaceuticals to targeted tissues within the body; (ii) crucibles (i.e. reaction vessels) for small-scale chemical reactions to occur within; (iii) computer logic or memory devices (i.e. bioelectronics); and (iv) antibiotics. For example, during the 1990s, M.R. Ghadiri created 'PEPTIDE NANOTUBES' made via the self-assembly of certain peptides into tubes of nanometre dimensions. These peptide nanotubes are 'membrane active' (i.e. insert one end of themselves into the outer membrane of a cell) and cause the cell (e.g. pathogenic bacteria) contents to 'leak out', which kills the bacteria.

molecular switch: A molecule that can be reversibly transformed between two or among more stable states.

molecular switches and timers: The importance of cytoplasm in the timing of the cell cycle during the early embryonic division of a frog has been demonstrated directly. Cell cycle timing is regulated by the cyclic accumulation of the cdc25 mitotic inducer, which, on reaching a critical level, results in the activation of the $p34^{cdc2}$ protein kinase. The accumulation of this inducer could play a part in coordinating cell division with growth. The M phase is regulated by a mechanism common to all eukaryotic cells. Entry into Mphase is determined by activation of the $p34^{cdc2}$ protein kinase, which requires $p34^{cdc2}$ dephosphorylation and association with cyclin.

molecular systemics: The process of using data on the molecular constitution of biological organisms' DNA, RNA or both to infer their correct scientific classification or taxonomy from an evolutionary point of view. The sequences of proteins and DNA are both used in phylogenetic inference.

molecular tinkering: (i) Molecular tinkering is the process of transforming pre-existing genes or parts of genes to produce new roles or functions, or of combining molecular systems to give new, often more complex, systems. (ii) An evolutionary process responsible for production of a protein with new properties by joining together pieces of several different genes.

molecular weight (m.w.): Numerically, the same as the relative molecular mass of a molecule expressed in Daltons. For example, a protein with a relative molecular mass of 20,000 has a molecular weight of 20,000.

molecular wires: Molecular chains that conduct electric current from one end to the other. They are the proposed building blocks for molecular electronic devices. Their typical diameters are less than three nanometres, while their lengths may be macroscopic, extending to centimetres or more.

molecular zippers: Molecular zippers are a class of transcriptional regulator proteins that binds to DNA at dyad-symmetric sites through a motif consisting of a 'leucine zipper' sequence that associates into noncovalent, parallel, α-helical dimers and a covalently connected basic region necessary for binding DNA. Many gene-regulating proteins 'zip' together into pairs. This linkage is critical to their ability to bind to DNA. The 'teeth' that join the molecules almost always consist of amino acid leucine.

molecule: Group of atoms joined together by covalent bonds. These bonds are formed as a consequence of the sharing or exchange of electrons among atoms.

molybdenum (Mb): An essential trace element in plant metabolism.

monintron gene: The gene that contains only one intron. Transcripts of such genes are not used for translation.

monitoring: Activities conducted to measure levels, concentrations or quantities of material and the use of these measurement results to evaluate potential exposures and doses and to determine existing environmental conditions, pollutant levels (rates) and effects on species in the environment.

monitoring protein level and its modification: It is important to monitor post-translational proteins and genetic network modifications (e.g. the phosphorylation state of a protein). It is possible to do two-dimensional protein gel analyses of proteins, followed by mass spectrometry analysis (MSA). MSA of peptide fragments helps in the identification of protein 'signatures'.

monoallelic: Applied to a polyploid in which all alleles at a particular locus are the same (in a tetraploid – *A1A1A1A1*) as opposed to DI-ALLELIC (in a tetraploid – *A1A1A1A2*), TRI-ALLELIC (in a tetraploid – *A1A1A2A3*), TETRA-ALLELIC (in a tetraploid – *A1A2A3A4*), etc.

monoallelic gene expression: Gene expression is termed 'monoallelic' when only one copy of the two copies of a gene is active while the other is not. Monoallelic gene expression is commonly initiated in the development of an organism and is stably maintained thereafter. Mammalian X inactivation, imprinting and allelic exclusion are some classic examples of monoallelic gene expression.

monoallelic methylation: A type of DNA methylation in which only one allele of a locus is methylated in a diploid organism.

monobrachial: Relating to one arm of a chromosome. A chromosome with a terminal centromere.

monocarpic: A plant flowering and setting seed only once before dying. *See* SEMELPAROUS.

monocentric chromosome: A chromosome with only one centromere.

monocentric crop plant: A crop plant species with only one centre of origin.

monochasium: A cymose inflorescence, characterized by the presence of a flower at the apex of the flower axis, with the branches arising singly. *Compare with* DICHASIUM *and* PLEIOCHASIUM.

monocistronic mRNA: An RNA molecule, mostly in eukaryotes, that contains information from only one cistron.

monoclonal antibodies (MAb): Discovered and developed in the 1970s by C Milstein and G. Kohler, monoclonal antibodies are the general name used for antibodies derived from a single source or clone of cells that recognize only one kind of antigen. Mab are made by fusing myeloma cancer cells (which multiply very fast) with antibody-producing cells, then spreading the resultant conjugate colony so thin that each cell can be grown into a whole, separate colony (i.e. cloning). In this way, one gets whole batches of the same (monoclonal) antibody, which are all specific to the same antigen. Monoclonal antibodies have found markets in diagnostic kits and show potential for use in drugs (e.g. to shrink tumours), imaging agents and in purification processes. One example of a diagnostic use is the invention in 1997 by B. Oesch of a monoclonal antibody-based rapid test to detect the prion (PrP 5c) that causes BOVINE SPONGIFORM ENCEPHALOPATHY (BSE) in cattle. *Contrasts* POLYCLONAL ANTIBODIES. *See* HYBRIDOMA.

monoclonal antibodies (MAb), preparation of: The procedure that involves injecting the mouse with an antigen, which provokes the production of antibodies by plasma cells which are concentrated in the spleen. The cells are taken from the spleen and mixed in a special selective growth medium with myeloma (tumour) cells. The fused cells (hybridoma) are then screened for antibody production and are then individually separated by means of a FLUORESCENCE-ACTIVATED CELL SORTER (FACS).

monoclonal antibodies (MAb), properties of: Monoclonal antibodies are very specific immunoglobulin molecules that display a wide range of biological activities. In addition to being used in diagnostics, the antigen-binding sites of antibody molecules have revealed great potential for developing bioactive peptides. Because of their specific ligand binding activity, they were well-thought-out as the magic bullets, as hypothesized by Paul Ehrlich. Hybridoma technology, developed by G. Kohler and C. Milestein (1975), used the fusion of myeloma and B cells to assist in the *in vitro* production of monoclonal antibodies. However, this technology was found to be not very helpful as most of the antibodies were of murine origin and had the problem of low immunogenicity. The application of recombinant DNA technology resulted in the development of chimeric and humanized antibodies with high efficiency and activity.

monoclonal antibodies (MAb), therapeutic use of: Even though antibodies act on wide varieties of pathways, therapeutic antibodies work in one of the following four ways: (i) as immunotoxicotherapy, where the antibody prevents or reverses the toxic effect of venom, toxin, drug or ligand; (ii) for destruction of target cells, where the antibodies are used to destroy the target cells, such as lymphocytes, cancer cells, etc.; (iii) for alteration of cell function; and finally, (iv) antibody-mediated drug delivery, where the drug is conjugated with the antibody for specific targeting. For the large-scale production of monoclonal antibodies, expression of monoclonal antibody genes is accomplished through recombinant DNA technology. More than 20 monoclonal antibodies have been approved for human uses.

monocot: Full form MONOCOTYLEDON. A plant with one cotyledon and flower parts usually in multiples of three, such as grasses and some flowering plants, such as grain crops, wheat, corn and barley. In addition, monocots are monosulcate, meaning that their pollen grains have one furrow or groove. *Contrasts* DICOTYLEDON (DICOT).

monocot epidermis: The stem's outer dermal tissue layer is the epidermis, a single layer of cells that thwarts damage caused by sunlight, pathogens and herbivores. In some stems, the dermal cells secrete a waxy substance that forms a cuticle, a protective covering, that helps the stem retain water. In monocot stems, there is no trichome (hairs) extending from the epidermis.

monocot hypodermis: In monocot stems, the hypodermis is a tightly packed layer of sclerenchyma cells located interior to the epidermis. It offers structural support to the stem.

monocot leaf: Monocot leaves have an upper and lower epidermis, both containing several small pores called STOMATA, which smooth gas exchange and allow water vapour to exit the leaf. The upper and lower epidermis are separated by ground tissue called MESOPHYLL, which contains parenchyma cells that carry out photosynthesis and sclerenchyma cells deliver structural support. In monocot leaves, the vascular structures form parallel veins that are connected to the stem's vascular structures.

monocot root: Monocot roots have a stringy structure, meaning they form a wide network of thin roots that initiate from the stem and stay close to the soil surface. The ground tissue of monocot roots, primarily composed of parenchyma cells, is divided by a ring of vascular tissue into the outer cortex and central pith.

monocot stem: The majority of the monocot stem is made of ground tissue, which primarily consists of parenchyma cells. Sclerenchyma cells are also found in regions that require extra strength. Monocot stems have vascular bundles that are composed of xylem and phloem and are scattered throughout the ground tissue.

monocot stomata: In monocot leaves, the upper and lower epidermis both contain stomata. Stomata (sing. stoma) are small epidermal pores that enable gases and water vapour to move between interior leaf structures and the surrounding air. In monocot leaves, the upper and lower epidermis both contain stomata. Stomata also occur in photosynthetic stems.

monocots versus dicots: Monocots and dicots have three key parts: roots, stems and leaves. Differences between monocots and dicots can be seen in the structure of three main parts: the roots, stems and leaves. The roots allow plants to absorb water and nutrients from the soil. Monocot roots are fibrous, meaning they form a wide network of thin roots that initiate from the stem and stay close to the soil surface. Dicot roots have a central 'taproot', meaning they form a single thick root with lateral branches that grows deep into the soil. Stems make plants stand tall, supporting their leaves and flowers. Vascular structures in the stem move water and nutrients upwards from the root to the leaves and transport food downwards from the leaves to the root. Monocot stems have bundles of vascular tissue scattered throughout, whereas vascular bundles in dicot stems are arranged in a ring. Leaves are the plant's food factories. Leaves receive water and nutrients from the root via vascular structures in the stem. They also absorb light and take in carbon dioxide. Parenchyma cells in the leaves carry out photosynthesis to produce sugar that the plant can break down for energy later. Pores called *stomata* allow carbon dioxide to enter the leaf and water vapour and oxygen to leave it. Monocot leaves have parallel vasculature, whereas dicot leaves have a net-like vasculature.

monocotyledon (abbrev. monocot): *See* MONOCOT.

monocropping: The high-yield agricultural practice of growing a single crop year after year on the same land in the absence of rotation through other crops. This practice gradually depletes the soil of certain nutrients and selects for both a highly competitive pest and weed community. Without balancing nutrient use and diversifying pest and weed communities, the productivity of monocultures is highly dependent on external inputs that may be harmful to the soil's fertility. Conversely, a well-designed crop rotation can reduce the need for synthetic fertilizers and herbicides by better using ecosystem services from a diverse set of crops. Additionally, crop rotations can improve soil structure and organic matter, which reduces erosion and increases farm system resilience. *See* MONOCULTURE.

monoculm mutant: A mutant type that shows only one culm instead of normally more (e.g. in wheat).

monoculture: The growth of the same plant species in close proximity to few or no other types of plants. It is the outcome of large-scale industrial farming systems that focus only on one crop. Such systems tend to eliminate biodiversity of plant life in an area and consequently lead to a loss of biological balance between pests and predators. This can result in intensified pest pressure, increased use of pesticides and therefore higher crop protection costs. *Contrasts* POLYCULTURE.

monocyclic disease: Having only one disease or life cycle per growing season. *Contrasts* POLYCYCLIC DISEASE.

monoecious: A category of plants possessing staminate (male) and pistilate (female) flowers separately on the same plant. Maize is a good example. The female flowers are somewhere located in the middle of the plant, whereas the male flowers (tassels) are located at the top of the plant. The male part is called ANDROECIUM, which consists of stamens and the female part is called GYNOECIUM, which consists of the pistil. *Contrasts* DIOECIOUS.

monoecism: The condition in plants that have male and female flowers separated on the same plant (e.g. in maize, cucumber, melon, pumpkin, coconuts, tomatoes, fig).

monofunctional alkylating agent: Alkylating agents carry one alkyl group in reactive form which is capable of being transferred to other molecules where electron density is high although actual number of such groups present in the molecule may be more than one. The drug reacts with just one strand of a pair, transfers one alkyl group separating it from its partner and eventually causing it and its attached sugar to break away from the DNA molecule. It may eventually cause permanent cell damage. For example, epoxides, dimethyl sulfate, diethyl sulfate, methyl methanesulfonate, ethyl methanesulfonate and diazomethane.

monogamy: A reproductive strategy in which one male and one female mate and reproduce exclusively with each other. *Contrasts* POLYGYNY *and* POLYANDRY.

monogenic: A trait governed by the alleles of one particular locus. *Compare with* DIGENIC, TRIGENIC, OLIGOGENIC *and* POLYGENIC.

monogenic disorder: A disorder caused by mutation of a single gene. *See* MUTATION *and* POLYGENIC DISORDER.

monogenic inheritance: A type of inheritance where it is controlled by alleles at a single locus. According to Mendelian genetics, each gene has two alleles that determine the character, and these alleles segregate independently during gamete formation to produce a distinct character. It gives, in case of complete dominance, 3:1 phenotypic ratio in F_2. *See* SINGLE-GENE RESISTANCE. *Contrasts* MULTIGENIC RESISTANCE, OLIGOGENIC RESISTANCE *and* POLYGENIC RESISTANCE.

monogenic resistance: Resistance determined by a single gene. This type of resistance develops in the plant against foreign pathogens.

monogenomatic: *See* MONOHAPLOID.

monogenotypic: Clone variety.

monohaploid: A haploid cell or individual possessing only one chromosome set in the nucleus.

monohybrid cross: A sexual cross in which the inheritance of just a single pair of alleles is followed (tall × dwarf). When an individual plant showing one phenotype is crossed with another having a contrasting phenotype for the same trait, the following generalizations were drawn from the results of monohybrid crosses: (i) The F_1 generation consisted of seeds all having the same phenotype as one of the parents; the phenotype that appears in F_1 generation is termed DOMINANT while other that does not is termed RECESSIVE, and the F_1 individual is called HYBRID. (ii) RECIPROCAL CROSSES between parents produced the same results. (iii) When plants of F_1 were allowed to undergo SELF-POLLINATION, the recessive trait, which had remained unexpressed in the F_1 generation, reappeared in the F_2 generation. The F_2 progeny consisted of dominant and recessive phenotypes in the ratio 3:1. This is known as a MONOHYBRID PHENOTYPIC RATIO. (iv) When plants of the F_2 generation were self-pollinated, the individuals with recessive phenotypes would always breed true. However, only one-third among the dominants were found to be pure and the remaining two-thirds repeated the ratio of 3 (dominant):1 (recessive) in the F_3 generation.

monohybrid heterosis: Hybrid vigour due to heterozygosity at a single locus. *Aka* SUPERDOMINANCE.

monohybrid segregation: A segregation pattern which is in accordance with a monogenic inheritance.

monoisodisomic: A cell or individual displaying monosomy for one chromosome but an isochromosome for one of the arms of the missing chromosome.

monoisosomic: A cell or individual displaying nullisomy of one chromosome but having an isochromosome for one arm of the missing chromosome pair.

monolayer: Describes the growth of eukaryotic cells in culture as a layer only one cell deep.

monomer: A molecule that can undergo polymerization thereby contributing constitutional units to the essential structure of a macromolecule. For example, amino acids (monomers) link together via condensation reactions (called peptide bonds) to yield polypeptides or proteins (polymers). A monomer is analogous to a link (monomer) in a metal chain (polymer). Carbohydrates, nucleic acids, proteins and lipids all are macromolecules formed from monomers through specific condensation reactions. *Contrasts* POLYMER.

monomorphic: Of one type, rather than several. *Compare with* DIMORPHIC *and* POLYMORPHIC.

monophyletic species: A group of species that share a common ancestry (i.e. being derived from a single interbreeding population).

monoploid: A haploid of a true diploid. It contains only the basic (x) chromosome number of a species, such that each type of chromosome appears only once in the nucleus. Maize is a true diploid, and thus the haploid of maize is called a monoploid. In the case of true diploids, the haploid and monoploid chromosome numbers are the same (e.g. in maize, x = n = 10), but in the case of polyploids, the monoploid and haploid numbers differ (e.g. in bread wheat, x = 7, n = 21, 2n = 42 or n = 3x and 2n = 6x = 42). Monoploids usually arise due to mutations. Monoploids may originate from the products of meiosis (e.g. in wasps, bees and ants, males are monoploids). Monoploids have a major role to play in modern approaches to animal and plant breeding.

monoploid number (x): The number of chromosomes in a single (non-homologous) set is called the monoploid number (x), which is different from the haploid number (n).

monoplontic: Refers to a haploid individual or monoploid phase of the life cycle.

monopodial: A mode of stem growth and branching in which the main axis is shaped by a single dominant meristem. *Contrasts* SYMPODIAL.

monosaccharide: A carbohydrate consisting of a simple sugar with the general formula $(CH_2O)_n$, where n = 3–8, carbon atoms not joined to other by glycosidic bonds. The word 'saccharide' has been derived from the Greek word 'Sakeharon', which means 'sugar'. On the basis of size, carbohydrates are grouped into three classes viz. monosaccharides, oligosaccharides and polysaccharides. Monosaccharides are simple sugars consisting of a single polyhydroxy aldehyde or ketone unit. Dextrose (6-carbon sugar D-glucose) is the most common monosaccharide. Monosaccharides of more than four carbons possess cyclic structures.

monosome: A chromosome that lacks its homologue in a diploid organism.

monosomic: A genome derived from a diploid but that has only one copy of one particular chromosome type, so that its chromosome number is 2n − 1. Monosomic series have developed and widely used mainly in wheat, rice, tomato, barley and several other crops. The aneuploids are being used in gene location studies. In rice 2n = 24 and x = 12 and thus we can have 12 monosomic lines. Monosomics are useful in locating newly discovered genes on specific chromosomes in plants.

monosomic alien addition lines (MAALs): A set of lines each one of which contains a different extra-chromosome from a related species.

monosomic analysis: A common procedure for gene mapping (e.g. in hexaploid wheat). When genes determining phenotypes of interest for which an aneuploid series is not available, crosses can be made to a monosomic series in a variety with a contrasting phenotypic pattern. The monosomics are used as female parents in order to ensure the majority (~72%) of progeny (the transmission of n − 1 gametes is about 75% by the egg cell, but only about 4% by the pollen) will be become monosomic. If the gene involved is both recessive and hemizygous-effective (an uncommon situation), direct phenotypic observations on F_1 monosomic progenies enable the cytogeneticist to locate the gene on a particular chromosome. Only monosomic individuals in the critical crosses are expected to exhibit the phenotype, whereas disomic sibs and monosomic (or disomic) progenies from all other crosses will display the dominant phenotype. If the plants of the monosomic series carry the dominant allele of a gene of interest, monosomic individuals in the critical cross display the recessive phenotype, whereas disomic sibs display the dominant phenotype.

monosomy: The condition in which one chromosome of a pair is missing. In humans, this would result in a total of 45 chromosomes. An example of monosomy is 45, X, *aka* TURNER SYNDROME. *See* CELL, CHROMOSOME, GENE EXPRESSION *and* TRISOMY.

monosomy, partial: When cells have one chromosome from a pair plus a portion of the second chromosome.

monospermous: Bearing one seed. *Aka* MONOSPERMIC.

monostromatic: Being a single cell thick, as in the alga Monostroma.

monotelic: A mitotic chromosome with one oriented and one unoriented centromere.

monotelocentric: A cell or individual lacking one chromosome pair but showing one telocentric chromosome for one arm of the two missing homologues.

monotelodisomic: A cell or individual lacking one chromosome pair but revealing two homologous telocentric chromosomes for one arm of the two missing homologues.

monotelomonoisosomic: A cell or individual lacking one chromosome pair but showing a telocentric chromosome for one arm of the missing homologous pair and an isochromosome for the other arm.

monotelotrisomic: A cell or individual displaying an additional telocentric chromosome to a certain pair of chromosomes.

monothecous: Having a sole compartment or cell. *Compare with* DITHECOUS.

monotypic: Containing only one taxon of the next lower rank. For example, a family with only one genus, or a genus that includes only a single species.

monovalent: Having one binding site; strictly, an atom with only one free electron available for binding in its highest energy shell.

monoxenic culture: A culture containing one species of an organism growing in the presence of one other species of organism.

monozygotic (MZ) twins: Twins developed from a single fertilized egg. MZ twins are identical genetically, but not epigenetically. *See* IDENTICAL TWINS.

Monte Carlo effect: The effect which is dependent upon template concentration; lower the abundance of any template, less likely its true abundance will be reflected in the amplified library.

morbid map: A diagram showing the chromosomal location of genes associated with disease.

morgan (M): Refers to a unit of relative distance between genes on a chromosome. One morgan (1 M) symbolizes a crossing-over value of 100%; a crossing-over value of 10% is a decimorgan (dM); 1% is a centimorgan (cM). Over short map distances, the distance between loci in morgans equals the recombination rate between them. Over long distances, the recombination rate is lower than the map distance because of multiple crossovers.

morph: Any relatively common heritable variation; any of the variant types of a polymorphic species.

morphogen: Signal molecule that can enforce a pattern on a field of cells by causing cells in different places to adopt different fates. A protein whose local concentration (or activity) causes the surrounding region to take up a particular structure or development fate. In each of these systems, the morphogen either is a transcription regulator or leads to the activation of a transcription factor in the localized region. Some well-known morphogens include cyclic adenosine monophosphate (cAMP), retinoic acid (RA), transforming growth factor β (TGFβ), bone marrow proteins (BMPs), bicoid (bcd) protein, decapentaplegic (dpp) protein and nanos protein.

morphogenesis: It refers to the creation of a mature form from an undifferentiated state to a differentiated state under *in vitro* or *in vivo* conditions. During plant tissue culture, for instance, the calli (undifferentiated mass of cells) subcultured on regeneration medium undergo morphogenesis, as a result of which the development of shoot buds, shoots and roots takes place. It reflects different aspects of cell behaviour, growth and division, movement relative to other cells, changes in shape and size, interaction with other cells and extracellular matrix and cell death. It is regarded as the 'third' component and the effector of developmental programmes. The basis of morphogenesis is cell behaviour. Morphogenesis means the formation of structure. Morphogenesis helps in the study of certain compounds important in development. The basis of morphogenesis is cell behaviour. This may be considered an isolated process or in terms of cells interacting with their environment. Morphogenesis is the process of the origin and development of a part, an organ or an organism. When this process is explained in molecular terms, it is known as molecular morphogenesis.

morphogenesis in *Bacillus subtilis*: The process of sporulation can be divided into a number of morphogenetically distinct stages which are dependent upon the expression of specific set of genes Gene expression is coordinated by a hierarchical cascade of transcriptional regulatory proteins. The network also includes post-translational regulation.

morphogenesis in filamentous phage particle: The pathway of filamentous phage particle morphogenesis has been worked out. The particle is secreted from the cell as it is formed through replacement of gene *V* protein bound to the newly synthesized single-stranded DNA by the membrane-bound gene *VIII major capsid protein*. For initiation of assembly process, morphological signal sequence in DNA located between genes *IV* and *II* is required. It is thought that the minor phage proteins encoded by genes *VII* and *IX* bind to the morphogenetic sequence and are responsible for initiating the assembly process.

morphogenesis in green algae: The adult green alga *Volvox carteri* consists of some 2000 small somatic cells located on the surface of a hollow sphere, and around 16 large reproductive cells just below, within a transparent extracellular matrix. By a series of predetermined cell divisions each reproductive cell will become an embryo that finally turns inside out by a process called INVERSION, to produce a juvenile with all its cells in the adult orientation. The discovery that a molecular motor of the kinesin family is involved in turning a multicellular green alga inside out might have implications for similar events in animal development.

morphogenesis in slime mould: Cyclic AMP (cAMP) is a morphogen in slime mould development. The life cycle begins with spores germinating in damp earth to yield amoeboid cells called *myxamoeba*, which feed on engulfing bacteria and reproduce by binary fission. The multiplying amoeba then forms a multicellular slug, which then undergoes complex morphogenesis to contract a fruiting body. Both aggregation and subsequent construction of the fruiting body seem to be governed by an intracellular signalling mechanism that uses cAMP as a chemical attractant. Within 6–8 h after onset of starvation, each amoeba can sense increase in local cAMP concentration (reception) and recreate a short pulse of cAMP in response (transmission). Membrane-bound cAMP receptor protein measures cAMP concentration and then triggers a burst of activity by a membrane-bound adenylate cyclase. Cells also possess a cAMP phosphodiesterase to destroy excess cellular cAMP. The first cell that spontaneously secretes a burst of cAMP defines the centre of each morphogenetic field. The cell that signals first is known as the founder cell, which effectively summons all others. cAMP not only brings together a mass of cells with radial symmetry but also allows them to organize into a multicellular organism with a distinct front and back end. Invariably, the appearances of enzymes at different levels of development were looked into for correlating their role in differentiation. One enzyme that was studied is trehalase, which breaks down trehalose. This enzyme appears after spore germination and disappears at the aggregation stage. It is assumed that trehalase is controlled by the vegetative genome. Another enzyme studied is UDP galactose polysaccharide transferase, which transfers galactosyl units for the synthesis of polysaccharide. This enzyme is observed to be key to the morphogenetic process of slime mould development. It is believed to be under the control of morphogens. One of the important factors in cell differentiation is the microenvironment. The microenvironmental pattern is most probably created by various cellular metabolites, which lead to the differentiation process.

morphogenesis in T_4 virus: T4 virus is made up of two parts: the head and the tail. The head is a bipyramidal hexagon, which consists of a coat and a core. The coat is made up of proteins, and the core is made up of a DNA molecule. The tail consists of the neck and collar. The inner cylinder is hollowed and surrounded by a sheath. On the other hand, there is a basal plate that bears six short spikes and long, thin, bent structures called tail fibres. Tail fibres and spikes help to fix the host. The first step in the life cycle of T4 virus is the degradation of bacterial DNA and many nucleoproteins. This will primarily be under the control of T4 DNA. Eighteen genes are known to synthesize the head of T4 virus. Gene *23* is the structural gene for the protein subunit of the head membrane. Mutants of this gene do not produce recognizable head membrane structures. The products of seven other genes are also required for the synthesis of a normal head membrane. *In vitro* complementation between lysate and head mutants has not been demonstrated, as might be expected. Any mutation in these genes results in aberrant assemblies of gene *23*. Gene *31* is responsible for solubility as it provides a solubility factor for proteins. Mutation in gene 66 causes shortening of the head, showing that this gene is responsible for elongation of the head. Mutation in gene *20* causes polyhead conditions or long tubular forms. Mutation in gene *21* induces the formation of an abnormal head that apparently contains no DNA. It has been proven experimentally that these genes are related to head structure. For tail, head and tail fibre, 21, 18 and 7 genes are, respectively, involved. The total number of genes in the morphogenetic pathway of the T4 particle is estimated to be 47.

morphogenetic: Refers to differentiation of tissues to form organs in an organism.

morphological mutations: The mutations that affect the visible characters or traits of an organism (e.g. size, shape, colour, etc.), plant or plant part, or bacterial or yeast colony.

morphological species concept: Organisms are classified in the same species if they appear similar.

morphology (Adj. morphologic): The word first used in print by the poet Johann Wolfgang von Goethe refers to the form and structure of an organism or any of their specific structural features. Currently, the term refers to the study of the shape, form and structure of an organism or a part thereof.

morphosis: A modification of the process of morphogenesis of an individual by environmental changes.

morphospecies: Species described wholly in terms of morphological characteristics.

mortality: The number of deaths of organisms because of some disorder. *In vitro* grown plants exhibit very high mortality upon their sudden transfer to soil (*ex vitro*). This is mainly because the *in vitro* grown plants are largely heterotrophs and grow under controlled temperature, light and humidity conditions. But when transferred to soil, these plants have to switch over to autotrophic mode of nutrition and are also subjected to several biotic/abiotic stresses. Hardening of *in vitro* grown plants prior to their transfer to soil reduces their mortality.

mosaic: (i) In plant pathology, a disease symptom (e.g. leaf showing non-uniform coloration, with intermingled normal, light green and yellowish patches) usually caused by a virus. *See* MOTTLE. (ii) In developmental biology, an organism made of two or more cell lines with different genotypes.

mosaic eggs: In the case of mosaic eggs, there is a difference in the pattern of cleavage of the ooplasm. The egg contains a mosaic of different cytoplasmic regions, each of which develops into a specific part of the embryo (e.g. frog eggs, ascidian eggs and many spirally cleaved eggs). Such eggs are unipotent.

mosaic evolution: Evolutionary rates may vary among different characters or character complexes because selection may act differently upon different functions.

mosaic gene: A discontinuous gene.

mosaicism: The occurrence in an individual of two or more genotypically distinct cell populations within the same tissue or individual that have a different genetic or chromosomal constitution.

mother cell: Special cells in the anther and ovule that produce pollen or egg cells.

mother plant: (i) The female ancestor of a hybrid and/or hybrid progeny. (ii) In horticulture, a mature plant from which cuttings are taken.

motif: A conserved region of a protein sequence alignment that generally associates with a particular function. Typically, it is conserved and is hence likely to be predictive of any subsequent occurrence of such a structural or biochemically functional region (e.g. the active site of an enzyme) in any other new protein sequence. The same motif can be identified in a variety of types of organisms. A motif is built from precise combinations of secondary structures (typically, α-helices and β-sheets). These are the definite sequences identified in some regulatory genes. These motifs code for conserved amino acid regions in proteins. In general, the term motif also applies to DNA and RNA elements associated with a particular function, but the more commonly used term in this case is element.

motile: Capable of self-propulsion by means of flagella, cilia or amoeboid or swimming movement.

motility: The ability of an organism to move independently, using metabolic energy. *Contrasts* SESSILITY.

motor protein: Protein that expends energy derived from nucleoside triphosphate hydrolysis to drive itself along a protein filament or another polymeric molecule.

mottle: *See* MOSAIC.

mould: Any microfungus with noticeable, abundant, or woven superficial growth (mycelium and/or spore masses) on various substrates, especially an economically important saprobe. Moulds generally grow on damp or decaying matter and on the surfaces of plant tissues.

movable genes: Movable genes are DNA elements that can transpose from one position to another in the genome of an organism. These genes were first discovered in maize by B. McClintock in 1950s. The existence of movable genes challenges the hypothesis of a fixed location of the gene in the chromosome.

movement proteins (MPs): Movement proteins are essential for cell-to-cell movement of plant viruses. These proteins have been shown to modify the gating function of plasmodesmata thereby allowing the virus particles or their nucleoprotein derivatives to spread to adjacent cells.

M phase: The period of cell cycle when mitosis or meiosis occurs.

M phase Cdk (M-Cdk): Complex moulded in vertebrate cells by an M-cyclin and the corresponding cyclin-dependent kinase (Cdk). The events of mitosis are prompted by M-Cdk, which is activated after completion of the S phase. *See* M-CYCLINS.

MRE11 complex: MRE11, RAD50 and NBS1 form a highly conserved protein complex that is involved in the detection, signalling and repair of DNA damage. MDC1-mediated focus formation by the MRE11 complex at sites of DNA damage is crucial for the efficient activation of the intra-S-phase checkpoint.

mRNA: *See* MESSENGER RNA.

mRNA cap: A structure that consists of 7^{m}GpppN (where 7^{m}G represents 7-methylguanylate, p represents a phosphate group and N represents any base), that is located at the 5′-end of eukaryotic mRNAs. In addition to its essential role of cap-dependent initiation of protein synthesis, the mRNA cap also functions as a protective group from 5′ to 3′ exonuclease cleavage.

mRNA cap 0: It consists of a guanine nucleotide connected to the mRNA via an unusual 5′ to 5′ triphosphate linkage. This guanosine at the 7 position is methylated directly after capping *in vitro* by a methyltransferase.

mRNA cap 1: It has a methylated 2′-hydroxy group on the first ribose sugar.

mRNA cap 2: It has methylated 2′-hydroxy groups on the first two ribose sugars.

mRNA decay pathways: Two general mRNA decay pathways exist. Both pathways are initiated by deadenylation by the Ccr4/Pop2/Not complex or possibly by the alternative deadenylases, Pan2/Pan3 and PARN. Poly(A) tail shortening can lead to either 3′–5′ exonucleolytic digestion by the exosome or decapping by the Dcp1/Dcp2 complex. Decapping is followed by 5′–3′ exonuclease digestion by Xrn1. The residual cap structure resulting from exosome digestion is cleaved by the scavenger decapping enzyme DcpS.

mRNA extraction: First, the mRNA is obtained and purified from the rest of the RNAs. Several methods exist for purifying RNA such as trizol extraction and column purification. Column purification is done by using oligomeric dT nucleotide-coated resins where only the mRNA having the poly(A) tail (consisting of a long sequence of adenine nucleotides) will bind. The rest of the RNAs are eluted out. The mRNA is eluted by using eluting buffer and some heat to separate the mRNA strands from oligo-dT.

mRNA helicase: Most mRNAs comprise secondary structure, yet their codons must be in single-stranded form to be translated. The ribosome itself functions as an mRNA helicase. The mRNA helicase activity is localized to the middle of the downstream tunnel, between the head and shoulder of the 30S subunit of the ribosome.

mRNA quantification: Methods used for mRNA quantification include: (i) Northern blotting, (ii) Reverse transcription quantitative polymerase chain reaction, and (iii) SAGE.

mRNA reading frame maintenance: The triplet-based genetic code necessitates that translating ribosomes uphold the reading frame of a messenger RNA faithfully to guarantee correct protein synthesis. However, in programmed ribosomal frameshifting, a specific insurrection of frame maintenance occurs, wherein the ribosome is forced to shift one nucleotide backward into an overlapping reading frame and to translate an entirely new sequence of amino acids.

mRNA 3′-end processing, regulation of: RNA polymerase II trusts on a repetitive sequence domain (characterized by amino acid sequence YSPTSPS) within its largest subunit to score transcription. Phosphorylation on serine-2/serine-5 of the carboxyl-terminal heptad repeats is well established. Threonine-4 phosphorylation was only detected after transcription end sites despite its functional implication in pausing, elongation, termination and messenger RNA processing. R.Y. Moreno *et al.* (2024) reveal that threonine-4 phosphorylation detection was blocked by flanking serine-5 phosphorylation at the onset of transcription, which can be removed selectively. Subsequent proteomic analyses recognized many proteins recruited to transcription via threonine-4 phosphorylation, which formerly were attributed to serine-2. Loss of threonine-4 phosphorylation significantly diminishes serine-2 phosphorylation, revealing a crosstalk between these two marks. Functional analysis of the threonine-4 phosphorylation emphasised its role in alternative 3′-end processing within pro-proliferative genes.

mRNA sequencing: Sensitively and accurately quantify gene expression, identify known and novel isoforms in the coding transcriptome, detect gene fusions and measure allele-specific expression.

mRNA stability: The steadiness of a given mRNA transcript is determined by the presence of sequences within an mRNA known as *cis*-elements, which can be bound by *trans*-acting RNA-binding proteins to inhibit or enhance mRNA decay.

MS medium: *In vitro* culture medium called MS medium after the description by T. Murashige and F.K. Skoog (1962).

MS2: A bacteriophage with an RNA genome and from which an RNA replicase has been isolated.

mSin3-interaction domains (SID): Interaction domains that are present on multiple transcriptional repressor proteins including TGFβ (transforming growth factor β) and Mad.

mt DNA: *See* MITOCHONDRIAL DNA.

mu particles: Bacteria-like particles found in the cytoplasm of *Paramecium* that have the male-killer phenotype.

Mu system of site-specific recombination: There are two tail fibre genes (*S* and *U*) in bacteriophage Mu. The *S* gene has a constant part (*Sc*) lying outside the invertible G region and alternative variable parts (*Sv* and *S'v*) which lie within it. Alternate versions of *U* (*U*, *U'*) also lie within the G region. Inversion of the G region is catalysed by protein Gin which facilitates recombination between the flanking gixR and gixL sites. The host protein, factor for inversion stimulation (FIS), stimulates inversion by binding to an enhancer outside the G region within the gin gene.

mulatto: The F_1 offspring of a pure Negro and a pure white.

mulch: A layer of material, such as organic matter or plastic, applied to the surface of the soil for purposes such as retention of water and inhibition of weeds.

mule: (i) A plant hybrid that is self-sterile and generally cross-sterile due to infertile pollen or undeveloped pistils. (ii) A domestic equine hybrid between a donkey and a horse. It is the offspring of a male donkey (a jack) and a female horse (a mare).

Muller-5 technique: Muller-5 *Drosophila* stock used in this procedure carries an X chromosome that contains *Bar* (*B*) and *white-apricot* (w^a) and two scute inversions. Muller-5 stock is used for detection of sex-linked recessive lethals. The precise genetic transcription of the stock is ln(1)scSIL sc8R + S, scS1 sc8 w^a *B*. If a lethal is induced in the wild-type male being tested, F_1 heterozygous female will carry the lethal on the X chromosome derived from the male parent. If this is so, wild-type males will not appear in F_2 progeny. *See* BASC TECHNIQUE.

Müllerian mimicry: An offensive/distasteful species that evolves to resemble another offensive/distasteful species. Both the species get amplified protection, because predators learn to avoid the common pattern more quickly.

multi-autotransformation (MAT) system: MAT system uses oncogenes of *Agrobacterium* (*ipt* or *rol* gene) for the selection of transgenic plants. Expression of the *ipt* gene causes abnormal shoot morphology called extreme shooty phenotype (ESP), which subsequently reverts into normal shoots with objective transgenes due to the excision of *ipt* gene. It is a positive selection system that gives the advantage of regeneration to the transgenic cells without killing the non-transgenic cells.

multi-autotransformation (MAT) vector: The MAT vector combines the use of the *ipt* or *rol* SMG with a marker removal system based on R/Rs site-specific recombination. MAT vector is a binary plasmid. which has a 'hit-and-run' cassette in which the chimeric *ipt* gene is inserted into the R/Rs system as a selectable marker. The CaMV 35S promoter drives the *gus*, *ipt* and *R* genes. The *nptII* is driven by the nopaline synthase promoter. The terminators of *gus*, *R* and *nptII* are derived from the nopaline synthase and that of the *ipt* gene is its native terminator. The *gus* and *nptII* genes are located outside the removable cassette.

multiallelic variation: Genes or DNA segments showing a high level of polymorphic variation, exemplified by a VARIABLE NUMBER OF TANDEM REPEATS (VNTRs) and microsatellites. *See* VARIABLE NUMBER OF TANDEM REPEAT (VNTR) LOCI *and* MICROSATELLITES.

multicellular: Any organism that is composed of many cells. *Contrasts* UNICELLULAR.

multicistronic message: An mRNA transcript with more than one cistron and thus encoding more than one polypeptide. These generally do not occur in eukaryotic organisms, due to differences in the mechanism of translation initiation.

multicloning sites (MCS): A short DNA sequence, found in most vectors in common use, that may contain many closely spaced, restriction-enzyme cleavage sites. *Aka* POLYLINKER.

multicopy genes: Gene duplication (or chromosomal duplication or gene amplification) is a major mechanism through which new genetic material is generated during molecular evolution. It can be defined as any duplication of a region of DNA that contains a gene. For example, sex of the liverwort *Marchantia polymorpha* is determined by the sex chromosomes Y and X, in male and female plant, respectively. Approximately half of the Y chromosome is made up of unique repeat sequences. K. Ishizaki *et al.* report (2002) that part of the Y chromosome, represented by a 90-kb insert of a genomic clone pMM2D3, contains five putative genes in addition to the *ORF162* gene, which is present also within the Y chromosome-specific repeat region. Sex of the liverwort *Marchantia* polymorpha is determined by the sex chromosomes Y and X, in male and female plant, respectively. Approximately half of the Y chromosome is made up of unique repeat sequences. Part of the Y chromosome, represented by a 90-kb insert of a genomic clone pMM2D3, contains five putative genes in addition to the *ORF162* gene, which is present also within the Y chromosome-specific repeat region.

multicopy plasmids: Plasmids present inside bacteria in quantities greater than one plasmid per (host) cell. Vectors for cloning DNA are usually multiple copies (10 to 20) per cell. *Contrasts* SINGLE-COPY PLASMIDS.

multicopy single-stranded DNA (msDNA): A type of extrachromosomal satellite DNA that comprises a single-stranded DNA molecule covalently linked via a 2′–5′ phosphodiester bond to an internal guanosine of an RNA molecule. The resultant DNA/RNA chimaera owns two stem-loops joined by a branch similar to the branches found in RNA splicing intermediates. The coding region for msDNA, called RETRON, also encodes a form of reverse transcriptase, which is essential for msDNA synthesis.

multidrug resistance (MDR) protein: Type of ABC transporter protein that can propel hydrophobic drugs (e.g. some anti-cancer drugs) out of the cytoplasm of eukaryotic cells.

multienzyme system: A sequence of related enzymes taking part in a particular metabolic pathway.

multi-epigenomics: Refers to biological studies, which leverages data from various epigenomics methods – EPIGENOMICS, EPITRANSCRIPTOMICS and EPIPROTEOMICS. *See details as separate items.*

multifactorial epigenomic profiling: S.J.A. Lochs *et al.* (2024) developed MULTIPLEXING ANTIBODIES by barcode identification (MAbID), a method for combined genomic profiling of histone modifications and chromatin-binding proteins in single cells, enabling researchers to study the interconnectivity between gene-regulatory mechanisms. They demonstrated MAbID's implementation in profiling multifactorial changes in chromatin signatures during *in vitro* neural differentiation and in primary mouse bone marrow tissue. MAbID holds the potential to gain unique insights into the interplay between gene regulatory mechanisms, especially for low-input samples and in single cells.

multifactorial inheritance: A pattern of inheritance exhibited by a number of conditions, such as heart disease, diabetes and certain birth defects. Multiple genetic or non-genetic factors – not just one – are involved in causing a multifactorial condition. *Aka* POLYGENIC INHERITANCE *or* MULTIGENIC INHERITANCE.

multifactorial or multigenic disorder: *See* POLYGENIC DISORDER.

multifactorial trait: A trait that is determined by the interaction of multiple genetic and environmental factors. *See* POLYGENIC TRAIT.

multiforked chromosome: When a bacterial chromosome has more than one replication fork, because a second replication initiation has occurred before the first cycle of replication has been completed.

multigene family: A multigene family comprises a group of nucleotide sequences or genes that displays four properties, namely, multiplicity, close linkage, sequence homology and related (or overlapping) phenotypic functions. Members of a multigene family are derived by duplication of an ancestral gene, followed by independent mutational events resulting in a series of independent genes either clustered together on a chromosome or dispersed throughout the genome.

multigene variety: A variety that carries a number of specific genes governing resistance to a particular pathogen.

multigenic inheritance: Inheritance determined by several genes with cumulative effect (e.g. ear length in corn).

multigenic perturbations: Transgenerational inheritance of DNA methylation statuses has been investigated through genome-wide DNA methylation contours of two *Arabidopsis thaliana* natural accessions to their F_1 offspring. The epialleles undergo stable transmission among plant populations. The combinatorial approach of construction of more advanced crosses (e.g. an F_2 generation or RILs) joined with QTL mapping methods have revealed the stability factor through sequence-independent chromatin inheritance (epigenetics) and *cis-* or *trans*-acting DNA-based factors.

multigenic resistance: Resistance deliberated by several genes. *See* POLYGENIC RESISTANCE. *Contrasts* MONOGENIC RESISTANCE (*aka* SINGLE-GENE RESISTANCE) *and* OLIGOGENIC RESISTANCE.

multigenic trait: A trait that is governed by many genes. *Contrasts* MONOGENIC TRAIT.

multigerm: An aggregate fruit containing numerous ovules.

multihybrid: An individual that is heterozygous for more than two genes. *Compare* MONOHYBRID *and* DIHYBRID.

multilayered genetic safeguards: Multilayered genetic safeguards include riboregulation, engineered addiction, auxotrophy and supplemental repressors.

multilevel evolution: Adaptive evolution occurring simultaneously at several levels of a biological hierarchy (e.g. nuclear and cytoplasmic genes).

multiline: A mixture of isoline each of which is different for a single gene controlling different forms of the same character (e.g. for different races of a pathogen or disease).

multiline cultivar: A mixture of cultivars that differ only in single, major genes for resistance to a pathogen, providing the opportunity to modify the mixture from year to year as required.

multiline variety: A composite (blended) population of several genetically related lines of a self-pollinated crop but bearing different genes (e.g. for resistance to pathogens). *Aka* MULTILINEAL VARIETY.

multilocation testing: A testing of breeder's strains and varieties at several geographically dissimilar sites in order to estimate the adaptive environmental retort and/or performance constancy.

multilocus enzyme electrophoresis (MLEE): Technique that enables the differentiation of organisms based on how far their intracellular enzymes migrate during electrophoresis.

multilocus probe: A DNA probe that hybridizes to a number of different sites in the genome of an organism.

multilocus sequence typing (MLST): Method that makes it possible to classify bacterial isolates based on the sequencing of six or seven well-conserved, housekeeping genes within the bacterial genome.

multimarker linkage disequilibrium mapping method: A method developed for fine mapping of QTL using dense marker map. The method compares the expected covariances between haplotype effects given a postulated QTL position to the covariances that are found in the data.

multimeric proteins: The proteins consisting of more than one subunit.

multinomial expansion: The terms generated when a mutinomial is raised to a power.

multintron gene: The gene that contains more than one intron. Transcripts of such genes are translated into proteins.

multinucleate: A cell possessing more than one nucleus per cell. *Contrasts* UNINUCLEATE.

multi-omics approach: Refers to biological studies that leverage data from various omics methods – GENOMICS, TRANSCRIPTOMICS, PROTEOMICS, METABOLOMICS, IONOMICS and MICROBIOME. *See details as separate items.*

multi-parent advanced generation intercross (MAGIC): C. Cavanagh *et al.* (2008) proposed MAGIC strategy to interrogate multiple alleles and to provide increased recombination and mapping resolution. *See* MULTI-PARENT ADVANCED GENERATION INTERCROSS (MAGIC) POPULATION.

multi-parent advanced generation intercross (MAGIC) population: The MAGIC populations serve a dual purpose – permanent mapping populations for precise QTL mapping and for direct and indirect use in variety development. Unlike a set of naturally diverse germplasm, this population is tailor-made for breeders with a combination of useful traits derived from multiple elite breeding lines. The MAGIC populations also present opportunities for studying the interactions of genome introgressions and chromosomal recombination.

multiparticulate virus: A plant virus in which the genome is split into more than one piece of nucleic acid encapsulated in more than one particle. *Aka* MULTIPARTITE VIRUS.

multipartite genome organization: The main chromosome having all genetic information is called the master chromosome. *See* MASTER CHROMOSOME.

multipass transmembrane protein: Membrane protein in which the polypeptide chain crosses the lipid bilayer more than once.

multipathotype tests: Seedlings of host plants are tested individually in the greenhouse for resistance against a range of pathotypes. These tests permit the postulation of some recessive genes in the host. Since these tests are usually performed on seedlings, important adult plant resistance genes may remain unidentified.

***Multiple Aleurone Layer* (*MAL*) gene:** A gene in corn (maize) that (when present in the DNA of a given plant) causes that plant to produce seed that contains higher-than-normal levels of calcium, magnesium, iron, zinc and manganese. These higher mineral levels are particularly useful for feeding of swine since traditional No. 2 yellow (dent) corn does not contain enough for optimal pig growth.

multiple alleles: The existence of more than two known allelic forms of a gene among the individuals of a population.

multiple allelism: A condition wherein a population of more than two alternate alleles represent the same locus in a given pair of chromosomes. Although only two actual alleles of a given gene can exist in a diploid cell, the total number of different allelic forms that might exist in a population is often quite large. The number of possible diploid combinations depends on the number of alleles in a series. If n is the number of alleles in a series, the number of diploid combinations (genotypes) is given by the expression $n(n + 1)/2$. For example, if the number of alleles is 5, the number of diploid genotypes in a population will be $\{5(5 + 1)\}/2 = 15$. Some examples of multiple allelism are self-incompatibility alleles in plants, the self-sterility multiple allelic system in *Nicotiana*, the ABO system of blood groups in man, eye colour mutants in *Drosophila*.

multiple allelism, conclusions from: Various conclusions have been drawn from studies on multiple alleles. (a) A gene exists in several alternate forms. (b) There are several mutable sites within a gene. (c) Members of an allelic series can show any type of dominance relationship with one another.

multiple cloning site: An artificially constructed region within a vector molecule which contains a number of closely spaced recognition sequences for restriction endonucleases. This serves as a convenient site into which foreign DNA may be inserted.

multiple comparisons: It is common practice for one study to involve the calculation of more than one P-value. For example, to test for treatment effects in three groups defined by disease aetiology, one may test the effects on four different patient response variables, or one may look for a significant difference in blood pressure at each of 24-hour measurements. When multiple statistical tests are done, the chances of at least one of them causing an assertion of an effect when there are no effects increases as the number of tests increases. This is called INFLATION OF TYPE I ASSERTION PROBABILITY α. When one wishes to control the overall type I probability, individual tests can be performed using a more stringent α level, or individual P-values can be adjusted upwards.

multiple convergence: When the desirable characters are distributed among a number of genotypes, a complex, cyclic crossing schedule is employed: $[(A \times B) \times (C \times D)] \times [(E \times F) \times (G \times H)]$ to form a complex hybrid. This approach leads to the convergence of several small streams of germplasm into one large stream for inbred improvement and their use in crosses.

multiple correlation coefficient: The degree of linear relationship of one quantitative variable with two or more simultaneously considered quantitative variables.

multiple correlations: Multiple correlations are an estimate of how well you can predict a variable using a linear function of other variables. It uses predictable variables to derive a conclusion.

multiple cropping: The practice of growing two or more crops on the same piece of land during 1 year instead of just one crop. When multiple crops are grown simultaneously, this is *aka* MULTICROPPING *or* INTERCROPPING. This cropping system helps farmers to double their crop productivity and their income. But, the selection of two or more crops for practicing multicropping mainly depends on the mutual benefit of the selected crops. In the Garhwal Himalaya of India, a practice called *barahnaja* involves sowing 12 or more crops on the same plot, including various types of beans, grains and millets, and harvesting them at different times. Threshing can be difficult in multiple cropping systems where crops are harvested together.

multiple DNA-protein interactions control transcription: The precise association of DNA-binding proteins with localized regions of DNA is vital for structured expression of the genome. For certain DNA transactions, the necessity for precision in localization and control is very high. Multiple DNA-protein interactions are also needed for controlled transcription of the eukaryotic genome. The basis of the exceptional precision of these DNA transactions is not immediately evident from the DNA-binding properties of the proteins that localize these and other such reactions. High-precision interactions involved in the initiation of transcription require DNA-binding proteins that localize. These reactions appear to use multiple DNA-protein and protein-protein interactions to generate a SPECIALIZED NUCLEOPROTEIN STRUCTURE (SNUP) in which the DNA is folded or wound. On the basis of prokaryotic work and additional inferences, snups seem to be responsible for these high-precision reactions.

multiple fruit: A bunch of fruits produced from more than one flower, appearing and harvested as a single fruit, often growing on a swollen axis, as with many species of the family Moraceae. *Compare with* AGGREGATE FRUIT.

multiple genes: Two or more genes at different loci that produce complementary or cumulative effects on a single, quantitative genetic trait. *See* POLYGENES.

multiple histone modifications, detection of: Unrestricted sequence alignment such as InsPecT, PTLMap, PILOT_PTLM, MODa and some specialized algorithms for top-down or middle-down tandem mass spectra are used for this purpose.

multiple origins of DNA replication: Replication occurs at many points along the eukaryotic chromosomes. The number of replicons is in thousands in a eukaryotic chromosome. Because of complexities of eukaryotic chromosomes, their rate of replication is sluggish compared to that of prokaryotes. If there were a single origin of replication in eukaryotes, it would require a little over 2 weeks to replicate one chromosome. For example, in reality, it takes only 4 min for a *Drosophila* cell to replicate its DNA, indicating the presence of 6000–7000 origins of replication.

multiple polyadenylation sites: A post-transcriptional step during RNA processing in which the precursor messenger RNA in eukaryotes matures to form a functional messenger RNA, about 200 adenosine nucleotides (polyA) are added in a polyadenylation reaction at the 3′ end. These poly(A) ribonucleotides are not encoded by the corresponding gene. In certain cases, there are multiple alternative polyadenylation sites in the primary transcript.

multiple population breeding system: The breeding population is subdivided into several smaller subpopulations that are bred for different objectives (e.g. different areas).

multiple regression: A regression that expresses the nature of the relationship between one (dependent) variable and two or more of the other (independent) variables. This can be linear or nonlinear.

multiple sequence alignment (MSA): A tool to determine levels of homology, and hence relatedness, between members of a series of globally related sequences. It is important for finding similar domains in a set of sequences and for doing phylogenetic analysis. If we have to find the evolution and the sequence conservation in such a group, we need to simultaneously compare several sequences. Here, MSA tools become very useful. MSA is very important for finding similar domains in a set of sequences and for doing phylogenetic analysis. Protein databases are often categorized by protein families. A protein family is a collection of proteins with similar three-dimensional shape, function or evolutionary history. The classification of a protein into a protein family provides hypotheses about its structure, function or evolutionary history.

multiple sequence alignment (MSA) programmes: (i) An alignment of three or more sequences with gaps inserted in the sequences such that residues with common structural positions and/or ancestral residues are aligned in the same column. Clustal W is one of the most widely used multiple sequence alignment programmes. (ii) A multiple alignment of k sequences is a rectangular array, consisting of characters taken from the alphabet A, that satisfies the following conditions: There are exactly k rows; ignoring the gap character, row number i is exactly the sequence sI; and each column contains at least one character different from '-'. In practice multiple sequence alignments include a cost/weight function, that defines the penalty for the insertion of gaps (the '-' character) and weights identities and conservative substitutions accordingly.

multiple sequence alignment (MSA), assumptions of: (i) All sequences are homologous. (ii) No duplicate sequences are present. (iii) In each column, amino acid residues are homologous. (iv) The alignment is optimal, with minimum gaps.

multiple sigma factors in bacteriophage SPO1: The control of gene expression through a cascade has been investigated in phage SPO1 infecting *Bacillus subtilis*. For the expression of the early genes of SPO1, bacterial RNA polymerase ($\alpha_2\beta\beta'\sigma^{55}$) is utilized because the promoters of the early genes like the promoters of host genes can be recognized by this RNA polymerase holoenzyme. The consensus sequence in SPO1 at −35 and −10 bp resembles those of *Escherichia coli* and *Bacillus subtilis*. For the expression of the middle and the late genes, however, expression of at least three regulator genes, namely, *28*, *33* and *34*, is required. The protein products of these genes are, respectively called gp28, gp33 and gp34. The gene *28* is an early gene whose expression is necessary for the expression of the middle genes. Similarly, genes *33* and *34* are the middle genes whose expression is pre-requisite for the expression of the late genes. It has been shown that gp28 replaces σ55 in the holoenzyme $\alpha_2\beta\beta'\sigma^{55}$ and, therefore, gp28 is also referred to as σgp28. This RNA polymerase with gp28 can then transcribe the middle genes. The next transition comprises the replacement of gp28 by gp33 and gp34, so that RNA polymerase may now transcribe the late genes. Thus, it is obvious that selective expression of genes can be caused by changes in sigma factor of RNA polymerase enzyme.

multiple sigma factors in *Escherichia coli*: For a long time, single sigma factor σ^{70} or σ^{D} was thought to occur in *E. coli*. However, new sigma factors in *E.*

coli were later discovered which operate under the conditions of heat shock (σ^{32} or σ^{H}), nitrogen starvation (σ^{54} or σ^{N}) or chemotaxis and flagellar structure (σ^{28} or σ^{F}).

multiple small RNAs as regulators: Small regulatory RNA molecules can act by pairing with their target messages, targeting themselves and the mRNAs for degradation. Multiple small RNAs are indispensable for regulation of the quorum-sensing systems of *Vibrio* species, including the regulation of virulence in *Vibrio cholerae*.

multiple targets to increase photosynthesis, manipulation of: Multiple targets should be identified and modified to enhance crop photosynthesis. One such target is ribulose bisphosphate carboxylase oxygenase (RuBisCO). This enzyme catalyses both carboxylation and oxygenation reactions. The majority of responses of photosynthesis to light, CO_2 and temperature are implanted in the kinetic properties of RuBisCO. Oxygenase activity of this enzyme can be reduced by increasing the concentration of CO_2 around the site of RuBisCO or by altering the kinetic properties of RuBisCO. RuBisCO manipulation may theoretically expand the range of kinetic properties and overcome the restraints that have been noticed for different kinetic parameters of RuBisCO.

multiple testing: If many significance tests are conducted, then some will reject the null hypothesis just by chance. Thus, significance levels need to be reduced when multiple tests are done.

multiple-copy plasmids: If the plasmid number per cell is more (up to 20), then these are called multicopy plasmids. Multicopy plasmids are able to replicate in the absence of protein synthesis.

multiple-factor hypothesis: H. Nilsson-Ehle (1909) and E.M. East (1916) postulated this hypothesis. They explained quantitative inheritance on the basis of action and segregation of a number of allelic pairs having duplicate and accumulative effects without complete dominance. Studies on ear length in corn by R.A. Emerson and E.M. East (1913), seed weight in beans by Johannsen, and grain colour in wheat by H. Nilsson-Ehle (1909) supported the multiple-factor hypothesis. G.U. Yule (1906) suggested that continuous quantitative variation might be produced by a multitude of individual genes, each with a small effect on the measured character.

multiplex automated genome engineering (MAGE): A highly efficient genome editing method that can generate a large and heterogeneous population of mutant bacterial genomes within days. Using oligonucleotide-mediated allelic replacement technology in a cyclic and automated manner, MAGE can simultaneously target and modify multiple genomic locations across a large population of cells.

multiplex sequencing: An approach to high-throughput sequencing that makes use of several pooled DNA samples run through gels simultaneously and then separated and analysed.

multiplexed base editing: Refers to simultaneously produce two distinct point mutations at different loci.

multiplexed orthogonal base editor (MOBE) systems: A suite of tools that leverages RNA aptamers to simultaneously produce two distinct point mutations at different loci. The researchers engineered the RNA aptamer to employ the cytidine or adenosine deaminase directly to the gRNA. A major restraint of base editor systems is attributed to the fact that the two types of base editors are hard to multiplex together due to guide RNA (gRNA) crosstalk, which results in frequent unwanted edits at on-target sites. To overcome this obstacle, researchers at the University of California, San Diego developed MOBE systems (M. Mukhopadhyay, 2024).

multiplexed reduced representation bisulfite sequencing (mRRBS): This method requires less input DNA, offers more flexibility in coverage and interrogated more CpG loci at a higher regional density.

multiplexing: A laboratory approach that performs multiple sets of reactions in parallel (simultaneously); greatly increasing speed and throughput. The application of nucleotide barcodes is followed by the subsequent pooling of multiple DNA samples that are sequenced as a single sample to increase throughput, take advantage of sequencer output, and decrease cost.

multiplication: The increase in number of individuals produced in the progeny from seed or by vegetative means.

multiplication population: Subpopulations of the general breeding population purposefully selected for different sets of traits or deployment destinations.

multiplication rule of probability: The probability of the joint occurrence of two or more independent events is the multiplication of their individual probabilities. For example, if an event occurs in n_1 ways, and another totally independent event occurs in n_2

ways, the number of ways both events can occur simultaneously is $n_1 \times n_2$.

multiplicational gene family: It comprises 10^1 to 10^4 copies of a gene, each 80–100 nucleotides long. Repeat units in this case are essentially identical.

multiplicity of infection: Synonym of virus load, the number of viral particles or infectious units per infected cell. Usually represented by the number of viral genomes – gcp/cell (genome-containing particles per cell) or TU/cell (transducing units per cell).

multipotency: Totipotent cells have total potential. They specialize into pluripotent cells that can give rise to most, but not all, of the tissues necessary for fetal development. Pluripotent cells undergo further specialization into multipotent cells that are committed to give rise to cells that have a particular function. For example, hematopoietic stem cells can differentiate into multiple types of blood cells, but not into muscle cells, skin cells, or cells of any other lineage. *Aka* PLURIPOTENCY.

multi-purpose vector system: This vector system enables screening of gene/cDNA libraries for functional genomics studies. The vector enables an all-in-one strategy for gene expression in mammalian, bacterial and insect cells and is also suitable for direct use *in vivo*.

multiregional model: The hypothesis that different hominins (e.g. *Homo erectus* and *Homo neanderthalensis*) found across the Old World evolved *in situ* into modern human populations. *Contrasts* OUT-OF-AFRICA MODEL.

multiseptate: Having many septa or crosswalls.

multisite mutant allele: A mutant allele differing from its wild-type form at two or more sites.

multispectral satellite imaging systems: Multispectral remote sensing enables analysts to differentiate objects that are hard to tell apart in the visible band. Some of the most important multispectral sensing systems are operated by government agencies as well as private commercial firms (C. Lavers, 2013). The US government programmes include (i) Landsat programme (Landsat Multispectral Scanner [MSS], Thematic Mapper [TM] and Enhanced Thematic Mapper Plus [ETM+], Landsat 8 satellite (operational land imager [OLI] and thermal infrared sensor [TIRS]); (ii) Advanced very high-resolution radiometer [AVHRR]) and (iii) moderate resolution imaging spectroradiometer [MODIS]. Several commercial systems, including Système Pour l'Observation de la Terre [SPOT], IKONOS, OrbView, GeoEye, DigitalGlobe's WorldView-3 sensor and Next Generation Satellite Imaging (NGSI). *See* Lavers, C. (2013), for details.

multi-tissue epigenetic clock: Accurate age estimators can be built on the basis of highly conserved CpGs. However, the most accurate clock results from applying elastic net regression to all CpGs. While the anti-ageing effect of calorie restriction could be detected with all types of epigenetic clocks, only ridge regression-based clocks replicated the finding of slow epigenetic ageing effects in dwarf mice.

multitude of genes: *See* POLYGENES.

multivalency: A property in which several histone modifications work together to increase the binding of reader proteins or the stability of a nucleosomal arrangement.

multivalent: An association of more than two chromosomes whose homologous regions are synapsed by pairs. For example, in autopolyploids or translocation heterozygotes. *Compare with* BIVALENT.

multivalent vaccines: Multiple foreign genes have been inserted into the Vaccinia virus. Vaccinia virus strain VP168 carries three foreign genes, namely, *InfHA*, *HSVgD* and *HBsAg*. Vaccinia virus recombinant VP168 responded by making antibodies to all three foreign antigens synthesized under Vaccinia virus regulation.

MUMmer: A system for rapidly aligning large DNA sequences to one another. It can align whole genomes to other genomes, large genome assemblies to one another, partial (draft) genomes sequences to one another or (with release 4) a set of reads to a genome.

mummy (in plants): (i) A dried, shrivelled fruit. (ii) Plant part or organ partially or wholly replaced with fungal structures.

mycorrhizae: Fungi that form symbiotic relationships with the roots of more developed plants.

Murashige and Skoog media: *See* MS MEDIUM.

***Mus musculus* genome:** *Mus musculus* has a genome size (almost the same as that of *Homo sapiens*) of 3,100,000,000 bp which comprises 30,000 genes, which is the same as that in human beings.

mutability: The ability of a gene to experience mutation.

mutable genes: A class of genes that spontaneously mutate very frequently.

mutable site: A site where a mutation occurs.

mutafacient: A gene or genetic element that determines or increases the chance of the occurrence of a mutation of anther gene or genetic element.

mutagenesis: (i) The process by which physical or chemical agents increase the rate of mutations. In nature, different alleles of a gene arise through spontaneous mutations and thus provide a basis for genetic variation which serves as raw material for evolution. (ii) The use of methods to physically change or 'mutate' the genetic sequence, without adding DNA from another organism. Various chemicals and ionizing radiation can be used to invoke these changes. SITE-DIRECTED MUTAGENESIS can also be used to invoke changes in specific genes. In plants, such agents are used to change a plant's genetic sequence, and the plant can pass on these new characteristics to its offspring.

mutagen: An agent that increases the mutation rate over the spontaneous mutation rate within an organism or cell. A mutagen can be a physical agent (e.g. UV light, X-rays, γ-rays, α-particles, β-particles, neutrons, etc.) or a chemical [e.g. base analogues, such as 5-bromouracil, 5-bromo deoxyuridine, 2-amino purine, 8-ethoxy caffeine, 1.3.7.9.-tetramethyluric acid, maleic hydrazide; antibiotics, such as azaserine, mitomycin C, streptomycin, streptonigrin, actinomycin D; alkylating agents, such as sulfur mustards (ethyl-2-chloroethyl sulfide), nitrogen mustards (2-chloroethyl-dimethyl amine), epoxides (ethylene oxide), ethyleneimines, sulfates, sulfonates, diazoalkanes, nitroso compounds (N-ethyl-N-nitroso urea); azide (sodium azide); hydroxylamine; nitrous acid; acridines (hydrocyclic dyes), such as acridine orange]. *See* TERATOGEN *and* MUTATION.

mutagenic: Substances and circumstances inducing mutations.

mutagenic DNA alterations: Any of those primary DNA alterations that do not prevent DNA replication and may give rise to gene mutations. They may arise due to mistakes in base pairing or may result from mistakes by the DNA polymerase.

mutagenic effectiveness: The ratio of factor mutations to dose (i.e. the frequency of a given mutation occurring at a given dose of mutagen employed).

mutagenicity: The capacity of a chemical or physical agent to cause permanent genetic alterations.

mutagenicity testing: The methodological assessment of the mutagenicity of chemical or physical agents.

mutagenize: Treatments that cause mutations.

mutagenized: Cells or individuals that were treated with chemical or physical mutagens.

mutant: A microorganism or cell, visible property, or eukaryotic individual that manifests new characteristics due to a change in its genetic material. A change from the normal to the exclusive or nonstandard resulting from a changed or new DNA sequence. The mutation in the DNA sequence can alter its meaning (i.e. its ability to produce the normal amount or normal kind of protein) such that the organism or cell itself is altered. Mutant organisms are used in basic studies related to gene expression and function. In plants, mutants do occur spontaneously but with a very low frequency such as one out of million. Whereas their number can be increased by treating seeds or plants with physical mutagen such as γ-irradiation or chemical mutagens like ethyl methanesulfonate (EMS), diethylsulfate (DES) and acridine orange. In plants some mutants have been released as crop varieties.

mutant allele: An allele differing from an allele found in the standard, or wild type.

mutant hunt: The process of accumulating different mutants showing abnormalities in a certain structure or function as a preparation for mutational dissection of that function.

mutant isoalleles: If the phenotype produced by different alleles is mutant, these alleles are called mutant isoalleles.

mutant site: A site on a chromosome at which a mutation can potentially occur or has occurred.

mutant strain: A strain of cells or individuals that, by one or more mutations, is distinguishable from the original strain.

mutase: An enzyme catalysing transposition of a functional group in the substrate. A substance is the substance that is acted upon by the enzyme. Intramolecular transfer of a chemical group from one position to another within the same molecule (i.e. the carbon atom). A well-known example of a mutase is phosphoglucomutase. It has a molecular weight of about 60,000 Daltons with about 600 amino acid residues (MONOMERS). The mutase can move a phosphate unit between positions 1 and 6. Here 1 refers to a carbon atom designated as '#1' and the 6 refers to a different carbon atom designated as '#6'.

mutasome: In SOS response, a complex of activated Umu D protein with Umu C and RecA proteins and DNA polymerase III. The mutasome is located at the lesion and may permit the polymerase to replicate past it.

mutation: A sudden heritable or genetic change in a nuclear, chloroplast or mitochondrial gene of an individual as a result of an alteration in a gene or chromosome, or of an increase in chromosome number. In RNA viruses, a mutation is a change to the genetic information that the RNA encodes. Similarly, in DNA viruses, a mutation is a change to the genetic information that the DNA encodes. Mutation does not include changes occurring during normal genetic recombination. The gene or chromosome set that results from such a process is also known as mutation. Mutation is thus a process as well as an effect. The term includes nucleotide substitutions, insertions, omissions, duplications, translocations, cross-linking, or other destruction within a gene, and also large chromosomal changes. A mutation can arise in a germ cell and be passed on to an individual's progeny, who will then carry it in every cell of their body. Most mutations do not have any effect, some may have positive effects and many have harmful effects. Loss-of-function mutations in a gene on the expressed allele will impair gene function, whereas a mutation in the silenced allele will have no apparent effect. Hence mutations in imprinted genes are associated with parent-of-origin effects on clinical phenotype. Mutations in genes that form various epigenetic mechanisms have an impact on their function. *See* GERMLINE MUTATION *and* SOMATIC MUTATION.

mutation breeding: Commonly used practice in plant breeding and other areas in which chemicals or radiation are applied to whole organisms (e.g. plants) or cells so that changes in the organism's DNA will occur. Mutations are evaluated for their beneficial effects, such as disease resistance. The identification and incorporation of useful mutations has been essential for traditional crop breeding. Several techniques in plant breeding have involved induced mutations, which were utilized by some plant breeders primarily during 1960s and 1970s to introduce desirable genes into the plants in which they were interested. Examples of desirable genes include those governing resistance to plant diseases, increased yield per acre/hectare or improvements in composition that were not present within the historic/natural germplasm of that plant species. These new-to-that-species genes were created by soaking its seeds or pollen in mutation-causing chemicals (i.e. mutagens) or by bombardment of seeds with ionizing radiation; followed by the grow-out of the resultant plants and the selection of the particular mutation (i.e. beneficial trait) desired by the plant breeder. That plant was then propagated via straightforward breeding to yield seeds that are still sown today.

mutation event: The actual occurrence of a mutation in time and space.

mutation frequency: The frequency at which a particular mutant is found in the population.

mutation hotspots during meiosis: Meiotic recombination is a genome-wide programme of exchanges between the paternal and maternal chromosomes that occurs in germ cells. R. Hinch *et al.* (2023) report that the mutation rate at meiotic recombination sites was several hundred-fold higher than the genome average. Some mutations were products of the meiotic recombination pathway, and others resulted from other DNA repair pathways.

mutation load: The total genetic burden in a population resulting from accumulated deleterious mutations.

mutation map: The frequency of mutations mapped along chromosomes and represented as a diagram.

mutation pressure: The continual production of an allele by mutations.

mutation rate (μ): The number of mutation events/gene/unit of time (e.g. per cell generation). The mutation rate is the likelihood of parentage when findings suggest otherwise. In humans, the mutation rate is 1 bp/10^9 bp per cell division; this corresponds to 10^{-6} mutations/gene/cell division and because there are 10^{16} divisions in a lifetime, 10^{10} mutations/gene/lifetime.

mutation spectrum: A term used to indicate that an organism may produce different kinds of mutants due to the differential action of various mutagens.

mutation theory of evolution: Hugo de Vries (1901) advocated that a single mutation can lead to radical changes in an organism which can transform one species to another. *Aka* DEVRIESISM.

mutation trend: A series of slight gene mutations in the same direction resulting in gradual intensification of a character change.

mutational dissection: The study of the components of a biological function through a study of mutations affecting that function.

mutational equilibrium: The equilibrium points at which the forward mutations balance the reverse mutations so that the allelic frequencies do not change as a consequence of mutations.

mutational heritability: The ratio between the mutational variance (V_M) and the environmental variance (V_E).

mutational hot spot: A site or region within a gene or genome that frequently mutates.

mutational load: Genetic load caused by mutation.

mutational site: The more or less defined position along a gene at which mutations occur.

mutational variance (V_M): The variance in a quantitative trait caused by new mutations in each generation.

mutations classified based on cause of mutations: (i) SPONTANEOUS MUTATIONS; (ii) INDUCED MUTATIONS. *See details as separate terms.*

mutations classified based on change in the sense of codon: (i) SAME-SENSE MUTATIONS, (ii) MISSENSE MUTATIONS and (iii) NONSENSE MUTATION. *See details as separate terms.*

mutations classified based on changes in chromosome number: Broadly speaking, such mutations are classified as (i) ANEUPLOIDS and (ii) EUPLOIDS. These terms refer to changes in chromosome number in a disomic cell/individual. *See details as separate terms.*

mutations classified based on changes in chromosome structure: (i) INTRACHROMOSOMAL CHANGES and (ii) INTERCHROMOSOMAL CHANGES.

mutations classified based on expression or effect on function: (i) MORPHOLOGICAL MUTATIONS; (ii) CONDITIONAL MUTATIONS; (iii) BIOCHEMICAL MUTATIONS; and (iv) POLAR MUTATIONS. *See details as separate terms.*

mutations classified based on location of genes: (i) NUCLEAR MUTATIONS and (ii) CYTOPLASMIC MUTATIONS. *See details as separate terms.*

mutations classified based on survival of the organism: (i) LETHAL MUTATIONS; (ii) VITAL MUTATIONS; and (iii) SUPERVITAL MUTATIONS. *See details as separate terms.*

mutations classified based on their size: (i) MICROMUTATIONS and (ii) MICROLESIONS. *See details as separate terms.*

mutations classified based on type of tissue in which it is induced: (i) SOMATIC MUTATIONS and (ii) GERMINAL MUTATIONS. *See details as separate terms.*

mutations in the promoter: Down mutations in the promoter decrease the transcription rate. Up mutations in the promoter increase the transcription rate compared to transcription rate of the normal promoter. Study of down and up mutations of the promoter suggests that the function of the −35 sequence is to deliver the signal for recognition by RNA polymerase while the −10 sequence permits the complex to convert from 'closed' to 'open' form. RNA polymerase may initially 'touch down' at the −35 sequence and then extend its contact over the −10 region. The enzyme covers around 60 bp of DNA. The initial 'closed' binary complex is converted by melting of a sequence of 12 bp that extends from the −10 region to the start point. The A-T rich base pair composition of the −10 sequence may be important for melting reaction. In certain genes, coding for small species of RNA (tRNAs, 5S rRNA, etc.), promoter sequences are located within the mature gene itself.

mutations, characteristics of: (i) Mutations are more or less permanent changes in kind, number or sequence of nucleotides in genetic material. The making or breaking of hydrogen bonds in a duplex DNA is a temporary, hence nonmutation change. (ii) Mutations are uncoded or unprogrammed changes. Methylation of cytosines or adenines is not a mutation because it is a programmed change since it occurs due to instructions coded by DNA. Mutations are relatively rare and unusual changes. (iii) Genetic material tends to be protected against the occurrence of uncoded and unusual changes by the double-helical structure and its bonds to proteins. (iv) Mutations are regarded as accidental, unintended, undirected, random events as we have no way of knowing when a mutation will occur, which individual, cell, cell organelle, chromosome or gene will mutate. (v) Mutations are unoriented with respect to purpose and result from deviation of a natural process.

mutations, classification of: Mutations are classified on the basis of (i) size of mutation, (ii) changes in chromosome structure, (iii) changes in chromosome number, (iv) cause of mutations, (v) type of damage, (vi) direction of mutations, (vii) sense of codon, (viii) types of suppressor mutations, (ix) types of tissues in which mutation is induced, (x) expression or effect on function, (xi) effect on survival of the organism, (xii) location of genes, (xiii) types of amino acids substitutes, (xiv) relevance to evolution and (xv) degree of character expressed.

mutator gene: A gene that increases the mutation rate of other genes in the same organism.

mutator mutations: Mutations of DNA polymerase that increase the overall mutation rate of another cell or organism.

muticous: Blunt, or without a definite point.

muton: A unit of mutation defined by the recombination test. Two mutations belong to the same muton if they do not show recombination. The smallest mutational unit in a double-stranded DNA molecule is a single nucleotide pair. While the muton is the unit of mutation, the muton is a larger unit that occurs within the recon. *See* RECON.

mutual translocation: Reciprocal transfer or crossing-over between the terminal portions of two non-homologous chromosomes.

mutualism: An association in which both symbionts benefit. There is not even one example of mutualism known where both partners benefit equally. In mutualism, the benefits may be nutritional or physical: (i) SYMBIOSIS BETWEEN ALGAE AND INVERTEBRATES: the algae provide the animals with organic compounds, which are products of photosynthesis, while the animals provide the algae with waste products such as nitrogenous compounds and carbon dioxide, which algae use in photosynthesis. (ii) BIOLOGICAL NITROGEN FIXATION: rhizobia normally live in the soil, but when there is limited soil nitrogen, legumes provide a home for symbiotic nitrogen-fixing bacteria. Rhizobia, in turn, fix nitrogen for the legumes. (iii) The association of microorganism in the digestive tract of ruminants. Microbes get nutrient-rich habitat, whereas ruminant get cellulose ferment. Another example is legume – *Rhizobium* symbiosis. *See* SYMBIOSIS.

mutually antagonistic chromatin marks: Histone variant H2A.Z and DNA methylation are mutually antagonistic chromatin marks. DNA methylation can alter chromatin structure and result in gene silencing by excluding H2A.Z and H2A.Z protects genes from DNA methylation.

mutually exclusive events: The set of events wherein only one can occur at a point in time. The blood group of a patient would be O, A, B or AB. These are mutually exclusive. Signs and symptoms such as pain, diarrhoea and vomiting are not mutually exclusive; they can occur together in a patient.

mutually exclusive histone modifications: Methylation of histone H3R2 by arginine methyltransferase PRMT6 and H3K4 by a MIXED-LINEAGE LEUKAEMIA (MLL) complex are mutually exclusive modifications. PTLMs of histones (or 'marks') are crucial in defining functional states, predominantly in promoter regions.

mutually exclusive mRNA splicing: A eukaryotic gene contains a large number of introns. Many of them are alternatively spliced. Four mechanisms of mutually exclusive splicing include stearic interference, spliceosomal incompatibility, disposal by nonsense-mediated decay and antagonism of repressor by Docker:selector pairing. Reverse splicing provides considerable insight into the catalytic flexibility of the spliceosome.

mutually exclusive patterns: In this type of alternative RNA splicing, the selection of exons of a pair is such that if one member of a pair is spliced in, other is necessarily to be excised out.

mutually exclusive RNA splicing: Characterized by splicing of exons in a coordinated manner such that two or more splicing events are not independent. As the name 'mutually exclusive' indicates, exactly one out of two exons (or one group out of two exon groups) is retained, while the other one is spliced out.

mycology: Branch of science specifically deals with the study of fungi. In agriculture, especially beneficial fungi such as mushrooms and mycorrhiza are dealt with in mycology. Mycology deals with the classification, isolation, modifications and commercial exploitation of various fungi.

mycoparasite: A fungus that attacks another fungus.

mycorrhiza: Fungus growing in symbiotic association with a wide variety of plants. Myco = fungus; rhiza = root. Mycorrhizal fungi are beneficial fungi that form symbiotic associations with a large number of field fruit and forest plants. These fungi help the plants in the uptake of inorganic elements such as phosphorus from soil, water absorption and also act as bioprotectants (protecting roots from pathogens). There are two main types of these fungi: endomycorrhizae (vesicular arbuscular mycorrhiza – VAM) and ectomycorrhizae. VAM form vesicles and arbuscules whereas the ectomycorrhizae do not form vesicles or arbuscules rather these fungi form mycelial mantle on the roots.

N

n, 2n: The symbol for gametic (haploid) and zygotic or somatic (diploid) chromosome numbers, respectively. *See* HAPLOID and DIPLOID.

N_3-CMC-enriched pseudouridine sequencing (CeU-seq): An approach used to map modified nucleotides that exploits the chemical properties of the modified nucleotide to interfere with reverse transcription. N3-carbodiimide-N-cyclohexyl-N′-(2-morpholinoethyl)carbodiimide metho-p-toluenesulfonate (CMC)-enriched pseudouridine sequencing (CeU-seq) was developed, which enhanced the sensitivity of pseudouridine mapping and led to an almost tenfold increase in the number of residues detectable.

N-acetylcysteine (NAC): An acetylated form of the amino acid L-cysteine ($C_5H_9NO_3S$). NAC is a precursor to glutathione, a powerful antioxidant in all cells. NAC is well captivated and is incorporated into glutathione once in the cells. It is capable of raising glutathione levels better than glutathione supplementation by itself. Glutathione is a tripeptide consisting of three amino acids, glycine, L-cysteine and L-glutamate. It decreases inflammation, increases immunity, helps in detoxification and more.

N-banding: A special chromosome staining method related to C banding, which reveals specific types of heterochromatins. The pattern of bands and interbands along a chromosome serves as a tool for chromosome or chromosome segment identification. The method has been successfully applied in wheat cytogenetic analysis.

N-end rule: According to the N-end rule, the amino acid at the amino or N-terminal end of a protein is a signal to proteases that control the average length of life of a protein. Almost complete predictability was attained in determining the life of the β-galactosidase protein based on the N-terminal amino acid. Protein lifespans range from 2 min for those with N-terminal arginine to greater than 20 h for those with N-terminal methionine.

N-ethylmaleimide-sensitive fusion (NSF) protein: A protein with ATPase activity that strips a complex of a v-SNARE and a t-SNARE.

N-linked oligosaccharide: Chain of sugars attached to a protein through the NH_2 group of the side chain of an asparagine residue.

N-segments: Sequences of nucleotides, added in a template-free fashion, at the joining junctions of heavy chain of antibodies.

Na^+-K^+ pump: Refers to Na^+-K^+ ATPase, a transmembrane carrier protein occurring in the plasma membrane of most animal cells that pumps Na^+ out of and K^+ into the cell, utilizing energy derived from ATP hydrolysis. *Aka* NA^+-K^+ ATPASE.

NADH dehydrogenase complex: The full form of NADH is 'nicotinamide adenine dinucleotide (NAD) + hydrogen (H)'. The first of the three electron-driven proton pumps in the mitochondrial respiratory chain. It receives electrons from NADH. NAD^+ and its phosphorylated form, $NADP^+$, are hydride-accepting coenzymes that play crucial roles in substrate oxidation reactions in the metabolism. In reduced forms, NADH and NADPH are hydride-donating coenzymes in substrate-reducing reactions.

NADH/NAD^+ ratio: The full form of NAD^+ is 'oxidized nicotinamide adenine dinucleotide'. The ratio of reduced NAD to oxidized NAD; changes in this ratio can promote or interfere with certain biochemical reactions that require either NAD^+ or NADH as co-factors.

naked gene: Pure, isolated DNA wanting of any proteins that may bind to it. A bare gene (a strand of DNA that codes for a protein) can be isolated from an organism or otherwise derived (e.g. synthesized from sequence data). Injecting the 'naked gene' for the relevant antigen of certain pathogens into some tissues in the (usual disease host) organism sometimes led to those (host organism) tissues taking up the 'naked gene' and expressing some of the (pathogen's) antigen(s), such that the (putative host organism's) immune system initiates an immune response (thereby resulting in vaccination against the disease conferred by the pathogen). When that happens, such naked genes are called DNA VACCINES.

naming a method: Editors of *Nature Methods* (2024, volume 21) deliver some guidance about how to best name methods and software tools. Many names, such as the acronyms SIMBA and TARDIS (TEMPORAL ANALYSIS OF RELATIVE DISTANCES), were inspired by pop culture references. A lot of methods start with a lowercase 'i', which we can only assume is inspired by the commercial popularity of the iPhone. And no one will be astonished that we have gotten a proliferation of technique names ending in '-omics' and '-seq'. A good method name should be attractive, unique and ideally reference the technique used. For example, 'Metabuli' is the name of a metagenomic taxonomic classifier tool, which is both unique as a word and references the technique of metagenomics. 'Molecular pixelation' is the name of a sequencing-based method to map the spatial organization of cell-surface proteins. While it does not specifically appeal to a particular technique, the name highlights the spatial and molecular aspects of the method and is a likeable phrase that should be easy to remember. Scientists need to come up with a method name that helps the technique or tool shine. *See* SIMBA, TEMPORAL ANALYSIS OF RELATIVE DISTANCES (TARDIS), METABULI *and* MOLECULAR PIXILATION (MPX).

nano-antibodies in recognition of mRNA modifications: The discovery of hcIgG antibodies resulted in the development of a set of commercial reagents for recognizing 2′-*O*-methyladenosine (2′-*O*-A^m), 2′-*O*-methylcytidine (2′-*O*-C^m), 2′-*O*-methylguanosine (2′-*O*-G^m) or 2′-*O*-methyluridine (2′-*O*-U^m), as well as comprehensively to all four 2′-*O*-methyl ribose moieties.

nanobiotechnology: The branch of nanotechnology with biological and biochemical applications or uses. Nanobiotechnology often studies existing elements of nature to fabricate new devices. *Aka* BIONANOTECHNOLOGY.

nanochips: Next-generation chips with higher storage capacity.

nanocrystal molecules: Coined by A.P. Alivisatos and P.G. Schultz, it is a term used to describe double-stranded DNA molecules that have several multi-atom clusters of gold attached to them. In 1996, these researchers were trying to generate nanometre-scale electrical circuits, semiconductors and so on. A separate methodology by C.A. Mirkin *et al.* utilized strands of DNA to reversibly assemble gold nanoparticles (nanometre-scale multi-atom particles) into supramolecular (many molecules) agglomerations, in which the gold particles are separated from each other by a distance of approximately 60 Å. The aggregation of these DNA-metal nanoparticles resulted in a visible colour change. In 1996, these researchers were working to produce simple and rapid tests that would indicate the presence of a virus (e.g. HIV-1 or HIV-2) via a visible colour change. Such a test would use two non-complementary DNA sequences, each with a gold nanoparticle attached to it (via a thiol group). The two sequences would be selected for their ability to fasten on to a target sequence in the desired virus, but these would be unable to combine, since these are non-complementary. When double-stranded DNA molecules possessing two sticky ends (complementary to the sequences attached to the virus) are added, the consequent colour change reflects the virus's presence.

nanomanipulator: A virtual reality device coupled to a scanning probe microscope, permitting the manipulation of atoms. Here individual molecules, DNA and atomic objects can be manipulated.

nanopipettes: Objects similar to pipettes with which controlled delivery of chemicals, reagents and light at nanometre-scale areas is accomplished.

nanopore: A device that can distinguish between DNA strands (molecules) that differ from each other by a single nucleotide (in the makeup of those molecular strands). Developed by D. Deamer and M. Akeson in 2001, nanopore consists of an artificial membrane (lipid bilayer) or ceramic objects such as zeolites, with a 'hole' (nanopore) punctured in that membrane by the protein alpha-haemolysin. Because a DNA molecule moving through such a nanopore temporarily blocks the nanopore (until it dissociates into a single DNA strand and 'slides' through), an electrical current/voltage applied to that nanopore varies (in amplitude, modulation, duration, etc.) as the DNA strand 'slides through', in a way that provides information about the nucleotides that make up that DNA strand. Nanopores are now in use for DNA sequencing. This technique has been developed using single-molecule sequencing and single real time sequencing, eliminating the need for clonal amplification. This reduces mistakes caused by PCR, simplifies library preparation and, most importantly, gives a much higher read length using higher-throughput platforms. Pacific Biosciences' platform uses SINGLE-MOLECULE REAL-TIME (SMRT) sequencing to give read lengths of around 1000 bases. Helicos Biosciences utilizes single-molecule sequencing and therefore does not require amplification prior to sequencing. Oxford Nanopore Technologies are currently developing silicon-based nanopores, which are subjected to a current that changes as DNA passes through the pore.

nanopore sequencing: A single molecule of DNA or RNA can be sequenced without the need for PCR amplification or the chemical labelling of the sample. Nanopore sequencing has the potential to offer relatively low-cost genotyping, high mobility for testing and rapid processing of samples with the ability to display results in real time.

nanopores in protein translocation: Nanopores are used in nucleic acid sequencing. Nanopore applications, however, extend far beyond nucleic acid sequencing. Nanopores have recently been used to identify and fingerprint proteins. Proteins, unlike DNA, do not have a uniform charge. The electrophoretic force cannot, in general, be used to translocate or linearize them. A. Sauciuc *et al.* (2023) show that the introduction of sets of charges in the lumen of the CytK nanopore spaced by ~1 nm creates an electroosmotic flow that induces the unidirectional transport of unstructured natural polypeptides against a strong electrophoretic force. Molecular dynamics simulations indicate that this electroosmotic-dominated force has a strength of ~20 pN at −100 mV, similar to the electric force on single-stranded DNA. Unfolded polypeptides produce current signatures as they traverse the nanopore, which may be used to identify proteins. This approach can be used to translocate and stretch proteins for enzymatic and non-enzymatic protein identification and sequencing.

nanoprobe: Nanoscale devices are used to understand the properties of materials and systems.

nanoreplicators: A nanobot system capable of replication.

***nanos* gene:** A gene whose product plays a role in the posterior structure morphogenesis.

nanospheres: An approach that utilizes a nanoparticle medium, core-shell type nanospheres, in conjunction with a pressurization technique during microchip electrophoresis. DNA fragments up to 15 kbp have been successfully analysed within 100 s without observing any saturation in migration rates. DNA fragments migrate in the medium while maintaining their characteristic molecular structure. To guarantee effective DNA loading and electro focusing in the nanosphere solution, scientists developed a double pressurization technique. Optimal pressure conditions and concentrations of packed nanospheres are important to attain improved DNA separations.

nanotechnology: An accurate molecule-by-molecule regulation of the products and by-products in the development of functional structures. From the Latin *nanus* = dwarf, it literally means DWARF TECHNOLOGY. The word was originally coined by Norio Taniguchi in 1974, to refer to high precision machining. However, R. Feynman and K.E. Drexler in the 1990s popularized the concept of nanotechnology as a new and developing technology in which humans deploy objects whose dimensions are approximately 1–100 nm. Theoretically, it is possible in the near future that a variety of man-made NANO-ASSEMBLERS or tiny (molecular) machines, smaller than a grain of sand, could manufacture things produced in factories today. The technology also incorporates biochips, biosensors and manipulating atoms and molecules to manufacture bigger but still microscopically functional structures and machines. Potential applications related to food include food packaging and processing to improve food safety and quality, and better nutrient and ingredient profiles to improve health.

narrow heritability: *See* HERITABILITY IN NARROW SENSE.

narrow-host-range plasmid: A plasmid capable of replicating only in one, or at most a few, different bacterial species.

narrowing the QTL down to candidate genes: Identifying the actual loci that affect a quantitative trait involves demonstrating causality using techniques like positional cloning followed by targeted gene replacement. The quest for individual genes within a **QTL** is assisted by the identification of a priori candidate genes using classical reverse genetics or bioinformatics. A functional relationship between the candidate gene and the QTL is demonstrated by using functional complementation. Once the QTL has been identified, molecular techniques can be employed to narrow the QTL down to candidate genes.

National Agricultural Research and Education System-Blended Learning Platform (NARES-BLP): NARES-BLP refers to the transformative shift in agricultural education and emphasizes the significance of BLENDED LEARNING PROGRAMS

(BLP). Blended learning involves a combination of online and face-to-face interactions between students and faculty. This platform emphasizes the importance of BLP technologies and tools, such as LEARNING MANAGEMENT SYSTEMS (LMS), video conferencing platforms and content creation tools, in the contemporary education system. NARES-BLP combines the best aspects of various teaching techniques and caters to diverse learning requirements. The Indian Council of Agricultural Research (ICAR), New Delhi is promoting blended learning through innovative in-house software.

National Center for Biotechnology Information (NCBI): NCBI was established in 1988 as a division of the National Library of Medicine (NLM) at the National Institutes of Health (NIH). It deals with a database of experimentally determined three-dimensional (3-D) biomolecular structures. Most 3-D structure data are obtained from X-ray crystallography and NMR spectroscopy. These provide a wealth of information regarding biological functions, mechanisms linked to function and the evolutionary history of and relationship between macromolecules. The addition of structure data to Entrez makes this information easily accessible to biologists and facilitates comparative analysis involving 3-D structure. Its mission is defined as development, distribution and maintenance of various molecular databases and computer software in order to support biological and biomedical studies at the molecular level. Further, it possesses (i) NCBI Sequence Submission System, (ii) Sequence Retrieval System and (iii) Sequence Analysis tools at NCBI.

National Institute of Health (NIH) guidelines: The Asilomar Conference resulted in a set of NIH guidelines in recombinant DNA technology. NIH has revised the document, 'Guidelines for Research Involving Recombinant DNA Molecules', several times since 1976. The NIH guidelines were modified in November 2013 at the latest.

National Institutions of Health (NIH): A non-regulatory agency that maintains oversight of research activities that the agency funds.

native: Naturally occurring in an area but not necessarily confined to it. *Compare with* ENDEMIC.

native ChIP: Native ChIP uses native chromatin obtained by the nuclease digestion of cell nuclei.

native DNA: Double-stranded DNA isolated from a cell with its hydrogen bonds between strands intact, as opposed to denatured DNA.

natural antisense transcript-derived siRNAs (nat-siRNA): Small RNAs formed from partially overlapping transcripts of antisense gene pairs surrounding an inducible and a constitutive gene.

natural cross-pollination: Cross-pollination that ensues naturally, in contrast to artificial or hand-pollination.

natural desmutagen: Antimutagenic agents that are extracted from the juices of various plants like cabbage, broccoli, green pepper, aubergine, apple, burdock, shallot, ginger, pineapple, mint leaf, radish, sweet potato, grape, Japanese ginger, cauliflower, beefsteak plant, enokidake mushroom and shimeji mushroom.

natural epialleles towards phenotypic variation: Plants possess dozens of natural epialleles contributing to phenotypic variation. The self-reinforcing loop between CHROMOMETHYLASE 3 (CMT3) and KRYPTONITE, involving DNA methylated in the CHG context and histone H3 lysine 9 methylation (H3K9me), maintains the natural epiallele in *Arabidopsis* and it mainly depends upon the CG dinucleotide symmetry.

natural epigenetic variation: The natural epigenetic variation residing in DNA methylation patterns has the potential to alter gene expression through natural epialleles which can be utilized for crop improvement. The natural sources of epigenetic variation include spontaneous changes, *cis* genetic changes including transposon insertions, structural rearrangements and *trans* genetic changes including siRNA signals. The naturally occurring variation linked with DNA methylation has the potential to generate epiallelic variation which could be used in plant research and used as an additional source of variation within a species for crop improvement in future.

natural farming: A chemical-free farming system rooted in Indian tradition enriched with a modern understanding of ecology, resource recycling and on-farm resource optimization. It is considered an agroecology-based diversified farming system that integrates crops, trees and livestock with functional biodiversity. It is largely based on on-farm biomass recycling with major stress on biomass mulching, the use of on-farm cow dung–urine formulations, maintaining soil aeration and the exclusion of all synthetic chemical inputs. Natural farming is expected to reduce dependency on purchased inputs. It is considered a cost-effective farming practice with scope for increasing employment and rural development.

natural host: A host in which the pathogenic microorganism (or parasite) is normally found and in which the pathogen can undergo its full development. *Aka* TYPICAL HOST.

natural hybrid: A hybrid taxon produced by chance in the wild.

natural inducer: A natural inducer is a protein manufactured in the notochord. It is released to the extracellular space and binds to the surface of ectodermal cells. This causes the stimulation of cAMP, causing the release of Ca^{2+}. This in turn activates some enzymes which alter the structure of chromosomes in the sense that the transcription of selected genes occurs.

natural inhibitor for cell death: An evolutionarily conserved molecule is used to make cholesterol which also acts as a defence against a cell-death mechanism called FERROPTOSIS. F.P. Freitas *et al.* (2024) and Y. Li *et al.* (2024) performed studies on a regulated form of cell death called ferroptosis, driven by an iron-dependent modification of lipids in cellular membranes. The results bring into sharp focus an unexpected hero, the molecule 7-DEHYDROCHOLESTEROL (7-DHC). *See* FERROPTOSIS.

natural killer cell (NK cell): Cytotoxic cell of the innate immune system that can kill virus-infected cells.

natural language processing (NLP): A subfield of linguistics, computer science, information engineering and artificial intelligence concerned with the interactions between computers and human (natural) languages, in particular on how to programme computers to process and analyse large amounts of natural language data. Challenges in natural language processing frequently involve speech recognition, natural language understanding and natural language generation.

natural morphogen: Retinoic acid (RA) is a natural morphogen and homeobox (*hox*) genes are among the candidate-responsive genes. This was shown in embryonal carcinoma human cell line NT2/D1. Nine genes of the *HOX-2* cluster are differentially activated when exposed to varying concentrations of retinoic acid in 3′ to 5′ directions. RA is a vitamin A derivative, which can regulate morphogenesis and differentiation in vertebrates. The *Drosophila* genome has been examined for RA receptor homologues. One such gene, XR2C, encodes a product with structural similarity to the human RA-responsive transcription factor, retinoid X receptor, which is encoded by ultraspiracle (usp), a locus required for both maternal and zygotic pattern formation.

natural phenomenon: A natural phenomenon is a non-artificial event in the physical sense, and therefore not produced by humans, although it may affect humans (e.g. birth, ageing and death). Common examples of natural phenomena include (but are not limited to) meteorological phenomena (like hurricanes, thunderstorms and tornadoes) and geological phenomena (like volcanic activity and earthquakes).

natural population change: A change in a population that transpires only due to the difference in birth and death rates, and not immigration or emigration.

natural populations: In natural populations, the relative frequency of alleles may vary greatly. In natural populations, much more is generally left to chance. Any genotype may mate with any genotype.

natural recombination: As a natural process, recombination takes place in sexual reproduction during meiosis.

natural resource: A source of material or energy that is pulled out from the environment.

natural selection: The concept developed by Charles Darwin in 1859; genes which produce characteristics that are more favourable in a particular environment will be more abundant in the next generation. This mechanism of evolution works in favour of those individuals that (for genetically based reasons) are better adapted to cope with environmental opportunities and constraints. Undesirable characteristics are thrown out of the gene pool. The more fit individuals have an improved chance of leaving descendants behind, eventually paving the way for genetically based changes in populations, or evolution.

natural selection in crop improvement: Natural selection is the process by which certain traits or alleles become more or less common in a population over time, driving evolution and shaping the genetic makeup of populations. Examples highlighting the impact of natural selection on genetic diversity and structure include studies on maize populations. They observed natural selection acting on genes related to local adaptation, resulting in allele fixation and reduced genetic diversity. Another study revealed evidence of natural selection on genes involved in herbivore defence, leading to allele fixation while maintaining genetic diversity within populations.

natural variation: Similar to genetic variation, it refers to the variation in the genetic and epigenetic material of a population or species, and includes the nuclear, mitochondrial, ribosomal genomes as well as the genomes of other organelles. New genetic variation is caused by epigenetic and/or genetic mutation. It may end up in the form of recombination, migration and/or alterations in the karyotype (the number, shape, size and internal arrangement of the chromosomes). Genetic drift is a statistical measure of the rate of genetic variation in a population.

natural variation in *Arabidopsis*: Over 750 natural accessions of *Arabidopsis thaliana* have been collected from around the world and are available from the two major seed stock centres, Arabidopsis Biological Resource Center (ABRC) and Nottingham Arabidopsis Stock Centre (NASC). These accessions are quite variable in terms of form and development (e.g. leaf shape, hairiness) and physiology (e.g. flowering time and disease resistance). Researchers around the world are using these differences in natural accessions to uncover the complex genetic interactions such as those underlying plant responses to the environment and the evolution of morphological traits. *FRI* and *FLC* loci controlling flowering time were mapped using linkage disequilibrium analysis.

natural world, changes in: The natural world continually changes, although some changes may be too slow to observe and the rates of change vary. For example, changes in the position and apparent shape of the moon, the evolution of the universe (CHEMICAL EVOLUTION) and the evolution of life (BIOLOGICAL EVOLUTION).

nature and nurture: The antithesis between genotype and environment.

nature of regulatory control: In prokaryotes, all the genes are always active, and in this case, regulatory mechanisms are designed to turn off those genes whose product is not required by the cell. In eukaryote, all genes are normally inactive unless turned on by gene regulatory mechanisms. In prokaryotes, the genes for enzymes involved in a particular pathway are linked and expressed in an operon fashion. They are under the control of the same promoter. In eukaryotes, genes are scattered among different chromosomes. Each is under the control of its own promoter.

nature of transforming principle: O.T. Avery, C.M. MacLeod and M. McCarty in 1944 demonstrated that DNA is the transforming principle. Their experiment elegantly demonstrated that genes, the units of heredity, are made up of DNA.

N_e: *See* EFFECTIVE POPULATION SIZE.

near cognate codon-anticodon pairing: Occurs when there is mispairing base pairing (bad fit) between the anticodon and codon at position 1 and complementary base paring at other positions (2 and 3) as, for example, between GAA/UUU.

nearest-neighbour analysis: A procedure of transferring radioactive atoms between adjacent nucleotides in DNA, used to demonstrate that the two strands of DNA run in opposite directions.

near-field scanning optical microscopy (NSOM): A microscopy technique for nanostructure investigation that breaks the far -field resolution limit by exploiting the properties of evanescent waves. In SNOM, the excitation laser light is focused through an aperture with a diameter smaller than the excitation wavelength, resulting in an evanescent field (or near-field) on the far side of the aperture. When the sample is scanned at a small distance below the aperture, the optical resolution of transmitted or reflected light is limited only by the diameter of the aperture. *See* SCANNING NEAR-FIELD OPTICAL MICROSCOPY (SNOM).

near-infrared (NIR) imaging system: Uses the NIR region spectrum to measure water content and its movement patterns in leaves and soil.

near-infrared (NIR) spectroscopy: Analytical instruments that shine light (possessing wavelengths between that of visible light and infrared light spectrum) on to samples (e.g. kernels of grain) and rapidly measure the reflected or transmitted (near-infrared) light in order to swiftly determine the amounts of protein, fat, moisture, lignans, etc. present in the sample. In certain samples, the near-infrared light causes cells (or specific molecules) to fluoresce (i.e. as a light of a very defined wavelength), which can subsequently be utilized for the measurement/identification of compounds within the sample. NIR is also being developed for use in the quantification of immunoassays and the detection of specific molecules (e.g. in the DNA sequencing process).

near-infrared infra-red (NIR) imaging system: Uses the NIR region spectrum to measure water content and its movement patterns in leaves and soil.

near-infrared spectroscopy: A spectroscopic method that utilizes the near-infrared region of the electromagnetic spectrum (from 780 nm to 2500 nm).

near-isogenic lines (NILs): A pair of lines that differ for small segments of one or more chromosomes carrying genes for the target trait. To develop NILs, a cross is made between a recurrent parent (desirable genotype lacking a particular trait) and a donor parent (carrying the gene of interest). F_1s are repeatedly backcrossed with recurrent parent to reconstitute its genotype. During backcrossing, only selected plants carrying the gene of interest are used in crossing. NILs thus produced are similar to recurrent parent but differ for the gene of interest. NILs can also be produced through repeated selfing.

necrosis (Adj. necrotic): Refers to cell death caused by physical injury to the cell (e.g. by exposure to toxin, exposure to ultraviolet light, and lack of oxygen, etc.). Pathogens, when infecting plant leaves, cause necrosis at the site of infection, which is seen in the form of brown or black spots on the green leaves. Necrosis of tissues also occurs when cultured on a medium containing toxins or antibiotics. During plant transformation, the target tissues are grown on a selection medium containing appropriate antibiotic such as hygromycin or kanamycin or herbicide such as bialaphos for two cycles of 2 weeks each. During incubation, the non-transformed cells exhibit necrosis and get eradicated.

nectar: A typically sweet, nutrient-rich fluid formed by the flowers of many plants and gathered by bees and other pollinators.

nectary (Adj. nectariferous): A specialized gland that secretes nectar.

needed for RDR2-independent DNA methylation (NERD) pathway: NERD is a Glycine-Tryptophan (G-W) repeat- and PHD finger-containing protein. *Arabidopsis* harbours a second siRNA-dependent DNA methylation pathway targeting a subset of non-conserved genomic loci. The activity of the NERD-dependent pathway differs from RdDM by the fact that it relies both on silencing-related factors previously implicated only in PTGS, including RdRP1/6 and ARGONAUTE2, and most likely on 21-nt siRNAs.

negative assortative mating: A form of non-random mating in which individuals who are not genetically alike for particular traits mate and those who are alike do not. The result is a progressive increase in the frequency of heterozygotes and a decrease in the homozygotes for the discriminated traits. *Compare with* POSITIVE ASSORTATIVE MATING and RANDOM MATING.

negative complementation: The complementation that occurs when interallelic complementation allows a mutant subunit to suppress the activity of a wild-type subunit.

negative conditional selectable marker genes (ncSMGs): These SMGs encode enzymes that metabolize a non-toxic substrate into a phytotoxin. Negative conditional SMGs are usually placed under the control of constitutive promoters because their lethality is only expressed when their substrates are provided.

negative control of gene expression: A kind of control of gene expression in which the active DNA-binding regulatory protein (repressor) turns the gene off.

negative cross-resistance: A type of resistance that occurs when a resistant biotype is more susceptible to other classes of herbicides than the susceptible biotype.

negative frequency-dependent selection: A kind of selection in which the fitness of a phenotype or genotype declines, as it becomes more common. This is an example of balancing selection.

negative hetero pyknosis: Deficient charging of heterochromatin with nucleic acid in meiotic and pre-meiotic divisions.

negative interference: The phenomenon whereby a crossover in a particular chromosomal region increases the occurrence of other apparent crossovers in the nearby regions. *Compare with* POSITIVE INTERFERENCE.

negative mass selection: Negative (or purifying) selection deters the spread of deleterious alleles When all undesired off-types are discarded in the grown crop population and the remaining individuals are proliferated further, the method is termed NEGATIVE MASS SELECTION. It is applied in the cross-pollinating of vegetable species, such as carrots, radishes or beetroots, in order to advance their uniformity. Negative mass selection is no longer an acceptable breeding method for highly advanced varieties. It is usually applied for the multiplication of established varieties or seed production, to eliminate diseased plants, casual hybrids or other defects.

negative non-conditional selectable marker genes (nncSMGs): These SMGs directly encode a phytotoxic protein, and their expression results in immediate lethality unless they are driven by a developmentally regulated or inducible promoter. Negative non-conditional SMGs are useful for selective GENETIC ABLATION experiments.

negative regulation: Negative feedback in biological systems mediated by allosteric regulatory enzymes.

negative regulators: The molecules function by switching off transcription or translation.

negative screening: A screening procedure designed to identify and eliminate the least desirable plants. *Compare with* POSITIVE SCREENING.

negative selectable marker genes (nSMGs): Those SMGs that confer selective disadvantage to transgenic cells/plants and cause the death of the transformed tissue. Negative SMGs are classified as CONDITIONAL or NON-CONDITIONAL. With negative selection markers, it is probable to accomplish genetically modified plants that no longer contain the marker genes essential for the transformation process. Once the transformed cells have been identified with the help of the marker genes, they are no longer required in the genetically modified plant. By using a 'chemical signal', plants carrying the marker gene can be separated. For example, the enzyme N-acetylphosphinothricin deacetylase (DEA) functions as a negative selection marker. It converts a non-toxic substance to the plant into one that is toxic. For example, using DEA, an inactive herbicide (N-acetylphosphinothricin) is converted into an active herbicide (phosphinothricin), which attacks the plant's central metabolism and kills it. If the transgenic progeny is treated with the inactive herbicide, the plants that carry the negative DEA selection marker will die. The plants not sensitive to the inactive herbicide remain unaffected; these are marker gene free.

negative staining: A stain that does not penetrate tissues, causing them to look bright against the stained background. It is done for studying some bacteria, which are protected by colourless capsules. During negative staining, the unstained (uncoloured) capsule is revealed in the presence of a stained background. This technique is used in the electron microscope in which a reverse, or negative, image of the object is formed.

negative supercoiling: Comprises the twisting of a duplex of DNA in space in the opposite sense to the turns of the strands occurring naturally in the double helix.

negative versus positive control: The product of a regulatory gene that also plays an important role in the regulation of gene expression. Accordingly, two types of controls are known to exist: NEGATIVE CONTROL and POSITIVE CONTROL.

negative-sense RNA (−RNA): The ribonucleic acid sequence complementary to the positive (or plus) sense sequence which is not translated into protein. *Contrasts with* POSITIVE-SENSE RNA (+RNA).

negative-sense single-stranded RNA viruses (−ssRNA viruses): Negative-strand RNA viruses (−ssRNA viruses) are a group of related viruses that have negative-sense, single-stranded genomes made of ribonucleic acid. They have genomes that act as complementary strands from which messenger RNA (mRNA) is synthesized by the viral enzyme RNA-dependent RNA polymerase (RdRp).

negative-stranded RNA viruses: Viruses having a genome that consists of single-stranded antisense RNA, that is, RNA that is the complement of the message sense. Examples of negative-stranded RNA viruses include measles, mumps, respiratory syncytial virus (RSV), parainfluenza viruses (PIV), human metapneumovirus, rabies, Ebola and influenza.

neighbour design: A design in which a concept of elements of a block being 'neighbours' occurs, so every pair of elements occurs as neighbours in a block the same number of times.

neighbour joining: A recursive distance-based phylogenetic reconstruction algorithm that considers its input as a distance matrix and generates an unrooted phylogenetic tree.

neighbour-balanced designs: These designs allow an adjustment to be made for the effect that a treatment may have on adjacent plots.

neighbourhood size: Proportional to the product of the rate of gene flow, σ^2, and the effective population size per unit area (i.e. the effective density), ρ, which determines the relative rate of gene flow and random drift in a spatially continuous population. In two dimensions, it is defined as Nb (or Nw) = $4\pi\rho\sigma^2$, where ρ is population density and σ is intergenerational dispersal.

nematodes: Non-segmented roundworms, some of which are microscopic, scrounging on plants or animals, or free living in soil or water.

neocentric activity: In plants, an aberrant behaviour during meiosis in which specific chromosome regions act as secondary sites of attachment for spindle fibres.

neocentromere: A functional ectopic centromere in a novel (non-centromeric) location that originates occasionally from non-centromeric regions of chromosomes. May lack specific classes of deoxyribonucleic acid (a satellite DNA) that are usually present in a centromere. *Aka* NEOCENTRIC.

neo-Darwinism: Modern evolutionary theory inspired by the unification of natural selection and Mendelism. A synonym of the MODERN SYNTHESIS. *See* SYNTHETIC THEORY OF EVOLUTION.

neofunctionalization: The gain of a novel function (or expression pattern) from one copy of the gene duplicate.

neomorph: (i) An anatomical feature not derived from a feature found in an ancestor, but instead a new development. (ii) A mutant gene with a different function from its non-mutant equivalent. (iii) An allele that results in a novel molecular function(s) or expression of the gene product.

neomycin: An aminoglycoside antibiotic ($C_{23}H_{46}N_6O_{13}$) produced by *Streptomyces fradiae* that interferes with ribosomal activity, thus producing errors in the reading of the mRNA.

neomycin phosphotransferase (NPT): An enzyme deployed in gene transfer experiments as a reporter for transient gene expression.

***neomycin phosphotransferase II* (*NPTII*) gene:** A marker gene that codes for the enzyme neomycin phosphotransferase II, which can inactivate the antibiotic kanamycin. The *NPTII* gene is commonly employed as a marker gene for genetically engineered plants. Neomycin phosphotransferase confers kanamycin resistance to cells expressing it (i.e. cells that contain the *NPTII* gene in addition to the other gene(s) inserted along with it). So those (engineered) cells will live in a laboratory container containing kanamycin.

***neomycin transferase*:** A bacterial gene encoding resistance to several common antibiotics (kanamycin, neomycin, G418) widely used as a selectable marker in eukaryotic cells.

neophyte: A plant that has recently been introduced to a geographic area. *Contrasts* ARCHAEOPHYTE.

neo-polyploid infertility: Plants often throw out offspring with different chromosome numbers (ploidy). This phenomenon can produce bigger flowers and fruits or provide environmental resilience. It has become a target in plant breeding for crop improvement. J. Westermann and co-workers in 2024 used a REVERSE GENOMICS OF ADAPTATION approach to identify what additional challenges to fertility might be significant in newly formed polyploids in *Arabidopsis arenosa*. They identified two enzymes – a kinase AGC1, which is important for pollen tube growth, and adenosine triphosphatase ACA8, a calcium transporter – both of which impact pollen tube structure. Homozygotes for tetraploid alleles at *AGC1.5* and *ACA8* loci produced pollen tubes with markedly improved growth compared with plants homozygous for diploid alleles, both in terms of growth rate and the prevention of premature bursting. The two loci thus additively associate with improved growth of pollen tubes after whole-genome duplication and may represent good candidates for improving the fertility of neo-polyploids in other species. Pollen tube tip growth has been identified as an important fertility challenge for neo-polyploid plants, as it provides insights into a naturally evolved multigenic solution.

neoschizomers: Pairs of restriction enzymes that recognize the same sequence (ACGT/TGCA) but cleave at different positions. For example, *Tai*I and *Mae*II restriction enzymes are neoschizomers.

Nernst equation: A quantitative expression that relates the equilibrium ratio of concentrations of an ion on either side of a permeable membrane to the voltage difference across the membrane.

nested block design (NBD): An experimental design in which the set of plots is divided into blocks, each of which is further partitioned into sub-blocks.

nested gene: A single gene that produces two or more nested products by moderating the end point of protein synthesis. This can happen by the leaky read-through of a termination codon or by cotranslational frameshifting. Similar strategies are used by eukaryotic RNA viruses and nested products. Such strategies can also be fashioned from eukaryotic genes by alternative RNA splicing or the use of alternative polyadenylation sites. Nested genes function as an example of a situation in which one gene resides within an intron of another gene.

nested PCR: A modification of PCR intended to reduce the contamination in products due to the amplification of unexpected primer binding sites. PCR itself is the process used to amplify DNA samples, via a temperature-mediated DNA polymerase. Conventional PCR requires primers complementary to the termini of the target DNA. A commonly occurring problem is primers binding to incorrect regions of the DNA, giving unexpected products. Nested polymerase chain reaction involves two sets of primers, used in two successive runs of polymerase chain reaction – the second set intended to amplify a secondary target within the first run product. The second-round amplification of an already PCR-amplified sequence uses a new pair of primers that are internal to the original primers, typically done when a single PCR reaction generates insufficient amounts of product.

nested PCR method: A very sensitive method for amplification of DNA, which takes part of the product of a single PCR reaction (after 30–35 cycles) and subjects it to a new round of PCR using a different set of PCR primers, which are nested within the region flanked by the original primer pair. Nested PCR involves two steps: (i) The target DNA undergoes the first run of polymerase chain reaction with the first set of primers, shown in green. The selection of alternative and similar primer binding sites gives a selection of products, only one containing the intended sequence. (ii) The product from the first reaction undergoes a second run with the second set of primers, shown in red. It is unlikely that any of the unwanted PCR products contain binding sites for both the new primers, ensuring the product from the second PCR has little contamination from unwanted products of primer dimers, hairpins and alternative primer target sequences.

nested-association mapping (NAM): An integrated mapping strategy that cost-effectively enables genome-wide high-resolution mapping. NAM greatly facilitates complex trait dissection in many species and is used as a large-scale maize mapping resource and a genomics and bioinformatics tool. It needs large amounts of SNPs and/or individuals, and it is sensitive to genetic heterogeneity.

nested-association mapping (NAM) population: It is an integrated multi-parent population approach that combines the advantages of linkage mapping and association mapping for high-resolution and high-power mapping of complex traits.

Net plant gene: A software developed for the prediction of exon/intron splice sites.

network biology: A biological network is a method of representing systems as complex sets of binary interactions or relations between various biological entities. Networks in biology include (i) protein-protein interaction networks, (ii) gene regulatory networks (DNA-protein interaction networks), (iii) gene co-expression networks (transcript-transcript association networks), (iv) metabolic networks, (v) signalling networks (cell signalling), (vi) neuronal network, (vii) food webs, (viii) between-species interaction networks, (ix) within-species interaction networks and (x) DNA-DNA chromatin networks.

***Neurospora*:** A pink mould, commonly found growing on old food. Being haploid, it is very useful in genetic studies. Working on biochemical mutants of *Neurospora crassa*, G. Beadle and E. Tatum in 1941 put forward ONE GENE–ONE ENZYME HYPOTHESIS.

neutral alleles: The alleles whose differential contribution to fitness is so small that their frequencies change more due to generations than to natural selection.

neutral drift: Incorporation of mutations that have little or no influence on the protein in its current environment. Neutral mutations, however, may have adaptive potential and permit adaptive measures to occur more frequently. *Aka* GENETIC DRIFT.

neutral evolution: The change and occasional fixation of alleles caused by the drift of alleles not correlated with reproductive success. Here, the population evolves without the influence of natural selection.

neutral gene hypothesis: The hypothesis that most genetic variation in natural populations is not maintained by selection and hence is neutral.

neutral mutations: (i) In evolutionary genetics, mutations that have no measurable effect on the Darwinian fitness of an organism that is these mutations are neither beneficial nor detrimental to the ability of an organism to survive and reproduce. (ii) In population genetics, mutations in which natural selection does not affect the spread of the mutation in a species are termed neutral mutations. (iii) In molecular genetics, changes in DNA sequence result in the substitution of an amino acid in a protein with another amino acid, both having a similar charge. *See* SILENT MUTATION.

neutral petites: These petites lack mitochondrial DNA. *Aka* VEGETATIVE PETITES.

neutral substitutions: The substitutions that have no effect on the phenotype or function of an individual. They are neutral to selection pressure. Mutations occurring in non-coding DNA are predominantly neutral.

neutralist: One who holds the belief that most molecular divergence and variation have an insignificant impact on fitness, at least at the time when it evolved.

neutralist-selectionist debate: Neutralists believe the driving forces of molecular evolution are random genetic drift and mutations, while selectionists explain molecular evolution with advantageous mutations and balancing selection. The molecular study of evolution has generated many new findings that have indicated the importance of mutation in the evolutionary change of DNA or protein molecules.

neutrality theory of evolution: A theory originally proposed by N. Kimura in 1969 suggests that much of the variation at the molecular level is neutral and due to the interaction between drift and mutation rather than being actively maintained by selection. Variation in state is not correlated with variation in reproductive success. All states are equally fit. Molecular evolution appears to display roughly a constant rate per unit of time. It is therefore said to show a MOLECULAR CLOCK, which seems to support the neutral theory of molecular evolution. However, there are many evolution-related observations that this theory does not explain, and controversy exists for acceptance of this theory. *Aka* NEUTRALITY THEORY OF PROTEIN EVOLUTION *or* NEUTRALISM.

new central dogma of molecular biology: The term was coined by S. Subramaniam in 1999. It is a restatement of the original 'central dogma' proposed in 1960 by F.H.C. Crick to include the fact that an organism's environment also impacts when and how some of its genes are expressed to cause certain proteins to be synthesized. Environmental factors impacting gene expression include temperature, sunlight, humidity, the presence of certain bacteria, the presence of SIGNAL TRANSDUCERS AND ACTIVATORS OF TRANSCRIPTION (STATs) and so on. The new central dogma also includes the fact that more than one protein can be manufactured from each gene in an organism's genome due to interactions between (i) genes, (ii) genes and their protein products (e.g. STATs) and (iii) genes and some environmental factors. Mechanistically, different proteins are formed via alternative splicing of the mRNA transcript. For example, a single intronic base substitution that is present within the *IKAP* gene (the allele responsible for the disease known as *Familial Dysautonomia*) affects the splicing of the IKAP transcript (i.e. the mRNA that determines which protein is subsequently manufactured by the cell), varying translation start or stop site (on the gene); or frameshifting (i.e. different set of triplet codons in the mRNA is translated). *Compare with* OLD CENTRAL DOGMA.

new central dogma of plant disease resistance: (i) The WRKY domain is a carboxy-terminal domain in R protein. WRKY proteins are **TFs** defined by a 60-amino acid region containing the conserved amino acid sequence WRKYGQK and a zinc-finger-like motif. WRKY proteins bind to the so-called W-boxes in DNA that contain the sequence motif (T)(T) TGAC(C/T). PATHOGEN-ASSOCIATED MOLECULAR PATTERNS (PAMP) recognition triggers immunity. These PAMP-triggered immune responses result in resistance. (ii) TTSS effectors suppress immunity. PAMP-triggered immune responses fail. The net result is susceptibility. (iii) When R proteins recognize effector activities, effector-triggered immune responses occur, and the result is resistance. (iv) When the effector recognized by the R protein is lost or modified, it results in susceptibility but with less virulence.

new chemical entity (NCE): A compound identified as a potential drug that is sent from research and development into clinical trials to determine its suitability.

new codon generation in immunoglobulin genes: Recombination at certain points between *V* and *J* sequences can generate new codons in the hypervariable regions. Recombination between different triplets can generate yet other new triplets.

new evolutionary synthesis: C.H. Waddington in 1953 and I.I. Schmalhausen in 1986 attempted to integrate embryology, ecology, genetics and palaeontology to create a new evolutionary synthesis that included developmental aspects. They deduced a similar framework whereby structures generated through developmental plasticity could become genetically fixed.

new insights into evolution: Developmental genetics and functional genomics will provide deeper insights into evolution, since DNA not only contains a detailed developmental programme but also vast information about the evolutionary history of organisms.

new patterns of regulation: Evolutionary loss of the ability of two species to hybridize is likely the result of the accumulation of incompatibilities between genes during embryonic development. Therefore, mammals seem to have undergone both rapid regulatory evolution and rapid rearrangement of genes. Thus, the gene rearrangements offer an important resource of accomplishing new patterns of regulation.

new phenotype of a differentiating cell: Cells with a given phenotype respond to the signal that converts them to a new phenotype. These cells may divide, preserving their cell identity, and they may influence other cells, or they may respond to some other signal, which again changes them to another distinct phenotype. A common event in developmental situations where a group of inducing cells transmit a signal that affects the future fate of responder cells. Some of these cells are altered to a phenotype, distinct from that of the cells that gave rise to them.

NextEra: A DNA library preparation kit provided by Illumina.

next-generation genome editing: A programmable RNA that bridges a genetic donor and a target could proclaim a harmless and more elastic approach to large-scale chromosome alterations. Jumping genes, such as the ones that give maize its distinctive colouration, provide a new gene-editing procedure.

This molecular peculiarity found in bacteria could grasp the key to redesigning genomes at will, enabling researchers to alter (insert, delete or flip) large segments of DNA. Guided by an RNA molecule called a 'bridge RNA' or 'seekRNA', the system has been shown to edit genes in a bacterium and test-tube reactions. According to H. Ledford in 2024, this system owing to its small size and its ability to make genetic changes that are thousands of bases long – much larger than is practical with the CRISPR-Cas9 genome editing system – without breaking DNA. This next-generation genome editing enzyme is also known as CRISPR: oddball 'jumping gene' enzyme. *See* BRIDGE RNA *and* SEEKRNA.

next-generation fertilizers: Governments and big agbiotech players are depending on plant-symbiotic bacteria to improve crop growth without the necessity for chemical fertilizers. Syngenta Biologicals and Intrinsyx Bio have arrived at a collaboration to exploit endophytes — naturally occurring symbiotic bacteria discovered living in trees flourishing in unusually harsh conditions. The endophytes developed by Intrinsyx benefit plant health not only by fixing nitrogen in root nodules, but also by improving phosphorous and potassium uptake, as well as enhancing crops' resistance to salinity and drought. Intrinsyx products consist of formulations that can be applied as seed.

next-generation sequencing (NGS): A massively parallel sequencing technology that offers ultra-high throughput, scalability and speed. The technology is used to determine the order of nucleotides in entire genomes or targeted regions of DNA or RNA. J. Eid *et al.* (2009) reported SINGLE-MOLECULE, REAL-TIME (SMRT) SEQUENCING data obtained from a DNA polymerase performing uninterrupted template-directed synthesis using four distinguishable fluorescently labelled DEOXYRIBONUCLEOSIDE TRIPHOSPHATES (dNTPs). They detected the time-based order of their enzymatic incorporation into a growing DNA strand with zero-mode waveguide nanostructure arrays, which provide optical observation volume confinement and enable parallel, simultaneous detection of thousands of single-molecule sequencing reactions. The conjugation of fluorophores to the terminal phosphate moiety of the dNTPs enables continuous observation of DNA synthesis over thousands of bases without steric hindrance. The data are reported directly on polymerase dynamics, displaying distinct polymerization states and pause sites corresponding to DNA secondary structure. Sequence data are aligned with the known reference sequence to assay biophysical parameters of polymerization for each template position. Consensus sequences are generated from the single-molecule reads at 15-fold coverage, showing a median precision of 99.3%, with no systematic error beyond fluorophore-dependent error rates. In genome sequencing, there is a quest for accuracy and speed. In third-generation sequencing, the current machine reads 3 bps. It sequenced the entire human genome in 3 min in 2013. It also claims an accuracy of greater than 99.9999%. *Aka* THIRD-GENERATION SEQUENCING or MASSIVELY PARALLEL SEQUENCING.

next-generation RNA sequencing (NGS) technologies: Any high-throughput sequencing technology used for RNA sequencing. Key RNA sequencing methods include MRNA SEQUENCING, TARGETED RNA SEQUENCING, ULTRA-LOW INPUT AND SINGLE-CELL RNA-SEQ, RNA EXOME CAPTURE SEQUENCING, TOTAL RNA SEQUENCING and SMALL RNA SEQUENCING. *See details as separate terms.*

next-generation sequencing (NGS) profiling of 6^{m}A RNA: NGS approaches have been used to map the locations of 6^{m}A and less-abundantly modified nucleosides.

NF-kB signalling: The cellular signalling pathway that leads to the activation of transcription factor nuclear factor kappa B (NF-kB) protein. NF-kB proteins are dimmers comprising of DNA-binding subunit and a transcription activation subunit.

niche: (i) The functional position of an organism in its environment constitutes the habitat in which the organism lives, the period during which it occurs and the resources it obtains there. (ii) The ecological role of a species; the set of resources it consumes and habitats it occupies. (iii) The set of ecological environments a species can survive and reproduce in. (iv) A species' niche includes all of its interactions with the biotic and abiotic factors of its environment.

nick: One of the phosphodiester bonds between two adjacent nucleotides in one strand of a double-stranded DNA molecule is ruptured. No base is removed from the strand; it is just opened at that point. Such damage in DNA is repaired by HOMOLOGY-DEPENDENT REPAIR (HDR).

nick translation: A tagging method for introducing labelled nucleotides with high specificity into a double-stranded DNA molecule, which involves making several small nicks in one strand of double-stranded DNA with DNase hydrolysis. Each break is extended by 5′→3′ exonuclease activity of DNA polymerase I and then repaired with the same enzyme. Such ^{32}P-labelled DNA is used by different DNA hybridization methods as a probe (e.g. Southern transfer). Nick translation is a misnomer as there is no translation (biosynthesis of protein) involved in this procedure.

nick-closing enzyme: A form of DNA polymerase that can restore base sequences. A nick is introduced adjacent to an incorrect base. The base is then removed, and the correct base is inserted.

nickase: Any enzyme that causes nicks (breaks in one strand) in the nucleic acid, allowing it to unwind. *See* CAS9 NICKASE.

nicked open circular DNA: The DNA of the plasmid has one strand cut.

nicking: Nuclease action to severe the sugar-phosphate backbone in one strand at one specific site.

nicking-closing enzyme: An enzyme involved in normal DNA replication. The enzyme in *Escherichia coli* is a monomer that uses a negatively supercoiled DNA as a template. Its function is to relax supercoiling stress that would otherwise build up in the DNA during unwinding for replication. *Aka* TOPOISOMERASE I. *See* SUPERCOILING.

nicotinamide adenine dinucleotide (NAD): A coenzyme that in different forms plays a very critical role in a wide range of cellular reactions. The conversion of NAD from its oxidized form (NAD^+) to its reduced form (NADH), and back, provides the cell with a mechanism for accepting and donating electrons. NAD^+/NADH plays a significant role in the reactions associated with glycolysis, oxidative phosphorylation and fermentation.

***nif* and *nod* genes:** Nodules containing bacteroides develop in the older parts of the root system. The vascular system of plant extends into the nodules and is how the nutrients pass in and out of the nodules. The *nif* genes code for enzyme nitrogenase. These genes are located on a very large plasmid pSym (symbiotic plasmid) in most of the fast-growing *Rhizobium* species. A close linkage has been demonstrated between the *nitrogenase* (*nif*) and *nodulation* (*nod*) genes on the pSym plasmid. Both bacterial and plant genes are involved in the formation of nodules.

***nif* genes:** Refer to the genetic designation of genes participating in the progression of nitrogen fixation. About 17 genes are organized in the *nif* operon. The *nif* genes produce proteins which fix the atmospheric nitrogen (N_2) as NH_4^+ and NO_3^- ions. Many soil bacteria also fix atmospheric nitrogen. There are many research activities dealing with the transfer of the bacterial system of nitrogen fixation into crop plants other than legumes. There is much interest in manipulation of *nif* genes from bacteria to allow plants to fix nitrogen.

***nif* genes, functions of:** *nif D* and *nif K* genes specify component I (or dinitrogenase). *nif H* gene encodes component II (or dinitrogenase reductase). The *nif B* gene forms part of a *nif A–nif B–fix N* transcriptional unit. Synthesis of the Fe-Mo co-factor of component I requires the products of the *nif E*, *nif N* and *nif B* genes.

NimaGen: A Dutch-based company that manufactures and supplies predesigned and customized tests, reagents and workflow solutions for DNA sequencing. Also, NimbleGen (Roche, Indianapolis, USA) has developed several array platforms for DNA methylation analysis as well as custom arrays to analyse DNA methylation on a genome-wide scale.

ninhydrin reaction: A colour reaction given by amino acids and peptides on heating with the chemical ninhydrin. The technique is widely used for the detection and quantitation (measurement) of amino acids and peptides. The concentration of amino acid in a solution (of hydrochloric acid) is proportional to the optical absorbance of the solution after heating it with ninhydrin. α-amino acids give an intense blue colour, and amino acids (e.g. proline) give a yellow colour. Ninhydrin reaction is used to determine the concentration of a protein or peptide and also obtain an idea of the type of protein or peptide that is present.

nitrates: Refers to nitrogen compounds that exist in a chemical form which plant roots are able to take in (i.e. utilized by the plant to make nitrogen-containing molecules such as proteins). Nitrates are produced from nitrogen taken out of (i) the atmosphere by nitrogen-fixing bacteria (living among the roots of legume plants such as the soybean, etc.), (ii) nitrites (in soil) by nitrate bacteria and (iii) the atmosphere by blue-green algae.

nitric oxide (NO): Gaseous signal molecule in both animals and plants. In animals, NO regulates smooth muscle contraction. For example, in plants NO is involved in responses to injury or infection.

nitrification: The oxidation of ammonium (NH_4) by action of nitrifying bacteria ultimately into nitrate (NO_3). Both ammonium and nitrate can be absorbed by plants.

nitrite reductase: An enzyme that catalyses the reduction of nitrite.

nitrites: Refers to nitrogen compounds that exist in a chemical form which plant roots are unable to take in. After conversion to nitrates via internal respiration by nitrate bacteria (in soil), the nitrates can be taken in by plant roots (i.e. utilized by the plant to make nitrogen-containing molecules such as proteins). Nitrites are made (via internal respiration) by nitrifying bacteria (e.g. in soil) from ammonia-containing substances (e.g. liquid wastes excreted by animals, decomposed animals and plants).

nitrogen: An element that is essential to all plants. It is found reduced and covalently bound in many organic compounds, and its chemical properties are especially important in the structure of proteins and nucleic acids. The deficiency of nitrogen in plants causes chlorosis and etiolation.

nitrogen base: One of the purines or pyrimidines that form part of the molecular structure of a nucleotide and hence a nucleic acid. *Aka* NITROGENOUS BASE.

nitrogen consumers: A crop plant in the crop rotation that takes nitrogen up from the soil, in contrast to plants (e.g. legumes) that deliver nitrogen to the soil by their nitrogen fixation activity.

nitrogen cycle: Conversions of nitrogen to organic compounds and back to nitrogen that occur in nature. This cycle is of immense importance in agriculture. In this cycle, atmospheric nitrogen is converted into ammonia (nitrogen fixation) then to nitrate (nitrification) and finally back to nitrogen gas (denitrification). This cycle is driven by different bacteria.

nitrogen fixation: Refers to the conversion of atmospheric nitrogen (N_2) into ammonia, a soluble, biologically available form (nitrates). Free nitrogen gas (N_2) constitutes about 80% of the atmosphere but it is not available to plants. The conversion is carried out by nitrogen-fixing organisms (e.g. *Rhizobium* bacteria) which live symbiotically in the roots of legume plants (e.g. lucernes or soybeans). This is one of nature's ways of fertilizing. Azotobacter (free-living) possesses the ability to fix atmospheric nitrogen into ammonia. These bacteria pay a high price, as they expend 16 ATP molecules to fix each nitrogen (N_2) molecule. It is also facilitated by lighting and photochemical fixation in the atmosphere. For example, the traditional varieties of soybeans typically leave ~40 lb of residual nitrogen per acre in fields at the end of the growing season. When enough nitrogen fixation does not occur or only non-legume plants are grown, the soil is not able to produce maximum crop yields and farmers may need to apply fixed nitrogen on to the field in the form of fertilizer (anhydrous ammonia, ammonium nitrate or sodium nitrate). Before its incorporation into a living system, N_2 must first be combined with the element hydrogen. The process of reduction of N_2 is referred to as 'nitrogen fixation'. Nitrogen fixation may be chemical or biological. Again, biological nitrogen fixation may be symbiotic or non-symbiotic.

nitrogen fixation (*nif/fix*) genes: The nif/fix genes are found in both free-living and symbiotic nitrogen-fixing bacteria and include the structural genes for nitrogenase and other regulatory genes. Nif/fix genes in *Rhizobium* are located on Sym plasmid (pSym). Nif/fix genes are organized into two groups – Cluster-I and Cluster-II. Cluster-I consists of about 12 *nif* genes (*nif HDKE*, *nif N*, *fix ABCX*, *nif AB* and *fix N*). Cluster-II possesses all fix genes, 11 in number (*fix LJ*, *fix K*, *Fix NOQP*, *fix GHFS*). There are 15 fix genes and 7 *nif* genes; gene *fixN* is present in both clusters.

nitrogen fixation by bacteria: (i) Plant geneticists in Hungary and England have identified the plant gene and protein that enable the plant to establish a relationship with nitrogen-fixing bacteria in the surrounding soil. (ii) Microbial geneticists at the University of Queensland have identified the bacterial gene that stimulates root nodule formation. (iii) Collaboration among molecular biologists in the European Union, USA and Canada yielded the complete genome sequence of one of the nitrogen-fixing bacteria species. (iv) Protein chemists have documented the precise structure of the bacterial enzyme that converts atmospheric nitrogen into a form the plant can use.

nitrogen fixation in cereals: In cereals, huge amounts of fertilizers are used because there is no symbiotic nitrogen fixation. Research is in progress to introduce genes into the wheat genome that may enable some variety of bacteria to fix nitrogen symbiotically in cereals. Gene complexes responsible for nodule (nod genes) formation and nitrogen fixation (nif genes) will have to be introduced into the cereals. Suitable plasmid and bacterial species will have to be discovered that can develop symbiotic relationship with cereals. This also requires genetic systems that can respond to each other.

nitrogen fixers: Microorganisms that convert atmospheric nitrogen into ammonia, using a complex enzyme system known as NITROGENASE.

nitrogen oxide (NO_x): A primary air pollutant produced from internal combustion engines and other industrial processes.

nitrogenase: This is the nitrogen-fixing enzyme complex that catalyses the reduction of molecular nitrogen in the nitrogen fixation process in which dinitrogen is reduced to ammonia. It consists of two proteins: protein I and protein II. Protein I is a dinitrogenase (Fe protein) which possesses a molecular weight of 200,000 kDa, iron content of 20.25 per molecule, sulfide groups/molecule 20 and a half-life in the air of 10 min. Protein II is a dinitrogenase reductase (Mo-Fe protein) and possesses a molecular weight of 60,000 kDa, iron content of 4 per molecule, sulfide groups/molecule 4, and a half-life in air of 0.5 min. This enzyme is destroyed by oxygen. Enzyme needs Mg^{++} to be active. It converts ATP into ADP while functioning. Nitrogenase is inhibited by ADP. It reduces N_2 (N≡N) and other small triple-bonded molecules. N≡N triple bond has a dissociation energy of 940 kJ.

nitrogenase activity: This is ensured by the spatial separation of the bacteroides inside the nodule structure and special oxygen-scavenging leghaemoglobin synthesized in the nodules. Although the rhizobial-legume symbiosis is a highly adapted and regulated interaction, it cannot be termed permanent or obligate. Both partners can live and propagate autonomously, and each host generation has to be populated by a new strain of free-living rhizobia.

nitrogenase complex: A complex of enzymes in nitrogen-fixing bacteria that catalyse the reduction of atmospheric N_2 to ammonia.

nitrogenase reaction: It also reduces H^+ to H_2. O_2 has a dissociation energy of 493 kJ for the double bond. Thus, nitrogen is extremely inert, and its activation is a very energy-demanding process. At least 8 electrons and 16 ATP molecules, 4 ATPs per pair of electrons, are required. The overall reaction is $N_2 + 8H^+ + 8e^- + 16ATP \rightarrow 2NH_3 + H_2 + 16ADP + 16 Pi$. In each cycle of electron transfer, dinitrogenase reductase (Fe protein) is reduced by ferredoxin (or flavodoxin) and binds two molecules of ATP. ATP binding lowers the reduction potential of dinitrogenase reductase (Fe protein) allowing it to interact with dinitrogenase (MoFe protein). After transferring the electron to dinitrogenase reductase (MoFe protein), the ATP is hydrolysed and dinitrogenase reductase dissociates from dinitrogenase. This step initiates a new cycle of reduction and ATP binding. When appropriately reduced, dinitrogenase then reduces N_2 to NH_3. Although only six electrons are necessary to reduce N_2 to two NH_3, eight electrons are required in the overall process as two electrons are lost as hydrogen (H_2).

nitrogenase system: A system of enzymes capable of reducing atmospheric nitrogen to ammonia in the presence of ATP.

nitrogenase, additional level of control on: The NifA protein itself is oxygen sensitive. Leghaemoglobin (Lb) genes are expressed in the prefixation zone, the interzone and the fixation zone. Leghaemoglobin proteins transport oxygen to sites of respiration, thus enabling ATP production in a low-oxygen environment. Senescence means deterioration with age. Cells lose the power of division and growth. Under conditions of excess nitrogen (+N), when the intracellular glutamine concentration is relatively high, GlnK is deuridylylated by uridyl transferase (UTase/UR) and interacts with the reduced form of NifL (NifL red) to promote the formation of a GlnK–NifL–NifA ternary complex in which NifA is inactive. Under nitrogen-limiting conditions (−N), when glutamine concentrations are low, GlnK is uridylylated by UTase/UR and is not competent to interact with NifL. Under such conditions, the binding of 2-oxoglutarate to NifA releases the activator from inhibition by NifL.

nitrogenous aromatic base: One of the purine or pyrimidine compounds that form part of the molecular structure of a nucleotide. These bases are adenine, guanine, cytosine, thymine and uracil.

nitrogenous bases: The DNA double helix looks like a twisted ladder due to the way the components of the nucleotides bind together. Hydrogen bonds join the nitrogenous bases of the two DNA strands, binding adenine (A) with thymine (T) and cytosine (C) with guanine (G) to form the base pairs that make up the steps of the ladder. These bases have different chemical structures that facilitate complementary base pairing. Cytosine, thymine and uracil are pyrimidines (containing one ring), whereas adenine and guanine are purines (containing two rings). Pyrimidines pair with purines – cytosine and guanine are joined by three hydrogen bonds, and adenine and thymine are joined by two hydrogen bonds. In double-stranded RNA, uracil pairs with adenine. *Aka* NITROGEN BASE.

nitroplast: A nitrogen-fixing organelle. In marine systems, some nitrogen-fixing bacteria are endosymbionts of microalgae, such as Candidatus *Atelocyanobacterium thalassa* (UCYN-A), a cyanobacterial symbiont of the unicellular algae *Braarudosphaera bigelowii*. T.H. Coale *et al.* (2024) report a close incorporation of the endosymbiont into the architecture and function of the host cell, which is representative of organelles. These discoveries display that UCYN-A has evolved from a symbiont to a eukaryotic organelle for nitrogen fixation – the nitroplast – thereby escalating a function believed to be entirely carried out by prokaryotic cells to eukaryotes.

nitrous acid (HNO_2): This chemical oxidatively deaminates the bases guanine, cytosine and adenine in DNA and RNA with decreased frequency. In this reaction, the amino group in these bases is replaced by oxygen. HNO_2 induces point mutations in T_4 phage and yeast. Its mutagenic effect is not reported for higher plants. Adenine is deaminated to HYPOXANTHINE, which then pairs preferentially with cytosine in the place of thymine. The result is the transition of A-T to G-C. Cytosine is deaminated to uracil, which now pairs with adenine in place of guanine. The result is the transition of G-C to A-T. Guanine is deaminated to xanthine, which still pairs with cytosine but by only two hydrogen bonds. Thus, the deamination of guanine is not mutagenic. Nitrous acid thus causes transitions A-T to G-C in both directions. Thymine, containing no amino group, is not altered by nitrous acid. Nitrous acid also causes INTERSTRAND CROSS-LINKING of DNA. The DNA strands fail to separate and there is no DNA duplication, which is lethal for phages. Blockage of replication may result in deletions. One cross-link is formed for four deaminations.

nitrous oxide (N_2O): Used as a polyploidizing agent to induce diploid (2n) pollen or unreduced gametes, particularly in tulip bulbs.

NK cell: *See* NATURAL KILLER CELL.

Nm-Judged-Universally sequencing (NJU-seq): A platform developed by Y. Tang in 2024 for high-throughput and site-specific detection of RNA 2′-O-methylation (Nm) which nominates thousands of mRNA Nm sites, with methylation ratios ranging from 1–30%.

no CRISPR: oddball 'jumping gene' enzyme: *See* NEXT-GENERATION GENOME EDITING.

no dosage compensation: In ZW females, the inadequate gene dosage of Z-linked genes is overcome in two ways. One, increased synthesis and storage products of maternally active genes. Two, a shift in the timing of action of zygotic genes such that they become effectively maternal genes during early development. In a sex-determination system where ZZ is male while ZW is female, no dosage compensation is observed for Z-linked genes.

no-go decay pathway: This pathway results in rapid mRNA cleavage and degradation. This pathway is different from other mRNA degradation pathways in yeast, in that stalling of the ribosomes by a stem-loop structure leads to the cleavage of the mRNA, a process that involves the Dom34 and Hbs1 proteins. This cleavage generates free ends that are subject to degradation: the enzyme Xrn1 chews up the mRNA from the 5′ end, whereas the multienzyme exosome complex degrades from the 3′ end. Dom34 and Hbs1 might interfere directly with the 'A site' of the stalled ribosome, possibly to promote the release of the ribosome and enable subsequent mRNA cleavage, or they may directly excite cleavage of the mRNA

no-tillage crop production: A practise of crop production in which the farmer utilizes virtually no mechanical cultivation. It involves only one pass over the field with a planter, instead of the conventional four passes per year with mechanical cultivator equipment plus one pass with planter used for traditional crop production. Reduction in field soil disturbance leaves more carbon in the soil thereby reducing 'greenhouse gases' in the atmosphere. It leaves more earthworms (*Eisenia foetida*) per unit area living in the topsoil. It reduces soil compaction or the reduction of interstitial spaces between individual soil particles, thereby, increasing the fertility of such 'no-till' farm fields. The plant residue remaining on the field's surface assists in controlling weeds and reduces soil erosion by 90–95% in comparison to traditional mechanical tillage. It also offers sites for insects to shelter and reproduce, leading to a need for increased insect control via methods such as inserting a *Bacillus thuringiensis* (*Bt*) gene into certain crop plants or utilizing integrated pest management. But, if a farmer needs to apply synthetic chemical pesticides, the plant residue remaining on the field's surface helps to cause breakdown (into substances such as carbon dioxide and water) of pesticides. That is because that plant residue helps to retain moisture in the field-surface environment, thereby enhancing the growth of the microorganisms that help break down pesticides. The use of the 'no-tillage crop production' practice helps farmers reduce the incidence of certain plant diseases such as white mould disease.

no-tillage/no-till farming: Planting crops directly into the residue of the previous year's crop. In addition to amplifying the benefits of conservation tillage, leaving crop residue untouched also helps to sequester carbon, a greenhouse gas, in the soil.

nod genes: A category of genes present within the DNA of certain soil-dwelling nitrogen-fixing bacteria. When nitrogen-fixing bacteria are in the presence of specific 'signalling molecules' (e.g. isoflavones or luteolin produced by the roots of legume plants), *nod* genes code for the bacterial production of specific chemical compounds which then trigger relevant plant genes (e.g. *NARK* gene in soybean plant) to cause the plant roots to create and/or grow nodules, which the bacteria subsequently move into and begin nitrogen fixation.

nod genes and the regulation of their expression: Genes directing specific nodulation events of a legume by a strain of *Rhizobium* are called nod formation genes which along with specificity genes (genes restrict a *Rhizobium* strain to a particular host plant) are borne on large (mega) plasmids called Sym plasmids (pSym). In *Rhizobium leguminosarum* bv *viciae*, nod genes are located between two clusters of *nif* genes, the genes for nitrogen fixation. Ten *nod* genes identified in *Rhizobium leguminosarum* are *nod M*, *nod L*, *nod E*, *nod F*, *nod D*, *nod A*, *nod B*, *nod C*, *nod I* and *nod J*. Five nod genes (*nod M*, *nod L*, *nod E*, *nod F* and *nod D*) located on one strand are transcribed in one direction, while the remaining five nod genes (*nod A*, *nod B*, *nod C*, *nod I* and *nod J*) located on the other DNA strand are transcribed in the opposite direction.

nod genes, functions of: The *nod ABC* genes are accountable for the synthesis of oligosaccharides, called nod factors, which induce root hair curling and trigger root cell division, ultimately leading to nodule formation. Nod factors consist of a backbone of N-acetylglucosamine to which different substituents are linked. Host specificity is determined by the precise structure of the nod factor of a given *Rhizobium* sp. *Nod ABC* genes direct the synthesis of the nod factor backbone, whereas variation components are synthesized under the direction of other *nod* genes. The *nod D* gene encodes a regulatory protein called *Nod* D that regulates the transcription of other *nod* genes. *nod F* gene encodes a specific acyl carrier protein utilized to acylate the nod factor by *nod* A. *Nod E* and *nod L* genes are accountable for determining host range (cross inoculation groups). The *nod M* gene is involved in nod factor synthesis. *Nod I* and *nod J* genes are membrane proteins that assist in exporting nod factors from the *Rhizobium* cells.

node: (i) An enlarged portion of a shoot at which leaves, branches or buds arise. (ii) A basic unit of a data structure, such as a linked list or tree data structure. Nodes contain data and also may link to other nodes. Links between nodes are often implemented by pointers.

nodulation: The process in which certain strains of soil-dwelling *Rhizobium* bacteria colonize the roots of specific plants (i.e. the legumes). As part of that process, (i) the *Rhizobium* bacteria are attracted to the vicinity of the plant's roots. For the soybean plant (*Glycine max* L.), that is accomplished by the plant synthesizing the signalling molecules known as isoflavones, which attract *Rhizobium japonicum* bacteria. For the lucerne plant, that is attained by the plant synthesizing luteolin molecules, which attract *Sinorhizobium meliloti* bacteria; (ii) certain genes (called *nod* genes) within the relevant *Rhizobium* bacteria are expressed (leading to the synthesis of specific chemical compounds); (iii) when the plant roots detect those chemical compounds, certain genes within those roots are expressed (leading to the formation of nodules on those roots) and (iv) the relevant *Rhizobium* bacteria move in and inhabit those plant root nodules, where the bacteria then fix nitrogen from the atmosphere, which converts that nitrogen into a chemical form (i.e. nitrates) that is available for use by plants (as fertilizer/plant food).

nodule: A small hard knot or an irregular, rounded lump or swelling on the roots of leguminous plants that contains nitrogen-fixing bacteria.

nodule bacteria: Refers to several species of nitrogen-fixing *Rhizobium* bacteria, which form ball-like nodules along legume roots.

nodule formation: Rhizobia are Gram-negative, aerobic, soil bacteria that can invade the root hairs of compatible host legumes and initiate the formation of nodules. Rhizobia exist as intracellular symbionts within the developed nodules, converting atmospheric nitrogen into ammonia for assimilation by the plant in exchange for plant-derived organic acids. Free rhizobia attaches to the root hair. Receptors help in the correct attachment of the bacteria with the root. *Rhizobium* contact stops root cell growth and the root hair curls. Infection threads of invading bacteria penetrate cortex cells causing them to divide. After bacteria are released from the infection thread, they are called bacteroides. Each host determines the number, size and shape of its enclosed bacteroides. Finally, the result is the formation of nodules, which serve as a site for nitrogen fixation. Signal exchange between the microsymbiont and the plant occurs through all stages of nodule development. Recognition of plant-derived flavonoids by rhizobial NodD proteins leads to the expression of rhizobial nod genes, culminating in the production of nod factors, lipochito-oligosaccharide signal molecules, that trigger nodule development in a host-specific manner. Bacteria fix atmospheric nitrogen to form ammonia in the nodules of the roots of the host. This ammonia is made use of by the host. To understand the genetics of nitrogen fixation, we have to analyse two genetic systems (organisms) that are involved in this process – host and parasite. Leghaemoglobin is a direct product of symbiosis, since its globin chains are synthesized by the plant and its haem groups are produced by the bacteroides. The pigment forms only after the symbiotic association has taken place that suggests that the plant genes that code for the globin part of the leghaemoglobin are expressed as a result of infection. The function of this pigment is to bind and store oxygen, which is required by the bacteroides. The binding process also helps to maintain a low concentration of free oxygen in the nodule. Nitrogenase is inactivated at high oxygen concentrations.

nodule formation, recognition and attachment in: In response to the variety of organic metabolites secreted by the roots of legume plants, the rhizobia migrate towards and grow in the rhizosphere and build up to high population density. It is considered that a series of flavonoid signals are there in organic metabolites that lead to the exchange of recognition signals thus attracting specific rhizobia species to specific legume root hairs. However, all species of *Rhizobium* (and *Bradyrhizobium*) possess a specific adhesion protein called rhicadhesin on their surface. Rhicadhesin is a calcium-binding protein and binds calcium complexes on the surface of root hairs. The rhicadhesin protein has been purified from *Rhizobium leguminosarum*, but its gene has not been cloned or otherwise identified. Recently, an *Rhizobium leguminosarum* bv. trifolii protein, RapA1, which has many of the properties attributed to rhicadhesin has been described. Both Rhicadhesin and RapA1 are secreted proteins that can bind calcium, bind at bacterial cell poles, mediate calcium-dependent agglutination and bind to root hairs. Rhicadhesin and RapA1 most likely are not the same proteins, as they are different sizes and RapA1 is found only in *Rhizobium leguminosarum*. Lectins, carbohydrate-containing proteins, also contribute to *Rhizobium*-Legume attachment. The plant then releases flavonoids which induce the expression of *nod* genes.

noise and gene expression: Noise or random fluctuations, in gene expression, may generate variability in cellular behaviour. To measure the noise intrinsic to eukaryotic gene expression, the differences in expression of two alleles in a diploid cell are measured. Such noise is gene specific and not dependent on the regulatory pathway or absolute rate of expression. The balance between the promoter activation and transcription influences the variability in messenger RNA levels. Both *cis*- and *trans*-acting mutations alter the noise of gene expression. These mutations suggest that noise is an evolvable trait and can be optimized to balance fidelity and diversity in eukaryotic gene expression.

nomenclature of restriction enzymes: Each enzyme is named after the bacterium from which it was isolated using a naming system based on bacterial genus, species and strain. For example, the name of the *Eco*RI restriction enzyme was derived from the genus *Escherichia*, species *coli* and strain R.

nominal scale: Assessment of a characteristic in terms of names only. The blood group is on a nominal scale since O, A, B and AB are just names with no order or no grading among them. The other type of scale for qualitative variables is ORDINAL, where grading is present such as good, better, best; poor, middle class, rich and so on.

nominal significance level: In the context of multiple comparisons involving multiple statistical tests, the seeming significance level α of each test is called the nominal significance level. The overall type I assertion probability for the study, the probability of at least one positive assertion when the true effect is zero, will be greater than α.

non-additive gene expression: Levels of gene expression in an auto- or allopolyploid are not equal to the sum of progenitors, suggesting gene activation includes dominance, overdominance or gene repression.

non-additive gene regulation in polyploids: The expression levels of duplicated genes in polyploids can show deviation from parental additivity (the arithmetic average of the parental expression levels). The non-additive expression has been widely observed in diverse polyploids and comprises at least three possible scenarios: (i) the total gene expression level in a polyploid is similar to that of one of its parents (expression-level dominance), (ii) total gene expression is lower or higher than in both parents (transgressive expression) and (iii) the relative contribution of the parental copies (homeologues) to the total gene expression is unequal (homeologue expression bias).

non-additive methylation in hybrids: The type of methylation that occurs in the hybrids more frequently when parental methylation levels are different. Non-additive methylation in the hybrid may be achieved in a number of ways, for instance, a non-additive increase in methylation can arise either through an increase in methylation of the low parent, the high parent or both parental epialleles. Non-additive methylation mechanisms in hybrids involve two types of *trans*-chromosomal interactions, which are named *TRANS*-CHROMOSOMAL METHYLATION and *TRANS*-CHROMOSOMAL DEMETHYLATION.

non-ambiguous code: One code word codes for only one amino acid. In only one case the code seems ambiguous in the sense that codon AUG specifies two kinds of **tRNA** ($tRNA_f^{met}$ at the initiation site and $tRNA^{met}$ in the reading frame).

non-AUG start codon: These codons are used at an astonishing frequency. These non-AUG initiation events are not simply errors but instead are used to generate or regulate proteins with key cellular functions, for example, during development or stress.

non-autonomous code: The nature of a genetic code where one codon codes for only one amino acid.

non-autonomous controlling elements: *See* NON-AUTONOMOUS TRANSPOSABLE ELEMENTS.

non-autonomous development: When a *Drosophila* imaginal disc from a mutant for eye colour develops into normal eyes in wild-type host.

non-autonomous transposable elements: A transposable element that does not produce the proteins required for transposition. Defective transposons that can transpose only when assisted by an autonomous controlling element of the same family for transposition. Mutation by these elements is stable (except when an autonomous element from the family is also present). *Aka* NON-AUTONOMOUS CONTROLLING ELEMENTS.

non-autonomous: Refers to biological units that cannot function by themselves; such units require the assistance of another unit, or 'helper'. *Contrasts* AUTONOMOUS.

non-biological vectors: Include the metal microparticles (coated with genes) which are 'shot' into cells by the Biolistic® gene gun to accomplish transfer of foreign DNA into a cell. *Aka* VEHICLE, DNA VECTOR or CLONING VECTOR.

non-canonical RdDM: The non-canonical RNA-directed DNA Methylation (RdDM) pathways involve 21–22 nt sRNAs from a variety of sources, allowing *de novo* DNA methylation to be initiated at many different types of loci. These 21–22 nt sRNAs are not specific to non-canonical RdDM and also function in other POST-TRANSCRIPTIONAL GENE SILENCING (PTGS) pathways.

non-CG methylation patterns in epigenetic landscape: Chromomethylase 2 (CMT2) is a functional methyltransferase *in vitro* and *in vivo* but is not well characterized. CMT2 preferentially binds histone mark H3K9me2 and methylates non-CG cytosines that are regulated by H3K9 methylation.

non-coding DNA: A certain portion of DNA that does not code for a gene product, such as a protein and/or character.

non-coding DNA hypothesis for sex determination: It has been proposed that key element lies in a non-coding DNA sequence, which serves as a high affinity binding site for a repressor-like molecule regulating the activity of major sex-determining gene. It has been argued that in several eukaryotes, certain DNA sequences that are sex determining are non-coding. In some species, these non-coding sequences are present in one sex and absent in the other, while in others their copy number or accessibility to regulatory molecules is significantly unequal between the two sexes. This inequality determines whether the embryo develops into a male or a female.

non-coding DNA strand: By convention, the base sequence for a transcribed region is that of messenger RNA transcribed from it. Its advantage is that the amino acid encoding codons and the start and stop codons are in their standard forms and thus easily recognizable.

non-coding region: Parts of a gene that include sequences that are transcribed but not translated; both 5′ and 3′ UNTRANSLATED REGIONS (UTRs), upstream promoter region and introns fall under the non-coding regions.

non-coding RNA (ncRNA): RNA molecules that do not contain protein-coding potential. ncRNAs can be highly abundant and functionally important RNA molecules that can epigenetically regulate gene expression or other cellular processes, such as nuclear organization and splicing. Most ancient ncRNAs probably encompass catalytic functions and are called RIBOZYMES (e.g. RNaseP). Others are tRNAs and ribonucleoprotein complexes such as **rRNA** within ribosomes or snRNPs of spliceosomes. Others function as RNA templates within telomerase which is related to reverse transcriptase. Most genomes are transcribed as non-coding RNA. Examples of ncRNAs include **micro RNAs (miRNAs)**, small interfering RNAs (siRNAs), **Piwi-interacting RNAs (piRNAs)**, large non-coding RNAs (lncRNAs), natural antisense transcripts (NATs) and large intergenic RNAs (lincRNAs).

non-coding RNA central dogma of molecular biology: S. Hughes and J.L. Jones described in 2007 that CpG methylated DNA (mDNA) is transcribed into pre-mRNA. Unmethylated DNA (uDNA) is transcribed into uRNA which, as a part of the spliceosome, is involved in the processing of pre-mRNA into mRNA. rDNA is transcribed into pre-RNA, which with the help of RNase MRP and snoRNP is processed into rRNAs, which are integral components of a ribosome. tDNA is transcribed into pre-rRNA, which with the help of RNase P, matures into tRNA. SRP and tmRNA are involved in translocating nascent proteins to the site of action.

non-coding RNAs (ncRNAs), classification of: The ncRNAs are chief players in gene regulation with absolutely no possibility to encode proteins. Based on their length, ncRNAs have been separated into three classes: small non-coding RNAs (sncRNAs; 18–30 nt), medium-sized ncRNAs (mncRNAs; 31–200 nt) and long non-coding RNAs (lncRNAs; >200 nt).

non-coding RNAs, different: All non-genetic RNAs except messenger RNA are non-coding RNA. *Aka* NONMESSENGER RNAS (NMRNAS) or FUNCTIONAL RNAS (FRNAS). The non-mRNAs have been given a variety of names. Non-coding RNAs include HETEROGENOUS NUCLEAR RNA (HNRNA), TRANSFER RNAS (TRNAS), RIBOSOMAL RNAS (RRNAS), SMALL NUCLEAR RNAS (SNRNAS), SMALL NUCLEOLAR RNAS (SNORNAS), TRANS-ACTIVATOR RESPONSIVE RNAS (TARRNAS), MICRO RNAS (MIRNAS), MEMORY RNAS (MEMRNAS), GUIDE RNAS (GRNAS), TELOMERASE RNAS (TELRNAS), TRANSFER-MESSENGER RNAS (TMRNAS), VAULT RNAS (VRNAS), SMALL INTERFERING RNAS (SIRNAS) and SCRAMBLED RIBOSOMAL RNAS (SCRRNAS). The term small RNAs (sRNAs) has been predominant in bacteria, whereas the term non-coding RNAs (ncRNAs) has been predominant in eukaryotes. *See details as separate terms.*

non-coding RNAs, expression profiling of: Numerous transcriptome- and epigenome-profiling procedures, such as genome-tiling microarray, high-throughput sequencing, BS-seq, histone mapping and small RNA profiling, are in use. The MASSIVELY PARALLEL SIGNATURE SEQUENCING (MPSS) technique was first used for genome-wide profiling of siRNAs. Micro RNA prediction, PhasiRNA detection and small RNA target prediction are some approaches.

non-coding RNAs, probing functional epigenetic roles for: CRISPR systems can be used to inquire ncRNA function by repurposing the sgRNAs that bind dCas9 as scaffolding molecules to trigger locus-specific regulation. Orthogonal integration of engineered sgRNA molecules presenting protein-binding RNA aptamers, for example, MS2 and PP7, has been recognized as an effective platform for parallel gene regulation in mammalian and yeast cells.

non-coding single-nucleotide polymorphism (ncSNP): Any single-nucleotide polymorphism that occurs in a non-coding region of the genome (e.g. an intron). NcSNPs are the most common types of SNPs in eukaryotes.

non-cognate codon-anticodon pairing: Occurs when there occurs mispairing base pairing (bad fit) between anticodon and codon at all three positions (1, 2 and 3) as, for example, between UGC/UUU.

non-composite transposons: Non-composite transposons also carry genes (e.g. for drug resistance) but do not terminate with IS elements. Transposition proteins are encoded in the central region. The ends are repeated sequences (but not IS elements). Non-composite transposons cause target site duplications (like composite transposons).

non-conjugative plasmids: These plasmids are incapable of initiating conjugation; hence, they can only be transferred with the assistance of conjugative plasmids, by 'accident'. An intermediate class of plasmids are mobilizable and carry only a subset of the genes required for transfer. They can 'parasitize' a conjugative plasmid, transferring at high frequency only in its presence.

non-conjunction: The failure of the metaphase chromosome to pair during meiosis.

non-conservative substitution: A mutation which results in the substitution of one amino acid within a polypeptide chain with an amino acid belonging to a different polarity/charge group.

non-covalent attraction: Chemical bond in which, in comparison with a COVALENT BOND, no electrons are shared. Non-covalent bonds are relatively weak interactions, but together they can sum to generate strong, highly specific force between molecules.

non-CP virus protein genes: Other than CP, those coding for potyvirus genome-linked proteins (*VPg*) are used for this purpose – VPg-proteinase (*Nla*), RNA-dependent RNA polymerase (*Nib*) and cylindrical inclusion (CI) proteins. These genes, when expressed in plants, confer resistance to the concerned potyvirus.

non-Cry genes for insect resistance: Several important insects are not susceptible to the currently available Cry proteins. For such insects, alternative insecticidal proteins will be needed. For example, the following proteins appear promising for some Lepidoptera and Coleoptera insects: (i) Cowpea trypsin inhibitor (*CpTI*) gene, (ii) Serine protease inhibitor (*aprotinin*) gene, (iii) Cysteine protease inhibitor (*CpI*) gene, (iv) Proteinase inhibitor II (*PiII*) and (v) Lectin.

non-cyclic photophosphorylation: Photosynthetic process that produces both ATP and NADPH in plants and cyanobacteria.

non-Darwinian evolution: A theory of evolution which considers natural selection incompetent to account for the survival of the fittest. The saltation hypothesis, punctuation equilibrium model and evolution by random walk are in contrast to the concept of DARWINIAN EVOLUTION.

non-degenerate code: When there is only one codon for each amino acid.

non-deterministic algorithm: An algorithm that, even for the same input, can exhibit different behaviours on different runs. *Compare with* DETERMINISTIC ALGORITHM.

non-disclosure agreement: In patenting of biological material, an agreement, common between entrepreneurs and potential investors, which permits a company to share protected information while preventing its release.

non-disjunction: During cell division, the separation fails to occur causing both sister chromatids or homologous chromosomes to be pulled to one pole of the cell. The failure of the homologous chromosomes to separate at anaphase-I of meiosis. Mitotic non-disjunction can occur due to the inactivation of either topoisomerase II, condensin or separase. Non-disjunction is of two types – PRIMARY NON-DISJUNCTION and SECONDARY NON-DISJUNCTION. *See details as separate terms.*

non-disjunction, primary: It may occur in an XX female and lead to the production of XX or no-X eggs (in addition to the normal X eggs), or it may occur in the first meiotic division in an XY male and result in XY and non-X sperms or (second division) in XX and non-X or XY and non-X sperm (in addition to normal Y or X sperms).

non-disjunction, secondary: It may occur in the second meiotic division in an XXY female, giving rise to eggs with XX, XY, X or Y chromosomal combinations.

non-DNA repair: Intrinsic features of the organism, like the structure of the code, intergenic and intragenic suppression and diploidy, in addition to normal repair mechanisms.

non-enveloped virus: A virus consisting of a nucleic acid core and protein capsid only.

non-essential amino acids: Amino acids of proteins that can be made (biochemically synthesized within the body) by humans and certain other vertebrate animals from simple chemical precursors (in contrast to the essential amino acids). These amino acids are thus not necessary in the diet (of humans and those other vertebrates). Non-essential amino acids include alanine, arginine, asparagine, aspartic acid, cysteine, glutamic acid, glutamine, glycine, proline, serine and tyrosine.

non-functional transcription complexes: J. Woodgate *et al.* (2024) show that translation, which is coupled to transcription in bacteria, actively dislodges stalled transcription elongation complexes (ECs) from the damaged DNA template. By contrast, paused, but otherwise elongation-competent, ECs are not dislodged by the ribosome. Instead, they are helped back into processive elongation. Coupled ribosomes functionally and kinetically discriminate between paused ECs and stalled ECs, ensuring the selective destruction of only the latter. This functional discrimination is controlled by the RNA polymerase's catalytic domain, the Trigger Loop. A coupled ribosome efficiently destroys misincorporated ECs that can cause conflicts with replication.

non-gel-based assays: Several non-gel-based methods are available for SNP detection. If SNP occurs at the 3′-end of a PCR primer binding site, it can be identified simply by the failure of amplification due to the mismatch b/w the primer sequence and the binding site in the template, although it may be difficult to discriminate this failure of PCR amplification due to SINGLE-NUCLEOTIDE POLYMORPHISM (SNP) from the PCR amplification failure due to other reasons. The common, non-gel-based assays for the detection of SNPs at the internal sites are based on the detection of a mismatch between the PCR product and an oligonucleotide used as a probe. The common methods used are TAQMAN ASSAY and MOLECULAR BEACON ASSAY. *See details as separate terms.*

non-genetic RNA: The RNA found in association with DNA that does not act as genetic material. Non-genetic RNA is made in a cell through the process of transcription. There are three major types of non-genetic RNA – MESSENGER RNA (MRNA), TRANSFER RNA (TRNA) and RIBOSOMAL RNA (RRNA). There are several other types of non-genetic RNAs present in smaller quantities as well, including SMALL NUCLEAR RNA (SNRNA), SMALL NUCLEOLAR RNA (SNORNA), MICRO RNAS (MIRNAS) the 4.5S SIGNAL RECOGNITION PARTICLE (SRP) RNA and so on. Non-genetic RNAs are classified into two types – NON-CODING RNA (NCRNA) and CODING RNA. mRNA is the only coding RNA. All non-coding RNAs are FUNCTIONAL RNAS. *Compare with* GENETIC RNA.

non-governmental organizations (NGOs): Charities and other not-for-profit organizations. *See* ENVIRONMENTAL NON-GOVERNMENTAL ORGANIZATIONS.

non-heme-iron proteins: Proteins containing iron but no porphyrin groups (within which iron atoms are held) in their structure.

non-heritable transgenesis: Refers to a type of genetic transformation that is not stably inherited and expressed from one parasitic nematode generation to the next.

non-histone protein modifications: High-mobility group (HMG) proteins, cohesins, condensins, replication proteins/factors, E2F protein family members, non-histone targets of histone modifiers, chromodomain proteins, DELTEX family ubiquitin ligases and ribosomal proteins are the non-histone proteins that affect/regulate the biological functions in which they are involved.

non-histone proteins: The proteins remaining in chromatin after the histones are removed. The scaffold structure is made of non-histone proteins. Some non-histone proteins are acidic.

non-homologous association: A pairing of chromosomes, which are noticeably not homologous. However, it is presumed that cryptically homologous segments enable the chromosome association.

non-homologous chromosomes: Chromosomes that are not of the same size and shape and contain different genes. The non-homologous chromosomes of a haploid chromosome set typically cannot pair with one another.

non-homologous DNA recombination: Recombination between two double-stranded molecules which have little or nucleotide sequence similarity.

non-homologous end joining (NHEJ) reaction: The predominant mechanism of recombinational DNA repair is non-homologous (illegitimate) end joining, where the entire reaction involves recognition, recruitment and physical alignment, processing of the termini, gap filling and ligation. A double-stranded break DNA repair mechanism that often introduces indels. NHEJ operates in the repair of DSBs, when the template strand is not available. NHEJ is thus an error-prone DSDB repair mechanism, based on the concerted activities of DNA-binding protein Ku and ligase LigD. This error-prone process often causes small insertions and deletions at the DSB site.

non-homologous gene displacement: A lateral gene transfer event in which a gene that governs a particular function is substituted by a non-homologous gene that performs a similar function.

non-homologous random recombination (NRR): NRR is based on DNase I fragmentation, blunt-end ligation and/or extension and capping using two asymmetrical DNA hairpins to stop the extension. This method potentially provides higher flexibility in modulating fragment size and crossover frequency, as well as the number of parental genes.

non-homologous recombination (NHR)-based methods: NHR-based methods include INCREMENTAL TRUNCATION FOR THE CREATION OF HYBRID ENZYMES (ITCHY), SEQUENCE HOMOLOGY INDEPENDENT PROTEIN RECOMBINATION (SHIPREC), EXON SHUFFLING and NRR.

non-homologous RNA recombination: Non-homologous RNA recombination transpires between two

RNA molecules and produces products that diverge noticeably from parental molecules.

non-host resistance: Incapability of a pathogen to infect a plant because the plant is not a host of the pathogen due to the lack of something in the plant that the pathogen requires, or the presence of substances that dissents with the pathogen.

non-identical DNA repeats: Non-identical (similar) DNA sequences are repeated in this case. For example, non-identical repeats of globin genes that produce α and β polypeptides of haemoglobin. Chromosome 16 holds the α gene family while chromosome 11 holds the β gene family. α and β genes are expressed at different developmental stages. Possible origins of gene families are mutation, transposition and duplication. Non-identical repeats also include PSEUDOGENES. Non-identical DNA sequences are *aka* NEARLY IDENTICAL DNA SEQUENCES.

non-infectious disease: A disease that is produced by an abiotic agent – by an environmental factor, but not by a pathogen. *See* ABIOTIC. *Contrasts* BIOTIC *and* INFECTIOUS.

non-invasive phenotyping: Real-time *in situ* monitoring of plant physiology is essential for establishing a phenotyping platform for precision agriculture. A key enabler for this monitoring is a device that can be non-invasively attached to plants and transduce their physiological status into digital data. Non-invasive imaging ranging from 400 to 2400 nm is used to define plant traits and link them to the abundant genomic data. The type of approach used depends on a number of factors and imaging methods can be combined to span broad scales. Platforms are broadly categorized by their adaptation to field conditions or indoor settings. Field applications (including aerial and ground-based cameras and sensors attached to satellites, aeroplanes, unmanned aerial vehicles, drones or tractors) provide remote sensing to estimate crop yield and monitor growing crops over large geographic areas. Indoor applications within greenhouses, controlled environment rooms and laboratories use mostly fixed systems and handheld devices to measure plant growth rate, estimate biomass, detect crop stress, measure physiological and biochemical traits, study the viability of seeds and characterize root systems for individual plants.

non-invasive phenotyping, all-organic transparent plant e-skin for: Y. Yang *et al.* (2024) reported an all-organic transparent plant e-skin by micropatterning poly(3,4-ethylenedioxythiophene) polystyrene sulfonate (PEDOT: PSS) on polydimethylsiloxane (PDMS) substrate. This plant e-skin is optically and mechanically invisible to plants with no observable adverse effects on plant health. They demonstrate the capabilities of these plant e-skins as strain and temperature sensors, with the application to *Brassica rapa* leaves for collecting corresponding parameters under normal and abiotic stress conditions. They also present a digital-twin interface to visualize the real-time plant surface environment, providing an intuitive and vivid platform for plant phenotyping.

non-ionizing radiations in induction of mutations: Non-ionizing radiations, such as Ultraviolet (UV) light, kill bacteria and induce mutations among the survivors. The first photochemical lesion found in the DNA of irradiated bacteria was thymine dimer. UV light produces thymine dimers mainly by linking adjacent thymine bases in the same strand of DNA via C-C bonds. Dimers containing thymine block DNA replication *in vitro* and *in vivo*.

non-linear tetrad: The tetrad in which the meiotic products are in no particular order.

non-long terminal repeat (non-LTR) retrotransposons: A retrotransposon that is not flanked by long terminal repeats.

non-Mendelian inheritance: An uncommon ratio of progeny phenotypes that does not reflect the simple operation of Mendel's principles. For example, mutant:wild-type ratios of 3:5, 5:3, 6:2 or 2:6 in segregating progeny indicate that GENE CONVERSION has occurred. In general, it refers to extrachromosomal and/or non-chromosomal inheritance. *Aka* NON-MENDELIAN RATIO.

non-overlapping code: When a DNA/mRNA molecule codes for as many amino acids as there are code words represented in the end-to-end sequence. For example, mRNA having sequence UUUCCC has two code words or codons, UUU and CCC.

non-parental ditype (NPD): A spore arrangement in Ascomycetes that contains only the two recombinant-type ascospores (assuming two segregating loci).

non-parentals: *See* RECOMBINANTS.

non-pathogenic: Unable to cause disease. *See* AVIRULENT.

non-pedigreed crop: A crop for which a crop certificate has not been issued or recognized by the official organization.

non-penetrative reproductive sex: Some animals use substitute methods of reproduction where penises do not deliver sperm into the vagina. For example, songbirds mate by means of a specific opening called the *cloaca*, which is also used for defecating waste. Sperm is transferred from the male birds to the females when they touch their cloacas together in a 'cloacal kiss'. N. J. Fasel *et al.* (2023) reported that during sex in serotine bats (*Eptesicus serotinus*), the male clutches the back of the female and bites the nape of her neck, apparently to hold on. Then he massages his erect penis around the female's tail membrane, probing for the vulva. Once this has been achieved, the male stays still, with the penis held firmly against the vulva. There is no penetration of the penis, but the sperm probably find their way into the vagina. The longest sex act continued for 12.7 h, but half of the recorded copulations lasted for 53 min or even less. After mating, the fur on the female's belly turns wet, which is considered an indication of the presence of semen.

non-permissive conditions: The conditions which do not allow lethal mutants to survive.

non-permissive host: An animal species whose cells may give rise to a tumour when infected with a particular virus. *Compare with* PERMISSIVE HOST.

non-persistent transmission: A type of virus transmission in which the virus is acquired and transmitted by the vector after short feeding times and is retained by the vector for only a short period. *See* STYLET-BORNE TRANSMISSION. *Contrasts* CIRCULATIVE TRANSMISSION, PERSISTENT TRANSMISSION *and* PROPAGATIVE TRANSMISSION.

non-polar group: A hydrophobic ('water hating') group on a molecule; usually hydrocarbon (composed of hydrogen and carbon atoms) in nature. These groups are more at home in a non-polar (oil-like) environment. Non-polar molecules lack any asymmetric accumulation of positive and negative charge and are generally insoluble in water. *Aka* APOLAR.

non-preference: Refers to a resistance mechanism where parasites favour some host genotypes more than others. The less preferred genotypes are classified as resistant.

non-radioactive probes: Probes that have been labelled with non-radioactive chemicals. As radioisotopes are generally hazardous to health, efforts have been made to produce non-radioactive probes using safe chemicals such as 'biotin' and 'digoxigenin'.

non-random assortment: Assortment or segregation of genes by a process that is not random. For example, linked genes show non-random assortment.

non-random mating: Mate selection based on one or more traits that are discriminated for or against. This is a mechanism of evolution. *See* POSITIVE ASSORTATIVE, NEGATIVE ASSORTATIVE *and* RANDOM MATING.

non-reciprocal recombination: Complementary recombination products resulting from a crossing-over event do not appear in a 1:1 ratio. *See* GENE CONVERSION. *Contrasts* RECIPROCAL RECOMBINATION.

non-recombinants: In mapping studies, offspring that have alleles arranged as in original parents.

non-recurrent apomixis: Refers to infrequent apomixis, usually instigated by haploid parthenogenesis.

non-recurrent parent: A parent that is not tangled in a backcross.

non-renewable resource: A resource present on Earth in finite quantities, such that its use depletes its future stock. Examples include metals and fossil fuels. *Aka* NON-RENEWABLE NATURAL RESOURCE. *Compare with* RENEWABLE RESOURCES.

non-repetitive DNA: Non-repetitive sequences occur once or a few times in the genome. Many of the sequences, which encode functional genes fall into this category. These sequences include but are not exclusively composed of protein-coding sequences. In reassociation studies, the complexity of a slow component corresponds with its physical size. This part of the genome is the solitary component of the prokaryotic DNA and is usually a major component in eukaryotes. *Aka* UNIQUE DNA or SINGLE-COPY DNA.

non-sampling error: The term describes variations in the estimations that may be triggered by population coverage limitations, data collection, processing and reporting methods.

non-selective herbicide: A chemical that kills all or nearly all the plants. *Contrasts* SELECTIVE HERBICIDE.

nonsense codon: A codon that does not specify any amino acid as no normal tRNA exists against a termination codon. It acts as a signal for the termination of translation. Also called a STOP CODON or NONSENSE CODON. The three nonsense codons are named AMBER (UAG), OCHRE (UAA) and OPAL (UGA) *See details as separate terms.*

nonsense intergenic suppressor: It is a gene coding for a mutant tRNA able to respond to one or more of the nonsense codons.

nonsense mutation: A mutation that alters a gene so that a nonsense codon is generated. Nonsense mutation prevents the synthesis of complete function proteins by premature termination of the polypeptide chain. *See* NONSENSE CODON.

nonsense suppressor: A mutation in tRNA that results in the insertion of an amino acid at the site of a stop codon and often reinstates enzyme activity in nonsense mutations and permits the completion of translation.

nonsense variant: A single base pair alteration which results in a premature stop codon.

nonsense-mediated decay (NMD): The pathway not only selectively degrades mRNAs harbouring premature termination codons (PTCs) but also regulates the abundance of a large number of cellular RNAs. In multicellular eukaryotes, long introns are documented through exon definition and most genes generate multiple mRNA variants through alternative RNA splicing. The NMD pathway may further profile the observed sets of variants by selectively degrading those mRNAs that contain premature termination codons, which are frequently generated in mammals. Thus, NMD functions in mRNA quality control. Independently of alternative RNA splicing, species with large intron numbers generally rely on NMD to compensate for suboptimal splicing efficiency and accuracy.

nonsense-mediated decay (NMD), inhibition of: NMD is a dynamically regulated process inhibited by amino acid deprivation, hypoxia and other cellular stresses commonly generated by the tumour microenvironment. This inhibition of NMD can result in the accumulation of misfolded, mutated and aggregated proteins.

non-sibling chromatid: A chromatid that is derived from the other homologue of the two homologous chromosomes or non-homologous chromosomes of the complement. *Aka* NON-SIBLING CHROMATID. *Contrasts* SISTER CHROMATIDS.

non-sister chromatids: The chromatids that result from partner chromosomes at pachytene or any distinct pair of chromosomes at mitosis. *See* NON-SIBLING CHROMATID.

non-specific DNA-protein interactions: Structural proteins that bind DNA are well-understood examples of non-specific DNA-protein interactions. Within eukaryotic nuclei chromosomes, DNA is held in complexes with basic structural proteins called histones. The histones organize the DNA into a compact structure called chromatin. In eukaryotes, multiple types of proteins are involved. The histones make a disk-shaped complex, called a nucleosome, which contains two complete turns of double-stranded DNA wrapped around its surface. These non-specific interactions are produced through basic residues in the histones making ionic bonds to the acidic sugar-phosphate backbone of the DNA and are therefore largely independent of the base sequence. Chemical modifications of these basic amino acid residues include methylation, phosphorylation, acetylation and so on. These chemical changes modify the strength of the interaction between the DNA and the histones, making the DNA more or less reachable to transcription factors and changing the rate of transcription. Other non-specific DNA-binding proteins in chromatin comprise the high-mobility group proteins, which bind to bent or distorted DNA. These proteins are central in bending arrays of nucleosomes and arranging them into the larger structures that ultimately make up chromosomes.

non-specific resistance: The type of host resistance not limited to specific races of the pathogen. *See* HORIZONTAL RESISTANCE.

non-standard amino acid (nsAA): An amino acid beyond the canonical 20 that are present in the genetic code. nsAAs generally fall into two classes: post-translational modifications (e.g. phosphoserine) and synthetic amino acids that do not exist in nature.

non-starch polysaccharides (NSP): The term refers to polysaccharide molecules (in plant seeds) other than starch. These include arabinoxylans, pectins, beta-glucans and alpha-galactosides (e.g. raffinose, stachyose and verbascose).

non-symbiotic (free-living) nitrogen fixation: Non-symbiotic nitrogen fixation takes place in (i) obligate anaerobes (e.g. *Clostridium pasteurianum*), (ii) facultative anaerobes (e.g. *Klebsiella*), (iii) photosynthetic bacteria (e.g. *Rhodobacter*), (iv) many cyanobacteria, (v) obligate aerobes (e.g. *Azotobacter*) and (vi) some methanogens.

non-synonymous mutation: A point mutation in a protein-coding region of a gene that changes a codon that results in a change in the amino acid sequence of the protein product.

non-synonymous single-nucleotide polymorphism (nsSNP): Any single-nucleotide polymorphism that exists in a coding region of a eukaryotic gene and changes the encoded amino acid. The nsSNPs may result in a non-functional protein, which may cause disease.

non-synonymous substitutions: The mutations that occur at non-degenerate sites. The rate of non-synonymous replacements for an evolving protein is indicative of the intensity of the selection pressure to maintain its structure. The non-synonymous substitution rate detected is dissimilar for different proteins.

non-synonymous substitution rate (K_a or D_n): The rate of non-synonymous substitutions in the gene that changes amino acid sequence in its protein product. *Compare with* SYNONYMOUS SUBSTITUTIONS RATE (K_s or D_s).

nonsynonymous variant: A single base pair alteration, which results in an altered amino acid.

non-target damage: Refers to damage from a pesticide to a non-target organism. Unintended (generally negative effects) short- or long-term impacts of genetically engineered (GE) crops on species (e.g. beneficial insects) on direct or indirect exposure to the GE crop. Concerns about non-target effects typically focus on unforeseen harms to other unintended species or environments, but non-target effects can also be neutral or beneficial. *See* BROAD-SPECTRUM PESTICIDE *and* NON-TARGET ORGANISM.

non-target genes: Genes in the recurrent parent that are not at the locus of the gene of interest. Basically, non-target genes are all those genes in the recurrent parent that the breeder would like to retain in the backcrossing programme.

non-target organism: Organisms that are not pests, but which may be affected by a pesticide treatment. During insect control, it is the major concern that beneficial insects like butterflies and honeybees should not get killed with the application of insecticides or by growing *Bt* transgenic plants. One of the major perceived and demonstrated benefits of using *Bt* in pest management programmes is the lack of impact on non-target organisms relative to the use of broad-spectrum commercial insecticides. However, continuous use may have potential environmental effects. *See* BROAD-SPECTRUM PESTICIDE *and* NON-TARGET DAMAGE.

non-targeted approach of transposon tagging: If the gene is to be isolated for a plant character for which mutants are not available, then a non-targeted approach is followed. Wild-type plant without transposon is crossed with wild-type plants having transposon. M_1 plants are selfed, and in M_2 desirable mutants are selected. A non-targeted approach is used to recover new mutations.

non-targeted versus targeted epigenetic diversity: Non-targeted epigenetic diversity deals with the random epigenetic variation existing in the genome. A priori knowledge about the epigenetic state of the target sequence is needed to induce target-specific epigenetic variations. This objective can be accomplished through target-specific epigenome editing methods that have been successfully used in animal systems to either add or remove the selected chromatin modifications at defined loci through various approaches.

non-traditional DNA cloning methods: Strategies that comprise non-traditional DNA cloning methods include POLYMERASE CHAIN REACTION (PCR)-MEDIATED CLONING, RECOMBINATIONAL CLONING, TOPO CLONING, SEAMLESS CLONING, LIGATION-INDEPENDENT CLONING, RESTRICTION AND LIGATION-INDEPENDENT CLONING, URACIL-DNA GLYCOSYLASE, RESTRICTION-FREE CLONING and IN-FUSION CLONING. *See details as separate terms.*

non-transcribed spacer (NTS): These spacers documented in a ribosomal gene family constitute only a part of the INTERGENIC SPACERS (IGS). The region between transcription units in a tandem gene cluster.

non-uniform resistance: The host prevents the establishment of certain races.

non-viral retroposons: Retroposons of non-viral origin, including (i) long interspersed repeated sequences and (ii) short interspersed repeated sequences.

non-viral retroelements: A mobile DNA element that transposes vis RNA intermediate that is copied into DNA by reverse transcription. Some non-viral retroelements are similar to retroviruses but lack genes for capsid proteins.

non-virulent: *See* AVIRULENT.

nopaline synthase (NOS) terminator: A termination codon (sequence of DNA) frequently utilized in the genetic engineering of plants to 'terminate' the expression of the inserted gene (i.e. to halt the synthesis of the desired protein in the plant, after the desired protein synthesis has occurred). The NOS terminator was originally extracted from the bacteria species *Agrobacterium tumefaciens*. *See* TERMINATOR.

nopaline: A rare derivate of an amino acid [(1,3-dicarboxypropyl)-L-arginine] produced in some crown galls of plants. The genes responsible for the synthesis of nopaline are part of the T-DNA of Ti plasmids of *Agrobacterium tumefaciens*.

NOR: *See* NUCLEOLAR ORGANIZER REGION.

norm: The account of the characteristics of a variety as supplied by the breeder. *See* VARIETY DESCRIPTION.

norm of reaction: The landscape of phenotypes produced by a given genotype under different environmental conditions. *See* RANGE OF REACTION.

normal curve: A bell-shaped curve that represents the dispersal of individuals segregating for a quantitative trait, such as height, yield, intelligence and so on. Usually, a curve is humped in the midway between two extremes, indicating that the majority of individuals in the population are grouped in this class called a MODE or MODEL CLASS. There may be two modes. The mode may be nearer to one end of the curve or the other, which is called a SKEWED CURVE.

normal distribution: Any of a family of bell-shaped frequency curves whose relative position and shape are demarcated on the basis of the mean and standard deviation. Most quantitative traits tend to follow a normal distribution curve. The central class, in which the mean, the mode and the median of the population coincide, is the high point of a balanced distribution of frequencies that falls off regularly and symmetrically in both directions. In any normal distribution (and this is true of the normal distribution of means to which standard error applies), about two-thirds of the individual observations have values within one standard deviation ($-\sigma$ and $+\sigma$), about 95% fall within two standard deviations (-2σ and $+2\sigma$) and over 99% within three standard deviations (-3σ and $+3\sigma$), above or below the mean. *Aka* NORMAL CURVE.

normal isoalleles: When the phenotype produced by alleles is wild type. *Aka* WILD-TYPE ISOALLELES.

normalizing selection: The removal by selection of all genes that are accountable for observed deviations from the normal (average) phenotype of a population.

normalized differentiation vegetation index (NDVI) satellite imagery in agriculture: NDVI is a remote sensing method for estimating crop health and biomass. The NDVI index measures the difference between visible and near-infrared reflectance of the vegetation. Crop reflectance depends on leaf area, chlorophyll content, age of leaves, canopy density and soil type. NDVI is often used with satellite imagery, which provides high-resolution images from space. The use of satellites has helped to make NDVI more accessible to farmers worldwide because it is accessible to anyone with an internet connection. Monitoring NDVI in agriculture for crop health screening has been around for decades. Still, it has only recently gained popularity due to the use of satellites and high-resolution aerial photography that provide frequent revisits. It gives an accurate estimate of plant biomass and leaf area index (LAI) without needing any ground data collection or manual interpretation.

northern blot(ting): Northern blotting, a procedure in which RNA fragments are separated by electrophoresis, was discovered by J. Alwine *et al.* (1977). It comprises the following steps: (i) RNAs are separated by gel electrophoresis and RNA bands are transferred to an appropriate membrane, for example, diazo benzyloxymethyl (DBM) paper or nylon membrane, for immobilization (baking). (ii) The RNAs are hybridized with radioactive single-stranded DNA probes. DBM paper, which is prepared from Whatman filter paper No. 540 after a series of reactions, is diazotized by incorporating the diazo group into a chemical compound through the treatment of an amine with nitrous acid. DBM is equally effective in binding to denatured DNA and is more effective than nitrocellulose filters for small DNA fragments. (iii) This specially prepared filter is engaged to blot RNA from agarose gel. Probe can be DNA or RNA, but the only condition is that it should be labelled. (iv) These bands are detected by autoradiography. D.B. Thomas in 1980 showed that mRNA bands could also be blotted directly on to a nitrocellulose membrane which can be hybridized with a labelled DNA or RNA probe. The single-stranded regions of probe are removed by nuclease so that quantitative estimation of hybridized mRNA can also be made. *See* SOUTHERN BLOT *and* WESTERN BLOT. *Aka* NORTHERN HYBRIDIZATION TECHNIQUE.

northern blots and RT-qPCR, limitations of: These two methods of RNA analysis are good for detecting whether a single gene is being expressed, but it quickly becomes impractical if many genes within the sample are being studied.

northern blotting, advantages of: (i) Northern blotting is used for observing protein products yielded from different mature mRNAs resulting from alternate RNA splicing of one precursor-mRNA molecule. (ii) It is used for the detection of RNA size. (iii) The use of probes with partial homology can be made. (iv) The quality and quantity of RNA can be measured on the gel itself prior to blotting. (v) The membranes can be stored for years after blotting for future re-probing.

northern blotting, problems of: The main problems with Northern blotting arise from the use of radioactive reagents which make the procedure time consuming and potentially dangerous. This method has lower-quality quantification than more modern methods because the quantification is done by measuring band strength in an image of a gel. Northern blotting is, however, still widely used procedure as it permits the discrimination of alternately spliced transcripts.

nopaline synthase (NOS) terminator: The part of the gene construct that signals the end of transcription. The gene constructs used in the plant transformation are comprised of promoter, marker gene, gene of interest and terminator. The NOS terminator possesses DNA sequences at the end and signals the end of transcription. The NOS terminator is a 127-bp sequence found in genetically modified plants.

NovaSeq: Whole-genome sequencing platform produced by Illumina.

novel activities/properties of proteins: (i) Point and deletion mutants and hybrid proteins are manufactured to obtain polypeptides with new properties. These proteins are either created individually by site-directed mutagenesis or produced as a large pool or library of millions of variants. The library is then screened or subjected to a special selection procedure to obtain the protein or proteins with the desired attributes. (ii) Proteins generally have stable and exclusive three-dimensional structures and are active at physiological temperatures of 37°C (98.6°F). In some instances, a protein may be cold sensitive – it may lose its conformation and activity at lower temperatures. The effect of temperature on protein stability can be modulated by protein engineering, in particular, by introducing amino acid replacements that enhance or destabilize the molecular packing interactions in the core of the protein structure. This has been achieved with the enzymes T4 lysozyme and staphylococcal nuclease. (iii) Protein engineering has been used to manufacture therapeutic proteins with improved properties such as increased solubility and stability. For example, insulin was engineered through mutagenesis to create monomeric forms that are fast acting (insulin lispro and insulin as part). On the other hand, another form (insulin glargine) was generated by mutagenesis to precipitate upon injection and give a sustained release of insulin. (iv) Proteins can be engineered to acquire new biological activities. For example, a catalytic antibody is a variant of an antibody. Antibodies are proteins that normally bind to a specific molecule but do not alter the bound molecule in any way. A catalytic antibody is one which has been changed by mutations to have a novel sequence that folds into a structure that catalyses a specific reaction (e.g. amide bond formation, ester hydrolysis and decarboxylation). Catalytic antibodies function like enzymes and are created to catalyse reactions for which there are no naturally occurring enzymes.

novel biosynthetic pathways: Development of novel pathways can be achieved by matching and mixing genes from different sources (even from unrelated metabolic routes) and by creating new biosynthetic functions by (i) random mutagenesis, (ii) recombination and (iii) selection. In this approach, no information on enzyme structure or catalytic mechanism involved in these pathways is needed. Metabolic engineering has been done for the biosynthesis of novel carotenoids and polyhydroxyalkanoates.

novel epialleles generation of: Procedures now available for producing new epialleles by changing the DNA methylation status. DNA methylation states at a large number of sites provide countless opportunities in crop genomes.

novel food: (i) A substance, including a microorganism, which does not have a history of safe use as a food. (ii) Food that has been manufactured, prepared, preserved or packaged by a process that has not been applied before to that food, causing the food to undergo a major change. (iii) A food derived from a plant, animal or microorganism that has been genetically modified.

novel joints: A new DNA sequence, thus, a new restriction fragment is created during any intrachromosomal amplification event. Cloned restriction fragments corresponding to amplified DNA and containing novel joints can be identified readily.

novel plant trait: A plant with characteristics not normally found in that species in which the new characteristic has been created through specific genetic manipulation, transformation, mutation and so on.

novel scientific concepts: Novel scientific concepts are new ideas that arise out of the interplay of three factors: an attempt to solve specific problems; the use of conceptual, analytical and material resources provided by the cognitive-social-cultural context of the problem; and dynamic processes of reasoning that extend ordinary cognition.

novel variant: A DNA alteration that has not been previously reported in a public database.

noxious weed: Of plants, containing harmful or unwholesome qualities. The term 'noxious' is applied in conjunction with 'weed' to specifically describe a plant that the law considers harmful to the environment. Each state/territory in Australia has specific legislation governing noxious weeds.

NPK analysis: The analysis of the ratio of nitrogen (N), phosphorus (P) and potassium (K) in an organic soil modification.

nu body: Nucleosome; the subunits of chromatin produced during chromosome coiling in eukaryotes with a roughly spherical shape, composed of a core octamer of histones and 146 nucleotide pairs of DNA. *Aka* NUCLEOSOME.

nucellar embryony: Parthenogenesis in which the embryo ascends directly from the nucellus.

nucellus: The tissue of the ovule of a seed plant that environs the female gametophyte. It is enclosed by integuments and is not of epidermal origin.

nuclear and mitochondrial codons: Four codons are different in mammalian nuclear (nu) and mitochondrial (mt) translational system: AUA = Met (mt), Ile (nu); AGA, AGG = Stop (mt), Arg (nu); UGA = Trp (mt), Stop (nu).

nuclear compaction: The packaging of DNA with the help of linker histone (e.g. H1) and core histones (H2A, H2B, H3 and H4) to form chromatin which allows the nearly 2 m of DNA material in each cell of the body to fit in a nucleus measuring just 4 to 8 μm in diameter. This process allows mitosis and meiosis and controls gene expression.

nuclear division cycle: The sequence of phases of the division of the nucleus.

nuclear division: The division of the cell nucleus by mitosis, meiosis or amitosis.

nuclear DNA (nDNA): DNA occurring in the nucleus of a eukaryotic cell.

nuclear dominance: Direct correlation between the differential heterochromatinization of NORs of one parental origin in hybrids, DNA methylation at cytosine residues and the switching off of rDNA units, clearly establishes the epigenomic origin of nucleolar dominance (W.S. Viegas *et al.*, 2007) Histone marks lead to marked modifications in chromatin organization patterns and to changes in the nuclear topology of specific chromatin domains. Disclosure of the 'histone code' associated with nucleolar dominance in hybrids was performed through identification of distinct modified histones on NORs from both parental species that associate with active or silent rDNA arrays.

nuclear dye: A nuclear stain.

nuclear envelope: The double membrane that separates the nucleus from the cytoplasm in the eukaryotic cell. A double membrane consists of an outer and inner membrane and is perforated by nuclear pores. Both membranes resemble the plasma membrane. These membranes possess small pores called nuclear pores. For communication of nucleus with cytoplasm.

nuclear epitranscriptome mapping methods: (i) RNA IMMUNOPRECIPITATION (RIP), (ii) METHYLATED RNA IMMUNOPRECIPITATION SEQUENCING (MERIP-SEQ), (iii) CHROMATIN IMMUNOPRECIPITATION SEQUENCING (CHIP-SEQ), (iv) ANTIBODY-MEDIATED CAPTURE (AMC), (v) MASSIVELY PARALLEL SEQUENCING (MPS), (vi) N_3-CMC-ENRICHED PSEUDOURIDINE SEQUENCING (CEU-SEQ), (vii) COVALENT TRAPPING OF PROTEINS FOR $5^{M}C$ MAPPING, (viii) 6-METHYLADENOSINE ($6^{M}A$) INDIVIDUAL NUCLEOTIDE CROSS-LINKING AND IMMUNOPRECIPITATION (MICLIP), (ix) 1-METHYLADENOSINE-SPECIFIC RNA IMMUNOPRECIPITATION WITH DEMETHYLASE-ASSISTED RNA SEQUENCING ($1^{M}A$-ID-SEQ) and (x) GLOBAL RNA MODIFICATION MAPPING STRATEGIES.

nuclear export of circular RNA (circRNA): Circular RNAs (circRNAs) have been implicated in a variety of functions in normal and cancerous cells. They are formed by the back-splicing of precursor mRNAs in the nucleus. circRNAs are predominantly localized in the cytoplasm, suggesting that they must be exported from the nucleus. L.H. Ngo *et al.* (2024) identified a pathway specific for the nuclear export of circular RNA. This pathway requires Ran-GTP, exportin-2 and IGF2BP1. Enhancing the nuclear Ran-GTP gradient by depletion or chemical inhibition of the major protein exporter CRM1 selectively increases the nuclear export of circRNAs, while reducing the nuclear Ran-GTP gradient selectively blocks circRNA export. Depletion or knock-out of exportin-2 specifically inhibits nuclear export of circRNA. Adaptors such as IGF2BP1 that bind directly to circular RNAs recruit Ran-GTP and exportin-2 to export circRNAs in a mechanism analogous to protein export, rather than mRNA export.

nuclear export signal: Cataloguing signal confined in the structure of molecules and complexes, such as RNA and new ribosomal subunits, which are transported from the nucleus to the cytosol through nuclear pore complexes.

nuclear fragmentation: The degeneration of the nucleus by partition of the nucleus into more or less different parts.

nuclear gene: A gene located in any of the nuclear chromosomes.

nuclear genome: In eukaryotes, the genetic information is coded in the nuclear DNA, as opposed to organellar genomes (MITOCHONDRIAL GENOME and CHLOROPLAST GENOME).

nuclear import of tRNAs in yeast: The transfer RNAs (tRNAs) were thought to cross the nuclear envelope to the cytosol only once after maturing in the nucleus. Several export mutants accumulate mature tRNAs in the nucleus even in the absence of transcription. Import requires energy but not the Ran cycle. The tRNAs shuttle between the nucleus and cytosol.

nuclear lamina: Stringy meshwork of proteins on the inner surface of the inner nuclear membrane. It is made up of a network of intermediate filaments cut from nuclear laminas. It is involved in most nuclear activities including DNA replication, RNA transcription, nuclear and chromatin organization, cell cycle regulation, cell development and differentiation, nuclear migration and apoptosis.

nuclear localization signal (NLS): Signal sequences or signal coverings found in proteins fated for the nucleus and which enable their selective transport into the nucleus from the cytosol through the nuclear pore complexes.

nuclear magnetic resonance (NMR): Resonant absorption of electromagnetic radiation at a specific frequency by atomic nuclei in a magnetic field, due to flipping of the orientation of their magnetic dipole moments. The NMR spectrum delivers information about the chemical environment of the nuclei. Two-dimensional NMR is used widely to work out the three-dimensional structure of small proteins.

nuclear male sterility (NMS): Refers to male sterility governed by nuclear genes. *Contrasts* CYTOPLASMIC MALE STERILITY GENE.

nuclear matrix: Protein molecules occurring in cancerous cells, but not in normal (i.e. non-mutated) cells.

nuclear membrane: The structure that separates the nucleus of eukaryotic cells from the cytoplasm. It includes two unit membranes each 10 nm thick, detached by a perinuclear space of 10–40 nm. At intervals, the two membranes are fused around the edges of circular pores, which enable the selective passage of materials into and out of the nucleus. *Aka* NUCLEAR ENVELOPE.

nuclear mutations: These mutations refer to mutations in the nuclear genes. In organisms having homogametic and heterogametic sexes, nuclear mutations may be autosomal, sex (or X)-linked, holandric (Y-linked) or X- and Y-linked (pseudoatutosomal).

nuclear pore complex (NPC): NPC is the bidirectional gate that mediates the exchange of macromolecules or their assemblies between nucleus and cytoplasm. NPCs consist of numerous copies of ~30 different proteins called nucleoporins (Nups). NPCs infiltrate the nuclear envelope (NE) and control the nucleocytoplasmic trafficking of macromolecules. The NPC arbitrates both exchange of proteins and RNA between the nucleus and the cytoplasm and physically interacts with chromatin to control transcription. J.H. Brickner in 2023 delivered new insight into the molecular basis for NPC-mediated epigenetic silencing by loading the replication processivity factor PROLIFERATING CELL NUCLEAR ANTIGEN (PCNA). Different NPCs facilitate different functions. Z. Li *et al.* (2023) report the cryo-electron microscopy structure of native pre-60S particles trapped in the channel of yeast NPCs. In addition to known assembly factors, multiple factors with export functions are also included in the structure. These factors in general bind to either the flexible regions or subunit interface of the pre-60S particle, and virtually provide many anchor sites for NPC binding. *In silico* analysis of the axial and radial distribution of pre-60S particles within the NPC shows that a single NPC can take up to four pre-60S particles simultaneously, and pre-60S particles are enriched in the inner ring regions close to the wall of the NPC with the solvent-exposed surface facing the centre of the nuclear pore. These data suggest a TRANSLOCATION MODEL for the export of pre-60S particles through the NPC.

nuclear pores: The nuclear membrane is not continuous. Nuclear pores disrupt the nuclear membrane. They regulate flow of material into and out of nucleus. There is evidence that the fibres of chromatin attach to the pores. This suggests that pores are involved in the control of DNA synthesis. The DNA being replicated is attached to the pores when pores are isolated. Nuclear pores thus act as selective barrier for certain substances to get in and get out of the nucleus. Discovered by E. Strasburger in 1875.

nuclear receptor superfamily: Intracellular receptors for hydrophobic signal molecules such as steroids and retinoic acid. The receptor-ligand complex acts as a transcription factor in the nucleus.

nuclear receptors: Receptors in a cell's outer membrane that serve to convey a 'signal' from outside the cell all the way into the cell's nucleus.

nuclear reprogramming: Restoration of the correct embryonic pattern of gene expression in a nucleus derived from a somatic cell and introduced into an oocyte.

nuclear RNA editing: The post-transcriptional modification of a glutamine codon CAA in apolipoprotien B (apoB) in mRNA generates a translational stop codon UAA in nucleus of mammals.

nuclear sap: A non-staining or slightly stainable liquid or semiliquid substance of the interphase nucleus in which the chromosomes and the nucleoli are found. In the nuclear sap, regulatory molecules, synthetic products, substrates and enzymes must pass on their way to and are found in the cytoplasm. Nuclear sap contains a wide variety of proteins including enzymes capable of anaerobic energy products (glycolysis). *Aka* NUCLEOPLASM.

nuclear speckles: The factors implicated in the transcription and processing of different classes of RNA are kept within specific nuclear bodies. One example is the nuclear speckle, characterized by high concentrations of protein and non-coding RNA regulators of pre-mRNA splicing. P. Bhat *et al.* (2024) demonstrated that the genes located near nuclear speckles display higher spliceosome concentrations, augmented spliceosome binding to their pre-mRNAs and higher co-transcriptional splicing levels than the genes positioned farther from nuclear speckles. Directed recruitment of a pre-mRNA to nuclear speckles is adequate to enhance mRNA splicing levels.

nuclear stain: Usually, basic dyes such as methylene blue, methyl green, crystal violet, green puron or azure B bromide that bind favourably to the cell nuclei (i.e. chromosomes and nucleolus) with a slight cytoplasmic effect.

nuclear system of heredity: The normal, Mendelian system in which inheritance is controlled by genes located in the nuclear chromosomes, as opposed to the corpuscular (=plastid) system and the cytoplasmic (molecular) system.

nuclear transfer (NT): Dolly, a female Finn-Dorset sheep, was cloned in 1996 by K. Campbell, I. Wilmut and colleagues at the Roslin Institute, part of the University of Edinburgh, Scotland, and the biotechnology company PPL Therapeutics, based near Edinburgh. They generated a new animal nearly identical to another one by injection of the nucleus from a cell of the donor animal into an enucleated oocyte of the recipient. It is possible to also delete or substitute genes (e.g. brought in from another species) as part of the nuclear transfer process, so nuclear transfer can be exploited to produce transgenic organisms or 'knock-out' organisms.

nuclear transplantation: The transfer of a nucleus into the cytoplasm of another cell.

nuclear transport: The movement of macromolecules into or out of the nucleus. The process is mediated by nuclear transport receptors.

nuclear-encoded plastid-localized RNA polymerase: Differential transcription of *rpoB-rpoC1-rpoC2* and other genes early in chloroplast development has been hypothesized to be caused by the increased expression of a nuclear-encoded plastid-localized RNA polymerase. It has been suggested that expression of the nuclear-encoded plastid-localized RNA polymerase is activated early in chloroplast development. This results in the differential transcription of the *rpoB* operon and other loci involved in plastid gene expression early in chloroplast development.

nuclease hypersensitive site (NHS): A region of eukaryotic chromosome that is explicitly vulnerable to nuclease attack because it is not wrapped as nucleosomes.

nuclease protection assays (NPA): An extremely sensitive method for the detection and quantitation of specific mRNAs. The basis of the NPA is solution hybridization of an antisense probe (radiolabelled or non-isotopic) to an RNA sample. After hybridization, single-stranded unhybridized probe and RNA are degraded by nucleases. The remaining protected fragments are separated through polyacrylamide gel electrophoresis.

nuclease-based genome editing: The use of engineered nucleases with a DNA-targeting mechanism (e.g. a ZINC-FINGER NUCLEASE (ZFN), TRANSCRIPTION ACTIVATOR-LIKE EFFECTOR NUCLEASE (TALEN) or a CLUSTERED REGULARLY INTERSPACED SHORT PALINDROMIC REPEAT (CRISPR)) for targeted mutagenesis *in vivo*.

nucleases: Those enzymes that cleave nucleic acids. Nucleases are universally distributed. A nuclease isolated from one organism may be different from that isolated from another organism. Some nucleases cleave DNAs while others cleave RNAs.

nucleases, functions of: (i) During DNA replication, DNA polymerase I acts to degrade primer through its exonuclease activity. (ii) During the process through which a circular molecule becomes linear, endonuclease activity occurs that produces nicks at two specific positions of the λ chromosome. (iii) During generalized recombination in prokaryotes and eukaryotes, endonucleases produce nicks. (iv) Endonucleases play an important role in dark repair of UV-induced DNA damage. The mechanism for excision of thymidine dimer employs an endonuclease. (v) The 3′ phosphate group during dark repair of UV-induced damage is removed by a 3′ endonuclease, and then 5′ endonuclease eliminates 6 or 7 nucleotide-long sections including the thymine dimers. (vi) Restriction endonucleases used in host restriction recognize a specific nucleotide sequence on DNA and cleave both strands of DNA at staggered positions. (vii) Exonuclase-3 nibbles double-stranded DNA from 3′→5′ directions thus exposing the 5′ single strands at each end. Thus, it can assist in transforming a linear DNA molecule into a circular DNA molecule. (viii) An RNase chews up the excessive transcripts up to the point where the poly(A) chain is to be added during RNA processing.

nucleation: (i) A critical stage in the assemblage of a polymeric structure, such as a microtubule, at which a small cluster of monomers aggregates in the correct arrangement to initiate rapid polymerization. More generally, nucleation is the rate-limiting step in an assembly process. (ii) Build-up of a histone modification and/or protein in a localized region of the genome.

nucleic acid hybridization: The process in which two complementary strands of nucleic acids – RNA, DNA or oligonucleotides – are joined. Nucleic acid hybridization is commonly used to identify a recombinant clone using complementary DNA or RNA probes. For example, to detect the presence of transgene say *cryl AC* in the putative transgenic plant, the radio labelled probe is allowed to hybridize with the genomic DNA of the putative transgenic plant on the filter membrane, using hybridization oven. If hybridization takes place, it clearly shows the stable integration of transgene in the plant genome.

nucleic acid sequencing: Refers to determining the arrangement of nucleotides in DNA or RNA. Nucleotide sequencing provides the most satisfying molecular map of a genome. *Aka* DNA/RNA SEQUENCING or NUCLEOTIDE SEQUENCING.

nucleic acid-protein relationship: Both DNA and RNA were formed abiotically. RNA was most probably the first to establish a functional relationship with proteins. The reasons for this are as follows: RNA is serving not only as a template for protein synthesis but also for code storage, for example, the TOBACCO MOSAIC VIRUS (TMV). Living systems would be intact even if their DNA is conceptually deleted.

nucleic acids: A biopolymer linear molecule composed of an approximately 200,000 nucleotides long chain of nucleotides. DEOXYRIBONUCLEIC ACID (DNA) and RIBONUCLEIC ACID (RNA), the two most important nucleic acids in living organisms and viruses, are made of thousands of four different deoxyribonucleotides and ribonucleotides, respectively. The bases involved are adenine, guanine, cytosine and thymine (uracil in RNA). Nucleic acids are either the specific (genetic) informational molecule (DNA), or act as an agent (RNA) in the machinery used by the cell for causing that information to be expressed (e.g. as a protein). Long molecules of nucleic acids are made up of smaller molecules, called NUCLEOTIDES, which are chemically linked together in a chain by phosphodiester bonds. DNA and RNA may be single- or double-stranded, linear or circular molecules. Nucleotide sequences contribute to transferring genetic information from cell to another or from one generation to another. DNA ensures that progeny resemble parents, whereas RNA plays important role in protein synthesis. In other words, RNA translates the information encoded in the DNA. *See* DEOXYRIBONUCLEIC ACID (DNA) *and* RIBONUCLEIC ACID (RNA).

nuclein: Swiss biologist F. Miescher (1868), while working on white blood cells, mined material from nucleus, which he termed as NUCLEIN, which was very rich in phosphorus. It was extracted from pus cells.

nucleobase: The nitrogen-containing biological compounds that form nucleosides, which in turn are components of nucleotides, with all of these monomers constituting the basic building blocks of nucleic acids.

nucleocytoplasmic interactions: Cellular development requires interaction between nucleus and cytoplasm – the activity of the nuclear genes during development is limited by the properties of the cytoplasm. Thus, the nucleus determines the phenotype. Phenotype can be regulated but not determined by the cytoplasm. The genes determine a cell's potential; the cytoplasm determines whether that potential will be reached.

nucleoid: An irregularly shaped region in prokaryotes, chloroplasts and mitochondria where 1-mm-long double-stranded DNA, nucleoid-associated proteins (NAPs) and RNA are concentrated. Unlike a nucleus, it is not bound by a membrane. *Aka* GENOPHORE. *See* NUCLEOID-ASSOCIATED PROTEINS.

nucleoid-associated proteins (NAPs): NAPs comprise various types of proteins found in the nucleoid region of prokaryotic cells which organize the structure of the chromosome and facilitate gene expression by binding chromosomal DNA. Some NAPs bind to non-specific sequences, introducing bends and folds to aid in condensing of the DNA, so it fits within the nucleoid. Other NAPs bind to specific sequences that facilitate DNA replication and RNA transcription.

nucleolar chromatin: The nucleolar chromatin contains gene for rRNA. The nucleolus is the largest structure in the nucleus of eukaryotic cells where it primarily serves as the site of ribosome synthesis and assembly. Nucleoli also have other important functions like assembly of signal recognition particles and playing a role in the cell's response to stress. Nucleoli are made of proteins and RNA and form around specific chromosomal regions.

nucleolar chromosome: The chromosome that possesses the nucleolus organizer region (NOR). *Aka* SATELLITE CHROMOSOME.

nucleolar constriction: The region of a chromosome that carries the nucleolus. This region is besides the centromeric region. It is observed as a secondary constriction along a particular chromosome and is not stained by standard chromosome techniques. Because of the presence of the nucleolar constriction, the chromosome arm appears divided into two parts; the terminal part is called 'SATELLITE', and the whole chromosome is called the 'SATELLITE CHROMOSOME'.

nucleolar dominance: An epigenetic phenomenon that occurs in genetic hybrids which describes the expression of 45S rRNA genes inherited from one progenitor due to the silencing of the other progenitor's rRNA genes. *Aka* AMPHIPLASTY.

nucleolar organizer region (NOR): A part of the acrocentric short arm that contains tandem copies of ribosomal or rRNA genes in large heteroclusters, present on the stalks of the short arms of chromosomes 13, 14, 15, 21 and 22. For the production of about 10 million ribosomes per cell in eukaryotes, rRNA genes exist in multiple copies in a tandem array of specific satellite chromosomes. The number of these genes may vary from 50 to 30,000 in a cell and this number may be equally dispersed on NORs. The DNA comprising these genes is called RIBOSOMAL DNA (RDNA), which is repetitive in nature. NORs are detected with silver staining. Lethal embryos are formed probably due to absence of nucleolar organizer. Nucleolar organizer plays a role in early development. *Aka* NUCLEOLAR ZONE *or* NUCLEAR DNA.

nucleolin: One of many non-ribosomal proteins. It is suggested to play a chief part in the regulation of rDNA transcription, perisomal synthesis, ribosomal assembly and maturation. It influences the nucleolar chromatin structure through its interaction with DNA and histones. Nucleolin is implicated in the cytoplasmic-nucleolar transport of preribosomal particles.

nucleolus: A clearly marked, often spherical area of the eukaryotic nucleus, composed of densely packed fibrils and granules. Its composition is similar to chromatin, except that it is highly rich in the rRNA and mRNA of ribosomal proteins. It is the site of the synthesis of ribosomal RNA. The assembly of ribosomes initiates in the nucleolus but is completed in the cytoplasm. The nucleolus is attached at the SECONDARY CONSTRICTION, that is called the NUCLEOLAR ORGANIZER REGION (NOR).

nucleolus-associated chromatin: It is composed of a chromatin fibre situated around the nucleolus and extending into it. Ribosomal RNA is transcribed from this chromatin. This chromatin contains multiple copies (900 in *Xenopus*) of genes coding for 18S, 28S and 5.8S rRNA.

nucleolytic reaction: A reaction that involves the hydrolysis of a phosphodiester bond in a nucleic acid.

nucleomorph: A highly reduced vestige of the nucleus. It occurs in many eukaryotic species that have gone through secondary symbioses in which one eukaryote became the endosymbiont of another eukaryote.

nucleophilic group: An electron-rich group with a strong propensity to contribute electrons to an electron-deficient nucleus.

nucleoplasm: A general name for the complex mixture of molecules that comprises the ground substance of the nucleus of a living cell.

nucleopore: One of a series of openings found in the cell's nuclear membrane. Nucleopores serve as channels for the selective transport of nucleic acids and proteins into and out of the cell nucleus.

nucleoporin: Any of a large number of different proteins that comprise nuclear pore complexes.

nucleoproteins: Complexes formed by interactions between nucleic acid and protein. These two substances are not linked by strong chemical bonds but are bound together by salt linkages and other weak bonds. Most viruses consist entirely of nucleoproteins, although some viruses also contain fatty substances. Nucleoproteins commonly occur in animal and plant cells and in bacteria.

nucleoside: A compound made up of a purine or pyrimidine base and a pentose sugar, but it lacks phosphate group. Purines include adenine (A) and guanine (G), whereas pyrimidines include thymine (T), cytosine (C) and uracil (U). Thus, nucleoside = base + sugar. *Compare with* NUCLEOTIDE.

nucleoside diphosphate kinases: Enzymes of this type are ubiquitous in nature. They catalyse reactions of following general type: $N_1TP + N_2DP \rightleftharpoons N_1DP + N_2TP$, where N_1 and N_2 are purine or pyrimidine ribonucleosides or deoxyribonucleosides. Nucleoside diphosphate kinases from many sources have been shown to function by means of the formation of a phosphoryl-enzyme intermediate.

nucleoside diphosphate sugar: A coenzyme-like carrier of a sugar molecule functioning in the enzymatic synthesis of polysaccharides and sugar derivatives.

nucleosome: Histones are relatively small proteins with a preponderance of the basic (positively charged) amino acids lysine and arginine, which makes sense because DNA is an acid. The histones' strong positive charge enables them to bind to and neutralize the negatively charged DNA throughout the chromatin. Histones together make up half of all chromatin protein by weight. All five types of histones – H1, H2A, H2B, H3 and H4 – appear throughout the chromatin of nearly all diploid eukaryotic cells. H1 histone is called LINKER HISTONE as it is found on the linker DNA – the DNA that connects two adjacent nucleosomes. Two copies of each of the four histones, H2A, H2B, H3 and H4, the core histones, form an octamer [(H2A-H2B) ×2 + (H3 × 2-H4 × 2)], which serves as a core particle in the formation of a nucleosome. Nucleosomes are 57 Å in height and 110 Å in diameter. The DNA double helix is wrapped around (~1.7 times) the histone octamer. With nuclease digestion, 146 bp of DNA are tightly associated with the nucleosome but ~200 bp of DNA in total are associated with the nucleosome. The difference is the linker DNA. Length of linker DNA can vary from 8–114 bp. The structure comprising histone proteins and DNA is the basic organizational unit in chromatin structure. Under an electron microscope, nucleosomes appear somewhat like a string of pearls. *Aka* Y BODY or NU BODY. *See* NU BODY *and* KARYOSOME.

nucleosome array: Methylation of histone H3 lysine 9 (H3K9) creates a binding site for heterochromatin protein 1 (HP1) (M. Lachner *et al.*, 2001). This epigenetic modification has a role in higher order chromatin organization by propagating heterochromatin-specific affinity in the histone H3 amino terminus. This histone in combination with other proteins compacts the 11-nm fibre into higher order structures such as the 30-nm fibre. H3 is accountable for the stable inheritance of the heterochromatic state. In the genome, thousands of nucleosomes are organized on a continuous DNA helix in linear strings separated by 10–60 bp of linker DNA. Thus, the lowest functional unit of chromatin is considered the NUCLEOSOMAL ARRAY.

nucleosome assembly: Sequences like CTG_nCAG_n have greater propensity for nucleosome assembly than any other DNA sequences because the DNA with such sequences have greater flexibility and, therefore, can easily be made curved than any other known sequences, they thus greatly favour nucleosomal assembly.

nucleosome barrier: RNA polymerase II transcribes RNA within a chromatin context, with nucleosomes acting as barriers to transcription.

nucleosome bead: An octamer of histone proteins plus 146 base pairs of DNA wrapped around the octamer in two turns. *Aka* CORE PARTICLE.

nucleosome binding proteins: The proteins that act to modulate the promoter chromatin architecture and transcription of target genes. For example, histone H1 and poly(ADP-ribose) polymerase-1 (PARP-1) exhibit a reciprocal pattern of chromatin binding at many RNA polymerase II-transcribed promoters.

nucleosome disassembly and reassembly: Although the nucleosomes are physical obstacles for the DNA-transcribing RNA polymerase II, it somehow passes through them while maintaining their structure. H. Ehara *et al.* (2022) obtained structural snapshots of the polymerase passaging through a nucleosome using cryo-electron microscopy and found that the polymerase forms a huge elongation complex with multiple transcription elongation factors, which mediates the disassembly of the downstream nucleosome and its subsequent reassembly behind the polymerase with the aid of the histone chaperone.

nucleosome dynamics: Nucleosome dynamics are governed by a complex interplay of histone composition, histone post-translational modifications, nucleosome occupancy and positioning within chromatin.

nucleosome occupancy: Often measured by the height of nucleosome signals after micrococcal nuclease

(MNase) digestion of chromatin followed by deep sequencing (MNase-seq).

nucleosome organization: It is critical for gene regulation. In living cells this organization is determined by multiple factors, including (i) the action of chromatin remodellers, (ii) competition with site-specific DNA-binding proteins and (iii) the DNA sequence preferences of nucleosomes themselves.

nucleosome phasing: Nucleosomes may be located in unique positions that are identical in a population of cells of the same kind, or they may occupy a small number of distinct positions. Such arrangements have been termed phasing.

nucleosome positioning: The term indicates the positions where nucleosomes are situated with respect to the genomic DNA sequence. Chromatin influences gene expression by restricting access of DNA-binding proteins to their related sites in the genome. A NUCLEOSOME-FREE REGION (NFR) is present within several hundred base pairs upstream of the translation start site. Many transcription factors bind within NFRs. Nucleosomes function largely to decouple the threshold of induction from the dynamic range.

nucleosome remodelling and deacetylase (NuRD) chromatin-remodelling complex: A multi-subunit chromatin remodelling complex containing two core subunits (Mi2 and HDAC1/2) with enzymatic functions. Mi2 catalyses ATP-dependent chromatin remodelling, while HDAC1/2 mediates histone/protein deacetylation.

nucleosome retention during transcription elongation: Cryo-electron microscopy structure of a mammalian Pol II-DSIF-SPT6-PAF1c-TFIIS-nucleosome complex stalled 54 base pairs within the nucleosome. The structure reported by M. Filipovski and co-workers in 2022 provides a mechanistic basis for nucleosome retention during transcription elongation where upstream DNA emerging from the Pol II cleft has rewrapped the proximal side of the nucleosome.

nucleosome-like structures: Nucleosome-like structures have been found to be associated with chromosome of archaebacterium. The major part of the chromosomal DNA is associated with protein particles with diameters of 9.4 nm arranged tandemly along the DNA fibres. Thus, the primary structure of the archaebacterium chromosome resembles that of eukaryote chromosome.

nucleosome-organizing proteins: A flexible, multivalent protein that can bridge nucleosome is HP1. HP1 is highly conserved dimeric protein that is present in different isoforms with slightly different functions and localization.

nucleosomes as transcription barriers: Nucleosomes present a strong barrier to transcription, raising important questions about the nature and the mechanisms of overcoming the barrier. DNA sequence, DNA-histone interactions and backtracking by RNA POLYMERASE II (RNA POL II) all contribute to formation of the barrier.

nucleotide: A chemical compound comprising a purine or pyrimidine base attached to five-carbon sugar, to which a mono-, di- or triphosphate is also attached. The monomeric unit of DNA and RNA. There are four types of deoxyribonucleotides in DNA: deoxyadenosine-5′-phosphate, deoxyguanosine-5′-phosphate, deoxycytidine-5′-phosphate and deoxythmidine-5′-phosphate. The prefix deoxy is removed while naming the nucleotides of RNA (ribonucleotides). Nucleotides are structural units of nucleic acids. Nucleotide = Base + Sugar + Phosphate. While POLYMERIZED NUCLEOTIDES are the structural units of a nucleic acid, FREE NUCLEOTIDES that are not an integral part of nucleic acids are also found in tissues and play important roles in the cell (e.g. ATP and cyclic AMP). *See* DEOXYRIBONUCLEOTIDES *and* RIBONUCLEOTIDES.

nucleotide addition cycle (NAC): C. Fang *et al.* (2024) report cryo-electron microscopy structures of RNA Polymerase IV (Pol IV) elongation complexes at four states of its NAC: post-translocation, guanosine triphosphate-bound, pretranslocation and backtracked states. The structures reveal that Pol IV maintains an open DNA cleft and kinked bridge helix in all NAC states, loosely interacts with the nucleoside triphosphate substrate and barely contacts proximal backtracked nucleotides.

nucleotide diversity (π): The coincidence that two randomly chosen copies of a nucleotide site will have different bases. It is similar to gene diversity and applies to single nucleotides.

nucleotide excision repair (NER): DNA repair mechanism in eukaryotes, which involves a large number of genes. A group of genes called *RAD* genes or *ERCC* (excision repair complementation competent) genes are involved in this process. This pathway is used by mammals to remove bulky DNA lesions such as those formed by UV light, environmental mutagens and some cancer chemotherapeutic adducts from DNA.

nucleotide pair: A pair of nucleotides present on complementary strands of DNA that are joined by hydrogen bonds. For example, nucleotides carrying adenine and thymine are joined by two hydrogen bonds (Adenine=Thymine) and those carrying guanine and cytosine are joined by three hydrogen bonds (Guanine≡Cytosine) in DNA.

nucleotide pair substitution: Refers to the replacement of a specific nucleotide pair by a different pair. Nucleotide pair substitution is often mutagenic. More commonly known as BASE PAIR SUBSTITUTION.

nucleotide sequence: The order of nucleotides along a DNA or RNA strand.

nucleotide sequence difference (K): The number of nucleotide substitutions per site in a pair of aligned sequences. Using RNA code, one can calculate how the homologous sites differentiated from each other in the course of evolution starting from the common ancestor.

nucleotide-binding site (NBS)-leucine-rich repeat (LRR) proteins: Structural features of these proteins are as follows: (i) They are among the largest proteins in plants and have 860–1900 amino acids. (ii) They have several domains: (a) AMINO-TERMINAL DOMAIN (involved in protein-protein interactions); (b) NBS DOMAIN (involved in specific binding and hydrolysis of ATP); (c) ARC DOMAIN (so called because of its presence in Apaf-1, R proteins and CED-4); (d) LRR DOMAIN (protein-protein interactions and ligand binding) and (e) CARBOXYL TERMINI in TNLs (TIR-NBS-LRR) are larger and more variable than CNLs (CC-NBS-LRR) and RNLs (RPW8-NBS-LRR).

nucleus: A large membrane-bound cellular organelle that contains the cell's hereditary information (DNA) organized in the form of chromosomes. Robert Brown first discovered the nucleus in 1833. Nuclear chromosomes decide what organism a fertilized egg will develop into. Structurally, the nucleus is composed of the nuclear envelope, chromatin, nucleolus, nuclear sap and SMALL NUCLEAR RIBONUCLEOPROTEINS (SNRNPS). It is the site of DNA replication, RNA synthesis, RNA processing and DNA recombination. RNA exits via the nuclear pores. The nucleus governs the cell's life activities (e.g. metabolism, growth and reproduction). The nucleus is a highly differentiated, relatively large organelle located in the cytoplasm of the cell. It is enclosed by a (nuclear) membrane that is quite similar to the plasma (cell) membrane, except that the nuclear membrane contains holes or pores. It is known for its high content of chromatin, which contains most of the cell's DNA. That chromatin is normally (when the cell is not in the process of dividing) distributed almost uniformly throughout the nucleus in a diffused manner.

nucleus breeding: A breeding scheme where populations in the breeding cycle are divided into (i) intensively managed nuclei with top-ranking genotypes and (ii) less intensively managed, genetically less advanced, main population.

nucleus-cytosol compartmentalization: The origin of the eukaryotic nucleus marked a seminal evolutionary transition. The nuclear envelope's budding function was to permit mRNA splicing, which is slow, to go to completion so that translation, which is fast, would occur only on mature mRNA with intact reading frames. The rapid, fortuitous spread of introns following the origin of mitochondria is cited as the selective pressure that led to nucleus-cytosol compartmentalization.

nucleus-independent inheritance of RNA interference: In *Caenorhabditis elegans* worms, epigenetic information is transmitted transgenerationally. To elucidate whether the effects transfer to the next generation inside or outside the nucleus, I. Rieger *et al.* (2023) used the tractability of gene-specific double-stranded RNA-induced silencing to demonstrate that RNA interference can be inherited independently of any nuclear factors via mothers that are genetically engineered to transmit only their ooplasm but not the oocytes' nuclei to the next generation. They characterize the mechanisms and, using RNA sequencing, chimeric worms and sequence polymorphism between different isolates, identify endogenous small RNAs which, similarly to exogenous siRNAs, are inherited in a nucleus-independent manner. These results might be regarded as partial vindication of discredited cytoplasmic inheritance theories from the 19th century, such as Darwin's 'pangenesis' theory.

null allele: A 'silent' allele without an obvious expression. It is generally thought equivalent to the complete deletion of the gene.

null hypothesis (H_0): The statistical hypothesis from which the expectations are predicted to test significance. If we wish to test whether observed numbers are equal to expected numbers, observed = expected, the null hypothesis, labelled H_0, is often used in the frequentist branch of statistical inference as a 'straw person'; classical statistics often assumes what one hopes does not happen (no effect of a treatment) and attempts to gather evidence against that assumption (i.e. tries to reject H_0). *Compare with* ALTERNATE HYPOTHESIS (H_1).

null hypothesis test: The standard hypothesis utilized in testing the statistical significance of the difference between the means of samples drawn from two populations. The null hypothesis states that there is no difference between the populations from which the samples are drawn. One can then determine the probability that one will find a difference equal to or greater than the one observed. If this probability is 0.05 or less, the null hypothesis is rejected, and the difference is said to be statistically significant.

null model: A model of no effect. It is used in statistics as the baseline against which a hypothesis is tested. *See* NULL ALLELE.

null mutation: A mutation that eliminates all enzymatic activity. Usually refers to deletion mutations.

nullihaploid: A cell or individual that holds a haploid chromosome set plus a missing single chromosome.

nulliplex: The condition in which a polyploid transmits a recessive gene at a particular locus in all homologues. *See* ALLOTETRAPLOID.

nullisome: A plant lacking both members of one specific pair of chromosomes. *Aka* NULLISOMIC.

nullisomic: A disomic (2n) individual/cell in which both members of one particular pair of homologous chromosomes are missing from the disomic chromosome complement (2n − 2). Nullisomics are valuable in discovering the position of newly found genes on specific chromosomes in plants. Nullisomics are generated as genetic stock for gene location studies.

nullisomic analysis: In nullisomic analysis, observations are recorded for phenotypic or other differences between the nullisomic for each chromosome and the disomic condition within the same variety. This method is applied for the localization of genes within a given genome. This method can only be employed in polyploids, while diploids commonly do not endure the loss of both homologous chromosomes.

nullisomic haploid: A haploid individual lacking one chromosome (e.g. x − 1, 2x − 1, 3x − 1). It is derived from disomic alien addition line.

nullisomic-tetrasomic: A cell or individual, typically an allopolyploid, which lacks one pair of chromosomes, which is partially compensated for by a tetrasomic (fourfold dosage) of another, usually homoeologous, chromosome. A whole series of nulli-tetrasomics has been produced in hexaploid wheat and effectively employed in numerous genetic and molecular studies.

numerator elements: Sex-determining elements found in X chromosomes of X/A ratio (e.g. in *Drosophila melanogaster* and *Caenorhabditis elegans*). They appear to have more than one numerator element (but possibly not a large number).

numeric aberration: The variation of the number of genomes or chromosomes, for example, ploidy variation, aneuploids (e.g. nullisomics, monosomics, trisomics, tetrasomics, etc.), substitutions or additions.

numeric constancy of chromosomes: The constant inheritance of the same number of chromosomes from generation to generation, that is facilitated by the mitotic and meiotic mechanisms.

numerical analysis: The study of algorithms that use numerical approximation (as opposed to symbolic manipulations) for mathematical analysis (as distinguished from discrete mathematics).

numerical method: In numerical analysis, a numerical method is a mathematical tool designed to solve numerical problems. The implementation of a numerical method with an appropriate convergence check in a programming language is called a NUMERICAL ALGORITHM.

numerical taxonomy: In general, any method of taxonomy using numerical measurements. In particular, it often refers to phenetic classification using large numbers of quantitatively measured characters.

numerous: Stamens are pronounced as numerous when there are more than twice as many sepals or petals, especially when there is no set number of them. *Compare with* INDEFINITE.

nurse cell: A cell connected by cytoplasmic bridges to an emerging oocyte and which thereby supplies it with ribosomes, mRNAs and proteins needed for the development of the early embryo.

nurse culture: In the culture of plant cells, the growth of a cell or cells on a contiguous culture of different origin that in turn is in contact with the tissue culture medium. The cultured cell or tissue may be separated from the feeder layer by a porous matrix such as filter paper or membranous filters. *See* FEEDER LAYER.

nurse plant: A plant, often a shrub, which affords shelter and shade to developing seedlings of other plants.

nursery: A place where plants are grown for sale, transplanting or experimentation. For example, young trees or other plants are raised up, either for propagation or for testing and observations.

NusA: A protein required in the elongation step of transcription.

nut: A hard, dry, indehiscent fruit containing only one seed.

nutation: The turning of a plant or plant organ towards light.

nutlet: (i) A small nut. (ii) One of the lobes or sections of the mature ovary of some members of the Boraginaceae, Verbenaceae and Lamiaceae.

nutraceuticals: Products isolated from foods that are sold in medicinal forms (biopharmaceuticals). Transgenic plants are now considered factories for producing a variety of nutraceuticals, such as provitamin A, sulfur-rich amino acids, lycopene, vitamin E, antioxidants and cholesterol-lowering margarine with plant sterols.

nutrient: Any chemical necessitated for the proper metabolism of organisms.

nutrient broth: Complex nutrient medium in liquid form. Most heterotrophic bacteria and fungi are routinely grown in complex mediums containing extracts from yeasts, plants or meat. In the complex mediums the energy, nitrogen, sulfur and carbon requirements of growing microorganisms are provided by protein. Vitamins and other organic growth factors are provided by yeast extracts or the meat extracts. Yeast extracts are rich in B vitamins. When agar is added to nutrient broth, it is called nutrient agar.

nutrient budget: A quantitative evaluation of the rates of nutrient input and output for an ecosystem, as well as the quantities occurring in and transferred within the system.

nutrient capital: The amount of nutrients that exists at a site in soil, living vegetation and dead organic matter.

nutrient cycling: Transfers and chemical conversions of nutrients in ecosystems, including recycling through decomposition.

nutrient medium: A solid or liquid combination of nutrients, usually including several salts, a carbohydrate (e.g. sucrose) and vitamins. Such a medium is often referred to as a basal medium and may be supplemented with growth hormones and occasionally with other defined or undefined substances.

nutrigenomics: The science of nutrigenomics seeks to provide a molecular understanding of how common dietary chemicals (i.e. nutrition) affect health by altering the expression and/or structure of an individual's genetic makeup.

nutrigenomics, tenets of: The conceptual basis for this new branch of genomic research can best be summarized by the following five tenets of nutrigenomics: (i) Under certain circumstances and in some individuals, diet can be a serious risk factor for a number of diseases. (ii) Common dietary chemicals can act on the human genome, either directly or indirectly, to alter gene expression or structure. (iii) The degree to which diet influences the balance between healthy and disease states may depend on an individual's genetic makeup. (iv) Some diet-regulated genes (and their normal, common variants) are likely to play a role in the onset, incidence, progression and/or severity of chronic diseases. (v) Dietary intervention based on knowledge of nutritional requirements, nutritional status and genotype (i.e. 'personalized nutrition') can be used to prevent, mitigate or cure chronic disease.

nutritional-deficient mutants: *See* AUXOTROPHS.

O

O protein: Phage protein that recognizes replication origin site Ori in phage. Four or more molecules are used for the formation of initial complex.

O-linked glycosylation: Addition of an oligosaccharide chain to a protein through the OH group of a serine or threonine side chain.

oat: A species of cereal grain used for food and as fodder for animals, especially poultry and horses.

object-oriented analysis and design (OOAD): An organizational method for investigating and scheming an application, system or business by using object-oriented programming; as well as applying pictorial modelling throughout the software development procedure to control stakeholder communication and product excellence.

object-oriented programming (OOP): A programming pattern based on the concept of 'objects', which can hold data in the form of fields (*aka* ATTRIBUTES *or* PROPERTIES) and code in the form of procedures (*aka* METHODS). A characteristic of objects is that an object's actions can contact and often amend the data fields of the object with which they are related (objects have a notion of 'this' or 'self'). In OOP, computer programmes are intended by making them out of objects that interact with one another. OOP languages are varied, but the most popular ones are class based, meaning that objects are examples of classes (which also govern their types).

object-relational database (ORD): Databases that syndicate the elements of object orientation and OOP languages with database capabilities. They offer more than persistent storage of programming language objects. Object databases outspread the functionality of object programming languages (e.g. C++, Smalltalk and Java) to deliver full-featured database programming competence. The consequence is a high degree of correspondence between the data model for the application and the data model of the database. Object-relational databases are employed in bioinformatics to map molecular biological objects (e.g. sequences, structures, maps and pathways) to their underlying illustrations (typically, within the rows and columns of relational database tables). This allows users to transact with the biological objects in a more spontaneous manner, as they would in the laboratory, without apprehension about the underlying data model of their illustration.

obligate aerobes: Organisms that need oxygen (O_2) for its growth. Obligate aerobes cannot grow in the absence of O_2. These organisms in the presence of O_2 produce more energy from nutrients. Since O_2 is less soluble in water, it becomes difficult for obligate aerobes to survive in stagnant water. Due to this problem, many aerobic bacteria adapt to less O_2 environment and can grow in the absence of O_2. These are called facultative anaerobes.

obligate anaerobes: Organisms that are unable to use oxygen (O_2) for their metabolism. These microbes grow only in the absence of oxygen. These organisms do not employ oxygen atoms existing in cellular materials; the atoms are generally attained from water. Genus *Clostridium* containing species that result in botulism and tetanus are the best instances of obligate anaerobes. *Compare with* FACULTATIVE ANAEROBES.

obligate apomictic: The plants that produce seed only through apomixis, without reversion to the sexual mode of reproduction.

obligate epialleles: The obligate epialleles have complete dependence on a genetic variant.

obligate parasite: An organism that cannot sustain the absenteeism of its host. *Aka* OBLIGATE PATHOGEN. *Compare with* FACULTATIVE PARASITE *and* FACULTATIVE PATHOGEN.

observation: A data collection approach in which the activities of subjects are visually scrutinized. The spectator tries to keep his attendance from intruding in or prompting any behaviours.

observational study: An investigation in which no experimental condition (e.g. treatment) is deployed by the investigator (i.e. randomization is not employed).

Such studies are often used to estimate physiognomies of subjects (means, proportions, etc.) and to evaluate relations between variables. They have known restrictions for therapeutic comparisons because of unidentified confounders.

occluding junction: Kind of cell junction that caps cells together in an epithelium, forming a barricade through which even small molecules cannot pass.

ochre codon: The nonsense codon UAA. One of the three nonsense codons that stop protein synthesis. *See* AMBER CODON *and* OPAL CODON.

ochre mutation: Amber mutations are the consequence of a base substitution that changes a codon specifying an amino acid into codon UAA, which signs chain termination.

ochre suppressor: A gene coding for a mutant tRNA, which retorts to the ochre (UAA) codon to permit continuation of protein synthesis.

octad: An ascus containing eight ascospores, formed in species in which the tetrad typically undergoes a post-meiotic mitotic division.

octanucleotide: DNA fragment made up of eight nucleotides.

octanucleotide motif (ONM): ONM sequence vital for viral protein, DNA and progeny virus biosynthesis at the origin of DNA replication of porcine circovirus type 2. This octanucleotide can be changed to an essential core element (ECE) characterized by AxTAxTAC. The sites stated by the indicated nucleotides are life threatening for viral DNA replication and stable infectious virus production, and they cannot be replaced by other bases, while the sites indicated by x can admit variable bases and produce stable progeny viruses.

octopine: A rare amino acid derivative which is produced by a certain type of crown-gall tissue. The genes responsible for the synthesis of octopine are part of the T-DNA from a Ti plasmid.

octoploid: Possessing eight chromosome sets of indistinguishable or dissimilar complements.

octavalent: Chromosome conformation comprises eight chromosomes.

odds: The probability of an event happens divided by the probability that it does not happen. An event that ensues 0.90 of the time has 9:1 odds of occurring, since $0.9/(1 - 0.9) = 9$.

odds ratio: The odds ratio for comparing two groups (A, B) on their probabilities of an outcome happening is the odds of the event taking place for group A divided by the odds that it ensues for group B. If P_A and P_B represent the probability of the consequence for the two groups of subjects, the A:B odds ratio is $[P_A/(1 - P_A)]/[P_B/(1 - P_B)]$. Odds ratios occur in the interval $[0, \infty]$. An odds ratio for a treatment is a degree of relative impact of that treatment on a binary outcome. As summary measures, odds ratios have benefits over risk ratios: they are not contingent on which of two possible consequences is labelled the 'event', and any odds ratio can apply to any likelihood of outcome in the reference group. Because of this, one frequently discovers that odds ratios for comparing treatments are relatively persistent across different types of patients. The same is not true of risk ratios or risk differences, as these are subject to the degree of risk in the reference group.

oedema (British word): A bulge or sweltering on leaves and other plant parts under conditions of high moisture and restricted transpiration. *See* INTUMESCENCE. Corresponding American word is EDEMA.

off-patent event: An existing event, originally patented by and commercially available through a PRP Holder, for which the patent protection has expired in a specific territory.

off-target effects: Unintended, short- or long-term consequences of an intervention on the genome of the organism in which the intended effect was incorporated.

off-target effects of genome editing: Cas9 cleavage at undesired locations due to gRNA targeting sequence with sufficient homology to recruit Cas9 to unintended genomic locations. *Aka* OFF-TARGET ACTIVITY.

off-target effects of RNA interference: Unintended and/or pleiotropic effects that occur as a result of sequence homology between novel, small RNAs in the GE crop and mRNA in the host crop plant or organisms in the environment. Off-target effects include gene silencing that leads to unexpected pollen lethality and unintended gene silencing in beneficial insects exposed to the crop plant.

office action: In patenting biological material, an official reply by a patent examiner concerning a patent application or amendment.

offspring: In biology, offspring are the young creation of living organisms, produced in case of asexual reproduction by a single organism or in the case of sexual reproduction by two organisms. Collective offspring may be known as a BROOD or PROGENY in a more general way.

***OH43* gene:** Gene in plants (e.g. maize) that governs manufacture of a seedcoat more resistant to tearing.

Greater tear resistance causes a lower incidence of fungi infestation, which causes less mycotoxin production in seed.

oil immersion: The oil that is placed between the lens of a microscope and the coverslip above a microscopical preparation. This transparent oil has high refractive index, which increases the numerical aperture of the objective lens.

oil quality in plants, modification of: Vegetable oils and fats institute an important component of human diet. Mammals (including humans) do not produce linoleic and linolenic fatty acids, which must be supplemented with diet. Various strategies have been used to modify fatty acids: (i) Introduction of a new enzyme (e.g. an acyl transferase or an acyl-ACP thioesterase and acyl-ACP desaturase), (ii) suppression of an enzyme activity (e.g. α acyl-ACP desaturase), (iii) site-directed mutagenesis to modify specificity of an enzyme (e.g. acyl-ACP desaturase) and (iv) construction of hybrid genes to generate novel enzyme activities.

oil-immersion objective: The objective lens system employed for highest resolution with the light microscope. The space between the coverslip over the object is examined and the lens is filled with a drop of oil with the same refractive index as the glass.

Okayama-Berg procedure: A technique of cDNA cloning by means of specialized vectors, favouring the production of full-size clones. This procedure permits cloning of full-length mRNA genes. The mRNA is extracted from postpolysomal supernatant of reticulocyte lysate of rabbits and made anaemic by phenylhydrazine injection.

Okazaki fragment: DNA polymerase on the leading strand produces the molecule in a 5′→3′ direction but polymerase on the lagging strand must go 'against the grain', making DNA from a series of RNA-primed fragments. Discontinuous DNA synthesis proceeds in steps of 200 base pairs in eukaryotes and 1000–2000 nucleotide long in prokaryotes. Each such fragment is called Okazaki fragment after its discoverer R. Okazaki and others in 1968. Completion of Okazaki fragment synthesis leaves a nick between the Okazaki fragment and the preceding RNA primer. DNA polymerase I extends the Okazaki fragment, while 5′→3′ exonuclease activity removes the RNA primer. This is called NICK TRANSLATION, which is a misnomer. DNA polymerase I dissociates after extending Okazaki fragment by 10–12 nucleotides. Then DNA ligase binds to the nick and joins Okazaki fragments together. DNA ligase uses NAD^+ as energy cofactor. The sealing is catalysed by the formation of phosphodiester bond, and then it gets dislocated. The lagging strand template loops back through the replisome so that the leading and lagging strands are synthesized in the same direction. This offers some hint of unit of packing of histone and DNA. *See* NICK TRANSLATION.

Okazaki fragment processing: During lagging strand chromatin replication, multiple Okazaki fragments (OFs) necessitate processing and nucleosome assembly; these two processes are linked. Using transmission electron microscopy and rapid degradation of DNA ligase Cdc9, G. Shi *et al.* (2024) observed flap structures accumulated on lagging strands, controlled by both Pol δ's strand displacement activity and Fen1's nuclease digestion. The distance between neighbouring flap structures exhibits a regular pattern, indicative of matured OF length. While *fen1*Δ or enhanced strand displacement activities by polymerase δ (Pol δ; *pol3exo*$^-$) minimally impact inter-flap distance, mutants influencing replication-coupled nucleosome assembly, such as *cac1*Δ and *mcm2-3A*, do meaningfully change it. Deletion of Pol32, a subunit of DNA Pol δ, expressively upsurges this distance. Mechanistically, Pol32 binds to histone H3-H4 and is perilous for nucleosome assembly on the lagging strand. Thus, Pol32 establishes a connection between nucleosome assembly and the processing of OFs on lagging strands.

oligodeoxyribonucleotide: A short, single-stranded DNA molecule (generally 15–50 nucleotides in length) which may be used as a primer or a hybridization probe. Oligodeoxyribonucleotides are synthesized chemically under automated conditions.

oligodeoxyribonucleotide-directed mutagenesis: An *in vitro* mutagenesis technique that involves the use of a synthetic oligodeoxynucleotide to introduce a predetermined nucleotide alteration into the gene to be mutated.

oligogene: A gene that exerts a marked phenotypic effect on characters that show typical qualitative inheritance.

oligogenic inheritance: Inheritance due to a small number of genes with discernible effects. *See* QUALITATIVE INHERITANCE.

oligogenic resistance: Resistance governed by one or a few genes.

oligomeric proteins: Large multipolypeptide proteins (e.g. haemoglobin).

oligomers: A short polymer of nucleotides, sugars or amino acids. Oligomer is a multimer with just a few subunits. Most oligomers possess identical subunits or repeating groups of non-identical subunits, generally in symmetric arrangements. In polymerase chain reactions, primers are oligomers consisting of 10–15 nucleotides.

oligonucleotide: A liner sequence of a few (10–15) nucleotides attached covalently by phosphodiester bonds. They are utilized as synthetic genes and DNA probes in site-directed mutagenesis, polymerase chain reactions, DNA sequencing and library construction. Oligonucleotide, composed of 25 or fewer nucleotides, is used as a DNA synthesis primer. Oligonucleotides are commonly used as primers during polymerase chain reaction for molecular profiling of the genotypes and in marker-assisted selection in plant breeding. Very short primers can give false positives, whereas the long primers sometimes do not amplify the desired DNA segment due to annealing problem. *See* OLIGODEOXYRIBONUCLEOTIDE *and* OLIGORIBONUCLEOTIDE.

oligonucleotide chips: These are synthesized *in situ* base by base, employing ~10^7 copies of selected oligonucleotides; each of 20–25 nucleotide length, based on prior sequence information on a 24H24 μm area of a glass surface by photolithography. This is a light-directed synthesis approach where a solid surface is derivatized with chemical linkers containing photolabile protecting groups. This is selectively activated by shining light through a photomask, exposing the 5′-OH group. The wafer is flooded with a modified nucleotide, causing the coupling of residues to the activated region of the chips. A second region of the chip is selectively activated with light using a different photomask, allowing the coupling of six residues to different activated region. A repeated series of steps enable parallel oligonucleotide synthesis at many addresses. Using complementorial synthesis strategy, a set of 4^K oligonucleotide of length K-mers can be generated in 4K synthesis cycles. K-mers are substrings of a biological sequence that are used in sequence analysis and computational genomics. Oligonucleotide chips engage multiple probes per gene.

oligonucleotide fingerprinting: The first microsatellite-based technique was derivative of restriction FRAGMENT LENGTH POLYMORPHISM (RFLP) analysis using microsatellite-complementary oligonucleotides as probes. Oligonucleotide printing comprises several steps: (i) isolation of high-molecular-weight genomic DNA, (ii) complete digestion of genomic DNA, (iii) electrophoretic separation of the restriction fragments, (iv) denaturation and immobilization of the separated DNA fragments, (v) hybridization of dried gel to labelled microsatellite-complementary oligonucleotide probes and (vi) detection of hybridizing fragments (fingerprints) by autoradiography or by chemiluminescence, and their documentation by photographing.

oligonucleotide ligation assay (OLA): A PCR-based method for single nucleotide polymorphism (SNP) typing. It is a ligase-mediated gene detection scheme which uses exact 3′ matching of a primer to one of the SNP alleles. If this transpires, the other labelled oligonucleotide (that binds to the nucleotide located immediately next to the SNP, on the other side) would be linked to the primer by ligase; the ensuing sample can then be verified for the presence of the label. This genotypic assay has been used to identify point mutations in DNA for a variety of diseases and to detect drug resistance-associated mutations in HIV-1 subtype B.

oligonucleotide microarrays: In oligonucleotide microarrays, the probes are short sequences premeditated to pair with the parts of the sequence of known or predicted open reading frames. Although oligonucleotide probes are often used in 'spotted' microarrays, the term 'oligonucleotide array' most often refers to a specific technique of manufacturing. One technique used to produce oligonucleotide arrays include photolithographic synthesis on a silica substrate where light and light-sensitive masking agents are used to 'build' a sequence one nucleotide at a time across the entire array. *Aka* *IN SITU* SYNTHESIZED ARRAYS. *Compare with* SPOTTED MICROARRAYS.

oligonucleotide probe design: The use of oligonucleotides for gene amplification, as diagnostic probes or antisense-based drugs, can be optimized by minimizing the possibility of non-specific hybridization. Oligonucleotides are also used in the blocking of the translation of transcripts produced by specific genes.

oligonucleotide probes: Short-chain fragments of DNA that are engaged in several gene analysis tests (e.g. the single base change in DNA that results in sickle-cell anaemia).

oligonucleotide therapy: Use of oligoribonucleotides and oligodeoxynucleotides as therapeutic agents. Three basic strategies used are an antisense inhibition of expression, triple helix formation and protein epitope targeting.

oligonucleotide-based insecticides: A natural method to control the number of insect pests based on the use of extracellular double-stranded DNA secreted by the plant *Pittosporum tobira*. Using a principle similar to one found in nature, V.V. Oberemok *et al.* (2022) showed that the topical application of artificially synthesized short antisense oligonucleotide insecticides (olinscides, DNA insecticides) is an effective and selective way to control the insect *Coccus hesperidum*.

oligonucleotide-directed mutagenesis (ODM): An *in vitro* mutagenesis technique that involves the use of a synthetic oligonucleotide to introduce a predetermined nucleotide alteration into the gene to be mutated.

oligonucleotide-mediated mutagenesis (OMM): Site-specific mutation with chemically synthesized oligonucleotide with homology to the target site (other than for the intended nucleotide modification).

oligonucleotides containing modified cytosines: Synthetic nucleotides used in epigenetics research.

oligopeptide: A few amino acids joined by peptide bonds. Peptides are the chains of amino acids. Amino acids join together through a substituted amide linkage, commonly called peptide bond. This bond results from the removal of water (dehydration). Tripeptides possess two peptide bonds. When a few amino acids get linked, the structure is called oligopeptide. Further when many amino acids join the resulting structure, it is called POLYPEPTIDE.

oligoribonuclease: A ribonuclease that degrades short oligoribonucleotides to monoribonucleotides.

oligoribonucleotide: An oligonucleotide made up of ribonucleotides. Oligoribonucleotides are synthesized on an applied biosystems DNA/RNA synthesizer using β-cyanoethyl phosphoramidite chemistry.

oligosaccharide: Carbohydrate consists of monosaccharide groups (usually 2–20) linked by glycosidic bonds (Ligo = few). Disaccharides are the most common oligosaccharides. Sucrose is a disaccharide of glucose and fructose, and lactose is a disaccharide of glucose and galactose. *See* COMPLEX OLIGOSACCHARIDE, HIGH-MANNOSE OLIGOSACCHARIDE, N-LINKED OLIGOSACCHARIDE *and* O-LINKED GLYCOSYLATION.

ome: In molecular biology, the suffix '-ome' is used to address the objects of study of fields such as the genome, proteome, transcriptome, metabolome and ionome; and refers to a totality of some sort. Relatively recently, fields included in this category are epigenome, epiproteome and epitranscriptome.

Omega RNA (ωRNA): ωRNA is a small RNA that directs IscB enzymes to cut specific DNA sequences. It plays the combined role of CRISPR RNA (crRNA) and TRANS-ACTIVATING CRISPR RNA (TRACRRNA) to guide DOUBLE-STRANDED DNA (dsDNA) cleavage. S.P. Nety and others in 2023 uncovered a potential *cis*-regulatory mechanism whereby a region of the programmable RNA-guided DNA endonuclease TnpB mRNA inhibits DNA cleavage by the RIBONUCLEOPROTEIN (RNP) COMPLEX. They further expand the characterization of TnpB by examining ωRNA processing and RNA-guided nuclease activity to reveal a new functionality, ωRNA biogenesis of TnpB; and characterizes additional members of this biotechnologically useful family of programmable enzymes.

O-micron DNA: An alternative name for the 2% plasmid of *Saccharomyces cerevisiae*.

omics: In biology, the suffix '-omics' refers to the sum of constituents within a cell. The omics sciences share the overarching aim of identifying, describing and quantifying the biomolecules and the molecular processes that contribute to the form and function of cells and tissues. It is used to address the studies in fields such as genomics, proteomics, transcriptomics, metabolomics and ionomics. Relatively recently, fields included in this category are epigenomics, epiproteomics and epitranscriptomics.

ommochromes: Molecules responsible for contributing brown pigment to eye colour in *Drosophila*. *Compare with* PTERIDINES.

omnipotency: *See* TOTIPOTENCY.

omnivorous: Parasites attacking numerous different hosts.

on-target activity: Cas9 cleavage at a desired location due to gRNA targeting sequence with sufficient homology to recruit Cas9 to desired genomic locations.

oncogenes: The genes that deliberate the ability to transform cells to a tumorigenic state. All human and subhuman species possess a complement of oncogenes that in their nonactivated state produce substances necessary for normal cell proliferation and cell-surface properties. Oncogenes were originally recognized in certain oncogenic retroviruses (*Aka* V-ONC) and their homologues also occur in eukaryotic cells (*Aka* C-ONC). Most of the oncogenes are mutant forms of normal PROTO-ONCOGENES implicated in the control of cell division, growth and development. In an attempt to locate the sequence responsible for oncogenicity, the proto-oncogene and oncogene are cleaved with restriction enzymes at the same site, the fragments are recombined and the recombinants confirmed for transforming activity. In this way, exact position has been narrowed to 350-bp region, which was sequenced. The cause of oncogenic activity has ultimately been attributed to a single base change (G → A) that causes substitution of valine or aspartate for glycine.

oncogenes, class I: These oncogenes (e.g. *src*, *yes*, *neu*, *abl*, *fps*, *fms*, *erb*, *ros*, *mos*, *fgr*) are involved in the synthesis of cell-surface receptor proteins. The cell-surface receptor proteins are triggered by growth factors formed by other cell types. A triggered cell proliferates as long as the second cell type manufactures its growth factor, in reply to some external stimulus. Class I oncogene products possess kinase activity and are localized in cytoplasm.

oncogenes, class II: These oncogenes (e.g. *HA-ras*, *kI-ras*, *N-ras*) are generally located in human tumours. Their protein products possess a common characteristic of controlling cellular metabolism.

oncogenes, class III: These oncogenes (e.g. *myc*, *myb*, *fos*, *ski*, *p53*) control nuclear activities (e.g. cell cycling).

oncogenes, class IV: These oncogenes (e.g. *cis*, *rel*, *B-lym*, *erb-A*, *ets*, *met*) are comparatively unrelated; some of them manufacture cell growth factors (proteins).

oncogenes, classes of: There are four main classes of oncogenes: Class I–IV. Viral oncogenes and cellular oncogenes are derived from cellular proto-oncogenes. *See* PROTO-ONCOGENE.

oncogenes, origin of: Viral oncogenes and cellular oncogenes are formed from cellular proto-oncogenes.

oncogenic virus: A DNA or RNA virus that causes cancer.

oncolytic herpes simplex virus (HSV) vectors: These vectors are promising therapeutic agents for cancer.

oncornavirus: An RNA virus that induces cancer.

one cistron–one polypeptide hypothesis: With the discovery of cistron by S. Benzer in 1962 and collinearity by C. Yanofsky in 1964, cistron is considered a part of the DNA molecule which codes for a single polypeptide. This leads to the proposal of one cistron–one polypeptide hypothesis. Some consider this hypothesis as an alternative name of ONE GENE–ONE POLYPEPTIDE HYPOTHESIS.

one enzyme–two functions concept: One enzyme, in some cases, may have two functions. For example, enzyme human paraoxonase 2 (pon2) has two functions: (i) enzymatic lactonase activity and (ii) reduction of intracellular oxidative stress. This observation introduced the one enzyme–two functions concept.

one gene–many proteins hypothesis: Alternative RNA splicing is a mechanism by which more than one protein can be made from a single gene. This observation yielded one gene–many proteins hypothesis. It means that the old paradigm (that one gene makes only one protein) is clearly in need of revision. The discovery of polyproteins gave rise to one gene–many protein hypothesis.

one gene–one antigen hypothesis: This hypothesis emphasizes that each of the genes in a tissue transplant, governing the host's reaction to it, is accountable for the production of a specific transplant antigen alone.

one gene–one chromomere hypothesis: B.H. Judd and others in 1972 proposed one gene–one chromomere (band) hypothesis in 1972, according to which there is one indispensable function articulated by genetic information contained in each chromomere. The ONE CHROMOMERE–ONE FUNCTION RELATIONSHIP holds regardless of the amount of DNA the chromomere contains. They thus projected that the chromomere is a relatively short length of structural DNA together with a variable length of regulatory DNA. Chromomere is one cistron within which much of regulatory DNA performs a regulatory function. Possibility also occurs that the condensed chromomeric DNA is entirely regulatory in function, whereas structural gene(s) are located in the interband region.

one gene–one enzyme hypothesis: G.W. Beadle and E.L. Tatum in 1941 induced nutritional mutants with X-rays in bread mould (*Aka Neurospora crassa*). The inability of the mutant to synthesize a substance was found to be due to absence or defect in one enzyme. Genetic analysis showed that the defect in the enzyme was due to a change in a single nuclear gene. This resulted in the proposal of one gene–one enzyme hypothesis. This hypothesis states that each gene controls the reproduction, function and specificity of a particular enzyme. Ample support to one gene–one enzyme hypothesis came from further studies on microorganisms like *Escherichia coli*, *Salmonella* and *Neurospora*.

one gene–one mRNA–one protein hypothesis: The ONE GENE–ONE RIBOSOME–ONE PROTEIN HYPOTHESIS was rejected when it was observed that RNA is translated into protein by ribosome, and this resulted in one gene–one mRNA–one protein hypothesis.

one gene–one polypeptide hypothesis: One gene controls the synthesis of only one polypeptide. This hypothesis is valid for all the protein–encoding genes. However, the one gene–one enzyme hypothesis could not be applied universally. Later studies on mutant haemoglobins, lactate dehydrogenase and tryptophan synthetase challenged the one gene–one enzyme hypothesis and lead to the one gene–one polypeptide relationship. One gene–one polypeptide hypothesis states that 'one gene controls the synthesis of one polypeptide'.

one gene–one primary cellular function hypothesis: Earlier hypotheses put forth to explain gene function were consequence of the investigations only on structural genes. In addition to structural genes which encode for polypeptides, there also occur other genes (e.g. tRNA genes and rRNA genes) which manufacture RNAs needed in protein biosynthesis; but they themselves do not function as a template during translation. The three types of RNAs (mRNAs, tRNAs and rRNAs) are considered primary cellular products of genes. A gene performs one specific primary cellular function. This gave birth to the one gene–one primary cellular function hypothesis. *See* PRIMARY CELLULAR FUNCTION.

one gene–one reaction hypothesis: This hypothesis states that every specific biochemical reaction is under the ultimate control of a different gene. This view was later refined as the one gene–one enzyme hypothesis.

one gene–one ribosome–one protein hypothesis: According to an early form of the theory of the information structure of protein synthesis, the RNA transcript was thought to offer the RNA moieties for newly formed ribosomes. Accordingly, each gene was envisioned to produce one specialized kind of ribosome, which in turn would perform the synthesis of one and only one kind of protein. F.H.C. Crick, L. Barnett, S. Brenner and R. Watts-Tobin in 1961 typified this scheme as the one gene–one ribosome–one protein hypothesis. Now we know that synthesis of new kinds of ribosomes is not a precondition for the synthesis of new kinds of proteins.

one-carbon metabolism: Network of biochemical reactions in which a chemical unit containing one-carbon atom (e.g. a methyl group) is transferred through several steps from a donor to another compound (e.g. DNA).

one-dimensional electrophoresis for DNA analysis: A simple way of separating fairly large fragments of DNA from one another, by size, is to use an agarose gel. DNA does not need a detergent, since it already has a large number of negative phosphate groups, evenly spaced. Thus, as with SDS-PAGE, the charge-to-mass ratio is constant.

one-dimensional electrophoresis for protein analysis: Various approaches include as follows: **SODIUM DODECYL SULFATE-POLYACRYLAMIDE GEL ELECTROPHORESIS** (SDS-PAGE), NATIVE GELS and **ISOELECTROFOCUSING** (IEF) gels.

1-methyladenosine (1^{m}A) in eukaryotic messenger RNA: 1^{m}A transpires on thousands of different gene transcripts in eukaryotic cells.

1-methyladenosine-specific RNA immunoprecipitation with demethylase-assisted RNA sequencing (1^{m}A-ID-seq): This procedure uses the 1^{m}A antibody to induce a mismatch or truncation during the reverse transcription process. More than 7000 1^{m}A modifications were recovered across the diverse mRNA transcripts in the cell.

one-point crossover: In genetic algorithms, a breeding technique that uses one randomly chosen point, interchanging the portions of the two breeding individuals to the right of that point.

one-sided alternative: A directional alternative hypothesis in the logic of asserting that the value can be only either more or less than the null value.

one-sided invasion (OSI) model: This is a non-conservative type of DNA recombination model, as the original parental sequences are lost during recombination. Thus, OSI model involves unlawful type of recombination. *Aka* SINGLE-STRAND INVASION MODEL.

one-sided test: Designed to test a directional hypothesis, yielding a one-sided *P*-value. For example, one might test the NULL HYPOTHESIS (H_0) that there is no difference in mortality between two treatments, with the ALTERNATIVE HYPOTHESIS (H_1) being that the new drug lowers mortality. *Aka* ONE-TAILED TEST. *See* TWO-SIDED TEST.

one-step gene editing: S. Tang and S.H. Sternberg in 2023 proposed retroelement-based one-step gene editing as a safer alternative to CRISPR-Cas approaches. Retroelements are endogenous DNA segments that use the host cell's RNA polymerase to generate an RNA copy of themselves, which is then turned into complementary DNA by a reverse transcriptase and incorporated into the genome elsewhere.

one-way analysis of variance: The one-way analysis of variance (ANOVA) enables us to compare several groups of observations, all of which are independent but possibly having a different mean for each group. A test of great reputation is whether or not all the means are equal. This is a technique that can be used to compare whether two samples' means are significantly different or not (using the F distribution). This technique can be used only for numerical response data, the 'Y', usually one variable, and numerical or (usually) categorical input data, the 'X', always one variable, hence 'one way'. The one-way ANOVA is employed to test for differences among at least three groups, since the two-group case can be covered by a t-test.

one-way classification: In a typical situation, it refers to the comparison of means of several univariate normal population, having the same unknown variance σ^2, on the basis of random samples selected from each population. The population means are denoted by $\mu_1, \mu_2, \ldots, \mu_k$ (if there are k populations).

ontogenesis: The progression of growth and development of an individual from zygote formation to maturity.

ontogenetic development: It deals with various processes involved in transformation of structures, eggs or some other rudimentary structures to complete adult as a result of an organism's experiences with its environment, rather than its genetic makeup. *Aka* ONTOGENY.

ontogenetic niche: The ecological niche occupied by an organism, at a given stage, in its ONTOGENY. In many organisms, this will change as the individual grows.

ontogeny: Growth; usually increase in size, or both size increase and differentiation of traits of the individual.

oocyte: A female gametocyte or germ cell that divides twice by mitosis and meiosis into two other oocytes or into two ootids; the developing egg. It is usually a large and immovable cell. It is involved in reproduction (*Aka* OVOCYTE).

oocyte microinjection: Frog oocytes are readily available in large numbers, as are the eggs and embryos. Amphibian embryogenesis occurs outside the mother so that access is not a problem, but breeding frogs is a problem. *Drosophila* has all the advantages of an extremely powerful genetic system.

oocyte, primary: *See* PRIMARY OOCYTE.

oocyte, secondary: *See* SECONDARY OOCYTE.

oogenesis: It is a process of development and maturation of ova from the primary germ cells produced by the meiosis in the cells of germinal epithelium of the ovary of the female organism. The process is virtually the same as spermatogenesis, but the end products are different. Oogenesis is broadly divided into three phases: multiplication phase, growth phase and maturation phase.

oogonia (sing. oogonium): Primordial germ cells that produce oocytes (by mitosis) from which the ovum and polar bodies develop by meiosis.

ooplasmic segregation: Massive reorganization of ooplasm is observed soon after fertilization. Ooplasmic segregation occurs with such a force that egg is temporarily distorted. Distortion takes the form of a constriction that spread from the animal pole to vegetal pole. During this process, the egg changes from a pear shape to a dumbbell shape and to a pear shape once again. Distinct morphogenic determinants are equally distributed during sea urchin cleavage. Cytoplasmic determinants play an important role in their development. Cleavage is involved in segregating these regions from one another.

opal codon: UGA, one of the three nonsense codons that cause termination of protein synthesis. *See* AMBER CODON *and* OCHRE CODON.

opal mutation: Amber mutation results from a base substitution that converts a codon specifying an amino acid into UGA, which gestures chain termination.

opal suppressor: A gene that codes for an altered transfer RNA such that its anticodon can recognize the opal codon and thus enables the continuation of protein synthesis. A suppressor of an opal mutation is a tRNA that is carrying the amino acid corresponding to the original codon or a neutral substitute. Some eukaryotic cells normally synthesize opal suppressor tRNAs. The role of these tRNAs is not clear, as they usually do not thwart normal termination of protein synthesis at an opal codon.

open chromatin: Loosely packed regions of the eukaryotic genome where the DNA regulatory elements can be easily assessed by transcriptional machinery.

open circle plasmid: One of the three conformations that a plasmid molecule may adopt. In an open circle, there has been either a single break or a number of staggered breaks made in the strands of the duplex. Such a molecule may not form supercoils and, therefore, is referred to as an OPEN CIRCLE or RELAXED CIRCLE.

open circular DNA (ocDNA): An open circular DNA molecule that has at least one nick. The ocDNA cannot be in the supercoiled form and has the same density as linear DNA in Caesium chloride/Ethidium bromide ($CsCl_2$/EB) density gradients.

open continuous culture: A cell culture in which influx of fresh medium is balanced by the discharge of an equivalent volume of culture.

open pollination: Natural, cross or random pollination.

open population: A population freely exposed to gene flow and subject to too much input of alien genes owing to immigration.

open promoter complex: The complex formed between *Escherichia coli* RNA polymerase and a promoter, in which the double helix is partially unwound in readiness for the start of RNA synthesis.

open reading frame (ORF): In a random sequence of nucleotides, 1 in every 20 codons in each reading frame is (on an average) a termination codon. ORF is a reading frame consisting of contiguous non-overlapping nucleotide codons that may or may not include a termination codon. The identification of an ORF is the first indication that a segment of DNA may be part of a functional gene. The term is deployed in investigations designed to find genes. *Compare with* GENE.

open reading frame (ORF) cDNA clones: Generation of expression-ready ORF cDNA clones often requires multistep cloning processes, as well as lengthy verification and sequence analysis. Gene *Copoeia* provides wide choices of expression systems, promoters, fusion tags and selection markers.

open system of gene expression profiles: An open system of gene expression analysis allows for the identification and analysis of genes not previously characterized. Such an 'open system' of gene expression is a serial analysis of gene expression (SAGE).

open-source software (OSS): A computer software in which source code is released under a license in which the copyright holder grants users the rights to study, change and allocate the software to anyone and for any purpose. Open-source software may be developed in a collaborative public manner. Open-source software is a prominent instance of open collaboration.

operating system (OS): A system software that manages computer hardware and software resources and delivers public services for computer programmes.

operational genes: Genes implicated in metabolic and other bordering processes. They are thought to be more disposed to lateral gene transfer than INFORMATIONAL GENES. *Contrasts* INFORMATIONAL GENES.

operational taxonomic unit (OTU): Any unit (e.g. an organism, gene, species or population) engaged for phylogenetic study.

operator: A region of DNA adjacent to structural genes that controls the expression of genes. A nucleotide sequence element to which a repressor protein attaches in order to prevent transcription of a gene or a group of genes that are cotranscribed (operon). Operator is the structural/biochemical 'switch' with which the operon is turned on or off. In *lactose* operon in *Escherichia coli*, the operator lies very close to the *lac z* gene. *Aka* O LOCUS. *See* OPERON.

operator mutations: Mutations in operator that regulate how much of the structural gene proteins are made, but which do not affect the amino acid sequence of the proteins themselves. These mutations are called REGULATORY MUTATIONS, as they are found to occur in the regulatory elements. The operator class of regulatory mutations controlled all the structural genes of an operon at the same time and in the same way. See OPERATOR-CONSTITUTIVE MUTATIONS (o^c).

operator-constitutive (o^c) mutation: Mutation in the operator which result in constitutive synthesis of all the gene products of the operon regardless of the presence or absence of the substrate. Constitutive mutants continually agitate the protein characteristic for that operon, as the operon unit cannot be turned off by the repressor molecule. The mutation affects the activity of all the genes of the operon in *cis*-arrangement, not in *trans*-arrangement.

operon: A cluster of genes that are cotranscribed under the governance of a single promoter-operator system. An operon comprises promoter (p), operator (o) and a number of structural genes. Operon model was given by F. Jacob and J. Monod in 1961. This model applies to the prokaryotic genes that are closely linked, cotranscribed and regulated coordinately.

operon concept: Metabolically related genes are linked together in prokaryotes; one such group of genes is termed as OPERON.

operon concept applicability to eukaryotes: The operon system of gene regulation has been evidenced not to be applicable to eukaryotes. Crossing-over and chromosome structural changes (e.g. translocations and inversions) disrupt clusters of genes that otherwise might become operons. In prokaryotes, metabolically related genes are located close together, whereas in eukaryotes, metabolically related genes may be dispersed throughout the genome. Moreover, compartmentalization exists in eukaryotic cells. Nucleus encompasses most of the genetic material. The compartmentalization has significant role in gene expression. Regulatory mechanisms in eukaryotes act, or might act, to regulate gene expression at various levels. Various potential control points recognized in eukaryotes include regulation at the level of gene structure, initiation of transcription, processing of transcript, transport of mRNA to cytoplasm, translation of mRNA and post-translational control. Evidences of controls operating at the above levels of gene expression are available. Models for controlling eukaryotic genes usually assume positive control. Two different classes of regulatory mechanisms may be recognized in eukaryotes. These are short-term and long-term controls. Short-term (reversible) regulation signifies the cell's retort to instabilities in environment.

opine synthetase: The enzyme system accountable for the synthesis of opines of a number of different types. The genetic information coding for opine synthetase is located in the T-DNA region of the Ti plasmid. These plasmids deliberate virulence to strains of *Agrobacterium*, which are in charge of the crown-gall disease. The T-DNA is transferred to the plant cells, where it is stably integrated into the nuclear genome, and the *opine synthetase* (*os*) gene is transcribed and translated. *See* OPINES.

opines: A group of low molecular weight compounds (in form of unusual amino acid derivatives) produced by plant cells of crown-gall tumours or hairy root tumours, transformed by pathogenic bacteria of the genus *Agrobacterium* and *Rhizobium* in order to feed the pathogen. Opines include agropines, nopalines and octopines and are characteristic of individual strains of *Agrobacterium*.

opportunism: According to this theory of evolution, what can happen usually does happen. The changes occur as they may and course of evolution follows opportunity rather than plan. Opportunity, however, is limited. Some species may get opportunity, whereas others may not. Thus, opportunism involves a strong component of chance.

opportunistic pathogens: Organisms that exist as a part of the normal body of microflora but which may become pathogenic under certain conditions (e.g. when the normal antimicrobial defence mechanism of the host has been impaired).

opportunistic selection: Mass-pedigree selection where bulk handling is terminated, whenever environment is conducive for some primary traits.

optic chiasma: A visible chiasma on meiotic chromosomes observed through a microscope.

optical density: The level of absorbance of light by a solution as a function of the concentration of a solute. *See* ABSORBANCE.

optimal alignment: The highest-scoring alignment found by an algorithm capable of producing multiple solutions. This is the best possible alignment that can be found, given any parameters supplied by the user to the sequence alignment programme.

optimal promoter: An optimal promoter is defined as a sequence consisting of the −35 hexamer, separated by 17 base pair from the −10 hexamer, lying 7 bp upstream of the start point.

optimization of existing biological functions: M. Stock and T.E. Gorochowski (2024) suggested an open-ended strategy to biological design, with the innovation of designed biology as its important goal. Rather than exclusively focusing on optimization towards a single best design, designing with novelty may move past the diminishing returns. Research from the artificial life community has established that embracing novelty can automatically produce innovative and unexpected resolutions to challenging problems beyond local optima. Synthetic biology thus offers a perfect playground to discover more creative methods to biological design.

optimum pH: The pH at which enzymes have maximal catalytic activity and the organism grows best. The optimum pH for plant tissue culture is around 5.8 that is adjusted using pH meter before autoclaving. The plants cultured on a medium possessing optimum pH exhibit the best growth.

optimum temperature: The temperature at which an organism grows the best. The optimum temperature varies with the type of organism. For example, most of the cultured plants grow best at a temperature of 25 ± 1°C in the incubation room. *Escherichia coli* grows best at 37°C, and *Agrobacterium tumefaciens* grows best at 28°C.

optogenetics: A synthetic biology technique that uses light to control genetic circuits for achieving spatiotemporal gene regulation of cells in living tissues.

orange flavour: Sweet orange (*Citrus sinensis*) displays restricted genetic diversity and high susceptibility to Huanglongbing (HLB), one of the most damaging diseases of citrus worldwide. Studies to understand the key compounds accountable for orange flavour and their genetic regulation has begun. Z. Fan and co-workers in 2024 have analysed juice samples (including oranges, mandarins, *Poncirus trifoliata* and hybrids) and noted distinct volatile compositions. Seven esters differentiated orange from mandarin flavour. Cluster analysis showed six esters with shared genetic control. Differential gene expression analysis identified *Clonorchis sinensis* alcohol acyltransferase 1 (CsAAT1) accountable for ester production in oranges. Its activity was authenticated through overexpression assays. Phylogeny discovered that the functional allele was inherited from pummelo. A single nucleotide polymorphism (SNP)-based DNA marker in the coding region precisely predicted phenotypes. This learning expands breeding options for orange-like cultivars.

orcein: Dye commonly employed in cytology, especially when liquified in acetic acid, as with aceto-orcein, for staining of squash preparations of the chromosomes.

order: One of the eight major hierarchical taxonomic ranks in Linnaean taxonomy. It is classified between family and class. In biological classification, the order is a taxonomic rank used in the classification of organisms and recognized by the nomenclature codes. *See* taxa.

ordered tetrads: In some fungi, the ascii are so narrow that ascospores cannot change their position during meiotic divisions and subsequent mitosis. An example of ordered tetrad is *Neurospora crassa*. *Compare with* UNORDERED TETRADS.

ordinal scale: A gauge that measures a polytomous attribute in a defined order, such as severity of disease into mild, moderate, serious and critical. Here, the distance between mild and moderate is undefined. Division of blood groups into O, A, B and AB is polytomous but not ordinal since these blood groups do not have any order; none is better or worse than the other.

ordinal variable: A categorical variable for which there is a definite ordering of the categories. For example, severity of lower back pain could be ordered as none, mild, moderate or severe and be coded using these names or using numeric codes such as 0, 1, 2, 10. Spacing between codes are not important.

organ: A collection of tissues that structurally form a functional unit specialized to perform a particular function. Heart, kidneys and lungs are examples of organs.

organ culture: *In vitro* culture of an organ for its multiplication or obtaining an entire organism. The root, shoot tip, embryo, leaf and so on, are the organs with specific functions in a plant. These can be cultured *in vitro* (e.g. root organ culture for production of secondary metabolites, shoot tip culture for micropropagation and embryo culture for obtaining hybrids).

organelle: A membrane-bound structure within eukaryotic cells. In the plant cell, for example, CHLOROPLAST, MITOCHONDRIA and NUCLEUS are the organelles that perform specific functions. Mitochondria possess its own DNA called MITOCHONDRIAL DNA (mtDNA), and this organelle is the site of respiration. Likewise, chloroplast possesses its own DNA called CHLOROPLAST DNA (cpDNA), and this organelle is the site of photosynthesis. For other organelles, *see* ENDOPLASMIC RETICULUM, LYSOSOMES, GOLGI BODIES, RIBOSOMES *and* LYSOSOMES.

organic: The matter from living or once-living things.

organic agriculture: A concept and practice of agricultural production that focuses on production without the use of synthetic inputs (e.g. fungicides, insecticides and herbicides). It uses fertilizers of organic origin (e.g. compost manure, green manure, bone meal) and places emphasis on techniques such as crop rotation and companion planting. It does not permit the use of transgenic organisms. The principles of organic agriculture were laid down in the 1940s by Sir Albert Howard, Lady Eve Balfour and others, and later modernized and interpreted in Britain by the SOIL ASSOCIATION. United States Department of Agriculture (USDA)s National 'Organic Program' has established a set of national standards for certified organic production, which are available online. An official label indicates certification by the USDA regulatory body (*Aka* ORGANIC FARMING).

organic compounds: Chemical compound that contains carbon and hydrogen. Chemically, the compounds can be divided into two classes: (i) inorganic compounds that lack carbon and in which ionic bonds play an important role (e.g. water, oxygen, salts, acid and bases) and (ii) organic compounds that always contain carbon, hydrogen and may also contain oxygen, nitrogen and various other elements. Carbon chains formulate the basis of many organic compounds in living cells. Typically, important organic compounds include cell wall constituents, polysaccharides, lipids, proteins and nucleic acids.

organic conservation breeding: Varieties that are of conventional origin are conserved and propagated by cultivation under organic standards. It is relevant to varieties that are older or of high importance, which should be conserved as a gene pool for future breeding efforts.

organic farming: *See* ORGANIC AGRICULTURE.

organic LED (OLED): LEDs based on carbon-based molecules.

organic matter: Amount of remainder after drying and incinerating of a sample of solid wastes or composts into an oven at 625°C for 3 h.

organic soil: A soil that is composed predominantly of organic matter. It usually refers to peat.

organism: Any living plant, animal, human, bacteria, fungus, virus etc. A living thing that has an organized structure, can react to stimuli, reproduce, grow, adapt and maintain homeostasis. Any biological unit which is capable of reproduction or transferring genetic material, and which is autonomous, that is able to exist without foreign help (e.g. microorganisms, fungi, plants, animals including humans). In certain international agreements like the CONVENTION ON BIOLOGICAL DIVERSITY (CBD), this term includes things (e.g. seeds, spores, eggs) possessing the possibility to grow into plants, animals, fungi and so on.

organismal classification: The distribution of organisms into hierarchical groups. Modern biological classifications are Linnaean and categorize organisms into species, genus, family, order, class, phylum, kingdom and certain intermediate categoric levels. Cladism, evolutionary classification and phenetic classification comprise the three methods of classification.

organismal evolution, steps in: Heterotrophic → Chemotrophic → Photoautotrophic → Photosynthetic → Respiration. *See* HETEROTROPHIC, CHEMOTROPHIC, PHOTOAUTOTROPHIC, PHOTOSYNTHETIC *and* RESPIRATION.

organization: Scientists have made the study of science manageable by organizing and classifying natural phenomena. For example, natural objects can be assembled in hierarchies (atoms, molecules, mineral grains, rocks, strata, hills, mountains and planets). Or objects can be arranged according to their complexity (prokaryotes, eukaryotes). Within eukaryotes, there are haploids and diploids; single celled and multi celled. Within diploid eukaryotes there are plants and animals; and within animals it is further split into fishes, amphibians, birds, reptiles and mammals. Organization occurs at micro as well as macro levels.

organization effect: An interaction among adjacent loci owing to some features of organization of the chromosomes.

organization of genome: Prokaryotic and eukaryotic chromosomes are similar with respect to two features: (i) unineme structure and (ii) semi-conservative DNA replication. Several differences in organization of prokaryotic and eukaryotic genomes exist with respect to (a) split versus non-split nature of gene; (b) coding versus non-coding DNA; (c) application of operon concept; (d) polycistronic versus unicistronic; (e) mRNAs, nature of regulatory control; (f) unicellularity versus multicellularity; (g) singularity versus plurality of chromosomes; (h) DNA packaging; (i) haploidy versus diploidy; (j) unique versus repeated sequences and (k) collinearity between gene and its product.

organized: Arranged into definite structures. Plant callus is considered as an unorganized mass of cells. Upon transfer to a differentiation medium, the cells start organizing into differentiated structures (e.g. embryoids, shoot buds and root primordia).

organizer: An inductor; a chemical substance in a living system that determines the fate of the development of certain cells or groups of cells. *See* SPEMANN'S ORGANIZER.

organogenesis: A process of differentiation during development by which organs are formed *de novo* or from pre-existing structures. Cells organize themselves into tissues and show natural organ form, function or both. During plant tissue culture, undifferentiated callus (homogeneous mass of cells) upon its subculturing on to differentiation medium (medium with auxins and cytokinins), undergoes organogenesis leading to the formation of organs (roots and shoots).

organogenic calli: In plant tissue culture, organogenesis is the formation of organs from the cultured explants (plant material; e.g. roots, leaves and flowers). The organogenesis process, is where the plant organs (shoots or roots), are developed.

organotypic: Resembling an organ *in vivo* in three-dimensional form, function or both. For example, a rudimentary organ in culture may differentiate in an organotypic manner; or a population of dispersed cells may become rearranged into an organotypic structure and may also function in an organotypic manner. This term is not used along with the word 'culture' but is used as a descriptive term.

Ori C: Replication origin of *Escherichia coli*. It contains four 9 bp non-contiguous repeats in 240 bp stretch of DNA.

Ori λ: Replication origin of λ phage. It contains four 19-bp repeats, each of which is an inverted repeat.

orientation: The movement of centromeres so that they lie axially with respect to the spindle.

origin of replication (Ori): A site within a DNA sequence of a chromosome, plasmid or non-integrated virus at which replication of the DNA is initiated.

origin of species: The term, coined by Charles Darwin, refers to the production of new types of organisms in descent from extant types. Natural selection is the process by which changes are favoured, leading to the production of new types.

origin of transfer (Ori-T): When gene transfer occurs through plasmids, one of the two strands of plasmid DNA is nicked at a site called Ori-T and this linear strand moves into the recipient bacteria.

origin recognition complex (ORC): Large protein complex that is bound to the DNA at origin of replication in eukaryotic chromosomes throughout the cell cycle.

origins of imprinting, theories on the: Genomic imprinting must have evolved because it deliberates a fitness benefit to the organisms (and their genes) that possess it. A widely acknowledged hypothesis for the evolution of genomic imprinting is the 'parental conflict hypothesis'. Several other hypotheses that suggest a co-adaptive reason for the evolution of genomic imprinting have been projected. Natural selection for genomic imprinting necessitates genetic variation in a population. A hypothesis for the origin of this genetic variation states that the host-defence system responsible for silencing foreign DNA elements (e.g. genes of viral origin) mistakenly silenced genes whose silencing turned out to be beneficial for the organism. There seems to be an over-representation of RETROTRANSPOSED GENES (i.e. the genes that are inserted into the genome by viruses) among imprinted genes. It has also been hypothesized that if the retrotransposed gene is introduced close to another imprinted gene, it may just attain this imprint. *See* PARENTAL CONFLICT HYPOTHESIS.

orphan crops: Those crops which are typically not traded internationally but which can play an important role in regional food security. Many of these crops have received minute attention from crop breeders or other research institutions.

orphan gene: A gene that has been identified by sequencing but its function still remains unknown.

orphones: The isolated individual genes found in isolated locations but related to members of a gene cluster.

ortet: The original plant from which a clone is started through rooted cuttings, grafting, tissue culture or other means of vegetative propagation (e.g. the original plus tree used to start a grafted clone for inclusion in a seed orchard). *See* PLUS TREE.

orthogenesis: (i) The mistaken idea that species are disposed to evolve in a fixed direction because of some inherent force driving them to do so. (ii) An inherent tendency for lineages to change in a particular direction.

orthogonal DNA polymerase (O-DNAP): Engineered error-prone DNA polymerase spotting a specific origin sequence to induce rapid mutagenesis of a gene of interest. Mutagenic orthogonal DNA polymerases selectively mutate the orthogonal replicon but not the genome.

orthogonal DNA replication system: The first orthogonal DNA replication system, OrthoRep, was reported in 2014 as a durable architecture for *in vivo* continuous evolution of user-defined genes. In OrthoRep, a linear DNA plasmid is replicated by a specific orthogonal DNA polymerase (O-DNA Pol or O-DNAP) through protein-primed replication. Orthogonal replication serves as a platform for *in vivo* continuous evolution and as a system whose replicative properties can be manipulated independently of the host's.

orthogonal hierarchical designs: Refer to randomized blocks, split-plots or split-split-plots.

orthogonal replication system in *Escherichia coli*: R. Tian and others in 2024 developed a stable orthogonal replication system in *Escherichia coli*. The orthogonal replicon can transport varied cargos of at least 16.5 kilobases and is not copied by host polymerases but is selectively copied by an orthogonal DNA polymerase (O-DNAP), which does not copy the genome. They designed mutant O-DNAPs that selectively upsurge the mutation rate of the orthogonal replicon by two to four orders of magnitude. This system displayed utility for accelerated continuous evolution by evolving a 150-fold increase in resistance to tigecycline (an antibiotic used to treat certain serious bacterial infections when other antibiotics may not work) in 12 days. And starting from a GREEN FLUORESCENT PROTEIN (GFP) variant, there was a 1000-fold increase in cellular fluorescence evolved in 5 days.

orthogonal translation systems (OTSs): Systems used in genetic code expansion to link incorporation of a NON-STANDARD AMINO ACID (NSAA) to a target codon. These systems comprise a tRNA that is selectively charged with a given nsAA by a COGNATE

AMINOACYL-TRNA SYNTHETASE (AARS). Other possible components of OTSs include mutant elongation factor Tu and ribosomes premeditated to accommodate the given nsAA.

orthologous: Chromosomes, genes, proteins, characters or structures in different species that have evolved from the same ancestor with convergent function. Normally, orthologous genes retain the same function throughout the course of evolution.

orthologous gene mapping: Marker clones, specifically cDNA clones of one crop plant, are being mapped on to the linkage maps of other crops. It is also called COMPARATIVE MAPPING, in which molecular markers are being used expansively for investigating the divergence and evolution of crop plants. This strategy is beneficial in several ways: (i) generation of saturated maps by mapping marker clones of one crop on to the linkage map of other crops, (ii) evolutionary history of various crop plants can be discovered and (iii) gene cloning from complex organisms. The first comparative maps in plants were generated by S.D. Tanksley *et al.* (1988) for tomato, pepper and potato genomes. Since then, comparative maps have been generated in several crops especially in grasses by the cross-hybridization of RFLP markers, the conserved gene order or synteny has been established in several families. Effectiveness of orthologous genes in different backgrounds has been demonstrated using gibberellic acid insensitive (*gai*) genes conferring dwarfness in crop plants. A *gai* gene construct from *Arabidopsis* introduced into a tall rice variety Basmati 370 lead to reduced height.

orthologous genes: Different genes retaining the same function during evolution (e.g. α-globin from humans and mice). *Contrasts* PARALOGOUS GENES.

orthologue: Orthologues are genes in different species that evolved from a common ancestral gene by speciation. Normally, orthologues retain the same function in the course of evolution. Identification of orthologues is critical for consistent forecast of gene function in newly sequenced genomes. *Contrast with* PARALOGUES.

***Oryza sativa* genome:** Rice (*Oryza sativa*) is the most important crop of human consumption and is a staple food for half of the world's population. It is the first cereal whose genome has been completely sequenced (S.A. Goff and others in 2002). The total genome of rice is 420 Mbp. Scientists selected the two varieties – Nippon bare and 93-11 (of the subspecies Japonica and Indica respectively) – to sequence the rice genome. Knowledge of the complete genetic code of rice will help breeders develop strains of the crop with specific characteristics (e.g. stress tolerance, disease resistance or high yield) much quicker than through traditional methods, which may require years of crossing to achieve the desired property. The sequencing of all the rice genes alone provides insufficient information on which all improvements in rice depend. Rice sequencing was done under the International Rice Genome Project. The two private Biotech companies Syngenta and Monsanto also sequenced the rice genome at their own level and submitted their data to GenBank. The researchers have also engaged their improved genome sequence to examine the evolutionary history of rice. In 2006, G.K. Wong and colleagues reported evidence in the rice DNA sequences for a whole-genome duplication event just before the grasses diverged from other flowering plants (about 55–70 million years ago). This genome duplication has likely played a role in the origin of the grasses, which then spread rapidly across the world to offer important sources of food that, among other things, possibly impacted human evolution. International Rice Genome Sequencing Project (2005) presents a map-based finished quality sequence that covers 95% of the 389 Mb genome, including virtually all the euchromatin and two complete centromeres. They found evidence for widespread and recurrent gene transfer from organelles to the nuclear chromosomes.

oryzalin: An agent that is effective for chromosome doubling of haploid apple shoots *in vitro*. Also, the shoot tips of diploid roses treated *in vitro* (at concentrations of 5–15 μM) produced tetraploid shoots in highest frequencies after exposure to 5 μM oryzalin for 14 days. Thin (1 mm) nodal sections were treated with 5 μM oryzalin for only 1 day.

oscillations in gene expression: Oscillations in gene expression regulate various cellular processes and so must be robust and tuneable. Interactions between both negative and positive feedback loops seem to ensure these features. Autonomous and self-sustained oscillator circuits mediating the periodic induction of specific target genes are minimal time-keeping devices found in the central and peripheral circadian clocks. These oscillators are important in controlling critical repair, metabolic and signalling pathways.

osmic acid (OsO_4): A fixing agent commonly utilized to prepare tissue samples for electron microscopy. *Aka* OSMIUM TETROXIDE.

osmolarity: A term used to designate the concentration of a solute in terms of the osmotic pressure it can exercise.

osmolyte: Osmolytes are osmolytic, active and neutral organic compounds, such as sugars (polyols), certain amino acids and quaternary ammonium compounds. Proline is the most extensively disseminated compatible osmolyte. There is a stout correlation between increased cellular proline levels, and the capacity to survive both water deficit and the effects of high environmental salinity.

osmosis: Movement of solvent across the membrane from an area possessing lower solute concentration to an area possessing higher solute concentration. When a plant cell is kept in hypertonic solution, then water from the vacuoles and the cytoplasm of a cell moves out and the process is called osmosis. It leads to the plasmolysis of the cell. It has a tendency to equalize the solute concentrations on the two sides.

osmotic pressure: The force with which a solvent moves from a solution possessing lower solute concentration to a solution of higher solute concentration. It is a pressure generated by the osmotic flow of water through a semipermeable membrane into an aqueous compartment containing solute at a higher concentration. Thus, the concentrated solution possesses higher osmotic pressure compared to dilute solutions.

osmoticum: Particles that cause osmosis.

osteogenesis imperfect: The most common and the mildest form of a group of rare disorders affecting the connective tissue, which are characterized by fragile bones that fracture easily (brittle bones).

outbreeding: A mating system branded by the breeding of genetically unrelated or dissimilar individuals. Since genetic diversity attempts to be enhanced and vigour or fitness of individuals can be enhanced by this process, it is often used to counter the detrimental effects of continuous inbreeding. *Contrasts* INBREEDING.

outcross: (i) Cross-pollination between plants of different genotypes. (ii) In hybrid seed production, the term is used to describe unintended pollination by an outside source of the same crop during. (iii) In biotechnology, it refers to the transfer of a given gene or genes (e.g. one synthesized by humans and introduced into a plant via genetic engineering) from a domesticated organism (e.g. crop plant) to a wild type (relative of crop).

outcrossing: The technique of crossing between different breeds. This is the practice of introducing distantly related genetic material into a breeding line, thereby increasing genetic diversity. *Aka* OUTBREEDING

outer membrane: (i) Outermost of the two membranes surrounding an organelle. (ii) The membrane adjacent to the cytosol.

outer nuclear membrane: The outermost of the two nuclear membranes. It is continuous with the endoplasmic reticulum and is speckled with ribosomes on its cytosolic face.

outgroup: An organism or gene from an evolutionary lineage that is separated from the lineages being studied prior to the appearance of their common ancestor.

output: In computer analysis, the outcome or result of the programme is referred to as the output.

output device: A device that transfers information from the computer to the outside world (e.g. monitor, printer).

outside markers: Loci on either side of another locus or specified region.

over seeding: Seeding into a prevailing crop stand or turf.

over winding: Over winding of double-stranded DNA results from the positive supercoiling, which puts further tension in the direction of the winding of two strands (about each other) in the duplex.

overdominance: Phenomenon in which the character of the heterozygote (*Aa*, F_1) is expressed more markedly in the phenotype than in that of either homozygote. Heterozygotes in this case have higher trait values (usually higher fitness). Overdominance can result in monohybrid heterosis when the hybrid vigour is attained by crossing parents differing in a single specified pair of allelic genes.

overdominance hypothesis: The theory that heterosis is caused by the heterozygous genotype (*Aa*) being superior to either of the homozygous parents (*AA* or *aa*). This hypothesis, given independently by E.M. East and G.H. Shull in 1908, attributes heterosis to the superior fitness of heterozygous genotypes over homozygous genotypes at an individual locus. *Aka* OVERDOMINANCE MODEL OF HETEROSIS.

overexpression vectors: *See* EXPRESSION VECTORS.

overfitting: In the framework of a prediction tool developed using a statistical model or an algorithmic technique (e.g. machine learning), there exists the tendency for the predicted values to be too extreme. Too extreme predictions make the calibration curve show symptoms of regression to the mean. A flattening of the curve should be less steep than 45°

the line of identify. When overfitting is present, low predicted values are too low and/or high predicted values are too high. Overfitting results from slicing the data into pieces that do not have huge denominators. The reason for overfitting is typically partaking in too many candidate features in a supervised learning (informed by Y) feature selection setting or estimating too many parameters in a pre-specified model. Each parameter estimated may be unbiased, but predictions are formed by putting all the parameters together. unless the model is intentionally underfitted using penalization (shrinkage, regularization), the combination of parameters exhibits the 'low values, too-low values, high values or too-high values' phenomenon. This is owing in part to sorting predicted values or selecting subjects with extreme predictions. It is possible that the overall mean predicted value is unbiased even with extreme overfitting. That is why it is imperative to estimate the entire calibration curve. *Aka* OVERINTERPRETATION.

overhangs: 3′ and 5′ **ssDNA** overhangs of dsDNA. Cleavage of DNA with many restriction endonucleases leaves such overhangs. *Aka* EXTENSIONS *or* STICKY ENDS.

overlap microtubule: In the mitotic or meiotic spindle, a microtubule interdigitating at the equator with the microtubules originating from the other pole.

overlapping clones: A collection of cloned sequences completed by generating randomly overlapping DNA fragments with infrequently cutting restriction enzymes. *See* GENOMIC LIBRARY.

overlapping code: When more amino acids are coded than there are code words represented in the end-to-end sequence. In case of non-overlapping codon, the mRNA sequence UUUCCC has two code words (UUU and CCC), but in case of overlapping code, the given sequence may generate four codons (UUU, UUC, UCC and CCC). Actually, the code is non-overlapping.

overlapping deletion: Compound deletion happening as two separate events, in which one deletion overlaps another.

overlapping generations: Population structure in which parents may still reproduce when their offspring are already reproductive.

overlapping genes: A special type of gene organization where one DNA sequence may code for different proteins. It materializes in the presence of two open reading frames in the same DNA sequence, which subsequently act as two functionally different genes. Thus, the overlapping genes share some of the same sequence. Start codons in different reading frames generate different polypeptides from the same DNA sequence. Such an arrangement of genes was first discovered in bacteriophage φX174 by B.G. Barrell and co-workers in 1976. Later, overlapping genes were also reported in eukaryotic multicellular organisms. *Aka* OVERLAPPING DNA SEGMENTS *and* OVERLAPPING READING FRAME.

overlapping inversion: A compound inversion resulting from a second inversion that includes a part of a previously inverted segment.

overpotting: A problem that describes a plant being cultured in a pot that is too large. When overpotted, a plant will expend the greatest part of its energy growing roots, thus reducing the amount of energy which otherwise should have been used to grow leaves and flowers. In addition, the larger pot size will hold more water than the plant can absorb and conditions of excessive moisture will leave the plant susceptible to potentially fatal diseases (e.g. crown rot, Pythium or root rot).

ovists: The people who supported this theory that miniature individuals were present in the egg.

ovulation: Release of an egg from the ovary.

ovule: A structure in angiosperms and gymnosperms that is an outgrowth of the ovary of a seed plant that encloses an embryo which, after fertilization, develops into a seed. It is analogous to the egg in animals.

ovum (pl. ova): A female sex cell or gamete. *See* EGG.

Oxford Nanopore Technologies (ONT): UK-based company that develops and sells nanopore sequencing products.

oxidant: The oxidizing agent is the entity that seeks and accepts electrons in an oxidation-reduction reaction. The oxidant is reduced by the end of the chemical reaction. That is, electron acceptance, by definition, is reduction.

oxidation (vb. oxidize): Oxidation is the removal of electrons (ė) from an atom or molecule associated with the release of energy. In many oxidation reactions, protons and electrons (hydrogen ions, H^+) are removed at the same time, which is equivalent to the removal of hydrogen atoms, and thus, these reactions are also called DEHYDROGENATION REACTIONS. Oxidation of a glucose molecule results in production of CO_2, H_2O and energy in the form of ATP. *Contrasts* REDUCTION

oxidation-reduction reaction: A reaction in which one compound is oxidized and the other is subsequently reduced. Oxidation and reduction reactions are always coupled. In this, reaction electrons are transferred from a donor to an acceptor molecule. Each time one substance is oxidized, another is simultaneously reduced. This type of coupled reaction is called 'oxidation-reduction' or a 'redox'.

oxidative burst: An early defence response against pathogens where 10–14 mol H_2O_2 is generated per cell per second. Here, the early oxidative burst (phase I) is common for compatible and incompatible interactions, whereas the later but sustained burst (phase II) is associated with a hypersensitive response/reaction.

oxidative damages to DNA: Base excision repair has an important role in preventing mutations associated with a common product of oxidative damage to DNA, 8-oxoguanine. 8-oxoguanine DNA glycosylases use an intricate series of steps to locate and excise 8-oxoguanine lesions efficiently.

oxidative deamination: A process in which an amino group in a molecule is replaced by an oxygen group. Some chemicals (e.g. nitrous acid) induce mutations in this way.

oxidative demethylation: The process of removing one or more methyl groups from a molecule, involving the oxidation (i.e. electron loss) of one or more atoms in the substrate.

oxidative phosphorylation: During oxidative phosphorylation, electrons derived from NADH and $FADH_2$ combine with O_2, and the energy released from these oxidation/reduction reactions is used to drive the synthesis of ATP from the ADP and inorganic (orthophosphate) phosphate (Pi). This process transpires in bacteria and mitochondria in which ATP formation is motivated by the transfer of electrons from food molecules to molecular oxygen. It involves the intermediate generation of a proton gradient (pH gradient) across a membrane and chemiosmotic coupling.

oxidative stress: The physiological stress/damage that is the consequence of the chemical breakdown of all or part of an organism via oxidation reaction(s). One common consequence of such oxidation reactions is the generation of reactive oxygen species (ROS, 'free radicals') within the organism's body that can adversely affect (i) endothelial function (i.e. the inner lining of blood vessels), (ii) platelet aggregation (e.g. inappropriate blood clotting/clumping), (iii) atherosclerosis (i.e. build-up of oxidized fatty deposits known as plaque on internal walls of arteries), (iv) myocardial function (e.g. heart failure) and (v) eye and kidney tissue (especially in diabetics). A key indicator of oxidative stress is the peroxidation of membrane lipids to form mono- and bifunctional aldehydes (e.g. 4-hydroxy2-nonenal, *Aka* HNE). Glutathione (GSH) can also lead to excessive oxidation and cell damage.

oxidizing agent (oxidant): The acceptor of electrons in the redox reaction. Ozone (O_3) and hydrogen peroxide (H_2O_2) are the common oxidizing agents and are widely used as antiseptic agents.

oxygen as a by-product of photosynthesis: Oxygen (O_2) molecules are a by-product of photosynthesis. They are produced during the light-dependent reactions that occur in the thylakoids of the chloroplast. This oxygen gas is released through the leaf's stomata into the surrounding air.

oxygenase: An enzyme catalysing a reaction in which oxygen is transferred into an acceptor molecule.

ozone layer: The region of the atmosphere, generally 11–26 km (7–16 miles) above Earth, where ozone occurs in high concentrations. The ozone layer absorbs ultraviolet radiation, shielding Earth from its damaging effects.

P: Probability value. A decimal fraction reveals the number of times an event will occur in a given number of trials. The probability of an alignment occurring with the score in question or better. The P value is calculated by relating the observed alignment score, S, to the expected distribution of HIGH-SCORING SEGMENT PAIRS (HSP) from comparisons of random sequences of the same length and composition as the query to the database. The most highly significant P values will be those close to 0. P values and E values are different ways of representing the significance of the alignment.

p: The relative frequency of a gene, when used together with 'q'. 'p' indicates the relative frequency of the dominant alleles, while 'q' indicates the relative frequency of the dominant alleles. 'p' denotes the short arm of a chromosome, while 'p' denotes the long arm of a chromosome.

P element: A transposon whose genes (within this transposon) counterattack rearrangement during the process (i.e. transposition) of the P element being incorporated into a new site within an organism's genome. In addition to 'transporting' genes to a new location(s) in the genome, the P element itself codes for transposase, an enzyme that makes transposition possible. Hybrid dysgenesis defects arise from the crossing of specific *Drosophila* strains. It occurs when the haploid genome of a male (P strain) possesses nearly 40 P elements/genome. P elements vary in length from 500 to 2900 bp and encode a transposase. P elements do not utilize an RNA intermediate during transposition. They can be inserted in many different positions. P elements code for a repressor, which makes them stable in the P strain in males (but unstable when crossed to the wild-type female as the female lacks a repressor in the cytoplasm). P elements are transposons that are carried in paternal (P) strains of *Drosophila melanogaster* (fly) but not in maternal (M) strains. When a P male is crossed with an M female, transposition is activated and the result is hybrid dysgenesis. The insertion of P elements at new sites in these crosses inactivates many genes and makes the cross infertile. Dysgenesis is principally a phenomenon of the germ cells. P-specific sequences can induce dysgenesis by insertional inactivation.

P I, P II: Symbols for the prophase of the first (P I) and second (P II) meiotic divisions, respectively.

P-DNA: J.F. Allemand *et al.* (1998) have shown that if DNA is artificially stretched, another structural form of the molecule is assumed, which is having structure with exposed bases. P-DNA thus shares features of the DNA structure proposed by L. Pauling and R.B. Corey (1953). It is longer than B-DNA. The phosphate groups are present inside the molecule. The nitrogenous bases are found closer to the external surface of the helix. Currently, this term is utilized for denoting a highly stretched and overwound form of DNA, with rises per residue along the helix axis of about 5.85 Å and 2.62 units per turn. *Aka* DNA, P-FORM.

P-element-mediated gene transfer: P element in *Drosophila* is 2.9 kb long and has 31-bp inverted terminal repeats. Probably this repeat is the site of action of transposase. The P element has been engineered into two types of elements to carry out *Drosophila* transformation: the helper element 'wings clipped', which encodes functional transposase and a series of elements that have functional ends but no functional transposase. The transposase gene has been replaced by genes for morphological markers (like the familiar *white* gene) and a MULTIPLE CLONING SITE (MCS) that allows the other DNA to be inserted. The cloning of the *rosy* gene in *Drosophila* and the precise location of the *rosy* DNA led to the choice of *rosy* for the first P-element-mediated transformation experiments. At present, the focus is on such basic genetic mechanisms as recombination and gene expression and the impact of position effects and transposable elements upon these mechanisms.

P-M system of hybrid dysgenesis in *Drosophila*: A system of hybrid dysgenesis where a female from maternal (M) strain when mated with a male from a paternal (P) strain, various germ-line abnormalities and characteristics of this syndrome. In reciprocal crosses, these abnormalities are not observed. P element is transposable element of 3-kb length found in certain strains of *Drosophila*.

P1, P2, P3, etc.: First, second, third and so on generation from a parent. Also used to designate different parents used in making hybrid or a series of hybrids.

P_1: The two parents of the F_1 generation.

P1-derived artificial chromosome (PAC): One type of vector used to clone DNA fragments (100 to 300 kb insert size; average, 150 kb) in *Escherichia coli* cells. Based on the bacteriophage (a virus) P1 genome. *See* CLONING VECTOR.

***p53* gene:** A tumour-suppressor gene that governs the passage of a given cell from the 'G1' phase to the 'S' (DNA synthesis) phase. The p53 protein that is coded for by the *p53* gene is a transcription factor (i.e. it reads its DNA to determine if it is damaged, and if 'yes', then acts to control cell division, while the *p53* gene codes for more production of additional p53 protein). If, in spite of the presence of p53 protein, a cell begins to divide uncontrollably following damage to its DNA, the *p53* gene can cause apoptosis, which is *aka* PROGRAMMED CELL DEATH (to prevent tumours). The *p53* gene encodes a gene regulatory protein that is activated by damage to DNA and is involved in blocking further progression through the cell cycle.

p53 protein: A tumour-suppressor protein, sometimes called the MASTER TRANSCRIPTION FACTOR or the GUARDIAN OF THE GENOME. Discovered in 1993 by A.J. Levine and colleagues, p53 is believed to be responsible for up to 50% of all human cancer tumours when the p53 gene is damaged or mutated. Amino acid sequence changes in p53 result from damage or mutation to the *p53* gene. The p53 protein has four domains, one of which (i.e. the core domain) binds to a specific sequence(s) of the cell's DNA in order to prevent the cell from dividing uncontrollably when the cell's DNA has been damaged until the damage to that DNA can be repaired. As the amount of DNA within a given (damaged) cell increases, the concentration of the p53 protein also increases. Because the p53 protein is a transcription factor (i.e. 'reads' DNA to determine if it is damaged, then acts to control cell division, while the p53 gene codes for the production of more p53), p53 is very efficient at preventing/inhibiting tumours.

p65 activation domain (p65AD): The principal transactivation domain of the 65 kDa polypeptide of the nuclear form of the NF-κB transcription factor.

PacBio sequencing: Pacific biosciences (PacBio) sequencing uses proprietary SINGLE MOLECULE REAL-TIME (SMRT) technology, which enables the real-time detection of nucleotide incorporation events during the elongation of the replicated strand from the non-amplified template.

pachytene: A third substage of prophase I of meiosis I during which homologous chromosomes are completely synapsed and undergo crossing-over. Crossing-over ensues between the non-sister chromatids of homologous chromosomes. The recombination that occurs during crossing-over in a heterozygous individual is responsible for part of the segregation detected in the progeny. *Aka* PACHYNEMA.

pachytene DNA: Some DNA synthesis takes place at the pachytene stage. This DNA is called pachytene DNA, which is involved exclusively with repair replication and thereby leads to crossing-over.

package: In recombinant DNA procedures, this refers to the step of incorporation of cosmid or other lambda vector DNA with an insert into a phage head for transduction of DNA into the host.

packaging: The act of establishing a viral head particle around a nucleic acid molecule or of filling a preformed virus with nucleic acid. *See IN VITRO* PACKAGING.

packing ratio: Refers to the length of DNA divided by the length into which it is packaged. For example, the shortest human chromosome contains 4.6×10^7 bp of DNA (about 10× the genome size of *Escherichia coli*). This is equivalent to 14,000 μm of extended DNA. In its most condensed state during mitosis, the chromosome is about 2 μm long. This gives a packing ratio of 7000 (14,000/2).

padlock capture: A method for simultaneously capturing and amplifying large numbers of regions of interest from whole genomic DNA. Each padlock probe has two complementary oligonucleotide sequences that flank a region of interest. The sequences are joined by a loop of DNA that ensures efficient joint hybridization and contains sequences for polymerase chain reaction (PCR) with universal primers.

padlock probes (PLPs): Padlock probes (PLPs) are long oligonucleotides whose ends are complementary to adjacent target sequences. Upon hybridization with the target, the two ends are brought into contact, allowing PLP circularization by ligation.

pair programming: An alert software development technique in which two programmers work together at one workstation. One, called the driver, writes code while the other, called the observer or navigator, reviews each line of code as it is typed in. The two programmers switch roles frequently.

pair-rule genes: The genes that influence the formation of body segments in *Drosophila*. The mutants have corresponding parts of the pattern deleted in every other segment. The afflicted segments may be even numbered or odd numbered. Mutations in pair-rule genes lead to corresponding deletions. Pair-rule genes are of two types – PRIMARY PAIR-RULE GENES and SECONDARY PAIR-RULE GENES. GAP PROTEINS regulate primary pair-rule genes. These genes are expressed in alternative parasegments. Expression patterns of these genes are alleviated by cross-regulation and assist in specifying the patterns produced by secondary pair-rule genes. Secondary pair-rule genes are activated by primary pair-rule proteins. Gap proteins are needed to establish the expression domains of secondary pair-rule genes. The dorsal-ventral polarity of the fly is established by a gradient of the nuclear protein dorsal. Dorsal acts both as an activator and repressor of transcription. Primary and secondary pair-rule genes together define the parasegments of the embryo. Different pair-rule genes respond to different gap protein concentrations. Mutations in gap genes cause embryos to develop with large gaps in an array of segments. Mutations in the pair-rule gene cause the organism to develop with only half the normal number of segments. Mutants in pair-rule genes have corresponding parts of the pattern deleted in every other segment. The afflicted segments may be even numbered or odd numbered. Examples of pair-rule genes in *Drosophila* include *runt*, *hairy*, *ftz (fushi tarazu)*, *even skipped*, *paired* and *odd paired*.

paired data: When each subject comprises two response measurements, there is a natural pairing to the data and the two responses are correlated. The correlation is a consequence of the fact that, generally, there is more variation between subjects than there is within subjects. Sometimes one can consider the difference or log ratio of the two responses for each subject and then analyse these 'effect measures' using an unpaired one-sample method such as the Wilcoxon signed-rank test or the paired t-test. One must be cautious that the effect measure is appropriately chosen so that it is independent of the baseline value.

pairing block: A definite chromosome segment that performs as a functional unit in meiotic chromosome pairing.

pairing of chromosomes: *See* SYNAPSIS.

pairing segments: The eukaryotic regions of chromosomes that synapse with a homologue in the first meiotic prophase.

palaeontology: The scientific study of fossils.

palea (pl. paleae): (i) The upper of two bracts enclosing a grass flower, major contributors to chaff in harvested grain. (ii) Chaffy scales on the receptacles of many Asteraceae. (iii) Chaffy scales on the stem of many ferns.

paleopolyploid: An ancient polyploid that undergoes chromosomal rearrangements, gene loss and mutations and eventually becomes diploidized such that it behaves cytogenetically as a diploid.

paleoproterozoic: Division of time from 2500 to 1600 mya.

palindrome: A sequence of letters (e.g. DAD, MADAM, MALAYALAM, REDIVIDER), words (e.g. AND MADAM DNA), phrases, nucleotides or nucleotide pairs that reads the same irrespective of which direction one starts from.

palindromic sequence: Nucleotide pairs that read the same regardless of which direction one starts from. Palindrome sequences are recognized by restriction endonucleases. That cleaves DNA from within. Palindromic sequences can form a cruciform structure due to the intrastrand base pairing of the self-complementary sequence. For example, *Eco*RI recognition site 5′-GAA•TTC-3′/ 3′- CTT•AAG-5′, where '•' indicates the point of symmetry.

palisade mesophyll: Unlike most monocot leaves, which have only one type of mesophyll, the ground tissue of dicot leaves is often arranged into two distinct types: palisade and spongy mesophyll. In the palisade mesophyll, parenchyma cells are tightly packed and their shape is usually polyhedral, elongated or lobed. This structure and the presence of chloroplasts facilitate photosynthesis.

palisade parenchyma: The tissue found underneath the upper epidermis of a leaf, composed of elongate, tubular cells organized upright in the manner of posts in a palisade fortification.

Pall's model of carcinogenesis: M.L. Pall in 1981 proposed a model that induced carcinogenesis is due to the production of tandem duplications of a proto-oncogene in the cell genome, producing a 5- to 10-fold increase in the proto-oncogene product. The initial event in carcinogenesis is proposed to be a mutation that produces a single tandem duplication of a proto-oncogene.

pan neutralist theory: This theory believes in the neutrality of most of the genetic variation and considers evolution as a random process. *Aka* NON-DARWINIAN EVOLUTION. *See* NEUTRALITY THEORY OF PROTEIN EVOLUTION.

pan selectionist theories: Theories that recognize natural selection as the only guiding force of evolution.

pan-edited cryptogenes: Incomplete genes whose transcripts are edited to produce translatable sequences. Several of the maxicircle genes represent cryptogenes. Three such pan-edited cryptogenes are known in *Trypanosoma brucei* maxicircle DNA, which lack more than 50% of the U residues present in mature mRNA.

pan-editing: Causes U insertions/deletions by editing sequence provided by external antisense guide RNA, which pairs with pre-edited mRNA in a ribonucleoprotein particle, the editosome and recognizes positions to be edited as mismatches. It also causes the insertion of multiple cytidine residues in several vertebrate mitochondrial mRNAs. Pre-mRNA lacking a stop codon is polyadenylated, with the first one or two adenylate residues providing the missing information by an unknown mechanism.

pandemic: An extensive and damaging outbreak of disease happening simultaneously in several countries, on multiple continents or worldwide.

pangene: The particle hypothesis by Charles Darwin in his pangenesis hypothesis to explain variation and evolution.

pangenesis: C. Darwin, in 1868, believed in the theory that very small, exact, but invisible copies of each body organ and component, called GEMMULES, were transported by the bloodstream to the sex organs and these were assembled into gametes. Upon fertilization, gemmules of the opposite sex mixed. During development, these 'gemmules' were then separated into different parts of the body. Thus, the individuals constituted a mixture of maternal and paternal organs. The pangenesis hypothesis was rejected on the basis of an experiment carried out by F. Galton in 1909. He transfused the blood between white and black rabbits but obtained no rabbits with intermediate skin colour in their progeny. This theory is now obsolete.

pangenome: In the fields of molecular biology and genetics, a pangenome or SUPRAGENOME, is the entire set of genes from all strains within a clade. The pangenome represents the entire set of genes within a species, consisting of a core genome – containing sequences shared between all individuals of the species – and the 'dispensable' genome. The human pangenome, which collated genetic sequences from 47 people of diverse ethnic backgrounds, could greatly expand the reach of personalized medicine. A draft human pangenome reference has been prepared by Wen-Wei Liao *et al.* (2023) and published in *Nature*.

panmictic population: Unstructured (random-mating) population.

panmictic varieties: *See* VARIETY.

panmictic: Describes a population in which every individual stands the same chance of mating with every other. In other words, where there is no population structure.

panmixis: A random-mating manner of sexual reproduction in which male and female gametes meet each other throughout a population incidentally (i.e. each mating between two individuals of a population has the same probability, as in a rye field).

panspermia theory: The hypothesis that organic molecules (e.g. RNA) or even living organisms (such as bacteria) came from outer space to Earth. *Aka* PANSPERMIA.

paper chromatography: A chromatographic method using absorbent paper by which minute amounts of material can be analysed. A paper strip with a drop of test material at the bottom is dipped into the carrier liquid (solvent) and removed when the solvent front almost reaches the top of the strip. Two-dimensional chromatograms can be produced using square paper and two different solvents. The paper is removed from the first carrier liquid, turned at right angles and dipped into the second. This gives a two-dimensional map of the constituents of the test drop. The identity of the constituents may be determined by measuring Rf values.

paper raft technique: A technique created to promote the development of single cells, which are taken from suspension cultures, in which cells are placed on to filter-paper squares set on actively growing callus (nurse tissue). Growth factors and nutrients from the callus tissue diffuse through the filter paper, promoting cell growth and development.

papovavirus: A class of animal viruses with small genomes (e.g. SV40 and polyoma). *See* VIRUS.

paracentric inversion: Paracentric means 'alongside of', 'beyond' centromere. Intrachromosomal rearrangements that result from a two-break event occurring in the same chromosome arm (short, p or long, q) are followed by the insertion of the intercalary segment after a 180° rotation. Because they do not alter chromosome arm ratios, paracentric inversions can remain undetected. However, in specialized chromosomes (such as polytene or salivary chromosomes of dipteran flies), paracentric inversions can be detected under a high-power light microscope.

paracodons: A second code exists that is imprinted in the structure of the aminoacyl synthetases (AASs). It works by matching the amino acid with the structural features of the tRNA. The chief attributes of paracodons are that they are non-degenerate, deterministic and older than the classical genetic code.

paracrine: In animals, a cell that produces hormones, growth factors or other signalling substances for which the target cells expressing the corresponding receptors are located in its vicinity or in a group adjacent to it. *See* AUTOCRINE *and* ENDOCRINE.

paracrine signalling: Short-range cell-cell communication process by means of secreted signal molecules that work on adjacent cells.

paradox: A statement that is seemingly contradictory or opposed to common sense and yet is true.

paragenetic: Changes within the eukaryotic chromosomes that affect the expression but not the constitution of the genes concerned. For example, V-type positions affect unstable loci in plants.

parallel computing: Type of computation in which several calculations or the execution of processes are carried out concurrently. Large problems can often be split into smaller ones, which can then be resolved simultaneously. There are numerous different kinds of parallel computing: bit-level, instruction-level, data-level and task parallelism.

parallel evolution: Evolution of similar attributes separately in two or more lineages of shared ancestry. The characteristics that were once different become similar after experiencing the same changes in different evolutionary lineages.

parallel mutations: Mutations that result in alike phenotypes but in different species or genotypes.

parallelism: Similarities arising in separate but related evolutionary lines due to similar selective factors. *See* CONVERGENT EVOLUTION *and* CONVERGENCE.

parallelism between RNA editing and self-splicing: There is parallelism between RNA editing (performed by gRNA) and self-splicing of introns in hnRNA (performed by ribozymes). A parallelism between the function of INTERNAL GUIDE SEQUENCE (IGS) in RNA splicing and the suggested role of GUIDE RNAS (GRNAS) in RNA editing in kinetoplastid mitochondria has been recognized. Scientists have recognized chimeric intermediate structures with gRNA covalently joined to the 3′ portion of the mRNA *in vitro*, indicating an evolutionary equivalence to the catalytic mechanism involved in RNA splicing. Like the self-splicing of introns, RNA editing also includes transesterification, but with the assistance of guide RNA. In the first step, the guide RNA lines up itself (by base pairing) with the unedited RNA, splits it into two and makes a new bond between one of the broken ends and the uridine at the tip of the tail of gRNA at its 3′-end.

parallelism in behaviour of chromosomes and Mendelian factors (genes): W.S. Sutton and Th. Boveri in 1903 independently discovered parallelism between the behaviour of chromosomes and Mendelian factors (genes) during meiosis and fertilization. Pairing and disjunction of homologous chromosomes at meiosis explain the mechanism of gene segregation. An independent assortment of genes can likewise be deduced from knowledge of chromosome behaviour as observed in microscopic preparations. Diploid cells contain two sets of chromosomes that are derived from two different parents. At meiosis, various pairs of chromosomes are sorted and distributed to gametes independent of each other. Gametes may thus contain any mixture of maternal chromosomes and paternal chromosomes. The number and morphology of the chromosomes are specific, distinct and ordinarily constant for each species. A firm basis for Mendel's laws of heredity exists in the behaviour of chromosomes during meiosis and fertilization. This paved the way for the CHROMOSOME THEORY OF INHERITANCE. Chromosome behaviour during meiosis is seen under the microscope, whereas the existence of genes is inferred from observation of the behaviour of traits in hybridization experiments.

paralogous genes: Genes that are homologous (having a common ancestry) and have become functionally diverged from each other after gene duplication events (e.g. α- and β-globins). *Contrasts* ORTHOLOGOUS GENES.

paralogous histones: Non-canonical histones that can replace the major core histone proteins and may have distinct gene regulatory functions.

paralogous traits: When characters are homologous with divergent functions.

paralogues: A pair of genes that derives from the same ancestral gene and now reside at different locations within the same genome.

paramecin: A toxin liberated by a 'killer' paramecium.

parameter(s): (i) Quantities such as selection coefficients or recombination rates that govern the behaviour of a model but do not evolve with it. *Contrasts* VARIABLES. (ii) In computer programming, a special kind of variable is used in a subroutine to refer to one of the pieces of data provided as input to the subroutine. These pieces of data are the value of the arguments (often called ACTUAL ARGUMENTS or ACTUAL PARAMETERS) with which the subroutine is going to be called/invoked. An ordered list of parameters is usually included in the definition of a subroutine, so that each time the subroutine is called, its arguments for that call are evaluated and the resulting values can be assigned to the corresponding parameters. (iii) An unknown quantity such as the population mean, population variance, difference in two means or regression coefficient. (iv) User-selectable values, characteristically revealed experimentally, that govern the boundaries of an algorithm or programme. For example, the selection of suitable input parameters rules the success of a search algorithm. Some of the most common search parameters in bioinformatics tools comprise the stringency of an alignment search tool and the weights (penalties) provided for mismatches and gaps.

parametric model: A model based on a mathematical function with a few unidentified parameters. Typically, the number of parameters in a parametric model does not increase with the sample size and a specific distribution is presumed for the dependent variable Y, which is conditional on X. *See* SEMIPARAMETRIC MODEL.

parametric test: A test that is based on specific norms about the distribution of the data or specific norms about model parameters. Examples include the t-test and the Pearson product-moment linear correlation test.

paramutable allele: A wobbly allele where the phenotypic significance is improved by the occurrence of a paramutagenic allele in a heterozygote gene. *See* PARAMUTAGENIC ALLELE.

paramutagenic allele: An allele that induces paramutation that leads to the heritable silenced state.

paramutation: A term given by R.A. Brink to explain a mutation in which one allele in a heterozygous condition changes its partner allele permanently. In this allelic interaction, PARAMUTAGENIC ALLELE suppresses PARAMUTABLE ALLELE in the heterozygote. In epigenetics, paramutation is a communication between two alleles of a single locus, causing a heritable change of one allele that is persuaded by the other allele. Paramutation is an allele-dependent transfer of epigenetic information, which results in the heritable silencing one allele by another. It is an epigenetic phenomenon involving changes in gene expression that are stably transmitted through mitosis as well as meiosis. These heritable changes are mediated by in *trans* interactions between homologous DNA sequences on different chromosomes. During these in *trans* interactions, epigenetic information is transferred from one allele of a gene to another allele of the same gene, resulting in a change in gene expression. Although paramutation was initially discovered in plants, it has recently been observed in mammals as well, suggesting that the mechanisms underlying paramutation might be evolutionarily conserved. Recent findings point to a crucial role for small RNAs in the paramutation process. *See* PARAMUTAGENIC ALLELE *and* PARAMUTABLE ALLELE.

paramutation models: Some models propose trans-communication between epialleles, including physical pairing of alleles and transfer of silencing complexes, histone modifiers, nucleosome repositioning factors and DNA-methylating activities on the paramutable locus. In some cases, the acquisition of DNA methylation accompanies the *trans*-inactivation of paramutable alleles, as found for the maize *r1*, *p1*, the petunia *A* loci and *FWA*.

paramutation violates Mendel's first principle: In epigenetics, paramutation is a communication between two alleles of a single locus, leading to a heritable change of one allele that is influenced by the other allele. Paramutation contradicts Mendel's first principle, which states that in the process of the formation of the gametes (egg or sperm) via meiosis, the allelic pairs separate, one going to each gamete, and that each allele remains completely unimpacted or uncontaminated by the other. In paramutation, an allele in one generation heritably influences the other allele in future generations, even if the allele triggering the change is itself not transmitted. What may be transmitted are patterns of DNA methylation or RNAs such as piRNAs, siRNAs, miRNAs or other regulatory RNAs. Through proper breeding, paramutation can produce isogenic sibling plants with drastically different phenotypes.

paramutation, discovery of: Paramutation was first revealed and studied in maize (*Zea mays* L.) by R.A. Brink at the University of Wisconsin–Madison in the 1950s. Brink noticed that definite weakly expressed alleles of the *red1 (r1)* locus in maize, which encodes a transcription factor that confers red pigment to maize kernels, can heritably change specific strongly expressed alleles to a weaker expression state. The weaker expression state assumed by the changed allele is heritable and can, in turn, alter the expression state of other active alleles in a process termed SECONDARY PARAMUTATION. Brink disclosed that the impact of the paramutagenic allele could continue for many generations. Interestingly, paramutation can occur in a single allele of a gene controlling a spectrum of phenotypes. At *r1* in maize, for example, the weaker expression state assumed by an allele following paramutation can range from completely colourless to nearly fully coloured kernels. This is an instance of exception to the general observation that continuous variation is governed by many genes. Allelic interactions similar to paramutation have now been reported in other organisms, including tomato, pea and mice.

paramutation, examples of: Two well-studied examples of paramutation are the booster 1 (*b1*) locus in maize and *DWARF1*(*D1*) gene in rice.

paramutation, molecular basis of: The molecular basis of paramutation is being unravelled, almost exclusively in maize. Paramutation may use common mechanisms with other epigenetic spectacles, such as gene silencing and genomic imprinting. In maize, paramutation appears to share many traits with the well understood RNA-directed DNA-methylation pathway in *Arabidopsis*, even though it has never been observed in the famous model plant. M. Alleman in 2006 reported that, in maize, 'paramutation is RNA-directed. Stability of the chromatin states associated with paramutation and transposon silencing requires the *mop1* gene, which encodes an RNA-dependent RNA polymerase'. Exactly how the RNA produced by this polymerase reasons paramutation in maize is as yet unstated, but like other epigenetic changes, it includes the covalent modification of DNA and/or the DNA-bound histone proteins without changing the DNA sequence itself. The *b1* locus encodes a basic helix-loop-helix transcription factor that activates the anthocyanin biosynthetic pathway and results in the presence of purple pigment throughout the plant. Two *b1* alleles are involved in paramutation, *B′* and *B-I*: these two variants have a 10–20-fold difference in transcription, but do not differ in DNA methylation in the promoter-proximal or coding regions. The paramutagenic *B′* state arises spontaneously from the unstable, highly expressed paramutable *B-I* state at frequencies around 0.1–10%. Unlike *B-I*, *B′* is extremely stable. The figure illustrates paramutation at *b1*; *B-I* (dark purple plant) is always changed to *B′* (light purple plant) in *B′*/*B-I* heterozygotes. The newly paramutated allele, *B′**, is indistinguishable from the *B′* parent; it always paramutates *B-I* in subsequent crosses. The full penetrance and extreme stability of *B′* enabled a genetic dissection of *b1* paramutation. This analysis identified mutations in other genes, as well as the minimal sequences that are required for paramutation at this locus. Fine-structure recombination mapping delimited a 6-kb region, which is located 100 kb upstream of the *b1* transcription start site, that was required for paramutation and *B-I* enhancer activity. In this region, *B-I* and *B′* have seven tandem repeats of an 853-bp sequence (black arrows) that is otherwise unique in the genome; neutral alleles only have one copy of this sequence. A comparison of recombinant alleles with a single copy and those with three or five tandem repeats showed that the repeats are required for both paramutation and high levels of transcription. DNA sequences are identical in *B-I* and *B′*, indicating that epigenetic mechanisms mediate the stable transcriptional silencing that is associated with *b1* paramutation. The tandem repeats are differentially methylated and have greater DNase I hypersensitivity in *B-I* relative to *B′*. The sequences required for paramutation are indicated in the figure by red rectangles. The *B′* and *B-I* states are represented by different numbers and sizes of ovals to symbolize the distinct epigenetic states within the tandem repeats.

paramutation-like effects in polyploids: Studies exploring paramutation that affects a transgenic selection reporter gene in tetraploid *Arabidopsis thaliana* have revealed that different types of small RNAs are derived from paramutable and paramutagenic epialleles. In addition, deletion of a repeat within the epiallele changes its paramutability. The temperature during the growth of the epiallelic hybrids determines the degree and timing of the allelic interaction. These results support a model for polyploidy-associated paramutation, with similarities as well as distinctions from other cases of paramutation.

paranemic joint: A region in which two complementary sequences of DNA are associated side by side instead of being intertwined in double helical structure. pBR322 is one of the standard plasmid helical structures.

parapatric: A geographic dispersal in which different types occur in different places and encounter only in a narrow zone.

parapatric speciation: In parapatric speciation, there is no specific extrinsic barrier to gene flow. The population is continuous, but none the less, the population does not mate randomly. Individuals are more likely to mate with their geographic neighbours than with individuals in a different part of the population's range. In this mode, divergence may happen because of reduced gene flow within the population and varying selection pressures across the population's range. Local selection pressures are strong enough to prevent homogenization by interbreeding.

paraphyletic group: A set of species comprising an ancestral species together with some, but not all, of its progenies. The species encompassed in this group are those that have sustained to resemble the ancestor; the excluded species have evolved fast and no longer resemble their ancestor.

pararetrovirus: A viral retroelement whose encapsidated genome is made of DNA. It is a retrovirus-like particle which cannot rearrange itself due to the absence of reverse transcriptase. It is thus a faulty retrovirus.

parasexual recombination: In bacteria, recombination transpires through parasexual means. The different forms of parasexual recombination in bacteria are transformation, transduction and conjugation. Parasexual recombination occurs in genetic systems that achieve genetic recombination by ameiotic exchange process.

parasexual reproduction: Any reproductive process that generates recombinants without the formation of gametes.

parasite: An organism that feeds on some other living host. Unlike saprophytes that live on dead organic matter, the parasites derive nutrients from the living host. There are several disease-causing bacteria and fungi that are parasites on humans. Likewise, tapeworms and hookworms are also parasites on human. In plants, cuscuta is good example of plant parasite. *Compare with* SAPROPHYTE *and* EPIPHYTE.

parasitic DNA: *See* SELFISH DNA.

parasitism: A symbiotic relationship between species where one organism, the parasite, lives on or inside another organism, the host, causing it some harm, and is adapted structurally to this way of life. The entomologist E.O. Wilson has characterized parasites as 'predators that eat prey in units of less than one'. In parasitism, one symbiont benefits at the expense of a host organism. For example, all host-parasite interactions (e.g. plant-pathogen interactions). Parasites use the host as a source or supply of food. Parasitic symbiosis varies in the degree to which the host is affected. Some parasites are so virulent that the host cells die shortly after the parasitism begins. Parasites that cause disease and the possible death of the host are called pathogens. The term parasite is more common for protozoa, worms and insects. Whereas microbial parasites are commonly called pathogens. Parasitism is a short-lived unstable symbiosis.

parasitology: The study of parasites, their hosts and the relationships between them. As a biological discipline, the scope of parasitology is not determined by the organism or environment in question but by their way of life.

parenchyma (Adj. parenchymatous): A versatile ground tissue made up of living primary cells that accomplishes a wide variety of structural and biochemical occupations in plants.

parent: Pertaining to an organism, cell or structure that produces another of its type.

parent-offspring analysis: The linear regression of the performance of offspring on that of the parents was projected as a procedure for estimating heritability.

parental conflict hypothesis: This hypothesis states that the inequality between parental genomes due to imprinting is a result of the differing interests of each parent in terms of the evolutionary fitness of their genes. The father is often more 'interested' in the growth of his offspring, at the expense of the mother. The mother's interest is often to conserve resources for her own survival while providing sufficient nourishment to current and subsequent litters. Accordingly, paternally expressed genes tend to be growth promoting whereas maternally expressed genes tend to be growth limiting. In support of this hypothesis, genomic imprinting has been found in all placental mammals, where post-fertilization offspring resource consumption at the expense of the mother is high; it has not been found in oviparous birds or monotremes (a class of oviparous mammals), where there is relatively little post-fertilization resource transfer and therefore less parental conflict. It is the maternal genome that controls much of the imprinting of both its own and the paternally derived genes in the zygote, making it difficult to explain why the maternal genes would willingly relinquish their dominance to that of the paternally derived genes in light of the conflict hypothesis.

parental ditype (PD): A spore arrangement in Ascomycetes that contains only the two non-recombinant-type ascospores.

parental generation: The generation including the immediate parents of the F_1 generation; the symbols P_2 and P_3 are generally used to designate grandparental and great-grandparental generations, respectively.

parental histone transfer and the replication fork: Replication of the eukaryotic genome must be tied to the spread of the epigenome encoded in the chromatin. N. Li *et al.* (2024) visualized cryo-electron microscopy assemblies of yeast (*Saccharomyces cerevisiae*) replisomes related to the FACT (facilitates chromatin transactions) complex (comprising Spt16 and Pob3) and an expelled histone hexamer. FACT is located at the forward-facing end of the replisome by communicating with the parental DNA duplex to seize the histones over the middle domain and the acidic carboxyl-terminal domain of Spt16. The H2A–H2B dimer chaperoned by the carboxyl-terminal domain of Spt16 is firmly tied to the H3–H4 tetramer, while the unoccupied H2A–H2B site is engaged by the histone-binding domain of Mcm2. The Mcm2 histone-binding domain envelopes around the DNA-binding surface of one H3–H4 dimer and outspreads across the tetramerization interface of the H3–H4 tetramer to the binding site of Spt16 middle domain before getting chaotic. This preparation leaves the residual DNA-binding surface of the other H3–H4 dimer available to additional interactions for further processing. The Mcm2 histone-binding domain and its downstream linker areas are nested on top of Tof1, repositioning the parental histones to the replisome forward-facing for transfer to the newly synthesized lagging-strand DNA. These findings provide vital mechanical perceptions into the mechanism of replication-coupled histone recycling for preserving epigenetic inheritance.

parental imprinting: The differential activation and deactivation of maternal and paternal genes. *See* GENE IMPRINTING.

parental type: Non-recombinants. An association of genetic markers found among the progeny of a cross is identical to an association of markers present in one or the other parent.

parietal: Attached to the marginal walls of a structure (e.g. ovules are attached to placentas on the wall of the ovary). *See* PLACENTATION.

paripatric speciation: A synonym of peripheral isolate speciation.

parsimony: A criterion used in cladistic tree-building. The best tree has the fewest changes in character states and the least homoplasy.

parthenocarpy: Production of fruit without fertilization and normally without seed. Banana is a good example of parthenocarpy. It is a triploid plant and does not produce viable gametes. Thus, the fruit develops without fertilization and lacks seeds. *Compare with* STENOSPERMOCARPY.

parthenogenesis: Development of an embryo from a sex cell of female gametophyte like an egg cell without pollination or fertilization. It is common, mainly in plants, where meiosis, with its chromosome pairing and disjunction, may be suppressed and diploid cells may give rise (without fertilization) to individuals of the next generation. Unless mutation occurs, the offspring produced carry the same genes that were present in the parent. If the parent is heterozygous so is the offspring.

parthenogenetic: An individual developed from an unfertilized egg. Examples are earthworms and some types of shrimp, parthenogenetic species of insects and lizards. *Aka* PARTHENOGENIC.

partial correlation: Shows the numerical relationship between a single Xi and Y in multiple correlations. The partial correlation demonstrates interaction between Xi variables and exemplifies the true relationship between the Xi and Y. *Contrasts* SIMPLE CORRELATION.

partial denaturation: The partial unwinding of the double helix means that these regions that remain intact are probably GC-rich, since G-C base pairs held together by three hydrogen bonds are more stable than A-T base pairs held together by two hydrogen bonds.

partial digest: A restriction digest that has not been allowed to go to completion and thus contains pieces of DNA restriction endonuclease sites that have not been cleaved. *Aka* INCOMPLETE DIGEST.

partial diploids: Bacteria containing two alleles for a particular locus owing to gene transfer by transduction or sexduction.

partial dominance: The appearance of a midway phenotype in individuals heterozygous for the gene concerned. It is generally believed to be a type of incomplete dominance, with the heterozygote approaching one homozygote more than the other gene. *Aka* SEMIDOMINANCE *and* INCOMPLETE DOMINANCE.

partial Latin square: An n × n array whose cells are either empty or contain a symbol from an alphabet of size n, such that each symbol happens at most once in each row or column.

partial pedigree: Some ancestors of a particular genotype are known, usually the female parents; partial pedigrees are most common where open-pollinated or polycross seed is used for progeny testing.

partial resistance: Resistance, which is expressed by the slower development of fewer pustules or lesions compared with normally susceptible varieties. *Contrasts* COMPLETE RESISTANCE.

partially balanced incomplete block design (PBIB): An experimental design where the residual variance of the difference between the candidates may assume one out of only two different values.

participatory plant breeding: Strategies that comprise intimate teamwork between researchers, farmers and other possible stakeholders to bring about plant genetic improvements within a species. It covers the entire research and development cycle of happenings associated with plant genetic improvement, such as identifying breeding objectives, generating genetic variability or diversity, selecting within variable populations to develop experimental materials, evaluating these materials, release of materials, diffusion, seed production and distribution. It may also comprise evaluating prevailing policy or legislative procedures or both and designing new ones where required.

participatory variety selection: Selection of fixed lines (including landraces) by farmers themselves in their target environments using their own selection criteria. It comprises four methodological steps: (i) situation analysis and identification of farmer's varietal requirements, (ii) search for appropriate genetic materials, (iii) farmer's investigation with new crop varieties in their own fields and with their own crop-management practices and (iv) wider distribution of farmer-preferred crop varieties.

particle acceleration method: *See* PARTICLE BOMBARDMENT METHOD.

particle bombardment method: A vectorless method for gene transfer into plants. This is also called a 'biolistic' approach or particle acceleration method that makes use of a particle gun machine. The process of transformation employs foreign DNA–coated minute 0.2–0.7 μm gold or tungsten particles to deliver into target plant cells. This method is being widely used because of its ability to deliver foreign DNA into regenerable cells, tissues or organs, irrespective of the monocots or dicots. Cotransformation with multiple genes or gene pyramiding is possible using the particle gun method. In this method of chloroplast engineering, cells are bombarded with DNA-coated tungsten particles. Here, the recipient for transformation is also a deletion mutant. The particle bombardment method has advantages over the glass bead method. (i) This is a more efficient method than the glass bead method, as it produces a far greater number of chloroplast transformants. (ii) Particle bombardment is a simpler method as we do not need to remove cell walls. (iii) By glass bead method we can produce transformants in *Chlamydomonas*. *Aka* PARTICLE GUN METHOD *or* PARTICLE ACCELERATION METHOD. *Compare with* GLASS BEAD METHOD.

particle gun: In genetic engineering, a gene gun or biolistic particle delivery system is a device used to deliver exogenous DNA (transgenes), RNA or protein to cells.

particle gun method: *See* PARTICLE BOMBARDMENT METHOD.

particulate inheritance: The model proposes that genetic information is transmitted from one generation to the next in discrete units (particles) so that the character of the offspring is not a smooth blend of essence from the parents. *Compare with* BLENDING INHERITANCE. *See* MENDEL'S PRINCIPLES.

particulate matter: Fine solid or liquid particles that adulterate the air and are supplemented to the atmosphere by natural and man-made processes at the Earth's surface. Examples of particulate matter are dust, smoke, soot, pollen and soil particles.

particulate nature of the gene: H.J. Muller in 1927 observed that mutation could be produced after being hit by a particle, which clearly indicated the particulate nature of the gene. At this stage of the gene concept, gene was regarded as a unit of mutation.

particulate radiation: Radiation that emits particles when passing through a medium.

particulate theory of inheritance: Gregor John Mendel's observation that one of the parental characteristics was absent in F_1 hybrid and reappeared in unchanged form in the F_2 generation. This leads to the conclusion that traits from the parental lines were transmitted as two discrete elements (particles) of particulate nature that retained their purity in the F_1 hybrids. *See* MENDEL'S PRINCIPLES. *Compare with* BLENDING INHERITANCE.

parvoviral vectors: Parvoviruses are small DNA viruses many of which exhibit onco-suppressive properties. Parvovirus-based vectors can be used to target the expression of therapeutic genes in tumours. Recombinant vectors derived from ADENO-ASSOCIATED VIRUSES (AAVs) have been successfully tested in many preclinical models of human diseases, including cancer.

Pascal's triangle: A technique for generating coefficients in a binomial distribution. Each term in the triangle is obtained by adding the numbers to the immediate left and right on the line above those terms.

passage: The transfer or transplantation of cells, with or without dilution, from one culture vessel to another. It is understood that any time cells are transferred from one vessel to another, a certain portion of the cells may be lost and, therefore, dilution of cells, whether deliberate or not, may occur. *See* SUBCULTURE.

passage number: The number of times the cells in the culture have been sub-cultured or passaged. In descriptions of this process, the ratio or dilution of the cells should be stated so that the relative cultural age can be ascertained.

passage time: The interval between successive subcultures or the culture period.

passenger DNA: Foreign DNA incorporated into a plasmid.

passivation: Covering the reactive surface with molecules or atoms so as to eliminate dangling bonds. A method used to protect nanoparticles.

passive DNA demethylation: In the absence of maintenance DNA methylation, $5^{m}C$ methylation marks are lost in newly synthesized DNA during DNA replication.

passive immunity: The protection against a given disease shaped by conditioning.

passive resistance: Inborn resistance that does not involve any self-protective reactions of the host to the existence of the pathogen.

passive transport: Transport of a solute across a membrane down its concentration gradient or its electrochemical gradient, utilizing only the energy stockpiled in the gradient. *Compare with* ACTIVE TRANSPORT.

passport data: Details about the source of an accession and any other pertinent information, including descriptors, that support the identification of the accession.

pasteurization: Moderate (mild) heat treatment given to liquid foods to kill pathogens and extend their shelf life. This is the technique given by Louis Pasteur to prevent food spoilage. It is now commonly used to kill spoilage and harmful bacteria in milk and in some alcoholic drinks. In the classic pasteurization treatment of milk, the milk was exposed to a temperature of about 63°C for 30 min. At present, milk pasteurization is done at 72°C, but only for 15 s. This is called HIGH-TEMPERATURE SHORT-TIME (HTST) pasteurization.

patch-clamp recording: Electrophysiological technique in which a tiny electrode tip is sealed on to a patch of cell membrane, thereby making it possible to record the flow of current through individual ion channels in the patch.

patent: In genetics, conferring the right or title to genes, gene variations or identifiable portions of sequenced genetic material to an individual or organization. Patents are approved for products and procedures that are considered new, workable and ingenious (i.e. novel, useful and inventive) or for valuable improvements to some prevailing invention. A patent is an exclusive right for a limited period of time (20 years) granted by the government to the patentee, subject to his full disclosure of the invention. An invention is developed with investment and ingenuity and deserves to be protected for the exploration of rewards for you. A patent is an exclusive right granted by a country to the owner of an invention to make, use, manufacture and market the invention, provided the invention satisfies certain conditions stipulated in the law. A patent in the law is a property right and hence, can be gifted, inherited, assigned, sold or licensed. As the right is conferred by the state, it can be revoked by the state under very special circumstances. The patent right is territorial in nature, meaning that a patent granted in India can be reinforced in India only. Every country is free to grant or refuse a patent on the basis of scrutiny by its patent office.

patent agent: An individual with methodological training who is proficiently representing an inventor in a patent trial.

patent attorney: An individual with legal training in patent law who is proficient in representing an inventor in patent suit and court case.

Patent Cooperation Treaty (PCT): This treaty implements the concept of a single international patent application, which has legal effect in the countries that are bound by the treaty and which are designed by the applicant. The PCT system consolidates and streamlines patenting procedures and reduces costs, providing applicants with a solid basis for important decision making.

patent infringement: When an unauthorized user copies or imitates a valid patent, the patent is said to have been infringed. The registered patentee or his assignee or a person authorized by him can initiate an infringement action in court.

patent office: The patent office in India is governed by the patents act 1970 (no. 39 of 1970) and the patents rules 1972, subsequently amended and is effective in the whole of India. This act was further amended in 1999 and 2002. The Patent Office, Nizam Palace, 5th-, 6th- and 7th Floors, 234/4, Acharya Jagdish Bose Road, Kolkata 700 020, is a subordinate office under the Department of Industrial Policy and Promotion, Ministry of Commerce & Industry, Government of India. The Patent Office has been functioning since 1912 to register and protect new inventions and technological improvements in India.

patent pool: A promise between two or more patent owners to authorization of one or more of their patents to one another or third parties.

patentable: For an invention to be patentable, it must satisfy three conditions – novelty, inventiveness (non-obvious) and usefulness. But it may not qualify for a patent in certain situations. Further, publication of an invention in any form by the inventor before filing a patent application would disqualify the invention from being patentable.

Patents (Amendment) Ordinance: An ordinance The Patents (Amendment) Ordinance was promulgated by the President of India for the purpose of bringing the existing patent law into conformity with the obligations undertaken in the agreement for the establishment of the World Trade Organization (WTO) and the agreement on Trade Related Aspects of Intellectual Property Rights (TRIPS). The ordinance came into effect on January 1, 2005. The domestic industry will not be able to manufacture the patented food, drugs and chemicals through alternate routes. A medicine or drug includes insecticides, germicides, fungicides, weedicides and all other substances included to be used for protection and preservation of plants. It also includes all the chemical substances that are ordinarily used as intermediates in the preparation or manufacture of any other medicine or drug. The product patent era will likely open new opportunities for pharmaceuticals, biotechnology and software companies.

paternal: Derived from the father.

paternal chromosome: The chromosome of a homologous pair that is inherited from the father.

paternal disomy: A condition where both copies of a gene at a locus in a diploid are inherited from the father; for example, in human patients with a disease known as BECKWITH WEIDEMANN SYNDROME (BWS), both copies of a region on the short arm of chromosome 11 (11p15.5) are inherited from the father.

paternal inheritance: Transmission of the cytoplasmic organelles from male parent to the offspring. First discovered by S.D. Russell in 1980 in an angiosperm *Plumbago zeylamica*.

paternal origin: Derived from the father.

paternal X chromosome inactivation: In rodents and marsupials, only the X chromosome of paternal origin (Xp) is silenced through early embryogenesis. This could be owing to a carryover outcome of the X chromosome's passage through the male germline, where it becomes momentarily silenced together with the Y chromosome, during MEIOTIC SEX CHROMOSOME INACTIVATION (MSCI).

paternal X chromosome: The X chromosome was contributed by father.

paternally expressed imprinted genes (PEGs): Genes inherited from the male parent that have been imprinted.

paternity: The distinctiveness of the father of an offspring.

path diagram: Pedigree preparation displaying only the straight line of descent from common ancestors.

path-coefficient analysis: A technique for analysing regular and irregular breeding systems. The path coefficient is a degree of the importance of a given path of impact from cause to effect. It represents the ratio of the standard deviation of the outcome to the total standard deviation.

pathogen: In the broadest sense, anything that can produce disease, though the term is most commonly used to refer specifically to an infectious microscopic organism such as a virus, bacterium, protozoan or another microbial agent that causes disease for a host organism by invading the host's tissues. Infection is not identical to disease because infection does not always lead to injury to the host. Bacteria *Escherichia coli* is a pathogen for humans and another bacteria *Xanthomonas oryzae*, is a pathogen of rice that reason bacterial blight disease of rice. Likewise, viruses and fungi also act as pathogens.

pathogen free: Free from specific organisms based on specific tests for the designated organisms.

pathogen genes: Generally dominant. Called *AVIRULENCE* (*AVR*) genes. They are adaptive and evolve continually. Pathogen genes evolve faster than the host genes. This leads to continually updating and replacing varieties with different genes or combinations of genes with resistance in response to ever-changing pathogen populations.

pathogen genomics: Approach using genomic sequencing to understand the evolution and spread of pathogens to help control the spread of an infectious disease.

pathogen or microbe-associated molecular pattern (PAMP or MAMP): Conserved (surface) structures produced by pathogens or microbes that are recognized, but not present in the host. PAMPs are defence elicitors that are evolutionary stable, forming a core component of the microorganism that cannot be sacrificed or even altered much without seriously impairing its viability. Examples of PAMPs include flagellin 22, EF-Tu and Chitin. *See* PATHOGEN OR MICROBE-ASSOCIATED MOLECULAR PATTERN.

pathogen or microbe-associated molecular pattern (PAMP) receptor FLS2: The best characterized plant PATTERN RECOGNITION RECEPTORS (PRR) for a PAMP is the *Arabidopsis thaliana* receptor-like kinase (RLK), FLS2 (flagellin sensing 2), which includes an extracellular LRR ligand-binding domain and an intracellular serine-threonine kinase and directly recognizes a conserved N-terminal fragment of bacterial flagellin FLG22. Stimulation of FLS2 by flagellin activates pathogenesis-related gene expression. The presence of FLS2 decreases host susceptibility to a pv. (pathovar) of *Pseudomonas syringae* when the pathogen is applied to leaves by spraying but not infiltration. This finding demonstrates a direct link between disease resistance and PAMP perception that had previously been lacking in plants.

pathogen or microbe-associated molecular pattern (PAMP)-triggered immunity (PTI): Basal defence mechanism effective against a broad range of pathogens induced after PAMP perception.

pathogen transmission and spread: Pathogens are transmitted, disseminated and spread by many factors, which include biotic, abiotic and environmental factors. Transmission usually implies active transfer by means of grafting, insects, mechanical factors, animals and so on. To disseminate or spread means to disperse or distribute.

pathogen virulence factors and innate immunity: All pathogens and their hosts engage in an arms race to gain the upper hand. One of the first lines of defence in plants is the cell wall. Pathogens use enzymes called polygalacturonases (PGs) to degrade it and at the same time, to suppress the plant's immune response. Plants, in turn, counteract by inhibiting PGs with inhibitory proteins (PGIPs) (E. Thynne and B. Kobe, 2024). Y. Xiao *et al.* (2024) show that interaction between *Phaseolus vulgaris* PGIP2 (PvPGIP2) and *Fusarium phyllophilum* polygalacturonase (FpPG) enhances substrate binding, resulting in inhibition of the enzyme activity of FpPG. This interaction promotes FpPG-catalysed production of long-chain immunoactive oligogalacturonides while diminishing immunosuppressive short oligogalacturonides. PvPGIP2 binding creates a substrate binding site on PvPGIP2-FpPG, forming a new polygalacturonase with boosted substrate binding activity and altered substrate preference. Structure-based engineering converts a putative PGIP that initially lacks FpPG-binding activity into an effective FpPG-interacting protein.

pathogen-free: Liberty from disease-causing organisms (bacteria, fungi, viruses, etc.).

pathogenicity: Ability to cause disease by overcoming the defences of the host. In general, there is inbuilt host resistance in plants, animals and humans towards pathogens. This kind of resistance is governed by specific genes called 'GENETIC RESISTANCE'. But the virulent pathogens possess the ability to adhere, penetrate or evade host defences and cause disease.

pathogenesis: The comprehensive order of proceedings starting with the arrival of the pathogen at the host surface to the accomplishment of the disease cycle.

pathogenesis-related (PR) proteins: Protective (i.e. disease-fighting) proteins that are produced within certain plants in response to the entry into plant of plant pathogens (bacteria, fungi, etc.) that infect and cause disease in plants. One PR protein is chitinase, a protein enzyme that degrades (breaks down) the chitin within the cell walls of pathogenic fungi. Production of PR proteins is often initiated by signalling molecules (e.g. harpin) produced by the pathogens.

pathogenic: Term used to describe a bacterium, virus or other microorganism that has the ability to cause disease.

pathogenic missense variants: The missense variants that are responsible for a disease.

pathogenicity: The capability of an organism to live at the cost of its host and to provoke disease.

pathogenicity islands: Adjoining sections of a pathogen's genome that comprise an unequal (relative to the number of base pairs) number of the factors that are sources of pathogenicity.

Pathogenwatch: Computer platform that makes it possible to compare pathogen genome sequences from around the world.

pathological: A condition of being diseased.

pathology: The study of diseases.

pathophysiology: Physiological processes or mechanisms whereby disease develops and progresses.

pathotype: A subdivision of a pathogen species categorized by its pattern of virulence or avirulence to a succession of differential host varieties.

pathovar (pv.): A subdivision of a plant-pathogenic bacterial species demarcated by host range. A pathovar for bacteria corresponds to forma specialis for fungi.

pathway: (i) In biochemistry, a chronological series of chemical reactions, each of which is reliant on previous ones in the pathway (e.g. the third reaction requires a chemical product produced by first/second chemical reactions), finally yields a valuable influence. For example, metabolism ensues via a very specific metabolic pathway. (ii) Bioinformatics endeavours to describe representations of key biological data types, algorithms and inference procedures, including sequences, structures, biological pathways and reactions. Representing and computing biological pathways necessitates ontologies for representing pathway knowledge, user interfaces to these databases, physicochemical possessions of enzymes and their substrates in pathways and pathway investigation of whole genomes, including recognizing common patterns across species and species differences.

pathway feedback mechanisms: Chemically based progressions (e.g. series of chemical reactions) that hamper (or upsurge rate of) a specified pathway. For example, when the body of bacteria requires catabolism (i.e. energy manufacture) to be decelerated, it utilizes the machinery of catabolite repression (to slow down catabolism via chemical/reaction means).

patient gene therapy: A type of gene therapy in which cells with healthy genes may be introduced into the affected tissue so that the healthy gene overcomes the defect without affecting the patient's inheritance.

patriclinous inheritance: An inheritance in which all offspring have the nucleus-based phenotype of the father.

patrilineal: Inherited from the father (e.g. the Y chromosome in mammals).

patromorphy: *See* MATROMORPHY.

pattern: A molecular biological design generally occurs at the level of the characters producing a gene or protein sequence. A pattern language must be definite in order to apply different criteria to different positions in a sequence. To enable a computer to carry out position-specific comparisons, a pattern-matching algorithm must permit alternative residues at a given position, repetitions of a residue, exclusion of alternative residues, weighting and ideally combinatorial representation.

pattern formation: Establishment of the basic plan of the organism. It is the procedure that causes cells to accept an accurate spatial organization. Spatial organization is a phenomenon that guarantees that all members of a species are morphologically alike. It involves organizing different types of cells into a spatial configuration. E- and N-cadherins mediate apical adhesion between retinal epithelial cells in *Drosophila*. Differential expression of N-cadherins within a subgroup of retinal cells (cone cells) causes them to form an overall shape that minimizes their surface contact with surrounding cells. The cells within this group pack together in the same way as soap bubbles do. The shaping of the cone cell group and packing of its components precisely imitate the physical tendency for the surfaces to be minimized. Thus, simple patterned expression of N-cadherin results in a complex spatial pattern of cells owing to cellular surface mechanics.

pattern formation, components of: (i) Regional specification. (ii) Morphogenesis.

pattern recognition receptors (PRR): Receptor proteins that can perceive PAMPs and trigger innate immunity in plants. A hallmark of PRRs is their sensitivity and specificity. Plants possessing the appropriate PRRs perceive a specific PAMP/MAMP at sub-nanomolar concentrations. Plants lacking the PRRs are completely blind to it. A given PAMP/MAMP is recognized through a specific conserved EPITOPE (ANTIGENIC DETERMINANT), such as the stretch of 22 amino acids (flg22) in the N-terminus of flagellin. Can pathogens avoid perception by the PRRs? It appears they can, but at a cost. Introduction of mutations into flagellin that make the molecule unrecognizable by FLS2 (a PRR) also renders the microbe motionless and reduces its virulence. The specificity of the PRR appears to be focused exactly on a highly conserved domain of the MAMP that is functionally important to the microbe.

pattern recognition receptors (PRR) (XA21)-Elicitor (Ax21) interaction: Activation of Xa21-based resistance is negatively regulated by *Xb24* and *Xb15* genes. The LRR domain of XA21 recognizes *Ax21* and this recognition activates downstream phosphorylation events. The interaction is extracellular Ax21 is a sulfated peptide that binds the rice PRR, XA21.

pattern recognition receptors (PRR)-mediated basal defences: Basal defences that are mediated by PRR proteins are suppressed by type III effectors such as AvrE1, AvrPto, HopAA1-1 and XopX. Through mitogen-activated protein kinase (MAPK) signalling, WORKY transcription factors activate the expression of the *FRK1* and *NHO1* genes. They act as regulators of many processes in plants.

pattern recognition: Concerned with the automatic discovery of regularities in data through the use of computer algorithms and with the use of these regularities to take actions such as classifying the data into different categories.

pattern searching: A procedure used for gene prediction. It involves scanning a nucleic acid or protein sequence for matches to short sequence patterns. These short patterns are known to be indicators of some biological function. The presence of a matching pattern in the target nucleic acid or protein is a signal of the same function for the target gene or protein sequence. Patterns most often examined in DNA sequences are recognition sites of restriction endonucleases, codons specifying the amino acid sequence of a protein, intron splice sites, promoter and binding sites for regulatory proteins. Patterns in protein sequences (motifs) can be used to predict the presence of active sites, the secondary structure of the protein and the presence of signals used to locate the protein in the cell.

patterns in leaves: Rather than being specified by genes alone, elaborate biological structures may arise through simple dynamical mechanisms. Patterns in leaves resemble the cracks in drying gel, suggesting that mechanical stresses might trigger the formation of the vein networks.

patterns of alternative RNA splicing: Five main types of alternative splicing events include constitutive splicing, mutually exclusive exons, cassette alternative exons, alternative 3′ splice sites, alternative 5′ splice site and intron retention.

patterns of speciation: Different models of speciation are covered under this term.

patulous: Spreading; standing at 45–50° to the axis. *See* ERECTO-PATENT.

pauciflor. Having few flowers per inflorescence. *Compare with* PLURIFLOR *and* UNIFLOR.

payoff: In an evolutionary context, the upsurge in fitness due to a competition.

PAZ domain: A conserved RNA-binding domain found in members of the Dicer and Argonaute protein families that preferentially interacts with the 3′-end of RNA.

pBluescript phagemid: pBluescript (pBS) or pBluescript II is a commercially available phagemid containing several useful sequences for use in cloning with bacteriophage. The sequences include a MULTIPLE CLONING SITE (MCS), an antibiotic resistance sequence to ampicillin and an *Escherichia coli* and f1 helper phage origin of replication. If the gene is disrupted by successful insertion of a DNA sequence, the bacteria exhibit a white coloration in blue-white screening, distinguishing successful recombination from those phagemids that were not altered. These recombinant plasmids can then be used in a variety of molecular techniques.

pBR322 vector: An artificial plasmid frequently used as a vector for cloning genes in Escherichia coli. In pBR322, the letter p stands for plasmid. BR denotes the name of two researchers, F. Bolivar and R. L. Rodriguez, who, with other scientists, constructed this plasmid in 1977. The number 322 denotes the order of synthesis, which distinguishes it from other plasmids synthesized in the same laboratory. The pBR322 vector has a size of 4.3 kb, about 20 copies per chromosome and carries ampicillin and tetracycline-resistance genes. It comprises a circle of double-stranded DNA.

Pearson correlation coefficient: Pearson's correlation coefficient is a statistical test that determines the connection between two continuous variables. Since experts build it based on covariance, they recognize it as the best approach to quantify the relationship among variables of interest.

pectin: A methylated polymer of galacturonic acid occurring in the middle lamella and the primary cell wall of plants; the jelly-forming material existing in fruit.

pectinate: Pinnately separated with narrow segments meticulously set like the teeth of a comb.

pedicel: A short slender stalk bearing an individual flower, inflorescence or spore.

pedigree: A record of the ancestry of an individual, family or strain. In humans, the family tree investigation is called pedigree analysis and it involves the careful compilation of phenotypic records of the family over several generations. Standard symbols are used to record the medical history and trace the inheritance of a disease or trait in a family. One of the modern applications of pedigree analysis is called 'genetic counselling'. In case of plants, the pedigree method is an important method for breeding self-pollinated crops like wheat. This involves the choice of parents, making hybrids and raising F_1 to F_8 generations and selecting the desirable plants while keeping a full pedigree record. *See* INHERIT.

pedigree analysis: A method based on information from various generations of individual to know the inheritance of a given trait. This method is largely used in those systems where classical crossing procedures are not applicable.

pedigree breeding: A scheme of breeding in which individual plants are carefully chosen in segregating generations from a cross depending on their desirability, judged individually and on the basis of a pedigree record. The advantages include the following: (i) if selection is effective, inferior genotypes may be rejected before lines are further assessed; (ii) selection in generation comprises a different environment, which offers good genetic variability and (iii) the genetic relationship of lines is assessed and can be utilized to maximize genetic variability.

pedigree selection: Selection procedure in a segregating population in which progenies of selected F_2 plants are reselected in succeeding generations until genetic purity is reached.

pedigreed seed: Seed of a named variety produced under the supervision of a seed certification agency to ensure genetic purity and quality of seed.

peduncle (Adj. pedunculate): The stalk of an inflorescence; part of an inflorescence, or a fructification.

peel-and-paste transposition: Insertion sequences are compact and pervasive transposable elements found in bacteria that encode only the genes necessary for their mobilization and maintenance. IS200- and IS605-family transposons undergo 'peel-and-paste' transposition catalysed by a TnpA transposase, but they also encode diverse, TnpB- and IscB-family proteins that are evolutionarily related to the CRISPR-associated effectors Cas12 and Cas9, respectively. TnpB and IscB function as RNA-guided DNA endonucleases. C. Meers *et al.* (2023) show that TnpB and IscB are essential to prevent permanent transposon loss as a consequence of the TnpA transposition mechanism. They selected a family of related insertion sequences from *Geobacillus stearothermophilus* that encode several TnpB and IscB orthologues. TnpA transposase was broadly active for transposon mobilization. The donor joints formed upon religation of transposon-flanking sequences were efficiently targeted for cleavage by RNA-guided TnpB and IscB nucleases and coexpression of TnpB and TnpA led to substantially greater transposon retention relative to conditions in which TnpA was expressed alone. TnpA and TnpB also stimulated recombination frequencies, surpassing rates observed with TnpB alone. RNA-guided DNA cleavage seems to have arisen as a primal biochemical activity to bias the selfish inheritance and spread of transposable elements, which was later co-opted during the evolution of CRISPR-Cas adaptive immunity for antiviral defence.

Pelham box: A conserved short DNA sequence (14 bp) in the 5′ non-coding region, immediately upstream from the TATA box, which serves as the promoter for transcription of heat shock genes.

pendulous: A hanging, for example, an ovule attached to a placenta on the top of the ovary. *Compare with* SUSPENDED.

penetrance: The ability of a gene to express itself in all the individuals that carry it in the appropriate genotype. It is the frequency with which a gene produces a recognizable effect in individuals who carry it. All the individuals with the same genotype may not reveal the same phenotype. Incomplete penetrance is implied if only a part of the population shows a character, whereas complete penetrance is implied if all individuals have the character. Penetrance represents the proportion of genotypes that show an expected phenotype under a defined set of environmental conditions. For example, if 100 females have two D mutations and 70 have symptoms of the disorder, the penetrance is 70%. *See* REDUCED PENETRANCE.

penetration: The early attack of a host by a pathogen. The phenotypic expression of an allele may be contingent on the growing condition or other factors, such as age or sex. Sometimes only a section of individuals with a certain genotype displays the 'expected' phenotype.

penetration and travel in nodule formation: After attachment, the root hair curls as a result of the action of substances excreted by the *Rhizobium* species called nod factors. After curling the root hair, the bacteria penetrate and enter the root hair and induce the plant to develop a cellulosic tube called an infection thread, which extends inwards to the root hair. The *Rhizobium* cells then spread within the infection thread, move into the underlying root cells and are released into the cytoplasm of the host cell through the action of an organizer produced by the interaction between the rhizobial polysaccharides and components of root cells. Nod factors now stimulate root cell division, eventually leading to the development of the root nodule.

penetration peg: A minute protuberance from a hypha, germ tube or appressorium that effects penetration of the host plant surface. *See* INFECTION PEG.

penetrative reproductive sex: In most mammals, this is the type of reproductive sex in which penises deliver sperm close to the eggs in the vagina, rectum or mouth during sex.

penicillin: An antibiotic formed by certain species of *Penicillium* and effective against Gram-positive bacteria.

Penicillium: Common mould species that grows in a diverse range of habitats.

pentaploid: Having two homologous sets and one non-homologous set of chromosomes (5n), as reported by T. N. Khoshoo and V.B. Sharma in 1959 while studying the cytology of *Allium neapolitanum* Cyr. Another example of a pentaploid is Kenai Birch (*Betula kenaica*), a medium-sized tree, in which 2n = 70.

pentasomic: The occurrence of five homologous chromosomes in a complement.

pentatricopeptide repeat (PPR) proteins: The PPR family of proteins was discovered as 'PPR cassette proteins' two decades ago by two independent research groups in France: one led by Alain Lecharny, Laboratoire de Biologie du Développement des Plantes, Université de Paris-Sud and the other led by N. Peeters, Station de Génétique, INRA, Route de St-Cyr, Versailles, Cedex, during protein exploration of the sequencing data of the *A. thaliana* genome. PPR proteins are encoded by nuclear genes and transported to the cell organelles to achieve numerous specific jobs, including RNA editing. The modular RNA-specific binding characteristic of the PPR proteins has been utilized for regulating organellar gene expression in some lower eukaryotes, mammals and plants.

pentatricopeptide repeat (PPR) protein classification based on function: On the basis of the function(s) performed, PPR proteins have been separated into two types: first, those proteins that are not implicated in RNA editing and second, those proteins that arbitrate RNA editing. RNA molecules comprising ligand-specific domains interact with PPR proteins.

pentatricopeptide repeat (PPR) protein classification based on motif structure: On the basis of their motif structure, the PPR proteins have been divided into two chief subfamilies – 'P' and 'PLS'. Based on the domain/motif structure, the P subfamily has been categorized into four subgroups, namely, P/DYW-like, P/RRM, P/SMR and P/PRORP and the PLS subfamily into three subgroups, namely, PLS/E, PLS/E+ and PLS/DYW among the above-mentioned plant species, with the objective of relating their structure with their function(s).

pentatricopeptide repeat (PPR) protein structure-function relationships: PPR proteins accomplish a variety of crucial functions in the cell nucleus. The difference in the PPR motifs is accountable to the variation in the structural domains, basic motif architecture and frequency of the number of PPR motifs per protein that contribute to a wide range of organellar activities. The structural studies of PPR proteins have revealed differences in the overall arrangement of PPR motifs, which have contributed to the understanding of PPR protein-RNA interaction. Overall, the PPR proteins that contain the DYW domain resemble in action the deaminase to perform the RNA-editing function. Exceptionally, the P subfamily PPR proteins without the DYW domain are also involved in the RNA-editing function by facilitating the interaction between the PPR protein and the DYW domain. Likewise, the PLS/E family also performs RNA editing without the presence of the DYW domain. Some PPRs with either an E1, E2 or E+ domain were also able to perform editing in the absence of the DYW domain.

pentatricopeptide repeat (PPR) protein-mediated programmable RNA editing (PRE): The engineered PPR proteins, because of their predictable recognition code for sequence-specific binding with RNA transcripts, led to the creation of DESIGNER PPR (DPPR) proteins. These dPPR proteins have the property to control the expression of organellar genes via modulating photosynthesis, respiration and organelle-dependent processes.

pentatricopeptide repeat (PPR) protein-mediated RNA editing: RNA editing changes the RNA sequence without changing the sequence or integrity of genomic DNA. RNA editing causes changes in seed sequences of MICRORNAS (MIRNAs) or in sequences of miRNA targeting sites; and influencing the stability of targeted RNAs through multiple mechanisms. The PPR proteins encoded by the nuclear genes accomplish RNA editing through interaction with mitochondrial and chloroplast transcripts. Each PPR motif formulates two α-helices, namely, helix A and helix B. The transcribed RNA transcript association facilitated by nucleotide interactions with this superhelix comprising PPR proteins achieves RNA editing. The presence of P, L, S, E and DYW domains in the PPR proteins experience RNA editing owing to the presence of the DYW domain. The resemblance of DYW domains with the cytidine deaminase domain helps perform C-to-U editing functions in the transcripts of two plant organelles – chloroplasts and mitochondria.

pentatricopeptide repeat (PPR) protein-RNA recognition code: PPR proteins are highly adaptable because of their widespread diversity in amino acid composition within the PPR motifs. This makes the PPR proteins suitable for recognizing different RNA sequences. The mechanism of RNA recognition by PPR proteins is based on the RNA recognition code specificity. The RNA recognition code comprises the association of the amino acids at positions 6 and 1 in the PPR protein that recognizes a specific RNA base.

pentose: A simple sugar (monosaccharide molecule) whose backbone structure comprises five carbon atoms. There are many different pentoses. Some examples of pentoses are ribose, deoxyribose, arabinose and xylose.

pentose phosphate pathway: A metabolic pathway that occurs along with glycolysis to produce pentoses and NADH. This is a catabolic pathway that begins with glucose and produces pentose phosphates. It is also called the HEXOSE MONOPHOSPHATE SHUNT, which operates simultaneously with glycolysis and results in the breakdown of five carbon sugars (pentoses) as well as glucose. This cycle produces intermediate pentoses for nucleic acid synthesis, glucose from CO_2 in photosynthesis and certain amino acids. Besides, this pathway also produces NDPH from $NADP^+$.

pepo: A type of berry formed from an inferior ovary and containing numerous seeds, usually large with a tough outer skin (e.g. a cucumber, pumpkin or watermelon).

peptidases: Enzymes that hydrolyse (cleave) peptide bonds. Polypeptides are broken down into smaller components by peptidases.

peptide: A short, linear chain synthesized by covalent joining of two or more amino acids linked through peptide bonds; dipeptide = two amino acids; oligopeptide = small number of amino acids, etc. It is an oligomer component of a polypeptide chain. *See* POLYPEPTIDE.

peptide aptamer: A short oligomer of amino acids, often fused to a protein scaffold.

peptide bond: A covalent bond between two amino acids in which the carboxyl group of one amino acid (HN_2-R1-HOOC) and the amino group of an adjacent amino acid (NH_2-R2-HOOC) react to form NH_2-R1-CO-NH-R2-COOH plus H_2O, where 'CO-NH' represents a peptide bond. PEPTIDYL TRANSFERASE is the enzyme that performs this biochemical reaction that forms the peptide bond, which holds the adjacent amino acid residues together to synthesize the polymer known as a POLYPEPTIDE. *Aka* AMIDE BOND. *See* OLIGOPEPTIDE.

peptide-bond formation in eubacteria: Eubacterial leucyl/phenylalanyl-tRNA protein transferase (LF-transferase) catalyses peptide-bond formation by using Leu-$tRNA^{Leu}$ (Phe-$tRNA^{Phe}$) and an amino-terminal Arg (or Lys) of a protein as donor and acceptor substrates, respectively. The protein-based mechanism for peptide-bond formation by LF-transferase is similar to the reverse reaction of the acylation step observed in the peptide hydrolysis reaction by serine proteases.

peptide fragment fingerprinting (PFF) strategy: This approach uses low-energy COLLISION-INDUCED FRAGMENTATION (CID) for peptide ion fragmentation in the gas phase of a mass spectrometry/mass spectrometry, also known as tandem mass spectrometry (abbreviated as MS/MS), instrument. Such fragmentation occurs mainly at peptide amide bonds and yields two types of fragments: the β ion, which preserves the N-terminus and the γ ion, which preserves the C-terminus. The fragmentation spectra obtained are compared with the search programmes containing the predicted spectra of peptides after *in silico* digestion of all available proteins in the database.

peptide map: Characteristic two-dimensional pattern (on paper or gel) formed by the separation of the mixture of peptides produced by the partial digestion of a protein.

peptide mapping: Refers to the characteristic pattern of peptides (i.e. pieces that make up a protein molecule) resulting from partial hydrolysis (cleavage, digestion) of a protein. The pattern (fingerprint) is obtained by separating the peptides via two-dimensional chromatography, in which the peptides are first subjected to chromatography using one solution which separates many, but not all, peptides. The chromatogram is then turned 90° and is again chromatographed using a second solution, which then separates all of the peptides; thereby producing the final 'fingerprint' of the protein. *Aka* FINGERPRINTING.

peptide mass fingerprinting (PMF) strategy: The most commonly used strategy to identify proteins and protein expression pattern in biological samples. The mass profile obtained after MALDI-TOF-MS is compared with the predicted mass values obtained from the *in silico* digestion of all proteins present in the database.

peptide microarrays: A collection of peptides displayed on a solid surface, usually a glass or plastic chip. Peptide chips are used to study the binding properties, functionality and kinetics of protein-protein interactions in general. *Aka* PEPTIDE CHIP *or* PEPTIDE EPITOPE MICROARRAY

peptide nucleic acid (PNA): In PNA, the backbone is polymeric N-(2-aminoethyl)glycine (AEG) and the N-acetic acids of the bases (N_9 for purines, N_1 for pyrimidines) are linked via amide bonds. AEG is formed in spark-tube experiments that also produces amino acids and may spontaneously polymerize at 100°. The N-acetic acids of the bases are also accessible in prebiotic syntheses. This suggested that PNA could have been early genetic material.

peptide pull-down assays: The pull-down assay is an *in vitro* method used to determine a physical interaction between two or more proteins. Pull-down assays are useful for both confirming the existence of a protein-protein interaction predicted by other research techniques (e.g. co-immunoprecipitation) and as an initial screening assay for identifying previously unknown protein-protein interactions.

peptide-mediated gene transfer: A range of protein and peptide sequences have been utilized to target, enhance or mediate the delivery of nucleic acids in a large variety of applications. Such proteins and peptides are regularly utilized along with cationic lipids. Targeting proteins and peptides enhances transfection efficiencies and provides targeted delivery.

peptidyl or **(P-) site:** The site on the ribosome to which the tRNA attached to the growing polypeptide is bound during translation.

peptidyl transferase: Peptide establishing enzyme, peptidyl transferase, is needed one copy for one 50S subunit between amino acids during process of translation. On completion of peptide-bond formation, the (A) site is engaged by F-met-amino acid-tRNA A1 and the (P) site is engaged by $tRNA_f^{Met}$. The translocation step follows. The $tRNA_f^{Met}$ is cleared from the ribosome. The F-met-amino acid 1-$tRNA^{A1}$ dipeptide is shifted from the (A) to the (P) site as the ribosome travels by the length of one codon along the mRNA in a $5' \rightarrow 3'$ direction, thus bringing a new codon to the (A) site. Thus, polypeptide assembly begins at the amino end and ends at the carboxyl end. mRNA is read in a $5' \rightarrow 3'$ direction.

peptidyltransferase centre (PTC): The catalytic centre of the ribosome. It is located on the large ribosomal subunit and catalyses peptide-bond formation and peptide release.

percentile: A percentile (or a centile) is a measure used in statistics indicating the value below which a given percentage of observations in a group of observations fall. For example, the 20th percentile is the value (or score) below which 20% of the observations may be found.

perennation: In biennials and herbaceous perennials, the aerial parts of the plant die down and the plants subsist by means of underground storage roots (e.g. carrot), rhizomes (e.g. couch grass, Solomon's seal), tubers (e.g. dahlia), bulbs (e.g. daffodil, snowdrop) or corms (e.g. crocus, gladiolus). *Contrasts* ANNUAL, BIENNIAL, MULTIVOLTINE *and* UNIVOLTINE.

perfect flower: A flower possessing all four types of essential structures (viz. pistil, stamens, sepals and petals). More precisely, a perfect flower possesses both essential floral parts (i.e. stamens and pistils). In crops like wheat, rice and soybean, the flowers are perfect. In contrast, in some cucurbits, fruits (papaya, date palm) and field crops like maize, the flowers are imperfect (i.e. possess only one sex), which leads to cross-pollination. *Contrasts* IMPERFECT FLOWERS.

perfect insecticide: A perfect pesticide is one that destroys a particular pest and is completely harmless to every other form of life.

perfect state: The sexual state in the life cycle of a fungus. *See* TELEOMORPH. *Contrasts* ANAMORPH, HOLOMORPH *and* IMPERFECT STATES.

pericentric heterochromatin: Major repressive heterochromatin domain in the nucleus, which is mainly composed of major satellite repeats.

pericentric inversion: A reversal of the normal order of genes in a chromosome segment involving parts of a chromosome on both sides of the centromere.

pericentromeres: Sites of constitutive heterochromatin that flank the central kinetochore-binding region of the centromere and are necessary for proper centromere function.

periodic dependence of transcription: Transcriptional elements perform vital role in gene expression. The transcriptional elements of the heat shock protein *hsp70* gene have been investigated in detail. The striking periodicity of around 10.5 bp seen for short insertions indicates that the best hsp70 transcription necessitates specific, rotational alignment of DNA-protein interaction sites.

periodic introns: Many of the inconsistent introns are periodic (i.e. they are relics of internal sequence repetitions within the ancestral gene). Some of these periodic-intron patterns are common between related genes.

periodicity: Refers to the number of base pairs per turn of the DNA double helix.

peripheral isolate speciation: A method of allopatric speciation in which the new species is formed from a small population isolated at the edge of an ancestral population's geographic range. Also called PERIPATRIC SPECIATION.

peripheral membrane protein: Protein that is committed to one face of a membrane by non-covalent interactions with other membrane proteins and which can be detached by relatively gentle treatments that leave the lipid bilayer intact.

peripheral: Any auxiliary or ancillary system connected to or integrated within a computer system and used to direct information to or recover information from the computer. An input manoeuvre sends data or instructions to the computer. An output device delivers output from the computer to the user and an input/output device accomplishes both functions.

permaculture: An approach to land management and settlement design that adopts arrangements observed in flourishing natural ecosystems. It includes a set of design principles derived from whole-systems thinking. It applies these principles in fields such as regenerative agriculture, town planning, rewilding and community resilience. The term was coined in 1978 by Bill Mollison and David Holmgren, who formulated the concept in opposition to modern industrialized methods, instead adopting a more traditional or 'natural' approach to agriculture. Permaculture has been criticized as being poorly defined and unscientific.

permanent cell line: A cell line that survives in culture through an unlimited number of generations.

permanent heterosis: A hybrid condition maintained by a balanced lethal system so that *Oenothera lamarckiana* is a permanent hybrid of the GAUDENS-VELANS COMPLEX.

permissive conditions: The conditions that allow conditional lethal mutants to survive.

permissive euchromatin: Describes most euchromatin, which is transcriptionally silent but potentially active. It is immediately responsive to specific signals, such as hormones. It includes luxury genes.

permissive host: A host cell in which a particular strain of virus can reproduce to yield progeny viruses. Cells that do not support such growth are defined as NON-PERMISSIVE.

permissive induction: When the responding cells are previously committed to a certain fate but require a particular signal to allow them to continue.

permissive locus: A locus for transgene integration that is permissive for high transgene expression levels and transgene stability.

permissive strain: The strain of a cell capable of acting as a permissive host.

permissive tissue interaction: Permissive interactions are those in which tissues do not differentiate in the same way due to a non-specific stimulus.

peroxidase: An enzyme that breaks down hydrogen peroxide. This enzyme breaks down hydrogen peroxide without producing oxygen. $H_2O_2 + 2H^+ \rightarrow 2H_2O$

peroxisome: Small membrane-bounded organelle that uses molecular oxygen to oxidize organic molecules. Contains some enzymes that produce and others that degrade hydrogen peroxide (H_2O_2).

peroxyacetyl nitrate (PAN): A secondary air pollutant formed when the exhaust of automobiles and other internal combustion engines occurs in the attendance of sunlight.

persistent: Staying attached to the plant beyond the usual time of falling. For example, sepals do not fall after flowering and flower parts linger through the maturity of fruit. *Compare* DECIDUOUS *and* CADUCOUS.

persistent modification: Non-heritable morphological or physiological modifications over a more or less long period of generation cycles persuaded by varying abiotic or biotic impacts.

persistent transmission: A type of virus transmission in which the virus is attained and spread by the vector after a relatively long feeding time and remains transmissible for a prolonged period while in association with its vector. *See* CIRCULATIVE TRANSMISSION *and* PROPAGATIVE TRANSMISSION. *Contrasts* NON-PERSISTENT TRANSMISSION *and* STYLET-BORNE TRANSMISSION.

pervasive transcription: Some transcripts either occur stably in cells (e.g. STABLE UNANNOTATED TRANSCRIPTS, SUTs) or become rapidly degraded by the RNA surveillance pathway (e.g. CRYPTIC UNSTABLE TRANSCRIPTS, CUTs). One characteristic of pervasive transcription is the extensive overlap of SUTs and CUTs. Transcription of SUTs and CUTs can be functional via mechanisms that involve the generated RNAs or their generation itself.

pest: Any organism that damages plants or plant products (e.g. seed).

PEST hypothesis: According to the PEST hypothesis, protein degradation is signalled by one of the four amino acids: proline (P), glutamic acid (E), serine (S) and threonine (T). Different proteins are programmes to survive different programmes in the cell.

pest management through RNA interference (RNAi): RNAi could be used selectively to kill pest species without adversely affecting non-target species by targeting the genes essential for pest insect and mite growth, development or reproduction. Insects and mites in which specific vital genes have been silenced include seven species of order Diptera (South American fruit fly, *Anastrepha fraterculus* (Wiedemann), Oriental fruit fly, *Bactrocera dorsalis* (Hendel), citrus fruit fly, *Bactrocera minax* (Enderlein), olive fruit fly, *Bactrocera oleae* (Rossi), pumpkin fruit fly, *Bactrocera tau* (Walker), Queensland fruit fly, *Bactrocera tryoni* (Froggatt), fruit fly *Drosophila suzuki* (Matsumura)), five species of order Hemiptera (whitefly, *Bemisia tabaci* (Gennadius), Asian citrus psyllid, *Diaphorina citri* (Kuwayama), green peach aphid, *Myzus persicae* (Sulzer), citrus mealybug, *Planococcus citri* (Risso) and grape mealybug, *Pseudococcus maritimus* (Ehrhorn)), one species of order Lepidoptera (light brown apple moth, *Epiphyas postvittana* (Walker)), one species of order Blattodea (Formosan subterranean termite, *Coptotermes formosanus* (Shiraki)), and two species of order Acari (citrus red mite, *Planococcus citri* (McGregor) and two-spotted spider mite, *Tetranychus urticae* (Koch)). These insects and mites infest a variety of fruit crops.

pest management using insects homozygous for a repressible dominant lethal: Transgenic insects, which are homozygous for a dominant, repressible, female-specific lethal gene system, are a major improvement to the SIT programmes. D.D. Thomas *et al.* (2000) demonstrated two methods for the genetic transformation of insects. The first system uses a sex-specific promoter or enhancer to drive the expression of a repressible transcription factor to control the expression of a toxic gene product while the second system uses non-sex-specific expression of the repressible transcription factor, which regulates a selectively lethal gene product. Both of these methods worked efficiently in *Drosophila melanogaster* and could be applied widely in other economically important insect pest species.

pest resistance: Plants with an inherent structural or chemical deterrent to insect, arthropod or fungal pests as a result of specific breeding or genetic engineering techniques.

pesticide: A wide-ranging class of crop protection products, including insecticides utilized to control insects, diseases, weeds, nematodes, fungi and other 'pests' on plants, fruits, vegetables and animals and in buildings. Both farmers and consumers use pesticides in the home or yard to eliminate termites and roaches, clean mould from shower curtains, keep at bay crab grass on the lawn, terminate fleas and ticks on pets, sanitize swimming pools and so on.

pesticide label: A legal document/text approved by the US Environmental Protection Agency that defines uses and provides other safety information for a commercial pesticide.

pesticide resistance: (i) A genetic change in response to selection by a pesticide resulting in the development of strains capable of surviving a dose lethal to a majority of individuals in a normal population. Resistance may develop in insects, weeds or pathogens. The existence of such resistant pests makes the pesticide less useful for handling pest populations. (ii) The development or selection of heritable traits (genes) in a pest population that enable the individuals expressing the trait to survive in the occurrence of levels of a pesticide (biological or chemical control agent) that would otherwise incapacitate or kill this pest.

pest-resistant GM crops: Pest-resistant GM crops (primarily cotton and maize) have been genetically modified so they are toxic to certain insects. They are often called *Bt* crops because the introduced genes were originally identified in a bacterial species called *Bacillus thuringiensis*. These bacteria produce a group of toxins called Cry toxins.

pest-resistant plants: Plants with the capability to endure, deter or repel pests and thereby thwart them from damaging the plants. Plant pests may include insects, nematodes, fungi, viruses, bacteria, weeds, etc.

pesticide treadmill: The inherent reliance of modern agriculture and public-health programmes on pesticides, often in increasing quantities, with the progression of time, to deal with pest problems.

pesticide-resistant insects: Insects with the mechanisms or skill to survive and reproduce in the presence of pesticides. These resistant pest variants will increase in frequency over time if pesticides persist in their environment.

petal: In a flower, one of the segments or divisions of the inner whorl of non-fertile parts surrounding the fertile organs, usually soft and conspicuously coloured. *Compare with* SEPAL.

petals and wings development: Both petals and wings display a marked asymmetry in cone growth, possibly in response to unknown long-range signals. In the *Antirrhinum* petal, such long-range signals could orient cell division, whereas in the *Drosophila* wing, the plane of division is random but cell positioning is oriented. In each case, the result is asymmetric clone growth along the main tissue axis.

petite yeast: A slow-growing strain of yeast (*Saccharomyces*) that lacks certain respiratory enzymes and forms unusually small colonies on agar. There are three types of petite mutants in yeast: SEGREGATIONAL, NEUTRAL and SUPPRESSIVE. *See details as separate terms.*

pH: A numeric scale used to specify the acidity or basicity (alkalinity) of an aqueous solution. It is roughly the negative of the logarithm to base 10 of the concentration, measured in units of moles per litre, of hydrogen ions (H^+). A measure of the acidity or basicity of a solution; on a scale of 0 (acidic) to 14 (basic): for example, lemon juice has a pH of 2.2 (acidic), water has a pH of 7.0 (neutral) and a solution of baking soda has a pH of 8.5 (basic).

pH buffers: The compounds that resist the change in pH. The organisms need the optimum pH of the nutrient broth for better growth. During culturing, bacteria excrete waste products that alter pH of the medium, which finally stops bacterial growth. To solve this problem, pH buffers are added to the culture medium. The commonly used buffer for bacterial culture media uses a mixture of K_2HPO_4 and KH_2PO_4.

PH domain: *See* PLECKSTRIN HOMOLOGY DOMAIN.

***Ph* locus:** A gene that controls homoeologous chromosome pairing in wheat and, similarly, in other allopolyploid plants; the *Ph* gene restricts pairing between homologous chromosomes in polyploidy. During evolution, a subtelomeric heterochromatin block became inserted into a group of cdc2-like genes on 5B; *Ph1* is involved in the regulation of chromatin condensation. It seems likely that this insertion event generated a functional and/or regulatory change in the 5B cdc2-like gene family.

PHABULOSA (PHB): A homeodomain leucine zipper III (HD-ZIPIII) transcription factor that mediates an auxin signalling loop to regulate vascular patterning in *Arabidopsis*.

phage: A submicroscopic organism that destroys bacteria. Phages are often viruses lacking cellular mechanisms of their own and so must infect a host cell to grow and reproduce. Bacteriophages are commonly employed as vectors for carrying foreign DNA into cells by genetic engineers. *Aka* BACTERIOPHAGE.

phage 'ghost': The phage protein shell left behind after the phage has injected its DNA into the host.

phage artificial chromosome (PAC) shuttle vector: A PAC shuttle vector creates a library representing two-fold coverage of the human genome. This could serve as an incredible resource to identify genes or sets of genes that cause disease.

phage artificial chromosomes (PACs): PACs are based on the P1 phage and have features of both P1 vectors and BACTERIAL ARTIFICIAL CHROMOSOMES (BACs). Similar to P1 vector, they contain a plasmid and a lytic replicon. Unlike P1 vectors, they do not need to be packaged into bacteriophage particles for transduction. Instead, they are introduced into *Escherichia coli* as circular DNA molecules through electroporation just as BACs are. They can transfer big amounts (about 100–300 kb) of other sequences for diverse bioengineering purposes.

phage display: A technique in which phage are engineered to fuse a foreign peptide or protein with their capsid (surface) proteins and hence display it on their cell surfaces. The immobilized phage may then be used as a screen to see what ligands bind to the expressed fusion protein exhibited (displayed) on the phage surface. This technology permits the manufacture of monoclonal antibodies for selected target antigen, without the use of any animal or hybridoma. DNA sequences coding for antibody V region are fused with DNA sequences coding for the amino terminus of the minor coat protein pIII of the phage. This fused DNA sequence, inserted in phage (f1, M13 or λ) genome produces a fused protein (carrying the antibody) that is exhibited on the surface of the phage (due to coat protein sequence).

phage induction: The transition from the prophage state to the lytic growth of a temperate phage in response to an external stimulus. Induction results in the initiation of transcription of phage genes, the excision of prophages from the host chromosome and the synthesis of phage genomes and capsid proteins.

phage therapy: Treatment to combat bacterial infection using phages, viruses that destroy bacteria.

phage typing: Means of classifying bacteria based on whether they get destroyed by specific phages.

phage vector: Phage vectors consist of an essentially complete phage genome into which is inserted DNA encoding the protein or peptide of interest.

phage-based recombination: Bacteria, such as *Escherichia coli*, encode their own homologous recombination systems. Viruses or phages that inhabit bacteria also often carry their own recombination functions, which can work with, or independently of, the bacterial recombination functions.

phagemid: Combining the desirable features of both plasmids and filamentous bacteriophages. (i) Phagemids contain the origin of replication from plasmid ColE 1, selectable markers for antibiotic resistance and the major intergenic (IG) region of a filamentous phage that is required for the initiation and termination of viral DNA synthesis and for the morphogenesis of phage particles. Important phagemids include PUC118 (a derivative of PUC18) and PUC119 (a derivative of PUC19). (ii) Phagemids contain an origin of replication (*ori*) for double-stranded replication as well as an flori to enable single-stranded replication and packaging into phage particles. They can carry inserts of size 35–50 kb and can be used as a type of cloning vector in combination with filamentous phage M13. Phagemids are also useful for generating templates for site-directed mutagenesis. *Aka* PHASMIDS.

phagocyte: An immune cell that engulfs and digests cells, cell debris, microorganisms and other foreign bodies in the bloodstream and tissues (PHAGOCYTOSIS). The ingested material is then degraded via enzymes. A whole class of cells is known to be PHAGOCYTIC. Monocytes, macrophages and neutrophils are examples of phagocytes.

phagocytosis: Eukaryotic cells can ingest substances through two mechanisms called phagocytosis and pinocytosis. In phagocytosis, particulate material is eaten up by a cell. Prominent in carnivorous cells, such as *Amoeba proteus* and in vertebrate macrophages and neutrophils. During phagocytosis, cellular projections called pseudopods engulf particles and bring them into cell. This mechanism is common in leukocytes (white blood cells, WBC) to destroy bacteria. *See* PHAGOCYTE, PINOCYTOSIS.

phagosome: Large intracellular membrane-bounded vesicle that is shaped as a consequence of phagocytosis.

phaneranthous: Showy (as in showy flowers of Dahlias, *Dahlia pinnata*) that advertise to pollinators, as opposed to APHANANTHOUS (unshowy) flowers of the river bushwillow, *Combretum erythrophyllum.*

phanerogam: (i) Gymnosperms and angiosperms, (ii) plants producing stamens and gynoecia and (iii) literally plants with conspicuous sexual reproductive organs. *Compare with* CRYPTOGAMS.

pharmacogenetics: The study of how genetics impacts an individual's response to medication. It deals with the variable response of humans to specific drugs. Pharmacogenetics helps in the classification of humans into responders and non-responders. Responders are further classified into two groups: those who derive benefit from drugs and those exhibiting side effects. The gene type of the individual plays an important role in drug response. The knowledge of pharmacogenetics would help drug companies design a drug to meet the individual's specific needs. Often used interchangeably with pharmacogenomics.

pharmacogenomics: A major objective of pharmacogenomics is the development of innovative classes of targeted drugs and vaccines designed to affect highly specific processes in the body while minimizing side effects. *See* PHARMACOGENETICS.

pharmacophore: The three-dimensional spatial arrangement of atoms, substituents, functional groups or chemical features that together are sufficient to describe the pharmacologically active components of a drug molecule or molecule series.

phase I studies: In toxicology, studies acquire preliminary information on dosage, absorption, metabolism and the relationship between toxicity and the dose schedule of treatment.

phase II studies: In toxicology, studies to investigate feasibility and estimate treatment activity and safety in diseases (e.g. tumour types) for which the treatment appears promising. This phase generates hypotheses for later testing.

phase III studies: In toxicology, comparative trial to find out the effectiveness and safety of a new treatment relative to standard therapy. These trials generally represent the most demanding proof of treatment efficacy (pivotal trials) and are considered the last stage before product licensing.

phase IV studies: Post-marketing studies of licensed products.

phase imaging: Imaging by mapping the phase of the cantilever oscillation in the TAPPING MODE, which is a dynamic AFM technique that images the sample topography by scanning the surface with an oscillating cantilever.

phase transition in development: The shift from one developmental phase to the next is under the control of spatial and temporal patterns of gene expression, so that selective activation or silencing of genes guides development through different phase changes. For example, the floral transition marks the beginning of reproductive development and the timing of this developmental shift is vital to determine the reproductive success of plant species. For that reason, plants control very precisely the time of flowering initiation in response to both endogenous and environmental factors, ensuring that the production of flowers and fruits takes place under optimal conditions.

phase, frame: Nucleotide triplets are 'in phase' with one another of they can be read in groups of three from an AUG codon and can, therefore, act as codons for amino acids during translation.

phase-contrast microscope: A sort of light microscope that makes use of the interference effects that transpire when light passes through material of different refractive indexes. Used to view living cells. Phase-contrast microscope is used to examine the cultured organisms without staining. A special condenser (phase contrast system) fitted with the microscope allows the examinations of structures inside the cells or to differentiate cell wall from the cytoplasm.

phased small interfering RNAs: A type of secondary siRNA that is processed from a miRNA-mediated cleavage of RNA transcripts, increasing silencing efficiency or simultaneously suppressing multiple target genes.

phased small RNA–mediated systemic signalling: SYSTEMIC ACQUIRED RESISTANCE (SAR) involves the generation of systemically transported signal that arms distal plant parts against secondary infections. M.B. Shine *et al.* (2022) show that two phased 21-nucleotide (nt) *trans*-acting small interfering RNA3a RNAs (tasi-RNA) derived from TAS3a and synthesized within 3 h of pathogen infection are the early mobile signal in SAR. *TAS3a* undergoes alternate polyadenylation, resulting in the generation of 555- and 367-nt transcripts. The 555-nt transcripts likely serve as the sole precursor for tasi-RNAs D7 and D8, which cleave auxin response factors (ARFs) 2, 3 and 4 to induce SAR. Conversely, increased expression of *ARF3* represses SAR. Tasi-ARFs are the early mobile signal in SAR.

PhaseTank approach: PhaseTank applies an algorithm that comes close to the relative sRNA production (RSRP) of each group.

phasiRNA detection: PhasiRNAs are sncRNAs with a length of 18–29 nt in plants. They are derived from *PHAS* loci which can be mRNA of protein-coding genes or lncRNAs in plants. Different methods available for computational detection of phasiRNA include ShortStack, PhaseTank and Unitas.

phasiRNA, biogenesis of: phasiRNA biogenesis is activated by an AGO RISC (RNA-induced silencing complex) complex, generally loaded with a 22-nt miRNA cleaving the phasiRNA precursor transcript. The 5′ fragment of the cleaved transcript is degraded whereas the 3′ fragment is transformed into a dsRNA by the RNA-dependent RNA polymerase6 (RDR6). The ds RNA is cleaved by a DCL yielding phasiRNAs. The phasiRNAs might be incorporated into AGOs in the RISC complex to target single-strand transcripts through base pairing.

phasmid: A hybrid molecule formed *in vivo* between a plasmid containing multiple *att* sites of bacteriophage λ and a bacteriophage λ derivative. The formation of a plasmid or its breakdown to release the plasmid and phage is controlled by the site-specific recombination system of λ. Phasmids can replicate as a plasmid (non-lytically) or as a phage (lytically). A number of sophisticated *in vivo* genetic manipulations can be accomplished using phasmids.

PHAVOLUTA (PHV): PHAVOLUTA, together with PHABULOSA, is sufficient to activate seed maturation genes in vegetative tissues.

phene: The phenotype of the plant, which is a product of the gene (gene (DNA) → transcription → RNA → processing → mRNA → translation → protein) and the interaction with the environment.

phenetic analyses: Computerized multivariate statistical methods for classifying cultivated or domesticated plants. It is also called numerical taxonomy for plants where (i) determining ancestor-descendant relationships may be trivial (i.e. domesticates are generally derived from wild plants), not vice versa; (ii) ancestor and descendant have diverged recently and are incompletely differentiated, so that their overall phenetic proximity is generally still strongly congruent with their degree of evolutionary divergence. Consequently, phenotypic traits which are similar in crops and their wild relatives may be generally homologous; and (iii) hybridization and introgression may be rampant. Phenetic analyses share with phylogenetic analyses the initial steps of collecting data regarding the taxa under study, grouping the measured traits into characters and character states according to known homologies. The lists of characters and character states comprise a data matrix, the variability of which is then used to calculate proximity measures.

phenetic classification: A method of classification in which the species are grouped together with other species that they most closely resemble phenotypically.

phenetic species concept. A concept of species according to which a species is a set of organisms that are phenotypically similar to one another. *Compare with* BIOLOGICAL SPECIES CONCEPT, CLADISTIC SPECIES CONCEPT, ECOLOGICAL SPECIES CONCEPT *and* RECOGNITION SPECIES CONCEPT.

phenocart: A device that automatically reads the vital signs of plants and uses a software package and GLOBAL POSITIONING SYSTEM (GPS) to organize the collected data. The age-old method of walking around a farm and examining thousands of plants by hand could soon be replaced by the phenocart.

phenocarting: The process of observing the physical characteristics of a plant, such as height and colour, in order to assess its health. It is an essential part of plant breeding, with farmers crossing different species and observing the results to determine if the new crops have the desired traits.

phenocopy: An abnormal individual resulting from special environmental conditions. It mimics a similar phenotype caused by gene mutation. Phenocopies are produced when an organism is exposed to a phenocopying agent during a sharply defined period of its development and the same effect can often be produced by very different agents. Changes in the environment can influence the expression of one or more genes. The most sensitive area is development, since relatively small changes that take place early in development can result in huge changes later. Certain drugs administered at the embryonic stage have been shown to produce effects that mimic those of known specific mutant alleles. The abnormal individual resulting from this treatment is called a phenocopy. Phenocopies are of two types – PHENOCOPYING ABNORMAL and PHENOCOPYING NORMAL. *See details as separate terms.*

phenocopying abnormal: In this case, phenocopy resembles a known mutation. For example, when *Drosophila* larvae are fed on silver nitrate mixed food, the adults emerged have straw body colour, which resemble *straw* mutant flies.

phenocopying normal: In this case, phenocopy resembles a normal individual whereas non-phenocopy individual is mutant. For example, the individuals suffering from diabetes when treated with insulin behave like normal individuals.

phenodeviants: Phenotypes that deviate from the population (or species) mean (or norm), owing to special gene combinations, for instance, extreme homozygosity.

phenogenetics: A branch of genetics that investigates the interaction of the genotype and its phenotypic manifestation.

phenogram: A branching illustration that relates entities by estimations of inclusive similarity. For example, diagrammatic arrangement of cluster analysis, taxa, germplasm collections or genetic markers. In CLUSTER ANALYSIS, taxa, germplasm collections or genetic markers are arranged in a hierarchy (called a 'phenogram' or 'dendrogram') by an agglomerative algorithm according to patterns occurring in a matrix of pair-wise genetic proximities. The hierarchies obtained from cluster analyses are highly dependent on both the proximity measure and the clustering algorithm used.

phenogroup: The antigenic determinants that are transmitted as a block from parents to offspring.

phenological isolation: Biological isolation of one species or group from other member of the same genus by reason of difference in time of flowering (early or late).

phenological synchrony: The harmonized development of parasites and their hosts based on degree-days and other environmental factors such that parasites can successfully infect their hosts. Phenological synchrony is exemplified by the production and release of ascospores of *Venturia inaequalis*, the apple scab pathogen, from fallen leaves from the previous season when new apple leaves are expanding.

phenology: The study dealing with the timing of seasonal biological phenomena or events, such as flowering, leaf emergence, fruit ripening and leaf fall.

phenome: All the phenotypic elements of an organism. *Compare with* PHENOTYPE.

phenomenon: Any state or process known through the senses rather than by intuition or reasoning is known as phenomenon. Thus, phenomenon is any observable occurrence.

phenomics: A technology that enables high-throughput and non-destructive phenotyping for crop improvement in response to various kinds of abiotic stresses. Phenomics refers to the relationship between genomics and phenotype/traits. Phenomics includes techniques such as infrared imaging, 3-D imaging, MRI, fluorescence imaging and spectral reflectance. *See* FORWARD PHENOMICS *and* REVERSE PHENOMICS.

phenomics tools, crop: Important crop phenomics tools include the following: (i) THREE-DIMENSIONAL (3D) IMAGING, (ii) FAR-INFRARED (FIR) IMAGING SYSTEM, (iii) NEAR INFRARED (NIR) IMAGING SYSTEM, (iv) FLUORESCENCE IMAGING SYSTEM and (v) MAGNETIC RESONANCE IMAGING (MRI) SYSTEM. *See details as separate terms.*

phenotype: The phenotype of an organism is its appearance (colour, form, size, behaviour, chemical composition and structure) and how we can observe its morphology, physiology and behaviour. The external appearance of an organism is contrasted with its genetic make-up or genotype for a particular character. The phenotype is determined by the genotype, environment and genotype-environment interaction and is represented by equation phenotype (P) = genotype (G) + environment (E) + genotype phenotype interaction (GEI). During the life of an individual, the phenotype may change; the genotype, however, remains constant. Organisms with similar phenotypes do not necessarily have identical genotypes. For instance, yellow peas can be either homozygous for the yellow allele or heterozygous for the yellow and green alleles. Phenotypes change continuously during development. *Compare with* GENOTYPE.

phenotypic: Relating to the visible and/or measurable characteristics of an organism (i.e. how it appears outwardly and physiologically) as compared with its genotype, or genetic characteristics.

phenotypic characters: Individual traits that can be observed in an organism in terms of appearance and behaviour which are consequence of the interaction between the organism's genetic make-up and its environment.

phenotypic contribution: Phenotypic variation in plant traits is strongly influenced by genetic and environmental factors. Over the lifespan of trees, developmental factors may also strongly influence leaf phenotypes.

phenotypic differentiation: The differentiation of phenotypes in separated gene pools during and after speciation.

phenotypic expression: The manifestation of a particular gene ultimately resulting in or contributing to in a particular phenotype.

phenotypic flexibility: The ability of a genotype to vary its expression in different environments. This is a character of considerable adaptive significance in most plants. Species in a genus, as also populations within a species, show differences in phenotypic flexibility indicating that this attribute is genetic-based and subjected to the control of the usual process of natural selection.

phenotypic mixing: A virus produced by the combination of the nucleic acid of one virus with the protein of another.

phenotypic plasticity. Phenotypic plasticity is the ability of one genotype to produce more than one phenotype when exposed to different environments. Phenotypic plasticity is more important for immobile organisms (e.g. plants) than mobile organisms (e.g. most animals), as mobile organisms can often move away from unfavourable environments. Phenotypic plasticity and exposure to hidden genetic variation both affect survival and evolution of new traits. Phenotypic variation is expressed by a single genotype in different environments. As an epigenetic mechanism that globally affects the relationship between genotype and phenotype, the yeast prion [PSI^+] expands the conceptual framework for phenotypic plasticity, provides a one-step mechanism for the acquisition of complex traits and affords a route to the genetic assimilation of initially transient epigenetic traits. *See* REACTION NORM.

phenotypic ratio: Ratio of different phenotypes in a segregating generation (e.g. phenotypic ratio in F_2 of a monohybrid cross in case of complete dominance is 3 tall:1 dwarf).

phenotypic segregation: The phenotypic difference patterns of cells or individuals in segregating populations. *Compare with* GENETIC SEGREGATION.

phenotypic selection: Development of a variety based on its physical appearance without considering its genetic constitution.

phenotypic sex determination: Sex determination by non-genetic means.

phenotypic value: The value observed when a quantitative trait, which is consequence of sum total of contributions from both genotype and environment, is measured on an individual.

phenotypic variance (V_P): The total variance observed for a character. It consists of experimental error, genotype × environment interaction and the genotypic variance.

phenotypic variation: Phenotypic variation observed in nature is due to combination of genetic and environmental (epigenetic) changes. It is the sum total of genotypic variation and environmental variation and their interaction. Mathematically, $V_P = V_G + V_E + V_{GE}$, where V_P = phenotypic variance, V_G = genotypic variance, V_E = environmental variance and V_{GE} = variance due to interaction between genotypic and environmental factors.

phenotyping: Refers to the set of observable physical appearance of an individual ensuing from the interaction of its genotype with the environment. Thus, phenotypes mirror the NATURE and the NURTURE of the organism.

phenylalanine (phe): An aromatic amino acid ($HO_2C\text{-}CH(NH_2)CH_2C_6H_5$). This essential amino acid is categorized non-polar because of the hydrophobic nature of the benzyl side chain. The codons for L-phenylalanine are UUU and UUC. L-Phenylalanine is one of the raw materials used to manufacture NutraSweet ® (NutraSweet Co.) synthetic sweetener.

phenylketonuria: An autosomal recessive metabolic disorder due to accumulation of phenylalanine because of defective enzyme phenylalanine hydroxylase. It is characterized by a serious mental and physical retardation.

***Pheres* gene:** A gene involved in embryonic development in *Arabidopsis* whose activity is modified by the MEDEA protein that binds to its promoter. *See* REGULATORY ELEMENTS.

ϕX174 phage: A bacteriophage which attacks *Escherichia coli*. The nucleic acid within the viral particle is a single-stranded DNA circle of 5375 bases. The replicative form is a double-stranded circle and was the first complete DNA molecule to be sequenced. Overlapping genes were first reported in ϕX174 phage.

phloem: In vascular plants, the xylem and phloem form continuous tubes that carry water, nutrients and other substances throughout the plant's roots, stem and leaves. Positive hydrostatic pressure in the phloem moves dissolved sugars and organic compounds from the leaves downward to the stem and roots via a process called TRANSLOCATION. The phloem of monocots and dicots contains conducting cells (sieve elements) and companion cells. Conducting cells have thin walls and they are alive in the mature plant, but they lack a nucleus and most other organelles.

phloem loading: Phloem loading is the process whereby carbohydrates enter the. sieve tubes at the source.

phloem necrosis: The death of phloem cells, often visible, caused by infection by systemic phloem pathogens such as phytoplasmas, spiroplasmas and viruses.

phloem unloading: Photoassimilate (sugars) removal from phloem and delivery to recipient sink cells.

phosphatase: An enzyme that catalyses the reaction involving the hydrolysis of esters of phosphoric acid and removes phosphate groups from a molecule.

phosphate bond: Usually the anhydride linkage between phosphate ions which may be an important source of energy.

phosphate homeostasis: An organism that can regulate its internal environment so as to maintain equilibrium has the property of homeostasis. Phosphate is an indispensable mineral nutrient for plant growth and development. Some miRNAs (e.g. miR-399) regulate phosphate homeostasis.

phosphate-sensing organelle: Inorganic phosphate (P_i) is one of the essential molecules for life. P_i starvation causes hyperproliferation in the digestive epithelium of *Drosophila melanogaster*, C. Xu *et al.* (2023) have determined that P_i starvation triggers the downregulation of the P_i transporter *PXo*. Immunostaining and ultrastructural analyses showed that PXo specifically marks non-canonical multilamellar organelles (PXo bodies). Further, by P_i imaging with a Förster resonance energy transfer (FRET)-based P_i sensor, it has been found that PXo restricts cytosolic P_i levels. PXo bodies require PXo for biogenesis and undergo degradation following P_i starvation. Proteomic and lipidomic characterization of PXo bodies unveiled their distinct feature as an intracellular P_i reserve. Therefore, P_i starvation triggers *PXo* downregulation and PXo body degradation as a compensatory mechanism to increase cytosolic P_i. Finally, connector of kinase to AP-1 (Cka), a component of the striatin-interacting phosphatase and kinase (STRIPAK) complex and c-Jun N-terminal kinase (JNK) signalling, was found as the mediator of *PXo* knockdown– or P_i starvation–induced hyperproliferation. Thus, PXo bodies were identified as a critical regulator of cytosolic P_i levels and identifies a P_i-dependent PXo–Cka–JNK signalling cascade controlling tissue homeostasis.

phosphatidylinositol 3-kinase (PI 3-kinase): A kinase involved in intracellular signalling pathways activated by a variety of cell-surface receptors. It phosphorylates inositol phospholipids at the 3 position of the inositol ring.

phosphatidylinositol: An inositol phospholipid.

***phosphinothricin acetyltransferase* (*pat*) gene:** A dominant gene isolated from the *Streptomyces viridochromogenes* bacterium that codes for the enzyme phosphinothricin acetyl transferase (PAT). When the *pat* gene is introduced into a plant's genome, it communicates resistance to glufosinate-ammonium-containing herbicides. The *pat* is also frequently employed by genetic engineers as a marker gene. In 1988, E. Strauch *et al.* successfully isolated the *pat* gene from bacteria and transferred it to plants by means of genetic engineering. The transgenic plants expressing the *pat* gene are able to degrade the herbicide phosphinotricin.

phosphodiester bond: The covalent bond between the 3′ hydroxyl in the sugar ring of one nucleotide and the 5′ phosphate group of the sugar ring of the adjacent nucleotide residue formed by DNA ligase within a nucleic acid 5′-Ribose-3′-O-P(O)$_2$-O-5′-Ribose-3′-, and so on, *Aka* PHOSPHODIESTER LINKAGE.

phosphodiesterases: A group of enzymes, called ligases, that inhibit apoptosis.

phosphoinositide: *See* INOSITOL PHOSPHOLIPID.

phospholipase C-β (PLC-β): Enzyme bound to the cytoplasmic face of the plasma membrane that converts membrane phosphatidylinositol 4,5-bisphosphate to diacylglycerol (which remains in the plasma membrane) and inositol 1,4,5-trisphosphate (IP3). It is activated by certain G proteins to trigger the inositol phospholipid signalling pathway.

phospholipase C-γ (PLC-γ): Like phospholipase C-β, an enzyme that cleaves inositol phospholipids to diacylglycerol and IP3 triggers the inositol phospholipid signalling pathway. Activated by certain receptor tyrosine kinases.

phospholipid: A lipid containing phosphate groups. In glycerophospholipids and some sphingolipids, a polar head group is joined to the hydrophobic moiety by a phosphodiester linkage and the resulting compound is called phospholipid.

phospholipid exchange protein: Water-soluble carrier protein that transfers a phospholipid molecule from one membrane to another.

phosphoprotein phosphatase: A type of enzyme that removes a phosphate group from a protein by hydrolysis.

phosphoproteins in gene regulation: Phosphoproteins play an important role in providing a high level of tissue specificity. The nuclear phosphoproteins possess most of the requirements of regulators of gene activity. These requirements include the following: (i) nuclear phosphoproteins are located on the active segment of chromatin, (ii) their distribution in various cell cycles is specific, (iii) phosphorylation of the hydroxyl group of serine residues reflects the intensity of RNA synthesis in this tissue, (iv) these proteins are species-specific, (v) transcription of a DNA-phosphoprotein complex is greater than that of DNA alone and (vi) some non-histone proteins that are synthesized in the S phase in Hela cells function to activate the transcription of histone genes in the G1 phase.

phosphorus (P): An element required by plants in the oxidized form. It is utilized in reactions, often involving ATP, in which energy is transferred.

phosphorylation: Addition of a phosphate group to an organic molecule. Phosphorylation leads to the formation of a phosphate derivative of an organic molecule by enzymatic transfer of phosphoryl group. Adenosine diphosphate phosphorylation by a kinase helps in the generation of ATP. The addition of P to ADP results in the formation of ATP as under: ADP + P + energy $\rightarrow$ ATP. With respect to proteins, certain amino acid side chains (serine, threonine and tyrosine) are subject to phosphorylation catalysed by protein kinases. This reaction modifies the phosphorylation status of a protein and may have dramatic influence on its biological properties. Phosphorylation serves as a common cellular control mechanism.

photo-cross-linking assisted 6^{m}A sequencing: This method is a variation of the 6^{m}A-seq method that was developed to map 6^{m}A at single-nucleotide resolution. It involves the cross-linking of the 6^{m}A-specific antibody to the RNA before the affinity purification of the RNA-antibody complexes, RNA isolation, reverse transcription and DNA sequencing.

photoautotroph (adj. **photoautotrophic**): The systems that can derive energy from the sun and synthesize most of their nutrition from simple inorganic molecules. Photoautotrophs evolved from chemoautotrophs. This great advance gave living systems independence from the environment. *See* PHOTOTROPH.

photoautotrophic: *See* PHOTOAUTOTROPH.

photobleaching: Photochemical reaction of fluorophores, light and oxygen that results in the intensity of the fluorescence emission declining with time.

photochemical reaction centre: The part of a photosystem that converts light energy into chemical energy.

photochemical reaction: A chemical reaction that necessitates sunlight.

photoinhibition: The slow down or discontinuing of a plant process by light (e.g. the germination of some seeds).

photomorphogenesis: In developmental biology, photomorphogenesis is light-mediated development, where plant growth patterns respond to the light spectrum. Light regulates almost all processes of growth and development in plants. For example, when a seedling gets exposed to light, the Cop1 protein is reduced in the nucleus and photosynthesis is initiated – the seedling becomes green; cryptochromes may interact with photoreceptor proteins, which can recognize blue light; thus, the interaction can 'switch off' the Cop1 protein. In studies focused on the roles of lncRNAs in response to light, many light-responsive long non-coding natural antisense transcripts (lncNATs) related to histone modifications have been identified.

photon: A single unit of light energy and other electromagnetic radiation. Photons empower photosynthesis in plant cells. Within the chloroplast's thylakoid membranes, pigments – including chlorophyll – absorb photons, initiating the light-dependent reactions of photosynthesis.

photoperiod: The optimal length or period of illumination in a 24-h period necessitated for the growth and maturation of a plant. Species have been categorized according to their day length requirements as short-day, long-day, intermediate-day or day-neutral. The photoperiod is separate from photosynthesis.

photoperiod pathway in flowering: Serves as a conduit for a large variety of environmental parameters that convert external information and integrate it into *FLOWERING LOCUS T* (*FT*) signal. These environmental signals merge to control the *FT* expression through numbers of transcription factors.

photoperiodic cycle: The cycling of day and night (light and darkness).

photoperiodic flowering: The flowering which occurs in response to the photoperiodic cycle.

photoperiodism in regional adaptation of crops: Plants measure the duration of metabolic activity to promote rapid growth in long days. Photoperiodism, the ability to measure day length, is critical for fitness and survival because of its influence on plant physiology and behaviour. The timing of numerous developmental processes is controlled by day length, including flowering, growth, dormancy and senescence. C.R. Buckley and M.J. Haydon (2024) support the view that photoperiodism has important implications for the regional adaptation of crops and migration of plant populations in response to climate change.

photoperiodism: The answer of a plant to periodic, often rhythmic, changes in either the intensity of light or the relative length of day.

photophosphorylation: The conversion of ADP into ATP is coupled to the light-dependent transfer of electrons in photosynthetic cells. The phosphorylation process occurs only in photosynthetic cells, which contain light-trapping pigments such as chlorophyll. The coupling of photosystems I and II with an electron transfer chain that transfers electrons from water (which is oxidized to form O_2) to $NADP^+$ (which is reduced to form NADPH). The transfer of electrons between photosystem II and photosystem I releases energy, which is preserved in the form of a transmembrane proton gradient and expended to synthesize ATP.

photoprotective heat dissipation: The significant input of photosystem II (PSII) to light-driven chemical energy renders PSII complex susceptible to photodamage. This light-induced damage, in general, occurs when the photons impinging on the leaf surface exceed the amount consumed by light-harvesting complex (LHC) of photosystems. To diminish such photodamage, a photoprotective machinery involving accessory photosynthetic pigments (xanthophylls, i.e. carotenoids) exists in photosystems, which disperses excess excitations towards heat dissipation rather than the photochemical reaction. Photoprotective heat dissipation does not keep on diverting energy from normal photosynthesis when there is no excess light. This characteristic should be of great value in improving photosynthetic efficiency in food crops.

photoreactivation: A light-induced reversal of ultraviolet light causing injury to cells. A large proportion of damaged cells are removed through light-activated enzyme photolyase which cleaves the thymidine dimers.

photoreceptor: A pigment that captivates light and is used in various plant metabolic processes that need light.

photoreceptor cell: A cell, functionally part of the nervous system, that reacts to the presence of light. It usually contains a pigment, called receptor, that undergoes a chemical change when light is absorbed. This chemical change kindles electrical fluctuations in the photoreceptor that, when passed along and processed by other neurons, form the basis of vision.

photorespiration: Photorespiration restricts plant carbon fixation by releasing CO_2 and using cellular resources to recycle the product of RuBisCO oxygenation, 2-phosphoglycolate, back into the Calvin cycle. This process confines agricultural productivity by dispersing energy and releasing CO_2. Photorespiration results in the loss of about one-third of the carbohydrate formed in C_3 plants. This physiological process can be viewed as a necessary evil, as this cyclic mechanism (known as C_2 CYCLE) is able to recuperate 75% of the carbon lost through RuBP oxidation. It also retains levels of toxic metabolite 2-phosphoglycolate produced by RuBisCO oxygenation reaction to a minimum and performs a role in N-metabolism. Rate of photorespiration upsurges with a rise in temperature in warm environments. Through removal or downregulation of an enzyme in the C_2 pathway, the photorespiratory C_2 metabolism downstream of RuBisCO can be hindered, but this will obstruct carbon recycling back to the C_3 cycle. Modelling studies on synthetic bypass routes of photorespiration are supportive in enhancing our understanding of photorespiration and dropping its impact on agricultural productivity. Photorespiration has been a major topic in plant science and researchers have employed multiple strategies in attempts to reduce photorespiration in C3 plants to improve yield. Yield is taking a systems approach to boosting yield in C3 plants with the goal of reducing photorespiration using yield targets such as C3003 from our 'Smart Carbon Grid' discovery platform.

photosensitizers: Molecules which have effective conversions to the triplet state after absorption of a photon. These triplet-state molecules either decrease other molecules to produce free radicals or they excite O_2, normally a triplet, to the excited singlet state. Singlet oxygen is very reactive and can cause significant cellular damage.

photosynthate: A product of photosynthesis.

photosynthesis: Refers to the synthesis of bioorganic compounds (molecules) using light energy as the power source. The synthesis of one molecule of hexose, glucose (a hexose sugar) and six oxygen molecules from six carbon dioxide and six water occurs via a complicated, multistep process involving reactions that occur both in the light (light reactions) and in the dark (dark reactions). In eukaryotic cells, the photosynthetic machinery essential to arrest light energy and subsequently exploit it. Photosynthetic machinery is contained in plant cell organelles called chloroplasts, which contain the molecule, called CHLOROPHYLL, that initially captures light energy. Chlorophyll appears green. Green plants manufacture carbohydrates from carbon dioxide and water, which are used as a hydrogen source. The synthesis reaction, which is light driven, liberates oxygen in this process. Other organisms use this oxygen to sustain life. From initial carbohydrates, plants subsequently also form other compounds (e.g. fatty acids). Plants are not the only users of photosynthesis technology. Other organisms such as green sulfur bacteria and purple bacteria also carry out photosynthesis, but they use other compounds besides water as a hydrogen source. All plants, except certain parasites, can perform photosynthesis. Virtually all other forms of life are directly or indirectly dependent on green plants for food. Photosynthesis is thus the foundation for almost all life on Earth. The light energy from sun is converted into chemical energy (i.e. synthesis of carbohydrates from CO_2 and H_2O). Photosynthetic organisms such as green plants capture solar energy and form ATP and NADPH to serve as energy sources for the synthesis of carbohydrates from CO_2 and H_2O, simultaneously they release O_2 into the atmosphere. Photosynthesis also ensues in a variety of bacteria and algae. The overall equation for photosynthesis in the presence of light is $6CO_2 + 6H_2O \rightarrow 6O_2 + C_6H_{12}O_6$.

photosynthetic: An organism that converts light energy to chemical energy in the form of sugars.

photosynthetic efficiency: Chlorophyll pigment is indispensable in thylakoid reactions of photosynthesis. Its concentration is regularly used as an index to evaluate the photosynthetic efficiency; more chlorophyll means more production of chemical energy (ATP and NADPH), which results in increased rate of photosynthesis. Despite the production of chemical energy, the efficiency, at which carbon-assimilating enzymes fix CO_2 and utilize chemical energy, helps significantly in evaluating the rate of photosynthesis. The photosynthetic efficiency is dependent on the wavelength of the light absorbed. Photosynthetically active radiation (400–700 nm) constitutes only 45% of the actual daylight. Therefore, the maximum theoretical efficiency of the photosynthesis process is approximately 11%.

photosynthetic efficiency, factors affecting: Photosynthesis is generally considered an extremely inefficient process. The biochemical events impacting photosynthetic efficiency in chloroplast are understood to such an extent that we can now translate light energy assimilation into carbon gain. In simple sense, the efficiency of photosynthesis signifies a balance between the energy offered for plant growth and the energy exploited in protecting the photosynthetic machinery against photo damage. Three constraints, namely, PHOTORESPIRATION, PHOTOPROTECTIVE HEAT DISSIPATION and PLANT CANOPY STRUCTURE influence the photosynthetic process in plants. These constraints have helped in identification of various physiological parameters that could be manipulated for crop enhancement.

photosynthetic efficiency, manipulation of genetic regulation of: Various genes/**quantitative trait loci (QTL)** from different plant species have displayed their role in refining photosynthetic efficiency. Modulation of Regulators of Complement Activation (RCA) genes could enhance photosynthetic rate and plant growth in soybean and other plants. The important major QTL associated with photosynthesis-related traits and candidate RCA genes can be effectively used in breeding programmes for photosynthesis enhancement. The markers linked to the desirable QTL could be further tested for **MAS** of genotypes with improved photosynthetic efficiency. Genetic regulation of photosynthetic-efficiency-related genes/QTL can be a gifted approach for crop improvement.

photosynthetic efficiency, strategies for improving: The possibilities for enhancing productive biomass and harvestable grain occur, but barriers (such as chlorophyll retention, radiation use, RuBisCO efficiency) forced by photosynthetic limits require to be overcome to enhance crop productivity. Even a trivial gain in the rate of net photosynthesis can be decoded into huge improvements in biomass and, hence, yield. The strategies used in such crop improvement programmes should focus on the introduction of stay-green trait, engineering of light-harvesting antenna size, modification of photorespiratory process or introduction of CARBON-CONCENTRATING MECHANISMS (CCM), exploitation of natural variation in RuBisCO and manipulation of RuBisCO biogenesis or its catalytic activity through upgraded thermotolerance of RuBisCO activase.

photosynthetic electron transfer: Light-driven reactions in photosynthesis in which electrons travel along the electron transport chain in the thylakoid membrane, producing ATP and NADPH.

photosynthetic improvement via phosphorus-use efficiency: Photosynthetic efficiency is now one of the few remaining key traits available to enhance crop yield potential further. The photosynthetic process is dependent on nutrient phosphorus (P). The genetic bases of photosynthesis and P-USE EFFICIENCY (PUE) in relation to yield have been investigated in several crop plants. QTL for both traits have been detected in soybean and three major QTL, viz. *q14-2*, *q15-2* and *q19-2*, have been observed to be related to both traits. A photosynthesis-specific QTL cluster, *q12-1*, has been detected under both normal and low-P conditions, suggesting that genes responsible for this region were not impacted by low-P stress and could be used in crop breeding programmes aimed at leveraging photosynthetic efficiency.

photosynthetic phosphorylation: Refers to the formation of ATP from the starting compounds ADP and inorganic phosphate (Pi), which is coupled to light-dependent electron movement in photosynthetic organisms. *Aka* PHOTOPHOSPHORYLATION.

photosystem: A cluster of chlorophyll and other molecules in a thylakoid that utilize the energy of light for photosynthesis, transmit the energy to photochemical reaction centres and interchange the excited electrons in a controlled manner to yield usable biochemical energy. Each photosystem encompasses hundreds of chlorophyll and carotenoid molecules functioning as antennae, while only a few chlorophyll molecules are working in the reaction centres.

phototroph: An organism that can use light energy to synthesize organic compounds such as carbohydrates. Considering the energy source, organisms are classified as phototrophs or chemotrophs. Phototrophs use light as their primary energy source. Photosynthetic organisms are phototrophs such as plants and blue-green algae. Chemotrophs depend on oxidation-reduction reactions of inorganic or organic compounds for their energy requirements.

phototropism: The specific movement (often termed bending) of an entire plant or just a plant part towards a light source.

phQTLdna: Refers to a QTL influencing a phenotype (ph), which can be demonstrated to be due to DNA sequence (dna).

phQTLepi: Refers to a QTL influencing a phenotype (ph), which can be demonstrated to be due to chromatin (epi).

phragmoplast: Structure made of microtubules and actin filaments that is formed in the prospective plane of division of a plant cell and directs the formation of cell plate.

phs1 **gene:** Pairing, synapsis and recombination are at the base of precise chromosome segregation during meiosis. Homologous pairing in maize is controlled by *phs1* gene. A mutation of this locus results in irregular departure of chromosomes at anaphase of meiosis and thereby loss of viability in pollen. In *phs1* mutant, synapsis is uncoupled from recombination and pairing. The protein encoded by the *phs1* gene possibly acts in a multistep process to coordinate pairing, recombination and synapsis.

phyletic evolution: The evolution by which a race or line is increasingly altered from its ancestral form without branching or separating into related parts. Thus, change occurs sequentially in a single line of ascent. *See* PHYLETIC GRADUALISM, PHYLETIC SUCCESSION *and* PHYLETIC SPECIATION.

phyletic gradualism: A theoretical model of evolution in which species change slowly at a more or less constant rate through time into other species. *See* PUNCTUATED EQUILIBRIUM.

phyletic speciation: The process of speciation caused by the gradual change in the genetic constitution of a population without the population splitting into demes and without any increase in the number of species produced by that population at any one time.

phyletic succession: The gradual replacement of one species by another without branching.

phyllody: The modification of floral organs into leaflike structures.

phylloplane: Using surface of a leaf as a habitat for organisms.

phylloplane-competent: Term applies to epiphytic microorganisms modified to live on the leaf surface.

phyllopodium: In ferns, a short outgrowth of the stem on which the frond is borne and which remains attached to the rhizome after the frond has been shed.

phyllosphere: The above ground surface of plants as a habitat for epiphytic microorganisms.

phyllotaxy: Relating to the arrangement of leaves on a stem with respect to one another.

phylogenetic: Pertaining to or based on phylogeny.

phylogenetic analysis: Determination of how the family might have been derived during evolution. Multiple sequence alignment methods are applied to a set of related sequences before a phylogenetic analysis can be performed. Closely related organisms have similar sequences and more distantly related organisms have more dissimilar sequences. Given a set of sequences, it should be possible to reconstruct the evolutionary relationships among genes and among organisms. This involves creating a branching structure, termed a phylogeny or tree that illustrates relationships between the sequences. Hence, a phylogenetic analysis of a family of related nucleic acids or protein sequences is a determination of how the family might have been derived during evolution.

phylogenetic analysis method: Phylogenetic analysis in general is a four-step method: (i) alignment strategy, (ii) determination of substitution model, (iii) tree building and (iv) tree evaluation.

phylogenetic anchor: The use of the phylogeny of a gene to deduce the organismal source of a small piece of DNA. It is used in METAGENOMICS.

phylogenetic development: It is the historical transformation of the form of life, starting with a simple form that might have been the first to appear and leading to complex ones that were the last to appear during evolution. *Aka* PHYLOGENY.

phylogenetic methods, classification of: There are two groups of methods of phylogenetic analysis – (i) DISTANCE METHODS and (ii) CHARACTER METHODS. *See details as separate terms.*

phylogenetic profiles: Phylogenetic profiles describe the pattern of presence or absence of a particular protein across a set of organisms whose genomes have been sequenced. If two proteins have the same phylogenetic profile (i.e. the same pattern of presence or absence in all surveyed genomes), it is inferred that the two proteins have a functional link. If there are 'n' fully sequenced genomes, there are up to '2^n' phylogenetic profiles.

phylogenetic studies: Evolutionary relationships within and between species, genera or larger taxonomic groupings. Molecular markers play an important role in phylogenetic studies for grouping individuals. Phylogenetic relationships involve studying similarities and differences among taxa using genetic markers. Phylogenetic trees have been established for many plant species on the basis of morphological markers, protein (isozyme) markers, chromosome homology and DNA markers.

phylogenetic trait analysis: A comparative method in which one constructs a phylogenetic tree, plots character states (traits) on the tree and infers transitions in character states form their position on the tree. Geographical locations of taxa can be plotted on to the tree to infer the location of ancestors.

phylogenetic tree: A diagram indicating the supposed sequence of historical relationship linking together species or high taxa. Phylogenetic tree is composed of nodes and branches representing *phylogenetic* relationships. The phylogenetic principles are based on the fact that all life forms have a common origin somewhere in the past, from which they evolved into the present-day species. Closely related species share a more recent common ancestor than distantly related species. Hence, the relationships between species can be represented by a PHYLOGENETIC TREE. In phylogenetic tree, the nodes represent taxonomic units, while the branches connecting them reflect their relationships in terms of descent. The TOPOLOGY is the pattern of branches found in a tree. The BRANCH LENGTH is commonly used to indicate some form of evolutionary distance represented by that branch. The actual, still existing taxonomic units or OPERATIONAL TAXONOMIC UNITS are represented by nodes on the tips of the branches, called EXTERNAL NODES. The other nodes are called INTERNAL NODES. *Aka* PHYLOGENY.

phylogenetics: Interpreting divergence from the perspective of evolutionary history, specifically in terms of ancestor-descendant (phylogenetic) relationships. Phylogenetics has its origin in the work of Willi Hennig, a German entomologist who first popularized cladistics. A phylogeny, synonymous with a phylogenetic tree, is a genealogical history of a taxon described in terms of ancestor-descendant relationships determined by proximity measures or by the distribution of various character states. Now there is a general consensus that classifications should be congruent with phylogenies whenever possible. In systematics, classifications are based solely on hierarchical branching patterns. A phylogenetic classification is developed by grouping taxa according to their patterns of phylogenetic divergence. Phylogenetics differs from phenetics especially by its character-based approaches, in which evolutionary change in each trait is assessed individually and phylogenetic relationships are estimated by the distribution of individual character states.

phylogenic development: Phylogenic development is the historical transformation of the form of life, starting with a simple form that might have been the first to appear and leading to complex ones that were the last to appear during evolution. *Aka* PHYLOGENY.

phylogenies, evaluation of: Even when alignments have been carefully crafted and conditions or scoring methods are carefully chosen, it is possible to generate a meaningless tree. Input data may simply not be robust (i.e. the data may itself contain more noise than an evolutionary signal). There are two ways to evaluate phylogenies: JUMBLING SEQUENCE ADDITION ORDER, BOOTSTRAPPING and JACKKNIFE REPLICATES. *See details as separate terms.*

phylogenomics: The amalgamation of genome analysis and evolutionary reconstructions into a single composite method. The term is also applied to the use of phylogenetic trees to forecast gene function.

phylogeny: A branching diagram or a tree showing the evolutionary relationships among various biological species or other entities based upon similarities and differences in their physical or genetic characteristics. All life on Earth is part of a single phylogenetic tree, indicating common ancestry.

phylogeny studies, tools for: A large number of computer tools are available for phylogeny studies. (i) Phylogeny.fr is a simple-to-use web service dedicated to reconstructing and analysing phylogenetic relationships between molecular sequences. It includes multiple alignments, phylogeny, a tree viewer and utility programmes. (ii) PHYML is a simple, fast and accurate algorithm to estimate maximum-likelihood phylogenies from DNA and protein sequences. (iii) ProtTest estimates the empirical model of amino acid substitution. (iv) Phylemon is a suite of web-based tools for molecular evolution, phylogenetics and phylogenomics. (v) POWER (PhylOgenetic Web

Repeater) allows users to carry out phylogenetic analysis on most programmes of the PHYLIP package repeatedly. POWER provides two pipelines to process the analysis. One of them includes MULTIPLE SEQUENCE ALIGNMENT (MSA) at the beginning of the pipeline, whereas the other begins phylogenetic analysis with aligned sequence. (vi) CVTree (Composition Vector) constructs whole-genome-based phylogenetic trees without sequence alignment by using a Composition Vector (CV) approach. It was first developed to infer the evolutionary relatedness of microbial organisms and then successfully applied to viruses, chloroplasts and fungi. (vii) Ribosomal RNA Analysis is the Ribosomal Database Project II. (viii) Ridom does ribosomal RNA analysis for clinically relevant bacteria. (ix) Rifle compares the restriction patterns of 16S rDNA.

phylogeny, assumptions of: (i) All sequences are homologous. (ii) No duplicate sequences are present. (iii) In each column, amino acid residues are homologous. (iv) The alignment is optimal, with minimum gaps. (v) No back mutation has occurred (some methods take this into account). (vi) All sequences are of the same length.

phylogeography: Interpretation of population history from the genealogy that attaches genes sampled from different geographic locations. The study is frequently based on genealogies inferred from mitochondrial DNA.

phylotype: The phylogenetic type of an uncultured organism as interpreted from investigation of its ribosomal RNA sequence.

phylum (pl. phyla): A taxonomic rank above 'kingdom' and below 'class'; it is equivalent to the botanical term 'division'. The segmentation of the animal kingdom into about 30 major groups, collectively known as PHYLA. The members of each phylum share the same basic structure and organization. For instance, fish, birds and human beings belong to one phylum – the Chordata – because all have spinal cords. *See* TAXON.

physical containment: The use of physical barriers to prevent the escape of genetically engineered organisms from the laboratory into the environment. Precautions employed include the maintenance of negative air pressure within the laboratory and the use of fume hoods with HEPA filters on their exhausts.

physical delivery for DNA-mediated gene transfer (DMGT): Physical delivery methods of gene transfer have been successfully used in the transfer of genes across reproductive barriers. Direct gene transfer, integration and expression of exogenous DNA have been reported to be very successful in producing transgenics in crops such as maize, rye, rice and wheat. The direct gene transfer methods include the physiochemical method, electroporation, microinjection (embryo microinjection, oocyte microinjection, fertilized egg microinjection), microprojectiles, liposome-mediated gene transfer, laser beam and pollen-aided direct gene transfer. Physiochemical and electroporation methods of gene transfer are suited to those species where regeneration of the whole plant from the protoplast is possible. *Aka* VECTORLESS *or* DIRECT GENE TRANSFER SYSTEMS.

physical interaction model of paramutation: In this model, the paramutagenic and paramutable repeats physically interact by means of a pairing protein complex. Subsequently, the epigenetic state of the paramutagenic repeats is transferred to the PARAMUTABLE REPEATS, changing the PARAMUTABLE REPEATS into PARAMUTAGENIC REPEATS.

physical map: The physical maps of DNA segments, chromosomal segments, whole chromosomes or whole genomes are based on physical distances. These physical distances are measured in terms of relative physical lengths or in terms of kilobase pairs (kbp) between genes or molecular markers or in terms of relative distances of these genes/markers from the centromeres. In a physical map, genes are depicted in the same order as they occur in the chromosome. Distance in physical maps is measured in base pairs. For the human genome, the lowest-resolution physical map is the banding patterns on the 24 different chromosomes, the highest-resolution map would be the complete nucleotide sequence of the chromosomes. These features of chromosomes are directly concerned with the genetic functions or production of a trait. Physical maps be defined cytologically or on the basis of molecular markers (i.e. probes). Cytologically based techniques include differential chromosome staining and somatic cell hybridization. Staining of chromosomes with Giemsa banding method provides landmarks for the markers in the form of light and dark bands. True physical maps can be made by the use of pulse field gel electrophoresis (PFGE). Megabase DNA fragments can be separated by this technique. It contrasts a GENETIC MAP, which is resolved using classical genetics. The two types of physical mapping are HETERODUPLEX MAPPING and RESTRICTION MAPPING. *See details as separate terms. See* GENETIC MAP.

physical mapping: The technologies to produce a physical map. Physical map is the linear arrangement of genes or other sequences on a chromosome (or part of it) as determined by techniques other than genetic recombination. Map distances are expressed as numbers of nucleotides (or nucleotide pairs) between identifiable genomic sites (e.g. markers, genes).

physical methods of gene transfer: These methods include MICROINJECTION, MACROINJECTION, MICROPROJECTILES (*aka* BIOLISTICS *or* GENE GUN), LASER BEAM and PROTOPLAST FUSION. *See details as separate terms.*

physical mutagens: Include ionizing radiations and non-ionizing radiations. Examples of ionizing radiations include heat or light from the sun, microwaves from an oven, X-rays from an X-ray tube and gamma rays from radioactive elements and those of non-ionizing radiations include UV light. Examples of non-ionizing radiations include Radiofrequency (RF) radiation used in many broadcast and communications applications, infrared radiation used in heat lamps and **ultraviolet (UV)** radiation from the sun and tanning beds. Radiations can change the genetic material.

physiochemical methods of gene transfer: These methods include DNA transfer by use of CALCIUM PHOSPHATE METHOD, DEAE-DEXTRAN, POLYCATION-DMSO, POLYETHYLENE GLYCOL (PEG), LIPOSOMES and PEPTIDES. These vectorless methods have been successfully tried in a wide range of plants and animals. *See details as separate terms.*

physiologic specialization: The presence of a number of races or forms of one species of pathogen based on their pathogenicity to different cultivars of a host species.

physiological changes in aged seeds: (i) Changes in moisture content of seeds, (ii) Decrease in germination percentage, (iii) Decrease in seedling length with the decrease in moisture content, (iv) Defective and stubby roots, (v) Loss of vigour of seedlings, (vi) Deterioration of seedcoat, (vii) Reduction in fruit size, and (viii) Some mutations and chromosomal aberrations.

physiological races: Pathogens of the same species with identical or similar morphology but diverging in certain cultural, physiological, biochemical, pathological or other characteristics (e.g. biotypes of a species that vary in the ability to infect particular varieties of the susceptible plant species).

physiological resistance: A type of resistance due to physiological or protoplasmical characters.

physiological suppression: A defect in one chemical pathway due to a mutation is circumvented by another mutation in another metabolic pathway. Thus, when one mutation shuts one pathway, another mutation opens another pathway leading to the same result.

physiology: The branch of biology dealing with the purposes and actions of living organisms and their parts, including all physical and chemical processes. The materials of physiology include all life (animals, plants, microorganisms and viruses).

phytase: It is an enzyme that catalyses the breakdown of phytin, the source of inorganic phosphorus in seed metabolism. An example of a heterologous enzyme produced in plants is phytase. Phytase is encoded by a gene from the fungus *Aspergillus niger*. It enhances phosphorus utilization by monogastric animals like chickens from their feed to the extent that phosphate supplements in feed become unnecessary. Transgenic tobacco seeds expressing the gene encoding phytase when fed to broiler chickens produced a gain in body weight.

phytoalexin: Substance manufactured by higher plants in response to a number of chemical, physical and biological provocations that impede the growth of certain microorganisms. Biological stimuli include the presence of infectious agents (e.g. fungus, bacteria) or their products. Phytoalexins possess antimicrobial (i.e. fungus-killing, bacteria-killing) properties, so they can help plants protect themselves against those microorganisms.

phytochemicals: Refer to certain biologically active chemical compounds that occur in fruits, vegetables, grains, herbs, flowers, bark and so on. Phytochemicals function to repel or control insects, prevent plant diseases and control fungi and adjacent weeds. Phytochemicals also sometimes have deliberate beneficial health effects on the animals (e.g. humans) that eat the plant (portions) containing such phytochemicals.

phytochemistry: The study of phytochemicals, which are chemicals derived from plants.

phytochrome: A photoreversible pigment that exists in every major taxonomic group of plants. It directs the course of plant growth, development and differentiation in a plant. The response is independent of photosynthesis; for example, in the photoperiod (length of light period), response exists in two interchangeable forms with respect to absorption, a red (Pr) and a far-red (Pr-r) form. While Pr phytochrome is accessible to orange-red light (600–680 nm) and inhibits flowering, Pf-r phytochrome is accessible to far-red light (700–760 nm) and induces flowering.

phytoestrogens: Compounds owning molecular structures somewhat similar to that of oestrogen and that naturally occur in all plants on earth. Consequently, all vegetable, fruit, cereal and legume plants comprise at least one form of 'phytoestrogen'. For example, flavones and flavanols are advantageous phytoestrogens (mostly red- and yellow-coloured pigments) found in coloured vegetables and fruits (red grapes, yellow grapefruit, oranges, etc.).

phytogenetics: Relating to studies on the inheritance, origin and evolution of plants. *Aka* PLANT GENETICS.

phytohormone: Plant hormone that control or regulates cell enlargement, cell division, root initiation and growth rate, germination, growth, metabolism or other physiological activities. The impact of phytochromes on the plant could be direct, via plant growth promotion or indirect, via improving plant nutrition for better development of the roots.

phytopathogenic: Capable of producing disease in plants.

phytopathology: The science of diagnosing and managing plant diseases. *See* PLANT PATHOLOGY.

phytoplasma: A plant-parasitic prokaryote, without cell wall, found in phloem tissue. It cannot yet be grown on artificial nutrient media. *Contrasts* MYCOPLASMA-LIKE ORGANISM (MLO).

phytoremediation: Refers to the use of specific green plants to remove, contain or render harmless environmental contaminants or pollutants from either soil (e.g. polluted fields) or water resources (e.g. polluted lakes). This includes the use of vegetation for *in situ* treatment of water, sediments, soils and air. For example, the Brazil water hyacinth (*Eichhornia crassipes*) naturally gathers in its tissue toxic metals such as lead, arsenic, cadmium, mercury, nickel and copper, and it has been utilized as a 'biofilter' (e.g. in India). Insertion of the *Escherichia coliform* bacteria gene *gsh 11* into the plant known as *Indian mustard* enables that plant to accumulate 40–90% higher amounts of cadmium (from cadmium-tainted soil) in its tissues than before. Specially selected or engineered plants can be used for extraction of toxic metals from soil or water, including removal of radioactive elements, removal of toxic organic compounds and, if possible, their mineralization.

phytosanitary certificate: An official paper that certifies that plant material has been inspected and found to be free of certain pathogens.

phytosterols: A group of phytochemicals (i.e. solid alcohols consisting of ring-structured molecules) that exist in seeds produced by certain plants (e.g. the soybean plant *Glycine max* L.). Human consumption of certain phytosterols can help to avert certain types of cancers and can assist in lowering total serum cholesterol and LOW-DENSITY LIPOPROTEINS (LDLP) levels, thereby reducing the risk of CORONARY HEART DISEASE (CHD). Some phytosterols (e.g. campesterol, stigmasterol, beta-sitosterol) hinder absorption of dietary cholesterol by the intestines and lower the body's recovery and recycle of cholesterol-containing bile salts, producing more cholesterol to be excreted from the body than before. In 2000, J. Judd fed phytosterols extracted from soybeans (*Glycine max* L.) to human volunteers who were consuming a 'low-fat' diet. Their total blood serum cholesterol and LDLP levels declined by >10% in a short time.

phytotoxic: Usually used to describe chemicals that are harmful to plants.

PI-derived artificial chromosomes (PACs): These are vectors for cloning large DNA fragments and incorporate features of both PI and F-factor systems and can be maintained in *Escherichia coli*. These vectors can accept inserts ranging from 100 to 300 kilobase pairs (kbp). Yeast artificial chromosomes can accept inserts ranging from 100 to 2000 kbp, whereas MAMMALIAN ARTIFICIAL CHROMOSOME (MAC) can clone fragments of more than 1000 kbp.

pico technology: Next smaller step after 'nanotechnology'.

picogram (pg): 10^{-12} g or 3.527×10^{-14} oz.

picorna virus: The smallest known RNA virus is known as picorna virus. These viruses result in common cold and hepatitis A in humans, one type of hoof and mouth disease in animals and at least one disease in maize. In 1994, A. Dasgupta discovered a cellular molecule within ordinary baker's yeast that thwarts picorna virus reproduction. This discovery may result in the creation of a treatment to cure one or more of the above-mentioned diseases after infection has begun.

pie diagram: A diagram showing the proportions of subjects in different groups or with different mutually exclusive characteristics by means of segments of a circular pie.

pigeon pea: Pigeon peas are grown as a commercial crop in India. Pigeon pea varieties are categorized into 'tall' and 'dwarf'. New hybrids are similar in height to Southern peas and beans. Pigeon peas must be grown as annuals in most parts of the USA since plants are killed by freezing temperatures. The plant is a vigorous, drought-tolerant legume which bears large pods that are easily harvested. This pea is a heavy bearer, yielding sweet-tasting peas. It is appropriate for early planting and will succeed under hot growing conditions.

PiggyBac (PB) transposon: During transposition, the PB transposase recognizes transposon-specific INVERTED TERMINAL REPEAT SEQUENCES (ITRs) located on both ends of the transposon vector and efficiently moves the contents from the original sites and efficiently integrates them into TTAA chromosomal sites. The powerful activity of the PB transposon system enables genes of interest between the two ITRs in the PB vector to be easily mobilized into target genomes. PB transposon is becoming a highly useful vector for genetic engineering of a wide variety of species, particularly insects.

pigment: Coloured molecule. A pigment that absorbs specific wavelengths of light. Chlorophyll absorbs blue and red light and reflects green light, so it appears green.

piliform: Having the shape of a pointed or close-fitting cap, a pileus.

pilus: A cytoplasmic bridge connecting bacteria exposed to UV irradiation, owing to the activation of cellular repair processes by visible light.

ping-pong mechanisms: Group-transfer reaction in which one or more products are released before all substrates get bound to the enzyme. Ping-pong bisubstrate reaction is represented by the equation which shows that as substrate A binds to the enzyme, enzyme-substrate complex EA forms. See Figure below. At this point, the intermediate state, E* forms. P is released from E*, then B binds to E*. B is converted to Q, which is released as the second product. E* becomes E and the process can be repeated. Often times, E* contains a fragment of the original substrate A. This fragment can alter the function of the enzyme, gets attached to substrate B or both. At the end of the reaction the enzyme must be found in its original form.

$$E \xrightarrow{A\,\downarrow} EA \rightleftharpoons E^*P \xrightarrow{P\,\uparrow} E^* \xrightarrow{B\,\downarrow} E^*B \rightleftharpoons EQ \xrightarrow{Q\,\uparrow} E$$

pink pigmented facultative methylotroph (PPFM): A type of bacteria that is naturally present in virtually all plants. PPFM produces cytokinin, which aids the cell division (growth) process in plants. PPFM also produces a chemical substance similar to vitamin B-12. In 1996, J. Polacco discovered that impregnation of aged seeds with PPFM improved the germination (sprouting) rate of those aged seeds.

pinocytosis: Type of endocytosis in which soluble materials are derived from the environment and incorporated into vesicles for digestion. Literally, CELL DRINKING. *See* FLUID-PHASE ENDOCYTOSIS *and* PHAGOCYTOSIS.

pipeline: Term used to describe an end-to-end process all the way from receipt of a sample through to getting validated data at the end.

piRNA-mediated transposon silencing: A few very large genes generate the small RNAs that silence parasitic DNA elements. RNAi pathway provides a crucial defence mechanism against transposons. A few trigger loci generate piRNAs, which target transposons elsewhere in the genome and prevent their transcription. Complementary binding of Aubergine- and Ago3-associated piRNA sequences results in their amplification, ensuring efficient silencing of transposons.

pistil: The ovule-bearing organ of the plant comprising of the ovary and its appendages (e.g. style, stigma).

pistillate flower: A flower possessing one or more pistils but no fertile stamens. Sometimes called a FEMALE FLOWER. *Contrasts* STAMINATE FLOWER.

pistillode: A sterile or rudimentary pistil that may appear in a STAMINATE FLOWER.

pistillody: The development of flowers pistils but without stamens.

pith: The parenchymatous tissue inhabiting the centre of the stem. The pith (or medulla) is composed of soft, spongy parenchyma cells with spaces between them. The pith is surrounded by a ring of vascular bundles, containing xylem and phloem. Its primary function is storing water and nutrients and transporting them throughout the plant.

Piwi proteins: The Piwi family proteins belong to the ARGONAUTE superfamily. The Piwi subfamily members bind to piRNAs. Tens of thousands of piRNA species, typically 24–32 nucleotide in length, have been found in mammals, zebrafish and *Drosophila*, but not in plants. A considerable portion of piRNAs maps to transposon-encoding regions. The expression of Piwi proteins and piRNAs is mainly restricted to the germline. Piwi-piRNA pathway has been implicated in a variety of germline functions. In mice, two Piwi proteins, MILI and MIWI2, are involved in piRNA generation in foetal male gonads and these piRNAs play important roles in silencing retrotransposons via DNA methylation. In male germ cells, *MILI*- or *MIWI2*-deficiency impairs *de novo* methylation of retrotransposons. *DNMT3L*-deficiency leaves the piRNA pathway largely intact. Piwi-piRNA pathway acts upstream of DNA methylation. piRNA-containing complexes are proposed to guide the *de novo* DNA-methylation machinery to transposon sequences. Piwi family proteins have been implicated in transposon control.

Piwi-interacting RNA (piRNA): The largest class of small non-coding RNA molecules expressed in animal cells; piRNAs are 26–34 nucleotides in length and form RNA-protein complexes through interactions with Piwi proteins; piRNAs differ from miRNAs and siRNA in both methods of biogenesis and function but are known to play a role in silencing retrotransposons in germ cells. Piwi and Aubergine bind RNAs that are predominantly antisense to transposons, whereas Ago3 complexes contain predominantly sense piRNAs.

P_L, P_R: Promoter leftward and promoter rightward are two strong promoters from bacteriophage λ. Transcription from both these promoters is under the control of *cI* gene.

plan of nature: The philosophical theory stating that nature is organized according to a plan. It has been influential in classification and is a type of idealism.

planosome: A supernumerary chromosome resulting from non-disjunction of a bivalent in meiosis.

plant: A living organism that grows from the ground using sunlight and water as food. Usually has leaves, roots. flowers and a stem. Plants are predominantly photosynthetic eukaryotes, forming the kingdom Plantae. Most of them are multicellular. Historically, the plant kingdom encompassed all living things that were not animals and included algae and fungi. All current definitions exclude the fungi and some of the algae. By one definition, plants form the clade Viridiplantae (Latin name for 'green plants') which consists of the green algae and the embryophyta or land plants. The latter include the hornworts, liverworts, mosses, lycophytes, ferns, conifers and other gymnosperms and the flowering plants. A definition based on genomes includes the Viridiplantae, along with the red algae and the glaucophytes, in the clade Archaeplastida.

plant airborne defence (PAD): Plants generate and release a wide array of VOLATILE ORGANIC COMPOUNDS (VOCs) when challenged by environmental stimuli such as biotic and abiotic stresses, which facilitate their reproduction, defence responses and plant-plant communication. According to S. Liu *et al.* (2023), once emitted, some VOCs can elicit defence signalling in neighbouring plants by interacting with specific receptor(s), a phenomenon referred to as airborne defence. *See* VOLATILE ORGANIC COMPOUNDS.

plant bioreactors: A plant which is designed to provide an effective environment for conversion of one material into another by appropriate biochemical reactions and this conversion is carried out by the action of biocatalysts like enzymes, microorganisms, cells of animals and plants or subcellular structures such as chloroplasts and mitochondria.

plant biotechnology research, current trends in: (i) Identification of genetically linked RGAs by BAC screening, (ii) Sequence analysis of a region surrounding the *Citrus tristeza* virus resistance gene, (iii) High efficiency transformation of *Penicillium nalgiovense*, (iv) Cloning and linkage mapping of resistance gene homologues in apple and (v) Male sterility through recombinant DNA technology. *See details as separate terms.*

plant breeder: A person or organization actively involved in the breeding and maintenance of varieties of plants, using a wide range of approaches of different scientific disciplines.

Plant Breeder's Rights (PBRs): The rights granted to a breeder under Protection of Plant Varieties and Farmers' Rights (PPV&FR) Act, 2001 is called the plant breeder's right. The PBR can be licensed, sold or inherited. It is applicable only in the country where it is established as per national law. This grant authorizes him to exclude others from producing, storing, offering for sale, commercializing, importing or exporting the propagating material of that variety for a period time within the jurisdictional limits of the country. The right of the owner of a variety may be analogous to copyright. The holder of PBR is free to inherent this right to successors or transfer or license it over to other parties. These rights, governed by Plant Breeder's Rights Acts stretch the plant breeder legal safety over the propagation of a cultivar and the exclusive rights to produce and to sell it, including the right to authorize others to produce and sell plants and reproductive material of a registered, deliberately bred variety. *Compare with* UPOV.

plant breeding: The purposeful manipulation of qualities in plants to create new varieties with a set of desired characteristics. The application of genetic principles and practices to the development of individuals or cultivars more suited to the needs of humans. It applies knowledge from agronomy, biochemistry, bioinformatics, biotechnology, botany, computer science, entomology, genetics, cytogenetics, epigenetics, molecular genetics, physiology, plant pathology or statistics.

plant canopy architecture: A mature, healthy plant normally possesses three or more layers of leaves. In those plants that have leaves roughly in a horizontal position, the topmost layer blocks most of the light at midday, whereas nearly 10% of light reaches the second layer and a meagre 1% is received by the layer below that. Thus, at least two-thirds of the energy blocked by the uppermost leaves remains under-utilized. A modification in plant architecture should enable the uppermost leaves to capture a smaller fraction of the light, thus permitting extra light to reach the leaves in the lower layers. This is possible when the uppermost leaves are vertical and the lowermost leaves are horizontal. For an erect leaf having a 75° angle with the horizontal one, the captured light energy per unit area would be just enough to saturate photosynthesis. Rest of the direct light would reach the leaves of lower layers of the canopy. Various models of canopy photosynthesis have been verified to see if a decrease in chlorophyll (Chl) content would result in an increase in the canopy photosynthesis in soybean. Various parameters have been used to test the relationship among leaf Chl content, leaf optical properties and biochemical capacity in 67 soybean accessions, along with a wild-type and Chl-deficient mutant. The canopy photosynthesis was conserved by reducing leaf chlorophyll and increasing more light interception by lower leaves of canopy, leading to a reduction in extra photon load on upper parts of the canopy and enrichment of the lower leaves of crop canopy with sufficient light to contribute to photosynthetic yield.

plant cell walls: Plants have walls of cellulose, as do green algae and certain protists, while fungi have cell walls of chitin.

plant cells: Plants are eukaryotes that are composed of several different types of cells, including parenchyma, sclerenchyma, collenchyma, xylem and phloem cells. Like other eukaryotes, plant cells have three main parts: the plasma membrane, the cytoplasm and the nucleus. Most plant cells also have a few unique structures, including chloroplasts, a central vacuole and a cell wall.

plant cloning vehicle: A vector used in gene manipulation to carry foreign DNA and insert it in an inheritable manner into plant cells. Possible vectors include the Ti plasmid of *Agrobacterium tumefaciens* and the DNA plant viruses.

plant communication: Plant communication encompasses communication using VOCs, electrical signalling and common mycorrhizal networks between plants and a host of other organisms such as soil microbes, other plants (of the same or other species), animals, insects and fungi. Plants communicate through a host of VOCs that can be separated into four broad categories, each the product of distinct chemical pathways: fatty acid derivatives, phenylpropanoids/benzenoids, amino acid derivatives and terpenoids. Scientists in Japan observed plants sending tiny, mist-like messages to each other through the air. They have seen visual evidence of plants talking to each other. That sweet, earthy smell that fills the air after cutting grass is actually a cacophony of screams in plant speak. For the first time, scientists have captured this fascinating conversation on camera. Masatsugu Toyota, a molecular biologist and the lead researcher from Saitama University found imaging technique a very powerful tool to prove that plants are very sensitive.

plant defence signalling: Pathogen recognition triggers energy-intensive defence systems. Successful defence depends on energy availability. K. Yamada and A. Mine in 2024 observe that sugar, especially glucose-6-phosphate (G6P), is critical in coordinating defence in *Arabidopsis*. Under sugar-sufficient conditions, phosphorylation levels of calcium-dependent protein kinase 5 (CPK5) are elevated by G6P-mediated suppression of protein phosphatases, enhancing defence responses before pathogen invasion. By monitoring sugar levels, plants determine defence levels and execute appropriate outputs against bacterial and fungal pathogens.

plant development: It is an overall term which refers to the various changes that occur in a plant during its life cycle.

plant disease: An abnormality in the structure and/or function of the host plant cells and/or tissue as a result of a continuous irritation caused by a pathogenic agent or an environmental factor. A disease is not static; it is a series of changes in the plant. All plants, to some extent, are subject to disease. Plant disease is the result of an infectious or biotic (a living component of an ecosystem) agent or a non-infectious or abiotic (non-living, physical and/or chemical component) factor.

plant diversification: Madagascar is a hotspot of biodiversity and unique species, both because of its long isolation from the mainland and its high rates of speciation. Landscape evolution has been hypothesized to contribute to plant diversification by creating barriers and habitat heterogeneity, providing opportunities for allopatric speciation. Using a process-explicit model, Y. Liu *et al.* (2024) tested this hypothesis by modelling topographic dynamics on Madagascar over the past 45 million years and found high plant species richness and diversification rates along the landward-shifting continental rift ESCARPMENT (a steep slope or long cliff that forms as a result of faulting or erosion and separates two relatively level areas having different elevations), with low north-south connectivity and high rates of local endemism. A relatively small-scale, transient landscape evolution can shape long-term biodiversity patterns. Habitat isolation and reconnection on a million-year timescale serves as an allopatric speciation pump creating the observed biodiversity.

plant DNA demethylation pathway: Active DNA demethylation involves the enzymatic removal of methylated cytosine. In plants, this process is initiated by a family of DNA glycosylases including DEMETER (DME), REPRESSOR OF SILENCING 1 (ROS1), DEMETER-LIKE 2 (DML2) and DEMETER-LIKE 3 (DML3).

plant DNA virus: A virus that contains DNA and infects plants. There are two groups of such viruses *Caulimoviruses* (with dsDNA) and *Geminivirus* (with ssDNA).

plant embryology: It is defined as the development and structure of microsporangium, megasporangium, male gametophyte, female gametophyte, fertilization, endosperm and embryo.

plant epigenome and genome evolution: Methylation of TRANSPOSABLE ELEMENTS (TEs) impacts gene expression. Methylation spreads from retroelements to flanking DNA. TE methylation may guide the resolution of polyploid events. Genic methylation is highly conserved over evolutionary time. With the beginning of high-throughput methods, it has now become a routine matter to quantify the genome-wide distribution of epigenetic marks and genome-wide patterns are providing further insights into evolutionary processes.

plant evolution: Evolution of eukaryotes from a presumed bacteria-like ancestor is considered one of the key events in evolutionary history. Eukaryotes reveal a distinct nucleus, organelles involved in energy metabolism (such as mitochondria and chloroplasts), extensive internal membranes and a cytoskeleton of protein fibres and filaments. Chloroplasts in green plants contain machinery for photosynthesis and algae originated as free-living bacteria related to the cyanobacteria whose chloroplastic DNA is more akin to free-living cyanobacteria DNA than to sequences from the plants in which the chloroplasts reside. The eukaryotic mitochondria providing machinery for ATP synthesis are endosymbionts like chloroplasts. Mitochondria were assimilated when aerobic eubacteria were engulfed by anaerobic host cells. As they conferred useful functions like aerobic respiration and photosynthesis, they were retained as endosymbionts. This must have transpired after the nucleus developed in the eukaryotic lineage. The origin of the eukaryotic nucleus is almost definitely autogenous and not a result of endosymbiosis. Mitochondria seem to have originated from an ancestor of the present-day purple photosynthetic bacteria that had lost its capacity for photosynthesis. All land plants most likely evolved from green algae or Chlorophyta. In the period before the Permian (the Carboniferous), the landscape was conquered by seedless ferns and their relatives. Vascular plants have been suggested to first appear in the Silurian (439–409 mya). After the Permian extinction, gymnosperms became more abundant. They evolved seeds and pollen. Angiosperms evolved from gymnosperms during the early Cretaceous about 140–125 mya. They further diversified and dispersed during the late Cretaceous (97.5–66.5 mya). Water lilies are one of the most ancient angiosperm plants. Presently, over three-fourths of all living plants are angiosperms.

plant gene expression control using epigenome editing: A large number of epigenome-editing tools based on SMALL INTERFERING RNAs (SIRNAs) have been developed for controlling plant gene expression in various plants.

plant genetic transformation: A way to insert DNA from another organism into the genome of a plant of interest. For example, insertion a gene called *Stilbene synthase* from grape into tomato to enable the production of the medicinal compound resveratrol in high levels in tomatoes.

plant genetics: The study of the genetics of traits in plants.

plant growth–promoting fungi: Rhizosphere fungi are capable of promoting plant growth when colonizing the plant root.

plant growth–promoting Rhizobacteria (PGPRs): Free-living bacteria that colonize roots or rhizosphere soil, some of them assault the tissues of plants and result in latent and asymptomatic infection and enhance plant growth and development directly or indirectly by averting deleterious effects of other microorganisms that are harmful to plants. Direct mechanisms include the production of stimulatory bacterial volatiles and phytohormones, dropping of the ethylene level in the plant (PHYTOSTIMULATORS), upgrading of the plant nutrient status (mobilize phosphates and micronutrients from insoluble sources, non-symbiotic nitrogen fixation) and stimulation of disease-resistance mechanisms (INDUCED SYSTEMIC RESISTANCE). Indirect mechanisms comprise the action of PGPRs like biocontrol agents, reducing diseases by the stimulation of other beneficial symbioses or controlling diseases largely by the production of antibiotics and antifungal metabolites (biopesticides), and, moreover, defending the plants by degrading xenobiotics like organic pollutants in contaminated soils (RHIZOREMEDIATORS).

plant growth: It is the irreversible, quantitative increase in size, mass and/or volume of a plant or its parts.

plant growth regulator: Signal molecule that helps coordinate growth and development. Examples are ethylene, auxins, gibberellins, cytokines, abscisic acid and the brassinosteroids. *Aka* PLANT HORMONE.

plant homeodomain (PHD) finger: A region within a protein capable of detecting methylated histones. Examples include the PHD finger of recombination activating gene 2 (RAG2) protein, which recognizes the trimethylation of lysine 4 on histone H3 (H3K4me3). RAG2 is not the catalytic component but is required for all known catalytic activities mediated by RAG1.

plant homeosis: Principles governing development in plants differ from those in animals reflecting their ancient evolutionary divergence. For example, animals have separate germline from soma quite early in development while in plants this decision is very late. Experiments on isolation and cloning of homeotic genes revealed that many homeotic genes encode transcription factors. For instance, *def* gene encodes DEFA protein, which resembles transcription factors like SERUM RESPONSE FACTOR (SRF) in mammals. Similarly, the product of *minichromosome maintenance* (*MCM 1*) gene in yeast, has been shown to be a transcription factor. Similar is the situation with the *agamous* (*ag*) gene in *Arabidopsis* producing a product AG. These genes seem to have a conserved domain in common with each other and include the gene products minichromosome maintenance1 (MCM 1), agamous (AG), deficiens A (DEF A) and SRF. Therefore, this domain has been given the name MADS-box.

plant hormones: Plant hormones are core internal chemical signals that impact plant development. Auxins endorse shoot growth through cell elongation and lateral root formation and help prevent senescence. They also constrain growth in lateral buds in favour of growth in apical meristem (apical dominance) and hinder cell elongation in roots. Cytokinins promote growth by stimulating cell division, including cytokinesis. Cytokinins also play a role in the deterrence of senescence, but unlike the auxins, promote the growth of lateral buds. Together, auxins and cytokinins aid in controlling a range of processes including meristem induction, apical dominance and root-shoot communication. Abscisic acid regulates various aspects of seed development and dormancy. It also arbitrates responses to water stress. Gibberellins control seed germination and stem elongation. Ethylene promotes maturation of the seedling and impacts the timing of leaf senescence. *Aka* PHYTOCHROME.

plant identification model: G. Xie *et al.* (2024) established a plant identification model for over 20,000 species (called the '20K model') engaging deep convolutional neural network algorithms. This model was skilled on a 15-year image data set from the Plant Photo Bank of China (PPBC, http://ppbc.cn). They accomplished the model via a branching neural network algorithm, which processes plant images from the training set. This algorithm captures fine-grained and contour features via its branching structures, making it predominantly operative for plant identification. The drill involved processing all images in the set through one cycle, termed an 'epoch'. The model experienced 90 epochs of exercise to ensure its strength and precision. *See* FLOWERMATE 2.0.

plant injury: An abrupt alteration of form or function caused by a discontinuous irritant. Plant injury includes insect, animal, physical, chemical or environmental agents.

plant introduction: The transport of a collection of seeds, plants or vegetative propagating materials from one ecological area into another.

plant metabolic engineering: A typical metabolic-engineering approach sets out with a focus on a particular metabolic intermediate or product. An assessment of knowledge on the pathway of biosynthesis is then performed and candidates for RATE-LIMITING STEPS are identified. These are often enzymes that catalyse reactions that are believed to be remote from equilibrium *in vivo* or that have known regulatory properties. These steps are then subjected to engineering through the introduction of a novel enzyme that increases the catalytic capacity at this step. Often, the enzyme that is chosen is 'heterologous' to that in the endogenous system and thus is known or postulated to be free of the regulatory controls found in the target organism.

plant metabolomics, challenges: (i) A typical plant metabolome exhibits enormous chemical diversity which needs multiple approaches for extraction. (ii) Individual metabolites exhibit a dynamic range of concentrations, so that special techniques are needed for separation of metabolites. (iii) There is a large variation in spatial/temporal distribution in metabolites, which means that metabolites differ in different organs and at different times (during lifespan and during different seasons/climates). (iv) Genomic information cannot be used for study of metabolome. These limitations require different strategies for the extraction, identification and quantification of plant metabolites.

plant mitochondrial genome: The plant mitochondrial genome is a circular double-stranded DNA molecule that encodes rRNAs, tRNAs, ribosomal proteins and a portion of the enzymes used in respiration. The mitochondrial DNA possesses direct repeats spread throughout the genome. Pairing and recombination among these direct repeats produces smaller circular DNA molecules.

plant molecular farming (PMF): This technique involves using genetically modified plants to produce substances that the plants typically do not produce naturally, such as industrial compounds or therapeutics.

plant nuclear DNA and organellar DNA separation: Human methyl-CG-binding domain 2 (MBD2) protein is used to fractionate angiosperm DNA into eukaryotic nuclear (methyl-CG-rich) versus organellar and prokaryotic (methyl-CG-poor) elements.

Plant Patent Act (PPA): In 1930, the PPA was enacted into law in the USA. The plant patent grants the breeder intellectual property protection exclusive right, for 17 years, to propagate the patented plant by asexual reproduction (e.g. via grafting). The purpose of the PPA was to encourage research investment in asexually reproduced plant species. Since 1930, more than 6000 plant patents have been issued by the Patent and Trademark Office, primarily for fruit trees, flowers, ornamental trees, grape and other horticultural species. *Aka* PLANT PROTECTION ACT (PPA).

plant pathogenesis: Refers to the process of disease development after infection of a host plant by a pathogen or parasite.

plant pathogenesis-related proteins (PPRP): Groups of proteins with different chemical properties produced in a cell within minutes or hours following inoculation, but all PPRPs being more or less toxic to pathogens.

plant pathology: The study of plant diseases and the abnormal conditions that constitute plant disorders. *See* PHYTOPATHOLOGY.

plant pests: Organisms that may directly or indirectly reason disease, spoilage or damage to plants, plant parts or processed plant materials. Common examples of plant pests include certain insects, mites, nematodes, fungi, moulds, viruses and bacteria.

plant phenomics: High-throughput phenotyping replaced the traditional plant phenotyping which is a high-throughput image analyses involving background estimation and plant segmentation in an automated phenotyping process. Unmanned Aerial Vehicle (UAV)-Based High-Throughput Phenotyping (HTP) tool in citrus has been developed. A novel **artificial intelligence (AI)**-based machine learning (ML) tool has been developed with the use for Gaussian mixture models for image refinement and background estimation with the use of Australian Plant Phenomics Facility Plant Accelerator (APPFPA). Plant phenomics has the potential to bridge between genotype to phenotype. It took the advantage of high-throughput phenotyping to keep the massive data in a place from high-dimensional imaging sensors to ease the identification of more complex phenotypic traits. *See* DEEP PLANT PHENOMICS (DPP), UNMANNED AIRCRAFT SYSTEMS (UAS) *and* ARTIFICIAL INTELLIGENCE (AI)-BASED UAS-BASED HIGH-THROUGHPUT PHENOTYPING (AUHTP) SYSTEM.

plant plasticity: Plasticity in plants supports homeostasis in response to variable environmental conditions. Plant flexibility is enabled by epigenetic processes that modify chromatin through self-motivated modifications in DNA methylation, histone variants, sRNAs and transposable elements.

plant quarantine: The isolation of newly imported plants to ensure that they are not introducing any foreign parasites. The term quarantine may refer to the quarantine station itself or to the process of testing and purifying the plant material.

Plant Regulomics Portal (PRP): PRP is an all-inclusive integrated regulatory information and analysis portal for plant genomes.

plant resistance (R) proteins: R proteins contain NUCLEOTIDE BINDING SITE (NBS) and LEUCINE-RICH REPEAT (LRR) domains.

plant stress memory: Plants have the ability to respond quickly to stressful situations and can also improve their defence systems by retaining and sustaining stressful memories, allowing for stronger or faster responses to repeated stressful situations.

plant stress recovery: Several studies evidence that plants are capable of efficient recovery even from relatively severe stress conditions. The recovery is dependent on various mechanisms including metal immobilization, sequestration and activation of repair systems.

plant stress response: Plant stress responses describe the suite of molecular and cellular processes that are triggered by the detection by the plant of some form of stress. Stresses can be abiotic, such as drought or excess light or biotic, such as herbivores or pathogens. miRNAs have been involved in cellular responses to oxidative stress. Some miRNAs in stress adaptation are conserved among plant species. For example, several N deficiency-related miRNAs in Arabidopsis, such as miR-160, miR-169, miR-171, miR-395, miR-397, miR-398, miR-399, miR-408 and miR-827, are involved in response to N deficiency in maize.

plant stress sensing: A rapid, non-destructive, method to detect both biotic and abiotic stressful conditions, utilized in precision agriculture and plant phenotyping for resistance breeding purposes. The processes involved are mainly based on the interaction between electromagnetic radiation and plants.

plant SUMOylation substrates, identification of: SUMOylation is an important type of post-translational modification in eukaryotic cells. This modification plays a critical role in plant development and stress responses, thus efficient identification of SUMOylation substrates in plant cells is a fundamental issue. R. Lai *et al.* (2023) initiated to establish a robust method to identify SUMOylation substrates in plant species, based on an approach using a plant cDNA library for substrate screening in a reconstituted reaction system in bacteria. SUMOylation has been shown to play critical roles in maize, including development and stress responses.

plant tissue culture: A technique of growing plant cells, tissues, organs or whole plants on synthetic medium under closely controlled and aseptic conditions. Plant tissue culture is an important area of plant biotechnology. Tissue culture methods can be exploited for creation, conservation and utilization of genetic variability for improvement of crop plants. Plant cells, tissues and organs are cultured on synthetic medium such as T. Murashige and F. Skoog in 1962 for obtaining desired growth. Micropropagation is exploited for commercial plant production, another culture for early release of varieties, embryo culture for developing hybrids otherwise not possible through conventional approaches. Similarly, there are other tissue culture techniques now being used in the industry and for producing transgenic plants.

plant transformation vectors: The plasmids that have been specifically designed to facilitate the generation of transgenic plants. The most commonly used plant transformation vectors are BINARY VECTORS because of their ability to replicate in both *Escherichia coli*, a common laboratory bacterium and *Agrobacterium tumefaciens*, a bacterium used to insert the recombinant DNA into plants. Plant transformation vectors contain three key elements: plasmid selection, plasmid replication and transfer DNA (T-DNA). TUMOUR-INDUCING PLASMID (PTI) and ROOT-INDUCING PLASMID (PRI) are most commonly used plant transformation vectors. *See details as separate terms.*

plant transposons: They have inverted repeat sequences and generate short direct target site repeats. Possible effects of transposons inserted into plant chromosome include the following: (i) activation or repression of adjacent genes by disrupting a cellular promoter or by action of transposon promoters, (ii) chromosome mutations such as duplications, deletions, inversions, translocations or breakage and (iii) disruption of genes to produce a null mutation (gene is non-functional).

plant variety: A group of individual plants that is uniform, stable and distinct genetically from other groups of individuals in the same species.

plant variety protection (PVP) requirements: PVP must meet requirements of DISTINCTIVENESS, UNIFORMITY AND STABILITY (DUS) besides novelty and a distinct denomination of the new variety.

Plant Variety Protection Act (PVPA): A law passed by the US Congress in 1970 that provides intellectual property protection (analogous to copyright protection) for new seed plants and seeds in America. Prior to 1970, breeders in the industry worked primarily with maize and sorghum, with some efforts directed to lucerne, cotton, sugarbeet and certain vegetables. The purpose of the PVPA was to encourage private plant breeding research and to offer better seed cultivars to farmers and gardeners.

Plant Variety Protection and Farmer's Rights (PVP&FR) Act: India has enacted The New Plant Variety Protection and Farmer's Rights Act in 2001. These criteria to arrive at a decision whether a plant variety is really new or not are distinctiveness (discountable from other varieties), uniformity (variation must be less than that within a species), stability (characteristics true to description after propagation at the end of each cycle), novelty and appropriate denomination (for identification). In India, PPV&FR Act, 2001 spells the legal provisions to this effect. The system of PVP is *aka* SUI GENERIS system which tends to provide a balance between the breeder's right and farmer's right.

plant variety rights (PVR): The legal rights of a plant breeder (a person, company or university) who has developed a new variety governed by the Plant Variety Rights. The registration of new varieties is now administered by Plant Breeders Rights.

plant water balance: The phytohormone ABSCISIC ACID (ABA) promotes plant water conservation by decreasing the aperture of stomatal pores in the epidermis through which water loss occurs. *Arabidopsis thaliana* plants harbouring transferred DNA insertional mutations in the sole prototypical heterotrimeric-binding protein (G protein) α subunit gene, *GAP1*, lack both ABA inhibition of guard cell inward K^+ channels and pH-independent ABA activation of anion channels. Stomatal opening in *gpa1* plants is insensitive to inhibition by ABA and the rate of water loss from *gpa1* mutants is greater than that from wild-type plants. Manipulation of G protein status in guard cells may provide a mechanism for controlling plant water balance.

plant's novel trait (PNT): A novel trait added to a plant (e.g. crop plant such as cotton, maize, soybean, etc.). Examples of novel traits are herbicide tolerance (via inserted CP4 *EPSPS* gene, *pat* gene, etc.), insect resistance (via inserted *Bt* gene, *Photorhabdus luminescens* gene, etc.) and resistance to aluminium toxicity (via inserted *CSb* gene, etc.). *Aka* PLANTS WITH NOVEL TRAITS.

plant-animal mutualism: One plant and one animal species develop a considerable dependence on each other such that elimination of one species leads to a marked numerical response in the other species.

plant-incorporated protectants (PIPs): Pesticidal substances produced by plants and the genetic material necessary for the plant to produce the substance. For example, scientists can take the gene for a specific *Bt* pesticidal protein and introduce the gene into the plant's genetic material.

plant-made pharmaceuticals (PMP): Chemicals having pharmaceutical applications produced in genetically engineered crops.

plant-microbe interactions: A complex, dynamic and continuous process that is as old as plant colonization on Earth. At the external face of the cell, microbial elicitors, PAMPs/MAMPs, are recognized by receptor proteins, PRRs. This interaction produces PAMP-TRIGGERED IMMUNITY (PTI). Intracellular perception of pathogen virulence factors (effectors) is explained by a direct recognition for protein-protein interactions of Avr and NBS-LRR, guard hypothesis or bait-and-switch model. This results in EFFECTOR TRIGGERED IMMUNITY (ETI).

plant-pathogen interaction studies in *Arabidopsis*: *Arabidopsis thaliana* has been successfully implemented in the study of the subdiscipline of plant pathology, that is, the interaction between plants and disease-causing pathogens. The use of *Arabidopsis* has led to many breakthroughs in the advancement of knowledge of how plants manifest plant disease resistance. The reason most plants are resistant to most pathogens is through non-host resistance. An example where *Arabidopsis* was used to determine the genes responsible for non-host resistance is *Blumeria graminis*, the causal agent of powdery mildew of grasses. Arabidopsis mutants were developed using the mutagen ethyl methanesulfonate and screened to determine which mutants had increased infection by *Blumeria graminis* (M. Stein *et al.*, 2006). The mutants with higher infection rates are referred to as PEN mutants due to the ability of *Blumeria graminis* to penetrate *Arabidopsis* to begin the disease process. The PEN genes were later mapped to identify the genes responsible for non-host resistance to *Blumeria graminis*.

plant-pollinator interactions: Colourful flower patterns are vital signals to attract pollinators. To generate such motifs, plants lay down boundaries dividing petals into subdomains where cells develop characteristic pigmentations, shapes and textures. While some transcription factors and biosynthetic pathways behind these characteristics are well known, the upstream processes restricting their activities to specific petal regions are being investigated. L. Riglet *et al.* (2024) unveil that the petal surface of *Hibiscus trionum*, an emerging model featuring a bullseye on its corolla, is prepatterned as the bullseye boundary position is stated long before it becomes visible. Pants can regulate boundary position during the prepatterning phase or modulate growth on either side of this boundary later in development to diverge bullseye proportions. Such modifications are functionally relevant, as buff-tailed bumblebees can dependably recognize food sources based on bullseye size and favour certain pattern proportions.

plant-specific RNA polymerases: Eukaryotic organisms possess three general multisubunit DNA-dependent RNA polymerases (RNAPs) with distinct functions (viz. RNAPI, RNAPII and RNAPIII). Higher plants possess two extra RNA polymerases, RNAPIV and RNAPV, that function mainly in RdDM pathway. RNAPIV functions upstream in RdDM to generate and amplify the small-RNA trigger for silencing, whereas RNAPV works downstream to transcribe non-coding RNAs that offer scaffolds for fascinating silencing complexes and could also be involved in the reinforcement of silencing through a positive-feedback loop.

PlantCFG: A comprehensive database developed by D. Liu, *et al.* (2024) for analysing candidate flowering genes in multiple plants. It consists of four main modules: (i) search, (ii) synteny, (iii) ABCDE model and (iv) evolution. The homepage enables users to quickly access the four main modules and external links, including FLOR-ID, TAIR, EnsemblPlants, InterPro and UniProt. In the 'search' module, users can retrieve candidate flowering genes using five search types: (i) species name, (ii) reported flowering genes, (iii) gene family, (iv) Pfam family and (v) sequence.

plantibodies: Antibodies synthesized in transgenic plants expressing the antibody-producing genes of an animal, for example, mouse, that had been previously injected with a pathogen (usually a virus) and that infects the plants.

Plantibodies™: A trademark owned by EPIcyte Pharmaceutical, Inc. It refers to antibodies (akin to mammalian ones) produced in plants that are genetically engineered to produce those specific antibodies. That process (genetically engineering plants to cause them to produce plantibodies) was invented during the 1990s by A. Hiatt and M. Hein. Although plants do not always glycosylate (i.e. attach oligosaccharide units to protein molecules such as these antibodies) in the same manner as animal cells, an antibody against HSV-2 pathogen expressed in genetically engineered soybean plants has proven comparable to that same antibody expressed in genetically engineered animal cells.

plantigens: Antigens (e.g. of pathogenic bacteria) produced in plants which are genetically engineered to produce those specific antigens. This process can be utilized to produce edible vaccines for the pathogenic bacteria possessing those antigens. Then people could be 'vaccinated' against disease merely by eating the genetically engineered plant (e.g. banana).

planting efficiency: *See* COLONY-FORMING EFFICIENCY.

plants resistant to environmental stress: Plants need to cope up with abiotic stresses such as drought, cold, heat and soils that are too acidic or salty to support plant growth. While plant breeders have successfully incorporated genetic resistance to biotic stresses into many crop plants through crossbreeding, their success at creating crops resistant to abiotic stresses has been more limited, largely because few crops have close relatives with genes for resistance to these stresses. Therefore, rDNA technology is being increasingly used to develop crops that can tolerate difficult growing conditions. Genetically modified tomato and canola plants that tolerate salt levels 300% greater than non-genetically modified varieties have been developed. Other researchers have identified many genes involved in cold, heat and drought tolerance found naturally in some plants and bacteria. Scientists in Mexico have produced maize and papaya that are tolerant to the high levels of aluminium that significantly impede crop plant productivity in many developing countries.

plants with enhanced photosynthesis cope with fluctuating environmental conditions: Proteome analyses in *Phaeodactylum tricornutum* revealed that incorporation of the PHOTOSYNTHETIC ELECTRON TRANSPORT CHAIN (PETC) caused increased photosynthesis in addition to maintenance of redox homeostasis. This was attributable to augmentation of cysteine residues in proteins in reply to the physiological state of the cell. The accelerated photosynthetic electron flow during PETC produces REACTIVE OXYGEN SPECIES (ROS), which function as molecular signals to switch on redox-regulatory networks to deal with unstable environmental conditions.

plants with enhanced photosynthesis display abiotic stress tolerance: Abiotic restrictions are one of the limiting factors that regulate crop productivity. As assimilate production from current photosynthesis and mobilization from stored reserves governs productivity. An increment/maintenance of rate of photosynthesis provides significant stress tolerance to crops and reduces yield losses. Transgenic maize having overexpressed *Zm*NF-YB16 protein displayed significant drought resistance via an increased rate of photosynthesis, which caused yield improvement under drought conditions. Changes in leaf metabolome to maintain photosynthetic rate in stressful environments pay towards inherent potential of abiotic stress tolerance in crops.

plants with enhanced photosynthesis produce heterotic biomass: In a large number of plant species, hybrid progeny of selected parental lines displays higher performance compared to their parents. Heterosis or hybrid vigour, is important in agriculture for improved yield of numerous important crops, such as maize, rice, sorghum, sunflower and canola. In progeny of crosses between *Arabidopsis thaliana* Columbia-0 and C24 accessions, heterosis was noticed in fully developed seeds and in seedlings sampled 3 days after sowing. Build-up of the heterotic biomass has been related with improved photosynthetic efficiency. Any activity connected with inhibition of photosynthesis diminishes heterosis and augmented light intensities enhance both photosynthesis and heterosis.

plants with increased quality of products, developing: One of the most successful research efforts to change the characteristics of a plant produce was carried out with tomatoes. Tomatoes need to be picked while still green so that they are firm enough to withstand mechanical handling and transport. Unfortunately, they do not develop the same flavour and texture of vine-ripened tomatoes. Softening of tomatoes and many other fruits is caused by the enzyme pectinase or polygalacturonase (PGA). This enzyme digests the pectin polysaccharide that cements the plant cells together. Softening of the fruit is caused, in part by this breakdown of pectin. In order to reduce the levels of PGA in ripening tomatoes, researchers placed the PGA gene in reverse orientation relative to the CaMV 35S promoter. This results in transcription of an antisense RNA that is complementary to the normal sense PGA mRNA. Although the exact mechanism is unknown, antisense RNA is able to arrest the translation of the endogenous PGA mRNA in the tomato fruit. Transgenic tomato plants that express an antisense PGA gene only have about 5% to 10% of normal PGA levels. Fruits of these plants have normal colour and flavour but they soften more slowly and can be picked and processed after they are ripe. They also have a higher content of soluble solids and are therefore better than normal tomatoes for processed tomato products. The gas hormone, ethylene is involved in the regulation of fruit ripening. Therefore, ripening can be slowed down by blocking or reducing ethylene production. This can be achieved by introducing ethylene forming gene(s) in a way that will suppress its own expression in the crop plants. Such fruits ripen very slowly (however, they can be ripened by ethylene application) and are very important for export to longer distances without spoilage as they show longer shelf life due to slow ripening. A notable example of this kind is the ‘Flavr Savr’ transgenic tomatoes, which were commercialized in the USA about 6 years ago. Transgenic lines of potato having increased levels of starch also have been developed by introducing a gene construct that expresses a gene from bacteria that produce an enzyme that enhances starch biosynthesis. A promoter from a potato gene that encodes the major protein in potato tubers has been used, so that the expression of the introduced gene is limited to the tuber. Tubers accumulate approximately 3–5% more starch than normal potatoes and when they are deep fried, they absorb less oil and yield chips having fewer calories. Golden rice contains beta-carotene to overcome vitamin A deficiency in regions where rice is the staple food. Canola containing high levels of oleic acids and laurate. Barley containing feed enzymes. Tomatoes which do not rot in room temperature. Other vegetables and fruits with delayed ripening as well as modified flavour characteristics.

plaque: A clear zone in the bacterial lawn resulting from lysis by phages. The plaque method is a common method for detecting and counting viruses. Bacteria and bacteriophage are co-cultured on solid medium. During incubation, all the bacteria in the area surrounding the original virus are destroyed. This produces a number of clearing (plaques) visible against a lawn of bacterial growth on the surface of agar.

The concentration of viral suspensions measured by the number of plaques are usually given in terms of PLAQUE-FORMING UNITS (pfu).

plaque-forming units (pfu): *See* PLAQUE.

plasma: The yellow fluid in which red blood cells are suspended. Pure samples of plasma are obtained by adding suitable agents to prevent coagulation and then sedimenting the red blood cells, usually by centrifugation.

plasma cells: From bone marrow, cells move to liver and spleen where some receptor molecules are attached to them. Cells increase in size and become plasma cells.

plasma gene: A self-replicating, cytoplasmically located gene.

plasma membrane: A thin structure that wholly environs the cell as a 'skin'. It may be observed with the aid of an electron microscope. The entire membrane seems to be about 100 Å thick and is composed of two dark lines, each about 30 Å thick which are, however, separated by a lighter area. This trilaminar 'sandwich' structure is known as the UNIT MEMBRANE. Besides the cell, membranes surround such systems as the endoplasmic reticulum, vacuoles, lysosomes, Golgi bodies, mitochondria, chloroplasts and the nucleus, to mention just a few. The plasma membrane and membranes in general function partially as a permeability barrier to the free movement of substances between the inside and exterior of the cell or organelles that they surround. Outer membrane of a cell is sometimes called the CELL MEMBRANE. The plasma membrane is a lipid bilayer, which is mostly composed of phospholipid molecules that have polar and non-polar parts. The polar part of each phospholipid is a phosphate-containing 'head' that is hydrophilic or attracted to water. The non-polar part has two 'tails' of fatty acid that are hydrophobic, meaning they do not dissolve in water. As a result, the phospholipid molecules arrange themselves with the heads facing outward towards the cytosol and extracellular fluid and the tails facing inward towards each other.

plasma membrane, functions of: It maintains a constant membrane potential, which is a very important phenomenon during fertilization. Due to the sodium pump, [Na^+] increases inside the cell. This results in inactive transportation. The fluid mosaic model explains the workings of the plasma membrane. Individual molecules of the phospholipid bilayer can move laterally, giving the bilayer fluidity, flexibility, high electric resistance and relative permeability. Mosaic membrane means that one side of the cell membrane is not the same as the other side. The plasma membrane is the main selective barrier separating cells from the outside world. Selective barriers work through transport mechanisms like simple diffusion, active transport, pinocytosis and phagocytosis. During simple diffusion, for a molecule to get into the cell, it should be water or lipid soluble. Simple diffusion works in the case of very small organisms. Active transport involves enzymes capable of utilizing energy made available by the breakdown of ATP. These enzymes may lie on the inner or outer surface of the cell membrane or both. Molecules move against the concentration gradient. Through active transport, complex macromolecules of biological significance like amino acids, lipids, nucleotides, sugars and phosphates are transported in addition to the small ions. Pinocytosis is a cell drinking process. Phagocytosis is a cell-eating process.

plasmagene: A self-replicating, cytoplasmically located gene showing no Mendelian inheritance. *Compare with* CHROMOGENE.

plasmalemma: A thin flexible membrane surrounding cytoplasm.

plasmid amplification: A method involving incubation with an inhibitor of protein synthesis aimed at increasing the copy number of certain types of plasmid in a bacterial culture.

plasmid copy number: One way to increase gene expression is to increase the number of transcripts. This is dependent, in part, on number of copies of the gene in a high–copy-number plasmid. Host cell genetic background can affect copy number.

plasmid incompatibility: The inability of two different plasmids to coexist in the same cell in absence of selection pressure. Twenty-five incompatibility groups of plasmids have been defined in *Escherichia coli*.

plasmid purification: The plasmid is often transformed into a bacterium like *Escherichia coli*. Ideally when the bacterium divides the plasmid should also be replicated. In the best-case scenario, each bacterial cell should have several copies of the plasmid. Kits are available to purify plasmid DNA, which are named by size of bacterial culture and corresponding plasmid yield. In increasing order, these are the MINIPREPARATION, MIDIPREPARATION, MAXIPREPARATION, MEGAPREPARATION and GIGAPREPARATION. *See details as separate terms.*

plasmid replicon: A plasmid replicates only when occurring within a host cell. Plasmid replication depends heavily on host replication machinery. Polymerase III is a major replication protein in bacteria. All plasmids examined to date replicate semi-conservatively but there are different replication patterns; some plasmids replicate unidirectionally while others replicate bidirectionally.

plasmid rescue: A technique used for transforming *Bacillus subtilis* in which an inactive donor plasmid interacts with a resident plasmid to effect a transformation. In practice, foreign DNA is ligated to monomeric vector DNA, and the recombinant inserted into cells containing a homologous plasmid. The resident plasmid results in transformation.

plasmid stability: Increases in the levels of gene expression of recombinant genes lead to reductions in cell growth rate.

plasmid vector: A plasmid or plasmid-like structure used as a carrier for alien DNA segments or genes.

plasmids: J. Lederberg (1952) gave the term plasmid. An extrachromosomal, small circular, independent, stable DNA within the cell with a self-duplicating and regulatory system. In addition to the chromosome, the bacteria often contain small circular, double-stranded DNA molecules called plasmids. These are extrachromosomal genetic elements and replicate independently of chromosomal DNA. This is in disparity to a similar genetic element, known as an EPISOME, that may occur independently of the chromosome or may become integrated into the host chromosome. Bacterial cells contain 1–10 plasmids, and each plasmid carries 5–100 genes that are generally not crucial for survival of bacterium but confer resistance to antibiotics and may be transferred by cell-to-cell contact (by conjugation via the sex pilus) or by viral-mediated transduction. These plasmids may be gained or lost without harming the cell. Plasmids may be advantageous to cells as they carry genes for antibiotic resistance, tolerance to heavy metals and the synthesis of enzymes. Plasmids can be transferred from one bacterium to another.

plasmids, characteristics of: (i) Size of plasmid in *Escherichia coli* is 30 μm, which is 20% of the main chromosome. (ii) DNA plasmids replicate in a semi-conservative manner. While plasmid genes control initiation of replication of plasmids, bacterial genes control polymerization and termination. When a plasmid is inserted into the bacterial chromosome, bacterial control of plasmid replication is absolute but when free, the control is less than absolute. (iii) Plasmids have the ability of transferring themselves from one bacterium to another. Transfer is facilitated by sex pili. In *Escherichia coli*, transfer is controlled by genes *tra-A*, *tra-B*, *tra-C* and so on. Mutations in these genes lead to failure of transfer. When transfer occurs, one of the two strands of plasmid DNA is nicked off at a site called origin of transfer (OriT) and linear strand thus formed moves to the recipient bacteria. The single strand in donor as well as recipient cell will then synthesize complementary strands so that duplex DNA plasmids are reconstituted in both the cells. (iv) Plasmid DNA may undergo recombination either with another plasmid DNA or with bacterial DNA. (v) In bacterial chromosome, there are INSERTION SEQUENCES (IS), which are homologous in plasmids and bacteria. These sequences help in recombination. There are at least 17 such sites in *Escherichia coli* chromosome. (vi) Genes of plasmid DNA have been mapped not by recombination but by complementation studies between strains having overlapping deletions. (vii) Usually plasmids have been discovered to be formed of DNA. However, RNA plasmids have also been reported in maize mitochondria.

plasmids, conformations of: Plasmid DNA may appear in one of five conformations, which (for a given size) run at different speeds in a gel during electrophoresis. The conformations of DNA in plasmids are listed here in order of electrophoretic mobility (speed for a given applied voltage) from slowest to fastest: (i) nicked open-circular DNA, (ii) relaxed circular DNA, (iii) linear DNA, (iv) supercoiled (or covalently closed circular) DNA and (v) supercoiled denatured DNA

plasmids, forms of: Known forms of plasmids include both linear and circular molecules. Plasmids are limited to sizes of 2.5–20 kb, in general. Plasmids have been identified in almost all strains of bacteria. Most, but not all, plasmids are dispensable. Some well-known examples of plasmids are SEX FACTORS (F and F′ factors), RESISTANCE TRANSFER FACTORS (RTFs), COLICIN FACTORS (COL FACTOR) and PHAGE. Modified plasmids are used for amplification of DNA segments in recombinant DNA experiments as acceptors and vectors of foreign DNA. It can carry genes into a host cell. Plasmids are now being used as vectors for gene cloning and transformation for developing transgenic organisms. *Aka* CIRCULAR MINICHROMOSOME.

plasmids, uses of: (i) Plasmids have been very valuable in creation of recombinant DNA molecules. Plasmids used in genetic engineering are called vectors. However, a plasmid can only contain inserts of about 1–10 kbp. To clone longer lengths of DNA, lambda phage with lysogeny genes deleted, cosmids, bacterial artificial chromosomes or yeast artificial chromosomes could be used. (ii) Plasmids serve as important tools in genetics and biotechnology labs, where they are commonly used to multiply (make many copies of) or express particular genes. Many plasmids are commercially available for such uses. (iii) Another major use of plasmids is to make large amounts of proteins. In this case, researchers grow bacteria containing a plasmid harbouring the gene of interest. Just as the bacteria produce to proteins to confer its antibiotic resistance, it can also be induced to produce large amounts of proteins from the inserted gene. This is a cheap and easy way of mass-producing a gene or the protein it then codes for, for example, insulin or even antibiotics. (iv) Plasmids are important to medical microbiology because plasmids can encode genes for antibiotic resistance or virulence factors. (v) Plasmids can also serve as markers of various bacterial strains when a typing system referred to as plasmid profiling or plasmid fingerprinting is used. (vi) Because many species of bacteria contain plasmids, plasmid profile typing has been used to investigate outbreaks of many bacterial diseases and to trace inter- and intraspecies spread of antibiotic resistance. (vii) Plasmids have a significant role in gene therapy.

plasmodesmata: Microscopic cytoplasmic bridges, lined with a plasma membrane, that connect adjacent cells in plants through a small pore in their cell walls. The plasmodesmata also occur in some algal cells. They offer major pathways of communication and transport between cells. *Aka* *PLASMODESMATA*.

plasmodium (pl. plasmodia): A naked, multinucleate mass of protoplasm moving and feeding in amoeboid fashion.

plasmogamy: Fusion of cytoplasm of two or more cells after karyogamy during the process of fertilization, during fungal life cycle. Fungal spores can be either asexual or sexual. Sexual spores result from sexual reproduction consisting of three phases: (i) plasmogamy, (ii) karyogamy and (iii) meiosis. Plasmogamy involves the penetration of a haploid nucleus of a donor cell into the cytoplasm of recipient cell. Subsequently, karyogamy takes place in which the haploid nuclei of donor and recipient cells fuse to form zygote.

plasmolysis: The consequence of placing plant cells in a hypertonic solution so that water is drawn out from the vacuole of the cell. The cytoplasm shrinks and the cell membrane is pulled away from the cell wall.

plasmon: All the extrachromosomal hereditary determinants.

plasmon mutation: A mutation that genetically changes the cytoplasm and/or its hereditary determinants.

plasmotype: The sum of the extrachromosomal hereditary determinants.

plastic plants: Some genetically engineered plants produce plastic within their stem structures which biodegrade in about 6 months. If the genes escape into the wild, through cross-pollination with wild relatives or by other means, then we face the prospect of natural areas littered with the plastic spines of decayed leaves. Plastic has the potential for disrupting entire food chains. It can be eaten by invertebrates, which are in turn eaten and so forth. If primary foods are inedible or poisonous, then whole food chains can die off.

plasticity: The ability of adult-derived stem cells to be capable of developing into cell types outside of the tissue of origin (e.g. human blood stem cells have been shown to differentiate into liver cells.

plasticity behaviour in plants: The dynamic changes in DNA methylation, histone variants, small RNAs, chromatin and transposable elements produce plasticity behaviour in plants. Plasticity in plants is crucial for sustaining homeostasis under varying environmental conditions through epigenetic processes.

plastid: A general name for a cell organelle of plants algae and a variety of single-celled eukaryotes. It is enclosed by a double membrane and contains a series of internal membranes and vesicles. In plants, plastids may differentiate into several forms (e.g. chloroplasts, leucoplasts, amyloplasts), depending upon which function they need to perform in the cell. Because there can be up to 10,000 plastids in a given plant cell, the insertion of a gene (e.g. via genetic engineering) into plastids can result in a higher yield (of the specific protein coded for by that gene) than is accomplished via insertion of the gene into the cell's nuclear DNA. They perform various physiological functions, such as the synthesis and storage of food.

plastid DNA: Organelle DNA that occurs in a plastid.

plastid genome organization: Most plastid genomes are circular and range in size from 120 to 217 kbp. Variation in genome size is related primarily to the size of an inverted repeat in the plastid genome and limited differences in coding capacity. Plastids contain from 22 to over 300 genome copies, depending on the developmental stage. The genome is membrane associated, complexed with protein and resembles bacterial nucleoids. Plastid DNA encodes approximately 135 genes that fall into three major functional categories: (i) genes encoding proteins and RNAs involved in transcription and translation of the plastid genome (RNA polymerase subunits, tRNAs, rRNAs, ribosomal proteins, initiation factor 1), (ii) genes encoding proteins of the photosynthetic apparatus (RuBisCO, PSI, PSII, ATP synthase, Cyt complex) and (iii) genes encoding proteins of the NADH oxidoreductase complex. Whereas plastids function in non-photosynthetic tissues, the plastid genome is clearly specialized for the production of the photosynthetic apparatus. Most plastid genes are organized in complex.

plastid inheritance: Inheritance which is governed by chloroplast genes.

plastidome mutation: Mutations that genetically change the DNA of the plastids of a plant cell. *Aka* PLASTOME MUTATION.

plastids, origin of: Plastids are believed to have originated from the engulfment of a photosynthetic bacterium by a eukaryotic cell that already contained mitochondria, which are the by-products of a separate endosymbiotic event. Plastids are thought to have originated from endosymbiotic cyanobacteria. They developed around 1.5×10^9 years ago and allowed eukaryotes to carry out oxygenic photosynthesis. Due to a split into three evolutionary lineages, the plastids are named differently: chloroplasts in green algae and plants, rhodoplasts in red algae and cyanelles in the glaucophytes. The plastids differ not only in their pigmentation but also in their ultrastructure. Many authors use terms plastid and chloroplast synonymously.

plastochron: The time gap between successive leaf initiation events.

plastogene: The hereditary determinants located in the plastid.

plastome: A plastome is the genome of a plastid, a type of organelle found in plants and in a variety of protoctists. The flowering plants are especially well represented in complete chloroplast genomes with more than 100 plastomes sequenced so far. The chloroplast genes are mappable by recombination analysis. Almost all chloroplast DNAs carefully examined so far exist as a single, more or less homogeneous, size class of circular molecules. Chloroplast genomes function in biogenesis of chloroplasts. Plastome encodes many of the gene products needed to perform photosynthesis. The plastome is replicated by nuclear-encoded proteins and its copy number appears to be highly controlled by the cell in a tissue-specific and developmental manner.

platform technology: Technology that has a common starting point but diverges once it is put into actual practice.

plating efficiency: This is a term which originally encompassed the terms, ATTACHMENT, SEEDING EFFICIENCY, CLONING EFFICIENCY, COLONY-FORMING EFFICIENCY and which is now better described by using one or more of them in its place as the term PLATING is not sufficiently descriptive of what is taking place. *See* ATTACHMENT, SEEDING, CLONING *and* COLONY-FORMING EFFICIENCY.

playback experiment: An experiment that describes the retrieval of DNA that has hybrid with RNA to check that it is non-repetitive by a further reassociation reaction.

pleckstrin homology domain (PH domain): Protein domain reported in intracellular signalling proteins by which they bind to inositol phospholipids phosphorylated by PI 3-kinase.

plectonemic winding: The interwinding of two strands in the classical double helix of DNA.

pleiochasium (pl. **pleiochasia**): An inflorescence in which numerous buds appear at the same time. *Compare with* MONOCHASIUM *and* DICHASIUM.

pleiotropic: With multiple effects of a given gene, usually in parts the organism not obviously related (e.g. flower colour and seedcoat colour).

pleiotropic genes: In general, one gene affects a single character but many cases are known in which a gene influences several unrelated characters. Pleiotropy helps in understanding various developmental processes. For example, biologist D. Ho in 1993 a single gene in barley (*Hordeum vulgare*) that controls the traits of the plant's height, drought resistance, strength and time to maturity.

pleiotropic mutation: A mutation that has effects on several different traits.

pleiotropy: The ability of a single gene to produce a complex phenotype (syndrome) that consists of two or more distinct characteristics. This phenomenon is important in development as it helps in understanding whether different phenotypic manifestations observed in an individual arise due to mutations in several different genes or a single mutation. Pleiotropy is of two types – SPURIOUS PLEIOTROPY and GENUINE PLEIOTROPY. *Aka* POLYPHENY. *See details as separate terms.*

pleomorphic: An organism able to adopt various shapes (and perhaps sizes).

pleomorphic bacteria: Some bacteria having the ability to alter their shape or size in response to environmental conditions.

ploidy: The occurrence of more than two complete sets of chromosomes in a genome. Natural polyploidy is the consequence of previous duplications of the whole genome or the introduction of complete genomes from another species during evolution.

ploidy-dependent changes in the epigenome of symbiotic cells: Transcriptional waves associate with growing ploidy levels and explored how the epigenome modifications might act during endoreduplication cycles. Differential DNA methylation was detected in only a small subset of symbiotic nodule-specific genes, including more than half of the nodule cysteine-rich (NCR) genes.

pluriflor: Showing many flowers per inflorescence. *Compare with* PAUCIFLOR *and* UNIFLOR.

pluripotency: (i) The ability to differentiate into multiple cell types. (ii) The potential ability of a cell to differentiate into all the three embryonic germ layers (endoderm, mesoderm and ectoderm) and to give rise to any embryonic tissue and any adult cell type. ESCs are pluripotent. *Aka* MULTIPOTENCY.

pluripotent stem cells: A stem cell that have the potential to differentiate into diverse types of cells. INDUCED PLURIPOTENT STEM CELLS (iPS cells or iPSCs) are a type of pluripotent stem cell artificially derived from an adult somatic cell, by inducing a 'forced' expression of certain genes.

pluripotent stem cells, uses of: Cell-based therapies from derivatives of pluripotent stem cells may prove effective in ameliorating the effects of some devastating diseases and injuries. But this will require ability to grow stem cells on a large scale. Human stem cell population proliferate more slowly than their murine counterparts and differentiate more readily and their cloning efficiency is very low. The addition of basic fibroblast growth factor (FGF) inhibits their tendency to differentiate and at the same time, improves cloning efficiency.

plus tree: A tree superior both to certain neighbouring trees and to the average of the stand.

plus/minus screening of cDNA libraries: mRNA is isolated from two populations of cells and reverse transcribed to produce a population of cDNAs. Random hexamers are oligonucleotide sequences of six bases which are synthesized entirely randomly to give a numerous range of sequences that have the potential to anneal at many random points on a DNA sequence and act as a primer to commence first-strand cDNA synthesis. Aliquots of the cDNAs can be converted to probe by random hexamer priming and used to screen duplicate lifts from a library.

plus-and-minus DNA sequencing: In this method of DNA sequencing, given by F. Sanger *et al.* (1977), DNA polymerase–produced single-stranded nucleotides are electrophoresed on polyacrylamide gels. Primer is elongated at the 3′ end. Method includes plus series of four experiments where the synthesis of new strand is stopped at various points by insertion of dideoxy analogues. In the minus series of four experiments, the oligonucleotide primer is elongated using only three of the deoxynucleoside triphosphates (e.g. dGTP, dCTP, dTTP are added, but dATP is missing). *Aka* ENZYMATIC METHOD *or* REPLICATION TERMINATION METHOD.

plus-end: The end of a microtubule or actin filament at which addition of monomers occurs most rapidly. *Aka* BARBED END.

pMB9: A small multicopy plasmid that carries a tetracycline-resistance gene.

PMC: *See* POLLEN MOTHER CELL.

pod: (i) A legume, the fruit of a leguminous plant, a dry fruit of a single carpel, splitting along two seams. (ii) A siliqua and silicula, the fruit of Brassicaceae, a dry fruit composed of two carpels separated by a partition.

point mutation: Refers to a change in single DNA nucleotide (e.g. adenine mutates to thymine or an insertion or deletion of a single nucleotide, especially in a region of DNA coding for protein). A mutation may cause an error in a codon of a DNA or RNA molecule. A mutation comprising a substitution of only one nucleotide in a DNA molecule. At 'hot spots' (i.e. certain locations on the DNA within some organisms), numerous point mutations can transpire. In the case of SINGLE-NUCLEOTIDE POLYMORPHISMS (SNPs), the same point mutation transpires at the same location (on the DNA within some organisms) across a population of individuals of that organism. *Compare with* FRAMESHIFT MUTATION.

point mutations in protein engineering: Many subtle variations in a particular protein can be generated by making amino acid replacements at specific positions in the polypeptide sequence. Each protein is unique by virtue of the sequence of its amino acids. At any location in the sequence, an amino acid can be replaced by another to generate a mutant protein that may have different characteristics by virtue of the single replaced amino acid. For example, enzyme pancreatic ribonuclease A that cleaves the covalent bonds that join RNA comprises 124 amino acids. If at position 119, the naturally occurring histidine is replaced with an alanine, the mutant protein histidine 119 → alanine is referred to as a H119A ribonuclease A. This mutant protein is expected to have little or no biological activity, because histidine 119 is important for that activity. If changes are also made at another position, for example, position 41 (where normally there is a lysine), then in principle there are 400 different variants that can be created by making all possible combinations of substitutions at just these two positions in the sequence. As more positions are varied, the number of combinations becomes enormous so that, with just six different positions subjected to all possible variations, 64 million different proteins can be generated. These proteins will share the identical sequence, except at the six varied positions.

point-accepted mutation (PAM): A unit introduced by M.O. Dayhoff *et al.* (1978) to quantify the amount of evolutionary change in a protein sequence. 1.0 PAM unit is the amount of evolution which will change, on average, 1% of amino acids in a protein sequence. A PAM(x) substitution matrix is a look-up table in which scores for each amino acid substitution have been calculated based on the frequency of that substitution in closely related proteins that have experienced a certain amount (x) of evolutionary divergence.

pointer: An object in many programming languages that stores a memory address. This can also be that of another value positioned in computer memory or in some cases, that of memory-mapped computer hardware. A pointer references a location in memory and obtaining the value stored at that location is known as DEREFERENCING THE POINTER. For example, a page number in a book's index could be considered a 'pointer' to the corresponding page; 'dereferencing' of such a pointer would be done by flipping to the page with the given page number and reading the text found on that page. The actual format and content of a pointer variable is reliant on the underlying computer architecture.

Poisson distribution: A frequency distribution for number of events per unit time, when the number of events is determined randomly and the probability of each event is low. The basis of a method whereby the distribution of a particular attribute in a population can be calculated from its mean occurrence in a random sample of the population, provided that the population is large and the probability that the attribute will occur is less than 0.1.

Pol I: *See* RNA POLYMERASE I.

Pol II: *See* RNA POLYMERASE II.

Pol III: *See* RNA POLYMERASE III.

polar: (i) At one end or pole of the cell, as for flagella. *Contrasts* PERITRICHOUS. (ii) In the electrical sense, a structure (e.g. a chemical bond, chemical group or molecule) with positive charge concerted towards one end and negative charge towards the other as a consequence of an uneven distribution of electrons. Polar molecules are most likely soluble in water.

polar body: The structure that is extruded from the oocyte during meiosis and contains one haploid set of chromosomes.

polar expansion theory: In this case, at the pole ends of the cell there is active growth of cell-surface material. Excess cell-surface accumulation of material deepens inwards. If cleavage were passive, injection of oil should block cytokinesis. But it was not observed to be so. Thus, this theory was rejected.

polar fibres: spindle fibres that extend from the two poles of a dividing cell.

polar gene conversion: A gradient of conversion frequency along the length of a gene.

polar mutations: Because transcription occurs only in one direction, sometimes a mutation in one gene reduces the expression of subsequent genes in the same transcription unit further down the line. Direction of transcription was determined by studies on polar mutants. For example, in the lactose operon of *Escherichia coli*, a polar mutation in *lacZ* gene influences *lacZ*, *lacY* and *lacA*; a polar mutation in *lacY* influences genes *lacY* and *lacA* only; and a polar mutation in gene *lacA* influences gene *lacA* only. These studies showed that direction of transcription is distally, from promoter to the operator and on to genes *lacZ*, *lacY* and *lacA* in that order.

polar nuclei: Two nuclei that migrate to the centre of the embryo sac and fuse with a male nucleus (sperm) to form the primary endosperm nucleus which divides and sometimes forms the endosperm.

polar plate: The functional centre for meiotic division of chromosomes. Spindle fibres from two poles meet on it.

polar relaxation theory: According to this theory, the surface of an egg is contractile. Evidence for this fact is provided by the observation that, just under the surface of the egg, there is an accumulation of actin. The surface somehow relaxes. This helps in the formation of cleavage furrows. Part of the surface that does not relax forms a cleavage.

polarity (genetic): (i) The distinction between the 5′ and the 3′ ends of nucleic acids. (ii) Having to do with the one way or unidirectionality of gene transcription in an operon unit. That is, the region near the operator is always transcribed before the more distant regions. By analogy, transcription begins at the left end of operon unit and proceeds towards the right end of the operon unit.

pole nucleus: One of the two haploid nuclei present in the centre of the embryo sac after division of the megaspore. They fuse to form a diploid definitive nucleus before fusing with the male gamete to form a triploid primary endosperm nucleus prior to double fertilization.

poles of embryo: In embryos, there are two centres of influence. One is VEGETAL POLE (V) and other is ANIMAL POLE (A). These two influences decrease to opposite poles. This is a double-gradient setup. According to this, micromeres have high vegetal and very little animal influence. Mesomeres have high animal influence. Is the gradient chemical in nature? If it is so, chemicals ought to mimic this effect. There are some chemicals that mimic the gradient. For example, if 16-cell embryo is treated with Li^{+} ions vegetalized embryo develops but if treatment is given with trypsin or Zn^{++} ions animalized embryo develops. This experiment shows that gradients are chemical in nature.

pollen: The cells that carry the male DNA of a plant. Together, the mass of microspores or pollen grains of angiosperms or microsporangia of gymnosperms produced within the anthers of a flowering plant. It is a highly specialized tissue whose function is the production of two sperm cells and their subsequent delivery through the style and ovary to the embryo sac cells where the double fertilization takes place. It is a highly reduced structure consisting of only three cells. The processes of pollen development, germination and fertilization involve the specific expression of a large number of genes.

pollen analysis: A method to study pollen grains, particularly their size, shape and surface. Since these pollen characters are highly specific for species, the method is widely used for taxonomic classifications.

pollen culture: The cultivation of pollen grains *in vitro* for producing haploid plants (generally referred to as another culture).

pollen embryogenesis: The process of developing embryos from cultured pollen. This is a unique phenomenon. In nature, after pollination the pollen germinates on the stigma to give pollen tube that helps in fertilization. But in tissue culture, particularly the growth regulators, change the mode of development of a pollen. The cultured pollen instead of forming pollen tube, develop into haploid pollen embryo that give rise to haploid pollen plant. This phenomenon is exploited commercially for production of haploid/doubled haploids for early release of varieties in rice, brassica, barley and wheat.

pollen escape: Escape of transgenic pollen that may pollinate a normal plant. In case of cross-pollinated crops, the transgenic pollen may pollinate the normal plants or the wild relatives of the transgenic crop. For example, in case of a transgenic variety is carrying herbicide resistance gene, then there is a risk that transgene may leak into weeds that can develop resistance to herbicide. There is no risk in case of transgenic vegetatively propagated plants that do not flower or the other crops/weeds are cross-incompatible.

pollen grain mitosis: In microsporogenesis, a nuclear division that ensues in the pollen grain after the formation of tetrads. It leads to appearance of a smaller generative nucleus and a larger vegetative nucleus.

pollen grain: A microspore in flowering plants that germinates to form the male gametophyte, a structure made up of the pollen grain plus a pollen tube.

pollen mass: Pollen grains adhering by a waxy texture or fine threads into a single body, called POLLINIUM (e.g. in orchids).

pollen mixing: Cross-incompatibility in interspecific crosses is related to proteins of the pistil that interact with proteins of the pollen to avert normal pollen tube germination and growth. This unfavourable reaction may be evaded in certain combinations by mixing pollen from a compatible species with pollen from an incompatible species.

pollen mother cell (PMC): The microsporocyte (2n), which experiences two meiotic divisions to produce four microspores, each microspore developing into a pollen grain.

pollen parent: The parent that contributes the pollen and that fertilizes the ovules of the other parent in the production of seed.

pollen pore: The pore present in the pollen, through which it ingresses the nutrition during its development. At the time of pollen germination, the pollen tube emerges from it.

pollen sac: A sac within the anther of a stamen within which microspores are formed.

pollen shedding: The status or process when the pollen grains are released from the anthers.

pollen sterility for hybrid seed production: To engineer pollen sterility, the *barnase* gene is placed under the control of a promoter that allows the gene to be expressed only during another development (i.e. in the male part of the flower). The *barnase* gene with its anther-specific promoter is stitched next to the transgene of interest, say, a gene coding for herbicide tolerance, also with its own promoter. Theoretically, there will be no fertile pollen from this transgenic crop. In the case of crops that are normally self-fertilized, there will be no seeds set. In out-crossing plants, the only fertile seeds set will be those fertilized by non-GM varieties nearby, which will not be herbicide tolerant; so, farmers who want the herbicide-tolerant trait will have to buy fresh seeds from the company every season. The problem is that such a male-sterile line by itself cannot be propagated, it does not breed true.

pollen sterility: Pollen that is not capable of fertilizing an egg cell.

pollen transfer: The mode by which pollen is blown out from the male to female organs (e.g. mediated by wind, insects and by hand).

pollen transmitting tissue: The tissue in the style of a flower through which the pollen tubes develop.

pollen tube: The structure formed from a germinating pollen grain and down which the two male gametes pass to the ovum.

pollen-aided direct gene transfer: This method of gene transfer was described by Y. Ohta in 1986, in maize. The exogenous DNA was mixed with pollen-recipient plants and applied on silks. The seeds thus obtained expressed genes from donor plants. The techniques of direct gene transfer are not yet successful to facilitate transfer of genes across a number of plant and animal species.

pollination: The transference of pollen from a male organ (such as an anther) to the receptive region of a female organ (such as a stigma) or the transference of pollen from anther to stigma or from a staminate cone to an ovulate cone. Pollen is most commonly transported by animals or by wind. Pollination can also be done in artificial culture (i.e. *in vitro* pollination).

pollination techniques: Particularly, in interspecific crosses different methods of pollination are functional in order to overcome PROGAMOUS INCOMPATIBILITY: (i) the grafted-ovary method, (ii) placental pollination, (iii) warm-water treatment, (iv) the mentor pollen technique and (v) hormone treatment.

pollinator: That parental individual, line, variety or species, which is used as a donor of pollen in a cross.

pollinium: *See* POLLEN MASS.

Polony sequencing: Polony sequencing, developed in George Church's laboratory in 2005 at Harvard, was among the first next-generation sequencing systems used to sequence a full genome. It combined an *in vitro* paired-tag library with emulsion PCR, an automated microscope and ligation-based sequencing chemistry to sequence an *Escherichia coli* genome at an accuracy of > 99.9999% and a cost of approximately one-tenth that of Sanger sequencing. The technology was licensed to Agencourt Biosciences, subsequently spun out into Agencourt Personal Genomics and ultimately incorporated into the Applied Biosystems SOLiD platform.

poly-dA/poly-dT technique: A method of inserting foreign DNA into a vehicle by making 5′ ploy-dA and 5′ ploy-dT tails on the vehicle and the foreign DNAs. Also feasible with poly-dG and poly-dC.

poly(A) polymerase: The enzyme responsible for polyadenylation of a eukaryotic mRNA molecule.

poly(A) site selection: In eukaryotes, tissue-specific selection (in certain transcription units) of alternative poly(A) sites and production of multiple messenger RNAs from one primary transcript.

poly(A) tail: A homopolymeric stretch of usually 25–200 adenine nucleotides that is added at the 3′-end of most eukaryotic mRNAs during processing, before its transport from the nucleus to the cytoplasm and subsequent translation at the ribosome. *Aka* POLY(A) TRACK.

poly(ADP-ribosyl)ation: Poly(ADP-ribosyl)ation frequently comprises the addition of long chains of ADP-ribose units, linked by glycosidic ribose-ribose bonds and is crucial for a wide range of processes, including DNA repair, regulation of chromosome structure, transcriptional regulation, mitosis and apoptosis.

polyacrylamide gel: A gel prepared by mixing a monomer (acrylamide) with a cross-linking agent

(N,N′-methylenebisacrylamide) in the attendance of a polymerizing agent. It results in the formation of an insoluble three-dimensional network of monomer chains, which become hydrated in water. There are tiny 'holes' in the gel (as in a plastic mesh bag) and with enough cross-linking the size of the holes begins to approach the size of the molecules that are to be separated. Since some molecules will be larger and some smaller, some of them will be able to pass through the gel matrix more easily than others. This is partially the basis for separation. If the gel is cross-linked enough and because of this the holes in that gel are smaller than the molecules to be separated, then the molecules will not be able to penetrate into the gel and no separation occurs. The charge on the molecule also performs a part in the separation. When run under conditions which denature proteins (i.e. in the presence of 2-mercaptoethanol, sodium dodecyl sulfate and possibly urea), molecules are separated mainly on the basis of size.

polyacrylamide gel electrophoresis (PAGE): A method used for separation of proteins and amino acids on the basis of their molecule size. The molecules move through the gel under the stimulus of an electric field. Polyacrylamide gel is engaged to separate biological molecules (proteins).

polyadenylation: Post-transcriptional addition of 50–250 adenine residues to the 3′ end of eukaryotic mRNA. The poly(A) tail can be used to separate eukaryotic mRNA from other RNA species with oligo(T) cellulose. The following consensus sequence is located 5–20 bp downstream from the ploy(A) addition site TTG/ANNNTTTTTT where N is any of the four standard DNA bases A, T, G and C.

polyadenylation editing: Addition of adenosine residues at end of several vertebrate mitochondrial pre-mRNAs to complete stop codons. Pre-mRNA lacking a stop codon is polyadenylated, with the first one or two adenylate residues providing the missing information.

polyadenylation signal sequence: A highly conserved sequence AAUAAA in pre-mRNA located 10–30 nucleotide 5′ to the cleavage site which defines the site of addition of poly(A) tail.

polyadenylation site: A site on the 3′-end of messenger RNA (mRNA) that signals the addition of a series of adenines during the RNA processing step and before the mRNA migrates to the cytoplasm. These so-called POLY(A) TAILS increase mRNA stability and allow one to isolate mRNA from cells by PCR amplification using poly(T) primers.

polyadenylation, functions of: Sometimes polyadenylation plays a role in the generation of translational STOP CODONS. Other functions of polyadenylation include providing for the stability of mRNAs, assisting in transport during processing and translation, and direct involvement in protein synthesis. Last activity was suggested by the observation that when mRNAs were being translated, originally short poly(A) tail became longer and, on the other hand, some other mRNAs underwent depolyadenylation when they were no longer being translated.

polyamidoamine (PAMAM)-mediated gene transfer: Dendrimers are another group of polymers that are composed of three-dimensional, branched structures known as dendrons. Among them, the polyamidoamine PAMAM family of dendrimers has been used for transfection. Since the sphere-shaped polycationic dendrimers are alike in proportion and shape to the histone clusters, they can compact the DNA to a small size and facilitate its entry into cells.

polybrid: A hybrid having more than two parental groups.

polycations: Large, positively charged molecules often used to mediate transfection by reducing charge repulsion between DNA and the cell membrane.

polycentric: Where chromosomes may sometimes contain more than two centromeres, as in *Luzula* plants and *Ascaris megalocephalus*, the dicentric and polycentric chromosomes form anaphase bridges.

polycistronic messenger RNA: A mRNA containing multiple ORFs is called polycistronic mRNA which often encodes proteins that perform related functions such as different steps in the biosynthesis of amino acids and nucleotides. Prokaryotic mRNA frequently contains two or more ORFs and hence can encode multiple polypeptide chains. Prokaryotic mRNAs are usually polycistronic. For example, lactose operon of *Escherichia coli*, from a single mRNA produces β-galactosidase, galactoside permease and thiogalactose transacetylase, which are used in breakdown of lactose. *Aka* POLYCISTRONIC TRANSCRIPTION UNIT. *Contrasts* MONOCISTRONIC MESSENGER RNA.

polycistronic versus unicistronic mRNAs: Messenger RNAs are polycistronic in prokaryotes – a mRNA may code for two or more different polypeptide chains. On the other hand, eukaryotic mRNAs are unicistronic – one mRNA codes for only one polypeptide chain.

polyclonal antibodies: A mixture of antibodies generated by different antibody-producing cells against more than one epitope of an antigen. *See* EPITOPE. *Contrasts* MONOCLONAL ANTIBODIES.

polycloning region: A segment of DNA in a cloning vector such as a plasmid that contains many different restriction-enzyme cleavage sites.

Polycomb complex: The Polycomb complex composition and localization changes dynamically at *FLC* during different phases of vernalization. Before the onset of cold, which triggers vernalization, the PRC2 core complex is already associated with chromatin over the length of the active *FLC* locus. The exon–intron structure is indicated beneath the chromatin fibre as black bars for each exon. Prolonged cold leads to the accumulation and nucleation of an alternative Polycomb complex containing PLANT HOMEODOMAIN (PHD) proteins (VIN3, VRN5) at a specific intragenic site near the beginning of the first intron. In plants returned to warm conditions, the cold-induced VIN3 PHD protein is lost. A modified PHD-PRC2 complex associates across the whole locus, inducing high levels of H3K27me3, which blanket the locus and provide repressive epigenetic stability (maintenance). *FLC* expression is progressively silenced by vernalization as plants are exposed to increasing periods of cold. This epigenetic process translates the prolonged exposure to cold into a stable silencing of *FLC* expression. This is maintained throughout the rest of development until it is reset in the embryo. The mitotic memory of vernalization involved changes in DNA methylation, a well-characterized epigenetic mark in plants. There is little DNA methylation at the *FLC* locus and it does not change with cold. H3K27me3 and the Polycomb protein Su(z)12 (PRC2) are required for the memory of *FLC* silencing. Reminiscent of Polycomb silencing in mammalian embryonic stem cells, *Arabidopsis* PRC2 is already localized at *FLC* before silencing. Two proteins, VRN5 and VIN3, containing PHD and FNIII domains then associate with PRC2 to trigger the silencing. An important part of the cold-induced epigenetic silencing at *FLC* is the cold-induced expression of one of the PHD proteins (VIN3). VIN3 heterodimerizes with the constitutively expressed PHD protein (VRN5). They associate with PRC2 at a site within the first intron of *FLC*, causing local accumulation of the histone modification H3K27me3, typical of Polycomb silencing. On transfer of plants back to warm temperatures, a PHD–PRC2 complex then spreads along the length of *FLC*, generating very high levels of H3K27me3 across the whole locus. These very high levels of H3K27me3 are required for the epigenetic maintenance of the silencing. LHP1, the *Arabidopsis* homologue of the metazoan HETEROCHROMATIN PROTEIN1 (HP1), is also required to maintain the epigenetic silencing and intriguingly binds to H3K27me3, rather than H3K9me3.

Polycomb group (PcG) proteins: A family of proteins first discovered in fruit flies that can remodel chromatin such that transcription factors cannot bind to promoter sequences in DNA. In Drosophila, the Trithorax group (trxG) and PcG proteins act antagonistically and interact with chromosomal elements, termed CELLULAR MEMORY MODULES (CMMs). PcG proteins possess a chromodomain-containing methylated histone H3 lysine 27 (H3K27me)-binding factor and function to maintain stable and heritable epigenetic repression.

Polycomb Repressive Complex 1 (PRC1): A conserved protein complexes that is composed in a hierarchical manner in which the catalytic core, RING1A/B, exclusively interacts with one of six Polycomb group RING finger (PCGF) proteins. PRC1 has role in transcription activation.

Polycomb Repressive Complex 2 (PRC2): A conserved protein complex that is involved in chromatin silencing through methylation of H3K27. They function as histone writers. *See* HISTONE WRITERS.

Polycomb/Trithorax system: Polycomb group (PcG) proteins and Trithorax group (TrxG) proteins are two classes of chromatin-modifying proteins that epigenetically regulate *Hox* gene expression. Both groups form multiprotein chromatin remodelling complexes that either repress (PcG) or activate transcription (TrxG). In *Drosophila*, the Trithorax group (trxG) and Polycomb group (PcG) proteins act antagonistically and interact with chromosomal elements, termed CELLULAR MEMORY MODULES (CMM). The trxG has been associated with gene activation and the H3K4me3 mark and PcG with gene repression and the H3K27me3 mark. The Polycomb/Trithorax system is conserved across the animal kingdom. The patterns of histone marks maintained by the Polycomb/Trithorax act as a kind of cellular GPS system specifying the anterior/posterior and proximal/distal coordinates of the cell within the body.

polycross: Open pollination of a group of genotypes (generally selected) in isolation from other compatible genotypes in such a way that it promotes random mating *inter se*. It is a widely used procedure for intercrossing parents by natural hybridization.

polycross progeny: Progeny from a selection, line or clone outcrossed to other selections growing in the same isolated polycross nursery.

polycross test: A progeny test to evaluate general combining ability from crosses among selected parents. Identities can be maintained only for the seed parents. A mixture of pollen is artificially applied to each female parent.

polyculture: Cropping systems in which different crop species are grown in mixtures in the same field at the same time. *Contrasts* MONOCULTURE.

polycyclic disease: A disease that completes two or more life cycles in 1 year. *Contrasts* MONOCYCLIC DISEASE.

polyembryony: (i) Refers to development of more than one embryo in a seed or multiple-embryo development in culture. Usually one seed carries one embryo but the seeds of certain plant species carry more than one embryo. For instance, in citrus, a seed carries three to four embryos. Among these, one embryo is zygotic, whereas others do develop from nuclear tissue. When seed is grown in soil, three to four plants arise from a single seed. (ii) The development of more than one embryo in an embryo sac. In plants usually there is one embryo in one seed as it can be seen during seed germination of wheat, rice, maize, soybean and cotton (one seed produces one plant), whereas some plant species exhibit polyembryony (e.g. citrus). In this case, from one seed, four to five seedlings arise of which one is zygotic and others are nucellar. The presence of more than one embryo in the seed is called polyembryony.

polyethylene: A plastic polymer of ethylene.

polyethylene glycol (PEG): A polymeric substance of molecular weight between 1000 and 6000. It is used for stimulation of protoplast fusion.

polyethylene glycol (PEG)-mediated gene transfer: PEG (15–25%) stimulates DNA uptake by endocytosis without any gross damage to protoplasts. The transformed protoplasts can be selected on a selection medium. PEG has long been known to promote uptake of naked DNA by protoplasts and is routinely used to deliver biologically active DNA into the plant nuclear genome. Chloroplast transformation can also be accomplished by chemical treatment of protoplasts with PEG in the presence of vector DNA.

polyethylene glycol (PEG)-stimulated DNA uptake: A method of gene transfer which involves use of 15.25% PEG, which stimulates uptake of DNA by endocytosis without any gross damage to protoplasts.

polyethylenimine (PEI)-mediated gene transfer: PEI is a polycationic organic macromolecule that has a high cationic charge density (also called a PROTON SPONGE). It condenses DNA into positively charged particle that interacts with the cell surfaces that are anionic in nature and gains entry into the cells by means of endocytosis.

polyetic: Which requiring more than 1 year to complete one life or disease cycle.

polyfunctional alkylating agent: Alkylating agents that carry more than two alkyl groups in reactive form which are capable of being transferred to other molecules where electron density is high. Prokaryotic mRNAs are usually polycistronic, for example, nitrogen mustard (IHN2) (Mustargen), Thio-TEPA, Chlorambucil (Leukeran), Cyclophosphaniide (Cytoxan) and Rusulfan (Myleran).

polygalacturonase-inhibiting proteins (PGIPs): Leucine-rich repeat (LRR) proteins implicated in the plant defence system. A number of PGIPs have been characterized from dicot species, whereas only limited data are available from monocots.

polygamodioecious: Bisexual and male flowers occur on some plants and bisexual and female flowers on others. *Compare with* ANDRODIOECIOUS, ANDROMONOECIOUS, DIOECIOUS, MONOECIOUS, POLYGAMOMONOECIOUS *and* POLYGAMOUS.

polygamomonoecious: Male, female and bisexual flowers occurring on the same plant. *Compare with* ANDRODIOECIOUS, ANDROMONOECIOUS, POLYGAMODIOECIOUS *and* POLYGAMOUS.

polygamous species: The species that produces bisexual, unisexual as well as neutral flowers. Such a condition is described as polygamous. Mango is an example.

polygenes: One of a group of genes that together contribute to a quantitative character. Individually each gene has little effect on the resulting phenotype, which instead requires the interaction of many genes. Initially, the term MULTIPLE FACTOR was used which was later replaced by the term POLYGENE by K. Mather in 1943. He defined polygenes as genes with a small effect on a particular character that can supplement each other to produce observable quantitative changes. Polygenes have been hypothesized to have two types of alleles. Those contributing to the traits are termed CONTRIBUTING ALLELES or EFFECTIVE ALLELES and those appearing not to do so are referred to as NON-CONTRIBUTING ALLELES, NON-EFFECTIVE ALLELES or INERT ALLELES. *Aka* MULTIPLE FACTORS. In molecular biology, they are *aka* QUANTITATIVE TRAIT LOCI (QTL).

polygenic: Pertaining to a trait controlled by many genes. *Aka* POLYGENY. *Contrasts* MONOGENIC.

polygenic balance: The harmonious balance, between the systems of polygenes in a species and the natural environment of that species, produced as a result of natural selection and enabling the organisms to develop normally and competitively.

polygenic disorder: Genetic disorder resulting from the combined action of alleles of more than one gene (e.g. heart disease is diabetes and some cancers). Although such disorders are inherited, they depend on the simultaneous presence of several alleles; thus, the hereditary patterns usually are more complex than those of single-gene disorders. *Compare with* SINGLE-GENE DISORDER.

polygenic inheritance: Inheritance involving alleles at many genetic loci, which interact with environmental factors. Polygenic refers to many genetic factors as opposed to the environmental component. *See* QUANTITATIVE INHERITANCE *and* MULTIPLE FACTOR HYPOTHESIS.

polygenic resistance: Resistance to parasites based on many genes (i.e. a series of genes governing a quantitative resistance). *See* MULTIGENIC RESISTANCE. *Contrasts* MONOGENIC RESISTANCE, OLIGOGENIC RESISTANCE *and* SINGLE-GENE RESISTANCE.

polygenic trait: An inherited trait that is determined by genes at two or more loci. Simple Mendelian rules of dominance do not apply to the complex interaction of genes that control quantitative traits. As a result, phenotypes may appear as apparent blends or intermediate expressions. Since multiple genes are involved, polygenic traits do not follow the patterns of Mendelian inheritance. Many polygenic traits are also influenced by the environment and are called MULTIFACTORIAL TRAITS. Human skin and hair colour are polygenic traits. For example, the level of protein produced in soybeans is controlled by five genes.

polygyny: Sexual reproduction strategy in which one male mates with several females. For example, lions, peacocks and gorillas all have polygynous mating systems. *Compare with* POLYANDRY *and* MONOGAMY.

polyhaploid: Haploid of a polyploid which contains more than one set (X) of chromosomes. The common bread wheat is allohexaploid with chromosome number, 2n = 6x = 42. Therefore, the haploids produced through wide crosses or another culture possess 21 chromosomes, n = 3x = 21 are called polyhaploids. This term is generally used for haploids of higher polyploids (>4x).

polyhybrid: Individuals that are heterozygous with respect to the alleles of many gene loci. *Aka* MULTIHYBRID.

polyhydroxyalkanoic acid (PHA): A 'family' of chemically related 'energy storage' substances (i.e. polyesters) that is naturally produced by certain bacteria (90 strains known). When PHA is detached from the bacteria and purified, this substance possesses physical properties resembling thermoplastics like polystyrene. PHA can rapidly be broken down by soil microorganisms. Thus, PHA is a biodegradable plastic. In 1994, researchers transferred genes for the production of one PHA into the weed plant *Arabidopsis thaliana* and the crop plant rapeseed (canola). In 1997, researchers transferred *phaB* and *phaC* genes into the crop plant cotton (*Gossypium hirsutum*), which caused those transformed plants to express (produce) PHA inside the fibres (seed hair cells) in the amount of 0.34% of the fibre weight. That PHA (inside those cotton fibres) resulted in a fabric (i.e. cottonPHA 'blend') possessing better insulation properties than traditional cotton fabric.

polyhydroxylbutylate (PHB): One of the polyhydroxyalkanoates (PHAs), polyhydroxylbutylate is an 'energy storage' substance that is naturally produced by certain bacteria, yeasts and plants. When removed from the bacteria and purified, this substance has physical properties quite similar to thermoplastics like polystyrene. PHB can quickly be broken down by soil microorganisms, so PHB is a biodegradable plastic. Three separate enzymes are utilized by the organism in order to make the PHB molecule. In 1994, researchers succeeded in transferring genes for PHB production into the weed plant *Arabidopsis thaliana* and the crop plant rapeseed (canola). Later in 1997, researchers transferred aceto-acetyl-CoA reductase (*phaB*) and PHB synthase (*phaC*) genes into cotton (*Gossypium hirsutum*), which caused those transformed plants to produce PHA inside the fibres (seed hair cells) in the amount of 0.34% of the fibre weight. That PHA inside those cotton fibres resulted in a fabric (i.e. cotton PHA 'blend') possessing better insulation properties than traditional cotton fabrics.

polykaryotic: Cells possessing many nuclei.

polylinker: A synthetic double-stranded short oligonucleotide consisting of a series of restriction sites. For special purposes, vectors are modified by inserting DNA fragments to create a series of restriction sites for different enzymes to facilitate its use in gene cloning.

polymer: Small organic molecules called monomers combine to form large molecules called polymers (macromolecules). The monomers join together through condensation reaction in which water is produced. Carbohydrates, proteins, lipids and nucleic acids are polymers. For example, DNA, a polynucleotide chain is made of monomers called nucleotides. Protein is a polypeptide chain made of monomers called amino acids.

polymerase: An enzyme that catalyses the assemblage of ribonucleotides into RNA (RNA POLYMERASE) and of deoxyribonucleotides into DNA (DNA POLYMERASE). There exists a wide variety of RNA and DNA polymerases which have a wide range of specific activities and which function efficiently under different conditions. In general, all polymerases necessitate templates upon which to build a new strand of DNA or RNA; however, DNA polymerases also require an RNA primer of 10–20 nucleotides to initiate the new strand, while RNA polymerases start synthesis at a specific promoter sequence.

polymerase chain reaction (PCR): *In vitro* method for DNA amplification using thermocycler. PCR is an *in vitro* method for the enzymatic synthesis of specific pieces of DNA that has revolutionized research in the area of life sciences. The process was developed by Kary Mullis (K.B. Mullis, US patent 4,683,195; K.B. Mullis, US patent 4,683,202) for which he was awarded 1993 Nobel Prize in Chemistry. PCR process requires two oligonucleotide primers that hybridize to opposite strands of the target DNA and flank the region to be amplified. A suitable DNA polymerase, the four DEOXYRIBONUCLEOSIDE TRIPHOSPHATES (dNTPs) and Mg^{2+} are also required. Amplification results from successive cycles of denaturation and oligonucleotide priming of the target DNA followed by the enzymatic synthesis of the strand of DNA that is complementary to the target DNA. To perform a PCR amplification, a mixture containing target DNA, oligonucleotide primers, dNTPs and a heat-stable DNA polymerase is heated to 90–95°C to denature the strands of the target DNA. PCR method for amplifying a DNA base sequence uses a heat-stable polymerase and two 20-base primers, one complementary to the (+) strand at one end of the sequence to be amplified and one complementary to the (–) strand at the other end. Heat-stable DNA polymerase, called *Taq* polymerase (from *Thermus aquaticus*) is used for multiple rounds of DNA synthesis to amplify that segment. The solution is cooled to a temperature that allows the primers to anneal to their complementary sequence on the target DNA and provide the 3′-OH required for DNA synthesis. The DNA polymerase then synthesizes a new DNA strand complementary to the target. In each cycle, the template concentration doubles with respect to the previous cycle. Therefore, for initial double-stranded DNA molecule there will be 2c molecules, where c = number of cycles. The newly synthesized strand initially extends a variable distance beyond the other primer-binding site on the complementary strand, creating additional primer-binding sites on the newly synthesized strand. This thermal cycling scheme is repeated numerous times with the DNA synthesized during the previous cycles serving as a template for each subsequent cycle. With each cycle the strands are denatured at high temperature, the primers annealed and then extended by the polymerase. The result is the doubling of the target DNA present with each cycle. This exponential accumulation of DNA of DNA sequences can theoretically produce over a million-fold amplification of the target in 20 cycles. In practice, the amplification efficiency is less than 100% and 30 to 40 cycles are commonly done.

polymerase chain reaction (PCR) amplification: *See* POLYMERASE CHAIN REACTION.

polymerase chain reaction (PCR) mutagenesis: A method for mutation induction where primers are premeditated to mismatch with their target and conditions are created so that annealing is still allowed. Amplification introduces the mutation into the amplified product.

polymerase chain reaction (PCR) mutagenesis and augmentation: Primer design can be engaged not only to intensify sequences but also to modify them. This has two major applications: the introduction of point mutations (a form of *in vitro* mutagenesis) and end modification of PCR products to enable further manipulation. In the first approach (PCR MUTAGENESIS) primers are designed to mismatch with their target and conditions are created so that annealing is still permitted. Amplification thus introduces the mutation into the amplified product, although only at the ends, which are specified by the primer. If a central replacement is required, the mutant product can itself be used as a large primer (a MEGAPRIMER) on the original template for extension or an overlapping pair of PCR reactions can be performed followed by combination of the two products, either by restriction digestion and ligation or by end-to-end recombination PCR. In the second approach, novel sequences are added to the 5′ primer ends which do not pair with the template DNA but permit extra sequences to be added on to each end of the PCR product. This approach has numerous applications: the addition of restriction endonuclease sites facilitates subcloning while the addition of universal primer-binding sites for sequencing and the addition of bacteriophage promoters facilitates *in vitro* transcription.

polymerase chain reaction (PCR) primers, designing of: PCR primers based upon protein sequence. PCR primers are based on multiple alignments. There are several excellent sites for designing PCR primers: (i) Primer3: The WWW primer tool is a very powerful PCR primer design programme permitting considerable control over the nature of the primers, including the size of the product desired, primer size and Tm range and presence/absence of a 3′-GC clamp. (ii) GeneFisher is an interactive PCR primer design and a very good site, allowing great control over primer design. (iii) PCR Now has been created to design REAL-TIME POLYMERASE CHAIN REACTION (RT-PCR) primers for any number of user-defined coding sequences. Great control over primer properties. (iv) Primer-BLAST was developed at National Centre of Biotechnology Information (NCBI) to help users make primers that are specific to the input PCR template. It uses Primer3 to design PCR primers and then submits them to BLAST search against a user-selected database. The blast results are then automatically analysed to avoid primer pairs that can cause amplification of targets other than the input template. (v) JOG 1:01 (JavaScript Oligonucleotide Generator) generates fixed length and composition or random length and composition oligonucleotides. If you have the protein sequence and want the DNA sequence, the best way is to reverse the translation of the protein. (a) European Molecular Biology Laboratory (EMBL): If you are interested in changing a specific amino acid into another, you should consult Reverse Translator (EMBL). (b) CODEHOP (the acronym is from COnsensus-DEgenerate Hybrid Oligonucleotide Primers) is used to design primers based upon multiple sequence alignments. (vi) The primaclade application accepts a multiple-species nucleotide alignment file as input and identifies a set of PCR primers that will bind across the alignment. (vii) PriFi: upload a file containing Fasta-formatted DNA sequences or alternatively, a *.aln file, select the control one wants over the primer design from an extensive list and press 'Find primers in alignment'. (viii) Genomic scale primers: See the JAVA page for additional downloadable programmes. (ix) The PCR Suite: this is a suite of four programmes based upon Primer3 for genomic primer design. All offer considerable control over primer properties: (a) Overlapping Primer: creates multiple overlapping PCR products in one sequence; (b) Genomic_Primers designs primers around exons in genomic sequence; (c) SNP_Primers designs primers around every SNP; and (d) cDNA_Primers – designs primers around open reading frames. (x) GenoFrag – a software package to design primers optimized for whole-genome scanning by long-range PCR.

polymerase chain reaction (PCR), applications of: PCR has medical applications, infectious disease applications, forensic applications and research applications.

polymerase chain reaction (PCR), drawbacks of: Standard PCR suffers from several drawbacks: (i) The sequence on the side of target DNA must be known. (ii) It works towards the target. So, this has been altered to anchored and inverse PCR. (iii) Contamination may occur in PCR for some reason which may generate problem both in research as well as in diagnostic applications.

polymerase chain reaction (PCR), forensic applications of: (i) In its most discriminating form, Genetic fingerprinting can uniquely discriminate any one person from the entire population of the world. Minute samples of DNA can be isolated from a crime scene and compared to that from suspects or from a DNA database of earlier evidence or convicts. Simpler versions of these tests are often used to rapidly rule out suspects during a criminal investigation. Evidence from decades-old crimes can be tested, confirming or exonerating the people originally convicted. (ii) Less discriminating forms of DNA fingerprinting can help in parental testing, where an individual is matched with their close relatives. DNA from unidentified human remains can be tested and compared with that from possible parents, siblings or children. Similar testing can be used to confirm the biological parents of an adopted (or kidnapped) child. The actual biological father of a newborn can also be confirmed (or ruled out).

polymerase chain reaction (PCR), procedure of: The reactants in PCR are DNA, primers (small pieces of DNA with base sequence homology to the ends of the DNA to be amplified), thermal stable DNA polymerase and deoxynucleotide triphosphates (dATP, dTTP, dGTP and dCTP). PCR procedure comprises the following steps: (i) Since the sequence of nucleotides on either side of the DNA sequence to be amplified is known, primers are constructed on either side of the sequence of interest. Once that is done, the sequence between the primers can be amplified. (ii) The primers and the ingredients for DNA replication are added to the sample. (iii) The mixture is then heated to 95°C for 20 s to denature the DNA, the step is called DENATURATION. (iv) The temperature is then lowered to 55°C for 20 s so that primers can anneal to their complementary sequences, the step is named ANNEALING. (v) The temperature is then raised to 72°C for 20 s for DNA replication, the step is called ELONGATION or EXTENSION. (vi) Then, a new cycle of replication is initiated. The various stages in the cycle are controlled by changes in temperature since the temperatures for denaturation, primer annealing and DNA replication are different. (vii) In about 20 cycles, a million copies of the DNA are made. In 30 cycles, a billion copies are made. (viii) The technique is made more efficient by using DNA polymerase from a hot springs bacterium, *Thermus aquaticus*, known as *Taq* polymerase, that can withstand the denaturing temperatures. Thus, after each cycle of replication, new components do not have to be added to the reaction mixture. Some machines can process 96 samples simultaneously. *Aka* MOLECULAR PHOTOCOPYING.

polymerase chain reaction (PCR), research applications of: (i) PCR permits rapid production of short pieces of DNA, even when nothing more than the sequence of the two primers is known. PCR thus augments many methods, such as generating hybridization probes for Southern or Northern blot hybridization. PCR supplies these techniques with large amounts of pure DNA, sometimes as a single strand, allowing analysis even from very small amounts of starting material. (ii) The task of DNA sequencing can also be aided by PCR. Known segments of DNA can easily be produced from a patient with a genetic disease mutation. Modifications to the amplification procedure can extract segments from a completely unknown genome or can generate just a single strand of an area of interest. (iii) PCR offers numerous applications to the more traditional process of DNA cloning. It can extract segments for insertion into a vector from a larger genome, which may be only available in small quantities. Using a single set of 'vector primers', it can even analyse or extract fragments that have already been inserted into vectors. Some alterations to the PCR protocol can generate mutations (general or site-directed) of an inserted fragment. (iv) Sequence-tagged site is a method where PCR is employed as an indicator that a particular segment of a genome is present in a particular clone. The Human Genome Project found this application dynamic to mapping the cosmid clones they were sequencing and to coordinating the results from different laboratories. (v) An exciting application of PCR comprises the phylogenic analysis of DNA from ancient sources, such as that the recovered bones of Neanderthals or from frozen tissues of Mammoths. In some cases, the highly degraded DNA from these sources might be reassembled during the early stages of amplification. (vi) A common application of PCR is the study of patterns of gene expression.

polymerase chain reaction (PCR), types of: Various types of PCR include the following: amplified fragment length polymorphism PCR (AFLP-PCR), allele-specific PCR, assembly PCR, asymmetric PCR, helicase-dependent amplification, hot start PCR, inverse PCR, *in situ* PCR, ISSR PCR, linear-after-the-exponential PCR (Late-PCR), long PCR, nested PCR, reverse transcription polymerase chain reaction (RT-PCR), real-time PCR and single-cell PCR. *See details as separate terms.*

polymerase chain reaction (PCR)-based molecular markers: Such markers are in demand because of their simplicity and requirement for only small quantities of sample DNA. Molecular markers are used for a variety of purposes, including chromosome mapping, DNA fingerprinting and genetic screening.

polymerase chain reaction (PCR)-based screens: Screening by PCR is an efficient way to identify pools containing specific clones. PCR-based screening is also a method to isolate novel cDNAs. The PCR primers can be designed and used to amplify reverse transcriptase products of mRNAs in an appropriate tissue.

polymerase chain reaction (PCR)-induced mutagenesis-restriction fragment length polymorphism (PIM-RFLP): L. Angelopoulou *et al.* (2023) present a novel strategy which is based on the well-documented robustness and simplicity of the classical PCR-RFLP approach. PIM-RFLP allows the assessment of the editing efficiency in detection of CRISPR-induced indels in pools of edited cells and the effective identification of fully mutated single-cell clones. This method is based on the creation by mutagenic PCR of a restriction-enzyme degenerate cleavage site in the PCR product of the wild-type allele, which can then be distinguished from the indel-containing alleles following the standard RFLP procedure. PIM-RFLP is highly accessible, can be executed in a single day and appears to outperform Sanger sequencing deconvolution algorithms in the detection of fully mutated clones.

polymerase chain reaction (PCR)-mediated cloning: The DNA fragment of interest and even the vector, can be amplified by the polymerase chain reaction (PCR) and ligated together, without the use of restriction enzymes. This is a rapid method for cloning genes, and is often used for projects that require higher throughput. It allows for the cloning of DNA fragments that are not available in large amounts.

polymerase chain reaction-restriction fragment length polymorphism (PCR-RFLP): Isolation of sufficient DNA for RFLP analysis is time consuming and labour intensive. However, PCR can be used to amplify very small amounts of DNA, usually in 2–3 h, to the levels required for RFLP analysis. Therefore, more samples can be analysed in a shorter time. *Aka* CLEAVED AMPLIFIED POLYMORPHIC SEQUENCE (CAPS) ASSAY.

polymerase-associated factor complex (PAFc): A protein complex that facilitates transcriptional initiation and elongation, in part through the recruitment of the mixed lineage leukaemia (MLL) compass complex and the RNF20–RAD6 complex, which leads to H3K4 trimethylation and H2AK119 monoubiquitylation.

polymeric effect: This type of gene interaction is shown between dominant genes in case of fruit shape in summer squash. Either dominant gene shows the same effect but interaction between dominant genes produced a novel phenotype. In this case we get three phenotypic classes in ratio 9:6:1.

polymeric genes: Genes with equivalent effects but collective action.

polymeric nanoparticles: Polymeric nanoparticles constructed from polycationic polymers are employed as efficient siRNA delivery systems which can shield and competently transport siRNAs to the target cells.

polymerism: A phenomenon in which non-allelomorphic genes having identical and cumulative action interact. The genes involved in such interactions cannot be identified individually. These interactions may be expressed statistically as components of genetic variance.

polymerize: The act or procedure of forming a polymer or polymeric compound.

polymery: The production of a trait via cooperation among several polymeric genes.

polymix breeding: This is an alternative method to full-sib crossing and testing. In this case, a pollen mix of many male parents rather than a single pollen for each cross is applied. Polymix breeding has some advantages: it is easy to implement, ensures good estimates of breeding values of the parent being pollinated and offers increased genetic gain opportunity because of the significantly increased number of effective parental combinations tested. It has, however, found limited use because in polymix breeding inbreeding and pedigree control is lost.

polymorphic: Of several different kinds (with respect to shape and/or size), hence polymorphic. *See* MONOMORPHIC (a single type) *and* DIMORPHIC (two types).

polymorphic DNA: A variable DNA sequence that can exist in a number of different but related forms.

polymorphic gene: Refers to a gene that has many different alleles, but none of them is predominant in the population. There is measurable variation in a phenotype. In biotechnology, when a marker locus is polymorphic, there are detectable differences in DNA sequence at that locus.

polymorphism: (i) In biochemistry, the term refers to the attribute of a chemical substance crystallizing (or simply existing) in two or more arrangements having different structures. For example, diamond and graphite are two different manifestations of the element carbon. (ii) In evolution, it is applied to a state in which a species of plant or animal is represented by several distinct, non-integrating forms or types unrelated to age or sex. The changes are often in coloration, though any property of the organism may be involved (e.g. nuclei shape for polymorphonuclear leukocytes). (iii) In population genetics, the presence of several forms of a trait or a gene in an interbreeding population. DNA sequences are mutating all the time. As two populations diverge over time, they each accumulate different mutations at different sites (e.g. when restriction sites in the vicinity of a given gene are compared from one genotype to another, one genotype may have the site and the other will not). (iv) In molecular genetics, DNA is a polymorphic compound because the polymer can take on different forms. A gene or unexpressed DNA variant that occurs in a population each of which has a frequency of 1% or greater in the population gene pool, too high to be explained by recurrent mutation. Polymorphisms in genes may result in protein polymorphisms. Also called GENETIC POLYMORPHISM. High polymorphism between two genotypes is evidence of genetic divergence. The most common type of polymorphism involves variation at a single nucleotide (also called a SINGLE-NUCLEOTIDE POLYMORPHISM or SNP). *See* BALANCED POLYMORPHISM *and* MUTATION.

polymorphism screening and genotyping of the mapping population: (i) Identify sufficient number of markers that reveal differences between parents (i.e. polymorphic markers). (ii) Sufficient polymorphism must exist between parents. (iii) Genotype the entire mapping population.

polymorphism, protein: A protein polymorphism is said to occur when the most common allele has a frequency of no greater than 99%.

polyneme hypothesis: The proposition that a chromatid contains more than one duplex DNA running in parallel. This hypothesis has been largely superseded by the present understanding that a single duplex of DNA is present in a chromatid, but that is elaborately packaged by supercoiling and looping out from a central axis. So, only at a superficial level can separate gene loci be regarded as being in purely linear order along a chromosome length.

polynemic chromosome: Describes metaphase chromosome and/or chromatids with more than two DNA helices. For example, polytene chromosomes found in *Drosophila* larval ganglionic cells.

polynucleotide: A linear sequence of many nucleotides.

polynucleotide kinase: Enzyme which catalyses the transfer of the terminal phosphate of ATP to 5′ hydroxyl termini of polynucleotides, either DNA or RNA. Usually derived from T4 bacteriophage.

polynucleotide phosphorylase (PNPase): A bifunctional enzyme with a phosphorolytic 3′ to 5′ exoribonuclease activity and a 3′-terminal oligonucleotide polymerase activity. It dismantles the RNA chain starting at the 3′ end and working towards the 5′ end. It also synthesizes long, highly heteropolymeric tails *in vivo*. It accounts for all of the observed residual polyadenylation in strains of *Escherichia coli* missing the normal polyadenylation enzyme. PNPase was discovered by M. Grunberg-Manago working in S. Ochoa's laboratory in 1955.

polypeptide: A long molecular chain of amino acids linked by peptide bonds. Via the synthesis of this chain, achieved by ribosomes, each polypeptide (protein) in nature is the ultimate expression product of a gene. All of the amino acids commonly found in proteins have an asymmetric carbon atom, except the amino acid glycine. Thus, the polypeptide is potentially chiral in nature. A protein is made up of one or several polypeptide chains. A small polypeptide chain usually consists of 100 or fewer amino acids. Titin, with its length of around 27,000–35,000 amino acids (depending on the splice isoform), is the largest known protein. *Aka* POLYPEPTIDE CHAIN. *See* OLIGOPEPTIDE *and* PROTEIN.

polypeptide backbone: The chain of repeating carbon and nitrogen atoms, linked by peptide bonds (−CONH−), in a polypeptide or protein. The side chains (called R groups or residues) of the amino acids project from this backbone.

polypeptide elongation: This process starts when acyl (A) site is occupied by an incoming amino acid–tRNA and peptide bond is formed between F-met (located on P site) and the incoming amino acid. The amino acid–tRNA that binds to A site is specifically determined by the triplet codon of the mRNA that occupies A site on the ribosome. N-terminal of first amino acid (Met) has formyl group. So, elongation cannot take place at this end. The other end (-COOH) is available for chain elongation. The factors required for elongation are Tu, Ts, G and GTP. Tu and Ts are involved in exchange reaction. One molecule of peptidyl transferase is required per 50S ribosome subunit. Thus, polypeptide assembly begins at the amino end and terminates at carboxyl end. mRNA is read in a 5′ → 3′ direction during protein synthesis.

polypeptide initiation. Translation initiation comprises reactions preceding the formation of the peptide bond between the first two amino acids of the protein, which is a relatively slow step in protein synthesis. For translation to be initiated, an INITIATION COMPLEX is formed.

polypeptide termination. As polypeptide chain grows, it remains linked covalently to tRNA and bound to the ribosome. When synthesized completely, polypeptide is released from both the components. When termination signal (codon UAA, UGA or UAG) is reached at A site of the 50S subunit, chain terminates. Release factors RF1, RF2, RF3, GTP and nonsense codon(s) are required at this step. RF1 is required for UAG and RF2 for UGA. UAA can accept either RF1 or RF2. RF3 activates RF1 and RF2.

polyphenism: A form of induced response in which the phenotypes are discrete.

polyphenol oxidase (PPO): One of the major enzymes accountable for browning of wheat food products. Wheat cultivars diverge in PPO activity and plant breeders prefer to select germplasm and cultivars with low PPO activity.

polyphenols: Phytochemicals function as antioxidants when consumed by humans. They are naturally found in coffee, certain types of grapes, certain red wines, green tea, cacao or cocoa, etc. Polyphenols are naturally produced within the beans of the cacao tree (*Theobroma cacao*) and thus are present in chocolate made from those beans. Polyphenols naturally produced in apples inhibit certain bacteria in the human mouth from producing the particular glucans that lead to a build-up of plaque on teeth. Prevention of such plaque formation may help prevent cavities from forming in teeth.

polyphyletic genes: A group of genes that do not have a common ancestor.

polyphyletic species: Designating a group of species arbitrarily classified together, some of the members of which have distinct evolutionary histories, not having descended from a common ancestor.

polyploids: Polyploids can be broadly classified as euploids and aneuploids. Euploids having an exact multiple of the basic genome may be autopolyploids or allopolyploids. Whereas aneuploids are those in which one or a few chromosomes of the genome are added and subtracted. Some important crops like potato, sweet potato and lucerne are natural autopolyploids and other crops like wheat brassica, cotton, oat, tobacco and sugarcane are natural allopolyploids. In plants, any sporophyte with cells containing three or more complete sets of chromosomes are called polyploids. Various combinations of words or numbers with '-ploid' indicate the number of haploid sets of chromosomes (e.g. triploid = 3 sets, tetraploid = 4 sets, pentaploid = 5 sets, hexaploid = 6 sets and so on). For example, approximately 70% of grass species and 20% of legumes are polyploid. In animals, polyploids are limited to groups that reproduce asexually or are HERMAPHRODITIC and PARTHENOGENETIC. Examples are earthworms, some types of shrimps, parthenogenetic species of insects and lizards. *See* HERMAPHRODITIC *and* PARTHENOGENETIC. *Contrasts* HAPLOID *and* DIPLOID. *See details as separate terms.*

polyploid genome architecture of sugarcane: A.L. Healey *et al.* (2024) suggest a major step towards advancing sugarcane biotechnology by generating a polyploid reference genome for R570, a typical modern cultivar derived from interspecific hybridization between the domesticated species (*Saccharum officinarum*) and the wild species (*Saccharum spontaneum*). In contrast to the existing single haplotype ('monoploid') representation of R570, this 8.7 billion base assembly encompasses a complete representation of unique DNA sequences across the nearly 12 chromosome copies in this polyploid genome. Using this highly contiguous genome assembly, they filled a previous unsized gap within an R570 physical genetic map to discover the likely causal genes underlying the single-copy *Bru1* brown rust resistance locus. This complex polyploid genome assembly with fine-grain descriptions of genome architecture and molecular targets for biotechnology is expected to fast-track molecular and transgenic breeding and adaptation of sugarcane to future environmental conditions.

polyploidization: Whole-genome duplication events that can occur through the doubling of the chromosomes in a single species (autopolyploidization) or through a cross between related species followed by chromosome doubling (allopolyploidization).

polyploidy: Although replication and cell division are coupled processes but replication does not necessarily cause cell division. It gives rise to endopolyploidy (where DNA and chromosome number are multiplied but no cell division takes place), polyploidy or polyteny. Polyploidy is a common phenomenon in the evolution of most species of plants but it is much rarer phenomenon in animals. Polyploidy has played role in evolution by having several copies of the same genome but it does not seem to have played role in causing evolutionary changes in genome size. *Compare with* POLYTENY.

polyploidy induction: Polyploidy is produced with colchicine, an alkaloid of the meadow saffron

(*Colchicum autumnale*) that impedes mitosis, the development of the nuclear spindle is hampered. The mitosis that takes place after treatment with colchicine is called C-MITOSIS. It also permits an easier detection and identification of chromosomes than a normal mitosis does. During the extended metaphase of C mitosis, the chromosomes form an X-shaped structure since the chromatids are still connected at the centromere though they may repel each other. After some time, the chromatids finally separate, but they do not segregate. They are enclosed by a new nuclear membrane and proceed to their interphase state. The number of chromosomes doubles – a diploid nucleus thus develops into a tetraploid.

polyprotein: A protein translated from an entire viral genome, which is then cleaved by proteases (protein-degrading enzymes) into the active protein products.

polyprotein genes: Many genes encode for one single large polypeptide which is, sometimes after translation, cleaved enzymatically into smaller subunits. Such polyprotein genes are also known in multicellular eukaryotes. Polyprotein genes also exist in some viruses. For example, ORF1ab (also ORF1a/b) refers jointly to two open reading frames (ORFs), ORF1a and ORF1b, that are conserved in the genomes of nidoviruses, a group of viruses that includes coronaviruses. The genes express large polyproteins that undergo proteolysis to form several non-structural proteins with various functions in the viral life cycle, including proteases and the components of the replicase-transcriptase complex (RTC). Polyprotein genes thus contradict the neoclassical view of the gene that each gene encodes for a single polypeptide.

polyribosome: If only one ribosome is attached to a single mRNA, protein synthesis will not be efficient. Instead, a single mRNA usually has several ribosomes attached to it, as many as 1 for every 90 nucleotides. One mRNA molecule simultaneously functions as a template for synthesis of a number of polypeptide chains at one time. This process enhances the efficiency of the translation machinery. *Aka* POLYSOME.

polysaccharide: A linear or branched polymer of more than 20 monosaccharides linked by glycosidic bonds. The polysaccharides are sugar polymers possessing 20 or more monosaccharide units and some possess hundreds and thousands of units. They include glycogen, starch, hyaluronic acid and cellulose. Cellulose is polysaccharide possess linear chains, whereas another polysaccharide glycogen is branched. Both cellulose and glycogen are composed of recurring units of D-glucose, but they differ in the type of glycosidic linkage that makes them different with respect to their properties and biological roles.

polysomic: An individual possessing either a single extra chromosome or one pair of extra chromosomes in a set.

polysomy: The reduplication of some but not all of the chromosomes of a set beyond the normal diploid number.

polyspermy: The entry of more than one spermatozoon into an egg cell, irrespective of whether or not the additional spermatozoon fertilizes.

polytene chromosome: A chromosome that consists of large numbers of parallel DNA strands (single chromatids) without separation into new chromosomes, making their structure (bends, bands, puffs or swells) under the high-powered light microscope clearly visible due to differential gene activation. For example, many-stranded giant chromosomes produced by repeated replication during synapsis in certain dipteran larval tissues, most prominently present in salivary glands, the gut and the Malpighian tubules of *Drosophila*, *Chironomus*, *Sciara* and *Rhyncosciara*. Also reported from suspensor cells of developing embryos in leguminous plants. *Aka* GIANT CHROMOSOME, POLYNEMIC CHROMOSOME *and* SALIVARY CHROMOSOME.

polytene chromosomes, bands of: Bands are visible as dense regions that contain the majority of DNA in tightly coiled form; bands of normal chromosomes are relatively much larger and are generated in the form of regions that retain a stain on certain chemical treatments.

polytene nucleus: The nucleus in which protein production and gene reduplication go side by side. A permanent stage in which the chromosomes are represented by many linearly stretched and paired thread, as in salivary cells of dipteran larvae.

polyteny: Multiplication of the number of DNA strands within a chromosome. This condition occurs in certain animal tissues (e.g. the salivary glands, the gut and the Malpighian tubules of Diptera). Polyteny increases amount of DNA manifold in somatic cells but not in germ cells. Thus, polyteny is not a general apparatus for increasing amount of DNA during evolution.

polytomy: A portion of a phylogenetic tree in which more than two branches emerge from a single node.

polytopic subspecies: When subspecies of a species differ in a single diagnostic character like colour, size or pattern, it may happen that several different and somewhat widely separated populations independently acquire an identical population.

polytypic species: A polytypic species has two or more genetically and phenotypically divergent subspecies, races or more generally speaking, populations that differ from each other so that a separate description is warranted.

polyunsaturated fatty acids (PUFA): Unsaturated fatty acids have more than one molecular double bond in their molecular 'backbone' (i.e. they contain at least two less than the maximum possible number of hydrogen atoms). Enzymes (e.g. Δ12 desaturase) present in some oilseed plants (soybean, canola, maize, etc.) convert some monounsaturated fatty acids (e.g. oleic acid) to some polyunsaturated fatty acids (e.g. linoleic acid), within their developing seeds. For example, soybean oil contains (historical average) 60% polyunsaturated fatty acids. Extensive research shows that PUFA impart diverse health benefits to humans that consume them. Some of the beneficial effects of PUFAs occur via PUFA interactions with several types of nuclear receptors (present in cells of some human tissues), which leads to modulation in the expression of certain genes in those cells.

pooled libraries: For long-term storage, phage particles or bacterial colonies on bacteriological plates are eluted in a mixed pool, which contains representatives of every different recombinant on the plate. *Aka* AMPLIFIED LIBRARIES.

population: (i) In statistics, a population is a set of similar items or events which is of interest for some question or experiment. A statistical population can be a group of existing objects or a hypothetical and potentially infinite group of objects conceived as a generalization from experience. A common aim of statistical analysis is to produce information about a chosen population. (ii) A group of organisms of the same species that are in close enough proximity to allow them to interbreed and share a gene pool (i.e. a more or less distinct group of individuals within a species who are reproductively isolated from other groups). In other words, they restrict their mate selection to members of their own population. This is usually due to geographic and/or social barriers to mating with outsiders. Members of a completely isolated small population tend to have similar genetic characteristics due to generations of inbreeding.

population biology: The study of populations of organisms, especially the regulation of population size and life history traits such as clutch size and extinction.

population bottleneck: A brief reduction in population size, which results in an eruption of random genetic drift. It may be consequence of a FOUNDER EFFECT.

population density: Number of cells/protoplasts/pollens per unit volume of medium or plated area. Population density is an important factor for obtaining better planting efficiency. For instance, the plant protoplasts are cultured at densities ranging from 1×10^5 to 5×10^5 protoplasts/petri dish. Lower and higher population densities give very poor plating efficiencies.

population doubling level: The total number of population doublings of a cell line or strain since its initiation *in vitro*. A formula to use for the calculation of 'population doublings' in a single passage is. number of population doublings = Log $(N/N_0) \times 3.33$ where. N equals the number of cells in the growth vessel at the end of a period of growth. N_0 stands for number of cells plated in the growth vessel. It is best to use the number of viable cells or number of attached cells for this determination. *Aka* CUMULATIVE POPULATION DOUBLINGS.

population doubling time: The interval calculated during the logarithmic phase of growth in which, for example, 1.0×10^6 cells increase to 2.0×10^6 cells. This term is not synonymous with CELL GENERATION TIME.

population ecology: A subfield of ecology that deals with the dynamics of species populations and how these populations interact with the environment. It is the study of how the population sizes of species change over time and space. *Aka* AUTOECOLOGY.

population epigenetics and epigenomics: The agricultural and evolutionary consequences of heritable epimutations in plant has been explored at population level. One major aspect of this work is the advancement of computational methods for the analysis of population-level epigenomic sequencing data. The goal is to develop effective ways to detect 'epimutable' regions in plant genomes and to characterize these regions in terms of epigenomic patterns of variation both within and across plant species.

population genetics: Study of the frequencies of genes and genotypes in a Mendelian population during transmission from one generation to the next. Population genetics studies inheritance patterns and the changes in frequencies of alleles in populations and variation across populations. It also identifies processes that result in evolution, including mutation, selection, inbreeding, assortative mating, gene flow and drift. *See* SYNTHETIC THEORY OF EVOLUTION.

population genomics: The large-scale application of genomic technologies to study populations of individuals. For example, population genomics research can be used to study human ancestry, migrations and health.

population selection: When populations involved in selection process are social groups.

population size: The number of individuals in a population that are included in reproduction during a certain generation.

population size effects on QTL mapping: Larger the population size, more accurate is the mapping study and the more likely it is to detect QTLs with smaller effects. Large population size provides statistical power, estimates of gene effects and confidence intervals of the locations of QTLs.

population structure: Any aberration from the ideal state of a single panmictic population. For example, individuals more likely mate with those in the same place or in the same microhabitat.

population structure in crop improvement: Refers to the distribution of genetic variation within and between populations, influenced by factors such as geographical distance, historical events and selective pressures. In crops, factors such as geographical distance, historical events and selective pressures contribute to population structure. Studies on pigeon pea, soybeans, maize and rapeseed have utilized population structure analysis to understand genetic diversity and structure. G. Ntsomboh-Ntsefong *et al.* (2023) provide evidence of population structure correlated with geographical distance, historical events such as founder effects and selective pressures such as adaptation and breeding programmes, indicating their role in genetic differentiation among populations.

population variance: The degree of distribution demarcated as the average of the squared deviations between the observed values of the elements of a population and the corresponding mean of those values.

population waves: Asymmetrical or periodic changes in the number of individuals in a population that are encompassed in reproduction during certain generations.

population-specific single-nucleotide polymorphism (population-specific SNP): Any single-nucleotide polymorphism that exists in one population and is absent in another.

porcine endogenous retrovirus (PERV): Endogenous retrovirus of pigs closely related to MLV. Some PERVs can infect human cells.

porosome: Permanent cup-shaped lipoprotein structure at the cell plasma membrane, where secretory vesicles transiently dock and fuse to release intra-vesicular contents from the cell.

porphyrins: A class of compounds derived from pyrrole nuclei which in metal chelate complexes form pigments important in many biological processes.

portmanteau of modulator-demodulator: A hardware device that adapts data into a format appropriate for a transmission medium so that it can be communicated from one computer to another (historically along telephone wires). A modem moderates one or more carrier wave signals to encode digital information for transmission and demodulates signals to decode the communicated information. The goal is to generate a signal that can be transmitted easily and decoded reliably to reproduce the original digital data. Modems can be used with almost any resources of transmitting analogue signals from light-emitting diodes to radio. A common form of modem is one that cracks the digital data of a computer into modulated electrical signal for transmission over telephone lines and demodulated by another modem at the receiver side to recuperate the digital data.

position effect: The alteration in the expression of a gene with respect to neighbouring genes which probably mirrors differences in the state of the chromatin along the chromosome. For example, a previously active gene may become inactive if positioned on a new site in the genome. *Aka* POSITIONAL EFFECT.

position effects on gene expression: (i) The yeast *ADE2* gene at its usual chromosomal location is expressed in all cells. When moved near the end of a yeast chromosome, which is incidentally folded into a form of heterochromatin, the gene is now no longer expressed in most cells of the population. The *ADE2* gene codes for one of the enzymes of adenine biosynthesis and the absence of the *ADE2* gene product results in the accumulation of a red pigment. Therefore, a colony of cells that expresses *ADE2* is white and one part of cells where the *ADE2* gene is not expressed is red. The white sectors that come out from the middle of the red colony grown on an agar surface signify the descendants of cells in which the *ADE2* gene has spontaneously become active. These white sectors are thought to be consequence of a heritable change to a less tightly packed state of chromatin near the *ADE2* gene in these cells. Although yeast chromosomes are too small to be seen under the light microscope, the chromatin structure at the ends of yeast chromosomes is thought to have many of the same structural features as the heterochromatin in the chromosomes of larger organisms. (ii) Position effects have also been detected for the *white* gene in the fruit fly *Drosophila*. The *white* gene governs eye pigment production and is named after the mutation that first identified it. Wild-type flies with a normal *white* gene (*white*$^+$) produce normal pigment, which gives them red eyes, but if the *white* gene is mutated and inactivated, the mutant flies (*white*$^-$) make no pigment and have *white* eyes. The flies in which a normal (*white*$^+$) gene has been moved near a region of heterochromatin, the eyes become mottled, with both red and white areas. The white patches characterize cells in which the *white*$^+$ gene has been silenced by the effects of the heterochromatin. In contrast, the red patches represent cells that express the *white*$^+$ gene because the heterochromatin did not spread across this gene at the given time, early in development, when the heterochromatin first formed. As in the yeast, the presence of large patches of red and white cells signifies that the state of transcriptional activity of the gene is inherited, once determined by its chromatin packaging in the early embryo.

position sensitive detector (PSD): *Aka* POST-SOURCE DECAY, in mass spectrometry.

position-effect variegation (PEV): Position effect is the effect on the expression of a gene when its location in a chromosome is changed, often by translocation. This has been well described in *Drosophila* with respect to eye colour. The phenotype is well characterized by unstable expression of a gene that results in the red eye coloration. In the mutant flies the eyes typically have a mottled appearance of white and red sectors. These phenotypes are often due to a chromosomal translocation such that the colour gene is now close to a region of heterochromatin. The heterochromatin can spread stochastically and switch off the colour gene resulting in the white eye sectors. PEV results in the variable silencing of a gene because of its proximity to heterochromatin. POSITION EFFECT is also used to describe the variation of expression exhibited by identical transgenes that insert into different regions of a genome. In this case the difference in expression is often due to enhancers that regulate neighbouring genes. These local enhancers can also affect the expression pattern of the transgene. Since each transgenic organism has the transgene in a different location each transgenic organism has the potential for a unique expression pattern.

position-specific scoring matrix (PSSM): The PSSM gives the log-odds score for finding a particular matching amino acid in a target sequence.

positional candidate gene approach (PCGA) for QTL mapping: PCGA involves integrated genome scan and candidate gene analyses, in which identification of CG is mainly based on physical linkage which provides the exact location of genes or genetic markers on chromosomes. PCGA identifies a gene within the vicinity of QTLs. In positional CG approach, a sequence comparison gives a better success with smaller confidence interval about QTLs.

positional candidate gene: A gene known to be located in the same region as a DNA marker that has been shown to be linked to a single-locus trait or to a quantitative trait locus and whose function suggests that it could be the source of genetic variation in the trait in question.

positional cloning: Method used to define the location of a gene on a chromosome and use this information to identify and clone the gene. The location of the gene is determined by linkage analysis of DNA from a large family containing afflicted and normal members to identify linkages between the transmission of the disease gene and observable genetic markers. This information is then used to screen (by chromosomal jumping and walking) the location of putative genes. The disease gene must be compared between the afflicted and normal family members and be shown to be different in the two groups. The full sequencing of the gene will then provide information regarding the characteristics and function of the gene product and a potential explanation for the cause of the disease. *See* MAP-BASED GENE CLONING.

positional gene identification: A laboratory approach used to locate the position of a disease-associated gene on a chromosome. Such a strategy can succeed even when nothing is known about the role of the gene's encoded protein in the disease. The technique typically relies on the use of known polymorphic markers whose inheritance can be traced through various members of families affected by the disease. *Aka* POSITIONAL CLONING.

positional homology: When an MSA is used for phylogenetic analysis, residues that are lined up in different sequences are thought to share a common ancestry (i.e. they are derived from a common ancestral residue).

positional information: The information that determines where a part of the body is to be formed. This information in the cell is present in form of localized substances in the cytoplasm. These cytoplasmic determinants and external forces initiate localized differences in development. Some cells or tissues can obtain information as to their positions because they are affected by such gradients. This positional information, in turn, can influence subsequent activity so that different groups of cells follow different developmental patterns, such as moving to new locations and/or causing the production of inducer substances that set up developmental gradients of their own.

positional value: A cell's interior record of its positional information.

positional/map-based cloning of QTLs: Studies in plants have succeeded in cloning QTLs using map-based methods. Successful map-based cloning requires that QTLs represent single genes that can be isolated in near-isogenic lines and that genotypes can be unambiguously inferred by progeny testing.

positions and patterns: One of the most obvious features of multicellular organisms is the special arrangement of their tissues and organs. That is, not only are these body parts different from each other, but they occupy localized positions that are fairly uniform to all individuals of the species. For such precise and repeatable patterns to occur, they must be preceded by a precisely ordered pattern of development. In a general way the initiation of pattern begins with the egg. Differences, such as those imposed by the uneven deposition of maternal substances in the egg cytoplasm or by external factors undoubtedly help to initiate localized differences in development.

positive and negative selection (PNS): A separation technique; a technique to speed up the task of selecting, from thousands of laboratory specimens, the few cells with precisely the desired genetic changes induced (via genetic engineering). The thousands of genetically altered cells are brought about and produced by genetic engineering experiments. Many genetic alterations are accomplished by injecting or flooding (specimen) cells with fragments of new genetic material (genes). A few cells are produced that have precisely the desired genetic changes among a large number of cells that do not have the desired changes. Sort of like a 'needle in a haystack'. By analogy, the few cells possessing the desired trait represent the needles while the multitude of cells not possessing the trait represent the hay. In order to isolate the few desired cells, the needles must be separated from the hay. PNS gets rid of the undesired cells and leaves only the cells possessing the desired genetic change. This is accomplished in the following way. The pieces of newly injected genetic material are composed not only of the desired sequence of DNA, but also another piece of DNA (known as a MARKER) which renders only those cells possessing the desired genetic change resistant to certain antibiotic drugs (such as neomycin) and certain antiviral drugs (e.g. Ganciclovir™). When all of the engineered cells are exposed to the drug (which normally kills all of the cells), only those cells possessing the desired genetic change (and the concomitant piece of DNA providing drug resistance) survive and hence are 'selected'. The other cells not having the drug resistance are selected against and die.

positive assortative mating: A form of non-random mating in which entities who are alike for particular traits mate and those who are not alike do not. The result is a progressive increase in the frequency of homozygotes and a decrease in the heterozygotes for the discriminated traits. *See* NEGATIVE ASSORTATIVE MATING, CONSANGUINEOUS MATING *and* RANDOM MATING.

positive conditional marker gene systems: SMGs of this category include antibiotics, herbicides, toxic and non-toxic drugs, metabolite analogue or a carbon source. All the commercially released transgenic lines have been selected through positive conditional selection systems.

positive control of *lac* operon: Positive control means activation of an operator site by a regulatory protein that is produced by a regulator site. The product of a regulatory gene functions to stimulate transcription. In addition to a negative control, there is also a positive control of gene action in the *lac* operon. A positive control in *lac* operon of *Escherichia coli* was detected by investigating the glucose effect. CYCLIC ADENOSINE MONOPHOSPHATE (cAMP) is the key metabolite that influences the glucose effect. Glucose blocks the activity of other operons through this metabolite. When glucose is used as an energy source, the cAMP level decreases, which prevents the functioning of other operons involved in the metabolism of other sugars like lactose, arabinose, galactose and maltose. When glucose levels decrease, cAMP levels increase and other sources of sugar can be utilized. cAMP is effective only when bound to CATABOLITE ACTIVATOR PROTEIN (CAP). When the cAMP-CAP complex binds to a specific site on the operon, it enhances the rate of transcription of the *lacZ-lacY-lacA* cluster. This system empowers the cells to utilize a large number of different sugars as a source of carbon. CAP-deficient strains and adenylate cyclase–deficient strains do not inspire glucose effect. Adenylate cyclase–deficient strains also synthesize low amounts of inducible enzymes. CAP-deficient strains produce low amounts of inducible enzymes. The positive and negative controls act together, but each is a separate regulatory system for operon transcription. cAMP is also called the second messenger. *Compare with* NEGATIVE CONTROL.

positive control: A sort of gene expression control strategy in which the active DNA-binding form of the regulatory protein switches the gene on.

positive dominant genes: Genes yielding, in hemizygous conditions, an intact multimeric protein in an optimal amount. As a result, they are capable of masking the defective action of mutant alleles and complementing themselves additively. Positive dominant loci represent the integrated form of a gene, which, through qualitative and quantitative degradation, gives rise to semi-dominant, codominant and negative dominant alleles.

positive eugenics: Enhancement of desirable characteristics through identification of 'good' genes and encouraging favourable mating by restricting childbearing to those best fitted to perpetuate the species. *Compare with* NEGATIVE EUGENICS.

positive frequency-dependent selection: In positive frequency-dependent selection, the fitness of a phenotype or genotype increases as it becomes more common.

positive interference: This type of interference exists when occurrence of one cross over reduces the probability of that a second will occur in the same chromosomal region. *Aka* INTERFERENCE. *Compare with* NEGATIVE INTERFERENCE.

positive mass selection: Positive (Darwinian) selection promotes the spread of beneficial alleles. Darwinian selection is the main mechanism that Darwin envisioned as giving rise to evolution, specific molecular genetic examples are very difficult to detect. Pioneering work by Z. Yang and R. Nielsen in 2000 provided a much more powerful methodology for detecting positive selection at the sequence level.

positive non-conditional selectable marker systems: SMGs of this category do not require use of any external selection agent but consist of a selection marker gene which encodes for a protein that can modulate morphogenesis/organogenesis either directly through synthesis/action of plant growth regulator or by acting as transcription factor.

positive predictive value: The probability that an individual with a positive test has or will develop, a particular disease or characteristic, that the test is designed to detect.

positive regulator proteins: The proteins required for the activation of transcription unit.

positive screening: A selection procedure in which the best individuals in a genetically varied population are identified and conserved to become the parents, either of the next screening generation or of a new cultivar. *Compare with* NEGATIVE SCREENING.

positive selectable marker genes (SMG): Those SMGs that promote the growth of only transformed cells/ tissue. Positive selection markers can be conditional or unconditional. Positive SMGs are divided into seven types (Type 1–7).

positive selectable marker genes, type 1: These SMGs cause chemical detoxification of phytotoxin. For example, *neo*, *nptII (aphA2)*, *nptI (aphA1)*, *aaC3*, *aaC4*, *aadA*, *SPT*, *hph (aphIV)*, *Ble*, *sulI*, *sat3* and *cat* genes.

positive selectable marker genes, type 2: These SMGs cause removal of phytotoxin from the sensitive cell compartment, for example, *pat*, *bar*, *EPSP synthase*, *aroA*, *cp4 epsps*, *gox*, *csr1-1*, *csr1-2*, *bnx*, *hemL* and *Cah* genes.

positive selectable marker genes, type 3: These SMGs cause overexpression of the sensitive or expression of an insensitive target molecule, for example, *DOG^{R}1*, *BADH*, *ocs*, *DHPS*, *TDC* and *DHFR* genes.

positive selectable marker genes, type 4: These SMGs confer pathogen resistance, for example, *xylA*, *manA (pmi)* and *uidA* (*gusA*) genes.

positive selectable marker genes, type 5: These SMGs provide heat tolerance, for example, *ipt*, *pga 22*, *rol*, *ESR1* and *CKI1 genes*.

positive selectable marker genes, type 6: These SMGs are responsible for carbohydrate metabolism and confer the ability to utilize a carbohydrate that otherwise is not metabolized, for example, *codA*, *aux2*, *tms2*, *dhlA* and *CYP105A* genes.

positive selectable marker genes, type 7: These SMGs are involved in growth regulator metabolism or regulation of the cell cycle and confer the ability to produce or increase sensitivity to cytokinin or auxin, for example, *Adh or Cue* genes.

positive selection: *See* DIRECTIONAL SELECTION.

positive supercoiling: Type of supercoiling that occurs in double-stranded cyclic DNA molecules bearing no breaks at all in either strand. If the double helix (of DNA) is wound further in the same direction as the natural winding of the two strands of the double helix molecule, then the circular duplex itself takes on superhelical turns. By analogy, supercoiling or superhelicity may be described as follows. In a piece of rope composed of two or three thinner strands, positive supercoiling become wound around each other to yield the finished rope. This is equivalent to the normal double-stranded DNA. If the ends of the rope are then joined or tied together and the resultant circle of rope is again wound in the same direction as the winding that produced the rope in the first place, supercoils will be formed and the rope will become a much thicker (supercoiled), but shorter, piece of rope. *Compare with* NEGATIVE SUPERCOILING.

positive transcription elongation factor b (P-TEFb) complex: A complex comprising CDK9 and cyclin T1 or T2 that phosphorylates the C-terminal domain of RNA polymerase II and the NELF–DSIF (negative elongation factor–DRB sensitivity-inducing factor) inhibitory complex, thereby facilitating the release of RNA polymerase II from a paused state into productive transcriptional elongation.

positive-negative selection: When the presence of a cassette is positively selected for, for example, by drug resistance and then negatively selected for, by eliminating cells that express a second selectable marker.

positive-sense RNA (+RNA): The RNA that can function straight as messenger RNA. *Contrasts* NEGATIVE-SENSE RNA (−RNA).

positive-sense single-stranded RNA viruses (+ssRNA viruses): Positive-strand RNA viruses (+ssRNA viruses) are a group of related viruses that have positive-sense, single-stranded genomes made of ribonucleic acid. The positive-sense genome can act as messenger RNA (mRNA) and can be directly translated into viral proteins by the host cell's ribosomes. Positive-strand RNA viruses encode an RNA-DEPENDENT RNA POLYMERASE (RDRP), which is used during replication of the genome to synthesize a negative-sense antigenome that is then used as a template to create a new positive-sense viral genome. Examples of +ssRNA viruses are polioviruses, rhinoviruses, coronaviruses, rubella, yellow fever virus, West Nile virus, dengue fever viruses, equine encephalitis viruses, hepatitis A, hepatitis C viruses and tobacco mosaic viruses.

***post hoc* test:** Researchers perform a *post hoc* test only after they have discovered a statistically relevant finding and need to identify where the differences actually originated.

post-adaptive mutation: A mutation that occurs as a direct response to an environment stress which enables the organism to adapt to that stress.

post-analytical: The phase of laboratory testing that occurs after the test result has been generated (e.g. test interpretation, test reporting and consultation). *See* PRE-ANALYTICAL.

post-bisulfite adapter tagging (PBAT): An improved single-cell single-base resolution method of DNA-methylation analysis. Bisulfite treatment is accomplished before adapter tagging by two rounds of random primer extension. The PBAT method can produce an extensive number of unamplified reads from as little as sub-nanogram quantities of DNA.

post-condition: In computer programming, a state or ground that must always be true just after the implementation of some segment of code or after a procedure in a formal specification. Post-conditions are occasionally verified using declarations within the code itself. Often, post-conditions are simply encompassed in the documentation of the affected sector of code.

post-consumer waste: Waste gathered after a consumer has disposed of something, for example, sweet wrappers or packaging from small electronic goods such as mobile phones or MP3 players.

post-embryonic development: The changes that occur after formation of the embryo are called POST-EMBRYONIC CHANGES and this period of development is called POST-EMBRYONIC PERIOD. Both these periods are included in development of an individual.

posterior. Situated towards the tail end, the back or the rear of the body. *Contrasts* ANTERIOR.

posterior odds ratio: A ratio of the likelihood of a hypothesis given a specific set of data relative to the likelihood of another hypothesis given the same data (Prob(H1 | D)/Prob(H2 | D)). This is the value that is estimated in BAYESIAN PHYLOGENETIC RECONSTRUCTION METHODS. *Aka* ODDS RATIO.

posterior probability: In a Bayesian context, this is the likelihood of an event after making use of the information in the data. In other words, it is the PRIOR PROBABILITY of an event after updating it with the data. Posterior probability can also be called

POST-TEST PROBABILITY if one parallels a diagnostic test with 'data'. *See* RECEIVER OPERATING CHARACTERISTIC (ROC) CURVE.

postharvest losses: Crop fatalities due to parasites that occur after harvest, usually in the store. These losses can be reduced or prevented by ensuring that the stored product is dry to prevent moulds from developing and that the product is stored in an airtight container that lacks oxygen to stop various animal pests from eating it.

postmeiotic segregation: Segregation of two strands of a duplex DNA that beat different in formation (created by heteroduplex formation during meiosis) when a subsequent replication allows the strands to separate.

postnatal diagnosis: Diagnosis of a genetic disorder in an individual after birth.

postponement of ageing: Following factors postpone ageing: Caloric restriction, replacement organs, new neurons, caloric reduction, immune system rejuvenation, superoxidant genes, protective hormones, blocking blood sugar, activation of telomeres and inactivation of p^{16INK4} pathway and protein named KLOTHO.

postreplication recombination (PRR)-mediated DNA repair: This repair process occurs mostly on DNA strand carrying gaps by inter- and intrastrand cross links. When such cross-linked strands replicate or are excised, gaps remain, which then enable the free end of a single-strand section to initiate an exchange with a similarly oriented strand on homologous DNA molecule.

postreplicative modification of DNA: The purines and pyrimidines are occasionally modified after DNA replication. The DNA of bacteriophage T2, T4 and T6 all contains 5-hydroxymethylcytosine ($5^{hm}C$) instead of cytosine. Most animal cells, bacterial and viral DNAs contain small amounts of the methylated bases 5-methylcytosine ($5^{m}C$) and 6-methyladenine ($6^{m}A$).

postreplicative repair: A DNA repair process initiated when DNA polymerase bypasses a damaged area. Enzymes in the *rec* system are used.

post-start G1: Cyclins of START PROMOTION FACTOR (SPF) are degraded during post-start G1 and p34cdc2 is phosphorylated. During the S phase, S-specific cyclin (A) is synthesized which somehow provides signal for completion of DNA replication and entry of cell to G2 phase.

post-termination complexes in eukaryotes: Eukaryotes do not encode a RIBOSOME RECYCLING FACTOR (RRF) homologue. In eukaryotes, a loosely associated subunit of initiation factors eIF3 promotes recycling of eukaryotic post-TCs. eIF3 is the principal factor that promotes splitting of post-termination ribosomes into 60S subunits and tRNA- and mRNA-bound 40S subunits. Its activity is enhanced by eIFs3j, eIF1 and eIF1A. eIF1 also mediates release of P site tRNA, whereas eIF3j ensures subsequent dissociation of mRNA. Initiation factor eIF3, in cooperation with its eIF3j subunit, eIF1 and eIF1A, split post-termination ribosomes into 60S subunits and tRNA- and mRNA-bound 40S subunits. eIF1 promotes subsequent release of P site deacylated tRNA, which is followed by dissociation of mRNA mediated by eIF3j.

post-termination complexes in prokaryotes: After translational termination, mRNA and P site deacylated tRNA remain associated with ribosomes in post-termination complexes (post-TCs), which must therefore be recycled by releasing mRNA and deacylated tRNA by dissociating ribosomes into subunits. Recycling of bacterial post-TCs requires elongation factor EF-G and a ribosome recycling factor (RRF).

post-transcriptional control: Any control on gene expression that is exerted at a stage after transcription has begun. These controls operate during RNA PROCESSING. Events of RNA processing – addition of cap at 3′ end, addition of poly(A) segment at 5′ end, removal of introns – occur post-transcriptionally in the heterogeneous RNA (hnRNA) in the nucleus. A mechanism allows further transcription when synthesis is needed and stop transcription whenever synthesis is not needed. This is achieved by a mechanism, which includes the presence of successive tryptophan codons at a specific site in *trpE* leader sequence. Similarly, in leader sequence of phenylalanine operon there are three consecutive phenylalanine codons. In leader sequence of leucine operon there are four consecutive leucine codons and in leader sequence of histidine operon there are eight consecutive histidine codons. These sequences are meant to give rise to mRNA for a small polypeptide about 15 amino acid residues long. Another example of post-transcriptional control of gene expression is the coat protein of RNA phage R17. It binds to a hairpin that encompasses the ribosome-binding site in the phage mRNA. Phage T_4 Reg A protein binds to the consensus sequence that includes the AUG initiation codon in several T_4 early mRNAs. T_4 DNA polymerase binds to a sequence in its mRNA that includes the SHINE-DALGARNO SEQUENCE needed for ribosome binding.

post-transcriptional control by leader sequences: In tryptophan operon of *Escherichia coli*, mRNA of gene *trpE* which codes for enzyme anthranilate synthetase and contains a sequence of 162 bases between its initiation codon (AUG) and the promoter-operator area of this gene. This sequence of 162 bases is called LEADER which determines post-transcriptional control. This post-transcriptional control is in addition to normal transcriptional control provided in the operon model.

post-transcriptional control in phage T4: Phage T4 Reg A protein binds to the consensus sequence that contains the AUG initiation codon in several T4 early mRNAs. T4 DNA polymerase binds to a sequence in its mRNA that includes the Shine-Dalgarno sequence required for ribosome binding.

post-transcriptional control in RNA phage R17: Coat protein of RNA phage R17 binds to a hairpin that includes the ribosome-binding site in the phage mRNA.

post-transcriptional control of gene regulation: Post-transcriptional regulation and stabilization might be implicated in precise regulation such as enzyme induction. In eukaryotes, mRNA production is not a simple step of transcribing RNA. The transcription units are actually much longer than the mature mRNA product. Several 'processing' steps transpire to primary RNA transcripts in the nucleus. These steps involve (i) addition of a methylated guanylate residues (mG), a 'cap', at the 5′-end of primary transcripts, (ii) addition of adenylic acid residues to a site in the primary transcript to create a 3′ poly(A) tail, (iii) removal of specified internal non-coding sequences (introns) and rejoining or splicing of the remaining coding RNA pieces (exons) to create a functional mRNA. Initiation site is around 30 nucleotides to the 3′ site of TATA.

post-transcriptional control through attenuators: In tryptophan operon, when tryptophan supply is satisfactory, transcription stops at about 147 bases after initiation so that no mRNA for *trpE* and other genes of operon are formed. But if tryptophan is in short supply, transcription continues non-stop. During adequate supply of tryptophan, mRNA for leader sequence is translated on ribosome. This synthesis of leader peptide results in the formation of hairpin loop. This peptide is *aka* ATTENUATOR, which is present in the sequences between 114 and 134 which stops further synthesis of mRNA. On other hand, if synthesis of mRNA is in short supply, two tryptophan codons cannot be translated due to lack of availability of amino acid tryptophan. The ribosome becomes stalled at this point and makes the formation of terminator hairpin loop impossible thus indirectly permitting transcription of *trpE* and other genes of the operon. This type of termination is possible only in those sequences where transcription and translation are coupled (i.e. this is possible only in prokaryotic genes).

post-transcriptional gene silencing (PTGS): Silencing of an endogenous gene was accomplished by the introduction of a homologous dsRNA, transgene or virus. In PTGS, the transcript of the silenced gene is synthesized but does not accumulate because it is rapidly degraded. PTGS is a more general term than RNA interference (RNAi), since it can be triggered by several different means. It was initially considered an unusual phenomenon limited to petunias and a few other plant species. While trying to deepen the purple colour of these flowers, C. Napoli *et al.* in 1990 introduced a pigment-producing gene under the control of a strong promoter. Instead of the expected deep purple colour, many of the flowers looked variegated or even white; the observed phenomenon was named 'cosuppression', since the expression of both the introduced gene and the homologous endogenous gene was suppressed. Meanwhile, it became one of the most important phenomena in molecular biology. It later became clear that PTGS occurs in both plants and animals and has roles in viral defence and transposon silencing mechanisms; most exciting is the emerging use of RNAi initiated by the introduction of double-stranded RNA (dsRNA) – as a tool to knock out expression of specific genes. *Compare with* TRANSCRIPTIONAL SILENCING. *See* RNA INTERFERENCE.

post-transcriptional gene silencing (PTGS) mechanism: PTGS/RNAi (RNA interference) is a mechanism at the cellular level triggered by DOUBLE-STRANDED RNA (DSRNA). RNAi is an emerging, biologically safe alternative to methods of insect pest control. This technique involves controlled expression of gene and regulation of protein function through small/short interfering RNA (siRNA) tools. siRNAs of 18–21 bp length are produced by cellular Dicer enzyme from exogenous dsRNA of a cell. These siRNAs are ultimately channelled to the RISC. This particular event blocks one or more steps of central dogma, resulting in either reduction (knockdown) or complete inhibition (knockout) of protein production of the target gene. PTGS is the result of mRNA of a particular gene being destroyed or blocked. The destruction of the mRNA prevents translation to form an active gene product (in most cases, a protein). A common mechanism of PTGS is RNAi. Other term related to PTGS is NONSENSE-MEDIATED DECAY (NMD). Ribozyme and antisense technology are the two approaches used to achieve precise PTGS in order to suppress a gene of interest. *Aka* RNA INTERFERENCE (RNAI) *or* RNA SILENCING.

post-transcriptional modification (PTCM or PTM): Changes that occur to a newly transcribed primary RNA transcript (hnRNA) after transcription has occurred and prior to its translation into a protein product. PTCMs in precursor mRNA include additions of caps and poly(A) tails, removal of introns, alternative RNA splicing and some adenine methylation in coding regions. Certain pore-mRNAs also experience RNA editing which was first discovered in mitochondria of trypanosomes. The processed/mature/functional mRNAs are then transferred to the cytoplasm and the tail is enzymatically detached to form the functional mRNAs. The poly(A) tail aids in the transfer of the complex and/or targets the complex to the cytoplasm. *See* RNA PROCESSING.

post-transcriptional processing: Addition or removal of nucleotides following RNA synthesis. In case of prokaryotes, primary mRNA is translated into protein without any modification. Whereas, in case of plants, animal and humans, the mRNAs are synthesized in the nucleus of the cell and the process of translation occurs in the cytoplasm. Therefore, mRNA transcripts undergo extensive post-transcriptional processing. Primary transcripts of structural genes are post-transcriptionally modified by the addition of 3′ poly (A) tail and 5′ cap. *Aka* RNA PROCESSING.

post-transcriptional regulatory element enhances expression of transgenes: The expression of genes delivered by retroviral vectors is often inefficient, a potential obstacle for their widespread use in human gene therapy. R. Zufferey *et al.* (1999) explored the possibility that the POST-TRANSCRIPTIONAL REGULATORY ELEMENT OF WOODCHUCK HEPATITIS VIRUS (WPRE) might help resolve this problem. Insertion of the WPRE in the 3′ untranslated region of coding sequences carried by either oncoretroviral or lentiviral vectors substantially increased their levels of expression in a transgene-, promoter- and vector-independent manner. The WPRE thus increased either luciferase or green fluorescent protein production five- to eightfold and effects of a comparable magnitude were observed with either the immediate-early cytomegalovirus or the herpesvirus thymidine kinase promoter and with both human immunodeficiency virus- and murine leukaemia virus-based vectors. The WPRE exerted this influence only when placed in the sense orientation, consistent with its predicted post-transcriptional mechanism of action. These results demonstrate that the WPRE significantly improves the performance of retroviral vectors and emphasize that post-transcriptional regulation of gene expression should be taken into account in the design of gene delivery systems.

post-translational control: A kind of control that functions after the protein synthesis has taken place. Examples are provided by synthesis of isoleucine from threonine, proteolytic control. of adrenocorticotropin or β-lipoprotein, rat kallikrein protein and rat preproinsulin. *Aka* FEEDBACK INHIBITION *or* END-PRODUCT INHIBITION.

post-translational gene regulation: (i) This level of control on gene regulation occurs after completion of translation. Stringent response and feedback (end-product) inhibition are mechanisms of post-translational gene regulation. (ii) The function and longevity of a protein may be affected by proteolytic enzymes that cleave polypeptide chain at specific places and also by kinases and methylases, and so on, that modify specific amino acid residues on specific polypeptide chain. It is only after such modification that active proteins or enzymes are synthesized.

post-translational modification (PTLM or PTM) in eukaryotes: (i) Removal or derivatization of amino acid residues or their modification following their incorporation into polypeptide. During protein synthesis, growing polypeptide starts folding to its nature configuration, which is facilitated by its interaction with endoplasmic reticulum (ER)-resident chaperone protein Hsp70. ER enzymes then initiate PTLM of the polypeptide, including attachment of 'core' carbohydrates and formation of disulfide bonds. (ii) Refers to one of many covalent protein modifications after translation. It is a vital mechanism regulating protein function. These modifications can be highly dynamic or have a function in epigenetic memory, for example, histone PTLM events involve phosphorylation, acetylation, glycosylation, carboxylation, hydroxylation, methylation, AMPylation, lipidation, ubiquitination, SUMOylation and so on, which are very important for protein function. Various variants of Eastern blotting are used to analyse specific types of post-translational modifications. Modifications in histones may lead to modifications of gene regulation and thereby may cause changes in phenotype, for example, post-translational modifications lead to isozymes. Various variants of eastern blotting are used to analyse specific types of PTLMs. *Aka* POST-TRANSLATIONAL REGULATION OF PROTEIN ACTIVITY. *See* EPIGENETIC MODIFICATIONS.

post-translational modification (PTLM or PTM) in prokaryotes: In prokaryotes, the first amino acid of a protein synthesized on ribosomes is formylated methionine. From recently synthesized polypeptide, N-formyl group is detached by deformylase. Then the elimination of first amino acid (methionine) occurs by methionine-specific amino-peptidase. Finally, terminal tRNA is detached and complete protein becomes available.

post-translational regulation of nitrogenase: Because of the metabolically demanding nature of nitrogen fixation, an additional layer of nitrogenase regulation is present in a few free-living diazotrophs. To prevent unproductive nitrogen fixation during energy-limiting or nitrogen-sufficient conditions, the nitrogenase complex is rapidly, reversibly inactivated *in vivo* by ADP-ribosylation of Fe protein.

postzygotic incompatibility: A condition prevailing in incompatible or wide crosses, where the zygote fails to develop, often due to nutritional reasons. In some cases, embryo culture can be utilized to rescue the embryo.

postzygotic isolating mechanism: Any one of several mechanisms that keep populations reproductively isolated from each other. In this case, a zygote is effectively formed but then it either flops to develop or it develops into an adult that is either sterile, non-viable or so weak that they do not survive, for example, donkeys and horses are post-zygotically isolated from one another, as a male donkey and a female horse can mate to produce a mule, but the mule is sterile.

potassium (K): An element that is essential for plant growth. Its deficit leads to reduced growth and to dark or blue-green coloration in the leaves.

potato epigenome: DNA methylation has a vital role in photoperiodic flowering of potato plant. A DNA-methylation inhibitor could promote tuber initiation in strict short-day genotypes.

potence: The property of a group of polygenes corresponding to the degree of dominance of a major gene.

potency: (i) Specifies the differentiation potential of the stem cell. Totipotent stem cells are produced from the fusion of an egg and sperm cell. (ii) The capacity of a gene for manifesting its presence. PARTIAL POTENCY is synonymous with INCOMPLETE DOMINANCE. *Compare with* EXPRESSIVITY *and* PENETRANCE.

potential heterozygosity (H): Having more than twice the normal haploid number of chromosomes.

potential reproducibility: The investigators have offered data management and analysis code that is fully self-contained and could be implemented by another person to obtain all their analytical results attained by the original researchers.

potentiation: When two or more factors, physical or chemical, given together cause an increase in mutation frequency beyond that each factor causes individually.

power: Likelihood of rejecting the null hypothesis for a set value of the unidentified effect. Power could also be termed the sensitivity of the statistical test in perceiving that effect. Power increases when the sample size and true unknown effect increase and when the intersubject variability decreases. In a two-group comparison, power generally increases as the allocation ratio gets closer to 1:1. For a given experiment it is required to use a statistical test expected to have extreme power (sensitivity). A less influential statistical test will have the same power as a better test that was applied after neglecting some of the observations, for example, testing for differences in the proportion of patients with hypertension in a 500-patient study may produce the same power as a 350-patient study which used blood pressure as a continuous variable.

poxviral vectors: Poxviruses are the largest and most complex viruses known. The most common poxviruses being vaccinia and variola virus (smallpox). Poxviral vectors are successful in immunogenetic therapy protocols, due to their strong immunogenicity.

pre- and post-genome era, differences between: The major difference between the pre- and post-genome era is that earlier the responses of cell or organisms could be studied on a limited extent (i.e. one gene or one pathway at a time), but now, expression of all the genes could be studied in totality at once which gives a deep insight into the behaviour of biological systems.

preadaptation: An adaptation evolved in one adaptive region, which demonstrates especially advantageous in an adjacent region and so enables the organism to radiate into it. In breeding, the pretreatment of plants under moderate climatic (light, temperature or nutrient) conditions slowly adapts them to stress conditions.

preadapted organism: An organism that possesses traits essential to a new environment before that environment is actually occupied, as air-breathing apparatus possessed by aquatic animals prior to their becoming terrestrial.

pre-adaptive mutation: A mutation that occurs before a specific environmental stress is encountered. When the stress occurs, the mutation that confers a selective advantage on the organism is selected for.

pre-analytical: The phase of laboratory testing that occurs before the test has been run (e.g. test requisition information, sample suitability and sample labelling/handling). *See* POST-ANALYTICAL.

prebiotic synthesis: The obviously occurring manufacture of organic compounds before there was life on Earth.

prebiotics: Chemical compounds more particularly, the oligosaccharides that stimulate the growth and colonization of probiotic bacteria. There are several prebiotic oligosaccharides such as fructam, fructo-oligosaccharides, (FOS), galacto-oligosaccharides and inulin available in Japan, USA and Europe.

prebreeding: All research and screening actions before a plant material arrives the directed breeding process (i.e. the development of germplasm to a state where it is viable for breeder's use). Primarily, it involves the evaluation of traits from exotic material and their introduction into more cultivated backgrounds.

precautionary principle: A principle associated with risk management. It states that where there are threats of serious or irreversible damage, lack of scientific certainty shall not be used as a reason for postponing cost-effective measures to prevent environmental degradation.

precautionary principle in genetic engineering: The precautionary principle states that no activity should be undertaken until it is certain that no detrimental effects will come as a result of that activity. By genetically modifying organisms, scientists are introducing an irreversible change into the natural environment. This change could potentially have drastic and far-reaching consequences, which could forever change the natural processes of nature. Only after it is proven that there are no harmful effects associated with the GMO should it be released for sale for human consumption.

precision: Gradation of absence of random error. The precision of a statistical estimator is associated with the expected error that ensues when approximating the infinite-data value. In other words, when you attempt to evaluate some quantity in a population, the precision is related to the error in the estimate. So, precision can be thought of as a 'margin of error' in approximating some unidentified value. Precision can be quantified by the width of a confidence (compatibility) interval and sometimes by a standard deviation of the estimator (standard error). For the confidence intervals, a 'margin for error' is calculated so that the quoted interval has a certain likelihood of encompassing the true value (e.g. population mean difference). Some authors define precision as the reciprocal of the variance of an estimate. By that definition, precision upsurges linearly as the sample size increases. If instead one expresses precision on the original scale of measurement instead of its square (i.e. if one uses the standard error or width of a confidence interval), precision rises as the square root of the sample size.

precision agriculture mapping: The Food and Agriculture Organization (FAO) of the United Nations, the world's population will reach 9.1 billion, 34% higher than today's population, by 2050. Due to this expected growth, there is pressure worldwide for higher agricultural production and reliable crop status information. To achieve these objectives, improved management of the world's agricultural resources is required, especially in developing countries. Utilizing HIGH-RESOLUTION, MULTISPECTRAL SATELLITE or AERIAL IMAGERY and ARTIFICIAL INTELLIGENCE (AI), MACHINE LEARNING (ML) and COMPUTER VISION (CV) ALGORITHMS, image data are collected and transferred into management solutions for crop health and improved production targets.

precision and accuracy: Precision and accuracy are two ways that scientists think about error. Accuracy refers to how close a measurement is to the true or accepted value. Precision refers to how close measurements of the same item are to each other. Precision is independent of accuracy. That means it is possible to be very precise but not very accurate and it is also possible to be accurate without being precise. The best quality scientific observations are both accurate and precise.

precision farming: An agricultural system which uses GLOBAL POSITIONING SYSTEM (GPS) involving satellites and sensors on the ground, along with rigorous information management tools to understand dissimilarities in resource conditions within fields. The information is utilized to more exactly apply fertilizers and other inputs and to more precisely predict crop yield.

precision guided sterile insect technique (pgSIT): In pgSIT, CRISPR gene-editing tool was employed to devise a technique of altering key genes that control insect sex determination and fertility. When pgSIT-derived eggs are introduced into targeted populations, only adult sterile males emerge, resulting in a novel, environmentally friendly, effective, species-specific, self-limiting, safe, scalable genetic population and thereby promising for a relatively low-cost method of controlling pest populations in the future.

precision insect pest control using microbes: Due to climate change, pollution and habitat loss, pollinators are under continuous threat. Bacteria are the ideal tools for developing precision bioinsecticides and preventing disease transmission by insects. In case of polyphagous insects, plant-based biocides are unlikely to offer much protection to other crops nearby. Also, it is hardly possible to engineer every species of vulnerable crop with a pest-specific biocide.

precision medicine: (i) An innovative approach to tailoring disease prevention and treatment that takes into account differences in people's genes, environments and lifestyles. (ii) The specific targeting of molecular abnormalities and the stratification of patients to specific drugs (epigenetic drugs). (iii) The specific targeting of molecular abnormalities and the stratification of patients to specific drugs. Can be used off the shelf. *Aka* PERSONALIZED MEDICINE.

precondition: In computer programming, a state or predicate that must always be factual just prior to the implementation of some section of code or before an operation in a proper specification. If a precondition is dishonoured, the effect of the section of code turns out to be undefined and thus may or may not carry out its projected work. Security glitches can arise due to incorrect preconditions.

precursor messenger RNA (pre-mRNA): An RNA molecule which is transcribed from chromosomal DNA in the nucleus of eukaryotic cells and then processed through splicing and other reactions to generate the mRNA which is transported the cytoplasm to directs protein synthesis there.

precursor micro RNA (pre-miRNA): An imperfect stem-loop RNA structure generated from a long transcript, known as primary miRNA, which is encoded by a non-protein-coding region in the genome. Dicer or Dicer-like enzyme cleaves a miRNA precursor and produces a mature miRNA.

precursor ribosomal RNA (pre-rRNA): The primary unprocessed transcript of a gene or a group of genes specifying rRNA molecules.

precursor ribosomal RNA (pre-rRNA) processing in eukaryotes: To form mature rRNA 18S, 5.8S and 28S, pre-rRNA 40S (in *Xenopus*) and 45S (in mammals) must go through a series of cleavages to remove the external and internal transcribed spacers (ETS/ITS). Eukaryotes have 5S rRNA as well which is transcribed from a divergent gene away from the nucleolus and is slightly processed.

precursor messenger RNA (pre-mRNA) processing in eukaryotes: *See* POST-TRANSCRIPTIONAL MODIFICATION (PTCM or PTM)

precursor ribosomal RNA (pre-rRNA) processing in prokaryotes: Prokaryotic cells synthesize three types of rRNA: 16S, 23S and 5S. Genomic arrangement of genes in the sequence 5′-16S, tRNAs, 23S, 5S, tRNA(s)-3′ occurs in a single operon. The entire operon is regularly transcribed as a unit. Cleavage enzymes are essential at a number of points even to separate several classes of pre-rRNAs. The rRNA transcript undergoes various vagaries either during or, less likely, after its synthesis but before its processing. To each transcript, around 110 methyl groups are added to the ribose moieties. These methyl groups are retained in the mature molecules. Thus, ribose methylation seems to act as a marker for the cleavage sites. The rRNA becomes bound to ribosomal proteins so that it is handled as ribonulceoprotein complex rather than as free rRNA.

precursor transfer RNA (pre-tRNA): The primary, unprocessed transcript of a gene or group of genes specifying tRNA molecules. *Aka* PRECURSOR SOLUBLE RNA (PRE-SOLUBLE RNA).

precursor transfer RNA (pre-tRNA) intron removal: Intervening sequences (IVS) are located in most of the tRNA precursors of eukaryotes and in advanced genera of prokaryotes, at a point, one site removed downstream from anticodon. However, in the gene for $tRNA^{Leu}$, the intron splits the anticodon sequence. For removal of intron from tRNA transcript, first step is cleavage of tRNA sections from the inserted sequence. These different tRNA precursors are separated by RNase E and III producing single tRNA precursors. The monomeric tRNA precursors interact with one or more RNases reducing the size of originally long tRNA precursor and permitting the mature tRNA precursor to assume its final secondary and tertiary structure. The introns of tRNA precursors are excised by precise endonucleolytic cleavage and ligation reactions catalysed by special splicing endonuclease and ligase activities. *Aka* PRE-SOLUBLE RNA INTRON REMOVAL.

precursor transfer RNA (pre-tRNA) processing: Different stages of tRNA maturation in fixed temporal order are as follows: (i) cleaving of the 5′ and 3′ ends in which RNase P acts on the 5′ terminus endonucleolytically and produces mature 5′-terminus by portraying mature tRNA sequence; (ii) the 3′-end maturation occurs by exonuclease activity by RNase D and 3′-CCA sequences of tRNA (as in prokaryotes); (iii) when -3′-CCA is not encoded in DNA, this triplet is enzymatically added before removal of intron (as is the case in eukaryotes; (iv) modification of the bases in mature coding region can occur on either the precursor molecular or cleavage products; and (v) removal of introns occurs when present. Introns are very short (13–60 nucleotides). The cleavage and splicing at intron junctions have been found to be highly dependent upon the presence of the -CCA. *Aka* PRE-SOLUBLE RNA PROCESSING.

predation: A biological interaction in which a predator kills and eats its prey.

predator: A biological interaction where one organism, the predator, kills and eats another organism, its prey. It is one of a family of common feeding behaviours that includes parasitism and micropredation (which usually do not kill the host) and parasitoidism (which always does, eventually). It is distinct from scavenging on dead prey, though many predators also scavenge; it overlaps with herbivory, as seed predators and destructive frugivores are predators.

predator-prey adaptations: Coexistence of two species delicately depends upon the genetic composition of the competing populations due to a counter evolutionary equilibrium.

predicting plant productivity. Plant productivity is predicted by using mathematical models constructed in view of different input components and their interactive influences on yield-related characteristics of the plant. Improvement in exactly forecasting plant biomass productivity is a vital component in nutrient management to assure a balance between nutrient stock and necessity under the climate change scenario.

predicting the human epigenome from DNA motifs: Epigram, an analysis pipeline that forecasts histone modification and DNA-methylation patterns from DNA motifs. The recognized cis elements denote interactions with the site-specific DNA-binding factors that launch and maintain epigenomic modifications. The *cis* elements in **ESCs** and four derived lineages were catalogued and noticed numerous motifs that have position preference, such as at the centre of H3K27ac or at the edges of H3K4me3 and H3K9me3, which provides mechanical vision about the shaping of the epigenome.

predictions: Reasoning including logical deduction from the hypothesis or theory.

predictive ecological risk assessment (PERA): PERA is the process by which future risks (harms, negative impacts) are estimated based on current knowledge and hypothesis-driven scientific research. Risk assessment frameworks typically involve logical steps of problem formulation, identification of potential hazards, identification of exposure pathways, risk characterization, prediction of the severity of harm (negligible, low, moderate or high) and an expression of uncertainty.

predictive metabolic engineering: The term refers to pathway manipulation that is based on the application of systems biology approaches (i.e. integrated metabolomics, proteomics and transcriptomics).

predictor, explanatory variable, risk factor, covariate, covariable and independent variable: Quantities which may be associated with better or worse outcome. Without further information, predictors (covariates) are taken to be measured at baseline. Time-dependent covariates are updated with post-baseline measurements. An external time-dependent covariate is one whose future values were already known at baseline, for example, in a crossover study, the new treatment assignment at 1 month (the crossover time) was already known at the time of randomization. Effects of external time-dependent covariates are easy to interpret. Internal time-dependent covariates (e.g. updated cholesterol measurements) may reflect changing subject condition. An especially difficult-to-interpret situation is a randomized trial in which one estimates the (supposedly constant) treatment effect after adjusting for internal time-dependent covariates.

predispose (n. **predisposition**)**:** To make disposed to infection and disease. The tendency of some non-genetic factors such as abiotic stress (water stress, temperature stress, etc.) to affect susceptibility of plants to diseases, hence associated with the disease triangle.

pre-empter stem: A configuration of leader transcript mRNA that does not terminate transcription in the attenuator-controlled amino acid operons.

preferential pairing: Chromosome pairing in allopolyploids in which the structurally most similar chromosomes preferentially pair with another.

preformation: The main feature of this idea of transmission of inheritance, given by J. Swammerdam in 1694, was a miniature man called HOMUNCULUS. The view advocated that the whole organism is already present in the sex cells and it merely gets bigger over time through the unfolding of pre-existing form. *See* THEORY OF ENCAPSULATION. *Contrasts* EPIGENESIS.

preformation theory: According to theory of preformation, things are always present in the egg in microscopic form. As development proceeds, these parts are enlarged.

preimplantation genetic disease diagnosis: Diagnosis of genetic disease in human embryo by *in vitro* fertilization prior to implantation.

pre-initiation complex (PIC): Pre-initiation complex formation is prerequisite for transcription initiation. The assembly of PIC begins with the binding of TATA-BINDING PROTEINS (TBP) or TFIID to the TATA box which greatly distorts its bound TATA-box DNA. This protein is the only GENERAL TRANSCRIPTION FACTOR (GTF) that binds to specific DNA segment and identify the transcription site. The regulatory proteins and other transcription factors exhibit certain distinct patterns or motifs in their structure that are important in their ability to bind to DNA. Various motifs of transcription factors that bind to DNA are ZINC FINGER, HELIX-TURN-HELIX, AMPHIPATHIC HELIX-LOOP-HELIX, LEUCINE ZIPPERS, ACID BLOBS and STEROID RECEPTORS.

pre-initiation complex (PIC) for transcription: RNA polymerase II (Pol II)-mediated transcription initiation requires assembly of a pre-initiation complex (PIC), during which the 14-subunit transcription factor IID (TFIID) recognizes core promoters and recruits TFIIA, TFIIB, TFIIE, TFIIF, TFIIH and RNA Pol II to sequentially assemble the core PIC (cPIC), intermediate PIC (mPIC) and holo PIC (hPIC).

pre-initiation complex (PIC), assembly of: Structural and biochemical analyses by X. Chen *et al.* (2022) have shown that the nucleosome makes multiple contacts with PIC-Mediator and may enhance transcription activity if the nucleosome is positioned 40 or 50 base pairs (but not 70 base pairs) downstream of TSS. This study provides a structural basis for understanding the regulatory function of the +1 nucleosome in transcription initiation.

premature cell division: Premature cell division relies on mechanisms similar to meiosis I, where genome haploidization is facilitated by homologous chromosome interactions, recognition and pairing.

premature centromere division: Premature centromere division represents a loss of control over the sequential separation and segregation of chromosome centromeres because the chromosomes are not attached at the centromere.

premature chromosome condensation: Results when an interphase cell fuses with a mitotic cell, causing the interphase cell to produce condensed chromosomes prematurely or it may be achieved following cell fusion mediated either by fusogenic viruses or by polyethylene glycol.

pre-meiotic male recombination: Recombination may rarely occur under natural conditions or may be induced with the help of a physical or a chemical agent in the pre-meiotic germ-line cells in males of *Drosophila melanogaster*.

prenatal: Existing or occurring before birth: prenatal medical care.

prenatal diagnosis: Diagnosis of a genetic disorder in an individual before birth.

prenylation: A covalent protein modification by attachment of an isoprenoid lipid group to it.

prepotency: The ability of a parent to impress its characteristics on offspring so they are more similar than usual.

pre-priming complex: The structure, consisting of a series of proteins attached to the origin of replication, which initiates DNA replication in *Escherichia coli*.

pre-priming protein: A protein that recognizes primer initiation sites.

preprophase band: Circumferential band of microtubules and actin filaments that is formed around a plant cell under the plasma membrane prior to mitosis and cell division.

pre-replication complex (pre-RC): A protein complex that is formed at the origin of replication during the initiation step of DNA replication. Eukaryotes have the most complex pre-RC. After ORC1-6 bind the origin of replication, Cdc6 is recruited. Cdc6 recruits the licensing factor Cdt1 and MCM2-7. Cdt1 binding and ATP hydrolysis by the ORC and Cdc6 load MCM2-7 on to DNA. There is a stoichiometric excess of the MCM proteins over the ORC and Cdc6 proteins, indicating that there may be multiple MCM heterohexamers bound to each origin of replication.

presence and absence hypothesis: The discredited idea that the normal phenotype is due to the presence of a factor and the mutant phenotype is its absence.

pre-setting: Describes the inheritance of transposable-element-mediated gene control that is independent of the transposable element. Usually confined to the first generation

pre-start G_1: During this subphase, the cells must attain a minimum size so that it can enter start stage. *G_1 cyclins* (C, D, E and G) are synthesized.

pre-symptomatic testing: A test for a genetic disease before it becomes apparent but which may develop in future if affected by certain factors.

prevalence: The proportion of individuals in a population having a disease.

prey: A family of common feeding behaviours that includes PARASITISM and MICROPREDATION (which usually do not kill the host) and PARASITOIDISM (which always does, eventually).

prezygotic incompatibility: In the case of incompatible or wide crosses, the inhibition of pollen germination or the prevention of pollen tube growth is among other possible barriers to plant fertilization. In some cases, such barrier can be overcome by *in vitro* pollination.

prezygotic isolation: Mechanisms that keep populations reproductively isolated from each other by preventing fertilization and zygote formation. Two mechanisms of prezygotic isolation are known. In one mechanism, the two species do not reach the stage of successful mating and thus no zygote is formation occurs. Examples of prezygotic isolation are the species that have different breeding seasons or courtship displays and which therefore never identify one another as potential mates. In the second mechanism, reproductive isolation sojourns production of an F_1 zygote by preventing cross-mating.

Pribnow box: The Pribnow box, *aka* the PRIBNOW-SCHALLER BOX, named after D. Pribnow in 1975 and H. Schaller *et al.* (1975), is a consensus sequence of TATAAT of six nucleotides (thymine, adenine, thymine, etc.) centred about 10 base pairs before the starting point of bacterial genes. It is an indispensable portion of a promoter site which is important in binding RNA polymerase for transcription to transpire in bacteria.

primary cellular function: Synthesis of one of the transcripts which are processed to have mRNA, tRNA or rRNA.

primary constriction: A constriction which is determined by and associated with the centromere region. *See* CENTROMERE.

primary contact: A condition where populations have been in contact throughout their divergence. *Contrasts* SECONDARY CONTACT.

primary culture: A culture started from cells, tissues or organs taken directly from organism. During plant tissue culture, the establishment of tissue culture is the first step. For this a suitable explant is aseptically cultured *in vitro* and incubated under optimum conditions for 4–6 weeks. The culture thus developed is called primary culture that is further subcultured (recultured) for further multiplication. A primary culture may be regarded as such until it is successfully subcultured for the first time. It then becomes a CELL LINE.

primary data: Data that are directly collected from the respondents. Primary data means original data that has been collected specially for the purpose in mind. It means someone collected the data from the original source first hand.

primary endosperm nucleus: The triploid nucleus ensuing from the fusion of a male gamete nucleus and two polar nuclei in the central cell of the embryo sac.

primary gene pool: Includes the cultivated species of a crop and related species from which valuable genes can be most willingly attained for breeding. In general, it refers to the sum total of all the genetic variation in the breeding population of a species and closely related species that commonly interbreed or can be regularly crossed with the species.

primary host: Occasionally refers to the TELIAL HOST of a heteroecious rust fungus and now and then used to refer to the most economically imperative host. *Contrasts* ALTERNATE HOST.

primary immune response: Adaptive immune response to an antigen that is attained on first encounter with that antigen.

primary induction: When a cell self-differentiates, for example, in frog, mesoderm self-differentiates.

primary infection: The first-time exposure of an organism to and infection by a pathogen. During a primary infection, organism's body has no innate defences against the pathogen, such as antibodies, which can be developed through exposure or vaccination. *Compare with* SECONDARY INFECTION.

primary inoculum: The propagules, such as spores and sclerotia that over winter or over summer that initiate an infection *See* INITIAL INOCULUM. *Contrasts* SECONDARY INOCULUM.

primary leaf: The first factual leaf that appears on a plant following the cotyledons.

primary metabolites: The molecules which are produced by plants such as carbohydrates, proteins, enzymes, sugars, flavours and dyes, which are further consumed in the plant metabolism for energy production. Primary metabolites are directly involved in normal growth, development and reproduction.

primary microRNA (pri-miRNA) processing: The cleavage site of pri-miRNA is identified by the ssRNA-dsRNA junction at the base of the miRNA hairpin, where Drosha cleaves ≈11 bp or one dsRNA helical turn, from the base. The ensuing pre-miRNA is a hairpin containing a 3′ two-nt overhang. After Drosha cleavage, the pre-miRNA exits the nucleus via transport by exportin-5, which binds the 3′ two-nt of the pre-miRNA.

primary non-disjunction: Non-disjunction is the fiasco of homologous chromosomes or sister chromatids to disperse accurately during cell division (mitosis/meiosis). The failure of homologous chromosomes to separate at anaphase I of meiosis I is a good example of primary non-disjunction, for example, in XX-XY system of sex determination (as in *Drosophila*), it is the production of eggs with two or no X chromosome(s) in an XX individual (female).

primary oocyte: The cell that undergoes meiosis in female animals.

primary Piwi-interacting RNAs (pri piRNAs): The product of piRNA precursor transcript processing. These piRNAs have a preference for a 5′ U. *Compare with* SECONDARY PIWI-INTERACTING RNAs.

primary pollutant: An air pollutant that is emitted directly into the atmosphere and is generally harmful to plants (e.g. SO_2 or NO_x). *Contrasts* SECONDARY POLLUTANT.

primary production: Productivity by autotrophic organisms (e.g. plants or algae), often estimated as biomass amassed over a unit of time or sometimes by the quantity of carbon fixed.

primary protein structure: Comprises the backbone and the sequence of the amino acids of a polypeptide chain.

primary resistance management method: Although transgenic crops expressing insecticidal toxins are widely used yet the economic benefits of these crops are lost if toxin resistance spreads through the pest population. The resistance management method is a high-dose/refuge strategy, requiring toxin-free crops as refuges near the insecticidal crops, as well as toxin doses sufficiently high than can kill wild-type insects and insects heterozygous for a resistance allele.

primary response: The immunological response to the first challenge with a particular antigen.

primary root: A root that grows straight from the radicle of an embryo rather than from a crown or node.

primary spermatocyte: The cell that undergoes meiosis in male animals.

primary storage: Often refers to simply as memory and is the only one directly reachable to the CENTRAL PROCESSING UNIT (CPU). The CPU uninterruptedly reads instructions stored there and implements them as required. Any data dynamically operated on are also stored there in unchanging manner. *Aka* MAIN MEMORY, INTERNAL MEMORY *or* PRIME MEMORY.

primary structure: Refers to the sequence of amino acids in a protein 'molecular' chain or to the linear sequence of nucleotides in a polynucleotide (RNA or DNA) 'molecular' chain. *Compare with* SECONDARY STRUCTURE, TERTIARY STRUCTURE *and* QUATERNARY STRUCTURE.

primary transcript: The immediate product of transcription of a gene or group of genes, which will subsequently be processed to give the mature (functional) transcript(s). *See* TRANSCRIPT.

primary trisomics: The trisomics in which the additional chromosome is a normal one.

primase: The RNA polymerase enzyme that synthesizes the primer needed to initiate replication of DNA polynucleotide.

primase-polymerase superfamily: A functionally diverse superfamily of replication and repair enzymes.

prime editing (PE): A 'search-and-replace' genome-editing technology in molecular biology by which the genome of living organisms may be modified. The technology directly writes new genetic information into a targeted DNA site. It is a gene-editing method that can perform targeted small insertions, deletions and base swapping in a precise way. It uses a fusion protein, consisting of a catalytically impaired Cas9 endonuclease fused to an engineered reverse transcriptase enzyme and a prime editing guide RNA (pegRNA), capable of identifying the target site and providing the new genetic information to replace the target DNA nucleotides. It mediates targeted insertions, deletions and base-to-base conversions without the need for double-strand breaks (DSBs) or donor DNA templates.

prime editing (PE) system: PE, developed by A.V. Anzalone *et al.* (2019), is a 'search-and-replace' genome-editing technology that can introduce 12 base-to-base conversions, as well as small insertions and deletions, without the need for DOUBLE-STRAND BREAKS (DSBs) or donor DNA templates. PE is based on targeted gRNA-templated reverse transcription and integration of the *de novo* synthesized single-stranded DNA. PEs consist of a nCas9 (H840A) conjugated with an engineered REVERSE TRANSCRIPTASE (RT) paired with a PRIME EDITING GUIDE RNA (PEGRNA) that both specifies the target site and encodes the desired edit.

prime editing guide RNA (pegRNA): A guide RNA capable of (i) identifying the target nucleotide sequence to be edited and (ii) encoding new genetic information that replaces the targeted sequence. The prime editor system collaborates with a pegRNA to facilitate a wide variety of precise genome edits in living cells. Towards the uncovering the molecular mechanism of pegRNA-guided reverse transcription by the prime editor, Y. Shuto *et al.* (2024) present cryo-electron microscopy structures of the SpCas9–M-MLV RTΔRNaseH–pegRNA–target DNA complex in multiple states. The termination structure divulges that M-MLV RT extends reverse transcription beyond the expected site, resulting in scaffold-derived incorporations that result in undesired edits at the target loci.

prime editing guide RNA-independent off-target effects: With all RNA-guided genome editors, guide-dependent off-target (OT) mutations can occur at other sites bearing similarity to the intended target. R. Gao *et al.* (2022) developed a cost-effective sensitive platform to profile guide-independent OT effects in human cells. They did not observe guide-independent OT mutations in the DNA or RNA of PRIME EDITOR 3 (PE3)-edited cells, or alterations to their telomeres, endogenous retroelements, alternative splicing events or gene expression. The results showed undetectable prime editing guide RNA–independent OT effects of PE3 in human cells, suggesting the high editing specificity of its reverse transcriptase moiety.

prime editing technology (PET): Prime editing (PE) technology enables precise alterations in the genetic code of a genome of interest. PE offers great potential for identifying major agronomically important genes in plants and editing them into superior variants, ideally targeting multiple loci simultaneously to realize the collective effects of the edits. A. Gupta *et al.* (2024) develop a modular assembly-based multiplex PE system in rice and demonstrate its efficacy in editing up to four genes – two for herbicide tolerance (*OsEPSPS1* and *OsALS1*) and two for *Xoo* resistance (*TFIIAγ5* and *OsSWEET11a*) in a single transformation experiment. This modular assembly-based system yielded high PE rates and streamlined the cloning of PE reagents, making it feasible for more labs to utilize PE for their editing experiments.

prime editing, improving: To identify cellular determinants of prime editing, J. Yan *et al.* (2024) develop scalable prime editing reporters and performed genome-scale CRISPR-interference screens from which a single factor arises as the strongest mediator of prime editing: the small RNA–binding exonuclease protection factor La. La promotes prime editing across approaches (PE2, PE3, PE4 and PE5) and edits (substitutions, insertions and deletions) endogenous loci and cell types but has no consistent effect on genome-editing approaches that rely on standard, unextended guide RNAs. They developed a novel prime editor protein (PE7) fused to the RNA-binding, N-terminal domain of La. This editor enhanced prime editing with expressed pegRNAs and ENGINEERED PEGRNAS (EPEGRNAS), as well as with synthetic pegRNAs optimized for La binding, thus providing crucial insights into how prime editing components interact with the cellular environment.

prime editors (PEs): PEs mediate genome modification without utilizing double-stranded DNA breaks or exogenous donor DNA as a template. PEs facilitate nucleotide substitutions or local insertions or deletions within the genome based on the template sequence encoded within the prime editing guide RNA (pegRNA).

prime editors, exonuclease-enhanced (Exo-PE): D. J.J. Truong *et al.* (2024) generated an enhanced fluorescence-activated cell sorting reporter cell line to develop an exonuclease-enhanced PE strategy ('Exo-PE') composed of an improved PE complex and an aptamer-recruited DNA exonuclease to remove the 5′ original DNA flap. Exo-PE achieved better overall editing efficacy than the reference PE2 strategy for insertions ≥30 base pairs in several endogenous loci and cell lines while maintaining the high editing precision of PE2. By enabling the precise incorporation of larger insertions, Exo-PE complements the growing palette of different PE tools and spurs additional refinements of the PE machinery.

primed *in situ* hybridization: Involves annealing of unlabelled primers followed by incubation with a reaction mixture that includes labelled nucleotides and DNA polymerase.

primed *in situ* labelling: A procedure that labels specific sequences in entire chromosomes by *in situ* DNA chain elongation or PCR.

primer: A single-stranded oligonucleotide which when attached to single-stranded template molecule acts as the starting point for complementary strand synthesis in the presence of DNA polymerase enzyme. It is short nucleic acid sequence containing a free 3′ hydroxyl group that forms base pairs with the complementary template strand and functions as the starting point for addition of nucleotides to copy the template strand. During PCR, synthetic primer is added to template DNA in the PCR tube in the presence of *Taq* polymerase, dNTPs and buffer solution. The primer binds to target DNA at 68°C and initiates chain elongation which finally results in the synthesis of a complementary strand. Appropriate primer also helps in checking the presence of transgene in the putative transgenic plant through PCR analysis.

primer annealing: A step in DNA replication of PCR in which the primers adhere to their complementary DNA sequences on the single-stranded DNA template.

primer dimers: Ensue when single-stranded primer oligonucleotides bind to each other rather than the DNA template.

primer extension: (i) A response in which DNA is reverse transcribed from an RNA template to which a specific oligonucleotide primer has been annealed. The new complementary DNA product is an extension of the primer, which is synthesized at the 3′ end of the primer in a direction outspreading towards the 5′ end of the RNA. This reaction is useful for discovering the extreme 5′ end of RNA molecules. (ii) A technique to detect the so-called SINGLE-NUCLEOTIDE POLYMORPHISMS (SNPs) in target DNA. In short, the target (e.g. a gene) is first amplified with specific primers in a conventional PCR and subsequently denatured. Then, a single-stranded primer is annealed to the single-stranded target DNA such that the primer ends exactly at the SNP site. After annealing, the duplex exposes a 3′-OH group for an extension catalysed by DNA polymerase in the presence of all four DIDEOXYNUCLEOSIDE TRIPHOSPHATES (DDNTPs) rather than DEOXYRIBONUCLEOSIDE TRIPHOSPHATES (DNTPs), each labelled with a specific fluorochrome. The matching ddNTP will then be incorporated and stops extension. The incorporated ddNTP is then identified by the specific fluorescence emission of its fluorochrome. A comparison with the wild-type sequence at the SNP site allows for identification of the type of SNP.

primer walking: A method for sequencing long (>1 kb) cloned pieces of DNA. The initial sequencing reaction reveals the sequence of the first few hundred nucleotides of the cloned DNA. On the basis of these data, a primer containing around 20 nucleotides and complementary to a sequence near the end of sequenced DNA is synthesized and is then used for sequencing the next few hundred nucleotides of the cloned DNA. This procedure is repeated until the complete nucleotide sequence of the cloned DNA is determined.

primer-independent DNA polymerases: M. Redrejo-Rodríguez *et al.* (2017) discovered a new group of DNA polymerases, called primer-independent PolB (piPolB), that might be a link between the previously known protein-primed and RNA/DNA-primed PolBs. PiPolBs are encoded by highly diverse mobile genetic elements, pipolins, integrated into the genomes of diverse bacteria and also present as circular plasmids in mitochondria. piPolB displays efficient DNA polymerization activity that can use undamaged and damaged templates and is endowed with proofreading and strand displacement capacities. Remarkably, the protein is also capable of template-dependent de novo DNA synthesis (i.e. DNA-priming activity), thereby breaking the long-standing dogma that replicative DNA polymerases require a pre-existing primer for DNA synthesis. piPolBs are suggested to be involved in self-replication of pipolins and may also contribute to bacterial DNA damage tolerance.

primers for DNA replication: A short oligoribonucleotide (of size ten ribonucleotides) synthesized by PRIMASE that is attached to a single-stranded DNA molecule in order to provide a site at which DNA replication of synthesis of OKAZAKI FRAGMENTS.

primers for PCR: Small segment of DNA nucleotides which binds to the single-stranded DNA template during PCR. These can be sequences specific for a gene or totally random, depending upon the objective of the experiment.

priming: (i) Phenomenon through which a transient biotic or abiotic stress cue leads to modified (typically faster or stronger) defence responses upon exposure to a recurring stress. The treatment of seeds with an osmotic solution, such as polyethylene glycol, which enables controlled hydration. The seed embryo grows to the point of germination and then is dried. (ii) Term is applied for more even and quick germination of certain vegetable seeds. (iii) In tobacco, this term is used to designate the harvesting of ripe tobacco leaves from the stalk as they ripen, beginning at the bottom and progressing upward. (iv) Term also used to describe for immunity and for responses to various abiotic stresses.

priming and stress memory: Experience and memory of environmental stimuli that indicate future stress can prepare (prime) organismic stress responses even in species lacking a nervous system. Such organisms prepare their phenotype for an improved response to future stress by 'priming'.

priming stress cue: Exposure to drug-associated environmental stimuli (cue), drug priming and stress have been thought of as the major risk factors contributing to relapse in abstinent subjects.

primitive: An organism or character judged to be less changed from a presumed ancestral state than other with which it is compared.

primitive data type: In computer science, primitive data types are a set of basic data types from which all other data types are constructed. Specifically, it often refers to the limited set of data representations in use by a particular processor, which all compiled programmes must use. Most processors support a similar set of primitive data types, although the specific representations vary. More generally, PRIMITIVE DATA TYPES may refer to the standard data types built into a programming language.

primordial germ cell (PGC): Cell reserved early in embryonic development that is a precursor to germ cells that produce gametes. Zygotically contributed *ovo* gene product is required for the survival of female germ cells in *Drosophila melanogaster*. The *ovo* locus appears to be involved in germ-line sex determination and is linked in some manner to sex determination in the soma. Egg and sperm cells of the mouse are derived from a founder population of primordial germ cells (PGCs) that are set aside early during embryogenesis. PGCs arise from the proximal epiblast. Embryonic germ cells show removal of methylation markers (imprints) of the *Igf2r* and *H19* genes, a property characteristic of the germ. Such model systems help in understanding germline epigenetic modification and mammalian gametogenesis.

primordial soup: Nickname given to the theory that life started in the oceans from the synthesis of organic molecules.

primordium (pl. primordia): The early cells that are reserved as the precursors of an organ to which they later give rise by mitotic development.

primosome: Primosome is a complex of two proteins, a primase and a helicase that initiates RNA primers on the lagging DNA strand during DNA replication. Primase synthesizes a short stretch of RNA, which is complementary to one of the DNA strands.

PrimPol damage tolerance pathway: During S phase, damage tolerance pathways allow lesion bypass to ensure efficient genome duplication. One such pathway is repriming, mediated by Primase-Polymerase (PrimPol) in human cells.

principle: A source or origin; that from which anything proceeds; fundamental substance or energy; primordial substance; ultimate element or cause; a basic generalization that is accepted as true and that can be used as a basis for reasoning or conduct. Examples: (i) BASIC NUCLEAR PRINCIPLES. The atom is composed of subatomic particles called PROTONS, NEUTRONS and ELECTRONS (ii) HARDY-WEINBERG PRINCIPLE.

principle of independent assortment (independent segregation): Different gene pairs segregate independently during meiosis, giving different traits an equal opportunity of occurring together. Mathematically, the principle of independent assortment is an extension of the principle of segregation which, in case of simple or complete dominance, can be expressed as $(a + b)^n$. Thus, for the F_2 progeny of a dihybrid cross, the phenotypic ratio will be $(3 + 1)^2$ (i.e. 9:3:3:1). The F_2 progeny produces two recombinant phenotypic classes in addition to the two parental ones in the F_2. When two gene pairs are showing independent assortment, it is inferred that the factors (genes) that control different traits and are located in non-homologous chromosomes. The principle of independent segregation can also be understood from backcrosses of dihybrids. Independent segregation is an important means for generating new genotypic and hence phenotypic combinations. Even when different gene pairs show incomplete dominance or codominance, segregation is independent and the genotypic ratio remains the same while the number of phenotypes increases. *Aka* INDEPENDENT SEGREGATION. *Compare with* LINKAGE.

principle of segregation: Deduced from G.J. Mendel's work involving monohybrid crosses. Accordingly, the two members of a gene pair separate from each other during meiosis, so that one-half of the gametes carry one member and the other half of the gametes carry the other. The principle of segregation was also deduced from Mendel's observations of phenotypic ratios in F_2 (3:1) and/or backcross (1:1) progeny.

prior odds: A ratio of the probability of one hypothesis (H_1) comparative to the probability of another hypothesis (H_2), denoted as $Prob(H_1)/Prob(H_2)$, prior to observing any data.

prior probability: The probability of an event as it could best be assessed before the experiment. In diagnostic testing this is called the PRETEST PROBABILITY. The prior probability can come from an objective model based on previously available information or it can be based on expert opinion. In some Bayesian analyses, prior probabilities are expressed as probability distributions which are flat lines, to reflect a complete absence of knowledge about an event. Such distributions are called non-informative, flat or reference distributions, and analyses based on them fully let the data 'speak for themselves'.

priority queue: An abstract data category which is like a regular queue or stack data structure but where additionally each element assumes a 'priority' associated with it. In a priority queue, an element with high priority is obliged before an element with low priority. In some applications, if two elements have the same priority, they are obliged according to the order in which they were arranged, while in other implementations, ordering of elements with the same priority is undefined.

privacy: In genetics, the right of people to restrict access to their genetic information.

privacy-preserving federated machine learning method (PPML-Omics): J. Zhou *et al.* (2024) proposed PPML-Omics by designing a decentralized differential private federated learning algorithm. They applied PPML-Omics to analyse data from three sequencing technologies and addressed the privacy concern in three major tasks of 'omic' data under three representative deep learning models. They examined privacy breaches in depth through privacy attack experiments and demonstrated that PPML-Omics could protect patients' privacy. In each of these applications, PPML-Omics was able to outperform methods of comparison under the same level of privacy guarantee, demonstrating the versatility of the method in simultaneously balancing the privacy-preserving capability and utility in omic data analysis.

private antigenic determinants: The determinants that are unique to an antigen.

private mutation: A rare DNA alteration which may be exclusively found in a single family or population.

pRNA: At least one species of DNA-containing phages, phi-29, uses a complex of six identical short RNA sequences as mechanical components (utilizing ATP for energy) of its DNA packaging machinery.

probability: Probability is the branch of mathematics concerning numerical descriptions of how likely an event is to occur or how likely it is that a proposition is true. The probability of an event is a number between 0 and 1, where, roughly speaking, 0 indicates impossibility of the event and 1 indicates certainty.

probability density function: The probability density is a statistical measurement that measures the likely outcome of a calculation over a given range, for example, when a random variable Y is continuous (i.e. it can take on every possible number within some interval), the probability density function is a function of y which is the limit, as the width δ of some interval goes to zero, of the probability that Y will be within the interval [y, y +δ], divided by δ. This is the first derivative (slope) of the cumulative probability distribution function for Y.

probability of ultimate fixation (U_p): The chance that a single copy of an allele will eventually fix throughout the entire population.

probability sample: A sample selected by a method such that each unit has a fixed and determined probability of selection.

proband: The affected individual by which a family with a genetic disorder is ascertained (in a family to present to medical attention). *Aka* PROPOSITUS *or* INDEX CASE.

probe: A radioactively or chemically labelled and defined molecule (DNA fragment, RNA fragment) that has some sequence homology to the gene that is represented in the library. It can be used in molecular hybridization to locate specific nucleic acid sequences by hybridization. Usually probe is employed for *ex situ* or *in situ* identification of specific, complementary nucleic acid sequences. *See* DNA PROBE.

probing the gene: DNA or protein probes are used for probing the gene.

probiotics: Living microorganisms that when ingested in certain numbers exert a beneficial effect beyond that of their nutrition value. Probiotics consist of either yeast or bacteria. Generally, probiotics belong to genus *Lactobacillus* and bifidobacteria. These microorganisms have a considerable safety record both within the fermented foods industry, where they have been used for many years and more recently, in probiotic foods. Though probiotic therapy dates back to at least 1905, there has been growing interest in the recent years. The exact mechanism of action of probiotics is still unclear. Initial theories revolved around 'competitive exclusion', in which 'good' bacteria eliminated or suppressed 'bad' bacteria. Alternative proposed mechanisms include production of antimicrobial factors, immunoregulatory effects, anti-inflammatory effects, anti-carcinogenic effects and so on.

problem: In algorithms, a problem can be defined as a real-world problem or real-world instance problem for which you need to develop a programme or set of instructions. An algorithm is a set of instructions.

problem formulation: The scoping phase of a risk assessment, in which the characteristics and use pattern of the product to be assessed are documented, as are the ecosystem or human population potentially at risk and the end points that will be the focus of the assessment.

problem of morphology: Each generation of adult has been produced by a series of developmental processes from an apparently structureless egg to the complex morphology of the adult.

problems associated with imprinting: Imprinting may cause problems in cloning, with clones having DNA that is not methylated in the correct position. It is possible that this is due to a lack of time for reprogramming to be completely achieved. When a nucleus is added to an egg during somatic cell nuclear transfer, the egg starts dividing in minutes, as compared to the days or months it takes for reprogramming during embryonic development. If time is the responsible factor, it may be possible to delay cell division in clones, giving time for proper reprogramming to occur.

procedural programming: A programming paradigm, derived from imperative programming, based on the concept of the procedure call. Procedures (a type of routine or subroutine) simply contain a series of computational steps to be carried out. Any given procedure might be called at any point during a programme's execution, including by other procedures or itself. The first major procedural programming languages appeared circa 1957–1964, including Fortran, ALGOL, COBOL, PL/I and BASIC. Pascal and C were published circa 1970–1972.

procedure: In computer programming, a subroutine is a sequence of programme instructions that performs a specific job, packaged as a unit. This unit can then be engaged in programmes wherever that particular job should be achieved. Subroutines may be defined within programmes or separately in libraries that can be engaged by many programmes. In different programming languages, a subroutine may be called ROUTINE, SUBPROGRAM, FUNCTION, METHOD or PROCEDURE. Technically, these terms all have different definitions. The generic, umbrella term CALLABLE UNIT is sometimes used.

process of risk assessment of transgenics: The process of risk assessment of transgenics consists of five components, namely, hazard identification, exposure assessment, effect's management, risk characterization and risk management.

process versus discard decision: Nuclear RNA complexity is 10–20 times greater than the complexity of the mRNA. According to this hypothetical possibility, there will be no processing of a particular transcript in one but successful processing takes place in another cell type. There is no evidence for existence of such a mechanism for nuclear RNA.

processed pseudogene: Processed pseudogenes are copies of messenger RNAs that have been reverse transcribed into DNA and inserted into the genome using the enzymatic activities of active L1 elements. Processed pseudogenes generally lack introns, end in a 3′ poly(A) and are flanked by target site duplications.

processing (P) bodies: Cytoplasmic ribonucleoprotein (RNP) granules primarily composed of translationally repressed mRNAs and proteins related to mRNA decay, suggesting roles in post-transcriptional regulation. Translation and mRNA degradation are influenced by a crucial transition where eukaryotic mRNAs exit translation and assemble an mRNP state that amasses into processing bodies (P-bodies).

Decapping activators (Dhh1p and Pat1p) work as translational repressors and facilitators of P body formation. This competition creates a magnificently balanced system setting the relative translation rate for an mRNA. mRNAs can be driven into either translation or repression by tipping the balance of this competition via any number of events. mRNA molecules within yeast P-bodies can also return to translation.

processing of 5S rRNA: The processing for maturation of 5S RNA occurs in two steps – the early steps and the late steps. In early steps, after transcription, the primary transcript is cleaved into two parts, one carries the genes for the minor rRNA and one or two tRNAs present on the intergenic spacer and other bears the major and the supplemental rRNAs, depending upon whether or not a tRNA ensues on the 3′-end. Neither RNase III nor E is involved in the initial enzymatic reaction. In late step, the 10S RNA comprises six stem-and-loop structures produced by RNase III. Then by action of RNase P, fragment 7S RNA is produced which contains 5S rRNA and a termination stem. Removal of this stem is achieved by RNase E.

processing of leaders of pre-mRNAs: In eukaryotes, a cap (methylated guanine) is added at 5′-end. Addition of cap (m^7GpppN_1) takes place at early stages of transcription in eukaryotes by RNA polymerase II. Cap is suggested to be derived by endonucleolytic cleavage and subsequent addition.

processing of other rRNAs: The primary transcript of the entire gene set is cleaved with RNase III or its equivalent, which breaks the precursor between the minor and major coding sectors.

processing of signals: Signals may be processed in three ways. (i) Certain chemical signals may penetrate the plasma membrane of the cell and interact with internal receptors (e.g. steroids, nitric oxide). Most signals, however, are hydrophilic molecules remaining outside the cell. These interact with transmembrane (membrane-spanning) or membrane-associated receptors and cause a change of receptor structure. The interaction may result in signal transport (i.e. the signalling molecule is internalized, either by carriage caused by the conformation change of the receptor, by the creation of a pore; for example, in the case of ion channels or by receptor-mediated endocytosis). (ii) The conformational change in the receptor may induce enzyme activity inside the cell which mediates downstream effects while the ligand remains on the outside (signal transduction). (iii) Physical stimuli may also interact with receptors or may mediate their effects directly. Light stimulates the G proteins linked to rhodopsin and cone opsin receptors when photons cause a conjugated light-sensitive molecule all-*cis* retinal to change to the all-*trans* conformation. Conversely, the response to heat shock and similar stresses is mediated directly by increases in denatured protein in the cell.

processing of the 5.8S rRNA: Processing of the supplemental species of rRNA is restricted to the nucleolus and involves small RNA U3 and another type of RNA, 7-1. During considerable part of its maturation, it is hydrogen bonded to the other pre-rRNAs. First product of this species is 12S RNA particle, trailed by the one sedimenting at 8S. Finally, 5.8S rRNA is formed.

processing of trains of pre-mRNAs: In eukaryotes, polyadenylation comprises two steps: cleavage and addition of poly(A) tail. Multiple copies of polyadenylation signal are of general existence. Tropomyosin of *Drosophila* has a cluster of five poly(A) signals followed by pair of about 250 residues downstream. First of these signals has a dual structure AAUAAUAAA. Different signals are implicated in addition to poly(A) tail in mRNA processing in different tissues.

processing the major species of rRNA: The primary transcript of shorter operons, 30S pre-rRNA is in the arrangement of two loops containing the minor and major species, respectively, each closed at one side by a powerfully base-paired stem. It is these stems that are cleaved by RNase III, the enzyme which specifically functions on double-stranded segments. The final trimming of 30S pre-rRNA occurs at the ribosome to produce 23S RNA.

processing unit: In algorithms, the input will be passed to the processing unit, producing the desired output.

processive enzyme: The ability of an enzyme to perform multiple catalytic reactions without releasing its substrate.

processivity of an enzyme: The extent to which an RNA or DNA polymerase adheres to a template before dissociating determines the average length (in kilobases) of the newly synthesized nucleic acid strands. It also applies to the action of exonucleases in digesting from the ends to the middle of a nucleic acid.

processivity of DNA replication: The frequency with which an enzyme dissociates from template-primer after addition of a nucleotide to the growing chain end.

prochoice determinant of cell fate: It is still unclear why a progenitor cell makes a choice to leave the progenitor pool and to adopt a specific cell fate. An initial decision in epithelial lineage determination involves the Notch signalling pathway. Notch gene from *Drosophila* is involved with cell fate choices in the neurogenic region of the blastoderm in developing nervous system and in the eye-antennal imaginal disc. cDNA clones were isolated from *Xenopus* embryos with *Notch* DNA in order to investigate whether cell-cell interactions in vertebrate embryos also depend on *Notch*-like molecules. *Xotch* was remarkably similar to *Notch* in both structural and developmental expression.

prochromosome: A heterochromatin block that is realized during the interphase of cell division and which is associated with the number of chromosomes per complement or less.

producer gene: A producer gene is comparable to a structural gene in prokaryotes. It produces precursor mRNA, which after several steps of processing becomes mRNA. Expression of a producer gene may be controlled by many RECEPTOR GENES.

producing mapping population: Various methods include selecting two genetically divergent parents, which show clear genetic differences for one or more traits of interest. Progenies used for genetic mapping are (i) second filial generation (F_2), (ii) backcross (BC), (iii) recombinant inbred lines (RILs), (iv) double haploids (DHs) and (v) near-isogenic lines (NILs).

producing transgenic plants using non-antibiotic resistance gene as marker: One potential alternative method to produce transformants without any antibiotic/herbicide marker gene is so-called POSITIVE SELECTION SYSTEM. In this approach, selection is based on the ability of plant cells to utilize non-metabolized carbohydrates. Only transformed cells are capable of utilizing mannose as a carbon source. Using same approach, selection can also be based on the ability of plant cells to autonomously produce growth regulators.

producing transgenic plants using plant-derived marker genes: A few plant genes can also be associated with antibiotic resistance. For instance, overexpression of an *Arabidopsis thaliana* ATP-BINDING CASSETTE (ABC) transporter (*At-WBC19*) gene confers high-level resistance to kanamycin in GM crops. The plant origin of these genes has been presented as a way to bypass some of the biosafety concerns associated with the use of ANTIBIOTIC RESISTANCE MARKER GENES (ARMGs) of microbial origin.

product of meiosis: One of the (usually four) cells formed by two meiotic divisions (meiosis I and meiosis II).

product rule of probability: This rule states that the probability of independent events occurring together is the product of the probabilities of the individual events.

production of first recombinant DNA molecule: P. Berg was among the first scientists to produce a recombinant DNA molecule in 1972. Stanley Cohen of Stanford University and Herbert Boyer of University of California, San Francisco (UCSF) applied for a patent on recombinant DNA technology in 1974, which was granted in 1980.

production of protease kallikrein in rat: Polypeptides exist that can be cut by proteases into different activities. This is proteolytic control, for example, same primary transcript may produce adrenocorticotropin or B lipoprotein. In case of rat *kallikrein* gene, the transcript is 867 nucleotides, including 24 in a leader and 48 in the train. The remaining 795 nucleotides encode a protein consisting of 265 amino acid residues. This protein is named preprokallikrein. Seventeen amino acids are removed enzymatically and become a proenzyme (prokallikrein), but before it becomes activated, the 11-amino acid 'pro' section is removed and catabolized. The active enzyme is only 237 amino acid long.

productive infection: An infection of animal cells by a virus which results in the structural gene binding with RNA polymerase.

proficiency testing (PT): The use of masked (blind) testing to assess whether a laboratory can perform a test correctly. Usually, the samples are provided by an organization independent of the laboratories performing the test. Also, there is a programme in which specimens are periodically sent to participating laboratories for analysis and/or identification, in which the results of each laboratory are compared with those of other participants and/or a target result. The results are then reported to participating laboratories and, possibly, also to others.

profile: (i) A table that lists the frequencies of each amino acid in each position of protein sequence. Frequencies are calculated from multiple alignments of sequences containing a domain of interest. (ii) A profile is a table of position-specific scores and gap penalties, representing a homologous family, that may be used to search sequence databases. In CLUSTAL-W-derived profiles those sequences that are more distantly related are assigned higher weights.

profile, improving sensitivity of: Attempts have been made to further improve the sensitivity of the profile by refining the procedures to construct a profile starting from a given multiple alignment. Other representations for sequence domains or motifs do not necessarily require the presence of a correct and complete multiple alignment, such as hidden Markov models. Attempts to further improve the sensitivity of a profile have been made through refining the methods to construct the profile, starting from a given multiple alignment. Other representations for sequence domains or motifs do not necessarily require the presence of a correct and complete multiple alignment, such as HIDDEN MARKOV MODELS. *See* POSITION-SPECIFIC SCORING MATRIX (PSSM).

proflavine: A chemical mutagen, one of the acridine dyes, frequently used to induce mutations for experimental purposes.

progenitor cell: Undifferentiated cell (i.e. immature cell), which has the potential to develop into any cell type.

progenote stage: DNA sequences of the most primitive unicellular eukaryotes may have initiated from primordial DNA sequences which had been randomly produced. Before a self-replicating cell could evolve but after the biological energy-generating system had evolved, DNA molecules were created in the primordial soup by random addition of the four nucleotides without the use of templates. The nucleotide sequences that coded for proteins were selected from the randomly produced nucleotide sequences in the primordial soup by natural selection. This represents the 'progenote' stage in the terminology of C. Woese and G.E. Fox (1977).

progeny: A genetic descendant or the descendants of a person, animal or plant; offspring.

progeny selection: Selection based on performance of progeny.

progeny testing: A test of the value of a genotype on the basis of the performance of its offspring generated in some definite system of mating.

progesterone: An endogenous steroid and progestogen sex hormone which plays a critical role in the menstrual cycle, pregnancy and embryogenesis in humans and other animal species.

prognosis: Prediction of the course and probable outcome of a disease.

programme: Established of instructions written in a precise sequence for the computer to accomplish a given task.

programme life cycle phase: Programme life cycle phases are the periods that a computer programme undergoes, from initial creation to deployment and execution. These phases comprise edit time, compile time, link time, distribution time, installation time, load time and run time.

programmable addition via site-specific targeting elements (PASTE): Tome Biosciences has launched a CRISPR tool for inserting large DNA sequences of around 36 kb into the genome at any position to correct genes *in vivo* in various dividing and non-dividing cell types. PASTE was developed in 2024 by Tome co-founders O. Abudayyeh and J. Gootenberg from the Massachusetts Institute of Technology. Investors include Andreessen Horowitz, Arch Venture Partners, Polaris and Fujifilm. PASTE uses a CRISPR-Cas9 nickase fused with two other enzymes: a reverse transcriptase and a serine integrase. The guide RNA-directed reverse transcriptase creates landing sites for the serine integrase, which inserts large sequences of DNA. This technology does not rely on repair responses to the DSBs that can result in insertions, deletions and off-target effects.

programmable endonucleolytic ribozyme: Ribozymes are RNA-based enzymes that typically catalyse self-cleavage or splicing reactions. Z.-X. Liu *et al.* (2024) discover hydrolytic endonucleolytic ribozymes (HYERs), which have the capability to precisely manipulate DNA. These ribozymes operate independently of proteins and can edit genes in both bacteria and human cells, offering a potential alternative to CRISPR and other gene-editing tools. A critical aspect of their function is that some HYERs adopt a homodimer structure that recognizes DNA through a complementary target recognition sequence and cleaves the DNA hydrolytically. Full complementarity between HYER TRSs and DNA targets is essential for efficient cleavage. Revised TRSs could redirect HYERs to cleave different DNA targets. Nine candidates (HYERs 1 to 9) were rationally selected for experimental characterization. Each HYER1 monomer contains a 6-nucleotide (nt) target recognition site (TRS) and a hydrolytic pocket for capturing and cleaving DNA substrates. Mutating the palindromes allowed HYER1 to form the heterodimer, which can carry two distinct TRSs and generate customizable dsDNA breaks with cleavage products bearing 5′ overhangs, blunt ends or 3′ overhangs. As an RNA-based DNA manipulation platform, HYER is highly programmable and holds the potential to be further developed and utilized in multiple genetic scenarios.

programmable RNA editing (PRE): Targeted RNA editing which relies on the delivery of exogenous proteins or chemically modified guide RNAs, which may lead to aberrant effector activity.

programmable RNA editing (PRE) in plants, safety aspects of: PPR-based programmable tools are safe, as dPPR proteins, built from consensus sequence, bind to the RNA target *in vivo*. The dPPR mimics the natural PPR proteins to allow PPR-RNA binding in an efficient manner. The absence of different components (adenosine deaminase acting on RNA (ADAR), Cas and gRNA) in dCas-based programmable RNA-editing tools prevents off-target effects.

programmable RNA editing (PRE) over genome editing, advantages of: The rewards of RNA editing over genome editing are several: (i) The biggest advantage is that RNA editing is temporary, unlike genome editing which is permanent. (ii) The pre-eminence of RNA editing lies in the half-life of the mRNA and the temporary nature of the editing strategy. (iii) The non-requirement of the HOMOLOGY-DIRECTED REPAIR (HDR) machinery and PROTOSPACER ADJACENT MOTIF (PAM) sequence at the target locus enhance the flexibility of RNA editing in targeting a particular transcript from the genome. (iv) The reversible nature of RNA editing would avoid the genomic off-target effects produced through NON-HOMOLOGOUS END JOINING (NHEJ).

programmable RNA editing (PRE), applications of: The programmable RNA-editing (PRE) technology should be able to overcome the faults ensuing from point mutation in plants as well as animals with high accuracy and proficiency. The combination of high-throughput detection and NGS-based techniques for transcriptome-wide documentation in the programmable approaches should be capable of specifically targeting the point mutations. The identification of PPR codes and customized dPPR motifs with programmed specificity could broaden prospect of dPPR towards practical applications in agricultural sciences. The dPPR proteins should be useful in studying the control of protein expression in the chloroplast and mitochondrial genetic systems more proficiently than natural PPR proteins in a programmable manner.

programmed cell death (PCD): The process in which healthy, normal cells kill themselves. It is a necessary part of shaping developing tissues and organs. PCD is a key mechanism for regulating cell numbers and connections and for sculpting tissues. PCD in unicellular organisms allows constant selection for the fittest cell in the colony, optimal adaptation of the cell number to the environment and tight regulation of cell cycle and cell differentiation that may be particularly useful when the cells are interacting.

programmed cell death (PCD) genes: Cell death (*ced*) genes control apoptosis. Loss-of-function mutations in the *ced-3* and *ced-4* genes and gain-of-function mutations in *ced-9* gene cause the failure of PCD. PCD also plays role in development of shape of different parts of the body. The *Caenorhabditis elegans* cell death genes *ced-3* and *ced-4* have been cloned and molecularly characterized. The *ced-4* gene encodes a 63 kDa protein with no significant similarities to other polypeptides. In contrast, *ced-3* gene encodes a protein that is similar to family of cysteine proteases, which includes INTERLEUKIN-1 β CONVERTING ENZYME (ICE). Overexpression of either ced-3 or ICE in mammalian cells causes apoptosis. Gene *ced-9* encodes a protein that is homologous to mammalian protein BCL-2. Expression of BCL-2 can inhibit cell death in nematode and can even partially substitute for the loss of ced-9 function.

programmed cell death (PCD) in development: About one-sixth of the cells that arise during *Caenorhabditis elegans* development are eliminated by PCD. During apoptosis, the cell shrinks and dies, after which it is engulfed and digested by neighbouring cell. The identity of the cells that die and the time at which they die is same from animal to animal. Purpose of cell death seems to be the elimination of the unneeded cells that could be detrimental to the organism. In plants, PCD is involved in the terminal differentiation of xylem vessels and may possibly be related to several other developmental processes. Inhibition of pathogen growth involves a rapid process of PCD localized at the site of invasion. The death of plant cells during this hypersensitive response (HR) process creates a zone of dead cells called *HR lesion* which limits the spread of pathogen, playing an important role in plant defence mechanism. Hence, PCD represents a potentially vulnerable target for pathogen signals that may act to either induce or suppress death by interfering with key signal transduction pathways. *See* APOPTOSIS.

programmed cell death (PCD), stages in: These stages are as follows: (i) decision stage (i.e. whether a given cell will die or adopt another fate), (ii) death of a cell, (iii) engulfment of the dead cell by phagocytes and (iv) degradation of the engulfed corpse. Mutations that affect each of these stages have been isolated. There seems to be a delicate balance between the activities of the proteins that promote and those that inhibit cell death.

programmed changes: An involuntary change where both the type of alteration and its place in the developmental programme are tightly regulated, for example, DNA methylation is a programmed change.

programmed non-reciprocal recombination: A type of programmed recombination that changes mating type switching in yeast and antigen switching in TRYPANOSOMES.

programmed recombination: The differentiation of some cell types is reliant on recombination. Two types of programmed recombination are recognized – programmed non-reciprocal recombination and site-specific recombination. *See* PROGRAMMED NON-RECIPROCAL RECOMBINATION *and* SITE-SPECIFIC RECOMBINATION.

programmed site-specific recombination: A type of programmed recombination that controls many aspects of vertebrate B-cell and T-cell differentiation.

programming language: A formal language, which encompasses a set of instructions that yield various kinds of output. Programming languages are employed in computer programming to apply algorithms.

programming language implementation: A system developed for effecting computer programmes. There are two general approaches to programming language application: interpretation and compilation.

programming language theory (PLT): A division of computer science that deals with the design, implementation, analysis, characterization and classification of programming languages and of their individual structures. PLT falls within the discipline of computer science, both reliant on and affecting mathematics, software engineering, linguistics and even cognitive science.

programmes and tools available for making searches: There are various programmes and tools available for making searches, algorithm analysis modelling and the computer graphics of the databases on genomes and proteomics like BLAST, FASTA, ENTREL, EDITVIEW, DATA-SEQUENCER, FACULTA, AUTOASSEMBLY, GLG Software and GENE EXPLORER 1.4. The first priority of any annotation software is to pinpoint the genes by looking for characteristic sequences at the beginning or ends of genes or by comparing new sequence to known genes. Additional computer programmes translate those genes into proteins and based on similarities to other protein to assign a function to each gene. NCBI's BLAST Programme compares the new genome data to that from other organisms, such as fruit fly or nematodes.

progressive heterosis: The additional hybrid vigour seen in double-cross tetraploid hybrids beyond that seen in their single-cross parents.

proinhibition: A plant defence compound that already exists before invasion by a pathogen.

prokaryote: A unicellular organism lacking a membrane-bound nucleus. Prokaryote is without a distinct nucleus, mitochondria or chloroplasts. They are the earliest and most primitive forms of life on Earth. This is supposed to be the result of endosymbiosis, with the outer membrane contributed by the eukaryotic cell and the inner membrane coming from the original prokaryote which was 'swallowed'. Prokaryotes are the single-cell organism such as bacteria in which nucleoid is not separated from the cytoplasm by a membrane. Therefore, prokaryotes are unicellular organisms with a single chromosome, no nuclear envelope and no membrane-bounded organelles. *Compare with* EUKARYOTE.

prokaryote and eukaryote genomes, similarities between: Prokaryotic and eukaryotic chromosomes are similar with respect to two features – unineme structure and semi-conservative DNA replication.

prokaryotes: Also spelled as PROCARYOTES. Simple organisms that lack a distinct nuclear membrane and other organelles. Many structural systems differentiate prokaryotes from eukaryotes, including the DNA arrangement, composition of membranes, the respiratory chain, the photosynthetic apparatus, ribosome size, the occurrence or absence of cytoplasmic streaming, the cell wall, flagella, the mode of sexual reproduction and the occurrence or absence of vacuoles. Some illustrative prokaryotes are the bacteria and blue-green algae. *Contrasts* EUKARYOTES (also spelled as EUCARYOTES).

prokaryotic: Also spelled as PROCARYOTIC. Pertaining to a one-celled organism lacking a true nucleus and nuclear membrane and having genetic material composed of a single circular piece of DNA. Prokaryotes, with the exception of spiroplasmas and mycoplasmas, have a rigid cell wall.

prokaryotic cell: A cell having no nuclear membrane and hence no nucleus.

prokaryotic cell reproduction: Most prokaryotic cells reproduce via cellular fission. During this process, the cell's one chromosome makes a copy of itself, so each new cell gets a copy of that chromosome.

prokaryotic chromosome: In prokaryotes, chromosomes are structures, located within the non-membrane-bound nucleoid, which contain most of the cell's genetic information or DNA. Most prokaryotic cells have one circular chromosome, composed of a long DNA double helix that is compacted by various NUCLEOID-ASSOCIATED PROTEINS (NAPs) and enzymes to fit within the nucleoid region of the cytoplasm. Since prokaryotic cells have only one chromosome, they are classified as haploid cells.

prokaryotic gene: It is a polymer of deoxyribonucleoside phosphates that includes the following regions: recognition region (~50 bp), transcription initiation site, 5′ untranslated region, translation initiation, coding region, translation stop site, 3′ untranslated region and transcription stop site.

prokaryotic genome: Structural organization of the prokaryotic genes within the genome was initially thought to be simple but discovery of overlapping genes in bacteriophages, animal viruses, bacterial IS and split genes in prokaryotic genomes suggested that organization of prokaryotic genome is fairly complex.

prokaryotic ribosome: It has sedimentation coefficient 70S. It contains 65% rRNA and 35% protein. Prokaryotic ribosomes have two subunits. Larger subunit (having sedimentation coefficient 50S) contains one molecule of 5S rRNA, one molecule of 23S rRNA and 34 different proteins. This ribosomal subunit has a fairly compact body (~150 × 200 × 200 Å) from which a central outgrowth and a stalk, stick out. This subunit is relatively more spherical than the smaller one. The partition between the head and body of the small subunit fit into the large subunit. There could be a space or tunnel between the two subunits. Smaller subunit (30S) has one molecule of 16S rRNA and 21 different proteins. This subunit has an elongated and asymmetrical shape (about 55 × 220 × 220 Å). It has a constricted region and cleft that separates the head from the base, while its platform projects out the base. The 30S and the 50S subunits may be held together by an association between discrete areas of these subunits.

prokaryotic ribosome-binding site: The leader sequence at 3′ end but 8–12 bases before the initiating codon has a sequence 5′ AGGAGGA 3′. This sequence is common to several *Escherichia coli* genes and was discovered by Shine and hence its name. This sequence pairs with the sequence 5′ACC′VC-C′UUA-OH3′ found at 3′ end of 16S rRNA and thus helps in mRNA binding to ribosome.

prokaryotic RNA polymerase operons: Three operons, namely σ, β and α, respectively contain genes for σ, β and β′ and α subunits of RNA polymerase.

prokaryotic translational release factors (RFs): In prokaryotes, RF1 and RF2 catalyse polypeptide release at UAG/UAA and UGA/UAA stop codons, respectively. RF3 is a GTPase found in a broad range of bacteria where it is thought to play a critical 'recycling' role in translation by facilitating the removal of RF1 and RF2 from the ribosome following peptide release.

prokaryotic versus eukaryotic genomes: Organization of prokaryotic and eukaryotic genomes diverges with respect to (i) split versus non-split nature of gene, (ii) coding versus non-coding DNA, (iii) applicability of operon concept, (iv) polycistronic versus unicistronic mRNAs (v) nature of regulatory control, (vi) unicellularity versus multicellularity, (vii) singularity versus plurality of chromosomes, (viii) DNA packaging, (ix) haploidy versus diploidy, (x) unique versus repeated sequences and (xi) collinearity between gene and its product.

proliferation: The growth or production of cells by multiplication of parts.

proliferating cell nuclear antigen (PCNA): A DNA clamp that acts as a processivity factor for DNA polymerase δ in eukaryotic cells and is essential for replication. PCNA is a homotrimer that forms a β subunit and achieves its processivity by encircling the DNA, where it acts as a scaffold to recruit proteins involved in DNA replication, DNA repair, chromatin remodelling and epigenetics. DNA pol α and pol δ in conjunction with PCNA appear to be essential for the complete replication of chromosomal DNA in eukaryotes. The observation that PCNA is relocated to sites of DNA damage suggests that it may also be involved in processes other than DNA replication.

proliferating cell nuclear antigen (PCNA) loading, mechanism of: PCNA is a ring-shaped clamp that encircles DNA and slides along the DNA duplex. The PCNA ring binds several DNA metabolic factors to recruit them to DNA and/or enhance their speed and efficiency on DNA. PCNA needs a multisubunit ATPase-driven 'clamp loader' to get onto DNA. Eukaryotes possess four distinct clamp loaders. Z. Yuan *et al.* (2024) resolve the structure of the Ctf18 alternative replication factor C (Ctf18-RFC) clamp loader, which specifically binds the DNA polymerase ε (Polε) that is devoted to synthesis of the leading strand of the duplex DNA genome. The structure discloses exclusive features in both Ctf18-RFC and Polε that have been fashioned over evolutionary time to safeguard their interdependent function.

proline (pro): A heterocyclic, non-polar imino acid, which is present in all proteins. The major pathway for proline synthesis, which occurs in the cytoplasm, is from glutamate, through gamma-glutamyl phosphate and glutamyl-gamma-semialdehyde, a two-step reaction that is catalysed by a single enzyme, D1-pyrroline-5-carboxylate synthetase.

proline-tryptophan-tryptophan-proline (PWWP) motif: A sequence of amino acids having the PWWP domain, which portrays proteins involved in DNA methylation, DNA repair and transcription regulation. Proteins with this PWWP motif are capable of distinguishing and binding methylated histones. Examples include the PWWP domain of the protein BRPF1, which recognizes the trimethylation of lysine 36 on histone H3 (H3K36me3).

prolog: A logic programming language related with AI and computational linguistics. Prolog has its roots in first-order logic, a formal logic, and unlike many other programming languages, Prolog is envisioned principally as a declarative programming language: the programme logic is articulated in terms of relations, represented as facts and rules. A computation is started by running a query over these relations.

propagule: Any material that functions in propagating an organism to the next stage in its life cycle, such as by dispersal.

prometaphase: Phase of mitosis immediately preceding metaphase in which the nuclear envelope breaks down and chromosomes first attach to the spindle for organization of the chromosomes on the metaphase plate.

promiscuous DNA: J. Ellis (1982–1983) defined promiscuous DNA as a nucleotide sequence which transpires in more than one of the three membrane-bound organellar genetic systems of eukaryotic cells and should be separated from a wide range of prokaryotic cells. The phenomenon of this DNA transfer is known as INTRACELLULAR PROMISCUITY and it reduces the separate identity of each genome. J.N. Timmis and N.S. Scott in 1983 observed that the nuclear DNA of spinach contains integrated sequences homologous with most of the chloroplast genome. Several lines of evidence have indicated that relocation of some organelle genes could have been more recent. S.L. Baldauf and J.D. Palmer in 1990 presented gene sequence and molecular phylogenetic evidence for the transfer of chloroplast *tufA* gene to the nucleus in the green algal ancestor of land plants. The *tufA* gene encodes chloroplast protein synthesis elongation factor Tu (EF-Tu). A 12-kilobase DNA sequence has been identified by D.B. Stern and D.M. Lonsdale in 1982 in the maize mitochondrial genome which is homologous to part of the inverted repeat of the maize chloroplast genome. In chloroplasts, the sequence contains a 16S rRNA gene and also the coding sequences for $tRNA^{Ile}$ and $tRNA^{Val}$.

promiscuous plasmid: A plasmid that is capable of promoting its own transfer to a wide range of Gram-negative bacteria and being stably maintained in these diverse hosts. These plasmids generally belong to the incompatibility classes known as P and Q.

promoter: A DNA sequence, upstream of a gene, at which RNA polymerase may bind for initiation of transcription. During plant transformation, promoter is a very powerful sequence in the gene construct which regulates the expression of transgene in the transgenic plant. CaMV 35S is constitutive promoter that causes the transgene to express in each and every cell. Further there are now seed-specific promoters and root-specific promoters that allow the expression of transgene in the specific tissues.

promoter and enhancer insertion: Activation of expression of a gene by integration of a nearby provirus, bringing expression of the gene under the control of regulatory elements in the LTR of the provirus.

promoter consensus sequences: The start point, in >90% cases, is a purine. Just upstream of the start point, a 6-bp region is documented in almost all the promoters. The centre of the hexamer generally is close to 10 bp upstream of the start point. This hexamer is, therefore, often named the −10 sequence. This −10 sequence is TATAAT; conservation of these bases varies from 45% to 96% (T_{80} A_{95} T_{45} A_{60} A_{50} T_{96}). At −35 sequence of promoter, consensus sequence is T_{82} T_{84} G_{78} A_{65} C_{54} A_{45}. Distance between −10 and −35 sequence in 90% of cases is between 16 and 18 bp. This distance between two sites may be important in appropriate separation for the geometry of RNA polymerase. Thus, an optimal promoter is defined as a sequence consisting of the −35 hexamer, separated by 17 bp from the −10 hexamer, lying 7 bp upstream of the start point.

promoter melting: This is a critical step in transcription initiation. The N-terminus of the β′ subunit (amino acids 1–314) and amino acids (94–507) of the σ subunit, together less than one-fifth of RNA polymerase holoenzyme, were able to melt an extended −10 promoter in a reaction unusually alike to that of true holoenzyme.

promoter mutants: These mutations are mapped to the left of the operator. Promoter mutants co-ordinately reduce proteins of *z-y-a* cluster. There mutants retort to *i* gene (repressor) control. That is, the three proteins are synthesized in low amounts in a promoter mutant cell when lactose is present. This signifies that the repressor control has not altered.

promoter single-nucleotide polymorphism: Any single-nucleotide polymorphism that ensues in the promoter sequence of a gene. If a pSNP prevents the binding of a transcription factor to its recognition sequence in the promoter, the promoter turns out to be partially dysfunctional. *Aka* PROMOTER SNP (PSNP).

promoter site: Defined by its recognition of eukaryotic RNA polymerase II for initiating transcription; its activity in a higher eukaryote; by experimental evidence or homology and sufficient similarity to an experimentally defined promoter; and by observed biological function.

promoters and enhancers in gene activation: Enhancer and promoter DNA recruit RNA polymerase II and genes to larger superstructures, sometimes referred to as transcription factories, hubs or condensates. M.E. Pownall *et al.* (2023) describe a new form of expansion microscopy, called *ChromExM*, which they used to expand chromatin in zebrafish embryos to visualize transcription at high resolution. This expanded view provides evidence that promoters and enhancers come together during gene activation, only to be 'kicked' apart soon after (T.J. Stasevich and H. Kimura, 2023). M.E. Pownall *et al.* (2023) also observed RNA polymerase II nanostructures engaged in active transcription and developed a 'kiss and kick' model of enhancer-promoter interactions in which transcription elongation leads to the physical separation of enhancers and promoters.

promoters of eukaryotic genes: RNA polymerase I, which transcribes rRNA genes, binds to promoter containing a CORE PROMOTER ELEMENT and an UPSTREAM CONTROL ELEMENT (UCE). The TATA BOX–BINDING PROTEIN (TBP), which is part of a larger complex called SL1, helps RNA polymerase I to recognize the core promoter. RNA polymerase II transcribes eukaryotic protein-encoding genes. The most common promoter element in eukaryotic protein-encoding genes is the TATA box, located at −35 to −20 to the site of transcription initiation. Its consensus sequence, $T_{82}A_{97}T_{93}A_{35}A_{43}/T_{27}A_{83}A_{50}/T_{37}$, is quite similar to the 10 region of the bacterial sigma 70 recognition site. Another promoter element is called the initiator (Inr). It has the consensus sequence PyPyAN(T/A)PyPy, where Py denotes pyrimidine (C or T), N = any and (T/A) means T or A. The base A at the third position is located at +1 (the transcriptional start site). TATA box and initiator are the core promoter elements. There are other elements often located within 100 or 200 bp of the transcriptional start site, such as CAAT box, having sequence GGT/CAATCT, lies between −70 and −80 and GC box which may be referred to as promoter-proximal elements. These elements may lead up to 200-fold increase in transcriptional activity. The protein, which interacts with the initiator and TATA box is known as the TATA BOX–BINDING PROTEIN (TBP), because the TATA box was discovered earlier than the initiator. It is a subunit of the general transcription factor TFIID. In eukaryotes, transcription by RNA polymerase II requires several different general transcription factors (GTFs) and, in most cases, the regulatory transcription factors. TBP recognizes not only the core promoter of protein genes, but also RNA promoters. The non-transcribed strand of the *Saccharomyces cerevisiae* U6 snRNA gene directs transcription initiation factor IIIB-requiring and accurately initiating transcription by RNA polymerase III. The non-transcribed strand promoter is much more extended than its duplex DNA counterpart, comprising the U6 gene TATA box, a downstream T7 tract and an upstream-lying segment. A requirement for placement of the 3′ end of the transcribed (template) strand within the confines of the transcription bubble is seen as indicating that the non-transcribed strand provides a scaffold for RNA polymerase recruitment but is deficient at a subsequent step of transcription initiation factor IIIB's direct involvement in promoter opening. Internal promoters are present in 5S rRNA, VA1, $tRNA_f^{met}$ genes. All these genes are transcribed by RNA polymerase III.

promoters of prokaryotic genes: A nucleotide sequence in the operon system that is recognized by RNA polymerase as the site at which to begin transcription of RNA. Promoters have shown consensus sequence which consists of −35 region (5′-TTGACA-3′) and the −10 region (5′-TATAAT-3′), called PRIBNOW BOX. Minimum size of promoter is 12 bp. The distance between −35 and −10 regions is important. Polymerase can travel along DNA if it is not blocked by a repressor bound to the operator between promoter site and structural gene cluster. Promoter controls the level of function of that gene. Promoter's influence on the timing/degree of gene expression is itself regulated by the molecules that bind to the promoter. The position and alignment of the promoter within a DNA molecule determines the start site of the new RNA. Other proteins (e.g. transcriptional activators such as sigma factor) are usually essential for an RNA polymerase to recognize a promoter. For example, the 'binding' of RNA polymerase causes transcription of RNA to begin and the 'binding' to promoter of other STATs (i.e. signal transducers and activators of transcription) can control the grade to which a given gene is expressed. A promoter is a region of DNA (deoxyribonucleic acid) which usually lies 'upstream' of the transcriptional initiation site of a gene. The promoter controls where (which portion of a plant, which organ within an animal, etc.) and when (which stage in the lifetime of an organism) the gene is expressed. For example, the promoter named 'Bce4' is 'seed specific' (i.e. it only 'promotes' the expression of a given gene's product: protein, fatty acid, amino acids, etc.) within a plant's seed. Promoters belong to two general classes. Some promoters can be recognized by RNA polymerase holoenzyme alone. Other promoters are not by themselves adequate to support transcription. Ancillary protein factors are needed for initiation to occur. These proteins recognize the sequence on DNA close to or overlap with, the sequence bound by RNA polymerase itself. In certain genes coding for small species of RNA (e.g. tRNAs, 5S rRNA), promoter sequences are located within the mature gene itself. The sequence of DNA needed for initiation reaction for transcription defines the promoter. Transcription enzyme, RNA polymerase binds at this site. The promoter contains the DNA sequence that can be modified through epigenetics to regulate gene expression. *See* TRANSCRIPTION.

promoters used in recombinant DNA technology: Commonly used promoters in recombinant DNA technology are T7 promoters, *lac* promoters (*bla* promoter) and cauliflower mosaic virus 35S promoter (for plant vectors).

promoters, terminators and other regulatory elements: (i) PromScan scans small genomes for potential factor-binding sites including IHF-binding sites. If a *.ptt file is included the results will indicate the position of the promoter relative to the nearest gene. (ii) Virtual Footprint offers two types of analyses: (a) Regulon analysis involves analysis of a whole prokaryotic genome with one regulator pattern and (b) Promoter analysis involves analysis of a promoter region with several regulator patterns. (iii) RibEx: (Riboswitch Explorer) scans <40kb DNA for potential genes (which are linked to BLASTP) and several hundred regulatory elements, including riboswitches.

pronuclear injection: The use of a fine needle to inject DNA into the nucleus of an unfertilized egg.

pronuclear transfer (PNT): In PNT method, developed by S. Mitalipov, both eggs are fertilized with the father's sperm first. Before the eggs have time to divide into early-stage embryos, the parents' chromosomes are removed from the mother's fertilized egg and dropped into the donor's.

pronuclei: Male and female nuclei prior to fertilization are called pronuclei. When pronuclei are near each other they seem to respond to each other's presence just before union of the pronuclei, DNA of the chromosomes are replicated in both pronuclei in preparation of mitotic division. In some species, pronuclei do not directly fuse rather they approach each other but each remains intact. Nuclear membranes of each pronucleus breakdown as first cleavage approaches. There is evidence to suggest that the male and female pronuclei are in communication with each other. Even though male and female pronuclei communicate, each is capable of independent activity in the absence of other. *See* NUCLEUS *and* TRANSGENIC.

proofreading: The ability of DNA polymerase to correct misincorporated nucleotides as a result of its exonuclease activity.

propagative transmission: A type of pathogen transmission characterized by a long period of attainment of the pathogen (usually a mollicute, e.g. phytoplasma or spiroplasma and sometimes a virus) by a vector (typically an insect), a latent period before the vector is able to communicate the pathogen and retaining of the pathogen by the vector for a long period because the pathogen reproduces or replicates in the vector. *See* CIRCULATIVE TRANSMISSION *and* PERSISTENT TRANSMISSION. *Contrasts* NON-PERSISTENT TRANSMISSION *and* STYLET-BORNE TRANSMISSION.

propagule: Any slice (seeds, spores, bulbils, etc.) of an organism skilled of independent growth of an organism.

proper accuracy scoring rule: When applied to forecasting categorical outcomes, a proper probability accuracy scoring rule is a quantity that is optimized when the predicted probabilities are the true outcome probabilities. Examples of proper accuracy scores include the Brier score, the logarithmic probability score and the log-likelihood from a correct statistical model. Examples of improper scoring rules (i.e. the rules that are optimized by a bogus model) are percentage classified correctly, sensitivity, specificity, precision, recall and the c-index (area under the RECEIVER OPERATING CHARACTERISTIC (ROC) CURVE).

proper block design: A block design is proper if all blocks have the same cardinality (i.e. contain the same number of plots). For binary designs, this just means that all blocks contain the same number of points.

prophage: The non-infectious form of a temperate bacteriophage in which the phage DNA has become incorporated into the lysogenic, host bacterial DNA.

prophase: The first stage of mitosis and meiosis. Chromosomes become visible. Each chromosome consists of two sister chromatids joined together by a centromere. Chromosomes shorten and thicken. Nucleoli start decreasing in size. Nuclear envelope begins to break down.

prophase I: Longest period of meiosis I during which chromosomes condense, homologous chromosomes pair, undergo crossing-over and then separate out but still close at the ends. It is divided into five substages LEPTOTENE, ZYGOTENE, PACHYTENE, DIPLOTENE and DIAKINESIS. *See details as separate terms.*

prophase II: A brief stage after completion of meiosis I. Meiosis II starts with prophase II. In some cell types, prophase II may be very short or even absent. During prophase II, centrioles move apart in pairs. Spindle is laid down. Nuclear envelope breaks down. Nucleoli begin to disappear.

proplastid: A small, colourless, double-membrane-bound organelle with diminutive internal structure that acts as a precursor in the development of all plastids (e.g. chloroplasts, chromoplasts and leucoplasts).

proportion: The measure of how big is the part of a whole. The whole is considered one. If 20% of a population have blood group A, the proportion is 0.20 (1/5).

proportional hazards: This supposition is fulfilled if two groups of patients are being compared and their hazard ratio is constant over time (though the immediate hazards may differ).

propositus: The member of a family who first comes to the attention of geneticist and through whom pedigree is discovered. Usually phenotype of the propositus is unusual or exceptional in some way. *Aka* PROBAND.

proprietary regulatory property (PRP): The data, dossiers and authorizations that enable the cultivation and sale of an event in any country in which it is approved for cultivation and allow the importation and use of material containing that event (seed product, grain or any product thereof regulated as a result of the event).

proprietary regulatory property (PRP) holder: an entity that owns or controls the PRP and any other relevant intellectual property rights for an off-patent event.

prospective study: One in which the investigation is first planned and then the subjects are enrolled. Prospective studies are usually featured by deliberate data collection.

protamine: A low molecular weight protein, lacking prosthetic group but rich in arginine content, found in association with nuclear DNA in certain cells.

protandrous: Having male sex organs which mature before the female ones (e.g. a flower sheds pollen before the stigma is receptive). *Compare with* PROTOGYNOUS.

protandry: Individuals are born as males, reproduce as males, then change sex and reproduce as females. In plants, individuals express male function prior to female function, producing pollen before being pollinated.

protease: Proteases signifies a whole class of protein-degrading enzymes. For example, the enzyme degrades proteins by hydrolysing some of their peptide bonds. *Aka* PROTEINASE *or* PROTEOLYTIC ENZYME.

protease inhibitors (PIs): Proteins that accrue in the seeds and storage organs of many plants. They protect plants against diverse insect, bacterial and fungal pathogens.

protease inhibitors, genes for: In cowpea (*Vigna unguiculata*), trypsin inhibitor (CpTI) was shown to be toxic to insects but cowpea seeds with high CpTI level are not toxic to humans. Gene for CpTI was used to produce transgenic plants of tobacco that showed insect resistance to Lepidoptera, Coleoptera and Orthoptera. Extensive field trials are necessary before these transgenic plants can be released to the farmers for cultivation.

proteasome: In protein degradation, ubiquitin-tagged proteins are identified by specific proteases in the cytosol which in turn cleave and degrade the tagged protein. These proteases are combined in form of a huge protein complex, called the proteasome. The proteasome (20S) encompasses 28 subunits and has a molecular weight of 700 kDa. Proteasome activates certain transcription factors that are implicated in instigating the cell to present antigens (i.e. from pathogens that invaded that cell) on the cell's surface and accomplish various other cellular functions. For example, the 26S proteasome degrades (breaks down) all ubiquitinated (ubiquitin-tagged) proteins in that cell.

protectant: An agent, usually a chemical, applied to a plant surface in advance of a pathogen attack to prevent infection.

protectant fungicide: A fungicide that retains itself on the surface where it is applied and stops infection often via the inhibition of spore germination and with no after-infection activity. *See* CONTACT FUNGICIDE. *Contrasts* SYSTEMIC FUNGICIDE.

protected polymorphism: When each allele will increase from very low frequency, we say that the polymorphism is protected.

protected variety: A variety that is released and granted a certificate of PVP under the legal statutes of a given country. The owner of a protected variety has the right to sell, offer, reproduce, import, export or use it for hybrid seed production.

protection: Various methods of plant disease management, including cultural practices that produce barriers or diminish the chance of infection, chemical protection, methods of biological control that protect plants and genetic resistance.

protection of mitochondria during endoplasmic reticulum (ER) stress: Cellular stress responses occur to maintain the status quo when malfunctions take place. When proteins misfold within the endoplasmic reticulum (ER) lumen, a crucial response is global repression of protein translation, which is mediated by protein kinase RNA-like endoplasmic reticulum kinase (PERK). This reaction benefits the overloaded ER. K.K. Brar *et al.* (2024) revealed that translational repression is actually not universal throughout the cell during ER stress. The mitochondrial protein ATPase family AAA domain-containing protein 3A (ATAD3A) binds PERK during stress, permitting key mitochondrial proteins to avoid repression. This mechanism offers a 'safe haven' for the protection of mitochondria during ER stress.

protein: Coined in 1838 by J. Berzelius, protein is any of a class of high molecular weight polymer compounds composed of a variety of α-amino acids linked by peptide linkages formed the enzyme peptidyl transferase. Protein synthesis is performed by ribosomes. Each protein is the ultimate expression product of a gene. After emerging from the cell's ribosome, proteins may also become phosphorylated, glycosylated, acetylated, farnesylated, ubiquitinated, sulfated or otherwise chemically modified. Proteins are the 'workhorses' of living systems and include enzymes, antibodies, receptors, peptide hormones and so on. Proteins in living organisms respond to environmental changes and other conditions by changing their location within cells, by getting cut into specific pieces, by deciding which other molecules they will bind to and so on. All of the amino acids commonly occurring in each and every one of the proteins have an asymmetric carbon atom, except the amino acid glycine. Thus, the protein is potentially chiral in nature. Proteins have many different functions: structure (collagen); movement (actin and myosin); catalysis (enzymes); transport (haemoglobin); regulation of cellular processes (insulin); and response to the stimuli (receptor proteins on surface of all cells). The information for making proteins is stored in the sequence of nucleotides in the DNA molecule.

protein amount polymorphism (PAP): A condition where there are at least two genetic variants (alleles) of some signs exist in the population with respect to amount of a protein.

protein biosynthesis: The production of protein from its constituent amino acids using mRNA as a template. This takes place in association with specific organelles known as ribosomes.

protein chip technology: A variety of bait proteins such as antibodies, are immobilized in an array format on to specially treated surface. The surface is then probed with the sample of interest and only the proteins that bind to the relevant antibodies remain bound to the chip. These are then placed in a chip reader which includes a mass spectrophotometer for analysis of the proteins. The protein chip is probed with fluorescently labelled proteins from two different cell states. Cell lysates are labelled by different fluorophores and mixed such that the colour acts as a reader for the change in abundance of the protein bound to the antibody. This technology allows high-throughput screening for gene expression and molecular interactions. *Aka* MICROARRAY TECHNOLOGY.

protein complexes in cell cycle: Time-dependent interaction network places both periodically and constitutively expressed proteins in a temporal cell cycle context, thereby revealing previously unknown components and modules. They discovered that most complexes consist of both periodically and constitutively expressed subunits, which suggested that the former controlled complex activity by a mechanism of just-in-time assembly. Consistent with this, they showed that additional regulation through target degradation and phosphorylation by Cdc28 (Cdk1) specifically affects the periodically expressed proteins.

Protein Data Bank (PDB): It is very large global repository for the processing and distribution of 3-D macromolecular structure data primarily determined experimentally. Depositors to PDB have derived the structures using a variety of tools and techniques such as X-ray crystallography, NMR, cryo-electron microscopy and theoretical modelling. This database provides access to the structural data as well as methods to visualize the structures and to download structural information. PDB provides pictorial summary of key information on each macromolecular structure. This information includes images of structure, annotated plots of each protein chain's secondary structure; ligand and protein; and DNA interaction. PDB structures can be displayed and manipulated in 3-D using Ras Mol/VRML browser. The former is also used for focusing on specific aspects of structure such as protein-ligand interaction.

protein degradation models: Two well-known models of protein degradation of proteins in the cell include N-END RULE and PEST HYPOTHESIS. *See details as separate terms.*

protein distribution, analysis of: Proteomics combines two-dimensional polyacrylamide gel electrophoresis (2D-PAGE) that separates, maps and quantifies proteins with mass spectroscopy (MS)-based sequencing techniques identifying both the amino acid sequences of proteins and their post-translation appendages. The approach is allied with database search algorithms to sequence and characterize individual proteins. For identification of unknown proteins, protein samples taken from different conditions are solubilized and applied to a 2-D gel electrophoresis where proteins are differentiated on the basis of their charge and mass ratio. Differences between the reference and altered states are measured by quantifying the ratios of spot intensities. Now protein spot can be excised and incubated with the sequence-specific proteases, such as trypsin, which reduces the protein to peptides that are more easily analysed by MS. The masses of the peptides are measured and protein is identified by comparison of resulting MASS FINGERPRINT to corresponding mass prediction contained in a database.

protein domain: A conserved, structurally well-defined sequence in a polypeptide chain. It has a unique folding and assumes a tertiary structure of its own and therefore performs a specific function. Larger proteins often have several of these domains. Each domain is connected to the next by short flexible regions of polypeptide chain and as a whole they determine the function of the protein. Different exons code for the different domains of a protein. Each domain is also found in other proteins. The origin of genes coding for such proteins may have been EXON SHUFFLING.

protein editing: Protein editing is achieved via site-directed mutagenesis. For example, this approach was effective in creating IgG with four amino acid substitutions.

protein electrophoresis: A procedure for analysing a mixture of proteins by separating them on the basis of physical characteristics such as size, shape or isoelectric point.

protein engineering of laundry detergent: The enzyme subtilisin is a protease (a protein-digesting protein) produced by bacteria and has a broad specificity for proteins that commonly soil clothing. To improve the efficiency of laundry detergents, detergent manufacturers added subtilisin. However, subtilisin is inactivated by bleach and it was found that this inactivation was due to the oxidation of the amino acid methionine at position 22 of the subtilisin molecule. Using site-directed mutagenesis of the subtilisin gene in *Escherichia coli*, this methionine was replaced by alanine subtilisin thus engineered the was the best in terms of activity and stability and now many laundry detergents contain cloned, genetically engineered subtilisin.

protein engineering: Production of recombinant proteins using recombinant DNA technology. It is also called ENZYME ENGINEERING if the protein is the enzyme. Using a recombinant DNA technology, a cloned gene can be expressed in variety of organisms to produce desired altered protein, which allows the production of novel proteins. It is also called METABOLIC ENGINEERING if aim is to modify the enzymes involved in the metabolism. Protein engineering leads to the production of superior enzymes and storage proteins using synthetic genes. First of all, a computer-aided protein model is prepared for specific function and then a synthetic gene is prepared, which produces the desired protein/enzyme in predictable manner. Technology has been advanced to incorporate non-standard amino acids into polypeptide through ribosome-based translation. Altered proteins are produced by site-directed mutagenesis. For this, the genetic code is expanded through creation of a 65th codon-anticodon pair from unnatural nucleoside bases having non-standard H-bonding patterns. This new codon-anticodon pair efficiently enables translation *in vitro* to yield polypeptides containing a non-standard amino acid. This is done in order to make the resultant proteins carry out novel functions. Genetic engineers employ recombinant DNA technology to change a particular nucleoside or triplet (codon) in the DNA (genes) of a cell. The resulting DNA is expected to code for the non-standard amino acid in the desired location in the protein produced by that cell.

protein engineering, applications of: (i) Novel activities or properties of proteins, (ii) Structure-function relationships and (iii) Minimalist proteins.

protein engineering, strategies for: There are two general strategies for protein engineering: (i) rational protein design and (ii) directed evolution. These techniques are not mutually exclusive; researchers often apply both these strategies. Other methods used for protein engineering are: (i) Transcription and translation systems from synthesized lengths of DNA or RNA with novel sequences, (ii) Chemical modification of 'normal' proteins and (iii) Solid-state polypeptide synthesis to form proteins.

protein expression analysis and mRNA profiling: Protein expression analysis is more complicating than mRNA profiling because proteins cannot be amplified like nucleic acids. Proteins are under-represented on standard 2-D gels. The recovery of membrane proteins can be improved by selecting standard detergents such as 3-[(3-cholamidopropyl)-dimethylammonio]-1-propanesulfonate (CHAPS) and selectively enriching the initial sample for membrane proteins (e.g. by preparing membrane fractions).

protein expression map: Prepared by associating the 2-D electrophoresis gel patterns of samples exposed to different physiological conditions or different treatments, it is likely to identify groups of proteins with related functions or whose expression is interdependent.

protein families: Sets of proteins that have a common evolutionary origin as revealed by their relatedness in function, which is usually reflected by similarities in sequence or in primary, secondary or tertiary structure. Families are subsets of proteins having related structure and function.

protein fingerprinting: A technique that produces a pattern of protein fragments (usually resolved on a two-dimensional electrophoretic gel) generated by cleavage with an enzyme such as trypsin. A protein having a repeated motif of amino acids with characteristic spacing of cysteines that may be involved in binding zinc, is characteristic of some proteins that bind DNA and/or RNA. It produces zinc-protein fingerprints.

protein folding: The complex interactions of a polypeptide molecular chain with its environment, itself and other protein units, which cause the polypeptide molecule to fold up into a highly organized, tightly packed, 3-D structure. The phenomenon was proven to occur spontaneously by C.B. Anfinsen during the 1960s, for protein molecules outside of living cells. This ability of polypeptide chains to fold into a great variety of topologies, combined with the large number of sequences (in the molecular chain) that can be derived from the 20 common amino acids in proteins, confers on protein molecules their great powers of recognition and selectivity. How a protein folds up determines its chemical function. During the 1990s, it was discovered that inside living cells, 'chaperone' molecules are needed for proper protein folding to transpire. These chaperones are protein molecules (e.g. certain heat shock proteins) that form a loosely bound complex to suppress incorrect protein folding as the protein molecule is emerging from the cell's ribosome. Protein folding is both complete and correct as soon as the newly formed protein molecule is released from the cell's ribosome. Cotranslational chaperones function in protein folding. RIBOSOME-ASSOCIATED TRIGGER FACTOR (RATF) is the first molecular chaperone encountered by nascent polypeptides in bacteria. RATF interacts with ribosomes and translating polypeptides in a dynamic reaction cycle. Ribosome binding stabilizes transcription factors (TFs) in an open, activated conformation. Activated RATF leaves the ribosome after a mean residence time of nearly 10s but may remain associated with the elongating nascent chain for up to 35 s, enabling entry of new RATF molecule at the ribosome docking site. The duration of nascent-chain interaction associates with the occurrence of hydrophobic motifs in translating polypeptides, reflecting a high aggregation tendency. Protein misfolding events during translation may provide a model for the regulation of nucleotide-independent chaperones.

protein glycosylation: Post-translational addition of oligosaccharide side chains to a protein.

protein ID (in GenBank): An identification number allocated to the amino acid sequence data entered within a sequence record. This sequence identifier uses the accession.version format. Each protein ID is made up of three letters, followed by five digits, a period and a version number. For example, in sequence record M12345, the protein ID for the sequence translation could be AAA35650.1. If the protein sequence data change in any way (even by only one amino acid), the version number in the Protein ID will be increased by an increment of one while the accession number base remains constant; for example, AAA12345.1 would become AAA12345.2. Each amino acid sequence change also results in the assignment of a new GI number to the altered protein translation.

protein identification strategies in proteomics: There are three main strategies: (i) THE PEPTIDE MASS FINGERPRINTING (PMF) STRATEGY, (ii) THE PEPTIDE FRAGMENT FINGERPRINTING (PFF) STRATEGY and (iii) *DE NOVO* STRATEGY. *See details as separate terms.*

protein interaction analysis: Refers to a number of different analyses/technologies applied to find out if a given (e.g. new) protein molecule interacts with a protein molecule whose function is already known (e.g. its use as a pharmaceutical). Through that analysis (e.g. inferring the 'new' protein's function by its action vis-à-vis the 'old' protein), useful information about the 'new' protein can be gathered.

protein interaction map of *Drosophila*: A two-hybrid-based protein-interaction map of the fly proteome has recently been prepared. A total of 10,623 predicted transcripts were isolated and screened against standard and normalized complementary DNA libraries to produce a draft map of 7048 proteins and 20,405 interactions. A computational method of rating two-hybrid interaction confidence was developed to refine this draft map to a higher confidence map of 4679 proteins and 4780 interactions. Statistical modelling of the network showed two levels of organization; a short-range organization, presumably corresponding to multiprotein complexes and a more global organization, presumably corresponding to intercomplex connections. The network recapitulated known pathways, extended pathways and uncovered previously unknown pathway components. This map serves as a starting point for systems biology modelling of multicellular organisms, including humans.

protein kinase C (PKC): Enzyme that transfers the terminal phosphate group of ATP to a specific amino acid of a target protein. Ca^{2+}-dependent protein kinase that, when activated by diacylglycerol and an increase in the concentration of Ca^{2+}, phosphorylates target proteins on specific serine and threonine residues.

protein kinases: Enzymes proficient in covalently bonding a phosphate group to certain amino acid residues in specific proteins. Protein kinases enact crucial part in the regulation of signalling within and between cells.

protein microarrays: Refers to a piece of glass, plastic or silicon on to which a number of proteins (or molecules of other chemical compounds that interact with proteins in a specific manner) have been placed. These microarrays (sometimes called BIOCHIPS) are employed to test (e.g. a single sample) for a wide variety of characteristics or effects (on or by the protein molecules in the sample that is exposed to that microarray).

protein module: According to Philip Ball in 2024, criticizes the view that a protein with a fixed shape binds to its target being similar to how key fits into a lock. Many proteins, he points out, have disordered domains – sections whose shape is not fixed, but changes constantly. This 'fuzziness and imprecision' is not sloppy design, but an essential feature of protein interactions. Being disordered makes proteins 'versatile communicators', able to respond rapidly to changes in the cell, binding to different partners and transmitting different signals depending on the circumstance. This also questions the classic views of evolution. Evolution is generally regarded as 'a slow affair of letting random mutations change one amino acid for another and observing the effect it produces'. But in fact, proteins are typically made up of several sections called modules – reshuffling, duplicating and tinkering with these modules is a common way to produce a useful new protein. *See* MODULE.

protein phosphatase: *See* PHOSPHOPROTEIN PHOSPHATASE.

protein phosphorylation/dephosphorylation: Phosphorylation (catalysed by kinases) and dephosphorylation (catalysed by phosphatases) are the critical events that regulate the cell cycle. They are used to control the activities of the regulatory circuit itself and activities of the substrates that execute the decisions of the regulatory circuit.

protein probes: If you have the protein product of the gene of interest, make an antibody against it and the antibody against the protein of interest is used to screen the library for the clone that is expressing the gene that codes for the protein. Clones can be detected based on their ability to confer a missing function on a mutant.

protein quantification: Measurement of abundance of different proteins in a protein profile.

protein sample preparation: The following are the steps: (i) extraction of membrane proteins, (ii) functional inactivation, (iii) protein degradation and (iv) sample contamination.

protein secretion: Proteins that are secreted from the cell must pass through the plasma membrane to the exterior. These proteins start their synthesis in the same way as membrane-associated proteins but pass entirely through the system instead of halting at some particular point within it. Interaction between the membrane systems of the cell and parts of the protein structure (sequences) or groups added covalently to it give signals. By means of such signals a protein is recognized by receptors located within the membrane. A travelling protein may then be secreted through the membrane or may be passed on to another membrane. Free polysomes synthesize proteins that are released directly into the cytosol. Some proteins have signals for targeting to organelles, such as the nucleus or mitochondria. Proteins synthesized on membrane-bound ribosomes pass into the ENDOPLASMIC RETICULUM (ER), along with the Golgi and them through the plasma membrane, unless they possess signals that cause retention at one of the steps on the pathway. Cotranslational transfer is the process through which membrane-bound polysomes synthesized proteins enter the ER when they are being synthesized. For entry of these proteins into the ER, leader sequence is essential at the N-terminal end of the protein. A protein that resides within a membrane is termed INTEGRAL MEMBRANE PROTEIN (IMP). Secretion of proteins is elucidated by signal hypothesis. Free ribosomes synthesize specific non-serum liver proteins. Leader sequences on mRNAs seem to decide which ribosomes shall translate them. In eukaryotes, ribosomes may be free or may be attached to the ER. Membrane-bound ribosomes synthesize proteins that are exported from the cell into the surrounding serum. 60S subunit of ribosome is attached to ER. The ER membrane forms a vesicle containing the protein. This vesicle then becomes secretary message whose membrane fuses with the cell membrane, emptying its contents outside the cell. In eukaryotes, the proteins that are designed for secretion are initially targeted to the ER. This needs an N-terminal hydrophobic signal peptide, which is recognized by a SIGNAL RECOGNITION PARTICLE (SRP). SRP consists of six proteins and a small cytoplasmic RNA (7S RNA) which is homologous to *Alu* element.

protein separation and identification: The following are the steps involved: (i) Production of a visual display of proteome by 2D-gel, (ii) Detection of low-abundance proteins, (iii) Detection of membrane proteins and (iv) Dynamic range of protein staining.

protein sequence phylogenetics: The technique used to determine the genetic relationship between various organisms. It is based on the determination of amino acid sequence of a comparable protein, often cytochrome C.

protein sequencing: A procedure used to know the sequential arrangement of amino acids in a protein molecule.

protein shuttling: There are certain proteins, which shuttle between nucleus and cytoplasm either constitutively or in response to regulatory signals, viz. nuclear proteins like hsc70 and cAMP-dependent protein kinase. Some of the ribosomal proteins are imported to associate with ribosomal RNA (rRNA) so that ribosomal subunits are assembled inside the nucleus and then exported through the pore to the cytoplasm; a process which is ATP dependent. Although many signal sequences for import are known but no signals for export are known.

protein signalling: The 'communication' by protein molecules to cells that governs their transportation and localization (i.e. destination in the cell). Discovered and delineated by Guenter Blobel during the 1970s, protein signalling, via a short sequence of amino acids attached to end of newly synthesized protein molecules, results in proteins travelling to the appropriate cell compartments (such as organelles) and/or out of the cell (i.e. secretion).

protein sorting: Protein sorting involves interaction of cell membranes either with fragments of newly produced protein or with a protein sequence added to this new protein. These protein sequences interacting with membranes, called SIGNAL SEQUENCES, TRANSIT PEPTIDES or LEADER SEQUENCES, are identified by receptors positioned within the membrane. The transfer of such proteins may be coupled with translation by COTRANSLATIONAL TRANSFER, in which proteins synthesized on ribosomes are attached to ER. Alternatively, this may take place after the translation, process called POST-TRANSLATIONAL TRANSFER in case the proteins are synthesized on free ribosomes. In eukaryotes, proteins can be targeted to any one of several intracellular organelles, as well as to the nucleus. Targeted proteins must have recognizable sequences or structures which enable them to be transported to the proper cellular compartment. *Aka* PROTEIN TRAFFICKING.

protein sorting signals: Amino acid sequences existing in transported proteins that selectively lead the distribution of the proteins to specific cellular compartments.

protein structure: A polypeptide chain may take on a certain structure in and of itself because of the amino acid monomers it contains and their specific location within the chain. The chain may furthermore interact with other polypeptide chains to make larger proteins known as OLIGOMERIC PROTEINS. The levels of protein structure normally encountered are PRIMARY STRUCTURE, SECONDARY STRUCTURE, TERTIARY STRUCTURE and QUATERNARY STRUCTURE.

protein synthesis: Chemical synthesis (manufacture) of a known protein molecule. Devised based upon the solid phase synthesis methodology developed by R.B. Merrifield in 1963, the desired proteins are assembled by repetitive coupling of the constituent amino acids to a growing polypeptide backbone, which itself is attached to a polymeric support (substrate). This procedure has been automated, so it is now possible to make proteins via automated synthesizers. For sake of discussion, the process of protein synthesis has been arbitrarily partitioned into three steps, TRANSLATION INITIATION, TRANSLATION ELONGATION and TRANSLATION TERMINATION, but they in fact form a continuum. Protein synthesis in *Escherichia coli* is well understood. *See details as separate terms.*

protein synthesis in cell organelles: In mitochondria and chloroplasts, following components of protein synthesis are found: DNA, DNA polymerase, RNA polymerase, ribosomes, tRNAs, initiation factors, elongation factors and termination factors. All these components are specific to cell organelles. Some components of the protein synthesis may be produced outside these organelles and then transported to the organelles.

protein synthesis in organelles versus cytoplasm: The components of the translation apparatus in chloroplasts and mitochondria differ from that in cytoplasm of eukaryotes in several features: (i) Ribosomes in these organelles are smaller in size (70S) than those in the cytoplasm (80S). (ii) The tRNAs are specific and differ in the number of tRNAs; in mitochondria this number is 22 as against 55 in cytoplasm. (iii) Initiation of translation occurs by formyl-methionyl tRNA both in chloroplasts and mitochondria, although no formylation of methionine transpires in cytoplasm. (iv) Translation in chloroplasts and mitochondria can be inhibited by chloramphenicol, as in bacteria since the 70S ribosomes are sensitive to chloramphenicol and not to cycloheximide. On the other hand, the translation in cytoplasm is inhibited by cycloheximide, since 80S ribosomes are sensitive to cycloheximide.

protein synthesis inhibition by micro RNAs: MicroRNA miR2 inhibits translation initiation without affecting mRNA stability. Inhibition of $m^7G_{ppp}G$ cap-mediated translation initiation is the mechanism of miR2 function.

protein synthesis machinery: In addition to mRNA, tRNAs, rRNAs and ribosomal proteins, some other factors are also essential for protein synthesis. Peptidyl transferase forms peptide bonds between amino acids. Protein factors catalyse partial reactions in the initiation, elongation and termination of peptides. Miscellaneous other factors such as ATP, GTP,

Mg^{++}, K^+ and NH_4^+ are vital for various biochemical reactions. The enzymes needed for protein synthesis include aminoacyl synthetases, phosphatase, transformylase, deformylase and methionine-specific peptidase.

protein targeting in eukaryotes: Proteins synthesized on cytoplasmic ribosomes of eukaryotic cells can either be retained in the cytoplasm or transferred to subcellular organelles or inserted into membranes or secreted outside the cell. All living cells have specific mechanisms that target proteins on the cell surface. In eukaryotes, the first part of this process involves recognition in the ER of amino-terminal signal sequences and translocation through SEC TRANSLOCONS, whereas subsequent targeting to different surface locations is promoted by internal sorting signals. This process is termed PROTEIN SORTING or PROTEIN TRAFFICKING. *See details as separate terms.*

protein targeting in prokaryotes: In prokaryotes, SIGNAL RECOGNITION PARTICLE (SRP) targets membrane proteins in the inner membrane. It binds translating ribosomes and screens the emerging nascent chain for a hydrophobic signal sequence, such as the transmembrane helix of inner membrane proteins (IMPs). If such a sequence emerges, the SRP binds tightly, enabling the SRP receptor to lock on it. This assembly delivers the ribosome-nascent chain complex to the protein translocation machinery in the membrane.

protein trafficking: Arrestin regulates almost all G protein–coupled receptor (GPCR)-mediated signalling and trafficking. Multidomain, spinophilin, antagonizes these multiple arrestin functions. Through blocking G PROTEIN RECEPTOR KINASE 2 (GRK2) association with receptor-Gβγ complexes, spinophilin reduces arrestin-stabilized receptor endocytosis and the acceleration of MITOGEN-ACTIVATED PROTEIN KINASE (MAPK) activity following endocytosis.

protein translation without ribosomes: Some short proteins are not synthesized on ribosomes. Examples of such proteins include some antibiotic polypeptides such as gramicidin and tyrocidine.

protein translocator: A membrane-bound protein that facilitates the transport of another protein transversely an organelle membrane.

protein *trans*-splicing: Intein-mediated protein splicing is a self-catalytic process in which the intervening intein sequence is removed from a precursor protein and the flanking extein segments are ligated with a native peptide bond. Splice junction proximal residues and internal residues within the intein direct these reactions. The protein *trans*-splicing postures problem for existing gene concept that start and end sites are not governed by genes.

protein turnover: Refers to the balance between protein synthesis and protein degradation. The turnover of some transcription factors is controlled through ubiquitin-dependent proteolysis. Usually, such processes are controlled by extracellular stimuli through phosphorylation of the target protein, which enables recognition by F box-containing E3 ubiquitin ligases.

protein-DNA interactions: Proteins interact with DNA through electrostatic interactions (salt bridges), dipolar interactions (hydrogen bonding, H-bonds), entropic effects (hydrophobic interactions) and dispersion forces (base stacking). These forces contribute in varying degrees to proteins binding in a sequence-specific or non-sequence-specific manner. Understanding how proteins interact with DNA, determining what proteins are present in these protein-DNA complexes and identifying the nucleic acid sequence (and possible structure) required to assemble these complexes are vital to understanding the role these complexes play in regulating cellular processes. DNA-protein interactions are a key mechanism in various biological processes, viz. site-specific recombination, DNA replication and initiation of transcription. The precise association of DNA-binding proteins with localized regions of DNA is vital for regulated expression of the genome. For certain DNA transactions, the requirement for precision in localization and control is extremely high. Multiple DNA-protein interactions are also required for controlled transcription of the eukaryotic genome. A number of laboratory techniques that have been developed to study the complex interactions of proteins with DNA, each with a unique history, varying utility and distinct strengths and weaknesses. Some of them are (i) CHROMATIN IMMUNOPRECIPITATION (CHIP) ASSAYS, (ii) DNA ELECTROPHORETIC MOBILITY SHIFT ASSAY (EMSA), (iii) DNA PULL-DOWN ASSAY, (iv) MICROPLATE CAPTURE AND DETECTION ASSAY and (v) REPORTER ASSAYS.

protein-encoding gene: The gene which on transcription yields messenger RNA (mRNA in prokaryotes) or precursor of mRNA (pre-mRNA in eukaryotes) which after various processing steps matures into functional mRNA. *Aka* STRUCTURAL GENE.

protein-protein interaction (PPI): Proteins do not function in isolation to perform a biological function in a cell. Protein-PPI is an important step in many cellular functions. Purification of entire multiprotein complexes can be done by affinity-based method. PPI is important to the regulation and execution of most biological processes. PPI occurs through a combination of hydrophobic bonding, van der Waals forces and salt bridges at specific binding domains on each protein. Protein-protein interactions are studied by direct DNA readout or yeast-two hybrid system. Yeast-two hybrid system was initially developed by S. Fields and C.-K. Song in 1989. It uses two fusion proteins to activate the transcription of a reporter gene in yeast. The first fusion protein contains a DNA-binding domain fused to a bait protein of interest, while the second is transcriptional activation domain fused to a second protein of interest. Specific interaction between the two chimeric proteins leads to transcriptional activation of the reporter genes which in easily scored by either colour-based assay or by auxotrophic complementation. Thus, study of protein-protein interactions is an important component of proteomics research. These interactions are being exploited for biotechnological applications. Methodologies to study protein-protein interactions include: (i) GENETIC METHODS, (ii) BIOINFORMATICS METHODS, (iii) AFFINITY-BASED BIOCHEMICAL METHODS and (iv) LIBRARY-BASED METHODS. *See details as separate terms.*

protein-protein interaction (PPI) analysis using affinity-based biochemical methods: These PPI approaches that use affinity-based biochemical methods include affinity chromatography, co-immunoprecipitation and GST pull-down.

protein-protein interaction (PPI) analysis using bioinformatics methods: Three PPI analysis methods that use bioinformatics methods include domain fusion (also called rosette-stone) method, comparative genomics method and phylogenetics profiling method.

protein-protein interaction (PPI) analysis using genetic methods: Different mutations are combined in the same cell or organism and resulting phenotype is observed. Different PPI analysis genetic methods used include suppressor mutations, synthetic lethal stress and dominant-negative approach.

protein-protein interaction (PPI) analysis using library-based methods: These methods are used for global analysis of binary interactions. The PPI methods using library-based methods include (i) standard expression libraries, (ii) phage interaction display and (iii) yeast-two hybrid system.

ProteinGPS™ engineering technology: ProteinGPS™ is the protein engineering system that quickly and efficiently designs proteins with improved characteristics. ProteinGPS™ instead relies on identifying key amino acid substitutions through bioinformatics-based mining of available sequence space and combining such substitutions in an information maximized variant data set (usually less than 100 unique gene variants). At that scale determining the activity for the commercially relevant function in an indicative assay can be readily performed.

proteinoids: Various amino acids in certain conditions of temperature and pH are referred to as 'PROTEINOIDS' or 'THERMAL PROTEINOIDS', which are thought to be involved in origin of life.

proteinoplasts: Differentiate from leucoplasts. Store and modify proteins.

proteins binding to particular DNA sequences: Some proteins have evolved to bind to particular DNA sequences. The most intensively studied of these are the various transcription factors (TFs), which are proteins that regulate transcription. Each TF binds to one particular set of DNA sequences and activates or inhibits the transcription of genes that have these sequences close to their promoters. The TFs do this in two ways. First, they can bind the RNA polymerase responsible for transcription, either directly or through other mediator proteins; this locates the polymerase at the promoter and allows it to begin transcription. Alternatively, TFs can bind enzymes that modify the histones at the promoter; this will change the accessibility of the DNA template to the polymerase. Such DNA targets occur throughout an organism's genome. Changes in the activity of one type of TF can affect thousands of genes. Consequently, these proteins are often the targets of the signal transduction processes that control responses to environmental changes or cellular differentiation and development. The specificity of these TFs' interactions with DNA come from the proteins making multiple contacts to the edges of the DNA bases, allowing them to read the DNA sequence. Most of these base interactions are made in the major groove, where the bases are most accessible.

proteogenomics: *See* PROTEOMICS IN MEDICINE AND HEALTH.

proteoglycan: Molecule made up of one or more glycosaminoglycan (GAG) chains attached to a core protein.

proteolysis: Degradation of a protein by hydrolysis at one or more of its peptide bonds.

proteolysis targeting chimera (PROTAC): A heterobifunctional molecule that contains three components: the protein-of-interest (POI) binding moiety, a linker and E3 ubiquitin ligase binding moiety) capable of triggering target degradation.

proteolytic enzymes: Enzymes which catalyse the hydrolysis (breakdown) of proteins or peptides. These enzymes thus destroy the structure and hence function of other proteins. These other proteins may or may not themselves be enzymes. *See* PROTEASE.

proteome: The entire collection of proteins in a cell. The proteome provides additional information not obtainable from transcriptome analysis. This is mainly because single mRNA (gene) can produce more than one protein due to post-translational processing. To study proteome, total protein of a cell or tissue is isolated and separated by two-dimensional electrophoresis. To identify the protein, it is treated with protease (to cut polypeptide). The resulting polypeptides are examined by mass spectrometry that determines the amino acid composition of each peptide. The proteome is a functional illustration of the genome that includes the types, functions and interactions of proteins that are present in a cell. A proteome is more complex than a genome because genome of an individual is constant but proteome is highly dynamic and specific to cell type, developmental stage and environment. Proteomes are expected to contribute towards understanding evolution at the molecular level.

proteome analysis: There are five main steps in proteome analysis: sample collection, handling and storage by assessing to relevant body fluid and tissue samples, separation of individual proteins by 2-D polyacrylamide gel electrophoresis (2-D PAGE) for separation of specific cell types from other types, producing a relatively pure cell sample for proteome analysis, identification by mass spectrometry or N-terminal sequencing of individual proteins recovered from the gel. Protein patterns on 2DE gels are analysed. Proteome maps are compared to detect proteins that are up or down regulated in disease, protein characterization is done using the protein prediction tools and storage, manipulation and comparison of data are done using bioinformatics. Public protein and databases are used for this purpose.

proteome analysis tools: Various tools of proteome analysis are (i) MASS SPECTROMETRY, (ii) TWO-DIMENSIONAL GEL ELECTROPHORESIS, (iii) ISOELECTRIC FOCUSING, (iv) POLYACRYLAMIDE GEL ELECTROPHORESIS, (v) COMPARATIVE 2-D GEL APPROACH, (vi) PROTEIN CHIP APPROACH, (vii) ENZYME-LINKED IMMUNOSORBENT ASSAY (ELISA) and (viii) LIQUID CHROMATOGRAPHY COUPLED WITH ELECTROSPRAY-IONIZATION TANDEM MASS SPECTROSCOPY. *See details as separate terms.*

proteome chip: A microarray (also called BIOCHIP) developed by M. Snyder *et al.* in 2001 reveals the following features. (i) It has a large number of known sequence protein molecules (e.g. all proteins present in a given organism) attached to its surface at known positions (i.e. specific 'addresses' on the microarray). (ii) It employs specific bioactive agents such as certain lipids or biotinylated calmodulin (i.e. calmodulin molecules to which a molecule of biotin is attached) in order to find out which of the protein molecules in #1 interacts with relevant bioactive agents. Because calmodulin is a well-known and very well-characterized calcium-binding protein (i.e. bioactive agent) involved in (known) cellular processes, the binding of calmodulin to specific protein molecules to the microarray/biochip offers critical information about the (cellular, protein-protein, etc.) functions and interactions of those protein molecules in the organism. (iii) It delivers a large amount of data concerning protein-protein interactions (e.g. via subsequent application to the microarray of dye-labelled streptavidin to identify the protein molecules via their addresses on the biochip) and protein-lipid interactions, all of which are required, in order to determine the organism's proteome.

proteome complexity: Proteome is more complex than a transcriptome as several different proteins may be manufactured from one transcript via one of the three processes: (i) Alternative splicing of the mRNA transcript. For example, a single intronic base substitution that is present within the *IκB kinase complex-associated protein* (*IKAP*) gene (i.e. the allele responsible) for the disease known as familial dysautonomia affects the splicing of the IKAP transcript (i.e. the mRNA segment that determines which specific protein is subsequently 'manufactured' by the ribosomes). (ii) Varying translation start or stop sites. (iii) Frameshifting (i.e. different set of triplet codons in the mRNA/transcript) is translated by the ribosome.

proteome researches: Using the more inclusive definition of proteomics, many different areas of study are now grouped under the title of proteomics.

proteomic studies on *Arabidopsis thaliana*: This is perhaps the only plant species which has been subjected to detailed proteomics study from all the three subcellular components including chloroplasts, mitochondria and nucleus.

proteomic studies on yeast: Complete genome sequence of yeast was worked out in 1996. There are 6200 genes predicted in this unicellular eukaryote. About 93% (5800) of proteins encoded by yeast genes have been elucidated.

proteomic studies, implications of: Nucleic acids are the only information carriers, therefore, the analysis of genes (by mutation) or of mRNA (by RNA interference or transcriptomic studies) can make us wiser about the gene function only indirectly. Proteins are actually functional molecules of the cell and are responsible for almost the entire biochemical activity of the cell which is achieved by proteins by interacting with a diverse spectrum of other molecules. Once function has been assigned to an individual protein, one can search for other proteins with related functions by seeking proteins whose amino acid sequences are similar to the original protein. This homology method is used widely to extend knowledge of protein function from one protein to its cousins, which are presently described from the same common ancestral protein. The powerful BLAST programme is used to extend experimental knowledge of protein function to new sequences in this way.

proteomic studies, model species for: *Escherichia coli*, yeast, humans, *Arabidopsis*, wheat barley, maize, soybean, pea, apricot, banana, rice, poplar have been used as model species for study of proteomes.

proteomics: The term proteomics was coined by V.C. Wasinger (1995) and derived from word 'proteome'. It involves large-scale analysis of total protein complement (proteome) of an organism. Analysis involves protein distribution, characterization and identification, protein-protein interactions and differential display proteomics. Modifications of the proteins that are not apparent from the DNA sequences, such as isoforms and post-translational modifications, can be determined only by proteomic methodologies. The localization of gene products which is often difficult to predict from the sequences can only be determined through proteomics. It checks out the functional relevance of the protein and how different proteins work together. There is no strict linear relationship between gene activity and the protein abundance in a cell.

proteomics in agriculture: With the completion of genome projects for *Arabidopsis thaliana*, *Oryza sativa* and several other plant species, an increasing number of whole-genome sequences are now available for plants. In this post-genomic era, a more thorough understanding of gene expression and function can be achieved through the characterization of the products of expression, the proteins, which are essential biological determinants of plant phenotypes. Proteomics offers a continually evolving set of novel techniques to study all facets of protein structure and function. The application of proteomics in plant pathology is becoming more commonplace with techniques such as two-dimensional gel electrophoresis (2-DE) and mass spectrometry (MS) being used to characterize cellular and extracellular virulence and pathogenicity factors produced by pathogens as well as to identify changes in protein levels in plant hosts upon infection by pathogenic organisms and symbiotic counterparts.

proteomics in genetically modified plant analysis: Concerns about the safety of genetically modified foods orbit around the idea of SUBSTANTIAL EQUIVALENCE (i.e. whether the process of genetic modification has presented any unanticipated or undesirable changes in the host plant). Such changes ensuing from the random integration of DNA into the plant genome and the influence it may have on endogenous genes, could consist of metabolic disruption and therefore changes in the concentrations of nutrients and possibly toxic compounds. Strategies for the comparative analysis of GM and non-GM crops have been often focused on specific compounds such as particular nutrients, metabolites or toxins. However, in order to fully characterize any changes that may have arisen, it is necessary to use global and non-targeted approaches, including transcriptional profiling, proteomics and metabolomics. Proteomics possibly has an important part to enact in food safety assessment because it also permits changes in protein abundance and post-translational modifications to be identified.

proteomics in medicine and health: (i) Proteomic technologies will play an important role in DRUG DISCOVERY, diagnostics and molecular medicine because there is a link between genes, proteins and disease. As researchers study defective proteins that cause particular diseases, their findings will help develop new drugs that either alter the shape of a defective protein or mimic a missing one. (ii) One of the most promising developments to come from the study of human genes and proteins are the documentation of potential NEW DRUGS for the treatment of disease. A molecule that fits the active site of an enzyme, but cannot be released by the enzyme, will inactivate the enzyme. This is the basis of new drug-discovery tools, which intend to find new drugs to inactivate proteins involved in disease. As genetic differences among individuals are understood, researchers expect to use these techniques to develop personalized drugs that are more effective for the individual. (iii) BIOMARKER is defined as a characteristic that is accurately measured and assessed as an indicator of normal biologic processes, pathogenic processes or pharmacologic responses to a therapeutic intervention. A number of techniques enable us to test for proteins produced during a particular disease, which assists to diagnose the disease quickly using the techniques WESTERN BLOT, IMMUNOHISTOCHEMICAL STAINING (IHS), ENZYME-LINKED IMMUNOSORBENT ASSAY (ELISA) or MASS SPECTROMETRY (MS). (iv) Identifying unique patterns of protein expression or biomarkers, associated with specific diseases is one of the most promising areas of CLINICAL PROTEOMICS. One of the first biomarkers used in disease diagnosis was prostate-specific antigen (PSA). Serum PSA levels are now commonly used in diagnosing prostate cancer in men. (v) SECRETOMICS is a subfield of proteomics that studies secreted proteins and secretion pathways using proteomic approaches, has recently emerged as an important tool for the discovery of biomarkers of disease. (vi) PROTEOGENOMICS includes use of proteomic technologies such as mass spectrometry is made for improving gene annotations. Parallel analysis of the genome and the proteome facilitates discovery of post-translational modifications and proteolytic events, especially when comparing multiple species. (vii) Already, many of the best-selling drugs today either act by targeting proteins or are proteins themselves. Advances in proteomics may help scientists eventually create PERSONALIZED MEDICATIONS for different individuals to be more effective and have fewer side effects. Current research is looking at protein families linked to diseases including cancer, diabetes and heart disease.

proteomics in secondary metabolite analysis: Secondary metabolites are not parts of central metabolism but they perform many useful functions, including defence against pathogens and attraction of pollinators. Secondary metabolites are utilized as drugs, dyes, fragrances, nutritional supplements and flavours. Generally, secondary metabolites are formed in low quantities. Proteomics has augmented discovery of plant secondary metabolism because it can identify not only enzymes but also regulatory proteins and the proteins involved in the shuttling of intermediates between compartments. Model organism commonly used for study of secondary metabolites is plant Madagascar periwinkle (*Catharanthus roseus*) which manufactures hundreds of alkaloids, including the potent antineoplastic drugs. Proteomic analysis of cells of this plant species under culture conditions known to influence alkaloid production has discovered five proteins whose abundance reflects the level of alkaloid accumulation, suggesting either a catalytic or regulatory role. Important strategy to upsurge levels of alkaloids in this system would be to identify and deploy transcription factors that control the expression of these genes. One such transcription factor is ORCA2, which was identified using yeast-two hybrid system.

proteomics methods in current research: Proteomic approaches used in current researches include the following: (i) Fluorescence two-dimensional differential gel electrophoresis (2-D DIGE) is used to quantify variation in the 2-D DIGE process and establish statistically valid thresholds for assigning quantitative changes between samples. (ii) Comparative proteomic analysis can help reveal the role of proteins in complex biological systems such as reproduction. For example, treatment with the insecticide triazophos results in an increase in the content of brown planthopper (*Nilaparvata lugens* (Stål)) male accessory gland proteins (Acps) that can be transferred to females via mating, thereby causing an increase in fecundity (i.e. birth rate) of females. To identify changes in the types of accessory gland proteins (Acps) and reproductive proteins that mated female planthoppers received from male planthoppers, researchers conducted a comparative proteomic analysis of mated *Nilaparvata lugens* females. The results showed that these proteins participate in the reproductive process of *Nilaparvata lugens* adult females and males. (iii) Proteome analysis of *Arabidopsis* peroxisomes has been established as the foremost unbiased method for identifying new peroxisomal proteins on a large scale. (iv) There are many strategies available for characterizing the human proteome, which is estimated to contain between 20,000 and 25,000 non-redundant proteins. The number of unique protein species likely increases to between 50,000 and 500,000, due to RNA splicing and proteolysis events, and when PTLM are also considered, the total number of unique human proteins is estimated to range in the low millions. (v) In addition, first promising attempts to decipher the proteome of animal tumours have recently been reported.

proteomics research: Proteomics, which focuses on gene products, is complementary to genomics and transcriptomics and is a thrust area of molecular biotechnological research. Verification of a gene product by methods of proteome analysis serves a very useful purpose for annotation of the genome. PTLM of proteins can be examined only by methods of proteome analysis. Sometimes results of transcriptome analysis do not correlate with those of proteome analysis, thus making it necessary to examine protein expression levels. Protein function is also regulated by proteolysis and recycling or sequestration of products in various cell compartments. Protein-PPIs also determine the function of gene and gene product, the protein.

proteomics study, impact of: The following are some methods utilized to determine which impact results from which protein: (i) Chemical genetics, to compare two same-species organisms (one of which has protein or a portion of protein, at least partially inactivated by a specific chemical). (ii) Gene expression analysis, to determine the protein(s) produced when a given gene is switched on; by measuring fluorescence of individual messenger RNA (mRNA) molecules (specific to which particular gene is switched on at the time), when that mRNA hybridizes (with DNA pieces corresponding to genes analysed, that were attached to hybridization surface on the biochip). (iii) Gene expression analysis, to determine impact when a given gene is knocked out/turned off. (iv) Protein interaction analysis, to determine if a newly discovered protein molecule interacts with a protein molecule whose function is already known (e.g. from previous research or use as a pharmaceutical). If the newly discovered protein molecule interacts with one whose function is already known, it generally has the same or similar function (in living cells) as the previously known protein molecule. Thus, the function of a newly discovered human protein can sometimes be inferred from a protein molecule discovered earlier in a microorganism (e.g. via Expressed Sequence Tags). (v) *In silico* biology (modelling), to compare computer-predicted events (e.g. the constituent peptides resulting from protein digestion) with actual or *in vitro* outcomes.

proteomics versus genomics: Proteomics refers to all proteins produced by a species, much as the genome is the entire set of genes. Unlike the genome, the proteome varies with time and is defined as the proteins present in one sample (tissue, organism or cell culture) at a certain point of time. Genomics deals with one genome per organism. There is a large number of proteomes in an organism. Genome is constant feature of an organism. Proteome varies with nature of tissue, the state of development, health or disease and effect of drug treatment. Therefore, proteomics is a much larger field than genomics. Proteomics has immense applications in disease identification. Proteomics enables correlations to be drawn between the range of proteins manufactured by a cell or tissue and the initiation or progression of a disease state. It permits discovery of new protein markers for diagnostic purposes and of novel molecular targets for drug discovery. The information provided by proteome research is complementary with the genetic information being generated by genomic research.

proteomics, analysers currently used in: (i) Quadropole (Q), (ii) Ion trap (IT), (iii) Time-of-flight (TOF), (iv) Fourier transform ion-cyclotron resonance (FT-ICR or FT-MS) and (v) Orbitrap (OT).

proteomics, branches of: (i) Sequence and structural proteomics, (ii) Expression proteomics, (iii) Interaction proteomics and (iv) Functional proteomics. *See details as separate terms.*

proteomics, broad areas of: There are three broad areas of proteomics: (i) expression genomics, (ii) cell map proteomics and (iii) structural proteomics.

protoandry: A state in hermaphroditic systems that is characterized by the development of male organs or maturation of their products before the appearance of the corresponding female product thus inhibiting self-fertilization and that is encountered commonly in mints, legumes and composites and among diverse groups of invertebrate animals. *Compare with* PROTOGYNY.

protocell: Believed to have given rise to first living cell on the Earth. It was thought to be heterotrophic (i.e. it obtained complex organic molecules from surroundings). When complex organic molecules began to disappear, only those systems survived that evolved metabolic pathway to change simpler organic molecules to complex molecules. Thus, auxototrophic mode of nutrition evolved from heterotrophic mode and origin of life can be traced back to auxotrophic protocells.

protoclonal variation: Variation among the somatic cells derived from protoplast culture or plants regenerated from cultured protoplasts. During plant tissue culture, the excessive cell division during callus multiplication leads to variation in the regenerated plants. When this variation results from the cultured protoplasts directly or via callus formation, more precisely it is called protoclonal variation. Using this technique, disease-resistant protoclones have been developed in potato and *Brassica napus*. This variation usually arises due to the changes in the base sequences of DNA, DNA amplification, duplication, deletions and so on during active cell division.

protoclone: A plant revived from a protoplast culture.

protofilament: A linear chain of protein subunits joined end to end, which contacts laterally with other protofilaments to make cytoskeletal components such as microtubules and intermediate filaments.

protogynous: Having female sex organs maturing before the male ones (e.g. a flower shedding pollen after the stigma has stopped to be receptive). *Compare with* PROTANDROUS.

protogyny: A state in hermaphroditic systems that is characterized by development of female organs or maturation of their products before the appearance of the corresponding male product thus inhibiting self-fertilization and that is encountered in apples, pears, figworts and among several groups of invertebrate animals. *Compare with* PROTOANDRY.

protomer: The asymmetric repeating unit (or units) from which an oligomeric protein is built up.

proton: Positively charged subatomic particle that formulates part of an atomic nucleus. Hydrogen has a nucleus composed of a single proton (H^+).

proton-motive force: Driving force that makes protons move across a membrane as a consequence of an electrochemical proton gradient.

proto-oncogene activation: Various mutagenic activities in proto-oncogenes can activate these genes: (i) A point mutation produced by a chemical or radiation carcinogen that can lead to change in the protein, thereby initiating cancerous growth. (ii) Induction of a chromosomal rearrangement that places in the proto-oncogene next to the regulatory area of an immunoglobulin gene. This may lead to inappropriate expression of the proto-oncogene. (iii) If the proto-oncogene is amplified either as repeat segments within the chromosome or extrachromosomally, the gene product will be overproduced. (iv) The retroviral transfer of a proto-oncogene picks up another animal cell. Here, transcription of proviral regulatory signals, not the signals associated with the proto-oncogenes in the cell's genome, abnormal regulation and/or abnormal product of v-onc expression produce the tumour phenotype.

proto-oncogenes: Normal cellular genes from which the viral oncogenes are derived. Proto-oncogenes do not produce cancer cells under normal circumstances, but they may do so if they are modified or changed by their incorporation into viral genomes. Almost every viral oncogene tested so far has revealed one or more closely related DNA sequences in uninfected host cells.

protoplasm: A general term, coined by J. E. Parkinje in 1840, that refers to the entire contents of a living cell or living substance. *See* CYTOPLASM.

protoplast: A cell without a cell wall, surrounded by a plasma membrane and capable of cell wall regeneration, growth and division upon culturing. Protoplasts are obtained from living cells by incubating them with cell wall–degrading enzyme such as cellulase, pectinase and cellulysin. Wall-less protoplasts are kept in osmoticum medium and used in basic as well as applied research such as somatic hybridization to develop somatic hybrids among remote plant species or for genetic transformation purposes. Under specific conditions (e.g. electroporation), certain DNA stretches (genes) synthesized by man can enter protoplasts. The cell then incorporates some or all of that DNA into its genetic complement (genome) and produces the product which the newly introduced gene codes for. In the case of plant protoplasts, whole plants can be regenerated from the (genetically engineered) protoplasts, resulting in plants that produce the product(s) for which the introduced gene(s) codes. Protoplasts are used for production of hybrid cells by protoplast fusion or for injection of foreign DNA. *Compare with* SPHEROPLAST.

protoplast culture: The isolation and culture of plant protoplasts by mechanical means or by enzymatic digestion of plant tissue, organs or cultures derived from these. Protoplasts are used for selection or hybridization at the cellular level and for a variety of other purposes.

protoplast fusion: A technique for producing hybrids between two cells which would not normally mate. The two cells may belong to the same or different species. The cell walls are removed from the two parental cells to create protoplasts and then fusion of the two cell membranes is promoted, usually by the addition of polyethylene glycol (PEG) of high molecular weight of 1500 at 25% concentration and Ca^{++} ions for 2–4 h. Alternately, protoplasts can also be fused using electric currents (electrofusion) generated by somatic cell hybridizer. Fusion results in the formation of somatic hybrid cell that regenerate into somatic hybrid plant. These fusogenic agents cause proteins to migrate from certain regions of the two cell membranes and allow areas of naked phospholipid to fuse. Subsequent regeneration of the cell wall enables the propagation of the hybrid organism. It is used for overcoming crossing barriers. If nuclear fusion does not follow cell fusion then HETEROKARYONS, rather than diploid organisms, are produced. In some crosses, the genetic contribution of the two parents to the stable hybrid. By protoplast fusion it is possible to transfer some useful genes from one species to another. Various enzymes used for breaking of plant cell walls are cellulase, pectinase, macerozyme and xylanase; from Gram-positive bacteria is lysozyme; and from fungal cells is Novozyme-234 which includes glucanase and chitinase.

protoplasts and somatic hybrids in crop improvement: Protoplast fusion and somatic hybrids can be used for (i) gene transfer and (ii) generation of cytoplasmic male sterility. Cytoplasmic hybrids (cybrids) have been utilized for transfer of cytoplasmic male sterility in *Petunia*, *Nicotiana* and *Brassica*.

protoporphyrinogen oxidase (protox or **PPO):** An enzyme used in photosynthesis and the target site of many herbicides in the cell membrane disrupter mode of action.

protospacer adjacent motif (PAM): A short (2–8 nucleotide) motif that occur on CRISPR (cr) RNA and are specifically recognized and targeted by CRISPR-Cas systems (except type III), as a means to discriminate self (CRISPR spacer) from non-self (invader protospacer) during by CRISPR-Cas systems (except type III), as a means to discriminate self (CRISPR spacer) from non-self (invader protospacer). PAMs follow the gRNA recognition sequence but are not in the gRNA. The PAM sequence varies between bacterial species. In *Streptococcus pyogenes*, it is NGG (which is more effective) or NAG (less effective).

protospacers: DNA sequences of 26–72 bp that are initially derived from invading phages and plasmids and that are embedded in the CRISPR loci in bacteria or archaea.

protostome: Member belonging to one of two large groups of bilaterian animals including ecdysozoans and lophotrochozoans (*see* DEUTEROSTOME) in which the initial embryonic opening develops into the mouth.

prototroph: A strain of organisms able to grow on a defined minimal medium from which they can synthesize all of the more complex biological molecules they require, as opposed to an AUXOTROPH.

protoxin: A chemical compound that becomes a toxin only after it is altered in some way. For example, the *Bt* protoxins (Cry9C, Cry1A (b), Cry1A (c), etc.) only become toxic after they are chemically altered by the alkaline environment inside the gut of certain insects.

protozoa: Free-living or parasitic, non-photosynthetic, single-celled, motile eukaryotic organisms, such as Paramecium and Amoeba. Free-living protozoa feed on bacteria or other microorganisms.

provirus: The DNA copy of retroviral genome that is integrated into host chromosomal DNA. The formation of complete viruses requires that DNA be transcribed back into the RNA that will serve as mRNA for viral protein synthesis for incorporation into new virions. But before transcription, viral DNA must get integrated into host DNA. Thus, the viral DNA in this integrated state is called provirus. *Aka* PROPHAGE.

provisional patent: A preliminary patent application filed without a formal patent claim, oath or declaration or any prior art statement. It provides the means to establish an early effective filing date in a subsequent non-provisional patent application and permitting the term PATENT PENDING to be applied.

proximal: Nearest to the point of attachment. *Contrasts* DISTAL.

proximal probes: Devices capable of fine positional control of atoms and molecules, enabling nanotechnology.

proximal proteins: With reference to assembly of complex structures (e.g. ribosomes), these are the proteins that bind early in the assembly process and whose presence in the partially assembled structure is required for binding of DISTAL PROTEINS.

proximal-to-distal patterning: What is the developmental mechanism that makes our upper arms different from our forearms or fingers? Two views of proximal-to-distal patterning in developing limb buds have been put forth. In the progress zone model, the fate of proximal (prospective upper arm) elements is specified before the fate of more distal elements, as the bud grows. Changes in cell fate occur in the progress zone, just next to the APICAL ECTODERMAL RIDGE (AER), a structure that forms from the ectodermal cells at the distal end of each limb bud and acts as a major signalling centre to ensure proper development of a limb. In the pre-specification, proximal-to-distal fates are already specified early on and the observed time sequence in skeletal development results from the selective expansion of these prespecified domains, along with cellular determination – the acquisition of definitive cell fate. How the two models can explain the effects of removing the AER? According to the progress zone model, the specification clock is arrested and distal specification never occurs. In the pre-specification model, the expansion of prespecified distal domains is aborted because of cell death. In both cases, the same results are expected.

proximity effect: Enzyme catalytic mechanism where free energy of activation is reduced by the prior bringing together of its reacting groups. During a biochemical reaction, reactants must come together with the proper spatial relationship. The enhancement in the rate of chemical reaction results from both proximity and orientation effects.

proximity ligation assay (PLA): A technology that extends the capabilities of traditional immunoassays to include direct detection of proteins, protein interactions, extracellular vesicles and post-translational modifications with high specificity and sensitivity.

prune: To purposely alter the shape and growth pattern of a plant by cutting stems, shoots or roots.

pSC101: A small, non-conjugative plasmid which encodes tetracycline resistance. Its origin is not clear as it is now thought to have come from a contaminant in a transformation experiment. It was one of the first vectors used in genetic engineering. The pSC101 tetracycline-resistance gene was used in the construction of pBR322.

pseudo-self-compatibility (PSC): Quantitative variation in the strength of the self-incompatibility (SI) response, such that matings between plants sharing active *S*-alleles produce some seed.

pseudoalleles: Such mutations as were allelic in complementation test and non-allelic on recombination test were termed pseudoalleles by Barbra McClintock in 1945. A. Chovnick in 1961 made similar observations at garnet locus of *Drosophila melanogaster*. It was suggested that such alleles (called pseudoalleles) seem to occupy complex locus which can be divided into subloci between which recombination can occur.

pseudoalleles *cis*-arrangement: Arrangement of two mutant alleles in one chromosome strand, with the wild-type alleles in the other.

pseudoalleles *trans*-arrangement: A mutant allele and a wild-type allele in each of the two chromosome strands.

pseudoautosomal gene: A gene that occurs in the pairing region of both the sex-determining heteromorphic (i.e. X and Y) chromosomes.

pseudoautosomal inheritance: Pattern of inheritance shown by genes that are located on the pairing region of the X and Y chromosomes in individuals having XY mechanism of sex determination. The X and Y chromosomes in humans contain regions (PAR1 and PR2) of sequence identity at the ends of their respective short and long arms. Recombination takes place in male meiosis between the X and Y within these regions of DNA sequence identity. Traditional patterns of sex-linked inheritance are sometimes not observed for alleles in these regions of X–Y sequence identity. The pattern of inheritance observed for alleles in the X–Y identical regions is referred to as pseudoautosomal inheritance, because it resembles the pattern seen for alleles located on autosomes. The areas of sequence identity on the X and Y are thus referred to as the PSEUDOAUTOSOMAL REGIONS (PAR). Human have 30 genes near the tips of these regions. Dyschondrosteosis, a dominantly inherited skeletal dysplasia of the forearm, is an example of a pseudoautosomal condition inherited in a dominant manner.

pseudoautosomal region: X and Y chromosomes have a short homologous segment which causes pairing. This region is called pseudoautosomal region and the genes located in this region show pseudoautosomal inheritance.

pseudobivalent: A bivalent-like association of two mitotic chromosomes owing to reciprocal chromatid exchange.

pseudocompatibility: The incidence of fertilization that normally is prohibited by incompatibility mechanisms. It is produced by specific environmental or genotypic conditions. For example, in rye, self-incompatibility can be broken by high temperature (about +30 °C) so that seeds are set.

pseudodiploid: This describes the condition where the number of chromosomes in a cell is diploid but, as a result of chromosomal rearrangements, the karyotype is abnormal and linkage relationships may be disrupted.

pseudodominance: A condition where a recessive gene occurring in a single dose expresses itself. Mostly pseudodominance is observed when the homologue portion corresponding to the recessive gene is deleted or naturally absent (i.e. recessive gene in hemizygous form expresses itself).

pseudoexon inclusion: Non-functional intronic sequences that can be activated for aberrant insertion in mRNA, which may cause disease.

pseudogamous heterosis: Increased vigour of maternal progeny owing to male parent effect on the endosperm.

pseudogamy: A kind of apomixis in which the diploid egg cell develops into the embryo without fertilization of the egg cell, although this happens only after fertilization of the polar nuclei with one of the sperm cells from the male gamete to make a normal triploid endosperm.

pseudogene: A nucleotide sequence that has similarity to a functional gene but within which the biological information has become scrambled so that the pseudogene is not itself functional. Such genes may appear due to mutations in initiation codons. Pseudogenes may also arise from reverse transcription of mRNA. The following are two important features of pseudogenes: (i) Most of the pseudogenes are more numerous than their genes and are therefore repetitive sequences. (ii) They are bordered by 6–21 base-long direct repeats. Pseudogenes are a common feature of many multigene families in higher eukaryotes. In fact, pseudogenes are ancient genes which have lost their function, and pseudogenes which have in the past been protein-coding genes usually contain many stop codons (i.e. their reading frame has been closed during the course of evolution). Gene-like sequences that lack regulatory sequences needed for transcription or translation or whose coding sequences are obviously defective due to frameshifts or premature stop codons. Pseudogenes are a type of junk DNA.

pseudoheterosis: Refers to lavishness. It designates hybrids between species, varieties or lines that exceed the parents in some traits; however, they occur neither through sheltering deleterious genes nor through balanced gene combinations.

pseudoisochromosome: A chromosome that reveals only equal ends as a consequence of interchanges.

pseudoknot: A feature of RNA tertiary structure best envisioned as two overlapping stemloops in which the loop of the first stemloop contributes as half of the stem in the second stemloop. Various tools used in pseudoknot studies include the following: (i) Kinefold RNA/DNA folding predictions including pseudoknots and entangled helices. (ii) pknotsRG tools for folding RNA secondary structures, including the class of simple recursive pseudoknots. (iii) HPknotter is used for detecting RNA H-type pseudoknots and offers a variety of tools, including pknotsRG, PNOTS and NUPACK. (iv) RNATOPS-W is a profile-based RNA structure search programme that can detect RNA pseudoknots in genomes. For input it requires a structure profile in the pasta format and genome sequences in the fasta format.

***Pseudomonas aeruginosa*:** A common rod-shaped Gram-negative bacterium that can cause infections in the blood, lungs (pneumonia) or other parts of the body after surgery. *Aka* PSEUDOMONAS spp.

***Pseudomonas fluorescens*:** A generally harmless soil microorganism (bacteria) that inhabits the roots of certain plants. At least one biotechnology company has introduced the gene for a protein that is toxic to insects (taken from *Bacillus thuringiensis*) into a *Pseudomonas fluorescens*. This was accomplished in order to confer insect resistance to the plants the roots of which the genetically engineered *Pseudomonas fluorescens* has occupied.

***Pseudomonas syringae*:** Bacterium that outbreaks several crop plants (e.g. rye, *Secale cereale*). In genetic engineering, a genetically modified strain of these bacteria absences a cell-surface protein that causes ice crystals to form. Spraying these bacteria on crops may avert freezing-related damage.

pseudo-overdominance model of heterosis: This model posits that complementation occurs for different recessive alleles that are present in close linkage but on opposite members of a pair of homologues such that overdominance appears to be operating.

pseudoresistance: Superficial resistance in an actually susceptible host due to chance, transitory (non-heritable) traits or environmental conditions.

pseudorevertant: A mutant virus or organism which has recovered a wild-type phenotype due to a second-site mutation (potentially located in a different region of the genome or involving a different polypeptide) which has eliminated the effect of the initial mutation.

pseudo-self-incompatibility: A partial seed setting following self-pollination in an otherwise self-incompatible plant.

pseudo-sequencing (Pseudo-seq) methods: Genome-wide, single-nucleotide resolution methods for ψ identification. Pseudo-seq accurately identified known modification sites as well as novel sites in ncRNAs and revealed hundreds of pseudouridylated sites in mRNAs. Such methods include N3-CMC-enriched pseudouridine sequencing method, Pre-enrichment of Ψ-containing RNA fragment method and the prediction of tRNA modifications.

pseudouridine (ψ): Pseudouridine (5-ribosyluracil) is an isomer of the nucleoside uridine in which the uracil is attached via a carbon-carbon instead of a nitrogen-carbon glycosidic bond. It is a ubiquitous constituent of structural RNA (transfer, ribosomal, small nuclear (snRNA) and small nucleolar (snoRNA)) and is present in coding RNA, across the three phylogenetic domains of life. Pseudouridine in rRNA and tRNA has been shown to fine-tune and stabilize the regional structure and help maintain their functions in mRNA decoding, ribosome assembly, processing and translation. Pseudouridine in snRNA has been shown to enhance spliceosomal RNA-pre-mRNA interaction to facilitate splicing regulation.

pseudouridylation: The isomerization of uridines within RNA to pseudouridine (ψ) by RNA enzymes known as pseudouridine (ψ) synthases.

PSI-BLAST and variants: The most commonly used programme for homology searches.

PSI-BLAST: Position-Specific Iterative BLAST. An iterative search using the BLAST algorithm. A profile is built after the initial search, which is then used in subsequent searches. The process may be repeated, if desired with new sequences found in each cycle used to refine the profile. Details can be found in this discussion of PSI-BLAST.

***Pst*I:** A type II restriction enzyme from bacterium *Providencia stuartii* 164. That recognizes the DNA sequence 5′-C T G C A↓G / 3′-G↑A C G T C where arrows indicate sites of cleavage.

psychobiology: The application of the principles of biology to the study of physiological, genetic and developmental mechanisms of behaviour in humans and other animals. *Aka* BEHAVIOURAL NEUROSCIENCE, BIOLOGICAL PSYCHOLOGY *or* BIOPSYCHOLOGY.

psychrophile: An organism that requires a temperature of 0°C (32°F) for growth.

pteridines: Molecules responsible for contributing red pigment for the eye colour to *Drosophila*. *Compare with* OMMOCHROMES.

***Pto*:** A specific resistance gene in tomato, which controls resistance against the bacterium *Pseudomonas syringae* pv. tomato. This gene encodes a cytosolic protein kinase. The *Pto* system remains the one of the best characterized specific resistance gene systems.

public antigenic determinants: The antigens that carry determinants common in all the antigens.

pUC18 vector: A plasmid cloning vector that has a size of 2.7 kb and nearly 100 copies per chromosome. It shows ampicillin resistance for selection and alpha complementing fragment of beta-galactosidase with in-frame polylinker for cloning.

puff: A structural modified region of a polytene chromosome. It originates from the despiralization of deoxyribonucleoprotein. *See* CHROMOSOME PUFF *and* BALBIANI RING.

puffing pattern: Puffs in polytene chromosomes appear as loose chromatin. Developmental changes in certain cell types are correlated with changes in puffing patterns. Puffs are the sites of active transcription. This was shown in polytene chromosome by isolating a unique species of high molecular weight RNA transcribed at a particular Balbiani ring. Puffing pattern is tissue and stage specific.

pulse-chase experiment: A method for following the passage of a substance through a biochemical or cellular pathway. Pulse-chase experiments are done by incubating cells very briefly with a radioactively labelled precursor (of some pathway or macromolecule) and then the fate of the label is followed during a subsequent incubating with a non-labelled precursor.

pulsed-field gel electrophoresis (PFGE): A technique for the electrophoretic separation of DNA molecules from the size of ordinary restriction fragments (<10 kb) to intact chromosomes of up to 20 Mb. The DNA is alternately subjected to two electrical fields at different angles for specific time intervals called pulse time. Under appropriate conditions, reorientation of DNA segments with different sizes and topology is different, which leads to the separation of the different segments.

pump: A transmembrane protein that energies the active transport of ions or small molecules across the lipid bilayer.

punctuated equilibrium model: A theoretical model of evolution in which species remain unchanged for long periods of time and then at times rapidly change as a result of major alterations in the environment and, subsequently, in natural selection. In some cases, the rapid change happens in quick bursts which are associated with speciation. *See* PHYLETIC GRADUALISM.

punctuation: A short period of rapid change breaking a long period of stasis in the fossil record.

punctuation codon: A codon that designates either the start or the end of a gene.

pungent: Having a sharp, hard point.

Punnett square: Developed in 1905 by British geneticist R.C. Punnett. A graphic representation of a particular cross accomplished to envisage the progeny of the cross. A lattice is used as a graphic representation of the progeny zygotes resulting from different gamete fusions in a specific cross. Punnett squares are commonly used by genetics counsellors to predict the odds of a couple passing on particular inherited traits.

pupa (pl. pupae; vb. pupate): The dormant stage between the larva and the adult of certain insects.

pure bred line (or strain): One derived from a line subjected to inbreeding. The population breeds true for a particular character or set of characters (e.g. tall → tall; dwarf → dwarf).

pure breeding: An individual or strain that is homozygous for a particular gene.

pure culture: A culture that contains only one form of species of a microorganism.

pure epialleles: Pure epialleles are independent of genetic variants.

pure line: A population that breeds true for a particular character or a set of characters (e.g. tall → tall; dwarf → dwarf). This is the progeny of a single homozygous individual produced through self-fertilization. A sound and scientific basis of pure line selection in self-pollinated crops was provided by Wilhelm L. Johannsen in 1900. He defined a 'pure line' as the descendants of single, homozygous self-fertilized organism and described the genetic mechanism by which different pure lines could be established. Further, Johannsen carried out his experiments on 'Princess' variety of beans (*Proteus vulgaris* L.) and gave the genetic basis of pure line theory that is now being used in breeding of self-pollinated crops. *Aka* PURE BREEDING LINE.

purifying selection: Selection against deleterious alleles.

purine (Pu): A basic nitrogenous two-ring heterocyclic compound occurring in nucleotides and nucleic acids as it contains fused pyrimidine and imidazole rings. Adenine (A) and guanine (G) are two purines in DNA and RNA.

purity testing: Resolving the gradation of contamination of seed lots with genetically non-identical, damaged or pest-infected seeds. The excellence of a seed lot is arbitrated by the relative percentage of various components. The quality is considered superior, if pure seed percentage is above 98 and the other-species seeds and inactive matter percentage are as low as possible. The percentage of seeds from other species should be practically insignificant or nil. Genetic purity of seeds refers to their trueness to type. If the seed possesses all the genetic qualities that the breeder has placed in the variety, it is said to be genetically pure. Genetic purity has a direct effect on ultimate yields. If there is any decline in the genetic makeup of the variety during the seed multiplication and distribution cycle, there will certainly be an equivalent decrease in its performance (e.g. yield and disease resistance).

puromycin: An antibiotic that inhibits polypeptide synthesis by competing with aminoacyl tRNAs for the ribosomal binding site 'A'.

putative transgenic: Assumed transgenic before their confirmation. During plant transformation, the *Agrobacterium-treated* or 'particle gun'–bombarded target tissues are grown on selection medium for two cycles of selection of 2 weeks each. Thus, selected tissues are regenerated into plants called putative transgenic plants. These plants are then characterized using molecular analysis to confirm the presence and expression of transgenes. Plants thus confirmed are called transgenic plants.

pycnotic: The concentration of the nucleus into a compact, strongly stained mass, occurring as the cell dies.

pyramiding of genes: The process of combining several disease-resistance or agronomically important genes from different sources into one genetic background (genotype). *See* GENE PYRAMIDING.

pyranosyl-RNA: It contains an altered ribose where RNA contains a 6-atom ring rather than a 5-atom ring of RNA's ribose. Because RNA is unstable and difficult to synthesize, the first genetic material may have used a simpler backbone than ribose.

pyridoxal phosphate: A major coenzyme which is involved in amino acid metabolism.

pyrimidine (Py): A basic, single six-membered ring structure, heterocyclic compound. The principal pyrimidines cytosine (C) and thymine (T) are present in DNA, while uracil (U) replaces thymine in RNA.

pyrimidine dimer: The compound formed when UV radiation induces the covalent joining in DNA of two thymine residues (forming thymine dimers) or two cytosine residues (forming cytosine dimers).

In RNA, uracil dimers are formed (in place of thymine dimers in DNA), where two uracil residues covalently join. The CYCLOBUTANE PYRIMIDINE DIMER (CPD) contains a four-membered ring arising from the coupling of the two double-bonded carbons of each of the pyrimidines. Such dimers interfere with base pairing during DNA replication, leading to mutations.

pyrimidine dimer: The compound that results when UV radiation induces covalent binding of two thymine residues (T-T dimer), two cytosine residues (C-C dimer) or a thymine and a cytosine residue (T-C dimer).

pyrophosphate cleavage: The enzymatic removal of two phosphate groups (designated as PPi) from ATP in one piece leaving AMP as another product. This cleavage releases more energy, which can be used in certain reactions that require more of a 'push' to get them going.

pyrosequencing: A method of DNA sequencing based on the 'sequencing by synthesis' principle. This method relies on the detection of pyrophosphate release on nucleotide incorporation, rather than chain termination with dideoxynucleotides. Pyrosequencing, a non-electrophoretic method for DNA sequencing, is a popular platform for DNA analysis (M. Ronaghi *et al.*, 1996, 1998 and 2001; M. Ronaghi in 2000). Advantage of this method are accuracy, ease of use and high flexibility for different applications. The technique dispenses with the need for labelled primers, labelled nucleotides and gel electrophoresis and it supports high levels of automation. The development of a chip-based format helps conduct pyrosequencing reactions on a microscale. Availability of cheap, high-throughput DNA sequencing chips helps to lift constraints on the cost and speed of doing experiments and paves the way towards a lab-on-a-chip solution for large-scale DNA sequencing. Currently, emphasis is on developing this platform for high-throughput microbial flora typing and genome sequencing.

pyrosequencing, commercialization of: Pyrosequencing has been commercialized by Biotage (for low-throughput sequencing) and 454 Life Sciences (for high-throughput sequencing). The latter platform sequences roughly 100 mb in a 7-hour run with a single machine. In the array-based method (commercialized by 454 Life Sciences), single-stranded DNA is annealed to beads and amplified via EmPCR. These DNA-bound beads are then placed into wells on a fibre-optic chip along with enzymes which produce light in the presence of ATP. When free nucleotides are washed over this chip, light is produced as ATP is generated when nucleotides join with their complementary base pairs. Addition of one (or more) nucleotide(s) results in a reaction that generates a light signal that is recorded by the CHARGE-COUPLED DEVICE (CCD) camera in the instrument. The signal strength is proportional to the number of nucleotides; for example, homopolymer stretches incorporated in a single-nucleotide flow.

pyruvate oxidation: Two molecules of pyruvate are converted into two molecules of acetyl-CoA. Pyruvate oxidation takes place in the mitochondrial matrix (in eukaryotes).

Python: An interpreted, high-level and general-purpose programming language created by Guido van Rossum. First released in 1991, Python's design philosophy underscores code readability with its notable use of significant whitespace. Its language constructs and object-oriented approach intend to help programmers write clear, logical code for small and large-scale projects.

Q

Q banding: A chromosomal staining technique using the fluorescence dye quinacrine mustard. Under UV light, a characteristic light and dark banding is induced.

q: The relative frequency of a gene, when used together with 'p'. (a) 'q' indicates the relative frequency of the recessive alleles while 'p' indicates the relative frequency of the recessive alleles. (b) 'q' denotes long arm of a chromosome while 'p' denotes short arm of a chromosome.

Q-beta (Qb): A bacteriophage that contaminates *Escherichia coli*. It contains a single-circular RNA genome. The viral RNA is termed the PLUS (+) STRAND because it can function as an mRNA molecule and direct synthesis of viral proteins. The '+' strand also acts as a template for synthesis of a complementary MINUS (−) STRAND. Qb is the best-known source of an RNA replicase. The resulting double-stranded molecule, together with Q-beta replicase, directs the synthesis of a large number of '+' strands. This autocatalytic replication can be harnessed to produce large quantities of desired RNA molecule using recombinant RNA technology.

Q-beta replicase: A viral RNA polymerase produced by a bacteriophage that infects *Escherichia coli* bacteria. Q-beta replicase can duplicate a naturally occurring almost any RNA sequence at a geometric rate.

Q-beta replicase technique: An RNA assay (test) that 'amplifies RNA probes' that a researcher is looking for. For example, by applying the Q-beta replicase technique to evaluate the presence of RNA specific to the AIDS virus, it is conceivable to sense an AIDS infection in a patient's blood sample long before that infection has advanced to the point where antibodies would occur in the blood.

QIAcube HT: Instrument used to perform nucleic acid purification.

quadratic selection gradient: The coefficient of a regression of squared deviation of a trait against fitness. Negative values designate stabilizing selection, which decreases the variance of the trait.

quadriallel analysis: An analysis based on double-crosses performed by crossing a number of homozygous lines in a diallel fashion and then crossing the F_1s so produced, according to a diallel scheme, with the constraint that in any double cross a homozygous line must not transpire more than once as a parent.

quadriduplex type: Autotetraploid. *Aka* QUADRUPLEX TYPE.

quadripartite structure: A structure having four corresponding parts.

quadriradial (Abbrev. qr): An interchange figure with four chromosome arms. *See* TRIRADIAL (TR).

quadrivalent: A chromosome conformation observable in late prophase and metaphase of the first meiotic division, where four chromosomes are connected by chiasmata. Quadrivalents can ensue in autotetraploids when four homologous chromosomes pair, or in diploids due to heterozygosity for a reciprocal translocation between two non-homologous chromosomes.

quadruplex: Refers to the inheritance of alleles in autotetraploids. For example, a genotype *AAAa* will yield gametes *AA* and *Aa* in the ratio 3:1.

quadruplex DNA: DNA can also occur in four-stranded form. Repeats of CGG in a single strand can produce quadruplex forms of DNA. In the presence of potassium, sodium and lithium ions, CGG repeats readily formulate quadruplex structures. Quadruplex DNA travels quicker on gel than other forms of DNA. Guanines are implicated in four-stranded structures with Hoogsteen hydrogen bonding. Segments of 2× GCGC tetrads bordered by two repeats of GGGG (G-QUARTETS) yield quadruplex strands. Many triplet repeats form hairpins and such multi-repeats can pleat into quadruplex. Telomeric DNA is an outstanding example of quadruplex DNA, which is found in the ends of eukaryotic chromosomes. Telomeric DNA shows short sequences made of Gs and Ts. They are highly distinctive of individual species. Such species-specific sequences are repeated hundred to thousand times in telomeric DNA. Eukaryotic chromosomal DNA is double-stranded, linear and runs from one end of the chromosome to the other end. It is anticipated that if the ends are open structures, they are susceptible for exonuclease digestion, but it is not actually the case. The ends are safe guarded by quadruplex structural organization, and furthermore, they are linked with specific telomeric DNA-binding proteins, which offer additional stability, protection and constancy.

qualitative character: A character in which variation is discontinuous and can be distinguished by visual observation. Individuals can be allocated to one of a small number of discrete classes like the red or white colour of flowers. A fewer number of major genes, with effect which is huge enough to produce distinct classes for each genotype and relatively little affected by the environment, regulate such traits. Data can be placed into a few distinct classes, and traits are not much affected by environment. Breeding for qualitative characters is relatively easier compared to quantitative traits. *Aka* QUALITATIVE TRAIT. *Compare with* QUANTITATIVE CHARACTER.

qualitative inheritance: An inheritance of a trait that contrasts markedly in its expression among individuals of a species. Variation is of discontinuous type. Such characters are generally governed by major (oligo) genes. *Aka* OLIGOGENIC INHERITANCE.

qualitative resistance: Resistance reactions that can be positioned in separate categories, usually governed by one or a few genes. *Contrasts* QUANTITATIVE RESISTANCE. *See* VERTICAL RESISTANCE.

qualitative variable: A variable that produces qualitative data and a trait that is evaluated in terms of characteristics, such as gender and grade of harshness of disease.

qualitative variation: Discrete variation where phenotypes can be divided into a fewer number of classes.

quality assurance: A laboratory programme that guarantees conception of reproducible results that are clinically valuable to patients and providers, in a timely fashion, through minimization of human error.

quality control: Intensively caring the notch of obedience to defined criteria, taking curative action when the system fails and recording all of these events to inform the total superiority of performance.

quality declared seed: A terminology presented by the Food and Agriculture Organization (FAO) USA for a seed system in which a proposed 10% of the seed formed and distributed is checked by a sovereign seed control agency and the rest by the seed-producing organization itself.

quantification of genetic variation: Measurement of amount of genetic variation present in a population. It depends upon determining power of the techniques used, intensifying the exertions made and variety of the strains examined.

quantification of global alterations of histone modifications: This method quantifies global histone PTLMs from colorectal cancer patients and healthy donors.

quantification of RNA modification $6^{m}A$: A computational method, XPORE, has been developed to recognize differential RNA modifications from nanopore direct RNA sequencing (RNA-seq) data. This method was assessed on transcriptome-wide $6^{m}A$ profiling data, demonstrating that xPore recognizes positions of $6^{m}A$ sites at single-base resolution, estimates the fraction of modified RNA species in the cell and enumerates the differential modification rate across conditions.

quantile (Q) plot: The quantile plot enables documentation of any individualities of the shape of the sample distribution, which might be symmetrical or skewed to higher or lower values.

quantile-quantile (QQ) plots: A graphical tool to help us assess if a set of data plausibly came from some theoretical distribution such as a normal or exponential.

quantitative allele-specific expression: Refers to the relative expression of two alleles in a diploid individual.

quantitative character: A character or trait in which difference is continuous so that its classification into separate classes is random. Likewise, eye colour in humans is from light blue to dark brown with all possible intermediate shades. These characters show a complete range of values from one extreme to another without noticeable discontinuity. Quantitative characters are not affected only by the deed of a few major genes. These characters are governed by large number of genes (polygenes) with small and similar effect. Rather environmental changes impact expression of such traits. Its basis lies in statistical models and methodology, albeit based on many strong assumptions. *Aka* QUANTITATIVE TRAIT. *Compare with* QUALITATIVE TRAIT.

quantitative complementation: A procedure for testing the candidacy of a gene at a QUANTITATIVE TRAIT

LOCUS (QTL). It is ignorant to the nature and position of the responsible sequence variant, a feature that has both its forte and feebleness, because whereas one interpretation of a positive result is allelism to the QTL, another is epistasis. The test is not complementation in its classical form because it is testing for an interaction between the null allele and the QTL, rather than for a main effect of either. For this reason, the QTL-KNOCKOUT INTERACTION TEST is preferable.

quantitative data: Collection of observations on features that could be numerically expressed for an individual such as haemoglobin level, blood pressure and blood glucose level.

quantitative genetics: The branch of genetics dealing with the inheritance of quantitative traits that display continuous variation, as opposed to qualitative traits. Since many of the critical targets of crop improvement in both plant and animal breeding are of this category, most practical crop improvement programmes include the application of quantitative genetics. *Aka* BIOMETRICAL GENETICS, STATISTICAL GENETICS *or* GENETICS OF COMPLEX TRAITS.

quantitative inheritance: Study of inheritance of those traits that exist in a range of phenotypes for a specific character. F. Galton in 1989 noted that the characters which showed continuous variation, appeared to be inherited, so that taller individuals for example, produced taller children on average. All quantitative characters are continuous over a range. Continuity of phenotypes in quantitative traits is result of two phenomena: Each genotype does not have a single phenotype but a wide phenotype range. So phenotypic differences between genotypic classes become blurred and thus it is not possible to assign a particular phenotype to a particular genotype and expression of quantitative traits depends on action and interaction of several genes, each having a small additive effect on the phenotype. Such characters are subject to considerable phenotypic modifications by environment. *Aka* POLYGENIC INHERITANCE.

quantitative mass spectrometry (qMS): A method in which bacterial cells are separately cultured in minimal media supplemented with ^{14}N- or ^{15}N-labelled ammonium sulfate as the only source of nitrogen. Using sucrose gradient ultracentrifugation, 70S ribosomes, 50S and 30S ribosomal subunits and pre-ribosomal particles are isolated from the cell lysate, trailed by rRNA extraction and purification. The 16S or 23S rRNA separated from mature ^{15}N-ribosomes is added as an external standard, and the combined mixture is treated with a nucleotide-specific ribonuclease T1, U2 or A before performing the LC-MS analysis.

quantitative models to predict the gene expression level: Significant correlation occurs between histone modification and gene expression levels. The histone modifications forecast gene expression driven by HIGH CG CONTENT PROMOTERS (HCPs) or LOW CG CONTENT PROMOTERS (LCPs). Likewise, the quantitative models relating H3K4me3 and H3K79me1 determine the gene expression levels in LCPs, whereas HCPs need H3K27ac and H4K20me1 to trigger the gene expression. Overall, the quantitative models have determined the association of histone modifications and gene expression to envisage the gene expression levels of a cell type.

quantitative polymerase chain reaction (qPCR): A method for detection and quantification of precise nucleic acid sequences. With qPCR, the amplification of target DNA can be observed in real time, as it is trending. *Aka* REAL-TIME PCR (RT-PCR).

quantitative proteomics for determining changes of histone modifications: The methods for protein quantification can also be engaged for determination of lively changes of histone marks. In these methods, an appropriate labelling method, such as *in vivo* labelling with stable isotope labelling by amino acids in cell culture (SILAC), *in vitro* labelling with TANDEM MASS TAGGING (TMT) or ISOBARIC TAGS FOR COMPARATIVE AND COMPLETE QUANTITATION (iTRAQ). iTRAQ is required for accurate quantification of histone marks.

quantitative resistance: Resistance reactions that display no distinct classes but vary continuously from resistant to susceptible. The consequence of few to many genes, the individual impacts of which may be trivial and difficult to detect. *Contrasts with* QUALITATIVE RESISTANCE. *See* HORIZONTAL RESISTANCE.

quantitative risk assessment: A practice for measuring estimations of the risk of mutation to an agent.

quantitative structure-activity relationship (QSAR): A computer modelling procedure that allows researchers to envisage the prospective activity of a new compound before that compound is actually created. QSAR is based on data from decades of research investigating the impact on 'activity' of the chemical structures of thousands of methodically studied molecules. For example, the biological activity of most antibiotics correlates with their tendency to dimerize.

quantitative trait: A trait for which there is an array of phenotypes differing by degree. Data cannot be placed into a few distinct classes. Most of the actual variations among organisms in a natural population are quantitative. Yield, height, weight, metabolic activity, reproductive rate and behaviour are some examples in crop plants. Various statistical parameters (viz. mean, mode, median, range, variance, standard deviation, coefficient of variation) are used to measure quantitative traits. A large number of genes with small additive effect are commonly implicated in determination of such traits. Expression of these traits is affected by environment. *Aka* POLYGENIC TRAIT.

quantitative trait loci (QTL): Quantitative trait locus is a segment of chromosome marked with molecular markers and has substantial impact on the presentation of a quantitative character. Widely held economically vital traits are inherited quantitatively and assumed to be governed by a large number of genes called POLYGENES whose exact number, mode of action and location are tough to be determined through Mendelian analysis. These are expected to be located on several chromosomes, if not the entire genome. The individual effect (of polygenes) is too small to be detected separately. The identification and location of such polygenes or QTLs, with the help of molecular markers through what is called QTL mapping, has the likelihood of providing analysis of quantitative characters with the comfort and accuracy now available for Mendelian characters.

quantitative trait loci (QTL) detection, factors influencing: The main factors that impact the detection of QTLs segregating in a population include genetic properties of QTLs that govern traits, environmental influences, population size and experimental error.

quantitative trait loci (QTL) mapping: Identification and location of polygenes or QTLs (Quantitative Trait Loci) with the help of molecular markers is called QTL mapping. Marker-facilitated genetic dissection of variation of complex phenotypes through appropriate experimental design and statistical analyses of segregating materials. This is done by demonstrating a statistically significant association between the quantitative trait phenotype and one or more genetic markers already located on a map. The joint analysis of genotype marker segregation and phenotypic values of individuals or lines enables the detection and location of loci affecting quantitative traits. Several recently developed PCR-based markers (e.g. RAPD, AFLP, SSR and STS) are used in QTL mapping. However, many of these markers are either dominant or anonymous, and their functions are not known. With the recent advances in the molecular marker technology now the quantitative traits can be mapped and tagged with molecular markers. It can provide answer to a number of fundamental questions like: what is the number of QTLs for a quantitative character? Are these QTLs the same for each character? Are these randomly distributed in the entire genome or there are some hot spots? QTL mapping is providing great help in MARKER-ASSISTED SELECTION in plant breeding.

quantitative trait loci (QTL) mapping, marker system for: Three important facets are considered while selecting a genetic marker system for QTL mapping the outbred nature of forest tree pedigrees and the probable comparative and candidate gene mapping. (a) Each parent of an outbreed pedigree is characteristically a different, highly heterozygous individual, where the transmission of up to four different alleles must be followed from the parents to progeny. Multiallelic codominant markers are best suited to perceive the maximum number of polymorphisms found in the heterozygous parents. (b) Comparative mapping, both within and among species, is an important tool for linking observations from different mapping experiments. Therefore, a subset of the markers used in a mapping experiment should be orthologous across pedigrees and species. (c) To recognize actual genes controlling a quantitative trait, genes with known or suggested function should be included in QTL mapping. Complete or partial cDNA sequences (ESTs) enable researchers to design ESTP markers that take into account all these aspects and can be used for genetic mapping of the entire genome and for measuring adaptive genetic diversity via QTL mapping analysis.

quantitative trait loci (QTL) mapping, principle of: Detection of QTLs is based upon the association of polymorphic markers with polymorphic genes controlling the trait of interest. Markers are employed to split the mapping population into different genotypic groups based on the presence or absence of a particular marker locus and to find out whether significant differences occur between groups with respect to the trait being studied.

quantitative trait loci (QTL) meta-analysis: A meta-analysis, with respect to QTL analysis, is performed when we have data concerning different populations. It determines whether QTL recognized for a given trait in one population correspond to those perceived in other populations, whether QTL locations identified in one species correspond to QTL of other types of loci spotted in corresponding regions of other species.

quantitative trait loci (QTL), analysis of: One should consider the whole picture of QTLs for single QTLs, multiple QTLs, separating to pyramiding single trait-to-trait complexes, single environments to multiple environments and mapping to cloning using ANOVA. QTL analysis could answer some important questions: how many genes are implicated in genetic control of a quantitative trait in a segregating population? Is it possible to separate closely linked QTLs into single units, and then pyramid favourable QTL alleles spread in genetic materials into a common background? Can we clone QTLs the same way as genes controlling qualitative traits? Mapping sections of the genome that comprise genes involved in specifying a quantitative trait are identified using molecular markers (tags) such as RESTRICTION FRAGMENT LENGTH POLYMORPHISM (RFLP), AMPLIFIED FRAGMENT LENGTH POLYMORPHISM (AFLP), SINGLE-NUCLEOTIDE POLYMORPHISM (SNPs), SIMPLE SEQUENCE LENGTH POLYMORPHISM (SSLP) and SIMPLE SEQUENCE REPEAT (SSR). Three widely used methods for detection of QTLs are SINGLE-MARKER ANALYSIS (SMA) *aka* SINGLE-POINT ANALYSIS (SPA), SIMPLE INTERVAL MAPPING (SIM) and COMPOSITE INTERVAL MAPPING (CIM). *See details as separate terms.*

quantitative trait loci (QTL), epigenetic modification of: Several DIFFERENTIALLY METHYLATED REGIONS (DMRs) have been revealed to behave as bona fide epi-QTL. These DMRs have been shown to explain for high heritability of two complex traits, flowering time and primary root.

quantitative trait loci sequencing (QTL-seq): A method that syndicates bulked segregant analysis (BSA) and high-throughput whole-genome re-sequencing to identify the major locus of a certain quantitative trait in a segregating population.

quantitative trait nucleotide (QTN): The DNA sequence polymorphism responsible for the QTL effect.

quantitative variable: A trait that is measured in terms of numeric, such as blood glucose level, a variable that produces quantitative data. *Aka* CONTINUOUS VARIABLE *or* DISCRETE VARIABLE.

quantitative variation: *See* CONTINUOUS VARIATION.

quantitative vertical resistance: An infrequent type of resistance. It occurs in cereals (e.g. Hessian fly resistance *Mayetiola destructor*) in wheat or rice blast. It is a puzzling term as its inheritance is qualitative while its properties are quantitative. It can easily be jumbled with horizontal resistance, and the best way to avoid it in a breeding programme is by selecting only parents that display the normal, qualitative or vertical resistance.

quantum dot: A 'molecular structure' that ranges between 1 and 100 nm in size, so it is midway between molecular and solid states. Quantum dots are made up of semiconductor materials, crystallites (grown via molecular beam epitaxy) and so on. Quantum dots can credibly be constructed to act as receptors (e.g. on 'biochips') for specific ligands in a way that would signal the presence of disease when a sample was passed over the quantum dot. That signal might be electronic, emission of specific wavelength, light and so on.

quantum evolution: Rapid evolutionary change, often assumed to be involved in the production of new higher categories.

quantum speciation: The budding off of a new and very different daughter species, with the 'founder effect' and random genetic drift, from a semi-isolated peripheral population of the ancestral species in a cross-fertilizing organism. *Aka* SALTATIONAL SPECIATION *or* ABRUPT SPECIATION.

quarantine: (i) The official imprisonment of plants subject to phytosanitary guidelines for visualization and research or for further inspection and/or testing. (ii) More general, a legal prohibition on the export or import of certain poisonous weeds or insects that may be attached to the plants. (iii) The legislative regulation on the transport of plants or plant parts to stop the spread of pests or pathogens. (iv) Keeping a person or living organism in isolation for a period (originally 40 days) after arrival in a new location to let the pre-existing disease symptoms appear, in case there was any disease present. (v) A procure used for restricting the sale or shipment of living organisms, usually to prevent disease or pest invasion of an area.

quarantine block: An isolated block in which all plats imported into the area are first grown to observe any disease or pest that may be carried by the plants in question and to prevent the spread of or disease or pest to plants grown in the locality.

quartet: The four intact nuclei and/or cells produced during meiosis. *See* TETRAD.

quartile: Refers to data divided into four equal parts. The three values – the 25th and 75th percentiles and the median – divide a variables distribution into four intervals containing equal numbers of observations, after the values are organized in an ascending order.

quasi-complete Latin square: It is called a Latin square because every entry occurs once in each row and once in each column. It is called quasi-complete because every entry is a horizontal neighbour to every other entry exactly twice, and every entry is a vertical neighbour to every other entry exactly twice.

quasidominance: Direct transmission, generation to generation, of a recessive trait leaving the impression of dominance. It occurs when the recessive gene is common or inbreeding is strong.

quasi-lethal: In F_2 progeny of Muller-5 method, four phenotypic classes of *Drosophila*, namely Muller-5 females, hetero (also called heterozygous) bar females, Bar-eyed males and normal males are typically present in 1:1:1:1 ratio, statistically. If the above-mentioned four phenotypic classes are statistically realized in ratio 4:4:4:1, lethal are called QUASI-LETHAL.

quasispecies: (i) A set of interrelated but dissimilar entities in nucleic acid clones during early evolution of life on Earth. (ii) A heterogeneous mixture of related genomes, as detected in most RNA viruses that do not contain a single genome species of defined sequence.

quaternary hybrid: A hybrid derived from four grand-parental types, all of which are different.

quaternary protein structure: Refers to how, in larger proteins, comprising of two or more individual polypeptide chains, the individual polypeptide chains are organized relative to each other. The association of multiple polypeptide subunits may be identical or different, held together by non-covalent bonds and frequently associate in a regular (symmetrical) arrangement, to form the final three-dimensional structure of a protein. At this level of protein structure, numerous individual molecules cooperate and assemble together to form a functional cluster. A classic example is haemoglobin, a complex of four myoglobin-like units. When the final protein is made of two, three, four . . . subunits, the protein is said to be a dimer, trimer, tetramer, . . . Quaternary structure in particular is the manner in which the subunits fit together. *See* PRIMARY STRUCTURE, SECONDARY STRUCTURE *and* TERTIARY STRUCTURE.

quelling: A form of POST-TRANSCRIPTIONAL GENE SILENCING (PTGS) or HOMOLOGY-DEPENDENT GENE SILENCING (HGDS) phenomena induced by sense transgenes in the fungus *Neurospora crassa* were termed quelling. Quelling was discovered during efforts to enhance the production of an orange pigment made by the gene *al1* of the fungus *Neurospora crassa*. One *Neurospora crassa* strain containing a wild-type *al1* gene (orange phenotype) was transformed with a plasmid containing a 1500 bp fragment of the coding sequence of the *al1* gene. A few transformants were firmly quelled and revealed albino phenotypes. In the *al1*-quelled strain, the level of unspliced *al1* mRNA was similar to that of the wild-type strain, whereas the native *al1* mRNA was highly reduced, indicating that quelling and not the rate of transcription impacted the level of mature mRNA in a homology-dependent way.

quenching: Refers to the process resulting in the de-excitation of a molecule from an excited singlet or triplet state back down to the ground state. For example, quenching of chlorophyll performs a vital protective role of carotenoids. If a triplet molecule is not rapidly quenched, it can react with oxygen to generate singlet oxygen. Singlet oxygen can result in much cellular damage.

query (sequence): A DNA, RNA or protein sequence employed to explore a sequence database in order to identify close or remote family members (homologues) of known function, or sequences with similar active sites or regions (analogues), from whom the function of the query may be inferred.

quick stop: *dna* mutants of *Escherichia coli* cease replication immediately when the temperature is increased to 42°C.

quiescence: The absence of growth, generally deducing the absence of environmental surroundings favouring growth. Although dormant seed is quiescent (inactive, tranquil), quiescence is distinguished from dormancy, which indicates the incapability to germinate even in the presence of environmental conditions favouring growth.

quiescent: Dormant or inactive; at rest.

quiescent culture: A culture of cells in which active cell growth and replication are minimal but in which cells continue to survive adequately. Confluent cell cultures may behave in this way.

quiescent transposable elements: The transposable elements that are in a state or period of inactivity or dormancy.

quinone (Q): Small, lipid-soluble, mobile electron carrier molecule found in the respiratory and photosynthetic electron transport chains.

quintile: Refers to data split into five equal parts. The four values – the 20th, 40th, 60th and 80th percentiles – divide a variables distribution into five intervals containing equal numbers of observations, after the values have been arranged in an ascending order.

quorum sensing (QS): (1) A mechanism that enables individual bacteria to sense the density of their population. (2) The ability of bacteria to interact with each other through a variety of mechanisms. QS allows a population of bacteria to behave more like a multicellular organism.

quorum sensing (QS) molecule: QS molecule in Gram-negative bacteria. N-Aceyl Homoserine Lactone (AHL) is a QS molecule. QS is a system of stimulus and response correlated to population density. Many species of bacteria use QS to coordinate gene expression according to the density of their local population. Ax21 (an elicitor) is a secreted peptide and a QS molecule, the production of which is regulated in a cell density-dependent manner. High level of conservation of *Ax21* in *Xanthomonas* suggests a crucial role of this protein in the biology of these pathogens. *Xanthomonas oryzae* pv. *oryzae* (*Xoo*) may employ QS to coordinate infection and production of virulence factors. Xa21 PATTERN RECOGNITION RECEPTOR (PRR) successfully eavesdrops (unauthorizedly intercepts) on the Ax21.

r: (i) The instantaneous rate of increase of a population or genotype. It is used as a measure of fitness. (Note. The Greek symbol for lambda is often used to define the finite rate of increase = e^r). (ii) In statistics, correlation coefficient. (iii) The portion of R plasmid carrying the gene for resistance to all drugs, with the probable exception of tetracycline.

R factors: Plasmids that carry genes that control resistance to various drugs. *Aka* R PLASMIDS.

R genes: Refers to genes within some plants that deliberate resistance to certain plant diseases through common signalling pathways involved in the investigation and stimulation of natural plant defence responses (e.g. SYSTEMIC ACQUIRED RESISTANCE, SAR). For example, the gene that codes for or causes the production of harpin protein is only present in a few bacteria (e.g. *Erwinia amylovora*), but R genes that respond to the presence of harpin are present within the genomes of plentiful species of plants. Thus, the spraying of man-made harpin protein on to any of those abundant species of (crop) plants results in those particular plants recruiting a protective/defensive response (cascade) against pathogenic bacteria, viruses, fungi and even some insects.

R line: The pollen parent line containing fertility-restoring genes, crossed with the A line in the production of hybrid seed.

R line (restorer): The pollen parents used in hybrid seed production. They restore the fertility of the A line (i.e. the male sterile female line); the B line or the maintainer, on the other hand, maintains the sterility of the A line on crossing (i.e. $A \times B \rightarrow A$), since B is isogenic to A.

R-loop profiling technique: Programmed constitutive heterochromatin silencing is essential for eukaryotic genome regulation. A large proportion of R-loops (RNA:DNA hybrids) were unexpectedly identified within *Arabidopsis* pericentromeric heterochromatin. Through a genome-wide R-loop profiling screen, J. Zhou *et al.* (2023) find that DDM1 (decrease in DNA methylation 1) is the primary restrictor of pericentromeric R-loops via its RNA:DNA helicase activity. Low levels of pericentromeric R-loops resolved by DDM1 cotranscriptionally can facilitate constitutive heterochromatin silencing. Furthermore, they demonstrate that DDM1 physically excludes histone H2A variant H2A.Z and promotes H2A.W deposition for faithful heterochromatin initiation soon after R-loop clearance. The dual functions of DDM1 in R-loop resolution and H2A.Z eviction are essential for sperm nuclei structure maintenance in mature pollen.

R-loop–dependent DNA looping: The nucleotide excision repair factor XPF interacts with the insulator binding protein CTCF and the cohesin subunits SMC1A and SMC3, leading to R-loop-dependent DNA looping upon transcription activation. To facilitate R-loop processing, XPF interacts and recruits with TOP2B on active gene promoters, leading to DSB accumulation and the activation of a DNA damage response (G. Chatzinikolaou *et al.*, 2023). Abrogation of TOP2B leads to the diminished recruitment of XPF, CTCF and the cohesin subunits to promoters of actively transcribed genes and R-loops and the concurrent impairment of CTCF-mediated DNA looping.

R-looping: The technique in which an RNA molecule is annealed to the complementary strand of a partially denatured DNA molecule. The DNA hybrid replaces the opposite DNA strand as a single-stranded bubble. The R-loop can be visualized under the electron microscope using the Klein-Schmidt technique. It was the R-looping technique that first revealed the presence of introns in eukaryotic genes.

r-oligo-capping: Cap structure of an mRNA is replaced with an oligoribonucleotide (r-oligo) to label the 5′-end of eukaryotic mRNAs. The method consists of removing the cap with tobacco acid pyrophosphatase (TAP) and ligating r-oligos to decapped mRNAs with T4 RNA ligase.

r selection: Selection acting on the measure of fitness r, contrasted with selection acting on the density-dependent measure, K. This concept is not used anymore in evolutionary biology. *Compare with* K-SELECTION.

R plasmids: Most widespread and well-studied group of bacterial plasmids conferring resistance to one or more transposons that bear various resistance genes that code for enzymes that are able to destroy and modify antibiotics. They are not usually integrated into the host chromosome. Some R plasmids possess only a single resistant gene, whereas others can have as many as eight. Many R plasmids are conjugative and possess drug-resistant genes as transposable elements. It comprises two components primarily – a RESISTANCE TRANSFER FACTOR (RTF) and resistance determinants. Plasmid R 100, for example, is a 94.3-kilobase-pair plasmid that carries resistant genes for sulfonamides streptomycin and spectinomycin, chloramphenicol, tetracycline and so on. It also carries genes conferring resistance to mercury. Many R plasmids are conjugative and possess drug-resistant genes as transposable elements, they play an important role in medical microbiology as their spread through natural populations can have profound consequences in the treatment of bacterial infections. *Aka* R FACTOR *and* RESISTANCE PLASMIDS.

R′ plasmid: A plasmid carrying selectable bacterial genes.

R programming language: A programming language and free software environment for statistical computing and graphics supported by the R Foundation for Statistical Computing. The R language is widely used among statisticians and data miners for developing statistical software and data analysis.

r selection: Selection occurring in an unpredictable environment where physically density-dependent factors are important. Traits for selection include small size, early maturation, single reproduction, large number of offspring and little parental care.

r^+m^-: Such bacterial strains commit suicide and they would commit suicide by restricting their own unmodified DNA.

R_0: The net reproductive rate or expected lifetime fecundity. It takes survival and fecundity into account. It is a measure of fitness when the population is stationary (i.e. not changing in size) or for species with non-overlapping generations.

R1, R2, R3, etc.: The first, second, third, etc. generations following any form of irradiation in mutation breeding.

Rab protein: Any of a large family of monomeric GTPases occurring in the plasma membrane and organelle membranes that are implicated in imparting specificity on vesicle docking.

Rabl configuration: In plants and insects, the term refers to chromosomes that are spatially arranged within the nucleus, with centromeres grouped on the nuclear membrane at one pole and telomeres attached to the nuclear membrane of the other hemisphere.

Rac prophage: *Escherichia coli* and other bacteria contain, in their chromosomes, remnants of viruses or prophages, such as Rac in *Escherichia coli*, that often are defective and contain only a few genes of the original virus. Two *Rac* genes, *recE* and *recT*, encode homologous recombination functions and are normally silent, but the *sbcA* mutation activates their constitutive expression.

race: (i) A genetically and, as a general rule, geographically discrete interbreeding division of a species. (ii) A population or group of populations divergent from other such populations of the same species by the frequencies of genes, chromosomal rearrangements or hereditary phenotypic characteristics. (iii) A race that has received a taxonomic name is a subspecies. For instance, populations of *Drosophila pseudoobscura* that inhabit different territories and differ in relative frequencies of the chromosomal types may be described as different races. (iv) A subgroup or biotype within a species or variety, distinguished from other races by virulence, symptom expression or host range, but not by morphology. (v) A group of pathogens of the same species, which are alike in physiological and pathogenic characteristics.

race-non-specific resistance: A resistance that is effective against all biotypes of the pathogen. *See* DURABLE RESISTANCE, GENERAL RESISTANCE *and* HORIZONTAL RESISTANCE. *Contrasts* SPECIFIC RESISTANCE *and* VERTICAL RESISTANCE.

race-non-specific type: Host-plant resistance that is operative against all races of a pathogen species. It is variable, sensitive to environmental changes and generally polygenically governed.

race-specific resistance (syn. **qualitative resistance**)**:** Resistance individually governed by single or very few genes in the plant – the specific resistance genes – and these are most usually operative against a limited proportion of the pathogen population. The proportion of the pathogen population proficient in causing disease on a resistant host is termed ‘virulent’ on the host. The portion incompetent of causing disease is termed 'avirulent' on that host. Correspondingly, the host is 'susceptible' or 'resistant', and the interaction is 'compatible' or 'incompatible', respectively. A specific avirulence gene in the pathogen matches a specific resistance gene in the plant. This relationship is termed the GENE-FOR-GENE HYPOTHESIS. *See* VERTICAL RESISTANCE.

race-specific type: Host-plant resistance that is effective against one or a few races of a pathogen species. Generally, it yields an immune or hypersensitive reaction and is governed by one or a few genes.

racemate: An equimolar (i.e. equal number of molecules) mixture of the D and L stereoisomers of an optically active compound. A solution of DEXTROROTARY (D) isomer (enantiomer) will rotate the plane in which the light was polarized a specific number of degrees to the right (dextro) while a solution containing the same number of LAEVOROTATORY (L) isomer molecules will rotate the plane in which the light was polarized the same number of degrees to the left (levo). The difference between D and L enantiomers is that the rotations of the plane of plane-polarized light are equal in magnitude but opposite in sign. Hence, a 50:50 mixture of both enantiomers (known as a RACEMIC MIXTURE) shows no optical activity. That is, a solution containing a 50:50 mixture of enantiomers will not rotate the plane of plane-polarized light when it is passed through the solution.

raceme (Adj. racemose): An indeterminate inflorescence in which the main axis produces a series of flowers on lateral stalks, the oldest at the base and the youngest at the top. *Compare with* SPIKE. *See* RACEMIFORM *and* RACEMOID.

rachis: The elongated main axis of an inflorescence.

radial-loop scaffold model: Certain non-histone proteins, including topoisomerase II, bind to chromatin every 60–100 kb and chain the supercoiled, nucleosome studded 300Å fibre into structural loops. Other non-histone proteins may, in turn, fold the loops into daisy-like rosettes. Some additional non-histone proteins may then wrap the rosette centres into a compressed bundle.

radiation absorbed dose (Abbrev. rad): A measure of the quantity of any ionizing radiation that is captivated by the tissue. One rad is equivalent to 100 ergs of energy riveted per gram tissue.

radiation genetics: A branch of genetics that deals with the effects of various types of radiation on chromosomes and genes.

radiation hybrid (RH) mapping: A hybrid cell encompassing small segments of irradiated human chromosomes. Maps of irradiation sites on chromosomes for the human, rat, mouse and other genomes offer important markers, enabling the construction of very precise SEQUENCE-TAGGED SITE (STS) maps crucial to studying multifactorial diseases. *See* SEQUENCE-TAGGED SITE.

radical amino acid substitutions: *See* DRASTIC AMINO ACID MUTATIONS.

radicle: The part of the plant embryo that grows into the primary root.

radioactive: Relating to a substance when a component chemical element is changing into another element through the release of radiant energy. Radioactivity is employed as a tool in research to tag or trace the movement of compounds. The presence of a compound containing a radioactive element is discovered by instruments that quantify the radiant energy produced or by radio-sensitive films.

radioactive isotope: An isotope with an unstable (atomic) nucleus that spontaneously discharges radiation as it decays. The radiation released includes alpha particles, nucleons, electrons and gamma rays.

radioactive marker: A radioactive atom that is incorporated into a chemical compound and whose radioactive emissions can be used to detect and follow the compound during a biochemical reaction.

radioactivity: The emission of energy due to changes in the nucleus of an atom. Such spontaneously released radiation is a characteristic of certain elements and at some levels can be harmful.

radiocarbon dating: A dating method for organic material that is relevant to about the last 70,000 years. It relies on the presumed constancy over time of atmospheric ^{14}C:^{12}C ratios and the known rate of decay of radioactive carbon, of which half vanishes in a period of about every 5730 years. *Aka* ^{14}C DATING.

radioimmunoassay (RIA): An assay system that uses the specificity of the antibody-antigen reaction of an immunoassay method with the sensitivity of detection by radioactive labelling. The radioactive atom compound is often ^{125}I. and it is attached to the antibody by the enzyme horseradish peroxidase. Any antigen-antibody complexes are detected by autoradiography.

radioisotope: In cytology, radioactive labelling of nucleic acids is generally accomplished by enzymatically incorporating nucleotides containing ^{32}P, ^{35}S, ^{125}I or ^{3}H; the labelled nucleotide is in essence identical to its unlabelled counterpart. The labelled probe is not susceptible to steric hindrance during hybridization and this may be the reason why radioactive probes can be more sensitive than non-radioactive probes. The particular isotope that is selected will depend upon the application, since there is an inverse relationship between sensitivity and resolution, such as the high emission energy (E_{max}) of ^{32}P makes it suitable for detecting sequences in less than 7 days but gives a wide scattering of silver grains in photographic emulsions during autoradiography and low resolution of the signal. In contrast, the weak beta emission of ^{3}H causes a high-resolution signal, but it can take at least 2 weeks to expose the photographic emulsion. Advantages of radioactive detection are higher sensitivity; non-radioactive methods necessitate elaborate detection methods (e.g. the supernatant and pellet cannot just be tested with a Geiger counter to see where the sample went; radioactivity permits highly accurate quantitation over numerous orders of magnitude).

radiolabelled probe: Probe labelled with radioisotope for detection through autoradiography. Nucleic acid (DNA, RNA) probes are labelled with ^{32}P, ^{35}S, ^{3}H or ^{125}I for detection of the matching sequences. Among these ^{32}P is the most commonly used. A radiolabelled probe (single-stranded) is allowed to hybridize with genomic/cDNA to locate the desired (matching) sequences in that organism.

radiolabelling: The technique by which a radioactive atom is attached to a chemical molecule or macromolecule (e.g. incorporation pf ^{32}P-dATP into DNA or ^{35}S into protein).

radiometric dating: A dating technique that uses the decay rate of radioactive isotopes to estimate the age of an object. *See* RADIOMETRIC DATING.

radiometry: A set of techniques for measuring electromagnetic radiation, including visible light. Radiometric techniques in optics characterize the distribution of the radiation's power in space, as opposed to photometric techniques, which characterize the light's interaction with the human eye. The fundamental difference between radiometry and photometry is that radiometry gives the entire optical radiation spectrum, while photometry is limited to the visible spectrum. Radiometry is distinct from quantum techniques such as photon counting. Landsat-simulating radiometer has been developed for agricultural remote sensing. Significant correlation was found in maize fields, little correlation occurred for small grain fields and sun angle had a minimal effect.

radiomimetic chemicals: The chemicals that mimic the effects of ionizing radiations on the genetic material.

radix: *Aka* BASE in digital numeral systems. The number of unique digits, including the digit zero, used to represent numbers in a positional numeral system. For example, in the decimal/denary system (the most common system in use today), the radix (base number) is 10 because it uses the ten digits from 0 through 9, and all other numbers are uniquely specified by positional combinations of these ten base digits; in the binary system that is the standard in computing, the radix is two, because it uses only two digits, 0 and 1, to uniquely specify each number.

rainforest: A moist temperate or tropical forest dominated by broad-leaved trees that form a continuous canopy.

Ramachandran plots: Proteins have definite conformation, which is determined by amino acid sequence.

ramet: An individual member of a clone.

Ran: Monomeric GTPase present in both cytosol and nucleus that is needed for the active transport of macromolecules into and out of the nucleus through nuclear pore complexes. Hydrolysis of GTP to GDP is considered to deliver the energy necessitated for this transport.

random: A word with many meanings. (i) In evolution mutations are random with respect to the needs of the organism in which they occur. (ii) They are not distributed at random along the DNA sequence (some parts of genomes are more mutable than others) and they are not always random with respect to environmental conditions (in bacteria and fungi environmental stimuli can increase the mutation rate). (iii) It is the lack of pattern or predictability in events. A random sequence of events, symbols or steps has no order and does not follow an intelligible pattern or combination. Individual random events are unpredictable. Each unit has the same chance of being picked up.

random amplified hybridized microsatellites (RAHMs): A new approach established to combine the several benefits of oligonucleotide fingerprinting with RAPD-PCR and MICROSATELLITE PRIMED-PCR (MP-PCR). In this approach, the genomic DNA is amplified with either a single arbitrary 10-mer primer as in RAPD or with a microsatellite complementary 15-mer/10-mer primer as in MP-PCR, and the PCR products are electrophoresed, Southern blotted and hybridized to ^{32}P or digoxigenin labelled microsatellite probe, such as $(CA)_8/(GA)_8$ $(GAC)_5$. Subsequent autoradiography discloses reproducible, probe-dependent fingerprints, polymorphic at the interspecific level. This offers for speed of the assay along with high sensitivity so that a high level of polymorphism is sensed.

random amplified microsatellite polymorphism (RAMP): PCR-based marker using a pair of primers for studying genetic polymorphism. In this approach, DNA amplification is achieved using a pair of primers of which one primer is an anchored microsatellite (SSR) and the other primer is an RAPD primer. RAMP amplified products determine length polymorphism that may occur either at the SSR target site itself or at the associated sequence between the binding sites of the two primers. RAMP is beneficial because undigested total genomic DNA is digested DNA that is used in AFLP. The RAPD primer-binding site actually obliges as an arbitrary end point for SSR-based specific amplified sequence.

random amplified polymorphic DNA (RAPD): RAPD is pronounced 'rapid'. It is a type of PCR reaction, but the segments of DNA that are amplified are random. The scientist performing RAPD generates several arbitrary, short primers (8–12 nucleotides), then proceeds with the PCR using a large template of genomic DNA, hoping that fragments will amplify. By resolving the resulting patterns, a semi-unique profile can be assembled from an RAPD reaction. RAPD markers are decamer (10-nucleotide length) DNA fragments from PCR amplification of random segments of genomic DNA with a single primer of arbitrary nucleotide sequence. No knowledge of the DNA sequence for the targeted gene is necessary, as the primers will bind somewhere in the sequence, but it is not sure exactly where. Unlike traditional PCR analysis, RAPD does not require any specific knowledge of the DNA sequence of the target organism: the identical 10-mer primers will or will not amplify a segment of DNA, depending on positions that are complementary to the primers' sequence. For example, no fragment is produced if primers are annealed too far apart or 3′ ends of the primers are not facing each other. Therefore, if a mutation has occurred in the template DNA at the site that was previously complementary to the primer, a PCR product will not be produced, resulting in a different pattern of amplified DNA segments on the gel. RAPD is an inexpensive yet powerful typing method for many bacterial species. Selecting the right sequence for the primer is very important because different sequences will produce different band patterns and possibly allow for more specific recognition of individual strains.

random amplified polymorphic DNA (RAPD) and inter simple sequence repeat (ISSR) markers, applications of: (i) Used to construct linkage maps in several species. (ii) Have high polymorphism and are simple. (iii) Non-reproducibility limits their application in mapping.

random amplified polymorphic DNA (RAPD) technique: A comparative investigation (among individuals, populations or species) of the DNA fragment length produced in controlled DNA synthesis reactions started with short sequences of DNA (primers). As a genetic mapping methodology, it exploits as its basis the fact that specific DNA sequences (polymorphic DNA) are repeated (i.e. appear in sequence) with a gene of interest. Thus, the polymorphic DNA sequences are linked to that specific gene. Their linked occurrence enables genetic mapping within a genome.

random amplified polymorphic DNA (RAPD), limitations of: As RAPD relies on a large, intact DNA template sequence, it has some limitations in the use of degraded DNA samples: (i) Its resolving power is considerably lesser than targeted, species-specific DNA comparison methods, such as short tandem repeats. (ii) Nearly all RAPD markers are dominant. That means it is not possible to distinguish whether a DNA segment is amplified from a locus that is heterozygous (1 copy) or homozygous (2 copies). Codominant RAPD markers, observed as different-sized DNA segments amplified from the same locus, are detected only rarely. (iii) PCR is an enzymatic reaction, therefore the quality and concentration of template DNA, concentrations of PCR components and the PCR cycling conditions may greatly influence the outcome. Thus, the RAPD technique is notoriously laboratory dependent and needs carefully developed laboratory protocols to be reproducible. (iv) Mismatches between the primer and the template may result in the total absence of PCR product as well as in a merely decreased amount of the product. Thus, the RAPD results can be difficult to interpret.

random amplified polymorphic DNA (RAPD), uses of: (i) RAPD analysis appears to offer a cost- and time-effective alternative to RESTRICTION FRAGMENT LENGTH POLYMORPHISM (RFLP) analysis.

(ii) RAPD is a favourite for comparing the DNA of biological systems that have not attracted the attention of the scientific community or in a system in which relatively few DNA sequences are compared (it is not appropriate for forming a DNA databank). (iii) RAPD has been used to characterize, and trace, the phylogeny of miscellaneous plant and animal species. (iv) RAPD can differentiate between genetically distinct individuals, although not necessarily in a reproducible way. It is used to analyse the genetic diversity of an individual by using random primers. (v) Molecular characterization of different strains of bacterial species from different sources has been done by using RAPD analysis.

random chimeragenesis on transient templates (RACHITT): RATCHITT is similar to DNA shuffling. It increases recombination frequencies through the use of small oligonucleotide linkers between low-homology regions.

random drift: The random change in genotype frequency caused by random variation in individual reproduction. Looking back in time, this process causes the coalescence of lineages. *Synonym of* GENETIC DRIFT.

random effects: Effects of a treatment in an experiment in which the treatments are given in relation to a whole range of possible treatment effects.

random epigenome editing: For this purpose, three methods have thus far been employed: (i) antimetabolite inhibitors, (ii) tissue culture-based approach and (iii) overexpression of effector molecules.

random error: An error triggered by sampling from a group rather than knowing the true value of a quantity such as the mean blood pressure for the entire group (e.g. healthy men over age 80). One can also state random errors in single measurements for individual subjects (e.g. the error in using a single blood pressure measurement to signify a subject's long-term blood pressure). These errors are small, and some are positive and some negative so that the long-term average is close to zero.

random gene amplification: A type of amplification that can occur randomly in the cell (e.g. in some types of cancer cells gene amplification is random).

random genetic drift: Variation in gene frequency from one generation to another owing to chance fluctuations.

random mating: For a given population, where an individual of one sex has an equal chance of mating with any individual of the opposite sex or insofar as the genotypes with respect to given genes are concerned. *Compare with* NON-RANDOM MATING.

random point mutagenesis: Genetic diversity introduced by random point mutagenesis using error-prone PCR (EP-PCR). This method is generally applicable to both single genes as well as to multiple contiguous genes constituting a metabolic pathway.

random polycross: A polycross scheme in which the pollination is arbitrary. This is possible with allogamous species, and with an autogamous species, which responds to a male gametocide or which has an easily controlled genetic male sterility. The advantage is that it makes very large numbers of crosses with very little labour. The disadvantage is that there is no control over the pollination and some parents may be more extensively represented than others.

random priming: A method of DNA labelling, introduced by A.P. Feinberg and B. Vogelstein in 1983, which relies on the ability of the KLENOW FRAGMENT of DNA polymerase to copy single-stranded DNA templates primed with random hexanucleotide mixtures.

random sample: A finite series of observations or individuals taken at random from the hypothetical infinitely large population of potential observations or individuals. A probability sample is a kind of random sample in which each possible subject has a known probability of being sampled, but the probabilities can vary. For example, one may wish to oversample African-Americans in a study to ensure good representation. In that case, one could sample African-Americans with the probability of 1.0 and others with a probability of 0.5.

random sampling: A sample is drawn from a population in such a way that every individual in the population has an equal chance of appearing in the sample. It ensures that the sample is representative and provides the necessary basis for virtually all forms of inference from sample to population, including informal inference, which is typical of re-randomization statistics. The popular methods of random sampling are simple, systematic, stratified, cluster and multistage.

random segregation: During meiosis, the two chromosomes of a pair are distributed randomly to the gametes, each gamete having an equal chance of receiving either chromosome.

random variable: A random variable is a variable in which the value is unknown. It can be discrete or continuous with any value given in a range.

random walks: A sequence of random changes; the total change is the sum of all the random steps.

randomization: The process of making assignments at random. In field trials, randomization of entries is needed to attain a valid (minimum) estimate of experimental error. Each entry must have the same chance of being assigned to any plot in a replication and an independent randomization is necessary for each replication.

randomized algorithm: In the randomized algorithm we use a random number so it gives immediate benefit. The random number helps in deciding the expected outcome.

randomized block: The entries to be tested are allocated at random to the plots within the block.

randomized block design: An experimental design that governs fertility variation in one direction only. A randomized block analysis of variance design (e.g. ONE-WAY BLOCKED ANALYSIS OF VARIANCE (ANOVA)) is fashioned by first grouping the experimental subjects into blocks; the subjects in each block are as similar as possible (e.g. littermates). There are as many subjects in each block as there are levels of the factor of interest. Randomly assigning a different level of the factor to each member of the block, such that each level exists once and only once per block; the blocks are presumed not to interact with the factor.

randomized complete block (RCB) design: Widely used in field experimentation, RCB design is an extension of the paired t-test. This design is suitable when quantitative data are collected (e.g. yield), and rigorous comparison between treatments is necessary. Two foundations of the design are replication (i.e. repetition) and randomization. RCB permits accommodation of any variability in the local environment and determination of the probability of the differences in results between treatments being real or simply due to chance. Three replications are the minimum but more are better. However, the statistical advantage gained is successively smaller with each added replicate. Each treatment must be included once in each block of replications. The treatment locations must be randomly assigned to plots within the block. The purpose of randomizing the locations is to avoid biasing the results. Two standard statistical tests are utilized together in analysis of data from this design: (i) an F-test, commonly called ANOVA, is employed to determine if there are significant differences between some of the treatments and (ii) if the F-test reveals that significant treatment differences do exist, then a 'means comparison test', such as the Dunn's test or Duncan's multiple range test, is engaged to determine which 'treatment means' (i.e. treatment averages) are really different from one another. There are two advantages to using this design: one is the capability to compare abundant treatments using one analysis (F-test), and the second is the capability to separate out differences between replicates owing to environmental gradients.

randomized controlled trial: *See* CLINICAL TRIAL.

randomized mutant library: A randomized mutant library is created by *de novo* synthesis of a gene. During synthesis, alternative nucleotides or codons are incorporated into the DNA sequence at specific positions. The expressed proteins can then be screened for variants, which exhibit favourable properties.

randomly amplified polymorphic DNA (RAPD): DNA-based marker with primer sequence ~ 10 nucleotides for fingerprinting. RAPDs were developed by J.G.K. Williams *et al.* (1990). It is a PCR-based technique, and it gives a large number of bands due to a smaller size of primers (~ 10 nucleotides). The polymorphism is resolved by the presence or absence of a particular amplified fragment. Now large number of designed random primers are commercially available for example 'operon primers' to carry out RAPD analysis in different plant species. Lack of reproducibility is the major problem in the use of RAPD primers. These markers have been used for tagging genes of agroeconomic importance and for constructing linkage maps in many plant species.

randomly reassociated DNA: The portion of DNA that anneals or renatures quickly after denaturation. The annealing comprises the retrieval of double strandedness by specific hydrogen bonding, and its rate in solution is a function partially of the frequency of a particular sequence in the mixture and partially of the frequency of the nature of the sequences involved. The DNA that reassociates most rapidly is the palindromic sequence material, which can hybridize with itself by the doubling over of a single-stranded molecule. For this reason, this DNA is called SNAP-BACK OR FOLD-BACK DNA. A fraction that reassociates somewhat more slowly than fold-back DNA, but still much faster than bulk eukaryotic DNA, highly repetitive and chiefly composed of centromeric DNA. DNA of intermediate speed of reassociation is termed MODERATELY REPETITIVE DNA and embraces material such as ribosomal and transfer RNA genes, as well as genes coding for histones, which are present in multiple copies. *See* ANNEALING OF DNA.

randomness: Absence of a systematic pattern. One might wish to study whether some hormone level varies methodically over the day as opposed to having a random pattern or whether events such as epileptic seizures are inclined to bunch or occur randomly in time. Sometimes the residuals in an ordinary regression model are plotted against the order in which subjects are amassed to make sure that the pattern is random. For example, there is no learning trend for the investigators.

range: The range is the difference between the lowest and highest values in a collection of data.

range of reaction: The range of all possible phenotypes that may advance by interaction with various environments from a given genotype. *Aka* NORM OF REACTION.

rapid amplification of cDNA ends (RACE): A technique that allows amplification of full-length cDNA from mRNA by extending to the 3′ or 5′ end, even without prior knowledge of the sequence.

rapid amplification of plasmid and phage DNA: This has been achieved by using Phi29 DNA polymerase and multiply primed rolling circle amplification. A simple method of using rolling circle amplification has been employed to amplify vector DNA such as M13 or plasmid DNA from single colonies or plaques. Using random primers and φ29 DNA polymerase, circular DNA templates can be amplified 10,000-fold in a few hours. This procedure removes the need for lengthy growth periods and traditional DNA isolation methods. Reaction products can be used directly for DNA sequencing after phosphatase treatment to inactivate unincorporated nucleotides. Amplified products can also be used for *in vitro* cloning, library construction and other molecular biology applications.

rapid DNA library construction: A rapid protocol for constructing plasmid libraries from small quantities of genomic/metagenomic DNA is a modification of the linker-amplified shotgun library (LASL) approach. Here linker amplification is combined with topoisomerase cloning. The technique is particularly well suited for functional screening, as it rapidly generates expression libraries with gene-sized inserts.

rapid expression of transgenes: Metabolic engineering of seed biosynthetic pathways has been done to diversify and improve crop product quality is a highly active research area. The validation of genes driven by seed-specific promoters is time consuming since the transformed plants must be grown to maturity before the gene function can be analysed. The genes driven by seed-specific promoters contained within complex constructs could be transiently expressed in the *Nicotiana benthamiana* leaf assay system by co-infiltrating the *Arabidopsis thaliana LEAFY COTYLEDON2* (*LEC2*) gene. Scientists first assembled an efficient transgenic docosahexaenoic acid (DHA) synthesis pathway using a traditional *Nicotiana benthamiana* CAULIFLOWER MOSAIC VIRUS (CaMV) 35S-driven leaf assay before using the LEC2-extended assay to rapidly validate a complex seed-specific construct containing the same genes before stable transformation in *Arabidopsis*. The LEC2-extended *Nicotiana benthamiana* assay allows the transient activation of seed-specific promoters in leaf tissue. The assay was used as a rapid preliminary screen of a complex seed-specific transgenic construct prior to stable transformation, a feature that will become increasingly useful as genetic engineering moves from the manipulation of single genes to the engineering of complex pathways.

rapid lysis (*r*) mutants: These mutants display a change in the pattern of lysis of *Escherichia coli* at the end of an infection by a T-even phage.

rapid sequencing: Comparative whole-genome analysis provides clues about molecular evolution. While a conserved sequence provides information about key motifs, the sequence differences provide information about the diversity of form and function. Bioinformatics provides facilities for storing and automation of the analysis of the sequences.

rare allele: At a given locus, the nucleotide seen least frequently in a given population. An allele is considered rare when it has a frequency of less than 1% in a given population.

rare base pairing: Pairing between two purines or pyrimidines in DNA.

rare codon recoding: Genetic encoding of non-canonical amino acids (ncAAs) is a vital technology at the boundary of chemistry and biology that augments our capability to manufacture proteins with characteristics that vary from natural proteins. The use of blank codons in earlier approaches has restricted the popularization and application of this technology. W. Ding *et al.* (2024) proposed a rare codon recoding approach that, in principle, unravels the bottleneck of ncAA integration and various functional proteins are synthesized with ncAAs in mammalian cells with the translation competence of natural proteins for a wide range of basic and applied advances for downstream applications.

rare single-nucleotide polymorphism (rare SNP): Any SINGLE-NUCLEOTIDE POLYMORPHISM (SNP) that transpires at an incidence of <1% in a population.

rare-cutter enzyme: (i) Restriction enzymes that recognize eight nucleotide or longer segments and generate DNA fragments ranging in size from 50 kb to over 9 mb (e.g. *Not*I, *Rsr*II, *Sfi*I and *Sty1*). Rare cutters are invaluable for long-range physical mapping of genomic DNA. (ii) A restriction enzyme that recognizes sites that are infrequently represented. *See* RESTRICTION-ENZYME CUTTING SITE.

Ras Mol: Programme that highlights key aspects of protein structure. It highlights the protein domains, PROSITE patterns and protein-ligand interactions, for interactive viewing in 3-D. Numerous links take the user to related sites. PDB structures can be displayed and manipulated in 3-D using the Ras Mol/VRML browser.

Ras proteins as switches: Ras proteins are the most celebrated members of a large family of GTP-binding proteins, called MONOMERIC GTPASES that assist in relaying signals from cell-surface receptors to the nucleus. Coded by the *ras* genes, Ras proteins were first identified in viruses that cause rat sarcomas. Ras proteins participate as a 'molecular switches' in the early steps of the signal transduction pathway that is associated with cell growth and differentiation. In its GTP complexed form, it is active, and in its GDP complex, it is inactive. 'On' and 'off' states of the switch are distinguished by conformational differences that span a length of more than 40 Å and are induced by γ-phosphate. The conformational differences also provide a structural basis for understanding biological and biochemical changes of the proteins due to oncogenic mutations, autophosphorylation and GTP hydrolysis. Cellular signalling pathways depend on the proper activation and inactivation of mediators. New structural information reveals an unusual mechanism by which one such mediator, Rap1, is switched off. GTPASE-ACTIVATING PROTEINS (GAPs) accelerate this intrinsically slow process. Transition to the active state requires the exchange of GDP for GTP, a process catalysed by GUANINE-NUCLEOTIDE-EXCHANGE FACTORS (GEFs).

rate: A ratio, such as a change per unit time. Rates are frequently the restrictions, which should not be confused with probabilities. The latter are controlled to be between 0 and 1, whereas there are no constraints on possible values for rates other than not being negative. A rate may also be a ratio such as falls per distance walked or bacteria per unit of surface area.

rate genes: Those genes that determine the rate of certain developmental processes.

rate-limiting step of a metabolic pathway: The relative importance of an enzyme in exerting control on a metabolic pathway is quantified by a control coefficient. The sum of all the control coefficients of the enzymes in a pathway is 1. The distribution of control can be highly variable, and that control can be redistributed according to environmental or developmental influences. The engineering of enzymes that catalyse reactions thought to be close to equilibrium often gives rise to large effects.

rational drug design: The development of drugs based on the three-dimensional molecular structure of a particular target. *Aka* STRUCTURE-BASED DRUG DESIGN.

rational protein design: The scientist uses detailed knowledge of the structure and function of the protein to make desired changes. This generally has the advantage of being inexpensive and technically easy, since site-directed mutagenesis techniques are well developed. However, its major drawback is that detailed structural knowledge of a protein is often unavailable, and even when it is available, it can be extremely difficult to predict the effects of various mutations. Computational protein design algorithms seek to identify novel amino acid sequences that are low in energy when folded to the pre-specified target structure. While the sequence-conformation space that needs to be searched is large, the most challenging requirement for computational protein design is a fast, yet accurate, energy function that can distinguish optimal sequences from similar suboptimal ones.

ratooning: Development of sprout or shoot from the root of a plant (e.g. in sugarcane) after it has been cropped.

***RAV6* epigenetic mutant in rice:** Heritable epigenetic variants of genes are termed epialleles. Epialleles may offer a new source of valuable traits for crop breeding. X. Zhang *et al.* (2015) identified *Epi-rav6*, a gain-of-function epiallele of rice (*Oryza sativa*) *RELATED TO ABSCISIC ACID INSENSITIVE3 (ABI3)/VIVIPAROUS1 (VP1) 6 (RAV6)*, which encodes a B3 DNA-binding domain-containing protein. The *Epi-rav6* plants display larger lamina inclination and smaller grain size. These agronomically relevant phenotypes are inherited in a semi-dominant manner. They did not find nucleotide sequence variation of *RAV6*. Instead, they found hypomethylation in the promoter region of *RAV6*, which resulted in ectopic expression of *RAV6* in *Epi-rav6* plants. Cytosine methylation of four CG and two CHG loci within a continuous 96-bp-region stages an important role in regulating *RAV6* expression. Treatment with brassinosteroid (BR) inhibitor reinstated the leaf angle defects of *Epi-rav6* plants. Thus, this work recognizes an exclusive rice epiallele, which may epitomize a common phenomenon in complex crop genomes. A semi-dominant mutant displays large lamina inclination and small seed size. A spontaneously occurring rice mutant with large leaf angle and small seeds was isolated from the *japonica* rice 'Zhonghua11'. This mutant was named *Epi-rav6* by X. Zhang *et al.* (2015), based on their characterization. Compared with the wild type (cv Zhonghua11), almost all leaf angles throughout all developmental stages were larger in *Epi-rav6* plants.

RAV6 regulates leaf angle by mediating brassinosteroids (BR) homeostasis: RAV6 (encoded by *Os02g45850* gene) is intimately associated with RAVL1 (encoded by *Os04g49230* gene) than other members of the RAV family. A comparison of deduced amino acid sequences proposed that RAV6 exhibits a high degree of sequence identity with RAVL1, especially in the B3 DNA-binding domain. These observations reveal that RAV6 and RAVL1 may share similar functions. RAVL1 maintains BR homeostasis via the coordinated activation of BR synthetic and receptor genes in rice. In addition, plants overexpressing *RAVL1* reveal increased lamina inclination, like the *Epi-rav6* mutants. Results suggest that RAV6 may also be involved in the BR pathway through activation of BR biosynthetic and signalling genes. X. Zhang *et al.* (2015) analysed the expression of representative genes encoding BR biosynthesis enzymes and the BR receptor, known as RAVL1 targets, in *Epi-rav6* mutants. Expression of BR synthetic genes, including *D2*, *D11* and *BRD1*, was intensely upregulated in the *Epi-rav6* mutant. Higher levels of induction for the BR receptor gene *BRI1* were also noticed in *Epi-rav6* mutants. RAV6 affects the expression of genes involved in BR signalling and BR biosynthesis. To further check that the large lamina inclination of *rav6* mutant results from BR overdose, X. Zhang *et al.* (2015) postulated that treatment with a BR inhibitor would reinstate, at least partially, the mutant phenotype. As expected, in the presence of propiconazole (Pcz), a BR biosynthesis inhibitor, the lamina inclination of the *Epi-rav6* mutant was restored. Taken together, these results advocate that RAV6 is implicated in BR-mediated developmental processes.

***RAV6*, cloning and characterization of:** To discover the molecular mechanism responsible for the *Epi-rav6* phenotype, X. Zhang *et al.* (2015) performed map-based cloning to isolate the causal gene. Using an F_2 population derived from a cross between the *Epi-rav6* mutant and *indica* var Huajingxian74, they mapped *Epi-rav6* to an 18.5-kb region on chromosome 2. This region contains only the promoter region of a putative gene, *Os02g45850*. They observed no nucleotide sequence difference between the mutant and the wild type in this region. They then investigated the expression level of *Os02g45850*, the only annotated gene in the mapped region. Compared with its low expression level in wild-type leaves, *Os02g45850* showed much higher expression in *Epi-rav6* (+/−) and *Epi-rav6* (−/−) leaves. The gradation of upregulation was much higher in homozygous than heterozygous lines.

raw score: Refers to the score of an alignment, S, calculated as the sum of substitution and gap scores. Substitution scores are given by a look-up table (*see* PAM, BLOSUM). Gap scores are typically calculated as the sum of G, the gap opening penalty and L, the gap extension penalty. For a gap of length n, the gap cost would be G+Ln. The choice of gap costs, G and L is experimental, but it is customary to choose a high value for G (10–15) and a low value for L (1–2).

ray: (i) Zygomorphic (ligulate) flowers in a radiate flowerhead, that is, ray florets/flowers (e.g. Asteraceae). (ii) Each of the branches of an umbel. (iii) The tissue that extends radially in the secondary xylem and phloem of a woody plant.

***RBS3* gene:** A gene that confers to any soybean plant (possessing that gene in its DNA) resistance to the antagonistic effects of the soil-borne fungus *Phialophora gregata*, which can result in the plant disease known as BROWN STEM ROT (BSR) in soybean plants.

reaction norm: (i) A property of genotype. How development maps the genotype into the phenotype as a function of the environment. (ii) Usually given for a single genotype across a range of environments. Populations and families can be described as having mean reaction norms. (iii) The set of phenotypes expressed by a single genotype.

reactive oxygen species (ROS): Highly reactive oxygen-containing free radicals that are generated during oxidative metabolism. ROS can react with and damage lipids, proteins and DNA in cells, causing oxidative stress. Common ROS include hydrogen peroxide, superoxide radicals and hydroxyl radicals.

reader/eraser paradox: Given that inhibition of readers (e.g. BET inhibitors, such as JQ1) blunts hypertrophy, it seems paradoxical that HDAC inhibitors, which would increase acetylation on histones and presumably BRD4 association with chromatin, decrease expression of hypertrophic genes. Conversely, increased HDAC activity, which would presumably promote BRD4 dissociation, increases the hypertrophic response. The answer to this conundrum may lie in recognizing that this biology is highly dynamic, not static. Both HATs and HDACs are targeted to transcribed regions of active genes, with the HATs activating and the HDACs resetting transcription. The binding of BRD4 to chromatin, the initiation of transcription and transcript elongation appears to require differential contact with combinations of acetylated histone lysines and an exchange of Brd4-interacting partners. In this way, HDAC-dependent release of BRD4 from chromatin after cell stress has been proposed as a requirement for proper pTEFb recruitment, with HDAC inhibitors blocking BRD4-dependent transcriptional elongation.

readers: Proteins that recognize and bind chromatin through DNA, RNA and histone modification recognition domains.

reading domains: Use of biological reading domains in place of histone modification antibodies helps in overcoming the problems such as high lot-to-lot variability, and the antibodies have insufficient specificity that arises due to the use of histone modification antibodies.

reading frame lengths (RFLs): This is the piece of DNA in between two successive stop codons. Computer analysis concluded that an RFL had an upper limit of 600 nucleotides. This helped in defining EXON LENGTH.

reading frame shift: A mutation that persuades the addition or deletion of a base within a coding sequence, resulting in the misreading of the coding frame during translation. The codon sequence that is determined by reading nucleotides in groups of three from some specific start codon. Because codons are nucleotide triplets, each mRNA has three reading frames (each nucleotide can participate in three codons, at the first, second and third base position). Duplex DNA strands have six reading frames, three in each strand. For example, GGGAAACCC, reading could start from the first letter, G and there will be three codons: GGG, AAA and CCC. If reading starts at G in the second position, the string will have two codons: GGA and AAC. If reading starts at the third base G, two codons will again result – GAA and ACC. The particular nucleotide sequence that begins at a specific point and is then partitioned into codons. The reading frame may be shifted by removing or adding a nucleotide(s). This would cause a new sequence of codons to be read. For example, the sequence CATGGT is normally read as the two codons: CAT and GGT. If another adenosine nucleotide (A) were inserted between the initial C and A, producing the sequence CAATGGT, then the reading frame would have been shifted in such a way that the two new (different) codons would be CAA and TGG, which would code for something completely different. *See* OPEN READING FRAME *and* UNIDENTIFIED READING FRAME.

reading mistake: The placement of an incorrect amino acid into a polypeptide chain during protein synthesis.

readthrough: In the presence of a suppressor-tRNA, an amino acid can be inserted into an elongating polypeptide chain in response to a stop codon. Termination is thus prevented and a longer-than-usual polypeptide is synthesized. This process is known as readthrough and the protein product is a readthrough protein.

reagent: Reagents can range from chemicals needed to perform a particular chemical reaction or constituents of a laboratory protocol to clones to be used in a large-scale gene expression study.

real inducer: An agent that actually induces gene expression. For example, allolactose is a real inducer of *lac* operon.

real-time PCR (RT-PCR): Although the traditional methods of quantitating mRNA, such as Northern blotting and *in situ* hybridization are fairly good, they do not match the ease and speed of RT-PCR. RT-PCR or reverse transcriptase PCR is semi-quantitative due to the necessity to load samples on a gel and the insensitivity of ethidium bromide. Thus, real-time PCR was developed out of the necessity to quantitate differences in mRNA expression in a relaxed and rapid manner, and due to the necessity to use small amounts of mRNA, such as those obtained by small tissue samples and LASER CAPTURE MICRODISSECTION (LCM) isolated cells. RT-PCR is categorized by the point in time during cycling when amplification of the PCR product of interest is first perceived rather than the amount of the PCR product of interest, which is amassed at the end point after PCR, which contains a large number of cycles. RT-PCR performs this act by monitoring the amount of fluorescence emitted during the PCR reaction, and this functions as an indicator of the amount of PCR amplification that transpires during each PCR cycle. Thus, in newer RT-PCR machines, one can visualize the progress of the reaction in 'real time'. RT-PCR also has a much wider dynamic range of up to 107-fold (compared to 1000-fold in conventional RT-PCR). This wide dynamic range also leads to a more accurate quantitation. RT-PCR is different from the other quantitative PCR as it quantitates the initial amount of the template instead of detecting the amount of the final amplified product.

real-time PCR primer design: QuantPrime is a flexible programme for reliable primer design for use in larger qPCR experiments. The flexible framework is also open for simple use in other quantification applications, such as hydrolysation probe design for qPCR and oligonucleotide probe design for quantitative *in situ* hybridization.

realized genetic gain: Refers to the observed difference between the mean phenotypic value of the offspring of the selected parents and the phenotypic value of the parental generation before selection.

realized heritability: Refers to heritability measured by a response to selection. It is the ratio of the single-generation progress of selection to the selection differential of the parents.

realized niche: Actual role an individual is known to play in an ecosystem.

reannealing: Spontaneous realignment of two single DNA strands to re-form a DNA double helix that had been denatured.

rearrangement: A structural alteration of a chromosome that results in an abnormal configuration, such as an inversion or translocation and may interrupt a gene sequence. Chromosomal rearrangements are typically identified on a karyotype.

reasoning system: In information technology, a reasoning system is a software system that generates conclusions from available knowledge using logical techniques such as deduction and induction. Reasoning systems play an important role in the implementation of artificial intelligence and knowledge-based systems.

reassociation kinetics of DNA: Reassociation refers to the pairing of complementary single DNA strands to make a double helix. Reassociation kinetics of DNA depends on sequence complexity. The rate of renaturation is dependent on the concentration of nucleotide strands and their sequences given that temperature of renaturation is kept constant.

rebuilding of chromatin structure: Certain proteins, termed epigenetic marks, are transferred from parent to daughter DNA strands in order to maintain the memory of transcriptional states. These pot-translationally modified proteins reinitiate the rebuilding of chromatin structure, which ultimately recruits RNA polymerase II to the newly replicated daughter strands. Pol II is believed to be recruited back to active genes only after chromatin is rebuilt. T.K. Fenstermaker *et al.* (2023) show that immediately after passage of the replication fork, Pol II in complex with other general transcription proteins and immature RNA reassociates with active genes on both leading and lagging strands of nascent DNA and rapidly resumes transcription. This suggests that the transcriptionally active Pol II complex is retained in close proximity to DNA, with a Pol II-Proliferating cell nuclear antigen (PCNA) interaction potentially underlying this retention. The Pol II machinery may not require epigenetic marks to be recruited to the newly synthesized DNA during the transition from DNA replication to the resumption of transcription.

Rec system: Genes that arbitrate recombination in *Escherichia coli* are designated as *rec*. The RecBCD complex of proteins helps to repair DNA by unwinding the DNA double helix in the region of a DOUBLE-STRAND BREAK (DSB) and chopping up the unwound strands. RecB and RecD are helicases that push on the DNA duplex together to unwind it. RecB also has a nuclease activity, which snips up the unwound DNA. RecC has been implicated in recognizing DNA sequences that direct the complex to slow down. The product of the *recA* locus of *Escherichia coli* performs dual activities: activating proteases and exchanging single strands of DNA molecules. The protease-activating activity controls the SOS response.

***recA* gene:** The gene that codes for RecA protein.

RecA protein: The prototype for a class of DNA-binding proteins that catalyse synapsis of DNA strands during genetic recombination. RecA protein is a product of the *RecA* locus central to recombination in *Escherichia coli*, and all organisms have RecA homologues. It is a protein with twin activities, acting as a protease and also able to exchange single strands of DNA (deoxyribonucleic acid) molecules. The protease activity controls the SOS RESPONSE. The nucleic acid handling facility (i.e. ability to exchange single strands of DNA) is implicated in recombination/repair pathways.

recapitulation: A partly or wholly erroneous hypothesis stating that an individual, during its development, passes through a series of stages corresponding to its successive evolutionary ancestors. According to the recapitulation hypothesis, an individual thus develops by 'climbing up its family tree'.

***recBC*:** The genetic notation of an *Escherichia coli* gene whose product mediates recombination. The enzyme encoded by the *recBC* gene is an ATP-dependent nuclease specific for double-stranded DNA. This enzyme degrades linear DNA that has been taken up by transformation.

receiver operating characteristic (ROC) curve: When an ordinal or continuous marker is engaged to diagnose a binary disease, an ROC curve can be drawn to study the judgement ability of the marker. The ROC curve is a plot of sensitivity versus one minus specificity of all possible dichotomizations of the marker as the cut points are varied. A major problem with the ROC curve is that it lures the researcher to publish cut points to somewhat arbitrarily classify patients as 'diseased' and 'normal'. In fact, when the diagnostic analysis is based on a cohort study, the marker's value can be converted into a post-test probability of disease enabling different physicians to use different cut points when the need arises (e.g. depending on available resources). Another advantage of the latter approach is that the current probability of disease also expresses the probability of an error. This area is identical to another more intuitive and easily computed measure of discrimination, the probability that in a randomly chosen pair of patients, one with and one without disease, the one with disease is the one with a higher value of the marker or post-test probability. This is also called the PROBABILITY OF CONCORDANCE between predicted and observed disease states. ROC curves play no part in formal decision making, as they ignore the utility (cost; loss) function or the cost of false positives and false negatives.

receptacle: The structure of a flower that bears the reproductive organs.

receptor: (i) Protein that binds a specific extracellular signal molecule (ligand) and initiates a response in the cell. Cell-surface receptors, such as the acetylcholine receptor and the insulin receptor, are located in the plasma membrane, with their ligand-binding site exposed to the external medium. Intracellular receptors, such as steroid hormone receptors, bind ligands that diffuse into the cell across the plasma membrane. (ii) In molecular genetics, a set of reactive chemical groups in the cell wall of a bacterium, which are complementary to a similar set in the tailpiece of a bacteriophage. (iii) A site that recognizes and binds an elicitor; any organ or molecular site that is sensitive to a distinct (specific) signal molecule. Usually, on the surface of a cell, receptors often bind to antibodies or hormones.

receptor element: A controlling element that can insert into a gene (making it a mutant) and can also exit from the gene thus making the mutation unstable. Both of these functions are non-autonomous, being under the influence of the regulator element. *See* CONTROLLING ELEMENT.

receptor gene: Receptor gene is comparable to an operator in bacterial operon. One such receptor site is supposed to be present adjacent to each producer gene. A specific receptor site is activated when a specific ACTIVATOR RNA or ACTIVATOR PROTEIN, a product of an integrator gene, complexes with it. *Aka* RECEPTOR SITE.

receptor-ligand model: *Avr* genes encode specific ligands and *R* genes encode receptors. Transgenic tobacco with a fusion of pathogen-induced tobacco *hsr203j* promoter and gene of the *Phytophthora cryptogea* coding cryptogein elicitor was developed. This fusion gene triggers an HR in transgenic tobacco upon pathogen infection and confers broad-spectrum fungal resistance.

receptor-mediated endocytosis: Refers to internalization of receptor-ligand complexes from the plasma membrane by endocytosis. It is utilized to take up some macromolecules, such as cholesterol-containing lipoproteins, from the extracellular fluid and is also a method of recycling receptor proteins once they have bound their ligands.

recessive: A quality found in the relationship between two versions of a gene. Individuals receive one version of a gene, called an ALLELE, from each parent. If the alleles are different, as in F_1, the dominant allele will be expressed, while the effect of the other allele, called RECESSIVE, is masked. In the case of a recessive genetic disorder, an individual must inherit two copies of the mutated allele in order for the disease to be present. Breeding for recessive traits is laborious and more time consuming. *Contrasts* DOMINANT.

recessive allele: Discovered by G.J. Mendel in the 1860s, this term refers to an allelic gene whose existence is obscured in the phenotype of a heterozygote by the DOMINANT ALLELE. In a heterozygote, the recessive allele does not produce a polypeptide; it is 'switched off'. In this case, only the dominant allele is the one producing the polypeptide chain (via cell's ribosome). *See* DOMINANT ALLELE *and* HEMIZYGOUS.

recessive epistasis: Either gene when homozygous recessive is epistatic to the other. Such genes are called COMPLEMENTARY GENES because two dominant genes complement each other to produce the wild-type phenotype.

recessive gene: (i) A gene that will be expressed only when there are two identical copies or, for a male, if one copy is present in the X chromosome. (ii) Any gene whose characteristic expression is dependent on the absence of a dominant gene.

recessive inheritance: *See* AUTOSOMAL RECESSIVE.

recessive lethal: An allele that is lethal when the cell is homozygous for it or hemizygous in males if genes lie in the X chromosome. *See* LETHAL.

recessive phenotype: When a pea plant showing one phenotype is crossed with a plant having another contrasting phenotype for the same trait, the phenotype that does not appear in F_1 generation is termed *recessive*.

recessive *S*-haplotype: Class II haplotypes in *Brassica rapa* include S44, S60, S40 and S29. There is a linear dominance relationship between class II haplotypes when in the heterozygous condition.

recessive trait: Refers to a trait that is expressed only when the genotype is homozygous. A trait that tends to be masked by other inherited traits yet persists in a population among heterozygous genotypes.

recessiveness: The situation where one member of the allelic pair is not expressed in the heterozygote. A recessive allele is expressed in the HOMOZYGOUS or HEMIZYGOUS (when only one member of the allelic pair is present) state. Recessiveness is the converse of dominance.

recipient cell: The cell that receives DNA during gene transfer between bacteria.

reciprocal altruism: An apparently altruistic behaviour done with the understanding that the recipient will reciprocate at some future date.

reciprocal chiasmata: Chiasmata formed between two chromatids in such a way that a double crossing-over occurs in which the other two chromatids are not involved.

reciprocal circulating chromatography (RCC): A method for RNA isolation that employs multiple-tip columns packed with solid-phase DNA probes to isolate multiple RNA species from a common sample of total RNAs.

reciprocal crosses: One of a pair of crosses in which the two opposite mating types are each attached with each of two different genotypes and mated with the reciprocal combination. For example, the male of genotype A is crossed with the female of genotype B (first cross), and the male of genotype B is crossed with the female of genotype A (reciprocal cross). Such crosses are used to detect sex linkage, maternal inheritance or cytoplasmic inheritance.

reciprocal full-sib selection: A method of interpopulation improvement for species in which the commercial product is a hybrid seed. A cycle of selection is accomplished in the fewest number of seasons by the use of plants from which both selfed and hybrid seed can be attained.

reciprocal genes: Non-allelic genes that reciprocate or complement one another gene.

reciprocal recombination: Two complementary recombination products resulting from a crossing-over event appear in 1:1 ratio. *Compare with* NON-RECIPROCAL RECOMBINATION.

reciprocal recurrent cross: A recurrent selection breeding system in which genetically different groups are mated, and in each selection cycle, individuals are mated from the different groups to test for combining ability.

reciprocal recurrent selection: A breeding method used to accomplish a build-up of genes that are valuable for specific traits but also for combining abilities. In practice, two populations constitute the basis of selection. Reciprocally, one population serves as a tester for the investigation of the selections deriving from the other population. From the first population, generally, a greater number of plants is selfed and at the same time crossed with numerous plants of the second population. the same procedure is repeated with the second population. During the second year, the progenies of the test-crosses are subjected to performance trials. the progenies of the cross of one female parent, derived from population 1, are combined with several male parents from population 2. Based on the performance testing, progenies from selfed seed, produced by the best plants, are grown during the third year. In the same year, the best progenies of selfings are crossed in many combinations. Seeds obtained from those crosses constitute the improved populations for growing during the fourth year. Within such populations, individual plants may be selfed again and crossed for operation in a new cycle.

reciprocal translocation: A chromosomal mutation in which parts of two different (non-homologous) chromosomes are reciprocally exchanged. In this case, in reality, two simple translocations are simultaneously accomplished.

reciprocally monophyletic: Two groups for which, at every locus, all genes within the group are more closely related to each other than they are to any organisms outside the group.

recircularization: The ligation of the two ends of a linear DNA molecule to re-form a circular DNA molecule. Recircularization is a common phenomenon in plasmids.

recoding: Certain directives in mRNA might lead to reprogramming of genetic information, which mRNA receives from DNA. This reprogramming of mRNA is reflected in the amino acid sequence of proteins. The set of instructions in mRNA, which bring about this recoding, is called RECODING SIGNALS. Two mechanisms for recoding have been proposed: first, change in a linear read-out of nucleotide sequence and, second, change in the meaning of the code.

recoding signals: The set of instructions in mRNA that commence programmed recoding are called recoding signals. Two mechanisms for recoding have been proposed: first, change in linear read-out of nucleotide sequence and, second, change in the meaning of the code. Second mechanism is the one in which the meaning of the code is altered has been observed in case of selenocysteine amino acid. This amino acid is not naturally occurring; therefore, no code has been assigned to this amino acid. It is believed that UGA stop codon with altered meaning acts as a codon for this amino acid.

recoding, applications of: (i) One can expand the genetic code of an organism to include new amino acids by adding new components to this template-directed biosynthetic machinery. (ii) Additional genetically encoded amino acids may offer evolutionary advantages to modern organisms and certainly extend our ability to manipulate the physicochemical and biological properties of proteins. (iii) UAAs with chemically reactive groups can be used as bio-orthogonal handles for the site-specific *in vitro*, and in some cases, intracellular modification of proteins; they can also be used to introduce a new or enhanced catalytic function into proteins. (iv) Genetically encoded UAAs can also allow the study of cellular processes, protein structure and function. (v) UAA mutagenesis is beginning to find many applications in the generation of therapeutic proteins, where the production of large quantities of homogenously modified protein is desired.

recognition and degradation of non-functional mRNAs: Eukaryotic mRNAs where translation termination occurs too soon, called NON-SENSE-MEDIATED DECAY, or fails to occur, called NON-STOP DECAY, are rapidly degraded. Yeast mRNAs with stalls in translation elongation are recognized and targeted for endonucleolytic cleavage, referred to as NO-GO DECAY. The cleavage triggered by no-go decay is dependent on translation and involves Dom34p and Hbs1p. Dom34p and Hbs1p are similar to the translation termination factors eRF1 and eRF3, indicating that these proteins might function in recognizing the stalled ribosome and triggering endonucleolytic cleavage.

recognition of transcription start signals: Initiation of transcription requires STARTISE, a point at which the first nucleotide is incorporated during transcription. Sigma (σ) factor recognizes start signals along the DNA molecules and also the correct strand of DNA to be used as a template. The presence of σ factor leads to very tight binding of holoenzyme to promoter regions of DNA, which contain start signals. Then very localized unwinding occurs, allowing synthesis of the RNA chain. The bases, which recognize RNA polymerase, are not themselves transcribed. Soon after this initial binding of the RNA polymerase recognition site, RNA polymerase disperses to the RNA polymerase binding site, which is A = T rich region. This binding region is called the TATA BOX.

recognition sequence: A specific palindromic sequence (typically of 4, 6 or 8 nucleotides) within a double-stranded DNA molecule, which is identified by a restriction endonuclease and at which the restriction endonuclease specifically slices the DNA molecule. So-called type II enzymes cut (and their corresponding modification enzymes methylate) within or very near the recognition sequence.

recognition site: A sequence of bases within the promoter region that serves to recognize RNA polymerase molecules. *See* PROMOTER *and* INITIATION SITE.

recognition species concept: A concept of species according to which a species is a set of organisms that recognize one another as potential mates; they have a shared mate recognition system. *Compare with* BIOLOGICAL SPECIES CONCEPT, CLADISTIC SPECIES CONCEPT, ECOLOGICAL SPECIES CONCEPT *and* PHENETIC SPECIES CONCEPT. *See details as separate terms.*

recognition system for recombination in immunoglobulin genes: Analysis of the DNA sequence in the vicinity of several V_L and V_H, and J_L and J_H sequences show a remarkable conservation of hepta- and decanucleotides. These conserved sequences are separated by a 10–11 or 21–23 bp non-homologous stretch of nucleotides. Presumably, similar sequences are noticed in the vicinity of D regions of the heavy chain genes. An immunoglobulin heavy chain variable region gene is produced from three sequences of DNA, V_H, D and J_H. These conserved hepta- and decanucleotides in the vicinity of V, D and J genes form a recognition system that is necessary for recombination.

recombinant: (i) Recombinants result from independent assortment or crossing-over. An individual or cell produced by combining genes from the recipient and donor, thus representing a new combination of genes. Morgan with his experiments with *Drosophila* was surprised to find significant numbers of flies with non-parental combinations of the eye and wing phenotypes of white eyes plus normal wings and red eyes plus miniature wings. The two non-parental combinations of linked genes are called recombinants. (ii) In case of plasmid DNA, the plasmid (vector) carrying genes from other organisms, inserted using restriction and ligase enzymes is called recombinant DNA.

recombinant chromosome: A structurally rearranged chromosome with a new segmental composition resulting from meiotic crossing-over involving a displaced segment and its normally located counterpart in certain types of structural (inversion or insertion) heterozygotes.

recombinant clone: Clone encompassing recombinant DNA molecules. *See* RECOMBINANT DNA TECHNOLOGY.

recombinant DNA (rDNA): A combination of DNA molecules of different origins that are linked together (spliced) *in vitro* using recombinant DNA technologies. For constructing a recombinant DNA molecule, DNA is treated from both the sources (donor and host) with the same restriction endonuclease. The ends of the cut have an overhanging piece of single-stranded DNA. These are called 'sticky ends' because they are able to base pair with any DNA molecule containing the complementary sticky end. In this case, both DNA preparations have complementary sticky ends and thus can pair with each other when mixed. DNA ligase covalently links the two into a molecule of recombinant DNA. Recombinant DNA is the basic requirement for transformation and recombinant DNA technology. P. Berg was among the first to produce a recombinant DNA molecule in 1972. P. Berg was awarded the Nobel Prize in 1980 for discovering a method for joining unrelated DNA molecules with terminal transferase. This pioneering work resulted in the development of a new technology, often called GENETIC ENGINEERING or GENE MANIPULATION, which has already had important practical applications. The recombinant DNA molecule is inserted into a host organism to produce new genetic combinations that are of value to science, medicine, agriculture and industry. *Aka* GENE CLONING *or* SPLICING.

Recombinant DNA Advisory Committee (RAC or RDAC): The former standing US national committee set up in 1974 by the US NIH to advise the NIH director on matters regarding policy and safety issues of recombinant DNA research and development. Over time, it was changed to become part of the American government's regulatory process for recombinant DNA research and product approval. The RAC was terminated by the director of the NIH in 1996 because the 'human health and environmental safety concerns expressed at the inception (of genetic engineering/biotechnology) had not materialized'.

recombinant DNA technology: S.N. Cohen *et al.* (1973) published a paper entitled 'Construction of Biologically Functional Bacterial Plasmids *in vitro*', which describes a method to isolate and amplify genes or DNA segments and introduce them into another cell with precision, creating a transgenic bacterium. P. Berg was awarded the Nobel Prize in 1980 for discovering a method for joining unrelated DNA molecules with terminal transferase. The laboratory manipulation of DNA in which DNA, or fragments of DNA from different sources, are cut and recombined using enzymes. This recombinant DNA is then inserted into a living organism. rDNA technology is usually used synonymously with genetic engineering. Under appropriate conditions, a recombinant DNA molecule can be inserted into a cell and replicated there, either autonomously or after it has become integrated into a cellular chromosome. It denotes procedures used in the manipulation of DNA in the laboratory including, but not limited to, cloning, polymerase chain reaction, restriction-enzyme digestion, ligation, nucleic acid replication, reverse transcription and ribonucleic acid synthesis. It differs from genetic recombination in that it does not occur through processes within the cell but is exclusively engineered *in vitro*. Recombinant DNA technology has assisted in protecting crops against detrimental pests, which pose a major challenge to fulfil the increasing demand for food. *Aka* IN VITRO RECOMBINATION. *See* GENETIC ENGINEERING.

recombinant DNA technology and protein engineering: The design and construction of new proteins or enzymes with novel or desired functions, has been achieved through the modification of amino acid sequences using recombinant DNA technology. The sizes and three-dimensional conformations of protein molecules are also manipulated by protein engineering. The basic techniques of genetic engineering are used to alter the genes that encode proteins, generating proteins with novel activities or properties. Such manipulations are frequently used to discover structure-function relationships, as well as to alter the activity, stability, localization and structure of proteins.

recombinant DNA technology experiments, guidelines for: P. Berg in 1975 discoursed that international research should continue but under rigorous guidelines. He was among the first to produce a recombinant DNA molecule in 1972 wrote a letter shortly afterwards, along with ten other researchers, to the journal *Science*. They counselled the National Institutes of Health (NIH) to regulate the use of recombinant DNA technology, and meanwhile, they urged scientists to pause most recombinant DNA experiments until they better understood whether the technique is safe. These concerns ultimately led to the 1975 Asilomar Conference, where one hundred scientists gathered to discuss the safety of manipulating DNA from different species. The meeting resulted in a set of NIH guidelines. NIH has revised the document, 'Guidelines for Research Involving Recombinant DNA Molecules', several times since 1976.

recombinant DNA technology in agriculture: The genetic manipulation of plants has been going on since prehistoric times when early farmers began carefully selecting and maintaining seeds from their best sow for the next season. Plant breeders have cross-fertilized related plants to provide next-generation plants with new characteristics such as higher yield, resistance to diseases and better nutrient content long before the science of genetics was developed. Recombinant DNA technology can be used for the insertion of genes in plants not only from related plant species but also from unrelated species such as microorganisms. This process of creation of transgenic plants is far more precise and selective than traditional breeding. The application of recombinant DNA technology is primarily useful for the production of transgenic plants with higher yield, nutritional values and increased resistance to stress and pests. Several commercially important transgenic crops (e.g. maize, soybean, tomato, cotton, potato, mustard and rice) have been genetically modified. In fact, in 2002, more than 5.5 million farmers worldwide cultivated about 58.7 million hectares (about 148 million acres) of crops that were genetically manipulated for herbicide tolerance, insect resistance, delayed fruit ripening and improved oil quality. Application of recombinant DNA technology has primarily assisted in producing three major types of transgenic plants that have improved performances such as (i) development of stress-tolerant plants, (ii) development of plants that have improved yield and (iii) transgenic plants as a source of biopharmaceuticals.

recombinant DNA technology in crop quality improvement: Recombinant DNA technology has played a very important role in transforming crop plants to improve the quality of the produce. At the international level, there are about 2–3 million deaths and 5 million blindness cases reported in children, mainly due to vitamin A deficiency. The need of the hour is to develop transgenic rice varieties that have a higher content of β-carotene level. (i) J.A. Paine *et al.* (2005) transformed the phytoene synthase (*psy)* from *Zea mays* L. and lycopene synthase gene (from *Narcissus pseudonarcissus*) in combination with the *Erwinia uredovora* carotene desaturase (*crtI*). They observed an increase in total carotenoids of up to 23-fold (37 μg/g of rice) compared to the original Golden Rice (having 1.6 μg/g of rice) and there was a preferential accumulation of β-carotene to about 84%. (ii) *Solanum tuberosum* is the fourth most important crop at the international level. Its carotene is also less to meet the daily requirements of the people who consume it as a staple crop. L.J.M. Ducreux *et al.* (2005) transformed potato and *Solanum phureja* plants with *CtrB* gene from *Erwinia carotovora*. The transformed plants had an enhanced carotene and level β-carotene level 5.6 and 0.0 to 35 and 11 µg/g DW, respectively and in *Solanum phureja* the carotene level increased from 20 to 78 µg/g DW. Similarly, transgenic plants have been developed for other nutritional parameters. (iii) Salt tolerance is an important characteristic required to overcome salinity-induced reduction in plant productivity. N. Sanan-Mishra *et al.* (2005) reported the isolation of a pea DNA helicase 45 (PDH45) that exhibited striking homology with the eukaryotic translation initiation factor eIF-4A. PDH45 mRNA is induced in pea seedlings in response to high salt, and its overexpression driven by a constitutive cauliflower mosaic virus-(35)S promoter in tobacco plants confers salinity tolerance, thus suggesting a previously undescribed pathway for manipulating stress tolerance in crop plants. The T(0) transgenic plants revealed high levels of PDH45 protein in normal and stress conditions, as compared with WT plants. The T(0) transgenics also showed tolerance to high salinity as tested by a leaf disk senescence assay. The T(1) transgenics were able to grow to maturity and set normal viable seeds under continuous salinity stress without any reduction in plant yield in terms of seed weight.

recombinant DNA technology in environment protection: A vast majority of applications of environmental biotechnology practice naturally occurring microorganisms (bacteria, fungi, etc.) to identify and filter manufacturing waste before it is introduced into the environment. Bioremediation programmes involving the use of microorganisms are currently in progress to clean up contaminated air, tracks of land, lakes and waterways. Recombinant technology helps in enhancing the efficacy of these processes so that their basic biological processes are more efficient and can degrade more complex chemicals and higher volumes of waste materials. Recombinant DNA technology also is being used in the development of bioindicators where bacteria have been genetically modified as 'bioluminescors' that give off light in response to several chemical pollutants. These are being used to measure the presence of some hazardous chemicals in the environment. Other genetic sensors that can be used to detect various chemical contaminants are also undergoing trials and include sensors that can be used to track how pollutants are naturally degrading in groundwater. For example, when the mercury resistance gene (*mer*) or the toluene degradation (*tol*) gene is linked to genes that code for bioluminescence within living bacterial cells, the biosensor cells can signal extremely low levels of inorganic mercury or toluene that occur in contaminated waters and soils by discharging visible light, which can be measured with fibre-optic fluorometers.

recombinant DNA technology patent: Stanley Cohen of Stanford University and Herbert Boyer of University of California, San Francisco (UCSF) applied for a patent on recombinant DNA technology in 1974, which was granted in 1980.

recombinant DNA technology, uses of: (i) Using recombinant DNA technology, genes in a single step have been transformed into different plants and animals, crossing all reproductive barriers. (ii) Plant diseases are one of the major factors that restrict crop productivity and enforce injudicious and large-scale use of pesticides that cause potentially harmful impacts on environmental safety and human health, besides the threat of pest resistance. (iii) Recombinant DNA technology has helped to protect crops against detrimental pests, which pose a major challenge to fulfil the increasing demand for food.

recombinant DNA technology (rDNA) in the pulp and paper industry: The application of recombinant DNA technology to problems in the pulp and paper industry are many. As a first step, these techniques are being used to provide much-needed fundamental information on the cellular and molecular mechanisms involved in the expression of extracellular enzymes that degrade lignocellulosic pulping wastes. The information gleaned from the studies on cellulolytic fungi and bacteria can be used to genetically engineer a yeast or bacterium capable of converting pulping wastes into ethanol and other useful by-products.

recombinant DNA, construction of: Steps involved in the construction of recombinant DNA are as follows: (i) Foreign (donor) DNA and plasmid vector are isolated. (ii) Donor and vector DNAs are cut with the same restriction endonuclease to produce STICKY ENDS. (iii) The donor and vector DNAs are joined together by mixing the DNAs together in a tube. When both DNAs have sticky ends generated with the same RE, the ends will match up. (iv) DNA ligase is the glue that holds the ends of the DNAs together by creating a phosphodiester bond between two DNA ends.

recombinant inbred line (RIL) population: A RIL population is developed using single seed descent from the F_2 generation. The result is a set of homogeneous, homozygous lines for which large amounts of seed can be produced for replicated trials. This type of population is often useful for mapping QTLs.

recombinant inbred lines (RILs): A kind of mapping population developed by continuous selfing of F_2 population until complete homozygosity is achieved. RILs are derived from a cross between two inbreds (true breeding) or pure lines, which should be diverse. RILs are generated usually using SINGLE SEED DESCENT (SSD) method and each RIL population is immortal. Because of the immortal nature of RILs, these are preferred over the other mapping populations such as F_2, F_3 and backcross populations. *See* MAPPING POPULATION.

recombinant inbred strains (RIS): Sets of animal strains that all are derived from the same two parental inbred strains and which each carry a specific combination of the parental genes; within each RI strain all animals are genetically identical.

recombinant joint: The point at which two recombining molecules of duplex DNA are connected (the edge of the heteroduplex region).

recombinant protein: The protein encoded by recombinant DNA that has been cloned in an expression vector that supports the expression of the gene and translation of messenger RNA.

recombinant RNA: A term used to describe RNA molecules joined *in vitro* by T_4 RNA ligase. This technique may be used to join an RNA sequence to a phage Q replicase template. The recombinant RNA molecule can be autocatalytically replicated by Q REPLICASE to produce large amounts of an RNA sequence of interest.

recombinant RNA technology: A technique that involves the construction of recombinant RNA and allows the production of large quantities of RNA by conferring on any RNA molecule the ability to be replicated in test tube reactions.

recombinant site-specifically modified histones: One approach for manufacturing site-specifically modified histones involves the ligation of protein fragments *in vitro*.

recombinant type: An arrangement of genetic markers, found among the progeny of a cross, that is unlike any association of markers present in the parents.

recombinant vaccine: A vaccine produced using recombinant DNA technology. Subunit vaccines that are produced by genetic engineering techniques in which other organisms are programmed to produce the desired antigenic fraction are called recombinant vaccines. For example, the vaccine against the hepatitis B virus consists of a portion of the viral protein coat that is produced by a genetically engineered yeast. These are safer vaccines because they cannot reproduce in the recipient.

recombinase: An enzyme that functions to 'cut open' the strand of DNA within a cell (e.g. to 'splice out' or 'splice in') a given gene. During 2000, N.-H. Chua and J. Zuo showed that activation of the gene for recombinase (via β oestradiol transcription factor) could be done to cause expression of recombinase in a manner that 'spliced out' (removed) antibiotic-resistance 'marker genes' from genetically engineered plants.

recombination: At a molecular level, recombination may be defined as the physical exchange of DNA sequences between chromatids of the homologous chromosomes. This exchange results in a duplex DNA in which two regions of opposite parental origins are connected by a stretch of hybrid (heteroduplex) DNA in which one strand is derived from each parent. Recombination results in the separation of linked genes and formation of new combinations of alleles of different genes located in the same chromosome not displayed by either parent. *Aka* GENERALIZED RECOMBINATION, GENETIC RECOMBINATION *or* DNA RECOMBINATION. *See* HOMOLOGOUS RECOMBINATION *and* CROSSING OVER.

recombination counting and ordering (RECORD) method: A method for the ordering of loci on genetic linkage maps. It minimizes the total number of recombination events; the search algorithm is a heuristic procedure, combining elements of branch and bound with local reshuffling. It does not require intensive calculations. The algorithm rapidly produces an optimal ordering as well as a series of near-optimal ones. The latter provides insight into the local certainty of ordering along the genetic map. Compared to JoinMap, RECORD is much faster and less sensitive to missing observations and scoring errors, since the optimization criterion is less dependent on the position of the erroneous markers. RECORD performs better in regions of the map with high marker density. *See* JOINMAP.

recombination factor: The gametic output in a hybrid of new combinations of two genes expressed as a fraction or percentage of the total gametic output of new and old combinations combined. The formulae for the gametic output of the double heterozygote *AaBb* in COUPLING PHASE and REPULSION PHASE, respectively are ½(1 − p) *A*½ ½p *Ab*, ½ *aB*, ½ (1 − p) *ab* and ½p *AB*, (1 − p) *Ab*, (1 − p) *aB*, ½p *ab*, where p represents the recombination fraction. Where there is no linkage, $p = 0.05$.

recombination fine mapping of QTLs: There are two approaches of this genomic method of QTL mapping: SUBSTITUTION MAPPING and INTROGRESSION LINE POPULATION. *See details as separate terms.*

recombination frequency: The proportion of recombinant progeny in the total progeny from a genetic cross. Such frequencies indicate relative distances between loci on a genetic map.

recombination hot spot: A localized region with an exceptionally high recombination rate.

recombination index: C.D. Darlington, in 1932, proposed a recombination index that consisted of the sum of the haploid number (n) and the average chiasma number per nucleus (x); this index (n+x) represents the average number of elements that segregate independently so that the number of possible gametes equals $2^{(n+x)}$.

recombination intermediates: Following reciprocal single-strand exchange, each DNA duplex has heteroduplex material covering the region from the initial site of exchange to the migration branch. In variants of the model in which some DNA is degraded and resynthesized, the initiating chromatid is the donor of genetic information. A property of circular DNA offers an approach to isolating recombination intermediates. Generally, the recombination in DNA duplexes is considered in linear molecule, but in many genomes (especially in viruses and plasmids), these molecules are circular.

recombination joint: The process of crossing-over starts with breakage at the corresponding points of the homologous strands of two paired DNA duplexes. The breakage allows movement of the free ends created by the nicks. Each strand leaves its partner and crosses over to pair with its complement in the other duplex. The reciprocal exchange creates a connection between the two DNA duplexes. Initially, this is sustained only by hydrogen bonding; at some point, it is made covalent by sealing the nicks at the sites of exchange. The connected pair of duplexes is called a JOINT MOLECULE. The point at which an individual strand of DNA crosses from one duplex to the other is called the RECOMBINATION JOINT. At the site of recombination, each duplex has a region consisting of one strand from each of the parental DNA molecules. This region is called HYBRID DNA or HETERODUPLEX DNA.

recombination landscape: The genomic landscape of recombination plays a role in shaping patterns of diversity along the genome and has an effect on the efficacy of selection.

recombination load: The loss of fitness because recombination breaks up associations between beneficial combinations of interacting alleles.

recombination nodules: Swellings along DNA strands and associated proteins during meiotic prophase pairing of chromosomes. The nodules can be identified under the electron microscope. It is suggested that recombination nodules are the sites of genetic crossing-over. When crossing-over occurs there is a breakage and reunion of chromatin strands. A small amount of DNA is synthesized at the breaks in short pieces of ~200 bp but repeated 700–2000 times per haploid genome. DNA synthesis inhibitors cause an increase in chromosome breakage. EARLY RECOMBINATION NODULES appear during zygotene coinciding with synapses. LATE RECOMBINATION NODULES occur during pachytene when synaptonemal complex formation is complete.

recombination rate: The proportion of recombinant gametes.

recombination repair: A mode of filling a gap in one strand of duplex DNA by retrieving a homologous single strand from another duplex. Usually the underlying mechanism behind homologous recombination and gene conversion.

recombination system: All factors that mediate and control the process of genetic recombination.

recombination theory of antibody diversity: According to this theory, there are as many different light and heavy chain genes as many light and heavy chains a cell can produce. The diverse types of antibodies are produced as a result of recombination between light and heavy chains.

recombination unit: On average, one centimorgan (cM) corresponds to roughly 1 million base pairs in the human genome. *See* MORGAN UNIT.

recombination, breakage and copying hypothesis of: Given by J. Bellings in 1933, this hypothesis proposed that the recombinant chromosome is formed by utilizing the physical section of parental chromosome and by copying the other. Thus, the recombinant chromosome retains a part of the parental chromosome due to the breakage and at the same time contains a part of the new chromosome synthesized by copying the other strand.

recombination, breakage and rejoining model of: This model assumes that recombinants arise as a result of breakage and rejoining of nucleic acid molecules. According to this theory proposed by C.D. Darlington in 1937, the two pieces of parental chromosomal information that are combined in the recombinant chromosome arise from the physical breaks in the parental chromosomes with subsequent physical exchange. The paired parental chromosomes break and rejoin at meiosis resulting in the formation of 'chiasmata' In this case, recombination does not depend upon the synthesis of new DNA molecules since the information entering the recombinants is merely transferred from the parents.

recombination, breakage and reunion hypothesis of: The classical and generally accepted model of crossing-over by physical breakage and crossways reunion of broken chromatids during meiosis. C.D. Darlington in 1937 proposed that the two pieces of parental chromosomal information that are combined in the recombinant chromosome arise from the physical breaks in the parental chromosomes with subsequent physical exchange. The paired parental chromosomes break and rejoin at meiosis, resulting in the formation of chiasmata. Recombination does not depend upon the synthesis of new DNA molecules since the information entering the recombinants is merely transferred from the two parents.

recombination, mechanism of: Theories explaining the mechanism of recombination may be divided into three general groups: breakage and copying, complete copy choice and breakage, and reunion. *See* BREAKAGE AND COPYING HYPOTHESIS, COMPLETE COPY CHOICE HYPOTHESIS *and* BREAKAGE-AND-REUNION HYPOTHESIS.

recombination, process of: DNA breakage and reunion are responsible for genetic recombination (crossing- over) in prokaryotes as well as eukaryotes. ENDONUCLEASE produces NICKS in single strands of DNA and a DNA LIGASE rejoins broken DNA strands. Eukaryotes mostly show reciprocal recombination products.

recombination-free (RF) cloning: A DNA cloning method developed from circular PCR. In this method, first, primers flanked by unique restriction enzymes were used to amplify the genes of interest. *Dpn*I recognizes the sequence 5′-GATC-3′; this sequence is very common within DNA sequences. The reason why *Dpn*I cleaves the parental plasmid but ignores the PCR products is that *Dpn*I only excises if the dATP is methylated. The polymerase recommended to be used in this method was the Pfu Turbo because this enzyme performs with high fidelity and does not add dATP on to the PCR products.

recombination-mediated genetic engineering: Recombineering (recombination-mediated genetic engineering) is a genetic and molecular biology technique based on homologous recombination systems, as opposed to the older or more common method of using restriction enzymes and ligases to combine DNA sequences in a specified order. *Aka* RECOMBINEERING.

recombinational cloning: This method uses *in vitro* site-specific recombination to accomplish the directional cloning of PCR products and the subsequent automatic subcloning of the DNA segment into new vector backbones at high efficiency. Three popularly used cloning systems are as follows: Gateway®, Creator™ and Echo Cloning™ systems. These systems use a site-specific recombinase to allow reliable transfer of a fragment from one vector to another without using restriction enzymes and ligases.

recombinational repair: Filling a gap in one strand of duplex DNA by retrieving a homologous single strand from another duplex. In case of recombinational repair, DNA strand exchange is promoted by RecA protein. Double-stranded DNA (dsDNA) pairs with the RecA presynaptic filament, which consists of RecA protein and single-stranded DNA (ssDNA), to produce heteroduplex DNA bound by RecA and the exchanged ssDNA. The DNA postreplication repair (PRR) system permits the cell to tolerate damage without actually repairing it. RecBCD is a DNA helicase comprising two motor subunits, RecB and RecD. *Aka* POSTREPLICATION REPAIR (PRR).

recombinator activity: Recombination does not occur in inactive chromosomes/heterochromatin. Recombination is not uniformly distributed throughout a given gene. If recombination is largely confined to structural genes, it is reasonable to equate promoter or operator sequences with recombinators. It is suggested that non-methylated promoters have recombinator activity and specifically initiate recombination. Hybrid DNA thus formed will be hemi-methylated, since homologous chromosome is very unlikely to carry an epigenetic defect at the same site. The maintenance enzyme will methylate the previously unmethylated site, thus eliminating the defect. Non-methylated recombinator could be a substrate for nuclease, which initiates recombination either by a single-strand or possibly a DSB. An enzyme that produces a double-stranded cut in the methylated site of a non-methylated sequence initiates recombination in *Neurospora*. The same methylated sequence is not a substrate for the enzyme.

recombinogen: An agent capable of inducing or enhancing recombination between linked genes..

recon: (i) A unit of recombination. It is defined by the recombination test. Two mutations belong to the same recon if they do not show recombination. A recon consists of several mutons. Although there is no lower limit on the size of recon, the smallest recombinational unit probably is two neighbouring nucleotide pairs. *See* MUTON. (ii) In cell culture, the viable cell reconstructed by the fusion of a karyoplast with a cytoplast.

RECON: A computer programme that allows *de novo* detection and classification of repeat family in genomic sequences.

RECON algorithm: This programme detects and groups repeats in the genome sequences by the blast itself and clusters them to a repeat family using multiple sequence alignment. This approach helps repeat annotation by determining repeat boundaries in genomic sequences and also enables the identification of new repeat elements. RECON is available from http://www.genetics.wustl.edu/eddy/recon/

reconstituting chemically orthogonal directed engineering (ReCODE™) technology: The Ambrx ReCODE™ technology (reconstituting chemically orthogonal directed engineering) confers unprecedented control over the site-directed placement of chemistry that is designed and demonstrated to be non-disruptive to protein function. Through directed evolution and selection, specialized ORTHOGONAL tRNA SYNTHETASES (RSo) are evolved to selectively and specifically amino-acylate a similarly ORTHOGONAL AMBER CODON-SUPPRESSING tRNA (tRNAo) in *Methanocaldococcus jannaschii* with unique, chemically specified amino acids. The cell's translational apparatus then incorporates the Ambrx amino acid into the elongating peptide sequence at positions designated by the amber codon. In the absence of the Ambrx aa-amino-acyl tRNAo, protein elongation is terminated at amber. Amino acids are designed to include side chains that are chemically reactive using coupling chemistry that preserves protein function and does not interact with native amino acids. The tRNAo and tRNAo synthetases are orthogonal because they do not interact with endogenous tRNA synthetases and tRNAs, respectively. The proprietary methods for the design, evolution and selection of amino acid and amino-acyl tRNA synthetase pairs have enabled the incorporation of over 30 different chemically specified amino acids into biosynthetic proteins in *Escherichia coli*, yeast and mammalian cell systems. Ambrx's initial focus has been in the *Escherichia coli* expression system. Ambrx has evolved the technology through several generations, driving amber suppression to high efficiency with yields that enable phase 3 and commercial-scale manufacturing. The components of this reconstituted system include the unique amino acid, the tRNAo to which it is conjugated by the tRNA synthetase (RSo) to create an amino-acyl tRNA that recognizes (suppresses) the stop codon TAG (amber) placed at any position in the protein of interest's coding sequence. This ReCODE™ technology has been demonstrated to confer engineering proteins for optimal performance, through site-directed placement of chemical reactivity during biosynthesis of the protein.

reconstructed cell: The viable cell reconstructed by the fusion of a karyoplast with a cytoplast.

record: A record is a basic data structure. Records in a database or spreadsheet are usually called ROWS. *Aka* STRUCTURE, STRUCT *or* COMPOUND DATA.

recreating mutant alleles in *Drosophila*: CRISPR/Cas9 and HDR have been used to develop a temperature-sensitive *transformer-2* (*tra-2*) allele in *Drosophila suzukii* by mutating the nucleotides orthologous to the *Drosophila melanogaster tra-2^{ts2}* allele. *Tra-2* is a sex-determination gene involved in regulating female-specific splicing of *doublesex* pre-mRNA. *Tra-2* is an attractive target to produce a male-only population of sterile insects for pest control using STERILITY-INDUCED TECHNIQUE (SIT). Conditional alleles of sex-determination genes could make SIT more efficient by reducing the size of breeding populations, eliminating radiation use and preventing the corelease of female siblings. Because *Drosophila suzukii* could not thrive at higher temperatures, this strategy for SIT may not be optimal for the control of these pests. However, the approach could be valuable for control of tropical insect pests.

recruitment heterochromatin protein 1: Tri-methylation of histone H3 lysine 9 is important for recruiting HETEROCHROMATIN PROTEIN 1 (HP1) to discrete regions of the genome, thereby regulating gene expression, chromatin packaging and heterochromatin formation.

recruitment of ATM and ATR: Ataxia-telangiectasia mutated (ATM), ataxia-telangiectasia and Rad3-related (ATR) and DNA-dependent protein kinase catalytic subunit (DNA-PKcs) are members of the phosphoinositide-3-kinase-related protein kinase (PIKK) family and are rapidly activated in response to DNA damage. ATM and DNA-PKcs respond mainly to DNA double-strand breaks, whereas ATR is activated by single-stranded DNA and stalled DNA replication forks.

recruitment of cohesion: M. Gullerova and N.J. Proudfoot in 2008 showed that DOUBLE-STRANDED (ds) RNA, which is generated by convergent genes in the G1 phase of the cell cycle in *Schizosaccharomyces pombe*, induces localized RNAi (Dicer and RITS)-dependent transient heterochromatin structures including histone H3 lysine 9 trimethylation marks and Swi6 association. Consequently, cohesin (protein) is recruited to these chromosomal positions through interaction with Swi6. The complex thus formed is called COHESION.

recruitment of the dosage compensation complex (DCC) to X chromosome: To achieve X chromosome dosage compensation, organisms must distinguish X chromosomes from autosomes. The DCC normally assembles along the entire X chromosome, but not all detached regions recruit complex, despite having genes known to be dosage compensated on the native X. DCC binds first to recruitment sites, then spreads to neighbouring X regions to accomplish chromosome-wide gene repression. From a large chromosomal domain, a 739-bp fragment was defined that functions *in vivo* as an X-recognition element to recruit DCC.

reculture: The process by which a cell monolayer or an explant is transferred, without subdivision into fresh medium. During tissue culture, the primary culture sometimes does not resume growth during the first 4–6 weeks of incubation. Under such a situation, the explants are recultured on fresh medium for more period to obtain growth. *See* PASSAGE.

recurrent backcrossing: The parent to which a hybrid is repetitively crossed in a backcross; it replaces the dragged alleles step by step with the alleles of the original variety. The parent used in such recurrent backcrosses is called RECURRENT PARENT.

recurrent full-sib selection: A method of intrapopulation improvement that involves the testing of paired-plant crosses. It is the only method of recurrent selection in which the seeds from two individuals, rather than one, are used for testing and to form the new population.

recurrent half-sib selection: A method of intrapopulation improvement that includes the evaluation of individuals through the use of their half-sib progeny. The general procedure for a cycle of selection is as follows: (i) to cross the plants being evaluated to a common tester, (ii) evaluate the half-sib progeny from each plant and (iii) intercross the elected individuals to form a new population.

recurrent mass selection: A breeding method designed to increase the levels of desirable qualities, which are quantitative variables, by changing the frequency of polygenes. In each screening generation, the best individuals are selected, and they become the parents of the next screening generation. The process is repeated for as many generations as necessary, but the rate of progress declines dramatically after a few generations.

recurrent neural network (RNN): A class of artificial neural networks where connections between nodes form a directed graph along a temporal sequence. This allows it to exhibit temporal dynamic behaviour. Unlike feedforward neural networks, RNNs can use their internal state (memory) to process sequences of inputs. This makes them applicable to tasks such as unsegmented, connected handwriting recognition or speech recognition.

recurrent parent: The parent that is repeatedly used in backcross breeding of crop plants. In plant breeding, backcross breeding is an important method for transferring useful genes into cultivated varieties from the wild species. To achieve this the cultivated variety is hybridized with wild species to obtain F_1 hybrid that is backcrossed with the cultivated variety to develop BC_1F_1 that is further crossed with cultivated variety to obtain BC_2F_1 and so on. In this case, the cultivated variety that is repeatedly used in crossing is called the recurrent parent.

recurrent reciprocal selection: A recurrent selection breeding system in which genetically different groups are maintained and, in each selection cycle, individuals are mated from the different groups to test for combining ability.

recurrent selection: A method designed to concentrate favourable genes scattered among a number of individuals. It is performed by repeated selection in each generation among the progeny produced by matings *inter se* of the selected individuals of the previous generation. In practice, plants from a population are selfed and, after the yield of the selfed seeds, the progenies of the phenotypically best individuals are grown in the second year. The best progenies are then crossed in as many combinations as possible and the seeds received thereby are grown in the third year as a population. Within the already improved population, selection and selfing can be carried out again. With this population, a second cycle of recurrent selection can be started.

recursion: An algorithmic procedure whereby an algorithm calls on itself to perform a calculation until the result exceeds a threshold, in which case the algorithm exits. Recursion is a powerful procedure with which to process data and is computationally quite efficient. Recursion occurs when a thing is defined in terms of itself or of its type. Recursion is used in a variety of disciplines ranging from linguistics to logic. The most common application of recursion is in mathematics and computer science, where a function being defined is applied within its own definition. While this apparently defines an infinite number of instances (function values), it is often done in such a way that no infinite loop or infinite chain of references can occur.

recursive algorithm: A recursive algorithm is based on recursion. In this case, a problem is broken into several subparts and called the same function again and again.

recycling: Procedure that is carried out without any chemical and biological treatment for some reusable materials such as paper, plastic, glass and can. These materials can be returned into economical processes.

recycling endosomes: Large intracellular membrane-bounded vesicle formed from a fragment of an endosome that is an intermediate stage on the passage of recycled receptors back to the cell membrane.

red biotechnology: The use of organisms for the improvement of medical processes. It includes the designing of organisms to manufacture pharmaceutical products like antibiotics and vaccines, the engineering of genetic cures through genomic manipulation and its use in forensics through DNA profiling.

Red Goo: Refers to designed nanotechnological objects for destruction, similar to Grey Goo, which is accidentally created.

red queen: Continual coevolution between two species (e.g. between host and parasite).

red-green colour blindness: The inability in man to distinguish red and green colours. One type of red-green colour blindness is determined by sex-linked gene located in non-homologous (differential) part of the X chromosome.

redifferentiation: The process of maturing dedifferentiated cells to perform specific functions and lose their capacity to divide again (e.g. formation of secondary cortex, xylem, phloem and cork).

rediploidization: In another culture, the haploidization of the genome by culturing pollen grains to haploid plantlets and its rediploidization after spontaneous or induced doubling of the chromosome set.

redirecting metabolic flow: In a branched biosynthetic pathway, there are several forks, where intermediates can enter alternative pathways. At such forks, resources (substrate, enzyme, transport system, ribosomes, etc.) are distributed equally or unequally to two or more metabolic pathways. When we are interested in one of the several possible products, at a fork the desired route can be made a priority, so that resources are minimized without decreasing cell viability. This can be achieved by (i) use of cloned genes suppressing feedback inhibition by the end products and (ii) elevation of the activity of rate-limiting enzyme.

redox gene regulation: A gene regulatory mechanism involving reductional-oxidation of a single conserved cysteine residue in the DNA-binding domain of the regulatory proteins.

redox pair: Pair of molecules in which one acts as an electron donor and one as an electron acceptor in an oxidation-reduction reaction; for example, NADH (electron donor) and NAD^+ (electron acceptor).

redox potential: The affinity of a redox pair for electrons, generally measured as the voltage difference between an equimolar mixture of the pair and a standard reference. $NADH/NAD^+$ has a low redox potential and O_2/H_2 has a high redox potential (high affinity for electrons).

redox reaction: A reaction in which one component becomes oxidized and the other reduced; an oxidation-reduction reaction. *See* OXIDATION REDUCTION REACTION.

redox/redox state: Shorthand for reduction/oxidation reactions. The term redox state is often used to describe the balance of NAD and NADH in a biological system such as a cell or organ. An abnormal redox state can develop in a variety of deleterious situations.

***Reduced height* (*Rht*) gene:** Short-straw mutant. There are two current widely used *Rht* mutant alleles, *Rht-B1b* and *Rht-D1b*. Both reduce plant height by 20% and increase seed yield by 5–10%.

reduced number of chromosomes: The gametic or haploid number of chromosomes. Some lower organisms, such as *Neurospora*, have reduced number of chromosomes in somatic cells.

reduced penetrance: A reduction in the number of individuals in any one phenotype through the number in the corresponding genotype is not reduced (i.e. the particular gene or genes involved, though present, are not always manifest).

reduced representation bisulfite sequencing (RRBS): An efficient and high-throughput technique for analysing the genome-wide methylation profiles on a single-nucleotide level. It combines restriction enzymes and bisulfite sequencing to enrich for areas of the genome with a high CYTOSINE-GUANINE DINUCLEOTIDE (CpG) content. Bisulfite treatment of DNA sequence converts unmethylated cytosines to uracil. To focus the assay and save costs, specific restriction enzymes that cut around CpG dinucleotides can reduce the genome to a portion specifically enriched in CpGs. This enriched sample is then sequenced to determine the methylation status of individual cytosines quantitatively.

reducers: One of three basic types of organisms, the other two being the producers and consumers. Reducers break down the organic chemicals of dead organisms, and they make these nutrients available for reuse by other organisms.

reducing sugar: Reducing sugar is a sugar that is capable of acting as a reducing agent because it has a free aldehyde group or a free ketone group. All monosaccharides are reducing sugars, along with some disaccharides, oligosaccharides and polysaccharides.

reductase: An enzyme responsible for reduction in an oxidation-reduction reaction.

reduction: Addition of electrons to an atom, as occurs during the addition of hydrogen to a molecule or the removal of oxygen from it. Opposite of OXIDATION. In a chemical reaction, the gain of (negatively charged) electrons by a chemical substance. When one substance is reduced by another, the other compound is oxidized (loses electrons) and is called the REDUCING AGENT.

reduction division: Nuclear division in which the chromosome number gets reduced to half of the somatic cell. Reduction division is very important for alternation of sporophytic and gametophytic generations. It takes place in the sex cells where meiosis occurs. During the meiosis-I, the chromosome number gets reduced in the gamete by one-half. Therefore, it is a reductional division whereas meiosis-II is an equational division.

reduction principle: If selection is the only evolutionary process acting, then the recombination rate will tend to decrease.

redundancy: A term used to describe the fact that some amino acids have more than one codon (that codes for production of that amino acid). There are approximately 64 possible codons available to code for 20 amino acids. Therefore, some amino acids will be specified by more than one codon. These (extra) codons are redundant.

redundant DNA: DNA that does not appear to be genetically active and hence is not translated or transcribed. It often consists of repeated sequences. *See* REPETITIVE DNA.

redundant genes: A group of genes that exhibit four properties, namely, multiplicity, close linkage, sequence homology and related or overlapping phenotypic functions. Presence of several copies of a single gene allows the organism to obtain large amounts of the gene product in short time interval. Examples of redundant genes are rRNA genes, tRNA genes, histone genes, antibody genes. These genes form a MULTIGENE FAMILY. *Aka* REPETITIVE GENES.

reduplication: Doubling of the genetic matter of a haploid chromosome set.

reduplication hypothesis: The discredited idea that gametes with parental gene combinations undergo a selective mitosis so that their frequencies are greater than the non-parental gene combinations. The reduplication hypothesis was meant to counter the idea of gene combinations generated by crossing-over.

refactoring: Reorganization of biological systems or pathways with the goal of improving the ease and predictability of future engineering efforts. This often entails the removal of native regulation, separation of overlapping genetic elements and the use of well-characterized regulatory elements.

reference: A value that enables a programme to indirectly access a particular datum, such as a variable's value or a record, in the computer's memory or in some other storage device. The reference is said to refer to the datum, and accessing the datum is called DEREFERENCING the reference.

reference counting: A programming technique of storing the number of references, pointers or handles to a resource (e.g. object, block of memory, disk space). In garbage collection algorithms, reference counts may be used to deallocate objects, which are no longer needed.

reference genome: The DNA sequence of an organism created from the combined DNA sequences of multiple individuals. It is considered to represent the common genetic information in humans and is useful for genetic analyses.

reference material: Material or substance one or more of whose property values are sufficiently homogeneous and well established to be used for calibration of a measuring system, the assessment of a measurement procedure or for assigning values to materials.

reference method: A thoroughly investigated method, in which exact and clear descriptions of the necessary conditions and procedures are given for the accurate determination of one or more property values; the documented accuracy and precision of the method are commensurate with the method's use for assessing the accuracy of other methods for measuring the same property values or for assigning reference method values to reference materials.

reference pangenome: The reference pangenome contains 47 phased, diploid assemblies from a cohort of genetically diverse individuals. These assemblies cover more than 99% of the expected sequence in each genome and are more than 99% accurate at the structural and base pair levels.

reference single-nucleotide polymorphism (refSNP, rSNP, rsID, SNP ID): Any single-nucleotide polymorphism at a specific site of a genome (or part of a genome, e.g. a BAC clone) that serves as reference point for the definition of other SNPs in its neighbourhood. The rsID number ('tag') is assigned to each refSNP at the time of its submission to the databanks.

referral: Individuals receive one version of a gene from each parent. Sometimes, a version of a gene is recessive. Recessive genes have a less powerful effect than dominant genes. If a dominant gene and a recessive gene are inherited, the effects of the recessive gene are not visible or are masked by the more powerful dominant gene's effects. If two recessive genes are inherited, then the effects of the recessive genes are visible.

refractive index: The relative velocity with which light passes through a substance. It is the measure of the light-bending ability of a medium. Refractive index of specimens can be changed by staining. After staining, when light rays pass through two materials (specimen and its medium) with different refractive indexes, the rays change direction (refract) from a straight path by changing an angle and thus increase the image's contrast between specimen and the medium.

refuge crop strategy: A strategy for management of *Bt* resistance in the field. The refuge crop strategy has been recommended for resistance management of *Bt* crops over prolonged period. Following this strategy, 25% area is cultivated with conventional (non-transgenic) crop. The conventional crop generates susceptible insects that mate with resistant ones developed in transgenic field so that resistant do not mate with resistant to produce super virulent strains.

regenerant: An entire plant grown from a single cell.

regeneration: (i) In plant tissue culture, a morphogenetic response to a stimulus that results in the production of roots, shoots and whole plants. The plant cells, tissues and organs cultured on growth medium supplemented with optimum levels of auxins and cytokinins exhibit regeneration of roots, shoots and entire plants during incubation for 4–6 weeks. (ii) It refers to the repair of damaged cells, tissues or organs to yield normal parts or complete organisms. Regeneration of certain cells and persistence of others involves the differentiation at the morphosis. Thus, regeneration is the replacement of a lost or damaged part of the body that comes as a result of some type of injury. In some species, regeneration power is very strong while in some species (e.g. human beings), the regeneration power is very poor.

regeneration of animal stem cells: Spermatogonia that have differentiated and are undergoing limited mitotic (transit-amplifying) divisions can repopulate the niche and revert to stem cell identity. Thus, in the appropriate microenvironment, transit-amplifying cells dedifferentiate, becoming functional stem cells during tissue regeneration. *Compare with* REGENERATION OF PLANT STEM CELLS.

regeneration of chloroplast transgenic plants through organogenesis: This method entails direct regeneration of putative chloroplast transgenic shoots from the bombarded leaf material. Achievement of HOMOPLASMY is not a prerequisite before regeneration, as HETEROPLASMIC SHOOT SECTIONS can be placed on selective shoot regeneration media for a second time to achieve homoplasmy.

regeneration of chloroplast transgenic plants through somatic embryogenesis: Somatic embryogenesis approach differs from organogenesis in that homoplasmy needs to be established prior to regeneration of transgenic plants. Once homoplasmy is established, plants can be regenerated that were homoplasmic.

regeneration of plant stem cells: Plant stem cells, contained in specialized structures, called MERISTEMS, have amazing regenerative powers. They enable plants to grow and produce new organs throughout lifetimes that can span hundreds of years. In *Arabidopsis*, shoot meristem contains some 100 cells and is less than 100 μm in diameter. The root meristem is similar in size. *Compare with* REGENERATION OF ANIMAL STEM CELLS.

regeneration protocols for chloroplast transgenic plants: ORGANOGENESIS and SOMATIC EMBRYOGENESIS are the two regeneration protocols for chloroplast transgenic plants.

regenerative agriculture: A broad term increasingly used to describe farming and ranching systems that are meant to restore the soil health and biological balance of an area in production through the use of organic, permaculture and other biologically integrated farming systems.

regenerative medicine: There are four phases in developing regenerative medicine – gene drugs, organ replacement, resetting the genetic clock and integration of non-biological substances with body. Anticipated benefits of regenerative medicines are in molecular medicines, microbial genomics, risk assessment, bioarchaeology, evolution, human migration, DNA forensics, agriculture, livestock breeding and bioprocessing.

region: An area of a chromosome lying between two adjacent landmarks. Regions are numbered outwards from the centromere on both chromosome arms.

regional climate outlook forums: Forums active in several parts of the world that routinely provide real-time regional climate outlook products.

regional specification: Process by which cells acquire positional information (i.e. instructions that govern their behaviour and allow them to form specific structures in appropriate positions, the basis of pattern formation). Thus, through regional specification, cells acquire positional values causing them to behave in a manner appropriate for their position in the organism as a whole. There are two types of positional specifications: (i) which controls differentiation and causes particular differentiated cell type to arise in particular position and (ii) which controls morphogenesis in cells, which are equivalent with respect to state of differentiation. *Aka* PATTERN FORMATION.

registered name: A cultivar name accepted by the relevant International Cultivar Registration Authority.

registered seed: The progeny of breeder or foundation so handled as closely as possible to maintain the genetic identity and purity of a variety. Registered seed is the source of certified seed. Registered seed must be approved and certified by an official seed certification agency.

registration: (i) The act of recording a new cultivar name with an International Cultivar Registration. Authority. (ii) Recording a new cultivar name with a statutory authority like the Plant Breeder's Rights Office. (iii) Recording a trademark with a trade marks office.

regression: The way in which a variable y depends on another variable x can be represented by a simple regression model, $y = \alpha + \beta x + \varepsilon$, where α is the y intercept (the value of Y where X = 0), β is the slope of the line and ε is a random error term. β is known as the REGRESSION COEFFICIENT.

regression analysis: The process of fitting a regression equation to a set of data by using the method of least squares. It includes the various statistical tests and estimates associated with the use of the fitted equation. An effective method for determining which factors have an effect on a certain variable of interest. The method of conducting a regression allows you to accurately establish the elements that are most important, which ones you may disregard and how these factors interact with one another. While there are many forms of regression analysis, they always focus on how independent variables impact a dependent variable.

regression coefficient (b): The change in the value of dependent variable corresponding to the unit change in the independent variable.

regression equation: The nature of relationship of one variable with one or more of others, generally expressed as a mathematical equation that best fits the data. It is also called REGRESSION MODEL.

regression line: A line that defines how much an increase or decrease in one factor may be expected from a unit increase in another. This is thus graphical presentation of linear regression.

regression testing: Rarely non-regression testing. A re-running functional and non-functional tests to ensure that previously developed and tested software still performs after a change. If not, that would be called a REGRESSION. Changes that may require regression testing include bug fixes, software enhancements, configuration changes and even substitution of electronic components. As regression test suites tend to grow with each found defect, test automation is frequently involved. Sometimes, a change impact analysis is performed to determine an appropriate subset of tests (non-regression analysis).

regression to the general mean: The phenomenon discovered by Sir Francis Galton in 1889 that in any series with complex phenomena that are dependent on many variables, where chance is involved, extreme outcomes tend to be followed by more moderate ones.

regressive variation: Variation in which the offspring tend to approach the mean of the population rather than to exceed the parental extremes.

regular patterned DNA lattices: Self-assembly of 2-D periodic lattices consisting of hundreds of thousands of DX tiles has been demonstrated. This is strong evidence of the scalability of this approach. In addition, DNA TRIPLE CROSSOVER MOLECULES (TX) have also been constructed from which we have also produced tiling lattices.

regulated enzymes: Regulated genes produce regulated enzymes. These enzymes are not needed by the cell at all times, like enzymes of metabolism of less common sugars (e.g. lactose, arabinose and galactose) and the enzymes of biosynthetic pathways of amino acids, purines and pyrimidines.

regulated gene: Regulated gene is the one whose expression is regulated. Expression of regulatory genes is subject to modification by chemical stimuli that appear in or disappear from cell's environment. Regulated genes produce regulated enzymes. These enzymes are not needed by the cell at all the times, like enzymes of metabolism of less common sugars such as lactose, arabinose and galactose, and the enzymes of biosynthetic pathways of amino acids, purines and pyrimidines.

regulated genes, categorization of: After identifying some set of regulated genes, the next step in expression profiling involves looking for patterns within the regulated set. Do the proteins made from these genes perform similar functions? Are they chemically similar? Do they reside in similar parts of the cell? Gene ontology analysis provides a standard way to define these relationships. Gene ontologies start with very broad categories (e.g. metabolic process) and break them down into smaller categories (e.g. carbohydrate metabolic process) and finally into quite restrictive categories (e.g. inositol and derivative phosphorylation). Genes have other attributes beside biological function, chemical properties and cellular location. One can compose sets of genes based on proximity to other genes, association with a disease and relationships with drugs or toxins. The molecular signatures database and the comparative toxicogenomic database are examples of resources to categorize genes in numerous ways.

regulated genes, patterns among: Regulated genes are categorized in terms of what they are and what they do, important relationships between genes may emerge. For example, we might see evidence that a certain gene creates a protein to make an enzyme that activates a protein to turn on a second gene on our list. This second gene may be a transcription factor that regulates yet another gene from our list. By observing these links, we may begin to suspect that they represent much more than chance associations in the results and that they are all on our list because of an underlying biological process. On the other hand, it could be that if one selected some genes at random, one might find many that seem to have something in common.

regulated pre-mRNA splicing by phosphorylation: RNA splicing is an essential step in the regulation of gene expression. It is controlled by the spliceosome, a mega-Dalton RNA-protein complex that assembles *de novo* on each pre-mRNA intron through an ordered assembly of intermediate complexes. Spliceosome activation is a major control step that requires substantial protein and RNA rearrangements leading to a catalytically active complex. SPLICING FACTOR 3B SUBUNIT 1 (SF3B1) protein – a subunit of the U2 small nuclear ribonucleoprotein – is phosphorylated during spliceosome activation. M. Hluchý *et al.* (2022) show that CYCLIN-DEPENDENT KINASE 11 (CDK11) associates with SF3B1 and phosphorylates threonine residues at its N terminus during spliceosome activation. The phosphorylation is important for the association between SF3B1 and U5 and U6 snRNAs in the activated spliceosome, termed the B^{ACT} COMPLEX, and the phosphorylation can be blocked by OTS964, a potent and selective inhibitor of CDK11.

regulation: A law made by a person or body that has been granted (delegated) law-making authority. A regulation is used both to indicate a specific type of delegated legislation as well as to refer generically to all forms of delegated legislation. The Department of Justice issues a special number to indicate that it is a regulation. More broadly, regulation may refer to all government intervention in the lives of citizens.

regulation eggs: The eggs in which case all the parts are equipotent and capable of regulating so as to replace any part that might be removed. It was concluded by H. Driesch in 1891 that early cleavage stage is capable of forming whole of the animal.

regulation of adaptation processes at the cellular level in plants: Under heat stress, aggregation and alteration in metabolites in chloroplasts may perform an important function in regulating the adaptation processes at the cellular levels in plants, which allows plants to activate heat stress responses at the optimal time to sustain photosynthesis. These metabolites can be osmolytes that accumulate in plants in response to environmental stresses (e.g. proline, glycine betaine), enzymic (e.g. SOD, catalase, peroxidase) and non-enzymic antioxidants (e.g. glutathione, ascorbic acid, tocopherols) and sugar molecules that are involved in signalling events for maintenance of cellular redox homeostasis under heat stress conditions. Multiple metabolic reprogramming in chloroplasts could play a crucial role in systemic acquired acclimatization against stress. This sort of metabolic reprogramming is important for plants to survive under stress conditions and to stop further impairment to the whole plant.

regulation of alternative RNA splicing by histone modifications: Most human genes are alternatively spliced in a cell or tissue-specific manner. Pre-mRNA splicing is influenced by RNA polymerase II elongation rate, chromatin remodellers and histone deacetylase inhibitors.

regulation of cap-dependent translation: Eukaryotic messenger RNAs contain a modified guanosine, termed a cap, at their 5′ ends. Translation of mRNAs requires the binding of an initiation factor, eIF4E, to the cap structure. family of proteins through a shared sequence regulate cap-dependent translation.

regulation of chromatin state: Histone acetylation controls gene expression, cell function and cell fate. P. Willnow and A.A. Teleman in 2024 investigate the profile of histone acetylation in the epithelial tissue of *Drosophila* wing disc. H3K18ac, H4K8ac and total lysine acetylation are enhanced in the outer rim of the disc. This acetylation profile is regulated by nuclear position, whereby nuclei incessantly travel from apical to basal positions within the epithelium and display increased levels of H3K18ac when they are in juxtaposition to the tissue surface. These surface nuclei have augmented levels of acetyl-CoA synthase, which produces the acetyl-CoA for histone acetylation. The carbon source for histone acetylation in the rim is fatty acid β-oxidation, which is also augmented in the rim. Inhibition of fatty acid β-oxidation results in decline in the H3K18ac levels in the genomic juxtaposition of genes involved in disc development.

regulation of chromatin structure: DNA methylation, histone modifications and RNA interference play critically important roles in regulating chromatin structure, thereby profoundly affecting transcription and other molecular events.

regulation of defence gene expression by NPR1: SAR is a ‘whole-plant’ resistance response that occurs following an earlier localized exposure to a pathogen. R genes/SAR activate SALICYLIC ACID (SA), which activates *Non-expressor of Pathogenesis-Related 1* gene (*NPR1**). Alternatively, INDUCED SYSTEMIC RESISTANCE (ISR) activates jasmonic acid (JA) and ethylene (ET), which activate NPR1. All this occurs in the cytoplasm. NPR1* enters into nucleus to complex with the transcription factors that bind specifically to the palindrome TGACGTCA (TGA). NPR1*-TGA complex binds the promoter to activate its downstream defence genes. Defences are thus activated.

regulation of flowering time by histone acetylation: The *Arabidopsis* autonomous floral-promotion pathway promotes flowering independently of the photoperiod and vernalization pathways by repressing *FLC*, a MADS-box transcription factor that blocks the transition from vegetative to reproductive development. *FLD*, one of six genes in the autonomous pathway, encodes a plant homologue of a protein found in histone deacetylases complexes in mammals. The *FLD* lesions undergo histone hyperacetylation in *FLC* chromatin and enhanced *FLC* expression, ultimately delay flowering. Multiple pathways correspond to *FLC* repression as all autonomous-pathway mutants do not undergo *FLC* hyperacetylation. The quantitative trait locus mapping showed that histone deacetylase HDAC6 controls chromatin remodelling capacity to regulate flowering.

regulation of flowering time by light quality: The environmental signals such as light and temperature influence flowering. The dense canopy plants receive reduced ratio of red to far-red incoming light, which lead to the emergence of warning mechanism 'shade-avoidance syndrome'. It results in the stem elongation at the expense of leaf expansion to increase the flowering. The phytochrome B (phyB) responds well in this mechanism to regulate flowering in response to the ratios of red to far-red light.

regulation of gene expression: Controlling the gene expression that involves transcription, translation and protein synthesis. In prokaryotes, gene expression is controlled usually at the level of transcription. Regulation of *lac* operon is mediated by binding of *lac* repressor to its operator. This binding prevents transcription, it is reversed by binding to *lac* repression of inducer the substrate for *lac* operon encoded enzymes.

regulation of transcription factor cooperativity: Pioneer transcription factors have the ability to access DNA in compacted chromatin. Multiple transcription factors can bind together to a regulatory element in a cooperative way, and cooperation between the pioneer transcription factors OCT4 (*aka* POU5F1) and SOX2 is important for pluripotency and reprogramming. K.K. Sinha *et al.* (2023) present cryo-electron microscopy structures of human OCT4 bound to a nucleosome containing human *LIN28B* or n*MATN1* DNA sequences, both of which bear multiple binding sites for OCT4. Binding of OCT4 induces changes to the nucleosome structure, repositions the nucleosomal DNA and facilitates cooperative binding of additional OCT4 and of SOX2 to their internal binding sites. The flexible activation domain of OCT4 contacts the N-terminal tail of histone H4, altering its conformation and thus promoting chromatin decompaction. Moreover, the DNA-binding domain of OCT4 engages with the N-terminal tail of histone H3, and post-translational modifications at H3K27 modulate DNA positioning and affect transcription factor cooperativity. Thus, the epigenetic landscape could regulate OCT4 activity to ensure proper cell programming.

regulation of transposable elements: Host genomes have evolved defence mechanisms, including chromatin modifications and RNA interference, to regulate transposable elements.

regulative development: In regulative development, the fate of each cell is determined by interactions with other cells. Each cell undergoes conditional specification (i.e. if it is removed from the embryo it will fail to fulfil its fate because it lacks the appropriate interactions with other cells). The remainder of the embryo, however, can regulate itself and replace the missing parts. This is because the appropriate interactions have yet to occur, and therefore, the fates of the remaining cells are not determined. The region of the embryo deriving interactions is termed organizer. It has the spatial property of being able to induce the formation of new structures when transplanted to another region of the embryo because it is the source of primary inductive signals. Mammals provide an extreme example of regulative development. There appears to be no intrinsic asymmetry in the egg, and in the first few divisions, any cell in the blastocyst can be isolated and will develop into a complete embryo.

regulator element: *See* CONTROLLING ELEMENT *and* RECEPTOR ELEMENT.

regulator(y) cell: Any cell that controls the fate of other cells.

regulator(y) gene: In the operon theory of gene regulation, a gene that is involved in switching on or off the transcription of structural genes. When transcribed, the regulator(y) gene produces a repressor protein, which switches off an operator gene and hence the operon that it controls. The regulator(y) gene is not part of the operon and may even be on a different chromosome. It has its own promoter. There are three groups of regulatory genes – MATERNAL GENES, SEGMENT GENES and HOMEOTIC GENES. The distinction between structural and regulatory genes is more or less arbitrary because the proteins (e.g. lac repressor of *Escherichia coli*) are the product of a structural gene; however, its function is regulatory in that it directly controls the expression of a set of structural genes coding for enzymes. A regulator(y) gene that can initiate or block the functions of other gene or a group of genes. Regulator(y) genes control the timing of production of a variety of chemicals in humans and other organisms. Shortly after conception, regulator genes work as master switches orchestrating the timely development of our body parts. They are also responsible for changes that occur in our bodies as we grow older. Small changes in regulatory genes cause large changes in phenotypes. *See details as separate terms.*

regulator(y) gene mutations: These mutations mapped very close to the operator of an operon. Mutations of regulatory gene resulted in constitutive synthesis of enzymes. Mutants of repressor gene were operationally different from operator mutants since they acted on all the three structural genes in the same way in both *cis*- and *trans*-arrangements in MERODIPLOIDS. Two types of *i* mutants in lactose operon have been isolated. The i^s mutants produced super-repressors that prevented induction of the proteins when substrate was present and the i^- mutants caused constitutive synthesis of the *i* gene mutants in haploids and merodiploids. The dominance-recessiveness relationship for *i* gene mutants observed in merodiploids is $i^s > i^+$; $i^+ > i^-$. Since *i* mutants acted in both *cis*- and *trans*-arrangement, this implied that *i* gene specified a product, which could diffuse from site of its synthesis to an altogether different region of the genome and thereby influence function of wild-type operator.

regulator(y) molecules: Proteins that themselves are specified by regulator(y) genes. *Aka* REPRESSORS.

regulator(y) region: Stretches of the DNA sequence, which control the activity of genes.

regulator(y) RNA: Any of the small RNA molecules that bind to messenger RNA or directly to DNA or both and regulate gene activity.

regulator(y) sequences: DNA sequences that direct transcription to occur in the desired tissues and at the correct stages of plant development.

regulators in cell differentiation: The continuous activity of positive and negative regulators is required to maintain differentiation in the course of normal development. There exist regulatory genes of high order termed differentiation genes, which are characterized by a system of mutual conditioning for susceptibility to environmental stimuli. Intracellular differentiation begins with a progenitor cell giving rise to two sibling cells that have different states of differentiation called REPLICATIVE DIFFERENTIATION.

regulatory change: Any change in the regulatory elements of a gene would be classified under regulatory change. The chief molecular barriers to development of hybrid organisms are regulatory. Regulatory genes code for RNAs or protein products whose function is to control the expression of other genes. Any change in these sequences would be classified under regulatory change. The chief molecular barriers to development of hybrid organisms are regulatory. Evolution at organismic level depends predominantly on regulatory changes and structural gene changes may have secondary role in organismal evolution. The changes in regulatory systems may be caused as a result of a number of gene rearrangements (e.g. inversions, translocations, duplications, fusions or fissions), thereby changing the order of genes or the chromosomes. This may alter the pattern(s) of gene expression.

regulatory code: The genome has more than one code for specifying life. Buried in DNA sequences is a regulatory code akin to the genetic code. The regulatory code is encoded in the arrangement of an enhancer's DNA-binding sites, in the spacing between binding sites or by the loss or gain of one or more of these sites.

regulatory controls in prokaryotes and eukaryotes: In prokaryotes, all the genes are always active and the regulatory mechanisms are designed to turn off those genes whose product is not required by the cell. Genes for enzymes involved in a particular pathway are linked in operon fashion. They are under the control of the same promoter. In eukaryote, all genes are normally inactive unless turned on by gene regulatory mechanisms. Genes are scattered among different chromosomes. Each is under the control of its own promoter. Barring some exceptions, operon concept does not apply to eukaryotes.

regulatory elements: DNA sequences to which regulatory proteins like transcription factors bind. They are classified as PROMOTERS, ENHANCERS, SILENCERS, CHROMATIN INSULATORS or CELLULAR MEMORY MODULES (CMMs). These elements are specific to certain types of genes are involved in regulating expression in response to changes in intra- or extracellular environment. It is in the upstream region of the promoter that regulatory elements, such as those involved in the transcriptional induction in response to heat shock or hormone/receptor complexes, are located.

regulatory elements in eukaryotic gene regulation: Regulatory elements are specific to certain types of genes, which are involved in regulating expression in response to changes in intra- or extracellular environment. It is in the upstream region of the promoter that regulatory elements, such as those involved in the transcriptional induction in response to heat shock or hormone/receptor complexes are located.

regulatory enzyme: A highly specialized enzyme having a regulatory (controlling) function through its capacity to undergo a change in its catalytic activity. There exist two major types of regulatory enzymes: (i) COVALENTLY MODULATED ENZYMES and (ii) ALLOSTERIC ENZYMES. *See details as separate terms.*

regulatory gene: A gene that codes for a repressor, which in turn regulates the transcription of structural genes in an operon by binding to the operator locus. *Compare with* STRUCTURAL GENE.

regulatory interactions between CNL proteins and pathogen virulence effectors: Regulatory interactions between three *Arabidopsis* CNL (CC-NBS-LRR) proteins (RPM1 and RPS2, RIN4 (RPM1-Interacting Protein 4)) and three pathogen virulence effectors [AvrB, AvrRpm1 and AvrRpt2] are understood. The protein RPM1 detects the phosphorylation of RIN4 by AvrB and AvrRpm1 and elicits the resistance response. This outcome can be blocked by AvrRpt2, a protease that cleaves RIN4. The disappearance of RIN4 is detected by RPS2, resulting in elicitation of the defence response.

regulatory mechanisms governing EMOs and GMOs: One important issue of EPIGENETICALLY MODIFIED ORGANISMS (EMOs) is regulatory, dealing with the extent to which the proponents of biotechnology will push back against EMOs. It is arguable that crops that have had their epigenome altered will be considered by both regulators and activists the same as GENETICALLY MODIFIED ORGANISMS (GMOs). One reason this may be the case is that GMOs and EMOs can use similar creation methods. For example, the genome-editing technique CRISPR has been modified so that researchers can make epigenetic alterations changes. While the US government has yet to weigh in on how it will see CRISPR-created EMOs, anti-biotechnology activists have made it clear that they want crops edited with CRISPR to be regulated the same as other biotechnologies.

regulatory networks, irreversibility in: Irreversibility, a transient perturbation which leaves a system in a new state, is a developing property in systems of interacting entities. This property has well-established consequences in statistical physics but remains underexplored in biological networks, especially for bacteria and other prokaryotes whose regulation of gene expression transpires predominantly at the transcriptional level. Focusing on the reconstructed regulatory network of *Escherichia coli*, Y. Zhao *et al.* (2024) investigate network responses to transient single-gene perturbations. They forecast irreversibility in numerous cases and find that the incidence of irreversibility rises with the proximity of the perturbed gene to positive circuits in the network. They suggest a relationship between the predicted irreversibility to transient perturbations and the evolutionary response to permanent perturbations.

regulatory protein: General name for any protein that binds to a specific DNA sequence to alter the expression of a gene or prevent the initiation of transcription. Also called REPRESSOR protein.

regulatory proteins in eukaryotes, working of: Regulatory proteins act in eukaryotes in a specific manner in the following ways: (i) The factor may be produced in certain tissues only. (ii) The factor may have to be activated by phosphorylation or some other modification. A control of this modification could provide specific regulation ability. (iii) The factor may be inactive. It becomes active only when it binds to a specific molecule. The availability of this molecule thus regulates the gene action of steroid receptors. (iv) The factor may be present in all types, but in non-target cell types, it is bound to an inhibitor, which may be a protein and is therefore inactive (e.g. factor NFKs; active kappa chains in B lymphocytes are bound to an inhibitory protein in cells other than B lymphocytes). (v) An inhibitory protein in the non-target tissue may occupy the target site of a factor in the promoter. As a result, the concerned factor is unable to bind to the promoter, and as a result, the transcription cannot be initiated by the CAAT-binding factor.

regulatory region or sequence: A DNA base sequence that controls gene expression.

regulatory science: The development and implementation of risk-analysis methods and the maximization the utility of risk analyses to inform regulatory decisions for biotechnology products, consistent with human health and environmental risk–benefit standards provided in relevant government statutes. Regulatory science includes the establishment of information and data quality standards and study guidelines and the generation of data and information to support risk analyses. It can include the development of risk-mitigation measures as well as the development and implementation of safety training and certification programmes to help ensure the intended benefits of products are realized and risks to workers, users and the environment are minimized. Individuals in government, industry, academia and non-governmental organizations that contribute to the advancement of regulatory science have degrees across disciplines in the natural, socio-economic and computational sciences, engineering and public policy.

regulatory sequence: A DNA sequence involved in regulating the expression of a gene (e.g. a promoter or operator region).

regulatory single-nucleotide polymorphism (regulatory SNP, rSNP): A relatively rare single-nucleotide polymorphism that affects the expression of a gene (or several genes). Usually this SNP is located in the promoter of the gene.

regulatory site: Site on an enzyme, other than the active site, that binds a molecule that affects enzyme activity.

reinforcement: An increase in reproductive isolation between incipient species by natural selection. Natural selection can directly favour only an increase in prezygotic isolation; reinforcement therefore amounts to selection for assortative mating between the incipiently speciating forms.

reiterated DNA: Nucleotide sequences that occur many times within a genome. Analysis of the two *Rhizobium* strains indicated that DNA reiteration is not confined to the chromosome or to some plasmids but is a property of the whole genome.

rejuvenation: Synonymous with dedifferentiation or treatment that leads to culture invigoration or revival. In horticulture (e.g. in fruit tree planting), reversion from adult to juvenile by restoration of juvenile vigour (growth) on a mature entity.

relation between probe and gene: The relation between a probe and the mRNA that it is expected to detect is not trivial. (i) Some mRNAs may cross-hybridize probes in the array that are supposed to detect another mRNA. (ii) mRNAs may experience amplification bias that is sequence or molecule specific. (iii) Probes that are designed to detect the mRNA of a particular gene may rely on genomic EST information that is incorrectly associated with that gene.

relational database: A digital database based on the relational model of data, as proposed by E. F. Codd in 1970. A software system used to maintain relational databases is a relational database management system (RDBMS). Many relational database systems have an option of using the SQL for querying and maintaining the database.

relational database management systems (RDBMS): A software system that includes a database architecture, query language, data loading and updating tools and other ancillary software that together allow the creation of a relational database application. An RDBMS stores data in a database consisting of one or more tables of rows and columns. The rows correspond to a record (tuple); the columns correspond to attributes (fields) in the record. In an RDBMS, a view, defined as a subset of the database that is the result of the evaluation of a query, is a table. RDBMSs use STRUCTURED QUERY LANGUAGE (SQL) for data definition, data management and data access and retrieval. Relational and object-relational databases are used extensively in bioinformatics to store sequences and other biological data.

relationship between genotype and environment: The equation 'phenotype = genotype × environment' describes the relationship between genotype and environment. The genotype is relatively stable throughout the life of an individual, whereas the phenotype changes continuously during development.

relative Darwinian fitness: The fitness divided by the mean fitness or by the fitness of a reference genotype. *See* ADAPTIVE FITNESS.

relative dating: The process of ordering fossils, rocks and geologic events from oldest to youngest. Because of the way sedimentary rocks form, lower layers in most series are older than higher layers, making it possible to determine which fossils found in those layers are oldest and which are youngest. By itself, relative dating cannot assign any absolute age to rocks or fossils.

relative rate test: A mathematical test for unequal rates of evolution.

relative risk: The ratio of the probabilities of two events. Unlike the odds ratios and hazard ratios, risk ratios are not capable of being constant but instead must depend on the base risk (e.g. the risk for a subject who does not have a risk factor). For example, a risk ratio of 2 may apply only to subjects with base risks <1/2. Also, unlike the odds ratio, the risk ratio depends greatly on which of two outcomes is labelled as the 'event'; a mortality ratio does not equal the survival ratio. The term relative risk is often inappropriately used to describe an odds ratio or a hazard ratio. *Aka* RISK RATIO.

relaxed control: A situation in which the numbers of a particular plasmid present in a bacterial cell are not closely regulated by the coupling of plasmid DNA synthesis to host cell DNA synthesis. In relaxed control, up to 100 copies of plasmid per cell may be found; this may be further increased using chloramphenicol.

relaxed DNA: The DNA is fully intact with both strands uncut but has been enzymatically 'relaxed' (supercoils removed). *See* SUPERCOIL.

relaxed mutant: A mutant that does not exhibit the STRINGENT RESPONSE under amino acid starvation.

relaxed plasmid: A plasmid whose replication is not linked to replication of the host genome and which therefore can exist as multiple copies within the cell.

relaxed replication control: The plasmids that are replicated independently of the replication of the bacterial chromosome. They have individual mechanisms for the initiation of replication (e.g. plasmids with high copy number are under relaxed control of replication).

relaxed response: In bacteria, under the conditions of amino acid starvation, the cell's continued production of tRNA and rRNA. *See* STRINGENT RESPONSE.

relaxed state of repair theory of ageing: The balance between the rate at which DNA damage occurs and DNA is repaired determines ageing. The germ cells may be preserved against the normal, time-dependent DNA damage. If it is so it is quite possible that ageing of somatic lines is caused from accumulation of DNA damage and relaxed state of DNA repair. This may lead to errors in transcription and translation processes, which would give rise to faulty enzymes and lead to decline in functional ability of the cell. DNA repair is important in determining longevity. In fact, repair capacity is correlated with maximum lifespan.

release factors (RFs): Proteins (RF1, RF2 and RF2) in prokaryotes responsible for the termination of translation and release of newly synthesized proteins when a nonsense (termination) codon appears in A site of the ribosome. Replaced by eRFs (eRF1, eRF1A and eRF3) in eukaryotes. *Aka* TERMINATION FACTORS.

reliability: It is the stability or consistency to repeat the same performance or result under consistent conditions. The performance could be poor, but the same performance every time means good reliability. Statistically, this means a smaller variance in repeated measurements. The range of reliability is 0 to 1.

reliability engineering: A subdiscipline of systems engineering that emphasizes dependability in the life cycle management of a product. Reliability describes the ability of a system or component to function under stated conditions for a specified period of time. Reliability is closely related to availability, which is typically described as the ability of a component or system to function at a specified moment or interval of time.

remediation: (i) The clean-up or containment (if chemicals are moving) of a hazardous waste disposal site to the satisfaction of the applicable regulatory agency, such as the Environmental Protection Agency (EPA). Such clean-up can sometimes be accomplished via use of microorganisms that have been adapted (naturally or via genetic engineering) to consume those chemical wastes present in the disposal site. (ii) Specific actions undertaken to deal with particular problems of environmental quality, such as the liming of acidic lakes and rivers to decrease their ecological damage. *Compare with* RESTORATION *and* RECLAMATION.

remote hybridization: Refers to crosses between two different species, genera or higher-ranking taxa, which can break species limits, increase genetic variation and combine the biological characteristics of existing species. *Aka* DISTANT HYBRIDIZATION.

removal of introns: Introns are *aka* INTERVENING SEQUENCES (IVS). When present, introns are very short (13–60 nucleotides). Introns in many eukaryotic genes begin with dinucleotide 5′-GT-3′ and with dinucleotide 5′-AG-3′. The cleavage and splicing at intron junctions have been found to be highly dependent upon the presence of the −CCA, at 3′-end. IVS are located in most of the tRNA precursors of eukaryotes and in advanced genera of prokaryotes at a point one site removed downstream from anticodon. However, in gene for $tRNA^{leu}$ the intron splits the anticodon sequence. Introns are removed by a two-step reaction. The first step is covalent modification of the substrate. RNA is cleaved at a precise 5′ site followed by covalent bond formation between 5′-end of the intervening sequence and adenosine residue near the 3′ splice site to produce a lariat molecule containing intervening sequence and 3′ exon. In the second step, the bipartite intermediate is subsequently resolved by cleavage at 3′ splice site followed by ligation of 5′ and 3′ exons via 3′,5′-phosphodiester bond. The phosphate at the 5′ splice site becomes incorporated into 2′-5′ phosphodiester bond in the branch structure and phosphate at 3′ splice site is used to form 3′,5′ phosphodiester bond linking the two exons in the produced RNA. The introns are deleted very precisely. *Aka* RNA SPLICING.

renaturation: The return to the natural structure of a protein or nucleic acid from a denatured (more random coil) state. For example, a protein may be denatured (lose its native (natural) structure) by exposure to surfactants such as SDS or to changes in the pH of the medium. If the surfactant is slowly removed, or the pH is slowly readjusted to the optimum for the protein, it will refold (snap) back into its original (native) form. *Compare with* DENATURATION.

renaturation kinetics: A technique that measures the rate of reassociation or reannealing of complementary single DNA strands derived from a single source that is used to indicate genome size and complexity. *See* COT PLOT.

renewable energy: Energy from renewable resources such as wind power, solar energy or biomass.

renewable resource: A resource that can be exploited without depletion because it is constantly replenished. This includes agricultural crops and fish, provided stocks are not overfished and are extended to cover the energy of solar radiation, wind, waves and tides.

Renner complexes: A specific gametic chromosome combination in evening primrose (*Oenothera* spp.).

Rep protein: An enzyme isolated from certain strains of *Escherichia coli* that opens the strands of the DNA duplex and therefore acts as a helicase. This enzyme also hydrolyses ATP.

repair: The correction of alterations in DNA structure before they are inherited as mutations.

repair DNA synthesis: Enzymatic excision and replacement of regions of damaged DNA. Repair of thymine dimers by UV irradiation is the best-understood example.

repair nuclease: A nuclease that recognizes a DNA strand with a missing base and cleaves out the damaged nucleotide, plus some neighbouring nucleotides, allowing later repair by insertion of an undamaged sequence by DNA polymerase and DNA ligase.

repair of DNA: The corrections of alterations in DNA structure before they are inherited as mutations.

repeat identification: Identification of repetitive sequences. Repetitive sequences occupy a large portion of most eukaryotic genomes and are divided into tandem (e.g. simple sequence repeats or SSRs) and interspersed repeats. Transposable elements (TEs), one of the interspersed repeats, are the most abundant in the repeat family.

repeat-associated siRNAs (rasiRNAs): Small, endogenous RNAs encoded by repeat sequences.

repeat-induced gene silencing (RIGS): A *cis*-TGS event induced by the insertion of multiple copies of a transgene at one locus. RIGS is abolished by reducing the copy number.

repeat-induced point (RIP) mutation: A phenomenon similar to METHYLATION INDUCED PREMEIOTICALLY (MIP) that is associated with point mutation in the duplicated sequences in the fungus *Neurospora crassa* whereby repeated sequences of DNA above a certain length are either excised through rearrangements (rearrangements induced pre-meiotically) or are modified and mutated (repeat-induced point mutation). The phenomenon is similar to MIP, which is associated with point mutation in the duplicated sequences in the fungus *Neurospora crassa*.

repeated blocks: An incidence structure of points and blocks has repeated blocks if there are two blocks incident with exactly the same points.

repeated DNA sequences and diseases: A number of human diseases are caused by having triplet repeat amplification such as in Huntington's disease where 11–34 repeats of CAG in the Huntington's disease gene is normal but ~50 to 100 results in the disease. Interspersed repeated DNA make up 25–40% of most mammalian genomes that are hundreds to thousands of base pairs long. Many interspersed repeated DNA sequences are TEs.

repeated gene family: It consists of many identical or nearly identical genes that cohabit a single haploid genome. In some cases, the genes are contained within tandemly repeated units.

repeated genes: Some genes may be present even in a haploid genome in many copies. These copies may be identical or only similar. The genes of ribosomal RNA are repeated in several tandem copies. Each one consists of one transcription unit, but the gene cluster is usually transmitted from one generation to the next as a single unit. Thus, the units of transmission and transcription are not always the same. The histone genes have been observed to be repeated in such tandem repeats in many higher eukaryotic organisms.

repeating unit: A tandem cluster is the length of the sequence that is repeated; appears circular on a restriction map.

RepeatMasker: Available at http://repeatmasker.org, this is a wildly used programme to find interspersed repeats (LINEs, SINEs, LTRs and DNA elements), simple sequence repeats (SSRs) and low complexity regions in the sequences using similarity search against well-defined repeat database. A repetitive element filter screens a sequence against a library of repetitive sequences and searches for low-complexity regions.

repeats in promoters: The nucleosome-free region directly upstream of genes (the promoter region) is enriched in repeats. One-fourth of all gene promoters contain tandem repeat sequences. Genes driven by these repeat-containing promoters showed a significantly higher rate of transcriptional divergence. Tandem repeats are variable elements in promoters that may facilitate evolutionary tuning of gene expression by affecting local chromatin structure.

repetition frequency: The integral number of copies of a given sequence present in the haploid genome; equals 1 for non-repetitive DNA, >2 for repetitive DNA.

repetitive DNA sequences: Repeat sequences and approximate repeats occur throughout the DNA of higher organisms (mammals). DNA is made up of copies of the same nucleotide sequence. Repeats in DNA sequences vary in size 5–100,000 nucleotides. Most repeats are located at telomeres and centromeres. Putative functions of repeated DNA are suggested to be many, including involvement in chromosome pairing, control of gene expression, processing of RNA (excision of introns) and participation in DNA replication. For example, (i) *Alu* sequences of about 300 characters in length appear over 600,000 times in human DNA, with about 87% homology to a consensus *Alu* string. (ii) Some short substrings, such as TATA-boxes, poly(A) and poly(TG), also appear more often than would be expected by chance. (iii) Repeat sequences may also occur within genes, as mutations or alterations to those genes. Repetitive sequences, especially mobile elements, have many applications in genetic research. (iv) DNA transposons and retroposons are used routinely for insertional mutagenesis, gene mapping, gene tagging and gene transfer in several model items. (v) Some repeat sequences may have unknown functions. (vi) Some mRNAs, tRNAs, 5S RNAs and histones are coded by repetitive sequences. (vii) In the human genome, 15% of DNA is repetitive. One-third is large piece repeats (10,000–300,000 bp) often copied from one chromosome location and 3% is single DNA short sequence repeats of -GTTAC-. Repeat sequences vary in size between 5 and 500 bp. *Aka* REPEATED DNA SEQUENCES, REPETITIOUS DNA SEQUENCES *or* REPETITIVE ELEMENTS.

repetitive element repression (RER): Repetitive elements such as retrotransposons, transposons, short interspersed nucleotide elements (SINEs), tandem repeats and microsatellites make up more than half of the genome. They are normally covered with epigenetic modifications that result in a compact DNA/protein structure termed heterochromatin. This compact structure prevents transcription of mobile repetitive elements or unequal recombination between sequences with similar DNA sequences that are located in different parts of the genome. RER is achieved through DNA methylation and histone modifications that depend on the class of repeat and the cell type. For example, in mouse embryonic stem cells (ESCs), intracisternal A particle (IAP) retrotransposons bear high levels of H4K20me3, whereas DNA transposons and pericentric satellite repeats have high levels of H3K9me3.

repetitive-and-unique sequence hypothesis: This hypothesis envisages that only a small proportion of DNA is meant for genes carrying the genetic information, and the rest of it has some other functions like control of gene activity. Now we know that genes in higher organisms are present both in repeated and unique sequences and amplification of all the genes is not necessarily found. Only a small proportion of DNA is unique and is meant for genes carrying the genetic information, and the rest of DNA is repetitive and has some other functions like controlling gene activity and serving as raw material for evolution.

replacement sites: In a gene, those sites at which mutations alter codons such that a different amino acid is incorporated during protein synthesis.

replacement vector: Bacteriophage λ vector in which cleavage sites flank a region containing genes not essential for the lytic cycle, and this region may be deleted and replaced by the DNA insert in the cloning process, and a larger sized DNA of 8–24 kb may be inserted.

replica plating: Inoculating a fresh plate by pressing it against a piece of velvet or similar material that has previously been pressed on to a plate with colonies of microorganisms. Replica plating is used to identify mutations. In this technique, a pad of sterile material such as latex, filter paper or velvet is pressed over the master plate and some of the cells from each colony adhere to the velvet. Further, the velvet is pressed down on to two or more sterile plates where colonies develop resembling to master plate. This method is used to select particular mutants from a plate.

replicase: An enzyme that copies any form of the genome (in the origin of life, the genome may not have been DNA- or RNA-based). An enzyme that unwinds double-helical DNA.

replicate: A more or less exact duplication or repetition of a test, an experiment or an experimental single plot to assure or to increase confidence in the resulting data.

replicate estimate: An estimate of the population quantity based on the replicate subsample using the same estimation methods used to compute the full sample estimate.

replicate sample: One of a set of subsamples, each obtained by deleting a number of observations in the original sample for the purpose of computing the appropriate variance based on the complex design of the survey.

replicated measurement t-test design: A well-suited design to field experiments when there is one treatment and a control condition or two treatments. When the measured data have been determined, the data can be analysed using an unpaired t-test. It is best to have between 10 and 20 observations for each treatment. Taking fewer than ten observations makes it difficult to detect the systematic effects of the treatment. A modification to the design that would improve the precision is the PAIRED T-TEST. The advantage of the method is that it detects differences between varieties A and B. The disadvantage is that it may involve more inconvenience to the breeder. The paired t-test is used when the design has been set up to link pairs of observations. Data may be taken as paired when it can be assumed that sources of experimental error (factors that influence the outcome of the experiment that are not part of the treatment) are the same for each of the pairs.

replicating form (RF): The double-stranded form of a single-stranded DNA virus (e.g. ϕX174).

replicating systems, origin of: Cross-catalytic self-replication of complementary templates can proceed with efficiency similar to that of autocatalytic self-replication of self-complementary templates. In 1960, it was proposed that the ancestor molecule was neither a DNA nor protein but RNA, a versatile messenger molecule. This concept got a further boost when it was discovered that RNA catalyses self-splicing replication by degrading the primer and filling in the gaps. May also be helpful in termination of DNA replication.

replication: (i) The copying of a nucleic acid molecule into a new nucleic acid molecule of a similar type (i.e. DNA → DNA or RNA → RNA). *See* DNA REPLICATION *and* RNA REPLICATION. (ii) The equal incorporation of all combinations two or more times in an experimental design, which is then said to be replicated.

replication error: Any modification that prevents or disturbs the DNA replication process.

replication eye: A region in which DNA has been replicated within a longer, unreplicated region. The term refers to the configuration of replicating DNA in eukaryotic chromosomes.

replication factor A (RFA): RFA is a heterotrimeric single-strand binding (ssb) protein, which is conserved in all eukaryotes. RFA function is critical not only for DNA replication but also for efficient DNA repair and recombination.

replication factor C (RFC): This protein is structurally, functionally and evolutionary related to γ complex. RFC is a five-subunit protein complex that is required for DNA replication. The subunits of this heteropentamer are named RFC1, RFC2, RFC3, RFC4 and RFC5 (in *Saccharomyces cerevisiae*). RFC is used in eukaryotic replication as a clamp loader, similar to the γ Complex in *Escherichia coli*. Its role as a clamp loader involves catalysing the loading of PCNA on to DNA. It binds to the 3′ end of the DNA and uses ATP hydrolysis to open the ring of PCNA so that it can encircle the DNA.

replication fork: Y-shaped structure found at the point where DNA is being synthesized. During DNA replication the parental DNA is unwound to allow the addition of nucleotides. The point at which replication occurs is called the replication fork. As the replication fork moves along the parental DNA, each of the unwound single strands combines with new nucleotides. The original strand and newly synthesized daughter strand are then rewind. The RNA-dependent DNA polymerases catalyse polymerization of DNA only in 5′→3′ direction. The replication fork looks like a fork in the road that is composed of a leading strand and a lagging strand of DNA. The leading strand in the replication fork grows in the 5′→3′ direction, and the lagging strand grows in the 3′←5′ direction.

replication fork assembly: The replication fork apparatus is known to be composed solely of DnaB helicase and DNA polymerase III holoenzyme.

replication fork progression: Chl1 helicase controls replication fork progression.

replication fork reversal: DNA replication is challenged by obstacles, including DNA damage, which impede synthesis and threaten genome stability. A common response to this replication stress is replication fork reversal, in which parental DNA reanneals and a new daughter strand duplex is formed. W. Liu *et al.* (2023) describe a mechanism dependent on the recombinase RAD51 by which cells can accomplish reversal without unloading the replication machinery, including the helicase complex, which cannot be reloaded once replication has started. This mechanism explains how reversal helps cells tolerate replication stress and rapidly resume DNA synthesis once the replication obstacle is removed.

replication number: The replication number of a treatment in a block design is the number of times the treatment occurs in the design. In the case of a BINARY DESIGN, this is equal to the number of blocks in which the treatment occurs. In a RESOLVABLE DESIGN, it is equal to the number of replicates or parallel classes in a resolution.

replication of foreign DNA segment in recipient DNA: There are four requirements: (i) The presence of a region on the DNA, which serves as point of origin of DNA replication. (ii) Controlling or regulating element is required for expression of the inserted gene. (iii) Move recombinant DNA into host cell that provides machinery for DNA replication (by transformation, electroporation or transfection). (iv) Select host cells that contain recombinant DNA. Gene cloning is based on host-vector system. A host (could be bacteria, fungus, higher organisms) should be *rec*⁻. A vector (could be plasmid, 4–5 kb; phage, 20–25 kb; cosmid, 40–45 kb; or YACs, 100–400 kb) must be *tra*⁻.

replication of virus: Reproduction of the original virus. This process can be viewed as occurring in stages, in which the first stage consists of the adsorption of the virus to the host cell; penetration of the virus (or its nucleic acid) into the cell, the taking over of the cell's biomachinery and harnessing of it to replicate viral nucleic acid along with the synthesis of other virus constituents; the correct assembly of the nucleic acids and other constituents into a functional virus; followed finally by release of the virus from the confines of the cell.

replication origin: A site on a DNA molecule where unwinding begins in order for replication to occur.

replication protein A (Abbrev. RPA): A eukaryotic single-stranded DNA-binding protein complex involved in DNA replication and repair.

replication stress: A complex phenomenon that has serious implications for genome stability, cell survival and human disease. Generation of aberrant replication fork structures containing single-stranded DNA activates the replication stress response, primarily mediated by the kinase ATM- AND RAD3-RELATED (ATR).

replication timing (RT): Refers to the order in which segments of DNA along the length of a chromosome are duplicated.

replication timing (RT) during early mammalian development: DNA replication enables genetic inheritance across the kingdoms of life. Replication occurs with a defined temporal order known as the RT programme, leading to the organization of the genome into early- or late-replicating regions. Using single-cell Repli-seq, T. Nakatani *et al.* (2024) generated genome-wide RT maps of mouse embryos from the zygote to the blastocyst stage. They showed that RT is initially not well defined but becomes defined progressively from the 4-cell stage, coinciding with the strengthening of the A and B compartments. RT depends on RNA polymerase II at zygotic genome activation.

replication-competent retroviral vectors: These retroviruses contain all necessary genes for virion synthesis and continue to propagate themselves once infection occurs.

replication-coupled DNA breaks, structure and repair of: R. Pavani *et al.* (2024) use a Cas9-nicking enzyme to produce site-specific nicks in the human genome. Depending on which DNA strand a nick is located in relative to an incoming replication fork, DSBs are produced with different end structures. Furthermore, these DSBs are processed differently compared with DSBs produced independently of replication. Using CRISPR-Cas9 nicking enzymes, they studied the interaction between the replication machinery and single-strand breaks, one of the most common forms of endogenous DNA damage. Replication forks breakdown at leading-strand nicks generates resected single-ended double-strand breaks (seDSBs) that are repaired by HOMOLOGOUS RECOMBINATION (HR).

replication-defective HSV vectors: These vectors are made by the deletion of one or more immediate-early genes (e.g. *ICP4*), which is then provided *in trans* by a complementing cell line.

replication-defective retroviral vectors: In these viruses, genes essential for DNA replication are removed. Replication-defective vectors are the most common choice in studies because the viruses have had the coding regions for the genes necessary for additional rounds of virion replication and packaging replaced with other genes or deleted. These vectors allow you to express your gene of interest from either a CYTOMEGALOVIRUS (CMV) promoter or the 5′-LTR promoter.

replication-defective virus: The virus that has lost one or more genes essential for completing the infective cycle.

replicative cell senescence: Phenomenon observed in primary cell cultures as they age, in which cell proliferation slows down and finally halts.

replicative DNA transposon: A DNA-based transposable element that moves itself to a new place in the genome and also leaves a copy in the original location.

replicative primases: DNA primase is an enzyme involved in the replication of DNA and is a type of RNA polymerase. Primase catalyses the synthesis of a short RNA (or DNA in some living organisms) segment called a PRIMER COMPLEMENTARY TO A SSDNA (SINGLE-STRANDED DNA) TEMPLATE.

replicative senescence: Perhaps cell division ceases when cells discern that their chromosomes have become too short. This process is called replicative senescence. This results in the activation of the p^{53}-dependent damage checkpoint, which induces growth arrest of the cells and structural changes may be seen. Functional interaction of Ataxia Telangiectasia Mutated (ATM) and telomeres *in vivo* in relation to acceleration of ageing have been evaluated. Impact of Atm deficiency as a function of progressive telomere attrition at both the cellular and whole-organism level was examined in mice doubly null for Atm and the telomerase RNA component (Terc). These compound mutants showed increased telomere erosion and genomic instability, yet they experienced a substantial elimination of T-cell lymphomas associated with Atm deficiency. A generalized proliferation defect was evident in all cell types and tissues examined, and this defect extended to tissue stem/progenitor cell compartments, thereby providing a basis for progressive multi-organ system compromise, accelerated ageing and premature death.

replicative transposition: In case of replicative transposition, element may be duplicated during the reaction via a DNA intermediate so that the transposing entity is a copy of the original element. at a new site. For example, Tn3, bacteriophage Mu.

replicator: A DNA segment that contains an origin of replication and is able to promote the replication of a plasmid DNA molecule in a host cell. The A, B1, B2 replicators bind DNA replication proteins while B3 element functions as replicator enhancer. Replicator usually refers to a DNA-based genome but can include prions and words passed on by cultural transmission.

replicon: A sequentially replicating segment of a nucleic acid, controlled by a subsegment known as a REPLICATOR. A single replicator is present in the bacterial chromosome, whereas the chromosomes of eukaryotes bear large numbers of replicons in series. Replicon is unit of DNA replication. It consists of origin and terminus. Origin is that sequence of DNA, which supports initiation of DNA replication. For example, *Ori*C in prokaryotes, *Ori*V, in eukaryotes, Oriλ in λ phage. TERMINUS is the point that stops the fork movement; hence DNA replication terminates. It lies exactly opposite to the point of origin. The whole of the *Escherichia coli* chromosome is one replicon. DNA replication occurs at many points along the eukaryotic chromosome, each replicating region, known as *replicon*, is generated into a bubble structure. The number of replicons is in the thousands in eukaryotic chromosomes. The termination point of DNA replication lies exactly opposite to the point of origin. The whole of the *Escherichia coli* chromosome is one replicon.

Repliscan: A tool for classifying RT regions.

Repli-seq assay: High-resolution Repli-Seq defines the temporal choreography of initiation, elongation and termination of replication in mammalian cells.

replisome: A discrete multiprotein structure required for DNA replication. Enzymes and proteins required to form a replisome are topoisomerases, including gyrase, unwinding protein, single-strand binding (ssb) protein, RNA polymerase, DNA polymerases I, II and III, polynucleotide ligase, exonucleases, endonuclease and pilot proteins. The DNA-replicating structure at the Y-junction consists of DNA polymerase III enzymes and a primosome (primase and DNA helicase).

replisome progression complex (RPC): Six conserved MINI-CHROMOSOME MAINTENANCE (MCM) proteins, Mcm2–7, are essential for the initiation and elongation of DNA replication, contain ATP binding pockets and can form a hexameric structure resembling that of known prokaryotic and viral helicases. MCM complex is part of the cellular machine responsible for the unwinding of DNA during S phase. Initiation factor Cdc45 and a four-subunit complex called GINS. The name GINS is an acronym created from the first letters of the Japanese numbers 5-1-2-3 (go-ichi-ni-san) in reference to the four protein subunits of the complex: Sld5, Psf1, Psf2 and Psf3. The Cdc45-MCM-GINS complex could constitute the core of a larger macromolecular structure.

replisome-RNA polymerase collision: Replication forks are impeded by DNA damage and protein-nucleic acid complexes such as transcribing RNA polymerase. Head-on collision of the replisome with RNA polymerase results in replication fork arrest. However, codirectional collision of the replisome with RNA polymerase has little or no effect on fork progression.

reporter assays, limitations of: Reporter Assays (i) Use exogenous DNA, (ii) Do not address changes due to genomic sequences near the promoter of interest and (iii) Artefacts due to gene dosage can occur.

reporter assays, strengths of: (i) *In vivo* monitoring, (ii) capture real-time data, (iii) powerful tool for mutational analysis of promoters and (iv) amendable to high-throughput screening.

reporter assays: Reporter assays provide a real-time *in vivo* read-out of translational activity for a promoter of interest. Reporter genes are fusions of a target promoter DNA sequence and a reporter gene DNA sequence. The promoter DNA sequence is customized by the researcher and the reporter gene DNA sequence codes for a protein with detectable properties such as firefly luciferase, Renilla luciferase or alkaline phosphatase. These genes produce enzymes only when the promoter of interest is activated. The enzyme, in turn, catalyses a substrate to produce either light, a colour change or other reaction that can be detected by spectroscopic instrumentation. The signal from the reporter gene is used as an indirect determinant for the translation of endogenous proteins driven from the same promoter.

reporter gene: A specific gene inserted into the DNA of a cell so that the cell will 'report' (to researchers) when signal transduction has occurred in that cell or when a (linked) gene was successfully expressed. The gene that codes for the production of the enzyme luciferase [which catalyses bioluminescence (light production)] is one of the most commonly used reporter genes. For example, when researchers are testing numerous candidate drugs for their ability to stop cells from (over-) producing a hormone or growth factor, the researchers need to quickly know when one of the candidate drugs has had the desired effect on the cell of interest. By prior insertion into that cell of a gene (e.g. which causes bioluminescence or a certain chemical to be produced by the cell when signal transduction has taken place), that cell 'reports' (when a candidate drug has had the desired effect on the cell) by producing the bioluminescence or chemical (coded for by the reporter gene), which can be rapidly detected by the researcher (e.g. via light sensors or biosensors placed adjacent to the cell). The use of a functional enzyme, such as beta-galactosidase, luciferase or chloramphenicol acetyltransferase, downstream of a gene, promoter or translational control element of interest to more easily identify the successful introduction of the gene into a host and to measure transcription and/or translation. *See* MARKER.

reporter marker gene (RMG): RMG produces pigmentation or fluorescence, which may be vital or non-vital. RMGs encode proteins whose activity can be easily quantified *in vitro* or imaged *in planta*. The marker gene that, when introduced, makes it possible to visually identify the cells or tissues expressing it. Thus, these marker genes can be used to monitor transgenic events and manually separate transgenic material from non-transformed material. REPORTER GENES (RGs) allow the investigation of regulatory, *cis*-acting sequences as well as *trans*-acting factors in transient and stable transformation experiments. RGs fall into two categories depending on whether they are conditional or non-conditional on the presence of external substrates. Some RGs can be adapted to also function as selectable marker genes through the development of novel substrates. *Aka* SCREENABLE MARKER GENE.

reporter plasmid: Reporter plasmid carries a gene whose expression is driven by the product of a regulatory gene carried by an effector plasmid.

representative sample: A sample that accurately reflects the distribution of relevant variables in the target population.

repressed allele: An allele that is in the 'off' state or is repressed.

repressed chromatin domains: Strikingly, repressed chromatin domains are preserved through local redeposition of parental nucleosomes.

repressible control: The end product, *aka* COREPRESSOR, acts to repress the production of the enzymes needed for catabolism. Transcription is initiated in the absence of the end product. In some enzyme systems and many other amino acids as well as purines and pyrimidines, enzymes are repressible in the sense that their production is repressed by a specific substance that is activated during the course of metabolism. Repressible systems are thus generally associated with biosynthetic processes. Repressible control produces an inactive repressor (occasionally called an APOREPRESSOR) that is activated by a substance called COREPRESSOR.

repressible enzyme system: A coordinated group of enzymes, involved in a synthetic pathway (anabolic), is repressible if excess quantities of the end product of the pathway lead to the termination of transcription of the genes for the enzymes. These systems are primarily prokaryotic operons. It is a way of shutting down the synthesis of an enzyme whose product is not required because so much of it is readily available to the cell. When that enzyme product is no longer available (e.g. because the cell has consumed that product), more of the enzyme is synthesized (to catalyse the production of more product).

repression: The inhibition of transcription (or translation) by the binding of a repressor protein to a specific site on the DNA (or mRNA) molecule. The repressor molecule is the product of a repressor gene.

repressive histone modifications: The methylation of residues H3K27 and H3K9 and the ubiquitination of H2AK119 are hallmarks of repressive chromatin and are often found at silent gene loci.

repressor: A protein that binds to the operator locus and thereby inhibits the transcription of adjacent genes of the operon by blocking the RNA polymerase from binding to the promoter to not allow it to transcribe the downstream genes. Repressor combines both with an inducer (or corepressor) and with an operator region (e.g. of DNA). Polypeptide size of lac repressor (monomer) is 360 amino acids. Repressors are themselves specified by regulatory genes. Regulation by repressor is called NEGATIVE REGULATION. Repressor binding is regulated by another small protein called EFFECTER.

Repressor of Silencing 1 (ROS1): A 5^{m}C DNA glycosylase/DNA demethylase that prevents DNA hypermethylation and transcriptional silencing in *Arabidopsis*.

repressor-operator binding: The gacS repressor that contains a helix-turn-helix motif binds to an operator within its own coding sequence. This operator is identical to the DNA sequence coding for residues 1–6 of the recognition helix of this repressor. Sometimes the products of regulator genes are affected by inducers in such a way that they cross-regulate the working of two operons simultaneously.

repriming: The act of priming something again. For example, repriming DNA synthesis is an intrinsic restart pathway that maintains efficient genome replication.

reproduce: To create another individual of the parental type that will, in turn, produce another.

reproducibility: In statistics, the closeness of the agreement between the results of measurements of the same analyte, where the measurements are carried out under changed conditions. Changed conditions may include principle or method of measurement, observer measuring instrument, location, conditions of use and time. Reproducibility may be expressed quantitatively in terms of dispersion characteristics of the results. For example, using samples from five individuals known to be homozygous for the disease F caused by mutation G, two technicians using three different sequencers on five different days can detect the presence of the mutation in all five samples in every case, so the reproducibility is 100%.

reproduction: (i) The process of forming new individuals of a species by sexual or asexual methods. Reproduction is a defining characteristic of all life, and every individual organism exists as the result of a reproductive event. There are two general methods by which reproduction takes place: sexual or asexual. *Aka* PROCREATION *or* BREEDING. (ii) In forestry, the process by which the forest is replaced or renewed, either through artificial reproduction, through seeding or planting or by natural reproduction, from natural seeding or sprouting.

reproductive biology: The branch of biology that studies the various types and mechanisms of reproduction used by living organisms, typically with special emphasis on cell division, fertility, endocrinology and/or the tissues, organs and systems involved in reproduction.

reproductive cells: The gametes and their immediate predecessors from which they are produced by division.

reproductive character displacement: The increased reproductive isolation between two closely related species when they live in the same geographic region (sympatry) as compared with when they live in separate geographic regions. A kind of character displacement in which the character concerned influences reproductive isolation, not ecological competition.

reproductive cloning: The cloning of an embryo for transplantation into a uterus to produce offspring genetically identical to the donor. Monozygotic (identical) twins are NATURAL CLONES.

reproductive cloning technology: This technology generates offspring that are genetically identical to the parent by the transfer of an entire donor nucleus into the enucleated cytoplasm of a host egg.

reproductive effort: Energy that is devoted to reproduction. There are several ways in which this may be defined.

reproductive growth (or **phase**)**:** Phase that involves the production of flowers, fruits and seeds.

reproductive isolating mechanisms: Methods that interfere with its interbreeding with organisms of other species.

reproductive isolation: Two populations or individuals of opposite sex are considered reproductively isolated from one another if they cannot together produce fertile offspring. *See* PREZYGOTIC ISOLATION *and* POSTZYGOTIC ISOLATION.

reproductive phase: At this stage, vegetative buds are transformed into floral buds. Androecium and gynoecium, male and female structures, respectively, are essential organs of a flower.

reproductive protein evolution: Sexual reproduction is a fundamental biological process among eukaryotes. Because of the significance of reproductive proteins to fitness, the diversity and rapid divergence of proteins acting at many stages of reproduction is surprising and suggests a role of adaptive diversification in reproductive protein evolution.

reproductive success: Number of successful gametes formed, thus number of offspring produced by a parent.

reproductive value: The expected contribution of organisms in that stage of life to the future lifetime reproductive success.

reprogramming: Conversion of a cell from one cell type to another. This involves a conversion first to a pluripotent state, then redifferentiating to the new cell type.

reprogramming to pluripotency: C.K. Wille *et al.* (2023) find that only the maintenance of elevated histone acetylation is essential for enhanced reprogramming and occurs at loci that are depleted for H3K79me2. Thus, DOT1L inhibition promotes hyperacetylation and hypertranscription pluripotent properties.

repulsion: An allelic arrangement in which each homologous chromosome has mutant and wild-type alleles. *Aka* *TRANS*-ARRANGEMENT.

repulsion linkage: The tendency of two dominant characters, inherited in an F_1 – one from one parent and the other from the other parent – to stay apart in the F_2 generation. For example, for genes and/or alleles A, a and B, b the repulsion heterozygote receives Ab from one parent and aB from the other, where A and B are dominant, and a and b are recessive. *Compare with* COUPLING LINKAGE.

repulsive regime tapping mode atomic force microscope (repulsive TM AFM): In repulsive TM AFM, imaging in the repulsive interaction regime is associated with the irreversible deformation of the molecules. This causes a significant loss in resolution and contrast.

repurposing the genome-editing tools: Repurposed CRISPR platforms provide a powerful tool to gain insights into the mechanisms regulating the epigenome.

requirements analysis: In systems engineering and software engineering, requirements analysis focuses on the tasks that determine the needs or conditions to meet the new or altered product or project, taking account of the possibly conflicting requirements of the various stakeholders, analysing, documenting, validating and managing software or system requirements.

research: Defined as the search for knowledge or any systematic investigation to establish facts. The term is also used to describe an entire collection of information about a particular subject. The researcher uses the scientific method. This research provides scientific information and theories for the explanation of the nature and the properties of the world around us. It makes practical applications possible. Scientific research is funded by public authorities, charitable organizations and private groups, including many companies. *Aka* SCIENTIFIC RESEARCH.

research in genetics works in a cascade: The genome project reads the sequence of a gene and annotates it. Structural genomics checks out the shape of the protein the gene dictates and proteomics checks out the functional relevance of the protein and how different proteins work together. This is precisely what is termed proteomics. There is no strict linear relationship between gene activity and protein abundance in a cell. The existence of an OPEN READING FRAME (ORF) in genomic data does not necessarily imply the existence of a functional gene. Modifications of the proteins that are not apparent from the DNA sequences, such as isoforms and posttranslational modifications, can be determined only by proteomic methodologies. The localization of gene products, which is often difficult to predict from the sequences can only be determined through proteomics. Proteomics involves large-scale analysis of the total protein complement (proteome) of an organism. Analysis involves protein distribution, characterization and identification, protein-protein interactions and differential display proteomics.

research priorities in epigenomics: **EPIC**'s goal is to facilitate the epigenomics community's search for answers to various fundamental biological questions: (i) What are the comparative roles of transcription factors and epigenetic regulators in the creation or maintenance of transcriptional states? (ii) How similar are the epigenomes of different species, and how can epigenomic states in model organisms be extrapolated to non-model plants, including crops that may be difficult to study? (iii) Is hybrid vigour an epigenetic phenomenon? (iv) How does the epigenome modify itself in response to environmental stresses (e.g. drought, salt, cold, heat, pollution, nutrient limitation) or pathogen attacks? (v) What is the molecular explanation for epigenetic inheritance and the continuance of acquired states? (vi) How much epigenetic variation exists within a population and is this variation associated with genetic variation? (vii) Can epigenomes be engineered or edited? (viii) How does the epigenome change during developmental transitions, such as when stem cells and their progeny differentiate into specific cell types? (ix) What are epigenetic changes that occur in shoot apical meristems as they transition from vegetative meristems to floral meristems, and how do photoperiod, light quality and temperature affect these transitions? (x) How are important facets of chromosome biology (e.g. centromere function, telomere function, replication, recombination and DNA repair) epigenetically regulated? (xi) How widespread is the crosstalk between DNA methylation and a histone PTLM, and through what mechanism does this crosstalk occur? (xii) How do different types of epigenetic modifications specify and communicate with one another? (xiii) What are the epigenetic significances of phenomena such as polyploidy, imprinting, paramutation, nucleolar dominance, apomixis and other common features of plant genomes?

research process, goal of the: To produce new knowledge, which takes three main forms – EXPLORATORY RESEARCH, CONSTRUCTIVE RESEARCH and EMPIRICAL RESEARCH – the boundaries between them may be fuzzy. *See details as separate terms.*

research reagents: Several proteins and enzymes are used in research activities could be produced in plants on a commercial scale. For example, (i) protein avidin is traditionally isolated from chicken egg; the avidin gene has been expressed in maize. (ii) β-glucuronidase (GUS) enzyme is being produced from transgenic maize.

resequencing: The sequencing of a specific exon or genomic region for the identification of DNA variants. Often performed to screen a large population for a gene of interest or to confirm a DNA variant identified by massively parallel sequencing.

resequencing by hybridization (RBH): A variant of the sequencing by hybridization technique that allows the detection of an SNP (or another type of mutation) in a target region of a genome. In short, an SNP is first detected in a genome-wide SNP mapping project, its linkage disequilibrium with a target region (e.g. conferring a disease phenotype) is determined and the region is amplified in a conventional polymerase chain reaction with bracketing primers. The amplification product or an oligonucleotide with the complementary sequence is then spotted on to a microarray together with thousands of other targets and hybridized to test samples of genomic DNA (also cDNA) from patients. Hybridizing probes are then re-sequenced and the SNP is verified.

resetting versus transgenerational inheritance: What is the mechanism that ensures that methylation marks at promoter DNA persist through meiosis, gametogenesis and early embryogenesis: Why some epigenetic marks are heritable whereas others are not? In part, this difference could be associated with the different types of epigenetic marks. For example, Polycomb marks could be more prone to resetting, whereas DNA methylation is more likely to be transmitted between generations. Many DNA methylation marks, including those that are set by viruses, are sometimes lost in the next generation. The persistent marks may have more or particular types of methylated C residues and be more difficult to erase than those that are reset. Alternatively, it could be that heritable epigenetic marks generate a signal that is transmitted from the maternal plant to the progeny, which guides reestablishment in the early zygote or meristematic cells of the next generation.

res-gene: Resistance gene.

residual: A statistical quantity that should be unrelated to certain other variables because their effects should have already been subtracted out. In ordinary multiple regression, the most commonly used residual is the difference between predicted and observed values.

residual proteins: These chromosomal proteins are the least soluble and they tend to remain behind after many extraction procedures.

residual reproductive value: The remaining contribution to lifetime reproductive success after the current activity has made its contribution.

residual risk: Another name for the negative predictive value. The risk of the disorder in an individual (or couple) with a negative test result.

residue: General term for the unit of a polymer. That portion of a sugar, amino acid or nucleotide that is retained as part of the polymer chain during the process of polymerization.

resilience: The ability of a population to persist in a given environment despite disturbance or reduced population size. Based upon the ability of individuals within the population to survive (fitness) and reproduce (fecundity) in a changed environment gene.

resinosis: A disease symptom in which there is a copious flow of resin from a wound or infection site of a conifer.

resistance: A host possessing properties that prevent or impede disease development. Plants, animals and humans possess varied degrees of resistance to biotic and abiotic stresses. In general, the resistance to diseases, insects and abiotic stresses is controlled by gene(s). The susceptible crop varieties can be made resistant by transferring genes for resistance through hybridization or a transgenic approach. In plants, resistance may be horizontal type or vertical type. The development of *Bt* cotton through a transgenic approach is a good example of developing insect resistance by introducing cry gene(s). There are different types of resistance: HYPERSENSITIVITY, SPECIFIC RESISTANCE, NON-UNIFORM RESISTANCE, MAJOR GENE RESISTANCE, VERTICAL RESISTANCE, FIELD RESISTANCE, NON-SPECIFIC RESISTANCE, UNIFORM RESISTANCE, MINOR GENE RESISTANCE and HORIZONTAL RESISTANCE. *See details as separate terms. Contrasts* SUSCEPTIBLE.

resistance against bacterial and fungal pathogens: Several transgenic plants in tobacco and tomato have been developed with resistance against bacterial and fungal pathogens. For example, tomato plants that have transferred gene *acetyl transferase* show resistance to wildfire caused by *Pseudomonas syringae*. Tomato transgenic plants having *Chitinase* gene from bacteria, show resistance to brown spot caused by *Alternaria longipes*. Potato transgenic plants having *osmotin* gene from potato show resistance to late blight caused by *Phytophthora infestans*.

resistance against fungal and bacterial pathogens, strategies for: In case of fungal and bacterial pathogens, resistance has been sought by expression of the following: (i) transgenes from non-plant sources and (ii) Transgenes from plant sources. Using these strategies, the transgenic plants showed increased resistance to the concerned diseases.

resistance against stress: A number of genes responsible for providing resistance against stress (e.g. heat, cold, salt, heavy metals, phytohormone, nitrogen) have been identified. Transgenic plants that have resistance to a variety of stresses may be produced in future. Resistance against chilling (1°C for 10 days) was introduced into tobacco plants by introducing a gene for glycerol-1-phosphate acyl transferase from *Arabidopsis*.

resistance against viral infection: If susceptible strain of a crop is inoculated with a mild strain of a virus, then the susceptible strain develops resistance against more virulent strains. This phenomenon is known as cross protection and has been used in tomato against potato spindle tuber viroid and in citrus against CITRUS TRISTEZA VIRUS. Cross protection has certain disadvantages – possibility of mutation in inducing mild virus strain; possibility of synergism between inducing virus and another unrelated virus; possibility of some yield losses due to mild strain also. These disadvantages can be overcome if a single gene imparting this benefit is used in production of transgenic plants. Three kinds of genes have been used for this purpose. Genes for virus coat or capsid protein (CP) from positive-strand RNA virus have been transferred to tobacco. In the transgenic plant, expression of CP was observed. When inoculated with TOBACCO MOSAIC VIRUS (TMV), information in transgenic plant was delayed. Gene for nucleocapsid (N) protein from TOMATO SPOTTED WILT VIRUS (TSWV) causes considerable yield losses in tomato, tobacco, lettuce, groundnut, pepper and so on. For producing transgenic tobacco plant, *N* gene was associated with CaMV 35S promoter and a leader sequence. The transgenic plant showed expression of *N* gene and significant resistance to TSWV. Satellite RNA are associated with specific strains of some plant RNA viruses although it is not necessary for their replication. The presence of satellite RNA leads to a reduction in the severity of disease symptoms and, therefore, has been used for developing resistance against specific viruses, such as transgenic plant of tobacco using DNA from RNA satellite of CUCUMBER

MOSAIC VIRUS (CMV) were produced, which led to a reduction in disease symptoms.

resistance breeding: Special crossing and selection methods used to improve the inherent capacity of a crop plant to prevent or retard the development of an infectious disease.

resistance evolution in a Cas9-based sex conversion-suppression gene drive: A suppression gene-drive system designed with a strategy to cause an all-male population to collapse has been developed, using *Drosophila melanogaster* as a model organism to experimentally evaluate the molecular basis of suppression gene-drive technology based on sex conversion using Cas-9 system. The *transformer* gene was targeted by a Cas-9-based gene-drive element, resulting in the conversion of females to males very efficiently. However, the gene drive was curbed due to the emergence of drive-resistant alleles at the target site. This occurred due to an increased rate of mutagenesis by NON-HOMOLOGOUS END JOINING (NHEJ). This simple homing CRISPR/Cas9-based gene-drive system is ineffective. Nevertheless, with adequate modifications and by adjusting the model for *Ceratitis capitata*, this sex conversion-based suppression gene-drive strategy can be applied successfully for the management of wild populations of *Ceratitis capitata*.

resistance gene analogues (RGA): With the recent molecular cloning of several plant disease resistance genes it has become apparent that resistance genes share certain homologies in conserved amino acid domain. PCR amplification of genomic DNA using degenerate primers on the basis of these conserved amino acid domains identified sequences with homologies to plant disease resistance genes (i.e. resistance gene analogues; RGAs); RGAs exist in large numbers in plant genomes and provide new possibilities for the investigation of resistance genetics in general and also for the analysis of certain plant disease resistances.

resistance gene enrichment sequencing (RenSeq): RenSeq is a NB-LRR (nucleotide-binding-site leucine-rich repeat) gene-targeted, resistance gene enrichment and sequencing method that enables the discovery and annotation of pathogen resistance gene family members in plant genome sequences. F. Jupe *et al.* (2013) applied RenSeq to the sequenced potato *Solanum tuberosum* clone DM (DM1–3 516 R44) and increased the number of identified NB-LRRs from 438 to 755. The majority of these identified R gene loci reside in poorly or previously unannotated regions of the genome.

resistance gene interactions: Some pairs of genes for resistance interact to produce a greater level of resistance than either one of them individually. Some gene interactions may suppress resistance.

resistance genes: The resistance genes could be monogenic, oligogenic or polygenic. Resistance genes encode predicted cytosolic resistance proteins of the NUCLEOTIDE-BINDING SITE (NBS) and LEUCINE-RICH REPEAT (LRR) class, with an N-terminal TOLL-LIKE INNATE IMMUNITY RECEPTORS (TIR) domain. LRR domain is the major determinant of specificity in the recognition of the corresponding Avr genes.

resistance in pest management: Refers to field strategies, which can maintain the insect resistance in the transgenic crop varieties for several years. United States EPA has mandated measures to reduce the risk of resistance development. Strategies are being formulated to delay the onset of resistance to *Bt*. Planting strategies typically manipulate transgenic plants in time and space in order to prevent immediate development of resistance. These include using rotations (in which transgenics may be alternated in time with non-transgenics), refugia (in which a portion of a field may be planted with non-transgenics) and mosaics (in which mixtures of transgenic and non-transgenic plants are grown together). Use of multiple transgene combinations (gene pyramiding) would arguably be more effective and easier to implement. Instead of directional selection driving insect populations towards resistance to *Bt*, other genes engineered into plants would give insect pests additional forces to reckon with. They would then have to overcome two or more insecticidal genes to become resistant, a less likely occurrence known as DISRUPTIONAL SELECTION. Besides helping to prevent the development of resistant insect biotypes, combining *Bt* with another gene or genes for insect resistance also increases the toxic effect to insects. For example, feeding assays have shown that mixtures of *Bt* and proteinase inhibitors have a synergistic interaction.

resistance to fungal and bacterial diseases: The disease resistance generated by employing pathogen genes is called PATHOGEN-DERIVED RESISTANCE (PDR). PDR has been realized in the cases of bacterial and fungal diseases. Ordinarily, hosts and pathogens show gene-for-gene relationship. This relationship is explained by gene-for-gene hypothesis put forth by

H.H. Flor (1955). Thus, for every gene for resistance possessed by the host, the pathogen has a corresponding gene for virulence. Disease resistance depends on the specific interaction between the products of the gene governing host resistance and those conditioning pathogenicity or virulence. Gene-for-gene relationship is classified into two types – incompatible reaction and compatible reaction.

resistance transfer factor (RTF): The segment of a conjugal R plasmid carrying the genetic factor that confers on the plasmid those properties needed for the transfer of plasmid from one cell to the other.

resistance, breakdown of: The inability of a plant to maintain resistance when attacked by a pest biotype that has a gene for virulence at every locus corresponding to a gene for resistance in the host.

resistant: State of host to exclude, hinder or resist the pathogen or other biotic and abiotic stresses. The development of resistant crop varieties is the major goal of plant breeding. A plant may be resistant to one pathogen or condition but not others. For instance, *Bt* cotton is resistant to American bollworm and transgenic papaya is resistant to a viral disease, papaya ringspot virus. Incorporation of resistance genes from other cultivated varieties or wild species through backcross breeding makes the recipient variety resistant to specific disease/insect or abiotic stress. Nowadays, the transgenic approach is being used to develop resistant transgenic crop varieties.

resistant insects: Insects that are difficult to kill or cannot be killed by insecticides/*Bt* transgenic crops. There is a strong risk that large-scale deployment of plants genetically engineered with *Bt* genes will cause insects to become resistant. Though, *Bt* has been used as a biological insecticide for over 30 years as a spray, however, spraying is not often practised by growers, because of its relatively high cost, critical timing of application and low efficacy. Even then, with *Bt* sprays prevalent for over three decades Bt microorganisms coexist naturally with herbivores, indicating that there has been little opportunity for insects to evolve resistance. This scenario may change rapidly because of the massive deployment of *Bt* crops in even more diverse geographic regions. Plants containing a single insect-resistant transgene will be planted commercially by growers. In transgenic plants, *Bt* insecticidal crystal proteins (ICPs) will be produced continually in plant tissue. This will in turn create strong selection pressure on both economically important and unimportant insects. If genes conferring resistance to *Bt* become fixed in insect populations, both sprays and transgenic plants would become ineffective as insect control agents.

resistant rootstocks: Some vegetatively propagated tree crops have superb agricultural or horticultural characteristics but are susceptible to various soil-borne parasites. They are then grafted on to resistant rootstocks (e.g. the grafting of classic, European, wine grapes on to American rootstocks to control Phylloxera). Many fruit trees (e.g. stone and pome fruits, citrus) and other high-yielding clones (e.g. rubber) are grafted on to resistant rootstocks for this reason.

resolution: The smallest distance by which two objects can be separated and still be resolved as separate objects. The resolving power can never be greater than the wavelength of light used. The image, which is seen through the eyepiece, is the aerial image formed by the microscope objective in the tube. It has a limit, where useful magnification ends and the empty magnification begins. The performance limit of the microscope is determined by the numerical aperture (NA) so that the total magnification of the microscope objective magnification multiplied by the eyepiece magnification cannot exceed 1000×.

resolvase: Enzyme actively involved in site-specific recombination between transposons present as direct repeats in a cointegrate structure.

resolving power: Ability of an experimental technique to distinguish between two genetic conditions (typically discussed when one condition is rare and of particular interest).

resonance contact scanning force microscope (RC SFM): A method used for high-speed imaging for real-time applications in which noise arising due to tip-surface interaction is reduced.

respiration: General term for a metabolic process in a cell involving the oxidative breakdown of sugars or other organic molecules and requiring the uptake of O_2 while producing CO_2 and H_2O as waste products. In this process, the energy (ATP-Adenosine triphosphate) is produced by the breakdown of glucose, which is further used by cells to perform various functions.

respiration in plants: Leaves possess stomata, which are tiny pores, for gaseous exchange. The oxygen consumed via stomata is used up by cells in the leaves to disintegrate glucose into water and carbon dioxide.

respiration, site of: Mitochondria is the site where cellular respiration takes place resulting in the production of ATP that serves as a source of energy for important metabolic processes.

respiratory chain: Electron transport chain in the inner mitochondrial membrane that receives high-energy electrons derived from the citric acid cycle and generates the proton gradient across the membrane that is used to power ATP synthesis.

respiratory control: Regulatory mechanism that controls the rate of electron transport in the respiratory chain according to need via a direct influence of the electrochemical proton gradient.

respiratory enzyme complex: Any of the major protein complexes of the mitochondrial respiratory chain that act as electron-driven proton pumps to generate the proton gradient across the inner membrane.

responding tissue: This is the tissue that responds to an inducing tissue. The ability of a cell tissue to respond in a particular way is known as competence. Competence is programmed. For example, changing Ca^{2+} ion concentration causes eggs to activate. This occurs in all the instructive tissue interactions.

response: In ecotoxicology, this refers to biological or ecological changes caused by exposure to an environmental stressor.

response to selection: The improvement of the inherent potential of a population through selection. *Aka* GENETIC ADVANCE.

response variable: *Aka* DEPENDENT VARIABLE *or* OUTCOME VARIABLE. Its value is predicted or its variation is explained by the explanatory variable. In an experimental study response, it is the outcome that is measured following manipulation of the explanatory variable.

responsible plant breeder: The plant breeder or breeding organization that is officially recognized as the maintainer of breeder seed reference samples and production for a variety.

rest period: Dormancy.

resting membrane potential: The membrane potential in equilibrium conditions in which there is no net flow of ions across the plasma membrane.

resting phase: A misnomer for INTERPHASE of mitosis and meiosis.

resting spore: A spore, often thick-walled, that can remain alive in a dormant state for some time, frequently after overwintering, as does an oospore or a teliospore, later germinating and capable of initiating infection.

restitution: The reunion of two broken surfaces of a fragmented chromosome, the original structure thereby being restored.

restitution nucleus: A nucleus with an unreduced chromosome number.

restitution type: Unreduced product of meiosis (i.e. chromosome number is not reduced during meiosis).

restoration ecology: Activities undertaken by ecologists to repair ecological damage, such as establishing vegetation on a degraded habitat, increasing the populations of endangered species and decreasing the area of threatened ecosystems.

restoration science: Studies relating to restoration activities that may be designed to replicate a pre-disturbance ecosystem or to create a new ecosystem where it had not previously occurred. Restoration ecology is the scientific study of repairing disturbed ecosystems through human intervention. According to T.S.P. Toma *et al.* (2024), restoration is expected to yield outcomes that mitigate and facilitate adaptation to climate change.

restoration: Establishment of a self-maintaining facsimile of a natural ecosystem on degraded land, as when abandoned farmland is converted back to a native prairie or forest. *Compare with* RECLAMATION *and* REMEDIATION.

restorer (R) line: A pollen parent line; it contains the restorer gene or genes, which restore cytoplasmic male sterile plants to pollen fertility; it is crossed with an A line in the production of hybrid seeds.

restorer gene: A gene and/or allele that is able to restore fertility of a sterile genotype. While genes for sterility frequently belong to the mitochondrial genome (i.e. cytoplasmic), the restorer genes are very often found to belong to the nuclear complement. They are used in hybrid variety production.

restricted maximum likelihood (REML): A particular form of maximum likelihood estimation that does not base estimates on a maximum likelihood fit of all the information, but instead uses a likelihood function calculated from a transformed set of data, so that nuisance parameters have no effect. *Aka* MULTI-TRAIT COVARIANCE ESTIMATION.

restricted transduction: *See* SPECIALIZED TRANSDUCTION.

restriction: The exclusion of foreign DNA from bacterial cells by restriction endonuclease-catalysed recognition and degradation.

restriction analysis: Determination of the number and size of the DNA fragments produced when a particular DNA molecule is cut with restriction endonucleases.

restriction and modification enzymes: Type II restriction endonuclease is responsible for restriction only and a separate enzyme is responsible for methylating the same target sequence. The structures of these proteins are also complicated. Best understood are the components of *Eco*RI system where restriction enzyme is a dimer of identical subunits and methylase is a monomer. The target sites for type II REs are the palindromic sequences of 4–6 bp. Because of symmetry, the bases to be methylated occur on both the strands. So, a target site could be fully methylated, hemi-methylated or non-methylated. A fully methylated site is not a target for restriction or methylation. A hemi-methylated site is not recognized by RE but may be converted by methylase into fully methylated condition. It occurs during the perpetuation of fully methylated DNA, which on replication produces hemi-methylated DNA. This is primarily the usual mode of action of methylase *in vivo*. A non-methylated target site may be a substrate for either restriction or modification *in vivo*. It is rare for unmodified DNA to survive by gaining the modification pattern of a new host. Most of the type II REs cleave DNA at an unmethylated target site by cleaving one bound in each strand. It could lead to blunt ends or sticky (also called cohesive ends).

restriction and modification in *Chlamydomonas*: In *Chlamydomonas*, methylation is involved in the maternal inheritance of chloroplast DNA. Chloroplast DNA from mating type plus (Mt^+) gametes is highly methylated. Chloroplast DNA from mating type minus (Mt^-) gametes is only slightly methylated. After zygote formation, the Mt^- chloroplast DNA is degraded. Thus, in *Chlamydomonas*, restriction modification clearly exists.

restriction digest: The results of the action of a restriction endonuclease on DNA.

restriction endonuclease digestion analysis: (i) WatCut provides restriction analysis coupled with where the sites are located within genes. (ii) Restriction Site Analysis provides one with a considerable choice of enzymes and output format, including pseudo-gel maps. (iii) Restriction-Enzyme Picker finds sets of four commercially available REs which together uniquely differentiate designated sequence groups from a supplied FASTA format sequence file for use in T-RFLP. (iv) NEBcutter provides opportunities to upload local files, choose from common vector sequences or enter GenBank accession numbers. Also includes the ability to map sites in genes. (v) *In silico* restriction digest of complete genomes allows *in silico* digestion of over 300 prokaryotic genomes and simulated pulsed-field gel electrophoretic separation of the fragments. (vi) Computation of size of DNA and protein fragments from their electrophoretic mobility. (vii) Sequence Extractor generates a clickable restriction map and PCR primer map of a DNA sequence.

restriction endonucleases, type I: These are complex multifunctional enzymes that recognize specific sites of size 15–16 bp in the host DNA and cleave DNA possibly random at least 1000 base pairs from the host specificity site from the 5′-end of the sequence 'TCA' located within the recognition site. These are multimeric and complex multifunctional enzymes as they undertake both endonuleolytic and methylation activities. Best known examples of type I RE are *Eco*B and *Eco*K, which are variants of the hsd system. They require AdoMet, ATP and Mg^{++} for restriction. Cleavage at these random sites follows a process of DNA translocation, which shows that these enzymes are also molecular motors. The recognition site is asymmetrical and is composed of two specific portions – one containing 3–4 nucleotides, and another containing 4–5 nucleotides – separated by a non-specific spacer of about 6–8 nucleotides. These enzymes are multifunctional and are capable of both restriction and modification activities, depending upon the methylation status of the target DNA. The cofactor S-Adenosyl methionine (AdoMet or SAM), adenosine triphosphate (ATP) and magnesium (Mg^{2+}) are required for their full activity. Best known examples of type I RE are *Eco*B and *Eco*K, which are variants of the hsd system. DNA sequences recognized in sB and sK strains are sB: T-G-A-N8-T-G-C-T; sK: A-A-C-N9-G-T-G-C.

restriction endonucleases, type II: These are separate endonucleases and methylases, which recognize a twofold symmetry sequence in DNA and generally cleave at the host specificity site. Type II restriction endonucleases have found applicability in the construction of recombinant DNA molecules. These enzymes are homodimers with recognition sites usually undivided and palindromic and 4–8 nucleotides in length. They recognize and cleave DNA at the same site, and they do not use adenosine triphosphate (ATP) or S-adenosine methionine for their activity – they usually require only Mg^{2+}

as a cofactor. Type II REs are remarkably stable and induce cleavage either, in most cases, within their recognition sequences or very close to them. The first type II enzyme to be isolated was *Hind*II in 1970. More than 350 different type II endonucleases with over 100 different recognition sequences are known. They require Mg^{2+} ions for cleavage. Type II restriction endonucleases recognize, bind and cleave at palindromic sequences.

restriction endonucleases, type IIB: These enzymes are multimers, containing more than one subunit. They cleave DNA on both sides of their recognition to cut out the recognition site. The recognition sequences can be symmetric or asymmetric. The recognition sequences can be continuous or discontinuous (two halves of the recognition sequence are separated). They cleave 10–12 base pairs away from the recognition sequence and produce a 3′-overhang. For example, *Bcg*I, *Bpl*I, *Bsp*24, *Bae*I and *Cje*I.

restriction endonucleases, type IIE: These enzymes cleave DNA following interaction with two copies of their recognition sequence. One recognition site acts as the target for cleavage, while the other acts as an allosteric effector that speeds up or improves the efficiency of enzyme cleavage. For example, *Nae*I.

restriction endonucleases, type IIF: Enzymes similar to type IIE enzymes, interact with two copies of their recognition sequence but cleave both sequences at the same time. For example, *Ngo*MIV.

restriction endonucleases, type IIG: These enzymes do have a single subunit, like classical type II restriction enzymes, but require the cofactor S-ADENOSYLMETHIONINE (SAM or ADOMET) to be active. Both restriction and modification activities reside in a single polypeptide. For example, *Eco*57I).

restriction endonucleases, type IIM: Enzymes that are able to recognize and cut methylated DNA. For example, *Dpn*I.

restriction endonucleases, type IIS: Enzymes that cleave DNA at a defined distance from their non-palindromic asymmetric recognition sites. These enzymes may function as dimers. For example, *Fok*I, *Bbv*I, *Bsr*I, *Ear*I, *Hph*I, *Alw*26, *Mbo*I, *Sfa*NI and *Tth*III.

restriction endonucleases, type IIT: Enzymes similar to type IIS, are composed of two different subunits with restriction and modification activities, respectively. Some recognize palindromic sequences while others have asymmetric recognition sites. *Bpu*10I is a heterodimeric enzyme, in which the two subunits form a single active site that recognizes asymmetric sequences. *Bsl*I is a heterotetramer, which recognizes palindromic sequences.

restriction endonucleases, type III: Type III REs are intermediate between the type I and type II enzymes; they cleave DNA in the immediate vicinity of their recognition sites. These are complex multifunctional enzymes, which recognize specific sites of size 5–6 bp in the host DNA and they cleave DNA 24–27 bp to 3′ side of the host specificity site. They require SAM or AdoMet, ATP and Mg^{++} for restriction. DNA sequences recognized in sP1 and s15 strains are: sP1: A-A-A-C-C; s15: C-A-G-C-A-G. DNA cleavage sites are 24–27 bp to 3′ side of the host specificity site. In type III REs, three modification and restriction enzymes recognized are *Eco*P1, *Eco*P15 and *Hinf*. Each enzyme consists of two subunits: R and MS. R subunit is responsible for restriction whereas MS is responsible for both modification and recognition. Modification (methylation) and restriction activities are performed simultaneously.

restriction endonucleases, type IV: McrA, McrBC and Mrr, encoded by genes *mcrA*, *mcrBC* and *mrr* are methylation-requiring systems that attack DNA only when it is methylated at specific positions. All three of the latter systems restrict DNA modified by CpG methyltransferase (M.Sss I) to contain methylcytosine in CpG dinucleotides. Mrr also attacks DNA with methyladenine in specific sequences. One or more of these sequences is present in the most commonly used strains of *Escherichia coli*.

restriction endonucleases, type V: Type V restriction enzymes utilize guide RNAs to target specific non-palindromic sequences found on invading organisms. They can cut DNA of variable length provided that a suitable guide RNA is provided. The flexibility and ease of use of these enzymes make them promising for future genetic engineering applications, for example, the cas9-gRNA complex from CLUSTERED REGULARLY INTERSPACED SHORT PALINDROMIC REPEATS (CRISPRs).

restriction endonucleases: A class of enzymes that cleave (cut) DNA at a specific and unique internal location along its length. These enzymes are naturally produced by bacteria that use them as a defence mechanism against viral infection. The enzymes chop up the viral nucleic acids and hence their function is destroyed. Discovered in the late 1970s by W. Arber, H. Smith and D. Nathans, they were awarded the Nobel Prize in 1978. Restriction endonucleases such as *Bam*HI, *Eco*RI, *Hin*dIII, *AluI* and *SalI* recognize and cut specific nucleotide sequences in DNA, which are four, six, eight or sixteen nucleotides long. *AluI* was the first restriction enzyme to be isolated from *Arthrobactor luteus*; *Hin*dIII was the third restriction enzyme to be isolated from *Haemophilus influenza* and *EcoRI* from *Escherichia coli*. Bacteria contain over 400 such enzymes that recognize and cut more than 100 different DNA sequences. They are important tools in genetic engineering, enabling the biotechnologist to splice new genes into the location(s) of a molecule of DNA where a restriction endonuclease has created a gap (via cleavage of the DNA RESTRICTION FRAGMENT LENGTH POLYMORPHISM (RFLP) technique. A 'genetic mapping' technique that analyses the specific sequence of bases (i.e. nucleotides) in a piece of DNA (from an organism). One type of restriction enzyme makes staggered cuts, and the second type results in blunt ends. Since the specific sequence of bases in DNA molecules is different for each species, strain, variety and individual (due to DNA polymorphism), RFLP can be utilized to 'map' those DNA molecules (for plant breeding purposes, for criminal investigation purposes, etc.). *Aka* RESTRICTION ENZYMES, RESTRICTION NUCLEASE *or* MOLECULAR SCISSORS.

restriction-enzyme cutting site: A specific nucleotide sequence of DNA at which a particular restriction enzyme cuts the DNA. Some sites occur frequently in DNA (e.g. every several hundred base pairs); others much less frequently (rare-cutter; e.g. every 10,000 base pairs).

restriction enzyme-mediated integration (REMI): REMI involves the transformation of cells with a mixture of plasmid DNA, linearized with a restriction enzyme, along with a restriction enzyme that is capable of generating compatible cohesive ends in the genome.

restriction enzymes, uses of: Over 90 restriction enzymes each with different site specificity have so far been isolated. These enzymes provide powerful tools for analysis of DNA organization, gene structure and gene regulation. Recombinant DNA molecules produced using these enzymes play an important role in genetic engineering and transfer of genes from one organism to another bypassing the limitations imposed by biological methods (sexual cycle, lack of pollination, fertilization, etc.).

restriction fragment length polymorphism (RFLP): A change in the DNA of an organism that changes how a restriction enzyme cuts the DNA into pieces. Two or more organisms can be compared based on the pattern of their DNA fragments when they are run on a gel (by electrophoresis). If the organisms have different spaces between their restriction enzyme sites, they produce different lengths of fragments when the restriction enzyme is used to cut up the DNA. When their fragments are run on a gel, they produce different patterns because the shorter fragments will move further than the longer ones. RFLP thus refers to the variation in the length of DNA fragments produced by a specific restriction endonuclease from genomic DNAs of two (or more) individuals of a species. An RFLP can originate from the mutation of a restriction enzyme target site (i.e. base sequence change generates or eliminates the restriction site). An RFLP can also originate from the insertion or deletion of a piece of DNA between the two larger sites. Unequal crossing or replication in which slippage has occurred, which creates the variation in a number of tandem DNA sequence repeats at the mini/microsatellites locus. RFLP markers exhibit codominance and are highly reliable. But RFLP analysis is laborious, time consuming and slow and thus not commonly used especially for the evaluation of large segregating breeding populations. Unlike several other molecular markers such as RAPD and microsatellites currently being used, RFLP is independent of PCR, and thus, it is more reliable, but this method is comparatively more laborious and time consuming.

restriction fragment length polymorphism (RFLP) fingerprinting: This technique of obtaining fingerprints is based on RFLP by introducing oligonucleotides complementary to microsatellites (e.g. $[GATA]_4$) as end-labelling probes.

restriction fragment length polymorphism (RFLP) markers: Any marker resulting from changes in length of genomic DNA with specific restriction endonucleases.

restriction fragment length polymorphism (RFLP), advantages of: (i) RFLPs are present everywhere in the genome and in living organisms. (ii) They show Mendelian inheritance (e.g. each band seen in a Southern blot indicates the presence of one or more restriction sites in a sequence; the sequence containing a restriction site is one allele, while the corresponding sequence missing the restriction site is the other allele; the 'phenotypes' of these alleles are the differences in banding patterns, due to presence or absence of bands). (iii) They show codominant expression. (iv) They have no pleiotropic effects. (v) They are independent of environmental effects. (vi) They are available at each developmental and/or physiological stage. (vii) Different loci within the genome can be identified by one DNA probe. (viii) Heterologous genes may also be used as probes. (ix) Any number of DNA probes can be established. (x) Probes are available for coding and silent genes (DNA sequences). (xi) Probes also show the variability of flanking DNA sequences. (xii) Several traits can be screened in the same experimental sample.

restriction fragment length polymorphism (RFLP), analysis of: Analyses of RFLPs are based on Southern blotting. It comprises the following steps: (i) RFLP analysis begins with a blood sample. (ii) DNA is extracted from nuclei of white blood cells and digested with a restriction endonuclease. (iii) The resulting DNA fragments are separated by gel electrophoresis, which separates them according to size. (iv) The RFLP is then detected by Southern blotting. First DNA in the gel is heated to denature (the two strands of DNA separate) and is blotted on a nylon membrane. (v) A probe, a radioactively labelled segment of single-stranded DNA that is complementary to the RFLP locus is applied to the membrane. (vi) The probe hybridizes with the fragment carrying the locus. A sheet of X-ray film placed over the membrane detects the radioactively tagged fragment and reveals the RFLP. (vii) In RFLP analysis, DNA samples from several individuals are often analysed at the same time.

restriction fragment length polymorphism (RFLP), applications of: (i) RFLPs are extremely useful in genetic improvement programmes. (ii) The RFLP probes are frequently used in genome mapping and in variation analysis (genotyping, forensics, paternity tests, hereditary disease diagnostics, etc.). (iii) The use of RFLPs as genetic markers in plant genetic analysis could also offer an alternative approach to specific problems, which previously have been unamenable to conventional techniques. (iv) RFLPs have potential usefulness in basic plant genetic studies. (v) Use of RFLPs can be made for studying varietal identification, identification and mapping of both quantitative and qualitative loci and early selection or recombinant lines and so on. (vi) For using RFLPs efficiently as genetic markers in these studies, knowledge of their individual genomic location or at least spatial arrangement of these is required. For this, genetic linkage map of RFLP should be available.

restriction fragment length polymorphism (RFLP), assumptions of: RFLP analysis is based on two assumptions: (i) All the bases are present in equal amounts in a genome. (ii) Random distribution of the bases occurs in the whole genome. On the basis of these assumptions, we can work out the expected molecular weight of the fragments and the average of probes showing polymorphism.

restriction fragment length polymorphism (RFLP), limitations of: There are several limitations of RFLPs: (i) Large amount of DNA is required for restriction/digestion and Southern blotting. (ii) The assay is time consuming and labour intensive. (iii) The requirement of radioactive isotopes makes the analysis relatively expensive and hazardous. (iv) Satellites and highly repetitive are inaccessible with naturally accruing probes usually. (v) RFLP analysis is unable to detect the single base change. RFLPs were actually used as a genetic marker for the diagnosis of human diseases/disorders.

restriction fragment length polymorphism (RFLP), method of: RFLP analysis comprises the following steps: (i) RFLP analysis begins with a blood sample. (ii) DNA is extracted from nuclei of white blood cells and digested with a restriction endonuclease. (iii) The resulting DNA fragments are separated by gel electrophoresis, which separates them according to size. (iv) The RFLP is then detected by Southern blotting. First DNA in the gel is heated to denature (the two strands of DNA separate) and is blotted on a nylon membrane. (v) A probe, a radioactively labelled segment of single-stranded DNA that is complementary to the RFLP locus is applied to the membrane. (vi) The probe hybridizes with the fragment carrying the locus. A sheet of X-ray film is placed over the membrane detects the radioactively tagged fragment and reveals the RFLP. (vii) In RFLP analysis, DNA samples from several individuals are often analysed at the same time.

restriction fragment length polymorphism (RFLP), salient features of: (i) Lack of dominance, (ii) multiple allelic forms, (iii) lack of pleiotropic effect on agronomic traits, (iv) codominant mode of inheritance, (v) no measurable effect on phenotype and (vi) no effect of environment on their evaluation.

restriction fragments: DNA fragments generated by digestion of a DNA preparation with one or more restriction endonucleases; usually separated by agarose gel electrophoresis and visualized by ethidium bromide staining under UV light (or alternatively subjected to Southern blot analysis).

restriction map: A linear sequence of restriction sites, each for a specific restriction endonuclease and distances among them are measured in terms of base pairs (bp) of DNA. Sites closer to 300 bp or less could thus be mapped. A restriction map of a chromosome (or a segment thereof) thus depicts the sites where one or more restriction endonucleases produce cuts in the DNA. A restriction map helps in selecting the correct restriction endonucleases for a particular cutting manipulation. The restriction map clearly indicates the recognition sequences (cutting sites) for different restriction enzymes on the DNA molecule. *Aka* CLEAVAGE MAP.

restriction mapping: Restriction mapping of a gene or a chromosome begins with an enzyme, which cleaves it at one or a few sites only. The fragments thus produced are separated through polyacrylamide (better for short fragments) and agarose (better for larger fragments) gel electrophoresis. The fastest-moving fragments in the gel are the shortest while the slow-moving ones are the longest. The electrophoretic technique now permits the separation of DNA fragments differing in length by one nucleotide only. Restriction mapping technique is an efficient way of finding differences in mtDNAs because the loss of a restriction site represents a mutational difference. Since mtDNA is known to be maternally inherited, the human mtDNA evolutionary tree data could lead to a speculation that all present-day humans might have shared a common ancestral mother.

restriction mapping methodology: One way of making a restriction map is to compare single-enzyme digests with double digests. In another method, a piece of DNA is labelled at each 5′-end with ^{32}P and cut to generate fragments with the label at one end. The longest fragment is isolated and digested with a second restriction enzyme, but the reaction is not allowed to go to completion; therefore, a population of variously sized labelled fragments is produced, each with one of the various restriction sites at the non-labelled end. These fragments are separated electrophoretically, and the distance of each restriction site from the labelled end can thus be deduced.

restriction mapping, approaches for: Several approaches are used for restriction mapping: (i) comparison of the products of restriction mapping with one enzyme, (ii) comparison of the products obtained after sequential digestion with two or more different restriction enzymes and (iii) comparison of the products of digestion of circular DNA under conditions, which permit only one cut by a restriction endonuclease (instead of its complete digestion). Followed by the digestion of linear DNA thus produced by another endonuclease, which again induced only one cut at a specific site. One approach in constructing a restriction map of a DNA molecule is to sequence the whole molecule and to run the sequence through a computer programme that will find the recognition sites that are present for every restriction enzyme known. To find the relative positions of restriction sites on a plasmid a technique involving single and double restriction digests is used. Based on the sizes of the resultant DNA fragments the positions of the sites can be inferred. Restriction maps are used as a reference to engineer plasmids or other relatively short pieces of DNA and sometimes for longer genomic DNA. Restriction mapping is a very useful technique when used for determining the orientation of an insert in a cloning vector, by mapping the position of an off-centre restriction site in the insert. The most common application of restriction mapping is presented: Determining the orientation of a cloned insert. This method requires that restriction maps of the cloning vector and the insert are already available. If a restriction site placed towards one end of the insert is known, the orientation can be determined by observing the size of the fragments in the gel. Oftentimes the orientation of inserts is important and this technique is used to screen for the correct orientation.

restriction point: Important checkpoint in the mammalian cell cycle. Passage through the restriction point commits the cell to enter S phase. It corresponds to the start of the yeast cell cycle.

restriction site: A certain nucleotide sequence within the double-stranded DNA. It is recognized by a restriction endonuclease. The enzyme cuts the double strand within the recognition sequence. The restriction sites are usually composed of four to six base pairs and are bilaterally symmetric. Both strands are cut either on exactly opposite positions (blunt ends) or alternated ones (sticky ends). The type of cutting depends on the enzyme used. *See* RECOGNITION SEQUENCE.

restriction site mapping: A technique in which genomic DNA is cleaved with restriction endonuclease into hundreds of fragments of varying size. The fragments are separated by gel electrophoresis and blot transferred to a cellulose nitrite film.

restriction-modification (R-M) system: A set of enzymes found in many bacteria and archaea that protects the host genome from genomic parasites. R-M systems consist of sequence-specific restriction endonucleases, which target invading DNA, and associated DNA methyltransferases with similar recognition sequences, which protect the host genome from the action of the endonucleases. Four types (I–IV) of restriction-modification (R-M) systems are known. *See details as separate terms.*

restriction-modification system (R-M), type I: It consists of three polypeptides: R (restriction), M (modification) and S (specificity). The resulting complex can both cleave and methylate DNA. Both reactions require ATP, and cleavage often occurs a considerable distance from the recognition site. The S subunit determines the specificity of both restriction and methylation.

restriction-modification system (R-M), type II: The methyltransferase and endonuclease are encoded as two separate proteins and act independently (there is no specificity protein). Both proteins recognize the same recognition site and therefore compete for activity. The methyltransferase acts as a monomer, methylating the duplex one strand at a time. The endonuclease acts as a homodimer, which facilitates the cleavage of both strands.

restriction-modification system (R-M), type III: Have R (restriction) and M (modification) proteins that form a complex of modification and cleavage. The M protein, however, can methylate on its own. Methylation also only occurs on one strand of the DNA, unlike most other known mechanisms. The heterodimer formed by the R and M proteins competes with itself by modifying and restricting the same reaction. This results in incomplete digestion.

restriction-modification system (R-M), type IV: They do not represent true R-M systems because they only contain a restriction enzyme and not a methylase. Unlike the other types, type IV restriction enzymes recognize and cut only modified DNA.

restrictive conditions: Environmental conditions under which a conditional mutant shows the mutant phenotype.

resynthesis: The artificial production of autopolyploids or allopolyploids of naturally occurring autopolyploid or allopolyploid plants by utilization of the presumable parental species (e.g. it was done in wheat and rapeseed).

retained introns: Retained introns may also result in the production of different mRNAs with corresponding different translated products. Intron retention, one form of alternative RNA splicing, is common in plants but rare in higher eukaryotes, because messenger RNAs with retained introns are subject to cellular restriction at the level of cytoplasmic control and expression.

reticulate: Having netlike markings.

reticulate evolution: Evolution in which lines combine as well as split.

retriever vectors: Vectors used to retrieve specific genes from a chromosome. Retriever vectors are being used for the isolation of genes from yeast. These vectors retrieve specific genes from the normal chromosome of an organism through recombination. These vectors can multiply in *Escherichia coli* and yeast, as well. The gene is removed by enzymes to produce a gapped vector that has flanking sequences. Through recombination, lost sequences are acquired from the host cell.

retroelement: An endogenous transposable element that moves along the genome via the transcription of an RNA intermediate. *Aka* RETROTRANSPOSABLE ELEMENT, RETROTRANSPOSON *or* RETROPOSON.

retrogene: In certain genes, there is reverse transcription of an mRNA and insertion of the product into the genome. This is a case of RNA to DNA flow of information. Some examples of retrogenes are *RPL36AL*, *HSPA2*, *SPIN2B* and *FAM50B*.

retrograde evolution: The mechanism of retrograde evolution has been proposed for biosynthetic pathways leading from one or more substrates via numerous enzymatic steps to more complex products. For example, nine enzymes are involved in histidine biosynthesis in bacteria like *Salmonella*, starting from ATP and phosphoribosylpyrophosphate (two substrates that are used in many other biochemical reactions) to synthesize histidine a linear biochemical pathway. Horowitz hypothesized that the most primitive bacteria had to take up histidine from the environment (as many higher organisms do) and that the first enzyme to evolve was the enzyme (E9) catalysing the last step in histidine biosynthesis. The organism possessing E9 had a strong selective advantage when histidine became scarce in the environment, since it could use the immediate precursor of histidine as a substrate. The next enzyme to evolve was presumably E8, which could have arisen by gene duplication of E9 and subsequent divergence and so on until a completely linear and streamlined pathway had been established in a retrograde manner. Recent amino acid sequence and structural comparisons of these enzymes indicate that the Horowitz model is not entirely correct, but there is evidence that two enzymes, His A and His F, with a β/α barrel structure have evolved by twofold gene duplication and gene fusion from a common half-barrel ancestor. *See* BARREL STRUCTURE.

retrograde regulation: Signalling pathways leading from the organelles (chloroplasts, mitochondria) to the nucleus are referred to as retrograde signalling and were first discovered in the yeast *Saccharomyces cerevisiae*. Retrograde signalling is broadly defined as cellular responses to alterations in the functional state of organelles. Retrograde signalling regulates nuclear gene expression depending on the developmental and functional state of plastids. Gene expression profiling, protein-interacting networks and changes in the metabolome of chloroplasts have shed light on systems' biological analysis of responses to retrograde signalling.

retrograde signalling: Signalling pathways leading from the organelles (chloroplasts, mitochondria) to the nucleus. It is broadly defined as cellular responses to alterations in the functional state of organelles. Retrograde signalling regulates nuclear gene expression depending on the developmental and functional state of plastids.

retron: Retron is a bacterial operon of unknown significance encoding reverse transcriptase and containing an untranslated region, which is first transcribed and then partially reverse transcribed to generate composite DNA-RNA structure. Retrons are not mobile because their products are unable to integrate, but they may be classified as retroelements because the sequence is mobilized by reverse transcription.

retroposon: A DNA transposable element that replicates through an RNA intermediate and does not have long terminal repeats. The RNA intermediate is then reverse-transcribed into DNA, which is inserted at a new site in the genome, which is inserted at a new site in the genome. Such as RETROTRANSPOSON.

retroposon microsatellite amplified polymorphism (REMAP): PCR-based marker using a pair of primers for studying genetic polymorphism. In this approach, DNA amplification is achieved using a thermocycler and a pair of primers, of which one primer is an anchored microsatellite (SSR) and the other primer is a retroposon LTR primer. REMAP amplified products resolve length polymorphism, which may be present either at the SSR target site itself or at the associated sequence between the binding sites of the two primers. REMAP is advantageous because undigested total genomic DNA is used as a template, rather than pre-amplified restriction digested DNA, which is used in AFLP.

retroregulation: The ability of a sequence downstream to regulate the translation of an mRNA.

retrosequence: Retrosequence is a mobile element similar in structure to retroposon. *Aka* PASSIVE RETROPOSON.

retrospective study: A study in which subjects were already enrolled before the study was designed, or the outcome of interest has occurred before the start of the study (an in a case–control study). Such studies often have difficulties such as absence of needed adjustment (confounder) variables and missing data.

retrotransposon: A transposon whose sequence shows homology with that of a retrovirus. Retrotransposons are common in eukaryotic organisms. Retrotransposons have RETROVIRUS LIKE ELEMENTS (LTR RETROTRANSPOSONS). LONG TERMINAL REPEATS (LTRs) regulate expression. Retrotransposon are exemplified by Yeast Ty-1-copia-like family and retroviral proviruses in vertebrates. *Aka* RETROPOSON.

retroviral infection: The presence of retroviral vectors, such as some viruses, which use their recombinant DNA to insert their genetic material into the chromosomes of the host's cells. The virus is then propagated by the host cell.

retroviral transduction: Retroviruses are found in many species including most mammals. Scattered in human and animal genomes are a class of repetitious genetic elements called ENDOGENOUS RETROVIRUSES (ERVs), which bear sequence homology to retroviral genomes. The ability of retroviruses to transpose within the host cell or to infect another host may have been designed for protein-coding, gene regulation and DNA repair by homologous and non-homologous recombination. Retroviruses act as natural genetic engineers, splicing genes from one species into another.

retroviral vectors: They are composed of a single RNA strand and use the enzyme reverse transcriptase to copy their genome into the DNA of the host cell's chromosomes. They are relatively genetically simple and have the ability to infect a wide variety of cell types with high efficiency. The genome of all retroviruses consists of two molecules of RNA, which are single stranded, (+) sense, and have 5′-cap and 3′-poly(A) (which is equivalent to mRNA). They vary in size from approximately 8 kb to 11 kb. Retroviral vectors can be either REPLICATION COMPETENT or REPLICATION DEFECTIVE. Certain retroviruses have been used by genetic engineers to carry new genes into cells. These molecules become part of that cell's protoplasm. Vector constructs in which the internal genes of a retrovirus are replaced by the gene of interest, flanked by the viral LTRs and packaging signals. After the transfection of helper cells, the vector is packaged into virus particles. Infection of target cells with these particles leads to the integration of the gene into cellular DNA as part of a provirus.

retrovirus: A type of virus that has RNA instead of DNA as its genetic material. Retroviruses replicate in a cell by first making a double-stranded DNA intermediate. It uses an enzyme called reverse transcriptase to become part of the host cells' DNA. This allows many copies of the virus to be made in the host cells.

reversal of central dogma: Crick's concept of central dogma was modified with increasing knowledge in the subject. In some plant viruses, for example, TMV, where RNA is the genetic material, genetic information flows from RNA to DNA with the help of an enzyme reverse transcriptase. Under laboratory conditions, proteins can be synthesized directly using DNA as a template. Genetic RNA can undergo self-replication. This concept was termed reverse transcription by H.M. Tamin and S. Mizutani in 1970.

reverse complement polymerase chain reaction (RC-PCR): Modification of polymerase chain reaction used to generate amplicon libraries for DNA sequencing.

reverse epimutation: Removal of an epigenetic mark from DNA or histone protein. For example, in DNA, if cytosine is replaced by 5-azacytidine, this site will remain unmethylated. *See* EPIMUTATIONS.

reverse genetics: From DNA sequence (gene) to a trait. Molecular marker and recombinant DNA technologies have now made it possible to change a particular trait by making a change in the base sequences of DNA (gene). This approach is called reverse genetics where we look for functions or traits controlled by a gene. Earlier geneticists and breeders were looking for genes or DNA sequences controlling specific traits of interest, which is called FORWARD GENETICS. In forward genetics, we start from phenotype, and in reverse genetics, we start from a DNA molecule (gene) to obtain the desired phenotype. There are several reverse genetics approaches: (i) An experimental molecular genetics technique that enables researchers to elucidate gene function by examining changes to phenotypes (of cells or organisms) caused by genetically engineering specific nucleic acid sequences (within DNA or RNA). *Compare with* MENDELIAN GENETICS *and* FORWARD GENETICS. (ii) This term is used to describe any of a variety of molecular methods that allow a wild-type allele of a gene to be targeted and replaced by an engineered mutant allele. *See* POSITIONAL CLONING. (iii) The use of recombinant DNA techniques to investigate gene function, usually by introducing specific mutations *in vitro* (e.g. by site-directed mutagenesis) and often to its original site in that host, by homogenization or transplacement. It is the opposite of conventional genetics where mutations are introduced at random into the whole genome followed by a lengthy procedure to select for the desired mutant.

reverse mutation: The production, by further mutation, of a premutation gene from a mutant gene. It restores the ability of the gene to produce a functional protein. In a strict sense, reversion is the correction of a mutation (i.e. it occurs at the same site). Reversion may be exact or equivalent. *See* REVERSION *and* BACK MUTATION. *Compare with* FORWARD MUTATION.

reverse phase chromatography (RPC): A method of separating a mixture of proteins or nucleic acids or other molecules by specific interactions of the molecules with a HYDROPHOBIC ('water hating') immobilized phase (i.e. stationary substrate), which interacts with hydrophobic regions of the protein (or nucleic acid) molecules to achieve (preferential) separation of the mixture.

reverse phenomics: It allows an in-depth mechanistic understanding of the basis of superiority of the traits shown to be valuable. It allows for advanced exploitation of these traits in future plant breeding schemes.

reverse RNA splicing: Introns are self-spliced in a very simple way. Catalytic steps of splicing can be efficiently reversed under appropriate conditions. Finding that reverse splicing occurs *in vivo* provides a piece of jigsaw that one day may allow us to reconstruct the paths that introns take when on the move. In group I introns, the full reversibility of the self-splicing reaction *in vitro*. It has implication for transposition of group I introns.

reverse tandem duplication: This occurs when the duplicate segment of a chromosome is present adjacent to the original sequences but in reverse order.

reverse transcriptase: A DNA polymerase that can synthesize a COMPLEMENTARY DNA (cDNA) strand using single-stranded RNA as a template as in case of retroviruses. This DNA synthesis is the reverse of normal transcription. *Aka* RNA-DEPENDENT DNA POLYMERASE. It is used for filling-in reactions, DNA sequencing and cDNA synthesis. It can also catalyse the degradation of RNA in an RNA-DNA hybrid and DNA polymerase then synthesizes a complementary strand of DNA creating double-stranded cDNA. Thus, this enzyme is very useful for developing a cDNA library. Reverse transcriptases closely resemble the DNA-directed DNA polymerases in that they require the same materials and conditions as the DNA polymerases (e.g. for RT-PCR). The exonucleolytic activity proceeds in a 5′→3′ and in a 3′→5′ direction. The RNA- or DNA-directed activity requires a template (DNA or RNA) and a primer. *Aka* RNA-DIRECTED DNA POLYMERASES.

reverse transcriptase polymerase chain reaction (RT-PCR): Procedure in which PCR amplification is carried out on DNA that is first generated by the conversion of mRNA to cDNA using reverse transcriptase.

reverse transcription: Some viruses produce enzymes called reverse transcriptase that reverse the transcription process by copying RNA back into a COMPLEMENTARY DNA (cDNA) sequence. This process produces a complementary DNA copy of an RNA molecule and is used by retroviruses and retrotransposons. In nature retroviruses use reverse transcription to insert themselves into an organism's genome. In the laboratory reverse transcription is used to translate unstable mRNA molecules into stable cDNA.

reverse transcription at low deoxyribonucleoside triphosphate concentrations-polymerase chain reaction (RTL-PCR): In this method, purified RNA is treated with DNase I to get rid of any DNA contaminant prior to RTL-P analysis and then transcribed into cDNA by RT at both low and high dNTP concentrations. At low dNTPs, the RT can stop one base prior to any 2′-*O*-methylated ribose, but it can read through the same methylated sites at high dNTPs.

reverse transcription in bacteria: Reverse transcriptase encoded by cellular genes has also been described in bacteria (myxobacteria and *Escherichia coli*). This enzyme is thus not unique to eukaryotes. This suggests that reverse transcriptase existed before retroviruses. Retrons with reverse transcriptase became RETROPOSONS, which with their ability to transpose became RETROTRANSPOSONS that with LTRs became RETROVIRUSES. They with the ability to form virus became PARARETROVIRUSES, which lost the ability to integrate and transpose, became DNA VIRUSES.

reverse transcription in cauliflower mosaic virus multiplication: Varmus proposed reverse transcription step in multiplication of CAULIFLOWER MOSAIC VIRUS (CaMV). Like retroviruses, CaMV has a capacity to produce a major RNA species, which contains all the viral information. CaMV uses primer for synthesis of DNA-LIKE RETROVIRUSES. But the genome of this virus is like that of HEPATITIS B VIRUS (HBV) (i.e. DNA).

reverse transcription in DNA viruses: The phenomenon of reverse transcription was first reported independently by H.M. Temin and S. Mizutani in 1970 in mice leukaemia virus and D. Baltimore in 1970 in Rous Sarcoma virus. Enzyme responsible for this type of transcription was named REVERSE TRANSCRIPTASE. The DNA product of RNA-directed DNA polymerase system has base sequence complementary to the viral RNA. Reverse transcription step has been proposed in replication of CaMV, a double-stranded DNA virus, which infects plants, on the basis of structural studies of encapsidated CaMV DNA and of intracellular DNA and RNA species judged to be intermediates in the replication of CaMV. W.S. Mason and J. Summers in 1982 proved that HBV replicates with the aid of reverse transcription. These viruses have double-stranded circular DNA. This reverse transcription in the virus occurs due to asymmetry of DNA strands. Synthesis of negative strand begins with DR1 sequence of PROGENOMIC RNA, and this initiation can occur either from 5′- or 3′-end of the RNA template. If it occurs from 3′-end then negative strand elongates without any interruption. However, if synthesis initiates from 5′-end, it needs transfer of nascent to the 3′-end of the same RNA molecule. Initiation of plus strand synthesis takes place from 3′-end of DR2 and the extension of plus strand takes place up to the terminal redundancy of minus strand to form the complete DNA molecule.

reverse transcription PCR (RT-PCR): RT-PCR is based on the process in which reverse transcriptase reverse transcribes RNA into DNA, and the enzyme reverse transcriptase was initially isolated from retroviruses. RT-PCR allows the formation of cDNA (complementary or copy DNA) from RNA, which stores the sequence of RNA (e.g. messenger RNA, mRNA) in the more stable form of nucleic acid, DNA. This reverse transcription from RNA into its reverse complement DNA (cDNA) is the first step of usually a two-step process of RT-PCR. Furthermore, by copying the RNA into DNA, one can then amplify the cDNA sequence by using primers specific for the DNA sequence. This amplification is the final second major step of the two-step process of RT-PCR. The exponential amplification of complementary sequence of mRNA or RNA sequences via reverse transcription polymerase chain reaction allow for a high sensitivity detection technique, where low copy number or less abundant RNA molecules can be detected. It is also used to clone mRNA sequences in the form of complementary DNA, allowing cDNA libraries to be created, which contain all the mRNA sequences of genes expressed in a cell. Furthermore, it allows the creation of cDNA constructs, which were cloned by RT-PCR and allow the expression of genes at the RNA and protein levels for further study. *Compare with* REAL-TIME PCR.

reverse transcription quantitative polymerase chain reaction (RT-qPCR): A more modern low-throughput approach for measuring mRNA abundance is reverse transcription quantitative polymerase chain reaction (RT-PCR followed with qPCR). RT-PCR first generates a DNA template from the mRNA by reverse transcription, which is called cDNA. This cDNA template is then used for qPCR where the change in fluorescence of a probe changes as the DNA amplification process progresses. With a carefully constructed standard curve qPCR can produce an absolute measurement such as number of copies of mRNA, typically in units of copies per nanolitre of homogenized tissue or copies per cell. qPCR is very sensitive (detection of a single mRNA molecule is possible) but can be expensive due to the fluorescent probes required.

reverse transcription without prior chemical treatment of the RNA: Chemical groups on the nucleobases (atoms 1, 2 and 6 for purines and 2, 3 and 4 for pyrimidines) hinder the base-pairing interaction with complementary bases. This causes attenuation of primed reverse transcription at such sites, causing a stop (e.g. 1^{m}A, 1^{m}G, 2^{m}A, 2^{m}G, 3^{m}U) or pause (e.g. 6^{m}A, D, 4^{m}C and 2′-*O*-Me nucleotides) on sequencing gels, one residue 3′ to the modified nucleobase; the abbreviations given here stand for 1^{m}A, 1-methyladenine; 1^{m}G, 1-methyl guanine; 2^{m}A, 2-methyladenine; 2^{m}G, 2-methylguanine; 3^{m}U, 3-methyluracil; 6^{m}A, 6-methyladenine; D, 4^{m}C, 4-methylcytosine; 2′-*O*-Me, 2′-O-methylation.

reverse transcription-chemical treatment of the RNA modification detection: Several different modifications that are indefinable to detection by reverse transcription can be identified following specific chemical treatment of the RNA. So far, detection methods have been established for some of the most common RNA modifications, such as I (inosine), Ψ (pseudouridine), 3^{m}C (3-methylcytosine), 5^{m}C (5-methylcytosine), 7^{m}G (7-methylguanine), ^{m}N (ribose 2′-hydroxyl group) and D (dihydrouridine).

reverse translation: A technique for isolating genes (or mRNAs) by their ability to hybridize with a short oligonucleotide sequence prepared by predicting the nucleic acid sequence from the known protein sequence.

reversibility of gene drives: An attractive feature of the proposed construct is that it is fully reversible. If one targets a gene that, when knocked out, is strongly deleterious, then there will be strong selection in favour of resistant allele sequences that are functional but are not recognized and are cut by the homing endonuclease.

reversible epigenetic modifications: All the epigenetic modifications, such as DNA modifications, histone modifications and RNA modifications are reversible.

reversible histone methylation: The discovery of enzymes that reverse the methylation of lysine and arginine challenges our current thinking on the unique nature of histone methylation and substantially increases the complexity of histone modification pathways.

reversible inhibition: This type of inhibition involves equilibrium between the enzyme and the inhibitor, the equilibrium constant (k_i) being a measure of affinity of the inhibition for the enzyme. Reversible enzyme inhibitors diminish on enzyme's activity by interacting reversibly with it. Some enzyme inhibitors are substances that structurally resemble their enzyme's substrate but either do not react or react very slowly. These substances are commonly used to probe the chemical and conformational nature of enzymes active site in effort to elucidate the enzyme catalytic mechanism.

reversible terminator sequencing (Illumina): Reversible terminator sequencing differs from the traditional Sanger method in that, instead of terminating the primer extension irreversibly using dideoxynucleotide, modified nucleotides are used in reversible termination. While many other techniques use emulsion PCR to amplify the DNA library fragments, reversible termination uses bridge PCR, improving the efficiency of this stage of the process. Reversible terminators can be grouped into two categories: 3′-O-blocked reversible terminators and 3′-unblocked reversible terminators.

reversion: Reversion of mutation is a change in DNA that either reverses the original alteration (EXACT REVERSION or TRUE REVERSION) or a different codon codes for the same amino acid or a different codon codes for a similar amino acid (EQUIVALENT REVERSION) or compensates for it by SECOND SITE REVERSION in the same gene. *See* REVERSE MUTATION *and* BACK MUTATION.

reversion of imprinted X chromosome: During mouse embryogenesis, reversion of imprinted X chromosome inactivation in the pluripotent inner cell mass of the female blastocyst is initiated by the repression of *Xist* from the paternal X chromosome.

revertant: A mutant gene, individual or strain that regains a former capability (e.g. production of a particular protein) by undergoing further mutation.

Review Committee on Genetic Manipulation (RCGM): This committee established under the Department of Biotechnology, Ministry of Science and Technology, Government of India monitors the safety of ongoing research projects and activities (e.g. small-scale field trials, import and export) involving genetically engineered organisms.

***Rhizobium*:** A genus of bacteria in the soil that live colonize the roots of leguminous plants to establish a symbiotic relationship. During the symbiosis, atmospheric nitrogen gas is converted into a form usable by the bacterium and the plant. For the legume soybean (*Glycine max* L.), the relevant strain of the bacteria is *Rhizobium japonicum*. For the legume lucerne, the relevant strain of the bacteria is *Sinorhizobium meliloti*.

***Rhizobium* nodules:** The galls on roots caused by *Rhizobium* spp.

rhizogenesis: Differentiation of roots from tissues, organs cultured *in vitro*. To obtain complete plants the shoot cultures are transferred to a medium supplemented with indolebutyric acid (IBA) and naphthalene acetic acid (NAA). During incubation, the shoots exhibit rhizogenesis (i.e. development of roots usually from the cut ends). Rhizogenesis is essential for transferring the propagules into the soil.

rhizome (Abbrev. rhiz.): A perennial underground stem usually growing horizontally. *See* STOLON.

rhizomorph: A macroscopic rope-like strand of compacted tissue formed by certain fungi.

rhizosphere: The area of the soil immediately surrounding the roots of a plant and is altered by the plant's root growth, nutrients, respiration and so on. Roots alter the nutrient status of the soil close to them by absorbing minerals and releasing other substances. This leads to an increase in the numbers of microorganisms and often alters the relative proportions of the different kinds of microorganisms present.

rhizosphere-competent: Used to describe microorganisms adapted to living in the rhizosphere of a plant.

rho (ρ) factor: A protein involved in (chemically) assisting *Escherichia coli* RNA polymerase in the termination of transcription at certain (ρ-dependent) sites on the DNA molecule.

rho (ρ)-dependent terminators: In case of rho-dependent terminators, there is a need for addition of rho (ρ) factor for termination *in vitro*. Rho factor is required for termination *in vivo* also. Most of the known rho-dependent terminators are found in phage genomes.

rho (ρ)-independent terminators: A DNA sequence signalling the termination of transcription, termination does not require the presence of rho protein. Rho-independent or simple terminator core enzyme can terminate *in vitro* at certain sites in the absence of any other factor. These terminators have two structural features – a hairpin in the secondary structure and a run of six uracils (Us) at the very end of the unit. Both features are needed for termination. The hairpins contain a G-C rich region near base of the stem. Probably all the hairpins that form in the RNA product cause the polymerase to slow or pause in RNA synthesis. Pausing creates an opportunity for termination to occur.

rho small GTP-binding protein (ROP) switch: M.R. Puli *et al.* (2024) identified a plant-specific RHO SMALL GTP-BINDING PROTEIN (ROP) SWITCH. They found that mutants lacking the ROP9 molecular switch exhibit enhanced water use efficiency (WUE) without yield penalties, both in greenhouse and large-scale field experiments. The ROP9 switch regulates the formation of reactive oxygen species, and its absence in *rop9* mutants leads to elevated stomata ROS levels and increased stomatal closure. The elevated stomatal closure observed in rop9 mutants is linked to an increase in ROS formation. This is very likely dependent on the RESPIRATORY BURST OXIDASE HOMOLOGUE (RBOH) NADPH oxidase and is not influenced by ABSCISIC ACID (ABA). Consistently, activated ROP9 can interact with respiratory BURST OXIDASE HOMOLOGUE B (RBOHB) in both yeast and plants. *See* WATER USE EFFICIENCY.

Ri plasmid: An *Agrobacterium* plasmid effectively used as a vector for gene transfer in plant cells.

RIBOmap: A new spatially resolved SINGLE-CELL TRANSLATOMICS technology developed by H. Zeng *et al.* (2023), accelerating our understanding of protein synthesis in the context of subcellular architecture, cell types and tissue anatomy. The pairwise SPATIAL TRANSLATOMIC and TRANSCRIPTOMIC MAPPING enabled us to systematically identify cell type- and tissue-region–specific translational regulation, paving the way for uncovering novel post-transcriptional gene regulation principles and mechanisms that shape the proteome for cellular and tissue functions. RIBOmap bypasses complicated polysome isolation steps and genetic manipulation, making it promising for studies in post hoc human tissue and disease samples. Integrating RIBOmap with other imaging-based measurements is expected to enable spatial multiomics mapping for a comprehensive understanding of biological systems.

ribonuclease (RNase): An enzyme that catalyses the degradation of RNA into smaller components. C.B. Afinsen, S. Moore and W.H. Stein were awarded the Nobel Prize in 1972 for the discovery of the chemical structure and activity of the enzyme ribonuclease. Ribonucleases can be ENDORIBONUCLEASES or EXORIBONUCLEASES. Some RNases cleave double-stranded RNA while some others cleave single-stranded RNA molecules. RNases play key roles in the maturation of all RNA molecules (both coding/messenger RNAs and non-coding RNAs). Ribonucleases cut an RNA molecule by hydrolysing one or more of its phosphodiester bonds. Active RNA degradation systems are the first defence against RNA viruses and provide the underlying machinery for more advanced cellular immune strategies such as RNA interference. There are many different RNASES. Some of the more important include RNASE A, which cleaves ssRNA 3′ of pyrimidines; RNASE T1, which cleaves single-stranded RNA at guanosine nucleotides; RNASE V1, which cleaves double-stranded RNA (helical regions); RNASE H, which degrades the RNA part of RNA:DNA hybrids. *See details as separate terms.*

ribonuclease E (RNase E): This enzyme has a key role in mRNA degradation and the processing of catalytic and structural RNAs in *Escherichia coli*. An evolutionarily conserved 17.4-kDa protein, named RraA (regulator of ribonuclease activity), binds to RNase E and inhibits RNase E endonucleolytic cleavages without altering cleavage site specificity or interacting detectably with substrate RNAs. Overexpression of RraA circumvents the effects of an autoregulatory mechanism that normally maintains the RNase E cellular level within a narrow range, resulting in the genome-wide accumulation of RNase E-targeted transcripts.

ribonuclease H (RNase H): a family of non-sequence-specific endonuclease enzymes that catalyse the cleavage of RNA in an RNA/DNA substrate via a hydrolytic mechanism. Members of the RNase H family can be found in nearly all organisms, from bacteria to archaea to eukaryotes.

ribonuclease III family: Members of the ribonuclease III (RNase III) family are DOUBLE-STRANDED RNA (dsRNA) specific endoribonucleases characterized by a signature motif in their active centres and a two-base 3′ overhang in their products. Bacterial RNase III is structurally simpler.

ribonuclease P (RNase P): A type of ribonuclease that cleaves RNA. RNase P is unique from other RNases in that it is a ribozyme – a ribonucleic acid that acts as a catalyst in the same way that a protein-based enzyme would. Its function is to cleave off an extra, or precursor, sequence of RNA on tRNA molecules. Further, RNase P is one of two known multiple turnover ribozymes in nature (the other being the ribosome), the discovery of which earned S. Altman and T. Cech the Nobel Prize in Chemistry in 1989.

ribonuclease T1: Enzyme that cuts nucleic acids at a very specific point on a nucleotide chain.

ribonuclease: *See* RNASE.

ribonucleic acid (RNA): A category of nucleic acids in which the component sugar is ribose and consisting of the four ribonucleotides containing cytosine, uracil, guanine or adenine. The three major types of RNA are (i) messenger RNA (mRNA), (ii) transfer RNA (tRNA) and (iii) ribosomal RNA (rRNA). In addition, there are non-coding RNAs that play a regulatory role in the cell. Like DNA, RNA is assembled as a linear chain of nucleotide bases but is usually single stranded. While DNA carries the genetic information of the cell, RNA has got various forms and functions; messenger RNA serves as template and tRNAs and rRNAs also play crucial role during protein synthesis. Some non-coding RNAs have a regulatory role in gene expression. RNA also constitutes the genome of RNA viruses, known as genetic RNA. Double-stranded RNA is also known to be the genetic material in some viruses.

ribonucleoprotein (RNP): A protein composed of pre-rRNAs and ribosomal as well as non-ribosomal protein components. One of the non-ribosomal proteins, the nucleolin, is considered to play a key role in the regulation of rDNA transcription, perisomal synthesis, ribosomal assembly and maturation.

ribonucleoproteins in pre-mRNA processing: One ribonucleoprotein (RNP) particle that contains a small RNA, U7, is necessary in processing the non-polyadenylated histone H_3 mRNA. Still another RNP that contains small RNA, U1, is involved in the formation of the poly(A) tail.

ribonucleoside: Composed of one ribose molecule plus either a purine or a pyrimidine base. There are four types of ribonucleosides – adenosine, cytidine, guanosine and uridine.

ribonucleotide: A nucleotide containing ribose as its pentose component. A ribonucleotide may have one of the four nitrogen bases – adenine, guanine, cytosine and uracil. One, two or three phosphate groups may be attached at the 5′ end of the ribose sugar molecule. Thus, ribonucleotides are named ADENOSINE 5′-MONOPHOSPHATE, GUANOSINE 5′-MONOPHOSPHATE, CYTIDINE 5′-MONOPHOSPHATE, URIDINE 5′-MONOPHOSPHATE, ADENOSINE 5′-DIPHOSPHATE, GUANOSINE 5′-DIPHOSPHATE, CYTIDINE 5′-DIPHOSPHATE, URIDINE 5′-DIPHOSPHATE, ADENOSINE 5′-TRIPHOSPHATE, GUANOSINE 5′-TRIPHOSPHATE, CYTIDINE 5′-TRIPHOSPHATE and URIDINE 5′-TRIPHOSPHATE. *See details as separate terms.*

ribonucleotide modifications, detection of: It utilizes one of three basic principles: (i) separation based on physicochemical characteristics, (ii) varying enzymatic turnover and (iii) varying chemical reactivity.

ribonucleotide triphosphates: RNA precursors (ATP, GTP, CTP and UTP) that form a Watson-Crick base pair with a DNA template base and its α-phosphate group is attached by a 3′-hydroxyl of the growing end of the RNA.

riboprobe: RNA probes. Riboprobes involve synthesis of labelled RNA with the help of RNA polymerase, using DNA probe as template in the presence of labelled ribonucleotide. Riboprobes being single-stranded offer several advantages over DNA probes.

ribose: A five-carbon-atom (pentose) monosaccharide (i.e. a sugar) present in ribonucleotide molecules and RNA. It is important to life because its closely related cousin deoxyribose forms a part of the molecules that constitute the backbone of deoxyribonucleic acids. Ribose sugar carry one more oxygen atom than deoxyribose. Ribose has an aldehyde group attached to the end of the chain in an open form. *Compare with* DEOXYRIBOSE.

ribose isomers: L-ribose and D-ribose are two optical isomers that can either form a pentagonal or hexagonal ring in aqueous solution. D-ribose is dextrorotatory. A beam of light is bent in the right direction when passed through its aqueous solution. While writing its structural formula, the hydroxyl groups are written on the right side of carbon atoms. On the other hand, L-ribose is laevorotatory. It bends the light rays to the left when passed through its aqueous solution. The hydroxyl groups are written on the left side of carbon atoms in the structural formula of L-ribose. Only D-ribose is found naturally in living organisms. The L-ribose is a synthetic molecule and does not occur in living structures.

ribosomal ambiguity (RAM) mutants: The mutants that allow incorrect tRNAs to be incorporated into the translation process.

ribosomal DNA (rDNA): DNA that codes for the rRNAs used in the production of ribosomes, the sites of cellular protein synthesis. In most eukaryotes, ribosomal DNA consists of tandemly repeated arrays of four or five genes located at the NUCLEOLUS ORGANIZER REGION (NOR) of one or more chromosomes.

ribosomal DNA internal transcribed spacers (rDNA-ITS): A multigene family with nuclear copies in eukaryotes that are arranged in tandem arrays in nucleolar organizer regions (NORs), generally at more than one chromosomal location. Each unit within a single array consists of genes coding for small (18S) and large (28S) rRNA subunits; the 5.8S nuclear rDNA gene lies embedded between these genes but separated by two internal transcribed spacers, ITS1 and ITS2. For example, the copy numbers of 18S–5.8S–28S rRNA genes in diploid genomes of *Quercus cerris*, *Quercus ilex*, *Quercus petraea*, *Quercus pubescens* and *Quercus robur* are estimated to be in the number of 1300–4000. The small subunit is highly conserved and has been used for the relationship between archaebacteria and eubacteria, while more conserved domains within the

28S region have been used to cover evolutionary time through the Palaeozoic and Mesozoic eras. The faster evolving ITS regions have been employed for population and congeneric phylogenies. Internal transcribed spacers represent a biparental nuclear mode of inheritance through several loci with several alleles per locus, are codominant, exhibit a high level of variability even within a single individual and are composed of non-coding sequences.

ribosomal elongation cycle: This cycle describes a series of reactions prolonging the nascent polypeptide chain by one amino acid and driven by two universal elongation factors termed EF-Tu and EF-G in bacteria. Extremely conserved LepA protein, present in all bacteria and mitochondria, is a third elongation factor required for accurate and efficient protein synthesis.

ribosomal RNA (rRNA): Any one of a number of specific RNA molecules that form part of the structure of a ribosome and participate in the synthesis of proteins. Often distinguished by their sedimentation coefficient (S) values. rRNAs make approximately two-thirds of the mass of the bacteria *Escherichia coli* ribosome and approximately one-half of the mass of mammalian ribosomes. Ribosomal RNA accounts for nearly 80% of the RNA content of the bacterial cell. RNA component plays catalytic roles in translation. Three sizes of rRNAs in bacteria include 5S, 16S and 23S. Four sizes of rRNAs in humans include 5S, 5.8S, 18S and 28S.

ribosomal RNA (rRNA) genes: The genes which on transcription yields precursor of rRNAs (pre-rRNA) which after various processing steps mature into functional rRNA. Ribosomes have two subunits: large subunit and small subunit. Large subunit has three rRNAs of size 28S, 5.8S and 5S. Small unit has 18S rRNA. There are four RNA genes. *18S rRNA*, *5.8S RNA* and *28S rRNA* genes are present in form of clusters, 18S rRNA-5.8S rRNA-28SrRNA, known as HETEROCLUSTERS. In eukaryotic cells, the genes for different rRNA molecules are transcribed as a single precursor. In most eukaryotes, the rRNA genes are found in clusters composed of 100 to 5000 repeats of these genes. *Aka* CLASS I GENES. Genes coding for rRNAs are present in NORs of the chromosome.

ribosomal RNA (rRNA) modification sites of a plant ribosome: Protein synthesis can be regulated with respect to changing environmental cues via the deposition of chemical modifications into rRNA. P. Cottilli *et al.* (2022) present the structure of a plant ribosome from tomato and a quantitative mass spectrometry analysis of its rRNAs. The study reveals fine features of the ribosomal proteins and 71 plant-specific rRNA modifications, and it re-annotates 30 rRNA residues in the available sequence. At the protein level, isoAsp is found in position 137 of uS11, and a zinc finger previously believed to be universal is missing from eL34, suggesting a lower effect of zinc deficiency on protein synthesis in plants. At the rRNA level, the plant ribosome differs markedly from its human counterpart with respect to the spatial distribution of modifications. Thus, it represents an additional layer of gene expression regulation, highlighting the molecular signature of a plant ribosome.

ribosomal RNA analysis tools: (i) Ridom is ribosomal RNA analysis tool for clinically relevant bacteria. (ii) Rifle compares restriction patterns of 16S rDNA amplicons against a database of theoretical restriction patterns generated from a 16S rDNA database.

ribosomal RNA gene family: A family of non-coding RNAs responsible for principal functions of the ribosome. rRNAs contribute directly to the catalytic properties of protein synthesis. Ribosomes have two subunits: large subunit and small subunit. Large subunit has three rRNAs of size 25S, 5.8S and 5S. Small unit has 18S rRNA. There are thus four RNA genes. *18S rRNA*, *5.8S RNA* and *28S rRNA* genes are present in form of clusters, 18S rRNA-5.8S rRNA-28SrRNA, known as HETEROCLUSTERS. In most eukaryotes, the rRNA genes are found in clusters composed of 100 to 5000 repeats of these genes and the genes for different rRNA molecules are transcribed as a single precursor. *Aka* CLASS I GENES.

ribosomal RNA modifications: Base methylation and pseudouridylation are two major types of ribosomal RNA modifications. Examples include isomerization of uridine to Ψ, base modification of C and G to 5^{m}C and 2^{m}G, respectively, 2′-O-methylation of uridine to 2′-O-methyluridine (^{m}U) and multiple modifications of Ψ to 1^{m}acp3Ψ. Ribose methylation has also been reported in ribosomal RNA.

ribosomal RNA processing in eukaryotes: At the nucleolar organizer region, 45S RNA is synthesized, which acts as a precursor molecule for the synthesis of 28S and 18S rRNAs. An endonuclease cleaves 45S RNA into two short precursor molecules of size 41S and 20S, which are cleaved into 28S and 18S rRNAs, respectively. The 41S RNA undergoes three sequential cleavage steps by exonuclease to yield 28S rRNA, whereas 20S RNA undergoes only one cleavage step by exonuclease to produce 18S rRNA.

ribosomal RNA transcription regulation: Ribosomal RNA (rRNA) transcription is regulated primarily at the level of initiation from rRNA promoters. The unusual kinetic properties of these promoters result in their specific regulation by two small molecule signals, ppGpp and the initiating NTP, that bind to RNA polymerase at all promoters.

ribosomal RNA transport: Cold treatment (10°C) of root meristems during a period longer than duration of the cell cycle in 20°C causes a blocking of most cells in G1. One of the causes of cell cycle blockade is the decrease of rRNA synthesis and inhibition of rRNA transport into cytoplasm. Transfer of seedlings to 20°C results in increase of rRNA synthesis and its transport to cytoplasm as well as transition of almost all cells from G1/ to S during 3 h. Synchronization of DNA replication, comprising cells within rows of cells, is preceded by similar transport pattern of rRNA to cytoplasm.

ribosomal RNA types in eukaryotes: Eukaryotes, on the other hand, have at least four flavours of rRNA: 5S, 5.8S, 18S and 28S rRNAs. The 5S, 5.8S and 28S rRNAs are found within the large ribosomal subunit, while the 18S rRNA is located in the small ribosomal subunit.

ribosomal RNA types in prokaryotes: Prokaryotes have three varieties of rRNA: 5S, 16S and 23S rRNAs. The 5S and 23S rRNAs are found within the large ribosomal subunit, while the 16S rRNA is located in the small ribosomal subunit.

ribosomal translocation: During the ribosomal translocation, the binding of elongation factor G (EF-G) to the pre-translational ribosome leads to a ratchet-like rotation of the 30S subunit relative to the 50S subunit in the direction of the mRNA movement. This rotation is accompanied by a 20 Å movement of the L1 stalk of the 50S subunit, implying that this region is involved in the translocation of deacylated tRNAs from the P to the E site. These ribosomal motions can occur only when the P-site tRNA is deacylated. Prior to peptidyl-transfer to the A-site tRNA or peptide removal, the presence of the charged P-site tRNA locks the ribosome and prohibits both of these motions.

ribosome: A complex ribonucleoprotein particle. Eukaryotic ribosomes contain 4 RNAs and at least 82 proteins, which is the 'machine' that translates mRNA into protein molecules. In eukaryotic cells, ribosomes are often in close proximity to the endoplasmic reticulum. Ribosomes coordinate the interplay of tRNAs, mRNAs and proteins in the complex process of protein synthesis (manufacture). RNA constitutes nearly two-thirds of the mass of these large (mega-Dalton) molecular assemblies, which are technically ribozymes (i.e. an enzyme in which the catalysis is performed by RNA). The formation of a ribosome (in the endoplasmic reticulum of a cell) from individual RNA and protein molecules is largely a self-assembly process, because all of the information needed for the correct assembly of this structure is contained in the primary structure of its (molecular) components. The assembly process is ordered and proceeds in stages. Many ribosomes (in a given cell) can simultaneously translate an mRNA molecule. The structure, consisting of a group of ribosomes bound to an mRNA molecule actively synthesizing protein, is called a POLYRIBOSOME or a POLYSOME. The ribosomes in this (polysome) unit operate independently of each other, each synthesizing a complete polypeptide (protein) MOLECULAR CHAIN. Ribosomes deposit proteins into the lumen of the rough ER, where they can be modified.

ribosome binding site: The nucleotide sequence that acts as the site for attachment of a ribosome to an mRNA molecule. *Aka* SHINE-DALGARNO SEQUENCE. The Shine-Dalgarno sequence in *Escherichia coli* is 5′AGGAGGU3′.

ribosome biosynthesis: The assembly of ribosome particles from RNA and protein components. This process is coordinated in such a way in eukaryotes and prokaryotes that neither excess ribosomal proteins nor ribosomal RNAs accumulate.

ribosome code: There is a possibility of ribosome code. In the budding yeast, 59 of the 79 cytoplasmic ribosomal proteins are encoded by two genes, stemming from an ancient genome duplication event. These paralogous genes are not functionally equivalent, suggesting the possible existence of a 'ribosome code'. Duplicated genes escape gene loss by conferring a dosage benefit or evolving diverged functions.

ribosome hibernation factors: To conserve energy during starvation and stress, many organisms use hibernation factor proteins to inhibit protein synthesis and protect their ribosomes from damage. In bacteria, two families of hibernation factors have been described, but the low conservation of these proteins and the huge diversity of species, habitats and environmental stressors have confounded their discovery.

By combining cryogenic electron microscopy, genetics and biochemistry, K. Helena-Bueno *et al.* (2024) identify Balon, a new hibernation factor in the cold-adapted bacterium *Psychrobacter urativorans*. Balon is a distant homologue of the archaeo-eukaryotic translation factor aeRF1. Balon occupies the ribosomal A site in both vacant and actively translating ribosomes in complex with EF-Tu, highlighting an unexpected role for EF-Tu in the cellular stress response. Unlike typical A-site substrates, Balon binds to ribosomes in an mRNA-independent manner, initiating a new mode of ribosome hibernation that can commence while ribosomes are still engaged in protein synthesis. Balon–EF-Tu-regulated ribosome hibernation has suggested to be a ubiquitous bacterial stress-response mechanism.

ribosome profiling: A strategy that is based on the deep sequencing of ribosome-protected mRNA fragments and enables genome-wide investigation of translation with subcodon resolution.

ribosome translocation, eukaryotic: During protein synthesis, coupled translocation of messenger RNA (mRNA) and transfer RNAs (tRNAs) advances the mRNA reading frame by one codon. In eukaryotes, translocation is accelerated and its fidelity is maintained by elongation factor 2 (eEF2). N. Milicevic *et al.* (2024) report ten high-resolution cryogenic electron microscopy (cryo-EM) structures of the elongating eukaryotic ribosome bound to the full translocation module consisting of mRNA, peptidyl-tRNA and deacylated tRNA, seven of which also contained ribosome-bound, naturally modified eEF2. This study recapitulates mRNA–tRNA2-growing peptide module progression through the ribosome, from the earliest states of eEF2 translocase accommodation until the very late stages of the process and shows an intricate network of interactions preventing the slippage of the translational reading frame. The accuracy of eukaryotic translocation relies on eukaryote-specific elements of the 80S ribosome, eEF2 and tRNAs. The sterically constrained environment imposed by diphthamide, a conserved eukaryotic post-translational modification in eEF2, not only stabilizes correct Watson-Crick codon-anticodon interactions but may also uncover erroneous peptidyl-tRNA and therefore contribute to higher accuracy of protein synthesis in eukaryotes.

ribosome, subunits of: Ribosomes contain a small subunit (SSU) and a large subunit (LSU). The LSU contains peptidyl transferase centre and SSU contains decoding centre. Each subunit is composed of one or more rRNAs and multiple proteins. It is the proteins in the ribosome that perform a largely structural function, not the RNA.

ribosomes, eukaryotic: Ribosomes in eukaryotes are different from those of prokaryotes. Size of eukaryotic ribosomes is 80S. It is made up of two subunits, 40S and 60S. 40S subunit constitutes of 18S rRNA (1900 bases) and 33 ribosomal proteins and 60S subunit is made up of 28S rRNA (4700 bases) + 5S rRNA (120 bases) + 5.8S rRNA (155 bases) and 45 ribosomal proteins.

ribosomes, prokaryotic: Prokaryotic ribosomes are made up of rRNA and various proteins. The size of the bacterial ribosome is 70S. It is made up of two subunits 30S and 50S. 30S subunit is made of 16S rRNA (1541 bases) and 21 ribosomal proteins. 50S subunit is made of 23S rRNA + 5S rRNA (2904 bases) and 31 ribosomal proteins ribosomal RNA Ribosomes are the sites where protein synthesis takes place. rRNA contributes to the binding of tRNA and mRNA to ribosome during protein synthesis.

ribosomes, tRNA binding sites on: The intact ribosome contains three tRNA binding sites that reach between two subunits; an A site where charged tRNA enters the ribosome, a P site that contains peptidyl-tRNA and an E site where deacylated tRNA exits the ribosome.

ribosubstitution method: Substituting a specific deoxynucleotide with a specific ribonucleotide in the DNA chain. The method facilitates a DNA chain to be cleaved at very specific points with an enzyme.

riboswitches: Riboswitches are natural RNA sensors that control gene expression via their capacity to bind small molecules (metabolites). They fold into RNA secondary structures whose conformation switches between an 'on' state and an 'off' state in response to ligand binding. They are often found at the 5′ end of the mRNA, in the untranslated region (the stretch that precedes the start codon): this way, they are the first domain to be synthesized and can therefore influence expression before the entire mRNA is created. Recently, a synthetic translational riboswitch controlled by the ligand theophylline was successfully employed in tobacco chloroplasts. *Aka* GENETIC SWITCHES.

riboswitch-mediated base-pairing changes: Bacteria make extensive use of riboswitches to sense metabolites and control gene expression and typically do so by modulating premature transcription termination or translation initiation. The most widespread riboswitch class known in bacteria responds to the coenzyme thiamine pyrophosphate (TPP), which is a derivative of vitamin B_1. Representatives of this class have also been identified in fungi and plants, where they are predicted to control messenger RNA splicing or processing.

ribozyme. An RNA molecule that possesses enzymatic (catalytic) activity. Discovered in 1982 by T. Cech and S. Altman, from a protozoan *Tetrahymena thermophila*, for which they were awarded the Nobel Prize in Chemistry in 1989. Such RNA molecules can specifically cleave (cut) other RNA molecules. The ribozyme (RNA) molecule and the other RNA molecule come together, whereupon the ribozyme molecule cuts the other RNA molecule at a specific defined (three-base) site. Because the ribozyme molecule acts as an enzyme in this reaction, the ribozyme molecule is not consumed or destroyed but goes on to similarly 'cut' other RNA molecules. In 2000, T. Steitz and P. Moore *et al.* proved that ribosomes (i.e. the cell's internal protein synthesis machinery) are functionally ribozymes. A good example is the hepatitis delta virus RNA, which is capable of self-cleavage and self-ligation in the absence of protein enzymes. D. Hershlag and T.R. Cech found in 1990 that DNA could be used as a weak substrate for the action of RNA enzymes as RNA catalyses the sequence-specific cleavage of single-stranded RNA. *See* RNA AS AN ENZYME.

ribozyme-mediated protection: Ribozymes are RNA molecules that exhibit enzyme activities. The ribozymes used to generate virus resistance are, in fact, hybrid RNA molecules consisting of TOBACCO RINGSPOT VIRUS (TobRV) satellite RNA endoribonuclease catalytic sequence linked to the antisense RNA of specific genes against which they are targeted. The strategy consists of (i) producing a ribozyme specific to a part of the target virus genome, (ii) producing cDNA of this ribozyme and (iii) integrating it into the host-plant genome. Transgenic tobacco plants expressing ribozyme against TMV showed some resistance to TMV infection. Ribozymes may be useful in producing resistance to several viruses within a group where conserved sequences exist. RNase P linked with external guide sequences is a potential alternative to ribozymes.

ribulose 1,5-bisphosphate carboxylase-oxygenase (RuBisCO): A CO_2-fixing enzyme; the key enzyme in photosynthesis. It is the most frequent protein on Earth. It has a unique double function of being both a carboxylase and an oxygenase. When acting as an oxygenase, it catalyses the light-dependent uptake of O_2 and the formation of CO_2 in a complicated process called PHOTORESPIRATION, which takes place concomitantly in three organelles – chloroplasts, peroxisomes and mitochondria. RuBisCO can be manipulated in order to create an artificial plant that can contribute to the reduction of CO_2 content in the atmosphere (i.e. decreasing the so-called greenhouse effect). This enzyme also catalyses a second wasteful competing reaction, which fixes oxygen instead of CO_2. This side reaction of RuBisCo reduces the efficiency of photosynthetic carbon fixation in C_3 plants in the process of photorespiration.

rice epigenome: Cytologically, thickly stained heterochromatin displayed less H3K4me2 and H3K4me3 and more DNA methylation marks than the less lightly stained euchromatin, whereas centromeres revealed an exclusive epigenetic composition. Most TEs showed highly methylated DNA but no H3K4 methylation, whereas more than half of protein-coding genes exhibited both methylated DNA and di- and/or trimethylated H3K4 epigenetic marks. Methylation of DNA but not H3K4 showed an association with repressed transcription. Rice epigenomic studies displayed both general and specific features, profiling additional chromatin modification profiles in a tissue-specific and cell-specific manner, which will help to largely improve the rice epigenomic architecture. Enormous available data divulge the enormous elasticity and possibility of diverse epigenomic landscapes. Cloning of several rice epialleles associated with important agronomic traits highlights the significance of epigenomic variation in rice plant growth, fitness and yield.

rice grassy stunt virus 1 (RGSV1): RGSV1 causes severe stunting, excessive tillering and pale green to yellow and narrow leaves with small rusty spots.

rice grassy stunt virus 2 (RGSV2): RGSV2 causes severe stunting, excessive tillering and yellow to orange and narrow leaves with small rusty spots.

rice plants subjected to the stress regenerate from tissue culture: The regenerated rice plants had changes to the genome-wide pattern of DNA methylation, including some at the promoters of genes. The changes are predominantly loss rather than gain of DNA methylation and they persist in the regenerated plants and their progeny. Such examples act as indicators rather than proof of transgenerational epigenetic change affecting an adaptive phenotype.

rice ragged stunt virus (RRSV): Symptoms are stunted plants, which remain dark green. Leaves are ragged and twisted. Vein swelling on leaf collar, leaf blades and leaf sheaths.

rice tungro virus: Causal agent is rice TUNGRO BACILLIFORM VIRUS (RTBV) and RICE TUNGRO SPHERICAL VIRUS (RTSV). The disease is transmitted by the green leafhopper *Nephotettix* spp. symptoms are yellow-to-yellow orange leaves, stunting and slightly reduced tillering.

rice whorl maggot disease: Causal agent is *Hydrellia philippina*. Symptoms are leaf margin feeding, which causes conspicuous damage and sometimes stunting of plants.

rifampicins: A group of antibiotics that inhibit the initiation of transcription in bacteria. It is a derivative of rifamycin SV, which specifically inhibits DNA-dependent bacteria and chloroplasts, but not mammalian RNA polymerase. It is also reported to inhibit the assembly of DNA and protein into mature virus particles. *Aka* RIFAMPIN *or* RIFAMYCIN.

right splicing junction: The boundary between the right end of an intron and the left end of the adjacent exon.

rigid populations: The populations that keep their genetic composition constant despite longitudinal and seasonal environmental changes.

***rII* locus of bacteriophage T4:** Complementation test determines if two mutations are in the same gene. Certain mutations in the *rII* region influence the length of the life cycle of T4 in the bacterial cell (r = rapid lysis). The *rII* mutants also make larger plaques than wild-type phages. The *rII* mutants and wild-type T4 phage grow equally well on *Escherichia coli* strain B, but on *Escherichia coli* K12(λ) strain, only wild type can multiply. Region *rII* of T4 phage consists of two genes, *rIIA* and *rIIB*. In a typical cross, *Escherichia coli* strain B bacteria were infected with two T4 phage particles each bearing independently isolated *rIIA* (or *rIIB*) mutation. As viruses multiplied, recombination occurred. The progeny was then grown on *Escherichia coli* K12(λ) strain (where only wild type can grow) to test for wild type. Normal particles were found in many large fractions of crosses, indicating that recombination had occurred within a gene.

ring bivalent: An association of two chromosomes with terminal chiasmata on both arms.

ring chromosome (r): Usually found in bacteria. Results when a chromosome breaks in two places and the ends of the chromosome arms fuse together to form a circular structure. An isochromosome may also form a ring in metaphase I of meiosis.

ring species: A situation in which two reproductively isolated populations living in the same region are connected by a geographic ring of populations that can interbreed. *See* REPRODUCTIVE ISOLATION.

Ringer solution: An aqueous solution of calcium chloride, sodium chloride and potassium chloride that provides a medium essentially isotonic to many animal tissues and is used topically to irrigate tissues or intravenously to replenish fluids and electrolytes.

ringspot: A disease symptom characterized by yellowish or necrotic rings enclosing green tissue, as in some plant diseases caused by viruses.

risk: Often used as another name for probability but a more accurate definition is the probability of an adverse event × the severity of the loss that experiencing that event would entail. The likelihood of a defined hazard being realized, which is the product of two probabilities: the probability of exposure, P(E) and the probability of the hazard resulting given that exposure has occurred, P(H/E). Risk (R) = P(E) × P(H/E).

risk analysis: Risk assessment, risk communication, risk management and policy relating to risk to human health and the environment, in the context of risks of concern to individuals; to public, private and non-governmental organizations; and to society at a local, regional, national or global level.

risk assessment: Determining the chance of a particular outcome by using a combination of family history and laboratory testing. An example of this is predicting breast cancer risk for a woman with a family history of breast cancer who has had BRCA gene testing. Functional genomics is now being applied in risk assessment where it quantifies DNA damage caused by radiation, mutagens and carcinogens.

risk communication: In genetics, a process in which a genetic counsellor or other medical professional interprets genetic test results and advises patients of the consequences for them and their offspring.

risk magnification: A treatment, even one for which there are no interactions with baseline covariates, which has a non-zero effect on a relative scale, by necessity must have different absolute effects. The variation of absolute differences is risk magnification due to baseline risk. Subjects that have baseline risks near 0 or 1 have nowhere to go; absolute risk differences are less restricted in the middle of the baseline risk distribution. Treatments have greater absolute benefit for sicker patients, up to a point, even if their relative effects are universal.

risk set: The set of patients in the study at a specified time.

risk-assessment end point: Societal, human health or environmental values that need to be managed or protected.

r^-m^-: Shorthand genetic nomenclature for a bacterial strain deficient in both the restriction and modification of DNA.

r^-m^+: Shorthand genetic nomenclature for a bacterial strain deficient in the restriction but not the modification of DNA. Such strains are useful hosts for the cloning of heterologous genes. (r^+m^- strains would commit suicide by restricting their own, unmodified, DNA).

RMBase (including RMBase v2:0): This database decodes the genome-wide landscape of RNA modifications recognized from high-throughput modification data produced by 18 independent studies. The current release of RMBase includes ~9500 Ψ modifications generated from Pseudo-seq and CeU-seq sequencing data, ~1000 5^mC predicted from Aza-IP data, ~124,200 6^mA modifications discovered from 6^mA-seq and ~1210 2′-*O*-Me identified from RiboMeth-seq data and public resources. RMBase v2.0 is a comprehensive database that assimilates epitranscriptome sequencing data for the exploration of PTCMs of RNAs and their relationships with miRNA binding events, disease-related SNPs and RNA-binding proteins (RBPs).

RNA accumulation studies in meiocytes: Currently there are two major approaches: (i) RNA-seq and (ii) microarray. The RNA-seq tech is more powerful for whole transcriptome analysis, the available analysers such as Illumina Genome Analyzer 2 and Roche 454 analyser made this approach more powerful.

RNA amount polymorphism: A condition where there are at least two genetic variants (alleles) of some sign that exist in the population with respect to the amount of RNA.

RNA analysis: Includes structural RNAs, RNA folding, pseudoknots, ribosomal RNAs and siRNAs.

RNA antiswitches: These molecules contain an aptamer domain that binds to a specific effector (small molecule ligand), as well as a sequence recognition domain complementary to a target mRNA. The binding of the effector changes the conformation of the switch and enables control of target gene expression.

RNA aptamers: RNA molecules that have high affinity and specificity for target molecules.

RNA as an enzyme: T.R. Cech *et al.* (1981), while working on an RNA from a protozoan *Tetrahymena thermophila* found that self-splicing of intron in 28S rRNA precursor occurs through TRANS-ESTERIFICATION mechanism. INTERVENING SEQUENCE (IVS) CYCLIZATION occurs. In this process, a guanosine nucleotide acts as a nucleophile and gets attached to the 5′-end of the existing intervening sequence via a normal 3′-5′ phosphodiester bond and then -OH of the 5′ exon would cause cleavage to produce 5′-phosphate and 3′-OH terminus. The ligation occurs without hydrolysis of ATP with the removal of covalent circular intron. The intron after its release undergoes reopening and refolding twice and ultimately loses 19 nucleotides. An RNA molecule that possesses enzymatic activity is known as RIBOZYME.

RNA as the genetic material, evidence of: TMV reconstitution experiment reported by H. Fraenkel-Conrat and B. Singer in 1957 provides evidence of RNA as the genetic material in TMV. TMV is one of the many viruses using RNA instead of DNA as its nucleic acid. There are different TMV strains with different protein envelopes. Protein and RNA of TMV could be separated. Without RNA, protein was not infective whereas pure TMV RNA was infective and caused the synthesis of viral particles. The type of reconstituted TMV depended upon RNA, not protein. RNA that serves as genetic material is termed GENETIC RNA. Single-stranded RNA is the genetic material in TMV, Influenza virus, Rous Sarcoma Virus, Poliomyelitis and bacteriophages MS2, F2, r17. Double-stranded RNA is the genetic material in wound tumour (plant) virus and retrovirus (animal virus). Examples of plant viruses that have RNA as the genetic material include CMV and TSWV. Human RNA viruses include Orthomyxoviruses, HEPATITIS C VIRUS (HCV), Ebola disease, SARS, influenza, polio measles and retrovirus, including adult HUMAN T-CELL LYMPHOTROPIC VIRUS TYPE 1 (HTLV-1) and human immunodeficiency virus (HIV). *See* POSITIVE-STRAND RNA VIRUSES *and* NEGATIVE-STRAND RNA VIRUSES.

RNA box: Refers to a short nucleic acid consensus sequence or motif. PILIPENKO'S BOX-A MOTIF is an

RNA box that may be involved in ribosome binding in some viral RNAs.

RNA capping: This first pre-mRNA processing step includes 5′-capping, which is a set of enzymatic reactions that add 7-methylguanosine ($7^{m}G$) to the 5′-end of pre-mRNA and thus protect the RNA from degradation by exonucleases. The $7^{m}G$ cap is then bound by cap binding complex heterodimer (CBC20/CBC80), which aids in mRNA export to cytoplasm and also protects the RNA from decapping.

RNA decay mechanisms: RNA decay plays a crucial role as a gene expression regulatory mechanism in eukaryotes. RNA can be degraded from the 3′-end by the exosome or monomeric ribonuclease DIS3L2. RNA can be degraded from the 5′-end by Xrn family of RNases.

RNA decoy: Decoy is something used to trick or mislead. Here is an example of a decoy in epigenetics. An endogenous lncRNA was first identified in *Arabidopsis thaliana*, which functioned as an endogenous target mimic (eTM) of miR399.

RNA editing: A process by which nucleotides not coded by the gene are introduced at specific positions in an mRNA molecule after transcription. It results in radical changes in the amino acid specified by a codon. Most of the editing events occur in the first two nucleotides of the coding sequences. RNA editing has been noticed in kinetoplastids, plant mitochondria, chloroplasts and mammalian nuclei. It has been shown that editing is ubiquitous in all gymnosperms and angiosperms except *Marchantia*. The editing requires at least two steps (i.e. specificity and biochemical modification). The nucleotides located immediately upstream of the edited cytosine are most often a pyrimidine (uridine) and very rarely guanidine. The three main biochemical processes involved may be deamination/amination, base exchange and nucleotide exchange. Editing is sequence-specific. It has been shown that editing mainly occurs in the mRNA fraction rather than rRNA or tRNA. It has been reported that the ACG codon appears at the 5′ terminus of chloroplast genes where initiation codon ATG would be expected. RNA editing also involves C-to-U change, multiple C-to-U or U-to-C changes, addition of G or C residues, A-to-G modification and conversion of U to A, U to G and A to G. RNA editing takes place with the help of GUIDE RNAs (gRNAs). The mRNAs are edited most frequently, whereas edits of tRNAs and rRNAs are extremely rare.

RNA editing adenosine to inosine: Primary transcripts of certain micro RNA (miRNA) genes are subject to RNA editing that converts adenosine to inosine. A-to-I RNA editing leads to transcriptome diversity.

RNA editing and RNA self-splicing, parallelism between: Parallelism between RNA editing (accomplished by gRNA) and self-splicing of introns in HETEROGENEOUS NUCLEAR RNA (hnRNA) (accomplished by ribozymes) was established by Thomas Cech in 1986 and 1987. A parallelism between the function of INTERNAL GUIDE SEQUENCE (IGS) in RNA splicing and the proposed role of gRNAs in RNA editing in kinetoplastid mitochondria was recognized subsequently by J.M. Mueller *et al.* (1993).

RNA editing and self-splicing: There is parallelism between RNA editing (done by gRNA) and self-splicing of introns in hnRNA (done by ribozymes) A parallelism between the function of IG-S in RNA splicing and the proposed role of guide RNAs in RNA editing in kinetoplastid mitochondria has been recognized. They identified a chimeric intermediate with gRNA covalently joined to 3′ portion of the mRNA *in vitro* suggesting an evolutionary analogy to the catalytic mechanism involved in RNA splicing. Like the self-splicing of introns, RNA editing also involves transesterification, but with the help of guide RNA. In the first step, the guide RNA aligns itself (by base pairing) with the unedited RNA, splits it into two and makes a new bond between one of the broken ends and the uridine at the tip of the tail of gRNA at its 3′-end.

RNA editing by base exchange: Exchange of base without cleavage of the sugar-phosphate backbone is another possible way of biochemical modification. Such transglycosylate reaction has been described in transfer RNAs. Pyrimidine replacement is also found in DNA repair metabolism. This process like deamination does not include a cut in the RNA chain.

RNA editing by deamination/amination: The simplest mechanism for C-to-U conversion is deamination at position 6 of cytosine leading to the formation of an uracil residue. This could be accomplished by cytidine deaminase, an enzyme of nucleotide metabolism. The reverse U-to-C modification found in plant mitochondria could be carried out by a CTP synthetase in an ATP-dependent reaction. However, the occurrence of reverse editing from U-to-C implies the presence of two different enzymatic activities and not exactly a reverse action of the same enzyme.

RNA editing by nucleotide exchange: Nucleotide deletions at RNA editing sites have been described in some wheat mitochondrial transcripts. Nucleotide exchange involves cleavage of the RNA chain, deletion of cytosine and addition of uracil, followed by religation.

RNA editing cytosine to uracil: All canonical transfer RNAs (tRNAs) have a uridine at position 8, involved in maintaining tRNA tertiary structure. However, the hyperthermophilic archeon *Methanopyrus kandleri* harbours 30 (out of 34) tRNA genes with cytidine at position 8. Cytosine-to-uracil (C-to-U) editing at this location in the tRNA's tertiary core has the cytidine deaminase domain linked to a tRNA-binding THUMP domain, named after THioUridine synthases, RNA Methyltransferases and Pseudo-uridine synthases. The presence of this C-to-U editing enzyme guarantees the proper folding and functionality of tRNAs. Three RNA editing mechanisms are potentially responsible for C-to-U conversion. Mechanism 1 involves deamination (or transamination) of the C-6 amide of cytosine to convert the base to uracil. Mechanism 2 involves transglycosylation of the ribosyl moiety to replace the base. Mechanism 3 involves deletion and insertion of a new nucleoside monophosphate.

RNA editing cytosine to uracil through base exchange: Exchange of base without cleavage of the sugar-phosphate backbone is another possible way of biochemical modification. Such TRANSGLYCOSYLATE REACTION has been described in transfer RNAs. Pyrimidine replacement is also found in DNA repair metabolism. This process like deamination does not include any cut in the RNA chain.

RNA editing cytosine to uracil through deamination/amination: The simplest mechanism for C-to-U conversion is deamination at position U of cytosine leading to uracil residue. This could be achieved by cytidine deaminase, an enzyme of nucleotide metabolism. The reverse U-to-C modification found in plant mitochondria could be carried out by a CTP synthetase in an ATP-dependent reaction. However, the existence of reverse editing from U to C implies the presence of two different enzymatic activities and not the reverse action of the same enzyme.

RNA editing cytosine to uracil through nucleotide exchange: Nucleotide deletions at RNA editing sites have been reported in wheat mitochondrial cDNA clones. These cDNAs were interpreted to represent single-nucleotide deleted intermediates of the editing process, and a deletion and insertion mechanism was proposed by these investigators. Nucleotide exchange involves cleavage of the RNA chain, deletion of a cytosine, addition of uracil, followed by the religation. The nucleotide exchange involves several steps and is similar to the one described for U addition and deletion in the editing of mRNA of trypanosomes. This mechanism is most complex and would probably need an apparatus composed of multiple factors.

RNA editing databases: RNA editing sites are already mapped on the reference human genome, and databases that source all the available and published data related to RNA editing can be easily found online.

RNA editing discovery: The term RNA editing was proposed by R. Benne *et al.* (1986). RNA editing was revealed in the mid-1980s, succeeding some surprising results attained by several groups of molecular biologists chiefly in the mitochondrial genes of Trypanosoma (which cause African sleeping sickness) and Leishmania (causing kala-azar): (i) Several genes having mutations, which were expected to make these genes inactive owing to the generation of premature stop signals, or due to loss of start signals, were still active. (ii) The sequences of messenger RNAs (mRNAs) at some positions were such that they could not have been derived like that from the DNA sequence of the corresponding genes.

RNA editing in chloroplasts: ACG codon appears at 5′-terminus of chloroplast genes where initiation codon ATG would be expected. Evidence of mRNA transcript from the *rpl2* gene in maize plastome shows that mRNA editing also occurs in chloroplasts. In organellar transcripts of higher plants, definite cytidine residues are changed into uridine residues. In many cases, editing leads to the restoration of conserved amino acid residues in a procedure that is essential for protein function in plastids.

RNA editing in different cell compartments: RNA editing has been observed to occur in chloroplasts, kinetoplastid mitochondria, plant mitochondria and nucleus.

RNA editing in kinetoplastid mitochondria: RNA editing has been divided into two classes – insertional editing and substitutional editing. Insertional editing involves the editing of kinetoplastid mitochondrial mRNA. It involves base pairing to specific gRNAs with poly(U) tails. The mechanism involves the identification of editing sites in the mRNA by mismatches with the gRNA, cleavage of the mRNA and religation of mRNA halves with added or deleted U residues by Terminal Uridylyl Transferase (TUTase). A similar mechanism underlies the insertion of C residues into slime mould mitochondrial pre-mRNA. The modification of apoB mRNA is different from insertional mRNA editing. No gRNA is implicated.

RNA editing in plant mitochondria: Mitochondria have adopted various modes of transcript maturation, such as RNA editing and *trans*-splicing. Transcripts of C-terminal gene modules undergo polyadenylation, and contiguous mRNAs are generated via the concatenation of separate module transcripts. The range of bases being edited varied from 0.8% to 5.8% of the total nucleotides present in a transcript. The mechanism of RNA editing in plant mitochondria requires at least two steps: (i) the specific identification of the cytosines to be edited and (ii) the biochemical modification. Several conclusions on the editing mechanism in plant mitochondria can be drawn from the information available to date. RNA editing changes more than 400 cytidines to uridines in the mRNAs of mitochondria in flowering plants. In other plants (e.g. ferns and mosses), RNA editing reactions, which change C to U or U to C are observed at almost equal frequencies.

RNA editing of micro RNAs: RNAs, including primary and precursor micro RNAs, can undergo PTCMs. Such modifications are produced by cellular stress, modifying the function of the affected RNAs. An important vasoactive micro RNA, miR-487b-3p is subject to A-to-I editing in the seed sequence of the micro RNA.

RNA editing, characteristics of: The following are the general characteristics of RNA editing: (i) Guide RNA editing is a post-transcription process. (ii) Editing frequently results in radical changes in amino acid(s) specified by a codon. (iii) Most editing events occur in the first two nucleotides of the codon. (iv) RNA editing efficiently modifies the transcripts such that a relatively homogenous population of polypeptide is produced. (v) RNA editing generally occurs in the coding sequences.

RNA editing, consequences of: RNA editing shows that genetic information resides not only in the genes per se but also in the apparatus that edits the transcript. RNA editing could be a relic of an RNA sequence correlation system that was utilized for polymerase-dependent nucleic acid synthesis evolved. RNA editing could regulate gene expression, perhaps, in a tissue-specific manner.

RNA editing, mechanisms of: Various hypotheses proposed for mechanisms of RNA editing are pseudogene hypothesis, variant of pseudogene hypothesis, edited template refractive to hybridization, cryptic signal sequences, self-replication edited RNA, alternative RNA/RNA heteroduplex model and multienzyme editing complex. Multienzyme editing complex hypothesis was proposed by L. Simpson in 1990 as a gRNA model of RNA editing and is the most viable proposal. The unedited transcript and gRNA hybridize and the mismatched nucleotides are edited using enzymes like ribonuclease, TUTase and RNA ligase. The enzyme complex moves along the hybrid in 3′→5′ direction.

RNA editing, patterns of: Various patterns of RNA editing are as follows: (i) Insertion/deletion of non-templated uridines, as in *Trypanosoma* mitochondrial transcripts. (ii) Incorporation of non-templated guanosines, as in paramyxoviral transcripts. (iii) Generation of a stop codon by conversion of a specific cytosine (C) to a uridine (U), as in apolipoprotein B transcripts in mammalian intestinal tissue. (iv) Change of multiple Cs to Us or Us to Cs, as in higher plant mitochondrial transcripts. This phenomenon often results in the synthesis of more than one functional transcript from a single ORF. RNA editing has also been shown to compensate for a frameshift mutation in the coding region.

RNA editing, significance of: RNA editing by addition and deletion of non-coded ribonucleotide residues to and from RNA transcripts converts them into correctly translatable mRNA. RNA editing thus offers yet another method of modulating genetic activity. RNA editing, which serves as a translational control mechanism, adds ribonucleotides for N-terminal amino acids in type II editing and creates the whole protein-encoding region in type III editing. It also regulates maxicircle gene expression at the developmental level. editing is ubiquitous in all gymnosperms and angiosperms except *Marchantia*. RNA editing shows that genetic information is located not only in the genes per se but also in the apparatus that edits the transcript. RNA editing may represent a mechanism that evolved to repair mutant genes and now has become entrenched in the gene expression system. RNA editing by host deaminases is an innate restriction process considered to counter the SARS-CoV-2 virus infection.

RNA editing, types of: There are four categories of RNA editing – SIMPLE EDITING, INSERTIONAL EDITING, PAN-EDITING and POLYADENYLATION EDITING. *See details as separate terms.*

RNA editing, wobble base improves precision in: The endogenous human ADENOSINE DEAMINASE ACTING ON RNA (ADAR) enzyme has been employed to perform adenosine-to-inosine (A-to-I) editing, where inosine resembles guanosine (G) in base pairing, enabling an eventual A-to-G edit. To achieve targeted editing, guide RNAs have been premeditated to produce double-stranded RNA structures with the target RNA, creating a site for ADAR recruitment. Thorsten Stafforst from the University of Tübingen, Germany explains that only one component, an RNA, is administered. This makes it predominantly straightforward to bring it into the clinic. The Stafforst group has developed various tools in the field. For example, L. Tang (2024) introduced *in silico* CLUSTER gRNAs, in which they offer design rules for optimizing editing outcomes.

RNA epigenetics: RNA modifications have emerged as a new layer in the transcriptional and epigenetic control of various biological processes, bringing about a new era of 'RNA epigenetics'.

RNA exome capture sequencing: Achieve cost-effective RNA exome analysis using sequence-specific capture of the coding regions of the transcriptome. Ideal for low-quality samples or limited starting material.

RNA folding: (i) LocARNA involves Multiple Alignment of RNAs. It is a tool for multiple alignment of RNA molecules. LocARNA requires only RNA sequences as input and will simultaneously fold and align the input sequences. LocARNA outputs a multiple alignment together with a consensus structure. For the folding, it makes use of a very realistic energy model. (ii) FOLDALIGN – folds and aligns RNA structures (make a fold alignment) based on a lightweight energy model and sequence similarity. The current version makes pairwise fold alignments. (iii) MFold – used for RNA folding. The data can be presented in a number of graphic formats. (iv) Vienna RNA secondary structure prediction – this site is useful for drawing tRNAs in cloverleaf format. (v) CONTRAfold achieves the highest single sequence prediction accuracies to date, outperforming currently available probabilistic and physics-based techniques. (vi) RADAR (**R**NA **D**ata **A**nalysis and **R**esearch) provides (a) constrained alignment of RNA secondary structures and (b) prediction of the consensus structure for a set of RNA sequences. RADAR performs many important RNA mining operations, including understanding the functionality of RNA sequences, the detection of structural RNA motifs and the clustering of RNA molecules.

RNA hybridization: When denatured DNA is mixed with complementary RNA, hybridization occurs. The rate of hybridization depends upon (i) concentration of DNA molecule (C_0), (ii) time given for renaturation (t) and extent of homology between molecules to be hybridized. More is the homology faster is rate of hybridization. The renaturation curve we get is known as the C_0t curve. Cot curves for several genomes plot a fraction of the DNA that has reassociated ($1 - C/C_0$) against the log of C_0t. The form of each curve is similar but the curve in each case is very different.

RNA immunoprecipitation protocol: This protocol uses a native approach without cross-linking or a method that uses formaldehyde cross-linking. iCLIP permits for high-resolution detection of single 6^mA residues and 6^mA clustering across the entire RNA).

RNA immunoprecipitation: An antibody-based approach engaged to map in vivo RNA-protein interactions. The RBP of interest is immunoprecipitated together with its related RNA for documentation of bound transcripts (mRNAs, ncRNAs or viral RNAs).

RNA import into mitochondria: RNA import into mammalian mitochondria is considered essential for replication, transcription and translation of the mitochondrial. Polynucleotide phosphorylase (PNPase), a 3′→5′ exoribonuclease and poly(A) polymerase, in the mitochondrial intermembrane space, a location lacking resident RNAs, has a role in regulating the import of nuclear-encoded RNAs into the mitochondrial matrix. Human mitochondrial mutations can be corrected by targeted RNA import. The RNA component of the human RNase P, appended to a non-imported RNA, directs the import of the resultant RNA fusion transcript into human mitochondria.

RNA in origin of life: There may have been a precursor molecule that had some of the properties of RNA (i.e. genetic information storage and chemical catalytic properties), but could have been existing prior to the synthesis of the RNA world. Two possible candidates are PYRANOSYL-RNA and PEPTIDE NUCLEIC ACID (PNA).

RNA insecticides: Insecticides made of RNA. According to E. Stokstad in 2024, RNA insecticides could offer a safer and more targeted weapon against crop pests.

RNA interference (RNAi): A reverse genetics process within living cells that moderates the activity of their genes. Historically, it was known by other names, including COSUPPRESSION, POST-TRANSCRIPTIONAL GENE SILENCING (PTGS) and QUELLING. The term RNAi is now used widely to describe RNA silencing in both plants and animals. RNAi can cause gene silencing both at the post-transcriptional and transcriptional level that is directed by small RNA molecules. Two types of small ribonucleic acid (RNA) molecules: MICRO RNA (miRNA) and SMALL INTERFERING RNA (siRNA) are central to RNA interference. RNAs are the direct products of genes, and these small RNAs can bind to other specific messenger RNA (mRNA) molecules and either increase or decrease their activity, for example by preventing an mRNA from producing a protein. RNA interference refers to the introduction of homologous double-stranded RNA (dsRNA) to specifically target a gene's product, resulting in null or hypomorphic phenotypes. The use of antisense RNA to interfere with a gene's activity in *Caenorhabditis elegans* was first utilized by S. Guo and K.J. Kemphues in 1995 to study *par-1*. A.Z. Fire *et al.* (1998) discovered that it is the presence of dsRNA, formed from the annealing of sense and antisense strands present *in vitro* RNA preparations, which is responsible for producing the interfering activity. Introduction of dsRNA into an adult worm results in the loss of the targeted endogenous mRNA from both the adult and its progeny. Two Americans, A.Z. Fire and C.C. Mello, won the Nobel Prize in Physiology or Medicine for the year 2005 for discovering a method of turning off selected genes, an important research tool *Aka* RNA SILENCING *or* RNA-MEDIATED INTERFERENCE.

RNA interference (RNAi) activity: Purified antisense and sense RNAs have only marginal interference activity. The sense-antisense mixture produced highly effective interference with endogenous gene activity. The mixture was at least two orders of magnitude more effective than either single strand alone in producing genetic interference. The potent interfering activity of the sense-antisense mixture reflected the formation of DOUBLE-STRANDED RNA (dsRNA). The degree of complementarity between the RNA silencing molecule and its cognate target determines the fate of the mRNA: blocked translation or immediate destruction.

RNA interference (RNAi) and dosage compensation: In mammals, dosage compensation is achieved by X-CHROMOSOME INACTIVATION (XCI) in the female. The non-coding *Xist* gene initiates silencing of the X chromosome, whereas its antisense partner *Tsix* blocks silencing.

RNA interference (RNAi) as an insect pest control method: RNAi is an emerging, biologically safe alternative among methods of insect pest control. RNAi studies with fruit flies (*Drosophila melanogaster*) revealed functional aspects of genes that encode proteins related to growth, reproduction, development and behaviour – encompassing almost all the aspects of an insect's life. RNAi technology shows very good potential as an environmentally safe alternative to bypass the harmful effects of chemical pesticides. RNAi is labour saving, cost effective and easy to perform. For the application of RNAi technology in IPM, there should be better stability in physiological condition, evade immune response, provide better cellular intake, escape endosomal degradation, resist renal clearance and ultimately establishes delivery of sufficient amount of dsRNA to generate RNAi that can retard the growth and survival of deadly insect pests in a cost-effective and environment safe approach. Although RNAi in the agriculture sector is still in a very nascent stage, it has undoubtedly bright future potential as this technology can easily be advantageous in the long run, limiting the chances of insect resistance.

RNA interference (RNAi) effector: Any RNA molecule that can mediate an inhibition in gene expression through RNA interference.

RNA interference (RNAi) in targeted genome elimination: The most exciting role for nuclear RNAi in genome stability is exemplified by targeted genome elimination in *Tetrahymena*. *Tetrahymena* species retain two nuclei, a germline micronucleus (Mic) and a somatic macronucleus (Mac). After zygote formation, a new Mac is generated by the deletion of ~6000 internal eliminated sequences (IES). These IES are enriched for H3K9 methylation before deletion and produce a population of 28-nt SCAN RNA (SCNRNA) that associates with the AGO Twi1p. An RNA helicase Ema1p facilitates the interaction between loaded Twi1p and chromatin by promoting base pairing with nascent transcripts, appropriating a cotranscriptional model.

RNA interference (RNAi) occurrence in the cell: Post-transcriptional RNA interference occurs in the cytoplasm, while transcriptional RNA interference occurs in the nucleus.

RNA interference (RNAi) phenotype: A phenotype generated as a result of the induction of RNA interference.

RNA interference (RNAi) proteins: Three nuclear-localized RNAi proteins have been implicated in generating and utilizing small RNAs in the RNA-directed DNA methylation pathway. RDR2, which is one of six RNA-dependent RNA polymerases encoded in the *Arabidopsis* genome, presumably produces double-stranded RNA from single-stranded templates. DCL3, which is one of four DICER-LIKE activities present in *Arabidopsis*, processes double-stranded RNA into 24-nucleotide 'heterochromatic' small RNAs. AGO4, which is one of ten ARGONAUTE proteins encoded in the *Arabidopsis* genome, is thought to be the small RNA-binding component of at least a subset of nuclear silencing effector complexes.

RNA interference (RNAi) regulation: RNAi itself is regulated during development. The non-LTR F-element possesses a short homologous retroelement suffix, which has both sense and antisense transcripts at all stages of Drosophila development. But suffix-specific siRNAs were detected among products of the expected RNAi only at a single stage.

RNA interference (RNAi) screening: The inhibition of gene expression by small non-coding RNA molecules. These molecules can hybridize with mRNA in the cell and direct the destruction of the message by Dicer, an endonuclease. Experimental screening is performed by exposing cells to many different RNAi sequences and assessing the effect on cell function or specific gene expression.

RNA interference (RNAi) technology: A mechanism that inhibits gene expression by inhibiting translation, causing the degradation of specific RNA molecules or hindering the transcription of specific genes. RNAi targets include RNA from viruses and transposons and also play a role in regulating development and genome maintenance. Small interfering RNA (siRNA) strands are key to the RNAi process and have complementary nucleotide sequences to the targeted RNA strand. Specific RNAi pathway proteins are guided by the siRNA to the targeted messenger RNA, where they cleave the target, breaking it down into smaller portions that can no longer be translated into protein. The RNAi pathway is initiated by the enzyme DICER, which cleaves long, double-stranded RNA molecules into short fragments of 20–25 bp. One of the two strands of each fragment, known as the GUIDE STRAND, is then incorporated into the RNA-INDUCED SILENCING COMPLEX (RISC) and pairs with complementary sequences. The outcome of this recognition event is post-transcriptional gene silencing.

RNA interference (RNAi) versus T-DNA-insertion mutagenesis: First, RNAi transgenes are dominant. Unlike recessive point or insertion mutations, RNAi genes can be combined readily with other mutations and reporter transgenes, since it is not necessary to screen another generation for progeny homozygous for all pre-existing mutations. Second, in cases where total inactivation of a gene might be lethal, selection for the drug-resistance marker linked to the transgene automatically selects for 'leaky' RNAi plants that are at least partially viable; the degree of growth impairment of RNAi plants may indicate the extent of silencing, hence the likelihood of a strong phenotype of interest.

RNA interference (RNAi), applications of: (i) RNAi has already proven to be an efficient and robust tool for functional genomics studies. (ii) Genetic analysis with RNAi is particularly valuable in studying organisms for which only a limited number of tools are available. (iii) Recently, it has been found that epigenetic regulation is caused by adjustments in the shape of complexes known as CHROMATIN, the bundles of DNA and certain fundamental proteins that make up the chromosome. By changing shape (i.e. becoming either more or less compact), chromatin can alter which genes are expressed. Small RNA responsible for RNAi wields tremendous control over chromatin's form. In doing so that can permanently shut down or delete sections of DNA. (iv) The discovery that synthetic double-stranded 21-nucleotide siRNA triggers gene-specific silencing in mammalian cells has further expanded the utility of RNAi in mammalian systems. Technology has been developed that allows the synthesis of siRNA from DNA templates *in vivo* to efficiently inhibit endogenous gene expression. (v) RNAi-based loss-of-function screens are tremendously powerful. (vi) By administering short, maybe chemically modified siRNAs, one might be able to reconstitute a functional siRNP complex *in vivo*. This will be in great demand as tools for functional genomics and as therapeutics, and RNAi will undoubtedly grow into one of the leading methodologies in the field. (vii) In any case, the discovery of RNAi adds a promising tool to the toolbox of molecular biologists. Introducing dsRNA corresponding to a particular gene will knock out the cell's own expression of that gene. This can be done in particular tissues at a chosen time. This often provides an advantage over conventional gene 'knockouts' where the missing gene is carried in the germline and thus whose absence may kill the embryo before it can be studied. (viii) It has been found out that human cells in culture can be protected from poliovirus and HIV-1 by siRNA with sequences that match RNA molecules encoded by viruses. (ix) Some viruses of both plants and animals have a genome of dsRNA. Many other viruses of both plants and animals have an RNA genome that in the host cell is briefly converted into dsRNA. So, RNAi may be a weapon to counter infections by these viruses by destroying their mRNAs and thus blocking the synthesis of essential viral proteins. (x) Transposons (one type of 'junk' DNA) may be transcribed into RNA molecules with regions that are double stranded. RNAi could then destroy these. (xi) In an attempt to use antisense RNA to investigate gene function in *Caenorhabditis elegans*, S. Guo and K.J. Kempheus in 1995 observed that injection of either antisense or sense RNAs into syncytial germline of worms was equally effective at silencing homologous target genes.

RNA interference (RNAi), biological role of: To date, some distinct roles for RNAi have emerged. (i) RNAi clearly acts as an antiviral defence. (ii) Genetic evidence has implicated RNAi as a GENO-PROTECTIVE MECHANISM. (iii) Recent findings have demonstrated a role for components of RNAi machinery in the regulation of cellular gene expression. (iv) RNAi is often triggered by and is used to silence repetitive transgenes. This raised the possibility that this same mechanism might function to silence endogenous repetitive sequences. RNAi evolved in part to protect organisms against both exogenous invaders and endogenous parasitic nucleic acids. Thus, the evolutionary conservation of RNAi mechanisms might indicate a critical role in maintaining genome stability. (v) Numerous clues have hinted that RNA interference and developmental control may be intimately linked.

RNA interference (RNAi), mechanism of: The RNAi included two steps – initiator and effector. In the initiator step, input dsRNA is digested into 21–26 small interfering RNA, which are also called guide RNAs. Biochemical analysis of the mechanism of RNAi action became possible with the development of a *Drosophila melanogaster in vitro* system that recapitulates many of the features of RNAi observed *in vivo*. The mechanism of RNAi involves two highly conserved enzyme systems, one is Dicer, in animals, including insects, and DICER-LIKE (DCL) elements in plants, and the second one is the RISC. Dicer family members are large, multidomain proteins that contain a putative RNA helicase domain, a PAZ (Piwi/Argonaute/Zwille) protein-protein interaction domain, two-tandem ribonuclease III (RNase III) and one or two dsRNA-binding domain(s). RISC consists of RNA and a major protein component (i.e. Argonaute), which is responsible for small interfering RNA (siRNA) binding as well as ribonuclease activity to cut the target mRNA. In the RNAi mechanism, dsRNA is processed by Dicer to convert it into 21–25 nt siRNA molecules, which are incorporated into RISC. RISC utilizes only the antisense strand of these small RNA molecules as a guide to search for its complementary base pairing with the target mRNA, which finally results in cleavage of target mRNA or translation repression. To induce RNAi in a system requires the introduction of homologous dsRNA corresponding to a vital gene(s) especially in the case of insects or intron-containing hairpin RNA (ihpRNA), which increases the gene silencing by 90–100%.

RNA interference (RNAi), processing pathways of: The two processing pathways of RNA interference are SHORT INTERFERING RNA (SIRNA) and MICRO RNAS (MIRNA).

RNA interference (RNAi), stages in: RNAi can be divided into four stages: double-stranded RNA cleavage, silencing complex formation, silencing complex activation and mRNA degradation. RNA interference is initiated by long double-stranded RNA molecules, which are processed into 21–26 nucleotides long RNAs by the small RNAs, often generated by the activity of an RNA-DEPENDENT RNA (RDR) POLYMERASE and a DICER (DCR) endoribonuclease, are bound by Argonaute (AGO) proteins and target cleavage of homologous mRNA transcripts. This RNase III protein is thought to act as a dimer that cleaves both strands of ds RNAs and leaves two-nucleotide, 3′ overhanging ends. These small interfering RNAs (siRNAs) are then incorporated into the RNA-INDUCED SILENCING COMPLEX (RISC), a protein-RNA complex and guide a nuclease, which degrades the target RNA.

RNA interference (RNAi)-based pesticides: Pest control agents that employ RNAi mechanisms as the basis of their action. They are viewed as environment-friendly and are a capable substitute for traditional chemical pesticides. The functioning substance in RNAi-based pesticides is double-stranded RNA (dsRNA) premeditated to match the nucleotide sequence of a target essential gene of the pest of concern. When taken up by the pest, this dsRNA exercises an RNAi effect and obstructs some dynamic biochemical/biological processes in the pest.

RNA interference (RNAi)-directed chromatin modifications: RNA interference (RNAi) acts on long double-stranded RNAs dsRNAs) in a variety of eukaryotes to generate small interfering RNAs that target homologous messenger RNA, resulting in their destruction. This process is widely used to 'knock-down' the expression of genes of interest to explore phenotypes. Short interfering RNAs (siRNAs) also induce DNA or chromatin modifications at the homologous genomic locus, which can result in transcriptional silencing or sequence elimination. RNAi-directed chromatin modification is coupled to transcription.

RNA interference-c (RNAi-c): Term refers to RNA interference as a result of transcript cleavage.

RNA interference-genomic modification (RNAi-gm): RNAi as a result of genomic DNA modification.

RNA interference (RNAi)-mediated epigenetic modifications in *Arabidopsis*: Synthesis of dsRNA from repeated DNA elements provides a substrate for Dicer-mediated cleavage and siRNA generation (DCL3 and other Dicers). RNA-directed RNA polymerases (RdRP, RDR2) and RNA polymerase IV (RNA Pol IV) may be directly involved in the synthesis of dsRNA or its amplification. siRNAs then load on to Argonaute proteins (e.g. AGO4), which is likely to help target cognate repeat sequences for DNA methylation and H3K9 methylation in association with other factors, including chromatin remodelling proteins and HDAC enzymes. Many of the silencing mutant screens in *Arabidopsis* have used inverted repeats introduced in *trans* to induce the silencing of endogenous or transgenic RGs, driven by tissue-specific promoters or promoters that responded to epigenetic signals. In each case, the promoter was targeted for silencing through a TGS pathway involving local chromatin changes at the promoter. Genes of note in the RNAi and TGS *Arabidopsis* pathways include DNA methyltransferases, particularly those related to the mammalian DNMT3 encoded by *DRM1* and *DRM2*. Also, there were components of the RNAi apparatus responsible for biogenesis and usage of 24-nt siRNA, such as *DICER-LIKE 3 (DCL3)* and *ARGONAUTE 4 (AGO4)*. Other mutants included those for the H3K9 methyltransferase genes *KYP/SUVH4*, *SUVH5*and *SUVH6* and the chromodomain-containing DNA methyltransferase gene *CMT3*. Apart from the DNA methyltransferase genes, the parallels with fission yeast are striking, namely, that the *Schizosaccharomyces pombe* RNA-INDUCED TRANSCRIPTIONAL SILENCING (RITS) complex contains both an Argonaute protein and the chromodomain protein Chp1, which depends on H3K9 methylation for its association with the chromosome. Unlike fission yeast, however, loss of heterochromatin-associated proteins that mediate TGS, like CMT3 or loss of the H3K9me2 mark does not result in loss of siRNA in *Arabidopsis*, and it is not yet clear if these DNA/chromatin-modifying proteins form a complex with AGO4. There is considerable redundancy in the proteins involved in *Arabidopsis* RNAi and TGS pathways. The pathway used to achieve heterochromatin formation depends on the underlying sequence – whether it is a transposon, retrotransposon or tandem repeat – resulting in both RNAi and RNAi-independent mechanisms of silencing. SUVH-4, -5 and -6 contain a SET- and RING-associated (SRA) domain that binds methylated DNA. DNA methyltransferase CMT3 has a chromodomain and a BAH domain that each recognizes K9me2. DNA methylation and H3K9me2 in heterochromatic regions act to reinforce each other, bypassing the role of RNAi in heterochromatin maintenance. Mutants in DNA polymerase epsilon and cullin 4, components of the Cryptic loci regulator 4 (Clr4)-containing Clr4 methyltransferase complex (CLRC) in *Schizosaccharomyces pombe*, relieve the silencing of some RNAi-sensitive transgenes in *Arabidopsis*.

RNA interference-r (RNAi-r): Term refers to RNA interference as a result of the repression of protein translation.

RNA ligase: This enzyme catalyses the formation of 3′→5′ phosphodiester bonds between 3′-OH and 5′-P groups of RNA molecules. This reaction is coupled to the pyrophosphorolysis of ATP to AMP.

RNA methylation: Methylation of RNA occurs at a variety of atoms, nucleotides, sequences and tertiary structures. Strongly related to other post-transcriptional modifications, methylation of different RNA species includes tRNA, rRNA, mRNA, tmRNA, snRNA, snoRNA, miRNA and viral RNA. Different catalytic strategies are employed for RNA methylation by a variety of RNA-methyltransferases.

Different functions of methyl groups in RNA include biophysical, biochemical and metabolic stabilization of RNA, quality control, resistance to antibiotics, mRNA reading frame maintenance, deciphering of normal and altered genetic code, selenocysteine incorporation, tRNA aminoacylation, ribotoxins, splicing, intracellular trafficking, immune response and others.

RNA methylation detection software tools: Such tools are pRNAm-PC, BisRNA, SRAMP, IR-NAm^{5}C-PseDNC, RAM-ESVM, 6^{m}Apred, RAMPred, RAM-PPS, Hmpre, RFAthM6A, M6ATH, iRNA-Methyl, pRNAm-PC, MethyRNA, RNA-MethylPred, iRNA-2OM and CAn.

RNA methylation in regulates gene expression: Messenger RNA methylation regulates gene expression post-transcriptionally. The first step in processing precursor messenger RNA is the addition of 7^{m}G cap at the 5′ end. 7^{m}G cap protects 5′ end of mRNA from degradation by ribonucleases. In the next step, some adenine bases are methylated to form N^6-methyladenosine (6^{m}A). 6^{m}A in mRNA accounts for the most abundant mRNA internal modification and has emerged as a widespread regulatory mechanism that controls gene expression in diverse physiological processes. Transcriptome-wide 6^{m}A mapping has revealed the distribution and pattern of 6^{m}A in cellular RNAs, referred to as the epitranscriptome. 6^{m}A in RNA plays a role in embryonic stem cell maintenance, differentiation, circadian rhythm modification heat shock response, meiotic progression and neuronal. 6^{m}A mRNA modification plays a role in cancer biology. 6^{m}A mRNA modification is critical for GLIOBLASTOMA STEM CELL (GSC) self-renewal and tumorigenesis.

RNA methylation in rRNA: Methylation on the base or the ribose is prevalent in eukaryotic ribosomal RNAs (rRNAs) and is thought to be crucial for ribosome biogenesis and function.

RNA methyltransferase complex: Methyltransferase-like 3 (METTL3) (*aka* Methyltransferase-like 14, METTL14) is a key component of the RNA methyltransferase complex. Knock-down of this component dramatically promotes human GSC growth, self-renewal and tumorigenesis.

RNA model of paramutation: In this model, the paramutagenic repeats are transcribed, while the paramutable repeats are not. The RNA derived from the paramutagenic repeats is made double stranded by an RNA-directed RNA polymerase and cleaved by a DCL protein into freely diffusible siRNAs. The siRNAs subsequently result in RNA-directed DNA methylation and chromatin silencing of the paramutable repeats.

RNA modification abundance: Cellular RNAs – ribosomal RNAs (rRNAs), transfer RNAs (tRNAs), messenger RNAs (mRNAs), small nuclear RNAs (snRNAs) and many others – have revealed more than 170 physically distinct PTCMs at thousands of sites. RNA contains diverse modifications that exert a significant impact on a variety of cellular processes by modifying gene expression. Methylation of RNA is the most common type of modification. Other well-known RNA modifications include conversion of adenosine-to-inosine (A-to-I) and guanosine to queuine (G-to-Q), pseudouridylation ((Ψ, an isomer of the nucleotide uridine), N1-methyladenosine (1^{m}A), N4-acetylcytidine (4^{ac}C), 2′-*O*-ribose methylation (2′-*O*-Me) and N7-methylguanosine (7^{m}G).

RNA modification analysis: RNA modifications are abundant in messenger RNAs (mRNAs), transfer RNAs (tRNAs) and ribosomal RNAs (rRNAs) and in regulatory non-coding RNAs (ncRNAs). RNA modifications are critical to development, including embryo development and stem cell fate determination. Understanding of role of different RNA modifications requires their precise detection and quantification.

RNA modification analysis methods: (i) Based on separation according to physicochemical features of RNA modification analysis include THIN LAYER CHROMATOGRAPHY (TLC), HIGH-PERFORMANCE LIQUID CHROMATOGRAPHY (HPLC) or mass spectrometry and sometimes on a combination of both LIQUID CHROMATOGRAPHY-MASS SPECTROSCOPY (LC-MS) and CAPILLARY ELECTROPHORESIS-LASER-INDUCED FLUORESCENCE (CE-LIF). Methods based on differential enzymatic turnover use ribonucleases such as RNase T1. (ii) Methods based on differential chemical reactivity use sodium bisulfite (BS; NaHSO3)-mediated conversion of cytosine to uracil.

RNA modification analysis using chromatography: The method is very effective when applied to short/small RNAs (sRNAs) (100–150 nt) such as tRNAs or other small non-coding RNAs (sncRNAs).

RNA modification analysis using cleavage methods: Such methods include RNase H-based method and RNA-cleaving deoxyribozyme (DNAzyme) cleavage (e.g. MC1 ribonuclease cleavage).

RNA modification analysis using deep sequencing: tRNAs contain modifications that can stall RT (e.g. methylation of adenine at position 58; $1^{m}A58$). Several experimental methods facilitated reverse transcription from such tRNA templates. First, the RT primer-binding site was moved closer to the anticodon stem-loop covering $1^{m}A58$ to exclude the blocking effect of this modification. Second, RT reaction was enabled using increased dNTP concentrations to overcome the transcriptional block. Third, RT reaction was primed using random hexamers creating a mixed rather than a tRNA-specific cDNA library. These adjustments in the RT reactions permitted them to amplify additional tRNAs, including $tRNA^{Val,3}$, $tRNA^{Val,4}$ and $tRNA^{Lys}$ from BS-treated total RNA preparations.

RNA modification analysis using dot blot method: This method is used for fast detection of total $6^{m}A$ levels in mRNA using a specific $6^{m}A$ antibody.

RNA modification analysis using ligation-based methods: Such methods have been developed on the basis of molecular recognition and enzymatic ligation of DNA oligonucleotides annealed to both modified and unmodified RNA, in regions flanking specific modification sites. These methods were successfully used to spot nucleotides 2′-*O*-Me, 6^{m}As and Ψs.

RNA modification analysis using liquid chromatography: Methods involve purification of the RNA molecule of interest (either cold or radioactively labelled), partial or complete digestion with nucleases and phosphatase treatment before separation on a chromatographic column. Mono and/or oligonucleotides (or nucleosides attained after removal of phosphate) are detected by UV absorption or estimation of radioactivity. The modified nucleotides are recognized on the basis of differences in their chromatographic mobility when compared with their unmodified counterparts.

RNA modification analysis using mass spectrometry: The principle forming the foundation of these methods involves the production of gas phase ions of the molecules of interest, either nucleosides or oligonucleotides and the separation of the generated ions according to their mass-to-charge ratios. The techniques that are most commonly used to generate gas phase ions include MATRIX ASSISTED LASER DESORPTION/IONIZATION (MALDI) and ELECTROSPRAY IONIZATION (ESI). TANDEM MASS SPECTROMETRY (TMS) technique is used to sequence the selected oligonucleotide.

RNA modification analysis using microarray and high-throughput sequencing platforms: Two major types of libraries that can be raised for RNA-seq include cDNA libraries and strand-specific RNA libraries.

RNA modification analysis using primer extension approach: Such method established for the precise mapping of 2′-*O*-Me residues in RNA engages primer extension by reverse transcriptase at low dNTP concentrations.

RNA modification analysis using reverse transcriptase approach: A synthetic DNA oligonucleotide designed to target a specific sequence is forged to cellular RNA without previous RNA purification. Primer elongation by reverse transcriptase permits both the sequencing of RNA species and the detection of specific reagent-dependent stops, which are suggestive of the presence and exact position of modified nucleotides.

RNA modification analysis using RiboMeth-seq method: This method detected known as well as novel modifications revealed partial modifications and unexpected communication between modification events and determined the order of modification at several sites during ribosome biogenesis. The RiboMeth-seq method also provides information on a subset of other modifications.

RNA modification analysis using thin-layer chromatography (TLC): Larger molecules are analysed after ribonuclease digestion and RNA fragment refinement. Complete digestion of RNA by numerous ribonucleases is accomplished before the analysis of the RNA digest products by two-dimensional separation on TLC plates. The products differ depending on the nuclease used. RNase T2 digestion generates 3′-monophosphate nucleosides; both nuclease P1 and the VENOM PHOSPHODIESTERASE (VPD) produce 5′-monophosphate nucleosides and piperidine produces both 2′- and 3′-monophosphate nucleosides.

RNA modification analysis using two-dimensional TLC-based technique: Identification of the chemical nature and exact location of RNA modifications characteristically trusts on two-dimensional thin layer chromatography (2D-TLC) of nucleotide digests, on HPLC coupled with MS or on the use of primer extension by reverse transcriptase.

RNA modification cofactors: Cofactors needed for RNA modification reactions are metabolites initiated from the central metabolism, such as

S-ADENOSYLMETHIONINE (SAM or AdoMet), flavin adenine dinucleotide (FAD), NICOTINAMIDE ADENINE DINUCLEOTIDE HYDROGEN (NADH), METHYLENE TETRAHYDROFOLATE (MTHF), ADENOSINE TRIPHOSPHATE (ATP), GUANOSINE TRIPHOSPHATE (GTP), ISOPENTENYL-PYROPHOSPHATE (IPP) and/or various amino acids.

RNA modification databases: Three important RNA modification databases are RMBase, including RMBase v2.0, MODOMICS and RNAMDB.

RNA modification detection using Maxam and Gilbert's approach: Using Gilbert's strategy, detection of $6^{m}A$, $5^{m}C$ and Ψ RNA modifications has been made possible.

RNA modification profiling: The existing approaches to spot and quantify RNA modification $3^{m}C$ levels can be placed into three classes: LC-MS/MS analysis, traditional primer extension assay and NGS-based methods for RNA modification profiling.

RNA modification profiling using LC-MS/MS analysis: While this approach is quantitative, it cannot easily determine which sites are $3^{m}C$-modified nor can it be used for the unbiased discovery of the modified subset of RNA species within a bulk population.

RNA modification profiling using miCLIP: This approach is used for high-resolution localization of $6^{m}A$ and $6^{m}A^{m}$ in eukaryotic RNA by individual-nucleotide resolution.

RNA modification profiling using multiplexed stable isotope labelling: STABLE ISOTOPE LABELLING by/with AMINO ACIDS IN CELL CULTURE (SILAC) is a technique based on MS that detects differences in protein abundance among samples using non-radioactive isotopic labelling.

RNA modification profiling using NGS-based methods: Two methods are as follows: (i) DM-tRNA-seq for mapping $1^{m}A$, $1^{m}G$, N2, N2-dimethylguanosine ($2{,}2^{m}G$) and $3^{m}C$ RNA modifications and (ii) AlkAniline-seq for mapping $7^{m}G$ and $3^{m}C$ RNA modifications.

RNA modification profiling using traditional primer extension assay: This method can be supportive in detecting sites of modification, but it can only be applied to candidate RNAs, and further, this method is not high-throughput nor can it discriminate between $3^{m}C$ and other RNA modifications that disturb base-pairing to terminate the reverse transcriptase. Another major limitation of this approach is that reverse transcription may also be arrested by RNA secondary structures and/or by other RNA modifications, both of which are very common in transfer RNAs (tRNAs).

RNA modification types: Chemical modifications of RNA have been classified into four types: (i) 5-ribosyluracil or pseudouridine (Ψ), (ii) simple modification of the bases, (iii) methylation of the ribose 2′-hydroxyl group (Nm) and (iv) complex or hypermodifications.

RNA modification versus RNA editing: The term RNA modification is used when the change is purely physical, in spite of the fact that the same enzymatic machinery can perform as an EDITOR or a MODIFICATOR, depending on the consequence of the enzymatic RNA alteration while the term RNA editing is used when the consequence of a given RNA change is to amend the genetic sense of the edited molecule.

RNA modifications: POST-TRANSCRIPTIONAL MODIFICATIONS (PTCMs) that are introduced at specific sites to enhance the structural and functional diversity of RNA in all organisms. Various types of RNAs in which RNA modification has been identified include messenger RNAs that convey genetic information; transfer RNAs that carry the building amino acids for protein synthesis, ribosomal RNAs that catalyse the formation of peptide bonds; LONG NON-CODING RNA (lncRNA) and SMALL NON-CODING RNA (sncRNA) that have manifold regulatory roles. Some instances of RNA modifications include RNA nucleotide methylation (e.g. N6-methyladenosine, $6^{m}A$), N1-methyladenosine ($1^{m}A$), 5-methylcytosine ($5^{m}C$), 5-hydroxymethylcytosine ($5^{hm}C$) and ribose 2′-hydroxyl group (Nm). Other RNA modifications include G-quadruplexes and RNA-loops (RGQs), Uridine transglycosylation and Guide-RNA-mediated RNA modifications. RNA modifications have the potential to alter the function of different RNA species such as messenger RNAs, transfer RNAs, ribosomal RNAs and small non-coding RNAs. RNA PROCESSING and RNA EDITING are major classes of RNA modifications.

RNA modifiers: Writes, readers and erasers of RNA modifications comprise RNA modifiers. In some RNA modifications, the modifiers have been identified. For example, (i) MT-A70 domain family proteins for mediating adenosine N^{6}-methylation, the removal of this RNA mark by members of ALPHA-KETOGLUTARATE-DEPENDENT DIOXYGENASE (AlkB) homologue domain family proteins and the detection of these $6^{m}A$-modified RNAs by YT521-B homology (YTH) domain family proteins and (ii) N6,2′-O-dimethyladenosine (6^mA^m) is present at the transcription start nucleotide of capped mRNAs in vertebrates. Cap-specific adenosine methyltransferase (CAPAM) is the RNA writer for this RNA modification.

RNA molecules, types of: *In vivo*, one might think of two classes of RNA molecules: (i) Unstable molecules such as mRNAs, whose function is to present a linear sequence of bases, are prevented from folding into compact internally base-paired structures by specific proteins whose function is to do just this. (ii) Stable molecules, such as rRNAs and tRNAs, fold into very characteristic secondary and tertiary structures that are indispensable for their particular functions.

RNA nanotechnology: RNA nanotechnology utilizes RNA as a programmable material to create self-assembling nanodevices for application in medicine and synthetic biology. The main challenge here is to develop advanced RNA ROBOTIC DEVICES that both sense, compute and motivate to obtain enhanced control over molecular processes.

RNA origami robot: *See* RNA NANOTECHNOLOGY.

RNA paired-end tag (RNA-PET): Simultaneous capture of RNAs with both a 5′ methyl cap and a poly(A) tail, which is indicative of a full-length RNA. This is then followed by sequencing a short tag from each end by high-throughput sequencing.

RNA phages: Phages whose genetic material is RNA (e.g. R17, f2, MS2 and QB). *Aka* RNA VIRUSES.

RNA phosphorylase: Enzyme RNA phosphorylase could link together ribonucleotides into a long RNA chain or break such a chain into smaller sections. In contrast to DNA polymerase, this enzyme could engage in polymerization activity in the absence of primer, and RNA synthesis proceeds easily no matter what proportions of the four ribonucleotides are used, as long as they are diphosphates rather than mono- or triphosphates.

RNA plasmid: Unusual genetic elements composed of single or double-stranded RNAs found in the plant mitochondria of some male sterile lines.

RNA polyadenylation: Second pre-mRNA processing step is 3′ cleavage and polyadenylation. This step occurs if POLYADENYLATION SIGNAL SEQUENCE (5′-AAUAAA-3′) is present in pre-mRNA, which is usually between protein-coding sequence and terminator. The pre-mRNA is first cleaved and then a series of ~200 adenines (A) are added to form a poly(A) tail, which protects the RNA from degradation. Poly(A) tail is bound by multiple POLY(A)-BINDING PROTEINS (PABP) necessary for mRNA export and translation re-initiation.

RNA polymerase core enzyme: In bacteria, it consists of five subunits of four types ($\alpha_2\beta\beta'\omega$) and is capable of elongating already initiated transcription by HOLOENZYME. *Aka* CORE ENZYME. *Compare with* RNA POLYMERASE HOLOENZYME.

RNA polymerase DNA-binding site: After initial binding to the recognition site, RNA polymerase diffuses to the RNAP binding site, which is A=T rich region. This binding region is called the TATA BOX. RNA polymerase tracks along DNA as template rotation generates twin domains of supercoiling. The binding sites at which RNAP forms a stable initiation complex with DNA lie within the promoter.

RNA polymerase holoenzyme: The version of *Escherichia coli* RNA polymerase that has subunit composition $\alpha_2\beta\beta'\omega\sigma$. It is involved in efficient recognition of promoter sequence and is capable of initiation of transcription.

RNA polymerase I (RNA Pol I): Synthesizes pre-rRNA 45S (35S in yeast), which matures into 28S, 18S and 5.8S rRNAs that form the major RNA components of the ribosome.

RNA polymerase I (RNA Pol I) specific promoters: Promoter sequences for RNA Pol I are located upstream (before start point of transcription).

RNA polymerase II (RNA Pol II): It transcribes all protein-coding genes but also some non-coding RNAs (e.g. 7SK). Transcription ends when the polymerase encounters a sequence called the *terminator*.

RNA polymerase II (RNA Pol II) core enzyme: The large pre-assembled complex of RNA polymerase II that is composed of five subunits of four types ($\alpha_2\beta\beta'\omega$) and is not capable of initiation of transcription but elongates already initiated transcription.

RNA polymerase II (RNA Pol II) holoenzyme: The large pre-assembled complex of RNA polymerase II that is composed of six subunits of five types ($\alpha_2\beta\beta'\omega\sigma$) and is involved in the efficient recognition of promoter sequences and is capable of initiation of transcription in the presence of five GENERAL TRANSCRIPTION FACTORS (GTFs) and mediators.

RNA polymerase II (RNA Pol II) in 'backtracked' state: Transcribing RNA polymerases oscillate between three stable states: pre-translocated state, post-translocated state and reverse translocated or backtracked state. The defining feature of the backtracked structure is a binding site for the first backtracked nucleotide. This binding site is occupied in case of nucleotide misincorporation in the RNA or damage to the DNA and is termed the 'P' site because it supports proofreading. The predominant mechanism of proofreading is the excision of a dinucleotide in the presence of the ELONGATION FACTOR SII (EFIIS).

RNA polymerase II (RNA Pol II) specific promoters: Promoters for RNA Pol II are present upstream and showed three regions at the start point. These three regions are TATA or 'Hogness box' (7-bp long), which is located at position −20, CAAT box, which is located between −70 to −80 and GC box, which is located between −60 and −100.

RNA polymerase II (RNA Pol II) transcription complex: A minimal RNA polymerase II transcription system comprises the polymerase and five GTFs TFIIB, -D, -E, -F and -H. The addition of mediator enables a response to regulatory factors. The GTFs are required for promoter recognition and the initiation of transcription. Following initiation, Pol II alone is capable of RNA-transcript elongation and of proofreading.

RNA polymerase II (RNA Pol II) transcription preinitiation complex (PIC): In PICs, the TFIIB linker and core domains are positioned over the central cleft and wall of the RNA Pol II. This positioning is not observed in the smaller RNA Pol II-TFIIB complex.

RNA polymerase II (RNA Pol II) ubiquitylation sites: Transcriptional arrest triggers ubiquitylation of RNA polymerase II. The yeast RNAPII ubiquitylation sites have been mapped. They play an important role in transcription elongation and the DNA damage response.

RNA polymerase III (RNA Pol III): It carries out about 10% of cellular transcription in eukaryotes. The genes transcribed by this enzyme are those for small non-coding RNA molecules such as 5S rRNA, tRNAs, the U6 snRNA found in spliceosomes and the 7S RNA component of signal recognition particle.

RNA polymerase III (RNA Pol III) specific promoters: Promoter sequences for RNA Pol III are located downstream with the exception of 7S genes of signal recognition particles where they are present both upstream and downstream.

RNA polymerase III (RNA Pol III) termination, mechanism of: J. Xie *et al.* (2022) combined high-resolution genome-wide analyses and *in vitro* transcription termination assays to revisit the mechanism of RNA polymerase III transcription termination in budding yeast. They showed that T-tracts are necessary but not always sufficient for termination and that secondary structures of the nascent RNAs are important auxiliary *cis*-acting elements. The helicase Sen1 plays a key role in a fail-safe termination pathway.

RNA polymerase IV (RNA Pol IV): Found in plants. Presumed maize RNA Pol IV is involved in paramutation, an inherited epigenetic change facilitated by an interaction between two alleles, as well as normal maize development. Presumed maize RNA Pol IV is involved in paramutation, an inherited epigenetic change facilitated by an interaction between two alleles, as well as normal maize development. It does not engage in the efficient RNA synthesis typical of the three major eukaryotic DNA-dependent RNA polymerases. It employs abnormal RNA polymerase activities to achieve genome-wide silencing, and its absence affects both maize development and heritable epigenetic changes. Also Synthesizes siRNA in plants micro RNAs.

RNA polymerase IV backtracking: RNA polymerase IV (RNA Pol IV) forms a complex with RNA-directed RNA polymerase 2 (RDR2) to generate double-stranded RNA (dsRNA) predecessors essential for plant gene silencing. Pol IV backtracks and carries its transcript's 3′ terminus to RDR2, which synthesizes dsRNA. Pol IV is incompetent in forward translocation and RNA cleavage. Pol IV transcription elongation is inclined to backtracking and incapable of RNA hydrolysis, safeguarding efficient dsRNA production by Pol IV–RDR2.

RNA polymerase recognition site: A site on the template strand that RNA polymerase initially recognizes and binds.

RNA polymerase secondary channel: High-resolution crystal structures have highlighted functionally important regions in multisubunit RNAPs, including the secondary channel, or pore, which is postulated to allow the diffusion of small molecules both into and out of the active centre of the enzyme. Regulatory factors and small molecules can exploit the secondary channel to gain access to the active site and modify the transcription properties of RNA polymerase.

RNA polymerase V (RNA Pol V): Synthesizes RNAs involved in siRNA-directed heterochromatin formation in plants. In addition to siRNAs, RNA-directed DNA methylation (RdDM) also requires non-coding RNA transcripts produced by RNA Pol V. When siRNAs are loaded on to the ARGONAUTE proteins AGO4 and/or AGO6, KTF1 binds the nascent non-coding RNA transcripts produced by RNA Pol V and is tethered to AGO4 by the base pairing between the non-coding RNA transcripts and the siRNAs. The interaction between AGO4 and the WG/GW motif of KTF1 reinforces the association between siRNAs and RNA Pol V transcripts. This step leads to the formation of a functional RdDM effector complex that directs the *de novo* DNA methyltransferase DRM2 to genomic loci being transcribed by RNA Pol V. AGO4 has a slicer activity by which the scaffold non-coding RNA transcripts are cleaved. The cleaved RNA fragments may be copied by RDR2 and then processed into 24-nt secondary siRNAs by DCL3. IDN2, a recently identified RdDM component, is an SGS3-like protein containing XS and XH domains separated by a coiled-coil region. The XS domain in IDN2 binds double-stranded RNA *in vitro*. RDM1 functions together with AGO4 and DRM2 as a part of the RdDM effector complex. RDM1 appears to bind single-stranded methylated DNA and may contribute to the recruitment of the RdDM effector complex to methylated DNA.

RNA polymerase V (RNA Pol V) structure: G. Xie *et al.* (2023) reported the structures of cauliflower RNA polymerase V in the free and elongation conformations. A conserved tyrosine residue of NRPE2 stacks with a double-stranded DNA branch of the transcription bubble to potentially attenuate elongation by inducing transcription stalling. The non-template DNA strand is captured by NRPE2 to enhance backtracking, thereby increasing 3′-5′ cleavage, which likely underpins Pol V's high fidelity. The structures also illuminate the mechanism of Pol V transcription stalling and enhanced backtracking, which may be important for Pol V's retention on chromatin to serve its function in tethering downstream factors for RNA-directed DNA methylation.

RNA polymerase in prokaryotes: An enzyme capable of synthesizing an RNA copy of a DNA template. In prokaryotes, there is only one type of RNA polymerase.

RNA polymerases and cancer: Oncogenic stimuli and tumour suppressors control RNA polymerase III-dependent transcription. Overexpression of a transcription factor of the RNA polymerase III apparatus transforms cells. Elevated tRNA synthesis can promote cellular transformation.

RNA polymerases in eukaryotes: An enzyme that catalyses the synthesis of RNA using a strand of DNA or RNA as a template. RNA polymerase copies one strand of DNA or RNA to make the complementary RNA strand. Using the rules of Watson-Crick base pairing, ribonucleotides (ATP, UTP, GTP and CTP) are added one at a time using as substrates ribonucleotide triphosphates and releasing pyrophosphate. There occur three types of RNA polymerases (RNA POLYMERASE I, RNA POLYMERASE II, RNA POLYMERASE III) in animals and five types of RNA polymerases (RNA POLYMERASE I, RNA POLYMERASE II, RNA POLYMERASE III, RNA POLYMERASE IV and RNA POLYMERASE V) in plants. *See details as separate terms.*

RNA primer (pRNA): Short stretch of RNA synthesized on a DNA template. It is required by DNA polymerases to start their DNA synthesis.

RNA primer (pRNA) in DNA replication: Required for initiation of DNA replication. First deoxynucleotide is always added to an RNA primer during DNA replication. For continuous DNA replication, only one primer molecule is required. DNA polymerase II does not initiate replication if RNA primer is missing. Synthesis of each OKAZAKI FRAGMENT during discontinuous DNA replication requires RNA primer, which is later degraded, replaced by deoxyribonucleotides. Gaps are filled in by DNA ligase.

RNA primer (pRNA) removal: The removal of RNA primers from Okazaki fragments to facilitate the union of two small Okazaki fragments takes place in two steps: (i) All except the last ribonucleosides are removed by RNase H, which acts endonucleolytically. (ii) The actual prN(p and N)$_2$ linkage is broken by DNA polymerase 5′→3′ exonuclease activity.

RNA primer (pRNA) synthesis: RNA primer is homogenous in length, but the sequence is not unique, not even at the RNA-DNA junction. RNA primer is required for DNA replication of only those DNA templates that do not have free ends. *Escherichia coli* primer synthesis seems to take place in a similar way. All known DNA polymerases require a pre-existing 'primer' from which to begin replication. A primer is an oligonucleotide that is base paired with a template strand and provides 3′-hydroxyl group upon which to start polymerizing a new DNA strand. *E. coli* uses these small RNAs as primers, which are synthesized by an enzyme called DnaG, coded by the *dnaG* gene. DnaG makes RNA primers (about ten nucleotides long) that are used by DNA polymerase III holoenzyme to start DNA synthesis.

RNA processing: Term collectively used to describe the sequence of events through which the primary transcript from a gene acquires its mature form. RNA processing may include modification, mainly through addition, deletion, cleavage and/or splicing, ribonucleotide modification, of precursor RNA transcripts so as to release mature transfer RNA (tRNA), ribosomal RNA (rRNA), messenger RNA (mRNA), small interfering RNAs (siRNA) and other species of RNA molecules from them. *Aka* RNA MATURATION. *See* PRE-MRNA PROCESSING, PRE-TRNA PROCESSING *and* PRE-RRNA PROCESSING.

RNA processing control: Control of gene expression by controlling how the RNA transcript is spliced or otherwise processed.

RNA quality-control systems: Eukaryotic cells contain numerous RNA quality-control systems that are important for shaping the transcriptome of eukaryotic cells. These systems not only prevent the accumulation of non-functional RNAs but also regulate normal mRNAs, repress viral and parasitic RNAs and potentially contribute to the evolution of new RNAs and hence proteins. RNA quality-control circuits depend on specific adaptor proteins that target aberrant RNAs for degradation as well as coupling of individual steps in mRNA biogenesis and function.

RNA recombination: RNA recombination is a mechanism universally used by all animal and plant RNA viruses. Genetic RNA recombination is a process of the exchange of genetic information between RNA molecules, and it slightly differs from a typical genetic DNA recombination. RNA recombination has been observed in riboviruses. RNA recombination was first discovered in poliovirus. The high frequency and widespread nature of RNA recombination indicate that this phenomenon plays a more significant role in the biology of RNA viruses than was previously recognized.

RNA repair: The recently discovered oxidative demethylation of 1-methyladenine ($1^{m}A$) and 3-methylcytosine ($3^{m}C$) by AlkB and human homologues expands the concept of alkylation repair, with respect to not only DNA but also RNA as well.

RNA replicase: A polymerase enzyme that catalyses the self-replication of single-stranded RNA.

RNA scaffold: Scaffold is a structure providing support. With nucleic acid duplication, fusion and recombination events, lncRNAs may attain multiple effector binding sites to become a scaffold.

RNA self-replication: The mechanism where RNA acts as a template for its own replication. This is considered to be the second modification to the original central dogma. The first modification is the ability of mRNA to act as a template for the synthesis of DNA. RNA self-replication is known to exist in RNA phages such as R17, f2, MS2 and QB. Its mode of replication is called RNA-DEPENDENT RNA SYNTHESIS.

RNA self-splicing: A set of linked cleavage-ligation reactions mediated by RNA itself and conserving the number of phosphodiester bonds. This process is observed in some nuclear pre-ribosomal RNAs and various mitochondrial chloroplast pre-mRNAs.

RNA sequence analysis: The determination of the base sequence in RNA using enzymes in a method directly analogous to the Maxam-Gilbert DNA technique, except that base-specific nucleases are employed, which cleave specifically next to guanosine, adenosine, cytidine and uridine.

RNA sequencing: The process of determining the order of ribonucleotides in RNA. It can be used to identify genetic mutations and changes in gene expression. It involves the isolation of RNA sequences, often with different purification techniques to isolate different fractions of RNA followed by high-throughput sequencing. With RNA sequencing, digital gene expression and RNA profiling are becoming increasingly popular. SINGLE-MOLECULE SEQUENCING is also joining this group. The frequency of a particular RNA sequence is counted and is used to infer the abundance of that specific transcript in the population. Since 10–40 million short sequences are obtained in a run, massive data become available with new machines. In order to detect rare transcripts, enrichment methods are used, where mainly the ribosomal RNA is eliminated before the samples are subjected to sequencing.

RNA sequencing methods: RNA sequencing methods include GENERATING DNA THROUGH REVERSE TRANSCRIPTION, ION EXCHANGE CHROMATOGRAPHY, USING RADIOACTIVE CHAIN TERMINATING RIBONUCLEOTIDES, DIRECT CHEMICAL METHOD and DIRECT RNA SEQUENCING. *See details as separate terms.*

RNA sequencing through reverse transcription: To sequence RNA, the usual method is first to reverse transcribe the sample to generate copy DNA fragments. This copy of DNA is then sequenced by employing one of many DNA-sequencing methods. Each molecule, with or without amplification, is then sequenced in a high-throughput manner to obtain short sequences from one end (single-end sequencing) or both ends (pair-end sequencing) using different NGS platforms (454 Roche, Illumina, SOLiD). The reads are typically 30–400 bp, depending on the DNA-sequencing technology used. The tetrabytes of data generated are stored in supercomputers.

RNA sequencing with chemical cleavage of ribonucleotides: A. Maxam and W. Gilbert in 1976–77 used four different base-specific chemical cleavage reactions and generated a means of directly sequencing RNA terminally labelled with ^{32}P.

RNA sequencing with radioactive chain-terminating ribonucleotides: The Sanger sequencing approach employs the 3′-deoxy analogues of the ribonucleoside triphosphates as specific chain terminators during RNA synthesis. The DNA molecules are radioactively labelled at 5′-end.

RNA sequencing-based RNA modification analysis: RNA-sequencing-based methods provide more accurate information about RNA modifications than do the complementary DNA-sequencing-based methods. RNA sequencing-based methods include A. Maxam and W. Gilbert's approach, demethylase-facilitated RNA-seq, bisulfite sequencing, adaptation of DNA bisulfite methylation mapping to RNA and 5^{m}C mapping and quantification in RNA.

RNA silencing: A group of related phenomena in diverse eukaryotes in which aberrant or dsRNA triggers a marked reduction in either transcription of the corresponding gene or direct degradation of the corresponding mRNA. Small interfering RNA (siRNA) (21–25 nucleotides) that is produced from long dsRNA by Dicer or DCL enzymes or chemically synthesized and can be recruited by multiple cellular proteins to form an RNA-induced silencing complex (RISC) that interferes with mRNA stability or mRNA translation. *See* RNA INTERFERENCE (RNAI).

RNA splicing: A complex and incompletely understood series of reactions occurring in the nucleus of eukaryotic cells in which pre-mRNA transcribed from chromosomal DNA is processed. In one of the steps, the intervening non-coding regions of the pre-mRNA, called INTRONS, are excised from the precursor and coding regions (EXONS) are covalently linked to produce an RNA molecule ready for transport to the cytoplasm. Because of splicing, eukaryotic DNA representing a gene encoding any given protein is usually much larger than the mature RNA.

RNA splicing mechanisms: Two types of splicing mechanisms, known as major and minor splicing mechanisms, have been identified. Accordingly, major and minor spliceosomes have been identified. These two splicing pathways are spatially separated in the cell and may have distinct functions.

RNA synthesis: Broadly speaking, this term has two meanings. (i) *In vivo* RNA synthesis using DNA as a template is termed transcription. (ii) RNA synthesized *in vitro* with or without a template.

RNA transcript: An RNA copy of a gene.

RNA transcriptase: The enzyme responsible for transcribing the information encoded in DNA into RNA. *Aka* RNA POLYMERASE.

RNA transposons: *Aka* RETROELEMENTS *or* RETROPOSONS. These transposons transpose via RNA intermediates. The RNA is copied by reverse transcriptase into DNA; the DNA integrates into the genome. Retroelements are found in all eukaryotes such as Tos in rice, copia in animals and Ty1 in yeast. Retroposons are non-LTR RNA transposons. Examples: Mammalian LINE repeats (long interspersed repetitive elements, L1s). Similar elements are found even in fungi. Mammalian SINE repeats (short interspersed repetitive elements, e.g. human Alu repeats).

RNA trimming: The removal of stretches of RNA at one or both ends of the primary transcript.

RNA turnover: mRNA turnover is determined by the rates of mRNA synthesis and degradation, which jointly adjust the level of gene expression. The degradation of mRNAs in eukaryotes occurs largely in the cytoplasm and begins with the removal of the poly(A) tail.

RNA vaccine: This type of vaccine uses a molecule called messenger RNA (mRNA) rather than part of an actual bacteria or virus. Messenger RNA is a type of RNA that is necessary for protein production. The protein thus produced stimulates an immune response.

RNA vectors: An RNA (ribonucleic acid) vehicle for transferring genetic information from one cell to another:

RNA viruses: Those viruses that contain RNA as genetic material.

RNA world: The stage before the evolution of the genetic code when RNA was responsible for both heredity and catalysis. Ribose is much more readily synthesized than deoxyribose under simulated prebiotic conditions. It is assumed that modern life forms arose from a molecular ancestor in which RNA molecules both stored genetic information and catalysed biochemical reactions. The realization that RNA combines both the features of DNA and proteins led to the 'RNA world' hypothesis for the origin of life. RNA enzymes like protein enzymes create a catalytic surface by folding into complex secondary and tertiary structures. DNA lacks 2′-OH group. It is less adept than RNA at folding. DNA precursors are always synthesized by reduction of RNA nucleoside diphosphates using ribonucleoside diphosphate reductase.

RNA world hypothesis: This hypothesis holds that during evolution the structural and enzymatic functions initially served by RNA were assumed by proteins, leading to the latter's domination of biological catalysis. This progression can still be seen in modern biology, where ribozymes, such as ribosome and RNase P, have evolved into protein-dependent RNA catalysis (RNPzymes). Similarly, group I introns use RNA-catalysed splicing reactions, but may function as RNPzymes bound to proteins that stabilize their catalytically active RNA structure.

RNA-binding proteins: There is a multitude of RBPs that play a role in the post-transcriptional regulation of gene expression. These proteins contain RNA-binding motifs.

RNA-coding triplets: Triplets (codons) found on mRNA that code for the amino acids – the codes being recognized by the anticodon present on tRNA charged with the specific amino acid. *See* GENETIC CODE.

RNA-dependent DNA polymerases: A group of enzymes that catalyse the formation of DNA molecules from DNA templates using RNA primers for initiation of DNA replication.

RNA-dependent RNA polymerase (Abbrev. RdRP or RDR): An enzyme that catalyses the replication of RNA from an RNA template. *Aka* RNA REPLICASE. This is in contrast to typical DNA-dependent RNA polymerases, which all organisms use to catalyse the transcription of RNA from a DNA template. RdRP is involved in RNA silencing in plants, worms, *Neurospora* and *Dictyostelium*, but not in *Drosophila* or human cells. RdRPs have been proposed to use cellular aberrant RNAs as templates and to copy them into cRNAs to form dsRNA. This newly synthesized dsRNA acts as a substrate for DCL enzymes. These enzymes are involved in RdDM. There are six RdRPs in Arabidopsis, RdRPs1-6. Three of RdRPs (RdRP3, 4 and 5) share an atypical DFDGD (aspartic acid-phenylalanine-aspartic acid-glycine-aspartic acid) amino acid motif in the catalytic domain. RdRP2 is the only one participating in 24-nt siRNA-mediated DNA methylation. *See* REVERSE TRANSCRIPTASE.

RNA-directed DNA methylation (RdDM) in plants: *Aka* DE NOVO METHYLATION as it is a conserved *de novo* DNA methylation mechanism in plants. RdDM was first found in transgenic potato containing viroid genes. It was subsequently recognized as a general transcriptional silencing mechanism in plants. It is involved in various epigenetic phenomena such as transgene silencing, transposon suppression, gene imprinting and paramutation.

RNA-directed DNA methylation (RdDM) pathway: Various components of the RdDM pathway include two plant-specific, atypical DNA-dependent RNA polymerases NRPD1 and NRPE1, which are similar to the largest subunits of canonical DNA-dependent RNA polymerases and function in *de novo* DNA methylation and transcriptional gene silencing. NRPD1 and NRPE1 are also multisubunit protein complexes and have been named (in the manner of DNA-dependent RNA polymerases) RNA Pol IV and RNA Pol V. These two enzymes have their own specific subunits but share some subunits with each other and share other subunits with RNA Pol II. The chromatin remodelling protein CLSY may facilitate RNA Pol IV transcription. RNA-dependent RNA polymerase (RDR2) converts the aberrant single-stranded RNAs to double-stranded RNAs, which are then cleaved into 24-nt short interfering (siRNAs) by the Dicer-like protein DCL3. The 24-nt siRNAs are bound by an ARGONAUTE protein (AGO4, AGO6 or AGO9). In intergenic non-coding (IGN) regions, RNA Pol V generates single-stranded scaffold RNA transcripts. Generation of RNA Pol V transcripts requires RDM4/DMS4, DRD1, DMS3 and RDM1. RDM1 may bind single-stranded methylated DNA and help recruit RNA Pol V and RNA Pol II to appropriate chromatin regions. DRD1, DMS3 and RDM1 (DDR) form a stable protein complex. KTF1 is an RNA-binding protein, which tethers AGO4 to nascent Pol V or Pol II RNA transcripts to form the RdDM effector complex. IDN2 may stabilize the base pairing between the nascent scaffold transcripts and 24-nt siRNAs. The effector complex directs the *de novo* DNA methyltransferase DRM2 to specific chromatin regions to catalyse new DNA methylation.

RNA-directed histone and DNA methylation: RdDM is an epigenetic process first elucidated in plants whereby small double-stranded RNAs (dsRNAs) are processed to guide methylation to complementary DNA loci. In the model plant organism *Arabidopsis thaliana*, these small dsRNAs may be generated from three sources: (i) viral replication intermediates, (ii) products of the endogenous RNA-dependent RNA polymerase and (iii) transcribed inverted repeats. These dsRNAs are then processed to direct histone 3 lysine 9 (H3K9) methylation via Ago4 and the SUVH (Suppressor of Variegation Homologue) histone methyltransferase family. This H3K9 dimethylation is then putatively bound by the cytosine methyltransferase CMT3, which methylates cytosines in a non-CG context. These marks, H3K9 dimethylation and cytosine methylation, are canonical marks of gene silencing. Thus, small double-stranded RNAs generated in diverse contexts bring about transcriptional silencing at specific loci.

RNA-directed RNA synthesis: Non-coding small RNAs regulate gene expression in all organisms, in some cases through direct association with RNA polymerase. For example, the DNA-dependent RNA polymerase uses bound 6S RNA as a template for RNA synthesis, producing 14- to 20-nucleotide RNA products (pRNA).

RNA-directed transcriptional gene silencing: The promoter-associated RNA model of RNA-mediated transcriptional gene silencing proposes that a variant species of mRNA, a promoter-associated mRNA, essentially an mRNA containing an extended 5′ UTR, is recognized by the antisense strand of siRNAs or possibly endogenous antisense RNAs during the act of transcription of the small RNA targeted promoter. The antisense strand of the siRNA might then guide a putative transcriptional silencing complex (composed of DNMT3A, Ago-1, HDAC-1 and possibly Ezh2) to the targeted promoter where silent-state histone modifications result in the initial gene-silencing event. The prolonged targeting of the promoter may result in heterochromatinization and/or DNA methylation of the small RNA-targeted loci.

RNA-DNA chimeras: A chimeric RNA/DNA oligonucleotide (chimera) was developed to correct point or frameshift mutations in episomal or chromosomal copies of genes.

RNA-DNA hybrid in biological functions: Formation of RNA-DNA hybrids is important in the transcription of DNA, reverse transcription of viral RNA and DNA replication.

RNA-driven hybridization reactions: The reactions that use an excess of RNA to react with all complementary sequences in a single-stranded preparation of DNA.

RNA-editing events: Molecular processes by which specific nucleotide sequences in an RNA molecule are changed, such as C-to-U and A-to-I editing by deaminases.

RNA-guided endonuclease (RGEN): Programmable nucleases composed of the Cas9 protein and a guide RNA.

RNA-induced silencing complex (RISC): RISC machinery consists of a large number of proteins to facilitate the complementary binding of siRNAs to polynucleotide strands of mRNA. A siRNA- or miRNA-protein complex that acts as an endonuclease and cleaves the complementary target mRNA or as a repressor and blocks the target mRNA translation. RISC is composed of a single-stranded guide RNA and an Argonaute protein. The Piwi domain, a highly conserved motif within Argonaute, has been shown to adopt an RNase H fold critical for the endonuclease activity of RISC.

RNA-induced transcriptional silencing (RITS): An AGO-containing RNP complex incorporating repeat-associated siRNA, which is implicated in chromatin silencing.

RNA-mediated interference (RNAi): Silencing of homologous gene expression by introducing long double-stranded RNA into cells of plants. This is also called RNA interference. Following this approach, a long double-stranded RNA molecule is introduced into the cell, which leads to sequence-specific degradation of the homologous gene transcripts. This technology has the potential to serve as an antiviral defence mechanism, for analysis of gene function and to suppress endogenous genes. RNAi has been used to produce low-caffeine coffee by suppressing the gene responsible for caffeine production.

RNA-PET: A paired-end tag (PET) sequencing method for full-length mRNA analysis using next-generation sequencer platforms such as the Illumina Genome Analyser and Applied Biosystems SOLiD. Simultaneous capture of RNAs with both a 5′ methyl cap and a poly(A) tail is indicative of a full-length RNA. This is then followed by sequencing a short tag from each end by high-throughput sequencing.

RNA-physical interaction model of paramutation: In this model, both RNA and physical pairing are required for the *trans*-inactivation. The paramutagenic as well as the paramutable repeats are transcribed and give rise to siRNAs, but only the paramutagenic state is susceptible to RdDM.

RNA-protein interactions: Proteins tend to interact with RNA where they form complex secondary structure elements (e.g. stem-loops and bulges). All ribonucleoproteins (e.g. ribosomes) result from RNA-protein interactions.

RNA-recognition motif (RRM) domain: The RRM domain contains two short consensus sequences embedded in a structurally conserved region of approximately 80 amino acids.

RNA-seq methodology: Technique that provides a means to examine the quantity and sequences of RNA in a sample using next-generation sequencing. RNA-seq, also called WHOLE TRANSCRIPTOME SHOTGUN SEQUENCING (WTSS) and dubbed 'a revolutionary tool for transcriptomics' refers to the use of high-throughput sequencing technologies to sequence cDNA in order to get information about a sample's RNA content. It involves the following steps: (i) RNA poly(A) library, (ii) Synthesis of cDNA, (iii) Sequencing and (iv) Transcriptome alignment. The creation of a library can change from platform to platform in high throughput sequencing, where each has several kits designed to build different types of libraries and adapt the resulting sequences to the specific requirements of their instruments. However, due to the nature of the template being analysed, there are commonalities within each technology. Due to the 5′ bias of randomly primed reverse transcription as well as secondary structures influencing primer binding sites, hydrolysis of RNA into 200–300 nucleotides prior to reverse transcription reduces both problems simultaneously. However, there are trade-offs with this method where although the overall body of the transcripts is efficiently converted to DNA, the 5′ and 3′ ends are less so. Once the cDNA is synthesized it can be further fragmented to reach the desired fragment length. The template is now ready to be prepared for the desired sequencing method. Due to the deep coverage and base-level resolution provided by next-generation sequencing instruments, RNA-seq provides researchers with efficient ways to measure transcriptome data experimentally, allowing them to get information, such as how different alleles of a gene are expressed, detecting post-transcriptional mutations or identifying gene fusions. *Aka* WHOLE TRANSCRIPTOME SHOTGUN SEQUENCING (WTSS).

RNA-transcript mapping: A procedure involving hybridization of the RNA to be analysed with a radioactively labelled and purified DNA probe under the conditions that favour RNA-DNA or DNA-DNA duplexes. After digestion with S1 nuclease, the nuclease-resistant fragments are resolved through polyacrylamide gel electrophoresis.

RNAi technology, challenges in using: Utilization of RNAi appears to be promising for plant protection. However, various key concerns, such as dsRNA stability, dsRNA uptake mechanisms, dsRNA production cost, off-target effects and RNAi resistance, need to be solved before its proficient practical applications are realized in the field. There is no doubt that a new era of RNAi-based pest control for fruit crop protection is right at the corner.

RNAMDB: serves as a focal point for information pertaining to naturally occurring RNA modifications, since its inception in 1994. This database uses an easy-to-use, searchable interface for obtaining detailed data on the currently known 109 RNA modifications.

RNase A: One of the hardiest enzymes in common laboratory usage and is specific for single-stranded RNAs. This endoribonuclease cleaves the 3′-end of unpaired C and U residues, ultimately forming a 3′-phosphorylated product via a 2′,3′-cyclic monophosphate intermediate. It does not require any cofactors for its activity (e.g. bovine pancreatic ribonuclease A).

RNase D: The enzyme responsible for processing pre-tRNA by cleaving at the 3′-termini of the mature tRNA sequences.

RNase H: An endoribonuclease that cleaves the RNA in a DNA/RNA duplex to produce ssDNA. It is a non-specific endonuclease and catalyses the cleavage of RNA via a hydrolytic mechanism, aided by an enzyme-bound divalent metal ion. RNase H leaves a 5′-phosphorylated product.

RNase I: This endoribonuclease cleaves the 3′-end of ssRNA at all dinucleotide bonds leaving a 5′-hydroxyl and 3′-phosphate, via a 2′,3′-cyclic monophosphate intermediate.

RNase II: An exoribonuclease responsible for the processive 3′→5′ degradation of single-stranded RNA.

RNase III: A type of endoribonuclease that cleaves rRNA (16S rRNA and 23S rRNA) from transcribed polycistronic RNA operon in prokaryotes. It also digests double-stranded RNA (dsRNA)-Dicer family of RNase, cutting pre-miRNA (60–70 bp long) at a specific site and transforming it in miRNA (22–30 bp), that is actively involved in the regulation of transcription and mRNA lifetime.

RNase L: An interferon-induced endoribonuclease that upon activation destroys all RNA within the cell.

RNase mapping: A method used to detect and quantify specific RNAs and to determine structural changes of RNA molecules by hybridization with SP6 single-stranded ^{32}P-RNA probes. An excess of ^{32}P-RNA complementary to test RNA is synthesized with SP6 RNA polymerase and hybridized to the RNA sample. Ribonuclease treatment digests the unhybridized single-stranded ^{32}P-probe. ^{32}P-RNA-RNA hybrids are ribonuclease resistant and can be detected and quantified by gel electrophoresis.

RNase P: This endoribonuclease specifically cleaves at the 5′-end of pre-RNA. It is a ribozyme. It is made of one 377-nucleotide RNA (H1-RNA) and 20-kDa ten proteins. Its function is to cleave off an extra, or precursor, sequence on tRNA molecules. It cleaves the 5′ leader sequence of all precursor tRNAs, producing a mature 5′-phosphate on the end of the tRNA and a hydroxyl group on the 3′-end of the leader sequence.

RNase PH: RNase PH is a member of the family of phosphorolytic 3′→5′ exoribonucleases that also include POLYNUCLEOTIDE PHOSPHORYLASE (PNPase). RNase PH is involved in the maturation of tRNA precursors and is especially important for the removal of nucleotide residues near the CCA acceptor end of the mature tRNAs.

RNase PhyM: A sequence-specific endoribonuclease enzyme for single-stranded RNAs. It cleaves the 3′-end of unpaired A and U residues.

RNase R: This exoribonuclease is a close homologue of RNase II, but it can, unlike RNase II, degrade RNA with secondary structures without the help of accessory factors.

RNase T: This exoribonuclease is the major contributor to the 3′→5′ maturation of many stable RNAs.

RNase T1: A sequence-specific endoribonuclease enzyme for single-stranded RNAs. It cleaves 3′-end of unpaired G residues.

RNase T2: A sequence-specific endoribonuclease enzyme for single-stranded RNAs. It cleaves 3′-end of all four residues, but preferentially 3′-end of As.

RNase U2: A sequence-specific endoribonuclease enzyme for single-stranded RNAs. It cleaves 3′-end of unpaired A residues.

RNase V: This endoribonuclease enzyme is specific for polyadenine and polyuridine RNA. Probably the activity named RNase V is due to residual RNase II activity or a related enzyme on ribosomes.

RNase V1: A non-sequence-specific endoribonuclease enzyme for double-stranded RNAs. It cleaves base-paired nucleotide residues.

RNase: Ribonuclease, an enzyme that hydrolyses RNA.

RNomics and gene regulation: RNomics majorly deals with the study of the structure and function of different RNA populations comprising total non-coding RNAs in the cell. Non-coding RNAs play very important roles in biological systems like DNA methylation, chromatin conformational changes and histone code. Regulatory RNA research has marked a new paradigm of RNA-directed gene expression regulation.

Robertsonian rearrangements: A chromosomal mutation due to centric fusion or centric fission (i.e. a reciprocal translocation with breakpoints within the centromeric regions). In wheat, ROBERTSONIAN TRANSLOCATIONS arise from centric misdivision of univalents at anaphase/telophase I, followed by segregation of the derived telocentric chromosomes to the same nucleus and fusion of the broken ends during the ensuing interkinesis.

robot: A virtual or mechanical artificial agent. In practice, it is usually an electro-mechanical machine, which is guided by computer or electronic programming and is thus able to do tasks on its own. Another common characteristic is that by its appearance or movements, a robot often conveys a sense that it has intent or agency of its own. The word robot can refer to both physical robots and virtual software agents, but the latter are usually referred to as bots.

robot, personality of a: Researchers are trying to create robots, which appear to have a personality. They use sounds, facial expressions and body language to try to convey an internal state, which may be joy, sadness or fear.

robot, speech recognition by: Speech would be the most natural way for the human to communicate, it is unnatural for the robot. Interpreting the continuous flow of sounds coming from a human (speech recognition) in real time is a difficult task for a computer, mostly because of the great variability of speech. The same word, spoken by the same person may sound different depending on local acoustics, volume, the previous word, whether or not the speaker has a cold and so on. It becomes even harder when the speaker has a different accent.

robotics: An interdisciplinary branch of engineering and science that includes mechanical engineering, electronic engineering, information engineering, computer science and others. Robotics involves design, construction, operation and use of robots, as well as computer systems for their perception, control, sensory feedback and information processing. The goal of robotics is to design intelligent machines that can help and assist humans in their day-to-day lives and keep everyone safe.

robotics, laws of: Three laws of robotics are as follows: (i) A robot may not injure a human being or, through inaction, allow a human being to come to harm. (ii) A robot must obey any orders given to it by human beings, except where such orders would conflict with the First Law. (iii) A robot must protect its own existence as long as such protection does not conflict with the First or Second Law.

robots, gestures by: A teacher also uses hand gestures to explain certain things in theory as well as practical classes. It is likely that gestures will make up a part of the interaction between humans and robots. A great many systems have been developed to recognize human hand gestures.

robust result: Getting largely the same result with a different analysis on the original data set.

robustness: In regard to methods of analysis, a measure of how dependent a result is on the assumptions of the method being used to obtain the result.

Roche 454 sequencing: This sequencing platform sequencing can sequence much longer reads than Illumina. Like Illumina, it does this by sequencing multiple reads at once by reading optical signals as bases are added. The DNA or RNA is fragmented into shorter reads of up to 1 kb. Generic adaptors are added to the ends and these are annealed to beads, one DNA fragment per bead. The fragments are then amplified by PCR using adaptor-specific primers. Each bead is then placed in a single well of a slide. So, each well will contain a single bead, covered in many PCR copies of a single sequence. The wells also contain DNA polymerase and sequencing buffers.

rod photoreceptor cell: Photoreceptor cell type in the retina that is responsible for non-colour vision in dim light.

Roentgen (r): The amount of ionizing radiation that produces about two ion pairs per cubic micrometre of matter, or about 1.6×10^{12} ion pairs per cubic centimetre.

rogue: To remove and destroy individual plants that are diseased, infested by insects or otherwise undesirable.

rolling circle amplification (RCA): Rapid amplification of plasmid and phage DNA has been achieved by using Phi29 (ϕ29) DNA polymerase and multiple-primed RCA in which oligonucleotide primers complementary to the amplification target circle are hybridized to the circle. The addition of DNA polymerase and DEOXYNUCLEOSIDE TRIPHOSPHATES (dNTPs) to the primed circle results in the extension of each primer and the displacement of each newly synthesized strand results from the elongation of the primer behind it. Secondary priming events can subsequently occur on the displaced product strands of the initial RCA step. RCA has been employed to amplify vector DNA such as M13 or plasmid DNA from single colonies or plaques. Using random primers and ϕ29 DNA polymerase, circular DNA templates can be amplified 10,000-fold in a few hours.

rolling circular DNA replication: A model of DNA replication that accounts for a circular DNA molecule producing linear daughter double helices. W. Gilbert and D. Dressler in 1968 proposed this model. Rolling circle is an alternative form of replication for circular DNA molecules in which an origin is nicked to provide a priming end. One strand of DNA is synthesized from this end, displacing the original partner strand, which is extruded as a tail. One round of replication (i.e. from origin to origin leads to the formation of one concatemer). As the replication is not terminated, a number of copies of concatemers are formed. Each concatemer has an ability to give rise to a copy of bacteriophage and thus a number of bacteriophages are produced. The rolling circle model has been valid in a number of organisms, including the late stages of growth of phage lambda, transfer of the *Escherichia coli* sex factor, replication of particular DNA sections in *Xenopus amphibians* that are specifically associated with the production of ribosomal RNA and replication of some single-stranded DNA phages such as ϕX174.

root: A unit of a plant's axial system, which is usually underground, does not bear leaves, tends to grow downward and is typically derived from the radicle of the embryo.

root cap: A cap of cells covering the apex of the growing point of a root and protecting it as it is forced through the soil.

root cutting: Cuttings consist of root only. other cuttings have no roots at the time they are made and inserted. *Contrasts* DIVISION.

root development: To understand the impairment of root development in *Arabidopsis*, four different phenotypic classes were studied. These results from single nuclear mutations in four different genes, namely *rhd1*, *rhd2*, *rhd3* and *rhd4*, showed that *rhd1* was involved in the proper initiation of root hair, while the other three were required for normal hair elongation. Auxin controls the growth of roots by modulating cellular responses to the phytohormone GIBBERELLIN (GA). GA promotes the growth of plants by opposing the effects of nuclear Della protein growth repressors, one of which is Arabidopsis RGA (for repressor of *gal-3*). GA opposes the action of several Della proteins by destabilizing them, reducing both the concentration of detectable Della proteins and their growth-restraining effects. They showed that auxin is necessary for GA-mediated control of root growth and that attenuation of AUXIN TRANSPORT or SIGNALLING delays the GA-induced disappearance of RGA from root cell nuclei.

root exudate: One of the various compounds that leak from growing and expanding sections of roots as well as from broken cells at exit points of lateral roots.

root graft: The fusion of roots of two adjacent plants so that their water- and food-conducting (vascular) systems become joined.

root hair: Root hairs extend from the epidermis of monocot and dicot roots. Each root hair grows laterally as an extension of a trichoblast, a type of epidermal cell located in the root's maturation zone. Root hairs enhance the root's total surface area to maximize water and mineral absorption from the surrounding soil. Their large central vacuoles allow them to collect water and minerals, which are passed into the root.

root microbiome: The dynamic community of microorganisms associated with plant roots.

root nodule: A small, gall-like growth on the roots of certain types of plants (e.g. legumes). The nodules develop as a result of infection of the root by bacteria.

root-inducing plasmids (pRi): *Agrobacterium rhizogenes* transfer a segment of its pRi to the plant genome. The transferred DNA contains genes, which are involved in the synthesis of plant hormones. These genes express in the plant cell and give rise to rooty tumours at the infection site. Transgenic plants can be readily regenerated from the rooty tumours, and the transferred DNA is transmitted to progeny plants. The Ri T-DNA region can be used for the construction of plant transformation vectors similar to the Ti-plasmid-derived vectors.

root-knot nematode: A nematode (*Meloidogyne naasi*) that induces small, gall-like growth on the roots of certain types of plants.

rootstock: (i) The part of a budded or grafted plant that supplies the root system, also simply called a STOCK. (ii) Plants selected to produce a root system with some specific attribute (e.g. a virus-free rootstock). *Contrasts* SCION.

Rosaceae: The rose family, of which the type of genus is *Rosa*.

rosette: (i) A disease symptom characterized by short, bunchy growth habit due to shortened internodes and no comparable reduction in leaf size. (ii) An arrangement of leaves radiating from a root crown near the earth. (iii) When parts are not whorled or opposite but appear so, due to the contractions of internodes (e.g. the petals in a double rose or a basal cluster of leaves, usually close to the ground, in some plants).

rot: The softening, discoloration and often disintegration of plant tissue as a result of fungal or bacterial infection.

R_0t value: The product of RNA concentration (R_0) and time (t) of incubation in an RNA-driven hybridization reaction.

rotation: The growth of different kinds of crops in succession in the same field.

rough endoplasmic reticulum (RER): Endoplasmic reticulum with ribosomes on its cytosolic surface. Involved in the synthesis of secreted and membrane-bound proteins. Found in animal cells and plant cells. It extends from the outer nuclear membrane throughout the cytoplasm. Rough ER is continuous with the outer nuclear membrane, and it has numerous, protein-synthesizing ribosomes attached to it. Ribosomes deposit proteins into the lumen of the rough ER, where they can be modified. Transport vesicles then take the modified proteins to the nearby Golgi body, where they are processed further and sorted for transport.

round-off error: Also, rounding error. The difference between the result produced by a given algorithm using exact arithmetic and the result produced by the same algorithm using finite-precision, rounded arithmetic. Rounding errors are due to inexactness in the representation of real numbers and the arithmetic operations done with them. This is a form of quantization error. When using approximation equations or algorithms, especially when using finitely many digits to represent real numbers (which in theory have infinitely many digits), one of the goals of numerical analysis is to estimate COMPUTATION ERRORS. Computation errors, also called NUMERICAL ERRORS, include both truncation errors and round-off errors.

router: A networking device that forwards data packets between computer networks. Routers perform the traffic-directing functions on the Internet. Data sent through the Internet, such as a web page or e-mail, are in the form of data packets. A packet is typically forwarded from one router to another router through the networks that constitute an internetwork (e.g. the Internet) until it reaches its destination node.

routine resequencing of megabase regions of genomic DNA: Understanding the disease susceptibilities and predisposition – basically sequence affected and unaffected people (e.g. isolate cancer cells, sequence tumour and normal tissue and get genotyping). Massively parallel DNA arrays are new sequencing technologies under development.

routing table: In computer networking, a ROUTING TABLE or ROUTING INFORMATION BASE (RIB), is a data table stored in a router or a network host that lists the routes to particular network destinations and, in some cases, metrics (distances) associated with those routes. The routing table contains information about the topology of the network immediately around it.

roving gene: *See* JUMPING GENES.

row-column design: An experimental design in which the set of plots has the structure of a rectangle, in which two plots may be in the same row, or in the same column or neither.

row-complete Latin square: A Latin square is row complete if each ordered pair of distinct symbols occurs precisely once in consecutive positions in a row of the square.

row-quasi-complete Latin square: A Latin square is row-quasi-complete if each unordered pair of distinct symbols occurs precisely twice in consecutive positions in a row of the square.

***Rps1c* gene:** A gene that confers to any soybean plant (possessing that gene in its DNA) resistance to several strains/races of phytophthora root rot (PRR) disease.

***Rps1k* gene:** A gene that confers to any soybean plant (possessing that gene in its DNA) resistance to 21 strains/races of PRR disease.

***Rps6* gene:** A gene that confers to any soybean plant (possessing that gene in its DNA) resistance to some strains/races of PRR disease.

RT/PCR reaction: A series of reactions that result in RNA being copied into DNA and then amplified. A single primer is used to make single-stranded cDNA copies from an RNA template under the direction of reverse transcriptase. A second primer complementary to this 'first strand' cDNA is added to the reaction mix along with *Taq* polymerase, resulting in the synthesis of double-stranded DNA. The reaction mix is then cycled (denaturation, annealing of primers,

Rubinstein–Taybi syndrome: Condition has symptoms that include intellectual disabilities. Its aetiology involves a mutation in CREB-binding protein (histone acetylation).

RuBisCO biogenesis or catalysis, manipulation of: Despite the crucial role of RuBisCO (ribulose 1,5-bisphosphate carboxylase-oxygenase) in CO_2 fixation in autotrophs, including food crops, the slow catalytic turnover rate and wasteful oxygenation reactions by this sluggish enzyme make it a good target to enhance photosynthetic efficiency. Because of its multimeric nature (i.e. L_8S_8 type in higher plants) and regulation by nuclear and chloroplast-encoded genes, its manipulation is extremely challenging.

ruderal: A plant that colonizes or occupies disturbed waste ground. *See* WEED.

rudiment: In the structure of a plant, an item that is at best hardly functional, either because it is immature and has not yet completed its development (e.g. a leaf still incompletely formed inside a bud) or because its role in the organism's morphology cannot be completed and therefore is futile (e.g. the leaf rudiment at the tip of a phyllode, that will be shed while immature, because the leaf function will be taken over by the phyllode). *Compare with* CATAPHYLL *and* VESTIGE.

rudimentary: Being of the nature of a rudiment; at most barely functional because incompletely developed; begun, but far from completed, either temporarily or permanently. *Compare with* VESTIGIAL.

rule: It is a principle or condition that customarily governs behaviour; a basic generalization that is accepted as true and that can be used as a basis for reasoning or conduct; the duration of a monarch's or government's power. Examples: Rules of probability, Rules of the road, British rule.

rule-based system: In computer science, a rule-based system is used to store and manipulate knowledge to interpret information in a useful way. It is often used in artificial intelligence applications and research. Normally, the term rule-based system is applied to systems involving human-crafted or curated rule sets. Rule-based systems constructed using automatic rule inference, such as rule-based machine learning, are normally excluded from this system type.

rules of thumb: A guide or approximation used to simplify decision making in complex circumstances.

run out: Separation of nucleic acid or protein molecules by gel electrophoresis.

run time: The final phase of a computer programme's life cycle, in which the code is being executed on the computer's CENTRAL PROCESSING UNIT (CPU) as machine code. In other words, 'run time' is the running phase of a programme. *Aka* RUN TIME *or* EXECUTION TIME.

run-time error: A run-time error is detected after or during the execution (running state) of a programme, whereas a compile-time error is detected by the compiler before the programme is ever executed. Type checking, register allocation, code generation and code optimization are typically done at compile time but may be done at run time depending on the particular language and compiler. Many other run-time errors exist and are handled differently by different programming languages, such as division by zero errors, domain errors, array subscript out-of-bounds errors, arithmetic underflow errors, several types of underflow and overflow errors, and many other run-time errors generally considered software bugs, which may or may not be caught and handled by any particular computer language.

runaway plasmid vector: A plasmid vector, which at a low temperature (30°C), is present in a moderate number of copies per cell because of some control on plasmid replication but at a higher temperature (35°C) all control on plasmid replication is lost and number of copies per cell increases continuously.

runaway process: Arises from the positive feedback between the evolution of a male trait and the female preference for that trait.

runaway replication: Uncontrolled replication of plasmid molecules, which multiply within the host building up to several thousand copies. This usually halts cell division but is a very powerful way of amplifying a gene product. Plasmid vectors that are temperature sensitive for a copy number control element have been constructed as runway replication vectors.

runner: A slender, horizontal stem that grows close to the soil surface. *See* STOLON.

runner plant: A new plant produced asexually on a runner or stolon.

runoff transcript: RNA that has been synthesized from plasmid DNA (usually by a bacteriophage RNA polymerase such as T7 or SP6) and which terminates at a specific 3′ site because of prior cleavage of the plasmid DNA with a restriction endonuclease.

rust: Plant disease caused by more than 7000 species of fungi of the phylum Basidiomycota. The rusts are a group of fungal diseases affecting the aerial parts of plants. Leaves are affected most commonly, but rust can also be found occasionally on stems and even flowers and fruit. The spore pustules produced by rusts vary in colour, according to the rust species and the type of spore that it is producing. *Puccinia* spp. attack small grain plants such as wheat, corn/maize, sorghum, oats, barley and rye. Its visual appearance is like that of rust on the surfaces of those plants.

RuvABC system: This enzyme complex helps to resolve the Holliday intermediate into recombination products. The binding of RuvABC complex causes the junction to adopt an open square-planar configuration. The bias towards resolution in one orientation or the other is determined by the way in which the Ruv proteins were positioned on the junction. Holliday junction resolution occurs as the resolvasome promotes branch migration.

S

s: The symbol representing the selection coefficient in a mathematical treatment of the effects of natural selection.

S phase: The phase of the cell cycle in which the chromosomes replicate by new DNA synthesis. S phase occurs between G1 and G2 phases. Both chromatids of a chromosome are joined together by a centromere. A diploid cell thus has double the amount of DNA. RNA and some proteins (histones) are synthesized in this phase. In G_1, the amount of DNA present is equal to 2C (diploid cells). *See* SYNTHETIC PHASE.

S phase, entry into: Excess cyclin E-Cdk2 accelerates entry into the S phase of the cell cycle and promotes polyploidy, which may contribute to genomic instability in cancer cells. Twenty amino acids in cyclin E have been identified as a CENTROSOMAL LOCALIZATION OF SIGNAL (CLS) essential for both centrosomal targeting and promoting DNA synthesis. Cyclin E has a molecular centrosomal-targeting domain essential for promoting S phase entry in a Cdk-2 independent manner.

S value (S): The unit of measurement of a sedimentation coefficient. *Aka* SVEDBERG UNIT *or* SEDIMENTATION COEFFICIENT.

S-adenosylmethionine (SAM) domain: Also written as S-adenosyl-L-methionine (SAM) domain. A protein domain containing an iron-sulfur (Fe-S) cluster that uses S-adenosylmethionine (SAM) to catalyse various radical reactions. *Aka* ADOMET DOMAIN.

***S*-allele:** An allele of a gene governing incompatibility in many allogamous (reproducing by cross-fertilization) plants. Alleles occurring in both style and pollen are referred to as matching *S*-alleles. *S*-alleles usually have their place in a series of multiple alleles.

***S*-locus:** A self-incompatibility (SI) locus that prevents self-fertilization, by a specific interaction between the *S*-gene product produced in the pistil and the *S*-gene products borne on or expressed by the male gametophyte, the pollen grain.

***S*-locus cysteine-rich (SCR) protein:** A small (<8 kDa) hydrophilic and positively charged peptide that is localized to the pollen coat.

***S*-locus glycoprotein (SLG):** This glycoprotein recognizes a pollen factor that leads to the rejection of self-pollen.

s-locus haplotype: Three tightly linked alleles [encoding *S*-locus glycoprotein (SLG), male-specific *S*-locus protein 11 (SP11) and female-specific *S*-locus receptor kinase (SRK)] arranged on chromosome as a heritable unit that controls self-incompatibility in *Brassica*.

***S*-locus protein 11 (SP11):** SP11 is found in the pollen coat and is involved in preventing self-pollination through a process known as SELF-INCOMPATIBILITY.

***S*-locus receptor kinase (SRK):** SRK is present on the stigma, the part of flowering plants that binds pollen. It allows stigma epidermal cells to discriminate between genetically related ('self') and genetically unrelated ('non-self') pollen in the self-incompatibility response of the Brassicaceae.

S_0: A symbol used to label the original selfed plant.

S1 nuclease: A 32-kDa endonuclease that specifically degrades (destroys or digests) single-stranded nucleic acids, including the single-stranded regions of duplex DNA, RNA or DNA/RNA. It preferentially targets DNA over RNA. This endonuclease can also introduce single-stranded nicks and breaks in duplex DNA, RNA and DNA/RNA. It has applications in nuclease protection assays, double-stranded DNA end blunting, RNA transcript mapping and cDNA hairpin removal. *Aka* S1 NUCLEASE.

S_1, S_2, S_3, etc.: The symbol used for continued selfing (self-fertilization) of plants. S_1 designates the generation obtained by selfing the parent plant (S_0), S_2 the generation obtained by selfing the S1 plant and so on.

S9 mix: A liver-derived supernatant used in the Ames test to activate or inactivate mutagens.

sacB: The *sacB* gene encodes the SacB protein, which converts sucrose into a toxic form that kills bacteria. This can be used in negative selection for the *sacB* gene.

***Saccharomyces cerevisiae* genome**: *Saccharomyces cerevisiae* (baker's yeast) has a genome size of 12,100,000. Number of genes identified is 6,300. The genome includes genes for transcription (1039), RNA metabolism (336), RNA processing (542), ribosome biogenesis and assembly (337), protein modification (361), Organelle organization and biogenesis (550) and macromolecular biogenesis (449), protein biosynthesis (442), RNA splicing (204), lipid metabolism and biogenesis (622), microtubule-related processes (247) and glycerophospholipid metabolism (64).

Saccharomyces: Genus of yeasts that reproduce asexually by budding or sexually by conjugation. Economically vital in brewing and baking, they are also widely engaged in genetic engineering and as simple model organisms in the study of eukaryotic cell biology.

saccharose: A sweet, crystalline substance, $C_{12}H_{22}O_{11}$, attained from the juice or sap of many plants (e.g. from sugarcane and sugarbeet).

safe gene drive: Offensive insect pest species can be contained by developing gene drive systems. The self-propagating gene drive technologies and the risk of their accidental spread to new areas might achieve better results. *See* GENE DRIVE.

Sal I box: A sequence, AGGTCGACCAGATNTCCG, that guides transcription termination of ribosomal DNA (mouse) transcription by RNA polymerase I. This sequence occurs eight times in the spacer region downstream of the 3′ end.

salicylic acid (SA): SA is a signalling molecule in SYSTEMIC ACQUIRED RESISTANCE (SAR) when it is activated in plants via spray application of COBRA R herbicide to soybean plants, spray application of harpin protein to various plants, chewing by insects on the leaves of tomato plants and/or the entry-into-plant of certain pathogenic bacteria/fungi and so on. *Salmonella*, is a genus of bacteria, consisting of more than 2400 serovars (strains/types) classified within two species, *Salmonella enterica* and *Salmonella bongori*. All of these serovars are potentially pathogenic (disease-causing) to humans. For example, some variants of *Salmonella typhimurium* can cause typhoid fever. The non-typhoid strains of *Salmonella* generally cause enterocolitis although that enterocolitis can lead to more serious systemic infections. *Salmonella enteritidis* and *Salmonella typhimurium* are increasingly causing outbreaks of foodborne illnesses (e.g. when foods are not washed or cooked thoroughly enough before consumption by humans).

salinization: The build-up of soluble salts in the soil surface; an important agricultural problem in drier regions.

salivary chromosomes: The giant chromosomes in the cells of the salivary glands of larval flies and other dipterans such as mosquitoes and *Drosophila* characteristically each show a highly specific pattern of contrasting bands. *See* GIANT CHROMOSOMES *and* POLYTENE CHROMOSOMES.

Salmon procedure: A procedure for producing haploids in hexaploid wheat. 'Salmon' is the name of an alloplasmic wheat variety carrying a 1RS.1BL chromosome translocation together with the cytoplasm of *Aegilops kotchyi*. Alloplasmic is defined as a part of the cytoplasm that is specialized to form cilia, flagella and similar structures. The interaction of cytoplasmically genetic determinants with, possibly, a parthenogenesis-inducing gene on chromosome arm 1RS of rye produces haploid progeny.

Salmonella: A type of rod-shaped Gram-negative bacteria that affects the intestinal tract. First discovered by D.E. Salmon in 1885, *Salmonella* commonly causes salmonellosis, a disease that has the symptoms of diarrhoea, fever and stomach cramps. Occasionally *Salmonella* can cause infection in urine, blood, bones, joints or the nervous system. *Salmonella* infections can be caused by the consumption of contaminated food or water and touching infected animals.

salt tolerance: Refers to the trait (of a plant) that allows a plant to grow/survive in soil that contains a high level of salts such as sodium chloride (NaCl). For example, in 2001, E. Blumwald and H.-X. Zhang inserted a salt-tolerance gene from *Arabidopsis thaliana* into a tomato plant (*Lycopersicon esculentum*) thus making that tomato plant resistant to salt concentrations up to 200 mM (which is far higher than it could previously survive). The *Arabidopsis* origin gene permits the tomato plant to take our salt from the soil, and then sequester and store the salt in vacuoles (i.e. small compartments) within its leaf cells. Using biotechnological approaches such as transformation with *Hal*2, *Gly*1, *Gly*2 and other genes is being attempted to develop transgenic salt tolerant crop varieties.

saltation: (i) A variation that exhibits a large effect. (ii) A major mutation in the asexual state of fungal growth, especially one taking place in *in vitro* culture.

saltation hypothesis: A hypothesis that explains evolution by stepwise changes in a population.

saltatory DNA replication: A sudden lateral amplification to produce a large number of copies of some DNA sequence.

***Sam-K* gene:** A gene naturally present within the *Escherichia coli* bacteriophage T3. If the *sam-k* gene is inserted via genetic engineering into a (fruit crop) plant's genome, that causes greatly reduced production of the chemical compound S-adenosylmethionine (SAM) in that plant's fruit. Because the SAM is normally converted (chemically) into l-aminocyclopropane-1-carboxylic acid (ACC) in the fruits of traditional varieties of (fruit crop) plants, such *sam-k* gene-containing plants produce fruits that ripen/soften far slower than fruit from traditional varieties of those plants, which can reduce spoilage/loss in the harvest and transport of such fruit. That is because ACC is required for fruits to produce ethylene, the plant hormone that triggers (over-) ripening/softening of fruit.

same-sense mutation: This is a type of point mutation in which the changed codon codes for the same amino acid as in the wild type. *Aka* NEUTRAL MUTATION *and* SILENT MUTATIONS.

sample: In statistics, a sample refers to a set of observations drawn from a population under study.

sample size: The number of experimental units on which observations are considered. It may be less than the number of observations in a data set, due to the possible multiplying effects of multiple variables and/or repeated measures within the experimental design.

sampling: The procedure by which an illustrative sample is taken from a seed lot or something else.

sampling accuracy: Refers to the extent to which sample statistics correctly estimate the population parameter. The terms BIASED and UNBIASED are typically used to describe the accuracy of sample statistics.

sampling design: Refers to the method used for selecting a sample.

sampling distribution: The distribution of a measurement based on all possible samples that can be drawn according to a specific sampling plan. The term usually refers to RANDOM SAMPLING.

sampling error: The error caused due to a sample giving a different result than the other sample, and neither possibly giving the same result as in the population.

sampling method: The procedure to choose or select a fraction out of the target population.

sampling precision: Proximity to a given standard.

sampling units: Non-overlapping groups of elements from the population.

SAND domain proteins: Protein domain SAND is named after a range of proteins in the protein family: Sp100, AIRE-1, NucP41/75 and DEAF-1. It is localized in the cell nucleus and has an important function in chromatin-dependent transcriptional control. It is found solely in eukaryotes.

Sanger sequencing: *See* DNA SEQUENCING BY CHAIN-TERMINATION METHOD.

sap transmission: The spread of pathogens, usually viruses, by rubbing sap from an infected plant on to a healthy plant to cause infection.

saponification value: Saponification value represents the number of milligrams of potassium hydroxide required to saponify 1 g of fat under the conditions specified. *Aka* SAPONIFICATION NUMBER *and* KOETTSTORFER NUMBER.

saprobe (Adj. saprobic): An organism that acquires nutrition from non-living organic matter. *See* SAPROPHYTE *and* SAPROTROPH.

saprophyte: An organism that feeds on dead organic matter. Most of the fungi are saprophytes as they obtain nutrition from decomposing dead organic matter. The microorganisms that feed on living organisms are generally called pathogens or parasites.

saprotroph (n. saprophyte; Adj. saprotrophic): An organism (such as a plant, a fungus or similar organism) that obtains nutrition from decaying non-living organic matter usually lacking chlorophyll. *See* SAPROBE *and* SAPROPHYTE. *Contrasts* PARASITE *and* EPIPHYTE.

sapwood: The physiologically active sector of wood contiguous to cambium. *Contrasts* HEARTWOOD.

sarment: A long, slender, prostrate stolon, commonly called a RUNNER.

sarmentose: Reproducing by sarments. Strawberry plants are the most familiar example. *See* SARMENT.

SAT-zone: Refers to the secondary constriction of a satellite chromosome.

satellite: A portion of chromosome appearing separated from the main body of the chromosome by secondary constrictions.

satellite association: All the chromosomes that have satellites tend to arrange themselves in a group during metaphase. This poses difficulty in assigning a specific number to a specific chromosome.

satellite band/DNA: When a eukaryotic DNA is centrifuged on a density gradient, two types of bands may be distinguished on the basis of optical absorbance. Most of the genome forms a rather broad peak centred on the buoyant density corresponding to the average G-C content of the genome. This is called MAIN BAND. Sometimes additional smaller peak(s) is/are seen at a different value. This material is called SATELLITE BAND and the DNA in this band is termed satellite DNA.

satellite chromosome: Each chromosome with a secondary constriction that divides a satellite part from the rest of the chromosome (arm).

satellite DNA (satDNA): A highly repetitive segment of middle repetitiveness and a segment of unique (or single copy) rest of the chromosome DNA. These segments constitute about 10%, 15% and 75%, respectively of total mouse DNA repeated million of times. The main characteristic of these satellites is that the size of the repeating unit is small (only 5–12 bp). The highly repetitive DNA occurs primarily around the centromeres and the telomeres. It is not known to be transcribed. Satellites are present in many eukaryotic genomes. Satellite DNA seems to enact no part in somatic functions but do play role in germ line. Satellite DNA fraction that has adequately different base composition from the bulk of the DNA in order to separate distinctly during caesium chloride density gradient centrifugation. It can be derived from nucleus, plastid or mitochondrial DNA. These are also called simple sequence DNA or SIMPLE SEQUENCE REPEATS (SSRs). This is also called satellite DNA because it migrates as 'satellite' bands during density gradient centrifugation. Satellite DNA does not encode RNAs or protein but it serves as good molecular marker in marker-aided selection in plant breeding. *See* MICROSATELLITE *and* MINISATELLITE.

satellite DNA and evolution: Satellite DNA is thought to be influential in pairing of chromosomes both in cell division and fusion of male and female gametes. Satellite DNA is not transcribed into RNA. There is clear distinction between satellite of one species and that of another no matter how closely related. This might represent a high rate of change of satellite sequences, which obviously must be faster than rate of speciation. A shift in the identity of satellite composition in a geographical isolated population of organisms could initiate incipient speciation through effective reproductive isolation.

satellite imagery: Employing satellite imagery in the agriculture sector is not new; organizations have been utilizing images from space to study land use ever since the first satellite of NASA's Landsat programme started beaming back pictures in 1972. The data now being generated can be integrated into all the additional innovations taking place on the farm. In India, agri-traders have been able to estimate state-by-state wheat yields 30 days ahead of harvest, giving them an advantage in terms of procurement and storage plans. Also, in India, insurance companies have started using satellite imagery to assess the extent and intensity of crop damage when resolving payment claims.

satellite males: Males that take up station in the vicinity of a calling or displaying male. The satellite male may attempt to intercept incoming females or to replace the displaying male when it leaves.

satellite region: A region of DNA that contains highly repetitive base sequences.

satellite RNA (satRNA): An RNA that is transmitted along with the virus but does not infect or damage the host is called satellite RNA.

satellite virus: Part of a defective virus associated functionally with another virus, which it depends on for replication. For example, part of a small virus associated with TOBACCO NECROSIS VIRUS (TNV) is reliant on the TNV genome for its replication.

saturated bonds: Describes a molecule containing carbon-carbon bonds that have only single covalent bonds, for example, in alkanes.

saturated linkage map: High-density linkage maps developed using molecular markers. Using different molecular markers, such as RESTRICTION FRAGMENT LENGTH POLYMORPHISM (RFLP), RANDOM AMPLIFIED POLYMORPHIC DNA (RAPD), AMPLIFIED FRAGMENT LENGTH POLYMORPHISM (AFLP) and SSRs, saturated linkage maps have been developed in several crop plants including tomato, rice, wheat, potato, barley, maize, Brassica and cotton. Saturated linkage maps are required for gene tagging, marker-aided selection and map-based gene cloning.

saturation density: The maximum cell number attainable, under specified culture conditions, in a culture vessel. This term is usually expressed as the number of cells per square centimetre in a monolayer culture or the number of cells per cubic centimetre in a suspension culture.

saturation hybridization: The experiment that has a large excess of one component, causing all complementary sequences in the other component to enter a duplex form.

saturation mapping: The enrichment of specific regions of an already-established genetic map with molecular markers such that marker density is extremely high. A saturated map is a prerequisite for map-based cloning of a gene of interest. Saturation mapping produces a high-resolution map and usually incorporates markers generated with different marker systems.

saturation mutagenesis: Induction and recovery of large numbers of mutations in one particular area of a genome, or in one biological function, in order to recognize all the genes in that area or influencing that function.

sB site: Host specificity site of *Escherichia coli* B strain.

SC_0 generation: Somaclones obtained from callus tissue cultures. SC is abbreviation of somaclones. SC_0 generation is composed of plants directly regenerated from callus cultures of plants. The variation in these plants may be due to genetic changes or due to epigenetic changes. SC_0 plants are characterized using morphological, biochemical and molecular methods to identify the variants.

SC_1 generation: A population of plants obtained through selfing of SC_0 plants. Genetic variation is heritable, whereas epigenetic variation is not heritable to following generations. Thus, all the variants of SC_0 generation do not breed true. Only the genetic variations in the SC_0 are carried forward to SC_1 generation.

scab: A roughened, crust-like diseased area appearing on the surface of a plant organ.

scabrid: Rough to the touch, with short hard protrusions or hairs. *Aka* SCABROUS.

scaffold: (i) In genomic mapping, a series of contigs that are in the right order but not necessarily connected in one continuous stretch of sequence. (ii) The eukaryotic chromosome structure seen after DNA and histones have been removed. Thus, a scaffold is made from non-histone proteins. Thirty non-histone proteins were found in scaffold. Non-histone proteins are also suggested to be responsible for chromosome structure. This model proposes that certain non-histone proteins, including topoisomerase II, bind to chromatin every 60–100 kb and tether the supercoiled, nucleosome-studded 300 Å fibre into structural loops. The removal of histones does not permit the chromatin to unfold completely. Other non-histone proteins may, in turn, gather the loops into daisy-like rosettes. Additional non-histone proteins may then compress the rosettes centres into a compact bundle. This proposal of looping and gathering is known as the RADIAL LOOP-SCAFFOLD MODEL OF COMPACTION, which offers a simple description of progressive chromosome compaction from interphase to metaphase chromosomes.

scaffold protein: Protein that systematizes assemblies of interacting intracellular signalling proteins into signalling complexes.

scalariform: Ladder-like in structure or appearance.

scald: A necrotic condition in which tissue is usually bleached and has the appearance of having been exposed to high temperatures.

scale: (i) A compressed epidermal outgrowth, such as those commonly found on the leaves and rhizomes of ferns. (ii) Refers to quantity, both relative and absolute. Thermometers, rulers and weighing devices help us to see that objects and energy vary in quantity.

scanning acoustic microscope (SAM): A pulse-echo method where an image is obtained by scanning a sample in a plane parallel to its surface (C-scan). A focused ultrasonic beam is transmitted from a transducer as a narrowband burst, through a fluid couplant (typically water) to a sample. *Aka* SELF-ASSEMBLED MONOLAYER. *See* COUPLANT.

scanning atom probe (SAP): SAP uses a nanoscopic probe that is scanned over a solid surface. The interaction between the probe and the surface may be a mechanical or electromagnetic force. The force signal is enhanced and then composed to a force diagram of the surface. Under ideal conditions, an atomic-scale resolution may be achieved.

scanning capacitance microscopy (SCM): SCM is used to map the local capacitance of a surface.

scanning electrochemical probe microscope (SEPM): A scanning probe technique that measures the local electrochemical activity of a sample in solution. The most common form of SEPM, feedback mode, measures the Faradaic current of a redox mediator that interacts with the sample, this leads to an inherent chemical selectivity.

scanning electron microscope (SEM): A type of electron microscope that gives three-dimensional (3D) images of a sample surface bombarded by an electron beam. SEM is used for studying surface topography and structures of biological specimens like hairs, flagella, stomata and embryoids. It can magnify the specimen from 1000 to 100,000× or even more.

scanning force microscopy (SFM): SFM acquires an image by scanning a sharp probe across a surface. This can be done by contacting the surface (contact mode) or by a variety of other scanning modes (intermittent contact and others are covered in more detail in separate laboratory modules). *Aka* ATOMIC FORCE MICROSCOPY.

scanning force spectroscopy (SFS): SFS was used to measure the mechanical properties of double-stranded RNA molecules in comparison with DNA. Similar to the B-S transition in DNA, RNA molecules are stretched from the assumed A′ conformation to a stretched conformation by applying a defined force (plateau force). The force depends on the G + C content of the RNA and is distinct from that required for the B-S transition of a homologous DNA molecule. *Aka* STATIC FORCE SPECTROSCOPY.

scanning hypothesis: A proposed system for the initiation of eukaryotic translation, in which the eukaryotic ribosome attaches to the 5′-terminal cap structure of the mRNA and then scans along the molecule until it reaches an initiation codon.

scanning ion-conductance microscopy (SICM): A scanning probe microscopy technique that uses an electrode as the probe tip. SICM allows for the determination of the surface topography of micrometre and even nanometre-range structures in aqueous media conducting electrolytes. The samples can be hard or soft, are generally non-conducting, and the non-destructive nature of the measurement allows for the observation of living tissues and cells, and biological samples in general.

scanning near-field optical microscopy (SNOM): A microscopy technique for nanostructure investigation that breaks the far-field resolution limit by exploiting the properties of evanescent waves. In SNOM, the excitation laser light is focused through an aperture with a diameter smaller than the excitation wavelength, resulting in an evanescent field (or near field) on the far side of the aperture. *See* NEAR-FIELD SCANNING OPTICAL MICROSCOPY (NSOM).

scanning probe electrochemistry (SPE): A technique within the broader class of SCANNING PROBE MICROSCOPY (SPM) that is used to measure the local electrochemical behaviour of liquid/solid, liquid/gas and liquid/liquid interfaces.

scanning probe microscope (SPM): Including AFMs and STMs, in which the effect of the interaction of a sharp probe with the sample, is measured to infer the atomic structure of the material.

scanning surface potential microscopy (SSPM): An electrical characterization method that produces maps of electrostatic potential across a sample surface. SSPM uses the principles of the Kelvin-probe method, which is a nulling technique and is therefore *aka* KELVIN-PROBE FORCE MICROSCOPY (KFM).

scanning thermal microscope (SThM): SThM was developed to probe thermal properties at the nanoscale level. SThM uses nanofabricated thermal probes to achieve unprecedented high spatial and thermal resolution and sensitivity with a unique signal detection scheme. SThM is available in two modes, TEMPERATURE CONTRAST MODE (TCM) and THERMAL CONDUCTIVITY CONTRAST MODE (CCM). TCM allows the user to measure the temperature variations on a sample surface, which has a heat source on it. CCM allows the user to measure variations of thermal conductivity on a sample surface.

scanning tunnelling microscopy (STM): Imaging technique using local temperature variations of the sample. *Aka* SCANNING THERMAL MICROSCOPY.

scanning tunnelling spectroscopy (STS): STS is an extension of STM and is used to provide information about the density of electrons in a sample as a function of their energy. In STM, a metal tip is moved over a conducting sample without making physical contact. A bias voltage applied between the sample and tip allows a current to flow between the two. This is a result of quantum tunnelling across a barrier; in this instance, the physical distance between the tip and the sample. The scanning tunnelling microscope is used to obtain TOPOGRAPHS, *aka* TOPOGRAPHIC MAPS of surfaces.

scanning vibrating electrode technique (SVET): SVET measures the local electric field above the surface of a sample in solution, allowing electrochemical processes to be visualized as they occur.

scarce RNA: mRNA that consists of a large number of individual RNA species, each present in very few very few copies per cell. *Aka* COMPLEX RNA.

scatter diagram: A diagram displaying values of one quantitative variable for different values of the other variable by plotting points. This is also called the (X, Y) PLOT.

scatterometers: Generally non-imaging sensors that are used to make precise quantitative measurements of the amount of energy backscattered from targets. The amount of energy backscattered depends on the surface properties (roughness) and the angle at which the microwave energy strikes the target.

Scatterometry measurements over ocean surfaces can be used to estimate wind speeds based on the sea surface roughness.

Schiff's reagent: A substance comprising of fuchsine bleached by sulfurous acid that produces a red colour upon reaction with an aldehyde. This reagent is employed for chromosome staining.

schizocarp: A dry fruit made from more than one carpel but breaking and splitting into individual carpels (called MERICARP) when ripe. *See* MERICARP.

science: A way of knowing about the natural world based on observations and experiments that can be confirmed or disproved by other scientists using accepted scientific methods.

scientific method: Refers to a body of techniques for investigating phenomena, acquiring new knowledge or correcting and integrating previous knowledge. A method of inquiry must be based on gathering observable, empirical and measurable evidence subject to specific principles of reasoning. A scientific method consists of the collection of data through observation and experimentation, and the formulation and testing of hypotheses. The scientific method attempts to minimize the influence of bias or prejudice in the experimenter when testing a hypothesis or a theory.

scientific method in discovery of DNA structure: Discovery of the structure of DNA is one of the best examples to illustrate the basic elements of the scientific method. Five basic elements of the discovery of DNA structure are DNA CHARACTERIZATIONS, DNA HYPOTHESES, DNA PREDICTIONS, DNA EXPERIMENTS and DNA ITERATIONS. *See details as separate terms.*

scientific method of research, limitations of: It is not possible for a scientist to record everything that took place in an experiment. He must select the facts he believes to be relevant to the experiment and report them. The problem is that parts of the theory itself need to be assumed to select and report the experimental conditions. The observations are hence sometimes described as being 'theory-laden'.

scientific method of research, misapplications of: (i) Lack of testing of hypothesis, (ii) error intrinsic to instruments, (iii) biasedness of observations, (iv) ruling out data that do not support the hypothesis and (v) failure to estimate quantitatively systematic errors.

scientific nomenclature: Scientific binomial method of classification. *See* SPECIES.

scion: The aerial portion of a graft combination, induced through various means to bond with a compatible understock or rootstock. *See* ROOTSTOCK.

sclerenchyma (Adj. sclerenchymatous): Sclerenchyma is a plant tissue providing mechanical stiffness and strength. Fibres and sclereids are the main types of sclerenchyma cells. Most sclerenchyma cells show intrusive growth. The cell walls of sclerenchyma have thickened secondary layers made from cellulose, hemicelluloses and lignin.

sclerophyll (Adj. sclerophyllous): Any structure stiffened with thick-walled cells (e.g. a plant with hard, stiff leaves).

sclerotium (pl. sclerotia): A vegetative inactive body of a fungus, composed of a compact mass of hyphae with or without host tissue, usually having a darkened rind.

scorable marker: *See* SCREENABLE MARKER.

scorch: Any symptom or structure that resembles the result of flame or fire on the affected part (e.g. often seen at the margins of leaves).

scorpioid: A cymose inflorescence, branching alternately on one side and then the other. *Compare with* HELICOID.

scrambled genome: Clitellates have the most scrambled genomes among animals studied so far. Freshwater worms and their land-based relatives have gone through massive genetic rearrangements. It helped the invertebrates exit the sea. There's a clear correlation between all these changes and a habitat shift. These worms have rejiggered their genomes so widely that the ribbons of genes consistent across other worms and many groups of ancestral invertebrates were largely unrecognizable. Everything broke and then rearranged completely randomly. Certain sets of genes holiday near one another on chromosomes even among distantly related animals. But new work reveals that the genes in these blocks became progressively scrambled in worm species that moved into fresh water and then on to land.

screenable marker: Marker gene that helps in screening cells/tissues/plants to identify transgenics through the expression of specific enzymes that produce distinctive phenotypes. During plant transformation, some commonly used detectable enzyme-producing genes include *nos, lux, cat, gus* and $m_2$fpu. In this case no selection pressure is applied on cells or regenerated shoots and only small pieces of cut plant tissues are observed for expression of marker gene. For example, *gus* gene produces an enzyme (β-glucoronidase) that acts on the substrate (X-gene) within the cells and converts it into a blue precipitate visible to naked eye. But this is a destructive method. Alternatively, the use of the green fluorescent protein gene (*m_2f-pu*) in the transgenic tissues gives green fluorescence under UV illumination and it is a non-destructive method.

screenable marker genes: *See* REPORTER MARKER GENES.

screening: Investigating the possessions and performance responses of individuals, lines, genotypes or other taxa under an assortment of conditions to assess the individuals or groups.

screening a library: Involves methods to identify the small number of clones bearing the DNA fragment/gene of interest among the numerous transformants within the library. If one or more details about their gene such as (i) the protein encoded by the gene they want to clone, (ii) the DNA sequence of a similar or homologous gene cloned from a different organism and (iii) the gene of interest has been mapped and is known to be closely linked to DNA markers, screening can be done. Screening is based on homology between a PROBE and one of the clones in the library. Gene expression can be used to synthesize a product in particular host cells (e.g. synthesis of human gene products in prokaryotic cells, which can be helpful in screening DNA library).

screening of cDNA libraries: The cloned bacteria are then screened through antibiotic selection. Once selected, stocks of the bacteria are created, which can later be grown and sequenced to compile the cDNA library. Full-length cDNA can be cloned. Various methods for screening of cDNA libraries that use DNA probes include (i) HOMOLOGY SCREENING, (ii) PCR-BASED SCREENS, (iii) SCREENING BY DATABASES, (iv) USING LABELLED OLIGONUCLEOTIDE PROBE BASED ON A KNOWN PEPTIDE SEQUENCE, (v) USING POLYNUCLEOTIDE KINASE AND Γ-^{32}P-LABELLED ATP TO RADIOLABEL OLIGO-NUCLEOTIDE PROBES and (vi) POSITIONAL CLONING. Another method of screening a cDNA library involves the use of polynucleotide kinase and γ-^{32}P-labelled ATP to radiolabel oligonucleotide probes. Screening a cDNA library can also be done based on their biological effects. *See details as separate terms.*

screening of cDNA library based on their biological effects: If a cDNA is thought to encode a soluble factor that has a known biological effect, and if that effect can be easily assayed, then the assay could be used to screen the library. Various approaches used to screen for desired cDNA clones based on their biological effects are FUNCTIONAL COMPLEMENTATION, EXPRESSION ON THE CELL SURFACE WITH ANTIBODY SCREENING, IMMUNOLOGICAL SCREENING, FUNCTIONAL CLONING OF RECEPTORS, PLUS/MINUS SCREENING AND DIFFERENTIAL DISPLAY, TWO-HYBRID SCREENING, PHAGE DISPLAY, CHEMICAL MODIFICATION OF SURFACE PEPTIDES, MULTI-PURPOSE VECTOR SYSTEM and TAGGING. *See details as separate terms.*

screening of cDNA library by databases: The rapid accumulation of sequence information and genetic data often allows scientists to bypass the steps required to isolate cDNAs. If partial protein sequence or partial cDNA sequence is available, searching databases may result in identifying candidate clones that can be ordered and tested to determine if they are the 'right' clone. Databases include the sequence of entire genomes as well as short sequences from cDNAs that serve to tag individual clones (ESTs = EXPRESSED SEQUENCE TAGS).

screening overkill: When screening a large population for horizontal resistance, there is a risk, in the early breeding cycles, that every individual will be killed and the entire breeding population lost. This can be prevented by using natural crop protection measures guaranteeing that the least susceptible plants produce at least a few seeds.

screening technique: A technique used to determine the genotype or phenotype of an organism.

screening the chloroplast transgenic plants: PCR analysis is used to screen the transgenic plants and distinguish true chloroplast transgenic plants from mutants or nuclear transgenic plants. Site-specific chloroplast integration of the transgene cassette is determined by using a set of primers, one of which anneals to the native chloroplast genome and the other anneals within the transgene cassette.

scrunching model of abortive transcription initiation: It invokes a flexible element in DNA. In each cycle of abortive initiation, RNA polymerase unwinds downstream DNA and pulls it into itself, accommodating the accumulated DNA as single-stranded bulges in the unwound region; upon release of the abortive RNA, RNA polymerase extrudes the internalized DNA.

scurf: Minute, loose, membranous scales observed on the surface of some plant parts (e.g. leaves).

seamless cloning: Seamless cloning methods typically combine attributes of more established cloning methods to create a unique solution to allow sequence-independent and scarless insertion of one or more fragments of DNA into a plasmid vector. Various commercial systems employ PCR to amplify the gene of interest, an exonuclease to chew back one strand of the insert and vector ends, and either a ligase, recombination event or *in vivo* repair to covalently join the insert to the vector through a true phosphodiester bond. The ability to quickly join a single insert to a plasmid at any sequence in the vector without a scar makes these technologies very appealing cloning methods.

seamless ligation cloning extract (SLiCE): This cloning method utilizes easy-to-generate bacterial cell extracts to assemble multiple DNA fragments into recombinant DNA molecules in a single *in vitro* recombination reaction. SLiCE facilitates seamless cloning by recombining short-end homologies (≥15 bp) with or without flanking heterologous sequences and provides an effective strategy for directional subcloning of DNA fragments from BACs or other sources.

search algorithm: Any algorithm that resolves the search problem, namely, to recover information stored within some data structure, or calculated in the search space of a problem domain, either with distinct or continuous values. Searching algorithms are employed for searching elements or groups of elements from a particular data structure. They can be of different types based on their method or the data structure in which the element should be instituted.

searching for other proteins with related functions: Once a function has been assigned to an individual protein, one can search for other proteins with related functions by seeking proteins whose amino acid sequences are similar to the original protein. This homology method is used widely to extend knowledge of protein function from one protein to its cousins, which are presently described from the same common ancestral protein. The powerful BLAST programme is used to extend experimental knowledge of protein function to new sequences in this way.

seasonal variation: Variation found among individuals of a species for a morphological biochemical or molecular trait cyclically in successive seasons through genetic reconstruction.

second code: For this, it is important to recognize that chromosomes are not merely DNA molecules but DNA-protein complexes, termed together as chromatin. Chromatin can be reversibly altered in many ways. For instance, the DNA can be methylated on its cytosines. The histone proteins that pack the DNA can also be chemically modified; the positions of histones and/or other chromatin proteins along the DNA can also be changed. Distinct histone protein variants can be engaged at specific sites along the DNA. These and other reversible modifications impact whether genes are turned on or off.

second genetic code: C. de Duve in 1988 proposed the existence of a second code, which is engraved in the structure of the aminoacyl synthetases (AASs). It matches the amino acid with the structural features of the tRNA, the phenomenon known as PARACODONS. The main features of this code are that it is non-degenerate, deterministic and older than the classical genetic code. Correct recognition of transfer RNAs by aminoacyl-tRNA synthetases is central to the maintenance of translational fidelity. The anticodon is indeed important for some of the 20 *Escherichia coli* isoaccepting groups. For many of the isoaccepting groups, the acceptor stem or position 73 (or both) is important as well. For accurate translation of genetic messages, the precision of two matchings is very important: first, that of amino acids with transfer RNA and, second, that of tRNA with mRNA. The later matching is a straightforward interaction of the codon of mRNA with anticodon in tRNA but the first matching is indirect and is mediated by specific enzymes, the AASs. These enzymes are specific for each amino acid, there being 20 AASs for 20 essential amino acids. Still lot of work is necessitated to properly define paracodons. *See* PARACODONS.

second meiotic division: Second meiotic division is in principle a mitotic (equational) division of chromosomes except that each dividing cell has only one set of homologous chromosomes.

second messenger: Small molecule that is formed in or released into the cytosol in response to an extracellular signal and helps to relay the signal to the interior of the cell. Examples include (i) cyclic AMP (cAMP), (ii) inositol triphosphate (IP3) and (iii) calcium ions (Ca^{2+}). *Aka* SMALL INTRACELLULAR MEDIATOR.

second polar event: The union of a male gamete with two haploid female nuclei at the time of the fertilization of the ovule. The resulting triploid nucleus divides to form the endosperm.

second site reversion: A second mutation that reverses the effect of a previous mutation in the same gene although without reinstating the original nucleotide sequence.

second wild ancestor of maize: It took more than a century for scientists to settle on the idea that it was domesticated about 9000 years ago in the lowlands of Mexico from a wild grass: a subspecies of teosinte called *Zea mays* ssp. *parviglumis*. N. Yang *et al.* (2023) report that between 15% and 20% of genes in all existing maize varieties come not from *Z. mays* ssp. *parviglumis*, but from a highland subspecies of teosinte called *Z. mays* ssp. *mexicana*, which hybridized with maize some 4000 years after people first domesticated the plant.

second-division restitution (SDR): SDR occurs when meiosis fails to separate sister chromatids. Due to premature cytokinesis before the second meiotic division takes place. Sister chromatids end up in the same nucleus. *Compare with* FIRST-DIVISION RESTITUTION.

second-division segregation (SDS): When two members of an allelic pair do not separate during meiosis I but rather at meiosis II. SDS produces 1:1:1:1 ratio during tetrad analysis. *Compare with* FIRST-DIVISION SEGREGATION.

second-generation molecular markers: DNA markers such as MICROSATELLITES (SSRs) and AMPLIFIED FRAGMENT LENGTH POLYMORPHISMS (AFLPs). *See details as separate terms.*

second-generation sequencing: *See* NEXT-GENERATION SEQUENCING.

secondary constriction: A non-centric constriction. For example, nucleolar organizing regions (NORs).

secondary contact: Contact between populations that had previously been geographically isolated (i.e. allopatric). *Contrasts* PRIMARY CONTACT.

secondary data: Data that have already been collected by someone, may be sorted, tabulated and has undergone a statistical treatment.

secondary DNA structure repeats: DNA often contains reiterated sequences of differing lengths. These include direct (e.g. GAAT-N6-GAAT) and inverted (GAAT-N6-ATTC) repeats. The later, if sufficiently close may form stable stem-loop structures. For secondary structures of RNA or DNA, highly recommend are Michael Zuker's sites: (i) MFold is used for RNA folding. (ii) Vienna RNA secondary structure prediction is useful for drawing tRNAs in cloverleaf format. (iii) pknotsRG is a series of three tools for folding RNA secondary structures, including the class of simple recursive pseudoknots. (iv) REPuter is used for fast computation of maximal repeats in complete genomes and gives an interesting graphical representation of repeats. (v) REPFIND, on sequences of less than 20 kb, it provides graphical and statistical analysis on direct repeats. (vi) EMBOSS (inverted, palindrome and tandem sequences) finds inverted and tandem repeats. (vii) CRISPRfinder presents a curious repeat structure found in many prokaryotic genomes. They show characteristics of both tandem and interspaced repeats.

secondary gene pool: Gene pool that consists of those species with a certain degree of hybridization barrier due to ploidy differences, chromosome alterations or incompatibility genes.

secondary growth: The growth in vascular plants from lateral meristems leading to wider stems and branches.

secondary immune response: Adaptive immune response to an antigen that is realized on a second or subsequent encounter with a given antigen. It is more rapid in onset, stronger and more specific than the primary immune response. *Compare with* PRIMARY IMMUNE RESPONSE.

secondary infection: Any infection triggered by inoculum produced as a result of a secondary or a subsequent infection and/or an infection caused by secondary inoculum. *Compare with* PRIMARY INFECTION.

secondary inoculum: The inoculum formed by infections that happened during the same growing season. *Contrasts* INITIAL INOCULUM *and* PRIMARY INOCULUM.

secondary metabolites: The metabolites that are produced by plants but not utilized by them, and thus not directly implicated in the normal growth, development and reproduction processes, but usually have important ecological functions. Antibiotics, pigments, alkaloids, caffeine, capsaicin, aroma and other metabolites of medicinal and aromatic value derived from plants are important secondary metabolites. *Compare with* PRIMARY METABOLITES.

secondary non-disjunction: A type of secondary non-disjunction that occurs during anaphase II. It is in reality a chromatid non-disjunction. In the XX-XY system of sex determination, the production of eggs with two XX chromosomes and a Y chromosome by an XXY individual.

secondary oocytes: The cells formed by meiosis I in female animals.

secondary organism (pathogen): An organism that reproduces in already diseased tissue but is not the primary pathogen.

secondary pairing: The association of bivalents in polyploids due to genetic, evolutionary or structural factors. For any reason, such bivalents give the impression of being in groups. Sometimes it seems that the bivalents of a certain genome are closer together than at random, or the bivalents of a certain genome occupy certain domains (spatial order) within the meiotic cell (prometaphase and metaphase).

secondary party: An entity that further develops or uses an off-patent event.

secondary phloem: The phloem formed by the vascular cambium in stems and branches of woody plants.

secondary piwi-interacting RNAs (spiRNAs): The product of a ping-pong amplification cycle that couples piRNA biogenesis to target silencing. These piRNAs are antisense to primary piRNAs and have a preference for an A at position 10. *Compare with* PRIMARY PIWI-INTERACTING RNAS.

secondary pollutant: An air pollutant that must be chemically produced from other air pollutants, such as ozone (O_3), which is a product of a photochemical reaction of exhaust products from combustion engines in the atmosphere. *Contrasts with* PRIMARY POLLUTANTS.

secondary protein structure: Refers to the shape (recurring arrangement in space in one dimension) of the individual polypeptide chain. In some cases, because of its primary structure, the chain may have an extended or longitudinally coiled conformation.

secondary reinforcement: The reinforcement of prezygotic barriers to hybridization after secondary contact between isolated populations. *See* DISRUPTIVE SELECTION *and* STABILIZING SELECTION.

secondary response: The immunological response to a second challenge with a particular antigen. Secondary responses give rise to much higher levels of antibody in a much shorter time than do the primary responses.

secondary root: A branch emerging from a primary root.

secondary spermatocytes: The cells formed by meiosis I in male animals.

secondary storage: This differs from PRIMARY STORAGE in that it is not directly reachable by the CENTRAL PROCESSING UNIT (CPU). The computer usually uses its input/output channels to access secondary storage and transfer the desired data to primary storage. Secondary storage is non-volatile (retaining data when power is shut off). Modern computer systems typically have two orders of magnitude more secondary storage than primary storage because secondary storage is less expensive. *Aka* EXTERNAL MEMORY *or* AUXILIARY STORAGE.

secondary structure: The folded, coiled or perverse shape of a polypeptide that results from hydrogen bonding between parts of a molecule. There are two main types of secondary structure: an α-helix and a β-pleated sheet. Local structure within a protein, which is deliberated by the nature of the side chains of adjacent amino acids (e.g. alpha helix, beta sheet and random coil). Local structure within an RNA molecule that is deliberated by base pairing of nucleotides, which are relatively closely positioned within the sequence (e.g. hairpins, stem-loop structures). *See* PRIMARY STRUCTURE, TERTIARY STRUCTURE *and* QUATERNARY STRUCTURE.

secondary trisomic: A trisomic having additional chromosome as isochromosome.

secondary xylem: The xylem formed by the vascular cambium in stems and branches of woody plants. *Aka* WOOD.

secretion of proteins synthesized on free ribosomes: Free ribosomes synthesize specific non-serum liver proteins. Leader sequences on mRNAs likely determine which ribosomes shall translate them.

secretion of proteins synthesized on membrane-bound ribosomes: In eukaryotes, ribosomes may lie free or attached to the endoplasmic reticulum (ER). Membrane-bound ribosomes synthesize proteins that are carried from the cell into the surrounding serum. 60S subunit of ribosome is attached to ER. The ER membrane is used to form a vesicle containing the protein. This vesicle then becomes secretary message whose membrane fuses with the cell membrane, discharging its contents outside the cell.

secretion system, type I (T1SS): It consists of only three protein subunits: the ABC protein, membrane fusion protein (MFP) and outer membrane protein (OMP). Type I secretion system carries various molecules, from ions, drugs, to proteins of various sizes (20–900 kDa). The molecules secreted vary in size from the small *Escherichia coli* peptide colicin V of 10 kDa to the large *Pseudomonas fluorescens* cell adhesion protein LapA of 900 kDa.

secretion system, type II (T2SS): T2SS relies on the Sec or Tat system for initial transport into the periplasm. Once there, they pass through the outer membrane via a multimeric (12–14 subunits) complex of pore-forming secretin proteins. In addition to the secretin protein, 10–15 other inner and outer membrane proteins comprise the full secretion apparatus.

secretion system, type III (T3SS): A bacterial system for transporting toxic proteins into the cells of their host. It is akin to a molecular syringe through which a Gram-negative bacterium can inject proteins into eukaryotic cells. The low Ca^{2+} concentration in the cytosol opens the gate that regulates T3SS.

secretion system, type IV (T4SS): This secretion system is homologous to conjugation machinery of bacteria. It was discovered in *Agrobacterium tumefaciens*, which uses this system to insert the T-DNA portion of the Ti plasmid into the plant host, which in turn reasons the affected area to develop into a crown gall (tumour).

secretion system, type V (T5SS): Also known as AUTOTRANSPORTER SYSTEM, type V secretion system encompasses the use of the Sec system for crossing the inner membrane. Proteins that use this pathway have the ability to form a β-barrel with their C-terminus, which inserts into the outer membrane, enabling the rest of the peptide (the passenger domain) to reach the outside of the cell.

secretion system, type VI (T6SS): T6SS gene clusters contain from 15 to >20 genes, two of which, *Hcp* and *VgrG*, have been displayed to be nearly universally secreted substrates of the system. Structural analysis of these and other proteins in this system bear a striking similarity to the tail spike of the T4 phage.

secretion vectors for use in *Bacillus subtilis*: *Bacillus* signal sequences are longer than those from Gram-negative bacteria and eukaryotes. Still these signal sequences can be used in an identical way to promote secretion of cloned genes' products.

secretomics: *See* PROTEOMICS IN MEDICINE AND HEALTH.

secretory proteins: Secretory proteins are transported in coated vesicles. A vesicle is generated by pinching off from a membrane, it fulfils its function by fusing with another membrane to release its contents. Some proteins have constitutive secretion while others are secreted through exocytosis. Proteins are imported into the cell by endocytosis. Exocytosis is the pathway for regulated secretion of proteins. The predominant protein in coat of vesicles is clathrin. The specificity of clathrin-coated vesicles presumably is conferred by other proteins located in coats.

secretory tissue: The tissues connected with the secretion of gums, resins, oils and other substances in plants.

secretory vesicle: Membrane-bounded organelle in which molecules destined for secretion are stored prior to release. Sometimes called SECRETORY GRANULE because darkly staining contents make the organelle visible as a small solid object.

section: A very thin slice of tissue, appropriate for observing under the microscope.

sector: An area of tissue whose phenotype is detectably different from the phenotype of the surrounding tissue.

securin-separase complex: Sister chromatids are held together by cohesins. At anaphase, separase is activated by degradation of its inhibitory partner, securing. Separase then cleaves cohesins, thus allowing sister-chromatid separation.

sedentary: Living in a fixed location. *Contrasts* MIGRATORY.

sedimentary rock: Rock formed from precipitated minerals such as calcite, or from lithified particles eroded from other rocks such as sandstone, shale and conglomerates.

sedimentation: A process by which mass eroded from elsewhere settles to the bottom of rivers, lakes or an ocean.

sedimentation coefficient (S): A value used to express the velocity with which a molecule or structure sediments when centrifuged in a dense solution. *Aka* UNIT.

seed: A mature ovule in the sexual reproductive cycle of gymnosperms and angiosperms consists of an embryonic plant together with a store of food, all surrounded by a protective coat, which can produce a new plant under suitable conditions.

seed alignment: Alignment that contains only one of each pair of homologues that are represented in a CLUSTAL-W-derived phylogenetic tree linked by a branch of length less than a distance of 0.2.

seed bank: (i) A place or storage in which seeds of rare plants or obsolete varieties are preserved, usually vacuum packed and under cold conditions to prolong their viability. (ii) An organization that preserves and maintains species and cultivars of plants for use by plant breeders. (iii) The collection of non-germinated seeds in a soil (weed seed bank).

seed breeder's rights: National and international rules and laws that offer plant breeders a legal means to apply for proprietary rights to cultivated plant varieties they have bred.

seed broadcasting: A method of sowing seeds in which the seeds are scattered randomly rather than planted in rows.

seed certification: A procedure developed as a way of guaranteeing that seeds have a high standard of purity and quality.

seed multiplication: All procedures mandatory to grow plants to maturity and produce seeds, including those practices necessary for harvesting, processing and preparing seeds for subsequent plantings.

seed protein quality in plants, modification of: There are two groups of approaches used to achieve this objective: (i) Introduction of an appropriate transgene. (ii) Modification of the endogenous protein-encoding gene. Cereal seeds are deficient in lysine while those of pulses are deficient in sulfur-containing amino acids (e.g. methionine and tryptophan). This limits their nutritional value for man since these amino acids are essential for man. Improvement of seed protein quality is an important objective. Gene encoding for sunflower *albumin 8* (*SFA8*) has been fused with *vicilin gene* promoter, which confers this gene with correct developmental and tissue-specific (seed-specific) expression and accumulation of protein in seeds of transgenic tobacco. The 2S seed protein of Amaranthus (Amal) is rich in methionine. Gene *ama1* has been transferred into potato, and the protein Ama1 was found to constitute about 0.3% of the total soluble protein.

seed region: The first 8–12 bp of a protospacer immediately adjacent to the PAM sequence that is required together with the PAM for successful targeting.

seed treatment: The application of a biological agent, chemical substance or physical treatment to seed to protect the seed or plant from pathogens or to stimulate germination or plant growth.

seed vigour: Seed properties that govern the potential for rapid uniform emergence and development of normal seedlings under both favourable and stress conditions.

seedborne pathogens: Pathogens carried on or in seeds, for example, in wheat, the streak mosaic virus of barley, the fungi, such as snow mould (*Fusarium nivale*), Septoria spike blotch (*Septoria nodorum*), *Helminthosporum* leaf blotch or spot blotch (*Cochliobolus sativus*, *Helminthosporium sativum*, syn *Bipolaris sorokiniana*, *Drechslera sorokiana*), loose smut (*Ustilago nuda* or *Ustilago tritici*), common smut or stinking smut (*Tilletia caries*) and dwarf smut (*Tilletia controversa*); in barley, the stripe mosaic virus, the fungi, such as snow mould (*Fusarium nivale*), leaf stripe disease (*Pyrenophora graminea*), *Helminthosporium* leaf blotch (*Helminthosporium gramineum*, syn *Drechslera graminea*), net blotch disease (*Pyrenophora teres*, syn *Helminthosporum teres* and syn *Drechslera teres*), loose smut (*Ustilago nuda*), black smut (*Ustilago nigra*) and hard smut (*Ustilago hordei*); in rye, the streak mosaic virus of barley, the fungi, such as snow mould (*Fusarium nivale*, syn *Griphosphaeria nivalis*), *Septoria* spike blotch (*Stagonospora nodorum*), stalk bunt (*Urocystis occulta* syn *Tuburcinia occulta*) and ergot (*Claviceps purpurea*); in oats, the streak mosaic virus of barley and the fungi, such as loose smut (*Ustilago avenae*); and in maize, the fungi, such as common smut (*Ustilago maydis*), seed rots (*Fusarium* spp. and *Penicillium* spp.) and seedling rots (*Pythium* spp., *Fusarium* spp., *Helminthosporum* spp., *Penicillium* spp., *Rhizopus* spp. *Rhizoctonia* spp. and *Deploida* spp.).

seeding efficiency: *See* ATTACHMENT EFFICIENCY.

seedling emergence mapping: Seedling emergence measurement and mapping is a common agricultural application of imagery collected using SMALL UNMANNED AERIAL SYSTEMS (sUAS). If crop germination is delayed or unsuccessful in portions of the field due to environmental factors, there is often a short window after plant emergence during which the producer could attempt replanting. At the early stages of crop development, the field can be mapped at very high resolution in order to view the seedlings and identify zones in which germination was unsuccessful. The required resolution depends upon the size of the plants' leaves just after emergence.

seedling resistance genes: The genes that confer resistance against a pathogen to a host at seedling stage.

seekRNA: IS*1111* and IS*110* insertion sequence (IS) family members encode a rare DEDD transposase type and display specific target site selection. Here, DEDD stands for four amino acids – Asp, Glu, Asp, Asp. The IS*1111* group comprises identifiable subterminal inverted repeats (sTIR) not found in the IS*110* type. IS in both these families contain a NON-CODING REGION (NCR) of substantial length and, as each individual IS or group of closely related IS selects a different site. It had been previously suggested that an NCR-derived RNA was involved in target selection. R. Siddiquee and co-workers in 2024 find that the NCR is usually downstream of the transposase gene in IS*1111* family IS and upstream in the IS*110* type. Four IS*1111* and one IS*110* family members that target different sequences are used to demonstrate that the NCR determines a short seeker RNA, called seekRNA, that co-purifies with the transposase. The seekRNA is indispensable for the transposition of the IS or a cargo bordered by IS ends from and to the preferred target. Short sequences matching both top and bottom strands of the target exist in the seekRNA but their direction in IS*1111* and IS*110* family IS reversed. Reprogramming the seekRNA and donor flank to target a different site has been demonstrated, indicating future biotechnological potential for these systems.

SEG: A programme for filtering low complexity regions in amino acid sequences. Residues that have been masked are represented as 'X' in an alignment. SEG filtering is performed by default in the blast subroutine of BLAST 2.0.

segment: (i) A part or subdivision of an organ (e.g. a petal is a segment of the corolla). (ii) A term sometimes used when the sepals and petals are indistinguishable.

segment genes: These genes are expressed after fertilization and decide the number of segments and their polarity. Mutation in a segmentation gene can change the segments concerning their number, polarity and so on. These genes are expressed after fertilization. Mutations in these genes alter the polarity of the segment. Segment genes are further classified GAP GENES, PAIR RULE GENES and SEGMENT POLARITY GENES. The functions of segmentation genes are to establish 'rules' by which segments form. *See details as separate terms.*

segment polarity genes: Mutations in segment polarity genes lead to loss of parts of the segment. Segment polarity genes are expressed in serial PARASEGMENTS and in the same relative position in each one and are responsible for the specification of cell types within the parasegments. These genes maintain and reinforce metameric pattern of parasegments by reciprocal signalling and establish the fate of the individual cells within each parasegment. These mutants have lost part of the 'P' compartment of each segment, and it has been replaced by a mirror-image duplication of the A compartment. Examples of segment polarity genes in *Drosophila* are *engrailed, wingless* and *gooseberry*.

segmental allopolyploid: An allopolyploid that contains some homologous segments between homoeologous chromosomes, in which meiotic pairing occurs between homologous chromosomes as well as homoeologous chromosomes.

segmentation gene: In eukaryotes, any of the genes that determine the number or polarity of body segments.

segments and parasegments: Mutations in segment genes alter the number or polarity of body segments. *Drosophila melanogaster* development can be considered in terms of two types of units – the segment and parasegment. The segment is a visible morphological structure. Primarily, there are three thoracic (T) and eight abdominal (A) segments. Another type of division has been suggested at blastoderm, when dividing lines can be seen between parasegments, each consisting of a P compartment succeeded by A compartment.

segregating population: A population, generally the progeny of a cross, in which genetic differences are noticeable, thus enabling identification of individuals having a desired trait and their selection for further breeding.

segregating sites: Sites that are polymorphic in a sample of sequences.

segregation: The movement of two homologous chromosomes during meiosis, one to each pole of the cell so that each gamete encompasses only one member of each pair of alleles. Mendel described segregation as the factors or genes making up pairs are separated out at the time of gamete formation in animals or spore formation in plants. In other words, only one gene or one factor of each pair goes into a given gamete or spore. The separation of two homologous chromosomes of each bivalent takes place at anaphase I of the meiosis. In breeding experiments, the F_1 hybrids are uniform and the F_2 generation is the first segregating generation. Segregating generations cannot be bulked to release as varieties in the self-pollinated crops.

segregation analysis: To compare the phenotypes of individuals in families to genotypes to determine the most likely mode of inheritance of a particular disease, or if a genetic variant is inherited by all individuals in a family with a particular disease. Can also refer to disjunction of homologous chromosomes at meiosis.

segregation distorter (SD): A factor that alters the segregation ratio in heterozygous *Drosophila*; SD^+ cells are eliminated during spermiogenesis in SD^+/SD males. It comprises both meiotic drive and differential survival of the haploid products of meiosis.

segregation distortion: Any systematic variation from equal representation of alleles or homologous chromosomes among the functional gametes. It leads to a meiotic drive. Deviation of segregation ratios from expected Mendelian ratios. During segregation analysis, distortion has been very commonly reported in a wide range of plants. Segregation distortion occurs either because of association with gametophytic/sterility genes, meiotic drive/preferential segregation as in the case of rice (Indica × Japonica crosses). It can also be specific concerning some markers in an otherwise normal mapping population.

segregation patterns and scoring methods: Segregation patterns for codominant and dominant markers in five types of mapping populations (F_2 population, **BC** populations derived by crossing F_1 to P_2, double haploids, recombinant inbred lines and near-isogenic line populations). The scoring method is different between codominant and dominant markers.

segregation ratio: The expected (Mendelian) or observed ratio between genotypes or phenotypes in the progeny of a cross.

segregational load: Genetic load exerted when the population is segregating fewer fit homozygotes because of heterozygote advantage.

segregational petites: These petites bear mutant nuclear gene(s).

segregative instability: The loss of plasmids due to defective partitioning is called segregative instability. Partitioning region *par* is essential for maintaining the stability of low copy number plasmids. ColE1 also contains *par* region that is deleted in pBR322, which is segregating randomly at cell division. Region *par* can be cloned into pBR322 to increase its stability.

selectable marker gene (SMG): SMGs are typically the genes for antibiotic resistance, which give the transformed organism (usually a single cell) the ability to live in the presence of an antibiotic. SMGs can be divided into several categories depending on whether they confer positive or negative selection and whether the selection is conditional or non-conditional in the presence of external substrates. Examples of SMGs that are frequently used to select transformed plant tissues include *nptII*, *hpt*, *bar* and *gox*, which confer resistance to kanamycin, hygromycin, phosphinothricin and glyphosate, respectively. SMGs are used in genetic engineering to facilitate the identification of cells that have incorporated another desirable trait that is not easy to identify in individual cells. *Aka* SELECTABLE MARKER. *See* SELECTIVE AGENT.

selectable marker gene (SMG) removal by cotransformation-based system: This method for SMG removal is based on the cotransformation of plants using two distinct transgene constructs present in the same transformed line of *A. tumefaciens*. One of these constructs contains the selectable marker transgene to be used, while the other includes the desired trait transgene, itself unlinked to any marker gene. Alternatively, these two transgenes are inserted into two different T-DNA elements present in the same 'super binary' plant transformation vector. Cotransformation methods for marker gene removal are based on the principle that a proportion of transformed plants carrying the SMG will also have integrated the required transgene at a second, unlinked insertion site. Marker genes can subsequently be removed from such plants by genetic segregation.

selectable marker gene (SMG) removal by Cre-*lox* recombinase system: The most common system used to mediate site-specific recombination in plants is the bacteriophage P1 Cre/*lox*. In this system, the plant should be previously transformed with an SMG cloned between two sequences of the *lox* gene, each with 34-bp repeats in direct orientation. In the second stage, the *Cre* gene should be introduced in this plant by a second transformation, either by sexual crossing or by transient expression. Once the *Cre* gene is expressed, the Cre enzyme catalyses the recombination between the *lox* repeat sequences, thus eliminating the marker gene in the secondary transformants.

selectable marker gene (SMG) removal by FLP-*FRT* recombinase system: The FLP recombinase, encoded by a gene of the *Saccharomyces cerevisiae* 2μ plasmid, catalyses the excision of sequences flanked by directly repeated *FRT* sites.

selectable marker gene (SMG) removal by genome-editing enzymes: Genome-editing enzymes – ZINC-FINGER NUCLEASES (ZFNs), meganucleases (also called HOMING NUCLEASES) and TRANSCRIPTION ACTIVATOR–LIKE EFFECTOR NUCLEASES (TALENs) – use the cell's natural repair system for the removal of SMGs from transgenics.

selectable marker gene (SMG) removal by intrachromosomal homologous recombination (ICR): Removal of nuclear marker genes in plants by this approach is based on intrachromosomal recombination between two directly repeated sequences flanking the marker gene to be excised.

selectable marker gene (SMG) removal by meganucleases: *Aka* HOMING NUCLEASES. Meganuclease engineering has been reported for transgene deletion. By using two I-*Sce*I cutting sites flanking an SMG in a transgene construct, it is possible to induce double-strand breaks on both restriction sites to release the SMG upon the expression of I-*Sce*I.

The cut ends were rejoined through a non-homologous end-joining (NHEJ) repair pathway.

selectable marker gene (SMG) removal by R-*Rs* recombinase system: The R recombinase of *Zygosaccharomyces rouxii* acts similarly to catalyse recombinatorial excision between directly repeated Rs sites. To construct this system, first, the R recombinase gene was completely re-synthesized to create a plant-adapted version of the gene. Next, the C-terminus of the R recombinase gene was fused to the ligand-binding domain (LBD) of the rat glucocorticoid receptor (GR) to achieve stringent post-translational regulation of recombinase activity.

selectable marker gene (SMG) removal by site-specific recombination (SSR) systems: This method for marker gene removal involved the heterologous expression of microbial recombinase enzymes in plants to excise marker transgenes that were flanked by microbial recombination sequences. Various microbial recombinase systems used for this purpose are CRE-LOX SYSTEM, FLP-FRT SYSTEM and R-RS SYSTEM.

selectable marker gene (SMG) removal by transcription activator–like effector nucleases (TALENs): Two identical sets of TALEN-binding sequences are designed to flank a SMG in a transformation vector. After the expression of TALEN, the double-strand break will be induced at both TALEN-binding sequences and release the SMG. This strategy can be devised for maker-free transgenic plant production or native chromosome fragment removal.

selectable marker gene (SMG) removal by transposon-based method: In this method, heterologous plant transposons are used for the removal of marker genes. In one such system, the maize Ac transposable element was engineered to contain the *isopentenyl transferase* (*ipt*) gene, conferring a selectable 'extreme shooty phenotype'. The Ac element encodes its transposase and so its excision conveniently removes this gene along with the *ipt* marker gene, thereby obviating the need for sexual reproduction steps in the procedure.

selectable marker gene (SMG) removal by zinc-finger nucleases (ZFNs): ZFNs were also deployed for gene excision in animal and plant cells. ZFNs can be used whereby two concurrent double-strand breaks introduced by ZFNs should be repaired such that the broken ends are religated by NHEJ and the intervening chromosomal segment is excised.

selectable marker gene (SMG) removal from the chloroplast genome by cotransformation-segregation: To determine whether targeted DNA insertion into the tobacco plastid genome can be obtained without physical linkage to an SMG, biolistic transformation of chloroplasts was carried out in tobacco leaf segments with a 1:1 mix of two independently targeted antibiotic resistance genes. Plastid transformants were selected by spectinomycin resistance due to expression of an integrated *aadA* gene. Integration of the unselected kanamycin resistance (*kan*) gene into the same plastid genome was established by Southern probing in ≈20% of the spectinomycin-selected clones. Efficient cotransformation facilitates targeted plastid genome modification without physical linkage to a marker gene.

selectable marker gene (SMG) removal from the chloroplast genome by homologous recombination systems: As the integration of foreign transgenes into the chloroplast genome takes place by homologous recombination, it is entirely logical to test this native plant mechanism as a means for the removal of marker genes from the chloroplast. Homoplastic marker-free plants are identified in this technique by PCR or by Southern blot analysis.

selectable marker gene (SMG) removal from the chloroplast genome by phage site-specific recombinases: Site-specific recombination systems including Cre-*lox* and φC31-*att* are used to remove an SMG inserted into plastid genomes.

selectable marker gene (SMG) removal from the chloroplast genome by transient co-integration of the marker gene: Marker-free chloroplast transformants were generated in tobacco using the reconstitution of wild-type pigmentation in combination with plastid transformation vectors, which prevent stable integration of the kanamycin selection marker. One benefit of this procedure using mutants is that marker-free plastid transformants can be produced directly in the first generation (T_0) without re-transformation or crossing.

selectable marker gene (SMG) removal from the chloroplast genome: Four protocols have been established to date for obtaining marker-free transplastomic plants. These protocols are EXCISION BY PHAGE SITE-SPECIFIC RECOMBINASES, HOMOLOGY-BASED EXCISION VIA DIRECT REPEATED SEQUENCES, COTRANSFORMATION–SEGREGATION and TRANSIENT

CO-INTEGRATION OF THE MARKER GENE. *See details as separate terms.*

selectable markers for mitochondrial transformation: Potential selectable markers exist for mitochondrial transformation in higher plants. The main challenge to mitochondrial transformation remains the identification of an acceptable selectable marker.

selectin: Member of a family of cell-surface carbohydrate-binding proteins that mediate transient, Ca^{2+}-dependent cell-cell adhesion in the bloodstream, for example between white blood cells and the endothelium of the blood vessel wall.

selection: (i) Differential (non-random) reproduction of different genotypes that induce evolutionary changes. Selection could be natural (SURVIVAL OF THE FITTEST) like in evolutionary processes where selection ensues if zygotic genotypes differ concerning fitness, or by man (ARTIFICIAL SELECTION). This includes the use of particular conditions (e.g. the presence of ampicillin) to allow survival only of cells with a particular phenotype (e.g. production of beta-lactamase). Artificial selection is also used in genetic engineering to identify a genetically modified organism due to its new characteristics (e.g. resistance to antibiotics). (ii) The stage of a genetic algorithm in which individual genomes are chosen from a population for later breeding (using the crossover operator).

selection at subindividual level: Selective processes discriminate between reproductive biological units lower than the individual organism in the hierarchy of organization.

selection coefficient (s): In population genetics, refers to differences in relative fitness. Selection coefficients are central to the quantitative description of evolution since fitness differences determine the change in genotype frequencies attributable to selection.

selection criteria: The precise characters and plant reactions on which the selection is focused during the breeding cycles.

selection differential (S): Difference between the mean value of the population and the mean value of the parents that give rise to the subsequent generation.

selection efficiency (SE): The number of independent transgenic events in 100 regenerated events. The difference 100-SE is often referred to as the ESCAPE RATE.

selection gain: In artificial selection, the difference in mean phenotypic value between the progeny of the selected parents and the parental generation.

selection gradient (β): The gradient of a regression of fitness on trait value.

selection index (pl. indices): Selection indices help to select the best individuals for the next breeding cycle based on observed phenotypic values for several traits of each candidate individual. They assign subjective economic weights to each trait and are relatively simple to analyse. Their shortcomings are that they need large amounts of information, economic weights are tough to assign, sampling error can be large and the statistical sampling properties of selection indices of the selection response are unknown except in the case of two traits.

selection intensity: Expressed as the ratio of the number of genotypes selected and the number of genotypes tested.

selection limit: The collapse of genetic variance in a population, so that no further selection response can be expected.

selection pressure: The effectiveness of natural selection in altering the genetic composition of a population over a series of generations.

selection pressure and the molecular clock: The selection pressure that maintains DNA coding sequences governs the distribution and spectrum of mutations observed. Since any amino acid change in that area might be deleterious, surviving mutations in coding DNA are base substitutions and are eliminated. Base substitutions occur more commonly at degenerate sites (sites where base substitutions will not alter the sense of the codon, often in the third position of the codon) than at non-degenerate sites (sites where missense mutations arise, causing amino acid replacements).

selection regime: The particular set of relative fitness for alternative genotypes within a population living in a particular environment.

selection response (R): The difference between the mean of the individuals selected to be parents and the mean of their offspring after selection. It is represented by the formula $R = h^2 \times S$ where h = heritability, S = selection coefficient and R = phenotypic difference between the mean of all selected fractions and the mean of total population.

selection sort: An in-place comparison sorting algorithm that has an $O(n^2)$ time complexity where n = number of items), which makes it inefficient on large lists, and generally performs worse than the similar insertion sort. Selection sort is noted for its simplicity and has performance advantages over more complicated algorithms in certain situations, particularly where auxiliary memory is limited. *See* TIME COMPLEXITY.

selection systems in chloroplast transformation: The bacterial *bar* gene, encoding PHOSPHINOTHRICIN ACETYLTRANSFERASE (PAT) and conferring herbicide resistance, has also been tested as a positive plastid-selectable marker. Expression of the *bar* gene in plastid conferred phosphinothricin resistance only when introduced by selection for a linked *aadA* gene. However, the *bar* gene was not found to be suitable for the direct selection of transplastomic lines, even when expressed at a higher level.

selection-mutation equilibrium: An equilibrium allelic frequency resulting from the balance between selection against an allele and mutation recreating this allele.

selectionism: The theory that some class of evolutionary events such as molecular or phenotypic changes have mainly been caused by natural selection.

selectionist: One who believes that most molecular divergence and variation is moulded by selection. *Contrasts* NEUTRALIST.

selective advantage: An advantage for survival of a genotype in a population, and for production of viable progeny as compared to other genotypes, which may display a selective disadvantage with respect to fitness and viability.

selective agent: An environmental or chemical mediator that imposes a lethal or sublethal stress on growing plants, or portions thereof in culture, enabling selection of resistant or tolerant individuals. Antibiotic or herbicides used to select the transgenic cells/plants *in vitro*. During plant transformation, the target tissues like meristem, callus, cotyledon and so on, are multicellular and transformation frequency is very low usually 2–5%. Therefore, *Agrobacterium* treated or particle gun-bombarded tissues contain mostly normal cells and few transformed cells. If such tissues are allowed to grow, then normal cells (non-transformed) mask the growth of the transgenic cells, which do not develop into transgenic plants. This problem has been overcome by the use of SMG(s) during transformation and then use of a selective agent to eliminate the non-transformed cells. For example, in rice transformation, the hygromycin resistance gene is introduced along with the gene of interest such as *Bt*. After transformation hygromycin (antibiotic) at the concentration of 30–50 mg/l is used in the tissue culture medium to select the transgenic cells/plants. Likewise, herbicide bialaphos is used as a selective agent if the *bar* gene is used as an SMG during transformation.

selective amplification: A type of amplification where a part of the genome has an increased copy number. Selective amplification is a structured developmental process occasionally utilized to increase the output of a high-demand gene.

selective amplification of microsatellite polymorphic loci (SAMPLs): A PCR-based marker for studying genetic polymorphism. It is the modification of AFLP in which an AFLP primer with three selective nucleotides is used in combination with a microsatellite primer. SAMPL primer that is 18–20 nucleotides long is based on sequences of two different adjacent SSRs and an associated intervening sequence, known to be found in compound repeats. Many different combinations of restriction endonucleases and primers can be used for the detection of limitless SSR based polymorphisms in the genome. In polyploid wheat where genome size is larger, AFLP results in crowded bands, which are difficult to score, SAMPL reduces the number of bands but resolves higher level of polymorphism.

selective breeding: Human-directed identification and breeding of plants or animals that possess useful traits compared to their ancestors. *See* ARTIFICIAL SELECTION. *Compare with* NATURAL SELECTION.

selective death: A fiasco to survive or reproduce. Also, a loss of fitness is characterizable to differences in genotype.

selective disadvantage: Substandard fitness of one genotype with others in the population.

selective fertilization: The non-random participation of male or female gametes or different genotypes in the formation of zygotes and/or hybrids.

selective gametocide: A treatment that incapacitates certain gametes, such as one that produces male sterility but does not affect the female gametes.

selective gene activation: The first step in differentiation might be expected to follow the selective activation of a battery of genes. It has been observed in the stage-specific puffing pattern of the giant chromosomes of insect larvae. Each stage of development is characterized by particular puffs in each larval tissue. The puff pattern can be modified in a predictable direction by insect hormones.

selective gene expression: Linker histone subtypes might exert selective gene expression. A linker histone H1b can specifically repress the expression of a master regulator of skeletal muscle differentiation, the *MyoD* gene, and thereby restrain the developmental decision to make muscle cells. Repression of the mouse *MyoD* gene by linker histone Hib and the homeodomain protein MSx1 has been observed. The MSx1 homeodomain protein forms a complex with the linker histone Hib, which binds to the enhancer of the *MyoD* gene. This blocks the expression of *MyoD* and prevents the differentiation of the muscle progenitor cells. Decreased production of H1b and Msx1 during development relieves the repression of *MyoD*, enabling skeletal muscle differentiation to proceed.

selective herbicide: A chemical that kills some plants but not others. *Contrasts* NON-SELECTIVE HERBICIDE.

selective mating: Non-random mating. Phenotypic differences influence the separation of the F_2 progeny into distinct types.

selective medium: A culture medium comprising substances that specifically impede the growth of some species of microorganisms over others. Following transformation, the tissues are grown on the selection medium containing an appropriate selective agent that can be antibiotic or herbicide. Two cycles of selection for 2 weeks each result in the elimination of non-transformed cells and transformed cells continue to grow on the selection medium. During plant transformation, the commonly used selective agents are hygromycin, kanamycin (antibiotics) and bialaphos (herbicide).

selective neutrality: The situation in which different alleles of a certain gene deliberate equal fitness.

selective pressures: Environmental forces such as scarcity of food or extreme temperatures that result in the survival of only certain organisms with characteristics that provide resistance.

selective sweep: Increase of neutral alleles by riding with a favourable mutation. This curves variation out of the region of the genome surrounding the favourable mutation.

selective system: Any experimental procedure that augments the recovery of specific genotypes.

selective value: A degree of the fitness of a gene within a genotype or of a genotype within a population. It is proportional to the probability of that gene or genotype surviving and is a function of gene/genotype frequencies.

selective whole-genome amplification (SWGA): SWGA technology preferentially amplifies the genomes of pathogens from complex mixtures of target and host DNA. SWGA does not require the separation of target DNA from background DNA, making it an attractive option for pathogens that cannot be amplified in culture.

selectivity: The selectivity of bioinformatics similarity search algorithms is defined as the significance threshold for reporting database sequence matches. For example, in BLAST searches, the parameter 'E' is interpreted as the upper bound on the expected frequency of chance occurrence of a match within the context of the entire database search. 'E' may be thought of as the number of matches that one expects to observe by chance alone during a database search.

selectivity filter: That part of an ion channel structure that determines which ions it can transport.

selector genes: The genes that cause cells to commit themselves to a particular developmental direction. For example, a homeotic gene influences the development of specific body segments in *Drosophila*.

selenocysteine insertion during protein biosynthesis: A complex between the non-coding Sec-insertion sequence (SECIS), SECIS-binding protein 2 (SBP2) and 40S ribosomal subunit enables Sec-specific elongation factor eEFSec to deliver Sec. T. Hilal *et al.* (2022) have shown that eEFSec and SBP2 do not interact directly but rather deploy their carboxyl-terminal domains to engage with the opposite ends of the SECIS. By using its Lys-rich and carboxyl-terminal segments, the ribosomal protein eS31 simultaneously interacts with Sec-specific transfer RNA (tRNASec) and SBP2, which further stabilizes the assembly. eEFSec is indiscriminate towards L-serine and facilitates its misincorporation at Sec UGA codons.

self-assembling DNA in patterns recognition: Self-assembling DNA can process information, but the computations have been limited to digital algorithms. A self-assembling DNA system has now been designed to perform complex pattern recognition (A. Phillips in 2024). Computations occur at multiple scales, involving molecules, cells, tissues, organs and beyond. For example, computational circuits are embedded in molecular pathways and formed by the neurons of the brain the immune system computes immune responses by processing a vast array of molecular inputs. Understanding such information processing could unlock strategies for treating disease and identify fundamental engineering principles of biology.

self-assembly: (i) Molecular organization without external stimulus. (ii) The capability of certain multimeric biological structures to come together from their component parts through random movements of the molecules and the formation of weak chemical bonds between surfaces with complementary shapes.

self-compatible: A plant that is capable of self-fertilization.

self-complementary nature of DNA: Represents the hydrogen bonding between the sequences of the bases on its intertwined polynucleotide chains (i.e. adenine in one strand pairs with thymine in the other and guanine in one strand pairs with cytosine in the other).

self-fertilization: A process in which sexual reproduction occurs as a result of the fusion of male and female gametes produced by the same individual. *Aka* SELFED. *Contrasts* CROSS-FERTILIZATION.

self-incompatibility: The phenomenon as a result of that a plant with normal pollen and ovules fails to set seed through self-pollination on account of some physiological hindrance that prevents fertilization. Plant is self-incompatible if pollen fails to set seed on the same plant and it is cross-incompatible when pollen is ineffective on another plant. The self-incompatibility acts either through the inhibition of pollen germination or growth of pollen tube. Thus, self-incompatibility restricts self-fertilization and encourages cross-fertilization is the most important way to enforce out-crossing. Self-incompatibility is widespread in different species of Cruciferae, Compositae, Rosaceae and Gramineae. It is exploited for hybrid seed production in crop plants. It shows its influence during the PROGAMIC STAGE OF DEVELOPMENT (i.e. before fertilization ensues, so that the chances of the egg cell being fertilized by foreign pollen remain undamaged).

self-incompatibility alleles: These are alleles of a multiple allelic series, which determine the compatibility of perfectly fertile male and female flowers. If a pollen grain bears a self-incompatibility (S) allele that is also present in the maternal parent, then it will not germinate but if that allele is not present in the maternal tissue, then the pollen grain produces a pollen tube containing the male nucleus, and this tube affects fertilization.

self-incompatibility mechanisms in higher plants: Pollen may nose dive to germinate on the stigma due to one or more of several reasons: (i) Pollen tube growth in the style may be repressed to the extent that pollen fails to reach the ovary. (ii) Pollen tubes of insufficient length may fail to penetrate the ovule. (iii) A male gamete that enters the embryo sac may fail to fuse with the egg cell. (iv) The relative length of stamens and style in a bisexual flower is associated with incompatibility. (5) Most plant species have homomorphic flowers in which the stamens and styles attain comparable lengths. Homomorphic flowers can be gametophytic or sporophytic. (vi) Some self-incompatibility species exhibit heteromorphic flowers in which the stamens and styles attain different heights. (vii) The presence of either pin or thrum flowers is termed DISTYLY.

self-incompatibility system and pollen features, correlations between: J.L. Brewbaker in 1957 revealed several remarkable correlations between a self-incompatibility system and pollen features. GAMETOPHYTIC SELF-INCOMPATIBILITY transpires nearly always in pollen with two nuclei; the stigma of the respective species is usually wet. SPOROPHYTIC SELF-INCOMPATIBILITY ensues in pollen with three nuclei; such pollen germinates badly under natural conditions and its life expectancy is short; the respective stigmata are dry; for the controlling mechanism, it was conceivable to identify glycoproteins at the surfaces of pollen and stigma.

self-incompatibility, genetic basis of: E.M. East and A.J. Mangelsdorf with their studies on *Nicotiana* in 1925 observed that self-incompatibility is caused by a gene (*SI*) with numerous alleles: *SI1*, *SI2*, *SI3*, *SI4* . . . *Sin*. A. Lundquist in 1954 and D. L. Hayman in 1956 discovered, independently of each other, a second self-incompatibility locus (*Z*) in grasses. In such cases, the pollen is only rejected if both plants contain the same alleles in both loci (*SI* and *Z*); such bifactorial systems were also later found in numerous monocot and dicot families.

self-management: The process by which computer systems manage their operation without human intervention.

self-pollination: The acceptance by stigmas of pollen from the same flower or flowers on the same plant, which means they are self-compatible. Also called SELFING. Self-pollination occurs in barley, chickpea, clover cowpea, crambe, crotalaria, guar, field bean, field pea, linseed, jute lentil, lespedeza, millet, mung bean, oats, groundnut, potato, rice, sesame, soybean, tobacco, tomato, vetch, wheat or wheatgrass. *Contrasts* CROSS-POLLINATION.

self-repair: Ability to heal itself without external influence.

self-replicating peptide: A peptide capable of templating its own synthesis autocatalytically. For example, a 32-residue 2-helical peptide based on the leucine-zipper domain of yeast transcription factor GCN4 can act as a self-replicating peptide by amide-bond condensation.

self-replication: (i) An ability of an assembler to replicate itself to a definite amount within a fixed period. (ii) The term is also used to describe a molecule or other structure that can cause its own replication.

self-splicing of rRNA introns: Group II introns are a type of RNA enzyme (ribozyme) that catalyses their own removal from RNA transcripts and addition into new genetic locations. Structural and functional analogies support the hypothesis that group II introns and the spliceosome share a common ancestor. The introns of *Tetrahymena* rRNA precursors are removed autocatalytically in a unique reaction mediated by the RNA molecule itself. Self-splicing of intron in 28S rRNA precursor occurs through a transesterification mechanism. In this process, a guanosine nucleotide acts as a nucleophile and gets attached to the 5′-end of the existing intervening sequence via a normal 3′-5′ phosphodiester bond and then −OH of the 5′ exon would cause cleavage to produce 5′-phosphate and 3′-OH terminus. The ligation occurs without hydrolysis of ATP with the removal of covalent circular intron. GMP initiates the process. A circular intron is created. 5′ splice-site selection by the self-splicing *Tetrahymena* ribosomal RNA precursor requires the formation of a particular RNA helix by base pairing between an intron 'guide sequence' and bases surrounding the 5′ splice junction. Part of the guide sequence subsequently holds the positions of the 3′-end of the cut 5′ exon to attack the 3′ splice junction.

self-sterile: Failure to complete fertilization and seed set after self-pollination.

self-sterility: Incapability of producing seed when self-pollinated. Several alleles, S_1, S_2, S_3, etc., are responsible for this phenomenon. *See* SELF-INCOMPATIBILITY.

selfing: When a pistil is fertilized with pollen from the same plant that bears the pistil. It also refers to seed resulting from such fertilization. *Compare with* NATURAL SELECTION.

selfish DNA: The term given in 1980 by W.F. Doolittle and C. Sapienza, and L.E. Orgel and F.H.C. Crick refers to DNA sequences, which seem to have slight or no function. The term includes all kinds of repetitive DNAs, introns, transposable elements and the DNA sequence present between the genes. Much of the DNA in higher organisms was considered junk. They suggested their very presence to be beneficial in facilitating gene rearrangements, thus increasing evolutionary adaptability or a storehouse from which new functional sequences could be recruited. Selfish DNA supports the process of species divergence. Selfish DNA consists of replicating units and it evolves by multiplication within a genome, exchange between genomes and random genetic drift at reproduction. Many species of repeated sequences dispersed throughout the genome have been mobile through evolutionary time. If mobility is a reality then some members may have important effects in the genome even if they have no normal function. Effect may include enhanced recombination and altering rate of gene expression. *Aka* PARASITIC DNA, NONSENSE DNA, JUNK DNA, IGNORANT DNA *or* INNOCENT DNA.

selfish DNA, function of: Molecular analysis is helping in understanding the function of selfish DNA in the genome. Selfish DNA has however been suggested to be involved in chromosome pairing, control of gene expression, processing of messenger RNA precursors and participation in DNA replication. But so far none of these functions has yet been established.

selfish gene: A gene that upsurges its transmission but decreases the fitness of the individual that carries it.

semaphyll: A structure such as a bract or a sepal (if the remainder of the perianth is inconspicuous), which has become modified to attract pollinators.

semelparity: Term applied when a plant flowers once and then dies. *Aka* 'BIG BANG' REPRODUCTION.

semi-allele: Mutant alleles that are allelic for the function but not the structure (i.e. they are not existing on the same chromosome position).

semi-arid: A region or land characterized by very little rainfall, usually from 250 to 5000 mm annually.

semi-conservative DNA replication: The mode of DNA replication in which each daughter double helix comprises one polynucleotide from the parent and the other one is newly synthesized. In *Vicia faba*, a semi-conservative mode of DNA replication was demonstrated by J.H. Taylor, P.S. Woods and W.L. Hughes in 1957. In *Escherichia coli*, the semi-conservative mode of DNA replication was shown by M. Messelson and F. Stahl in 1958.

semi-discontinuous DNA replication: POLYOMA VIRUS (PyV) replicates discontinuously on both arms of replication forks.

semi-dominance: There is a blending of the two alleles that results in a third phenotype different from the ones of the parents. For example, the cross between a white flower and a red flower generates a pink flower. *Aka* INCOMPLETE DOMINANCE *or* PARTIAL DOMINANCE.

semi-dwarf: In wheat breeding, this term labels individuals or a variety showing intermediate stem length compared to very short mutant lines. During the 1970s, semi-dwarf wheat genotypes generated the so-called 'GREEN REVOLUTION' in agriculture, mainly of third-world countries. Several alleles of *Rht* (*reduced height*) loci produced shorter straw length with better lodging resistance as well as higher spikelet fertility and thus yield. Several dwarf (*Rht*) mutants produced reduced amounts of certain gibberellins within the meristematic cells, which reduce plant height in wheat, rice, maize and thale cress sp. The genes modifying those gibberellic acids yield several proteins that regulate different cell activities. Some of the *Rht* genes that displayed a specific gibberellic acid insensitivity have been sequenced and are available for use in genetic engineering experiments.

semigamy: Male gamete penetrates the egg cell but does not fuse with the egg nucleus. Here segregation of maternal and paternal genes takes place but there is no fertilization.

semi-Latin squares: An (n × n) row-and-column design for k_n treatments where every row and every column of the design comprises a complete set of treatments, and every row-by-column intersection comprises a block of k plots. In 1937, F. Yates gave an example of a 7 × 7 semi-Latin square for 14 varieties, where the 14 varieties were divided into seven pairs and each pair was assigned to a different letter of a 7 × 7 Latin square. Each row-by-column intersection of the Latin square confined a block of two plots, and the two varieties of each pair were randomly allocated to pairs of plots within individual blocks. This design was of the split plot type and some treatment differences were estimated entirely within blocks, whereas other treatment differences were estimated entirely between blocks. Often, accuracy is greater for plots-within-blocks comparisons than for plots-between-blocks comparisons. F. Yates suggested that a better plan would be to assign the varieties to the two sets of letters of a 7 × 7 Graeco-Latin square where the row-by-column intersections of the superimposed squares would represent blocks of size two; each of the seven varieties allotted to one set of letters would occur once in the same block with each of the seven varieties assigned to the other set of letters. F. Yates demonstrated that, where all variety comparisons are equally important, this is the best choice of design for comparisons in the plots-within-blocks stratum. Finally, he proposed that hyper-Graeco-Latin square designs with row-by-column intersections split into three or more plots could be constructed from three or more orthogonal squares.

semi-lethal: In F_2 progeny of the Muller-5 method, four phenotypic classes, namely Muller-5 females, hetero bar females, Bar-eyed males and normal males are normally present in 1:1:1:1 ratio, statistically. If the above-mentioned four phenotypic classes are present in ratio 2:2:2:1, lethal is termed SEMI-LETHAL.

semi-lethal mutation: A mutation that in the homozygous condition produces more than 50% mortality but not complete mortality. *See* VITALITY MUTATION.

seminiferous: Seed-bearing cone consisting of seed-bearing, overlapping scales surrounding a central axis (e.g. a seed-bearing spike of a grass; a pistil as a seed-bearing organ of a flower).

semiparametric model: 'Parametric' assumptions may be made about some aspects of a model, while other components may be estimated 'non-parametrically'. In the Cox regression procedure, a parametric model for the relative hazard is overlaid on a non-parametric estimate of baseline hazard. Like the proportional odds ordinal logistic model, the Cox semiparametric (proportional hazards) model is fully parametric on the right-hand side, and non-parametric on the left-hand (dependent variable Y) side. These types of semiparametric models essentially have an intercept for each distinct value of Y occurring in the data, allowing for the estimation of the distribution of Y in a way that is very similar to the empirical cumulative distribution function, a non-parametric distribution estimator.

semi-species: The populations that have acquired some but not yet all features of a species rank. These are border cases between species and subspecies.

semi-sterility: A situation in which half or more of all zygotes are inviable.

sen DNA: The amplified circular and plasmid-like section of mitochondrial DNA found in senescent *Podospora* cultures.

senesce: To weaken, as with maturation, age or disease stress.

senescence: Overall effect of all metabolic changes in an organism, which lead to a decline in active cellular functions. Senescence starts at different times in different organisms. Its expression varies as a result of gene-environmental interactions. In plants, senescence is the phase of growth that extends from full maturity to death. It is characterized by an accumulation of metabolic products, an increased respiratory rate and a loss in dry weight. *See* IN VITRO SENESCENCE.

senescent cells: Animal cells that have nearly reached the limit of lifespan (usually around 50 doublings) in cell culture and are beginning to show signs of impending death.

sense codon: A codon (triplet) that codes for a particular amino acid during protein synthesis. There is a total of 20 amino acids and 64 possible codons. Out of 64 codons, 61 are sense codons and 3 are nonsense codons. Sense codons code for amino acids, whereas nonsense codons do not. The language of mRNA is in the form of codons such as AAA, GGC and AUG. The sequence of codons on mRNA determines the sequence of amino acids in the resulting protein. *Aka* SENSE WORD.

sense DNA strand: Strand complementary to the antisense DNA strand. *Aka* NON-CODING STRAND, NON-TRANSCRIBING STRAND, NON-TEMPLATE STRAND *or* CRICK STRAND. *Compare with* ANTISENSE DNA STRAND.

sense mutation: A mutation that changes a termination (stop) codon into one that codes for an amino acid, usually resulting in an elongated protein.

sense RNA: The RNA formed on the template or antisense strand of the gene.

sense-posttranscriptional gene silencing (sense-PTGS): Sense-PTGS is the PTGS mediated by sense transgenes that lead to RNA degradation and DNA methylation of the transcribed region.

sensitive: A descriptive term for referring to stigmas that, in response to touch, close the two lobes of the stigma together, thereby ending the receptivity of the stigma, at least for the time that the lobes are closed together. *Mimulus* is perhaps the best-known example.

sensitive development period: The period of development during which the action of a gene is sensitive to the influence of external conditions.

sensitivity: (i) The proportion of individuals with the abnormality or disease that will yield a positive test. (ii) The sensitivity of bioinformatics similarity search algorithms centres around two areas – how best the method can perceive biologically meaningful relationships between two related sequences in the presence of mutations and sequencing errors and how the exploratory nature of the algorithm marks the probability that a matching sequence will not be spotted. At the user's discretion, the speed of most similarity search programmes can be forfeited in exchange for greater sensitivity – with stress on detecting lower-scoring matches.

sensitivity analysis: A method used in modelling that identifies critical variables that significantly affect the system under study, and then changes their values to determine the type and extent of their effect.

sensitivity and specificity: One way to quantify the utility of a diagnostic test when both the disease and the test are binary. The sensitivity is the probability that a patient with a disease will have a positive test, and specificity is the probability that a patient without a disease will have a negative test. Because these probabilities are conditional on the outcome, they are more useful for retrospective case-control studies. In general, it is more natural and useful to study variations in post-test probabilities of disease given different test results and different patient pre-test characteristics because (i) in general both the sensitivity and specificity will vary with the type of patient being diagnosed, (ii) sensitivity increases with the severity of the disease present unless the disease is all-or-nothing, (iii) specificity can vary with gradations in preclinical amount of disease and (iv) many diagnostic tests are based on continuous rather than binary measurements. Multivariable models are very useful for estimating post-test probabilities. The calibration and discrimination of the post-test probabilities can be quantified.

sensitivity of reporter gene assays: It is the ratio between reporter gene and background signal.

sensor gene: A eukaryotic regulatory gene that is sensitive to external/developmental signals. A sensor gene controls the activity of an integrator gene, which can be transcribed only when the sensor site is activated. The sensor sites are also regulatory sequences that are sensitive to external stimuli, like temperature or light or internal developmental signals like hormones or chemical inducers after binding with some regulatory proteins.

sensor region: In the Britten-Davidson model of eukaryote gene regulation proposed by J. Britten and E.H. Davidson in 1969, the sensor region responds to a single component in the environment (perhaps a hormone) and activates integrator genes, which activate receptor genes, which in turn activate producer (structural) genes.

sensor site: A constituent of the eukaryotic gene regulatory system. A sensor site controls the activity of an integrator gene, which can be transcribed only when the sensor site is activated. A sensor site retorts to particular internal signals or external stimuli. A specific sensor site thus turns on a particular INTEGRATOR GENE or a set of INTEGRATOR GENES.

sensory bias: An inborn fondness for particular male traits, which evolved not as a result of the sexual selection caused by that preference.

sepal: In a flower, one of the segments or divisions of the outer whorl of non-fertile parts surrounding the fertile organ, usually green. *Compare with* PETAL.

separase: Anaphase I in meiosis I is triggered by the cleavage of arm Rec8 by an enzyme named separase, allowing the homologous pairs to disengage and move to opposite poles.

separate creation: The theory that species have separate origins and never change after their origin. Most versions of the theory of separate creation are religiously inspired and suggest that the origin of species occurs by supernatural action.

septate: With cross walls; having septa. *See* SEPTUM. *Contrasts* COENOCYTIC *and* NON-SEPTATE.

septum: (i) A dividing wall; in fungi, a cross wall. (ii) A partition (e.g. the membranous wall that separates the two valves of the pod of Brassicaceae).

sequence alignment: A way of positioning the primary sequence of two or more macromolecules (DNA, RNA or protein) such that individual residues in each sequence are lined up with related residues in the other sequence(s). Gaps can be introduced into one or more regions of each sequence to facilitate the lining up of residues. Which individual residues are lined up with which other ones depends on the purpose of the alignment as well as the method used to create the alignment?

sequence alignment searches: Sequence alignment methods lead to identification of similar sequences. Multiple sequence alignment methods are applied to a set of related sequences before a phylogenetic analysis can be performed. Closely related organisms have similar sequences and more distantly related organisms have more dissimilar sequences. Given a set of sequences, it should be possible to reconstruct the evolutionary relationships among genes and among organisms. This involves creating a branching structure, termed a phylogeny or tree.

sequence analysis: Sequence determines shape and shape determines function. This holds for proteins as well as nucleic acids. The similarity in the sequence has both quantitative and qualitative aspects: Measurement of similarity gives a quantitative answer, saying that two sequences show a certain degree of similarity. An alignment is a mutual agreement of two sequences, which is a sort of qualitative answer; it exhibits where the two sequences are similar and where they differ. An optimal alignment, of course, is one that exhibits the most correspondences and least differences. Sequence alignment can be useful for discovering information related to functions, structure and evolution.

sequence analysis of a region surrounding the *Citrus tristeza* virus (CTV) resistance gene: CTV is the major virus pathogen causing significant economic damage to citrus worldwide, and a single dominant gene, *Ctv*, provides broad spectrum resistance to CTV in *Poncirus trifoliata* L. Raf. *Ctv* was physically mapped to a 282-kb region using a *Poncirus trifoliata* bacterial artificial chromosome library. This region was completely sequenced by Z.-N. Yang *et al.* (2003) to about 8x coverage using a shotgun sequencing strategy and primer walking for gap closure. Sequence analysis predicts 22 putative genes, two mutator-like transposons and eight retrotransposons. This sequence analysis also revealed some interesting features of this region of the *Poncirus trifoliata* genome: a disease-resistance gene cluster with seven members and eight retrotransposons clustered in a 125-kb gene-poor region. Comparative sequence analysis suggests that six genes in the Ctv region have significant sequence similarity with their orthologues in bacterial artificial chromosome clones F7H2 and F21T11 from *Arabidopsis* chromosome I. However, the analysis of gene collinearity between *Poncirus trifoliata* and *Arabidopsis* indicates that *Arabidopsis* genome sequence information may be of limited use for positional gene cloning in *Poncirus trifoliata* and citrus. Analysis of candidate genes for Ctv has been performed.

sequence analysis, models for: There are four major variants of biologically motivated inexact matching problems – GLOBAL ALIGNMENT, LOCAL ALIGNMENT, END FREE-SPACE ALIGNMENT and GAP PENALTY. *See details as separate terms.*

sequence assembly: In bioinformatics, sequence assembly refers to aligning and merging fragments from a longer DNA sequence in order to reconstruct the original sequence. This is needed as

DNA-sequencing technology might not be able to 'read' whole genomes in one go, but rather reads small pieces of between 20 and 30,000 bases, depending on the technology used.

sequence capture: The massively parallel replacement for PCR. The process of simultaneously isolating thousands or millions of regions of the genome prior to massively parallel sequencing. *Aka* TARGETED SEQUENCE CAPTURE *or* TARGETED GENOMIC ENRICHMENT.

sequence characterized amplified regions (SCARs): SCAR primers identify loci, which are amplified in RAPD/arbitrarily primed polymerase chain reaction (AP-PCR), DNA amplification fingerprinting (DAF) or amplified fragment length polymorphism (AFLP). SCARs are PCR-based DNA markers amplified by PCR using specific 15–30 bp primers, designed from nucleotide sequences established in cloned RAPD fragments linked to a trait of interest. For this RAPD fragments are cloned and their termini sequenced. These are similar to sequence-tagged sites (STS) and are dominant markers. Variation in the sequence is resolved by the presence or absence of the band. A high level of polymorphism may be detected by digesting PCR-amplified products with different restriction enzymes. SCARs and STSs are dominant markers (i.e. presence or absence of bands tells the polymorphism). SCARs do not face the problem of low reproducibility generally encountered with RAPDs. Obtaining a codominant marker may be an additional advantage of converting RAPDs into SCARs.

sequence complexity of a genome: The components of complex genomes differ not only in repetition frequency (highly repetitive, moderately repetitive, single copy) but also in sequence complexity. Complexity (denoted by N) is the number of base pairs of unique or non-repeating DNA in a given segment of DNA or component of the genome. Denaturation, renaturation and hybridization techniques are used to govern the sequence intricacy of any genome.

sequence divergence: Variations in the DNA or protein sequences of homologous genes in different species due to the autonomous accrual of mutations and natural selection since these species share a common ancestor.

sequence homology-independent protein recombination (SHIPREC): The SHIPREC method generates variants with a single crossover without requiring sequence homology between the input sequences.

sequence map: A pictorial representation of the sequence of amino acids in a peptide or of bases in an oligonucleotide.

sequence motif: A short conserved amino acid or nucleotide sequence design that signifies a specific functional site of a molecule.

sequence polymorphism: *See* POLYMORPHISM.

sequence similarity: To understand sequence similarity, one needs to understand the various terms, namely, identical, similar, homologous, analogous, orthologous and paralogous.

sequence-tag-based technologies: Since sequencing the whole transcriptome is difficult. Sequence tags (300–800 bp long) from each mRNA molecule are used to study transcriptome. EXPRESSED SEQUENCE TAGS (ESTs), SERIAL ANALYSIS OF GENE EXPRESSION (SAGE), MASSIVELY PARALLEL SIGNATURE SEQUENCING (MPSS), CAP-SELECTED ANALYSIS OF GENE EXPRESSION (CAGE), RNA SEQUENCING (RNA-SEQ), HIGH-RESOLUTION TILING ARRAYS, CHROMATIN IMMUNOPRECIPITATION and CHIP-CHIP and CHIP-SEQ) are sequence-tag based technologies that are used for transcriptome analysis. *See details as separate terms.*

sequence-tagged microsatellite (STM) marker: A locus-specific, codominant molecular marker used for DNA fingerprinting or genetic and physical mapping that is generated by sequencing microsatellite-flanking sequences, designing specific PCR primers towards these flanks and using them to amplify repeat regions of genomic DNA. In most cases, only one amplification product can be detected, which represents an STM marker.

sequence-tagged sequence (STS) marker: Unlike arbitrary primers, STS markers rely on some degree of sequence knowledge. A unique (single-copy) DNA sequence is used as a mapping landmark on a chromosome. Markers based on STS are codominant. They tend to be more reproducible because longer primer sequences are used. Require the same basic laboratory protocols and equipment as standard PCR.

sequence-tagged sites (STS): A relatively short, easily PCR-amplified sequence (200 to 500 bp) that can be specifically amplified by PCR and detected in the presence of all other genomic sequences and whose location in the genome is mapped. The DNA sequence of an STS may contain repetitive elements, sequences that appear elsewhere in the genome, but as long as the sequences at both ends of the site are unique and conserved, researchers can uniquely identify this portion of the genome using tools usually present in any laboratory. STS include such markers as MICROSATELLITES – SSRs, SEQUENCE-TAGGED MICROSATELLITE SITE (STMS) or SIMPLE SEQUENCE REPEAT POLYMORPHISMS (SSRPs), SEQUENCE CHARACTERIZED AMPLIFIED REGIONS (SCARs), CLEAVED AMPLIFIED POLYMORPHIC SEQUENCES (CAPs) and INTER-SIMPLE SEQUENCE REPEAT (ISSRs). ESTs are STSs derived from cDNAs. STSs act as physical markers for genomic mapping and cloning.

sequence-tagged sites (STS), uses of: (i) STS markers have been established for RFLP markers linked with bacterial blight rest genes *Xa5, Xa13, Xa21* in rice and stem rust rest gene in barley. (ii) STSs can be easily spotted by PCR using specific primers. For this reason, they are valuable for constructing genetic and physical maps from sequence data reported from many different laboratories. They serve as landmarks on the developing physical map of a genome. (iii) When STS loci contain genetic polymorphisms, they become valuable genetic markers (i.e. loci that can be used to distinguish individuals). They are used in shotgun sequencing, specifically to aid sequence assembly.

sequence-specific DNA-protein interactions: Sequence-specific DNA-binding proteins generally interact with the major groove of B-DNA, because it exposes more functional groups that identify a base pair. However, there also exist some known minor groove DNA-binding ligands such as Netropsin, Distamycin, Hoechst 33258 and Pentamidine.

sequencing: Determination of the order of nucleotides (base sequences) in a DNA or RNA molecule or the order of amino acids in a protein.

sequencing by binding on the ONSO system: The ONSO system is an integrated hardware and software short-read sequencing platform focused on delivering best-in-class accuracy, resulting in over 90 issued patents and numerous pending applications in the US and abroad. The ONSO system utilizes SBB chemistry to free researchers from the limitations of existing short-read sequencing platforms, providing ≥90% of bases at Q40+ accuracy.

sequencing by oligonucleotide ligation and detection (SOLiD): SOLiD is an enzymatic method of sequencing that uses DNA ligase. Emulsion PCR is used to immobilize/amplify a ssDNA primer-binding region (known as an adapter), which has been conjugated to the target sequence (i.e. the sequence that is to be sequenced) on a bead. These beads are then deposited on to a glass surface. Once bead deposition has occurred, a primer of length N is hybridized to the adapter. Then the beads are exposed to a library of 8-mer probes, which have different fluorescent dyes at the 5′ end and a hydroxyl group at the 3′ end. Bases 1 and 2 are complementary to the nucleotides to be sequenced while bases 3–5 are degenerate and bases 6–8 are inosine bases. Only a complementary probe will hybridize to the target sequence, adjacent to the primer. DNA ligase is then used to join the 8-mer probe to the primer. A phosphorothioate linkage between bases 5 and 6 allows the fluorescent dye to be cleaved from the fragment using silver ions. This cleavage allows fluorescence to be measured (four different fluorescent dyes are used, all of which have different emission spectra) and also generates a 5′-phosphate group, which can undergo further ligation. Once the first round of sequencing is completed, the extension product is melted off and then a second round of sequencing is performed with a primer of length N − 1. Many rounds of sequencing using shorter primers each time (i.e. N − 2 and N − 3) and measuring the fluorescence ensures that the target is sequenced. SOLiD technique is highly accurate (at 99.999% with a sixth primer and inexpensive). It can complete a single run in 7 days and in that time can produce 30 Gb of data. Main disadvantage is that read lengths are short, making it unsuitable for many applications.

sequencing gel: A long polyacrylamide salt gel that has sufficient resolving power to separate single-stranded fragments of DNA or RNA, which differ in length by only a single nucleotide. Electrophoresis is carried out at high voltage and with the gel in a vertical position.

sequencing of bisulfite-converted DNA: Methylation-specific restriction enzyme digestion has been combined with high-throughput sequencing methods such as high-throughput deep sequencing, Roche/454 sequencing, AB SOLiD sequencing and Illumina sequencing are such methods, which are described here briefly.

sequential epitope: *See* LINEAR EPITOPE.

sequon: A (potential) site on a protein molecule's 'backbone' where a sugar molecule (or a chain of sugar molecules, i.e. an oligosaccharide) may be attached. For example, proteins with asparagine (Asn) residues in the tripeptide motif Asn-Xaa-Ser/Thr that are N-glycosylated with preformed oligosaccharide units, or other residues may also be involved in the determination of a glycosylation site, or the structure of the protein itself may inhibit the glycosylation of certain polypeptides.

serial analysis of gene expression (SAGE): ESTs evaluate only a limited number of genes at a time. The bottleneck was broken by a method known as SAGE, given by V.E. Velculescu *et al.* (1995). This method allows a rapid, detailed analysis of thousands of transcripts. It is based on two principles: (i) A short nucleotide sequences tag (9–11 bp) can uniquely identify a transcript, provided it is isolated from a defined position within the transcript. For example, a sequence as short as 9 bp can distinguish four transcripts given a random nucleotide distribution at the tagged site whereas the number of transcripts is far less than this. (ii) Concatenation of short sequence tags allows the efficient analysis of transcripts by serially sequencing multiple tags within a single clone. SAGE involves the synthesis of double-stranded cDNA from mRNA by means of a biotinylated oligo(dT) primer. The cDNA is then cleaved with a 4 bp restriction enzyme AE. The 3′ portion of the cleaved cDNA is then isolated by binding to streptavidin beads. The cDNA is then divided in half and ligated to one of the two linkers containing the type II shift cleavage (IIs) restriction site. The linkers are so designed that cleavage of the ligation products with the TE results in the release of the linker with a short piece of cDNA (9bp-Tag). After blunt ends are created, the two pools of released tags are ligated to each other. Ligated tags then serve as the template for PCR amplification with primers specific to each linker. This provides orientation and punctuation of 4 bp per ditag sequence. Cleavage of the PCR product with the AE allows the isolation of ditags that could then be concatenated by ligation, cloned and sequenced. This method provides quantitative information on the abundance of transcripts and to recognize novel expressed genes. It is an ‘open system’ of gene expression analysis to our set of transcript profiling tools. SAGE supplements the expression analyses data generated by microarrays. Such open-system approaches allow for the identification and analysis of genes not previously characterized. SAGE analyses have already been used to study gene expression in plant systems.

serial analysis of gene expression (SAGE) methodology: SAGE involves the synthesis of double-stranded cDNA from mRNA by means of a biotinylated oligo(dT) primer. The cDNA is then cleaved with a 4 bp restriction enzyme (anchoring enzyme, AE). The 3′ portion of the cleaved cDNA is then isolated by binding to streptavidin beads. The cDNA is then divided in half and ligated to one of two linkers containing type II shift cleavage (IIs) restriction site. The linkers are so designed that cleavage of the ligation products with the tagging enzyme (TE) results in the release of the linker with a short piece of cDNA (9bp-Tag). After blunt ends are created, the two pools of released tags are ligated to each other. Ligated tags then serve as templates for PCR amplification with a primer specific to each linker. This provides orientation and punctuation of 4 bp per ditag sequence. Cleavage of the polymerase chain reaction (PCR) product with the AE allows the isolation of ditags that could then be concatenated by ligation, cloned and sequenced. This method provides quantitative information on plenty of transcripts and identifies novel expressed genes.

serial analysis of gene expression (SAGE), applications of: (i) Analysis of yeast transcriptome, (ii) gene expression profiles in normal and cancer cells, (iii) insights into p53-mediated apoptosis, (iv) 14-3-3 sigma is a p53-regulated inhibitor of G2/M progression, (v) identification of Myc as a target of the APC pathway, (vi) PPARd is an APC-regulated target of non-steroidal anti-inflammatory drugs, (vii) identification and classification of p53-regulated genes, (viii) analysis of human transcriptomes, (ix) serial microanalysis of renal transcriptomes, (x) genes expressed in human tumour endothelium, (xi) ferredoxin reductase affects 5-FU apoptosis, (xii) analysis of colorectal metastases (PRL-3), (xiii) characterization of gene expression in colorectal adenomas and cancers, (xiv) using the transcriptome to analyse the genome (LongSAGE) and (xv) DEC1 is a downstream target of TGF-beta.

serial analysis of gene expression (SAGE), efficiency of: This technique is influential but not very expedient for the comparison of many different samples and for the study of rare transcripts. ‘Tag-based’ technologies like SAGE, which can provide a relative measure of the cellular concentration of different messenger RNAs, can be used. The great advantage of tag-based methods is the ‘open architecture’, allowing for the exact measurement of any transcript, with a known or unknown sequence. SAGE is a method for comprehensive analysis of gene expression patterns with digital analysis. Because SAGE does not require a pre-existing clone, it can be used to identify and quantitate new genes as well as known genes. SAGE is less expensive to conduct and is 26 times more sensitive to ESTs in detecting rarely expressed transcripts. It also allows the digitalization of data for comparison across different experiment. Disadvantages of SAGE include ambiguity in identifying transcripts owing to short tag size and its linear sequencing process compared with MPSS and microarrays.

serial analysis of gene expression (SAGE), principle of: (i) A short sequence tag (10–14 bp) contains sufficient information to uniquely identify a transcript provided that the tag is obtained from a unique position within each transcript. (ii) Sequence tags can be linked together to from long serial molecules that can be cloned and sequenced. (iii) Quantitation of the number of times a particular tag is observed provides the expression level of the corresponding transcript.

serial dilution: Multistep process of diluting a sample several times. During the isolation of microbes, it is not possible to study samples that have suspension containing billion (10^9) cells/ml. The practical solution is to make serial dilutions. Usually, serial 100-fold or 10-fold dilutions are made (e.g. for 10-fold dilutions, 1 ml of suspension is added to 9 ml medium, and this process is repeated). After six cycles, it gets diluted to a million-fold (10^6).

serine (ser): A non-essential amino acid, with chemical formula $C_3H_7NO_3$, a biosynthetic precursor of several metabolites, including cysteine, glycine and choline. Some mammals synthesize D-serine within their brains and it acts as a neurotransmitter there.

serine phosphorylation: Serine is one of the building blocks of protein that can be modified by the addition of a chemical tag, in this case, a phosphate group.

serine protease: A protease that has a reactive serine in the active site of some proteins.

sessile: (i) Organisms generally, lacking motility or means of self-locomotion; immobile or incapable of movement. Sessile organisms may move via external forces such as wind or water currents but are more often permanently fixed to a solid object such as a rock, soil or another organism. (ii) In botany, the property of a plant or plant part about a leaf, leaflet, flower, floret, fruit, ascocarp, basidiocarp and so on, without a stalk, petiole, pedicel, stipe or stem; (of nematodes) permanently attached; not capable of moving about.

sessility: The state of organisms that do not possess a means of self-locomotion and are normally immobile.

set: An abstract data type that can stock unique values, without any specific order. It is a computer implementation of the mathematical concept of a finite set. Unlike most other collection types, rather than recovering a specific element from a set, one typically tests a value for membership in a set.

SET domain: A 120-amino acid signature domain for histone lysine methyltransferases (KMTs) that is conserved in Suppressor of variegation 3–9, Enhancer of zeste and Trithorax.

70S initiation complex: The final complex formed at the second stage of initiation of polypeptide synthesis in prokaryotes. It consists of the small and large subunits of the ribosome, mRNA and initiator tRNA charged with $_f^{Met}$. Three sites exist on 50S subunit – one for attachment of aminoacyl, second for growing polypeptide chain attached to tRNA and third for exit of tRNA.

7-dehydrocholesterol (7-DHC) and ferroptosis sensitivity: 7-dehydrocholesterol dictates ferroptosis sensitivity. Y. Li *et al.* (2024) identified that the enzymes involved in distal cholesterol biosynthesis have pivotal yet opposing roles in regulating ferroptosis through dictating the level of 7-dehydrocholesterol (7-DHC) – an intermediate metabolite of distal cholesterol biosynthesis that is synthesized by sterol C5-desaturase (SC5D) and metabolized by 7-DHC reductase (DHCR7) for cholesterol synthesis. The pathway components, including genes *MSMO1*, *CYP51A1*, *EBP* and *SC5D*, function as potential suppressors of ferroptosis, whereas *DHCR7* functions as a pro-ferroptotic gene.

7-dehydrocholesterol (7-DHC): F.P. Freitas *et al.* (2024) identify proferroptotic activity of 7-dehydrocholesterol reductase (DHCR7) and an unexpected prosurvival function of its substrate, 7-DHC. 7-DHC effectively shields (phospho)lipids from autoxidation and subsequent fragmentation. The accumulation of 7-DHC is capable of inducing a shift towards a ferroptosis-resistant state in these tumours ultimately resulting in a more aggressive phenotype. This observation suggests an antiferroptotic activity of 7-DHC as a cell-intrinsic mechanism that could be exploited by cancer cells to escape ferroptosis.

severity values: A means to measure accruing opportunities for pathogen infection (e.g. infection periods), concerning a predetermined edge that, when met, generates a disease-management activity (e.g. a fungicide application); used in disease estimation or forecasting.

Sewall Wright effect: *See* RANDOM GENETIC DRIFT.

sex: Contrasting features observed in male and female individuals of the same species.

sex allocation: The allocation of reproductive effort to male versus female function in hermaphrodites and to male versus female offspring in species with separate sexes.

sex cell: A gamete, either a sperm or an ovum. Sex cells are produced by the meiosis process. *Compare with* SOMATIC CELL.

sex chromatin: Barr body; facultative heterochromatin of one of the two X chromosomes of normal mammalian females but not of normal males.

sex chromosomes: Chromosomes that are different in two sexes that determine the sex. Presence or absence of such chromosomes is related to the sex of the bearer. Sex chromosomes are not similarly distributed in both sexes. These chromosomes play a role in sex determination and are inherited differently by the two sexes. For example, in mammals, males carry an X and a Y chromosome while females carry two X chromosomes. In birds and butterflies, males carry two Z chromosomes, and females carry a Z and a W chromosome. *Silene latifolia* is a dioecious plant species (with separate male and female individuals), which harbours XY sex chromosomes and constitutes an important model for sex determination in plants. D.S. Guttman and D. Charlesworth in 1998 proposed that the Y chromosome of *Silene latifolia* carries two loci involved in sex determination: the first involved in the suppression of female organ development and the second involved in the activation of male organ development. F. Monéger (2007) discovered *SlSTM* and *SlCUC* as strong candidate genes for their involvement in sex determination in this plant species. *Aka* ALLOSOME. *Compare with* AUTOSOME.

sex chromosomes of ape: Apes possess two sex chromosomes—the male-specific Y chromosome and the X chromosome, which occurs in both males and females. The Y chromosome is decisive for male reproduction. Using the methodology developed for the telomere-to-telomere (T2T) human genome, K.D. Markova *et al.* (2024) formed gapless assemblies of the X and Y chromosomes for five great apes—bonobo (*Pan paniscus*), chimpanzee (*Pan troglodytes*), western lowland gorilla (*Gorilla gorilla gorilla*), Bornean orang-utan (*Pongo pygmaeus*) and Sumatran orang-utan (*Pongo abelii*)—and a lesser ape, the siamang gibbon (*Symphalangus syndactylus*). Compared with the X chromosomes, the ape Y chromosomes diverge greatly in size and have low alignability and high levels of structural rearrangements—owing to the accumulation of lineage-specific ampliconic regions, palindromes, transposable elements and satellites. Many Y chromosome genes enlarge in multi-copy families and some evolve under purifying selection. Thus, the Y chromosome displays dynamic evolution, whereas the X chromosome is more stable.

sex correction: Normal individuals with abnormal chromosome constitution. For example, normal individuals with chromosome constitution 46+XX were produced from a loss of Y chromosome from a 47+XXY complement.

sex determination: Mechanisms responsible for determining the sex of a zygote during development.

sex factor: An episome of bacteria (e.g. the F plasmid in *Escherichia coli*) that when integrated into the chromosome causes chromosome breakage at the junction and facilitates genetic exchange. Also exists autonomously in the F^+ bacterial cell. *Aka* SEX PARTICLE *or* SEX PLASMID. *Aka* F-FACTOR, FERTILITY FACTOR.

sex factor affinity (sfa): A region on the bacterial chromosome that has a specific affinity for the sex factor. It is a region consisting of sex factor DNA left behind when the sex factor recombined out of the bacterial chromosome. *Aka* SEX FERTILITY AFFINITY.

sex linkage: Genes located and inherited on a sex chromosome.

sex mosaic: Combination of male and female features in the body of an individual.

sex pilus: A tube structure present on the exterior of F^+ bacteria, thought to be involved in the transfer of DNA during conjugation.

sex ratio: The ratio of the number of males and the number of females (may also be presented as per cent) at fertilization (equals primary sex ratio). Usually, this is 1.1 in sexual species, but under certain conditions, sex ratios can be strongly female biased.

sex reversal: Transformation of one sex into another. Sex reversal can be achieved by deploying a male-determining factor. There are diverse primary signals that control sex in insects. A. Meccariello *et al.* (2019) identified the male-determining factor, *Maleness-on-the-Y* (*MoY*) in *Ceratitis capitata*, which can be used in genetic control methods. Since this factor is preserved functionally in other fruit flies, it can assist as a new tool to operate their reproduction. A male-determining factor linked to the Y chromosome has been identified in many insects. In case of *Bactrocera dorsalis*, a target site in the transformer gene *Bdtra* was targeted and CRISPR/Cas9-mediated knockout of an autosomal-derived micro RNA miR-1-3p caused the transformation of XY males into phenotypic females (W. Peng *et al.*, 2020). Micro RNA miR-1-3p has thus been demonstrated to play a role in male sex determination in *B. dorsalis*.

sex reversal by a gene-edited mutation: CRISPR/Cas9 homology-directed repair gene-editing procedure was employed to create temperature-sensitive point mutation in the *transformer-2* (*tra-2*) sex determination gene in *Drosophila suzukii* for achieving sex reversal from female to male adults. *Ds-tra-2* ^{ts2} mutants, when cultured at permissive temperatures (16–20°C), developed as normal fertile XX and XY adults, but when cultured at restrictive temperatures (26–29°C), the individuals developed into females, which were sterile with incomplete pseudomale phenotype and males with normal morphology, but sterile. Ds-TRA-2^{ts2} protein was thus found to be functionally temperature dependent as evidenced by the up- and downregulation of expression of the female-specific *Ds-Yolk protein 1* (*Ds-Yp1*) gene with temperature shifts. It was suggested that gene-edited mutation could be temperature dependent and this would assist in generation of conditional mutations in insects for future functional genomics analysis. Furthermore, temperature-dependent creation of sterile male populations will help improve the efficiency of STERILE INSECT TECHNIQUES (SIT).

sex-biased inheritance: Passage of genes to the next generation through one sex or the other (e.g. maternal inheritance or sex linkage). *See* MATERNAL INHERITANCE *and* SEX LINKAGE.

sex-controlled: A trait whose appearance is controlled by the type of sex of the individual (e.g. in asparagus, hops and hemp). *See* SEX-INFLUENCED.

sex-influenced genes: A gene is so termed when the dominance-recessiveness relationship between two alleles in a heterozygote is influenced by the sex of the individual. *Aka* SEX-CONTROLLED GENES.

sex-influenced traits: These are those characters in which the dominance of an autosomal gene depends on the sex of the individual. In the case of sex-influenced traits, the heterozygotes express differently in males and females because of differences in the male and female hormones. Thus, a trait may be dominant in one sex and recessive in the other. Sex-influenced traits are *aka* SEX-CONTROLLED TRAITS. This phenomenon is *aka* SEX-INFLUENCED DOMINANCE. Some examples of sex-influenced traits are baldness in men, horns in sheep, gout, spina bifida, white forelock in humans, short index finger, fissure of upper lip, cleft palate, absence of upper lateral incisor and enlargement of terminal joints of fingers.

sex-lethal gene: *Sxl* gene in *Drosophila*, located on the X chromosome, is a sex switch, guiding development controlled by the ratio of numerator and the denominator elements that act to influence the genic balance equation (X/A).

sex-limited characters: A trait controlled by a gene located on a sex (X) chromosome in organisms having XY system of sex determination. The characters are governed by autosomal genes whose expression is determined by the presence or absence of one of the sex hormones. The phenotypic effect is thus limited to one sex or the other. Thus, one sex is uniform in expression of a particular trait and yet the same genes produce a different phenotype or do not express at all in the other sex in its progeny. Some examples of sex-limited traits are spots on the body of Ayrshire cattle, feathering in domestic fowl, colour of clover butterfly, milk production in mammals, horns in Rambowillet sheep, egg production in chickens, age of onset of menstruation, width of pelvis, stuttering and distribution of body hairs.

sex-limited gene: The autosomal gene whose expression is limited to only one of the two sexes because of the presence or absence of sex hormones. It refers to a gene that is located in differential region of the X chromosome.

sex-linked gene: Refers to a gene that is differentially expressed in two sexes because of the presence or absence of sex hormones.

sex-linked inheritance: It has been seen that in *Drosophila* and man there are regions in X and Y chromosomes that do not have any homology. These are called DIFFERENTIAL REGIONS of X and Y chromosomes. There are regions in X and Y chromosomes that have homology and are known as pairing regions of X and Y chromosomes. This fact needs to be kept in mind for understanding sex-linked inheritance. Sex-linked inheritance is of three types: X-LINKED (CRISS-CROSS) INHERITANCE, Y-LINKED (HOLANDRIC)

INHERITANCE and X- AND Y-LINKED (PSEUDOAUTOSOMAL) INHERITANCE. X-linked inheritance is *aka* CRISS-CROSS INHERITANCE *or* SKIP-GENERATION INHERITANCE. *See details as separate terms.*

sex-linked trait: As related to genetics, the term refers to characteristics (or traits) that are influenced by genes carried on the sex chromosomes. In humans, the term often refers to traits or disorders influenced by genes on the X chromosome, as it contains many more genes than the smaller Y chromosome. *See* SEX CHROMOSOMES.

sex-promoting alleles: Certain alleles (e.g. *xa*, *xb*, *xc*) are concerned with sex determination. Females are always heterozygous, males homozygous or hemizygous.

sex-ratio phenotype: A mutant trait in *Drosophila* whereby females produce mostly, if not only, daughters.

sexduction: Incorporation of bacterial chromosomal genes in the fertility plasmid, with subsequent transfer to a recipient cell in conjugation. *Aka* F-DUCTION *or* F-MEDIATED TRANSDUCTION.

sexual: Relating to processes of reproduction of individuals in which meiosis and fertilization are included leading to genetic recombination.

sexual conjugation: An infrequent occurrence in which two adjacent bacteria stretch out portions of their cell membranes to touch one another, fuse and then pass transposons, jumping genes or plasmids to each other.

sexual dimorphism: The condition where sexes of the same species exhibit different morphological characteristics, particularly characteristics not directly involved in reproduction (e.g. in *Melandrium alba*, hops, hemp). *Aka* SEX DIMORPHISM.

sexual PCR: Term given by Willem Stemmer (1994) is a form of PCR in which similar, but not identical, DNA sequences are reassembled to obtain novel juxtapositions, simulating the result of genetic recombination. The result is the creation of an array of related genes, which may possess improved characteristics. By repeated rounds of recombination, selection and PCR-based amplification vastly improved gene products, such as enzymes with greater activity, may be generated and selected. The intellectual property surrounding 'DNA shuffling', is often referred to as 'molecular breeding' or 'sexual PCR'. *Aka* MOLECULAR DIVERSITY.

sexual reproduction: In organisms that reproduce sexually, the gametes (reproductive cells like eggs and sperm) are haploid cells that are produced by female and male PRIMORDIAL GERM CELLS (PMCs). This means they have only one copy of each chromosome instead of homologous pairs. Gametes are produced via meiosis, a process in which an original diploid cell divides into four haploid daughter cells. Individuals produced by sexual reproduction differ from either of their parents and each other. Sexual reproduction is a rule rather exception in humans, whereas plants reproduce sexually or asexually or by both methods. *Contrasts* ASEXUAL REPRODUCTION.

sexual selection: A selection on mating behaviour, either by competition among members of one sex (usually males) for reaching to members of the other sex or by choice of members of one sex (usually females) for certain members of the other sex. In sexual selection, individuals are favoured by their fitness relative to other members of the same sex, whereas natural selection works on the fitness of a genotype relative to the whole population. *Compare with* NATURAL SELECTION.

sexual spore: A spore produced during the sexual cycle.

sexually compatible: Able to be cross-mated or cross-fertile.

sexually dimorphic: When males and females of a species have considerably different appearances, which may include size, colouration or other features, such as special plumage.

shade avoidance and transition to flowering: The transition to flowering in plants is regulated by environmental factors such as temperature and light. Plants grown under dense canopies or at high density perceive a decrease in the ratio of red-to-far-red incoming light triggering a series of responses collectively known as 'shade-avoidance syndrome'. During shade avoidance, stems elongate at the expense of leaf expansion, and flowering is accelerated. Phytochrome B has the most significant role in shade avoidance responses. PFT1, a nuclear protein, induces flowering in response to suboptimal light conditions PFT1 functions downstream of phyB to regulate the expression of flowering locus T (*FT*).

shallow learning (SL): SL implies a maximum of two layers residing in between the input and output neural network to authenticate the prediction.

sharon phages: *See* CHARON PHAGES.

shear: (i) It is the movement of one layer relative to the parallel adjacent layer. (ii) To fragment DNA molecules into smaller pieces. DNA is very susceptible to hydrodynamic shear forces. Forcing a DNA solution through a hypodermic needle will fragment it into smaller pieces. The size of all the fragments obtained is inversely proportional to the diameter of the needle's bore. The actual sites at which the shear force breaks a DNA molecule are approximated by random shear and then cloned by either tailing their ends or using linkers to create a complete gene library of an organism.

shelf life: The term or period for which a stored commodity (vegetables or food) remains effective, useful or suitable for consumption.

shift: A chromosome aberration in which a chromosome segment is transposed to a new location in the same or a different chromosome. The shift is a sample type of translocation.

shifting balance theory (SBT): A theory developed by Sewall Wright, in which species evolve towards the best among many alternative adaptive peaks.

shifty codon: The codons that supposedly increase frameshifts. The homopolymer codons AAA, CCC, GGG and TTT are called 'shifty' codons.

shifty transfer RNA: A normal tRNA that promotes frameshifting at a site of non-triplet movement, called a SHIFTY CODON. Such tRNAs tend to cause two-base and four-base translocation errors by mistranslating certain non-cognate codons.

Shine-Dalgarno sequence: A short sequence many prokaryotic ORFs contain upstream on the 5′-end of start codon called RIBOSOME BINDING SITE that eases the binding by a ribosome. This sequence typically located 3–9 bp on the 5′ side of the start codon is complementary to the sequence located near 3′-end of one of the RNA components (16S rRNA). In prokaryotes, part or all of the polypurine sequence AGGAGG are located on mRNA just upstream of an AUG initiation codon. The ribosome binding site base pairs with 16S rRNA component thereby aligning the ribosome with the beginning of the ORF.

Shivalik-valley (Chandigarh)-based startup Ag-Next: This is also helpful in the improvement of the quality of agricultural produce through innovative AI-driven hardware, software and analytics. The use of sensor-based technologies for precision agriculture, advanced farming systems, robotics and unmanned vehicle technologies in farm and food sector for predictive analytics and policy interventions would help promote AI in Indian Agriculture.

shoot: The aerial part of a plant, including a stem and all of its dependent parts (leaves, flowers, etc).

shoot apex (TIP) culture: The shoot apex is the tip of the stem, which contains shoot apical meristematic cells and the region of growth plus one to the numerous primordial leaves, usually measuring from 0.1 to 1.0 mm in length. In cases where more mature leaves are included, the structure can measure up to several centimetres in length.

shoot apical meristem: Undifferentiated tissue, located within the shoot tip, generally appearing as a shiny dome-like structure distal to the youngest leaf primordium and measuring less than 0.1 mm in length when excised. In instances where more mature leaves are included, the structure can measure up to several centimetres in length.

shoot tip culture: *In vitro* culturing of shoot tip consisting of apical meristem covered with several primordial leaves and measuring from 0.5–1.5 mm in length. Shoot tip culturing is commonly used to establish shoot cultures *in vitro* for micropropagation of plants. The shoot tip is already a growing point of the plant, which easily starts growing in tissue culture after 1–2 weeks of incubation on a medium with low auxin and high cytokinins.

short hairpin RNA (shRNA): A sequence of small RNA that styles a tight hairpin turn, which can be employed to produce a siRNA to silence target gene expression via RNA interference. *Aka* SMALL HAIRPIN RNA.

short interfering ribonucleoprotein (siRNP) assembly: Three distinct complexes (R1, R2 and R3) accumulate on siRNAs *in vitro* in *Drosophila*. All three complexes require Dicer-2 (Dcr-2), which directly contacts siRNAs in the ATP-independent complex.

short interfering RNA (siRNA) design: A siRNA guides sequence-specific degradation of the homologous mRNA, thus producing 'knock-down' cells. SiRNA Selector scans a target gene for candidate siRNA sequences that satisfy user-adjustable rules. The programme evaluates siRNA functionality and specificity.

short interfering RNA types: Include HETEROCHROMATIC SIRNAS (HC-SIRNAS), NATURAL ANTISENSE TRANSCRIPT-DERIVED SIRNAS (NAT-SIRNAS), *TRANS*-ACTING SMALL INTERFERING RNAS (TASIRNAS) and SHORT HAIRPIN RNA (SHRNA). *See details as separate terms.*

short interspersed nuclear elements (SINEs): In many mammals, a large portion of the middle repetitive DNA is composed of many copies of a short sequence spread throughout the genome. For example, in human beings, there are about 500,000 copies of a 300-bp sequence that constitutes 5–6% of the total human genome. It exists as one copy for every 4 kb of the human genome. Because this sequence is cleaved by the restriction endonuclease Alu1; it is called the Alu family. Alu insertions cause many mutations resulting in disease. For example, an Alu insert is the cause of a severe form of haemophilia A (factor VIII). In case of plants, similar types of repeated and dispersed sequences are present in species that have large genomes (e.g. maize, rye and wheat). *Aka* SHORT-PERIOD INTERSPERSION OF DNA SEQUENCES.

short tandem repeats (STRs): Sequences in which repeating units of 1–6 base pairs form an array of up to 100 base pairs long. Variations in STR length are associated with, and often the causative agents of, gene expression changes present in some hereditary conditions. C.A. Horton *et al.* (2023) show that STRs exert their effects by directly binding transcription factor proteins. TF-preferred STRs need not resemble known binding sites. This binding can be explained and predicted by simple additive models in which repeated instances of low-affinity binding sites seem to have large effects. This explained how STRs might influence gene expression in both normal and diseased states.

short/small interfering RNAs (siRNAs): Small RNAs that are generated from double-stranded RNAs by Dicer or Dicer-like enzymes or chemically synthesized and serve as guides for the cleavage of homologous mRNAs in RNA interference or for the addition of chromatin modifications including histone and DNA methylation at homologous genomic sequences in transcriptional gene silencing.

short/small non-coding RNAs: Small (20–25 nts long) non-coding RNAs (ncRNAs) are a large family of endogenously expressed transcripts, 18–200 nt long, that play a crucial role in regulating cell function. Small RNAs are classified as (i) MICRO RNAS (MIRNAS), (ii) SMALL INTERFERING RNAS (SIRNAS), (iii) PIWI-INTERACTING RNAS (PIRNAS), (iv) TINY NON-CODING RNAS (TNCRNAS), (v) SMALL MODULATORY RNA (SMRNA), (vi) PHASED SIRNAS (PHASIRNAS), (vii) *TRANS*-ACTING SIRNAS (TA-SIRNAS), (viii) TELOMERASE RNAS (TELRNAS) and (ix) GUIDE RNAS (GRNAS). RNA-induced silencing complex (RISC) that interferes with mRNA stability or mRNA translation. The process that both naturally occurs in cells and is being used as a technology to switch certain genes off.

short-day plants: Plants that flower when exposed to a dark period longer than the critical length (light period shorter than the critical length).

short-period interspersion: A pattern in a genome in which moderately repetitive DNA sequences of 300 bp. alternate with non-repetitive sequences of 1000 bp.

short-period interspersion of DNA sequences: The prominent arrangement is short-period interspersion. This arrangement is characterized by repeated sequences of 100–200 bp in length interspersed among single-copy sequences that are 1000–2000 bp in length. This arrangement is found in animals, fungi and plants.

short-stem transfer RNA: Messenger RNA has 64 possible triplet sequences, or codons, three of which usually terminate protein synthesis. However, some organisms can use all codons to specify amino acids, thanks in part to a surprising feature of a transfer RNA (P.V. Baranov and J.F. Atkins, 2023). When a protein is synthesized, a chain of amino acid residues is built on the basis of messenger RNA instructions, and this construction must be brought to a stop appropriately. A. Kachale *et al.* (2023) report the discovery of an unusual mechanism that breaks with the conventional way in which the mRNA decoding process occurs. They analyse the in-frame stop codons in 7259 predicted protein-coding genes of a previously undescribed trypanosomatid, *Blastocrithidia nonstop*. They reveal that in this species in-frame stop codons are under-represented in genes expressed at high levels and that UAA serves as the only termination codon. Whereas new $tRNAs^{Glu}$ fully cognate to UAG and UAA evolved to reassign these stop codons, the UGA reassignment followed a different path through shortening the anticodon stem of $tRNA^{Trp}{}_{CCA}$ from five to four-base pairs (bp). The canonical 5-bp $tRNA^{Trp}$ recognizes UGG as dictated by the genetic code, whereas its shortened 4-bp variant incorporates tryptophan also into in-frame UGA.

short-term gene regulation: It signifies cells' response to oscillations in the environment. This is a reversible type of gene regulation. Short-term regulation comprises changes in activities or concentrations of enzymes as levels of a particular substrate or hormone increase or decrease as the cell cycle is reversed. This is the only type of gene regulation occurring in viruses and bacteria. Short-term regulation is an important feature of both developing and fully differentiated eukaryotic cells; even when a cell is undergoing differentiation, it responds to its environment.

ShortStack approach: In this method, SMALL RNA (sRNAs) are grouped depending on the alignment localization of the reads. sRNAs of the cluster within the user-defined size range and threshold are annotated as Dicer-derived. The primary sRNA size of the cluster is denoted 'DicerCall'. A phase score is calculated for the phase size of the DicerCall to evaluate the level of phasing of the loci.

shot-hole: A symptom in which small lesions fall out of leaves, giving leaf the appearance of being hit by buckshot (small lead shot for shotgun shells).

shotgun cloning: A technique for obtaining the desired gene that involves 'chopping up' the entire genetic complement of a cell using restriction enzymes, then attaching each (resultant) DNA fragment to a vector and transferring it into a bacterium and finally screening those (engineered) bacteria to locate the bacteria that are producing the desired product (e.g. a protein).

shotgun cloning of *Bacillus subtilis* genome in *Escherichia coli*: In this technique, *Escherichia coli* is used as an alternate intermediate host and bifunctional vectors, which replicate in both *Escherichia coli* and *Bacillus subtilis*. Plasmid pHV33 has been used for this purpose to construct a *Bacillus subtilis* bank in *Escherichia coli*.

shotgun experiment: Cloning of an entire genome in the form of randomly produced fragments.

shotgun library: Initially genome is fragmented using a nebulizer (shotgun). Fragmented DNA is coupled with vectors in the presence of DNA ligase. Recombinant vectors are then transformed into *Escherichia coli* and cultured in a selectable medium. From the colonies grown in the culture, DNA is isolated from the vector and sequenced one by one. The key requirement of shotgun approach to genome sequencing is that it must be possible to identify overlaps between all the individual sequences (DNA fragments) that are generated, and this identification process must be accurate and unambiguous so that the correct genome sequence is obtained.

shotgun sequencing: The genome is broken into pieces, each one of them is sequenced and then a powerful computer is used to assemble these pieces into the correct order by sequence overlaps. R.D. Fleischmann *et al.* (1995) used this method for mapping the genome of the bacterium *Haemophilus influenza*. This method abolishes the necessity for time-consuming mapping. However, the problem in shotgun sequencing arises due to the occurrence of repetitive nucleotide sequences, which make up large portion of vertebrate genomes. Only 5% of genome codes for the protein, and the rest is repetitive. Suppose there is a DNA fragment, when we perform random fragmentation and then try to assemble, it may assemble incorrectly and sometimes intervening information is lost. If that part contains an important gene, it changes the phenotypic trait of that organism. Sometimes called WHOLE-GENOME SHOTGUN SEQUENCING.

shrub: A woody perennial plant without a single main trunk, branching freely but is generally smaller than a tree.

shuttle breeding: A programme that shuttles seed between two (or more) locations to be grown at each. This method was introduced by 1970 Nobel Laureate Peace N.E. Borlaug and is a proven contributor to the success of CIMMYT (Mexico). Shuttle breeding enables two breeding cycles per year instead of one (e.g. a winter cycle in the northern desert of Sonora and a summer crop in the central Mexican highlands). This method not only fast-forward selection, but also exposes test varieties to radically different day lengths, temperatures, altitudes and diseases. The resulting plants are broadly adapted. They grow well in numerous environments.

shuttle mutagenesis: In this process of mutagenesis, first a gene is cloned and then mutated through the use of transposons and the mutated gene is then transferred back in the original organism.

shuttle vector: A plasmid (vector) that can exist and replicate in the cells of more than one organism. Some shuttle vectors can exist and replicate in *Escherichia coli* and yeast. pJDB219 is a yeast cloning vector, and it is also a shuttle vector. Shuttle vectors consisting of BPV and pBR322 sequences and capable of replication in both bacterial and mouse cells have also been constructed for genetic engineering. Using these vectors, manipulation of large DNA inserts (up to 3 Mb pairs) can be done.

SI mapping: A method for determining the transcriptional organization of a gene. A laboratory method is used for locating the start and end points of transcripts and for mapping introns.

SI system: System International d' Unites. An international system of units for the measurement of lengths, volume and mass. It is based on the metric system.

sib: Short for SIBLING; one of two or more offspring with both parents in common but deriving from different gametes.

sib mating: A form of inbreeding in which progeny of the same parents (siblings) are intercrossed.

siblicide: The killing of one offspring by its sibling.

sibling species: Morphologically similar or identical populations that are reproductively isolated. These species may be sympatric or allopatric. These species may exit at the same time or at different times. The well-studied classical pairs of sibling species in *Drosophila* are *Drosophila pseudoobscura* and *Drosophila persimilis*, and *Drosophila melanogaster* and *Drosophila simulans*.

sibling technique: A method for the separation of bacteria encompassing a specific cloned gene (e.g. in shotgun cloning experiments) whose phenotype cannot be easily identified on individual colonies but can be spotted with great sensitivity in a large population where only a small minority of the cells contains the desired gene.

sibship: The children of one generation in a single family.

side chain: A chemical group that is attached to a core part of the molecule called the 'main chain' or backbone. The side chain is a hydrocarbon branching element of a molecule that is attached to a larger hydrocarbon backbone. It is one factor in determining a molecule's properties and reactivity.

sieve element: A tube-shaped living cell in the phloem, functioning in the transport of dissolved organic substances in the plant. *Aka* SIEVE TUBE ELEMENT.

sieve plate: Perforated wall area between two phloem cells through which their protoplasts are interconnected.

sigma (σ) factor: A protein component of RNA polymerase that is a must for initiation of transcription in prokaryotes. It is not required for elongation of the transcript. In prokaryotes, the sequence of a promoter is recognized by a Sigma (σ) factor of the RNA polymerase. *Escherichia coli* has five sigma factors. Different sigma factors recognize different promoter sequences. (i) SIGMA 70 regulates the expression of most genes. (ii) SIGMA 32 regulates the expression of heat-shock proteins. (iii) SIGMA 28 regulates the expression of flagellar operon (involved in cell motion). (iv) SIGMA 38 regulates gene expression against external stresses. (v) SIGMA 54 regulates gene expression for nitrogen metabolism.

sigma adaptation: Transcriptional activator PafBC in the presence of single-stranded DNA activates transcription by reprogramming the canonical −10 and −35 promoter specificity of RNA polymerase associated with the housekeeping sigma subunit. The structure reveals a unique mode of promoter recognition, termed 'sigma adaptation'. PafBC inserts between DNA and sigma factor to mediate recognition of hybrid promoters lacking the −35 but featuring the canonical −10 and a PafBC-specific −26 element. Sigma adaptation may constitute a more general mechanism of transcriptional control in mycobacteria.

sigma chain: One of the polypeptide chains comprising RNA polymerase, which is essential for the recognition of signals in the DNA for the start of transcription. It is not required for elongation of the transcript. *Aka* SIGMA FACTOR.

sigmoid (curve): S-shaped curve (e.g. the growth curve of a plant).

sign: A warning of disease from direct observation of a pathogen or its parts. *Contrasts* SYMPTOM.

sign test: A non-parametric test based on signs (negative or positive) of differences between two sets of observations. This test is used to compare two groups for their central tendency.

signal: Cells reply to their environment by rearranging their structure, regulating the activity of proteins and changing patterns of gene expression. The provocation for such responses may be a small molecule, a macromolecule or a physical agent, such as light.

signal codons: The codons that code for either start or stop signal in protein synthesis.

signal hypothesis: Ribosomes are either free in the cytoplasm or associated with membranes. The choice is resolved by the type of protein being synthesized. Membrane-bound ribosomes, polysomes, synthesize proteins that enter the membrane or in eukaryotes are passed into membrane-bound organelles (e.g. the Golgi apparatus, mitochondria, chloroplasts, vacuoles) or transported outside the cell membrane. The procedure for membrane attachment is explained by the signal hypothesis, which was developed by G. Blobel and D.D. Sabatini in 1971 and C. Milstein and their colleagues in 1972. The process is relevant for both prokaryotes and eukaryotes. The signal for membrane insertion is coded into the first one to three dozen amino acids of membrane-bound proteins. This signal peptide takes part in a chain of events resulting in membrane attachment by the ribosome and membrane insertion of the protein. The first step ensues when the signal peptide becomes reachable outside of the ribosome. It is documented by a ribonucleoprotein particle called the SIGNAL RECOGNITION PARTICLE (SRP). The SRP contains six proteins and a small cytoplasmic RNA (7S RNA), which is homologous to the Alu element. It binds to the signal sequence as it emerges from the ribosome and stalls protein synthesis until the complex reaches the ER. Here, the SRP binds to its receptor, a dimeric DOCKING PROTEIN situated on the ER membrane. The SRP has three functional domains, which sponsor signal binding, receptor binding and elongation arrest, respectively. The ribosome then contacts its receptor, a trimeric transmembrane protein called Sec61, and protein synthesis resumes. The polypeptide is fed into the ER lumen, a process termed VECTORIAL DISCHARGE or COTRANSLATIONAL IMPORT, through the Sec61 complex (some proteins require an additional component called TRAM). After synthesis, STP exchange releases the ribosome from the SRP, and the SRP itself is released from its receptor back into the cytoplasm. During this time, protein synthesis is paused. The ribosome is brought in direct contact with the membrane. Other proteins of the membrane help anchor the ribosome. Protein synthesis then resumes, with the nascent protein passing directly into the membrane or passing into the membrane shortly after being synthesized. Once through the membrane, the SIGNAL PEPTIDE is cleaved from the protein by an enzyme called SIGNAL PEPTIDASE.

signal molecule: Extracellular or intracellular molecule that cues the response of a cell to the behaviour of other cells or objects in the environment. Such a signalling phenomenon has been observed in *Agrobacterium*.

signal patch: Protein sorting signal that consists of a specific three-dimensional arrangement of atoms on the folded protein's surface.

signal peptidase: An enzyme that is present within the lumen of the ER, which proteolytically cleaves a secreted protein at the site of a signal sequence.

signal peptide: Protein sorting requires a specific amino acid sequence (of 15–30 amino acids) the newly synthesized protein, which interacts with membranes. These amino acid sequences are predictable by specific receptors located within the membrane. Signal peptide takes part in a chain of events leading to membrane attachment by the ribosome and membrane insertion of the protein. The signal peptide does not appear to have a consensus sequence like the transcription or translation recognition boxes. Rather, similarities (at least for the ER and bacterial membrane-bound proteins) include a positively charged (basic) amino acid (commonly lysine or arginine) near the beginning (N-terminal end) followed by about a dozen hydrophobic (nonpolar) amino acids, commonly alanine, isoleucine, leucine, phenylalanine and valine. The signal peptide sequence of the bovine prolactin protein reported by N.L. Sasavage *et al.* (1982) is: NH_2-Met Asp Ser Lys Gly Ser Ser Gln Lys Ser Arg Leu Leu Leu Leu Leu Val Val Ser Asn Leu Leu Leu Cys Gln Gly Val Val Ser Thr Pro Val . . . Asn Asn Cys-COOH. The amino acids represented in bold separate the signal peptide from the rest of the protein, which consists of 199 amino acid residues. The amino acid sequences that target proteins for secretion from cells or for integration into cell membranes. The signal sequence is presented at the ribosomal tunnel exit in an exposed position ready for lodging in the hydrophobic groove of the rearranged SRP54 M domain. Upon ribosome binding, the SRP 54 NG domain also experiences a conformational rearrangement, priming it for the subsequent docking reaction with the NG domain of the SRP receptor. Later in mature proteins, the sequence vanishes. It is removed by a protease. *See* SIGNAL RECOGNITION PARTICLE. *Aka* TRANSIT PEPTIDE, LEADER SEQUENCES *and* SIGNAL SEQUENCE.

signal pulses: Noise limits the efficiency of information transfer. But correlations of the intervals between signal pulses can reduce low-frequency noise and thereby increase the transfer of information.

signal recognition particle (SRP): It consists of a ribonucleoprotein particle that comprises six different proteins and a 7S RNA, which is about 300 nucleotides long. The complex of SRP, ribosome and signal peptide then pass or disperse to a membrane in which the SRP binds to a receptor on the membrane, called a DOCKING PROTEIN (DP) or SIGNAL RECOGNITION PARTICLE RECEPTOR (SRPR). Signal sequences target proteins for secretion from cells for integration into cell membranes. As nascent proteins arise from the ribosome, signal sequences are recognized by the SRP, which subsequently associates with its signal receptor (SR). In this complex, SRP and SR stimulate each other's GTP activity, and GTP hydrolysis ensures unidirectional targeting of cargo through a translocation pore in the membrane. To define the mechanism of reciprocal activation, they determined the 1.9 Å structure of the complex formed between these two GTPases. The two partners form a quasi-two-fold symmetrical heterodimer.

signal recognition particle RNA (SRP RNA): Non-coding RNAs that are implicated in protein translocation to ER. The RNA component of the SRP in the eukaryotes is called 4.5S RNA. The SRP RNA is an RNA-protein complex present in the cytoplasm of cells that binds to the mRNA of the proteins that are destined for secretion from the cell.

signal sequence: A short sequence added to the amino-terminal end of a polypeptide chain that forms an amphipathic helix allowing the nascent polypeptide to migrate through membranes such as the ER or the cell membrane. It is cleaved from the polypeptide after the protein has crossed the membrane. *Aka* LEADER SEQUENCE.

signal transducers: The DNA damage checkpoints employ damage sensor proteins to detect DNA damage and to initiate signal transduction cascades that employ Chk1 and Chk2, Serine/ Threonine kinases and Cdc25 phosphatases. These signal transducers activate p53 and inactivate cyclin-dependent kinases to inhibit cell cycle progression from G1 to S, (the G1/S checkpoint), DNA replication (intra-S-checkpoint) or G2 to mitosis (the G2/M checkpoint).

signal transducers and activators of transcription (STATs): Molecules that reason signal transduction to happen (i.e. when a hormone or other chemical 'binds' to it), or molecules that cause transcription to transpire (i.e. when transcription factor(s) 'bind' to it). STATs can be attached to solid surfaces (e.g. in a bioassay or biosensor) for use in such research applications as high-throughput screening. Upon cytokine treatment, members of STATs family of proteins are phosphorylated on tyrosine and serine sites within the carboxyl-terminal region in cells. In response to cytokine treatment, Stat3 is also acetylated on a single lysine residue, Lys^{685}. Histone acetyltransferase p300-mediated Stat3 acetylation on Lys^{685} was reversible by type 1 histone deacetylase (HDAC). Use of a prostate cancer cell line (PC3) that lacks Stat3 and PC3 cells expressing wild-type Stat3 or a Stat3 mutant containing a Lys^{685}-to-Arginine substitution revealed that Lys^{685} acetylation was critical for Stat3 to form stable dimmers required for cytokine-stimulated DNA binding and transcriptional regulation, to enhance transcription of cell growth-related genes and to promote cell cycle progression in response to treatment with on costatin M. STATs carry signals from cytokines and other extracellular stimuli into the cell nucleus. STATs each carry a Src homology2 (SH2) domain that recruits STATs into cytokine receptors at the cell surface that have been stimulated by ligand and phosphorylated. STAT dimmers form through the association of their two phosphorylated SHA domains, and then move to the nucleus where they bind to the DNA and switch on the expression of specific target genes. STATs after phosphorylation by kinases move to the nucleus, bind to specific DNA elements and direct transcription of interferon genes whose protein products interfere with viral replication.

signal transduction: Refers to the reception and conversion of a chemical message (e.g. hormone) by a cell. The process by which extracellular signal (electrical, chemical or mechanical) is reflected in the cellular response. Living cells possess the unique ability to receive, amplify and act on signals received from the outside environment. The signal represents information that is detected by specific receptors and converted to the cellular response that involves a chemical process. This conversion of information into a chemical change is called signal transduction. For instance, when plant tissues are exposed to specific growth regulators in the culture, the growth pattern changes accordingly. If a callus is transferred to a medium containing IBA, the callus produces roots. Likewise, if transferred to a medium containing cytokinin (kinetin), then callus produces shoots. In this case, hormones act as chemical signals. Signal transduction pathways occur before most of the mechanisms of regulation gene expression. In eukaryotes, a cascade of molecules causing the activation of one or more specific transcription factors is involved. Among oncoproteins, G-PROTEINS are represented by Ras proteins. Since G-proteins transduce signals, they are also described as TRANSDUCERS.

signal transduction knowledge environment (STKE): STKE database organizes information on the signalling circuitry that regulates cellular function. Brief information on the following aspects of signal transduction is available on the database: (i) toll-like receptor, (ii) cell-cell communication and signal transduction, (iii)chemotaxing, (iv) cytokines, (v) visual pathway, (vi) adrenergic and myocytic signalling pathways, (vii) apoptosis and (viii) circadian oscillators.

signalling: The communication (signalling) that ensues between and within cells of an organism (e.g. via hormones and nitric acid). Such signalling directs certain cells to grow, change or produce specific proteins at specific times.

signalling and repair of DNA damage: As a signal, DNA damage activates a series of biochemical reactions in response to ionizing radiation insult, triggering a variety of cellular processes.

signalling between plastids and the nucleus: Plastids perform versatile functions. Plastid genomes encode fewer than 100 proteins. The vast majority of plastid proteins are encoded by nuclear genes. Approximately 3000 proteins are transcribed in the nucleus, translated into the cytoplasm and then imported into plastids. After endosymbiosis, most genes were either lost or transferred to the nucleus, where they acquired the regulatory sequences for expression, as well as sequences that encode a transit peptide for targeting proteins to the plastid where they are imported and cleaved. Once these proteins reach their proper compartment within the plastid, many are incorporated into multisubunit complexes that include components encoded in plastid genomes. Thus, signalling between plastids and the nucleus is required to maintain plastid biological functions. Signalling between chloroplasts and the nucleus is bidirectional.

signalling molecule: A molecule used to communicate with cells, or to deliver a signal to other organisms (e.g. a signal by the soybean plant to attract beneficial *Rhizobium* bacteria to colonize the roots of that soybean plant). For example, the young offspring of fleas can remain immature (larvae) for up to 2 years in the absence of a food source, until carbon dioxide molecules and heat from a nearby mammal (potential host/food source) signal them to mature into adults to prey on the mammal. Another example includes the larvae of North American tree frogs that are signalled by chemicals released into a pond's water when the first such frog larva is killed by a predatory dragonfly nymph (i.e. when those dragonflies first reach each year at a given pond to prey on the frog larvae). That chemical signal reasons all of the North American tree frog larvae in that pond to subsequently grow tails that are twice as large as were grown by them before that chemical signal, to facilitate their escape from the dragonfly nymphs.

signalling specificity: Signalling specificity is complicated in eukaryotic cells because signal transduction pathways that respond to different stimuli and perform distinct functions often share the same pathway components. In haploid budding yeast, for example, the mitogen-activated protein kinase (MAPK) cascades that regulate both mating and filamentous growth share the same three activating kinases – Ste20 MAPKKKK or PAK-type kinase, Ste11 PAPKKK and Ste7MAPKK (which are activated serially) – as well as Ste12 transcription factor. The same Ste7 MAPKK activates two separate MAPKs: Fus3, which is essential for the mating programme, and Kss1, which is essential for the filamentous growth programme. Fus3 and Kss1 are both activated in response to mating pheromone by the MAPK scaffold protein Ste5; Kss1 also can be activated by cell-surface proteins such as mucin (Msb2) and other conditions that invasive growth. Signalling specificity is possible in part because MAPKs choose different substrates. For example, during mating, Fus3 phosphorylates the cell cycle regulator Far1, whereas both Fus3 and Kss1 can phosphorylate Ste12. During filamentous growth, Kss1 regulates Tec1 transcription factor indirectly through Ste12, which is guided by Tec1 to the promoters of filamentous growth genes.

signal sequence receptors: The receptors that seem to determine the fate of the entry of different proteins into different membrane-bound compartments.

significance: In statistical treatments, the probability value of >0.05 is termed not significant, those ≤0.05 but >0.01 are highly significant. Significant probability values indicate that the results, although possible, deviate too greatly from the expectancy to be acceptable when chance alone is operating.

significance level: A pre-set value of α (usually 0.05) against which P-values are judged in order to reject H_0 (*see* TYPE I ERROR). Sometimes a P-value itself is called the significance level.

signs and symptoms of plant disorders: The signs and symptoms of plant disorders are the appearance or manifestation of changes in the normal form and/or function of the plant. Signs and symptoms are usually the first indication you will notice in plant problems. Signs are the appearance and/or physical evidence of the causal factor of the plant abnormality. Signs are the physical evidence of damage caused by biotic or abiotic agents such as the pathogen itself, pests, spores, fruiting bodies, chemical residue, bacterial ooze and so forth. Symptoms are the visible response of a plant to biotic and/or abiotic factors that result in a change or abnormality in the plant. Symptoms can take the form as galls, chlorosis, ringspots, wilt, rot and so on.

SILAC-based peptide/protein quantification: SILAC is an abbreviation for Stable Isotope Labelling by Amino Acids in Cell Culture. In this method, the 'heavy' and 'light' ratios of the modified peptide, its unmodified counterpart and modified protein are produced in HPLC/MS/MS analysis.

silencer sequence: The region of a protein (usually N-terminal) responsible for cotranslational insertion into membranes of the ER.

silencers: DNA sequence that helps to inhibit transcription, located at a long distance from the core promoter.

silencing: In this context, complete or partial shut-off of transgene expression owing to factors acting at the locus of vector integration and/or the structure of the locus. Shutdown of transcription of a gene usually occurs by methylation of cytosine (C) residues.

silent alleles: *See* AMORPHIC ALLELES.

silent epialleles: These genome-wide maps provide strong evidence for the existence of silent epialleles in plant genomes, which, once identified, can be targeted for reactivation leading to increased phenotypic variation.

silent epigenetic changes: Epigenetic changes and mutations in structural genes may result in divergent effects. The effect of epigenetic changes will primarily be on the regulation of gene activity, rather than on the integrity of coding sequences. The effects of epigenetic changes may be less harmful than those of mutations in structural genes. Genes that are inactive in somatic cells may often have a cluster of 5-METHYLCYTOSINE (5^mC) rather than one 5^mC in the adjacent promoter region. Thus, the expression of the gene depends not on the loss of a single methyl group but on the sequential loss of several methyl groups. Many silent epigenetic defects are required to experience a deleterious phenotypic effect. Changes associated with ageing accumulate more and more rapidly as the end of the normal span is approached.

silent gene: A gene without a promoter that can be activated by an insertion sequence placed near an unexpressed coding sequence. The insertion sequence offers a promoter that directs transcription into sequences adjacent to its insertion site.

silent information regulator proteins: A complex of *trans*-acting silencing proteins involved in establishing and maintaining heterochromatin in budding yeast.

silent mutation: A mutation in a gene that leads to no detectable change in the biological characteristics of that gene's product (e.g. a protein) due to the degeneracy of the genetic code. Mutation do occur spontaneously or can be triggered by using different kinds of physical and chemical mutagens, such as X-rays, UV rays, γ-rays, ethyl methanesulfonate (EMS), diethyl sulfate (DES), acridine orange (AO). *See* SAMESENSE MUTATION.

silent sites: The sites in a gene describe those positions at which mutations do not alter the product.

silica gel: A substance that captivates water vapour and is stored in air-tight vessels to keep the contents dry. It is particularly useful for the long-term storage of seeds in genetic conservation.

silver staining: Via silver nitrate ($AgN0_3$) and an appropriate cell pretreatment, interphase nuclei can be rendered visible without phase contrast microscopy. They look yellow to light brown and the nucleoli seem brown. At metaphase, silver nitrate stains nucleolar organizer regions that have been active in the preceding interphase. Although the quantity of silver nitrate has some relation to the quantity of activity or number of active rDNA copies, it cannot be used for an accurate quantitative study.

silviculture: Deals with the use of the science of forestry and the farming of woodlands for commercial purposes and wildlife conservation.

SIMBA: H. Chen *et al.* (2024) offer SIMBA, a graph embedding procedure that together embeds single cells and their defining features, such as genes, chromatin-accessible regions and DNA sequences, into a common concealed space. By promoting the co-embedding of cells and features, SIMBA permits the study of cellular heterogeneity, clustering-free marker discovery, gene regulation inference, batch effect removal and omics data integration. SIMBA delivers a single framework that enables diverse single-cell problems to be formulated in a unified way and thus streamlines the development of new analyses and extensions to new single-cell modalities. SIMBA is implemented as a comprehensive Python library (https://simba-bio.readthedocs.io).

similar species: The degree to which two species or populations share identities.

similarity: The extent to which nucleotide or protein sequences are related. The extent of similarity between two sequences can be based on per cent sequence identity and/or conservation. In BLAST, similarity refers to a positive matrix score.

similarity search: Given a newly sequenced gene, two main approaches are employed to predict structure and function from the amino acid sequence. HOMOLOGY METHODS are the most commanding and are based on the detection of significant extended sequence similarity to a protein of known structure or of a sequence pattern characteristic of a protein family. STATISTICAL METHODS are less efficacious but more universal and are based on the derivation of structural preference values for single residues, pairs of residues, short oligopeptides or short sequence patterns. The transfer of structure and function information to a potentially homologous protein is forthright when the sequence similarity is high and extended in length, but the assessment of the structural significance of sequence similarity can be problematic when sequence similarity is weak or restricted to a short region. *Aka* HOMOLOGY SEARCH.

simple: Undivided or unsegmented (e.g. a leaf not divided into leaflets; however, that a simple leaf may still be complete, saw-toothed or lobed) or an unbranched hair or inflorescence.

simple gene: A continuous sequence in a nucleic acid that postulates a particular polypeptide or functional RNA. This condition occurs in the prokaryotes and viruses.

simple incomplete Trojan squares: With a simple factorial treatment structure, they can be created by omitting the complete first row from a standardized Trojan square. Assuming that a Trojan square is constructed by using standardized orthogonal Latin squares, the first row of the square will contain blocks that each comprise only a single level of the letters factor. Omitting the first row of a standardized Trojan square will bounce a design that has full factorial balance for the letters and squares factors in both the blocks-within-columns and the plots-within-blocks strata. Simple incomplete Trojan designs possess a simple factorial ANALYSIS OF VARIANCE (ANOVA) in every layer of the analysis and, if the factorial influences of the design are related to real factorial treatment effects, the design possesses a balanced factorial treatment structure.

simple interval mapping (SIM): SIM is a method of QTL mapping, which makes use of linkage maps and analyses intervals between adjacent pairs of linked markers along chromosomes simultaneously, instead of analysing single markers. The use of linked markers for analysis compensates for recombination between the markers and the QTL and is considered statistically more powerful compared with single-point analysis (SPA). MapMaker/QTL and QGene are used to conduct SIM.

simple linear regression: A regression study in which the mean value of a dependent variable y is presumed to be related to a single independent variable x by the expression $y = a + \beta x$ where a = intercept and β the slope of the regression. For example, regression of height on age.

simple modification of RNA bases: Modifications of this type include simple base modifications such as methylation (e.g. $3^{m}U$, $7^{m}G$), deamination (e.g. inosine), reduction (e.g. dihydrouridine), thiolation (e.g. $s^{2}U$) or alkylation (e.g. $tRNA^{Thr}$) of the base or methylation of the ribose sugar moiety.

simple protein: A protein that produces only amino acids upon hydrolysis (i.e. cleavage of the protein molecule into fragments) and does not have other molecular constituents such as lipids or polysaccharide attachments.

simple random sampling: A sampling procedure wherein the target population is treated as a unitary whole and each element possesses an equal probability of being selected for the sample.

simple regression: Relationship between one independent and one dependent variable. It could be linear or non-linear.

simple RNA editing: Produces C→U transitions in pre-mRNAs by specific cytidine deaminases, A→G transitions in introns by dsRNA deaminase and C→U transitions in plant organelles by an unknown mechanism.

simple sequence DNA: Stretches of DNA that consist of only one, or a few tandemly repeated nucleotides, for example, poly (dA) X poly (dT) or poly (dG-dT) X poly (dC-dA). These two types of simple sequence have been shown to be repetitive and interspersed in many eukaryotic genomes. *See* SATELLITE DNA.

simple sequence gene family: The simple sequence family includes segments of DNA derived from 10^{2}–10^{7} repetitions of a short essential sequence, generally 6–15 nucleotides in length. The degree of homology among the repeat units ranges between 80% and 100%.

simple sequence repeats (SSRs): SSRs are tandem repeats of DNA sequences of only a few base pairs (1–6 bp) in length, the most plentiful being the dinucleotide repeats. The term microsatellite was introduced to characterize the simple sequence stretches augmented by PCR. SSRs are multilocus probes creating complex banding patterns and are usually non-species specific happening ubiquitously. They both belong to the repetitive DNA family and GINGER PRINTS generated by these probes are *aka* OLIGONUCLEOTIDE FINGERPRINTS. The DNA sequences flanking SSRs are known to be conserved. The length differences are attributed to the variation in the number of repeat units at a particular SSR locus, possibly caused by slippage during replication. Different experiments have discovered that SSRs are codominant like RFLPs when used as sequence-tagged microsatellites but dominant when used as fingerprinting markers. Many alleles occur in a population and the level of heterozygosity is extremely high. Markers are inherited in Mendelian fashion. *Aka* MICROSATELLITES, SIMPLE SEQUENCE REPEAT MARKER *and* SHORT TANDEM REPEATS. *See* SATELLITE DNA.

simple sequence repeats (SSRs), applications of: (i) The rapid identification of molecules that may be targeted by researchers and the pharmaceutical industry for the development of better drug therapies. (ii) Guiding physicians to identify patients who will experience optimal benefit and minimal risk from a specific therapy. (iii) The delineation of molecular signatures that may indicate an increased risk to develop an autoimmune disease later on, or may predict a more severe disease course that requires early and/or aggressive therapy. Technological advances provide opportunities for a long-anticipated personalized medicine approach to the treatment of autoimmune diseases. The ever-increasing numbers of novel, innovative therapies will need to be harnessed wisely to achieve optimal long-term outcomes in as many patients as possible while complying with the demands of health authorities and healthcare providers for evidence-based, economically sound prescriptions of novel, expensive drugs.

simple sequence repeats (SSRs), limitations of: SSRs do have some limitations: (i) Use of SSR marker system is hampered by the requirement for sequence information from flanking regions, from which primers are designed for POLYMERASE CHAIN REACTION (PCR) amplification. (ii) Discovery and characterization of a large number of SSRs is therefore time consuming and expensive for many taxa.

simple sequence repeats (SSRs), uses of (i) SSRs have been utilized for constructing appropriate primers for amplification of the SSR loci using PCR. Any such primer or a pair of primers, when engaged to amplify a particular SSR locus in several genotypes divulges SSR polymorphism in the form of differences in length of the amplified product, each band signifying an allele at that locus. (ii) SSRs can be used for linkage analysis. (iii) SSRs, being small in size, are more mutual, clonable and docile to PCR analysis. Recently, more than 300 loci representing microsatellites, were mapped each in human and mouse genomes. Similar efforts are also being made in several crop plants. (iv) Loci representing microsatellites can be mapped. (v) New microsatellites can be cloned directly from total genomic DNA libraries or libraries or libraries enriched for specific microsatellites. For this purpose, genomic libraries with relatively small clones may be generated in sequencing vectors by ligating them with the genomic DNA restricted with frequent cutting restriction endonucleases (REs). The library can be screened for microsatellites by hybridization with repetitive oligonucleotide probes, and the positive clones identified on screening can then be sequenced. (vi) Owing to their ubiquity, hypervariability, abundance and genome-wide distribution, SSR loci represent a new generation of powerful genetic markers for eukaryotes.

simple sequences (SSs): *See* MICROSATELLITES.

simple tandem repeats (STRs): *See* MICROSATELLITES.

simple transcription units: Those simple transcription units from which only one functional mRNA is moulded from the transcription unit. The primary transcript may or may not need poly(A) addition/splicing). For instance, histone genes do not necessitate poly(A) addition or splicing to their RNAs. In protein IX mRNA in adenovirus, α-interferon mRNAs and many yeast mRNAs, poly(A) is added but splicing is not compulsory, and α-and β-globin RNAs in many cell mRNAs need poly(A) addition as well as splicing.

simple translocation: Involves transference of part of a chromosome to a non-homologous chromosome.

simplest living system: The simplest living system must be capable of replication. First living molecule was almost certainly a nucleic acid. First living molecule might have been an RNA replicase that catalysed its own replication without help from a protein. This hypothetical RNA would have functioned as both genetic material and the replication enzyme.

simplex family: A family in which only one family member is affected by a disease or disorder.

simulated annealing: An optimization algorithm that selects random deviations that improve the desired trait.

simulation: A way of forecasting the result of an action based on modelling with a computer. The imitation may be other DETERMINISTIC (thus, the same action always gives the same result) or MONTE CARLO (different runs can give different results because random sampling is involved in the simulation programme).

simulation experiment: An intended research to examine some phenomenon of interest about the model (e.g. consequence of perturbations, parameter sensitivity).

simultaneous control of different operons: Occasionally the outcomes of regulatory genes are influenced by inducers in such a way that they cross-regulate activity of two operons simultaneously. For example, products of regulatory gene i_{II} are affected by inducer i_{II} but control operon I. Products of regulatory gene i_{I} are affected by inducer i_{I} but control the operon II.

simultaneous selection: In a selection process, where several traits are targeted in the same generation.

simultaneous single-molecule imaging: An ultrasensitive single-molecule imaging technology capable of detecting and quantifying 5^{hm}C and 5^{m}C from trace amounts of DNA.

single cross: A cross between two parental plants and/or lines.

single primer amplification reaction (SPAR): A variant of melting profile (MP)-PCR technique that involves single di-, tri-, tetra- and pentanucleotide repeats to amplify genomic DNA across a panel of eukaryotes.

single random sampling: A study method in which every sample point has an identical chance of being sampled.

single replication: An experimental plot design in which each variant occurs in one replication, either randomized or non-randomized.

single-base extension (SBE): A technique for the detection of SINGLE-NUCLEOTIDE POLYMORPHISM (SNP) that uses primers ending directly adjacent to the SNP mismatch and is used to incorporate the complementary, fluorescently labelled ddNTP in a conventional PCR with *Taq* polymerase. After incorporation of the ddNTP, the reaction is terminated and the incorporated nucleotide can be detected by laser-induced fluorescence. In one single reaction, several SNPs can be discovered if primers with different 5′ tails and different fluorochromes are used.

single-cell analysis of epigenetic phenomenon: Various aspects investigated under this heading include single-cell chromatin accessibility and conformation assays (using single-cell combinational indexing-based coassay (sci-CAR) chromVAR package, EpiQ™ chromatin analysis kit and EpiDiff software), histone modifications and transcription factor binding, protein-DNA interactions, chromatin structure and chromosome organization, assaying genome-wide nucleosome occupancy, assessment of chromosome conformation, single-cell transcriptome and methylome sequencing, single-site methylation within the promoter, multi-omics of single cells and single-cell epigenomics.

single-cell analysis, challenges for: Two major challenges are faced: (i) Incompleteness of the data both for DNA and RNA analysis and (ii) Generation in the heterogeneity of gene expression from identical DNA sequences.

single-cell DNA methylation sequencing method: The highly innovative single-cell WGBS library preparation and Illumina next-generation sequencing technology have been combined to visualize genomic methylation states at single-cell resolution. Single-cell whole-genome bisulfite sequencing enables the discovery of cellular heterogeneity typically masked by standard, bulk methylation sequencing.

single-cell epigenomics: Different cells of an organism possess the same genetic material, but they still function differently. Single-cell RNA sequencing has shown how heterogeneous the transcriptome of an individual cell might be with a homogeneous cell population or tissue. Studying epigenomics in single cells confers some advantages such as detection of subpopulations expressing specific combinations of genes, finding that transcription occurs in bursts rather than in a continuous fashion, and discovery of the relationship between variation in DNA sequence and heterogeneity of gene expression.

single-cell gene expression profiling: Simultaneous visualization of gene expression profiling inside single cell with high spatial and temporal resolution. A key goal of biology is to relate the expression of specific genes to a particular cellular phenotype. However, current assays for gene expression destroy the structural context. By combining advances in computational fluorescence microscopy with multiplex probe design, researchers devised technology in which the expression of many genes can be visualized simultaneously inside single cells with high spatial and temporal resolution. Analysis of 11 genes in serum-stimulated cultured cells revealed unique patterns of gene expression within individual cells. Using the nucleus as the substrate for parallel gene analysis, J.M. Levsky *et al.* (2002) provided a platform for the fusion of genomics and cell biology under the term CELLULAR GENOMICS.

single-cell joint profiling of multiple epigenetic proteins and gene transcription: H. Xiong *et al.* (2024) developed uCoTarget, utilizing a split-pool barcoding strategy for realizing ultra-high-throughput single-cell joint profiling of multiple epigenetic proteins. and demonstrated that uCoTarget enables simultaneous detection of five histone modifications (H3K27ac, H3K4me3, H3K4me1, H3K36me3 and H3K27me3) in 19,860 single cells. They applied uCoTarget to the *in vitro* generation of HAEMATOPOIETIC STEM/PROGENITOR CELLS (HSPCs) from human embryonic stem cells, presenting multimodal epigenomic profiles in 26,418 single cells. uCoTarget revealed the establishment of the pairing of HSPC enhancers (H3K27ac) and promoters (H3K4me3) and RUNX1 engagement priming for H3K27ac activation along the HSPC path. They then developed uCoTargetX, an expansion of uCoTarget to simultaneously measure transcriptome and multiple epigenome targets. These methods thus enabled generalizable, versatile multimodal profiles for reconstructing comprehensive epigenome and transcriptome landscapes and analysing the regulatory interplay at the single-cell level.

single-cell methylation and transcriptome sequencing (scM&T-seq): This is a method for parallel single-cell genome-wide methylome and transcriptome sequencing, enabling the discovery of associations between transcriptional and epigenetic variation. Profiling of mouse embryonic stem cells confirmed known links between DNA methylation and transcription. This method reveals novel associations between heterogeneously methylated distal regulatory elements and the transcription of key pluripotency genes.

single-cell PCR (scPCR): The invention of the polymerase chain reaction (PCR) has altered the way molecular biology can be studied. With PCR, it is now possible to amplify and examine minute quantities of rare genetic material: the limit of this exploration being the single cell. Single-cell PCR has applications in many areas, especially in the field of prenatal diagnostics. In prenatal diagnosis, single-cell PCR has made possible preimplantation genetic analysis and the use of foetal cells enriched from the blood of pregnant women for the assessment of single-gene Mendelian disorders. Sc PCR has not only proven its usefulness in diagnostics but also lately has been very useful to basic scientists investigating immunological, neurological and developmental problems. Recent advances in molecular biology techniques such as whole-genome amplification (WGA) procedures and single-cell complementary DNA arrays will permit the genetic analysis of single cells to become a more common practice, thus creating new avenues for diagnosis and research.

single-cell proteins (SCPs): Proteins derived from single-celled organisms, such as yeast, fungi, algae and bacteria, which are grown on various carbon sources for synthesis, with a high protein content. Yeast is an example. SCPs are generally utilized in respect of those organisms that are edible by domesticated animals or humans. SINGLE-DOMAIN ANTIBODIES (dAbs) V_H 'heavy chains' (a portion of antibody molecules) yielded by genetically engineered *Escherichia coli* cells that function to bind antigens like antibodies or MONOCLONAL ANTIBODIES (MAbs). Similar to MAbs, dAbs can be formed in large quantities, to be used as human or animal therapeutics (e.g. to combat diseases).

single-cell reduced representation bisulfite sequencing (scRRBS): A protocol that uses one or multiple restriction enzymes on the genomic DNA to produce sequence-specific fragmentation. This methylome analysis technique enables single-cell and single-base resolution DNA methylation analysis based on scRRBS.

single-cell RNA sequencing (scRNA-seq), detection and analysis of variants in: G. Quinones-Valdez *et al.* (2022) present scAllele, an integrative approach that detects single-nucleotide variants, insertions, deletions and their allelic linkage with splicing patterns in scRNA-seq. They demonstrate that scAllele achieves better performance in identifying nucleotide variants than other commonly used tools. In addition, the read-specific variant calls by scAllele enable allele-specific splicing analysis, a unique feature not offered by other methods.

single-cell RNA sequencing (scRNA-seq): scRNA-seq methods stand as a front-line technology widely employed in biological and biomedical research. Existing scRNA-seq methods trust on reverse transcription (RT) and second-strand synthesis (SSS) to convert RNA to cDNA before amplification. These procedures frequently suffer from inadequate RT/SSS efficiency, which reduces the sensitivity of RNA detection. *See* LINEARLY AMPLIFIED SINGLE-STRANDED RNA-DERIVED TRANSCRIPTOME SEQUENCING (LAST-SEQ).

single-chain variable fragment (scFV): A fusion protein in which the epitope-binding regions of the heavy and light chains of an antibody are connected by a short linker peptide and are expressed in soluble form in cells.

single-channel microarray detection: In single-channel microarrays or one-colour microarrays, the arrays provide intensity data for each probe or probe set indicating a relative level of hybridization with the labelled target. However, they do not truly indicate abundance levels of a gene but rather relative abundance when compared to other samples or conditions when processed in the same experiment. Each RNA molecule encounters protocol and batch-specific bias during the amplification, labelling and hybridization phases of the experiment making comparisons between genes for the same microarray uninformative. The comparison of two conditions for the same gene requires two separate single-dye hybridizations. One strength of the single-dye system lies in the fact that an aberrant sample cannot affect the raw data derived from other samples, because each array chip is exposed to only one sample (as opposed to a two-colour system in which a single low-quality sample may drastically impinge on overall data precision even if the other sample was of high quality). Another benefit is that data are more easily compared to arrays from different experiments so long as batch effects have been accounted for.

single-copy artificial chromosomes: *See* HUMAN ARTIFICIAL CHROMOSOMES (HACs).

single-copy gene: A gene or DNA sequence that occurs only once in a haploid genome. Most of the structural genes (coding for proteins) signify single-copy genes.

single-copy nuclear (scn) DNA: scnDNA refers to DNA sequences that occur only once in the genome. These sequences typically include structural genes, which are responsible for coding proteins. In contrast, repetitive DNA consists of sequences that are repeated multiple times throughout the genome. Single-copy DNA is unique and thus does not have such repetitions.

single-copy plasmid: Plasmids maintained in bacterial cells in a ratio of one plasmid for each host chromosome. The advantage of using a single-copy plasmid rather than simply integrating the device into the genome is that a separate plasmid offers some isolation from the host and makes moving the device between different host strains slightly easier.

single-copy probe: A radioactive-labelled DNA or RNA sequence with exceedingly precise activity. It is used to recognize complementary regions by applying the northern and Southern hybridization techniques or colony hybridization.

single-copy region: A unique DNA region present only once in the haploid genome. Most of the structural (protein-coding) genes are single-copy genes.

single-copy sequences: These sequences comprise but do not exclusively consist of protein-coding sequences. A length of DNA with no repetitive nucleotide sequences is termed UNIQUE DNA, for which 1–3 repeats may be found. *Aka* UNIQUE SEQUENCES.

single-cross hybrid: The first generation (F_1) of a cross between two specified inbred lines.

single-cross parent: The F_1 offspring of two inbred parents, which in turn is utilized as a parent, usually with another single-cross parent, to produce a DOUBLE-CROSS HYBRID.

single-feature polymorphisms (SFPs): SFPs are recognized in transcript profiling data by envisioning differences in hybridization signals in diverse plant varieties. The polymorphisms existing in DNA are transcribed into the messenger RNA and can theoretically impact hybridization to the microarrays or GeneChip probes if occurring in a region complementary to the probe. Polymorphisms created during mRNA processing, such as alternative splicing and polyadenylation, could also impact the hybridization of the target RNA.

single-gene character: Any genetic trait whose inheritance is controlled by a single gene. The inheritance of a single-gene character follows Mendel's principles of inheritance.

single-gene disorder: A genetic disorder caused by alterations in a single gene, such as sickle-cell disease, retinoblastoma and Duchenne muscular dystrophy. *Compare with* POLYGENIC DISORDER, COMPLEX DISORDER *and* MULTIFACTORIAL DISORDER.

single-gene perturbations: The function of a single gene or a small set of genes displays variability to reveal the epigenome consequences. The comparative analysis of genome-wide DNA methylation contours of wild-type *A. thaliana* plants with mutants lacking genes shows the DNA methylation importance at several hundred genes located in *trans* to the conditioning loci. The unbiased search for genetic loci that impact DNA methylation and chromatin modifications has certain limitations, including the requirement of the successive perturbation of all genes and combinations followed by a global epigenomic evaluation of each perturbation.

single-gene resistance: Resistance deliberated by a single gene. *See* MONOGENIC RESISTANCE. *Contrasts with* MULTIGENIC RESISTANCE, OLIGOGENIC RESISTANCE *and* POLYGENIC RESISTANCE.

single-guide RNA (sgRNA): A chimeric RNA composed of crRNA and tracrRNA connected by a short RNA linker. sgRNA provides both targeting specificity and scaffolding/binding ability for Cas9 nuclease, which does not exist in nature.

single-hit single-nucleotide polymorphism (single-hit SNP): Any SNP for which each allele is present in only one sample from a distinct population.

single-marker analysis (SMA): The statistical methods of QTL mapping used for SMA include t-tests, ANOVA and linear regression. Linear regression is most commonly used because the coefficient of determination (R^2) from the marker explains the phenotypic variation arising from the QTL linked to the marker. *Aka* SINGLE-POINT ANALYSIS (SPA).

single-molecule fluorescence resonance energy transfer (smFRET): A single-molecule fluorescence microscopy technique in which the proximity or environment of donor (e.g. Cy3) and acceptor (e.g. Cy5) fluorophores is monitored over time.

single-molecule real-time (SMRT) detection of 6^{m}A mRNA using reverse transcription (RT): SMRT®) technology to track the activity of RT enzymes synthesizing cDNA on thousands of single RNA templates concurrently in real time with single-nucleotide turnover resolution using arrays of Zero-mode waveguides (ZMWs).

single-molecule real-time (SMRT) sequencing: A parallelized single-molecule DNA-sequencing method. SMRT sequencing utilizes a ZMW. This method was given by M.J. Levene *et al.* (2003). A single DNA polymerase enzyme is affixed at the bottom of a ZMW with a single molecule of DNA as a template. The ZMW is a structure that creates an illuminated observation volume that is small enough to observe only a single nucleotide of DNA being incorporated by DNA polymerase. Each of the four DNA bases is attached to one of four different fluorescent dyes. When a nucleotide is incorporated by the DNA polymerase, the fluorescent tag is cleaved off and diffuses out of the observation area of the ZMW where its fluorescence is no longer observable. A detector detects the fluorescent signal of the nucleotide incorporation, and the base call is made according to the corresponding fluorescence of the dye. *See* ZERO-MODE WAVEGUIDE.

single-molecule real-time bisulfite sequencing (SMRT-BS): This method comprises five steps: (i) bisulfite conversion of genomic DNA, (ii) amplification of bisulfite-treated DNA using region-specific primers coupled with universal oligonucleotide tags, (iii) second-round PCR with universal primers tagged with barcodes, (iv) amplicon pooling and SMRT sequencing and (v) and data analysis.

single-molecule sequencing: Refers to techniques that can read the base sequence directly from individual strands of DNA or RNA present in a sample of interest.

single-molecule sequencing (SMS) technologies: Third-generation DNA-sequencing technologies that are under development are expected to significantly reduce the cost and effort of sequencing in comparison with second-generation technologies – despite the fact that cost reductions continue to be made for second-generation approaches. Various single-molecule DNA methylation sequencing technologies include NANOPORE SEQUENCING, SINGLE-MOLECULE REAL-TIME (SMRT) SEQUENCING, SIMULTANEOUS SINGLE-MOLECULE IMAGING and DNA FIBRE-BASED FLUORESCENCE *IN SITU* HYBRIDIZATION (DNA FIBRE FISH).

single-nucleotide polymorphism (SNP): Pronounced 'snip'. DNA sequence differences between individuals at certain nucleotide positions in the genome arise when a single nucleotide (A, T, C or G) in the genome sequence is altered. For example, two sequenced DNA fragments from different individuals, AAGCCTA to AAGCTTA, contain a difference in a single nucleotide. In this case, we say that there are two alleles: C and T. Such genetic variations trigger differences in our susceptibility to, or protection from all kinds of diseases. The severity of illness and the way our body retorts to treatments are also indicators of genetic variations. For example, a single-base mutation in the *Apolipoprotein E* gene is related to a higher risk for Alzheimer's disease. This type of variation is common and can be characterized in populations. SNPs occur within DNA at the rate of approximately one in every 1300 bp in most organisms; SNPs usually occur in the same genomic location in different individuals. The human genome should be approximately 3,000,000 SNPs in the entire genome.

SNPs are useful markers for gene mapping studies. SNPs involve either gel-based assays or non-gel-based assays. SNPs approach relies on the capacity to distinguish a perfect match from a single-base mismatch. These are more stable and the relative fidelity of their inheritance is higher than other marker systems like microsatellites and AFLPs. SNPs have also been constructed in humans. Besides, SNPs have also been discovered in large numbers in crops like maize, wheat and barley. *Aka* SIMPLE-NUCLEOTIDE POLYMORPHISM *and* SINGLE-NUCLEOTIDE VARIANT (SNV).

single-nucleotide polymorphism (SNP) cluster: Accumulation of SNPs in relatively small genomic regions. Such clusters can extend over 1 kb (or more), are relatively rare but very old, and originate from ancestral chromosome fragments inherited in the extant species.

single-nucleotide polymorphism database (dbSNP): A database for SNPs, short deletions and insertion polymorphisms.

single-nucleotide polymorphism (SNP) density: The frequency of SNPs per unit length of genomic DNA (usually SNPs per kilobase). SNP density is different for different regions of genomes.

single-nucleotide polymorphism (SNP) frequency: The frequency with which SNPs occur along a defined stretch of DNA or a chromosome (usually expressed as SNP per kilobase). SNP frequency varies between genomic regions in the same individual and between related individuals.

single-nucleotide polymorphism (SNP) map: The linear arrangement of SNPs along a specific region in two homologous genomes. A group of known/detailed SNPs, superimposed on to the genome map of an organism (e.g. to facilitate genetic/population studies, such as of genetically related disease susceptibility).

single-nucleotide polymorphism (SNP) scanning: The *in silico* search for SNPs in a sequenced stretch of DNA (e.g. a BAC clone, a genomic segment or, in extreme cases, whole genomes).

single-nucleotide polymorphism (SNP) scoring: The search for SNPs in two (or more) DNA sequences (in extreme cases, genomes) and their characterization and use for genome analysis.

single-nucleotide polymorphism (SNP), abundance of: As of 26 June 2012, dbSNP listed 187,852,828 SNPs in humans. (i) rs6311 and rs6313 are SNPs in the *HTR2A* gene on human chromosome 13. (ii) A SNP in the F5 gene causes a hypercoagulability disorder with the variant Factor V Leiden. (iii) rs3091244 is an example of a triallelic SNP in the CRP gene on human chromosome 1. (iv) TAS2R38 codes for PTC tasting ability and contains six annotated SNPs.

single-nucleotide polymorphism (SNP), analysis of: Analytical procedures used to discover novel SNPs and detect known SNPs include (i) DNA sequencing, (ii) capillary electrophoresis, (iii) mass spectrometry, (iv) single-strand conformation polymorphism (SSCP), (v) electrochemical analysis, (vi) denaturing HPLC and gel electrophoresis, (vii) RFLP and (viii) hybridization analysis.

single-nucleotide polymorphism (SNP), characteristics of: (i) Numerous. (ii) Can be quite polymorphic. (iii) Should be highly transferable. (iv) Can be restricted to coding DNA. (v) Can be codominant. (vi) Several SNP genotyping technologies have been developed. Suitability of SNP technology is determined in terms of (a) sensitivity, (b) Reproducibility, (c) accuracy and (d) capability of multiplexing. No single protocol meets all research needs.

single-nucleotide polymorphism (SNP), databases: As for genes, bioinformatics databases also occur for SNPs. dbSNP is a SNP database from the National Center for Biotechnology Information (NCBI). SNPedia is a wiki-style database supporting personal genome annotation, interpretation and analysis. The OMIM database describes the relation between polymorphisms and diseases, the Human Gene Mutation Database delivers gene mutations producing or related with human inherited diseases and functional SNPs and GWAS Central enables users to visually interrogate the factual summary-level association data in one or more genome-wide association studies.

single-nucleotide polymorphism (SNP), detection of: Identifying single nucleotide polymorphism among alleles within or between populations. Several applications of microarrays make use of SNP detection, including genotyping, forensic analysis, measuring predisposition to disease, identifying drug candidates, evaluating germline mutations in individuals or somatic mutations in cancers, assessing loss of heterozygosity, or genetic linkage analysis.

single-nucleotide polymorphism (SNP), examples of: Variations (in individual nucleotides) that ensue within DNA at the rate of approximately 1 in every 1300 base pairs in most organisms (approximately one in every 100 base pairs in humans' DNA). SNPs usually arise in the same genomic location (e.g. on the organism's DNA) in different individuals. These variations account for diversity within a given species (e.g. black cattle and white cattle and different strains/serotypes within a given bacteria species), and some genetic diseases (e.g. cystic fibrosis) and familial dysautonomia Duchenne muscular dystrophy each] are due to one SNP and (c) the body's response to certain pharmaceuticals and food ingredients (e.g. the diuretic drug thiazide works to control hypertension in 60% of US African Americans, but only 8% of US Caucasian people, due to one SNP). Certain pharmaceuticals do not have the desired effect in some groups of humans possessing certain specific 'grouped SNPs' known as HAPLOTYPES. Because those groupings of SNPs are linked (i.e. tend to travel together as a group within the genetics of a given population), they can collectively confer a given MULTIPLE-SNP-TRAIT to an identifiable subpopulation of individuals. For example, the pharmaceuticals acetaminophen, aspirin and Valium remain in the bodies of women (who constitute a haplotype for that pharmacogenomic trait) longer than in men. Methods utilized to identify SNPs include examination of the DNA of populations of individuals with and without a given (genetically related) disease and with and without a given trait. SNP MAPPING is a genetic mapping technique that utilizes the fact that individual nucleotides (within a DNA molecule) can exist in different forms (for a particular site/location on that DNA molecule), which enables such SNPs to be utilized as markers. One example would be to track a given SNP versus the occurrence of genetically related disease in a given human population.

single-nucleotide polymorphism (SNP), importance of: Variations in human DNA sequences can impact how humans develop diseases and retort to pathogens, chemicals, drugs, vaccines and other agents. SNPs are also considered the key enablers in understanding the notion of personalized medicine. However, their greatest relevance in biomedical research is for comparing regions of the genome between cohorts (such as with matched cohorts with and without a disease) in genome-wide association studies. The study of SNPs is also vital in crop and livestock breeding programmes. SNPs are usually biallelic and thus easily assayed. A single SNP may cause a Mendelian disease. For complex diseases, SNPs do not generally function individually; rather, they work in harmonization with other SNPs to manifest a disease condition. *See* SNP GENOTYPING for details on the various methods used to identify SNPs.

single-nucleotide polymorphism (SNP), limitations of: (i) These are biallelic which means in a population only two alleles are present. (ii) These are population specific and can be used only on one population.

single-nucleotide polymorphism (SNP), salient features of: (i) Two of every three SNPs involve the replacement of cytosine (C) with thymine (T). (ii) SNPs can occur in both coding (gene) and NCRs of the genome. Many SNPs do not affect cell function and they don't form a mutant phenotype. (iii) When they are present in coding regions, they may present at the third position of the codon hence no effect on the phenotype. But when a nucleotide is present at first and second position they may lead to a change in the sequence of amino acid and hence leads to the mutant phenotype. (iv) When the SNP is present in the coding region, they may show cent per cent transmission with the trait. But when they are away from the gene or are present in the non-coding region, they don't show cent per cent transmission with the trait. (v) SNPs are sequence-based molecular markers. SNPs show intraspecific diversity all individuals of the same species have the same DNA sequence but the differences occur at certain loci due to SNPs. In case of human beings, 99.9% of DNA is the same but the difference is due to 0.1% SNPs. These SNPs make us individuals rather than identical clones.

single-nucleotide polymorphism (SNP), types of: SNPs may be located within coding sequences of genes, non-coding regions of genes, or in the intergenic regions (regions between genes). SNPs within a coding sequence do not essentially change the amino acid sequence of the protein that is produced, due to degeneracy of the genetic code. A SNP in which both alleles produce the same polypeptide sequence is called a SYNONYMOUS POLYMORPHISM (or SILENT MUTATION). If a different polypeptide sequence is produced, the polymorphism is a REPLACEMENT POLYMORPHISM. A replacement polymorphism change may be either missense, which results in a different amino acid, or nonsense, which results in a premature stop codon. Over half of all known disease mutations come from replacement polymorphisms.

SNPs that are not in protein-coding regions may still affect gene splicing, transcription factor binding, messenger RNA degradation or the sequence of non-coding RNA. Gene expression affected by this type of SNP is referred to as an eSNP (expression SNP) and may be upstream or downstream from the gene.

single-nucleotide polymorphism (SNP), uses of: (i) These are the most abundant type of molecular markers present and are spread on the whole of the genome. and can be used to create high-density maps and also can be used to tag the gene of interest. (ii) These are codominant markers and can be used to distinguish heterozygotes from the homozygotes. (iii) Cultivar identification and evaluation of diversity identify the level of inter-varietal variation for each genome. In wheat, one SNP per 20 bp and in the maize genome, one SNP per 70 bp have been recorded in certain regions of these genomes. (iv) Functional proteomics identifies the SNPs that modify proteins and protein active sites. (v) Gene therapy involves the insertion of normal genes into germ cells or fertilized cells to create a beneficial genetic change.

single-nucleotide polymorphic marker: Most commonly, SNPs are found in the DNA between genes. They can act as biological markers, helping scientists locate genes that are associated with disease. When SNPs occur within a gene or in a regulatory region near a gene, they may play a more direct role in disease by affecting the gene's function.

single-pass sequencing: Rapid sequencing of large segments of the genome of an organism by isolating as many expressed (cDNA) sequences as possible and performing single sequencer runs on their 5′ or 3′ ends. Single-pass sequencing characteristically causes individual, error-prone sequencing reads of 400–700 bases, depending on the type of sequencer used. However, if many of these are produced from several clones from different tissues, they may be overlapped and assembled to remove the errors and generate a contiguous sequence for the entire expressed gene.

single-pass transmembrane protein: Membrane protein in which the polypeptide chain crosses the lipid bilayer only once.

single-plant comparisons: A method to guess crop losses by comparing yields of individual plants or components differentially affected by yield constraints.

single-plant plot: Design used for the replicated assessment of experimental lines or cultivars by the, for example, honeycomb design. The number of plants evaluated is equal to the number of replications in the experiment. The plots are organized systematically to allow comparison of a plant of one line with adjacent plants of other lines.

single-polypeptide nuclear RNA polymerase: Transcription of some mRNAs in humans and rodents is arbitrated by SINGLE-POLYPEPTIDE NUCLEAR RNA POLYMERASE IV (spRNAPIV). It is expressed from an alternative transcript of the mitochondrial RNA polymerase gene *POLRMT*. The spRNAPIV lacks 262 amino-terminal amino acids of mitochondrial RNA polymerase, including the mitochondrial-targeting signal and localizes into the nucleus.

single-row plot: An experimental field design in which a single-plant row of different lengths embodies one plot.

single-seed descent (SSD): Derivation of plants by employing a selection process in which F_2 plants and their progeny are advanced by single seeds until genetic purity is achieved. Single-seed descent methods (single-seed, single-hill, multiple-seed) are easy means to continue populations during inbreeding. Natural selection cannot influence the population unless genotypes differ in their ability to produce viable seeds. Artificial selection is based on the phenotype of individual plants, not on the progeny performance.

single-seed descent (SSD) method: SSD method is the modification of bulk procedure, which was first proposed by C.H. Goulden in 1939 and later modified by C.A. Brim in 1966. Instead of bulking whole seed lot of selected plants, a single seed is selected randomly from each selected plant to make bulk in each generation. This approach assumes a rapid segregation advanced from F_1 to F_6 with limited space and effort, without performing any selection. This method is now generally used for developing populations for gene mapping and tagging.

single-sequence repeat (SSR) DNA marker technique: A genetic mapping technique that utilizes the fact that microsatellite sequences appear repeatedly in sequence, within the DNA molecule, in a manner that allows them to be utilized as markers.

single-strand annealing (SSA): A process initiated when a double-strand break is made between two repeated sequences oriented in the same direction.

single-strand annealing (SSA) model of recombination: This model was projected to pronounce extrachromosomal recombination in somatic plant cells. This is a non-conservative type of DNA recombination.

single-strand assimilation: The ability of RecA protein to cause a single strand of DNA to displace its homologous strand in a duplex, that is, the single strand is assimilated into the duplex.

single-strand binding (ssb) protein: One of the proteins that attach to single-stranded DNA in the region of the replication fork, preventing reannealing of unreplicated DNA. These proteins are ubiquitous and essential for a wide variety of DNA metabolic processes, including DNA replication, recombination, damage detection and repair. SSBs have multiple roles in binding and sequestering ssDNA, detecting DNA damage, stimulating nucleases, helicases and strand-exchange proteins, activating transcription and mediating protein-protein interactions.

single-strand conformation polymorphism (SSCP): This molecular tool enables the detection of polymorphism due to alterations of one or more base pairs in the PCR products and banks on the secondary structure being different for single strands derived from PCR products that differ by one or more nucleotides at an internal site within the strand. In order to detect such differences, PCR products are devalued and electrophoretically separated in neutral acrylamide gels. PCR products that do not differ in fragment length have been shown to exhibit SSCP in several studies. Sequence information requires designing the primers and radioactive labelling, and autoradiography is used to detect SSCP variants. This makes this technique relatively unsuitable for routine mapping/tagging studies. SSCP trusts on secondary and tertiary structural differences between denatured and rapidly cooled amplified DNA fragments that differ slightly in their DNA sequences. Different SSCP alleles are determined on non-denaturing acrylamide gels, usually at low temperatures. The ability to identify alleles depends on the conditions of electrophoresis and this requires DNA sequence data.

single-strand DNA-binding proteins: A distinct group of DNA-binding proteins are the DNA-binding proteins that specifically bind single-stranded DNA. In humans, replication protein A is the best-understood member of this family and is used in processes where the double helix is separated, including DNA replication, recombination and DNA repair. These binding proteins seem to stabilize single-stranded DNA and protect it from forming stem-loops or being degraded by nucleases. *Aka* SSB PROTEINS.

single-strand exchange: A reaction in which one of the strands of a duplex DNA leaves its former partner and instead pairs with the complementary strand in another molecule, displacing its homologue in the second duplex.

single-stranded antisense or guide strand: The strand of the siRNA complementary to the target transcript and the favoured strand for loading into effective RISCs for sequence-specific cleavage. Functional analysis using RNAi, an assessment of protein function through the use of RNAi-based methods, resources and technologies.

single-stranded DNA (ssDNA): Although DNA mostly is double stranded, single-stranded DNA has also been found to occur. Single-stranded DNA molecules have 5′→3′ polarity. When ssDNA is held in an aqueous solution, it shows random coiled features, for each of the nucleotides have bases linked to their sugars by glycosidic bonds at C1′ position in such a way their sugars show C2-endo and C2-Exo puckering, hence they show anti-configuration. The ssDNA strands also display duplex or hairpin or stem-loop structural features wherever the sequences are complementary to each other. Most of the ssDNA genomes, either linear or circular, are generally packed into viral proteinaceous coats or capsids. In host cells, they reveal supercoiled states associated with proteins that provide stability. Replication of ssDNA is through ds replicative form of DNA. Many viruses have single-stranded forms of DNA. Many viruses have single-stranded DNAs as their genome. They can be either circular or linear (e.g. ϕX174 and M13).

single-stranded DNA break (ssDB): Cleavage of the DNA such that the backbone of one of the strands is nicked while the other one remains intact.

single-stranded DNA library preparation: Single-stranded library preparation is performed by ligating a biotinylated adapter oligonucleotide to the 3′-ends of heat-denatured DNA. The resulting strands are then immobilized on streptavidin-coated beads and copied with a polymerase. A second adapter is attached by blunt-end ligation, and library preparation is completed by PCR amplification.

single-stranded DNA recombineering (SSDR): Recombineering based on single-stranded linear DNA fragments. SSDR only requires expression of the phage-derived ssDNA-binding protein Beta or RecT.

single-stranded linear T-DNA transfer intermediate: The presence of a single-stranded linear T-DNA molecule called T-strand was detected and confirmed. T-strand is produced at about one copy per bacterium and corresponds to the bottom strand of the nopaline T-DNA region such that its 5′ and 3′ ends map to the right and left T-DNA border repeats.

single-stranded M13 DNA isolation using phenol: This semi-automated isolation procedure is used in Sequenase[TM]-catalysed dye-terminator reactions. The phage particles are precipitated with **PEG**, collected by centrifugation and the pellet is resuspended in buffer. The phage protein coat is denatured and removed by one phenol and two ether extractions. After ethanol precipitation, the dried DNA pellet is resuspended in a buffer, and the concentration and purity crudely are assessed by agarose gel electrophoresis against known standards.

single-stranded RNA (ssRNA): (i) RNA produced by transcription is ssRNA. (ii) ssRNA is the simplest form of genetic molecule and constitutes the genome in some viruses. There are two kinds of ssRNA viruses: NEGATIVE-SENSE SINGLE-STRANDED RNA VIRUSES and POSITIVE-SENSE SINGLE-STRANDED RNA VIRUSES. *See details as separate terms.*

single-stranded RNA (ssRNA) viruses: Those viruses that possess single-stranded RNA (ssRNA) as genetic material. Examples of ssRNA viruses are Tobacco Mosaic Virus (TMV), Influenza Virus, Rous Sarcoma Virus, Poliomyelitis and bacteriophages MS2, F2 and r17.

singleton variable: A variable that is referenced only once and may be used as a dummy argument in a function call, or when its address is assigned to another variable, which subsequently accesses its assigned storage. Singleton variables sometimes arise because of a mistake made in assigning a value to a variable and forgetting to use it later or mistyping one instance of the variable name. Some compilers and lint-like tools flag incidences of singleton variables.

singularity versus plurality of chromosomes: The main genetic element contained in a prokaryotic cell is a single, circular 1 mm long double-stranded DNA molecule called NUCLEOID. In addition to the main chromosome, there may be one or more small-sized circular double-stranded molecules, called plasmids. Plasmids are negatively supercoiled DNA molecules that usually carry non-essential genes (e.g. drug resistance). Plasmids are replicating entities independently of the bacterial chromosome when lying free in the bacterial cell. Some plasmids can become integrated into the main bacterial chromosome. In this case, plasmid DNA will replicate along with main bacterial chromosome. Plasmids have been found to be very useful in the construction of recombinant DNA molecules. In addition to nucleoids and plasmids, some bacterial cells carry bacteriophages. In eukaryotes, genetic material is divided into of several large chromosomes and sometimes a few small (supernumerary) chromosomes. In eukaryotes, DNA is coiled around nucleosome particles. In eukaryotes, in addition to the nucleus, genetic material is present in cell organelles like mitochondria, chloroplasts and kinetoplasts.

sink A term used for storage organs (e.g. grain, fruit, tubers, roots as in carrot, stem as in sugarcane) of a plant into which products of photosynthesis are transferred and stored.

siRNA Target Designer: SiRNA that guides sequence-specific degradation of the homologous mRNA, thus producing 'knock-down' cells.

sister-chromatid cohesion during DNA replication termination: Cohesion is established during DNA replication but how sister DNAs become trapped inside cohesin rings was made known when G. Cameron *et al.* (2024) demonstrated that cohesins are pushed along DNA by the replication machinery (replisome) until they meet another replisome. Whereas replisomes disassemble upon fork convergence during DNA replication termination, cohesins persist, establishing cohesion at replication termination sites. Disrupting replisome disassembly after fork convergence in living cells hinders cohesion, underscoring a crucial connection between sister-chromatid cohesion and the termination of DNA replication. Additionally, they display that **CMG** (CDC45–MCM2-7–GINS) helicase disassembly during replication termination is vital for proper cohesion in budding yeast. This study supports a model wherein sister-chromatid cohesion is established during DNA replication termination.

sister-chromatid cohesion: Eukaryotic genomes are organized by loop extrusion and sister-chromatid cohesion, both mediated by the multimeric cohesin protein complex. Using quantitative super-resolution imaging, F. Ochs *et al.* (2024) identified two discrete populations of chromatin-bound cohesin in postreplicative human cells. Whereas most complexes appear dimeric, cohesin that localized to sites of sister-chromatid cohesion and associated with sororin was exclusively monomeric. The monomeric stoichiometry of sororin:cohesin complexes demonstrates that sister-chromatid cohesion is conferred by individual cohesin rings, a key prediction of the proposal that cohesion arises from the co-entrapment of sister DNAs.

sister-chromatid cohesion, establishment of: Two newly duplicated copies of genomic DNA are held together by the ring-shaped COHESIN complex to ensure the faithful inheritance of the genome during cell division. Cohesin mediates sister-chromatid cohesion by topologically entrapping two sister DNAs during DNA replication. Y. Murayama *et al.* (2024) studied the interplay between cohesin and replication by reconstituting a functional replisome using purified proteins. Once DNA is encircled before replication, the cohesin ring accommodates replication in its entirety, from initiation to termination, leading to topological capture of newly synthesized DNA. Replisome-associated cohesion establishment factors Chl1 helicase and Ctf4 promote cohesin loading specifically during continuing replication. Bubble DNA, which mimics the state of DNA unwinding, induces topological cohesin loading, which is further promoted by Chl1. These results show how cohesin initially responds to replication and provide a molecular model for the establishment of sister-chromatid cohesion.

sister-chromatid differential (SCD) staining: Differential staining with fluorochromes and conventional dyes of sister chromatids, when the DNA has been substituted for two rounds of replication with 5′ bromodeoxyuridine or one round followed by another in thymidine.

sister-chromatid exchange (SCE): An event, similar to crossing-over, that can ensue between sister chromatids at mitosis and meiosis. Such events lead to homogenization. It may be detected in harlequin chromosomes. *See* HARLEQUIN CHROMOSOMES.

sister chromatids: The two identical chromatids derived from the replication of one chromosome and joined together by a centromere are called SISTER CHROMATIDS.

site and timing-specific gene targeting: The efficient introduction of somatic mutations in a given gene, at a given time, in a specific cell type, facilitates studies of gene function and the generation of animal models for human diseases. A conditional site-specific recombination system has been developed in mice using a new version of the Cre/*lox* system.

site(s): Sites in sequences can be located either in DNA (e.g. binding sites, cleavage sites) or in proteins. To find a site in DNA, ambiguity symbols are used to permit several different symbols at one position. Proteins need a different mechanism, however (*see* PATTERN). Restriction enzyme cleavage sites, for example, have the following properties: limited length (typically, fewer than 20 base pairs), definition of the cleavage site and its appearance (3′, 5′ overhang or blunt) and definition of the binding site.

site-determined selection: A technique to select varieties (also population varieties) that are optimally modified to specific environmental conditions. The parental lines are also selected with the final growing site of the target crop. Crosses are carried out at a central location where the F_1 (up to F_4) generations are also grown. The F_5 to F_n plants are assessed and selected at different locations through pedigree selection (i.e. a combination of natural and artificial selection). Thus, environmental factors control that characteristics are expressed and which are not. This may influence the breeder's selection of promising phenotypes. This method is frequently practiced in cereal breeding but is also relevant to other crops.

site-directed: Any biological activity specific to a specific location.

site-directed mutagenesis (SDM): The introduction of a mutation, usually a point mutation or an insertion, into a particular location in a cloned DNA fragment. A technique that can be employed to produce a protein that differs slightly in its structure from the protein normally produced (by an organism or cell). A single mutation (in the cell's DNA) is caused by hybridizing the region in a codon to be mutated with a short, synthetic oligonucleotide. This causes the codon to code for a different specific amino acid in the protein gene product. SDM holds the potential to enable man to create modified (engineered) proteins that have desirable properties not currently available in the proteins produced by existing organisms. This mutated fragment may be used to 'knock out' a gene in the organism of interest by homologous recombination.

site-directed nuclease (SDN): Engineered DNA nucleases that are programmed to specific sites within the genome where they cleave a DNA chain by separating nucleotides.

sites on the 50S subunit: Three sites exist on the 50S ribosomal subunit: (i) Acyl site (A site) required for the attachment of aminoacyl (amino acid tRNA), (ii)

A peptidyl site (P site) required for the growing peptide chain attached to tRNA. F-met-tRNA$_f^{met}$ sits in the P site and (c) The exit site (E-site) is the binding site for tRNA that is released after peptide bond formation.

site-specific epigenome editing: The site-specific approaches of epigenome editing are classified into three categories: (i) targeting of short interfering RNAs (siRNAs), (ii) targeting of lncRNAs and (c3) engineered DNA-binding proteins attached with effector molecules.

site-specific epigenome-editing tools: Classified according to the source of the DNA-binding domain. For example, (i) MEGANUCLEASES, *aka* HOMONUCLEASES-BASED GENOME-EDITING TOOLS, (ii) ZINC-FINGER PROTEIN (ZFP)-BASED EPIGENOME-EDITING TOOLS, (iii) TRANSCRIPTION ACTIVATOR–LIKE EFFECTOR (TALE)-BASED EPIGENOME-EDITING TOOLS and (iv) CLUSTERED REGULARLY INTERSPACED SHORT PALINDROMIC REPEAT (CRISPR)/DCAS-BASED EPIGENOME-EDITING TOOLS.

site-specific mutagenesis: The use of recombinant DNA technology to generate specific deletions, insertions or substitutions *in vitro* at a particular site in the gene. The technique enables the production of proteins having any desired amino acid at any position.

site-specific recombination (SSR): A specialized recombination in which a plasmid is inserted into or is spliced out of a host chromosome. This type of recombination that occurs in two specific sequences that require not be homologous. The proteins mediating this process are site-specific recombinases, which identify short, specific DNA sequences in the donor and recipient molecules, and interaction between the proteins facilitates recombination. Examples of SSR are (i) F-factor and *Escherichia coli* chromosome and (ii) l phage and *Escherichia coli* chromosome.

site-specific recombination, flippase (FLP)-flippase recombination target (FRP) system of: The FLP-FRP system of yeast 2μ plasmid is also believed to function in resolving the plasmid multimers and may enact a part in the increase of copy number. Unlike Tn3 resolvase and Cre, however, FLP recombinase falls under the λ integrase family and catalyses inversions and fusions as well as resolutions. Of these systems, FLP-FRP and Cre-*loxP* have been widely used as tools for genome manipulation and gene regulation.

site-specific recombineering: A recombination engineering system that allows efficient manipulation of genomic DNA at predetermined locations. It does not require extensive sequence similarity and relies on-site-specific recombinases that catalyse reciprocal recombination of DNA at short sequences.

6-methyladenosine (6^{m}A) individual nucleotide cross-linking and immunoprecipitation (miCLIP): This is the preferred method for mapping 6^{m}A. UV light covalently crosslinks 6^{m}A-specific antibodies at 6^{m}A residues in purified RNA. Crosslinked antibody-RNA complexes are then proteolysed to produce peptide-nucleotide adducts that induce terminations or nucleotide misincorporations adjacent to 6^{m}A during RT.

6-methyladenosine (6^{m}A) methylome of mRNA: Detections of revocable 6^{m}A on *Arabidopsis* mRNA and mapping of 6^{m}A methylomes in mammals and yeast have discovered probable regulatory functions of this RNA modification.

6-methyladenosine (6^{m}A) methyltransferases: Those enzymes that generate N6-methyladenine (EC 2.1.1.72). Found in the three existing types of bacterial restriction-modification systems (in type I system the A-MTase is the product of the *hsdM* gene. Responsible for the methylation of specific DNA sequences to prevent the host from digesting its genome via its restriction enzymes. Have the same sequence specificity as their corresponding restriction enzymes. Contain a conserved motif Asp/Asn-Pro-Pro-Tyr/Phe in their N-terminal section. Folds into two domains – an N-terminal catalytic domain, which contains the catalytic and cofactor-binding sites, and a C-terminal DNA recognition domain.

6-methyladenosine (6^{m}A) RNA and gene expression control: Methyltransferase-like enzymes (METTL3 and METTL14) and Wilms tumour 1-associated protein (WTAP) are recognized as constituents of a 6^{m}A methyltransferase complex. FTO and ALKBH5 (α-ketoglutarate-dependent dioxygenase alkB homologue 5) are the two 6^{m}A demethylating enzymes yet recognized. But based on the studies of the mRNA targeting specificity and tissue-specific expression patterns of these enzymes, it is highly probable that many additional 6^{m}A demethylases exist. 6^{m}A is present in a large subset of the transcriptome in specific regions of mRNA. mRNA may undergo this post-transcriptional modification to regulate its fate and function.

6-methyladenosine (6^{m}A) RNA marks on primary microRNAs for processing: METTL3 methylates pri-miRNAs, marking them competent for recognition and processing by DGCR8.

6-methyladenosine ($6^{m}A$) RNA regulates UV-induced DNA damage response: $6^{m}A$ in RNA is rapidly (within 2 min) and transiently induced at DNA damage sites in response to ultraviolet (UV) irradiation. This modification occurs on numerous poly(A)$^{+}$ transcripts and is regulated by METTL3.

sK site: Host specificity site of *Escherichia coli* K strain.

skew: Distortion in the shape of the normal distribution to near one side or the other.

skewed distribution: A distribution is skewed if one of its tails is longer than the other. It is right skewed when mean > median > mode and left skewed when mean < median < mode.

skip-generation inheritance: A male transmits his sex-linked genes to his grandsons through his daughters, never to or through his sons. *See* SEX-LINKED INHERITANCE *and* CRISS-CROSS INHERITANCE.

skotomorphogenesis: The development of a seedling in the dark, which is characterized by etiolation. Upon exposure to light, the seedling switches rapidly to PHOTOMORPHOGENESIS (light development).

slab gel sequencing systems: These systems make use of ultrathin (75 μm) slab gels and involve running of at least 96 lanes per gel, the runs being completed twice daily (Jun and Yoshihide, 1999). In these systems, automation in sample loading of sequencing gels has also been achieved, by using a plexiglass block that has wells that are same distance apart as the comb teeth cut in drawing samples by capillary action. Each well in plexiglass block is filled with a sample (PCR dideoxy-reaction mixture), so that when the porous membrane comb is lowered on to the sample wells in the plexiglass block, the samples are drawn up automatically into the comb teeth by capillary action. Using this approach of employing porous combs, automated loading of up to 192, 384 or 480 samples per gel has been achieved. The porous comb with the samples is placed between the glass plates of the gel apparatus above the flat surface of the polymerized gel and the samples are driven from the comb into the gel by electrophoresis.

sleeping beauty (SB) transposon system: A transposon related to mariner, originally isolated from fish. SB transposon system composed of an SB transposase and a transposon that was designed to insert specific sequences of DNA into genomes of vertebrate animals.

slipped DNA: In slipped DNA, one of the strands contains sequence repeats. Slipped DNA arises during replication by DNA polymerases, where it can produce greater length of DNA in one strand and deletions in the other. Slipped strands have been identified and isolated. Nine such loci in humans have been identified that are found to be highly unstable. The repeats found are CTGnCAGn, CTGn CAGn. Single-stranded oligos of CCG's or CGG's can form intermolecular duplexes or intramolecular duplexes. Such stable slipped segments can expand and become heritable. They can affect gene expression, protein bindings, transcription initiation and possibly replication. Slipped DNA is found upstream of important regulatory sites.

slip-strand mispairing (SSM): A procedure in which a DNA polymerase adds too many or too few copies of a repetitive sequence during replication.

sloppy copiers: The sloppy copiers DNA polymerases are *aka* ERROR-PRONE DNA POLYMERASES (EP POLS). In *Escherichia coli* two EP pols: umuD'2C (PolV) and DinB (PolIV) are well studied. EP pols have the following properties: (i) an ability to copy damaged DNA with high efficiency, (ii) poor accuracy in nucleotide incorporation with base substitution error frequencies of $\sim 10^{-1}$–10^{-3}, (iii) a tendency to form base mispairs rather than correct Watson-Crick base pairs and (iv) a propensity to catalyse incorporation using aberrant DNA primer ends, including base mismatches, misaligned primer-template and DNA damage sites.

slot blot procedure: Slot blot procedure consists of seven steps: (i) To each sample, equal amounts of DNA are added to the wells of a 96-well plate and volume is kept as small as possible. (ii) 100 μl of 0.1 N NaOH is added to each well. Plate is incubated at 65°C for 30 min. (iii) 100 μl of 2 M ammonium acetate is added to each well. (iv) Nylon membrane(s) is prepared, cut to the suitable size and labelled. Membrane is placed in double-distilled water (ddH_2O). ddH_2O is poured out and 20× SSC (saline-sodium citrate) buffer added. The slot blot device is set up. Slot blot device is closed. (v) Samples are transferred to the well of the slot blot apparatus. Samples are suck into wells by putting on house vacuum to the slot blot device. (vi) Filter is removed and baked in a vacuum oven for 2 h. (vii) Hybridization is done with the suitable probe using the Southern hybridization technique.

slow component of eukaryotic genome: The last fraction (III) to renature represents 45% of total DNA. It extends over C_0t range from $\sim 10^2$–10^4 with $C_0t_{1/2}$ value of 630.

slow rusting (genotype): A genotype in which rust develops slowly but never reaches an appropriately high degree of severity. This is a type of 'partial' or 'incomplete' resistance. This trait is quantitatively inherited.

slow-stop: *dna* mutants of *Escherichia coli* current round of bacterial replication but cannot initiate another at 42°C.

sludge: Solid collected from sewage. After primary sewage treatment that removes large floating materials, sand, oil and grease, the sewage is passed through sedimentation tanks where more solid material gets settled. This solid material settling on the bottom of the tanks is called sludge. It may be disposed of in a landfill, but if organic, can be used as a beneficial soil amendment.

***Sma*I:** A type II restriction enzyme from the bacterium *Serratia macesccens* that recognizes the DNA sequence 5′-C C C↓G G G / 3′-G G G↑C C C, where arrows indicate sites of cleavage.

small activating RNA (saRNA): miRNAs that can activate gene expression by binding to promoter sequences.

small cytoplasmic ribonucleoproteins (scRNPs): Complexes of scRNA (small cytoplasmic RNAs) and proteins found in the cytoplasm. An example is SIGNAL RECOGNITION PARTICLE (SRP).

small cytoplasmic RNAs (scRNAs): Any of various low-molecular mass RNA molecules (100–300 nt) found in the cytoplasm as components of small cytoplasmic ribonucleoprotein.

small interfering RNA (siRNA) from transposons: Transposons in *Arabidopsis* are a chief source of 24-nt siRNA. *Arabidopsis* centromeric 180-bp satellite repeats, which are organized in tens of thousands of tandem copies on either side of Athila LTR retroelements, are also transcribed and processed by RNAi. This processing relies on chromatin remodelling by the SWI2/SNF2 ATPase DDM1, dsRNA production from RNA polymerase IV transcripts by RDR2 and dicing by DCL3. Silencing is contingent on H3K9me2 and the associated DNA methyltransferase CMT3. Retrotransposon insertions into the repeats can silence the adjacent repeats, and this activity is contingent on other mechanisms including MET1, DDM1 and the histone deacetylase HDA6. DDM1 has an exquisite specificity for transposons and repeats and must somehow recognize these as being different from genes. Loss of DNA methylation in *met1* and *ddm1* mutants is epigenetically inherited in crosses to wild-type plants and only those transposons that retain siRNA can be remethylated. Thus, RNAi has a vital role in the initiation of silencing in *Arabidopsis*. In *Arabidopsis*, two novel RNA polymerases (RNA polymerase IV and RNA polymerase V) were recovered. It is not yet known what template is used by RNA polymerase IV, but RNA polymerase V is responsible for intergenic transcription of repeats and recruits AGO4 via glycine and tryptophan (GW) repeats in the carboxy-terminal domain and in accessory elongation factors. Only the largest subunits are unique to RNA polymerase IV and RNA polymerase V, which use many of the same small subunits as RNA polymerase II. Additional SWI2/SNF2 chromatin remodellers, namely, CLSY2, were recovered, which may modify local chromatin structure to ease processivity of RNA polymerases. It is therefore likely that they facilitate transcription by RNA polymerase IV. A similar role can be proposed for DDM1. The requirement for DDM1 (also a chromatin remodeller) in silencing transposons is far more severe than that of RNA polymerase IV or the other SWI2/SNF2 proteins. Thus, the role of RNAi-mediated heterochromatic silencing in plants is well documented. RNAi-mediated heterochromatic silencing mechanism in plants is more complicated than in *S. pombe*, largely because of DNA methylation. Parallels with fission yeast may become clearer in mutants that have lost RNAi-independent DNA methylation, such as *ddm1*. For example, in pollen grains, 24-nt siRNAs guide RNA-dependent DNA methylation independently of DDM1. 21-nt siRNA are responsible for transposon silencing in sperm cells. Plants do not possess the piRNA pathway.

small interfering RNA (siRNA) interference pathways: After entry into the cytoplasm, siRNA is either loaded on to RISC directly or utilize a dicer-mediated process. After RISC loading, the passenger strand departs, thereby commencing the RNA interference process via target mRNA cleavage and degradation.

small interfering RNA (siRNA) modifications: The known modifications in sRNA species have been primarily limited to 2′-O-Me in plants and inosine in mammals.

small interfering RNA (siRNA) sequencing strategies: Genome-wide profiling of sRNAs was first accomplished using MASSIVELY PARALLEL SIGNATURE SEQUENCING (MPSS) approaches. With the advent of next-generation high-throughput sequencing technologies, high-resolution sRNA maps have been prepared using the 454-sequencing and the Illumina sequencing system, which are ideally suited for sRNA discovery and profiling. The core methodologies of genome-wide profiling of sRNAs using high-throughput-sequencing approaches are sRNA purification and adapter ligation.

small interfering RNA (siRNA) target prediction: Because of differences among animal and plant miRNA target sites, different tools are required to predict the target sites. Some such tools are psRNATarget, Targetfinder and comTar.

small interfering RNAs (siRNAs): Models of post-transcriptional gene silencing (PTGS) indicate that 21–24 nucleotide dsRNAs mediate PTGS. Introduction of siRNAs can persuade PTGS (e.g. in mammalian cells). siRNAs are seemingly produced *in vivo* by cleavage of dsRNA introduced directly or via a transgene or virus; amplification by an RNA-dependent RNA polymerase (RdRP) may transpire in some organisms. siRNAs are incorporated into the RISC, guiding the complex to the homologous endogenous mRNA where the complex cleaves the transcript. RNAs are indeed the mediators of sequence-specific mRNA degradation. Chemically synthesized 21- and 22-nucleotide RNA duplexes are accomplished of guiding target RNA cleavage. These small RNAs were therefore named short interfering RNAs (siRNAs) and miRNA cleaving complexes were referred to as small or short interfering ribonucleoprotein particles (siRNPs). *Aka* SMALL RNA *or* SHORT SILENCING RNA (sRNA).

small interfering RNAs (siRNAs) for functional analysis: RNA duplexes generated to structurally resemble naturally occurring siRNAs. Can be produced by (i) chemical synthesis, (ii) through the action of endonucleases on dsRNA (sometimes termed esiRNAs) or (iii) coexpression of single-stranded RNAs (from independent, tandem or convergent expression cassettes) to generate a siRNA. Care should be made to clearly define the method of production.

small interfering RNAs (siRNAs) in heterochromatinization through methylation: siRNAs are involved in formation of heterochromatin by DNA methylation. Small RNA-silencing pathways require the biogenesis of a small RNA from either long ssRNA or dsRNA. The resultant small RNA is loaded on to effector complexes that contain a member of the Argonaute/Piwi (AGO) family of proteins, which bind to the small RNA (~22–28 nt in size) using their conserved middle (MID) and PIWI-Argonaute-Zwille (PAZ) domains. RNAi-mediated silencing transpires through multiple mechanisms. In the nucleus, RNAi promotes DNA and chromatin modification to induce heterochromatin formation and transcriptional gene silencing (TGS). In addition, it cotranscriptionally degrades RNAs that are transcribed in heterochromatic domains by a process called cotranscriptional gene silencing (CTGS). In the cytoplasm, RNAi mediates the degradation of target mRNAs or their translational repression (PTGS). Plants have methylated DNA in most heterochromatin areas. Genetic screens for mutants that relieve RNA-mediated TGS have recovered mutants in H3K9-specific lysine methyltransferases (KMTs) and RNAi components. The requirement for DNA methyltransferases, SWI/SNF remodelling complexes and two novel RNA polymerases in heterochromatin formation has been revealed.

small interfering RNAs (siRNAs) processing pathway: The following are three steps involved in this pathway: (i) ATP-dependent processive double-stranded RNA cleavage into double-stranded fragments 21–25 nucleotides long. They contain 5′-phosphate and 3′-hydroxyl termini and a 2′ additional overhanging nucleotide on their 3′-ends. The fragments thus generated are called siRNA molecules. This step is carried by Dicer. (ii) siRNAs are incorporated into a protein complex (RISC), which is inactive in this form to conduct RNAi. This step comprises unwinding of the siRNA duplex and remodelling of the complex to generate an active form of RISC. (iii) The recognition and cleavage of mRNA complementary to the siRNA strand present in RISC. *Compare with* MICRORNAS (MIRNA) PROCESSING PATHWAY.

small intracellular mediator: *See* SECOND MESSENGER.

small molecules in gene regulation: Small molecules that target specific DNA sequences have the potential to control gene expression. For example, pyrrole-imidazole polyamides are cell permeable and can inhibit the transcription of specific genes.

small molecules interacting with RNA (smiRNA): A 24-nt small RNA derived from SMI transcript from SMI genomic region.

small nuclear ribonucleoprotein particles (snurps), function of: Snurps play role in nuclear function. Both their protein components and the sequences of their RNA are highly conserved across all higher eukaryotic species. Five snurps (i.e. U1, U2, U4, U5, U6) reside in the nucleoplasm and are often found associated with heterogeneous nuclear RNA (hnRNA) are therefore thought to contribute to various aspects of pre-mRNA processing. The major chemical constituent of snurps is uracil residues and they are thus named U1 to U12. U1 to U4 are most abundantly found in all eukaryotic cells. Since U3 snRNP has been localized to the nucleolus, it is expected to participate in ribosome biogenesis.

small nuclear ribonucleoproteins (snRNPs or snurps): The introns of nuclear mRNA precursor transcripts are spliced out in a two-step reaction performed by complex ribonucleoprotein particles called SPLICEOSOMES. They also attach new non-coding segments to the leading and trailing ends of the mRNA. Snurps help to remove worthless introns from the message issued by a cell's genes. Without snurps cellular activity would come to a halt. snRNPs are components of the spliceosome, the intron-removing device in eukaryotic nuclei. The major snRNPs, U_1, U_2, U_4 and U_6 and U_5, have to be transported to nucleus for the cytoplasm where they are synthesized. The U1 and U2 snRNPs form separate particles, while the U4, U5 and U6 components join together in the U4/U6.U5 tri-snRNP. In nucleus they splice pre-mRNAs. 5′ Cap of snRNA molecule is involved in nucleo-cytoplasmic transport of Usn RNPs. One step in the nuclear uptake of U_1 snRNPs involves the 3^{m}G Cap structure. snRNAs are complexed with proteins to make small nuclear ribonucleoproteins. There are three possibilities of their mechanism of action: (i) They help to organize the intron in a configuration that encourages self-splicing. (ii) In some organisms, precursor mRNA constructs lariat structure and splicing of intron takes place without the help of protein or other RNA factors. (iii) splicing may provide missing intron sequences necessary for spontaneous splicing. *Aka* SMALL RIBONUCLEOPROTEINS *or* SMALL NUCLEORIBOPROTEIN STRUCTURES (SNURPS).

small nuclear RNA (snRNA) gene family: snRNA genes are of 11–13 types. Major 5–6 types of snRNA genes (U1, U2, U3, U4, U5 and U6) exist in many copies and are regularly transcribed. These small nuclear RNAs, called *U RNAs*, institute spliceosome formed during splicing of introns from precursors of messenger RNA. The rest snRNA genes are less common. Suggested number of genes for each snRNA is 2000.

small nuclear RNAs (snRNAs): snRNA is used to refer to a number of small RNA molecules found in the nucleus. These RNA molecules are important in a number of processes including RNA splicing (removal of the introns from hnRNA) and maintenance of the telomeres, or chromosome ends. They are always found associated with specific proteins and the complexes are referred to as SMALL NUCLEAR RIBONUCLEOPROTEINS (snRNPs) or sometimes as snurps. It involves five snRNAs and their associated proteins. These ribonucleoproteins form a large (60S) complex, called SPLICEOSOME. Then, after a two-step enzymatic reaction, the intron is removed and two neighbouring exons are joined together. Small nuclear RNAs (snRNAs), called U1, U2, U3, U4, U5 and U6, which range in size from 100–215 nucleotides, are present in cells of all vertebrate species. They possess unusual trimethylated ($m_3^{2,2,7}G$) cap structure at 5′ ends, except for U6. Snurps, namely U1, U2, U4, U5 and U6, reside in the nucleoplasm and are often found associated with HETEROGENEOUS NUCLEAR RNA (hnRNA) are therefore thought to contribute to various aspects of pre-mRNA processing. The major chemical constituent of snurps is uracil and they are thus named U1 to U12. U1 to U4 are most abundantly found in all eukaryotic cells. No intervening sequences are observed in U1, U2 and U3 genes. These genes are dispersed throughout the genome. Putative adenylation signal (AATAAA) and the termination signal (TTT, at the 3′ end of the RNA) are not found for the U1 gene. U1 RNA is also not polyadenylated.

small nucleolar organizer RNAs (snoRNAs): Those non-coding RNAs that are implicated in modification of other RNAs. Before ribosome assembly, rRNA transcripts are processed to yield mature rRNA in all organisms. These processing steps include covalent post-transcriptional modifications such as base methylation, methylation of the 2′-hydroxyl group of sugar residues and conversion of uridine into pseudouridine by base rotation. These modifications are performed by protein enzymes that modify pre-rRNAs at specific sites. snoRNAs associate with specific proteins to form SMALL NUCLEOLAR RIBOPROTEINS (snoRNPs). More than 150 snoRNAs are known in different eukaryotic lineages. They are found in cell's nucleolus where it processes and methylates rRNA. It seems that they build up a complex, the snorposome, which carries out rRNA maturation.

small nucleolar riboproteins (snorps): Snurps-like particles identified in nucleolus.

small nucleolar RNA (snoRNAs): A class of small RNA molecules that functions in ribosome biogenesis in the nucleolus by guiding the assembly of macromolecular complexes on the target RNA to allow site-specific modifications or processing reactions to occur. They guide chemical modification (e.g. methylation or pseudouridylation) of eukaryotic rRNAs.

small organic molecule: Carbon-containing compounds that usually have a molecular weight of less than 2000 g mol^{-1}. *Aka* SMALL MOLECULE.

small population size effect: Rapid changes in gene pool frequencies that can occur in small populations. *See* GENETIC DRIFT, FOUNDER PRINCIPLE *and* BOTTLENECK EFFECT.

small RNA chaperons: Small U-RNA molecules involved in ribosome biosynthesis.

small RNA sequencing: Isolate and sequence small RNA species, such as micro RNA, to understand the role of non-coding RNA in gene silencing and post-transcriptional regulation of gene expression.

small temporal RNAs (stRNAs): Over the years, structural and regulatory RNAs have been identified. The first 22-nucleotide RNAs, *lin-4* and *let-7*, are expressed during the first and the forth larval instars of *Caenorhabditis elegans*, respectively. These RNAs downregulate target mRNAs through UNTRANSLATED REGION (UTR) elements that are complementary to each regulatory RNA to specify the temporal expression of cell fates.

smallest eukaryote: *Ostreococcus tauri* is a photosynthetic picoeukaryote, which is classified as green algae. Electron microscopy reveals their extreme simple ultrastructure. Mean length and width are 0.97 and 0.70 μm, respectively, and the DNA content per cell is 33.31 fg.

smallest known gene: MICROCIN C7 (Mcc C7), a modified linear heptapeptide that inhibits protein synthesis in Enterobacteriaceae, consists of Acetyl Met-Arg-Thr-Gly-Asn-Ala-Asp. It is synthesized from a 21-bp open reading frame gene called *mccA*. This is one of the smallest genes so far identified by J.E. Gonzalez-Pastor, J.L.S. Millan and F. Moreno in 1994.

smart breeding: A breeding method that uses molecular biological methods. It works in a way similar to traditional breeding. However, the gene or gene variant responsible for a specific trait can be accurately identified through molecular biological procedures (DNA-sequencing, PCR). In this way it is possible to test the offspring of a cross for the existence of the crossed gene, even before the actual trait is signalled by a changed external appearance. Only those plants that contain the desired gene grow. The purpose of this method is to introduce into crop plants genes from wild populations, which deliberate features of interest to breeders (disease resistance, fruit colour, sugar content, etc.).

smart decision support system (SDSS) in agriculture: SDSS acquires the input parameters based on real-time monitoring to optimize the yield that realizes sustainability by improving per hectare production and lessening water seepage wastage in agribusiness. The proposed model comprises three basic units including an intelligent sensor module, smart irrigation system and controlled fertilizer module. The system has integrated sensors, cloud employing decision support layers and networking-based DSS to recommend cautions for optimum sustainable yield.

smart genes: Term was coined to describe the genes to combinations of signals sent from one gene to another in control networks. The brain of these smart genes is complicated assembly of proteins, known as TRANSCRIPTIONAL COMPLEX, which is composed of transcription factors. These transcriptions factors are DNA-binding proteins, which can discriminate between distinctive DNA sequences and thus help RNA polymerase to transcribe gene.

smart materials: Products with ability to respond to the environment, such as shape transformation.

smart transcription factors: These are the proteinous transcription factors that are smart enough to make their contact with proteins through their phosphorylation to initiate transcription. Smart transcription factor first gets phosphorylated and then binds to co-activator binding protein. Then this complex binds to DNA element, which initiates transcription. For example, smart transcription factor cAMP RESPONSE ELEMENT BINDING (CREB) protein is first phosphorylated and then it binds to coactivator CREB BINDING PROTEIN (CBP). This complex finally binds to DNA element known as cAMP-regulated enhancer and this leads to initiation of transcription.

smear: A thin film containing microorganisms spread on slide for microscopy. For microscopic studies, the microorganisms are usually fixed by heating or treating with chemicals. In heat fixation, a drop of liquid containing microorganism is spread in the form of thin film (smear) on the microscopic slide and allowed to air dry followed by passing the slide through flame. Heat causes denaturation of microbial proteins and the coagulated proteins stick the microbes to slide.

Smith model: Selection and random genetic drift are two main forces influencing selection response in recurrent selection programmes. The precise evaluation of both these forces enables a better comparison of the efficiency of different recurrent selection schemes. The population diallel analysis proposed by J.J. Hammond and C.O. Gardner (1974), and the model proposed by O.S. Smith (1979), with full thought of inbreeding depression due to random genetic drift, can also be used for quantitative analysis. The impact of random genetic drift is expected to be large, principally in studies with many selection cycles and/or high rates of inbreeding. Therefore, the extension of the population diallel permits a better analysis of the selection retort in recurrent selection.

smog: Air pollution consisting of smoke and fog, which occurs in large urban and industrial areas, is mainly caused by the action of sunlight on burned fuels, mostly from car exhausts. Smog can cause eye irritations and breathing problems and damage plant life.

smoltification: The process through which a juvenile salmon becomes physiologically ready to enter salt water within its migratory life history.

smooth endoplasmic reticulum (SER): SER is a network of tubules, which gets its name from the lack of ribosomes on its surface. The SER detoxifies drugs and poisons; stores ions and produces lipids, hormones and other cellular products, which are distributed throughout the cell by transport vesicles.

smooth method: A statistical method to remove genotyping errors from genetic linkage data during the mapping process. The programme SMOOTH calculates the difference between the observed and predicted values of data points based on data points of neighbouring loci in a given marker order. Highly improbable data points are removed by the programme in an iterative process with a mapping algorithm that recalculates the map after cleaning. The method is able to detect a high amount of scoring errors and demonstrates that the programme enables mapping software to successfully construct a very accurate high density.

SMsplice: A fully interpretable model of pre-mRNA splicing, described by K. Mccue and C.B. Burge (2024), that combines models of core splice sites (SS) motifs, splicing regulatory elements (SREs) and exonic and intronic length preferences. They learn models that predict SS locations with 83–86% accuracy in fish, insects and plants and about 70% in mammals. Learned SRE motifs include both known SPLICING FACTOR (SF) binding motifs and unfamiliar motifs.

smut: A plant disease produced by a fungus of the order Ustilaginales. The indicators include the formation of masses of black, shootlike spores. Infected plants frequently show some degree of distortion.

SNAp REceptors (SNAREs): Large family of transmembrane proteins existing in organelle membranes and the vesicles derived from them. They are implicated in guiding vesicles to their correct destinations. They occur in pairs – a v-SNARE in the vesicle membrane that binds precisely to a complementary t-SNARE in the target membrane.

snowball sampling: A singular non-probability method employed when the desired sample characteristic is rare. It may be extremely difficult or cost-prohibitive to locate respondents in these situations. Snowball sampling trusts on recommendations from initial subjects to generate additional subjects. While this technique can involve dramatically lower search costs, it comes at the cost of familiarizing bias because the technique itself diminishes the likelihood that the sample will represent a good cross section from the population.

snRNP action, mechanism of: snRNP stands for SMALL RIBONUCLEOPROTEIN. There are three possibilities of the mechanism of action of snRNPs: (i) they help to arrange the intron in a configuration that encourages self-splicing; (ii) in some organisms, precursor mRNA formulates lariat structure and splicing of intron ensues without the help of protein or other RNA factors; (iii) splicing may offer missing intron sequences necessary for spontaneous splicing.

social Darwinism: A doctrine that applies the principles of selection to the structure of society, asserting that social structure is determined by how well people are suited to living conditions.

sociobiology: The study of evolution of social behaviour in animals.

sodium bisulfite treatment: Method to detect 5-methylcytosine in DNA. Following treatment of DNA, all unmethylated cytosine are converted to uracil. Subsequent PCR amplification displays these bases as thymine. Methyl group at C5 of cytosine inhibits the hydrolytic deamination of cytosine to uracil.

sodium dodecyl sulfate polyacrylamide gel electrophoresis (SDS-PAGE): SDS denatures proteins by binding to the hydrophobic regions and essentially coating the linear protein sequence with a set of SDS molecules. The SDS is negatively charged and thus becomes the dominant charge of the complex. The number of SDS molecules that bind is simply proportional to the size of the protein. Therefore, the charge to mass ratio should not change with size. In solution (water), in principle all different-sized proteins covered with SDS would run at about the same mobility. However, the proteins are not run through water. Instead they are run through an inert polymer, polyacrylamide. The density and pore size of this polymer can be varied. Thus, the size of molecules that can pass through the matrix can be varied. This determines in what molecular weight range the gel will have the highest resolving power. SDS-PAGE separates proteins according to molecular mass irrespective of charge. The basis of the technique is the exposure of denatured Pr^- to the detergent SDS, which binds stoichiometrically to the polypeptide backbone and caries a large negative SDS molecules on each polypeptide dwarfs any intrinsic charge carried by the proteins themselves, and stoichiometric binding means that large proteins bind to more SDS than do the smaller proteins.

sodium dodecyl sulfate polyacrylamide gel electrophoresis (SDS-PAGE), important features of: SDS-PAGE has two important consequences: (i) All protein-SDS complexes have essentially the same charge density. (ii) The relative differences in mass between proteins are maintained in SDS-protein complexes. The gel enhances the size-dependent separation by sieving the proteins as they migrate. Sieving effect depends on the pore size of gel which in turn depends on gel concentration. Since polyacrylamide gels are made up of gelling agent acrylamide and cross-linking agent bis-acrylamide, so pore size also depends on monomer by mass, that is represented by bis-acrylamide. The mass of the protein can be estimated by including in one of the lanes of the gel a series of protein markers whose masses are known. *See* POLYACRYLAMIDE GEL ELECTROPHORESIS.

sodium flux in plants: Plants have evolved several approaches to cope with salinity, which in most species involve efficient Na^+ management. F. Gámez-Arjona *et al.* (2024) believe that the Ca^{2+}-sensor protein SALT OVERLY SENSITIVE 3 (SOS3), a critical determinant of the salt tolerance of *Arabidopsis*, is the missing molecular link that co-regulates in reverse the two major Na^+ transport systems operating in vascular plants, the SOS1- and HKT1-like proteins that mediate Na^+ loading and unloading at the xylem, respectively. Under salt stress, SOS3 promotes the recruitment of SOS1 to the plasma membrane and its activation by the SOS2/SOS3 kinase complex, while simultaneously SOS3 interacts with and commands the degradation of HKT1;1. Thus, SOS3 functions as the molecular switch shifting the balance from Na^+ retention in the root towards delivery to the shoot.

soft computing: Soft computing is dedicated to system solutions based on soft computing techniques. It provides rapid dissemination of important results in soft computing technologies, a fusion of research in evolutionary algorithms and genetic programming, neural science and neural net systems, fuzzy set theory and fuzzy systems and chaos theory and chaotic systems.

soft rot: A softening, discoloration and often disintegration of plant tissue as a result of fungal or bacterial infection.

soft selection: Selection that favours a particular number or percentage of the population regardless of their absolute characteristics, frequency and density dependent. *Aka* RANK-ORDER SELECTION. *See* HARD SELECTION.

software: A collection of data or computer instructions that tell the computer how to work. This is in contrast to physical hardware, from which the system is built and actually performs the work. In computer science and software engineering, computer software is all information processed by computer systems, programmes and data. Computer software includes computer programmes, libraries and related non-executable data, such as online documentation or digital media. Software is in *contrast to* PHYSICAL HARDWARE, from which the system is built and actually performs the work. Computer hardware and software require each other and neither is realistically usable on its own. *Aka* COMPUTER SOFTWARE.

software agent: A computer programme that functions for a user or other programme in a relationship of agency, which derives from the Latin 'agere' (which means 'to do'). An agreement to perform on one's behalf. Such 'action on behalf of' infers the authority to decide which, if any, action is appropriate.

Agents are colloquially known as BOTS, from ROBOT. They may be personified, as and when execution pairs with a robot body, or as software such as a chatbot executing on a phone (e.g. Siri) or other computing device. Software agents may be self-governing or work in coordination with other agents or people. Software agents interacting with people (e.g. chatbots, human-robot interaction environments) may own human-like qualities such as natural language understanding and speech, personality or embody humanoid form.

software construction: In software engineering, it is the comprehensive creation of working, meaningful software using an amalgamation of coding, verification, unit testing, integration testing and debugging. It is associated with all the other software engineering disciplines, most powerfully to software design and software testing.

software deployment: All of the activities that make a software system available for use.

software design: The process by which an agent creates a specification of a software artefact, intended to accomplish goals, using a set of primitive components and subject to constraints. Software design may refer to either 'all the activity involved in conceptualizing, framing, implementing, commissioning and ultimately modifying complex systems' or 'the activity following requirements specification and before programming, as . . . [in] a stylized software engineering process'.

software development: In software engineering, a software development process is the process of dividing software development work into distinct phases to improve design, product management and project management. The methodology may comprise the predefinition of specific deliverables and items that are created and completed by a project team to develop or maintain an application. Most modern development processes can be vaguely described as AGILE. Other methodologies include waterfall, prototyping, iterative and incremental development, spiral development, rapid application development and extreme programming. *Aka* SOFTWARE DEVELOPMENT LIFE CYCLE (SDLC).

software engineering: The systematic application of engineering methods to the development of software. Software engineering is a computing science.

software maintenance: In software engineering, software maintenance refers to the modification of a software product after delivery to correct faults, to improve performance or other features.

software prototyping: The activity of creating prototypes of software applications (i.e. incomplete versions of the software programme being developed). This activity can transpire in software development and is comparable to prototyping as known from other fields, such as mechanical engineering or manufacturing. A prototype typically simulates only a few aspects of and may sometimes be completely unlike from the final product.

software requirements specification (SRS): Refers to a description of a software system to be developed. The specification of software requirements lays out functional and non-functional requirements, and it may contain a set of use cases that describe user interactions that the software must provide to the user for perfect interaction.

software testing: An investigation directed to provide stakeholders with information about the quality of the software product or service under test. Software testing can also offer an objective and independent view of the software to enable the business community to appreciate and understand the risks of software implementation. Test techniques include the process of executing a programme or application with the commitment of finding software bugs (errors or other defects) and verifying that the software product is fit for use.

software toolbox for epigenome mapping: The availability of new high-throughput methods accelerated the demand for efficient software tools to process and analyse the overwhelming flowing incomprehensible raw data. The most popular examples of such tools, with high profile examples, include Bowtie for NGS data, MACS for ChIP-seq data and DEseq for RNA-seq data.

software tools commonly used in biotechnology: For sequence assembly, prediction of coding regions, annotation, detecting repeats and splice sites, software programmes used include: TRNASCAN, FGENEH, GENSCAN, Genquest, MUMmer, REPEATMASKER, Annotator, PSI-BLAST and variants, Genefinder, Grail X GRAIL/Grailcnt, Net plant gene and AAT.

soil association: *See* ORGANIC AGRICULTURE.

soil biodiversity: It is extraordinarily difficult to enumerate soil biodiversity. M.A. Anthony *et al.* (2023) estimate that about 59% of all species on Earth live in soil, estimate researchers who reviewed global biodiversity data. This would make the ground the planet's single most biodiverse habitat. It is home to 99% of Enchytraeidae worms, 90% of fungi, 86% of plants and more than 50% of bacteria – but only 3% of mammals live in it.

soil conservation practices: *See* MINIMAL TILLAGE PRACTICES.

soil drench: An application of a solution or suspension of a chemical to the soil, especially application of a pesticide to manage soilborne pathogens.

soil inhabitant: An organism that maintains its population in soil over a long period of time.

soil invader: An organism whose population in soil reduces in several months to years.

soil microflora: The extraction of large amounts of DNA and RNA from soil and the accurate taxonomic and functional characterization of these can increase our knowledge of the composition of soil microflora, of pathogenicity, of trophic interactions between soil microorganisms and of the production of bioactive molecules by soil microorganisms. Of the various genetic approaches used to estimate the microbial composition and diversity, the PCR amplification of conserved and variable regions of 16S rRNA gene may give either a rough overview of taxonomically distant groups or a deeper insight into selected ecophysiological groups.

soil pasteurization: The procedure used to free soil of selected harmful microorganisms using heat. *Aka* SOIL STERILIZATION.

soil plant analysis development (SPAD) meter: SPAD meter is commonly used to measure leaf N status in a non-destructive manner. As leaf chlorophyll status impacts crop productivity, association of leaf N with crop yield was studied in rice. Here SPAD reading was significantly correlated with number of kernels per row, grain yield/hectare, number of kernel rows/ear and ear weight (B. Ghimire, D. Timsin and J. Nepal in 2015). In maize, SPAD chlorophyll readings taken at a late developmental stage showed a high positive genetic correlation with single-plant yield. The SPAD meter has also been used in vegetables (cabbage, onion and carrots) and the results showed the use of SPAD to monitor tissue N was cost-effective.

soil sterilization: *See* SOIL PASTEURIZATION.

soilborne: An organism carried on or beneath the soil surface.

solar energy in crop improvement: Solar power is a renewable energy source that has numerous benefits for agriculture, such as (i) reduced energy costs, (ii) increased sustainability, (iii) reliable power supply, (iv) improved water management, (v) increased crop yield, (vi) government incentives and (vii) long-term investment.

solarization: A disease control exercise in which moist planting area of the soil is covered with polyethylene sheeting and exposed to sunlight, thereby heating the soil and inhibiting or killing soilborne plant pathogens.

solenoid: A higher order structure in a eukaryotic chromosome in which 100 fibres of nucleosomes are stacked with 5–6 nucleosomes/helix. H1 histone helps in solenoid formation, which gives a package ratio of 1:50. One model of additional compaction past nucleosomal winding proposes that the 110-Å nucleosomal chromatin supercoils into a 300-Å superhelix called solenoid. Nucleosomes, connected by about 20–60 bp of linker DNA, form a 10-nm beads-on-a-string array, which can be compacted further into a 30-nm chromatin fibre. The 30-nm fibre is a second level of chromatin organization, which provides an ~100-fold compaction of the DNA. This helical structure has six nucleosomes per turn. H1 histone seems to play vital role in formation of this superhelix. The 300-Å chromatin fibre condenses into a chromatid of compact metaphase chromosome with the help of non-histone proteins. Two chromatids when join together by a centromere form a complete chromosome. H1 stabilized the nucleosomes and helped in the formation of solenoid structure of 300 Å in diameter by making H1 polymers.

solid state fermentation (SSF): SSF consists of the microbial growth and product formation on solid particles in the absence (or near absence) of free water.

solid wastes: Extremely variable municipal wastes that include discarded food, garden discards, newspapers, bottles, cans, construction debris, old cars and disused furniture.

solid-phase synthesis: Chemical synthesis method using a solid support, such as a plastic bead.

solitary: Single, of flowers that grow one plant per year, one in each axil, or widely separated on the plant; not grouped in an inflorescence.

soluble N-ethylmaleimide-sensitive factor activating protein receptors (SNAREs): Large family of transmembrane proteins occurring in organelle membranes and the vesicles derived from them. They are implicated in guiding vesicles to their correct destinations. They exist in pairs – a v-SNARE in the vesicle membrane that binds specifically to a complementary t-SNARE in the target membrane.

soluble RNA (sRNA): *See* TRANSFER RNA.

solute: A solid substance that get dissolved in the solvent. During the preparation of NaCl solution the solid NaCl is called solute, whereas the water in which NaCl dissolves is called solvent. Polarity of water makes it an excellent dissolving medium (SOLVENT). Several polar compounds like NaCl dissociate in water to produce individual molecules. The positive part of solute gets attracted towards the negative part of the water molecule and vice versa and thus the solute rapidly dissolves in the solvent.

solvent: The liquid in which a solute is dissolved. *See* SOLUTE.

solvent accessibility: The surface area (typically measured in square angstroms) of a biological molecule, usually a protein, which is exposed to solvent in its native, folded form. Determining the solvent accessibility of a protein helps define which amino acids in its molecular sequence are on the exterior of the molecule and thus available to participate in interactions with other molecules.

soma: The cellular parts of a multicellular organism that do not directly produce gametes. *Contrasts* GERMLINE.

somaclonal variation: Phenotypic variation may be genetic or epigenetic among the callus-derived plants. Occurrence of somaclonal variation among the tissue culture regenerated plants possesses much significance for breeding crop plants particularly the vegetatively propagated crops. Occurrence of callus phase during tissue culture lead to somaclonal variation. Various causes of somaclonal variation include; chromosomal changes both in number and structure, point mutations, DNA amplification and virus elimination.

somaclone: Tissue culture regenerated plant showing phenotypic variation. For instance, if chrysanthemum with red flowers is cultured and among the regenerants another white colour appears, then this plant with changed colour is called somaclone, which is identical to the parent plant from which they are produced and cultured. Some somaclones in crop plants have been released as new crop varieties.

somatic: Refers to diploid body cells, normally with one set of chromosomes coming from the male parent and one set from the female parent. In case of plants, the whole plant body is somatic except the gametic cells. Accordingly, roots, hypocotyls, cotyledons, leaves, stems, flower petals and so on are somatic tissues. Pollen and egg cells are gametic cells. The human body is somatic and only the sperm and egg cells are gametic.

somatic cell: All body cell except the gametes. In a growing plant, cells present in roots, cotyledons, stem, leaves, flower petals and so on, all are somatic cells, which possess the sporophytic (zygotic) number of chromosome (2n), whereas the egg cell and pollen are the gametic cells that contain half the chromosome (gametic) number (n) of the mother plant. When cell cultures are established from somatic cells, these are called somatic cell cultures. *Compare with* SEX CELL.

somatic cell fusion: The fusion of somatic cells (usually protoplasts) by different *in vitro* methods.

somatic cell gene therapy: Altered gene is inserted into the somatic cells and this leads to a change only in the individual receiving the treatment. Yet, most governments have limited all gene therapy experiments to somatic cell gene therapy because the alteration performed in germ gene therapy would change future generations. *See* GENE THERAPY.

somatic cell genetic mutation: A change in the genetic structure that is neither inherited nor passed to offspring. *Aka* ACQUIRED MUTATIONS. *See* GERMLINE GENETIC MUTATION.

somatic cell genetics: Genetics of somatic cells. Somatic cells are cultured *in vitro* in the form of cell suspension cultures and callus tissue cultures or dividing somatic cells can be excised from actively growing root tips. After preparation of slides, genetic studies with respect to chromosome number and structures are made using compound microscope. Further genomic DNA can be isolated and genetic profiling can be done using PCR-based markers like RAPD and microsatellites.

somatic cell hybrid: The cell or plant resulting from the fusion of animal cells or plant protoplasts, respectively, derived from somatic cells that differ genetically. *See* HYBRID CELL *and* HETEROKARYON.

somatic cell hybridization: A technique that permits hybridization between genetically differing somatic cells of same or different species.

somatic cell mutation and recombination (SCMR): A test that detects induction of mutations and recombination in somatic cells (like eyes and wings of *Drosophila*). It is used to test mutagenicity of individual chemical compounds and mixtures.

somatic cell nuclear transfer (SCNT): A cloning technique where the nucleus from an unfertilized egg is removed and replaced with the nucleus from a somatic cell. This technique can be used both for reproductive cloning and therapeutic cloning. Transplantation of a diploid nucleus from a somatic cell to an enucleated egg cell, artificially mimicking fertilization and potentiating development. The resulting egg will carry the full complement of genetic material of the host organism. This is how Dolly the cloned sheep was produced; she was genetically identical to her 'mother'. *See* THERAPEUTIC CLONING.

somatic cell nuclear transfer (SCNT) technology: This technology has recently been used to generate animals with a common genetic composition. A pluripotent ESC (SCNT-hES-1) line has been derived from a cloned human blastocyst. The SCNT-hES-1 cells displayed typical ESC morphology and cell-surface markers and were capable of differentiating into embryoid bodies *in vitro* and of forming teratomas *in vivo* containing cell derivatives from all three embryonic germ layers in SEVERE COMBINED IMMUNODEFICIENT (SCID) mice. After continuous proliferation for more than 70 passages, SCNT-hES-1 cells maintained normal karyotes and were genetically identical to the somatic nuclear donor cells.

somatic cell nuclear transfer (SCNT) technology, issues concerning: SCNT could be used to clone a human as well. SCNT uses embryos in the process. Many people believe it is unethical to create embryos and then destroy them in order to harvest the stem cells.

somatic cell variant: A somatic cell with unique features not shared by others, as selected for in a screening trial that may follow a mutational event.

somatic crossing-over: Crossing-over in somatic cells during mitosis such that parent cells heterozygous for a given allele, instead of giving rise to two identical heterozygous daughter cells, give rise to non-identical daughter cells, one of which is homozygous for one of these alleles, the other being homozygous for the other allele.

somatic differentially methylated regions (somatic DMRs): DMRs that acquire their parental-specific DNA methylation in somatic diploid cells.

somatic diversity theory: A theory that account for the high degree of antibody diversity. It suggests that mutation of a basic immunoglobulin gene is responsible for all the different types of immunoglobulins produced by B lymphocytes. Only a small number ofV sequences are transmitted from parents to offspring, perhaps as few as three V_k sequences, five V_γ sequences and a few V_H sequences. These sequences then undergo thousands of different base pair changes in individual lymphocyte lines to generate the observed diversity. One lymphocyte clone will, for example, carry sequences $V_{\delta 1a}$ and $V_{\gamma 11c}$, another $V_{\delta 1b}$ and $V_{\delta 11e}$ and so on. Two of these sequences are then selected for expression in a given lymphocyte clone. *See* GERMLINE THEORY.

somatic doubling: An interruption in the mitotic process that produces a cell with twice the normal chromosome number.

somatic embryo: Development of embryo from a somatic cell in culture.

somatic embryogenesis: The process of embryo development from somatic cells in culture. Somatic embryogenesis is a unique phenomenon under which the cultured somatic cells, tissues or organs exhibit the formation of embryoids (somatic embryos) similar to zygotic embryos, but these differ genetically, being mother type. This phenomenon occurs under the influence of growth regulators being used in tissue culture. Somatic embryos are being used for, mass cloning of plants, production of synthetic seeds and for genetic transformation of plants.

somatic hybrid cell lines: Hybrid cells created by fusing cells from the organism of interest with an immortalized cell line.

somatic hybridization: The fusion of genetically different somatic cells by different means, usually to bypass the natural crossing (incompatibility) barriers.

somatic hypermutation: The mutation of the immunoglobulin variable region in mature B cells during an immune response. It results in affinity maturation of the antibody response. Like class switch recombination, it requires activation-induced cytidine deaminase.

somatic mosaicism: Accumulating mutations in somatic cells. Mosaic brain mutations have been to a diversity of other neurological disorders, including schizophrenia and Alzheimer's disease. Somatic mosaicism is likely playing a role in these conditions.

somatic mutation: Alterations of the DNA that occur after conception in any of the cells of the body except the germ cells and which therefore cannot be passed on to offspring, unless vegetative methods of propagation are available or vegetative cells are ordained to become sex cells.

somatic mutation theory: A theory to explain the high degree of antibody variability. Accordingly, mutation of a basic immunoglobulin gene accounts for all the different types of immunoglobulins produced by B lymphocytes. *See* GERMLINE THEORY.

somatic mutation theory of ageing: L. Szilard (1959) gave somatic mutation theory to explain basis of ageing. Mutations occur randomly and spontaneously, destroy genes and part of chromosomes in postmitotic cells during the lifespan of organisms and the mutation load is gradually increased, which decreases the production of functional proteins. This theory is supported by death of rats exposed to whole-body irradiations that were too low to produce syndromes died earlier, fall out in brain cells with age, ageing in fibrous tissue, reduction of lifespan in ice by induced mutations. The limitation of this theory, however, is that there is no method of accurately measuring somatic mutations. It is not possible to drive statements like 'DNA damage does/does not accumulate during cellular ageing', 'DNA repair capacity increases/decreases/remains unaltered during ageing', or 'the defect in DNA repair capacity is – is not – the cause of cellular ageing'.

somatic nuclear RNA interference: While the germlines of metazoans have a clear role in nuclear RNAi, some evidence shows that TGS also occurs in somatic cells. Feeding *C. elegans* with dsRNA targeting an endogenous gene triggers H3K9 methylation at the target locus in somatic cells that is dependent on the nuclear RNAi pathway nuclear RNAi defective (NRDE) and on the RdRP RRF-1.

somatic pairing: The phenomenon of close association of homologue was in somatic cells, as seen in dipteran giant chromosomes.

somatic regeneration: A means by which plants can regenerate bipolar structures from a somatic cell. During the process of cell differentiation, the explant responds to endogenous stimuli, which trigger the induction of a signalling response and, consequently, modify the gene programme of the cell.

somatic segregation: The formation by mitosis of cells of cells differing from one another, either through mutation or somatic crossover in the nucleus or through an unequal assortment of cytoplasmic determinants such as is not required by normal differentiation.

somatic selection: Selection among cells within a tissue or organ.

somatic sexual development: A process that determines phenotypic sex of an individual.

somatic stress memory: Stress memory whose duration is limited to one generation of organisms. It may be mitotically heritable, but often lasts only a fraction of the lifespan of the organism.

somatic variants: Regenerated plants (i.e. clones) produced from cells that originally were derived from the same plant but are not genetically identical. Such plants (clones) are called SPORTS or SOMATIC VARIANTS because they differ (genetically) from the 'parent' plant. Sometimes such somatic variants have been developed by man to become a new plant variety (the nectarine is an example of this).

somatoplastic sterility: The failure of zygotes during embryonic stages due to disturbances in embryo-endosperm relationships.

sorting: A process employed to sort and/or separate different cells, particles, chromosomes and so on. Some automated methods of cell sorting in use for different applications include (i) utilizing controlled electrical fields to gather specific cell types on to electrodes in biochips, (ii) fluorescence-activated cell (FAC) sorter machines or (iii) magnetic particles (e.g. attached to antibodies). The ploidy level can also be determined by cell sorting equipment.

sorting algorithm: An algorithm that places elements of a list in a certain specific order. The most frequently used orders include NUMERICAL ORDER and LEXICOGRAPHICAL ORDER. Effective sorting is imperative for optimizing the efficiency of other algorithms (such as search and merge algorithms) that require input data to be in sorted lists. Sorting is also often advantageous for canonicalizing data and for producing human-readable output. More formally, the output of any sorting algorithm must satisfy two conditions: (i) The output should be in non-decreasing order (each element is no smaller than the previous element according to the desired total order) and (ii) The output should be a permutation (a reordering, yet retaining all of the original elements) of the input. Further, the input data is frequently stored in an array, which permits random access, rather than a list, which only enables sequential access. Though many algorithms can be functional to either type of data after suitable modification. Generally, sorting algorithms are engaged to sort groups of data in an increasing or decreasing manner.

sorting signal: Amino acid sequence that guides the delivery of a protein to a specific location outside the cytosol.

SOS: Acronym for SAVE OUR SOULS.

SOS box: The region of the promoters of various genes that is recognized by the LexA repressor. Release of repression results in the induction of the SOS RESPONSE.

SOS genes: In *Escherichia coli*, any group of genes that are stimulated upon DNA damage by physical and chemical genotoxic agents, leading to the interaction of the LexA repressor. SOS gene are implicated in DNA repair, mutagenesis and cell division, as well as a whole set of plasmid-encoded colicins.

SOS regulon: The SOS repair system is a complex one with over 20 genes involved. Two of these genes are the important regulator genes *lexA* and *recA*. SOS repair is inducible and consists of several other operons called SOS regulon controlled by a common element. *See* SOS.

SOS repair system: The repair systems (recA, uvr) stimulated by the presence of single-stranded DNA that usually occurs from postreplicative gaps caused by various types of DNA damage. The RecA protein, stimulated by single-stranded DNA, is involved in the inactivation of the LexA repressor, thereby inducing the SOS response. *Aka* SOS RESPONSE.

SOS response: A global response to DNA damage recognized from 'switching on' of genetic repair machinery in which the cell cycle is arrested and DNA repair and mutagenesis are induced usually by radiation.

sounds emitted by plants under stress: Stressed plants show altered phenotypes, including changes in colour, smell and shape. I. Khait *et al.* (2023) show that stressed plants emit airborne sounds that can be recorded from a distance and classified. They recorded ultrasonic sounds emitted by tomato and tobacco plants inside an acoustic chamber, and in a greenhouse, while monitoring the plant's physiological parameters. They developed machine learning models that succeeded in identifying the condition of the plants, including dehydration level and injury, based solely on the emitted sounds. These informative sounds may also be detectable by other organisms. This work opens avenues for understanding plants and their interactions with the environment and may have significant impact on agriculture.

source: Part or organ of plant where metabolites such as sucrose are synthesized.

source clone: A source that originates from a single plant or explant within a clone – as from a specific virus tested (SVT) or specific pathogen tested (SPT) individual explant or plant.

source code: In computing, source code is any collection of code, with or without comments, written using a human-readable programming language, usually as plain text. The source code of a programme is specially designed to ease the work of computer programmers, who specify the actions to be performed by a computer mostly by writing source code, which is often transformed by an assembler or compiler into binary machine code that can be implemented by the computer. The machine code might then be stored for execution at a later time. Alternatively, source code may be interpreted and thus immediately executed.

source plant: A mother plant or donor plant from which an explant used to initiate a culture is taken.

Southern blot(ting): This technique, devised by E.M. Southern (1975), is a frequently used method for the documentation of DNA fragments that are complementary to a known DNA sequence. DNA-DNA hybridization provides the basis of Southern hybridization. Various steps in Southern hybridization include denaturation, blotting, baking, pretreatment, hybridization and autoradiography. The restriction fragments of DNA present in polyacrylamide gel (for smaller fragments up to 20 kb), agarose gel (for larger fragments, greater than 20 kb), agarose gel with pulsed electrical fields (PEF) or field inversion or pulsed field gel electrophoresis (PFGE) (for very large fragments up to 1000–2000 kb) separated following electrophoresis are denatured into single-stranded DNA by alkali treatment. Mechanical shearing or REs can also produce fragments of DNA. This step is known as DENATURATION. Gel reveals varying fragments depending upon their size. Since DNA is negatively charged, it migrates to positive pole. They are then transferred to a nitrocellulose filter membrane. The DNA becomes stuck in nitrocellulose membrane. This step is known as BLOTTING. It takes several hours but bands retain their positions on the filter and there is a minimum loss in their resolution. The nitrocellulose membrane is removed and the DNA is permanently immobilized on the membrane by baking it at 80°C *IN VACUO*. This step is called BAKING. Single-stranded DNA shows high affinity towards nitrocellulose filter (RNA lacks it). The baked membrane is treated with a solution containing 0.2% each of Ficoll (a polymer of sucrose), polyvinylpyrrolidone and bovine serum albumin. This is called PRETREATMENT. It prevents non-specific binding of radioactive probes. Marker fragments of known sizes are positioned in a separate lane to interpolate sizes of the unknown fragments. The pre-treated membrane is placed in a solution of radioactive single-stranded DNA or oligodeoxynucleotide, called probe. The probe hybridizes with complementary DNA and this step is known as HYBRIDIZATION reaction. After hybridization, the membrane is washed to remove the unbound probe. The membrane is now placed in close contact with an X-ray film and incubated to permit images due to radioactive probe. The film is developed to show distinct bands in the gel, which show complementarity with the probe This step is called AUTORADIOGRAPHY. *Compare with* NORTHERN BLOTTING *and* WESTERN BLOTTING. *Aka* SOUTHERN HYBRIDIZATION.

Southern and Northern blots, differences between: Southern and Northern blots diverge from each in certain ways: (i) Southern blotting separates DNA while Northern blotting separates RNA. (ii) In case of Southern blotting, DNA has to be denatured before blotting. This step is not needed in case of Northern blotting. (iii) Nitrocellulose membrane filter is normally used in Southern blotting but nitrocellulose membrane filter is not engaged in Northern blotting as RNA does not bind to it. (iv) In Southern blotting, hybridization with probe produces DNA/DNA hybrids while in Northern blotting, hybridization with probe produces RNA/DNA hybrids.

Southern blot(ting), applications of: This technique has several applications. It enables a comparison between the genome of a particular organism and that of an available gene or gene fragment (the probe). It can state whether an organism contains a particular gene and deliver information about the organization and restriction map of that gene or specific DNA fragments. It has been used to detect RFLP and VARIABLE NUMBER OF TANDEM REPEAT (VNTR) Polymorphism. VNTR is the basis of DNA fingerprinting. It is also engaged in detection and identification of transferred genes in the transgenic individuals. Now nylon membranes have been developed. These membranes are physically strong. Both DNA and RNA become cross-linked by a brief exposure to UV light. These membranes save baking time. Same membrane blot can be utilized for search with more than one probe. It is reusable. When gels are stained with intercalating dye ethidium bromide, it provides visible fluorescence on illumination of gel with ultraviolet (UV) light. As little as 0.05 μg DNA band can be detected.

***SP11D*:** A dominant *SP11* locus.

SP11-methylation-inducing (SMI) genomic region: A region that can generate non-coding transcript of about 2 kb in length.

***SP11R*:** A recessive *SP11* locus.

SP6 RNA polymerase: A bacteriophage RNA polymerase that is commonly used to transcribe plasmid DNA into RNA. The plasmid must contain an SP6 promoter upstream of the relevant sequence.

spacer: Short (30–45 nt) DNA sequences derived from target genetic elements; located in the CRISPR array and flanked by repeat sequences.

spacer DNA: Sections of non-transcribed DNA between transcribed segments. Term generally applies to immunoglobulin genes. These segments of DNA are removed before transcription of such genes begins.

spacer region: A sequence of nucleotides in the DNA between coding genes.

spacer sequence: A portion of DNA sequences that does not code any RNA. It is rich in adenine and thymine and is found between the transcriptional DNA segments.

spacer tRNA gene: In *Escherichia coli*, any gene coding for transfer RNA and located in the spacer sequence between the genes coding for 16S and 23S ribosomal RNA or downstream from the 5S rRNA genes.

spatial gene expression: Place (cell, tissue of organ) where a gene is expressed. Sea urchin genes coding for calcium-binding proteins of about 15 kDa are also specifically expressed in dorsal ectoderm cells at gastrulation.

spatial organization: The phenomenon, which guarantees that all members of a species, are morphologically similar. Two components of pattern formation are REGIONAL SPECIFICATION and MORPHOGENESIS. The pattern involving spatial arrangement of tissues and organs are formed through a complex network of reactions and interactions influenced by genes.

spatial transcriptomics: Use of transformative technology is largely limited to detecting gene expression. Since the life of an RNA molecule is rich and dynamic, starting from DNA transcription to nascent RNA synthesis, splicing, export, translation and degradation, it is desirable to not only detect the presence of a given RNA molecule but also to have information on its stage of life. H. Zeng *et al.* (2023) developed a spatial transcriptomic approach that is specific for mRNAs that are undergoing translation, enabling the study of their biological function *in situ*. This ribosome-bound messenger RNA imaging technique is called RIBOMAP and applied it in single cells *in situ* to profile translation events with spatial coordinates. Thousands of genes were simultaneously mapped in intact cells and tissues at molecular resolution, revealing the regulatory principles that specify the location and efficiency of protein production for functionally relevant gene programmes across different cell types and tissue regions.

Spatial-CUT&Tag Method: Named for pixel by pixel on a frozen tissue section without dissociation, which resolved spatially distinct and cell-type-specific chromatin modifications in mouse embryonic organogenesis and postnatal brain development.

spatial-temporal reasoning: An area of artificial intelligence that draws from the fields of computer science, cognitive science and cognitive psychology. The theoretic goal – on the cognitive side – involves representing and reasoning spatial-temporal knowledge in mind. The applied goal – on the computing side – involves developing high-level control systems of automata for navigating and understanding time and space.

spatially resolved epigenomics: An approach combining *in situ* tagmentation and transcription with MERFISH enables spatial profiling of the epigenome in tissues with single-cell resolution (R. Strack, 2022).

spatio-temporal regulation of mRNAs: Micro RNAs contribute in spatio-temporal regulation of messenger RNA and protein synthesis. Aberrant miRNA expression results in developmental abnormalities and diseases, such as cardiovascular disorders and cancer.

Spearman rank correlation coefficient: The statistical parameter usually calculated on occasions when it is not convenient, economical or even possible to give actual values to variables, but only to assign a rank order to instances of each variable. It may also be a better indicator that a relationship exists between two variables when the relationship is non-linear.

special creation: The discredited idea that all living things occupying the Earth were specially created by God.

special nucleotides: A number of special nucleotides and oligonucleotides play a vital role in the regulation of metabolism in all types of cells. Some such nucleotides are adenosine 3′,5′-cyclic monophosphate (cyclic AMP), guanosine 3′,5′-cyclic monophosphate (cyclic GMP), guanosine-5′-diphosphate 3′-diphosphate (ppGpp), guanosine-5′-triphosphate 3′-diphosphate (pppGpp), diadenosine 5′,5″-P1, P4-tetraphosphate (A5′pppp5′A or Ap4A), pppA′p5′A2′p5′A (Two-five A″).

specialized genes: The genes whose activity is under the control of constitutive as well as regulatory factors. They are active only in specialized cells to form specialized functions. A distant enhancer element is activated by protein factor binding and is transmitted to the promoter and a signal is transmitted to specialized promoter, where a number of different constitutive (C) and regulatory (R) transcription factors are bound. The interaction between these bound factors transmits a signal to RNA polymerase II, which is held at the TATA box by another factor, and initiation occurs. Different combinations of factor-binding sites generate promoters and enhancers with different properties. This model derived from studies on specialized genes cannot be directly applied to all types of mammalian genes.

specialized nucleoprotein structures (snups): The DNA-binding proteins that localize these reactions appear to use multiple DNA-protein and protein-protein interactions to generate a snup structure in which the DNA is folded or wound. On the basis of prokaryotic work and additional inferences, snups seem to be responsible for high-precision reactions (transactions), such as site-specific recombination, initiation of DNA replication and initiation of transcription. *Compare with* SPECIALIZED RIBONUCLEOPROTEIN STRUCTURES (SNURPS).

specialized nucleoriboprotein structures (snurps): These are small ribonucleoprotein particles that assist in removal of introns from the messages issued by a cell's genes.

specialized packaging of sperm DNA: R. Dubruille *et al.* (2023) show that paternal chromosomes undergo a radical conformational change during sperm development, compacting 20- to 200-fold. Compaction is achieved by the replacement of canonical DNA packaging proteins with sperm-specialized versions. This specialized packaging renders paternal DNA biologically inert and unable to replicate or divide postfertilization in the zygote. Without this epigenetic distinction, the paternal DNA shatters before the first zygotic division. The specialized packaging of sperm DNA thus preserves genome stability in the fruit fly zygote.

specialized pre-mRNA processing: Spliced leader (SL) *trans*-splicing in *C. elegans* attaches a 22-nucleotide (nt) exon on to the 5′-end of many mRNAs. A particular class of SL, SL2, splices mRNAs of downstream operon genes. The UAYYUU motif hybridizes with the 5′ splice site on the SL2 RNA to recruit the SL to the pre-mRNA 5′ splice site of introns, effectively reversing the roles of snRNP and pre-mRNA in *trans*-splicing.

specialized transduction: Specialized transduction requires a phage that uses the lysogenic cycle for reproduction. In the lysogenic cycle, phage DNA can exist as a prophage integrated into the bacterial chromosome. Occasionally when the prophage exits it can take adjacent bacterial genes with it. The phage DNA directs synthesis of new phages. The phage particles carry phage DNA and bacterial DNA. The cell lyses, releasing the phages. A phage carrying bacterial DNA infects another cell. The joined phage and bacterial DNA circularize. Along with the prophage, bacterial DNA integrates with the recipient chromosome by a cross-over event. This forms a recombinant cell. In specialized transduction, only certain bacterial genes can be transferred. These genes must exist on either side of the prophage. *Compare with* GENERALIZED TRANSDUCTION.

speciation: A process whereby, over time, one species transforms through single evolutionary line into a different species (the process known as ANAGENESIS, PHYLETIC SPECIATION or SUCCESSIVE SPECIATION) or one species diverges to become two or more species (the process known as CLADOGENESIS or TRUE SPECIATION). Changes in related organisms ensue to the point where they are different enough to be considered separate species. This occurs when populations of one species are separated and adapt to their new environment or conditions (physiological, geographic or behavioural).

speciation reversal: The merging of two previously distinct lineages. After up to 2 million years of separate evolution, two types of common raven have been 'caught in the act' of consolidation.

species (Abbrev. sp.): (i) A group of actually or potentially interbreeding natural populations that are reproductively isolated from other similar groups (E. Mayr in 1910). (ii) A largest and most conclusive reproductive community of sexual and cross-fertilizing individuals, which share a common gene pool (Th. Dobzhansky in 1937). (iii) A lineage evolving separately from others and with its own evolutionary role and tendencies (E.H. Simpson in 1949). (iv) From development point of view, species is a group of individuals having complementary receptors on sperm and eggs. Most specific level in the binomial classification. The system of naming (nomenclature) of an organism was established in 1735 by Carolus Linnaeus. Scientific names are Latinized because Latin was traditionally used language by scholars at that time. According to this nomenclature each organism is assigned two names – the genus is the first name and is always capitalized; the specific epithet (species name) written in lowercase letters. Both these names are italicized. For example, the scientific name of rice is *Oryza sativa* L. in which *Oryza* is genus and *sativa* a species and L. stands for authority (i.e. Carolus Linnaeus).

species boundaries: Species boundaries emphasize the importance of reproductive isolation or lack of genetic exchange that separates species. By this account, crossing species boundaries would involve the transfer of genetic materials between populations of organisms that do not interbreed. Some individuals argue that crossing species boundaries is unnatural, immoral and in violation of God's laws, which presumes that species boundaries are fixed and readily delineated. In fact, species boundaries are simply a matter of a naming convention, and there are no truly fixed boundaries to cross.

species concept: An important classificatory category, that can be variously defined by the biological species concept, cladistic species concept, ecological species concept, phenetic species concept and recognition species concept. A particular species is referred to by a Linnaean binomial, such as *Homo sapiens* for human beings.

species group: A group of closely related species, usually with partially overlapping ranges.

species hybridization: The mating of individuals from two different species. Based on species hybridization, several wild species are characterized in the lineage of the cultivated crop plants (e.g. of wheat). Induced species hybridization is also practiced for introgression of desirable genes into the cultivated crop plants.

species recognition: The exchange of appropriate (species specific) stimuli and responses between individuals, particularly during courtship.

species selection: Selection between species ascending from differences in the rate of speciation and/or extinction of lineages.

species swarm: A large number of closely related species. A species swarm can be the result of a population splitting into a number of founder populations that form new species.

species tree: A phylogenetic tree revealing the relationships among species. It *contrasts* GENE TREES, which may include procedures such as gene duplication and lateral gene transfer.

species, cautions in defining: Generally, a species is referred to as a group, or populations of individuals, sharing common features and/or ancestry, generally the smallest group that can be readily and consistently recognized; often, a group of individuals capable of interbreeding and producing fertile offspring. Species is the basic unit of classification, the category of taxa of the lowest principal rank in the nomenclatural hierarchy. Strict assignment to a species is not always possible, as it is subject to particular contexts, and the species concept under consideration. Sibling species, interspecies crosses, intergeneric crosses in several cases pose challenge in defining a species. It is commonly assumed that members of one species are reproductively isolated from members of all other species (i.e. they cannot mate with them to produce fertile offspring). However, we must be cautious in defining species with this criterion because members of very closely related species can sometimes produce offspring together, and a small fraction of those may be fertile to some degree. This is the case with mules, which are the product of mating between horses and donkeys. About one out of 10,000 mules is fertile. This suggests that some species differences are a matter of degree.

species, origin of: The production of new types of organisms in descent from extant types. Natural selection is the process by which changes are favoured leading to the production of new types. The term was introduced by Charles Darwin.

specific combining ability (sca): The performance of specific combinations of genetic strains in crosses in relation to the average performance of all combinations. *Compare with* GENERAL COMBINING ABILITY (GCA).

specific epithet: Follows the name of the genus and is the second word of a botanical binomial system. The generic name and specific epithet together constitute the name of a species (i.e. the specific epithet is not the species name).

specific modifier: A gene that has the specific and perhaps exclusive function of modifying the expression of a gene at another locus.

specific resistance: A resistance that is operative against some biotypes or races of the pathogen, but not others, usually inherited monogenically and expressed qualitatively. *See* VERTICAL RESISTANCE. *Contrasts* DURABLE RESISTANCE, GENERAL RESISTANCE, HORIZONTAL RESISTANCE *and* RACE-NON-SPECIFIC RESISTANCE.

specific-locus test: A system for detecting recessive mutations in diploids. Normal individuals treated with a mutagen are mated to the testers that are homozygous for the recessive alleles at a number of specific loci. The progeny is then screened for presence of recessive phenotypes.

specificity: The proportion of individuals without the abnormality or disease that will have a negative test.

specificity factors: Proteins that reversibly associate with the core component of RNA polymerase and determine, which promoters the polymerase will bind to.

specificity in eukaryotic transcription: Cells regulate the genetic machinery for synthesis of ubiquitous housekeeping proteins but impart specificity in the synthesis of other proteins characterizing the cell type or its differentiated state. Specificity in the regulation of gene transcription basically rests on the specificity of interaction of *cis*-acting elements with *trans*-acting factors. Mechanisms in eukaryotic cells exist that permit degeneracy in DNA-protein interactions, at the same time ensuring specificity through auxiliary DNA-protein and protein-protein contacts, such that a finite number of *trans*-acting factors can differentially interact with different architectures of *cis*-acting elements. This can be illustrated with examples of B-cell specificity in immunoglobulin heavy chain gene expression and specificity in liver gene expression.

speckels in RNA splicing: Pre-mRNA splicing is a predominantly cotranscriptional event, which involves a large number of essential splicing factors within the mammalian cell nucleus. Most splicing factors are concentrated in 30–40 distinct domains called SPECKLETS. Speckles are highly dynamic structures that respond specifically to activation of nearby genes. These dynamic events are dependent on RNA polymerase II transcription and are sensitive to inhibitors of protein kinases and Ser/Thr phosphatases. When single genes were transcriptionally activated in living cells splicing factors left speckles in peripheral extensions and accumulated at new sites of transcription. One function of speckles is to supply splicing factors to active genes.

spectral data: Spectral analysis involves the calculation of waves or oscillations in a set of sequenced data. These data may be observed as a function of one or more independent variables such as the three Cartesian spatial coordinates or time. The spatial or temporal observation interval is assumed to be constant.

spectral karyotype (SKY): An image of all chromosomes of a complement, each labelled with a different fluorescence colour. This technique is valuable for identifying structural chromosome abnormalities or alien chromosomes. *See* CHROMOSOME.

spectral reflectance: Spectral reflectance is the ratio of upwelling to down-welling radiant fluxes for given surface, illumination and observation properties, including the sensor's spectral sampling properties. Similar to the human eye's brightness and chromatic adaptation, such normalization by a down-welling flux helps the interpretability of reflectance as opposed to the radiance reflected by a given target. Information provided by analysis of remotely sensed multispectral reflectance data of agricultural crops can be applied to improve crop type inventory and crop yield assessment.

spectrometry: The measurement of the absorption or emission of light by a substance at a specific wavelength.

spectrophotometer: An optical instrument that measures the concentration of a compound that has been dissolved in a solvent (water, alcohol, etc.). The instrument shines a light through the solution, measures the fraction of the light that is absorbed by the solution and calculates the concentration from that absorbance value.

spectroradiometer: A machine that operates almost like the visible region spectrophotometers, are designed to measure the spectral density of illuminants. Applications may include evaluation and categorization of lighting for sales by the manufacturer, or for the customers to confirm the lamp they decided to purchase is within their specifications. It has four components: (i) The light source shines on to or through the sample. (ii) The sample transmits or reflects light. (iii) The detector detects how much light was reflected from or transmitted through the sample. (iv) The detector then converts how much light the sample transmitted or reflected into a number.

spectroradiometry: It is the measurement of absolute radiometric quantities in narrow bands of wavelength.

speech recognition: An interdisciplinary subfield of computational linguistics that develops methodologies and technologies that enables the recognition and translation of spoken language into text by computers. It incorporates knowledge and research in the linguistics, computer science and electrical engineering fields. *Aka* AUTOMATIC SPEECH RECOGNITION (ASR), COMPUTER SPEECH RECOGNITION or SPEECH TO TEXT (STT).

speed breeding: Speed breeding is a revolutionary plant breeding technology that greatly reduces generation time by using optimal day length, light intensity, light quality and temperature to improve biomass accumulation and stimulate flowering and seed production. A controlled environment also allows the use of immature seed for propagation to minimize generation time. So far, speed breeding techniques have been well established in many plant species, including long-day plants, day-neutral plants and a limited number of short-day plants.

speltoid: A mutation that arises spontaneously (e.g. in wheat). This mutant resembles spelt wheat with respect to certain features.

sperm: The male reproductive cell, or gamete, in anisogamous forms of sexual reproduction (forms in which there is a larger, female reproductive cell and a smaller, male one). Animals produce motile sperm with a tail known as a flagellum, which are known as SPERMATOZOA (sing. SPERMATOZOON) while some red algae and fungi produce non-motile sperm cells, known as SPERMATIA. *Compare with* EGG.

sperm and ovum, differences between: In spermatogenesis, the four products of meiosis are viable while in oogenesis only one product is functional and the remaining three products are degenerated. In size, sperm is smaller than ovum. Sperm is motile while ovum is immotile. Process of spermatogenesis is completed within the testis and usually mature sperms come out of it. In case of oogenesis, second maturation stage occurs outside the ovary at the time of fertilization. Sperms are produced in million while ovum is only one.

sperm attachment: Attachment of sperm with egg is the first step in fertilization. Attachment involves binding of sperm and egg. When sperm bind to surface of egg, cortical granules fuse with plasma membrane and cortical granules open up. This is called cortical reaction or fertilization wave as a result of which Ca^{++} is responsible for everything else that occurred during fertilization. Protease separates plasma membrane and vitelline layer. As a result, vitelline layer is lifted up and fertilization envelope is formed. Fertilization envelope is derived from vitelline layer. Protease also breaks sperm receptors and removes them. As a result of this, other sperm will be released from the egg. Polyspermy is thus prevented.

sperm DNA modifications: Sperm DNA has been found to be methylation but it has so far not been demonstrated to play role in epigenetic inheritance.

sperm middle piece: Middle part mainly consists of mitochondria. Mitochondria carry oxidative enzymes, which provide energy to flagellum, required for propulsion of flagellum. Middle part is surrounded a condensed sheath of cytoplasm called manchette.

sperm miRNAs: Sperm contain miRNAs and alterations in sperm miRNAs have been detected. Zygotic injection of miR-19b (the expression of which is upregulated in the sperm of mice consuming a high-fat–high-sugar diet) induced obesity and glucose intolerance in the offspring, recapitulating paternal phenotypes. Epigenetic changes induced by miR-19b injection were sustained in the germline.

sperm nucleus (pl. nuclei): In angiosperms, the two male gametes formed by division of the generative cell migrate down the pollen tube behind the vegetative nucleus. When the pollen tube enters the embryo sac, the tip of the tube bursts to release the generative nuclei; one of them fuses with the egg nucleus to form the zygote, and the other typically fuses with the polar nuclei or definitive nucleus to form the primary endosperm nucleus.

sperm RNA modifications: Modifications are found in sperm RNA. Full map of RNA modifications in sperm RNAs will be of big help to understand the role of tsRNAs and miRNAs in the inheritance of acquired traits.

sperm RNAs in oocytes: Sperm is believed to pass on only DNA to oocyte during fertilization. Now it is known that human male gametes pass over more to the oocyte than just the haploid male genome. Paternal messenger RNAs are also delivered to the egg at fertilization. These sperm RNAs are suggested to be important in early zygotic and embryonic development. Significant biological functions assigned to sperm RNAs include their involvement in non-Mendelian inheritance in mammals, such as the paramutation phenomenon in mice, and their contribution to the intergenerational inheritance of paternally acquired traits, including mental and nutritional stresses.

sperm tail: Tail contains thin layer of cytoplasm. It consists of ten pairs of longitudinal fibres in mammals. It consists of nine pairs. It is a wedge-shaped structure.

sperm transfer RNA-derived small RNAs (tsRNAs): Presence of tsRNAs has been detected through RNA sequencing. Sperm tsRNAs mainly result from the 5′-end of tRNAs, ranging in size from 29–34 nt and are more plentiful than miRNAs, instituting the majority of small ncRNAs in sperm. Sperm tsRNAs can act as sensitive markers of environmental exposure.

sperm typing: The technique that analyses genotype of each meiotic product directly and does not require study of diploid offspring.

sperm-borne mitochondrial RNAs: Spermatozoa dock a complex and environment-sensitive pool of small non-coding RNAs (sncRNAs), which impacts offspring development and adult phenotypes. A. Tomar *et al.* (2024) engaged two diverse patterns of presumption acute high-fat diet to dissect epididymal versus testicular contributions to the sperm sncRNA pool and offspring health. They reveal that epididymal spermatozoa, but not developing germ cells, are sensitive to the environment and identify mitochondrial tRNAs (mt-tRNAs) and their fragments (mt-tsRNAs) as sperm-borne factors. Upregulation of mt-tsRNAs is downstream of mitochondrial dysfunction. Single-embryo transcriptomics of genetically hybrid two-cell embryos revealed sperm-to-oocyte transfer of mt-tRNAs at fertilization and suggested their participation in the control of early-embryo transcription. mt-tRNAs are diet induced and sperm borne and demonstrate, in a physiological setting, father-to-offspring transfer of sperm mitochondrial RNAs at fertilization.

sperm-mediated gene transfer (SMGT): SMGT is based on the spontaneous ability of sperm cells to bind and internalize exogenous DNA and carry it to oocyte during fertilization produces genetically modified animals with high efficiency. The SMGT procedure involves (i) collection of sperm cells from many different members of the species, (ii) evaluation of fertility and ability to take up foreign DNA, (iii) removal of seminal fluid and mix with transgene and (iv) transfer of sperm to recipient via artificial insemination. A variant of this method is TESTIS-MEDIATED GENE TRANSFER (TMGT).

sperm-mediated transfection: Introduction of DNA into an oocyte by first mixing it with sperm, which then is used for fertilization.

spermatids: End product of meiosis. Formation of spermatids starts from the primary germ cells or primordial germ cells of testes and is differentiated into three phases – multiplication phase, growth phase and maturation phase.

spermatocyte: The cell that undergoes meiosis to produce four spermatids.

spermatogenesis: The process responsible for formation of mature male gametes (sperms) from an undifferentiated germline cell, as a consequence of meiosis, is known as spermatogenesis. Its end product is spermatids. Germinal epithelium cells present in the testes possess the power to undergo this process. The diploid epithelium cells when undergo spermatogenesis undergo reduction in chromosome number. The process of maturation of reproductive cells occurs in testes. The process includes formation of spermatids from the germinal cells of epithelium through meiosis and metamorphosis of spermatids into sperms, known as SPERMIOGENESIS or SPERMATELIOSIS.

spermatogenesis and oogenesis, significance of: Reduction of chromosome number to half helps in restoring the parental combination of diploid (2n) chromosomes after fusion of gametes at the time of fertilization. In embryo, sex cells have the same origin. Original cells are PRIMORDIAL GERM CELLS (PGCs), which are set apart very early during development. They arise from endoderm close to kidney. They migrate to gonads, then called spermatogonia and oogonia. Primordial germ cells migrate to gonads. They circulate through blood by ambeodic movement.

spermatogonium: A cell type in the testis of male vertebrates that produces primary spermatocytes by mitosis.

spermiogenesis: It is gradual differentiation of stationary rounded spermatids into an active and motile sperm cell. Spermatids of a typical animal cell contain mitochondria, Golgi bodies and centriole, cytoplasm and a nucleus with n number of chromosomes. Spermatid is not functional as such as it is to reach the ovum and, therefore, has to be motile. So, it reduces in weight and loses all the superfluous structures to become motile. Is Y chromosome needed for spermiogenesis? How can the genes of the Y chromosome be needed for spermiogenesis when this process occurs after both meiotic divisions, a time when half of spermatids possess X chromosome and the other half possess Y chromosome? X and Y chromosomes are separated in first division yet spermatozoa differentiate from secondary spermatocyte receiving X not Y, produce normal sperm. This is explained on the basis of storage of stable mRNAs produced by gene or genes of the Y chromosome. The protein synthesis going on during spermiogenesis appears to use mRNA templates that were made in primary spermatocyte stage before the time X and Y chromosomes were separated. *Aka* SPERMATELIOSIS.

spheroplast: A bacterial or plant cell from which most of the cell wall has been degraded or removed. *See* PROTOPLAST.

spherosomes: Vesicles in plant cells that appear to perform functions similar to those of lysosomes in animal cells.

spike: A botanical term use to denote an unbranched inflorescence in which flowers (or SPIKELETS of grasses) are attached directly, without petioles, to a central stem (e.g. wheat, gladiolus). *Compare with* RACEME.

spikelet: A spikelike appendage made of one or more reduced flowers and associated bracts; unit of inflorescence in grasses; a small spike.

spin-polarized scanning tunnelling microscopy (SP STM): SP STM allows to record constant current images, which represent the electron density near the sample surface and simultaneously the spin polarization of the electron density, which is related to the magnetic moment.

spindle: The microtubule apparatus that controls chromosome movement during anaphase stages of mitosis and meiosis.

spindle assembly checkpoint: This checkpoint senses the completion of chromosome alignment at metaphase. It differs from the G1 and G2 checkpoints in that its default setting is 'on' as cells enter mitosis. It shuts off only when every chromosome is properly attached to the spindle.

spindle attachment: That point of a chromatid to which the spindle fibre appears to be attached at metaphase and anaphase and that, at anaphase starts to move towards the pole before the rest of the chromatid. *Aka* CENTROMERE, INSERTION REGION *or* KINETOCHORE.

spindle fibres: Aggregates of microtubules that move chromosomes during cell division.

spindle poison: Any poison negatively affecting the formation or function of the spindle of the dividing cell.

spindle pole: One of the two opposite poles to which the spindle fibres pull the chromosomes or chromatids during anaphase.

spindle-attachment checkpoint: Checkpoint that operates during mitosis to guarantee that all chromosomes are properly attached to the spindle before sister-chromatid separation starts.

spintronics: Electronic devices exploiting the spin of electrons, in addition to charge.

spiral: An arrangement where plant parts are arranged in a succession of curves like the thread of a screw or coiled in a cylindrical or conical manner.

spiral cleavage: The cleavage process in molluscs and some invertebrates whereby the spindle at mitosis is tipped in relation to the original egg axis.

spiral model: A risk-driven software of development process model. Based on the unique risk patterns of a given project, the spiral model guides a team to adopt elements of one or more process models, such as incremental, waterfall or evolutionary prototyping.

splice form: By using alternative splicing, a single message precursor from DNA can generate an entire family of mRNAs and proteins. This can be utilized to create specificity in cell-cell or cell-ligand interactions. A cell may produce a given protein, but it will be a different splice form of the protein than that produced by an adjacent cell. In this manner, the two cells have the potential to interact differently with other cells or molecules. Two places where this has been extremely important is in the production of cell-surface specificity proteins in the immune and nervous systems.

splice site (SS): Introns are removed from primary transcripts by cleavage at conserved sequences called 'splice sites'. These sites are found at the 5′ and 3′ ends of introns. Most commonly, the RNA sequence that is removed begins with the dinucleotide GU at its 5′ end and ends with AG at its 3′ end.

splice variants: Refers to all possible mature messenger RNAs resulting from alternative splicing of one precursor messenger RNA molecule.

splice-site compatibility: The phenomenon that results in differential use of splice sites as certain splice-site combinations are specifically preferred over the others and different mRNAs are formed. Differential transfactors also influence splice-site compatibility as they have ability to regulate the selection of differential splice-site utilization in a positive or negative manner.

splice-site mutations: Nucleotide signals at the splice sites of pre-mRNA direct the enzymatic machinery. If a mutation alters one of these signals, then the intron is not removed and it remains as part of the final mRNA molecule. The translation of this mRNA sequence alters the amino acid sequence of the protein product.

spontaneous mutation rate: Refers to frequency of mutations that occur naturally and in absence of an artificial mutagenic treatment.

splice-site recognition: Serine and arginine-rich (SR) proteins have important roles in facilitating splice-site recognition. For example, they recruit the U1 snRNP to the 5′ splice site and the U2AF complex and U2 snRNP to the 3′ splice site by binding to an ESE and directly interacting with protein targets. Inhibition of splice-site recognition can be achieved in many ways. First, when splicing silencers are located close to splice sites or to splicing enhancers, inhibition can transpire by sterically blocking the access of snRNPs or of positive regulatory factors. Some silencers can be located over 100–200 bp away from enhancers. Such splicing inhibitors act by masking splice-site recognition through multimerization along the RNA.

splice-site variant (splicing variant): A nucleotide change that alters or removes the specific sequence necessary for excising out of an intron. Variants that alter this sequence can result in the retention of an intron of the loss of an exon in the mature messenger RNA, thereby resulting in an altered protein product. May be denoted splice-3 if in the 3′ end of a splice site, or splice-5 if at the 5′ end of a splice site.

spliced leader (SL): The SL is a gene that generates a functional ncRNA that is composed of two regions: an intronic region of unknown function (SLi) and an exonic region (SLe), which is transferred to the 5′ end of independent transcripts yielding mature mRNAs, in a process known as spliced leader *trans*-splicing (SLTS).

spliced leader RNA (SL RNA): Spliced leader exonic (SLe) region addition could lead to increased mRNA stability, differential recruitment of the translational machinery, modification of the 5′ region or a combination of these effects.

spliceosome: A dynamic structure assembled on pre-mRNA templates in a stepwise fashion and then dissociates as the splicing reaction is completed. Even a small error in splicing is not tolerable as it results into non-functional polypeptide. Thus, RNA processing involves high degree of specificity. Specialized nuceoriboprotein structures (snurps)-like particles identified in cytoplasm and nucleolus are, respectively, named small cytoplasmic ribonucleoproteins (scRNPs or scyrps) and small nucleolar ribonucleoproteins (snorps). The multiple snRNP nature of spliceosome also raises the possibility that different introns may be recognized by different combinations of snRNPs. Five types of small nuclear RNAs (snRNAs) (U1, U2, U4, U5 and U6) constitute a spliceosome. snurps help to eliminate worthless introns from the messages issued by a cell's genes. Without snurps cellular activity would come to a standstill. J.A. Steitz *et al.* discovered snurps in 1978. These small complexes of RNA and a variety of proteins are critical components of a sophisticated molecular assemblage called spliceosome.

spliceosomal B complex: The pre-catalytic spliceosome (B complex) is preceded by its precursor spliceosome (pre-B complex) and followed by the activated spliceosome (B^{act} complex). The pre-B-to-B and B-to-B^{act} transitions are driven by the ATPase/helicases Prp28 and Brr2.

spliceosome activation: Catalytic activation of the spliceosome is critically dependent on its association with the NineTeen Complex (NTC) named after its core E3 ubiquitin ligase subunit PRP19.

spliceosome assembly: Intricate RNA-RNA and RNP networks that serve to align the reactive groups of the pre-mRNA for catalysis and are formed and repeatedly rearranged during spliceosome assembly. Exon-intron 5′ splice site is recognized by U1 snRNP. The precursor RNA is probably also associated with hnRNP core proteins. U5 snRNP binds to the 3′ splice site at this stage. Incubation of substrate RNA in the presence of ATP leads to the rapid formation of a U2 snRNP complex. Further incubation produces assembly of the spliceosome containing at least U2, U4, U5 and U6 snRNAs. The U4 and U6 snRNAs can be associated in one small ribonucleoprotein (snRNP) particle. The two RNAs characteristics of the intermediate in splicing, the 5′ exon and the lariat intervening sequence terminating in the 3′ exon RNAs, are found solely in the spliceosome.

spliceosome disassembly: M.K. Vorländer *et al.* (2024) describe 2.6–3.2 Å resolution cryo-electron microscopy structures of nematode and human terminal intron-lariat spliceosomes. They reveal how four disassembly factors and the conserved RNA helicase DHX15 recruit spliceosome disassembly. The disassembly factors investigate large inner and outer spliceosome surfaces to perceive the release of ligated mRNA. Two of these factors, TFIP11 and C19L1, and three general spliceosome subunits, SYF1, SYF2 and SDE2, then dock and stimulate DHX15 on the catalytic U6 snRNA to initiate disassembly. U6 thus controls both the start and end of pre-mRNA splicing.

spliceosome remodelling: Spliceosome remodelling, executed by conserved adenosine triphosphatase (ATPase)/helicases including Prp2, enables pre-mRNA splicing. These conserved mechanisms explain the coupling of spliceosome remodelling to pre-mRNA splicing.

splicing: A process involving the excision of introns and joining exons during maturation of pre-mRNA. Eukaryotic genes contain both introns (non-coding) and exons (coding region). This is necessary, because the eukaryotic genes if transferred to bacteria as such, then bacterium will not be able to remove the introns and therefore will not be able to make correct protein. *See* GENE SPLICING *and* RNA SPLICING.

splicing and polyadenylation: Pattern of alternative RNA splicing and alternative cleavage and polyadenylation were powerfully correlated across tissues, suggesting synchronized regulation of both splicing and polyadenylation processes.

splicing in mRNA cytoplasmic localization: An important newly discovered activity for splicing is regulation of MESSENGER RIBONUCLEOPROTEIN (mRNP) complex assembly and organization for mRNA cytoplasmic localization.

splicing junctions: In RNA molecules, the sequences of nucleotides immediately flanking the exon-intron boundaries.

splicing-dependent microarray (SDM). A microarray method in which fluorescent probes are used to specifically detect pre-mRNA, mRNA and total RNA obtained from a cell culture. This can be used to study changes in splicing that occur *in vivo*.

SPlinted Ligation Adapter Tagging (SPLAT): SPLAT is based on initial bisulfite treatment of the DNA, followed by adaptor tagging of single-stranded DNA fragments, and enables WGBS using low quantities of the DNA samples. If SEQUENCE-TAGGED SITE (STS) does not display polymorphism, it is usually converted into SPLAT. Discrete STS products from different genotypes are themselves individually sequenced. Any difference thus revealed can be sequenced from nuclear ribosomal DNA. They can be used in the production of internal primers. In general, single-polymorphic amplified sequences are dominant markers.

split gene: A discontinuous gene. In split genes, coding sequences (exons) are split by non-coding sequences (introns). Based on the number of introns in a split gene, the genes are of two types. Of the first are the MONINTRON GENES as they have only one intron. Such gene have been reported in case of $tRNA^{tyr}$ gene in *Xenopus* and $tRNA^{leu}$ gene in *Drosophila*. The transcripts of such genes are not used for translation and they as such perform a biological function. The importance of this intron is reflected by the fact that removal of this intron leads to lack of processing and loss of function of the transcript. Of the second type are the MULTINTRON GENES as they have several introns. Transcripts of multintron genes are translated into proteins (e.g. rat muscle α-actin gene has six exons). Size of introns is usually much larger than the exons. Introns are spliced out during processing of the pre-mRNA. Collagen gene has 52 introns.

split plot design: An experimental design that permits the testing of two factors in combination – one factor (the main effect) serves as a replication for the second factor (the split effect). There are many split plot design options, but the basic principle involves assigning one set of treatments to the main plots that are arranged in RANDOMIZED COMPLETE BLOCKS; the second set of treatments is assigned to subplots within each main plot. The statistical analysis is similar to that used with the randomized complete block design (i.e. an ANOVA set up for a split plot is followed by a means comparison test such as the Dunn's test or Duncan's multiple range test). *Aka* NESTED BLOCK DESIGN (NBD) *and* SPLIT DESIGN.

split promoter: The intergenic promoter of eukaryotic genes encoding transfer RNAs that consist of two separated regions within the gene's coding sequence.

split versus non-split gene: In prokaryote, nature of gene is non-split, and no processing is required for a transcript to become mRNA. In eukaryotes, most of the genes are split (again, with some exceptions known for example histone genes, interferon genes) in nature (i.e. coding sequences are separated by non-coding sequences). Transcript requires to be processed in the nucleus before finished mRNA is obtained. Non-split gene in eukaryotes are known as EXONIC GENES.

spontaneous generation theory: The theory, now known to be invalid, that plants, animals and microorganisms suddenly arise from non-living materials under certain environmental conditions.

spontaneous haploids: The spontaneous existence of haploids, usually by asexual processes.

spontaneous mutation: The only operational definition of spontaneous mutations is that these mutations are generated in natural or laboratory populations under normal growth conditions. The frequency may, therefore, depend on the growth medium, temperature, pH and so on, which are normally employed. These are those mutations whose cause is not defined. These may originate due to errors in DNA replication, due to environmental mutagens, misalignment mutagenesis, rare base pairing between two purines or two pyrimidines (tautomerism) or transposable genetic elements. The frequency of spontaneous mutations may depend on the growth medium, temperature, pH and so on. In plants, the frequency of such mutations is very low usually one out of millions. In horticultural plants, many cultivars have been developed from bud mutants – product of spontaneous mutation. These mutations are common in the microorganisms.

spontaneous mutations due to transposition: Transposons are segments of DNA that can move around to different positions in the genome of a single cell. In the process, they may cause mutations or can increase (or decrease) the amount of DNA in the genome. These mobile segments of DNA are sometimes called JUMPING GENES.

spontaneous mutations, mechanism of: Mechanisms of spontaneous mutations include changes in electronic structure of bases, rare base pairing between two purines or two pyrimidines, errors in DNA replication, agents residing inside the cell, transposition and misalignment mutagenesis.

spontaneous protoplast fusion: Protoplasts fuse through their plasmodesmata.

spontaneous univalent: This is the spontaneous production of an aberration in which one member of a homologous pair of chromosomes in the cell or an individual is lacking.

spontaneous variation: The variation in plants derived from tissue culture not exposed to mutagens but occurring as a result of the culture conditions.

sporadic cancer: Cancer that occurs randomly and is not inherited from parents; caused by DNA changes in one cell that grows and divides, spreading throughout the body. *See* HEREDITARY CANCER.

sporadic case: The occurrence of a genetic disorder in one individual of a family due to a new mutation or by chance. These disorders have a low likelihood of recurring in the family. *Aka* ISOLATED CASE.

sporadic disease: Disease that occurs, occasionally in a population. In case of humans, the disease like tetanus that occurs only occasionally is called sporadic disease. Likewise, there are sporadic diseases of plants and animals. Some insect pests are also sporadic in nature.

sporangiophore: An organ bearing sporangia (e.g. the cones of Equisetum, an example of a living fossil).

sporangium (pl. sporangia): A structure in which spores are formed and the mature spores are released.

spore: In biology, a spore is a unit of sexual or asexual reproduction that may be adapted for dispersal and for survival, often for extended periods of time, in unfavourable conditions. Spores form part of the life cycles of many plants, algae, fungi and protozoa. Spore is a thick walled, resistant, non-growing structures produced by bacteria and fungi during unfavourable conditions. Spore is a haploid propagule, produced by meiosis in diploid cells of a sporophyte that can germinate to produce a multicellular gametophyte. A minute reproductive unit in fungi and lower plant forms is also called spore. *See* SPORULATION.

spore dormancy: DNA in dormant spores of *Bacillus* sp. is well protected against damage resulting from treatments, such as desiccation, heat, oxidizing agents, UV and γ radiations. The DNA-binding α/β-type SMALL ACID-SOLUBLE PROTEINS (SASPs) are a major factor in the resistance and long-term survival of spores of *Bacillus* species by protecting spore DNA against such damages.

spore-specific repair: A type of DNA repair which in *Bacillus subtilis* restores the spore-specific UV photoproduct 5-thyminyl-5,6-dihydrothymine into two residues in site leaving the DNA backbone intact.

sporophyte: In alternation of generations, the whole plant body is sporophyte possessing diploid chromosome number and producing MEIOSPORES upon meiosis. The sporophytes possess diploid chromosome number in comparison to gametophytes, which possess haploid chromosome number. Unlike some lower plant in higher plants, the sporophyte phase is much longer in comparison to gametophytic phase that starts with the formation of gametes and ends with the development of zygote in the flowers. *Compare with* GAMETOPHYTE.

sporophytic: Of the spore-bearing diploid portion of a plant.

sporophytic self-incompatibility (SSI): Refers to SI based on the genotypic and phenotypic relationship between the female and male reproductive system. Alleles in cells of the pistil determine its receptivity to pollen. The phenotype of the pollen, expressed as its inability to effect fertilization, may be determined by the maternal plant, known as SPOROPHYTIC INCOMPATIBILITY.

sport: A naturally occurring variant of a species, not usually occurring in a population or group of plants. A plant that has mutated spontaneously so that it differs from its parent plant.

sporulate: To produce spores.

sporulation: The process of forming spores. Sporulation is common process in bacteria and fungi. During unfavourable conditions (stress conditions) bacteria for instance sporulate to produce endospores, the most resistant structures (non-growing) that can survive up to hundreds of years, withstanding moisture stress, temperature stress, toxic metals and radiations. On the onset of favourable conditions endospores germinate to produce new vegetative cells. These endospores are sometimes also called spores.

sporulation in bacteria: Loss of the sigma activity of RNA polymerase of *Bacillus subtilis* has been observed during sporulation. Alterations in RNA polymerase are involved in the initiation of sporulation in *Bacillus subtilis*. In its RNA polymerase, at least three sigma factors, namely, σ^{43} (43,000 Da), σ^{37} (37,000 Da) and σ^{29} (29,000 Da), have been identified. σ^{43} is involved in transcription during vegetative phase, while σ^{37} and σ^{29} are involved during sporulation. At the beginning of sporulation, σ^{43} is replaced by σ^{37}, perhaps due to the activity of any of the five *spo O* genes, which either removes σ^{43} or modifies the core enzyme. The displaced σ^{43} can be recovered from the sporulating cells. Therefore, $\alpha_2\beta\beta'\omega\sigma^{37}$ initiates the transcription of the gene needed for sporulation, another protein σ^{29} appears and RNA polymerase $\alpha_2\beta\beta'\omega\sigma^{29}$ functions for transcription. Several other sigma factors have also been isolated and characterized in *Bacillus subtilis.*

spot: An indication of disease characterized by a limited necrotic area, as on leaves, flowers and stems.

spot test: In mutagenicity testing, an *in vivo* procedure for detecting mutagens and carcinogens. The method involves placing the agent in the centre of the Petri dish containing the minimal medium that is seeded with auxotrophic bacteria. Revertants from auxotrophy to prototrophy are identified as colonies in a ring around the spot where the potential mutagen was deposited. This test permits characterization of mutagens inducing base pair substitutions, frameshift mutations or large deletions.

spotted microarrays: In spotted microarrays, the probes are oligonucleotides, cDNA or small fragments of PCR products that correspond to mRNAs. The probes are synthesized prior to deposition on the array surface and are then ‘spotted’ on to glass. A common method utilizes an array of fine pins or needles controlled by a robotic arm that is dipped into wells containing DNA probes and then depositing each probe at designated locations on the array surface. The consequent 'grid' of probes represents the nucleic acid profiles of the prepared probes and is ready to receive complementary cDNA or cRNA 'targets' derived from experimental or clinical samples.

spotting-microarray technology: In this microarray technology, PCR products from cDNA clones are spotted on to glass slides using robotic suspensors. Each spot contains a specific probe for a gene.

spreading: Standing out at right angles to an axis (e.g. in leaves or hairs).

spreading of RNA silencing: It occurs from 5′ to 3′ direction along the transgene without any effect on transgene methylation.

spreading position effect: Phenomenon in which the effect of an inversion or translocation in inactivating adjacent gene sequences affects more than one such sequence.

springwood: The part of the yearly xylem growth ring in woody plants formed early in the growing season, consisting typically of cells that are larger than those formed later in the season. *Contrasts* SUMMERWOOD.

spur: (i) A short shoot. (ii) A conical or tubular outgrowth from the base of a perianth segment, often containing nectar.

spurious pleiotropism: When one effect causes other thigs to happen. It may be brought about in two ways. Either the gene does two (or more) things directly, but by means of the same mechanism, or else it has one effect which in turn causes two things to happen.

spurious transcription: Refers to both cryptic transcription and the ectopic and/or extemporaneous expression of annotated transcripts in cells where the corresponding genes are normally silenced.

spurious transcriptional activation: Spurious transcriptional activation by unliganded nuclear receptors is inhibited by histone lysine methylation. This inhibitory histone modification code is efficiently countered by ligand-dependent recruitment of histone lysine demethylase, including lysine-specific demethylase 1 (LSD1), which appears to be used for this purpose by a number of signal-regulated transcriptional activators. Histone methyltransferase (HMT)-dependent mechanisms (e.g. H3K9 HMTs) impose ligand-dependency for gene activation by nuclear receptors.

square design: A design having equally many points and blocks.

squash and dot hybridization: In this *in situ* hybridization method, one root tip from a germinated seed may be quashed on to nitrocellulose filter and filters are hybridized with radioactively labelled repeated DNA probe derived from the alien species in question. After washing the filters in appropriate buffers, they are utilized for exposure of X-ray film, which on developing yields the desired information about the existence of alien genetic material.

Src family: Family of cytoplasmic tyrosine kinases (pronounced 'sark') that relate with the cytoplasmic domains of some enzyme-linked receptors (e.g. the T cell antigen receptor) that lack intrinsic tyrosine kinase activity. They transmit a signal onwards by phosphorylating the receptor itself and other signalling proteins.

Src homology domain 2 (SH_2 domain): A protein domain existing in many signalling proteins. This domain binds a short amino acid sequence containing a phosphotyrosine.

***Sst*I:** A type II restriction enzyme produced by bacterium *Streptomyces stanford* that recognizes the DNA sequence 5′-G A G C T↓C 3′ / 3′-C↑T C G A G 5′ where arrows indicate sites of cleavage.

stabilizing selection: A type of selection that removes individuals from both ends of a phenotypic distribution divided by deviation of a sample of means.

stability: Resistance to environmental changes that leads to reduced genotype by environment interactions and increased fitness.

stability against environmental changes: Resistance to environmental changes that leads to reduced genotype by environment interactions and increased fitness.

stabilizing selection: A type of selection that removes individuals from a population (i.e. from both ends of a phenotypic distribution divided by deviation of a sample of means. In the case of polygenic traits that are expressed as a continuum of phenotypes, such as human stature, it would be selection for people who are in the middle of the range and against those who are very tall or very short. The result would be fewer people who are at the extremes in height. In the case of a trait controlled by only two alleles, it would be selection against both homozygotes and the heterozygote. An example is nature selecting for people who are heterozygous for the gene locus. The theorized competitive disadvantage of unnecessary virulence genes is that the races with excess genes would have decreased fitness relative to races with fewer virulence genes, so a 'super-race' would be less likely to appear in multiline crops.

stable DNA methylation changes: Global DNA methylation was stable over 18 months in blood at −20°C and −80°C and DNA at 4°C and −80°C. However, at 18 months, DNA methylation from DNA stored at −20°C relatively decreased by 6.1% compared to baseline.

stable expression: Permanent kind of expression of transgene(s) that is heritable to following generations. During plant transformation, the introduced transgene(s) when gets integrated with the host genome and expressed along with the other native genes, the phenomenon is called stable gene expression. This is actually required to produce stable transgenics. Such transgene(s) are heritable to following generations through sexual reproduction.

stable messenger RNA: Stable mRNA is responsible for differentiation. Support to this view comes from the observations that during the formation of fibre cells two changes take place: reduction in the quantity of mRNA and mRNA becomes long-lifed. The red blood cells have no nucleus but protein synthesis is there. This may be due to the presence of stable mRNA molecules. Cells save lot of energy by synthesizing stable mRNA.

stable RNA: RNA molecules, such as ribosomal RNA and transfer RNA (but not messenger RNA, normally), which are subject to rapid turnover in the cell. The stabilization of these RNAs is probably the result of complexing with a protein.

stable transgene expression: Stable transgene expression requires an efficient promoter expression vector that resists gene silencing. Stability of transgene expression can be enhanced following various approaches: (i) integration of multiple copies, (ii) insertion of repeated sequences in concatemeric arrangement at one locus and (iii) length of homology, especially by the position of the interacting sequences.

stabilization of the mitotic spindle: Clathrin has an established function in the generation of vesicles that transfer membrane and proteins around the cell. The formation of clathrin-coated vesicles occurs continuously in non-dividing cells, but is shut off during mitosis, when clathrin concentrates at the spindle apparatus. Clathrin stabilizes fibres of the mitotic spindle to avoid congression of chromosomes. Clathrin are bound to the spindle directly by the amino-terminal domain of clathrin heavy chain.

stack: An abstract data type that functions as a collection of elements, with two main principal operations: push, which adds an element to the collection, and pop, which removes the most recently added element that was not yet removed. The order in which elements come off a stack has given rise to its alternative name, LIFO (last in, first out). Additionally, a peek operation may yield access to the top without modifying the stack. The name 'stack' for this type of structure comes from the analogy to a set of physical items stacked on top of each other. This structure makes it easy to take an item off the top of the stack, while getting to an item deeper in the stack may require taking off multiple other items first.

stacked genes: Refers to the insertion of two or more (synthetic) genes into the genome of an organism. One example would be of a plant into which a gene from *Bacillus thuringiensis* (*Bt*) and a gene for resistance to a specific herbicide have been inserted.

stacked traits: In plant biotechnology, stacked traits refer to the scenario where two or more new traits have been introduced into the plant to produce improvements in the field or improvements in yield.

stacking: The packing of the flattish nitrogen bases at the centre of the DNA double helix.

stacking forces: Weak hydrophobic forces between two adjacent base pairs of double-helical DNA. *Aka* HYDROPHOBIC INTERACTIONS.

stage-wise selection: Often used to denote the selection of individuals in stages. Individuals are preselected (the first stage) followed by reselection (the second stage) of the best individuals based on another testing method. For example, phenotypic selection is followed by clonal test of those preselected based on phenotype. *Aka* STEPWISE SELECTION.

staggered cuts: Scissions (cuts) made in duplex DNA when the two strands of DNA that make up the duplex DNA are cleaved at different points near each other by REs. What is produced is a single-stranded structure (in which the single strands are several nucleotide-bases long) with a double-stranded core section. This core section is much longer than the single-stranded region.

staggered extension process (StEP): StEP generates recombination events through PCR amplification of a template with very short extension times, whereby partially elongated oligonucleotides can anneal.

staining: Colouring a cell, tissue or organism for microscopy. In plants for mitotic and meiotic studies, the dividing cells are stained with an acetocarmine stain that is easily taken up by DNA causing the chromatin (chromosomes) to appear darkly stained in comparison to the cytoplasm and other cell organelles. In the case of bacteria, 'Gram stain' developed by Danish Bacteriologist, Hans Christian Gram in 1884, is a very common stain that classifies bacteria into gram-positive and Gram-negative types.

stalk: The supporting structure of an organ, usually narrower in diameter than the organ itself.

stalled replication forks: Stalled replication forks are characterized by extensive ssDNA, generated by polymerase-helicase uncoupling or nucleolytic processing, which is very unstable and therefore needs to be protected.

stamen: A male structure of a flower, composed of a pollen-bearing anther (microsporangia) and a filament or stalk.

staminal flower: A flower bearing stamens but not functional pistils (i.e. a male flower). *Aka* STAMINATE FLOWER.

staminode: A sterile stamen, often rudimentary, sometimes petal-like. Commonly has a role in attracting pollinators that feed on the staminodes.

staminophore: A structure around the apex of eucalypt, myrtaceae hypanthia that supports the stamens.

stand-alone BLAST search: This search can be performed at any local computer by installing the BLAST software. Stand-alone BLAST reduces searching time significantly by avoiding online communication and allows batch blast (submit multiple queries at once) against local databases downloaded from GenBank or created by user. Stand-alone BLAST document including installation can be found at NCBI's FTP site (ftp://ftp.ncbi.nlm. nih. gov/blast/documents/blast.txt).

stand-alone rolling circle amplification (RCA) with capillary electrophoresis (CE): This method is used for specific and sensitive detection of small RNAs in biological samples. To enhance the overall reaction efficiency and simplify the procedure, RCA is not preceded with ligation, and a preformed circular probe is employed as the template for the target small RNA-primed isothermal amplification. The long RCA product was digested and analysed by CE. Two DNA polymerases, the Phi29 and Bst, were compared for their detection performance. Bst is superior in the aspects of specificity, procedure simplicity and reproducibility, while Phi29 leads to a 5-fold lower detection limit and can detect as low as 35 amol (attomole, an SI unit of amount of substance equal to 10^{-18} moles) of the target small RNA.

standard: The large posterior petal of pea flowers.

standard check: An appropriate and generally satisfactory base for comparisons. However, check varieties have a variable but short lifespan. With each change of check variety there is a break in the continuity of comparisons over time. A standard check variety is also a single genotype with homeostatic properties that may yield a response at variance with other genotypes under test.

standard check, alternatives to: Sometimes there are conditions where the standard check varieties are not relevant to the breeder's needs (e.g. regional or international nurseries). In such cases, alternatives can be (i) the use of existing entries or the addition of new entries in a nursery or (ii) the use of derived statistics generated by the nursery entries; the latter can be realized by so-called floating checks. *See* FLOATING CHECK.

standard deviation (SD, s or σ): A measure of the variability in a population of items; a set of 'n' measurements ($x_1, x_2, x_3, \ldots, x_n$) is equal to the positive square root of the variance of the measurements. The square root of the variance (σ^2). The standard deviation has a simple interpretation only if the data distribution is Gaussian (normal), and in that restrictive case the mean ±1.96 standard deviations is expected to cover 0.95 of the distribution of the measurement. This interpretation does not apply superbly to asymmetric (skewed) distributions and is not robust to outliers. It is also used to quantify the amount of variation or dispersion of a set of data values around the mean. *See* VARIANCE.

standard error (SE): The approximate standard deviation of a statistical sample population. The SE is a statistical term that measures the accuracy with which a sample distribution represents a population by using standard deviation. A measure of variation of a population of means.

standard error of difference in means (SEM): The SE for the difference between two means is larger than the SE of either mean. It quantifies uncertainty. The uncertainty of the difference between two means is greater than the uncertainty in either mean. So, the SE of the difference is greater than either SEM but is less than their sum.

standard error of sample mean (s_x): The SE of the mean, or simply SE, indicates how different the population mean is likely to be from a sample mean.

standard error of the mean (SE_m): It is a kind of measure of dispersion. SE is the standard deviation of the sampling distribution of a statistic. It measures the accuracy with which a sample represents a population. In statistics, the sample mean deviates from the actual mean of a population; this deviation is known as STANDARD ERROR. The SE uses statistics (sample data) but standard deviation uses parameters (population data). SE of the mean is calculated by dividing the standard deviation by the square root of the sample size. SE goes to zero as the sample size reaches ∞.

standard free-energy change (ΔG°): Free-energy change (ΔG) of two reacting molecules at standard temperature and pressure when all components are present at a concentration of 1 mole per litre.

standard genetic code: The set of rules used by living cells to translate information encoded within genetic material into proteins. *See* CANONICAL CODE.

standard neutral model: The simplest form of the neutral theory, in which mutation and random genetic drift act in a single panmictic population of constant size.

standard siRNA structures: siRNAs consist of two RNA strands, an antisense (or guide) strand and a sense (or passenger) strand, which form a duplex 19 to 25 bp in length with 3′ dinucleotide overhangs.

standard specimen: A representative specimen of a cultivar or other taxon, which establishes how the name of that taxon should be used.

standard type: The randomly specified genotype used as a basis of comparison for genetic studies. The genes of a standard type are usually symbolized by a '+' sign. All genes deviating from those of standard type are designated by names or symbols. This term is synonymously used with WILD TYPE.

staple crops: The most common crops in people's diets, such as rice, wheat and maize (corn), which provide 60% of the world's food energy intake. Typically, staple crops are well adapted to the climate in which they are grown, and many are tolerant of drought, pests or soil low in nutrients.

starch: A homopolysaccharide, consisting of glucose molecules, which is the major storage biomolecule of plants and the major nutritionally important substance in the human diet. It is synthesized in large amounts in leaves during the day and degraded during the subsequent night. Plants produce starch in two different molecular forms, amylopectin and amylose. The starch content in traditional maize kernels averages 72–76% amylopectin and 24–28% amylose. Starch is broken down by amylases to yield glucose, which can be used as an energy source. The analogous polymer used by mammalian systems is called GLYCOGEN or, in old usage, ANIMAL STARCH. Two crystalline forms are found in starch molecules, A and B, the helices in the A-form being more densely packed than in the B-form.

starch (or amylum): A polymeric carbohydrate consisting of a large number of glucose units joined by glycosidic bonds. The principal reserve food materials of the green plants, frequently found in colourless plastid.

starch gel electrophoresis: A method for high-resolution separation of soluble proteins using starch as a matrix. It was introduced by O. Smithies in 1955. M. Poulik (a collaborator of O. Smithies) in 1957 introduced discontinuous buffer systems and enzyme-histochemical staining of electrophoretically separated enzymes, which were later called ZYMOGRAMS. The problem with this method is the fact that, due to the presence of various anionic residues, such as glucuronic acid in the purest starch fractions, the measured mobilities are always biased by the electro-osmotic flow of water in the gels. Since the gels have net negative charges, the water moves relative to the stationary gels to compensate. In addition, the concentration of glucuronic acid residues varies in starch preparations from batch to batch. Thus, reproducibility is often severely compromised.

starch quality in plants, modification of: The constituents of starch grain are amylose and amylopectin. Amylose forms the core of the starch grain. It is linear or slightly branched and is made up of 295–335 glucose units. Amylose synthesis is caused by GRANULE-BOUND STARCH SYNTHASES (GBSS). Amylopectin is highly branched and is the major component of starch granules. Amylopectin is synthesized by SOLUBLE STARCH SYNTHASES (SSS). Modified forms of starch have varied applications in food and other industries. Genes encoding the main enzymes of starch biosynthesis have been cloned in potato to produce transgenic lines.

starLink: A brand of genetically modified maize containing two modifications: a gene for resistance to glufosinate, and a variant of the *B. thuringiensis* (*Bt*) protein called Cry9C. Cry9C had not been used in a GM crop prior to StarLink, causing heightened regulatory scrutiny. This brand of transgenic maize was approved for animal feed only, but which also has been found in the human food supply.

start codon: A codon that codes for initiation of protein synthesis. In both eukaryotes and prokaryotes, mostly start codon is AUG, but in some cases GUG also works as start codon. *Aka* INITIATION CODON.

start G_1: This subphase is an important control point. During this phase, cells should get better supply of nutrients. During start, the cells are committed to enter S phase. G_1 cyclins bind to cell division cycle protein Cdc2 (or p34cdc2) to produce START PROMOTION FACTOR (SPF), which is an active kinase. Once the SPF is formed the cells are committed to enter S phase.

start point: The position on DNA corresponding to the first base incorporated into RNA. *Aka* START SITE *or* STARTISE.

start-transfer signal: Short amino acid sequence that allows a polypeptide chain to start being translocated across the ER membrane through a PROTEIN TRANSLOCATOR. Multipass membrane proteins have both N-terminal (signal sequence) and internal start-transfer signals.

startise: Initiation of transcription requires startise, a point at which first nucleotide is incorporated during transcription. Sigma factor recognizes start signals along the DNA molecules and also the correct strand of DNA to be used as template. Presence of σ factor leads to very tight binding of holoenzyme to promoter regions of DNA, which contain start signals. Then very localized unwinding occurs, allowing RNA synthesis. RNA polymerase binding sites are not themselves transcribed. Soon after this initial binding of RNA polymerase at the recognition site, RNA polymerase diffuses to RNA polymerase binding site, which is A=T rich region. This binding region is referred to as TATA BOX or START POINT.

stasipatric speciation: A rapid speciation that may occur among small breeding populations that are not completely isolated genetically or spatially. Such speciation may occur either parapatrically or sympatrically.

stasis: A long period without evolutionary change. *See* PUNCTUATION.

state: In information technology and computer science, a system is described as stateful if it is premeditated to remember preceding actions or user interactions; the remembered information is called the STATE OF THE SYSTEM.

State Biosafety Coordination Committee (SBCC): SBCC is a monitoring committee at State level. It has powers to (i) inspect, investigate and to take punitive action in case of violations of statutory provisions through the State Pollution Control Board (SPCB) or the Directorate of Health and so on, (ii) review periodically the safety and control measures established at various institutions handling GE organisms, and (iii) act as a nodal agency at the State level to assess the damage, if any, due to the release of GE organisms and to take on-site control measures.

statement: In computer programming, a statement is a syntactic unit of an imperative programming language that expresses some act to be carried out. A programme written in such a language is formed by a sequence of one or more statements. A statement may have internal components (e.g. expressions).

static view of epigenome dynamics: Current QTL approaches including linkage and association mapping lead to the detection of stable DNA sequence variants displaying phenotypic variation.

stationary frequency distribution: The representation of a probability density function showing the way the probability in a given situation is distributed over the possible events.

stationery phase: During bacterial growth in the liquid medium, it is the period when number of dividing cells is almost equal to the number of dying cells. *See* BACTERIAL GROWTH CURVE.

statistic: A function of the 'n' sample observations. A quantity (such as mean, median or standard deviation) that is calculated from a sample of data to extract information about unknown values in a corresponding population.

statistical analysis: Analysing collected data for summarizing information to make it more usable and/or drawing generalizations about a population based on a sample drawn from that population.

statistical and hidden Markov Models-based approaches: GCG provides several tools that help to identify protein-encoding sequences by statistics that measure codon usage (CODON-PREFERENCE) and non-random use of particular nucleotides in the third position of each codon (ESTCODE). These statistical methods are imprecise but can help identify possible coding regions in large chunks of genomic DNA sequence. Standard learning approaches for pattern recognition include neutral networks and hidden Markov models (HMMs). HMMGENE and GENEMARK are popular gene recognition programmes based on HMMs.

statistical distribution: The array of frequencies of different quantitative or qualitative classes in a population.

statistical genetics: A scientific field concerned with the development and application of statistical methods for drawing inferences from genetic data. *See* BIOMETRICAL GENETICS.

statistical inference: Statistical inference occurs when you use sample data to generate an inference or conclusion. Statistical inference can include regression, confidence intervals or hypothesis tests.

statistical model: A model with identified parameters that comprises a model for the data through a probability distribution and favours additivity of effects. Examples of statistical models include ordinary linear regression with an assumption of a Gaussian distribution for the residuals, logistic regression, Cox proportional hazards regression, longitudinal models, quantile regression, ridge regression, lasso and elastic net.

statistical parameters: Mean, mode, median, range, variance, standard deviation and coefficient of variation are some of the statistical parameters used to measure quantitative traits.

statistical power: (i) The chance that the null hypothesis (H_0) will be rejected when the data are produced by a different model. (ii) The probability of rejection of the H_0 when it is in fact false. Mathematically, statistical power is equal to $1 - \beta$, where β is the probability of a type II error.

statistical proteins: In the absence of a genetic code, proteins synthesized will have a random sequence of amino acids.

statistical significance: A result is statistically significant if the chance of wrongly rejecting a true null hypothesis (type I error) is less than the prefixed level such as 5% or 1%. The implication is that chance differences in samples would produce that kind of result in less than 5 times out of 100. Thus, chance is not an explanation for that result, and it is most likely real.

statistical test: A type of statistical procedure that is applied to data to determine whether or not the results are statistically significant.

statistics: The branch of mathematics dealing with the collection, organization, presentation, analysis and interpretation of data.

statolith: (i) A sotolith is a crystalline particle of calcium carbonate and a protein adhering to the gelatinous membrane of the maculae of the utricle and the saccule. (ii) In botany, an inclusion (e.g. starch grains) in plant cells is involved in geotropic responses. Their sedimentation transmits the gravitropic signal by activating mechano-sensing channels. The gravitropic signal then leads to the reorientation of auxin efflux carriers and subsequent redistribution of auxin streams in the root cap and root as a whole. The changed relations in the concentration of auxin lead to differential growth of the root tissues. Taken together, the root then turns following the gravity stimuli.

stay-green trait: 'Stay green' is an important trait for the genetic improvement of several crops; it allows plants to keep their leaves on the active photosynthetic level under stress conditions. However, chlorophyll degradation during leaf senescence is essential for the mobilization of mineral nutrients, especially nitrogen, to young, developed sinks. This catabolic pathway is disturbed when mutations occur in genes encoding enzymes, such as Mg-dechelatase and chl *b* reductase leading to greenness of tissue. Stay greens are broadly classified into two types – FUNCTIONAL STAY GREENS, where the whole senescence condition is delayed, and COSMETIC STAY GREENS, which possess mutations in genes responsible for chlorophyll catabolism. Physiologically, delayed senescence has a crucial effect on carbon capture and nutrient mobilization to develop sinks and is considered a valuable trait in imparting tolerance to abiotic stresses. Various routes for introducing the stay-green trait into plants to enhance photosynthesis include (i) knock out of chlorophyll-degradation genes, (ii) manipulating endogenous cytokinin and senescence genes, (iii) overexpression of genes regulating chlorophyll biosynthesis and (iv) development of varieties/cultivars with delayed senescence.

stearic fit model: Any early hypothesis to explain the relationship between an amino acid and its codon on the basis of the stearic relationship between the two.

stele: The stele (or vascular cylinder) is in the central region of the root, where the xylem and phloem develop. In monocot roots, the stele contains an exterior ring of phloem and an interior ring of xylem arranged around the central pith.

stem: The plant axis, either aerial or subterranean, that bears nodes, leaves, branches and flowers.

stem cell: A type of undifferentiated or partially undifferentiated cell that is capable of differentiating into any of the other types of specialized cells and also capable of dividing to produce more of the same type of stem cell. Stem cells are the earliest type of cell in a cell lineage. Stem cells reside in specialized niches. Stem cell research is gaining importance in medical biotechnology because stem cells exhibit frequent divisions in the culture and exhibit totipotency and are thus being exploited for cloning purposes. The stem cells are taken from early stages of developing embryos. In animals and humans, stem cells are found in different tissues and have the potential to differentiate mainly into the cells of the tissue to which they belong. But the embryonic stem cells (ESCs) are extraordinary being totipotent being used in cloning.

stem cell competence: Some plant cells, like the totipotent meristems, are more versatile than animal cells. Recent discoveries, however, show that the difference in the retention of competence between animals and plant cells is not as great as once thought. Adult humans have a reservoir of cells that retain some ability to become other cell types. Cells derived from foetal tissue have been used to generate so-called ESCs. Recent studies suggest that adult stem cells may be more versatile than had been previously thought. MULTIPOTENT ADULT PROGENITOR CELLS (MAPCs) have been derived from the bone marrow of adult mice. These cells appear to be able to differentiate into virtually all cell types of mouse when injected into mouse blastocysts. These MAPCs have also been injected into living adult mice and have differentiated into liver, lung and intestine tissue. The fact that the same principles and many of the same genes direct the development of such different and diverse animals has generated renewed interest and study of how developmental systems evolve. Given the striking similarity of genes used, how do they manifest differences across animals arise and evolve? This question will keep biologists busy for many years to come.

stem cell differentiation: The continuation of mammalian species requires the formation and development of the sexually dimorphic germ cells. Cultured ESCs are generally considered pluripotent rather than totipotent because of the failure to detect germline cells under differentiating conditions. Totipotent stem cells can become any cell in the human body. Pluripotent stem cells can become almost any cell in the human body, but they cannot become placental tissue needed for development in the human uterus. Multipotent stem cells can become only a certain type of cell, such as blood cells. Mammalian ESCs in culture can develop into oogonia that enter meiosis, recruit adjacent cells to form follicle-like structures and later develop into blastocysts. Oogenesis in culture should contribute to various areas, including nuclear transfer and manipulation of the germline, and advance studies on fertility treatment and germ and somatic cell interaction and differentiation. The pluripotency genes are generally hypomethylated in stem cells and gain methylation during cell differentiation. *Oct4* and *Nanog* are two such pluripotency genes, and they are essential for the pluripotency of ES cells. Transcription factor Oct4 is sufficient to reprogramme human neural stem cells to pluripotency. *De novo* DNA methylation and silencing of the pluripotency genes contribute to the loss of pluripotency in differentiated cells. Genome-wide DNA methylation analysis indicated widespread differences in DNA methylation between the genomes of ES cells and differentiated cells.

stem cell fusion: Stem cells adopt the functional features of other lineages by means of cell-fusion-mediated acquisition of the lineage-specific determinants (chromosomal DNA) rather than by signal-mediated differentiation.

stem cell maintenance and division: Neural-derived *Drosophila* insulin-like peptides (DILPs) directly regulate germline stem cell division rate, demonstrating that signals mediating the ovarian response to nutritional input can modify stem cell activity in a niche-dependent manner DILPs have a role in controlling germline cyst growth and vitellogenesis.

stem cell mobilization: A key process in organ homeostasis is the mobilization of stem cells out of their niches. Telomere length as well as the catalytic component of telomerase, Tert, are critical determinants in the mobilization of epidermal stem cells. Telomere shortening inhibited the mobilization of stem cells out of their niche, impaired hair growth and resulted in the suppression of stem cell proliferative capacity *in vitro*. In contrast, Tert overexpression in the absence of changes in telomere length promoted stem cell mobilization, hair growth and stem cell proliferation *in vitro*. The effects of telomeres and telomerase on stem biology anticipate their role in cancer and ageing.

stem cell plasticity: Somatic stem cells have been claimed to possess an unexpectedly broad differentiation potential, referred to as plasticity, which could be induced by exposing stem cells to the extracellular developmental signals of other lineages in mixed-cell cultures. Stem cell plasticity was found to be a true characteristic of neural stem cells and the conversion of NSCs to unanticipated cell types can be accomplished without cell fusion.

stem cell pluripotency: Pluripotent stem cells can produce theoretically every cell type in the animal body and are derived from embryonic (not adult) tissues. Three types of mammalian pluripotent stem cell lines have been isolated. ES cells are derived from the INNER CELL MASS (ICM) of the preimplantation embryo. EMBRYONIC GERM CELLS (EGCs) are derived from PRIMORDIAL GERM CELLS (PGCs) isolated from the embryonic gonad. EMBRYONIC CARCINOMA CELLS (ECCs) are derived from PGCs in the embryonic gonad but usually are detected as components of testicular tumours in the adult. All of the three pluripotent stem cell types are usually derived by culture on layers of mitotically inactive fibroblasts, termed feeder layers.

stem cell totipotency: Most stem cells are not totipotent. Instead, they are partially committed but remain undifferentiated. Upon appropriate stimulation, they are capable of regenerating mature cell types. In nodal point in adult melanocyte stem cell differentiation, Pax3 simultaneously functions to initiate a melanogenic cascade while acting downstream to prevent terminal differentiation. Pax3 activates the expression of Mitf, a transcription factor critical for melanogenesis, while at the same time, it competes with Mitf for occupancy of an enhancer required for the expression of dopachrome tautomerase, an enzyme that functions in melanin synthesis. Pax3-expressing melanoblasts are thus committed but undifferentiated until Pax3-mediated repression is relieved by activated β-catenin. Thus, a stem cell transcription factor can both determine cell fate and simultaneously maintain an undifferentiated state, leaving a cell poised to differentiate in response to external stimuli. Neural stem cells can form blood and muscle tissue, mesenchymal stem cells can produce differentiated cell types in the brain and skin stem cells can make neurons, glia, smooth muscle and adipocytes.

stem cell-based therapy: The ability of pluripotent stem cells to give rise to the wide array of differentiated derivatives attests to their usefulness for purposes of cell-based therapy. In the presence of factors that inhibit their differentiation has typically been directed by manipulating their environment by trial and error (e.g. by growing stem cells at high density, by growing them on different types of feeder cells, by addition of growth factors and by growth on crude or defined extracellular matrix substrates). In these conditions, the types of differentiated cell types formed can be varied and haphazard. When grown in suspension, pluripotent stem cells will form embryoid bodies, structures originally described in teratomas and which resemble the early preimplant embryo.

stem line: A term referring to the most basic clone of a tumour cell population. The stem line is always listed first in the nomenclature string.

stem rust (in wheat): A hazardous fungus (*Puccinia graminis* f. sp. tritici) causing wheat yield loss in many countries of the world. Recently, a new and virulent race, previously found in East Africa and Yemen, has spread to major wheat-growing areas in Iran. Afghanistan, India, Pakistan, Turkmenistan, Uzbekistan and Kazakhstan are most threatened countries by this fungus. As much as 80% of all wheat varieties planted in Asia and Africa are susceptible to wheat stem rust. The spores of wheat rust are mostly flown by wind over long distances. The fungus is spreading rapidly and could utterly lower wheat production in countries at direct risk. The new race first emerged in Uganda in 1999 and is therefore called Ug99. Subsequently, it spread to Kenya and Ethiopia. In 2007, Ug99 had affected wheat fields in Yemen. Disease surveillance and wheat breeding are already under way to monitor the fungus and to develop Ug99-resistant varieties.

stem-loop structure: A feature of RNA secondary structure, in which two complementary, inverted sequences, which are separated by a short-intervening sequence within a single strand of RNA base pair to form a 'stem' with a 'loop' at one end. Similar to a hairpin, but these usually have very small loops and longer stems.

stenospermocarpy: The development or production of fruit that is seedless or has minute seeds because of the abortion of seed development, notably in many table grapes. *Compare with* PARTHENOCARPY.

step allelomorphism: The occurrence of a series of multiple allelomorphs with overlapping effects, which can supposedly be related to a linear order in the distribution of units of change within the gene.

stepladder gel: A DNA-sequencing gel. The numerous bands in each lane give the impression of a stepladder.

stepped cline: A cline with a sudden change in gene or character frequency.

stepwise regression: It is a method of selecting the 'best' set of independent variables for a regression equation. Variables are introduced one at a time, with the criterion for accepting a variable based on the correlation of that variable and y in the presence of the other variables already in the model.

stereomicroscope: A microscope having a set of optics for each eye so arranged that the sets view the object from slightly different directions and make it appear in the three dimensions.

sterile fungus: A fungus that is not known to produce any kind of spores.

sterile: Environment that is devoid of any living microorganisms or spores. For example, a hypodermic needle that has been sterilized (e.g. by heating it) and is free of living microorganisms is said to be sterile. Sterile conditions are the prerequisite of plant and animal tissue culture. These are maintained by autoclaving of culture medium at 121°C at 15 pounds/square inch pressure for 20 min and the surface sterilization of inoculum either with mercuric hypochlorite, sodium hypochlorite solution, 70% ethanol, use of Laminar Air Flow Cabinets for inoculation. Plant, animal, insect or human do not reproduce sexually when they are unable to produce functional gametes, bear pollen or a flower that does not bear seed. For example, a bull that has been castrated is rendered sterile. Male sterility is exploited in plant breeding for production of hybrid seeds and in entomology for insect control.

sterile insect technique (SIT) area-wide integrated pest management (SIT AW-IPM): SIT has been applied effectively as a component of SIT AW-IPM for *Ceratitis capitata* since 1970s. But to increase the efficacy of AW-IPM, there is a need to improve biological strategies. Transgenic approaches should widen the applicability of such programmes to different pest species. To achieve this objective, two major strategies are followed: First approach is to cause sterility without interfering with spermatogenesis to maintain males and their sperm as competitive as possible. The system has been tested successfully in *Drosophila melanogaster*. Second, a strain of *Ceratitis capitata* was engineered to carry a sperm marking system with two fluorescent markers. It was possible to distinguish the wild-type males from engineered released males during the monitoring process. Also, the mating success of sperm-marked, fluorescent males with wild females can be monitored and their spermathecae could also be examined. Further studies on *Ceratitis* reproductive biology can be made using this easily recognizable sperm marking process as well as SIT-based AW-IPM programmes can be optimized.

sterile insect techniques (SIT): Genetics-based tools used for the management of insect pests include (i) INDUCED STERILITY, (ii) PRECISION-GUIDED STERILE INSECT TECHNIQUE (PGSIT), (iii) SEX REVERSAL BY MANIPULATING A MALE-DETERMINING FACTOR, (iv) AREA-WIDE INTEGRATED PEST MANAGEMENT (SIT-BASED AW-IPM) and (v) IMPROVED SIT. *See details as separate terms.*

sterility: The failure to complete fertilization and obtain seed as a result of defective pollen or ovules, or other aberrations.

sterility gene: A gene that causes sterility in an individual organism.

sterility test in *Drosophila*: Groups of one- to two-day-old males are fed with food having selected concentrations of the compound to be tested. After an overnight feeding, each group of males with about twice the number of females is placed in separate culture bottles. After 2–3 days, males are given a new set of virgin females. This is continued for 10–12 days to raise 4–5 broods. Number of progenies in different broods is compared with control. If some reduction in progeny (number of flies) is observed then a dominant lethal test is carried out.

sterilization (adj. **sterilized**)**:** The total devastation of living organisms or contaminating agents by various means, including heat, chemicals or irradiation. See AUTOCLAVE.

stickiness: The genetically or artificially induced agglutination of chromosomes. Usually it results in different sorts of chromosome aberrations.

sticky association: Any association of two or more chromosomes at mitosis or meiosis due to non-specific sticky adhesions rather than to homology between/among the chromosomes.

sticky end: The terminus of a DNA molecule that has either a 3′ or 5′ overhang, and that typically results from a cut by a restriction endonuclease. Such termini are capable of specific ligation reactions with other termini, which have complementary overhangs. A sticky end can be 'blunt ended' either by the removal of an overhang, or a 'filling in' reaction, which adds additional nucleotides complementary to the overhang. Sticky ends produced by the restriction enzymes of type II (e.g. *Eco*RI, are 5′ G↓AATTC / 3′ CTTAA↑G). *Aka* ADHESIVE ENDS *and* COHESIVE ENDS. *Compare with* BLUNT END.

stigma: The pollen-receptive surface of a female reproductive organ, carpel or group of fused carpels, usually sticky; usually a point or small head at the summit of the style that receives pollen and on which the pollen germinates.

stimulatory G protein (Gs): G protein that, when activated, activates the enzyme adenylyl cyclase and thus stimulates the production of cyclic AMP.

stimulon: A set of genes that retort to a single stimulus (regulon, operon). Stimulus may overlap if they contain an operon with multiple promoters that have different regulatory properties.

stimulus for cleavage: It is not known how exactly the contractile ring is formed. Still an experiment indicates that part of the mitotic apparatus is required for cell division. Even if chromosomes from metaphase cells are removed, cleavage furrow is still formed. This shows that spindle apparatus has material to initiate cleavage and chromosomes are not necessary for cytokinesis to occur. The spindle has microtubules, which interact with microfilaments (actin). Cleavage results in formation of blastula. Cells rearrange and embryo gastrulates. In gastrulation, one obvious result is the formation of gut. So, this is the time for cell movement and cell interaction.

stippling: A series of small dots or speckles in which chlorophyll is absent.

stipule: The small, leaflike appendage at the base of a leaf petiole, usually occurring in pairs.

stochastic: A process with an indeterminate or random element as opposed to a DETERMINISTIC PROCESS that has no random element.

stochastic DNA rearrangements: Random DNA arrangements continue taking place without any need so that a small fraction of the cell population carrying rearranged genes always exist. One of such examples involves the inversion of DNA elements carrying the promoter for one of the flagellar protein operons in *S. typhimurium*. In one orientation, the element encourages the transcription of a gene encoding one flagellar protein and the repressor of a distant gene encoding an alternative flagellar protein. Inversion of the element blocks transcription of both, the first flagellar protein and repressor gene of the second flagellar protein, thus enabling expression of the second flagellar protein.

stock: An artificial mating group. *See* ROOTSTOCK.

stock culture: A culture maintained in a culture collection for future investigation or reference.

stock plant: A mother plant set aside for cuttings to reproduce the plant and/or variety.

stock seeds: The supply of seeds, tubers or other propagules reserved for planting.

stoichiometry of histone modifications: Several MS-based strategies have been developed to determine the stoichiometry of PTLMs: Absolute quantification (AQUA) of modification states, SILAC-based peptide/protein quantification, Isotope labelling *in vitro*, Label-free quantification. These strategies are applied for studying both histone and non-histone PTLMs.

stolon: In general, a slender, horizontal stem that grows close to the soil surface. In fungi, a hypha that grows horizontally along the surface. *See* RUNNER *and* RHIZOME.

stoma: A structure composed of two guard cells and the opening between them in the epidermis of a leaf or stem, functioning in gas exchange between tissues and the atmosphere. Guard cells regulate the pore size.

stomatal closure response: The response by plants to environmental stress (such as drought) that reduces stomatal pore size. This leads to reduced water transpiration rates and restricted access of CO_2 to the leaves.

stomatal patterning: Coordinated spacing and patterning of stomata allow efficient gas exchange between plants and the atmosphere. Three ERECTA (ER)-family leucine-rich repeat-receptor-like kinases (LRR-RLKs) together control stomatal patterning, with specific family members regulating the specification of stomatal stem cell fate and the differentiation of guard cells. Loss-of-function mutations in all three ER-family genes cause stomatal clustering. Genetic interactions with a known stomatal patterning mutant with *too many mouths* (*tmm*) revealed stoichiometric epistasis and combination-specific neomorphism. Their findings suggest that the negative regulation of ER-family RLKs by TMM, which is an LRR receptor-like protein, is critical for proper stomatal differentiation.

stone cell: A reduced form of sclerenchyma cells with highly thickened, lignified cellular walls that form small bundles of durable layers of tissue and are developed in the leaves of most halophytes. These cells form the tissue of nut shells and the stones of drupes. *Aka* SCLEREID CELL.

stop codon: A codon for which there is no corresponding tRNA molecule to insert an amino acid into the polypeptide chain. Instead, a termination factor is typically inserted, causing the ribosome to be disassembled, terminate translation and the completely synthesized protein to be released from the ribosome. There are three stop codons – UAA, UAG and UGA. *Aka* TERMINATION CODON *or* NONSENSE CODON.

stop-gain variant: A DNA sequence change that results in a new stop codon is generated.

stop-lost variant: A DNA sequence change that results in a loss of a stop codon.

stop-transfer signal: Hydrophobic amino acid sequence that stops translocation of a polypeptide chain through the ER membrane, thus affixing the protein chain in the membrane.

storage: Computer data storage is a technology comprising of computer components and recording media that are utilized to recall digital data. It is a core function and fundamental component of computers.

storage protein genes: Structural genes coding the synthesis of storage proteins. The storage proteins are encoded by multigene families. For example, in wheat, each homoeologous chromosome group contains structural genes that have similar positions within each genome. The differences in the strength and elasticity of dough are under control of several storage protein loci (the high- and low-molecular-weight (HMW and LMW), glutenins (Glu-loci) and the gliadins (Gli-loci). The HMW glutenins seem to be responsible for dough strength, the LMW glutenins and gliadins for dough extensibility and elasticity. Through electrophoretic methods, it has been possible to identify five loci controlling these proteins. They are located on the homoeologous chromosomes 1 and 6. Alleles at these loci are highly variable and correlate with differences between good and poor bread or biscuit-making quality. Globulins are major storage proteins in legumes. They are soluble in water. Eighteen genes belonging to three families of globulin genes are known in pear. Homologies in genes coding for proteins and globulins in different crop plants have also been demonstrated. Their multiplicity suitably fulfils the demand for rapid synthesis of pear protein in the developing seeds. Prolamines that are major storage proteins in cereals and are soluble in aqueous alcohol. For each of the 10 prolamines families, 3–10 genes are reported in maize. More than a dozen genes for prolamine represent several families in wheat. The tissue-specific expression of the DP's genes is illustrated by the following, for example, when wheat glutenin gene on barley hordein gene (with respective regulatory sequences) were fused with *chloramphenicol acetylase* (*CAT*) gene in tobacco. CAT activity was found only in the endosperm. Another reporter gene used for the same purpose is β-*glucuronidase* (*GUS*), which also showed endospermic expression with wheat glutenin gene with its regulatory sequences. These regulatory sequences are called *endospermic box*. There are some mechanisms responsible for tissue-specific regulation.

storage protein variation: The storage proteins of seeds are an important group of plant proteins, which provide nitrogen and amino acids in the germinating seedlings and constitute a major portion of proteins for human and livestock consumption. Storage proteins occur predominantly in the endosperm in the cereals. Variation in storage proteins is genetically controlled. Different mutants are known to affect substantial changes in the amino acid content in maize, sorghum and barley. In the opaque and floury mutants of maize, prolamine is reduced and lysine is increased. Genetic variants in maize and barley result in change in protein composition with decrease in weight and starch content. A wrinkled mutation in peas affects legumin content. The structural gene for subunit of polypeptides of legumin has been located on chromosome 7 and closely linked to *wrinkled* mutation. Mineral deficiency and amounts of hormones during synthesis also influence the variation in storage proteins. Abscisic acid stimulates the accumulation of 7S protein in French bean and 42S protein in rapeseed.

storage proteins: The proteins that are stored in a cell tissue or an organism for some future use. Storage proteins have been reported both in plant and animal systems. In plants, seed storage proteins have been well studied. The quiescent seeds can remain dormant for months together. As the seeds enter the dormant phase, seeds accumulate large amounts of plant hormone ABSCISIC ACID (ABA). This hormone is involved in dropping (abscises) of leaves of deciduous trees in the fall. ABA is also involved in blocking precocious germination of seeds. Accumulation of storage proteins occurs little during cell division phase but legumin accumulation occurs during this phase. Differential rate of accumulation has been shown for each of the storage proteins during seed development. Accumulation of a protein is a reflection of the equilibrium established between rate of synthesis and rate of degradation. Protein accumulation in legume seeds has sigmoidal kinetics followed by dry seed weight gain. Synthesis of seed proteins also serves as a model for regulation of gene expression. Multigene families encode these proteins. About 30–100 zein genes encode the prolamine fraction (zein) in maize. Variation of storage proteins is clearly controlled by genetic and non-genetic factors.

strain: (i) A distinct taxonomic form of an organism or virus within a species that differs from other forms of the species biologically, physically or chemically. (ii) The term is most commonly used to identify particular types of bacteria and viruses. (iii) The term is also used to label a population of cells derived from a single cell. (iv) In plant breeding, a group of individuals developed through breeding or natural process that in the case of plants becomes a variety after it is named and identified for commercial cultivation. During crop breeding, a plant breeder tests the different strains (material from breeding experiments) for 3 years and in different locations including regional trials during the third year. Based on 3-year data, a superior strain is recommended for adaptive trials at farmer's fields in comparison to the check variety. Till this stage, it is called 'strain'. In case it performs superior to check variety, it is released as variety. *See* CELL STRAIN.

strain building: The improvement of cross-fertilizing plants by any one of a number of methods of selection.

strand: Each of the two helices in a double-helical molecule (e.g. DNA).

strand displacement: It is a mode of replication of some viruses in which a new DNA strand grows by displacing the previous (homologous) strand of the duplex.

strand displacement cascades: The cascades of strand displacement reactions can be used for either computational or structural purposes. An individual strand displacement reaction involves revealing a new sequence in response to the presence of some initiator strand.

strand transfer repair: A DNA repair mechanism in which a recombination-like process forms gaps opposite the induced lesion in a replicating template strand. This type of repair is induced by irradiation. If other DNA repair systems are defective, this repair is known to occur in *Escherichia coli*. A gap of 10^3 or more base pairs opposite the lesion through a recombination-like process is created. These gaps are then filled in via strand transfer from the previously replicated normal strand. During this process, the thymine dimer becomes randomly distributed through old or newly made strands by reciprocal exchanges. In this way, damage is reduced.

strand transferase: Any of the enzymes that encourage homologous pairing by forming a complex between a double-stranded DNA molecule and a homologous single-stranded DNA molecule.

strand-directed mismatch repair: *See* MISMATCH REPAIR.

strand-resolved mutagenicity of DNA damage and repair: Damage of DNA base can produce strand-phased mutation patterns and multiallelic difference through the process of lesion segregation. C.J. Anderson *et al.* (2024) exploited these properties to show how strand-asymmetric processes, such as replication and transcription, shape DNA damage and repair. Despite distinct mechanisms of leading and lagging strand replication, they report identical fidelity and damage tolerance for both strands. For small alkylation adducts of DNA, they support a model in which the same translesion polymerase is recruited on-the-fly to both replication strands, unambiguously contrasting the strand asymmetric tolerance of bulky UV-induced adducts. The accumulation of multiple distinct mutations at the site of persistent lesions offers the means to quantify the relative efficiency of repair processes genome wide and at single-base resolution. At multiple scales, they show that DNA damage-induced mutations are largely shaped by the influence of DNA accessibility on repair efficiency, rather than gradients of DNA damage. Finally, they disclose specific genomic conditions that can actively drive oncogenic mutagenesis by demeaning the fidelity of nucleotide excision repair.

strategy: (i) It is comparable to a phenotype or reaction norm. It refers to the morphology or behaviour expressed by an individual under a variety of circumstances. (ii) The terms strategy and tactic have frequently been applied to alternative methods of maximizing fitness. For example, SEMELPARITY and ITEROPARITY may be called STRATEGIES (or TACTICS). The two terms are, strictly not synonyms and apply specifically to warfare, not biology. In a life history context, they are usually regarded as SYNONYMOUS.

stratification: The process by which seed dormancy is broken and hence germination percentages improved.

stratified mass selection: Mass selection in which the population is separated into subpopulations that are grown under different environmental conditions (i.e. in different fields or in different parts of a field). Plants for next-generation seed are selected from different subpopulations.

stratified random sample: A sample attained by separating the population elements into non-overlapping groups, called STRATA, and then selecting a simple random sample within each stratum.

stratified sampling: A sampling procedure in which the sample space is separated into groups or strata and each stratum is sampled randomly. It guarantees that the sample contains values over the full range of the sample space, for example, Latin hypercube sampling.

streak plate method: A routine method of obtaining a pure microbial culture. It is the most commonly used method for bacterial culture in Petri dishes. The bacterial population is picked up from the broth using a sterile loop that is moved back and forth on the surface of solid medium in the Petri dish. The bacterial population keeps reducing towards the end. This process is repeated several times and after incubation for sufficient time, the individual colonies can be picked up for further investigations.

stream: A sequence of data elements made available over time. A stream can be supposed as items on a conveyor belt being processed one at a time rather than in large batches.

streptavidin: A protein produced by *Streptomyces avidinii*. It displays affinity to the vitamin 'biotin'. It is engaged for the detection of biotinylated hybrid DNA.

***Streptomyces coelicolor*:** A species of the Streptomyces family, a commonly existing family of bacteria that are highly adapted to living in soil. These bacteria are highly productive natural factories, which make over half of the naturally derived antibiotics in current use. Streptomyces can make over 6,000 different chemical products that include antibacterial agents (e.g. tetracyclines, erythromycin, rifamycin, kanamycin and vancomycin), antifungals (candicidin and nystatin), anti-cancer drugs (bleomycin, daunorubicin and doxorubicin), antihelminthics (avermectin) and immunosuppressants (FK506 and rapamycin). The *S. coelicolor* genome contains ~8000 genes.

streptomycin: An antibiotic that prevents the elongation of protein biosynthesis. It is produced by some strains of *Streptomyces*. In genetic experiments, streptomycin resistance is used as a selection marker.

stress adaptation: Refers to the ability of the young to adapt to changing stress in the environment. The origin of these stressors can be physical (pollution, nutrient status, infections) or psychological (trauma, parenting, home environment).

stress granules: Assemblies of untranslated MESSENGER RIBONUCLEOPROTEINS (mRNPs) that form from mRNAs stalled in translation initiation. Stress granules form through interactions between mRNA-binding proteins that link together populations of mRNPs.

Stress Knowledge Map (SKM): A freely available resource (https://skm.nib.si) comprising two complementary knowledge graphs that pronounce the present knowledge of biochemical, signalling and regulatory molecular interactions in plants in the form of a highly curated model of plant stress signalling (PSS; 543 reactions) and a large comprehensive knowledge network (488,390 interactions). Both graphs have been built by domain experts through systematic curation of varied literature and database resources. SKM offers a single-entry point for investigations of plant stress response and related growth trade-offs, as well as interactive investigations of existing knowledge. PSS is also outlined as a qualitative and quantitative model for systems biology and thus symbolizes a starting point for a plant digital twin. C. Bleker *et al.* (2024) describe the landscapes of SKM and show, through two case studies, how it can be engaged for complex analyses, including systematic hypothesis generation and design of authentication experiments, or to advance visions into experimental annotations in plant biology.

stress memory: Describes the phenomenon through which information on a past stress cue is retained and results in a modified response upon recurring stress or a sustained response after the priming stress cue. *See* PRIMING *and* SOMATIC STRESS MEMORY.

stress proteins: These proteins are produced by many organisms (plant, bacteria and mammal) cells when cells are stressed by environmental conditions such as certain chemicals, pathogens or heat. When maize (*Z. mays* L.) is stressed during its growing season by high night-time temperatures, the plant shifts from its normal synthesis of ('immune system' defence) chitinase to the synthesis of heat-shock (i.e. stress) proteins, instead. Two stress proteins, known as HSP90 and HSP70, guarantee that certain crucial proteins in cells are folded into the conformation needed by that cell. *Aka* HEAT-SHOCK PROTEINS.

stress response: A specific response by the plant to a stimulus that disturbs or interferes with the normal physiological equilibrium.

stress tolerance: Stress tolerance is the ability to withstand adverse events and stressful situations without developing physical or emotional symptoms, by actively and positively coping with stress.

stress-induced endoreplication: Damage to appropriate control of the cell cycle can result in whole-genome doubling (WGD) and aneuploidy. C. McKenney *et al.* (2024) discovered the effects of ribotoxic stress (interference with translation by the antibiotic anisomycin) and other pressures on cell cycle control in mammalian cells in culture. Ribotoxic stress ensues when ribosome activity is reduced during protein elongation, which causes ribosome collisions. In response to ribosome crashes, activation of the mitogen-activated protein kinase kinase kinase (MAP3K) ZAKα and the alleged stress-activated protein kinases (SAPKs) downstream results in widespread G2 cell cycle exit. Mechanically, G2 cell cycle exit ensues by premature activation of the anaphase-promoting complex or cyclosome (APC/C) in G2. They observed that early APC/C reactivation happens by determined concurrent inhibition of cyclin-dependent kinase 1 (CDK1) and CDK4/6 in G2 by SAPKs. When ribosome activity is re-established and stress signalling diminishes, the cells that have already replicated their DNA are ready to resume the cell cycle. This causes a second round of DNA replication to occur, thus completing the WGD process. In this context, p53 function is not needed, as cells deficient for the p53 gene are still able to undergo this process. SAPK-mediated WGD may represent an important source of genetic instability during ageing and tumorigenesis.

stress-responsive genes: Several constitutive stress-responsive genes that are differentially upregulated due to perturbation of different stress signals, for example, a gene (*AT1G68440*) involved in phenylpropanoid metabolism and polyamine catabolism as responsive to seven different stress signals. Fourteen genes were identified as responsive to stresses, such as heat, dehydration and phosphate stress. Examples of these genes are *LEA2* (*late embryogenesis-abundant protein 2*), *LEA5* (*late embryogenesis-abundant protein 5*) and *dehydrin*.

stressor: Any agent or actor with the potential to alter a component of the ecosystem.

striate (n. **striations**)**:** Marked with delicate parallel lines, ridges or grooves.

STRING: A biological database and web resource of known and predicted protein-protein interactions. It is conventionally a sequence of characters, either as an exact constant or as some kind of variable. The latter may permit its elements to be mutated and the length changed, or it may be fixed (after creation). A string is generally classified as a data type and is often implemented as an array data structure of bytes (or words) that stores a sequence of elements, typically characters, using some character encoding. String may also represent more general arrays or other sequence (or list) data types and structures.

stringency: Reaction conditions (e.g. temperature, salt and pH) that dictate the annealing of single-stranded DNA or RNA molecules to produce double-stranded DNA/DNA, RNA/RNA or DNA/RNA hybrids. At high stringency, duplexes form only between strands with perfect one-to-one complementarity. Lower stringency allows annealing between strands with some degree of mismatch between bases.

stringent factor: A protein that catalyses the formation of an unusual nucleotide (guanosine tetraphosphate) during the stringent response under amino acid starvation.

stringent plasmid: A plasmid present at a low copy number in the cell.

stringent replication: The limitation of single-copy plasmids to replication with the bacterial chromosome.

stringent response (SR): A broadly conserved bacterial stress response that controls adaptation to nutrient deprivation and is activated by a number of different starvation and stress signals. Stringent response ensues when prokaryotic cells are starved to amino acids when an uncharged tRNA finds its way into the A site of the ribosome. This event transpires under general amino acid starvation. It results in an idling reaction by the ribosome, which entails the production of the odd nucleotide guanine tetraphosphate (3′-ppGpp-5′). This is a post-translational control of gene regulation.

stroma (pl. stromata): (i) The connective tissue in which a glandular or other epithelium is embedded. (ii) The large interior space of a chloroplast, containing enzymes that incorporate CO_2 into sugars.

strong overlap: The number of bases matching between the two clones (e.g. BAC clones) determined by using the strictest matching criteria.

strong promoter: A promoter whose copy number is under strict control and, therefore, has only one or two copies per chromosome (i.e. a low copy number plasmid; e.g. the F plasmid).

structural alleles: Alleles defined on the basis of the recombination test were termed structural alleles. Different alleles of a gene may differ in the amount of genetic material. T.H. Morgan and A.H. Sturtevant in 1923 studied *bar* locus in *Drosophila melanogaster*. A *bar* mutation consists of insertion of a *bar* gene (duplication) at that locus (16A). When this repeated segment is lost, the mutant revertas back to wild type. *Ultrabar* has two repeats. Duplication of 16A reduces the number of facets in the eye: B^+/B^+, 779; B^+/B, 358, B/B, 68, *Ultrabar*, 45. *Ultrabar* arises due to unequal pairing and crossing-over. From these observations it was the conclusion that a gene does not necessarily represent a constant amount of genetic material and its alleles may differ from each other in this regard. When such recombined sections appear in *cis*-position, the phenotype produced may be different than when they are in *trans*-position.

structural and regulatory genes, distinction between: The distinction between structural and regulatory genes is more or less arbitrary. For example, lac repressor of *Escherichia coli* is the product of a structural gene; however, its function is regulatory in that it directly controls the expression of a set of structural genes coding for enzymes. Best criterion to distinguish between structural and regulatory genes will be on the basis of the function their gene products perform.

structural biology: The branch of molecular biology, biochemistry and biophysics concerned with the molecular structure of biological macromolecules, especially proteins and nucleic acids, how they acquire the structures they have and how alterations in their structures affect their function.

structural chromosomal change: Change in the genetic structure of one or more chromosomes. May be intraradial or extraradial concerning arms; internal or external with regard to chromosomes and eccentric or dicentric concerning the possession of the centromere or to the direction of a segment in relation to the centromere.

structural classification of proteins (SCOP): SCOP database has been created by manual inspection and abetted by a battery of automated methods. This database provides a detailed and comprehensive description of the structural and evolutionary relationship between all proteins whose structure is known including all entries in the Protein Data Bank. It is available as a set of tightly linked hypertext documents, which make the large database comprehensible and accessible.

structural DNA nanotechnology (SDN): Focuses on synthesizing and characterizing nucleic acid complexes and materials that assemble into a static, equilibrium end state.

structural gene: A gene that codes for any RNA (ribonucleic acid) or protein product other than a regulator molecule (e.g. repressor). The product of a structural gene performs a functional or structural role in the cell. In a strict sense, this term refers to the genes that govern the molecular organization (amino acid sequence) of the proteins.

structural gene changes in evolution: Through electrophoretic analysis of structural genes, it has been estimated that about twenty electrophoretically detectable allelic substitutions per hundred loci accumulate before reproductive isolation is completed. As a general rule, only those structural genes whose products have some important regulatory function contribute significantly to the process of speciation. Otherwise, the organismal evolution and structural gene evolution go on at virtually independent rates. Speciation can occur even without the alteration of structural genes. And, therefore, for the evolution of new species, the evolution of new control mechanisms seems to be more important than the evolution of new proteins.

structural gene mutations: Mutations in structural genes may result in amino acid changes in the protein.

structural genomics, applications of: (i) Structural genomics applies high-speed techniques to make a systematic survey of all protein structures, cataloguing the common ways in which proteins fold. This information leads to computer programmes capable of predicting the shape and function of any protein from the simple linear sequence of As, Gs, Cs and Ts in genes. (ii) The complete spectrum of protein structures can be surveyed in the same way as DNA sequencing machines are used to decode entire genomes. The approach is expected to extend the genomics revolution from a catalogue of genes to a catalogue of the 3D shapes of the proteins for which the genes code. (iii) In the wake of these sequencing efforts, plant research enters an exciting period in which genome-wide approaches are becoming an integral part of plant biology (C. Somerville and S. Somerville in 1999), with potentially highly rewarding but as yet unpredictable biotechnological applications. This is reflected by the current frenzy with which new agricultural biotechnical companies are being founded and the rate at which existing companies are investing in the development of tools to exploit and further expand this wealth of information. (iv) Access to information from structural genomics and new tools that exploit it will profoundly alter the ways we select and approach questions in biology. This will in turn directly affect the application of genetic methods for improving economically important species.

structural genomics: It represents an initial phase of genome analysis and involves the construction of high-resolution genetic, physical and transcript maps of an organism. The effort to determine the 3D structures of large numbers of proteins using both experimental techniques and computer simulation. The complete genome sequences available from model organisms of yeast, *Arabidopsis and Drosophila* will rapidly help in knowing the gene function or molecular biology of other non-model organisms of complex genomes.

structural heterozygosity: Heterozygosity for chromosome mutations.

structural hybridity: Heterozygosity for a chromosomal rearrangement.

structural process in research: Generally, research follows a certain stepwise process. Though step order may vary depending on the subject matter and researcher, the following steps are usually part of most formal research, both basic and applied: (i) define the question, (ii) form hypothesis, (iii) conceptual definition, (iv) operational definition, (v) performing experiments and gathering of data, (vi) analysis of data, (vii) interpretation of data and concluding, (viii) publish results, (ix) test, revising of hypothesis and (x) retest (frequently done by other scientists), conclusions and iterations, if necessary.

structural proteins: Proteins that play a role in giving shape or pattern to a chromosome, a cell, a tissue, an organ or an organism (e.g. tubulin, collagen, carrier protein haemoglobin).

structural proteomics: It deals with the preparation of structural maps of protein and protein domain three-dimensional configuration. It deals with structure-function relationships, drug discovery, prediction of three-dimensional structure from genomic sequences and identification of structurally related orthologues.

structural proteomics, history of: (i) The first protein (insulin 51 amino acids) sequence was completed in 1956. (ii) The first RNA (yeast $tRNA^{Ala}$) sequence was completed in 1966. (iii) The first DNA (*Escherichia coli* lac operator) sequence was reported in 1973. (iv) Protein sequencing by Edman degradation provided a crucial link between the activity of a protein and genetic basis of a particular phenotype. (v) A dedicated protein database was formed in 1986 (i.e. SWISS-PROT). TrEMBL database contains automatic translations of nucleotide sequences in EMBL and is not manually curated. (vi) The study of 3-D protein structure is done using crystallography and nuclear magnetic resonance. The Protein Data Bank (PDB) was the first protein structure database.

structural RNAs: (i) TRNAs - provides secondary structure diagrams of the tRNA molecules. It can also be accessed here or here. Identifies tRNA genes without introns or with introns at canonical or non-canonical positions. (a) ARWEN is a programme to detect tRNAs in metazoan mitochondrial DNA sequences. (b) Rfam is a collection of RNA families, each represented by multiple sequence alignments, consensus secondary structures and covariance models. (c) TFAM, unlike other tRNA gene-finders, uses information from the total sequences of tRNAs and not just their anticodons to predict their function. Therefore, TFAM has an advantage in predicting initiator tRNAs, the amino acid charging identity of non-standard tRNAs such as suppressors and the former identity of pseudo-tRNAs. New TFAM models are available including a proteobacterial model for the bacterial glycosylated isoleucine tRNAs, making it now possible for TFAM to correctly classify all tRNA genes for some bacterial taxa. (ii) TM-RNAs – BRUCE detects transfer-messenger RNAs, which function to free ribosome-stalled mRNAs. (iii) SMALL NUCLEOLAR RNAs (snoRNAs) – can be detected with Snoscan for methylation-guide for snoRNAs and snoGPS for pseudouridylation-guide snoRNAs. (iv) RNAmmer 1.2 – predicts 5s/8s, 16s/18s and 23s/28s ribosomal RNA in full genome sequences. (v) Ribosomal RNA BLAST – BLASTN searches against the small (SSU) &/or large (LSU) subunit RNA databases Unfortunately the input is limited to a maximum of 10,000 nucleotides. This site provides comprehensive, quality and regularly updated databases of aligned small (16S/18S, SSU) and large subunit (23S/28S, LSU) ribosomal RNA (rRNA) sequences for all three domains of life.

structural stability: It may arise by deletion, insertion or rearrangement of DNA. To avoid this, it is desirable to minimize the expression of cloned genes until the organism is introduced into the final vessels. This can be done by using controllable promoters. Another way is the use R1-derived runaway replication mutants. Better still, both techniques can be combined.

structure and function relationship: A relationship exists between the way organisms and objects look (feel, smell, sound and taste) and the things they do. Structure and function of cell, DNA, chloroplasts, mitochondria.

structure prediction: Algorithms that envisage the secondary, tertiary and even quaternary structure of proteins from their sequences. Determining protein structure from a sequence has been called 'the second half of the genetic code' since it is the higher-level folded structure of a protein that governs its function as a gene product. So far, most structure prediction methods have been only partly successful and typically work best for certain well-defined classes of proteins.

structure-based combinatorial protein engineering (SCOPE): Multiple PCR primers are designed, based on protein primary, secondary and tertiary structure, and used to create crossover gene libraries from distantly related proteins. SCOPE provides an effective means for the creation of functional gene libraries from gene families that share both low- and high-sequence homology.

structure-function relationships: With over 180 genomes sequenced, including human and mouse, the amino acid sequence of a particular protein is now generally available from many different organisms. This advance has created a situation in which protein sequences are available even when there is no information on the biological activity of the protein. Thus, the focus has shifted to understanding the function of all of these proteins. Techniques such as the yeast two-hybrid system have been applied to identify protein interactions, and crystallographic structures are being determined to elucidate biological information about all the proteins in an entire organism. Still, the chemical basis for activity is often not completely understood, even after the biological activity and high-resolution structure have been determined. The construction of mutant proteins can elucidate the role of a particular amino acid at a specific position in the sequence.

structured coalescent: The COALESCENT PROCESS is a description of population evolution and underlies much of modern population genetics. The Wright-Fisher model provides the simplest explanation of the evolutionary change in a population of identical individuals. Structured coalescent is an extension to the coalescent process in which lineages move from one place to another as they trace backwards in time while they coalesce.

structured population: *See* POPULATION STRUCTURE.

structured storage: A NoSQL (originally referring to 'non-SQL' or 'non-relational') database offers a mechanism for storage and retrieval of data that is modelled in means other than the tabular relations used in relational databases. Such databases have occurred since the late 1960s, but the name 'NoSQL' was only coined in the early 21st century, prompted by the needs of Web 2.0 companies. NoSQL databases are increasingly engaged in big data and real-time web applications. NoSQL systems are also called 'Not only SQL' to highlight that they may support SQL-like query languages or sit alongside SQL databases in polyglot-persistent architectures.

struggle for existence: Charles Darwin suggested that there is competition among the individuals that are born to obtain what they need to live such as food or a place to live, hide or breed. The individuals possessing useful variations are selected for and those wanting useful variation are selected against.

Student's t-distribution: A continuous probability distribution that generalizes the standard normal distribution. Like the latter, it is symmetric around zero and bell shaped.

Student's t-test: A statistical procedure to discover the significant difference between two sample means from two normally distributed populations with unknown variances. This can include correlated means, correlation, independent proportions or independent means. In 1908, W.S. Gosset, an Englishman publishing under the pseudonym Student, developed the t-test and t-distribution. *Aka* T-TEST.

stunting: The reduction in height of a vertical axis of an individual resulting from a progressive reduction in the length of successive internodes or a decrease in their number.

style: An elongated part of a carpel or a group of fused carpels between the ovary and the stigma through which the pollen tube grows. The pollen must grow through the style in form of a tube to fertilize the plant.

stylet-borne transmission: A type of virus transmission in which the virus is acquired and transmitted by the vector after short feeding times and is retained by the vector for only a short period. *See* NON-PERSISTENT TRANSMISSION. *Contrasts* CIRCULATIVE TRANSMISSION, PERSISTENT TRANSMISSION *and* PROPAGATIVE TRANSMISSION.

subclone: A method in which smaller DNA fragments are cloned from a large insert, which has already been cloned in a vector.

subcloning: Transplantation of a fragment of DNA from one vector to another.

subcloning of DNA without dephosphorylation or gel electrophoresis: Conventional subcloning into plasmid vectors often involves dephosphorylation, gel electrophoresis, DNA extraction and purification to isolate the target insert and the cleaved plasmid. This is not only time consuming but very often problematic as well. These steps can be circumvented by mixing digested donor and recipient plasmids for ligation.

subculture: Tissue culture process by which the cultured tissue is first subdivided and then transferred to fresh culture medium. Tissue cultures are maintained and multiplied through periodic subculturing. For instance, during micropropagation plants shoot clumps from cultured bottles is subdivided into 5–10 parts and then subcultured into 5–10 bottles containing fresh medium. The incubation period varying from 2–3 weeks comprises one cycle. Therefore, subculturing after every cycle leads to a huge multiplication in a year. *See* PASSAGE.

subculturing: Subculturing, also called PASSAGING CELLS, is the removal of the medium and transfer of cells from a previous culture into fresh growth medium, a procedure that enables the further propagation of the cell line or cell strain.

subdividing genomic DNA: The process of subdividing genomic DNA into clonable elements and inserting them into hosts is called creating a library, a clone bank or a gene bank. Thus, a complete library of host cells will contain all of the genomic DNA of the source organism.

subfunctionalization: Functional (or expression) divergence of gene duplicate from the ancestral gene (e.g. tissue-specific expression of gene duplicate).

subgenomic RNA: A piece of viral RNA, shorter than the entire genome of the virus, found in cells infected by the virus and sometimes encapsidated.

subgenus: A category of supplementary taxa intermediate between genus and section. The name of a subgenus is a singular noun, always has an uppercase initial letter and is used in combination with the generic name (e.g. *Primula* subgenus *Primula*).

subinert chromosome: A chromosome with a slight genetic effect is said to be inert. *See* INERT GENE.

sublethal mutation: *See* VITALITY MUTATION.

sublining: Refers to splitting a breeding population into numerous smaller populations. All controlled crosses for forward selection are made within a subline, leading to inbreeding within sublines.

submedian centromere: Centromere located not quite in the middle of the chromosome.

submerged fermentation (SmF): Characterized by the cultivation of microorganisms in a liquid medium.

submetacentric chromosome: A chromosome where the centromere is nearer one end than the other, resulting in the arms not being of equal length.

submicroscopic: Objects that are too small to be resolved and made visible by an ordinary light microscope.

subpetiolate: A leaf possessing an extremely short petiole and may appear sessile.

subphases of G_1 phase: G_1 phase is divided into three subphases – pre-start G_1, start G_1 and post-start G_1 stages.

subpopulation: Either an identifiable fraction or subdivision of a population or specific biotype within a natural population.

subroutine: In computer programming, a subroutine is a sequence of programme instructions that accomplishes a specific task, packaged as a unit. This unit can then be used in programmes wherever that particular task should be performed. Subroutines may be defined within programmes, or separately in libraries that can be utilized by many programmes. In different programming languages, a subroutine may be called a ROUTINE, SUBPROGRAMME, FUNCTION, METHOD or PROCEDURE. Technically, all these terms have different definitions. The generic, umbrella term CALLABLE UNIT is sometimes used.

subsexual reproduction: In English, the word 'asyndetic' means 'not connected by a conjunction'. Asyndetic meiosis is the meiotic cell with one or two bivalents, cells with more than two bivalents being exceedingly rare. The term 'subsexual reproduction' refers to a partially asyndetic meiosis that leads to a restitution nucleus with limited crossing-over, which in turn leads to genetic variability.

subshrub: Also, undershrub. A small shrub that may have partially herbaceous stems, but generally a woody plant less than 1 m (3.3 ft) high.

subspecies (ssp.): (i) A species forming a population or group of populations that are distinguishable from other members of the species and often partially reproductively isolated from them. (ii) One of the several geographic or similar subdivisions between which interbreeding takes place at a reduced level.

substances that maintain gradients: ANIMALIZED SUBSTANCES seem to include a nucleotide, a small polypeptide or an amino acid. VEGETALIZED SUBSTANCES include aromatic amino acids like tyrosine and tryptophan. Purification of vegetalized and animalized substances and then getting the work done can prove its role. In animalization, ectodermal structures are more pronounced. Larvae are devoid of mesodermal and endodermal differentiation. Examples of animalized substances are Zn^{++} ions and trypsin. In vegetablization, exaggeration of mesodermal and endodermal differentiation at the expense of endodermal structures is observed. An example of a vegetalized substance is Li^{+}.

substantial equivalence: (i) A principle inherent in the safety assessment process that compares a genetically modified food with a conventional non-modified food with a long history of safe use. If the modified food has essentially all the characteristics of the non-modified food concerning food and feed value it is said to be substantially equivalent. (ii) The concept that two genetically different plant lines are deemed the same based on composition and safety.

substantial equivalence, principle of: Any genetically engineered (GE) product that is found to be 'substantially equivalent' to its unmodified counterpart would be regarded as safe and fit for human consumption.

substitution: In genetics, a mutation that occurs when a base in the DNA sequence is randomly swapped for a different base is known as a SUBSTITUTION MUTATION. A substitution mutation will only change the amino acid for the triplet (group of three consecutive bases) where the mutation occurs. It will not have a knock-on effect further along the gene/polypeptide. *See* MUTATION.

substitution haploid: A haploid individual having one chromosome replaced by that of a related species. For example, 2x − 1 + 1, 3x − 1 + 1. They are derived from alien substitution lines.

substitution line: A line in which one or more chromosomes are substituted by one or more chromosomes of a donor variety or species. It is used for genetic analysis and gene transfer.

substitution load: The total loss of mean fitness triggered when favourable alleles are substituted gradually by selection rather than instantaneously.

substitution mapping of QTLs: A method for fine mapping of QTLs previously localized to chromosomal regions. It uses tomato as a model system. Substitution mapping is analogous to deletion mapping of genetic loci in humans, in which altered phenotype is associated with loss of a specific chromosomal region. However, substitution mapping utilizes meiotic recombinants, rather than deletions. By prior QTL analysis, an RFLP is identified, which lies near a QTL of interest. Different chromosomal segments carrying the RFLP are identified, and their regions of overlap are determined using all available genetic markers.

substitution matrix: (i) A model of protein evolution at the sequence level dealing with the development of a set of widely used substitution matrices. These are regularly called Dayhoff, mutation data matrix (MDM), BLOSUM or per cent accepted mutation (PAM) matrices. They are derived from global alignments of closely related sequences. Matrices for greater evolutionary distances are extrapolated from those for lesser distances. (ii) A table displaying the probability of change taking place between different macromolecular residues (nucleotides or amino acids).

substitutional editing: No guide RNA is implicated in this type of editing. The phosphodiester backbone of the RNA is not cut and the new base is synthesized *in situ*. Substitutional editing blurs the distinction between RNA editing and RNA modification.

substitutional load: The cost to a population of replacing one allele with another in the course of evolutionary change.

substrain: A substrain can be derived from a strain by isolating a single cell or groups of cells having properties or markers not shared by all cells of the parent strain.

substrate: (i) In biochemical reactions, substrate is the substance acted upon by an enzyme. For example, the enzyme amylase catalyses the breakdown of starch molecules into glucose molecules. In this instance, starch is the substrate of the enzyme amylase. Similarly, sucrose is the substrate of an enzyme sucrase that catalyses the hydrolysis of sucrose to glucose and fructose. Likewise, the enzyme luciferase acts on substrate luciferin and the X-gluc acts as the substrate for enzyme β-glucuronidase to produce a blue colour in the GUS assay. (ii) In chromatography, the substance (usually solid or gel) that attracts and non-covalently binds (interacts) with one or more of the molecules in a solution that is passed over that substrate (e.g. in a chromatography column). This preferential binding (interaction with the substrate) allows one or more of the solution's molecular ingredients to be separated from the other(s).

substrate race: A local race selected to agree in its colouration with that of the substrate, for example, a black race on a lava flow.

substrate subspecies: A population whose colouration agrees with the soil or rocky substrate on which it lives, for example, a pale-coloured population living on a sandy desert.

substratum: Solid surface to which a cell adheres.

subtelocentric chromosome: A chromosome whose centromere is positioned between its middle and its end but closer to the middle.

subtelomeric region: The chromosomal region just proximal to the telomere (end of the chromosome) composed of highly polymorphic repetitive DNA sequences that are typically situated adjacent to gene-rich areas.

subtend: To stand below or close to, as in a bract at the base of a flower.

subtraction library: A cDNA library that only contains cDNAs uniquely expressed in a given cell or tissue (e.g. T cells and B cells will express many common RNAs, as well as a very small percentage, which will be unique for T cells and B cells, respectively). To make a T cell subtraction library, the cDNA from a T cell library is hybridized with a vast excess of B-cell RNA. The commonly expressed genes will result in RNA-cDNA hybrids, which can be removed (or subtracted) to leave only T cell-specific cDNAs.

subtractive hybridization: A form of differential hybridization, where a probe is designed in which the signal representing sequences common to the two transcript pools of interest are removed. The product is a probe population enriched for sequences favourably present in one of the pools. It is considered a more sensitive technique than differential hybridization.

subunit: Part of a multi-component complex where subunit refers to one protein of a protein complex or one polypeptide chain of a multichain protein. For example, oligomeric protein haemoglobin is a tetramer consisting of two α-subunits (α1 and α2) and two β-subunits (β1 and β2).

subunit vaccine: A vaccine presenting a (protein) antigen to the immune system rather than the entire pathogen (viral particles, bacteria). Recombinant vaccines make use of only those antigenic fragments of microorganisms that best stimulate an immune response. *See* RECOMBINANT VACCINES.

subvital: Characterized by a reduction in vitality; said of a gamete or a zygote. The effect is less than semi-lethal.

subvital gene: A gene that causes the death of some proportion (but not all) of the individuals that express it.

succession: The sequence of transient communities occurring in an area before the climax.

succulent: Juicy or fleshy fruit, vegetables and meat or a plant with a fleshy habit.

sucker: (i) A shoot of more or less subterranean origin. (ii) An erect shoot arising from a bud on a root or a rhizome, sometimes at a distance from the stem of the plant.

sucrose: Disaccharide comprising of one glucose unit and one fructose unit. Sucrose is the major form in which glucose is transported between plant cells.

sucrose gradient centrifugation: Procedure used for separation of DNA, RNA or proteins based on size and conformation.

sudden speciation: New species evolve suddenly through individuals. *Aka* SPECIATION BY SALTATION. Processes may be genetical (through macromutations) or chromosomal (through hybridization and/or polyploidy). The plausibility of macromutations as a mechanism is less.

sugar: Small carbohydrates with a monomer unit of general formula $(CH_2O)_n$. Examples of monosaccharides include glucose, fructose and mannose, and those of disaccharides include sucrose (one molecule of glucose linked with one molecule of fructose and one of together) and lactose (one molecule of glucose linked with one molecule of galactose).

sugar-phosphate backbone: The DNA double helix looks like a twisted ladder due to the way the components of the nucleotides bind together. Covalent bonds join the sugar of one nucleotide to the phosphate group of the next nucleotide along each DNA strand, forming the sugar-phosphate backbone that makes up the sides of the ladder through phosphodiester bonds.

suicide seeds: *See* GENETIC USE RESTRICTION TECHNOLOGY (GURT).

suitability for food processing: Bruise-resistant tomatoes were developed, which express antisense RNA against POLYGALACTURONASE (PG) that attacks pectin in the cell walls of the ripening fruit. Tomato exhibiting delayed ripening was developed by using antisense RNA against enzymes involved in ethylene production. This increases the shelf life of tomato.

sulfhydryl (thiol, SH): Chemical group containing sulfur and hydrogen found in the amino acids (e.g. methionine and cysteine incorporated in proteins) and other molecules. Two sulfhydryls in cysteines can covalently bind to produce a disulfide bond.

sulfonamide: Bacteriostatic compound that interferes with folic acid synthesis. These are also called sulfa drugs, which were among the first synthetic antimicrobial drugs used to treat microbial diseases. In the recent years, the antibiotics have almost totally replaced the sulfa drugs. The action of these drugs is due to their structural similarity to PARA-AMINOBENZOIC ACID (PABA). Most widely used sulfa drug is a combination of trimethoprim and sulfamethoxazole (TMP-SMZ). This combination has a broader spectrum of action and prevents the development of resistant strains.

sulfosate: An active ingredient in some herbicides. It kills plants (e.g. weeds) by inhibiting the crucial plant enzyme EPSP Synthase. Chemically, sulfosate is a trimethylsulfonium salt of the same organic acid as glyphosate, so sulfosate can be applied over crops (e.g. soybeans) that have been GE to be tolerant to glyphosate-based herbicides.

sulfur (S): A yellow mineral; an element that is needed by the plant. It is found covalently bound, especially in proteins, where it stabilizes plant structures. Deficiency of S causes chlorosis and etiolation. It is also used as a dust or wet table powder for control of various diseases and insects, including powdery mildew, rust and mites.

sulfur cycle: Chemical conversions of sulfur in the environment, mostly caused by microorganisms. The sulfur cycle plays important roles in nutrition and in energy generation in the form of ATP. Sulfur is a component of some enzymes and cofactors. Some autotrophic prokaryotes generate ATP by oxidizing reduced forms of sulfur. The main loop of the cycle consists of reducing the sulfate ions (SO_4^{2-}) to hydrogen sulfide gas (H_2S) and reoxidizing it to sulfate.

sulfur dioxide (SO_2): The principal air pollutant abundantly produced in industrial processes and coal burning that results in interveinal necrosis on broadleaved plants and tip necrosis on conifers.

sulfuration: A treatment with sulfur, with fumes of burning sulfur, sulfur dioxide or with sulfites as done in fumigation, bleaching or preserving. It is used in greenhouses for reducing and/or preventing leaf diseases (e.g. in cereals).

sum of pairs method: Sums the substitution score of all possible pair-wise combinations of sequence characters in one column of a multiple sequence alignment.

sum of squares (SS): The sum of squares of the deviations of individual values from the arithmetic mean. SS is used to compute variance. SS_G refers to the sum of squares of deviations due to genetic factors; SS_E refers to the SS due to environmental differences, and SS_m is sometimes used for the total sum of squares, $SS_G + SS_E$.

sum rule of probability: This rule states that the probability of either of two mutually exclusive events occurring is the sum of their individual probabilities.

summerwood: The part of the yearly xylem growth ring in woody plants formed late in the growing season and consisting of cells smaller than those of springwood. *Contrasts* SPRINGWOOD.

SUMO: An acronym for Small Ubiquitin-related MOdifier protein. SUMO is a member of a ubiquitin-like protein superfamily that is covalently attached to target proteins as a POST-TRANSLATIONAL MODIFICATION (PTLM) to alter the localization, stability and/or function of the target protein in response to changes in the cellular environment.

SUMOylation: A posttranslational modification (PTLM) that plays a role in various cellular processes. SMALL UBIQUITIN-LIKE MODIFIER (SUMO) proteins are a family of covalently linked small proteins (consisting of ~100 amino acids, or 12 kDa) that are enzymatically attached to specific proteins (in a process analogous to ubiquitylation). SUMOylation influences the function of proteins by inhibiting the usual interaction between the target of SUMOylation and another protein. The interaction involves the provision of a new binding surface or the initiation of conformational changes in the target protein. SUMOylated proteins are not destined for destruction. Chemically, SUMOylation involves the formation of a peptide bond between a lysine on the target protein and a C-terminal glycine on the SUMO protein that is generated by the cleavage of the last four amino acids in the SUMO protein.

sunscald: The injury of plant tissues burned or scorched by direct sun. *Aka* SUNBURN.

super dominance: *See* TRUE OVERDOMINANCE.

super elongation complex (SEC): An assembly of transcription elongation factors that aids in the reinitiation of transcription by the proximal promoter–paused RNA polymerase II. Tudor domains: Regions within proteins that are capable of detecting and binding methylated histones. An example is the Tudor domain of 53BP1, which recognizes the dimethylation of lysine 20 on histone H4 (H4K20me2).

super suppressor: A mutation that can suppress a variety of other mutations; characteristically, a nonsense suppressor.

super-multigene family: Genes that have considerable homology but do not perform the same function.

super-repressed: Uninducible.

super-suppressor mutation: A mutation that can suppress a diversity of other mutations; typically, a nonsense suppressor.

supercoil: Double-stranded circular DNA that is twisted about itself. Commonly observed with plasmids and circular viral DNA genomes (such as that of hepatitis B virus). A nick in one strand of the plasmid may remove the twist, resulting in a relaxed, circular DNA molecule. A complete break in the DNA puts the plasmid in a linear form. Supercoils, relaxed circular DNA and linear DNA all have different migration properties in agarose gels, even though they contain the same number of base pairs.

supercoiled denatured DNA: DNA is like SUPERCOILED DNA, but has unpaired regions that make it slightly less compact; this can result from excessive alkalinity during plasmid preparation.

supercoiled DNA: Region of DNA in which the double helix is further twisted on itself. DNA is fully intact with both strands uncut, and with a twist built in, resulting in a compact form. It is also called SUPERHELICITY, which gives a peculiar twisted appearance of DNA. In bacteria and viruses, DNA is often circular. In eukaryotes, large loops of DNA occur in such a way that each loop represents a circle. A DNA molecule without supercoiling is said to be RELAXED or have ZERO SUPERCOILING. Such organization often puts an additional constraint on double-helical structure and the DNA becomes supercoiled. One supercoil is added every time the duplex thread is twisted about its axis. The supercoiling places the DNA molecule under torsion, which can be released by a break in one of the two strands. The supercoils may be negative or positive. The NEGATIVE SUPERCOILS are found *in vivo* when they are in a direction opposite to the clockwise turns of right-handed DNA. This will actually relieve the torsion and will become underwound or even single stranded in a region. This helps in unwinding of DNA for replication and so on. The POSITIVE SUPERCOILS are in a clockwise direction in right-handed DNA. This leads to overwinding (superhelical), which can be created *in vitro*, but does not occur in nature. Supercoiling is often controlled by some enzymes and is described by parameters such as linking number, twisting number and writhing number. *Aka* COVALENTLY CLOSED-CIRCULAR DNA.

supercoiling: The coiling of a closed duplex DNA molecule in space so that it crosses over its own axis. *Aka* SUPERHELICITY.

supercoiling in bacterial chromosome: A folded genome of *Escherichia coli* reveals about 100 loops, folds or domains. These folds are stabilized by RNA molecules. Each fold in the chromosome is independently supercoiled. The formation of loops and supercoils within the loops packs the 1100 μm long *Escherichia coli* chromosome into a structure of 1–2 μm diameter.

superfemale: An abnormal type, almost completely sterile, in *Drosophila* with an X chromosome/autosome ratio greater than 1 (e.g. 3X/2A = 1.5).

supergene: A group of neighbouring genes on a chromosome that tend to be inherited together and sometimes are functionally related. This chromosome segment is protected from the occurrence of crossing-over and so is transmitted intact from generation to generation, like a recon. Advantageous grouping of genes within an inversion; the entire gene complex is inherited as a whole.

superinfection: Phage infection of a cell that already docks a prophage.

superior ovary: In plants, an ovary borne above the level of attachment of the other floral parts, or above the base of a hypanthium. *Compare with* INFERIOR OVARY *and* HALF-INFERIOR OVARY.

superlattices: Materials that exhibit double periodicity. A nanoparticle can be crystallized into a superlattice in which there will be a periodicity within the nanoparticle (gold lattice for example) and a periodicity of the particles themselves. Such superlattices are known in conventional materials also, in addition to multi-layered films.

superlethal mutation: *See* VITALITY MUTATION.

supermale: An abnormal male individual that carries higher doses of male determiners. It is almost completely sterile, in *Drosophila* with the X chromosome/autosome ratio less than 0.5 (e.g. 1X/3A = 0.33).

***SUPERMAN* gene:** A floral homeotic gene, mutations of which affect flower development in *Arabidopsis*. Hypermethylated *SUPERMAN* alleles *clarkkent* (*clk*) were used for genetic screens that led to the isolation of mutations in genes *KRYPTONITE* (*KYP*), *CHROMOMETHYLASE3* (*CMT3*) and *domain-rearranged methyltransferases* (*DRMs*).

supernatural phenomenon: The phenomena that cannot be described by natural laws, cannot be tested by scientific methodology and are therefore outside the realm of science.

supernumerary chromosomes: Chromosomes frequently occur in varying numbers, in addition to the characteristic constant complement of chromosomes. *Aka* ACCESSORY CHROMOSOMES.

superoxide dismutase (SOD): An enzyme that destroys superoxide free radicals. Obligate anaerobes in the presence of oxygen, produce superoxide free radicals (anions), which are toxic to cellular components, to hydrogen peroxide. Therefore, microorganisms, attempting to grow in O_2 produce an enzyme SOD to neutralize the superoxide free radicals. SOD plays a critical role in the defence of cells against the toxic effects of oxygen radicals. SOD competes with nitric oxide (NO) for superoxide anion, which inactivates NO to form peroxynitrite.

superoxide free radical: Toxic form of oxygen produced during aerobic respiration. *See* SUPEROXIDE DISMUTASE.

superspecies: A monophyletic group of completely or principally allopatric species that are too discrete to be included in a single species.

supervital mutations: Mutations that increase the survival of the individuals carrying them in the appropriate environment or under stress.

superweed: A colloquial, derogatory term referring to a plant pest resistant to an herbicide.

supplementary genes: Two dominant genes when present together produce a novel phenotype. These genes yield 9:3:4 phenotypic ratio in F_2 progeny.

support-vector machines (SVMs): In machine learning, SVMs, also called SUPPORT-VECTOR NETWORKS (SVNs), are supervised learning models with associated learning algorithms that analyse data used for classification and regression analysis.

suppression: A mutation in a gene that reverses the effect of a previous mutation in a different gene. This term is aka RESCUE of an allele by means of another gene. Suppression also serves as a DNA PROTECTIVE MECHANISM or DNA SAFEGUARDING MECHANISM.

suppression gene interaction: In case of inhibitory genes, one gene when dominant is epistatic to the other but other when recessive homozygous is epistatic to the first. In F_2, 13 mutant and 3 coloured individuals are observed.

suppression of endogenous genes: In many crops, certain quality-related traits can be improved by reducing or eliminating the expression of specific genes. The level of endogenous gene expression can be reduced by the following four approaches: (i) antisense gene, (ii) ribozyme, (iii) gene disruption and (iv) over-transcription leading to cosuppression.

suppression of transposition by DNA methylation: In plants, high DNA methylation generally occurs in these transposable elements. Methylation of symmetric cytosines mainly relies on the methylase MET1 and chromatin remodeller DDM1. Methylation at CHG and CHH sites mainly depends on the methylases CMT3 and DRM2, respectively. Loss of DNA methylation in mutants of these genes leads to the release of transcriptional silencing of transposable elements and other repetitive DNA sequences. Mutation of *DDM1*causes a variety of developmental abnormalities. One of the *ddm1*-induced abnormalities in the *clam* mutant is caused by the transposition of an endogenous CACTA family transposon *CAC1* from its original site to the *DWF4* locus.

suppressive petites: These petites bear mutant mitochondrial DNA.

suppressive rescue: A transferred gene is tagged with an amber suppressor gene to isolate a human oncogene.

suppressive soil: A soil in which various diseases are naturally occurring at levels lower than expected due to biological factors in the soil. It serves as an example of natural biological control.

suppressor gene: A gene that can inverse the effect of a specific type of mutation in other genes, such as a premature termination sequence.

suppressor mutation: A mutation that completely or moderately restores a function lost by a primary mutation. It is located at a site in the gene different from the site of the primary mutation.

Suppressor of variegation 3–9 (Su(var)3–9): Originally identified in genetic screens for position effect variegation in *Drosophila melanogaster* and later shown to encode a histone H3 lysine 9 (H3K9) methylating enzyme.

suppressor rescue: A transferred gene is tagged with an *amber* suppressor gene to isolate a human oncogene.

suppressor T (T_S) cells: T cells that regulate the immune response by interacting with the B cells, turning it off when the antigen is absent. Most immunologists feel that T_S cells are not a distinct population but represent suppressive activity by the T_H and T_C cell populations.

suppressor tRNA: Any tRNA species that, due to mutation in a transfer RNA gene, displays changed coding specificity and is able to translate nonsense or missense codons. Nonsense and missense suppressor tRNAs are frequently altered in the anticodon loop or in an adjacent region. Frameshift suppressor tRNA rewards for frameshift mutations by reading non-triplet codons.

suppressor, extragenic: *See* EXTRAGENIC SUPPRESSOR.

suppressor, intragenic: *See* INTRAGENIC SUPPRESSOR.

suppressor-sensitive mutation: A mutation whose phenotype is repressed in a genotype that also carries an intergenic suppressor of that mutation, for example, amber (UAG), ochre (UAA) or opal (UGA).

suprachiasmatic nucleus: A bilateral structure located in the anterior part of the hypothalamus. It is the central pacemaker of the circadian timing system and regulates most circadian rhythms in the body.

supramolecular assembly: Refers to a very large molecular structure.

supramolecular complex: A double-crossover (DX) structure that consists of five DNA single strands that form two double-helical domains, on the top and the bottom in this image. There are two crossover points where the strands cross from one domain into the other.

supra-operon control: The synchronized control of a gene cluster with multiple operator and promoter sites that consists, in spirit, of several interacting operons controlling different biochemical pathways.

surface or substrate-dependent cells or cultures: *See* ANCHORAGE DEPENDENT CELLS.

surrogate genetics: A branch of genetics that deals with the introduction of a manipulated DNA into a living nucleus where its expression can be monitored. Effects of directed DNA changes, changes in RNA transcript and changes in protein encoded by it can all be studied.

surveillance: Systematic collection, analysis, interpretation and dissemination of data (generated by the laboratory and private and public domain literature) related to the biotechnology field to assist in the planning and implementation of research, evaluation and management of risks and public health interventions and programmes (if needed).

survey: A survey is an investigation of the characteristics of a given population by means of collecting data from a sample of that population through a questionnaire and estimating their characteristics through the systematic use of the statistical methodology.

survival analysis: A branch of statistics dealing with the analysis of the time until an event such as death. Survival analysis is distinguished by its emphasis on estimating the time course of events and in dealing with censoring. *See* COX MODEL.

survival factor: Extracellular signal required for a cell to survive. In the absence of such a signal, the cell will undergo apoptosis and die.

survival function: The probability of being free of the event at a specified time.

survival of the fittest: The term refers to the survival of only those organisms best fit to acquire and use resources, resulting in the evolution of organisms that are best adapted to the environment. Darwin used metaphorically to describe 'natural selection'. The phrase was invented by the 19th-century philosopher H. Spencer. Survival alone is insufficient for evolution – it's reproduction – passing on of genes that really counts. Most modern biologists no longer use this term when describing or discussing natural selection. This phenomenon is, however, the foundation of DARWIN'S THEORY OF NATURAL SELECTION.

survival time: Interval between the time origin and the occurrence of the event or censoring.

suscept: An abbreviated term for a susceptible plant.

susceptibility: The antithesis of RESISTANCE. Lack of resistance to a disease. Resistance and susceptibility of crop varieties to specific diseases or insect depends upon the genotype and environment. Susceptible varieties can be made resistant by transferring resistance genes from wild species through backcross breeding. Resistant varieties after a few years' cultivation become susceptible due to the development of new virulent races of pathogens/insects. The transgenic approach is increasingly becoming important to develop resistance in otherwise susceptible but well-adopted varieties. Nutrient deficiency and excess also can greatly affect the susceptibility of plants to disease and disorders.

susceptible: Being subject to infection or injury by a pathogen. *Contrasts* RESISTANT.

suspension culture: A type of culture in which the single cells or aggregates of a few cells multiply while suspended in a liquid medium. Both animal and plant cells can be grown in the liquid medium. For raising the cell suspension culture, agitation of the liquid medium is required to provide proper aerobic conditions to the growing cells. In laboratories usually the batch-type suspension cultures are grown by keeping the flasks containing cell cultures on reciprocating shakers. Whereas, in the industry continuous type of suspension cultures are grown by adjusting the inflow of the fresh medium and outflow of the spent medium.

suspensor cells: The group or chain of cells formed by the zygote that pushes the developing proembryo towards the centre of the ovule in contact with the nutrient supply.

sustainability: The adoption of practices that allow for the long-term maintenance of the productive capacity, the viability and quality of life, and conservation of the environment and resource base.

sustainable agriculture: A systematic approach to agriculture that concentrates on guaranteeing the long-term productivity of human and natural resources for meeting food, textile and industrial needs. It can be based on an understanding of ecosystem services. It is an integrated system of plant and animal production practices having a site-specific application that will, over the long term satisfy human food or industrial needs, improve environmental quality and the natural resource base upon which the agricultural economy depends, make the most efficient use of non-renewable resources and on-farm resources and integrate natural biological cycles and controls, sustain the economic viability of farm operations and augment the quality of life for farmers and society as a whole.

sustainable agriculture methods: Elements of sustainable agriculture include PERMACULTURE, AGROFORESTRY, MIXED FARMING, MULTIPLE CROPPING and CROP ROTATION.

sustainable development: All aspects of development that meet the needs of the present without compromising the ability of future generations to meet their own needs.

sustainable green revolution: The Green Revolution has encountered major problems of equity, stability and sustainability, calling for a 'new phase' of agricultural research and development.

SV40 early gene promoter: The early gene promoter of mammalian virus SV40 has three regions: (i) Enhancer (150 bp), (ii) 21-bp repeats (60 bp) and (iii) TATA region. For efficient transcription, the regulatory protein binds to each of the three and then they contact each other. Insertion of 5 or 15 bp between (i) enhancer and (ii) 21-bp repeat segment decreases transcription more drastically than does insertion of 10 or 21 bp. Similar results have been found for insertions between the middle segment and TATA. Here DNA bound protein interact when lying at adjacent sites. DNA probably loops to allow the interaction to take place. Multiple protein-protein interactions have a key role in the generation of an active transcriptional initiation complex at a promoter, which is a very large specific nucleoprotein structure composed of multiple sequence elements recognized by different cooperating factors.

Svedberg unit (S): A unit of sedimentation rate during centrifugation.

swilase: *See* DNA GYRASE.

switch gene: A gene that causes the epigenotype to switch to a different developmental pathway. *See* OLIGOGENE.

switch proteins: Refers to certain protein molecules that signal a plant when environmental conditions are so dry (or cold, etc.) that the plant needs to protect itself (via extreme measures) to survive.

switch-key of (epi)genetic changes: Plant tissue culture environment is considered a switch-key of (epi) genetic changes.

switching on target genes: Signal transduction pathways in living systems are activated by information in the environment; by encoding and transmitting this information, they produce 'music' that is interpreted by cells and results in direct changes to cellular behaviour. A combination of experiments and computer modelling have been used to make sense of a particular signal transduction pathway and illustrate how the pathway can be manipulated to switch on different target genes.

swivel: The place in DNA at which replication starts.

swivelase: *See* DNA GYRASE.

SWR1 chromatin remodelling complex: SWI2/SNF2-related 1 chromatin remodelling complex (SWR1-C) belongs to the INO80 chromatin remodelling family and mainly catalyses the exchange of H2A-H2B with the H2A.Z-H2B dimer.

symbiogenesis: *See* ENDOSYMBIOTIC THEORY.

symbiont: Any organism involved in any type of symbiosis with another organism, either of the same or a different species. Organisms live together in physical association and are mutually interdependent. Legume–*Rhizobium* symbiosis is a common example that contributes a lot to agriculture. In this case, plant provides shelter to the *Rhizobium*, which fixes atmospheric nitrogen for the plant. Host plant and *Rhizobium* are symbionts.

symbiosis: The living together of two different kinds of organisms that may, but does not necessarily, benefit each organism. Examples: (i) Symbiosis is demonstrated in the beneficial relationship between mycorrhizal fungi and the roots of over 85% of the plants in nature. The relationship between mycorrhizal fungi and the host roots of the plant results in increased surface area for the absorption of nutrients and water. In return, the fungi attain carbohydrates (simple sugars) from the plant. (ii) The nitrogen-fixing nodules on the roots of legumes formed by bacteria of the genus *Rhizobium* and the symbiotic relationship of certain fungi and a photosynthetic partner, either an alga or a cyanobacterium, as in lichens.

symbiosis, types of: (i) COMMENSALISM, (ii) MUTUALISM and (iii) PARASITISM. *See details as separate terms.*

symbiotic theory of origin of eukaryotic cell: L. Marguilis gave this theory of the origin of the eukaryotic cell, in 1970. A prokaryotic cell having a flexible cell membrane ingested bacterial cell, which evolved into mitochondria. Spirochetes became ingested and converted into nuclear fibres, centrioles and flagella of a primitive unicellular eukaryote. Blue-green algae cells became incorporated into colourless unicellular eukaryotes to form a simple green flagellate. This is the most favoured theory.

symbolic computation: In mathematics and computer science, computer algebra (also called ALGEBRAIC COMPUTATION), is a scientific area that refers to the study and development of algorithms and software for manipulating mathematical expressions and other mathematical objects. Although computer algebra could be considered a subfield of scientific computing, they are generally considered distinct fields because scientific computing is typically based on numerical computation with approximate floating-point numbers, while symbolic computation underscores exact computation with expressions containing variables that have no given value and are manipulated as symbols.

symmetric cell division: A type of cell division that produces identical cell types, is required for cells to proliferate and contribute to growth. *Aka* PROLIFERATIVE CELL DIVISION. *Compare with* ASYMMETRIC CELL DIVISION.

symmetric distribution: One in which values to the left of the mean by a certain amount are just as likely to be observed as values to the right of the mean by the same amount. For symmetric distributions, the population mean and median are identical and the distance between the 25th and 50th percentiles equals the distance between the 50th and 75th percentiles. *See* QUARTILES.

symmetric RNA structures: Small RNA molecules containing a 19-bp duplex with 2-nt overhangs at each 3′-end. This is the standard siRNA structure.

symmetrical: Capable of being split into at least two equal, mirror-image halves (e.g. zygomorphic) or having rotational symmetry (e.g. regular or actinomorphic). *Compare with* IRREGULAR *and* ASYMMETRICAL.

symmetrical sequences: DNA fragments that display the same sequence on both DNA strands.

sympatric: The populations or species that inhabit, at least in part, the same geographic region. *Compare with* ALLOPATRIC.

sympatric hybridization: Hybridization between individuals of populations within the range and habitat of the parent species. This type of hybridization may be frequent in parasites.

sympatric speciation: Speciation in which the evolution of reproductive isolating mechanisms ensues within the range and habitat of the parent species. This type of speciation may be common in parasites.

sympatric species: Species that are reproductively isolated but occupy a common geographical area.

sympatry: Coexistence in the same place. *Compare with* ALLOPATRY.

symplastic coupling: A special type of cell-cell interaction in which two or more plant cells, coupled by plasmodesmata, can exchange macromolecules such as introduced dyes, proteins or RNA. Cells interlinked in this way can form a network of cells that behave as a supracellular domain.

symptomless carrier: A plant that, although infected with a pathogen (usually a virus), produces no obvious symptoms.

symptoms: An indication of disease by reaction of the host (e.g. pustules, canker, leaf spot, wilt) on diseased organisms. Different diseases produce specific symptoms, which are being used in disease diagnosis. The virus generally causes yellowing of leaves (mosaic) or curling. *Agrobacterium* results in gall formation in dicot plants generally at root-shoot junctions. The fungus *Colletotrichum falcatum* causes redness in sugarcane pith, called red rot disease of sugarcane.

Syn 0, Syn 1, Syn 2: Symbols for designating the original populations (Syn 0), first synthetic progeny (progeny of Syn 0) and second synthetic progeny (progeny of Syn 1).

synanthous: A type of growth in which new leaves and flowers appear and die back at the same time. *See* HYSTERANTHOUS *and* PROTERANTHOUS.

synapomorphy: A shared, derived character state indicating that two species belong to the same group.

synapsis: (i) In genetic recombination, the initial formation of base pairs between complementary DNA strands in different DNA molecules that exist at sites of crossing-over between chromosomes. (ii) The pairing of homologous chromosomes in the pachytene stage of prophase I of meiosis I. Two homologous duplex molecules of DNA must be brought into close contact so that the corresponding sequences can be exchanged. Homologous chromosomes (each consisting of the two sister chromatids produced by the prior replication) approach one another. They become laterally associated in the form of a SYNAPTONEMAL COMPLEX (SC). Each chromosome at this stage appears as a mass of chromatin bounded by a lateral element, which in this case has a striated structure. The two lateral elements are separated from each other by a central element. The triplet of parallel dense strands lies in a single plane that curves and twists along its axis. The distance between the homologous chromosomes is considerable in molecular terms, more than 200 nm. The generation of the SC coincides with the presumed time of crossing-over, although there is no direct evidence that recombination occurs at the stage of synapsis.

synaptic remodelling: The process of changes in synapse number (e.g. gain or loss) is referred to as synaptic remodelling.

synaptic signalling: A type of cell-cell communication that occurs across chemical synapses in the nervous system.

synaptomere: Any of the specific points where the SC formation starts.

synaptonemal complex (SC): Ribbonlike structures observed in electron micrographs of nuclei in the synaptic stages of meiosis. The tripartite ribbon represents a system that promotes the synapsis of homologous chromosomes. The SC seems to be the essential prerequisite for homologous chromosome pairing and crossing-over. *Aka* SYNAPTINEMAL COMPLEX.

synchronization of polymerization on leading and lagging strands: A hallmark feature of DNA replication is the coordination between the continuous polymerization of nucleotides on the leading strand and the discontinuous synthesis of DNA on the lagging strand. This synchronization requires a precisely timed series of enzymatic steps that regulate the synthesis of an RNA primer, the recycling of the lagging-strand DNA polymerase and the production of an OKAZAKI FRAGMENT.

synchronized cells: All cells that are in the same stage, for example, mitotic selection where up to 99% of the mitotic cells can be obtained.

synchronous culture: A cell culture in which all the cells are at the same stage of cell cycle at the same time. During cell division, the cell passes through different stages like 'G' phase, 'S' phase and the dividing phase. A synchronous culture is one in which all the cells are either at the 'G' phase or 'S' phase or at the dividing phase.

syncytial endosperm: A mass of cytoplasm with several nuclei, not divided into separate cells.

syncytium (pl. syncytia): Mass of cytoplasm comprising many nuclei enclosed by a single plasma membrane. Typically, syncytium is the result either of cell fusion or of a series of incomplete division cycles in which the nuclei divide but the cell does not.

syndrome: A group of symptoms that consistently occur together, or a condition characterized by a set of associated symptoms. Specific symptoms that are associated with a disease. A specific group of symptoms or signs accompany a particular disease.

synergid (pl. synergidae): One of two haploid cells that lie inside the embryo sac, beside the ovum. They nourish the ovum.

synergism (Adj. synergistic): A greater-than-additive effect of two or more interacting factors or genes. During plant tissue culture, the growth regulators exhibit synergism and thus enhance the growth *in vitro*. For instance, if two cytokinins – kinetin and benzylaminopurine – are added together, in several instances, the shoot regeneration is better than when either is used alone. *Contrasts* ANTAGONISM.

synergistic gene expression: Gene expression is supposed to be synergistic when the combined quantitative effect of two transcriptional activation elements is more than the sum of their expression. GAL4 is a sequence-specific DNA-binding protein that activates transcription in yeast. GR is a nuclear receptor superfamily protein that mediates the diverse actions of glucocorticoids as a ligand-dependent transcription factor. A synergistic effect of GR and GAL4 was observed. *Aka* SYNERGISTIC TRANSCRIPTION.

syngamy: (i) The union of the nuclei of sex cells (gametes) in reproduction. (ii) The union of two genomes that leads to a doubling of the ploidy level.

syngeneic: Genetically identical members of the same species.

synkaryon: A hybrid cell having a nucleus as a consequence of the fusion of two genetically different nuclei, sometimes from two different species.

synonymous: A word/code that has the same meaning as another word.

synonymous codons: Different genetic codons that specify the same amino acid in a degenerate code. For example, GCU, GCC, GCA and GCG are synonymous codons as all of them code for alanine. *Aka* SAME-SENSE CODONS.

synonymous mutation: A point mutation (change in a single nucleotide) that does not change the amino acid for which the DNA triplet codes.

synonymous single-nucleotide polymorphism: Abbreviated as SYNONYMOUS SNP (or SYNSNP). Any SNP that occurs in an exon but does not change the amino acid composition of the encoded protein.

synonymous substitutions: Those mutations that ensue at the degenerate sites where substitutions do not change the sense of the codon (e.g. often in the third position of the codon). Such substitutions are neutral to selection pressure. The rate of synonymous substitutions in a gene is independent of selective pressure.

synonymous substitutions rate (K_s or D_s): The rate of synonymous substitutions that do not change amino acid sequence. *Compare with* NON-SYNONYMOUS SUBSTITUTIONS RATE.

synonymous variant: A nucleotide change in the DNA that does not alter the protein sequence. *Aka* SILENT ALLELE *and* CODING SYNONYMOUS.

syntax: The syntax of a computer language is the set of rules that defines the combinations of symbols that are considered to be correctly structured statements or expressions in that language. This applies both to programming languages, where the document represents source code, and to mark-up languages, where the document represents data.

syntax error: An error in the syntax of a sequence of characters or tokens that is intended to be written in compile-time. A programme will not compile until all syntax errors are corrected. For interpreted languages, however, a syntax error may be detected during programme execution, and an interpreter's error messages might not differentiate syntax errors from errors of other kinds. There is some disagreement as to just what errors are syntax errors. Some think that the use of an uninitialized variable's value in Java code is a syntax error, but many others disagree and classify this as a (static) semantic error. *See* SEMANTICS.

syntelic chromosomes: When kinetochores of sister chromatids attach to microtubules coming from the same pole, the chromosomes are said to be syntelic, and no tension is generated.

syntenic: Genetic loci located in the same chromosome.

syntenic genes: Genes that are believed to be linked (i.e. reside on the same chromosome) because of their concordant association with a specific marker gene known to be located on a particular chromosome.

synteny: (i) Chromosomal association of genes (linkage) established in somatic cell culture. (ii) In biotechnology, the conservation of the genome organization (gene order) on the chromosomes, which is called collinearity) among phylogenetically related species. It allows understanding of the evolutionary relationships between species. Conserved clustering of genes/unique sequences in different species belonging to the same group. *See* COMPARATIVE GENOMICS, LINKAGE *and* CONSERVED SEQUENCE.

synteny mapping: The localization of DNA sequences (e.g. whole chromosome arms, large genomic DNA regions, genes) in a target genome that are syntenic to regions in a reference genome.

synteny test: A test that by observing concordance in hybrid cell lines determines whether two loci belong to the same group.

synteny-based positional cloning: Any positional cloning of a gene (or genes) in the genome of one organism (say 'A') for which mapping information from the genome of another organism (say 'B') is used. This approach is based on synteny.

synthases: Enzymes that catalyse condensation reactions in which no nucleoside triphosphate (ATP and GTP, etc.) is required as a source of energy. Citrate synthase catalyses condensation reactions that yield a product, which is more complex than the precursors. *Compare with* SYNTHETASES.

synthesis of 28S and 18S rRNA: At the nucleolar organizer region, 45S RNA is synthesized, which acts as a precursor molecule for the synthesis of 28S and 18S rRNAs. An endonuclease cleaves 45S RNA into two short precursor molecules of size 41S and 20S, which are cleaved into 28S and 18S rRNAs, respectively. The 41S RNA undergoes three sequential cleavage steps by exonuclease to yield 28S rRNA, whereas 20S RNA undergoes only one cleavage step by exonuclease to produce 18S rRNA.

synthesis of gene from a true precursor tRNA: DNA duplex that gives rise to this tRNA precursor is synthesized in the form of 26 small (oligonucleotide) segments. These were then arranged into six DNA duplex fragments that have single-stranded ends. These six fragments gave rise to the presumed gene for *Escherichia coli* tyrosine suppressor tRNA precursor. This gene, however, still lacked promoter region and other sequences essential for processing.

synthesis of genes using PCR: Crude synthetic oligonucleotides are sometimes used for rapid generation of DNA fragments through polymerase chain reaction (PCR). The full-length double-stranded DNA could be cloned in a plasmid/phage vector and multiplied in *Escherichia coli*. Overlapping oligonucleotides could be extended through several rounds of PCR. Full-length double-stranded DNA could then be amplified using short primers complementary to the 5′ends of the two strands.

synthesis of lagging strand: By extending RNA primers, the lagging-strand polymerase restarts at short intervals and produces Okazaki fragments. At least in prokaryotic systems, this directionality problem is solved by the formation of a loop in the lagging strand of the replication fork to reorient the lagging-strand DNA polymerase so that it advances in parallel with the leading-strand polymerase. The replication loop grows and shrinks during each cell cycle of OKAZAKI FRAGMENT synthesis.

synthesis of siRNA: (i) RNA Pol IV was proposed to synthesize RNA transcripts by using the RdDM target loci as templates. (ii) The single-stranded RNA is converted into double-stranded RNA by the RNA-dependent-RNA polymerase RDR2. (iii) The double-stranded RNA is cleaved into 24-nt primary siRNAs by a Dicer-like protein DCL3. (iv) The maturation of siRNAs requires HUA ENHANCER 1 (HEN1), a conserved *S*-adenosyl-L-methionine-dependent RNA methyltransferase that methylates the 2′-OH group on the 3′-terminal nucleotides of siRNAs.

synthetases: Enzymes that catalyse condensation reactions, which use ATP or another nucleoside triphosphate as a source of energy. The conversion of succinyl-CoA into succinate is catalysed by an enzyme called succinyl-CoA synthetase. This enzyme uses nucleoside triphosphate in this reaction. *Compare with* SYNTHASES.

synthetic (variety): A variety produced by crossing *inter se* several genotypes selected for good combining ability in all possible hybrid combinations, with subsequent maintenance of the variety by open pollination. Usually, the first generation of a synthetic variety is obtained by a polycross involving a certain number of components with a good GCA. The components are maintained by identical reproduction, either by vegetative propagation (clones) or by continued sib mating (inbred populations).

synthetic amphiploid: An artificially produced amphiploidy.

synthetic auxins: Chemically synthesized auxins such as 2,4-dichlorophenoxy acetic acid (2,4-D) and 2,4,5-trichlorophenoxyacetic acid (2,4,5-T) stimulate unrestrained DNA and protein synthesis and are effective herbicides. 2,4-D is still used as an herbicide, but the use of 2,4,5-T has been banned because it disposed to be contaminated with carcinogenic dioxin. At extremely low concentrations, 2,4-D is used to promote the growth of orchid seedlings or *in vitro* cell cultures.

synthetic biology: An interdisciplinary branch of biology and engineering combining various disciplines from within these domains, including biotechnology, evolutionary biology, molecular biology, systems biology, biophysics, computer engineering and genetic engineering. Engineering principles are applied to reduce genetics into DNA 'parts' so that those parts can be understood in terms of how they can be combined to build desired functions in living cells. Through this process, it is possible to change the parts of a living thing and assemble new organisms from parts of DNA from more than one source organism or to build synthetic DNA from molecules. It deals with things like making new cells and enzymes to make organisms stronger or fight against diseases.

synthetic chromosome: An artificial chromosome synthesized from simple chemical building blocks.

Owing to limitations in the length of DNA that is amenable to direct chemical synthesis, the construction of synthetic chromosomes is a hierarchical process, in which synthetic oligonucleotides are assembled into larger DNA segments in a stepwise manner using *in vitro* and *in vivo* assembly methods.

synthetic enhancers, cell-type-directed design of: Transcriptional enhancers act as docking stations for combinations of transcription factors and thereby regulate spatio-temporal activation of their target genes. I.I. Taskiran *et al.* (2024) show that deep learning models, can be used to efficiently design synthetic, cell-type-specific enhancers, starting from random sequences, and that this optimization process allows detailed tracing of enhancer features at single-nucleotide resolution. They characterize enhancer codes through the strength, combination and arrangement of transcription factor activator and transcription factor repressor motifs. They successfully design human enhancers, which adhere to enhancer rules similar to those of *Drosophila* enhancers. Enhancer design guided by deep learning leads to better understanding of how enhancers work and shows that their code can be exploited to manipulate cell states.

synthetic gene: A gene that is synthesized in the laboratory. Nowadays, the gene(s) can be synthesized chemically in the laboratory by joining specific nucleotides as per sequence of gene. For example, a synthetic *Bt* gene is 500 times more effective than the natural *Bt* genes from *B. thuringiensis.* During chemical synthesis bases are closer to prokaryotes or eukaryotes. *See* GeneDesign.

synthetic genetic circuit: A circuit that consists of genetic components optimized to function in their natural context, not in the context of the synthetic circuit. Such a circuit is constructed to regulate a cellular function.

synthetic genome: D.G. Gibson *et al.* (2008) synthesized a 582,970-bp *Mycoplasma genitalium* genome. To identify the genome as synthetic, they inserted 'watermarks' at intergenic sites known to tolerate transposon insertions. Overlapping 'cassettes' of 5–7 kb assembled from chemically synthesized oligonucleotides were joined by *in vitro* recombination. The complete synthetic genome was assembled by transformation-associated recombination cloning in yeast, then isolated and sequenced.

synthetic intelligence (SI): An alternative term for artificial intelligence that emphasizes that the intelligence of machines need not be an imitation or in any way artificial; it can be a genuine form of intelligence.

synthetic lethal: Lethal chromosomes derived from normally viable chromosomes by recombination, as a result of crossing-over.

synthetic lethality: A condition in which a combination of mutations in two or more genes leads to the death of a cell or organism, whereas each mutation does not.

synthetic mammalian oscillator: A synthetic mammalian oscillator based on an auto-regulated sense-antisense transcription control circuit encoding a positive and a time-delayed negative feedback loop, enabling autonomous, self-sustained and tuneable oscillatory gene expression.

synthetic medium: A chemically defined substrate upon which microorganisms are grown.

synthetic peptides: Short peptides (2 to 30-amino acid long) synthesized through chemical reactions in the laboratory.

synthetic phase: A stage in interphase during which DNA synthesis occurs. At the end of this phase, the amount of DNA doubles. *Aka* S phase.

synthetic plasmid: A circular DNA molecule that is constructed from a replicon and one or two gene sequences. The synthetic plasmids have a genetic makeup that differs from wild-type plasmids. They are used as cloning vehicles.

synthetic polyribonucleotides: RNA made *in vitro* without a nucleic acid template.

synthetic populations: Populations derived from inter se crossing in the diallel fashion of a number (generally 4–8) of inbreds of known high per se performance and combining ability and subsequently maintained by open population. Composites are also synthetic populations but involve any number and any kind of parental components – inbreds, varieties, single crosses, double-crosses and so on – of known performance.

synthetic seed: Somatic embryos or shoot buds encapsulated in a synthetic seedcoat, which is capable of producing normal plants. Synthetic seeds serve an efficient delivery system for micropropagation. The somatic embryos or shoot buds are encapsulated in calcium alginate beads enriched with nutrient medium. Synthetic seeds are then given to growers instead of the soil-grown plants that require more labour, space and transport. *See* artificial seed.

synthetic shuffling: A variation of the original family (DNA) shuffling format that does not require naturally occurring segments of homology for recombination.

synthetic theory of evolution: The 20th-century conception of evolution is caused by a number of complex and often interacting processes. Th. Dobzhansky published *Genetics and Origin of Species* in 1937. This is essentially a combination of Darwin's concept of natural selection, Mendel's basic genetics, reproductive isolation along with the facts and theories of population genetics and molecular biology.

synthetic vaccines: The vaccines that do not contain intact viruses or complete polypeptides but merely small peptides that have been synthesized in the laboratory to mimic a very small region of the outer coat of the virus. These peptides elicit antibodies capable of neutralizing the virus. *Aka* SYNTHETIC PEPTIDE VACCINES.

synthetic variety: A variety that originates as a known mixture of cross-fertilizing genotypes; advanced generations of open-pollinated seed mixtures of a group of strains, clones or inbreds among them, maintained by open pollination for a specific number of generations. The component units are propagated and the synthetics are constituted at regular intervals.

synthetic yeast genome: A yeast strain, synthesized by Y. Zhao *et al.* (2023), D. Schindler *et al.* (2023) and G.J. Taghon and E.A. Strychalski (2023) with over 50% synthetic DNA replicates and is able to bud normally, splitting into two cells. Standard brewer's yeast (*Saccharomyces cerevisiae*) stores its genetic blueprint across 16 chromosomes. In the new strain, 6.5 of those chromosomes were edited and synthesized in the laboratory – and an extra one was stitched together from edited bits of the yeast's genetic code.

synthetics: Populations comprising of seed mixture of genetically known strains, clones, inbreds or hybrids among them.

system console: The system console, computer console, root console, operator's console or simply console is the text entry and display device for system administration messages, particularly those from the BIOS or boot loader, the kernel, from the init (short for initialization) system and the system logger. It is a physical device comprising of a keyboard and a screen, and traditionally is a text terminal, but may also be a graphical terminal. System consoles are generalized to computer terminals, which are abstracted, respectively, by virtual consoles and terminal emulators. Today communication with system consoles is generally done abstractly, via the standard streams (stdin, stdout and stderr), but there may be system-specific interfaces, for example, those used by the system kernel.

systematic identification of common variants in genes: The approach is to get gene variants sequenced of 100 random individuals and move from family-based linkage analyses. Testing disease susceptibility against all common variants a well-characterized clinical group with a comprehensive DNA array does this simultaneously by genotyping. This is done by characterizing SNPs associated with a given human disease. Use of DNA array technology for human SNP is already under way. This aids in the identification of susceptibility genes.

systematic mutation: Mutation of major effect presumed to give rise to new species or higher categories at a single step.

systematic pressure: One of the non-random evolutionary pressures – selection, mutation or migration.

systematic sampling: A method of selecting a sample by a systematic method as opposed to random sampling, for example, selecting each third plant from a sample.

systematics: An extension of taxonomy, it is the scientific classification of living organisms. The scientific study of biodiversity. It is concerned with the discovery and naming of new species of organisms (nomenclature) and arranging these taxa into classification schemes (taxonomy). A large part of modern systematics is concerned with understanding the evolutionary relationships between various taxa (phylogenetics) using methods of comparative biology (e.g. physiology, behaviour, biochemistry, morphology, genetics) and statistical analysis.

systemic: Pertaining to a disease in which the pathogen (or a single infection) spreads generally throughout the plant; pertaining to chemicals that spread internally through the plant.

systemic acquired resistance (SAR): A mechanism of induced defence that confers long-lasting protection against a broad spectrum of microorganisms. SAR requires the signal molecule SA and is associated with the accumulation of pathogenesis-related proteins, which are thought to contribute to resistance. The induction of disease resistance in areas of a plant remote from the site of initial infection; it is an important element in plants' strategies to protect themselves from disease attack. The gene *DIR1* is essential to the SAR response. *See* INDUCED SYSTEMIC RESISTANCE.

systemic fungicide: A fungicide that is absorbed into plant tissue and may offer some curative or after-infection activity; includes locally systemic fungicides, xylem-mobile (upwards moving) and amphimobile (move in phloem upwards as well as downwards in the plant). *Contrasts* CONTACT FUNGICIDE *and* PROTECTANT FUNGICIDE.

systemic infection: Type of infection in which a pathogen spreads throughout the host body. In plants, viral diseases, in general, are systemic in nature where the virus particles spread throughout the plant body. Such kinds of diseases cause heavy losses and are difficult to control. Meristematic tissues usually remain free even from such diseases. Therefore, meristem tip culture in plants helps in producing disease-free plants.

systemic pesticide: An agent that is distributed throughout the plant. It protects the entire host against pests for a certain time.

systems: A system as a whole that is composed of parts arranged in an orderly manner according to some scheme or plan. In science, systems involve matter, energy and information that move through defined pathways. In animals, the digestive system, reproductive system and circulatory system are some of the examples.

systems biology: (i) A branch of biology concerned with the computational and mathematical analysis of complex biological systems. (ii) It is an interdisciplinary field, which combines elements of systems theory and applied mathematics with theoretical biology, with a primary aim to discover and model the emergent properties of interacting biological entities. (iii) The simultaneous study of complex interactions of multiple levels of biological information including DNA, RNA, proteins and biochemicals.

systems biology approach: This approach aims to identify metabolic switches and transcriptional regulators and reveal the intrinsic biological networks that occur within cells. Mutants and metabolically altered plants are analysed using transcriptomics, proteomics and metabolomics to understand how different intrinsic networks are influenced. This systems view gives the ease of integration of various approaches to understand their structure and function. While the traditional approach is experimental, bioinformatics offers a synthetic approach. The synthetic approach is much faster and in certain cases, is as reliable as the experimental approach. It's in stark contrast to decades of reductionist biology, which involves taking the pieces apart.

systems increasing homozygosity in a breeding population: Mating systems increasing homozygosity in a breeding population are inbreeding, genetic drift and selection pressure.

t: (i) The ratio of observed deviation to its estimated standard deviation. (ii) Symbol for translocation.

T antigen: An enzyme encoded by the mammalian virus SV40, which is essential for viral replication. It is also essential for cellular transformation and was named T for transforming antigens.

T cell: A type of lymphocyte that carries the T-cell receptor (TcR) and is responsible for cell-mediated immunity. T cells act against foreign cells/tissues/organisms. These cells use antigen receptors to recognize and destroy the foreign body. These cells are produced by the thymus gland. Further, the T cells have four types: helper T (T_H) cells, cytotoxic T (T_C), delayed hypersensitivity T (T_D) cells and suppressor T (T_S) cells.

T chromosome: A chromosome in which a terminal (T) region shows neocentric activity.

T I, T II: Symbols, respectively, for telophase of first (T I) and second (T II) meiotic division.

T phages: T phages (T1–T7) are virulent phages that contain double-stranded DNA and affect its antigenic properties.

T-cell receptor genes: T-cell receptor genes synthesize T-cell receptors and are more complex in structure than immunoglobulin genes.

T-cell receptors (TcRs): Surface receptors of T cells that enable the T cells to recognize host cells that have been infected. Proteins present on cell membrane that bind to special molecules to facilitate their entry into the cell.

T-DNA: Complete Ti plasmid is not found in plant tumour cells, only 20kb DNA of Ti plasmid called T-DNA is found integrated into plant nuclear DNA. Genes of T-DNA are in eukaryotic in origin and have been captured by Ti plasmid during evolution.

T-DNA transfer intermediate: T-DNA transfer intermediate can be a DOUBLE-STRANDED CIRCULAR, DOUBLE-STRANDED LINEAR or SINGLE-STRANDED LINEAR MOLECULE. However, there are arguments against circular or linear double-stranded T-DNA.

T-DNA transfer mechanism: There are seven steps in the transfer of the T-DNA molecule from *Agrobacterium* to the plant cell: (i) recognition of a susceptible plant cell; (ii) induction of *vir*-gene expression; (iii) production of a transferable T-DNA copy; (iv) transfer of T-DNA complex through the bacterial; (v) transfer of T-DNA complex through the bacterial and the plant cell membrane; (vi) transfer of T-DNA complex through the plant cytoplasm and through the nuclear membrane and (vii) stable integration of the T-DNA element as a linear fragment into the plant nuclear genome.

t-score: A t-score in a t-distribution refers to the number of standard deviations a sample is away from the average.

t-SNARE: *See* SOLUBLE N-ETHYLMALEIMIDE-SENSITIVE FACTOR ACTIVATING PROTEIN RECEPTORS (SNAP RECEPTORS),

t-test: A test of statistical significance that is used for comparing two means when the sample size is small (less than 30). It is of two types – (i) STUDENT'S T-TEST is used with paired observations and (ii) FISHER'S T-TEST is used when observations are not paired.

T_0 generation: T_0 generation is the term used for a population of transgenic plants directly regenerated from tissue cultures. T_0 plants possess transgene in hemizygous condition and are being grown in the transgenic glass house till maturity for molecular characterization.

T_1 generation: In plant genetics, this term refers to the progeny resulting from self-pollination of the primary transformant regenerated from tissue culture. This is a segregating population of transgenic plants developed after selfing the T_0 plants. T_1 generation is also grown in the transgenic glasshouse. With respect to transgene, this population is comprised of three kinds of plants, viz., homozygous, hemizygous and null types. Further, selfing helps in isolating true breeding lines.

T4 phage: A type of bacteriophage used as a source of commonly used ligase, DNA polymerase and polynucleotide kinase.

T4 polynucleotide kinase: It catalyses the transfer and exchange of inorganic phosphate (Pi) from the γ position of ATP to the 5′-hydroxyl terminus of polynucleotides (double- and single-stranded DNA and RNA) and nucleoside 3′-monophosphates. Polynucleotide kinase also catalyses the removal of 3′-phosphoryl groups from 3′-phosphoryl polynucleotides, deoxynucleoside 3′-monophosphates and deoxynucleoside 3′-diphosphates.

T7 RNA polymerase: A bacteriophage RNA polymerase that is generally used to transcribe plasmid DNA into RNA. The plasmid must contain a T7 promoter upstream of the relevant sequence.

tachytelic evolution: An accelerated evolution that typically occurs when an organism enters a new adaptive zone and initiates adaptive radiation to fill the available niches. Tachytelic evolution at a much faster rate than HOROTELIC EVOLUTION. *See* HOROTELIC EVOLUTION.

TACTAAC box: A consensus DNA sequence in eukaryotic genes surrounding the lariat branch point of eukaryotic pre-mRNA, which has a role in intron removal.

tag single-nucleotide polymorphism (tagSNP): Any one of two (or more) strongly associated (i.e. commonly inherited) SINGLE-NUCLEOTIDE POLYMORPHISM (SNP) loci that are characteristic of a specific haplotype. tagSNPs distinguish different haplotypes segregating in a population.

tail: Single-stranded DNA extension added by terminal deoxynucleotidyl transferase.

tailing: Non-transcriptive addition of a homopolymeric nucleotide to the 3′ end of a restriction fragment by use of enzyme terminal transferase.

tailored chloroplasts: The ability to transform chloroplasts by homologous transformation and drive high-level expression of transgenes in chloroplasts, coupled with their maternal mode of inheritance in most species of interest, make chloroplasts a prime target for biotechnological improvement of crop plants. However, commercial varieties harbouring transgenic chloroplasts have not been generated yet. One promising target for modifying chloroplast functions is photorespiration.

tandem DNA repeats: A series of repeated sequences, arrayed next to each other.

tandem duplication: A duplication mutation in which the duplicated DNA is found next to the original DNA.

tandem genes: Genes that are arranged tandemly in the chromosomes, especially multicopy genes where copies are tandemly duplicated. *See* TANDEM DUPLICATION.

tandem mass spectrometry (MS/MS): Tandem analysis is achieved in three steps: (i) Selection of the oligonucleotide of interest by setting a mass window such that all oligonucleotides of different masses are omitted from the mix. (ii) Fragmentation of the selected oligonucleotide by collision with an inert gas (e.g. argon). (iii) Mass analysis of resultant fragments followed by determination of the RNA oligonucleotide sequence and modification sites.

tandem repeats: A chromosomal mutation in which several copies of chromosome segments lie adjacent to each other, with the same gene order. The DNA, which codes for the rRNA, contains many tandem repeats. *Aka* TANDEM ARRAYS. *See* PHYSICAL MAP.

tandem repeats, types of: Tandem repeats can be of three types – (i) perfect tandem repeats, (ii) imperfect (interrupted by several nonrepeat nucleotides) and (iii) compound repeats. They are well-distributed throughout a genome. MICROSATELLITES are *aka* SIMPLE-SEQUENCE REPEATS (SSRs), SIMPLE TANDEM REPEATS (STRs) and SIMPLE SEQUENCES (SSs).

tandem satellites: Two short segments of chromosomes separated from each other by a constriction and from the main body of the chromosome by a second constriction.

tandem selection: In the case of successive multiple selection cycles, the selection concerns other traits in the first few generations than in later generations.

tandemly repeated DNA: A tandem repeat is a segment of DNA containing a collection of similar or identical tandemly repeating units that are adjacent to each other with no intervening DNA. They comprise 10–15% of mammalian genomes. Tandemly repeated DNA consists of rows of many copies of the same sequence. The repeated unit ranges from 1 to 2000 bp in length. Often the repeat is less than 10 bp and is referred to as SSR DNA or SATELLITE DNA (due to centrifugation 'satellite' bands). Centromeres and telomeres are rich in SSR DNA. At a given site, the amount of SSR DNA may vary greatly. In DNA minisatellites, the satellites may vary between 100 and 100,000 bp in length.

tandon: A multiple-copy tandemly repeated gene.

tapetal cell layer: The innermost nutritive cell layer in the anther, which surrounds the developing POLLEN MOTHER CELLS (PMCs) and/or microspores supplying nutrition and enzymes required for microsporogenesis and pollen maturation.

tapetum: A layer of cells, rich in nutrients, which surrounds the spore mother cells. In case of higher plants, the pollens are enclosed in the tapetal layer of cells in the anther. These cells are somatic cells (mother type) and it is also called nurse layer, which feeds the developing pollen. Tapetum can be degenerated by introduction of *barnase* gene into plants through transformation. Thus, developed transgenic plants exhibit male sterility and are used in hybrid seed production.

tapping mode atomic force microscope (TM AFM): The probe tip is oscillated at a fixed frequency, close to the resonance frequency and scanned. Amplitude of the oscillation changes with the topography of the sample, which is used for imaging. TM AFM has two tip sample interaction regimes, ATTRACTIVE TM AFM and REPULSIVE TM AFM. *See details as separate terms.*

taproot: The primary single dominant plunging root of a plant from which smaller lateral roots branch out.

***Taq* DNA ligase:** Enzyme catalyses the formation of a phosphodiester bond between juxtaposed 5′-phosphate and 3′-hydroxyl termini of two adjacent oligonucleotides, which are hybridized to a complementary target DNA. The ligation will occur only if the oligonucleotides are perfectly paired to the complementary target DNA and have no gaps between them; therefore, a single-base substitution can be detected. *Taq* DNA ligase is active at elevated temperatures (45–65°C)

***Taq* DNA polymerase:** *Taq* DNA polymerase is an 832-amino acid single subunit enzyme with a MW of 94 kDa. It was first isolated from the thermophilic bacteria *Thermus aquaticus*. It catalyses the incorporation of dNTPs into DNA and requires a DNA template, a primer terminus and the divalent cation Mg^{2+}. It contains a polymerization dependent 5′-3′ exonuclease activity. It does not have a 3′-5′ exonuclease and thus no proofreading function. Despite this, the enzyme synthesizes DNA *in vitro* with reasonable fidelity. Commonly used to catalyse PCR reactions due to its heat resistance repetitively through high temperatures during the denaturation step. It can tolerate the high-temperature denaturation step and is stable throughout all the reaction cycles of PCR.

TaqMan assay: Genotyping assay used to amplify and detect specific alleles in genomic DNA. In this assay, an oligonucleotide probe is labelled with fluorescent reporter molecule at 5′-end and a quencher molecule at 3′-end. This probe is called TAQMAN PROBE because it is denatured by endonuclease activity of *Taq* polymerase. After the digestion of PCR, the reporter is released, resulting in a rise in fluorescent signal. However, when due to the presence of an SNP, the probe mismatches with the template leading to a failure in duplex formation, no such degradation occurs, no reporter is released and no rise in fluorescent signals occurs. Different combinations of reporters and quenchers will permit multiplexing so that as many as six SNPs are scored in a single PCR reaction.

taraxum type: Diplospory where the spore mother cell enters the meiotic prophase, but because of synapsis there is no pairing and the univalents remain scattered over the whole spindle. the first meiosis results in a restitution nucleus. The second meiosis results in an unreduced dyad.

target: A protein or nucleic acid (DNA or RNA) whose activity can be modified by an external stimulus, such as a drug or another biological molecule (e.g. hormone, antibody and neurotransmitter).

target identification: The process of finding cellular macromolecules that bind to a small molecule or peptide.

target locus: Genomic target of the gRNA. The sequence includes the nearly 20 unique base target specified by the gRNA plus the genomic PROTOSPACER ADJACENT MOTIF (PAM) sequence.

target nucleic acid: DNA or RNA to be hybridized with the labelled probe.

target population: The entire group of individuals a breeder is interested in. The group about which a researcher wishes to draw conclusions by using several methods.

target region amplification polymorphism (TRAP) marker: A novel marker technique for plant genotyping. A rapid and efficient PCR-based technique was developed, which uses bioinformatics tools and EXPRESSED-SEQUENCE TAG (EST) database to generate polymorphic markers around targeted candidate gene sequences. This technique uses two primers to generate markers. One of the primers, the FIXED PRIMER, is designed from the targeted EST/gene sequence in the database. The second primer, the ARBITRARY PRIMER, is an arbitrary sequence with either an AT- or GC-rich core to anneal with an intron or exon, respectively. The TRAP technique is valuable in genotyping germplasm collections and in tagging genes governing desirable agronomic traits of crop plants.

target sequence: Genomic target of the gRNA targeting sequence; the 20 nucleotides that are incorporated into the gRNA plus the PAM sequence; not to be confused with the gRNA targeting sequence.

target site: Restriction endonucleases bind to DNA at a specific sequence known as their RECOGNITION SITE and cut the DNA within the sequence known as the target site. For type II restriction endonuclease, the kind most commonly used in genetic engineering experiments, the recognition site and the target site are one and the same thing.

target site duplication: A sequence of DNA that is duplicated when a transposable element inserts, usually found at each end of the insertion.

target theory: A theory that predicts response curves based on the number of events required to cause the mutational phenomenon in a gene. This theory was used to estimate size of the gene.

target tissue: Tissue that is used for genetic manipulation using *Agrobacterium* and 'particle gun' methods. Any tissue which can be regenerated into plants can be used as target tissue. In general, the embryogenic calli or meristems that can be easily regenerated into plants, are being commonly and widely used as target tissues. These tissues are treated with *Agrobacterium* or bombarded using the 'particle gun' method. Further, these tissues are grown on a selection medium and finally transferred to a regeneration medium for obtaining complete plants.

target-directed miRNA degradation (TDMD) model: Micro RNAs help to regulate many genes in animal cells. Each miRNA associates with an AGO protein, forming a complex in which the miRNA pairs with a MESSENGER RNA (mRNA) target and AGO recruits the factors that accelerate mRNA decay.

target-ligand interaction screening: An HTS methodology utilized to screen a large number of candidates (e.g. compounds) based upon their interaction (e.g. chemical 'binding') to a preselected 'target' (e.g. molecule present within a cell membrane, molecule placed on a biochip or other bioassay to facilitate HTS, a molecule present on the surface of a nematode utilized in HTS, etc.).

targeted epigenome editing: For accomplishing this task, much effort is needed before successfully using epigenome editing technology for crop improvement: (a) The scientist has to ensure that the trait to be improved is epigenetically inherited. (b) Genome and epigenome maps of the plant species in question should be available. (c) Particular epigenetic change(s) must be unambiguously related with phenotypic change. (d) The target site in the genome has to be carefully selected to ensure the desired epigenetic change. Then comes the final decision of selecting, designing and constructing the epigenome editing tool.

targeted gene modification: In this method of mutation induction, genes are introduced by any one of the traditional gene transfer methods. This is followed by site-specific mutations.

targeted gene replacement: In this method of gene replacement, homologous recombination replaces a resident gene with a foreign gene.

targeted gene transfer: The gene transfer at homologous sites in the host genome.

targeted mutagenesis: Deliberate change in the genetic structure directed at a specific site on the chromosome. Used in research to determine the targeted region's function. *See* MUTATION *and* POLYMORPHISM.

targeted RNA delivery into mitochondria: It has been shown that the tRNA import machinery from the *Leishmania* can be engineered on to human mitochondria and the mitochondrial translational defects can be nullified to some extent. RNA can be genetically engineered to be targeted into the mitochondria to nullify the tRNA mutations that lead to human disorders.

targeted RNA sequencing: Analyse gene expression in a focused set of genes of interest. Targeted RNA-Seq can be achieved via either enrichment or amplicon-based approaches.

targeted sequencing: Targeted sequencing with HiFi reads allows you to sequence only the genomic regions you select easily and cost-effectively at scale. Hybrid capture and amplicon workflows are used, giving all the benefits you expect from HiFi reads including accurate haplotype resolution and comprehensive detection of all variant types. The length and accuracy of HiFi reads allow you to access genomic dark regions such as areas of high homology, GC-rich and repetitive regions and to resolve complex gene families like HLA and pharmacogenes like *CYP2D6*.

targeting of proteins into nucleus: Yeast nucleus contains a discrete set of proteins that are synthesized in cytoplasm. A segment of *MAT*α*-2* gene product, which was 13 amino acid long and was sufficient to localized β-galactosidase activity in the nucleus.

targeting-induced local lesions in genomes (TILLING): A technique to screen a population that has either been deliberately mutagenized or possesses natural biodiversity (Eco-TILLING) to identify those plants with a change in a specific gene of interest. This technique uses high-throughput methods convenient to plant breeders, including a reverse genetics method that combines random chemical mutagenesis with PCR-based screening of gene regions of interest. TILLING is an influential reverse genetics method in plant functional genomics and breeding to identify mutagenized individuals with improved behaviour for a trait of interest. Two variations of TILLING are Eco-TILLING and double-stranded Eco-TILLING (DEco-TILLING). It provides a range of allele types, including missense and knockout mutations. By comparing the phenotypes of isogenic genotypes differing in single sequence motifs, TILLING provides direct proof of the function of both induced and natural polymorphisms without the use of transgenic modifications.

targetome: An ANTISENSE OLIGONUCLEOTIDE (ASO) consisting of a target-site-specific domain and an ADAR-recruiting domain directs the adenosine deaminase acting on RNA (ADAR) to a target site in the transcriptome, leading to site-specific A-to-I editing of the target RNA.

targets: Biomolecules that can bind active agents are called targets. They can be receptors, enzymes or ion channels. If agent and target interact with each other, the term agent-target-specific effect is used. The identification of targets is very important in biomedical and pharmaceutical research because a specific interaction can help to understand basic biomolecular processes. This is essential to identify new points of application.

TAT: A protein that functions as the main activator of HIV gene expression in cells. This protein complexes with TAR (a 60-nucleotide sequence found in all viral messenger ribonucleic acid) to mediate synthesis of proteins (in an infected cell) necessary for HIV to reproduce.

TATA box: In molecular biology, the TATA box consisting of six nucleotides TATAAT (also called the GOLDBERG-HOGNESS BOX) is a sequence of DNA found in the core promoter region of genes in archaea and eukaryotes. The bacterial homologue of the TATA box is called the Pribnow box, discovered by D. Pribnow in 1975 in promoters of prokaryotic genes, which has a shorter consensus sequence. TATA box is found in both prokaryotes and eukaryotes; however, in the two it is spacing from the transcriptional start site that differs. The TATA box is a common promoter sequence recognized by RNA polymerase II. *See* PRIBNOW BOX.

TATA homology: An adenine-thymidine-rich (gene) sequence present 20–30 nucleotides 'upstream' of the TRANSCRIPTION START SITE (TSS) on most eukaryotic protein coding genes. It is required for correct expression of the gene. Blocking this sequence of the gene may inhibit ability of the AIDS virus to reproduce.

TATA-binding protein (TBP): TBP is essential for all transcription. It detects in eukaryotic cell nuclei and eukaryotes are fundamentally homologous.

tautomer: An isomer of an organic substance, differing from other forms of the substance in the position of a hydrogen atom and a double bond.

tautomeric shift: Reversible shift of proton position in a molecule. Based on nucleic acids shift between keto and enol forms or between amino and imino forms.

tautomerism: The presence of two or more chemical compounds that are capable of facile interconversion, in many cases merely exchanging a hydrogen atom between two other atoms, to either of which it forms a covalent bond. Tautomeric bases show unusual pairing. Tautomers are rare and short lived.

tautomerization: The spontaneous isomerization of a nitrogen base from its normal keto (or amino) form to an alternative hydrogen-bonding enol (or imino) form.

taxa (sing. **taxon**)**:** A unit of classification (e.g. species, genus). Separate sources of sequences are referred to as *taxa* defined as phylogenetically distinct units on the tree. The tree is composed of nodes representing the taxa and the branches representing the relationship among the taxa. The lengths of branches are often drawn proportional to the number of sequence changes in the branch and hence can represent the divergence. A CLADE is a monophyletic taxon.

taxis: A change of direction of locomotion in a motile cell, made in response to certain types of external stimulus (e.g. temperature, light and nutrients). *See* CHEMOTAXIS.

taxonomy (Adj. taxonomic): Taxonomy is from the Greek for orderly arrangement. In biology, taxonomy is the scientific study of naming, defining (circumscribing) and classifying groups of biological organisms based on shared characteristics. Organisms are grouped into *taxa* (sing. *taxon*) and these groups are given a taxonomic rank; groups of a given rank can be aggregated to form a more inclusive group of higher rank, thus creating a taxonomic hierarchy. The principal ranks in modern use are domain, kingdom, phylum (division is sometimes used in botany in place of phylum), class, order, family, genus and species. It also helps to identify an unknown organism and then to include the new organism in the group of other organisms that have similar characteristics. Nowadays, MOLECULAR TAXONOMY is an upcoming field for precise identification and classification of the organisms. *Aka* SYSTEMATICS.

TCA cycle: *See* CITRIC ACID CYCLE *and* KREBS CYCLE.

technology protection system: *See* GENETIC USE RESTRICTION TECHNOLOGY (GURT).

tegmen: The inner layer of the testa (seedcoat). It develops from the inner integument of the ovule.

TEL: A DNA segment containing telomeric sequences. TEL DNA is required for replication of linear chromosomes. TEL DNA is highly conserved among divergent species. TEL sequence in yeast is $(C_{1-3}A)_n$, in Tetrahymena $(C_4A_2)_n$ and in mammals, birds and reptiles $(C_3T_2)_n$.

teleomorph: The sexual form in the life cycle of a fungus. *See* PERFECT STATE. *Contrasts* ANAMORPH, HOLOMORPH *and* IMPERFECT STATE.

telial host: Primary host in heteroecious rusts. The host plant on which a heteroecious rust fungus produces telia (spore-producing body of some rust fungi) and, sometimes, uredinia (a usually reddish or black mass of hyphae and spores of a rust fungus forming pustules that rupture the host's cuticle). *Contrasts* AECIAL HOST.

teliospore: A thick-walled resting or overwintering spore made by the rust fungi (Pucciniales) and smut fungi (Ustilaginomycotina) in which karyogamy occurs. It germinates to form a promycelium (basidium) in which meiosis occurs. *Aka* TELEUTOSPORE *or* TELEUTOSPORODESM.

telium (pl. telia): A fruiting body (sorus) of a rust fungus that makes teliospores.

telocentric chromosome: A chromosome with the centromere located very close to one end. *See* TELOCHROMOSOME.

telochromomere: A chromomere that is terminally located.

telochromosome (Abbrev. telosome): (i) A chromosome with a terminally located centromere. (ii) The length of telomeric DNA is maintained by the enzyme telomerase, but in addition, six telomere-associated proteins – TRF1, TRF2, POT1, RAP1, TIN2 and TPP1 in mammalian cells – have been shown to form a complex known as the TELOSOME or SHELTERIN COMPLEX, which is essential for telomere function.

teloisodisomic: A cell or individual in which one chromosome pair is lacking but telocentric for one arm and the missing pair and an isochromosome for the same arm occur.

teloisotrisomic: An individual lacking in one chromosome but which has a telocentric chromosome and an isochromosome for the same arm of the missing chromosome.

telomerase: A reverse transcriptase contains an RNA molecule that acts as the template for the tandem repeat at the telomere. It synthesizes the telomere to maintain its length after each cell division. Telomerase extends old 3′-end of G-rich strand, offering a longer template for the new 5′-end and adds specific DNA sequence repeats ('TTAGGG' in all vertebrates) 100 to 1000 times to the 3′-end of DNA strands in the telomere regions at the ends of chromosomes. It is active in young cells and gametes, inactive in differentiated somatic cells and reactivated in malignant cells. Telomerase can add one base at a time to the telomeric end of a chromosome. This maintenance work is required for cells to escape from replicative senescence. Thus, telomerase is responsible for telomere maintenance. Most somatic cells have little or no telomerase and the number of divisions they can undergo is limited. *Aka* TELOMERASE TERMINAL TRANSFERASE.

telomerase RNA: The replication of the ends of a linear genome is a difficult task for eukaryotic cells because of the inability of conventional DNA polymerase to replicate linear ends correctly. Eukaryotic telomerase activity is due to a ribonucleoprotein, which consists of an RNA moiety with a secondary structure highly conserved among eukaryotes and of several protein subunits, of which the catalytic one is a conserved reverse transcriptase. The telomerase reverse transcriptase subunit uses the RNA moiety as a primer for the extension of linear DNA during telomerase replication. RNA component of telomerase is 159 nucleotide long. It contains a short sequence complementary to one or more repeats from C-rich strand of telomere. It is used as a template for synthesis of a repeat sequence of G-rich strand. By repeatedly extending the G-rich strand, the shortening effect of DNA replication at the telomere can be counter-balanced. Its RNA component binds to the single-stranded end of the telomere. An associated reverse transcriptase activity is able to maintain the length and structure of telomere.

telomerase RNA component (TERC): An ncRNA found in eukaryotes that is a component of telomerase, the enzyme used to extend telomeres. TERC serves as a template for telomere replication (reverse transcription) by telomerase. Telomerase RNAs differ greatly in sequence and structure between vertebrates, ciliates and yeasts, but they share a 5′ pseudoknot structure close to the template sequence. The vertebrate telomerase RNAs have a 3′ H/ACA snoRNA-like domain.

telomerase RNA processing. Chromosomal ends contain telomeric repeats, which are added to the 3′-end of DNA molecules by telomerase enzyme complex. Telomerase is a ribonucleoprotein as it consists of a REVERSE TRANSCRIPTASE ENZYME and TELOMERASE RNA, as a template for telomere synthesis.

telomere: One of the two terminal chromomeres of a chromosome, associated with a typical repeat DNA sequence TTAGGG that is replicated by ribonucleoprotein telomerase enzyme in a special way. A telomere enacts a part in the spatial orientation of chromosomes within the interphase nucleus. Chromosomes normally lose around 100 bp from the telomere every time the cell divides. This specialized structure is involved in the replication and stability of the end of a linear DNA molecule. *See* DNA REPLICATION.

telomere erosion theory of ageing: Telomeres are special functional complexes at the end of eukaryotic chromosomes. The telomeres get shortened with each cell division like the burning of candlewick. This is due to the end replication problem. This problem arises from the inability of normal mechanisms for replicating chromosomes to do a complete job. When DNA polymerase copies the two parental DNA strands in a chromosome, they leave each daughter strand shortened at one tip. If cells do not compensate for this flaw in replication mechanism, chromosomes will inexorably shrink. Eventually erosion would eliminate telomeres and critical genes in some generation of cells. Telomerase adds DNA to chromosomes at the telomeres before replication begins. This added DNA consists of one or more telomeric subunits. Many human cells lack telomerase. These cells are not able to correct end replication problems and chromosome shorten.

telomere fusion: The end-to-end joining of two chromosomes to constitute one PSEUDODICENTRIC CHROMOSOME by terminal rearrangement. If one centromere of the pseudodicentric is suppressed, telomere fusion may result in a stable chromosome.

telomere mapping: The localization of telomeres on the ends of chromosomes by, for example, fluorescent *in situ* hybridization and the identification of other sequence elements within the telomeric (and also subtelomeric) region. TELOMERES are assemblies consisting of protein and DNA sequences (that do not code for proteins), which are located at the ends of chromosomes.

telomere position effect (TPE): In *Saccharomyces cerevisiae*, TPE can result in the reversible silencing of a gene near a telomere by a mechanism that depends both on telomere length and on the distance to the gene.

telomere proteins: Telomeric DNA end-binding proteins have generally been found to inhibit rather than stimulate the action of the chromosome end-replicating enzyme, telomerase. TPP1 and POT1 form a complex with telomeric DNA that increases the activity and processivity of the human telomerase core enzyme. POT1-TPP1 switches from inhibiting telomerase access to the telomere, as a component of sheltrin, to serving as a processivity factor for telomerase during telomere extension.

telomere repeat binding (TRB) proteins in plant development: Plant-specific transcriptional regulators, called TELOMERE REPEAT BINDING proteins (TRBs), unite two DNA-binding domains, the GH1 domain, which binds to linker DNA and is shared with H1 histones, and the Myb/SANT domain, which precisely recognizes the telobox DNA-binding site motif. TRB1, TRB2 and TRB3 proteins employ Polycomb group complex 2 (PRC2) to deposit H3K27me3 and JMJ14 to eliminate H3K4me3 at gene promoters containing telobox motifs to repress transcription. S. Amiard *et al.* (2024) reveal that TRB4 and TRB5, two related paralogues belonging to a separate TRB clade conserved in spermatophytes, regulate the transcription of several hundred genes involved in developmental responses to environmental signals. TRB4 binds to several thousand sites in the genome, mainly at transcription start sites and promoter regions of transcriptionally active and H3K4me3-marked genes, but, unlike TRB1, it is not enriched at H3K27me3-marked gene bodies. TRB4 can physically interact with the catalytic components of PRC2, SWINGER and CURLY LEAF (CLF). TRB4 and TRB5 are essential for distinctive phenotypic traits detected in *clf* mutant plants and thus function as transcriptional activators of several hundred CLF-controlled genes, including key flowering genes. TRB4 shares multiple target genes with TRB1 and physically and genetically interacts with members of both TRB clades. TRB proteins participate in both positive and negative interactions with other members of the family to regulate plant development through both PRC2-dependent and -independent mechanisms.

telomere replication problem: The DNA synthesis ensues exclusively in 5′ to 3′ direction. Initiation of DNA replication requires a short RNA primer. In linear chromosome, the extreme 5′ end is not replicated. As a result, the newly synthesized chromosomal DNA shortens at the end. If this process occurs enough times, the chromosome length reduces. In all the eukaryotes, one strand of telomeric DNA is rich in G nucleotides while other is rich in C nucleotides. The G-rich strand is always the one that contributes the 3′ end of the DNA molecule, which extends 12–16 nucleotides beyond the terminus of the C-rich strand. Nucleosomes are not present in telomeric regions of chromosomes. The repeat sequences act as binding sites for telomere-specific proteins, such as TRF1 in humans. The bound proteins are presumed to act as a cap. Thus, ends of the chromosomes are prevented from being degraded or fusing with other chromosomes.

telomere terminal transferase: A ribonucleoprotein enzyme that can produce the tandem repeat units of the telomere 3′ end of telomeric primers without template. The RNA component of the enzyme contains the complementary sequence of the telomeric repeats that it synthesizes.

telomeres as epigenetic agents: Telomere position effect is a phenomenon that refers to the ability of mammalian telomeres to silence subtelomeric genes. Telomere shortening in humans is associated with a decrease in TPE while TPE increases with telomere elongation.

telomere-to-telomere (T2T) assembly: Telomere-to-Telomere (T2T) Consortium presents a complete 3.055 billion–base pair sequence of a human genome, T2T-CHM13, that includes gapless assemblies for all chromosomes except Y, corrects errors in the prior references and introduces nearly 200 million bp of sequence containing 1956 gene predictions, 99 of which are predicted to be protein coding.

telomeric association: Fusion of chromosomes by their telomeres, which predisposes a cell to genetic instability.

telomeric chromatin organization: H. Hu *et al.* (2023) revealed the structural basis of telomeric nucleosome recognition by shelterin factor TRF1. Shelterin and nucleosomes are the key players that organize mammalian chromosome ends into the protective telomere caps. Cryo-electron microscopy structures of a human telomeric nucleosome both unbound and bound to the shelterin factor TRF1 revealed that TRF1 binds unwrapped nucleosomal DNA ends by engaging both the nucleosomal DNA and the histone octamer. Unexpectedly, TRF1 binding shifts the register of the nucleosomal DNA by 1 bp. Phosphorylation of the TRF1 C-terminus and a non-canonical DNA-binding surface on TRF1 are critical for its association with telomeric nucleosomes. These insights into shelterin-chromatin interactions have crucial implications for understanding telomeric chromatin organization and other roles of shelterin at telomeres including replication and transcription.

telomeric C-strand: The telomeric DNA strand consisting of CCCTAA repeats which pair with the telomerase-synthesized TTAGGG repeats.

telomeric DNA: Telomeric DNA shows some characteristic features. (a) Telomeric DNA is heterochromatic and contains direct tandemly repeated sequences. In yeast, it has been shown that different strains contain different lengths of telomeres and that the length is under genetic control. (b) Telomeric DNA of the chromosome is to seal the end, making degradation by enzymes less likely. (c) Telomeres carry multiple copies of a simple repeat. The exact sequence of the repeat varies from organism to organism. Most repeat sequences can be described by consensus sequence 5′-$C_{(1–8)}A$ / $T_{(1–4)}$-3′. One version of this consensus sequence 5′-C_3TA_2-3′ is found at telomeres in distantly related organisms, viz. protozoan *Tetrahymena* and all vertebrates tested to date. In some organisms, telomeric repeat is precise (e.g. C_4A_2 for *Trypanosoma* and C_4A_4 for *Oxytricha*). The repeat is more heterogenous in others. (d) Telomeric sequences have precise orientation in that C-rich strand always run 5′®3′ from the end toward the inside of the DNA molecule. Single-base gaps or nicks are present in the C-rich strand. € Amount of telomeric DNA varies from telomere to telomere. Virtually for all organisms the amount of telomeric DNA varies from telomere to telomere. (f) In some organisms, telomere length may vary during development. For example, the length of terminal restriction fragments from human sex chromosomes are 5 kb longer in sperm DNA than in DNA extracted from blood from the same person. (g) Subtelomeric regions of chromosome usually contain middle repetitive elements. In addition to simple repeats found at the very ends of the chromosomes, subtelomeric regions of chromosome usually contain middle repetitive elements called TELOMERE-ASSOCIATED SEQUENCES. They have been studied in great detail in *Saccharomyces*. The internal tracts of $C_{(1-3)}A$ sequences in yeast chromosome could serve as backup telomeres. Many of the features of telomeric DNA are conserved among eukaryotes.

telomeric length: Telomerase is known to preferentially elongate the shortest telomeres in a cell. Short telomeres are elongated more frequently than long telomeres, arguing that telomeres switch between two states, one that allows telomere extension and the one that does not.

telomeric repeat-containing RNA (TERRA): Mammalian telomeres are transcribed into long non-coding RNA, called TERRA. TERRA molecules are heterogenous in length, are transcribed from several subtelomeric loci toward chromosome ends. Suppressors with morphogenetic defects in genitalia (SMG) proteins, which are enriched at telomeres *in vivo*, negatively regulate TERRA association with chromatin and protect chromosome ends from telomere loss. Thus, telomeres are actively transcribed into TERRA and SMG factors represent the molecular link between TERRA regulation and the maintenance of telomere integrity.

telophase: The fourth and final phase of mitosis and the two divisions of meiosis, during which the spindle disappears, nucleoli reappear, the nuclear membrane starts to develop around the two groups of daughter chromosomes and/or chromatids and the chromosomes return to their extended state, in which they are no longer visible. The nuclei then enter a resting stage as they were before the occurrence of the division.

telophase I: It is the stage of meiosis I in which the chromosomes at each pole of spindle elongate and form a nucleus with a nuclear envelope and nucleoli. The spindle and astral rays gradually disappear.

telophase II: The concluding stage of meiosis II, which is characterized by the disappearance of spindle apparatus and reappearance of the nuclei, four in number, which carry haploid number of chromosomes.

telosomic trisomy: The extra chromosome is a telocentric chromosome fragment homologous to one arm of a chromosome pair in the standard complement.

telotertiary compensating trisomic: A compensating trisomic in which the lost chromosome is remunerated for by one telocentric chromosome and one tertiary chromosome.

telotrisomic: In allopolyploids (e.g. hexaploid wheat), a cell or individual with one lost chromosome but having a telocentric and an isochromosome for the same arm of the missing chromosome.

teminism: *See* REVERSE TRANSCRIPTION.

temperate phages: The phages that have a dual existence such that they may either perpetuate through a lytic cycle (lysis) or may take the form of a prophage by integration of their DNA with the DNA of bacterial cell (lysogenic pathway) and phage thus lives in harmony with the bacterial cell.

temperate symbiotic phage: A bacterial virus that tends to take up residence in the host cell without destroying it.

temperature gradient capillary electrophoresis (TGCE): A molecular procedure that is utilized to distinguish heteroduplex from homoduplex DNA molecules and thus be applied to the detection of various types of DNA polymorphisms. Unlike most SINGLE-NUCLEOTIDE POLYMORPHISM (SNP) detection technologies, TGCE can be utilized even in the absence of prior knowledge of the sequences of the underlying polymorphisms. TGCE is both sensitive and reliable in detecting SNPs, small Indel (insertion/deletion) polymorphisms (IDPs) and SSRs and engaging this technique it is possible to detect a single SNP in amplicons of over 800 bp and 1-bp IDPs in amplicons of approximately 500 bp. Genotyping data obtained via TGCE are consistent with data obtained via gel-based detection technologies. For genetic mapping experiments, TGCE has a number of benefits over alternative heteroduplex-detection technologies, such as celery endonuclease (CELI) and denaturing high-performance liquid chromatography (DHPLC). Multiplexing can enhance TGCE's throughput to 12 markers on 94 recombinant inbreds per day. Given its ability to efficiently and reliably identify a variety of subtle DNA polymorphisms that exist at high frequency in genes, TGCE reveals great promise for discovering polymorphisms and conducting genetic mapping and genotyping experiments.

***temperature-sensitive* (*ts*) mutant:** Organism or cell carrying a genetically modified protein (or RNA molecule) that performs normally at one temperature but is abnormal at another (usually higher) temperature.

temperature-sensitive mutation: An organism with a mutant allele that expresses normal at a permissive temperature but mutant at a restrictive temperature.

template: In general terms, it is a mould or pattern that can be copied or its shape reproduced. When used with reference to molecular biology. For example, a single-stranded DNA, complementary to nascent RNA or DNA strand, serves to specify the nucleotide sequence of the nascent strand. *Aka* TEMPLATE STRAND.

template DNA and its mRNA transcript, relationship between: From a given piece of double-helical DNA, mRNA that will be synthesized will have a predicted sequence. With respect to a particular gene (say *x*), RNA polymerase binds at the promoter. mRNA is synthesized using one strand (called antisense strand in 3′←5′ direction) of DNA as a template. Its complementary strand (called sense strand in 5′→3′ direction) is not used as a template with respect to gene *x* but this strand in opposite 3′←5′ direction may be an antisense strand with respect to some other gene (say *y*). Terms sense and antisense DNA strand are thus defined with reference to a particular gene. A bacterial mRNA often consists of several coding regions each of which has its own initiation codons. These mRNAs may encode for more than one polypeptide chains. Such molecules are termed polycistronic messenger RNAs. Each eukaryotic mRNA usually codes only for a single polypeptide chain. Such molecules are known as UNICISTRONIC MRNAS.

template method pattern: The template method pattern suggests that you break down an algorithm into a series of steps, turn these steps into methods and put a series of calls to these methods inside a single template method.

temporal and spatial regulation of expression of introduced genes: *In vivo* electroporation is a powerful technique for the introduction of genes into organisms. Temporal and spatial regulation of expression of introduced genes, or of RNAi, would further enhance the utility of this method. Conditional regulation of gene expression was demonstrated from electroporated plasmids in the postnatal rat retina and the embryonic mouse brain. For temporal regulation, Cre/loxP-mediated inducible expression vectors were used in combination with a vector expressing a conditionally active form of Cre recombinase, which is activated by 4-hydroxytamoxifen. The onset of gene expression was regulated by the timing of 4-hydroxytamoxifen administration. For spatial regulation, transgenes were expressed by using promoters specific for rod photoreceptors, bipolar cells, amacrine cells, Müller glia or progenitor cells. Combinations of these constructs will facilitate a variety of experiments, including cell-type-specific gene misexpression, conditional RNAi and fate mapping of progenitor and precursor cells.

temporal gene: Any gene that governs the relative expression of structural genes in a particular organ or at a particular developmental time. Temporal genes are divided into two classes – CLASS I TEMPORAL GENES are *trans*-acting that lie at a distance from the structural genes whose expression they influence and CLASS II TEMPORAL GENES are *cis*-acting that lie in close proximity to the structural genes.

temporal gene expression: Time during development when a gene is expressed. Temporal expression of actin genes has been observed during sea urchin development.

temporal sequences of gene expression during phage infection: Regulation of gene expression during the life cycles of virulent bacteriophages is quite unlike from the reversible on-off switches, characteristic of bacterial operons. In phage-infected bacteria, the viral genes are expressed in genetically pre-programmed sequences or cascades. *Bacillus subtilis* phage SP01 exhibits a slightly more complex pathway of sequential gene expression, including three sets of genes – the early, middle and late – in reference to their time of expression during the phage reproductive cycle.

temporary resistance: Vertical resistances are temporary in action as they stop functioning on the appearance of a matching vertical pathotype. They are within the capacity for microevolutionary change of the parasite.

temporary wilt: The wilting due to the occurrence of insufficient soil water from which a plant can recuperate when water is supplied.

tendril: Any slender organ modified from a stem, leaf, leaflet or stipule that is used by climbing plants via clinging to an object.

tenoplast: *Aka* VACUOLAR MEMBRANE.

tentative consensus sequence (TCS): The identification of a sequence from an EST cluster that represents part or all of a complete gene. TCs are usually determined by clustering ESTs allowing for sequencing errors, artefacts such as chimeric clones, and naturally occurring biological phenomena such as alternative splicing. Creation of a cluster allows one to generate a consensus sequence and then identify a long open reading frame, which would suggest the possibility of that consensus representing a bona fide gene.

tentative human consensus sequences (THCS): A consensus sequence generated from human EST fragments. THCs may be validated by comparison against databases of known human gene sequences and human genomic sequences, or by identification of the ORFs or other sequence features contained within the consensus as belonging to a known human gene product.

10T1/2 cell line: A contact-inhibited fibroblastic cell line derived from a mouse embryo that is commonly used for toxicity screening, transfection, transformation, tumourigenicity and cloning studies.

teosinte: A wild plant (*Zea diploperennis*) native to Mexico, which is related to domesticated corn/maize (*Zea mays* L).

teosinte pollen drive: Selfish genetic elements contribute to hybrid incompatibility and bias or 'drive' their own transmission. Chromosomal drive typically functions in asymmetric female meiosis, whereas gene drive is normally post-meiotic and typically found in males. Using single-molecule and single-pollen genome sequencing, B. Berube *et al.* (2024) describe *Teosinte Pollen Drive*, an instance of gene drive in hybrids between maize (*Zea mays* ssp. *mays*) and teosinte *mexicana* (*Zea mays* ssp. *mexicana*) that depends on RNA interference (RNAi). 22-nucleotide small RNAs from a non-coding RNA hairpin in *mexicana* rely on *Dicer-like 2* (*Dcl2*) and target *Teosinte Drive Responder 1* (*Tdr1*), which encodes a lipase needed for pollen viability. *Dcl2*, *Tdr1* and the hairpin are in tight pseudolinkage on chromosome 5, but only when transmitted through the male.

tepal: A segment of a perianth, either sepal or petal that is usually used when all perianth segments are indistinguishable in appearance.

teratogen (adj. teratogenic): Any agent that causes an abnormality following fetal exposure during pregnancy.

teratogenesis: Nongenetic process that leads to the production of phenocopies. *See* PHENOCOPY.

teratogenic: Causing a serious deformity in fetal development.

teratoma: A tumour composed of a chaotic array of different tissue types.

terminal affinity: The property by which chromosomes are held together end to end from diplotene until first metaphase of meiosis by a terminal chiasma or are brought together in this way at metaphase.

terminal association: In meiosis, the achiasmatic (non-chiasmatic) association of homologous chromosomes just by unspecific end-to-end attachments.

terminal CCA: All 3′ end of all tRNAs there is a sequence CCA and to the adenine moiety of this sequence a specific amino acid is attached.

terminal cell differentiation: It means the end product of cell differentiation, for example, nerve cells and skin cells.

terminal centromere: A chromosome showing a terminal centromere. *Aka* T-END *or* TELOSOME.

terminal chiasma: The association of homologous chromosomes at meiosis, involving only the chromosome ends. This may be the result of a crossing-over event at the chromosomal extremity, or of terminalization of chiasma, which earlier were positioned some distance away from the chromosome terminus.

terminal deletion: A deletion involving the end of a chromosome.

terminal deoxynucleotidyl transferase (TdT): A specialized DNA polymerase expressed in immature, pre-B, pre-T lymphoid cells and acute lymphoblastic leukaemia/lymphoma cells. It adds N-nucleotides to the V, D and J exons during antibody gene recombination, enabling the phenomenon of junctional diversity. TdT functions by catalysing the polymerization of deoxynucleoside triphosphate to the 3′-hydroxyl group of a preformed polynucleotide chain. TdT is a non-template-directed DNA polymerase. It was first detected in the thymus gland. Terminal transferase is useful for at least two procedures – labelling the 3′-ends of DNA and adding complementary homopolymeric tails to DNA. *Aka* DNA NUCLEOTIDYLEXOTRANSFERASE (DNTT) *or* TERMINAL TRANSFERASE.

terminal gene: A gene situated at the extreme end of the chromosome telomere.

terminal inverted repeat (TIR): Repeats that flank most DNA transposons and lie in an inverted orientation. *Aka* INVERTED TERMINAL REPEAT (ITR).

terminal redundancy: The repetition of the same sequence at both ends of, for example, a phage genome.

terminal transferase: An enzyme that makes DNA without using a template.

terminalization: Refers to the movement of the chiasma toward the ends of the bivalent during meiosis.

terminase: An enzyme of phage lambda, which produces the staggered cuts at the 'cos' sites during packaging.

termination codon: One of the three codons – UAG, UAA or UGA – found in DNA or mRNA molecules (genes) that cause termination of protein synthesis. Each one of these sequences causes the termination of the peptide chain and its release in free form. *Aka* NONSENSE CODON *or* TERMINATOR SEQUENCE. *See* STOP CODON *and* NONSENSE CODON. *Compare with* START CODON *and* INITIATION CODON.

termination factors (TFs): Proteins required to achieve release of newly synthesized polypeptide chain from tRNA. *Aka* RELEASE FACTORS (RFs).

termination of DNA replication: In bacteria, the termination sites are located between 23 and 29 min (TerA, TerD and TerE), 33 and 36 min (TerB and TerC) and at 48 min (TerF). These sites are spread over a long distance (1 min is ~50 kb). Termination site has been hypothesized to be T-shaped. Replication forks meeting at the top of the T are arrested (that is, the clockwise fork will pass through sites TerE, TerD and TerA, but will stop at TerC, TerB or TerF). A protein called Tus binds to the Ter sites, and this binding stops DnaB (helicase) action.

termination of transcription in prokaryotes: Termination (as well as initiation) of transcription requires breaking of hydrogen bonds and additional proteins to interact with core enzyme. Accountability for termination lies with the sequences already transcribed by RNA polymerase. With regard to requirement of rho (ρ) factors in termination, terminators are of two types – rho-independent and rho-dependent. Rho-independent or simple terminator core enzyme can terminate *in vitro* at certain sites in the absence of any other factor. These terminators have two structural features, a hairpin in secondary structure and a run of U's at the very end of the unit. Both features are required for termination. The hairpins contain a G-C-rich region near base of the stem. Probably all hairpins that form in the RNA product cause the polymerase to slow or pause in RNA synthesis. Pausing creates an opportunity for termination to occur. Rho-independent terminators include palindromic regions that form hairpins of varying lengths from 7 to 20 bp. The stem-loop structure is followed by a run of U residues. The string of U residues probably provides the signal that enables RNA polymerase to dissociate from the template when it pauses at the hairpin. In case of rho-dependent terminators, there is a need for addition of rho (ρ) factor for termination *in vitro*. Rho factor is needed for termination *in vivo* also. Most of the known rho-dependent terminators are found in phage genomes.

termination of translation in prokaryotes: As polypeptide chain grows, it remains linked covalently to tRNA and bound to the ribosome. When complete, polypeptide is released from both the components. When termination signal (codon UAA, UGA or UAG) is reached at A-site of the 50S subunit, chain terminates. Release factors (RF1, RF2 and RF3), GTP and nonsense codon(s) are required at this step. RF1 is needed for UAG and RF2 for UGA. UAA can accept either RF1 or RF2. RF3 activates RF1 and RF2.

termination sequence: *See* TERMINATOR.

terminator: A sequence downstream from the 3′ end of an open reading frame that serves to halt transcription by the RNA polymerase. In bacteria these are commonly sequences that are palindromic and thus capable of forming hairpins. Many prokaryotic and eukaryotic terminators necessitate a hairpin to build in the secondary structure of the RNA being transcribed. This shows that termination depends on the RNA product and is not determined simply by scrutiny of the DNA sequence during transcription. At some terminators, termination event can be prevented by specific ancillary sequences (factors) that interact with RNA polymerase. Sometimes termination needs the action of a protein, such as rho (ρ) factor in *Escherichia coli*. *Aka* TERMINATOR SEQUENCE. *See* nopaline synthase (NOS) TERMINATOR.

terminator gene: A specific genetic sequence inserted by scientists into a seed's DNA that renders the seed and the crop it produces sterile. A *ribosomal inhibitor protein* (*RIP*) gene is the lethal gene, the SAPONIN 4 RIP being particularly preferred. RIP directly interferes in the expression of all proteins in a plant cell, without being toxic to other organisms. Expression of RIP in the cells of the embryo would be entirely preventing germination of the seed. *Aka* TECHNOLOGY PROTECTION SYSTEM. *See* TERMINATOR SEEDS.

terminator seeds: A descriptive term used for seeds that have been genetically engineered to produce a crop whose first generation produces sterile seeds, thereby stopping a second generation from being grown from seeds saved from the first. It has been designed as a way to build patent protection directly into a high-value, genetically engineered crop variety and thus recover high research investment costs. *See* TERMINATOR GENES.

terminator sequence: A sequence in DNA that signals the termination of transcription to RNA polymerase.

terminator stem: A configuration of the leader transcript that signals transcription termination in attenuator-controlled amino acid operons.

terminator technology: A biotechnological technique to block seed germination through the destruction of an embryo in developing seed. This technology has been developed jointly by USDA and Delta and Pine Land Company in the USA. This technology makes use of blocking sequences. When these sequences are present, these do not allow the production of RIBOSOME INACTIVATING PROTEIN (RIP) and embryo develops normally. But the excision of the 'blocking sequence' results in the production of RIP, which degenerates embryo to make seed unfit for germination. The standardization of 'terminator technology' though not actually used so far, reveals the amazing opportunities that molecular biology provides in the production of transgenic plants with fascinating characteristics. *See* GENETIC USE RESTRICTION TECHNOLOGY (GURT).

terminator technology, components of: The terminator technology consists of three genes Late Embryogenesis Abundant (LEA) promoter, recombinase and excision sequences, with their on/off switches, blocker and inducer.

territory: A defined region.

tertiary gene pool: The tertiary gene pool comprises those species where hybridization with the cultivated species is either not possible or hybrids tends to be anomalous, lethal, or completely sterile. This gene pool generally includes distant relatives in other genera. Gene transfer from a species of the tertiary gene pool to the cultivated species requires special crossing and embryo rescue techniques in order to get viable hybrids. *Aka* GENE POOL 3 (GP-3).

tertiary monosomic: An individual in which two arms of two non-homologous chromosomes missing but the remaining two arms are joined to constitute a tertiary chromosome.

tertiary nucleic acid structure: Refers to higher-ordered structures conferred on nucleic acids by interactions between nucleotides, which are not closely positioned within the sequence (primary structure) of the molecule. In nucleic acids, folding in a molecule is achieved by recognition among complementary regions.

tertiary protein structure: Refers to how the polypeptide chain (the primary structure) is bent and folded in three-dimensional space in order to form the normal tightly folded and compact structure. Folding of a protein chain occurs via interactions of its side chain molecules, including formation of disulfide bonds between cysteine residues. *See* PRIMARY STRUCTURE, SECONDARY STRUCTURE *and* QUATERNARY STRUCTURE.

tertiary trisomic: A chromosome occurring in addition to the normal diploid complement but as a result of a reciprocal interchange between two standard chromosomes. It generally occurs in the progeny of a translocation heterozygote.

test cross: Refers to a cross between a heterozygote of an unknown genotype and an individual homozygous for the recessive genes in question. If the unknown is heterozygous, then approximately 50% of the offspring should display the recessive phenotype (e.g. in a trihybrid cross, $AbC/aBc \times abc/abc$, where A/a, B/b and C/c represent three pairs of alleles, and the slashes separate the contributions from the two parents of each individual). In general, each cross contributes to the solution of an experimental question by using more or less defined crossing partners. In other test crosses, promising parental genotypes are crossed with a number of other known genotypes. The progeny is grown separately and assessed for the desired traits. The measurements can be utilized for the determination of GENERAL COMBINING ABILITY (gca) and SPECIFIC COMBINING ABILITY (sca), and these are required if effective crosses are to be made. In inbreeding, it is not enough merely to select the best varieties, but parents with good crossing ability must also be available so that a number of important traits are combined and passed on to progeny. It is especially critical for asexually propagated plant or hybrids. Test crosses necessitate extensive testing facilities and are time-consuming. It is used to determine the genotype (homozygous or heterozygous) of an individual of the dominant phenotype. It is also used to construct linkage maps.

test for allelism: It is a fundamental problem in heredity to understand the genes that contribute to one particular biological process. Such an analysis begins with a collection of related mutant phenotypes centred on the process. For example, if a geneticist were interested in the genes determining flower colour, genetic dissection would begin with isolating a set of different flower colours. An important task is to determine how many different genes are represented by the mutations that determine the related phenotypes, because this number defines the set of genes that affect the process under study. Hence it is necessary to have a test for ascertaining whether the mutations are alleles of one gene or different genes. The allelism test having the widest application is the COMPLEMENTATION TEST.

test mating: *See* TEST CROSS.

test of significance: A test designed to assess significance to distinguish differences due to sampling errors from differences indicating discrepancies between observation and hypothesis. *See* SIGNIFICANCE TEST.

test systems: The organisms and methodology used to test mutagenicity and/or carcinogenicity of environmental chemical and physical agents.

testa (pl. testae): The seedcoat.

tester plant: Plants of like kind and similar physiological conditions used in experiments to measure performance or quality characters.

Tet-assisted bisulfite sequencing (TAB-seq): This method employs a clever use of the Tet enzyme to distinguish between $5^{m}C$ and $5^{hm}C$ using bisulfite.

Tet-on and Tet-off: Tet-off activates gene expression in the absence of Doxycycline (Dox), whereas Tet-on activates gene expression in the presence of Dox. Dox is also used in 'Tet-on' and 'Tet-off' tetracycline-controlled transcriptional activation to regulate transgene expression in organisms and cell cultures. A method of inducible gene expression where transcription is reversibly turned on or off in the presence of the antibiotic tetracycline or one of its derivatives (e.g. doxycycline). In nature, the Ptet promoter expresses TetR, the repressor and TetA, the protein that pumps tetracycline antibiotic out of the cell. The difference between 'Tet-on' and 'Tet-off' is not whether a transactivator turns a gene on or off, as the name might suggest; rather, both proteins activate expression.

tetra-allelic: In tetraploids, when multiple allele loci all have different alleles (e.g. four alleles, $A_1A_2A_3A_4$).

tetra-allelic single-nucleotide polymorphism (tetra-allelic SNP): Any single-nucleotide polymorphism of which four different alleles occur in a population.

tetracycline: An antibacterial antibiotic from *Streptomyces* spp.

tetrad: A group of four. (1) Usually refers to four pollen grains that remain fused together through maturity (e.g. in the Epacridaceae). (2) Four homologous chromatids in a bundle during the first meiotic prophase and metaphase. In meiosis, the four haploid cells that result from a single diploid cell during gametogenesis.

tetrad analysis: Term is generally applied to Ascomycetes. The use of tetrads to study the behaviour of chromosomes and genes in crossing over during meiosis. It is particularly used in gene mapping of fungi.

tetrad analysis in *Neurospora*: To know the genotype of a particular ascospore arising from crosses between strains of *Neurospora crassa*, ascospores must be isolated and cultured separately. Because of ORDERED TETRADS in *Neurospora*, the distance between *a* gene locus and its centromere can be determined.

tetrad divisions: The first and second meiotic divisions that the spore mother cell undergoes to form four spores.

tetramer: Structure resulting from association of four subunits.

tetranucleotide hypothesis: P.A. Levene in 1909 put forward a hypothesis that DNA is a linear, single-stranded polynucleotide consisting of four repeating bases (adenine, thymine, guanine and cytosine) linked to each other by a deoxyribose phosphate ester backbone.

tetraploid: A polyploid cell, tissue or organism with four sets of chromosomes (4n) in the nucleus.

tetraploidization: The mitotic or meiotic process that produces tetraploids.

tetrasome: A chromosome occurring four times.

tetrasomic: Having one or more chromosomes of a complement embodied four times in each nucleus. *Aka* TETRASOMY.

tetraspore: The asexual spore of red algae. It is so named because each sporangium produces just four spores. *See* RHODOPHYCEAE.

tetratype (TT): A spore arrangement in Ascomycetes that consists of two parental and two recombinant spores, indicating a single crossover between two linked loci.

tetravalent: An association at first meiotic division of four chromosomes that are completely or partially homologous. *See* QUADRIVALENT.

tetrazolium test: A quick test to determine seed viability. Tetrazolium is a group of chemicals that have the capability to accept hydrogen atoms from dehydrogenase enzymes during the respiration process in viable seeds. This ability is the basis of the test during which the tetrazolium undergoes a colour change, usually from colourless to red (formazan). The method was developed in Germany in the early 1940s by G. Lakon. The test is highly valued throughout the world for estimating seed viability. This test can be finished in only a few hours.

thale cress: Common name for *Arabidopsis thaliana*.

thallus: The vegetative body of a fungus or a 'simple plant' (e.g. moss).

The Arabidopsis Information Resource (TAIR): TAIR maintains a database of genetic and molecular biology data for the model higher plant *Arabidopsis thaliana*. Data available from TAIR includes the complete genome sequence along with gene structure, gene product information, gene expression, DNA and seed stocks, genome maps, genetic and physical markers, publications and information about the Arabidopsis research community. Gene product function data are updated every week from the latest published research literature and community data submissions. TAIR also provides extensive link-outs from our data pages to other *Arabidopsis* resources.

theistic evolution: A view in which religious teachings are seen as compatible with biological evolution. Species share a common ancestry and change through natural selection and other evolutionary processes but with divine guidance and intervention.

theobromine: A mutagenically active purine analogue.

theories of ageing: Various theories are: (i) SOMATIC MUTATION THEORY; (ii) RELAXED STATE OF REPAIR THEORY; (iii) GENE REGULATION THEORY; (iv) FREE RADICAL THEORY; (v) TELOMERE EROSION THEORY; (vi) MITOCHONDRIAL DNA DAMAGE THEORY; (vii) TISSUE REGENERATION THEORY; and (viii) EPIGENETIC DEFECTS THEORY. *See details as separate terms.*

theory: A scientific theory summarizes a hypothesis or group of hypotheses that have been supported with repeated testing. A theory is valid as long as there is no evidence to dispute it. Therefore, theories can be disproven. Basically, if evidence accumulates to support a hypothesis, then the hypothesis can become accepted as a good explanation of a phenomenon. One definition of a theory is to say it is an accepted hypothesis. Theories are analytical tools for understanding, explaining and making predictions about a given subject matter. Examples: Darwin's theory of natural selection, Lamarck's theory of inheritance of acquired traits, H. DeVries' theory of mutation, Modern synthetic theory of evolution; Big bang theory; Newton's theory of gravitation; Atomic theory; Cell theory; Germplasm theory; Einstein's theory of relativity; Particulate theory of inheritance. Some of these theories have not stood test of time and have been rejected. Some theories have been modified.

theory of encapsulation: Both C. Bonnet and L. Spallanzani, during 1765–1768, accepted the preformation theory. According to their version of this theory, the germs of all living things were created by God in the beginning and were encapsulated within the first female of each species. Thus, the new individual present in each egg was not formed *de novo* but developed as the result of an expansion of parts the delineation of which had been laid down within the germ by God at the creation. It was assumed that the semen provided a stimulus for this expansion.

theory of epigenesis. This theory was put forward by C.F. Wolf in 1759. He studied the development of adult embryos and said that he never saw any miniature chick in the eggs. He saw only certain granules, which by arranging and rearranging formed the adult chick. Thus, adult organization is formed gradually. Although his idea did not seem to be very strong initially yet his idea is correct even today. According to this theory, development is result of action and interaction among different parts. Each generation of adults has been produced by a series of developmental processes from an apparently structureless egg to the complex morphology of adults. This is referred to as PROBLEM OF MORPHOLOGY.

theory of evolution: Scientific theory about the origins of life on Earth and how it has changed over time.

theory of inheritance of acquired characteristics: Jean-Baptiste Lamarck (1809) tried to explain the extraordinary ability of hereditary factors to respond directly to the environment. He assumed that each hereditary particle had a spiritual conscious-like property that could absorb and interpret messages from outside. He believed that variations in organisms are induced in response to an urgent need and regarded these variations as heritable. This theory could not be proven experimentally. *Aka* THEORY OF INHERITANCE OF ACQUIRED TRAITS. *See* LAMARCKISM.

theory of preformation: According to theory of preformation, things are already present in the egg in microscopic, transparent and folded form. As development proceeds, these parts are enlarged, become dense, visible and unfolded. Pupal development, flower buds and vegetative buds exemplify this theory.

theory of reversibility: The theory that states that a structure or adaptation, which has been lost cannot be regained in its exact original form.

therapeutic cloning: Therapeutic cloning involves creating a cloned embryo for the sole purpose of producing embryonic stem cells with the same DNA as the donor cell. These stem cells can be used in experiments aimed at understanding disease and developing new treatments for disease.

therapeutic proteins, enzymes and diagnostics: Transgenic plants can produce a variety of proteins that can be used in diagnostics for detecting human diseases and therapeutics for curing human and animal diseases in large scale with low cost. The monoclonal antibodies, blood plasma proteins, peptide hormones and cytokinins are being produced in transgenic plants and their parts such as tobacco leaves, potato tubers, sugarcane stem and maize seed endosperm). Plants are a wonderful source of cheap chemical factories that require only water, minerals, sunlight and carbon dioxide to produce thousands of sophisticated chemical molecules with different structures. Given the right genes, plants can function as bioreactors to produce modified or new compounds such as amino acids, proteins, vitamins, plastics, pharmaceuticals (peptides and proteins), drugs and enzymes for food industry and so on.

thermocycler: A laboratory apparatus most commonly used to amplify segments of DNA via the POLYMERASE CHAIN REACTION (PCR). Thermal cyclers may also be used in laboratories to facilitate other temperature-sensitive reactions, including restriction enzyme digestion or rapid diagnostics. The device has a thermal block with holes where tubes holding the reaction mixtures can be inserted. The cycler then raises and lowers the temperature of the block in discrete, pre-programmed steps.

thermoduric: An organism that can survive high temperatures but does not necessarily grow at such temperatures.

thermophile: A thermophile is an organism – a type of extremophile – that thrives at relatively high temperatures, between 41 and 122°C (106 and 252°F). Thermophiles grow best at high temperature ranging from 50°C to 60°C. Many of these cannot grow at temperature below 45°C. Thermophiles play an important role in making organic compost in which temperature rises up to 60°C. Some microbes can grow at 80°C or even more and these are called hyperthermophiles or extreme thermophiles. Most of these microorganisms live in hot springs. The bacteria *Thermus aquaticus* (source of *Taq* polymerase) is a good example that grows in hot springs.

thermoregulation of a gene: Regulation of expression is controlled by temperature. Histone-like bacterial proteins seem to play role in this process.

thermotherapy: The use of high temperature to reduce or eliminate pathogens in plant tissue. Thermotherapy is often used on plants before meristem culture to produce pathogen-free plants.

θ: The product $4N_e\mu$, which measures the relative rates of random drift ($1/2N_e$) and mutation (μ) and hence determines the variation maintained in a balance between these processes.

θ structure: A θ like structure seen as an intermediate in the replication of a circular DNA molecule.

θ-mediated end joining: The polymerase theta-mediated end joining pathway has an important role in resolving these breaks and is essential in cancers defective in other DNA repair pathways, thus making it an emerging therapeutic target. It requires annealing of 2–6 nucleotides of complementary sequence, microhomologies, that are adjacent to the broken ends, followed by initiation of end-bridging DNA synthesis by polymerase θ. S.S.O. Carvajal-Garcia *et al.* (2023) demonstrated that exonucleolytic digestion of unpaired 3′ tails is required before polymerase θ can initiate synthesis, then a switch to a more accurate, processive and strand-displacing polymerase to complete repair. The replicative polymerase, polymerase δ, is required for both steps; its 3′ to 5′ exonuclease activity for flap trimming, then its polymerase activity for extension and completion of repair.

thin layer chromatography (TLC): A chromatographic method in which a glass plate is covered with a thin layer of inert, absorbent material (e.g. cellulose or silica gel) and the materials to be analysed are spotted near the lower edge of the plate. The base of the plate is then placed in a solvent, which rises up the plate capillary action, separating the constituents of the mixtures. The principals involved are similar to those of paper chromatography and, like paper chromatography, two-dimensional methods can also be applied.

thin layer gel filtration: A technique in which molecules are separated on a flat plate (typically 20 × 20 cm) in which a layer of xerogel 0.5 × 1.0 mm thick is spread.

thioester bond: High-energy bond created by a condensation reaction between an acid (acyl) group and a thiol group (−SH). Such bonds are seen, for example, in acetyl-CoA and in many enzyme-substrate complexes.

thiogalactoside transacetylase: Product of gene *a* of *lac* operon. Thiogalactoside acetylase, *aka* PROTEIN 'a', is implicated at some point in lactose metabolism. Function of this enzyme is obscure. Acetylation reaction conducted by this enzyme may act as a detoxifying mechanism.

thiol: *See* SULFHYDRYL.

third-generation molecular markers: DNA markers include EXPRESSED-SEQUENCE TAGS (ESTs), SINGLE-NUCLEOTIDE POLYMORPHISMS (SNPs) and MINIATURE INVERTED-REPEAT TRANSPOSABLE ELEMENTS (MITEs). *See* MOLECULAR MARKERS.

third-generation programming language: A third-generation programming language (3GL) is a high-level computer programming language that tends to be more machine-independent and programmer-friendly than the machine code of the first generation and assembly languages of the second generation, while having a less specific focus to the fourth and fifth generations. Examples of common and historical third-generation programming languages are ALGOL, BASIC, C, COBOL, Fortran, Java and Pascal.

third-generation sequencing: Third-generation sequencing technologies offer the capability for single-molecule real time sequencing of longer reads and detection of DNA modification without the aforementioned assay. PacBio SMRT technology and Oxford Nanopore can use unaltered DNA to detect methylation. *See* NANOPORES.

30-nm chromatin fibre: The substructure of chromatin that consists of a possible helix array of nucleosomes in a fibre approximately 30 nm in diameter.

30S initiation complex: The complex first formed during initiation of polypeptide synthesis in prokaryotes. It comprises mRNA, 30S subunit of ribosome, a special initiating species of aminoacyl~tRNA charged with f^{Met}, GTP and three initiation factors (IF_1, IF_2 and IF_3).

^{32}P (Phosphorous-32): ^{32}P is a radioactive isotope that is commonly deployed to label biological molecules. The marker is particularly helpful in the identification of malignant tumours because cancer cells tend to accumulate more phosphate than normal cells. ^{32}P label of DNA along with ^{35}S-label of protein was used by A. Hershey and M. Chase experiment in 1952 that DNA is the genetic material to demonstrate that DNA is the genetic material in T2 bacteriophage.

^{35}S: ^{35}S is a low-energy beta emitter that is often used as a radioactive label for biological molecules. The ^{35}S-labelled protein may be used as a molecular probe in a number of situations. Pulse-chase labelling of antigens with [^{35}S]methionine is used to determine the relative half-life of a protein. ^{35}S-label of protein along with ^{32}P label of DNA was used by A. Hershey and M. Chase experiment in 1952 that DNA is the genetic material to demonstrate that DNA is the genetic material in T2 bacteriophage.

Thomas circle: A double-stranded DNA fragment from eukaryotic chromosomes that can form circular structures after treatment with exonucleases. The formation of Thomas circles is assumed to be due to the presence of repetitious DNA sequences in the fragments.

thousand-grain weight (TGW): Equals 1000-grain weight expressed in grams. It is a standard measure for seed weight, and thus indirectly for seed size. From a seed lot, 1000 seeds are randomly sampled and the weight is determined in grams.

threading DNA: a surface-independent tip-based method developed by A.C. Payne *et al.* (2013) for stretching and depositing single and double-stranded DNA molecules. DNA is stretched into air at a liquid-air interface and can be subsequently deposited on to a dry substrate isolated from solution. A. Singh further expanded this technology in 2023 to make it more sensitive to biologically relevant sequences, including those containing POST-TRANSLATIONAL MODIFICATIONS (PTLMs). This involved chemically linking a second DNA oligomer, called the threading DNA, to the other end of the peptide.

three-breed crossing: A system of breeding in which breed A is crossed with B and the F_1 with C. This hybrid is then crossed with B and the product is crossed with A to produce a hybrid, which crosses with C. The next cross is with B and then A and so on.

three-dimensional (3D) diffusion. It includes diffusion both from outside to inside and inside to outside. Proteins are not randomly distributed. Some proteins may be found inside while others may be found outside the plasma membrane. Proteins can pass through if they are lipid soluble. Different cell types are recognized on the basis of types of proteins found on the surfaces, which serve as markers for different cell types.

three-dimensional (3D) genome *in situ*: 3D mapping and accelerated super-resolution imaging of the human genome using *in situ* sequencing.

three-dimensional (3D) imaging: Captures digital photos of the sides and top (of shoot mass, shape, leaf number, leaf health and colour) and combines them into a 3D image.

three-parent embryos: Human embryos have been produced containing DNA from three people. For this, chromosomes are taken from one zygote and put into a zygote stripped of its original chromosomes but left with its original mitochondria. The resulting embryos contained the nuclear DNA from two traditional parents and mitochondrial DNA from a third. Two techniques, namely, MATERNAL SPINDLE TRANSFER and PRONUCLEAR TRANSFER, have been developed to have three-parent embryos.

three-point test cross: Cross involving one parent with three heterozygous gene pairs (*ABC/abc*) and the other (tester) with three homozygous recessive gene pairs (*abc/abc*). These crosses are performed to establish linkage of a gene with already mapped genes.

3′ end: Describes the different, complementary ends of a DNA or RNA single strand. It ends with an OH group.

3′ splice site: The 3′ end of an intron where cleavage takes place in RNA splicing.

3′ untranslated (3′ UTR) region: Sequence of nucleotides at the 3′ end of mRNA; does not code for the amino acids of a protein but affects both the stability of the mRNA and its translation.

3′-5′ exoribonucleases: Enzymes that remove nucleotides from the 3′ end of RNA.

3′-end processing of pre-mRNAs: During 3′-end processing, histone pre-mRNAs are cleaved five nucleotides after a conserved stem loop by an endonuclease dependent on the U7 SMALL NUCLEAR RIBONUCLEOPROTEIN (snRNP). The upstream cleavage product is degraded by a 5′-3′ exonuclease, also dependent on the U7 snRNP.

3′-O-blocked reversible terminators: The mechanism uses a sequencing by synthesis approach, elongating the primer in a stepwise manner. Sequencing primers and templates are fixed to a solid support. The support is exposed to each of the four DNA bases, which have a different fluorophore attached (to the nitrogenous base) in addition to a 3′-O-azidomethyl group. Only the correct base anneals to the target and is subsequently ligated to the primer. The solid support is then imaged and nucleotides that have not been incorporated are washed away and the fluorescent branch is cleaved using TCEP (tris(2-carboxyethyl)phosphine). TCEP also removes the 3′-O-azidomethyl group, regenerating 3′-OH and the cycle can be repeated.

3′-OH terminus: The end of a polynucleotide that terminates with a hydroxyl group attached to the 3′ carbon of the sugar.

3-phosphoglyceraldehyde (3-PGAL): Isomer of phosphoglycerate (PGA) formed during the Calvin cycle.

3′-unblocked reversible terminators: The reversible termination group of 3′-unblocked reversible terminators is linked to both the base and the fluorescence group, which now acts as part of the termination group as well as a reporter. This method differs from the 3′-O-blocked reversible terminators method in three ways: the 3′-position is not blocked (i.e. the base has free 3′-OH). The fluorophore is the same for all four bases. Each modified base is flowed in sequentially rather than at the same time. The main disadvantage of these techniques lies with their poor read length. To prevent incorporation of two nucleotides in a single step, a block is put in place. Noise can also be created if the fluorophore is unsuccessfully attached or removed. This technique was pioneered by Illumina, with their HiSeq and MiSeq platforms. HiSeq is the cheapest of the second-generation sequencers ($0.02 per million bases). It also has a high data output of 600 Gb per run, which takes around 8 days to complete.

three-way cross: A first-generation hybrid between a single-cross and an inbred (or a pure) line variety. It produces a THREE-WAY HYBRID progeny.

threonine (thr): An aliphatic, polar alpha-amino acid ($HO_2CCH(NH_2)CH(OH)CH_3$). It has four codons (ACU, ACA, ACC and ACG). Together with serine and tyrosine, threonine is one of three alcohol groups bearing amino acids commonly found in proteins. It is biosynthesized from aspartic acid and is a precursor of isoleucine in microorganisms. The threonine residue is susceptible to numerous POSTTRANSLATIONAL MODIFICATIONS (PTLMs).

threonine phosphorylation: Threonine is one of the building blocks of protein that can be modified by the addition of a chemical tag, in this case, a phosphate group.

threshold: A term used in studying the effect of radiation. Below a certain dose, if there is a threshold, there is no measurable effect. No threshold exists for most genetic defects.

threshold character: A term used for those phenotypic traits whose segregating distributions are phenotypically discontinuous but whose inheritance is multigenic as if they are quantitative characters. Threshold characters may segregate into many discontinuous phenotypic classes.

threshold model: A model that states that a discrete trait is present only when the quantity of an underlying continuous trait is greater than some threshold.

threshold value: A critical value on an underlying scale of liability above which individuals express a trait or disease.

throwbacks: In some cases of gene-gene interaction, F_1 hybrids have phenotype unlike their immediate parents but like their remote ancestors. *Aka* AVITISM *or* REVERSION. For example, in a cross between two white-flowered sweet pea plants, F_1 hybrid had a purple colour unlike their immediate parental phenotype, white flower colour. At the molecular level, this phenomenon is explained based on complementation between two dominant genes. Another example of avitism is provided by coat colour in mice. When black mice are crossed with albino mice, F_1 hybrid mice show agouti pattern. This phenotype results due to interaction between two dominant genes.

thylakoid: Flattened sac of the membrane in a chloroplast that contains chlorophyll and other pigments and carries out the light-trapping reactions of photosynthesis. Stacks of thylakoids occur form the grana of chloroplasts. The thylakoid membrane contains the pigment chlorophyll, which is stored in protein complexes called PHOTOSYSTEMS I and II. These chlorophyll molecules absorb light energy from the sun, which the cell alters into chemical energy through a series of chemical reactions. *See* STROMA.

thymidine: The deoxyribonucleoside that contains the pyrimidine thymine.

thymidylic acid: The deoxyribonucleotide that contains the pyrimidine thymine and one phosphate group is attached to the sugar moiety.

thymine (T): A nitrogenous base ($C_5H_6N_2O_2$), one member of the base pair AT (adenine-thymine) in DNA not in RNA, the other three bases being adenine, guanine and cytosine. It acts as an indispensable growth factor for microorganisms. Its hydrogen-bonding counterpart in RNA is uracil (U). *Aka* 2,6-DIOXY,5-METHYLPYRIMIDINE.

thymine dimer (T-T): The first photochemical lesion found in the DNA of irradiated bacteria was a thymine dimer. UV light produces thymine dimers mainly by linking adjacent thymine bases in the same strand of DNA via C-C bonds. Dimers containing thymine block DNA replication *in vitro* and *in vivo*.

thymine DNA glycosylase (TDG): A G/T mismatch DNA glycosylase in mammals has a weak 5^{m}C DNA glycosylase activity *in vitro*.

Ti plasmid: Tumour-inducing plasmid of *Agrobacterium tumefaciens*, which is used as the basis for a series of cloning vectors for higher plants and is now widely being used to transform plants. The bacterium possesses a plasmid called Ti plasmid, which is about 200 kb and carries t-DNA (transfer DNA). During the co-cultivation of this bacterium with plant tissues, the t-DNA gets excised and transferred to plant cells in the form of a single strand. Any gene that is to be transferred to the plant, is first placed in the t-DNA region of Ti plasmid. *Aka* TUMOUR-INDUCING PLASMID.

Ti plasmid *vir* region: Genes in the virulence region are grouped into numerous operons, *virABCDEFG*, which code for the enzymes responsible for mediating transduction of T-DNA to plant cells. *virA* codes for a receptor that reacts to the presence of phenolic compounds such as acetosyringone, syringealdehyde or acetovanillone, which leak out of damaged plant tissues. *virB* encodes proteins that produce a pore/pilus-like structure. *virC* binds the overdrive sequence. *virD1* and *virD2* produce endonucleases, which target the direct repeat borders of the T-DNA segment; virD4 produces the coupling protein. *virE* gene product binds to the T-strand protecting it from nuclease attack and intercalates with lipids to form channels in the plant membranes through which the T-complex passes, beginning with the right border. *virG* codes for a protein that activates *vir*-gene expression after binding to a consensus sequence, once it has been phosphorylated by virA.

Ti plasmid-based transformation vectors: There are two classes of Ti plasmid-based transformation vectors of *Agrobacterium tumefaciens*: COINTEGRATE TI VECTORS and BINARY TI VECTORS. *See details as separate terms.*

Tier 1 cell type: Tier 1 cell types are the highest priority set and comprise three widely studied cell lines: K562 erythroleukemia cells; GM12878, a B-lymphoblastoid cell line that is also part of the 1000 Genomes project (http://1000genomes.org) 55; and the H1 embryonic stem cell (H1 hESC) line.

Tier 2 cell type: The second-priority set of cell types in the ENCODE project, which includes HeLa-S3 cervical carcinoma cells, HepG2 hepatoblastoma cells and primary (non-transformed) human umbilical vein endothelial cells (HUVECs).

Tier 3 cell type: Any other ENCODE cell types not included in tier 1 or tier 2.

Tier system: In mutagenicity testing, a test system in which potentially mutagenic chemicals detected in a relatively simple test are subjected subsequently to biologically more complex tests until they are either eliminated from further testing or bioassayed in full-scale animal systems.

tight junction: Cell-cell junction that caps adjacent epithelial cells together, stopping the passage of most dissolved molecules from one side of the epithelial sheet to the other.

tile-based nanostructures: Short oligonucleotides, known as MOTIFS, assemble to make a tile and tiles assemble to make a complete lattice.

tiling array: A subtype of microarray containing small probes that are designed to cover the entire genome or contigs of the genome in an unbiased manner. These arrays can be used coupled with chromatin immunoprecipitation (ChIP-chip), with methyl-DNA immunoprecipitation (MeDIP-chip) and in DNase chip studies. The purpose is to empirically detect expression of transcripts or alternatively splice forms that may not have been previously known or predicted.

tiling microarrays: *See* DNA OLIGONUCLEOTIDE MICROARRAYS.

tiling path: The ordered assemblage of **BAC** clones using sequence overlaps of neighbouring clones such that they completely cover the entire corresponding section of the chromosome.

tiling-array analysis: Commercial chips containing 25–60 nt oligonucleotide probes designed to continuously cover a genomic region that is used to produce unbiased maps of histone modifications following CHROMATIN IMMUNOPRECIPITATION (ChIP-chip), or of DNA methylation following methylated DNA immunoprecipitation (MeDIP-chip), or of gene expression following cDNA hybridization (RNA-chip).

tillage: Practice of preparing the ground for planting and controlling weeds between plantings by turning or aerating the soil. Conventional tillage can lead to an increased risk of erosion; therefore, conservation tillage has been increasingly adopted to preserve soil, a non-renewable resource.

tiller: A lateral shoot, culm or stalk ascending from a crown bud. Tillers are common in grasses.

time of segregation: Segregation of allelic members is complete before the gametes are formed. To be precise, we say that segregation occurs during meiosis.

time-dependent covariate: *See* PREDICTOR.

time-resolved electrochemical technology for plant root environment *in situ* chemical sensing (TETRIS): Traditional single-point measurements fail to capture dynamic chemical responses of plants, which are complex, nonequilibrium biological systems. P. Coatsworth *et al.* (2024) report TETRIS, a real-time chemical phenotyping system for continuously monitoring chemical signals in the often-neglected plant root environment. TETRIS consists of low-cost, highly scalable screen-printed electrochemical sensors for monitoring concentrations of salt, pH and H_2O_2 in the root environment of whole plants, where multiplexing allowed for parallel sensing operation. TETRIS was used to measure ion uptake in tomato, kale and rice and detected differences between nutrient and heavy metal ion uptake. Modulation of ion uptake with ion channel blocker LaCl3 was monitored by TETRIS and machine learning used to predict ion uptake. TETRIS has the potential to overcome the urgent 'bottleneck' in high-throughput screening in producing high-yielding plant varieties with improved resistance against stress.

***timing of CAB expression 1* (*TOC1*) gene:** Gene responsible for encoding Timing of CAB expression 1 (TOC1) protein in *Arabidopsis thaliana*.

tip: The terminal node on a phylogenetic tree.

tissue: A section of an organ that consists of a largely homogeneous population of cell types. Since many organs are multifunctional, they have developed highly specialized cell types to perform different functions. Identifying the section of an organ that is homogeneous for a particular cell type ensures that the gene expression profiles extracted from those cells will accurately resemble the class of cells that make up the tissue. Organs are then formed by the functional grouping together of multiple tissues.

tissue culture: A process of growing a plant through tissue culture from cells rather than seeds. The growth and maintenance of cells from higher organisms are done in laboratory in a sterile test tube or petri dish environment, which contains the nutrients essential for cell growth. Many times, the organs are also used for tissue culture. The media used for the growth of the culture is broth and agar. The maintenance or growth of tissue *in vitro* is done in a way that may permit further differentiation, preservation or regeneration. Tissue culture in plants is an important aspect being exploited in plant breeding for creation, conservation and utilization of genetic variability for the improvement of crop plants. Various aspects like micropropagation, somaclonal variation, embryo culture, another culture, pollen culture and protoplast culture are now routinely and widely being used in plant breeding programmes. Besides, tissue culture is the backbone of plant transformation for developing transgenic plants. Likewise, animal tissue culture is being done for basic and applied research. *Aka* MICROPROPAGATION.

tissue differentiation: The specification and proper arrangements of new cell types during tissue differentiation require the coordinated regulation of gene expression and precise interactions between neighbouring cells. Of the many growth factors involved in these events, Wnts are particularly interesting regulators, because a key component of their signalling pathway, β-catenin, also functions as a component of the cadherin complex, which controls cell-cell adhesion and influences cell migration. They have provided evidence of possible interrelations between Wnt and other growth factor signalling, β-catenin functions and cadherin-mediated adhesion.

tissue engineering: Grafting of artificial skin derived from cultured cells. Animal and human skin can be made artificially from the cultured cells. Of the few cell types that can be cultured, keratinocytes are the most important and these constitute 90% of epidermis of skin. The cultured keratinocytes must come from the healthy skin of same person to prevent rejection.

tissue regeneration theory of ageing: According to this theory, there is diminished regenerative potential of tissues. With age, resident precursor cells (satellite cells) had a markedly impaired propensity to proliferate and to produce myoblasts necessary for muscle regeneration. This was due to insufficient upregulation of the Notch-ligand Delta and, thus, diminished activation of Notch in aged, regenerating muscle. Inhibition of the Notch impaired regeneration of young muscle, whereas forced activation of Notch restored the regenerative potential to old muscle. Thus, Notch signalling is a key determinant of muscle regenerative potential that declines with age.

tissue specificity: The question of tissue specificity can be illustrated with the following example: HNF1 and HNF4 are present in the liver and kidney, but the genes that have binding sites for these factors in the kidney are not active or at least not as active as in the liver. This situation suggests that negative regulation could be one mechanism operating. HNF1β expression is correlated with the repression of a subset of liver genes. Another interesting feature of the liver transcription factors is their relation to homeotic genes in *Drosophila*. HNF1, through its DNA-binding region, is distantly related to homeobox proteins. In HNF3, 86 out of 110 amino acids in the DNA-binding region are identical to those in the product of *forkhead*, a *Drosophila* homeotic gene. HNF4 has high homology in its DNA-binding, ligand-binding or dimerization domains to the *Drosophila* homologue than to any mammalian nuclear receptor. There is evidence for the presence of these liver transcription factors in rat embryos at mid-gestation. These factors may, therefore, play a role in endodermal differentiation decision in the primitive gut in a fashion analogous to the effect of homeobox proteins on anterior-posterior axis patterning in *Drosophila*. The CYP2B1/B2 family of the cytochrome P-450 gene superfamily, induced in rat liver by the prototype drug phenobarbitone, may set a new paradigm for tissue specificity of gene expression.

tissue-specific gene expression: Expression of a gene in specific tissues. During plant genetic engineering, using tissue-specific promoters, gene expression can be achieved in desired tissues rather than all the cells of plant. For example, the use of TA 29 promoter allows the *barnase* gene to express only in anthers. Likewise, there are now seed-specific, root-specific and fruit-specific promoters that allow the transgene to express only in the desired tissue/organ. Galactose 4 (GAL4), a yeast regulator, when expressed in particular tissue *Drosophila* larvae, stimulates tissue-specific transcription of a *Drosophila* promoter linked to GAL4 binding sites. For this purpose, they made use of effector and reporter genes.

tissue-specific transgene expression: Single-cell mouse embryos were infected *in vitro* with recombinant lentiviral vectors to generate transgenic mice carrying the *green fluorescent protein* (*gfp*) gene driven by a ubiquitously expressing promoter. Eighty per cent of founder mice carried at least one copy of the transgene, and 90% of these expressed GFP at high levels. Progeny inherited the transgene(s) and displayed green Fluorescence. Mice generated using lentiviral vectors with muscle-specific and T lymphocyte-specific promoters expressed high levels of GFP only in the appropriate cell types. They also generated transgenic rats that express GFP at high levels, suggesting that this technique can be used to produce other transgenic animal species.

titre of library: Concentration of a substance in solution or the strength of such a substance determined by titration. Calculating the titre allows researchers to approximate how many infectious viral particles were successfully created in the library.

T_m: Refers to the (1) midpoint of the temperature range over which DNA is melted or denatured by heat. (2) temperature at which a duplex nucleic acid molecule is 50% melted into single strands. T_m is dependent upon the number and proportion of G-C base pairs besides the ionic conditions. T_m is a measure of the thermal firmness of a nucleic acid probe: target sequence hybrid.

Tn transposons: In some of the Tn elements, the IS elements at both the ends are identical (e.g. IS1 in Tn9), but in others, these elements are closely related but not identical (e.g. IS10L and IS10R in Tn10). When two IS elements of a Tn element are identical, either element can have transposition (Tn9), but when they are non-identical, transposition will depend upon one of the two elements (IS10R and IS10L) for Tn10, IS10R is active; IS10L shows only 1–10% activity.

Tn10 transposon: Tn10 is a composite transposon. It is 9.3 kb long and contains 6.5 kb of central DNA with genes including tetracycline resistance. 1.4 kb IS elements ($IS10_L$ and $IS10_R$) at each end, in an inverted orientation. Transposition of composite transposons results from the IS elements, which supply transposase and its recognition signals, the IRs. Transposons, like IS elements, produce target site duplications (e.g. a 9-bp duplication for Tn10).

Tn3 transposon: It is a non-composite transposon. Tn3's length is about 5 kb, with 38-bp inverted terminal repeats. It has three genes in its central region: (a) *bla* encodes β-lactamase, which breaks down ampicili, (b) *tnpA* encodes transposase, needed for insertion into a new site and (c) *tnpB* encodes resolvase, involved in recombinational events needed for transposition (not found in all transposons). Tn3 produces a 5-bp duplication upon insertion.

Tn5 transposon: This transposon carries a gene encoding resistance to antibiotics kanamycin and meomycin, as well as genes involved in its own transposition.

tobacco mosaic virus (TMV): One of the smallest viruses that is made up of some 2200 chains of identical polypeptides and a molecule of RNA. All of the genetic/heredity information of the Tobacco Mosaic Virus is confined in its RNA. The first discovery of a self-assembling, active biological structure occurred in 1955, when H. Fraenkel-Conrat and R. Williams showed that TMV reassembles into functioning, infectious virus particles after the TMV has been separated into its components via immersion in concentrated acetic acid. The TMV virus infects the leaves of tomato and tobacco plants, triggering disease. Tobacco plants can be genetically engineered to resist TMV infection. A tomato plant, genetically engineered to resist TMV infection, has been made commercially available since 1992.

tobacco mosaic virus infectivity: Tobacco mosaic virus (TMV) is one of the many viruses using RNA instead of DNA as its nucleic acid. There are different TMV strains with different protein envelopes. Protein and RNA of TMV could be separated. Without RNA, protein was not infective whereas pure TMV RNA was infective and caused the synthesis of viral particles. H. Fraenkel-Conrat and B.B. Singer observed in 1957 that the type of reconstituted TMV depended upon RNA not protein.

toehold-mediated strand displacement: An enzyme-free molecular tool to exchange one strand of DNA or RNA (output) with another strand (input). It is based on the hybridization of two complementary strands of DNA or RNA via Watson-Crick base pairing (A-T/U and C-G) and makes use of a process called branch migration.

toeprinting assay: Method to investigate the interaction of messenger RNA with ribosomes or other RNA-binding proteins. It differs from standard DNA FOOTPRINTING ASSAY. Originally called EXTENSION INHIBITION ASSAY. *Compare with* DNA FOOTPRINTING.

tolerable mutations: The mutations that affect the function of the gene to such a degree that they do not have a drastic effect on the individual and are tolerable. Such mutations accompany the process of evolution and play an important role in speciation. Tolerable mutations are of two types – neutral and favourable.

tomato epigenome: Single-base resolution methylomes of wild-type tomato fruits at four different developmental stages, immature (17 DPA), mature green (39 DPA), breaker (42 DPA) and red ripe (52 DPA), fruits (42 DPA) of tomato ripening mutants, *rin* and *Cnr* and tomato leaves were published. Tomatoes exposed to the methyltransferase inhibitor 5-5-azaC ripen prematurely. Chilling induces methylation changes in tomato fruits.

top cross: A cross between a selection, line, clone and so on and a common pollen parent, which may be a variety, inbred line, single cross and so on. The common pollen parent is called the TOP CROSS or TESTER parent. In maize, a top cross is commonly called an INBRED-VARIETY CROSS. Usually, top cross is used in order to test the gca.

top cross hybrid: The first generation of a cross between an inbred line and an open-pollinated variety.

top cross progeny: Progeny from outcrossed seed of selections, clones or lines to a common pollen.

top dressing: The additional fertilizer application when crop plants are already established, usually before flowering. Top dressing comprises material (e.g. compost or manure) that is applied to the surface around the plant to aid in drainage, decrease erosion, prevent moisture loss and keep weeds down.

top-down and bottom-up design: Top-down and bottom-up are both strategies of information processing and knowledge ordering, used in a variety of fields including software, humanistic and scientific theories (*see* SYSTEMICS) and management and organization. In practice, they can be seen as a style of thinking, teaching or leadership. A top-down approach (*Aka* STEPWISE DESIGN and STEPWISE REFINEMENT and in some cases used as a synonym of DECOMPOSITION) is essentially the breaking down of a system to gain insight into its compositional sub-systems in a reverse engineering fashion. A bottom-up approach is the piecing together of systems to give rise to more complex systems, thus making the original systems sub-systems of the emergent system. Bottom-up processing is a type of information processing based on incoming data from the environment to form a perception.

top-down approach: Making nanosystems, materials and devices starting from bulk materials.

top-down approach in nanobiotechnology: Top-down systems are where microscopic elements or small numbers of atoms or molecules fashion elegant patterns. It is currently being used in computer chip manufacturing. It aims for miniaturization and lithography/etching proximity. Top-down approach also shows that if desired the two pieces of DNA can be joined covalently to form a single double helix.

top-down proteomics: Top-down proteomics is a method of protein identification that either uses an ion trapping mass spectrometer to store an isolated protein ion for mass measurement and tandem MS (MS/MS) analysis.

topo cloning: Topo cloning method uses topoisomerase instead of ligase and cloning may be done more rapidly without the need for restriction digest of the vector or insert. In TOPO cloning method a linearized vector is activated by attaching topoisomerase I to its ends and this 'TOPO-activated' vector may then accept a PCR product by ligating both the 5′-ends of the PCR product, releasing the topoisomerase and forming a circular vector in the process. To harness the relegating activity of topoisomerase, TOPO vectors are provided linearized with topoisomerase I covalently bound to each 3′ phosphate. This enables the vectors to readily ligate DNA sequences with compatible ends. The ligation is complete in only 5 min at room temperature. Three types of TOPO® cloning known are: TOPO® TA cloning® of Taq-amplified DNA, Zero Blunt® TOPO® cloning of blunt-end DNA and directional TOPO® cloning of blunt-end DNA.

topoisomerase: An enzyme that can relieve or create supercoiling in DNA by creating transitory breaks in one (type I topoisomerase) or both (type II topoisomerase) strands of the helical backbone. It is important during DNA replication to relieve the negative supercoiling that occurs when the two strands of the DNA double helix are separated. It makes reversible cuts in a double-helical DNA molecule for the purpose of removing knots or unwinding excessive twists. *Aka* DNA TOPOISOMERASE.

topological isomers: Forms of DNA with the same sequence but differing in their linkage number (coiling). *Aka* TOPOISOMERS.

topological operon: Analogous to the control of multiple genes by a single ON/OFF regulatory switch in the bacterial operon system (e.g. the *lac* operon), the term topological operon was coined by M. Levo *et al.* (2022) to describe the co-regulation of multiple genes by shared enhancers in eukaryotic genomes.

topologically associated domains (TADs): Large genomic regions promoting regulatory interactions by forming higher-order chromatin structures separated by boundary regions.

torpedo model of transcriptional termination: This model postulates that poly(A) site cleavage offers a defenceless RNA 5′-end that is degraded by 5′→3′

exonuclease activities (torpedoes) and so causes dissociation of RNA polymerase II from the DNA template. The 3′-end of the cleavage RNA is subjected to degradation by the exosome (Exo).

torpedo stage: During this developmental stage, the embryo elongates into the cellular endosperm and the internal layers of the hypocotyl and radicle differentiate to produce the vascular tissue. Lipid deposition into the cotyledons begins at this stage. Organelle differentiation transpires leading to the greening of the embryo.

total RNA sequencing: Accurately measure gene and transcript abundance and detect both known and novel features in coding and multiple forms of non-coding RNA.

totipotency: Ability of a cell to unfold its morphogenetic potential when provided with suitable cultural conditions. A. Hansen in 1879 first hypothesized that plants are totipotent, mentioning the power of regeneration of Begonia that can develop new plantules of leaf cuttings, a characteristic still used by gardeners today. The science of tissue culture is based on the concept of 'totipotency' given by G. Haberlandt in 1902. For instance, in case of plants, all the plant cells come from a single cell (i.e. zygote through successive mitotic divisions). In mitosis, daughter cells are considered to be similar to the parent cell. Therefore, G. Haberlandt conceived the idea that if the zygote has the potential to produce complete plant then why cannot the individual cell, which is the mitotic derivative of zygote. It took about 50 years to prove this concept. Today, it is true and has been well documented with most of the plant species. Even in humans, the 'stem cells' have been found to be totipotent.

totipotency and development: Multicellular organisms are made up of diverse types of cells, which can be easily recognized in microscope. With a few exceptions, gene content of all the cells of an organism is the same. The genes coding for these specialized proteins are present in all the cells of the embryo. Some mechanisms must result in expression of specific genes in the appropriate cell type and their repression in all other cell types.

totipotency and pluripotency, connection between: L. Li *et al.* (2023) found that SOX2 regulates the pluripotency programme by opening global enhancers and then occupying pre-accessible enhancers opened by early-stage-expressing transcription factors (TFs) TFAP2C and NR5A2. SOX2 then redistributes and opens enhancers or primes enhancers for future activation when cells acquire naive and formative pluripotency. These data revealed flexible pioneer TF-chromatin interactions and a transitionary 'pre-pluripotency' state that connects totipotency and pluripotency.

totipotent cell: The state of an undifferentiated cell (or nucleus) which when isolated can give rise to any and all adult cell types, as *compared* with a DIFFERENTIATED CELL whose fate is determined. *Aka* TOTIPOTENT NUCLEUS *and* TOTIPOTENTIAL CELL. *See* STEM CELL.

totipotent stem cells: Embryonic stem cells that are present during the first few cell divisions postfertilization and can form any of the different types of cells in the body. Multipotent stem cells are adult stem cells that can form other cell types but have limited potency.

touch receptor differentiation: The extent of gene regulation in cell differentiation is poorly understood. Genes have been identified that are known to be expressed in touch receptors, a previously uncloned gene (*nec-17*) that is needed for maintaining touch receptor differentiation and more than 50 previously unknown mec-3-dependent genes. These genes are randomly distributed in the genome and under-represented both for genes that are co-expressed in operons and for multiple members of gene families. Using regions 5′ of the start codon of the first 20 genes, they have also identified an over-represented heptanucleotide, AATGCAT, which is required for the expression of touch receptor genes.

toxicity: The measurements of a substance to interfere with the vital processes of an organism's life.

toxicogenomics: A fusion of genomics and toxicology disciplines intended to identify, classify and manage the latent (inherent susceptibility), incipient and overt adverse (toxic) effects on genome structure and expression levels (RNA, protein, cell/tissue/organ type) as a consequence of an organism's exposure to environmental substances (contaminants such as chemicals, drugs and micro/multicellular organisms and/or components) and stressors (for example, quality of air, climate, soil, solar radiation and water). It combines genome-wide mRNA expression profiling with protein expression patterns using bioinformatics to understand the role of gene-environment interactions in disease and dysfunction.

toxicology: The science of the harmful effects of chemicals (including drugs) on living biological systems. It seeks to determine the mechanisms by which chemicals produce adverse effects in cells and organisms. *Compare with* ENVIRONMENTAL TOXICOLOGY *and* ECOTOXICOLOGY.

toxin: Poisonous protein that can either be natural (produced by plants, animals or bacteria) or manufactured. There are several bacteria and fungi, which produce toxin and kill the other organisms. For example, bacteria *Bacillus thuringiensis* produces a delta-endotoxin that kills insects belonging to order Lepidoptera. This ability of *Bacillus thuringiensis* has been exploited to produce transgenic *Bt* crops carrying *Bt* genes, which produce the same delta-endotoxin that is produced by the bacteria.

toxin-antitoxin behaviour: Antibiotic resistance, virulence and other plasmids in bacteria use toxin-antitoxin gene pairs to ensure their persistence during host replication. The toxin-antitoxin system eliminates plasmid-free cells that emerge as a result of segregation or replication defects and contributes to intra- and interspecies plasmid dissemination. Chromosomal homologues of toxin-antitoxin genes are widely distributed in pathogenic and other bacteria and induce reversible cell cycle arrest or programmed cell death in response to starvation or other adverse conditions. The dissection of the interaction of the toxins with intracellular targets and the elucidation of the tertiary structures of toxin-antitoxin complexes has provided exciting insights into toxin-antitoxin behaviour.

***tra* genes:** A group of genes carried by an F plasmid and coding for proteins involved in DNA transfer during conjugation. *tra*-gene-encoded proteins are involved in bacterial conjugation: pili assembly and production (*traA*, *traB*, *traE*, *traC*, *traF*, *traG*, *traH*, *traK*, *traL*, *traQ*, *traU*, *traV*, *traW*), inner membrane proteins (*traB*, *traE*, *traG*, *traL*, *traP*), periplasmic proteins (*traC*, *traF*, *traH*, *traK*, *traU*, *traW*), DNA transfer (*traC*, *traD*, *traI*, *traM*, *traY*), surface exclusion proteins (*traS*, *traT*) and mating pair stabilization (*traN*, *traG*).

tracer: A chemically or radioactivity-labelled nucleic acid component involved in a reassociation reaction in amounts too small to influence the progress of reaction but it can be traced in a biochemical process or its presence readily located in a cell or tissue.

tracheid: A cell with strengthened walls that functions to transport fluid within plants.

tracking chromatin state changes: Interactions between biomolecules underlie all cellular processes and ultimately control cell fate. Perturbation of native interactions through mutation, changes in expression levels or external stimuli leads to altered cellular physiology and can result in either disease or therapeutic effects. Mapping these interactions and determining how they respond to stimulus is the genesis of many drug development efforts, leading to new therapeutic targets and improvements in human health. C.P. Seath *et al.* 2023 describe a method for the traceless incorporation of iridium-photosensitizers into the nuclear micro-environment using engineered split inteins. These Ir-catalysts can activate diazirine warheads through Dexter energy transfer to form reactive carbenes within an approximately 10 nm radius, cross-linking with proteins in the immediate micro-environment (a process termed μMap) for analysis using quantitative chemoproteomics. This NANOSCALE PROXIMITY-LABELLING method can reveal the critical changes in interactomes in the presence of cancer-associated mutations, as well as treatment with small-molecule inhibitors. Understanding nuclear protein-protein interactions is expected to have a significant effect on the field of epigenetic drug discovery in both academia and industry.

trade-off: A change in one trait that increases fitness causes a change in the other trait that decreases fitness. The two traits are fitness components. A trade-off is often seen in the form of a negative correlation between two fitness components: (i) egg size (larger eggs have better survival) and (ii) egg number.

traditional breeding: Modification of plants and animals through selective breeding. Practices used in traditional plant breeding may include aspects of biotechnology such as tissue culture and mutational breeding. *See* TRADITIONAL BREEDING METHODS.

traditional breeding methods: Refers to some or most procedures/techniques/technologies utilized by crop plant breeders prior to 'genetic engineering'.

traditional DNA cloning: Usually refers to the use of restriction endonucleases to generate DNA fragments with specific complementary end sequences that can be joined together with a DNA ligase, prior to transformation.

trailer segment: The untranslated segment that lies downstream of the nonsense codon of an mRNA molecule. *Aka* TRAIN SEQUENCE.

train sequence: A region of the non-coding sequence following the 3′-end of gene proper (mature gene, antisense strand). Train contains signals associated with termination of transcription. Typically, there occurs a series of T-A bp, in prokaryotes usually six or more in number, while in metazoans four times seems to suffice. A stem-and-loop assembly could conceivably be formed in the transcript.

trait: A phenotypic trait, simply trait, or character state is a distinct variant of a phenotypic characteristic of an organism. Biological property, characteristic or simply character are the terms synonymously used for trait. Traits may be measured in terms of colour, shape, pattern, molecular structure or statistical parameters. It may be either inherited or determined environmentally, but typically occurs as a combination of the two. Some important plant traits are upright leaves, drought tolerance, *Bt* resistance. Many traits are the consequence of the expression of a single gene, but some are polygenic. For example, the level of protein content in soybeans is controlled by five genes. A trait can be influenced by the environment. The term phenotype is sometimes used as a synonym for trait in common use, but strictly speaking, does not indicate the trait, but the state of that trait (e.g. the trait eye colour has the phenotypes blue, brown and hazel). Qualitative traits have been found to be more useful for understanding the mechanics of inheritance.

trait analysis by association, evolution and linkage (TASSEL): TASSEL is a software package used for genome wide association mapping to evaluate trait associations, evolutionary patterns and linkage disequilibrium. Strengths of this software include: (a) the opportunity for a number of new and powerful statistical approaches to association mapping such as a GENERAL LINEAR MODEL (GLM) and MIXED LINEAR MODEL (MLM). MLM is an implementation of the technique, which reduces Type I error in association mapping with complex pedigrees, families, founding effects and population structure, (b) An ability to handle a wide range of indels (insertion & deletions). Most software ignore this type of polymorphism. In some species (like maize), this is the most common type of polymorphism.

trait-level genetic use restriction technology (T-GURT): In the case of T-GURTs, one or more genes conferring a single trait are switched on or off through application of chemical inducers. T-GURT thus modifies a crop in such a way that the genetic enhancement engineered into the crop does not function until the crop plant is treated with an activator chemical that is sold by the biotechnology company. T-GURTs are not intended to affect the viability of seeds. Traits of interest that could be controlled by T-GURTs include male sterility, pest resistance, stress tolerance, nutrient production, seed germination or flower development. The technology is restricted at the trait level, hence the term T-GURT.

traitor technology: This technology offers an opportunity to use a number of commercial characteristics into a plant variety, which company to choose either to activate or deactivate through external chemical sprays or soaking of platform. The Gene Giants want to tie their traitor seed to their property chemical so that one is useless without the other. This technology described in Zeneca's new plant killer patent is to create plant varieties, which must be exposed to chemicals in order to germinate and grow properly.

traits involved in the biological control of pests: Involved in biological control are those key traits of natural enemy that determine the main events in their life, also called LIFE HISTORY TRAITS, which determine the reproductive potential (searching behaviour, fecundity, sex ratio) and the environmental sensitivity (temperature tolerance, starvation resistance, diapause) of a natural enemy. Knowledge of these traits on molecular level will helps us innovate effective strategies for rearing, sampling and release of natural enemies. Likewise, knowledge on the genetic variation of those traits in insect-pest species is expected to be helpful in determining the most efficient way of targeting them.

***trans*:** As used in molecular biology, an interaction that involves two sites, which are located on separate molecules.

***trans* face:** Face of a Golgi stack at which material leaves the organelle for the cell surface or another cell compartment. It is adjacent to the *trans* Golgi network.

***trans* fatty acids:** One of the two isomeric forms that fatty acids can occur in. *Trans* fatty acids naturally exist in some meat and dairy products (which constitute approximately 5% of the average American diet).

***trans* Golgi network (TGN):** Network of interconnected cisternae and tubules at the trans face of the Golgi apparatus, through which material is transferred out of the Golgi.

***trans*-acting:** Refers to mutations of (e.g. a repressor gene) that act through a diffusible protein product. This is the normal mode of action of most recessive mutations. *Compare with* *CIS*-ACTING.

***trans*-acting function:** The ability of a DNA sequence or transcript to regulate the expression of one or more genes on different chromosomes or to regulate mature RNAs in the cytoplasm. *Aka* GENE REGULATION *IN TRANS*.

***trans*-acting gene:** A gene acting on or cooperating with another gene located in a different chromosome.

***trans*-acting gene regulation:** The gene regulation where the regulator was produced from one gene, and subsequently diffused across the cell to bind to suitable binding sites at operators for other genes.

***trans*-acting ncRNAs:** *Trans*-acting ncRNAs possess well-defined mechanisms that can regulate transcription or translation, they generally lack the ability to directly sense cellular signals.

***trans*-acting protein:** A *trans*-acting protein has an exceptional feature of acting on or impacting only the molecule of DNA (deoxyribonucleic acid) from which it was expressed.

***trans*-acting small interfering RNAs (tasiRNAs):** A class of siRNAs that repress gene expression through PTGS in land plants. Precursor transcripts from TAS loci are polyadenylated and converted to double-stranded RNA and are then processed into 21-nucleotide-long RNA duplexes with overhangs. These segments are incorporated into RISC and direct the sequence-specific cleavage of target mRNA.

***trans*-activating CRISPR RNA (tracrRNA):** A small RNA encoded upstream of the CRISPR locus in type II CRISPR systems, with a 24-nucleotide sequence complementary to repeats of the CRISPR RNA (crRNA) precursor transcripts. Essential for the processing of pre-crRNA to mature crRNA.

***trans*-arrangement:** Linkage of the dominant allele of one pair and the recessive allele of another on the same chromosome. *Aka* *TRANS*-CONFIGURATION.

***trans*-configuration:** A double heterozygote where each of the dominant alleles are present in the opposite homologue.

***trans*-dominant mutation:** A mutation in a regulatory gene that can affect the expression of a structural gene on another chromosome.

***trans*-heterozygote:** A heterozygote of two linked genes with a mutant and wild-type allele linked, $a+/+b$. This is a more appropriate term than the 'repulsion phase' of linkage, since like alleles do not repel.

***trans*-hydrogenation:** Sequential transfer of electrons in cellular oxidation cycles.

***trans*-inactivation:** It involves inactivation of genes encoded on one T-DNA following transformation with a second T-DNA encoding a different marker gene.

***trans*-position:** When two mutations are located in two different homologous chromosomes. For example, mutations in repressor sequence that act through a diffusible protein product.

***trans*-regulatory elements (TREs):** TREs are DNA sequences encoding upstream regulators (i.e. *trans*-acting factors), which may modify or regulate the expression of distant genes. *Trans*-acting factors interact with *cis*-regulatory elements to regulate gene expression.

***trans*-regulatory hubs (TRHs):** Clusters of interchromosomal associations between *CIS*-REGULATORY DOMAINS (CRDs).

trans-specific evolution: The evolution of groups above the species level. *Aka* MACROEVOLUTION.

***trans*-splicing:** (i) A reconstruction of a larger transcript by joining the 5′ and 3′ portions of a transcript from two independent viruses (co-administered to the same tissue) that have formed intermolecular concatemers. (ii) A type of eukaryotic gene regulation where exon RNA is synthesized as a part of a separate transcript. (iii) Intermolecular splicing that happens in trypanosomes and worms where a short sequence called spliced leader (SL RNA) is linked to the 5′-end of many pre-mRNAs.

***trans*-tagging:** The obstructive activity of a functional gene by introducing a foreign DNA sequence.

***trans*-transcriptional gene silencing (*trans*-TGS):** *Trans*-TGS is a unidirectional event impacting an active locus (i.e. a locus that does not undergo *cis*-TGS spontaneously) induced by allelic, ectopic or extrachromosomal homologous sequences. Paramutation is a form of *trans*-TGS activity.

transalkylation: The alkylation of bases in DNA does not seem to be direct but transfer of the alkyl group from phosphate to base. This type of alkylation of bases in DNA is known as TRANSALKYLATION.

transaminase: A large group of enzymes whose function is to transfer the amino group from any one of at least 12 amino acids to α-keto acid to form another amino acid. *Aka* AMINOTRANSFERASES.

transamination: Enzymatic transfer of an amino group from amino acid to some other organic acid. During this catabolic reaction, the α-amino group is transferred to the α-carbon atom of α-ketoglutarate, leaving behind the corresponding α-keto acid analogue of the amino acid. There is no net deamination in these reactions because the α-ketoglutarate becomes aminated when an amino acid is deaminated.

Transamination is catalysed by enzymes called TRANSAMINASES or AMINOTRANSFERASES.

transcellular transport: Transport of solutes, such as nutrients, across an epithelium, by employing membrane transport proteins in the apical and basal appearances of the epithelial cells.

transcribed spacer: Part of an rRNA transcription unit that is transcribed but discarded during maturation; that is, it does not give rise to part of rRNA.

transcript: Initial eukaryotic RNA product is longer as it contains introns and exons. The primary transcript undergoes RNA processing (removal of introns) and other modifications as well to produce active RNA species.

transcript analysis: Experiments designed to determine which regions of a DNA molecule are transcribing into RNA at a given time, under defined conditions and/or at particular developmental stage.

transcript imaging (TI): A technique based on AMPLIFIED FRAGMENT LENGTH POLYMORPHISM (AFLP) analysis applied to cDNA. TI offers a quantified view of all the transcripts in a sample on an electronic image where approximately 20,000 AFLP bands, each representing a transcript, can be identified and quantitatively measured; differences between strains, samples, or treatments can be recognized; and the differential bands can be sequenced. The sequence of the band leads to the identification of the transcript.

transcript profiling (TP): TP measures the mRNA levels of all expressed genes relative to one another. A technique that evaluates transcriptional activity in the context of known biological pathways. Transcript profiling technology enables quantitative measurement of the transcriptional activity of potentially thousands of genes in biological samples.

transcript profiling (TP), analysis of: (i) Identification of the genes and biological pathways associated with clinically defined conditions. (ii) Large-scale gene function studies. (iii) Transcription profiling studies in biopsies for prostate cancer.

transcript profiling (TP), applications: (i) Gene expression. (ii) Microarray approach. (iii) Measure of expression. (iv) Single-nucleotide variation (SNV) discovery. (v) Post-transcriptional single-nucleotide variants. (vi) Fusion gene detection.

transcript profiling (TP), limitations of: The information gathered when sequencing a sample's transcriptome in this way has many of the same limitations as other RNA expression analysis pipelines. Mainly, the information gathered is tissue-specific. Gene expression is not uniform throughout an organism's cells, it is strongly dependent on the tissue type being measured. Also, the information gathered is time-dependent: During a cell's lifetime and context, its gene expression levels change. Because of this, care must be taken when drawing conclusions from the sequencing.

transcript profiling (TP), techniques of: Various techniques of transcript profiling include (a) cDNA-AFLP transcript profiling technique; (b) meiome and (c) GeneCalling.

transcript responsiveness: Responsiveness of genes to manipulation of transcription factors in **ESCs** is associated with histone modifications and tissue specificity.

transcript-based cloning (TBC): An innovation of gene identification and isolation based on DNA microarrays and the instability of RNA transcripts from genes damaged by mutations. The technique works well with relatively simple genomes, in both the model systems and crop species (e.g. barley) that have a much larger and more complex genome. Favourite genes can be isolated in one-tenth of the time and for one-tenth of the cost of existing methods. TBC relies on taking 'snapshots' of gene activity in normal plants and plants with genetic mutations. By comparing 'snapshots', it is possible to recognize and separate the gene affected by the mutation.

transcription: A biological process in which DNA is read by an enzyme called DNA-DEPENDENT RNA POLYMERASE, which uses one strand of DNA as a template to produce a complementary RNA molecule. Transcription is the first step in expression of a gene. *Aka* DNA TRANSCRIPTION. In this process, three major types of RNA are produced. MESSENGER RNAs (mRNAs) that encode amino acids, TRANSFER RNAs (tRNAs) read the information encoded on mRNA and transfer appropriate amino acid to growing polypeptide chain and RIBOSOMAL RNA (rRNAs), which are the constituents of ribosomal-site of protein synthesis.

transcription activation: Transcriptional activators are required to turn on the expression of genes in a eukaryotic cell. Activators bound to the enhancer can facilitate either the recruitment of RNA polymerase II to the promoter or its elongation.

transcriptional activation domains, identification of: Gene expression in *Arabidopsis* is controlled by more than 1,900 transcription factors (TFs). Activator TFs contain activation domains (ADs) that engage coactivator complexes. M. Morffy *et al.* (2024) use a yeast library strategy to experimentally recognize *Arabidopsis* ADs on a proteome-wide scale, and find that more than half of the *Arabidopsis* TFs contain an AD. They develop a neural network to precisely envisage ADs and to identify sequence features that are necessary to recruit coactivator complexes.

transcription activator-like effector (TALE) DNA-binding domain: TALEs are found in bacterial plant pathogens and are involved in regulating the genes of the host plant in order to facilitate bacterial virulence, proliferation and dissemination. They contain a central region of tandem 33–35 residue repeats and each repeat region encodes a single DNA base in the TALE's binding site.

transcription activator-like effector nucleases (TALENs): Artificial restriction enzymes generated by fusing a TAL effector DNA-binding domain to a DNA cleavage domain. TRANSCRIPTION ACTIVATOR-LIKE EFFECTORS (TALEs) can be quickly engineered to bind practically any desired DNA sequence. By combining such an engineered TALE with a DNA cleavage domain, one can engineer restriction enzymes that are specific for any desired DNA sequence. When these restriction enzymes are introduced into cells, they can be used for genome-editing *in situ*. TALENs are fusions of the *Fok*I cleavage domain and DNA-binding domains derived from TALE proteins. TALEs contain multiple 33–35 amino acid repeat domains that each recognizes a single-base pair. Like ZFNs, TALENs induce targeted DSBs that activate DNA damage response pathways and enable custom alterations.

transcription activator-like effectors (TALEs): DNA-binding proteins with a modular structure derived from *Xanthomonas* spp. (a plant pathogen). Each module is composed of nearly 34 amino acids and recognizes a single nucleotide. The base specificity is determined by the amino acids at positions 12 and 13 (known as repeat variable diresidues) in each module.

transcription and recombination: Effect of RNA polymerase II-dependent transcription on recombination has been reported in *Saccharomyces cerevisiae*. Stimulation in recombination is mediated by events initiating within the integrated plasmid sequences.

transcription and repair link: The general transcription factor TFIIH forms an essential part of both the basal transcriptional apparatus of the cell, called *transcriptome* and the complex of repair proteins, called REPAIROSOME. In the transcription complex, the TFIIH core is associated with a cyclin-dependent kinase complex (CDK7/cyclin H and the assembly factor MAT1 in mammals), whereas in repair, it is associated with other repair proteins.

transcription and translation in eukaryotes: In eukaryotes, nuclear membrane separates the site where mRNA is synthesized (nucleus) and where protein is to be synthesized (cytoplasm). Heterogeneous nuclear RNA (hnRNA) includes pre-mRNA, which is processed to form mRNA. The mature mRNA is then transported to cytoplasm in order for the translation process to start. Thus, transcription and translation are separate processes in eukaryotes.

transcription and translation in prokaryotes: In prokaryotes, transcription and translation are integrated processes as there is no separation between the sites of transcription and protein synthesis. Here, structural gene directly produces mRNA, which start getting translated even before the transcription has terminated. There is no processing of mRNA required in prokaryotes before it is used as a template for translation.

transcription attenuation: Inhibition of gene expression in bacteria by premature termination of transcription.

transcription bubble: A region of locally melted DNA in the process of being transcribed by RNA polymerase.

transcription coexpression: Transcription coexpression of interacting gene products is required for complex molecular processes. *Cis*-regulatory elements orchestrate coexpression. Syntactical rules governing this regulatory function are flexible but become highly constrained evolutionarily once they are established in a particular element.

transcription elongation: Elongation of transcription takes place in alternating laps of monotonous and inchworm-like movement with the flexible DNA polymerase configuration being subject to direct sequence control. During this step, RNA polymerase catalyses nucleophilic attack by the 3′-OH group of the growing RNA chain on α phosphorous of the ribonucleoside triphosphate, resulting in the formation of a new phosphodiester linkage and release of pyrophosphate. Twelve nucleotides of the template strand base pair with newly synthesized RNA to form a hybrid RNA:DNA helix. When a new nucleotide joins the 3′-end of the growing RNA chain and lengthens the RNA: DNA hybrid by following steps: (i) The length of the RNA:DNA hybrid must be reduced by melting one bp at the upstream end of the transcription bubble. (ii) The DNA bp must be melted at the downstream end of bubble. (iii) The DNA bp must be reformed at the upstream end of bubble. (iv) RNAP must advance along the DNA by one nucleotide, positioning its active site over the next unpaired nucleotide on the template strand. Elongation lasts till a termination complex is formed. The mechanism of substrate loading in multisubunit RNA polymerase is vital for understanding the general principles of transcription.

transcription elongation complex: The mechanism of substrate loading in multisubunit RNA polymerase is vital for understanding the general principles of transcription.

transcription elongation tuning: RNA chain elongation is a highly processive and accurate process that is finely regulated by numerous intrinsic and extrinsic signals. G. Bar-Nahum *et al.* (2005) describe a general mechanism that that governs RNAP movement and response to regulatory inputs such as pause, terminators and elongation factors. *Escherichia coli* RNA polymerase moves by a complex Brownian ratchet mechanism, which acts prior to phosphodiester bond formation. The incoming substrate and flexible F bridge domain of the catalytic centre serve as two separate ratchet devices that function in concert to drive forward translocation. The adjacent G loop domain controls F bridge motion, thus keeping the proper balance between productive and inactive states of the elongation complex. This balance is critical for cell viability since it determines the rate, processivity and fidelity of transcription.

transcription factor- and microRNA-mediated gene regulation: Gene regulatory factors that control the expression of genomic information come in a variety of flavours, with transcription factors and microRNAs representing the most numerous gene regulatory factors in multicellular genomes. There are common principles of transcription factor- and microRNA-mediated gene regulatory events.

transcription factor binding sites: Several sites are available: (i) PLACE (Plant *cis*-acting regulatory DNA elements). (ii) PlnTFDB (Plant Transcriptional Factor Database). (iii) DBD: Transcription factor prediction database. (iv) rVista 2.0 – High-throughput discovery of functional regulatory elements in sequence alignments. Excluding up to 95% false-positive transcription factor binding sites predictions while maintaining high sensitivity of the search. (v) TESS (Transcription Element Search System) is a tool for predicting transcription factor binding sites in DNA sequences. It can identify binding sites using site or consensus strings. This resource is best for scanning short DNA sequences.

transcription factor CCAAT-enhancer-binding proteins (C/EBPs): Transcription factors of family C/EBP use a bipartite structural motif to bind DNA. The C/EBP family consists of six transcription factors (α, β, γ, δ, ε and ζ). Two protein chains dimerize through a set of amphipathic α-helices termed the leucine zippers. 'SCISSORS GRIP' MODEL has been proposed to describe how C/EBP binds DNA in the major groove.

transcription factors (TFs): The proteins that regulate the activation of transcription in the eukaryotic nucleus. Transcription factors localize to regions of promoter and enhancer sequence elements either through direct binding to DNA or through binding other DNA-bound proteins. They act by promoting the formation of the PREINITIATION COMPLEX (PIC) that recruits and activates RNA polymerase. TF consists of at least two functional domains – a DNA-BINDING DOMAIN (DBD) and an activating domain (AD). Three classes of transcription factors known are GENERAL TRANSCRIPTION FACTORS, UPSTREAM ENHANCER FACTORS and INDUCIBLE TRANSCRIPTION FACTORS. The transcription initiation of most structural genes requires the presence of seven GENERAL TRANSCRIPTION FACTORS (GTFs), named TFIIA, TFIIB, TFIID, TFIIE, TFIIF, TFIIH and TFIIJ, many of which are multiprotein complexes. These GTFs combine in an ordered pathway with promoter and RNA polymerase near the transcription start site to form a so-called PREINITIATION COMPLEX. TFIIF accelerates RNA chain growth uniformly and TFIIS helps in elongation by relieving obstacles. TFIIS causes hydrolytic cleavage at 3′-end of RNA chains, which are blocked. Its action is similar to cleavage by GreA and GreB in *Escherichia coli*. Rpb9 is small subunit of yeast RNA Pol II participating in elongation. Proteins are directly involved in regulation of transcription initiation by binding to the control elements, and there are ubiquitous as well as cell- and tissue-specific transcription factors. Several families of TFs include helix-loop-helix proteins, helix-turn-helix proteins, leucine zipper proteins and zinc-finger proteins. TFs interpret the genetic regulatory information, such as in transcriptional enhancers and promoters. Salient properties of transcriptional factors are: (i) They are composed of functional modules. (ii) They regulate transcription via recruitment of coactivators and corepressors. (iii) They can be regulated by posttranslational modifications. (iv) They are often members of multiprotein family. (v) Chromatin is an integral component in the function of transcriptional factors. (vi) Recognition sites for transcriptional factors tend to be located in clusters. *Aka* TRANSCRIPTIONAL FACTORS.

transcription factors (TFs), potential of: TFs (i) cure diseases (e.g. by blocking the deleterious effects of certain disease causing genes); (ii) assist farmers in crop protection (e.g. by switching on the genes that cause crop plants to initiate 'cold hardening' or certain types of insect resistance mechanisms); (iii) improve human health (e.g. PUFA modulation of genes, modulation of genes by some vitamins, etc.); (iv) Nowadays the transcription factors are used to upregulate the transcription of gene(s) controlling abiotic stresses such as drought/salt through genetic transformation.

transcription in bacteriophages: When a temperate bacteriophage such as lambda (λ) infects a bacterium, it can follow either of two pathways of development. A temperate phage can either enter the lytic cycle, during which it reproduces and lyses the host cell, just like a virulent phage or enter the lysogenic pathway, during which its chromosome is inserted into the chromosome of the host and thereafter is replicated like any other segment of that chromosome. When integrated in the chromosome of the host cell, the temperate phage chromosome is called a PROPHAGE. In a lysogenic bacterium, the genes of the prophage that encode products involved in the lytic pathway must not be expressed.

transcription in eukaryotes: Transcription apparatus of eukaryotic cells is more complex than that of prokaryotes. Eukaryotes require transcription factors to first bind to the promoter region and then help recruit the appropriate polymerase. RNA polymerase I is located in the nucleolus, a specialized nuclear substructure in which ribosomal RNA (rRNA) is transcribed, processed and assembled into ribosomes. RNA polymerase II accounts for synthesis of HETEROGENEOUS NUCLEAR RNA (hnRNA), including the precursor of mRNA. RNA polymerase III transcribes a variety of structural RNAs that includes the 5S pre-rRNA, transfer pre-RNAs (pre-tRNAs) and small nuclear pre-RNAs. In plants, two additional RNA polymerases, RNA Pol IV and RNA Pol V have been found. In eukaryotes, there are certain post-transcriptional processes that occur before translation begins.

transcription in nonbacterial viruses: In cells infected by viruses that encapsidate DNA, synthesis of viral mRNA must precede production of proteins. The majority of DNA including the parvo-, papova-, hepadna-, adeno- and herpesviruses, depend on cellular RNA polymerase II, the enzyme that synthesizes cellular mRNA. The retroviruses, which establish themselves as DNA proviruses early in their infection cycles, also use the cellular RNA polymerase II system.

transcription in prokaryotes: (i) Four types of ribonucleotides of A, U, C and G (ii) Template strand – two complementary strands of double-helical DNA separate and only one of them ($3' \leftarrow 5'$), in relation to a particular gene, is utilized as a template for RNA synthesis. This strand of DNA is known as ANTISENSE STRAND. RNA chain grows in $5' \rightarrow 3'$ direction. In prokaryotes, there is only one species of DNA-dependent RNA polymerase. A unit of DNA that acts as a template for RNA transcription is called an OPERON in prokaryotes. An operon contains a promoter, an operator and structural gene(s). (iii) DNA-dependent RNA polymerase – this enzyme is composed of five polypeptides (σ_2, β, β', ω and σ) of five different types. This means five different genes (σ, β, β', ω and σ) are required to produce *Escherichia coli* polymerase. Active form of enzyme is known as HOLOENZYME. Holoenzyme comprises a core enzyme and a sigma (σ) factor. Core enzyme has four types of subunits (σ_2, β, β' and ω). No covalent bond is formed between various chains. Aggregation results from formation of secondary bonds. Core enzyme catalyses formation of internucleotide 3′-5′ phosphodiester bonds equally well in the absence or presence of σ factor. Only holoenzyme can initiate transcription; but then σ factor is released, leaving the core enzyme to undertake elongation. It also requires PRIBNOW BOX. Only holoenzyme can initiate transcription. Sigma factor and core enzyme recycle at different points in transcription. Rho factor plays role in termination of transcription. This figure also shows role of nusA protein in transcription elongation. Transcription and translation processes are integrated processes in prokaryotes. Translation on mRNA begins before transcription is complete. After translation, mRNA is released and degraded.

transcription in viruses: The viral polymerases are diverse, including some forms that can use RNA as a template instead of DNA. This happens in negative-strand RNA viruses and dsRNA viruses, both of which occur for a portion of their life cycle as double-stranded RNA. However, some positive-strand RNA viruses, such as polio, also comprise these RNA-DEPENDENT RNA POLYMERASES.

transcription initiation sites: To the GCCAAT recognition motif, often referred the CCAAT box, DNA-binding proteins, collectively designated as CCAAT-BINDING TRANSCRIPTION FACTORS (CTF) bind that act as a bona fide sequence-specific transcription initiation proteins.

transcription initiation, intermediate complexes in: *See* INTERMEDIATE COMPLEXES IN TRANSCRIPTION INITIATION.

transcription initiation, structural visualization of: X. Chen and others (2023) determined the structures of 16 de novo transcribing complexes, TC2 to TC17, containing nascent RNAs of 2 to 17 nucleotides, respectively. Connecting these structures revealed an abrupt transition from initially transcribing complexes (in TC2 to TC9, GTFs remained bound to the promoter and Pol II) to early elongation complexes (in TC10 to TC17, GTFs dissociated from the promoter). This transition was accompanied by remarkable conformational and compositional changes, including transcription bubble collapse and GTF dissociation, which allowed RNA Pol II to escape the promoter for product RNA synthesis.

transcription itself regulates gene expression: The process of transcribing DNA can itself regulate gene expression in yeast: in this case concerned it is the key act of reading the DNA. Process of intergenic transcription and gene regulation is given here in yeast, as an example. During yeast growth on minimal medium lacking the amino acids serine and glycine, transcription of *SER3* messenger RNA is required to produce the SER3 protein, which is essential for serine biosynthesis. SER3 transcription activator(s) bind to the *SER3* promoter sequence, allowing RNA polymerase II to transcribe the gene. When yeast cells grow on rich medium, SER3 protein is not required. Repression of the *SER3* gene is achieved by activation of the intergenic *SRG1* promoter located upstream of *SER3*. After the initiation of transcription at *SRG1*, the RNA polymerase II machinery runs over the *SER3* gene promoter, thereby preventing the binding of SER3 transcriptional activators and thus silencing the *SER3* promoter. Transcription of *SER3* gene is tightly repressed during growth in rich medium.

transcription memory: (i) Epigenetic transcriptional memory, whereby the rate or strength of the expression of genes in response to a stimulus is enhanced by previous exposure to that stimulus. (ii) Altered gene expression responses in a primed plant; either a sustained change after a short stimulus or a modified gene activation pattern upon a recurring stress.

transcription of centromeric repeats: Heterochromatin in eukaryotic genomes controls diverse chromosomal processes including transcriptional silencing. However, in fission yeast RNA polymerase II transcription of centromeric repeats is vital for RNA interference-mediated heterochromatin assembly. RNA polymerase II transcription during S phase is related with the loading of RNA interference and heterochromatin factors such as the Ago1 subunit of the RITS complex and the Clr4 methyltransferase complex subunit Rik1.

transcription of Line-1 elements: LINE-1 (L1) elements are retrotransposons that comprise large fractions of mammalian genomes. Transcription through L1 open reading frames is inefficient owing to an elongation defect, inhibiting the robust expression of L1 RNA and proteins, the substrate and enzyme(s) for retrotransposition. This elongation defect probably controls L1 transposition frequency in mammalian cells. This transcriptional defect has been bypassed by synthesizing the open reading frames of L1 from synthetic oligonucleotides, altering 24% of the nucleic acid sequence without changing the amino acid sequence. Such resynthesis leads to greatly enhanced steady state L1 RNA and protein levels.

transcription of mammalian genes: RNA polymerase II appears to be responsible for the transcription of all protein-encoding genes in mammalian cells. Two types of control regions are involved in transcription initiation – promoters and enhancers. Two types of control elements recognized in the promoter region are constitutive elements and regulatory elements.

transcription pausing: The phenomenon where in both prokaryotic and eukaryotic RNA polymerases display slow genetic transcription at transcription pause sites where particular regions of a DNA molecule are being transcribed (i.e. RNA chain elongation does not transpire at a persistent rate). Pausing may be necessary for proper control of transcription and happens immediately after transcription of regions with dyad symmetry by formation of a hairpin near the 3′ end of growing transcript.

transcription proofreading mechanism: Mistakes can happen as RNA polymerase copies DNA into transcripts. A proofreading mechanism that eliminates incorrect RNA is triggered by the erroneous RNA itself. Correction of misincorporated errors at the growing end of the transcribed RNA is inspired by the misincorporated nucleotide.

transcription regulators: These proteins, perform two functions – recognize specific sequences of enhancer or other regulatory elements and bind to them, initiate transcription of regulated genes by activating the promoter. In yeast, it has been demonstrated that the DNA binding and gene activation activities of a yeast transcription regulator can be separated. Yeast GAL4 regulator is believed to consist of two domains, the AMINO DOMAIN, which recognizes sequences in the GALACTOSE UPSTREAM ACTIVATING SEQUENCE (UAS_G) and the CARBOXYL DOMAIN, which when bound to DNA through the amino domain, contacts another protein to facilitate positive control.

transcription silencing control by H3K9me: Histone H3 lysine-9 methylation (H3K9me) regulates transcription silencing. Cells delimit H3K9me domains to avoid silencing of essential genes. Using *Arabidopsis* genetic systems that induce H3K9me2 in genes and transposons *de novo*, K. Yabe *et al.* (2024) display that H3K9me2 accumulation inconsistently also results in the deposition of the euchromatic mark H3K36me3 by a SET domain Histone-lysine N-methyltransferase, ASHH3. ASHH3-induced H3K36me3 deliberates anti-silencing by averting the demethylation of H3K4me1 by LDL2, which arbitrates transcriptional silencing downstream of H3K9me2. Thus, H3K9me2 not only enables but also arranges silencing by actuating antagonistic silencing and anti-silencing pathways, providing insights into the molecular basis underlying proper partitioning of chromatin domains and the creation of metastable epigenetic variation.

transcription start site (TSS): The position in a gene at which the mRNA synthesis starts. The segment from this point downstream to the translation initiation site is called the 5′ UNTRANSLATED REGION (5′ UTR).

transcription termination, intrinsic: Cellular RNA polymerases in both bacteria and eukaryotes can terminate their transcription through a factor-independent termination pathway, called INTRINSIC TERMINATION TRANSCRIPTION, in which RNA polymerase recognizes terminator sequences, stops nucleotide addition and releases nascent RNA spontaneously. L. You *et al.* (2023) report a set of single-particle cryo-electron microscopy structures of *Escherichia coli* transcription intrinsic termination complexes representing key intermediate states of the event. The structures show that RNA polymerase pauses at terminator sequences, the terminator RNA hairpin folds inside RNA polymerase and RNA polymerase rewinds the transcription bubble to release RNA and then DNA. These macromolecular snapshots define a structural mechanism for bacterial intrinsic termination and a pathway for RNA release and DNA collapse that is relevant for factor-independent termination by all RNA polymerases. Transcription terminators prevent synthesis of unnecessary long transcripts to avoid energy drain on a cell. Undesirable secondary structures formed may reduce efficiency of translation. Promoter occlusion (transcription of cloned gene from promoter may interfere with transcription of other essential regulatory genes) may occur. Inclusion of terminator solves the problem.

transcription terminator: Transcription termination process in prokaryotes comprises formation of 'hairpin structure'. DNA sequence that delivers the signal to cease or stop transcription is known as a TERMINATOR. Terminators require that all the hydrogen bonds holding RNA-DNA together must be broken. Termination of transcription transpires because the elongation complex is less stable when transcribing certain specific DNA sequences. Terminators may be either rho-dependent or rho-independent. Termination of transcription is an important regulatory step closely related to transcriptional interference, and even transcriptional initiation. DNA sequence 3′-AAAAAAT-5′ in sense strand of DNA is terminator sequence. Probably, RHO (ρ) FACTOR recognizes this sequence. At some terminators, termination event can be prevented by specific ancillary sequences (factors) that interact with RNA polymerase. Responsibility for termination lies with the sequences already transcribed by RNA polymerase.

transcription unit: A group of genes that code for functionally associated RNA molecules or protein molecules. A sequence of DNA transcribed into a single RNA, starting at the promoter and ending at the terminator. This group of genes is thus transcribed together as a unit. *Aka* TRANSCRIPTIONAL UNIT.

transcription unit design: It refers to whether one or more mRNAs encoding different proteins are produced from a primary transcript. Two types of transcriptional units (simple and complex) are known. *See* SIMPLE TRANSCRIPTION UNIT *and* COMPLEX TRANSCRIPTION UNIT.

transcription-coupled alternative RNA splicing: Two models have been proposed to explain the role of RNA polymerase II in the regulation of alternative RNA splicing. The first model is known as the RECRUITMENT MODEL, in which RNA polymerase II and transcription factors interact, directly or indirectly, with splicing factors, thereby increasing or decreasing the efficiency of splicing. A second, kinetic model proposes that the rate of transcription elongation influences the inclusion of alternative exons by affecting whether the splicing machinery is recruited sufficiently quickly for spliceosome assembly and splicing to occur.

transcription-coupled nucleotide excision repair (TC-NER): In both prokaryotes and eukaryotes, transcribing DNA is repaired at a faster rate than non-transcribing DNA. The process of repair of transcribing DNA is known as transcription-coupled nucleotide excision repair (TC-NER). In eukaryotes, initial damage is recognized by stalled RNA polymerase. All the proteins involved in GGR except XPC-HR23B are required for TC-NER. It requires additional proteins, including CSA and CSB. Cells use TC-NER to efficiently eliminate DNA lesions.

transcription-replication conflicts: An important advance in cancer therapy has been the development of POLY(ADP-RIBOSE) POLYMERASE (PARP) inhibitors for the treatment of HOMOLOGOUS RECOMBINATION (HR) deficient cancers. PARP inhibitors trap PARPs on DNA. The trapped PARPs are thought to block replisome progression, resulting in DNA double-strand breaks that require HR for repair. M. Petropoulos *et al.* (2024) show that PARP1 functions together with TIMELESS and TIPIN to protect the replisome in early S phase from transcription-replication conflicts.

transcription-replication interaction profiles (TRIPs): Refers to the transcriptional response of each gene to its own replication, which they term the transcription-replication interaction profile which is based on the analysing single-bacterium RNA-sequencing data. A.W. Pountain *et al.* (2024) found that the response to the universal perturbation of chromosomal replication integrates biological regulatory factors with biophysical molecular events on the chromosome to reveal the local regulatory context of a gene. Whereas the TRIPs of many genes conform to a gene dosage-dependent pattern, others diverge in distinct ways and this is shaped by factors such as intra-operon position and repression state.

transcription-replication interactions: Bacterial genome regulation takes place by transcription-replication interactions, which means that novel genome-wide classification of regulatory modes is based on the transcriptional response of each gene to its own replication. By revealing the underlying mechanistic drivers of gene expression heterogeneity, A.W. Pountain *et al.* (2024) provide a quantitative, biophysical framework for modelling replication-dependent expression dynamics.

transcription-translation reaction: An *in vitro* series of reactions, involving the synthesis (transcription) of an mRNA from a plasmid (usually with T7 or SP6 RNA polymerase), followed by use of the mRNA to programme translation in a cell-free system such as a rabbit reticulocyte lysate. The polypeptide product of translation is usually labelled with [^{35}S]-methionine and examined in an SDS-PAGE gel with or without prior immunoprecipitation. This series of reactions permits the synthesis of a polypeptide from DNA *in vitro*.

transcription, fidelity of: During transcription elongation, fidelity of transcription refers to the hydrolytic reaction stimulated by misincorporated nucleotides proofread most of the misincorporation events and thus serves as an inherent mechanism of transcriptional fidelity.

transcription, requirements for: Process of transcription requires: (i) Four types of ribonucleotides containing A, U, C and G. (ii) Two complementary strands of double-helical DNA separate and at a given time only one of them (3′←5′) is utilized as a template for RNA synthesis. This is called the TEMPLATE STRAND, CODING STRAND or ANTISENSE STRAND. RNA chain grows in 5′→3′ direction. The other strand that is not being used as template is called NON-TEMPLATE STRAND, NON-CODING STRAND or SENSE STRAND. (iii) TRANSCRIBING ENZYME (DNA-dependent RNA polymerase).

transcriptional activation, rate-limiting step of: The rate-limiting step of transcriptional activation in eukaryotes, and thus the critical point for gene regulation is unknown. Random removal and reformation of promoter nucleosomes can account for stochastic and kinetic properties of *PHO5* expression. Disassembly of promoter nucleosomes could be the rate-limiting step for *PHO* expression. SWR1 complex catalyses an ATP-dependent histone exchange with specificity for the histone variant H2A.Z. This provides a new theme in chromatin remodelling mechanisms as an explanation of how histone variants are incorporated into the nucleosome outside of the S phase.

transcriptional activator: Any DNA-binding protein that stimulates transcription by relating with DNA sequence motifs (transcription control regions), RNA polymerase, or both. Some transcriptional activators have two surfaces: (i) the DNA-binding region that places the protein on DNA and (ii) the activating region that networks with a target protein to endorse initiation of transcription.

transcriptional antitermination: A mechanism that enables transcription termination signals of an operon to be overcome.

transcriptional apparatus in Archaea: The transcriptional apparatus in Archaea can be described as a simplified version of its eukaryotic RNA polymerase II counterpart, comprising of an RNAPII-like enzyme as well as two general transcription factors, the TATA-binding protein (TBP) and the eukaryotic TFIIB ortholog TFB.

transcriptional bursting: RNA polymerase II initiates transcription either randomly or in bursts. M. Oehler *et al.* (2023) show that transcription factor WHITE COLLAR COMPLEX (WCC) of *Neurospora* is not only an activator but also represses transcription by recruiting HISTONE DEACETYLASE 3 (HDA3). Bursts of transcription are governed by a long-lived refractory state established and maintained by WCC and HDA3 at the core promoter, at an upstream activating sequence. Thus, transcription factor-mediated repression also influences transcriptional bursting. T. Fukaya in 2023 found that transcriptional bursting regulates gene expression during cell fate determination and disease processes in mammals, plays a critical role in dosage compensation mechanisms of X-linked genes during sex specification, as well as in establishing the temporal and spatial patterns of gene expression during Drosophila embryogenesis, serves as a major source of random effects in gene expression and contribute to cellular variability within a population of cells.

transcriptional complexity: Transcribed portions of several organisms are larger and more complex than expected, and many functional properties of transcripts are based not on coding sequences but on regulatory sequences in untranslated regions or non-coding RNAs. Alternative start and polyadenylation sites and regulation of intron splicing add extra dimensions to the enrichment of transcriptional output.

transcriptional control: In general, genes of eukaryotes are apparently 'turned off' and must be activated. Differentiation and growth/development takes place only in eukaryotes. At any one time, in a eukaryote cell, only 2–15% of total genome is expressed. It means some sort of regulation is occurring. Histones act as repressors of gene activity. Nonhistones are acidic in nature and are thought to act as controlling elements for gene transcription. Expression of mammalian genes may be meticulously controlled by repressors acting on the translation of messenger RNA. Eukaryotic DNA used for RNA polymerase recognition, binding and transcription shows the presence of the 'TATA' box as in prokaryotes. The TATA box lies exactly where the transcription begins. In addition, a number of promoters show a 9-nucleotide sequence GGC/TCAATCT, called the 'CAAT' box, which lies about 40 nucleotides upstream from TATA box. Other sequences drawn in gene regulation in eukaryotes are enhancer sequences, which are most probably involved in recognition and binding of RNA polymerase to DNA. Enhancers also seem to play crucial role in cell differentiation. Regulation of protein synthesis occurs by regulating the synthesis of mRNA. It is the common term for all *cis*-acting DNA regulatory sequences that regulate transcription of a particular gene. An example of transcriptional control of gene activity is provided by the gene coding for enzyme β-galactosidase, which degrades lactose to form galactose and glucose. The synthesis of this enzyme is induced when lactose is present in the medium and glucose is absent. When the β-galactosidase (*lac z*) gene is induced, mRNA transcribed from the gene lac *z* appears and rate of β-galactosidase synthesis increases to 1000 folds and is maintained as long as the inducing substance is present.

transcriptional derepression: When a substrate binds a repressor protein and causes it to undergo a conformational change. If the repressor is bound upstream of a gene, such as in an operator sequence, then it would be repressing the gene's expression.

transcriptional elements: Transcriptional elements possess a consensus sequence. They enact vital part in gene expression. For example, maximal transcription of *heat shock protein 70* (*hsp70*) gene necessitates two copies of a 14-bp heat shock element (HSE) with consensus sequence CNNGAANNTTCNNG.

transcriptional enhancer: Any of a class of *cis*-acting DNA sequences (usually 200 bp long) that are able to activate the transcription by RNA polymerase II of cellular and viral genes in a manner relatively independent of its position and orientation. Transcription activation can transpire over distances >6 kb from either the 5′ or 3′ end of a gene. Many enhancers contain an octamer sequence motif ATTTGCATT. Frequently, enhancers occur near Z-DNA and are localized in introns. *See* ENHANCERS.

transcriptional factors (TFs): *See* TRANSCRIPTION FACTORS.

transcriptional forests: Genomic mapping of the transcriptome reveals transcriptional forests, with overlapping transcription on both strands, separated by deserts in which few transcripts are observed. The data thus obtained provide a comprehensive platform for the comparative analysis of mammalian transcriptional regulation in differentiation and development.

transcriptional gene silencing (TGS): A form of gene silencing identified by run-on experiments showing that transcription initiation is blocked at the DNA level in the nucleus. TGS often begins when transcription factors (proteins that regulate gene activity) by recognizing and binding to specific sequences to temporarily block RNA production. Silencing of transgenes or endogenous gene achieved by the formation of a repressive chromatin environment at a locus targeted by small RNA making it unapproachable to transcriptional machinery. TGS is usually associated with DNA methylation and often triggered by siRNAs. TGS is the result of histone modifications, creating an environment of heterochromatin around a gene that makes it unreachable to RNA polymerase, transcription factors, etc.). Two prevailing models occur for silencing: (i) stearic hindrance in silenced chromatin inhibits the binding of upstream activator protein or polymerase or (ii) silencing primarily blocks steps downstream of the transcription preinitiation complex formation. Results of recent investigations contradict both the models. Transcriptional silencing at several *URA3* transgenes, and the naturally silenced endogenous *HMRa* and *HMR*α mating type genes, has been revealed to act downstream of gene activator protein through binding to strongly reduce the occupancy of TFIIB, RNA polymerase II and TFIIE at the silenced promoters. Certain terms related to specific topics of transcriptional gene silencing include GENOMIC IMPRINTING, PARAMUTATION, TRANSPOSON SILENCING, TRANSGENE SILENCING, POSITION EFFECT and RNA-DIRECTED DNA METHYLATION. *See details as separate terms. See* CIS-TGS *and* TRANS-TGS. *Compare with* POST-TRANSCRIPTIONAL GENE SILENCING (PTGS).

transcriptional gene silencing through convergent transcription: Convergent transcription induces TGS in *trans* for both fission yeast and mammalian cells. Introduction of convergent transcription plasmids into either *Schizosaccharomyces pombe* or mammalian cells allows the production of dsRNA from inserted gene fragments, resulting in TGS of endogenous genes.

transcriptional hub: Why would two distinct genes on separate chromosomes and from different nuclear locations unite in response to signals for gene expression? They might be seeds for formation of transcriptional hubs. In the absence of oestrogen, two genes (*TEF1* and *GREB1*) that are activated by this hormone and found on different chromosomes, reside in different locations within the nucleus.

transcriptional memory: Sustained differential response in gene expression after an exogenous cue. Transcriptional memory can be evident from either sustained change in the expression (activation or repression) or from a modified response after a second cue.

transcriptional pausing: Transcriptional pausing by RNA polymerase plays an important role in the regulation of gene expression. Defined, sequence-specific pause sites have been identified biochemically. Single-molecule studies have also shown that bacterial RNA polymerase pauses frequently during transcription elongation. Elongation may be impeded by pause sites, which induces temporary reversible block to nucleotide addition or by arrest or dead ends that stops the transcription, which could only be resumed by factors like Gre A and Gre B. Gre A and Gre B are elongation factors involved in transcription elongation including suppression of transcription arrest, enhancement of transcription fidelity and facilitating transcription from abortive initiation to productive elongation. During transcription, RNA polymerase moves processively along a DNA template, producing a complementary

RNA. This stage in early elongation appears to be an important broadly used target of gene regulation. *Aka* TRANSCRIPTIONAL STUTTERING.

transcriptional processivity: The affinity of RNA polymerase for template DNA is stated as transcriptional processivity. The interactions that occur during elongation process come under this phenomenon. Repressors bind to selected sets of genes at sites called SILENCERS. They interfere with functioning of activators and slow down transcription. Three types of factors play role in transcriptional processivity: (i) BASAL FACTORS respond to activators and position RNA polymerase at the start of the protein-encoding region of a gene and send the enzyme on its way. (ii) COACTIVATORS are the adaptor molecules that integrate signals from activators and repressors and relay the result to basal factors. (iii) ACTIVATORS bind to genes at sites called ENHANCERS. They help to determine which genes will be switched on and they speed up the rate of transcription.

transcriptional regulation of nitrogenase: Nitrogen fixation is regulated at the transcriptional level in response to environmental oxygen and ammonium levels. Nitrogenase components are oxygen labile, bacteria repress transcription when oxygen levels are high. Nitrogenase synthesis is also repressed when the cellular level of fixed nitrogen is sufficiently high. Legume nodules consist of five distinct regions from free end to attachment end – (i) nodule meristem, (ii) prefixation zone, (iii) interzone, (iv) nitrogen fixation zone and (v) senescence zone. An oxygen barrier is present in the nodule parenchyma surrounding the nodule vascular bundle that reduces oxygen access to the central tissue of the nodule. This oxygen barrier is interrupted in the meristem, an oxygen gradient forms that extends from the distal to the proximal end of the nodule. In the first cell layer of the interzone, the low oxygen concentration leads to the events described in the region known as green box. Here low oxygen concentrations activate the bacterial transmembrane oxygen sensor protein FixL, which in turn phosphorylates and thereby activates the transcriptional activator FixJ. The activated FixJ protein (FixJ*) induces transcription of *nifA* and *fixK*. Protein products of *nifA* and *fixK* genes induce the transcription of different genes that are encoding proteins involved in the process of nitrogen fixation. NifA induces transcription of *nifHDK* and FixK induces transcription of *fixNOOP* genes.

transcriptional regulators: DNA-binding transcriptional regulators infer the genome's regulatory cypher by binding to specific sequences to promote or repress gene expression. Transcriptional regulators may function as therapeutic targets.

transcriptional regulatory networks: The gene regulatory networks, which consist of physical and functional interactions between transcription factors (TFs) and their target genes.

transcriptional repression or activation, choice of: A factor MeCP2 implicated in autism spectrum disorders can both suppress and activate the expression of genes involved in neurodevelopment. MeCP2 may play a more complex role, coordinating either transcriptional repression or activation in the hypothalamus, depending on the molecular contact.

transcriptional repression: Transcriptional repressors are usually viewed as proteins that bind to promoters in a way that impedes subsequent binding of RNA polymerase.

transcriptional start site (TSS): The nucleotide of a gene or cistron at which transcription (RNA synthesis) starts; the most common triplet at which transcription begins in *Escherichia coli* is CAT. Primer extension identifies the transcriptional start site.

transcriptional terminators: (i) TransTerm searches for rho-independent terminators in the vicinity of annotated genes. (ii) FindTerm (Softberry Inc.) can also be used for mapping rho-independent terminators. (iii) RibEx: Riboswitch Explorer scans <40 kb DNA for potential genes (which are linked to BLASTP) and several hundred regulatory elements, including riboswitches.

transcriptional unit: *See* TRANSCRIPTION UNIT.

transcriptions silencing and paramutation: The genetic effects of transcriptional silencing are similar to the process of paramutation, an allelic interaction where one paramutagenic allele suppresses another paramutable allele in the heterozygote. The result of paramutation in most cases is that the paramutable allele becomes paramutagenic and the change is heritable through meiosis but ultimately reversible. There is no change in DNA sequence and the effect appears to be related to gene copy number and chromosome position. In some cases, paramutation is associated with a change in the amount of DNA methylation.

transcriptome: (i) Refers to the entire set of all gene transcripts (i.e. mRNA segments resulting from the transcription process) in a given organism. (ii) Total RNA complements generally described for a specific tissue or an organ under specific conditions. (iii)

The transcriptome of the plant as a whole would consist of mRNA available in all kinds of cells in a plant species in time and space. Transcriptomics helps to understand the role of transcriptome in an organism's structure, growth, health, disease (and/or that organism's resistance to disease) and so on. Those roles are predominantly due to the impact of each protein molecule (i.e. resulting from the mRNA segments being translated in cells' ribosomes), which is in turn due to the protein molecule's composition and its tertiary conformation (which determines the protein's impact in the organism's tissues, metabolism, etc.). Transcriptome is dynamic as it depends on the organism, cell type, developmental stage and environment.

transcriptome analysis: Transcriptional profiling of a cell. Many of the non-coding regions, previously thought to be functionally inert, are actually transcriptionally active genomic sections with various features. Once mRNA is isolated, it is used to develop a cDNA library. This library is utilized for a variety of studies. All these activities of mRNA can be collectively addressed by several available high-throughput technologies, which are classified into a number of different categories: (i) cDNA-AFLP. (ii) Sequence tag-based technologies. (iii) Microarray-based technologies. (iv) GeneCalling. Transcriptome analysis includes the following three activities: (i) complete transcript cataloguing, (ii) dynamic transcript profiling and (iii) analysis of gene expression regulatory network.

transcriptome assembly: *De novo* transcriptome assembly is the *de novo* sequence assembly method of creating a transcriptome without the aid of a reference genome. Once RNA is extracted and purified from cells, it is sent to a high-throughput sequencing facility, where it is first reverse-transcribed to create a cDNA library. This cDNA can then be fragmented into various lengths depending on the platform used for sequencing. Each of the following platforms utilizes a different type of technology to sequence millions of short reads: 454 Sequencing, Illumina and SOLiD.

transcriptome complexity: Transcriptome is more complex than a genome. More than one protein can result from each gene in an organism's genome, due to interactions (a) between genes, (b) between genes and their (protein) products and (c) between genes and some environmental factors. Mechanistically, this results in different proteins being produced (during the translation process). Microarrays are used to study transcriptomes, but it is possible only for those organisms that possess few genes. Microarray for all human genes could be carried by ten glass slides of 18 mm × 18 mm. For example, for studying changes in transcriptome as a result of a disease, a microarray is prepared with a cDNA library from normal tissue. Hybridization with labelled cDNA from the cancerous tissue, reveals which genes are up or downregulated. An alternative to microarray is provided by DNA chips.

transcriptome diversity: A-to-I RNA editing results in transcriptome diversity whose importance has been recognized in normal brain function.

transcriptome mapping: The procedure to create a genetic map with cDNA fragments that differ in size in the respective parents and segregate in the progeny. Usually, the resultant map divulges maternal or paternal inheritance of the fragments. Since the cDNA fragments are displayed on northern-type gels, interesting fragments can be directly isolated and sequenced. The sequence can be annotated such as to identify the underlying gene.

transcriptome sequencing: Transcriptome sequencing employs high-throughput sequencing technologies to get access to almost all transcripts of specific tissues or cells in a certain state by comprehensive and rapid cDNA sequencing, becoming the basis as well as the starting point for the research of gene expression.

transcriptomics: This area of research investigates transcripts including the number, type and modification, many of which can impact phenotype. Transcriptomics studies mRNA from each possible tissue, at each developmental stage and under each of a number of possible environments (stresses). A variety of small non-coding RNA sequences (ncRNA) and the LARGE INTERGENIC NON-CODING RNA SEQUENCES (lincRNA) are known, and their role in gene regulation is examined. The ncRNAs control gene expression through mRNA degradation and the lincRNAs presumably control gene expression through their association with CHROMATIN-MODIFYING COMPLEXES (CMCs). A study of transcriptome also elucidates the complex interactions, which generally take place among transcripts, before they are translated. Transcriptomics also analyses global gene expression and alterations produced by genetic effects, environmental effects, drugs or chemicals, or tissue location. It is the application of micro- or macroarrays and sequence-based methods to conduct expression profiling to determine genome-wide gene expression. Transcriptomics is usually done using RNA-sequencing technology.

transcriptomics tools: (i) Expression Profiler Next Generation is an open, extensible web-based collaborative platform for microarray gene expression. (ii) GEPAS (Gene Expression Pattern Analysis Suite) is useful for the analysis of microarray gene expression data. (iii) DNA arrays: Analysis Tools include pro-processing, viewers, unsupervised clustering, differential gene expression, supervised classification and data mining modules. (iv) NIA Array Analysis Tool is used for microarray data analysis. (v) M@CBETH (MicroArray Classification Benchmarking Tool on Host server) offers the microarray community a simple tool for making optimal two-class predictions by using randomizations of the benchmarking data set. (vi) VAMPIRE microarray analysis suite is a statistical framework that models the dependence of measurement variance on the level of gene expression. (vii) MIDAW (MIcroarray Data Analysis Web tool) performs data normalization (global, lowess & variance stabilizing procedures), descriptive analysis (boxplot, density estimation, cluster analysis), feature extraction (hypothesis testing, principal component, partial least square analysis) and discriminant analysis (Prediction Analysis of microarray, PAM). Results are displayed in output pages where figures and results can be easily downloaded. There are tutorial pages for each session of data processing. (viii) GEMS (Gene Expression Mining Server) is a simple but promising new approach for biclustering based on a Gibbs sampling paradigm.

transcytosis: The process involving uptake of material at one face of a cell by endocytosis, its transfer across a cell in vesicles and its discharge from another face by exocytosis.

transdifferentiation: A process invoked to explain how tissue-specific adult stem cells can generate cells of other tissues. Transdifferentiation is a poorly understood process. New results challenge its existence as it has been demonstrated that transplanted bone marrow cells can turn into unexpected lineages including myocytes, hepatocytes, neurons and many others. It has been demonstrated that mouse bone marrow cells can fuse spontaneously with embryonic stem cells in culture *in vitro* that contains interleukin-3.

transducin: Retinal rod outer segments contain a light-activated cyclic GMP phosphodiesterase. Transducin, a multisubunit protein, is the amplified signal carrier in this activation process. A photolysed rhodopsin molecule triggers the exchange of GTP for GDP in hundreds of molecules of transducin. GTP-transducin then switches on the phosphodiesterase. Amplifiers like transducin may be present in many kinds of cells, as exemplified by the G-protein in the hormone-stimulated adenylate cyclase system. G-protein involved in yeast mating-type switching is known as transducin.

transducing particle: A defective phage, carrying a part of the host's genome.

transduction: A virus that infects bacteria picks up some bacterial DNA from one host and transfers it to the next host, which may incorporate the DNA if it survives the infection. Recombination in bacteria whereby DNA is transferred by a phage from one cell to another. Clear implication of DNA as the genetic material was derived from transduction experiments by N. Zinder and J. Lederberg in 1952 on mouse typhoid bacterium *Salmonella typhimurium*. Their experiment involved process of transduction in which a bacterium-infecting virus (bacteriophage) assists as the vector transferring DNA from one bacterial cell to another. They used two strains of bacteria, LA2 (phe^+ trp^+ met^- his^-) and LA22 (phe^- trp^- met^+ his^+), each placed in one of the two arms of a Davis-U-tube separated by a sintered glass filter. This filter prevents the transfer of bacterial cells but permits the free passage of growth medium between the two arms through the application of suction and pressure. Phage P_{22} is capable of GENERALIZED TRANSDUCTION. As a result, a 'filterable agent' (FA) arose in connection with LA2 that could produce phototrophs in LA22. This FA was phage P_{22}. It was found that 'phe^+ trp^+ factor' could pass through the filters that hold bacteria but allow viruses to pass through. The 'phe^+ trp^+ factor' was located inside the virus. Production of recombinant viral DNA was consequence of transduction. Transduction is defined as the process of the transfer of genetic material from one genotype to another through agency of bacteriophages. Transduction is one of three mechanisms of gene transfer between the bacteria. In this process, a bacteriophage infects one bacteria cell (say donor), then multiplies in it. During this multiplication, some pieces of bacterial DNA get packed inside the phage coats. These newly formed phages when infect another bacterial cell (say recipient) transfer the bacterial DNA from donor into recipient cell.

Transduction is of two types – GENERALIZED TRANSDUCTION and SPECIALIZED TRANSDUCTION. *See details as separate terms.*

transduction mapping: The use of transduction to determine the relative positions of two or more genes on a bacterial DNA molecule.

transesterification reactions in RNA recombination: Replicase-driven template-switching in RNA recombination may be caused by various transesterification reactions, which are either performed by RNA molecules themselves or are promoted by some proteins.

transfarmers: The animals that are used to study the production of a pharmaceutical protein or antibiotics in their milk. Ultimately, all transfarmers are female because males cannot produce milk. Transfarming results in easy product purification, low production costs and eliminates the need for drug refrigeration. *Aka* TRANSPHARMERS.

transfectant: A genetic transformation of a cell by free (naked) DNA.

transfection: When DNA from an animal virus or bacteriophage is used in transformation, intact virus particles are formed inside the recipient bacterial cells. In this case, there is no need for donor DNA to become integrated into the host chromosome. When such bacteria come in contact with the animal host, which the virus is able to infect, they release the virus particles causing the infection of the host animal. Transgenic animals are produced through transfection of specialized cells or embryo since only eggs or embryos can develop into whole animal. This term is equivalent to 'transformation' – a term commonly used for plants. *Compare* with TRANSFORMATION.

transfer DNA (T-DNA): The transferred DNA of the TUMOUR-INDUCING (TI) PLASMID of some species of bacteria such as *Agrobacterium tumefaciens* and *Agrobacterium rhizogenes* (actually a Ri plasmid). The T-DNA is transferred from bacterium into the host plant's nuclear DNA genome. T-DNA contains genes that *inter alia* enable the infected plant cells to make two unusual compounds, NOPALINE and OCTOPINE, both characteristic of transformed cells; the cells form a gall that hosts the bacterium; this DNA transfer mechanism is exploited in the genetic engineering of plants. *Aka* TRANSFERRED DNA.

transfer operon (*tra*): Sequence of the loci that impart the male (F-pili-producing) phenotype on a bacterium. The male cell can transfer the F plasmid to an F^- cell.

transfer reaction: A reaction that attaches an amino acid with its specific tRNA. This reaction is conducted by a specific aminoacyl synthetase.

transfer RNA (tRNA): A non-coding RNA to which amino acids must attach prior to their incorporation into polypeptides. The secondary structure of tRNA has a typical clover-leaf configuration. It is further folded through alternative hydrogen-bonding geometries (including Hoogsteen base pairs) into an L-shaped tertiary structure. tRNA comprises several unusual (modified) bases. The principal features of tRNA clover-leaf comprise an accepter stem, three stem loops, which are referred to as ψ loop, the D loop, the anticodon loop and a variable loop. Transfer RNA is the direct interface between the amino acid sequence of a protein and the information in DNA. Transfer RNA brings the amino acids to the ribosomes to make proteins. The tRNA amounts to 10–15% of total RNA in the cell. Cells produce at least 40 different kinds of tRNAs. Each of these is a copy of a specific tRNA gene. First, PRECURSORS OF tRNAs (pre-tRNAs) are synthesized, which after certain modifications become tRNAs. They contain a large number of unusual bases. The tRNAs interact with specific amino acids. tRNAs have various LOOPS that serve useful functions. Some sequences in tRNA are complementary and they give stem-like structures. It is a small molecule of 70–90 ribonucleotides. The 3′-end of tRNA carries CCA sequence. It is with adenine of this sequence that an amino acid attaches itself with tRNA. tRNA contains an anticodon, which permits temporary complementary pairing with codons on mRNA. tRNAs are of several different kinds and each type attaches with it a specific amino acid and brings it to the ribosome. The partial pairing of some complementary sequences gives tRNA a clover-leaf structure. Each species of tRNA has a specific anticodon. A specific tRNA attaches to a specific amino acid and then unites that amino acid with a specific codon on the mRNA. tRNA has an anticodon, which is a sequence of three bases complementary to an mRNA codon. There is at least one tRNA for each of the 20 amino acids. In protein synthesis, each codon in the messenger RNA pairs with the appropriate tRNA's anticodon and the amino acids are arranged in order to make the protein. *Aka* ADAPTOR RNA *or* SOLUBLE RNA. *See* TRANSCRIPTION.

transfer RNA (tRNA) acceptance: During transfer RNA selection, a cognate codon-anticodon interaction triggers a series of events that ultimately lead to the acceptance of that tRNA into the ribosome for peptide-bond formation. High fidelity discrimination between the cognate tRNA and near- and cognate ones relies on their differential dissociation rates from the ribosome and on specific acceleration of forward rate constants by cognate species.

transfer RNA (tRNA) CCA-adding polymerase: Transfer RNA nucleotidyltransferases (CCA-adding polymerases) mature the essential 3′-CCA terminus of transfer RNA without any nucleic acid template. These enzymes are responsible for the maturation or repair of the functional 3′-end of tRNAs by means of the addition of the essential nucleotides CCA.

transfer RNA (tRNA) chaperone: Some modification enzymes and RNA-binding proteins possibly function as RNA chaperones improving the efficiency of RNA processing. Enzyme tRNA pseudouridine synthase TruB is a tRNA chaperone.

transfer RNA (tRNA) deacylase: The enzyme responsible for cleaving the bond between a tRNA molecule and the growing polypeptide during translation.

transfer RNA (tRNA) defect, diseases owing to: A single change in a person's tRNA gene can contribute to a range of life-shortening risk factors, including high blood pressure, high cholesterol and other metabolic disorders.

transfer RNA (tRNA) display: A method developed by D.L. Dunkelmann *et al.* (2024) that enables direct, rapid and scalable selection for orthogonal synthetases that selectively acylate their cognate orthogonal tRNAs with non-canonical monomers (ncMs) in *Escherichia coli*, independent of whether the ncMs are ribosomal substrates. Using tRNA display, they directly select orthogonal synthetases that specifically acylate their cognate orthogonal tRNA with eight non-canonical amino acids and eight ncMs, including several β-amino acids, α,α-disubstituted-amino acids and β-hydroxy acids. They demonstrate the genetically encoded, site-specific cellular incorporation of β-amino acids and α,α-disubstituted amino acids into a protein and thereby expand the chemical scope of the genetic code to new classes of monomers.

transfer RNA (tRNA) gene family: Each type of tRNA gene contains ten to several hundred copies per haploid genome. The tRNA genes also occur in the form of heteroclusters. These heteroclusters may be tandemly repeated or 50–60 sites in different chromosomes. Ten to several hundred genes for each tRNA occur in each haploid genome.

transfer RNA (tRNA) gene: A gene that on transcription yields precursor of tRNA (pre-tRNA) which after various processing steps matures into a functional tRNA.

transfer RNA (tRNA) identity: Correct recognition of tRNA by aminoacyl-tRNA synthetases is essential to the maintenance of translational fidelity. The hypothesis that synthetases recognize anticodon nucleotides was not accepted until relatively recently. Implementation of the new methodology has clearly demonstrated that anticodon is indeed important for 7 of the 20 *Escherichia coli* isoaccepting groups. For many of the isoaccepting groups, the acceptor stem or position 73 (or both) is important as well.

transfer RNA (tRNA) isoacceptors: Differential species of tRNA that bind the same amino acid but, because of coding degeneracy, have differing anticodon triplets.

transfer RNA (tRNA) maturation, different stages of Different stages of tRNA maturation in fixed temporal order include (i) trimming of the 5′ and 3′ ends, (ii) modification of the bases and (iii) removal of introns.

transfer RNA (tRNA) modifications: Transfer RNA modifications include 5-methyluridine (5^mU), N3-methylcytidine (3^mC), 5-methylcytidine (5^mC), N7-methylguanosine (7^mG) and 7-(4,5-cis-dihydroxy-1-cyclopenten-3-yl-aminomethyl)-7-deazaguanosine (Q).

transfer RNA (tRNA) nucleotidal transferase: The enzyme responsible for the post-transcriptional attachment of the nucleotide sequence 5′-CCA-3′ to the 3′-end of a tRNA molecule.

transfer RNA (tRNA) precursors: A form of tRNA that is about 20% larger than the mature molecules and contains extra sequences at both the 5′ and 3′ ends. In prokaryotes, they are removed by specific enzymes, RNaseP and RNaseQ, which cleave, respectively, at the 5′ and 3′ ends.

transfer RNA (tRNA) releasing factor: A factor catalysing the release from ribosome. It is distinct from the G factor in that it loses its activity upon heating and does not require GTP for its action. It may play a role in removing tRNA from the ribosome after the polypeptide chain termination factor has released the completed chain from the ribosome-bound tRNA.

transfer RNA (tRNA)-specific adenosine deaminase (TadA): Adenine base editors (ABEs) are precise gene-editing agents that convert A:T pairs into G:C through a deoxyinosine intermediate. Existing

ABEs function most effectively when the target A is in a TA context. Y.L. Wu *et al.* (2024) evolved the *Escherichia coli* transfer RNA-specific adenosine deaminase (TadA) to generate TadA8r, which extends potent deoxyadenosine deamination to RA (R = A or G) and is faster in processing GA than TadA8.20 and TadA8e, the two most active TadA variants reported so far. ABE8r, comprising TadA8r and a *Streptococcus pyogenes* Cas9 nickase, expands the editing window at the protospacer adjacent motif-distal end and outperforms ABE7.10, ABE8.20 and ABE8e in correcting disease-associated G:C-to-A:T transitions in the human genome, with a controlled off target profile.

transfer RNA (tRNA) transport: tRNA nuclear transport in *Xenopus lovis* oocyte resembles a carrier-mediated translocation process rather than diffusion through a simple pore or channel. tRNA transport is saturable by tRNA, with a maximal rate measured to be in the mature oocyte. Many tRNA species share a common carrier system. tRNA nuclear transport is sharply dependent on temperature, with an optimal rate observed at 31 °C.

transfer RNA (tRNA), types of: Based on the size of the extra or variable loop, there are two types of tRNAs – Class I tRNAs and Class II tRNAs. *See* TRNA, CLASS I TRNAS *and* CLASS II TRNAS.

transfer RNAs (tRNAs), class I: Have only 3–5 bases in their extra loop. They represent nearly 75% of all tRNAs.

transfer RNAs (tRNAs), class II: Have a larger loop with 13–21 bases in this extra loop and 5 bp in the stem.

transfer-messenger RNA (tmRNA): tmRNA, or 10Sa RNA, is a stable cytoplasmic RNA found in a wide variety of Eubacteria, but with no known homologous archaeal or eukaryotic counterpart. tm RNA contains a $tRNA^{ala}$ –like structure formed by base pairing between its 5′ and 3′ ends and an internal reading frame encoding a short peptide tag. It is involved in a newly discovered translational mechanism, *trans*-translation, in which a single peptide is synthesized from two different mRNA molecules. In this process, tmRNA acts as both tRNA and mRNA to rescue ribosomes stalled on a truncated mRNA lacking a stop codon. Thus, tmRNA displays a double function: like $tRNA^{ala}$, it can be charged by the corresponding alanyl-tRNA synthetase and, like mRNA, its reading frame can be translated by the ribosome. When a truncated mRNA lacking a stop codon is translated, it stalls the ribosome at its 3′ end. Alanine-charged tmRNA enters the ribosome and its alanine is added to the C-terminus of the truncated peptide. Thus, a chimeric peptide whose C-terminal tag is a protease degradation signal is formed, tmRNA is released and ribosome recycling is made possible. tmRNAs help in bacterial translation and act as a quality control factor.

transferases: Enzymes that catalyse the transfer of functional groups to molecules from other molecules.

transferring nucleic acids by conjugating mitochondrial targeting sequences (MTS): The MTS is recognized by the mitochondrial import complexes: TRANSLOCASES OF THE OUTER MEMBRANE (TOM) and TRANSLOCASES OF THE INNER MEMBRANE (TIM) and mediates mitochondrial localization and subsequently deliver mitochondrial proteins to the matrix compartment. Thus, the MTS and mitochondrial protein import complexes have been explored for nucleic acid import to mitochondria.

transform: In oncology, to convert a normal cell into a cancerous one.

transformant: A cell that has been transformed by the uptake of naked DNA.

transformation: (i) Transformation is the progression in which genetic material is transferred from one bacterium to another without involving any intermediate organism or medium. Genetic alteration of a cell resulting from the direct uptake and incorporation of exogenous DNA from its surroundings and taken up through the cell membrane(s). (ii) Insertion of new genetic material into nonbacterial cells, including animal and plant cells (iii) Transformation has a special meaning in relation to animal cells, indicating progression to a cancerous state, the term is avoided for animal cells when describing the introduction of exogenous genetic material. (iv) The transfer of genetic information by DNA separated from the cell. (v) With respect to cloning of DNA, it refers to the transformation of bacteria (usually to specific antibiotic resistance) due to the uptake of foreign DNA. (vi) With respect to eukaryotic cells, usually means conversion to less-restrained or unrestrained growth. (vii) In plant cell culture, the introduction and stable genomic integration of foreign DNA into a plant cell by any means, resulting in a genetic modification. This definition is the traditional microbiological definition. (viii) The procedure in which free DNA is transferred directly into a competent recipient cell. The direct transfer of genetic material from donor to recipient. The acquisition (e.g.

by bacteria cells) of new genetic markers (new traits coded for by the new DNA) via the process of transformation. (ix) With respect to animal cell culture, *see* IN VITRO TRANSFORMATION, IN VITRO NEOPLASTIC TRANSFORMATION *and* TRANSFECTION. (x) In oncogenesis, transformation means cancerous growth of cells. In genetic engineering experiments, the foreign DNA/plasmids are introduced into competent cells through the method of electroporation or simple transformation. In the case of plants, cloned genes are introduced at the cellular or protoplast level using various vector/vectorless methods. Bacterial transformation helps in gene cloning, whereas in case of plants transformation helps in the generation of transgenic crop varieties such as *Bt* cotton. Transformation is achieved by using vector and vectorless gene transfer methods.

transformation by plasmid rescue: If plasmid DNA is linearized by restriction endonuclease cleavage, no transformation of *Bacillus subtilis* occurs. If the recipient carries a homologous plasmid, and if a restriction endonuclease cut occurs within the homologous moiety, then this same marker transforms efficiently. This requires *Bacillus subtilis recE* gene. This is known as transformation by plasmid rescue.

transformation efficiency (TE): (i) The number of independent transgenic events obtained from 100 explants. (ii) In biolistic transformation, TE is often given as the number of transgenic events per bombarded plate. (iii) The number of bacterial and/or plant cells that uptake and express plasmid DNA and/or a foreign gene, respectively, divided by the mass of plasmid used.

transformation mapping: A method of gene mapping in bacteria through conjugation.

transformation of cells with free DNA: Genes can also be transferred into the cells from solutions of free DNA (i.e. by transformation). In mammalian cells, it has been possible to transform TK^- (thymidine kinase deficient) mouse cell into TK^+ cells by transformation, where HAT (hypoxanthine-aminopterin-thymidine) medium was used for screening. The transforming potency of DNA could be increased by the digestion of DNA with a restriction endonuclease and then fractionating fragments by size. This type of transformation can rarely be brought about in fertilized mouse eggs also and it was shown that rabbit β globin was produced by the mouse developed from a transformed egg and that the rabbit β globin gene thus introduced is inherited to the next generation.

transformation of heterochromatin into transcriptionally active state: Histone exchange is a mechanism through which heterochromatin is transformed into a transcriptionally active state. RNA levels at the transcriptional site increase immediately after the induction of transcription and the rate of synthesis slows over time.

transformation of protoplasts: Transformation of protoplasts involves POLYETHYLENE GLYCOL (PEG) induction of DNA uptake in protoplasts and subsequent regeneration of bacterial cell walls. This procedure yields up to 80% transformants making the method suitable for the introduction of cryptic plasmids. Plasmids isolated from *Staphylococcus aureus* that code for resistance to tetracycline (pT127) or chloramphenicol (pC194) can be transformed into *Bacillus subtilis* where they replicate and express antibiotic resistance normally.

transformation, eukaryotic: Conversion of eukaryotic cells to a condition of unrestrained growth in culture, resembling or identical to the tumourigenic condition.

transformation-competent viruses: These viruses can transform the infected cells and may replicate via DNA or RNA intermediate and can transfer genetic information horizontally or vertically.

transformation-incompetent viruses: The viruses that re able to replicate but not transform the infected cells. They may replicate via DNA or RNA intermediate and can transfer genetic information horizontally or vertically.

transformed cells: Cells that have become capable of sustained proliferation *in vitro*.

transformer: An allele in fruit flies that converts chromosomal females into sterile males.

transformer architecture: The transformer architecture uses an encoder-decoder structure that does not rely on recurrence and convolutions to generate an output. The encoder maps an input sequence to a series of continuous representations. The transformer transforms an input sequence into an output sequence. This includes speech recognition, text-to-speech transformation and so on.

transforming growth factor (TGF): A growth factor that is involved in proliferation and cell differentiation.

transforming principle: Substance or an agent responsible for transforming *Streptococcus* bacteria from one serotype (nonvirulent) to another (virulent). O. Avery, C. MacLeod and M. McCarty in

1944 identified the chemical nature of the transforming principle as DNA. It was thus shown that DNA is the unit of heredity of which genes are made.

transforming protein engineering with Infologs™: Using independently designed synthetic genes where substitutions are systematically incorporated (Infologs™) leads to uniform sampling, systematic variance and unrestricted information-rich results. Wheat GST with the ability to detoxify a panel of common herbicides was designed using this patented DNA2.0 bioengineering method. The relative functional contribution of 60 amino acid substitutions against 14 herbicides was quantified using only 96 infologs and dramatically improved by a small set (16) of second-generation infologs.

transformism: The evolutionary theory of Lamarck in which changes occur within a lineage of populations, but in which lineages do not split (i.e. no speciation occurs, at least not in the sense of the cladistic species concept) and do not go extinct.

transgene: (i) A gene that has been isolated from one organism and transferred to a different organism by any of a number of genetic engineering techniques from one species to another by passing all reproductive barriers. (ii) Artificially reconstituted gene, which can be expressed either pre- or post-integration into the genome; often encodes a therapeutic or a marker gene. A gene used for transformation that comes from *trans* species or *trans* organism. Conventional breeding allows the transfer of gene(s) only from cross-compatible-related species, whereas the transgenic approach has broadened the available gene pool for a breeder, as the transgene(s) cloned from viruses, bacteria, fungus, insects, animals, humans can be introduced into crop plants. The transgenes cloned from prokaryotes need some modifications for their better expression in eukaryotes. For example, the bacterial genes are rich in A and T bases, whereas plant genes are rich in G and C bases. Therefore, a change from AT rich to GC rich, is generally required for bacterial transgenes to express them into plants.

transgene silencing: Observations of unpredictable transgene silencing and variable expression have pioneered research into a novel type of epigenetic regulation, termed homology-dependent gene silencing. Plant gene silencing was originally thought to be a quirk of transformation procedures but is now recognized to be a facet of vitally important gene regulatory systems, present in all organisms. Monocot plants, especially the grasses, play a foremost role in the agricultural economy of all nations, and their biotechnological manipulation offers great potential for both developed and developing countries.

transgene validation: Validation that a transgene insertion designed to express the wild-type gene product (what we refer to as the parent transgene insertion) is fully functional. The gold standard for validating parent transgene insertions is rescue of null mutants (either isolated from forward genetic screens, generated by gene knockout consortia, or made using CRISPR/Cas9).

transgene-induced gene silencing: (i) The most effective way to induce transgenic gene silencing is to introduce INVERTED-REPEAT (IR) DOUBLE-STRANDED RNA (dsRNA) or ARTIFICIAL MICRORNA (amiRNA) instead of a transgenic sense or antisense strand of genes. (ii) The number of copies of a transgene that integrate into the genome of a transformed plant and the position of the integration site cannot be predicted, regardless of whether *Agrobacterium*-mediated transfer or direct gene transfer are used. Therefore, one or multiple intact or rearranged copies can integrate at one or multiple unlinked loci.

transgeneration: A heritable change taking place at a single gene locus, presumably caused by chemical alteration of the individual gene – a mutation proper, point mutation, or genovariation.

transgenerational and intergenerational stress memory (TISM): A stress imprint that extends from one stressed generation of organisms to at least the first stress-free offspring generation. In this review, we use the term intergenerational memory when only the first stress-free generation has a detectable memory effect and TRANSGENERATIONAL MEMORY if memory is detectable after at least two stress-free generations. As the progeny develops on the mother plant, INTERGENERATIONAL MEMORY may be mediated by the conditions in which the seed grows and by cues introduced into the seed or embryo by the mother plant. Transgenerational memory, by contrast, probably has an epigenetic basis.

transgenerational epigenetic effect: One example related to defence priming offers good evidence for a transgenerational epigenetic effect that could be adaptive. The progeny of *Arabidopsis* plants infected with bacteria are more resistant to secondary infection with an oomycete (fungi-like unicellular organisms) than the control progeny of unprimed plants.

This effect was obvious in the first-generation progeny. However, oomycete resistance also occurred in the second-generation progeny that were produced on plants that were not themselves primed by infection. This observation rules out inherited resistance being due to the transport of a physiological or biochemical factor from parents to seed – the most likely explanation is therefore epigenetic. Chromatin analysis of defence genes has strengthened the notion that inherited priming is because of epigenetic mechanisms. Défense genes that are upregulated following primed resistance were associated with the enrichment of acetylated histones, a known activating epigenetic mark, in the promoter region. In contrast, downregulated genes following primed resistance had higher levels of the repressive mark H3K27me3. However, this is not a simple mechanism because plants that were defective for DNA methylation at CpHpG sites mimicked the transgenerational priming. Transgenerational priming may be mediated by hypomethylation of DNA at these CpHpG sites. There may be cascades of epigenetic regulation in which stress leads to a loss of repressive marks that, in turn, trigger activating epigenetic marks.

transgenerational epigenetic inheritance (TEI): Transmission of modifications in gene expression patterns that are transmitted from one generation to the next via germ lines. These modifications are not due to changes in the DNA sequence but to changes in the epigenetic state of the genes.

transgenerational epigenetic regulation in response to stress: Stress-induced responses are mostly transient. However, the virus-induced gene silencing illustrates the potential for transgenerational epigenetic effects in plants if a stress induces siRNAs or chromatin changes. In dandelions (*Taraxacum officinale*), the genome-wide pattern of DNA methylation is modified if the parental plants are exposed to environmental stress and the progeny shows modifications of root/shoot biomass ratio, P content, leaf morphology and stress tolerance relative to the control. DNA methylation is suggested to be involved because chemical suppression of DNA methyltransferase activity blocks the transgenerational effects.

transgenerational epigenetics: Epigenetics study that gives an exhaustive investigation of the legacy of epigenetic phenomena between generations. This epigenetic data can be exchanged through various components including DNA methylation, histone changes or RNA and the impacts can endure for different generations.

transgenerational immune priming: Immune responses induced in the progeny of infected/elicited plants.

transgenerational induced resistance: Transgenerational responses to stress have been reported in both plants and animals, ranging from maladaptive pathological effects of environmental pollutants to adaptive immunological traits that increase disease resistance. The latter response has been referred to as TRANSGENERATIONAL ACQUIRED RESISTANCE (TAR) or TRANSGENERATIONAL INDUCED RESISTANCE (T-IR), which is typically based on a sensitization, or 'priming', of the immune system, mediating a faster and/or stronger immune response.

transgenerational inheritance: Transmission of epigenetic information that is passed on to gametes without alteration of the DNA sequence.

transgenerational priming: Priming is a phenomenon in which exposure to one stimulus influences how a person responds to a subsequent, related stimulus. Defence genes that are upregulated following primed resistance were linked with the enrichment of acetylated histones, which is a known activating epigenetic mark.

transgenerational stress memory: A stress imprint that extends from one stressed generation of organisms to at least the first stress-free offspring generation. This memory is detectable if it is detectable after at least two stress-free generations. As the progeny develops on the mother plant, intergenerational memory may be mediated by the conditions in which the seed grows and by cues introduced into the seed or embryo by the mother plant. It probably has an epigenetic basis. *Compare with* INTERGENERATIONAL STRESS MEMORY.

transgenes from nonplant sources: Nonplant sources of genes for improving plant resistance include (i) Genes encoding insensitive target enzymes; (ii) genes specifying toxin inactivation; (iii) expression of antibacterial peptides; (iv) expression of bacterial lysozymes; and (v) genes specifying artificially PROGRAMMED CELL DEATH (PCD).

transgenes from plant sources: In case of fungal and bacterial pathogens, plant resistance has been sought by expression of the following transgenes using plant genes: (i) expression of heterologous phytoalexins; (ii) genes encoding ribosome-activating proteins; (iii) expression of heterologous thionins; (iv) ectopic (out of the natural sources) expression of pathogenesis-related proteins; and (v) ectopic expression of chitinases.

transgenesis: The process of transferring an exogenous DNA segment or gene (transgene) to an animal so that it is able to pass the transgene on to all of the offspring. The whole process involving the introduction, integration, expression of transgene(s) into a cell and development of transgenic organism. Transgenesis is the process of developing transgenic organism. Transgene(s) is introduced at a cellular level using vector or vectorless systems, which randomly integrates with the recipient genome and starts expressing along with the other native genes. Transgenic cell/tissue is made to regenerate into entire organism through cultural process, which is finally subjected to molecular analysis for confirmation. This term also denotes genetic transformation of whole, live parasitic nematodes accompanied by promoter-regulated, tissue-appropriate transgene expression. *See* TRANSFORMATION.

transgenic: A **genetically modified organism (GMO)** whose genetic material has been altered using recombinant DNA technology so that it exhibits a new property. An organism carrying a transgene. For example, *Bt* cotton is transgenic as this carry an additional cry gene of bacterium *Bacillus thuringiensis*, which is stably integrated and expresses along with the other native cotton genes. *Bt* cotton has been developed through the process of transformation. Transgenics have been produced in large number especially in plants and in some animals, the world over. Transgenic crop varieties such as *Bt* cotton and *Bt* maize have been released for commercial cultivation in many countries. Transgenic varieties carry transgene(s) governing resistance to various biotic and abiotic stresses and for the improvement of nutritional/processing quality. Transgenic organisms are also called GMOs or LIVING-MODIFIED ORGANISMS (LMOs).

transgenic allelic complementation for candidate gene verification: This method provides proof that a candidate gene is actually controlling natural variation. This is a definitive step to confirm a QTL by introducing alleles into QTL parental accessions and null knockout mutants. Such a transgenic approach identifies alleles that have significantly different effects on the phenotype.

transgenic animal/plant: An animal or a plant that contains an alien or modified DNA (gene) into its germ line, incorporated by biotechnological means and which is more or less stably inherited.

transgenic crops expressing insecticidal toxins: With the insertion of genes in crop plants from other organisms, crops resistant to pests and diseases can be developed. Genes for various purposes such as for producing an insecticidal protein or functioning as an endotoxin, avoiding diseases that rot the root or for tolerance to certain herbicides, has been incorporated into the plant genome. Pests themselves can be genetically transformed and it is possible to release insects, which are genetically changed to make them sterile or die when a certain lethal gene is expressed. Genetic engineering can be employed to enhance natural plant resistance to pests and diseases. Crop protection through engineered pest-resistant genes is also now possible. Transgenic crops expressing insecticidal toxins have been developed for insect-pest management. One approach for achieving this objective is by expression of *Bacillus thuringiensis* (*Bt*) genes that produce proteins conferring tolerance to herbicides. Also, insecticides need to be sprayed on *Bt* crops less frequently, in less quantity and in a more targeted manner, sparing beneficial insects and spiders. But the insects have the tendency to develop resistance to engineered crops and thus continual research is needed.

transgenic crops with improved nutrition quality: Transgenic crops have been developed by introducing genes involved in the metabolism of vitamins, minerals and amino acids. Few examples of genetic modification of nutritional quality include vitamin A deficiency and seed protein quality. Vitamin A deficiency can cause night blindness and skin disorders, among others. The staple food rice is extremely low in vitamin A and therefore the improvement of vitamin A content is very essential. In a remarkable example of genetic engineering, Prof Ingo Potrykus and Dr Peter Beyer developed genetically engineered rice (popularly known as 'Golden Rice'), which is enriched in pro-vitamin A by introducing three genes involved in the biosynthetic pathway for carotenoid, the precursor for vitamin A. The seeds of Golden Rice are yellow in colour because pro-vitamin A is produced in the entire grain. The nutritional quality of cereals and legumes is poor because of deficiency of the essential amino acids (i.e. lysine in cereals and methionine and tryptophan in pulses). Two genetic engineering approaches have been employed to improve the seed protein quality. In the first case, a transgene (e.g. gene for protein containing sulfur-rich amino acids) was incorporated into pea plant (which is deficient in methionine and cysteine, but rich in lysine) under the regulation of seed-specific promoter. In the second approach, the endogenous genes were modified so as to increase the essential amino acids like lysine in the seed proteins of cereals.

transgenic disease models: Transgenic disease models are created to mimic human diseases so scientists can learn more about the diseases including symptoms, pathways, causes and research treatments, as well as possible cures. The most well-known disease models are ALZHEIMER'S MOUSE, ONCOMOUSE™, AIDS MOUSE, ANDi and SUPER FLY.

transgenic enhancement of plant nutrients: The principal strategy for transgenic enhancement of plant nutrients involves increasing the expression of anabolic biosynthetic genes and second, feedback metabolic loops tend to maintain homeostatic levels of nutrients. Plants might respond to an increase in nutrient production induced by overexpression of biosynthetic genes by activating a degradation pathway that either negates nutrient accumulation or converts nutrients into undesirable metabolites.

transgenic event: The unique recombinant DNA insertion event that took place in one plant cell, which was then used to generate a transgenic plant. The selected Event is used for breeding and development of commercial varieties in a crop.

transgenic organism: (i) A genetically engineered organism to which gene(s) from unrelated or related species or families have been introduced and have become an integral part of its genome. (ii) An organism whose gamete (sperm/egg) contains genetic material originally derived from an organism other than the parents, or in addition to the parental genetic material. (iii) A plant or animal that has stably incorporated one or more genes from another cell or organism and can pass them on to successive generations.

transgenic plant as a source of biopharmaceuticals: Plants are among the most effective bioreactors, which produce quantities of material with sunlight and soil-based nutrients as inputs. Efforts are in progress to replace the traditional fermentation procedure for the production of biopharmaceuticals with plant-based production. The advantages of using plants are the ability to increase production at low cost by planting more acres, rather than building fermentation capacity, lower capital and operating cost, simplified downstream processing and so on. Therapeutic drugs to treat cancer, infectious diseases, autoimmune diseases, cardiovascular diseases and other conditions and several vaccines can potentially be grown in plants. Plant transgenic technology is being used to produce a plant that will generate a seed that expresses a desired THERAPEUTIC PROTEIN. This seed can proliferate under the right growing conditions to yield plants and seed stock for manufacturing the desired protein. The desired protein can be extracted from the seed to make a biopharmaceutical. Plant-based therapeutics are expected to be much more cost effective. For example, Dow Plant Pharmaceuticals is using maize to grow pharmaceuticals by designing and selecting the plant, which will contain the active pharmaceutical within the endosperm seed compartment. Benefits of producing pharmaceuticals in maize include long-term storage advantage, easier purification in view of a limited number of soluble seed proteins in maize seed, low microbial load, low proteolytic activity and specialized promoters to enable expression of the protein in specific parts of the plants.

transgenic plants: Transgenic plants are those which carry additional, stably integrated and expressed foreign gene(s), usually from a distantly related (*trans*) species/organisms. The whole process of introduction, integration and expression of foreign gene(s) in the host is called GENETIC TRANSFORMATION. Now there are a large number of plant species where transgenic plants have been successfully produced. Initially, transgenic plant production was restricted to dicotyledons, but now it has been extended to several monocotyledons also. The advancement in the molecular biology and genetic engineering specifically in pathogenicity and disease resistance offer a promising alternative for producing new disease-resistant crop varieties. By adopting such technology, it has become possible to have a direct access to an ocean of genes (i.e. genes can be transferred across species, genera or even from animals to plants and vice versa). This is also a single-step process and eliminates the requirement for repeated backcrosses, as in conventional plant breeding, for transfer of a particular trait from wild species. This also does not have any harmful impact on environment and beneficial insects. Recombinant DNA technology has been effectively used to transform plants for various traits like disease, insect, herbicide and stress resistance and for improving the nutritional quality of the produce. For example, the plant product known as

'*Bt* corn' is a transgenic plant because it has a gene from the bacterium *Bacillus thuringiensis* (*Bt*). That gene produces a protein with pesticidal properties that, when incorporated into a plant, allows the plant to produce this protein, thus transferring the bacteria's natural defence to the plant. Scientists are now trying to produce transgenic plants that can be used for the production of edible vaccines, good quality polyester and so on.

transgenic plants in dicotyledons: Transgenic dicotyledonous plants have been engineered for (i) herbicide resistance; (ii) insect resistance; (iii) resistance against viral infection; virus-resistant tobacco was the first transgenic crop commercialized in China in 1990; (iv) resistance against bacterial and fungal pathogens; (v) resistance against stress; (vi) suitability for food processing, (vii) male sterility and fertility restoration.

transgenic plants in monocotyledons: Transgenic monocotyledonous plants have been engineered: (i) With methods for uptake of DNA, it has now been possible to produce transgenic plants in some cereals. A rice transgenic plant carried a functional gene for aminoglycoside phosphotransferase II along with a reporter gene for neomycin phosphotransferase. The transgenic rice plants possessed a bacterial *hph* gene, encoding hygromycin B resistance along with the reporter gene encoding β-glucuronidase (GUS). (ii) A reporter gene for neomycin phosphotransferase (NPT II) associated with 35S promoter gene of the cauliflower mosaic virus (CaMV) was used for production of transgenic plants in maize. *Kurstaki* gene from *Bacillus thuringiensis* attached to CaMV 355 promoter was used to produce transgenic plant, which enhanced the production of toxins, 1000 times. The toxin controlled European maize borer. (iii) Resistance gene (*bar* gene) against herbicide PPT along with *gus* marker gene associated with CaMV 35S promoter + *Adh* I intron (to enhance activity of 35S Promoter) was introduced into wheat. Safety of grains for human consumption will have to be ensured before successful commercial use. (iv) Gene for iron deficiency has been identified and soon we will have a designer rice having more iron to serve the iron-deficient persons. Transgenic plants can also be used for the production of antibiotics for immunotherapy. (v) In 1994, a delayed ripening variety of tomato, Flavr Savr, was released in USA.

transgenic RNAi plants: Insects pose a huge threat to crop production. Variations in their genomes cause selection to favour those who can survive pesticides and *Bt* crops. Silencing specific genes by RNA interference is a desirable natural solution to this problem. Potent silencing effects have been produced by using target double-stranded RNAs through an effective vector system. Transgenic plants expressing RNAi vectors as well as dsRNA-containing crop sprays have been successful in the laboratory.

transgenic stacking: Introducing multiple transgenes into a host organism.

transgenome: A genetic element transferred by chromosome-mediated gene transfer.

transgression: (i) In legal terms, the act or process of breaking a law or moral rule. (ii) Gene transfer from one species to another according to natural laws is transgression.

transgressive: A phenomenon in which some individual in the progeny shows characters exceeding in value those shown by either of the parents. The transgressive inheritance is usually seen in crosses involving quantitative characters.

transgressive breeding: Isolation of gene combinations (recombinants) possessing new characters or a new intensity of traits surpassing both the parents in F_2 onwards. Development of productive variety through 'assembling productivity alleles' is an example of transgressive breeding.

transgressive segregation: A plant breeding (propagation) technique, in which genetically very different members of the same species, are mated with each other. The offspring of that mating can be more healthy, productive (e.g. fast growing) and uniform than their parents, a phenomenon known as HYBRID VIGOUR. In this case, some individuals in the F_2 of a cross that fall beyond the parental range for a quantitative trait.

transgressive variation: The production of F_2 phenotypes that exceed the parental extremes.

transient cellular events: M.C. Huppertz ***et al.*** (2024) present a method in which modified proteins can stably record biological activity during a particular time period in which cells are exposed to a fluorescent substrate. They altered the HaloTag method so that a tag was split between two proteins of interest, thus monitoring their interaction. *See* HALOTAG METHOD.

transient diploid: The stage of the life cycle of predominantly haploid fungi (and algae) during which meiosis occurs.

transient epigenetic regulation in response to stress: There are many instances of effects in plants that may have an epigenetic basis, including those that are induced by stress. One of these candidate epigenetic phenomena is well-known to gardeners who 'harden off' plants by a mild exposure to low temperature so that they are then protected against later freezing temperatures. Of course, this process is required to be repeated in each generation. A longer-lived 'epigenetic memory' occurs following exposure of plants to various pathogens. The initial exposure to the pathogen may activate defence systems involving hormones including salicylic acid (SA) and genes that interfere with growth of the pathogen. The primary responses are, however, transient, but if the plants are then exposed to a second pathogen, the genes are activated more rapidly than observed previously, after the first inoculation. A similar defence priming effect is produced by β-amino butyric acid. This 'defence priming memory' persists for 28 days or even longer. The possibility that pathogen-induced changes are epigenetic in nature is further strengthened by the extensive and dynamic modifications to the methylation of genomic DNA in *Arabidopsis* infected with bacteria or treated with SA. The key modifications may involve loss of DNA methylation at defence gene loci so that they are more easily activated at the transcriptional level.

transient expression: The temporary or unstable expression of a gene or genes shortly after the transformation of a host cell (i.e. cells in which the transgene has not been physically integrated into the genome, called 'stable transformation', but is just carried as an episome that can be lost). For example, during plant transformation, the introduced transgenes like *GUS* can express (produce protein) even without integrating into the genome. This expression is temporary and not stable. Transient expression is used for basic studies relating to promoter analysis or checking the DNA coating on gold particles and so on.

transient knockdown: The binding of an oligonucleotide to the active gene or its transcripts causes decreased expression through a variety of processes. Binding can occur either through the blocking of transcription (in the case of gene-binding), the degradation of the mRNA transcript (e.g. by small interfering RNA (siRNA) or RNase-H dependent antisense), or through the blocking of either mRNA translation, pre-mRNA splicing sites, or nuclease cleavage sites used for maturation of other functional RNAs, including miRNA (e.g. by Morpholino oligos or other RNase-H independent antisense).

transient peptide: The short leader sequence cleaved from proteins that are imported into cellular organelles by post-translational passage of the membrane.

transient polymorphism: Some populations contain detrimental mutations, which are gradually eliminated by selection. Selection in such cases acts to decrease polymorphism.

transit peptide: A peptide that, when fused to a protein, functions to transport that protein between compartments within eukaryotic cells. Once inside the 'destination compartment', the transit peptide is cleaved off the protein and that protein is then free to perform its designated task.

transition: The change from purine to purine (A → G or G → A) or pyrimidine to pyrimidine (T → C or C → T). Transitions may arise due to PROTON MIGRATION within bases and they show unusual pairing tautomeric adenine (A^*) pairs with cytosine and this mispairing leads to A.T → G.C transition. Tautomeric guanine (G^*) may pair with thymine. This mispair leads to G.C → A.T transition. Similarly, T^*G forbidden base pairing leads to T.A → C.G transition and C*=A forbidden base pairing leads to C.G → T.A transition. Proton migration within base pairs can set off a chain of modifications that ends in DNA mutation.

transition from RNA to DNA world: Evolution occurred in several steps: first, from RNA to RNA; second, RNA to protein world and finally, to the present world based on DNA. The DNA world itself probably originated in two steps: first, the U-DNA world, following the invention of ribonucleotide reductase and second, the T-DNA world, with the independent invention of at least two thymidylate synthases.

transition from totipotency to pluripotency: Genetic and epigenetic mechanisms regulate the transition from the totipotent zygote to pluripotent cells. These pluripotent cells can be propagated infinitely *in vitro*, underpinned by a unique epigenetic state. Regenerating totipotency during development of germ cells contains re-expression of pluripotency-specific genes and extensive erasure of epigenetic modifications. Increasing knowledge of key underlying mechanisms heightens prospects for creating pluripotent cells directly from adult somatic cells.

transition from transcription initiation to elongation mode: Once the primer reaches 10-nucleotide length, the RNA holoenzyme undergoes a transition from the initiation to the elongation mode and the transcription complex moves away from the promoter along the DNA template. Transition to elongation stage of transcription is associated with a conformational change in holoenzyme that releases the σ factor. An accessory protein NusA binds to the core enzyme and prevents σ factor from reassociating during elongation. NusA-core enzyme elongates the RNA chain.

transition state: Structure that is formed transiently during a chemical reaction and has the highest free energy of any reaction intermediate. Its formation is a rate-limiting step in the reaction.

transition to flowering, epigenetic control of: The reproductive success is based on the floral transition timing and flowering genes in angiosperms. The *FLC* and *FT* act as the regulators of the flowering genes in *Arabidopsis*. A number of chromatin-modifying components are involved in the activation or repression of *FLC* expression. The *FLC* expression activators include 'active' chromatin modifications including acetylation of core histone tails, histone H3 lysine 4 (H3K4) methylation, H2B monoubiquitination, H3 lysine 36 (H3K36) di- and tri-methylation and deposition of the histone variant H2A.Z, while *FLC* expression suppression include 'repressive' histone modifications associated with histone deacetylation, H3K4 demethylation, histone H3 lysine-9 (H3K9) and H3 lysine 27 (H3K27) methylation and histone arginine methylation. In addition, the Polycomb group gene-mediated transcriptional-silencing mechanism represses both *FT* and *FLC* expression. The regulation of flowering through *FLC* expression and chromatin modifications control the various developmental genes in plants.

transitional phase of gene concept: This phase of gene concept deals with pseudoalleles. Two mutations *Star* (*S*) and *asteriod* (*ast*) showed no complementation and *S* and *ast* mutations were concluded as allelic. E.B. Lewis in 1941, however, showed that *S* and *ast* showed 0.02% recombination.

transitions through tautomerism: Tautomeric adenine (A*) pairs with cytosine and this mispairing lead to A.T®G.C transition. Tautomeric guanine (G*) may pair with thymine. This mispairing leads to the G.C®A.T transition. T*=G forbidden base pairing leads to T.A®C.G transition and C*=A forbidden base pairing leads to C.G®T.A transition.

transitive gene silencing (TGS): Process in which RNA-dependent RNA polymerase (RdRp) amplifies the RNAi signal throughout the plant, silencing genes in other plant tissues and organs. RNAi can produce unexpected pleiotropic effects, such as reduced pollen viability, even when other aspects of plant growth seem to be normal. Vertical gene flow of an RNAi-mediated pollen lethality phenotype to native plants could alter fitness, plant community composition and biodiversity.

translatable exon: *See* EXON. *Compare with* UNTRANSLATABLE EXON.

translation: It is the conversion of the genetic information contained in mRNA into protein synthesis. This is the process in which the genetic information carried by mRNA molecule specifies the sequence of amino acids in the growing polypeptide chain during protein synthesis. In other words, protein synthesis is called translation because it involves decoding the 'language' of nucleic acids and translating that information into the 'language' of proteins. Translation is a biochemical process by which ribosomes in the cytoplasm or endoplasmic reticulum synthesize proteins following the transcription of DNA to RNA in the cell's nucleus. For sake of discussion, translation is divided into initiation, elongation and termination phases. Eukaryotic genes may be regulated at the transcriptional, post-transcriptional, translational or post-translational level. Four major steps involved in the process of translation include charging of tRNA, initiation, elongation and termination. The machinery responsible for translating the language of mRNA to the language of proteins is composed of four primary components – mRNA, ribosome, rRNA and aminoacyl-tRNA synthetase. *Aka* RNA TRANSLATION.

translation elongation, requirements of prokaryotic: Polypeptide chain is elongated by covalent attachment of successive amino acid units each carried to ribosomes by their respective tRNAs, which successively base pair with a corresponding codon in mRNA at A-site. Elongation is promoted by elongation factors. Elongation requires (a) the initiation complex; (b) the next aminoacyl-tRNA, specified by the next codon in the mRNA; (c) a set of three soluble cytosolic proteins called ELONGATION FACTORS (EF-Tu, EF-Ts and EF-G) and (d) GTP.

translation elongation, steps in prokaryotic: Three steps are recognized in translation elongation. In the first step, the next aminoacyl-tRNA is first bound to a complex of a complex of EF-Tu containing a molecule of bound GTP. The resulting aminoacyl-tRNA-EF-Tu. GTP complex is then bound to the A-site of the 70S initiation complex. The GTP is hydrolysed, an EF-Tu. GDP complex is regenerated. In the second step, a new peptide bond is formed between the amino acids bound by their tRNAs to the A and P sites on the ribosome. The amino acid-tRNA that binds to the (A) site is specifically determined by the triplet codon of the mRNA that occupies the (A) site on the ribosome.

translation initiation in eukaryotes: A complex process in which initiator tRNA, 40S and 60S ribosomal subunits are assembled by eukaryotic initiation factors (eIFs) into an 80S ribosome at the initiation codon of mRNA. In eukaryotes, only AUG is recognized as initiation codon. Also, methionine is not formylated in eukaryotes $tRNA_m$ is present in eukaryotes compared to $tRNA_f$ in prokaryotes. Newly synthesized bacterial proteins start with formylmethionine and in eukaryotes, proteins start with methionine. In eukaryotes, there are at least eight initiation factors (eIF1, eIF2, eIF3, eIF4A, eIF4B, eIF4C, eIF4D and eIF5). Their roles are not yet clear. *See* INTERMEDIATE COMPLEXES IN EUKARYOTIC TRANSLATION INITIATION.

translation initiation in prokaryotes: Three steps are involved in this process: (i) The 30S initiation complex is formed. It requires mRNA, 30S subunit of ribosome, a special initiating species of aminoacyl~tRNA (that seemingly starts all polypeptide chain and three initiation protein factors – IF1, IF2, IF3. IF3 prevents the 30S and 50S subunits from combining prematurely. Binding of the mRNA to the 30S subunit then takes place in such a way that the initiation codon (AUG) binds to a precise location on the 30S subunit. The initiating AUG is guided to the correct position on the 30S subunit by an initiating signal called the SHINE-DALGARNO SEQUENCE in the mRNA. The first amino acid incorporated in prokaryotes is formylated N-methionine. At this stage, formation of 30 initiation complex is complete. (ii) The complex consisting of the 30S subunit, IF3 and mRNA now forms a still larger complex by binding IF2, which already is bound to GTP and the initiating fMet-$tRNA_f^{Met}$. The anticodon of this tRNA pairs correctly with the initiation codon in this step. (iii) This large complex combines with the 50S ribosomal subunit; simultaneously, the GTP molecule bound to IF2 is hydrolysed to GDP and Pi (which are released). IF3 and IF2 also depart from the ribosome. The 70S initiation complex is thus formed. Precise function of IF1 is not known. However, it has been demonstrated to play role in initial phase of protein synthesis in *Escherichia coli*. Studies reveal RNA chaperone activity of IF1. During translation initiation, the mRNA bearing the code for polypeptide to be formed binds to smaller subunit of ribosome, this is followed by binding of initiating amino acyl-tRNA and large subunit to form initiation complex. The initiating aminoacyl-tRNA base pairs with mRNA codon AUG at P site, which signals the beginning of polypeptide chain.

translation initiation sites in eukaryotes, choice of: A mature mRNA may contain more than one translation initiation codons. More of the same α-amylase enzyme is produced in rat salivary glands than in liver. DNAs complementary to α-amylase mRNA from two tissues were sequenced. The 5′ nucleotides of the two mRNAs were found to differ while the remainder of the mRNA chains was identical. The 'cap' sites for the two mRNAs lie approximately 2.8 kb apart, whereas poly(A) site is the same. The 'cap' sites always seem to be start sites for RNA syntheses. α-amylase production would appear to be controlled by selecting specific transcriptional start sites.

translation preinitiation steps: Activation and charging of transfer RNAs, binding of charged tRNA to the ribosome, loading of mRNA to the ribosome are important translation preinitiation steps.

translation regulation through ribosomal RNA pseudouridylation: Pseudouridine is enriched in ribosomal, spliceosomal, transfer and messenger RNA and thus integral to the central dogma. Y. Zhao *et al.* (2023) studied the translation of a hypopseudouridylated ribosome initiated by the INTERNAL RIBOSOME ENTRY SITE (IRES) elements analysed eight cryo-electron microscopy structures of the ribosome bound with the Taura syndrome virus IRES in multiple functional states. They found widespread loss of pseudouridine-mediated interactions through water and long-range base pairings. In the presence of the translocase, eukaryotic elongation factor 2 and guanosine 5′-triphosphate hydrolysis, the hypopseudouridylated ribosome favours a rare unconducive conformation for decoding that is partially recouped in the ribosome population that remains modified at the P-site uridine. The structural principles learned establish the link between functional defects and modification loss and are likely applicable to other pseudouridine-associated processes.

translation repression: This type of repression occurs when a translationally 'active' mRNA structure or initiation complex is rendered 'inactive' by an mRNA conformational change and/or binding of a *trans*-acting repressor. Another translation repression process inhibits the translation of messenger RNA into protein by blocking ribosome recruitment or function.

translation termination: Refers to the incorporation of the last amino acid into a polypeptide chain and the release of the complete polypeptide chain from the ribosomes during protein biosynthesis.

translation termination in eukaryotes: Translation termination in eukaryotes occurs in response to a stop codon in the ribosomal A-site and requires two release factors (RFs), eRF1 and eRF3, which bind to the A-site as an eRF1/eRF3/GTP complex with eRF1 responsible for codon recognition.

translation termination in prokaryotes: The successful completion of the polypeptide chain is signalled by a termination codon in the mRNA. The polypeptide chain is then released with the help of proteins called RELEASE FACTORS. Once a termination codon occupies the ribosomal A-site, three termination or release factors (RF1, RF2 and RF3) contribute to (i) the hydrolysis of the terminal peptidyl-tRNA bond; (ii) release of the free polypeptide and the last tRNA, now uncharged, from the P site and (iii) the dissociation of the 70S ribosome into its 30S and 50S subunits, ready to start a new cycle of polypeptide synthesis. RF1 recognizes the termination codons UAG and UAA and RF2 recognizes UGA and UAA. The class II RF3 is a GTPase that removes class I RFs (RF1 and RF2) from ribosome after release of nascent polypeptide. Termination of polypeptide synthesis is signalled by one of three termination codons in the mRNA (UAA, UAG, UGA), immediately following the last amino acid codon. RF1 is required for UAG and RF2 for UGA. UAA can accept either RF1 or RF2. RF3 activates RF1 and RF2.

translation, proofreading in: Protective role of proofreading in translation is evident from an instance in which a small mistake has a catastrophic effect. Transfer of information from the DNA sequence of a gene to the corresponding amino acid sequence of a protein requires transcription of the sequence into messenger RNA, then translation of that RNA into an amino acid sequence at ribosomes, through the agency of amino acid-specific transfer RNAs. All these steps are prone to error.

translational amplification: The use of the same mRNA molecule many times, allowing the synthesis of large amounts of polypeptide from a single mRNA and, therefore, from a single gene product.

translational control: Control of gene expression by the selection of which mRNAs in the cytoplasm are translated into proteins by ribosomes.

translational gene silencing: When gene silencing takes place through suppression of translation.

translational initiation in eukaryotic gene regulation: Assembly of the eIF4E/eIF4G complex has a chief role in the regulation of gene expression at the level of translational initiation. This complex is regulated by the eIF4E-binding proteins (E4-BPs), which compete with eIF4G for binding to eIF4E and have tumour-suppressor activity.

translational initiation sequences in eukaryotes: Eukaryotic translation and transcription is an asynchronous process. Eukaryotic translation initiation takes place on ribosomes bound to the endoplasmic reticulum in the cytoplasm.

translational initiation sequences in prokaryotes: In addition to AUG initiation codon, three structural features have been recognized on mRNA. (a) Shine-Dalgarno (S-D) sequence is nine nucleotides long, having the consensus sequence UAAGGAGGU. (b) In the polycistronic messages are required one or more termination codons in the ribosome-binding site. (c) Ribosome-binding site of genes encoding highly expressed proteins contains all or part of sequence PuPuUUUPuPu. Another feature is triplet immediately preceding the AUG start codon (e.g. for β-galactosidase mRNA translation), most favoured triplets are UAU or CUU.

translational reading frame slippage prevention: Coupled translocation of mRNA and tRNAs requires advancement of the mRNA reading frame by one codon. In eukaryotes, translocation is accelerated and its fidelity is maintained by elongation factor 2 (eEF2). N. Milicevic *et al.* (2024) show an intricate network of interactions preventing the slippage of the translational reading frame. The accuracy of eukaryotic translocation relies on eukaryote-specific elements of the 80S ribosome, eEF2 and tRNAs. The sterically constrained environment imposed by diphthamide, a conserved eukaryotic post-translational modification in eEF2, not only stabilizes correct Watson-Crick codon-anticodon interactions but may also uncover erroneous peptidyl-tRNA and therefore contribute to higher accuracy of protein synthesis in eukaryotes.

translational reading frame: In translation, a mature mRNA strand is translated in the 5′→3′ direction by a ribosome to produce a polypeptide chain. Reading frames in the 3′→5′ direction on the mRNA strand may be disregarded during translation. There are therefore three possible reading frames for an mRNA strand that can be used during translation, although usually only one reading frame is used. Each reading frame corresponds to a different starting nucleotide.

translational recoding: A process that causes ribosome slippage along the mRNA, thereby changing the amino acid sequence of the synthesized protein. This phenomenon has been reported in chloroplasts of *Oenothera* and tobacco. *Aka* RIBOSOMAL FRAMESHIFTING.

translatome: The collection of **open reading frames (ORFs)** that are being actively translated in a cell or organism.

translatomics: The study of all ORFs that are being actively translated in a cell or organism.

translesion synthesis (TLS): Refers to DNA replication past the point of lesion. TLS is *aka* MUTAGENIC REPAIR (MR) or ERROR-PRONE REPAIR (EPR). In humans, four ERROR-PRONE DNA POLYMERASES are known.

translocase of the inner membrane (TIM) complexes: Protein translocators positioned in the mitochondrial inner membrane. TIM23 complex mediates the transport of proteins into the matrix and the insertion of some proteins into the inner membrane. TIM22 complex mediates the insertion of a subgroup of proteins into the inner membrane.

translocase of the outer membrane (TOM) complex: A multisubunit protein complex that transports proteins across the mitochondrial outer membrane.

translocase: Elongation factor G (EF-G) in prokaryotes is necessary for proper translocation at the ribosome during the translation process. This elongation factor is replaced by eEF2 in eukaryotes.

translocation: (i) Genetic mutation in which a section of a chromosome 'breaks off' and moves to a new position in that or a different chromosome. They may drastically alter size of a chromosome as well as the position of its centromere. Translocation is utilized as an important tool in gene transfers and production of duplications and deficiencies. *See* INTERCALARY TRANSLOCATION, SIMPLE TRANSLOCATION *and* RECIPROCAL TRANSLOCATION. (ii) The process by which a newly synthesized protein is directed toward a specific cellular compartment (i.e. the nucleus, the endoplasmic reticulum). (iii) Transport of water, nutrients, chemicals, or food materials from site of synthesis or storage to other part of plant where they are required for metabolic activities.

translocation of chromosome: A rearrangement in which part of a chromosome is removed by breakage and then becomes attached elsewhere in the same chromosome or to some other chromosome.

translocation of gene: Appearance of a new copy of a gene at the location in the genome elsewhere from the original copy.

translocation of protein: The movement of a protein from the site of synthesis to the site of action.

translocation of ribosome: It is movement a ribosome from one codon to the next along mRNA after the addition of each amino acid to the polypeptide chain.

translocation shift: Transfer of an interior segment of a chromosome close to the interior of a non-homologous chromosome.

translocation tester set: A series of more or less defined homozygous reciprocal translocations. They may be utilized in test crosses to identify accessory or unknown chromosomes by homologous pairing in MI of meiosis (e.g. this sort of tester set was produced in diploid rye and barley).

translocation-and-pause cycles: Translation occurs through successive translocation-and-pause cycles. The distribution of pause lengths, with a median of 2.8S, indicates that at least two rate-limiting processes control each pause. Each translocation step measures three bases (i.e. one codon) and occurs in less than 0.1 s. Translocation and RNA unwinding are strictly coupled ribosomal functions.

translucent: So clear that light may pass through but not transparent.

transmembrane proteins. The protein molecules that extend from one side of a cell membrane to the other side of that membrane. For example, G-proteins are transmembrane proteins that accomplish signal transduction (i.e. convey 'signal' from outside the cell to one or more internal cell parts). EPIDERMAL GROWTH FACTOR (EGF) receptors bind to EGF molecules (e.g. passing-by in the blood), then both enter the cell (through the cell membrane) together, where the EGF stimulates growth/division of that cell.

transmembrane region: The region of a transmembrane protein that actually spans the membrane. Transmembrane regions are usually hydrophobic in order to be thermodynamically compatible with the lipid bilayer portion of the membrane. They may consist of either alpha-helical or beta-strand secondary structure elements, but in either case the external residues (the ones facing the membrane) are invariably hydrophobic while the internal residues may be hydrophilic (as in the case of a pore or channel) or polar. One common transmembrane structural domain is the seven-helix bundle seen in numerous channel proteins.

transmission disequilibrium test (TdT): A statistical test that detects associations between genetic markers and disease alleles by looking for marker alleles that are transmitted in excess to affected offspring.

transmission electron microscope (TEM): Electron microscope for detailed study of thin sections specimen with magnification ranging from 10,000 to 100,000×, even more. *See* ELECTRON MICROSCOPE.

transmission genetics: The study of the mechanisms involved in the transmission of a gene from one generation to the next. *Aka* CLASSICAL GENETICS *or* MENDELIAN GENETICS.

transmission of epigenetic changes: Epigenetic changes can be reversible. For example, the controlling element Ac (activator) in maize is capable of transposition. A derivative that has lost transposase activity has been shown to have methylated cytosine in CCGG, CAG and CTG sequence. This is inherited through sexual crosses but can revert to active Ac either in the germline or somatic tissues, with the loss of methylation but without a change in base sequence.

transmit (n. transmission): To spread or transfer of an infectious pathogen from plant to plant in the same generation or from one plant generation to another.

transpiration: Loss of water from plants via stomata to the outside atmosphere. The plants growing in the field absorb water from soil through roots. Some of the water is lost through the stomata of leaves that is called transpiration. During transplanting of seedlings, transpiration losses sometimes cause dehydration. and the seedlings do not establish. To minimize the transpiration loss, some antitranspirants, like glycerol and so on, are applied to the leaf surfaces. Another way of minimizing transpiration loss is to create high humidity in the green house by using humidifiers. *Compare with* EVAPOTRANSPIRATION.

transplacement of gene: A technique originally developed in yeast for exchanging a DNA sequence on a recombinant plasmid for the corresponding sequence in one of the host's chromosomes. Thus, a wild-type copy involves specific protein 'carriers' located within the cell membrane.

transplantation: The implanting of cells, tissues, or organs that have been retrieved from a living or deceased donor into a recipient.

transplastomic plants: Plants carrying transgenic plastome. Attempts are now being made to develop transgenic plants with engineered chloroplast genome through chloroplast transformation. The foreign genes transferred to chloroplast have shown up to 40% foreign proteins in the total cellular protein. Since the chloroplast possesses maternal inheritance, therefore, transgenes do not go to pollen and hence there is no risk regarding the transgene escape via pollen. Therefore, transplastomic plants are self-contained and thus considered environmental friendly. Therefore, transplastomic plants may be a solution for warranting biocontainment.

transport vesicles: Transport vesicles fuse with and arise from the edges of the Golgi body's cisternae. Found in animal and plant cells. Ribosomes deposit proteins into the lumen of the rough endoplasmic reticulum, where they can be modified. Transport vesicles then take the modified proteins to the nearby Golgi body, where they are processed further and sorted for transport. Transport vesicles carry the fully processed proteins throughout the cell or to the plasma membrane for export. Transport vesicles also distribute lipids, hormones and other cellular products from the smooth ER throughout the cell.

transposable DNA/genes: Stretch of DNA that as discrete genetic and physical entities can move from one position to another in the same or a different chromosome or a different genome through a process of excision and integration. *Aka* JUMPING DNA/GENE, TRANSPOSABLE GENETIC ELEMENT, TRANSPOSON *and* TRANSPOSABLE ELEMENT.

transposable DNA: *See* TRANSPOSABLE ELEMENTS.

transposable element cycling: The transformation of a transposable element from an epigenetically inactive state to an active state, and back to an epigenetically inactive state, over the course of development or over multiple generations.

transposable elements (TEs): Discovered in 1950 by geneticist B. McClintock in corn (maize) plants (*Zea mays* L.); and in bacteria a decade later by J. Lederberg. Transposons generally consist of more than 2000 nucleotides, which can either carry genes along one organism's genome, or even into another organism's genome (e.g. via sexual conjugation, in bacteria). By such sexual conjugation, transposons can transport genes that confer new phenotypic properties (e.g. resistance to certain antibiotics, for a given bacterial cell). A DNA sequences that can 'jump' from one location in the genome to another. They include retrotransposons that are mobile following a 'copy and paste' mechanism and require an RNA intermediate and DNA transposons that move using a 'cut-and-paste' mechanism. The process is called TRANSPOSITION. The movement of a transposable element can generate mutations or chromosomal rearrangements and thus affect the expression of other genes. Move directly from one site in the genome to another (do not need other vectors. The insertion site of an element does not have extensive homology to the transposon. They exist as multiple copies dispersed in the genome. They do not rely on any relationship between sequences at the donor and recipient sites (i.e. it is a non-homologous recombination). Internal counterpart to vectors that move sequences between genomes (phages and plasmids). May provide a major source of mutations in the genome. They contain inverted repeats at element termini. A short, direct repeat of genomic DNA often flanks the transposon. The elements found in prokaryotes as well as eukaryotes can get inserted into structural genes and block their action. On excision of a transposable element, the function of the gene is restored. It appears as if a non-functional allele has reversed to a functional one. Some of these elements are autonomous. Probably they encode a transposase. Based on size, orientation and nature of terminal repeats, transposons can be classified into four types: transposons with LONG TERMINAL DIRECT REPEATS (LTDRs), transposons with LONG TERMINAL INVERTED REPEATS (LTIRs), transposons with SHORT TERMINAL INVERTED REPEATS (STIRs) and transposons WITHOUT TERMINAL REPEATS (without TR). *Aka* TRANSPOSONS, JUMPING DNA/GENES, INSERTION SEQUENCES (IS) *or* TRANSPOSABLE GENETIC ELEMENTS. *See* TRANSPOSITION.

transposable elements as potential vectors for transformation: The successful genetic transformation of *Drosophila melanogaster* using transposable vectors such as P elements indicated that other insect species could be transformed genetically in the same manner. Although *Drosophila* P elements have not been successful as vectors in non-drosophilids, the demonstration of introduction of exogenous DNA sequences into *Drosophila* germ lines have raised the possibilities of using new techniques with other insect species. Transposable elements analogous to the P elements of *Drosophila* are under investigation.

transposable elements in evolution: Transposable elements can amplify a small genetic disturbance into an earthquake. Transposons are certainly a stuff of remodelling and building of organismic evolution.

transposable genetic element: *See* TRANSPOSABLE ELEMENTS *and* TRANSPOSON.

transposase: An enzyme (ribonucleoprotein) that is coded by the transposase gene present in the transposon and catalyses its transposition. For example, *Drosophila* transposon, known as the P element, carries a gene for the synthesis of transposase. A significant portion of genomes of higher plants and vertebrates consists of transposable elements and their derivatives. Autonomous DNA-type transposons encode a transposase that enables them to mobilize to a new chromosomal position in the host genome by a cut-and-paste mechanism. As this is potentially mutagenic, the host limits transposition through epigenetic change silencing and heterochromatin formation. A transposase from *Arabidopsis thaliana* that they named *day sleeper* is essential for normal plant growth; it shares several characteristics with *haT* (*hobo, Activator, Tam3*) family of transposases. *Day sleeper* was isolated as a factor binding to a motif (Kubox) present in upstream region of the *Arabidopsis* DNA repair gene *Ku70*. This motif is also present in the upstream region of many other plant genes. Plants lacking *day sleeper* or overexpressing this gene do not develop in a normal manner. *Day sleeper* overexpression results in altered expression of many genes. Transposase-like genes can be essential for plant development and can also regulate GLOBAL GENE EXPRESSION. Thus, transposase can become domesticated by the host to fulfil important cellular functions.

transposase activation, molecular basis of: Although some transposases act alone by themselves, many trust on devoted AAA+ ATPase subunits that control site selectivity and catalytic function. Using IS21 as a model transposase system, Á. de la Gándara *et al.* (2024) discover that an ATPase regulator employs nucleotide-controlled assembly and DNA deformation to allow structure-based site selectivity, transposase employment and activation and integration. The IstB ATPase self-assembles into an autoinhibited pentamer of dimers that firmly curves target DNA into a half-coil. Two of these decamers dimerize, which stabilizes the target nucleic acid into a kinked S-shaped configuration that occupies the IstA transposase at the interface between the two IstB oligomers to form almost 1 MDa transpososome complex. Specific interactions inspire regulator ATPase activity and activate a large conformational change on the transposase that places the catalytic site to achieve DNA strand transfer. *See* TRANSPOSOME.

transposase gene: A transposon gene that codes for enzyme transposase that catalyses transposition. Transposase binds to specific DNA sequences called INVERTED REPEATS (IRs) because the two regions are upside-down and backward to each other and cut DNA to remove the gene. The transposon can enter several locations. When it invades a gene it usually inactivates the gene by interrupting the coding sequence and the protein that the gene codes for.

transposase silencing through RNA interference: Germline-limited transposase genes play key roles in this process of genome-wide DNA excision, which suggest that transposases function in large eukaryotic genomes containing thousands of active transposons. Transposase gene expression occurs during germline-soma differentiation and that silencing of transposase by RNA interference leads to abnormal DNA rearrangement in the offspring.

transposition: The process whereby a transposon or an insertion sequence inserts itself into a new site on the same or another DNA molecule. Commonly, this process involves TRANSPOSABLE ELEMENTS. Here, movement of a gene or set of genes occurs from one site in the genome to another site without a reciprocal exchange. Different transposons may transpose by different mechanisms: ILLEGITIMATE RECOMBINATION, TRANSPOSITION IMMUNITY and TRANSPOSON TAGGING. *See* TRANSPOSONS. *See details as separate terms.*

transposition elements in humans: Two types of transposable elements that occur in humans are (i) LONG INTERSPERSED NUCLEAR ELEMENTS (LINEs) and (ii) SHORT INTERSPERSED NUCLEAR ELEMENTS (SINEs). *See details as separate terms.*

transposition immunity: Phenomenon through which some transposons have the ability to prevent others of the same type from transposing to the same DNA molecule.

transposition pesudoalleles: Pseudoalleles are very closely linked. Same as *TRANS*-HETEROZYGOTE.

transposition, mechanisms of: There are three mechanisms of transposition: (i) CONSERVATIVE TRANSPOSITION, (ii) REPLICATIVE TRANSPOSITION and (iii) RETROTRANSPOSITION. *See details as separate terms.*

transposition, types of: On the basis of the process of transposition, two main classes of mobile genetic elements are (i) those which transpose directly from DNA to DNA and (ii) retroposons which involve an RNA intermediate.

transpositional suppression: Transposons are known to cause vast genomic alterations. In plants, transposon mobilization is strongly repressed by DNA methylation. W. Fan *et al.* (2023) suggest that transposon RNA is marked by 6^{m}A RNA methylation and can be localized in stress granules (SGs). SG-localized *At*ALKBH9B selectively demethylates a heat-activated retroelement, *Onsen*, and thereby releases it from spatial confinement, allowing for its mobilization. m^{6}A RNA methylation contributes to transpositional suppression by inhibiting virus-like particle assembly and extrachromosomal DNA production.

transposon activation: Transposable elements (TE) become active and/or mobile in the genomes of replicatively and stress-induced senescent mammalian cells.

transposon burst: Bursts of transposon activity make use of master-slave and random mechanisms. Significant increases in transposon activity have been noted in plants after allopolyploidization, during chromosome rearrangements in *Drosophila* and after species hybridization.

transposon elements in prokaryotes: TEs of prokaryotes are of three types: insertion sequence (IS) elements, transposons (Tn) and bacteriophage Mu.

transposon families in maize: Each family of controlling elements in maize has two classes: autonomous and nonautonomous.

transposon immunity: A plasmid containing a transposon (say Tn3) is immune to further insertions of the transposon (Tn3). This phenomenon works in *cis*.

transposon mutagenesis: Transposable elements in prokaryotes and eukaryotes when inserted in a given locus can control expression of the locus and cause large-scale rearrangements. When introduced into an operon, they not only abolish the function of the gene but also lead to reduction in the activities of all other genes downstream in the direction of transcription (5′→3′). Such mutations are *aka* POLAR MUTATIONS.

transposon reactivation: The act or process of making transposon active again or becoming active again. Reactivated transposons harm the host in multiple ways.

transposon silencing: Transposon silencing is a form of transcriptional gene silencing targeting transposons. Transcriptional gene silencing is a product of histone modifications that stop the transcription of a particular area of DNA. Nearly a century ago, two geneticists described 'rogue' pea plants with an unorthodox pattern of inheritance. More than a century ago, William Bateson and Caroline Pellew found that crossing inferior rogues with normal plants always produced rogue offspring, suggesting that the rogue appearance was a dominant trait. The real shock came when rogue progeny was crossed back to normal plants. Following the principles of Mendelian inheritance, these crosses should have produced a mix of normal and rogue plants, but instead they produced only rogue plants. This phenomenon, later dubbed 'paramutation', allowed the rogues to break the rules by acting 'epigenetically' – inducing heritable changes in gene expression without DNA mutations. In one-sided interactions between gene pairs, or alleles, only 'paramutagenic' alleles can attenuate and eventually silence, the expression of 'paramutable' alleles.

transposon silencing, maintenance of: Given the damage that transposons can cause by inserting themselves into essential genes, it is not surprising that organisms have evolved enduring mechanisms to keep jumping genes in their place. This study contributes a valuable framework for identifying the factors that regulate the enigmatic epigenetic processes that defend the genome against invasive elements – and helps explain how these changes can continue and be transmitted to the next generation. The gene, *Mediator of paramutation1 (Mop1)*, encodes an RNA-processing enzyme called RNA-DEPENDENT RNA POLYMERASE 2 (RDR2) that is required to make the small RNAs needed to maintain silencing of a transposon (called *MuDR*). While *Mop1* maintains *MuDR* silencing, the authors show that a second gene is required to establish heritability of silencing. *MuDR* can be heritably silenced by a paramutagenic gene called *Mu killer (Muk)* – an inverted-repeat variant of the *MuDR* transposon. *MuDR* encodes two genes – *mudrA* causes excision of the element and, with *mudrA*, *mudrB* helps reinsert it. *Muk* produces hairpin double-stranded RNAs that generate rapid processing of full-length *mudrA* into small RNAs, leading to the destruction of *mudrA* transcripts and the methylation and silencing of *MuDR*. *Muk*-induced *MuDR* silencing begins with *mudrA*, then spreads to the adjacent *mudrB* gene.

transposon tagging: Blocking activity of a functional gene by inserting a foreign DNA sequence. Specific genes are tagged. Crosses are designed such that one parent is homozygous recessive for targeted gene while other is homozygous dominant. Transposon tagging is particularly effective when transposons are transferred to nonhost plant species (heterologous system). Transposon tagging has the advantage of greater probability of hitting the gene compared to the T-DNA tagging. Transposon tagging has two approaches: TARGETED APPROACH and NON-TARGETED APPROACH.

transposon-arrayed gene knock out (TAG-KO): A highly efficient method for gene discovery and gene function assignment in the rice blast fungus *Magnaporthe grisea*.

transposon-based vectors: Transposon-based vectors have been prepared using P ELEMENT, SLEEPING BEAUTY (SB), PIGGYBAC (PB) and AC-DS TRANSPOSON SYSTEMS.

transposon-mediated gene transfer: Genes can be transferred from one species to another by use of a transposon. Examples are P ELEMENT in *Drosophila* and DS ELEMENT in maize.

transposons: Segment of DNA (gene) that can move from one position to another in the genome. Transposons, also called JUMPING GENES, TRANSPOSABLE ELEMENTS or TRANSPOSABLE GENETIC ELEMENTS, were discovered by Barbara McClintock in 1950 in maize but now have been found in virtually all organisms. The transposons are small segments (700–40,000 bp) of DNA, which can move or jump (be transposed) from one place on the chromosome to another on the same or a different chromosome. This movement is called transposition, which is a random process. Jumping of these transposons causes mutation and observable changes in the phenotype because the expression of genes in both the regions (donor and recipient) gets changed. The transposons of maize (Ac and Ds) were earlier called the Ac-Ds system (activator–dissociation system). Ac measures more than 4500 bp and Ds about 400 bp. Each possesses genes including the gene for transposase enzyme responsible for transposition. Part of this region can be deleted to use the transposon as a cloning vector. The transposons of *Drosophila* are called P elements. *See* TRANSPOSABLE ELEMENTS.

transposons with long terminal direct repeats (LTDRs): LTDRs are known as type 1 transposons, examples of LTDR transposons are (i) *copia*-like elements in *Drosophila*; intracistronic A type particles (IAPs) in mice and (ii) Ty elements in yeast.

transposons with long terminal inverted repeats (LTIRs): LTIR transposons contain LTIRs. Such transposons include FB and TE elements of *Drosophila melanogaster*. *Aka* TYPE II TRANSPOSONS.

transposons with short terminal inverted repeats (STIRs): STIR transposons are found in animals as well as plants. Such transposons in animal systems include P-M and I-R elements in *Drosophila melanogaster*, Piggy Back and Sleeping Beauty DNA transposons and Te1 element in nematode *Caenorhabditis elegans*. Plant STIRs include Ac-Ds elements in *Zea mays* L., Spm (suppressor mutation) in maize, Tam1 and Tam3 elements in *Antirrhinum majus* and a putative transposable element in soybean.

transposons without terminal repeats: This type of transposons is Alu sequences. Now they are better known as RETROPOSONS and are classified as PSEUDOGENES. These transposons are incapable of transposition in the manner transposons do. *Aka* TYPE IV TRANSPOSONS.

transposons, biological significance of: Mutation of genes is caused by insertion of transposons. Regulation of gene expression occurs by transposon insertion. Transposable element-mediated rearrangements include deletions, duplications and inversions. Transposons play a role in the evolution of new genomes and lead to the evolution of new functions during evolution.

transposons, composite: A composite transposon consists of two inverted repeats from two separate transposons moving together as one unit and carrying the DNA between them.

transposons, non-composite: Unlike the composite transposon, non-composite transposons do not contain IS elements at their ends but have the repeated sequences at their ends that are required for transposition.

transpososome: A complex structure that functions together with an enzyme called transposase to enable mobile genetic elements to move around in the genome. The IS110 and IS1111 systems need only a single protein, which is less than half the size of many of the Cas enzymes used in CRISPR genome-editing systems. The IS110 system positions out as an innovative and sophisticated way via which mobile DNA elements can hitchhike around the genome. So far, IS*110* family members do not appear to function well in mammalian cells. Scientists are now attempting to engineer them to function well in the mammalian cells.

transsexual: A person who has adopted the appearance of the opposite sex with the aid of surgical procedures and hormone therapy.

transvection: An epigenetic phenomenon that involves the interaction between two homologous chromosomes, resulting in either gene activation or repression at an allele.

transversion: A point mutation changing from a purine into a pyrimidine, or vice versa (A→T, A→C, G→T, G→C) or vice versa (T→A, C→A, T→G, C→G). Transversions may arise by mistake pairing between synadenine produced by proton migration and rotation and common adenine and between neoguanine produced by shifting of the glycosidic bond from normal form to extracyclic nitrogen and common guanine. Methylated guanine undergoes tautomeric shifts, pairs with thymine, and leads to G.C → T.A transversions. Transversions occur by apposition of two errors – proton migration in one and 180° rotation around the glycosidic bond (syn) in the other. Tautomeric synadenine (A*) may pair with syn A, which leads to A.T → T.A transversion. Similarly, tautomeric neoguanine (G*) may also pair with syn G, which leads to G.C → C.G transversion. Rare base pairing between two purines or two pyrimidines can cause spontaneous production of transversion. These A.G and T.C pairs use one or two hydrogen bonds. In this case, they have slightly wrong distances or angles for two sugar (C)-base (N) bonds. Since the DNA backbone is not completely stiff, it can accumulate such distortions at least occasionally, as a mistake. *Compare with* TRANSITION.

transversions through tautomerism: Proton migration within base pairs can set off a chain of modifications that end in DNA mutation. Methylated guanine undergoes tautomeric shifts, pairs with thymine, and leads to G.C®T.A transversions. Transversions occur by apposition of two errors – proton migration in one and 180° rotation around glycosidic bond (syn) in the other tautomeric synadenine (A*) may pair with synA, which leads to A.T®T.A transversion. Similarly, tautomeric neoguanine (G*) may also pair with synG, which leads to G.C®C.G transversion.

TRANSWITCH®: A 'sense' technology employed to 'turn off' or 'suppress' a gene (e.g. the one that reasons tomato to ripen) that causes an unwanted effect (e.g. premature softening of tomato). TRANSWITCH® and its registered trademark are owned by DNA Plant Technology Corp.

trap crop: A crop planted around a field to guard the inner crop from diseases transmitted by aerial vectors; host crop of a parasitic plant, (e.g. witchweed, *Striga* spp.), that is planted to stimulate seed germination and later sacrificed by ploughing under before the parasitic plant produces new seeds.

traptamer: N.S. Vallina *et al.* (2024) use the RNA ORIGAMI METHOD to prototype an RNA robotic device, named the 'Traptamer', which mechanically traps the fluorescent aptamer, iSpinach. The Traptamer senses two RNA key strands, acts as a Boolean AND gate and reversibly controls the fluorescence of the iSpinach aptamer. Traptamer structure at 5.45 Å resolution reveals the mechanical mode of distortion of the iSpinach motif. and suggests a general approach to distorting RNA motifs and a path forward to build sophisticated RNA machines that through sensing, computing and actuation modules can be used to precisely control RNA functionalities in cellular systems.

treadmilling: The process by which a polymeric protein filament is maintained at constant length by the addition of protein subunits at one end and loss of subunits at the other.

trenching: The physical separation of soil in a vertical plane to sever grafted roots between trees.

tri-allelic single-nucleotide polymorphism (tri-allelic SNP): Any single-nucleotide polymorphism of which three different alleles exist in a population.

triacylglycerol: Molecule constituted of three fatty acids esterified to glycerol. The main constituent of fat droplets in animal tissues (where the fatty acids are saturated) and of vegetable oils (where the fatty acids are mainly unsaturated). *Aka* TRIGLYCERIDE.

tricarboxylic acid (TCA) cycle: A series of biochemical reactions that serve to generate energy from the metabolism of acetyl-CoA, which in turn is derived from the metabolism of sugars, fats and proteins. *See* CITRIC ACID CYCLE.

trichogyne: A female receptive hypha.

trichome: Most dicot leaves have trichomes, hairs that extend from the epidermal cells of the upper and lower epidermis. Trichomes can also be present on dicot stems, as well as monocot leaves and stems. Depending on the type of plant, trichomes can be unicellular or multicellular, branched or unbranched and glandular or non-glandular. They protect the plant from damage, which can be caused by herbivores, UV light and extreme temperatures.

trifoliate: A compound leaf having three leaflets. For example, a clover leaf.

trigeneric hybrid: A spontaneous or experimental hybrid consisting of three genomes of different genera.

trigger protein: *See* U PROTEIN.

triggering stress cue: The response to stressful stimuli is elaborated and triggered by the stress system. Priming modifies responses to a triggering stress cue.

trihybrid: An individual heterozygous for three pairs of genes (e.g. $+++/abc$).

trihybrid ratio: Three contrasting pairs of traits and hence three gene pairs are considered at one time. In this case also the PRINCIPLE OF INDEPENDENT SEGREGATION is valid according to Mendel's principles.

triisopropylbenzene sulfonyl chloride (TPS): A chemical agent responsible for condensation reaction in the chemical synthesis of oligonucleotide fragments. The reaction is known as a TPS-MEDIATED CONDENSATION REACTION.

triisosomic: In allopolyploids, such as hexaploid wheat, when a cell or individual absences one chromosome pair while three homologous isosomes for the same arm are present.

trimeric GTP-binding protein: *See* GTP-BINDING PROTEIN.

trimerous: In three parts, particularly with respect to flowers; having three parts in each whorl.

trimming of the 5′ and 3′ ends: For this, RNase P acts on the 5′ terminus endonucleolytically and produces mature 5′ terminus by characterizing mature tRNA sequence. The 3′ end maturation occurs by exonuclease activity by RNase D and 3′-CCA sequences of tRNA (as in prokaryotes). When 3′-CCA is not encoded in DNA, this triplet is enzymatically added post-transcriptionally. But this triplet addition has to transpire before removal of intron. The second situation occurs in eukaryotes.

trinucleotide repeat: A sequence of three nucleotides that is repeated. Such regions of the genome may be prone to unstable expansion of the number of repeats with subsequent generations and result in characteristic group of genetic disorders such as Fragile X Syndrome, Spinal and Bulbar Muscular Atrophy, Myotonic Dystrophy.

trioecious: Species that have male, female and hermaphroditic flowers on different individuals.

tripeptidyl peptidase 1 (TPP1 or simply T) glutamate (E) and leucine (L)-rich patch (TEL

patch): A small group of specific amino acids on the surface of the shelterin protein TPP1. Tripeptidyl peptidase 1 directly recruits telomerase to telomeres and then stimulates its activity.

triple X: Trisomy for X chromosome.

triple-helical DNA: Three strands of DNA, which are complementary to each other, have tendency for triple helix formations. Homo- or hetero-polypurine or polypyrimidine tracts can adopt triple strand conformations. Triplex strands can impact transcription, replication and gene expression and they even present specific protein binding. Triplex DNA structures can be either intermolecular or intramolecular forms. Triple-helical DNA may occur during recombination and DNA repair. *Aka* H-DNA *or* TRIPLEX DNA.

triple-helical DNA in repressor-specific repression: Triple-helical complexes can repress transcription primarily by hindering promoter DNA assembly into initiation complexes rather than by occluding Sp1 binding. Triple-helix-induced repression most likely includes changes in DNA flexibility. These intermolecular triplex structures may function as antisense inhibitors of gene expression and may have a role in recombination.

triplet binding assay: An experimental technique that enables the coding specificity of a triplet of nucleotide to be determined.

triplet code: A group of three successive nucleotide in RNA or DNA that, in the genetic code, specifies a particular amino acid in the synthesis of polypeptide chains.

triplex: The condition in autotetraploidy in which at a locus three genes are dominant and one recessive.

triplicate genes: Three genes, any one of which produces the same phenotypic effect. For example, red seedcoat colour in wheat is produced by the R_1, R_2 and R_3 genes.

triplinerved: Leaves having three main nerves with the lateral nerves arising from the mid nerve above the base of the leaf.

triploid fish: Fish are ordinarily diploid. By exposing fish eggs to certain specific combinations of temperature and pressure, immediately after fertilization of those eggs, scientists can cause the resultant fish to become triploid. Triploid fish are unable to reproduce. This sterility is desired by man, to prevent certain fish (e.g. those that have been genetically engineered) from mating with wild fish. Such induced (triploid) sterility also prevents the (genetically engineered) fish from wasting energy on the act of reproduction, so they grow faster and larger. *See* TRIPOID FRUITS.

triploid fruits: Refers to organisms that possess three sets of chromosomes, instead of the normal two. Conversion of a diploid (i.e. two sets of chromosomes) organism to triploid has been done by man. Triploids are characteristically sterile. A practical application of sterility associated with triploidy is the production of 'seedless' varieties of grapes, seedless watermelon, banana and so on. This type of transfer of energy use from reproduction to growth also holds true for 'seedless' grapes, watermelons and so on. *See* TRIPLOID FISH.

triploid fusion nucleus: The result of the fusion of one male gamete (pollen) with two haploid female nuclei (polar cells) in the second polar event. This nucleus divides to produce the cells of the endosperm.

triploidy: A state in which three chromosome sets are present.

triradial chromosome configuration: A chromosomal pairing arrangement of three chromosomes in which the homologous chromosomes are organized like a star. *See* QUADRIRADIAL (qr).

Tris-ethylenediamine tetra-acetic acid (Tris-EDTA or TE) buffer: TE buffer is the most commonly used solution in molecular biology. TE is a component of Tris (a common pH buffer) and EDTA (molecule that chelate cations). Tris-EDTA buffer solubilizes DNA and RNA while protecting it from degradation.

trisomic: A genome that is diploid (2n) but that contains an extra chromosome (2n + 1), homologous with one of the existing pairs, so that one kind of chromosome is present in triplicate. Trisomics have played an important role in gene location programmes.

trisomic analysis: A procedure for mapping gene loci on individual chromosomes by comparing disomic and trisomic segregation patterns of a series of individuals.

trisomic series: A complete set of trisomics of a given plant species, in which all different chromosomes of the complement are available as trisomics in different individual lines (e.g. trisomic series have been established in in barley, rye, tomato and maize).

trisomy: Condition of a diploid cell or individual having three chromosomes of a particular type. Trisomy for chromosome 21 is the cause of Down syndrome. Trisomy for chromosome 18 and trisomy for chromosome 13 cause very serious birth defects.

tritelosomic: An individual that lacks one chromosome pair but possesses three additional, homologous telocentric chromosomes.

triticale: An allopolyploid obtained by combining the chromosomes of wheat (*Triticum*) and rye (*Secale*) first bred in laboratories during the late 19th century in Scotland and Germany to produce a new species *Triticosecale wittmack*.

tritium: A radioactive isotope of hydrogen.

trivalent: Three synapsed homologous chromosomes.

TRNASCAN: Detects for RNA coding regions.

Trojan gene: A gene that drives a population to extinction during the process of spread as a result of destructive, self-reinforcing cycles of natural selection.

Trojan squares: Semi-Latin square designs based on sets of mutually orthogonal superimposed Latin squares, which have been shown to be maximally efficient semi-Latin square designs for pairwise treatment comparisons in the plots-within-blocks stratum. Trojan squares of size (n × n)/k are based on k sets of n × n orthogonal Latin squares and have a natural factorial treatment structure; the 'k' squares and 'n' letters of the design represent either pseudo-factors for an unstructured treatment set or real factors for a factorial treatment set. Trojan squares have generalized balance and have a particularly simple analysis of variance.

trophophyll: A vegetative, nutrient-manufacturing leaf or microphyll whose key function is photosynthesis. It is never specialized or modified for some other function. *Compare with* SPOROPHYLL.

true breeding: A condition in which a group of identical individuals when selfed always yield offspring of the same phenotype. The alleles for this type of plants are homozygous.

true heritability: *See* HERITABILITY IN BROAD SENSE.

true negative: Does not have the abnormality or disease and is correctly classified by the test.

true overdominance hypothesis: The allozyme loci have a direct effect on the quantitative trait. Heterosis ($F_1 > P$) for a single locus. In contrast, pseudodominance may occur due to repulsion phase-linkage of more than one gene/locus mimicking the true overdominance, while SUPER DOMINANCE relates to heterosis accruing from heterozygotes that carry divergent alleles (e.g. A1A4 > A1A3 > Q1A2).

true positive: Does have the abnormality or disease and is correctly classified by the test.

true speciation: *See* CLADOGENESIS.

truncate: Cut off squarely; having an abruptly transverse end.

truncate distribution: An atypical sample of a population in which certain types of matings or individuals have been omitted.

truncation selection: A breeding technique in which individuals in whom quantitative expression of a phenotype is above or below a certain value (the truncation point) are selected as parents for the next generation.

trunk: The upright, large and typically woody main stem of a tree.

truss: A compact cluster of flowers or fruits arising from one centre; evident in many rhododendrons.

trypanosome surface antigen switching: About 1000 genes for 100 different variable surface glycoproteins (VSGs) are found scattered on different chromosomes in the genome of trypanosome. Diversity rests on changing expression from one pre-existing gene to another. Each VSG is coded by a single basic copy gene, which may be telomeric or internal in location and there may be several isogenes for the same VSG or similar VSGs. The copy of the gene, which is active, is called an EXPRESSION-LINKED COPY (ELC) and is located on an expression site. The creation of an ELC may comprise the transfer of a basic copy gene to the expression site or vice versa. Almost all switches in VSG type include the replacement of the ELC by a pre-existing silent copy.

tryptophan (trp): A heterocyclic, nonpolar, alpha, essential amino acid having chemical formula $C_{11}H_{12}N_2O_2$. It is a precursor of important biochemical molecules such as indoleacetic acid, serotonin and nicotinic acid. L-tryptophan is used as a common feed additive for livestock to ensure that their diet includes an adequate amount of this essential amino acid.

tryptophan (*trp*) operon in *Escherichia coli* and *Salmonella*: The regulation of the tryptophan operon of *Escherichia coli* is perhaps the best-known repressible system. In this operon, there are five different structural genes – *trp A*, *trp B*, *trp C*, *trp D* and *trp E* – which code for different enzymes meant for the synthesis of tryptophan from anthranilate. Genes *trpE* and *trpD* code for two polypeptides of anthranilate synthetase, *trpC* codes for indole-glycerol-synthetase, *trpB* codes for tryptophan synthetase β chain while *trpA* codes for tryptophan synthetase α chain. Flanking these five structural genes are promoter-operator-leader-attenuator (*P-O-L-a*) sequences

(162 nucleotides) on the upstream and *t* (36 nucleotides) and *t′* (250 nucleotides) terminator sequences on the downstream. In this operon, operator *trpO* is positioned within the promoter.

Tsix RNA: A non-coding RNA. Its gene is antisense to, and overlaps with, the *Xist* gene. Tsix transcription across Xist is thought to repress Xist upregulation and XCI initiation. The future Xi chromosome downregulates Tsix and upregulates Xist. The future active X chromosome downregulates Xist and sustains Tsix transcription.

Tubby proteins: Dysfunction of tubby protein results in maturity-onset obesity in mice. Tubby has been implicated as a transcription regulator. Tubby functions in signal transduction from heterotrimeric GTP-BINDING PROTEIN (G-PROTEIN)-COUPLED RECEPTORS. Receptor-mediated activation of G-protein αq (Gαq) releases tubby from the plasma membrane through the action of phospholipase C-β, triggering translocation of tubby to the cell nucleus. The localization of TUBBY-LIKE PROTEIN 3 (TULP3) is similarly regulated. Tubby proteins function as membrane-bound transcription regulators that translocate to the nucleus in response to phosphoinositide hydrolysis, providing a direct link between G-protein signalling and the regulation of gene expression.

tuberoid: An alternative name for an underground storage organ caused by the swelling of a root. It exists in many orchids.

tuberous: Similar to a tuber or producing tubers. *Aka* TUBULAR.

tubulin: Tubulin is organized into microtubules. They are fairly rigid, long and cylindrical. They give shape to the cell and are responsible for movement of the cell through cilia, flagella or pseudopodia. Tublins, each one having a bead-like structure, are arranged in the form of a long chain to form microtubule through polymerization. The polymerization is stabilized through D_2O and is disturbed by temperature shock and colchicine and becomes tubulin molecules.

Tudor domain: A conserved protein motif that is able to recognize symmetrically dimethylated arginines. Tudor domain-containing proteins (TDRDs) have been shown to participate in the Piwi pathway to suppress retrotransposons via DNA methylation.

tumour: (i) An abnormal benign or malignant mass of tissue that is not inflammatory, arises without obvious cause from cells of pre-existent tissue and possesses no physiological function. (ii) An abnormal swelling or localized outgrowth, often roughly spherical, produced by a plant as a result of attack by a fungus, bacterium, nematode, insect or other organism. *See* GALL *and* KNOT.

tumour progression: The process by which an initial mildly disordered cell behaviour gradually evolves into a full-blown cancer.

tumour-inducing (Ti) plasmid: A circular plasmid that often, but not always, is a part of the genetic equipment that *Agrobacterium tumefaciens* uses to transduce its genetic material to plants. The plasmid has 196 genes that code for 195 proteins. There is one structural RNA. The plasmid is 206,479 nucleotide long. The Ti plasmid is lost when *Agrobacterium* is grown above 28°C. Such cured bacteria do not induce crown galls (i.e. they become avirulent). Ti plasmid consists of a T-DNA region, which contains genes for the synthesis of opines needed by the bacteria for its metabolism. The Ti plasmids are classified into different types based on the type of opine produced by their genes: OCTOPINE PLASMID, NOPALINE PLASMID, SUCCINAMOPINE PLASMID and LEUCINOPINE PLASMID. Opine genes are responsible for causing galls on the plants these infect. pTi also contains the virulence region, which activates the T-DNA region to enable transfer to the plant tissues. The Ti plasmid is disarmed by removing the genes responsible for opine synthesis, and the gene of interest can be put in the T-DNA region. *Aka* TI PLASMID.

tumour-suppressive mechanisms: Evolution has installed in the proliferative programmes of mammalian cells a variety of innate tumour-suppressive mechanisms that trigger apoptosis or senescence, should proliferation become aberrant. These contingent processes rely on a series of sensors and transducers that act in a coordinated network to target the machinery responsible for apoptosis and cell cycle arrest of different points.

tumour-suppressor gene (TSG): A TSG, or anti-oncogene, is a gene that regulates a cell during cell division and replication. If TSG mutates, it results in a loss or reduction in its function and the cell grows uncontrollably, it will result in cancer. Examples are p16 and p53, dubbed the GUARDIAN OF THE GENOME.

tumour-suppressor protein: The principal tumour-suppressor protein is nuclear transcription factor p53, which accumulates in cells in response to DNA damage, oncogene activation and other stresses. It acts as a nuclear transcription factor that transactivates genes involved in apoptosis, cell cycle regulation and numerous other processes.

tumourigenesis: A process responsible for induction of tumours.

tunnelling: A truly quantum mechanical effect due to the wave nature of matter by which a particle can cross over a barrier if its wavefunction has a finite probability of existence.

turgidity: A state of being rigid or swollen as a consequence of internal water pressure.

Turing machines: First described by Alan Turing in 1936, Turing machines are simple abstract computational devices intended to help investigate the extent and limitations of what can be computed. Turing's 'automatic machines', as he termed them in 1936, were specifically devised for the computing of real numbers.

Turing test: A test of a machine's ability to exhibit intelligent behaviour equivalent to, or indistinguishable from, that of a human, developed by Alan Turing in 1950. Turing proposed that a human evaluator would judge natural language conversations between a human and a machine designed to generate human-like responses. The evaluator would be aware that one of the two partners in conversation is a machine and all participants would be separated from one another. The conversation would be limited to a text-only channel such as a computer keyboard and screen so the result would not depend on the machine's ability to render words as speech. If the evaluator cannot reliably tell the machine from the human, the machine is said to have passed the test. The test results do not depend on the machine's ability to give correct answers to questions, only how closely its answers resemble those a human would give.

Turner's syndrome: A series of abnormalities in humans due to monosomy for the X chromosome. Individuals are phenotypically female but are sterile. It has 45 chromosomes; chromatin negative.

turning CO_2 into a cheap, clean fuel: Carbon efficiency is one of the most pressing problems of carbon dioxide electroreduction today. Z. Zhang *et al.* (2023) published their efficient process for converting CO_2 into fuel. A 2018 study called CO_2 a 'notoriously inert molecule'. It was declared the invisible gas was 'far more pernicious' to work with than researchers had thought. The breakthrough follows an almost century-long effort to turn CO_2 into a cheap, clean fuel. Researchers at the Massachusetts Institute of Technology exposed CO_2 to catalysts and then electrolysis that turns the gas into a powder called sodium formate, which can be safely stored for decades. The resulting powder closely resembles a commercial product that has been safely used for years to melt ice on highways and airports. It has been stored for 2000 h in tanks without a hint of corrosion. This powder could produce electricity for homes, he said and 'nothing goes into the atmosphere'.

turnover number: Number of substrate molecules, an enzyme molecule alters into a product per second. The metabolic reactions catalysed by enzymes display much higher (up to 10 billion times) reaction rates. The turnover number is known by the number of substrate molecules that act on an enzyme molecule per unit time. For example, DNA polymerase 1 that plays a role in DNA synthesis has a turnover number of 15.

turnover rate: The rate at which a population of molecules is degraded in the cell.

tussock: A dense bunch of vegetation, usually well separated from neighbouring tussocks, as observed, for example, in some grasses. *Compare with* SWARD.

twilight length, plant growth and flowering time: D. Mehta *et al.* (2024) demonstrated that the length of twilight impacts growth and flowering time via the *Late Elongated Hypocotyl* (*LHY*)/*Circadian Clock Associated 1* (*CCA1*) clock genes in *Arabidopsis*. Plants have been found to be more sensitive to twilight length compared to equivalent deviations exclusively in photoperiods. Transcriptome and proteome analyses have displayed that twilight length influences reactive oxygen species metabolism, photosynthesis and carbon metabolism. Genetic analyses have revealed a twilight sensing pathway from the photoreceptors *PHY E*, *PHY B*, *PHY D* and *CRY2* via *LHY/CCA1* to flowering modulation via the *GIGANTEA-FLOWERING TIME* (*GI-FT*) pathway.

twin species: Two species that are phenotypically very similar but separated from each other by reproductive isolation. *See* SIBLING SPECIES.

twin spot: A pair of mutant sectors within wild-type tissue, produced by a mitotic crossover in an individual of appropriate heterozygous genotype.

twisting number: In double-helical DNA, this is the number of base pairs divided by the number of base pairs per turn of the double helix.

two genes-one polypeptide hypothesis: The lambda chain appears to be encoded by two separate germline genes – a specificity region gene and a common region gene – which are expressed as a single, continuous polypeptide chain. Antibody light chains seem to be an exception to the rule of one gene-one polypeptide chain hypothesis. There is, however, still a debate about whether two genes-one polypeptide chain is fact or fiction.

two-celled stage: The embryo after the first mitotic division produces a smaller apical cell and larger basal cell. The apical cell gives rise to all of the embryo except the root. The basal cell gives rise to the radicle and the suspensor.

two-channel microarray: Two-channel microarrays are characteristically hybridized with cDNA prepared from two samples to be compared (e.g. diseased tissue versus healthy tissue) and that are labelled with two different fluorophores. Fluorescent dyes commonly used for cDNA labelling include cyanine dye Cy3, which has a fluorescence emission wavelength of 570 nm (corresponding to the green part of the light spectrum) and cyanine dye Cy5 with a fluorescence emission wavelength of 670 nm (corresponding to the red part of the light spectrum). The two Cy-labelled cDNA samples are mixed and hybridized into a single microarray that is then scanned in a microarray scanner to envision the fluorescence of the two fluorophores after excitation with a laser beam of a defined wavelength. Relative intensities of each fluorophore may then be used in ratio-based analysis to identify upregulated and downregulated genes. *Aka* TWO-COLOUR MICROARRAYS.

two-channel versus one-channel microarrays: Relative intensities of each fluorophore (Cy3 and Cy5) are utilized in ratio-based analysis to identify upregulated and downregulated genes. In SINGLE-CHANNEL MICROARRAYS or ONE-COLOUR MICROARRAYS, the arrays provide intensity data for each probe or probe set indicating a relative level of hybridization with the labelled target. However, they do not indicate abundance levels of a gene but rather relative abundance when compared to other samples or conditions when processed in the same experiment. Each RNA molecule faces protocol and batch-specific bias during the amplification, labelling and hybridization phases of the experiment making comparisons between genes for the same microarray uncommunicative. The comparison of two conditions for the same gene needs two separate single-dye hybridizations. One strength of the single-dye system lies in the fact that an aberrant sample cannot influence the raw data derived from other samples, because each array chip is exposed to only one sample (as opposed to a TWO-COLOUR SYSTEM in which a single low-quality sample may drastically impinge on overall data precision even if the other sample was of high quality). Another benefit is that data are more easily comparable to arrays from different experiments so far as batch effects have been accounted for. *See* TWO-CHANNEL MICROARRAY.

two-dimensional (2D) diffusion: In this case, diffusion takes place from inside to inside or outside to outside the plasma membrane. Diffusion cannot take place from outside to inside or inside to outside.

two-dimensional (2D) gel electrophoresis: 2D gel electrophoresis is a powerful and widely used method for the analysis of complex protein mixtures extracted from cells, tissues, or other biological samples. This technique separates proteins in two steps based on two independent properties: the first step is ISOELECTRIC FOCUSING (IEF), which separates proteins according to their isoelectric points; the second step is SDS-polyacrylamide gel electrophoresis (SDS-PAGE), which separates proteins according to their molecular weights. In this way, complex mixtures consisting of thousands of different proteins can be resolved and the relative amount of each protein can be determined. A fixed-time gel run (i.e. with the appropriate gel and the appropriate electrical field applied to the gel) leaves a scientist with approximately 1000 'spots' (of protein molecules) on the gel. Each 'spot' is a collection of the molecules of one protein (or of several proteins with similar molecular weights) from the original sample (mixture). To identify the protein(s) in the 'spots', the scientist stains them and then assesses the entire gel with an electronic image scanner (or he assesses it visually). From the pattern (coupled with intensity) of the 'spots', two such gels could be utilized to confirm if two organisms were the same species/strain/variety, or to determine the differences (in gene expression) between samples of diseased versus healthy tissues.

two-dimensional (2D) PAGE: Two-dimensional polyacrylamide gel electrophoresis. The most common technique for protein separation. Proteins are separated in one dimension according to their size and in the second dimension, according to their charge (that is, their isoelectric point). After separation, the gel is stained so that protein spots can be seen.

two-dimensional (2D) PAGE correlative analysis: Two-dimensional correlation analysis is a method for visualizing the relationships among the variables in multivariate data and their temporal behaviour by applying the complex cross-correlation function. This function measures correlations that occur at the same rate or frequency concerning the data acquisition time. The complex cross-correlation function yields real and imaginary components that contain information about the phase behaviour of the variables. The real component provides information about mutually dependent in-phase variations.

two-dimensional (2D) paper fraction technique: Method developed by F. Sanger and colleagues in 1965 to determine the sequence of nucleic acids. It used a combination of radioactive labelling and electrophoresis and laid the groundwork for sequencing RNA.

two-hybrid screening for isolating cDNA: This screening method takes advantage of a specific protein-protein interaction that occurs between two proteins each of which is itself a hybrid protein. The entire assay relies on the ability of one part of each of the hybrid proteins to form a specific interaction that is reasonably stable under physiological conditions with the other. Same as the yeast-two-hybrid system.

two-hybrid system: Procedure for recognizing interacting proteins consuming genetically engineered yeast cells.

two-locus gametophytic self-incompatibility: If the alleles expressed at these two loci in the pollen grain both match the corresponding alleles in the pistil, the pollen grain will be recognized as incompatible.

two-out-of-three reading frame: The hypothesis according to which codons may be read by trusting mainly on the Watson-Crick pairs formed with the first two codon positions. Mispaired nucleotides in the third position of codon and anticodon, called wobble positions, make only a marginal contribution to the total stability of the reading interactions.

two-point cross: A cross involving two loci.

two-point test cross: A test cross involving two loci.

two-ranked: Refers to leaves arranged in two rows in the same plane, on opposite sides of the branch. *See* DISTICHOUS.

two-RNA-polymerase machine: The plant-specific RNA Pol IV forms a complex with RDR2 (an RNA-dependent RNA polymerase) to produce double-stranded precursors of siRNA essential for genomic DNA methylation.

two-sided alternative: An alternative hypothesis stipulates that the parameter value can be higher or lower than the null value. Both directions are admissible. This type of alternative is set up when it is not known that the value will be higher or lower than the specified value.

two-sided test: A test that is non-directional and that leads to a two-sided P-value. If the null hypothesis H_0 is that two treatments have the same mortality outcome, a two-sided alternative is that the mortality difference is nonzero. Two-sided P-values are larger than one-sided P-values (they are double if the distribution of the test statistic is symmetric). They can be thought of as a multiplicity adjustment that would allow a claim to be made that treatment lowers or raises mortality. *Compare with* ONE-SIDED TEST.

two-state model for splicing: A model for the catalytically activated spliceosome in which specific conformations exist in equilibrium that promote either 5′ splice site (ss) cleavage or exon ligation. This equilibrium can be modulated by mutations in core spliceosome components, the pre-mRNA, or other factors.

two-strand double crossover: A double crossover that involves only two of the four chromatids of a tetrad comprising one bivalent.

two-tailed test: A two-tailed test is a statistical test in which the critical region of a distribution is two-sided and tests whether a sample is greater than or less than a certain range of values. If the sample being tested falls into either of the critical areas, the alternative hypothesis is accepted instead of the null hypothesis. The two-tailed test gets its name from testing the area under both tails of a normal distribution.

two-way analysis of variance: A method of studying the impacts of two factors separately (their main effects) and (sometimes) together (their interaction effect).

two-way classification: The classification of a set of observations based on two criteria. The 'two-way' comes because each item is classified in two ways, for example, classification by gender and political party. *Compare with* ONE-WAY CLASSIFICATION.

2n: The symbol for zygotic (diploid) chromosome numbers. *Compare with* 'n'.

Ty element: A yeast transposon that has interspersed repeat DNA. They are classic retrotransposons, with a reverse transcriptase activity. They transpose via an RNA intermediate. It is 5.9 kb including two terminal direct repeats of 334 bp, the long terminal repeats (LTR) or deltas (δ). Each delta contains a promoter and transposase recognition sequences. Two ORFs in the mRNA, designated TyA and TyB, encoding two different proteins.

type: An item, usually a herbarium specimen, to which the name of a taxon is permanently attached (i.e. a designated representative of a plant name). This is important in determining the priority of names existing for a particular taxon.

type genus: In nomenclature, a single genus on which a taxonomic family is based.

type I error: In statistics, refers to the rejection of a true hypothesis.

type II error: In statistics, refers to the acceptance of a false hypothesis.

type number: The most frequently occurring chromosome number in a particular taxonomic group (= modal number).

type specimen: The actual physical, a stuffed or dried specimen that a scientist who describes and names a previously unknown species places in a museum or other recognized repository in order to have the right to name that newly discovered species. This 'officially deposited specimen' is essential for three purposes: (a) So that comparisons can later be made if there is ever a doubt whether another 'new' species is simply a member of this same species (and thus already named). (b) So that taxonomists who determine and keep the official scientific names by which scientists must refer to each of the world's organisms can name each of the newly discovered species in accordance with the complex rules of the International Codes for Nomenclature. Examples of such names include *Arabidopsis thaliana*, *Escherichia coli* and *Agrobacterium tumefaciens*. (c) So that patent claims for genetically engineered organisms can later be imposed.

type theory: In mathematics, logic and computer science, a type of theory is any of a class of formal systems, some of which can serve as alternatives to set theory as a foundation for all mathematics. In type theory, every 'term' has a 'type' and operations are restricted to terms of a certain type.

typological thinking: The thinking that organisms of a species conform to a specific norm. According to this view, variation is considered abnormal.

typology: (i) The definition of classificatory groups by phenetic similarity to a 'type' specimen. A species, for example, might be defined as all individuals less than x phenetic units from the species' type. (ii) The theory that distinct 'types' exist in nature, perhaps because they are part of some plan of nature. The type of the species is then the most important form of it and variants around that type are noise, or 'mistakes'. Neo-Darwinism opposes typology, because in a gene pool, no one variant is any more important than any others.

tyrosinase: An enzyme that converts tyrosine to dopa and oxidizes this to dopa quinone.

tyrosine (tyr): A phenolic α-amino acid with chemical formula $C_9H_{11}NO_3$. It is a precursor of the hormones such as epinephrine, norepinephrine, thyroxine and triiodothyronine. It is also a precursor of the molecule, known as MELANIN, which is the pigment of a suntan.

tyrosine kinase inhibitors (TKI): Refers to various chemical compounds that obstruct the activity of tyrosine kinase enzyme (inside the body). One example of TKI is GENISTEIN. Because the activity of tyrosine kinase triggers cancerous (tumour) cells to metastasize (spread/grow), consumption by humans of relevant TKI acts to help prevent (spreading of) certain cancers.

tyrosine kinase receptors: Tyrosine kinase receptors are transmembrane proteins that transduce signals controlling cell growth, survival, motility and differentiation. The binding of a ligand to a receptor, which then forms a dimmer, activates these signals. This dimmer adds phosphate groups to itself, resulting in the creation of docking sites that bind to downstream signal transduction molecules containing SRC HOMOLOGY 2 (SH2) domains. Upon ligand binding, several signalling pathways can be activated, each one consisting of a chain of signalling molecules that indirectly alter the expression of target genes in the nucleus. Erb-4, a member of EPITHELIAL GROWTH FACTOR (EGF) tyrosine kinase receptor family, also activates gene expression more directly. The Erb-4 undergoes proteolysis within its plasma membrane domain – a process called RIP for intramembrane proteolysis – and its intracellular portion moves to the nucleus, where it may affect the transcription of target genes.

tyrosine phosphorylation: Tyrosine is one of the building blocks of proteins that can be modified by the addition of a chemical tag, in this case, a phosphate group.

tyrosinosis: A human disease caused by a metabolic defect and characterized by the excretion of p-hydroxyphenyl pyruvic acid and tyrosine into urine.

U

U orientation: Expressive of the condition in which the vector and the inserted fragment of **DNA** have the opposite orientation.

U protein: Theoretical protein believed to control the transition of cells from the G_0 to G_1 phase of the cell cycle and thus surely into the S phase. The concentration of this unstable protein is expected to exceed a threshold level for triggering the progression through the cell cycle. This is supposed to occur if the cell ensues to have adequate access to growth factors or to nutrients. *Aka* TRIGGER PROTEIN.

US Patent and Trademark Office (USPTO): The USPTO is a Washington, DC-based American Government agency responsible for common patent protection matters for all of America's 50 states and its territorial possessions. The USPTO permits the patenting of new and unique microbes, plants and animals, as well as novel and exclusive methods to produce biotechnology advances.

U1 small nuclear RNA (U1snRNA): One step in the nuclear uptake of U1 snRNPs involves the m^3G cap structure. U1 small nuclear RNA (U1 RNA) is essential for the processing of **pre-mRNA** The U1 snRNP particle, which contained U1 RNA, plus a set of seven to nine proteins, appears to function in the recognition of 5′-splice sites of pre-mRNA.

U2 small nuclear RNA (U2snRNA): The conserved branch site UACAAAC is known to form base pairs with the complementary sequence GUAGUA in U2 snRNA during mRNA splicing in yeast. The GUAGUA element is conserved in mammalian U2 snRNA.

U3 small nuclear RNA (U3snRNA): U3 snRNA is found in the nucleolus, the site of rRNA synthesis. All are very abundant and there may be as many as 1 million copies per nucleus. U3 snRNP base sequences are highly conserved between organisms, and all seem to contain unusual trimethylguanosine structures at their 5′ ends. The role of these RNAs is not known, but one suggestion is that they are involved in ribosome biogenesis.

U4 small nuclear RNA (U4snRNA): U4 snRNA is a constituent of a base-paired di-snRNP complex that is crucial for pre-mRNA splicing of the major class of metazoan nuclear introns.

U5 small nuclear RNA (U5snRNA): U5 snRNA constitutes a coaxially stacked series of helices that venture into the active site of the spliceosome. Loop 1, which caps this series of helices, forms 4–5 base pairs with the 5′-exon during the two chemical reactions of splicing.

U6 small nuclear RNA (U6snRNA): U6 is one of the five snRNAs that are required for splicing of nuclear precursor mRNA. U6 snRNAs do not possess unusual trimethylated ($3m^{2,2,7}G$) cap structure at 5′ ends. Two domains in U6 RNA have been identified, each of which is obligatory for *in vitro* splicing. Single-nucleotide replacements in these two domains block splicing either at the first or the second step.

U7 small nuclear RNA (U7snRNA): Human U7 snRNA plays a role in the 3′ end maturation of histone pre-mRNAs.

U8 small nuclear RNA (U8snRNA): U8 snRNA functions in the modification of other small nuclear RNAs.

U9 small nuclear RNA (U9snRNA): Involved in ribosome biogenesis.

U10 small nuclear RNA (U10snRNA): Involved in ribosome biogenesis.

U11 small nuclear RNA (U11snRNA): Perhaps involved in the maturation and processing of cellular RNAs.

U12 small nuclear RNA (U12snRNA) U12 snRNA is analogous to U2 snRNA of the U2-dependent spliceosome and is indispensable for the splicing of U12-dependent introns in metazoan cells. The essential region of U12 snRNA, which base pairs to the branch site of minor class introns, is well studied.

UACUAAC box: In yeast introns, a conserved sequence that specifies the branch point for the lariat and to which U2 snRNA binds by complementary base pairing during RNA splicing.

UAG stop codon: Amber codon. One of the three translation termination codons.

ubiquitin ligase: Any one of a large number of enzymes that adhere ubiquitin to a protein, thus marking it for degradation in a proteasome. The process catalysed by a ubiquitin ligase is termed UBIQUITYLATION.

ubiquitin: A small, highly conserved regulatory protein ubiquitous in all eukaryotic cells, including the proteasomes, which plays a significant role in 'tagging' with lysine of other proteins fated for degradation (via proteolytic cleavage). Such proteins are then fragmented and removed or recycled because they are damaged or no longer required by the body. Such 'tagged' protein molecules are said to have been ubiquitinated.

ubiquitination: A **post-translational modification (PTLM)** process in which carboxylic acid of the terminal glycine from the di-glycine motif in the activated ubiquitin forms an amide bond to the epsilon (ε) amine of the lysine in the modified protein.

ubiquitylation: The process by which ubiquitins control the activity of larger proteins by reversibly marking them – either for destruction by the proteasome or for transport to different cellular compartments. This process involves a cascade of three types of enzymes, known as E1, E2 and E3. The E1 enzyme activates ubiquitin and transfers it to an E2 enzyme, which interacts with a specific E3 partner to mediate the ligation of the ubiquitin moiety to the target protein. There are hundreds of E3 ubiquitin ligases, each of which (with help from various E2 conjugating enzymes) selects the specific protein that receives the ubiquitin tag, thus determining the destination (e.g. the proteasome or the nucleus) of the protein. A major class of E3 ubiquitin ligases is the RING (named for 'really interesting new gene') group of proteins, which function as scaffolds that juxtapose the target protein and the catalytic ubiquitin E2 enzyme.

Ug99 fungus: A specific race of wheat stem rust *Puccinia graminis* f. sp. *tritici* that was first identified in Uganda (Ug) in 1999.

UGDase flipping mechanism: Flipping is important for the productive binding of uracil in the active site of UDGases. The enzyme probably does not recognize spontaneously flipped uracils. It catalyses a local falsification of the DNA backbone which causes the base to flip out into the enzyme.

ultimate causation: The evolutionary determination of the state of a trait at the level of population or lineage through adaptive evolution or drift.

ultracentrifugation: Process in which a very high-speed centrifuge is used to precipitate large biological molecules from solution or separate/purify them by their diverse rates of sedimentation.

ultracentrifuge: A high-speed centrifuge that can achieve revolving speeds up to 85,000 rpm and centrifugal fields up to 500,000 times gravity. A centrifuge is a device that rotates ultracentrifugation tubes around very rapidly, like a merry-go-round, to force the heavier suspended materials (in the solutions in the tubes) to the bottoms of those test tubes before the lighter material.

ultrafiltration: A mixture separation procedure that utilizes the capability of synthetic semipermeable membranes (possessing suitable physical and chemical properties) to discriminate between molecules in the mixture, principally on the basis of their molecular size and shape. Invented and developed by R.J. Taylor in the 1950s and 1960s, ultrafiltration is characteristically exploited for the separation/purification of relatively high molecular-weight solutes (e.g. proteins, gums, polymers and other complex organic molecules) and colloidally discrete substances (e.g. minerals and microorganisms) from their solvents (e.g. water).

ultrahigh vacuum (UHV), variable temperature scanning probe microscope (UHV VT SPM): A state-of-the-art investigational scheme that can be utilized for surface-science research on the nanometre gauge. It comprises an Atomic Force and Scanning Tunnelling Microscopes talented for Atomic Resolution (AFM, STM), Low Energy Electron Diffraction (LEED) for surface depiction and ion beam setup for surface popping. This system allows the user to discover the properties of various phenomena at clean environment at variable temperature.

ultrahigh vacuum (UHV): A necessity of most of the high-quality scientific instrumentation, which considers vacuum better than 10^{-8} torr. Torr is a non-SI unit of pressure defined as 1/760 of an atmosphere. It was so named after Evangelista Torricelli, an Italian physicist and mathematician who discovered the principle of the barometer in 1644.

ultralow background DNA cloning system: This cloning system is founded on yeast-based *in vivo* cloning by virtually entirely abolishing the production of by-products and using the technique to clone commonly used *Escherichia coli* vectors, predominantly those deficient in yeast replication origins and possessing an ampicillin resistance gene (*Amp*^{r}).

ultralow input and single-cell RNA-Seq: This method uses deep RNA-Seq to study the signals and behaviour of a cell in the context of its surrounding environment. This method is principally valuable for biologists studying processes such as differentiation, proliferation and tumorigenesis.

ultrametric tree: A phylogenetic tree in which all the branch lengths are forced to be equidistant from the root. *Aka* DENDROGRAM.

ultrasonic atomic force microscopy (UAFM): In the presence of ultrasonic vibration, hard surfaces can become concave and dented with the tip of a soft cantilever, due to its inertia. Ultrasound diminishes or even abolishes friction and hence modifies the tip-nanoparticle surface interactions in AFM manipulation. The subsurface sensitivity of the technique makes achievable the purposed management of subsurface nanoscale features by ultrasonic actuation.

ultrasonic force microscopy (UFM): UFM is based on atomic force microscopy. A tip on the end of a cantilever is in contact with a sample surface at a very high frequency in the megahertz range (10^6–10^8 Hz). The force between the tip and the sample is reliant in a nonlinear way on the position of the tip, and this is done to induce a low frequency kilohertz tip vibration by moderating the amplitude of the megahertz vibration at this low frequency. Then by scanning the tip laterally across the sample and monitoring this kilohertz vibration, one can make an image of the elastic properties of the sample on nanometre length scales.

ultrastructure: The detailed structure of a sample (e.g. a cell, tissue or organ) that can be detected only through electron microscopy. *Aka* FINE STRUCTURE.

ultraviolet (UV) fluorescent microscopy: Ultraviolet microscope in which the UV light is focused on biological material that shines in a specific colour when so irradiated. This may happen due to staining in or association with fluorescently tagged antibody. *See* FLUORESCENCE MICROSCOPY.

ultraviolet (UV) light rays: The fraction of the electromagnetic spectrum with wavelengths ranging between 100 nm and 400 nm, between ionizing radiation (X-rays) and visible light. UV is absorbed by DNA and is highly mutagenic to unicellular organisms and to the epidermal cells of multicellular organisms. UV light is used in tissue culture for its mutagenic and bactericidal properties. UV rays are produced from mercury vapour lamps or tubes. *Aka* ULTRAVIOLET RADIATION.

ultraviolet (UV) spectroscopy: The determination of the amount of UV radiation absorbed by a sample for recognizing the chemical compounds and to measure the concentrations of nucleic acid solutions.

ultraviolet (UV) visualization: A procedure employed to perceive the position of proteins, nucleic acids and other compounds that absorb UV light or fluorescence in the presence of UV light, on paper, gel or thin layer chromatograms or electrophoretic gels. In some cases, the matrix is sprayed with a suitable reagent prior to inspection to empower visualization of the compounds.

unassigned reading frames (URFs): URFs are those sequences that have correct initiation codons and are uninterrupted by stop codons. The occurrence of long URFs has been reported. Probably these genes are in search of a function. Some URFs that occur within an intron may be involved in specifying proteins important in the splicing of introns. *Aka* UNIDENTIFIED READING FRAMES.

unattached kinetochores: Kinetochores that are not attached to microtubules prevent chromosome missegregation via the spindle assembly checkpoint.

unbalanced diallelic: A genotype involving a multiple allelic locus in autotetraploid where two alleles are found to occur an unequal number of times.

unbalanced experimental design: An experiment or set of data in which all treatments or treatment combinations are not equally epitomized. A common source of unbalanced experiments is inadequate mortality among entries in a test.

unbalanced translocation: A form of chromosomal translocation in which the absence of a chromosomal segment causes a deleterious genetic effect.

uncertainty: A situation in which an individual decision maker is not conscious of all possible outcomes of his decision and/or the likelihoods with which those consequences may ensue given their decision.

uncoating: The decomposition of a viral capsid. Provides informal and simplified description of the way a virus infectious material enters the cell, usually appearing in light science material for the general public.

uncoiling of chromomeres: Condition most readily observed in the puffing of polytene chromosomes and the looping out of chromatin coils in lampbrush chromosomes.

uncompetitive inhibition: A type of enzyme inhibition in which an inhibitor binds to the enzyme-substrate complex and thereby declines its apparent K_m and its apparent V_{max} values by the same factor. In uncompetitive inhibition, the inhibitor binds directly to the enzyme-substrate complex but not to the free enzyme. The inhibitor, which need not resemble the substrate, seemingly distorts the active site, thereby rendering the enzyme catalytically inactive. Actually, uncompetitive inhibition is substantial only for multi-substrate enzymes.

uncondensed chromatin: Chromatin that is not extremely compressed and is therefore more probable to be transcriptionally active. A robust relationship has been observed between gene activity and chromatin decondensation. Both the puffs and interbands of polytene chromosomes are uncondensed and thus are the sites of strong and modest transcriptional activity, respectively.

unconscious selection: Indirect selection by breeders, sometimes called PARALLEL SELECTION, when, in addition to a target characteristic, another trait is also betrothed to the next generation. It can be a morphological, biochemical or physiological characteristic genetically associated with the target characteristic.

uncultured microbes: Microbes that have never been grown successfully in isolation in the laboratory.

undefined medium: A growth medium in which all the components cannot be defined. In plant tissue culture, T. Murashige and F. Skoog medium is a defined medium as we know the concentration of each and every component. Sometimes, we add other growth adjuvants such as coconut milk and potato extract into this medium then it becomes an undefined medium because we do not exactly know what coconut milk or potato extract contain.

under-replication: Certain heterochromatic chromosomal areas and ribosomal DNAs that reveal a slower rate of replication as compared to the rest of the genomic DNA.

underdominance: A state in which the phenotypic expression of the heterozygote (e.g. *Aa*) is less than either of the homozygote (*AA* or *aa*).

understanding gene structure: The structure of gene can be understood in terms of its two functions – recombination and mutation. The concept of gene structure has been changing with the developments in the subject. This changing scenario can be discussed under three phases – classical, transitional and modern.

understorey: Plant life growing beneath the forest canopy.

underwinding: Underwinding of DNA is produced by negative supercoiling because the double helix is itself coiled in the opposite sense from the intertwining of the strands.

undifferentiated: This is the condition wherein the cell in culture lacks the specialized structures and/or function of the cell type *in vivo*. In the case of plants, the organs such as leaves, embryos and cotyledons cultured on a medium supplemented with higher levels of auxins, such as 2, 4-D (2–4mg/l), undergo the process of dedifferentiation to form a homogeneous mass of undifferentiated cells called callus comprised parenchymatous cells. In plant cells, prevailing in a state of cell development characterized by an isodiametric cell shape, with very little or no vacuole and a large nucleus, and exemplified by cells comprising an apical meristem or embryo. In animal cells, this is the situation wherein the cell in culture lacks the specialized structure and/or function of the cell type *in vivo*.

undifferentiated plastids: Undifferentiated plastids (ETIOPLASTS, PROPLASTIDS) may develop into any of the several plastids – CHLOROPLASTS, CHROMOPLASTS, GERONTOPLASTS and LEUCOPLASTS.

undulipodium: An intracellular projection of a eukaryotic cell containing a microtubule array, used for locomotion or for propelling fluids.

unequal bivalent: The one which is composed of a pair of chromosomes of different lengths, one of which consequently contains a segment that is lacking in the other.

unequal crossing over (UEC): A crossing over after inappropriate pairing between chromosome homologues that are not flawlessly aligned due to the existence of repetitive DNA. UEC is thus a type of gene duplication or deletion event that deletes a sequence in one strand and replaces it with a duplication from its sister chromatid in mitosis or from its homologous chromosome during meiosis. The consequence is, for example, one crossing over chromatid with one copy of the segment and another with three copies. Since homology is required for this process, the underlying mechanism is normal homologous recombination.

unfolded protein response: Cellular retort prompted by a build-up of misfolded proteins in the endoplasmic reticulum. It may happen due to amplified transcription of endoplasmic reticulum chaperones and degradative enzymes. Protein misfolding is involved in ageing. By utilizing the unfolded protein response, several hundred yeast genes have been identified with a role in endoplasmic reticulum folding and methodically categorized their functional independencies by determining unfolded protein response levels in double mutants.

unicellular: An organism made of only one cell, for example, yeast. *Contrasts* MULTICELLULAR.

unicellularity versus multicellularity: Prokaryotes are unicellular organisms where the genome is not separated from the cytoplasm. Eukaryotes may be unicellular or multicellular. Genomes in the eukaryotic cells are separated from the cytoplasm by nuclear membrane.

monocistronic mRNA: When one messenger RNA holds genetic information only from one cistron (gene) and is used as a template to produce only one polypeptide chain under defined conditions. Eukaryotic mRNAs are monocistronic.

unidentified reading frame (URF): An OPEN READING FRAME (ORF) recognized from a DNA sequence for which no genetic function is known.

unidirectional import: Nuclear proteins, like other proteins, are synthesized in the cytoplasm and are later imported to the nucleus. This transportation is unidirectional (i.e. they cannot exit the nuclear pore; e.g. SV40 T-antigens and *Xenopus* nucleoplasmin). Unidirectional export through the nuclear pore of TRANSFER RNA (tRNA), SMALL NUCLEAR RNA (snRNA) and PROCESSED MESSENGER RNA (mRNA) is known. Importins play a role in the import process.

unidirectional replication: Movement of a single replication fork from a given origin. An example of unidirectional replication is the replication of mitochondrial DNA by D-loops in vertebrates.

unifactorial: (i) In genetics, monogenic (i.e. having or being characters or a mode of inheritance dependent on a single genetic locus). (ii) In general, having, involving, concerned with or produced by a single element or cause.

uniflagellate. Having one flagellum.

uniflor: With a single flower. *Compare with* PAUCIFLOR (few) *and* PLURIFLOR (many).

uniform crossover: In genetic algorithms, a breeding procedure in which it is arbitrarily obvious for each of a breeding pair of individuals whether they should be exchanged.

uniform resistance: The type of host resistance that is comparable for all races of the pathogen, rather than being good only for some races.

Uniform Resource Locator (URL): Colloquially web address. A reference to a web resource that specifies its location on a computer network and a mechanism for retrieving it. A URL is an explicit type of Uniform Resource Identifier (URI), although many people use these two terms interchangeably. URLs occur most commonly to reference web pages (http) but are also used for file transfer (ftp), e-mail (mailto), database access (JDBC) and many other applications.

uniformitarianism: The postulation that the same natural processes functioned in the past that are known to be operating now.

uniformity: Refers to the condition of a population or group in which all the individuals are genetically identical. It is a typical property of clonal varieties (e.g. potato). In general, lack of diversity within and between plant species is evident in current cultivars.

uniformity trial: An experiment in which a single variety is raised under unvarying cultural conditions in a large experimental area to investigate the shape of variation for soil fertility.

UniGene database: A public database, preserved by the National Centre for Biotechnology Information (NCBI), which brings together sets of GenBank sequences that signify the transcription products of distinct genes.

unigene-derived microsatellite (UGMS) markers: A specific type of molecular marker having the advantage of investigative variation in the expressed component of the genome with unique identity and positions.

unigenes: A nonredundant set of genes that is delineated after clustering (computational) analysis of sequences produced through an EST or a genome sequencing project.

unilateral inheritance: Inheritance that is related to linkage in sex chromosomes.

unilinear inheritance: The transfer of a donor chromosome to a suitable recipient (e.g. bacteria).

unineme structure of chromosomes: A single unceasing length of DNA double helix makes one chromosome. Genes are arranged in prokaryotic and eukaryotic chromosomes in linear order. T_4 phage and *E. coli* established this view, respectively, for viruses and prokaryotes. The unineme structure of rat lampbrush chromosomes were anticipated by O.L. Miller in 1965 which was later supported by electron microscope studies of O.L. Miller and B. Beatty in 1969. They demonstrated that after digestion the remaining thickness of lateral loops was about 20 Å. *Aka* UNINEME HYPOTHESIS.

uninucleate: Having one nucleus. *Contrasts* MULTINUCLEATE.

Union for Protection of New Varieties of Plants (UPOV): A group of the world's countries that have together agreed to mutually guard the intellectual property (of owners, breeders) that is inherent in new plant varieties developed by man. These intellectual property protectors are often collectively referred to as 'Breeder's Rights'. Established in 1961, the secretariat for UPOV is in Geneva, Switzerland.

uniparental disomy: A situation where both chromosomes in a cell are derived from a single parent and there is no contribution of chromosomal material from the other parent. Uniparental disomy may impact the whole chromosome (complete) or part of a chromosome. If the chromosomal region contains an imprinted gene, then uniparental disomy will be associated with alterations in imprinted gene expression.

uniparental inheritance: A form of inheritance in which only one parent delivers genes to the progeny. A phenomenon, generally displayed by extranuclear genes, in which all progeny resembles the phenotype of only one parent.

unipartite structure: Refers to single units.

unipolar: A cell spindle having only one pole. *Aka* SPINDLE.

uniporter: Carrier protein that transfers a single solute from one side of the membrane to the other.

unique clone: A sequence that has no match in GenBank or other public database.

unique DNA: Mainstream of the structural genes are unique DNA sequences. The part played by noncoding unique DNA sequences in evolution is still very incomprehensible. A major portion of these unique DNA sequences do not code for proteins and are not transcribed. For example, regulatory sequences such as promoters and other sequences of yet unknown function. Introns, leader sequences and intercistronic sequences, are transcribed in association with structural genes. Any mutation that interferes with the correct removal of intron sequences may have a substantial influence on the regulation of gene expression. *Aka* NONREPETITIVE DNA *or* SINGLE-COPY DNA.

unique event polymorphism (UEP): An allele for which all copies are derived from just one mutational event.

unique features of broken chromosome end: Barbra McClintock in 1951 noticed that if two chromosomes were broken in a cell, the end of one could attach to the other and vice versa. She never observed the attachment of a broken end to the end of an unbroken chromosome. Thus, the ends of broken chromosomes are sticky, whereas the normal end is not sticky, suggesting the ends of chromosomes have unique protective features.

unique versus repeated sequences: In bacteria, most of the genome comprises unique sequence DNA, representative of genes and regulatory elements. Some genes and other sequences are repetitive, but their copy number (or repetition frequency) is usually low (i.e. less than 10). Examples of such repetitious sequences consist of the ribosomal RNA genes (there are seven in *E. coli*) and transposable elements such as IS elements. Infrequently, certain sequence motifs may be moderately repetitive. For example, in the 1.8 Mbp *Haemophilus influenzae* genome, there are around 1500 copies of the 30-bp DNA uptake site. Similarly, the *E. coli* genome comprises many copies of two repetitive elements ENTEROBACTERIAL REPEATED INTERNAL CONSENSUS (ERIC) and REPEATED EXTRAGENIC PALINDROME (REP). Together, however, repetitive DNA comprises less than 1% of bacterial genomes and genome size is a direct indicator of genome complexity. On the other hand, eukaryotes carry a very large portion of repetitive sequences of DNA in their genomes.

unisexual: Individuals possessing only male or only female reproductive organs. When male and female flowers are on separate individuals, they are called dioecious (dioicous). *See* SEXUAL REPRODUCTION IN PLANTS. *Compare with* MONOECIOUS.

unisexual flowers: Only one kind of reproductive structure is present in a flower. The unisexual flower may be staminate (stamens present) or pistilate (gynoecium present).

unisite mutant allele: A mutant allele differing from its wild-type form at only one site.

unit: A measure of the biological activity of a substance, as estimated by various consistent assays.

unit character: A character or trait governed in diploids by one pair of alleles. *Contrast with* QUANTITATIVE *or* POLYGENIC CHARACTERS.

unit factor: The term used by Gregor J. Mendel for the gene. *Aka* ELEMENT, FACTOR *or* GENE.

unitary matrix: A recording system in which only identical characters obtain a positive score. *Aka* IDENTITY MATRIX.

Unitas approach: A tool for annotating small RNAs. This tool is better able to detect tRNA-derived fragments. It can help researchers identify cryptic RNAs and their functions.

univalent: A single chromosome is noticed during meiosis when bivalents also occur. It has no pairing mate.

univalent shift: A natural modification in monosomy from one chromosome to another. It transpires by partial asynapsis or desynapsis during meiosis. *Aka* UNIVALENT SWITCH.

universal codon: A triplet codon codes for the same amino acid in diverse forms of life (that is, bacteria, wheat, *Drosophila* and man).

universal codon, exceptions to: A triplet codon codes for the same amino acid in diverse forms of life (viz. bacteria, wheat, *Drosophila*, man). Mitochondrial genetic systems endure relatively recurrent deviations in codon assignment. This tolerance of mitochondrial genomes for changes in codon assignment occurs by RNA editing.

universal donor: A person with group O blood, whose erythrocytes, therefore, bear neither A nor B antigens and whose blood can be donated to any member of the blood types O, A, B and AB, if necessary.

universal genetic code: A triplet codon codes for the same amino acid in different forms of life (e.g. bacteria, wheat, Drosophila and humans). *See* CANONICAL CODE.

universal homology: A homologous characteristic occurring in all cellular organisms.

universal nucleoside: 1-(2'deoxy-β-D-ribofuranosyl)-3-nitropyrrole, designated 'M' is a universal nucleoside that maximizes stacking while minimizing hydrogen-binding interactions without sterically disrupting a DNA helix. The universal nucleoside is used at ambiguous sites in DNA primers.

universal recipient: A person with group AB blood, who can receive blood from any members of the blood types O, A, B and AB, if necessary.

unknome: Represents the poorly understood genes. Unknome demonstrates the importance of poorly understood genes, provides a resource to accelerate future research and highlights a need to support database curation to ensure that mis-annotation does not erode our awareness of our ignorance. The unknome can be observed at http://unknome.org, with the entire database available to download as SQLite Version 3 files. Data from the functional screens occur in the main text or the supplementary data sets S2 and S3 Data. Code for the functional assays exists at https://github.com/tjs23/unknome.

Unknome database: A tool for annotating small RNAs. This tool is better able to detect tRNA-derived fragments. It can help researchers identify cryptic RNAs and their functions. This publicly available and customizable 'Unknome database' ranks proteins based on how little is known about them.

unmatched *S* gene: An *S* gene is any gene governing the specificity of incompatible reactions in plants. Unmatched S gene in a given pollination is any *S* gene that is signified either in style or pollen but not in both.

unmethylated site: When neither of the complementary strands of DNA carries methyl group at a site. *Aka* HOMODUPLEX UNMODIFIED DNA.

unmixed families: In a genetic code dictionary box, a group of four codons share their first two bases.

unnatural amino acid: An amino acid that is not naturally encoded or found in the genetic code of an organism.

unordered tetrads: In some fungi, ascospores can change their position. An example of unordered tetrads is baker's yeast. *Compare with* ORDERED TETRADS.

unreduced gametes: Such gametes are not produced from common meiosis and thus carry the number of chromosomes per cell that is typical of a sporophyte. Such gametes impulsively appear as a result of irregular division in anaphase I of meiosis which may cause spontaneous (meiotic) polyploidization. In rye, such gametes have been used for the production of tetraploids via VALENCE CROSSES.

unrooted tree: A phylogenetic tree in which the root is not revealed, mostly because it is unknown.

unsaturated: Refers to a molecule that comprises one or more double or triple carbon-carbon bonds, such as those seen in isoprene or benzene.

unscheduled DNA synthesis: DNA synthesis that transpires at times other than S phase of the cell cycle. Most of such synthesis is DNA repair or recombination related. In some case (e.g. in mitochondria), synthesis of DNA in the cytoplasm also occurs nonstop.

unstable equilibrium: A frequency equilibrium between alleles in a population that is readily upset because of a selective disadvantage of the heterozygotes.

unstable gene: A gene that mutates frequently.

unstable mutation: A mutation that has a small frequency of reversion. For example, a mutation caused by the insertion of a controlling element, whose successive exit yields a reversion.

unsupervised learning: A form of self-organized Hebbian learning that assists find earlier unknown patterns in data sets without pre-existing labels. It is *aka* SELF-ORGANIZATION which enables modelling probability densities of given inputs. It is considered one of the main three categories of machine learning, along with SUPERVISED LEARNING and REINFORCEMENT LEARNING. SEMI-SUPERVISED LEARNING is also known which is a hybridization of supervised and unsupervised techniques.

untranscribed spacer: The DNA sequence that resides between two coding sequences and is not transcribed by RNA polymerase. Although such sequences evidently do exist, some spacers detected in the past have been demonstrated to be transcribed; such spacers are exemplified by the spacers occurring between the multiple copies of ribosomal RNA genes.

untranslatable exon: *See* EXON.

untranslated regions (UTRs): The regions on either side of the coding sequence of a strand of mRNA which typically include regulatory sequences. The region on the 5′ end of the gene is called the 5′UTR, and the region on the 3′ end of the gene is called the 3′UTR. The UTRs may contain important translational regulatory elements involved in pre-mRNA processing, mRNA stability and translation initiation. *Aka* NONTRANSLATED SEQUENCES.

untwisting enzyme: An enzyme that conserves or reinstates the helical structure of the double-stranded DNA during replication of circular DNA molecules. These enzymes produce a nick that enables free rotation and then reseal the nicked DNA strand once the supercoils are relaxed.

unusual bases: Other bases, in addition to four standard bases (adenine, cytosine, guanine and uracil), are found mainly in tRNAs. *Aka* MODIFIED BASES (e.g. inosine, pseudouridine, ribothymidine, methylguanosine, methylinosine).

unusual chromosomes: The chromosomes which show adaptational forms of the normal chromosomes (e.g. lampbrush chromosomes, polytene chromosomes) or these may be permanently specialized structures (e.g. B CHROMOSOMES and SEX CHROMOSOMES).

unwinding protein: The protein that binds to and unwinds the DNA helix at the replicating fork.

up mutations: Mutations that cause enhanced transcription levels are known as up mutations. These mutations generally involve only a single base pair, showing that RNA polymerase reaction with these sites must be highly specific.

up promoter mutations: Mutations in the promoter that enhance the frequency of transcription initiation.

upgrading regulation: A mechanism in dosage compensation where a single X chromosome in one sex elevates expression of its genes to equalize the expression of two X chromosomes in the other sex. For example, in *Drosophila*, hypertranscription exists.

uploading: In computer networks, it refers to transferring data to a remote system such as a server or another client so that the remote system can store a copy. *Contrasts* DOWNLOADING.

upper tier: In an eight-celled embryo, the term refers to the upper four cells produced by mitotic division of the apical cell. These cells produce the apical meristem and cotyledons.

upper/lower epidermis: Leaves have two outer dermal tissue layers, the upper and lower epidermis, each is composed of a single layer of cells that averts damage triggered by sunlight, pathogens and herbivores. The monocot leaf epidermis contains silica, which offers a deterrent to herbivores.

upregulation of the mitochondrial respiratory chain enzyme-associated genes in cotton: HIGH TEMPERATURE (HT)-INDUCED PROMOTER METHYLATION MODIFICATIONS resulted in the upregulation of the mitochondrial respiratory chain enzyme-associated genes – *GhNDUS7*, *GhCOX6A*, *GhCX5B2* and *GhATPBM* – ultimately encouraging a series of redox progressions to form ATP for normal anther development under HT stress.

upstream: In genetics, the direction which on DNA molecule is opposite to the direction RNA polymerase moves during transcription. This also is the direction opposite to the direction in which ribosomes move during translation. The +1 position of DNA (gene) is the first transcribed base; nucleotides upstream from this position are called –1, –2, –3, etc. *Compare with* DOWNSTREAM.

upstream activating sequence (UAS): This is a DNA segment required for activating the transcription of eukaryotic genes.

upstream enhancer factors: These proteins bind to specific DNA sequences upstream of the initiation site to stimulate or repress transcription initiation by the GTF-RNA polymerase II complex. The binding of upstream factors to DNA is unregulated (i.e. they adhere to any available DNA containing target sequences). The upstream transcription factors that are present in a cell vary with its developmental stage and its needs; their synthesis is also controlled.

uracil (U): A pyrimidine base, with the chemical name 2,6-dioxypyrimidine, which is found in RNA. Uracil is capable of forming a complementary base pair with adenine.

uracil-DNA glycosylase (UDGase) cloning: The *ung* gene codes for uracil-DNA glycosylase (UDG) which is a good device to produce cohesive ends. Primers are synthesized with four dUMPs on the 3′-end for amplifying the vector and target DNA. UDG cleaves the N-glycosylic bonds between other deoxyriboses and uracil in DOUBLE-STRANDED DNA (dsDNA), so the adjacent oligonucleotides are unstable and drift away from the dsDNA, which results in the formation of long tails on both the 5′-ends of the PCR products. By introducing dUMP instead of dTMP into the primers, scientists have successfully developed novel procedures for vector construction. *Aka* URACIL-N-GLYCOSYLASE UNG.

urea: An organic compound with the chemical formula $CO(NH_2)_2$. This is the most concentrated form of solid nitrogen fertilizer, and it is marketed in the prilled form. It is sometimes used for aerial top dressing. In the soil, urea changes to ammonium carbonate which may temporarily result in a harmful local high pH.

urease: An enzyme that converts urea into ammonium carbonate, occurring in microbes (i.e. bacteria, fungi).

urediniospore: The asexual, dikaryotic, often rust-coloured spore of a rust fungus, formed in a structure called UREDINIUM, a 'repeating stage' of a heteroecious rust fungus, which is capable of infecting the host species on which it is formed. *Aka* UREDIOSPORE *or* UREDOSPORE.

uredinium: The fruiting body (sorus) of a rust fungus that forms urediniospores. *Aka* UREDIUM or UREDINIA.

uridine: The ribonucleoside that contains the pyrimidine uracil (U).

uridine 5′-diphosphate (UDP): A ribonucleotide that contains pyrimidine and two phosphate groups are attached at 5′ end of the ribose sugar moiety.

uridine 5′-monophosphate (UMP): A ribonucleotide that contains pyrimidine and one phosphate group is attached at 5′ end of the ribose sugar moiety.

uridine 5′-triphosphate (UTP): A ribonucleotide that contains pyrimidine and three phosphate groups are attached at 5′ end of the ribose sugar moiety.

uridylate-rich small nuclear ribonucleic acids (UsnRNAs): An important component of U small nuclear ribonucleoproteins. At least 12 species of UsnRNAs (U1–U12) are known. The U RNAs (U1–U7) complexed with certain proteins are used in pre-mRNA processing. *Aka* U RNAs. *See details of different U RNAs as separate terms.*

uridylate-rich small nuclear ribonucleoproteins (UsnRNPs): Nuclear particles, each consisting of one or two U-RNAs and several proteins, involved in splicing and other transcript processing events. They regulate sex-specific alternative splicing of transcripts required for sexual differentiation. U snRNPs are classified based on U snRNA(s) present. At least five different kinds of snRNPs join the spliceosome to participate in splicing. They can be envisioned by gel electrophoresis and are known individually as U1, U2, U4, U5 and U6. UsnNPs control sex-specific alternative splicing of transcripts essential for sexual differentiation.

uridylate-rich small nuclear RNAs (UsnRNAs): UsnRNAs include the major or abundant (U1–U6) and the minor or low abundant (U7–U14). UsnRNAs are a distinct class of evolutionarily highly conserved, metabolically stable molecules that serve as crucial functional components (in the form of RNP particles, the UsnRNPs) or nuclear RNP complexes in which splicing of pre-mRNA (U1, U2, U4–U6 RNAs) or processing of pre-rRNA (U3, U8, U13, U14 RNAs) transpires.

uridylic acid: The ribonucleotide that contains the pyrimidine uracil.

use and disuse doctrine: The discredited idea that the more a structure is used, the more prominent it becomes in future generations and that the less it is used, the less prominence it assumes in later generations. *Aka* THEORY OF ACQUIRED CHARACTERISTICS.

user: An individual who uses a computer or network service. Users of computer systems and software products mostly lack the technical know-how essential to fully understand how they work. Power users make use of advanced features of programmes, though they are not necessarily capable of computer programming and system administration.

user agent: Software (or software agent) that acts on behalf of a user, such as a web browser that 'retrieves, renders and facilitates end user interaction with Web content'. For example, an e-mail reader is a mail user agent.

user interface (UI): The space where interactions between humans and machines take place. The goal of this interaction is to enable effective operation and control of the machine from the human end, while the machine concurrently feeds back information that assists the operators' decision making process.

Examples of this broad concept of user interfaces include the interactive aspects of computer operating systems, hand tools, heavy machinery operator controls and process controls. The design deliberations applicable when making user interfaces are connected with or involve such disciplines as ergonomics (process of designing or arranging workplaces, products and systems so that they suit the people who use them) and psychology.

user interface (UI) design: The design of user interfaces for machines and software, such as computers, home appliances, mobile devices and other electronic devices, with an emphasis on usability and the user experience. The goal of UI design is to enable the user's interaction as simple and efficient as possible, in terms of achieving user goals (user-centred design). *Aka* USER INTERFACE ENGINEERING.

utility function: A measure of the usefulness/merits/demerits of the model.

utricle: A small cell, sac or bladder-like protuberance in an animal or plant. *See* PERIGYNIUM.

UV-induced mutations: Mutations are produced if the UV light-induced lesions are not repaired by the repair mechanism of the cell. Unrepaired pyrimidine dimers cause mutations by inducing formation of daughter-strand gaps. One gap is formed opposite each lesion. These gaps if not repaired cannot be mended by repair replication, because each is located opposite a noncoding lesion.

uvrABC endonuclease: A multienzyme complex in bacteria involved in DNA repair by nucleotide excision repair, and it is, therefore, sometimes called an EXCINUCLEASE. This UvrABC repair process, sometimes called the SHORT-PATCH PROCESS, involves the removal of 12 nucleotides where a genetic mutation has arisen followed by a DNA polymerase, replacing these aberrant nucleotides with the correct nucleotides and completing the DNA repair.

V gene segment: Gene segment coding for most of the variable (V) area of the polypeptide chains of immunoglobulins and T cell receptors.

V-chromosomes: Chromosomes with two arms (i.e. metacentric chromosomes).

V-J joining: The process of joining a variable gene and a joining gene as the first step of the formation in functioning immunoglobulin gene.

v-onc: The version of an oncogene carried by a transforming retrovirus.

V(D)J joining: Recombination progression by which different gene segments (V, D and J) are joined together to produce a functional gene for the synthesis of a polypeptide chain of an immunoglobulin or T cell receptor.

V(D)J recombination: Developing B lymphocytes assemble immunoglobulin genes from widely scattered gene segments, using a somatic DNA rearrangement through the process of V(D)J recombination, which is catalysed by a 'recombinase' enzymatic machinery. An immunoglobulin heavy chain variable region gene is produced from three sequences of DNA, V_H, D and J_H. Conserved hepta- and decanucleotides in the vicinity of V, D and J genes generate a recognition system required for recombination. These conserved sequences are separated by a 10–11 or 21–23 bp non-homologous stretch of nucleotides. Presumably, similar sequences occur in the vicinity of D regions of the heavy chain genes. Histone H3 trimethylation at lysine 4 (H3K4me3) is involved in V(D)J recombination. V(D)J recombination thus assembles antigen receptor genes from component gene segments.

vaccination: Administering vaccines for conferring immunity. Vaccination is one of man's most significant inventions. It has eliminated smallpox and brought under control such diseases as diphtheria, poliomyelitis and measle. *See* IMMUNIZATION.

vaccine: A preparation that contains an agent or its component (antigen), administered to excite an immune response that will protect a person from illness due to that agent. A THERAPEUTIC VACCINE is given after disease has started and is intended to reduce or arrest the progress of the disease. A PREVENTIVE VACCINE or PROPHYLACTIC VACCINE is intended to thwart disease from starting. Agents used in vaccines may be whole-killed (inactive), live-attenuated (weakened) or artificially manufactured. Vaccines can be made using the recombinant DNA process. Now using a transgenic approach edible vaccines have been produced in fruits like banana and apple. Eating these transgenic fruits will result in vaccination.

vacuole: A cytoplasmic empty organelle surrounded by a membrane called tonoplast and contains sugar, organic acids, inorganic salts, proteins and pigments. It temporarily stores water, food, waste or salt and also pigments in certain plant cells. The solution in the vacuole is isotonic with the cytoplasm. In most plant cells, there is a single large vacuole that fills more than one-third of the cell's volume. Some bacterial cells contain gas vacuoles. Vacuoles help in maintaining water balance.

vagility: Refers to the ability of organisms to spread throughout a given habitat.

valence cross: Crossing between individuals of different ploidy levels. For example, the method used in the production of tetraploid rye varieties involves crossing between a tetraploid female and a diploid male. The tetraploid F_1 seeds are identified by green-grained xenia on a pale-grained mother plant, resulting from the fusion of reduced gametes of the mother plant (carrying a recessive allele for pale seed colour) with unreduced gametes of the male parent (carrying the dominant allele for green seed colour).

validated target: A gene or gene product that is a known target of a drug or chemical.

validating the hypothesis: Sometimes experimenters may make methodical errors during their experiments, unconsciously veering from a scientific method (pathological science) for various reasons, or in rare cases deliberately falsify their results. It is a

Glossary of Agricultural Biotechnology (G.S. Miglani, P. Chhuneja, S.S. Gosal)

common practice for other scientists to attempt to repeat the experiments to duplicate the results, thus further validating the hypothesis.

validation: The action (or process) of proving that a procedure, process, system, equipment or method used works as expected and achieves the intended result (can be performed at both analytical and clinical levels).

validation of histone modification antibodies: Authentication of commercial antibodies is essential to produce the high-quality data sets desired by the modENCODE and ENCODE consortia. All commercial histone antibodies should be authenticated by at least two independent methods.

validity: The extent to which a measurement precisely replicates the theoretical meaning of a variable.

valine (val): An essential, aliphatic, nonpolar amino acid ($HO_2CCH(NH_2)CH(CH_3)_2$) considered indispensable for normal growth of animals. It is biosynthesized from pyruvic acid.

value-enhanced grains: Those grains that possess innovative characteristics that are economically valuable, such as higher-than-normal protein content, better quality protein, higher-than-normal oil content, etc. For instance, (i) high-oil corn (maize) possesses a kernel oil content of 5.8% or greater versus oil content of 3.5% or less for traditional No. 2 yellow corn, (ii) glutamate dehydrogenase (GDH) corn (maize) possesses a kernel protein content that tends to be around 10% greater than the protein content of traditional corn (maize) varieties and (iii) high-amylose corn possesses a kernel amylose content of 50% or more of the total kernel starch.

valvate: Meeting of sepals and petals in bud edge to edge but not overlapping.

valve: A slice of an organ that trashes or splits open, such as the teeth-like portions of a pericarp in a split (dehisced) capsule or pod when ripe.

Van der Waals forces: The comparatively feeble forces of desirability between molecules that donate to intermolecular bonding (i.e. binding together two or more adjacent molecules). Historically, van der Waals forces were always believed to be weaker than the hydrogen bond forces responsible for intramolecular bonding. However, in 1995, Alfred French discovered that van der Waals forces are chiefly accountable for holding together a mass of cellulose molecules, in which hydrogen bonding plays a lesser role. *Aka* VAN DER WAALS ATTRACTION.

var: An abbreviation of VARIETAS.

variability: The totality of different genetic or phenotypic properties within different taxa.

variable: (i) In genetics, a variable is any characteristic, number or quantity that can be measured or counted. A variable may also be called a data item. Age, sex, milk yield, fat yield and type of feed are some of the examples of variables. It is called a variable because the value may vary between data units in a population and may change in value over time or condition. (ii) In computer programming, a variable or scalar, is a storage location (identified by a memory address) paired with an associated symbolic name (an identifier), which includes some known or unknown quantity of information referred to as a value. The variable name is the normal way to reference the stored value, in addition to referring to the variable itself, depending on the context. This separation of name and content enables the name to be utilized independently of the exact information it represents. The identifier in computer source code can be bound to a value during run time, and the value of the variable may therefore change during the course of programme execution. (iii) A changeable factor is believed to impact a natural phenomenon of interest or that can be installed during an experiment. *Contrasts* PARAMETERS.

variable (V) region: This region of an immunoglobulin chain is coded by the V genelets and varies extensively when different chains are compared, as the result of multiple (different) genomic copies and changes introduced during the construction of an active immunoglobulin. It comprises the antigen-binding site.

variable expression: A dissimilarity in phenotype between impacted members of the same family (e.g. individuals carrying identical mutations). It transpires in many dominant conditions and may be related to diminished penetrance. *Aka* VARIABLE EXPRESSIVITY.

variable numbers of tandem repeats (VNTRs): DNA sequence blocks of 2 to 60 base pairs which are repeated from 2 to >20 times end to end in different individuals. VNTRs exist in satellite, microsatellite and minisatellite sequences. Presumably, variability is produced by unequal crossing over. This polymorphism renders VNTRs very valued DNA markers used in genomic mapping, linkage analysis and DNA fingerprinting.

variance (σ^2): A statistic providing an unbiased estimate of population variability and is calculated as the average squared deviation about the mean of a set of data. It is the square of the standard deviation. To find the variance of the set of numbers 4, 6 and 8, the first to be calculated is the mean, which is 6. Then are calculated sum of the squared deviations from the mean $(4-6)^2 + (6-6)^2 + (8-6)^2$, which comes to 8, which is divided by (n – 1) (which is 2 in this case). The minus 1 is a kind of 'penalty' that corrects for estimating the population mean with the sample mean. The variance of the three numbers is 8/2 = 4. The variance of a set of identical numbers (such as 6, 6 and 6) is zero. The more variable the set of numbers, the higher is the variance. *Compare with* STANDARD DEVIATION.

variance balance block design: A block design is variance balanced if the variances of the best estimators of all the normalized treatment disparities are alike.

variance ratio: The ratio of treatment or replication variance to the error variance.

variance, components of: A contribution to the phenotypic variance due to a definite type of variation. It includes various types of variances such as ENVIRONMENTAL VARIANCE, ADDITIVE VARIANCE, DOMINANCE VARIANCE and INTERACTION VARIANCE.

variance, environmental: The variance ensuing from environmental or nongenetic causes. *See* ENVIRONMENTAL VARIANCE.

variance, genetic: The variance resulting from genetic causes. *See* GENETIC VARIANCE.

variance, types of: Variance can be divided into PHENOTYPIC VARIANCE, GENOTYPIC VARIANCE and ENVIRONMENTAL VARIANCE. *See details as separate terms.*

variant: (i) A culture exhibiting a stable phenotypic change whether genetic or epigenetic in origin. (ii) A plant or group of plants showing some measure of difference from the characteristics associated with a particular taxon. Prolonged *in vitro* culturing particularly involving the callus phase results in variant cells/plants. Genetic variation is heritable whereas epigenetic variation is generally not heritable through the sexual mode of reproduction. All mutants are variants but all variants are not mutants.

variant of uncertain significance (VUS): A genetic variant that has been identified through genetic testing but whose significance to the function or health of an organism is not known. Two related terms are GENE OF UNCERTAIN SIGNIFICANCE (GUS) and INSIGNIFICANT MUTATION. *See details as separate terms.*

variant sequence: An alteration of the nucleotide sequence from the reference genome. Alterations may be benign (polymorphisms), pathogenic (mutation) or of unknown significance.

variate: A specific numerical value of a variable.

variation: (i) Differences in pattern or function between individuals or other taxa. (ii) The cause of the differences between individuals. This is the reason that brothers and sisters who do resemble each other are still unique individuals. (iii) To understand the concept of organic evolution and the statistical nature of the world, one needs to understand that all organisms and objects have distinctive properties. Variation is an important characteristic of natural populations. Variation exists at various levels. Interspecific variation, intraspecific variation, polymorphism at phenotypic, chromosomal, protein and DNA levels. Occurrence of differences among individuals due to genetic or environmental causes. Variation is a common phenomenon at the level of cell/organism. Many times, except the clones, different cells/organisms exhibit a great deal of variability which may be attributed to genetic factors or environmental conditions. The occurrence of variation in a cell line used for commercial production is considered an undesirable phenomenon, whereas occurrence of variation is considered a desirable phenomenon for improvement purposes as superior variants can be released as better strains/varieties.

variation, classical view: The view that genetic variation is typically the consequence of harmful mutations. This contrasts the BALANCED VIEW that most variation is sustained by balancing selection.

variegated: Sporadically marked with blotches or patches of another colour.

variegation: Appearance of differently coloured zones in the leaves and sometimes the stems of plants due to several causes. It may be an inherited trait or may arise due to viral infection. *See* POSITION-EFFECT VARIEGATION (PEV).

varietal hybrid: The progeny emerging from the mating of two varieties.

varietal revamping: Reestablishment of a variety (released one) that has run out of cultivation (deteriorated due to various reasons). It can be rectified by secondary selection or other means and then again can be revamped under cultivation.

varietas: *See* VARIETY.

variety (Adj. varietal): A taxonomic rank occurring below species and between the ranks of subspecies and form. Often, VARIETY is in common usage and abbreviated as VAR. A plant type within a species that is true to type and possesses recognizable characteristics, described by a third italicized word in a scientific name. Breeding varieties can be classified according to the manner of propagation, such as CLONE

VARIETIES (maintained by vegetative propagation), LINE VARIETIES (maintained by self-fertilization), PANMICTIC VARIETIES (propagated by cross-fertilization) or HYBRID VARIETIES (produced by directed crosses). A variety is considered a unit with respect to its suitability for being propagated unchanged.

variety blend: A mechanical mixture of seed from two or more varieties.

variety description: A document in which a plant breeder lays down the distinctive characteristics of a variety.

variety in endogenous gene expression using epigenome-editing technology. Transgenic epimutagenesis, founded on ectopic expression of human TEN-ELEVEN TRANSLOCATION METHYLCYTOSINE DIOXYGENASE 1 (TET1), can yield hypomethylated plants that when crossed with wild type can be used to generate EPIGENETIC RECOMBINANT INBRED LINES (epiRILs) with new phenotypic variation.

variety-level genetic use restriction technology (V-GURT): V-GURT produces sterile seeds, so the seed from this crop could not be used as seeds, but only for sale as food or fodder. Consequentially, resistance to the introduction of GURT technology into developing countries is strong. The technology is restricted at the plant variety level, hence the term V-GURT. Manufacturers of genetically enhanced crops would use this technology to protect their products from unauthorized use.

vascular: Refers to the conducting tissues (xylem and phloem) of vascular plants.

vascular bundle: A bundle of vascular tissue in the primary stems of vascular plants, consisting of specialized conducting cells for the transport of water (XYLEM) and assimilate (phloem). Vascular bundles transport water, nutrients and other substances throughout the plant. Within each vascular bundle, the phloem is located exterior to the xylem. In monocot stems, the vascular bundles are scattered throughout the ground tissue. In monocot leaves, the vascular structures form parallel veins that are connected to the stem's vascular structures.

vascular cambium: A cylinder of meristematic cells (lateral meristem) that organizes secondary phloem to the exterior and secondary xylem (wood) to the interior of a branch or trunk of a woody plant.

vascular cylinder: The cylinder of vascular tissue found in stems or roots of plants. *See* STELE.

vascular plant tissues: Flowering plants, *aka* ANGIOSPERMS, are vascular plants – meaning that they have a transportation system that enables water, minerals and food to pass throughout the plant. They are composed of three types of tissue: DERMAL TISSUE, VASCULAR TISSUE and GROUND TISSUE. There are two main types of vascular tissue in plants, the XYLEM and PHLOEM, which together form vascular bundles throughout the plant's interior.

vascular plants: Division of plants that has vascular tissues, which serve to transport fluids and nutrients throughout the plant. Vascular plants include the clubmosses, horsetails, ferns, gymnosperms (including conifers) and angiosperms (flowering plants).

vascular wilt disease: A disease of the xylem that interrupts the normal uptake of water and minerals, leading to wilting and yellowing of foliage.

vasculum: A stiff container used by botanists to keep field samples viable for transportation. The main purpose of the vasculum is to transport plants without crushing them and by maintaining a cool, humid environment. *Aka* BOTANICAL BOX.

vault RNA: The vault RNA is ribonucleoprotein particle occurring in the cytoplasm of eukaryotes. It is associated with the nuclear pore complex and sticks to the nuclear membrane. The vault RIBONUCLEOPROTEIN (RNP) has a molecular weight of 13 MDa, a diameter ranging from 26 nm to 35 nm and a cyclic eightfold symmetry. It consists of one vault RNA and three proteins that together make a complex architecture. The function of this RNP is not yet clear but its structure suggests a role in subcellular transport.

vector: In molecular biology, vector is a DNA molecule (often plasmid or virus) that is used as a vehicle to carry a particular DNA segment into a host cell as part of a cloning or recombinant DNA technique. Vector could be a plasmid, or the genome of a bacteriophage or virus, Charon phages, cosmids and BACTERIAL ARTIFICIAL CHROMOSOMES (BACs), that is self-replicating and that can be engaged as agents to transfer DNA segments from host into target cells of another host. These vectors are of prokaryotic origin. Among eukaryotic vectors are yeast vectors (yeast artificial chromosome), animal vectors (Simian virus 40, a mammalian virus) and plant vectors (Ti plasmid of *Agrobacterium tumefaciens*). *See* CLONING VECTOR and EXPRESSION VECTOR.

Vector Alignment Search Tool (VAST): It is a computer algorithm developed at NCBI and used to identify similar protein 3D structures. The 'structure neighbours' for every structure in MMDB are pre-computed and accessible via links on MMDB structure summary pages. The neighbours can be used to identify distant homologs that cannot be recognized by sequence comparison alone.

vector amplification: The plasmid is often transformed into a bacterium (e.g. *Escherichia coli*) perfectly when the bacterium divides the plasmid should also be replicated. Each bacterial cell possesses several copies of the plasmid. After a good number of bacterial colonies have grown, they are miniprepped to harvest the plasmid DNA. To ensure growth of only transformed bacteria (which carry the desired plasmids to be harvested), a marker gene is used in the destination vector for selection. The typical marker genes include those for antibiotic resistance or nutrient biosynthesis, for example, the marker gene could be for resistance to the antibiotic ampicillin. If the bacteria that were supposed to pick up the desired plasmid had picked up the desired gene then they would also contain the marker gene. Now, the bacteria that picked up the plasmid would be able to grow in ampicillin, whereas the bacteria that did not pick up the desired plasmid would be susceptible to devastation by the ampicillin. Therefore, successfully transformed bacteria would be efficiently selected.

vector constructs for RNA interference in plants: Commonwealth Scientific and Industrial Research Organisation (CSIRO, Australia) has designed the pHANNIBAL and pKANNIBAL vectors for the transformation of dicotyledonous plants. These plasmids are comprised cauliflower mosaic virus 35S RNA promoter (CaMV 35S), an intron and several sites for target gene insertion in sense and antisense orientations at both ends of the intron. The pHANNIBAL and pKANNIBAL vectors contain bacterial genes of ampicillin and kanamycin resistance, respectively. American company Invitrogen designed plasmids pHELLSGATE, pWATERGATE and pSTARGATE. The pHELLSGATE vector contains a CaMV 35S promoter, while the pWATERGATE vector contains a light-inducible promoter of a **ribulose-1,5-bisphosphate carboxylase (RuBisCO)** small subunit. The pSTARGATE vector has been specifically designed for the transformation of monocotyledonous plants and contains a promoter of the maize ubiquitin gene and intron, as well as a selective marker of hygromycin resistance. Vectors of the pOpOff series are comprised a pOp6 promoter and an LhGR transcription factor.

vector designing: A vector is comprised of a promoter, multiple cloning sites, termination signal, antibiotic resistance gene, replication origin (Ori) and antibiotic resistance gene(s). Using complex cloning strategies, completely customized vectors can be established when needed. Plasmid vectors can be designed by choosing from modular, easily commutable fragments encoding regulatory and functional elements or from custom-made elements. These individual modules can be attained via PCR amplification in some cases, or they can be produced synthetically. Synthetic vector elements can be adapted to include practically any known regulatory or functional elements to fit your exclusive requirements. Plasmids can be assembled using the GeneAssembler® synthesis platform. This advanced assembly skill can build vectors rapidly and dependably. In addition, sequence optimization further improves plasmid design, serving to accommodate your design features and achieve maximum expression in your host. While designing a vector a few points are kept in mind. Vector should have maximum genetic constancy in the host organism, minimal vector size, addition/deletion of immune-relevant sequence motifs, optimized cell/tissue-specific promoters and meaningfully abridged homology to wild-type sequences to decrease the risk of homologous recombination events.

vectorial discharge: A step involved in protein secretion in eukaryotes where a polypeptide is fed into endoplasmic reticulum lumen through Sec61 complex. *Aka* COTRANSLATIONAL IMPORT.

vectorless methods of genes transfer: Vectorless methods do not use any vector system for gene transfer. These methods are classified as PHYSIOCHEMICAL, PHYSICAL and ELECTRICAL. *Aka* DIRECT GENE TRANSFER SYSTEMS (DGTS) or DNA-MEDIATED GENE TRANSFER (DMGT). *See details as separate terms. See* VECTOR METHODS.

vectors used in construction of genomic library: (i) PLASMIDS have a capacity of 15 kb. (ii) PHAGE (LAMBDA) have a capacity of 25 kb. (iii) COSMIDS or FOSMIDS have a capacity of 35–45 kb. (iv) BACTERIAL ARTIFICIAL CHROMOSOMES (BAC)(P-1 derived) have a capacity of 50–300 kb. (v) yeast artificial chromosomes (YAC) have a capacity of 300–more than 1500 kb. (vi) human artificial chromosomes (HAC) have a capacity of >2000 kb.

vegetative: Referring to somatic or asexual portions of a plant that do not participate in sexual reproduction.

vegetative cell: A cell that is not a sex cell. *Aka* SOMATIC CELL.

vegetative cell nucleus: Nucleus of the terminally differentiated vegetative cell which does not contribute genetic information to subsequent generations.

vegetative growth (or phase): The period of growth between germination and flowering.

vegetative growth regulation: Q. Wang *et al.* (2024) have described a new procedure of daylength measurement that perceives the duration of photosynthetic output and is essential for rapid growth during long summer days. Plants use multiple, independent methods to measure daylength that can differentially influence discrete developmental processes. Whereas flowering time principally relies on light signalling, vegetative growth is governed by photosynthetic and metabolic signals that become modified according to photoperiod. Photosynthetic rheostat of vegetative growth is partly dependent on the production of myo-inositol, encoded by gene *MYO-INOSITOL-1-PHOSPHATE SYNTHASE 1* (*MIPS1*), which functions as a forerunner for many other molecules involved in plant growth.

***vegetative insecticidal protein* (*Vip*) gene:** The genes in Bt strains that produced Vip proteins during the vegetative growth phase and these proteins do not share any structural and sequence homology with Cry proteins. Therefore, these are considered a tremendous complement or supplement source of Cry toxins in resistance management and crop protection.

vegetative nucleus: The large nucleus occurring within the pollen grain of angiosperms along with other one or two smaller generative nuclei. After germination of the pollen grain this nucleus travels to the tip of the pollen tube where it may be termed the tube nucleus.

vegetative phase: During this phase, the sporophyte continues to grow vigorously for some time and bear only the leaves and vegetative buds. The vegetative phase is dominant in life history of an angiosperm. *Aka* VEGETATIVE LAG PHASE.

vegetative propagation: Plant propagation through vegetative means without the true botanical seeds. A form of asexual reproduction in plants in which cuttings, bulbs, tubers and other vegetative plant parts are utilized to produce new plants (e.g. cloning of potato or strawberry). It is important to maintain and multiply especially those plants which do not produce seeds (e.g. banana). Even in seed crops like eucalyptus, citrus and mango, vegetative propagation ensures true to type uniform plants. Conventionally, it is done through budding, grafting, layering bulbs, tubers, corms, etc. Using tissue culture, micropropagation ensures true to type, rapid and mass multiplication of plants.

vegetative to inflorescence meristem switch: The switch in apical meristem from a vegetative to an inflorescence meristem and then floral meristem could be a two-phase transition or directly form vegetative meristem to floral meristem could be a single-phase transition depending on critical cues, both environmental and genetic. The *Arabidopsis* mutants reveal four partially redundant pathways involved in floral induction: PHOTOPERIOD PATHWAY, AUTONOMOUS PATHWAY, VERNALIZATION PATHWAY and GIBBERELLIN PATHWAY. Most of the genetic elements involved in these pathways are conserved.

vehicle: The host organism used for the replication or expression of a cloned gene or other sequence. The term vehicle is little used and is often confused with vector. *Compare with* VECTOR.

vehicle plasmid: A plasmid containing a piece of passenger DNA; used in recombinant DNA work.

vein: A strand of vascular tissue in the leaves of vascular plants. *Aka* NERVE.

vein banding: A symptom of virus disease in which regions along veins are either greener or markedly yellower than tissue between veins.

vein clearing: The fading of green colour in or around leaf veins.

veinlet: A small vein; the ultimate visible division of a vein.

velocity sedimentation analysis: A method that measures the mass of a molecule or particle by determining the rate at which the molecule sediments through a dense solution when subjected to a high centrifugal force.

venation: The pattern of veins in a leaf.

Venereal Disease Research Laboratory (VDRL) test: Immunological test to detect the presence of *Treponema pallidum* the causal agent of syphilis. It is a rapid screening test to detect the syphilis disease, which is a serious Pelvic Inflammatory Disease in women. This test detects the antibodies against the causal bacterium *Treponema pallidum*. This test detects only about 70–80% of primary syphilis cases.

verification: (i) Process of matching an unknown but keyed specimen with known material in the herbarium. (ii) In statistics, the process of subjecting hypotheses to empirical tests to authenticate whether or not the given theory is reinforced.

verminator technology: Verminator technology comprises 'a gene switch', which is inducible by external application of a chemical inducer which controls expression of a gene product, which, in turn, impacts the expression of a second gene in the genome. This technology has been developed to prevent plant growth, rather than killing the seed. The technology utilizes a gene from the fat tissue of a rat, which is capable of blocking the normal plant growth, unless the blocking process is deactivated by a chemical tetracycline. The technique has been labelled as 'VERMINATOR' by Rural Advancement Foundation International (RAFI) and appears to be wider and more flexible than the 'TERMINATOR', though intended to serve the same purpose.

vernalization: Process of treating the germinating seeds with low temperatures so as to induce flowering at a particular preferred time in the genetic model plant, *Arabidopsis thaliana* (thale cress) in which a gene, *Flc*, suppresses the formation of flowers during cold periods. Another gene, *Vrn2*, triggers that suppression. However, in spring the suppression is raised in a way that the gene 'remembers' the previous cold period exposure, because the gene *Vrn2* itself is cold-insensitive. There is a similar gene, reported in fruit fly, *Drosophila melanogaster*, which also assists as CHEMICAL MEMORY. *Aka* JAROVIZATION.

vernalization pathway in flowering: In many plant species, flowering is promoted by a period of exposure to low temperature through a process known as vernalization. Triggers stable repression of FLC. This pathway is operated through gibberellins.

vernalization responses: Genetically, flowers are just an adaptation of normal leaf and stem components in plants. The combination of floral, stem and leaf genes is generally required for the formation of new shoots during plant development. The floral regulation depends upon the endogenous signals; (age and developmental condition) and environmental conditions (day length, light quality and temperature) in plants. The diverse vernalization responses across angiosperms is controlled by EPIGENETIC MECHANISMS (DNA methylation, histone modifications and small RNAs).

vernation: (i) The arrangement of unexpanded leaves in a bud. (ii) The order in which leaves unfold from a bud.

versatile (of anthers): Swinging freely about the point of attachment to the filament.

version (in GenBank): Similar to the Protein ID for protein sequences, 'version' is a nucleotide sequence identification number assigned to each GenBank sequence. The format for this sequence identifier is accession.version (e.g. M12345.1). Whenever the author of a particular sequence record modifies the sequence data in any way (even if just a single nucleotide is altered), the version number will be increased by an increment of one while the accession number base remains constant. For example, M12345.1 would become M12345.2. Each sequence modification also leads to the assignment of a new GI number (link to GI entry). Whenever an NCBI sequence database is searched, only the most recent version of a record is retrieved. NCBI' s Sequence Revision History page is used to view the various GI numbers, version numbers or update dates related with a particular GenBank record.

vertical descent: The evolution of species by a branching pattern.

vertical DNA transfer: The transfer of DNA from parent organisms to the offspring via reproduction.

vertical evolution: *See* VERTICAL DESCENT.

vertical gene transfer: Out-crossing. Referred to as the transfer of genetic information, including any genetic mutations, from a parent to their offspring. It could happen by artificial means, asexual reproduction or sexual reproduction. This procedure is a typical mechanism that occurs in groups of related species. Humans, plants and animals transmit their genes by vertical transmission. In contrast, the bacteria can transfer their genes not only to their offspring but also laterally to other microbes that is called horizontal gene transfer that may occur through conjugation and transduction.

vertical inheritance: Refers to the transmission of traits from parent to offspring. *Aka* VERTICAL TRANSMISSION.

vertical resistance (VR): A resistance that is operative against some biotypes or races of the pathogen, but not others, generally inherited monogenically and expressed qualitatively. It obeys H.H. Flor's gene-for-gene relationship. *Aka* SINGLE-GENE RESISTANCE *or* RACE-SPECIFIC RESISTANCE. *Contrasts* GENERAL RESISTANCE, HORIZONTAL RESISTANCE *and* RACE-NONSPECIFIC RESISTANCE.

verticillate: Arranged in one or more whorls (i.e. several similar parts arranged at the same point of the axis; e.g. leaf arrangement). *Compare with* PSEUDOVERTICILLATE where several similar parts appear whorled or verticillate but are not actually so.

vertifolia effect: The loss of general (durable, horizontal, host-nonspecific) resistance in a cultivar after several generations of selection during which a major gene deliberates resistance to the dominant race or biotype of the pathogen. Vertifolia effect was first detected in the potato cultivar Vertifolia with late blight resistance.

very short patch (VSP) repair: In *Escherichia coli*, a process that repairs T/G mismatches reproducing original CC(AT)GG sites. This mechanism has perhaps evolved to minimize C→T transitions that are consequence of deamination of methylated cytosines.

vesicle: (i) A small structure or vacuole within or sometimes external to a cell, consisting of fluid enclosed by a lipid bilayer. (ii) A thin sac in which zoospores are differentiated and released. (iii) The bulbous head terminating the conidiophores of *Aspergillus*, a structure formed by endomycorrhizal fungi within living cells of the root.

vesicle SNARE: The SNAREs (soluble N-ethylmaleimide-sensitive factor attachment protein receptor) mediate the release of synaptic vesicles containing neurotransmitters in neurons. These neuronal SNAREs are the targets of the neurotoxins responsible for botulism and tetanus produced by certain bacteria.

vesicular transport: A mechanism that enables a protein molecule to pass between compartments within eukaryotic cells. The compartment wall (membrane) holds a 'sensor' (receptor) that senses the occurrence of correct protein (e.g. after that protein has been synthesized in the cell's ribosomes), then swells outward along with that protein molecule. The membrane swelling containing protein then breaks off and carries the protein to its destination in another compartment in the cell.

vessel: A capillary tube shaped from a succession of open-ended cells in the water-conducting tissue of a plant.

vestige: A small part or amount of something larger, stronger, or more important that still exists from something that existed in the past. A famous example of vestige is a Drosophila mutant with rudimentary wings, called *vestigial*.

vestigial: Any structure that has been greatly abridged in size and function over evolutionary time to the level that now it appears to have little or no function.

vestigiality: The retaining during the course of evolution of genetically determined structures or characteristics that have lost some or all of their family functions in a certain species.

viability: (i) The likelihood that a fertilized egg will live and progress into an adult organism. (ii) The term is also often useful to plant germination tests with comparisons across phenotypic classes under specific environmental conditions.

viable: (i) The state of being flourishing. (ii) Able to germinate, as seeds, fungal spores, sclerotia, etc. (iii) Capable of growth.

vibrio: Comma-shaped bacterium.

vigour test: This test uses seed to determine its possibility to develop into normal, healthy plants under a wide range of field conditions. Stress tests, accelerated ageing and other techniques are employed to deliver a more sensitive index of seed quality. Poor emergence, dormancy, seed deterioration and other quality aspects can be identified through vigour test.

villous: (i) Plentiful in or enclosed with long, soft, straight hairs. (ii) Any structure that is bushy with soft hairs.

Vilmorin principle: Given by Louis de Vilmorin in 1886, this principle refers to the selection carried out on the basis of progeny test (plant-to-row) performance. Pure line breeding is based on this principle.

vine: Scandent plant, a member of the genus *Vitis*, having climbing habit by means of trailing or twining stems or runners.

viral genome methylation as epigenetic defence: Geminiviruses encapsidate single-stranded DNA genomes that replicate in plant cell nuclei through double-stranded DNA intermediates with which bind the cellular histone proteins to form MINICHROMOSOMES. Geminiviruses are targeted by RNA silencing machinery and encode suppressor proteins such as AL2 and L2 that act against this defence. These related proteins can overcome silencing by multiple mechanisms, one of which involves interaction with and inhibit adenosine kinase (ADK), a cellular enzyme related with the methyl cycle that generates S-adenosylmethionine, an essential methyltransferase cofactor. P. Raja *et al.* (2008) hypothesized that the viral genome is targeted by small-RNA-directed methylation. They revealed that *Arabidopsis* plants with mutations in genes encoding cytosine or histone H3 lysine 9 (H3K9) methyltransferases, RNA-directed methylation pathway components or ADK are hypersensitive to geminivirus infection. They also showed that viral DNA and associated histone H3 are methylated in infected plants and that cytosine methylation levels get significantly diminished in viral DNA isolated from methylation-deficient mutants. Beet curly top virus L2 mutant DNA occurring in tissues that have recovered from infection is hypermethylated and that host recovery needs AGO4, a component of the RNA-directed methylation pathway. They thus propose that plants employ chromatin methylation as a defence mechanism against DNA viruses, which geminiviruses counter by inhibiting global methylation.

viral infectivity experiment: A. Hershey and M. Chase (1952) performed experiments on bacteriophage T_2. This phage is relatively simple in molecular structure mostly protein with DNA contained inside the protein sheath of its head. Knowing that phosphorus is not present in proteins but is an integral part of DNA and sulfur does not occur in DNA but is present in proteins, they labelled phage DNA with radioisotope of phosphorous ^{32}P and phage proteins with that of sulfur ^{35}S in different cultures. They observed that after infection label ^{32}P went inside the bacteria, whereas label ^{35}S was left outside, showing that the phage phosphorus-containing DNA enters the cell while residue containing at least 80% of the sulfur-containing protein of phage remains at the cell surface. This residue consists of material forming the protective membrane of the resting phage particle and enacts no further part in infection after the attachment of phage to the bacterium. Viral infectivity experiment thus demonstrated that genetic material of phage T_2 is DNA, not protein.

viral recombination: Viral recombination is not restricted to simple pairing between two homologous chromosomes and to only one set of recombinational actions in the zygotes. Since many copies of viral chromosomes exist in any one cell, genetic exchange may transpire between members of plentiful populations of chromosomes. Viral chromosome multiplication is geometric or clonal. Same viral chromosomes may pair and exchange genetic material more than once and may possibly mate with more than one chromosome at a time. Virus must exchange genetic material in a host cytoplasm. Thus, viral recombination is a population phenomenon whose events are described statistically.

viral resistance: Resistance in a plant to a viral disease(s). Viral diseases cause huge crop losses. Yellow bean mosaic of pulses, tomato mosaic, chilli mosaic, grassy stunt of sugarcane, p PVX, PVY of potato are serious viral diseases. Using transgenic approach there are now several strategies to develop viral resistance in crops. Some of these include coat protein-mediated resistance (CPMR), dysfunctional movement of protein, antisense RNA, satellite RNAs, plantibodies and use of RIBOSOME-INACTIVATING PROTEIN (RIP) gene. Development and commercialization of transgenic papaya in USA is the best example of the development of viral resistance – resistance against PAPAYA RING SPOT VIRUS (PRSV), through transgenic approach.

viral retroelement: A virus whose genome replication process involves reverse transcription.

viral RNA replication: Replication of phage RNA occurs by viral RNA replicase (RNA-dependent RNA polymerase). The enzyme is made of four subunits – three (α, γ, δ) are host polypeptides while one is a virus-specified polypeptide. The γ and the δ subunits function as the protein synthesis elongation factors EF-Tu and EF-Ts, respectively. The α subunit is a component of the 30S ribosomal subunit.

viral suppressor of RNA silencing (VSR): VSRs can act by binding to siRNA or proteins of the RNA silencing pathway, directly or indirectly, effectively sequestering them to prevent the viral RNA from being targeted. Suppressor 19-kDa protein (p19) from tombusviruses forms a dimeric clamp around the ds form of siRNA. The two p19 amino termini are positioned precisely around the two ends of the RNA duplex structure so that the two strands of the duplex cannot separate and release the active ss form. VSRs include P38 from carmoviruses that binds the **AGO** protein itself. P0 from poleroviruses is an atypical F box protein that directs AGO degradation through an autophagy.

viral transactivating protein: A specific protein employed by a lytic virus to switch on the cascade of gene regulation by which that virus controls a healthy cell to disrupt its molecular machinery to produce virus components. This transactivating protein is vital to the whole lytic cycle of the virus and therefore a potential target for therapeutic intervention.

viral vector: These vectors are used to effectively transfer genetic material into a host in order to modify its cells or tissue. This will allow for the manipulation of these cells and tissues so that they may be used to express genes differently.

viremia: Virus in blood.

virescence (Adj. virescent): A process or situation in which normally white or coloured plant tissues (such as flower petals) become green due to abnormal development of flowers in which all organs are green and partly or wholly transformed into small leaf-like structures.

virion: A non-replicating fully developed intact virus particle. It is a nucleic acid core enclosed by a protein coat called CAPSID. Besides, it contains some enzyme molecules, structural proteins and a surrounding membrane called ENVELOPE. Sometimes, other proteins form structures called SPIKES extended from the envelope.

virogene: The complete viral genome.

viroid: Any of a class of plant pathogenic agents consisting of an infectious, single-stranded, free RNA molecule. Small pieces of naked RNA (300–400 nucleotides long) lacking protein coat. These nucleotides are internally paired to give three-dimensional structure of viroid. This RNA does not code for any protein. For example, POTATO SPINDLE TUBER VIROID (PSTV) is a common disease of potato caused by viroid.

virtual libraries: The creation and storage of vast collections of molecular structures in an electronic database. These databases may be interrogated for subsets that display specific physicochemical features, or may be 'virtually screened' for their ability to bind a drug target. This process should be performed before the synthesis and testing of the molecules themselves.

virtual machine (VM): An imitation of a computer system. VMs are based on computer architectures and attempt to offer the same functionality as a physical computer. Their implementations may comprise specialized hardware, software or a combination of both.

virulence: Degree of pathogenicity of a pathogen. Different strains of the same pathogen possess different kinds of virulence. Some strains are mild, whereas others are virulent and cause symptoms (disease) very rapidly. The virulence of the pathogen is governed by genes called virulence genes or *vir* genes.

virulence (*avr*): The relative capability of a microorganism to overcome the resistance of a host. *Compare with* AVIRULENCE.

virulence factors: Bacterial flagella and Pili are examples of virulence factors. Bacterial flagella are hair-like structures used for locomotion. Pili are appendage required for bacterial conjugation.

virulence gene: Gene that triggers an organism's ability to cause disease.

virulence plasmids: These bacterial plasmids cause pathogenicity on the host bacterium. They make the bacterium more pathogenic as the bacterium is better able to resist the host defence or to produce toxins. For example, Ti-plasmids of *Agrobacterium tumefaciens* induce crown gall disease in angiospermic plants. Enterotoxigenic strains of *Escherichia coli* cause travellers' diarrhoea because of a plasmid that codes for an enterotoxin which persuades extensive secretion of water and salts into the bowel.

virulent: A pathogen capable of causing disease by aggressively interfering with the immune system of the host. (i) A disease-causing bacterium leading to the death of the host. (ii) A bacteriophage that follows the lytic mode of infection and destroys its host bacterial cell. *Contrasts* AVIRULENT.

virus: A submicroscopic, filterable agent containing nucleic acid draped in the protein coat. Viruses were first recognized in 1892 by D. Ivanovsky. According to Sir Peter Medawar's definition given in 1960, a virus is a piece of bad news wrapped in a protein. A kind of intracellular parasite particle consisting of nucleic acid (RNA or DNA) encoding for a small number of genes, enclosed in a protein coat and capable of replicating within a host cell and spreading from cell to cell. Viruses are very small (25–300 nm) particles (i.e. one-tenth to one-third of the smallest bacteria). Viruses lack structures such as ribosomes, nucleus and even the cytoplasm. Viruses contain fewer genes. For example, the largest virus T_4 contains only 77 genes as compared to human being (about 30,000 genes) and rice (about 40,000 genes).

virus coat protein (CP) gene: CP gene is attached to a strong promoter. This has been attained in tobacco, tomato, lucerne, sugarbeet, potato and so on. The first type of such plant was developed in 1986 by A. Powell *et al.* Most likely, the resistance is generated by CP due to the blocking of the process of uncoating of virus particles and delay of systemic spread of the viruses.

virus free: *See* DISEASE FREE.

virus resistance: Some methods engaged to engineer plants for virus resistance include: (i) virus coat protein (CP) gene, (ii) non-CP protein gene, (iii) cDNA of satellite, (iv) defective viral genome and (v) ribozyme-mediated protection. All these strategies are based on genes derived from pathogenic viruses themselves.

virus-free: Free from specific viruses based on tests designed to detect the presence of the organism in question.

virus-induced gene silencing (VIGS): A kind of POST-TRANSCRIPTIONAL GENE SILENCING (PTGS) that is triggered by viruses rather than transgenes. VIGS is linked with the processing of double-stranded RNA encoded by a virus. This viral mechanism explains the possibility for endogenous genome elements to initiate persistent gene silencing if they are activated by stress or other external stimuli. RNA silencing is part of the antiviral defence systems. RNA silencing functions when plant cells recognize the viral RNAs as foreign and copy it with an RNA-DEPENDENT RNA POLYMERASE (RDRP) into a DOUBLE-STRANDED (DS) RNA. The dsRNA is then processed by a Dicer nuclease into 21–24-nt siRNAs. In plants, the Dicer proteins are referred to as DICER-LIKE (DCL) although strictly they are not DCL – they are, in fact, Dicer proteins. The small RNAs initially appear in a double-strand form but eventually its single-stranded (ss) version is then incorporated into an effector ribonucleoprotein complex that also contains an AGRONAUTE (AGO) nuclease. AGO has structural similarity to ribonuclease H, and the siRNA serves as a guide for AGO. The siRNA base pairs with other RNAs in the cell and thereby guides AGO to its target for degradation or to be translationally suppressed. The most frequent target of viral siRNAs would, of course, be the viral RNA itself. In an uninfected cell, there are no viral siRNAs, hence, AGO proteins have no guide RNA to target viral RNAs. Following infection, the first few rounds of virus replication would be unconstrained by RNA silencing. Once the viral RNA has accumulated to a substantial level it could prime AGO and the rate of viral RNA accumulation would slow down. The steady-state levels are also impacted by the extent to which the viral RNA can elude the silencing machinery by, for example, encapsidation or through the action of VIRAL SUPPRESSOR OF RNA SILENCING (VSR).

virus-resistant crops: Plants having the ability to endure plant viral diseases. Developed via traditional breeding or genetic engineering (e.g. papaya ringspot virus-resistant papaya). *See* BREEDING.

visible: Mutation recognizable by their phenotypic expression, in contrast to vitality mutations and lethal factors.

vision processing unit (VPU): A type of microprocessor designed to accelerate machine vision tasks, such as image and video processing, object detection, feature extraction, and machine learning inference. VPUs are designed to handle real-time, high-volume data streams efficiently and with low power consumption. Examples of VPUs include the Movidius Myriad X and Myriad 2 from Intel Corporation, which are used in various applications including Google Project Tango, Google Clips, and DJI drones.

visualization: A process of presenting abstract scientific data in the form of images that can help in understanding the meaning of the data.

vital mutations: These mutations do not impact the survival of the individuals in which they occur. Some mutations, particularly relating to height, shape and size of leaves, fall in this category.

vital staining: A stain that is accomplished with the ability to enter and stain a living cell without killing it. These stains have thus been useful for both diagnostic and surgical techniques in a variety of medical specialities.

vitality mutation: Any mutation the effect of which, in contrast to visible and lethal factors, is difficult to determine but that changes the viability of the carrier genotype if present in an effective dosage.

vitamins: Vitamins are biomolecules, required to grow and develop by our body. There are 13 vitamins your body needs, for example, vitamins A, B, C, D, K, B-6, B-12, etc.

vitrification: Abnormal looking generates of shoot cultures. During plant tissue culture, the shoot cultures of many plant species (e.g. chrysanthemum, strawberry and carnations) exhibit abnormal growth such as little leaves and spindly shoots. These shoots are called vitrified and the major cause of vitrification is the high humidity in the culture vessel. Vitrification is undesirable phenomenon and can be reduced by controlling the humidity levels using porous plugs and bottom cooling.

viviparous: (i) Seeds or fruits which germinate before being shed from the parent plant. (ii) The development of plantlets on non-floral organs (e.g. leaves).

V_{max}: Defined as maximal velocity of the enzymatic reaction. It occurs at high substrate concentrations when the enzyme is saturated, that is, when it is entirely in the ES form: $Vmax = k_2[E]_T$. This equation is basic equation of enzyme kinetics.

volatile communication in plants: Plants are continually exposed to volatile organic compounds (VOCs) that are released during plant-plant communication, within-plant self-signalling and plant–microbe interactions. Understanding VOC perception and downstream signalling is vital for understanding the mechanisms behind information exchange in plants. Using the hormone-like function of volatile terpenoids in reproductive organ development as a system with a visual marker for communication, S.A. Stirling and co-workers (2024) demonstrate that a petunia Karrikin insensitive receptor, PhKAI2ia, stereospecifically perceives the (-)-germacrene D signal, triggering a KAI2-mediated signalling cascade and affecting plant fitness. This study illumines the involvement of a KAI2ia-dependent signalling pathway in volatile communication.

volatile organic compounds (VOCs): Aphid infestation can stimulate emission of VOCs from host plants, and METHYL SALICYLATE (MeSA) is considered one of the major components of VOCs. Q. Gong *et al.* (2023) found that MeSA can enhance plant defence by repelling or inhibiting the survival rate of insect pests as well as by attracting their natural enemies. MeSA can also function as a mobile signal, not only conferring within-plant and long-distance SYSTEMIC ACQUIRED RESISTANCE (SAR) but also establishing interplant communication under microbial pathogen infection. MeSA plays an important role in plant AIRBORNE DEFENCE (AD). They investigated virulence approaches of the CMV1a protein (CMV1a) from CUCUMBER MOSAIC VIRUS (CMV) and identified the transcription factor NAC2 from *Nicotiana benthamiana* as an interactor of CMV1a and demonstrated that NAC2 is essential for plant antiviral defence through virus infection assays on NAC2 knockout plants. The lack of MeSA emission is responsible for NAC2-mediated attractiveness to aphids. The aphid preference for *nac2* plants still exists after gaseous MeSA treatment, approving the role of NAC2 in volatile MeSA-induced plant antixenosis resistance to aphids.

volunteer: A self-set plant or a plant seeded by chance.

von Baer laws: The four laws given by Karl Ernst von Baer in 1828, which are called von Baer's laws of embryology, can be stated in form of following generalizations: (i) More general characters of a large group of animals appear earlier in their embryo than the more special characters. (ii) From the most general forms the fewer general forms are developed and so on, until finally the most special forms appear. (iii) Every embryo of a given animal form, instead of passing through the other forms, becomes separated from them. (iv) Fundamentally, embryos of higher forms never resemble any other form but only its embryo. These empirical laws retain their validity even today.

voucher specimen: An herbarium specimen that describes the morphology and identification of chromosome, ecological and anatomical studies, plant breeding (parents and progeny) and taxonomy.

VP64: A transcription activator composed of four tandem copies of the herpes simplex virus VP16 activation domain connected by Gly-Ser linkers. VP64 is often fused to DNA-binding proteins for targeted transcription activation.

VSG switching: *Trypanosome brucei* is the causal agent of African sleeping sickness in humans and one of the sources of nagana in cattle. This protozoan parasite eludes the host immune system by antigenic variation, a periodic switching of its VARIANT SURFACE GLYCOPROTEIN (VSG) COAT. Introduction of a DNA DOUBLE-STRAND BREAK (DSB) adjacent to the roughly 70 bp repeats upstream of the transcribed *VSG* gene upsurges switching *in vitro*. A DSB is a natural intermediate of VSG gene conversion and that *VSG* switching is the consequence of this DSB by break-induced replication. Antigenic switching is induced by a single I-*Sce*I-produced DSB.

vybrid: Result of a cross between two open-pollinated varieties. The first and subsequent generation progenies obtained from crosses of heterozygous facultative apomicts. These crosses produce two types of plants – maternals and sexuals, out of which maternals are selected so that mean values of hybrids remain same over generations.

W

W: The symbol representing adaptive value (or relative fitness) in a mathematical treatment of natural selection.

W chromosome: X chromosome in organisms where the female is the heterogametic sex.

Wahlund effect: The decreased frequency of homozygotes in the progeny after mating made among individuals of two previously isolated subpopulations.

walling-off: The separation of diseased tissues from healthy ones by BARRIER TISSUES formed by a diseased plant.

water-shoot: An erect, strong-growing or epicormic shoot arising from near the base of a shrub or tree, but distinct from a sucker.

water-soaked: A disease symptom in which plant tissues or lesions seem wet, dark, usually sunken and translucent.

water as a reactant of photosynthesis: Water (H_2O) is a reactant of photosynthesis and six water molecules are required for each set of photosynthesis reactions. These water molecules travel from the soil surrounding the plant, enter the roots through the root hairs and passage through the xylem in the roots and stem to a leaf. Once inside the leaf, the water molecules travel into the chloroplasts inside the palisade mesophyll cells. During the light-dependent reactions that transpire in the thylakoids of the chloroplast, the water molecules are torn apart to convert $NADP^+$ to NADPH, to help in the production of ATP, and to produce oxygen as a by-product of photosynthesis.

water sprout: A small, rapidly growing shoot or branch on a large stem, developing from an adventitious tissue.

Watson-Crick base pairing: The normal base pairing, viz. adenine (A) pairs with thymine (T) and guanine (G) pairs with cytosine (C) in DNA and adenine (A) pairs with uracil (U) and guanine (G) pairs with cytosine (C) in RNA.

Watson-Crick model of DNA structure: DNA molecule is double-stranded and coiled like a rope. The two DNA strands are complementary; adenine (A) pairs with thymine (T) with two hydrogen bonds and guanine (G) pairs with cytosine (C) with three hydrogen bonds. Two strands of DNA run antiparallel, one strand is called 5'→3' and the other one 3'←5'. Each strand is a chain of deoxyribonucleotides held together by PHOSPHODIESTER BONDS between bases. The purines and pyrimidines are joined with deoxyribose by β-GLYCOSIDIC BONDS. The intertwining of two strands of DNA is just like a staircase in which railings are made up of sugar-phosphate backbones and stairs are made up of bases. The bases stack on the top of one another in a twisted structure of double helix and HYDROPHOBIC INTERACTIONS keep nucleotide pairs at their positions. The double-helical model of DNA proposed by J.D. Watson and F.H.C. Crick in 1953 revealed important inferences in suggesting explanations for a few important biological phenomena: How is biological information stored in DNA? How does DNA replicate?

wave of advance: The idea that a favourable allele advances behind a moving cline.

***waxy hull-less barley* (*WHB*):** A barley mutant that is rich in soluble fibre and low in fat content. These characteristics render it a nutritionally treasured ingredient for food products. It has been revealed that the soluble fibre beta-glucan reduces cholesterol and lowers blood glucose and insulin in response to a meal.

***waxy* maize:** Maize that yields kernels in which the starch is at least 99% amylopectin, versus the average of 72–76% amylopectin in common maize.

***waxy* wheat:** Varieties of wheat that have higher amylopectin content and, thus, a lower amylose content in the starch within their seeds. For instance, bread flour made from waxy seeds contains 0–3% amylose versus 24–27% amylose in bread flour made from traditional varieties.

weak AI: Artificial intelligence that is focused on one narrow task.

weather: The environmental conditions at a particular time. Typical characteristics to describe weather are precipitation (rain or snow), temperature (hot or cold), humidity (moist or dry), wind and atmospheric pressure.

web crawler: An Internet bot that systematically browses the World Wide Web, characteristically for web indexing (web spidering). *Aka* SPIDER, SPIDERBOT *or simply* CRAWLER.

webserver: A web server is a software and hardware that uses Hypertext Transfer Protocol (HTTP) and other protocols to reply to client requests through the World Wide Web. The main job of a web server is to reveal website content through storing, processing and delivering webpages to users.

weed: (i) Any plant growing where it is not required; commonly associated with disturbed habitats. *See* RUDERAL. (ii) An unwelcome plant that grows among agricultural crops. (iii) A naturalized, exotic or ecologically 'out of balance' indigenous species outside of the agricultural or garden context, which, as a consequence of attack, adversely affects the survival or regeneration of indigenous species in natural or partly natural vegetation communities.

weed management: Managing weeds in ornamental plant production, whether in field soil, greenhouses or outdoor containers, can be difficult but is indispensable to successful production. Weeds not only compete with the crop for plant nutrients and sunlight but are also unsightly and do not encounter clean nursery quality standards.

weed management methods: Various methods used for weed management include prevention, cultivation, cover crops, mowing, flaming, hand removal, mulches, soil solarization, transplants, herbicides, preplant herbicides, preemergence herbicides and postemergence herbicides.

weed mapping: A commonly used application of remote sensing in agriculture and drones deliver advantages in this application due to the high degree of flexibility in spatial resolution. Successful mapping of weeds necessitates the agronomist to work closely with the image analyst to guarantee understanding of the form and phenology of the weeds and how they are different from the crop species.

weight matrix: The density of binding sites in a gene or sequence can be engaged to derive a ratio of density for each element in a pattern of interest. The combined individual density ratios of all elements are then utilized together to build a scoring profile, known as a weight matrix. This profile can be employed to test the estimate of the identification of the pattern selected and the ability of the algorithm to differentiate it from a nonpattern sequence.

Western blotting: An immunological technique, developed by H. Towbin *et al.* (1979), is used to detect specific protein in the total soluble protein. Gene expression in putative transgenic plants is detected by using this technique. For this, total soluble proteins are isolated from the transgenics. These are separated by gel electrophoresis and then blotted to filter paper and specific ones are spotted by labelled antibodies. If specific antibodies hybridize, the transgene is considered expressing. Even the level of expression can be quantified by the amount of protein produced. *Aka* WESTERN HYBRIDIZATION. *Compare with* SOUTHERN BLOTTING, NORTHERN BLOTTING *and* EASTERN BLOTTING.

wet nanotechnology: Bionanotechnology where nanosystems function in the presence of water. The nanomaterials are membranes, enzymes and such biological entities. All life is based on such technologies.

what is inherited? It is the genes and not the traits that are inherited. No trait is simple enough to be controlled by a single gene. When we talk about the inheritance of a trait, we refer to the genetic differences that determine the phenotypic differences existing between two parental lines for a particular trait. The development of a trait is a consequence of complex interactions involving products of many genes.

wheat epigenome: Epigenetic variations, documented in the epigenome of allohexaploid 16-GB *Triticum aestivum*, impacted important traits like height, yield and flowering time. In detailed analyses of the open chromatin DNA methylome, seven major chromatin marks (namely, H3K4me3, a modification that generally exists in promoters; H3K27me3, associated with repression; H3K9ac and H3K27ac, associated with active regulation, mostly in promoters and enhancers; H3K36me3, associated with transcribed regions; H3K9me2, responsible for the repression of TEs and H3K4me1 is preferentially associated with enhancers in vertebrates, but in gene-body regions in *Arabidopsis thaliana*) and transcriptomic data produced for seedlings of wheat. Subgenome-divergent regulation in its massive allohexaploid genome contributed to its genome plasticity and domestication.

Whitehouse-Hastings model: In homologous chromatids, strands of different polarity are broken. The complementary nucleotide chains, one from each

DNA molecule, break nearly but not exactly at homologous points. The parts of the two broken chains between the two breakage points uncoil from their complementary chains and then coil around each other so that complementary base pairing transpires between their terminal portions to make a crossover molecule. New nucleotide chains are synthesized alongside the old chains that have not crossed over. The new chains uncoil from their complements, peel back and unite by complementary base pairing to make the other crossover molecule. A fundamental feature of this theory is thus the incidence of HYBRID DNA in the region of cross over. Proposed by H.L. Whitehouse and P.G. Hastings during 1963–1967.

whole exome sequencing (WES): A laboratory procedure that is employed to mark the nucleotide sequence primarily of the exonic (or protein-coding) regions of an individual's genome and related sequences, representing nearly 1% of the complete DNA sequence. *Compare with* SEQUENCING.

whole genome amplification (WGA): A methodology for non-selective amplification of the whole genome sequence, which L. Zhang *et al.* first reported in 1992. In eukaryotes, genome duplication can ensue if the cell flops to undergo mitosis after DNA replication. This is seen, for example, in mammalian liver cells, which are tetraploid and most obviously in the salivary glands of dipterans (e.g. *Drosophila*), where several rounds of DNA replication transpire to produce giant polytene chromosomes containing up to 2^{10} (= 1,024) chromatids. Its chief purpose is to amplify and reflect the whole genome of trace tissues and single cells without sequence bias and to provide sufficient DNA templates for subsequent multigene and multilocus analysis, along with comprehensive genome research.

whole genome bisulfite sequencing: A next-generation sequencing procedure engaged to determine the DNA methylation status of single cytosines by treating the DNA with sodium bisulfite before high-throughput DNA sequencing. The DNA methylation status of various genes can divulge information regarding gene regulation and transcriptional activities.

whole-genome duplication (WGD): An incident that generates an organism with additional copies of the entire genome of a species.

whole genome sequencing (WGS): DNA sequencing method that intends to sequence all of the bases of an individual's genome, including the exome and noncoding regions. WGS is the mapping out of a person's unique DNA. The genome of the test sample is sequenced and computationally compared to a reference genome to identify many types of genetic variation. The unique combination of length and accuracy allows HiFi reads to determine difficult repeats, segmental duplications and centromeres to generate more contiguous and phased assemblies. With PacBio HiFi sequencing, this method delivers a more complete and more accurate genome with long, single-molecule reads and 90% of bases at ≥Q30.

whole plant from a single somatic cell: Many plants have evolved repair mechanisms and reproductive strategies that enable them to yield offspring even without the union of sperm and egg. Some plants can reproduce from an outgrowth of stems, roots and bulbs, but others are even more fundamental, able to create new embryos from single somatic cells. In the laboratory, the herbicide 2,4-dichlorophenoxyacetic acid (sold as a weed killer, called 2,4-D) can stimulate cells in culture to elongate, build new cells in culture to elongate, build a new cell wall and start dividing to form embryos. The herbicide is a synthetic analogue of the plant hormones called auxins, which controls everything from the plant's response to light and gravity to the ripening of fruit. Auxins might also be important in natural somatic embryogenesis.

why DNA has thymine in place of uracil: It seems the original DNA genomes contained uracil. Spontaneous deamination of cytosine to uracil takes place. Harmful influences of these mutations are prevented by the enzyme uracil-DNA glycosylase, which excises uracil (but not thymine) from DNA. Replacement of uracil by thymine transpired only once after the initial conversion of RNA to DNA genomes. Thus, the increasingly efficient repair enzymes could recognize the hydrolysis products of cytosine in DNA.

Wi-Fi: A family of wireless networking technologies, based on the IEEE 802.11 family of standards, which are commonly utilized for local area networking of devices and Internet access. Wi-Fi is a trademark of the non-profit Wi-Fi Alliance, which limits the use of the term Wi-Fi Certified to products that successfully complete interoperability certification testing.

wide hybridization: Interspecific and intergeneric hybridization is a first step to introducing alien variation and to transferring wanted genes from wild species to cultivated species. Barriers to wide hybridization are incompatibility between the parental species, the viability of the F_1 hybrids and sterility of the F_1 hybrids or their progeny.

wild species: Related species of crop plants that are not grown commercially. Crop breeding has resulted in the development and release of improved varieties in various crops which are being grown commercially by the farmers. Wild species of these crops carry genes for resistance to various biotic and abiotic stresses. Therefore, in breeding such genes are transferred from wild species into cultivated varieties through backcross breeding. For example, rice (*Oryza sativa* L.) has more than 20 wild species, such as *Oryza nivara, Oryza barthii, Oryza longistaminata* and *Oryza rufipogon*.

wild type: The genotype or phenotype, out of a set of genotypes or phenotypes of a species that is largely found in nature (i.e. normal; unmutated, non-transformed) organism. During mutation breeding of crop plants, the wild type is the normal variety used for mutagenesis to develop mutants. In the case of plant transformation, the original variety (non-transformed) is called the wild type, which serves as a control to compare the transgenic variety developed from this original variety. The expression is mainly utilized to differentiate rare mutant forms of a species from the lab stock of normal individuals. *Contrasts* MUTANT *and* VARIANT.

wild-type isoalleles: If the phenotype produced by different alleles is wild type, these alleles are called wild-type isoalleles. *See* NORMAL ISOALLELES.

wilt: The drooping of leaves and stems resulting from lack of water (i.e. inadequate water supply or excessive transpiration). It is a vascular disease that disturbs normal water uptake.

wilting coefficient: The level of soil moisture at which water becomes unavailable to plants and causes permanent wilting ensures.

win QTL cartographer (winQTLCart): Windows QTL Cart maps quantitative trait loci (QTL) in cross populations from inbred lines. WinQTLCart comprises a powerful graphic tool for presenting and summarizing mapping results and can import and export data in a variety of formats.

windbreak: A row or other grouping of trees or shrubs used to offer protection against the impacts of high-velocity winds.

winter-type: Concerning growth habit, the term refers to plants germinating in autumn and requiring vernalization during the wintertime for flower initiation during the following season.

WIPONET: A global intellectual Property information network is being developed by WIPO. It will encourage international cooperation by facilitating the digital exchange of intellectual property information between intellectual property offices of its member States. WIPONET is of particular significance to developing countries.

wobble base pairing: The pairing of mRNA codon with tRNA anticodon in which the first two bases of a codon obey the standard G-C and A-U base-pairing convention and the third base has abnormal base pairing. *Aka* WOBBLE HYPOTHESIS.

wobble modification: Antibody production is a well-regulated process. S. Giguère *et al.* (2024) reveal that TRANSFER RNA (tRNA) and CODON USAGE are congruently optimized as B cells differentiate from a naïve phenotype to plasma cells. The tRNA pool in plasma cells is enriched in inosine (I^{34})-modifiable anticodons, which enables one tRNA to decode multiple synonymous codons by 'wobble' (non–Watson-Crick) base-pairing. Codon usage and cell-specific tRNA pools may serve as an attractive future target for both the exogenous production of pharmaceutical antibodies and the rational design of vaccines. This codon optimization is accompanied by an increase in tRNAs with complementary modifications, which enhances antibody biosynthesis by ANTIBODY-SECRETING CELLS (ASCs). R.A. Alvarez and L.K. James (2024) suggest wobble modification as a mechanism for DECODING THE LANGUAGE OF IMMUNITY.

wobble rules: There are 32 tRNAs (including one for initiation) in all. They can complement all 61 sense codons. The genetic code is degenerate as a given amino acid may have more than one codon. The lesser importance of the third position in the genetic code ties in with two truths about tRNAs. First, although there would appear to be a demand for 62 tRNAs, there are in reality only about 32 different tRNAs in *Escherichia coli* cells. Second, a rare base such as inosine can occur in the anticodon, usually in the position that is complementary to the third position of the codon. These two facts resulted in the concept that some kind of conservation of tRNAs is taking place and that rare bases may be involved. Since the first position of the anticodon (5') is not as constrained as the other two positions, a given base at that position may pair with any of the several bases in the third position of the anticodon. F.H.C. Crick (1966) characterized this ability as a wobble. The genetic code dictionary shows the possible pairings that would produce a tRNA system compatible with the known code. For example, if an isoleucine tRNA has the anticodon 3'-UAI-5', it is compatible with three codons for the amino acid: 5'-AUU-3', 5'-AUC-3' and 5'-AUA-3'. This means that inosine is the first (5') position of the anticodon can recognize U, C or A in the third (3') position of the anticodon, and thus one tRNA complements all three codons for isoleucine. *Aka* WOBBLE HYPOTHESIS.

wood: The secondary xylem formed by the vascular cambium in stems and branches of woody plants.

woody: Hard and lignified; not herbaceous.

working collection: An assemblage of germplasm saved under short-term storage conditions, commonly exploited by breeders or researchers.

working copy: A gene that functions the way it is intended to.

world collection: A global collection of samples of a species or genera. It is a synchronized act of several countries and institutions under the International Plant Genetic Resources Institute (IPGRI).

World Intellectual Property Organization (WIPO): An international organization dedicated to assisting to guarantee that the rights of creators and owners of intellectual property are safeguarded worldwide and that inventors and authors are, thus, recognized and rewarded for their ingenuity. This international protection acts as an offshoot of human creativity, pushing forward the boundaries of science and technology and enriching the world of literature and the arts.

wound: An injury to plant tissues that oftentimes breaches barriers (i.e. cuticle, bark, cell walls) that might otherwise eliminate pathogens. Some pathogens (e.g. viruses) can enter plants only through a wound. A wound may happen from natural growth processes, physical and chemical agents, animals (especially insects), and many human agricultural activities (e.g. pruning).

Wright's effect: Nonadaptive differentiation into different types due to 'drift' followed by random fixation in small, isolated populations.

Wright's inbreeding coefficient: The proportion of loci at which an individual organism is homozygous. *See* INBREEDING COEFFICIENT.

Wright-Castle estimator: A method for approximating the number of genes that affect a quantitative trait. It is based on comparability between the variance in an F_2 population and the difference in means between the parent populations.

Wright-Fisher model: A normal model of random genetic drift, in which each gene is drawn at random from 2N genes in the previous generation.

writers: Enzymes that add epigenetic marks to DNA, RNA and histones.

writhing number: The number of times a duplex axis crosses over itself in space.

WRKY transcription factors: These TFs bind to W-box *cis*-elements and activate or repress transcription of immune-responsive genes. They carry a 60 amino acid domain containing WRKY domain, named after the almost invariant amino acid sequence, Tryptophan (W)-Arginine (R)-Lysine (K)-Tyrosine (Y), at the N-terminus. These TFs are one of the largest families of transcriptional regulators in plants with 72 representatives in *Arabidopsis*, and more than 100 members in rice, soybean or popular.

WW domain: A protein domain that binds to proline-rich regions.

X

x: Basic chromosome number.

X chromosome: A sex chromosome that normally occurs as paired (XX homogametic sex) in each female cell and as single (i.e. unpaired) in each male cell in those species in which the male characteristically has two dissimilar sex chromosomes (XY in heterogametic sex) (e.g. *Drosophila*, humans). *See* Y CHROMOSOME and SEX CHROMOSOME.

X chromosome counting: The process by which a cell governs how many, if any, X chromosomes are required to be inactivated to attain a ratio of one active X chromosome per diploid set of autosomes. When a cell inactivates too many or too few X chromosomes, this characteristically reflects a fault in the counting mechanism. However, failure to maintain the correct active X chromosome to autosome ratio might take place due to glitches at other stages in XCI and might not necessarily signify a malfunction in the counting mechanisms.

X chromosome inactivation (XCI): A mechanism by which cells with two X chromosomes inactivate one to produce the same number of active X chromosomes as male XY cells. In mice, two forms of XCI occur: IMPRINTED INACTIVATION and RANDOM INACTIVATION. Imprinted XCI inactivates the PATERNAL X CHROMOSOME (Xp) in all cells. Random XCI takes place in embryonic cells after the Xp chromosome is reactivated, inactivates either the paternal or the maternal X chromosome with equal frequency. XCI is accomplished through the formation of heterochromatin, and the highly clumped and condensed inactive X chromosome is termed a BARR BODY.

X chromosome inactivation (XCI) choice: The procedure a cell uses to ascertain which X chromosomes are to be inactivated. Failed or skewed counting refers to the phenomenon when one X chromosome is inactivated more frequently than another. Skewing can be a consequence of failures that take place prior to XCI, and, as such, it does not necessarily symbolize malfunctions in the choice mechanisms. Choice can only be observed if counting is successful and not bypassed by any other mechanism. *Aka* X CHROMOSOME CHOICE.

X chromosome inactivation (XCI) initiation: The first steps in XCI include counting, choice and the initial stages of reduction and prevention of the transcription of genes on the X chromosome.

X chromosome-pairing region (Xpr): An X-linked region that is positioned 200-kb upstream from *Xist* and is implicated in X chromosome pairing prior to XCI.

X- and Y-linked inheritance: When a trait is controlled by genes that are located in the pairing region of the X and Y chromosomes, such genes are also called PSEUDOAUTOSOMAL GENES, and the type of inheritance shown is called PSEUDOAUTOSOMAL INHERITANCE. In *Drosophila*, the *bobbed* (*b*) gene, which produces shorter and slender bristles gene, is an example. In the F_1 generation, the results in the reciprocal crosses are the same. X- and Y-linkages are revealed only in the F_2 generation.

X inactivation: The genetic inactivation of all X chromosomes in excess of one takes place on a random basis in each cell at an early stage of embryogenesis in mammals.

X inactivation centre (XIC): A part of the X chromosome is thought to contain the elements needed for counting, choice and the initiation of inactivation. X CHROMOSOME INACTIVATION (XCI) possibly begins at the XIC and spreads to the rest of the chromosome. Inactivation was initially proposed to spread from either a single locus or multiple loci, but later studies identified a single XIC within a 680-kb to 1.2-Mb region. Subsequent experiments revealed that a 450-kb portion of the XIC, and an even smaller 80-kb sequence that contained Xist, Tsix and Xite, could function as XICs in multicopy arrays. The transgenes used in these studies, however, might not be in a position to determine all XCI initiation elements that exist in the endogenous XIC. Methylated CpG islands suggest an important role for late DNA replication in propagating X-inactivation from the cell to its progeny. X-inactivation is an important research tool for identifying the factors that programme and maintain developmental processes. Enough evidence is available for the existence and location of the XIC in the mouse X chromosome. It is located somewhere between T16H and HD3.

***X inactive specific transcript* (*Xist*) gene:** The gene responsible for X-inactivation. *Xist* produces a long RNA transcript that accurately sticks along one of every pair of X-chromosomes, switching most genes off.

X inactive specific transcript (Xist) RNA: A noncoding RNA is produced from the *Xist* gene located on the X chromosome and is upregulated on the future INACTIVE X (XI) CHROMOSOME when random X chromosome inactivation (XCI) is initiated. Once upregulated, Xist coats the entire Xi chromosome.

***X-inactivation intergenic transcription element* (*Xite*) gene:** The *Xite* gene codes for Xite RNA, which docks intergenic transcription start sites and DNase I hypersensitive sites with allelic differences. At the onset of XCI, deleting Xite downregulates Tsix in *cis* and skews XCI ratios, suggesting that *Xite* promotes Tsix promotes persistence on the active X chromosome.

X-inactivation intergenic transcription element (Xite) RNA: A noncoding RNA coded by a gene that contains an enhancer for Tsix that continues its expression during differentiation.

X-linked disease: A genetic disease caused by an allele at a locus on the X chromosome.

X-linked dominant: An inheritance pattern for which the causative gene of the given disorder is located on the X chromosome. The phenotype can be expressed by both males and females, and affected males tend to have more severe symptoms than affected females.

X-linked genes: Genes located on the X chromosome are said to be X-linked. X-linked genes have distinctive inheritance patterns because they are present in different numbers in females (XX) and males (XY). *See* SEX-LINKED GENES.

X-linked recessive: An inheritance pattern for which the causative gene of the given disorder is located on the X chromosome. As males have got only one X chromosome (hemizygous, XY), a single gene change on the X chromosome is enough to cause the disorder or trait. Since women have two X chromosomes, it is uncommon for women to have X-linked disorders, though not impossible. Carrier females may express mild symptoms of the trait due to differences in X chromosome inactivation. Examples of such diseases are haemophilia and Duchenne muscular dystrophy.

X-linked traits: Traits genes that are located in the X chromosome. Such traits display differences in reciprocal crosses, giving different phenotypic ratios. Males that carry (inherit) X-linked recessive character due to a hemizygous condition, because the Y chromosome does not contain a counterpart allele. Males are most often affected by X-LINKED DISORDERS. Affected males transmit the X-linked gene to all their daughters but none to their sons. Some examples are white eyes in *Drosophila*, yellow bodies in *Drosophila*, haemophilia, Duchenne muscular dystrophy and colour-blindness in men

X-ray crystallography: An X-ray diffraction technique is used to determine the shape, size and three-dimensional structure of a crystalline compound. There are several *Bt* genes, like *Cry 1AC*, *Cry 1Ab*, *Cry 9C*and *Cry 2A*. The word 'cry' means crystal. Different strains of *Bacillus thuringiensis* produce different types of crystal proteins. The shape and size of the crystal protein help distinguishing one strain from the other.

X-ray diffraction analysis: A method for determining the structure of a molecule by analysing the angles at which X-rays are deflected by a crystal of the compound.

X-ray diffraction pattern: The pattern of spots on an X-ray sensitive film that is produced by passing X-rays through a crystal.

X-rays: Electromagnetic radiation with wavelengths in the range of approximately 0.1–10 nm, between UV radiation and γ-rays and capable of penetrating solids.

x, 2x, 3x, etc.: Symbols for the number of genomes in the gametic nucleus of species in a genus with a polyploid series.

X/A ratio: The ratio between the number of X chromosomes (X) and the number of sets of autosomes (A) that determines sex in *Drosophila*. For example, X/A 1.0 is female; 0.5 is male; 1.33 is metafemale; 0.33 is metamale.

X0: A female genotype with only one X chromosome. X0 mice are viable, fertile and otherwise normal, whereas X0 humans are infertile and develop other health problems related to this genotype.

X_1, X_2, X_3, etc.: Symbols denoting first, second, third, etc. generations from an irradiated ancestral plant (X_0).

xanthine oxidase: An enzyme accountable for the synthesis of free radicals in the body.

xanthophyll: A yellow-coloured, oxygen-containing carotenoid ($C_{40}H_{56}O_2$), photosynthetic pigment found in chloroplasts (e.g. the lutein of leaves). Xanthophylls do not need light for synthesis, so xanthophylls are present in all young and etiolated leaves. Xanthophylls in leaves function as accessory pigments, capturing certain wavelengths of sunlight not absorbed by chlorophylls and thereby increasing the overall absorptance of the visible spectrum of sunlight.

xenia effect: A condition in which the genotype of the pollen influences the developing embryo of the maternal tissue (endosperm) of the fruit to produce a visible effect on the seed.

xenobiotic compounds: The compounds (veterinary drugs, agrochemical herbicides, etc.) that are fashioned for use in an ecosystem comprising more than one species. For example, herbicides supposed to kill weeds leave commercial crops undamaged, or veterinary drugs that are supposed to kill parasitic worms leave the host livestock unharmed.

xenobiotics: Chemical compounds with a structure or substituent on their structure that is not found, is not produced by and habitually cannot be destroyed by living organisms.

xenogamy: The transfer of pollen grains from the anther of one flower to the stigma of another flower of a different plant. *Aka* CROSS-POLLINATION. *Compare with* GEITONOGAMY.

xeromorph: A plant with structural properties (e.g. hard or succulent leaves) or functional adaptations that avert water loss by evaporation. Xeromorphs are generally associated with arid habitats but are not necessarily drought-tolerant. *Compare with* XEROPHYTE.

xerophyte: A plant normally living in a dry habitat, characteristically displaying xeromorphic or succulent adaptation. Xeromorphs can tolerate long periods of drought. *Compare with* XEROMORPH.

X-inactivation: (i) A dosage compensation mechanism in which one of two X chromosomes in the somatic cells of female mammals is epigenetically silenced. (ii) The genetic inactivation of all X chromosomes in excess of one, occurs on a random basis in each cell at an early stage of embryogenesis.

Xis protein: A protein required for excisive site-specific recombination between phages and *E. coli* between prophage *attL* and *attR* sites. This protein is also needed for the formation of a localized nucleoprotein structure on *attR* DNA. *Aka* EXCISIVE PROTEIN.

Xist RNA: X inactive specific transcript; the noncoding RNA transcribed from the X inactivation centre (Xic) that binds along the entire chromosome from which it is transcribed to mediate X chromosome inactivation in placental mammals.

***Xma*I:** A type II restriction enzyme from the bacterium *Xanthomonas malvaceous* that recognizes the DNA sequence 5′-C↓C C G G G / 3′-G G G C C↑C, where arrows indicate sites of cleavage.

XO condition: Having an X chromosome but no Y chromosome, as the female in poultry and some other birds.

XO system of sex determination: C.E. McClung observed that somatic cells in female grasshoppers carry a different number of chromosomes than do the corresponding cells in the male. He called the extra chromosome in the female an X BODY. E.B. Wilson in 1909 reported different numbers of chromosomes in diploid cells of male and female proteron; males carried 13, and females carried 14 chromosomes in unreduced cells. Males form two types of gametes (HETEROGAMETIC SEX). Some gametes carried 6 chromosomes, while others carried 7. All the eggs carried seven chromosomes (HOMOGAMETIC SEX). Eggs fertilized with sperm carrying seven chromosomes produced females. It was thus hypothesized that females carried 12 autosomes and 2 X chromosomes, whereas males carried 12 autosomes and 1 X chromosome. In domestic fowl, fume (a moth), and butterflies, the female is a HETEROGAMETIC SEX (XO), and the male is a HOMOGAMETIC SEX (XX).

XXY syndrome: Klinefelter syndrome, 47 chromosomes, chromatin positive.

XY type system of sex determination: The XY system of sex determination is found in higher animals and some plants (e.g. *Melandrium album*). In humans, the presence of the Y chromosome determines maleness. The number of chromosomes in the somatic cells of males and females is the same. Males carry autosomes and one X and one Y chromosome (heterogametic sex).

xylem: In vascular plants, xylem and phloem form continuous tubes that carry water, nutrients and other substances throughout the plant's roots, stem and leaves. The evaporation of water at the leaf's stomata (transpiration) creates negative hydrostatic pressure in the xylem that transports water and dissolved minerals from the roots upward to the stem and leaves. The xylem contains several types of cells, including tracheids, vessel elements, parenchyma and fibres. Tracheids and vessel elements are thick-walled cells that are dead at maturity, and they form side by side, connecting together to form tubes.

Y

Y chromosome: A sex chromosome other than the X chromosome in heterogametic sex in the XY system of sex-determination (e.g. males of *Drosophila* and humans). The Y chromosome is not embodied in the homogametic sex and only once in the heterogametic sex. In mammals, Y chromosome prompts maleness.

Y chromosome shredding: An innovative gene-drive approach for converting XY males into fertile XO females, which is termed Y chromosome deletion, has been established by T.A. Prowse *et al.* (2019) using ORTHOGONAL PROGRAMMABLE ENDONUCLEASES (Y-CHOPE). This methodology combines a programmable endonuclease that 'shreds' the Y chromosome. Y chromosomes can be eradicated with very high efficacy from embryonic stem cells by CRISPR/Cas12a system. A Y-shredding drive system leads to a diminution of the pool of XY males that could impact population eradication through limitation in mating.

Y RNA: Y RNAs are an emergent class of highly conserved, small ncRNAs. Y RNAs and their fragments are not just random degradation products but are themselves bioactive molecules. Most of the reported roles of Y RNAs involve nuclear functions. Some of the Y RNA fragments have been annotated as a new kind of microRNA. Accumulating evidence is available to suggest functional involvement of Y RNAs in diseases such as cancer and immune-related pathologies.

Y-junction: The point of active DNA replication where the double helix opens up so that each strand can synchronize to serve as a template for replication. *Aka* REPLICATION FORK.

Y-ligation-based block shuffling (YLBS): YLBS is a universal method that permits modules of variable size to be integrated into sequences through repeated cycles of ligation of sequence blocks with a stem and two branches, formed by two unique single-stranded DNA molecules.

Y-linked genes: It refers to genes located in the Y chromosome. *See* HOLANDRIC GENES.

Y-linked traits: The traits controlled by genes located in the Y chromosomes (e.g. hairy pinna in man). *Aka* HOLANDRIC TRAITS.

Yates's correction: A correction pragmatic in the calculation of normal deviates or chi-squares (χ^2's) to permit for the discrepancy arising by the observations being discontinuous while tables of the normal deviate and χ^2 are calculated on the supposition of continuity in the variate. In chi-square analysis, the reduction of ½ from the absolute value of (observed – expected) deviation, when the expected number of either of the classes is small (usually less than 5), is termed Yates correction. This correction factor offers the accuracy of chi-square determinations. *See the* CHI-SQUARE TEST.

yeast artificial chromosome (YAC): YAC is a genetically engineered chromosome produced from the DNA of the yeast, *Saccharomyces cerevisiae*, which is then incorporated into a bacterial plasmid. By inserting large fragments of DNA, the inserted sequences can be cloned and physically mapped using the process of CHROMOSOME WALKING. A YAC consists of six components: (i) telomeres (TEL), which are ends of chromosomes involved in the replication and stability of linear DNA; (ii) origin of replication (*ori*) sequences necessary for the replication in yeast cells; (iii) a yeast centromere (CEN), which is a specialized chromosomal region where spindle fibres attach during mitosis; (iv) selectable markers for identification in yeast cells; (v) ampicillin resistance (*Amp*^{r}) gene for selective amplification and (vi) recognition sites for restriction enzymes. YAC can carry DNA inserts from 100 kbp to 1000 kbp. *See* CLONING VECTOR *and* COSMID.

yeast artificial chromosome (YAC) clone: YAC cloning systems allow the cloning of DNA stretches of 50 to well over 2000 kb.

yeast artificial chromosome (YAC) library: YAC libraries are engaged for cloning extremely large DNA fragments (of more than 1 Mb) and are valuable for cloning large genes (such as the 250-kb cystic fibrosis gene) and for creating libraries of large overlapping clones, such as for individual chromosomes isolated from organisms (chromosomal libraries). These have been employed extensively for mapping genomes of complex organisms.

yeast artificial mitochondria (YAM): Mouse mitochondrial genomes were cloned and stably preserved in *Escherichia coli* at low copy number. One full and three deleted mouse mitochondrial genome clones carrying a selection marker, γ-ori and yeast *COX2* gene bordered by duplicated sequences have been produced. These mouse mitochondrial genome clones were stably sustained as circular monomers being transformed into yeast mitochondria. The exogenous sequences used for cloning and screening were detached by homologous recombination in yeast mitochondria. YAM carried only engineered mouse mitochondrial genomes in their mitochondria.

yeast autonomously replicating sequences (yARS): In yeast, yARS have core consensus sequence 5′-(A/T)TTTAT(A/G)TTT(A/T)-3′. The functioning of ARS requires additional near-matching sequences in the 3′ bordering region. ARS contain two essential domains: one 11-bp core consensus and the other broad A+T-rich 3′ flanking sequence. ARS elements can enable effective circular plasmid DNA replication.

yeast centromere plasmid (YCp): Plasmid which carries functional yeast centromere. Most of the plasmids carrying ARS were unstable in yeast. One was stable as it behaved like a MINICHROMOSOME. Centromeres from chromosomes IV and XI of yeast conferred on plasmids similar properties to centromere of Chromosome III. The isolation of plasmids bearing centromere functional in yeast permits detailed molecular analysis of events occurring at mitosis and meiosis.

yeast centromeres: Yeast centromeres include two highly conserved nucleotide sequences each only 10–15 bp long, separated by around 90 bp of AT-rich DNA. Short stretch of 120 nucleotides is adequate to specify a centromere. Centromeric sequences of different yeast chromosomes are closely related. The centromeres of higher eukaryotic organisms are much larger and more complex than those of yeast.

yeast chromosomal proteins: Yeast chromosomal proteins in many ways resemble those of higher eukaryotes. Histone H1 is apparently absent in yeast chromosomes.

yeast cloning vectors: The yeasts, and especially *Saccharomyces cerevisiae*, are preferred organisms in which to clone and express DNA. They are eukaryotes and so can splice out introns, the noncoding sequences in between the coding sequences of many eukaryotic genes.

yeast episomal plasmid (YEp): Many strains of yeast contain a 2 μ (6.3 kb) long plasmid that has no known function. It replicates under nuclear control. Two plasmids pJDB219 and pJDB248 were recovered from yeast cells. YEp is a cloning vehicle engaged for introduction of genes and pieces of genetic material into certain yeast strains at high copy number. YEp can replicate in both *Escherichia coli* and certain yeast strains.

yeast extract: The concentrates of the soluble fraction of yeast and more particularly *Saccharomyces cerevisiae*. As food ingredients, yeast extracts are described for their nutritional composition and flavouring properties. They are produced principally from autolysis.

yeast genome: *Saccharomyces cerevisiae* is a haploid eukaryote having 12,068 kb DNA consisting of 6,340 genes, including 5,885 potential protein-encoding genes, 140 rRNA genes, 40snRNA genes and 275 tRNA genes.

yeast integrative plasmid (YIp): A yeast vector that trusts on its integration into a host chromosome for replication.

yeast mating type switching: A notable attribute in yeast is the ability of some strains (which carry dominant *HO* allele and not those which carry recessive *HO*allele) to switch their mating type from 'a' to 'α' and vice versa. As a result, an *HO* haploid strain, irrespective of its original mating type, will soon possess cells of both mating types giving diploids (*MATa*/*MATα*). This is achieved by the presence of a *MAT* (*a* or *α*) locus with an active cassette (type *a* or α) associated with two silent cassettes at loci *HML(α)* on the left and *HMR(a)* on the right. The switching is affected by the activity of an endonuclease coded by the dominant gene *HO*.

yeast model species: Eukaryote, single-celled (unicellular) fungi that reproduce by fission or budding to produce two daughter cells of equal sizes. Fission yeast *Schizosaccharomyces pombe* and budding yeast *Saccharomyces cerevisiae* are genetically best-known species. Yeast is an important eukaryotic microorganism serving as model organism for molecular genetics/epigenetics research (i.e. especially cell cycle research).

yeast protein RAP1: The yeast protein RAP1 was earlier delineated as a transcriptional regulator that binds *in vitro* to sequences occurring in a number of apparently unrelated genomic loci. These sequences comprise the silencers at the transcriptionally repressed mating-type genes, the promoters of many genes important for cell growth and the poly[(cytosine)1-3adenine] repeats of telomeres. RAP1 may be engaged in telomere formation *in vivo*.

yeast replicating plasmids (YRp): These plasmids contain autonomous replicating sequences (ARS) derived from yeast chromosome. These vectors can replicate independently of the yeast chromosome; however, they tend to be unstable and may be lost during budding.

yeast transcription activator GCN4: 'bZIP', a DNA-binding protein that includes a basic region that contacts DNA and an adjacent 'leucine zipper' that stimulates protein dimerization. A peptide model for basic regions of yeast transcription activator GCN4 has been developed in which leucine zipper has been replaced by disulfide bond.

yeast transcription regulator: Yeast transcription regulator executes two functions: (i) DNA binding and (ii) gene activation. These two functions of a yeast transcription regulator can be separated. The yeast GAL4 regulator consists of two domains – the AMINO DOMAIN recognizes sequences in the GALACTOSE UPSTREAM ACTIVATING SEQUENCE (UAS_G) and the CARBOXYL DOMAIN, when bound to DNA through the amino domain, contacts another protein to mediate positive control.

yeast: A fungus of the family Saccharomycetaceae that man uses specifically in the making of alcoholic liquors and as a leavening (fermenting) agent in bread making. Some strains of yeast cells are also commonly engaged in bioprocesses, because they are relatively simple to engineer genetically (via RECOMBINANT DNA) and relatively easy to propagate (via FERMENTATION) to yield desired products (e.g. proteins). Yeast is common contaminant in plant tissue culture. Yeast can occur in the form of single cells, reproducing asexually by fission or by budding. Some yeast species are pathogenic. Examples of such yeast species that result in human disease are *Candida albicans*, *Candida tropicalis*, *Candida glabrata*, *Candida parapsilosis* and *Cryptococcus neoformans*. *See* LEAVENING AGENT.

yeast two-hybrid (Y2H) system: Yeast two-hybrid is an *in vivo* system for protein-protein interaction studies. Protein-protein interactions are studied by direct DNA readout or yeast two-hybrid system. Y2H system was initially developed by S. Fields and O. Song in 1989. A transcription factor generally comprises a DNA-BINDING DOMAIN (DBD) and an ACTIVATION DOMAIN (AD), the former helps in DNA binding and the later facilitates the activation of a gene lying downstream. It uses two fusion proteins to activate the transcription of a reporter gene in yeast. If one wants to study interaction between two proteins X and Y, two hybrid gene constructs are prepared, one having a hybrid gene coding for fusion protein DBD + X along with a reporter gene and the other coding for the fusion protein AD + Y. Yeast cells are cotransformed with both hybrid constructs, so that X and Y interact physically. This will bring DBD and AD in close proximity. Specific interaction between the two chimeric proteins results in transcriptional activation of the reporter genes which is easily detected by either colour-based assay or by auxotrophic complementation. Thus, Y2H system is used for investigation of protein-protein interactions using genetically engineered yeast cells.

yeast vectors: All yeast vectors can replicate in *Escherichia coli*, often at high copy number and all use markers *Leu2*$^+$, *His*$^+$, *Ura3*$^+$ and *Trp1*$^+$ that can be selected readily in yeast. Further, all of them contain unique target sites for a number of restriction endonucleases.

yield: The quantity of an agricultural crop (e.g. a grain, fruit or vegetable) yielded in a season. It can be measured in pounds or bushels per acre, or kilograms or metric tons per hectare.

yield drag: A negative effect on grain yield connected with crop plants that possess a specific gene or reveal a specific trait.

yield lag: A proportionate decrease in yield observed in some hybrids or varieties compared with that observed in the most recently produced hybrids or varieties.

yield traits: Crops for feed or food are based on plants that possess a certain set of characteristics or traits that can include, for example, how much seed or biomass yield they will produce under typical field conditions. Historically, crop yield is determined by traits such as grain/seed size, grain/seed number, number of tillers/branches, fruit yield, harvest index among others in different crops that will bring step-change improvements to crop yield by enabling crops to fix and seize more carbon and then target that additional carbon to the seed.

yolk: Nutritional reserves rich in lipids, proteins and polysaccharides, occurring in the eggs of many animals whose primary role is to supply food for the development of the embryo.

Young's modulus: A modulus relating tensile (or compressive) stress to strain.

Z

z: The natural logarithm of the ratio of two estimated standard deviations.

Z chromosome: Used to denote the Y chromosome, where the female is the heterogametic sex.

Z DNA: *See* DNA, Z-FORM.

Z RNA: Z-RNA is a left-handed alternative conformation for the RNA double helix. Just like Z-DNA, Z-RNA is preferred by a sequence composed of purine/pyrimidine repeats and especially CG repeats.

z-score: A measurement of the distance between the mean and data point of a variable. It is measured in standard deviation units. More technically, it is an amount of how many standard deviations below or above the population mean a raw score is located. A z-score can be placed on a normal distribution curve. Z-scores range from –3 standard deviations (which would fall to the far left of the normal distribution curve) up to +3 standard deviations (which would fall to the far right of the normal distribution curve). *Aka* STANDARD SCORE.

z-test: A test of significance which is used to compare two means when the sample size is large (more than 30). To use a z-test, you need to know the differences in variances and have a large sample size. The test statistic is assumed to have a normal distribution.

zeaxanthin: A carotenoid (i.e. 'light-harvesting' compound used in photosynthesis) that is naturally synthesized in Brussels sprouts, summer squash, maize, avocado, green beans and dark green leafy vegetables. It is a phytochemical/nutraceutical whose intake by humans has been shown to lower the risk of age-related macular degeneration, a leading reason for blindness in elderly people.

zebra chromosome: A chromosome so named because it has a striped genomic *in situ* hybridization pattern.

zein: A storage prolamin and alcohol-soluble protein of maize occurring in the endosperm. It is composed of ten consecutive helical segments that are arranged over each other in an antiparallel fashion. The van der Waals forces and intramolecular hydrogen bonding interactions seem to stabilize the molecular structure of zein. It is made of two major proteins, with a molecular weight of 19,000 and 21,000 Da.

zero selection: When selection intensity is the highest and a proportion of the population rejected 100% (vice versa in the case of no selection; i.e. lowest selection and no rejection at all). In this case, the entire population is carried forward.

zero tillage: A method to recover soil betterment where the new crop is planted in the chaff of the previous crop with less soil disturbance than even with minimum cultivation.

zero time-binding DNA: DNA enters the duplex form at the start of a reassociation reaction. This is the consequence of intramolecular reassociation of inverted repeats.

zero-mode waveguide structures (ZMWs): Sub-wavelength optical nanostructures, 50–200 nm in size, that direct light energy into a volume that is small in all dimensions compared to the wavelength of the light. ZMWs are utilized as instruments for focal volume quarantine.

zetta: Zetta means 10^{21}, which refers to the typical number of distinct designed parts in a product, made by the systems we visualize (molecular, mature or molecular-manufacturing based nanotechnology). The term ZETTA TECHNOLOGY refers to the implemented technology and its products, rather than to intermediate steps on the pathway.

zinc-finger nickases (ZFNickases): Zinc-Finger nucleases that contain inactivating mutations in one of the two *Fok*I cleavage domains. ZFNickases make only single-strand DNA breaks and induce HOMOLOGY-DEPENDENT RECOMBINATION (HDR) without activating the mutagenic NONHOMOLOGOUS END-JOINING (NHEJ) pathway.

***ZFY* gene:** ZFY symbolizes the zinc finger on the Y chromosome. Originally, it was believed to be the human male sex-switch gene, located in the short arm of the Y chromosome.

Zika virus: First isolated in the Zika Forest of Uganda in 1947, the virus is transmitted between humans and animals by mosquitoes. In most cases, it causes a very mild infection with no long-term harm. But it can be dangerous in pregnancy as it can lead to birth defects, in particular causing babies to have abnormally small heads (microcephaly).

zinc fertilizers: Compounds of zinc in various formulations applied to supply zinc as crop nutrition, including zinc sulfates $ZnSO_4 \cdot 7H_2O$, (21% Zn) and $ZnSO_4 \cdot H_2O$ (33% Zn), Zn-EDTA chelate (12% Zn), zinc oxide ZnO (78% Zn), zinc frits (variable Zn content) and natural Zn chelates.

zinc finger (ZF): Crystallographic structure (PDB 1R4O) of a dimer of the zinc finger containing DNA-BINDING DOMAIN (DBD) of the glucocorticoid receptor bound to DNA. The zinc-finger domain is generally between 23 and 28 amino acids long and is stabilized by coordinating Zinc ions with regularly spaced zinc-coordinating residues (either histidines or cysteines). The most common class of zinc finger (Cys_2His_2) coordinates a single zinc ion and consists of a recognition helix and a 2-strand beta-sheet. In transcription factors, these domains often occur in arrays (usually separated by short linker sequences), and adjacent fingers are spaced at 3-bp intervals when bound to DNA. During plant transformation, these zinc fingers serve to activate transcription factors governing the expression of gene(s) related to drought tolerance.

zinc-finger motif: A finger-like loop of peptides enclosing a bound zinc ion at one end, characteristically part of a larger protein, called ZINC-FINGER PROTEIN, which plays a crucial role in regulating the gene transcription.

zinc-finger nucleases (ZFNs): Programmable nucleases comprising the DNA-binding domain of a zinc-finger protein and nonspecific DNA-cleaving nuclease domain of the *Fok*I restriction endonuclease. ZFN dimers produce targeted DNA DOUBLE-STRANDED BREAKS (DSBs) that excite DNA damage response pathways. The binding specificity of the designed zinc-finger domain guides the ZFN to a specific genomic site. ZFNs are used for the development of tools for genome and epigenome editing.

zinc-finger protein (ZFP): A protein containing a DNA-binding domain with a finger-like overhang that is characterized by the coordination of zinc ion(s) to stabilize its fold. There is collinearity between the protein sequence of the zinc finger and its target DNA sequence, with each finger mainly recognizing three base pairs. ZFP acts to either repress or activate transcription (i.e. of the gene the 'finger' touches within a DNA molecule). Thus, they could potentially be engaged in functional genomics to investigate the specific function of a given gene.

zipper trichomes: Self-pollinating plants are treasured for agriculture because favourable traits can easily be transmitted to offspring. Domesticated tomatoes accomplish this structurally by establishing a cone around the male anthers, which guarantees that pollen can effortlessly reach the stigma. The anther cone is seized and organized by a thick web of hairs, called zipper trichomes. M. Wu *et al.* (2024) recognize three HOMEODOMAIN-LEUCINE ZIPPER IV (HD-Zip IV) genes that co-ordinately endorse the establishment of interlocking trichomes at the anther edge to hold together the neighbouring anthers, thus producing a closed anther cone and cleistogamous flower morphology necessary for stringent self-pollination. These HD-Zip IV genes also determine style length by controlling the transition from cell division to endoreduplication. The expression of these HD-Zip IV genes and their downstream gene, *Style 2.1*, was consecutively modified to outline the cleistogamous morphology during tomato evolution and domestication. These observations offer insights into the molecular basis of cleistogamy in modern tomato and determine goals for enhancing fruit set and preventing pollen contamination in genetically modified crops.

zombie seeds: Another term given by environmentalists for terminator seeds.

zona pellucida: This is the site where a developing embryo is located before implantation. To achieve implantation, the embryo must first hatch so that it can establish direct membrane-to-membrane contact with cells in the uterine wall. Embryonic and fetal development within a uterus is a fundamental characteristic of all mammals except the primitive egg-laying platypus.

zonate: (i) Occurrence of light and dark circular bands or rings, characteristically on leaves or flowers. (ii) Relating to a target-like development of a tree canker, characterized by successive, perennial rings of callus, referring to any indication appearing in concentric rings.

zone electrophoresis: A kind of electrophoresis in which the mobile substance passes through a support medium such as paper, cellulose, acetate, starch gel, silica gel, polyurethane foam or polyacrylamide polymers. The sample is applied to the support medium in the form of a narrow band approximately midway between the buffer reservoirs housing the electrodes. During electrophoresis, each component migrates towards the cathode or anode, depending on its charge, at a characteristic rate. Using this method, charged molecules that are separated into discrete units can be divided into sections and individual components may be separately eluted.

zone of integration: The area of contact between two populations of opposing phenotypes in which a mixture of phenotypes occurs through interbreeding. This area is sometimes called the ZONE OF HYBRIDIZATION.

zoo blot: It describes the practice of the Southern blotting method to assess the ability of a DNA probe from one species to hybridize with the DNA from the genomes of a variety of other species.

zoonotic infection: Broadcast of an infectious agent from an animal reservoir to humans.

zooplankton: Tiny animals that exist in the water column of lakes and oceans.

zoosporangium (pl. zoosporangia): A sporangium, or spore case, that holds zoospores.

zoospore: A spore of a fungus or fungus-like organism with flagella, capable of movement in water.

zygomere: Hypothetical pairing sites located along the chromosomes. For example, in barley, the zygomeres are lying proximal, as opposed to rye in which the zygomeres are lying terminal along the chromosome.

zygomorphic: Bilaterally symmetrical (i.e. symmetrical about one vertical plane only). It applies to flowers in which the perianth segments within each whorl vary in size and shape. *Contrasts with* ACTINOMORPHIC *and* IRREGULAR.

zygomycete: An unceremonious term utilized for a member of the Zygomycota.

zygomycota: A fungus group categorized by nonseptate hyphae, sexual spores and sporangiospores formed in a sporangium; common postharvest pathogens. *See* ZYGOMYCETE.

zygonema: See ZYGOTENE.

zygosome: A newly fused chromosome pair produced by synapsis (ZYGOTENE). *Aka* MIXOCHROMOSOME.

zygospore: A thick-walled resistant spore developing from a zygote produced from the fusion of isogametes. It represents the diploid reproductive stage in the life cycle of many fungi and protists.

zygote: (Gr. *zygon*, a yoke) A fertilized egg resulting from the fusion of the male (sperm) and female (egg) sex cells. The zygote divides mitotically to form the embryo of the seed. The union of a sperm nucleus with the two polar nuclei of the embryo sac produces a TRIPLOID (3x) cell that divides mitotically to produce the endosperm of the seed.

zygotene: The second stage of the prophase-I of meiosis-I, following leptotene, during which homologous chromosomes are connected side by side to initiate synapsis. For example, a diploid crop species with 2n = 20 chromosomes would reveal ten pairs of homologous chromosomes.

zygotene chromosomal bouquet: During meiosis, chromosomal pairing necessitates microtubule-dependent telomere securing and rotation on the nuclear envelope, a process that also performs chromosome homology searches. Formation of the structure involves telomere pulling towards the centrosome. The formation of this structure known as zygotene cilium was called the zygotene chromosomal bouquet by A. Mytlis *et al.* (2022). This structure helps to guarantee chromosomal bouquet formation and germ cell morphogenesis in zebrafish and mouse oocytes. The cilium announces the centrosome, enabling it to resist telomere pulling, and is indispensable for the bouquet and synaptonemal complex formation and fruitful oogenesis.

zygotene DNA (zygDNA): DNA of the homologous chromosomes associate with each other and pair, via the synaptonemal complex. Some DNA synthesis takes place at the zygotene stage. *Aka* zDNA. *See* PACHYTENE DNA.

zygotene RNA: RNA transcribed from zygotene DNA.

zygotic checkboard: *See* PUNNET SQUARE.

zygotic embryo: An embryo that emerges from a union of male and female gametes.

zygotic genes: These genes are expressed in the zygote. Broadly speaking, zygotic genes are classified as SEGMENT GENES and HOMEOTIC SELECTOR or HOMEOTIC GENES.

zygotic induction: When a prophage passages from an F^+ cell into an F^- cell during CONJUGATION it may initiate vegetative growth.

zygotic lethal: A lethal gene whose effect is in the embryo, larval or adult, in contrast to the gametic lethal affecting a gamete.

zygotic mutation: A mutation produced in the zygote shortly after fertilization.

zygotic selection: The process that causes differential mortality of an organism at any stage (other than gametes) during its life cycle.

zyme systems: Chemical reactions characterized by the existence of an inactive precursor of an enzyme which is activated via another enzyme that normally eliminates an extra piece of peptide chain at a physiologically apposite time and place. *See* ZYMOGENS.

zymogens: The enzymatically inactive precursors of certain proteolytic enzymes. The enzymes are inactive because they contain an extra piece of peptide chain. When this peptide is removed by another proteolytic enzyme, the zymogen is thus converted into the normal, active enzyme. The function of zymogens probably is to protect the cell, its machinery and/or the place of manufacture within the cell from the potentially harmful or lethal effects of an active, proteolytic enzyme. In other words, the approach is to activate the enzyme only when, and especially where, it is needed. *See* ENZYME SYSTEM.

zymogram: An electrophoretic strip (as of starch gel) or a photographic representation of it displaying the pattern of separated enzymes and particularly isoenzymes after electrophoresis.

zymography: An electrophoretic procedure for the recognition of hydrolytic enzymes. Zymography can also be applied to other types of enzymes such as xylanases, lipases and chitinases. Three types of zymography procedures include *IN-GEL* ZYMOGRAPHY (IGZ), *IN SITU* ZYMOGRAPHY (ISZ) and *IN VIVO* ZYMOGRAPHY (IVZ). *See details as separate terms.*

BIBLIOGRAPHY

Agricultural Biotechnology Glossary. U.S. Department of Agriculture. Available at: https://www.usda.gov/topics/biotechnology/biotechnology-glossary.

Agricultural Biotechnology: Informing the Dialogue. (2003) Cornell University College of Agriculture and Life Sciences; Ithaca, NY. *An Electronic Journal of the U.S. Department of State* 8(3), 36–38.

Alberts, B., Johnson, A., Lewis, J., Raff, M., Roberts K. and Walter, P. (2002) *Molecular Biology of the Cell*. 4th edn. Glossary. Garland Science, New York.

Bailey, R. (2019) *Cell Biology Glossary*. ThoughtCo. https://www.thoughtco.com/cell-biology-glossary-373293.

BioInformatics Glossary. Bio-Informatics Services. Available at: https://www.biosyn.com/bioinformatics.aspx.

Biology Dictionary: Molecular Biology. Available at: https://biologydictionary.net/category/molecular-biology/.

Biology Terms Dictionary. Available at: https://www.genscript.com/biology-glossary/8779/biotechnology.

Biostatistics Terminology. (2009) Available at: https://www.cram.com/flashcards/biostatistics-terminology-871485.

Biotechnology and Biosafety Glossary of Terms. Ministry of Environment, Forest and Climate Change, Government of India & CABI. Available at: https://www.whatisbiotechnology.org/index.php/glossary.

Cassells, A. and Gahan, P.B. (2006) *Dictionary of Plant Tissue Culture*. Taylor & Francis, Boca Raton, Florida.

Cell Biology. (2010) Available at: https://www.vocabulary.com/lists/29530.

Chahal, G.S. and Gosal, S.S. (2002). Glossary. In: *Principles and procedures of plant breeding: biotechnological and conventional approaches*. Alpha Science International Ltd., Pangbourne, UK.

Choudhary, A.K. and Vijayakumar, A.G. (2012) *Glossary of Plant Breeding: A Perspective*. LAP Lambert Academic Publishing, Germany.

Cooper, J. (2007) *Basic Terminology and Definitions in Plant Pathology*. WSU Extension, SJC. Available at: https://s3.wp.wsu.edu/uploads/sites/2054/2014/04/TermsandDefinitions_001.pdf.

Cytogenetics Glossary. Available at: https://quizlet.com/636132949/cytogenetics-glossary-flash-cards/.

Datta, S. (2018) *Unit I: Biostatistics and Computer Application Glossary* (Alphabetically arranged). https://www.researchgate.net/publication/327885091_Biostatistics_and_Computer_application_Glossary_Alphabetically_arranged.

Definitions of Evolutionary Terms. National Academy of Sciences. Available at: https://www.nationalacademies.org/evolution/definitions.

The Department of Energy (DOE). Human Genome Program. Available at: http://www.ornl.gov/sci/techresources/Human_Genome/glossary/.

Ebert, D. *Glossary in Evolutionary Biology*. Available at: https://www.evolution.unibas.ch/teaching/entstehung_des_lebens/pdf/glossary_evolutionary-biology.pdf.

Evolution Glossary. Available at: http://evolution-textbook.org/content/free/glossary/glossary.html.

Glossary. (2007) Government of Canada. Available at: https://www.canada.ca/en/health-canada/services/science-research/emerging-technology/biotechnology/about-biotechnology/glossary-biotechnology.html.

Glossary. (2011) ENVIS Centre on Environmental Biotechnology. Hosted by University of Kalyani, West Bengal. Sponsored by Ministry of Environment, Forests & Climate Change, Govt. of India. Available at: http://deskuenvis.nic.in/glossary.asp.

Glossary Evolution. (2001) Available at: https://www.pbs.org/wgbh/evolution/library/glossary/index.html.

Glossary of artificial intelligence. Available at: https://en.wikipedia.org/wiki/Glossary_of_artificial_intelligence.

A glossary of bioinformatics terms. Available at: https://mipt.ru/dbmp/student/files/bioinformatics/books/glossary_bioinf.php.

Glossary of biotechnology terms. (1986) *High Technology Law Journal* 1, 253–258.

Glossary of Botanical Terms. Available at: https://en.wikipedia.org/wiki/Glossary_of_botanical_terms.

Glossary of computer science. (2023) Available at: https://en.wikipedia.org/wiki/Glossary_of_computer_science.

Glossary of Genetic Terminology. The Genetics Center Incorporation. Available at: http://www.thegeneticscenter.com/glossary.htm.

Glossary of the Human Genome Project. http://www.ornl.gov/sci/techresources/Human_Genome/glossary/.

Glossary of Nanotechnology Terms. Available at: https://www.scribd.com/document/175704490/Glossary-of-Nanotechnology-Terms.

Glossary of Probability and Statistics. Available at: https://en.wikipedia.org/wiki/Glossary_of_probability_and_statistics.

Glossary of Scientific Terms. FSD Society. Available at: https://www.fshdsociety.org/glossary-of-scientific-terms/.

Glossary of Statistical Terms. UC Berkeley Statistics Department. Available at: https://www.stat.berkeley.edu/~stark/SticiGui/Text/gloss.htm.

Glossary of Terms. Available at: https://www.palomar.edu/anthro/synthetic/glossary.htm.

Glossary of Terms and Abbreviations. Iowa Institute of Human Genetics. Available at: https://medicine.uiowa.edu/humangenetics/resources/glossary-terms-and-abbreviations.

Glossary of Terms Regarding Evolution. Tought Co. Available at: https://www.thoughtco.com/glossary-of-evolution-terms-1224596.

Glossary. The MindFuel Foundation. Available at: https://wonderville.org/glossary/artificial-intelligence.

Glossary: Cells. Available at: https://www.visiblebody.com/learn/biology/cells/glossary.

Glossary: Dicot Plants. Available at: https://www.visiblebody.com/learn/biology/monocot-dicot/dicot-glossary.

Glossary: DNA and Chromosomes. Available at: https://www.visiblebody.com/learn/biology/dna-chromosomes/glossary.

Glossary: Monocot Plant. Available at: https://www.visiblebody.com/learn/biology/monocot-dicot/monocot-glossary.

Glossary: Photosynthesis. Available at: https://www.visiblebody.com/learn/biology/photosynthesis/glossary.

Harrell, F.E., Jr. (2022) *Glossary of Statistical Terms*. https://hbiostat.org/glossary/.

He, M. and Petoukhov, S. (2011) *Mathematics of Bioinformatics: Theory, Practice and Applications*. Wiley, pp. 270–295.

Illustrated Glossary of Plant Pathology. Available at: https://www.apsnet.org/edcenter/resources/illglossary/Pages/default.aspx.

Kaushansky, K. (2002) Glossary of molecular biology terminology. *Hematology American Society Hematology Education Program* 2002(1), 490–509. doi:10.1182/asheducation-2002.1.490.

Khandekar, J.D. (1997) Glossary: common terms in molecular biology. Part 1 (a–f). *Postgraduate Medical Journal* 73, 317–318.

King, R.C., Stansfield, W.D. and Mulligan, P.K. (2006) *A Dictionary of Genetics*. 7th edn. Oxford University Press, New York.

Krebs, J.R. and Davies. N.B. (1993) *An Introduction to Behavioural Ecology*. 3rd edn. Blackwell, UK.

Lackie, J.M. (2013) *The Dictionary of Cell and Molecular Biology*. 5th edn. Academic Press, New York.

Lavers, C. (2013) *The Origins of High-Resolution Civilian Satellite Imaging – Part 2: Civilian Imagery Programs and Providers*. Available at: https://www.directions-mag.com/article/1646.

Miglani, G.S. (1996) *A Language of Genetics: From Microbes to Man*. Department of Genetics, Punjab Agricultural University, Ludhiana, India.

Miglani, G.S. (1998a) Glossary. In: *Basic Genetics*. Narosa Publishing House, New Delhi, India, pp. 508–569

Miglani, G.S. (1998b) *Dictionary of Plant Genetics and Molecular Biology*. Howarth Press, New York.

Miglani, G.S. (2002) Glossary. In: *Advanced Genetics*. 1st edn. Alpha Science International Pvt. Ltd., Pangbourne, UK, pp. 569–625.

Miglani, G.S. (2007a) Glossary. In *Advanced Genetics*. 2nd edn. Alpha Science International Pvt. Ltd., Pangbourne, UK, pp. 620–701.

Miglani, G.S. (2007b) Glossary. In: *Developmental Genetics*. Anshan Ltd., UK, pp. 553–663.

Miglani, G.S. (2008a) Glossary. In: *Fundamentals of genetics*. Alpha Science International Pvt. Ltd., Pangbourne, UK, pp. G.1–G.66.

Miglani, G.S. (2008b) Mendelian genetics. In: Dashek, W.V. and Harrison, M. (eds) *Plant Cell Biology*. Science Publishers, Enfield, New Hampshire, pp. 259–348.

Miglani, G.S. (2011) Scientific method of research. In: Trivedi, P.C. (ed.) *Plant, Environment and Sustainability*. Agrobios, Jodhpur, India. pp. 307–324.

Miglani, G.S. (2013a) Glossary. In: *Gene regulation*. Alpha Science International Pvt. Ltd., Pangbourne, UK, pp. G.1–.43.

Miglani, G.S. (2013b) Glossary. In: *Genetic material*. Alpha Science International Pvt. Ltd., Pangbourne, UK, pp. G.1–G.57.

Miglani, G.S. (2013c) Plastid development and genomics. In: Dashek, W. and Miglani, G.S. (eds) *Plant Cells and their Organelles*. Wiley, West Sussex, UK.

Miglani, G.S. (2014) Glossary. In: *Gene expression*. Alpha Science International Pvt. Ltd., Pangbourne, UK, pp. G.1–G.43.

Miglani, G.S. (2015) Glossary. In: *Essentials of Molecular Genetics*. Alpha Science International Pvt. Ltd., Oxford, UK, pp. G.1–G.58.

Miglani, G.S. (2016) Glossary. In: *Genetic engineering: principles, procedures and consequences*. Alpha Science International Pvt. Ltd., Oxford, UK, pp. G.1–G.62.

Miglani, G.S. (2017) Genome editing in crop improvement: present scenario and future prospects. *Journal of Crop Improvement* 31(4), 553–559.

Miglani, G.S. (2019) Glossary. In: *Genome editing: a comprehensive treatise*. Alpha Science International Pvt. Ltd., Oxford, UK, pp. G.1–G.25.

Miglani, G.S. (2023a) Glossary. In: *Epigenomics*. I.K. International Pvt. Ltd., New Delhi, India, pp. 461–505.

Miglani, G.S. (2023b). Glossary. In: *Fundamentals of epigenetics*. I.K. International Pvt. Ltd., New Delhi, India, pp. 659–697.

Miglani, G.S. and Kaur, M. (2021a) Artificial intelligence – boon for agriculture. In: Shekhawat, G.S. (ed.) *Agricultural Innovations and Sustainability. Festschrift volume in honour of Prof. P.C. Trivedi*. Agrobios Research, Jodhpur, India, pp. 1–35.

Miglani, G.S. and Kaur, M. (2021b) Flower development and regulation of flowering. In: Meena, P. (ed.) *Plant Physiology, Development and Adoptions. Festschrift volume in honour of Prof. P.C. Trivedi*. Agrobios Research, Jodhpur, India, pp. 1–27.

Miglani, G.S. and Kaur, M. (2024) Glossary. In: *Epigenetics and biological processes*. I.K. International Pvt. Ltd., New Delhi, India, pp. 623–677.

Miglani, G.S., Kaur, A. and Kaur, L. (2020a) Plant gene expression control using genome- and epigenome-editing technologies. *Journal of Crop Improvement* 34(1), 1–63.

Miglani, G.S., Singh, S., Sharma, M. and Arora, S. (2020b) Engineered drives for plant pest management. In: Trivedi, P.C. (ed.) *Biotechnology for Plant Disease Diagnostics and Management*. Pointer, Jaipur, India, pp. 68–92.

Miglani, G.S., Kaur, R., Sharma, P. and Gupta, N. (2021) Leveraging photosynthetic efficiency toward improving crop yields. *Journal of Crop Improvement* 35(3), 361–402.

Miglani, G.S., Singh, S., Kaur S. and Kaur, M. (2022a) Current scenario of RNA interference-based control of insect and mite pests of fruit crops. In: Tanda, A.S. (ed.) *Molecular Advances in Insect Resistance of Field Crops: Modern and Applied Approaches*. Springer Nature, Cham, Switzerland, pp. 43–86.

Miglani, G.S., Singh, S., Li. Z. and Sandhu, R.K. (2022b) Intervention of modern genetic tools for managing insect-pests of fruit crops. In: Chakravarthy, A.K. (ed.) *Modern Genetic Tools for Pest Management in Agro-ecosystems*. Springer, Singapore, pp. 483–513.

Miglani, G.S., Kaur, M., Bhandari, G. and Kaur, G. (2024a) Epigenetic architecture of complex traits in plants. In: Kumar, N. (ed.) *Plant Mutagenesis and Crop Improvement*. CRC Press, pp. 29–57.

Miglani, G.S., Nisa, W.U. and Sandhu, S.K. (2024b) Cryptic genetic variation in evolution and crop improvement. In: Kumar, N. (ed.) *Plant Mutagenesis and Crop Improvement*. CRC Press, pp. 162–179.

Monday, P. (2009) *Glossary of Terms for Molecular Technology*. Available at: https://molecular-biology.coe.hawaii.edu/glossary/.

Monocots, Dicots, and Their Tissues. Available at: https://www.visiblebody.com/learn/biology/monocot-dicot/overview.

Nagpur, T., Oswalt, D.L. and Singh, F. (1992) *A Glossary for Crop Improvement*. Skill Development Series No. 1. Human Resource Development Program. International Crop Research Institute for the Semi-Arid Tropics, Patancheru, India.

National Center for Biotechnology Information. *BLAST Glossary*. Available at: http://www.ncbi.nlm.nih.gov/Education/BLASTinfo/glossary2.html.

Nill, K. (2021) *Glossary of Biotechnology & Agrobiotechnology Terms*. 5th edn. CRC Press.

Paneda, B. and Mejia, M. (2007) *Glossary: Multi-Institutional Distance Learning Course on the Ex Situ Conservation of Plant Genetic Resources.* Centro Internacional de Agricultura Tropical (CIAT) Publication No. 360. CIAT, Cali, Colombia.

Pradeep, T. (2007) Glossary of nano terms. In: *NANO: The Essentials: Understanding Nanoscience and Nanotechnology*. 1st edn. McGraw-Hill, New Delhi, India. Available at: https://www.accessengineeringlibrary.com/content/book/9780071548298/back-matter/glossary1.

Preparing for Future Products of Biotechnology. (2017) National Academies Press (US), Washington, DC. Available at: https://www.ncbi.nlm.nih.gov/books/NBK442195/.

Protein folds glossary from the Structural Classification of Proteins website (SCOP). Available at: https://www.ebi.ac.uk/pdbe/scop/.

Reif, J.H. (2002) *DNA Lattices: A Programmable Method for Molecular Scale Patterning and Computation.* Duke University, Durham, North Carolina. Available at: https://users.cs.duke.edu.

Reif, J.H., LaBean, T.H. and Seeman, N.C. (2001) *Challenges and Applications for Self-Assembled DNA Nanostructures.* DIMACS Series in Discrete Mathematics and Theoretical Computer Science, Leiden, The Netherlands, (June), ed. A. Condon. Lecture Notes in Computer Science 2000, Springer. Available at: https://users.cs.duke.edu/~reif/paper/DNAlattice/DNAlattice.pdf.

Rieger, R., Michaelis, A. and Green. M.M. (2013) *A Glossary of Genetics and Cytogenetics.* Springer Berlin, Heidelberg, Germany.

Roff, D.A. (1992) *The Evolution of Life Histories: Theory and Analysis.* Chapman & Hall, London.

Schaeffer, W.I., Mueller, S., Renfroe, M., Shay, J.W. and Vaughn, J. (1990) Terminology associated with cell, tissue and organ culture, molecular biology and molecular genetics. *In Vitro Cellular & Developmental Biology* 26, 97–101.

Schlegel, R.H.J. (2010) *Dictionary of Plant Breeding*. 2nd edn. CRC Press, Boca Raton, Florida.

Sharma, D. (2022) *50 Statistics Terms to Know (With Definitions)*. https://www.indeed.com/career-advice/career-development/list-of-statistics-terms.

Shih, W. and Wong, W. (2021) *DNA Nanotechnology Tools: From Design to Applications.* Available at: https://wyss.harvard.edu/technology/dna-nanotechnology-tools-from-design-to-applications/.

SMART database. Available at: http://smart.embl-heidelberg.de/help/smart_glossary.shtml.

Stearns, S.C. and Hoeckstra, R.F. (2005) *Evolution: An Introduction*. Oxford University Press, Oxford, UK.

Talking Glossary of Genomic and Genetic Terms. National Human Genome Research Institute. https://www.genome.gov/genetics-glossary..

UCMP glossary: cell biology. (2009) Available at: https://ucmp.berkeley.edu/glossary/gloss4cell.html.

Vanangamudi, K., Natarajan, K., Sheela, K.R.V.S., Renuka, R., Kayalvizhi, G., Sujatha, E. and Saravanan, T. (2013) *Agros Dictionary of Plant Sciences (Including Genetics, Plant Breeding, Seed Science and Technology and Agriculture Biotechnology)*. Agrobios, Jodhpur, India.

What is biotechnology? Available at: https://www.whatisbiotechnology.org/index.php/glossary.

Zaid, A., Hughes, H.G., Porceddu, E. and Nicholas, F. (1999) *Glossary of Biotechnology and Genetic Engineering*. FAO Research and Technology Paper 7. Rome, Italy.